Prepare for Class "Read the Book"

Feature	Description	Benefit	Page
Every chapter begins with....			
Chapter Opening Article & Project	Each chapter begins with a current article and ends with a related project.	The Article describes a real situation. The Project lets you apply what you learned to solve a related problem.	400, 501
Internet-based Projects	The projects allow for the integration of spreadsheet technology that students will need to be a productive member of the workforce.	The projects allow the opportunity for students to collaborate and use mathematics to deal with issues that come up in their lives.	400, 501
Every section begins with....			
Learning Objectives	Each section begins with a list of objectives. Objectives also appear in the text where the objective is covered.	These focus your studying by emphasizing what's most important and where to find it.	421
Most sections contain...			
PREPARING FOR THIS SECTION	Most sections begin with a list of key concepts to review with page numbers.	Ever forget what you've learned? This feature highlights previously learned material to be used in this section. Review it, and you'll always be prepared to move forward.	421
Now Work the 'Are You Prepared?' Problems	Problems that assess whether you have the prerequisite knowledge for the upcoming section.	Not sure you need the Preparing for This Section review? Work the 'Are You Prepared?' problems. If you get one wrong, you'll know exactly what you need to review and where to review it!	421, 432
"Now Work" PROBLEMS	These follow most examples and direct you to a related exercise.	We learn best by doing. You'll solidify your understanding of examples if you try a similar problem right away, to be sure you understand what you've just read.	428, 430
WARNING	Warnings are provided in the text.	These point out common mistakes and help you to avoid them.	453, 454
Explorations and **Seeing the Concept**	These represent graphing utility activities to foreshadow a concept or solidify a concept just presented.	You will obtain a deeper and more intuitive understanding of theorems and definitions.	245, 426
In Words	These provide alternative descriptions of select definitions and theorems.	Does math ever look foreign to you? This feature translates math into plain English.	430
Calculus Icon	These appear next to information essential for the study of calculus.	Pay attention – if you spend extra time now, you'll do better later!	429
NEW! Showcase **EXAMPLES**	These examples provide "how-to" instruction by offering a guided, step-by-step approach to solving a problem.	With each step presented on the left and the mathematics displayed on the right, students can immediately see how each step is employed.	332–333
NEW! Model It! Examples and Problems	Marked with ●. These are examples and problems that require you to build a mathematical model from either a verbal description or data. The homework Model It! problems are marked by purple numbers.	It is rare for a problem to come in the form, "Solve the following equation". Rather, the equation must be developed based on an explanation of the problem. These problems require you to develop models that will allow you to describe the problem mathematically and suggest a solution to the problem.	320, 351–360

Practice "Work the Problems"

Feature	Description	Benefit	Page
"Assess Your Understanding" contains a variety of problems at the end of each section.			
'Are You Prepared?' Problems	These assess your retention of the prerequisite material you'll need. Answers are given at the end of the section exercises. This feature is related to the Preparing for This Section feature.	Do you always remember what you've learned? Working these problems is the best way to find out. If you get one wrong, you'll know exactly what you need to review and where to review it!	421, 432
Concepts and Vocabulary	These short-answer questions, mainly Fill-in-the-Blank and True/False items, assess your understanding of key definitions and concepts in the current section.	It is difficult to learn math without knowing the language of mathematics. These problems test your understanding of the formulas and vocabulary.	432
Skill Building	Correlated to section examples, these problems provide straightforward practice.	It's important to dig in and develop your skills. These problems provide you with ample practice to do so.	432–434
NEW! Mixed Practice	These problems offer comprehensive assessment of the skills learned in the section by asking problems that relate to more than one concept or objective. These problems may also require you to utilize skills learned in previous sections.	Learning mathematics is a building process. Many concepts are interrelated. These problems help you see how mathematics builds on itself and also see how the concepts tie together.	434–435
Applications and Extensions	These problems allow you to apply your skills to real-world problems. These problems also allow you to extend concepts learned in the section.	You will see that the material learned within the section has many uses in everyday life.	435-437
Explaining Concepts: Discussion and Writing	"Discussion and Writing" problems are colored red. These support class discussion, verbalization of mathematical ideas, and writing and research projects.	To verbalize an idea, or to describe it clearly in writing, shows real understanding. These problems nurture that understanding. Many are challenging but you'll get out what you put in.	437
NEW! Interactive Exercises	In selected exercise sets, applets are provided to give a "hands-on" experience.	The applets allow students to interact with mathematics in an active learning environment. By exploring a variety of scenarios, the student is able to visualize the mathematics and develop a deeper conceptual understanding of the material.	257
"Now Work" PROBLEMS	Many examples refer you to a related homework problem. These related problems are marked by a pencil and yellow numbers.	If you get stuck while working problems, look for the closest Now Work problem and refer back to the related example to see if it helps.	421, 432
Graphing Calculator	These optional problems require the use of a graphing utility, and are marked by a special icon and green numbers.	Your instructor will usually provide guidance on whether or not to do these problems. If so, these problems help to verify and visualize your analytical results.	427

Review "Study for Quizzes and Tests"

Feature	Description	Benefit	Page
Chapter Reviews at the end of each chapter contain...			
"Things to Know"	A detailed list of important theorems, formulas, and definitions from the chapter.	Review these and you'll know the most important material in the chapter!	494–495
"You Should Be Able to..."	Contains a complete list of objectives by section, examples that illustrate the objective, and practice exercises that test your understanding of the objective.	Do the recommended exercises and you'll have mastery over the key material. If you get something wrong, review the suggested page numbers and try again.	495–496
Review Exercises	These provide comprehensive review and practice of key skills, matched to the Learning Objectives for each section.	Practice makes perfect. These problems combine exercises from all sections, giving you a comprehensive review in one place.	496–499
CHAPTER TEST	About 15–20 problems that can be taken as a Chapter Test. Be sure to take the Chapter Test under test conditions—no notes!	Be prepared. Take the sample practice test under test conditions. This will get you ready for your instructor's test. If you get a problem wrong, watch the Chapter Test Prep video.	500
CUMULATIVE REVIEW	These problem sets appear at the end of each chapter, beginning with Chapter 2. They combine problems from previous chapters, providing an ongoing cumulative review.	These are really important. They will ensure that you are not forgetting anything as you go. These will go a long way toward keeping you constantly primed for the final exam.	500–501
CHAPTER PROJECTS	The Chapter Project applies what you've learned in the chapter. Additional projects are available on the Instructor's Resource Center (IRC).	The Project gives you an opportunity to apply what you've learned in the chapter to solve a problem related to the opening article. If your instructor allows, these make excellent opportunities to work in a group, which is often the best way of learning math.	501–502
NEW! Internet-based Projects	In selected chapters, a web-based project is given.	The projects allow the opportunity for students to collaborate and use mathematics to deal with issues that come up in their lives.	501–502

To the Student

As you begin, you may feel anxious about the number of theorems, definitions, procedures, and equations. You may wonder if you can learn it all in time. Don't worry, your concerns are normal. This textbook was written with you in mind. If you attend class, work hard, and read and study this book, you will build the knowledge and skills you need to be successful. Here's how you can use the book to your benefit.

Read Carefully

When you get busy, it's easy to skip reading and go right to the problems. Don't. . . the book has a large number of examples and clear explanations to help you break down the mathematics into easy-to-understand steps. Reading will provide you with a clearer understanding, beyond simple memorization. Read before class (not after) so you can ask questions about anything you didn't understand. You'll be amazed at how much more you'll get out of class if you do this.

Use the Features

I use many different methods in the classroom to communicate. Those methods, when incorporated into the book, are called "features." The features serve many purposes, from providing timely review of material you learned before (just when you need it), to providing organized review sessions to help you prepare for quizzes and tests. Take advantage of the features and you will master the material.

To make this easier, I've provided a brief guide to getting the most from this book. Refer to the "Prepare for Class," "Practice," and "Review" pages on the inside front cover of this book. Spend fifteen minutes reviewing the guide and familiarizing yourself with the features by flipping to the page numbers provided. Then, as you read, use them. This is the best way to make the most of your textbook.

Please do not hesitate to contact me, through Pearson Education, with any questions, suggestions, or comments that would improve this text. I look forward to hearing from you, and good luck with all of your studies.

Best Wishes!

Michael Sullivan

STUDY SMARTER

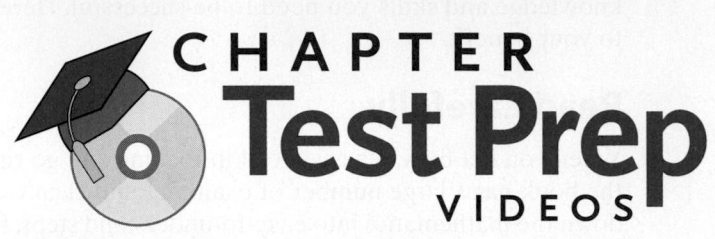

CHAPTER
Test Prep
VIDEOS

Step-by-step solutions on video for all chapter test exercises from the text

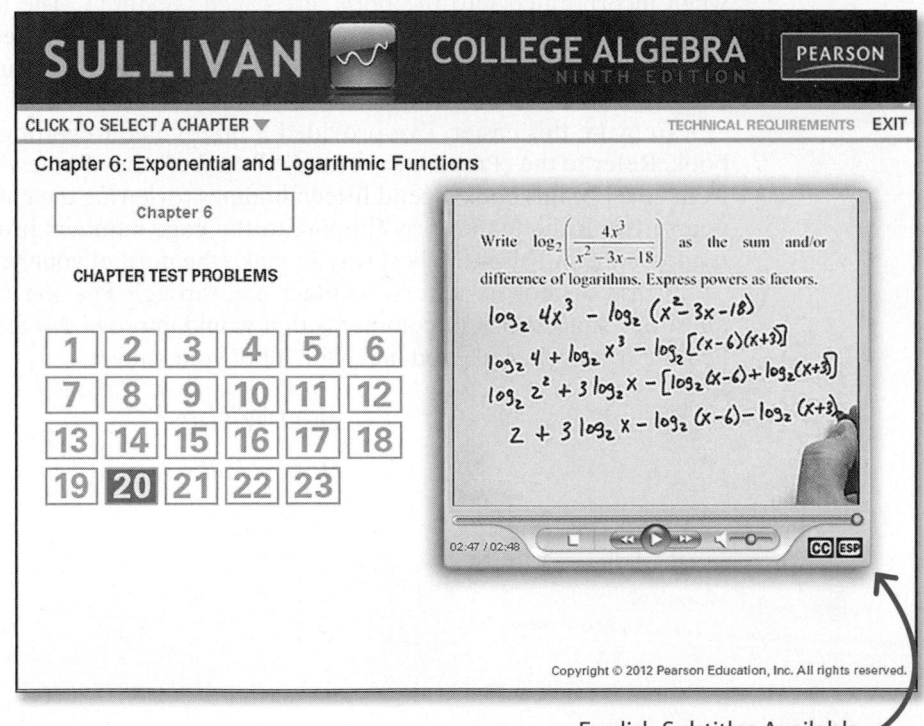

English Subtitles Available

CHAPTER TEST PREP VIDEOS ARE ACCESSIBLE THROUGH THE FOLLOWING:

Dedicated to the Memory of Mary

COLLEGE ALGEBRA

NINTH EDITION

COLLEGE ALGEBRA

NINTH EDITION

Michael Sullivan
Chicago State University

Prentice Hall

Boston Columbus Indianapolis New York San Francisco Upper Saddle River
Amsterdam Cape Town Dubai London Madrid Milan Munich Paris Montréal Toronto
Delhi Mexico City São Paulo Sydney Hong Kong Seoul Singapore Taipei Tokyo

Editor in Chief: *Anne Kelly*
Sponsoring Editor: *Dawn Murrin*
Assistant Editor: *Joseph Colella*
Executive Marketing Manager: *Roxanne McCarley*
Marketing Assistant: *Katherine Minton*
Senior Managing Editor: *Karen Wernholm*
Associate Managing Editor: *Tamela Ambush*
Senior Production Project Manager: *Peggy McMahon*
Production Editor: *Bob Walters, Prepress Management, Inc.;*
 Sherry Berg, Nesbitt Graphics, Inc.
Senior Design Supervisor: *Andrea Nix*
Art Direction and Cover Design: *Barbara T. Atkinson*
Interior Design: *Tamara Newnam*
Image Manager/Image Management Services: *Rachel Youdelman*
Photo Researcher: *Caroline Commins*
Permissions Project Supervisor: *Michael Joyce*
Media Producer: *Vicki Dreyfus*
Senior Author Support/Technology Specialist: *Joe Vetere*
Manufacturing Manager: *Evelyn Beaton*
Senior Manufacturing Buyer: *Carol Melville*
Composition: *MPS Limited, a Macmillan Company*
Technical Illustrations: *Precision Graphics and Laserwords*
Printer/Binder: *Courier/Kendallville*
Cover Printer: *Lehigh-Phoenix Color/Hagerstown*
Text Font: *Times Ten Roman*

Credits and acknowledgments borrowed from other sources and reproduced, with permission, in this textbook appear on page xxx.

Many of the designations used by manufacturers and sellers to distinguish their products are claimed as trademarks. Where those designations appear in this book, and Pearson was aware of a trademark claim, the designations have been printed in initial caps or all caps.

Library of Congress Cataloging-in-Publication Data

Sullivan, Michael, 1942–
 College algebra / Michael Sullivan.—9th ed.
 p. cm.
 Includes index.
 ISBN 978-0-321-71681-1
 1. Algebra—Textbooks. I. Title.
 QA154.3.S763 2012
 512.9—dc22

 2010047534

1 2 3 4 5 6 7 8 9 10—CRK—14 13 12 11 10

Prentice Hall
is an imprint of

www.pearsonhighered.com

ISBN-10: 0-321-71681-7
ISBN-13: 978-0-321-71681-1

Table of Contents

Three Distinct Series

Students have different goals, learning styles, and levels of preparation. Instructors have different teaching philosophies, styles, and techniques. Rather than write one series to fit all, the Sullivans have written three distinct series. All share the same goal—to develop a high level of mathematical understanding and an appreciation for the way mathematics can describe the world around us. The manner of reaching that goal, however, differs from series to series.

Contemporary Series, Ninth Edition

The Contemporary Series is the most traditional in approach yet modern in its treatment of precalculus mathematics. Graphing utility coverage is optional and can be included or excluded at the discretion of the instructor: *College Algebra, Algebra & Trigonometry, Trigonometry, Precalculus.*

Enhanced with Graphing Utilities Series, Fifth Edition

This series provides a more thorough integration of graphing utilities into topics, allowing students to explore mathematical concepts and foreshadow ideas usually studied in later courses. Using technology, the approach to solving certain problems differs from the Contemporary Series, while the emphasis on understanding concepts and building strong skills does not: *College Algebra, Algebra & Trigonometry, Trigonometry, Precalculus.*

Concepts through Functions Series, Second Edition

This series differs from the others, utilizing a functions approach that serves as the organizing principle tying concepts together. Functions are introduced early in various formats. This approach supports the Rule of Four, which states that functions are represented symbolically, numerically, graphically, and verbally. Each chapter introduces a new type of function and then develops all concepts pertaining to that particular function. The solutions of equations and inequalities, instead of being developed as stand-alone topics, are developed in the context of the underlying functions. Graphing utility coverage is optional and can be included or excluded at the discretion of the instructor: *College Algebra; Precalculus, with a Unit Circle Approach to Trigonometry; Precalculus, with a Right Triangle Approach to Trigonometry.*

Preface to the Instructor

As a professor of mathematics at an urban public university for 35 years, I understand the varied needs of college algebra students. Students range from being underprepared, with little mathematical background and a fear of mathematics, to being highly prepared and motivated. For some, this is their final course in mathematics. For others, it is preparation for future mathematics courses. I have written this text with both groups in mind.

A tremendous benefit of authoring a successful series is the broad-based feedback I receive from teachers and students who have used previous editions. I am sincerely grateful for their support. Virtually every change to this edition is the result of their thoughtful comments and suggestions. I hope that I have been able to take their ideas and, building upon a successful foundation of the eighth edition, make this series an even better learning and teaching tool for students and teachers.

Features in the Ninth Edition

Rather than provide a list of features here, that information can be found on the endpapers in the front of this book.

This places the features in their proper context, as building blocks of an overall learning system that has been carefully crafted over the years to help students get the most out of the time they put into studying. Please take the time to review this and to discuss it with your students at the beginning of your course. My experience has been that when students utilize these features, they are more successful in the course.

New to the Ninth Edition

- **Chapter Projects**, which apply the concepts of each chapter to a real-world situation, have been enhanced to give students an up-to-the-minute experience. Many projects are new and Internet-based, requiring the student to research information online in order to solve problems.

- **Author Solves It MathXL Video Clips**—author Michael Sullivan works by section through MathXL exercises typically requested by students for more explanation or tutoring. These videos are a result of Sullivan's experiences in teaching online.

- **Showcase Examples** are used to present examples in a guided, step-by-step format. Students can immediately see how each of the steps in a problem are employed. The "How To" examples have a two-column format in which the left column describes the step in solving the problem and the right column displays the algebra complete with annotations.

- **Model It** examples and exercises are clearly marked with a ● icon. These examples and exercises are meant to develop the student's ability to build models from both verbal descriptions and data. Many of the problems involving data require the students to first determine the appropriate model (linear, quadratic, and so on) to fit to the data and justify their choice.

- **Exercise Sets** at the end of each section remain classified according to purpose. The "*Are You Prepared?*" exercises have been expanded to better serve the student who needs a just-in-time review of concepts utilized in the section. The *Concepts and Vocabulary* exercises have been updated. These fill-in-the-blank and True/False problems have been written to serve as reading quizzes. *Mixed Practice* exercises have been added where appropriate. These problems offer a comprehensive assessment of the skills learned in the section by asking problems that relate to more than one objective. Sometimes these require information from previous sections so students must utilize skills learned throughout the course. *Applications and Extension* problems have been updated and many new problems involving sourced information and data have been added to bring relevance and timeliness to the exercises. The *Explaining Concepts: Discussion and Writing* exercises have been updated and reworded to stimulate discussion of concepts in online discussion forums. These can also be used to spark classroom discussion. Finally, in the **Annotated Instructor's Edition,** I have preselected problems that can serve as sample homework assignments. These are indicated by a blue underline, and they are assignable in MyMathLab® if desired.

- The **Chapter Review** now identifies Examples to review for each objective in the chapter.

Changes in the Ninth Edition

- **CONTENT**
 - **Chapter 3, Section 3** A new objective "Use a graph to locate the absolute maximum and the absolute minimum" has been added. The Extreme Value Theorem is also cited here.
 - **Chapter 4, Section 3** A new objective "Find a quadratic function given its vertex and one point" has been added.
 - **Chapter 5, Section 1** A new objective "Build cubic models from data" has been added.
 - **Chapter 5, Section 5** Descartes' Rule of Signs has been removed as its value is redundant to the information collected from other sources.
 - **Chapter 6, Section 3** The definition of an exponential function has been broadened.

- **ORGANIZATION**
 - ○ **Chapter R, Section 5** The objective "Complete the Square" has been relocated to here from Chapter 1.

Using the Ninth Edition Effectively with Your Syllabus

To meet the varied needs of diverse syllabi, this book contains more content than is likely to be covered in a *College Algebra* course. As the chart illustrates, this book has been organized with flexibility of use in mind. Within a given chapter, certain sections are optional (see the detail following the flow chart) and can be omitted without loss of continuity.

Chapter R Review
This chapter consists of review material. It may be used as the first part of the course or later as a just-in-time review when the content is required. Specific references to this chapter occur throughout the book to assist in the review process.

Chapter 1 Equations and Inequalities
Primarily a review of Intermediate Algebra topics, this material is prerequisite for later topics. The coverage of complex numbers and quadratic equations with a negative discriminant is optional and may be postponed or skipped entirely without loss of continuity.

Chapter 2 Graphs
This chapter lays the foundation for functions. Section 2.5 is optional.

Chapter 3 Functions and Their Graphs
Perhaps the most important chapter. Section 3.6 is optional.

Chapter 4 Linear and Quadratic Functions
Topic selection depends on your syllabus. Sections 4.2 and 4.4 may be omitted without a loss of continuity.

Chapter 5 Polynomial and Rational Functions
Topic selection depends on your syllabus.

Chapter 6 Exponential and Logarithmic Functions
Sections 6.1–6.6 follow in sequence. Sections 6.7, 6.8, and 6.9 are optional.

Chapter 7 Analytic Geometry
Sections 7.1–7.4 follow in sequence.

Chapter 8 Systems of Equations and Inequalities
Sections 8.2–8.7 may be covered in any order, but each requires Section 8.1. Section 8.8 requires Section 8.7.

Chapter 9 Sequences; Induction; The Binomial Theorem
There are three independent parts: Sections 9.1–9.3; Section 9.4; and Section 9.5.

Chapter 10 Counting and Probability
The sections follow in sequence.

Acknowledgments

Textbooks are written by authors, but evolve from an idea to final form through the efforts of many people. It was Don Dellen who first suggested this book and series to me. Don is remembered for his extensive contributions to publishing and mathematics.

Thanks are due to the following people for their assistance and encouragement to the preparation of this edition:

- From Pearson Education: Anne Kelly for her substantial contributions, ideas, and enthusiasm; Roxanne McCarley, who is a huge fan and supporter; Dawn Murrin, for her unmatched talent at getting the details right; Bob and Carol Walters for their superb organizational skills in directing production; Peggy McMahon for stepping in and directing the final stages of production; Chris Hoag for her continued support and genuine interest; Greg Tobin for his leadership and commitment to excellence; and the Pearson Math and Science Sales team, for their continued confidence and personal support of our books.

- As this book went to press, Bob Walters, Production Manager, passed away after a long and valiant battle fighting lung disease. He was an old and dear friend—a true professional in every sense of the word.

- Accuracy checkers: C. Brad Davis, who read the entire manuscript and accuracy checked answers. His attention to detail is amazing; Timothy Britt, for creating the Solutions Manuals and accuracy checking answers.

- *Reviewers*: Larissa Williamson, University of Florida; Richard Nadel, Florida International University; Robin Steinberg, Puma CC; Mike Rosenthal, Florida International University; Gerardo Aladro, Florida International University; Tammy Muhs, University of Central Florida; Val Mohanakumar, Hillsborough CC.

Finally, I offer my grateful thanks to the dedicated users and reviewers of my books, whose collective insights form the backbone of each textbook revision.

My list of indebtedness just grows and grows. And, if I've forgotten anyone, please accept my apology. Thank you all.

James Africh, College of DuPage
Steve Agronsky, Cal Poly State University
Grant Alexander, Joliet Junior College
Dave Anderson, South Suburban College
Richard Andrews, Florida A&M University
Joby Milo Anthony, University of Central Florida
James E. Arnold, University of Wisconsin-Milwaukee
Adel Arshaghi, Center for Educational Merit
Carolyn Autray, University of West Georgia
Agnes Azzolino, Middlesex County College
Wilson P Banks, Illinois State University
Sudeshna Basu, Howard University
Dale R. Bedgood, East Texas State University
Beth Beno, South Suburban College
Carolyn Bernath, Tallahassee Community College
Rebecca Berthiaume, Edison State College
William H. Beyer, University of Akron
Annette Blackwelder, Florida State University
Richelle Blair, Lakeland Community College
Kevin Bodden, Lewis and Clark College
Barry Booten, Florida Atlantic University
Larry Bouldin, Roane State Community College
Bob Bradshaw, Ohlone College
Trudy Bratten, Grossmont College
Tim Bremer, Broome Community College
Tim Britt, Jackson State Community College
Michael Brook, University of Delaware
Joanne Brunner, Joliet Junior College
Warren Burch, Brevard Community College
Mary Butler, Lincoln Public Schools
Melanie Butler, West Virginia University
Jim Butterbach, Joliet Junior College
William J. Cable, University of Wisconsin-Stevens Point
Lois Calamia, Brookdale Community College
Jim Campbell, Lincoln Public Schools
Roger Carlsen, Moraine Valley Community College
Elena Catoiu, Joliet Junior College
Mathews Chakkanakuzhi, Palomar College
Tim Chappell, Penn Valley Community College
John Collado, South Suburban College
Alicia Collins, Mesa Community College
Nelson Collins, Joliet Junior College
Jim Cooper, Joliet Junior College
Denise Corbett, East Carolina University
Carlos C. Corona, San Antonio College
Theodore C. Coskey, South Seattle Community College
Donna Costello, Plano Senior High School
Paul Crittenden, University of Nebraska at Lincoln
John Davenport, East Texas State University
Faye Dang, Joliet Junior College
Antonio David, Del Mar College
Stephanie Deacon, Liberty University
Duane E. Deal, Ball State University

Jerry DeGroot, Purdue North Central
Timothy Deis, University of Wisconsin-Platteville
Joanna DelMonaco, Middlesex Community College
Vivian Dennis, Eastfield College
Deborah Dillon, R. L. Turner High School
Guesna Dohrman, Tallahassee Community College
Cheryl Doolittle, Iowa State University
Karen R. Dougan, University of Florida
Jerrett Dumouchel, Florida Community College
 at Jacksonville
Louise Dyson, Clark College
Paul D. East, Lexington Community College
Don Edmondson, University of Texas-Austin
Erica Egizio, Joliet Junior College
Jason Eltrevoog, Joliet Junior College
Christopher Ennis, University of Minnesota
Kathy Eppler, Salt Lake Community College
Ralph Esparza, Jr., Richland College
Garret J. Etgen, University of Houston
Scott Fallstrom, Shoreline Community College
Pete Falzone, Pensacola Junior College
W.A. Ferguson, University of Illinois-Urbana/Champaign
Iris B. Fetta, Clemson University
Mason Flake, student at Edison Community College
Timothy W. Flood, Pittsburg State University
Robert Frank, Westmoreland County
 Community College
Merle Friel, Humboldt State University
Richard A. Fritz, Moraine Valley
Community College
Dewey Furness, Ricke College
Randy Gallaher, Lewis and Clark College
Tina Garn, University of Arizona
Dawit Getachew, Chicago State University
Wayne Gibson, Rancho Santiago College
Robert Gill, University of Minnesota Duluth
Nina Girard, University of Pittsburgh at Johnstown
Sudhir Kumar Goel, Valdosta State University
Adrienne Goldstein, Miami Dade College,
 Kendall Campus
Joan Goliday, Sante Fe Community College
Lourdes Gonzalez, Miami Dade College, Kendall Campus
Frederic Gooding, Goucher College
Donald Goral, Northern Virginia Community College
Sue Graupner, Lincoln Public Schools
Mary Beth Grayson, Liberty University
Jennifer L. Grimsley, University of Charleston
Ken Gurganus, University of North Carolina
James E. Hall, University of Wisconsin-Madison
Judy Hall, West Virginia University
Edward R. Hancock, DeVry Institute of Technology
Julia Hassett, DeVry Institute-Dupage
Christopher Hay-Jahans, University of South Dakota

Michael Sullivan
Chicago State University

STUDENT RESOURCES

Available to students are the following supplements:

- *Student Solutions Manual (ISBN 10: 0321716876; ISBN 13: 9780321716873)*
 Fully worked solutions to odd-numbered exercises.

- *Algebra Review (ISBN 10: 0131480065; ISBN 13: 9780131480063)*
 Four chapters of Intermediate Algebra Review. Perfect for a slower-paced course or for individual review.

- *Videos on DVD for College Algebra 9e (ISBN 10: 0321716965; ISBN 13: 9780321716965)*
 The Videos on DVD contain short video clips of Michael Sullivan III working key book examples. Chapter Test Prep Videos (also included) provide fully worked solutions to the Chapter Test exercises. The Chapter Test Prep Videos are also available within MyMathLab® or on YouTube™ (go to www.youtube.com/SullivanColAlgebra9e). Videos have optional subtitles.

MathXL® Online Course (access code required)

MathXL® is a powerful online homework, tutorial, and assessment system that accompanies Pearson Education's textbooks in mathematics or statistics. With MathXL, instructors can:

- Create, edit, and assign online homework and tests using algorithmically generated exercises correlated at the objective level to the textbook.

- Create and assign their own online exercises and import TestGen tests for added flexibility.

- Maintain records of all student work tracked in MathXL's online gradebook.

With MathXL, students can:

- Take chapter tests in MathXL and receive personalized study plans based on their test results.

- Use the study plan to link directly to tutorial exercises for the objectives they need to study and retest.

- Access supplemental animations and video clips directly from selected exercises.

MathXL is available to qualified adopters. For more information, visit our website at **www.mathxl.com**, or contact your Pearson sales representative.

MyMathLab® Online Course (access code required)

MyMathLab® is a text-specific, easily customizable online course that integrates interactive multimedia instruction with textbook content. MyMathLab gives you the tools you need to deliver all or a portion of your course online, whether your students are in a lab setting or working from home.

- **Interactive homework exercises**, correlated to your textbook at the objective level, are algorithmically generated for unlimited practice and mastery. Most exercises are free-response and provide guided solutions, sample problems, and tutorial learning aids for extra help.

- **Personalized homework** assignments that you can design to meet the needs of your class. MyMathLab tailors the assignment for each student based on their test or quiz scores. Each student receives a homework assignment that contains only the problems they still need to master.

- **Personalized Study Plan,** generated when students complete a test or quiz or homework, indicates which topics have been mastered and links to tutorial exercises for topics students have not mastered. You can customize the Study Plan so that the topics available match your course content.

- **Multimedia learning aids**, such as video lectures and podcasts, animations, and a complete multimedia textbook, help students independently improve their understanding and performance. You can assign these multimedia learning aids as homework to help your students grasp the concepts.

- **Homework and Test Manager** lets you assign homework, quizzes, and tests that are automatically graded. Select just the right mix of questions from the MyMathLab exercise bank, instructor-created custom exercises, and/or TestGen® test items.

- **Gradebook,** designed specifically for mathematics and statistics, automatically tracks students' results, lets you stay on top of student performance, and gives you control over how to calculate final grades. You can also add offline (paper-and-pencil) grades to the gradebook.

- **MathXL Exercise Builder** allows you to create static and algorithmic exercises for your online assignments. You can use the library of sample exercises as an easy starting point, or you can edit any course-related exercise.

- **Pearson Tutor Center** (**www.pearsontutorservices.com**) access is automatically included with MyMathLab. The Tutor Center is staffed by qualified math instructors who provide textbook-specific tutoring for students via toll-free phone, fax, email, and interactive Web sessions.

- **NEW Resources for Sullivan, College Algebra, 9e**

 - **Author Solves It videos** feature Mike Sullivan III working by section through MathXL exercises typically requested by students for more explanation or tutoring. These videos are a result of Sullivan's experiences in teaching online.

 - **Sample homework assignments**, preselected by the author, are indicated by a blue underline within the end-of-section exercise sets in the Annotated Instructor's Edition and are assignable in MyMathLab.

 - **Chapter Project MathXL Exercises** allow instructors to assign problems based on the new Internet-based Chapter Projects.

Students do their assignments in the Flash®-based MathXL Player, which is compatible with almost any browser (Firefox®, Safari™, or Internet Explorer®) on almost any platform (Macintosh® or Windows®). MyMathLab is powered by CourseCompass™, Pearson Education's online teaching and learning environment, and by MathXL®, our online homework, tutorial, and assessment system. MyMathLab is available to qualified adopters. For more information, visit www.mymathlab.com or contact your Pearson representative.

Applications Index

Acoustics
amplifying sound, 498
loudness of sound, 449
whispering galleries, 520–21, 538

Aeronautics
Challenger disaster, 485

Agriculture
crop allocation, 632
farm management, 627
field enclosure, 612
minimizing cost, 627
watering a field, 102

Air travel
parking at O'Hare International
 Airport, 242

Archaeology
age of ancient tools, 478–79
age of fossil, 484
age of tree, 484
date of prehistoric man's
 death, 498

Architecture
brick staircase, 652, 676
Burj Khalifa building, 31
floor design, 650–51, 676
football stadium seating, 652
mosaic design, 652, 676
Norman window, 37, 307
parabolic arch, 307
racetrack design, 523
special window, 307, 315
stadium construction, 652
window design, 307
window dimensions, 102

Art
framing a painting, 147

Astronomy
distance from Earth to its moon, 29
distances of planets from Sun, 646
light-year, 29
planetary orbits, 520
 Earth, 523
 Jupiter, 523
 Mars, 523
 Neptune, 539
 Pluto, 523, 539

Aviation
orbital launches, 553

Biology
alcohol and driving, 445, 450

bacterial growth, 477–78, 490–91
 E-coli, 233, 273
blood types, 684–85
bone length, 315–16
cricket chirp rate and temperature, 309
healing of wounds, 435, 449
maternal age versus Down syndrome, 288
yeast biomass as function of time, 489–90

Business
advertising, 287, 316
automobile production, 407–8, 570
blending coffee, 141
candy bar size, 103
car rentals, 279
checkout lines, 703
cigarette exports, 491
clothing store, 705
commissions, 315
cookie orders, 632
copying machines, 147
cost
 of can, 364, 367
 of charter bus, 147
 of commodity, 408
 of manufacturing, 29, 141, 340, 373,
 620–21
 marginal, 299, 315
 minimizing, 315, 627
 of production, 233, 408, 596, 633
 of transporting goods, 243
cost equation, 180, 192
cost function, 280
 average, 217
Dell personal computer price and
 demand, 491–92
demand
 for candy, 192
 for jeans, 287
 for PCs, 491
demand equation, 315, 398
depreciation, 400
discount pricing, 91, 92, 408
drive-thru rate
 at Burger King, 431
 at Citibank, 435, 449
 at McDonald's, 435
equipment depreciation, 662
expense computation, 142
Jiffy Lube's car arrival rate, 435, 449
managing a meat market, 627
mixing candy, 141
mixing nuts, 141
orange juice production, 570
precision ball bearings, 29
price markup, 91
 of new car, 129
price *vs.* quantity demanded, 281
product design, 628

production scheduling, 627
product promotion, 181
profit, 596
 cigar company, 256
 on figurines, 634
 maximizing, 625–26, 627–28
profit function, 213
rate of return on, 474
restaurant management, 554
revenue, 141, 299, 312–13
 advertising and, 287
 airline, 628
 of clothing store, 585
 daily, 299
 from digital music, 256
 maximizing, 299, 306
 monthly, 299
 theater, 555
revenue equation, 192
revenue function, 196
RV rental, 317
salary, 652
 gross, 212
 increases in, 662, 676
sales
 commission on, 128–29
 of movie theater ticket, 541, 546, 553
 net, 156
salvage value, 498
straight-line depreciation, 276–77, 280
supply and demand, 277–78, 280
tax, 373
theater attendance, 92
toy truck manufacturing, 620–21
transporting goods, 621
truck rentals, 180, 281
unemployment, 706
wages
 of car salesperson, 180
 hourly, 89, 91

Calculus
area under a curve, 257
area under graph, 232
Simpson's rule, 307

Carpentry
pitch, 181

Chemistry, 91
alpha particles, 536
decomposition reactions, 485
drug concentration, 366
gas laws, 193
mixing acids, 146
pH, 448
purity of gold, 142
radioactive decay, 484, 491, 498
radioactivity from Chernobyl, 485

xxv

Photo Credits

COLLEGE ALGEBRA

NINTH EDITION

Review

Outline

A Look Ahead ▷

Chapter R, as the title states, contains review material. Your instructor may choose to cover all or part of it as a regular chapter at the beginning of your course or later as a just-in-time review when the content is required. Regardless, when information in this chapter is needed, a specific reference to this chapter will be made so you can review.

R.1 Real Numbers

PREPARING FOR THIS BOOK *Before getting started, read "To the Student" on Page ii at the front of this book.*

OBJECTIVES 1 Work with Sets (p. 2)
2 Classify Numbers (p. 4)
3 Evaluate Numerical Expressions (p. 8)
4 Work with Properties of Real Numbers (p. 9)

1 Work with Sets

A **set** is a well-defined collection of distinct objects. The objects of a set are called its **elements.** By **well-defined,** we mean that there is a rule that enables us to determine whether a given object is an element of the set. If a set has no elements, it is called the **empty set,** or **null set,** and is denoted by the symbol \varnothing.

For example, the set of **digits** consists of the collection of numbers 0, 1, 2, 3, 4, 5, 6, 7, 8, and 9. If we use the symbol D to denote the set of digits, then we can write

$$D = \{0, 1, 2, 3, 4, 5, 6, 7, 8, 9\}$$

In this notation, the braces $\{\ \}$ are used to enclose the objects, or **elements,** in the set. This method of denoting a set is called the **roster method.** A second way to denote a set is to use **set-builder notation,** where the set D of digits is written as

$$D = \{\quad x \quad | \quad x \text{ is a digit}\}$$

Read as "*D* is the set of all *x* such that *x* is a digit."

EXAMPLE 1 **Using Set-builder Notation and the Roster Method**

(a) $E = \{x | x \text{ is an even digit}\} = \{0, 2, 4, 6, 8\}$
(b) $O = \{x | x \text{ is an odd digit}\} = \{1, 3, 5, 7, 9\}$

Because the elements of a set are distinct, we never repeat elements. For example, we would never write $\{1, 2, 3, 2\}$; the correct listing is $\{1, 2, 3\}$. Because a set is a collection, the order in which the elements are listed is immaterial. $\{1, 2, 3\}$, $\{1, 3, 2\}$, $\{2, 1, 3\}$, and so on, all represent the same set.

If every element of a set A is also an element of a set B, then we say that A is a **subset** of B and write $A \subseteq B$. If two sets A and B have the same elements, then we say that A **equals** B and write $A = B$.

For example, $\{1, 2, 3\} \subseteq \{1, 2, 3, 4, 5\}$ and $\{1, 2, 3\} = \{2, 3, 1\}$.

DEFINITION
If A and B are sets, the **intersection** of A with B, denoted $A \cap B$, is the set consisting of elements that belong to both A and B. The **union** of A with B, denoted $A \cup B$, is the set consisting of elements that belong to either A or B, or both.

EXAMPLE 2 **Finding the Intersection and Union of Sets**

Let $A = \{1, 3, 5, 8\}$, $B = \{3, 5, 7\}$, and $C = \{2, 4, 6, 8\}$. Find:

(a) $A \cap B$ (b) $A \cup B$ (c) $B \cap (A \cup C)$

Solution (a) $A \cap B = \{1, 3, 5, 8\} \cap \{3, 5, 7\} = \{3, 5\}$
(b) $A \cup B = \{1, 3, 5, 8\} \cup \{3, 5, 7\} = \{1, 3, 5, 7, 8\}$
(c) $B \cap (A \cup C) = \{3, 5, 7\} \cap [\{1, 3, 5, 8\} \cup \{2, 4, 6, 8\}]$
$= \{3, 5, 7\} \cap \{1, 2, 3, 4, 5, 6, 8\} = \{3, 5\}$

Now Work PROBLEM 13

Usually, in working with sets, we designate a **universal set** U, the set consisting of all the elements that we wish to consider. Once a universal set has been designated, we can consider elements of the universal set not found in a given set.

DEFINITION If A is a set, the **complement** of A, denoted \overline{A}, is the set consisting of all the elements in the universal set that are not in A.*

EXAMPLE 3 **Finding the Complement of a Set**

If the universal set is $U = \{1, 2, 3, 4, 5, 6, 7, 8, 9\}$ and if $A = \{1, 3, 5, 7, 9\}$, then $\overline{A} = \{2, 4, 6, 8\}$.

It follows from the definition of complement that $A \cup \overline{A} = U$ and $A \cap \overline{A} = \varnothing$. Do you see why?

Now Work PROBLEM 17

Figure 1

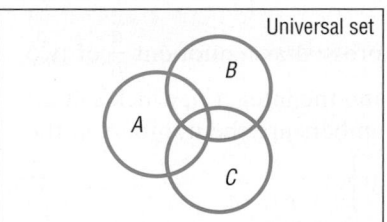

It is often helpful to draw pictures of sets. Such pictures, called **Venn diagrams,** represent sets as circles enclosed in a rectangle, which represents the universal set. Such diagrams often help us to visualize various relationships among sets. See Figure 1.

If we know that $A \subseteq B$, we might use the Venn diagram in Figure 2(a). If we know that A and B have no elements in common, that is, if $A \cap B = \varnothing$, we might use the Venn diagram in Figure 2(b). The sets A and B in Figure 2(b) are said to be **disjoint.**

Figure 2

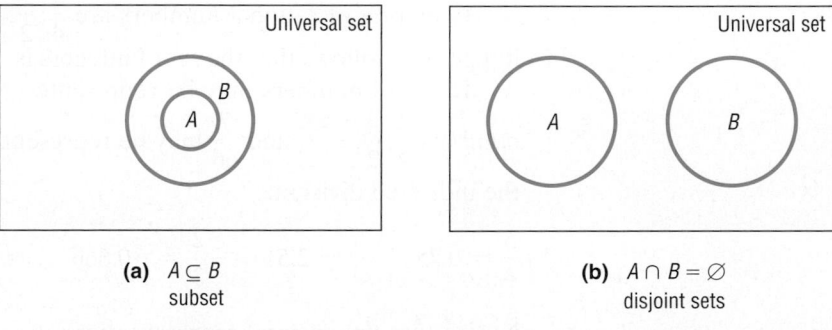

(a) $A \subseteq B$
subset

(b) $A \cap B = \varnothing$
disjoint sets

Figures 3(a), 3(b), and 3(c) use Venn diagrams to illustrate the definitions of intersection, union, and complement, respectively.

Figure 3

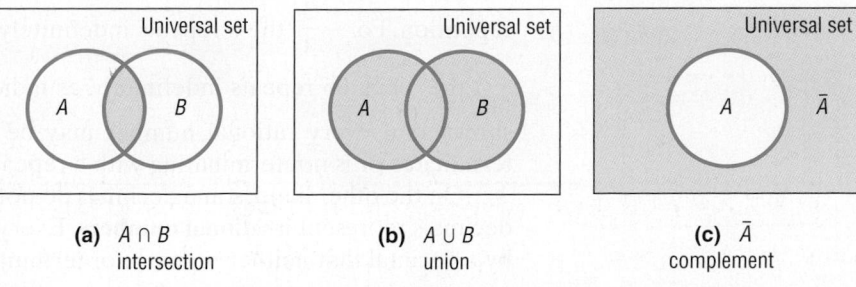

(a) $A \cap B$
intersection

(b) $A \cup B$
union

(c) \overline{A}
complement

*Some books use the notation A' for the complement of A.

2 Classify Numbers

It is helpful to classify the various kinds of numbers that we deal with as sets. The **counting numbers,** or **natural numbers,** are the numbers in the set $\{1, 2, 3, 4 \dots\}$. (The three dots, called an **ellipsis,** indicate that the pattern continues indefinitely.) As their name implies, these numbers are often used to count things. For example, there are 26 letters in our alphabet; there are 100 cents in a dollar. The **whole numbers** are the numbers in the set $\{0, 1, 2, 3, \dots\}$, that is, the counting numbers together with 0. The set of counting numbers is a subset of the set of whole numbers.

DEFINITION The **integers** are the set of numbers $\{\dots, -3, -2, -1, 0, 1, 2, 3, \dots\}$.

These numbers are useful in many situations. For example, if your checking account has $10 in it and you write a check for $15, you can represent the current balance as $-\$5$.

Each time we expand a number system, such as from the whole numbers to the integers, we do so in order to be able to handle new, and usually more complicated, problems. The integers allow us to solve problems requiring both positive and negative counting numbers, such as profit/loss, height above/below sea level, temperature above/below 0°F, and so on.

But integers alone are not sufficient for *all* problems. For example, they do not answer the question "What part of a dollar is 38 cents?" To answer such a question, we enlarge our number system to include *rational numbers.* For example, $\dfrac{38}{100}$ answers the question "What part of a dollar is 38 cents?"

DEFINITION A **rational number** is a number that can be expressed as a quotient $\dfrac{a}{b}$ of two integers. The integer a is called the **numerator,** and the integer b, which cannot be 0, is called the **denominator.** The rational numbers are the numbers in the set $\left\{ x \mid x = \dfrac{a}{b}, \text{ where } a, b \text{ are integers and } b \neq 0 \right\}$.

Examples of rational numbers are $\dfrac{3}{4}, \dfrac{5}{2}, \dfrac{0}{4}, -\dfrac{2}{3}$, and $\dfrac{100}{3}$. Since $\dfrac{a}{1} = a$ for any integer a, it follows that the set of integers is a subset of the set of rational numbers.

Rational numbers may be represented as **decimals.** For example, the rational numbers $\dfrac{3}{4}, \dfrac{5}{2}, -\dfrac{2}{3}$, and $\dfrac{7}{66}$ may be represented as decimals by merely carrying out the indicated division:

$$\frac{3}{4} = 0.75 \qquad \frac{5}{2} = 2.5 \qquad -\frac{2}{3} = -0.666\dots = -0.\overline{6} \qquad \frac{7}{66} = 0.1060606\dots = 0.1\overline{06}$$

Notice that the decimal representations of $\dfrac{3}{4}$ and $\dfrac{5}{2}$ terminate, or end. The decimal representations of $-\dfrac{2}{3}$ and $\dfrac{7}{66}$ do not terminate, but they do exhibit a pattern of repetition. For $-\dfrac{2}{3}$, the 6 repeats indefinitely, as indicated by the bar over the 6; for $\dfrac{7}{66}$, the block 06 repeats indefinitely, as indicated by the bar over the 06. It can be shown that every rational number may be represented by a decimal that either terminates or is nonterminating with a repeating block of digits, and vice versa.

On the other hand, some decimals do not fit into either of these categories. Such decimals represent **irrational numbers.** Every irrational number may be represented by a decimal that neither repeats nor terminates. In other words, irrational numbers cannot be written in the form $\dfrac{a}{b}$, where a, b are integers and $b \neq 0$.

Irrational numbers occur naturally. For example, consider the isosceles right triangle whose legs are each of length 1. See Figure 4. The length of the hypotenuse is $\sqrt{2}$, an irrational number.

Also, the number that equals the ratio of the circumference C to the diameter d of any circle, denoted by the symbol π (the Greek letter pi), is an irrational number. See Figure 5.

Figure 4

Figure 5 $\pi = \dfrac{C}{d}$

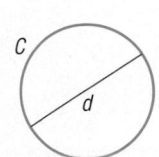

DEFINITION The set of **real numbers** is the union of the set of rational numbers with the set of irrational numbers.

Figure 6 shows the relationship of various types of numbers.*

Figure 6

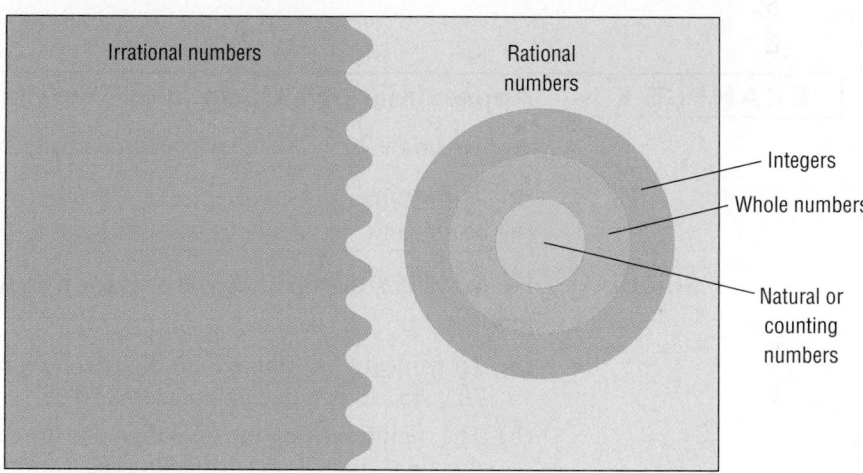

Real numbers

EXAMPLE 4 **Classifying the Numbers in a Set**

List the numbers in the set

$$\left\{ -3, \frac{4}{3}, 0.12, \sqrt{2}, \pi, 10, 2.151515\ldots \text{(where the block 15 repeats)} \right\}$$

that are

(a) Natural numbers (b) Integers (c) Rational numbers
(d) Irrational numbers (e) Real numbers

Solution (a) 10 is the only natural number.

(b) -3 and 10 are integers.

(c) -3, 10, $\dfrac{4}{3}$, 0.12, and 2.151515… are rational numbers.

(d) $\sqrt{2}$ and π are irrational numbers.

(e) All the numbers listed are real numbers.

 Now Work PROBLEM 23

* The set of real numbers is a subset of the set of complex numbers. We discuss complex numbers in Chapter 1, Section 1.3.

Approximations

Every decimal may be represented by a real number (either rational or irrational), and every real number may be represented by a decimal.

In practice, the decimal representation of an irrational number is given as an approximation. For example, using the symbol ≈ (read as "approximately equal to"), we can write

$$\sqrt{2} \approx 1.4142 \qquad \pi \approx 3.1416$$

In approximating decimals, we either *round off* or *truncate* to a given number of decimal places.* The number of places establishes the location of the *final digit* in the decimal approximation.

> **Truncation:** Drop all the digits that follow the specified final digit in the decimal.
>
> **Rounding:** Identify the specified final digit in the decimal. If the next digit is 5 or more, add 1 to the final digit; if the next digit is 4 or less, leave the final digit as it is. Then truncate following the final digit.

| EXAMPLE 5 | **Approximating a Decimal to Two Places** |

Approximate 20.98752 to two decimal places by

(a) Truncating
(b) Rounding

Solution For 20.98752, the final digit is 8, since it is two decimal places from the decimal point.

(a) To truncate, we remove all digits following the final digit 8. The truncation of 20.98752 to two decimal places is 20.98.

(b) The digit following the final digit 8 is the digit 7. Since 7 is 5 or more, we add 1 to the final digit 8 and truncate. The rounded form of 20.98752 to two decimal places is 20.99. ↵

| EXAMPLE 6 | **Approximating a Decimal to Two and Four Places** |

Number	Rounded to Two Decimal Places	Rounded to Four Decimal Places	Truncated to Two Decimal Places	Truncated to Four Decimal Places
(a) 3.14159	3.14	3.1416	3.14	3.1415
(b) 0.056128	0.06	0.0561	0.05	0.0561
(c) 893.46125	893.46	893.4613	893.46	893.4612 ↵

══════ Now Work PROBLEM 27

Calculators

Calculators are finite machines. As a result, they are incapable of displaying decimals that contain a large number of digits. For example, some calculators are capable of displaying only eight digits. When a number requires more than eight digits,

* Sometimes we say "correct to a given number of decimal places" instead of "truncate."

the calculator either truncates or rounds. To see how your calculator handles decimals, divide 2 by 3. How many digits do you see? Is the last digit a 6 or a 7? If it is a 6, your calculator truncates; if it is a 7, your calculator rounds.

There are different kinds of calculators. An **arithmetic** calculator can only add, subtract, multiply, and divide numbers; therefore, this type is not adequate for this course. **Scientific** calculators have all the capabilities of arithmetic calculators and also contain **function keys** labeled ln, log, sin, cos, tan, x^y, inv, and so on. As you proceed through this text, you will discover how to use many of the function keys. **Graphing** calculators have all the capabilities of scientific calculators and contain a screen on which graphs can be displayed.

For those who have access to a graphing calculator, we have included comments, examples, and exercises marked with a ▦, indicating that a graphing calculator is required. We have also included an appendix that explains some of the capabilities of a graphing calculator. The ▦ comments, examples, and exercises may be omitted without loss of continuity, if so desired.

Operations

In algebra, we use letters such as x, y, a, b, and c to represent numbers. The symbols used in algebra for the operations of addition, subtraction, multiplication, and division are $+$, $-$, \cdot, and $/$. The words used to describe the results of these operations are **sum, difference, product,** and **quotient.** Table 1 summarizes these ideas.

Table 1

Operation	Symbol	Words
Addition	$a + b$	Sum: a plus b
Subtraction	$a - b$	Difference: a minus b
Multiplication	$a \cdot b, (a) \cdot b, a \cdot (b), (a) \cdot (b),$ $ab, (a)b, a(b), (a)(b)$	Product: a times b
Division	a/b or $\dfrac{a}{b}$	Quotient: a divided by b

In algebra, we generally avoid using the multiplication sign \times and the division sign \div so familiar in arithmetic. Notice also that when two expressions are placed next to each other without an operation symbol, as in ab, or in parentheses, as in $(a)(b)$, it is understood that the expressions, called **factors,** are to be multiplied.

We also prefer not to use mixed numbers in algebra. When mixed numbers are used, addition is understood; for example, $2\dfrac{3}{4}$ means $2 + \dfrac{3}{4}$. In algebra, use of a mixed number may be confusing because the absence of an operation symbol between two terms is generally taken to mean multiplication. The expression $2\dfrac{3}{4}$ is therefore written instead as 2.75 or as $\dfrac{11}{4}$.

The symbol $=$, called an **equal sign** and read as "equals" or "is," is used to express the idea that the number or expression on the left of the equal sign is equivalent to the number or expression on the right.

EXAMPLE 7 **Writing Statements Using Symbols**

(a) The sum of 2 and 7 equals 9. In symbols, this statement is written as $2 + 7 = 9$.

(b) The product of 3 and 5 is 15. In symbols, this statement is written as $3 \cdot 5 = 15$.

Now Work PROBLEM 39

3 Evaluate Numerical Expressions

Consider the expression $2 + 3 \cdot 6$. It is not clear whether we should add 2 and 3 to get 5, and then multiply by 6 to get 30; or first multiply 3 and 6 to get 18, and then add 2 to get 20. To avoid this ambiguity, we have the following agreement.

In Words
Multiply first, then add.

We agree that whenever the two operations of addition and multiplication separate three numbers, the multiplication operation will always be performed first, followed by the addition operation.

For $2 + 3 \cdot 6$, we have

$$2 + 3 \cdot 6 = 2 + 18 = 20$$

EXAMPLE 8 **Finding the Value of an Expression**

Evaluate each expression.

(a) $3 + 4 \cdot 5$ (b) $8 \cdot 2 + 1$ (c) $2 + 2 \cdot 2$

Solution (a) $3 + 4 \cdot 5 = 3 + 20 = 23$ (b) $8 \cdot 2 + 1 = 16 + 1 = 17$
$\qquad\qquad\quad \uparrow$ \uparrow
\qquad Multiply first Multiply first

(c) $2 + 2 \cdot 2 = 2 + 4 = 6$

Now Work PROBLEM 51

To first add 3 and 4 and then multiply the result by 5, we use parentheses and write $(3 + 4) \cdot 5$. Whenever parentheses appear in an expression, it means "perform the operations within the parentheses first!"

EXAMPLE 9 **Finding the Value of an Expression**

(a) $(5 + 3) \cdot 4 = 8 \cdot 4 = 32$
(b) $(4 + 5) \cdot (8 - 2) = 9 \cdot 6 = 54$

When we divide two expressions, as in

$$\frac{2 + 3}{4 + 8}$$

it is understood that the division bar acts like parentheses; that is,

$$\frac{2 + 3}{4 + 8} = \frac{(2 + 3)}{(4 + 8)}$$

Rules for the Order of Operations

1. Begin with the innermost parentheses and work outward. Remember that in dividing two expressions the numerator and denominator are treated as if they were enclosed in parentheses.
2. Perform multiplications and divisions, working from left to right.
3. Perform additions and subtractions, working from left to right.

EXAMPLE 10 **Finding the Value of an Expression**

Evaluate each expression.

(a) $8 \cdot 2 + 3$ (b) $5 \cdot (3 + 4) + 2$

(c) $\dfrac{2 + 5}{2 + 4 \cdot 7}$ (d) $2 + [4 + 2 \cdot (10 + 6)]$

Solution

(a) $8 \cdot 2 + 3 = 16 + 3 = 19$

 ↑
 Multiply first

(b) $5 \cdot (3 + 4) + 2 = 5 \cdot 7 + 2 = 35 + 2 = 37$

 ↑ ↑
 Parentheses first Multiply before adding

(c) $\dfrac{2 + 5}{2 + 4 \cdot 7} = \dfrac{2 + 5}{2 + 28} = \dfrac{7}{30}$

(d) $2 + [4 + 2 \cdot (10 + 6)] = 2 + [4 + 2 \cdot (16)]$

 $= 2 + [4 + 32] = 2 + [36] = 38$

Figure 7

Be careful if you use a calculator. For Example 10(c), you need to use parentheses. See Figure 7.* If you don't, the calculator will compute the expression

$$2 + \frac{5}{2} + 4 \cdot 7 = 2 + 2.5 + 28 = 32.5$$

giving a wrong answer.

Now Work PROBLEMS **57** AND **65**

4 Work with Properties of Real Numbers

The equal sign is used to mean that one expression is equivalent to another. Four important properties of equality are listed next. In this list, a, b, and c represent real numbers.

1. The **reflexive property** states that a number always equals itself; that is, $a = a$.
2. The **symmetric property** states that if $a = b$ then $b = a$.
3. The **transitive property** states that if $a = b$ and $b = c$ then $a = c$.
4. The **principle of substitution** states that if $a = b$ then we may substitute b for a in any expression containing a.

Now, let's consider some other properties of real numbers.

EXAMPLE 11 **Commutative Properties**

(a) $3 + 5 = 8$ (b) $2 \cdot 3 = 6$

 $5 + 3 = 8$ $3 \cdot 2 = 6$

 $3 + 5 = 5 + 3$ $2 \cdot 3 = 3 \cdot 2$

This example illustrates the **commutative property** of real numbers, which states that the order in which addition or multiplication takes place will not affect the final result.

* Notice that we converted the decimal to its fraction form. Consult your manual to see how your calculator does this.

Commutative Properties

$$a + b = b + a \tag{1a}$$
$$a \cdot b = b \cdot a \tag{1b}$$

Here, and in the properties listed next and on pages 11–13, a, b, and c represent real numbers.

EXAMPLE 12 **Associative Properties**

(a) $2 + (3 + 4) = 2 + 7 = 9$ (b) $2 \cdot (3 \cdot 4) = 2 \cdot 12 = 24$
$\quad (2 + 3) + 4 = 5 + 4 = 9$ $\quad (2 \cdot 3) \cdot 4 = 6 \cdot 4 = 24$
$\quad 2 + (3 + 4) = (2 + 3) + 4$ $\quad 2 \cdot (3 \cdot 4) = (2 \cdot 3) \cdot 4$

The way we add or multiply three real numbers will not affect the final result. Expressions such as $2 + 3 + 4$ and $3 \cdot 4 \cdot 5$ present no ambiguity, even though addition and multiplication are performed on one pair of numbers at a time. This property is called the **associative property.**

Associative Properties

$$a + (b + c) = (a + b) + c = a + b + c \tag{2a}$$
$$a \cdot (b \cdot c) = (a \cdot b) \cdot c = a \cdot b \cdot c \tag{2b}$$

Distributive Property

$$a \cdot (b + c) = a \cdot b + a \cdot c \tag{3a}$$
$$(a + b) \cdot c = a \cdot c + b \cdot c \tag{3b}$$

The **distributive property** may be used in two different ways.

EXAMPLE 13 **Distributive Property**

(a) $2 \cdot (x + 3) = 2 \cdot x + 2 \cdot 3 = 2x + 6$ Use to remove parentheses.
(b) $3x + 5x = (3 + 5)x = 8x$ Use to combine two expressions.
(c) $(x + 2)(x + 3) = x(x + 3) + 2(x + 3) = (x^2 + 3x) + (2x + 6)$
$\qquad = x^2 + (3x + 2x) + 6 = x^2 + 5x + 6$

Now Work PROBLEM 87

The real numbers 0 and 1 have unique properties, called the *identity properties.*

EXAMPLE 14 **Identity Properties**

(a) $4 + 0 = 0 + 4 = 4$ (b) $3 \cdot 1 = 1 \cdot 3 = 3$

Identity Properties

$$0 + a = a + 0 = a \qquad \text{(4a)}$$

$$a \cdot 1 = 1 \cdot a = a \qquad \text{(4b)}$$

We call 0 the **additive identity** and 1 the **multiplicative identity.**

For each real number a, there is a real number $-a$, called the **additive inverse** of a, having the following property:

Additive Inverse Property

$$a + (-a) = -a + a = 0 \qquad \text{(5a)}$$

EXAMPLE 15 **Finding an Additive Inverse**

(a) The additive inverse of 6 is -6, because $6 + (-6) = 0$.

(b) The additive inverse of -8 is $-(-8) = 8$, because $-8 + 8 = 0$.

The additive inverse of a, that is, $-a$, is often called the *negative* of a or the *opposite* of a. The use of such terms can be dangerous, because they suggest that the additive inverse is a negative number, which may not be the case. For example, the additive inverse of -3, or $-(-3)$, equals 3, a positive number.

For each *nonzero* real number a, there is a real number $\dfrac{1}{a}$, called the **multiplicative inverse** of a, having the following property:

Multiplicative Inverse Property

$$a \cdot \frac{1}{a} = \frac{1}{a} \cdot a = 1 \qquad \text{if } a \neq 0 \qquad \text{(5b)}$$

The multiplicative inverse $\dfrac{1}{a}$ of a nonzero real number a is also referred to as the **reciprocal** of a.

EXAMPLE 16 **Finding a Reciprocal**

(a) The reciprocal of 6 is $\dfrac{1}{6}$, because $6 \cdot \dfrac{1}{6} = 1$.

(b) The reciprocal of -3 is $\dfrac{1}{-3}$, because $-3 \cdot \dfrac{1}{-3} = 1$.

(c) The reciprocal of $\dfrac{2}{3}$ is $\dfrac{3}{2}$, because $\dfrac{2}{3} \cdot \dfrac{3}{2} = 1$.

With these properties for adding and multiplying real numbers, we can define the operations of subtraction and division as follows:

DEFINITION The **difference** $a - b$, also read "a less b" or "a minus b," is defined as

$$a - b = a + (-b) \qquad \text{(6)}$$

To subtract b from a, add the opposite of b to a.

DEFINITION If b is a nonzero real number, the **quotient** $\dfrac{a}{b}$, also read as "a divided by b" or "the ratio of a to b," is defined as

$$\frac{a}{b} = a \cdot \frac{1}{b} \qquad \text{if } b \neq 0 \tag{7}$$

EXAMPLE 17 **Working with Differences and Quotients**

(a) $8 - 5 = 8 + (-5) = 3$

(b) $4 - 9 = 4 + (-9) = -5$

(c) $\dfrac{5}{8} = 5 \cdot \dfrac{1}{8}$

For any number a, the product of a times 0 is always 0; that is,

> **In Words**
> The result of multiplying by zero is zero.

Multiplication by Zero

$$a \cdot 0 = 0 \tag{8}$$

For a nonzero number a,

Division Properties

$$\frac{0}{a} = 0 \qquad \frac{a}{a} = 1 \quad \text{if } a \neq 0 \tag{9}$$

NOTE Division by 0 is not defined. One reason is to avoid the following difficulty: $\dfrac{2}{0} = x$ means to find x such that $0 \cdot x = 2$. But $0 \cdot x$ equals 0 for all x, so there is no unique number x such that $\dfrac{2}{0} = x$. ∎

Rules of Signs

$$a(-b) = -(ab) \qquad (-a)b = -(ab) \qquad (-a)(-b) = ab$$
$$-(-a) = a \qquad \frac{a}{-b} = \frac{-a}{b} = -\frac{a}{b} \qquad \frac{-a}{-b} = \frac{a}{b} \tag{10}$$

EXAMPLE 18 **Applying the Rules of Signs**

(a) $2(-3) = -(2 \cdot 3) = -6$ (b) $(-3)(-5) = 3 \cdot 5 = 15$

(c) $\dfrac{3}{-2} = \dfrac{-3}{2} = -\dfrac{3}{2}$ (d) $\dfrac{-4}{-9} = \dfrac{4}{9}$

(e) $\dfrac{x}{-2} = \dfrac{1}{-2} \cdot x = -\dfrac{1}{2}x$

Cancellation Properties

$$ac = bc \quad \text{implies} \quad a = b \quad \text{if } c \neq 0$$

$$\frac{ac}{bc} = \frac{a}{b} \qquad\qquad\qquad \text{if } b \neq 0, c \neq 0 \qquad \textbf{(11)}$$

EXAMPLE 19 **Using the Cancellation Properties**

(a) If $2x = 6$, then

$$2x = 6$$
$$2x = 2 \cdot 3 \qquad \text{Factor 6.}$$
$$x = 3 \qquad \text{Cancel the 2's.}$$

NOTE We follow the common practice of using slash marks to indicate cancellations. ∎

(b) $\dfrac{18}{12} = \dfrac{3 \cdot \cancel{6}}{2 \cdot \cancel{6}} = \dfrac{3}{2}$

\qquad Cancel the 6's.

In Words
If a product equals 0, then one or both of the factors is 0.

Zero-Product Property

$$\text{If } ab = 0, \text{ then } a = 0, \text{ or } b = 0, \text{ or both.} \qquad \textbf{(12)}$$

EXAMPLE 20 **Using the Zero-Product Property**

If $2x = 0$, then either $2 = 0$ or $x = 0$. Since $2 \neq 0$, it follows that $x = 0$.

Arithmetic of Quotients

$$\frac{a}{b} + \frac{c}{d} = \frac{ad}{bd} + \frac{bc}{bd} = \frac{ad + bc}{bd} \qquad \text{if } b \neq 0, d \neq 0 \qquad \textbf{(13)}$$

$$\frac{a}{b} \cdot \frac{c}{d} = \frac{ac}{bd} \qquad\qquad\qquad\qquad \text{if } b \neq 0, d \neq 0 \qquad \textbf{(14)}$$

$$\frac{\dfrac{a}{b}}{\dfrac{c}{d}} = \frac{a}{b} \cdot \frac{d}{c} = \frac{ad}{bc} \qquad\qquad \text{if } b \neq 0, c \neq 0, d \neq 0 \qquad \textbf{(15)}$$

EXAMPLE 21 **Adding, Subtracting, Multiplying, and Dividing Quotients**

(a) $\dfrac{2}{3} + \dfrac{5}{2} = \dfrac{2 \cdot 2}{3 \cdot 2} + \dfrac{3 \cdot 5}{3 \cdot 2} = \dfrac{2 \cdot 2 + 3 \cdot 5}{3 \cdot 2} = \dfrac{4 + 15}{6} = \dfrac{19}{6}$

$\qquad\quad$ ↑ By equation (13)

(b) $\dfrac{3}{5} - \dfrac{2}{3} = \dfrac{3}{5} + \left(-\dfrac{2}{3} \right) = \dfrac{3}{5} + \dfrac{-2}{3}$

\quad ↑ $\qquad\qquad\quad$ ↑

\quad By equation (6) By equation (10)

$$= \frac{3 \cdot 3 + 5 \cdot (-2)}{5 \cdot 3} = \frac{9 + (-10)}{15} = \frac{-1}{15} = -\frac{1}{15}$$

↑ By equation (13)

NOTE Slanting the cancellation marks in different directions for different factors, as shown here, is a good practice to follow, since it will help in checking for errors. ∎

(c) $\dfrac{8}{3} \cdot \dfrac{15}{4} = \dfrac{8 \cdot 15}{3 \cdot 4} = \dfrac{2 \cdot 4 \cdot 3 \cdot 5}{3 \cdot 4 \cdot 1} = \dfrac{2 \cdot 5}{1} = 10$

↑ By equation (14) ↑ By equation (11)

(d) $\dfrac{\frac{3}{5}}{\frac{7}{9}} = \dfrac{3}{5} \cdot \dfrac{9}{7} = \dfrac{3 \cdot 9}{5 \cdot 7} = \dfrac{27}{35}$

↑ By equation (14)

↑ By equation (15)

NOTE In writing quotients, we shall follow the usual convention and write the quotient in lowest terms. That is, we write it so that any common factors of the numerator and the denominator have been removed using the cancellation properties, equation (11). As examples,

$$\dfrac{90}{24} = \dfrac{15 \cdot 6}{4 \cdot 6} = \dfrac{15}{4}$$

$$\dfrac{24x^2}{18x} = \dfrac{4 \cdot 6 \cdot x \cdot x}{3 \cdot 6 \cdot x} = \dfrac{4x}{3} \qquad x \neq 0$$ ∎

▬▬▬ **Now Work** PROBLEMS 67, 71, AND 81

Sometimes it is easier to add two fractions using *least common multiples* (LCM). The LCM of two numbers is the smallest number that each has as a common multiple.

EXAMPLE 22 **Finding the Least Common Multiple of Two Numbers**

Find the least common multiple of 15 and 12.

Solution To find the LCM of 15 and 12, we look at multiples of 15 and 12.

$$15, \ 30, \ 45, \ 60, \ 75, \ 90, \ 105, \ 120, \dots$$
$$12, \ 24, \ 36, \ 48, \ 60, \ 72, \ 84, \ 96, \ 108, \ 120, \dots$$

The *common* multiples are in blue. The *least* common multiple is 60.

EXAMPLE 23 **Using the Least Common Multiple to Add Two Fractions**

Find: $\dfrac{8}{15} + \dfrac{5}{12}$

Solution We use the LCM of the denominators of the fractions and rewrite each fraction using the LCM as a common denominator. The LCM of the denominators (12 and 15) is 60. Rewrite each fraction using 60 as the denominator.

$$\dfrac{8}{15} + \dfrac{5}{12} = \dfrac{8}{15} \cdot \dfrac{4}{4} + \dfrac{5}{12} \cdot \dfrac{5}{5}$$
$$= \dfrac{32}{60} + \dfrac{25}{60}$$
$$= \dfrac{32 + 25}{60}$$
$$= \dfrac{57}{60}$$
$$= \dfrac{19}{20}$$

▬▬▬ **Now Work** PROBLEM 75

Historical Feature

The real number system has a history that stretches back at least to the ancient Babylonians (1800 BC). It is remarkable how much the ancient Babylonian attitudes resemble our own. As we stated in the text, the fundamental difficulty with irrational numbers is that they cannot be written as quotients of integers or, equivalently, as repeating or terminating decimals. The Babylonians wrote their numbers in a system based on 60 in the same way that we write ours based on 10. They would carry as many places for π as the accuracy of the problem demanded, just as we now use

$$\pi \approx 3\frac{1}{7} \quad \text{or} \quad \pi \approx 3.1416 \quad \text{or} \quad \pi \approx 3.14159$$

$$\text{or} \quad \pi \approx 3.14159265358979$$

depending on how accurate we need to be.

Things were very different for the Greeks, whose number system allowed only rational numbers. When it was discovered that $\sqrt{2}$ was not a rational number, this was regarded as a fundamental flaw in the number concept. So serious was the matter that the Pythagorean Brotherhood (an early mathematical society) is said to have drowned one of its members for revealing this terrible secret.

Greek mathematicians then turned away from the number concept, expressing facts about whole numbers in terms of line segments.

In astronomy, however, Babylonian methods, including the Babylonian number system, continued to be used. Simon Stevin (1548–1620), probably using the Babylonian system as a model, invented the decimal system, complete with rules of calculation, in 1585. [Others, for example, al-Kashi of Samarkand (d. 1429), had made some progress in the same direction.] The decimal system so effectively conceals the difficulties that the need for more logical precision began to be felt only in the early 1800s. Around 1880, Georg Cantor (1845–1918) and Richard Dedekind (1831–1916) gave precise definitions of real numbers. Cantor's definition, although more abstract and precise, has its roots in the decimal (and hence Babylonian) numerical system.

Sets and set theory were a spin-off of the research that went into clarifying the foundations of the real number system. Set theory has developed into a large discipline of its own, and many mathematicians regard it as the foundation upon which modern mathematics is built. Cantor's discoveries that infinite sets can also be counted and that there are different sizes of infinite sets are among the most astounding results of modern mathematics.

R.1 Assess Your Understanding

Concepts and Vocabulary

1. The numbers in the set $\left\{ x \mid x = \dfrac{a}{b}, \text{where } a, b \text{ are integers and } b \neq 0 \right\}$, are called _____ numbers.

2. The value of the expression $4 + 5 \cdot 6 - 3$ is _____.

3. The fact that $2x + 3x = (2 + 3)x$ is a consequence of the _____ Property.

4. "The product of 5 and $x + 3$ equals 6" may be written as _____.

5. *True or False* Rational numbers have decimals that either terminate or are nonterminating with a repeating block of digits.

6. *True or False* The Zero-Product Property states that the product of any number and zero equals zero.

7. *True or False* The least common multiple of 12 and 18 is 6.

8. *True or False* No real number is both rational and irrational.

Skill Building

In Problems 9–20, use U = universal set = $\{0, 1, 2, 3, 4, 5, 6, 7, 8, 9\}$, A = $\{1, 3, 4, 5, 9\}$, B = $\{2, 4, 6, 7, 8\}$, and C = $\{1, 3, 4, 6\}$ to find each set.

9. $A \cup B$

10. $A \cup C$

11. $A \cap B$

12. $A \cap C$

13. $(A \cup B) \cap C$

14. $(A \cap B) \cup C$

15. \overline{A}

16. \overline{C}

17. $\overline{A \cap B}$

18. $\overline{B \cup C}$

19. $\overline{A} \cup \overline{B}$

20. $\overline{B} \cap \overline{C}$

In Problems 21–26, list the numbers in each set that are (a) Natural numbers, (b) Integers, (c) Rational numbers, (d) Irrational numbers, (e) Real numbers.

21. $A = \left\{ -6, \dfrac{1}{2}, -1.333\ldots \text{(the 3's repeat)}, \pi, 2, 5 \right\}$

22. $B = \left\{ -\dfrac{5}{3}, 2.060606\ldots \text{(the block 06 repeats)}, 1.25, 0, 1, \sqrt{5} \right\}$

23. $C = \left\{ 0, 1, \dfrac{1}{2}, \dfrac{1}{3}, \dfrac{1}{4} \right\}$

24. $D = \{ -1, -1.1, -1.2, -1.3 \}$

25. $E = \left\{ \sqrt{2}, \pi, \sqrt{2} + 1, \pi + \dfrac{1}{2} \right\}$

26. $F = \left\{ -\sqrt{2}, \pi + \sqrt{2}, \dfrac{1}{2} + 10.3 \right\}$

In Problems 27–38, approximate each number (a) rounded and (b) truncated to three decimal places.

27. 18.9526 **28.** 25.86134 **29.** 28.65319 **30.** 99.05249 **31.** 0.06291 **32.** 0.05388

33. 9.9985 **34.** 1.0006 **35.** $\dfrac{3}{7}$ **36.** $\dfrac{5}{9}$ **37.** $\dfrac{521}{15}$ **38.** $\dfrac{81}{5}$

In Problems 39–48, write each statement using symbols.

39. The sum of 3 and 2 equals 5.

40. The product of 5 and 2 equals 10.

41. The sum of x and 2 is the product of 3 and 4.

42. The sum of 3 and y is the sum of 2 and 2.

43. The product of 3 and y is the sum of 1 and 2.

44. The product of 2 and x is the product of 4 and 6.

45. The difference x less 2 equals 6.

46. The difference 2 less y equals 6.

47. The quotient x divided by 2 is 6.

48. The quotient 2 divided by x is 6.

In Problems 49–86, evaluate each expression.

49. $9 - 4 + 2$

50. $6 - 4 + 3$

51. $-6 + 4 \cdot 3$

52. $8 - 4 \cdot 2$

53. $4 + 5 - 8$

54. $8 - 3 - 4$

55. $4 + \dfrac{1}{3}$

56. $2 - \dfrac{1}{2}$

57. $6 - [3 \cdot 5 + 2 \cdot (3 - 2)]$

58. $2 \cdot [8 - 3(4 + 2)] - 3$

59. $2 \cdot (3 - 5) + 8 \cdot 2 - 1$

60. $1 - (4 \cdot 3 - 2 + 2)$

61. $10 - [6 - 2 \cdot 2 + (8 - 3)] \cdot 2$

62. $2 - 5 \cdot 4 - [6 \cdot (3 - 4)]$

63. $(5 - 3)\dfrac{1}{2}$

64. $(5 + 4)\dfrac{1}{3}$

65. $\dfrac{4 + 8}{5 - 3}$

66. $\dfrac{2 - 4}{5 - 3}$

67. $\dfrac{3}{5} \cdot \dfrac{10}{21}$

68. $\dfrac{5}{9} \cdot \dfrac{3}{10}$

69. $\dfrac{6}{25} \cdot \dfrac{10}{27}$

70. $\dfrac{21}{25} \cdot \dfrac{100}{3}$

71. $\dfrac{3}{4} + \dfrac{2}{5}$

72. $\dfrac{4}{3} + \dfrac{1}{2}$

73. $\dfrac{5}{6} + \dfrac{9}{5}$

74. $\dfrac{8}{9} + \dfrac{15}{2}$

75. $\dfrac{5}{18} + \dfrac{1}{12}$

76. $\dfrac{2}{15} + \dfrac{8}{9}$

77. $\dfrac{1}{30} - \dfrac{7}{18}$

78. $\dfrac{3}{14} - \dfrac{2}{21}$

79. $\dfrac{3}{20} - \dfrac{2}{15}$

80. $\dfrac{6}{35} - \dfrac{3}{14}$

81. $\dfrac{\frac{5}{18}}{\frac{11}{27}}$

82. $\dfrac{\frac{5}{21}}{\frac{2}{35}}$

83. $\dfrac{1}{2} \cdot \dfrac{3}{5} + \dfrac{7}{10}$

84. $\dfrac{2}{3} + \dfrac{4}{5} \cdot \dfrac{1}{6}$

85. $2 \cdot \dfrac{3}{4} + \dfrac{3}{8}$

86. $3 \cdot \dfrac{5}{6} - \dfrac{1}{2}$

In Problems 87–98, use the Distributive Property to remove the parentheses.

87. $6(x + 4)$

88. $4(2x - 1)$

89. $x(x - 4)$

90. $4x(x + 3)$

91. $2\left(\dfrac{3}{4}x - \dfrac{1}{2}\right)$

92. $3\left(\dfrac{2}{3}x + \dfrac{1}{6}\right)$

93. $(x + 2)(x + 4)$

94. $(x + 5)(x + 1)$

95. $(x - 2)(x + 1)$

96. $(x - 4)(x + 1)$

97. $(x - 8)(x - 2)$

98. $(x - 4)(x - 2)$

Explaining Concepts: Discussion and Writing

99. Explain to a friend how the Distributive Property is used to justify the fact that $2x + 3x = 5x$.

100. Explain to a friend why $2 + 3 \cdot 4 = 14$, whereas $(2 + 3) \cdot 4 = 20$.

101. Explain why $2(3 \cdot 4)$ is not equal to $(2 \cdot 3) \cdot (2 \cdot 4)$.

102. Explain why $\dfrac{4 + 3}{2 + 5}$ is not equal to $\dfrac{4}{2} + \dfrac{3}{5}$.

103. Is subtraction commutative? Support your conclusion with an example.

104. Is subtraction associative? Support your conclusion with an example.

105. Is division commutative? Support your conclusion with an example.

106. Is division associative? Support your conclusion with an example.

107. If $2 = x$, why does $x = 2$?

108. If $x = 5$, why does $x^2 + x = 30$?

109. Are there any real numbers that are both rational and irrational? Are there any real numbers that are neither? Explain your reasoning.

110. Explain why the sum of a rational number and an irrational number must be irrational.

111. A rational number is defined as the quotient of two integers. When written as a decimal, the decimal will either repeat or terminate. By looking at the denominator of the rational number, there is a way to tell in advance whether its decimal representation will repeat or terminate. Make a list of rational numbers and their decimals. See if you can discover the pattern. Confirm your conclusion by consulting books on number theory at the library. Write a brief essay on your findings.

112. The current time is 12 noon CST. What time (CST) will it be 12,997 hours from now?

113. Both $\dfrac{a}{0}(a \neq 0)$ and $\dfrac{0}{0}$ are undefined, but for different reasons. Write a paragraph or two explaining the different reasons.

R.2 Algebra Essentials

OBJECTIVES **1** Graph Inequalities (p. 18)
2 Find Distance on the Real Number Line (p. 19)
3 Evaluate Algebraic Expressions (p. 20)
4 Determine the Domain of a Variable (p. 21)
5 Use the Laws of Exponents (p. 21)
6 Evaluate Square Roots (p. 23)
7 Use a Calculator to Evaluate Exponents (p. 24)
8 Use Scientific Notation (p. 24)

The Real Number Line

Real numbers can be represented by points on a line called the **real number line.** There is a one-to-one correspondence between real numbers and points on a line. That is, every real number corresponds to a point on the line, and each point on the line has a unique real number associated with it.

Pick a point on the line somewhere in the center, and label it O. This point, called the **origin,** corresponds to the real number 0. See Figure 8. The point 1 unit to the right of O corresponds to the number 1. The distance between 0 and 1 determines the **scale** of the number line. For example, the point associated with the number 2 is twice as far from O as 1. Notice that an arrowhead on the right end of the line indicates the direction in which the numbers increase. Points to the left of the origin correspond to the real numbers -1, -2, and so on. Figure 8 also shows the points associated with the rational numbers $-\dfrac{1}{2}$ and $\dfrac{1}{2}$ and with the irrational numbers $\sqrt{2}$ and π.

Figure 8
Real number line

DEFINITION The real number associated with a point P is called the **coordinate** of P, and the line whose points have been assigned coordinates is called the **real number line.**

━━━▶ **Now Work** PROBLEM 11

Figure 9

Negative real numbers Zero Positive real numbers

The real number line consists of three classes of real numbers, as shown in Figure 9.

1. The **negative real numbers** are the coordinates of points to the left of the origin O.
2. The real number **zero** is the coordinate of the origin O.
3. The **positive real numbers** are the coordinates of points to the right of the origin O.

Multiplication Properties of Positive and Negative Numbers

1. The product of two positive numbers is a positive number.
2. The product of two negative numbers is a positive number.
3. The product of a positive number and a negative number is a negative number.

1 Graph Inequalities

Figure 10

(a) $a < b$

(b) $a = b$

(c) $a > b$

An important property of the real number line follows from the fact that, given two numbers (points) a and b, either a is to the left of b, or a is at the same location as b, or a is to the right of b. See Figure 10.

If a is to the left of b, we say that "a is less than b" and write $a < b$. If a is to the right of b, we say that "a is greater than b" and write $a > b$. If a is at the same location as b, then $a = b$. If a is either less than or equal to b, we write $a \leq b$. Similarly, $a \geq b$ means that a is either greater than or equal to b. Collectively, the symbols $<, >, \leq$, and \geq are called **inequality symbols.**

Note that $a < b$ and $b > a$ mean the same thing. It does not matter whether we write $2 < 3$ or $3 > 2$.

Furthermore, if $a < b$ or if $b > a$, then the difference $b - a$ is positive. Do you see why?

EXAMPLE 1 **Using Inequality Symbols**

(a) $3 < 7$ (b) $-8 > -16$ (c) $-6 < 0$

(d) $-8 < -4$ (e) $4 > -1$ (f) $8 > 0$

In Example 1(a), we conclude that $3 < 7$ either because 3 is to the left of 7 on the real number line or because the difference, $7 - 3 = 4$, is a positive real number.

Similarly, we conclude in Example 1(b) that $-8 > -16$ either because -8 lies to the right of -16 on the real number line or because the difference, $-8 - (-16) = -8 + 16 = 8$, is a positive real number.

Look again at Example 1. Note that the inequality symbol always points in the direction of the smaller number.

An **inequality** is a statement in which two expressions are related by an inequality symbol. The expressions are referred to as the **sides** of the inequality. Inequalities of the form $a < b$ or $b > a$ are called **strict inequalities,** whereas inequalities of the form $a \leq b$ or $b \geq a$ are called **nonstrict inequalities.**

Based on the discussion so far, we conclude that

$a > 0$ is equivalent to a is positive

$a < 0$ is equivalent to a is negative

We sometimes read $a > 0$ by saying that "a is positive." If $a \geq 0$, then either $a > 0$ or $a = 0$, and we may read this as "a is nonnegative."

Now Work PROBLEMS 15 AND 25

| EXAMPLE 2 | **Graphing Inequalities** |

(a) On the real number line, graph all numbers x for which $x > 4$.

(b) On the real number line, graph all numbers x for which $x \leq 5$.

Figure 11

Solution (a) See Figure 11. Notice that we use a left parenthesis to indicate that the number 4 is *not* part of the graph.

Figure 12

(b) See Figure 12. Notice that we use a right bracket to indicate that the number 5 *is* part of the graph.

Now Work PROBLEM 31

2 Find Distance on the Real Number Line

Figure 13

The *absolute value* of a number a is the distance from 0 to a on the number line. For example, -4 is 4 units from 0, and 3 is 3 units from 0. See Figure 13. Thus, the absolute value of -4 is 4, and the absolute value of 3 is 3.

A more formal definition of absolute value is given next.

DEFINITION The **absolute value** of a real number a, denoted by the symbol $|a|$, is defined by the rules

$$|a| = a \quad \text{if } a \geq 0 \qquad \text{and} \qquad |a| = -a \quad \text{if } a < 0$$

For example, since $-4 < 0$, the second rule must be used to get $|-4| = -(-4) = 4$.

| EXAMPLE 3 | **Computing Absolute Value** |

(a) $|8| = 8$ (b) $|0| = 0$ (c) $|-15| = -(-15) = 15$

Look again at Figure 13. The distance from -4 to 3 is 7 units. This distance is the difference $3 - (-4)$, obtained by subtracting the smaller coordinate from the larger. However, since $|3 - (-4)| = |7| = 7$ and $|-4 - 3| = |-7| = 7$, we can use absolute value to calculate the distance between two points without being concerned about which is smaller.

DEFINITION If P and Q are two points on a real number line with coordinates a and b, respectively, the **distance between P and Q**, denoted by $d(P, Q)$, is

$$d(P, Q) = |b - a|$$

Since $|b - a| = |a - b|$, it follows that $d(P, Q) = d(Q, P)$.

EXAMPLE 4 | **Finding Distance on a Number Line**

Let P, Q, and R be points on a real number line with coordinates -5, 7, and -3, respectively. Find the distance

(a) between P and Q　　　　　(b) between Q and R

Solution See Figure 14.

Figure 14

(a) $d(P, Q) = |7 - (-5)| = |12| = 12$

(b) $d(Q, R) = |-3 - 7| = |-10| = 10$

━━━**Now Work** PROBLEM 37

3 Evaluate Algebraic Expressions

Remember, in algebra we use letters such as x, y, a, b, and c to represent numbers. If the letter used is to represent *any* number from a given set of numbers, it is called a **variable**. A **constant** is either a fixed number, such as 5 or $\sqrt{3}$, or a letter that represents a fixed (possibly unspecified) number.

Constants and variables are combined using the operations of addition, subtraction, multiplication, and division to form *algebraic expressions*. Examples of algebraic expressions include

$$x + 3 \qquad \frac{3}{1 - t} \qquad 7x - 2y$$

To evaluate an algebraic expression, substitute for each variable its numerical value.

EXAMPLE 5 | **Evaluating an Algebraic Expression**

Evaluate each expression if $x = 3$ and $y = -1$.

(a) $x + 3y$　　(b) $5xy$　　(c) $\dfrac{3y}{2 - 2x}$　　(d) $|-4x + y|$

Solution (a) Substitute 3 for x and -1 for y in the expression $x + 3y$.

$$x + 3y = 3 + 3(-1) = 3 + (-3) = 0$$
$$\uparrow$$
$$x = 3, y = -1$$

(b) If $x = 3$ and $y = -1$, then

$$5xy = 5(3)(-1) = -15$$

(c) If $x = 3$ and $y = -1$, then

$$\frac{3y}{2 - 2x} = \frac{3(-1)}{2 - 2(3)} = \frac{-3}{2 - 6} = \frac{-3}{-4} = \frac{3}{4}$$

(d) If $x = 3$ and $y = -1$, then

$$|-4x + y| = |-4(3) + (-1)| = |-12 + (-1)| = |-13| = 13$$

━━━**Now Work** PROBLEMS 39 AND 47

4 Determine the Domain of a Variable

In working with expressions or formulas involving variables, the variables may be allowed to take on values from only a certain set of numbers. For example, in the formula for the area A of a circle of radius r, $A = \pi r^2$, the variable r is necessarily restricted to the positive real numbers. In the expression $\dfrac{1}{x}$, the variable x cannot take on the value 0, since division by 0 is not defined.

DEFINITION The set of values that a variable may assume is called the **domain of the variable.**

EXAMPLE 6 **Finding the Domain of a Variable**

The domain of the variable x in the expression

$$\frac{5}{x - 2}$$

is $\{x | x \neq 2\}$, since, if $x = 2$, the denominator becomes 0, which is not defined.

EXAMPLE 7 **Circumference of a Circle**

In the formula for the circumference C of a circle of radius r,

$$C = 2\pi r$$

the domain of the variable r, representing the radius of the circle, is the set of positive real numbers. The domain of the variable C, representing the circumference of the circle, is also the set of positive real numbers.

In describing the domain of a variable, we may use either set notation or words, whichever is more convenient.

Now Work PROBLEM 57

5 Use the Laws of Exponents

Integer exponents provide a shorthand device for representing repeated multiplications of a real number. For example,

$$3^4 = 3 \cdot 3 \cdot 3 \cdot 3 = 81$$

Additionally, many formulas have exponents. For example,

- The formula for the horsepower rating H of an engine is

$$H = \frac{D^2 N}{2.5}$$

where D is the diameter of a cylinder and N is the number of cylinders.

- A formula for the resistance R of blood flowing in a blood vessel is

$$R = C\frac{L}{r^4}$$

where L is the length of the blood vessel, r is the radius, and C is a positive constant.

DEFINITION If a is a real number and n is a positive integer, then the symbol a^n represents the product of n factors of a. That is,

$$a^n = \underbrace{a \cdot a \cdot \,...\, \cdot a}_{n\text{ factors}} \tag{1}$$

Here it is understood that $a^1 = a$.

Then $a^2 = a \cdot a$, $a^3 = a \cdot a \cdot a$, and so on. In the expression a^n, a is called the **base** and n is called the **exponent,** or **power.** We read a^n as "a raised to the power n" or as "a to the nth power." We usually read a^2 as "a squared" and a^3 as "a cubed."

In working with exponents, the operation of *raising to a power* is performed before any other operation. As examples,

$$4 \cdot 3^2 = 4 \cdot 9 = 36 \qquad 2^2 + 3^2 = 4 + 9 = 13$$

$$-2^4 = -16 \qquad 5 \cdot 3^2 + 2 \cdot 4 = 5 \cdot 9 + 2 \cdot 4 = 45 + 8 = 53$$

Parentheses are used to indicate operations to be performed first. For example,

$$(-2)^4 = (-2)(-2)(-2)(-2) = 16 \qquad (2 + 3)^2 = 5^2 = 25$$

DEFINITION If $a \neq 0$, we define

$$a^0 = 1 \quad \text{if } a \neq 0$$

DEFINITION If $a \neq 0$ and if n is a positive integer, then we define

$$a^{-n} = \frac{1}{a^n} \quad \text{if } a \neq 0$$

Whenever you encounter a negative exponent, think "reciprocal."

EXAMPLE 8 **Evaluating Expressions Containing Negative Exponents**

(a) $2^{-3} = \dfrac{1}{2^3} = \dfrac{1}{8}$ (b) $x^{-4} = \dfrac{1}{x^4}$ (c) $\left(\dfrac{1}{5}\right)^{-2} = \dfrac{1}{\left(\dfrac{1}{5}\right)^2} = \dfrac{1}{\dfrac{1}{25}} = 25$

----Now Work PROBLEMS 75 AND 95

The following properties, called the **Laws of Exponents,** can be proved using the preceding definitions. In the list, a and b are real numbers, and m and n are integers.

THEOREM **Laws of Exponents**

$$a^m a^n = a^{m+n} \qquad (a^m)^n = a^{mn} \qquad (ab)^n = a^n b^n$$

$$\frac{a^m}{a^n} = a^{m-n} = \frac{1}{a^{n-m}} \quad \text{if } a \neq 0 \qquad \left(\frac{a}{b}\right)^n = \frac{a^n}{b^n} \quad \text{if } b \neq 0$$

EXAMPLE 9 | **Using the Laws of Exponents**

(a) $x^{-3} \cdot x^5 = x^{-3+5} = x^2 \quad x \neq 0$ (b) $(x^{-3})^2 = x^{-3 \cdot 2} = x^{-6} = \dfrac{1}{x^6} \quad x \neq 0$

(c) $(2x)^3 = 2^3 \cdot x^3 = 8x^3$ (d) $\left(\dfrac{2}{3}\right)^4 = \dfrac{2^4}{3^4} = \dfrac{16}{81}$

(e) $\dfrac{x^{-2}}{x^{-5}} = x^{-2-(-5)} = x^3 \quad x \neq 0$

━━━ **Now Work** PROBLEM 77

EXAMPLE 10 | **Using the Laws of Exponents**

Write each expression so that all exponents are positive.

(a) $\dfrac{x^5 y^{-2}}{x^3 y} \quad x \neq 0, \quad y \neq 0$ (b) $\left(\dfrac{x^{-3}}{3y^{-1}}\right)^{-2} \quad x \neq 0, \quad y \neq 0$

Solution (a) $\dfrac{x^5 y^{-2}}{x^3 y} = \dfrac{x^5}{x^3} \cdot \dfrac{y^{-2}}{y} = x^{5-3} \cdot y^{-2-1} = x^2 y^{-3} = x^2 \cdot \dfrac{1}{y^3} = \dfrac{x^2}{y^3}$

(b) $\left(\dfrac{x^{-3}}{3y^{-1}}\right)^{-2} = \dfrac{(x^{-3})^{-2}}{(3y^{-1})^{-2}} = \dfrac{x^6}{3^{-2}(y^{-1})^{-2}} = \dfrac{x^6}{\dfrac{1}{9}y^2} = \dfrac{9x^6}{y^2}$

━━━ **Now Work** PROBLEM 87

6 Evaluate Square Roots

A real number is squared when it is raised to the power 2. The inverse of squaring is finding a **square root.** For example, since $6^2 = 36$ and $(-6)^2 = 36$, the numbers 6 and -6 are square roots of 36.

The symbol $\sqrt{}$, called a **radical sign,** is used to denote the **principal,** or nonnegative, square root. For example, $\sqrt{36} = 6$.

In Words
$\sqrt{36}$ means "give me the nonnegative number whose square is 36."

DEFINITION | If a is a nonnegative real number, the nonnegative number b such that $b^2 = a$ is the **principal square root** of a, and is denoted by $b = \sqrt{a}$.

The following comments are noteworthy:

1. Negative numbers do not have square roots (in the real number system), because the square of any real number is *nonnegative*. For example, $\sqrt{-4}$ is not a real number, because there is no real number whose square is -4.
2. The principal square root of 0 is 0, since $0^2 = 0$. That is, $\sqrt{0} = 0$.
3. The principal square root of a positive number is positive.
4. If $c \geq 0$, then $(\sqrt{c})^2 = c$. For example, $(\sqrt{2})^2 = 2$ and $(\sqrt{3})^2 = 3$.

EXAMPLE 11 | **Evaluating Square Roots**

(a) $\sqrt{64} = 8$ (b) $\sqrt{\dfrac{1}{16}} = \dfrac{1}{4}$ (c) $\left(\sqrt{1.4}\right)^2 = 1.4$

Examples 11(a) and (b) are examples of square roots of perfect squares, since

$64 = 8^2$ and $\dfrac{1}{16} = \left(\dfrac{1}{4}\right)^2$.

Consider the expression $\sqrt{a^2}$. Since $a^2 \geq 0$, the principal square root of a^2 is defined whether $a > 0$ or $a < 0$. However, since the principal square root is nonnegative, we need an absolute value to ensure the nonnegative result. That is,

$$\sqrt{a^2} = |a| \qquad a \text{ any real number} \qquad \text{(2)}$$

EXAMPLE 12 **Using Equation (2)**

(a) $\sqrt{(2.3)^2} = |2.3| = 2.3$ (b) $\sqrt{(-2.3)^2} = |-2.3| = 2.3$

(c) $\sqrt{x^2} = |x|$

━━━━ Now Work PROBLEM 83

7 Use a Calculator to Evaluate Exponents

Your calculator has either a caret key, $\boxed{\wedge}$, or an $\boxed{x^y}$ key, which is used for computations involving exponents.

EXAMPLE 13 **Exponents on a Graphing Calculator**

Evaluate: $(2.3)^5$

Figure 15

Solution Figure 15 shows the result using a TI-84 graphing calculator.

━━━━ Now Work PROBLEM 113

8 Use Scientific Notation

Measurements of physical quantities can range from very small to very large. For example, the mass of a proton is approximately 0.00000000000000000000000000167 kilogram and the mass of Earth is about 5,980,000,000,000,000,000,000,000 kilograms. These numbers obviously are tedious to write down and difficult to read, so we use exponents to rewrite each.

DEFINITION When a number has been written as the product of a number x, where $1 \leq x < 10$, times a power of 10, it is said to be written in **scientific notation.**

In scientific notation,

$$\text{Mass of a proton} = 1.67 \times 10^{-27} \text{ kilogram}$$

$$\text{Mass of Earth} = 5.98 \times 10^{24} \text{ kilograms}$$

Converting a Decimal to Scientific Notation

To change a positive number into scientific notation:

1. Count the number N of places that the decimal point must be moved to arrive at a number x, where $1 \leq x < 10$.

2. If the original number is greater than or equal to 1, the scientific notation is $x \times 10^N$. If the original number is between 0 and 1, the scientific notation is $x \times 10^{-N}$.

| EXAMPLE 14 | **Using Scientific Notation** |

Write each number in scientific notation.

(a) 9582 (b) 1.245 (c) 0.285 (d) 0.000561

Solution (a) The decimal point in 9582 follows the 2. Count left from the decimal point

$$9\ 5\ 8\ 2\ .$$
$$\qquad\quad 3\ \ 2\ \ 1$$

stopping after three moves, because 9.582 is a number between 1 and 10. Since 9582 is greater than 1, we write

$$9582 = 9.582 \times 10^3$$

(b) The decimal point in 1.245 is between the 1 and 2. Since the number is already between 1 and 10, the scientific notation for it is $1.245 \times 10^0 = 1.245$.

(c) The decimal point in 0.285 is between the 0 and the 2. We count

$$0\ .\ 2\ 8\ 5$$
$$\qquad\quad 1$$

stopping after one move, because 2.85 is a number between 1 and 10. Since 0.285 is between 0 and 1, we write

$$0.285 = 2.85 \times 10^{-1}$$

(d) The decimal point in 0.000561 is moved as follows:

$$0\ .\ 0\ 0\ 0\ 5\ 6\ 1$$
$$\qquad\quad 1\ \ 2\ \ 3\ \ 4$$

As a result,

$$0.000561 = 5.61 \times 10^{-4}$$

━━━━━**Now Work** PROBLEM **119**

| EXAMPLE 15 | **Changing from Scientific Notation to Decimals** |

Write each number as a decimal.

(a) 2.1×10^4 (b) 3.26×10^{-5} (c) 1×10^{-2}

Solution (a) $2.1 \times 10^4 = 2\ .\ 1\ 0\ 0\ 0 \times 10^4 = 21{,}000$
$$\qquad\qquad\qquad\qquad\quad 1\ \ 2\ \ 3\ \ 4$$

(b) $3.26 \times 10^{-5} = 0\ 0\ 0\ 0\ 0\ 3\ .\ 2\ 6 \times 10^{-5} = 0.0000326$
$$\qquad\qquad\qquad\qquad\quad 5\ \ 4\ \ 3\ \ 2\ \ 1$$

(c) $1 \times 10^{-2} = 0\ 0\ 1\ .\ 0 \times 10^{-2} = 0.01$
$$\qquad\qquad\qquad\quad 2\ \ 1$$

━━━━━**Now Work** PROBLEM **127**

EXAMPLE 16 **Using Scientific Notation**

(a) The diameter of the smallest living cell is only about 0.00001 centimeter (cm).* Express this number in scientific notation.

(b) The surface area of Earth is about 1.97×10^8 square miles.† Express the surface area as a whole number.

Solution (a) 0.00001 cm = 1×10^{-5} cm because the decimal point is moved five places and the number is less than 1.

(b) 1.97×10^8 square miles = 197,000,000 square miles.

⌐

━━━ **Now Work** PROBLEM 153

COMMENT *On a calculator, a number such as 3.615×10^{12} is usually displayed as* $\boxed{3.615E12.}$ ∎

Powers of Ten, Philip and Phylis Morrison.
†*1998 Information Please Almanac.*

Historical Feature

The word *algebra* is derived from the Arabic word *al-jabr*. This word is a part of the title of a ninth century work, "Hisâb al-jabr w'al-muqâbalah," written by Mohammed ibn Músâ al-Khwârizmî. The word *al-jabr* means "a restoration," a reference to the fact that, if a number is added to one side of an equation, then it must also be added to the other side in order to "restore" the equality. The title of the work, freely translated, is "The Science of Reduction and Cancellation." Of course, today, algebra has come to mean a great deal more.

R.2 Assess Your Understanding

Concepts and Vocabulary

1. A(n) _____ is a letter used in algebra to represent any number from a given set of numbers.

2. On the real number line, the real number zero is the coordinate of the _____.

3. An inequality of the form $a > b$ is called a(n) _____ inequality.

4. In the expression 2^4, the number 2 is called the _____ and 4 is called the _____.

5. In scientific notation, 1234.5678 = _____.

6. **True or False** The product of two negative real numbers is always greater than zero.

7. **True or False** The distance between two distinct points on the real number line is always greater than zero.

8. **True or False** The absolute value of a real number is always greater than zero.

9. **True or False** When a number is expressed in scientific notation, it is expressed as the product of a number $x, 0 \le x < 1$, and a power of 10.

10. **True or False** To multiply two expressions having the same base, retain the base and multiply the exponents.

Skill Building

11. On the real number line, label the points with coordinates $0, 1, -1, \dfrac{5}{2}, -2.5, \dfrac{3}{4}$, and 0.25.

12. Repeat Problem 11 for the coordinates $0, -2, 2, -1.5, \dfrac{3}{2}, \dfrac{1}{3}$, and $\dfrac{2}{3}$.

In Problems 13–22, replace the question mark by <, >, or =, whichever is correct.

13. $\dfrac{1}{2}$? 0

14. 5 ? 6

15. -1 ? -2

16. -3 ? $-\dfrac{5}{2}$

17. π ? 3.14

18. $\sqrt{2}$? 1.41

19. $\dfrac{1}{2}$? 0.5

20. $\dfrac{1}{3}$? 0.33

21. $\dfrac{2}{3}$? 0.67

22. $\dfrac{1}{4}$? 0.25

In Problems 23–28, write each statement as an inequality.

23. x is positive

24. z is negative

25. x is less than 2

26. y is greater than -5

27. x is less than or equal to 1

28. x is greater than or equal to 2

In Problems 29–32, graph the numbers x on the real number line.

29. $x \geq -2$

30. $x < 4$

31. $x > -1$

32. $x \leq 7$

In Problems 33–38, use the given real number line to compute each distance.

$$
\begin{array}{ccccc}
A & B & C & D & E \\
\end{array}
$$

$$-4 \;-3 \;-2 \;-1 \;\;0 \;\;1 \;\;2 \;\;3 \;\;4 \;\;5 \;\;6$$

33. $d(C, D)$

34. $d(C, A)$

35. $d(D, E)$

36. $d(C, E)$

37. $d(A, E)$

38. $d(D, B)$

In Problems 39–46, evaluate each expression if $x = -2$ and $y = 3$.

39. $x + 2y$

40. $3x + y$

41. $5xy + 2$

42. $-2x + xy$

43. $\dfrac{2x}{x - y}$

44. $\dfrac{x + y}{x - y}$

45. $\dfrac{3x + 2y}{2 + y}$

46. $\dfrac{2x - 3}{y}$

In Problems 47–56, find the value of each expression if $x = 3$ and $y = -2$.

47. $|x + y|$

48. $|x - y|$

49. $|x| + |y|$

50. $|x| - |y|$

51. $\dfrac{|x|}{x}$

52. $\dfrac{|y|}{y}$

53. $|4x - 5y|$

54. $|3x + 2y|$

55. $||4x| - |5y||$

56. $3|x| + 2|y|$

In Problems 57–64, determine which of the value(s) (a) through (d), if any, must be excluded from the domain of the variable in each expression:

(a) $x = 3$ (b) $x = 1$ (c) $x = 0$ (d) $x = -1$

57. $\dfrac{x^2 - 1}{x}$

58. $\dfrac{x^2 + 1}{x}$

59. $\dfrac{x}{x^2 - 9}$

60. $\dfrac{x}{x^2 + 9}$

61. $\dfrac{x^2}{x^2 + 1}$

62. $\dfrac{x^3}{x^2 - 1}$

63. $\dfrac{x^2 + 5x - 10}{x^3 - x}$

64. $\dfrac{-9x^2 - x + 1}{x^3 + x}$

In Problems 65–68, determine the domain of the variable x in each expression.

65. $\dfrac{4}{x - 5}$

66. $\dfrac{-6}{x + 4}$

67. $\dfrac{x}{x + 4}$

68. $\dfrac{x - 2}{x - 6}$

In Problems 69–72, use the formula $C = \dfrac{5}{9}(F - 32)$ for converting degrees Fahrenheit into degrees Celsius to find the Celsius measure of each Fahrenheit temperature.

69. $F = 32°$

70. $F = 212°$

71. $F = 77°$

72. $F = -4°$

In Problems 73–84, simplify each expression.

73. $(-4)^2$

74. -4^2

75. 4^{-2}

76. -4^{-2}

77. $3^{-6} \cdot 3^4$

78. $4^{-2} \cdot 4^3$

79. $(3^{-2})^{-1}$

80. $(2^{-1})^{-3}$

81. $\sqrt{25}$

82. $\sqrt{36}$

83. $\sqrt{(-4)^2}$

84. $\sqrt{(-3)^2}$

In Problems 85–94, simplify each expression. Express the answer so that all exponents are positive. Whenever an exponent is 0 or negative, we assume that the base is not 0.

85. $(8x^3)^2$

86. $(-4x^2)^{-1}$

87. $(x^2 y^{-1})^2$

88. $(x^{-1} y)^3$

89. $\dfrac{x^2 y^3}{xy^4}$

90. $\dfrac{x^{-2} y}{xy^2}$

91. $\dfrac{(-2)^3 x^4 (yz)^2}{3^2 xy^3 z}$

92. $\dfrac{4x^{-2}(yz)^{-1}}{2^3 x^4 y}$

93. $\left(\dfrac{3x^{-1}}{4y^{-1}}\right)^{-2}$

94. $\left(\dfrac{5x^{-2}}{6y^{-2}}\right)^{-3}$

In Problems 95–106, find the value of each expression if $x = 2$ and $y = -1$.

95. $2xy^{-1}$

96. $-3x^{-1}y$

97. $x^2 + y^2$

98. x^2y^2

99. $(xy)^2$

100. $(x + y)^2$

101. $\sqrt{x^2}$

102. $(\sqrt{x})^2$

103. $\sqrt{x^2 + y^2}$

104. $\sqrt{x^2} + \sqrt{y^2}$

105. x^y

106. y^x

107. Find the value of the expression $2x^3 - 3x^2 + 5x - 4$ if $x = 2$. What is the value if $x = 1$?

108. Find the value of the expression $4x^3 + 3x^2 - x + 2$ if $x = 1$. What is the value if $x = 2$?

109. What is the value of $\dfrac{(666)^4}{(222)^4}$?

110. What is the value of $(0.1)^3(20)^3$?

In Problems 111–118, use a calculator to evaluate each expression. Round your answer to three decimal places.

111. $(8.2)^6$

112. $(3.7)^5$

113. $(6.1)^{-3}$

114. $(2.2)^{-5}$

115. $(-2.8)^6$

116. $-(2.8)^6$

117. $(-8.11)^{-4}$

118. $-(8.11)^{-4}$

In Problems 119–126, write each number in scientific notation.

119. 454.2

120. 32.14

121. 0.013

122. 0.00421

123. 32,155

124. 21,210

125. 0.000423

126. 0.0514

In Problems 127–134, write each number as a decimal.

127. 6.15×10^4

128. 9.7×10^3

129. 1.214×10^{-3}

130. 9.88×10^{-4}

131. 1.1×10^8

132. 4.112×10^2

133. 8.1×10^{-2}

134. 6.453×10^{-1}

Applications and Extensions

In Problems 135–144, express each statement as an equation involving the indicated variables.

135. Area of a Rectangle The area A of a rectangle is the product of its length l and its width w.

136. Perimeter of a Rectangle The perimeter P of a rectangle is twice the sum of its length l and its width w.

137. Circumference of a Circle The circumference C of a circle is the product of π and its diameter d.

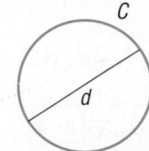

138. Area of a Triangle The area A of a triangle is one-half the product of its base b and its height h.

139. Area of an Equilateral Triangle The area A of an equilateral triangle is $\dfrac{\sqrt{3}}{4}$ times the square of the length x of one side.

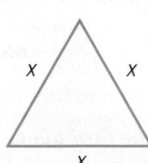

140. Perimeter of an Equilateral Triangle The perimeter P of an equilateral triangle is 3 times the length x of one side.

141. Volume of a Sphere The volume V of a sphere is $\dfrac{4}{3}$ times π times the cube of the radius r.

142. Surface Area of a Sphere The surface area S of a sphere is 4 times π times the square of the radius r.

143. Volume of a Cube The volume V of a cube is the cube of the length x of a side.

144. Surface Area of a Cube The surface area S of a cube is 6 times the square of the length x of a side.

145. Manufacturing Cost The weekly production cost C of manufacturing x watches is given by the formula $C = 4000 + 2x$, where the variable C is in dollars.
 (a) What is the cost of producing 1000 watches?
 (b) What is the cost of producing 2000 watches?

146. Balancing a Checkbook At the beginning of the month, Mike had a balance of $210 in his checking account. During the next month, he deposited $80, wrote a check for $120, made another deposit of $25, wrote two checks: one for $60 and the other for $32. He was also assessed a monthly service charge of $5. What was his balance at the end of the month?

In Problems 147 and 148, write an inequality using an absolute value to describe each statement.

147. x is at least 6 units from 4.

148. x is more than 5 units from 2.

149. U.S. Voltage In the United States, normal household voltage is 110 volts. It is acceptable for the actual voltage x to differ from normal by at most 5 volts. A formula that describes this is

$$|x - 110| \leq 5$$

 (a) Show that a voltage of 108 volts is acceptable.
 (b) Show that a voltage of 104 volts is not acceptable.

150. Foreign Voltage In other countries, normal household voltage is 220 volts. It is acceptable for the actual voltage x to differ from normal by at most 8 volts. A formula that describes this is

$$|x - 220| \leq 8$$

 (a) Show that a voltage of 214 volts is acceptable.
 (b) Show that a voltage of 209 volts is not acceptable.

151. Making Precision Ball Bearings The FireBall Company manufactures ball bearings for precision equipment. One of its products is a ball bearing with a stated radius of 3 centimeters (cm). Only ball bearings with a radius within 0.01 cm of this stated radius are acceptable. If x is the radius of a ball bearing, a formula describing this situation is

$$|x - 3| \leq 0.01$$

 (a) Is a ball bearing of radius $x = 2.999$ acceptable?
 (b) Is a ball bearing of radius $x = 2.89$ acceptable?

152. Body Temperature Normal human body temperature is 98.6°F. A temperature x that differs from normal by at least 1.5°F is considered unhealthy. A formula that describes this is

$$|x - 98.6| \geq 1.5$$

 (a) Show that a temperature of 97°F is unhealthy.
 (b) Show that a temperature of 100°F is not unhealthy.

153. Distance from Earth to Its Moon The distance from Earth to the Moon is about 4×10^8 meters.* Express this distance as a whole number.

154. Height of Mt. Everest The height of Mt. Everest is 8848 meters.* Express this height in scientific notation.

155. Wavelength of Visible Light The wavelength of visible light is about 5×10^{-7} meter.* Express this wavelength as a decimal.

156. Diameter of an Atom The diameter of an atom is about 1×10^{-10} meter.* Express this diameter as a decimal.

157. Diameter of Copper Wire The smallest commercial copper wire is about 0.0005 inch in diameter.† Express this diameter using scientific notation.

158. Smallest Motor The smallest motor ever made is less than 0.05 centimeter wide.† Express this width using scientific notation.

159. Astronomy One light-year is defined by astronomers to be the distance that a beam of light will travel in 1 year (365 days). If the speed of light is 186,000 miles per second, how many miles are in a light-year? Express your answer in scientific notation.

160. Astronomy How long does it take a beam of light to reach Earth from the Sun when the Sun is 93,000,000 miles from Earth? Express your answer in seconds, using scientific notation.

161. Does $\dfrac{1}{3}$ equal 0.333? If not, which is larger? By how much?

162. Does $\dfrac{2}{3}$ equal 0.666? If not, which is larger? By how much?

Explaining Concepts: Discussion and Writing

163. Is there a positive real number "closest" to 0?

164. Number game I'm thinking of a number! It lies between 1 and 10; its square is rational and lies between 1 and 10. The number is larger than π. Correct to two decimal places (that is, truncated to two decimal places) name the number. Now think of your own number, describe it, and challenge a fellow student to name it.

165. Write a brief paragraph that illustrates the similarities and differences between "less than" ($<$) and "less than or equal to" (\leq).

166. Give a reason why the statement $5 < 8$ is true.

Powers of Ten, Philip and Phylis Morrison.
†*1998 Information Please Almanac.*

R.3 Geometry Essentials

Figure 16

1 Use the Pythagorean Theorem and Its Converse

The *Pythagorean Theorem* is a statement about *right triangles*. A **right triangle** is one that contains a **right angle,** that is, an angle of 90°. The side of the triangle opposite the 90° angle is called the **hypotenuse;** the remaining two sides are called **legs.** In Figure 16 we have used c to represent the length of the hypotenuse and a and b to represent the lengths of the legs. Notice the use of the symbol \ulcorner to show the 90° angle. We now state the Pythagorean Theorem.

PYTHAGOREAN THEOREM

In a right triangle, the square of the length of the hypotenuse is equal to the sum of the squares of the lengths of the legs. That is, in the right triangle shown in Figure 16,

$$c^2 = a^2 + b^2 \qquad\qquad (1)$$

A proof of the Pythagorean Theorem is given at the end of this section.

EXAMPLE 1 **Finding the Hypotenuse of a Right Triangle**

In a right triangle, one leg has length 4 and the other has length 3. What is the length of the hypotenuse?

Solution Since the triangle is a right triangle, we use the Pythagorean Theorem with $a = 4$ and $b = 3$ to find the length c of the hypotenuse. From equation (1), we have

$$c^2 = a^2 + b^2$$
$$c^2 = 4^2 + 3^2 = 16 + 9 = 25$$
$$c = \sqrt{25} = 5$$

Now Work PROBLEM 13

The converse of the Pythagorean Theorem is also true.

CONVERSE OF THE PYTHAGOREAN THEOREM

In a triangle, if the square of the length of one side equals the sum of the squares of the lengths of the other two sides, the triangle is a right triangle. The 90° angle is opposite the longest side.

A proof is given at the end of this section.

EXAMPLE 2 **Verifying That a Triangle Is a Right Triangle**

Show that a triangle whose sides are of lengths 5, 12, and 13 is a right triangle. Identify the hypotenuse.

Solution We square the lengths of the sides.

$$5^2 = 25, \qquad 12^2 = 144, \qquad 13^2 = 169$$

Figure 17

Notice that the sum of the first two squares (25 and 144) equals the third square (169). Hence, the triangle is a right triangle. The longest side, 13, is the hypotenuse. See Figure 17.

━━━ **Now Work** PROBLEM 21

EXAMPLE 3 | **Applying the Pythagorean Theorem**

The tallest building in the world is Burj Khalifa in Dubai, United Arab Emirates, at 2717 feet and 160 floors. The observation deck is 1450 feet above ground level. How far can a person standing on the observation deck see (with the aid of a telescope)? Use 3960 miles for the radius of Earth.

Source: Wikipedia 2010

Solution From the center of Earth, draw two radii: one through Burj Khalifa and the other to the farthest point a person can see from the observation deck. See Figure 18. Apply the Pythagorean Theorem to the right triangle.

Since 1 mile $= 5280$ feet, then 1450 feet $= \dfrac{1450}{5280}$ mile. So we have

$$d^2 + (3960)^2 = \left(3960 + \frac{1450}{5280}\right)^2$$
$$d^2 = \left(3960 + \frac{1450}{5280}\right)^2 - (3960)^2 \approx 2175.08$$
$$d \approx 46.64$$

A person can see almost 47 miles from the observation tower.

Figure 18

━━━ **Now Work** PROBLEM 53

2 Know Geometry Formulas

Certain formulas from geometry are useful in solving algebra problems.

For a rectangle of length *l* and width *w*,

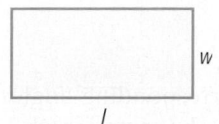

$$\text{Area} = lw \qquad \text{Perimeter} = 2l + 2w$$

For a triangle with base *b* and altitude *h*,

$$\text{Area} = \frac{1}{2}bh$$

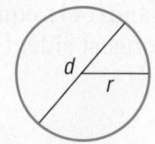

For a circle of radius r (diameter $d = 2r$),

$$\text{Area} = \pi r^2 \qquad \text{Circumference} = 2\pi r = \pi d$$

For a closed rectangular box of length l, width w, and height h,

$$\text{Volume} = lwh \qquad \text{Surface area} = 2lh + 2wh + 2lw$$

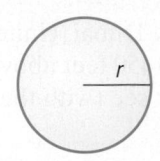

For a sphere of radius r,

$$\text{Volume} = \frac{4}{3}\pi r^3 \qquad \text{Surface area} = 4\pi r^2$$

For a right circular cylinder of height h and radius r,

$$\text{Volume} = \pi r^2 h \qquad \text{Surface area} = 2\pi r^2 + 2\pi rh$$

Now Work PROBLEM 29

EXAMPLE 4 | **Using Geometry Formulas**

A Christmas tree ornament is in the shape of a semicircle on top of a triangle. How many square centimeters (cm²) of copper is required to make the ornament if the height of the triangle is 6 cm and the base is 4 cm?

Solution See Figure 19. The amount of copper required equals the shaded area. This area is the sum of the areas of the triangle and the semicircle. The triangle has height $h = 6$ and base $b = 4$. The semicircle has diameter $d = 4$, so its radius is $r = 2$.

Figure 19

$$\text{Area} = \text{Area of triangle} + \text{Area of semicircle}$$
$$= \frac{1}{2}bh + \frac{1}{2}\pi r^2 = \frac{1}{2}(4)(6) + \frac{1}{2}\pi \cdot 2^2 \qquad b = 4; h = 6; r = 2$$
$$= 12 + 2\pi \approx 18.28 \text{ cm}^2$$

About 18.28 cm² of copper is required.

Now Work PROBLEM 47

3 Understand Congruent Triangles and Similar Triangles

Throughout the text we will make reference to triangles. We begin with a discussion of *congruent* triangles. According to dictionary.com, the word **congruent** means coinciding exactly when superimposed. For example, two angles are congruent if they have the same measure and two line segments are congruent if they have the same length.

In Words

Two triangles are congruent if they have the same size and shape.

DEFINITION Two triangles are **congruent** if each of the corresponding angles is the same measure and each of the corresponding sides is the same length.

In Figure 20, corresponding angles are equal and the lengths of the corresponding sides are equal: $a = d$, $b = e$, and $c = f$. We conclude that these triangles are congruent.

Figure 20 Congruent triangles

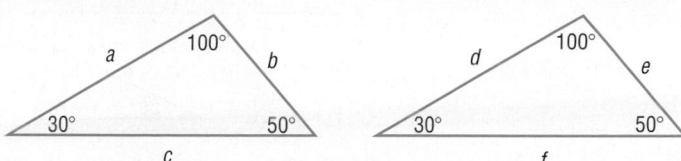

It is not necessary to verify that all three angles and all three sides are the same measure to determine whether two triangles are congruent.

Determining Congruent Triangles

1. **Angle–Side–Angle Case** Two triangles are congruent if two of the angles are equal and the lengths of the corresponding sides between the two angles are equal.

 For example, in Figure 21(a), the two triangles are congruent because two angles and the included side are equal.

2. **Side–Side–Side Case** Two triangles are congruent if the lengths of the corresponding sides of the triangles are equal.

 For example, in Figure 21(b), the two triangles are congruent because the three corresponding sides are all equal.

3. **Side–Angle–Side Case** Two triangles are congruent if the lengths of two corresponding sides are equal and the angles between the two sides are the same.

 For example, in Figure 21(c), the two triangles are congruent because two sides and the included angle are equal.

Figure 21

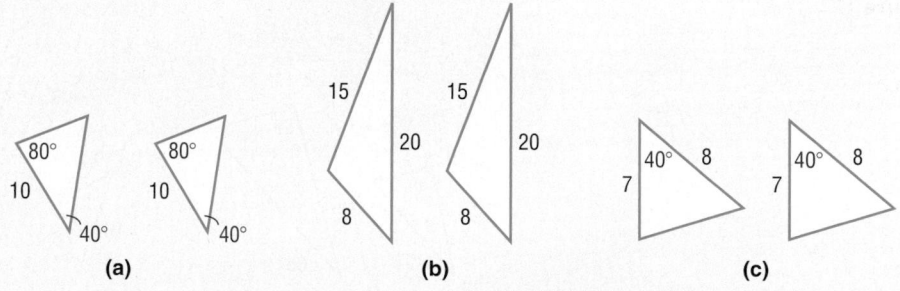

We contrast congruent triangles with *similar* triangles.

DEFINITION

Two triangles are **similar** if the corresponding angles are equal and the lengths of the corresponding sides are proportional.

In Words
Two triangles are similar if they have the same shape, but (possibly) different sizes.

For example, the triangles in Figure 22 are similar because the corresponding angles are equal. In addition, the lengths of the corresponding sides are proportional because each side in the triangle on the right is twice as long as each corresponding side in the triangle on the left. That is, the ratio of the corresponding sides is a constant: $\dfrac{d}{a} = \dfrac{e}{b} = \dfrac{f}{c} = 2$.

Figure 22

a 80° b
30° 70°
c

d = 2e 80°
e = 2b
30° 70°
f = 2c

It is not necessary to verify that all three angles are equal and all three sides are proportional to determine whether two triangles are congruent.

Determining Similar Triangles

1. **Angle–Angle Case** Two triangles are similar if two of the corresponding angles are equal.
 For example, in Figure 23(a), the two triangles are similar because two angles are equal.

2. **Side–Side–Side Case** Two triangles are similar if the lengths of all three sides of each triangle are proportional.
 For example, in Figure 23(b), the two triangles are similar because

$$\frac{10}{30} = \frac{5}{15} = \frac{6}{18} = \frac{1}{3}.$$

3. **Side–Angle–Side Case** Two triangles are similar if two corresponding sides are proportional and the angles between the two sides are equal.
 For example, in Figure 23(c), the two triangles are similar because $\frac{4}{6} = \frac{12}{18} = \frac{2}{3}$ and the angles between the sides are equal.

Figure 23

80°
80°

35° 35°

(a)

15
5 30
10 6 18

(b)

12 18
120° 120°
4 6

(c)

EXAMPLE 5 **Using Similar Triangles**

Given that the triangles in Figure 24 are similar, find the missing length x and the angles A, B, and C.

Figure 24

60° 6 30°
3
90°

B x
C
5
A

Solution Because the triangles are similar, corresponding angles are equal. So $A = 90°$, $B = 60°$, and $C = 30°$. Also, the corresponding sides are proportional. That is, $\dfrac{3}{5} = \dfrac{6}{x}$. We solve this equation for x.

$$\frac{3}{5} = \frac{6}{x}$$

$$5x \cdot \frac{3}{5} = 5x \cdot \frac{6}{x} \qquad \text{Multiply both sides by } 5x.$$

$$3x = 30 \qquad \text{Simplify.}$$

$$x = 10 \qquad \text{Divide both sides by 3.}$$

The missing length is 10 units.

━━━━━**Now Work** PROBLEM **41**

Proof of the Pythagorean Theorem Begin with a square, each side of length $a + b$. In this square, form four right triangles, each having legs equal in length to a and b. See Figure 25. All these triangles are congruent (two sides and their included angle are equal). As a result, the hypotenuse of each is the same, say c, and the pink shading in Figure 25 indicates a square with an area equal to c^2.

Figure 25

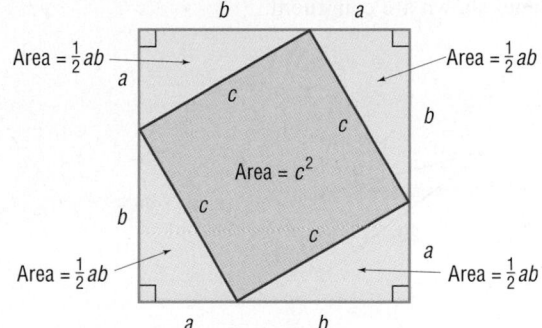

The area of the original square with sides $a + b$ equals the sum of the areas of the four triangles (each of area $\dfrac{1}{2}ab$) plus the area of the square with side c. That is,

$$(a + b)^2 = \frac{1}{2}ab + \frac{1}{2}ab + \frac{1}{2}ab + \frac{1}{2}ab + c^2$$

$$a^2 + 2ab + b^2 = 2ab + c^2$$

$$a^2 + b^2 = c^2$$

Figure 26

(a)

(b) $c^2 = a^2 + b^2$

The proof is complete. ∎

Proof of the Converse of the Pythagorean Theorem Begin with two triangles: one a right triangle with legs a and b and the other a triangle with sides a, b, and c for which $c^2 = a^2 + b^2$. See Figure 26. By the Pythagorean Theorem, the length x of the third side of the first triangle is

$$x^2 = a^2 + b^2$$

But $c^2 = a^2 + b^2$. Then,

$$x^2 = c^2$$

$$x = c$$

The two triangles have the same sides and are therefore congruent. This means corresponding angles are equal, so the angle opposite side c of the second triangle equals $90°$.
 The proof is complete. ∎

R.3 Assess Your Understanding

Concepts and Vocabulary

1. A(n) _____ triangle is one that contains an angle of 90 degrees. The longest side is called the _____.

2. For a triangle with base b and altitude h, a formula for the area A is _____.

3. The formula for the circumference C of a circle of radius r is _____.

4. Two triangles are _____ if corresponding angles are equal and the lengths of the corresponding sides are proportional.

5. *True or False* In a right triangle, the square of the length of the longest side equals the sum of the squares of the lengths of the other two sides.

6. *True or False* The triangle with sides of length 6, 8, and 10 is a right triangle.

7. *True or False* The volume of a sphere of radius r is $\frac{4}{3}\pi r^2$.

8. *True or False* The triangles shown are congruent.

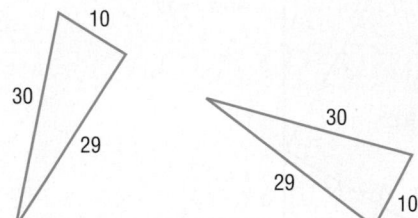

9. *True or False* The triangles shown are similar.

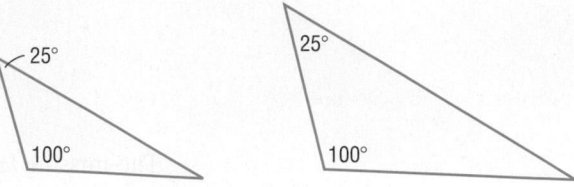

10. *True or False* The triangles shown are similar.

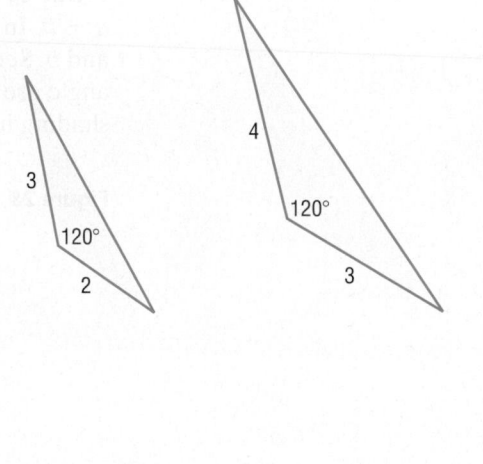

Skill Building

In Problems 11–16, the lengths of the legs of a right triangle are given. Find the hypotenuse.

11. $a = 5$, $b = 12$
12. $a = 6$, $b = 8$
13. $a = 10$, $b = 24$
14. $a = 4$, $b = 3$
15. $a = 7$, $b = 24$
16. $a = 14$, $b = 48$

In Problems 17–24, the lengths of the sides of a triangle are given. Determine which are right triangles. For those that are, identify the hypotenuse.

17. 3, 4, 5
18. 6, 8, 10
19. 4, 5, 6
20. 2, 2, 3
21. 7, 24, 25
22. 10, 24, 26
23. 6, 4, 3
24. 5, 4, 7

25. Find the area A of a rectangle with length 4 inches and width 2 inches.

26. Find the area A of a rectangle with length 9 centimeters and width 4 centimeters.

27. Find the area A of a triangle with height 4 inches and base 2 inches.

28. Find the area A of a triangle with height 9 centimeters and base 4 centimeters.

29. Find the area A and circumference C of a circle of radius 5 meters.

30. Find the area A and circumference C of a circle of radius 2 feet.

31. Find the volume V and surface area S of a rectangular box with length 8 feet, width 4 feet, and height 7 feet.

32. Find the volume V and surface area S of a rectangular box with length 9 inches, width 4 inches, and height 8 inches.

33. Find the volume V and surface area S of a sphere of radius 4 centimeters.

34. Find the volume V and surface area S of a sphere of radius 3 feet.

35. Find the volume V and surface area S of a right circular cylinder with radius 9 inches and height 8 inches.

36. Find the volume V and surface area S of a right circular cylinder with radius 8 inches and height 9 inches.

In Problems 37–40, find the area of the shaded region.

37.

2
2

38.

2
2

39.

2
2

40.

2
2

In Problems 41–44, each pair of triangles is similar. Find the missing length x and the missing angles A, B, and C.

41.

60° 2 90°
4
30°
B x A
8
C

42.

16 30°
75°
12
75°
x
B A
6 C

43.

60°
20
95° 25°
45
A
30
B C
x

44.

10 125°
50°
50
8 B
A 5°
x
C

Applications and Extensions

45. How many feet does a wheel with a diameter of 16 inches travel after four revolutions?

46. How many revolutions will a circular disk with a diameter of 4 feet have completed after it has rolled 20 feet?

47. In the figure shown, *ABCD* is a square, with each side of length 6 feet. The width of the border (shaded portion) between the outer square *EFGH* and *ABCD* is 2 feet. Find the area of the border.

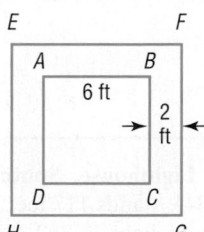
E F
A B
6 ft
2 ft
D C
H G

48. Refer to the figure. Square *ABCD* has an area of 100 square feet; square *BEFG* has an area of 16 square feet. What is the area of the triangle *CGF*?

A B E
G F
D C

49. Architecture A **Norman window** consists of a rectangle surmounted by a semicircle. Find the area of the Norman window shown in the illustration. How much wood frame is needed to enclose the window?

6'
4'

50. Construction A circular swimming pool, 20 feet in diameter, is enclosed by a wooden deck that is 3 feet wide. What is the area of the deck? How much fence is required to enclose the deck?

3'
20'

51. How Tall Is the Great Pyramid? The ancient Greek philosopher Thales of Miletus is reported on one occasion to have visited Egypt and calculated the height of the Great Pyramid of Cheops by means of shadow reckoning. Thales knew that each side of the base of the pyramid was 252 paces and that his own height was 2 paces. He measured the length of the pyramid's shadow to be 114 paces and determined the length of his shadow to be 3 paces. See the illustration. Using similar triangles, determine the height of the Great Pyramid in terms of the number of paces.

Source: www.anselm.edu/homepage/dbanach/thales.htm. This site references another source: Selections, from Julia E. Diggins, *String, Straightedge, and Shadow,* Viking Press, New York, 1965, Illustrations by Corydon Bell.

52. The Bermuda Triangle Karen is doing research on the Bermuda Triangle which she defines roughly by Hamilton, Bermuda; San Juan, Puerto Rico; and Fort Lauderdale, Florida. On her atlas Karen measures the straight-line distances from Hamilton to Fort Lauderdale, Fort Lauderdale to San Juan, and San Juan to Hamilton to be approximately 57 millimeters (mm), 58 mm, and 53.5 mm respectively. If the actual distance from Fort Lauderdale to San Juan is 1046 miles, approximate the actual distances from San Juan to Hamilton and from Hamilton to Fort Lauderdale.

Source: Reprinted with permission from Red River Press, Inc., Winnipeg, Canada.

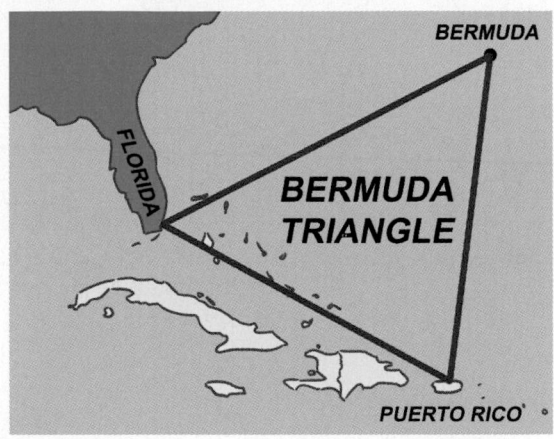

Source: www.en.wikipedia.org/wiki/Bermuda_Triangle.

In Problems 53–55, use the facts that the radius of Earth is 3960 miles and 1 mile = 5280 feet.

53. How Far Can You See? The conning tower of the U.S.S. *Silversides,* a World War II submarine now permanently stationed in Muskegon, Michigan, is approximately 20 feet above sea level. How far can you see from the conning tower?

54. How Far Can You See? A person who is 6 feet tall is standing on the beach in Fort Lauderdale, Florida, and looks out onto the Atlantic Ocean. Suddenly, a ship appears on the horizon. How far is the ship from shore?

55. How Far Can You See? The deck of a destroyer is 100 feet above sea level. How far can a person see from the deck?

How far can a person see from the bridge, which is 150 feet above sea level?

56. Suppose that m and n are positive integers with $m > n$. If $a = m^2 - n^2$, $b = 2mn$, and $c = m^2 + n^2$, show that a, b, and c are the lengths of the sides of a right triangle. (This formula can be used to find the sides of a right triangle that are integers, such as 3, 4, 5; 5, 12, 13; and so on. Such triplets of integers are called **Pythagorean triples.**)

Explaining Concepts: Discussion and Writing

57. You have 1000 feet of flexible pool siding and wish to construct a swimming pool. Experiment with rectangular-shaped pools with perimeters of 1000 feet. How do their areas vary? What is the shape of the rectangle with the largest area? Now compute the area enclosed by a circular pool with a perimeter (circumference) of 1000 feet. What would be your choice of shape for the pool? If rectangular, what is your preference for dimensions? Justify your choice. If your only consideration is to have a pool that encloses the most area, what shape should you use?

58. The Gibb's Hill Lighthouse, Southampton, Bermuda, in operation since 1846, stands 117 feet high on a hill 245 feet high, so its beam of light is 362 feet above sea level. A brochure states that the light itself can be seen on the horizon about 26 miles distant. Verify the correctness of this information. The brochure further states that ships 40 miles away can see the light and planes flying at 10,000 feet can see it 120 miles away. Verify the accuracy of these statements. What assumption did the brochure make about the height of the ship?

R.4 Polynomials

We have described algebra as a generalization of arithmetic in which letters are used to represent real numbers. From now on, we shall use the letters at the end of the alphabet, such as x, y, and z, to represent variables and the letters at the beginning of the alphabet, such as a, b, and c, to represent constants. In the expressions $3x + 5$ and $ax + b$, it is understood that x is a variable and that a and b are constants, even though the constants a and b are unspecified. As you will find out, the context usually makes the intended meaning clear.

1 Recognize Monomials

DEFINITION

A **monomial** in one variable is the product of a constant and a variable raised to a nonnegative integer power. A monomial is of the form

$$ax^k$$

NOTE The nonnegative integers are the integers 0, 1, 2, 3, ∎

where a is a constant, x is a variable, and $k \geq 0$ is an integer. The constant a is called the **coefficient** of the monomial. If $a \neq 0$, then k is called the **degree** of the monomial.

EXAMPLE 1 **Examples of Monomials**

Monomial	Coefficient	Degree	
(a) $6x^2$	6	2	
(b) $-\sqrt{2}x^3$	$-\sqrt{2}$	3	
(c) 3	3	0	Since $3 = 3 \cdot 1 = 3x^0, x \neq 0$
(d) $-5x$	-5	1	Since $-5x = -5x^1$
(e) x^4	1	4	Since $x^4 = 1 \cdot x^4$

EXAMPLE 2 **Examples of Nonmonomial Expressions**

(a) $3x^{1/2}$ is not a monomial, since the exponent of the variable x is $\frac{1}{2}$ and $\frac{1}{2}$ is not a nonnegative integer.
(b) $4x^{-3}$ is not a monomial, since the exponent of the variable x is -3 and -3 is not a nonnegative integer.

Now Work PROBLEM 7

2 Recognize Polynomials

Two monomials with the same variable raised to the same power are called **like terms.** For example, $2x^4$ and $-5x^4$ are like terms. In contrast, the monomials $2x^3$ and $2x^5$ are not like terms.

We can add or subtract like terms using the Distributive Property. For example,

$$2x^2 + 5x^2 = (2 + 5)x^2 = 7x^2 \quad \text{and} \quad 8x^3 - 5x^3 = (8 - 5)x^3 = 3x^3$$

The sum or difference of two monomials having different degrees is called a **binomial.** The sum or difference of three monomials with three different degrees is called a **trinomial.** For example,

$x^2 - 2$ is a binomial.
$x^3 - 3x + 5$ is a trinomial.
$2x^2 + 5x^2 + 2 = 7x^2 + 2$ is a binomial.

DEFINITION

A **polynomial** in one variable is an algebraic expression of the form

$$a_n x^n + a_{n-1} x^{n-1} + \cdots + a_1 x + a_0 \tag{1}$$

> **In Words**
> A polynomial is a sum of monomials.

where $a_n, a_{n-1}, \ldots, a_1, a_0$ are constants,* called the **coefficients** of the polynomial, $n \geq 0$ is an integer, and x is a variable. If $a_n \neq 0$, it is called the **leading coefficient,** $a_n x^n$ is called the **leading term,** and n is the **degree** of the polynomial.

The monomials that make up a polynomial are called its **terms.** If all the coefficients are 0, the polynomial is called the **zero polynomial,** which has no degree.

Polynomials are usually written in **standard form,** beginning with the nonzero term of highest degree and continuing with terms in descending order according to degree. If a power of x is missing, it is because its coefficient is zero.

EXAMPLE 3

Examples of Polynomials

Polynomial	Coefficients	Degree
$-8x^3 + 4x^2 - 6x + 2$	$-8, 4, -6, 2$	3
$3x^2 - 5 = 3x^2 + 0 \cdot x + (-5)$	$3, 0, -5$	2
$8 - 2x + x^2 = 1 \cdot x^2 + (-2)x + 8$	$1, -2, 8$	2
$5x + \sqrt{2} = 5x^1 + \sqrt{2}$	$5, \sqrt{2}$	1
$3 = 3 \cdot 1 = 3 \cdot x^0$	3	0
0	0	No degree

Although we have been using x to represent the variable, letters such as y or z are also commonly used.

$3x^4 - x^2 + 2$ is a polynomial (in x) of degree 4.
$9y^3 - 2y^2 + y - 3$ is a polynomial (in y) of degree 3.
$z^5 + \pi$ is a polynomial (in z) of degree 5.

Algebraic expressions such as

$$\frac{1}{x} \quad \text{and} \quad \frac{x^2 + 1}{x + 5}$$

* The notation a_n is read as "a sub n." The number n is called a **subscript** and should not be confused with an exponent. We use subscripts to distinguish one constant from another when a large or undetermined number of constants is required.

are not polynomials. The first is not a polynomial because $\frac{1}{x} = x^{-1}$ has an exponent that is not a nonnegative integer. Although the second expression is the quotient of two polynomials, the polynomial in the denominator has degree greater than 0, so the expression cannot be a polynomial.

→ **Now Work** PROBLEM 17

3 Add and Subtract Polynomials

Polynomials are added and subtracted by combining like terms.

EXAMPLE 4 **Adding Polynomials**

Find the sum of the polynomials:

$$8x^3 - 2x^2 + 6x - 2 \quad \text{and} \quad 3x^4 - 2x^3 + x^2 + x$$

Solution We shall find the sum in two ways.

Horizontal Addition: The idea here is to group the like terms and then combine them.

$$(8x^3 - 2x^2 + 6x - 2) + (3x^4 - 2x^3 + x^2 + x)$$
$$= 3x^4 + (8x^3 - 2x^3) + (-2x^2 + x^2) + (6x + x) - 2$$
$$= 3x^4 + 6x^3 - x^2 + 7x - 2$$

Vertical Addition: The idea here is to vertically line up the like terms in each polynomial and then add the coefficients.

$$
\begin{array}{cccccc}
 & x^4 & x^3 & x^2 & x^1 & x^0 \\
 & & 8x^3 & - 2x^2 & + 6x & - 2 \\
+ & 3x^4 & - 2x^3 & + x^2 & + x & \\
\hline
 & 3x^4 & + 6x^3 & - x^2 & + 7x & - 2
\end{array}
$$

We can subtract two polynomials horizontally or vertically as well.

EXAMPLE 5 **Subtracting Polynomials**

Find the difference: $(3x^4 - 4x^3 + 6x^2 - 1) - (2x^4 - 8x^2 - 6x + 5)$

Solution *Horizontal Subtraction:*

$$(3x^4 - 4x^3 + 6x^2 - 1) - (2x^4 - 8x^2 - 6x + 5)$$
$$= 3x^4 - 4x^3 + 6x^2 - 1 + \underbrace{(-2x^4 + 8x^2 + 6x - 5)}$$

Be sure to change the sign of each
term in the second polynomial.

$$= \underset{\uparrow}{(3x^4 - 2x^4)} + (-4x^3) + (6x^2 + 8x^2) + 6x + (-1 - 5)$$

Group like terms.

$$= x^4 - 4x^3 + 14x^2 + 6x - 6$$

COMMENT Vertical subtraction will be used when we divide polynomials. ∎

Vertical Subtraction: We line up like terms, change the sign of each coefficient of the second polynomial, and add.

$$
\begin{array}{c}
\overset{x^4 \quad\quad x^3 \quad\quad x^2 \quad\quad x^1 \quad\quad x^0}{3x^4 - 4x^3 + 6x^2 \quad\quad - 1} = \\
- \ [2x^4 \quad\quad - 8x^2 - 6x + 5] = +
\end{array}
\quad
\begin{array}{c}
\overset{x^4 \quad\quad x^3 \quad\quad x^2 \quad\quad x^1 \quad\quad x^0}{3x^4 - 4x^3 + 6x^2 \quad\quad - 1} \\
\underline{-2x^4 \quad\quad + 8x^2 + 6x - 5} \\
x^4 - 4x^3 + 14x^2 + 6x - 6
\end{array}
$$

The choice of which of these methods to use for adding and subtracting polynomials is left to you. To save space, we shall most often use the horizontal format.

➤ **Now Work** PROBLEM 29

4 Multiply Polynomials

Two monomials may be multiplied using the Laws of Exponents and the Commutative and Associative Properties. For example,

$$(2x^3) \cdot (5x^4) = (2 \cdot 5) \cdot (x^3 \cdot x^4) = 10x^{3+4} = 10x^7$$

Products of polynomials are found by repeated use of the Distributive Property and the Laws of Exponents. Again, you have a choice of horizontal or vertical format.

EXAMPLE 6

Multiplying Polynomials

Find the product: $(2x + 5)(x^2 - x + 2)$

Solution

Horizontal Multiplication:

$$(2x + 5)(x^2 - x + 2) = \underset{\uparrow}{2x(x^2 - x + 2) + 5(x^2 - x + 2)}$$
Distributive Property

$$= \underset{\uparrow}{(2x \cdot x^2 - 2x \cdot x + 2x \cdot 2) + (5 \cdot x^2 - 5 \cdot x + 5 \cdot 2)}$$
Distributive Property

$$= \underset{\uparrow}{(2x^3 - 2x^2 + 4x) + (5x^2 - 5x + 10)}$$
Law of Exponents

$$= \underset{\uparrow}{2x^3 + 3x^2 - x + 10}$$
Combine like terms.

Vertical Multiplication: The idea here is very much like multiplying a two-digit number by a three-digit number.

$$
\begin{array}{r}
x^2 - x + 2 \\
2x + 5 \\
\hline
2x^3 - 2x^2 + 4x \\
(+) 5x^2 - 5x + 10 \\
\hline
2x^3 + 3x^2 - x + 10
\end{array}
$$

This line is $2x(x^2 - x + 2)$.

This line is $5(x^2 - x + 2)$.

Sum of the above two lines.

➤ **Now Work** PROBLEM 45

5 Know Formulas for Special Products

Certain products, which we call **special products,** occur frequently in algebra. We can calculate them easily using the **FOIL** (*First, Outer, Inner, Last*) method of multiplying two binomials.

$$(ax + b)(cx + d) = ax(cx + d) + b(cx + d)$$

$$= \overbrace{ax \cdot cx}^{\text{First}} + \overbrace{ax \cdot d}^{\text{Outer}} + \overbrace{b \cdot cx}^{\text{Inner}} + \overbrace{b \cdot d}^{\text{Last}}$$

$$= acx^2 + adx + bcx + bd$$

$$= acx^2 + (ad + bc)x + bd$$

EXAMPLE 7 **Using FOIL**

(a) $(x - 3)(x + 3) = x^2 + 3x - 3x - 9 = x^2 - 9$

(b) $(x + 2)^2 = (x + 2)(x + 2) = x^2 + 2x + 2x + 4 = x^2 + 4x + 4$

(c) $(x - 3)^2 = (x - 3)(x - 3) = x^2 - 3x - 3x + 9 = x^2 - 6x + 9$

(d) $(x + 3)(x + 1) = x^2 + x + 3x + 3 = x^2 + 4x + 3$

(e) $(2x + 1)(3x + 4) = 6x^2 + 8x + 3x + 4 = 6x^2 + 11x + 4$

Now Work PROBLEMS 47 AND 55

Some products have been given special names because of their form. The following special products are based on Examples 7(a), (b), and (c).

Difference of Two Squares

$$(x - a)(x + a) = x^2 - a^2 \qquad \textbf{(2)}$$

Squares of Binomials, or Perfect Squares

$$(x + a)^2 = x^2 + 2ax + a^2 \qquad \textbf{(3a)}$$
$$(x - a)^2 = x^2 - 2ax + a^2 \qquad \textbf{(3b)}$$

EXAMPLE 8 **Using Special Product Formulas**

(a) $(x - 5)(x + 5) = x^2 - 5^2 = x^2 - 25$ — Difference of two squares

(b) $(x + 7)^2 = x^2 + 2 \cdot 7 \cdot x + 7^2 = x^2 + 14x + 49$ — Square of a binomial

(c) $(2x + 1)^2 = (2x)^2 + 2 \cdot 1 \cdot 2x + 1^2 = 4x^2 + 4x + 1$ — Notice that we used 2x in place of x in formula (3a).

(d) $(3x - 4)^2 = (3x)^2 - 2 \cdot 4 \cdot 3x + 4^2 = 9x^2 - 24x + 16$ — Replace x by 3x in formula (3b).

Now Work PROBLEMS 65, 67, AND 69

Let's look at some more examples that lead to general formulas.

EXAMPLE 9 **Cubing a Binomial**

(a) $(x + 2)^3 = (x + 2)(x + 2)^2 = (x + 2)(x^2 + 4x + 4)$ Formula (3a)

$$= (x^3 + 4x^2 + 4x) + (2x^2 + 8x + 8)$$
$$= x^3 + 6x^2 + 12x + 8$$

(b) $(x - 1)^3 = (x - 1)(x - 1)^2 = (x - 1)(x^2 - 2x + 1)$ Formula (3b)

$$= (x^3 - 2x^2 + x) - (x^2 - 2x + 1)$$
$$= x^3 - 3x^2 + 3x - 1$$

Cubes of Binomials, or Perfect Cubes

$$(x + a)^3 = x^3 + 3ax^2 + 3a^2x + a^3 \qquad \textbf{(4a)}$$
$$(x - a)^3 = x^3 - 3ax^2 + 3a^2x - a^3 \qquad \textbf{(4b)}$$

Now Work PROBLEM 85

EXAMPLE 10 **Forming the Difference of Two Cubes**

$$(x - 1)(x^2 + x + 1) = x(x^2 + x + 1) - 1(x^2 + x + 1)$$
$$= x^3 + x^2 + x - x^2 - x - 1$$
$$= x^3 - 1$$

EXAMPLE 11 **Forming the Sum of Two Cubes**

$$(x + 2)(x^2 - 2x + 4) = x(x^2 - 2x + 4) + 2(x^2 - 2x + 4)$$
$$= x^3 - 2x^2 + 4x + 2x^2 - 4x + 8$$
$$= x^3 + 8$$

Examples 10 and 11 lead to two more special products.

Difference of Two Cubes

$$(x - a)(x^2 + ax + a^2) = x^3 - a^3 \qquad \textbf{(5)}$$

Sum of Two Cubes

$$(x + a)(x^2 - ax + a^2) = x^3 + a^3 \qquad \textbf{(6)}$$

6 Divide Polynomials Using Long Division

The procedure for dividing two polynomials is similar to the procedure for dividing two integers.

EXAMPLE 12 **Dividing Two Integers**

Divide 842 by 15.

Solution

$$
\begin{array}{r}
56 \quad \leftarrow \text{Quotient} \\
\text{Divisor} \rightarrow \quad 15\overline{)842} \quad \leftarrow \text{Dividend} \\
\underline{75} \quad \leftarrow 5 \cdot 15 \ (\text{subtract}) \\
92 \\
\underline{90} \quad \leftarrow 6 \cdot 15 \ (\text{subtract}) \\
2 \quad \leftarrow \text{Remainder}
\end{array}
$$

So, $\dfrac{842}{15} = 56 + \dfrac{2}{15}$.

In the long division process detailed in Example 12, the number 15 is called the **divisor,** the number 842 is called the **dividend,** the number 56 is called the **quotient,** and the number 2 is called the **remainder.**

To check the answer obtained in a division problem, multiply the quotient by the divisor and add the remainder. The answer should be the dividend.

$$
\boxed{(\text{Quotient})(\text{Divisor}) + \text{Remainder} = \text{Dividend}}
$$

For example, we can check the results obtained in Example 12 as follows:

$$
(56)(15) + 2 = 840 + 2 = 842
$$

To divide two polynomials, we first must write each polynomial in standard form. The process then follows a pattern similar to that of Example 12. The next example illustrates the procedure.

EXAMPLE 13 **Dividing Two Polynomials**

Find the quotient and the remainder when

$$
3x^3 + 4x^2 + x + 7 \quad \text{is divided by} \quad x^2 + 1
$$

Solution

Each polynomial is in standard form. The dividend is $3x^3 + 4x^2 + x + 7$, and the divisor is $x^2 + 1$.

NOTE Remember, a polynomial is in standard form when its terms are written according to descending degree. ∎

STEP 1: Divide the leading term of the dividend, $3x^3$, by the leading term of the divisor, x^2. Enter the result, $3x$, over the term $3x^3$, as follows:

$$
\begin{array}{r}
3x \qquad\qquad\quad \\
x^2 + 1 \overline{)3x^3 + 4x^2 + x + 7}
\end{array}
$$

STEP 2: Multiply $3x$ by $x^2 + 1$ and enter the result below the dividend.

$$
\begin{array}{r}
3x \qquad\qquad\quad \\
x^2 + 1 \overline{)3x^3 + 4x^2 + x + 7} \\
\underline{3x^3 \qquad\quad + 3x} \quad \leftarrow 3x \cdot (x^2 + 1) = 3x^3 + 3x
\end{array}
$$

Notice that we align the 3x term under the x to make the next step easier.

STEP 3: Subtract and bring down the remaining terms.

$$
\begin{array}{r}
3x \qquad\qquad\quad \\
x^2 + 1 \overline{)3x^3 + 4x^2 + x + 7} \\
\underline{3x^3 \qquad\quad + 3x} \quad \leftarrow \text{Subtract (change the signs and add).} \\
4x^2 - 2x + 7 \quad \leftarrow \text{Bring down the } 4x^2 \text{ and the 7.}
\end{array}
$$

Step 4: Repeat Steps 1–3 using $4x^2 - 2x + 7$ as the dividend.

$$
\begin{array}{r}
3x + 4 \\
x^2 + 1 \overline{)\smash{3x^3 + 4x^2 + x + 7}} \\
\underline{3x^3 + 3x } \\
4x^2 - 2x + 7 \\
\underline{4x^2 + 4} \\
-2x + 3
\end{array}
$$

← Divide $4x^2$ by x^2 to get 4.

← Multiply $x^2 + 1$ by 4; subtract.

Since x^2 does not divide $-2x$ evenly (that is, the result is not a monomial), the process ends. The quotient is $3x + 4$, and the remainder is $-2x + 3$.

✓**Check:** (Quotient)(Divisor) + Remainder

$$
\begin{aligned}
&= (3x + 4)(x^2 + 1) + (-2x + 3) \\
&= 3x^3 + 3x + 4x^2 + 4 + (-2x + 3) \\
&= 3x^3 + 4x^2 + x + 7 = \text{Dividend}
\end{aligned}
$$

Then

$$
\frac{3x^3 + 4x^2 + x + 7}{x^2 + 1} = 3x + 4 + \frac{-2x + 3}{x^2 + 1}
$$

The next example combines the steps involved in long division.

EXAMPLE 14

Dividing Two Polynomials

Find the quotient and the remainder when

$$
x^4 - 3x^3 + 2x - 5 \quad \text{is divided by} \quad x^2 - x + 1
$$

Solution

In setting up this division problem, it is necessary to leave a space for the missing x^2 term in the dividend.

$$
\begin{array}{r}
x^2 - 2x - 3 \\
x^2 - x + 1 \overline{)\smash{x^4 - 3x^3 + 2x - 5}} \\
\underline{x^4 - x^3 + x^2 } \\
-2x^3 - x^2 + 2x - 5 \\
\underline{-2x^3 + 2x^2 - 2x } \\
-3x^2 + 4x - 5 \\
\underline{-3x^2 + 3x - 3} \\
x - 2
\end{array}
$$

← Quotient

← Divisor

← Dividend

Subtract →

Subtract →

Subtract →

← Remainder

✓**Check:** (Quotient)(Divisor) + Remainder

$$
\begin{aligned}
&= (x^2 - 2x - 3)(x^2 - x + 1) + x - 2 \\
&= x^4 - x^3 + x^2 - 2x^3 + 2x^2 - 2x - 3x^2 + 3x - 3 + x - 2 \\
&= x^4 - 3x^3 + 2x - 5 = \text{Dividend}
\end{aligned}
$$

As a result,

$$
\frac{x^4 - 3x^3 + 2x - 5}{x^2 - x + 1} = x^2 - 2x - 3 + \frac{x - 2}{x^2 - x + 1}
$$

The process of dividing two polynomials leads to the following result:

THEOREM Let Q be a polynomial of positive degree and let P be a polynomial whose degree is greater than or equal to the degree of Q. The remainder after dividing P by Q is either the zero polynomial or a polynomial whose degree is less than the degree of the divisor Q.

━━━━━▶ **Now Work** PROBLEM 93

7 Work with Polynomials in Two Variables

A **monomial in two variables** x and y has the form $ax^n y^m$, where a is a constant, x and y are variables, and n and m are nonnegative integers. The **degree** of a monomial is the sum of the powers of the variables.

For example,

$$2xy^3, \quad x^2 y^2, \quad \text{and} \quad x^3 y$$

are monomials, each of which has degree 4.

A **polynomial in two variables** x and y is the sum of one or more monomials in two variables. The **degree of a polynomial** in two variables is the highest degree of all the monomials with nonzero coefficients.

EXAMPLE 15 | **Examples of Polynomials in Two Variables**

$$3x^2 + 2x^3 y + 5 \qquad \pi x^3 - y^2 \qquad x^4 + 4x^3 y - xy^3 + y^4$$

Two variables, degree is 4. Two variables, degree is 3. Two variables, degree is 4.

Multiplying polynomials in two variables is handled in the same way as polynomials in one variable.

EXAMPLE 16 | **Using a Special Product Formula**

To multiply $(2x - y)^2$, use the Squares of Binomials formula (3b) with $2x$ instead of x and y instead of a.

$$(2x - y)^2 = (2x)^2 - 2 \cdot y \cdot 2x + y^2$$
$$= 4x^2 - 4xy + y^2$$

━━━━━▶ **Now Work** PROBLEM 79

R.4 Assess Your Understanding

Concepts and Vocabulary

1. The polynomial $3x^4 - 2x^3 + 13x^2 - 5$ is of degree _____. The leading coefficient is _____.

2. $(x^2 - 4)(x^2 + 4) = $ _____.

3. $(x - 2)(x^2 + 2x + 4) = $ _____.

4. *True or False* $4x^{-2}$ is a monomial of degree -2.

5. *True or False* The degree of the product of two nonzero polynomials equals the sum of their degrees.

6. *True or False* $(x + a)(x^2 + ax + a) = x^3 + a^3$.

Skill Building

In Problems 7–16, tell whether the expression is a monomial. If it is, name the variable(s) and the coefficient and give the degree of the monomial. If it is not a monomial, state why not.

7. $2x^3$
8. $-4x^2$
9. $\dfrac{8}{x}$
10. $-2x^{-3}$
11. $-2xy^2$

12. $5x^2y^3$
13. $\dfrac{8x}{y}$
14. $-\dfrac{2x^2}{y^3}$
15. $x^2 + y^2$
16. $3x^2 + 4$

In Problems 17–26, tell whether the expression is a polynomial. If it is, give its degree. If it is not, state why not.

17. $3x^2 - 5$
18. $1 - 4x$
19. 5
20. $-\pi$
21. $3x^2 - \dfrac{5}{x}$

22. $\dfrac{3}{x} + 2$
23. $2y^3 - \sqrt{2}$
24. $10z^2 + z$
25. $\dfrac{x^2 + 5}{x^3 - 1}$
26. $\dfrac{3x^3 + 2x - 1}{x^2 + x + 1}$

In Problems 27–46, add, subtract, or multiply, as indicated. Express your answer as a single polynomial in standard form.

27. $(x^2 + 4x + 5) + (3x - 3)$

28. $(x^3 + 3x^2 + 2) + (x^2 - 4x + 4)$

29. $(x^3 - 2x^2 + 5x + 10) - (2x^2 - 4x + 3)$

30. $(x^2 - 3x - 4) - (x^3 - 3x^2 + x + 5)$

31. $(6x^5 + x^3 + x) + (5x^4 - x^3 + 3x^2)$

32. $(10x^5 - 8x^2) + (3x^3 - 2x^2 + 6)$

33. $(x^2 - 3x + 1) + 2(3x^2 + x - 4)$

34. $-2(x^2 + x + 1) + (-5x^2 - x + 2)$

35. $6(x^3 + x^2 - 3) - 4(2x^3 - 3x^2)$

36. $8(4x^3 - 3x^2 - 1) - 6(4x^3 + 8x - 2)$

37. $(x^2 - x + 2) + (2x^2 - 3x + 5) - (x^2 + 1)$

38. $(x^2 + 1) - (4x^2 + 5) + (x^2 + x - 2)$

39. $9(y^2 - 3y + 4) - 6(1 - y^2)$

40. $8(1 - y^3) + 4(1 + y + y^2 + y^3)$

41. $x(x^2 + x - 4)$

42. $4x^2(x^3 - x + 2)$

43. $-2x^2(4x^3 + 5)$

44. $5x^3(3x - 4)$

45. $(x + 1)(x^2 + 2x - 4)$

46. $(2x - 3)(x^2 + x + 1)$

In Problems 47–64, multiply the polynomials using the FOIL method. Express your answer as a single polynomial in standard form.

47. $(x + 2)(x + 4)$
48. $(x + 3)(x + 5)$
49. $(2x + 5)(x + 2)$

50. $(3x + 1)(2x + 1)$
51. $(x - 4)(x + 2)$
52. $(x + 4)(x - 2)$

53. $(x - 3)(x - 2)$
54. $(x - 5)(x - 1)$
55. $(2x + 3)(x - 2)$

56. $(2x - 4)(3x + 1)$
57. $(-2x + 3)(x - 4)$
58. $(-3x - 1)(x + 1)$

59. $(-x - 2)(-2x - 4)$
60. $(-2x - 3)(3 - x)$
61. $(x - 2y)(x + y)$

62. $(2x + 3y)(x - y)$
63. $(-2x - 3y)(3x + 2y)$
64. $(x - 3y)(-2x + y)$

In Problems 65–88, multiply the polynomials using the special product formulas. Express your answer as a single polynomial in standard form.

65. $(x - 7)(x + 7)$
66. $(x - 1)(x + 1)$
67. $(2x + 3)(2x - 3)$
68. $(3x + 2)(3x - 2)$

69. $(x + 4)^2$
70. $(x + 5)^2$
71. $(x - 4)^2$
72. $(x - 5)^2$

73. $(3x + 4)(3x - 4)$
74. $(5x - 3)(5x + 3)$
75. $(2x - 3)^2$
76. $(3x - 4)^2$

77. $(x + y)(x - y)$ **78.** $(x + 3y)(x - 3y)$ **79.** $(3x + y)(3x - y)$ **80.** $(3x + 4y)(3x - 4y)$

81. $(x + y)^2$ **82.** $(x - y)^2$ **83.** $(x - 2y)^2$ **84.** $(2x + 3y)^2$

85. $(x - 2)^3$ **86.** $(x + 1)^3$ **87.** $(2x + 1)^3$ **88.** $(3x - 2)^3$

In Problems 89–104, find the quotient and the remainder. Check your work by verifying that

$$(Quotient)(Divisor) + Remainder = Dividend$$

89. $4x^3 - 3x^2 + x + 1$ divided by $x + 2$

90. $3x^3 - x^2 + x - 2$ divided by $x + 2$

91. $4x^3 - 3x^2 + x + 1$ divided by x^2

92. $3x^3 - x^2 + x - 2$ divided by x^2

93. $5x^4 - 3x^2 + x + 1$ divided by $x^2 + 2$

94. $5x^4 - x^2 + x - 2$ divided by $x^2 + 2$

95. $4x^5 - 3x^2 + x + 1$ divided by $2x^3 - 1$

96. $3x^5 - x^2 + x - 2$ divided by $3x^3 - 1$

97. $2x^4 - 3x^3 + x + 1$ divided by $2x^2 + x + 1$

98. $3x^4 - x^3 + x - 2$ divided by $3x^2 + x + 1$

99. $-4x^3 + x^2 - 4$ divided by $x - 1$

100. $-3x^4 - 2x - 1$ divided by $x - 1$

101. $1 - x^2 + x^4$ divided by $x^2 + x + 1$

102. $1 - x^2 + x^4$ divided by $x^2 - x + 1$

103. $x^3 - a^3$ divided by $x - a$

104. $x^5 - a^5$ divided by $x - a$

Explaining Concepts: Discussion and Writing

105. Explain why the degree of the product of two nonzero polynomials equals the sum of their degrees.

106. Explain why the degree of the sum of two polynomials of different degrees equals the larger of their degrees.

107. Give a careful statement about the degree of the sum of two polynomials of the same degree.

108. Do you prefer adding two polynomials using the horizontal method or the vertical method? Write a brief position paper defending your choice.

109. Do you prefer to memorize the rule for the square of a binomial $(x + a)^2$ or to use FOIL to obtain the product? Write a brief position paper defending your choice.

R.5 Factoring Polynomials

> **OBJECTIVES** **1** Factor the Difference of Two Squares and the Sum and Difference of Two Cubes (p. 50)
> **2** Factor Perfect Squares (p. 51)
> **3** Factor a Second-Degree Polynomial: $x^2 + Bx + C$ (p. 52)
> **4** Factor by Grouping (p. 53)
> **5** Factor a Second-Degree Polynomial: $Ax^2 + Bx + C, A \neq 1$ (p. 54)
> **6** Complete the Square (p. 56)

Consider the following product:

$$(2x + 3)(x - 4) = 2x^2 - 5x - 12$$

The two polynomials on the left side are called **factors** of the polynomial on the right side. Expressing a given polynomial as a product of other polynomials, that is, finding the factors of a polynomial, is called **factoring**.

We shall restrict our discussion here to factoring polynomials in one variable into products of polynomials in one variable, where all coefficients are integers. We call this **factoring over the integers.**

Any polynomial can be written as the product of 1 times itself or as -1 times its additive inverse. If a polynomial cannot be written as the product of two other polynomials (excluding 1 and -1), then the polynomial is said to be **prime.** When a polynomial has been written as a product consisting only of prime factors, it is said to be **factored completely.** Examples of prime polynomials (over the integers) are

$$2, \quad 3, \quad 5, \quad x, \quad x+1, \quad x-1, \quad 3x+4 \quad x^2+4$$

The first factor to look for in a factoring problem is a common monomial factor present in each term of the polynomial. If one is present, use the Distributive Property to factor it out. Continue factoring out monomial factors until none are left.

COMMENT Over the real numbers, $3x + 4$ factors into $3(x + \frac{4}{3})$. It is the noninteger $\frac{4}{3}$ that causes $3x + 4$ to be prime over the integers. In most instances, we will be factoring over the integers. ∎

EXAMPLE 1		**Identifying Common Monomial Factors**		

Polynomial	Common Monomial Factor	Remaining Factor	Factored Form
$2x + 4$	2	$x + 2$	$2x + 4 = 2(x + 2)$
$3x - 6$	3	$x - 2$	$3x - 6 = 3(x - 2)$
$2x^2 - 4x + 8$	2	$x^2 - 2x + 4$	$2x^2 - 4x + 8 = 2(x^2 - 2x + 4)$
$8x - 12$	4	$2x - 3$	$8x - 12 = 4(2x - 3)$
$x^2 + x$	x	$x + 1$	$x^2 + x = x(x + 1)$
$x^3 - 3x^2$	x^2	$x - 3$	$x^3 - 3x^2 = x^2(x - 3)$
$6x^2 + 9x$	$3x$	$2x + 3$	$6x^2 + 9x = 3x(2x + 3)$

Notice that, once all common monomial factors have been removed from a polynomial, the remaining factor is either a prime polynomial of degree 1 or a polynomial of degree 2 or higher. (Do you see why?)

━━━━━▶ **Now Work** PROBLEM 5

1 Factor the Difference of Two Squares and the Sum and Difference of Two Cubes

When you factor a polynomial, first check for common monomial factors. Then see whether you can use one of the special formulas discussed in the previous section.

Difference of Two Squares	$x^2 - a^2 = (x - a)(x + a)$
Perfect Squares	$x^2 + 2ax + a^2 = (x + a)^2$
	$x^2 - 2ax + a^2 = (x - a)^2$
Sum of Two Cubes	$x^3 + a^3 = (x + a)(x^2 - ax + a^2)$
Difference of Two Cubes	$x^3 - a^3 = (x - a)(x^2 + ax + a^2)$

EXAMPLE 2		**Factoring the Difference of Two Squares**

Factor completely: $x^2 - 4$

Solution Notice that $x^2 - 4$ is the difference of two squares, x^2 and 2^2.

$$x^2 - 4 = (x - 2)(x + 2)$$

EXAMPLE 3

Factoring the Difference of Two Cubes

Factor completely: $x^3 - 1$

Solution Because $x^3 - 1$ is the difference of two cubes, x^3 and 1^3,

$$x^3 - 1 = (x - 1)(x^2 + x + 1)$$

EXAMPLE 4

Factoring the Sum of Two Cubes

Factor completely: $x^3 + 8$

Solution Because $x^3 + 8$ is the sum of two cubes, x^3 and 2^3,

$$x^3 + 8 = (x + 2)(x^2 - 2x + 4)$$

EXAMPLE 5

Factoring the Difference of Two Squares

Factor completely: $x^4 - 16$

Solution Because $x^4 - 16$ is the difference of two squares, $x^4 = (x^2)^2$ and $16 = 4^2$,

$$x^4 - 16 = (x^2 - 4)(x^2 + 4)$$

But $x^2 - 4$ is also the difference of two squares. Then,

$$x^4 - 16 = (x^2 - 4)(x^2 + 4) = (x - 2)(x + 2)(x^2 + 4)$$

══➤ **Now Work** PROBLEMS **15** AND **33**

2 Factor Perfect Squares

When the first term and third term of a trinomial are both positive and are perfect squares, such as x^2, $9x^2$, 1, and 4, check to see whether the trinomial is a perfect square.

EXAMPLE 6

Factoring a Perfect Square

Factor completely: $x^2 + 6x + 9$

Solution The first term, x^2, and the third term, $9 = 3^2$, are perfect squares. Because the middle term $6x$ is twice the product of x and 3, we have a perfect square.

$$x^2 + 6x + 9 = (x + 3)^2$$

EXAMPLE 7

Factoring a Perfect Square

Factor completely: $9x^2 - 6x + 1$

Solution The first term, $9x^2 = (3x)^2$, and the third term, $1 = 1^2$, are perfect squares. Because the middle term, $-6x$, is -2 times the product of $3x$ and 1, we have a perfect square.

$$9x^2 - 6x + 1 = (3x - 1)^2$$

EXAMPLE 8

Factoring a Perfect Square

Factor completely: $25x^2 + 30x + 9$

Solution The first term, $25x^2 = (5x)^2$, and the third term, $9 = 3^2$, are perfect squares. Because the middle term, $30x$, is twice the product of $5x$ and 3, we have a perfect square.

$$25x^2 + 30x + 9 = (5x + 3)^2$$

══➤ **Now Work** PROBLEMS **25** AND **99**

If a trinomial is not a perfect square, it may be possible to factor it using the technique discussed next.

3 Factor a Second-Degree Polynomial: $x^2 + Bx + C$

The idea behind factoring a second-degree polynomial like $x^2 + Bx + C$ is to see whether it can be made equal to the product of two, possibly equal, first-degree polynomials.

For example, we know that

$$(x + 3)(x + 4) = x^2 + 7x + 12$$

The factors of $x^2 + 7x + 12$ are $x + 3$ and $x + 4$. Notice the following:

$$x^2 + 7x + 12 = (x + 3)(x + 4)$$

12 is the product of 3 and 4
7 is the sum of 3 and 4

In general, if $x^2 + Bx + C = (x + a)(x + b) = x^2 + (a + b)x + ab$, then $ab = C$ and $a + b = B$.

To factor a second-degree polynomial $x^2 + Bx + C$, find integers whose product is C and whose sum is B. That is, if there are numbers a, b, where $ab = C$ and $a + b = B$, then

$$x^2 + Bx + C = (x + a)(x + b)$$

EXAMPLE 9 **Factoring a Trinomial**

Factor completely: $x^2 + 7x + 10$

Solution First, determine all pairs of integers whose product is 10 and then compute their sums.

Integers whose product is 10	1, 10	−1, −10	2, 5	−2, −5
Sum	11	−11	7	−7

The integers 2 and 5 have a product of 10 and add up to 7, the coefficient of the middle term. As a result,

$$x^2 + 7x + 10 = (x + 2)(x + 5)$$

EXAMPLE 10 **Factoring a Trinomial**

Factor completely: $x^2 - 6x + 8$

Solution First, determine all pairs of integers whose product is 8 and then compute each sum.

Integers whose product is 8	1, 8	−1, −8	2, 4	−2, −4
Sum	9	−9	6	−6

Since −6 is the coefficient of the middle term,

$$x^2 - 6x + 8 = (x - 2)(x - 4)$$

EXAMPLE 11 **Factoring a Trinomial**

Factor completely: $x^2 - x - 12$

Solution First, determine all pairs of integers whose product is -12 and then compute each sum.

Integers whose product is −12	1, −12	−1, 12	2, −6	−2, 6	3, −4	−3, 4
Sum	−11	11	−4	4	−1	1

Since -1 is the coefficient of the middle term,

$$x^2 - x - 12 = (x + 3)(x - 4)$$

EXAMPLE 12 **Factoring a Trinomial**

Factor completely: $x^2 + 4x - 12$

Solution The integers -2 and 6 have a product of -12 and have the sum 4. So,

$$x^2 + 4x - 12 = (x - 2)(x + 6)$$

To avoid errors in factoring, always check your answer by multiplying it out to see if the result equals the original expression.
When none of the possibilities works, the polynomial is prime.

EXAMPLE 13 **Identifying a Prime Polynomial**

Show that $x^2 + 9$ is prime.

Solution First, list the pairs of integers whose product is 9 and then compute their sums.

Integers whose product is 9	1, 9	−1, −9	3, 3	−3, −3
Sum	10	−10	6	−6

Since the coefficient of the middle term in $x^2 + 9 = x^2 + 0x + 9$ is 0 and none of the sums equals 0, we conclude that $x^2 + 9$ is prime.

Example 13 demonstrates a more general result:

THEOREM Any polynomial of the form $x^2 + a^2$, a real, is prime.

Now Work PROBLEMS 39 AND 83

4 Factor by Grouping

Sometimes a common factor does not occur in every term of the polynomial, but in each of several groups of terms that together make up the polynomial. When this happens, the common factor can be factored out of each group by means of the Distributive Property. This technique is called **factoring by grouping.**

EXAMPLE 14 **Factoring by Grouping**

Factor completely by grouping: $(x^2 + 2)x + (x^2 + 2) \cdot 3$

Solution Notice the common factor $x^2 + 2$. By applying the Distributive Property, we have

$$(x^2 + 2)x + (x^2 + 2) \cdot 3 = (x^2 + 2)(x + 3)$$

Since $x^2 + 2$ and $x + 3$ are prime, the factorization is complete.

The next example shows a factoring problem that occurs in calculus.

EXAMPLE 15 | **Factoring by Grouping**

Factor completely by grouping: $3(x - 1)^2(x + 2)^4 + 4(x - 1)^3(x + 2)^3$

Solution Here, $(x - 1)^2(x + 2)^3$ is a common factor of $3(x - 1)^2(x + 2)^4$ and of $4(x - 1)^3(x + 2)^3$. As a result,

$$3(x - 1)^2(x + 2)^4 + 4(x - 1)^3(x + 2)^3 = (x - 1)^2(x + 2)^3[3(x + 2) + 4(x - 1)]$$
$$= (x - 1)^2(x + 2)^3[3x + 6 + 4x - 4]$$
$$= (x - 1)^2(x + 2)^3(7x + 2)$$

EXAMPLE 16 | **Factoring by Grouping**

Factor completely by grouping: $x^3 - 4x^2 + 2x - 8$

Solution To see if factoring by grouping will work, group the first two terms and the last two terms. Then look for a common factor in each group. In this example, we can factor x^2 from $x^3 - 4x^2$ and 2 from $2x - 8$. The remaining factor in each case is the same, $x - 4$. This means that factoring by grouping will work, as follows:

$$x^3 - 4x^2 + 2x - 8 = (x^3 - 4x^2) + (2x - 8)$$
$$= x^2(x - 4) + 2(x - 4)$$
$$= (x - 4)(x^2 + 2)$$

Since $x^2 + 2$ and $x - 4$ are prime, the factorization is complete.

━━━━━━━**Now Work** PROBLEMS **51** AND **127**

5 Factor a Second-Degree Polynomial: $Ax^2 + Bx + C, A \neq 1$

To factor a second-degree polynomial $Ax^2 + Bx + C$, when $A \neq 1$ and $A, B,$ and C have no common factors, follow these steps:

> ### Steps for Factoring $Ax^2 + Bx + C$, when $A \neq 1$ and $A, B,$ and C Have No Common Factors
>
> **STEP 1:** Find the value of AC.
> **STEP 2:** Find a pair of integers whose product is AC that add up to B. That is, find a and b so that $ab = AC$ and $a + b = B$.
> **STEP 3:** Write $Ax^2 + Bx + C = Ax^2 + ax + bx + C$.
> **STEP 4:** Factor this last expression by grouping.

EXAMPLE 17 | **Factoring a Trinomial**

Factor completely: $2x^2 + 5x + 3$

Solution Comparing $2x^2 + 5x + 3$ to $Ax^2 + Bx + C$, we find that $A = 2, B = 5,$ and $C = 3$.

STEP 1: The value of AC is $2 \cdot 3 = 6$.

STEP 2: Determine the pairs of integers whose product is $AC = 6$ and compute their sums.

Integers whose product is 6	1, 6	−1, −6	2, 3	−2, −3
Sum	7	−7	5	−5

STEP 3: The integers whose product is 6 that add up to $B = 5$ are 2 and 3.

$$2x^2 + 5x + 3 = 2x^2 + 2x + 3x + 3$$

STEP 4: Factor by grouping.

$$2x^2 + 2x + 3x + 3 = (2x^2 + 2x) + (3x + 3)$$
$$= 2x(x + 1) + 3(x + 1)$$
$$= (x + 1)(2x + 3)$$

As a result,

$$2x^2 + 5x + 3 = (x + 1)(2x + 3)$$

EXAMPLE 18 **Factoring a Trinomial**

Factor completely: $2x^2 - x - 6$

Solution Comparing $2x^2 - x - 6$ to $Ax^2 + Bx + C$, we find that $A = 2, B = -1$, and $C = -6$.

STEP 1: The value of AC is $2 \cdot (-6) = -12$.

STEP 2: Determine the pairs of integers whose product is $AC = -12$ and compute their sums.

Integers whose product is −12	1, −12	−1, 12	2, −6	−2, 6	3, −4	−3, 4
Sum	−11	11	−4	4	−1	1

STEP 3: The integers whose product is −12 that add up to $B = -1$ are −4 and 3.

$$2x^2 - x - 6 = 2x^2 - 4x + 3x - 6$$

STEP 4: Factor by grouping.

$$2x^2 - 4x + 3x - 6 = (2x^2 - 4x) + (3x - 6)$$
$$= 2x(x - 2) + 3(x - 2)$$
$$= (x - 2)(2x + 3)$$

As a result,

$$2x^2 - x - 6 = (x - 2)(2x + 3)$$

Now Work PROBLEM 57

SUMMARY

Type of Polynomial	Method	Example
Any polynomial	Look for common monomial factors. (Always do this first!)	$6x^2 + 9x = 3x(2x + 3)$
Binomials of degree 2 or higher	Check for a special product: Difference of two squares, $x^2 - a^2$ Difference of two cubes, $x^3 - a^3$ Sum of two cubes, $x^3 + a^3$	$x^2 - 16 = (x - 4)(x + 4)$ $x^3 - 64 = (x - 4)(x^2 + 4x + 16)$ $x^3 + 27 = (x + 3)(x^2 - 3x + 9)$
Trinomials of degree 2	Check for a perfect square, $(x \pm a)^2$ Factoring $x^2 + Bx + C$ (p. 52) Factoring $Ax^2 + Bx + C$ (p. 54)	$x^2 + 8x + 16 = (x + 4)^2$ $x^2 - 10x + 25 = (x - 5)^2$ $x^2 - x - 2 = (x - 2)(x + 1)$ $6x^2 + x - 1 = (2x + 1)(3x - 1)$
Four or more terms	Grouping	$2x^3 - 3x^2 + 4x - 6 = (2x - 3)(x^2 + 2)$

6 Complete the Square

The idea behind completing the square in one variable is to "adjust" an expression of the form $x^2 + bx$ to make it a perfect square. Perfect squares are trinomials of the form

$$x^2 + 2ax + a^2 = (x + a)^2 \text{ or } x^2 - 2ax + a^2 = (x - a)^2$$

For example, $x^2 + 6x + 9$ is a perfect square because $x^2 + 6x + 9 = (x + 3)^2$. And $p^2 - 12p + 36$ is a perfect square because $p^2 - 12p + 36 = (p - 6)^2$.

So how do we "adjust" $x^2 + bx$ to make it a perfect square? We do it by adding a number. For example, to make $x^2 + 6x$ a perfect square, add 9. But how do we know to add 9? If we divide the coefficient on the first-degree term, 6, by 2, and then square the result, we obtain 9. This approach works in general.

> ### Completing the Square
>
> Identify the coefficient of the first-degree term. Multiply this coefficient by $\dfrac{1}{2}$ and then square the result. That is, determine the value of b in $x^2 + bx$ and compute $\left(\dfrac{1}{2}b\right)^2$.

EXAMPLE 19 **Completing the Square**

Determine the number that must be added to each expression to complete the square. Then factor the expression.

Start	Add	Result	Factored Form
$y^2 + 8y$	$\left(\dfrac{1}{2} \cdot 8\right)^2 = 16$	$y^2 + 8y + 16$	$(y + 4)^2$
$x^2 + 12x$	$\left(\dfrac{1}{2} \cdot 12\right)^2 = 36$	$x^2 + 12x + 36$	$(x + 6)^2$
$a^2 - 20a$	$\left(\dfrac{1}{2} \cdot (-20)\right)^2 = 100$	$a^2 - 20a + 100$	$(a - 10)^2$
$p^2 - 5p$	$\left(\dfrac{1}{2} \cdot (-5)\right)^2 = \dfrac{25}{4}$	$p^2 - 5p + \dfrac{25}{4}$	$\left(p - \dfrac{5}{2}\right)^2$

Notice that the factored form of a perfect square is either

$$x^2 + bx + \left(\frac{b}{2}\right)^2 = \left(x + \frac{b}{2}\right)^2 \text{ or } x^2 - bx + \left(\frac{b}{2}\right)^2 = \left(x - \frac{b}{2}\right)^2$$

── **Now Work** PROBLEM 69

Are you wondering why we call making an expression a perfect square "completing the square"? Look at the square in Figure 27. Its area is $(y + 4)^2$. The yellow area is y^2 and each orange area is $4y$ (for a total area of $8y$). The sum of these areas is $y^2 + 8y$. To complete the square, we need to add the area of the green region: $4 \cdot 4 = 16$. As a result, $y^2 + 8y + 16 = (y + 4)^2$.

Figure 27

R.5 Assess Your Understanding

Concepts and Vocabulary

1. If factored completely, $3x^3 - 12x = $ _____.

2. If a polynomial cannot be written as the product of two other polynomials (excluding 1 and -1), then the polynomial is said to be _____.

3. *True or False* The polynomial $x^2 + 4$ is prime.

4. *True or False* $3x^3 - 2x^2 - 6x + 4 = (3x - 2)(x^2 + 2)$.

Skill Building

In Problems 5–14, factor each polynomial by removing the common monomial factor.

5. $3x + 6$

6. $7x - 14$

7. $ax^2 + a$

8. $ax - a$

9. $x^3 + x^2 + x$

10. $x^3 - x^2 + x$

11. $2x^2 - 2x$

12. $3x^2 - 3x$

13. $3x^2y - 6xy^2 + 12xy$

14. $60x^2y - 48xy^2 + 72x^3y$

In Problems 15–22, factor the difference of two squares.

15. $x^2 - 1$

16. $x^2 - 4$

17. $4x^2 - 1$

18. $9x^2 - 1$

19. $x^2 - 16$

20. $x^2 - 25$

21. $25x^2 - 4$

22. $36x^2 - 9$

In Problems 23–32, factor the perfect squares.

23. $x^2 + 2x + 1$

24. $x^2 - 4x + 4$

25. $x^2 + 4x + 4$

26. $x^2 - 2x + 1$

27. $x^2 - 10x + 25$

28. $x^2 + 10x + 25$

29. $4x^2 + 4x + 1$

30. $9x^2 + 6x + 1$

31. $16x^2 + 8x + 1$

32. $25x^2 + 10x + 1$

In Problems 33–38, factor the sum or difference of two cubes.

33. $x^3 - 27$

34. $x^3 + 125$

35. $x^3 + 27$

36. $27 - 8x^3$

37. $8x^3 + 27$

38. $64 - 27x^3$

In Problems 39–50, factor each polynomial.

39. $x^2 + 5x + 6$

40. $x^2 + 6x + 8$

41. $x^2 + 7x + 6$

42. $x^2 + 9x + 8$

43. $x^2 + 7x + 10$

44. $x^2 + 11x + 10$

45. $x^2 - 10x + 16$

46. $x^2 - 17x + 16$

47. $x^2 - 7x - 8$

48. $x^2 - 2x - 8$

49. $x^2 + 7x - 8$

50. $x^2 + 2x - 8$

In Problems 51–56, factor by grouping.

51. $2x^2 + 4x + 3x + 6$

52. $3x^2 - 3x + 2x - 2$

53. $2x^2 - 4x + x - 2$

54. $3x^2 + 6x - x - 2$

55. $6x^2 + 9x + 4x + 6$

56. $9x^2 - 6x + 3x - 2$

In Problems 57–68, factor each polynomial.

57. $3x^2 + 4x + 1$

58. $2x^2 + 3x + 1$

59. $2z^2 + 5z + 3$

60. $6z^2 + 5z + 1$

61. $3x^2 + 2x - 8$

62. $3x^2 + 10x + 8$

63. $3x^2 - 2x - 8$

64. $3x^2 - 10x + 8$

65. $3x^2 + 14x + 8$

66. $3x^2 - 14x + 8$

67. $3x^2 + 10x - 8$

68. $3x^2 - 10x - 8$

In Problems 69–74, determine the number that should be added to complete the square of each expression. Then factor each expression.

69. $x^2 + 10x$

70. $p^2 + 14p$

71. $y^2 - 6y$

72. $x^2 - 4x$

73. $x^2 - \frac{1}{2}x$

74. $x^2 + \frac{1}{3}x$

Mixed Practice

In Problems 75–122, factor completely each polynomial. If the polynomial cannot be factored, say it is prime.

75. $x^2 - 36$

76. $x^2 - 9$

77. $2 - 8x^2$

78. $3 - 27x^2$

79. $x^2 + 11x + 10$

80. $x^2 + 5x + 4$

81. $x^2 - 10x + 21$

82. $x^2 - 6x + 8$

83. $4x^2 - 8x + 32$

84. $3x^2 - 12x + 15$

85. $x^2 + 4x + 16$

86. $x^2 + 12x + 36$

87. $15 + 2x - x^2$

88. $14 + 6x - x^2$

89. $3x^2 - 12x - 36$

90. $x^3 + 8x^2 - 20x$

91. $y^4 + 11y^3 + 30y^2$

92. $3y^3 - 18y^2 - 48y$

93. $4x^2 + 12x + 9$

94. $9x^2 - 12x + 4$

95. $6x^2 + 8x + 2$

96. $8x^2 + 6x - 2$

97. $x^4 - 81$

98. $x^4 - 1$

99. $x^6 - 2x^3 + 1$

100. $x^6 + 2x^3 + 1$

101. $x^7 - x^5$

102. $x^8 - x^5$

103. $16x^2 + 24x + 9$

104. $9x^2 - 24x + 16$

105. $5 + 16x - 16x^2$

106. $5 + 11x - 16x^2$

107. $4y^2 - 16y + 15$

108. $9y^2 + 9y - 4$

109. $1 - 8x^2 - 9x^4$

110. $4 - 14x^2 - 8x^4$

111. $x(x + 3) - 6(x + 3)$

112. $5(3x - 7) + x(3x - 7)$

113. $(x + 2)^2 - 5(x + 2)$

114. $(x - 1)^2 - 2(x - 1)$

115. $(3x - 2)^3 - 27$

116. $(5x + 1)^3 - 1$

117. $3(x^2 + 10x + 25) - 4(x + 5)$

118. $7(x^2 - 6x + 9) + 5(x - 3)$

119. $x^3 + 2x^2 - x - 2$

120. $x^3 - 3x^2 - x + 3$

121. $x^4 - x^3 + x - 1$

122. $x^4 + x^3 + x + 1$

Applications and Extensions

△ *In Problems 123–132, expressions that occur in calculus are given. Factor completely each expression.*

123. $2(3x + 4)^2 + (2x + 3) \cdot 2(3x + 4) \cdot 3$

124. $5(2x + 1)^2 + (5x - 6) \cdot 2(2x + 1) \cdot 2$

125. $2x(2x + 5) + x^2 \cdot 2$

126. $3x^2(8x - 3) + x^3 \cdot 8$

127. $2(x + 3)(x - 2)^3 + (x + 3)^2 \cdot 3(x - 2)^2$

128. $4(x + 5)^3(x - 1)^2 + (x + 5)^4 \cdot 2(x - 1)$

129. $(4x - 3)^2 + x \cdot 2(4x - 3) \cdot 4$

130. $3x^2(3x + 4)^2 + x^3 \cdot 2(3x + 4) \cdot 3$

131. $2(3x - 5) \cdot 3(2x + 1)^3 + (3x - 5)^2 \cdot 3(2x + 1)^2 \cdot 2$

132. $3(4x + 5)^2 \cdot 4(5x + 1)^2 + (4x + 5)^3 \cdot 2(5x + 1) \cdot 5$

133. Show that $x^2 + 4$ is prime.

134. Show that $x^2 + x + 1$ is prime.

Explaining Concepts: Discussion and Writing

135. Make up a polynomial that factors into a perfect square.

136. Explain to a fellow student what you look for first when presented with a factoring problem. What do you do next?

R.6 Synthetic Division

OBJECTIVE 1 Divide Polynomials Using Synthetic Division (p. 58)

1 Divide Polynomials Using Synthetic Division

To find the quotient as well as the remainder when a polynomial of degree 1 or higher is divided by $x - c$, a shortened version of long division, called **synthetic division,** makes the task simpler.

To see how synthetic division works, we use long division to divide the polynomial $2x^3 - x^2 + 3$ by $x - 3$.

$$
\begin{array}{r}
2x^2 + 5x \;+\; 15 \quad\;\; \leftarrow \text{Quotient} \\
x - 3 \overline{)\,2x^3 - \;\;x^2 \qquad\quad + \;3\;} \\
\underline{2x^3 - 6x^2\qquad\qquad} \\
5x^2 \qquad\qquad \\
\underline{5x^2 - 15x\qquad} \\
15x + \;\;3 \\
\underline{15x - 45} \\
48 \quad \leftarrow \text{Remainder}
\end{array}
$$

✓ **Check:** (Divisor) · (Quotient) + Remainder

$$
\begin{aligned}
&= (x - 3)(2x^2 + 5x + 15) + 48 \\
&= 2x^3 + 5x^2 + 15x - 6x^2 - 15x - 45 + 48 \\
&= 2x^3 - x^2 + 3
\end{aligned}
$$

The process of synthetic division arises from rewriting the long division in a more compact form, using simpler notation. For example, in the long division above, the terms in blue are not really necessary because they are identical to the terms directly above them. With these terms removed, we have

$$
\begin{array}{r}
2x^2 + 5x \;+\; 15\qquad \\
x - 3 \overline{)\,2x^3 - \;\;x^2 \qquad\quad + \;3\;} \\
- 6x^2\qquad\qquad \\
\overline{5x^2\qquad\qquad} \\
- 15x \\
\overline{15x\qquad} \\
- 45 \\
\overline{48}
\end{array}
$$

Most of the x's that appear in this process can also be removed, provided that we are careful about positioning each coefficient. In this regard, we will need to use 0 as the coefficient of x in the dividend, because that power of x is missing. Now we have

$$
\begin{array}{r}
2x^2 + 5x + 15\qquad \\
x - 3 \overline{)\,2 \quad - 1 \qquad 0 \qquad 3\;} \\
- 6\qquad\qquad\quad \\
\overline{5\qquad\qquad\quad} \\
- 15\quad\;\; \\
\overline{15\qquad} \\
- 45 \\
\overline{48}
\end{array}
$$

We can make this display more compact by moving the lines up until the numbers in blue align horizontally.

$$
\begin{array}{rcccc}
& 2x^2 + 5x + 15 & & & \text{Row 1} \\
x - 3 \overline{)\,2} & -1 & 0 & 3 & \text{Row 2} \\
& -6 & -15 & -45 & \text{Row 3} \\
\hline
\bigcirc & 5 & 15 & 48 & \text{Row 4}
\end{array}
$$

Because the leading coefficient of the divisor is always 1, we know that the leading coefficient of the dividend will also be the leading coefficient of the synthetic quotient. So we place the leading coefficient of the quotient, 2, in the circled position. Now, the first three numbers in row 4 are precisely the coefficients of the quotient, and the last

number in row 4 is the remainder. Thus, row 1 is not really needed, so we can compress the process to three rows, where the bottom row contains both the coefficients of the quotient and the remainder.

$$
\begin{array}{r}
x - 3\overline{)\,2 \quad -1 \quad \ \ 0 \quad \ \ 3\,} \quad \text{Row 1} \\
\underline{-6 - 15 - 45} \quad \text{Row 2 (subtract)} \\
2 \quad \ \ 5 \quad \ 15 \quad \ 48 \quad \text{Row 3}
\end{array}
$$

Recall that the entries in row 3 are obtained by subtracting the entries in row 2 from those in row 1. Rather than subtracting the entries in row 2, we can change the sign of each entry and add. With this modification, our display will look like this:

$$
\begin{array}{r}
x - 3\overline{)\,2 \quad -1 \quad \ \ 0 \quad \ \ 3\,} \quad \text{Row 1} \\
\underline{6 \quad \ 15 \quad \ 45} \quad \text{Row 2 (add)} \\
2 \quad \ \ 5 \quad \ 15 \quad \ 48 \quad \text{Row 3}
\end{array}
$$

Notice that the entries in row 2 are three times the prior entries in row 3. Our last modification to the display replaces the $x - 3$ by 3. The entries in row 3 give the quotient and the remainder, as shown next.

$$
\begin{array}{r}
3\overline{)\,2 \quad -1 \quad \ \ 0 \quad \ \ 3\,} \quad \text{Row 1} \\
\underline{6 \quad \ 15 \quad \ 45} \quad \text{Row 2 (add)} \\
2 \quad \ \ 5 \quad \ 15 \quad \ 48 \quad \text{Row 3}
\end{array}
$$

$$\underbrace{\qquad}_{\text{Quotient}} \qquad \underbrace{\ }_{\text{Remainder}}$$

$$2x^2 + 5x + 15 \qquad 48$$

Let's go through an example step by step.

EXAMPLE 1 **Using Synthetic Division to Find the Quotient and Remainder**

Use synthetic division to find the quotient and remainder when

$$x^3 - 4x^2 - 5 \quad \text{is divided by} \quad x - 3$$

Solution **STEP 1:** Write the dividend in descending powers of x. Then copy the coefficients, remembering to insert a 0 for any missing powers of x.

$$1 \quad -4 \quad 0 \quad -5 \quad \text{Row 1}$$

STEP 2: Insert the usual division symbol. In synthetic division, the divisor is of the form $x - c$, and c is the number placed to the left of the division symbol. Here, since the divisor is $x - 3$, we insert 3 to the left of the division symbol.

$$3\overline{)\,1 \quad -4 \quad 0 \quad -5\,} \quad \text{Row 1}$$

STEP 3: Bring the 1 down two rows, and enter it in row 3.

$$
\begin{array}{r}
3\overline{)\,1 \quad -4 \quad 0 \quad -5\,} \quad \text{Row 1} \\
\downarrow \qquad\qquad\qquad\qquad \text{Row 2} \\
\underline{} \\
1 \qquad\qquad\qquad\qquad\quad \text{Row 3}
\end{array}
$$

STEP 4: Multiply the latest entry in row 3 by 3, and place the result in row 2, one column over to the right.

$$
\begin{array}{r}
3\overline{)\,1 \quad -4 \quad 0 \quad -5\,} \quad \text{Row 1} \\
3 \qquad\qquad\qquad \text{Row 2} \\
\underline{} \\
1 \qquad\qquad\qquad\qquad\quad \text{Row 3}
\end{array}
$$

STEP 5: Add the entry in row 2 to the entry above it in row 1, and enter the sum in row 3.

$$
\begin{array}{r}
3\overline{)\,1 \quad -4 \quad 0 \quad -5\,} \quad \text{Row 1} \\
3 \qquad\qquad\qquad \text{Row 2} \\
\underline{} \\
1 \quad -1 \qquad\qquad\quad \text{Row 3}
\end{array}
$$

STEP 6: Repeat Steps 4 and 5 until no more entries are available in row 1.

$$3)\overline{1 \quad -4 \quad 0 \quad -5} \quad \text{Row 1}$$
$$3 \quad -3 \quad -9 \quad \text{Row 2}$$
$$1 \quad -1 \quad -3 \quad -14 \quad \text{Row 3}$$

STEP 7: The final entry in row 3, the -14, is the remainder; the other entries in row 3, the $1, -1,$ and $-3,$ are the coefficients (in descending order) of a polynomial whose degree is 1 less than that of the dividend. This is the quotient. Thus,

$$\text{Quotient} = x^2 - x - 3 \qquad \text{Remainder} = -14$$

✓ **Check:** $(\text{Divisor})(\text{Quotient}) + \text{Remainder}$

$$= (x - 3)(x^2 - x - 3) + (-14)$$
$$= (x^3 - x^2 - 3x - 3x^2 + 3x + 9) + (-14)$$
$$= x^3 - 4x^2 - 5 = \text{Dividend}$$

Let's do an example in which all seven steps are combined.

EXAMPLE 2

Using Synthetic Division to Verify a Factor

Use synthetic division to show that $x + 3$ is a factor of

$$2x^5 + 5x^4 - 2x^3 + 2x^2 - 2x + 3$$

Solution The divisor is $x + 3 = x - (-3)$, so we place -3 to the left of the division symbol. Then the row 3 entries will be multiplied by -3, entered in row 2, and added to row 1.

$$-3)\overline{2 \quad 5 \quad -2 \quad 2 \quad -2 \quad 3} \quad \text{Row 1}$$
$$-6 \quad 3 \quad -3 \quad 3 \quad -3 \quad \text{Row 2}$$
$$2 \quad -1 \quad 1 \quad -1 \quad 1 \quad 0 \quad \text{Row 3}$$

Because the remainder is 0, we have

$$(\text{Divisor})(\text{Quotient}) + \text{Remainder}$$
$$= (x + 3)(2x^4 - x^3 + x^2 - x + 1) = 2x^5 + 5x^4 - 2x^3 + 2x^2 - 2x + 3$$

As we see, $x + 3$ is a factor of $2x^5 + 5x^4 - 2x^3 + 2x^2 - 2x + 3$.

As Example 2 illustrates, the remainder after division gives information about whether the divisor is, or is not, a factor. We shall have more to say about this in Chapter 5.

Now Work PROBLEMS **7** AND **17**

R.6 Assess Your Understanding

Concepts and Vocabulary

1. To check division, the expression that is being divided, the dividend, should equal the product of the _____ and the _____ plus the _____.

2. To divide $2x^3 - 5x + 1$ by $x + 3$ using synthetic division, the first step is to write _____$\overline{)\qquad\qquad}$.

3. *True or False* In using synthetic division, the divisor is always a polynomial of degree 1, whose leading coefficient is 1.

4. *True or False* $-2)\overline{5 \quad 3 \quad 2 \quad 1}$ means $\dfrac{5x^3 + 3x^2 + 2x + 1}{x + 2} = 5x^2 - 7x + 16 + \dfrac{-31}{x + 2}$.
$$-10 \quad 14 \quad -32$$
$$5 \quad -7 \quad 16 \quad -31$$

Skill Building

In Problems 5–16, use synthetic division to find the quotient and remainder when:

5. $x^3 - x^2 + 2x + 4$ is divided by $x - 2$

6. $x^3 + 2x^2 - 3x + 1$ is divided by $x + 1$

7. $3x^3 + 2x^2 - x + 3$ is divided by $x - 3$

8. $-4x^3 + 2x^2 - x + 1$ is divided by $x + 2$

9. $x^5 - 4x^3 + x$ is divided by $x + 3$

10. $x^4 + x^2 + 2$ is divided by $x - 2$

11. $4x^6 - 3x^4 + x^2 + 5$ is divided by $x - 1$

12. $x^5 + 5x^3 - 10$ is divided by $x + 1$

13. $0.1x^3 + 0.2x$ is divided by $x + 1.1$

14. $0.1x^2 - 0.2$ is divided by $x + 2.1$

15. $x^5 - 1$ is divided by $x - 1$

16. $x^5 + 1$ is divided by $x + 1$

In Problems 17–26, use synthetic division to determine whether $x - c$ is a factor of the given polynomial.

17. $4x^3 - 3x^2 - 8x + 4;\ \ x - 2$

18. $-4x^3 + 5x^2 + 8;\ \ x + 3$

19. $3x^4 - 6x^3 - 5x + 10;\ \ x - 2$

20. $4x^4 - 15x^2 - 4;\ \ x - 2$

21. $3x^6 + 82x^3 + 27;\ \ x + 3$

22. $2x^6 - 18x^4 + x^2 - 9;\ \ x + 3$

23. $4x^6 - 64x^4 + x^2 - 15;\ \ x + 4$

24. $x^6 - 16x^4 + x^2 - 16;\ \ x + 4$

25. $2x^4 - x^3 + 2x - 1;\ \ x - \dfrac{1}{2}$

26. $3x^4 + x^3 - 3x + 1;\ \ x + \dfrac{1}{3}$

Applications and Extensions

27. Find the sum of $a, b, c,$ and d if

$$\frac{x^3 - 2x^2 + 3x + 5}{x + 2} = ax^2 + bx + c + \frac{d}{x + 2}$$

Explaining Concepts: Discussion and Writing

28. When dividing a polynomial by $x - c$, do you prefer to use long division or synthetic division? Does the value of c make a difference to you in choosing? Give reasons.

R.7 Rational Expressions

> **OBJECTIVES** **1** Reduce a Rational Expression to Lowest Terms (p. 62)
> **2** Multiply and Divide Rational Expressions (p. 63)
> **3** Add and Subtract Rational Expressions (p. 64)
> **4** Use the Least Common Multiple Method (p. 66)
> **5** Simplify Complex Rational Expressions (p. 68)

1 Reduce a Rational Expression to Lowest Terms

If we form the quotient of two polynomials, the result is called a **rational expression.** Some examples of rational expressions are

(a) $\dfrac{x^3 + 1}{x}$ (b) $\dfrac{3x^2 + x - 2}{x^2 + 5}$ (c) $\dfrac{x}{x^2 - 1}$ (d) $\dfrac{xy^2}{(x - y)^2}$

Expressions (a), (b), and (c) are rational expressions in one variable, x, whereas (d) is a rational expression in two variables, x and y.

Rational expressions are described in the same manner as rational numbers. In expression (a), the polynomial $x^3 + 1$ is called the **numerator,** and x is called the **denominator.** When the numerator and denominator of a rational expression contain no common factors (except 1 and -1), we say that the rational expression is **reduced to lowest terms,** or **simplified.**

The polynomial in the denominator of a rational expression cannot be equal to 0 because division by 0 is not defined. For example, for the expression $\dfrac{x^3 + 1}{x}$, x cannot take on the value 0. The domain of the variable x is $\{x | x \neq 0\}$.

A rational expression is reduced to lowest terms by factoring the numerator and the denominator completely and canceling any common factors using the Cancellation Property:

$$\frac{a\cancel{c}}{b\cancel{c}} = \frac{a}{b} \qquad \text{if } b \neq 0, c \neq 0 \tag{1}$$

EXAMPLE 1

Reducing a Rational Expression to Lowest Terms

Reduce to lowest terms: $\dfrac{x^2 + 4x + 4}{x^2 + 3x + 2}$

Solution

Begin by factoring the numerator and the denominator.

$$x^2 + 4x + 4 = (x + 2)(x + 2)$$
$$x^2 + 3x + 2 = (x + 2)(x + 1)$$

WARNING Apply the Cancellation Property only to rational expressions written in factored form. Be sure to cancel only common factors! ∎

Since a common factor, $x + 2$, appears, the original expression is not in lowest terms. To reduce it to lowest terms, use the Cancellation Property:

$$\frac{x^2 + 4x + 4}{x^2 + 3x + 2} = \frac{\cancel{(x + 2)}(x + 2)}{\cancel{(x + 2)}(x + 1)} = \frac{x + 2}{x + 1} \qquad x \neq -2, -1$$

EXAMPLE 2

Reducing Rational Expressions to Lowest Terms

Reduce each rational expression to lowest terms.

(a) $\dfrac{x^3 - 8}{x^3 - 2x^2}$

(b) $\dfrac{8 - 2x}{x^2 - x - 12}$

Solution

(a) $\dfrac{x^3 - 8}{x^3 - 2x^2} = \dfrac{\cancel{(x - 2)}(x^2 + 2x + 4)}{x^2\cancel{(x - 2)}} = \dfrac{x^2 + 2x + 4}{x^2} \qquad x \neq 0, 2$

(b) $\dfrac{8 - 2x}{x^2 - x - 12} = \dfrac{2(4 - x)}{(x - 4)(x + 3)} = \dfrac{2(-1)\cancel{(x - 4)}}{\cancel{(x - 4)}(x + 3)} = \dfrac{-2}{x + 3} \qquad x \neq -3, 4$

Now Work PROBLEM 5

2 Multiply and Divide Rational Expressions

The rules for multiplying and dividing rational expressions are the same as the rules for multiplying and dividing rational numbers. If $\dfrac{a}{b}$ and $\dfrac{c}{d}$, $b \neq 0$, $d \neq 0$, are two rational expressions, then

$$\frac{a}{b} \cdot \frac{c}{d} = \frac{ac}{bd} \qquad \text{if } b \neq 0, d \neq 0 \tag{2}$$

$$\frac{\dfrac{a}{b}}{\dfrac{c}{d}} = \frac{a}{b} \cdot \frac{d}{c} = \frac{ad}{bc} \qquad \text{if } b \neq 0, c \neq 0, d \neq 0 \tag{3}$$

In using equations (2) and (3) with rational expressions, be sure first to factor each polynomial completely so that common factors can be canceled. Leave your answer in factored form.

EXAMPLE 3 **Multiplying and Dividing Rational Expressions**

Perform the indicated operation and simplify the result. Leave your answer in factored form.

(a) $\dfrac{x^2 - 2x + 1}{x^3 + x} \cdot \dfrac{4x^2 + 4}{x^2 + x - 2}$ (b) $\dfrac{\dfrac{x + 3}{x^2 - 4}}{\dfrac{x^2 - x - 12}{x^3 - 8}}$

Solution (a) $\dfrac{x^2 - 2x + 1}{x^3 + x} \cdot \dfrac{4x^2 + 4}{x^2 + x - 2} = \dfrac{(x - 1)^2}{x(x^2 + 1)} \cdot \dfrac{4(x^2 + 1)}{(x + 2)(x - 1)}$

$$= \dfrac{(x - 1)^2(4)\cancel{(x^2 + 1)}}{x\cancel{(x^2 + 1)}(x + 2)\cancel{(x - 1)}}$$

$$= \dfrac{4(x - 1)}{x(x + 2)} \qquad x \neq -2, 0, 1$$

(b) $\dfrac{\dfrac{x + 3}{x^2 - 4}}{\dfrac{x^2 - x - 12}{x^3 - 8}} = \dfrac{x + 3}{x^2 - 4} \cdot \dfrac{x^3 - 8}{x^2 - x - 12}$

$$= \dfrac{x + 3}{(x - 2)(x + 2)} \cdot \dfrac{(x - 2)(x^2 + 2x + 4)}{(x - 4)(x + 3)}$$

$$= \dfrac{\cancel{(x + 3)}\,\cancel{(x - 2)}(x^2 + 2x + 4)}{\cancel{(x - 2)}(x + 2)(x - 4)\cancel{(x + 3)}}$$

$$= \dfrac{x^2 + 2x + 4}{(x + 2)(x - 4)} \qquad x \neq -3, -2, 2, 4$$

Now Work PROBLEMS **17** AND **25**

3 Add and Subtract Rational Expressions

The rules for adding and subtracting rational expressions are the same as the rules for adding and subtracting rational numbers. So, if the denominators of two rational expressions to be added (or subtracted) are equal, we add (or subtract) the numerators and keep the common denominator.

In Words
To add (or subtract) two rational expressions with the same denominator, keep the common denominator and add (or subtract) the numerators.

If $\dfrac{a}{b}$ and $\dfrac{c}{b}$ are two rational expressions, then

$$\dfrac{a}{b} + \dfrac{c}{b} = \dfrac{a + c}{b} \qquad \dfrac{a}{b} - \dfrac{c}{b} = \dfrac{a - c}{b} \qquad \text{if } b \neq 0 \qquad \textbf{(4)}$$

EXAMPLE 4 **Adding and Subtracting Rational Expressions with Equal Denominators**

Perform the indicated operation and simplify the result. Leave your answer in factored form.

(a) $\dfrac{2x^2 - 4}{2x + 5} + \dfrac{x + 3}{2x + 5} \qquad x \neq -\dfrac{5}{2}$ (b) $\dfrac{x}{x - 3} - \dfrac{3x + 2}{x - 3} \qquad x \neq 3$

Solution (a) $\dfrac{2x^2 - 4}{2x + 5} + \dfrac{x + 3}{2x + 5} = \dfrac{(2x^2 - 4) + (x + 3)}{2x + 5}$

$$= \dfrac{2x^2 + x - 1}{2x + 5} = \dfrac{(2x - 1)(x + 1)}{2x + 5}$$

(b) $\dfrac{x}{x-3} - \dfrac{3x+2}{x-3} = \dfrac{x - (3x+2)}{x-3} = \dfrac{x - 3x - 2}{x-3}$

$$= \dfrac{-2x-2}{x-3} = \dfrac{-2(x+1)}{x-3}$$

EXAMPLE 5 **Adding Rational Expressions Whose Denominators Are Additive Inverses of Each Other**

Perform the indicated operation and simplify the result. Leave your answer in factored form.

$$\dfrac{2x}{x-3} + \dfrac{5}{3-x} \qquad x \neq 3$$

Solution Notice that the denominators of the two rational expressions are different. However, the denominator of the second expression is the additive inverse of the denominator of the first. That is,

$$3 - x = -x + 3 = -1 \cdot (x-3) = -(x-3)$$

Then

$$\dfrac{2x}{x-3} + \dfrac{5}{3-x} = \dfrac{2x}{x-3} + \dfrac{5}{-(x-3)} = \dfrac{2x}{x-3} + \dfrac{-5}{x-3}$$

$$\underset{\substack{\uparrow \\ 3-x = -(x-3)}}{} \qquad \underset{\substack{\uparrow \\ \frac{a}{-b} = \frac{-a}{b}}}{}$$

$$= \dfrac{2x + (-5)}{x-3} = \dfrac{2x-5}{x-3}$$

Now Work PROBLEMS **37 AND 43**

If the denominators of two rational expressions to be added or subtracted are not equal, we can use the general formulas for adding and subtracting rational expressions.

$$\dfrac{a}{b} + \dfrac{c}{d} = \dfrac{a \cdot d}{b \cdot d} + \dfrac{b \cdot c}{b \cdot d} = \dfrac{ad + bc}{bd} \qquad \text{if } b \neq 0, d \neq 0 \qquad \textbf{(5a)}$$

$$\dfrac{a}{b} - \dfrac{c}{d} = \dfrac{a \cdot d}{b \cdot d} - \dfrac{b \cdot c}{b \cdot d} = \dfrac{ad - bc}{bd} \qquad \text{if } b \neq 0, d \neq 0 \qquad \textbf{(5b)}$$

EXAMPLE 6 **Adding and Subtracting Rational Expressions with Unequal Denominators**

Perform the indicated operation and simplify the result. Leave your answer in factored form.

(a) $\dfrac{x-3}{x+4} + \dfrac{x}{x-2} \qquad x \neq -4, 2$ (b) $\dfrac{x^2}{x^2-4} - \dfrac{1}{x} \qquad x \neq -2, 0, 2$

Solution (a) $\dfrac{x-3}{x+4} + \dfrac{x}{x-2} = \dfrac{x-3}{x+4} \cdot \dfrac{x-2}{x-2} + \dfrac{x+4}{x+4} \cdot \dfrac{x}{x-2}$

$$\underset{\substack{\uparrow \\ (5a)}}{}$$

$$= \dfrac{(x-3)(x-2) + (x+4)(x)}{(x+4)(x-2)}$$

$$= \dfrac{x^2 - 5x + 6 + x^2 + 4x}{(x+4)(x-2)} = \dfrac{2x^2 - x + 6}{(x+4)(x-2)}$$

(b) $\dfrac{x^2}{x^2-4} - \dfrac{1}{x} = \dfrac{x^2}{x^2-4} \cdot \dfrac{x}{x} - \dfrac{x^2-4}{x^2-4} \cdot \dfrac{1}{x} = \dfrac{x^2(x) - (x^2-4)(1)}{(x^2-4)(x)}$

\uparrow
(5b)

$= \dfrac{x^3 - x^2 + 4}{(x-2)(x+2)(x)}$

Now Work PROBLEM 47

4 Use the Least Common Multiple Method

If the denominators of two rational expressions to be added (or subtracted) have common factors, we usually do not use the general rules given by equations (5a) and (5b). Just as with fractions, we apply the **least common multiple (LCM) method.** The LCM method uses the polynomial of least degree that has each denominator polynomial as a factor.

The LCM Method for Adding or Subtracting Rational Expressions

The Least Common Multiple (LCM) Method requires four steps:

STEP 1: Factor completely the polynomial in the denominator of each rational expression.

STEP 2: The LCM of the denominators is the product of each of these factors raised to a power equal to the greatest number of times that the factor occurs in the polynomials.

STEP 3: Write each rational expression using the LCM as the common denominator.

STEP 4: Add or subtract the rational expressions using equation (4).

We begin with an example that only requires Steps 1 and 2.

EXAMPLE 7

Finding the Least Common Multiple

Find the least common multiple of the following pair of polynomials:

$$x(x-1)^2(x+1) \quad \text{and} \quad 4(x-1)(x+1)^3$$

Solution

STEP 1: The polynomials are already factored completely as

$$x(x-1)^2(x+1) \quad \text{and} \quad 4(x-1)(x+1)^3$$

STEP 2: Start by writing the factors of the left-hand polynomial. (Or you could start with the one on the right.)

$$x(x-1)^2(x+1)$$

Now look at the right-hand polynomial. Its first factor, 4, does not appear in our list, so we insert it.

$$4x(x-1)^2(x+1)$$

The next factor, $x-1$, is already in our list, so no change is necessary. The final factor is $(x+1)^3$. Since our list has $x+1$ to the first power only, we replace $x+1$ in the list by $(x+1)^3$. The LCM is

$$4x(x-1)^2(x+1)^3$$

Notice that the LCM is, in fact, the polynomial of least degree that contains $x(x - 1)^2(x + 1)$ and $4(x - 1)(x + 1)^3$ as factors.

Now Work PROBLEM 53

EXAMPLE 8 **Using the Least Common Multiple to Add Rational Expressions**

Perform the indicated operation and simplify the result. Leave your answer in factored form.

$$\frac{x}{x^2 + 3x + 2} + \frac{2x - 3}{x^2 - 1} \qquad x \neq -2, -1, 1$$

Solution **STEP 1:** Factor completely the polynomials in the denominators.

$$x^2 + 3x + 2 = (x + 2)(x + 1)$$

$$x^2 - 1 = (x - 1)(x + 1)$$

STEP 2: The LCM is $(x + 2)(x + 1)(x - 1)$. Do you see why?

STEP 3: Write each rational expression using the LCM as the denominator.

$$\frac{x}{x^2 + 3x + 2} = \frac{x}{(x + 2)(x + 1)} = \frac{x}{(x + 2)(x + 1)} \cdot \frac{x - 1}{x - 1} = \frac{x(x - 1)}{(x + 2)(x + 1)(x - 1)}$$

↑ Multiply numerator and denominator by x − 1 to get the LCM in the denominator.

$$\frac{2x - 3}{x^2 - 1} = \frac{2x - 3}{(x - 1)(x + 1)} = \frac{2x - 3}{(x - 1)(x + 1)} \cdot \frac{x + 2}{x + 2} = \frac{(2x - 3)(x + 2)}{(x - 1)(x + 1)(x + 2)}$$

↑ Multiply numerator and denominator by x + 2 to get the LCM in the denominator.

STEP 4: Now we can add by using equation (4).

$$\frac{x}{x^2 + 3x + 2} + \frac{2x - 3}{x^2 - 1} = \frac{x(x - 1)}{(x + 2)(x + 1)(x - 1)} + \frac{(2x - 3)(x + 2)}{(x + 2)(x + 1)(x - 1)}$$

$$= \frac{(x^2 - x) + (2x^2 + x - 6)}{(x + 2)(x + 1)(x - 1)}$$

$$= \frac{3x^2 - 6}{(x + 2)(x + 1)(x - 1)} = \frac{3(x^2 - 2)}{(x + 2)(x + 1)(x - 1)}$$

EXAMPLE 9 **Using the Least Common Multiple to Subtract Rational Expressions**

Perform the indicated operation and simplify the result. Leave your answer in factored form.

$$\frac{3}{x^2 + x} - \frac{x + 4}{x^2 + 2x + 1} \qquad x \neq -1, 0$$

Solution **STEP 1:** Factor completely the polynomials in the denominators.

$$x^2 + x = x(x + 1)$$

$$x^2 + 2x + 1 = (x + 1)^2$$

STEP 2: The LCM is $x(x + 1)^2$.

STEP 3: Write each rational expression using the LCM as the denominator.

$$\frac{3}{x^2 + x} = \frac{3}{x(x+1)} = \frac{3}{x(x+1)} \cdot \frac{x+1}{x+1} = \frac{3(x+1)}{x(x+1)^2}$$

$$\frac{x+4}{x^2 + 2x + 1} = \frac{x+4}{(x+1)^2} = \frac{x+4}{(x+1)^2} \cdot \frac{x}{x} = \frac{x(x+4)}{x(x+1)^2}$$

STEP 4: Subtract, using equation (4).

$$\frac{3}{x^2 + x} - \frac{x+4}{x^2 + 2x + 1} = \frac{3(x+1)}{x(x+1)^2} - \frac{x(x+4)}{x(x+1)^2}$$

$$= \frac{3(x+1) - x(x+4)}{x(x+1)^2}$$

$$= \frac{3x + 3 - x^2 - 4x}{x(x+1)^2}$$

$$= \frac{-x^2 - x + 3}{x(x+1)^2}$$

Now Work PROBLEM 63

5 Simplify Complex Rational Expressions

When sums and/or differences of rational expressions appear as the numerator and/or denominator of a quotient, the quotient is called a **complex rational expression.*** For example,

$$\frac{1 + \dfrac{1}{x}}{1 - \dfrac{1}{x}} \quad \text{and} \quad \frac{\dfrac{x^2}{x^2 - 4} - 3}{\dfrac{x-3}{x+2} - 1}$$

are complex rational expressions. To **simplify** a complex rational expression means to write it as a rational expression reduced to lowest terms. This can be accomplished in either of two ways.

Simplifying a Complex Rational Expression

METHOD 1: Treat the numerator and denominator of the complex rational expression separately, performing whatever operations are indicated and simplifying the results. Follow this by simplifying the resulting rational expression.

METHOD 2: Find the LCM of the denominators of all rational expressions that appear in the complex rational expression. Multiply the numerator and denominator of the complex rational expression by the LCM and simplify the result.

We use both methods in the next example. By carefully studying each method, you can discover situations in which one method may be easier to use than the other.

EXAMPLE 10 **Simplifying a Complex Rational Expression**

Simplify: $\dfrac{\dfrac{1}{2} + \dfrac{3}{x}}{\dfrac{x+3}{4}}$ $x \neq -3, 0$

* Some texts use the term **complex fraction.**

Solution *Method 1:* First, we perform the indicated operation in the numerator, and then we divide.

$$\frac{\dfrac{1}{2}+\dfrac{3}{x}}{\dfrac{x+3}{4}} = \frac{\dfrac{1\cdot x + 2\cdot 3}{2\cdot x}}{\dfrac{x+3}{4}} = \frac{\dfrac{x+6}{2x}}{\dfrac{x+3}{4}} = \frac{x+6}{2x}\cdot\frac{4}{x+3}$$

$\underset{\text{Rule for adding quotients}}{\uparrow}$ $\underset{\text{Rule for dividing quotients}}{\uparrow}$

$$= \frac{(x+6)\cdot 4}{2\cdot x\cdot (x+3)} = \frac{2\cdot 2\cdot (x+6)}{2\cdot x\cdot (x+3)} = \frac{2(x+6)}{x(x+3)}$$

$\underset{\text{Rule for multiplying quotients}}{\uparrow}$

Method 2: The rational expressions that appear in the complex rational expression are

$$\frac{1}{2}, \quad \frac{3}{x}, \quad \frac{x+3}{4}$$

The LCM of their denominators is $4x$. We multiply the numerator and denominator of the complex rational expression by $4x$ and then simplify.

$$\frac{\dfrac{1}{2}+\dfrac{3}{x}}{\dfrac{x+3}{4}} = \frac{4x\cdot\left(\dfrac{1}{2}+\dfrac{3}{x}\right)}{4x\cdot\left(\dfrac{x+3}{4}\right)} = \frac{4x\cdot\dfrac{1}{2}+4x\cdot\dfrac{3}{x}}{\dfrac{4x\cdot(x+3)}{4}}$$

$\underset{\substack{\text{Multiply the} \\ \text{numerator and} \\ \text{denominator by } 4x.}}{\uparrow}$ $\underset{\substack{\text{Use the Distributive Property} \\ \text{in the numerator.}}}{\uparrow}$

$$= \frac{2\cdot 2x\cdot\dfrac{1}{2}+4x\cdot\dfrac{3}{x}}{\dfrac{4x\cdot(x+3)}{4}} = \frac{2x+12}{x(x+3)} = \frac{2(x+6)}{x(x+3)}$$

$\underset{\text{Simplify}}{\uparrow}$ $\underset{\text{Factor}}{\uparrow}$

EXAMPLE 11 **Simplifying a Complex Rational Expression**

Simplify: $\dfrac{\dfrac{x^2}{x-4}+2}{\dfrac{2x-2}{x}-1}$ $x \neq 0, 2, 4$

Solution We will use Method 1.

$$\frac{\dfrac{x^2}{x-4}+2}{\dfrac{2x-2}{x}-1} = \frac{\dfrac{x^2}{x-4}+\dfrac{2(x-4)}{x-4}}{\dfrac{2x-2}{x}-\dfrac{x}{x}} = \frac{\dfrac{x^2+2x-8}{x-4}}{\dfrac{2x-2-x}{x}}$$

$$= \frac{\dfrac{(x+4)(x-2)}{x-4}}{\dfrac{x-2}{x}} = \frac{(x+4)(x-2)}{x-4}\cdot\frac{x}{x-2}$$

$$= \frac{(x+4)\cdot x}{x-4}$$

Now Work PROBLEM 73

Application

EXAMPLE 12 **Solving an Application in Electricity**

An electrical circuit contains two resistors connected in parallel, as shown in Figure 28. If the resistance of each is R_1 and R_2 ohms, respectively, their combined resistance R is given by the formula

Figure 28

$$R = \dfrac{1}{\dfrac{1}{R_1} + \dfrac{1}{R_2}}$$

Express R as a rational expression; that is, simplify the right-hand side of this formula. Evaluate the rational expression if $R_1 = 6$ ohms and $R_2 = 10$ ohms.

Solution We will use Method 2. If we consider 1 as the fraction $\dfrac{1}{1}$, then the rational expressions in the complex rational expression are

$$\frac{1}{1}, \quad \frac{1}{R_1}, \quad \frac{1}{R_2}$$

The LCM of the denominators is $R_1 R_2$. We multiply the numerator and denominator of the complex rational expression by $R_1 R_2$ and simplify.

$$\frac{1}{\dfrac{1}{R_1} + \dfrac{1}{R_2}} = \frac{1 \cdot R_1 R_2}{\left(\dfrac{1}{R_1} + \dfrac{1}{R_2}\right) \cdot R_1 R_2} = \frac{R_1 R_2}{\dfrac{1}{R_1} \cdot R_1 R_2 + \dfrac{1}{R_2} \cdot R_1 R_2} = \frac{R_1 R_2}{R_2 + R_1}$$

Thus,

$$R = \frac{R_1 R_2}{R_2 + R_1}$$

If $R_1 = 6$ and $R_2 = 10$, then

$$R = \frac{6 \cdot 10}{10 + 6} = \frac{60}{16} = \frac{15}{4} \quad \text{ohms}$$

♪

R.7 Assess Your Understanding

Concepts and Vocabulary

1. When the numerator and denominator of a rational expression contain no common factors (except 1 and −1), the rational expression is in _____ _____.

2. LCM is an abbreviation for _____ _____.

3. **True or False** The rational expression $\dfrac{2x^3 - 4x}{x - 2}$ is reduced to lowest terms.

4. **True or False** The LCM of $2x^3 + 6x^2$ and $6x^4 + 4x^3$ is $4x^3(x + 1)$.

Skill Building

In Problems 5–16, reduce each rational expression to lowest terms.

5. $\dfrac{3x + 9}{x^2 - 9}$

6. $\dfrac{4x^2 + 8x}{12x + 24}$

7. $\dfrac{x^2 - 2x}{3x - 6}$

8. $\dfrac{15x^2 + 24x}{3x^2}$

9. $\dfrac{24x^2}{12x^2 - 6x}$

10. $\dfrac{x^2 + 4x + 4}{x^2 - 4}$

11. $\dfrac{y^2 - 25}{2y^2 - 8y - 10}$

12. $\dfrac{3y^2 - y - 2}{3y^2 + 5y + 2}$

13. $\dfrac{x^2 + 4x - 5}{x^2 - 2x + 1}$

14. $\dfrac{x - x^2}{x^2 + x - 2}$

15. $\dfrac{x^2 + 5x - 14}{2 - x}$

16. $\dfrac{2x^2 + 5x - 3}{1 - 2x}$

In Problems 17–34, perform the indicated operation and simplify the result. Leave your answer in factored form.

17. $\dfrac{3x + 6}{5x^2} \cdot \dfrac{x}{x^2 - 4}$

18. $\dfrac{3}{2x} \cdot \dfrac{x^2}{6x + 10}$

19. $\dfrac{4x^2}{x^2 - 16} \cdot \dfrac{x^3 - 64}{2x}$

20. $\dfrac{12}{x^2 + x} \cdot \dfrac{x^3 + 1}{4x - 2}$

21. $\dfrac{4x - 8}{-3x} \cdot \dfrac{12}{12 - 6x}$

22. $\dfrac{6x - 27}{5x} \cdot \dfrac{2}{4x - 18}$

23. $\dfrac{x^2 - 3x - 10}{x^2 + 2x - 35} \cdot \dfrac{x^2 + 4x - 21}{x^2 + 9x + 14}$

24. $\dfrac{x^2 + x - 6}{x^2 + 4x - 5} \cdot \dfrac{x^2 - 25}{x^2 + 2x - 15}$

25. $\dfrac{\dfrac{6x}{x^2 - 4}}{\dfrac{3x - 9}{2x + 4}}$

26. $\dfrac{\dfrac{12x}{5x + 20}}{\dfrac{4x^2}{x^2 - 16}}$

27. $\dfrac{\dfrac{8x}{x^2 - 1}}{\dfrac{10x}{x + 1}}$

28. $\dfrac{\dfrac{x - 2}{4x}}{\dfrac{x^2 - 4x + 4}{12x}}$

29. $\dfrac{\dfrac{4 - x}{4 + x}}{\dfrac{4x}{x^2 - 16}}$

30. $\dfrac{\dfrac{3 + x}{3 - x}}{\dfrac{x^2 - 9}{9x^3}}$

31. $\dfrac{\dfrac{x^2 + 7x + 12}{x^2 - 7x + 12}}{\dfrac{x^2 + x - 12}{x^2 - x - 12}}$

32. $\dfrac{\dfrac{x^2 + 7x + 6}{x^2 + x - 6}}{\dfrac{x^2 + 5x - 6}{x^2 + 5x + 6}}$

33. $\dfrac{\dfrac{2x^2 - x - 28}{3x^2 - x - 2}}{\dfrac{4x^2 + 16x + 7}{3x^2 + 11x + 6}}$

34. $\dfrac{\dfrac{9x^2 + 3x - 2}{12x^2 + 5x - 2}}{\dfrac{9x^2 - 6x + 1}{8x^2 - 10x - 3}}$

In Problems 35–52, perform the indicated operation and simplify the result. Leave your answer in factored form.

35. $\dfrac{x}{2} + \dfrac{5}{2}$

36. $\dfrac{3}{x} - \dfrac{6}{x}$

37. $\dfrac{x^2}{2x - 3} - \dfrac{4}{2x - 3}$

38. $\dfrac{3x^2}{2x - 1} - \dfrac{9}{2x - 1}$

39. $\dfrac{x + 1}{x - 3} + \dfrac{2x - 3}{x - 3}$

40. $\dfrac{2x - 5}{3x + 2} + \dfrac{x + 4}{3x + 2}$

41. $\dfrac{3x + 5}{2x - 1} - \dfrac{2x - 4}{2x - 1}$

42. $\dfrac{5x - 4}{3x + 4} - \dfrac{x + 1}{3x + 4}$

43. $\dfrac{4}{x - 2} + \dfrac{x}{2 - x}$

44. $\dfrac{6}{x - 1} - \dfrac{x}{1 - x}$

45. $\dfrac{4}{x - 1} - \dfrac{2}{x + 2}$

46. $\dfrac{2}{x + 5} - \dfrac{5}{x - 5}$

47. $\dfrac{x}{x + 1} + \dfrac{2x - 3}{x - 1}$

48. $\dfrac{3x}{x - 4} + \dfrac{2x}{x + 3}$

49. $\dfrac{x - 3}{x + 2} - \dfrac{x + 4}{x - 2}$

50. $\dfrac{2x - 3}{x - 1} - \dfrac{2x + 1}{x + 1}$

51. $\dfrac{x}{x^2 - 4} + \dfrac{1}{x}$

52. $\dfrac{x - 1}{x^3} + \dfrac{x}{x^2 + 1}$

In Problems 53–60, find the LCM of the given polynomials.

53. $x^2 - 4, \quad x^2 - x - 2$

54. $x^2 - x - 12, \quad x^2 - 8x + 16$

55. $x^3 - x, \quad x^2 - x$

56. $3x^2 - 27, \quad 2x^2 - x - 15$

57. $4x^3 - 4x^2 + x, \quad 2x^3 - x^2, \quad x^3$

58. $x - 3, \quad x^2 + 3x, \quad x^3 - 9x$

59. $x^3 - x, \quad x^3 - 2x^2 + x, \quad x^3 - 1$

60. $x^2 + 4x + 4, \quad x^3 + 2x^2, \quad (x + 2)^3$

In Problems 61–72, perform the indicated operations and simplify the result. Leave your answer in factored form.

61. $\dfrac{x}{x^2 - 7x + 6} - \dfrac{x}{x^2 - 2x - 24}$

62. $\dfrac{x}{x - 3} - \dfrac{x + 1}{x^2 + 5x - 24}$

63. $\dfrac{4x}{x^2 - 4} - \dfrac{2}{x^2 + x - 6}$

64. $\dfrac{3x}{x - 1} - \dfrac{x - 4}{x^2 - 2x + 1}$

65. $\dfrac{3}{(x - 1)^2(x + 1)} + \dfrac{2}{(x - 1)(x + 1)^2}$

66. $\dfrac{2}{(x + 2)^2(x - 1)} - \dfrac{6}{(x + 2)(x - 1)^2}$

67. $\dfrac{x + 4}{x^2 - x - 2} - \dfrac{2x + 3}{x^2 + 2x - 8}$

68. $\dfrac{2x - 3}{x^2 + 8x + 7} - \dfrac{x - 2}{(x + 1)^2}$

69. $\dfrac{1}{x} - \dfrac{2}{x^2 + x} + \dfrac{3}{x^3 - x^2}$

70. $\dfrac{x}{(x - 1)^2} + \dfrac{2}{x} - \dfrac{x + 1}{x^3 - x^2}$

71. $\dfrac{1}{h}\left(\dfrac{1}{x + h} - \dfrac{1}{x}\right)$

72. $\dfrac{1}{h}\left[\dfrac{1}{(x + h)^2} - \dfrac{1}{x^2}\right]$

In Problems 73–84, perform the indicated operations and simplify the result. Leave your answer in factored form.

73. $\dfrac{1 + \dfrac{1}{x}}{1 - \dfrac{1}{x}}$

74. $\dfrac{4 + \dfrac{1}{x^2}}{3 - \dfrac{1}{x^2}}$

75. $\dfrac{2 - \dfrac{x+1}{x}}{3 + \dfrac{x-1}{x+1}}$

76. $\dfrac{1 - \dfrac{x}{x+1}}{2 - \dfrac{x-1}{x}}$

77. $\dfrac{\dfrac{x+4}{x-2} - \dfrac{x-3}{x+1}}{x+1}$

78. $\dfrac{\dfrac{x-2}{x+1} - \dfrac{x}{x-2}}{x+3}$

79. $\dfrac{\dfrac{x-2}{x+2} + \dfrac{x-1}{x+1}}{\dfrac{x}{x+1} - \dfrac{2x-3}{x}}$

80. $\dfrac{\dfrac{2x+5}{x} - \dfrac{x}{x-3}}{\dfrac{x^2}{x-3} - \dfrac{(x+1)^2}{x+3}}$

81. $1 - \dfrac{1}{1 - \dfrac{1}{x}}$

82. $1 - \dfrac{1}{1 - \dfrac{1}{1-x}}$

83. $\dfrac{2(x-1)^{-1} + 3}{3(x-1)^{-1} + 2}$

84. $\dfrac{4(x+2)^{-1} - 3}{3(x+2)^{-1} - 1}$

In Problems 85–92, expressions that occur in calculus are given. Reduce each expression to lowest terms.

85. $\dfrac{(2x+3)\cdot 3 - (3x-5)\cdot 2}{(3x-5)^2}$

86. $\dfrac{(4x+1)\cdot 5 - (5x-2)\cdot 4}{(5x-2)^2}$

87. $\dfrac{x\cdot 2x - (x^2+1)\cdot 1}{(x^2+1)^2}$

88. $\dfrac{x\cdot 2x - (x^2-4)\cdot 1}{(x^2-4)^2}$

89. $\dfrac{(3x+1)\cdot 2x - x^2\cdot 3}{(3x+1)^2}$

90. $\dfrac{(2x-5)\cdot 3x^2 - x^3\cdot 2}{(2x-5)^2}$

91. $\dfrac{(x^2+1)\cdot 3 - (3x+4)\cdot 2x}{(x^2+1)^2}$

92. $\dfrac{(x^2+9)\cdot 2 - (2x-5)\cdot 2x}{(x^2+9)^2}$

Applications and Extensions

93. **The Lensmaker's Equation** The focal length f of a lens with index of refraction n is

$$\frac{1}{f} = (n-1)\left[\frac{1}{R_1} + \frac{1}{R_2}\right]$$

where R_1 and R_2 are the radii of curvature of the front and back surfaces of the lens. Express f as a rational expression. Evaluate the rational expression for $n = 1.5$, $R_1 = 0.1$ meter, and $R_2 = 0.2$ meter.

94. **Electrical Circuits** An electrical circuit contains three resistors connected in parallel. If the resistance of each is R_1, R_2, and R_3 ohms, respectively, their combined resistance R is given by the formula

$$\frac{1}{R} = \frac{1}{R_1} + \frac{1}{R_2} + \frac{1}{R_3}$$

Express R as a rational expression. Evaluate R for $R_1 = 5$ ohms, $R_2 = 4$ ohms, and $R_3 = 10$ ohms.

Explaining Concepts: Discussion and Writing

95. The following expressions are called **continued fractions:**

$$1 + \frac{1}{x}, \quad 1 + \cfrac{1}{1 + \cfrac{1}{x}}, \quad 1 + \cfrac{1}{1 + \cfrac{1}{1 + \cfrac{1}{x}}}, \quad 1 + \cfrac{1}{1 + \cfrac{1}{1 + \cfrac{1}{1 + \cfrac{1}{x}}}}, \ldots$$

Each simplifies to an expression of the form

$$\frac{ax + b}{bx + c}$$

Trace the successive values of $a, b,$ and c as you "continue" the fraction. Can you discover the patterns that these values follow? Go to the library and research Fibonacci numbers. Write a report on your findings.

96. Explain to a fellow student when you would use the LCM method to add two rational expressions. Give two examples of adding two rational expressions, one in which you use the LCM and the other in which you do not.

97. Which of the two methods given in the text for simplifying complex rational expressions do you prefer? Write a brief paragraph stating the reasons for your choice.

R.8 *n*th Roots; Rational Exponents

PREPARING FOR THIS SECTION *Before getting started, review the following:*

• Exponents, Square Roots (Section R.2, pp. 21–24)

Now Work the **'Are You Prepared?'** problems on page 78.

OBJECTIVES **1** Work with *n*th Roots (p. 73)
 2 Simplify Radicals (p. 74)
 3 Rationalize Denominators (p. 75)
 4 Simplify Expressions with Rational Exponents (p. 76)

1 Work with *n*th Roots

DEFINITION

The **principal *n*th root of a real number *a*,** $n \geq 2$ an integer, symbolized by $\sqrt[n]{a}$, is defined as follows:

$$\sqrt[n]{a} = b \quad \text{means} \quad a = b^n$$

where $a \geq 0$ and $b \geq 0$ if *n* is even and *a*, *b* are any real numbers if *n* is odd.

In Words
The symbol $\sqrt[n]{a}$ means "give me the number, which when raised to the power n, equals a."

Notice that if *a* is negative and *n* is even then $\sqrt[n]{a}$ is not defined. When it is defined, the principal *n*th root of a number is unique.

The symbol $\sqrt[n]{a}$ for the principal *n*th root of *a* is called a **radical;** the integer *n* is called the **index,** and *a* is called the **radicand.** If the index of a radical is 2, we call $\sqrt[2]{a}$ the **square root** of *a* and omit the index 2 by simply writing \sqrt{a}. If the index is 3, we call $\sqrt[3]{a}$ the **cube root** of *a*.

EXAMPLE 1 **Simplifying Principal *n*th Roots**

(a) $\sqrt[3]{8} = \sqrt[3]{2^3} = 2$ (b) $\sqrt[3]{-64} = \sqrt[3]{(-4)^3} = -4$

(c) $\sqrt[4]{\dfrac{1}{16}} = \sqrt[4]{\left(\dfrac{1}{2}\right)^4} = \dfrac{1}{2}$ (d) $\sqrt[6]{(-2)^6} = |-2| = 2$

These are examples of **perfect roots,** since each simplifies to a rational number. Notice the absolute value in Example 1(d). If *n* is even, the principal *n*th root must be nonnegative.

In general, if $n \geq 2$ is an integer and *a* is a real number, we have

$$\sqrt[n]{a^n} = a \qquad \text{if } n \geq 3 \text{ is odd} \tag{1a}$$
$$\sqrt[n]{a^n} = |a| \qquad \text{if } n \geq 2 \text{ is even} \tag{1b}$$

Now Work PROBLEM **7**

Radicals provide a way of representing many irrational real numbers. For example, there is no rational number whose square is 2. Using radicals, we can say that $\sqrt{2}$ is the positive number whose square is 2.

Figure 29

```
5ˣ√16
         1.741101127
```

EXAMPLE 2 — Using a Calculator to Approximate Roots

Use a calculator to approximate $\sqrt[5]{16}$.

Solution Figure 29 shows the result using a TI-84 plus graphing calculator.

━━ **Now Work** PROBLEM 101

2 Simplify Radicals

Let $n \geq 2$ and $m \geq 2$ denote positive integers, and let a and b represent real numbers. Assuming that all radicals are defined, we have the following properties:

Properties of Radicals

$$\sqrt[n]{ab} = \sqrt[n]{a}\,\sqrt[n]{b} \tag{2a}$$

$$\sqrt[n]{\frac{a}{b}} = \frac{\sqrt[n]{a}}{\sqrt[n]{b}} \qquad b \neq 0 \tag{2b}$$

$$\sqrt[n]{a^m} = \left(\sqrt[n]{a}\right)^m \tag{2c}$$

When used in reference to radicals, the direction to "simplify" will mean to remove from the radicals any perfect roots that occur as factors.

EXAMPLE 3 — Simplifying Radicals

(a) $\sqrt{32} = \underset{\uparrow}{\sqrt{16 \cdot 2}} = \underset{\uparrow}{\sqrt{16} \cdot \sqrt{2}} = 4\sqrt{2}$

$$ Factor out 16, (2a)
$$ a perfect square.

(b) $\sqrt[3]{16} = \underset{\uparrow}{\sqrt[3]{8 \cdot 2}} = \underset{\uparrow}{\sqrt[3]{8} \cdot \sqrt[3]{2}} = \sqrt[3]{2^3} \cdot \sqrt[3]{2} = 2\sqrt[3]{2}$

$$ Factor out 8, (2a)
$$ a perfect cube.

(c) $\sqrt[3]{-16x^4} = \underset{\uparrow}{\sqrt[3]{-8 \cdot 2 \cdot x^3 \cdot x}} = \underset{\uparrow}{\sqrt[3]{(-8x^3)(2x)}}$

$$ Factor perfect Group perfect
$$ cubes inside radical. cubes.

$$= \underset{\uparrow}{\sqrt[3]{(-2x)^3 \cdot 2x}} = \sqrt[3]{(-2x)^3} \cdot \sqrt[3]{2x} = -2x\sqrt[3]{2x}$$

$$ (2a)

(d) $\sqrt[4]{\dfrac{16x^5}{81}} = \sqrt[4]{\dfrac{2^4 x^4 x}{3^4}} = \sqrt[4]{\left(\dfrac{2x}{3}\right)^4 \cdot x} = \sqrt[4]{\left(\dfrac{2x}{3}\right)^4} \cdot \sqrt[4]{x} = \left|\dfrac{2x}{3}\right| \sqrt[4]{x}$

━━ **Now Work** PROBLEMS 11 AND 17

Two or more radicals can be combined, provided that they have the same index and the same radicand. Such radicals are called **like radicals.**

EXAMPLE 4 — Combining Like Radicals

(a) $-8\sqrt{12} + \sqrt{3} = -8\sqrt{4 \cdot 3} + \sqrt{3}$

$ = -8 \cdot \sqrt{4}\,\sqrt{3} + \sqrt{3}$

$ = -16\sqrt{3} + \sqrt{3} = -15\sqrt{3}$

(b) $\sqrt[3]{8x^4} + \sqrt[3]{-x} + 4\sqrt[3]{27x} = \sqrt[3]{2^3x^3x} + \sqrt[3]{-1 \cdot x} + 4\sqrt[3]{3^3x}$

$$= \sqrt[3]{(2x)^3} \cdot \sqrt[3]{x} + \sqrt[3]{-1} \cdot \sqrt[3]{x} + 4\sqrt[3]{3^3} \cdot \sqrt[3]{x}$$

$$= 2x\sqrt[3]{x} - 1 \cdot \sqrt[3]{x} + 12\sqrt[3]{x}$$

$$= (2x + 11)\sqrt[3]{x}$$

----Now Work PROBLEM 33

3 Rationalize Denominators

When radicals occur in quotients, it is customary to rewrite the quotient so that the new denominator contains no radicals. This process is referred to as **rationalizing the denominator.**

The idea is to multiply by an appropriate expression so that the new denominator contains no radicals. For example:

If a Denominator Contains the Factor	Multiply by	To Obtain a Denominator Free of Radicals
$\sqrt{3}$	$\sqrt{3}$	$(\sqrt{3})^2 = 3$
$\sqrt{3} + 1$	$\sqrt{3} - 1$	$(\sqrt{3})^2 - 1^2 = 3 - 1 = 2$
$\sqrt{2} - 3$	$\sqrt{2} + 3$	$(\sqrt{2})^2 - 3^2 = 2 - 9 = -7$
$\sqrt{5} - \sqrt{3}$	$\sqrt{5} + \sqrt{3}$	$(\sqrt{5})^2 - (\sqrt{3})^2 = 5 - 3 = 2$
$\sqrt[3]{4}$	$\sqrt[3]{2}$	$\sqrt[3]{4} \cdot \sqrt[3]{2} = \sqrt[3]{8} = 2$

In rationalizing the denominator of a quotient, be sure to multiply both the numerator and the denominator by the expression.

EXAMPLE 5

Rationalizing Denominators

Rationalize the denominator of each expression:

(a) $\dfrac{1}{\sqrt{3}}$ (b) $\dfrac{5}{4\sqrt{2}}$ (c) $\dfrac{\sqrt{2}}{\sqrt{3} - 3\sqrt{2}}$

Solution

(a) The denominator contains the factor $\sqrt{3}$, so we multiply the numerator and denominator by $\sqrt{3}$ to obtain

$$\frac{1}{\sqrt{3}} = \frac{1}{\sqrt{3}} \cdot \frac{\sqrt{3}}{\sqrt{3}} = \frac{\sqrt{3}}{(\sqrt{3})^2} = \frac{\sqrt{3}}{3}$$

(b) The denominator contains the factor $\sqrt{2}$, so we multiply the numerator and denominator by $\sqrt{2}$ to obtain

$$\frac{5}{4\sqrt{2}} = \frac{5}{4\sqrt{2}} \cdot \frac{\sqrt{2}}{\sqrt{2}} = \frac{5\sqrt{2}}{4(\sqrt{2})^2} = \frac{5\sqrt{2}}{4 \cdot 2} = \frac{5\sqrt{2}}{8}$$

(c) The denominator contains the factor $\sqrt{3} - 3\sqrt{2}$, so we multiply the numerator and denominator by $\sqrt{3} + 3\sqrt{2}$ to obtain

$$\frac{\sqrt{2}}{\sqrt{3} - 3\sqrt{2}} = \frac{\sqrt{2}}{\sqrt{3} - 3\sqrt{2}} \cdot \frac{\sqrt{3} + 3\sqrt{2}}{\sqrt{3} + 3\sqrt{2}} = \frac{\sqrt{2}(\sqrt{3} + 3\sqrt{2})}{(\sqrt{3})^2 - (3\sqrt{2})^2}$$

$$= \frac{\sqrt{2}\sqrt{3} + 3(\sqrt{2})^2}{3 - 18} = \frac{\sqrt{6} + 6}{-15} = -\frac{6 + \sqrt{6}}{15}$$

----Now Work PROBLEM 47

4 Simplify Expressions with Rational Exponents

Radicals are used to define rational exponents.

DEFINITION

If a is a real number and $n \geq 2$ is an integer, then

$$a^{1/n} = \sqrt[n]{a} \tag{3}$$

provided that $\sqrt[n]{a}$ exists.

Note that if n is even and $a < 0$ then $\sqrt[n]{a}$ and $a^{1/n}$ do not exist.

EXAMPLE 6

Writing Expressions Containing Fractional Exponents as Radicals

(a) $4^{1/2} = \sqrt{4} = 2$ 　　　　　　(b) $8^{1/2} = \sqrt{8} = 2\sqrt{2}$

(c) $(-27)^{1/3} = \sqrt[3]{-27} = -3$ 　　(d) $16^{1/3} = \sqrt[3]{16} = 2\sqrt[3]{2}$

DEFINITION

If a is a real number and m and n are integers containing no common factors, with $n \geq 2$, then

$$a^{m/n} = \sqrt[n]{a^m} = (\sqrt[n]{a})^m \tag{4}$$

provided that $\sqrt[n]{a}$ exists.

We have two comments about equation (4):

1. The exponent $\dfrac{m}{n}$ must be in lowest terms and n must be positive.

2. In simplifying the rational expression $a^{m/n}$, either $\sqrt[n]{a^m}$ or $(\sqrt[n]{a})^m$ may be used, the choice depending on which is easier to simplify. Generally, taking the root first, as in $(\sqrt[n]{a})^m$, is easier.

EXAMPLE 7

Using Equation (4)

(a) $4^{3/2} = (\sqrt{4})^3 = 2^3 = 8$ 　　(b) $(-8)^{4/3} = (\sqrt[3]{-8})^4 = (-2)^4 = 16$

(c) $(32)^{-2/5} = (\sqrt[5]{32})^{-2} = 2^{-2} = \dfrac{1}{4}$ 　(d) $25^{6/4} = 25^{3/2} = (\sqrt{25})^3 = 5^3 = 125$

━━━ **Now Work** PROBLEM 55

It can be shown that the Laws of Exponents hold for rational exponents. The next example illustrates using the Laws of Exponents to simplify.

EXAMPLE 8

Simplifying Expressions Containing Rational Exponents

Simplify each expression. Express your answer so that only positive exponents occur. Assume that the variables are positive.

(a) $(x^{2/3}y)(x^{-2}y)^{1/2}$ 　　(b) $\left(\dfrac{2x^{1/3}}{y^{2/3}}\right)^{-3}$ 　　(c) $\left(\dfrac{9x^2y^{1/3}}{x^{1/3}y}\right)^{1/2}$

Solution (a) $(x^{2/3}y)(x^{-2}y)^{1/2} = (x^{2/3}y)[(x^{-2})^{1/2}y^{1/2}]$

$$= x^{2/3}yx^{-1}y^{1/2}$$

$$= (x^{2/3} \cdot x^{-1})(y \cdot y^{1/2})$$

$$= x^{-1/3}y^{3/2}$$

$$= \frac{y^{3/2}}{x^{1/3}}$$

(b) $\left(\dfrac{2x^{1/3}}{y^{2/3}}\right)^{-3} = \left(\dfrac{y^{2/3}}{2x^{1/3}}\right)^3 = \dfrac{(y^{2/3})^3}{(2x^{1/3})^3} = \dfrac{y^2}{2^3(x^{1/3})^3} = \dfrac{y^2}{8x}$

(c) $\left(\dfrac{9x^2y^{1/3}}{x^{1/3}y}\right)^{1/2} = \left(\dfrac{9x^{2-(1/3)}}{y^{1-(1/3)}}\right)^{1/2} = \left(\dfrac{9x^{5/3}}{y^{2/3}}\right)^{1/2} = \dfrac{9^{1/2}(x^{5/3})^{1/2}}{(y^{2/3})^{1/2}} = \dfrac{3x^{5/6}}{y^{1/3}}$

─────**Now Work** PROBLEM 71

The next two examples illustrate some algebra that you will need to know for certain calculus problems.

EXAMPLE 9 **Writing an Expression as a Single Quotient**

Write the following expression as a single quotient in which only positive exponents appear.

$$(x^2 + 1)^{1/2} + x \cdot \frac{1}{2}(x^2 + 1)^{-1/2} \cdot 2x$$

Solution $(x^2 + 1)^{1/2} + x \cdot \dfrac{1}{\cancel{2}}(x^2 + 1)^{-1/2} \cdot \cancel{2}x = (x^2 + 1)^{1/2} + \dfrac{x^2}{(x^2 + 1)^{1/2}}$

$$= \dfrac{(x^2 + 1)^{1/2}(x^2 + 1)^{1/2} + x^2}{(x^2 + 1)^{1/2}}$$

$$= \dfrac{(x^2 + 1) + x^2}{(x^2 + 1)^{1/2}}$$

$$= \dfrac{2x^2 + 1}{(x^2 + 1)^{1/2}}$$

─────**Now Work** PROBLEM 77

EXAMPLE 10 **Factoring an Expression Containing Rational Exponents**

Factor: $\dfrac{4}{3}x^{1/3}(2x + 1) + 2x^{4/3}$

Solution Begin by writing $2x^{4/3}$ as a fraction with 3 as the denominator.

$$\frac{4}{3}x^{1/3}(2x + 1) + 2x^{4/3} = \frac{4x^{1/3}(2x + 1)}{3} + \frac{6x^{4/3}}{3} = \frac{4x^{1/3}(2x + 1) + 6x^{4/3}}{3}$$

$$\underset{\text{Add the two fractions}}{\uparrow}$$

$$= \frac{2x^{1/3}[2(2x + 1) + 3x]}{3} = \frac{2x^{1/3}(7x + 2)}{3}$$

$$\underset{2 \text{ and } x^{1/3} \text{ are common factors}}{\uparrow} \qquad \underset{\text{Simplify}}{\uparrow}$$

─────**Now Work** PROBLEM 89

Historical Note

The radical sign, $\sqrt{}$, was first used in print by Christoff Rudolff in 1525. It is thought to be the manuscript form of the letter *r* (for the Latin word *radix* = *root*), although this is not quite conclusively confirmed. It took a long time for $\sqrt{}$ to become the standard symbol for a square root and much longer to standardize $\sqrt[3]{}$, $\sqrt[4]{}$, $\sqrt[5]{}$, and so on. The indexes of the root were placed in every conceivable position, with

$$\sqrt[3]{8}, \quad \sqrt{\text{③}8}, \quad \text{and} \quad \underset{3}{\sqrt{}}8$$

all being variants for $\sqrt[3]{8}$. The notation $\sqrt{}\,\sqrt{16}$ was popular for $\sqrt[4]{16}$. By the 1700s, the index had settled where we now put it.

The bar on top of the present radical symbol, as follows,

$$\sqrt{a^2 + 2ab + b^2}$$

is the last survivor of the **vinculum**, a bar placed atop an expression to indicate what we would now indicate with parentheses. For example,

$$\overline{ab + c} = a(b + c)$$

R.8 Assess Your Understanding

'Are You Prepared?' *Answers are given at the end of these exercises. If you get a wrong answer, read the pages in red.*

1. $(-3)^2 =$ _____; $-3^2 =$ _____ (pp. 21–24)

2. $\sqrt{16} =$ _____; $\sqrt{(-4)^2} =$ _____ (pp. 21–24)

Concepts and Vocabulary

3. In the symbol $\sqrt[n]{a}$, the integer n is called the _____.

4. *True or False* $\sqrt[5]{-32} = -2$

5. We call $\sqrt[3]{a}$ the _____ _____ of a.

6. *True or False* $\sqrt[4]{(-3)^4} = -3$

Skill Building

In Problems 7–42, simplify each expression. Assume that all variables are positive when they appear.

7. $\sqrt[3]{27}$

8. $\sqrt[4]{16}$

9. $\sqrt[3]{-8}$

10. $\sqrt[3]{-1}$

11. $\sqrt{8}$

12. $\sqrt[3]{54}$

13. $\sqrt[3]{-8x^4}$

14. $\sqrt[4]{48x^5}$

15. $\sqrt[4]{x^{12}y^8}$

16. $\sqrt[5]{x^{10}y^5}$

17. $\sqrt[4]{\dfrac{x^9y^7}{xy^3}}$

18. $\sqrt[3]{\dfrac{3xy^2}{81x^4y^2}}$

19. $\sqrt{36x}$

20. $\sqrt{9x^5}$

21. $\sqrt{3x^2}\,\sqrt{12x}$

22. $\sqrt{5x}\,\sqrt{20x^3}$

23. $\left(\sqrt{5}\,\sqrt[3]{9}\right)^2$

24. $\left(\sqrt[3]{3}\,\sqrt{10}\right)^4$

25. $\left(3\sqrt{6}\right)\left(2\sqrt{2}\right)$

26. $\left(5\sqrt{8}\right)\left(-3\sqrt{3}\right)$

27. $3\sqrt{2} + 4\sqrt{2}$

28. $6\sqrt{5} - 4\sqrt{5}$

29. $-\sqrt{18} + 2\sqrt{8}$

30. $2\sqrt{12} - 3\sqrt{27}$

31. $\left(\sqrt{3} + 3\right)\left(\sqrt{3} - 1\right)$

32. $\left(\sqrt{5} - 2\right)\left(\sqrt{5} + 3\right)$

33. $5\sqrt[3]{2} - 2\sqrt[3]{54}$

34. $9\sqrt[3]{24} - \sqrt[3]{81}$

35. $\left(\sqrt{x} - 1\right)^2$

36. $\left(\sqrt{x} + \sqrt{5}\right)^2$

37. $\sqrt[3]{16x^4} - \sqrt[3]{2x}$

38. $\sqrt[4]{32x} + \sqrt[4]{2x^5}$

39. $\sqrt{8x^3} - 3\sqrt{50x}$

40. $3x\sqrt{9y} + 4\sqrt{25y}$

41. $\sqrt[3]{16x^4y} - 3x\sqrt[3]{2xy} + 5\sqrt[3]{-2xy^4}$

42. $8xy - \sqrt{25x^2y^2} + \sqrt[3]{8x^3y^3}$

In Problems 43–54, rationalize the denominator of each expression. Assume that all variables are positive when they appear.

43. $\dfrac{1}{\sqrt{2}}$

44. $\dfrac{2}{\sqrt{3}}$

45. $\dfrac{-\sqrt{3}}{\sqrt{5}}$

46. $\dfrac{-\sqrt{3}}{\sqrt{8}}$

47. $\dfrac{\sqrt{3}}{5 - \sqrt{2}}$

48. $\dfrac{\sqrt{2}}{\sqrt{7} + 2}$

49. $\dfrac{2 - \sqrt{5}}{2 + 3\sqrt{5}}$

50. $\dfrac{\sqrt{3} - 1}{2\sqrt{3} + 3}$

51. $\dfrac{5}{\sqrt[3]{2}}$

52. $\dfrac{-2}{\sqrt[3]{9}}$

53. $\dfrac{\sqrt{x+h}-\sqrt{x}}{\sqrt{x+h}+\sqrt{x}}$

54. $\dfrac{\sqrt{x+h}+\sqrt{x-h}}{\sqrt{x+h}-\sqrt{x-h}}$

In Problems 55–66, simplify each expression.

55. $8^{2/3}$

56. $4^{3/2}$

57. $(-27)^{1/3}$

58. $16^{3/4}$

59. $16^{3/2}$

60. $25^{3/2}$

61. $9^{-3/2}$

62. $16^{-3/2}$

63. $\left(\dfrac{9}{8}\right)^{3/2}$

64. $\left(\dfrac{27}{8}\right)^{2/3}$

65. $\left(\dfrac{8}{9}\right)^{-3/2}$

66. $\left(\dfrac{8}{27}\right)^{-2/3}$

In Problems 67–74, simplify each expression. Express your answer so that only positive exponents occur. Assume that the variables are positive.

67. $x^{3/4}x^{1/3}x^{-1/2}$

68. $x^{2/3}x^{1/2}x^{-1/4}$

69. $(x^3y^6)^{1/3}$

70. $(x^4y^8)^{3/4}$

71. $\dfrac{(x^2y)^{1/3}(xy^2)^{2/3}}{x^{2/3}y^{2/3}}$

72. $\dfrac{(xy)^{1/4}(x^2y^2)^{1/2}}{(x^2y)^{3/4}}$

73. $\dfrac{(16x^2y^{-1/3})^{3/4}}{(xy^2)^{1/4}}$

74. $\dfrac{(4x^{-1}y^{1/3})^{3/2}}{(xy)^{3/2}}$

Applications and Extensions

In Problems 75–88, expressions that occur in calculus are given. Write each expression as a single quotient in which only positive exponents and/or radicals appear.

75. $\dfrac{x}{(1+x)^{1/2}}+2(1+x)^{1/2}\quad x>-1$

76. $\dfrac{1+x}{2x^{1/2}}+x^{1/2}\quad x>0$

77. $2x(x^2+1)^{1/2}+x^2\cdot\dfrac{1}{2}(x^2+1)^{-1/2}\cdot2x$

78. $(x+1)^{1/3}+x\cdot\dfrac{1}{3}(x+1)^{-2/3}\quad x\neq-1$

79. $\sqrt{4x+3}\cdot\dfrac{1}{2\sqrt{x-5}}+\sqrt{x-5}\cdot\dfrac{1}{5\sqrt{4x+3}}\quad x>5$

80. $\dfrac{\sqrt[3]{8x+1}}{3\sqrt[3]{(x-2)^2}}+\dfrac{\sqrt[3]{x-2}}{24\sqrt[3]{(8x+1)^2}}\quad x\neq2,\ x\neq-\dfrac{1}{8}$

81. $\dfrac{\sqrt{1+x}-x\cdot\dfrac{1}{2\sqrt{1+x}}}{1+x}\quad x>-1$

82. $\dfrac{\sqrt{x^2+1}-x\cdot\dfrac{2x}{2\sqrt{x^2+1}}}{x^2+1}$

83. $\dfrac{(x+4)^{1/2}-2x(x+4)^{-1/2}}{x+4}\quad x>-4$

84. $\dfrac{(9-x^2)^{1/2}+x^2(9-x^2)^{-1/2}}{9-x^2}\quad -3<x<3$

85. $\dfrac{\dfrac{x^2}{(x^2-1)^{1/2}}-(x^2-1)^{1/2}}{x^2}\quad x<-1\ \text{ or }\ x>1$

86. $\dfrac{(x^2+4)^{1/2}-x^2(x^2+4)^{-1/2}}{x^2+4}$

87. $\dfrac{\dfrac{1+x^2}{2\sqrt{x}}-2x\sqrt{x}}{(1+x^2)^2}\quad x>0$

88. $\dfrac{2x(1-x^2)^{1/3}+\dfrac{2}{3}x^3(1-x^2)^{-2/3}}{(1-x^2)^{2/3}}\quad x\neq-1,\ x\neq1$

In Problems 89–98, expressions that occur in calculus are given. Factor each expression. Express your answer so that only positive exponents occur.

89. $(x+1)^{3/2}+x\cdot\dfrac{3}{2}(x+1)^{1/2}\quad x\geq-1$

90. $(x^2+4)^{4/3}+x\cdot\dfrac{4}{3}(x^2+4)^{1/3}\cdot2x$

91. $6x^{1/2}(x^2+x)-8x^{3/2}-8x^{1/2}\quad x\geq0$

92. $6x^{1/2}(2x+3)+x^{3/2}\cdot8\quad x\geq0$

93. $3(x^2 + 4)^{4/3} + x \cdot 4(x^2 + 4)^{1/3} \cdot 2x$

94. $2x(3x + 4)^{4/3} + x^2 \cdot 4(3x + 4)^{1/3}$

95. $4(3x + 5)^{1/3}(2x + 3)^{3/2} + 3(3x + 5)^{4/3}(2x + 3)^{1/2}$ $\quad x \ge -\dfrac{3}{2}$

96. $6(6x + 1)^{1/3}(4x - 3)^{3/2} + 6(6x + 1)^{4/3}(4x - 3)^{1/2}$ $\quad x \ge \dfrac{3}{4}$

97. $3x^{-1/2} + \dfrac{3}{2}x^{1/2}$ $\quad x > 0$

98. $8x^{1/3} - 4x^{-2/3}$ $\quad x \ne 0$

In Problems 99–106, use a calculator to approximate each radical. Round your answer to two decimal places.

99. $\sqrt{2}$
100. $\sqrt{7}$
101. $\sqrt[3]{4}$
102. $\sqrt[3]{-5}$

103. $\dfrac{2 + \sqrt{3}}{3 - \sqrt{5}}$
104. $\dfrac{\sqrt{5} - 2}{\sqrt{2} + 4}$
105. $\dfrac{3\sqrt[3]{5} - \sqrt{2}}{\sqrt{3}}$
106. $\dfrac{2\sqrt{3} - \sqrt[3]{4}}{\sqrt{2}}$

107. Calculating the Amount of Gasoline in a Tank A Shell station stores its gasoline in underground tanks that are right circular cylinders lying on their sides. See the illustration. The volume V of gasoline in the tank (in gallons) is given by the formula

$$V = 40h^2\sqrt{\frac{96}{h} - 0.608}$$

where h is the height of the gasoline (in inches) as measured on a depth stick.
(a) If $h = 12$ inches, how many gallons of gasoline are in the tank?
(b) If $h = 1$ inch, how many gallons of gasoline are in the tank?

108. Inclined Planes The final velocity v of an object in feet per second (ft/sec) after it slides down a frictionless inclined plane of height h feet is

$$v = \sqrt{64h + v_0^2}$$

where v_0 is the initial velocity (in ft/sec) of the object.
(a) What is the final velocity v of an object that slides down a frictionless inclined plane of height 4 feet? Assume that the initial velocity is 0.
(b) What is the final velocity v of an object that slides down a frictionless inclined plane of height 16 feet? Assume that the initial velocity is 0.
(c) What is the final velocity v of an object that slides down a frictionless inclined plane of height 2 feet with an initial velocity of 4 ft/sec?

Problems 109–112 require the following information.

Period of a Pendulum *The period T, in seconds, of a pendulum of length l, in feet, may be approximated using the formula*

$$T = 2\pi\sqrt{\frac{l}{32}}$$

In Problems 109–112, express your answer both as a square root and as a decimal.

109. Find the period T of a pendulum whose length is 64 feet.

110. Find the period T of a pendulum whose length is 16 feet.

111. Find the period T of a pendulum whose length is 8 inches.

112. Find the period T of a pendulum whose length is 4 inches.

Explaining Concepts: Discussion and Writing

113. Give an example to show that $\sqrt{a^2}$ is not equal to a. Use it to explain why $\sqrt{a^2} = |a|$.

'Are You Prepared?' Answers

1. $9; -9$
2. $4; 4$

Equations and Inequalities

1

Outline

Financing a Purchase

Whenever we make a major purchase, such as an automobile or a house, we often need to finance the purchase by borrowing money from a lending institution, such as a bank. Have you ever wondered how the bank determines the monthly payment? How much total interest will be paid over the course of the loan? What roles do the rate of interest and the length of the loan play?

 —See the Internet-based Chapter Project I—

A Look Ahead ▷ Chapter 1, Equations and Inequalities, reviews many topics covered in Intermediate Algebra. If your instructor decides to exclude complex numbers from the course, don't be alarmed. The book has been designed so that the topic of complex numbers can be included or excluded without any confusion later on.

1.1 Linear Equations

PREPARING FOR THIS SECTION *Before getting started, review the following:*
- Properties of Real Numbers (Section R.1, pp. 9–13)
- Domain of a Variable (Section R.2, p. 21)

Now Work the 'Are You Prepared?' problems on page 90.

OBJECTIVES 1 Solve a Linear Equation (p. 84)
 2 Solve Equations That Lead to Linear Equations (p. 86)
 3 Solve Problems That Can Be Modeled by Linear Equations (p. 87)

An **equation in one variable** is a statement in which two expressions, at least one containing the variable, are equal. The expressions are called the **sides** of the equation. Since an equation is a statement, it may be true or false, depending on the value of the variable. Unless otherwise restricted, the admissible values of the variable are those in the domain of the variable. These admissible values of the variable, if any, that result in a true statement are called **solutions,** or **roots,** of the equation. To **solve an equation** means to find all the solutions of the equation.

For example, the following are all equations in one variable, x:

$$x + 5 = 9 \qquad x^2 + 5x = 2x - 2 \qquad \frac{x^2 - 4}{x + 1} = 0 \qquad \sqrt{x^2 + 9} = 5$$

The first of these statements, $x + 5 = 9$, is true when $x = 4$ and false for any other choice of x. That is, 4 is a solution of the equation $x + 5 = 9$. We also say that 4 **satisfies** the equation $x + 5 = 9$, because, when we substitute 4 for x, a true statement results.

Sometimes an equation will have more than one solution. For example, the equation

$$\frac{x^2 - 4}{x + 1} = 0$$

has $x = -2$ and $x = 2$ as solutions.

Usually, we will write the solution of an equation in set notation. This set is called the **solution set** of the equation. For example, the solution set of the equation $x^2 - 9 = 0$ is $\{-3, 3\}$.

Some equations have no real solution. For example, $x^2 + 9 = 5$ has no real solution, because there is no real number whose square when added to 9 equals 5.

An equation that is satisfied for every value of the variable for which both sides are defined is called an **identity.** For example, the equation

$$3x + 5 = x + 3 + 2x + 2$$

is an identity, because this statement is true for any real number x.

One method for solving an equation is to replace the original equation by a succession of *equivalent equations* until an equation with an obvious solution is obtained.

For example, all the following equations are equivalent.

$$2x + 3 = 13$$
$$2x = 10$$
$$x = 5$$

We conclude that the solution set of the original equation is $\{5\}$.

How do we obtain equivalent equations? In general, there are five ways.

Procedures That Result in Equivalent Equations

1. Interchange the two sides of the equation:

$$\text{Replace } 3 = x \text{ by } x = 3$$

2. Simplify the sides of the equation by combining like terms, eliminating parentheses, and so on:

$$\text{Replace } (x + 2) + 6 = 2x + (x + 1)$$
$$\text{by } x + 8 = 3x + 1$$

3. Add or subtract the same expression on both sides of the equation:

$$\text{Replace } 3x - 5 = 4$$
$$\text{by } (3x - 5) + 5 = 4 + 5$$

4. Multiply or divide both sides of the equation by the same nonzero expression:

$$\text{Replace } \frac{3x}{x - 1} = \frac{6}{x - 1} \quad x \neq 1$$
$$\text{by } \frac{3x}{x - 1} \cdot (x - 1) = \frac{6}{x - 1} \cdot (x - 1)$$

5. If one side of the equation is 0 and the other side can be factored, then we may use the Zero-Product Property* and set each factor equal to 0:

$$\text{Replace } x(x - 3) = 0$$
$$\text{by } x = 0 \text{ or } x - 3 = 0$$

WARNING Squaring both sides of an equation does not necessarily lead to an equivalent equation. ∎

Whenever it is possible to solve an equation in your head, do so. For example,

The solution of $2x = 8$ is $x = 4$.
The solution of $3x - 15 = 0$ is $x = 5$.

Now Work PROBLEM 9

EXAMPLE 1 **Solving an Equation**

Solve the equation: $3x - 5 = 4$

Solution Replace the original equation by a succession of equivalent equations.

$$3x - 5 = 4$$
$$(3x - 5) + 5 = 4 + 5 \quad \text{Add 5 to both sides.}$$
$$3x = 9 \quad \text{Simplify.}$$
$$\frac{3x}{3} = \frac{9}{3} \quad \text{Divide both sides by 3.}$$
$$x = 3 \quad \text{Simplify.}$$

The last equation, $x = 3$, has the single solution 3. All these equations are equivalent, so 3 is the only solution of the original equation, $3x - 5 = 4$.

* The Zero-Product Property says that if $ab = 0$ then $a = 0$ or $b = 0$ or both equal 0.

✓**Check:** It is a good practice to check the solution by substituting 3 for x in the original equation.

$$3x - 5 = 4$$
$$3(3) - 5 \overset{?}{=} 4$$
$$9 - 5 \overset{?}{=} 4$$
$$4 = 4$$

The solution checks. The solution set is $\{3\}$.

➤ **Now Work** PROBLEM 23

Steps for Solving Equations

STEP 1: List any restrictions on the domain of the variable.
STEP 2: Simplify the equation by replacing the original equation by a succession of equivalent equations following the procedures listed earlier.
STEP 3: If the result of Step 2 is a product of factors equal to 0, use the Zero-Product Property and set each factor equal to 0 (procedure 5).
STEP 4: Check your solution(s).

1 Solve a Linear Equation

Linear equations are equations such as

$$3x + 12 = 0 \qquad -2x + 5 = 0 \qquad \frac{1}{2}x - \sqrt{3} = 0$$

DEFINITION

A **linear equation in one variable** is equivalent to an equation of the form

$$\boxed{ax + b = 0}$$

where a and b are real numbers and $a \neq 0$.

Sometimes, a linear equation is called a **first-degree equation**, because the left side is a polynomial in x of degree 1.

It is relatively easy to solve a linear equation. The idea is to *isolate* the variable:

$$ax + b = 0 \qquad a \neq 0$$
$$ax = -b \qquad \text{Subtract } b \text{ from both sides.}$$
$$x = \frac{-b}{a} \qquad \text{Divide both sides by } a, a \neq 0.$$

The linear equation $ax + b = 0$, $a \neq 0$, has the single solution given by the formula $x = -\dfrac{b}{a}$.

EXAMPLE 2

Solving a Linear Equation

Solve the equation: $\dfrac{1}{2}(x + 5) - 4 = \dfrac{1}{3}(2x - 1)$

Solution To clear the equation of fractions, multiply both sides by 6, the least common multiple of the denominators of the fractions $\dfrac{1}{2}$ and $\dfrac{1}{3}$.

$$\frac{1}{2}(x+5) - 4 = \frac{1}{3}(2x-1)$$

$$6\left[\frac{1}{2}(x+5) - 4\right] = 6\left[\frac{1}{3}(2x-1)\right] \qquad \text{Multiply both sides by 6, the LCM of 2 and 3.}$$

$$3(x+5) - 6 \cdot 4 = 2(2x-1) \qquad \begin{array}{l}\text{Use the Distributive Property on the left and}\\ \text{the Associative Property on the right.}\end{array}$$

$$3x + 15 - 24 = 4x - 2 \qquad \text{Use the Distributive Property.}$$

$$3x - 9 = 4x - 2 \qquad \text{Combine like terms.}$$

$$3x - 9 + 9 = 4x - 2 + 9 \qquad \text{Add 9 to each side.}$$

$$3x = 4x + 7 \qquad \text{Simplify.}$$

$$3x - 4x = 4x + 7 - 4x \qquad \text{Subtract } 4x \text{ from each side.}$$

$$-x = 7 \qquad \text{Simplify.}$$

$$x = -7 \qquad \text{Multiply both sides by } -1.$$

✔**Check:** $\dfrac{1}{2}(x+5) - 4 = \dfrac{1}{2}(-7+5) - 4 = \dfrac{1}{2}(-2) - 4 = -1 - 4 = -5$

$\dfrac{1}{3}(2x-1) = \dfrac{1}{3}[2(-7) - 1] = \dfrac{1}{3}(-14 - 1) = \dfrac{1}{3}(-15) = -5$

Since the two expressions are equal, the solution $x = -7$ checks and the solution set is $\{-7\}$.

━━━━ **Now Work** PROBLEM 33

EXAMPLE 3 **Solving a Linear Equation Using a Calculator**

Solve the equation: $\quad 2.78x + \dfrac{2}{17.931} = 54.06$

Round the answer to two decimal places.

Solution To avoid rounding errors, solve for x before using the calculator.

$$2.78x + \frac{2}{17.931} = 54.06$$

$$2.78x = 54.06 - \frac{2}{17.931} \qquad \text{Subtract } \frac{2}{17.931} \text{ from each side.}$$

$$x = \frac{54.06 - \dfrac{2}{17.931}}{2.78} \qquad \text{Divide each side by 2.78.}$$

Now use your calculator. The solution, rounded to two decimal places, is 19.41.

✔**Check:** Store the unrounded solution 19.40592134 in memory and proceed to evaluate $2.78x + \dfrac{2}{17.931}$.

$$(2.78)(19.40592134) + \frac{2}{17.931} = 54.06$$

━━━━ **Now Work** PROBLEM 65

2 Solve Equations That Lead to Linear Equations

EXAMPLE 4 **Solving an Equation That Leads to a Linear Equation**

Solve the equation: $(2y + 1)(y - 1) = (y + 5)(2y - 5)$

Solution

$$(2y + 1)(y - 1) = (y + 5)(2y - 5)$$
$$2y^2 - y - 1 = 2y^2 + 5y - 25 \qquad \text{Multiply and combine like terms.}$$
$$-y - 1 = 5y - 25 \qquad \text{Subtract } 2y^2 \text{ from each side.}$$
$$-y = 5y - 24 \qquad \text{Add 1 to each side.}$$
$$-6y = -24 \qquad \text{Subtract 5y from each side.}$$
$$y = 4 \qquad \text{Divide both sides by } -6.$$

✔**Check:** $(2y + 1)(y - 1) = [2(4) + 1](4 - 1) = (8 + 1)(3) = (9)(3) = 27$

$(y + 5)(2y - 5) = (4 + 5)[2(4) - 5] = (9)(8 - 5) = (9)(3) = 27$

Since the two expressions are equal, the solution $y = 4$ checks.
The solution set is $\{4\}$.

EXAMPLE 5 **Solving an Equation That Leads to a Linear Equation**

Solve the equation: $\dfrac{3}{x - 2} = \dfrac{1}{x - 1} + \dfrac{7}{(x - 1)(x - 2)}$

Solution First, notice that the domain of the variable is $\{x | x \neq 1, x \neq 2\}$. Clear the equation of fractions by multiplying both sides by the least common multiple of the denominators of the three fractions, $(x - 1)(x - 2)$.

$$\frac{3}{x - 2} = \frac{1}{x - 1} + \frac{7}{(x - 1)(x - 2)}$$

$$(x - 1)(x - 2)\frac{3}{x - 2} = (x - 1)(x - 2)\left[\frac{1}{x - 1} + \frac{7}{(x - 1)(x - 2)}\right] \quad \begin{array}{l}\text{Multiply both sides by}\\ (x - 1)(x - 2). \text{ Cancel}\\ \text{on the left.}\end{array}$$

$$3x - 3 = (x - 1)(x - 2)\frac{1}{x - 1} + (x - 1)(x - 2)\frac{7}{(x - 1)(x - 2)} \quad \begin{array}{l}\text{Use the Distributive}\\ \text{Property on each side;}\\ \text{cancel on the right.}\end{array}$$

$$3x - 3 = (x - 2) + 7 \qquad \text{Combine like terms.}$$
$$3x - 3 = x + 5 \qquad \text{Add 3 to each side.}$$
$$2x = 8 \qquad \text{Subtract x from each side.}$$
$$x = 4 \qquad \text{Divide by 2.}$$

✔**Check:** $\dfrac{3}{x - 2} = \dfrac{3}{4 - 2} = \dfrac{3}{2}$

$$\frac{1}{x - 1} + \frac{7}{(x - 1)(x - 2)} = \frac{1}{4 - 1} + \frac{7}{(4 - 1)(4 - 2)} = \frac{1}{3} + \frac{7}{3 \cdot 2} = \frac{2}{6} + \frac{7}{6} = \frac{9}{6} = \frac{3}{2}$$

Since the two expressions are equal, the solution $x = 4$ checks.
The solution set is $\{4\}$.

Now Work PROBLEM 59

EXAMPLE 6 **An Equation with No Solution**

Solve the equation: $\dfrac{3x}{x - 1} + 2 = \dfrac{3}{x - 1}$

Solution First, notice that the domain of the variable is $\{x \mid x \neq 1\}$. Since the two quotients in the equation have the same denominator, $x - 1$, we can simplify by multiplying both sides by $x - 1$. The resulting equation is equivalent to the original equation, since we are multiplying by $x - 1$, which is not 0. (Remember, $x \neq 1$.)

$$\frac{3x}{x-1} + 2 = \frac{3}{x-1}$$

$$\left(\frac{3x}{x-1} + 2\right) \cdot (x-1) = \frac{3}{x-1} \cdot (x-1) \qquad \text{Multiply both sides by } x - 1; \text{ cancel on the right.}$$

$$\frac{3x}{x-1} \cdot (x-1) + 2 \cdot (x-1) = 3 \qquad \text{Use the Distributive Property on the left side; cancel on the left.}$$

$$3x + 2x - 2 = 3 \qquad \text{Simplify.}$$

$$5x - 2 = 3 \qquad \text{Combine like terms.}$$

$$5x = 5 \qquad \text{Add 2 to each side.}$$

$$x = 1 \qquad \text{Divide both sides by 5.}$$

The solution appears to be 1. But recall that $x = 1$ is not in the domain of the variable. The equation has no solution.

---**Now Work** PROBLEM 49

EXAMPLE 7	**Converting to Fahrenheit from Celsius**

In the United States we measure temperature in both degrees Fahrenheit (°F) and degrees Celsius (°C), which are related by the formula $C = \frac{5}{9}(F - 32)$. What are the Fahrenheit temperatures corresponding to Celsius temperatures of 0°, 10°, 20°, and 30°C?

Solution We could solve four equations for F by replacing C each time by 0, 10, 20, and 30. Instead, it is much easier and faster first to solve the equation $C = \frac{5}{9}(F - 32)$ for F and then substitute in the values of C.

$$C = \frac{5}{9}(F - 32)$$

$$9C = 5(F - 32) \qquad \text{Multiply both sides by 9.}$$

$$9C = 5F - 160 \qquad \text{Use the Distributive Property.}$$

$$5F - 160 = 9C \qquad \text{Interchange sides.}$$

$$5F = 9C + 160 \qquad \text{Add 160 to each side.}$$

$$F = \frac{9}{5}C + 32 \qquad \text{Divide both sides by 5.}$$

Now do the required arithmetic.

$$0°C: \quad F = \frac{9}{5}(0) + 32 = 32°F$$

$$10°C: \quad F = \frac{9}{5}(10) + 32 = 50°F$$

$$20°C: \quad F = \frac{9}{5}(20) + 32 = 68°F$$

NOTE The icon ● is a Model It! icon. It indicates that the discussion or problem involves modeling. ∎

$$30°C: \quad F = \frac{9}{5}(30) + 32 = 86°F$$

3 Solve Problems That Can Be Modeled by Linear Equations

Although each situation has its unique features, we can provide an outline of the steps to follow to solve applied problems.

NOTE It is a good practice to choose a variable that reminds you of the unknown. For example, use *t* for time. ∎

Steps for Solving Applied Problems

STEP 1: Read the problem carefully, perhaps two or three times. Pay particular attention to the question being asked in order to identify what you are looking for. If you can, determine realistic possibilities for the answer.

STEP 2: Assign a letter (variable) to represent what you are looking for, and, if necessary, express any remaining unknown quantities in terms of this variable.

STEP 3: Make a list of all the known facts, and translate them into mathematical expressions. These may take the form of an equation (or, later, an inequality) involving the variable. The equation (or inequality) is called the **model.** If possible, draw an appropriately labeled diagram to assist you. Sometimes a table or chart helps.

STEP 4: Solve the equation for the variable, and then answer the question, usually using a complete sentence.

STEP 5: Check the answer with the facts in the problem. If it agrees, congratulations! If it does not agree, try again.

 EXAMPLE 8 **Investments**

A total of $18,000 is invested, some in stocks and some in bonds. If the amount invested in bonds is half that invested in stocks, how much is invested in each category?

Step-by-Step Solution

Step 1: Determine what you are looking for.

We are being asked to find the amount of two investments. These amounts must total $18,000. (Do you see why?)

Step 2: Assign a variable to represent what you are looking for. If necessary, express any remaining unknown quantities in terms of this variable.

If *x* equals the amount invested in stocks, then the rest of the money, $18,000 - x$, is the amount invested in bonds.

Step 3: Translate the English into mathematical statements. It may be helpful to draw a figure that represents the situation. Sometimes a table can be used to organize the information. Use the information to build your model.

Set up a table:

Amount in Stocks	Amount in Bonds	Reason
x	18,000 − x	Total invested is $18,000

We also know that:

Total amount invested in bonds is one-half that in stocks

$$18,000 - x \qquad = \qquad \frac{1}{2}(x)$$

Step 4: Solve the equation and answer the original question.

$$18,000 - x = \frac{1}{2}x$$

$$18,000 = x + \frac{1}{2}x \qquad \text{Add } x \text{ to both sides.}$$

$$18,000 = \frac{3}{2}x \qquad \text{Simplify.}$$

$$\left(\frac{2}{3}\right)18,000 = \left(\frac{2}{3}\right)\left(\frac{3}{2}x\right) \qquad \text{Multiply both sides by } \frac{2}{3}.$$

$$12,000 = x \qquad \text{Simplify.}$$

So, $12,000 is invested in stocks and $18,000 - $12,000 = $6000 is invested in bonds.

Step 5: Check your answer with the facts presented in the problem.

The total invested is $12,000 + $6000 = $18,000, and the amount in bonds, $6000, is half that in stocks, $12,000.

━━━ **Now Work** PROBLEM 83

EXAMPLE 9 **Determining an Hourly Wage**

Shannon grossed $435 one week by working 52 hours. Her employer pays time-and-a-half for all hours worked in excess of 40 hours. With this information, can you determine Shannon's regular hourly wage?

Solution **STEP 1:** We are looking for an hourly wage. Our answer will be expressed in dollars per hour.

STEP 2: Let x represent the regular hourly wage; x is measured in dollars per hour.

STEP 3: Set up a table:

	Hours Worked	Hourly Wage	Salary
Regular	40	x	$40x$
Overtime	12	$1.5x$	$12(1.5x) = 18x$

The sum of regular salary plus overtime salary will equal $435. From the table, $40x + 18x = 435$.

STEP 4:
$$40x + 18x = 435$$
$$58x = 435$$
$$x = 7.50$$

Shannon's regular hourly wage is $7.50 per hour.

STEP 5: Forty hours yields a salary of $40(7.50) = \$300$, and 12 hours of overtime yields a salary of $12(1.5)(7.50) = \$135$, for a total of $435.

 Now Work PROBLEM 85

SUMMARY **Steps for Solving a Linear Equation**

To solve a linear equation, follow these steps:

STEP 1: List any restrictions on the variable.

STEP 2: If necessary, clear the equation of fractions by multiplying both sides by the least common multiple (LCM) of the denominators of all the fractions.

STEP 3: Remove all parentheses and simplify.

STEP 4: Collect all terms containing the variable on one side and all remaining terms on the other side.

STEP 5: Simplify and solve.

STEP 6: Check your solution(s).

Historical Feature

Solving equations is among the oldest of mathematical activities, and efforts to systematize this activity determined much of the shape of modern mathematics.

Consider the following problem and its solution using only words: Solve the problem of how many apples Jim has, given that

"Bob's five apples and Jim's apples together make twelve apples" by thinking,

"Jim's apples are all twelve apples less Bob's five apples" and then concluding,

"Jim has seven apples."

The mental steps translated into algebra are

$$5 + x = 12$$
$$x = 12 - 5$$
$$= 7$$

The solution of this problem using only words is the earliest form of algebra. Such problems were solved exactly this way in Babylonia in 1800 BC. We know almost nothing of mathematical work before this date, although most authorities believe the sophistication of the earliest known texts indicates that a long period of previous development must have occurred. The method of writing out equations in words persisted for thousands of years, and although it now seems extremely cumbersome, it was used very effectively by many generations of mathematicians. The Arabs developed a good deal of the theory of cubic equations while writing out all the equations in words. About AD 1500, the tendency to abbreviate words in the written equations began to lead in the direction of modern notation; for example, the Latin word *et* (meaning *and*) developed into the plus sign, +. Although the occasional use of letters to represent variables dates back to AD 1200, the practice did not become common until about AD 1600. Development thereafter was rapid, and by 1635 algebraic notation did not differ essentially from what we use now.

1.1 Assess Your Understanding

'Are You Prepared?' *Answers are given at the end of these exercises. If you get a wrong answer, read the pages listed in red.*

1. The fact that $2(x + 3) = 2x + 6$ is because of the _____ Property. (pp. 9–13)

2. The fact that $3x = 0$ implies that $x = 0$ is a result of the _____ Property. (pp. 9–13)

3. The domain of the variable in the expression $\dfrac{x}{x - 4}$ is _____. (p. 21)

Concepts and Vocabulary

4. **True or False** Multiplying both sides of an equation by any number results in an equivalent equation.

5. An equation that is satisfied for every value of the variable for which both sides are defined is called a(n) _____.

6. An equation of the form $ax + b = 0$ is called a(n) _____ equation or a(n) _____ _____ equation .

7. **True or False** The solution of the equation $3x - 8 = 0$ is $\dfrac{3}{8}$.

8. **True or False** Some equations have no solution.

Skill Building

In Problems 9–16, mentally solve each equation.

9. $7x = 21$

10. $6x = -24$

11. $3x + 15 = 0$

12. $6x + 18 = 0$

13. $2x - 3 = 0$

14. $3x + 4 = 0$

15. $\dfrac{1}{3}x = \dfrac{5}{12}$

16. $\dfrac{2}{3}x = \dfrac{9}{2}$

In Problems 17–64, solve each equation.

17. $3x + 4 = x$

18. $2x + 9 = 5x$

19. $2t - 6 = 3 - t$

20. $5y + 6 = -18 - y$

21. $6 - x = 2x + 9$

22. $3 - 2x = 2 - x$

23. $3 + 2n = 4n + 7$

24. $6 - 2m = 3m + 1$

25. $2(3 + 2x) = 3(x - 4)$

26. $3(2 - x) = 2x - 1$

27. $8x - (3x + 2) = 3x - 10$

28. $7 - (2x - 1) = 10$

29. $\dfrac{3}{2}x + 2 = \dfrac{1}{2} - \dfrac{1}{2}x$

30. $\dfrac{1}{3}x = 2 - \dfrac{2}{3}x$

31. $\dfrac{1}{2}x - 5 = \dfrac{3}{4}x$

32. $1 - \dfrac{1}{2}x = 6$

33. $\dfrac{2}{3}p = \dfrac{1}{2}p + \dfrac{1}{3}$

34. $\dfrac{1}{2} - \dfrac{1}{3}p = \dfrac{4}{3}$

35. $0.9t = 0.4 + 0.1t$

36. $0.9t = 1 + t$

37. $\dfrac{x + 1}{3} + \dfrac{x + 2}{7} = 2$

38. $\dfrac{2x + 1}{3} + 16 = 3x$

39. $\dfrac{2}{y} + \dfrac{4}{y} = 3$

40. $\dfrac{4}{y} - 5 = \dfrac{5}{2y}$

41. $\dfrac{1}{2} + \dfrac{2}{x} = \dfrac{3}{4}$

42. $\dfrac{3}{x} - \dfrac{1}{3} = \dfrac{1}{6}$

43. $(x + 7)(x - 1) = (x + 1)^2$

44. $(x + 2)(x - 3) = (x + 3)^2$

45. $x(2x - 3) = (2x + 1)(x - 4)$

46. $x(1 + 2x) = (2x - 1)(x - 2)$

47. $z(z^2 + 1) = 3 + z^3$

48. $w(4 - w^2) = 8 - w^3$

49. $\dfrac{x}{x - 2} + 3 = \dfrac{2}{x - 2}$

50. $\dfrac{2x}{x + 3} = \dfrac{-6}{x + 3} - 2$

51. $\dfrac{2x}{x^2 - 4} = \dfrac{4}{x^2 - 4} - \dfrac{3}{x + 2}$

52. $\dfrac{x}{x^2 - 9} + \dfrac{4}{x + 3} = \dfrac{3}{x^2 - 9}$

53. $\dfrac{x}{x + 2} = \dfrac{3}{2}$

54. $\dfrac{3x}{x - 1} = 2$

55. $\dfrac{5}{2x - 3} = \dfrac{3}{x + 5}$

56. $\dfrac{-4}{x + 4} = \dfrac{-3}{x + 6}$

57. $\dfrac{6t + 7}{4t - 1} = \dfrac{3t + 8}{2t - 4}$

58. $\dfrac{8w + 5}{10w - 7} = \dfrac{4w - 3}{5w + 7}$

59. $\dfrac{4}{x-2} = \dfrac{-3}{x+5} + \dfrac{7}{(x+5)(x-2)}$ **60.** $\dfrac{-4}{2x+3} + \dfrac{1}{x-1} = \dfrac{1}{(2x+3)(x-1)}$ **61.** $\dfrac{2}{y+3} + \dfrac{3}{y-4} = \dfrac{5}{y+6}$

62. $\dfrac{5}{5z-11} + \dfrac{4}{2z-3} = \dfrac{-3}{5-z}$ **63.** $\dfrac{x}{x^2-1} - \dfrac{x+3}{x^2-x} = \dfrac{-3}{x^2+x}$ **64.** $\dfrac{x+1}{x^2+2x} - \dfrac{x+4}{x^2+x} = \dfrac{-3}{x^2+3x+2}$

In Problems 65–68, use a calculator to solve each equation. Round the solution to two decimal places.

65. $3.2x + \dfrac{21.3}{65.871} = 19.23$ **66.** $6.2x - \dfrac{19.1}{83.72} = 0.195$

67. $14.72 - 21.58x = \dfrac{18}{2.11}x + 2.4$ **68.** $18.63x - \dfrac{21.2}{2.6} = \dfrac{14}{2.32}x - 20$

Applications and Extensions

In Problems 69–74, solve each equation. The letters a, b, and c are constants.

69. $ax - b = c, \quad a \neq 0$ **70.** $1 - ax = b, \quad a \neq 0$ **71.** $\dfrac{x}{a} + \dfrac{x}{b} = c, \quad a \neq 0, b \neq 0, a \neq -b$

72. $\dfrac{a}{x} + \dfrac{b}{x} = c, \quad c \neq 0$ **73.** $\dfrac{1}{x-a} + \dfrac{1}{x+a} = \dfrac{2}{x-1}$ **74.** $\dfrac{b+c}{x+a} = \dfrac{b-c}{x-a}, \quad c \neq 0, a \neq 0$

75. Find the number a for which $x = 4$ is a solution of the equation

$$x + 2a = 16 + ax - 6a$$

76. Find the number b for which $x = 2$ is a solution of the equation

$$x + 2b = x - 4 + 2bx$$

Problems 77–82 list some formulas that occur in applications. Solve each formula for the indicated variable.

77. Electricity $\dfrac{1}{R} = \dfrac{1}{R_1} + \dfrac{1}{R_2}$ for R

78. Finance $A = P(1 + rt)$ for r

79. Mechanics $F = \dfrac{mv^2}{R}$ for R

80. Chemistry $PV = nRT$ for T

81. Mathematics $S = \dfrac{a}{1-r}$ for r

82. Mechanics $v = -gt + v_0$ for t

85. Finance A total of $20,000 is to be invested, some in bonds and some in certificates of deposit (CDs). If the amount invested in bonds is to exceed that in CDs by $3000, how much will be invested in each type of investment?

84. Finance A total of $10,000 is to be divided between Sean and George, with George to receive $3000 less than Sean. How much will each receive?

85. Computing Hourly Wages Sandra, who is paid time-and-a-half for hours worked in excess of 40 hours, had gross weekly wages of $442 for 48 hours worked. What is her regular hourly rate?

86. Computing Hourly Wages Leigh is paid time-and-a-half for hours worked in excess of 40 hours and double-time for hours worked on Sunday. If Leigh had gross weekly wages of $456 for working 50 hours, 4 of which were on Sunday, what is her regular hourly rate?

87. Computing Grades Going into the final exam, which will count as two tests, Brooke has test scores of 80, 83, 71, 61, and 95. What score does Brooke need on the final in order to have an average score of 80?

88. Computing Grades Going into the final exam, which will count as two-thirds of the final grade, Mike has test scores of 86, 80, 84, and 90. What score does Mike need on the final in order to earn a B, which requires an average score of 80? What does he need to earn an A, which requires an average of 90?

89. Business: Discount Pricing A builder of tract homes reduced the price of a model by 15%. If the new price is $425,000, what was its original price? How much can be saved by purchasing the model?

90. Business: Discount Pricing A car dealer, at a year-end clearance, reduces the list price of last year's models by 15%. If a certain four-door model has a discounted price of $8000, what was its list price? How much can be saved by purchasing last year's model?

91. Business: Marking up the Price of Books A college book store marks up the price that it pays the publisher for a book by 35%. If the selling price of a book is $92.00, how much did the bookstore pay for this book?

92. **Personal Finance: Cost of a Car** The suggested list price of a new car is $18,000. The dealer's cost is 85% of list. How much will you pay if the dealer is willing to accept $100 over cost for the car?

93. **Business: Theater Attendance** The manager of the Coral Theater wants to know whether the majority of its patrons are adults or children. One day in July, 5200 tickets were sold and the receipts totaled $29,961. The adult admission is $7.50, and the children's admission is $4.50. How many adult patrons were there?

94. **Business: Discount Pricing** A wool suit, discounted by 30% for a clearance sale, has a price tag of $399. What was the suit's original price?

95. **Geometry** The perimeter of a rectangle is 60 feet. Find its length and width if the length is 8 feet longer than the width.

96. **Geometry** The perimeter of a rectangle is 42 meters. Find its length and width if the length is twice the width.

97. **Sharing the Cost of a Pizza** Judy and Tom agree to share the cost of an $18 pizza based on how much each ate. If Tom ate $\frac{2}{3}$ the amount that Judy ate, how much should each pay?

[**Hint:** Some pizza may be left.]

Tom's portion

Judy's portion

Explaining Concepts: Discussion and Writing

98. **What Is Wrong?** One step in the following list contains an error. Identify it and explain what is wrong.

$$
\begin{array}{lr}
x = 2 & \textbf{(1)} \\
3x - 2x = 2 & \textbf{(2)} \\
3x = 2x + 2 & \textbf{(3)} \\
x^2 + 3x = x^2 + 2x + 2 & \textbf{(4)} \\
x^2 + 3x - 10 = x^2 + 2x - 8 & \textbf{(5)} \\
(x - 2)(x + 5) = (x - 2)(x + 4) & \textbf{(6)} \\
x + 5 = x + 4 & \textbf{(7)} \\
1 = 0 & \textbf{(8)}
\end{array}
$$

99. The equation

$$
\frac{5}{x + 3} + 3 = \frac{8 + x}{x + 3}
$$

has no solution, yet when we go through the process of solving it we obtain $x = -3$. Write a brief paragraph to explain what causes this to happen.

100. Make up an equation that has no solution and give it to a fellow student to solve. Ask the fellow student to write a critique of your equation.

'Are You Prepared?' Answers

1. Distributive 2. Zero-Product 3. $\{x \mid x \neq 4\}$

1.2 Quadratic Equations

PREPARING FOR THIS SECTION *Before getting started, review the following:*

- Factoring Polynomials (Section R.5, pp. 49–55)
- Zero-Product Property (Section R.1, p. 13)
- Square Roots (Section R.2, pp. 23–24)
- Complete the Square (Section R.5, p. 56)

Now Work the 'Are You Prepared?' problems on page 101.

OBJECTIVES **1** Solve a Quadratic Equation by Factoring (p. 93)
 2 Solve a Quadratic Equation by Completing the Square (p. 95)
 3 Solve a Quadratic Equation Using the Quadratic Formula (p. 96)
 4 Solve Problems That Can Be Modeled by Quadratic Equations (p. 99)

Quadratic equations are equations such as

$$
\begin{array}{c}
2x^2 + x + 8 = 0 \\
3x^2 - 5x + 6 = 0 \\
x^2 - 9 = 0
\end{array}
$$

DEFINITION

A **quadratic equation** is an equation equivalent to one of the form

$$ax^2 + bx + c = 0 \qquad\qquad (1)$$

where a, b, and c are real numbers and $a \neq 0$.

A quadratic equation written in the form $ax^2 + bx + c = 0$ is said to be in **standard form.**

Sometimes, a quadratic equation is called a **second-degree equation,** because the left side is a polynomial of degree 2. We shall discuss three ways of solving quadratic equations: by factoring, by completing the square, and by using the quadratic formula.

1 Solve a Quadratic Equation by Factoring

When a quadratic equation is written in standard form $ax^2 + bx + c = 0$, it may be possible to factor the expression on the left side into the product of two first-degree polynomials. Then, by using the Zero-Product Property and setting each factor equal to 0, we can solve the resulting linear equations and obtain the solutions of the quadratic equation.

EXAMPLE 1

Solving a Quadratic Equation by Factoring

Solve the equation:

(a) $x^2 + 6x = 0$

(b) $2x^2 = x + 3$

Solution

(a) The equation is in the standard form specified in equation (1). The left side may be factored as

$$x^2 + 6x = 0$$
$$x(x + 6) = 0 \quad \text{Factor.}$$

Using the Zero-Product Property, set each factor equal to 0 and then solve the resulting first-degree equations.

$$x = 0 \quad \text{or} \quad x + 6 = 0 \quad \text{Zero-Product Property}$$
$$x = 0 \quad \text{or} \quad x = -6 \quad \text{Solve.}$$

The solution set is $\{0, -6\}$.

(b) Place the equation $2x^2 = x + 3$ in standard form by adding $-x - 3$ to both sides.

$$2x^2 = x + 3$$
$$2x^2 - x - 3 = 0 \quad \text{Add } -x - 3 \text{ to both sides.}$$

The left side may now be factored as

$$(2x - 3)(x + 1) = 0 \quad \text{Factor.}$$

so that

$$2x - 3 = 0 \quad \text{or} \quad x + 1 = 0 \quad \text{Zero-Product Property}$$
$$x = \frac{3}{2} \qquad\qquad x = -1 \quad \text{Solve.}$$

The solution set is $\left\{ -1, \frac{3}{2} \right\}$.

When the left side factors into two linear equations with the same solution, the quadratic equation is said to have a **repeated solution.** This solution is also called a **root of multiplicity 2,** or a **double root.**

| EXAMPLE 2 | **Solving a Quadratic Equation by Factoring** |

Solve the equation: $9x^2 - 6x + 1 = 0$

Solution This equation is already in standard form, and the left side can be factored.

$$9x^2 - 6x + 1 = 0$$
$$(3x - 1)(3x - 1) = 0$$

so

$$x = \frac{1}{3} \quad \text{or} \quad x = \frac{1}{3}$$

This equation has only the repeated solution $\frac{1}{3}$. The solution set is $\left\{\frac{1}{3}\right\}$.

━━━━━**Now Work** PROBLEMS 11 AND 21

The Square Root Method

Suppose that we wish to solve the quadratic equation

$$x^2 = p \tag{2}$$

where $p \geq 0$ is a nonnegative number. Proceed as in the earlier examples.

$$x^2 - p = 0 \qquad \text{Put in standard form.}$$
$$(x - \sqrt{p})(x + \sqrt{p}) = 0 \qquad \text{Factor (over the real numbers).}$$
$$x = \sqrt{p} \quad \text{or} \quad x = -\sqrt{p} \qquad \text{Solve.}$$

We have the following result:

> If $x^2 = p$ and $p \geq 0$, then $x = \sqrt{p}$ or $x = -\sqrt{p}$. $\tag{3}$

When statement (3) is used, it is called the **Square Root Method.** In statement (3), note that if $p > 0$ the equation $x^2 = p$ has two solutions, $x = \sqrt{p}$ and $x = -\sqrt{p}$. We usually abbreviate these solutions as $x = \pm\sqrt{p}$, read as "x equals plus or minus the square root of p."

For example, the two solutions of the equation

$$x^2 = 4$$

are

$$x = \pm\sqrt{4} \qquad \text{Use the Square Root Method.}$$

and, since $\sqrt{4} = 2$, we have

$$x = \pm 2$$

The solution set is $\{-2, 2\}$.

| EXAMPLE 3 | **Solving a Quadratic Equation Using the Square Root Method** |

Solve each equation.

(a) $x^2 = 5$ (b) $(x - 2)^2 = 16$

Solution (a) Use the Square Root Method to get

$$x^2 = 5$$
$$x = \pm\sqrt{5} \qquad \text{Use the Square Root Method.}$$
$$x = \sqrt{5} \quad \text{or} \quad x = -\sqrt{5}$$

The solution set is $\{-\sqrt{5}, \sqrt{5}\}$.

(b) Use the Square Root Method to get

$$(x - 2)^2 = 16$$
$$x - 2 = \pm\sqrt{16} \qquad \text{Use the Square Root Method.}$$
$$x - 2 = \pm 4$$
$$x - 2 = 4 \quad \text{or} \quad x - 2 = -4$$
$$x = 6 \quad \text{or} \qquad x = -2$$

The solution set is $\{-2, 6\}$.

──── **Now Work** PROBLEM 31

2 Solve a Quadratic Equation by Completing the Square

EXAMPLE 4

Solving a Quadratic Equation by Completing the Square

Solve by completing the square: $x^2 + 5x + 4 = 0$

Solution Always begin this procedure by rearranging the equation so that the constant is on the right side.

$$x^2 + 5x + 4 = 0$$
$$x^2 + 5x = -4$$

Since the coefficient of x^2 is 1, we can complete the square on the left side by adding $\left(\dfrac{1}{2} \cdot 5\right)^2 = \dfrac{25}{4}$. Of course, in an equation, whatever is added to the left side must also be added to the right side. So add $\dfrac{25}{4}$ to *both* sides.

$$x^2 + 5x + \frac{25}{4} = -4 + \frac{25}{4} \qquad \text{Add } \frac{25}{4} \text{ to both sides.}$$
$$\left(x + \frac{5}{2}\right)^2 = \frac{9}{4} \qquad \text{Factor.}$$
$$x + \frac{5}{2} = \pm\sqrt{\frac{9}{4}} \qquad \text{Use the Square Root Method.}$$
$$x + \frac{5}{2} = \pm\frac{3}{2}$$
$$x = -\frac{5}{2} \pm \frac{3}{2}$$
$$x = -\frac{5}{2} + \frac{3}{2} = -1 \quad \text{or} \quad x = -\frac{5}{2} - \frac{3}{2} = -4$$

The solution set is $\{-4, -1\}$.

──── THE SOLUTION OF THE EQUATION IN EXAMPLE **4** ALSO CAN BE OBTAINED BY FACTORING. REWORK EXAMPLE **4** USING THIS TECHNIQUE.

The next example illustrates an equation that cannot be solved by factoring.

EXAMPLE 5

Solving a Quadratic Equation by Completing the Square

Solve by completing the square: $2x^2 - 8x - 5 = 0$

Solution First, rewrite the equation so that the constant is on the right side.

$$2x^2 - 8x - 5 = 0$$
$$2x^2 - 8x = 5$$

Next, divide both sides by 2 so that the coefficient of x^2 is 1. (This enables us to complete the square at the next step.)

$$x^2 - 4x = \frac{5}{2}$$

Finally, complete the square by adding 4 to both sides.

$$x^2 - 4x + 4 = \frac{5}{2} + 4$$

$$(x - 2)^2 = \frac{13}{2}$$

$$x - 2 = \pm\sqrt{\frac{13}{2}} \qquad \text{Use the Square Root Method.}$$

$$x - 2 = \pm\frac{\sqrt{26}}{2} \qquad \sqrt{\frac{13}{2}} = \frac{\sqrt{13}}{\sqrt{2}} = \frac{\sqrt{13}}{\sqrt{2}} \cdot \frac{\sqrt{2}}{\sqrt{2}} = \frac{\sqrt{26}}{2}$$

$$x = 2 \pm \frac{\sqrt{26}}{2}$$

The solution set is $\left\{ 2 - \frac{\sqrt{26}}{2}, 2 + \frac{\sqrt{26}}{2} \right\}$.

NOTE If we wanted an approximation, say rounded to two decimal places, of these solutions, we would use a calculator to get $\{-0.55, 4.55\}$. ∎

⟶ **Now Work** PROBLEM 35

3 Solve a Quadratic Equation Using the Quadratic Formula

We can use the method of completing the square to obtain a general formula for solving any quadratic equation

$$ax^2 + bx + c = 0 \qquad a \neq 0$$

NOTE There is no loss in generality to assume that $a > 0$, since if $a < 0$ we can multiply by -1 to obtain an equivalent equation with a positive leading coefficient. ∎

As in Examples 4 and 5, rearrange the terms as

$$ax^2 + bx = -c \quad a > 0$$

Since $a > 0$, we can divide both sides by a to get

$$x^2 + \frac{b}{a}x = -\frac{c}{a}$$

Now the coefficient of x^2 is 1. To complete the square on the left side, add the square of $\frac{1}{2}$ of the coefficient of x; that is, add

$$\left(\frac{1}{2} \cdot \frac{b}{a}\right)^2 = \frac{b^2}{4a^2}$$

to both sides. Then

$$x^2 + \frac{b}{a}x + \frac{b^2}{4a^2} = \frac{b^2}{4a^2} - \frac{c}{a}$$

$$\left(x + \frac{b}{2a}\right)^2 = \frac{b^2 - 4ac}{4a^2} \qquad \frac{b^2}{4a^2} - \frac{c}{a} = \frac{b^2}{4a^2} - \frac{4ac}{4a^2} = \frac{b^2 - 4ac}{4a^2} \quad \textbf{(4)}$$

Provided that $b^2 - 4ac \geq 0$, we can now use the Square Root Method to get

$$x + \frac{b}{2a} = \pm\sqrt{\frac{b^2 - 4ac}{4a^2}}$$

$$x + \frac{b}{2a} = \frac{\pm\sqrt{b^2 - 4ac}}{2a} \qquad \text{The square root of a quotient equals the quotient of the square roots.}$$
$$\text{Also, } \sqrt{4a^2} = 2a \text{ since } a > 0.$$

$$x = -\frac{b}{2a} \pm \frac{\sqrt{b^2 - 4ac}}{2a} \quad \text{Add } -\frac{b}{2a} \text{ to both sides.}$$

$$= \frac{-b \pm \sqrt{b^2 - 4ac}}{2a} \quad \text{Combine the quotients on the right.}$$

What if $b^2 - 4ac$ is negative? Then equation (4) states that the left expression (a real number squared) equals the right expression (a negative number). Since this occurrence is impossible for real numbers, we conclude that if $b^2 - 4ac < 0$ the quadratic equation has no *real* solution. (We discuss quadratic equations for which the quantity $b^2 - 4ac < 0$ in detail in the next section.)

THEOREM **Quadratic Formula**

Consider the quadratic equation

$$ax^2 + bx + c = 0 \quad a \neq 0$$

If $b^2 - 4ac < 0$, this equation has no real solution.
If $b^2 - 4ac \geq 0$, the real solution(s) of this equation is (are) given by the **quadratic formula:**

$$x = \frac{-b \pm \sqrt{b^2 - 4ac}}{2a} \tag{5}$$

The quantity $b^2 - 4ac$ is called the **discriminant** of the quadratic equation, because its value tells us whether the equation has real solutions. In fact, it also tells us how many solutions to expect.

Discriminant of a Quadratic Equation

For a quadratic equation $ax^2 + bx + c = 0$:

1. If $b^2 - 4ac > 0$, there are two unequal real solutions.
2. If $b^2 - 4ac = 0$, there is a repeated solution, a root of multiplicity 2.
3. If $b^2 - 4ac < 0$, there is no real solution.

When asked to find the real solutions, of a quadratic equation, always evaluate the discriminant first to see if there are any real solutions.

EXAMPLE 6 **Solving a Quadratic Equation Using the Quadratic Formula**

Use the quadratic formula to find the real solutions, if any, of the equation

$$3x^2 - 5x + 1 = 0$$

Solution The equation is in standard form, so we compare it to $ax^2 + bx + c = 0$ to find a, b, and c.

$$3x^2 - 5x + 1 = 0$$
$$ax^2 + bx + c = 0 \quad a = 3, b = -5, c = 1$$

With $a = 3, b = -5$, and $c = 1$, evaluate the discriminant $b^2 - 4ac$.

$$b^2 - 4ac = (-5)^2 - 4(3)(1) = 25 - 12 = 13$$

Since $b^2 - 4ac > 0$, there are two real solutions, which can be found using the quadratic formula.

$$x = \frac{-b \pm \sqrt{b^2 - 4ac}}{2a} = \frac{-(-5) \pm \sqrt{13}}{2(3)} = \frac{5 \pm \sqrt{13}}{6}$$

The solution set is $\left\{ \dfrac{5 - \sqrt{13}}{6}, \dfrac{5 + \sqrt{13}}{6} \right\}$.

EXAMPLE 7

Solving a Quadratic Equation Using the Quadratic Formula

Use the quadratic formula to find the real solutions, if any, of the equation

$$\frac{25}{2}x^2 - 30x + 18 = 0$$

Solution The equation is given in standard form. However, to simplify the arithmetic, clear the fractions.

$$\frac{25}{2}x^2 - 30x + 18 = 0$$

$$25x^2 - 60x + 36 = 0 \quad \text{Clear fractions; multiply by 2.}$$

$$ax^2 + bx + c = 0 \quad \text{Compare to standard form.}$$

With $a = 25, b = -60$, and $c = 36$, evaluate the discriminant.

$$b^2 - 4ac = (-60)^2 - 4(25)(36) = 3600 - 3600 = 0$$

The equation has a repeated solution, which is found by using the quadratic formula.

$$x = \frac{-b \pm \sqrt{b^2 - 4ac}}{2a} = \frac{60 \pm \sqrt{0}}{50} = \frac{60}{50} = \frac{6}{5}$$

The solution set is $\left\{ \dfrac{6}{5} \right\}$.

EXAMPLE 8

Solving a Quadratic Equation Using the Quadratic Formula

Use the quadratic formula to find the real solutions, if any, of the equation

$$3x^2 + 2 = 4x$$

Solution The equation, as given, is not in standard form.

$$3x^2 + 2 = 4x$$

$$3x^2 - 4x + 2 = 0 \quad \text{Put in standard form.}$$

$$ax^2 + bx + c = 0 \quad \text{Compare to standard form.}$$

With $a = 3, b = -4$, and $c = 2$, we find

$$b^2 - 4ac = (-4)^2 - 4(3)(2) = 16 - 24 = -8$$

Since $b^2 - 4ac < 0$, the equation has no real solution.

Now Work PROBLEMS **45** AND **55**

EXAMPLE 9 **Solving a Quadratic Equation Using the Quadratic Formula**

Find the real solutions, if any, of the equation: $9 + \dfrac{3}{x} - \dfrac{2}{x^2} = 0,\ x \neq 0$

Solution In its present form, the equation

$$9 + \frac{3}{x} - \frac{2}{x^2} = 0$$

is not a quadratic equation. However, it can be transformed into one by multiplying each side by x^2. The result is

$$9x^2 + 3x - 2 = 0$$

Although we multiplied each side by x^2, we know that $x^2 \neq 0$ (do you see why?), so this quadratic equation is equivalent to the original equation.

Using $a = 9$, $b = 3$, and $c = -2$, the discriminant is

$$b^2 - 4ac = 3^2 - 4(9)(-2) = 9 + 72 = 81$$

Since $b^2 - 4ac > 0$, the new equation has two real solutions.

$$x = \frac{-b \pm \sqrt{b^2 - 4ac}}{2a} = \frac{-3 \pm \sqrt{81}}{2(9)} = \frac{-3 \pm 9}{18}$$

$$x = \frac{-3 + 9}{18} = \frac{6}{18} = \frac{1}{3} \quad \text{or} \quad x = \frac{-3 - 9}{18} = \frac{-12}{18} = -\frac{2}{3}$$

The solution set is $\left\{ -\dfrac{2}{3}, \dfrac{1}{3} \right\}$.

SUMMARY Procedure for Solving a Quadratic Equation

To solve a quadratic equation, first put it in standard form:

$$ax^2 + bx + c = 0$$

Then:

STEP 1: Identify a, b, and c.

STEP 2: Evaluate the discriminant, $b^2 - 4ac$.

STEP 3: (a) If the discriminant is negative, the equation has no real solution.

 (b) If the discriminant is zero, the equation has one real solution, a repeated root.

 (c) If the discriminant is positive, the equation has two distinct real solutions.

If you can easily spot factors, use the factoring method to solve the equation. Otherwise, use the quadratic formula or the method of completing the square.

 4 Solve Problems That Can Be Modeled by Quadratic Equations

Many applied problems require the solution of a quadratic equation. Let's look at one that you will probably see again in a slightly different form if you study calculus.

 EXAMPLE 10 **Constructing a Box**

From each corner of a square piece of sheet metal, remove a square of side 9 centimeters. Turn up the edges to form an open box. If the box is to hold 144 cubic centimeters (cm³), what should be the dimensions of the piece of sheet metal?

Solution Use Figure 1 as a guide. We have labeled by x the length of a side of the square piece of sheet metal. The box will be of height 9 centimeters, and its square base will measure $x - 18$ on each side. The volume V (Length \times Width \times Height) of the box is therefore

$$V = (x - 18)(x - 18) \cdot 9 = 9(x - 18)^2$$

Figure 1

Volume $= 9(x - 18)(x - 18)$

Since the volume of the box is to be 144 cm³, we have

$$9(x - 18)^2 = 144 \qquad \textit{V = 144}$$
$$(x - 18)^2 = 16 \qquad \textit{Divide each side by 9.}$$
$$x - 18 = \pm 4 \qquad \textit{Use the Square Root Method.}$$
$$x = 18 \pm 4$$
$$x = 22 \quad \text{or} \quad x = 14$$

✔**Check:** If we begin with a piece of sheet metal 22 centimeters by 22 centimeters, cut out a 9 centimeter square from each corner, and fold up the edges, we get a box whose dimensions are 9 by 4 by 4, with volume $9 \times 4 \times 4 = 144$ cm³, as required.

Discard the solution $x = 14$ (do you see why?) and conclude that the sheet metal should be 22 centimeters by 22 centimeters.

═════**Now Work** PROBLEM **97**

Historical Feature

P roblems using quadratic equations are found in the oldest known mathematical literature. Babylonians and Egyptians were solving such problems before 1800 BC. Euclid solved quadratic equations geometrically in his *Data* (300 BC), and the Hindus and Arabs gave rules for solving any quadratic equation with real roots. Because negative numbers were not freely used before AD 1500, there were several different types of quadratic equations, each with its own rule. Thomas Harriot (1560–1621) introduced the method of factoring to obtain solutions, and François Viète (1540–1603) introduced a method that is essentially completing the square.

Until modern times it was usual to neglect the negative roots (if there were any), and equations involving square roots of negative quantities were regarded as unsolvable until the 1500s.

Historical Problems

1. *One of al-Khwǎrízmí solutions* Solve $x^2 + 12x = 85$ by drawing the square shown. The area of the four white rectangles and the yellow square is $x^2 + 12x$. We then set this expression equal to 85 to get the equation $x^2 + 12x = 85$. If we add the four blue squares, we will have a larger square of known area. Complete the solution.

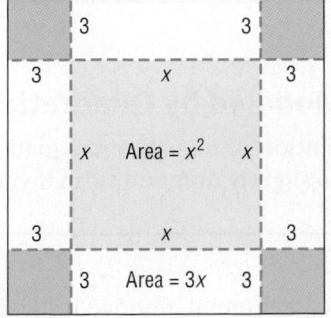

2. *Viète's method* Solve $x^2 + 12x - 85 = 0$ by letting $x = u + z$. Then

$$(u + z)^2 + 12(u + z) - 85 = 0$$
$$u^2 + (2z + 12)u + (z^2 + 12z - 85) = 0$$

Now select z so that $2z + 12 = 0$ and finish the solution.

3. *Another method to get the quadratic formula* Look at equation (4) on page 96. Rewrite the right side as $\left(\dfrac{\sqrt{b^2 - 4ac}}{2a}\right)^2$ and then subtract it *from* each side. The right side is now 0 and the left side is a difference of two squares. If you factor this difference of two squares, you will easily be able to get the quadratic formula, and, moreover, the quadratic expression is factored, which is sometimes useful.

1.2 Assess Your Understanding

'Are You Prepared?' *Answers are given at the end of these exercises. If you get a wrong answer, read the pages listed in red.*

1. Factor: $x^2 - 5x - 6$ (pp. 49–55)
2. Factor: $2x^2 - x - 3$ (pp. 49–55)
3. The solution set of the equation $(x - 3)(3x + 5) = 0$ is
 _____ . (p. 13)

4. **True or False** $\sqrt{x^2} = |x|$. (pp. 23–24)
5. Complete the square of $x^2 + 5x$. Factor the new expression. (p. 56)

Concepts and Vocabulary

6. The quantity $b^2 - 4ac$ is called the _____ of a quadratic equation. If it is _____, the equation has no real solution.

7. **True or False** Quadratic equations always have two real solutions.

8. **True or False** If the discriminant of a quadratic equation is positive, then the equation has two solutions that are negatives of one another.

Skill Building

In Problems 9–28, solve each equation by factoring.

9. $x^2 - 9x = 0$
10. $x^2 + 4x = 0$
11. $x^2 - 25 = 0$
12. $x^2 - 9 = 0$

13. $z^2 + z - 6 = 0$
14. $v^2 + 7v + 6 = 0$
15. $2x^2 - 5x - 3 = 0$
16. $3x^2 + 5x + 2 = 0$

17. $3t^2 - 48 = 0$
18. $2y^2 - 50 = 0$
19. $x(x - 8) + 12 = 0$
20. $x(x + 4) = 12$

21. $4x^2 + 9 = 12x$
22. $25x^2 + 16 = 40x$
23. $6(p^2 - 1) = 5p$
24. $2(2u^2 - 4u) + 3 = 0$

25. $6x - 5 = \dfrac{6}{x}$
26. $x + \dfrac{12}{x} = 7$
27. $\dfrac{4(x - 2)}{x - 3} + \dfrac{3}{x} = \dfrac{-3}{x(x - 3)}$
28. $\dfrac{5}{x + 4} = 4 + \dfrac{3}{x - 2}$

In Problems 29–34, solve each equation by the Square Root Method.

29. $x^2 = 25$
30. $x^2 = 36$
31. $(x - 1)^2 = 4$

32. $(x + 2)^2 = 1$
33. $(2y + 3)^2 = 9$
34. $(3z - 2)^2 = 4$

In Problems 35–40, solve each equation by completing the square.

35. $x^2 + 4x = 21$
36. $x^2 - 6x = 13$
37. $x^2 - \dfrac{1}{2}x - \dfrac{3}{16} = 0$

38. $x^2 + \dfrac{2}{3}x - \dfrac{1}{3} = 0$
39. $3x^2 + x - \dfrac{1}{2} = 0$
40. $2x^2 - 3x - 1 = 0$

In Problems 41–64, find the real solutions, if any, of each equation. Use the quadratic formula.

41. $x^2 - 4x + 2 = 0$
42. $x^2 + 4x + 2 = 0$
43. $x^2 - 4x - 1 = 0$

44. $x^2 + 6x + 1 = 0$
45. $2x^2 - 5x + 3 = 0$
46. $2x^2 + 5x + 3 = 0$

47. $4y^2 - y + 2 = 0$
48. $4t^2 + t + 1 = 0$
49. $4x^2 = 1 - 2x$

50. $2x^2 = 1 - 2x$
51. $4x^2 = 9x$
52. $5x = 4x^2$

53. $9t^2 - 6t + 1 = 0$
54. $4u^2 - 6u + 9 = 0$
55. $\dfrac{3}{4}x^2 - \dfrac{1}{4}x - \dfrac{1}{2} = 0$

56. $\frac{2}{3}x^2 - x - 3 = 0$

57. $\frac{5}{3}x^2 - x = \frac{1}{3}$

58. $\frac{3}{5}x^2 - x = \frac{1}{5}$

59. $2x(x + 2) = 3$

60. $3x(x + 2) = 1$

61. $4 - \frac{1}{x} - \frac{2}{x^2} = 0$

62. $4 + \frac{1}{x} - \frac{1}{x^2} = 0$

63. $\frac{3x}{x - 2} + \frac{1}{x} = 4$

64. $\frac{2x}{x - 3} + \frac{1}{x} = 4$

In Problems 65–70, find the real solutions, if any, of each equation. Use the quadratic formula and a calculator. Express any solutions rounded to two decimal places.

65. $x^2 - 4.1x + 2.2 = 0$

66. $x^2 + 3.9x + 1.8 = 0$

67. $x^2 + \sqrt{3}x - 3 = 0$

68. $x^2 + \sqrt{2}x - 2 = 0$

69. $\pi x^2 - x - \pi = 0$

70. $\pi x^2 + \pi x - 2 = 0$

In Problems 71–76, use the discriminant to determine whether each quadratic equation has two unequal real solutions, a repeated real solution, or no real solution, without solving the equation.

71. $2x^2 - 6x + 7 = 0$

72. $x^2 + 4x + 7 = 0$

73. $9x^2 - 30x + 25 = 0$

74. $25x^2 - 20x + 4 = 0$

75. $3x^2 + 5x - 8 = 0$

76. $2x^2 - 3x - 7 = 0$

Mixed Practice

In Problems 77–90, find the real solutions, if any, of each equation. Use any method.

77. $x^2 - 5 = 0$

78. $x^2 - 6 = 0$

79. $16x^2 - 8x + 1 = 0$

80. $9x^2 - 12x + 4 = 0$

81. $10x^2 - 19x - 15 = 0$

82. $6x^2 + 7x - 20 = 0$

83. $2 + z = 6z^2$

84. $2 = y + 6y^2$

85. $x^2 + \sqrt{2}x = \frac{1}{2}$

86. $\frac{1}{2}x^2 = \sqrt{2}x + 1$

87. $x^2 + x = 4$

88. $x^2 + x = 1$

89. $\frac{x}{x - 2} + \frac{2}{x + 1} = \frac{7x + 1}{x^2 - x - 2}$

90. $\frac{3x}{x + 2} + \frac{1}{x - 1} = \frac{4 - 7x}{x^2 + x - 2}$

Applications and Extensions

91. Pythagorean Theorem How many right triangles have a hypotenuse that measures $2x + 3$ meters and legs that measure $2x - 5$ meters and $x + 7$ meters? What are the dimensions of the triangle(s)?

92. Pythagorean Theorem How many right triangles have a hypotenuse that measures $4x + 5$ inches and legs that measure $3x + 13$ inches and x inches? What are the dimensions of the triangle(s)?

93. Dimensions of a Window The area of the opening of a rectangular window is to be 143 square feet. If the length is to be 2 feet more than the width, what are the dimensions?

94. Dimensions of a Window The area of a rectangular window is to be 306 square centimeters. If the length exceeds the width by 1 centimeter, what are the dimensions?

95. Geometry Find the dimensions of a rectangle whose perimeter is 26 meters and whose area is 40 square meters.

96. Watering a Field An adjustable water sprinkler that sprays water in a circular pattern is placed at the center of a square field whose area is 1250 square feet (see the figure). What is

the shortest radius setting that can be used if the field is to be completely enclosed within the circle?

97. Constructing a Box An open box is to be constructed from a square piece of sheet metal by removing a square of side 1 foot from each corner and turning up the edges. If the box is to hold 4 cubic feet, what should be the dimensions of the sheet metal?

98. Constructing a Box Rework Problem 97 if the piece of sheet metal is a rectangle whose length is twice its width.

99. Physics A ball is thrown vertically upward from the top of a building 96 feet tall with an initial velocity of

80 feet per second. The distance s (in feet) of the ball from the ground after t seconds is $s = 96 + 80t - 16t^2$.

(a) After how many seconds does the ball strike the ground?

(b) After how many seconds will the ball pass the top of the building on its way down?

100. Physics An object is propelled vertically upward with an initial velocity of 20 meters per second. The distance s (in meters) of the object from the ground after t seconds is $s = -4.9t^2 + 20t$.

(a) When will the object be 15 meters above the ground?

(b) When will it strike the ground?

(c) Will the object reach a height of 100 meters?

101. Reducing the Size of a Candy Bar A jumbo chocolate bar with a rectangular shape measures 12 centimeters in length, 7 centimeters in width, and 3 centimeters in thickness. Due to escalating costs of cocoa, management decides to reduce the volume of the bar by 10%. To accomplish this reduction, management decides that the new bar should have the same 3 centimeter thickness, but the length and width of each should be reduced an equal number of centimeters. What should be the dimensions of the new candy bar?

102. Reducing the Size of a Candy Bar Rework Problem 101 if the reduction is to be 20%.

103. Constructing a Border around a Pool A circular pool measures 10 feet across. One cubic yard of concrete is to be used to create a circular border of uniform width around the pool. If the border is to have a depth of 3 inches, how wide will the border be? (1 cubic yard = 27 cubic feet) See the illustration.

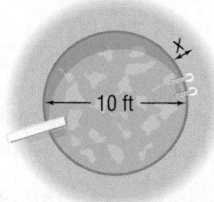

104. Constructing a Border around a Pool Rework Problem 103 if the depth of the border is 4 inches.

105. Constructing a Border around a Garden A landscaper, who just completed a rectangular flower garden measuring 6 feet by 10 feet, orders 1 cubic yard of premixed cement, all of which is to be used to create a border of uniform width around the garden. If the border is to have a depth

of 3 inches, how wide will the border be? (1 cubic yard = 27 cubic feet)

106. Dimensions of a Patio A contractor orders 8 cubic yards of premixed cement, all of which is to be used to pour a patio that will be 4 inches thick. If the length of the patio is specified to be twice the width, what will be the patio dimensions? (1 cubic yard = 27 cubic feet)

107. Comparing TVs The screen size of a television is determined by the length of the diagonal of the rectangular screen. Traditional TVs come in a 4 : 3 format, meaning the ratio of the length to the width of the rectangular screen is 4 to 3. What is the area of a 37-inch traditional TV screen? What is the area of a 37-inch LCD TV whose screen is in a 16 : 9 format? Which screen is larger?

[**Hint:** If x is the length of a 4 : 3 format screen, then $\frac{3}{4}x$ is the width.]

traditional LCD
4:3 16:9

108. Comparing TVs Refer to Problem 107. Find the screen area of a traditional 50-inch TV and compare it with a 50-inch Plasma TV whose screen is in a 16 : 9 format. Which screen is larger?

109. The sum of the consecutive integers $1, 2, 3, \ldots, n$ is given by the formula $\frac{1}{2}n(n + 1)$. How many consecutive integers, starting with 1, must be added to get a sum of 666?

110. Geometry If a polygon of n sides has $\frac{1}{2}n(n - 3)$ diagonals, how many sides will a polygon with 65 diagonals have? Is there a polygon with 80 diagonals?

111. Show that the sum of the roots of a quadratic equation is $-\frac{b}{a}$.

112. Show that the product of the roots of a quadratic equation is $\frac{c}{a}$.

113. Find k such that the equation $kx^2 + x + k = 0$ has a repeated real solution.

114. Find k such that the equation $x^2 - kx + 4 = 0$ has a repeated real solution.

115. Show that the real solutions of the equation $ax^2 + bx + c = 0$ are the negatives of the real solutions of the equation $ax^2 - bx + c = 0$. Assume that $b^2 - 4ac \geq 0$.

116. Show that the real solutions of the equation $ax^2 + bx + c = 0$ are the reciprocals of the real solutions of the equation $cx^2 + bx + a = 0$. Assume that $b^2 - 4ac \geq 0$.

Explaining Concepts: Discussion and Writing

117. Which of the following pairs of equations are equivalent? Explain.
(a) $x^2 = 9$; $x = 3$
(b) $x = \sqrt{9}$; $x = 3$
(c) $(x - 1)(x - 2) = (x - 1)^2$; $x - 2 = x - 1$

118. Describe three ways that you might solve a quadratic equation. State your preferred method; explain why you chose it.

119. Explain the benefits of evaluating the discriminant of a quadratic equation before attempting to solve it.

120. Create three quadratic equations: one having two distinct solutions, one having no real solution, and one having exactly one real solution.

121. The word *quadratic* seems to imply four (*quad*), yet a quadratic equation is an equation that involves a polynomial of degree 2. Investigate the origin of the term *quadratic* as it is used in the expression *quadratic equation*. Write a brief essay on your findings.

'Are You Prepared?' Answers

1. $(x - 6)(x + 1)$ **2.** $(2x - 3)(x + 1)$ **3.** $\left\{-\dfrac{5}{3}, 3\right\}$ **4.** True **5.** $x^2 + 5x + \dfrac{25}{4} = \left(x + \dfrac{5}{2}\right)^2$

1.3 Complex Numbers; Quadratic Equations in the Complex Number System*

PREPARING FOR THIS SECTION *Before getting started, review the following:*

- Classification of Numbers (Section R.1, pp. 4–5)
- Rationalizing Denominators (Section R.8, p. 45)

Now Work the 'Are You Prepared?' problems on page 111.

OBJECTIVES 1 Add, Subtract, Multiply, and Divide Complex Numbers (p. 105)
2 Solve Quadratic Equations in the Complex Number System (p. 109)

Complex Numbers

One property of a real number is that its square is nonnegative. For example, there is no real number x for which

$$x^2 = -1$$

To remedy this situation, we introduce a new number called the *imaginary unit*.

DEFINITION

The **imaginary unit,** which we denote by *i*, is the number whose square is -1. That is,

$$i^2 = -1$$

This should not surprise you. If our universe were to consist only of integers, there would be no number x for which $2x = 1$. This unfortunate circumstance was remedied by introducing numbers such as $\dfrac{1}{2}$ and $\dfrac{2}{3}$, the *rational numbers*. If our universe were to consist only of rational numbers, there would be no x whose square equals 2. That is, there would be no number x for which $x^2 = 2$. To remedy this, we introduced numbers such as $\sqrt{2}$ and $\sqrt[3]{5}$, the *irrational numbers*. The *real numbers*, you will recall, consist of the rational numbers and the irrational numbers. Now, if our universe were to consist only of real numbers, then there would be no number x whose square is -1. To remedy this, we introduce a number i, whose square is -1.

*This section may be omitted without any loss of continuity.

In the progression outlined, each time we encountered a situation that was unsuitable, we introduced a new number system to remedy this situation. The number system that results from introducing the number i is called the **complex number system.**

DEFINITION

Complex numbers are numbers of the form $a + bi$, where a and b are real numbers. The real number a is called the **real part** of the number $a + bi$; the real number b is called the **imaginary part** of $a + bi$; and i is the imaginary unit, so $i^2 = -1$.

For example, the complex number $-5 + 6i$ has the real part -5 and the imaginary part 6.

When a complex number is written in the form $a + bi$, where a and b are real numbers, we say it is in **standard form.** However, if the imaginary part of a complex number is negative, such as in the complex number $3 + (-2)i$, we agree to write it instead in the form $3 - 2i$.

Also, the complex number $a + 0i$ is usually written merely as a. This serves to remind us that the real numbers are a subset of the complex numbers. The complex number $0 + bi$ is usually written as bi. Sometimes the complex number bi is called a **pure imaginary number.**

1 Add, Subtract, Multiply, and Divide Complex Numbers

Equality, addition, subtraction, and multiplication of complex numbers are defined so as to preserve the familiar rules of algebra for real numbers. Two complex numbers are equal if and only if their real parts are equal and their imaginary parts are equal. That is,

Equality of Complex Numbers

$$a + bi = c + di \qquad \text{if and only if } a = c \text{ and } b = d \qquad \textbf{(1)}$$

Two complex numbers are added by forming the complex number whose real part is the sum of the real parts and whose imaginary part is the sum of the imaginary parts. That is,

Sum of Complex Numbers

$$(a + bi) + (c + di) = (a + c) + (b + d)i \qquad \textbf{(2)}$$

To subtract two complex numbers, use this rule:

Difference of Complex Numbers

$$(a + bi) - (c + di) = (a - c) + (b - d)i \qquad \textbf{(3)}$$

EXAMPLE 1 **Adding and Subtracting Complex Numbers**

(a) $(3 + 5i) + (-2 + 3i) = [3 + (-2)] + (5 + 3)i = 1 + 8i$

(b) $(6 + 4i) - (3 + 6i) = (6 - 3) + (4 - 6)i = 3 + (-2)i = 3 - 2i$

⌐

Now Work PROBLEM 13

Products of complex numbers are calculated as illustrated in Example 2.

EXAMPLE 2 **Multiplying Complex Numbers**

$$(5 + 3i) \cdot (2 + 7i) = 5 \cdot (2 + 7i) + 3i(2 + 7i) = 10 + 35i + 6i + 21i^2$$

<div style="text-align:center">↑ ↑</div>

<div style="text-align:center">Distributive Property Distributive Property</div>

$$= 10 + 41i + 21(-1)$$

<div style="text-align:center">↑</div>
<div style="text-align:center">$i^2 = -1$</div>

$$= -11 + 41i$$

Based on the procedure of Example 2, the **product** of two complex numbers is defined as follows:

Product of Complex Numbers

$$(a + bi) \cdot (c + di) = (ac - bd) + (ad + bc)i \qquad (4)$$

Do not bother to memorize formula (4). Instead, whenever it is necessary to multiply two complex numbers, follow the usual rules for multiplying two binomials, as in Example 2, remembering that $i^2 = -1$. For example,

$$(2i)(2i) = 4i^2 = -4$$

$$(2 + i)(1 - i) = 2 - 2i + i - i^2 = 3 - i$$

➤ **Now Work** PROBLEM 19

Algebraic properties for addition and multiplication, such as the commutative, associative, and distributive properties, hold for complex numbers. The property that every nonzero complex number has a multiplicative inverse, or reciprocal, requires a closer look.

DEFINITION If $z = a + bi$ is a complex number, then its **conjugate,** denoted by \bar{z}, is defined as

$$\bar{z} = \overline{a + bi} = a - bi$$

For example, $\overline{2 + 3i} = 2 - 3i$ and $\overline{-6 - 2i} = -6 + 2i$.

EXAMPLE 3 **Multiplying a Complex Number by Its Conjugate**

Find the product of the complex number $z = 3 + 4i$ and its conjugate \bar{z}.

Solution Since $\bar{z} = 3 - 4i$, we have

$$z\bar{z} = (3 + 4i)(3 - 4i) = 9 - 12i + 12i - 16i^2 = 9 + 16 = 25$$

The result obtained in Example 3 has an important generalization.

THEOREM The product of a complex number and its conjugate is a nonnegative real number. That is, if $z = a + bi$, then

$$z\bar{z} = a^2 + b^2 \qquad (5)$$

Proof If $z = a + bi$, then

$$z\bar{z} = (a + bi)(a - bi) = a^2 - (bi)^2 = a^2 - b^2 i^2 = a^2 + b^2 \qquad \blacksquare$$

To express the reciprocal of a nonzero complex number z in standard form, multiply the numerator and denominator of $\dfrac{1}{z}$ by \bar{z}. That is, if $z = a + bi$ is a nonzero complex number, then

$$\frac{1}{a + bi} = \frac{1}{z} = \frac{1}{z} \cdot \frac{\bar{z}}{\bar{z}} = \frac{\bar{z}}{z\bar{z}} = \frac{a - bi}{a^2 + b^2}$$

$$\uparrow$$
$$\text{Use (5).}$$

$$= \frac{a}{a^2 + b^2} - \frac{b}{a^2 + b^2}i$$

EXAMPLE 4

Writing the Reciprocal of a Complex Number in Standard Form

Write $\dfrac{1}{3 + 4i}$ in standard form $a + bi$; that is, find the reciprocal of $3 + 4i$.

Solution The idea is to multiply the numerator and denominator by the conjugate of $3 + 4i$, that is, by the complex number $3 - 4i$. The result is

$$\frac{1}{3 + 4i} = \frac{1}{3 + 4i} \cdot \frac{3 - 4i}{3 - 4i} = \frac{3 - 4i}{9 + 16} = \frac{3}{25} - \frac{4}{25}i$$

To express the quotient of two complex numbers in standard form, multiply the numerator and denominator of the quotient by the conjugate of the denominator.

EXAMPLE 5

Writing the Quotient of Two Complex Numbers in Standard Form

Write each of the following in standard form.

(a) $\dfrac{1 + 4i}{5 - 12i}$

(b) $\dfrac{2 - 3i}{4 - 3i}$

Solution (a) $\dfrac{1 + 4i}{5 - 12i} = \dfrac{1 + 4i}{5 - 12i} \cdot \dfrac{5 + 12i}{5 + 12i} = \dfrac{5 + 12i + 20i + 48i^2}{25 + 144}$

$$= \frac{-43 + 32i}{169} = -\frac{43}{169} + \frac{32}{169}i$$

(b) $\dfrac{2 - 3i}{4 - 3i} = \dfrac{2 - 3i}{4 - 3i} \cdot \dfrac{4 + 3i}{4 + 3i} = \dfrac{8 + 6i - 12i - 9i^2}{16 + 9}$

$$= \frac{17 - 6i}{25} = \frac{17}{25} - \frac{6}{25}i$$

Now Work PROBLEM 27

EXAMPLE 6

Writing Other Expressions in Standard Form

If $z = 2 - 3i$ and $w = 5 + 2i$, write each of the following expressions in standard form.

(a) $\dfrac{z}{w}$

(b) $\overline{z + w}$

(c) $z + \bar{z}$

Solution (a) $\dfrac{z}{w} = \dfrac{z \cdot \overline{w}}{w \cdot \overline{w}} = \dfrac{(2 - 3i)(5 - 2i)}{(5 + 2i)(5 - 2i)} = \dfrac{10 - 4i - 15i + 6i^2}{25 + 4}$

$\qquad\qquad = \dfrac{4 - 19i}{29} = \dfrac{4}{29} - \dfrac{19}{29}i$

(b) $\overline{z + w} = \overline{(2 - 3i) + (5 + 2i)} = \overline{7 - i} = 7 + i$

(c) $z + \overline{z} = (2 - 3i) + (2 + 3i) = 4$

The conjugate of a complex number has certain general properties that we shall find useful later.

For a real number $a = a + 0i$, the conjugate is $\overline{a} = \overline{a + 0i} = a - 0i = a$. That is,

THEOREM The conjugate of a real number is the real number itself.

Other properties of the conjugate that are direct consequences of the definition are given next. In each statement, z and w represent complex numbers.

THEOREM The conjugate of the conjugate of a complex number is the complex number itself.

$$\overline{(\overline{z})} = z \tag{6}$$

The conjugate of the sum of two complex numbers equals the sum of their conjugates.

$$\overline{z + w} = \overline{z} + \overline{w} \tag{7}$$

The conjugate of the product of two complex numbers equals the product of their conjugates.

$$\overline{z \cdot w} = \overline{z} \cdot \overline{w} \tag{8}$$

We leave the proofs of equations (6), (7), and (8) as exercises.

Powers of i

The **powers of i** follow a pattern that is useful to know.

$$i^1 = i \qquad\qquad\qquad i^5 = i^4 \cdot i = 1 \cdot i = i$$
$$i^2 = -1 \qquad\qquad\qquad i^6 = i^4 \cdot i^2 = -1$$
$$i^3 = i^2 \cdot i = -1 \cdot i = -i \qquad i^7 = i^4 \cdot i^3 = -i$$
$$i^4 = i^2 \cdot i^2 = (-1)(-1) = 1 \qquad i^8 = i^4 \cdot i^4 = 1$$

And so on. The powers of i repeat with every fourth power.

EXAMPLE 7 **Evaluating Powers of i**

(a) $i^{27} = i^{24} \cdot i^3 = (i^4)^6 \cdot i^3 = 1^6 \cdot i^3 = -i$

(b) $i^{101} = i^{100} \cdot i^1 = (i^4)^{25} \cdot i = 1^{25} \cdot i = i$

EXAMPLE 8

Writing the Power of a Complex Number in Standard Form

Write $(2 + i)^3$ in standard form.

Solution

Use the special product formula for $(x + a)^3$.

$$(x + a)^3 = x^3 + 3ax^2 + 3a^2x + a^3$$

NOTE Another way to find $(2 + i)^3$ is to multiply out $(2 + i)^2 (2 + i)$. ∎

Using this special product formula,

$$(2 + i)^3 = 2^3 + 3 \cdot i \cdot 2^2 + 3 \cdot i^2 \cdot 2 + i^3$$
$$= 8 + 12i + 6(-1) + (-i)$$
$$= 2 + 11i.$$

━━━**Now Work** PROBLEM **41**

2 Solve Quadratic Equations in the Complex Number System

Quadratic equations with a negative discriminant have no real number solution. However, if we extend our number system to allow complex numbers, quadratic equations will always have a solution. Since the solution to a quadratic equation involves the square root of the discriminant, we begin with a discussion of square roots of negative numbers.

DEFINITION

If N is a positive real number, we define the **principal square root of** $-N$, denoted by $\sqrt{-N}$, as

WARNING In writing $\sqrt{-N} = \sqrt{N}i$ be sure to place i outside the $\sqrt{}$ symbol. ∎

$$\boxed{\sqrt{-N} = \sqrt{N}\,i}$$

where i is the imaginary unit and $i^2 = -1$.

EXAMPLE 9

Evaluating the Square Root of a Negative Number

(a) $\sqrt{-1} = \sqrt{1}\,i = i$
(b) $\sqrt{-4} = \sqrt{4}\,i = 2i$
(c) $\sqrt{-8} = \sqrt{8}\,i = 2\sqrt{2}\,i$

EXAMPLE 10

Solving Equations

Solve each equation in the complex number system.

(a) $x^2 = 4$ (b) $x^2 = -9$

Solution

(a) $x^2 = 4$
$$x = \pm\sqrt{4} = \pm 2$$
The equation has two solutions, -2 and 2. The solution set is $\{-2, 2\}$.

(b) $x^2 = -9$
$$x = \pm\sqrt{-9} = \pm\sqrt{9}\,i = \pm 3i$$
The equation has two solutions, $-3i$ and $3i$. The solution set is $\{-3i, 3i\}$.

━━━**Now Work** PROBLEMS **49** AND **53**

WARNING When working with square roots of negative numbers, do not set the square root of a product equal to the product of the square roots (which can be done with positive numbers). To see why, look at this calculation: We know that $\sqrt{100} = 10$. However, it is also true that $100 = (-25)(-4)$, so

$$10 = \sqrt{100} = \sqrt{(-25)(-4)} \neq \sqrt{-25}\,\sqrt{-4} = \left(\sqrt{25}\,i\right)\left(\sqrt{4}\,i\right) = (5i)(2i) = 10i^2 = -10$$

\uparrow

Here is the error. ∎

Because we have defined the square root of a negative number, we can now restate the quadratic formula without restriction.

THEOREM **Quadratic Formula**

In the complex number system, the solutions of the quadratic equation $ax^2 + bx + c = 0$, where a, b, and c are real numbers and $a \neq 0$, are given by the formula

$$x = \frac{-b \pm \sqrt{b^2 - 4ac}}{2a} \tag{9}$$

EXAMPLE 11 **Solving a Quadratic Equation in the Complex Number System**

Solve the equation $x^2 - 4x + 8 = 0$ in the complex number system.

Solution Here $a = 1, b = -4, c = 8$, and $b^2 - 4ac = 16 - 4(1)(8) = -16$. Using equation (9), we find that

$$x = \frac{-(-4) \pm \sqrt{-16}}{2(1)} = \frac{4 \pm \sqrt{16}\,i}{2} = \frac{4 \pm 4i}{2} = 2 \pm 2i$$

The equation has two solutions $2 - 2i$ and $2 + 2i$. The solution set is $\{2 - 2i, 2 + 2i\}$.

✓**Check:** $2 + 2i$: $(2 + 2i)^2 - 4(2 + 2i) + 8 = 4 + 8i + 4i^2 - 8 - 8i + 8$
$$= 4 - 4 = 0$$

$2 - 2i$: $(2 - 2i)^2 - 4(2 - 2i) + 8 = 4 - 8i + 4i^2 - 8 + 8i + 8$
$$= 4 - 4 = 0$$

━━━━**Now Work** PROBLEM 59

The discriminant $b^2 - 4ac$ of a quadratic equation still serves as a way to determine the character of the solutions.

Character of the Solutions of a Quadratic Equation

In the complex number system, consider a quadratic equation $ax^2 + bx + c = 0$ with real coefficients.

1. If $b^2 - 4ac > 0$, the equation has two unequal real solutions.
2. If $b^2 - 4ac = 0$, the equation has a repeated real solution, a double root.
3. If $b^2 - 4ac < 0$, the equation has two complex solutions that are not real. The solutions are conjugates of each other.

The third conclusion in the display is a consequence of the fact that if $b^2 - 4ac = -N < 0$ then, by the quadratic formula, the solutions are

$$x = \frac{-b + \sqrt{b^2 - 4ac}}{2a} = \frac{-b + \sqrt{-N}}{2a} = \frac{-b + \sqrt{N}i}{2a} = \frac{-b}{2a} + \frac{\sqrt{N}}{2a}i$$

and

$$x = \frac{-b - \sqrt{b^2 - 4ac}}{2a} = \frac{-b - \sqrt{-N}}{2a} = \frac{-b - \sqrt{N}i}{2a} = \frac{-b}{2a} - \frac{\sqrt{N}}{2a}i$$

which are conjugates of each other.

EXAMPLE 12 **Determining the Character of the Solutions of a Quadratic Equation**

Without solving, determine the character of the solutions of each equation.

(a) $3x^2 + 4x + 5 = 0$ (b) $2x^2 + 4x + 1 = 0$ (c) $9x^2 - 6x + 1 = 0$

Solution (a) Here $a = 3, b = 4$, and $c = 5$, so $b^2 - 4ac = 16 - 4(3)(5) = -44$. The solutions are two complex numbers that are not real and are conjugates of each other.

(b) Here $a = 2, b = 4$, and $c = 1$, so $b^2 - 4ac = 16 - 8 = 8$. The solutions are two unequal real numbers.

(c) Here $a = 9, b = -6$, and $c = 1$, so $b^2 - 4ac = 36 - 4(9)(1) = 0$. The solution is a repeated real number, that is, a double root.

━━━ **Now Work** PROBLEM 73

1.3 Assess Your Understanding

'Are You Prepared?' *Answers are given at the end of these exercises. If you get a wrong answer, read the pages listed in red.*

1. Name the integers and the rational numbers in the set $\left\{-3, 0, \sqrt{2}, \dfrac{6}{5}, \pi\right\}$. (pp. 4–5)

2. **True or False** Rational numbers and irrational numbers are in the set of real numbers. (pp. 4–5)

3. Rationalize the denominator of $\dfrac{3}{2 + \sqrt{3}}$. (p. 45)

Concepts and Vocabulary

4. In the complex number $5 + 2i$, the number 5 is called the _____ part; the number 2 is called the _____ part; the number i is called the _____ _____.

5. The equation $x^2 = -4$ has the solution set _____.

6. **True or False** The conjugate of $2 + 5i$ is $-2 - 5i$.

7. **True or False** All real numbers are complex numbers.

8. **True or False** If $2 - 3i$ is a solution of a quadratic equation with real coefficients, then $-2 + 3i$ is also a solution.

Skill Building

In Problems 9–46, write each expression in the standard form $a + bi$.

9. $(2 - 3i) + (6 + 8i)$

10. $(4 + 5i) + (-8 + 2i)$

11. $(-3 + 2i) - (4 - 4i)$

12. $(3 - 4i) - (-3 - 4i)$

13. $(2 - 5i) - (8 + 6i)$

14. $(-8 + 4i) - (2 - 2i)$

15. $3(2 - 6i)$

16. $-4(2 + 8i)$

17. $2i(2 - 3i)$

18. $3i(-3 + 4i)$

19. $(3 - 4i)(2 + i)$

20. $(5 + 3i)(2 - i)$

21. $(-6 + i)(-6 - i)$

22. $(-3 + i)(3 + i)$

23. $\dfrac{10}{3 - 4i}$

24. $\dfrac{13}{5 - 12i}$

25. $\dfrac{2 + i}{i}$

26. $\dfrac{2 - i}{-2i}$

27. $\dfrac{6 - i}{1 + i}$

28. $\dfrac{2 + 3i}{1 - i}$

29. $\left(\dfrac{1}{2} + \dfrac{\sqrt{3}}{2}i\right)^2$

30. $\left(\dfrac{\sqrt{3}}{2} - \dfrac{1}{2}i\right)^2$

31. $(1 + i)^2$

32. $(1 - i)^2$

33. i^{23} **34.** i^{14} **35.** i^{-15} **36.** i^{-23} **37.** $i^6 - 5$

38. $4 + i^3$ **39.** $6i^3 - 4i^5$ **40.** $4i^3 - 2i^2 + 1$ **41.** $(1 + i)^3$ **42.** $(3i)^4 + 1$

43. $i^7(1 + i^2)$ **44.** $2i^4(1 + i^2)$ **45.** $i^6 + i^4 + i^2 + 1$ **46.** $i^7 + i^5 + i^3 + i$

In Problems 47–52, perform the indicated operations and express your answer in the form $a + bi$.

47. $\sqrt{-4}$ **48.** $\sqrt{-9}$ **49.** $\sqrt{-25}$

50. $\sqrt{-64}$ **51.** $\sqrt{(3 + 4i)(4i - 3)}$ **52.** $\sqrt{(4 + 3i)(3i - 4)}$

In Problems 53–72, solve each equation in the complex number system.

53. $x^2 + 4 = 0$ **54.** $x^2 - 4 = 0$ **55.** $x^2 - 16 = 0$ **56.** $x^2 + 25 = 0$

57. $x^2 - 6x + 13 = 0$ **58.** $x^2 + 4x + 8 = 0$ **59.** $x^2 - 6x + 10 = 0$ **60.** $x^2 - 2x + 5 = 0$

61. $8x^2 - 4x + 1 = 0$ **62.** $10x^2 + 6x + 1 = 0$ **63.** $5x^2 + 1 = 2x$ **64.** $13x^2 + 1 = 6x$

65. $x^2 + x + 1 = 0$ **66.** $x^2 - x + 1 = 0$ **67.** $x^3 - 8 = 0$ **68.** $x^3 + 27 = 0$

69. $x^4 = 16$ **70.** $x^4 = 1$ **71.** $x^4 + 13x^2 + 36 = 0$ **72.** $x^4 + 3x^2 - 4 = 0$

In Problems 73–78, without solving, determine the character of the solutions of each equation in the complex number system.

73. $3x^2 - 3x + 4 = 0$ **74.** $2x^2 - 4x + 1 = 0$ **75.** $2x^2 + 3x = 4$

76. $x^2 + 6 = 2x$ **77.** $9x^2 - 12x + 4 = 0$ **78.** $4x^2 + 12x + 9 = 0$

79. $2 + 3i$ is a solution of a quadratic equation with real coefficients. Find the other solution.

80. $4 - i$ is a solution of a quadratic equation with real coefficients. Find the other solution.

In Problems 81–84, $z = 3 - 4i$ and $w = 8 + 3i$. Write each expression in the standard form $a + bi$.

81. $z + \bar{z}$ **82.** $w - \bar{w}$ **83.** $z\bar{z}$ **84.** $\overline{z - w}$

Applications and Extensions

85. Electrical Circuits The impedance Z, in ohms, of a circuit element is defined as the ratio of the phasor voltage V, in volts, across the element to the phasor current I, in amperes, through the elements. That is, $Z = \dfrac{V}{I}$. If the voltage across a circuit element is $18 + i$ volts and the current through the element is $3 - 4i$ amperes, determine the impedance.

86. Parallel Circuits In an ac circuit with two parallel pathways, the total impedance Z, in ohms, satisfies the formula $\dfrac{1}{Z} = \dfrac{1}{Z_1} + \dfrac{1}{Z_2}$, where Z_1 is the impedance of the first pathway

and Z_2 is the impedance of the second pathway. Determine the total impedance if the impedances of the two pathways are $Z_1 = 2 + i$ ohms and $Z_2 = 4 - 3i$ ohms.

87. Use $z = a + bi$ to show that $z + \bar{z} = 2a$ and $z - \bar{z} = 2bi$.

88. Use $z = a + bi$ to show that $\bar{\bar{z}} = z$.

89. Use $z = a + bi$ and $w = c + di$ to show that $\overline{z + w} = \bar{z} + \bar{w}$.

90. Use $z = a + bi$ and $w = c + di$ to show that $\overline{z \cdot w} = \bar{z} \cdot \bar{w}$.

Explaining Concepts: Discussion and Writing

91. Explain to a friend how you would add two complex numbers and how you would multiply two complex numbers. Explain any differences in the two explanations.

92. Write a brief paragraph that compares the method used to rationalize the denominator of a radical expression and the method used to write the quotient of two complex numbers in standard form.

93. Use an Internet search engine to investigate the origins of complex numbers. Write a paragraph describing what you find and present it to the class.

94. Explain how the method of multiplying two complex numbers is related to multiplying two binomials.

95. What Went Wrong? A student multiplied $\sqrt{-9}$ and $\sqrt{-9}$ as follows:

$$\sqrt{-9} \cdot \sqrt{-9} = \sqrt{(-9)(-9)}$$
$$= \sqrt{81}$$
$$= 9$$

The instructor marked the problem incorrect. Why?

'Are You Prepared?' Answers

1. Integers: $\{-3, 0\}$; rational numbers: $\left\{-3, 0, \dfrac{6}{5}\right\}$ **2.** True **3.** $3(2 - \sqrt{3})$

1.4 Radical Equations; Equations Quadratic in Form; Factorable Equations

PREPARING FOR THIS SECTION *Before getting started, review the following:*

- Square Roots (Section R.2, pp. 23–24)
- Factoring Polynomials (Section R.5, pp. 49–55)
- *n*th Roots; Rational Exponents (Section R.8, pp. 73–75)

Now Work the '**Are You Prepared?**' problems on page 117.

OBJECTIVES 1 Solve Radical Equations (p. 113)
2 Solve Equations Quadratic in Form (p. 114)
3 Solve Equations by Factoring (p. 116)

1 Solve Radical Equations

When the variable in an equation occurs in a square root, cube root, and so on, that is, when it occurs in a radical, the equation is called a **radical equation.** Sometimes a suitable operation will change a radical equation to one that is linear or quadratic. A commonly used procedure is to isolate the most complicated radical on one side of the equation and then eliminate it by raising each side to a power equal to the index of the radical. Care must be taken, however, because apparent solutions that are not, in fact, solutions of the original equation may result. These are called **extraneous solutions.** Therefore, we need to check all answers when working with radical equations.

EXAMPLE 1 **Solving a Radical Equation**

Find the real solutions of the equation: $\sqrt[3]{2x - 4} - 2 = 0$

Solution The equation contains a radical whose index is 3. Isolate it on the left side.

$$\sqrt[3]{2x - 4} - 2 = 0$$
$$\sqrt[3]{2x - 4} = 2$$

Now raise each side to the third power (the index of the radical is 3) and solve.

$$\left(\sqrt[3]{2x - 4}\right)^3 = 2^3 \quad \text{Raise each side to the power 3.}$$
$$2x - 4 = 8 \quad \text{Simplify.}$$
$$2x = 12 \quad \text{Add 4 to both sides.}$$
$$x = 6 \quad \text{Divide both sides by 2.}$$

✓**Check:** $\sqrt[3]{2(6) - 4} - 2 = \sqrt[3]{12 - 4} - 2 = \sqrt[3]{8} - 2 = 2 - 2 = 0.$

The solution set is $\{6\}$.

Now Work PROBLEM 7

EXAMPLE 2 **Solving a Radical Equation**

Find the real solutions of the equation: $\sqrt{x - 1} = x - 7$

Solution Square both sides since the index of a square root is 2.

$$\sqrt{x - 1} = x - 7$$
$$\left(\sqrt{x - 1}\right)^2 = (x - 7)^2 \quad \text{Square both sides.}$$
$$x - 1 = x^2 - 14x + 49 \quad \text{Remove parentheses.}$$
$$x^2 - 15x + 50 = 0 \quad \text{Put in standard form.}$$

$$(x - 10)(x - 5) = 0 \qquad \text{Factor.}$$
$$x = 10 \quad \text{or} \quad x = 5 \qquad \text{Apply the Zero-Product Property and solve.}$$

✓**Check:**

$$x = 10: \quad \sqrt{x - 1} = \sqrt{10 - 1} = \sqrt{9} = 3 \text{ and } x - 7 = 10 - 7 = 3$$
$$x = 5: \quad \sqrt{x - 1} = \sqrt{5 - 1} = \sqrt{4} = 2 \text{ and } x - 7 = 5 - 7 = -2$$

The solution $x = 5$ is extraneous; the only solution of the equation is $x = 10$. The solution set is $\{10\}$.

━━━━**Now Work** PROBLEM 19

Sometimes we need to raise each side to a power more than once in order to solve a radical equation.

EXAMPLE 3

Solving a Radical Equation

Find the real solutions of the equation: $\sqrt{2x + 3} - \sqrt{x + 2} = 2$

Solution First, we choose to isolate the more complicated radical expression (in this case, $\sqrt{2x + 3}$) on the left side.

$$\sqrt{2x + 3} = \sqrt{x + 2} + 2$$

Now square both sides (the index of the radical on the left is 2).

$$\left(\sqrt{2x + 3}\right)^2 = \left(\sqrt{x + 2} + 2\right)^2 \qquad \text{Square both sides.}$$
$$2x + 3 = \left(\sqrt{x + 2}\right)^2 + 4\sqrt{x + 2} + 4 \qquad \text{Remove parentheses.}$$
$$2x + 3 = x + 2 + 4\sqrt{x + 2} + 4 \qquad \text{Simplify.}$$
$$2x + 3 = x + 6 + 4\sqrt{x + 2} \qquad \text{Combine like terms.}$$

Because the equation still contains a radical, isolate the remaining radical on the right side and again square both sides.

$$x - 3 = 4\sqrt{x + 2} \qquad \text{Isolate the radical on the right side.}$$
$$(x - 3)^2 = \left(4\sqrt{x + 2}\right)^2 \qquad \text{Square both sides.}$$
$$x^2 - 6x + 9 = 16x + 32 \qquad \text{Remove parentheses.}$$
$$x^2 - 22x - 23 = 0 \qquad \text{Put in standard form.}$$
$$(x - 23)(x + 1) = 0 \qquad \text{Factor.}$$
$$x = 23 \quad \text{or} \quad x = -1$$

The original equation appears to have the solution set $\{-1, 23\}$. However, we have not yet checked.

✓**Check:**

$$x = 23: \quad \sqrt{2x + 3} - \sqrt{x + 2} = \sqrt{2(23) + 3} - \sqrt{23 + 2} = \sqrt{49} - \sqrt{25} = 7 - 5 = 2$$
$$x = -1: \quad \sqrt{2x + 3} - \sqrt{x + 2} = \sqrt{2(-1) + 3} - \sqrt{-1 + 2} = \sqrt{1} - \sqrt{1} = 1 - 1 = 0$$

The equation has only one solution, 23; the solution -1 is extraneous. The solution set is $\{23\}$.

━━━━**Now Work** PROBLEM 29

2 Solve Equations Quadratic in Form

The equation $x^4 + x^2 - 12 = 0$ is not quadratic in x, but it is quadratic in x^2. That is, if we let $u = x^2$, we get $u^2 + u - 12 = 0$, a quadratic equation. This equation can be solved for u and, in turn, by using $u = x^2$, we can find the solutions x of the original equation.

In general, if an appropriate substitution u transforms an equation into one of the form

$$au^2 + bu + c = 0 \qquad a \neq 0$$

then the original equation is called an **equation of the quadratic type** or an **equation quadratic in form.**

The difficulty of solving such an equation lies in the determination that the equation is, in fact, quadratic in form. After you are told an equation is quadratic in form, it is easy enough to see it, but some practice is needed to enable you to recognize such equations on your own.

EXAMPLE 4 | **Solving an Equation Quadratic in Form**

Find the real solutions of the equation: $(x + 2)^2 + 11(x + 2) - 12 = 0$

Solution For this equation, let $u = x + 2$. Then $u^2 = (x + 2)^2$, and the original equation,

$$(x + 2)^2 + 11(x + 2) - 12 = 0$$

becomes

$$u^2 + 11u - 12 = 0 \quad \text{Let } u = x + 2. \text{ Then } u^2 = (x + 2)^2.$$
$$(u + 12)(u - 1) = 0 \quad \text{Factor.}$$
$$u = -12 \quad \text{or} \quad u = 1 \quad \text{Solve.}$$

But we want to solve for x. Because $u = x + 2$, we have

$$x + 2 = -12 \quad \text{or} \quad x + 2 = 1$$
$$x = -14 \qquad\qquad x = -1$$

✓ **Check:** $x = -14$: $\quad (-14 + 2)^2 + 11(-14 + 2) - 12$
$$= (-12)^2 + 11(-12) - 12 = 144 - 132 - 12 = 0$$
$x = -1$: $\quad (-1 + 2)^2 + 11(-1 + 2) - 12 = 1 + 11 - 12 = 0$

The original equation has the solution set $\{-14, -1\}$. ♩

EXAMPLE 5 | **Solving an Equation Quadratic in Form**

Find the real solutions of the equation: $(x^2 - 1)^2 + (x^2 - 1) - 12 = 0$

Solution For the equation $(x^2 - 1)^2 + (x^2 - 1) - 12 = 0$, we let $u = x^2 - 1$ so that $u^2 = (x^2 - 1)^2$. Then the original equation,

$$(x^2 - 1)^2 + (x^2 - 1) - 12 = 0$$

becomes

$$u^2 + u - 12 = 0 \quad \text{Let } u = x^2 - 1. \text{ Then } u^2 = (x^2 - 1)^2.$$
$$(u + 4)(u - 3) = 0 \quad \text{Factor.}$$
$$u = -4 \quad \text{or} \quad u = 3 \quad \text{Solve.}$$

But remember that we want to solve for x. Because $u = x^2 - 1$, we have

$$x^2 - 1 = -4 \quad \text{or} \quad x^2 - 1 = 3$$
$$x^2 = -3 \qquad\qquad x^2 = 4$$

The first of these has no real solution; the second has the solution set $\{-2, 2\}$.

✓ **Check:** $x = -2$: $\quad (4 - 1)^2 + (4 - 1) - 12 = 9 + 3 - 12 = 0$
$x = 2$: $\quad (4 - 1)^2 + (4 - 1) - 12 = 9 + 3 - 12 = 0$

The original equation has the solution set $\{-2, 2\}$. ♩

| EXAMPLE 6 | **Solving an Equation Quadratic in Form** |

Find the real solutions of the equation: $x + 2\sqrt{x} - 3 = 0$

Solution For the equation $x + 2\sqrt{x} - 3 = 0$, let $u = \sqrt{x}$. Then $u^2 = x$, and the original equation,

$$x + 2\sqrt{x} - 3 = 0$$

becomes

$$u^2 + 2u - 3 = 0 \quad \text{Let } u = \sqrt{x}. \text{ Then } u^2 = x.$$
$$(u + 3)(u - 1) = 0 \quad \text{Factor.}$$
$$u = -3 \quad \text{or} \quad u = 1 \quad \text{Solve.}$$

Since $u = \sqrt{x}$, we have $\sqrt{x} = -3$ or $\sqrt{x} = 1$. The first of these, $\sqrt{x} = -3$, has no real solution, since the principal square root of a real number is never negative. The second, $\sqrt{x} = 1$, has the solution $x = 1$.

✓**Check:** $1 + 2\sqrt{1} - 3 = 1 + 2 - 3 = 0$

The original equation has the solution set $\{1\}$.

ANOTHER METHOD FOR SOLVING EXAMPLE **6** WOULD BE TO TREAT IT AS A RADICAL EQUATION. SOLVE IT THIS WAY FOR PRACTICE.

The idea should now be clear. If an equation contains an expression and that same expression squared, make a substitution for the expression. You may get a quadratic equation.

Now Work PROBLEM 51

3 Solve Equations by Factoring

We have already solved certain quadratic equations using factoring. Let's look at examples of other kinds of equations that can be solved by factoring.

| EXAMPLE 7 | **Solving an Equation by Factoring** |

Solve the equation: $x^4 = 4x^2$

Solution Begin by collecting all terms on one side. This results in 0 on one side and an expression to be factored on the other.

$$x^4 = 4x^2$$
$$x^4 - 4x^2 = 0$$
$$x^2(x^2 - 4) = 0 \quad \text{Factor.}$$
$$x^2 = 0 \quad \text{or} \quad x^2 - 4 = 0 \quad \text{Apply the Zero-Product Property.}$$
$$x^2 = 4$$
$$x = 0 \quad \text{or} \quad x = -2 \quad \text{or} \quad x = 2$$

The solution set is $\{-2, 0, 2\}$.

✓**Check:** $x = -2$: $(-2)^4 = 16$ and $4(-2)^2 = 16$ So -2 is a solution.
 $x = 0$: $0^4 = 0$ and $4 \cdot 0^2 = 0$ So 0 is a solution.
 $x = 2$: $2^4 = 16$ and $4 \cdot 2^2 = 16$ So 2 is a solution.

EXAMPLE 8	**Solving an Equation by Factoring**

Solve the equation: $x^3 - x^2 - 4x + 4 = 0$

Solution Do you recall the method of factoring by grouping? (If not, review pp. 53–54.) Group the terms of $x^3 - x^2 - 4x + 4 = 0$ as follows:

$$(x^3 - x^2) - (4x - 4) = 0$$

Factor out x^2 from the first grouping and 4 from the second.

$$x^2(x - 1) - 4(x - 1) = 0$$

This reveals the common factor $(x - 1)$, so we have

$$(x^2 - 4)(x - 1) = 0$$
$$(x - 2)(x + 2)(x - 1) = 0 \qquad \text{Factor again.}$$
$$x - 2 = 0 \quad \text{or} \quad x + 2 = 0 \quad \text{or} \quad x - 1 = 0 \quad \text{Set each factor equal to 0.}$$
$$x = 2 \qquad\qquad x = -2 \qquad\qquad x = 1 \quad \text{Solve.}$$

The solution set is $\{-2, 1, 2\}$.

✓**Check:**

$x = -2$: $(-2)^3 - (-2)^2 - 4(-2) + 4 = -8 - 4 + 8 + 4 = 0$ -2 is a solution.

$x = 1$: $1^3 - 1^2 - 4(1) + 4 = 1 - 1 - 4 + 4 = 0$ 1 is a solution.

$x = 2$: $2^3 - 2^2 - 4(2) + 4 = 8 - 4 - 8 + 4 = 0$ 2 is a solution.

════── **Now Work** PROBLEM **79**

1.4 Assess Your Understanding

'Are You Prepared?' *Answers are given at the end of these exercises. If you get a wrong answer, read the pages listed in red.*

1. ***True or False*** The principal square root of any nonnegative real number is always nonnegative. (pp. 23–24)

2. $\sqrt[3]{-8} = $ _____ (pp. 73–75)

3. Factor $6x^3 - 2x^2$ (pp. 49–55)

Concepts and Vocabulary

4. When an apparent solution does not satisfy the original equation, it is called a(n) _____ solution.

5. If u is an expression involving x, the equation $au^2 + bu + c = 0, a \neq 0$, is called a(n) equation _____ _____ _____.

6. ***True or False*** Radical equations sometimes have extraneous solutions.

Skill Building

In Problems 7–40, find the real solutions of each equation.

7. $\sqrt{2t - 1} = 1$

8. $\sqrt{3t + 4} = 2$

9. $\sqrt{3t + 4} = -6$

10. $\sqrt{5t + 3} = -2$

11. $\sqrt[3]{1 - 2x} - 3 = 0$

12. $\sqrt[3]{1 - 2x} - 1 = 0$

13. $\sqrt[4]{5x - 4} = 2$

14. $\sqrt[5]{2x - 3} = -1$

15. $\sqrt[5]{x^2 + 2x} = -1$

16. $\sqrt[4]{x^2 + 16} = \sqrt{5}$

17. $x = 8\sqrt{x}$

18. $x = 3\sqrt{x}$

19. $\sqrt{15 - 2x} = x$

20. $\sqrt{12 - x} = x$

21. $x = 2\sqrt{x - 1}$

22. $x = 2\sqrt{-x - 1}$

23. $\sqrt{x^2 - x - 4} = x + 2$

24. $\sqrt{3 - x + x^2} = x - 2$

25. $3 + \sqrt{3x + 1} = x$

26. $2 + \sqrt{12 - 2x} = x$

27. $\sqrt{2x + 3} - \sqrt{x + 1} = 1$

28. $\sqrt{3x + 7} + \sqrt{x + 2} = 1$

29. $\sqrt{3x + 1} - \sqrt{x - 1} = 2$

30. $\sqrt{3x - 5} - \sqrt{x + 7} = 2$

31. $\sqrt{3 - 2\sqrt{x}} = \sqrt{x}$

32. $\sqrt{10 + 3\sqrt{x}} = \sqrt{x}$

33. $(3x + 1)^{1/2} = 4$

34. $(3x - 5)^{1/2} = 2$

35. $(5x - 2)^{1/3} = 2$

36. $(2x + 1)^{1/3} = -1$

37. $(x^2 + 9)^{1/2} = 5$

38. $(x^2 - 16)^{1/2} = 9$

39. $x^{3/2} - 3x^{1/2} = 0$

40. $x^{3/4} - 9x^{1/4} = 0$

In Problems 41–72, find the real solutions of each equation.

41. $x^4 - 5x^2 + 4 = 0$

42. $x^4 - 10x^2 + 25 = 0$

43. $3x^4 - 2x^2 - 1 = 0$

44. $2x^4 - 5x^2 - 12 = 0$

45. $x^6 + 7x^3 - 8 = 0$

46. $x^6 - 7x^3 - 8 = 0$

47. $(x + 2)^2 + 7(x + 2) + 12 = 0$

48. $(2x + 5)^2 - (2x + 5) - 6 = 0$

49. $(3x + 4)^2 - 6(3x + 4) + 9 = 0$

50. $(2 - x)^2 + (2 - x) - 20 = 0$

51. $2(s + 1)^2 - 5(s + 1) = 3$

52. $3(1 - y)^2 + 5(1 - y) + 2 = 0$

53. $x - 4x\sqrt{x} = 0$

54. $x + 8\sqrt{x} = 0$

55. $x + \sqrt{x} = 20$

56. $x + \sqrt{x} = 6$

57. $t^{1/2} - 2t^{1/4} + 1 = 0$

58. $z^{1/2} - 4z^{1/4} + 4 = 0$

59. $4x^{1/2} - 9x^{1/4} + 4 = 0$

60. $x^{1/2} - 3x^{1/4} + 2 = 0$

61. $\sqrt{5x^2 - 6} = x$

62. $\sqrt[4]{4 - 5x^2} = x$

63. $x^2 + 3x + \sqrt{x^2 + 3x} = 6$

64. $x^2 - 3x - \sqrt{x^2 - 3x} = 2$

65. $\dfrac{1}{(x + 1)^2} = \dfrac{1}{x + 1} + 2$

66. $\dfrac{1}{(x - 1)^2} + \dfrac{1}{x - 1} = 12$

67. $3x^{-2} - 7x^{-1} - 6 = 0$

68. $2x^{-2} - 3x^{-1} - 4 = 0$

69. $2x^{2/3} - 5x^{1/3} - 3 = 0$

70. $3x^{4/3} + 5x^{2/3} - 2 = 0$

71. $\left(\dfrac{v}{v + 1}\right)^2 + \dfrac{2v}{v + 1} = 8$

72. $\left(\dfrac{y}{y - 1}\right)^2 = 6\left(\dfrac{y}{y - 1}\right) + 7$

In Problems 73–88, find the real solutions of each equation by factoring.

73. $x^3 - 9x = 0$

74. $x^4 - x^2 = 0$

75. $4x^3 = 3x^2$

76. $x^5 = 4x^3$

77. $x^3 + x^2 - 20x = 0$

78. $x^3 + 6x^2 - 7x = 0$

79. $x^3 + x^2 - x - 1 = 0$

80. $x^3 + 4x^2 - x - 4 = 0$

81. $x^3 - 3x^2 - 4x + 12 = 0$

82. $x^3 - 3x^2 - x + 3 = 0$

83. $2x^3 + 4 = x^2 + 8x$

84. $3x^3 + 4x^2 = 27x + 36$

85. $5x^3 + 45x = 2x^2 + 18$

86. $3x^3 + 12x = 5x^2 + 20$

87. $x(x^2 - 3x)^{1/3} + 2(x^2 - 3x)^{4/3} = 0$

88. $3x(x^2 + 2x)^{1/2} - 2(x^2 + 2x)^{3/2} = 0$

In Problems 89–94, find the real solutions of each equation. Use a calculator to express any solutions rounded to two decimal places.

89. $x - 4x^{1/2} + 2 = 0$

90. $x^{2/3} + 4x^{1/3} + 2 = 0$

91. $x^4 + \sqrt{3}x^2 - 3 = 0$

92. $x^4 + \sqrt{2}x^2 - 2 = 0$

93. $\pi(1 + t)^2 = \pi + 1 + t$

94. $\pi(1 + r)^2 = 2 + \pi(1 + r)$

Mixed Practice

95. If $k = \dfrac{x + 3}{x - 3}$ and $k^2 - k = 12$, find x.

96. If $k = \dfrac{x + 3}{x - 4}$ and $k^2 - 3k = 28$, find x.

Applications

97. Physics: Using Sound to Measure Distance The distance to the surface of the water in a well can sometimes be found by dropping an object into the well and measuring the time elapsed until a sound is heard. If t_1 is the time (measured in seconds) that it takes for the object to strike the water, then t_1 will obey the equation $s = 16t_1^2$, where s is the distance (measured in feet). It follows that $t_1 = \dfrac{\sqrt{s}}{4}$. Suppose that t_2 is the time that it takes for the sound of the impact to reach your ears. Because sound waves are known to travel at a speed of approximately 1100 feet per second, the time t_2 to travel the distance s will be $t_2 = \dfrac{s}{1100}$. See the illustration.

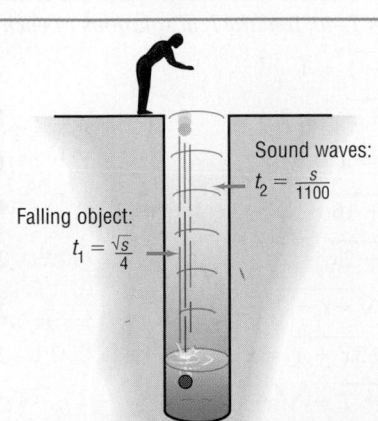

Sound waves:
$t_2 = \dfrac{s}{1100}$

Falling object:
$t_1 = \dfrac{\sqrt{s}}{4}$

Now $t_1 + t_2$ is the total time that elapses from the moment that the object is dropped to the moment that a sound is heard. We have the equation

$$\text{Total time elapsed} = \frac{\sqrt{s}}{4} + \frac{s}{1100}$$

Find the distance to the water's surface if the total time elapsed from dropping a rock to hearing it hit water is 4 seconds.

98. **Crushing Load** A civil engineer relates the thickness T, in inches, and height H, in feet, of a square wooden pillar to its crushing load L, in tons, using the model $T = \sqrt[4]{\dfrac{LH^2}{25}}$. If a square wooden pillar is 4 inches thick and 10 feet high, what is its crushing load?

99. **Foucault's Pendulum** The period of a pendulum is the time it takes the pendulum to make one full swing back and forth. The period T, in seconds, is given by the formula $T = 2\pi\sqrt{\dfrac{l}{32}}$, where l is the length, in feet, of the pendulum. In 1851, Jean Bernard Leon Foucault demonstrated the axial rotation of Earth using a large pendulum that he hung in the Panthéon in Paris. The period of Foucault's pendulum was approximately 16.5 seconds. What was its length?

Explaining Concepts: Discussion and Writing

100. Make up a radical equation that has no solution.

101. Make up a radical equation that has an extraneous solution.

102. Discuss the step in the solving process for radical equations that leads to the possibility of extraneous solutions. Why is there no such possibility for linear and quadratic equations?

103. **What Went Wrong?** On an exam, Jane solved the equation $\sqrt{2x + 3} - x = 0$ and wrote that the solution set was $\{-1, 3\}$. Jane received 3 out of 5 points for the problem. Jane asks you why she received 3 out of 5 points. Provide an explanation.

'Are You Prepared?' Answers

1. True 2. -2 3. $2x^2(3x - 1)$

1.5 Solving Inequalities

PREPARING FOR THIS SECTION *Before getting started, review the following:*

• Algebra Essentials (Section R.2, pp. 17–26)

Now Work the 'Are You Prepared?' problems on page 127.

OBJECTIVES **1** Use Interval Notation (p. 120)
 2 Use Properties of Inequalities (p. 121)
 3 Solve Inequalities (p. 123)
 4 Solve Combined Inequalities (p. 124)

Suppose that a and b are two real numbers and $a < b$. We shall use the notation $a < x < b$ to mean that x is a number *between* a and b. The expression $a < x < b$ is equivalent to the two inequalities $a < x$ and $x < b$. Similarly, the expression $a \le x \le b$ is equivalent to the two inequalities $a \le x$ and $x \le b$. The remaining two possibilities, $a \le x < b$ and $a < x \le b$, are defined similarly.

Although it is acceptable to write $3 \ge x \ge 2$, it is preferable to reverse the inequality symbols and write instead $2 \le x \le 3$ so that, as you read from left to right, the values go from smaller to larger.

A statement such as $2 \le x \le 1$ is false because there is no number x for which $2 \le x$ and $x \le 1$. Finally, never mix inequality symbols, as in $2 \le x \ge 3$.

1 Use Interval Notation

Let a and b represent two real numbers with $a < b$.

<table>
<tr>
<td>**DEFINITION**</td>
<td>

An **open interval,** denoted by **(a, b),** consists of all real numbers x for which $a < x < b$.

A **closed interval,** denoted by **[a, b],** consists of all real numbers x for which $a \leq x \leq b$.

The **half-open,** or **half-closed, intervals** are **(a, b],** consisting of all real numbers x for which $a < x \leq b$, and **[a, b),** consisting of all real numbers x for which $a \leq x < b$.

</td>
</tr>
</table>

In each of these definitions, a is called the **left endpoint** and b the **right endpoint** of the interval.

The symbol ∞ (read as "infinity") is not a real number, but a notational device used to indicate unboundedness in the positive direction. The symbol $-\infty$ (read as "negative infinity") also is not a real number, but a notational device used to indicate unboundedness in the negative direction. Using the symbols ∞ and $-\infty$, we can define five other kinds of intervals:

$[a, \infty)$	Consists of all real numbers x for which $x \geq a$
(a, ∞)	Consists of all real numbers x for which $x > a$
$(-\infty, a]$	Consists of all real numbers x for which $x \leq a$
$(-\infty, a)$	Consists of all real numbers x for which $x < a$
$(-\infty, \infty)$	Consists of all real numbers

Note that ∞ and $-\infty$ are never included as endpoints, since neither is a real number.

Table 1 summarizes interval notation, corresponding inequality notation, and their graphs.

Table 1

Interval	Inequality	Graph
The open interval (a, b)	$a < x < b$	
The closed interval $[a, b]$	$a \leq x \leq b$	
The half-open interval $[a, b)$	$a \leq x < b$	
The half-open interval $(a, b]$	$a < x \leq b$	
The interval $[a, \infty)$	$x \geq a$	
The interval (a, ∞)	$x > a$	
The interval $(-\infty, a]$	$x \leq a$	
The interval $(-\infty, a)$	$x < a$	
The interval $(-\infty, \infty)$	All real numbers	

EXAMPLE 1 **Writing Inequalities Using Interval Notation**

Write each inequality using interval notation.

(a) $1 \leq x \leq 3$ (b) $-4 < x < 0$ (c) $x > 5$ (d) $x \leq 1$

Solution (a) $1 \leq x \leq 3$ describes all numbers x between 1 and 3, inclusive. In interval notation, we write $[1, 3]$.

(b) In interval notation, $-4 < x < 0$ is written $(-4, 0)$.

(c) $x > 5$ consists of all numbers x greater than 5. In interval notation, we write $(5, \infty)$.

(d) In interval notation, $x \leq 1$ is written $(-\infty, 1]$.

EXAMPLE 2

Writing Intervals Using Inequality Notation

Write each interval as an inequality involving x.

(a) $[1, 4)$ (b) $(2, \infty)$ (c) $[2, 3]$ (d) $(-\infty, -3]$

Solution

(a) $[1, 4)$ consists of all numbers x for which $1 \leq x < 4$.

(b) $(2, \infty)$ consists of all numbers x for which $x > 2$.

(c) $[2, 3]$ consists of all numbers x for which $2 \leq x \leq 3$.

(d) $(-\infty, -3]$ consists of all numbers x for which $x \leq -3$.

━━━ **Now Work** PROBLEMS **11, 23,** AND **31**

2 Use Properties of Inequalities

The product of two positive real numbers is positive, the product of two negative real numbers is positive, and the product of 0 and 0 is 0. For any real number a, the value of a^2 is 0 or positive; that is, a^2 is nonnegative. This is called the **nonnegative property.**

In Words

The square of a real number is never negative.

Nonnegative Property

For any real number a,

$$a^2 \geq 0 \tag{1}$$

If we add the same number to both sides of an inequality, we obtain an equivalent inequality. For example, since $3 < 5$, then $3 + 4 < 5 + 4$ or $7 < 9$. This is called the **addition property** of inequalities.

In Words

The addition property states that the sense, or direction, of an inequality remains unchanged if the same number is added to each side.

Addition Property of Inequalities

For real numbers a, b, and c,

$$\text{If } a < b, \text{ then } a + c < b + c. \tag{2a}$$
$$\text{If } a > b, \text{ then } a + c > b + c. \tag{2b}$$

Figure 2 illustrates the addition property (2a). In Figure 2(a), we see that a lies to the left of b. If c is positive, then $a + c$ and $b + c$ each lie c units to the right of a and b, respectively. Consequently, $a + c$ must lie to the left of $b + c$; that is, $a + c < b + c$. Figure 2(b) illustrates the situation if c is negative.

Figure 2

(a) If $a < b$ and $c > 0$, then $a + c < b + c$.

(b) If $a < b$ and $c < 0$, then $a + c < b + c$.

━━━ DRAW AN ILLUSTRATION SIMILAR TO FIGURE **2** THAT ILLUSTRATES THE ADDITION PROPERTY **(2b).**

| EXAMPLE 3 | **Addition Property of Inequalities** |

(a) If $x < -5$, then $x + 5 < -5 + 5$ or $x + 5 < 0$.

(b) If $x > 2$, then $x + (-2) > 2 + (-2)$ or $x - 2 > 0$. ⏎

━━━ **Now Work** PROBLEM 39

| EXAMPLE 4 | **Multiplying an Inequality by a Positive Number** |

Express as an inequality the result of multiplying each side of the inequality $3 < 7$ by 2.

Solution Begin with

$$3 < 7$$

Multiplying each side by 2 yields the numbers 6 and 14, so we have

$$6 < 14$$ ⏎

| EXAMPLE 5 | **Multiplying an Inequality by a Negative Number** |

Express as an inequality the result of multiplying each side of the inequality $9 > 2$ by -4.

Solution Begin with

$$9 > 2$$

Multiplying each side by -4 yields the numbers -36 and -8, so we have

$$-36 < -8$$ ⏎

In Words

Multiplying by a negative number reverses the inequality.

Note that the effect of multiplying both sides of $9 > 2$ by the negative number -4 is that the direction of the inequality symbol is reversed.

Examples 4 and 5 illustrate the following general **multiplication properties** for inequalities:

In Words

The multiplication properties state that the sense, or direction, of an inequality *remains the same* if each side is multiplied by a *positive* real number, whereas the direction is *reversed* if each side is multiplied by a *negative* real number.

Multiplication Properties for Inequalities

For real numbers a, b, and c,

If $a < b$ and if $c > 0$, then $ac < bc$. If $a < b$ and if $c < 0$, then $ac > bc$.	**(3a)**
If $a > b$ and if $c > 0$, then $ac > bc$. If $a > b$ and if $c < 0$, then $ac < bc$.	**(3b)**

| EXAMPLE 6 | **Multiplication Property of Inequalities** |

(a) If $2x < 6$, then $\dfrac{1}{2}(2x) < \dfrac{1}{2}(6)$ or $x < 3$.

(b) If $\dfrac{x}{-3} > 12$, then $-3\left(\dfrac{x}{-3}\right) < -3(12)$ or $x < -36$.

(c) If $-4x < -8$, then $\dfrac{-4x}{-4} > \dfrac{-8}{-4}$ or $x > 2$.

(d) If $-x > 8$, then $(-1)(-x) < (-1)(8)$ or $x < -8$. ⏎

━━━ **Now Work** PROBLEM 45

Reciprocal Property for Inequalities

$$\text{If } a > 0, \text{ then } \frac{1}{a} > 0 \qquad \text{If } \frac{1}{a} > 0, \text{ then } a > 0 \qquad \textbf{(4a)}$$

$$\text{If } a < 0, \text{ then } \frac{1}{a} < 0 \qquad \text{If } \frac{1}{a} < 0, \text{ then } a < 0 \qquad \textbf{(4b)}$$

3 Solve Inequalities

An **inequality in one variable** is a statement involving two expressions, at least one containing the variable, separated by one of the inequality symbols $<$, \leq, $>$, or \geq. To **solve an inequality** means to find all values of the variable for which the statement is true. These values are called **solutions** of the inequality.

For example, the following are all inequalities involving one variable x:

$$x + 5 < 8 \qquad 2x - 3 \geq 4 \qquad x^2 - 1 \leq 3 \qquad \frac{x + 1}{x - 2} > 0$$

As with equations, one method for solving an inequality is to replace it by a series of equivalent inequalities until an inequality with an obvious solution, such as $x < 3$, is obtained. We obtain equivalent inequalities by applying some of the same properties as those used to find equivalent equations. The addition property and the multiplication properties form the bases for the following procedures.

Procedures That Leave the Inequality Symbol Unchanged

1. Simplify both sides of the inequality by combining like terms and eliminating parentheses:

$$\text{Replace} \qquad x + 2 + 6 > 2x + 5(x + 1)$$
$$\text{by} \qquad x + 8 > 7x + 5$$

2. Add or subtract the same expression on both sides of the inequality:

$$\text{Replace} \qquad 3x - 5 < 4$$
$$\text{by} \qquad (3x - 5) + 5 < 4 + 5$$

3. Multiply or divide both sides of the inequality by the same *positive* expression:

$$\text{Replace} \qquad 4x > 16 \quad \text{by} \quad \frac{4x}{4} > \frac{16}{4}$$

Procedures That Reverse the Sense or Direction of the Inequality Symbol

1. Interchange the two sides of the inequality:

$$\text{Replace} \qquad 3 < x \quad \text{by} \quad x > 3$$

2. Multiply or divide both sides of the inequality by the same *negative* expression:

$$\text{Replace} \qquad -2x > 6 \quad \text{by} \quad \frac{-2x}{-2} < \frac{6}{-2}$$

As the examples that follow illustrate, we solve inequalities using many of the same steps that we would use to solve equations. In writing the solution of an

inequality, we may use either set notation or interval notation, whichever is more convenient.

EXAMPLE 7 | **Solving an Inequality**

Solve the inequality: $3 - 2x < 5$
Graph the solution set.

Solution
$$3 - 2x < 5$$
$$3 - 2x - 3 < 5 - 3 \quad \text{Subtract 3 from both sides.}$$
$$-2x < 2 \quad \text{Simplify.}$$
$$\frac{-2x}{-2} > \frac{2}{-2} \quad \text{Divide both sides by } -2. \text{ (The sense}$$
$$\text{of the inequality symbol is reversed.)}$$
$$x > -1 \quad \text{Simplify.}$$

Figure 3

The solution set is $\{x \mid x > -1\}$ or, using interval notation, all numbers in the interval $(-1, \infty)$. See Figure 3 for the graph.

EXAMPLE 8 | **Solving an Inequality**

Solve the inequality: $4x + 7 \geq 2x - 3$
Graph the solution set.

Solution
$$4x + 7 \geq 2x - 3$$
$$4x + 7 - 7 \geq 2x - 3 - 7 \quad \text{Subtract 7 from both sides.}$$
$$4x \geq 2x - 10 \quad \text{Simplify.}$$
$$4x - 2x \geq 2x - 10 - 2x \quad \text{Subtract 2x from both sides.}$$
$$2x \geq -10 \quad \text{Simplify.}$$
$$\frac{2x}{2} \geq \frac{-10}{2} \quad \text{Divide both sides by 2. (The direction}$$
$$\text{of the inequality symbol is unchanged.)}$$
$$x \geq -5 \quad \text{Simplify.}$$

Figure 4

The solution set is $\{x \mid x \geq -5\}$ or, using interval notation, all numbers in the interval $[-5, \infty)$. See Figure 4 for the graph.

Now Work PROBLEM 53

4 Solve Combined Inequalities

EXAMPLE 9 | **Solving a Combined Inequality**

Solve the inequality: $-5 < 3x - 2 < 1$
Graph the solution set.

Solution Recall that the inequality

$$-5 < 3x - 2 < 1$$

is equivalent to the two inequalities

$$-5 < 3x - 2 \quad \text{and} \quad 3x - 2 < 1$$

We solve each of these inequalities separately.

$$-5 < 3x - 2 \qquad\qquad\qquad 3x - 2 < 1$$

$$-5 + 2 < 3x - 2 + 2 \quad \text{Add 2 to both sides.} \quad 3x - 2 + 2 < 1 + 2$$

$$-3 < 3x \qquad\qquad \text{Simplify.} \qquad\qquad 3x < 3$$

$$\frac{-3}{3} < \frac{3x}{3} \qquad\qquad \text{Divide both sides by 3.} \qquad \frac{3x}{3} < \frac{3}{3}$$

$$-1 < x \qquad\qquad \text{Simplify.} \qquad\qquad x < 1$$

The solution set of the original pair of inequalities consists of all x for which

$$-1 < x \quad \text{and} \quad x < 1$$

Figure 5

This may be written more compactly as $\{x \mid -1 < x < 1\}$. In interval notation, the solution is $(-1, 1)$. See Figure 5 for the graph.

Observe in the preceding process that the solution of the two inequalities required exactly the same steps. A shortcut to solving the original inequality algebraically is to deal with the two inequalities at the same time, as follows:

$$-5 < \quad 3x - 2 \quad < 1$$

$$-5 + 2 < 3x - 2 + 2 < 1 + 2 \quad \text{Add 2 to each part.}$$

$$-3 < \quad 3x \quad < 3 \quad \text{Simplify.}$$

$$\frac{-3}{3} < \quad \frac{3x}{3} \quad < \frac{3}{3} \quad \text{Divide each part by 3.}$$

$$-1 < \quad x \quad < 1 \quad \text{Simplify.}$$

EXAMPLE 10 **Solving a Combined Inequality**

Solve the inequality: $-1 \leq \dfrac{3 - 5x}{2} \leq 9$

Graph the solution set.

Solution

$$-1 \leq \quad \frac{3 - 5x}{2} \quad \leq 9$$

$$2(-1) \leq 2\left(\frac{3 - 5x}{2}\right) \leq 2(9) \quad \text{Multiply each part by 2 to remove the denominator.}$$

$$-2 \leq \quad 3 - 5x \quad \leq 18 \quad \text{Simplify.}$$

$$-2 - 3 \leq 3 - 5x - 3 \leq 18 - 3 \quad \text{Subtract 3 from each part to isolate the term containing } x.$$

$$-5 \leq \quad -5x \quad \leq 15 \quad \text{Simplify.}$$

$$\frac{-5}{-5} \geq \quad \frac{-5x}{-5} \quad \geq \frac{15}{-5} \quad \text{Divide each part by } -5 \text{ (reverse the sense of each inequality symbol).}$$

$$1 \geq \quad x \quad \geq -3 \quad \text{Simplify.}$$

$$-3 \leq \quad x \quad \leq 1 \quad \text{Reverse the order so that the numbers get larger as you read from left to right.}$$

Figure 6

The solution set is $\{x \mid -3 \leq x \leq 1\}$, that is, all x in the interval $[-3, 1]$. Figure 6 illustrates the graph.

━ Now Work PROBLEM **73**

| EXAMPLE 11 | Using the Reciprocal Property to Solve an Inequality |

Solve the inequality: $(4x - 1)^{-1} > 0$
Graph the solution set.

Solution Since $(4x - 1)^{-1} = \dfrac{1}{4x - 1}$ and since the Reciprocal Property states that when $\dfrac{1}{a} > 0$ then $a > 0$, we have

$$(4x - 1)^{-1} > 0$$

$$\dfrac{1}{4x - 1} > 0$$

$$4x - 1 > 0 \quad \text{Reciprocal Property}$$

$$4x > 1$$

$$x > \dfrac{1}{4}$$

Figure 7

The solution set is $\left\{x \middle| x > \dfrac{1}{4}\right\}$, that is, all x in the interval $\left(\dfrac{1}{4}, \infty\right)$. Figure 7 illustrates the graph.

Now Work PROBLEM 83

| EXAMPLE 12 | Creating Equivalent Inequalities |

If $-1 < x < 4$, find a and b so that $a < 2x + 1 < b$.

Solution The idea here is to change the middle part of the combined inequality from x to $2x + 1$, using properties of inequalities.

$$
\begin{aligned}
-1 <\ & x & < 4 \\
-2 <\ & 2x & < 8 \quad \text{Multiply each part by 2.} \\
-1 <\ & 2x + 1 & < 9 \quad \text{Add 1 to each part.}
\end{aligned}
$$

Now we see that $a = -1$ and $b = 9$.

Now Work PROBLEM 91

Application

| EXAMPLE 13 | Physics: Ohm's Law |

In electricity, Ohm's law states that $E = IR$, where E is the voltage (in volts), I is the current (in amperes), and R is the resistance (in ohms). An air-conditioning unit is rated at a resistance of 10 ohms. If the voltage varies from 110 to 120 volts, inclusive, what corresponding range of current will the air conditioner draw?

Solution The voltage lies between 110 and 120, inclusive, so

$$
\begin{aligned}
110 \leq\ & E & \leq 120 \\
110 \leq\ & IR & \leq 120 \quad \text{Ohm's law, } E = IR \\
110 \leq\ & I(10) & \leq 120 \quad R = 10 \\
\dfrac{110}{10} \leq\ & \dfrac{I(10)}{10} & \leq \dfrac{120}{10} \quad \text{Divide each part by 10.} \\
11 \leq\ & I & \leq 12 \quad \text{Simplify.}
\end{aligned}
$$

The air conditioner will draw between 11 and 12 amperes of current, inclusive.

1.5 Assess Your Understanding

'Are You Prepared?' *Answers are given at the end of these exercises. If you get a wrong answer, read the pages listed in red.*

1. Graph the inequality: $x \geq -2$. (pp. 17–26)

2. *True or False* $-5 > -3$ (pp. 17–26)

Concepts and Vocabulary

3. If each side of an inequality is multiplied by a(n) _____ number, then the sense of the inequality symbol is reversed.

4. A(n) _____ _____, denoted $[a, b]$, consists of all real numbers x for which $a \leq x \leq b$.

5. The _____ _____ state that the sense, or direction, of an inequality remains the same if each side is multiplied by a positive number, while the direction is reversed if each side is multiplied by a negative number.

In Problems 6–9, assume that $a < b$ and $c < 0$.

6. *True or False* $a + c < b + c$

7. *True or False* $a - c < b - c$

8. *True or False* $ac > bc$

9. *True or False* $\dfrac{a}{c} < \dfrac{b}{c}$

10. *True or False* The square of any real number is always nonnegative.

Skill Building

In Problems 11–16, express the graph shown in blue using interval notation. Also express each as an inequality involving x.

11.

12.

13.

14.

15.

16.

In Problems 17–22, an inequality is given. Write the inequality obtained by:

(a) *Adding 3 to each side of the given inequality.*
(b) *Subtracting 5 from each side of the given inequality.*
(c) *Multiplying each side of the given inequality by 3.*
(d) *Multiplying each side of the given inequality by −2.*

17. $3 < 5$ **18.** $2 > 1$ **19.** $4 > -3$ **20.** $-3 > -5$ **21.** $2x + 1 < 2$ **22.** $1 - 2x > 5$

In Problems 23–30, write each inequality using interval notation, and illustrate each inequality using the real number line.

23. $0 \leq x \leq 4$ **24.** $-1 < x < 5$ **25.** $4 \leq x < 6$ **26.** $-2 < x < 0$

27. $x \geq 4$ **28.** $x \leq 5$ **29.** $x < -4$ **30.** $x > 1$

In Problems 31–38, write each interval as an inequality involving x, and illustrate each inequality using the real number line.

31. $[2, 5]$ **32.** $(1, 2)$ **33.** $(-3, -2)$ **34.** $[0, 1)$

35. $[4, \infty)$ **36.** $(-\infty, 2]$ **37.** $(-\infty, -3)$ **38.** $(-8, \infty)$

In Problems 39–52, fill in the blank with the correct inequality symbol.

39. If $x < 5$, then $x - 5$ _____ 0.

40. If $x < -4$, then $x + 4$ _____ 0.

41. If $x > -4$, then $x + 4$ _____ 0.

42. If $x > 6$, then $x - 6$ _____ 0.

43. If $x \geq -4$, then $3x$ _____ -12.

44. If $x \leq 3$, then $2x$ _____ 6.

45. If $x > 6$, then $-2x$ _____ -12.

46. If $x > -2$, then $-4x$ _____ 8.

47. If $x \geq 5$, then $-4x$ _____ -20.

48. If $x \leq -4$, then $-3x$ _____ 12.

49. If $2x > 6$, then x _____ 3.

50. If $3x \leq 12$, then x _____ 4.

51. If $-\dfrac{1}{2}x \leq 3$, then x _____ -6.

52. If $-\dfrac{1}{4}x > 1$, then x _____ -4.

In Problems 53–88, solve each inequality. Express your answer using set notation or interval notation. Graph the solution set.

53. $x + 1 < 5$

54. $x - 6 < 1$

55. $1 - 2x \leq 3$

56. $2 - 3x \leq 5$

57. $3x - 7 > 2$

58. $2x + 5 > 1$

59. $3x - 1 \geq 3 + x$

60. $2x - 2 \geq 3 + x$

61. $-2(x + 3) < 8$

62. $-3(1 - x) < 12$

63. $4 - 3(1 - x) \leq 3$

64. $8 - 4(2 - x) \leq -2x$

65. $\frac{1}{2}(x - 4) > x + 8$

66. $3x + 4 > \frac{1}{3}(x - 2)$

67. $\frac{x}{2} \geq 1 - \frac{x}{4}$

68. $\frac{x}{3} \geq 2 + \frac{x}{6}$

69. $0 \leq 2x - 6 \leq 4$

70. $4 \leq 2x + 2 \leq 10$

71. $-5 \leq 4 - 3x \leq 2$

72. $-3 \leq 3 - 2x \leq 9$

73. $-3 < \frac{2x - 1}{4} < 0$

74. $0 < \frac{3x + 2}{2} < 4$

75. $1 < 1 - \frac{1}{2}x < 4$

76. $0 < 1 - \frac{1}{3}x < 1$

77. $(x + 2)(x - 3) > (x - 1)(x + 1)$

78. $(x - 1)(x + 1) > (x - 3)(x + 4)$

79. $x(4x + 3) \leq (2x + 1)^2$

80. $x(9x - 5) \leq (3x - 1)^2$

81. $\frac{1}{2} \leq \frac{x + 1}{3} < \frac{3}{4}$

82. $\frac{1}{3} < \frac{x + 1}{2} \leq \frac{2}{3}$

83. $(4x + 2)^{-1} < 0$

84. $(2x - 1)^{-1} > 0$

85. $0 < \frac{2}{x} < \frac{3}{5}$

86. $0 < \frac{4}{x} < \frac{2}{3}$

87. $0 < (2x - 4)^{-1} < \frac{1}{2}$

88. $0 < (3x + 6)^{-1} < \frac{1}{3}$

Applications and Extensions

In Problems 89–98, find a and b.

89. If $-1 < x < 1$, then $a < x + 4 < b$.

90. If $-3 < x < 2$, then $a < x - 6 < b$.

91. If $2 < x < 3$, then $a < -4x < b$.

92. If $-4 < x < 0$, then $a < \frac{1}{2}x < b$.

93. If $0 < x < 4$, then $a < 2x + 3 < b$.

94. If $-3 < x < 3$, then $a < 1 - 2x < b$.

95. If $-3 < x < 0$, then $a < \frac{1}{x + 4} < b$.

96. If $2 < x < 4$, then $a < \frac{1}{x - 6} < b$.

97. If $6 < 3x < 12$, then $a < x^2 < b$.

98. If $0 < 2x < 6$, then $a < x^2 < b$.

99. What is the domain of the variable in the expression $\sqrt{3x + 6}$?

100. What is the domain of the variable in the expression $\sqrt{8 + 2x}$?

101. A young adult may be defined as someone older than 21, but less than 30 years of age. Express this statement using inequalities.

102. Middle-aged may be defined as being 40 or more and less than 60. Express this statement using inequalities.

103. Life Expectancy The Social Security Administration determined that an average 30-year-old male in 2005 could expect to live at least 46.60 more years and an average

30-year-old female in 2005 could expect to live at least 51.03 more years.

(a) To what age can an average 30-year-old male expect to live? Express your answer as an inequality.

(b) To what age can an average 30-year-old female expect to live? Express your answer as an inequality.

(c) Who can expect to live longer, a male or a female? By how many years?

Source: Social Security Administration, Period Life Table, 2005

104. General Chemistry For a certain ideal gas, the volume V (in cubic centimeters) equals 20 times the temperature T (in degrees Celsius). If the temperature varies from 80° to 120° C inclusive, what is the corresponding range of the volume of the gas?

***105. Real Estate** A real estate agent agrees to sell an apartment complex according to the following commission schedule: $45,000 plus 25% of the selling price in excess of $900,000. Assuming that the complex will sell at some price between $900,000 and $1,100,000 inclusive, over what range

does the agent's commission vary? How does the commission vary as a percent of selling price?

106. **Sales Commission** A used car salesperson is paid a commission of $25 plus 40% of the selling price in excess of owner's cost. The owner claims that used cars typically sell for at least owner's cost plus $200 and at most owner's cost plus $3000. For each sale made, over what range can the salesperson expect the commission to vary?

107. **Federal Tax Withholding** The percentage method of withholding for federal income tax (2010) states that a single person whose weekly wages, after subtracting withholding allowances, are over $693, but not over $1302, shall have $82.35 plus 25% of the excess over $693 withheld. Over what range does the amount withheld vary if the weekly wages vary from $700 to $900 inclusive?

 Source: Employer's Tax Guide. Internal Revenue Service, 2010.

108. **Exercising** Sue wants to lose weight. For healthy weight loss, the American College of Sports Medicine (ACSM) recommends 200 to 300 minutes of exercise per week. For the first six days of the week, Sue exercised 40, 45, 0, 50, 25, and 35 minutes. How long should Sue exercise on the seventh day in order to stay within the ACSM guidelines?

109. **Electricity Rates** Commonwealth Edison Company's charge for electricity in January 2010 was 9.44¢ per kilowatt-hour. In addition, each monthly bill contains a customer charge of $12.55. If last year's bills ranged from a low of $76.27 to a high of $248.55, over what range did usage vary (in kilowatt-hours)?

 Source: Commonwealth Edison Co., Chicago, Illinois, 2010.

110. **Water Bills** The Village of Oak Lawn charges homeowners $37.62 per quarter-year plus $3.86 per 1000 gallons for water usage in excess of 10,000 gallons. In 2010 one homeowner's quarterly bill ranged from a high of $122.54 to a low of $68.50. Over what range did water usage vary?

 Source: Village of Oak Lawn, Illinois, January 2010.

111. **Markup of a New Car** The markup over dealer's cost of a new car ranges from 12% to 18%. If the sticker price is $18,000, over what range will the dealer's cost vary?

112. **IQ Tests** A standard intelligence test has an average score of 100. According to statistical theory, of the people who take the test, the 2.5% with the highest scores will have scores of more than 1.96σ above the average, where σ (sigma, a number called the **standard deviation**) depends on the nature of the test. If $\sigma = 12$ for this test and there is (in principle) no upper limit to the score possible on the test, write the interval of possible test scores of the people in the top 2.5%.

113. **Computing Grades** In your Economics 101 class, you have scores of 68, 82, 87, and 89 on the first four of five tests. To get a grade of B, the average of the first five test scores must be greater than or equal to 80 and less than 90.
 (a) Solve an inequality to find the range of the score that you need on the last test to get a B.
 (b) What score do you need if the fifth test counts double?

What do I need to get a B?

114. **"Light" Foods** For food products to be labeled "light," the U.S. Food and Drug Administration requires that the altered product must either contain one-third or fewer calories than the regular product or it must contain one-half or less fat than the regular product. If a serving of Miracle Whip® Light contains 20 calories and 1.5 grams of fat, then what must be true about either the number of calories or the grams of fat in a serving of regular Miracle Whip®?

115. **Arithmetic Mean** If $a < b$, show that $a < \dfrac{a+b}{2} < b$. The number $\dfrac{a+b}{2}$ is called the **arithmetic mean** of a and b.

116. Refer to Problem 115. Show that the arithmetic mean of a and b is equidistant from a and b.

117. **Geometric Mean** If $0 < a < b$, show that $a < \sqrt{ab} < b$. The number \sqrt{ab} is called the **geometric mean** of a and b.

118. Refer to Problems 115 and 117. Show that the geometric mean of a and b is less than the arithmetic mean of a and b.

119. **Harmonic Mean** For $0 < a < b$, let h be defined by
 $$\frac{1}{h} = \frac{1}{2}\left(\frac{1}{a} + \frac{1}{b}\right)$$
 Show that $a < h < b$. The number h is called the **harmonic mean** of a and b.

120. Refer to Problems 115, 117, and 119. Show that the harmonic mean of a and b equals the geometric mean squared divided by the arithmetic mean.

121. **Another Reciprocal Property** Prove that if $0 < a < b$, then $0 < \dfrac{1}{b} < \dfrac{1}{a}$.

Explaining Concepts: Discussion and Writing

122. Make up an inequality that has no solution. Make up one that has exactly one solution.

123. The inequality $x^2 + 1 < -5$ has no real solution. Explain why.

124. Do you prefer to use inequality notation or interval notation to express the solution to an inequality? Give your reasons. Are there particular circumstances when you prefer one to the other? Cite examples.

125. How would you explain to a fellow student the underlying reason for the multiplication properties for inequalities (page 122) that is, the sense or direction of an inequality remains the same if each side is multiplied by a positive real number, whereas the direction is reversed if each side is multiplied by a negative real number.

'Are You Prepared?' Answers

1. ![number line from -4 to 0 marked at -2]

2. False

1.6 Equations and Inequalities Involving Absolute Value

PREPARING FOR THIS SECTION *Before getting started, review the following:*

- Algebra Essentials (Chapter R, Section R.2, pp. 17–26)

Now Work the *'Are You Prepared?'* problems on page 132.

OBJECTIVES **1** Solve Equations Involving Absolute Value (p. 130)
2 Solve Inequalities Involving Absolute Value (p. 130)

1 Solve Equations Involving Absolute Value

Recall that, on the real number line, the absolute value of a equals the distance from the origin to the point whose coordinate is a. For example, there are two points whose distance from the origin is 5 units, -5 and 5. So the equation $|x| = 5$ will have the solution set $\{-5, 5\}$. This leads to the following result:

THEOREM If a is a positive real number and if u is any algebraic expression, then

$$|u| = a \quad \text{is equivalent to} \quad u = a \quad \text{or} \quad u = -a \qquad \textbf{(1)}$$

EXAMPLE 1 **Solving an Equation Involving Absolute Value**

Solve the equations:

(a) $|x + 4| = 13$ (b) $|2x - 3| + 2 = 7$

Solution (a) This follows the form of equation (1), where $u = x + 4$. There are two possibilities.

$$x + 4 = 13 \quad \text{or} \quad x + 4 = -13$$
$$x = 9 \quad \text{or} \quad x = -17$$

The solution set is $\{-17, 9\}$.

(b) The equation $|2x - 3| + 2 = 7$ is not in the form of equation (1). Proceed as follows:

$$|2x - 3| + 2 = 7$$
$$|2x - 3| = 5 \qquad \text{Subtract 2 from each side.}$$
$$2x - 3 = 5 \quad \text{or} \quad 2x - 3 = -5 \qquad \text{Apply (1).}$$
$$2x = 8 \quad \text{or} \quad 2x = -2$$
$$x = 4 \quad \text{or} \quad x = -1$$

The solution set is $\{-1, 4\}$.

Now Work PROBLEM 9

2 Solve Inequalities Involving Absolute Value

EXAMPLE 2 **Solving an Inequality Involving Absolute Value**

Solve the inequality: $|x| < 4$

Figure 8

Less than 4 units
from origin 0

Solution We are looking for all points whose coordinate x is a distance less than 4 units from the origin. See Figure 8 for an illustration. Because any number x between -4 and 4 satisfies the condition $|x| < 4$, the solution set consists of all numbers x for which $-4 < x < 4$, that is, all x in the interval $(-4, 4)$.

THEOREM If a is a positive number and if u is an algebraic expression, then

$$|u| < a \quad \text{is equivalent to} \quad -a < u < a \qquad \text{(2)}$$

$$|u| \le a \quad \text{is equivalent to} \quad -a \le u \le a \qquad \text{(3)}$$

In other words, $|u| < a$ is equivalent to $-a < u$ and $u < a$.

See Figure 9 for an illustration of statement (3).

Figure 9

$|u| \le a, a > 0$

EXAMPLE 3 **Solving an Inequality Involving Absolute Value**

Solve the inequality: $|2x + 4| \le 3$
Graph the solution set.

Solution

$$|2x + 4| \le 3$$ *This follows the form of statement (3); the expression $u = 2x + 4$ is inside the absolute value bars.*

$$-3 \le \quad 2x + 4 \quad \le 3$$ *Apply statement (3).*

$$-3 - 4 \le 2x + 4 - 4 \le 3 - 4$$ *Subtract 4 from each part.*

$$-7 \le \quad 2x \quad \le -1$$ *Simplify.*

$$\frac{-7}{2} \le \quad \frac{2x}{2} \quad \le \frac{-1}{2}$$ *Divide each part by 2.*

$$-\frac{7}{2} \le \quad x \quad \le -\frac{1}{2}$$ *Simplify.*

Figure 10

The solution set is $\left\{ x \middle| -\dfrac{7}{2} \le x \le -\dfrac{1}{2} \right\}$, that is, all x in the interval $\left[-\dfrac{7}{2}, -\dfrac{1}{2} \right]$. See Figure 10 for the graph of the solution set.

EXAMPLE 4 **Solving an Inequality Involving Absolute Value**

Solve the inequality: $|1 - 4x| < 5$
Graph the solution set.

Solution

$$|1 - 4x| < 5$$ *This expression follows the form of statement (2); the expression $u = 1 - 4x$ is inside the absolute value bars.*

$$-5 < \quad 1 - 4x \quad < 5$$ *Apply statement (2).*

$$-5 - 1 < 1 - 4x - 1 < 5 - 1$$ *Subtract 1 from each part.*

$$-6 < \quad -4x \quad < 4$$ *Simplify.*

$$\frac{-6}{-4} > \quad \frac{-4x}{-4} \quad > \frac{4}{-4}$$ *Divide each part by -4, which reverses the sense of the inequality symbols.*

$$\frac{3}{2} > \quad x \quad > -1$$ *Simplify.*

$$-1 < \quad x \quad < \frac{3}{2}$$ *Rearrange the ordering.*

Figure 11

The solution set is $\left\{ x \middle| -1 < x < \dfrac{3}{2} \right\}$, that is, all x in the interval $\left(-1, \dfrac{3}{2} \right)$. See Figure 11 for the graph of the solution set.

➤**Now Work** PROBLEM 39

<div style="border:1px solid">**EXAMPLE 5**</div> **Solving an Inequality Involving Absolute Value**

Solve the inequality: $|x| > 3$
Graph the solution set.

Solution

We are looking for all points whose coordinate x is a distance greater than 3 units from the origin. Figure 12 illustrates the situation. We conclude that any number x less than -3 or greater than 3 satisfies the condition $|x| > 3$. Consequently, the solution set consists of all numbers x for which $x < -3$ or $x > 3$, that is, all x in $(-\infty, -3) \cup (3, \infty)$.*

Figure 12

THEOREM

If a is a positive number and u is an algebraic expression, then

$	u	> a$ is equivalent to $u < -a$ or $u > a$	**(4)**
$	u	\geq a$ is equivalent to $u \leq -a$ or $u \geq a$	**(5)**

Figure 13

See Figure 13 for an illustration of statement (5).

<div style="border:1px solid">**EXAMPLE 6**</div> **Solving an Inequality Involving Absolute Value**

Solve the inequality: $|2x - 5| > 3$
Graph the solution set.

Solution

$$|2x - 5| > 3$$ This follows the form of statement (4); the expression $u = 2x - 5$ is inside the absolute value bars.

$$2x - 5 < -3 \quad \text{or} \quad 2x - 5 > 3$$ Apply statement (4).

$$2x - 5 + 5 < -3 + 5 \quad \text{or} \quad 2x - 5 + 5 > 3 + 5$$ Add 5 to each part.

$$2x < 2 \quad \text{or} \quad 2x > 8$$ Simplify.

$$\frac{2x}{2} < \frac{2}{2} \quad \text{or} \quad \frac{2x}{2} > \frac{8}{2}$$ Divide each part by 2.

$$x < 1 \quad \text{or} \quad x > 4$$ Simplify.

Figure 14

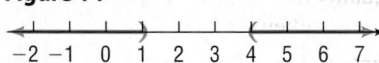

The solution set is $\{x \mid x < 1 \text{ or } x > 4\}$, that is, all x in $(-\infty, 1) \cup (4, \infty)$. See Figure 14 for the graph of the solution set.

WARNING A common error to be avoided is to attempt to write the solution $x < 1$ or $x > 4$ as the combined inequality $1 > x > 4$, which is incorrect, since there are no numbers x for which $1 > x$ and $x > 4$. ∎

Now Work PROBLEM 43

*Recall that the symbol \cup stands for the union of two sets. Refer to page 2 if necessary.

1.6 Assess Your Understanding

'Are You Prepared?' *Answers are given at the end of these exercises. If you get a wrong answer, read the pages listed in red.*

1. $|-2| =$ _____ (pp. 17–26)

2. *True or False* $|x| \geq 0$ for any real number x. (pp. 17–26)

Concepts and Vocabulary

3. The solution set of the equation $|x| = 5$ is $\{$ _____ $\}$.

4. The solution set of the inequality $|x| < 5$ is $\{x \mid$ _____ $\}$.

5. *True or False* The equation $|x| = -2$ has no solution.

6. *True or False* The inequality $|x| \geq -2$ has the set of real numbers as its solution set.

Skill Building

In Problems 7–34, solve each equation.

7. $|2x| = 6$

8. $|3x| = 12$

9. $|2x + 3| = 5$

10. $|3x - 1| = 2$

11. $|1 - 4t| + 8 = 13$

12. $|1 - 2z| + 6 = 9$

13. $|-2x| = |8|$

14. $|-x| = |1|$

15. $|-2|x = 4$

16. $|3|x = 9$

17. $\dfrac{2}{3}|x| = 9$

18. $\dfrac{3}{4}|x| = 9$

19. $\left|\dfrac{x}{3} + \dfrac{2}{5}\right| = 2$

20. $\left|\dfrac{x}{2} - \dfrac{1}{3}\right| = 1$

21. $|u - 2| = -\dfrac{1}{2}$

22. $|2 - v| = -1$

23. $4 - |2x| = 3$

24. $5 - \left|\dfrac{1}{2}x\right| = 3$

25. $|x^2 - 9| = 0$

26. $|x^2 - 16| = 0$

27. $|x^2 - 2x| = 3$

28. $|x^2 + x| = 12$

29. $|x^2 + x - 1| = 1$

30. $|x^2 + 3x - 2| = 2$

31. $\left|\dfrac{3x - 2}{2x - 3}\right| = 2$

32. $\left|\dfrac{2x + 1}{3x + 4}\right| = 1$

33. $|x^2 + 3x| = |x^2 - 2x|$

34. $|x^2 - 2x| = |x^2 + 6x|$

In Problems 35–62, solve each inequality. Express your answer using set notation or interval notation. Graph the solution set.

35. $|2x| < 8$

36. $|3x| < 15$

37. $|3x| > 12$

38. $|2x| > 6$

39. $|x - 2| + 2 < 3$

40. $|x + 4| + 3 < 5$

41. $|3t - 2| \le 4$

42. $|2u + 5| \le 7$

43. $|2x - 3| \ge 2$

44. $|3x + 4| \ge 2$

45. $|1 - 4x| - 7 < -2$

46. $|1 - 2x| - 4 < -1$

47. $|1 - 2x| > 3$

48. $|2 - 3x| > 1$

49. $|-4x| + |-5| \le 1$

50. $|-x| - |4| \le 2$

51. $|-2x| > |-3|$

52. $|-x - 2| \ge 1$

53. $-|2x - 1| \ge -3$

54. $-|1 - 2x| \ge -3$

55. $|2x| < -1$

56. $|3x| \ge 0$

57. $|5x| \ge -1$

58. $|6x| < -2$

59. $\left|\dfrac{2x + 3}{3} - \dfrac{1}{2}\right| < 1$

60. $3 - |x + 1| < \dfrac{1}{2}$

61. $5 + |x - 1| > \dfrac{1}{2}$

62. $\left|\dfrac{2x - 3}{2} + \dfrac{1}{3}\right| > 1$

Applications and Extensions

63. Body Temperature "Normal" human body temperature is 98.6°F. If a temperature x that differs from normal by at least 1.5° is considered unhealthy, write the condition for an unhealthy temperature x as an inequality involving an absolute value, and solve for x.

64. Household Voltage In the United States, normal household voltage is 110 volts. However, it is not uncommon for actual voltage to differ from normal voltage by at most 5 volts. Express this situation as an inequality involving an absolute value. Use x as the actual voltage and solve for x.

65. Reading Books A Gallup poll conducted May 20–22, 2005, found that Americans read an average of 13.4 books per year. Gallup is 99% confident that the result from this poll is off by fewer than 1.35 books from the actual average x. Express this situation as an inequality involving absolute

value, and solve the inequality for x to determine the interval in which the actual average is likely to fall.

Note: In statistics, this interval is called a 99% **confidence interval.**

66. Speed of Sound According to data from the Hill Aerospace Museum (Hill Air Force Base, Utah), the speed of sound varies depending on altitude, barometric pressure, and temperature. For example, at 20,000 feet, 13.75 inches of mercury, and –12.3°F, the speed of sound is about 707 miles per hour, but the speed can vary from this result by as much as 55 miles per hour as conditions change.

(a) Express this situation as an inequality involving an absolute value.

(b) Using x for the speed of sound, solve for x to find an interval for the speed of sound.

67. Express the fact that x differs from 3 by less than $\dfrac{1}{2}$ as an inequality involving an absolute value. Solve for x.

68. Express the fact that x differs from -4 by less than 1 as an inequality involving an absolute value. Solve for x.

69. Express the fact that x differs from -3 by more than 2 as an inequality involving an absolute value. Solve for x.

70. Express the fact that x differs from 2 by more than 3 as an inequality involving an absolute value. Solve for x.

In Problems 71–76, find a and b.

71. If $|x - 1| < 3$, then $a < x + 4 < b$.

72. If $|x + 2| < 5$, then $a < x - 2 < b$.

73. If $|x + 4| \leq 2$, then $a \leq 2x - 3 \leq b$.

74. If $|x - 3| \leq 1$, then $a \leq 3x + 1 \leq b$.

75. If $|x - 2| \leq 7$, then $a \leq \dfrac{1}{x - 10} \leq b$.

76. If $|x + 1| \leq 3$, then $a \leq \dfrac{1}{x + 5} \leq b$.

77. Show that: if $a > 0, b > 0$, and $\sqrt{a} < \sqrt{b}$, then $a < b$.

 [**Hint:** $b - a = (\sqrt{b} - \sqrt{a})(\sqrt{b} + \sqrt{a})$.]

78. Show that $a \leq |a|$.

79. Prove the *triangle inequality* $|a + b| \leq |a| + |b|$.

 [**Hint:** Expand $|a + b|^2 = (a + b)^2$, and use the result of Problem 78.]

80. Prove that $|a - b| \geq |a| - |b|$.

 [**Hint:** Apply the triangle inequality from Problem 79 to $|a| = |(a - b) + b|$.]

81. If $a > 0$, show that the solution set of the inequality

$$x^2 < a$$

consists of all numbers x for which

$$-\sqrt{a} < x < \sqrt{a}$$

82. If $a > 0$, show that the solution set of the inequality

$$x^2 > a$$

consists of all numbers x for which

$$x < -\sqrt{a} \quad \text{or} \quad x > \sqrt{a}$$

In Problems 83–90, use the results found in Problems 81 and 82 to solve each inequality.

83. $x^2 < 1$

84. $x^2 < 4$

85. $x^2 \geq 9$

86. $x^2 \geq 1$

87. $x^2 \leq 16$

88. $x^2 \leq 9$

89. $x^2 > 4$

90. $x^2 > 16$

91. Solve $|3x - |2x + 1|| = 4$.

92. Solve $|x + |3x - 2|| = 2$.

Explaining Concepts: Discussion and Writing

93. The equation $|x| = -2$ has no solution. Explain why.

94. The inequality $|x| > -0.5$ has all real numbers as solutions. Explain why.

95. The inequality $|x| > 0$ has as solution set $\{x | x \neq 0\}$. Explain why.

'Are You Prepared?' Answers

1. 2 **2.** True

1.7 Problem Solving: Interest, Mixture, Uniform Motion, Constant Rate Job Applications

OBJECTIVES **1** Translate Verbal Descriptions into Mathematical Expressions (p. 135)

 2 Solve Interest Problems (p. 136)

 3 Solve Mixture Problems (p. 137)

 4 Solve Uniform Motion Problems (p. 138)

 5 Solve Constant Rate Job Problems (p. 140)

Applied (word) problems do not come in the form "Solve the equation...." Instead, they supply information using words, a verbal description of the real problem. So, to solve applied problems, we must be able to translate the verbal description into the language of mathematics. We do this by using variables to represent unknown quantities and then finding relationships (such as equations) that involve these variables. The process of doing all this is called **mathematical modeling.**

Any solution to the mathematical problem must be checked against the mathematical problem, the verbal description, and the real problem. See Figure 15 for an illustration of the **modeling process.**

Figure 15

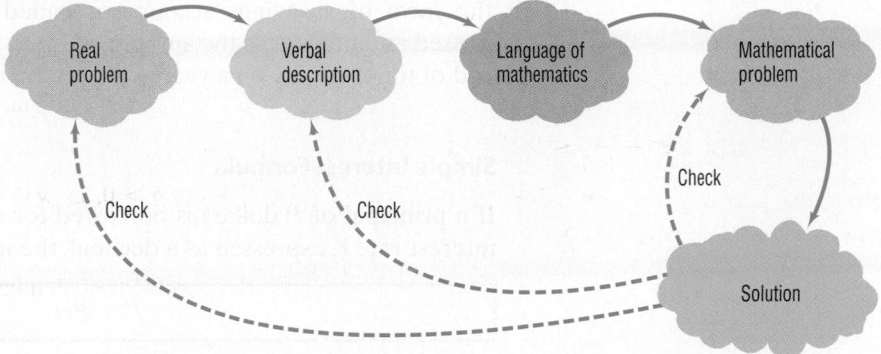

1 Translate Verbal Descriptions into Mathematical Expressions

EXAMPLE 1 **Translating Verbal Descriptions into Mathematical Expressions**

(a) For uniform motion, the constant speed of an object equals the distance traveled divided by the time required.

Translation: If r is the speed, d the distance, and t the time, then $r = \dfrac{d}{t}$.

(b) Let x denote a number.

The number 5 times as large as x is $5x$.

The number 3 less than x is $x - 3$.

The number that exceeds x by 4 is $x + 4$.

The number that, when added to x, gives 5 is $5 - x$.

Now Work PROBLEM 7

Always check the units used to measure the variables of an applied problem. In Example 1(a), if r is measured in miles per hour, then the distance d must be expressed in miles and the time t must be expressed in hours. It is a good practice to check units to be sure that they are consistent and make sense.

The steps to follow for solving applied problems, given earlier, are repeated next:

Steps for Solving Applied Problems

STEP 1: Read the problem carefully, perhaps two or three times. Pay particular attention to the question being asked in order to identify what you are looking for. If you can, determine realistic possibilities for the answer.

STEP 2: Assign a letter (variable) to represent what you are looking for, and, if necessary, express any remaining unknown quantities in terms of this variable.

STEP 3: Make a list of all the known facts, and translate them into mathematical expressions. These may take the form of an equation or an inequality involving the variable. If possible, draw an appropriately labeled diagram to assist you. Sometimes a table or chart helps.

STEP 4: Solve the equation for the variable, and then answer the question.

STEP 5: Check the answer with the facts in the problem. If it agrees, congratulations! If it does not agree, try again.

2 Solve Interest Problems

Interest is money paid for the use of money. The total amount borrowed (whether by an individual from a bank in the form of a loan or by a bank from an individual in the form of a savings account) is called the **principal.** The **rate of interest,** expressed as a percent, is the amount charged for the use of the principal for a given period of time, usually on a yearly (that is, on a per annum) basis.

Simple Interest Formula

If a principal of P dollars is borrowed for a period of t years at a per annum interest rate r, expressed as a decimal, the interest I charged is

$$I = Prt \tag{1}$$

Interest charged according to formula (1) is called **simple interest.** When using formula (1), be sure to express r as a decimal.

EXAMPLE 2 **Finance: Computing Interest on a Loan**

Suppose that Juanita borrows $500 for 6 months at the simple interest rate of 9% per annum. What is the interest that Juanita will be charged on the loan? How much does Juanita owe after 6 months?

Solution The rate of interest is given per annum, so the actual time that the money is borrowed must be expressed in years. The interest charged would be the principal, $500, times the rate of interest $(9\% = 0.09)$ times the time in years, $\frac{1}{2}$:

$$\text{Interest charged} = I = Prt = (500)(0.09)\left(\frac{1}{2}\right) = \$22.50$$

After 6 months, Juanita will owe what she borrowed plus the interest:

$$\$500 + \$22.50 = \$522.50$$

EXAMPLE 3 **Financial Planning**

Candy has $70,000 to invest and wants an annual return of $2800, which requires an overall rate of return of 4%. She can invest in a safe, government-insured certificate of deposit, but it only pays 2%. To obtain 4%, she agrees to invest some of her money in noninsured corporate bonds paying 7%. How much should be placed in each investment to achieve her goal?

Solution **STEP 1:** The question is asking for two dollar amounts: the principal to invest in the corporate bonds and the principal to invest in the certificate of deposit.

STEP 2: We let x represent the amount (in dollars) to be invested in the bonds. Then $70,000 - x$ is the amount that will be invested in the certificate. (Do you see why?)

STEP 3: We set up a table:

	Principal ($)	Rate	Time (yr)	Interest ($)
Bonds	x	7% = 0.07	1	0.07x
Certificate	70,000 − x	2% = 0.02	1	0.02(70,000 − x)
Total	70,000	4% = 0.04	1	0.04(70,000) = 2800

Since the total interest from the investments is equal to $0.04(70,000) = 2800$, we have the equation

$$0.07x + 0.02(70,000 - x) = 2800$$

(Note that the units are consistent: the unit is dollars on each side.)

STEP 4: $0.07x + 1400 - 0.02x = 2800$
$$0.05x = 1400$$
$$x = 28,000$$

Candy should place \$28,000 in the bonds and $70,000 - \$28,000 = \$42,000$ in the certificate.

STEP 5: The interest on the bonds after 1 year is $0.07(\$28,000) = \1960; the interest on the certificate after 1 year is $0.02(\$42,000) = \840. The total annual interest is \$2800, the required amount. ↵

Now Work PROBLEM 17

3 Solve Mixture Problems

Oil refineries sometimes produce gasoline that is a blend of two or more types of fuel; bakeries occasionally blend two or more types of flour for their bread. These problems are referred to as **mixture problems** because they combine two or more quantities to form a mixture.

EXAMPLE 4 **Blending Coffees**

The manager of a Starbucks store decides to experiment with a new blend of coffee. She will mix some B grade Colombian coffee that sells for \$5 per pound with some A grade Arabica coffee that sells for \$10 per pound to get 100 pounds of the new blend. The selling price of the new blend is to be \$7 per pound, and there is to be no difference in revenue from selling the new blend versus selling the other types. How many pounds of the B grade Colombian and A grade Arabica coffees are required?

Solution Let x represent the number of pounds of the B grade Colombian coffee. Then $100 - x$ equals the number of pounds of the A grade Arabica coffee. See Figure 16.

Figure 16

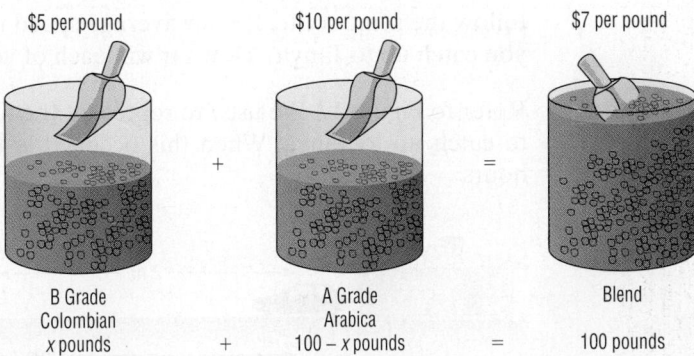

$5 per pound $10 per pound $7 per pound

B Grade A Grade Blend
Colombian Arabica 100 pounds
x pounds + $100 - x$ pounds =

Since there is to be no difference in revenue between selling the A and B grades separately versus the blend, we have

$$\left\{\begin{matrix}\text{Price per pound}\\ \text{of B grade}\end{matrix}\right\}\left\{\begin{matrix}\text{\# Pounds}\\ \text{B grade}\end{matrix}\right\} + \left\{\begin{matrix}\text{Price per pound}\\ \text{of A grade}\end{matrix}\right\}\left\{\begin{matrix}\text{\# Pounds}\\ \text{A grade}\end{matrix}\right\} = \left\{\begin{matrix}\text{Price per pound}\\ \text{of blend}\end{matrix}\right\}\left\{\begin{matrix}\text{\# Pounds}\\ \text{blend}\end{matrix}\right\}$$

$$\$5 \quad\cdot\quad x \quad + \quad \$10 \quad\cdot (100 - x) \quad = \quad \$7 \quad\cdot\quad 100$$

We have the equation

$$5x + 10(100 - x) = 700$$
$$5x + 1000 - 10x = 700$$
$$-5x = -300$$
$$x = 60$$

The manager should blend 60 pounds of B grade Colombian coffee with $100 - 60 = 40$ pounds of A grade Arabica coffee to get the desired blend.

✓**Check:** The 60 pounds of B grade coffee would sell for $(\$5)(60) = \300, and the 40 pounds of A grade coffee would sell for $(\$10)(40) = \400; the total revenue, \$700, equals the revenue obtained from selling the blend, as desired.

➤ **Now Work** PROBLEM 21

4 Solve Uniform Motion Problems

Objects that move at a constant speed are said to be in **uniform motion.** When the average speed of an object is known, it can be interpreted as its constant speed. For example, a bicyclist traveling at an average speed of 25 miles per hour is in uniform motion.

Uniform Motion Formula

If an object moves at an average speed (rate) r, the distance d covered in time t is given by the formula

$$d = rt \tag{2}$$

That is, Distance = Rate · Time.

EXAMPLE 5

Physics: Uniform Motion

Tanya, who is a long-distance runner, runs at an average speed of 8 miles per hour (mi/hr). Two hours after Tanya leaves your house, you leave in your Honda and follow the same route. If your average speed is 40 mi/hr, how long will it be before you catch up to Tanya? How far will each of you be from your home?

Solution

Refer to Figure 17. We use t to represent the time (in hours) that it takes the Honda to catch up to Tanya. When this occurs, the total time elapsed for Tanya is $t + 2$ hours.

Figure 17

Set up the following table:

	Rate mi/hr	Time hr	Distance mi
Tanya	8	$t + 2$	$8(t + 2)$
Honda	40	t	$40t$

Since the distance traveled is the same, we are led to the following equation:

$$8(t + 2) = 40t$$
$$8t + 16 = 40t$$
$$32t = 16$$
$$t = \frac{1}{2} \text{ hour}$$

It will take the Honda $\frac{1}{2}$ hour to catch up to Tanya. Each will have gone 20 miles.

✓**Check:** In 2.5 hours, Tanya travels a distance of $(2.5)(8) = 20$ miles. In $\frac{1}{2}$ hour, the Honda travels a distance of $\left(\frac{1}{2}\right)(40) = 20$ miles. ↵

| EXAMPLE 6 | **Physics: Uniform Motion** |

A motorboat heads upstream a distance of 24 miles on a river whose current is running at 3 miles per hour (mi/hr). The trip up and back takes 6 hours. Assuming that the motorboat maintained a constant speed relative to the water, what was its speed?

Solution See Figure 18. We use r to represent the constant speed of the motorboat relative to the water. Then the true speed going upstream is $r - 3$ mi/hr, and the true speed going downstream is $r + 3$ mi/hr. Since Distance = Rate × Time, then Time = $\dfrac{\text{Distance}}{\text{Rate}}$. Set up a table.

Figure 18

24 miles

r − 3 mi/hr

r + 3 mi/hr

	Rate mi/hr	Distance mi	Time = $\dfrac{\text{Distance}}{\text{Rate}}$ hr
Upstream	$r - 3$	24	$\dfrac{24}{r - 3}$
Downstream	$r + 3$	24	$\dfrac{24}{r + 3}$

Since the total time up and back is 6 hours, we have

$$\frac{24}{r - 3} + \frac{24}{r + 3} = 6$$

$$\frac{24(r + 3) + 24(r - 3)}{(r - 3)(r + 3)} = 6 \qquad \text{Add the quotients on the left.}$$

$$\frac{48r}{r^2 - 9} = 6 \qquad \text{Simplify.}$$

$$48r = 6(r^2 - 9) \qquad \text{Multiply both sides by } r^2 - 9.$$

$$6r^2 - 48r - 54 = 0 \qquad \text{Place in standard form.}$$

$$r^2 - 8r - 9 = 0 \qquad \text{Divide by 6.}$$

$$(r - 9)(r + 1) = 0 \qquad \text{Factor.}$$

$$r = 9 \quad \text{or} \quad r = -1 \qquad \text{Apply the Zero-Product Property and solve.}$$

We discard the solution $r = -1$ mi/hr, so the speed of the motorboat relative to the water is 9 mi/hr. ↵

━━ **Now Work** PROBLEM 27

5 Solve Constant Rate Job Problems

This section involves jobs that are performed at a **constant rate.** Our assumption is that, if a job can be done in t units of time, then $\dfrac{1}{t}$ of the job is done in 1 unit of time.

EXAMPLE 7 **Working Together to Do a Job**

At 10 AM Danny is asked by his father to weed the garden. From past experience, Danny knows that this will take him 4 hours, working alone. His older brother, Mike, when it is his turn to do this job, requires 6 hours. Since Mike wants to go golfing with Danny and has a reservation for 1 PM, he agrees to help Danny. Assuming no gain or loss of efficiency, when will they finish if they work together? Can they make the golf date?

Solution Set up Table 2. In 1 hour, Danny does $\dfrac{1}{4}$ of the job, and in 1 hour, Mike does $\dfrac{1}{6}$ of the job. Let t be the time (in hours) that it takes them to do the job together. In 1 hour, then, $\dfrac{1}{t}$ of the job is completed. We reason as follows:

$$\left(\begin{array}{c}\text{Part done by Danny}\\\text{in 1 hour}\end{array}\right) + \left(\begin{array}{c}\text{Part done by Mike}\\\text{in 1 hour}\end{array}\right) = \left(\begin{array}{c}\text{Part done together}\\\text{in 1 hour}\end{array}\right)$$

From Table 2,

$$\frac{1}{4} + \frac{1}{6} = \frac{1}{t}$$

$$\frac{3}{12} + \frac{2}{12} = \frac{1}{t}$$

$$\frac{5}{12} = \frac{1}{t}$$

$$5t = 12$$

$$t = \frac{12}{5}$$

Table 2

	Hours to Do Job	Part of Job Done in 1 Hour
Danny	4	$\dfrac{1}{4}$
Mike	6	$\dfrac{1}{6}$
Together	t	$\dfrac{1}{t}$

Working together, the job can be done in $\dfrac{12}{5}$ hours, or 2 hours, 24 minutes. They should make the golf date, since they will finish at 12:24 PM. ◢

Now Work PROBLEM 33

1.7 Assess Your Understanding

Concepts and Vocabulary

1. The process of using variables to represent unknown quantities and then finding relationships that involve these variables is referred to as _____ _____.

2. The money paid for the use of money is _____.

3. Objects that move at a constant speed are said to be in _____ _____.

4. *True or False* The amount charged for the use of principal for a given period of time is called the rate of interest.

5. *True or False* If an object moves at an average speed r, the distance d covered in time t is given by the formula $d = rt$.

6. Suppose that you want to mix two coffees in order to obtain 100 pounds of a blend. If x represents the number of pounds of coffee A, write an algebraic expression that represents the number of pounds of coffee B.

Applications and Extensions

In Problems 7–16, translate each sentence into a mathematical equation. Be sure to identify the meaning of all symbols.

7. **Geometry** The area of a circle is the product of the number π and the square of the radius.

8. **Geometry** The circumference of a circle is the product of the number π and twice the radius.

9. **Geometry** The area of a square is the square of the length of a side.

10. **Geometry** The perimeter of a square is four times the length of a side.

11. **Physics** Force equals the product of mass and acceleration.

12. **Physics** Pressure is force per unit area.

13. **Physics** Work equals force times distance.

14. **Physics** Kinetic energy is one-half the product of the mass and the square of the velocity.

15. **Business** The total variable cost of manufacturing x dishwashers is $150 per dishwasher times the number of dishwashers manufactured.

16. **Business** The total revenue derived from selling x dishwashers is $250 per dishwasher times the number of dishwashers sold.

17. **Financial Planning** Betsy, a recent retiree, requires $6000 per year in extra income. She has $50,000 to invest and can invest in B-rated bonds paying 15% per year or in a certificate of deposit (CD) paying 7% per year. How much money should be invested in each to realize exactly $6000 in interest per year?

18. **Financial Planning** After 2 years, Betsy (see Problem 17) finds that she will now require $7000 per year. Assuming that the remaining information is the same, how should the money be reinvested?

19. **Banking** A bank loaned out $12,000, part of it at the rate of 8% per year and the rest at the rate of 18% per year. If the interest received totaled $1000, how much was loaned at 8%?

20. **Banking** Wendy, a loan officer at a bank, has $1,000,000 to lend and is required to obtain an average return of 18% per year. If she can lend at the rate of 19% or at the rate of 16%, how much can she lend at the 16% rate and still meet her requirement?

21. **Blending Teas** The manager of a store that specializes in selling tea decides to experiment with a new blend. She will mix some Earl Grey tea that sells for $5 per pound with some Orange Pekoe tea that sells for $3 per pound to get 100 pounds of the new blend. The selling price of the new blend is to be $4.50 per pound, and there is to be no difference in revenue from selling the new blend versus selling the other types. How many pounds of the Earl Grey tea and Orange Pekoe tea are required?

22. **Business: Blending Coffee** A coffee manufacturer wants to market a new blend of coffee that sells for $3.90 per pound by mixing two coffees that sell for $2.75 and $5 per pound, respectively. What amounts of each coffee should be blended to obtain the desired mixture?

 [**Hint:** Assume that the total weight of the desired blend is 100 pounds.]

23. **Business: Mixing Nuts** A nut store normally sells cashews for $9.00 per pound and almonds for $3.50 per pound. But at the end of the month the almonds had not sold well, so, in order to sell 60 pounds of almonds, the manager decided to mix the 60 pounds of almonds with some cashews and sell the mixture for $7.50 per pound. How many pounds of cashews should be mixed with the almonds to ensure no change in the profit?

24. **Business: Mixing Candy** A candy store sells boxes of candy containing caramels and cremes. Each box sells for $12.50 and holds 30 pieces of candy (all pieces are the same size). If the caramels cost $0.25 to produce and the cremes cost $0.45 to produce, how many of each should be in a box to make a profit of $3?

25. **Physics: Uniform Motion** A motorboat can maintain a constant speed of 16 miles per hour relative to the water. The boat makes a trip upstream to a certain point in 20 minutes; the return trip takes 15 minutes. What is the speed of the current? See the figure.

26. **Physics: Uniform Motion** A motorboat heads upstream on a river that has a current of 3 miles per hour. The trip upstream takes 5 hours, and the return trip takes 2.5 hours. What is the speed of the motorboat? (Assume that the motorboat maintains a constant speed relative to the water.)

27. **Physics: Uniform Motion** A motorboat maintained a constant speed of 15 miles per hour relative to the water in going 10 miles upstream and then returning. The total time for the trip was 1.5 hours. Use this information to find the speed of the current.

28. **Physics: Uniform Motion** Two cars enter the Florida Turnpike at Commercial Boulevard at 8:00 AM, each heading for Wildwood. One car's average speed is 10 miles per hour more than the other's. The faster car arrives at Wildwood at 11:00 AM, $\frac{1}{2}$ hour before the other car. What was the average speed of each car? How far did each travel?

29. **Moving Walkways** The speed of a moving walkway is typically about 2.5 feet per second. Walking on such a moving walkway, it takes Karen a total of 40 seconds to travel 50 feet with the movement of the walkway and then back again against the movement of the walkway. What is Karen's normal walking speed?

 Source: Answers.com

30. **Moving Walkways** The Gare Montparnasse train station in Paris has a high-speed version of a moving walkway. If he walks while riding this moving walkway, Jean Claude can travel 200 meters in 30 seconds less time than if he stands still on the moving walkway. If Jean Claude walks at a normal rate of 1.5 meters per second, what is the speed of the Gare Montparnasse walkway?

Source: Answers.com

31. **Tennis** A regulation doubles tennis court has an area of 2808 square feet. If it is 6 feet longer than twice its width, determine the dimensions of the court.

Source: United States Tennis Association

32. **Laser Printers** It takes an HP LaserJet 1300 laser printer 10 minutes longer to complete a 600-page print job by itself than it takes an HP LaserJet 2420 to complete the same job by itself. Together the two printers can complete the job in 12 minutes. How long does it take each printer to complete the print job alone? What is the speed of each printer?

Source: Hewlett-Packard

33. **Working Together on a Job** Trent can deliver his newspapers in 30 minutes. It takes Lois 20 minutes to do the same route. How long would it take them to deliver the newspapers if they work together?

34. **Working Together on a Job** Patrice, by himself, can paint four rooms in 10 hours. If he hires April to help, they can do the same job together in 6 hours. If he lets April work alone, how long will it take her to paint four rooms?

35. **Enclosing a Garden** A gardener has 46 feet of fencing to be used to enclose a rectangular garden that has a border 2 feet wide surrounding it. See the figure.
 (a) If the length of the garden is to be twice its width, what will be the dimensions of the garden?
 (b) What is the area of the garden?
 (c) If the length and width of the garden are to be the same, what would be the dimensions of the garden?
 (d) What would be the area of the square garden?

36. **Construction** A pond is enclosed by a wooden deck that is 3 feet wide. The fence surrounding the deck is 100 feet long.
 (a) If the pond is square, what are its dimensions?
 (b) If the pond is rectangular and the length of the pond is to be three times its width, what are its dimensions?
 (c) If the pond is circular, what is its diameter?
 (d) Which pond has the most area?

37. **Football** A tight end can run the 100-yard dash in 12 seconds. A defensive back can do it in 10 seconds. The tight end catches a pass at his own 20-yard line with the defensive back at the 15-yard line. (See the figure.) If no other players are nearby, at what yard line will the defensive back catch up to the tight end?

 [**Hint:** At time $t = 0$, the defensive back is 5 yards behind the tight end.]

38. **Computing Business Expense** Therese, an outside salesperson, uses her car for both business and pleasure. Last year, she traveled 30,000 miles, using 900 gallons of gasoline. Her car gets 40 miles per gallon on the highway and 25 in the city. She can deduct all highway travel, but no city travel, on her taxes. How many miles should Therese be allowed as a business expense?

39. **Mixing Water and Antifreeze** How much water should be added to 1 gallon of pure antifreeze to obtain a solution that is 60% antifreeze?

40. **Mixing Water and Antifreeze** The cooling system of a certain foreign-made car has a capacity of 15 liters. If the system is filled with a mixture that is 40% antifreeze, how much of this mixture should be drained and replaced by pure antifreeze so that the system is filled with a solution that is 60% antifreeze?

41. **Chemistry: Salt Solutions** How much water must be evaporated from 32 ounces of a 4% salt solution to make a 6% salt solution?

42. **Chemistry: Salt Solutions** How much water must be evaporated from 240 gallons of a 3% salt solution to produce a 5% salt solution?

43. **Purity of Gold** The purity of gold is measured in karats, with pure gold being 24 karats. Other purities of gold are expressed as proportional parts of pure gold. Thus, 18-karat gold is $\frac{18}{24}$, or 75% pure gold; 12-karat gold is $\frac{12}{24}$, or 50% pure gold; and so on. How much 12-karat gold should be mixed with pure gold to obtain 60 grams of 16-karat gold?

44. **Chemistry: Sugar Molecules** A sugar molecule has twice as many atoms of hydrogen as it does oxygen and one more atom of carbon than oxygen. If a sugar molecule has a total of 45 atoms, how many are oxygen? How many are hydrogen?

45. **Running a Race** Mike can run the mile in 6 minutes, and Dan can run the mile in 9 minutes. If Mike gives Dan a head start of 1 minute, how far from the start will Mike pass Dan? How long does it take? See the figure.

46. Range of an Airplane An air rescue plane averages 300 miles per hour in still air. It carries enough fuel for 5 hours of flying time. If, upon takeoff, it encounters a head wind of 30 mi/hr, how far can it fly and return safely? (Assume that the wind remains constant.)

47. Emptying Oil Tankers An oil tanker can be emptied by the main pump in 4 hours. An auxiliary pump can empty the tanker in 9 hours. If the main pump is started at 9 AM, when should the auxiliary pump be started so that the tanker is emptied by noon?

48. Cement Mix A 20-pound bag of Economy brand cement mix contains 25% cement and 75% sand. How much pure cement must be added to produce a cement mix that is 40% cement?

49. Emptying a Tub A bathroom tub will fill in 15 minutes with both faucets open and the stopper in place. With both faucets closed and the stopper removed, the tub will empty in 20 minutes. How long will it take for the tub to fill if both faucets are open and the stopper is removed?

50. Using Two Pumps A 5-horsepower (hp) pump can empty a pool in 5 hours. A smaller, 2-hp pump empties the same pool in 8 hours. The pumps are used together to begin emptying this pool. After two hours, the 2-hp pump breaks down. How long will it take the larger pump to empty the pool?

51. A Biathlon Suppose that you have entered an 87-mile biathlon that consists of a run and a bicycle race. During your run, your average speed is 6 miles per hour, and during your bicycle race, your average speed is 25 miles per hour. You finish the race in 5 hours. What is the distance of the run? What is the distance of the bicycle race?

52. Cyclists Two cyclists leave a city at the same time, one going east and the other going west. The westbound cyclist bikes 5 mph faster than the eastbound cyclist. After 6 hours they are 246 miles apart. How fast is each cyclist riding?

53. Comparing Olympic Heroes In the 1984 Olympics, C. Lewis of the United States won the gold medal in the 100-meter race with a time of 9.99 seconds. In the 1896 Olympics, Thomas Burke, also of the United States, won the gold medal in the 100-meter race in 12.0 seconds. If they ran in the same race repeating their respective times, by how many meters would Lewis beat Burke?

54. Constructing a Coffee Can A 39-ounce can of Hills Bros.® coffee requires 188.5 square inches of aluminum. If its height is 7 inches, what is its radius? **[Hint:** The surface area S of a right cylinder is $S = 2\pi r^2 + 2\pi rh$, where r is the radius and h is the height.]

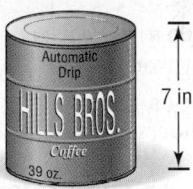

7 in.

Explaining Concepts: Discussion and Writing

55. Critical Thinking You are the manager of a clothing store and have just purchased 100 dress shirts for $20.00 each. After 1 month of selling the shirts at the regular price, you plan to have a sale giving 40% off the original selling price. However, you still want to make a profit of $4 on each shirt at the sale price. What should you price the shirts at initially to ensure this? If, instead of 40% off at the sale, you give 50% off, by how much is your profit reduced?

56. Critical Thinking Make up a word problem that requires solving a linear equation as part of its solution. Exchange problems with a friend. Write a critique of your friend's problem.

57. Critical Thinking Without solving, explain what is wrong with the following mixture problem: How many liters of 25% ethanol should be added to 20 liters of 48% ethanol to obtain a solution of 58% ethanol? Now go through an algebraic solution. What happens?

58. Computing Average Speed In going from Chicago to Atlanta, a car averages 45 miles per hour, and in going from Atlanta to Miami, it averages 55 miles per hour. If Atlanta is halfway between Chicago and Miami, what is the average speed from Chicago to Miami? Discuss an intuitive solution. Write a paragraph defending your intuitive solution. Then solve the problem algebraically. Is your intuitive solution the same as the algebraic one? If not, find the flaw.

59. Speed of a Plane On a recent flight from Phoenix to Kansas City, a distance of 919 nautical miles, the plane arrived 20 minutes early. On leaving the aircraft, I asked the captain, "What was our tail wind?" He replied, "I don't know, but our ground speed was 550 knots." How can you determine if enough information is provided to find the tail wind? If possible, find the tail wind. (1 knot = 1 nautical mile per hour)

CHAPTER REVIEW

Things to Know

Quadratic formula (pp. 97 and 110)

If $ax^2 + bx + c = 0$, $a \neq 0$, then $x = \dfrac{-b \pm \sqrt{b^2 - 4ac}}{2a}$

If $b^2 - 4ac < 0$, there are no real solutions.

Discriminant (pp. 97 and 110)

If $b^2 - 4ac > 0$, there are two distinct real solutions.

If $b^2 - 4ac = 0$, there is one repeated real solution.

If $b^2 - 4ac < 0$, there are no real solutions, but there are two distinct complex solutions that are not real; the complex solutions are conjugates of each other.

Interval notation (p. 120)

$[a, b]$	$\{x \mid a \leq x \leq b\}$	$[a, \infty)$	$\{x \mid x \geq a\}$
$[a, b)$	$\{x \mid a \leq x < b\}$	(a, ∞)	$\{x \mid x > a\}$
$(a, b]$	$\{x \mid a < x \leq b\}$	$(-\infty, a]$	$\{x \mid x \leq a\}$
(a, b)	$\{x \mid a < x < b\}$	$(-\infty, a)$	$\{x \mid x < a\}$
		$(-\infty, \infty)$	All real numbers

Properties of inequalities

Addition property (p. 121)

If $a < b$, then $a + c < b + c$

If $a > b$, then $a + c > b + c$

Multiplication properties (p. 122)

(a) If $a < b$ and if $c > 0$, then $ac < bc$ (b) If $a > b$ and if $c > 0$, then $ac > bc$

If $a < b$ and if $c < 0$, then $ac > bc$ If $a > b$ and if $c < 0$, then $ac < bc$

Reciprocal properties (p. 123)

If $a > 0$, then $\dfrac{1}{a} > 0$ If $a < 0$, then $\dfrac{1}{a} < 0$

If $\dfrac{1}{a} > 0$, then $a > 0$ If $\dfrac{1}{a} < 0$, then $a < 0$

Absolute value

If $|u| = a, a > 0$, then $u = -a$ or $u = a$ (p. 130)

If $|u| \leq a, a > 0$, then $-a \leq u \leq a$ (p. 131)

If $|u| \geq a, a > 0$, then $u \leq -a$ or $u \geq a$ (p. 132)

Objectives

Section	You should be able to . . .	Examples	Review Exercises
1.1	**1** Solve a linear equation (p. 84)	1–3	1–6, 11–12
	2 Solve equations that lead to linear equations (p. 86)	4–6	7, 8, 36
	3 Solve problems that can be modeled by linear equations (p. 87)	8	82, 101
1.2	**1** Solve a quadratic equation by factoring (p. 93)	1, 2	10, 13, 14, 33–35
	2 Solve a quadratic equation by completing the square (p. 95)	4, 5	9, 10, 13–16, 19, 20, 33–35
	3 Solve a quadratic equation using the quadratic formula (p. 96)	6–9	9, 10, 13–16, 19, 20, 33–35
	4 Solve problems that can be modeled by quadratic equations (p. 99)	10	84, 90, 95, 96, 100, 102
1.3	**1** Add, subtract, multiply, and divide complex numbers (p. 105)	1–5	61–70
	2 Solve quadratic equations in the complex number system (p. 109)	9–12	71–78
1.4	**1** Solve radical equations (p. 113)	1–3	17, 18, 23–30, 37, 38
	2 Solve equations quadratic in form (p. 114)	4–6	21, 22, 31, 32
	3 Solve equations by factoring (p. 116)	7, 8	43–46
1.5	**1** Use interval notation (p. 120)	1, 2	47–60
	2 Use properties of inequalities (p. 121)	3–6	47–60
	3 Solve inequalities (p. 123)	7, 8	47, 48
	4 Solve combined inequalities (p. 124)	9, 10	49–52
1.6	**1** Solve equations involving absolute value (p. 130)	1	39–42
	2 Solve inequalities involving absolute value (p. 130)	2–6	53–60
1.7	**1** Translate verbal descriptions into mathematical expressions (p. 135)	1	79, 80
	2 Solve interest problems (p. 136)	2, 3	81, 82
	3 Solve mixture problems (p. 137)	4	82, 93, 94, 97
	4 Solve uniform motion problems (p. 138)	5, 6	83, 85–89, 104, 105
	5 Solve constant rate job problems (p. 140)	7	91, 92, 99, 103

Review Exercises

In Problems 1–46, find the real solutions, if any, of each equation. (Where they appear, a, b, m, and n are positive constants.)

1. $2 - \dfrac{x}{3} = 8$

2. $\dfrac{x}{4} - 2 = 4$

3. $-2(5 - 3x) + 8 = 4 + 5x$

4. $(6 - 3x) - 2(1 + x) = 6x$

5. $\dfrac{3x}{4} - \dfrac{x}{3} = \dfrac{1}{12}$

6. $\dfrac{4 - 2x}{3} + \dfrac{1}{6} = 2x$

7. $\dfrac{x}{x - 1} = \dfrac{6}{5}$ $x \neq 1$

8. $\dfrac{4x - 5}{3 - 7x} = 2$ $x \neq \dfrac{3}{7}$

9. $x(1 - x) = 6$

10. $x(1 + x) = 6$

11. $\dfrac{1}{2}\left(x - \dfrac{1}{3}\right) = \dfrac{3}{4} - \dfrac{x}{6}$

12. $\dfrac{1 - 3x}{4} = \dfrac{x + 6}{3} + \dfrac{1}{2}$

13. $(x - 1)(2x + 3) = 3$

14. $x(2 - x) = 3(x - 4)$

15. $2x + 3 = 4x^2$

16. $1 + 6x = 4x^2$

17. $\sqrt[3]{x^2 - 1} = 2$

18. $\sqrt{1 + x^3} = 3$

19. $x(x + 1) + 2 = 0$

20. $3x^2 - x + 1 = 0$

21. $x^4 - 5x^2 + 4 = 0$

22. $3x^4 + 4x^2 + 1 = 0$

23. $\sqrt{2x - 3} + x = 3$

24. $\sqrt{2x - 1} = x - 2$

25. $\sqrt[4]{2x + 3} = 2$

26. $\sqrt[5]{3x + 1} = -1$

27. $\sqrt{x + 1} + \sqrt{x - 1} = \sqrt{2x + 1}$

28. $\sqrt{2x - 1} - \sqrt{x - 5} = 3$

29. $2x^{1/2} - 3 = 0$

30. $3x^{1/4} - 2 = 0$

31. $x^{-6} - 7x^{-3} - 8 = 0$

32. $6x^{-1} - 5x^{-1/2} + 1 = 0$

33. $x^2 + m^2 = 2mx + (nx)^2$ $n \neq 1$

34. $b^2x^2 + 2ax = x^2 + a^2$ $b \neq 1$

35. $10a^2x^2 - 2abx - 36b^2 = 0$

36. $\dfrac{1}{x - m} + \dfrac{1}{x - n} = \dfrac{2}{x}$ $x \neq 0, x \neq m, x \neq n$

37. $\sqrt{x^2 + 3x + 7} - \sqrt{x^2 - 3x + 9} + 2 = 0$

38. $\sqrt{x^2 + 3x + 7} - \sqrt{x^2 + 3x + 9} = 2$

39. $|2x + 3| = 7$

40. $|3x - 1| = 5$

41. $|2 - 3x| + 2 = 9$

42. $|1 - 2x| + 1 = 4$

43. $2x^3 = 3x^2$

44. $5x^4 = 9x^3$

45. $2x^3 + 5x^2 - 8x - 20 = 0$

46. $3x^3 + 5x^2 - 3x - 5 = 0$

In Problems 47–60, solve each inequality. Express your answer using set notation or interval notation. Graph the solution set.

47. $\dfrac{2x - 3}{5} + 2 \leq \dfrac{x}{2}$

48. $\dfrac{5 - x}{3} \leq 6x - 4$

49. $-9 \leq \dfrac{2x + 3}{-4} \leq 7$

50. $-4 < \dfrac{2x - 2}{3} < 6$

51. $2 < \dfrac{3 - 3x}{12} < 6$

52. $-3 \leq \dfrac{5 - 3x}{2} \leq 6$

53. $|3x + 4| < \dfrac{1}{2}$

54. $|1 - 2x| < \dfrac{1}{3}$

55. $|2x - 5| \geq 9$

56. $|3x + 1| \geq 10$

57. $2 + |2 - 3x| \leq 4$

58. $\dfrac{1}{2} + \left|\dfrac{2x - 1}{3}\right| \leq 1$

59. $1 - |2 - 3x| < -4$

60. $1 - \left|\dfrac{2x - 1}{3}\right| < -2$

In Problems 61–70, use the complex number system and write each expression in the standard form a + bi.

61. $(6 + 3i) - (2 - 4i)$

62. $(8 - 3i) + (-6 + 2i)$

63. $4(3 - i) + 3(-5 + 2i)$

64. $2(1 + i) - 3(2 - 3i)$

65. $\dfrac{3}{3 + i}$

66. $\dfrac{4}{2 - i}$

67. i^{50}

68. i^{29}

69. $(2 + 3i)^3$

70. $(3 - 2i)^3$

In Problems 71–78, solve each equation in the complex number system.

71. $x^2 + x + 1 = 0$ **72.** $x^2 - x + 1 = 0$ **73.** $2x^2 + x - 2 = 0$ **74.** $3x^2 - 2x - 1 = 0$

75. $x^2 + 3 = x$ **76.** $2x^2 + 1 = 2x$ **77.** $x(1 - x) = 6$ **78.** $x(1 + x) = 2$

90 mi/hr

5 mi/hr

150 mi

79. Translate the following statement into a mathematical expression: The perimeter p of a rectangle is the sum of two times the length l and two times the width w.

80. Translate the following statement into a mathematical expression: The total cost C of manufacturing x bicycles in one day is $50,000 plus $95 times the number of bicycles manufactured.

81. Banking A bank lends out $9000 at 7% simple interest. At the end of 1 year, how much interest is owed on the loan?

82. Financial Planning Steve, a recent retiree, requires $5000 per year in extra income. He has $70,000 to invest and can invest in A-rated bonds paying 8% per year or in a certificate of deposit (CD) paying 5% per year. How much money should be invested in each to realize exactly $5000 in interest per year?

83. Lightning and Thunder A flash of lightning is seen, and the resulting thunderclap is heard 3 seconds later. If the speed of sound averages 1100 feet per second, how far away is the storm?

84. Physics: Intensity of Light The intensity I (in candlepower) of a certain light source obeys the equation $I = \dfrac{900}{x^2}$, where x is the distance (in meters) from the light. Over what range of distances can an object be placed from this light source so that the range of intensity of light is from 1600 to 3600 candlepower, inclusive?

85. Extent of Search and Rescue A search plane has a cruising speed of 250 miles per hour and carries enough fuel for at most 5 hours of flying. If there is a wind that averages 30 miles per hour and the direction of the search is with the wind one way and against it the other, how far can the search plane travel before it has to turn back?

86. Extent of Search and Rescue If the search plane described in Problem 85 is able to add a supplementary fuel tank that allows for an additional 2 hours of flying, how much farther can the plane extend its search?

87. Rescue at Sea A life raft, set adrift from a sinking ship 150 miles offshore, travels directly toward a Coast Guard station at the rate of 5 miles per hour. At the time that the raft is set adrift, a rescue helicopter is dispatched from the Coast Guard station. If the helicopter's average speed is 90 miles per hour, how long will it take the helicopter to reach the life raft?

88. Physics: Uniform Motion Two bees leave two locations 150 meters apart and fly, without stopping, back and forth between these two locations at average speeds of 3 meters per second and 5 meters per second, respectively. How long is it until the bees meet for the first time? How long is it until they meet for the second time?

89. Physics: Uniform Motion A Metra commuter train leaves Union Station in Chicago at 12 noon. Two hours later, an Amtrak train leaves on the same track, traveling at an average speed that is 50 miles per hour faster than the Metra train. At 3 PM the Amtrak train is 10 miles behind the commuter train. How fast is each going?

90. Physics An object is thrown down from the top of a building 1280 feet tall with an initial velocity of 32 feet per second. The distance s (in feet) of the object from the ground after t seconds is $s = 1280 - 32t - 16t^2$.
(a) When will the object strike ground?
(b) What is the height of the object after 4 seconds?

91. Working Together to Get a Job Done Clarissa and Shawna, working together, can paint the exterior of a house in 6 days. Clarissa by herself can complete this job in 5 days less than Shawna. How long will it take Clarissa to complete the job by herself?

92. Emptying a Tank Two pumps of different sizes, working together, can empty a fuel tank in 5 hours. The larger pump can empty this tank in 4 hours less than the smaller one. If the larger pump is out of order, how long will it take the smaller one to do the job alone?

93. Chemistry: Salt Solutions How much water should be added to 64 ounces of a 10% salt solution to make a 2% salt solution?

94. Chemistry: Salt Solutions How much water must be evaporated from 64 ounces of a 2% salt solution to make a 10% salt solution?

95. Geometry The hypotenuse of a right triangle measures 13 centimeters. Find the lengths of the legs if their sum is 17 centimeters.

96. Geometry The diagonal of a rectangle measures 10 inches. If the length is 2 inches more than the width, find the dimensions of the rectangle.

97. Chemistry: Mixing Acids A laboratory has 60 cubic centimeters (cm^3) of a solution that is 40% HCl acid. How many cubic centimeters of a 15% solution of HCl acid should be mixed with the 60 cm^3 of 40% acid to obtain a solution of 25% HCl? How much of the 25% solution is there?

98. Framing a Painting An artist has 50 inches of oak trim to frame a painting. The frame is to have a border 3 inches wide surrounding the painting.
 (a) If the painting is square, what are its dimensions? What are the dimensions of the frame?
 (b) If the painting is rectangular with a length twice its width, what are the dimensions of the painting? What are the dimensions of the frame?

99. Using Two Pumps An 8-horsepower (hp) pump can fill a tank in 8 hours. A smaller, 3-hp pump fills the same tank in 12 hours. The pumps are used together to begin filling this tank. After four hours, the 8-hp pump breaks down. How long will it take the smaller pump to fill the tank?

100. Pleasing Proportion One formula stating the relationship between the length l and width w of a rectangle of "pleasing proportion" is $l^2 = w(l + w)$. How should a 4 foot by 8 foot sheet of plasterboard be cut so that the result is a rectangle of "pleasing proportion" with a width of 4 feet?

101. Finance An inheritance of $900,000 is to be divided among Scott, Alice, and Tricia in the following manner: Alice is to receive $\frac{3}{4}$ of what Scott gets, while Tricia gets $\frac{1}{2}$ of what Scott gets. How much does each receive?

102. Business: Determining the Cost of a Charter A group of 20 senior citizens can charter a bus for a one-day excursion trip for $15 per person. The charter company agrees to reduce the price of each ticket by 10¢ for each additional passenger in excess of 20 who goes on the trip, up to a maximum of 44 passengers (the capacity of the bus). If the final bill from the charter company was $482.40, how many seniors went on the trip, and how much did each pay?

103. Utilizing Copying Machines A new copying machine can do a certain job in 1 hour less than an older copier. Together they can do this job in 72 minutes. How long would it take the older copier by itself to do the job?

104. Evening Up a Race In a 100-meter race, Todd crosses the finish line 5 meters ahead of Scott. To even things up, Todd suggests to Scott that they race again, this time with Todd lining up 5 meters behind the start.
 (a) Assuming that Todd and Scott run at the same pace as before, does the second race end in a tie?
 (b) If not, who wins?
 (c) By how many meters does he win?
 (d) How far back should Todd start so that the race ends in a tie?

After running the race a second time, Scott, to even things up, suggests to Todd that he (Scott) line up 5 meters in front of the start.
 (e) Assuming again that they run at the same pace as in the first race, does the third race result in a tie?
 (f) If not, who wins?
 (g) By how many meters?
 (h) How far ahead should Scott start so that the race ends in a tie?

105. Physics: Uniform Motion A man is walking at an average speed of 4 miles per hour alongside a railroad track. A freight train, going in the same direction at an average speed of 30 miles per hour, requires 5 seconds to pass the man. How long is the freight train? Give your answer in feet.

CHAPTER TEST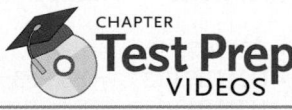

The Chapter Test Prep Videos are step-by-step test solutions available in the Video Resources DVD, in **MyMathLab**, or on this text's You**Tube** Channel. Flip back to the Student Resources page to see the exact web address for this text's YouTube channel.

In Problems 1–7, find the real solutions, if any, of each equation.

1. $\dfrac{2x}{3} - \dfrac{x}{2} = \dfrac{5}{12}$ **2.** $x(x - 1) = 6$ **3.** $x^4 - 3x^2 - 4 = 0$ **4.** $\sqrt{2x - 5} + 2 = 4$

5. $|2x - 3| + 7 = 10$ **6.** $3x^3 + 2x^2 - 12x - 8 = 0$ **7.** $3x^2 - x + 1 = 0$

In Problems 8–10, solve each inequality. Express your answer using interval notation. Graph the solution set.

8. $-3 \le \dfrac{3x - 4}{2} \le 6$ **9.** $|3x + 4| < 8$ **10.** $2 + |2x - 5| \ge 9$

11. Write $\dfrac{-2}{3 - i}$ in the standard form $a + bi$.

12. Solve the equation $4x^2 - 4x + 5 = 0$ in the complex number system.

13. Blending Coffee A coffee house has 20 pounds of a coffee that sells for $4 per pound. How many pounds of a coffee that sells for $8 per pound should be mixed with the 20 pounds of $4-per-pound coffee to obtain a blend that will sell for $5 per pound? How much of the $5-per-pound coffee is there to sell?

CHAPTER PROJECTS

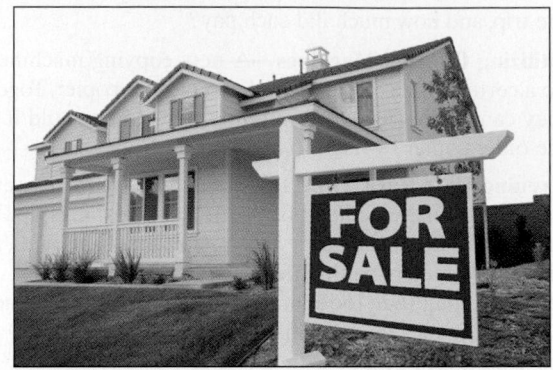

((•)) **Internet-based Project**

I. Financing a Purchase At some point in your life you are likely going to need to borrow money to finance a purchase. For example, most of us will finance the purchase of a car or a home. What is the mathematics behind financing a purchase? When we borrow money from a bank, the bank uses a rather complex equation (or formula) to determine how much you need to pay each month to repay the loan. There are a number of variables that determine the monthly payment. These variables include the amount borrowed, the interest rate, and the length of the loan. The interest rate is determined based on current economic conditions, the length of the loan, the type of item being purchased, and your credit history. To learn how banks judge your credit worthiness, read the article "How Credit Scores Work" at *http://money.howstuffworks.com /personal-finance/debt-management/credit-score.htm*

The formula below gives the monthly payment *P* required to pay off a loan amount *L* at an annual interest rate *r*,

expressed as a decimal, but usually given as a percent. The time *t*, measured in months, is the length of the loan. For example, a 30-year loan requires $12 \times 30 = 360$ monthly payments.

$$P = L\left[\frac{\dfrac{r}{12}}{1 - \left(1 + \dfrac{r}{12}\right)^{-t}}\right]$$

P = monthly payment
L = loan amount
r = annual rate of interest
 expressed as a decimal
t = length of loan, in months

1. Interest rates change daily. Many websites post current interest rates on loans. Go to *www.bankrate.com* (or some other website that posts lender's interest rates) and find the current best interest rate on a 48-month new car purchase loan. Use this rate to determine the monthly payment on a $20,000 automobile loan.

2. Determine the total amount paid for the loan by multiplying the loan payment by the term of the loan.

3. Determine the total amount of interest paid by subtracting the loan amount from the total amount paid from question 2.

4. More often than not, we decide how much of a payment we can afford and use that information to determine the loan amount. Suppose you can afford a monthly payment of $500. Use the interest rate from question 1 to determine the maximum amount you can borrow. If you have $5000 to put down on the car, what is the maximum value of a car you can purchase?

5. Repeat questions 1 through 4 using a 60-month new car purchase loan, a 48-month used car purchase loan, and a 60-month used car purchase loan.

6. We can use the power of a spreadsheet, such as Excel, to amortize the loan. A loan amortization schedule is a list of the monthly payments, a breakdown of interest and principal, along with a current loan balance. Create a loan amortization schedule for each of the four loan scenarios discussed above using the following as a guide. You may want to use an Internet search engine to research specific keystrokes for creating an amortization schedule in a spreadsheet. We supply a sample spreadsheet with formulas included as a guide. Use the spreadsheet to verify your results from questions 1 through 5.

Loan Information		Payment Number	Payment Amount	Interest	Principal	Balance	Total Interest Paid
Loan Amount	$20,000.00	1	=PMT(B3/12,B5,−B2,0)	=B2*B3/12	=E2−F2	=B2−G2	=B2*B3/12
Annual Interest Rate	0.05	2	=PMT(B3/12,B5,−B2,0)	=H2*B3/12	=E3−F3	=H2−G3	=I2+F3
Length of Loan (years)	4	3	=PMT(B3/12,B5,−B2,0)	=H3*B3/12	=E4−F4	=H3−G4	=I3+F4
Number of Payments	=B4*12	·	·	·	·	·	·
		·	·	·	·	·	·
		·	·	·	·	·	·

7. Go to an online automobile website such as *www.cars.com, www.vehix.com,* or *www.autobytel.com.* Research the types of vehicles you can afford for a monthly payment of $500. Decide on a vehicle you would purchase based on your analysis in questions 1–6. Be sure to justify your decision and include the impact the term of the loan has on your decision. You might consider other factors in your decision such as expected maintenance costs and insurance costs.

Citations:

Obringer, Lee Ann. "How Credit Scores Work." 16 July 2002. HowStuffWorks.com. <http://money.howstuffworks.com/personal-finance/ debt-management/credit-score.htm> 15 March 2010; Excel ©2010 Microsoft Corporation. Used with permission from Microsoft.

The following project is also available on the Instructor's Resource Center (IRC):

II. Project at Motorola How Many Cellular Phones Can I Make? An industrial engineer uses a model involving equations to be sure production levels meet customer demand.

Graphs

2

Outline

The First Modern Olympics: Athens, 1896

The birth of the modern Olympic Games

By John Gettings—"I hereby proclaim the opening of the first International Olympic Games at Athens." With these words on April 6, 1896, King George I of Greece welcomed the crowd that had gathered in the newly reconstructed Panathenean Stadium to the modern-day Olympic Summer Games.

The event was the idea of Baron Pierre de Coubertin of France who traveled the world to gather support for his dream to have nations come together and overcome national disputes, all in the name of sport.

The program for the Games included track and field, fencing, weightlifting, rifle and pistol shooting, tennis, cycling, swimming, gymnastics, and wrestling. Although 14 nations participated, most of the athletes were Greek.

The Games reached their high point on Day 5 with the first modern-day marathon. The idea to hold an event to commemorate the Ancient Olympic games was suggested by a friend of de Coubertin and was met with great anticipation. The race was run from Marathon to Athens (estimated at 22–26 miles), watched by more than 100,000 people, and won by a Greek runner, Spiridon Louis. *Gettings, The First Modern Olympics, Athens, 1986, © 2000–2010 Pearson Education, publishing as Infoplease. Reprinted with permission.*

—See the Internet-based Chapter Project—

◁ **A Look Back** In Chapter R we reviewed algebra essentials and geometry essentials and in Chapter 1 we studied equations in one variable.

A Look Ahead ▷ Here we connect algebra and geometry using the rectangular coordinate system to graph equations in two variables. The idea of using a system of rectangular coordinates dates back to ancient times, when such a system was used for surveying and city planning. Apollonius of Perga, in 200 BC, used a form of rectangular coordinates in his work on conics, although this use does not stand out as clearly as it does in modern treatments. Sporadic use of rectangular coordinates continued until the 1600s. By that time, algebra had developed sufficiently so that René Descartes (1596–1650) and Pierre de Fermat (1601–1665) could take the crucial step, which was the use of rectangular coordinates to translate geometry problems into algebra problems, and vice versa. This step was important for two reasons. First, it allowed both geometers and algebraists to gain new insights into their subjects, which previously had been regarded as separate, but now were seen to be connected in many important ways. Second, these insights made the development of calculus possible, which greatly enlarged the number of areas in which mathematics could be applied and made possible a much deeper understanding of these areas.

2.1 The Distance and Midpoint Formulas

PREPARING FOR THIS SECTION *Before getting started, review the following:*

- Algebra Essentials (Chapter R, Section R.2, pp. 17–26)
- Geometry Essentials (Chapter R, Section R.3, pp. 30–35)

Now Work the 'Are You Prepared?' problems on page 154.

OBJECTIVES **1** Use the Distance Formula (p. 151)
2 Use the Midpoint Formula (p. 153)

Figure 1

Figure 2

Figure 3

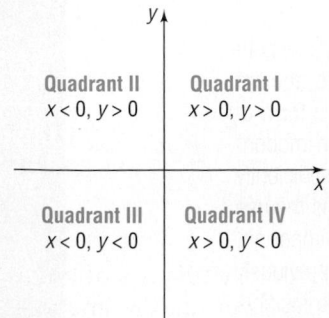

Quadrant II
$x < 0, y > 0$

Quadrant I
$x > 0, y > 0$

Quadrant III
$x < 0, y < 0$

Quadrant IV
$x > 0, y < 0$

Rectangular Coordinates

We locate a point on the real number line by assigning it a single real number, called the *coordinate of the point*. For work in a two-dimensional plane, we locate points by using two numbers.

We begin with two real number lines located in the same plane: one horizontal and the other vertical. The horizontal line is called the **x-axis,** the vertical line the **y-axis,** and the point of intersection the **origin O.** See Figure 1. We assign coordinates to every point on these number lines using a convenient scale. We usually use the same scale on each axis, but in applications, different scales appropriate to the application may be used.

The origin O has a value of 0 on both the x-axis and y-axis. Points on the x-axis to the right of O are associated with positive real numbers, and those to the left of O are associated with negative real numbers. Points on the y-axis above O are associated with positive real numbers, and those below O are associated with negative real numbers. In Figure 1, the x-axis and y-axis are labeled as x and y, respectively, and we have used an arrow at the end of each axis to denote the positive direction.

The coordinate system described here is called a **rectangular** or **Cartesian*** **coordinate system.** The plane formed by the x-axis and y-axis is sometimes called the **xy-plane,** and the x-axis and y-axis are referred to as the **coordinate axes.**

Any point P in the xy-plane can be located by using an **ordered pair** (x, y) of real numbers. Let x denote the signed distance of P from the y-axis (*signed* means that, if P is to the right of the y-axis, then $x > 0$, and if P is to the left of the y-axis, then $x < 0$); and let y denote the signed distance of P from the x-axis. The ordered pair (x, y), also called the **coordinates** of P, then gives us enough information to locate the point P in the plane.

For example, to locate the point whose coordinates are $(-3, 1)$, go 3 units along the x-axis to the left of O and then go straight up 1 unit. We **plot** this point by placing a dot at this location. See Figure 2, in which the points with coordinates $(-3, 1)$, $(-2, -3)$, $(3, -2)$, and $(3, 2)$ are plotted.

The origin has coordinates $(0, 0)$. Any point on the x-axis has coordinates of the form $(x, 0)$, and any point on the y-axis has coordinates of the form $(0, y)$.

If (x, y) are the coordinates of a point P, then x is called the **x-coordinate,** or **abscissa,** of P and y is the **y-coordinate,** or **ordinate,** of P. We identify the point P by its coordinates (x, y) by writing $P = (x, y)$. Usually, we will simply say "the point (x, y)" rather than "the point whose coordinates are (x, y)."

The coordinate axes divide the xy-plane into four sections called **quadrants,** as shown in Figure 3. In quadrant I, both the x-coordinate and the y-coordinate of all points are positive; in quadrant II, x is negative and y is positive; in quadrant III, both x and y are negative; and in quadrant IV, x is positive and y is negative. Points on the coordinate axes belong to no quadrant.

Now Work PROBLEM 13

*Named after René Descartes (1596–1650), a French mathematician, philosopher, and theologian.

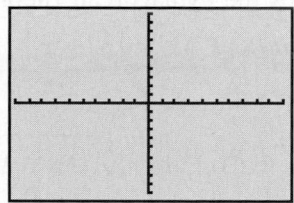 COMMENT On a graphing calculator, you can set the scale on each axis. Once this has been done, you obtain the **viewing rectangle.** See Figure 4 for a typical viewing rectangle. You should now read Section 1, *The Viewing Rectangle,* in the Appendix. ∎

Figure 4

1 Use the Distance Formula

If the same units of measurement, such as inches, centimeters, and so on, are used for both the *x*-axis and *y*-axis, then all distances in the *xy*-plane can be measured using this unit of measurement.

EXAMPLE 1 | **Finding the Distance between Two Points**

Find the distance *d* between the points $(1, 3)$ and $(5, 6)$.

Solution First plot the points $(1, 3)$ and $(5, 6)$ and connect them with a straight line. See Figure 5(a). We are looking for the length *d*. We begin by drawing a horizontal line from $(1, 3)$ to $(5, 3)$ and a vertical line from $(5, 3)$ to $(5, 6)$, forming a right triangle, as shown in Figure 5(b). One leg of the triangle is of length 4 (since $|5 - 1| = 4$), and the other is of length 3 (since $|6 - 3| = 3$). By the Pythagorean Theorem, the square of the distance *d* that we seek is

$$d^2 = 4^2 + 3^2 = 16 + 9 = 25$$
$$d = \sqrt{25} = 5$$

Figure 5

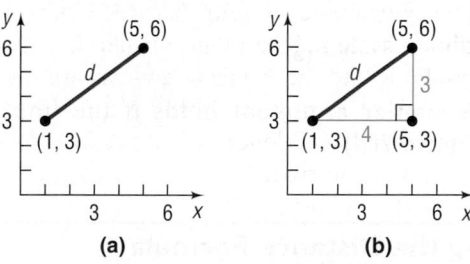

(a) (b)

The **distance formula** provides a straightforward method for computing the distance between two points.

THEOREM | **Distance Formula**

The distance between two points $P_1 = (x_1, y_1)$ and $P_2 = (x_2, y_2)$, denoted by $d(P_1, P_2)$, is

$$d(P_1, P_2) = \sqrt{(x_2 - x_1)^2 + (y_2 - y_1)^2} \qquad \text{(1)}$$

In Words

To compute the distance between two points, find the difference of the x-coordinates, square it, and add this to the square of the difference of the y-coordinates. The square root of this sum is the distance.

Proof of the Distance Formula Let (x_1, y_1) denote the coordinates of point P_1 and let (x_2, y_2) denote the coordinates of point P_2. Assume that the line joining P_1 and P_2 is neither horizontal nor vertical. Refer to Figure 6(a) on page 152. The coordinates of P_3 are (x_2, y_1). The horizontal distance from P_1 to P_3 is the absolute

value of the difference of the x-coordinates, $|x_2 - x_1|$. The vertical distance from P_3 to P_2 is the absolute value of the difference of the y-coordinates, $|y_2 - y_1|$. See Figure 6(b). The distance $d(P_1, P_2)$ that we seek is the length of the hypotenuse of the right triangle, so, by the Pythagorean Theorem, it follows that

$$[d(P_1, P_2)]^2 = |x_2 - x_1|^2 + |y_2 - y_1|^2$$
$$= (x_2 - x_1)^2 + (y_2 - y_1)^2$$
$$d(P_1, P_2) = \sqrt{(x_2 - x_1)^2 + (y_2 - y_1)^2}$$

Figure 6

(a) (b)

Now, if the line joining P_1 and P_2 is horizontal, then the y-coordinate of P_1 equals the y-coordinate of P_2; that is, $y_1 = y_2$. Refer to Figure 7(a). In this case, the distance formula (1) still works, because, for $y_1 = y_2$, it reduces to

$$d(P_1, P_2) = \sqrt{(x_2 - x_1)^2 + 0^2} = \sqrt{(x_2 - x_1)^2} = |x_2 - x_1|$$

Figure 7

(a) (b)

A similar argument holds if the line joining P_1 and P_2 is vertical. See Figure 7(b). ∎

EXAMPLE 2 **Using the Distance Formula**

Find the distance d between the points $(-4, 5)$ and $(3, 2)$.

Solution Using the distance formula, equation (1), the distance d is

$$d = \sqrt{[3 - (-4)]^2 + (2 - 5)^2} = \sqrt{7^2 + (-3)^2}$$
$$= \sqrt{49 + 9} = \sqrt{58} \approx 7.62$$

─────**Now Work** PROBLEMS **17** AND **21**

The distance between two points $P_1 = (x_1, y_1)$ and $P_2 = (x_2, y_2)$ is never a negative number. Furthermore, the distance between two points is 0 only when the points are identical, that is, when $x_1 = x_2$ and $y_1 = y_2$. Also, because $(x_2 - x_1)^2 = (x_1 - x_2)^2$ and $(y_2 - y_1)^2 = (y_1 - y_2)^2$, it makes no difference whether the distance is computed from P_1 to P_2 or from P_2 to P_1; that is, $d(P_1, P_2) = d(P_2, P_1)$.

The introduction to this chapter mentioned that rectangular coordinates enable us to translate geometry problems into algebra problems, and vice versa. The next example shows how algebra (the distance formula) can be used to solve geometry problems.

| EXAMPLE 3 | **Using Algebra to Solve Geometry Problems** |

Consider the three points $A = (-2, 1)$, $B = (2, 3)$, and $C = (3, 1)$.

(a) Plot each point and form the triangle ABC.
(b) Find the length of each side of the triangle.
(c) Verify that the triangle is a right triangle.
(d) Find the area of the triangle.

Solution

(a) Figure 8 shows the points A, B, C and the triangle ABC.
(b) To find the length of each side of the triangle, use the distance formula, equation (1).

Figure 8

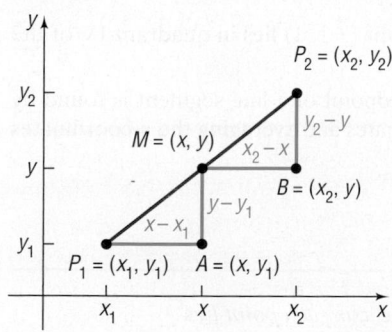

$$d(A, B) = \sqrt{[2 - (-2)]^2 + (3 - 1)^2} = \sqrt{16 + 4} = \sqrt{20} = 2\sqrt{5}$$
$$d(B, C) = \sqrt{(3 - 2)^2 + (1 - 3)^2} = \sqrt{1 + 4} = \sqrt{5}$$
$$d(A, C) = \sqrt{[3 - (-2)]^2 + (1 - 1)^2} = \sqrt{25 + 0} = 5$$

(c) To show that the triangle is a right triangle, we need to show that the sum of the squares of the lengths of two of the sides equals the square of the length of the third side. (Why is this sufficient?) Looking at Figure 8, it seems reasonable to conjecture that the right angle is at vertex B. We shall check to see whether

$$[d(A, B)]^2 + [d(B, C)]^2 = [d(A, C)]^2$$

Using the results in part (b),

$$[d(A, B)]^2 + [d(B, C)]^2 = \left(2\sqrt{5}\right)^2 + \left(\sqrt{5}\right)^2$$
$$= 20 + 5 = 25 = [d(A, C)]^2$$

It follows from the converse of the Pythagorean Theorem that triangle ABC is a right triangle.

(d) Because the right angle is at vertex B, the sides AB and BC form the base and height of the triangle. Its area is

$$\text{Area} = \frac{1}{2}(\text{Base})(\text{Height}) = \frac{1}{2}\left(2\sqrt{5}\right)\left(\sqrt{5}\right) = 5 \text{ square units}$$

Now Work PROBLEM 29

2 Use the Midpoint Formula

We now derive a formula for the coordinates of the **midpoint of a line segment.** Let $P_1 = (x_1, y_1)$ and $P_2 = (x_2, y_2)$ be the endpoints of a line segment, and let $M = (x, y)$ be the point on the line segment that is the same distance from P_1 as it is from P_2. See Figure 9. The triangles P_1AM and MBP_2 are congruent. Do you see why? $d(P_1, M) = d(M, P_2)$ is given; $\angle AP_1M = \angle BMP_2$* and $\angle P_1MA = \angle MP_2B$. So, we have angle–side–angle. Because triangles P_1AM and MBP_2 are congruent, corresponding sides are equal in length. That is,

Figure 9

$$x - x_1 = x_2 - x \quad \text{and} \quad y - y_1 = y_2 - y$$
$$2x = x_1 + x_2 \qquad\qquad 2y = y_1 + y_2$$
$$x = \frac{x_1 + x_2}{2} \qquad\qquad y = \frac{y_1 + y_2}{2}$$

*A postulate from geometry states that the transversal $\overline{P_1P_2}$ forms congruent corresponding angles with the parallel line segments $\overline{P_1A}$ and \overline{MB}.

THEOREM **Midpoint Formula**

The midpoint $M = (x, y)$ of the line segment from $P_1 = (x_1, y_1)$ to $P_2 = (x_2, y_2)$ is

> **In Words**
> To find the midpoint of a line segment, average the x-coordinates and average the y-coordinates of the endpoints.

$$M = (x, y) = \left(\frac{x_1 + x_2}{2}, \frac{y_1 + y_2}{2} \right) \qquad (2)$$

EXAMPLE 4 **Finding the Midpoint of a Line Segment**

Find the midpoint of the line segment from $P_1 = (-5, 5)$ to $P_2 = (3, 1)$. Plot the points P_1 and P_2 and their midpoint.

Solution Apply the midpoint formula (2) using $x_1 = -5$, $y_1 = 5$, $x_2 = 3$, and $y_2 = 1$. Then the coordinates (x, y) of the midpoint M are

Figure 10

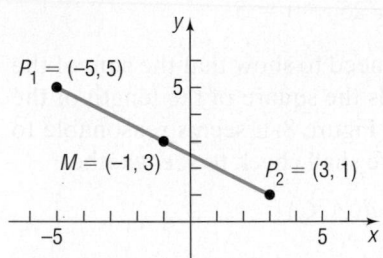

$$x = \frac{x_1 + x_2}{2} = \frac{-5 + 3}{2} = -1 \quad \text{and} \quad y = \frac{y_1 + y_2}{2} = \frac{5 + 1}{2} = 3$$

That is, $M = (-1, 3)$. See Figure 10.

━━━━ **Now Work** PROBLEM 35

2.1 Assess Your Understanding

'Are You Prepared?' *Answers are given at the end of these exercises. If you get a wrong answer, read the pages listed in red.*

1. On the real number line the origin is assigned the number _____. (p. 17)

2. If -3 and 5 are the coordinates of two points on the real number line, the distance between these points is _____. (pp. 19–20)

3. If 3 and 4 are the legs of a right triangle, the hypotenuse is _____. (p. 30)

4. Use the converse of the Pythagorean Theorem to show that a triangle whose sides are of lengths 11, 60, and 61 is a right triangle. (pp. 30–31)

5. The area A of a triangle whose base is b and whose altitude is h is $A =$ _____. (p. 31)

6. **True or False** Two triangles are congruent if two angles and the included side of one equals two angles and the included side of the other (pp. 32–33).

Concepts and Vocabulary

7. If (x, y) are the coordinates of a point P in the xy-plane, then x is called the _____ of P and y is the _____ of P.

8. The coordinate axes divide the xy-plane into four sections called _____.

9. If three distinct points P, Q, and R all lie on a line and if $d(P, Q) = d(Q, R)$, then Q is called the _____ of the line segment from P to R.

10. **True or False** The distance between two points is sometimes a negative number.

11. **True or False** The point $(-1, 4)$ lies in quadrant IV of the Cartesian plane.

12. **True or False** The midpoint of a line segment is found by averaging the x-coordinates and averaging the y-coordinates of the endpoints.

Skill Building

In Problems 13 and 14, plot each point in the xy-plane. Tell in which quadrant or on what coordinate axis each point lies.

13. (a) $A = (-3, 2)$
 (b) $B = (6, 0)$
 (c) $C = (-2, -2)$
 (d) $D = (6, 5)$
 (e) $E = (0, -3)$
 (f) $F = (6, -3)$

14. (a) $A = (1, 4)$
 (b) $B = (-3, -4)$
 (c) $C = (-3, 4)$
 (d) $D = (4, 1)$
 (e) $E = (0, 1)$
 (f) $F = (-3, 0)$

15. Plot the points $(2, 0)$, $(2, -3)$, $(2, 4)$, $(2, 1)$, and $(2, -1)$. Describe the set of all points of the form $(2, y)$, where y is a real number.

16. Plot the points $(0, 3)$, $(1, 3)$, $(-2, 3)$, $(5, 3)$, and $(-4, 3)$. Describe the set of all points of the form $(x, 3)$, where x is a real number.

In Problems 17–28, find the distance $d(P_1, P_2)$ between the points P_1 and P_2.

17. **18.** **19.** **20.**

21. $P_1 = (3, -4)$; $P_2 = (5, 4)$

22. $P_1 = (-1, 0)$; $P_2 = (2, 4)$

23. $P_1 = (-3, 2)$; $P_2 = (6, 0)$

24. $P_1 = (2, -3)$; $P_2 = (4, 2)$

25. $P_1 = (4, -3)$; $P_2 = (6, 4)$

26. $P_1 = (-4, -3)$; $P_2 = (6, 2)$

27. $P_1 = (a, b)$; $P_2 = (0, 0)$

28. $P_1 = (a, a)$; $P_2 = (0, 0)$

In Problems 29–34, plot each point and form the triangle ABC. Verify that the triangle is a right triangle. Find its area.

29. $A = (-2, 5)$; $B = (1, 3)$; $C = (-1, 0)$

30. $A = (-2, 5)$; $B = (12, 3)$; $C = (10, -11)$

31. $A = (-5, 3)$; $B = (6, 0)$; $C = (5, 5)$

32. $A = (-6, 3)$; $B = (3, -5)$; $C = (-1, 5)$

33. $A = (4, -3)$; $B = (0, -3)$; $C = (4, 2)$

34. $A = (4, -3)$; $B = (4, 1)$; $C = (2, 1)$

In Problems 35–42, find the midpoint of the line segment joining the points P_1 and P_2.

35. $P_1 = (3, -4)$; $P_2 = (5, 4)$

36. $P_1 = (-2, 0)$; $P_2 = (2, 4)$

37. $P_1 = (-3, 2)$; $P_2 = (6, 0)$

38. $P_1 = (2, -3)$; $P_2 = (4, 2)$

39. $P_1 = (4, -3)$; $P_2 = (6, 1)$

40. $P_1 = (-4, -3)$; $P_2 = (2, 2)$

41. $P_1 = (a, b)$; $P_2 = (0, 0)$

42. $P_1 = (a, a)$; $P_2 = (0, 0)$

Applications and Extensions

43. If the point $(2, 5)$ is shifted 3 units to the right and 2 units down, what are its new coordinates?

44. If the point $(-1, 6)$ is shifted 2 units to the left and 4 units up, what are its new coordinates?

45. Find all points having an x-coordinate of 3 whose distance from the point $(-2, -1)$ is 13.
(a) By using the Pythagorean Theorem.
(b) By using the distance formula.

46. Find all points having a y-coordinate of -6 whose distance from the point $(1, 2)$ is 17.
(a) By using the Pythagorean Theorem.
(b) By using the distance formula.

47. Find all points on the x-axis that are 6 units from the point $(4, -3)$.

48. Find all points on the y-axis that are 6 units from the point $(4, -3)$.

49. The midpoint of the line segment from P_1 to P_2 is $(-1, 4)$. If $P_1 = (-3, 6)$, what is P_2?

50. The midpoint of the line segment from P_1 to P_2 is $(5, -4)$. If $P_2 = (7, -2)$, what is P_1?

51. Geometry The **medians** of a triangle are the line segments from each vertex to the midpoint of the opposite side (see the figure). Find the lengths of the medians of the triangle with vertices at $A = (0, 0)$, $B = (6, 0)$, and $C = (4, 4)$.

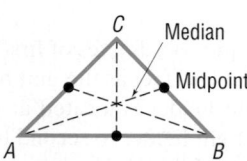

52. Geometry An **equilateral triangle** is one in which all three sides are of equal length. If two vertices of an equilateral triangle are $(0, 4)$ and $(0, 0)$, find the third vertex. How many of these triangles are possible?

53. Geometry Find the midpoint of each diagonal of a square with side of length s. Draw the conclusion that the diagonals of a square intersect at their midpoints.

[**Hint:** Use $(0, 0)$, $(0, s)$, $(s, 0)$, and (s, s) as the vertices of the square.]

54. Geometry Verify that the points $(0, 0)$, $(a, 0)$, and $\left(\dfrac{a}{2}, \dfrac{\sqrt{3}\,a}{2}\right)$ are the vertices of an equilateral triangle. Then show that the midpoints of the three sides are the vertices of a second equilateral triangle (refer to Problem 52).

*In Problems 55–58, find the length of each side of the triangle determined by the three points P_1, P_2, and P_3. State whether the triangle is an isosceles triangle, a right triangle, neither of these, or both. (An **isosceles triangle** is one in which at least two of the sides are of equal length.)*

55. $P_1 = (2, 1);$ $P_2 = (-4, 1);$ $P_3 = (-4, -3)$

56. $P_1 = (-1, 4);$ $P_2 = (6, 2);$ $P_3 = (4, -5)$

57. $P_1 = (-2, -1);$ $P_2 = (0, 7);$ $P_3 = (3, 2)$

58. $P_1 = (7, 2);$ $P_2 = (-4, 0);$ $P_3 = (4, 6)$

59. Baseball A major league baseball "diamond" is actually a square, 90 feet on a side (see the figure). What is the distance directly from home plate to second base (the diagonal of the square)?

60. Little League Baseball The layout of a Little League playing field is a square, 60 feet on a side. How far is it directly from home plate to second base (the diagonal of the square)?

Source: Little League Baseball, Official Regulations and Playing Rules, 2010.

61. Baseball Refer to Problem 59. Overlay a rectangular coordinate system on a major league baseball diamond so that the origin is at home plate, the positive x-axis lies in the direction from home plate to first base, and the positive y-axis lies in the direction from home plate to third base.
 (a) What are the coordinates of first base, second base, and third base? Use feet as the unit of measurement.
 (b) If the right fielder is located at $(310, 15)$, how far is it from the right fielder to second base?
 (c) If the center fielder is located at $(300, 300)$, how far is it from the center fielder to third base?

62. Little League Baseball Refer to Problem 60. Overlay a rectangular coordinate system on a Little League baseball diamond so that the origin is at home plate, the positive x-axis lies in the direction from home plate to first base, and the positive y-axis lies in the direction from home plate to third base.
 (a) What are the coordinates of first base, second base, and third base? Use feet as the unit of measurement.
 (b) If the right fielder is located at $(180, 20)$, how far is it from the right fielder to second base?
 (c) If the center fielder is located at $(220, 220)$, how far is it from the center fielder to third base?

63. Distance between Moving Objects A Dodge Neon and a Mack truck leave an intersection at the same time. The Neon heads east at an average speed of 30 miles per hour, while the truck heads south at an average speed of 40 miles per hour. Find an expression for their distance apart d (in miles) at the end of t hours.

64. Distance of a Moving Object from a Fixed Point A hot-air balloon, headed due east at an average speed of 15 miles per hour and at a constant altitude of 100 feet, passes over an intersection (see the figure). Find an expression for the distance d (measured in feet) from the balloon to the intersection t seconds later.

65. Drafting Error When a draftsman draws three lines that are to intersect at one point, the lines may not intersect as intended and subsequently will form an **error triangle.** If this error triangle is long and thin, one estimate for the location of the desired point is the midpoint of the shortest side. The figure shows one such error triangle.

Source: www.uwgb.edu/dutchs/STRUCTGE/sl00.htm

 (a) Find an estimate for the desired intersection point.
 (b) Find the length of the median for the midpoint found in part (a). See Problem 51.

66. Net Sales The figure illustrates how net sales of Wal-Mart Stores, Inc., have grown from 2002 through 2008. Use the midpoint formula to estimate the net sales of Wal-Mart Stores, Inc., in 2005. How does your result compare to the reported value of $282 billion?

Source: Wal-Mart Stores, Inc., 2008 Annual Report

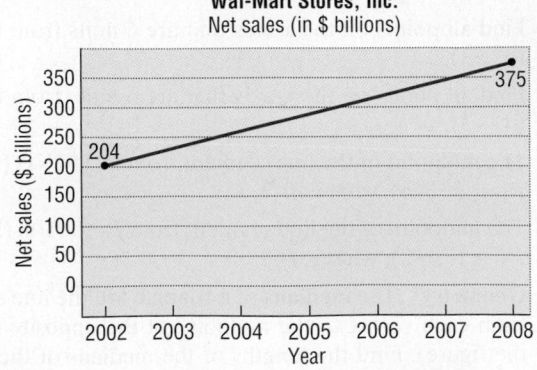

67. Poverty Threshold Poverty thresholds are determined by the U.S. Census Bureau. A poverty threshold represents the minimum annual household income for a family not to be considered poor. In 1998, the poverty threshold for a family of four with two children under the age of 18 years was $16,530. In 2008, the poverty threshold for a family of four with two children under the age of 18 years was $21,834.

Assuming poverty thresholds increase in a straight-line fashion, use the midpoint formula to estimate the poverty threshold of a family of four with two children under the age of 18 in 2003. How does your result compare to the actual poverty threshold in 2003 of $18,660?

Source: U.S. Census Bureau

Explaining Concepts: Discussion and Writing

68. Write a paragraph that describes a Cartesian plane. Then write a second paragraph that describes how to plot points in the Cartesian plane. Your paragraphs should include the terms "coordinate axes," "ordered pair," "coordinates," "plot," "*x*-coordinate," and "*y*-coordinate."

'Are You Prepared?' Answers

1. 0 **2.** 8 **3.** 5 **4.** $11^2 + 60^2 = 121 + 3600 = 3721 = 61^2$ **5.** $A = \dfrac{1}{2}bh$ **6.** True

2.2 Graphs of Equations in Two Variables; Intercepts; Symmetry

PREPARING FOR THIS SECTION *Before getting started, review the following:*

- Solving Linear Equations (Section 1.1, pp. 82–87)
- Solving Equations by Factoring (Section 1.2, pp. 93–94)

Now Work the 'Are You Prepared?' problems on page 164.

OBJECTIVES **1** Graph Equations by Plotting Points (p. 157)
 2 Find Intercepts from a Graph (p. 159)
 3 Find Intercepts from an Equation (p. 160)
 4 Test an Equation for Symmetry with Respect to the *x*-Axis, the *y*-Axis, and the Origin (p. 160)
 5 Know How to Graph Key Equations (p. 163)

1 Graph Equations by Plotting Points

An **equation in two variables,** say *x* and *y*, is a statement in which two expressions involving *x* and *y* are equal. The expressions are called the **sides** of the equation. Since an equation is a statement, it may be true or false, depending on the value of the variables. Any values of *x* and *y* that result in a true statement are said to **satisfy** the equation.

For example, the following are all equations in two variables *x* and *y*:

$$x^2 + y^2 = 5 \qquad 2x - y = 6 \qquad y = 2x + 5 \qquad x^2 = y$$

The first of these, $x^2 + y^2 = 5$, is satisfied for $x = 1$, $y = 2$, since $1^2 + 2^2 = 1 + 4 = 5$. Other choices of *x* and *y*, such as $x = -1$, $y = -2$, also satisfy this equation. It is not satisfied for $x = 2$ and $y = 3$, since $2^2 + 3^2 = 4 + 9 = 13 \neq 5$.

The **graph of an equation in two variables** *x* and *y* consists of the set of points in the *xy*-plane whose coordinates (x, y) satisfy the equation.

Graphs play an important role in helping us to visualize the relationships that exist between two variables or quantities. Figure 11 on page 158 shows the relation between the level of risk in a stock portfolio and the average annual rate of return. From the graph, we can see that, when 30% of a portfolio of stocks is invested in foreign companies, risk is minimized.

Figure 11
Source: T. Rowe Price

| EXAMPLE 1 | **Determining Whether a Point Is on the Graph of an Equation** |

Determine if the following points are on the graph of the equation $2x - y = 6$.

(a) $(2, 3)$ (b) $(2, -2)$

Solution

(a) For the point $(2, 3)$, check to see if $x = 2, y = 3$ satisfies the equation $2x - y = 6$.

$$2x - y = 2(2) - 3 = 4 - 3 = 1 \neq 6$$

The equation is not satisfied, so the point $(2, 3)$ is not on the graph of $2x - y = 6$.

(b) For the point $(2, -2)$,

$$2x - y = 2(2) - (-2) = 4 + 2 = 6$$

The equation is satisfied, so the point $(2, -2)$ is on the graph of $2x - y = 6$.

Now Work PROBLEM 11

| EXAMPLE 2 | **Graphing an Equation by Plotting Points** |

Graph the equation: $y = 2x + 5$

Solution

We want to find all points (x, y) that satisfy the equation. To locate some of these points (and get an idea of the pattern of the graph), assign some numbers to x and find corresponding values for y.

Figure 12

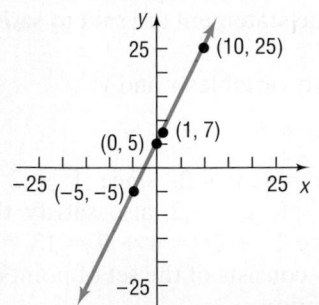

If	Then	Point on Graph
$x = 0$	$y = 2(0) + 5 = 5$	$(0, 5)$
$x = 1$	$y = 2(1) + 5 = 7$	$(1, 7)$
$x = -5$	$y = 2(-5) + 5 = -5$	$(-5, -5)$
$x = 10$	$y = 2(10) + 5 = 25$	$(10, 25)$

By plotting these points and then connecting them, we obtain the graph of the equation (a *line*), as shown in Figure 12.

| EXAMPLE 3 | **Graphing an Equation by Plotting Points** |

Graph the equation: $y = x^2$

Solution

Table 1 provides several points on the graph. In Figure 13 we plot these points and connect them with a smooth curve to obtain the graph (a *parabola*).

Table 1

x	$y = x^2$	(x, y)
-4	16	$(-4, 16)$
-3	9	$(-3, 9)$
-2	4	$(-2, 4)$
-1	1	$(-1, 1)$
0	0	$(0, 0)$
1	1	$(1, 1)$
2	4	$(2, 4)$
3	9	$(3, 9)$
4	16	$(4, 16)$

Figure 13

The graphs of the equations shown in Figures 12 and 13 do not show all points. For example, in Figure 12, the point $(20, 45)$ is a part of the graph of $y = 2x + 5$, but it is not shown. Since the graph of $y = 2x + 5$ could be extended out as far as we please, we use arrows to indicate that the pattern shown continues. It is important when illustrating a graph to present enough of the graph so that any viewer of the illustration will "see" the rest of it as an obvious continuation of what is actually there. This is referred to as a **complete graph.**

One way to obtain a complete graph of an equation is to plot a sufficient number of points on the graph until a pattern becomes evident. Then these points are connected with a smooth curve following the suggested pattern. But how many points are sufficient? Sometimes knowledge about the equation tells us. For example, we will learn in the next section that, if an equation is of the form $y = mx + b$, then its graph is a line. In this case, only two points are needed to obtain the graph.

One purpose of this book is to investigate the properties of equations in order to decide whether a graph is complete. Sometimes we shall graph equations by plotting points. Shortly, we shall investigate various techniques that will enable us to graph an equation without plotting so many points.

Two techniques that sometimes reduce the number of points required to graph an equation involve finding *intercepts* and checking for *symmetry*.

COMMENT Another way to obtain the graph of an equation is to use a graphing utility. Read Section 2, Using a Graphing Utility to Graph Equations, in the Appendix. ■

Figure 14

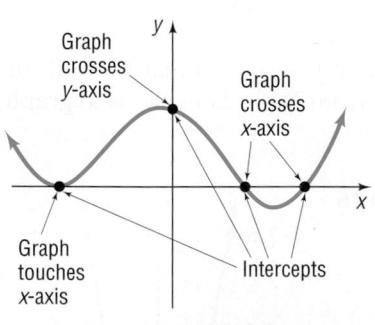

Graph crosses y-axis

Graph crosses x-axis

Graph touches x-axis

Intercepts

2 Find Intercepts from a Graph

The points, if any, at which a graph crosses or touches the coordinate axes are called the **intercepts.** See Figure 14. The x-coordinate of a point at which the graph crosses or touches the x-axis is an **x-intercept,** and the y-coordinate of a point at which the graph crosses or touches the y-axis is a **y-intercept.** For a graph to be complete, all its intercepts must be displayed.

EXAMPLE 4

Finding Intercepts from a Graph

Find the intercepts of the graph in Figure 15. What are its x-intercepts? What are its y-intercepts?

Solution The intercepts of the graph are the points

Figure 15

$$(-3, 0), \quad (0, 3), \quad \left(\frac{3}{2}, 0\right), \quad \left(0, -\frac{4}{3}\right), \quad (0, -3.5), \quad (4.5, 0)$$

The x-intercepts are $-3, \frac{3}{2}$, and 4.5; the y-intercepts are $-3.5, -\frac{4}{3}$, and 3.

In Example 4, you should notice the following usage: If we do not specify the type of intercept (x- versus y-), then report the intercept as an ordered pair. However, if the type of intercept is specified, then report the coordinate of the specified intercept. For x-intercepts, report the x-coordinate of the intercept; for y-intercepts, report the y-coordinate of the intercept.

 Now Work PROBLEM 39(a)

3 Find Intercepts from an Equation

The intercepts of a graph can be found from its equation by using the fact that points on the x-axis have y-coordinates equal to 0 and points on the y-axis have x-coordinates equal to 0.

COMMENT For many equations, finding intercepts may not be so easy. In such cases, a graphing utility can be used. Read the first part of Section 3, *Using a Graphing Utility to Locate Intercepts and Check for Symmetry*, in the Appendix, to find out how to locate intercepts using a graphing utility. ∎

Procedure for Finding Intercepts

1. To find the x-intercept(s), if any, of the graph of an equation, let $y = 0$ in the equation and solve for x, where x is a real number.
2. To find the y-intercept(s), if any, of the graph of an equation, let $x = 0$ in the equation and solve for y, where y is a real number.

EXAMPLE 5

Finding Intercepts from an Equation

Find the x-intercept(s) and the y-intercept(s) of the graph of $y = x^2 - 4$. Then graph $y = x^2 - 4$ by plotting points.

Solution To find the x-intercept(s), let $y = 0$ and obtain the equation

$$x^2 - 4 = 0 \quad \text{$y = x^2 - 4$ with $y = 0$}$$
$$(x + 2)(x - 2) = 0 \quad \text{Factor.}$$
$$x + 2 = 0 \quad \text{or} \quad x - 2 = 0 \quad \text{Zero-Product Property}$$
$$x = -2 \quad \text{or} \quad x = 2 \quad \text{Solve.}$$

The equation has two solutions, -2 and 2. The x-intercepts are -2 and 2.

To find the y-intercept(s), let $x = 0$ in the equation.

$$y = x^2 - 4$$
$$= 0^2 - 4 = -4$$

The y-intercept is -4.

Since $x^2 \geq 0$ for all x, we deduce from the equation $y = x^2 - 4$ that $y \geq -4$ for all x. This information, the intercepts, and the points from Table 2 enable us to graph $y = x^2 - 4$. See Figure 16.

Table 2

x	$y = x^2 - 4$	(x, y)
-3	5	$(-3, 5)$
-1	-3	$(-1, -3)$
1	-3	$(1, -3)$
3	5	$(3, 5)$

Figure 16

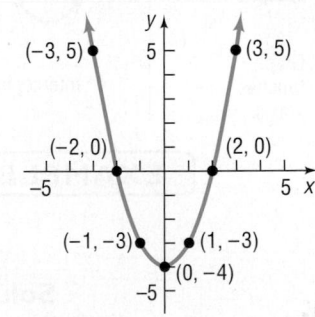

Now Work PROBLEM 21

4 Test an Equation for Symmetry with Respect to the *x*-Axis, the *y*-Axis, and the Origin

We just saw the role that intercepts play in obtaining key points on the graph of an equation. Another helpful tool for graphing equations involves symmetry, particularly symmetry with respect to the x-axis, the y-axis, and the origin.

DEFINITION A graph is said to be **symmetric with respect to the *x*-axis** if, for every point (x, y) on the graph, the point $(x, -y)$ is also on the graph.

Figure 17 illustrates the definition. When a graph is symmetric with respect to the *x*-axis, notice that the part of the graph above the *x*-axis is a reflection or mirror image of the part below it, and vice versa.

Figure 17
Symmetry with respect to the *x*-axis

EXAMPLE 6 **Points Symmetric with Respect to the *x*-Axis**

If a graph is symmetric with respect to the *x*-axis and the point $(3, 2)$ is on the graph, then the point $(3, -2)$ is also on the graph.

Figure 18
Symmetry with respect to the *y*-axis

DEFINITION A graph is said to be **symmetric with respect to the *y*-axis** if, for every point (x, y) on the graph, the point $(-x, y)$ is also on the graph.

Figure 18 illustrates the definition. When a graph is symmetric with respect to the *y*-axis, notice that the part of the graph to the right of the *y*-axis is a reflection of the part to the left of it, and vice versa.

EXAMPLE 7 **Points Symmetric with Respect to the *y*-Axis**

If a graph is symmetric with respect to the *y*-axis and the point $(5, 8)$ is on the graph, then the point $(-5, 8)$ is also on the graph.

DEFINITION A graph is said to be **symmetric with respect to the origin** if, for every point (x, y) on the graph, the point $(-x, -y)$ is also on the graph.

Figure 19
Symmetry with respect to the origin

Figure 19 illustrates the definition. Notice that symmetry with respect to the origin may be viewed in three ways:

1. As a reflection about the *y*-axis, followed by a reflection about the *x*-axis

2. As a projection along a line through the origin so that the distances from the origin are equal

3. As half of a complete revolution about the origin

EXAMPLE 8 **Points Symmetric with Respect to the Origin**

If a graph is symmetric with respect to the origin and the point $(4, 2)$ is on the graph, then the point $(-4, -2)$ is also on the graph.

━━━━━**Now Work** PROBLEMS 29 AND 39(b)

When the graph of an equation is symmetric with respect to a coordinate axis or the origin, the number of points that you need to plot in order to see the pattern is reduced. For example, if the graph of an equation is symmetric with respect to the *y*-axis, then, once points to the right of the *y*-axis are plotted, an equal number of points on the graph can be obtained by reflecting them about the *y*-axis. Because of this, before we graph an equation, we first want to determine whether it has any symmetry. The following tests are used for this purpose.

Tests for Symmetry

To test the graph of an equation for symmetry with respect to the

x-Axis Replace y by $-y$ in the equation and simplify. If an equivalent equation results, the graph of the equation is symmetric with respect to the x-axis.

y-Axis Replace x by $-x$ in the equation and simplify. If an equivalent equation results, the graph of the equation is symmetric with respect to the y-axis.

Origin Replace x by $-x$ and y by $-y$ in the equation and simplify. If an equivalent equation results, the graph of the equation is symmetric with respect to the origin.

EXAMPLE 9 **Testing an Equation for Symmetry**

Test $y = \dfrac{4x^2}{x^2 + 1}$ for symmetry.

Solution *x-Axis:* To test for symmetry with respect to the x-axis, replace y by $-y$. Since $-y = \dfrac{4x^2}{x^2 + 1}$ is not equivalent to $y = \dfrac{4x^2}{x^2 + 1}$, the graph of the equation is not symmetric with respect to the x-axis.

y-Axis: To test for symmetry with respect to the y-axis, replace x by $-x$. Since $y = \dfrac{4(-x)^2}{(-x)^2 + 1} = \dfrac{4x^2}{x^2 + 1}$ is equivalent to $y = \dfrac{4x^2}{x^2 + 1}$, the graph of the equation is symmetric with respect to the y-axis.

Origin: To test for symmetry with respect to the origin, replace x by $-x$ and y by $-y$.

$$-y = \frac{4(-x)^2}{(-x)^2 + 1} \qquad \text{Replace x by } -x \text{ and y by } -y.$$

$$-y = \frac{4x^2}{x^2 + 1} \qquad \text{Simplify.}$$

$$y = -\frac{4x^2}{x^2 + 1} \qquad \text{Multiply both sides by } -1.$$

Since the result is not equivalent to the original equation, the graph of the equation $y = \dfrac{4x^2}{x^2 + 1}$ is not symmetric with respect to the origin. ⨩

Seeing the Concept

Figure 20 shows the graph of $y = \dfrac{4x^2}{x^2 + 1}$ using a graphing utility. Do you see the symmetry with respect to the y-axis?

Figure 20

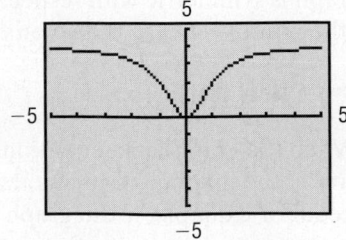

Now Work PROBLEM 59

5 Know How to Graph Key Equations

The next three examples use intercepts, symmetry, and point plotting to obtain the graphs of key equations. It is important to know the graphs of these key equations because we use them later. The first of these is $y = x^3$.

EXAMPLE 10

Graphing the Equation $y = x^3$ by Finding Intercepts, Checking for Symmetry, and Plotting Points

Graph the equation $y = x^3$ by plotting points. Find any intercepts and check for symmetry first.

Solution
First, find the intercepts. When $x = 0$, then $y = 0$; and when $y = 0$, then $x = 0$. The origin $(0, 0)$ is the only intercept. Now test for symmetry.

x-Axis: Replace y by $-y$. Since $-y = x^3$ is not equivalent to $y = x^3$, the graph is not symmetric with respect to the x-axis.

y-Axis: Replace x by $-x$. Since $y = (-x)^3 = -x^3$ is not equivalent to $y = x^3$, the graph is not symmetric with respect to the y-axis.

Origin: Replace x by $-x$ and y by $-y$. Since $-y = (-x)^3 = -x^3$ is equivalent to $y = x^3$ (multiply both sides by -1), the graph is symmetric with respect to the origin.

To graph $y = x^3$, we use the equation to obtain several points on the graph. Because of the symmetry, we only need to locate points on the graph for which $x \geq 0$. See Table 3. Since $(1, 1)$ is on the graph, and the graph is symmetric with respect to the origin, the point $(-1, -1)$ is also on the graph. Plot the points from Table 3 and use the symmetry. Figure 21 shows the graph.

Figure 21

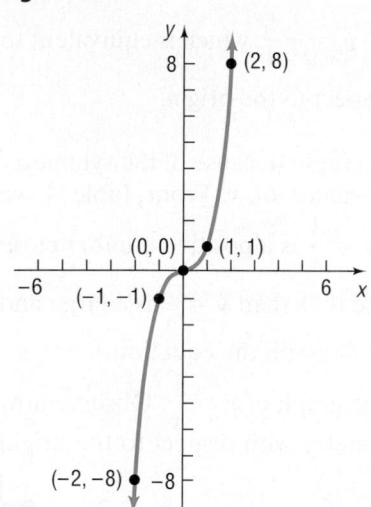

Table 3

x	$y = x^3$	(x, y)
0	0	$(0, 0)$
1	1	$(1, 1)$
2	8	$(2, 8)$
3	27	$(3, 27)$

EXAMPLE 11

Graphing the Equation $x = y^2$

(a) Graph the equation $x = y^2$. Find any intercepts and check for symmetry first.
(b) Graph $x = y^2$, $y \geq 0$.

Solution
(a) The lone intercept is $(0, 0)$. The graph is symmetric with respect to the x-axis. (Do you see why? Replace y by $-y$.) Figure 22 shows the graph.

(b) If we restrict y so that $y \geq 0$, the equation $x = y^2$, $y \geq 0$, may be written equivalently as $y = \sqrt{x}$. The portion of the graph of $x = y^2$ in quadrant I is therefore the graph of $y = \sqrt{x}$. See Figure 23.

Figure 24

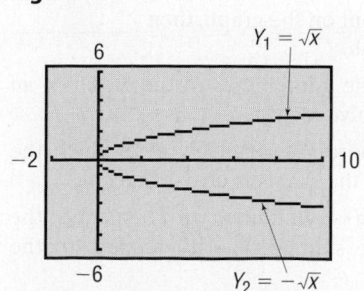

Figure 22
$x = y^2$

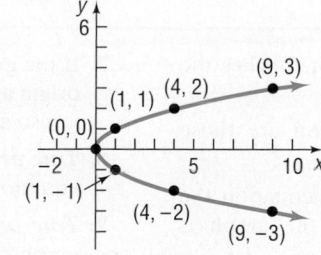

Figure 23
$y = \sqrt{x}$

 COMMENT To see the graph of the equation $x = y^2$ on a graphing calculator, you will need to graph two equations: $Y_1 = \sqrt{x}$ and $Y_2 = -\sqrt{x}$. We discuss why in Chapter 3. See Figure 24. ∎

EXAMPLE 12

Graphing the Equation $y = \dfrac{1}{x}$

Graph the equation $y = \dfrac{1}{x}$. Find any intercepts and check for symmetry first.

Solution Check for intercepts first. If we let $x = 0$, we obtain 0 in the denominator, which makes y undefined. We conclude that there is no y-intercept. If we let $y = 0$, we get the equation $\dfrac{1}{x} = 0$, which has no solution. We conclude that there is no x-intercept.

The graph of $y = \dfrac{1}{x}$ does not cross or touch the coordinate axes.

Next check for symmetry:

x-Axis: Replacing y by $-y$ yields $-y = \dfrac{1}{x}$, which is not equivalent to $y = \dfrac{1}{x}$.

y-Axis: Replacing x by $-x$ yields $y = \dfrac{1}{-x} = -\dfrac{1}{x}$, which is not equivalent to $y = \dfrac{1}{x}$.

Origin: Replacing x by $-x$ and y by $-y$ yields $-y = -\dfrac{1}{x}$, which is equivalent to $y = \dfrac{1}{x}$. The graph is symmetric with respect to the origin.

Now, set up Table 4, listing several points on the graph. Because of the symmetry with respect to the origin, we use only positive values of x. From Table 4 we infer that if x is a large and positive number, then $y = \dfrac{1}{x}$ is a positive number close to 0. We also infer that if x is a positive number close to 0, then $y = \dfrac{1}{x}$ is a large and positive number. Armed with this information, we can graph the equation.

Figure 25 illustrates some of these points and the graph of $y = \dfrac{1}{x}$. Observe how the absence of intercepts and the existence of symmetry with respect to the origin were utilized.

Table 4

x	$y = \dfrac{1}{x}$	(x, y)
$\dfrac{1}{10}$	10	$\left(\dfrac{1}{10}, 10\right)$
$\dfrac{1}{3}$	3	$\left(\dfrac{1}{3}, 3\right)$
$\dfrac{1}{2}$	2	$\left(\dfrac{1}{2}, 2\right)$
1	1	$(1, 1)$
2	$\dfrac{1}{2}$	$\left(2, \dfrac{1}{2}\right)$
3	$\dfrac{1}{3}$	$\left(3, \dfrac{1}{3}\right)$
10	$\dfrac{1}{10}$	$\left(10, \dfrac{1}{10}\right)$

Figure 25

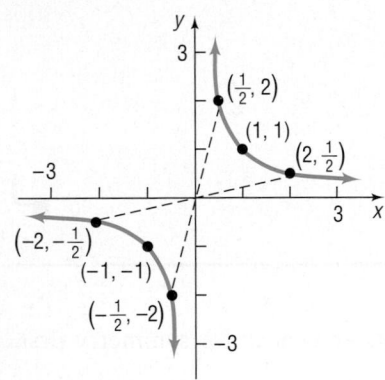

COMMENT Refer to Example 2 in the Appendix, Section 3, for the graph of $y = \dfrac{1}{x}$ using a graphing utility. ∎

2.2 Assess Your Understanding

'Are You Prepared?' *Answers are given at the end of these exercises. If you get a wrong answer, read the pages listed in red.*

1. Solve the equation $2(x + 3) - 1 = -7$. (pp. 82–85)

2. Solve the equation $x^2 - 9 = 0$. (pp. 93–94)

Concepts and Vocabulary

3. The points, if any, at which a graph crosses or touches the coordinate axes are called _____.

4. The x-intercepts of the graph of an equation are those x-values for which _____.

5. If for every point (x, y) on the graph of an equation the point $(-x, y)$ is also on the graph, then the graph is symmetric with respect to the _____.

6. If the graph of an equation is symmetric with respect to the y-axis and -4 is an x-intercept of this graph, then _____ is also an x-intercept.

7. If the graph of an equation is symmetric with respect to the origin and $(3, -4)$ is a point on the graph, then _____ is also a point on the graph.

8. *True or False* To find the y-intercepts of the graph of an equation, let $x = 0$ and solve for y.

9. *True or False* The y-coordinate of a point at which the graph crosses or touches the x-axis is an x-intercept.

10. *True or False* If a graph is symmetric with respect to the x-axis, then it cannot be symmetric with respect to the y-axis.

Skill Building

In Problems 11–16, determine which of the given points are on the graph of the equation.

11. Equation: $y = x^4 - \sqrt{x}$
Points: $(0,0); (1,1); (-1,0)$

12. Equation: $y = x^3 - 2\sqrt{x}$
Points: $(0,0); (1,1); (1,-1)$

13. Equation: $y^2 = x^2 + 9$
Points: $(0,3); (3,0); (-3,0)$

14. Equation: $y^3 = x + 1$
Points: $(1,2); (0,1); (-1,0)$

15. Equation: $x^2 + y^2 = 4$
Points: $(0,2); (-2,2); \left(\sqrt{2}, \sqrt{2}\right)$

16. Equation: $x^2 + 4y^2 = 4$
Points: $(0,1); (2,0); \left(2, \dfrac{1}{2}\right)$

In Problems 17–28, find the intercepts and graph each equation by plotting points. Be sure to label the intercepts.

17. $y = x + 2$

18. $y = x - 6$

19. $y = 2x + 8$

20. $y = 3x - 9$

21. $y = x^2 - 1$

22. $y = x^2 - 9$

23. $y = -x^2 + 4$

24. $y = -x^2 + 1$

25. $2x + 3y = 6$

26. $5x + 2y = 10$

27. $9x^2 + 4y = 36$

28. $4x^2 + y = 4$

In Problems 29–38, plot each point. Then plot the point that is symmetric to it with respect to (a) the x-axis; (b) the y-axis; (c) the origin.

29. $(3,4)$

30. $(5,3)$

31. $(-2,1)$

32. $(4,-2)$

33. $(5,-2)$

34. $(-1,-1)$

35. $(-3,-4)$

36. $(4,0)$

37. $(0,-3)$

38. $(-3,0)$

In Problems 39–50, the graph of an equation is given. (a) Find the intercepts. (b) Indicate whether the graph is symmetric with respect to the x-axis, the y-axis, or the origin.

39.

40.

41.

42.

43.

44.

45.

46.

47.

48.

49.

50.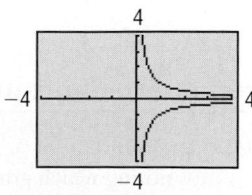

In Problems 51–54, draw a complete graph so that it has the type of symmetry indicated.

51. y-axis

52. x-axis

53. Origin

54. y-axis

In Problems 55–70, list the intercepts and test for symmetry.

55. $y^2 = x + 4$

56. $y^2 = x + 9$

57. $y = \sqrt[3]{x}$

58. $y = \sqrt[5]{x}$

59. $x^2 + y - 9 = 0$

60. $x^2 - y - 4 = 0$

61. $9x^2 + 4y^2 = 36$

62. $4x^2 + y^2 = 4$

63. $y = x^3 - 27$

64. $y = x^4 - 1$

65. $y = x^2 - 3x - 4$

66. $y = x^2 + 4$

67. $y = \dfrac{3x}{x^2 + 9}$

68. $y = \dfrac{x^2 - 4}{2x}$

69. $y = \dfrac{-x^3}{x^2 - 9}$

70. $y = \dfrac{x^4 + 1}{2x^5}$

In Problems 71–74, draw a quick sketch of each equation.

71. $y = x^3$

72. $x = y^2$

73. $y = \sqrt{x}$

74. $y = \dfrac{1}{x}$

75. If $(3, b)$ is a point on the graph of $y = 4x + 1$, what is b?

76. If $(-2, b)$ is a point on the graph of $2x + 3y = 2$, what is b?

77. If $(a, 4)$ is a point on the graph of $y = x^2 + 3x$, what is a?

78. If $(a, -5)$ is a point on the graph of $y = x^2 + 6x$, what is a?

Applications and Extensions

79. Given that the point $(1, 2)$ is on the graph of an equation that is symmetric with respect to the origin, what other point is on the graph?

80. If the graph of an equation is symmetric with respect to the y-axis and 6 is an x-intercept of this graph, name another x-intercept.

81. If the graph of an equation is symmetric with respect to the origin and -4 is an x-intercept of this graph, name another x-intercept.

82. If the graph of an equation is symmetric with respect to the x-axis and 2 is a y-intercept, name another y-intercept.

83. Microphones In studios and on stages, cardioid microphones are often preferred for the richness they add to voices and for their ability to reduce the level of sound from the sides and rear of the microphone. Suppose one such cardioid pattern is given by the equation $(x^2 + y^2 - x)^2 = x^2 + y^2$.
(a) Find the intercepts of the graph of the equation.
(b) Test for symmetry with respect to the x-axis, y-axis, and origin.

Source: www.notaviva.com

84. Solar Energy The solar electric generating systems at Kramer Junction, California, use parabolic troughs to heat a heat-transfer fluid to a high temperature. This fluid is used to generate steam that drives a power conversion system to produce electricity. For troughs 7.5 feet wide, an equation for the cross-section is $16y^2 = 120x - 225$.

(a) Find the intercepts of the graph of the equation.
(b) Test for symmetry with respect to the x-axis, y-axis, and origin.

Source: U.S. Department of Energy

Explaining Concepts: Discussion and Writing

85. (a) Graph $y = \sqrt{x^2}$, $y = x$, $y = |x|$, and $y = (\sqrt{x})^2$, noting which graphs are the same.
(b) Explain why the graphs of $y = \sqrt{x^2}$ and $y = |x|$ are the same.
(c) Explain why the graphs of $y = x$ and $y = (\sqrt{x})^2$ are not the same.
(d) Explain why the graphs of $y = \sqrt{x^2}$ and $y = x$ are not the same.

86. Explain what is meant by a complete graph.

87. Draw a graph of an equation that contains two x-intercepts; at one the graph crosses the x-axis, and at the other the graph touches the x-axis.

88. Make up an equation with the intercepts $(2, 0)$, $(4, 0)$, and $(0, 1)$. Compare your equation with a friend's equation. Comment on any similarities.

89. Draw a graph that contains the points $(-2, -1)$, $(0, 1)$, $(1, 3)$, and $(3, 5)$. Compare your graph with those of other students. Are most of the graphs almost straight lines? How many are "curved"? Discuss the various ways that these points might be connected.

90. An equation is being tested for symmetry with respect to the x-axis, the y-axis, and the origin. Explain why, if two of these symmetries are present, the remaining one must also be present.

91. Draw a graph that contains the points $(-2, 5)$, $(-1, 3)$, and $(0, 2)$ that is symmetric with respect to the y-axis. Compare your graph with those of other students; comment on any similarities. Can a graph contain these points and be symmetric with respect to the x-axis? the origin? Why or why not?

Interactive Exercises

Ask your instructor if the applets below are of interest to you.

92. y-axis Symmetry *Open the y-axis symmetry applet.* Move point A around the Cartesian Plane with your mouse. How are the coordinates of point A and the coordinates of point B related?

93. x-axis Symmetry *Open the x-axis symmetry applet.* Move point A around the Cartesian Plane with your mouse. How

are the coordinates of point A and the coordinates of point B related?

94. Origin Symmetry *Open the origin symmetry applet.* Move point A around the Cartesian Plane with your mouse. How are the coordinates of point A and the coordinates of point B related?

'Are You Prepared?' Answers

1. $\{-6\}$ **2.** $\{-3, 3\}$

2.3 Lines

OBJECTIVES **1** Calculate and Interpret the Slope of a Line (p. 167)
 2 Graph Lines Given a Point and the Slope (p. 170)
 3 Find the Equation of a Vertical Line (p. 170)
 4 Use the Point–Slope Form of a Line; Identify Horizontal Lines (p. 171)
 5 Find the Equation of a Line Given Two Points (p. 172)
 6 Write the Equation of a Line in Slope–Intercept Form (p. 172)
 7 Identify the Slope and y-Intercept of a Line from Its Equation (p. 173)
 8 Graph Lines Written in General Form Using Intercepts (p. 174)
 9 Find Equations of Parallel Lines (p. 175)
 10 Find Equations of Perpendicular Lines (p. 176)

In this section we study a certain type of equation that contains two variables, called a *linear equation,* and its graph, a *line.*

Figure 26

1 Calculate and Interpret the Slope of a Line

Consider the staircase illustrated in Figure 26. Each step contains exactly the same horizontal **run** and the same vertical **rise.** The ratio of the rise to the run, called the *slope,* is a numerical measure of the steepness of the staircase. For example, if the run is increased and the rise remains the same, the staircase becomes less steep. If the run is kept the same, but the rise is increased, the staircase becomes more steep. This important characteristic of a line is best defined using rectangular coordinates.

DEFINITION Let $P = (x_1, y_1)$ and $Q = (x_2, y_2)$ be two distinct points. If $x_1 \neq x_2$, the **slope m** of the nonvertical line L containing P and Q is defined by the formula

$$m = \frac{y_2 - y_1}{x_2 - x_1} \qquad x_1 \neq x_2 \tag{1}$$

If $x_1 = x_2$, L is a **vertical line** and the slope m of L is **undefined** (since this results in division by 0).

Figure 27(a) on page 168 provides an illustration of the slope of a nonvertical line; Figure 27(b) illustrates a vertical line.

Figure 27

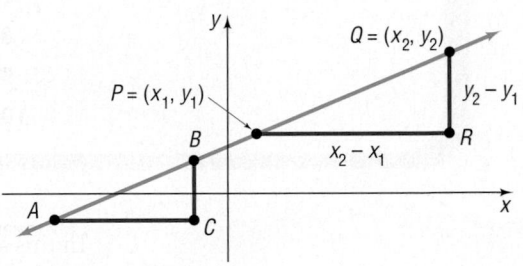

(a) Slope of L is $m = \dfrac{y_2 - y_1}{x_2 - x_1}$ **(b)** Slope is undefined; L is vertical

As Figure 27(a) illustrates, the slope m of a nonvertical line may be viewed as

$$m = \frac{y_2 - y_1}{x_2 - x_1} = \frac{\text{Rise}}{\text{Run}} \quad \text{or} \quad m = \frac{y_2 - y_1}{x_2 - x_1} = \frac{\text{Change in } y}{\text{Change in } x} = \frac{\Delta y}{\Delta x}$$

That is, the slope m of a nonvertical line measures the amount y changes when x changes from x_1 to x_2. The expression $\dfrac{\Delta y}{\Delta x}$ is called the **average rate of change** of y, with respect to x.

Two comments about computing the slope of a nonvertical line may prove helpful:

1. Any two distinct points on the line can be used to compute the slope of the line. (See Figure 28 for justification.)

Figure 28
Triangles ABC and PQR are similar (equal angles), so ratios of corresponding sides are proportional. Then

Slope using P and $Q = \dfrac{y_2 - y_1}{x_2 - x_1} =$

$\dfrac{d(B, C)}{d(A, C)} =$ Slope using A and B

Since any two distinct points can be used to compute the slope of a line, the average rate of change of a line is always the same number.

2. The slope of a line may be computed from $P = (x_1, y_1)$ to $Q = (x_2, y_2)$ or from Q to P because

$$\frac{y_2 - y_1}{x_2 - x_1} = \frac{y_1 - y_2}{x_1 - x_2}$$

EXAMPLE 1

Finding and Interpreting the Slope of a Line Given Two Points

The slope m of the line containing the points $(1, 2)$ and $(5, -3)$ may be computed as

$$m = \frac{-3 - 2}{5 - 1} = \frac{-5}{4} = -\frac{5}{4} \quad \text{or as} \quad m = \frac{2 - (-3)}{1 - 5} = \frac{5}{-4} = -\frac{5}{4}$$

For every 4-unit change in x, y will change by -5 units. That is, if x increases by 4 units, then y will decrease by 5 units. The average rate of change of y with respect to x is $-\dfrac{5}{4}$.

Now Work PROBLEMS **11** AND **17**

EXAMPLE 2 **Finding the Slopes of Various Lines Containing the Same Point (2, 3)**

Compute the slopes of the lines L_1, L_2, L_3, and L_4 containing the following pairs of points. Graph all four lines on the same set of coordinate axes.

$$L_1: \quad P = (2,3) \qquad Q_1 = (-1,-2)$$
$$L_2: \quad P = (2,3) \qquad Q_2 = (3,-1)$$
$$L_3: \quad P = (2,3) \qquad Q_3 = (5,3)$$
$$L_4: \quad P = (2,3) \qquad Q_4 = (2,5)$$

Solution Let m_1, m_2, m_3, and m_4 denote the slopes of the lines L_1, L_2, L_3, and L_4, respectively. Then

Figure 29

$$m_1 = \frac{-2-3}{-1-2} = \frac{-5}{-3} = \frac{5}{3} \qquad \text{A rise of 5 divided by a run of 3}$$

$$m_2 = \frac{-1-3}{3-2} = \frac{-4}{1} = -4$$

$$m_3 = \frac{3-3}{5-2} = \frac{0}{3} = 0$$

m_4 is undefined because $x_1 = x_2 = 2$

The graphs of these lines are given in Figure 29.

Figure 29 illustrates the following facts:

1. When the slope of a line is positive, the line slants upward from left to right (L_1).
2. When the slope of a line is negative, the line slants downward from left to right (L_2).
3. When the slope is 0, the line is horizontal (L_3).
4. When the slope is undefined, the line is vertical (L_4).

Seeing the Concept

On the same screen, graph the following equations:

$Y_1 = 0$ Slope of line is 0.

$Y_2 = \dfrac{1}{4}x$ Slope of line is $\dfrac{1}{4}$.

$Y_3 = \dfrac{1}{2}x$ Slope of line is $\dfrac{1}{2}$.

$Y_4 = x$ Slope of line is 1.

$Y_5 = 2x$ Slope of line is 2.

$Y_6 = 6x$ Slope of line is 6.

Figure 30

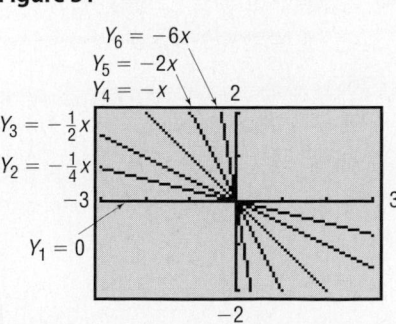

See Figure 30.

Seeing the Concept

On the same screen, graph the following equations:

$Y_1 = 0$ Slope of line is 0.

$Y_2 = -\dfrac{1}{4}x$ Slope of line is $-\dfrac{1}{4}$.

$Y_3 = -\dfrac{1}{2}x$ Slope of line is $-\dfrac{1}{2}$.

$Y_4 = -x$ Slope of line is −1.

$Y_5 = -2x$ Slope of line is −2.

$Y_6 = -6x$ Slope of line is −6.

Figure 31

See Figure 31.

Figures 30 and 31 on page 169 illustrate that the closer the line is to the vertical position, the greater the magnitude of the slope.

2 Graph Lines Given a Point and the Slope

EXAMPLE 3 **Graphing a Line Given a Point and a Slope**

Draw a graph of the line that contains the point $(3, 2)$ and has a slope of:

(a) $\dfrac{3}{4}$ (b) $-\dfrac{4}{5}$

Solution (a) Slope $= \dfrac{\text{Rise}}{\text{Run}}$. The fact that the slope is $\dfrac{3}{4}$ means that for every horizontal movement (run) of 4 units to the right there will be a vertical movement (rise) of 3 units. Look at Figure 32. If we start at the given point $(3, 2)$ and move 4 units to the right and 3 units up, we reach the point $(7, 5)$. By drawing the line through this point and the point $(3, 2)$, we have the graph.

Figure 32

(b) The fact that the slope is

$$-\frac{4}{5} = \frac{-4}{5} = \frac{\text{Rise}}{\text{Run}}$$

means that for every horizontal movement of 5 units to the right there will be a corresponding vertical movement of -4 units (a downward movement). If we start at the given point $(3, 2)$ and move 5 units to the right and then 4 units down, we arrive at the point $(8, -2)$. By drawing the line through these points, we have the graph. See Figure 33.

Alternatively, we can set

$$-\frac{4}{5} = \frac{4}{-5} = \frac{\text{Rise}}{\text{Run}}$$

Figure 33

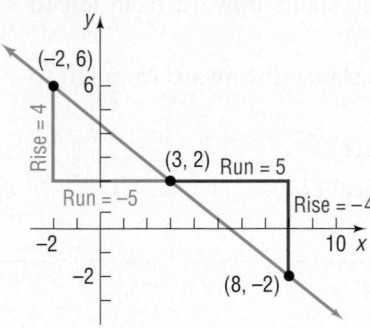

so that for every horizontal movement of -5 units (a movement to the left) there will be a corresponding vertical movement of 4 units (upward). This approach brings us to the point $(-2, 6)$, which is also on the graph shown in Figure 33.

Now Work PROBLEM 23

3 Find the Equation of a Vertical Line

EXAMPLE 4 **Graphing a Line**

Graph the equation: $x = 3$

Solution To graph $x = 3$, we find all points (x, y) in the plane for which $x = 3$. No matter what y-coordinate is used, the corresponding x-coordinate always equals 3. Consequently, the graph of the equation $x = 3$ is a vertical line with x-intercept 3 and undefined slope. See Figure 34.

Figure 34

As suggested by Example 4, we have the following result:

THEOREM **Equation of a Vertical Line**

A vertical line is given by an equation of the form

$$x = a$$

where a is the x-intercept.

> COMMENT To graph an equation using a graphing utility, we need to express the equation in the form y = {expression in x}. But x = 3 cannot be put in this form. To overcome this, most graphing utilities have special commands for drawing vertical lines. DRAW, LINE, PLOT, and VERT are among the more common ones. Consult your manual to determine the correct methodology for your graphing utility. ∎

Figure 35

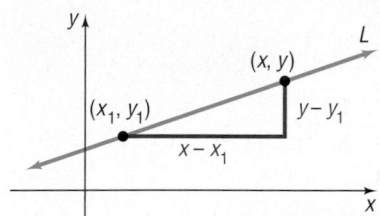

4 Use the Point–Slope Form of a Line; Identify Horizontal Lines

Let L be a nonvertical line with slope m and containing the point (x_1, y_1). See Figure 35. For any other point (x, y) on L, we have

$$m = \frac{y - y_1}{x - x_1} \quad \text{or} \quad y - y_1 = m(x - x_1)$$

THEOREM **Point–Slope Form of an Equation of a Line**

An equation of a nonvertical line with slope m that contains the point (x_1, y_1) is

$$y - y_1 = m(x - x_1) \qquad\qquad \textbf{(2)}$$

EXAMPLE 5 **Using the Point–Slope Form of a Line**

Figure 36

An equation of the line with slope 4 and containing the point $(1, 2)$ can be found by using the point–slope form with $m = 4$, $x_1 = 1$, and $y_1 = 2$.

$$y - y_1 = m(x - x_1)$$
$$y - 2 = 4(x - 1) \qquad m = 4, x_1 = 1, y_1 = 2$$
$$y = 4x - 2 \qquad \text{Solve for y.}$$

See Figure 36 for the graph.

━━━━ **Now Work** PROBLEM 45

EXAMPLE 6 **Finding the Equation of a Horizontal Line**

Find an equation of the horizontal line containing the point $(3, 2)$.

Solution Because all the y-values are equal on a horizontal line, the slope of a horizontal line is 0. To get an equation, we use the point–slope form with $m = 0$, $x_1 = 3$, and $y_1 = 2$.

Figure 37

$$y - y_1 = m(x - x_1)$$
$$y - 2 = 0 \cdot (x - 3) \qquad m = 0, x_1 = 3, \text{ and } y_1 = 2$$
$$y - 2 = 0$$
$$y = 2$$

See Figure 37 for the graph.

As suggested by Example 6, we have the following result:

THEOREM **Equation of a Horizontal Line**

A horizontal line is given by an equation of the form

$$y = b$$

where b is the y-intercept.

5 Find the Equation of a Line Given Two Points

EXAMPLE 7 **Finding an Equation of a Line Given Two Points**

Find an equation of the line containing the points $(2, 3)$ and $(-4, 5)$. Graph the line.

Solution First compute the slope of the line.

$$m = \frac{5 - 3}{-4 - 2} = \frac{2}{-6} = -\frac{1}{3}$$

Figure 38

Use the point $(2, 3)$ and the slope $m = -\dfrac{1}{3}$ to get the point–slope form of the equation of the line.

$$y - 3 = -\frac{1}{3}(x - 2)$$

See Figure 38 for the graph.

In the solution to Example 7, we could have used the other point, $(-4, 5)$, instead of the point $(2, 3)$. The equation that results, although it looks different, is equivalent to the equation that we obtained in the example. (Try it for yourself.)

 Now Work PROBLEM 37

6 Write the Equation of a Line in Slope–Intercept Form

Another useful equation of a line is obtained when the slope m and y-intercept b are known. In this event, we know both the slope m of the line and a point $(0, b)$ on the line; then we use the point–slope form, equation (2), to obtain the following equation:

$$y - b = m(x - 0) \quad \text{or} \quad y = mx + b$$

THEOREM **Slope–Intercept Form of an Equation of a Line**

An equation of a line with slope m and y-intercept b is

$$y = mx + b \qquad (3)$$

Now Work PROBLEM 51 (EXPRESS ANSWER
IN SLOPE–INTERCEPT FORM)

Figure 39 $y = mx + 2$

$Y_5 = -3x + 2$ $Y_4 = 3x + 2$
$Y_3 = -x + 2$ $Y_2 = x + 2$

$Y_1 = 2$

Seeing the Concept

To see the role that the slope m plays, graph the following lines on the same screen.

$$Y_1 = 2$$
$$Y_2 = x + 2$$
$$Y_3 = -x + 2$$
$$Y_4 = 3x + 2$$
$$Y_5 = -3x + 2$$

See Figure 39. What do you conclude about the lines $y = mx + 2$?

Figure 40 $y = 2x + b$

$Y_4 = 2x + 4$ $Y_2 = 2x + 1$
$Y_1 = 2x$
$Y_3 = 2x - 1$
$Y_5 = 2x - 4$

Seeing the Concept

To see the role of the y-intercept b, graph the following lines on the same screen.

$$Y_1 = 2x$$
$$Y_2 = 2x + 1$$
$$Y_3 = 2x - 1$$
$$Y_4 = 2x + 4$$
$$Y_5 = 2x - 4$$

See Figure 40. What do you conclude about the lines $y = 2x + b$?

7 Identify the Slope and y-Intercept of a Line from Its Equation

When the equation of a line is written in slope–intercept form, it is easy to find the slope m and y-intercept b of the line. For example, suppose that the equation of a line is

$$y = -2x + 7$$

Compare it to $y = mx + b$.

$$y = -2x + 7$$
$$y = \quad mx \; + \; b$$

The slope of this line is -2 and its y-intercept is 7.

Now Work PROBLEM 71

EXAMPLE 8 **Finding the Slope and y-Intercept**

Find the slope m and y-intercept b of the equation $2x + 4y = 8$. Graph the equation.

Solution To obtain the slope and y-intercept, write the equation in slope–intercept form by solving for y.

$$2x + 4y = 8$$
$$4y = -2x + 8$$
$$y = -\frac{1}{2}x + 2 \quad y = mx + b$$

Figure 41

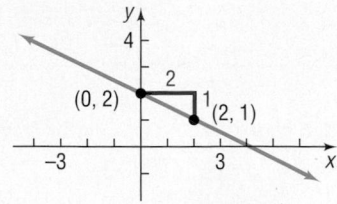

The coefficient of x, $-\dfrac{1}{2}$, is the slope, and the y-intercept is 2. Graph the line using the fact that the y-intercept is 2 and the slope is $-\dfrac{1}{2}$. Then, starting at the point $(0, 2)$, go to the right 2 units and then down 1 unit to the point $(2, 1)$. See Figure 41.

Now Work PROBLEM 77

8 Graph Lines Written in General Form Using Intercepts

Refer to Example 8. The form of the equation of the line $2x + 4y = 8$ is called the *general form*.

DEFINITION

The equation of a line is in **general form*** when it is written as

$$Ax + By = C \qquad\qquad (4)$$

where A, B, and C are real numbers and A and B are not both 0.

If $B = 0$ in (4), then $A \neq 0$ and the graph of the equation is a vertical line: $x = \dfrac{C}{A}$. If $B \neq 0$ in (4), then we can solve the equation for y and write the equation in slope–intercept form as we did in Example 8.

Another approach to graphing the equation (4) would be to find its intercepts. Remember, the intercepts of the graph of an equation are the points where the graph crosses or touches a coordinate axis.

EXAMPLE 9

Graphing an Equation in General Form Using Its Intercepts

Graph the equation $2x + 4y = 8$ by finding its intercepts.

Solution
To obtain the x-intercept, let $y = 0$ in the equation and solve for x.

$$2x + 4y = 8$$
$$2x + 4(0) = 8 \quad \text{Let } y = 0.$$
$$2x = 8$$
$$x = 4 \quad \text{Divide both sides by 2.}$$

The x-intercept is 4 and the point $(4, 0)$ is on the graph of the equation.
To obtain the y-intercept, let $x = 0$ in the equation and solve for y.

$$2x + 4y = 8$$
$$2(0) + 4y = 8 \quad \text{Let } x = 0.$$
$$4y = 8$$
$$y = 2 \quad \text{Divide both sides by 4.}$$

The y-intercept is 2 and the point $(0, 2)$ is on the graph of the equation.
Plot the points $(4, 0)$ and $(0, 2)$ and draw the line through the points. See Figure 42.

Figure 42

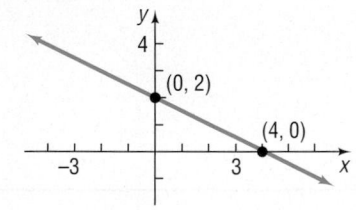

Now Work PROBLEM 91

Every line has an equation that is equivalent to an equation written in general form. For example, a vertical line whose equation is

$$x = a$$

can be written in the general form

$$1 \cdot x + 0 \cdot y = a \quad A = 1, B = 0, C = a$$

A horizontal line whose equation is

$$y = b$$

can be written in the general form

$$0 \cdot x + 1 \cdot y = b \quad A = 0, B = 1, C = b$$

*Some books use the term **standard form**.

Lines that are neither vertical nor horizontal have general equations of the form

$$Ax + By = C \quad A \neq 0 \text{ and } B \neq 0$$

Because the equation of every line can be written in general form, any equation equivalent to equation (4) is called a **linear equation.**

9 Find Equations of Parallel Lines

Figure 43

When two lines (in the plane) do not intersect (that is, they have no points in common), they are said to be **parallel.** Look at Figure 43. There we have drawn two parallel lines and have constructed two right triangles by drawing sides parallel to the coordinate axes. The right triangles are similar. (Do you see why? Two angles are equal.) Because the triangles are similar, the ratios of corresponding sides are equal.

THEOREM Criterion for Parallel Lines

Two nonvertical lines are parallel if and only if their slopes are equal and they have different y-intercepts.

The use of the words "if and only if" in the preceding theorem means that actually two statements are being made, one the converse of the other.

If two nonvertical lines are parallel, then their slopes are equal and they have different y-intercepts.

If two nonvertical lines have equal slopes and they have different y-intercepts, then they are parallel.

EXAMPLE 10 **Showing That Two Lines Are Parallel**

Show that the lines given by the following equations are parallel:

$$L_1: \quad 2x + 3y = 6, \qquad L_2: \quad 4x + 6y = 0$$

Solution

To determine whether these lines have equal slopes and different y-intercepts, write each equation in slope–intercept form:

Figure 44

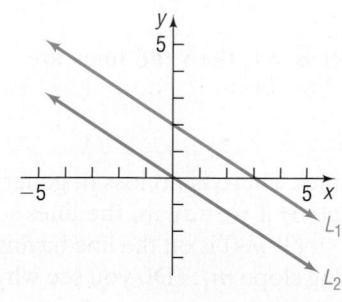

$$L_1: \quad 2x + 3y = 6 \qquad\qquad L_2: \quad 4x + 6y = 0$$
$$3y = -2x + 6 \qquad\qquad\qquad 6y = -4x$$
$$y = -\frac{2}{3}x + 2 \qquad\qquad\qquad y = -\frac{2}{3}x$$

$$\text{Slope} = -\frac{2}{3}; \, y\text{-intercept} = 2 \qquad \text{Slope} = -\frac{2}{3}; \, y\text{-intercept} = 0$$

Because these lines have the same slope, $-\dfrac{2}{3}$, but different y-intercepts, the lines are parallel. See Figure 44.

EXAMPLE 11 **Finding a Line That Is Parallel to a Given Line**

Find an equation for the line that contains the point $(2, -3)$ and is parallel to the line $2x + y = 6$.

Solution

Since the two lines are to be parallel, the slope of the line that we seek equals the slope of the line $2x + y = 6$. Begin by writing the equation of the line $2x + y = 6$ in slope–intercept form.

$$2x + y = 6$$
$$y = -2x + 6$$

Figure 45

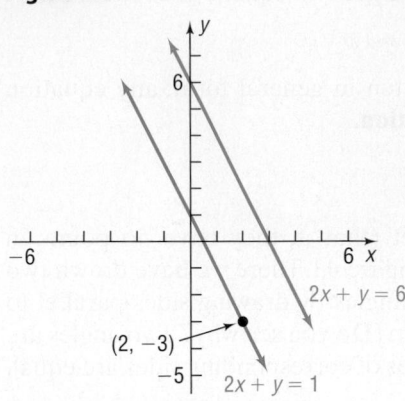

$2x + y = 6$

$(2, -3)$

$2x + y = 1$

The slope is -2. Since the line that we seek also has slope -2 and contains the point $(2, -3)$, use the point–slope form to obtain its equation.

$$y - y_1 = m(x - x_1) \quad \text{Point–slope form}$$
$$y - (-3) = -2(x - 2) \quad m = -2, x_1 = 2, y_1 = -3$$
$$y + 3 = -2x + 4 \quad \text{Simplify.}$$
$$y = -2x + 1 \quad \text{Slope–intercept form}$$
$$2x + y = 1 \quad \text{General form}$$

This line is parallel to the line $2x + y = 6$ and contains the point $(2, -3)$. See Figure 45.

━━━ **Now Work** PROBLEM 59

Figure 46

$90°$

10 Find Equations of Perpendicular Lines

When two lines intersect at a right angle (90°), they are said to be **perpendicular.** See Figure 46.

The following result gives a condition, in terms of their slopes, for two lines to be perpendicular.

THEOREM

Criterion for Perpendicular Lines

Two nonvertical lines are perpendicular if and only if the product of their slopes is -1.

Here we shall prove the "only if" part of the statement:

If two nonvertical lines are perpendicular, then the product of their slopes is -1.

In Problem 128 you are asked to prove the "if" part of the theorem; that is:

If two nonvertical lines have slopes whose product is -1, then the lines are perpendicular.

Figure 47

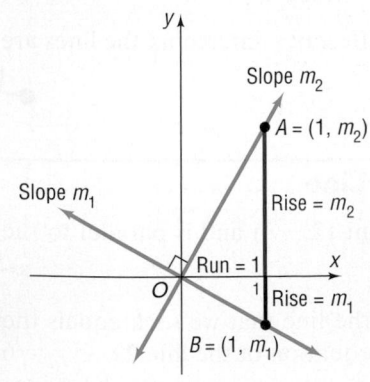

Slope m_2

$A = (1, m_2)$

Slope m_1

Rise $= m_2$

Run $= 1$

Rise $= m_1$

$B = (1, m_1)$

Proof Let m_1 and m_2 denote the slopes of the two lines. There is no loss in generality (that is, neither the angle nor the slopes are affected) if we situate the lines so that they meet at the origin. See Figure 47. The point $A = (1, m_2)$ is on the line having slope m_2, and the point $B = (1, m_1)$ is on the line having slope m_1. (Do you see why this must be true?)

Suppose that the lines are perpendicular. Then triangle OAB is a right triangle. As a result of the Pythagorean Theorem, it follows that

$$[d(O, A)]^2 + [d(O, B)]^2 = [d(A, B)]^2 \tag{5}$$

Using the distance formula, the squares of these distances are

$$[d(O, A)]^2 = (1 - 0)^2 + (m_2 - 0)^2 = 1 + m_2^2$$
$$[d(O, B)]^2 = (1 - 0)^2 + (m_1 - 0)^2 = 1 + m_1^2$$
$$[d(A, B)]^2 = (1 - 1)^2 + (m_2 - m_1)^2 = m_2^2 - 2m_1 m_2 + m_1^2$$

Using these facts in equation (5), we get

$$\left(1 + m_2^2\right) + \left(1 + m_1^2\right) = m_2^2 - 2m_1m_2 + m_1^2$$

which, upon simplification, can be written as

$$m_1m_2 = -1$$

If the lines are perpendicular, the product of their slopes is -1. ∎

You may find it easier to remember the condition for two nonvertical lines to be perpendicular by observing that the equality $m_1m_2 = -1$ means that m_1 and m_2 are negative reciprocals of each other; that is, either $m_1 = -\dfrac{1}{m_2}$ or $m_2 = -\dfrac{1}{m_1}$.

EXAMPLE 12 **Finding the Slope of a Line Perpendicular to Another Line**

If a line has slope $\dfrac{3}{2}$, any line having slope $-\dfrac{2}{3}$ is perpendicular to it.

EXAMPLE 13 **Finding the Equation of a Line Perpendicular to a Given Line**

Find an equation of the line that contains the point $(1, -2)$ and is perpendicular to the line $x + 3y = 6$. Graph the two lines.

Solution First write the equation of the given line in slope–intercept form to find its slope.

$$x + 3y = 6$$
$$3y = -x + 6 \qquad \text{\textit{Proceed to solve for y.}}$$
$$y = -\frac{1}{3}x + 2 \qquad \text{\textit{Place in the form y = mx + b.}}$$

The given line has slope $-\dfrac{1}{3}$. Any line perpendicular to this line will have slope 3.

Because we require the point $(1, -2)$ to be on this line with slope 3, use the point–slope form of the equation of a line.

$$y - y_1 = m(x - x_1) \qquad \text{\textit{Point–slope form}}$$
$$y - (-2) = 3(x - 1) \qquad \text{\textit{m = 3, x}}_1 \text{\textit{ = 1, y}}_1 \text{\textit{ = -2}}$$

To obtain other forms of the equation, proceed as follows:

$$y + 2 = 3(x - 1)$$
$$y + 2 = 3x - 3 \qquad \text{\textit{Simplify.}}$$
$$y = 3x - 5 \qquad \text{\textit{Slope–intercept form}}$$
$$3x - y = 5 \qquad \text{\textit{General form}}$$

Figure 48 shows the graphs.

Figure 48

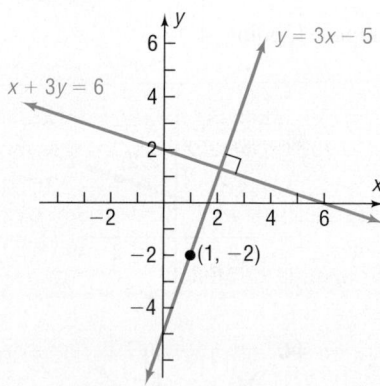

── **Now Work** PROBLEM 65

WARNING Be sure to use a square screen when you graph perpendicular lines. Otherwise, the angle between the two lines will appear distorted. A discussion of square screens is given in Section 5 of the Appendix. ∎

2.3 Assess Your Understanding

Concepts and Vocabulary

1. The slope of a vertical line is _____ ; the slope of a horizontal line is _____ .

2. For the line $2x + 3y = 6$, the x-intercept is _____ and the y-intercept is _____ .

3. A horizontal line is given by an equation of the form _____ , where b is the _____ .

4. *True or False* Vertical lines have an undefined slope.

5. *True or False* The slope of the line $2y = 3x + 5$ is 3.

6. *True or False* The point $(1, 2)$ is on the line $2x + y = 4$.

7. Two nonvertical lines have slopes m_1 and m_2, respectively. The lines are parallel if _____ and the _____ are unequal; the lines are perpendicular if _____ .

8. The lines $y = 2x + 3$ and $y = ax + 5$ are parallel if $a = $ _____ .

9. The lines $y = 2x - 1$ and $y = ax + 2$ are perpendicular if $a = $ _____ .

10. *True or False* Perpendicular lines have slopes that are reciprocals of one another.

Skill Building

In Problems 11–14, (a) find the slope of the line and (b) interpret the slope.

11.

12.

13.

14.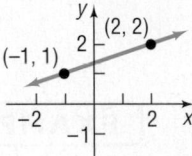

In Problems 15–22, plot each pair of points and determine the slope of the line containing them. Graph the line.

15. $(2, 3)$; $(4, 0)$

16. $(4, 2)$; $(3, 4)$

17. $(-2, 3)$; $(2, 1)$

18. $(-1, 1)$; $(2, 3)$

19. $(-3, -1)$; $(2, -1)$

20. $(4, 2)$; $(-5, 2)$

21. $(-1, 2)$; $(-1, -2)$

22. $(2, 0)$; $(2, 2)$

In Problems 23–30, graph the line containing the point P and having slope m.

23. $P = (1, 2)$; $m = 3$

24. $P = (2, 1)$; $m = 4$

25. $P = (2, 4)$; $m = -\dfrac{3}{4}$

26. $P = (1, 3)$; $m = -\dfrac{2}{5}$

27. $P = (-1, 3)$; $m = 0$

28. $P = (2, -4)$; $m = 0$

29. $P = (0, 3)$; slope undefined

30. $P = (-2, 0)$; slope undefined

In Problems 31–36, the slope and a point on a line are given. Use this information to locate three additional points on the line. Answers may vary.

[**Hint:** *It is not necessary to find the equation of the line. See Example 3.*]

31. Slope 4; point $(1, 2)$

32. Slope 2; point $(-2, 3)$

33. Slope $-\dfrac{3}{2}$; point $(2, -4)$

34. Slope $\dfrac{4}{3}$; point $(-3, 2)$

35. Slope -2; point $(-2, -3)$

36. Slope -1; point $(4, 1)$

In Problems 37–44, find an equation of the line L.

37.

38.

39.

40.

41.

L is parallel to $y = 2x$

42.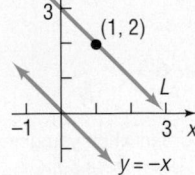

L is parallel to $y = -x$

43.

L is perpendicular to $y = 2x$

44.

L is perpendicular to $y = -x$

In Problems 45–70, find an equation for the line with the given properties. Express your answer using either the general form or the slope–intercept form of the equation of a line, whichever you prefer.

45. Slope = 3; containing the point $(-2, 3)$

46. Slope = 2; containing the point $(4, -3)$

47. Slope = $-\dfrac{2}{3}$; containing the point $(1, -1)$

48. Slope = $\dfrac{1}{2}$; containing the point $(3, 1)$

49. Containing the points $(1, 3)$ and $(-1, 2)$

50. Containing the points $(-3, 4)$ and $(2, 5)$

51. Slope = -3; y-intercept = 3

52. Slope = -2; y-intercept = -2

53. x-intercept = 2; y-intercept = -1

54. x-intercept = -4; y-intercept = 4

55. Slope undefined; containing the point $(2, 4)$

56. Slope undefined; containing the point $(3, 8)$

57. Horizontal; containing the point $(-3, 2)$

58. Vertical; containing the point $(4, -5)$

59. Parallel to the line $y = 2x$; containing the point $(-1, 2)$

60. Parallel to the line $y = -3x$; containing the point $(-1, 2)$

61. Parallel to the line $2x - y = -2$; containing the point $(0, 0)$

62. Parallel to the line $x - 2y = -5$; containing the point $(0, 0)$

63. Parallel to the line $x = 5$; containing the point $(4, 2)$

64. Parallel to the line $y = 5$; containing the point $(4, 2)$

65. Perpendicular to the line $y = \dfrac{1}{2}x + 4$; containing the point $(1, -2)$

66. Perpendicular to the line $y = 2x - 3$; containing the point $(1, -2)$

67. Perpendicular to the line $2x + y = 2$; containing the point $(-3, 0)$

68. Perpendicular to the line $x - 2y = -5$; containing the point $(0, 4)$

69. Perpendicular to the line $x = 8$; containing the point $(3, 4)$

70. Perpendicular to the line $y = 8$; containing the point $(3, 4)$

In Problems 71–90, find the slope and y-intercept of each line. Graph the line.

71. $y = 2x + 3$

72. $y = -3x + 4$

73. $\dfrac{1}{2}y = x - 1$

74. $\dfrac{1}{3}x + y = 2$

75. $y = \dfrac{1}{2}x + 2$

76. $y = 2x + \dfrac{1}{2}$

77. $x + 2y = 4$

78. $-x + 3y = 6$

79. $2x - 3y = 6$

80. $3x + 2y = 6$

81. $x + y = 1$

82. $x - y = 2$

83. $x = -4$

84. $y = -1$

85. $y = 5$

86. $x = 2$

87. $y - x = 0$

88. $x + y = 0$

89. $2y - 3x = 0$

90. $3x + 2y = 0$

In Problems 91–100, (a) find the intercepts of the graph of each equation and (b) graph the equation.

91. $2x + 3y = 6$

92. $3x - 2y = 6$

93. $-4x + 5y = 40$

94. $6x - 4y = 24$

95. $7x + 2y = 21$

96. $5x + 3y = 18$

97. $\dfrac{1}{2}x + \dfrac{1}{3}y = 1$

98. $x - \dfrac{2}{3}y = 4$

99. $0.2x - 0.5y = 1$

100. $-0.3x + 0.4y = 1.2$

101. Find an equation of the x-axis.

102. Find an equation of the y-axis.

In Problems 103–106, the equations of two lines are given. Determine if the lines are parallel, perpendicular, or neither.

103. $y = 2x - 3$
$y = 2x + 4$

104. $y = \dfrac{1}{2}x - 3$
$y = -2x + 4$

105. $y = 4x + 5$
$y = -4x + 2$

106. $y = -2x + 3$
$y = -\dfrac{1}{2}x + 2$

In Problems 107–110, write an equation of each line. Express your answer using either the general form or the slope–intercept form of the equation of a line, whichever you prefer.

107.

108.

109.

110.

Applications and Extensions

111. Geometry Use slopes to show that the triangle whose vertices are $(-2, 5)$, $(1, 3)$, and $(-1, 0)$ is a right triangle.

112. Geometry Use slopes to show that the quadrilateral whose vertices are $(1, -1)$, $(4, 1)$, $(2, 2)$, and $(5, 4)$ is a parallelogram.

113. Geometry Use slopes to show that the quadrilateral whose vertices are $(-1, 0)$, $(2, 3)$, $(1, -2)$, and $(4, 1)$ is a rectangle.

114. Geometry Use slopes and the distance formula to show that the quadrilateral whose vertices are $(0, 0)$, $(1, 3)$, $(4, 2)$, and $(3, -1)$ is a square.

115. Truck Rentals A truck rental company rents a moving truck for one day by charging $29 plus $0.20 per mile. Write a linear equation that relates the cost C, in dollars, of renting the truck to the number x of miles driven. What is the cost of renting the truck if the truck is driven 110 miles? 230 miles?

116. Cost Equation The **fixed costs** of operating a business are the costs incurred regardless of the level of production. Fixed costs include rent, fixed salaries, and costs of leasing machinery. The **variable costs** of operating a business are the costs that change with the level of output. Variable costs include raw materials, hourly wages, and electricity. Suppose that a manufacturer of jeans has fixed daily costs of $500 and variable costs of $8 for each pair of jeans manufactured. Write a linear equation that relates the daily cost C, in dollars, of manufacturing the jeans to the number x of jeans manufactured. What is the cost of manufacturing 400 pairs of jeans? 740 pairs?

117. Cost of Driving a Car The annual fixed costs for owning a small sedan are $1289, assuming the car is completely paid for. The cost to drive the car is approximately $0.15 per mile. Write a linear equation that relates the cost C and the number x of miles driven annually.

Source: www.pacebus.com

118. Wages of a Car Salesperson Dan receives $375 per week for selling new and used cars at a car dealership in Oak Lawn, Illinois. In addition, he receives 5% of the profit on any sales that he generates. Write a linear equation that represents Dan's weekly salary S when he has sales that generate a profit of x dollars.

119. Electricity Rates in Illinois Commonwealth Edison Company supplies electricity to residential customers for a monthly customer charge of $10.55 plus 9.44 cents per kilowatt-hour for up to 600 kilowatt-hours.

(a) Write a linear equation that relates the monthly charge C, in dollars, to the number x of kilowatt-hours used in a month, $0 \le x \le 600$.
(b) Graph this equation.
(c) What is the monthly charge for using 200 kilowatt-hours?
(d) What is the monthly charge for using 500 kilowatt-hours?
(e) Interpret the slope of the line.

Source: Commonwealth Edison Company, January, 2010.

120. Electricity Rates in Florida Florida Power & Light Company supplies electricity to residential customers for a monthly customer charge of $5.69 plus 8.48 cents per kilowatt-hour for up to 1000 kilowatt-hours.

(a) Write a linear equation that relates the monthly charge C, in dollars, to the number x of kilowatt-hours used in a month, $0 \le x \le 1000$.
(b) Graph this equation.
(c) What is the monthly charge for using 200 kilowatt-hours?
(d) What is the monthly charge for using 500 kilowatt-hours?
(e) Interpret the slope of the line.

Source: Florida Power & Light Company, February, 2010.

121. Measuring Temperature The relationship between Celsius (°C) and Fahrenheit (°F) degrees of measuring temperature is linear. Find a linear equation relating °C and °F if 0°C corresponds to 32°F and 100°C corresponds to 212°F. Use the equation to find the Celsius measure of 70°F.

122. Measuring Temperature The Kelvin (K) scale for measuring temperature is obtained by adding 273 to the Celsius temperature.

(a) Write a linear equation relating K and °C.
(b) Write a linear equation relating K and °F (see Problem 121).

123. Access Ramp A wooden access ramp is being built to reach a platform that sits 30 inches above the floor. The ramp drops 2 inches for every 25-inch run.

(a) Write a linear equation that relates the height y of the ramp above the floor to the horizontal distance x from the platform.
(b) Find and interpret the x-intercept of the graph of your equation.
(c) Design requirements stipulate that the maximum run be 30 feet and that the maximum slope be a drop of 1 inch for each 12 inches of run. Will this ramp meet the requirements? Explain.
(d) What slopes could be used to obtain the 30-inch rise and still meet design requirements?

Source: www.adaptiveaccess.com/wood_ramps.php

124. Cigarette Use A report in the Child Trends DataBase indicated that, in 1996, 22.2% of twelfth grade students reported daily use of cigarettes. In 2006, 12.2% of twelfth grade students reported daily use of cigarettes.

(a) Write a linear equation that relates the percent y of twelfth grade students who smoke cigarettes daily to the number x of years after 1996.

(b) Find the intercepts of the graph of your equation.

(c) Do the intercepts have any meaningful interpretation?

(d) Use your equation to predict the percent for the year 2016. Is this result reasonable?

Source: www.childtrendsdatabank.org

125. Product Promotion A cereal company finds that the number of people who will buy one of its products in the first month that it is introduced is linearly related to the amount of money it spends on advertising. If it spends \$40,000 on advertising, then 100,000 boxes of cereal will be sold, and if it spends \$60,000, then 200,000 boxes will be sold.

(a) Write a linear equation that relates the amount A spent on advertising to the number x of boxes the company aims to sell.

(b) How much advertising is needed to sell 300,000 boxes of cereal?

(c) Interpret the slope.

126. Show that the line containing the points (a, b) and (b, a), $a \neq b$, is perpendicular to the line $y = x$. Also show that the midpoint of (a, b) and (b, a) lies on the line $y = x$.

127. The equation $2x - y = C$ defines a **family of lines,** one line for each value of C. On one set of coordinate axes, graph the members of the family when $C = -4$, $C = 0$, and $C = 2$. Can you draw a conclusion from the graph about each member of the family?

128. Prove that if two nonvertical lines have slopes whose product is -1 then the lines are perpendicular. [**Hint:** Refer to Figure 47 and use the converse of the Pythagorean Theorem.]

Explaining Concepts: Discussion and Writing

129. Which of the following equations might have the graph shown? (More than one answer is possible.)

(a) $2x + 3y = 6$

(b) $-2x + 3y = 6$

(c) $3x - 4y = -12$

(d) $x - y = 1$

(e) $x - y = -1$

(f) $y = 3x - 5$

(g) $y = 2x + 3$

(h) $y = -3x + 3$

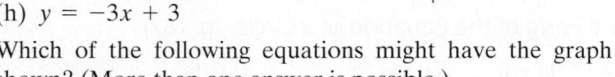

130. Which of the following equations might have the graph shown? (More than one answer is possible.)

(a) $2x + 3y = 6$

(b) $2x - 3y = 6$

(c) $3x + 4y = 12$

(d) $x - y = 1$

(e) $x - y = -1$

(f) $y = -2x - 1$

(g) $y = -\dfrac{1}{2}x + 10$

(h) $y = x + 4$

131. The figure shows the graph of two parallel lines. Which of the following pairs of equations might have such a graph?

(a) $x - 2y = 3$
 $x + 2y = 7$

(b) $x + y = 2$
 $x + y = -1$

(c) $x - y = -2$
 $x - y = 1$

(d) $x - y = -2$
 $2x - 2y = -4$

(e) $x + 2y = 2$
 $x + 2y = -1$

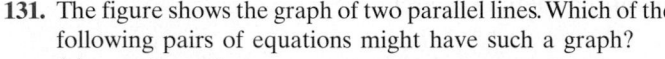

132. The figure shows the graph of two perpendicular lines. Which of the following pairs of equations might have such a graph?

(a) $y - 2x = 2$
 $y + 2x = -1$

(b) $y - 2x = 0$
 $2y + x = 0$

(c) $2y - x = 2$
 $2y + x = -2$

(d) $y - 2x = 2$
 $x + 2y = -1$

(e) $2x + y = -2$
 $2y + x = -2$

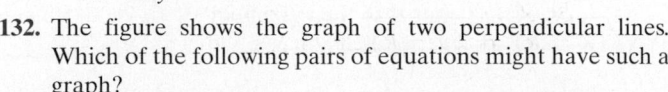

133. m is for Slope The accepted symbol used to denote the slope of a line is the letter m. Investigate the origin of this symbolism. Begin by consulting a French dictionary and looking up the French word *monter*. Write a brief essay on your findings.

134. Grade of a Road The term *grade* is used to describe the inclination of a road. How does this term relate to the notion of slope of a line? Is a 4% grade very steep? Investigate the grades of some mountainous roads and determine their slopes. Write a brief essay on your findings.

135. Carpentry Carpenters use the term *pitch* to describe the steepness of staircases and roofs. How does pitch relate to slope? Investigate typical pitches used for stairs and for roofs. Write a brief essay on your findings.

136. Can the equation of every line be written in slope–intercept form? Why?

137. Does every line have exactly one x-intercept and one y-intercept? Are there any lines that have no intercepts?

138. What can you say about two lines that have equal slopes and equal y-intercepts?

139. What can you say about two lines with the same x-intercept and the same y-intercept? Assume that the x-intercept is not 0.

140. If two distinct lines have the same slope, but different x-intercepts, can they have the same y-intercept?

141. If two distinct lines have the same y-intercept, but different slopes, can they have the same x-intercept?

142. Which form of the equation of a line do you prefer to use? Justify your position with an example that shows that your choice is better than another. Have reasons.

143. What Went Wrong? A student is asked to find the slope of the line joining $(-3, 2)$ and $(1, -4)$. He states that the slope is $\dfrac{3}{2}$. Is he correct? If not, what went wrong?

Interactive Exercises

Ask your instructor if the applet below is of interest to you.

144. Slope *Open the slope applet.* Move point B around the Cartesian plane with your mouse.
 (a) Move B to the point whose coordinates are $(2, 7)$. What is the slope of the line?
 (b) Move B to the point whose coordinates are $(3, 6)$. What is the slope of the line?
 (c) Move B to the point whose coordinates are $(4, 5)$. What is the slope of the line?
 (d) Move B to the point whose coordinates are $(4, 4)$. What is the slope of the line?
 (e) Move B to the point whose coordinates are $(4, 1)$. What is the slope of the line?
 (f) Move B to the point whose coordinates are $(3, -2)$. What is the slope of the line?
 (g) Slowly move B to a point whose *x*-coordinate is 1. What happens to the value of the slope as the *x*-coordinate approaches 1?
 (h) What can be said about a line whose slope is positive? What can be said about a line whose slope is negative? What can be said about a line whose slope is 0?
 (i) Consider the results of parts (a)–(c). What can be said about the steepness of a line with positive slope as its slope increases?
 (j) Move B to the point whose coordinates are $(3, 5)$. What is the slope of the line? Move B to the point whose coordinates are $(5, 6)$. What is the slope of the line? Move B to the point whose coordinates are $(-1, 3)$. What is the slope of the line?

2.4 Circles

PREPARING FOR THIS SECTION *Before getting started, review the following:*

• Completing the Square (Chapter R, Section R.5, p. 56) • Square Root Method (Section 1.2, pp. 94–95)

Now Work the '**Are You Prepared?**' problems on page 185.

OBJECTIVES **1** Write the Standard Form of the Equation of a Circle (p. 182)
 2 Graph a Circle (p. 183)
 3 Work with the General Form of the Equation of a Circle (p. 184)

1 Write the Standard Form of the Equation of a Circle

One advantage of a coordinate system is that it enables us to translate a geometric statement into an algebraic statement, and vice versa. Consider, for example, the following geometric statement that defines a circle.

DEFINITION A **circle** is a set of points in the *xy*-plane that are a fixed distance *r* from a fixed point (h, k). The fixed distance *r* is called the **radius,** and the fixed point (h, k) is called the **center** of the circle.

Figure 49

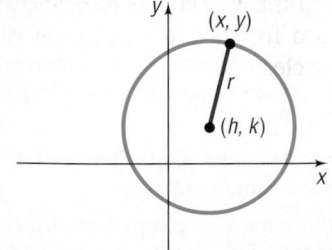

Figure 49 shows the graph of a circle. To find the equation, let (x, y) represent the coordinates of any point on a circle with radius *r* and center (h, k). Then the distance between the points (x, y) and (h, k) must always equal *r*. That is, by the distance formula

$$\sqrt{(x - h)^2 + (y - k)^2} = r$$

or, equivalently,

$$(x - h)^2 + (y - k)^2 = r^2$$

DEFINITION The **standard form of an equation of a circle** with radius *r* and center (h, k) is

$$(x - h)^2 + (y - k)^2 = r^2 \tag{1}$$

THEOREM

The standard form of an equation of a circle of radius r with center at the origin $(0, 0)$ is

$$x^2 + y^2 = r^2$$

DEFINITION

If the radius $r = 1$, the circle whose center is at the origin is called the **unit circle** and has the equation

$$x^2 + y^2 = 1$$

See Figure 50. Notice that the graph of the unit circle is symmetric with respect to the x-axis, the y-axis, and the origin.

Figure 50
Unit circle $x^2 + y^2 = 1$

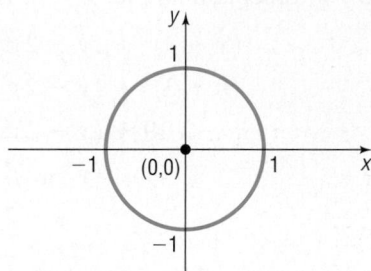

EXAMPLE 1

Writing the Standard Form of the Equation of a Circle

Write the standard form of the equation of the circle with radius 5 and center $(-3, 6)$.

Solution Using equation (1) and substituting the values $r = 5$, $h = -3$, and $k = 6$, we have

$$(x - h)^2 + (y - k)^2 = r^2$$
$$(x + 3)^2 + (y - 6)^2 = 25$$

Now Work PROBLEM 7

2 Graph a Circle

EXAMPLE 2

Graphing a Circle

Graph the equation: $(x + 3)^2 + (y - 2)^2 = 16$

Solution Since the equation is in the form of equation (1), its graph is a circle. To graph the equation, compare the given equation to the standard form of the equation of a circle. The comparison yields information about the circle.

Figure 51

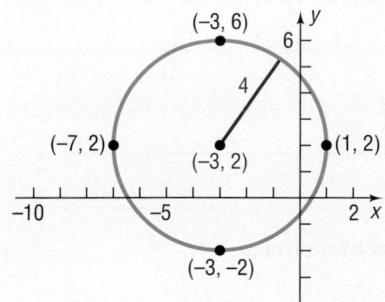

$$(x + 3)^2 + (y - 2)^2 = 16$$
$$(x - (-3))^2 + (y - 2)^2 = 4^2$$
$$\uparrow \qquad\qquad \uparrow \qquad\quad \uparrow$$
$$(x - h)^2 + (y - k)^2 = r^2$$

We see that $h = -3$, $k = 2$, and $r = 4$. The circle has center $(-3, 2)$ and a radius of 4 units. To graph this circle, first plot the center $(-3, 2)$. Since the radius is 4, we can locate four points on the circle by plotting points 4 units to the left, to the right, up, and down from the center. These four points can then be used as guides to obtain the graph. See Figure 51.

Now Work PROBLEMS 23(a) AND (b)

EXAMPLE 3

Finding the Intercepts of a Circle

For the circle $(x + 3)^2 + (y - 2)^2 = 16$, find the intercepts, if any, of its graph.

Solution

This is the equation discussed and graphed in Example 2. To find the x-intercepts, if any, let $y = 0$. Then

$$(x + 3)^2 + (y - 2)^2 = 16$$
$$(x + 3)^2 + (0 - 2)^2 = 16 \qquad y = 0$$
$$(x + 3)^2 + 4 = 16 \qquad \text{Simplify.}$$
$$(x + 3)^2 = 12 \qquad \text{Simplify.}$$
$$x + 3 = \pm\sqrt{12} \qquad \text{Apply the Square Root Method.}$$
$$x = -3 \pm 2\sqrt{3} \qquad \text{Solve for } x.$$

The x-intercepts are $-3 - 2\sqrt{3} \approx -6.46$ and $-3 + 2\sqrt{3} \approx 0.46$.
To find the y-intercepts, if any, let $x = 0$. Then

$$(x + 3)^2 + (y - 2)^2 = 16$$
$$(0 + 3)^2 + (y - 2)^2 = 16$$
$$9 + (y - 2)^2 = 16$$
$$(y - 2)^2 = 7$$
$$y - 2 = \pm\sqrt{7}$$
$$y = 2 \pm \sqrt{7}$$

The y-intercepts are $2 - \sqrt{7} \approx -0.65$ and $2 + \sqrt{7} \approx 4.65$.
Look back at Figure 51 to verify the approximate locations of the intercepts.

 Now Work PROBLEM 23(c)

3 Work with the General Form of the Equation of a Circle

If we eliminate the parentheses from the standard form of the equation of the circle given in Example 2, we get

$$(x + 3)^2 + (y - 2)^2 = 16$$
$$x^2 + 6x + 9 + y^2 - 4y + 4 = 16$$

which, upon simplifying, is equivalent to

$$x^2 + y^2 + 6x - 4y - 3 = 0 \tag{2}$$

It can be shown that any equation of the form

$$x^2 + y^2 + ax + by + c = 0$$

has a graph that is a circle, or a point, or has no graph at all. For example, the graph of the equation $x^2 + y^2 = 0$ is the single point $(0, 0)$. The equation $x^2 + y^2 + 5 = 0$, or $x^2 + y^2 = -5$, has no graph, because sums of squares of real numbers are never negative.

DEFINITION

When its graph is a circle, the equation

$$x^2 + y^2 + ax + by + c = 0$$

is referred to as the **general form of the equation of a circle.**

Now Work PROBLEM 13

If an equation of a circle is in the general form, we use the method of completing the square to put the equation in standard form so that we can identify its center and radius.

EXAMPLE 4 **Graphing a Circle Whose Equation Is in General Form**

Graph the equation $x^2 + y^2 + 4x - 6y + 12 = 0$

Solution Group the terms involving x, group the terms involving y, and put the constant on the right side of the equation. The result is

$$(x^2 + 4x) + (y^2 - 6y) = -12$$

Next, complete the square of each expression in parentheses. Remember that any number added on the left side of the equation must also be added on the right.

$$(x^2 + 4x + 4) + (y^2 - 6y + 9) = -12 + 4 + 9$$

$$\left(\frac{4}{2}\right)^2 = 4 \qquad \left(\frac{-6}{2}\right)^2 = 9$$

$$(x + 2)^2 + (y - 3)^2 = 1 \quad \text{Factor.}$$

Figure 52

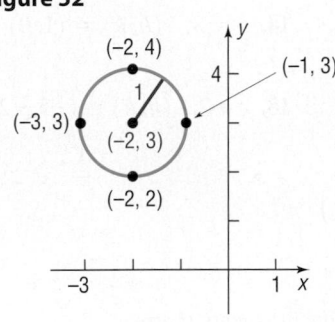

This equation is the standard form of the equation of a circle with radius 1 and center $(-2, 3)$.

To graph the equation use the center $(-2, 3)$ and the radius 1. See Figure 52.

───── **Now Work** PROBLEM 27

 EXAMPLE 5 **Using a Graphing Utility to Graph a Circle**

Graph the equation: $x^2 + y^2 = 4$

Figure 53

Solution This is the equation of a circle with center at the origin and radius 2. To graph this equation, solve for y.

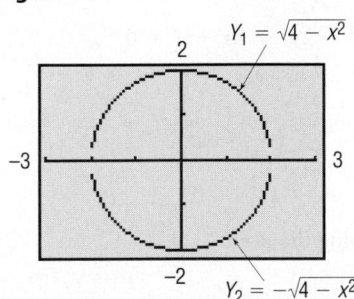

$$x^2 + y^2 = 4$$

$$y^2 = 4 - x^2 \qquad \text{Subtract } x^2 \text{ from each side.}$$

$$y = \pm\sqrt{4 - x^2} \qquad \text{Apply the Square Root Method to solve for } y$$

There are two equations to graph: first graph $Y_1 = \sqrt{4 - x^2}$ and then graph $Y_2 = -\sqrt{4 - x^2}$ on the same square screen. (Your circle will appear oval if you do not use a square screen.) See Figure 53.

2.4 Assess Your Understanding

Are You Prepared? *Answers are given at the end of these exercises. If you get a wrong answer, read the pages listed in red.*

1. To complete the square of $x^2 + 10x$, you would (*add/subtract*) the number _____. (p. 56)

2. Use the Square Root Method to solve the equation $(x - 2)^2 = 9$. (pp. 94–95)

Concepts and Vocabulary

3. *True or False* Every equation of the form
$$x^2 + y^2 + ax + by + c = 0$$
has a circle as its graph.

4. For a circle, the _____ is the distance from the center to any point on the circle.

5. *True or False* The radius of the circle $x^2 + y^2 = 9$ is 3.

6. *True or False* The center of the circle
$$(x + 3)^2 + (y - 2)^2 = 13$$
is $(3, -2)$.

Skill Building

In Problems 7–10, find the center and radius of each circle. Write the standard form of the equation.

7.

8.

9.

10.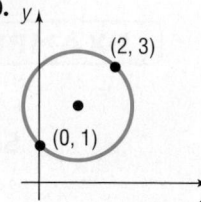

In Problems 11–20, write the standard form of the equation and the general form of the equation of each circle of radius r and center (h, k). Graph each circle.

11. $r = 2$; $(h, k) = (0, 0)$

12. $r = 3$; $(h, k) = (0, 0)$

13. $r = 2$; $(h, k) = (0, 2)$

14. $r = 3$; $(h, k) = (1, 0)$

15. $r = 5$; $(h, k) = (4, -3)$

16. $r = 4$; $(h, k) = (2, -3)$

17. $r = 4$; $(h, k) = (-2, 1)$

18. $r = 7$; $(h, k) = (-5, -2)$

19. $r = \dfrac{1}{2}$; $(h, k) = \left(\dfrac{1}{2}, 0\right)$

20. $r = \dfrac{1}{2}$; $(h, k) = \left(0, -\dfrac{1}{2}\right)$

In Problems 21–34, (a) find the center (h, k) and radius r of each circle; (b) graph each circle; (c) find the intercepts, if any.

21. $x^2 + y^2 = 4$

22. $x^2 + (y - 1)^2 = 1$

23. $2(x - 3)^2 + 2y^2 = 8$

24. $3(x + 1)^2 + 3(y - 1)^2 = 6$

25. $x^2 + y^2 - 2x - 4y - 4 = 0$

26. $x^2 + y^2 + 4x + 2y - 20 = 0$

27. $x^2 + y^2 + 4x - 4y - 1 = 0$

28. $x^2 + y^2 - 6x + 2y + 9 = 0$

29. $x^2 + y^2 - x + 2y + 1 = 0$

30. $x^2 + y^2 + x + y - \dfrac{1}{2} = 0$

31. $2x^2 + 2y^2 - 12x + 8y - 24 = 0$

32. $2x^2 + 2y^2 + 8x + 7 = 0$

33. $2x^2 + 8x + 2y^2 = 0$

34. $3x^2 + 3y^2 - 12y = 0$

In Problems 35–42, find the standard form of the equation of each circle.

35. Center at the origin and containing the point $(-2, 3)$

36. Center $(1, 0)$ and containing the point $(-3, 2)$

37. Center $(2, 3)$ and tangent to the x-axis

38. Center $(-3, 1)$ and tangent to the y-axis

39. With endpoints of a diameter at $(1, 4)$ and $(-3, 2)$

40. With endpoints of a diameter at $(4, 3)$ and $(0, 1)$

41. Center $(-1, 3)$ and tangent to the line $y = 2$

42. Center $(4, -2)$ and tangent to the line $x = 1$

In Problems 43–46, match each graph with the correct equation.

(a) $(x - 3)^2 + (y + 3)^2 = 9$ (b) $(x + 1)^2 + (y - 2)^2 = 4$ (c) $(x - 1)^2 + (y + 2)^2 = 4$ (d) $(x + 3)^2 + (y - 3)^2 = 9$

43.

44.

45.

46.

Applications and Extensions

47. Find the area of the square in the figure.

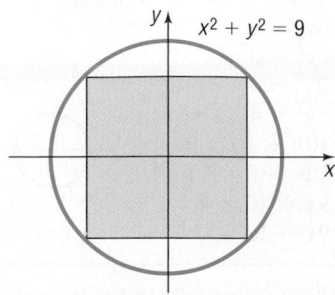

48. Find the area of the blue shaded region in the figure, assuming the quadrilateral inside the circle is a square.

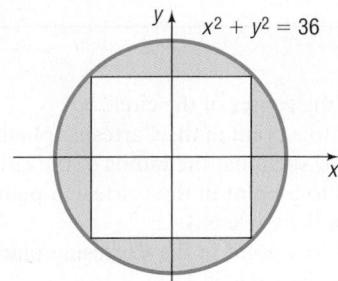

49. Ferris Wheel The original Ferris wheel was built in 1893 by Pittsburgh, Pennsylvania, bridge builder George W. Ferris. The Ferris wheel was originally built for the 1893 World's Fair in Chicago, but was also later reconstructed for the 1904 World's Fair in St. Louis. It had a maximum height of 264 feet and a wheel diameter of 250 feet. Find an equation for the wheel if the center of the wheel is on the y-axis.

Source: inventors.about.com

50. Ferris Wheel In 2008, the Singapore Flyer opened as the world's largest Ferris wheel. It has a maximum height of 165 meters and a diameter of 150 meters, with one full rotation taking approximately 30 minutes. Find an equation for the wheel if the center of the wheel is on the y-axis.

Source: Wikipedia

51. Weather Satellites Earth is represented on a map of a portion of the solar system so that its surface is the circle with equation $x^2 + y^2 + 2x + 4y - 4091 = 0$. A weather satellite circles 0.6 unit above Earth with the center of its circular orbit at the center of Earth. Find the equation for the orbit of the satellite on this map.

52. The **tangent line** to a circle may be defined as the line that intersects the circle in a single point, called the **point of tangency.** See the figure.

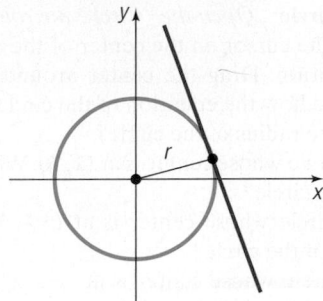

If the equation of the circle is $x^2 + y^2 = r^2$ and the equation of the tangent line is $y = mx + b$, show that:

(a) $r^2(1 + m^2) = b^2$

[**Hint:** The quadratic equation $x^2 + (mx + b)^2 = r^2$ has exactly one solution.]

(b) The point of tangency is $\left(\dfrac{-r^2 m}{b}, \dfrac{r^2}{b} \right)$.

(c) The tangent line is perpendicular to the line containing the center of the circle and the point of tangency.

53. The Greek Method The Greek method for finding the equation of the tangent line to a circle uses the fact that at any point on a circle the lines containing the center and the tangent line are perpendicular (see Problem 52). Use this method to find an equation of the tangent line to the circle $x^2 + y^2 = 9$ at the point $(1, 2\sqrt{2})$.

54. Use the Greek method described in Problem 53 to find an equation of the tangent line to the circle $x^2 + y^2 - 4x + 6y + 4 = 0$ at the point $(3, 2\sqrt{2} - 3)$.

55. Refer to Problem 52. The line $x - 2y + 4 = 0$ is tangent to a circle at $(0, 2)$. The line $y = 2x - 7$ is tangent to the same circle at $(3, -1)$. Find the center of the circle.

56. Find an equation of the line containing the centers of the two circles

$$x^2 + y^2 - 4x + 6y + 4 = 0$$

and

$$x^2 + y^2 + 6x + 4y + 9 = 0$$

57. If a circle of radius 2 is made to roll along the x-axis, what is an equation for the path of the center of the circle?

58. If the circumference of a circle is 6π, what is its radius?

Explaining Concepts: Discussion and Writing

59. Which of the following equations might have the graph shown? (More than one answer is possible.)
(a) $(x - 2)^2 + (y + 3)^2 = 13$
(b) $(x - 2)^2 + (y - 2)^2 = 8$
(c) $(x - 2)^2 + (y - 3)^2 = 13$
(d) $(x + 2)^2 + (y - 2)^2 = 8$
(e) $x^2 + y^2 - 4x - 9y = 0$
(f) $x^2 + y^2 + 4x - 2y = 0$
(g) $x^2 + y^2 - 9x - 4y = 0$
(h) $x^2 + y^2 - 4x - 4y = 4$

60. Which of the following equations might have the graph shown? (More than one answer is possible.)
(a) $(x - 2)^2 + y^2 = 3$
(b) $(x + 2)^2 + y^2 = 3$
(c) $x^2 + (y - 2)^2 = 3$
(d) $(x + 2)^2 + y^2 = 4$
(e) $x^2 + y^2 + 10x + 16 = 0$
(f) $x^2 + y^2 + 10x - 2y = 1$
(g) $x^2 + y^2 + 9x + 10 = 0$
(h) $x^2 + y^2 - 9x - 10 = 0$

61. Explain how the center and radius of a circle can be used to graph the circle.

62. What Went Wrong? A student stated that the center and radius of the graph whose equation is $(x + 3)^2 + (y - 2)^2 = 16$ are $(3, -2)$ and 4, respectively. Why is this incorrect?

Interactive Exercises

Ask your instructor if the applets below are of interest to you.

63. Center of a Circle *Open the "Circle: the role of the center" applet.* Place the cursor on the center of the circle and hold the mouse button. Drag the center around the Cartesian plane and note how the equation of the circle changes.
(a) What is the radius of the circle?
(b) Draw a circle whose center is at $(1, 3)$. What is the equation of the circle?
(c) Draw a circle whose center is at $(-1, 3)$. What is the equation of the circle?
(d) Draw a circle whose center is at $(-1, -3)$. What is the equation of the circle?
(e) Draw a circle whose center is at $(1, -3)$. What is the equation of the circle?
(f) Write a few sentences explaining the role the center of the circle plays in the equation of the circle.

64. Radius of a Circle *Open the "Circle: the role of the radius" applet.* Place the cursor on point B, press and hold the mouse button. Drag B around the Cartesian plane.

(a) What is the center of the circle?
(b) Move B to a point in the Cartesian plane directly above the center such that the radius of the circle is 5.
(c) Move B to a point in the Cartesian plane such that the radius of the circle is 4.
(d) Move B to a point in the Cartesian plane such that the radius of the circle is 3.
(e) Find the coordinates of two points with integer coordinates in the fourth quadrant on the circle that result in a circle of radius 5 with center equal to that found in part (a).
(f) Use the concept of symmetry about the center, vertical line through the center of the circle, and horizontal line through the center of the circle to find three other points with integer coordinates in the other three quadrants that lie on the circle of radius five with center equal to that found in part (a).

'Are You Prepared?' Answers

1. add; 25 **2.** $\{-1, 5\}$

2.5 Variation

OBJECTIVES **1** Construct a Model Using Direct Variation (p. 189)
2 Construct a Model Using Inverse Variation (p. 189)
3 Construct a Model Using Joint Variation or Combined Variation (p. 190)

 When a mathematical model is developed for a real-world problem, it often involves relationships between quantities that are expressed in terms of proportionality:

Force is proportional to acceleration.

When an ideal gas is held at a constant temperature, pressure and volume are inversely proportional.

The force of attraction between two heavenly bodies is inversely proportional to the square of the distance between them.

Revenue is directly proportional to sales.

Each of the preceding statements illustrates the idea of **variation,** or how one quantity varies in relation to another quantity. Quantities may vary *directly, inversely,* or *jointly.*

1 Construct a Model Using Direct Variation

DEFINITION

Let x and y denote two quantities. Then y **varies directly** with x, or y is **directly proportional to** x, if there is a nonzero number k such that

$$y = kx$$

The number k is called the **constant of proportionality.**

Figure 54
$y = kx; k > 0, x \geq 0$

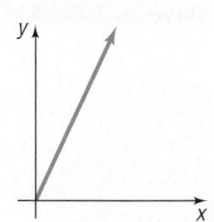

The graph in Figure 54 illustrates the relationship between y and x if y varies directly with x and $k > 0$, $x \geq 0$. Note that the constant of proportionality is, in fact, the slope of the line.

If we know that two quantities vary directly, then knowing the value of each quantity in one instance enables us to write a formula that is true in all cases.

EXAMPLE 1 **Mortgage Payments**

The monthly payment p on a mortgage varies directly with the amount borrowed B. If the monthly payment on a 30-year mortgage is \$6.65 for every \$1000 borrowed, find a formula that relates the monthly payment p to the amount borrowed B for a mortgage with these terms. Then find the monthly payment p when the amount borrowed B is \$120,000.

Solution Because p varies directly with B, we know that

$$p = kB$$

Figure 55

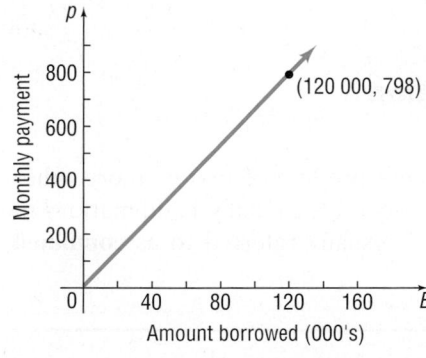

for some constant k. Because $p = 6.65$ when $B = 1000$, it follows that

$$6.65 = k(1000)$$
$$k = 0.00665 \quad \text{Solve for } k.$$

Since $p = kB$,

$$p = 0.00665B$$

In particular, when $B = \$120,000$,

$$p = 0.00665(\$120,000) = \$798$$

Figure 55 illustrates the relationship between the monthly payment p and the amount borrowed B.

━━━**Now Work** PROBLEMS 3 AND 21

2 Construct a Model Using Inverse Variation

Figure 56
$y = \dfrac{k}{x}; k > 0, x > 0$

DEFINITION

Let x and y denote two quantities. Then y **varies inversely** with x, or y is **inversely proportional to** x, if there is a nonzero constant k such that

$$y = \frac{k}{x}$$

The graph in Figure 56 illustrates the relationship between y and x if y varies inversely with x and $k > 0$, $x > 0$.

| EXAMPLE 2 | **Maximum Weight That Can Be Supported by a Piece of Pine** |

Figure 57

See Figure 57. The maximum weight W that can be safely supported by a 2-inch by 4-inch piece of pine varies inversely with its length l. Experiments indicate that the maximum weight that a 10-foot-long 2-by-4 piece of pine can support is 500 pounds. Write a general formula relating the maximum weight W (in pounds) to length l (in feet). Find the maximum weight W that can be safely supported by a length of 25 feet.

Solution Because W varies inversely with l, we know that

$$W = \frac{k}{l}$$

for some constant k. Because $W = 500$ when $l = 10$, we have

$$500 = \frac{k}{10}$$

$$k = 5000$$

Since $W = \frac{k}{l}$,

$$W = \frac{5000}{l}$$

Figure 58

In particular, the maximum weight W that can be safely supported by a piece of pine 25 feet in length is

$$W = \frac{5000}{25} = 200 \text{ pounds}$$

Figure 58 illustrates the relationship between the weight W and the length l.

━━━━━━➤ **Now Work** PROBLEM 31

3 Construct a Model Using Joint Variation or Combined Variation

When a variable quantity Q is proportional to the product of two or more other variables, we say that Q **varies jointly** with these quantities. Finally, combinations of direct and/or inverse variation may occur. This is usually referred to as **combined variation.**

| EXAMPLE 3 | **Loss of Heat Through a Wall** |

The loss of heat through a wall varies jointly with the area of the wall and the difference between the inside and outside temperatures and varies inversely with the thickness of the wall. Write an equation that relates these quantities.

Solution Begin by assigning symbols to represent the quantities:

L = Heat loss T = Temperature difference
A = Area of wall d = Thickness of wall

Then

$$L = k\frac{AT}{d}$$

where k is the constant of proportionality.

In direct or inverse variation, the quantities that vary may be raised to powers. For example, in the early seventeenth century, Johannes Kepler (1571–1630) discovered that the square of the period of revolution T around the Sun varies directly with the cube of its mean distance a from the Sun. That is, $T^2 = ka^3$, where k is the constant of proportionality.

| EXAMPLE 4 | **Force of the Wind on a Window** |

The force F of the wind on a flat surface positioned at a right angle to the direction of the wind varies jointly with the area A of the surface and the square of the speed v of the wind. A wind of 30 miles per hour blowing on a window measuring 4 feet by 5 feet has a force of 150 pounds. See Figure 59. What is the force on a window measuring 3 feet by 4 feet caused by a wind of 50 miles per hour?

Solution Since F varies jointly with A and v^2, we have

$$F = kAv^2$$

Figure 59

where k is the constant of proportionality. We are told that $F = 150$ when $A = 4 \cdot 5 = 20$ and $v = 30$. Then

$$150 = k(20)(900) \qquad F = kAv^2, F = 150, A = 20, v = 30$$

$$k = \frac{1}{120}$$

Since $F = kAv^2$,

$$F = \frac{1}{120}Av^2$$

For a wind of 50 miles per hour blowing on a window whose area is $A = 3 \cdot 4 = 12$ square feet, the force F is

$$F = \frac{1}{120}(12)(2500) = 250 \text{ pounds}$$

——— **Now Work** PROBLEM **39**

2.5 Assess Your Understanding

Concepts and Vocabulary

1. If x and y are two quantities, then y is directly proportional to x if there is a nonzero number k such that _____.

2. *True or False* If y varies directly with x, then $y = \dfrac{k}{x}$, where k is a constant.

Skill Building

In Problems 3–14, write a general formula to describe each variation.

3. y varies directly with x; $y = 2$ when $x = 10$

4. v varies directly with t; $v = 16$ when $t = 2$

5. A varies directly with x^2; $A = 4\pi$ when $x = 2$

6. V varies directly with x^3; $V = 36\pi$ when $x = 3$

7. F varies inversely with d^2; $F = 10$ when $d = 5$

8. y varies inversely with \sqrt{x}; $y = 4$ when $x = 9$

9. z varies directly with the sum of the squares of x and y; $z = 5$ when $x = 3$ and $y = 4$

10. T varies jointly with the cube root of x and the square of d; $T = 18$ when $x = 8$ and $d = 3$

11. M varies directly with the square of d and inversely with the square root of x; $M = 24$ when $x = 9$ and $d = 4$

12. z varies directly with the sum of the cube of x and the square of y; $z = 1$ when $x = 2$ and $y = 3$

13. The square of T varies directly with the cube of a and inversely with the square of d; $T = 2$ when $a = 2$ and $d = 4$

14. The cube of z varies directly with the sum of the squares of x and y; $z = 2$ when $x = 9$ and $y = 4$

Applications and Extensions

In Problems 15–20, write an equation that relates the quantities.

15. **Geometry** The volume V of a sphere varies directly with the cube of its radius r. The constant of proportionality is $\dfrac{4\pi}{3}$.

16. **Geometry** The square of the length of the hypotenuse c of a right triangle varies jointly with the sum of the squares of the lengths of its legs a and b. The constant of proportionality is 1.

17. **Geometry** The area A of a triangle varies jointly with the lengths of the base b and the height h. The constant of proportionality is $\dfrac{1}{2}$.

18. **Geometry** The perimeter p of a rectangle varies jointly with the sum of the lengths of its sides l and w. The constant of proportionality is 2.

19. **Physics: Newton's Law** The force F (in newtons) of attraction between two bodies varies jointly with their masses m and M (in kilograms) and inversely with the square of the distance d (in meters) between them. The constant of proportionality is $G = 6.67 \times 10^{-11}$.

20. **Physics: Simple Pendulum** The **period** of a pendulum is the time required for one oscillation; the pendulum is usually referred to as **simple** when the angle made to the vertical is less than 5°. The period T of a simple pendulum (in seconds) varies directly with the square root of its length l (in feet). The constant of proportionality is $\dfrac{2\pi}{\sqrt{32}}$.

21. **Mortgage Payments** The monthly payment p on a mortgage varies directly with the amount borrowed B. If the monthly payment on a 30-year mortgage is $6.49 for every $1000 borrowed, find a linear equation that relates the monthly payment p to the amount borrowed B for a mortgage with the same terms. Then find the monthly payment p when the amount borrowed B is $145,000.

22. **Mortgage Payments** The monthly payment p on a mortgage varies directly with the amount borrowed B. If the monthly payment on a 15-year mortgage is $8.99 for every $1000 borrowed, find a linear equation that relates the monthly payment p to the amount borrowed B for a mortgage with the same terms. Then find the monthly payment p when the amount borrowed B is $175,000.

23. **Physics: Falling Objects** The distance s that an object falls is directly proportional to the square of the time t of the fall. If an object falls 16 feet in 1 second, how far will it fall in 3 seconds? How long will it take an object to fall 64 feet?

24. **Physics: Falling Objects** The velocity v of a falling object is directly proportional to the time t of the fall. If, after 2 seconds, the velocity of the object is 64 feet per second, what will its velocity be after 3 seconds?

25. **Physics: Stretching a Spring** The elongation E of a spring balance varies directly with the applied weight W (see the figure). If $E = 3$ when $W = 20$, find E when $W = 15$.

26. **Physics: Vibrating String** The rate of vibration of a string under constant tension varies inversely with the length of the string. If a string is 48 inches long and vibrates 256 times per second, what is the length of a string that vibrates 576 times per second?

27. **Revenue Equation** At the corner Shell station, the revenue R varies directly with the number g of gallons of gasoline sold. If the revenue is $47.40 when the number of gallons sold is 12, find a linear equation that relates revenue R to the number g of gallons of gasoline. Then find the revenue R when the number of gallons of gasoline sold is 10.5.

28. **Cost Equation** The cost C of roasted almonds varies directly with the number A of pounds of almonds purchased. If the cost is $23.75 when the number of pounds of roasted almonds purchased is 5, find a linear equation that relates the cost C to the number A of pounds of almonds purchased. Then find the cost C when the number of pounds of almonds purchased is 3.5.

29. **Demand** Suppose that the demand D for candy at the movie theater is inversely related to the price p.
 (a) When the price of candy is $2.75 per bag, the theater sells 156 bags of candy. Express the demand for candy in terms of its price.
 (b) Determine the number of bags of candy that will be sold if the price is raised to $3 a bag.

30. **Driving to School** The time t that it takes to get to school varies inversely with your average speed s.
 (a) Suppose that it takes you 40 minutes to get to school when your average speed is 30 miles per hour. Express the driving time to school in terms of average speed.
 (b) Suppose that your average speed to school is 40 miles per hour. How long will it take you to get to school?

31. **Pressure** The volume of a gas V held at a constant temperature in a closed container varies inversely with its pressure P. If the volume of a gas is 600 cubic centimeters (cm^3) when the pressure is 150 millimeters of mercury (mm Hg), find the volume when the pressure is 200 mm Hg.

32. Resistance The current i in a circuit is inversely proportional to its resistance Z measured in ohms. Suppose that when the current in a circuit is 30 amperes the resistance is 8 ohms. Find the current in the same circuit when the resistance is 10 ohms.

33. Weight The weight of an object above the surface of Earth varies inversely with the square of the distance from the center of Earth. If Maria weighs 125 pounds when she is on the surface of Earth (3960 miles from the center), determine Maria's weight if she is at the top of Mount McKinley (3.8 miles from the surface of Earth).

34. Intensity of Light The intensity I of light (measured in foot-candles) varies inversely with the square of the distance from the bulb. Suppose that the intensity of a 100-watt light bulb at a distance of 2 meters is 0.075 foot-candle. Determine the intensity of the bulb at a distance of 5 meters.

35. Geometry The volume V of a right circular cylinder varies jointly with the square of its radius r and its height h. The constant of proportionality is π. See the figure. Write an equation for V.

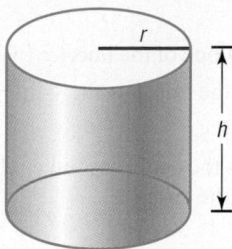

36. Geometry The volume V of a right circular cone varies jointly with the square of its radius r and its height h. The constant of proportionality is $\dfrac{\pi}{3}$. See the figure. Write an equation for V.

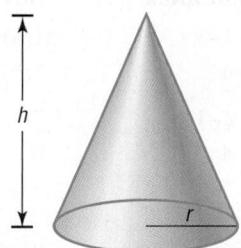

37. Weight of a Body The weight of a body above the surface of Earth varies inversely with the square of the distance from the center of Earth. If a certain body weighs 55 pounds when it is 3960 miles from the center of Earth, how much will it weigh when it is 3965 miles from the center?

38. Force of the Wind on a Window The force exerted by the wind on a plane surface varies jointly with the area of the surface and the square of the velocity of the wind. If the force on an area of 20 square feet is 11 pounds when the wind velocity is 22 miles per hour, find the force on a surface area of 47.125 square feet when the wind velocity is 36.5 miles per hour.

39. Horsepower The horsepower (hp) that a shaft can safely transmit varies jointly with its speed (in revolutions per minute, rpm) and the cube of its diameter. If a shaft of a certain material 2 inches in diameter can transmit 36 hp at 75 rpm, what diameter must the shaft have in order to transmit 45 hp at 125 rpm?

40. Chemistry: Gas Laws The volume V of an ideal gas varies directly with the temperature T and inversely with the pressure P. Write an equation relating V, T, and P using k as the constant of proportionality. If a cylinder contains oxygen at a temperature of 300 K and a pressure of 15 atmospheres in a volume of 100 liters, what is the constant of proportionality k? If a piston is lowered into the cylinder, decreasing the volume occupied by the gas to 80 liters and raising the temperature to 310 K, what is the gas pressure?

41. Physics: Kinetic Energy The kinetic energy K of a moving object varies jointly with its mass m and the square of its velocity v. If an object weighing 25 kilograms and moving with a velocity of 10 meters per second has a kinetic energy of 1250 joules, find its kinetic energy when the velocity is 15 meters per second.

42. Electrical Resistance of a Wire The electrical resistance of a wire varies directly with the length of the wire and inversely with the square of the diameter of the wire. If a wire 432 feet long and 4 millimeters in diameter has a resistance of 1.24 ohms, find the length of a wire of the same material whose resistance is 1.44 ohms and whose diameter is 3 millimeters.

43. Measuring the Stress of Materials The stress in the material of a pipe subject to internal pressure varies jointly with the internal pressure and the internal diameter of the pipe and inversely with the thickness of the pipe. The stress is 100 pounds per square inch when the diameter is 5 inches, the thickness is 0.75 inch, and the internal pressure is 25 pounds per square inch. Find the stress when the internal pressure is 40 pounds per square inch if the diameter is 8 inches and the thickness is 0.50 inch.

44. Safe Load for a Beam The maximum safe load for a horizontal rectangular beam varies jointly with the width of the beam and the square of the thickness of the beam and inversely with its length. If an 8-foot beam will support up to 750 pounds when the beam is 4 inches wide and 2 inches thick, what is the maximum safe load in a similar beam 10 feet long, 6 inches wide, and 2 inches thick?

Explaining Concepts: Discussion and Writing

45. In the early 17th century, Johannes Kepler discovered that the square of the period T of the revolution of a planet around the Sun varies directly with the cube of its mean distance a from the Sun. Go to the library and research this law and Kepler's other two laws. Write a brief paper about these laws and Kepler's place in history.

46. Using a situation that has not been discussed in the text, write a real-world problem that you think involves two variables that vary directly. Exchange your problem with another student's to solve and critique.

47. Using a situation that has not been discussed in the text, write a real-world problem that you think involves two variables that vary inversely. Exchange your problem with another student's to solve and critique.

48. Using a situation that has not been discussed in the text, write a real-world problem that you think involves three variables that vary jointly. Exchange your problem with another student's to solve and critique.

CHAPTER REVIEW

Things to Know

Formulas

Distance formula (p. 151)	$d = \sqrt{(x_2 - x_1)^2 + (y_2 - y_1)^2}$
Midpoint formula (p. 154)	$(x, y) = \left(\dfrac{x_1 + x_2}{2}, \dfrac{y_1 + y_2}{2} \right)$
Slope (p. 167)	$m = \dfrac{y_2 - y_1}{x_2 - x_1}$ if $x_1 \neq x_2$; undefined if $x_1 = x_2$
Parallel lines (p. 175)	Equal slopes ($m_1 = m_2$) and different y-intercepts ($b_1 \neq b_2$)
Perpendicular lines (p. 176)	Product of slopes is -1 ($m_1 \cdot m_2 = -1$)
Direct variation (p. 189)	$y = kx$
Inverse variation (p. 189)	$y = \dfrac{k}{x}$

Equations of Lines and Circles

Vertical line (p. 171)	$x = a$; a is the x-intercept
Horizontal line (p. 172)	$y = b$; b is the y-intercept
Point–slope form of the equation of a line (p. 171)	$y - y_1 = m(x - x_1)$; m is the slope of the line, (x_1, y_1) is a point on the line
Slope–intercept form of the equation of a line (p. 172)	$y = mx + b$; m is the slope of the line, b is the y-intercept
General form of the equation of a line (p. 174)	$Ax + By = C$; A, B not both 0
Standard form of the equation of a circle (p. 182)	$(x - h)^2 + (y - k)^2 = r^2$; r is the radius of the circle, (h, k) is the center of the circle
Equation of the unit circle (p. 183)	$x^2 + y^2 = 1$
General form of the equation of a circle (p. 184)	$x^2 + y^2 + ax + by + c = 0$, with restrictions on $a, b,$ and c

Objectives

Section	You should be able to ...	Examples	Review Exercises
2.1	**1** Use the distance formula (p. 151)	1–3	1(a)–6(a), 48, 49(a), 50
	2 Use the midpoint formula (p. 153)	4	1(b)–6(b), 50
2.2	**1** Graph equations by plotting points (p. 157)	1–3	7
	2 Find intercepts from a graph (p. 159)	4	8
	3 Find intercepts from an equation (p. 160)	5	9–16
	4 Test an equation for symmetry with respect to the x-axis, the y-axis, and the origin (p. 160)	6–9	9–16
	5 Know how to graph key equations (p. 163)	10–12	45, 46
2.3	**1** Calculate and interpret the slope of a line (p. 167)	1, 2	1(c)–6(c), 1(d)–6(d), 49(b), 51
	2 Graph lines given a point and the slope (p. 170)	3	47
	3 Find the equation of a vertical line (p. 170)	4	29
	4 Use the point–slope form of a line; identify horizontal lines (p. 171)	5, 6	27, 28
	5 Find the equation of a line given two points (p. 172)	7	30–32
	6 Write the equation of a line in slope–intercept form (p. 172)	8	27, 28, 30–36
	7 Identify the slope and y-intercept of a line from its equation (p. 173)	8	37–40
	8 Graph lines written in general form using intercepts (p. 174)	9	41–44
	9 Find equations of parallel lines (p. 175)	10, 11	33, 34
	10 Find equations of perpendicular lines (p. 176)	12, 13	35, 36
2.4	**1** Write the standard form of the equation of a circle (p. 182)	1	17–20, 50
	2 Graph a circle (p. 183)	2, 3	21–26
	3 Work with the general form of the equation of a circle (p. 184)	4	23–26

Section	You should be able to . . .	Examples	Review Exercises
2.5	1 Construct a model using direct variation (p. 189)	1	52, 53, 55
	2 Construct a model using inverse variation (p. 189)	2	54
	3 Construct a model using joint or combined variation (p. 190)	3, 4	55

Review Exercises

In Problems 1–6, find the following for each pair of points:

(a) The distance between the points
(b) The midpoint of the line segment connecting the points
(c) The slope of the line containing the points
(d) Interpret the slope found in part (c)

1. $(0, 0); (4, 2)$

2. $(0, 0); (-4, 6)$

3. $(1, -1); (-2, 3)$

4. $(-2, 2); (1, 4)$

5. $(4, -4); (4, 8)$

6. $(-3, 4); (2, 4)$

7. Graph $y = x^2 + 4$ by plotting points.

8. List the intercepts of the graph below.

In Problems 9–16, list the intercepts and test for symmetry with respect to the x-axis, the y-axis, and the origin.

9. $2x = 3y^2$

10. $y = 5x$

11. $x^2 + 4y^2 = 16$

12. $9x^2 - y^2 = 9$

13. $y = x^4 + 2x^2 + 1$

14. $y = x^3 - x$

15. $x^2 + x + y^2 + 2y = 0$

16. $x^2 + 4x + y^2 - 2y = 0$

In Problems 17–20, find the standard form of the equation of the circle whose center and radius are given.

17. $(h, k) = (-2, 3); r = 4$

18. $(h, k) = (3, 4); r = 4$

19. $(h, k) = (-1, -2); r = 1$

20. $(h, k) - (2, -4); r = 3$

In Problems 21–26, find the center and radius of each circle. Graph each circle. Find the intercepts, if any, of each circle.

21. $x^2 + (y - 1)^2 = 4$

22. $(x + 2)^2 + y^2 = 9$

23. $x^2 + y^2 - 2x + 4y - 4 = 0$

24. $x^2 + y^2 + 4x - 4y - 1 = 0$

25. $3x^2 + 3y^2 - 6x + 12y = 0$

26. $2x^2 + 2y^2 - 4x = 0$

In Problems 27–36, find an equation of the line having the given characteristics. Express your answer using either the general form or the slope–intercept form of the equation of a line, whichever you prefer.

27. Slope $= -2$; containing the point $(3, -1)$

28. Slope $= 0$; containing the point $(-5, 4)$

29. Vertical; containing the point $(-3, 4)$

30. x-intercept $= 2$; containing the point $(4, -5)$

31. y-intercept $= -2$; containing the point $(5, -3)$

32. Containing the points $(3, -4)$ and $(2, 1)$

33. Parallel to the line $2x - 3y = -4$; containing the point $(-5, 3)$

34. Parallel to the line $x + y = 2$; containing the point $(1, -3)$

35. Perpendicular to the line $x + y = 2$; containing the point $(4, -3)$

36. Perpendicular to the line $3x - y = -4$; containing the point $(-2, 4)$

In Problems 37–40, find the slope and y-intercept of each line. Graph the line, labeling any intercepts.

37. $4x - 5y = -20$

38. $3x + 4y = 12$

39. $\frac{1}{2}x - \frac{1}{3}y = -\frac{1}{6}$

40. $-\frac{3}{4}x + \frac{1}{2}y = 0$

In Problems 41–44, find the intercepts and graph each line.

41. $2x - 3y = 12$

42. $x - 2y = 8$

43. $\frac{1}{2}x + \frac{1}{3}y = 2$

44. $\frac{1}{3}x - \frac{1}{4}y = 1$

45. Sketch a graph of $y = x^3$.

46. Sketch a graph of $y = \sqrt{x}$.

47. Graph the line with slope $\dfrac{2}{3}$ containing the point $(1, 2)$.

48. Show that the points $A = (3, 4)$, $B = (1, 1)$, and $C = (-2, 3)$ are the vertices of an isosceles triangle.

49. Show that the points $A = (-2, 0)$, $B = (-4, 4)$, and $C = (8, 5)$ are the vertices of a right triangle in two ways:
 (a) By using the converse of the Pythagorean Theorem
 (b) By using the slopes of the lines joining the vertices

50. The endpoints of the diameter of a circle are $(-3, 2)$ and $(5, -6)$. Find the center and radius of the circle. Write the standard equation of this circle.

51. Show that the points $A = (2, 5)$, $B = (6, 1)$, and $C = (8, -1)$ lie on a line by using slopes.

52. Mortgage Payments The monthly payment p on a mortgage varies directly with the amount borrowed B. If the monthly payment on a 30-year mortgage is $854.00 when $130,000 is borrowed, find an equation that relates the monthly payment p to the amount borrowed B for a mortgage with the same terms. Then find the monthly payment p when the amount borrowed B is $165,000.

53. Revenue Function At the corner Esso station, the revenue R varies directly with the number g of gallons of gasoline sold. If the revenue is $46.67 when the number of gallons sold is 13, find an equation that relates revenue R to the number g of gallons of gasoline. Then find the revenue R when the number of gallons of gasoline sold is 11.2.

54. Weight of a Body The weight of a body varies inversely with the square of its distance from the center of Earth. Assuming that the radius of Earth is 3960 miles, how much would a man weigh at an altitude of 1 mile above Earth's surface if he weighs 200 pounds on Earth's surface?

55. Kepler's Third Law of Planetary Motion Kepler's Third Law of Planetary Motion states that the square of the period of revolution T of a planet varies directly with the cube of its mean distance a from the Sun. If the mean distance of Earth from the Sun is 93 million miles, what is the mean distance of the planet Mercury from the Sun, given that Mercury has a "year" of 88 days?

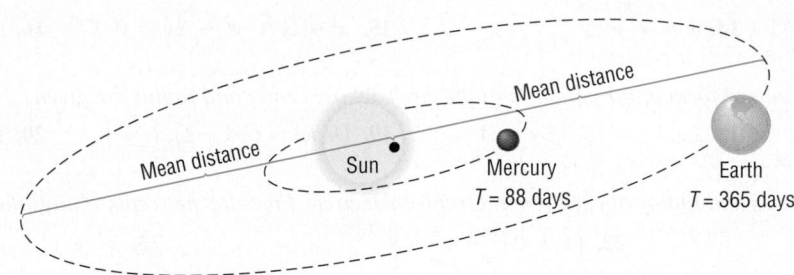

56. Create four problems that you might be asked to do given the two points $(-3, 4)$ and $(6, 1)$. Each problem should involve a different concept. Be sure that your directions are clearly stated.

57. Describe each of the following graphs in the xy-plane. Give justification.
 (a) $x = 0$ (b) $y = 0$ (c) $x + y = 0$ (d) $xy = 0$ (e) $x^2 + y^2 = 0$

58. Suppose that you have a rectangular field that requires watering. Your watering system consists of an arm of variable length that rotates so that the watering pattern is a circle. Decide where to position the arm and what length it should be so that the entire field is watered most efficiently. When does it become desirable to use more than one arm?

 [**Hint:** Use a rectangular coordinate system positioned as shown in the figures. Write equations for the circle(s) swept out by the watering arm(s).]

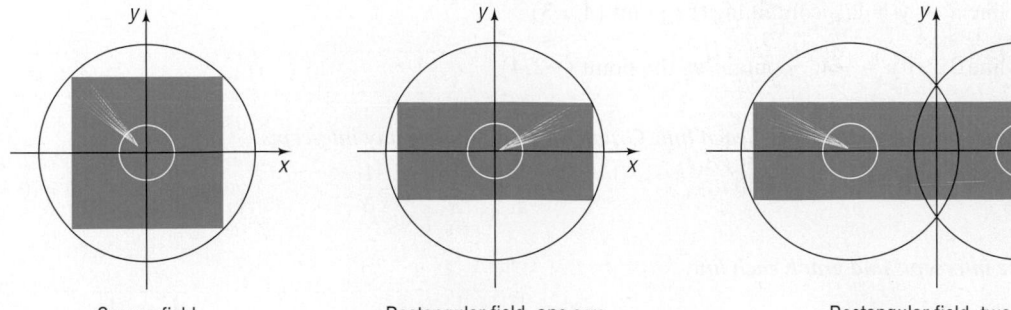

Square field Rectangular field, one arm Rectangular field, two arms

CHAPTER TEST

The Chapter Test Prep Videos are step-by-step test solutions available in the Video Resources DVD, in **MyMathLab**, or on this text's **You Tube** Channel. Flip back to the Student Resources page to see the exact web address for this text's YouTube channel.

In Problems 1–3, use $P_1 = (-1, 3)$ and $P_2 = (5, -1)$.

1. Find the distance from P_1 to P_2.

2. Find the midpoint of the line segment joining P_1 and P_2.

3. (a) Find the slope of the line containing P_1 and P_2.
 (b) Interpret this slope.

4. Graph $y = x^2 - 9$ by plotting points.

5. Sketch the graph of $y^2 = x$.

6. List the intercepts and test for symmetry: $x^2 + y = 9$.

7. Write the slope–intercept form of the line with slope -2 containing the point $(3, -4)$. Graph the line.

8. Write the general form of the circle with center $(4, -3)$ and radius 5.

9. Find the center and radius of the circle $x^2 + y^2 + 4x - 2y - 4 = 0$. Graph this circle.

10. For the line $2x + 3y = 6$, find a line parallel to it containing the point $(1, -1)$. Also find a line perpendicular to it containing the point $(0, 3)$.

11. **Resistance due to a Conductor** The resistance (in ohms) of a circular conductor varies directly with the length of the conductor and inversely with the square of the radius of the conductor. If 50 feet of wire with a radius of 6×10^{-3} inch has a resistance of 10 ohms, what would be the resistance of 100 feet of the same wire if the radius is increased to 7×10^{-3} inch?

CUMULATIVE REVIEW

In Problems 1–8, find the real solution(s) of each equation.

1. $3x - 5 = 0$

2. $x^2 - x - 12 = 0$

3. $2x^2 - 5x - 3 = 0$

4. $x^2 - 2x - 2 = 0$

5. $x^2 + 2x + 5 = 0$

6. $\sqrt{2x + 1} = 3$

7. $|x - 2| = 1$

8. $\sqrt{x^2 + 4x} = 2$

In Problems 9 and 10, solve each equation in the complex number system.

9. $x^2 = -9$

10. $x^2 - 2x + 5 = 0$

In Problems 11–14, solve each inequality. Graph the solution set.

11. $2x - 3 \leq 7$

12. $-1 < x + 4 < 5$

13. $|x - 2| \leq 1$

14. $|2 + x| > 3$

15. Find the distance between the points $P = (-1, 3)$ and $Q = (4, -2)$. Find the midpoint of the line segment from P to Q.

16. Which of the following points are on the graph of $y = x^3 - 3x + 1$?
 (a) $(-2, -1)$ (b) $(2, 3)$ (c) $(3, 1)$

17. Sketch the graph of $y = x^3$.

18. Find the equation of the line containing the points $(-1, 4)$ and $(2, -2)$. Express your answer in slope–intercept form.

19. Find the equation of the line perpendicular to the line $y = 2x + 1$ and containing the point $(3, 5)$. Express your answer in slope–intercept form and graph the line.

20. Graph the equation $x^2 + y^2 - 4x + 8y - 5 = 0$.

CHAPTER PROJECT

Internet-based Project

Predicting Olympic Performance Measurements of human performance over time sometimes follow a strong linear relationship for reasonably short periods. In 2004 the Summer Olympic Games returned to Greece, the home of both the ancient Olympics and the first modern Olympics. The following data represent the winning times (in hours) for men and women in the Olympic marathon.

Year	Men	Women
1984	2.16	2.41
1988	2.18	2.43
1992	2.22	2.54
1996	2.21	2.43
2000	2.17	2.39

Source: www.hickoksports.com/history/olmtandf.shtml

1. Treating year as the independent variable and the winning value as the dependent variable, find linear equations relating these variables (separately for men and women) using the data for the years 1992 and 1996. Compare the equations and comment on any similarities or differences.

2. Interpret the slopes in your equations from part 1. Do the y-intercepts have a reasonable interpretation? Why or why not?

3. Use your equations to predict the winning time in the 2004 Olympics. Compare your predictions to the actual results (2.18 hours for men and 2.44 hours for women). How well did your equations do in predicting the winning times?

4. Repeat parts 1 to 3 using the data for the years 1996 and 2000. How do your results compare?

5. Would your equations be useful in predicting the winning marathon times in the 2104 Summer Olympics? Why or why not?

6. Pick your favorite Winter Olympics event and find the winning value (that is distance, time, or the like) in two Winter Olympics prior to 2006. Repeat parts 1 to 3 using your selected event and years and compare to the actual results of the 2006 Winter Olympics in Torino, Italy.

Functions and Their Graphs

3

Outline

Choosing a Cellular Telephone Plan

Most consumers choose a cellular telephone provider first, and then select an appropriate plan from that provider. The choice as to the type of plan selected depends upon your use of the phone. For example, is text messaging important? How many minutes do you plan to use the phone? Do you desire a data plan to browse the Web? The mathematics learned in this chapter can help you decide the plan best-suited for your particular needs.

 —*See the Internet-based Chapter Project*—

◁ A Look Back So far, our discussion has focused on techniques for graphing equations containing two variables.

A Look Ahead ▷ In this chapter, we look at a special type of equation involving two variables called a *function*. This chapter deals with what a function is, how to graph functions, properties of functions, and how functions are used in applications. The word function apparently was introduced by René Descartes in 1637. For him, a function simply meant any positive integral power of a variable *x*. Gottfried Wilhelm Leibniz (1646–1716), who always emphasized the geometric side of mathematics, used the word function to denote any quantity associated with a curve, such as the coordinates of a point on the curve. Leonhard Euler (1707–1783) employed the word to mean any equation or formula involving variables and constants. His idea of a function is similar to the one most often seen in courses that precede calculus. Later, the use of functions in investigating heat flow equations led to a very broad definition, due to Lejeune Dirichlet (1805–1859), which describes a function as a correspondence between two sets. It is his definition that we use here.

3.1 Functions

PREPARING FOR THIS SECTION *Before getting started, review the following:*

- Intervals (Section 1.5, pp. 120–121)
- Solving Inequalities (Section 1.5, pp. 123–126)
- Evaluating Algebraic Expressions, Domain of a Variable (Chapter R, Section R.2, pp. 20–23)

Now Work the 'Are You Prepared?' problems on page 210.

OBJECTIVES **1** Determine Whether a Relation Represents a Function (p. 200)
2 Find the Value of a Function (p. 203)
3 Find the Domain of a Function Defined by an Equation (p. 206)
4 Form the Sum, Difference, Product, and Quotient of Two Functions (p. 208)

1 Determine Whether a Relation Represents a Function

Often there are situations where one variable is somehow linked to the value of another variable. For example, an individual's level of education is linked to annual income. Engine size is linked to gas mileage. When the value of one variable is related to the value of a second variable, we have a *relation*. A **relation** is a correspondence between two sets. If x and y are two elements in these sets and if a relation exists between x and y, then we say that x **corresponds** to y or that y **depends on** x, and we write $x \rightarrow y$.

There are a number of ways to express relations between two sets. For example, the equation $y = 3x - 1$ shows a relation between x and y. It says that if we take some number x, multiply it by 3, and then subtract 1 we obtain the corresponding value of y. In this sense, x serves as the **input** to the relation and y is the **output** of the relation. We can also express this relation as a graph as shown in Figure 1.

Figure 1

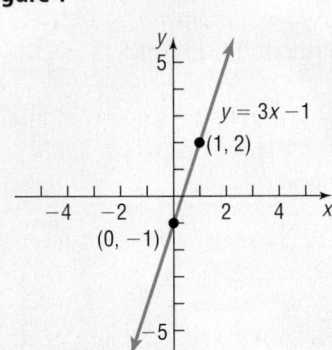

Not only can a relation be expressed through an equation or graph, but we can also express a relation through a technique called *mapping*. A **map** illustrates a relation by using a set of inputs and drawing arrows to the corresponding element in the set of outputs. **Ordered pairs** can be used to represent $x \rightarrow y$ as (x, y).

EXAMPLE 1 **Maps and Ordered Pairs as Relations**

Figure 2 shows a relation between states and the number of representatives each state has in the House of Representatives. The relation might be named "number of representatives."

Figure 2

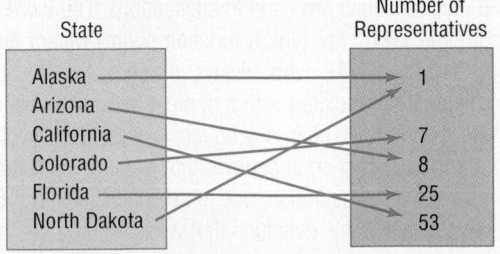

In this relation, Alaska corresponds to 1, Arizona corresponds to 8, and so on. Using ordered pairs, this relation would be expressed as

$\{(\text{Alaska}, 1), (\text{Arizona}, 8), (\text{California}, 53), (\text{Colorado}, 7), (\text{Florida}, 25), (\text{North Dakota}, 1)\}$

One of the most important concepts in algebra is the *function*. A function is a special type of relation. To understand the idea behind a function, let's revisit the relation presented in Example 1. If we were to ask, "How many representatives does Alaska have?," you would respond "1". In fact, each input *state* corresponds to a single output *number of representatives*.

Let's consider a second relation where we have a correspondence between four people and their phone numbers. See Figure 3. Notice that Colleen has two telephone numbers. If asked, "What is Colleen's phone number?", you cannot assign a single number to her.

Let's look at one more relation. Figure 4 is a relation that shows a correspondence between *animals* and life *expectancy*. If asked to determine the life expectancy of a dog, we would all respond "11 years." If asked to determine the life expectancy of a rabbit, we would all respond "7 years."

Figure 3

Figure 4

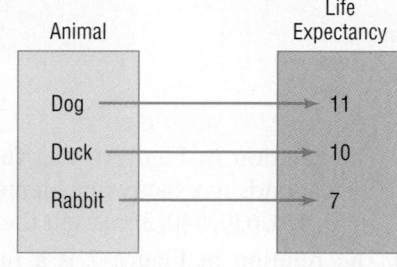

Notice that the relations presented in Figures 2 and 4 have something in common. What is it? The common link between these two relations is that each input corresponds to exactly one output. This leads to the definition of a *function*.

Figure 5

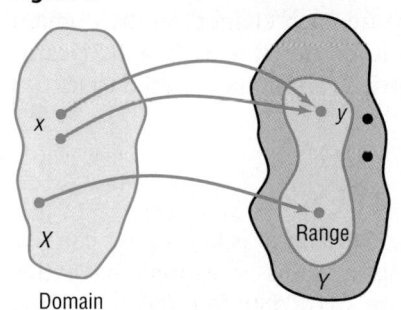

Domain

DEFINITION

Let X and Y be two nonempty sets.* A **function** from X into Y is a relation that associates with each element of X exactly one element of Y.

The set X is called the **domain** of the function. For each element x in X, the corresponding element y in Y is called the **value** of the function at x, or the **image** of x. The set of all images of the elements in the domain is called the **range** of the function. See Figure 5.

Since there may be some elements in Y that are not the image of some x in X, it follows that the range of a function may be a subset of Y, as shown in Figure 5.

Not all relations between two sets are functions. The next example shows how to determine whether a relation is a function.

EXAMPLE 2

Determining Whether a Relation Represents a Function

Determine which of the following relations represent a function. If the relation is a function, then state its domain and range.

(a) See Figure 6. For this relation, the domain represents the number of calories in a sandwich from a fast-food restaurant and the range represents the fat content (in grams).

Figure 6
Source: Each company's Web site

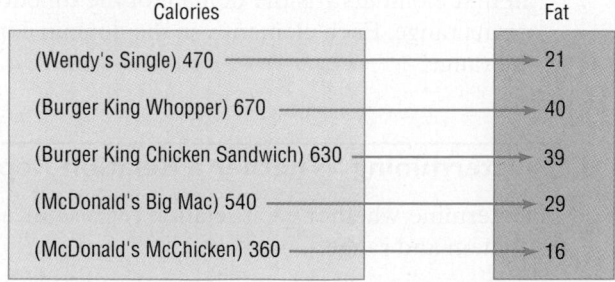

* The sets X and Y will usually be sets of real numbers, in which case a (real) function results. The two sets can also be sets of complex numbers, and then we have defined a complex function. In the broad definition (due to Lejeune Dirichlet), X and Y can be any two sets.

(b) See Figure 7. For this relation, the domain represents gasoline stations in Collier County, Florida, and the range represents the price per gallon of unleaded regular in July 2010.

(c) See Figure 8. For this relation, the domain represents the weight (in carats) of pear-cut diamonds and the range represents the price (in dollars).

Figure 7

Gas Station Price of regular per gallon

Figure 8

Source: Used with permission of Diamonds.com

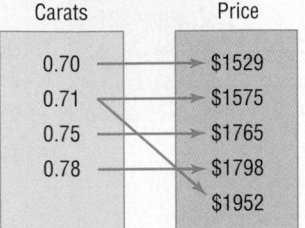

Carats Price

Solution

(a) The relation in Figure 6 is a function because each element in the domain corresponds to exactly one element in the range. The domain of the function is $\{470, 670, 630, 540, 360\}$, and the range of the function is $\{31, 40, 39, 29, 16\}$.

(b) The relation in Figure 7 is a function because each element in the domain corresponds to exactly one element in the range. The domain of the function is $\{\text{Mobil, Shell, Sunoco, 7-Eleven}\}$. The range of the function is $\{2.69, 2.71, 2.72\}$. Notice that it is okay for more than one element in the domain to correspond to the same element in the range (Shell and 7-Eleven each sell gas for $2.72 a gallon).

(c) The relation in Figure 8 is not a function because each element in the domain does not correspond to exactly one element in the range. If a 0.71-carat diamond is chosen from the domain, a single price cannot be assigned to it.

Now Work PROBLEM 15

The idea behind a function is its predictability. If the input is known, we can use the function to determine the output. With "nonfunctions," we don't have this predictability. Look back at Figure 7. The inputs are $\{410, 580, 540, 750, 600, 430\}$. The correspondence is *number of fat grams*, and the outputs are $\{19, 29, 24, 33, 23\}$. If asked, "How many grams of fat are in a 410-calorie sandwich?," we can use the correspondence to answer "19." Now consider Figure 8. If asked, "What is the price of a 0.71-carat diamond?," we could not give a single response because two outputs result from the single input "0.71." For this reason, the relation in Figure 8 is not a function.

We may also think of a function as a set of ordered pairs (x, y) in which no ordered pairs have the same first element and different second elements. The set of all first elements x is the domain of the function, and the set of all second elements y is its range. Each element x in the domain corresponds to exactly one element y in the range.

In Words

For a function, no input has more than one output. The domain of a function is the set of all inputs; the range is the set of all outputs.

EXAMPLE 3 | **Determining Whether a Relation Represents a Function**

Determine whether each relation represents a function. If it is a function, state the domain and range.

(a) $\{(1, 4), (2, 5), (3, 6), (4, 7)\}$

(b) $\{(1, 4), (2, 4), (3, 5), (6, 10)\}$

(c) $\{(-3, 9), (-2, 4), (0, 0), (1, 1), (-3, 8)\}$

Solution (a) This relation is a function because there are no ordered pairs with the same first element and different second elements. The domain of this function is $\{1, 2, 3, 4\}$, and its range is $\{4, 5, 6, 7\}$.

(b) This relation is a function because there are no ordered pairs with the same first element and different second elements. The domain of this function is $\{1, 2, 3, 6\}$, and its range is $\{4, 5, 10\}$.

(c) This relation is not a function because there are two ordered pairs, $(-3, 9)$ and $(-3, 8)$, that have the same first element and different second elements. ⌐

In Example 3(b), notice that 1 and 2 in the domain each have the same image in the range. This does not violate the definition of a function; two different first elements can have the same second element. A violation of the definition occurs when two ordered pairs have the same first element and different second elements, as in Example 3(c).

Now Work PROBLEM 19

Up to now we have shown how to identify when a relation is a function for relations defined by mappings (Example 2) and ordered pairs (Example 3). But relations can also be expressed as equations. We discuss next the circumstances under which equations are functions.

To determine whether an equation, where y depends on x, is a function, it is often easiest to solve the equation for y. If any value of x in the domain corresponds to more than one y, the equation does not define a function; otherwise, it does define a function.

EXAMPLE 4 **Determining Whether an Equation Is a Function**

Determine if the equation $y = 2x - 5$ defines y as a function of x.

Solution The equation tells us to take an input x, multiply it by 2, and then subtract 5. For any input x, these operations yield only one output y. For example, if $x = 1$, then $y = 2(1) - 5 = -3$. If $x = 3$, then $y = 2(3) - 5 = 1$. For this reason, the equation is a function. ⌐

EXAMPLE 5 **Determining Whether an Equation Is a Function**

Determine if the equation $x^2 + y^2 = 1$ defines y as a function of x.

Solution To determine whether the equation $x^2 + y^2 = 1$, which defines the unit circle, is a function, solve the equation for y.

$$x^2 + y^2 = 1$$
$$y^2 = 1 - x^2$$
$$y = \pm\sqrt{1 - x^2}$$

For values of x between -1 and 1, two values of y result. For example, if $x = 0$, then $y = \pm 1$, so two different outputs result from the same input. This means that the equation $x^2 + y^2 = 1$ does not define a function. ⌐

Now Work PROBLEM 33

2 Find the Value of a Function

Functions are often denoted by letters such as f, F, g, G, and others. If f is a function, then for each number x in its domain the corresponding image in the range is designated by the symbol $f(x)$, read as "f of x" or as "f at x." We refer to $f(x)$ as the **value of f at the number x;** $f(x)$ is the number that results when x is given and the function f is applied; $f(x)$ is the output corresponding to x or the image of x; $f(x)$

does *not* mean "*f* times *x*." For example, the function given in Example 4 may be written as $y = f(x) = 2x - 5$. Then $f\left(\dfrac{3}{2}\right) = -2$.

Figure 9 illustrates some other functions. Notice that, in every function, for each *x* in the domain there is one value in the range.

Figure 9

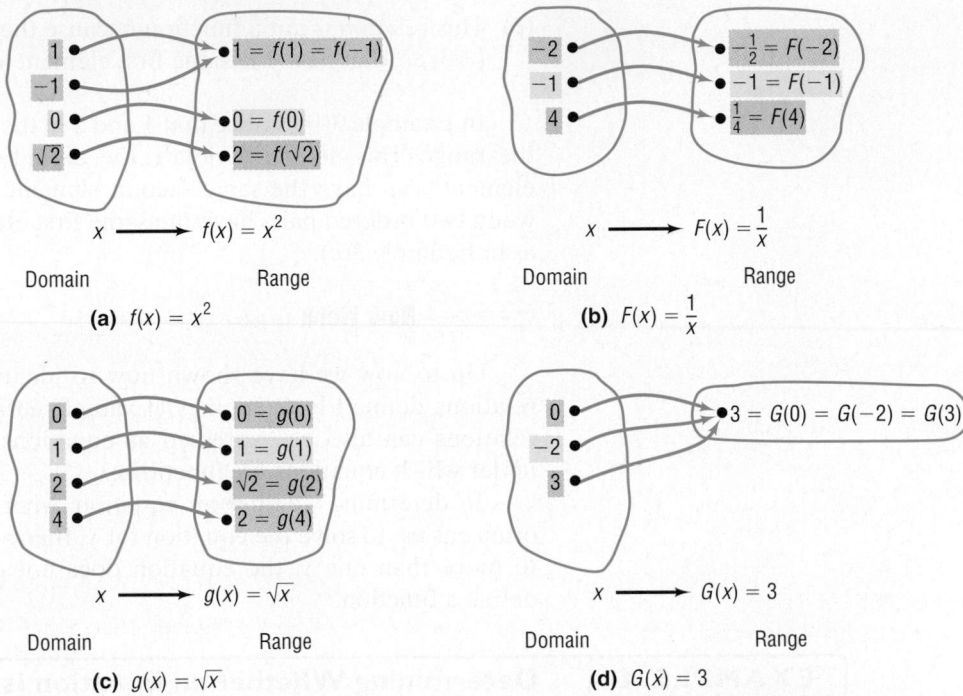

(a) $f(x) = x^2$

(b) $F(x) = \dfrac{1}{x}$

(c) $g(x) = \sqrt{x}$

(d) $G(x) = 3$

Figure 10

Input *x*

Output $y = f(x)$

Sometimes it is helpful to think of a function *f* as a machine that receives as input a number from the domain, manipulates it, and outputs a value. See Figure 10.

The restrictions on this input/output machine are as follows:

1. It only accepts numbers from the domain of the function.

2. For each input, there is exactly one output (which may be repeated for different inputs).

For a function $y = f(x)$, the variable *x* is called the **independent variable,** because it can be assigned any of the permissible numbers from the domain. The variable *y* is called the **dependent variable,** because its value depends on *x*.

Any symbols can be used to represent the independent and dependent variables. For example, if *f* is the *cube function*, then *f* can be given by $f(x) = x^3$ or $f(t) = t^3$ or $f(z) = z^3$. All three functions are the same. Each tells us to cube the independent variable to get the output. In practice, the symbols used for the independent and dependent variables are based on common usage, such as using *C* for cost in business.

The independent variable is also called the **argument** of the function. Thinking of the independent variable as an argument can sometimes make it easier to find the value of a function. For example, if *f* is the function defined by $f(x) = x^3$, then *f* tells us to cube the argument. Thus, $f(2)$ means to cube 2, $f(a)$ means to cube the number *a*, and $f(x + h)$ means to cube the quantity $x + h$.

EXAMPLE 6 **Finding Values of a Function**

For the function *f* defined by $f(x) = 2x^2 - 3x$, evaluate

(a) $f(3)$ (b) $f(x) + f(3)$ (c) $3f(x)$ (d) $f(-x)$

(e) $-f(x)$ (f) $f(3x)$ (g) $f(x + 3)$ (h) $\dfrac{f(x + h) - f(x)}{h}$ $h \neq 0$

Solution (a) Substitute 3 for x in the equation for f, $f(x) = 2x^2 - 3x$, to get

$$f(3) = 2(3)^2 - 3(3) = 18 - 9 = 9$$

The image of 3 is 9.

(b) $f(x) + f(3) = (2x^2 - 3x) + (9) = 2x^2 - 3x + 9$

(c) Multiply the equation for f by 3.

$$3f(x) = 3(2x^2 - 3x) = 6x^2 - 9x$$

(d) Substitute $-x$ for x in the equation for f and simplify.

$$f(-x) = 2(-x)^2 - 3(-x) = 2x^2 + 3x \quad \text{Notice the use of parentheses here.}$$

(e) $-f(x) = -(2x^2 - 3x) = -2x^2 + 3x$

(f) Substitute $3x$ for x in the equation for f and simplify.

$$f(3x) = 2(3x)^2 - 3(3x) = 2(9x^2) - 9x = 18x^2 - 9x$$

(g) Substitute $x + 3$ for x in the equation for f and simplify.

$$\begin{aligned} f(x + 3) &= 2(x + 3)^2 - 3(x + 3) \\ &= 2(x^2 + 6x + 9) - 3x - 9 \\ &= 2x^2 + 12x + 18 - 3x - 9 \\ &= 2x^2 + 9x + 9 \end{aligned}$$

(h) $$\frac{f(x + h) - f(x)}{h} = \frac{[2(x + h)^2 - 3(x + h)] - [2x^2 - 3x]}{h}$$
$$\uparrow$$
$$\scriptstyle f(x + h) = 2(x + h)^2 - 3(x + h)$$

$$= \frac{2(x^2 + 2xh + h^2) - 3x - 3h - 2x^2 + 3x}{h} \quad \text{Simplify.}$$

$$= \frac{2x^2 + 4xh + 2h^2 - 3h - 2x^2}{h} \quad \begin{array}{l}\text{Distribute and combine} \\ \text{like terms.}\end{array}$$

$$= \frac{4xh + 2h^2 - 3h}{h} \quad \text{Combine like terms.}$$

$$= \frac{h(4x + 2h - 3)}{h} \quad \text{Factor out } h.$$

$$= 4x + 2h - 3 \quad \text{Divide out the } h\text{'s.}$$

Notice in this example that $f(x + 3) \neq f(x) + f(3)$, $f(-x) \neq -f(x)$, and $3f(x) \neq f(3x)$.

The expression in part (h) is called the **difference quotient** of f, an important expression in calculus.

Now Work PROBLEMS 39 AND 75

Most calculators have special keys that allow you to find the value of certain commonly used functions. For example, you should be able to find the square function $f(x) = x^2$, the square root function $f(x) = \sqrt{x}$, the reciprocal function $f(x) = \dfrac{1}{x} = x^{-1}$, and many others that will be discussed later in this book (such as $\ln x$ and $\log x$). Verify the results of Example 7, which follows, on your calculator.

EXAMPLE 7 **Finding Values of a Function on a Calculator**

(a) $f(x) = x^2$ $f(1.234) = 1.234^2 = 1.522756$

(b) $F(x) = \dfrac{1}{x}$ $F(1.234) = \dfrac{1}{1.234} \approx 0.8103727715$

(c) $g(x) = \sqrt{x}$ $g(1.234) = \sqrt{1.234} \approx 1.110855526$

COMMENT Graphing calculators can be used to evaluate any function that you wish. Figure 11 shows the result obtained in Example 6(a) on a TI-84 Plus graphing calculator with the function to be evaluated, $f(x) = 2x^2 - 3x$, in Y_1.

Figure 11

Implicit Form of a Function

COMMENT The explicit form of a function is the form required by a graphing calculator. ∎

In general, when a function f is defined by an equation in x and y, we say that the function f is given **implicitly.** If it is possible to solve the equation for y in terms of x, then we write $y = f(x)$ and say that the function is given **explicitly.** For example,

Implicit Form	**Explicit Form**
$3x + y = 5$	$y = f(x) = -3x + 5$
$x^2 - y = 6$	$y = f(x) = x^2 - 6$
$xy = 4$	$y = f(x) = \dfrac{4}{x}$

SUMMARY Important Facts about Functions

(a) For each x in the domain of a function f, there is exactly one image $f(x)$ in the range; however, an element in the range can result from more than one x in the domain.

(b) f is the symbol that we use to denote the function. It is symbolic of the equation (rule) that we use to get from an x in the domain to $f(x)$ in the range.

(c) If $y = f(x)$, then x is called the independent variable or argument of f, and y is called the dependent variable or the value of f at x.

3 Find the Domain of a Function Defined by an Equation

Often the domain of a function f is not specified; instead, only the equation defining the function is given. In such cases, we agree that the **domain of f** is the largest set of real numbers for which the value $f(x)$ is a real number. The domain of a function f is the same as the domain of the variable x in the expression $f(x)$.

EXAMPLE 8 **Finding the Domain of a Function**

Find the domain of each of the following functions:

(a) $f(x) = x^2 + 5x$ (b) $g(x) = \dfrac{3x}{x^2 - 4}$

(c) $h(t) = \sqrt{4 - 3t}$ (d) $F(x) = \dfrac{\sqrt{3x + 12}}{x - 5}$

Solution

(a) The function tells us to square a number and then add five times the number. Since these operations can be performed on any real number, we conclude that the domain of f is the set of all real numbers.

(b) The function g tells us to divide $3x$ by $x^2 - 4$. Since division by 0 is not defined, the denominator $x^2 - 4$ can never be 0, so x can never equal -2 or 2. The domain of the function g is $\{x | x \neq -2, x \neq 2\}$.

(c) The function h tells us to take the square root of $4 - 3t$. But only nonnegative numbers have real square roots, so the expression under the square root (the radicand) must be nonnegative (greater than or equal to zero). This requires that

$$4 - 3t \geq 0$$
$$-3t \geq -4$$
$$t \leq \frac{4}{3}$$

The domain of h is $\left\{ t \,\middle|\, t \leq \frac{4}{3} \right\}$ or the interval $\left(-\infty, \dfrac{4}{3} \right]$.

(d) The function F tells us to take the square root of $3x + 12$ and divide this result by $x - 5$. This requires that $3x + 12 \geq 0$, so $x \geq -4$, and also that $x - 5 \neq 0$, so $x \neq 5$. Combining these two restrictions, the domain of F is $\{x | x \geq -4, x \neq 5\}$.

> **In Words**
>
> The domain of g found in Example 8(b) is $\{x | x \neq -2, x \neq 2\}$. This notation is read, "The domain of the function g is the set of all real numbers x such that x does not equal -2 and x does not equal 2."

For the functions that we will encounter in this book, the following steps may prove helpful for finding the domain of a function that is defined by an equation and whose domain is a subset of the real numbers.

Finding the Domain of a Function Defined by an Equation

1. Start with the domain as the set of real numbers.
2. If the equation has a denominator, exclude any numbers that give a zero denominator.
3. If the equation has a radical of even index, exclude any numbers that cause the expression inside the radical to be negative.

Now Work PROBLEM 51

If x is in the domain of a function f, we shall say that f **is defined at x,** or $f(x)$ **exists.** If x is not in the domain of f, we say that f **is not defined at x,** or $f(x)$ **does not exist.** For example, if $f(x) = \dfrac{x}{x^2 - 1}$, then $f(0)$ exists, but $f(1)$ and $f(-1)$ do not exist. (Do you see why?)

We have not said much about finding the range of a function. We will say more about finding the range when we look at the graph of a function in the next section. When a function is defined by an equation, it can be difficult to find the range. Therefore, we shall usually be content to find just the domain of a function when the function is defined by an equation. We shall express the domain of a function using inequalities, interval notation, set notation, or words, whichever is most convenient.

When we use functions in applications, the domain may be restricted by physical or geometric considerations. For example, the domain of the function f defined by $f(x) = x^2$ is the set of all real numbers. However, if f is used to obtain the area of a square when the length x of a side is known, then we must restrict the domain of f to the positive real numbers, since the length of a side can never be 0 or negative.

<div>
<table>
<tr><td>**EXAMPLE 9**</td><td>**Finding the Domain in an Application**</td></tr>
</table>
</div>

Express the area of a circle as a function of its radius. Find the domain.

Solution See Figure 12. The formula for the area A of a circle of radius r is $A = \pi r^2$. If we use r to represent the independent variable and A to represent the dependent variable, the function expressing this relationship is

Figure 12

$$A(r) = \pi r^2$$

In this setting, the domain is $\{r \mid r > 0\}$. (Do you see why?)

Observe in the solution to Example 9 that the symbol A is used in two ways: It is used to name the function, and it is used to symbolize the dependent variable. This double use is common in applications and should not cause any difficulty.

━━━━▶ **Now Work** PROBLEM 89

4 Form the Sum, Difference, Product, and Quotient of Two Functions

Next we introduce some operations on functions. We shall see that functions, like numbers, can be added, subtracted, multiplied, and divided. For example, if $f(x) = x^2 + 9$ and $g(x) = 3x + 5$, then

$$f(x) + g(x) = (x^2 + 9) + (3x + 5) = x^2 + 3x + 14$$

The new function $y = x^2 + 3x + 14$ is called the *sum function f + g*. Similarly,

$$f(x) \cdot g(x) = (x^2 + 9)(3x + 5) = 3x^3 + 5x^2 + 27x + 45$$

The new function $y = 3x^3 + 5x^2 + 27x + 45$ is called the *product function f · g*. The general definitions are given next.

DEFINITION If f and g are functions:
The **sum $f + g$** is the function defined by

$$(f + g)(x) = f(x) + g(x)$$

> **In Words**
> Remember, the symbol ∩ stands for intersection. It means you should find the elements that are common to two sets.

The domain of $f + g$ consists of the numbers x that are in the domains of both f and g. That is, domain of $f + g$ = domain of f ∩ domain of g.

DEFINITION The **difference $f - g$** is the function defined by

$$(f - g)(x) = f(x) - g(x)$$

The domain of $f - g$ consists of the numbers x that are in the domains of both f and g. That is, domain of $f - g$ = domain of f ∩ domain of g.

DEFINITION The **product $f \cdot g$** is the function defined by

$$(f \cdot g)(x) = f(x) \cdot g(x)$$

The domain of $f \cdot g$ consists of the numbers x that are in the domains of both f and g. That is, domain of $f \cdot g$ = domain of f ∩ domain of g.

DEFINITION

The **quotient** $\dfrac{f}{g}$ is the function defined by

$$\left(\frac{f}{g}\right)(x) = \frac{f(x)}{g(x)} \qquad g(x) \neq 0$$

The domain of $\dfrac{f}{g}$ consists of the numbers x for which $g(x) \neq 0$ and that are in the domains of both f and g. That is,

$$\text{domain of } \frac{f}{g} = \{x \mid g(x) \neq 0\} \cap \text{ domain of } f \cap \text{ domain of } g$$

EXAMPLE 10 **Operations on Functions**

Let f and g be two functions defined as

$$f(x) = \frac{1}{x + 2} \quad \text{and} \quad g(x) = \frac{x}{x - 1}$$

Find the following, and determine the domain in each case.

(a) $(f + g)(x)$ (b) $(f - g)(x)$ (c) $(f \cdot g)(x)$ (d) $\left(\dfrac{f}{g}\right)(x)$

Solution The domain of f is $\{x \mid x \neq -2\}$ and the domain of g is $\{x \mid x \neq 1\}$.

(a) $(f + g)(x) = f(x) + g(x) = \dfrac{1}{x + 2} + \dfrac{x}{x - 1}$

$$= \frac{x - 1}{(x + 2)(x - 1)} + \frac{x(x + 2)}{(x + 2)(x - 1)} = \frac{x^2 + 3x - 1}{(x + 2)(x - 1)}$$

The domain of $f + g$ consists of those numbers x that are in the domains of both f and g. Therefore, the domain of $f + g$ is $\{x \mid x \neq -2, x \neq 1\}$.

(b) $(f - g)(x) = f(x) - g(x) = \dfrac{1}{x + 2} - \dfrac{x}{x - 1}$

$$= \frac{x - 1}{(x + 2)(x - 1)} - \frac{x(x + 2)}{(x + 2)(x - 1)} = \frac{-(x^2 + x + 1)}{(x + 2)(x - 1)}$$

The domain of $f - g$ consists of those numbers x that are in the domains of both f and g. Therefore, the domain of $f - g$ is $\{x \mid x \neq -2, x \neq 1\}$.

(c) $(f \cdot g)(x) = f(x) \cdot g(x) = \dfrac{1}{x + 2} \cdot \dfrac{x}{x - 1} = \dfrac{x}{(x + 2)(x - 1)}$

The domain of $f \cdot g$ consists of those numbers x that are in the domains of both f and g. Therefore, the domain of $f \cdot g$ is $\{x \mid x \neq -2, x \neq 1\}$.

(d) $\left(\dfrac{f}{g}\right)(x) = \dfrac{f(x)}{g(x)} = \dfrac{\dfrac{1}{x + 2}}{\dfrac{x}{x - 1}} = \dfrac{1}{x + 2} \cdot \dfrac{x - 1}{x} = \dfrac{x - 1}{x(x + 2)}$

The domain of $\dfrac{f}{g}$ consists of the numbers x for which $g(x) \neq 0$ and that are in the domains of both f and g. Since $g(x) = 0$ when $x = 0$, we exclude 0 as well as -2 and 1 from the domain. The domain of $\dfrac{f}{g}$ is $\{x \mid x \neq -2, x \neq 0, x \neq 1\}$.

Now Work PROBLEM 63

In calculus, it is sometimes helpful to view a complicated function as the sum, difference, product, or quotient of simpler functions. For example,

$$F(x) = x^2 + \sqrt{x} \text{ is the sum of } f(x) = x^2 \text{ and } g(x) = \sqrt{x}.$$

$$H(x) = \frac{x^2 - 1}{x^2 + 1} \text{ is the quotient of } f(x) = x^2 - 1 \text{ and } g(x) = x^2 + 1.$$

SUMMARY

Function A relation between two sets of real numbers so that each number x in the first set, the domain, has corresponding to it exactly one number y in the second set.

A set of ordered pairs (x, y) or $(x, f(x))$ in which no first element is paired with two different second elements.

The range is the set of y values of the function that are the images of the x values in the domain.

A function f may be defined implicitly by an equation involving x and y or explicitly by writing $y = f(x)$.

Unspecified domain If a function f is defined by an equation and no domain is specified, then the domain will be taken to be the largest set of real numbers for which the equation defines a real number.

Function notation $y = f(x)$

f is a symbol for the function.

x is the independent variable or argument.

y is the dependent variable.

$f(x)$ is the value of the function at x, or the image of x.

3.1 Assess Your Understanding

'Are You Prepared?' *Answers are given at the end of these exercises. If you get a wrong answer, read the pages listed in red.*

1. The inequality $-1 < x < 3$ can be written in interval notation as _____. (pp. 120–121)

2. If $x = -2$, the value of the expression $3x^2 - 5x + \dfrac{1}{x}$ is _____. (pp. 20–23)

3. The domain of the variable in the expression $\dfrac{x - 3}{x + 4}$ is _____. (pp. 20–23)

4. Solve the inequality: $3 - 2x > 5$. Graph the solution set. (pp. 123–126)

Concepts and Vocabulary

5. If f is a function defined by the equation $y = f(x)$, then x is called the _____ variable and y is the _____ variable.

6. The set of all images of the elements in the domain of a function is called the _____.

7. If the domain of f is all real numbers in the interval $[0, 7]$ and the domain of g is all real numbers in the interval $[-2, 5]$, the domain of $f + g$ is all real numbers in the interval _____.

8. The domain of $\dfrac{f}{g}$ consists of numbers x for which $g(x)$ ____ 0 that are in the domains of both ____ and ____.

9. If $f(x) = x + 1$ and $g(x) = x^3$, then _____ $= x^3 - (x + 1)$.

10. **True or False** Every relation is a function.

11. **True or False** The domain of $(f \cdot g)(x)$ consists of the numbers x that are in the domains of both f and g.

12. **True or False** The independent variable is sometimes referred to as the argument of the function.

13. **True or False** If no domain is specified for a function f, then the domain of f is taken to be the set of real numbers.

14. **True or False** The domain of the function $f(x) = \dfrac{x^2 - 4}{x}$ is $\{x | x \neq \pm 2\}$.

Skill Building

In Problems 15–26, determine whether each relation represents a function. For each function, state the domain and range.

15.

16.

17.

18.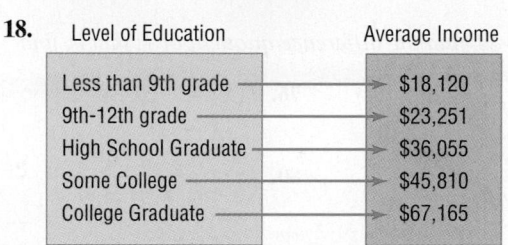

19. $\{(2, 6), (-3, 6), (4, 9), (2, 10)\}$

20. $\{(-2, 5), (-1, 3), (3, 7), (4, 12)\}$

21. $\{(1, 3), (2, 3), (3, 3), (4, 3)\}$

22. $\{(0, -2), (1, 3), (2, 3), (3, 7)\}$

23. $\{(-2, 4), (-2, 6), (0, 3), (3, 7)\}$

24. $\{(-4, 4), (-3, 3), (-2, 2), (-1, 1), (-4, 0)\}$

25. $\{(-2, 4), (-1, 1), (0, 0), (1, 1)\}$

26. $\{(-2, 16), (-1, 4), (0, 3), (1, 4)\}$

In Problems 27–38, determine whether the equation defines y as a function of x.

27. $y = x^2$

28. $y = x^3$

29. $y = \dfrac{1}{x}$

30. $y = |x|$

31. $y^2 = 4 - x^2$

32. $y = \pm\sqrt{1 - 2x}$

33. $x = y^2$

34. $x + y^2 = 1$

35. $y = 2x^2 - 3x + 4$

36. $y = \dfrac{3x - 1}{x + 2}$

37. $2x^2 + 3y^2 = 1$

38. $x^2 - 4y^2 = 1$

In Problems 39–46, find the following for each function:

(a) $f(0)$ (b) $f(1)$ (c) $f(-1)$ (d) $f(-x)$ (e) $-f(x)$ (f) $f(x + 1)$ (g) $f(2x)$ (h) $f(x + h)$

39. $f(x) = 3x^2 + 2x - 4$

40. $f(x) = -2x^2 + x - 1$

41. $f(x) = \dfrac{x}{x^2 + 1}$

42. $f(x) = \dfrac{x^2 - 1}{x + 4}$

43. $f(x) = |x| + 4$

44. $f(x) = \sqrt{x^2 + x}$

45. $f(x) = \dfrac{2x + 1}{3x - 5}$

46. $f(x) = 1 - \dfrac{1}{(x + 2)^2}$

In Problems 47–62, find the domain of each function.

47. $f(x) = -5x + 4$

48. $f(x) = x^2 + 2$

49. $f(x) = \dfrac{x}{x^2 + 1}$

50. $f(x) = \dfrac{x^2}{x^2 + 1}$

51. $g(x) = \dfrac{x}{x^2 - 16}$

52. $h(x) = \dfrac{2x}{x^2 - 4}$

53. $F(x) = \dfrac{x - 2}{x^3 + x}$

54. $G(x) = \dfrac{x + 4}{x^3 - 4x}$

55. $h(x) = \sqrt{3x - 12}$

56. $G(x) = \sqrt{1 - x}$

57. $f(x) = \dfrac{4}{\sqrt{x - 9}}$

58. $f(x) = \dfrac{x}{\sqrt{x - 4}}$

59. $p(x) = \sqrt{\dfrac{2}{x - 1}}$

60. $q(x) = \sqrt{-x - 2}$

61. $P(t) = \dfrac{\sqrt{t - 4}}{3t - 21}$

62. $h(z) = \dfrac{\sqrt{z + 3}}{z - 2}$

In Problems 63–72, for the given functions f and g, find the following. For parts (a)–(d), also find the domain.

(a) $(f + g)(x)$ (b) $(f - g)(x)$ (c) $(f \cdot g)(x)$ (d) $\left(\dfrac{f}{g}\right)(x)$

(e) $(f + g)(3)$ (f) $(f - g)(4)$ (g) $(f \cdot g)(2)$ (h) $\left(\dfrac{f}{g}\right)(1)$

63. $f(x) = 3x + 4;$ $g(x) = 2x - 3$

64. $f(x) = 2x + 1;$ $g(x) = 3x - 2$

65. $f(x) = x - 1;$ $g(x) = 2x^2$

66. $f(x) = 2x^2 + 3;$ $g(x) = 4x^3 + 1$

67. $f(x) = \sqrt{x}$; $g(x) = 3x - 5$

68. $f(x) = |x|$; $g(x) = x$

69. $f(x) = 1 + \dfrac{1}{x}$; $g(x) = \dfrac{1}{x}$

70. $f(x) = \sqrt{x - 1}$; $g(x) = \sqrt{4 - x}$

71. $f(x) = \dfrac{2x + 3}{3x - 2}$; $g(x) = \dfrac{4x}{3x - 2}$

72. $f(x) = \sqrt{x + 1}$; $g(x) = \dfrac{2}{x}$

73. Given $f(x) = 3x + 1$ and $(f + g)(x) = 6 - \dfrac{1}{2}x$, find the function g.

74. Given $f(x) = \dfrac{1}{x}$ and $\left(\dfrac{f}{g}\right)(x) = \dfrac{x + 1}{x^2 - x}$, find the function g.

⚠ In Problems 75–82, find the difference quotient of f; that is, find $\dfrac{f(x + h) - f(x)}{h}$, $h \neq 0$, for each function. Be sure to simplify.

75. $f(x) = 4x + 3$

76. $f(x) = -3x + 1$

77. $f(x) = x^2 - x + 4$

78. $f(x) = 3x^2 - 2x + 6$

79. $f(x) = \dfrac{1}{x^2}$

80. $f(x) = \dfrac{1}{x + 3}$

81. $f(x) = \sqrt{x}$
[**Hint:** Rationalize the numerator.]

82. $f(x) = \sqrt{x + 1}$

Applications and Extensions

83. If $f(x) = 2x^3 + Ax^2 + 4x - 5$ and $f(2) = 5$, what is the value of A?

84. If $f(x) = 3x^2 - Bx + 4$ and $f(-1) = 12$, what is the value of B?

85. If $f(x) = \dfrac{3x + 8}{2x - A}$ and $f(0) = 2$, what is the value of A?

86. If $f(x) = \dfrac{2x - B}{3x + 4}$ and $f(2) = \dfrac{1}{2}$, what is the value of B?

87. If $f(x) = \dfrac{2x - A}{x - 3}$ and $f(4) = 0$, what is the value of A? Where is f not defined?

88. If $f(x) = \dfrac{x - B}{x - A}$, $f(2) = 0$ and $f(1)$ is undefined, what are the values of A and B?

89. Geometry Express the area A of a rectangle as a function of the length x if the length of the rectangle is twice its width.

90. Geometry Express the area A of an isosceles right triangle as a function of the length x of one of the two equal sides.

91. Constructing Functions Express the gross salary G of a person who earns \$10 per hour as a function of the number x of hours worked.

92. Constructing Functions Tiffany, a commissioned salesperson, earns \$100 base pay plus \$10 per item sold. Express her gross salary G as a function of the number x of items sold.

93. Population as a Function of Age The function

$$P(a) = 0.015a^2 - 4.962a + 290.580$$

represents the population P (in millions) of Americans that are a years of age or older.

(a) Identify the dependent and independent variables.
(b) Evaluate $P(20)$. Provide a verbal explanation of the meaning of $P(20)$.
(c) Evaluate $P(0)$. Provide a verbal explanation of the meaning of $P(0)$.

94. Number of Rooms The function

$$N(r) = -1.44r^2 + 14.52r - 14.96$$

represents the number N of housing units (in millions) that have r rooms, where r is an integer and $2 \leq r \leq 9$.

(a) Identify the dependent and independent variables.
(b) Evaluate $N(3)$. Provide a verbal explanation of the meaning of $N(3)$.

95. Effect of Gravity on Earth If a rock falls from a height of 20 meters on Earth, the height H (in meters) after x seconds is approximately

$$H(x) = 20 - 4.9x^2$$

(a) What is the height of the rock when $x = 1$ second? $x = 1.1$ seconds? $x = 1.2$ seconds? $x = 1.3$ seconds?
(b) When is the height of the rock 15 meters? When is it 10 meters? When is it 5 meters?
(c) When does the rock strike the ground?

96. Effect of Gravity on Jupiter If a rock falls from a height of 20 meters on the planet Jupiter, its height H (in meters) after x seconds is approximately

$$H(x) = 20 - 13x^2$$

(a) What is the height of the rock when $x = 1$ second? $x = 1.1$ seconds? $x = 1.2$ seconds?
(b) When is the height of the rock 15 meters? When is it 10 meters? When is it 5 meters?
(c) When does the rock strike the ground?

97. Cost of Trans-Atlantic Travel A Boeing 747 crosses the Atlantic Ocean (3000 miles) with an airspeed of 500 miles per hour. The cost C (in dollars) per passenger is given by

$$C(x) = 100 + \frac{x}{10} + \frac{36,000}{x}$$

where x is the ground speed (airspeed \pm wind).
(a) What is the cost per passenger for quiescent (no wind) conditions?
(b) What is the cost per passenger with a head wind of 50 miles per hour?
(c) What is the cost per passenger with a tail wind of 100 miles per hour?
(d) What is the cost per passenger with a head wind of 100 miles per hour?

98. Cross-sectional Area The cross-sectional area of a beam cut from a log with radius 1 foot is given by the function $A(x) = 4x\sqrt{1 - x^2}$, where x represents the length, in feet, of half the base of the beam. See the figure. Determine the cross-sectional area of the beam if the length of half the base of the beam is as follows:
(a) One-third of a foot
(b) One-half of a foot
(c) Two-thirds of a foot

$$A(x) = 4x\sqrt{1 - x^2}$$

99. Economics The **participation rate** is the number of people in the labor force divided by the civilian population (excludes military). Let $L(x)$ represent the size of the labor force in year x and $P(x)$ represent the civilian population in year x. Determine a function that represents the participation rate R as a function of x.

100. Crimes Suppose that $V(x)$ represents the number of violent crimes committed in year x and $P(x)$ represents the number of property crimes committed in year x. Determine a function T that represents the combined total of violent crimes and property crimes in year x.

101. Health Care Suppose that $P(x)$ represents the percentage of income spent on health care in year x and $I(x)$ represents income in year x. Determine a function H that represents total health care expenditures in year x.

102. Income Tax Suppose that $I(x)$ represents the income of an individual in year x before taxes and $T(x)$ represents the individual's tax bill in year x. Determine a function N that represents the individual's net income (income after taxes) in year x.

103. Profit Function Suppose that the revenue R, in dollars, from selling x cell phones, in hundreds, is $R(x) = -1.2x^2 + 220x$. The cost C, in dollars, of selling x cell phones is $C(x) = 0.05x^3 - 2x^2 + 65x + 500$.
(a) Find the profit function, $P(x) = R(x) - C(x)$.
(b) Find the profit if $x = 15$ hundred cell phones are sold.
(c) Interpret $P(15)$.

104. Profit Function Suppose that the revenue R, in dollars, from selling x clocks is $R(x) = 30x$. The cost C, in dollars, of selling x clocks is $C(x) = 0.1x^2 + 7x + 400$.
(a) Find the profit function, $P(x) = R(x) - C(x)$.
(b) Find the profit if $x = 30$ clocks are sold.
(c) Interpret $P(30)$.

105. Some functions f have the property that $f(a + b) = f(a) + f(b)$ for all real numbers a and b. Which of the following functions have this property?
(a) $h(x) = 2x$ (b) $g(x) = x^2$
(c) $F(x) = 5x - 2$ (d) $G(x) = \dfrac{1}{x}$

Explaining Concepts: Discussion and Writing

106. Are the functions $f(x) = x - 1$ and $g(x) = \dfrac{x^2 - 1}{x + 1}$ the same? Explain.

107. Investigate when, historically, the use of the function notation $y = f(x)$ first appeared.

108. Find a function H that multiplies a number x by 3, then subtracts the cube of x and divides the result by your age.

'Are You Prepared?' Answers

1. $(-1, 3)$ **2.** 21.5 **3.** $\{x | x \neq -4\}$ **4.** $\{x | x < -1\}$

3.2 The Graph of a Function

PREPARING FOR THIS SECTION *Before getting started, review the following:*

- Graphs of Equations (Section 2.2, pp. 157–159)
- Intercepts (Section 2.2, pp. 159–160)

Now Work the '**Are You Prepared?**' problems on page 218.

OBJECTIVES 1 Identify the Graph of a Function (p. 214)
2 Obtain Information from or about the Graph of a Function (p. 215)

In applications, a graph often demonstrates more clearly the relationship between two variables than, say, an equation or table would. For example, Table 1 shows the average price of gasoline at a particular gas station in Texas (for the years 1980–2009 adjusted for inflation, based on 2008 dollars). If we plot these data and then connect the points, we obtain Figure 13.

Table 1

Year	Price	Year	Price	Year	Price
1980	3.41	1990	2.25	2000	1.85
1981	3.26	1991	1.90	2001	1.40
1982	3.15	1992	1.82	2002	1.86
1983	2.51	1993	1.70	2003	1.79
1984	2.51	1994	1.85	2004	2.13
1985	2.46	1995	1.68	2005	2.60
1986	1.63	1996	1.87	2006	2.62
1987	1.90	1997	1.65	2007	3.29
1988	1.77	1998	1.50	2008	2.10
1989	1.83	1999	1.73	2009	2.45

Source: http://www.randomuseless.info/gasprice/gasprice.html

Figure 13
Average retail price of gasoline (2008 dollars)

Source: http://www.randomuseless.info/gasprice/gasprice.html

We can see from the graph that the price of gasoline (adjusted for inflation) fell from 1980 to 1986 and rose rapidly from 2003 to 2007. The graph also shows that the lowest price occurred in 2001. To learn information such as this from an equation requires that some calculations be made.

Look again at Figure 13. The graph shows that for each date on the horizontal axis there is only one price on the vertical axis. The graph represents a function, although the exact rule for getting from date to price is not given.

When a function is defined by an equation in x and y, the **graph of the function** is the graph of the equation, that is, the set of points (x, y) in the xy-plane that satisfy the equation.

1 Identify the Graph of a Function

Not every collection of points in the xy-plane represents the graph of a function. Remember, for a function, each number x in the domain has exactly one image y in the range. This means that the graph of a function cannot contain two points with the same x-coordinate and different y-coordinates. Therefore, the graph of a function must satisfy the following **vertical-line test.**

THEOREM **Vertical-line Test**

A set of points in the xy-plane is the graph of a function if and only if every vertical line intersects the graph in at most one point.

| **EXAMPLE 1** | **Identifying the Graph of a Function** |

Which of the graphs in Figure 14 are graphs of functions?

Figure 14

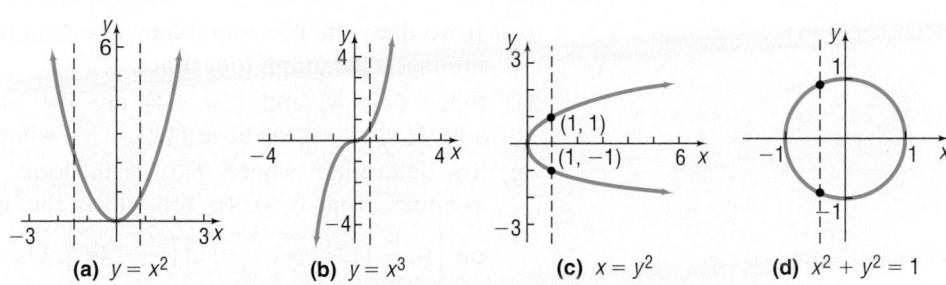

(a) $y = x^2$ (b) $y = x^3$ (c) $x = y^2$ (d) $x^2 + y^2 = 1$

Solution The graphs in Figures 14(a) and 14(b) are graphs of functions, because every vertical line intersects each graph in at most one point. The graphs in Figures 14(c) and 14(d) are not graphs of functions, because there is a vertical line that intersects each graph in more than one point. Notice in Figure 14(c) that the input 1 corresponds to two outputs, −1 and 1. This is why the graph does not represent a function.

Now Work PROBLEM 15

2 Obtain Information from or about the Graph of a Function

If (x, y) is a point on the graph of a function f, then y is the value of f at x; that is, $y = f(x)$. Also if $y = f(x)$, then (x, y) is a point on the graph of f. For example, if $(-2, 7)$ is on the graph of f, then $f(-2) = 7$, and if $f(5) = 8$, then the point $(5, 8)$ is on the graph of $y = f(x)$. The next example illustrates how to obtain information about a function if its graph is given.

| **EXAMPLE 2** | **Obtaining Information from the Graph of a Function** |

Figure 15

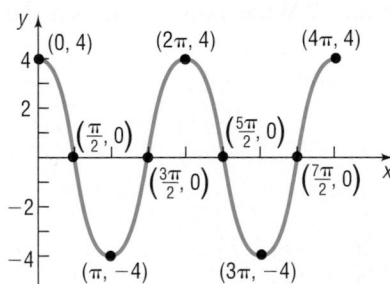

Let f be the function whose graph is given in Figure 15. (The graph of f might represent the distance y that the bob of a pendulum is from its *at-rest* position at time x. Negative values of y mean that the pendulum is to the left of the at-rest position, and positive values of y mean that the pendulum is to the right of the at-rest position.)

(a) What are $f(0)$, $f\left(\dfrac{3\pi}{2}\right)$, and $f(3\pi)$?

(b) What is the domain of f?

(c) What is the range of f?

(d) List the intercepts. (Recall that these are the points, if any, where the graph crosses or touches the coordinate axes.)

(e) How many times does the line $y = 2$ intersect the graph?

(f) For what values of x does $f(x) = -4$?

(g) For what values of x is $f(x) > 0$?

Solution (a) Since $(0, 4)$ is on the graph of f, the y-coordinate 4 is the value of f at the x-coordinate 0; that is, $f(0) = 4$. In a similar way, we find that when $x = \dfrac{3\pi}{2}$, then $y = 0$, so $f\left(\dfrac{3\pi}{2}\right) = 0$. When $x = 3\pi$, then $y = -4$, so $f(3\pi) = -4$.

(b) To determine the domain of f, we notice that the points on the graph of f have x-coordinates between 0 and 4π, inclusive; and for each number x between 0 and 4π, there is a point $(x, f(x))$ on the graph. The domain of f is $\{x | 0 \leq x \leq 4\pi\}$ or the interval $[0, 4\pi]$.

(c) The points on the graph all have y-coordinates between -4 and 4, inclusive; and for each such number y, there is at least one number x in the domain. The range of f is $\{y | -4 \leq y \leq 4\}$ or the interval $[-4, 4]$.

(d) The intercepts are the points

$$(0, 4), \left(\frac{\pi}{2}, 0\right), \left(\frac{3\pi}{2}, 0\right), \left(\frac{5\pi}{2}, 0\right), \quad \text{and} \quad \left(\frac{7\pi}{2}, 0\right)$$

(e) If we draw the horizontal line $y = 2$ on the graph in Figure 15, we find that it intersects the graph four times.

(f) Since $(\pi, -4)$ and $(3\pi, -4)$ are the only points on the graph for which $y = f(x) = -4$, we have $f(x) = -4$ when $x = \pi$ and $x = 3\pi$.

(g) To determine where $f(x) > 0$, look at Figure 15 and determine the x-values from 0 to 4π for which the y-coordinate is positive. This occurs on $\left[0, \frac{\pi}{2}\right) \cup \left(\frac{3\pi}{2}, \frac{5\pi}{2}\right) \cup \left(\frac{7\pi}{2}, 4\pi\right]$. Using inequality notation, $f(x) > 0$ for

$$0 \le x < \frac{\pi}{2} \text{ or } \frac{3\pi}{2} < x < \frac{5\pi}{2} \quad \text{or} \quad \frac{7\pi}{2} < x \le 4\pi.$$

When the graph of a function is given, its domain may be viewed as the shadow created by the graph on the x-axis by vertical beams of light. Its range can be viewed as the shadow created by the graph on the y-axis by horizontal beams of light. Try this technique with the graph given in Figure 15.

─────**Now Work** PROBLEMS 9 AND 13

EXAMPLE 3

Obtaining Information about the Graph of a Function

Consider the function: $f(x) = \dfrac{x + 1}{x + 2}$

(a) Find the domain of f.

(b) Is the point $\left(1, \dfrac{1}{2}\right)$ on the graph of f?

(c) If $x = 2$, what is $f(x)$? What point is on the graph of f?

(d) If $f(x) = 2$, what is x? What point is on the graph of f?

(e) What are the x-intercepts of the graph of f (if any)? What point(s) are on the graph of f?

Solution

(a) The domain of f is $\{x | x \ne -2\}$.

(b) When $x = 1$, then

$$f(x) = \frac{x + 1}{x + 2}$$

$$f(1) = \frac{1 + 1}{1 + 2} = \frac{2}{3}$$

The point $\left(1, \dfrac{2}{3}\right)$ is on the graph of f; the point $\left(1, \dfrac{1}{2}\right)$ is not.

(c) If $x = 2$, then

$$f(x) = \frac{x + 1}{x + 2}$$

$$f(2) = \frac{2 + 1}{2 + 2} = \frac{3}{4}$$

The point $\left(2, \dfrac{3}{4}\right)$ is on the graph of f.

(d) If $f(x) = 2$, then

$$f(x) = 2$$

$$\frac{x + 1}{x + 2} = 2$$

$$x + 1 = 2(x + 2) \quad \text{Multiply both sides by } x + 2.$$
$$x + 1 = 2x + 4 \quad \text{Remove parentheses.}$$
$$x = -3 \quad \text{Solve for } x.$$

If $f(x) = 2$, then $x = -3$. The point $(-3, 2)$ is on the graph of f.

(e) The x-intercepts of the graph of f are the real solutions of the equation $f(x) = 0$ that are in the domain of f. The only real solution of the equation $f(x) = \dfrac{x + 1}{x + 2} = 0$, is $x = -1$, so -1 is the only x-intercept. Since $f(-1) = 0$, the point $(-1, 0)$ is on the graph of f.

―――Now Work PROBLEM 25

EXAMPLE 4

Average Cost Function

The average cost \overline{C} of manufacturing x computers per day is given by the function

$$\overline{C}(x) = 0.56x^2 - 34.39x + 1212.57 + \frac{20{,}000}{x}$$

Determine the average cost of manufacturing:

(a) 30 computers in a day
(b) 40 computers in a day
(c) 50 computers in a day
(d) Graph the function $\overline{C} = \overline{C}(x), 0 < x \le 80$.
(e) Create a TABLE with TblStart $= 1$ and ΔTbl $= 1$. Which value of x minimizes the average cost?

Solution

(a) The average cost of manufacturing $x = 30$ computers is

$$\overline{C}(30) = 0.56(30)^2 - 34.39(30) + 1212.57 + \frac{20{,}000}{30} = \$1351.54$$

(b) The average cost of manufacturing $x = 40$ computers is

$$\overline{C}(40) = 0.56(40)^2 - 34.39(40) + 1212.57 + \frac{20{,}000}{40} = \$1232.97$$

(c) The average cost of manufacturing $x = 50$ computers is

$$\overline{C}(50) = 0.56(50)^2 - 34.39(50) + 1212.57 + \frac{20{,}000}{50} = \$1293.07$$

(d) See Figure 16 for the graph of $\overline{C} = \overline{C}(x)$.
(e) With the function $\overline{C} = \overline{C}(x)$ in Y_1, we create Table 2. We scroll down until we find a value of x for which Y_1 is smallest. Table 3 shows that manufacturing $x = 41$ computers minimizes the average cost at $\$1231.74$ per computer.

Figure 16

Table 2

Table 3

―――Now Work PROBLEM 31

SUMMARY

Graph of a Function The collection of points (x, y) that satisfies the equation $y = f(x)$.

Vertical Line Test A collection of points is the graph of a function provided that every vertical line intersects the graph in at most one point.

3.2 Assess Your Understanding

'Are You Prepared?' *Answers are given at the end of these exercises. If you get a wrong answer, read the pages listed in red.*

1. The intercepts of the equation $x^2 + 4y^2 = 16$ are _____. (pp. 159–160)

2. **True or False** The point $(-2, -6)$ is on the graph of the equation $x = 2y - 2$. (pp. 157–159)

Concepts and Vocabulary

3. A set of points in the xy-plane is the graph of a function if and only if every _____ line intersects the graph in at most one point.

4. If the point $(5, -3)$ is a point on the graph of f, then $f(___) = ____$.

5. Find a so that the point $(-1, 2)$ is on the graph of $f(x) = ax^2 + 4$.

6. **True or False** A function can have more than one y-intercept.

7. **True or False** The graph of a function $y = f(x)$ always crosses the y-axis.

8. **True or False** The y-intercept of the graph of the function $y = f(x)$, whose domain is all real numbers, is $f(0)$.

Skill Building

9. Use the given graph of the function f to answer parts (a)–(n).

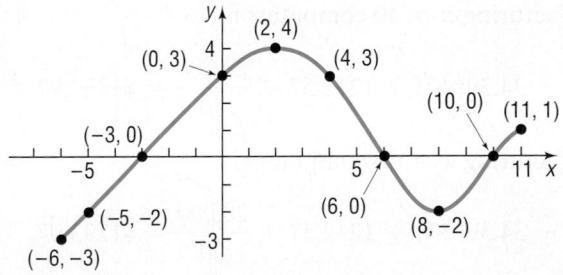

10. Use the given graph of the function f to answer parts (a)–(n).

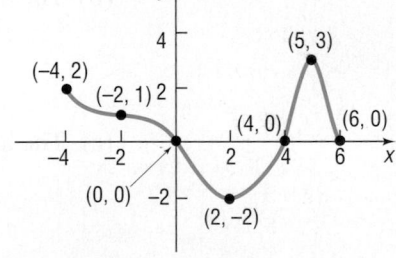

(a) Find $f(0)$ and $f(-6)$.
(b) Find $f(6)$ and $f(11)$.
(c) Is $f(3)$ positive or negative?
(d) Is $f(-4)$ positive or negative?
(e) For what values of x is $f(x) = 0$?
(f) For what values of x is $f(x) > 0$?
(g) What is the domain of f?
(h) What is the range of f?
(i) What are the x-intercepts?
(j) What is the y-intercept?
(k) How often does the line $y = \dfrac{1}{2}$ intersect the graph?
(l) How often does the line $x = 5$ intersect the graph?
(m) For what values of x does $f(x) = 3$?
(n) For what values of x does $f(x) = -2$?

(a) Find $f(0)$ and $f(6)$.
(b) Find $f(2)$ and $f(-2)$.
(c) Is $f(3)$ positive or negative?
(d) Is $f(-1)$ positive or negative?
(e) For what values of x is $f(x) = 0$?
(f) For what values of x is $f(x) < 0$?
(g) What is the domain of f?
(h) What is the range of f?
(i) What are the x-intercepts?
(j) What is the y-intercept?
(k) How often does the line $y = -1$ intersect the graph?
(l) How often does the line $x = 1$ intersect the graph?
(m) For what value of x does $f(x) = 3$?
(n) For what value of x does $f(x) = -2$?

In Problems 11–22, determine whether the graph is that of a function by using the vertical-line test. If it is, use the graph to find:

(a) *The domain and range*
(b) *The intercepts, if any*
(c) *Any symmetry with respect to the x-axis, the y-axis, or the origin*

11.

12.

13.

14.

15.

16.

17.

18.

19.

20.

21.

22.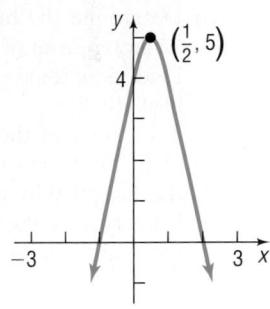

In Problems 23–28, answer the questions about the given function.

23. $f(x) = 2x^2 - x - 1$
 (a) Is the point $(-1, 2)$ on the graph of f?
 (b) If $x = -2$, what is $f(x)$? What point is on the graph of f?
 (c) If $f(x) = -1$, what is x? What point(s) are on the graph of f?
 (d) What is the domain of f?
 (e) List the x-intercepts, if any, of the graph of f.
 (f) List the y-intercept, if there is one, of the graph of f.

24. $f(x) = -3x^2 + 5x$
 (a) Is the point $(-1, 2)$ on the graph of f?
 (b) If $x = -2$, what is $f(x)$? What point is on the graph of f?
 (c) If $f(x) = -2$, what is x? What point(s) are on the graph of f?
 (d) What is the domain of f?
 (e) List the x-intercepts, if any, of the graph of f.
 (f) List the y-intercept, if there is one, of the graph of f.

25. $f(x) = \dfrac{x + 2}{x - 6}$
 (a) Is the point $(3, 14)$ on the graph of f?
 (b) If $x = 4$, what is $f(x)$? What point is on the graph of f?
 (c) If $f(x) = 2$, what is x? What point(s) are on the graph of f?
 (d) What is the domain of f?
 (e) List the x-intercepts, if any, of the graph of f.
 (f) List the y-intercept, if there is one, of the graph of f.

26. $f(x) = \dfrac{x^2 + 2}{x + 4}$
 (a) Is the point $\left(1, \dfrac{3}{5}\right)$ on the graph of f?

(b) If $x = 0$, what is $f(x)$? What point is on the graph of f?
(c) If $f(x) = \dfrac{1}{2}$, what is x? What point(s) are on the graph of f?
(d) What is the domain of f?
(e) List the x-intercepts, if any, of the graph of f.
(f) List the y-intercept, if there is one, of the graph of f.

27. $f(x) = \dfrac{2x^2}{x^4 + 1}$
 (a) Is the point $(-1, 1)$ on the graph of f?
 (b) If $x = 2$, what is $f(x)$? What point is on the graph of f?
 (c) If $f(x) = 1$, what is x? What point(s) are on the graph of f?
 (d) What is the domain of f?
 (e) List the x-intercepts, if any, of the graph of f.
 (f) List the y-intercept, if there is one, of the graph of f.

28. $f(x) = \dfrac{2x}{x - 2}$
 (a) Is the point $\left(\dfrac{1}{2}, -\dfrac{2}{3}\right)$ on the graph of f?
 (b) If $x = 4$, what is $f(x)$? What point is on the graph of f?
 (c) If $f(x) = 1$, what is x? What point(s) are on the graph of f?
 (d) What is the domain of f?
 (e) List the x-intercepts, if any, of the graph of f.
 (f) List the y-intercept, if there is one, of the graph of f.

Applications and Extensions

29. Free-throw Shots According to physicist Peter Brancazio, the key to a successful foul shot in basketball lies in the arc of the shot. Brancazio determined the optimal angle of the arc from the free-throw line to be 45 degrees. The arc also depends on the velocity with which the ball is shot. If a player shoots a foul shot, releasing the ball at a 45-degree angle from a position 6 feet above the floor, then the path of the ball can be modeled by the function

$$h(x) = -\frac{44x^2}{v^2} + x + 6$$

where h is the height of the ball above the floor, x is the forward distance of the ball in front of the foul line, and v is the initial velocity with which the ball is shot in feet per second. Suppose a player shoots a ball with an initial velocity of 28 feet per second.

(a) Determine the height of the ball after it has traveled 8 feet in front of the foul line.

(b) Determine the height of the ball after it has traveled 12 feet in front of the foul line.

(c) Find additional points and graph the path of the basketball.

(d) The center of the hoop is 10 feet above the floor and 15 feet in front of the foul line. Will the ball go through the hoop? Why or why not? If not, with what initial velocity must the ball be shot in order for the ball to go through the hoop?

Source: The Physics of Foul Shots, Discover, Vol. 21, No. 10, October 2000

30. Granny Shots The last player in the NBA to use an underhand foul shot (a "granny" shot) was Hall of Fame forward Rick Barry who retired in 1980. Barry believes that current NBA players could increase their free-throw percentage if they were to use an underhand shot. Since underhand shots are released from a lower position, the angle of the shot must be increased. If a player shoots an underhand foul shot, releasing the ball at a 70-degree angle from a position 3.5 feet above the floor, then the path of the ball can be modeled by the function $h(x) = -\frac{136x^2}{v^2} + 2.7x + 3.5$,

where h is the height of the ball above the floor, x is the forward distance of the ball in front of the foul line, and v is the initial velocity with which the ball is shot in feet per second.

(a) The center of the hoop is 10 feet above the floor and 15 feet in front of the foul line. Determine the initial velocity with which the ball must be shot in order for the ball to go through the hoop.

(b) Write the function for the path of the ball using the velocity found in part (a).

(c) Determine the height of the ball after it has traveled 9 feet in front of the foul line.

(d) Find additional points and graph the path of the basketball.

Source: The Physics of Foul Shots, Discover, Vol. 21, No. 10, October 2000

31. Motion of a Golf Ball A golf ball is hit with an initial velocity of 130 feet per second at an inclination of 45° to the horizontal. In physics, it is established that the height h

of the golf ball is given by the function

$$h(x) = \frac{-32x^2}{130^2} + x$$

where x is the horizontal distance that the golf ball has traveled.

(a) Determine the height of the golf ball after it has traveled 100 feet.

(b) What is the height after it has traveled 300 feet?

(c) What is the height after it has traveled 500 feet?

(d) How far was the golf ball hit?

(e) Use a graphing utility to graph the function $h = h(x)$.

(f) Use a graphing utility to determine the distance that the ball has traveled when the height of the ball is 90 feet.

(g) Create a TABLE with TblStart = 0 and ΔTbl = 25. To the nearest 25 feet, how far does the ball travel before it reaches a maximum height? What is the maximum height?

(h) Adjust the value of ΔTbl until you determine the distance, to within 1 foot, that the ball travels before it reaches a maximum height.

32. Cross-sectional Area The cross-sectional area of a beam cut from a log with radius 1 foot is given by the function $A(x) = 4x\sqrt{1 - x^2}$, where x represents the length, in feet, of half the base of the beam. See the figure.

$$A(x) = 4x\sqrt{1 - x^2}$$

(a) Find the domain of A.

(b) Use a graphing utility to graph the function $A = A(x)$.

(c) Create a TABLE with TblStart = 0 and ΔTbl = 0.1 for $0 \le x \le 1$. Which value of x maximizes the cross-sectional area? What should be the length of the base of the beam to maximize the cross-sectional area?

33. Cost of Trans-Atlantic Travel A Boeing 747 crosses the Atlantic Ocean (3000 miles) with an airspeed of 500 miles per hour. The cost C (in dollars) per passenger is given by

$$C(x) = 100 + \frac{x}{10} + \frac{36,000}{x}$$

where x is the ground speed (airspeed ± wind).
(a) Use a graphing utility to graph the function $C = C(x)$.
(b) Create a TABLE with TblStart = 0 and ΔTbl = 50.
(c) To the nearest 50 miles per hour, what ground speed minimizes the cost per passenger?

34. Effect of Elevation on Weight If an object weighs m pounds at sea level, then its weight W (in pounds) at a height of h miles above sea level is given approximately by

$$W(h) = m \left(\frac{4000}{4000 + h} \right)^2$$

(a) If Amy weighs 120 pounds at sea level, how much will she weigh on Pike's Peak, which is 14,110 feet above sea level?
(b) Use a graphing utility to graph the function $W = W(h)$. Use $m = 120$ pounds.

(c) Create a Table with TblStart = 0 and ΔTbl = 0.5 to see how the weight W varies as h changes from 0 to 5 miles.
(d) At what height will Amy weigh 119.95 pounds?
(e) Does your answer to part (d) seem reasonable? Explain.

35. The graph of two functions, f and g, is illustrated. Use the graph to answer parts (a)–(f).

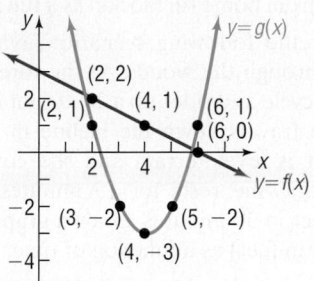

(a) $(f + g)(2)$ (b) $(f + g)(4)$
(c) $(f - g)(6)$ (d) $(g - f)(6)$
(e) $(f \cdot g)(2)$ (f) $\left(\dfrac{f}{g} \right)(4)$

Explaining Concepts: Discussion and Writing

36. Describe how you would proceed to find the domain and range of a function if you were given its graph. How would your strategy change if you were given the equation defining the function instead of its graph?

37. How many x-intercepts can the graph of a function have? How many y-intercepts can the graph of a function have?

38. Is a graph that consists of a single point the graph of a function? Can you write the equation of such a function?

39. Match each of the following functions with the graph that best describes the situation.
(a) The cost of building a house as a function of its square footage
(b) The height of an egg dropped from a 300-foot building as a function of time
(c) The height of a human as a function of time
(d) The demand for Big Macs as a function of price
(e) The height of a child on a swing as a function of time

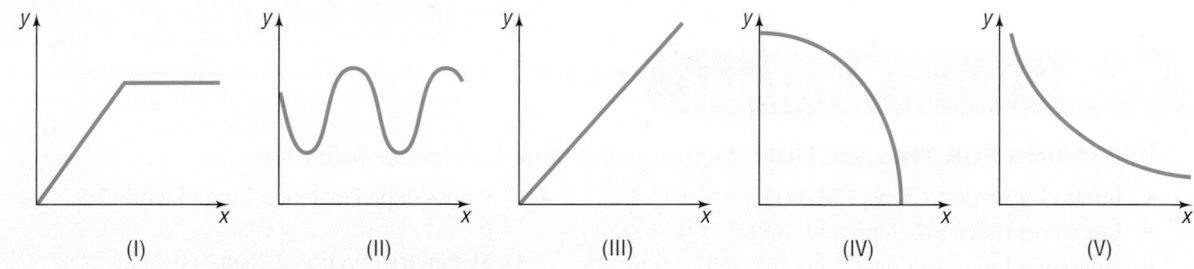

(I) (II) (III) (IV) (V)

40. Match each of the following functions with the graph that best describes the situation.
(a) The temperature of a bowl of soup as a function of time
(b) The number of hours of daylight per day over a 2-year period
(c) The population of Florida as a function of time
(d) The distance travelled by a car going at a constant velocity as a function of time
(e) The height of a golf ball hit with a 7-iron as a function of time

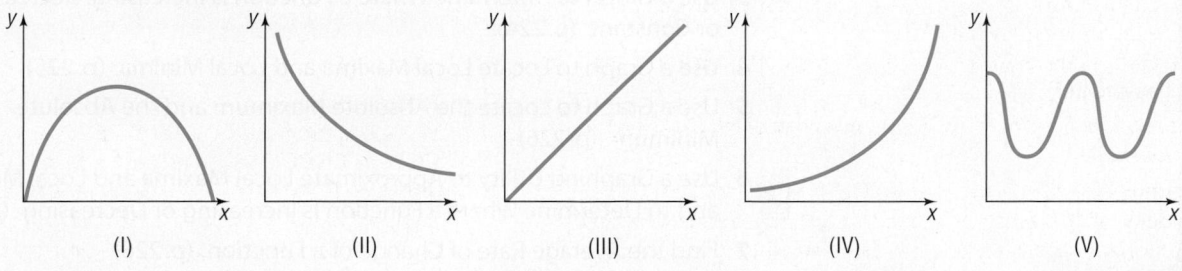

(I) (II) (III) (IV) (V)

41. Consider the following scenario: Barbara decides to take a walk. She leaves home, walks 2 blocks in 5 minutes at a constant speed, and realizes that she forgot to lock the door. So Barbara runs home in 1 minute. While at her doorstep, it takes her 1 minute to find her keys and lock the door. Barbara walks 5 blocks in 15 minutes and then decides to jog home. It takes her 7 minutes to get home. Draw a graph of Barbara's distance from home (in blocks) as a function of time.

42. Consider the following scenario: Jayne enjoys riding her bicycle through the woods. At the forest preserve, she gets on her bicycle and rides up a 2000-foot incline in 10 minutes. She then travels down the incline in 3 minutes. The next 5000 feet is level terrain and she covers the distance in 20 minutes. She rests for 15 minutes. Jayne then travels 10,000 feet in 30 minutes. Draw a graph of Jayne's distance traveled (in feet) as a function of time.

43. The following sketch represents the distance *d* (in miles) that Kevin was from home as a function of time *t* (in hours). Answer the questions based on the graph. In parts (a)–(g), how many hours elapsed and how far was Kevin from home during this time?

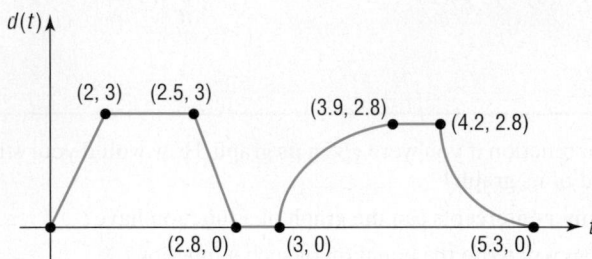

(a) From $t = 0$ to $t = 2$
(b) From $t = 2$ to $t = 2.5$

(c) From $t = 2.5$ to $t = 2.8$
(d) From $t = 2.8$ to $t = 3$
(e) From $t = 3$ to $t = 3.9$
(f) From $t = 3.9$ to $t = 4.2$
(g) From $t = 4.2$ to $t = 5.3$
(h) What is the farthest distance that Kevin was from home?
(i) How many times did Kevin return home?

44. The following sketch represents the speed *v* (in miles per hour) of Michael's car as a function of time *t* (in minutes).

(a) Over what interval of time was Michael traveling fastest?
(b) Over what interval(s) of time was Michael's speed zero?
(c) What was Michael's speed between 0 and 2 minutes?
(d) What was Michael's speed between 4.2 and 6 minutes?
(e) What was Michael's speed between 7 and 7.4 minutes?
(f) When was Michael's speed constant?

45. Draw the graph of a function whose domain is $\{x | -3 \le x \le 8, \ x \ne 5\}$ and whose range is $\{y | -1 \le y \le 2, \ y \ne 0\}$. What point(s) in the rectangle $-3 \le x \le 8, -1 \le y \le 2$ cannot be on the graph? Compare your graph with those of other students. What differences do you see?

46. Is there a function whose graph is symmetric with respect to the *x*-axis? Explain.

'Are You Prepared?' Answers

1. $(-4, 0), (4, 0), (0, -2), (0, 2)$ **2.** False

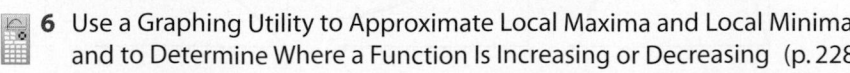

3.3 Properties of Functions

PREPARING FOR THIS SECTION *Before getting started, review the following:*

- Intervals (Section 1.5, pp. 120–121)
- Intercepts (Section 2.2, pp. 159–160)
- Slope of a Line (Section 2.3, pp. 167–169)

- Point–Slope Form of a Line (Section 2.3, p. 171)
- Symmetry (Section 2.2, pp. 160–162)

Now Work the 'Are You Prepared?' problems on page 230.

OBJECTIVES **1** Determine Even and Odd Functions from a Graph (p. 223)

2 Identify Even and Odd Functions from the Equation (p. 224)

3 Use a Graph to Determine Where a Function Is Increasing, Decreasing, or Constant (p. 224)

4 Use a Graph to Locate Local Maxima and Local Minima (p. 225)

5 Use a Graph to Locate the Absolute Maximum and the Absolute Minimum (p. 226)

6 Use a Graphing Utility to Approximate Local Maxima and Local Minima and to Determine Where a Function Is Increasing or Decreasing (p. 228)

7 Find the Average Rate of Change of a Function (p. 228)

To obtain the graph of a function $y = f(x)$, it is often helpful to know certain properties that the function has and the impact of these properties on the way that the graph will look.

1 Determine Even and Odd Functions from a Graph

The words *even* and *odd*, when applied to a function f, describe the symmetry that exists for the graph of the function.

A function f is even, if and only if, whenever the point (x, y) is on the graph of f then the point $(-x, y)$ is also on the graph. Using function notation, we define an even function as follows:

DEFINITION A function f is **even** if, for every number x in its domain, the number $-x$ is also in the domain and

$$f(-x) = f(x)$$

A function f is odd, if and only if, whenever the point (x, y) is on the graph of f then the point $(-x, -y)$ is also on the graph. Using function notation, we define an odd function as follows:

DEFINITION A function f is **odd** if, for every number x in its domain, the number $-x$ is also in the domain and

$$f(-x) = -f(x)$$

Refer to page 162, where the tests for symmetry are listed. The following results are then evident.

THEOREM A function is even if and only if its graph is symmetric with respect to the y-axis. A function is odd if and only if its graph is symmetric with respect to the origin.

EXAMPLE 1 **Determining Even and Odd Functions from the Graph**

Determine whether each graph given in Figure 17 is the graph of an even function, an odd function, or a function that is neither even nor odd.

Figure 17

(a) (b) (c)

Solution (a) The graph in Figure 17(a) is that of an even function, because the graph is symmetric with respect to the y-axis.

(b) The function whose graph is given in Figure 17(b) is neither even nor odd, because the graph is neither symmetric with respect to the y-axis nor symmetric with respect to the origin.

(c) The function whose graph is given in Figure 17(c) is odd, because its graph is symmetric with respect to the origin.

Now Work PROBLEMS 21(a), (b), AND (d)

2 Identify Even and Odd Functions from the Equation

EXAMPLE 2

Identifying Even and Odd Functions Algebraically

Determine whether each of the following functions is even, odd, or neither. Then determine whether the graph is symmetric with respect to the *y*-axis, or with respect to the origin.

(a) $f(x) = x^2 - 5$ (b) $g(x) = x^3 - 1$
(c) $h(x) = 5x^3 - x$ (d) $F(x) = |x|$

Solution

(a) To determine whether f is even, odd, or neither, replace x by $-x$ in $f(x) = x^2 - 5$. Then

$$f(-x) = (-x)^2 - 5 = x^2 - 5 = f(x)$$

Since $f(-x) = f(x)$, we conclude that f is an even function, and the graph of f is symmetric with respect to the *y*-axis.

(b) Replace x by $-x$ in $g(x) = x^3 - 1$. Then

$$g(-x) = (-x)^3 - 1 = -x^3 - 1$$

Since $g(-x) \neq g(x)$ and $g(-x) \neq -g(x) = -(x^3 - 1) = -x^3 + 1$, we conclude that g is neither even nor odd. The graph of g is not symmetric with respect to the *y*-axis nor is it symmetric with respect to the origin.

(c) Replace x by $-x$ in $h(x) = 5x^3 - x$. Then

$$h(-x) = 5(-x)^3 - (-x) = -5x^3 + x = -(5x^3 - x) = -h(x)$$

Since $h(-x) = -h(x)$, h is an odd function, and the graph of h is symmetric with respect to the origin.

(d) Replace x by $-x$ in $F(x) = |x|$. Then

$$F(-x) = |-x| = |-1| \cdot |x| = |x| = F(x)$$

Since $F(-x) = F(x)$, F is an even function, and the graph of F is symmetric with respect to the *y*-axis.

――― **Now Work** PROBLEM 33

3 Use a Graph to Determine Where a Function Is Increasing, Decreasing, or Constant

Consider the graph given in Figure 18. If you look from left to right along the graph of the function, you will notice that parts of the graph are going up, parts are going down, and parts are horizontal. In such cases, the function is described as *increasing, decreasing,* or *constant*, respectively.

Figure 18

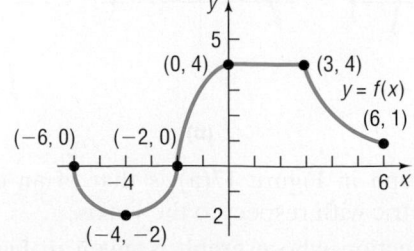

EXAMPLE 3

Determining Where a Function Is Increasing, Decreasing, or Constant from Its Graph

Where is the function in Figure 18 increasing? Where is it decreasing? Where is it constant?

Solution

To answer the question of where a function is increasing, where it is decreasing, and where it is constant, we use strict inequalities involving the independent variable x, or we use open intervals* of x-coordinates. The function whose graph is given in Figure 18 is increasing on the open interval $(-4, 0)$ or for $-4 < x < 0$. The function is decreasing on the open intervals $(-6, -4)$ and $(3, 6)$ or for $-6 < x < -4$ and $3 < x < 6$. The function is constant on the open interval $(0, 3)$ or for $0 < x < 3$. ↵

More precise definitions follow:

DEFINITIONS

A function f is **increasing** on an open interval I if, for any choice of x_1 and x_2 in I, with $x_1 < x_2$, we have $f(x_1) < f(x_2)$.

A function f is **decreasing** on an open interval I if, for any choice of x_1 and x_2 in I, with $x_1 < x_2$, we have $f(x_1) > f(x_2)$.

A function f is **constant** on an open interval I if, for all choices of x in I, the values $f(x)$ are equal. ↵

Figure 19 illustrates the definitions. The graph of an increasing function goes up from left to right, the graph of a decreasing function goes down from left to right, and the graph of a constant function remains at a fixed height.

Figure 19

(a) For $x_1 < x_2$ in I, $f(x_1) < f(x_2)$; f is increasing on I

(b) For $x_1 < x_2$ in I, $f(x_1) > f(x_2)$; f is decreasing on I

(c) For all x in I, the values of f are equal; f is constant on I

Now Work PROBLEMS **11, 13, 15,** AND **21(C)**

4 Use a Graph to Locate Local Maxima and Local Minima

Suppose f is a function defined on an open interval containing c. If the value of f at c is greater than or equal to the values of f on I, then f has a *local maximum* at c[†]. See Figure 20(a).

If the value of f at c is less than or equal to the values of f on I, then f has a *local minimum* at c. See Figure 20(b).

Figure 20

f has a local maximum at c.

f has a local minimum at c.

(a) **(b)**

* The open interval (a, b) consists of all real numbers x for which $a < x < b$.
[†] Some texts use the term *relative* instead of *local*.

DEFINITIONS

A function f has a **local maximum** at c if there is an open interval I containing c so that for all x in I, $f(x) \leq f(c)$. We call $f(c)$ a **local maximum value of** f.

A function f has a **local minimum** at c if there is an open interval I containing c so that, for all x in I, $f(x) \geq f(c)$. We call $f(c)$ a **local minimum value of** f.

If f has a local maximum at c, then the value of f at c is greater than or equal to the values of f near c. If f has a local minimum at c, then the value of f at c is less than or equal to the values of f near c. The word *local* is used to suggest that it is only near c, that is, in some open interval containing c, that the value $f(c)$ has these properties.

Figure 21

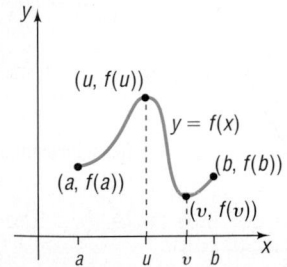

EXAMPLE 4

Finding Local Maxima and Local Minima from the Graph of a Function and Determining Where the Function Is Increasing, Decreasing, or Constant

Figure 21 shows the graph of a function f.

(a) At what value(s) of x, if any, does f have a local maximum? List the local maximum values.

(b) At what value(s) of x, if any, does f have a local minimum? List the local minimum values.

(c) Find the intervals on which f is increasing. Find the intervals on which f is decreasing.

Solution

The domain of f is the set of real numbers.

WARNING The y-value is the local maximum value or local minimum value and it occurs at some x-value. For example, in Figure 21, we say f has a local maximum at 1 and the local maximum value is 2. ∎

(a) f has a local maximum at 1, since for all x close to 1, we have $f(x) \leq f(1)$. The local maximum value is $f(1) = 2$.

(b) f has local minima at -1 and at 3. The local minima values are $f(-1) = 1$ and $f(3) = 0$.

(c) The function whose graph is given in Figure 21 is increasing for all values of x between -1 and 1 and for all values of x greater than 3. That is, the function is increasing on the intervals $(-1, 1)$ and $(3, \infty)$ or for $-1 < x < 1$ and $x > 3$. The function is decreasing for all values of x less than -1 and for all values of x between 1 and 3. That is, the function is decreasing on the intervals $(-\infty, -1)$ and $(1, 3)$ or for $x < -1$ and $1 < x < 3$. ↵

 Now Work PROBLEMS **17** AND **19**

Figure 22

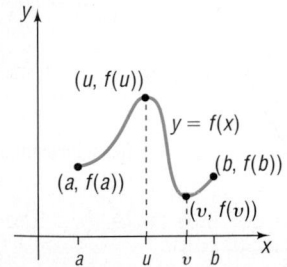

domain: $[a, b]$
for all x in $[a, b]$, $f(x) \leq f(u)$
for all x in $[a, b]$, $f(x) \geq f(v)$
absolute maximum: $f(u)$
absolute minimum: $f(v)$

5 Use a Graph to Locate the Absolute Maximum and the Absolute Minimum

Look at the graph of the function f given in Figure 22. The domain of f is the closed interval $[a, b]$. Also, the largest value of f is $f(u)$ and the smallest value of f is $f(v)$. These are called, respectively, the *absolute maximum* and the *absolute minimum* of f on $[a, b]$.

DEFINITION Let f denote a function defined on some interval I. If there is a number u in I for which $f(x) \leq f(u)$ for all x in I, then $f(u)$ is the **absolute maximum of** f on I and we say **the absolute maximum of** f **occurs at** u.

If there is a number v in I for which $f(x) \geq f(v)$ for all x in I, then $f(v)$ is the **absolute minimum of** f on I and we say **the absolute minimum of** f **occurs at** v. ↵

The absolute maximum and absolute minimum of a function f are sometimes called the **extreme values** of f on I.

The absolute maximum or absolute minimum of a function f may not exist. Let's look at some examples.

EXAMPLE 5 **Finding the Absolute Maximum and the Absolute Minimum from the Graph of a Function**

For each graph of a function $y = f(x)$ in Figure 23 on the following page, find the absolute maximum and the absolute minimum, if they exist.

Figure 23

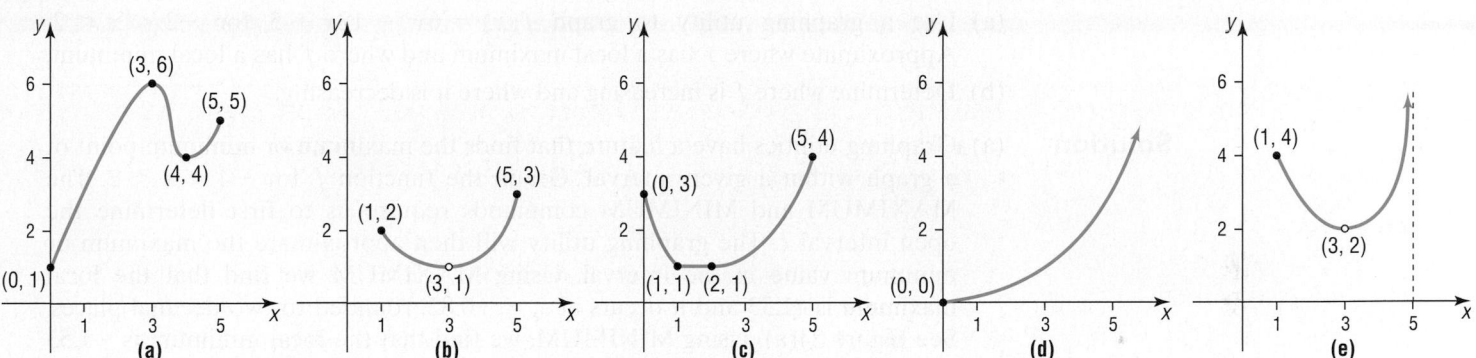

(a) (b) (c) (d) (e)

Solution (a) The function f whose graph is given in Figure 23(a) has the closed interval $[0, 5]$ as its domain. The largest value of f is $f(3) = 6$, the absolute maximum. The smallest value of f is $f(0) = 1$, the absolute minimum.

(b) The function f whose graph is given in Figure 23(b) has the domain $\{x | 1 \le x \le 5, x \ne 3\}$. Note that we exclude 3 from the domain because of the "hole" at $(3, 1)$. The largest value of f on its domain is $f(5) = 3$, the absolute maximum. There is no absolute minimum. Do you see why? As you trace the graph, getting closer to the point $(3, 1)$, there is no single smallest value. [As soon as you claim a smallest value, we can trace closer to $(3, 1)$ and get a smaller value!]

(c) The function f whose graph is given in Figure 23(c) has the interval $[0, 5]$ as its domain. The absolute maximum of f is $f(5) = 4$. The absolute minimum is 1. Notice that the absolute minimum 1 occurs at any number in the interval $[1, 2]$.

(d) The graph of the function f given in Figure 23(d) has the interval $[0, \infty)$ as its domain. The function has no absolute maximum; the absolute minimum is $f(0) = 0$.

(e) The graph of the function f in Figure 23(e) has the domain $\{x | 1 \le x < 5, x \ne 3\}$. The function f has no absolute maximum and no absolute minimum. Do you see why? ↵

In calculus, there is a theorem with conditions that guarantee a function will have an absolute maximum and an absolute minimum.

THEOREM **Extreme Value Theorem**

If f is a continuous function* whose domain is a closed interval $[a, b]$, then f has an absolute maximum and an absolute minimum on $[a, b]$.

━━━━━▶ **Now Work** PROBLEM **45**

* Although it requires calculus for a precise definition, we'll agree for now that a continuous function is one whose graph has no gaps or holes and can be traced without lifting the pencil from the paper.

6 Use a Graphing Utility to Approximate Local Maxima and Local Minima and to Determine Where a Function Is Increasing or Decreasing

To locate the exact value at which a function f has a local maximum or a local minimum usually requires calculus. However, a graphing utility may be used to approximate these values by using the MAXIMUM and MINIMUM features.

EXAMPLE 6

Using a Graphing Utility to Approximate Local Maxima and Minima and to Determine Where a Function Is Increasing or Decreasing

(a) Use a graphing utility to graph $f(x) = 6x^3 - 12x + 5$ for $-2 < x < 2$. Approximate where f has a local maximum and where f has a local minimum.

(b) Determine where f is increasing and where it is decreasing.

Solution

(a) Graphing utilities have a feature that finds the maximum or minimum point of a graph within a given interval. Graph the function f for $-2 < x < 2$. The MAXIMUM and MINIMUM commands require us to first determine the open interval I. The graphing utility will then approximate the maximum or minimum value in the interval. Using MAXIMUM we find that the local maximum is 11.53 and it occurs at $x = -0.82$, rounded to two decimal places. See Figure 24(a). Using MINIMUM, we find that the local minimum is -1.53 and it occurs at $x = 0.82$, rounded to two decimal places. See Figure 24(b).

Figure 24

(a) (b)

(b) Looking at Figures 24(a) and (b), we see that the graph of f is increasing from $x = -2$ to $x = -0.82$ and from $x = 0.82$ to $x = 2$, so f is increasing on the intervals $(-2, -0.82)$ and $(0.82, 2)$ or for $-2 < x < -0.82$ and $0.82 < x < 2$. The graph is decreasing from $x = -0.82$ to $x = 0.82$, so f is decreasing on the interval $(-0.82, 0.82)$ or for $-0.82 < x < 0.82$.

Now Work PROBLEM 53

7 Find the Average Rate of Change of a Function

In Section 2.3, we said that the slope of a line could be interpreted as the average rate of change. To find the average rate of change of a function between any two points on its graph, calculate the slope of the line containing the two points.

DEFINITION

If a and b, $a \neq b$, are in the domain of a function $y = f(x)$, the **average rate of change of f** from a to b is defined as

$$\text{Average rate of change} = \frac{\Delta y}{\Delta x} = \frac{f(b) - f(a)}{b - a} \qquad a \neq b \qquad \textbf{(1)}$$

The symbol Δy in (1) is the "change in y," and Δx is the "change in x." The average rate of change of f is the change in y divided by the change in x.

| **EXAMPLE 7** | **Finding the Average Rate of Change** |

Find the average rate of change of $f(x) = 3x^2$:

(a) From 1 to 3 (b) From 1 to 5 (c) From 1 to 7

Solution (a) The average rate of change of $f(x) = 3x^2$ from 1 to 3 is

$$\frac{\Delta y}{\Delta x} = \frac{f(3) - f(1)}{3 - 1} = \frac{27 - 3}{3 - 1} = \frac{24}{2} = 12$$

(b) The average rate of change of $f(x) = 3x^2$ from 1 to 5 is

$$\frac{\Delta y}{\Delta x} = \frac{f(5) - f(1)}{5 - 1} = \frac{75 - 3}{5 - 1} = \frac{72}{4} = 18$$

(c) The average rate of change of $f(x) = 3x^2$ from 1 to 7 is

$$\frac{\Delta y}{\Delta x} = \frac{f(7) - f(1)}{7 - 1} = \frac{147 - 3}{7 - 1} = \frac{144}{6} = 24$$

Figure 25

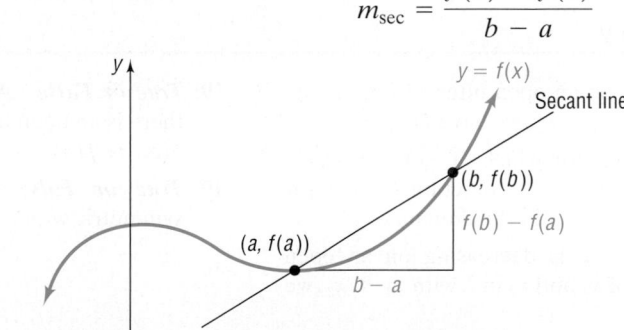

See Figure 25 for a graph of $f(x) = 3x^2$. The function f is increasing for $x > 0$. The fact that the average rate of change is positive for any $x_1, x_2, x_1 \neq x_2$ in the interval $(1, 7)$ indicates that the graph is increasing on $1 < x < 7$. Further, the average rate of change is consistently getting larger for $1 < x < 7$, indicating that the graph is increasing at an increasing rate.

Now Work PROBLEM 61

The Secant Line

The average rate of change of a function has an important geometric interpretation. Look at the graph of $y = f(x)$ in Figure 26. We have labeled two points on the graph: $(a, f(a))$ and $(b, f(b))$. The line containing these two points is called the **secant line**; its slope is

$$m_{\text{sec}} = \frac{f(b) - f(a)}{b - a}$$

Figure 26

| **THEOREM** | **Slope of the Secant Line** |

The average rate of change of a function from a to b equals the slope of the secant line containing the two points $(a, f(a))$ and $(b, f(b))$ on its graph.

| **EXAMPLE 8** | **Finding the Equation of a Secant Line** |

Suppose that $g(x) = 3x^2 - 2x + 3$.

(a) Find the average rate of change of g from -2 to 1.

(b) Find an equation of the secant line containing $(-2, g(-2))$ and $(1, g(1))$.

(c) Using a graphing utility, draw the graph of g and the secant line obtained in part (b) on the same screen.

Solution (a) The average rate of change of $g(x) = 3x^2 - 2x + 3$ from -2 to 1 is

$$\text{Average rate of change} = \frac{g(1) - g(-2)}{1 - (-2)}$$

$$= \frac{4 - 19}{3} \qquad \begin{array}{l} g(1) = 3(1)^2 - 2(1) + 3 = 4 \\ g(-2) = 3(-2)^2 - 2(-2) + 3 = 19 \end{array}$$

$$= -\frac{15}{3} = -5$$

Figure 27

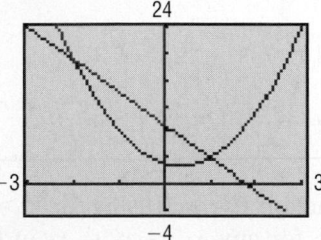

(b) The slope of the secant line containing $(-2, g(-2)) = (-2, 19)$ and $(1, g(1)) = (1, 4)$ is $m_{\text{sec}} = -5$. We use the point–slope form to find an equation of the secant line.

$$\begin{array}{ll} y - y_1 = m_{\text{sec}}(x - x_1) & \text{Point–slope form of the secant line} \\ y - 19 = -5(x - (-2)) & x_1 = -2, y_1 = g(-2) = 19, m_{\text{sec}} = -5 \\ y - 19 = -5x - 10 & \text{Simplify.} \\ y = -5x + 9 & \text{Slope–intercept form of the secant line} \end{array}$$

(c) Figure 27 shows the graph of g along with the secant line $y = -5x + 9$.

Now Work PROBLEM 67

3.3 Assess Your Understanding

'Are You Prepared?' *Answers are given at the end of these exercises. If you get a wrong answer, read the pages listed in red.*

1. The interval $(2, 5)$ can be written as the inequality _____. (pp. 120–121)

2. The slope of the line containing the points $(-2, 3)$ and $(3, 8)$ is ___. (pp. 167–169)

3. Test the equation $y = 5x^2 - 1$ for symmetry with respect to the x-axis, the y-axis, and the origin. (pp. 160–162)

4. Write the point–slope form of the line with slope 5 containing the point $(3, -2)$. (p. 171)

5. The intercepts of the equation $y = x^2 - 9$ are _____. (pp. 159–160)

Concepts and Vocabulary

6. A function f is _____ on an open interval I if, for any choice of x_1 and x_2 in I, with $x_1 < x_2$, we have $f(x_1) < f(x_2)$.

7. A(n) _____ function f is one for which $f(-x) = f(x)$ for every x in the domain of f; a(n) _____ function f is one for which $f(-x) = -f(x)$ for every x in the domain of f.

8. *True or False* A function f is decreasing on an open interval I if, for any choice of x_1 and x_2 in I, with $x_1 < x_2$, we have $f(x_1) > f(x_2)$.

9. *True or False* A function f has a local maximum at c if there is an open interval I containing c so that for all x in I, $f(x) \leq f(c)$.

10. *True or False* Even functions have graphs that are symmetric with respect to the origin.

Skill Building

In Problems 11–20, use the graph of the function f given.

11. Is f increasing on the interval $(-8, -2)$?

12. Is f decreasing on the interval $(-8, -4)$?

13. Is f increasing on the interval $(2, 10)$?

14. Is f decreasing on the interval $(2, 5)$?

15. List the interval(s) on which f is increasing.

16. List the interval(s) on which f is decreasing.

17. Is there a local maximum value at 2? If yes, what is it?

18. Is there a local maximum value at 5? If yes, what is it?

19. List the number(s) at which f has a local maximum. What are the local maximum values?

20. List the number(s) at which f has a local minimum. What are the local minimum values?

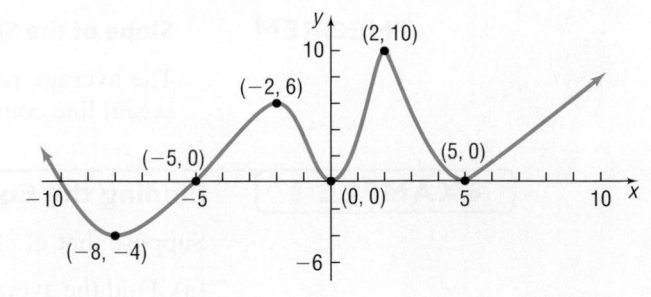

In Problems 21–28, the graph of a function is given. Use the graph to find:

 (a) *The intercepts, if any*
 (b) *The domain and range*
 (c) *The intervals on which it is increasing, decreasing, or constant*
 (d) *Whether it is even, odd, or neither*

21.

22.

23.

24.

25.

26.

27.

28.
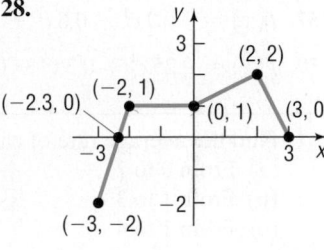

In Problems 29–32, the graph of a function f is given. Use the graph to find:

 (a) *The numbers, if any, at which f has a local maximum value. What are the local maximum values?*
 (b) *The numbers, if any, at which f has a local minimum value. What are the local minimum values?*

29.

30.

31.

32.
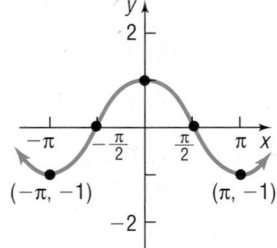

In Problems 33–44, determine algebraically whether each function is even, odd, or neither.

33. $f(x) = 4x^3$

34. $f(x) = 2x^4 - x^2$

35. $g(x) = -3x^2 - 5$

36. $h(x) = 3x^3 + 5$

37. $F(x) = \sqrt[3]{x}$

38. $G(x) = \sqrt{x}$

39. $f(x) = x + |x|$

40. $f(x) = \sqrt[3]{2x^2 + 1}$

41. $g(x) = \dfrac{1}{x^2}$

42. $h(x) = \dfrac{x}{x^2 - 1}$

43. $h(x) = \dfrac{-x^3}{3x^2 - 9}$

44. $F(x) = \dfrac{2x}{|x|}$

In Problems 45–52, for each graph of a function $y = f(x)$, find the absolute maximum and the absolute minimum, if they exist.

45.

46.

47.

48.

49.

50.

51.

52.

In Problems 53–60, use a graphing utility to graph each function over the indicated interval and approximate any local maximum values and local minimum values. Determine where the function is increasing and where it is decreasing. Round answers to two decimal places.

53. $f(x) = x^3 - 3x + 2$ $(-2, 2)$

54. $f(x) = x^3 - 3x^2 + 5$ $(-1, 3)$

55. $f(x) = x^5 - x^3$ $(-2, 2)$

56. $f(x) = x^4 - x^2$ $(-2, 2)$

57. $f(x) = -0.2x^3 - 0.6x^2 + 4x - 6$ $(-6, 4)$

58. $f(x) = -0.4x^3 + 0.6x^2 + 3x - 2$ $(-4, 5)$

59. $f(x) = 0.25x^4 + 0.3x^3 - 0.9x^2 + 3$ $(-3, 2)$

60. $f(x) = -0.4x^4 - 0.5x^3 + 0.8x^2 - 2$ $(-3, 2)$

61. Find the average rate of change of $f(x) = -2x^2 + 4$
 (a) From 0 to 2
 (b) From 1 to 3
 (c) From 1 to 4

62. Find the average rate of change of $f(x) = -x^3 + 1$
 (a) From 0 to 2
 (b) From 1 to 3
 (c) From −1 to 1

63. Find the average rate of change of $g(x) = x^3 - 2x + 1$
 (a) From −3 to −2
 (b) From −1 to 1
 (c) From 1 to 3

64. Find the average rate of change of $h(x) = x^2 - 2x + 3$
 (a) From −1 to 1
 (b) From 0 to 2
 (c) From 2 to 5

65. $f(x) = 5x - 2$
 (a) Find the average rate of change from 1 to 3.
 (b) Find an equation of the secant line containing $(1, f(1))$ and $(3, f(3))$.

66. $f(x) = -4x + 1$
 (a) Find the average rate of change from 2 to 5.
 (b) Find an equation of the secant line containing $(2, f(2))$ and $(5, f(5))$.

67. $g(x) = x^2 - 2$
 (a) Find the average rate of change from −2 to 1.
 (b) Find an equation of the secant line containing $(-2, g(-2))$ and $(1, g(1))$.

68. $g(x) = x^2 + 1$
 (a) Find the average rate of change from −1 to 2.
 (b) Find an equation of the secant line containing $(-1, g(-1))$ and $(2, g(2))$.

69. $h(x) = x^2 - 2x$
 (a) Find the average rate of change from 2 to 4.
 (b) Find an equation of the secant line containing $(2, h(2))$ and $(4, h(4))$.

70. $h(x) = -2x^2 + x$
 (a) Find the average rate of change from 0 to 3.
 (b) Find an equation of the secant line containing $(0, h(0))$ and $(3, h(3))$.

Mixed Practice

71. $g(x) = x^3 - 27x$
 (a) Determine whether g is even, odd, or neither.
 (b) There is a local minimum value of −54 at 3. Determine the local maximum value.

72. $f(x) = -x^3 + 12x$
 (a) Determine whether f is even, odd, or neither.
 (b) There is a local maximum value of 16 at 2. Determine the local minimum value.

73. $F(x) = -x^4 + 8x^2 + 8$
 (a) Determine whether F is even, odd, or neither.
 (b) There is a local maximum value of 24 at $x = 2$. Determine a second local maximum value.

 (c) Suppose the area under the graph of F between $x = 0$ and $x = 3$ that is bounded below by the x-axis is 47.4 square units. Using the result from part (a), determine the area under the graph of F between $x = -3$ and $x = 0$ bounded below by the x-axis.

74. $G(x) = -x^4 + 32x^2 + 144$
 (a) Determine whether G is even, odd, or neither.
 (b) There is a local maximum value of 400 at $x = 4$. Determine a second local maximum value.

 (c) Suppose the area under the graph of G between $x = 0$ and $x = 6$ that is bounded below by the x-axis is 1612.8 square units. Using the result from part (a), determine the area under the graph of G between $x = -6$ and $x = 0$ bounded below by the x-axis.

Applications and Extensions

75. Minimum Average Cost The average cost per hour in dollars, \overline{C}, of producing x riding lawn mowers can be modeled by the function

$$\overline{C}(x) = 0.3x^2 + 21x - 251 + \frac{2500}{x}$$

(a) Use a graphing utility to graph $\overline{C} = \overline{C}(x)$.
(b) Determine the number of riding lawn mowers to produce in order to minimize average cost.
(c) What is the minimum average cost?

76. Medicine Concentration The concentration C of a medication in the bloodstream t hours after being administered is modeled by the function

$$C(t) = -0.002x^4 + 0.039t^3 - 0.285t^2 + 0.766t + 0.085$$

(a) After how many hours will the concentration be highest?
(b) A woman nursing a child must wait until the concentration is below 0.5 before she can feed him. After taking the medication, how long must she wait before feeding her child?

77. E-coli Growth A strain of E-coli Beu 397-recA441 is placed into a nutrient broth at $30°$ Celsius and allowed to grow. The data shown in the table are collected. The population is measured in grams and the time in hours. Since population P depends on time t and each input corresponds to exactly one output, we can say that population is a function of time; so $P(t)$ represents the population at time t.

(a) Find the average rate of change of the population from 0 to 2.5 hours.
(b) Find the average rate of change of the population from 4.5 to 6 hours.
(c) What is happening to the average rate of change as time passes?

Time (hours), t	Population (grams), P
0	0.09
2.5	0.18
3.5	0.26
4.5	0.35
6	0.50

78. e-Filing Tax Returns The Internal Revenue Service Restructuring and Reform Act (RRA) was signed into law by President Bill Clinton in 1998. A major objective of the RRA was to promote electronic filing of tax returns. The data in the table that follows, show the percentage of individual income tax returns filed electronically for filing years 2000–2008. Since the percentage P of returns filed electronically depends on the filing year y and each input corresponds to exactly one output, the percentage of returns filed electronically is a function of the filing year; so $P(y)$ represents the percentage of returns filed electronically for filing year y.

(a) Find the average rate of change of the percentage of e-filed returns from 2000 to 2002.
(b) Find the average rate of change of the percentage of e-filed returns from 2004 to 2006.
(c) Find the average rate of change of the percentage of e-filed returns from 2006 to 2008.
(d) What is happening to the average rate of change as time passes?

Year	Percentage of returns e-filed
2000	27.9
2001	31.1
2002	35.9
2003	40.6
2004	47.0
2005	51.8
2006	54.5
2007	58.0
2008	59.8

SOURCE: Internal Revenue Service

79. For the function $f(x) = x^2$, compute each average rate of change:
(a) From 0 to 1
(b) From 0 to 0.5
(c) From 0 to 0.1
(d) From 0 to 0.01
(e) From 0 to 0.001
(f) Use a graphing utility to graph each of the secant lines along with f.
(g) What do you think is happening to the secant lines?
(h) What is happening to the slopes of the secant lines? Is there some number that they are getting closer to? What is that number?

80. For the function $f(x) = x^2$, compute each average rate of change:
(a) From 1 to 2
(b) From 1 to 1.5
(c) From 1 to 1.1
(d) From 1 to 1.01
(e) From 1 to 1.001
(f) Use a graphing utility to graph each of the secant lines along with f.
(g) What do you think is happening to the secant lines?
(h) What is happening to the slopes of the secant lines? Is there some number that they are getting closer to? What is that number?

⚖ *Problems 81–88 require the following discussion of a secant line. The slope of the secant line containing the two points $(x, f(x))$ and $(x + h, f(x + h))$ on the graph of a function $y = f(x)$ may be given as*

$$m_{sec} = \frac{f(x + h) - f(x)}{(x + h) - x} = \frac{f(x + h) - f(x)}{h} \qquad h \neq 0$$

*In calculus, this expression is called the **difference quotient of** f.*

(a) *Express the slope of the secant line of each function in terms of x and h. Be sure to simplify your answer.*
(b) *Find m_{sec} for $h = 0.5, 0.1$, and 0.01 at $x = 1$. What value does m_{sec} approach as h approaches 0?*
(c) *Find the equation for the secant line at $x = 1$ with $h = 0.01$.*
📱 (d) *Use a graphing utility to graph f and the secant line found in part (c) on the same viewing window.*

81. $f(x) = 2x + 5$ **82.** $f(x) = -3x + 2$ **83.** $f(x) = x^2 + 2x$ **84.** $f(x) = 2x^2 + x$

85. $f(x) = 2x^2 - 3x + 1$ **86.** $f(x) = -x^2 + 3x - 2$ **87.** $f(x) = \dfrac{1}{x}$ **88.** $f(x) = \dfrac{1}{x^2}$

Explaining Concepts: Discussion and Writing

89. Draw the graph of a function that has the following properties: domain: all real numbers; range: all real numbers; intercepts: $(0, -3)$ and $(3, 0)$; a local maximum value of -2 is at -1; a local minimum value of -6 is at 2. Compare your graph with those of others. Comment on any differences.

90. Redo Problem 89 with the following additional information: increasing on $(-\infty, -1), (2, \infty)$; decreasing on $(-1, 2)$. Again compare your graph with others and comment on any differences.

91. How many x-intercepts can a function defined on an interval have if it is increasing on that interval? Explain.

92. Suppose that a friend of yours does not understand the idea of increasing and decreasing functions. Provide an explanation, complete with graphs, that clarifies the idea.

93. Can a function be both even and odd? Explain.

94. Using a graphing utility, graph $y = 5$ on the interval $(-3, 3)$. Use MAXIMUM to find the local maximum values on $(-3, 3)$. Comment on the result provided by the calculator.

95. A function f has a positive average rate of change on the interval $[2, 5]$. Is f increasing on $[2, 5]$? Explain.

96. Show that a constant function $f(x) = b$ has an average rate of change of 0. Compute the average rate of change of $y = \sqrt{4 - x^2}$ on the interval $[-2, 2]$. Explain how this can happen.

'Are You Prepared?' Answers

1. $2 < x < 5$ **2.** 1 **3.** symmetric with respect to the y-axis **4.** $y + 2 = 5(x - 3)$ **5.** $(-3, 0), (3, 0), (0, -9)$

3.4 Library of Functions; Piecewise-defined Functions

PREPARING FOR THIS SECTION *Before getting started, review the following:*

- Intercepts (Section 2.2, pp. 159–160)
- Graphs of Key Equations (Section 2.2: Example 3, p. 158; Example 10, p. 163; Example 11, p. 163; Example 12, p. 164)

Now Work the 'Are You Prepared?' problems on page 241.

> **OBJECTIVES** **1** Graph the Functions Listed in the Library of Functions (p. 234)
> **2** Graph Piecewise-defined Functions (p. 239)

1 Graph the Functions Listed in the Library of Functions

First we introduce a few more functions, beginning with the *square root function*.

On page 163, we graphed the equation $y = \sqrt{x}$. Figure 28 shows a graph of the function $f(x) = \sqrt{x}$. Based on the graph, we have the following properties:

Figure 28

Properties of $f(x) = \sqrt{x}$

1. The domain and the range are the set of nonnegative real numbers.
2. The x-intercept of the graph of $f(x) = \sqrt{x}$ is 0. The y-intercept of the graph of $f(x) = \sqrt{x}$ is also 0.
3. The function is neither even nor odd.
4. The function is increasing on the interval $(0, \infty)$.
5. The function has an absolute minimum of 0 at $x = 0$.

EXAMPLE 1

Graphing the Cube Root Function

(a) Determine whether $f(x) = \sqrt[3]{x}$ is even, odd, or neither. State whether the graph of f is symmetric with respect to the y-axis or symmetric with respect to the origin.

(b) Determine the intercepts, if any, of the graph of $f(x) = \sqrt[3]{x}$.

(c) Graph $f(x) = \sqrt[3]{x}$.

Solution

(a) Because

$$f(-x) = \sqrt[3]{-x} = -\sqrt[3]{x} = -f(x)$$

the function is odd. The graph of f is symmetric with respect to the origin.

(b) The y-intercept is $f(0) = \sqrt[3]{0} = 0$. The x-intercept is found by solving the equation $f(x) = 0$.

$$f(x) = 0$$
$$\sqrt[3]{x} = 0 \quad f(x) = \sqrt[3]{x}$$
$$x = 0 \quad \text{Cube both sides of the equation.}$$

The x-intercept is also 0.

(c) Use the function to form Table 4 and obtain some points on the graph. Because of the symmetry with respect to the origin, we find only points (x, y) for which $x \geq 0$. Figure 29 shows the graph of $f(x) = \sqrt[3]{x}$.

Table 4

x	$y = f(x) = \sqrt[3]{x}$	(x, y)
0	0	$(0, 0)$
$\frac{1}{8}$	$\frac{1}{2}$	$\left(\frac{1}{8}, \frac{1}{2}\right)$
1	1	$(1, 1)$
2	$\sqrt[3]{2} \approx 1.26$	$\left(2, \sqrt[3]{2}\right)$
8	2	$(8, 2)$

Figure 29

From the results of Example 1 and Figure 29, we have the following properties of the cube root function.

Properties of $f(x) = \sqrt[3]{x}$

1. The domain and the range are the set of all real numbers.
2. The x-intercept of the graph of $f(x) = \sqrt[3]{x}$ is 0. The y-intercept of the graph of $f(x) = \sqrt[3]{x}$ is also 0.
3. The graph is symmetric with respect to the origin. The function is odd.
4. The function is increasing on the interval $(-\infty, \infty)$.
5. The function does not have any local minima or any local maxima.

EXAMPLE 2	**Graphing the Absolute Value Function**

(a) Determine whether $f(x) = |x|$ is even, odd, or neither. State whether the graph of f is symmetric with respect to the y-axis or symmetric with respect to the origin.

(b) Determine the intercepts, if any, of the graph of $f(x) = |x|$.

(c) Graph $f(x) = |x|$.

Solution

(a) Because

$$f(-x) = |-x|$$
$$= |x| = f(x)$$

the function is even. The graph of f is symmetric with respect to the y-axis.

(b) The y-intercept is $f(0) = |0| = 0$. The x-intercept is found by solving the equation $f(x) = 0$ or $|x| = 0$. So the x-intercept is 0.

(c) Use the function to form Table 5 and obtain some points on the graph. Because of the symmetry with respect to the y-axis, we need to find only points (x, y) for which $x \geq 0$. Figure 30 shows the graph of $f(x) = |x|$.

Table 5

x	$y = f(x) = \|x\|$	(x, y)
0	0	(0, 0)
1	1	(1, 1)
2	2	(2, 2)
3	3	(3, 3)

Figure 30

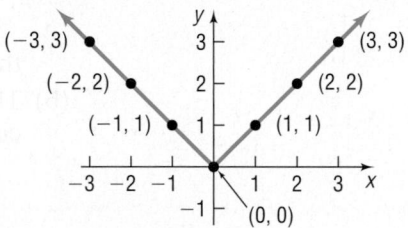

From the results of Example 2 and Figure 30, we have the following properties of the absolute value function.

Properties of $f(x) = |x|$

1. The domain is the set of all real numbers. The range of f is $\{y | y \geq 0\}$.

2. The x-intercept of the graph of $f(x) = |x|$ is 0. The y-intercept of the graph of $f(x) = |x|$ is also 0.

3. The graph is symmetric with respect to the y-axis. The function is even.

4. The function is decreasing on the interval $(-\infty, 0)$. It is increasing on the interval $(0, \infty)$.

5. The function has an absolute minimum of 0 at $x = 0$.

 Seeing the Concept

Graph $y = |x|$ on a square screen and compare what you see with Figure 30. Note that some graphing calculators use abs(x) for absolute value.

Below is a list of the key functions that we have discussed. In going through this list, pay special attention to the properties of each function, particularly to the shape of each graph. Knowing these graphs along with key points on each graph will lay the foundation for further graphing techniques.

Figure 31 Constant Function

Constant Function

$f(x) = b$	b is a real number

See Figure 31.

The domain of a **constant function** is the set of all real numbers; its range is the set consisting of a single number b. Its graph is a horizontal line whose y-intercept is b. The constant function is an even function.

Identity Function

$$f(x) = x$$

See Figure 32.

The domain and the range of the **identity function** are the set of all real numbers. Its graph is a line whose slope is 1 and whose y-intercept is 0. The line consists of all points for which the x-coordinate equals the y-coordinate. The identity function is an odd function that is increasing over its domain. Note that the graph bisects quadrants I and III.

Square Function

$$f(x) = x^2$$

See Figure 33.

The domain of the **square function** f is the set of all real numbers; its range is the set of nonnegative real numbers. The graph of this function is a parabola whose intercept is at $(0, 0)$. The square function is an even function that is decreasing on the interval $(-\infty, 0)$ and increasing on the interval $(0, \infty)$.

Cube Function

$$f(x) = x^3$$

See Figure 34.

The domain and the range of the **cube function** are the set of all real numbers. The intercept of the graph is at $(0, 0)$. The cube function is odd and is increasing on the interval $(-\infty, \infty)$.

Square Root Function

$$f(x) = \sqrt{x}$$

See Figure 35.

The domain and the range of the **square root function** are the set of nonnegative real numbers. The intercept of the graph is at $(0, 0)$. The square root function is neither even nor odd and is increasing on the interval $(0, \infty)$.

Cube Root Function

$$f(x) = \sqrt[3]{x}$$

See Figure 36.

The domain and the range of the **cube root function** are the set of all real numbers. The intercept of the graph is at $(0, 0)$. The cube root function is an odd function that is increasing on the interval $(-\infty, \infty)$.

Figure 32 Identity Function

Figure 33 Square Function

Figure 34 Cube Function

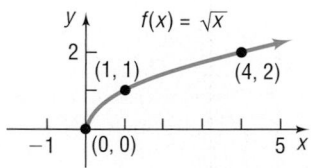

Figure 35 Square Root Function

Figure 36 Cube Root Function

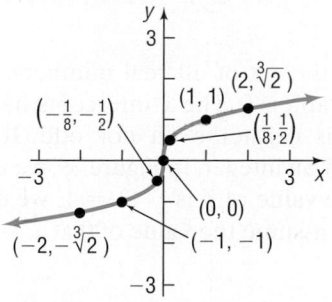

Reciprocal Function

$$f(x) = \frac{1}{x}$$

Figure 37 Reciprocal Function

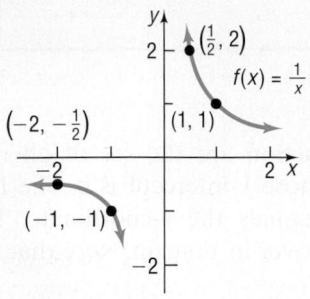

Refer to Example 12, page 164, for a discussion of the equation $y = \frac{1}{x}$. See Figure 37.

The domain and the range of the **reciprocal function** are the set of all nonzero real numbers. The graph has no intercepts. The reciprocal function is decreasing on the intervals $(-\infty, 0)$ and $(0, \infty)$ and is an odd function.

Absolute Value Function

$$f(x) = |x|$$

Figure 38 Absolute Value Function

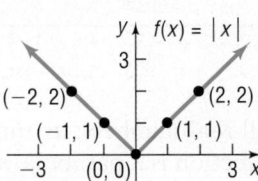

See Figure 38.

The domain of the **absolute value function** is the set of all real numbers; its range is the set of nonnegative real numbers. The intercept of the graph is at $(0, 0)$. If $x \geq 0$, then $f(x) = x$, and the graph of f is part of the line $y = x$; if $x < 0$, then $f(x) = -x$, and the graph of f is part of the line $y = -x$. The absolute value function is an even function; it is decreasing on the interval $(-\infty, 0)$ and increasing on the interval $(0, \infty)$.

The notation int(x) stands for the largest integer less than or equal to x. For example,

$$\text{int}(1) = 1, \quad \text{int}(2.5) = 2, \quad \text{int}\left(\frac{1}{2}\right) = 0, \quad \text{int}\left(-\frac{3}{4}\right) = -1, \quad \text{int}(\pi) = 3$$

This type of correspondence occurs frequently enough in mathematics that we give it a name.

DEFINITION

Greatest Integer Function

$$f(x) = \text{int}(x)^* = \text{greatest integer less than or equal to } x$$

We obtain the graph of $f(x) = \text{int}(x)$ by plotting several points. See Table 6. For values of x, $-1 \leq x < 0$, the value of $f(x) = \text{int}(x)$ is -1; for values of x, $0 \leq x < 1$, the value of f is 0. See Figure 39 for the graph.

Table 6

x	y = f(x) = int(x)	(x, y)
-1	-1	$(-1, -1)$
$-\dfrac{1}{2}$	-1	$\left(-\dfrac{1}{2}, -1\right)$
$-\dfrac{1}{4}$	-1	$\left(-\dfrac{1}{4}, -1\right)$
0	0	$(0, 0)$
$\dfrac{1}{4}$	0	$\left(\dfrac{1}{4}, 0\right)$
$\dfrac{1}{2}$	0	$\left(\dfrac{1}{2}, 0\right)$
$\dfrac{3}{4}$	0	$\left(\dfrac{3}{4}, 0\right)$

Figure 39 Greatest Integer Function

The domain of the **greatest integer function** is the set of all real numbers; its range is the set of integers. The y-intercept of the graph is 0. The x-intercepts lie in the interval $[0, 1)$. The greatest integer function is neither even nor odd. It is constant on every interval of the form $[k, k + 1)$, for k an integer. In Figure 39, we use a solid dot to indicate, for example, that at $x = 1$ the value of f is $f(1) = 1$; we use an open circle to illustrate that the function does not assume the value of 0 at $x = 1$.

* Some books use the notation $f(x) = [x]$ instead of int(x).

Figure 40 $f(x) = \text{int}(x)$

(a) Connected mode

(b) Dot mode

 Although a precise definition requires the idea of a limit, discussed in calculus, in a rough sense, a function is said to be **continuous** if its graph has no gaps or holes and can be drawn without lifting a pencil from the paper on which the graph is drawn. We contrast this with a *discontinuous* function. A function is **discontinuous** if its graph has gaps or holes so that its graph cannot be drawn without lifting a pencil from the paper.

From the graph of the greatest integer function, we can see why it is also called a **step function.** At $x = 0$, $x = \pm 1$, $x = \pm 2$, and so on, this function is discontinuous because, at integer values, the graph suddenly "steps" from one value to another without taking on any of the intermediate values. For example, to the immediate left of $x = 3$, the y-coordinates of the points on the graph are 2, and at $x = 3$ and to the immediate right of $x = 3$, the y-coordinates of the points on the graph are 3. So, the graph has gaps in it.

COMMENT When graphing a function using a graphing utility, you can choose either the **connected mode,** in which points plotted on the screen are connected, making the graph appear without any breaks, or the **dot mode,** in which only the points plotted appear. When graphing the greatest integer function with a graphing utility, it may be necessary to be in the dot mode. This is to prevent the utility from "connecting the dots" when $f(x)$ changes from one integer value to the next. See Figure 40. ∎

The functions discussed so far are basic. Whenever you encounter one of them, you should see a mental picture of its graph. For example, if you encounter the function $f(x) = x^2$, you should see in your mind's eye a picture like Figure 33.

Now Work PROBLEMS **9** THROUGH **16**

2 Graph Piecewise-defined Functions

Sometimes a function is defined using different equations on different parts of its domain. For example, the absolute value function $f(x) = |x|$ is actually defined by two equations: $f(x) = x$ if $x \geq 0$ and $f(x) = -x$ if $x < 0$. For convenience, these equations are generally combined into one expression as

$$f(x) = |x| = \begin{cases} x & \text{if } x \geq 0 \\ -x & \text{if } x < 0 \end{cases}$$

When a function is defined by different equations on different parts of its domain, it is called a **piecewise-defined** function.

EXAMPLE 3

Analyzing a Piecewise-defined Function

The function f is defined as

$$f(x) = \begin{cases} -2x + 1 & \text{if } -3 \leq x < 1 \\ 2 & \text{if } x = 1 \\ x^2 & \text{if } x > 1 \end{cases}$$

(a) Find $f(-2)$, $f(1)$, and $f(2)$. (b) Determine the domain of f.
(c) Locate any intercepts. (d) Graph f.
(e) Use the graph to find the range of f. (f) Is f continuous on its domain?

Solution

(a) To find $f(-2)$, observe that when $x = -2$ the equation for f is given by $f(x) = -2x + 1$. So

$$f(-2) = -2(-2) + 1 = 5$$

When $x = 1$, the equation for f is $f(x) = 2$. That is,

$$f(1) = 2$$

When $x = 2$, the equation for f is $f(x) = x^2$. So

$$f(2) = 2^2 = 4$$

(b) To find the domain of f, look at its definition. Since f is defined for all x greater than or equal to -3, the domain of f is $\{x | x \geq -3\}$, or the interval $[-3, \infty)$.

(c) The y-intercept of the graph of the function is $f(0)$. Because the equation for f when $x = 0$ is $f(x) = -2x + 1$, the y-intercept is $f(0) = -2(0) + 1 = 1$. The x-intercepts of the graph of a function f are the real solutions to the equation $f(x) = 0$. To find the x-intercepts of f, solve $f(x) = 0$ for each "piece" of the function and then determine if the values of x, if any, satisfy the condition that defines the piece.

$$f(x) = 0 \qquad\qquad f(x) = 0 \qquad\qquad f(x) = 0$$
$$-2x + 1 = 0 \quad -3 \leq x < 1 \qquad 2 = 0 \quad x = 1 \qquad x^2 = 0 \quad x > 1$$
$$-2x = -1 \qquad\qquad\qquad \text{No solution} \qquad\qquad x = 0$$
$$x = \frac{1}{2}$$

The first potential x-intercept, $x = \dfrac{1}{2}$, satisfies the condition $-3 \leq x < 1$, so $x = \dfrac{1}{2}$ is an x-intercept. The second potential x-intercept, $x = 0$, does not satisfy the condition $x > 1$, so $x = 0$ is not an x-intercept. The only x-intercept is $\dfrac{1}{2}$. The intercepts are $(0, 1)$ and $\left(\dfrac{1}{2}, 0\right)$.

(d) To graph f, we graph "each piece." First we graph the line $y = -2x + 1$ and keep only the part for which $-3 \leq x < 1$. Then we plot the point $(1, 2)$ because, when $x = 1$, $f(x) = 2$. Finally, we graph the parabola $y = x^2$ and keep only the part for which $x > 1$. See Figure 41.

(e) From the graph, we conclude that the range of f is $\{y | y > -1\}$, or the interval $(-1, \infty)$.

(f) The function f is not continuous because there is a "jump" in the graph at $x = 1$.

Figure 41

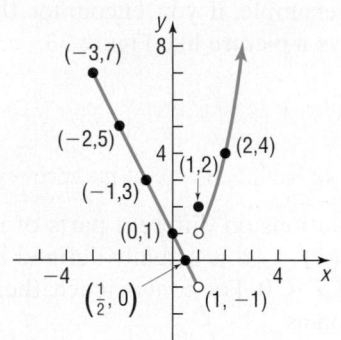

━ **Now Work** PROBLEM 29

EXAMPLE 4 **Cost of Electricity**

In the summer of 2009, Duke Energy supplied electricity to residences of Ohio for a monthly customer charge of $4.50 plus 4.2345¢ per kilowatt-hour (kWhr) for the first 1000 kWhr supplied in the month and 5.3622¢ per kWhr for all usage over 1000 kWhr in the month.

(a) What is the charge for using 300 kWhr in a month?

(b) What is the charge for using 1500 kWhr in a month?

(c) If C is the monthly charge for x kWhr, develop a model relating the monthly charge and kilowatt-hours used. That is, express C as a function of x.

Source: Duke Energy, 2009.

Solution (a) For 300 kWhr, the charge is $4.50 plus 4.2345¢ = $0.042345 per kWhr. That is,

$$\text{Charge} = \$4.50 + \$0.042345(300) = \$17.20$$

(b) For 1500 kWhr, the charge is $4.50 plus 4.2345¢ per kWhr for the first 1000 kWhr plus 5.3622¢ per kWhr for the 500 kWhr in excess of 1000. That is,

$$\text{Charge} = \$4.50 + \$0.042345(1000) + \$0.053622(500) = \$73.66$$

(c) Let x represent the number of kilowatt-hours used. If $0 \leq x \leq 1000$, the monthly charge C (in dollars) can be found by multiplying x times $0.042345 and adding the monthly customer charge of $4.50. So, if $0 \leq x \leq 1000$, then $C(x) = 0.042345x + 4.50$.

For $x > 1000$, the charge is $0.042345(1000) + 4.50 + 0.053622(x - 1000)$, since $x - 1000$ equals the usage in excess of 1000 kWhr, which costs $0.053622 per kWhr. That is, if $x > 1000$, then

$$C(x) = 0.042345(1000) + 4.50 + 0.053622(x - 1000)$$
$$= 46.845 + 0.053622(x - 1000)$$
$$= 0.053622x - 6.777$$

The rule for computing C follows two equations:

Figure 42

$$C(x) = \begin{cases} 0.042345x + 4.50 & \text{if } 0 \le x \le 1000 \\ 0.053622x - 6.777 & \text{if } x > 1000 \end{cases}$$ The Model

See Figure 42 for the graph.

3.4 Assess Your Understanding

'Are You Prepared?' *Answers are given at the end of these exercises. If you get a wrong answer, read the pages listed in red.*

1. Sketch the graph of $y = \sqrt{x}$. (p. 163)

2. Sketch the graph of $y = \dfrac{1}{x}$. (pp. 164–165)

3. List the intercepts of the equation $y = x^3 - 8$. (pp. 159–160)

Concepts and Vocabulary

4. The function $f(x) = x^2$ is decreasing on the interval _____.

5. When functions are defined by more than one equation, they are called _____ functions.

6. *True or False* The cube function is odd and is increasing on the interval $(-\infty, \infty)$.

7. *True or False* The cube root function is odd and is decreasing on the interval $(-\infty, \infty)$.

8. *True or False* The domain and the range of the reciprocal function are the set of all real numbers.

Skill Building

In Problems 9–16, match each graph to its function.

A. *Constant function*
E. *Square root function*

B. *Identity function*
F. *Reciprocal function*

C. *Square function*
G. *Absolute value function*

D. *Cube function*
H. *Cube root function*

9. **10.** **11.** **12.**

13. **14.** **15.** **16.**

In Problems 17–24, sketch the graph of each function. Be sure to label three points on the graph.

17. $f(x) = x$

18. $f(x) = x^2$

19. $f(x) = x^3$

20. $f(x) = \sqrt{x}$

21. $f(x) = \dfrac{1}{x}$

22. $f(x) = |x|$

23. $f(x) = \sqrt[3]{x}$

24. $f(x) = 3$

25. If $f(x) = \begin{cases} x^2 & \text{if } x < 0 \\ 2 & \text{if } x = 0 \\ 2x + 1 & \text{if } x > 0 \end{cases}$

find: (a) $f(-2)$ (b) $f(0)$ (c) $f(2)$

26. If $f(x) = \begin{cases} -3x & \text{if } x < -1 \\ 0 & \text{if } x = -1 \\ 2x^2 + 1 & \text{if } x > -1 \end{cases}$

find: (a) $f(-2)$ (b) $f(-1)$ (c) $f(0)$

27. If $f(x) = \begin{cases} 2x - 4 & \text{if } -1 \le x \le 2 \\ x^3 - 2 & \text{if } 2 < x \le 3 \end{cases}$

find: (a) $f(0)$ (b) $f(1)$ (c) $f(2)$ (d) $f(3)$

28. If $f(x) = \begin{cases} x^3 & \text{if } -2 \le x < 1 \\ 3x + 2 & \text{if } 1 \le x \le 4 \end{cases}$

find: (a) $f(-1)$ (b) $f(0)$ (c) $f(1)$ (d) $f(3)$

In Problems 29–40:

(a) *Find the domain of each function.* (b) *Locate any intercepts.* (c) *Graph each function.*

(d) *Based on the graph, find the range.* (e) *Is f continuous on its domain?*

29. $f(x) = \begin{cases} 2x & \text{if } x \neq 0 \\ 1 & \text{if } x = 0 \end{cases}$

30. $f(x) = \begin{cases} 3x & \text{if } x \neq 0 \\ 4 & \text{if } x = 0 \end{cases}$

31. $f(x) = \begin{cases} -2x + 3 & \text{if } x < 1 \\ 3x - 2 & \text{if } x \geq 1 \end{cases}$

32. $f(x) = \begin{cases} x + 3 & \text{if } x < -2 \\ -2x - 3 & \text{if } x \geq -2 \end{cases}$

33. $f(x) = \begin{cases} x + 3 & \text{if } -2 \leq x < 1 \\ 5 & \text{if } x = 1 \\ -x + 2 & \text{if } x > 1 \end{cases}$

34. $f(x) = \begin{cases} 2x + 5 & \text{if } -3 \leq x < 0 \\ -3 & \text{if } x = 0 \\ -5x & \text{if } x > 0 \end{cases}$

35. $f(x) = \begin{cases} 1 + x & \text{if } x < 0 \\ x^2 & \text{if } x \geq 0 \end{cases}$

36. $f(x) = \begin{cases} \dfrac{1}{x} & \text{if } x < 0 \\ \sqrt[3]{x} & \text{if } x \geq 0 \end{cases}$

37. $f(x) = \begin{cases} |x| & \text{if } -2 \leq x < 0 \\ x^3 & \text{if } x > 0 \end{cases}$

38. $f(x) = \begin{cases} 2 - x & \text{if } -3 \leq x < 1 \\ \sqrt{x} & \text{if } x > 1 \end{cases}$

39. $f(x) = 2 \operatorname{int}(x)$

40. $f(x) = \operatorname{int}(2x)$

In Problems 41–44, the graph of a piecewise-defined function is given. Write a definition for each function.

41.

42.

43.

44.
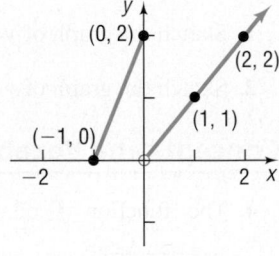

45. If $f(x) = \operatorname{int}(2x)$, find

(a) $f(1.2)$ (b) $f(1.6)$ (c) $f(-1.8)$

46. If $f(x) = \operatorname{int}\left(\dfrac{x}{2}\right)$, find

(a) $f(1.2)$ (b) $f(1.6)$ (c) $f(-1.8)$

Applications and Extensions

47. **Cell Phone Service** Sprint PCS offers a monthly cellular phone plan for $39.99. It includes 450 anytime minutes and charges $0.45 per minute for additional minutes. The following function is used to compute the monthly cost for a subscriber:

$$C(x) = \begin{cases} 39.99 & \text{if } 0 \leq x \leq 450 \\ 0.45x - 162.51 & \text{if } x > 450 \end{cases}$$

where *x* is the number of anytime minutes used. Compute the monthly cost of the cellular phone for use of the following number of anytime minutes:

(a) 200 (b) 465 (c) 451

Source: Sprint PCS

48. **Parking at O'Hare International Airport** The short-term (no more than 24 hours) parking fee *F* (in dollars) for parking *x* hours at O'Hare International Airport's main parking garage can be modeled by the function

$$F(x) = \begin{cases} 3 & \text{if } 0 < x \leq 3 \\ 5 \operatorname{int}(x + 1) + 1 & \text{if } 3 < x < 9 \\ 50 & \text{if } 9 \leq x \leq 24 \end{cases}$$

Determine the fee for parking in the short-term parking garage for

(a) 2 hours (b) 7 hours (c) 15 hours
(d) 8 hours and 24 minutes

Source: O'Hare International Airport

49. **Cost of Natural Gas** In April 2009, Peoples Energy had the following rate schedule for natural gas usage in single-family residences:

Monthly service charge	$15.95
Per therm service charge	
1st 50 therms	$0.33606/therm
Over 50 therms	$0.10580/therm
Gas charge	$0.3940/therm

(a) What is the charge for using 50 therms in a month?
(b) What is the charge for using 500 therms in a month?
(c) Develop a model that relates the monthly charge *C* for *x* therms of gas.
(d) Graph the function found in part (c).

Source: Peoples Energy, Chicago, Illinois, 2009

50. **Cost of Natural Gas** In April 2009, Nicor Gas had the following rate schedule for natural gas usage in single-family residences:

Monthly customer charge	$8.40
Distribution charge	
1st 20 therms	$0.1473/therm
Next 30 therms	$0.0579/therm
Over 50 therms	$0.0519/therm
Gas supply charge	$0.43/therm

(a) What is the charge for using 40 therms in a month?
(b) What is the charge for using 150 therms in a month?
(c) Develop a model that gives the monthly charge *C* for *x* therms of gas.
(d) Graph the function found in part (c).

Source: Nicor Gas, Aurora, Illinois, 2009

51. **Federal Income Tax** Two 2009 Tax Rate Schedules are given in the accompanying table. If x equals taxable income and y equals the tax due, construct a function $y = f(x)$ for Schedule X.

REVISED 2009 TAX RATE SCHEDULES										
Schedule X—Single						**Schedule Y-1—Married Filing jointly or qualifying Widow(er)**				
If Taxable Income Is Over	But Not Over	The Tax Is This Amount		Plus This %	Of the Excess Over	If Taxable Income Is Over	But Not Over	The Tax Is This Amount	Plus This %	Of The Excess Over
$0	$8,350	–	+	10%	$0	$0	$16,700	–	+ 10%	$0
8,350	33,950	$835.00	+	15%	8,350	16,700	67,900	$1,670.00	+ 15%	16,700
33,950	82,250	4,675.00	+	25%	33,950	67,900	137,050	9,350.00	+ 25%	67,900
82,250	171,550	16,750.00	+	28%	82,250	137,050	208,850	26,637.50	+ 28%	137,050
171,550	372,950	41,754.00	+	33%	171,550	208,850	372,950	46,741.50	+ 33%	208,850
372,950	–	108,216.00	+	35%	372,950	372,950	–	100,894.50	+ 35%	372,950

Source: Internal Revenue Service

52. **Federal Income Tax** Refer to the revised 2009 tax rate schedules. If x equals taxable income and y equals the tax due, construct a function $y = f(x)$ for Schedule Y-1.

53. **Cost of Transporting Goods** A trucking company transports goods between Chicago and New York, a distance of 960 miles. The company's policy is to charge, for each pound, $0.50 per mile for the first 100 miles, $0.40 per mile for the next 300 miles, $0.25 per mile for the next 400 miles, and no charge for the remaining 160 miles.
 (a) Graph the relationship between the cost of transportation in dollars and mileage over the entire 960-mile route.
 (b) Find the cost as a function of mileage for hauls between 100 and 400 miles from Chicago.
 (c) Find the cost as a function of mileage for hauls between 400 and 800 miles from Chicago.

54. **Car Rental Costs** An economy car rented in Florida from National Car Rental® on a weekly basis costs $95 per week. Extra days cost $24 per day until the day rate exceeds the weekly rate, in which case the weekly rate applies. Also, any part of a day used counts as a full day. Find the cost C of renting an economy car as a function of the number x of days used, where $7 \leq x \leq 14$. Graph this function.

55. **Minimum Payments for Credit Cards** Holders of credit cards issued by banks, department stores, oil companies, and so on, receive bills each month that state minimum amounts that must be paid by a certain due date. The minimum due depends on the total amount owed. One such credit card company uses the following rules: For a bill of less than $10, the entire amount is due. For a bill of at least $10 but less than $500, the minimum due is $10. A minimum of $30 is due on a bill of at least $500 but less than $1000, a minimum of $50 is due on a bill of at least $1000 but less than $1500, and a minimum of $70 is due on bills of $1500 or more. Find the function f that describes the minimum payment due on a bill of x dollars. Graph f.

56. **Interest Payments for Credit Cards** Refer to Problem 55. The card holder may pay any amount between the minimum due and the total owed. The organization issuing the card charges the card holder interest of 1.5% per month for the first $1000 owed and 1% per month on any unpaid balance over $1000. Find the function g that gives the amount of interest charged per month on a balance of x dollars. Graph g.

57. **Wind Chill** The wind chill factor represents the equivalent air temperature at a standard wind speed that would produce the same heat loss as the given temperature and wind speed. One formula for computing the equivalent temperature is

$$W = \begin{cases} t & 0 \leq v < 1.79 \\ 33 - \dfrac{(10.45 + 10\sqrt{v} - v)(33 - t)}{22.04} & 1.79 \leq v \leq 20 \\ 33 - 1.5958(33 - t) & v > 20 \end{cases}$$

where v represents the wind speed (in meters per second) and t represents the air temperature (°C). Compute the wind chill for the following:
 (a) An air temperature of 10°C and a wind speed of 1 meter per second (m/sec)
 (b) An air temperature of 10°C and a wind speed of 5 m/sec
 (c) An air temperature of 10°C and a wind speed of 15 m/sec
 (d) An air temperature of 10°C and a wind speed of 25 m/sec
 (e) Explain the physical meaning of the equation corresponding to $0 \leq v < 1.79$.
 (f) Explain the physical meaning of the equation corresponding to $v > 20$.

58. **Wind Chill** Redo Problem 57(a)–(d) for an air temperature of -10°C.

59. **First-class Mail** In 2009 the U.S. Postal Service charged $1.17 postage for first-class mail retail flats (such as an 8.5″ by 11″ envelope) weighing up to 1 ounce, plus $0.17 for each additional ounce up to 13 ounces. First-class rates do not apply to flats weighing more than 13 ounces. Develop a model that relates C, the first-class postage charged, for a flat weighing x ounces. Graph the function.

Source: United States Postal Service

Explaining Concepts: Discussion and Writing

In Problems 60–67, use a graphing utility.

60. Exploration Graph $y = x^2$. Then on the same screen graph $y = x^2 + 2$, followed by $y = x^2 + 4$, followed by $y = x^2 - 2$. What pattern do you observe? Can you predict the graph of $y = x^2 - 4$? Of $y = x^2 + 5$?

61. Exploration Graph $y = x^2$. Then on the same screen graph $y = (x - 2)^2$, followed by $y = (x - 4)^2$, followed by $y = (x + 2)^2$. What pattern do you observe? Can you predict the graph of $y = (x + 4)^2$? Of $y = (x - 5)^2$?

62. Exploration Graph $y = |x|$. Then on the same screen graph $y = 2|x|$, followed by $y = 4|x|$, followed by $y = \frac{1}{2}|x|$. What pattern do you observe? Can you predict the graph of $y = \frac{1}{4}|x|$? Of $y = 5|x|$?

63. Exploration Graph $y = x^2$. Then on the same screen graph $y = -x^2$. What pattern do you observe? Now try $y = |x|$ and $y = -|x|$. What do you conclude?

64. Exploration Graph $y = \sqrt{x}$. Then on the same screen graph $y = \sqrt{-x}$. What pattern do you observe? Now try $y = 2x + 1$ and $y = 2(-x) + 1$. What do you conclude?

65. Exploration Graph $y = x^3$. Then on the same screen graph $y = (x - 1)^3 + 2$. Could you have predicted the result?

66. Exploration Graph $y = x^2$, $y = x^4$, and $y = x^6$ on the same screen. What do you notice is the same about each graph? What do you notice that is different?

67. Exploration Graph $y = x^3$, $y = x^5$, and $y = x^7$ on the same screen. What do you notice is the same about each graph? What do you notice that is different?

68. Consider the equation

$$y = \begin{cases} 1 & \text{if } x \text{ is rational} \\ 0 & \text{if } x \text{ is irrational} \end{cases}$$

Is this a function? What is its domain? What is its range? What is its *y*-intercept, if any? What are its *x*-intercepts, if any? Is it even, odd, or neither? How would you describe its graph?

69. Define some functions that pass through $(0, 0)$ and $(1, 1)$ and are increasing for $x \geq 0$. Begin your list with $y = \sqrt{x}$, $y = x$, and $y = x^2$. Can you propose a general result about such functions?

'Are You Prepared?' Answers

1.

2.

3. $(0, -8), (2, 0)$

3.5 Graphing Techniques: Transformations

OBJECTIVES 1 Graph Functions Using Vertical and Horizontal Shifts (p. 244)

2 Graph Functions Using Compressions and Stretches (p. 247)

3 Graph Functions Using Reflections about the *x*-Axis and the *y*-Axis (p. 250)

At this stage, if you were asked to graph any of the functions defined by $y = x$, $y = x^2$, $y = x^3$, $y = \sqrt{x}$, $y = \sqrt[3]{x}$, $y = \frac{1}{x}$, or $y = |x|$, your response should be, "Yes, I recognize these functions and know the general shapes of their graphs." (If this is not your answer, review the previous section, Figures 32 through 38.)

Sometimes we are asked to graph a function that is "almost" like one that we already know how to graph. In this section, we develop techniques for graphing such functions. Collectively, these techniques are referred to as **transformations.**

1 Graph Functions Using Vertical and Horizontal Shifts

EXAMPLE 1

Vertical Shift Up

Use the graph of $f(x) = x^2$ to obtain the graph of $g(x) = x^2 + 3$.

Solution Begin by obtaining some points on the graphs of *f* and *g*. For example, when $x = 0$, then $y = f(0) = 0$ and $y = g(0) = 3$. When $x = 1$, then $y = f(1) = 1$ and

$y = g(1) = 4$. Table 7 lists these and a few other points on each graph. Notice that each y-coordinate of a point on the graph of g is 3 units larger than the y-coordinate of the corresponding point on the graph of f. We conclude that the graph of g is identical to that of f, except that it is shifted vertically up 3 units. See Figure 43.

Figure 43

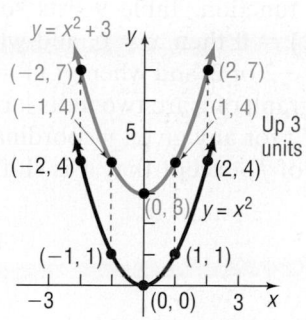

Table 7

x	$y = f(x)$ $= x^2$	$y = g(x)$ $= x^2 + 3$
−2	4	7
−1	1	4
0	0	3
1	1	4
2	4	7

EXAMPLE 2

Vertical Shift Down

Use the graph of $f(x) = x^2$ to obtain the graph of $g(x) = x^2 - 4$.

Solution

Table 8 lists some points on the graphs of f and g. Notice that each y-coordinate of g is 4 units less than the corresponding y-coordinate of f.

To obtain the graph of g from the graph of f, subtract 4 from each y-coordinate on the graph of f. So the graph of g is identical to that of f, except that it is shifted down 4 units. See Figure 44.

Figure 44

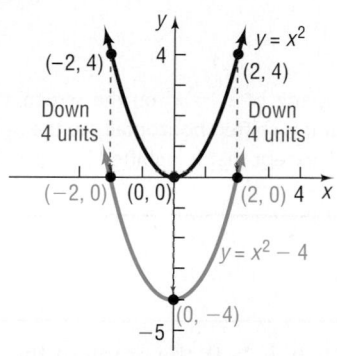

Table 8

x	$y = f(x)$ $= x^2$	$y = g(x)$ $= x^2 - 4$
−2	4	0
−1	1	−3
0	0	−4
1	1	−3
2	4	0

Figure 45

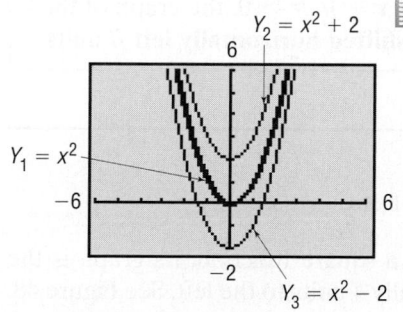

Exploration

On the same screen, graph each of the following functions:

$$Y_1 = x^2$$
$$Y_2 = x^2 + 2$$
$$Y_3 = x^2 - 2$$

Figure 45 illustrates the graphs. You should have observed a general pattern. With $Y_1 = x^2$ on the screen, the graph of $Y_2 = x^2 + 2$ is identical to that of $Y_1 = x^2$, except that it is shifted vertically up 2 units. The graph of $Y_3 = x^2 - 2$ is identical to that of $Y_1 = x^2$, except that it is shifted vertically down 2 units.

We are led to the following conclusions:

> If a positive real number k is added to the output of a function $y = f(x)$, the graph of the new function $y = f(x) + k$ is the graph of f **shifted vertically up** k units.

> If a positive real number k is subtracted from the output of a function $y = f(x)$, the graph of the new function $y = f(x) - k$ is the graph of f **shifted vertically down** k units.

Now Work PROBLEM **39**

| EXAMPLE 3 | **Horizontal Shift to the Right** |

Use the graph of $f(x) = x^2$ to obtain the graph of $g(x) = (x - 2)^2$.

Solution The function $g(x) = (x - 2)^2$ is basically a square function. Table 9 lists some points on the graphs of f and g. Note that when $f(x) = 0$ then $x = 0$, and when $g(x) = 0$, then $x = 2$. Also, when $f(x) = 4$, then $x = -2$ or 2, and when $g(x) = 4$, then $x = 0$ or 4. Notice that the x-coordinates on the graph of g are two units larger than the corresponding x-coordinates on the graph of f for any given y-coordinate. We conclude that the graph of g is identical to that of f, except that it is shifted horizontally 2 units to the right. See Figure 46.

Figure 46

Table 9

x	$y = f(x)$ $= x^2$	$y = g(x)$ $= (x - 2)^2$
-2	4	16
0	0	4
2	4	0
4	16	4

Figure 47

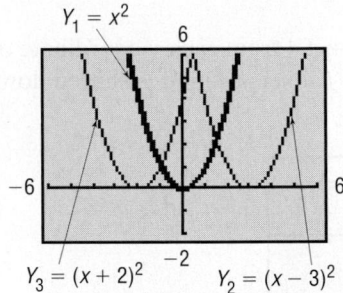

$Y_3 = (x + 2)^2$ $Y_2 = (x - 3)^2$

Exploration

On the same screen, graph each of the following functions:

$$Y_1 = x^2$$
$$Y_2 = (x - 3)^2$$
$$Y_3 = (x + 2)^2$$

Figure 47 illustrates the graphs.

You should have observed the following pattern. With the graph of $Y_1 = x^2$ on the screen, the graph of $Y_2 = (x - 3)^2$ is identical to that of $Y_1 = x^2$, except that it is shifted horizontally to the right 3 units. The graph of $Y_3 = (x + 2)^2$ is identical to that of $Y_1 = x^2$, except that it is shifted horizontally to the left 2 units.

We are led to the following conclusion.

In Other Words
If a positive number h is subtracted from x in $y = f(x)$, the graph of the new function $y = f(x - h)$ is the graph of $y = f(x)$ shifted horizontally right h units. If h is added to x, shift horizontally left h units.

If the argument x of a function f is replaced by $x - h$, $h > 0$, the graph of the new function $y = f(x - h)$ is the graph of f **shifted horizontally right** h units. If the argument x of a function f is replaced by $x + h$, $h > 0$, the graph of the new function $y = f(x + h)$ is the graph of f **shifted horizontally left** h units.

| EXAMPLE 4 | **Horizontal Shift to the Left** |

Use the graph of $f(x) = x^2$ to obtain the graph of $g(x) = (x + 4)^2$.

Solution Again, the function $g(x) = (x + 4)^2$ is basically a square function. Its graph is the same as that of f, except that it is shifted horizontally 4 units to the left. See Figure 48.

Figure 48

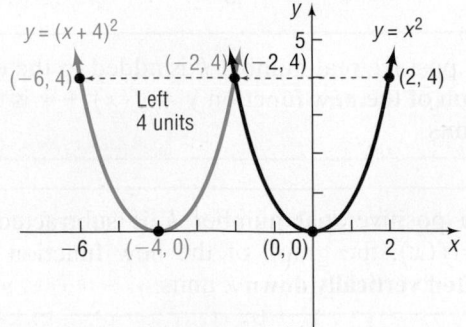

━━━━**Now Work** PROBLEM 43

Notice the distinction between vertical and horizontal shifts. The graph of $f(x) = \sqrt{x} + 3$ is obtained by shifting the graph of $y = \sqrt{x}$ *up* 3 units, because we evaluate the square root function first and then add 3. The graph of $g(x) = \sqrt{x + 3}$ is obtained by shifting the graph of $y = \sqrt{x}$ *left* 3 units, because we add 3 to x before we evaluate the square root function.

Vertical and horizontal shifts are sometimes combined.

EXAMPLE 5 **Combining Vertical and Horizontal Shifts**

Graph the function: $f(x) = (x + 3)^2 - 5$

Solution We graph f in steps. First, notice that the rule for f is basically a square function, so begin with the graph of $y = x^2$ as shown in Figure 49(a). Next, to get the graph of $y = (x + 3)^2$, shift the graph of $y = x^2$ horizontally 3 units to the left. See Figure 49(b). Finally, to get the graph of $y = (x + 3)^2 - 5$, shift the graph of $y = (x + 3)^2$ vertically down 5 units. See Figure 49(c). Note the points plotted on each graph. Using key points can be helpful in keeping track of the transformation that has taken place.

Figure 49

(a) (b) (c)

 ✓**Check:** Graph $Y_1 = f(x) = (x + 3)^2 - 5$ and compare the graph to Figure 49(c).

In Example 5, if the vertical shift had been done first, followed by the horizontal shift, the final graph would have been the same. Try it for yourself.

Now Work PROBLEM 45

2 Graph Functions Using Compressions and Stretches

EXAMPLE 6 **Vertical Stretch**

Use the graph of $f(x) = |x|$ to obtain the graph of $g(x) = 2|x|$.

Solution To see the relationship between the graphs of f and g, form Table 10, listing points on each graph. For each x, the y-coordinate of a point on the graph of g is 2 times as large as the corresponding y-coordinate on the graph of f. The graph of $f(x) = |x|$ is vertically stretched by a factor of 2 to obtain the graph of $g(x) = 2|x|$ [for example, $(1, 1)$ is on the graph of f, but $(1, 2)$ is on the graph of g]. See Figure 50.

Table 10

x	y = f(x) = \|x\|	y = g(x) = 2\|x\|
−2	2	4
−1	1	2
0	0	0
1	1	2
2	2	4

Figure 50

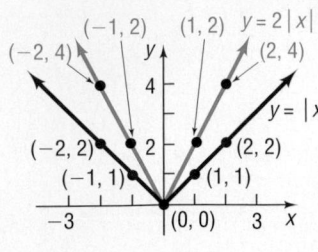

EXAMPLE 7 **Vertical Compression**

Use the graph of $f(x) = |x|$ to obtain the graph of $g(x) = \frac{1}{2}|x|$.

Solution For each x, the y-coordinate of a point on the graph of g is $\frac{1}{2}$ as large as the corresponding y-coordinate on the graph of f. The graph of $f(x) = |x|$ is vertically compressed by a factor of $\frac{1}{2}$ to obtain the graph of $g(x) = \frac{1}{2}|x|$ [for example, $(2, 2)$ is on the graph of f, but $(2, 1)$ is on the graph of g]. See Table 11 and Figure 51.

Table 11

x	y = f(x) = \|x\|	y = g(x) = \frac{1}{2}\|x\|
−2	2	1
−1	1	$\frac{1}{2}$
0	0	0
1	1	$\frac{1}{2}$
2	2	1

Figure 51

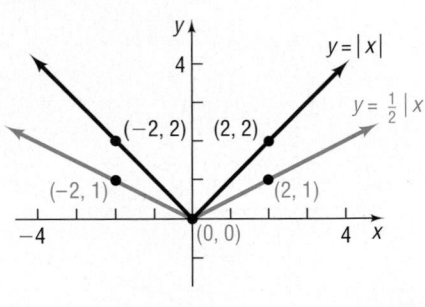

When the right side of a function $y = f(x)$ is multiplied by a positive number a, the graph of the new function $y = af(x)$ is obtained by multiplying each y-coordinate on the graph of $y = f(x)$ by a. The new graph is a **vertically compressed** (if $0 < a < 1$) or a **vertically stretched** (if $a > 1$) version of the graph of $y = f(x)$.

Now Work PROBLEM 47

What happens if the argument x of a function $y = f(x)$ is multiplied by a positive number a, creating a new function $y = f(ax)$? To find the answer, look at the following Exploration.

Exploration

On the same screen, graph each of the following functions:

$$Y_1 = f(x) = \sqrt{x} \qquad Y_2 = f(2x) = \sqrt{2x} \qquad Y_3 = f\left(\frac{1}{2}x\right) = \sqrt{\frac{1}{2}x} = \sqrt{\frac{x}{2}}$$

Create a table of values to explore the relation between the x- and y-coordinates of each function.

Result You should have obtained the graphs in Figure 52. Look at Table 12(a). Notice that $(1, 1)$, $(4, 2)$, and $(9, 3)$ are points on the graph of $Y_1 = \sqrt{x}$. Also, $(0.5, 1)$, $(2, 2)$, and $(4.5, 3)$ are points on the graph of $Y_2 = \sqrt{2x}$. For a given y-coordinate, the x-coordinate on the graph of Y_2 is $\frac{1}{2}$ of the x-coordinate on Y_1.

Figure 52

Table 12

(a)

(b)

We conclude that the graph of $Y_2 = \sqrt{2x}$ is obtained by multiplying the x-coordinate of each point on the graph of $Y_1 = \sqrt{x}$ by $\frac{1}{2}$. The graph of $Y_2 = \sqrt{2x}$ is the graph of $Y_1 = \sqrt{x}$ *compressed* horizontally.

Look at Table 12(b). Notice that (1, 1), (4, 2), and (9, 3) are points on the graph of $Y_1 = \sqrt{x}$. Also notice that (2, 1), (8, 2), and (18, 3) are points on the graph of $Y_3 = \sqrt{\frac{x}{2}}$. For a given y-coordinate, the x-coordinate on the graph of Y_3 is 2 times the x-coordinate on Y_1. We conclude that the graph of $Y_3 = \sqrt{\frac{x}{2}}$ is obtained by multiplying the x-coordinate of each point on the graph of $Y_1 = \sqrt{x}$ by 2. The graph of $Y_3 = \sqrt{\frac{x}{2}}$ is the graph of $Y_1 = \sqrt{x}$ *stretched* horizontally.

Based on the results of the Exploration, we have the following result:

If the argument x of a function $y = f(x)$ is multiplied by a positive number a, the graph of the new function $y = f(ax)$ is obtained by multiplying each x-coordinate of $y = f(x)$ by $\frac{1}{a}$. A **horizontal compression** results if $a > 1$, and a **horizontal stretch** occurs if $0 < a < 1$.

EXAMPLE 8

Graphing Using Stretches and Compressions

The graph of $y = f(x)$ is given in Figure 53. Use this graph to find the graphs of

(a) $y = 2f(x)$ 　　　　　　　　　　　　　(b) $y = f(3x)$

Solution　(a) The graph of $y = 2f(x)$ is obtained by multiplying each y-coordinate of $y = f(x)$ by 2. See Figure 54.

(b) The graph of $y = f(3x)$ is obtained from the graph of $y = f(x)$ by multiplying each x-coordinate of $y = f(x)$ by $\frac{1}{3}$. See Figure 55.

Figure 53

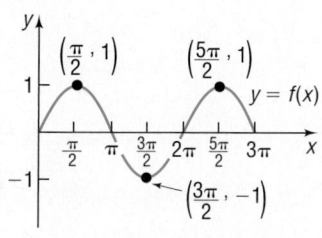

$y = f(x)$

Figure 54

$y = 2f(x)$

Figure 55

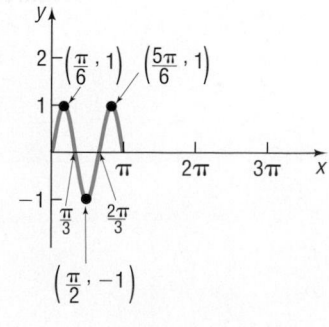

$y = f(3x)$

━━ **Now Work** PROBLEMS **63(e)** AND **(g)**

3 Graph Functions Using Reflections about the *x*-Axis and the *y*-Axis

EXAMPLE 9	**Reflection about the *x*-Axis**

Graph the function: $f(x) = -x^2$

Solution

Begin with the graph of $y = x^2$, as shown in black in Figure 56. For each point (x, y) on the graph of $y = x^2$, the point $(x, -y)$ is on the graph of $y = -x^2$, as indicated in Table 13. Draw the graph of $y = -x^2$ by reflecting the graph of $y = x^2$ about the *x*-axis. See Figure 56.

Figure 56

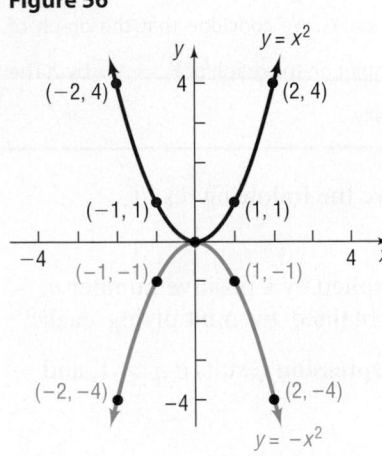

Table 13

x	y = x²	y = −x²
−2	4	−4
−1	1	−1
0	0	0
1	1	−1
2	4	−4

> When the right side of the function $y = f(x)$ is multiplied by -1, the graph of the new function $y = -f(x)$ is the **reflection about the *x*-axis** of the graph of the function $y = f(x)$.

━━━**Now Work** PROBLEM 49

EXAMPLE 10	**Reflection about the *y*-Axis**

Graph the function: $f(x) = \sqrt{-x}$

Solution

First, notice that the domain of f consists of all real numbers x for which $-x \geq 0$ or, equivalently, $x \leq 0$. To get the graph of $f(x) = \sqrt{-x}$, begin with the graph of $y = \sqrt{x}$, as shown in Figure 57. For each point (x, y) on the graph of $y = \sqrt{x}$, the point $(-x, y)$ is on the graph of $y = \sqrt{-x}$. Obtain the graph of $y = \sqrt{-x}$ by reflecting the graph of $y = \sqrt{x}$ about the *y*-axis. See Figure 57.

Figure 57

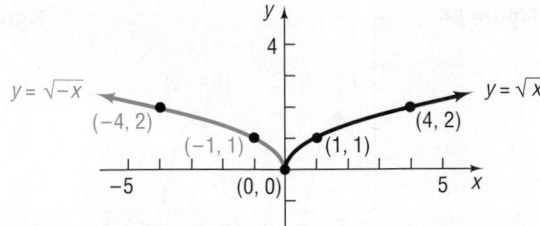

> When the graph of the function $y = f(x)$ is known, the graph of the new function $y = f(-x)$ is the **reflection about the *y*-axis** of the graph of the function $y = f(x)$.

SUMMARY OF GRAPHING TECHNIQUES

To Graph:	Draw the Graph of f and:	Functional Change to $f(x)$
Vertical shifts		
$y = f(x) + k,\quad k > 0$	Raise the graph of f by k units.	Add k to $f(x)$.
$y = f(x) - k,\quad k > 0$	Lower the graph of f by k units.	Subtract k from $f(x)$.
Horizontal shifts		
$y = f(x + h),\quad h > 0$	Shift the graph of f to the left h units.	Replace x by $x + h$.
$y = f(x - h),\quad h > 0$	Shift the graph of f to the right h units.	Replace x by $x - h$.
Compressing or stretching		
$y = af(x),\quad a > 0$	Multiply each y-coordinate of $y = f(x)$ by a. Stretch the graph of f vertically if $a > 1$. Compress the graph of f vertically if $0 < a < 1$.	Multiply $f(x)$ by a.
$y = f(ax),\quad a > 0$	Multiply each x-coordinate of $y = f(x)$ by $\dfrac{1}{a}$. Stretch the graph of f horizontally if $0 < a < 1$. Compress the graph of f horizontally if $a > 1$.	Replace x by ax.
Reflection about the x-axis		
$y = -f(x)$	Reflect the graph of f about the x-axis.	Multiply $f(x)$ by -1.
Reflection about the y-axis		
$y = f(-x)$	Reflect the graph of f about the y-axis.	Replace x by $-x$.

EXAMPLE 11 **Determining the Function Obtained from a Series of Transformations**

Find the function that is finally graphed after the following three transformations are applied to the graph of $y = |x|$.

1. Shift left 2 units
2. Shift up 3 units
3. Reflect about the y-axis

Solution
1. Shift left 2 units: Replace x by $x + 2$. $y = |x + 2|$
2. Shift up 3 units: Add 3. $y = |x + 2| + 3$
3. Reflect about the y-axis: Replace x by $-x$. $y = |-x + 2| + 3$

━━━━━ **Now Work** PROBLEM **27**

EXAMPLE 12 **Combining Graphing Procedures**

Graph the function $f(x) = \dfrac{3}{x - 2} + 1$. Find the domain and the range of f.

Solution It is helpful to write f as $f(x) = 3\left(\dfrac{1}{x - 2}\right) + 1$. Now use the following steps to obtain the graph of f:

STEP 1: $y = \dfrac{1}{x}$ Reciprocal function

STEP 2: $y = 3 \cdot \left(\dfrac{1}{x}\right) = \dfrac{3}{x}$ Multiply by 3; vertical stretch of the graph of $y = \dfrac{1}{x}$ by a factor of 3.

STEP 3: $y = \dfrac{3}{x - 2}$ Replace x by $x - 2$; horizontal shift to the right 2 units.

STEP 4: $y = \dfrac{3}{x - 2} + 1$ Add 1; vertical shift up 1 unit.

See Figure 58.

Figure 58

(a) $y = \dfrac{1}{x}$ Multiply by 3; Vertical stretch (b) $y = \dfrac{3}{x}$ Replace x by $x - 2$; Horizontal shift right 2 units (c) $y = \dfrac{3}{x-2}$ Add 1; Vertical shift up 1 unit (d) $y = \dfrac{3}{x-2} + 1$

The domain of $y = \dfrac{1}{x}$ is $\{x | x \neq 0\}$ and its range is $\{y | y \neq 0\}$. Because we shifted right 2 units and up 1 unit to obtain f, the domain of f is $\{x | x \neq 2\}$ and its range is $\{y | y \neq 1\}$. ↵

Other orderings of the steps shown in Example 12 would also result in the graph of f. For example, try this one:

STEP 1: $y = \dfrac{1}{x}$ Reciprocal function

STEP 2: $y = \dfrac{1}{x - 2}$ Replace x by $x - 2$; horizontal shift to the right 2 units.

STEP 3: $y = \dfrac{3}{x - 2}$ Multiply by 3; vertical stretch of the graph of $y = \dfrac{1}{x - 2}$ by a factor of 3.

STEP 4: $y = \dfrac{3}{x - 2} + 1$ Add 1; vertical shift up 1 unit.

Hint: Although the order in which transformations are performed can be altered, you may consider using the following order for consistency:

1. Reflections
2. Compressions and stretches
3. Shifts

EXAMPLE 13 **Combining Graphing Procedures**

Graph the function $f(x) = \sqrt{1 - x} + 2$. Find the domain and the range of f.

Solution Because horizontal shifts require the form $x - h$, we begin by rewriting $f(x)$ as

$$f(x) = \sqrt{1 - x} + 2 = \sqrt{-(x - 1)} + 2.$$ Now use the following steps:

STEP 1: $y = \sqrt{x}$ Square root function

STEP 2: $y = \sqrt{-x}$ Replace x by −x; reflect about the y-axis.

STEP 3: $y = \sqrt{-(x - 1)} = \sqrt{1 - x}$ Replace x by x − 1; horizontal shift to the right 1 unit.

STEP 4: $y = \sqrt{1 - x} + 2$ Add 2; vertical shift up 2 units.

See Figure 59.

Figure 59

The domain of f is $(-\infty, 1]$ and the range is $[2, \infty)$.

Now Work PROBLEM 55

3.5 Assess Your Understanding

Concepts and Vocabulary

1. Suppose that the graph of a function f is known. Then the graph of $y = f(x - 2)$ may be obtained by a(n) _____ shift of the graph of f to the _____ a distance of 2 units.

2. Suppose that the graph of a function f is known. Then the graph of $y = f(-x)$ may be obtained by a reflection about the _____-axis of the graph of the function $y = f(x)$.

3. Suppose that the graph of a function g is known. The graph of $y = g(x) + 2$ may be obtained by a _____ shift of the graph of g _____ a distance of 2 units.

4. *True or False* The graph of $y = -f(x)$ is the reflection about the x-axis of the graph of $y = f(x)$.

5. *True or False* To obtain the graph of $f(x) = \sqrt{x + 2}$, shift the graph of $y = \sqrt{x}$ horizontally to the right 2 units.

6. *True or False* To obtain the graph of $f(x) = x^3 + 5$, shift the graph of $y = x^3$ vertically up 5 units.

Skill Building

In Problems 7–18, match each graph to one of the following functions:

A. $y = x^2 + 2$ B. $y = -x^2 + 2$ C. $y = |x| + 2$ D. $y = -|x| + 2$

E. $y = (x - 2)^2$ F. $y = -(x + 2)^2$ G. $y = |x - 2|$ H. $y = -|x + 2|$

I. $y = 2x^2$ J. $y = -2x^2$ K. $y = 2|x|$ L. $y = -2|x|$

7.

8.

9.

10.

11.

12.

13.

14.

15.

16.

17.

18.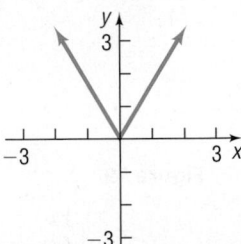

In Problems 19–26, write the function whose graph is the graph of $y = x^3$, but is:

19. Shifted to the right 4 units

20. Shifted to the left 4 units

21. Shifted up 4 units

22. Shifted down 4 units

23. Reflected about the y-axis

24. Reflected about the x-axis

25. Vertically stretched by a factor of 4

26. Horizontally stretched by a factor of 4

In Problems 27–30, find the function that is finally graphed after each of the following transformations is applied to the graph of $y = \sqrt{x}$ in the order stated.

27. (1) Shift up 2 units
 (2) Reflect about the x-axis
 (3) Reflect about the y-axis

28. (1) Reflect about the x-axis
 (2) Shift right 3 units
 (3) Shift down 2 units

29. (1) Reflect about the x-axis
 (2) Shift up 2 units
 (3) Shift left 3 units

30. (1) Shift up 2 units
 (2) Reflect about the y-axis
 (3) Shift left 3 units

31. If $(3, 6)$ is a point on the graph of $y = f(x)$, which of the following points must be on the graph of $y = -f(x)$?
 (a) $(6, 3)$
 (b) $(6, -3)$
 (c) $(3, -6)$
 (d) $(-3, 6)$

32. If $(3, 6)$ is a point on the graph of $y = f(x)$, which of the following points must be on the graph of $y = f(-x)$?
 (a) $(6, 3)$
 (b) $(6, -3)$
 (c) $(3, -6)$
 (d) $(-3, 6)$

33. If $(1, 3)$ is a point on the graph of $y = f(x)$, which of the following points must be on the graph of $y = 2f(x)$?
 (a) $\left(1, \dfrac{3}{2}\right)$
 (b) $(2, 3)$
 (c) $(1, 6)$
 (d) $\left(\dfrac{1}{2}, 3\right)$

34. If $(4, 2)$ is a point on the graph of $y = f(x)$, which of the following points must be on the graph of $y = f(2x)$?
 (a) $(4, 1)$
 (b) $(8, 2)$
 (c) $(2, 2)$
 (d) $(4, 4)$

35. Suppose that the x-intercepts of the graph of $y = f(x)$ are -5 and 3.
 (a) What are the x-intercepts of the graph of $y = f(x + 2)$?
 (b) What are the x-intercepts of the graph of $y = f(x - 2)$?
 (c) What are the x-intercepts of the graph of $y = 4f(x)$?
 (d) What are the x-intercepts of the graph of $y = f(-x)$?

36. Suppose that the x-intercepts of the graph of $y = f(x)$ are -8 and 1.
 (a) What are the x-intercepts of the graph of $y = f(x + 4)$?
 (b) What are the x-intercepts of the graph of $y = f(x - 3)$?
 (c) What are the x-intercepts of the graph of $y = 2f(x)$?
 (d) What are the x-intercepts of the graph of $y = f(-x)$?

37. Suppose that the function $y = f(x)$ is increasing on the interval $(-1, 5)$.
 (a) Over what interval is the graph of $y = f(x + 2)$ increasing?
 (b) Over what interval is the graph of $y = f(x - 5)$ increasing?
 (c) What can be said about the graph of $y = -f(x)$?
 (d) What can be said about the graph of $y = f(-x)$?

38. Suppose that the function $y = f(x)$ is decreasing on the interval $(-2, 7)$.
 (a) Over what interval is the graph of $y = f(x + 2)$ decreasing?
 (b) Over what interval is the graph of $y = f(x - 5)$ decreasing?
 (c) What can be said about the graph of $y = -f(x)$?
 (d) What can be said about the graph of $y = f(-x)$?

In Problems 39–62, graph each function using the techniques of shifting, compressing, stretching, and/or reflecting. Start with the graph of the basic function (for example, $y = x^2$) and show all stages. Be sure to show at least three key points. Find the domain and the range of each function.

39. $f(x) = x^2 - 1$

40. $f(x) = x^2 + 4$

41. $g(x) = x^3 + 1$

42. $g(x) = x^3 - 1$

43. $h(x) = \sqrt{x - 2}$

44. $h(x) = \sqrt{x + 1}$

45. $f(x) = (x - 1)^3 + 2$

46. $f(x) = (x + 2)^3 - 3$

47. $g(x) = 4\sqrt{x}$

48. $g(x) = \frac{1}{2}\sqrt{x}$

49. $f(x) = -\sqrt[3]{x}$

50. $f(x) = -\sqrt{x}$

51. $f(x) = 2(x + 1)^2 - 3$

52. $f(x) = 3(x - 2)^2 + 1$

53. $g(x) = 2\sqrt{x - 2} + 1$

54. $g(x) = 3|x + 1| - 3$

55. $h(x) = \sqrt{-x} - 2$

56. $h(x) = \frac{4}{x} + 2$

57. $f(x) = -(x + 1)^3 - 1$

58. $f(x) = -4\sqrt{x - 1}$

59. $g(x) = 2|1 - x|$

60. $g(x) = 4\sqrt{2 - x}$

61. $h(x) = 2 \operatorname{int}(x - 1)$

62. $h(x) = \operatorname{int}(-x)$

In Problems 63–66, the graph of a function f is illustrated. Use the graph of f as the first step toward graphing each of the following functions:

(a) $F(x) = f(x) + 3$

(b) $G(x) = f(x + 2)$

(c) $P(x) = -f(x)$

(d) $H(x) = f(x + 1) - 2$

(e) $Q(x) = \frac{1}{2}f(x)$

(f) $g(x) = f(-x)$

(g) $h(x) = f(2x)$

63.

64.

65.

66.

Mixed Practice

In Problems 67–74, complete the square of each quadratic expression. Then graph each function using the technique of shifting. (If necessary, refer to Chapter R, Section R.5 to review completing the square.)

67. $f(x) = x^2 + 2x$

68. $f(x) = x^2 - 6x$

69. $f(x) = x^2 - 8x + 1$

70. $f(x) = x^2 + 4x + 2$

71. $f(x) = 2x^2 - 12x + 19$

72. $f(x) = 3x^2 + 6x + 1$

73. $f(x) = -3x^2 - 12x - 17$

74. $f(x) = -2x^2 - 12x - 13$

Applications and Extensions

75. The equation $y = (x - c)^2$ defines a *family of parabolas,* one parabola for each value of c. On one set of coordinate axes, graph the members of the family for $c = 0$, $c = 3$, and $c = -2$.

76. Repeat Problem 75 for the family of parabolas $y = x^2 + c$.

77. Thermostat Control Energy conservation experts estimate that homeowners can save 5% to 10% on winter heating bills by programming their thermostats 5 to 10 degrees lower while sleeping. In the given graph, the temperature T (in degrees Fahrenheit) of a home is given as a function of time t (in hours after midnight) over a 24-hour period.

(a) At what temperature is the thermostat set during daytime hours? At what temperature is the thermostat set overnight?

(b) The homeowner reprograms the thermostat to $y = T(t) - 2$. Explain how this affects the temperature in the house. Graph this new function.

(c) The homeowner reprograms the thermostat to $y = T(t + 1)$. Explain how this affects the temperature in the house. Graph this new function.

Source: Roger Albright, 547 Ways to Be Fuel Smart, 2000

78. Digital Music Revenues The total projected worldwide digital music revenues R, in millions of dollars, for the years 2005 through 2010 can be estimated by the function

$$R(x) = 170.7x^2 + 1373x + 1080$$

where x is the number of years after 2005.

(a) Find $R(0)$, $R(3)$, and $R(5)$ and explain what each value represents.

(b) Find $r = R(x - 5)$.

(c) Find $r(5)$, $r(8)$, and $r(10)$ and explain what each value represents.

(d) In the model r, what does x represent?

(e) Would there be an advantage in using the model r when estimating the projected revenues for a given year instead of the model R?

Source: eMarketer.com, May 2006

79. Temperature Measurements The relationship between the Celsius (°C) and Fahrenheit (°F) scales for measuring temperature is given by the equation

$$F = \frac{9}{5}C + 32$$

The relationship between the Celsius (°C) and Kelvin (K) scales is $K = C + 273$. Graph the equation $F = \frac{9}{5}C + 32$ using degrees Fahrenheit on the y-axis and degrees Celsius on the x-axis. Use the techniques introduced in this section to obtain the graph showing the relationship between Kelvin and Fahrenheit temperatures.

80. Period of a Pendulum The period T (in seconds) of a simple pendulum is a function of its length l (in feet) defined by the equation

$$T = 2\pi\sqrt{\frac{l}{g}}$$

where $g \approx 32.2$ feet per second per second is the acceleration of gravity.

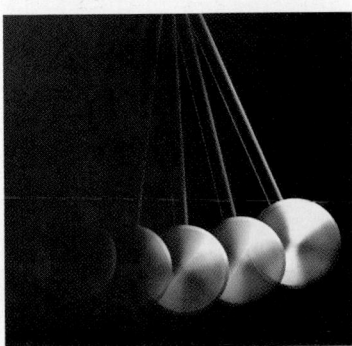

(a) Use a graphing utility to graph the function $T = T(l)$.

(b) Now graph the functions $T = T(l + 1)$, $T = T(l + 2)$, and $T = T(l + 3)$.

(c) Discuss how adding to the length l changes the period T.

(d) Now graph the functions $T = T(2l)$, $T = T(3l)$, and $T = T(4l)$.

(e) Discuss how multiplying the length l by factors of 2, 3, and 4 changes the period T.

81. Cigar Company Profits The daily profits of a cigar company from selling x cigars are given by

$$p(x) = -0.05x^2 + 100x - 2000$$

The government wishes to impose a tax on cigars (sometimes called a *sin tax*) that gives the company the option of either paying a flat tax of $10,000 per day or a tax of 10% on profits. As chief financial officer (CFO) of the company, you need to decide which tax is the better option for the company.

(a) On the same screen, graph $Y_1 = p(x) - 10,000$ and $Y_2 = (1 - 0.10)p(x)$.

(b) Based on the graph, which option would you select? Why?

(c) Using the terminology learned in this section, describe each graph in terms of the graph of $p(x)$.

(d) Suppose that the government offered the options of a flat tax of $4800 or a tax of 10% on profits. Which would you select? Why?

82. The graph of a function f is illustrated in the figure.

(a) Draw the graph of $y = |f(x)|$.

(b) Draw the graph of $y = f(|x|)$.

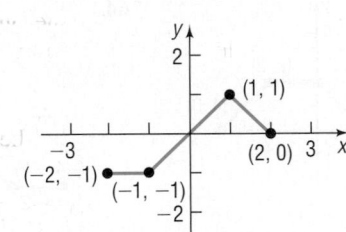

83. The graph of a function f is illustrated in the figure.

(a) Draw the graph of $y = |f(x)|$.

(b) Draw the graph of $y = f(|x|)$.

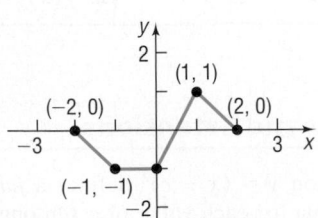

84. Suppose $(1, 3)$ is a point on the graph of $y = f(x)$.

(a) What point is on the graph of $y = f(x + 3) - 5$?

(b) What point is on the graph of $y = -2f(x - 2) + 1$?

(c) What point is on the graph of $y = f(2x + 3)$?

85. Suppose $(-3, 5)$ is a point on the graph of $y = g(x)$.

(a) What point is on the graph of $y = g(x + 1) - 3$?

(b) What point is on the graph of $y = -3g(x - 4) + 3$?

(c) What point is on the graph of $y = g(3x + 9)$?

Explaining Concepts: Discussion and Writing

86. Suppose that the graph of a function f is known. Explain how the graph of $y = 4f(x)$ differs from the graph of $y = f(4x)$.

87. Suppose that the graph of a function f is known. Explain how the graph of $y = f(x) - 2$ differs from the graph of $y = f(x - 2)$.

△ 88. The area under the curve $y = \sqrt{x}$ bounded below by the x-axis and on the right by $x = 4$ is $\dfrac{16}{3}$ square units. Using the ideas presented in this section, what do you think is the area under the curve of $y = \sqrt{-x}$ bounded below by the x-axis and on the left by $x = -4$? Justify your answer.

Interactive Exercises: Exploring Transformations

Ask your instructor if the applets below are of interest to you.

89. Vertical Shifts *Open the vertical shift applet.* Use your mouse to grab the slider and change the value of k. Note the role k plays in the graph of $g(x) = f(x) + k$, where $f(x) = x^2$.

90. Horizontal Shifts *Open the horizontal shift applet.* Use your mouse to grab the slider and change the value of h. Note the role h plays in the graph of $g(x) = f(x - h)$, where $f(x) = x^2$.

91. Vertical Stretches *Open the vertical stretch applet.* Use your mouse to grab the slider and change the value of a. Note the role a plays in the graph of $g(x) = af(x)$, where $f(x) = |x|$.

92. Horizontal Stretches *Open the horizontal stretch applet.*
(a) Use your mouse to grab the slider and change the value of a. Note the role a plays in the graph of $g(x) = f(ax) = \sqrt{ax}$, where $f(x) = \sqrt{x}$. What happens to the points on the graph of g when $0 < a < 1$? What happens to the points on the graph when $a > 1$?
(b) To further understand the concept of horizontal compressions, fill in the spreadsheet to the right of the graph as follows:
(i) What x-coordinate is required on the graph of $g(x) = \sqrt{2x}$, if the y-coordinate is to be 1?

(ii) What x-coordinate is required on the graph of $g(x) = \sqrt{2x}$, if the y-coordinate is to be 2?
(iii) What x-coordinate is required on the graph of $g(x) = \sqrt{2x}$, if the y-coordinate is to be 3?
(iv) What x-coordinate is required on the graph of $g(x) = \sqrt{\dfrac{1}{2}x}$, if the y-coordinate is to be 1?
(v) What x-coordinate is required on the graph of $g(x) = \sqrt{\dfrac{1}{2}x}$, if the y-coordinate is to be 2?
(vi) What x-coordinate is required on the graph of $g(x) = \sqrt{\dfrac{1}{2}x}$, if the y-coordinate is to be 3?

93. Reflection about the y-axis *Open the reflection about the y-axis applet.* Move your mouse to grab the slide and change the value of a from 1 to -1.

94. Reflection about the x-axis *Open the reflection about the x-axis applet.* Move your mouse to grab the slide and change the value of a from 1 to -1.

3.6 Mathematical Models: Building Functions

> **OBJECTIVE 1** Build and Analyze Functions (p. 257)

1 Build and Analyze Functions

Real-world problems often result in mathematical models that involve functions. These functions need to be constructed or built based on the information given. In building functions, we must be able to translate the verbal description into the language of mathematics. We do this by assigning symbols to represent the independent and dependent variables and then by finding the function or rule that relates these variables.

EXAMPLE 1 **Finding the Distance from the Origin to a Point on a Graph**

Let $P = (x, y)$ be a point on the graph of $y = x^2 - 1$.

(a) Express the distance d from P to the origin O as a function of x.

(b) What is d if $x = 0$?

(c) What is d if $x = 1$?

(d) What is d if $x = \dfrac{\sqrt{2}}{2}$?

(e) Use a graphing utility to graph the function $d = d(x)$, $x \geq 0$. Rounded to two decimal places, find the value(s) of x at which d has a local minimum. [This gives the point(s) on the graph of $y = x^2 - 1$ closest to the origin.]

Figure 60

Solution

(a) Figure 60 illustrates the graph of $y = x^2 - 1$. The distance d from P to O is

$$d = \sqrt{(x - 0)^2 + (y - 0)^2} = \sqrt{x^2 + y^2}$$

Since P is a point on the graph of $y = x^2 - 1$, substitute $x^2 - 1$ for y. Then

$$d(x) = \sqrt{x^2 + (x^2 - 1)^2} = \sqrt{x^4 - x^2 + 1}$$

The distance d is expressed as a function of x.

(b) If $x = 0$, the distance d is

$$d(0) = \sqrt{0^4 - 0^2 + 1} = \sqrt{1} = 1$$

(c) If $x = 1$, the distance d is

$$d(1) = \sqrt{1^4 - 1^2 + 1} = 1$$

(d) If $x = \dfrac{\sqrt{2}}{2}$, the distance d is

Figure 61

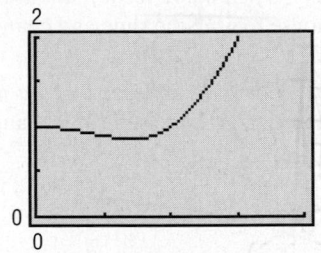

$$d\left(\frac{\sqrt{2}}{2}\right) = \sqrt{\left(\frac{\sqrt{2}}{2}\right)^4 - \left(\frac{\sqrt{2}}{2}\right)^2 + 1} = \sqrt{\frac{1}{4} - \frac{1}{2} + 1} = \frac{\sqrt{3}}{2}$$

(e) Figure 61 shows the graph of $Y_1 = \sqrt{x^4 - x^2 + 1}$. Using the MINIMUM feature on a graphing utility, we find that when $x \approx 0.71$ the value of d is smallest. The local minimum is $d \approx 0.87$ rounded to two decimal places. Since $d(x)$ is even, by symmetry, it follows that when $x \approx -0.71$ the value of d is also a local minimum. Since $(\pm 0.71)^2 - 1 \approx -0.50$, the points $(-0.71, -0.50)$ and $(0.71, -0.50)$ on the graph of $y = x^2 - 1$ are closest to the origin.

━━━━**Now Work** PROBLEM 1

EXAMPLE 2 **Area of a Rectangle**

Figure 62

A rectangle has one corner in quadrant I on the graph of $y = 25 - x^2$, another at the origin, a third on the positive y-axis, and the fourth on the positive x-axis. See Figure 62.

(a) Express the area A of the rectangle as a function of x.

(b) What is the domain of A?

(c) Graph $A = A(x)$.

(d) For what value of x is the area largest?

Solution

(a) The area A of the rectangle is $A = xy$, where $y = 25 - x^2$. Substituting this expression for y, we obtain $A(x) = x(25 - x^2) = 25x - x^3$.

(b) Since (x, y) is in quadrant I, we have $x > 0$. Also, $y = 25 - x^2 > 0$, which implies that $x^2 < 25$, so $-5 < x < 5$. Combining these restrictions, we have the domain of A as $\{x | 0 < x < 5\}$, or $(0, 5)$ using interval notation.

(c) See Figure 63 for the graph of $A = A(x)$.

(d) Using MAXIMUM, we find that the maximum area is 48.11 square units at $x = 2.89$ units, each rounded to two decimal places. See Figure 64.

Figure 63

Figure 64

Now Work PROBLEM 7

EXAMPLE 3

Making a Playpen*

A manufacturer of children's playpens makes a square model that can be opened at one corner and attached at right angles to a wall or, perhaps, the side of a house. If each side is 3 feet in length, the open configuration doubles the available area in which the child can play from 9 square feet to 18 square feet. See Figure 65.

Now suppose that we place hinges at the outer corners to allow for a configuration like the one shown in Figure 66.

Figure 65

Figure 66

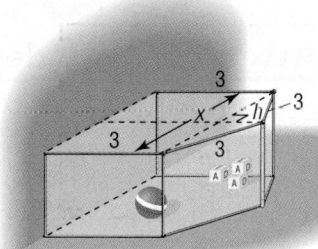

(a) Build a model that expresses the area A of the configuration shown in Figure 66 as a function of the distance x between the two parallel sides.

(b) Find the domain of A.

(c) Find A if $x = 5$.

 (d) Graph $A = A(x)$. For what value of x is the area largest? What is the maximum area?

Solution

(a) Refer to Figure 66. The area A we seek consists of the area of a rectangle (with width 3 and length x) and the area of an isosceles triangle (with base x and two equal sides of length 3). The height h of the triangle may be found using the Pythagorean Theorem.

$$h^2 + \left(\frac{x}{2}\right)^2 = 3^2$$

$$h^2 = 3^2 - \left(\frac{x}{2}\right)^2 = 9 - \frac{x^2}{4} = \frac{36 - x^2}{4}$$

$$h = \frac{1}{2}\sqrt{36 - x^2}$$

* Adapted from *Proceedings, Summer Conference for College Teachers on Applied Mathematics* (University of Missouri, Rolla), 1971.

The area A enclosed by the playpen is

$$A = \text{area of rectangle} + \text{area of triangle} = 3x + \frac{1}{2}x\left(\frac{1}{2}\sqrt{36 - x^2}\right)$$

The area A expressed as a function of x is

$$A(x) = 3x + \frac{x\sqrt{36 - x^2}}{4} \qquad \text{The Model}$$

(b) To find the domain of A, notice that $x > 0$, since x is a length. Also, the expression under the square root must be positive, so

$$36 - x^2 > 0$$
$$x^2 < 36$$
$$-6 < x < 6$$

Combining these restrictions, the domain of A is $0 < x < 6$, or $(0, 6)$ using interval notation.

Figure 67

(c) If $x = 5$, the area is

$$A(5) = 3(5) + \frac{5}{4}\sqrt{36 - (5)^2} \approx 19.15 \text{ square feet}$$

If the length of the playpen is 5 feet, its area is 19.15 square feet.

(d) See Figure 67. The maximum area is about 19.82 square feet, obtained when x is about 5.58 feet.

3.6 Assess Your Understanding

Applications and Extensions

1. Let $P = (x, y)$ be a point on the graph of $y = x^2 - 8$.
 (a) Express the distance d from P to the origin as a function of x.
 (b) What is d if $x = 0$?
 (c) What is d if $x = 1$?
 (d) Use a graphing utility to graph $d = d(x)$.
 (e) For what values of x is d smallest?

2. Let $P = (x, y)$ be a point on the graph of $y = x^2 - 8$.
 (a) Express the distance d from P to the point $(0, -1)$ as a function of x.
 (b) What is d if $x = 0$?
 (c) What is d if $x = -1$?
 (d) Use a graphing utility to graph $d = d(x)$.
 (e) For what values of x is d smallest?

3. Let $P = (x, y)$ be a point on the graph of $y = \sqrt{x}$.
 (a) Express the distance d from P to the point $(1, 0)$ as a function of x.
 (b) Use a graphing utility to graph $d = d(x)$.
 (c) For what values of x is d smallest?

4. Let $P = (x, y)$ be a point on the graph of $y = \frac{1}{x}$.

 (a) Express the distance d from P to the origin as a function of x.
 (b) Use a graphing utility to graph $d = d(x)$.
 (c) For what values of x is d smallest?

5. A right triangle has one vertex on the graph of $y = x^3$, $x > 0$, at (x, y), another at the origin, and the third on the positive y-axis at $(0, y)$, as shown in the figure. Express the area A of the triangle as a function of x.

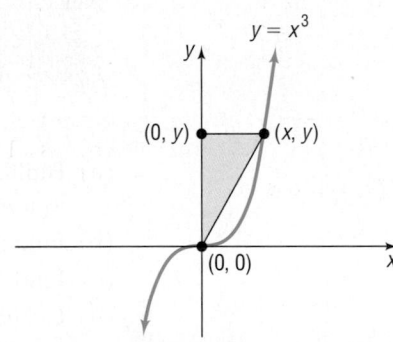

6. A right triangle has one vertex on the graph of $y = 9 - x^2$, $x > 0$, at (x, y), another at the origin, and the third on the positive x-axis at $(x, 0)$. Express the area A of the triangle as a function of x.

7. A rectangle has one corner in quadrant I on the graph of $y = 16 - x^2$, another at the origin, a third on the positive y-axis, and the fourth on the positive x-axis. See the figure.

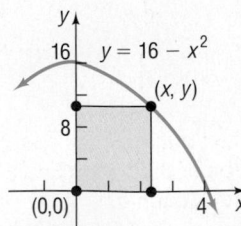

(a) Express the area A of the rectangle as a function of x.

(b) What is the domain of A?

(c) Graph $A = A(x)$. For what value of x is A largest?

8. A rectangle is inscribed in a semicircle of radius 2. See the figure. Let $P = (x, y)$ be the point in quadrant I that is a vertex of the rectangle and is on the circle.

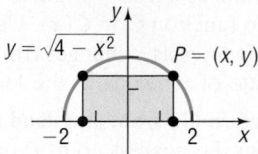

(a) Express the area A of the rectangle as a function of x.

(b) Express the perimeter p of the rectangle as a function of x.

(c) Graph $A = A(x)$. For what value of x is A largest?

(d) Graph $p = p(x)$. For what value of x is p largest?

9. A rectangle is inscribed in a circle of radius 2. See the figure. Let $P = (x, y)$ be the point in quadrant I that is a vertex of the rectangle and is on the circle.

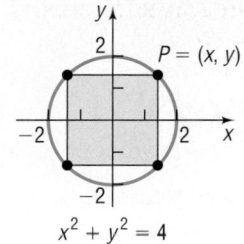

(a) Express the area A of the rectangle as a function of x.

(b) Express the perimeter p of the rectangle as a function of x.

(c) Graph $A = A(x)$. For what value of x is A largest?

(d) Graph $p = p(x)$. For what value of x is p largest?

10. A circle of radius r is inscribed in a square. See the figure.

(a) Express the area A of the square as a function of the radius r of the circle.

(b) Express the perimeter p of the square as a function of r.

11. Geometry A wire 10 meters long is to be cut into two pieces. One piece will be shaped as a square, and the other piece will be shaped as a circle. See the figure.

(a) Express the total area A enclosed by the pieces of wire as a function of the length x of a side of the square.

(b) What is the domain of A?

(c) Graph $A = A(x)$. For what value of x is A smallest?

12. Geometry A wire 10 meters long is to be cut into two pieces. One piece will be shaped as an equilateral triangle, and the other piece will be shaped as a circle.

(a) Express the total area A enclosed by the pieces of wire as a function of the length x of a side of the equilateral triangle.

(b) What is the domain of A?

(c) Graph $A = A(x)$. For what value of x is A smallest?

13. A wire of length x is bent into the shape of a circle.

(a) Express the circumference C of the circle as a function of x.

(b) Express the area A of the circle as a function of x.

14. A wire of length x is bent into the shape of a square.

(a) Express the perimeter p of the square as a function of x.

(b) Express the area A of the square as a function of x.

15. Geometry A semicircle of radius r is inscribed in a rectangle so that the diameter of the semicircle is the length of the rectangle. See the figure.

(a) Express the area A of the rectangle as a function of the radius r of the semicircle.

(b) Express the perimeter p of the rectangle as a function of r.

16. Geometry An equilateral triangle is inscribed in a circle of radius r. See the figure. Express the circumference C of the circle as a function of the length x of a side of the triangle.

[**Hint:** First show that $r^2 = \dfrac{x^2}{3}$.]

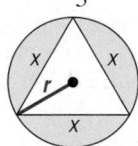

17. Geometry An equilateral triangle is inscribed in a circle of radius r. See the figure in Problem 16. Express the area A within the circle, but outside the triangle, as a function of the length x of a side of the triangle.

18. Uniform Motion Two cars leave an intersection at the same time. One is headed south at a constant speed of 30 miles per hour, and the other is headed west at a constant speed of 40 miles per hour (see the figure). Build a model that expresses the distance d between the cars as a function of the time t.

[**Hint:** At $t = 0$, the cars leave the intersection.]

19. **Uniform Motion** Two cars are approaching an intersection. One is 2 miles south of the intersection and is moving at a constant speed of 30 miles per hour. At the same time, the other car is 3 miles east of the intersection and is moving at a constant speed of 40 miles per hour.
 (a) Build a model that expresses the distance d between the cars as a function of time t.

 [**Hint:** At $t = 0$, the cars are 2 miles south and 3 miles east of the intersection, respectively.]
 (b) Use a graphing utility to graph $d = d(t)$. For what value of t is d smallest?

20. **Inscribing a Cylinder in a Sphere** Inscribe a right circular cylinder of height h and radius r in a sphere of fixed radius R. See the illustration. Express the volume V of the cylinder as a function of h.

 [**Hint:** $V = \pi r^2 h$. Note also the right triangle.]

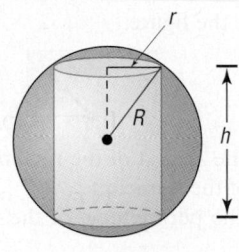

Sphere

21. **Inscribing a Cylinder in a Cone** Inscribe a right circular cylinder of height h and radius r in a cone of fixed radius R and fixed height H. See the illustration. Express the volume V of the cylinder as a function of r.

 [**Hint:** $V = \pi r^2 h$. Note also the similar triangles.]

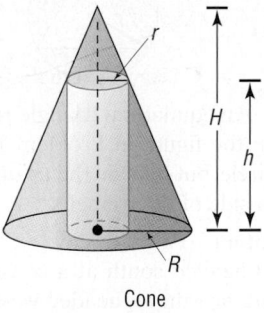

Cone

22. **Installing Cable TV** MetroMedia Cable is asked to provide service to a customer whose house is located 2 miles from the road along which the cable is buried. The nearest connection box for the cable is located 5 miles down the road. See the figure.

(a) If the installation cost is $500 per mile along the road and $700 per mile off the road, build a model that expresses the total cost C of installation as a function of the distance x (in miles) from the connection box to the point where the cable installation turns off the road. Give the domain.
(b) Compute the cost if $x = 1$ mile.
(c) Compute the cost if $x = 3$ miles.
(d) Graph the function $C = C(x)$. Use TRACE to see how the cost C varies as x changes from 0 to 5.
(e) What value of x results in the least cost?

23. **Time Required to Go from an Island to a Town** An island is 2 miles from the nearest point P on a straight shoreline. A town is 12 miles down the shore from P. See the illustration.

(a) If a person can row a boat at an average speed of 3 miles per hour and the same person can walk 5 miles per hour, build a model that expresses the time T that it takes to go from the island to town as a function of the distance x from P to where the person lands the boat.
(b) What is the domain of T?
(c) How long will it take to travel from the island to town if the person lands the boat 4 miles from P?
(d) How long will it take if the person lands the boat 8 miles from P?

24. **Filling a Conical Tank** Water is poured into a container in the shape of a right circular cone with radius 4 feet and height 16 feet. See the figure. Express the volume V of the water in the cone as a function of the height h of the water.

 [**Hint:** The volume V of a cone of radius r and height h is $V = \dfrac{1}{3}\pi r^2 h$.]

25. **Constructing an Open Box** An open box with a square base is to be made from a square piece of cardboard 24 inches on a side by cutting out a square from each corner and turning up the sides. See the figure.

(a) Express the volume V of the box as a function of the length x of the side of the square cut from each corner.

(b) What is the volume if a 3-inch square is cut out?

(c) What is the volume if a 10-inch square is cut out?

(d) Graph $V = V(x)$. For what value of x is V largest?

26. **Constructing an Open Box** An open box with a square base is required to have a volume of 10 cubic feet.

(a) Express the amount A of material used to make such a box as a function of the length x of a side of the square base.

(b) How much material is required for a base 1 foot by 1 foot?

(c) How much material is required for a base 2 feet by 2 feet?

(d) Use a graphing utility to graph $A = A(x)$. For what value of x is A smallest?

CHAPTER REVIEW

Library of Functions

Constant function (p. 236)

$$f(x) = b$$

The graph is a horizontal line with y-intercept b.

Identity function (p. 237)

$$f(x) = x$$

The graph is a line with slope 1 and y-intercept 0.

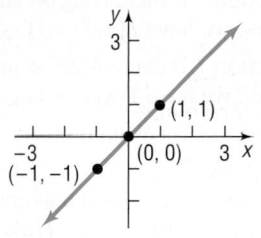

Square function (p. 237)

$$f(x) = x^2$$

The graph is a parabola with intercept at $(0, 0)$.

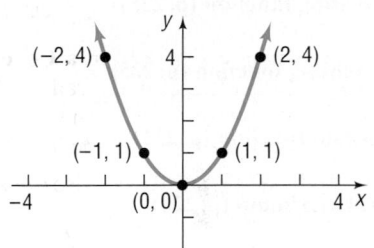

Cube function (p. 237)

$$f(x) = x^3$$

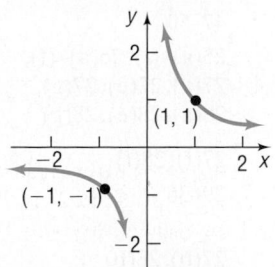

Square root function (p. 237)

$$f(x) = \sqrt{x}$$

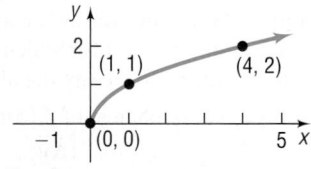

Cube root function (p. 237)

$$f(x) = \sqrt[3]{x}$$

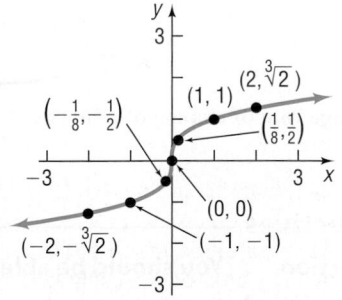

Reciprocal function (p. 238)

$$f(x) = \frac{1}{x}$$

Absolute value function (p. 238)

$$f(x) = |x|$$

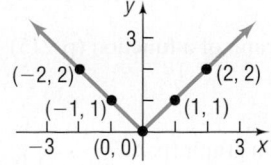

Greatest integer function (p. 238)

$$f(x) = \text{int}(x)$$

Things to Know

Function (pp. 201–203)	A relation between two sets so that each element x in the first set, the domain, has corresponding to it exactly one element y in the second set. The range is the set of y values of the function for the x values in the domain. A function can also be characterized as a set of ordered pairs (x, y) in which no first element is paired with two different second elements.
Function notation (pp. 203–206)	$y = f(x)$ f is a symbol for the function. x is the argument, or independent variable. y is the dependent variable. $f(x)$ is the value of the function at x, or the image of x. A function f may be defined implicitly by an equation involving x and y or explicitly by writing $y = f(x)$.
Difference quotient of f (pp. 205 and 234)	$\dfrac{f(x + h) - f(x)}{h} \quad h \neq 0$
Domain (pp. 206–208)	If unspecified, the domain of a function f defined by an equation is the largest set of real numbers for which $f(x)$ is a real number.
Vertical-line test (p. 214)	A set of points in the plane is the graph of a function if and only if every vertical line intersects the graph in at most one point.
Even function f (p. 223)	$f(-x) = f(x)$ for every x in the domain ($-x$ must also be in the domain).
Odd function f (p. 223)	$f(-x) = -f(x)$ for every x in the domain ($-x$ must also be in the domain).
Increasing function (p. 225)	A function f is increasing on an open interval I if, for any choice of x_1 and x_2 in I, with $x_1 < x_2$, we have $f(x_1) < f(x_2)$.
Decreasing function (p. 225)	A function f is decreasing on an open interval I if, for any choice of x_1 and x_2 in I, with $x_1 < x_2$, we have $f(x_1) > f(x_2)$.
Constant function (p. 225)	A function f is constant on an open interval I if, for all choices of x in I, the values of $f(x)$ are equal.
Local maximum (p. 226)	A function f has a local maximum at c if there is an open interval I containing c so that, for all x in I, $f(x) \leq f(c)$.
Local minimum (p. 226)	A function f has a local minimum at c if there is an open interval I containing c so that, for all x in I, $f(x) \geq f(c)$.
Absolute maximum and Absolute minimum (p. 226)	Let f denote a function defined on some interval I. If there is a number u in I for which $f(x) \leq f(u)$ for all x in I, then $f(u)$ is the absolute maximum of f on I and we say the absolute maximum of f occurs at u. If there is a number v in I for which $f(x) \geq f(v)$, for all x in I, then $f(v)$ is the absolute minimum of f on I and we say the absolute minimum of f occurs at v.
Average rate of change of a function (p. 228)	The average rate of change of f from a to b is $$\frac{\Delta y}{\Delta x} = \frac{f(b) - f(a)}{b - a} \quad a \neq b$$

Objectives

Section		You should be able to . . .	Examples	Review Exercises
3.1	1	Determine whether a relation represents a function (p. 200)	1–5	1, 2
	2	Find the value of a function (p. 203)	6, 7	3–8, 23, 24, 71, 72
	3	Find the domain of a function defined by an equation (p. 206)	8, 9	9–16
	4	Form the sum, difference, product, and quotient of two functions (p. 208)	10	17–22
3.2	1	Identify the graph of a function (p. 214)	1	47–50
	2	Obtain information from or about the graph of a function (p. 215)	2–4	25(a)–(f), 26(a)–(f), 27(a), 27(e), 27(g), 28(a), 28(e), 28(g)
3.3	1	Determine even and odd functions from a graph (p. 223)	1	27(f), 28(f)
	2	Identify even and odd functions from the equation (p. 224)	2	29–36
	3	Use a graph to determine where a function is increasing, decreasing, or constant (p. 224)	3	27(b), 28(b)

Section	You should be able to . . .	Examples	Review Exercises
	4 Use a graph to locate local maxima and local minima (p. 225)	4	27(c), 28(c)
	5 Use a graph to locate the absolute maximum and the absolute minimum (p. 226)	5	27(d), 28(d)
	6 Use a graphing utility to approximate local maxima and local minima and to determine where a function is increasing or decreasing (p. 228)	6	37–40, 74(d), 75(b)
	7 Find the average rate of change of a function (p. 228)	7, 8	41–46
3.4	**1** Graph the functions listed in the library of functions (p. 234)	1, 2	51–54
	2 Graph piecewise-defined functions (p. 239)	3, 4	67–70
3.5	**1** Graph functions using vertical and horizontal shifts (p. 244)	1–5	25(f), 26(f), 26(g) 55, 56, 59–66
	2 Graph functions using compressions and stretches (p. 247)	6–8	25(g), 26(h), 57, 58, 65, 66
	3 Graph functions using reflections about the x-axis or y-axis (p. 250)	9–10	25(h), 57, 61, 62, 66
3.6	**1** Build and analyze functions (p. 257)	1–3	73–75

Review Exercises

In Problems 1 and 2, determine whether each relation represents a function. For each function, state the domain and range.

1. $\{(-1, 0), (2, 3), (4, 0)\}$

2. $\{(4, -1), (2, 1), (4, 2)\}$

In Problems 3–8, find the following for each function:

(a) $f(2)$ (b) $f(-2)$ (c) $f(-x)$ (d) $-f(x)$ (e) $f(x - 2)$ (f) $f(2x)$

3. $f(x) = \dfrac{3x}{x^2 - 1}$

4. $f(x) = \dfrac{x^2}{x + 1}$

5. $f(x) = \sqrt{x^2 - 4}$

6. $f(x) = |x^2 - 4|$

7. $f(x) = \dfrac{x^2 - 4}{x^2}$

8. $f(x) = \dfrac{x^3}{x^2 - 9}$

In Problems 9–16, find the domain of each function.

9. $f(x) = \dfrac{x}{x^2 - 9}$

10. $f(x) = \dfrac{3x^2}{x - 2}$

11. $f(x) = \sqrt{2 - x}$

12. $f(x) = \sqrt{x + 2}$

13. $h(x) = \dfrac{\sqrt{x}}{|x|}$

14. $g(x) = \dfrac{|x|}{x}$

15. $f(x) = \dfrac{x}{x^2 + 2x - 3}$

16. $F(x) = \dfrac{1}{x^2 - 3x - 4}$

In Problems 17–22, find $f + g, f - g, f \cdot g$, and $\dfrac{f}{g}$ for each pair of functions. State the domain of each of these functions.

17. $f(x) = 2 - x$; $g(x) = 3x + 1$

18. $f(x) = 2x - 1$; $g(x) = 2x + 1$

19. $f(x) = 3x^2 + x + 1$; $g(x) = 3x$

20. $f(x) = 3x$; $g(x) = 1 + x + x^2$

21. $f(x) = \dfrac{x + 1}{x - 1}$; $g(x) = \dfrac{1}{x}$

22. $f(x) = \dfrac{1}{x - 3}$; $g(x) = \dfrac{3}{x}$

In Problems 23 and 24, find the difference quotient of each function f; that is, find

$$\frac{f(x + h) - f(x)}{h} \qquad h \neq 0$$

23. $f(x) = -2x^2 + x + 1$

24. $f(x) = 3x^2 - 2x + 4$

25. Using the graph of the function f shown:

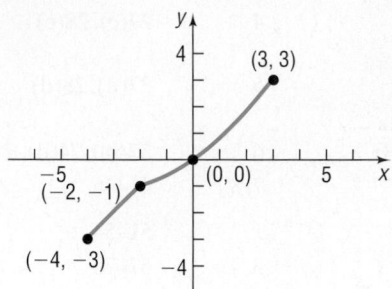

(a) Find the domain and the range of f.
(b) List the intercepts.
(c) Find $f(-2)$.
(d) For what value of x does $f(x) = -3$?
(e) Solve $f(x) > 0$.
(f) Graph $y = f(x - 3)$.
(g) Graph $y = f\left(\dfrac{1}{2}x\right)$.
(h) Graph $y = -f(x)$.

26. Using the graph of the function g shown:

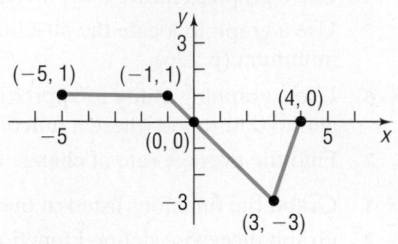

(a) Find the domain and the range of g.
(b) Find $g(-1)$.
(c) List the intercepts.
(d) For what value of x does $g(x) = -3$?
(e) Solve $g(x) > 0$.
(f) Graph $y = g(x - 2)$.
(g) Graph $y = g(x) + 1$.
(h) Graph $y = 2g(x)$.

In Problems 27 and 28, use the graph of the function f to find:

(a) The domain and the range of f.
(b) The intervals on which f is increasing, decreasing, or constant.
(c) The local minimum values and local maximum values.
(d) The absolute maximum and absolute minimum.
(e) Whether the graph is symmetric with respect to the x-axis, the y-axis, or the origin.
(f) Whether the function is even, odd, or neither.
(g) The intercepts, if any.

27.

28.

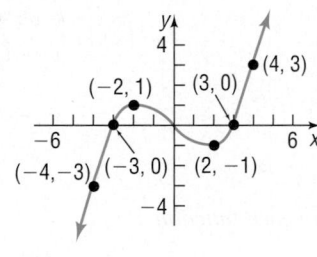

In Problems 29–36, determine (algebraically) whether the given function is even, odd, or neither.

29. $f(x) = x^3 - 4x$

30. $g(x) = \dfrac{4 + x^2}{1 + x^4}$

31. $h(x) = \dfrac{1}{x^4} + \dfrac{1}{x^2} + 1$

32. $F(x) = \sqrt{1 - x^3}$

33. $G(x) = 1 - x + x^3$

34. $H(x) = 1 + x + x^2$

35. $f(x) = \dfrac{x}{1 + x^2}$

36. $g(x) = \dfrac{1 + x^2}{x^3}$

In Problems 37–40, use a graphing utility to graph each function over the indicated interval. Approximate any local maximum values and local minimum values. Determine where the function is increasing and where it is decreasing.

37. $f(x) = 2x^3 - 5x + 1 \quad (-3, 3)$

38. $f(x) = -x^3 + 3x - 5 \quad (-3, 3)$

39. $f(x) = 2x^4 - 5x^3 + 2x + 1 \quad (-2, 3)$

40. $f(x) = -x^4 + 3x^3 - 4x + 3 \quad (-2, 3)$

In Problems 41 and 42, find the average rate of change of f:

(a) From 1 to 2 *(b) From 0 to 1* *(c) From 2 to 4*

41. $f(x) = 8x^2 - x$

42. $f(x) = 2x^3 + x$

In Problems 43–46, find the average rate of change from 2 to 3 for each function f. Be sure to simplify.

43. $f(x) = 2 - 5x$

44. $f(x) = 2x^2 + 7$

45. $f(x) = 3x - 4x^2$

46. $f(x) = x^2 - 3x + 2$

In Problems 47–50, is the graph shown the graph of a function?

47.

48.

49.

50.

In Problems 51–54, sketch the graph of each function. Be sure to label at least three points.

51. $f(x) = |x|$

52. $f(x) = \sqrt[3]{x}$

53. $f(x) = \sqrt{x}$

54. $f(x) = \dfrac{1}{x}$

In Problems 55–66, graph each function using the techniques of shifting, compressing or stretching, and reflections. Identify any intercepts on the graph. State the domain and, based on the graph, find the range.

55. $F(x) = |x| - 4$

56. $f(x) = |x| + 4$

57. $g(x) = -2|x|$

58. $g(x) = \dfrac{1}{2}|x|$

59. $h(x) = \sqrt{x - 1}$

60. $h(x) = \sqrt{x} - 1$

61. $f(x) = \sqrt{1 - x}$

62. $f(x) = -\sqrt{x + 3}$

63. $h(x) = (x - 1)^2 + 2$

64. $h(x) = (x + 2)^2 - 3$

65. $g(x) = 3(x - 1)^3 + 1$

66. $g(x) = -2(x + 2)^3 - 8$

In Problems 67–70,

(a) Find the domain of each function.
(b) Locate any intercepts.
(c) Graph each function.
(d) Based on the graph, find the range.
(e) Is f continuous on its domain?

67. $f(x) = \begin{cases} 3x & \text{if } -2 < x \le 1 \\ x + 1 & \text{if } x > 1 \end{cases}$

68. $f(x) = \begin{cases} x - 1 & \text{if } -3 < x < 0 \\ 3x - 1 & \text{if } x \ge 0 \end{cases}$

69. $f(x) = \begin{cases} x & \text{if } -4 \le x < 0 \\ 1 & \text{if } x = 0 \\ 3x & \text{if } x > 0 \end{cases}$

70. $f(x) = \begin{cases} x^2 & \text{if } -2 \le x \le 2 \\ 2x - 1 & \text{if } x > 2 \end{cases}$

71. A function f is defined by

$$f(x) = \frac{Ax + 5}{6x - 2}$$

If $f(1) = 4$, find A.

72. A function g is defined by

$$g(x) = \frac{A}{x} + \frac{8}{x^2}$$

If $g(-1) = 0$, find A.

73. Page Design A page with dimensions of $8\frac{1}{2}$ inches by 11 inches has a border of uniform width x surrounding the printed matter of the page, as shown in the figure.
(a) Develop a model that expresses the area A of the printed part of the page as a function of the width x of the border.
(b) Give the domain and the range of A.
(c) Find the area of the printed page for borders of widths 1 inch, 1.2 inches, and 1.5 inches.
(d) Graph the function $A = A(x)$.

$8\frac{1}{2}$ in.

11 in.

74. Constructing a Closed Box A closed box with a square base is required to have a volume of 10 cubic feet.
(a) Build a model that expresses the amount A of material used to make such a box as a function of the length x of a side of the square base.
(b) How much material is required for a base 1 foot by 1 foot?
(c) How much material is required for a base 2 feet by 2 feet?
(d) Graph $A = A(x)$. For what value of x is A smallest?

75. A rectangle has one vertex in quadrant I on the graph of $y = 10 - x^2$, another at the origin, one on the positive x-axis, and one on the positive y-axis.
(a) Express the area A of the rectangle as a function of x.
(b) Find the largest area A that can be enclosed by the rectangle.

CHAPTER TEST

CHAPTER
Test Prep
VIDEOS

The Chapter Test Prep Videos are step-by-step test solutions available in the Video Resources DVD, in *MyMathLab*, or on this text's You**Tube** Channel. Flip back to the Student Resources page to see the exact web address for this text's YouTube channel.

1. Determine whether each relation represents a function. For each function, state the domain and the range.
 (a) $\{(2, 5), (4, 6), (6, 7), (8, 8)\}$
 (b) $\{(1, 3), (4, -2), (-3, 5), (1, 7)\}$
 (c)

 (d)

 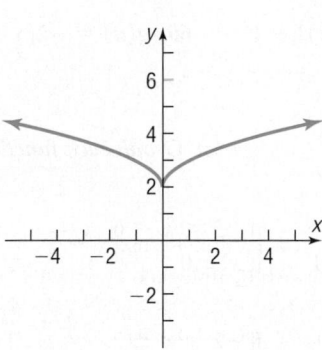

In Problems 2–4, find the domain of each function and evaluate each function at $x = -1$.

2. $f(x) = \sqrt{4 - 5x}$

3. $g(x) = \dfrac{x + 2}{|x + 2|}$

4. $h(x) = \dfrac{x - 4}{x^2 + 5x - 36}$

5. Using the graph of the function f:

 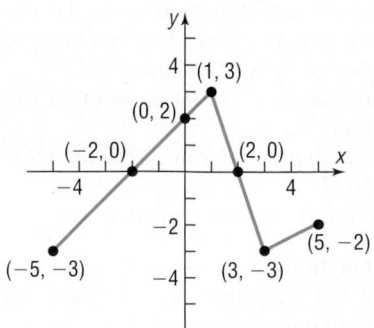

 (a) Find the domain and the range of f.
 (b) List the intercepts.
 (c) Find $f(1)$.
 (d) For what value(s) of x does $f(x) = -3$?
 (e) Solve $f(x) < 0$.

6. Use a graphing utility to graph the function $f(x) = -x^4 + 2x^3 + 4x^2 - 2$ on the interval $(-5, 5)$. Approximate any local maximum values and local minimum values rounded to two decimal places. Determine where the function is increasing and where it is decreasing.

7. Consider the function $g(x) = \begin{cases} 2x + 1 & \text{if } x < -1 \\ x - 4 & \text{if } x \geq -1 \end{cases}$
 (a) Graph the function.
 (b) List the intercepts.
 (c) Find $g(-5)$.
 (d) Find $g(2)$.

8. For the function $f(x) = 3x^2 - 2x + 4$, find the average rate of change of f from 3 to 4.

9. For the functions $f(x) = 2x^2 + 1$ and $g(x) = 3x - 2$, find the following and simplify:
 (a) $f - g$
 (b) $f \cdot g$
 (c) $f(x + h) - f(x)$

10. Graph each function using the techniques of shifting, compressing or stretching, and reflections. Start with the graph of the basic function and show all stages.
 (a) $h(x) = -2(x + 1)^3 + 3$
 (b) $g(x) = |x + 4| + 2$

11. The variable interest rate on a student loan changes each July 1 based on the bank prime loan rate. For the years 1992–2007, this rate can be approximated by the model $r(x) = -0.115x^2 + 1.183x + 5.623$, where x is the number of years since 1992 and r is the interest rate as a percent.
 (a) Use a graphing utility to estimate the highest rate during this time period. During which year was the interest rate the highest?
 (b) Use the model to estimate the rate in 2010. Does this value seem reasonable?

 Source: U.S. Federal Reserve

12. A community skating rink is in the shape of a rectangle with semicircles attached at the ends. The length of the rectangle is 20 feet less than twice the width. The thickness of the ice is 0.75 inch.
 (a) Build a model that expresses the ice volume, V, as a function of the width, x.
 (b) How much ice is in the rink if the width is 90 feet?

 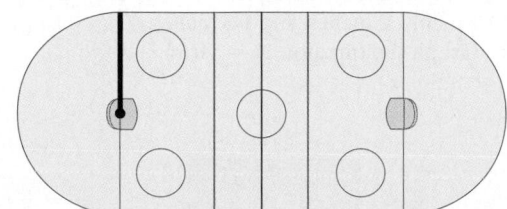

CUMULATIVE REVIEW

In Problems 1–6, find the real solutions of each equation.

1. $3x - 8 = 10$

2. $3x^2 - x = 0$

3. $x^2 - 8x - 9 = 0$

4. $6x^2 - 5x + 1 = 0$

5. $|2x + 3| = 4$

6. $\sqrt{2x + 3} = 2$

In Problems 7–9, solve each inequality. Graph the solution set.

7. $2 - 3x > 6$ **8.** $|2x - 5| < 3$ **9.** $|4x + 1| \geq 7$

10. (a) Find the distance from $P_1 = (-2, -3)$ to $P_2 = (3, -5)$.
 (b) What is the midpoint of the line segment from P_1 to P_2?
 (c) What is the slope of the line containing the points P_1 and P_2?

In Problems 11–14, graph each equation.

11. $3x - 2y = 12$ **12.** $x = y^2$

13. $x^2 + (y - 3)^2 = 16$ **14.** $y = \sqrt{x}$

15. For the equation $3x^2 - 4y = 12$, find the intercepts and check for symmetry.

16. Find the slope–intercept form of the equation of the line containing the points $(-2, 4)$ and $(6, 8)$.

In Problems 17–19, graph each function.

17. $f(x) = (x + 2)^2 - 3$

18. $f(x) = \dfrac{1}{x}$

19. $f(x) = \begin{cases} 2 - x & \text{if } x \leq 2 \\ |x| & \text{if } x > 2 \end{cases}$

CHAPTER PROJECTS

 Internet-based Project

I. Choosing a Cellular Telephone Plan Collect information from your family, friends, or consumer agencies such as Consumer Reports. Then decide on a cellular telephone provider, choosing the company that you feel offers the best service. Once you have selected a service provider, research the various types of individual plans offered by the company by visiting the provider's website.

1. Suppose you expect to use 400 anytime minutes without a texting or data plan. What would be the monthly cost of each plan you are considering?

2. Suppose you expect to use 600 anytime minutes with unlimited texting, but no data plan. What would be the monthly cost of each plan you are considering?

3. Suppose you expect to use 500 anytime minutes with unlimited texting and an unlimited data plan. What would be the monthly cost of each plan you are considering?

4. Suppose you expect to use 500 anytime minutes with unlimited texting and 20 MB of data. What would be the monthly cost of each plan you are considering?

5. Build a model that describes the monthly cost C as a function of the number of anytime minutes used m assuming unlimited texting and 20 MB of data each month for each plan you are considering.

6. Graph each function from Problem 5.

7. Based on your particular usage, which plan is best for you?

8. Now, develop an Excel spreadsheet to analyze the various plans you are considering. Suppose you want a plan that offers 700 anytime minutes with additional minutes costing $0.40 per minute that costs $39.99 per month. In addition, you want unlimited texting, which costs an additional $20 per month, and a data plan that offers up to 25 MB of data each month, with each additional MB costing $0.20. Because cellular telephone plans cost structure is based on piecewise-defined functions, we need "if-then" statements within Excel to analyze the cost of the plan. Use the Excel spreadsheet below as a guide in developing your worksheet. Enter into your spreadsheet a variety of possible minutes and data used to help arrive at a decision regarding which plan is best for you.

9. Write a paragraph supporting the choice in plans that best meets your needs.

10. How are "if/then" loops similar to a piecewise-defined function?

	A	B	C	D
1				
2	Monthly Fee	$ 39.99		
3	Alloted number of anytime minutes	700		
4	Number of anytime minutes used:	700		
5	Cost per additional minute	$ 0.40		
6	Monthly cost of text messaging:	$ 20.00		
7	Monthly cost of data plan	$ 9.99		
8	Alloted data per month (MB)	25		
9	Data used	30		
10	Cost per additional MB of data	$ 0.20		
11				
12	Cost of phone minutes	=IF(B4<B3,B2,B2+B5*(B4-B3))		
13	Cost of data	=IF(B9<B8,B7,B7+B10*(B9-B8))		
14				
15	Total Cost	=B6+B12+B13		
16				

The following projects are available on the Instructor's Resource Center (IRC):

II. Project at Motorola: *Wireless Internet Service* Use functions and their graphs to analyze the total cost of various wireless Internet service plans.

III. Cost of Cable When government regulations and customer preference influence the path of a new cable line, the Pythagorean Theorem can be used to assess the cost of installation.

IV. Oil Spill Functions are used to analyze the size and spread of an oil spill from a leaking tanker.

Citation: Excel © 2010 Microsoft Corporation. Used with permission from Microsoft.

Linear and Quadratic Functions

4

Outline

The Beta of a Stock

Investing in the stock market can be rewarding and fun, but how does one go about selecting which stocks to purchase? Financial investment firms hire thousands of analysts who track individual stocks (equities) and assess the value of the underlying company. One measure the analysts consider is the *beta* of the stock. **Beta** measures the relative risk of an individual company's equity to that of a market basket of stocks, such as the Standard & Poor's 500. But how is beta computed?

 —See the Internet-based Chapter Project—

◁ **A Look Back** Up to now, our discussion has focused on graphs of equations and functions. We learned how to graph equations using the point-plotting method, intercepts, and the tests for symmetry. In addition, we learned what a function is and how to identify whether a relation represents a function. We also discussed properties of functions, such as domain/range, increasing/decreasing, even/odd, and average rate of change.

A Look Ahead ▷ Going forward, we will look at classes of functions. In this chapter, we focus on linear and quadratic functions, their properties, and applications.

4.1 Linear Functions and Their Properties

PREPARING FOR THIS SECTION *Before getting started, review the following:*

- Lines (Section 2.3, pp. 167–175)
- Graphs of Equations in Two Variables; Intercepts; Symmetry (Section 2.2, pp. 157–164)
- Linear Equations (Section 1.1, pp. 82–87)

- Functions (Section 3.1, pp. 200–208)
- The Graph of a Function (Section 3.2, pp. 214–217)
- Properties of Functions (Section 3.3, pp. 222–230)

Now Work the 'Are You Prepared?' problems on page 278.

OBJECTIVES **1** Graph Linear Functions (p. 272)
2 Use Average Rate of Change to Identify Linear Functions (p. 272)
3 Determine Whether a Linear Function Is Increasing, Decreasing, or Constant (p. 275)
4 Build Linear Models from Verbal Descriptions (p. 276)

1 Graph Linear Functions

In Section 2.3 we discussed lines. In particular, for nonvertical lines we developed the slope–intercept form of the equation of a line $y = mx + b$. When we write the slope–intercept form of a line using function notation, we have a *linear function*.

DEFINITION

A **linear function** is a function of the form

$$f(x) = mx + b$$

The graph of a linear function is a line with slope m and y-intercept b. Its domain is the set of all real numbers.

Functions that are not linear are said to be **nonlinear.**

EXAMPLE 1

Graphing a Linear Function

Graph the linear function: $f(x) = -3x + 7$

Figure 1

Solution

This is a linear function with slope $m = -3$ and y-intercept $b = 7$. To graph this function, we plot the point $(0, 7)$, the y-intercept, and use the slope to find an additional point by moving right 1 unit and down 3 units. See Figure 1.

Alternatively, we could have found an additional point by evaluating the function at some $x \neq 0$. For $x = 1$, we find $f(1) = -3(1) + 7 = 4$ and obtain the point $(1, 4)$ on the graph.

Now Work PROBLEMS 13(a) AND (b)

2 Use Average Rate of Change to Identify Linear Functions

Look at Table 1, which shows certain values of the independent variable x and corresponding values of the dependent variable y for the function $f(x) = -3x + 7$. Notice that as the value of the independent variable, x, increases by 1 the value of the dependent variable y decreases by 3. That is, the average rate of change of y with respect to x is a constant, -3.

Table 1

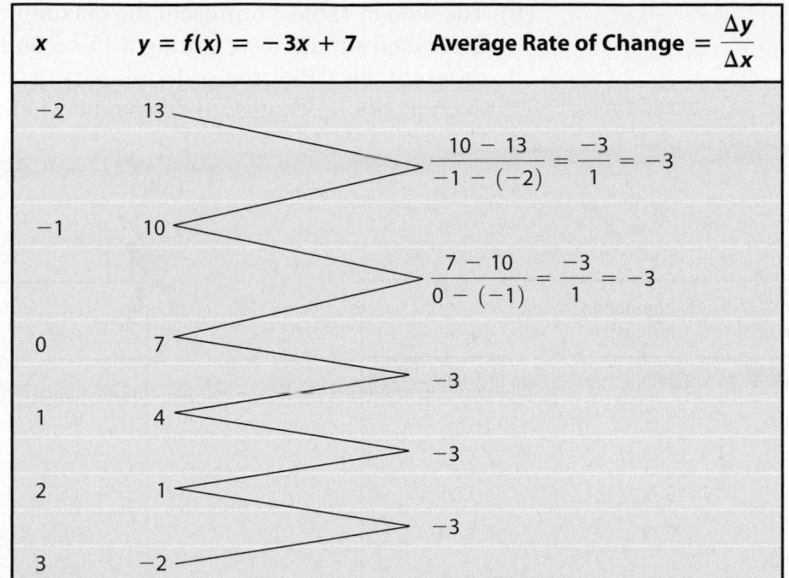

x	y = f(x) = −3x + 7	Average Rate of Change = $\dfrac{\Delta y}{\Delta x}$
−2	13	
		$\dfrac{10 - 13}{-1 - (-2)} = \dfrac{-3}{1} = -3$
−1	10	
		$\dfrac{7 - 10}{0 - (-1)} = \dfrac{-3}{1} = -3$
0	7	
		−3
1	4	
		−3
2	1	
		−3
3	−2	

It is not a coincidence that the average rate of change of the linear function $f(x) = -3x + 7$ is the slope of the linear function. That is, $\dfrac{\Delta y}{\Delta x} = m = -3$. The following theorem states this fact.

THEOREM

Average Rate of Change of a Linear Function

Linear functions have a constant average rate of change. That is, the average rate of change of a linear function $f(x) = mx + b$ is

$$\frac{\Delta y}{\Delta x} = m$$

Proof The average rate of change of $f(x) = mx + b$ from x_1 to x_2, $x_1 \neq x_2$, is

$$\frac{\Delta y}{\Delta x} = \frac{f(x_2) - f(x_1)}{x_2 - x_1} = \frac{(mx_2 + b) - (mx_1 + b)}{x_2 - x_1}$$

$$= \frac{mx_2 - mx_1}{x_2 - x_1} = \frac{m(x_2 - x_1)}{x_2 - x_1} = m \qquad \blacksquare$$

Based on the theorem just proved, the average rate of change of the function $g(x) = -\dfrac{2}{5}x + 5$ is $-\dfrac{2}{5}$.

━━━ **Now Work** PROBLEM 13(c)

As it turns out, only linear functions have a constant average rate of change. Because of this, we can use the average rate of change to determine whether a function is linear or not. This is especially useful if the function is defined by a data set.

EXAMPLE 2

Using the Average Rate of Change to Identify Linear Functions

(a) A strain of E-coli Beu 397-recA441 is placed into a Petri dish at 30° Celsius and allowed to grow. The data shown in Table 2 on page 274 are collected. The population is measured in grams and the time in hours. Plot the ordered pairs (x, y) in the Cartesian plane and use the average rate of change to determine whether the function is linear.

(b) The data in Table 3 represent the maximum number of heartbeats that a healthy individual should have during a 15-second interval of time while exercising for different ages. Plot the ordered pairs (x, y) in the Cartesian plane and use the average rate of change to determine whether the function is linear.

Table 2

Time (hours), x	Population (grams), y	(x, y)
0	0.09	(0, 0.09)
1	0.12	(1, 0.12)
2	0.16	(2, 0.16)
3	0.22	(3, 0.22)
4	0.29	(4, 0.29)
5	0.39	(5, 0.39)

Table 3

Age, x	Maximum Number of Heartbeats, y	(x, y)
20	50	(20, 50)
30	47.5	(30, 47.5)
40	45	(40, 45)
50	42.5	(50, 42.5)
60	40	(60, 40)
70	37.5	(70, 37.5)

Source: American Heart Association

Solution Compute the average rate of change of each function. If the average rate of change is constant, the function is linear. If the average rate of change is not constant, the function is nonlinear.

(a) Figure 2 shows the points listed in Table 2 plotted in the Cartesian plane. Notice that it is impossible to draw a straight line that contains all the points. Table 4 displays the average rate of change of the population.

Figure 2

Table 4

Time (hours), x	Population (grams), y	Average Rate of Change = $\dfrac{\Delta y}{\Delta x}$
0	0.09	
		$\dfrac{0.12 - 0.09}{1 - 0} = 0.03$
1	0.12	
		0.04
2	0.16	
		0.06
3	0.22	
		0.07
4	0.29	
		0.10
5	0.39	

Because the average rate of change is not constant, we know that the function is not linear. In fact, because the average rate of change is increasing as the value of the independent variable increases, the function is increasing at an increasing rate. So not only is the population increasing over time, but it is also growing more rapidly as time passes.

(b) Figure 3 shows the points listed in Table 3 plotted in the Cartesian plane. We can see that the data in Figure 3 lie on a straight line. Table 5 contains the average rate of change of the maximum number of heartbeats. The average rate of change of the heartbeat data is constant, -0.25 beat per year, so the function is linear.

Figure 3

Table 5

Age, x	Maximum Number of Heartbeats, y	Average Rate of Change $= \dfrac{\Delta y}{\Delta x}$
20	50	
		$\dfrac{47.5 - 50}{30 - 20} = -0.25$
30	47.5	
		-0.25
40	45	
		-0.25
50	42.5	
		-0.25
60	40	
		-0.25
70	37.5	

Now Work PROBLEM 21

3 Determine Whether a Linear Function Is Increasing, Decreasing, or Constant

Look back at the Seeing the Concept on page 169. When the slope m of a linear function is positive $(m > 0)$, the line slants upward from left to right. When the slope m of a linear function is negative $(m < 0)$, the line slants downward from left to right. When the slope m of a linear function is zero $(m = 0)$, the line is horizontal.

THEOREM

Increasing, Decreasing, and Constant Linear Functions

A linear function $f(x) = mx + b$ is increasing over its domain if its slope, m, is positive. It is decreasing over its domain if its slope, m, is negative. It is constant over its domain if its slope, m, is zero.

EXAMPLE 3

Determining Whether a Linear Function Is Increasing, Decreasing, or Constant

Determine whether the following linear functions are increasing, decreasing, or constant.

(a) $f(x) = 5x - 2$ (b) $g(x) = -2x + 8$

(c) $s(t) = \dfrac{3}{4}t - 4$ (d) $h(z) = 7$

Solution

(a) For the linear function $f(x) = 5x - 2$, the slope is 5, which is positive. The function f is increasing on the interval $(-\infty, \infty)$.

(b) For the linear function $g(x) = -2x + 8$, the slope is -2, which is negative. The function g is decreasing on the interval $(-\infty, \infty)$.

(c) For the linear function $s(t) = \dfrac{3}{4}t - 4$, the slope is $\dfrac{3}{4}$, which is positive. The function s is increasing on the interval $(-\infty, \infty)$.

(d) We can write the linear function h as $h(z) = 0z + 7$. Because the slope is 0, the function h is constant on the interval $(-\infty, \infty)$.

Now Work PROBLEM 13 (d)

4 Build Linear Models from Verbal Descriptions

When the average rate of change of a function is constant, we can use a linear function to model the relation between the two variables. For example, if your phone company charges you $0.07 per minute to talk regardless of the number of minutes used, we can model the relation between the cost C and minutes used x as the linear function $C(x) = 0.07x$, with slope $m = \dfrac{0.07\,\text{dollar}}{1\,\text{minute}}$.

Modeling with a Linear Function

If the average rate of change of a function is a constant m, a linear function f can be used to model the relation between the two variables as follows:

$$f(x) = mx + b$$

where b is the value of f at 0, that is, $b = f(0)$.

EXAMPLE 4

Straight-line Depreciation

Book value is the value of an asset that a company uses to create its balance sheet. Some companies depreciate their assets using straight-line depreciation so that the value of the asset declines by a fixed amount each year. The amount of the decline depends on the useful life that the company places on the asset. Suppose that a company just purchased a fleet of new cars for its sales force at a cost of $28,000 per car. The company chooses to depreciate each vehicle using the straight-line method over 7 years. This means that each car will depreciate by $\dfrac{\$28{,}000}{7} = \4000 per year.

(a) Write a linear function that expresses the book value V of each car as a function of its age, x.
(b) Graph the linear function.
(c) What is the book value of each car after 3 years?
(d) Interpret the slope.
(e) When will the book value of each car be $8000?
 [**Hint:** Solve the equation $V(x) = 8000$.]

Solution

(a) If we let $V(x)$ represent the value of each car after x years, then $V(0)$ represents the original value of each car, so $V(0) = \$28{,}000$. The y-intercept of the linear function is $28,000. Because each car depreciates by $4000 per year, the slope of the linear function is -4000. The linear function that represents the book value V of each car after x years is

$$V(x) = -4000x + 28{,}000$$

(b) Figure 4 shows the graph of V.
(c) The book value of each car after 3 years is

$$V(3) = -4000(3) + 28{,}000$$
$$= \$16{,}000$$

Figure 4

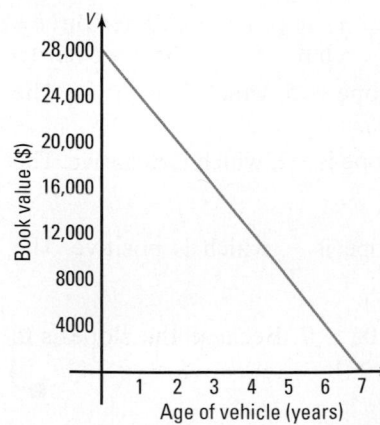

(d) Since the slope of $V(x) = -4000x + 28{,}000$ is -4000, the average rate of change of book value is $-\$4000/\text{year}$. So for each additional year that passes the book value of the car decreases by $4000.

(e) To find when the book value will be $8000, solve the equation

$$V(x) = 8000$$
$$-4000x + 28{,}000 = 8000$$
$$-4000x = -20{,}000 \quad \text{Subtract 28,000 from each side.}$$
$$x = \frac{-20{,}000}{-4000} = 5 \quad \text{Divide by } -4000.$$

The car will have a book value of $8000 when it is 5 years old.

━━━━**Now Work** PROBLEM 45

EXAMPLE 5 **Supply and Demand**

The **quantity supplied** of a good is the amount of a product that a company is willing to make available for sale at a given price. The **quantity demanded** of a good is the amount of a product that consumers are willing to purchase at a given price. Suppose that the quantity supplied, S, and quantity demanded, D, of cellular telephones each month are given by the following functions:

$$S(p) = 60p - 900$$
$$D(p) = -15p + 2850$$

where p is the price (in dollars) of the telephone.

(a) The **equilibrium price** of a product is defined as the price at which quantity supplied equals quantity demanded. That is, the equilibrium price is the price at which $S(p) = D(p)$. Find the equilibrium price of cellular telephones. What is the **equilibrium quantity,** the amount demanded (or supplied) at the equilibrium price?

(b) Determine the prices for which quantity supplied is greater than quantity demanded. That is, solve the inequality $S(p) > D(p)$.

(c) Graph $S = S(p)$, $D = D(p)$ and label the equilibrium price.

Solution (a) To find the equilibrium price, solve the equation $S(p) = D(p)$.

$$60p - 900 = -15p + 2850 \quad \begin{array}{l} S(p) = 60p - 900; \\ D(p) = -15p + 2850 \end{array}$$
$$60p = -15p + 3750 \quad \text{Add 900 to each side.}$$
$$75p = 3750 \quad \text{Add 15p to each side.}$$
$$p = 50 \quad \text{Divide each side by 75.}$$

The equilibrium price is $50 per cellular phone. To find the equilibrium quantity, evaluate either $S(p)$ or $D(p)$ at $p = 50$.

$$S(50) = 60(50) - 900 = 2100$$

The equilibrium quantity is 2100 cellular phones. At a price of $50 per phone, the company will produce and sell 2100 phones each month and have no shortages or excess inventory.

(b) The inequality $S(p) > D(p)$ is

$$60p - 900 > -15p + 2850 \quad S(p) > D(p)$$
$$60p > -15p + 3750 \quad \text{Add 900 to each side.}$$
$$75p > 3750 \quad \text{Add 15p to each side.}$$
$$p > 50 \quad \text{Divide each side by 75.}$$

If the company charges more than $50 per phone, quantity supplied will exceed quantity demanded. In this case the company will have excess phones in inventory.

(c) Figure 5 shows the graphs of $S = S(p)$ and $D = D(p)$ with the equilibrium point labeled.

Figure 5

Price ($)

──Now Work PROBLEM 39

4.1 Assess Your Understanding

'Are You Prepared?' *Answers are given at the end of these exercises. If you get a wrong answer, read the pages listed in red.*

1. Graph $y = 2x - 3$. (pp. 157–164)

2. Find the slope of the line joining the points $(2, 5)$ and $(-1, 3)$. (pp. 167–175)

3. Find the average rate of change of $f(x) = 3x^2 - 2$, from 2 to 4. (pp. 222–230)

4. Solve: $60x - 900 = -15x + 2850$. (pp. 82–87)

5. If $f(x) = x^2 - 4$, find $f(-2)$. (pp. 200–208)

6. **True or False** The graph of the function $f(x) = x^2$ is increasing on the interval $(0, \infty)$. (pp. 214–217)

Concepts and Vocabulary

7. For the graph of the linear function $f(x) = mx + b$, m is the _____ and b is the _____.

8. For the graph of the linear function $H(z) = -4z + 3$, the slope is _____ and the y-intercept is _____.

9. If the slope m of the graph of a linear function is _____, the function is increasing over its domain.

10. **True or False** The slope of a nonvertical line is the average rate of change of the linear function.

11. **True or False** If the average rate of change of a linear function is $\frac{2}{3}$, then if y increases by 3, x will increase by 2.

12. **True or False** The average rate of change of $f(x) = 2x + 8$ is 8.

Skill Building

In Problems 13–20, a linear function is given.

(a) *Determine the slope and y-intercept of each function.*
(b) *Use the slope and y-intercept to graph the linear function.*
(c) *Determine the average rate of change of each function.*
(d) *Determine whether the linear function is increasing, decreasing, or constant.*

13. $f(x) = 2x + 3$

14. $g(x) = 5x - 4$

15. $h(x) = -3x + 4$

16. $p(x) = -x + 6$

17. $f(x) = \frac{1}{4}x - 3$

18. $h(x) = -\frac{2}{3}x + 4$

19. $F(x) = 4$

20. $G(x) = -2$

In Problems 21–28, determine whether the given function is linear or nonlinear. If it is linear, determine the slope.

21.

x	y = f(x)
−2	4
−1	1
0	−2
1	−5
2	−8

22.

x	y = f(x)
−2	1/4
−1	1/2
0	1
1	2
2	4

23.

x	y = f(x)
−2	−8
−1	−3
0	0
1	1
2	0

24.

x	y = f(x)
−2	−4
−1	0
0	4
1	8
2	12

25.

x	y = f(x)
−2	−26
−1	−4
0	2
1	−2
2	−10

26.

x	y = f(x)
−2	−4
−1	−3.5
0	−3
1	−2.5
2	−2

27.

x	y = f(x)
−2	8
−1	8
0	8
1	8
2	8

28.

x	y = f(x)
−2	0
−1	1
0	4
1	9
2	16

Applications and Extensions

29. Suppose that $f(x) = 4x - 1$ and $g(x) = -2x + 5$.
 (a) Solve $f(x) = 0$. (b) Solve $f(x) > 0$.
 (c) Solve $f(x) = g(x)$. (d) Solve $f(x) \le g(x)$.
 (e) Graph $y = f(x)$ and $y = g(x)$ and label the point that represents the solution to the equation $f(x) = g(x)$.

30. Suppose that $f(x) = 3x + 5$ and $g(x) = -2x + 15$.
 (a) Solve $f(x) = 0$. (b) Solve $f(x) < 0$.
 (c) Solve $f(x) = g(x)$. (d) Solve $f(x) \ge g(x)$.
 (e) Graph $y = f(x)$ and $y = g(x)$ and label the point that represents the solution to the equation $f(x) = g(x)$.

31. In parts (a)–(f), use the following figure.

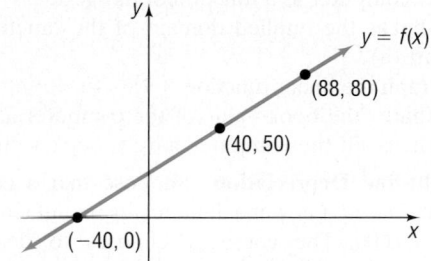

 (a) Solve $f(x) = 50$. (b) Solve $f(x) = 80$.
 (c) Solve $f(x) = 0$. (d) Solve $f(x) > 50$.
 (e) Solve $f(x) \le 80$. (f) Solve $0 < f(x) < 80$.

32. In parts (a)–(f), use the following figure.

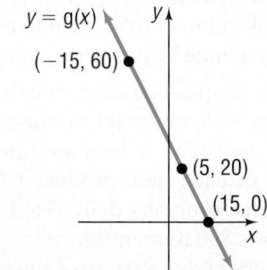

 (a) Solve $g(x) = 20$. (b) Solve $g(x) = 60$.
 (c) Solve $g(x) = 0$. (d) Solve $g(x) > 20$.
 (e) Solve $g(x) \le 60$. (f) Solve $0 < g(x) < 60$.

33. In parts (a) and (b) use the following figure.

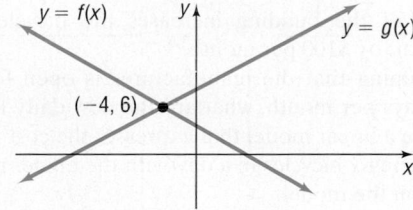

 (a) Solve the equation: $f(x) = g(x)$.
 (b) Solve the inequality: $f(x) > g(x)$.

34. In parts (a) and (b), use the following figure.

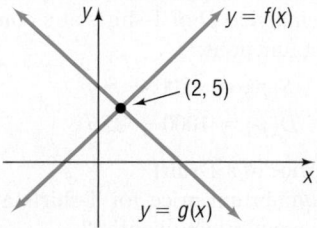

 (a) Solve the equation: $f(x) = g(x)$.
 (b) Solve the inequality: $f(x) \le g(x)$.

35. In parts (a) and (b), use the following figure.

 (a) Solve the equation: $f(x) = g(x)$.
 (b) Solve the inequality: $g(x) \le f(x) < h(x)$.

36. In parts (a) and (b), use the following figure.

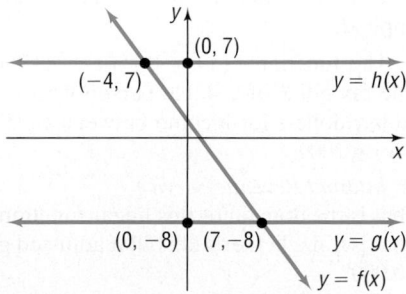

 (a) Solve the equation: $f(x) = g(x)$.
 (b) Solve the inequality: $g(x) < f(x) \le h(x)$.

37. Car Rentals The cost C, in dollars, of renting a moving truck for a day is modeled by the function $C(x) = 0.25x + 35$, where x is the number of miles driven.
 (a) What is the cost if you drive $x = 40$ miles?
 (b) If the cost of renting the moving truck is \$80, how many miles did you drive?
 (c) Suppose that you want the cost to be no more than \$100. What is the maximum number of miles that you can drive?
 (d) What is the implied domain of C?

38. Phone Charges The monthly cost C, in dollars, for international calls on a certain cellular phone plan is modeled by the function $C(x) = 0.38x + 5$, where x is the number of minutes used.
(a) What is the cost if you talk on the phone for $x = 50$ minutes?
(b) Suppose that your monthly bill is $29.32. How many minutes did you use the phone?
(c) Suppose that you budget yourself $60 per month for the phone. What is the maximum number of minutes that you can talk?
(d) What is the implied domain of C if there are 30 days in the month?

39. Supply and Demand Suppose that the quantity supplied S and quantity demanded D of T-shirts at a concert are given by the following functions:

$$S(p) = -200 + 50p$$
$$D(p) = 1000 - 25p$$

where p is the price of a T-shirt.
(a) Find the equilibrium price for T-shirts at this concert. What is the equilibrium quantity?
(b) Determine the prices for which quantity demanded is greater than quantity supplied.
(c) What do you think will eventually happen to the price of T-shirts if quantity demanded is greater than quantity supplied?

40. Supply and Demand Suppose that the quantity supplied S and quantity demanded D of hot dogs at a baseball game are given by the following functions:

$$S(p) = -2000 + 3000p$$
$$D(p) = 10,000 - 1000p$$

where p is the price of a hot dog.
(a) Find the equilibrium price for hot dogs at the baseball game. What is the equilibrium quantity?
(b) Determine the prices for which quantity demanded is less than quantity supplied.
(c) What do you think will eventually happen to the price of hot dogs if quantity demanded is less than quantity supplied?

41. Taxes The function $T(x) = 0.15(x - 8350) + 835$ represents the tax bill T of a single person whose adjusted gross income is x dollars for income between $8350 and $33,950, inclusive, in 2009.
Source: Internal Revenue Service
(a) What is the domain of this linear function?
(b) What is a single filer's tax bill if adjusted gross income is $20,000?
(c) Which variable is independent and which is dependent?
(d) Graph the linear function over the domain specified in part (a).
(e) What is a single filer's adjusted gross income if the tax bill is $3707.50?

42. Luxury Tax In 2002, major league baseball signed a labor agreement with the players. In this agreement, any team whose payroll exceeded $136.5 million in 2006 had to pay a luxury tax of 40% (for second offenses). The linear function $T(p) = 0.40(p - 136.5)$ describes the luxury tax T of a team whose payroll was p (in millions of dollars).
Source: Major League Baseball

(a) What is the implied domain of this linear function?
(b) What was the luxury tax for the New York Yankees whose 2006 payroll was $171.1 million?
(c) Graph the linear function.
(d) What is the payroll of a team that pays a luxury tax of $11.7 million?

The point at which a company's profits equal zero is called the company's **break-even point**. *For Problems 43 and 44, let R represent a company's revenue, let C represent the company's costs, and let x represent the number of units produced and sold each day.*
(a) *Find the firm's break-even point; that is, find x so that $R = C$.*
(b) *Find the values of x such that $R(x) > C(x)$. This represents the number of units that the company must sell to earn a profit.*

43. $R(x) = 8x$
$C(x) = 4.5x + 17,500$

44. $R(x) = 12x$
$C(x) = 10x + 15,000$

45. Straight-line Depreciation Suppose that a company has just purchased a new computer for $3000. The company chooses to depreciate the computer using the straight-line method over 3 years.
(a) Write a linear model that expresses the book value V of the computer as a function of its age x.
(b) What is the implied domain of the function found in part (a)?
(c) Graph the linear function.
(d) What is the book value of the computer after 2 years?
(e) When will the computer have a book value of $2000?

46. Straight-line Depreciation Suppose that a company has just purchased a new machine for its manufacturing facility for $120,000. The company chooses to depreciate the machine using the straight-line method over 10 years.
(a) Write a linear model that expresses the book value V of the machine as a function of its age x.
(b) What is the implied domain of the function found in part (a)?
(c) Graph the linear function.
(d) What is the book value of the machine after 4 years?
(e) When will the machine have a book value of $72,000?

47. Cost Function The simplest cost function is the linear cost function, $C(x) = mx + b$, where the y-intercept b represents the fixed costs of operating a business and the slope m represents the cost of each item produced. Suppose that a small bicycle manufacturer has daily fixed costs of $1800 and each bicycle costs $90 to manufacture.
(a) Write a linear model that expresses the cost C of manufacturing x bicycles in a day.
(b) Graph the model.
(c) What is the cost of manufacturing 14 bicycles in a day?
(d) How many bicycles could be manufactured for $3780?

48. Cost Function Refer to Problem 47. Suppose that the landlord of the building increases the bicycle manufacturer's rent by $100 per month.
(a) Assuming that the manufacturer is open for business 20 days per month, what are the new daily fixed costs?
(b) Write a linear model that expresses the cost C of manufacturing x bicycles in a day with the higher rent.
(c) Graph the model.
(d) What is the cost of manufacturing 14 bicycles in a day?
(e) How many bicycles can be manufactured for $3780?

49. Truck Rentals A truck rental company rents a truck for one day by charging $29 plus $0.07 per mile.
(a) Write a linear model that relates the cost C, in dollars, of renting the truck to the number x of miles driven.
(b) What is the cost of renting the truck if the truck is driven 110 miles? 230 miles?

50. Long Distance A phone company offers a domestic long distance package by charging $5 plus $0.05 per minute.
(a) Write a linear model that relates the cost C, in dollars, of talking x minutes.
(b) What is the cost of talking 105 minutes? 180 minutes?

Mixed Practice

51. Developing a Linear Model from Data The following data represent the price p and quantity demanded per day q of 24" LCD monitor.

Price, p (in dollars)	Quantity Demanded, q
150	100
200	80
250	60
300	40

(a) Plot the ordered pairs (p, q) in a Cartesian plane.
(b) Show that quantity demanded q is a linear function of the price p.
(c) Determine the linear function that describes the relation between p and q.
(d) What is the implied domain of the linear function?
(e) Graph the linear function in the Cartesian plane drawn in part (a).
(f) Interpret the slope.
(g) Interpret the values of the intercepts.

52. Developing a Linear Model from Data The following data represent the various combinations of soda and hot dogs that Yolanda can buy at a baseball game with $60.

Soda, s	Hot Dogs, h
20	0
15	3
10	6
5	9

(a) Plot the ordered pairs (s, h) in a Cartesian plane.
(b) Show that the number of hot dogs purchased h is a linear function of the number of sodas purchased s.
(c) Determine the linear function that describes the relation between s and h.
(d) What is the implied domain of the linear function?
(e) Graph the linear function in the Cartesian plane drawn in part (a).
(f) Interpret the slope.
(g) Interpret the values of the intercepts.

Explaining Concepts: Discussion and Writing

53. Which of the following functions might have the graph shown? (More than one answer is possible.)
(a) $f(x) = 2x - 7$
(b) $g(x) = -3x + 4$
(c) $H(x) = 5$
(d) $F(x) = 3x + 4$
(e) $G(x) = \dfrac{1}{2}x + 2$

54. Which of the following functions might have the graph shown? (More than one answer is possible.)
(a) $f(x) = 3x + 1$
(b) $g(x) = -2x + 3$
(c) $H(x) = 3$
(d) $F(x) = -4x - 1$
(e) $G(x) = -\dfrac{2}{3}x + 3$

55. Under what circumstances is a linear function $f(x) = mx + b$ odd? Can a linear function ever be even?

56. Explain how the graph of $f(x) = mx + b$ can be used to solve $mx + b > 0$.

'Are You Prepared?' Answers

1.

2. $\dfrac{2}{3}$ **3.** 18 **4.** $\{50\}$ **5.** 0 **6.** True

4.2 Linear Models: Building Linear Functions from Data

PREPARING FOR THIS SECTION *Before getting started, review the following:*

- Rectangular Coordinates (Section 2.1, pp. 150–151)
- Functions (Section 3.1, pp. 200–208)
- Lines (Section 2.3, pp. 167–175)

Now Work the **'Are You Prepared?'** problems on page 285.

OBJECTIVES **1** Draw and Interpret Scatter Diagrams (p. 282)

2 Distinguish between Linear and Nonlinear Relations (p. 283)

 3 Use a Graphing Utility to Find the Line of Best Fit (p. 284)

1 Draw and Interpret Scatter Diagrams

In Section 4.1, we built linear models from verbal descriptions. Linear models can also be constructed by fitting a linear function to data. The first step is to plot the ordered pairs using rectangular coordinates. The resulting graph is called a **scatter diagram.**

EXAMPLE 1 **Drawing and Interpreting a Scatter Diagram**

In baseball, the on-base percentage for a team represents the percentage of time that the players safely reach base. The data given in Table 6 represent the number of runs scored y and the on-base percentage x for teams in the National League during the 2008 baseball season.

Table 6

Team	On-Base Percentage, x	Runs Scored, y	(x, y)
Atlanta	34.5	753	(34.5, 753)
St. Louis	35.0	779	(35.0, 779)
Colorado	33.6	747	(33.6, 747)
Houston	32.3	712	(32.3, 712)
Philadelphia	33.2	799	(33.2, 799)
San Francisco	32.1	640	(32.1, 640)
Pittsburgh	32.0	735	(32.0, 735)
Florida	32.6	770	(32.6, 770)
Chicago Cubs	35.4	855	(35.4, 855)
Arizona	32.7	720	(32.7, 720)
Milwaukee	32.5	750	(32.5, 750)
Washington	32.3	641	(32.3, 641)
Cincinnati	32.1	704	(32.1, 704)
San Diego	31.7	637	(31.7, 637)
NY Mets	34.0	799	(34.0, 799)
Los Angeles	33.3	700	(33.3, 700)

Source: Based on data from http://www.baseball-reference.com. A Sports Reference, LLC, web site.

(a) Draw a scatter diagram of the data, treating on-base percentage as the independent variable.

 (b) Use a graphing utility to draw a scatter diagram.

(c) Describe what happens to runs scored as the on-base percentage increases.

Solution (a) To draw a scatter diagram, plot the ordered pairs listed in Table 6, with the on-base percentage as the *x*-coordinate and the runs scored as the *y*-coordinate. See Figure 6(a). Notice that the points in the scatter diagram are not connected.

(b) Figure 6(b) shows a scatter diagram using a TI-84 Plus graphing calculator.

(c) We see from the scatter diagrams that, as the on-base percentage increases, the trend is that the number of runs scored also increases.

Figure 6

(a)

(b)

Now Work PROBLEM 11(a)

2 Distinguish between Linear and Nonlinear Relations

Notice that the points in Figure 6 do not follow a perfect linear relation (as they do in Figure 3 in Section 4.1). However, the data do exhibit a linear pattern. There are numerous explanations as to why the data are not perfectly linear, but one easy explanation is the fact that other variables besides on-base percentage play a role in determining runs scored, such as number of home runs hit.

Scatter diagrams are used to help us to see the type of relation that exists between two variables. In this text, we will discuss a variety of different relations that may exist between two variables. For now, we concentrate on distinguishing between linear and nonlinear relations. See Figure 7.

Figure 7

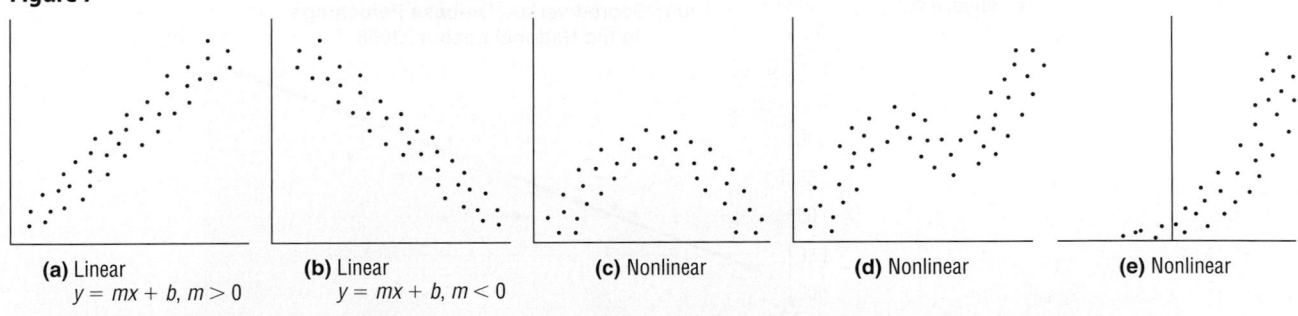

(a) Linear
$y = mx + b, m > 0$

(b) Linear
$y = mx + b, m < 0$

(c) Nonlinear

(d) Nonlinear

(e) Nonlinear

EXAMPLE 2 **Distinguishing between Linear and Nonlinear Relations**

Determine whether the relation between the two variables in Figure 8 is linear or nonlinear.

Figure 8

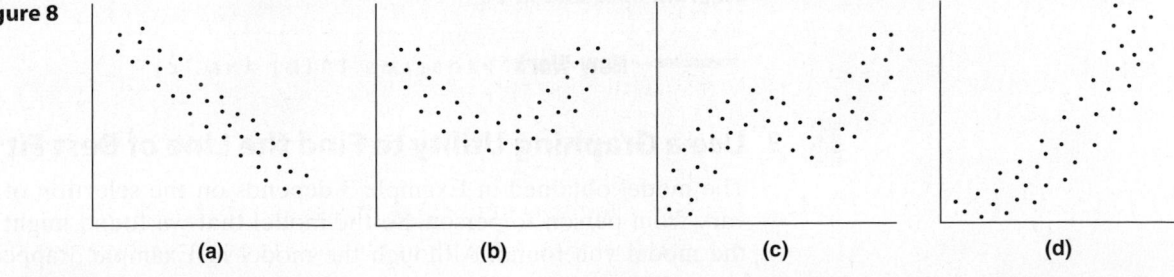

(a)

(b)

(c)

(d)

Solution (a) Linear (b) Nonlinear (c) Nonlinear (d) Nonlinear

 Now Work PROBLEM 5

In this section we study data whose scatter diagrams imply that a linear relation exists between the two variables.

Suppose that the scatter diagram of a set of data appears to be linearly related as in Figure 7(a) or (b). We might want to model the data by finding an equation of a line that relates the two variables. One way to obtain a model for such data is to draw a line through two points on the scatter diagram and determine the equation of the line.

EXAMPLE 3 **Finding a Model for Linearly Related Data**

Use the data in Table 6 from Example 1 to:

(a) Select two points and find an equation of the line containing the points.
(b) Graph the line on the scatter diagram obtained in Example 1(a).

Solution (a) Select two points, say $(32.7, 720)$ and $(35.4, 855)$. The slope of the line joining the points $(32.7, 720)$ and $(35.4, 855)$ is

$$m = \frac{855 - 720}{35.4 - 32.7} = \frac{135}{2.7} = 50$$

The equation of the line with slope 50 and passing through $(32.7, 720)$ is found using the point–slope form with $m = 50$, $x_1 = 32.7$, and $y_1 = 720$.

$$y - y_1 = m(x - x_1) \qquad \text{Point–slope form of a line}$$
$$y - 720 = 50(x - 32.7) \qquad x_1 = 32.7,\ y_1 = 720,\ m = 50$$
$$y - 720 = 50x - 1635$$
$$y = 50x - 915 \qquad \text{The Model}$$

(b) Figure 9 shows the scatter diagram with the graph of the line found in part (a).

Figure 9

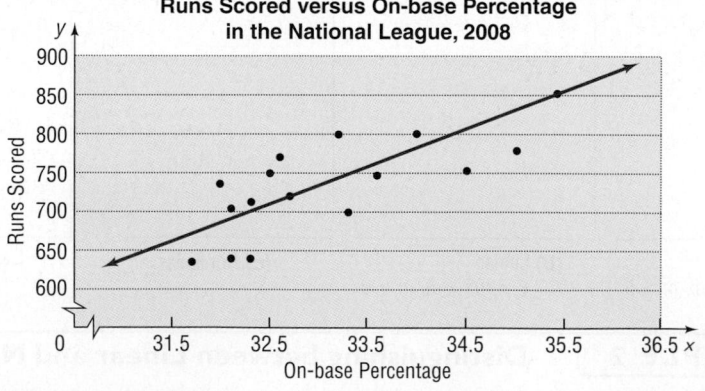

Runs Scored versus On-base Percentage
in the National League, 2008

Select two other points and complete the solution. Graph the line on the scatter diagram obtained in Figure 6.

 Now Work PROBLEMS 11(b) AND (c)

3 Use a Graphing Utility to Find the Line of Best Fit

The model obtained in Example 3 depends on the selection of points, which will vary from person to person. So the model that we found might be different from the model you found. Although the model in Example 3 appears to fit the data

well, there may be a model that "fits it better." Do you think your model fits the data better? Is there a *line of best fit*? As it turns out, there is a method for finding a model that best fits linearly related data (called the **line of best fit**).*

EXAMPLE 4

Finding a Model for Linearly Related Data

Use the data in Table 6 from Example 1.

(a) Use a graphing utility to find the line of best fit that models the relation between on-base percentage and runs scored.

(b) Graph the line of best fit on the scatter diagram obtained in Example 1(b).

(c) Interpret the slope.

(d) Use the line of best fit to predict the number of runs a team will score if their on-base percentage is 34.1.

Solution

Figure 10

(a) Graphing utilities contain built-in programs that find the line of best fit for a collection of points in a scatter diagram. Upon executing the LINear REGression program, we obtain the results shown in Figure 10. The output that the utility provides shows us the equation $y = ax + b$, where a is the slope of the line and b is the y-intercept. The line of best fit that relates on-base percentage to runs scored may be expressed as the line

$$y = 40.85x - 617.66 \quad \text{The Model}$$

Figure 11

(b) Figure 11 shows the graph of the line of best fit, along with the scatter diagram.

(c) The slope of the line of best fit is 40.85, which means that, for every 1 percent increase in the on-base percentage, runs scored increase 40.85, on average.

(d) Letting $x = 34.1$ in the equation of the line of best fit, we obtain $y = 40.85(34.1) - 617.66 \approx 775$ runs.

Now Work PROBLEMS 11(d) AND (e)

Does the line of best fit appear to be a good fit? In other words, does the line appear to accurately describe the relation between on-base percentage and runs scored? And just how "good" is this line of best fit? Look again at Figure 10. The last line of output is $r = 0.751$. This number, called the **correlation coefficient**, r, $-1 \leq r \leq 1$, is a measure of the strength of the linear relation that exists between two variables. The closer that $|r|$ is to 1, the more perfect the linear relationship is. If r is close to 0, there is little or no linear relationship between the variables. A negative value of r, $r < 0$, indicates that as x increases y decreases; a positive value of r, $r > 0$, indicates that as x increases y does also. The data given in Table 6, having a correlation coefficient of 0.751, are indicative of a linear relationship with positive slope.

4.2 Assess Your Understanding

'Are You Prepared?' *Answers are given at the end of these exercises. If you get a wrong answer, read the pages listed in red.*

1. Plot the points $(1, 5)$, $(2, 6)$, $(3, 9)$, $(1, 12)$ in the Cartesian plane. Is the relation $\{(1, 5), (2, 6), (3, 9), (1, 12)\}$ a function? Why? (pp. 150 and 200–208)

2. Find an equation of the line containing the points $(1, 4)$ and $(3, 8)$. (pp. 167–175)

Concepts and Vocabulary

3. A _____ is used to help us to see the type of relation, if any, that may exist between two variables.

4. *True or False* The correlation coefficient is a measure of the strength of a linear relation between two variables and must lie between -1 and 1, inclusive.

* We shall not discuss the underlying mathematics of lines of best fit in this book.

Skill Building

In Problems 5–10, examine the scatter diagram and determine whether the type of relation is linear or nonlinear.

5.

6.

7.

8.

9.

10.

In Problems 11–16,

(a) *Draw a scatter diagram.*
(b) *Select two points from the scatter diagram and find the equation of the line containing the points selected.*
(c) *Graph the line found in part (b) on the scatter diagram.*
(d) *Use a graphing utility to find the line of best fit.*
(e) *Use a graphing utility to draw the scatter diagram and graph the line of best fit on it.*

11.

x	3	4	5	6	7	8	9
y	4	6	7	10	12	14	16

12.

x	3	5	7	9	11	13
y	0	2	3	6	9	11

13.

x	−2	−1	0	1	2
y	−4	0	1	4	5

14.

x	−2	−1	0	1	2
y	7	6	3	2	0

15.

x	−20	−17	−15	−14	−10
y	100	120	118	130	140

16.

x	−30	−27	−25	−20	−14
y	10	12	13	13	18

Applications and Extensions

17. Candy The following data represent the weight (in grams) of various candy bars and the corresponding number of calories.

Candy Bar	Weight, x	Calories, y
Hershey's Milk Chocolate®	44.28	230
Nestle's Crunch®	44.84	230
Butterfinger®	61.30	270
Baby Ruth®	66.45	280
Almond Joy®	47.33	220
Twix® (with Caramel)	58.00	280
Snickers®	61.12	280
Heath®	39.52	210

Source: Megan Pocius, Student at Joliet Junior College

(a) Draw a scatter diagram of the data treating weight as the independent variable.
(b) What type of relation appears to exist between the weight of a candy bar and the number of calories?
(c) Select two points and find a linear model that contains the points.

(d) Graph the line on the scatter diagram drawn in part (a).
(e) Use the linear model to predict the number of calories in a candy bar that weighs 62.3 grams.
(f) Interpret the slope of the line found in part (c).

18. Raisins The following data represent the weight (in grams) of a box of raisins and the number of raisins in the box.

Weight (in grams), w	Number of Raisins, N
42.3	87
42.7	91
42.8	93
42.4	87
42.6	89
42.4	90
42.3	82
42.5	86
42.7	86
42.5	86

Source: Jennifer Maxwell, Student at Joliet Junior College

(a) Draw a scatter diagram of the data treating weight as the independent variable.
(b) What type of relation appears to exist between the weight of a box of raisins and the number of raisins?
(c) Select two points and find a linear model that contains the points.
(d) Graph the line on the scatter diagram drawn in part (b).
(e) Use the linear model to predict the number of raisins in a box that weighs 42.5 grams.
(f) Interpret the slope of the line found in part (c).

19. **Video Games and Grade-Point Average** Professor Grant Alexander wanted to find a linear model that relates the number of hours a student plays video games each week, h, to the cumulative grade-point average, G, of the student. He obtained a random sample of 10 full-time students at his college and asked each student to disclose the number of hours spent playing video games and the student's cumulative grade-point average.

Hours of Video Games per Week, h	Grade-point Average, G
0	3.49
0	3.05
2	3.24
3	2.82
3	3.19
5	2.78
8	2.31
8	2.54
10	2.03
12	2.51

(a) Explain why the number of hours spent playing video games is the independent variable and cumulative grade-point average is the dependent variable.
(b) Use a graphing utility to draw a scatter diagram.
(c) Use a graphing utility to find the line of best fit that models the relation between number of hours of video game playing each week and grade-point average. Express the model using function notation.
(d) Interpret the slope.
(e) Predict the grade-point average of a student who plays video games for 8 hours each week.
(f) How many hours of video game playing do you think a student plays whose grade-point average is 2.40?

20. **Height versus Head Circumference** A pediatrician wanted to find a linear model that relates a child's height, H, to head circumference, C. She randomly selects nine children from her practice, measures their height and head circumference, and obtains the data shown. Let H represent the independent variable and C the dependent variable.
(a) Use a graphing utility to draw a scatter diagram.
(b) Use a graphing utility to find the line of best fit that models the relation between height and head circumference. Express the model using function notation.
(c) Interpret the slope.
(d) Predict the head circumference of a child that is 26 inches tall.

(e) What is the height of a child whose head circumference is 17.4 inches?

Height, H (inches)	Head Circumference, C (inches)
25.25	16.4
25.75	16.9
25	16.9
27.75	17.6
26.5	17.3
27	17.5
26.75	17.3
26.75	17.5
27.5	17.5

Source: Denise Slucki, Student at Joliet Junior College

21. **Demand for Jeans** The marketing manager at Levi-Strauss wishes to find a function that relates the demand D for men's jeans and p, the price of the jeans. The following data were obtained based on a price history of the jeans.

Price ($/Pair), p	Demand (Pairs of Jeans Sold per Day), D
20	60
22	57
23	56
23	53
27	52
29	49
30	44

(a) Does the relation defined by the set of ordered pairs (p, D) represent a function?
(b) Draw a scatter diagram of the data.
(c) Using a graphing utility, find the line of best fit that models the relation between price and quantity demanded.
(d) Interpret the slope.
(e) Express the relationship found in part (c) using function notation.
(f) What is the domain of the function?
(g) How many jeans will be demanded if the price is $28 a pair?

22. **Advertising and Sales Revenue** A marketing firm wishes to find a function that relates the sales S of a product and A, the amount spent on advertising the product. The data are obtained from past experience. Advertising and sales are measured in thousands of dollars.

Advertising Expenditures, A	Sales, S
20	335
22	339
22.5	338
24	343
24	341
27	350
28.3	351

(a) Does the relation defined by the set of ordered pairs (A, S) represent a function?
(b) Draw a scatter diagram of the data.
(c) Using a graphing utility, find the line of best fit that models the relation between advertising expenditures and sales.
(d) Interpret the slope.
(e) Express the relationship found in part (c) using function notation.
(f) What is the domain of the function?
(g) Predict sales if advertising expenditures are $25,000.

Explaining Concepts: Discussion and Writing

23. **Maternal Age versus Down Syndrome** A biologist would like to know how the age of the mother affects the incidence rate of Down syndrome. The data to the right represent the age of the mother and the incidence rate of Down syndrome per 1000 pregnancies.

 Draw a scatter diagram treating age of the mother as the independent variable. Would it make sense to find the line of best fit for these data? Why or why not?

24. Find the line of best fit for the ordered pairs $(1, 5)$ and $(3, 8)$. What is the correlation coefficient for these data? Why is this result reasonable?

25. What does a correlation coefficient of 0 imply?

26. Explain why it does not make sense to interpret the y-intercept in Problem 17.

27. Refer to Problem 19. Solve $G(h) = 0$. Provide an interpretation of this result. Find $G(0)$. Provide an interpretation of this result.

Age of Mother, x	Incidence of Down Syndrome, y
33	2.4
34	3.1
35	4
36	5
37	6.7
38	8.3
39	10
40	13.3
41	16.7
42	22.2
43	28.6
44	33.3
45	50

Source: Hook, E.B., *Journal of the American Medical Association,* 249, 2034-2038, 1983.

'Are You Prepared?' Answers

1.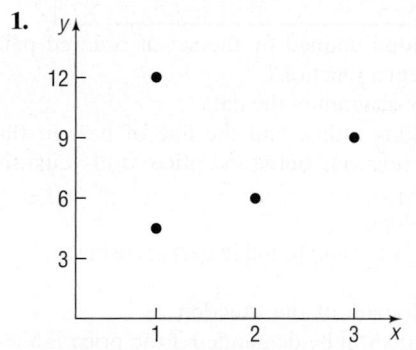

No, because the input, 1, corresponds to two different outputs.

2. $y = 2x + 2$

4.3 Quadratic Functions and Their Properties

PREPARING FOR THIS SECTION *Before getting started, review the following:*

- Intercepts (Section 2.2, pp. 159–160)
- Graphing Techniques: Transformations (Section 3.5, pp. 244–253)
- Completing the Square (Section R.5, p. 56)
- Quadratic Equations (Section 1.2, pp. 92–99)

Now Work the 'Are You Prepared?' problems on page 297.

OBJECTIVES
1 Graph a Quadratic Function Using Transformations (p. 290)
2 Identify the Vertex and Axis of Symmetry of a Quadratic Function (p. 292)
3 Graph a Quadratic Function Using Its Vertex, Axis, and Intercepts (p. 292)
4 Find a Quadratic Function Given Its Vertex and One Other Point (p. 295)
5 Find the Maximum or Minimum Value of a Quadratic Function (p. 296)

Quadratic Functions

Here are some examples of quadratic functions.

$$F(x) = 3x^2 - 5x + 1 \quad g(x) = -6x^2 + 1 \quad H(x) = \frac{1}{2}x^2 + \frac{2}{3}x$$

DEFINITION

A **quadratic function** is a function of the form

$$f(x) = ax^2 + bx + c$$

where a, b, and c are real numbers and $a \neq 0$. The domain of a quadratic function is the set of all real numbers.

In Words

A quadratic function is a function defined by a second-degree polynomial in one variable.

Many applications require a knowledge of quadratic functions. For example, suppose that Texas Instruments collects the data shown in Table 7, which relate the number of calculators sold to the price p (in dollars) per calculator. Since the price of a product determines the quantity that will be purchased, we treat price as the independent variable. The relationship between the number x of calculators sold and the price p per calculator is given by the linear equation

$$x = 21,000 - 150p$$

Table 7

Price per Calculator, p (Dollars)	Number of Calculators, x
60	12,000
65	11,250
70	10,500
75	9,750
80	9,000
85	8,250
90	7,500

Then the revenue R derived from selling x calculators at the price p per calculator is equal to the unit selling price p of the calculator times the number x of units actually sold. That is,

$$R = xp$$
$$R(p) = (21,000 - 150p)p \qquad x = 21,000 - 150p$$
$$= -150p^2 + 21,000p$$

So the revenue R is a quadratic function of the price p. Figure 12 illustrates the graph of this revenue function, whose domain is $0 \leq p \leq 140$, since both x and p must be nonnegative.

Figure 12

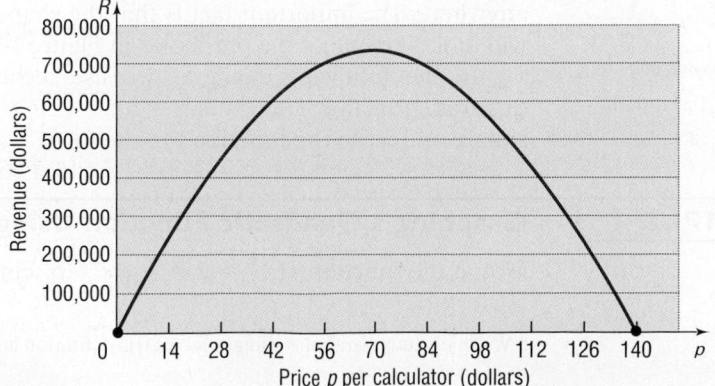

Price p per calculator (dollars)

Figure 13
Path of a cannonball

A second situation in which a quadratic function appears involves the motion of a projectile. Based on Newton's Second Law of Motion (force equals mass times acceleration, $F = ma$), it can be shown that, ignoring air resistance, the path of a projectile propelled upward at an inclination to the horizontal is the graph of a quadratic function. See Figure 13 for an illustration.

1 Graph a Quadratic Function Using Transformations

We know how to graph the square function $f(x) = x^2$. Figure 14 shows the graph of three functions of the form $f(x) = ax^2, a > 0$, for $a = 1$, $a = \dfrac{1}{2}$, and $a = 3$. Notice that the larger the value of a, the "narrower" the graph is, and the smaller the value of a, the "wider" the graph is.

Figure 15 shows the graphs of $f(x) = ax^2$ for $a < 0$. Notice that these graphs are reflections about the x-axis of the graphs in Figure 14. Based on the results of these two figures, we can draw some general conclusions about the graph of $f(x) = ax^2$. First, as $|a|$ increases, the graph becomes "taller" (a vertical stretch), and as $|a|$ gets closer to zero, the graph gets "shorter" (a vertical compression). Second, if a is positive, the graph opens "up," and if a is negative, the graph opens "down."

Figure 14 **Figure 15**

The graphs in Figures 14 and 15 are typical of the graphs of all quadratic functions, which we call **parabolas.*** Refer to Figure 16, where two parabolas are pictured. The one on the left **opens up** and has a lowest point; the one on the right **opens down** and has a highest point. The lowest or highest point of a parabola is called the **vertex.** The vertical line passing through the vertex in each parabola in Figure 16 is called the **axis of symmetry** (usually abbreviated to **axis**) of the parabola. Because the parabola is symmetric about its axis, the axis of symmetry of a parabola can be used to find additional points on the parabola.

Figure 16

Graphs of a quadratic function,
$f(x) = ax^2 + bx + c, a \neq 0$

Axis of symmetry Vertex is highest point

Vertex is lowest point Axis of symmetry

(a) Opens up **(b)** Opens down
$a > 0$ $a < 0$

The parabolas shown in Figure 16 are the graphs of a quadratic function $f(x) = ax^2 + bx + c, a \neq 0$. Notice that the coordinate axes are not included in the figure. Depending on the values of a, b, and c, the axes could be placed anywhere. The important fact is that the shape of the graph of a quadratic function will look like one of the parabolas in Figure 16.

In the following example, we use techniques from Section 3.5 to graph a quadratic function $f(x) = ax^2 + bx + c, a \neq 0$. In so doing, we shall complete the square and write the function f in the form $f(x) = a(x - h)^2 + k$.

<hr>

EXAMPLE 1 **Graphing a Quadratic Function Using Transformations**

Graph the function $f(x) = 2x^2 + 8x + 5$. Find the vertex and axis of symmetry.

<hr>

* We shall study parabolas using a geometric definition later in this book.

Solution Begin by completing the square on the right side.

$$f(x) = 2x^2 + 8x + 5$$

$$= 2(x^2 + 4x) + 5 \qquad \text{Factor out the 2 from } 2x^2 + 8x.$$

$$= 2(x^2 + 4x + 4) + 5 - 8 \qquad \text{Complete the square of } x^2 + 4x \text{ by adding 4.}$$
$$\text{Notice that the factor of 2 requires that 8 be}$$
$$= 2(x + 2)^2 - 3 \qquad \text{added and subtracted.}$$

The graph of f can be obtained from the graph of $y = x^2$ in three stages, as shown in Figure 17. Now compare this graph to the graph in Figure 16(a). The graph of $f(x) = 2x^2 + 8x + 5$ is a parabola that opens up and has its vertex (lowest point) at $(-2, -3)$. Its axis of symmetry is the line $x = -2$.

Figure 17

(a) $y = x^2$ Multiply by 2; Vertical stretch (b) $y = 2x^2$ Replace x by $x + 2$; Shift left 2 units (c) $y = 2(x + 2)^2$ Subtract 3; Shift down 3 units (d) $y = 2(x + 2)^2 - 3$

Now Work PROBLEM 23

The method used in Example 1 can be used to graph any quadratic function $f(x) = ax^2 + bx + c, a \neq 0$, as follows:

$$f(x) = ax^2 + bx + c$$

$$= a\left(x^2 + \frac{b}{a}x\right) + c \qquad \text{Factor out } a \text{ from } ax^2 + bx.$$

$$= a\left(x^2 + \frac{b}{a}x + \frac{b^2}{4a^2}\right) + c - a\left(\frac{b^2}{4a^2}\right) \qquad \text{Complete the square by adding } \frac{b^2}{4a^2}.$$
$$\text{Look closely at this step!}$$

$$= a\left(x + \frac{b}{2a}\right)^2 + c - \frac{b^2}{4a} \qquad \text{Factor.}$$

$$= a\left(x + \frac{b}{2a}\right)^2 + \frac{4ac - b^2}{4a} \qquad c - \frac{b^2}{4a} = c \cdot \frac{4a}{4a} - \frac{b^2}{4a} = \frac{4ac - b^2}{4a}$$

Based on these results, we conclude the following:

If $h = -\dfrac{b}{2a}$ and $k = \dfrac{4ac - b^2}{4a}$, then

$$\boxed{f(x) = ax^2 + bx + c = a(x - h)^2 + k \qquad \qquad (1)}$$

The graph of $f(x) = a(x - h)^2 + k$ is the parabola $y = ax^2$ shifted horizontally h units (replace x by $x - h$) and vertically k units (add k). As a result, the vertex is at (h, k), and the graph opens up if $a > 0$ and down if $a < 0$. The axis of symmetry is the vertical line $x = h$.

For example, compare equation (1) with the solution given in Example 1.

$$f(x) = 2(x + 2)^2 - 3$$
$$= 2(x - (-2))^2 + (-3)$$
$$= a(x - h)^2 + k$$

We conclude that $a = 2$, so the graph opens up. Also, we find that $h = -2$ and $k = -3$, so its vertex is at $(-2, -3)$.

2 Identify the Vertex and Axis of Symmetry of a Quadratic Function

We do not need to complete the square to obtain the vertex. In almost every case, it is easier to obtain the vertex of a quadratic function f by remembering that its x-coordinate is $h = -\dfrac{b}{2a}$. The y-coordinate k can then be found by evaluating f at $-\dfrac{b}{2a}$. That is, $k = f\left(-\dfrac{b}{2a}\right)$.

Properties of the Graph of a Quadratic Function

$$f(x) = ax^2 + bx + c \qquad a \neq 0$$

$$\text{Vertex} = \left(-\frac{b}{2a}, f\left(-\frac{b}{2a}\right)\right) \quad \text{Axis of symmetry: the line } x = -\frac{b}{2a} \quad \textbf{(2)}$$

Parabola opens up if $a > 0$; the vertex is a minimum point.
Parabola opens down if $a < 0$; the vertex is a maximum point.

EXAMPLE 2

Locating the Vertex without Graphing

Without graphing, locate the vertex and axis of symmetry of the parabola defined by $f(x) = -3x^2 + 6x + 1$. Does it open up or down?

Solution

For this quadratic function, $a = -3$, $b = 6$, and $c = 1$. The x-coordinate of the vertex is

$$h = -\frac{b}{2a} = -\frac{6}{-6} = 1$$

The y-coordinate of the vertex is

$$k = f\left(-\frac{b}{2a}\right) = f(1) = -3 + 6 + 1 = 4$$

The vertex is located at the point $(1, 4)$. The axis of symmetry is the line $x = 1$. Because $a = -3 < 0$, the parabola opens down.　♩

3 Graph a Quadratic Function Using Its Vertex, Axis, and Intercepts

The location of the vertex and intercepts of a quadratic function, $f(x) = ax^2 + bx + c$, $a \neq 0$, along with knowledge as to whether the graph opens up or down, usually provides enough information to graph it.

The y-intercept is the value of f at $x = 0$; that is, the y-intercept is $f(0) = c$.

The x-intercepts, if there are any, are found by solving the quadratic equation

$$ax^2 + bx + c = 0$$

This equation has two, one, or no real solutions, depending on whether the discriminant $b^2 - 4ac$ is positive, 0, or negative. Depending on the value of the discriminant, the graph of f has x-intercepts, as follows:

The x-Intercepts of a Quadratic Function

1. If the discriminant $b^2 - 4ac > 0$, the graph of $f(x) = ax^2 + bx + c$ has two distinct x-intercepts so it crosses the x-axis in two places.

2. If the discriminant $b^2 - 4ac = 0$, the graph of $f(x) = ax^2 + bx + c$ has one x-intercept so it touches the x-axis at its vertex.

3. If the discriminant $b^2 - 4ac < 0$, the graph of $f(x) = ax^2 + bx + c$ has no x-intercepts so it does not cross or touch the x-axis.

Figure 18 illustrates these possibilities for parabolas that open up.

Figure 18
$f(x) = ax^2 + bx + c, \ a > 0$

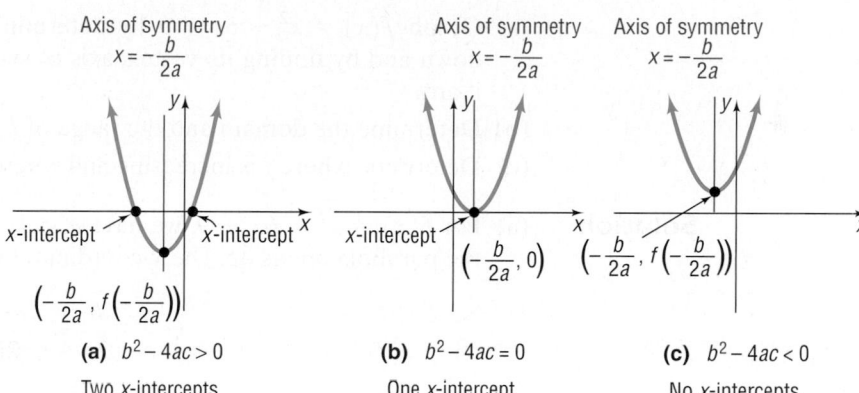

(a) $b^2 - 4ac > 0$
Two x-intercepts

(b) $b^2 - 4ac = 0$
One x-intercept

(c) $b^2 - 4ac < 0$
No x-intercepts

EXAMPLE 3

Graphing a Quadratic Function Using Its Vertex, Axis, and Intercepts

(a) Use the information from Example 2 and the locations of the intercepts to graph $f(x) = -3x^2 + 6x + 1$.

(b) Determine the domain and the range of f.

(c) Determine where f is increasing and where it is decreasing.

Solution

(a) In Example 2, we found the vertex to be at $(1, 4)$ and the axis of symmetry to be $x = 1$. The y-intercept is found by letting $x = 0$. The y-intercept is $f(0) = 1$. The x-intercepts are found by solving the equation $f(x) = 0$. This results in the equation

$$-3x^2 + 6x + 1 = 0 \quad a = -3, b = 6, c = 1$$

The discriminant $b^2 - 4ac = (6)^2 - 4(-3)(1) = 36 + 12 = 48 > 0$, so the equation has two real solutions and the graph has two x-intercepts. Using the quadratic formula, we find that

$$x = \frac{-b + \sqrt{b^2 - 4ac}}{2a} = \frac{-6 + \sqrt{48}}{-6} = \frac{-6 + 4\sqrt{3}}{-6} \approx -0.15$$

and

$$x = \frac{-b - \sqrt{b^2 - 4ac}}{2a} = \frac{-6 - \sqrt{48}}{-6} = \frac{-6 - 4\sqrt{3}}{-6} \approx 2.15$$

The x-intercepts are approximately -0.15 and 2.15.

The graph is illustrated in Figure 19. Notice how we used the y-intercept and the axis of symmetry, $x = 1$, to obtain the additional point $(2, 1)$ on the graph.

Figure 19

(b) The domain of f is the set of all real numbers. Based on the graph, the range of f is the interval $(-\infty, 4]$.

(c) The function f is increasing on the interval $(-\infty, 1)$ and decreasing on the interval $(1, \infty)$.

 Graph the function in Example 3 by completing the square and using transformations. Which method do you prefer?

Now Work PROBLEM 31

If the graph of a quadratic function has only one x-intercept or no x-intercepts, it is usually necessary to plot an additional point to obtain the graph.

EXAMPLE 4 **Graphing a Quadratic Function Using Its Vertex, Axis, and Intercepts**

(a) Graph $f(x) = x^2 - 6x + 9$ by determining whether the graph opens up or down and by finding its vertex, axis of symmetry, y-intercept, and x-intercepts, if any.

(b) Determine the domain and the range of f.

(c) Determine where f is increasing and where it is decreasing.

Solution (a) For $f(x) = x^2 - 6x + 9$, we have $a = 1$, $b = -6$, and $c = 9$. Since $a = 1 > 0$, the parabola opens up. The x-coordinate of the vertex is

$$h = -\frac{b}{2a} = -\frac{-6}{2(1)} = 3$$

Figure 20

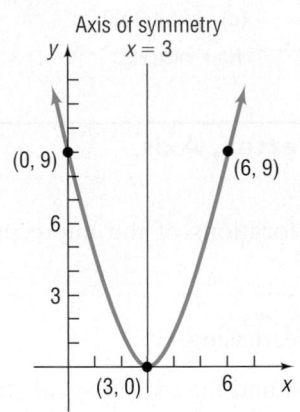

The y-coordinate of the vertex is

$$k = f(3) = (3)^2 - 6(3) + 9 = 0$$

So the vertex is at $(3, 0)$. The axis of symmetry is the line $x = 3$. The y-intercept is $f(0) = 9$. Since the vertex $(3, 0)$ lies on the x-axis, the graph touches the x-axis at the x-intercept. By using the axis of symmetry and the y-intercept at $(0, 9)$, we can locate the additional point $(6, 9)$ on the graph. See Figure 20.

(b) The domain of f is the set of all real numbers. Based on the graph, the range of f is the interval $[0, \infty)$.

(c) The function f is decreasing on the interval $(-\infty, 3)$ and increasing on the interval $(3, \infty)$.

Now Work PROBLEM 37

EXAMPLE 5 **Graphing a Quadratic Function Using Its Vertex, Axis, and Intercepts**

(a) Graph $f(x) = 2x^2 + x + 1$ by determining whether the graph opens up or down and by finding its vertex, axis of symmetry, y-intercept, and x-intercepts, if any.

(b) Determine the domain and the range of f.

(c) Determine where f is increasing and where it is decreasing.

Solution (a) For $f(x) = 2x^2 + x + 1$, we have $a = 2$, $b = 1$, and $c = 1$. Since $a = 2 > 0$, the parabola opens up. The x-coordinate of the vertex is

$$h = -\frac{b}{2a} = -\frac{1}{4}$$

NOTE In Example 5, since the vertex is above the x-axis and the parabola opens up, we can conclude that the graph of the quadratic function will have no x-intercepts. ∎

The y-coordinate of the vertex is

$$k = f\left(-\frac{1}{4}\right) = 2\left(\frac{1}{16}\right) + \left(-\frac{1}{4}\right) + 1 = \frac{7}{8}$$

So the vertex is at $\left(-\frac{1}{4}, \frac{7}{8}\right)$. The axis of symmetry is the line $x = -\frac{1}{4}$. The y-intercept is $f(0) = 1$. The x-intercept(s), if any, obey the equation $2x^2 + x + 1 = 0$. Since the discriminant $b^2 - 4ac = (1)^2 - 4(2)(1) = -7 < 0$, this equation has no real solutions, and therefore the graph has no x-intercepts.

We use the point $(0, 1)$ and the axis of symmetry $x = -\frac{1}{4}$ to locate the additional point $\left(-\frac{1}{2}, 1\right)$ on the graph. See Figure 21.

Figure 21

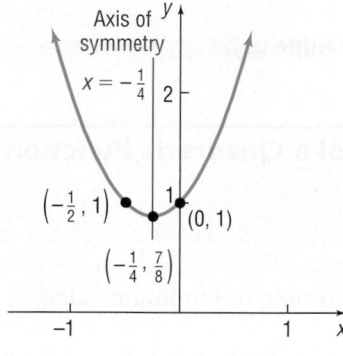

(b) The domain of f is the set of all real numbers. Based on the graph, the range of f is the interval $\left[\frac{7}{8}, \infty\right)$.

(c) The function f is decreasing on the interval $\left(-\infty, -\frac{1}{4}\right)$ and is increasing on the interval $\left(-\frac{1}{4}, \infty\right)$.

━━━ **Now Work** PROBLEM 41

4 Find a Quadratic Function Given Its Vertex and One Other Point

Given the vertex (h, k) and one additional point on the graph of a quadratic function $f(x) = ax^2 + bx + c, a \neq 0$, we can use

$$f(x) = a(x - h)^2 + k \qquad \textbf{(3)}$$

to obtain the quadratic function.

EXAMPLE 6

Finding the Quadratic Function Given Its Vertex and One Other Point

Determine the quadratic function whose vertex is $(1, -5)$ and whose y-intercept is -3. The graph of the parabola is shown in Figure 22.

Solution

The vertex is $(1, -5)$, so $h = 1$ and $k = -5$. Substitute these values into equation (3).

$$f(x) = a(x - h)^2 + k \quad \text{Equation (3)}$$
$$f(x) = a(x - 1)^2 - 5 \quad h = 1, k = -5$$

Figure 22

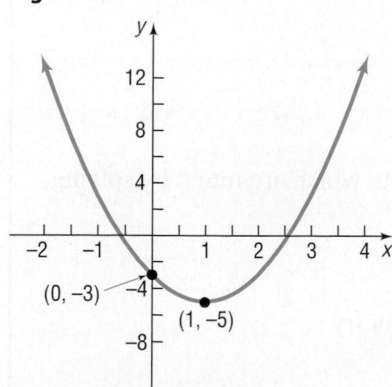

To determine the value of a, we use the fact that $f(0) = -3$ (the y-intercept).

$$f(x) = a(x - 1)^2 - 5$$
$$-3 = a(0 - 1)^2 - 5 \quad x = 0, y = f(0) = -3$$
$$-3 = a - 5$$
$$a = 2$$

The quadratic function whose graph is shown in Figure 22 is

$$f(x) = a(x - h)^2 + k = 2(x - 1)^2 - 5 = 2x^2 - 4x - 3$$

━━━ **Now Work** PROBLEM 47

5 Find the Maximum or Minimum Value of a Quadratic Function

The graph of a quadratic function

$$f(x) = ax^2 + bx + c \qquad a \neq 0$$

is a parabola with vertex at $\left(-\dfrac{b}{2a}, f\left(-\dfrac{b}{2a}\right)\right)$. This vertex is the highest point on the graph if $a < 0$ and the lowest point on the graph if $a > 0$. If the vertex is the highest point $(a < 0)$, then $f\left(-\dfrac{b}{2a}\right)$ is the **maximum value** of f. If the vertex is the lowest point $(a > 0)$, then $f\left(-\dfrac{b}{2a}\right)$ is the **minimum value** of f.

EXAMPLE 7 | **Finding the Maximum or Minimum Value of a Quadratic Function**

Determine whether the quadratic function

$$f(x) = x^2 - 4x - 5$$

has a maximum or minimum value. Then find the maximum or minimum value.

Solution Compare $f(x) = x^2 - 4x - 5$ to $f(x) = ax^2 + bx + c$. Then $a = 1$, $b = -4$, and $c = -5$. Since $a > 0$, the graph of f opens up, so the vertex is a minimum point. The minimum value occurs at

$$x = -\frac{b}{2a} = -\frac{-4}{2(1)} = \frac{4}{2} = 2$$

$$\uparrow$$
$$a = 1, b = -4$$

The minimum value is

$$f\left(-\frac{b}{2a}\right) = f(2) = 2^2 - 4(2) - 5 = 4 - 8 - 5 = -9$$

♩

━━━━▶ **Now Work** PROBLEM **55**

SUMMARY Steps for Graphing a Quadratic Function $f(x) = ax^2 + bx + c, a \neq 0$

Option 1

STEP 1: Complete the square in x to write the quadratic function in the form $f(x) = a(x - h)^2 + k$.

STEP 2: Graph the function in stages using transformations.

Option 2

STEP 1: Determine whether the parabola opens up $(a > 0)$ or down $(a < 0)$.

STEP 2: Determine the vertex $\left(-\dfrac{b}{2a}, f\left(-\dfrac{b}{2a}\right)\right)$.

STEP 3: Determine the axis of symmetry, $x = -\dfrac{b}{2a}$.

STEP 4: Determine the y-intercept, $f(0)$, and the x-intercepts, if any.
 (a) If $b^2 - 4ac > 0$, the graph of the quadratic function has two x-intercepts, which are found by solving the equation $ax^2 + bx + c = 0$.
 (b) If $b^2 - 4ac = 0$, the vertex is the x-intercept.
 (c) If $b^2 - 4ac < 0$, there are no x-intercepts.

STEP 5: Determine an additional point by using the y-intercept and the axis of symmetry.

STEP 6: Plot the points and draw the graph.

4.3 Assess Your Understanding

'Are You Prepared?' *Answers are given at the end of these exercises. If you get a wrong answer, read the pages listed in red.*

1. List the intercepts of the equation $y = x^2 - 9$. (pp. 159–160)
2. Find the real solutions of the equation $2x^2 + 7x - 4 = 0$. (pp. 92–99)
3. To complete the square of $x^2 - 5x$, you add the number _____ . (p. 56)
4. To graph $y = (x - 4)^2$, you shift the graph of $y = x^2$ to the _____ a distance of _____ units. (pp. 244–253)

Concepts and Vocabulary

5. The graph of a quadratic function is called a(n) _____ .
6. The vertical line passing through the vertex of a parabola is called the _____ .
7. The *x*-coordinate of the vertex of $f(x) = ax^2 + bx + c$, $a \neq 0$, is _____ .
8. *True or False* The graph of $f(x) = 2x^2 + 3x - 4$ opens up.
9. *True or False* The *y*-coordinate of the vertex of $f(x) = -x^2 + 4x + 5$ is $f(2)$.
10. *True or False* If the discriminant $b^2 - 4ac = 0$, the graph of $f(x) = ax^2 + bx + c, a \neq 0$, will touch the *x*-axis at its vertex.

Skill Building

In Problems 11–18, match each graph to one the following functions.

11. $f(x) = x^2 - 1$
12. $f(x) = -x^2 - 1$
13. $f(x) = x^2 - 2x + 1$
14. $f(x) = x^2 + 2x + 1$

15. $f(x) = x^2 - 2x + 2$
16. $f(x) = x^2 + 2x$
17. $f(x) = x^2 - 2x$
18. $f(x) = x^2 + 2x + 2$

A

B

C

D

E

F

G

H
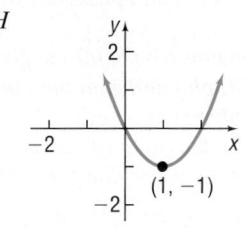

In Problems 19–30, graph the function f by starting with the graph of $y = x^2$ and using transformations (shifting, compressing, stretching, and/or reflection).

[Hint: If necessary, write *f* in the form $f(x) = a(x - h)^2 + k$.]

19. $f(x) = \dfrac{1}{4}x^2$
20. $f(x) = 2x^2 + 4$
21. $f(x) = (x + 2)^2 - 2$
22. $f(x) = (x - 3)^2 - 10$

23. $f(x) = x^2 + 4x + 2$
24. $f(x) = x^2 - 6x - 1$
25. $f(x) = 2x^2 - 4x + 1$
26. $f(x) = 3x^2 + 6x$

27. $f(x) = -x^2 - 2x$
28. $f(x) = -2x^2 + 6x + 2$
29. $f(x) = \dfrac{1}{2}x^2 + x - 1$
30. $f(x) = \dfrac{2}{3}x^2 + \dfrac{4}{3}x - 1$

In Problems 31–46, (a) graph each quadratic function by determining whether its graph opens up or down and by finding its vertex, axis of symmetry, y-intercept, and x-intercepts, if any. (b) Determine the domain and the range of the function. (c) Determine where the function is increasing and where it is decreasing.

31. $f(x) = x^2 + 2x$
32. $f(x) = x^2 - 4x$
33. $f(x) = -x^2 - 6x$
34. $f(x) = -x^2 + 4x$

35. $f(x) = x^2 + 2x - 8$
36. $f(x) = x^2 - 2x - 3$
37. $f(x) = x^2 + 2x + 1$
38. $f(x) = x^2 + 6x + 9$

39. $f(x) = 2x^2 - x + 2$
40. $f(x) = 4x^2 - 2x + 1$
41. $f(x) = -2x^2 + 2x - 3$
42. $f(x) = -3x^2 + 3x - 2$

43. $f(x) = 3x^2 + 6x + 2$
44. $f(x) = 2x^2 + 5x + 3$
45. $f(x) = -4x^2 - 6x + 2$
46. $f(x) = 3x^2 - 8x + 2$

In Problems 47–52, determine the quadratic function whose graph is given.

47.

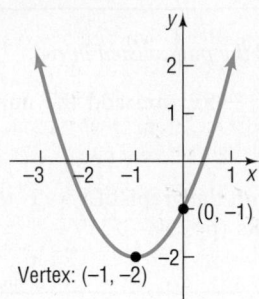

Vertex: (−1, −2)
(0, −1)

48.

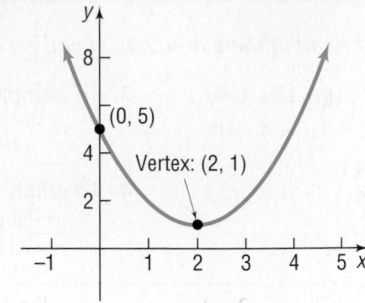

(0, 5)
Vertex: (2, 1)

49.

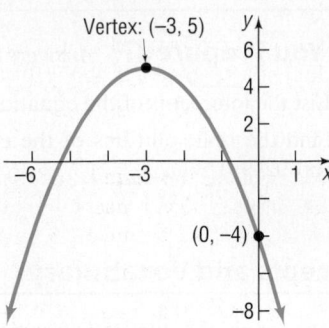

Vertex: (−3, 5)
(0, −4)

50.

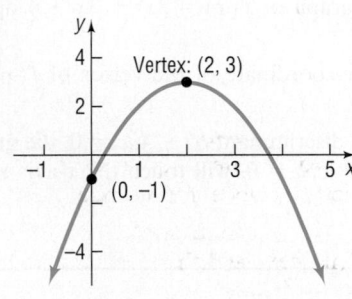

Vertex: (2, 3)
(0, −1)

51.

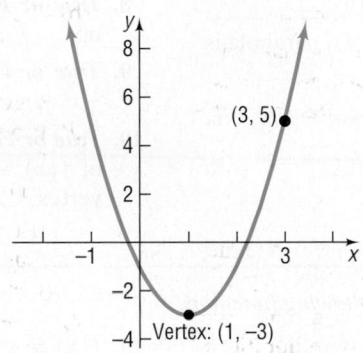

(3, 5)
Vertex: (1, −3)

52.

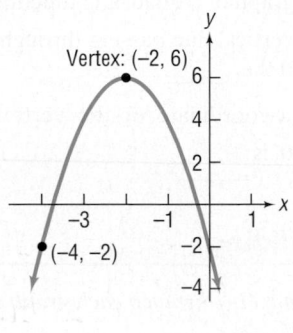

Vertex: (−2, 6)
(−4, −2)

In Problems 53–60, determine, without graphing, whether the given quadratic function has a maximum value or a minimum value and then find the value.

53. $f(x) = 2x^2 + 12x$ **54.** $f(x) = -2x^2 + 12x$ **55.** $f(x) = 2x^2 + 12x - 3$ **56.** $f(x) = 4x^2 - 8x + 3$

57. $f(x) = -x^2 + 10x - 4$ **58.** $f(x) = -2x^2 + 8x + 3$ **59.** $f(x) = -3x^2 + 12x + 1$ **60.** $f(x) = 4x^2 - 4x$

Applications and Extensions

61. The graph of the function $f(x) = ax^2 + bx + c$ has vertex at $(0, 2)$ and passes through the point $(1, 8)$. Find a, b, and c.

62. The graph of the function $f(x) = ax^2 + bx + c$ has vertex at $(1, 4)$ and passes through the point $(-1, -8)$. Find a, b, and c.

In Problems 63–68, for the given functions f and g,
(a) Graph f and g on the same Cartesian plane.
(b) Solve $f(x) = g(x)$.
(c) Use the result of part (b) to label the points of intersection of the graphs of f and g.
(d) Shade the region for which $f(x) > g(x)$, that is, the region below f and above g.

63. $f(x) = 2x - 1$; $g(x) = x^2 - 4$ **64.** $f(x) = -2x - 1$; $g(x) = x^2 - 9$

65. $f(x) = -x^2 + 4$; $g(x) = -2x + 1$ **66.** $f(x) = -x^2 + 9$; $g(x) = 2x + 1$

67. $f(x) = -x^2 + 5x$; $g(x) = x^2 + 3x - 4$ **68.** $f(x) = -x^2 + 7x - 6$; $g(x) = x^2 + x - 6$

Answer Problems 69 and 70 using the following: A quadratic function of the form $f(x) = ax^2 + bx + c$ with $b^2 - 4ac > 0$ may also be written in the form $f(x) = a(x - r_1)(x - r_2)$, where r_1 and r_2 are the x-intercepts of the graph of the quadratic function.

69. (a) Find a quadratic function whose x-intercepts are −3 and 1 with $a = 1$; $a = 2$; $a = -2$; $a = 5$.
 (b) How does the value of a affect the intercepts?
 (c) How does the value of a affect the axis of symmetry?
 (d) How does the value of a affect the vertex?
 (e) Compare the x-coordinate of the vertex with the midpoint of the x-intercepts. What might you conclude?

70. (a) Find a quadratic function whose x-intercepts are −5 and 3 with $a = 1$; $a = 2$; $a = -2$; $a = 5$.
 (b) How does the value of a affect the intercepts?
 (c) How does the value of a affect the axis of symmetry?
 (d) How does the value of a affect the vertex?
 (e) Compare the x-coordinate of the vertex with the midpoint of the x-intercepts. What might you conclude?

71. Suppose that $f(x) = x^2 + 4x - 21$
 (a) What is the vertex of f?
 (b) What are the x-intercepts of the graph of f?
 (c) Solve $f(x) = -21$ for x. What points are on the graph of f?
 (d) Use the information obtained in parts (a)–(c) to graph $f(x) = x^2 + 4x - 21$.

72. Suppose that $f(x) = x^2 + 2x - 8$
 (a) What is the vertex of f?
 (b) What are the x-intercepts of the graph of f?
 (c) Solve $f(x) = -8$ for x. What points are on the graph of f?
 (d) Use the information obtained in parts (a)–(c) to graph $f(x) = x^2 + 2x - 8$.

73. Find the point on the line $y = x$ that is closest to the point $(3, 1)$.

 [**Hint:** Express the distance d from the point to the line as a function of x, and then find the minimum value of $[d(x)]^2$.

74. Find the point on the line $y = x + 1$ that is closest to the point $(4, 1)$.

75. Maximizing Revenue Suppose that the manufacturer of a gas clothes dryer has found that, when the unit price is p dollars, the revenue R (in dollars) is

$$R(p) = -4p^2 + 4000p$$

What unit price should be established for the dryer to maximize revenue? What is the maximum revenue?

76. Maximizing Revenue The John Deere company has found that the revenue, in dollars, from sales of riding mowers is a function of the unit price p, in dollars, that it charges. If the revenue R is

$$R(p) = -\frac{1}{2}p^2 + 1900p$$

what unit price p should be charged to maximize revenue? What is the maximum revenue?

77. Minimizing Marginal Cost The **marginal cost** of a product can be thought of as the cost of producing one additional unit of output. For example, if the marginal cost of producing the 50th product is $6.20, it cost $6.20 to increase production from 49 to 50 units of output. Suppose the marginal cost C (in dollars) to produce x thousand mp3 players is given by the function

$$C(x) = x^2 - 140x + 7400$$

 (a) How many players should be produced to minimize the marginal cost?

 (b) What is the minimum marginal cost?

78. Minimizing Marginal Cost (See Problem 77.) The marginal cost C (in dollars) of manufacturing x cell phones (in thousands) is given by

$$C(x) = 5x^2 - 200x + 4000$$

 (a) How many cell phones should be manufactured to minimize the marginal cost?

 (b) What is the minimum marginal cost?

79. Business The monthly revenue R achieved by selling x wristwatches is figured to be $R(x) = 75x - 0.2x^2$. The monthly cost C of selling x wristwatches is $C(x) = 32x + 1750$.

 (a) How many wristwatches must the firm sell to maximize revenue? What is the maximum revenue?

 (b) Profit is given as $P(x) = R(x) - C(x)$. What is the profit function?

 (c) How many wristwatches must the firm sell to maximize profit? What is the maximum profit?

 (d) Provide a reasonable explanation as to why the answers found in parts (a) and (c) differ. Explain why a quadratic function is a reasonable model for revenue.

80. Business The daily revenue R achieved by selling x boxes of candy is figured to be $R(x) = 9.5x - 0.04x^2$. The daily cost C of selling x boxes of candy is $C(x) = 1.25x + 250$.

 (a) How many boxes of candy must the firm sell to maximize revenue? What is the maximum revenue?

 (b) Profit is given as $P(x) = R(x) - C(x)$. What is the profit function?

 (c) How many boxes of candy must the firm sell to maximize profit? What is the maximum profit?

 (d) Provide a reasonable explanation as to why the answers found in parts (a) and (c) differ. Explain why a quadratic function is a reasonable model for revenue.

81. Stopping Distance An accepted relationship between stopping distance, d (in feet), and the speed of a car, v (in mph), is $d = 1.1v + 0.06v^2$ on dry, level concrete.

 (a) How many feet will it take a car traveling 45 mph to stop on dry, level concrete?

 (b) If an accident occurs 200 feet ahead of you, what is the maximum speed you can be traveling to avoid being involved?

 (c) What might the term $1.1v$ represent?

 Source: www2.nsta.org/Energy/fn_braking.html

82. Birthrate of Unmarried Women In the United States, the birthrate B of unmarried women (births per 1000 unmarried women) for women whose age is a is modeled by the function $B(a) = -0.27a^2 + 14.23a - 120.16$.

 (a) What is the age of unmarried women with the highest birthrate?

 (b) What is the highest birthrate of unmarried women?

 (c) Evaluate and interpret $B(40)$.

 Source: United States Statistical Abstract, 2009

83. Find a quadratic function whose x-intercepts are -4 and 2 and whose range is $[-18, \infty)$.

84. Find a quadratic function whose x-intercepts are -1 and 5 and whose range is $(-\infty, 9]$.

85. Let $f(x) = ax^2 + bx + c$, where a, b, and c are odd integers. If x is an integer, show that $f(x)$ must be an odd integer.

 [**Hint:** x is either an even integer or an odd integer.]

Explaining Concepts: Discussion and Writing

86. Make up a quadratic function that opens down and has only one x-intercept. Compare yours with others in the class. What are the similarities? What are the differences?

87. On one set of coordinate axes, graph the family of parabolas $f(x) = x^2 + 2x + c$ for $c = -3$, $c = 0$, and $c = 1$. Describe the characteristics of a member of this family.

88. On one set of coordinate axes, graph the family of parabolas $f(x) = x^2 + bx + 1$ for $b = -4$, $b = 0$, and $b = 4$. Describe the general characteristics of this family.

89. State the circumstances that cause the graph of a quadratic function $f(x) = ax^2 + bx + c$ to have no x-intercepts.

90. Why does the graph of a quadratic function open up if $a > 0$ and down if $a < 0$?

91. Can a quadratic function have a range of $(-\infty, \infty)$? Justify your answer.

92. What are the possibilities for the number of times the graphs of two different quadratic functions intersect?

'Are You Prepared?' Answers

1. $(0, -9)$, $(-3, 0)$, $(3, 0)$ **2.** $\left\{-4, \frac{1}{2}\right\}$ **3.** $\frac{25}{4}$ **4.** right; 4

4.4 Build Quadratic Models from Verbal Descriptions and from Data

PREPARING FOR THIS SECTION *Before getting started, review the following:*

- Problem Solving (Section 1.7, pp. 134–140)
- Linear Models: Building Linear Functions from Data (Section 4.2, pp. 282–285)

Now Work the '**Are You Prepared?**' problems on page 305.

OBJECTIVES **1** Build Quadratic Models from Verbal Descriptions (p. 300)
 2 Build Quadratic Models from Data (p. 304)

In this section we will first discuss models in the form of a quadratic function when a verbal description of the problem is given. We end the section by fitting a quadratic function to data, which is another form of modeling.

When a mathematical model is in the form of a quadratic function, the properties of the graph of the quadratic function can provide important information about the model. In particular, we can use the quadratic function to determine the maximum or minimum value of the function. The fact that the graph of a quadratic function has a maximum or minimum value enables us to answer questions involving **optimization,** that is, finding the maximum or minimum values in models.

1 Build Quadratic Models from Verbal Descriptions

In economics, revenue R, in dollars, is defined as the amount of money received from the sale of an item and is equal to the unit selling price p, in dollars, of the item times the number x of units actually sold. That is,

$$R = xp$$

The Law of Demand states that p and x are related: As one increases, the other decreases. The equation that relates p and x is called the **demand equation.** When the demand equation is linear, the revenue model is a quadratic function.

EXAMPLE 1 **Maximizing Revenue**

The marketing department at Texas Instruments has found that, when certain calculators are sold at a price of p dollars per unit, the number x of calculators sold is given by the demand equation

$$x = 21,000 - 150p$$

(a) Find a model that expresses the revenue R as a function of the price p.

(b) What is the domain of R?

(c) What unit price should be used to maximize revenue?

(d) If this price is charged, what is the maximum revenue?

(e) How many units are sold at this price?

(f) Graph R.

(g) What price should Texas Instruments charge to collect at least $675,000 in revenue?

Solution (a) The revenue R is $R = xp$, where $x = 21,000 - 150p$.

$$R = xp = (21,000 - 150p)p = -150p^2 + 21,000p \quad \text{The Model}$$

(b) Because x represents the number of calculators sold, we have $x \geq 0$, so $21,000 - 150p \geq 0$. Solving this linear inequality, we find that $p \leq 140$. In addition, Texas Instruments will only charge a positive price for the calculator, so $p > 0$. Combining these inequalities, the domain of R is $\{p \mid 0 < p \leq 140\}$.

(c) The function R is a quadratic function with $a = -150, b = 21,000$, and $c = 0$. Because $a < 0$, the vertex is the highest point on the parabola. The revenue R is a maximum when the price p is

$$p = -\frac{b}{2a} \underset{\substack{\uparrow \\ a = -150, \, b = 21,000}}{=} -\frac{21,000}{2(-150)} = -\frac{21,000}{-300} = \$70.00$$

(d) The maximum revenue R is

$$R(70) = -150(70)^2 + 21,000(70) = \$735,000$$

(e) The number of calculators sold is given by the demand equation $x = 21,000 - 150p$. At a price of $p = \$70$,

$$x = 21,000 - 150(70) = 10,500$$

calculators are sold.

(f) To graph R, plot the intercept $(140, 0)$ and the vertex $(70, 735000)$. See Figure 23 for the graph.

Figure 23

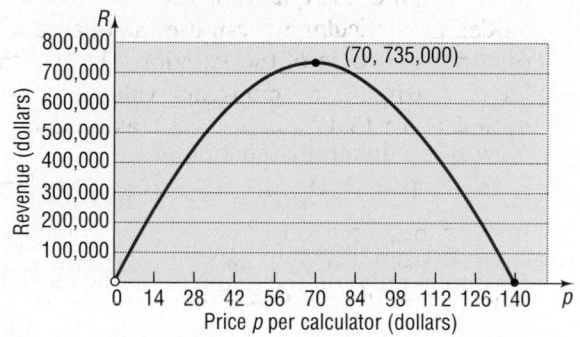

Price p per calculator (dollars)

(g) Graph $R = 675,000$ and $R(p) = -150p^2 + 21,000p$ on the same Cartesian plane. See Figure 24. We find where the graphs intersect by solving

$$675,000 = -150p^2 + 21,000p$$

$$150p^2 - 21,000p + 675,000 = 0 \quad \text{Add } 150p^2 - 21,000p \text{ to both sides.}$$

$$p^2 - 140p + 4500 = 0 \quad \text{Divide both sides by 150.}$$

$$(p - 50)(p - 90) = 0 \quad \text{Factor.}$$

$$p = 50 \text{ or } p = 90 \quad \text{Use the Zero-Product Property.}$$

Figure 24

Price p per calculator (dollars)

The graphs intersect at $(50, 675,000)$ and $(90, 675,000)$. Based on the graph in Figure 24, Texas Instruments should charge between \$50 and \$90 to earn at least \$675,000 in revenue.

Now Work PROBLEM 3

| EXAMPLE 2 | **Maximizing the Area Enclosed by a Fence** |

A farmer has 2000 yards of fence to enclose a rectangular field. What are the dimensions of the rectangle that encloses the most area?

Figure 25

Solution Figure 25 illustrates the situation. The available fence represents the perimeter of the rectangle. If x is the length and w is the width, then

$$2x + 2w = 2000 \qquad (1)$$

The area A of the rectangle is

$$A = xw$$

To express A in terms of a single variable, solve equation (1) for w and substitute the result in $A = xw$. Then A involves only the variable x. [You could also solve equation (1) for x and express A in terms of w alone. Try it!]

$$2x + 2w = 2000$$
$$2w = 2000 - 2x$$
$$w = \frac{2000 - 2x}{2} = 1000 - x$$

Then the area A is

$$A = xw = x(1000 - x) = -x^2 + 1000x$$

Now, A is a quadratic function of x.

$$A(x) = -x^2 + 1000x \qquad a = -1, b = 1000, c = 0$$

Figure 26

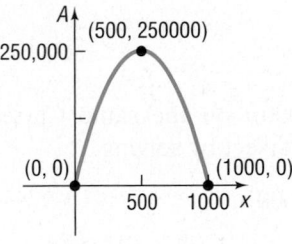

Figure 26 shows the graph of $A(x) = -x^2 + 1000x$. Since $a < 0$, the vertex is a maximum point on the graph of A. The maximum value occurs at

$$x = -\frac{b}{2a} = -\frac{1000}{2(-1)} = 500$$

The maximum value of A is

$$A\left(-\frac{b}{2a}\right) = A(500) = -500^2 + 1000(500) = -250{,}000 + 500{,}000 = 250{,}000$$

The largest rectangle that can be enclosed by 2000 yards of fence has an area of 250,000 square yards. Its dimensions are 500 yards by 500 yards.

───── **Now Work** PROBLEM 7

| EXAMPLE 3 | **Analyzing the Motion of a Projectile** |

A projectile is fired from a cliff 500 feet above the water at an inclination of 45° to the horizontal, with a muzzle velocity of 400 feet per second. In physics, it is established that the height h of the projectile above the water can be modeled by

$$h(x) = \frac{-32x^2}{(400)^2} + x + 500$$

where x is the horizontal distance of the projectile from the base of the cliff. See Figure 27.

Figure 27

(a) Find the maximum height of the projectile.

(b) How far from the base of the cliff will the projectile strike the water?

Solution (a) The height of the projectile is given by a quadratic function.

$$h(x) = \frac{-32x^2}{(400)^2} + x + 500 = \frac{-1}{5000}x^2 + x + 500$$

We are looking for the maximum value of h. Since $a < 0$, the maximum value is obtained at the vertex, whose x-coordinate is

$$x = -\frac{b}{2a} = -\frac{1}{2\left(-\frac{1}{5000}\right)} = \frac{5000}{2} = 2500$$

The maximum height of the projectile is

$$h(2500) = \frac{-1}{5000}(2500)^2 + 2500 + 500 = -1250 + 2500 + 500 = 1750 \text{ ft}$$

(b) The projectile will strike the water when the height is zero. To find the distance x traveled, solve the equation

$$h(x) = \frac{-1}{5000}x^2 + x + 500 = 0$$

The discriminant of this quadratic equation is

$$b^2 - 4ac = 1^2 - 4\left(\frac{-1}{5000}\right)(500) = 1.4$$

Then

$$x = \frac{-b \pm \sqrt{b^2 - 4ac}}{2a} = \frac{-1 \pm \sqrt{1.4}}{2\left(-\frac{1}{5000}\right)} \approx \begin{cases} -458 \\ 5458 \end{cases}$$

Discard the negative solution. The projectile will strike the water at a distance of about 5458 feet from the base of the cliff.

Now Work PROBLEM 11

Seeing the Concept

Graph

$$h(x) = \frac{-1}{5000}x^2 + x + 500$$

$$0 \leq x \leq 5500$$

Use MAXIMUM to find the maximum height of the projectile, and use ROOT or ZERO to find the distance from the base of the cliff to where it strikes the water. Compare your results with those obtained in Example 3.

EXAMPLE 4 **The Golden Gate Bridge**

The Golden Gate Bridge, a suspension bridge, spans the entrance to San Francisco Bay. Its 746-foot-tall towers are 4200 feet apart. The bridge is suspended from two huge cables more than 3 feet in diameter; the 90-foot-wide roadway is 220 feet above the water. The cables are parabolic in shape* and touch the road surface at the center of the bridge. Find the height of the cable above the road at a distance of 1000 feet from the center.

Solution See Figure 28 on page 304. Begin by choosing the placement of the coordinate axes so that the x-axis coincides with the road surface and the origin coincides with the center of the bridge. As a result, the twin towers will be vertical (height $746 - 220 = 526$ feet above the road) and located 2100 feet from the center. Also, the cable, which has the shape of a parabola, will extend from the towers, open up, and have its vertex at $(0, 0)$. The choice of placement of the axes enables us to identify the equation of the parabola as $y = ax^2, a > 0$. Notice that the points $(-2100, 526)$ and $(2100, 526)$ are on the graph.

* A cable suspended from two towers is in the shape of a **catenary**, but when a horizontal roadway is suspended from the cable, the cable takes the shape of a parabola.

Figure 28

Based on these facts, we can find the value of a in $y = ax^2$.

$$y = ax^2$$
$$526 = a(2100)^2 \quad x = 2100, y = 526$$
$$a = \frac{526}{(2100)^2}$$

The equation of the parabola is

$$y = \frac{526}{(2100)^2} x^2$$

The height of the cable when $x = 1000$ is

$$y = \frac{526}{(2100)^2}(1000)^2 \approx 119.3 \text{ feet}$$

The cable is 119.3 feet above the road at a distance of 1000 feet from the center of the bridge.

 Now Work PROBLEM 13

2 Build Quadratic Models from Data

In Section 4.2, we found the line of best fit for data that appeared to be linearly related. It was noted that data may also follow a nonlinear relation. Figures 29(a) and (b) show scatter diagrams of data that follow a quadratic relation.

Figure 29

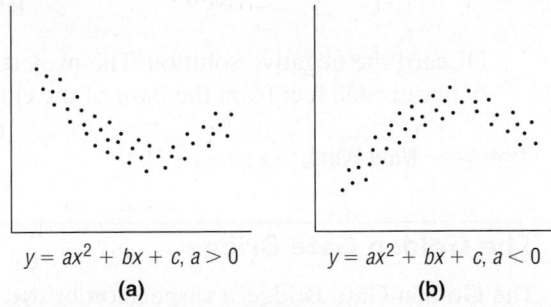

$y = ax^2 + bx + c, a > 0$
(a)

$y = ax^2 + bx + c, a < 0$
(b)

EXAMPLE 5

Fitting a Quadratic Function to Data

The data in Table 8 represent the percentage D of the population that is divorced for various ages x in 2007.

(a) Draw a scatter diagram of the data treating age as the independent variable. Comment on the type of relation that may exist between age and percentage of the population divorced.

(b) Use a graphing utility to find the quadratic function of best fit that models the relation between age and percentage of the population divorced.

(c) Use the model found in part (b) to approximate the age at which the percentage of the population divorced is greatest.

(d) Use the model found in part (b) to approximate the highest percentage of the population that is divorced.

(e) Use a graphing utility to draw the quadratic function of best fit on the scatter diagram.

Table 8

Age, x	Percentage Divorced, D
22	0.8
27	2.8
32	6.4
37	8.7
42	12.3
50	14.5
60	13.8
70	9.6
80	4.9

Source: United States Statistical Abstract, 2009

Solution

(a) Figure 30 shows the scatter diagram, from which it appears the data follow a quadratic relation, with $a < 0$.

(b) Upon executing the QUADratic REGression program, we obtain the results shown in Figure 31. The output of the utility shows us the equation $y = ax^2 + bx + c$. The quadratic function of best fit that models the relation between age and percentage divorced is

$$D(x) = -0.0136x^2 + 1.4794x - 26.3412 \quad \text{The Model}$$

where a represents age and D represents the percentage divorced.

(c) Based on the quadratic function of best fit, the age with the greatest percentage divorced is

$$-\frac{b}{2a} = -\frac{1.4794}{2(-0.0136)} \approx 54 \text{ years}$$

(d) Evaluate the function $D(x)$ at $x = 54$.

$$D(54) = -0.0136(54)^2 + 1.4794(54) - 26.3412 \approx 13.9 \text{ percent}$$

According to the model, 54-year-olds have the highest percentage divorced at 13.9 percent.

(e) Figure 32 shows the graph of the quadratic function found in part (b) drawn on the scatter diagram.

Look again at Figure 31. Notice that the output given by the graphing calculator does not include r, the correlation coefficient. Recall that the correlation coefficient is a measure of the strength of a linear relation that exists between two variables. The graphing calculator does not provide an indication of how well the function fits the data in terms of r since a quadratic function cannot be expressed as a linear function.

Figure 30

Figure 31

```
QuadReg
y=ax²+bx+c
a=-.0136351781
b=1.47938947
c=-26.34117462
```

Figure 32

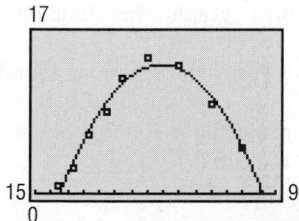

Now Work PROBLEM 25

4.4 Assess Your Understanding

'Are You Prepared?' *Answers are given at the end of these exercises. If you get a wrong answer, read the pages listed in red.*

1. Translate the following sentence into a mathematical equation: The total revenue R from selling x hot dogs is $3 times the number of hot dogs sold. (pp. 134–140)

2. Use a graphing utility to find the line of best fit for the following data: (pp. 282–285)

x	3	5	5	6	7	8
y	10	13	12	15	16	19

Applications and Extensions

3. **Maximizing Revenue** The price p (in dollars) and the quantity x sold of a certain product obey the demand equation

$$p = -\frac{1}{6}x + 100$$

 (a) Find a model that expresses the revenue R as a function of x. (Remember, $R = xp$.)
 (b) What is the domain of R?
 (c) What is the revenue if 200 units are sold?
 (d) What quantity x maximizes revenue? What is the maximum revenue?
 (e) What price should the company charge to maximize revenue?

4. **Maximizing Revenue** The price p (in dollars) and the quantity x sold of a certain product obey the demand equation

$$p = -\frac{1}{3}x + 100$$

 (a) Find a model that expresses the revenue R as a function of x.
 (b) What is the domain of R?
 (c) What is the revenue if 100 units are sold?
 (d) What quantity x maximizes revenue? What is the maximum revenue?
 (e) What price should the company charge to maximize revenue?

5. **Maximizing Revenue** The price p (in dollars) and the quantity x sold of a certain product obey the demand equation
$$x = -5p + 100 \qquad 0 < p \le 20$$
 (a) Express the revenue R as a function of x.
 (b) What is the revenue if 15 units are sold?
 (c) What quantity x maximizes revenue? What is the maximum revenue?
 (d) What price should the company charge to maximize revenue?
 (e) What price should the company charge to earn at least $480 in revenue?

6. **Maximizing Revenue** The price p (in dollars) and the quantity x sold of a certain product obey the demand equation
$$x = -20p + 500 \qquad 0 < p \le 25$$
 (a) Express the revenue R as a function of x.
 (b) What is the revenue if 20 units are sold?
 (c) What quantity x maximizes revenue? What is the maximum revenue?
 (d) What price should the company charge to maximize revenue?
 (e) What price should the company charge to earn at least $3000 in revenue?

7. **Enclosing a Rectangular Field** David has 400 yards of fencing and wishes to enclose a rectangular area.
 (a) Express the area A of the rectangle as a function of the width w of the rectangle.
 (b) For what value of w is the area largest?
 (c) What is the maximum area?

8. **Enclosing a Rectangular Field** Beth has 3000 feet of fencing available to enclose a rectangular field.
 (a) Express the area A of the rectangle as a function of x, where x is the length of the rectangle.
 (b) For what value of x is the area largest?
 (c) What is the maximum area?

9. **Enclosing the Most Area with a Fence** A farmer with 4000 meters of fencing wants to enclose a rectangular plot that borders on a river. If the farmer does not fence the side along the river, what is the largest area that can be enclosed? (See the figure.)

x x

$4000 - 2x$

10. **Enclosing the Most Area with a Fence** A farmer with 2000 meters of fencing wants to enclose a rectangular plot that borders on a straight highway. If the farmer does not fence the side along the highway, what is the largest area that can be enclosed?

11. **Analyzing the Motion of a Projectile** A projectile is fired from a cliff 200 feet above the water at an inclination of 45° to the horizontal, with a muzzle velocity of 50 feet per second. The height h of the projectile above the water is modeled by

$$h(x) = \frac{-32x^2}{(50)^2} + x + 200$$

 where x is the horizontal distance of the projectile from the face of the cliff.
 (a) At what horizontal distance from the face of the cliff is the height of the projectile a maximum?
 (b) Find the maximum height of the projectile.
 (c) At what horizontal distance from the face of the cliff will the projectile strike the water?
 (d) Using a graphing utility, graph the function h, $0 \le x \le 200$.
 (e) Use a graphing utility to verify the solutions found in parts (b) and (c).
 (f) When the height of the projectile is 100 feet above the water, how far is it from the cliff?

12. **Analyzing the Motion of a Projectile** A projectile is fired at an inclination of 45° to the horizontal, with a muzzle velocity of 100 feet per second. The height h of the projectile is modeled by

$$h(x) = \frac{-32x^2}{(100)^2} + x$$

 where x is the horizontal distance of the projectile from the firing point.
 (a) At what horizontal distance from the firing point is the height of the projectile a maximum?
 (b) Find the maximum height of the projectile.
 (c) At what horizontal distance from the firing point will the projectile strike the ground?
 (d) Using a graphing utility, graph the function h, $0 \le x \le 350$.

(e) Use a graphing utility to verify the results obtained in parts (b) and (c).

(f) When the height of the projectile is 50 feet above the ground, how far has it traveled horizontally?

13. Suspension Bridge A suspension bridge with weight uniformly distributed along its length has twin towers that extend 75 meters above the road surface and are 400 meters apart. The cables are parabolic in shape and are suspended from the tops of the towers. The cables touch the road surface at the center of the bridge. Find the height of the cables at a point 100 meters from the center. (Assume that the road is level.)

14. Architecture A parabolic arch has a span of 120 feet and a maximum height of 25 feet. Choose suitable rectangular coordinate axes and find the equation of the parabola. Then calculate the height of the arch at points 10 feet, 20 feet, and 40 feet from the center.

15. Constructing Rain Gutters A rain gutter is to be made of aluminum sheets that are 12 inches wide by turning up the edges 90°. See the illustration.

(a) What depth will provide maximum cross-sectional area and hence allow the most water to flow?

(b) What depths will allow at least 16 square inches of water to flow?

16. Norman Windows A **Norman window** has the shape of a rectangle surmounted by a semicircle of diameter equal to the width of the rectangle. See the figure. If the perimeter of the window is 20 feet, what dimensions will admit the most light (maximize the area)?

[**Hint:** Circumference of a circle $= 2\pi r$; area of a circle $= \pi r^2$, where r is the radius of the circle.]

17. Constructing a Stadium A track and field playing area is in the shape of a rectangle with semicircles at each end. See the figure. The inside perimeter of the track is to be 1500 meters. What should the dimensions of the rectangle be so that the area of the rectangle is a maximum?

18. Architecture A special window has the shape of a rectangle surmounted by an equilateral triangle. See the figure. If the perimeter of the window is 16 feet, what dimensions will admit the most light?

[**Hint:** Area of an equilateral triangle $= \left(\dfrac{\sqrt{3}}{4}\right)x^2$, where x is the length of a side of the triangle.]

19. Chemical Reactions A self-catalytic chemical reaction results in the formation of a compound that causes the formation ratio to increase. If the reaction rate V is modeled by

$$V(x) = kx(a - x), \qquad 0 \le x \le a$$

where k is a positive constant, a is the initial amount of the compound, and x is the variable amount of the compound, for what value of x is the reaction rate a maximum?

20. Calculus: Simpson's Rule The figure shows the graph of $y = ax^2 + bx + c$. Suppose that the points $(-h, y_0)$, $(0, y_1)$, and (h, y_2) are on the graph. It can be shown that the area enclosed by the parabola, the x-axis, and the lines $x = -h$ and $x = h$ is

$$\text{Area} = \frac{h}{3}(2ah^2 + 6c)$$

Show that this area may also be given by

$$\text{Area} = \frac{h}{3}(y_0 + 4y_1 + y_2)$$

21. Use the result obtained in Problem 20 to find the area enclosed by $f(x) = -5x^2 + 8$, the x-axis, and the lines $x = -1$ and $x = 1$.

22. Use the result obtained in Problem 20 to find the area enclosed by $f(x) = 2x^2 + 8$, the x-axis, and the lines $x = -2$ and $x = 2$.

23. Use the result obtained in Problem 20 to find the area enclosed by $f(x) = x^2 + 3x + 5$, the x-axis, and the lines $x = -4$ and $x = 4$.

24. Use the result obtained in Problem 20 to find the area enclosed by $f(x) = -x^2 + x + 4$, the x-axis, and the lines $x = -1$ and $x = 1$.

25. Life Cycle Hypothesis An individual's income varies with his or her age. The following table shows the median income *I* of males of different age groups within the United States for 2006. For each age group, let the class midpoint represent the independent variable, *x*. For the class "65 years and older," we will assume that the class midpoint is 69.5.

Age	Class Midpoint, x	Median Income, I
15–24 years	19.5	$10,964
25–34 years	29.5	$32,131
35–44 years	39.5	$42,637
45–54 years	49.5	$45,693
55–64 years	59.5	$41,477
65 years and older	69.5	$23,500

Source: U.S. Census Bureau

(a) Use a graphing utility to draw a scatter diagram of the data. Comment on the type of relation that may exist between the two variables.
(b) Use a graphing utility to find the quadratic function of best fit that models the relation between age and median income.
(c) Use the function found in part (b) to determine the age at which an individual can expect to earn the most income.
(d) Use the function found in part (b) to predict the peak income earned.
(e) With a graphing utility, graph the quadratic function of best fit on the scatter diagram.

26. Height of a Ball A shot-putter throws a ball at an inclination of 45° to the horizontal. The following data represent the height of the ball *h* at the instant that it has traveled *x* feet horizontally.

Distance, x	Height, h
20	25
40	40
60	55
80	65
100	71
120	77
140	77
160	75
180	71
200	64

(a) Use a graphing utility to draw a scatter diagram of the data. Comment on the type of relation that may exist between the two variables.
(b) Use a graphing utility to find the quadratic function of best fit that models the relation between distance and height.
(c) Use the function found in part (b) to determine how far the ball will travel before it reaches its maximum height.
(d) Use the function found in part (b) to find the maximum height of the ball.
(e) With a graphing utility, graph the quadratic function of best fit on the scatter diagram.

Mixed Practice

27. Which Model? The following data represent the square footage and rents (dollars per month) for apartments in the Del Mar area of San Diego, California.

Square Footage, x	Rent per Month, R
686	1600
770	1665
817	1750
800	1685
809	1700
901	1770
803	1725

Source: apartments.com

(a) Using a graphing utility, draw a scatter diagram of the data treating square footage as the independent variable. What type of relation appears to exist between square footage and rent?
(b) Based on your response to part (a), find either a linear or quadratic model that describes the relation between square footage and rent.
(c) Use your model to predict the rent of an apartment in San Diego that is 850 square feet.

28. Which Model? An engineer collects the following data showing the speed *s* of a Toyota Camry and its average miles per gallon, *M*.

Speed, s	Miles per Gallon, M
30	18
35	20
40	23
40	25
45	25
50	28
55	30
60	29
65	26
65	25
70	25

(a) Using a graphing utility, draw a scatter diagram of the data treating speed as the independent variable. What type of relation appears to exist between speed and miles per gallon?

(b) Based on your response to part (a), find either a linear or quadratic model that describes the relation between speed and miles per gallon.

(c) Use your model to predict the miles per gallon for a Camry that is traveling 63 miles per hour.

29. Which Model? The following data represent the percentage of the U.S. population whose age is x who do not have a high school diploma as of March 2005.

Age, a	Percentage without a High School Diploma, P
30	13.3
40	11.6
50	10.9
60	13.7
70	22.3
80	30.2

Source: U.S. Census Bureau

(a) Using a graphing utility, draw a scatter diagram of the data treating age as the independent variable. What type of relation appears to exist between age and percentage of the population without a high school diploma?

(b) Based on your response to part (a), find either a linear or quadratic model that describes the relation between age and percentage of the population that do not have a high school diploma.

(c) Use your model to predict the percentage of 35-year-olds that do not have a high school diploma.

30. Which Model? A cricket makes a chirping noise by sliding its wings together rapidly. Perhaps you have noticed that the number of chirps seems to increase with the temperature. The following data list the temperature (in degrees Fahrenheit) and the number of chirps per second for the striped ground cricket.

Temperature, x	Chirps per Second, C
88.6	20.0
93.3	19.8
80.6	17.1
69.7	14.7
69.4	15.4
79.6	15.0
80.6	16.0
76.3	14.4
75.2	15.5

Source: Pierce, George W. *The Songs of Insects.* Cambridge, MA Harvard University Press, 1949, pp. 12 – 21

(a) Using a graphing utility, draw a scatter diagram of the data treating temperature as the independent variable. What type of relation appears to exist between temperature and chirps per second?

(b) Based on your response to part (a), find either a linear or quadratic model that best describes the relation between temperature and chirps per second.

(c) Use your model to predict the chirps per second if the temperature is 80°F.

Explaining Concepts: Discussion and Writing

31. Refer to Example 1 on page 300. Notice that if the price charged for the calculators is $0 or $140 the revenue is $0. It is easy to explain why revenue would be $0 if the price charged is $0, but how can revenue be $0 if the price charged is $140?

'Are You Prepared?' Answers

1. $R = 3x$ **2.** $y = 1.7826x + 4.0652$

4.5 Inequalities Involving Quadratic Functions

PREPARING FOR THIS SECTION *Before getting started, review the following:*

• Solve Inequalities (Section 1.5, pp. 123–126) • Use Interval Notation (Section 1.5, pp. 120–121)

Now Work the 'Are You Prepared?' problems on page 312.

OBJECTIVE 1 Solve Inequalities Involving a Quadratic Function (p. 309)

1 Solve Inequalities Involving a Quadratic Function

In this section we solve inequalities that involve quadratic functions. We will accomplish this by using their graphs. For example, to solve the inequality

$$ax^2 + bx + c > 0 \qquad a \neq 0$$

graph the function $f(x) = ax^2 + bx + c$ and, from the graph, determine where it is above the x-axis, that is, where $f(x) > 0$. To solve the inequality $ax^2 + bx + c < 0$, $a \neq 0$, graph the function $f(x) = ax^2 + bx + c$ and determine where the graph is below the x-axis. If the inequality is not strict, include the x-intercepts, if any, in the solution.

| EXAMPLE 1 | **Solving an Inequality** |

Solve the inequality $x^2 - 4x - 12 \leq 0$ and graph the solution set.

Solution Graph the function $f(x) = x^2 - 4x - 12$. The intercepts are

Figure 33

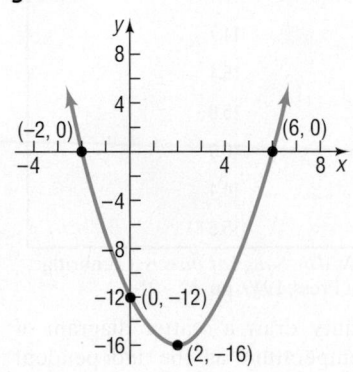

$$y\text{-intercept:} \qquad\qquad\qquad f(0) = -12 \quad \text{Evaluate } f \text{ at } 0.$$

$$x\text{-intercepts (if any):} \quad x^2 - 4x - 12 = 0 \quad \text{Solve } f(x) = 0.$$

$$(x - 6)(x + 2) = 0 \quad \text{Factor.}$$

$$x - 6 = 0 \quad \text{or} \quad x + 2 = 0 \quad \text{Apply the Zero-Product Property.}$$

$$x = 6 \quad \text{or} \qquad x = -2$$

The y-intercept is -12; the x-intercepts are -2 and 6.

The vertex is at $x = -\dfrac{b}{2a} = -\dfrac{-4}{2} = 2$. Since $f(2) = -16$, the vertex is $(2, -16)$.

See Figure 33 for the graph.

Figure 34

The graph is below the x-axis for $-2 < x < 6$. Since the original inequality is not strict, include the x-intercepts. The solution set is $\{x \mid -2 \leq x \leq 6\}$ or, using interval notation, $[-2, 6]$. See Figure 34 for the graph of the solution set.

⟵ Now Work PROBLEM 9

| EXAMPLE 2 | **Solving an Inequality** |

Solve the inequality $2x^2 < x + 10$ and graph the solution set.

Solution **Method 1** Rearrange the inequality so that 0 is on the right side.

$$2x^2 < x + 10$$

$$2x^2 - x - 10 < 0 \qquad \text{Subtract } x + 10 \text{ from both sides.}$$

This inequality is equivalent to the one that we want to solve.

Next graph the function $f(x) = 2x^2 - x - 10$ to find where $f(x) < 0$. The intercepts are

Figure 35

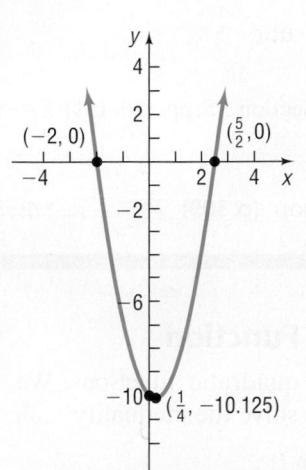

$$y\text{-intercept:} \qquad\qquad\qquad f(0) = -10 \qquad\qquad \text{Evaluate } f \text{ at } 0.$$

$$x\text{-intercepts (if any):} \quad 2x^2 - x - 10 = 0 \qquad \text{Solve } f(x) = 0.$$

$$(2x - 5)(x + 2) = 0 \qquad \text{Factor.}$$

$$2x - 5 = 0 \quad \text{or} \quad x + 2 = 0 \quad \text{Apply the Zero-Product Property.}$$

$$x = \frac{5}{2} \quad \text{or} \quad x = -2$$

The y-intercept is -10; the x-intercepts are -2 and $\dfrac{5}{2}$.

The vertex is at $x = -\dfrac{b}{2a} = -\dfrac{-1}{4} = \dfrac{1}{4}$. Since $f\left(\dfrac{1}{4}\right) = -10.125$, the vertex is $\left(\dfrac{1}{4}, -10.125\right)$. See Figure 35 for the graph.

The graph is below the x-axis ($f(x) < 0$) between $x = -2$ and $x = \dfrac{5}{2}$. Since the inequality is strict, the solution set is $\left\{ x \,\middle|\, -2 < x < \dfrac{5}{2} \right\}$ or, using interval notation, $\left(-2, \dfrac{5}{2} \right)$.

Method 2 If $f(x) = 2x^2$ and $g(x) = x + 10$, the inequality that we want to solve is $f(x) < g(x)$. Graph the functions $f(x) = 2x^2$ and $g(x) = x + 10$. See Figure 36. The graphs intersect where $f(x) = g(x)$. Then

$$
\begin{aligned}
2x^2 &= x + 10 && f(x) = g(x) \\
2x^2 - x - 10 &= 0 \\
(2x - 5)(x + 2) &= 0 && \text{Factor.} \\
2x - 5 = 0 \quad &\text{or} \quad x + 2 = 0 && \text{Apply the Zero-Product Property.} \\
x = \tfrac{5}{2} \quad &\text{or} \quad x = -2
\end{aligned}
$$

Figure 36

The graphs intersect at the points $(-2, 8)$ and $\left(\dfrac{5}{2}, \dfrac{25}{2} \right)$. To solve $f(x) < g(x)$, we need to find where the graph of f is below the graph of g. This happens between the points of intersection. Since the inequality is strict, the solution set is $\left\{ x \,\middle|\, -2 < x < \dfrac{5}{2} \right\}$ or, using interval notation, $\left(-2, \dfrac{5}{2} \right)$.

See Figure 37 for the graph of the solution set.

Figure 37

(number line from −4 to 4 with open interval from −2 to 2)

━━━━**Now Work** PROBLEMS **5** AND **13**

| EXAMPLE 3 | **Solving an Inequality** |

Solve the inequality $x^2 + x + 1 > 0$ and graph the solution set.

Figure 38

(parabola graph with points (1, 3), (−1, 1), (0, 1), and vertex $\left(-\tfrac{1}{2}, \tfrac{3}{4} \right)$)

Solution Graph the function $f(x) = x^2 + x + 1$. The y-intercept is 1; there are no x-intercepts (Do you see why? Check the discriminants). The vertex is at $x = -\dfrac{b}{2a} = -\dfrac{1}{2}$. Since $f\left(-\dfrac{1}{2} \right) = \dfrac{3}{4}$, the vertex is at $\left(-\dfrac{1}{2}, \dfrac{3}{4} \right)$. The points $(1, 3)$ and $(-1, 1)$ are also on the graph. See Figure 38.

The graph of f lies above the x-axis for all x. The solution set is the set of all real numbers. See Figure 39.

Figure 39

(number line from −4 to 4)

━━━━**Now Work** PROBLEM **17**

4.5 Assess Your Understanding

'Are You Prepared?' *Answers are given at the end of these exercises. If you get a wrong answer, read the pages listed in red.*

1. Solve the inequality $-3x - 2 < 7$ (pp. 123–126)

2. Write $(-2, 7]$ using inequality notation. (pp. 120–121)

Skill Building

In Problems 3–6, use the figure to solve each inequality.

3. **4.** **5.** **6.**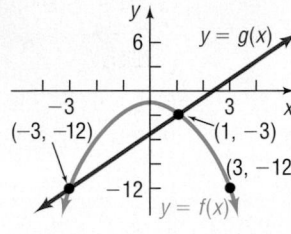

3.
(a) $f(x) > 0$
(b) $f(x) \leq 0$

4.
(a) $g(x) < 0$
(b) $g(x) \geq 0$

5.
(a) $g(x) \geq f(x)$
(b) $f(x) > g(x)$

6.
(a) $f(x) < g(x)$
(b) $f(x) \geq g(x)$

In Problems 7–22, solve each inequality.

7. $x^2 - 3x - 10 < 0$
8. $x^2 + 3x - 10 > 0$
9. $x^2 - 4x > 0$
10. $x^2 + 8x > 0$

11. $x^2 - 9 < 0$
12. $x^2 - 1 < 0$
13. $x^2 + x > 12$
14. $x^2 + 7x < -12$

15. $2x^2 < 5x + 3$
16. $6x^2 < 6 + 5x$
17. $x^2 - x + 1 \leq 0$
18. $x^2 + 2x + 4 > 0$

19. $4x^2 + 9 < 6x$
20. $25x^2 + 16 < 40x$
21. $6(x^2 - 1) > 5x$
22. $2(2x^2 - 3x) > -9$

Mixed Practice

23. What is the domain of the function $f(x) = \sqrt{x^2 - 16}$?

24. What is the domain of the function $f(x) = \sqrt{x - 3x^2}$?

In Problems 25–32, use the given functions f and g.

(a) *Solve* $f(x) = 0$.
(b) *Solve* $g(x) = 0$.
(c) *Solve* $f(x) = g(x)$.
(d) *Solve* $f(x) > 0$.

(e) *Solve* $g(x) \leq 0$.
(f) *Solve* $f(x) > g(x)$.
(g) *Solve* $f(x) \geq 1$.

25. $f(x) = x^2 - 1$
$g(x) = 3x + 3$

26. $f(x) = -x^2 + 3$
$g(x) = -3x + 3$

27. $f(x) = -x^2 + 1$
$g(x) = 4x + 1$

28. $f(x) = -x^2 + 4$
$g(x) = -x - 2$

29. $f(x) = x^2 - 4$
$g(x) = -x^2 + 4$

30. $f(x) = x^2 - 2x + 1$
$g(x) = -x^2 + 1$

31. $f(x) = x^2 - x - 2$
$g(x) = x^2 + x - 2$

32. $f(x) = -x^2 - x + 1$
$g(x) = -x^2 + x + 6$

Applications and Extensions

33. Physics A ball is thrown vertically upward with an initial velocity of 80 feet per second. The distance s

$s = 80t - 16t^2$

96 ft

(in feet) of the ball from the ground after t seconds is $s(t) = 80t - 16t^2$.
(a) At what time t will the ball strike the ground?
(b) For what time t is the ball more than 96 feet above the ground?

34. Physics A ball is thrown vertically upward with an initial velocity of 96 feet per second. The distance s (in feet) of the ball from the ground after t seconds is $s(t) = 96t - 16t^2$.
(a) At what time t will the ball strike the ground?
(b) For what time t is the ball more than 128 feet above the ground?

35. Revenue Suppose that the manufacturer of a gas clothes dryer has found that, when the unit price is p dollars, the revenue R (in dollars) is

$$R(p) = -4p^2 + 4000p$$

(a) At what prices p is revenue zero?

(b) For what range of prices will revenue exceed $800,000?

36. Revenue The John Deere company has found that the revenue from sales of heavy-duty tractors is a function of the unit price p, in dollars, that it charges. If the revenue R, in dollars, is

$$R(p) = -\frac{1}{2}p^2 + 1900p$$

(a) At what prices p is revenue zero?

(b) For what range of prices will revenue exceed $1,200,000?

37. Artillery A projectile fired from the point $(0,0)$ at an angle to the positive x-axis has a trajectory given by

$$y = cx - (1 + c^2)\left(\frac{g}{2}\right)\left(\frac{x}{v}\right)^2$$

where

$x =$ horizontal distance in meters

$y =$ height in meters

$v =$ initial muzzle velocity in meters per second (m/sec)

$g =$ acceleration due to gravity $= 9.81$ meters per second squared (m/sec^2)

$c > 0$ is a constant determined by the angle of elevation.

A howitzer fires an artillery round with a muzzle velocity of 897 m/sec.

(a) If the round must clear a hill 200 meters high at a distance of 2000 meters in front of the howitzer, what c values are permitted in the trajectory equation?

(b) If the goal in part (a) is to hit a target on the ground 75 kilometers away, is it possible to do so? If so, for what values of c? If not, what is the maximum distance the round will travel?

Source: www.answers.com

38. Runaway Car Using Hooke's Law, we can show that the *work* done in compressing a spring a distance of x feet from its at-rest position is $W = \frac{1}{2}kx^2$, where k is a stiffness constant depending on the spring. It can also be shown that the work done by a body in motion before it comes to rest is given by $\tilde{W} = \frac{w}{2g}v^2$, where $w =$ weight of the object (lb), $g =$ acceleration due to gravity (32.2 ft/sec^2), and $v =$ object's velocity (in ft/sec). A parking garage has a spring shock absorber at the end of a ramp to stop runaway cars. The spring has a stiffness constant $k = 9450$ lb/ft and must be able to stop a 4000-lb car traveling at 25 mph. What is the least compression required of the spring? Express your answer using feet to the nearest tenth.

[**Hint:** Solve $W > \tilde{W}, x \geq 0$].

Source: www.sciforums.com

Explaining Concepts: Discussion and Writing

39. Show that the inequality $(x - 4)^2 \leq 0$ has exactly one solution.

40. Show that the inequality $(x - 2)^2 > 0$ has one real number that is not a solution.

41. Explain why the inequality $x^2 + x + 1 > 0$ has all real numbers as the solution set.

42. Explain why the inequality $x^2 - x + 1 < 0$ has the empty set as solution set.

43. Explain the circumstances under which the x-intercepts of the graph of a quadratic function are included in the solution set of a quadratic inequality.

'Are You Prepared?' Answers

1. $\{x | x > -3\}$ or $(-3, \infty)$

2. $-2 < x \leq 7$

CHAPTER REVIEW

Things to Know

Linear function (p. 272)

$f(x) = mx + b$ 　　Average rate of change $= m$

The graph is a line with slope m and y-intercept b.

Quadratic function (pp. 289–293)

$f(x) = ax^2 + bx + c, a \neq 0$ 　　The graph is a parabola that opens up if $a > 0$ and opens down if $a < 0$.

Vertex: $\left(-\frac{b}{2a}, f\left(-\frac{b}{2a}\right)\right)$

Axis of symmetry: $x = -\frac{b}{2a}$

y-intercept: $f(0) = c$

x-intercept(s): If any, found by finding the real solutions of the equation $ax^2 + bx + c = 0$

Objectives

Section		You should be able to ...	Examples	Review Exercises
4.1	1	Graph linear functions (p. 272)	1	1(a)–6(a), 1(b)–6(b)
	2	Use average rate of change to identify linear functions (p. 272)	2	7, 8
	3	Determine whether a linear function is increasing, decreasing, or constant (p. 275)	3	1(d)–6(d)
	4	Build linear models from verbal descriptions (p. 276)	4, 5	37, 38
4.2	1	Draw and interpret scatter diagrams (p. 282)	1	46(a), 47(a)
	2	Distinguish between linear and nonlinear relations (p. 283)	2	46(b), 47(a)
	3	Use a graphing utility to find the line of best fit (p. 284)	4	46(c)
4.3	1	Graph a quadratic function using transformations (p. 290)	1	9–14
	2	Identify the vertex and axis of symmetry of a quadratic function (p. 292)	2	15–24
	3	Graph a quadratic function using its vertex, axis, and intercepts (p. 292)	3–5	15–24
	4	Find a quadratic function given its vertex and one other point (p. 295)	6	35, 36
	5	Find the maximum or minimum value of a quadratic function (p. 296)	7	25–30, 39–44
4.4	1	Build quadratic models from verbal descriptions (p. 300)	1–4	39–45
	2	Build quadratic models from data (p. 304)	5	47
4.5	1	Solve inequalities involving a quadratic function (p. 309)	1–3	31–34

Review Exercises

In Problems 1–6:

(a) *Determine the slope and y-intercept of each linear function.*
(b) *Find the average rate of change of each function.*
(c) *Graph each function. Label the intercepts.*
(d) *Determine whether the function is increasing, decreasing, or constant.*

1. $f(x) = 2x - 5$

2. $g(x) = -4x + 7$

3. $h(x) = \dfrac{4}{5}x - 6$

4. $F(x) = -\dfrac{1}{3}x + 1$

5. $G(x) = 4$

6. $H(x) = -3$

In Problems 7 and 8, determine whether the function is linear or nonlinear. If the function is linear, state its slope.

7.

x	y = f(x)
−1	−2
0	3
1	8
2	13
3	18

8.

x	y = g(x)
−1	−3
0	4
1	7
2	6
3	1

In Problems 9–14, graph each quadratic function using transformations (shifting, compressing, stretching, and/or reflecting).

9. $f(x) = (x - 2)^2 + 2$

10. $f(x) = (x + 1)^2 - 4$

11. $f(x) = -(x - 4)^2$

12. $f(x) = (x - 1)^2 - 3$

13. $f(x) = 2(x + 1)^2 + 4$

14. $f(x) = -3(x + 2)^2 + 1$

In Problems 15–24, (a) graph each quadratic function by determining whether its graph opens up or down and by finding its vertex, axis of symmetry, y-intercept, and x-intercepts, if any. (b) Determine the domain and the range of the function. (c) Determine where the function is increasing and where it is decreasing.

15. $f(x) = (x - 2)^2 + 2$

16. $f(x) = (x + 1)^2 - 4$

17. $f(x) = \dfrac{1}{4}x^2 - 16$

18. $f(x) = -\dfrac{1}{2}x^2 + 2$

19. $f(x) = -4x^2 + 4x$

20. $f(x) = 9x^2 - 6x + 3$

21. $f(x) = \dfrac{9}{2}x^2 + 3x + 1$

22. $f(x) = -x^2 + x + \dfrac{1}{2}$

23. $f(x) = 3x^2 + 4x - 1$

24. $f(x) = -2x^2 - x + 4$

In Problems 25–30, determine whether the given quadratic function has a maximum value or a minimum value, and then find the value.

25. $f(x) = 3x^2 - 6x + 4$

26. $f(x) = 2x^2 + 8x + 5$

27. $f(x) = -x^2 + 8x - 4$

28. $f(x) = -x^2 - 10x - 3$

29. $f(x) = -3x^2 + 12x + 4$

30. $f(x) = -2x^2 + 4$

In Problems 31–34, solve each quadratic inequality.

31. $x^2 + 6x - 16 < 0$

32. $3x^2 - 2x - 1 \geq 0$

33. $3x^2 \geq 14x + 5$

34. $4x^2 < 13x - 3$

In Problems 35 and 36, find the quadratic function for which:

35. Vertex is $(-1, 2)$; contains the point $(1, 6)$

36. Vertex is $(3, -4)$; contains the point $(4, 2)$

37. Comparing Phone Companies Marissa must decide between one of two companies as her long-distance phone provider. Company A charges a monthly fee of $7.00 plus $0.06 per minute, while Company B does not have a monthly fee, but charges $0.08 per minute.
 (a) Find a linear function that relates cost, C, to total minutes on the phone, x, for each company.
 (b) Determine the number of minutes x for which the bill from Company A will equal the bill from Company B.
 (c) Over what interval of minutes x will the bill from Company B be less than the bill from Company A?

38. Sales Commissions Bill was just offered a sales position for a computer company. His salary would be $15,000 per year plus 1% of his total annual sales.
 (a) Find a linear function that relates Bill's annual salary, S, to his total annual sales, x.
 (b) In 2010, Bill had total annual sales of $1,000,000. What was Bill's salary?
 (c) What would Bill have to sell to earn $100,000?
 (d) Determine the sales required of Bill for his salary to exceed $150,000.

39. Demand Equation The price p (in dollars) and the quantity x sold of a certain product obey the demand equation

$$p = -\frac{1}{10}x + 150 \qquad 0 \leq x \leq 1500$$

 (a) Express the revenue R as a function of x.
 (b) What is the revenue if 100 units are sold?
 (c) What quantity x maximizes revenue? What is the maximum revenue?
 (d) What price should the company charge to maximize revenue?

40. Landscaping A landscape engineer has 200 feet of border to enclose a rectangular pond. What dimensions will result in the largest pond?

41. Enclosing the Most Area with a Fence A farmer with 10,000 meters of fencing wants to enclose a rectangular field and then divide it into two plots with a fence parallel to one of

the sides. See the figure. What is the largest area that can be enclosed?

42. Architecture A special window in the shape of a rectangle with semicircles at each end is to be constructed so that the outside dimensions are 100 feet in length. See the illustration. Find the dimensions of the rectangle that maximizes its area.

43. Minimizing Marginal Cost Callaway Golf Company has determined that the marginal cost C of manufacturing x Big Bertha golf clubs may be expressed by the quadratic function

$$C(x) = 4.9x^2 - 617.4x + 19,600$$

 (a) How many clubs should be manufactured to minimize the marginal cost?
 (b) At this level of production, what is the marginal cost?

44. A rectangle has one vertex on the line $y = 10 - x$, $x > 0$, another at the origin, one on the positive x-axis, and one on the positive y-axis. Express the area A of the rectangle as a function of x. Find the largest area A that can be enclosed by the rectangle.

45. Parabolic Arch Bridge A horizontal bridge is in the shape of a parabolic arch. Given the information shown in the figure, what is the height h of the arch 2 feet from shore?

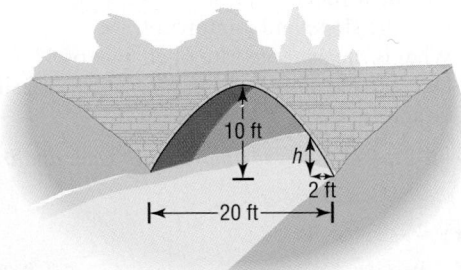

46. Bone Length Research performed at NASA, led by Dr. Emily R. Morey-Holton, measured the lengths of the right humerus and right tibia in 11 rats that were sent to space on Spacelab Life Sciences 2. The data on page 316 were collected.
 (a) Draw a scatter diagram of the data treating length of the right humerus as the independent variable.
 (b) Based on the scatter diagram, do you think that there is a linear relation between the length of the right humerus and the length of the right tibia?
 (c) Use a graphing utility to find the line of best fit relating length of the right humerus and length of the right tibia.

(d) Predict the length of the right tibia on a rat whose right humerus is 26.5 millimeters (mm).

Right Humerus (mm), x	Right Tibia (mm), y
24.80	36.05
24.59	35.57
24.59	35.57
24.29	34.58
23.81	34.20
24.87	34.73
25.90	37.38
26.11	37.96
26.63	37.46
26.31	37.75
26.84	38.50

Source: NASA Life Sciences Data Archive

47. Advertising A small manufacturing firm collected the following data on advertising expenditures A (in thousands of dollars) and total revenue R (in thousands of dollars).
 (a) Draw a scatter diagram of the data. Comment on the type of relation that may exist between the two variables.

Advertising	Total Revenue
20	$6101
22	$6222
25	$6350
25	$6378
27	$6453
28	$6423
29	$6360
31	$6231

 (b) The quadratic function of best fit to these data is
$$R(A) = -7.76A^2 + 411.88A + 942.72$$
 Use this function to determine the optimal level of advertising.
 (c) Use the function to predict the total revenue when the optimal level of advertising is spent.
 (d) Use a graphing utility to verify that the function given in part (b) is the quadratic function of best fit.
 (e) Use a graphing utility to draw a scatter diagram of the data and then graph the quadratic function of best fit on the scatter diagram.

CHAPTER TEST

CHAPTER **Test Prep** VIDEOS

The Chapter Test Prep Videos are step-by-step test solutions available in the Video Resources DVD, in *MyMathLab*, or on this text's You Tube™ Channel. Flip back to the Student Resources page to see the exact web address for this text's YouTube channel.

1. For the linear function $f(x) = -4x + 3$,
 (a) Find the slope and y-intercept.
 (b) What is the average rate of change of f?
 (c) Determine whether f is increasing, decreasing, or constant.
 (d) Graph f.

In Problems 2 and 3, find the intercepts of each quadratic function.

2. $f(x) = 3x^2 - 2x - 8$

3. $G(x) = -2x^2 + 4x + 1$

4. Given that $f(x) = x^2 + 3x$ and $g(x) = 5x + 3$, solve $f(x) = g(x)$. Graph each function and label the points of intersection.

5. Graph $f(x) = (x - 3)^2 - 2$ using transformations.

6. For the quadratic function $f(x) = 3x^2 - 12x + 4$,
 (a) Determine whether the graph opens up or down.
 (b) Determine the vertex.
 (c) Determine the axis of symmetry.
 (d) Determine the intercepts.
 (e) Use the information from parts (a)–(d) to graph f.

7. Determine whether $f(x) = -2x^2 + 12x + 3$ has a maximum or minimum value. Then find the maximum or minimum value.

8. Solve $x^2 - 10x + 24 \geq 0$.

9. **RV Rental** The weekly rental cost of a 20-foot recreational vehicle is $129.50 plus $0.15 per mile.
 (a) Find a linear function that expresses the cost C as a function of miles driven m.
 (b) What is the rental cost if 860 miles are driven?
 (c) How many miles were driven if the rental cost is $213.80?

CUMULATIVE REVIEW

1. Find the distance between the points $P = (-1, 3)$ and $Q = (4, -2)$. Find the midpoint of the line segment P to Q.

2. Which of the following points are on the graph of $y = x^3 - 3x + 1$?
 (a) $(-2, -1)$
 (b) $(2, 3)$
 (c) $(3, 1)$

3. Solve the inequality $5x + 3 \geq 0$ and graph the solution set.

4. Find the equation of the line containing the points $(-1, 4)$ and $(2, -2)$. Express your answer in slope–intercept form and graph the line.

5. Find the equation of the line perpendicular to the line $y = 2x + 1$ and containing the point $(3, 5)$. Express your answer in slope–intercept form and graph the line.

6. Graph the equation $x^2 + y^2 - 4x + 8y - 5 = 0$.

7. Does the following relation represent a function? $\{(-3, 8), (1, 3), (2, 5), (3, 8)\}$.

8. For the function f defined by $f(x) = x^2 - 4x + 1$, find:
 (a) $f(2)$
 (b) $f(x) + f(2)$
 (c) $f(-x)$
 (d) $-f(x)$
 (e) $f(x + 2)$
 (f) $\dfrac{f(x + h) - f(x)}{h}$ $h \neq 0$

9. Find the domain of $h(z) = \dfrac{3z - 1}{6z - 7}$.

10. Is the following graph the graph of a function?

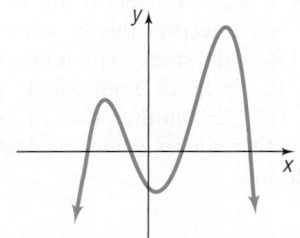

11. Consider the function $f(x) = \dfrac{x}{x + 4}$.
 (a) Is the point $\left(1, \dfrac{1}{4}\right)$ on the graph of f?
 (b) If $x = -2$, what is $f(x)$? What point is on the graph of f?
 (c) If $f(x) = 2$, what is x? What point is on the graph of f?

12. Is the function $f(x) = \dfrac{x^2}{2x + 1}$ even, odd, or neither?

13. Approximate the local maximum values and local minimum values of $f(x) = x^3 - 5x + 1$ on $(-4, 4)$. Determine where the function is increasing and where it is decreasing.

14. If $f(x) = 3x + 5$ and $g(x) = 2x + 1$,
 (a) Solve $f(x) = g(x)$. (b) Solve $f(x) > g(x)$.

15. For the graph of the function f,

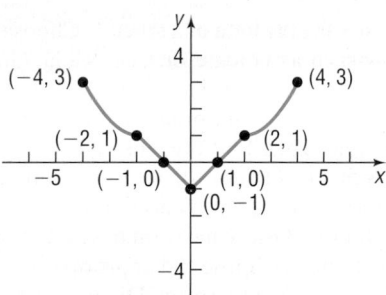

(a) Find the domain and the range of f.
(b) Find the intercepts.
(c) Is the graph of f symmetric with respect to the x-axis, the y-axis, or the origin?
(d) Find $f(2)$.
(e) For what value(s) of x is $f(x) = 3$?
(f) Solve $f(x) < 0$.
(g) Graph $y = f(x) + 2$.
(h) Graph $y = f(-x)$.
(i) Graph $y = 2f(x)$.
(j) Is f even, odd, or neither?
(k) Find the interval(s) on which f is increasing.

CHAPTER PROJECTS

the percentage change in the stock you chose as the dependent variable. The easiest way to draw a scatter diagram in Excel is to place the two columns of data next to each other (for example, have the percentage change in the S&P500 in column F and the percentage change in the stock you chose in column G). Then highlight the data and select the Scatter Diagram icon under Insert. Comment on the type of relation that appears to exist between the two variables.

3. **Finding beta.** To find beta requires that we find the line of best fit using least-squares regression. The easiest approach is to click inside the scatter diagram. Across the top of the screen you will see an option entitled "Chart Layouts." Select the option with a line drawn on the scatter diagram and *fx* labeled on the graph. The line of best fit appears on the scatter diagram. See below.

Internet-based Project

I. The Beta of a Stock You want to invest in the stock market, but are not sure which stock to purchase. Information is the key to making an informed investment decision. One piece of information that many stock analysts use is the beta of the stock. Go to Wikipedia (*http://en.wikipedia.org/wiki/Beta_%28finance%29*) and research what beta measures and what it represents.

1. **Approximating the beta of a stock.** Choose a well-known company such as Google or Coca-Cola. Go to a website such as Yahoo! Finance (*http://finance.yahoo.com/*) and find the weekly closing price of the company's stock for the past year. Then find the closing price of the Standard & Poor's 500 (S&P500) for the same time period. To get the historical prices in Yahoo! Finance click the price graph, choose Basic Chart, then scroll down and select Historical Prices. Choose the appropriate time period and select Weekly. Finally, select Download to Spreadsheet. Repeat this for the S&P500 and copy the data into the same spreadsheet. Finally, rearrange the data in chronological order. Be sure to expand the selection to sort all the data. Now, using the adjusted close price, compute the percentage change in price for each week using the formula% change $= \dfrac{P_1 - P_o}{P_o}$. For example, if week 1 price is in cell D1 and week 2 price is in cell D2, then % change $= \dfrac{D2 - D1}{D1}$. Repeat this for the S&P500 data.

2. **Using Excel to draw a scatter diagram.** Treat the percentage change in the S&P500 as the independent variable and

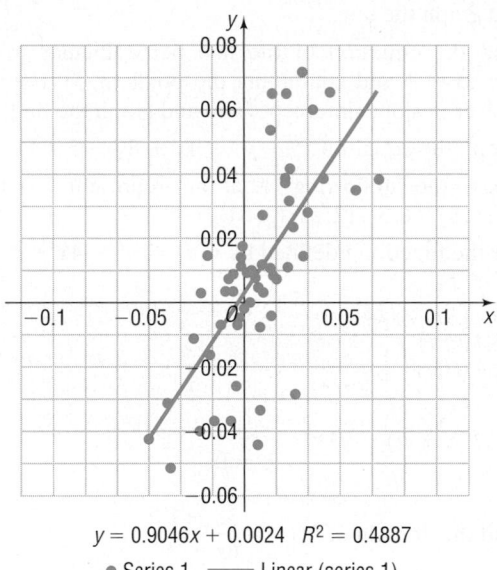

$$y = 0.9046x + 0.0024 \quad R^2 = 0.4887$$

● Series 1 ——— Linear (series 1)

The line of best fit for this data is $y = 0.9046x + 0.0024$. You may click on Chart Title or either axis title and insert the appropriate names. The beta is the slope of the line of best fit, 0.9046. We interpret this by saying if the S&P500 increases by 1%, this stock will increase by 0.9%, on average. Find the beta of your stock and provide an interpretation. NOTE: Another way to use Excel to find the line of best fit requires using the Data Analysis Tool Pack under add-ins.

The following projects are available on the Instructor's Resource Center (IRC):

II. Cannons A battery commander uses the weight of a missile, its initial velocity, and the position of its gun to determine where the missile will travel.

III. First and Second Differences Finite differences provide a numerical method that is used to estimate the graph of an unknown function.

IV. CBL Experiment Computer simulation is used to study the physical properties of a bouncing ball.

5

Polynomial and Rational Functions

Outline

Day Length

Day length refers to the time each day from the moment the upper limb of the sun's disk appears above the horizon during sunrise to the moment when the upper limb disappears below the horizon during sunset. The length of a day depends upon the day of the year as well as the latitude of the location. Latitude gives the location of a point on Earth north or south of the equator. In the Internet Project at the end of this chapter, we use information from the chapter to investigate the relation between the length of day and latitude for a specific day of the year.

 —See the Internet-based Chapter Project I—

◁ **A Look Back** In Chapter 3, we began our discussion of functions. We defined domain, range, and independent and dependent variables, found the value of a function, and graphed functions. We continued our study of functions by listing the properties that a function might have, such as being even or odd, and created a library of functions, naming key functions and listing their properties, including their graphs.

In Chapter 4, we discussed linear functions and quadratic functions, which belong to the class of *polynomial functions*.

A Look Ahead ▷ In this chapter, we look at two general classes of functions, polynomial functions and rational functions, and examine their properties. Polynomial functions are arguably the simplest expressions in algebra. For this reason, they are often used to approximate other, more complicated functions. Rational functions are ratios of polynomial functions.

5.1 Polynomial Functions and Models

PREPARING FOR THIS SECTION *Before getting started, review the following:*

- Polynomials (Chapter R, Section R.4, pp. 39–47)
- Using a Graphing Utility to Approximate Local Maxima and Local Minima (Section 3.3, p. 228)
- Intercepts of a Function (Section 3.2, pp. 215–217)
- Graphing Techniques: Transformations (Section 3.5, pp. 244–251)
- Intercepts (Section 2.2, pp. 159–160)

Now Work the 'Are You Prepared?' problems on page 337.

OBJECTIVES 1 Identify Polynomial Functions and Their Degree (p. 320)
2 Graph Polynomial Functions Using Transformations (p. 324)
3 Identify the Real Zeros of a Polynomial Function and Their Multiplicity (p. 325)
4 Analyze the Graph of a Polynomial Function (p. 332)
5 Build Cubic Models from Data (p. 336)

1 Identify Polynomial Functions and Their Degree

In Chapter 4, we studied the linear function $f(x) = mx + b$, which can be written as

$$f(x) = a_1 x + a_0$$

and the quadratic function $f(x) = ax^2 + bx + c, a \neq 0$, which can be written as

$$f(x) = a_2 x^2 + a_1 x + a_0 \qquad a_2 \neq 0$$

Each of these functions is an example of a *polynomial function*.

DEFINITION

A **polynomial function** is a function of the form

$$f(x) = a_n x^n + a_{n-1} x^{n-1} + \cdots + a_1 x + a_0 \qquad \textbf{(1)}$$

In Words
A polynomial function is the sum of monomials.

where $a_n, a_{n-1}, \ldots, a_1, a_0$ are real numbers and n is a nonnegative integer. The domain of a polynomial function is the set of all real numbers.

A polynomial function is a function whose rule is given by a polynomial in one variable. The **degree** of a polynomial function is the largest power of x that appears. The zero polynomial function $f(x) = 0 + 0x + 0x^2 + \cdots + 0x^n$ is not assigned a degree.

Polynomial functions are among the simplest expressions in algebra. They are easy to evaluate: only addition and repeated multiplication are required. Because of this, they are often used to approximate other, more complicated functions. In this section, we investigate properties of this important class of functions.

EXAMPLE 1 | **Identifying Polynomial Functions**

Determine which of the following are polynomial functions. For those that are, state the degree; for those that are not, tell why not.

(a) $f(x) = 2 - 3x^4$ (b) $g(x) = \sqrt{x}$ (c) $h(x) = \dfrac{x^2 - 2}{x^3 - 1}$

(d) $F(x) = 0$ (e) $G(x) = 8$ (f) $H(x) = -2x^3(x - 1)^2$

Solution
(a) f is a polynomial function of degree 4.

(b) g is not a polynomial function because $g(x) = \sqrt{x} = x^{\frac{1}{2}}$, so the variable x is raised to the $\dfrac{1}{2}$ power, which is not a nonnegative integer.

(c) h is not a polynomial function. It is the ratio of two distinct polynomials, and the polynomial in the denominator is of positive degree.

(d) F is the zero polynomial function; it is not assigned a degree.

(e) G is a nonzero constant function. It is a polynomial function of degree 0 since $G(x) = 8 = 8x^0$.

(f) $H(x) = -2x^3(x-1)^2 = -2x^3(x^2 - 2x + 1) = -2x^5 + 4x^4 - 2x^3$. So H is a polynomial function of degree 5. Do you see a way to find the degree of H without multiplying out?

Now Work PROBLEMS 15 AND 19

We have already discussed in detail polynomial functions of degrees 0, 1, and 2. See Table 1 for a summary of the properties of the graphs of these polynomial functions.

Table 1

Degree	Form	Name	Graph
No degree	$f(x) = 0$	Zero function	The *x*-axis
0	$f(x) = a_0,\quad a_0 \neq 0$	Constant function	Horizontal line with *y*-intercept a_0
1	$f(x) = a_1x + a_0,\quad a_1 \neq 0$	Linear function	Nonvertical, nonhorizontal line with slope a_1 and *y*-intercept a_0
2	$f(x) = a_2x^2 + a_1x + a_0,\quad a_2 \neq 0$	Quadratic function	Parabola: graph opens up if $a_2 > 0$; graph opens down if $a_2 < 0$

One objective of this section is to analyze the graph of a polynomial function. If you take a course in calculus, you will learn that the graph of every polynomial function is both smooth and continuous. By **smooth,** we mean that the graph contains no sharp corners or cusps; by **continuous,** we mean that the graph has no gaps or holes and can be drawn without lifting pencil from paper. See Figures 1(a) and (b).

Figure 1

(a) Graph of a polynomial function: smooth, continuous

(b) Cannot be the graph of a polynomial function

Power Functions

We begin the analysis of the graph of a polynomial function by discussing *power functions,* a special kind of polynomial function.

DEFINITION

In Words
A power function is defined by a single monomial.

A **power function of degree n** is a monomial function of the form

$$f(x) = ax^n \tag{2}$$

where a is a real number, $a \neq 0$, and $n > 0$ is an integer.

Examples of power functions are

$$f(x) = 3x \qquad f(x) = -5x^2 \qquad f(x) = 8x^3 \qquad f(x) = -5x^4$$
$$\text{degree 1} \qquad\quad \text{degree 2} \qquad\quad \text{degree 3} \qquad\quad \text{degree 4}$$

The graph of a power function of degree 1, $f(x) = ax$, is a straight line, with slope a, that passes through the origin. The graph of a power function of degree 2, $f(x) = ax^2$, is a parabola, with vertex at the origin, that opens up if $a > 0$ and down if $a < 0$.

If we know how to graph a power function of the form $f(x) = x^n$, a compression or stretch and, perhaps, a reflection about the x-axis will enable us to obtain the graph of $g(x) = ax^n$. Consequently, we shall concentrate on graphing power functions of the form $f(x) = x^n$.

We begin with power functions of even degree of the form $f(x) = x^n, n \geq 2$ and n even. The domain of f is the set of all real numbers, and the range is the set of nonnegative real numbers. Such a power function is an even function (do you see why?), so its graph is symmetric with respect to the y-axis. Its graph always contains the origin and the points $(-1, 1)$ and $(1, 1)$.

If $n = 2$, the graph is the familiar parabola $y = x^2$ that opens up, with vertex at the origin. If $n \geq 4$, the graph of $f(x) = x^n, n$ even, will be closer to the x-axis than the parabola $y = x^2$ if $-1 < x < 1, x \neq 0$, and farther from the x-axis than the parabola $y = x^2$ if $x < -1$ or if $x > 1$. Figure 2(a) illustrates this conclusion. Figure 2(b) shows the graphs of $y = x^4$ and $y = x^8$ for comparison.

Figure 2

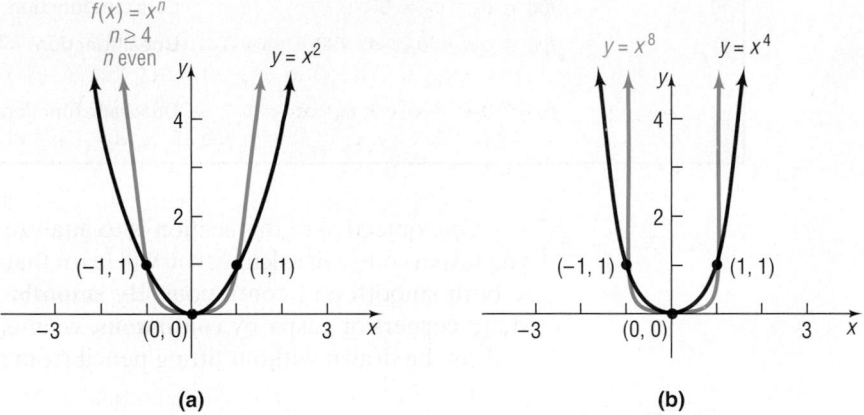

(a)

(b)

From Figure 2, we can see that as n increases the graph of $f(x) = x^n, n \geq 2$ and n even, tends to flatten out near the origin and to increase very rapidly when x is far from 0. For large n, it may appear that the graph coincides with the x-axis near the origin, but it does not; the graph actually touches the x-axis only at the origin (see Table 2). Also, for large n, it may appear that for $x < -1$ or for $x > 1$ the graph is vertical, but it is not; it is only increasing very rapidly in these intervals. If the graphs were enlarged many times, these distinctions would be clear.

Table 2

	$x = 0.1$	$x = 0.3$	$x = 0.5$
$f(x) = x^8$	10^{-8}	0.0000656	0.0039063
$f(x) = x^{20}$	10^{-20}	$3.487 \cdot 10^{-11}$	0.000001
$f(x) = x^{40}$	10^{-40}	$1.216 \cdot 10^{-21}$	$9.095 \cdot 10^{-13}$

Seeing the Concept

Graph $Y_1 = x^4, Y_2 = x^8$, and $Y_3 = x^{12}$ using the viewing rectangle $-2 \leq x \leq 2, -4 \leq y \leq 16$. Then graph each again using the viewing rectangle $-1 \leq x \leq 1, 0 \leq y \leq 1$. See Figure 3. TRACE along one of the graphs to confirm that for x close to 0 the graph is above the x-axis and that for $x > 0$ the graph is increasing.

Figure 3

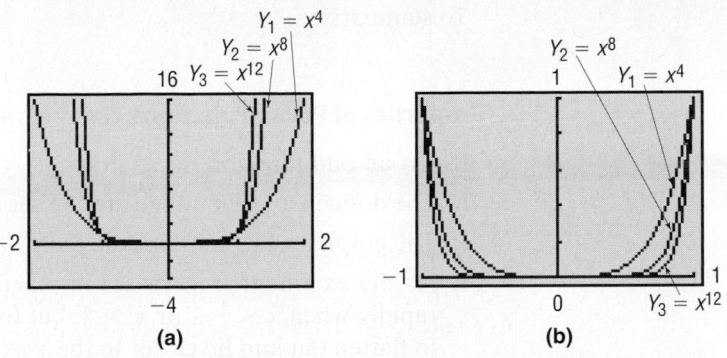

(a) (b)

Properties of Power Functions, $f(x) = x^n$, n Is a Positive Even Integer

1. f is an even function, so its graph is symmetric with respect to the y-axis.
2. The domain is the set of all real numbers. The range is the set of nonnegative real numbers.
3. The graph always contains the points $(-1, 1)$, $(0, 0)$, and $(1, 1)$.
4. As the exponent n increases in magnitude, the function increases more rapidly when $x < -1$ or $x > 1$; but for x near the origin, the graph tends to flatten out and lie closer to the x-axis.

Figure 4

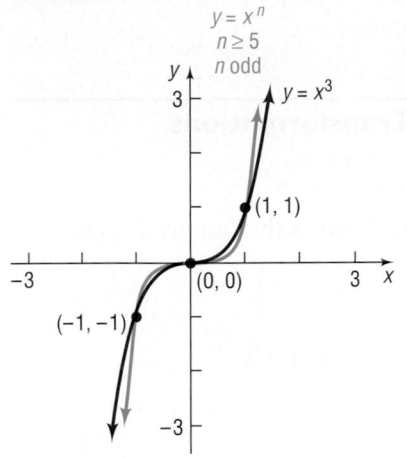

Now we consider power functions of odd degree of the form $f(x) = x^n$, $n \geq 3$ and n odd. The domain and the range of f are the set of real numbers. Such a power function is an odd function (do you see why?), so its graph is symmetric with respect to the origin. Its graph always contains the origin and the points $(-1, -1)$ and $(1, 1)$.

The graph of $f(x) = x^n$ when $n = 3$ has been shown several times and is repeated in Figure 4. If $n \geq 5$, the graph of $f(x) = x^n$, n odd, will be closer to the x-axis than that of $y = x^3$ if $-1 < x < 1$ and farther from the x-axis than that of $y = x^3$ if $x < -1$ or if $x > 1$. Figure 4 also illustrates this conclusion.

Figure 5 shows the graph of $y = x^5$ and the graph of $y = x^9$ for further comparison.

Figure 5

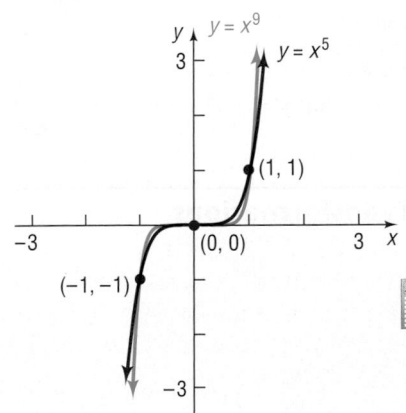

It appears that each graph coincides with the x-axis near the origin, but it does not; each graph actually crosses the x-axis at the origin. Also, it appears that as x increases the graph becomes vertical, but it does not; each graph is increasing very rapidly.

Seeing the Concept

Graph $Y_1 = x^3$, $Y_2 = x^7$, and $Y_3 = x^{11}$ using the viewing rectangle $-2 \leq x \leq 2$, $-16 \leq y \leq 16$. Then graph each again using the viewing rectangle $-1 \leq x \leq 1$, $-1 \leq y \leq 1$. See Figure 6. TRACE along one of the graphs to confirm that the graph is increasing and crosses the x-axis at the origin.

Figure 6

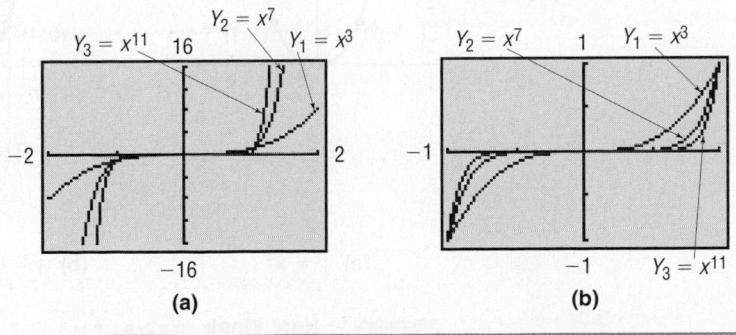

(a) (b)

To summarize:

Properties of Power Functions, $f(x) = x^n$, n Is a Positive Odd Integer

1. f is an odd function, so its graph is symmetric with respect to the origin.
2. The domain and the range are the set of all real numbers.
3. The graph always contains the points $(-1, -1)$, $(0, 0)$, and $(1, 1)$.
4. As the exponent n increases in magnitude, the function increases more rapidly when $x < -1$ or $x > 1$; but for x near the origin, the graph tends to flatten out and lie closer to the x-axis.

2 Graph Polynomial Functions Using Transformations

The methods of shifting, compression, stretching, and reflection studied in Section 3.5, when used with the facts just presented, will enable us to graph polynomial functions that are transformations of power functions.

| EXAMPLE 2 | **Graphing a Polynomial Function Using Transformations** |

Graph: $f(x) = 1 - x^5$

Solution It is helpful to rewrite f as $f(x) = -x^5 + 1$. Figure 7 shows the required stages.

Figure 7

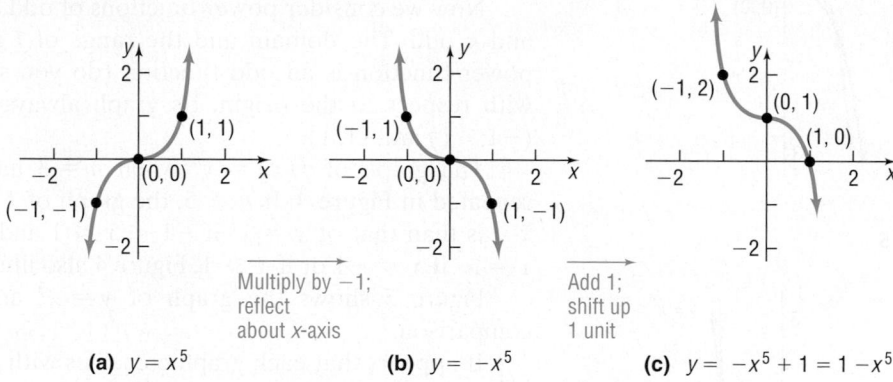

(a) $y = x^5$ (b) $y = -x^5$ (c) $y = -x^5 + 1 = 1 - x^5$

Multiply by -1; reflect about x-axis

Add 1; shift up 1 unit

| EXAMPLE 3 | **Graphing a Polynomial Function Using Transformations** |

Graph: $f(x) = \dfrac{1}{2}(x - 1)^4$

Solution Figure 8 shows the required stages.

Figure 8

(a) $y = x^4$ (b) $y = (x - 1)^4$ (c) $y = \dfrac{1}{2}(x - 1)^4$

Replace x by $x - 1$; shift right 1 unit

Multiply by $\dfrac{1}{2}$; compression by a factor of $\dfrac{1}{2}$

Now Work PROBLEMS 27 AND 33

3 Identify the Real Zeros of a Polynomial Function and Their Multiplicity

Figure 9 shows the graph of a polynomial function with four *x*-intercepts. Notice that at the *x*-intercepts the graph must either cross the *x*-axis or touch the *x*-axis. Consequently, between consecutive *x*-intercepts the graph is either above the *x*-axis or below the *x*-axis.

Figure 9

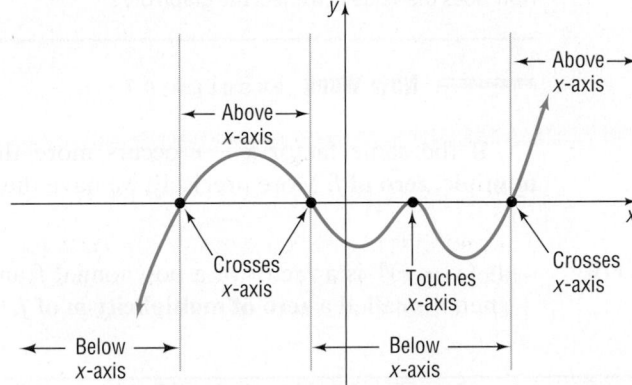

If a polynomial function *f* is factored completely, it is easy to locate the *x*-intercepts of the graph by solving the equation $f(x) = 0$ and using the Zero-Product Property. For example, if $f(x) = (x - 1)^2(x + 3)$, then the solutions of the equation

$$f(x) = (x - 1)^2(x + 3) = 0$$

are identified as 1 and -3. That is, $f(1) = 0$ and $f(-3) = 0$.

DEFINITION

> If *f* is a function and *r* is a real number for which $f(r) = 0$, then *r* is called a **real zero** of *f*.

As a consequence of this definition, the following statements are equivalent.

1. *r* is a real zero of a polynomial function *f*.
2. *r* is an *x*-intercept of the graph of *f*.
3. $x - r$ is a factor of *f*.
4. *r* is a solution to the equation $f(x) = 0$.

So the real zeros of a polynomial function are the *x*-intercepts of its graph, and they are found by solving the equation $f(x) = 0$.

EXAMPLE 4 **Finding a Polynomial Function from Its Zeros**

(a) Find a polynomial function of degree 3 whose zeros are -3, 2, and 5.

(b) Use a graphing utility to graph the polynomial found in part (a) to verify your result.

Solution (a) If *r* is a real zero of a polynomial function *f*, then $x - r$ is a factor of *f*. This means that $x - (-3) = x + 3$, $x - 2$, and $x - 5$ are factors of *f*. As a result, any polynomial function of the form

$$f(x) = a(x + 3)(x - 2)(x - 5)$$

where *a* is a nonzero real number, qualifies. The value of *a* causes a stretch, compression, or reflection, but does not affect the *x*-intercepts of the graph. Do you know why?

Figure 10

 (b) We choose to graph f with $a = 1$. Then

$$f(x) = (x + 3)(x - 2)(x - 5) = x^3 - 4x^2 - 11x + 30$$

Figure 10 shows the graph of f. Notice that the x-intercepts are $-3, 2,$ and 5.

Seeing the Concept

Graph the function found in Example 4 for $a = 2$ and $a = -1$. Does the value of a affect the zeros of f? How does the value of a affect the graph of f?

 Now Work PROBLEM 41

If the same factor $x - r$ occurs more than once, r is called a **repeated,** or **multiple, zero of** f. More precisely, we have the following definition.

DEFINITION If $(x - r)^m$ is a factor of a polynomial f and $(x - r)^{m+1}$ is not a factor of f, then r is called a **zero of multiplicity m of f.** *

EXAMPLE 5 **Identifying Zeros and Their Multiplicities**

For the polynomial

$$f(x) = 5(x - 2)(x + 3)^2\left(x - \frac{1}{2}\right)^4$$

2 is a zero of multiplicity 1 because the exponent on the factor $x - 2$ is 1.

-3 is a zero of multiplicity 2 because the exponent on the factor $x + 3$ is 2.

$\dfrac{1}{2}$ is a zero of multiplicity 4 because the exponent on the factor $x - \dfrac{1}{2}$ is 4.

 Now Work PROBLEM 49(a)

Suppose that it is possible to factor completely a polynomial function and, as a result, locate all the x-intercepts of its graph (the real zeros of the function). These x-intercepts then divide the x-axis into open intervals and, on each such interval, the graph of the polynomial will be either above or below the x-axis. Let's look at an example.

EXAMPLE 6 **Graphing a Polynomial Using Its x-Intercepts**

For the polynomial: $f(x) = x^2(x - 2)$

(a) Find the x- and y-intercepts of the graph of f.

(b) Use the x-intercepts to find the intervals on which the graph of f is above the x-axis and the intervals on which the graph of f is below the x-axis.

(c) Locate other points on the graph and connect all the points plotted with a smooth, continuous curve.

Solution (a) The y-intercept is $f(0) = 0^2(0 - 2) = 0$. The x-intercepts satisfy the equation

$$f(x) = x^2(x - 2) = 0$$

from which we find

$$x^2 = 0 \quad \text{or} \quad x - 2 = 0$$
$$x = 0 \quad \text{or} \qquad x = 2$$

The x-intercepts are 0 and 2.

* Some books use the terms **multiple root** and **root of multiplicity m.**

(b) The two x-intercepts divide the x-axis into three intervals:

$$(-\infty, 0) \qquad (0, 2) \qquad (2, \infty)$$

Since the graph of f crosses or touches the x-axis only at $x = 0$ and $x = 2$, it follows that the graph of f is either above the x-axis $[f(x) > 0]$ or below the x-axis $[f(x) < 0]$ on each of these three intervals. To see where the graph lies, we only need to pick one number in each interval, evaluate f there, and see whether the value is positive (above the x-axis) or negative (below the x-axis). See Table 3.

(c) In constructing Table 3, we obtained three additional points on the graph: $(-1, -3)$, $(1, -1)$, and $(3, 9)$. Figure 11 illustrates these points, the intercepts, and a smooth, continuous curve (the graph of f) connecting them.

Figure 11

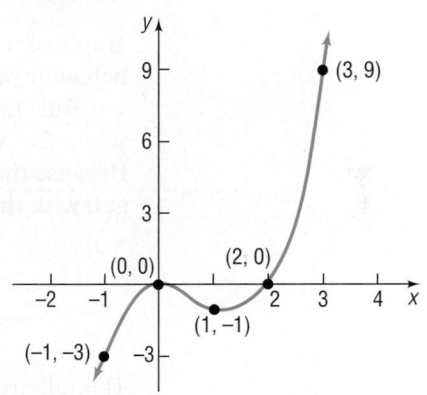

Table 3

	0	2	
Interval	$(-\infty, 0)$	$(0, 2)$	$(2, \infty)$
Number chosen	-1	1	3
Value of f	$f(-1) = -3$	$f(1) = -1$	$f(3) = 9$
Location of graph	Below x-axis	Below x-axis	Above x-axis
Point on graph	$(-1, -3)$	$(1, -1)$	$(3, 9)$

Look again at Table 3. Since the graph of $f(x) = x^2(x - 2)$ is below the x-axis on both sides of 0, the graph of f *touches* the x-axis at $x = 0$, a *zero of multiplicity 2*. Since the graph of f is below the x-axis for $x < 2$ and above the x-axis for $x > 2$, the graph of f *crosses* the x-axis at $x = 2$, a *zero of multiplicity 1*.

This suggests the following results:

If r Is a Zero of Even Multiplicity

The sign of $f(x)$ does not change from one side to the other side of r.

The graph of f **touches** the x-axis at r.

If r Is a Zero of Odd Multiplicity

The sign of $f(x)$ changes from one side to the other side of r.

The graph of f **crosses** the x-axis at r.

Now Work PROBLEM 49(b)

Behavior Near a Zero

The multiplicity of a zero can be used to determine whether the graph of a function touches or crosses the x-axis at the zero. However, we can learn more about the behavior of the graph near its zeros than just whether the graph crosses or touches the x-axis. Consider the function $f(x) = x^2(x - 2)$ whose graph is drawn in Figure 11. The zeros of f are 0 and 2. Table 4 on page 328 shows the values of $f(x) = x^2(x - 2)$ and $y = -2x^2$ for x near 0. Figure 12 shows the points $(-0.1, -0.021)$, $(-0.05, -0.0051)$, and so on, that are on the graph of $f(x) = x^2(x - 2)$ along with the graph of $y = -2x^2$ on the same Cartesian plane. From the table and graph, we can see that the points on the graph of $f(x) = x^2(x - 2)$ and the points on the

Table 4

x	$f(x) = x^2(x - 2)$	$y = -2x^2$
−0.1	−0.021	−0.02
−0.05	−0.005125	−0.005
−0.03	−0.001827	−0.0018
−0.01	−0.000201	−0.0002
0	0	0
0.01	−0.000199	−0.0002
0.03	−0.001773	−0.0018
0.05	−0.004875	−0.005
0.1	−0.019	−0.02

Figure 12

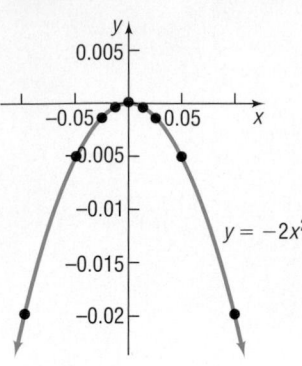

graph of $y = -2x^2$ are indistinguishable near $x = 0$. So $y = -2x^2$ describes the behavior of the graph of $f(x) = x^2(x - 2)$ near $x = 0$.

But how did we know that the function $f(x) = x^2(x - 2)$ behaves like $y = -2x^2$ when x is close to 0? In other words, where did $y = -2x^2$ come from? Because the zero, 0, comes from the factor x^2, we evaluate all factors in the function f at 0 with the exception of x^2.

$$f(x) = x^2(x - 2)$$
$$\approx x^2(0 - 2)$$
$$= -2x^2$$

The factor x^2 gives rise to the zero, so we keep the factor x^2 and let $x = 0$ in the remaining factors to find the behavior near 0.

This tells us that the graph of $f(x) = x^2(x - 2)$ will behave like the graph of $y = -2x^2$ near $x = 0$.

Now let's discuss the behavior of $f(x) = x^2(x - 2)$ near $x = 2$, the other zero. Because the zero, 2, comes from the factor $x - 2$, we evaluate all factors of the function f at 2 with the exception of $x - 2$.

$$f(x) = x^2(x - 2)$$
$$\approx 2^2(x - 2)$$
$$= 4(x - 2)$$

The factor $x - 2$ gives rise to the zero, so we keep the factor $x - 2$ and let $x = 2$ in the remaining factors to find the behavior near 2.

So the graph of $f(x) = x^2(x - 2)$ will behave like the graph of $y = 4(x - 2)$ near $x = 2$. Table 5 verifies that $f(x) = x^2(x - 2)$ and $y = 4(x - 2)$ have similar values for x near 2. Figure 13 shows the points $(1.9, -0.361)$, $(1.99, -0.0396)$, and so on, that are on the graph of $f(x) = x^2(x - 2)$ along with the graph of $y = 4(x - 2)$ on the same Cartesian plane. We can see that the points on the graph of $f(x) = x^2(x - 2)$ and the points on the graph of $y = 4(x - 2)$ are indistinguishable near $x = 2$. So $y = 4(x - 2)$, a line with slope 4, describes the behavior of the graph of $f(x) = x^2(x - 2)$ near $x = 2$.

Table 5

x	$f(x) = x^2(x - 2)$	$y = 4(x - 2)$
1.9	−0.361	−0.4
1.99	−0.0396	−0.04
1.999	−0.003996	−0.004
2	0	0
2.001	0.004004	0.004
2.01	0.0404	0.04
2.1	0.441	0.4

Figure 13

Figure 14

Figure 14 illustrates how we would use this information to begin to graph $f(x) = x^2(x - 2)$.

- The multiplicity of a real zero determines whether the graph crosses or touches the x-axis at the zero.
- The behavior of the graph near a real zero determines how the graph touches or crosses the x-axis.

Now Work PROBLEM 49(c)

Turning Points

 Look again at Figure 11 on page 327. We cannot be sure just how low the graph actually goes between $x = 0$ and $x = 2$. But we do know that somewhere in the interval $(0, 2)$ the graph of f must change direction (from decreasing to increasing). The points at which a graph changes direction are called **turning points.** In calculus, techniques for locating them are given. So we shall not ask for the location of turning points in our graphs. Instead, we will use the following result from calculus, which tells us the maximum number of turning points that the graph of a polynomial function can have.

THEOREM **Turning Points**

If f is a polynomial function of degree n, then the graph of f has at most $n - 1$ turning points.

If the graph of a polynomial function f has $n - 1$ turning points, the degree of f is at least n.

For example, the graph of $f(x) = x^2(x - 2)$ shown in Figure 11 is the graph of a polynomial function of degree 3 and has $3 - 1 = 2$ turning points: one at $(0, 0)$ and the other somewhere between $x = 0$ and $x = 2$.

Based on the theorem, if the graph of a polynomial function has three turning points, then the degree of the function must be at least 4.

Exploration

A graphing utility can be used to locate the turning points of a graph. Graph $Y_1 = x^2(x - 2)$. Use MINIMUM to find the location of the turning point for $0 < x < 2$. See Figure 15.

Figure 15

Now Work PROBLEM 49(d)

EXAMPLE 7 **Identifying the Graph of a Polynomial Function**

Which of the graphs in Figure 16 could be the graph of a polynomial function? For those that could, list the real zeros and state the least degree the polynomial can have. For those that could not, say why not.

Figure 16

(a) (b) (c) (d)

Solution

(a) The graph in Figure 16(a) cannot be the graph of a polynomial function because of the gap that occurs at $x = -1$. Remember, the graph of a polynomial function is continuous—no gaps or holes.

(b) The graph in Figure 16(b) could be the graph of a polynomial function because the graph is smooth and continuous. It has three real zeros, at -2, at 1, and at 2. Since the graph has two turning points, the degree of the polynomial function must be at least 3.

(c) The graph in Figure 16(c) cannot be the graph of a polynomial function because of the cusp at $x = 1$. Remember, the graph of a polynomial function is smooth.

(d) The graph in Figure 16(d) could be the graph of a polynomial function. It has two real zeros, at -2 and at 1. Since the graph has three turning points, the degree of the polynomial function is at least 4.

Now Work PROBLEM 61

End Behavior

One last remark about Figure 11. For very large values of x, either positive or negative, the graph of $f(x) = x^2(x - 2)$ looks like the graph of $y = x^3$. To see why, we write f in the form

$$f(x) = x^2(x - 2) = x^3 - 2x^2 = x^3\left(1 - \frac{2}{x}\right)$$

Now, for large values of x, either positive or negative, the term $\dfrac{2}{x}$ is close to 0, so for large values of x

$$f(x) = x^3 - 2x^2 = x^3\left(1 - \frac{2}{x}\right) \approx x^3$$

The behavior of the graph of a function for large values of x, either positive or negative, is referred to as its **end behavior.**

THEOREM

End Behavior

For large values of x, either positive or negative, the graph of the polynomial function

$$f(x) = a_n x^n + a_{n-1} x^{n-1} + \cdots + a_1 x + a_0$$

resembles the graph of the power function

$$y = a_n x^n$$

In Words
The end behavior of a polynomial function resembles that of its leading term.

For example, if $f(x) = -2x^3 + 5x^2 + x - 4$, then the graph of f will behave like the graph of $y = -2x^3$ for very large values of x, either positive or negative. We can see that the graphs of f and $y = -2x^3$ "behave" the same by considering Table 6 and Figure 17.

Table 6

x	f(x)	y = -2x³
10	-1,494	-2,000
100	-1,949,904	-2,000,000
500	-248,749,504	-250,000,000
1,000	-1,994,999,004	-2,000,000,000

Figure 17

Notice that, as x becomes a larger and larger positive number, the values of f become larger and larger negative numbers. When this happens, we say that f is **unbounded in the negative direction.** Rather than using words to describe the behavior of the graph of the function, we explain its behavior using notation. We can symbolize "the value of f becomes a larger and larger negative number as x becomes a larger and larger positive number" by writing $f(x) \rightarrow -\infty$ as $x \rightarrow \infty$ (read "the values of f approach negative infinity as x approaches infinity"). In calculus, **limits** are used to convey these ideas. There we use the symbolism $\lim_{x \to \infty} f(x) = -\infty$, read "the limit of $f(x)$ as x approaches infinity equals negative infinity," to mean that $f(x) \rightarrow -\infty$ as $x \rightarrow \infty$.

When the value of a limit equals infinity, we mean that the values of the function are unbounded in the positive or negative direction and call the limit an **infinite limit.** When we discuss limits as x becomes unbounded in the negative direction or unbounded in the positive direction, we are discussing **limits at infinity.**

Look back at Figures 2 and 4. Based on the preceding theorem and the previous discussion on power functions, the end behavior of a polynomial function can only be of four types. See Figure 18.

Figure 18

End behavior of $f(x) = a_n x^n + a_{n-1} x^{n-1} + \cdots + a_1 x + a_0$

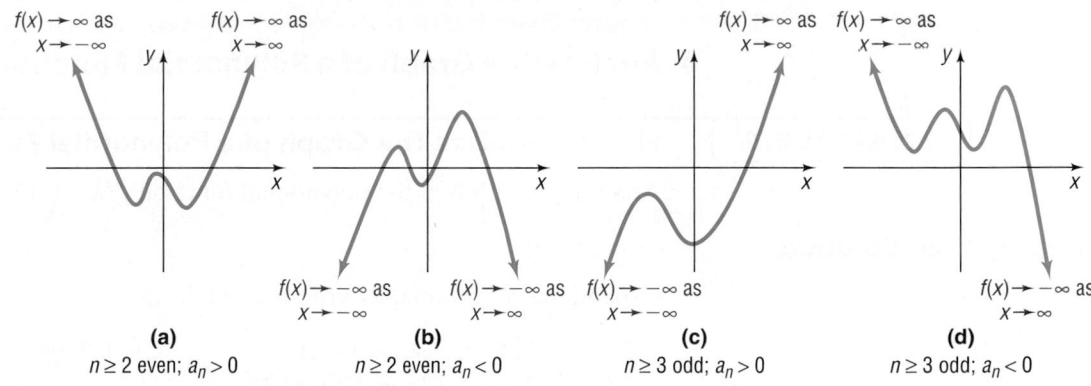

(a)
$n \geq 2$ even; $a_n > 0$

(b)
$n \geq 2$ even; $a_n < 0$

(c)
$n \geq 3$ odd; $a_n > 0$

(d)
$n \geq 3$ odd; $a_n < 0$

For example, if $f(x) = -2x^4 + x^3 + 4x^2 - 7x + 1$, the graph of f will resemble the graph of the power function $y = -2x^4$ for large $|x|$. The graph of f will behave like Figure 18(b) for large $|x|$.

Now Work PROBLEM **49(e)**

EXAMPLE 8

Identifying the Graph of a Polynomial Function

Which of the graphs in Figure 19 could be the graph of

$$f(x) = x^4 + 5x^3 + 5x^2 - 5x - 6?$$

Figure 19

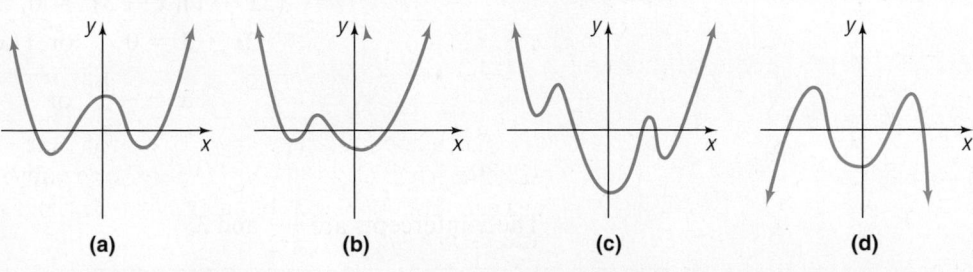

(a)

(b)

(c)

(d)

Solution The y-intercept of f is $f(0) = -6$. We can eliminate the graph in Figure 19(a), whose y-intercept is positive.

We don't have any methods for finding the x-intercepts of f, so we move on to investigate the turning points of each graph. Since f is of degree 4, the graph of f has at most 3 turning points. We eliminate the graph in Figure 19(c) since that graph has 5 turning points.

Now we look at end behavior. For large values of x, the graph of f will behave like the graph of $y = x^4$. This eliminates the graph in Figure 19(d), whose end behavior is like the graph of $y = -x^4$.

Only the graph in Figure 19(b) could be (and, in fact, is) the graph of $f(x) = x^4 + 5x^3 + 5x^2 - 5x - 6$. ↵

━━━**Now Work** PROBLEM 65

SUMMARY Graph of a Polynomial Function $f(x) = a_n x^n + a_{n-1} x^{n-1} + \cdots + a_1 x + a_0 \quad a_n \neq 0$

Degree of the polynomial function f: n
Graph is smooth and continuous.
Maximum number of turning points: $n - 1$
At a zero of even multiplicity: The graph of f touches the x-axis.
At a zero of odd multiplicity: The graph of f crosses the x-axis.
Between zeros, the graph of f is either above or below the x-axis.
End behavior: For large $|x|$, the graph of f behaves like the graph of $y = a_n x^n$.

4 Analyze the Graph of a Polynomial Function

EXAMPLE 9

How to Analyze the Graph of a Polynomial Function

Analyze the graph of the polynomial function $f(x) = (2x + 1)(x - 3)^2$.

Step-by-Step Solution

Step 1: Determine the end behavior of the graph of the function.

Expand the polynomial to write it in the form

$$f(x) = a_n x^n + a_{n-1} x^{n-1} + \cdots + a_1 x + a_0$$
$$f(x) = (2x + 1)(x - 3)^2$$
$$= (2x + 1)(x^2 - 6x + 9)$$
$$= 2x^3 - 12x^2 + 18x + x^2 - 6x + 9 \quad \text{Multiply.}$$
$$= 2x^3 - 11x^2 + 12x + 9 \quad \text{Combine like terms.}$$

The polynomial function f is of degree 3. The graph of f behaves like $y = 2x^3$ for large values of $|x|$.

Step 2: Find the x- and y-intercepts of the graph of the function.

The y-intercept is $f(0) = 9$. To find the x-intercepts, we solve $f(x) = 0$.

$$f(x) = 0$$
$$(2x + 1)(x - 3)^2 = 0$$
$$2x + 1 = 0 \quad \text{or} \quad (x - 3)^2 = 0$$
$$x = -\frac{1}{2} \quad \text{or} \quad x - 3 = 0$$
$$x = 3$$

The x-intercepts are $-\dfrac{1}{2}$ and 3.

Step 3: Determine the zeros of the function and their multiplicity. Use this information to determine whether the graph crosses or touches the x-axis at each x-intercept.

The zeros of f are $-\dfrac{1}{2}$ and 3. The zero $-\dfrac{1}{2}$ is a zero of multiplicity 1, so the graph of f crosses the x-axis at $x = -\dfrac{1}{2}$. The zero 3 is a zero of multiplicity 2, so the graph of f touches the x-axis at $x = 3$.

Step 4: Determine the maximum number of turning points on the graph of the function.

Because the polynomial function is of degree 3 (Step 1), the graph of the function will have at most $3 - 1 = 2$ turning points.

Step 5: Determine the behavior of the graph of f near each x-intercept.

The two x-intercepts are $-\dfrac{1}{2}$ and 3.

$$\text{Near } -\frac{1}{2}: \quad f(x) = (2x + 1)(x - 3)^2$$

$$\approx (2x + 1)\left(-\frac{1}{2} + 3\right)^2$$

$$= (2x + 1)\left(\frac{25}{4}\right)$$

$$= \frac{25}{2}x + \frac{25}{4} \qquad \text{A line with slope } \frac{25}{2}$$

$$\text{Near 3}: \quad f(x) = (2x + 1)(x - 3)^2$$

$$\approx (2 \cdot 3 + 1)(x - 3)^2$$

$$= 7(x - 3)^2 \qquad \text{A parabola that opens up}$$

Step 6: Put all the information from Steps 1 through 5 together to obtain the graph of f.

Figure 20(a) illustrates the information obtained from Steps 1 through 5. We evaluate f at -1, 1, and 4 to help establish the scale on the y-axis. The graph of f is given in Figure 20(b).

Figure 20

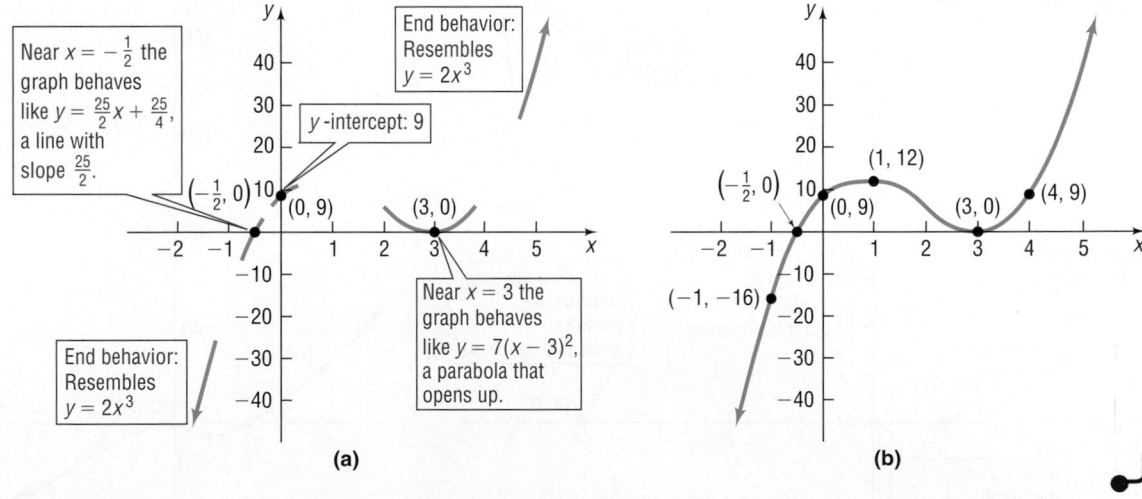

(a) (b)

SUMMARY Analyzing the Graph of a Polynomial Function

STEP 1: Determine the end behavior of the graph of the function.

STEP 2: Find the x- and y-intercepts of the graph of the function.

STEP 3: Determine the zeros of the function and their multiplicity. Use this information to determine whether the graph crosses or touches the x-axis at each x-intercept.

STEP 4: Determine the maximum number of turning points on the graph of the function.

STEP 5: Determine the behavior of the graph near each x-intercept.

STEP 6: Use the information in Steps 1 through 5 to draw a complete graph of the function.

EXAMPLE 10

Analyzing the Graph of a Polynomial Function

Analyze the graph of the polynomial function

$$f(x) = x^2(x - 4)(x + 1)$$

Solution

STEP 1: End behavior: the graph of f resembles that of the power function $y = x^4$ for large values of $|x|$.

STEP 2: The y-intercept is $f(0) = 0$. The x-intercepts satisfy the equation

$$f(x) = x^2(x - 4)(x + 1) = 0$$

So

$$x^2 = 0 \quad \text{or} \quad x - 4 = 0 \quad \text{or} \quad x + 1 = 0$$
$$x = 0 \quad \text{or} \quad x = 4 \quad \text{or} \quad x = -1$$

The x-intercepts are $-1, 0,$ and 4.

STEP 3: The intercept 0 is a zero of multiplicity 2, so the graph of f will touch the x-axis at 0; 4 and -1 are zeros of multiplicity 1, so the graph of f will cross the x-axis at 4 and -1.

STEP 4: The graph of f will contain at most three turning points.

STEP 5: The three x-intercepts are $-1, 0,$ and 4.

Near -1: $\quad f(x) = x^2(x - 4)(x + 1) \approx (-1)^2(-1 - 4)(x + 1) = -5(x + 1)$ \qquad A line with slope -5

Near 0: $\quad f(x) = x^2(x - 4)(x + 1) \approx x^2(0 - 4)(0 + 1) = -4x^2$ \qquad A parabola opening down

Near 4: $\quad f(x) = x^2(x - 4)(x + 1) \approx 4^2(x - 4)(4 + 1) = 80(x - 4)$ \qquad A line with slope 80

STEP 6: Figure 21(a) illustrates the information obtained from Steps 1–5. The graph of f is given in Figure 21(b). Notice that we evaluated f at $-2, -\dfrac{1}{2}, 2,$ and 5 to help establish the scale on the y-axis.

Figure 21

(a)

(b)

Exploration

Graph $Y_1 = x^2(x - 4)(x + 1)$. Compare what you see with Figure 21(b). Use MAXIMUM/MINIMUM to locate the three turning points.

Now Work PROBLEM 69

For polynomial functions that have noninteger coefficients and for polynomials that are not easily factored, we utilize the graphing utility early in the analysis. This is because the amount of information that can be obtained from algebraic analysis is limited.

 EXAMPLE 11

How to Use a Graphing Utility to Analyze the Graph of a Polynomial Function

Analyze the graph of the polynomial function

$$f(x) = x^3 + 2.48x^2 - 4.3155x + 2.484406$$

Step-by-Step Solution

Step 1: Determine the end behavior of the graph of the function.

The polynomial function f is of degree 3. The graph of f behaves like $y - x^3$ for large values of $|x|$.

Step 2: Graph the function using a graphing utility.

See Figure 22 for the graph of f.

Figure 22

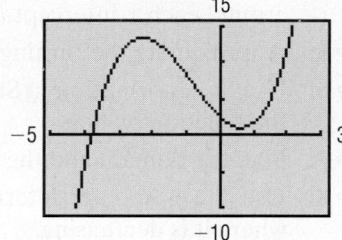

Step 3: Use a graphing utility to approximate the x- and y-intercepts of the graph.

The y-intercept is $f(0) = 2.484406$. In Examples 9 and 10, the polynomial function was factored, so it was easy to find the x-intercepts algebraically. However, it is not readily apparent how to factor f in this example. Therefore, we use a graphing utility's ZERO (or ROOT or SOLVE) feature and find the lone x-intercept to be -3.79, rounded to two decimal places.

Step 4: Use a graphing utility to create a TABLE to find points on the graph around each x-intercept.

Table 7 shows values of x on each side of the x-intercept. The points $(-4, -4.57)$ and $(-2, 13.04)$ are on the graph.

Table 7

Step 5: Approximate the turning points of the graph.

From the graph of f shown in Figure 22, we can see that f has two turning points. Using MAXIMUM, one turning point is at $(-2.28, 13.36)$, rounded to two decimal places. Using MINIMUM, the other turning point is at $(0.63, 1)$, rounded to two decimal places.

Step 6: Use the information in Steps 1 through 5 to draw a complete graph of the function by hand.

Figure 23 shows a graph of f using the information in Steps 1 through 5.

Figure 23

End behavior:
Resembles $y = x^3$

$(-2.28, 13.36)$
$(-2, 13.04)$
$(0, 2.484406)$
$(-3.79, 0)$
$(-4, -4.57)$
$(0.63, 1)$

End behavior:
Resembles $y = x^3$

Step 7: Find the domain and the range of the function.

The domain and the range of f are the set of all real numbers.

Step 8: Use the graph to determine where the function is increasing and where it is decreasing.

Based on the graph, f is increasing on the intervals $(-\infty, -2.28)$ and $(0.63, \infty)$. Also, f is decreasing on the interval $(-2.28, 0.63)$.

Using a Graphing Utility to Analyze the Graph of a Polynomial Function

STEP 1: Determine the end behavior of the graph of the function.

STEP 2: Graph the function using a graphing utility.

STEP 3: Use a graphing utility to approximate the x- and y-intercepts of the graph.

STEP 4: Use a graphing utility to create a TABLE to find points on the graph around each x-intercept.

STEP 5: Approximate the turning points of the graph.

STEP 6: Use the information in Steps 1 through 5 to draw a complete graph of the function by hand.

STEP 7: Find the domain and the range of the function.

STEP 8: Use the graph to determine where the function is increasing and where it is decreasing.

━━━━━ **Now Work** PROBLEM 87

5 Build Cubic Models from Data

In Section 4.2 we found the line of best fit from data, and in Section 4.4 we found the quadratic function of best fit. It is also possible to find polynomial functions of best fit. However, most statisticians do not recommend finding polynomials of best fit of degree higher than 3.

Data that follow a cubic relation should look like Figure 24(a) or (b).

Figure 24

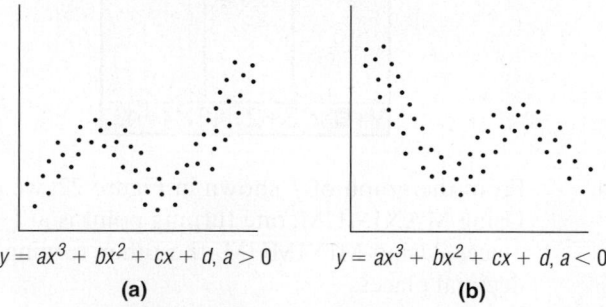

$y = ax^3 + bx^2 + cx + d, a > 0$ $y = ax^3 + bx^2 + cx + d, a < 0$

(a) (b)

EXAMPLE 12 **A Cubic Function of Best Fit**

The data in Table 8 represent the weekly cost C (in thousands of dollars) of printing x thousand textbooks.

(a) Draw a scatter diagram of the data using x as the independent variable and C as the dependent variable. Comment on the type of relation that may exist between the two variables x and C.

(b) Using a graphing utility, find the cubic function of best fit $C = C(x)$ that models the relation between number of texts and cost.

(c) Graph the cubic function of best fit on your scatter diagram.

(d) Use the function found in part (b) to predict the cost of printing 22 thousand texts per week.

Table 8

Number of Textbooks, x	Cost, C
0	100
5	128.1
10	144
13	153.5
17	161.2
18	162.6
20	166.3
23	178.9
25	190.2
27	221.8

Solution

(a) Figure 25 shows the scatter diagram. A cubic relation may exist between the two variables.

(b) Upon executing the CUBIC REGression program, we obtain the results shown in Figure 26. The output that the utility provides shows us the equation $y = ax^3 + bx^2 + cx + d$. The cubic function of best fit to the data is $C(x) = 0.0155x^3 - 0.5951x^2 + 9.1502x + 98.4327$.

(c) Figure 27 shows the graph of the cubic function of best fit on the scatter diagram. The function fits the data reasonably well.

Figure 25

Figure 26

```
CubicReg
y=ax³+bx²+cx+d
a=.0154590051
b=-.5951424724
c=9.150171681
d=98.43272255
```

Figure 27

(d) Evaluate the function $C(x)$ at $x = 22$.

$$C(22) = 0.0155(22)^3 - 0.5951(22)^2 + 9.1502(22) + 98.4327 \approx 176.8$$

The model predicts that the cost of printing 22 thousand textbooks in a week will be 176.8 thousand dollars, that is $176,800.

5.1 Assess Your Understanding

'Are You Prepared?' *Answers are given at the end of these exercises. If you get a wrong answer, read the pages listed in red.*

1. The intercepts of the equation $9x^2 + 4y = 36$ are _____. (pp. 159–160)

2. Is the expression $4x^3 - 3.6x^2 - \sqrt{2}$ a polynomial? If so, what is its degree? (pp. 39–47)

3. To graph $y = x^2 - 4$, you would shift the graph of $y = x^2$ _____ a distance of _____ units. (pp. 244–251)

4. Use a graphing utility to approximate (rounded to two decimal places) the local maximum value and local minimum value of $f(x) = x^3 - 2x^2 - 4x + 5$, for $-3 < x < 3$. (p. 228)

5. **True or False** The x-intercepts of the graph of a function $y = f(x)$ are the real solutions of the equation $f(x) = 0$. (pp. 215–217)

6. If $g(5) = 0$, what point is on the graph of g? What is the corresponding x-intercept of the graph of g? (pp. 215–217)

Concepts and Vocabulary

7. The graph of every polynomial function is both _____ and _____.

8. If r is a real zero of even multiplicity of a function f, then the graph of f _____ (crosses/touches) the x-axis at r.

9. The graphs of power functions of the form $f(x) = x^n$, where n is an even integer, always contain the points _____, _____, and _____.

10. If r is a solution to the equation $f(x) = 0$, name three additional statements that can be made about f and r assuming f is a polynomial function.

11. The points at which a graph changes direction (from increasing to decreasing or decreasing to increasing) are called _____.

12. The graph of the function $f(x) = 3x^4 - x^3 + 5x^2 - 2x - 7$ will behave like the graph of _____ for large values of $|x|$.

13. If $f(x) = -2x^5 + x^3 - 5x^2 + 7$, then $\lim_{x \to -\infty} f(x) = $ _____ and $\lim_{x \to \infty} f(x) = $ _____.

14. Explain what the notation $\lim_{x \to \infty} f(x) = -\infty$ means.

Skill Building

In Problems 15–26, determine which functions are polynomial functions. For those that are, state the degree. For those that are not, tell why not.

15. $f(x) = 4x + x^3$

16. $f(x) = 5x^2 + 4x^4$

17. $g(x) = \dfrac{1 - x^2}{2}$

18. $h(x) = 3 - \dfrac{1}{2}x$

19. $f(x) = 1 - \dfrac{1}{x}$

20. $f(x) = x(x - 1)$

21. $g(x) = x^{3/2} - x^2 + 2$

22. $h(x) = \sqrt{x}(\sqrt{x} - 1)$

23. $F(x) = 5x^4 - \pi x^3 + \dfrac{1}{2}$

24. $F(x) = \dfrac{x^2 - 5}{x^3}$

25. $G(x) = 2(x - 1)^2(x^2 + 1)$

26. $G(x) = -3x^2(x + 2)^3$

In Problems 27–40, use transformations of the graph of $y = x^4$ or $y = x^5$ to graph each function.

27. $f(x) = (x + 1)^4$

28. $f(x) = (x - 2)^5$

29. $f(x) = x^5 - 3$

30. $f(x) = x^4 + 2$

31. $f(x) = \dfrac{1}{2}x^4$

32. $f(x) = 3x^5$

33. $f(x) = -x^5$

34. $f(x) = -x^4$

35. $f(x) = (x - 1)^5 + 2$

36. $f(x) = (x + 2)^4 - 3$

37. $f(x) = 2(x + 1)^4 + 1$

38. $f(x) = \dfrac{1}{2}(x - 1)^5 - 2$

39. $f(x) = 4 - (x - 2)^5$

40. $f(x) = 3 - (x + 2)^4$

In Problems 41–48, form a polynomial function whose real zeros and degree are given. Answers will vary depending on the choice of a leading coefficient.

41. Zeros: $-1, 1, 3$; degree 3

42. Zeros: $-2, 2, 3$; degree 3

43. Zeros: $-3, 0, 4$; degree 3

44. Zeros: $-4, 0, 2$; degree 3

45. Zeros: $-4, -1, 2, 3$; degree 4

46. Zeros: $-3, -1, 2, 5$; degree 4

47. Zeros: -1, multiplicity 1; 3, multiplicity 2; degree 3

48. Zeros: -2, multiplicity 2; 4, multiplicity 1; degree 3

In Problems 49–60, for each polynomial function:

(a) List each real zero and its multiplicity.
(b) Determine whether the graph crosses or touches the x-axis at each x-intercept.
(c) Determine the behavior of the graph near each x-intercept (zero).
(d) Determine the maximum number of turning points on the graph.
(e) Determine the end behavior; that is, find the power function that the graph of f resembles for large values of $|x|$.

49. $f(x) = 3(x - 7)(x + 3)^2$

50. $f(x) = 4(x + 4)(x + 3)^3$

51. $f(x) = 4(x^2 + 1)(x - 2)^3$

52. $f(x) = 2(x - 3)(x^2 + 4)^3$

53. $f(x) = -2\left(x + \dfrac{1}{2}\right)^2(x + 4)^3$

54. $f(x) = \left(x - \dfrac{1}{3}\right)^2(x - 1)^3$

55. $f(x) = (x - 5)^3(x + 4)^2$

56. $f(x) = (x + \sqrt{3})^2(x - 2)^4$

57. $f(x) = 3(x^2 + 8)(x^2 + 9)^2$

58. $f(x) = -2(x^2 + 3)^3$

59. $f(x) = -2x^2(x^2 - 2)$

60. $f(x) = 4x(x^2 - 3)$

In Problems 61–64, identify which of the graphs could be the graph of a polynomial function. For those that could, list the real zeros and state the least degree the polynomial can have. For those that could not, say why not.

61.

62.

63.

64.

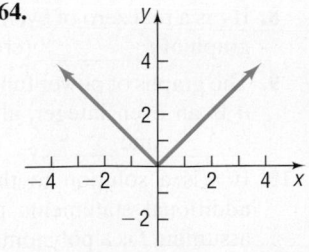

In Problems 65–68, construct a polynomial function that might have the given graph. (More than one answer may be possible.)

65. **66.** **67.** **68.**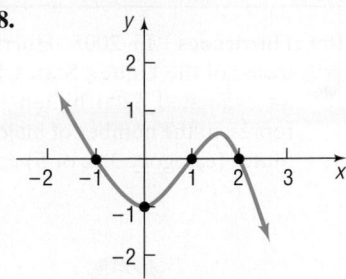

In Problems 69–86, analyze each polynomial function by following Steps 1 through 6 on page 333.

69. $f(x) = x^2(x - 3)$

70. $f(x) = x(x + 2)^2$

71. $f(x) = (x + 4)(x - 2)^2$

72. $f(x) = (x - 1)(x + 3)^2$

73. $f(x) = -2(x + 2)(x - 2)^3$

74. $f(x) = -\dfrac{1}{2}(x + 4)(x - 1)^3$

75. $f(x) = (x + 1)(x - 2)(x + 4)$

76. $f(x) = (x - 1)(x + 4)(x - 3)$

77. $f(x) = x^2(x - 2)(x + 2)$

78. $f(x) = x^2(x - 3)(x + 4)$

79. $f(x) = (x + 1)^2(x - 2)^2$

80. $f(x) = (x + 1)^3(x - 3)$

81. $f(x) = x^2(x - 3)(x + 1)$

82. $f(x) = x^2(x - 3)(x - 1)$

83. $f(x) = (x + 2)^2(x - 4)^2$

84. $f(x) = (x - 2)^2(x + 2)(x + 4)$

85. $f(x) = x^2(x - 2)(x^2 + 3)$

86. $f(x) = x^2(x^2 + 1)(x + 4)$

In Problems 87–94, analyze each polynomial function f by following Steps 1 through 8 on page 336.

87. $f(x) = x^3 + 0.2x^2 - 1.5876x - 0.31752$

88. $f(x) = x^3 - 0.8x^2 - 4.6656x + 3.73248$

89. $f(x) = x^3 + 2.56x^2 - 3.31x + 0.89$

90. $f(x) = x^3 - 2.91x^2 - 7.668x - 3.8151$

91. $f(x) = x^4 - 2.5x^2 + 0.5625$

92. $f(x) = x^4 - 18.5x^2 + 50.2619$

93. $f(x) = 2x^4 - \pi x^3 + \sqrt{5}x - 4$

94. $f(x) = -1.2x^4 + 0.5x^2 - \sqrt{3}x + 2$

Mixed Practice

In Problems 95–102, analyze each polynomial function by following Steps 1 through 6 on page 333.
[Hint: You will need to first factor the polynomial].

95. $f(x) = 4x - x^3$

96. $f(x) = x - x^3$

97. $f(x) = x^3 + x^2 - 12x$

98. $f(x) = x^3 + 2x^2 - 8x$

99. $f(x) = 2x^4 + 12x^3 - 8x^2 - 48x$

100. $f(x) = 4x^3 + 10x^2 - 4x - 10$

101. $f(x) = -x^5 - x^4 + x^3 + x^2$

102. $f(x) = -x^5 + 5x^4 + 4x^3 - 20x^2$

In Problems 103–106, construct a polynomial function f with the given characteristics.

103. Zeros: $-3, 1, 4$; degree 3; y-intercept: 36

104. Zeros: $-4, -1, 2$; degree 3; y-intercept: 16

105. Zeros: -5 (multiplicity 2); 2 (multiplicity 1); 4 (multiplicity 1); degree 4; contains the point $(3, 128)$

106. Zeros: -4 (multiplicity 1); 0 (multiplicity 3); 2 (multiplicity 1); degree 5; contains the point $(-2, 64)$

107. $G(x) = (x + 3)^2(x - 2)$
(a) Identify the x-intercepts of the graph of G.
(b) What are the x-intercepts of the graph of $y = G(x + 3)$?

108. $h(x) = (x + 2)(x - 4)^3$
(a) Identify the x-intercepts of the graph of h.
(b) What are the x-intercepts of the graph of $y = h(x - 2)$?

Applications and Extensions

109. Hurricanes In 2005, Hurricane Katrina struck the Gulf Coast of the United States, killing 1289 people and causing an estimated $200 billion in damage. The following data represent the number of major hurricane strikes in the United States (category 3, 4, or 5) each decade from 1921 to 2000.

Decade, x	Major Hurricanes Striking United States, H
1921–1930, 1	5
1931–1940, 2	8
1941–1950, 3	10
1951–1960, 4	8
1961–1970, 5	6
1971–1980, 6	4
1981–1990, 7	5
1991–2000, 8	5

Source: National Oceanic & Atmospheric Administration

(a) Draw a scatter diagram of the data. Comment on the type of relation that may exist between the two variables.
(b) Use a graphing utility to find the cubic function of best fit that models the relation between decade and number of major hurricanes.
(c) Use the model found in part (b) to predict the number of major hurricanes that struck the United States between 1961 and 1970.
(d) With a graphing utility, draw a scatter diagram of the data and then graph the cubic function of best fit on the scatter diagram.
(e) Concern has risen about the increase in the number and intensity of hurricanes, but some scientists believe this is just a natural fluctuation that could last another decade or two. Use your model to predict the number of major hurricanes that will strike the United States between 2001 and 2010. Does your result appear to agree with what these scientists believe?
(f) From 2001 through 2009, 10 major hurricanes struck the United States. Does this support or contradict your prediction in part (e)?

110. Cost of Manufacturing The following data represent the cost C (in thousands of dollars) of manufacturing Chevy Cobalts and the number x of Cobalts produced.
(a) Draw a scatter diagram of the data using x as the independent variable and C as the dependent variable. Comment on the type of relation that may exist between the two variables C and x.
(b) Use a graphing utility to find the cubic function of best fit $C = C(x)$.
(c) Graph the cubic function of best fit on the scatter diagram.
(d) Use the function found in part (b) to predict the cost of manufacturing 11 Cobalts.
(e) Interpret the y-intercept.

Number of Cobalts Produced, x	Cost, C
0	10
1	23
2	31
3	38
4	43
5	50
6	59
7	70
8	85
9	105
10	135

111. Temperature The following data represent the temperature T (°Fahrenheit) in Kansas City, Missouri, x hours after midnight on May 15, 2010.

Hours after Midnight, x	Temperature (°F), T
3	45.0
6	44.1
9	51.1
12	57.9
15	63.0
18	63.0
21	59.0
24	54.0

Source: The Weather Underground

(a) Draw a scatter diagram of the data. Comment on the type of relation that may exist between the two variables.
(b) Find the average rate of change in temperature from 9 AM to 12 noon.
(c) What is the average rate of change in temperature from 3 PM to 6 PM?
(d) Decide on a function of best fit to these data (linear, quadratic, or cubic) and use this function to predict the temperature at 5 PM.
(e) With a graphing utility, draw a scatter diagram of the data and then graph the function of best fit on the scatter diagram.
(f) Interpret the y-intercept.

112. Future Value of Money Suppose that you make deposits of $500 at the beginning of every year into an Individual Retirement Account (IRA) earning interest r. At the beginning of the first year, the value of the account will be $500; at the beginning of the second year, the value of the account, will be

$$\underbrace{\$500 + \$500r}_{\text{Value of 1st deposit}} + \underbrace{\$500}_{\text{Value of 2nd deposit}} = \$500(1 + r) + \$500 = 500r + 1000$$

(a) Verify that the value of the account at the beginning of the third year is $T(r) = 500r^2 + 1500r + 1500$.

(b) The account value at the beginning of the fourth year is $F(r) = 500r^3 + 2000r^2 + 3000r + 2000$. If the annual rate of interest is $5\% = 0.05$, what will be the value of the account at the beginning of the fourth year?

113. **A Geometric Series** In calculus, you will learn that certain functions can be approximated by polynomial functions. We will explore one such function now.

(a) Using a graphing utility, create a table of values with
$$Y_1 = f(x) = \frac{1}{1 - x} \text{ and } Y_2 = g_2(x) = 1 + x + x^2 + x^3$$
for $-1 < x < 1$ with $\Delta\text{Tbl} = 0.1$.

(b) Using a graphing utility, create a table of values with
$$Y_1 = f(x) = \frac{1}{1 - x} \text{ and } Y_3 = g_3(x) = 1 + x + x^2 + x^3 + x^4$$
for $-1 < x < 1$ with $\Delta\text{Tbl} = 0.1$.

(c) Using a graphing utility, create a table of values with
$$Y_1 = f(x) = \frac{1}{1 - x} \text{ and } Y_4 = g_4(x) = 1 + x + x^2 + x^3 + x^4 + x^5$$
for $-1 < x < 1$ with $\Delta\text{Tbl} = 0.1$.

(d) What do you notice about the values of the function as more terms are added to the polynomial? Are there some values of x for which the approximations are better?

Explaining Concepts: Discussion and Writing

114. Can the graph of a polynomial function have no y-intercept? Can it have no x-intercepts? Explain.

115. Write a few paragraphs that provide a general strategy for graphing a polynomial function. Be sure to mention the following: degree, intercepts, end behavior, and turning points.

116. Make up a polynomial that has the following characteristics: crosses the x-axis at -1 and 4, touches the x-axis at 0 and 2, and is above the x-axis between 0 and 2. Give your polynomial to a fellow classmate and ask for a written critique.

117. Make up two polynomials, not of the same degree, with the following characteristics: crosses the x-axis at -2, touches the x-axis at 1, and is above the x-axis between -2 and 1. Give your polynomials to a fellow classmate and ask for a written critique.

118. The graph of a polynomial function is always smooth and continuous. Name a function studied earlier that is smooth and not continuous. Name one that is continuous, but not smooth.

119. Which of the following statements are true regarding the graph of the cubic polynomial $f(x) = x^3 + bx^2 + cx + d$? (Give reasons for your conclusions.)
(a) It intersects the y-axis in one and only one point.
(b) It intersects the x-axis in at most three points.
(c) It intersects the x-axis at least once.
(d) For $|x|$ very large, it behaves like the graph of $y = x^3$.
(e) It is symmetric with respect to the origin.
(f) It passes through the origin.

120. The illustration shows the graph of a polynomial function.

(a) Is the degree of the polynomial even or odd?
(b) Is the leading coefficient positive or negative?
(c) Is the function even, odd, or neither?
(d) Why is x^2 necessarily a factor of the polynomial?
(e) What is the minimum degree of the polynomial?
(f) Formulate five different polynomials whose graphs could look like the one shown. Compare yours to those of other students. What similarities do you see? What differences?

121. Design a polynomial function with the following characteristics: degree 6; four distinct real zeros, one of multiplicity 3; y-intercept 3; behaves like $y = -5x^6$ for large values of $|x|$. Is this polynomial unique? Compare your polynomial with those of other students. What terms will be the same as everyone else's? Add some more characteristics, such as symmetry or naming the real zeros. How does this modify the polynomial?

Interactive Exercises

Ask your instructor if the applet exercise below is of interest to you.

Multiplicity and Turning Points Open the Multiplicity applet. On the screen you will see the graph of $f(x) = (x + 2)^a x^b(x - 2)^c$ where $a = \{1, 2, 3\}, b = \{1, 2, 3\}$, and $c = \{1, 2, 3, 4\}$.

1. Grab the slider for the exponent a and move it from 1 to 2 to 3. What happens to the graph as the value of a changes? In particular, describe the behavior of the graph around the zero -2.

2. On the same graph, grab the slider for the exponent a and move it to 1. Grab the slider for the exponent b and move it from 1 to 2 to 3. What happens to the graph as the value of b changes? In particular, describe the behavior of the graph around the zero 0.

3. On the same graph, grab the slider for the exponent b and move it to 1. Grab the slider for the exponent c and move it from 1 to 2 to 3 to 4. What happens to the graph as the value of c changes? In particular, describe the behavior of the graph around the zero 2.

4. Experiment with the graph by adjusting a, b, and c. Based on your experiences conjecture the role the exponent plays in the behavior of the graph around each zero of the function.

5. Obtain a graph of the function for the values of a, b, and c in the following table. Conjecture a relation between the degree of a polynomial and the number of turning points after completing the table. In the table, a can be $1, 2,$ or 3; b can be $1, 2,$ or 3; and c can be $1, 2, 3,$ or 4.

Values of a, b, and c	Degree of Polynomial	Number of Turning Points
$a = 1, b = 1, c = 1$	3	
$a = 1, b = 1, c = 2$	4	
$a = 1, b = 1, c = 3$	5	
$a = 1, b = 1, c = 4$		
$a = 1, b = 2, c = 1$		
$a = 1, b = 2, c = 2$		
$a = 1, b = 2, c = 3$		
$a = 1, b = 2, c = 4$		
$a = 1, b = 3, c = 1$		
$a = 1, b = 3, c = 2$		
$a = 1, b = 3, c = 3$		
$a = 1, b = 3, c = 4$		
$a = 2, b = 1, c = 1$		
$a = 2, b = 1, c = 2$		
$a = 2, b = 1, c = 3$		
$a = 2, b = 1, c = 4$		
$a = 2, b = 2, c = 1$		
$a = 2, b = 2, c = 2$		
$a = 3, b = 3, c = 4$		

'Are You Prepared?' Answers

1. $(-2, 0), (2, 0), (0, 9)$ **2.** Yes; 3 **3.** Down; 4 **4.** Local maximum value 6.48 at $x = -0.67$; local minimum value -3 at $x = 2$

5. True **6.** $(5, 0)$; 5

5.2 Properties of Rational Functions

PREPARING FOR THIS SECTION *Before getting started, review the following:*

- Rational Expressions (Chapter R, Section R.7, pp. 62–69)
- Polynomial Division (Chapter R, Section R.4, pp. 44–47)
- Graph of $f(x) = \dfrac{1}{x}$ (Section 2.2, Example 12, p. 164)
- Graphing Techniques: Transformations (Section 3.5, pp. 244–251)

Now Work the 'Are You Prepared?' problems on page 350.

OBJECTIVES **1** Find the Domain of a Rational Function (p. 343)
　　　　　　　　2 Find the Vertical Asymptotes of a Rational Function (p. 346)
　　　　　　　　3 Find the Horizontal or Oblique Asymptote of a Rational Function (p. 347)

Ratios of integers are called *rational numbers*. Similarly, ratios of polynomial functions are called *rational functions*. Examples of rational functions are

$$R(x) = \frac{x^2 - 4}{x^2 + x + 1} \qquad F(x) = \frac{x^3}{x^2 - 4} \qquad G(x) = \frac{3x^2}{x^4 - 1}$$

DEFINITION

A **rational function** is a function of the form

$$R(x) = \frac{p(x)}{q(x)}$$

where p and q are polynomial functions and q is not the zero polynomial. The domain of a rational function is the set of all real numbers except those for which the denominator q is 0.

1 Find the Domain of a Rational Function

EXAMPLE 1

Finding the Domain of a Rational Function

(a) The domain of $R(x) = \dfrac{2x^2 - 4}{x + 5}$ is the set of all real numbers x except -5; that is, the domain is $\{x \mid x \neq -5\}$.

(b) The domain of $R(x) = \dfrac{1}{x^2 - 4}$ is the set of all real numbers x except -2 and 2; that is, the domain is $\{x \mid x \neq -2, x \neq 2\}$.

(c) The domain of $R(x) = \dfrac{x^3}{x^2 + 1}$ is the set of all real numbers.

(d) The domain of $R(x) = \dfrac{x^2 - 1}{x - 1}$ is the set of all real numbers x except 1; that is, the domain is $\{x \mid x \neq 1\}$.

Although $\dfrac{x^2 - 1}{x - 1}$ reduces to $x + 1$, it is important to observe that the functions

$$R(x) = \frac{x^2 - 1}{x - 1} \quad \text{and} \quad f(x) = x + 1$$

are not equal, since the domain of R is $\{x \mid x \neq 1\}$ and the domain of f is the set of all real numbers.

Now Work PROBLEM 15

If $R(x) = \dfrac{p(x)}{q(x)}$ is a rational function and if p and q have no common factors, then the rational function R is said to be in **lowest terms.** For a rational function $R(x) = \dfrac{p(x)}{q(x)}$ in lowest terms, the real zeros, if any, of the numerator in the domain of R are the x-intercepts of the graph of R and so will play a major role in the graph of R. The real zeros of the denominator of R [that is, the numbers x, if any, for which $q(x) = 0$], although not in the domain of R, also play a major role in the graph of R.

We have already discussed the properties of the rational function $y = \dfrac{1}{x}$. (Refer to Example 12, page 164). The next rational function that we take up is

$$H(x) = \frac{1}{x^2}.$$

EXAMPLE 2

Graphing $y = \dfrac{1}{x^2}$

Analyze the graph of $H(x) = \dfrac{1}{x^2}$.

Solution

The domain of $H(x) = \dfrac{1}{x^2}$ is the set of all real numbers x except 0. The graph has no y-intercept, because x can never equal 0. The graph has no x-intercept because the equation $H(x) = 0$ has no solution. Therefore, the graph of H will not cross or touch either of the coordinate axes. Because

$$H(-x) = \frac{1}{(-x)^2} = \frac{1}{x^2} = H(x)$$

H is an even function, so its graph is symmetric with respect to the y-axis.

Table 9 shows the behavior of $H(x) = \dfrac{1}{x^2}$ for selected positive numbers x. (We will use symmetry to obtain the graph of H when $x < 0$.) From the first three rows of Table 9, we see that, as the values of x approach (get closer to) 0, the values of $H(x)$ become larger and larger positive numbers, so H is unbounded in the positive direction. We use limit notation, $\lim\limits_{x \to 0} H(x) = \infty$, read "the limit of $H(x)$ as x approaches zero equals infinity," to mean that $H(x) \to \infty$ as $x \to 0$.

Look at the last four rows of Table 9. As $x \to \infty$, the values of $H(x)$ approach 0 (the end behavior of the graph). In calculus, this is symbolized by writing $\lim\limits_{x \to \infty} H(x) = 0$. Figure 28 shows the graph. Notice the use of red dashed lines to convey the ideas discussed above.

Table 9

x	$H(x) = \dfrac{1}{x^2}$
$\dfrac{1}{2}$	4
$\dfrac{1}{100}$	10,000
$\dfrac{1}{10,000}$	100,000,000
1	1
2	$\dfrac{1}{4}$
100	$\dfrac{1}{10,000}$
10,000	$\dfrac{1}{100,000,000}$

Figure 28

$H(x) = \dfrac{1}{x^2}$

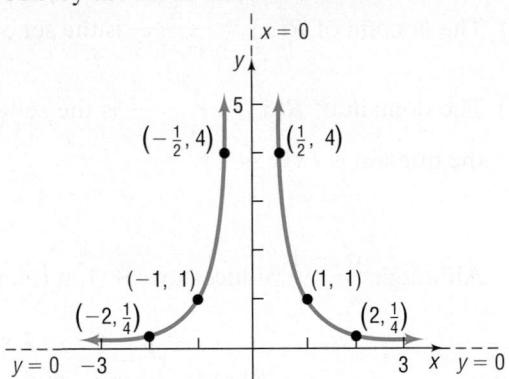

EXAMPLE 3

Using Transformations to Graph a Rational Function

Graph the rational function: $R(x) = \dfrac{1}{(x-2)^2} + 1$

Solution

The domain of R is the set of all real numbers except $x = 2$. To graph R, start with the graph of $y = \dfrac{1}{x^2}$. See Figure 29 for the steps.

Figure 29

Replace x by $x - 2$; shift right 2 units

Add 1; shift up 1 unit

(a) $y = \dfrac{1}{x^2}$ (b) $y = \dfrac{1}{(x-2)^2}$ (c) $y = \dfrac{1}{(x-2)^2} + 1$

━━━➤ **Now Work** PROBLEM 33

Asymptotes

Let's investigate the roles of the vertical line $x = 2$ and the horizontal line $y = 1$ in Figure 29(c).

First, we look at the end behavior of $R(x) = \dfrac{1}{(x-2)^2} + 1$. Table 10(a) shows the values of R at $x = 10, 100, 1000, 10,000$. Notice that, as x becomes unbounded in the positive direction, the values of R approach 1, so $\lim\limits_{x \to \infty} R(x) = 1$. From Table 10(b) we see that, as x becomes unbounded in the negative direction, the values of R also approach 1, so $\lim\limits_{x \to -\infty} R(x) = 1$.

Even though $x = 2$ is not in the domain of R, the behavior of the graph of R near $x = 2$ is important. Table 10(c) shows the values of R at $x = 1.5$, $1.9, 1.99, 1.999,$ and 1.9999. We see that, as x approaches 2 for $x < 2$, denoted $x \to 2^-$, the values of R are increasing without bound, so $\lim\limits_{x \to 2^-} R(x) = \infty$. From Table 10(d), we see that, as x approaches 2 for $x > 2$, denoted $x \to 2^+$, the values of R are also increasing without bound, so $\lim\limits_{x \to 2^+} R(x) = \infty$.

Table 10

x	R(x)
10	1.0156
100	1.0001
1000	1.000001
10,000	1.00000001

(a)

x	R(x)
−10	1.0069
−100	1.0001
−1000	1.000001
−10,000	1.00000001

(b)

x	R(x)
1.5	5
1.9	101
1.99	10,001
1.999	1,000,001
1.9999	100,000,001

(c)

x	R(x)
2.5	5
2.1	101
2.01	10,001
2.001	1,000,001
2.0001	100,000,001

(d)

The vertical line $x = 2$ and the horizontal line $y = 1$ are called *asymptotes* of the graph of R.

DEFINITION

Let R denote a function.

If, as $x \to -\infty$ or as $x \to \infty$, the values of $R(x)$ approach some fixed number L, then the line $y = L$ is a **horizontal asymptote** of the graph of R. [Refer to Figures 30(a) and (b).]

If, as x approaches some number c, the values $|R(x)| \to \infty [R(x) \to -\infty$ or $R(x) \to \infty]$, then the line $x = c$ is a **vertical asymptote** of the graph of R. [Refer to Figures 30(c) and (d).]

Figure 30

(a) End behavior:
As $x \to \infty$, the values of $R(x)$ approach L [$\lim\limits_{x \to \infty} R(x) = L$]. That is, the points on the graph of R are getting closer to the line $y = L$; $y = L$ is a horizontal asymptote.

(b) End behavior:
As $x \to -\infty$, the values of $R(x)$ approach L [$\lim\limits_{x \to -\infty} R(x) = L$]. That is, the points on the graph of R are getting closer to the line $y = L$; $y = L$ is a horizontal asymptote.

(c) As x approaches c, the values of $|R(x)| \to \infty$ [$\lim\limits_{x \to c^-} R(x) = \infty$; $\lim\limits_{x \to c^+} R(x) = \infty$]. That is, the points on the graph of R are getting closer to the line $x = c$; $x = c$ is a vertical asymptote.

(d) As x approaches c, the values of $|R(x)| \to \infty$ [$\lim\limits_{x \to c^-} R(x) = -\infty$; $\lim\limits_{x \to c^+} R(x) = \infty$]. That is, the points on the graph of R are getting closer to the line $x = c$; $x = c$ is a vertical asymptote.

Figure 31

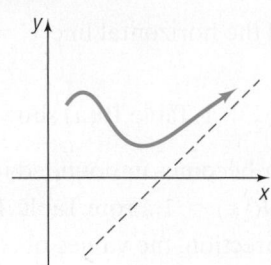

A horizontal asymptote, when it occurs, describes the **end behavior** of the graph as $x \to \infty$ or as $x \to -\infty$. **The graph of a function may intersect a horizontal asymptote.**

A vertical asymptote, when it occurs, describes the behavior of the graph when x is close to some number c. **The graph of a rational function will never intersect a vertical asymptote.**

There is a third possibility. If, as $x \to -\infty$ or as $x \to \infty$, the value of a rational function $R(x)$ approaches a linear expression $ax + b$, $a \neq 0$, then the line $y = ax + b$, $a \neq 0$, is an **oblique asymptote** of R. Figure 31 shows an oblique asymptote. An oblique asymptote, when it occurs, describes the end behavior of the graph. **The graph of a function may intersect an oblique asymptote.**

─────**Now Work** PROBLEM 25

2 Find the Vertical Asymptotes of a Rational Function

The vertical asymptotes of a rational function $R(x) = \dfrac{p(x)}{q(x)}$, in lowest terms, are located at the real zeros of the denominator of $q(x)$. Suppose that r is a real zero of q, so $x - r$ is a factor of q. As x approaches r, symbolized as $x \to r$, the values of $x - r$ approach 0, causing the ratio to become unbounded, that is, $|R(x)| \to \infty$. Based on the definition, we conclude that the line $x = r$ is a vertical asymptote.

THEOREM

Locating Vertical Asymptotes

A rational function $R(x) = \dfrac{p(x)}{q(x)}$, *in lowest terms*, will have a vertical asymptote $x = r$ if r is a real zero of the *denominator* q. That is, if $x - r$ is a factor of the denominator q of a rational function $R(x) = \dfrac{p(x)}{q(x)}$, in lowest terms, R will have the vertical asymptote $x = r$.

WARNING If a rational function is not in lowest terms, an application of this theorem may result in an incorrect listing of vertical asymptotes. ∎

EXAMPLE 4

Finding Vertical Asymptotes

Find the vertical asymptotes, if any, of the graph of each rational function.

(a) $F(x) = \dfrac{x + 3}{x - 1}$ (b) $R(x) = \dfrac{x}{x^2 - 4}$

(c) $H(x) = \dfrac{x^2}{x^2 + 1}$ (d) $G(x) = \dfrac{x^2 - 9}{x^2 + 4x - 21}$

Solution

(a) F is in lowest terms and the only zero of the denominator is 1. The line $x = 1$ is the vertical asymptote of the graph of F.

WARNING In Example 4(a), the vertical asymptote is x = 1. Do not say that the vertical asymptote is 1. ∎

(b) R is in lowest terms and the zeros of the denominator $x^2 - 4$ are -2 and 2. The lines $x = -2$ and $x = 2$ are the vertical asymptotes of the graph of R.

(c) H is in lowest terms and the denominator has no real zeros, because the equation $x^2 + 1 = 0$ has no real solutions. The graph of H has no vertical asymptotes.

(d) Factor the numerator and denominator of $G(x)$ to determine if it is in lowest terms.

$$G(x) = \frac{x^2 - 9}{x^2 + 4x - 21} = \frac{(x + 3)(x - 3)}{(x + 7)(x - 3)} = \frac{x + 3}{x + 7} \qquad x \neq 3$$

The only zero of the denominator of $G(x)$ in lowest terms is -7. The line $x = -7$ is the only vertical asymptote of the graph of G.

As Example 4 points out, rational functions can have no vertical asymptotes, one vertical asymptote, or more than one vertical asymptote.

Exploration

Graph each of the following rational functions:

$$R(x) = \frac{1}{x-1} \qquad R(x) = \frac{1}{(x-1)^2} \qquad R(x) = \frac{1}{(x-1)^3} \qquad R(x) = \frac{1}{(x-1)^4}$$

Each has the vertical asymptote $x = 1$. What happens to the value of $R(x)$ as x approaches 1 from the right side of the vertical asymptote; that is, what is $\lim\limits_{x \to 1^+} R(x)$? What happens to the value of $R(x)$ as x approaches 1 from the left side of the vertical asymptote; that is, what is $\lim\limits_{x \to 1^-} R(x)$? How does the multiplicity of the zero in the denominator affect the graph of R?

──**Now Work** PROBLEM 47 (FIND THE VERTICAL ASYMPTOTES, IF ANY.)

3 Find the Horizontal or Oblique Asymptote of a Rational Function

The procedure for finding horizontal and oblique asymptotes is somewhat more involved. To find such asymptotes, we need to know how the values of a function behave as $x \to -\infty$ or as $x \to \infty$. That is, we need to find the end behavior of the rational function.

If a rational function $R(x)$ is **proper,** that is, if the degree of the numerator is less than the degree of the denominator, then as $x \to -\infty$ or as $x \to \infty$ the value of $R(x)$ approaches 0. Consequently, the line $y = 0$ (the x-axis) is a horizontal asymptote of the graph.

THEOREM If a rational function is proper, the line $y = 0$ is a horizontal asymptote of its graph.

EXAMPLE 5 | **Finding a Horizontal Asymptote**

Find the horizontal asymptote, if one exists, of the graph of

$$R(x) = \frac{x - 12}{4x^2 + x + 1}$$

Solution Since the degree of the numerator, 1, is less than the degree of the denominator, 2, the rational function R is proper. The line $y = 0$ is a horizontal asymptote of the graph of R.

To see why $y = 0$ is a horizontal asymptote of the function R in Example 5, we investigate the behavior of R as $x \to -\infty$ and $x \to \infty$. When $|x|$ is very large, the numerator of R, which is $x - 12$, can be approximated by the power function $y = x$, while the denominator of R, which is $4x^2 + x + 1$, can be approximated by the power function $y = 4x^2$. Applying these ideas to $R(x)$, we find

$$R(x) = \frac{x - 12}{4x^2 + x + 1} \underset{\substack{\uparrow \\ \text{For } |x| \text{ very large}}}{\approx} \frac{x}{4x^2} = \frac{1}{4x} \underset{\substack{\uparrow \\ \text{As } x \to -\infty \text{ or } x \to \infty}}{\to} 0$$

This shows that the line $y = 0$ is a horizontal asymptote of the graph of R.

If a rational function $R(x) = \dfrac{p(x)}{q(x)}$ is **improper,** that is, if the degree of the numerator is greater than or equal to the degree of the denominator, we use long division to write the rational function as the sum of a polynomial $f(x)$ (the quotient) plus a proper rational function $\dfrac{r(x)}{q(x)}$ ($r(x)$ is the remainder). That is, we write

$$R(x) = \frac{p(x)}{q(x)} = f(x) + \frac{r(x)}{q(x)}$$

where $f(x)$ is a polynomial and $\dfrac{r(x)}{q(x)}$ is a proper rational function. Since $\dfrac{r(x)}{q(x)}$ is proper, $\dfrac{r(x)}{q(x)} \to 0$ as $x \to -\infty$ or as $x \to \infty$. As a result,

$$R(x) = \frac{p(x)}{q(x)} \to f(x) \qquad \text{as } x \to -\infty \text{ or as } x \to \infty$$

The possibilities are listed next.

1. If $f(x) = b$, a constant, the line $y = b$ is a horizontal asymptote of the graph of R.

2. If $f(x) = ax + b, a \ne 0$, the line $y = ax + b$ is an oblique asymptote of the graph of R.

3. In all other cases, the graph of R approaches the graph of f, and there are no horizontal or oblique asymptotes.

We illustrate each of the possibilities in Examples 6, 7, and 8.

EXAMPLE 6 **Finding a Horizontal or Oblique Asymptote**

Find the horizontal or oblique asymptote, if one exists, of the graph of

$$H(x) = \frac{3x^4 - x^2}{x^3 - x^2 + 1}$$

Solution Since the degree of the numerator, 4, is greater than the degree of the denominator, 3, the rational function H is improper. To find a horizontal or oblique asymptote, we use long division.

$$
\begin{array}{r}
3x + 3 \\
x^3 - x^2 + 1 \overline{)3x^4 \qquad\quad - x^2} \\
3x^4 - 3x^3 \qquad\quad + 3x \\
\overline{3x^3 - x^2 - 3x} \\
3x^3 - 3x^2 \qquad + 3 \\
\overline{2x^2 - 3x - 3}
\end{array}
$$

As a result,

$$H(x) = \frac{3x^4 - x^2}{x^3 - x^2 + 1} = 3x + 3 + \frac{2x^2 - 3x - 3}{x^3 - x^2 + 1}$$

As $x \to -\infty$ or as $x \to \infty$,

$$\frac{2x^2 - 3x - 3}{x^3 - x^2 + 1} \approx \frac{2x^2}{x^3} = \frac{2}{x} \to 0$$

As $x \to -\infty$ or as $x \to \infty$, we have $H(x) \to 3x + 3$. We conclude that the graph of the rational function H has an oblique asymptote $y = 3x + 3$.

EXAMPLE 7 **Finding a Horizontal or Oblique Asymptote**

Find the horizontal or oblique asymptote, if one exists, of the graph of

$$R(x) = \frac{8x^2 - x + 2}{4x^2 - 1}$$

Solution Since the degree of the numerator, 2, equals the degree of the denominator, 2, the rational function R is improper. To find a horizontal or oblique asymptote, we use long division.

$$\begin{array}{r} 2 \\ 4x^2 - 1 \overline{)8x^2 - x + 2} \\ \underline{8x^2 \qquad - 2} \\ -x + 4 \end{array}$$

As a result,

$$R(x) = \frac{8x^2 - x + 2}{4x^2 - 1} = 2 + \frac{-x + 4}{4x^2 - 1}$$

Then, as $x \to -\infty$ or as $x \to \infty$,

$$\frac{-x + 4}{4x^2 - 1} \approx \frac{-x}{4x^2} = \frac{-1}{4x} \to 0$$

As $x \to -\infty$ or as $x \to \infty$, we have $R(x) \to 2$. We conclude that $y = 2$ is a horizontal asymptote of the graph. ↵

In Example 7, notice that the quotient 2 obtained by long division is the quotient of the leading coefficients of the numerator polynomial and the denominator polynomial $\left(\frac{8}{4}\right)$. This means that we can avoid the long division process for rational functions where the numerator and denominator *are of the same degree* and conclude that the quotient of the leading coefficients will give us the horizontal asymptote.

Now Work PROBLEMS 43 AND 45

| EXAMPLE 8 | **Finding a Horizontal or Oblique Asymptote** |

Find the horizontal or oblique asymptote, if one exists, of the graph of

$$G(x) = \frac{2x^5 - x^3 + 2}{x^3 - 1}$$

Solution Since the degree of the numerator, 5, is greater than the degree of the denominator, 3, the rational function G is improper. To find a horizontal or oblique asymptote, we use long division.

$$\begin{array}{r} 2x^2 - 1 \\ x^3 - 1 \overline{)2x^5 - x^3 \qquad + 2} \\ \underline{2x^5 \qquad - 2x^2} \\ -x^3 + 2x^2 + 2 \\ \underline{-x^3 \qquad + 1} \\ 2x^2 + 1 \end{array}$$

As a result,

$$G(x) = \frac{2x^5 - x^3 + 2}{x^3 - 1} = 2x^2 - 1 + \frac{2x^2 + 1}{x^3 - 1}$$

Then, as $x \to -\infty$ or as $x \to \infty$,

$$\frac{2x^2 + 1}{x^3 - 1} \approx \frac{2x^2}{x^3} = \frac{2}{x} \to 0$$

As $x \to -\infty$ or as $x \to \infty$, we have $G(x) \to 2x^2 - 1$. We conclude that, for large values of $|x|$, the graph of G approaches the graph of $y = 2x^2 - 1$. That is, the graph of G will look like the graph of $y = 2x^2 - 1$ as $x \to -\infty$ or $x \to \infty$. Since $y = 2x^2 - 1$ is not a linear function, G has no horizontal or oblique asymptote. ♩

SUMMARY Finding a Horizontal or Oblique Asymptote of a Rational Function

Consider the rational function

$$R(x) = \frac{p(x)}{q(x)} = \frac{a_n x^n + a_{n-1} x^{n-1} + \cdots + a_1 x + a_0}{b_m x^m + b_{m-1} x^{m-1} + \cdots + b_1 x + b_0}$$

in which the degree of the numerator is n and the degree of the denominator is m.

1. If $n < m$ (the degree of the numerator is less than the degree of the denominator), then R is a proper rational function, and the graph of R will have the horizontal asymptote $y = 0$ (the x-axis).

2. If $n \geq m$ (the degree of the numerator is greater than or equal to the degree of the denominator), then R is improper. Here long division is used.

 (a) If $n = m$ (the degree of the numerator equals the degree of the denominator), the quotient obtained will be the number $\dfrac{a_n}{b_m}$, and the line $y = \dfrac{a_n}{b_m}$ is a horizontal asymptote.

 (b) If $n = m + 1$ (the degree of the numerator is one more than the degree of the denominator), the quotient obtained is of the form $ax + b$ (a polynomial of degree 1), and the line $y = ax + b$ is an oblique asymptote.

 (c) If $n \geq m + 2$ (the degree of the numerator is two or more greater than the degree of the denominator), the quotient obtained is a polynomial of degree 2 or higher, and R has neither a horizontal nor an oblique asymptote. In this case, for very large values of $|x|$, the graph of R will behave like the graph of the quotient.

Note: The graph of a rational function either has one horizontal or one oblique asymptote or else has no horizontal and no oblique asymptote. ∎

5.2 Assess Your Understanding

'Are You Prepared?' *Answers are given at the end of these exercises. If you get a wrong answer, read the pages listed in red.*

1. *True or False* The quotient of two polynomial expressions is a rational expression. (pp. 62–69)

2. What are the quotient and remainder when $3x^4 - x^2$ is divided by $x^3 - x^2 + 1$. (pp. 44–47)

3. Graph $y = \dfrac{1}{x}$. (p. 164)

4. Graph $y = 2(x + 1)^2 - 3$ using transformations. (pp. 244–251)

Concepts and Vocabulary

5. *True or False* The domain of every rational function is the set of all real numbers.

6. If, as $x \to -\infty$ or as $x \to \infty$, the values of $R(x)$ approach some fixed number L, then the line $y = L$ is a _____ _____ of the graph of R.

7. If, as x approaches some number c, the values of $|R(x)| \to \infty$, then the line $x = c$ is a _____ _____ of the graph of R.

8. For a rational function R, if the degree of the numerator is less than the degree of the denominator, then R is _____.

9. *True or False* The graph of a rational function may intersect a horizontal asymptote.

10. *True or False* The graph of a rational function may intersect a vertical asymptote.

11. If a rational function is proper, then _____ is a horizontal asymptote.

12. *True or False* If the degree of the numerator of a rational function equals the degree of the denominator, then the ratio of the leading coefficients gives rise to the horizontal asymptote.

Skill Building

In Problems 13–24, find the domain of each rational function.

13. $R(x) = \dfrac{4x}{x - 3}$

14. $R(x) = \dfrac{5x^2}{3 + x}$

15. $H(x) = \dfrac{-4x^2}{(x - 2)(x + 4)}$

16. $G(x) = \dfrac{6}{(x + 3)(4 - x)}$

17. $F(x) = \dfrac{3x(x - 1)}{2x^2 - 5x - 3}$

18. $Q(x) = \dfrac{-x(1 - x)}{3x^2 + 5x - 2}$

19. $R(x) = \dfrac{x}{x^3 - 8}$

20. $R(x) = \dfrac{x}{x^4 - 1}$

21. $H(x) = \dfrac{3x^2 + x}{x^2 + 4}$

22. $G(x) = \dfrac{x - 3}{x^4 + 1}$

23. $R(x) = \dfrac{3(x^2 - x - 6)}{4(x^2 - 9)}$

24. $F(x) = \dfrac{-2(x^2 - 4)}{3(x^2 + 4x + 4)}$

In Problems 25–30, use the graph shown to find

 (a) *The domain and range of each function* (b) *The intercepts, if any* (c) *Horizontal asymptotes, if any*

 (d) *Vertical asymptotes, if any* (e) *Oblique asymptotes, if any*

25.

26.

27.

28.

29.

30.

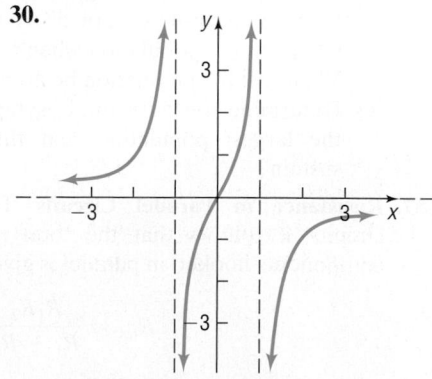

In Problems 31–42, graph each rational function using transformations.

31. $F(x) = 2 + \dfrac{1}{x}$

32. $Q(x) = 3 + \dfrac{1}{x^2}$

33. $R(x) = \dfrac{1}{(x - 1)^2}$

34. $R(x) = \dfrac{3}{x}$

35. $H(x) = \dfrac{-2}{x + 1}$

36. $G(x) = \dfrac{2}{(x + 2)^2}$

37. $R(x) = \dfrac{-1}{x^2 + 4x + 4}$

38. $R(x) = \dfrac{1}{x - 1} + 1$

39. $G(x) = 1 + \dfrac{2}{(x - 3)^2}$

40. $F(x) = 2 - \dfrac{1}{x + 1}$

41. $R(x) = \dfrac{x^2 - 4}{x^2}$

42. $R(x) = \dfrac{x - 4}{x}$

In Problems 43–54, find the vertical, horizontal, and oblique asymptotes, if any, of each rational function.

43. $R(x) = \dfrac{3x}{x + 4}$

44. $R(x) = \dfrac{3x + 5}{x - 6}$

45. $H(x) = \dfrac{x^3 - 8}{x^2 - 5x + 6}$

46. $G(x) = \dfrac{x^3 + 1}{x^2 - 5x - 14}$

47. $T(x) = \dfrac{x^3}{x^4 - 1}$

48. $P(x) = \dfrac{4x^2}{x^3 - 1}$

49. $Q(x) = \dfrac{2x^2 - 5x - 12}{3x^2 - 11x - 4}$

50. $F(x) = \dfrac{x^2 + 6x + 5}{2x^2 + 7x + 5}$

51. $R(x) = \dfrac{6x^2 + 7x - 5}{3x + 5}$

52. $R(x) = \dfrac{8x^2 + 26x - 7}{4x - 1}$

53. $G(x) = \dfrac{x^4 - 1}{x^2 - x}$

54. $F(x) = \dfrac{x^4 - 16}{x^2 - 2x}$

Applications and Extensions

55. Gravity In physics, it is established that the acceleration due to gravity, g (in meters/sec²), at a height h meters above sea level is given by

$$g(h) = \frac{3.99 \times 10^{14}}{(6.374 \times 10^6 + h)^2}$$

where 6.374×10^6 is the radius of Earth in meters.
(a) What is the acceleration due to gravity at sea level?
(b) The Willis Tower in Chicago, Illinois, is 443 meters tall. What is the acceleration due to gravity at the top of the Willis Tower?
(c) The peak of Mount Everest is 8848 meters above sea level. What is the acceleration due to gravity on the peak of Mount Everest?
(d) Find the horizontal asymptote of $g(h)$.
(e) Solve $g(h) = 0$. How do you interpret your answer?

56. Population Model A rare species of insect was discovered in the Amazon Rain Forest. To protect the species, environmentalists declared the insect endangered and transplanted the insect into a protected area. The population P of the insect t months after being transplanted is

$$P(t) = \frac{50(1 + 0.5t)}{2 + 0.01t}$$

(a) How many insects were discovered? In other words, what was the population when $t = 0$?
(b) What will the population be after 5 years?
(c) Determine the horizontal asymptote of $P(t)$. What is the largest population that the protected area can sustain?

57. Resistance in Parallel Circuits From Ohm's law for circuits, it follows that the total resistance R_{tot} of two components hooked in parallel is given by the equation

$$R_{tot} = \frac{R_1 R_2}{R_1 + R_2}$$

where R_1 and R_2 are the individual resistances.

(a) Let $R_1 = 10$ ohms, and graph R_{tot} as a function of R_2.
(b) Find and interpret any asymptotes of the graph obtained in part (a).
(c) If $R_2 = 2\sqrt{R_1}$, what value of R_1 will yield an R_{tot} of 17 ohms?

Source: en.wikipedia.org/wiki/Series_and_parallel_circuits

58. Newton's Method In calculus you will learn that, if

$$p(x) = a_n x^n + a_{n-1} x^{n-1} + \cdots + a_1 x + a_0$$

is a polynomial function, then the *derivative* of $p(x)$ is

$$p'(x) = na_n x^{n-1} + (n - 1)a_{n-1} x^{n-2} + \cdots + 2a_2 x + a_1$$

Newton's Method is an efficient method for approximating the x-intercepts (or real zeros) of a function, such as $p(x)$. The following steps outline Newton's Method.

STEP 1: Select an initial value x_0 that is somewhat close to the x-intercept being sought.

STEP 2: Find values for x using the relation

$$x_{n+1} = x_n - \frac{p(x_n)}{p'(x_n)} \quad n = 1, 2, \ldots$$

until you get two consecutive values x_n and x_{n+1} that agree to whatever decimal place accuracy you desire.

STEP 3: The approximate zero will be x_{n+1}.

Consider the polynomial $p(x) = x^3 - 7x - 40$.
(a) Evaluate $p(5)$ and $p(-3)$.
(b) What might we conclude about a zero of p? Explain.
(c) Use Newton's Method to approximate an x-intercept, r, $-3 < r < 5$, of $p(x)$ to four decimal places.
(d) Use a graphing utility to graph $p(x)$ and verify your answer in part (c).
(e) Using a graphing utility, evaluate $p(r)$ to verify your result.

Explaining Concepts: Discussion and Writing

59. If the graph of a rational function R has the vertical asymptote $x = 4$, the factor $x - 4$ must be present in the denominator of R. Explain why.

60. If the graph of a rational function R has the horizontal asymptote $y = 2$, the degree of the numerator of R equals the degree of the denominator of R. Explain why.

61. Can the graph of a rational function have both a horizontal and an oblique asymptote? Explain.

62. Make up a rational function that has $y = 2x + 1$ as an oblique asymptote. Explain the methodology that you used.

'Are You Prepared?' Answers

1. True **2.** Quotient: $3x + 3$; remainder: $2x^2 - 3x - 3$ **3.** **4.**

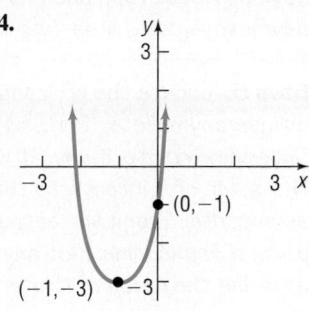

5.3 The Graph of a Rational Function

PREPARING FOR THIS SECTION *Before getting started, review the following:*

- Intercepts (Section 2.2, pp. 159–160)

Now Work the **'Are You Prepared?'** problem on page 365.

OBJECTIVES **1** Analyze the Graph of a Rational Function (p. 353)
 2 Solve Applied Problems Involving Rational Functions (p. 364)

1 Analyze the Graph of a Rational Function

We commented earlier that calculus provides the tools required to graph a polynomial function accurately. The same holds true for rational functions. However, we can gather together quite a bit of information about their graphs to get an idea of the general shape and position of the graph.

EXAMPLE 1

How to Analyze the Graph of a Rational Function

Analyze the graph of the rational function: $R(x) = \dfrac{x - 1}{x^2 - 4}$

Step-by-Step Solution

Step 1: Factor the numerator and denominator of R. Find the domain of the rational function.

$$R(x) = \frac{x - 1}{x^2 - 4} = \frac{x - 1}{(x + 2)(x - 2)}$$

The domain of R is $\{x \,|\, x \neq -2, x \neq 2\}$.

Step 2: Write R in lowest terms.

Because there are no common factors between the numerator and denominator, R is in lowest terms.

Step 3: Locate the intercepts of the graph. Determine the behavior of the graph of R near each x-intercept using the same procedure as for polynomial functions. Plot each x-intercept and indicate the behavior of the graph near it.

Since 0 is in the domain of R, the y-intercept is $R(0) = \dfrac{1}{4}$. The x-intercepts are found by determining the real zeros of the numerator of R that are in the domain of R. By solving $x - 1 = 0$, the only real zero of the numerator is 1, so the only x-intercept of the graph of R is 1. We analyze the behavior of the graph of R near $x = 1$:

$$\text{Near 1:} \quad R(x) = \frac{x - 1}{(x + 2)(x - 2)} \approx \frac{x - 1}{(1 + 2)(1 - 2)} = -\frac{1}{3}(x - 1)$$

Plot the point $(1, 0)$ and draw a line through $(1, 0)$ with a negative slope. See Figure 32(a) on page 355.

Step 4: Locate the vertical asymptotes. Graph each vertical asymptote using a dashed line.

The vertical asymptotes are the zeros of the denominator with the rational function in lowest terms. With R written in lowest terms, we find that the graph of R has two vertical asymptotes: the lines $x = -2$ and $x = 2$.

Step 5: Locate the horizontal or oblique asymptote, if one exists. Determine points, if any, at which the graph of R intersects this asymptote. Graph the asymptotes using a dashed line. Plot any points at which the graph of R intersects the asymptote.

Because the degree of the numerator is less than the degree of the denominator, R is proper and the line $y = 0$ (the x-axis) is a horizontal asymptote of the graph. To determine if the graph of R intersects the horizontal asymptote, solve the equation $R(x) = 0$:

$$\frac{x - 1}{x^2 - 4} = 0$$

$$x - 1 = 0$$

$$x = 1$$

The only solution is $x = 1$, so the graph of R intersects the horizontal asymptote at $(1, 0)$.

Step 6: Use the zeros of the numerator and denominator of R to divide the x-axis into intervals. Determine where the graph of R is above or below the x-axis by choosing a number in each interval and evaluating R there. Plot the points found.

The zero of the numerator, 1, and the zeros of the denominator, -2 and 2, divide the x-axis into four intervals:

$$(-\infty, -2) \quad (-2, 1) \quad (1, 2) \quad (2, \infty)$$

Now construct Table 11.

Table 11

Interval	$(-\infty, -2)$	$(-2, 1)$	$(1, 2)$	$(2, \infty)$
Number chosen	-3	0	$\dfrac{3}{2}$	3
Value of R	$R(-3) = -0.8$	$R(0) = \dfrac{1}{4}$	$R\left(\dfrac{3}{2}\right) = -\dfrac{2}{7}$	$R(3) = 0.4$
Location of graph	Below x-axis	Above x-axis	Below x-axis	Above x-axis
Point on graph	$(-3, -0.8)$	$\left(0, \dfrac{1}{4}\right)$	$\left(\dfrac{3}{2}, -\dfrac{2}{7}\right)$	$(3, 0.4)$

Figure 32(a) shows the asymptotes, the points from Table 11, the y-intercept, the x-intercept, and the behavior of the graph near the x-intercept, 1.

Step 7: Analyze the behavior of the graph of R near each asymptote and indicate this behavior on the graph.

- Since $y = 0$ (the x-axis) is a horizontal asymptote and the graph lies below the x-axis for $x < -2$, we can sketch a portion of the graph by placing a small arrow to the far left and under the x-axis.
- Since the line $x = -2$ is a vertical asymptote and the graph lies below the x-axis for $x < -2$, we place an arrow well below the x-axis and approaching the line $x = -2$ from the left $\left(\lim\limits_{x \to -2^-} R(x) = -\infty\right)$.
- Since the graph is above the x-axis for $-2 < x < 1$ and $x = -2$ is a vertical asymptote, the graph will continue on the right of $x = -2$ at the top $\left(\lim\limits_{x \to -2^+} R(x) = +\infty\right)$. Similar explanations account for the other arrows shown in Figure 32(b).

Step 8: Use the results obtained in Steps 1 through 7 to graph R.

Figure 32(c) shows the graph of R.

Figure 32

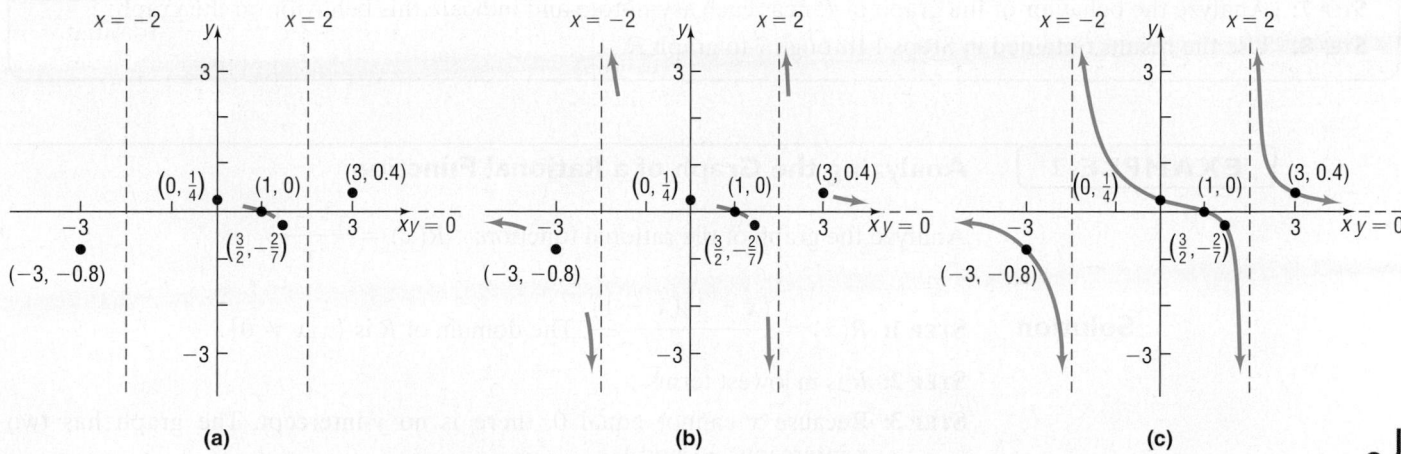

(a) (b) (c)

![calculator icon] **Exploration**

Graph $R(x) = \dfrac{x-1}{x^2-4}$

Result The analysis just completed in Example 1 helps us to set the viewing rectangle to obtain a complete graph. Figure 33(a) shows the graph of $R(x) = \dfrac{x-1}{x^2-4}$ in connected mode, and Figure 33(b) shows it in dot mode. Notice in Figure 33(a) that the graph has vertical lines at $x = -2$ and $x = 2$. This is due to the fact that, when the graphing utility is in connected mode, it will connect the dots between consecutive pixels and vertical lines may occur. We know that the graph of R does not cross the lines $x = -2$ and $x = 2$, since R is not defined at $x = -2$ or $x = 2$. So, when graphing rational functions, dot mode should be used to avoid extraneous vertical lines that are not part of the graph. See Figure 33(b).

Figure 33

Connected mode
(a)

Dot mode
(b)

![pencil icon] **Now Work** PROBLEM **7**

SUMMARY Analyzing the Graph of a Rational Function R

STEP 1: Factor the numerator and denominator of R. Find the domain of the rational function.

STEP 2: Write R in lowest terms.

STEP 3: Locate the intercepts of the graph. The x-intercepts are the zeros of the numerator of R that are in the domain of R. Determine the behavior of the graph of R near each x-intercept.

STEP 4: Determine the vertical asymptotes. Graph each vertical asymptote using a dashed line.

STEP 5: Determine the horizontal or oblique asymptote, if one exists. Determine points, if any, at which the graph of R intersects this asymptote. Graph the asymptote using a dashed line. Plot any points at which the graph of R intersects the asymptote.

(Continued)

STEP 6: Use the zeros of the numerator and denominator of R to divide the x-axis into intervals. Determine where the graph of R is above or below the x-axis by choosing a number in each interval and evaluating R there. Plot the points found.

STEP 7: Analyze the behavior of the graph of R near each asymptote and indicate this behavior on the graph.

STEP 8: Use the results obtained in Steps 1 through 7 to graph R.

EXAMPLE 2 **Analyzing the Graph of a Rational Function**

Analyze the graph of the rational function: $R(x) = \dfrac{x^2 - 1}{x}$

Solution **STEP 1:** $R(x) = \dfrac{(x + 1)(x - 1)}{x}$. The domain of R is $\{x \mid x \neq 0\}$.

STEP 2: R is in lowest terms.

STEP 3: Because x cannot equal 0, there is no y-intercept. The graph has two x-intercepts: -1 and 1.

$$\text{Near } -1: \quad R(x) = \frac{(x + 1)(x - 1)}{x} \approx \frac{(x + 1)(-1 - 1)}{-1} = 2(x + 1)$$

$$\text{Near } 1: \quad R(x) = \frac{(x + 1)(x - 1)}{x} \approx \frac{(1 + 1)(x - 1)}{1} = 2(x - 1)$$

Plot the point $(-1, 0)$ and indicate a line with positive slope there. Plot the point $(1, 0)$ and indicate a line with positive slope there.

STEP 4: The real zero of the denominator with R in lowest terms is 0, so the graph of R has the line $x = 0$ (the y-axis) as a vertical asymptote. Graph $x = 0$ using a dashed line.

STEP 5: Since the degree of the numerator, 2, is one greater than the degree of the denominator, 1, the rational function will have an oblique asymptote. To find the oblique asymptote, we use long division.

NOTE Because the denominator of the rational function is a monomial, we can also find the oblique asymptote as follows:

$$\frac{x^2 - 1}{x} = \frac{x^2}{x} - \frac{1}{x} = x - \frac{1}{x}$$

Since $\dfrac{1}{x} \to 0$ as $x \to \infty$, $y = x$ is the oblique asymptote. ∎

$$\begin{array}{r} x \\ x \overline{)\, x^2 - 1\,} \\ \underline{x^2 } \\ -1 \end{array}$$

The quotient is x, so the line $y = x$ is an oblique asymptote of the graph. Graph $y = x$ using a dashed line.

To determine whether the graph of R intersects the asymptote $y = x$, we solve the equation $R(x) = x$.

$$R(x) = \frac{x^2 - 1}{x} = x$$

$$x^2 - 1 = x^2$$

$$-1 = 0 \quad \text{Impossible}$$

We conclude that the equation $\dfrac{x^2 - 1}{x} = x$ has no solution, so the graph of R does not intersect the line $y = x$.

COMMENT Notice that R is an odd function and so its graph is symmetric with respect to the origin. This observation reduces the work involved in graphing R. ∎

STEP 6: The zeros of the numerator are -1 and 1; the zero of the denominator is 0. Use these values to divide the x-axis into four intervals:

$$(-\infty, -1) \quad (-1, 0) \quad (0, 1) \quad (1, \infty)$$

Now construct Table 12. Plot the points from Table 12. You should now have Figure 34(a).

Table 12

	-1	0	1	$\rightarrow x$
Interval	$(-\infty, -1)$	$(-1, 0)$	$(0, 1)$	$(1, \infty)$
Number chosen	-2	$-\dfrac{1}{2}$	$\dfrac{1}{2}$	2
Value of R	$R(-2) = -\dfrac{3}{2}$	$R\left(-\dfrac{1}{2}\right) = \dfrac{3}{2}$	$R\left(\dfrac{1}{2}\right) = -\dfrac{3}{2}$	$R(2) = \dfrac{3}{2}$
Location of graph	Below x-axis	Above x-axis	Below x-axis	Above x-axis
Point on graph	$\left(-2, -\dfrac{3}{2}\right)$	$\left(-\dfrac{1}{2}, \dfrac{3}{2}\right)$	$\left(\dfrac{1}{2}, -\dfrac{3}{2}\right)$	$\left(2, \dfrac{3}{2}\right)$

STEP 7: Since the graph of R is below the x-axis for $x < -1$ and is above the x-axis for $x > 1$, and since the graph of R does not intersect the oblique asymptote $y = x$, the graph of R will approach the line $y = x$ as shown in Figure 34(b).

Since the graph of R is above the x-axis for $-1 < x < 0$, the graph of R will approach the vertical asymptote $x = 0$ at the top to the left of $x = 0$ $\left[\lim\limits_{x \to 0^-} R(x) = \infty\right]$; since the graph of R is below the x-axis for $0 < x < 1$, the graph of R will approach the vertical asymptote $x = 0$ at the bottom to the right of $x = 0$ $\left[\lim\limits_{x \to 0^+} R(x) = -\infty\right]$. See Figure 34(b).

STEP 8: The complete graph is given in Figure 34(c).

Figure 34

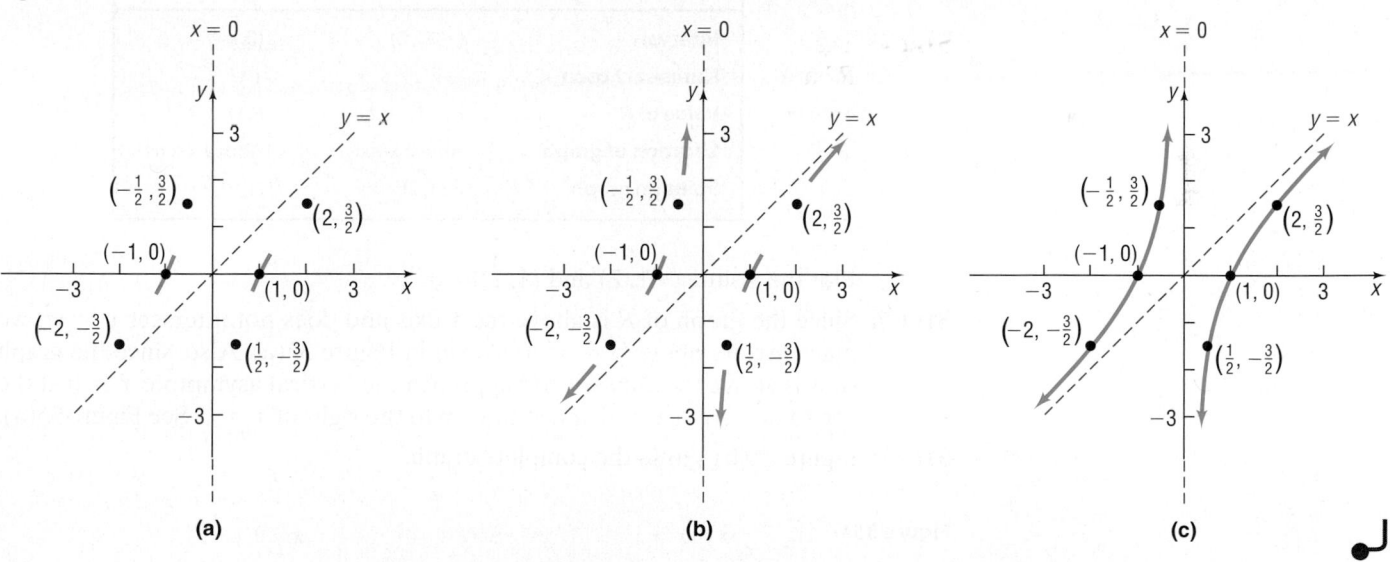

(a) (b) (c)

Seeing the Concept

Graph $R(x) = \dfrac{x^2 - 1}{x}$ and compare what you see with Figure 34(c). Could you have predicted from the graph that $y = x$ is an oblique asymptote? Graph $y = x$ and ZOOM-OUT. What do you observe?

Now Work PROBLEM 15

EXAMPLE 3 **Analyzing the Graph of a Rational Function**

Analyze the graph of the rational function: $R(x) = \dfrac{x^4 + 1}{x^2}$

Solution **STEP 1:** R is completely factored. The domain of R is $\{x \mid x \neq 0\}$.

STEP 2: R is in lowest terms.

STEP 3: There is no y-intercept. Since $x^4 + 1 = 0$ has no real solutions, there are no x-intercepts.

STEP 4: R is in lowest terms, so $x = 0$ (the y-axis) is a vertical asymptote of R. Graph the line $x = 0$ using dashes.

STEP 5: Since the degree of the numerator, 4, is two more than the degree of the denominator, 2, the rational function will not have a horizontal or oblique asymptote. We use long division to find the end behavior of R.

$$\begin{array}{r} x^2 \\ x^2 \overline{)x^4 + 1} \\ \underline{x^4 } \\ 1 \end{array}$$

The quotient is x^2, so the graph of R will approach the graph of $y = x^2$ as $x \to -\infty$ and as $x \to \infty$. The graph of R does not intersect $y = x^2$. Do you know why? Graph $y = x^2$ using dashes.

STEP 6: The numerator has no real zeros, and the denominator has one real zero at 0. We divide the x-axis into the two intervals

$$(-\infty, 0) \qquad (0, \infty)$$

and construct Table 13.

Table 13

	$(-\infty, 0)$	$(0, \infty)$
Interval	$(-\infty, 0)$	$(0, \infty)$
Number chosen	-1	1
Value of R	$R(-1) = 2$	$R(1) = 2$
Location of graph	Above x-axis	Above x-axis
Point on graph	$(-1, 2)$	$(1, 2)$

Plot the points $(-1, 2)$ and $(1, 2)$.

STEP 7: Since the graph of R is above the x-axis and does not intersect $y = x^2$, we place arrows above $y = x^2$ as shown in Figure 35(a). Also, since the graph of R is above the x-axis, it will approach the vertical asymptote $x = 0$ at the top to the left of $x = 0$ and at the top to the right of $x = 0$. See Figure 35(a).

STEP 8: Figure 35(b) shows the complete graph.

NOTE Notice that R in Example 3 is an even function. Do you see the symmetry about the y-axis in the graph of R? ∎

Figure 35

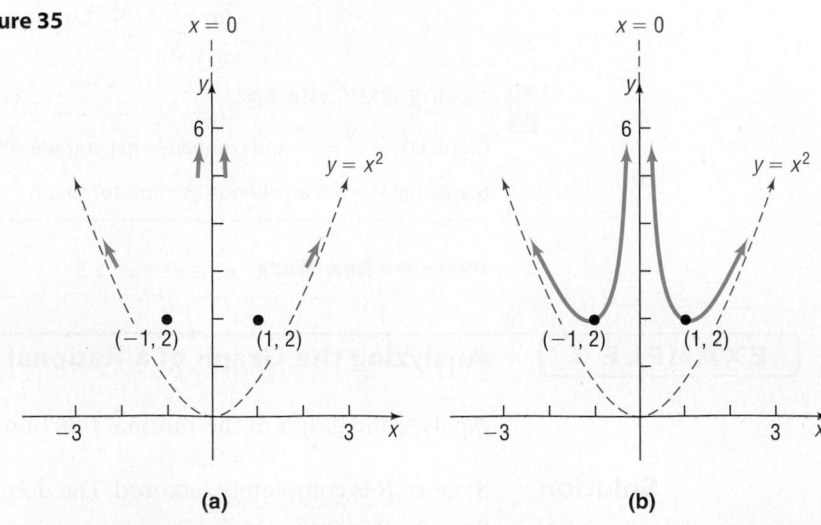

(a)　　　　　(b)

Seeing the Concept

Graph $R(x) = \dfrac{x^4 + 1}{x^2}$ and compare what you see with Figure 35(b). Use MINIMUM to find the two turning points. Enter $Y_2 = x^2$ and ZOOM-OUT. What do you see?

Now Work PROBLEM 13

EXAMPLE 4 **Analyzing the Graph of a Rational Function**

Analyze the graph of the rational function: $R(x) = \dfrac{3x^2 - 3x}{x^2 + x - 12}$

Solution **STEP 1:** Factor R to get

$$R(x) = \frac{3x(x - 1)}{(x + 4)(x - 3)}$$

The domain of R is $\{x \mid x \neq -4, x \neq 3\}$.

STEP 2: R is in lowest terms.

STEP 3: The y-intercept is $R(0) = 0$. Plot the point $(0, 0)$. Since the real solutions of the equation $3x(x - 1) = 0$ are $x = 0$ and $x = 1$, the graph has two x-intercepts, 0 and 1. We determine the behavior of the graph of R near each x-intercept.

$$\text{Near } 0: \quad R(x) = \frac{3x(x - 1)}{(x + 4)(x - 3)} \approx \frac{3x(0 - 1)}{(0 + 4)(0 - 3)} = \frac{1}{4}x$$

$$\text{Near } 1: \quad R(x) = \frac{3x(x - 1)}{(x + 4)(x - 3)} \approx \frac{3(1)(x - 1)}{(1 + 4)(1 - 3)} = -\frac{3}{10}(x - 1)$$

Plot the point $(0, 0)$ and show a line with positive slope there. Plot the point $(1, 0)$ and show a line with negative slope there.

STEP 4: R is in lowest terms. The real solutions of the equation $(x + 4)(x - 3) = 0$ are $x = -4$ and $x = 3$, so the graph of R has two vertical asymptotes, the lines $x = -4$ and $x = 3$. Graph these lines using dashes.

STEP 5: Since the degree of the numerator equals the degree of the denominator, the graph has a horizontal asymptote. To find it, form the quotient of the leading coefficient of the numerator, 3, and the leading coefficient of the denominator, 1. The graph of R has the horizontal asymptote $y = 3$.

To find out whether the graph of R intersects the asymptote, solve the equation $R(x) = 3$.

$$R(x) = \frac{3x^2 - 3x}{x^2 + x - 12} = 3$$
$$3x^2 - 3x = 3x^2 + 3x - 36$$
$$-6x = -36$$
$$x = 6$$

The graph intersects the line $y = 3$ at $x = 6$, and $(6, 3)$ is a point on the graph of R. Plot the point $(6, 3)$ and graph the line $y = 3$ using dashes.

STEP 6: The real zeros of the numerator, 0 and 1, and the real zeros of the denominator, -4 and 3, divide the x-axis into five intervals:

$$(-\infty, -4) \quad (-4, 0) \quad (0, 1) \quad (1, 3) \quad (3, \infty)$$

Construct Table 14. Plot the points from Table 14. Figure 36(a) shows the graph we have so far.

Table 14

Interval	$(-\infty, -4)$	$(-4, 0)$	$(0, 1)$	$(1, 3)$	$(3, \infty)$
Number chosen	-5	-2	$\dfrac{1}{2}$	2	4
Value of R	$R(-5) = 11.25$	$R(-2) = -1.8$	$R\left(\dfrac{1}{2}\right) = \dfrac{1}{15}$	$R(2) = -1$	$R(4) = 4.5$
Location of graph	Above x-axis	Below x-axis	Above x-axis	Below x-axis	Above x-axis
Point on graph	$(-5, 11.25)$	$(-2, -1.8)$	$\left(\dfrac{1}{2}, \dfrac{1}{15}\right)$	$(2, -1)$	$(4, 4.5)$

STEP 7: • Since the graph of R is above the x-axis for $x < -4$ and only crosses the line $y = 3$ at $(6, 3)$, as x approaches $-\infty$ the graph of R will approach the horizontal asymptote $y = 3$ from above ($\lim\limits_{x \to -\infty} R(x) = 3$).

• The graph of R will approach the vertical asymptote $x = -4$ at the top to the left of $x = -4$ ($\lim\limits_{x \to -4^-} R(x) = +\infty$) and at the bottom to the right of $x = -4$ ($\lim\limits_{x \to -4^+} R(x) = -\infty$).

• The graph of R will approach the vertical asymptote $x = 3$ at the bottom to the left of $x = 3$ ($\lim\limits_{x \to 3^-} R(x) = -\infty$) and at the top to the right of $x = 3$ ($\lim\limits_{x \to 3^+} R(x) = +\infty$).

• We do not know whether the graph of R crosses or touches the line $y = 3$ at $(6, 3)$. To see whether the graph, in fact, crosses or touches the line $y = 3$, we plot an additional point to the right of $(6, 3)$. We use $x = 7$ to find $R(7) = \dfrac{63}{22} < 3$. The graph crosses $y = 3$ at $x = 6$. Because $(6, 3)$ is the only point where the graph of R intersects the asymptote $y = 3$, the graph must approach the line $y = 3$ from below as $x \to \infty$ ($\lim\limits_{x \to \infty} R(x) = 3$). See Figure 36(b).

STEP 8: The complete graph is shown in Figure 36(c).

Figure 36

(a)

(b)

(c)

Exploration

Graph $R(x) = \dfrac{3x^2 - 3x}{x^2 + x - 12}$

Result Figure 37 shows the graph in connected mode, and Figure 38(a) shows it in dot mode. Neither graph displays clearly the behavior of the function between the two x-intercepts, 0 and 1. Nor do they clearly display the fact that the graph crosses the horizontal asymptote at $(6, 3)$. To see these parts better, we graph R for $-1 \le x \le 2$ [Figure 38(b)] and for $4 \le x \le 60$ [Figure 39(b)].

Figure 37

Connected mode

Figure 38

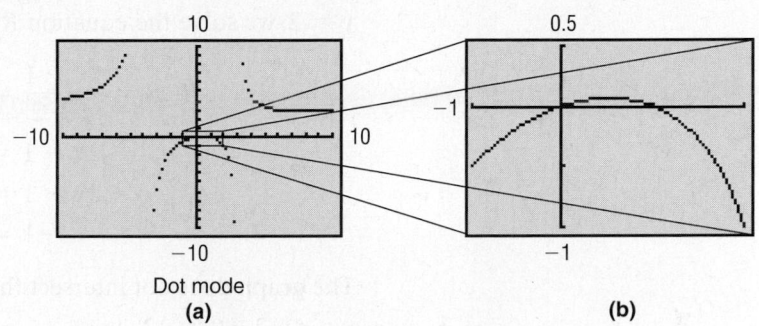

Dot mode
(a)

(b)

Figure 39

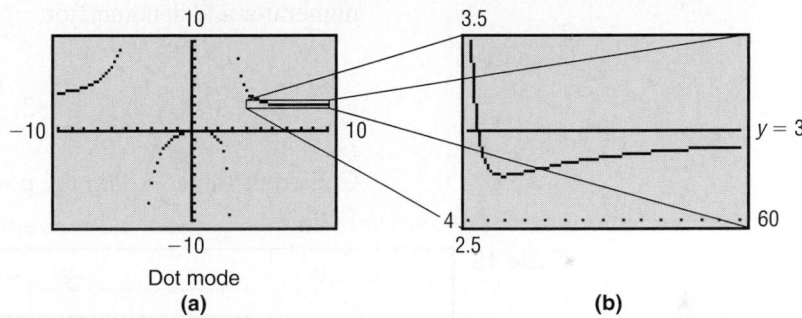

Dot mode
(a)

(b)

The new graphs reflect the behavior produced by the analysis. Furthermore, we observe two turning points, one between 0 and 1 and the other to the right of 6. Rounded to two decimal places, these turning points are $(0.52, 0.07)$ and $(11.48, 2.75)$.

EXAMPLE 5 | **Analyzing the Graph of a Rational Function with a Hole**

Analyze the graph of the rational function: $\quad R(x) = \dfrac{2x^2 - 5x + 2}{x^2 - 4}$

Solution **STEP 1:** Factor R and obtain

$$R(x) = \frac{(2x - 1)(x - 2)}{(x + 2)(x - 2)}$$

The domain of R is $\{x \mid x \ne -2, x \ne 2\}$.

STEP 2: In lowest terms,

$$R(x) = \frac{2x - 1}{x + 2} \qquad x \ne -2, x \ne 2$$

STEP 3: The y-intercept is $R(0) = -\dfrac{1}{2}$. Plot the point $\left(0, -\dfrac{1}{2}\right)$.
The graph has one x-intercept: $\dfrac{1}{2}$.

Near $\dfrac{1}{2}$: $\quad R(x) = \dfrac{2x - 1}{x + 2} \approx \dfrac{2x - 1}{\dfrac{1}{2} + 2} = \dfrac{2}{5}(2x - 1)$

Plot the point $\left(\dfrac{1}{2}, 0\right)$ showing a line with positive slope.

STEP 4: Since $x + 2$ is the only factor of the denominator of $R(x)$ *in lowest terms*, the graph has one vertical asymptote, $x = -2$. However, the rational function is undefined at both $x = 2$ and $x = -2$. Graph the line $x = -2$ using dashes.

STEP 5: Since the degree of the numerator equals the degree of the denominator, the graph has a horizontal asymptote. To find it, form the quotient of the leading coefficient of the numerator, 2, and the leading coefficient of the denominator, 1. The graph of R has the horizontal asymptote $y = 2$. Graph the line $y = 2$ using dashes.

To find out whether the graph of R intersects the horizontal asymptote $y = 2$, we solve the equation $R(x) = 2$.

$$R(x) = \frac{2x - 1}{x + 2} = 2$$
$$2x - 1 = 2(x + 2)$$
$$2x - 1 = 2x + 4$$
$$-1 = 4 \qquad \textit{Impossible}$$

The graph does not intersect the line $y = 2$.

STEP 6: Look at the factored expression for R in Step 1. The real zeros of the numerator and denominator, -2, $\frac{1}{2}$, and 2, divide the x-axis into four intervals:

$$(-\infty, -2) \qquad \left(-2, \frac{1}{2}\right) \qquad \left(\frac{1}{2}, 2\right) \qquad (2, \infty)$$

Construct Table 15. Plot the points in Table 15.

Table 15

	-2	$1/2$	2	

Interval	$(-\infty, -2)$	$\left(-2, \frac{1}{2}\right)$	$\left(\frac{1}{2}, 2\right)$	$(2, \infty)$
Number chosen	-3	-1	1	3
Value of R	$R(-3) = 7$	$R(-1) = -3$	$R(1) = \frac{1}{3}$	$R(3) = 1$
Location of graph	Above x-axis	Below x-axis	Above x-axis	Above x-axis
Point on graph	$(-3, 7)$	$(-1, -3)$	$\left(1, \frac{1}{3}\right)$	$(3, 1)$

STEP 7: • From Table 15 we know that the graph of R is above the x-axis for $x < -2$.

• From Step 5 we know that the graph of R does not intersect the asymptote $y = 2$. Therefore, the graph of R will approach $y = 2$ from above as $x \rightarrow -\infty$ and will approach the vertical asymptote $x = -2$ at the top from the left.

• Since the graph of R is below the x-axis for $-2 < x < \frac{1}{2}$, the graph of R will approach $x = -2$ at the bottom from the right.

• Finally, since the graph of R is above the x-axis for $x > \frac{1}{2}$ and does not intersect the horizontal asymptote $y = 2$, the graph of R will approach $y = 2$ from below as $x \rightarrow \infty$. See Figure 40(a).

NOTE The coordinates of the hole were obtained by evaluating R in lowest terms at 2. R in lowest terms is $\frac{2x - 1}{x + 2}$, which, at $x = 2$, is $\frac{2(2) - 1}{2 + 2} = \frac{3}{4}$. ∎

STEP 8: See Figure 40(b) for the complete graph. Since R is not defined at 2, there is a hole at the point $\left(2, \frac{3}{4}\right)$.

Figure 40

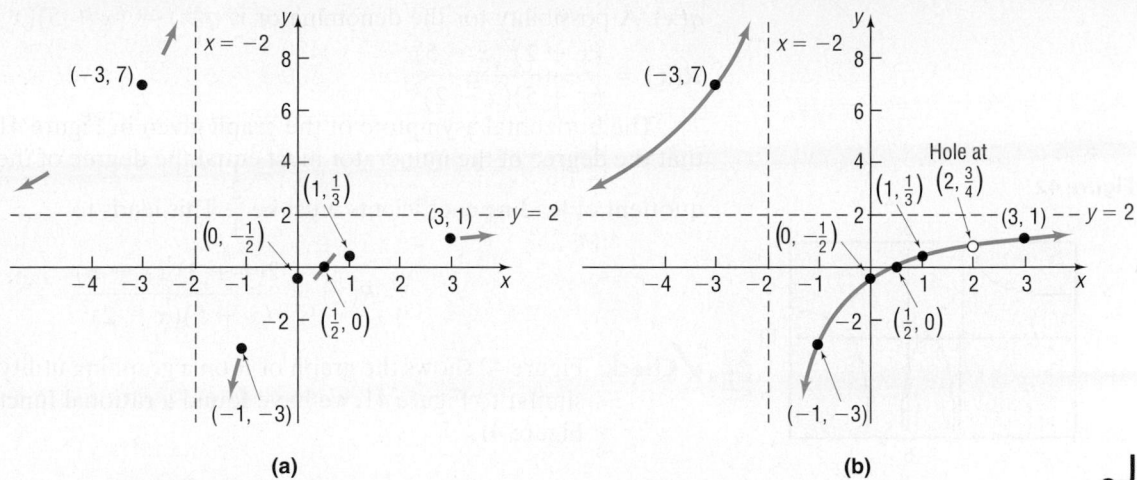

(a) (b)

Exploration

Graph $R(x) = \dfrac{2x^2 - 5x + 2}{x^2 - 4}$. Do you see the hole at $\left(2, \dfrac{3}{4}\right)$? TRACE along the graph. Did you obtain an ERROR at $x = 2$? Are you convinced that an algebraic analysis of a rational function is required in order to accurately interpret the graph obtained with a graphing utility?

As Example 5 shows, **the zeros of the denominator of a rational function give rise to either vertical asymptotes or holes in the graph.**

Now Work PROBLEM 33

EXAMPLE 6 **Constructing a Rational Function from Its Graph**

Find a rational function that might have the graph shown in Figure 41.

Figure 41

Solution The numerator of a rational function $R(x) = \dfrac{p(x)}{q(x)}$ in lowest terms determines the x-intercepts of its graph. The graph shown in Figure 41 has x-intercepts -2 (even multiplicity; graph touches the x-axis) and 5 (odd multiplicity; graph crosses the x-axis). So one possibility for the numerator is $p(x) = (x + 2)^2(x - 5)$.

The denominator of a rational function in lowest terms determines the vertical asymptotes of its graph. The vertical asymptotes of the graph are $x = -5$ and $x = 2$. Since $R(x)$ approaches ∞ to the left of $x = -5$ and $R(x)$ approaches $-\infty$ to the right of $x = -5$, we know that $(x + 5)$ is a factor of odd multiplicity in $q(x)$. Also, $R(x)$ approaches $-\infty$ on both sides of $x = 2$, so $(x - 2)$ is a factor of even multiplicity in

$q(x)$. A possibility for the denominator is $q(x) = (x + 5)(x - 2)^2$. So far we have

$$R(x) = \frac{(x + 2)^2(x - 5)}{(x + 5)(x - 2)^2}.$$

The horizontal asymptote of the graph given in Figure 41 is $y = 2$, so we know that the degree of the numerator must equal the degree of the denominator and the quotient of leading coefficients must be $\frac{2}{1}$. This leads to

$$R(x) = \frac{2(x + 2)^2(x - 5)}{(x + 5)(x - 2)^2}$$

Figure 42

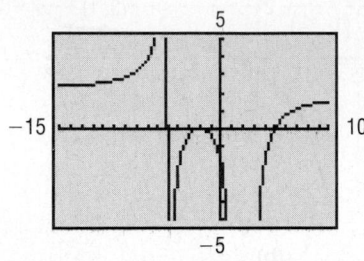

✔**Check:** Figure 42 shows the graph of R on a graphing utility. Since Figure 42 looks similar to Figure 41, we have found a rational function R for the graph in Figure 41.

Now Work PROBLEM 45

2 Solve Applied Problems Involving Rational Functions

EXAMPLE 7 **Finding the Least Cost of a Can**

Reynolds Metal Company manufactures aluminum cans in the shape of a cylinder with a capacity of 500 cubic centimeters $\left(\frac{1}{2} \text{ liter}\right)$. The top and bottom of the can are made of a special aluminum alloy that costs 0.05¢ per square centimeter. The sides of the can are made of material that costs 0.02¢ per square centimeter.

(a) Express the cost of material for the can as a function of the radius r of the can.
(b) Use a graphing utility to graph the function $C = C(r)$.
(c) What value of r will result in the least cost?
(d) What is this least cost?

Solution (a) Figure 43 illustrates the components of a can in the shape of a right circular cylinder. Notice that the material required to produce a cylindrical can of height h and radius r consists of a rectangle of area $2\pi rh$ and two circles, each of area πr^2. The total cost C (in cents) of manufacturing the can is therefore

Figure 43

$$C = \text{Cost of the top and bottom} + \text{Cost of the side}$$

$$= \underbrace{2(\pi r^2)}_{\substack{\text{Total area} \\ \text{of top and} \\ \text{bottom}}} \underbrace{(0.05)}_{\substack{\text{Cost/unit} \\ \text{area}}} + \underbrace{(2\pi rh)}_{\substack{\text{Total} \\ \text{area of} \\ \text{side}}} \underbrace{(0.02)}_{\substack{\text{Cost/unit} \\ \text{area}}}$$

$$= 0.10\pi r^2 + 0.04\pi rh$$

But we have the additional restriction that the height h and radius r must be chosen so that the volume V of the can is 500 cubic centimeters. Since $V = \pi r^2 h$, we have

$$500 = \pi r^2 h \quad \text{so} \quad h = \frac{500}{\pi r^2}$$

Figure 44

Substituting this expression for h, the cost C, in cents, as a function of the radius r is

$$C(r) = 0.10\pi r^2 + 0.04\pi r \cdot \frac{500}{\pi r^2} = 0.10\pi r^2 + \frac{20}{r} = \frac{0.10\pi r^3 + 20}{r}$$

(b) See Figure 44 for the graph of $C = C(r)$.
(c) Using the MINIMUM command, the cost is least for a radius of about 3.17 centimeters.
(d) The least cost is $C(3.17) \approx 9.47$¢.

Now Work PROBLEM 55

5.3 Assess Your Understanding

'Are You Prepared?' *The answer is given at the end of these exercises. If you get a wrong answer, read the pages listed in red.*

1. Find the intercepts of the graph of the equation $y = \dfrac{x^2 - 1}{x^2 - 4}$. (pp. 159–160)

Concepts and Vocabulary

2. If the numerator and the denominator of a rational function have no common factors, the rational function is

___ _____ _____.

3. The graph of a rational function never intersects a _____ asymptote.

4. *True or False* The graph of a rational function sometimes intersects an oblique asymptote.

5. *True or False* The graph of a rational function sometimes has a hole.

6. $R(x) = \dfrac{x(x - 2)^2}{x - 2}$

(a) Find the domain of R.
(b) Find the x-intercepts of R.

Skill Building

In Problems 7–44, follow Steps 1 through 8 on pages 355–356 to analyze the graph of each function.

7. $R(x) = \dfrac{x + 1}{x(x + 4)}$

8. $R(x) = \dfrac{x}{(x - 1)(x + 2)}$

9. $R(x) = \dfrac{3x + 3}{2x + 4}$

10. $R(x) = \dfrac{2x + 4}{x - 1}$

11. $R(x) = \dfrac{3}{x^2 - 4}$

12. $R(x) = \dfrac{6}{x^2 - x - 6}$

13. $P(x) = \dfrac{x^4 + x^2 + 1}{x^2 - 1}$

14. $Q(x) = \dfrac{x^4 - 1}{x^2 - 4}$

15. $H(x) = \dfrac{x^3 - 1}{x^2 - 9}$

16. $G(x) = \dfrac{x^3 + 1}{x^2 + 2x}$

17. $R(x) = \dfrac{x^2}{x^2 + x - 6}$

18. $R(x) = \dfrac{x^2 + x - 12}{x^2 - 4}$

19. $G(x) = \dfrac{x}{x^2 - 4}$

20. $G(x) = \dfrac{3x}{x^2 - 1}$

21. $R(x) = \dfrac{3}{(x - 1)(x^2 - 4)}$

22. $R(x) = \dfrac{-4}{(x + 1)(x^2 - 9)}$

23. $H(x) = \dfrac{x^2 - 1}{x^4 - 16}$

24. $H(x) = \dfrac{x^2 + 4}{x^4 - 1}$

25. $F(x) = \dfrac{x^2 - 3x - 4}{x + 2}$

26. $F(x) = \dfrac{x^2 + 3x + 2}{x - 1}$

27. $R(x) = \dfrac{x^2 + x - 12}{x - 4}$

28. $R(x) = \dfrac{x^2 - x - 12}{x + 5}$

29. $F(x) = \dfrac{x^2 + x - 12}{x + 2}$

30. $G(x) = \dfrac{x^2 - x - 12}{x + 1}$

31. $R(x) = \dfrac{x(x - 1)^2}{(x + 3)^3}$

32. $R(x) = \dfrac{(x - 1)(x + 2)(x - 3)}{x(x - 4)^2}$

33. $R(x) = \dfrac{x^2 + x - 12}{x^2 - x - 6}$

34. $R(x) = \dfrac{x^2 + 3x - 10}{x^2 + 8x + 15}$

35. $R(x) = \dfrac{6x^2 - 7x - 3}{2x^2 - 7x + 6}$

36. $R(x) = \dfrac{8x^2 + 26x + 15}{2x^2 - x - 15}$

37. $R(x) = \dfrac{x^2 + 5x + 6}{x + 3}$

38. $R(x) = \dfrac{x^2 + x - 30}{x + 6}$

39. $f(x) = x + \dfrac{1}{x}$

40. $f(x) = 2x + \dfrac{9}{x}$

41. $f(x) = x^2 + \dfrac{1}{x}$

42. $f(x) = 2x^2 + \dfrac{16}{x}$

43. $f(x) = x + \dfrac{1}{x^3}$

44. $f(x) = 2x + \dfrac{9}{x^3}$

In Problems 45–48, find a rational function that might have the given graph. (More than one answer might be possible.)

45.

46.

47.

48.

Applications and Extensions

49. Drug Concentration The concentration C of a certain drug in a patient's bloodstream t hours after injection is given by

$$C(t) = \frac{t}{2t^2 + 1}$$

(a) Find the horizontal asymptote of $C(t)$. What happens to the concentration of the drug as t increases?
(b) Using your graphing utility, graph $C = C(t)$.
(c) Determine the time at which the concentration is highest.

50. Drug Concentration The concentration C of a certain drug in a patient's bloodstream t minutes after injection is given by

$$C(t) = \frac{50t}{t^2 + 25}$$

(a) Find the horizontal asymptote of $C(t)$. What happens to the concentration of the drug as t increases?
(b) Using your graphing utility, graph $C = C(t)$.
(c) Determine the time at which the concentration is highest.

51. Minimum Cost A rectangular area adjacent to a river is to be fenced in; no fence is needed on the river side. The enclosed area is to be 1000 square feet. Fencing for the side parallel to the river is $5 per linear foot, and fencing for the other two sides is $8 per linear foot; the four corner posts are $25 apiece. Let x be the length of one of the sides perpendicular to the river.
(a) Write a function $C(x)$ that describes the cost of the project.
(b) What is the domain of C?
(c) Use a graphing utility to graph $C = C(x)$.
(d) Find the dimensions of the cheapest enclosure.
Source: www.uncwil.edu/courses/math111hb/PandR/rational/rational.html

52. Doppler Effect The Doppler effect (named after Christian Doppler) is the change in the pitch (frequency) of the sound from a source (s) as heard by an observer (o) when one or both are in motion. If we assume both the source and the observer are moving in the same direction, the relationship is

$$f' = f_a\left(\frac{v - v_o}{v - v_s}\right)$$

where f' = perceived pitch by the observer
 f_a = actual pitch of the source
 v = speed of sound in air (assume 772.4 mph)
 v_o = speed of the observer
 v_s = speed of the source

Suppose that you are traveling down the road at 45 mph and you hear an ambulance (with siren) coming toward you from the rear. The actual pitch of the siren is 600 hertz (Hz).
(a) Write a function $f'(v_s)$ that describes this scenario.
(b) If $f' = 620$ Hz, find the speed of the ambulance.
(c) Use a graphing utility to graph the function.
(d) Verify your answer from part (b).
Source: www.kettering.edu/~drussell/

53. Minimizing Surface Area United Parcel Service has contracted you to design a closed box with a square base that has a volume of 10,000 cubic inches. See the illustration.

(a) Express the surface area S of the box as a function of x.
(b) Using a graphing utility, graph the function found in part (a).
(c) What is the minimum amount of cardboard that can be used to construct the box?
(d) What are the dimensions of the box that minimize the surface area?
(e) Why might UPS be interested in designing a box that minimizes the surface area?

54. Minimizing Surface Area United Parcel Service has contracted you to design an open box with a square base that has a volume of 5000 cubic inches. See the illustration.

(a) Express the surface area S of the box as a function of x.
(b) Using a graphing utility, graph the function found in part (a).
(c) What is the minimum amount of cardboard that can be used to construct the box?
(d) What are the dimensions of the box that minimize the surface area?
(e) Why might UPS be interested in designing a box that minimizes the surface area?

55. Cost of a Can A can in the shape of a right circular cylinder is required to have a volume of 500 cubic centimeters. The top and bottom are made of material that costs 6¢ per square centimeter, while the sides are made of material that costs 4¢ per square centimeter.

(a) Express the total cost C of the material as a function of the radius r of the cylinder. (Refer to Figure 43.)
(b) Graph $C = C(r)$. For what value of r is the cost C a minimum?

56. Material Needed to Make a Drum A steel drum in the shape of a right circular cylinder is required to have a volume of 100 cubic feet.

(a) Express the amount A of material required to make the drum as a function of the radius r of the cylinder.
(b) How much material is required if the drum's radius is 3 feet?
(c) How much material is required if the drum's radius is 4 feet?
(d) How much material is required if the drum's radius is 5 feet?
(e) Graph $A = A(r)$. For what value of r is A smallest?

Explaining Concepts: Discussion and Writing

57. Graph each of the following functions:

$$y = \frac{x^2 - 1}{x - 1} \qquad y = \frac{x^3 - 1}{x - 1}$$

$$y = \frac{x^4 - 1}{x - 1} \qquad y = \frac{x^5 - 1}{x - 1}$$

Is $x = 1$ a vertical asymptote? Why not? What is happening for $x = 1$? What do you conjecture about $y = \dfrac{x^n - 1}{x - 1}$, $n \geq 1$ an integer, for $x = 1$?

58. Graph each of the following functions:

$$y = \frac{x^2}{x - 1} \qquad y = \frac{x^4}{x - 1} \qquad y = \frac{x^6}{x - 1} \qquad y = \frac{x^8}{x - 1}$$

What similarities do you see? What differences?

59. Write a few paragraphs that provide a general strategy for graphing a rational function. Be sure to mention the following: proper, improper, intercepts, and asymptotes.

60. Create a rational function that has the following characteristics: crosses the x-axis at 2; touches the x-axis

at -1; one vertical asymptote at $x = -5$ and another at $x = 6$; and one horizontal asymptote, $y = 3$. Compare your function to a fellow classmate's. How do they differ? What are their similarities?

61. Create a rational function that has the following characteristics: crosses the x-axis at 3; touches the x-axis at -2; one vertical asymptote, $x = 1$; and one horizontal asymptote, $y = 2$. Give your rational function to a fellow classmate and ask for a written critique of your rational function.

62. Create a rational function with the following characteristics: three real zeros, one of multiplicity 2; y-intercept 1; vertical asymptotes, $x = -2$ and $x = 3$; oblique asymptote, $y = 2x + 1$. Is this rational function unique? Compare your function with those of other students. What will be the same as everyone else's? Add some more characteristics, such as symmetry or naming the real zeros. How does this modify the rational function?

63. Explain the circumstances under which the graph of a rational function will have a hole.

'Are You Prepared?' Answer

1. $\left(0, \dfrac{1}{4}\right)$, $(1, 0)$, $(-1, 0)$

5.4 Polynomial and Rational Inequalities

PREPARING FOR THIS SECTION *Before getting started, review the following:*

- Solving Linear Inequalities (Section 1.5, pp. 123–125)
- Solving Quadratic Inequalities (Section 4.5, pp. 309–311)

Now Work the 'Are You Prepared?' problems on page 371.

OBJECTIVES **1** Solve Polynomial Inequalities (p. 368)

2 Solve Rational Inequalities (p. 369)

1 Solve Polynomial Inequalities

In this section we solve inequalities that involve polynomials of degree 3 and higher, along with inequalities that involve rational functions. To help understand the algebraic procedure for solving such inequalities, we use the information obtained in the previous three sections about the graphs of polynomial and rational functions. The approach follows the same methodology that we used to solve inequalities involving quadratic functions.

EXAMPLE 1

Solving a Polynomial Inequality Using Its Graph

Solve $(x + 3)(x - 1)^2 > 0$ by graphing $f(x) = (x + 3)(x - 1)^2$.

Solution

Graph $f(x) = (x + 3)(x - 1)^2$ and determine the intervals of x for which the graph is above the x-axis. These values of x result in $f(x)$ being positive. Using Steps 1 through 6 on page 333, we obtain the graph shown in Figure 45.

Figure 45

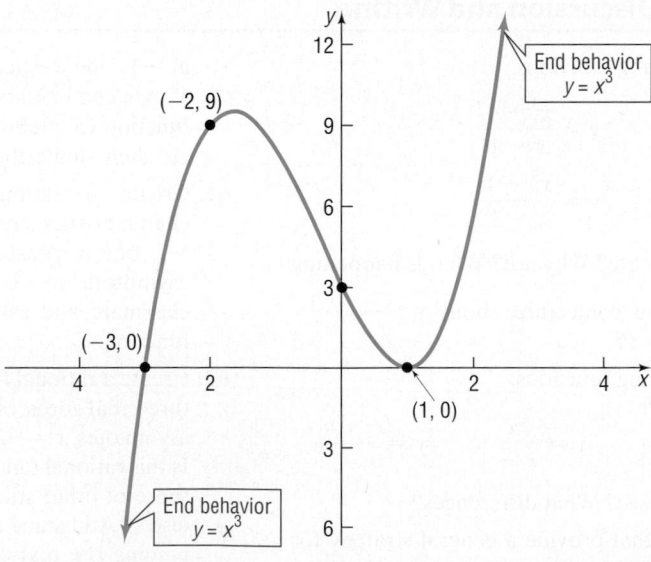

From the graph, we can see that $f(x) > 0$ for $-3 < x < 1$ or $x > 1$. The solution set is $\{x | -3 < x < 1 \text{ or } x > 1\}$ or, using interval notation, $(-3, 1) \cup (1, \infty)$.

Now Work PROBLEM **9**

The results of Example 1 lead to the following approach to solving polynomial and rational inequalities algebraically. Suppose that the polynomial or rational inequality is in one of the forms

$$f(x) < 0 \qquad f(x) > 0 \qquad f(x) \leq 0 \qquad f(x) \geq 0$$

Locate the zeros of f if f is a polynomial function, and locate the zeros of the numerator and the denominator if f is a rational function. If we use these zeros to divide the real number line into intervals, we know that on each interval the graph of f is either above the x-axis $[f(x) > 0]$ or below the x-axis $[f(x) < 0]$. In other words, we have found the solution of the inequality.

| EXAMPLE 2 | **How to Solve a Polynomial Inequality Algebraically** |

Solve the inequality $x^4 > x$ algebraically, and graph the solution set.

Step-by-Step Solution

Step 1: Write the inequality so that a polynomial expression f is on the left side and zero is on the right side.

Rearrange the inequality so that 0 is on the right side.

$$x^4 > x$$
$$x^4 - x > 0 \quad \text{Subtract } x \text{ from both sides of the inequality.}$$

This inequality is equivalent to the one we wish to solve.

Step 2: Determine the real zeros (x-intercepts of the graph) of f.

Find the real zeros of $f(x) = x^4 - x$ by solving $x^4 - x = 0$.

$$x^4 - x = 0$$
$$x(x^3 - 1) = 0 \quad \text{Factor out } x.$$
$$x(x - 1)(x^2 + x + 1) = 0 \quad \text{Factor the difference of two cubes.}$$
$$x = 0 \quad \text{or} \quad x - 1 = 0 \quad \text{or} \quad x^2 + x + 1 = 0 \quad \text{Set each factor equal to zero and solve.}$$
$$x = 0 \quad \text{or} \quad x = 1$$

The equation $x^2 + x + 1 = 0$ has no real solutions. Do you see why?

Step 3: Use the zeros found in Step 2 to divide the real number line into intervals.

Use the real zeros to separate the real number line into three intervals:

$$(-\infty, 0) \qquad (0, 1) \qquad (1, \infty)$$

Step 4: Select a number in each interval, evaluate f at the number, and determine whether f is positive or negative. If f is positive, all values of f in the interval are positive. If f is negative, all values of f in the interval are negative.

Select a test number in each interval found in Step 3 and evaluate $f(x) = x^4 - x$ at each number to determine if $f(x)$ is positive or negative. See Table 16.

Table 16

Interval	$(-\infty, 0)$	$(0, 1)$	$(1, \infty)$
Number chosen	-1	$\dfrac{1}{2}$	2
Value of f	$f(-1) = 2$	$f\left(\dfrac{1}{2}\right) = -\dfrac{7}{16}$	$f(2) = 14$
Conclusion	Positive	Negative	Positive

NOTE If the inequality is not strict (\leq or \geq), include the solutions of $f(x) = 0$ in the solution set. ∎

Since we want to know where $f(x)$ is positive, we conclude that $f(x) > 0$ for all numbers x for which $x < 0$ or $x > 1$. Because the original inequality is strict, numbers x that satisfy the equation $x^4 = x$ are not solutions. The solution set of the inequality $x^4 > x$ is $\{x \mid x < 0 \text{ or } x > 1\}$ or, using interval notation, $(-\infty, 0) \cup (1, \infty)$.

Figure 46 shows the graph of the solution set.

Figure 46

$-2\ -1\ \ 0\ \ 1\ \ 2$

───────**Now Work** PROBLEM 21

2 Solve Rational Inequalities

Just as we presented a graphical approach to help us understand the algebraic procedure for solving inequalities involving polynomials, we present a graphical

approach to help us understand the algebraic procedure for solving inequalities involving rational expressions.

EXAMPLE 3

Solving a Rational Inequality Using Its Graph

Solve $\dfrac{x-1}{x^2-4} \geq 0$ by graphing $R(x) = \dfrac{x-1}{x^2-4}$.

Solution

Graph $R(x) = \dfrac{x-1}{x^2-4}$ and determine the intervals of x such that the graph is above or on the x-axis. Do you see why? These values of x result in $R(x)$ being positive or zero. We graphed $R(x) = \dfrac{x-1}{x^2-4}$ in Example 1, Section 5.3 (pp. 353–354). We reproduce the graph in Figure 47.

Figure 47

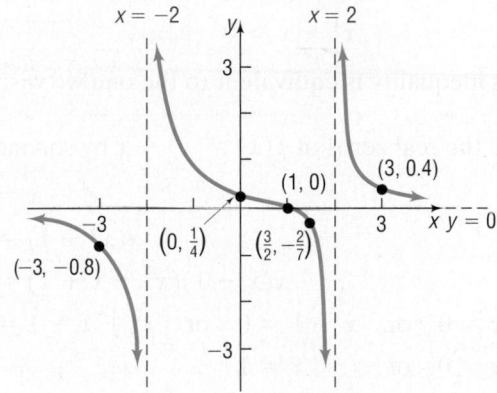

From the graph, we can see that $R(x) \geq 0$ for $-2 < x \leq 1$ or $x > 2$. The solution set is $\{x \mid -2 < x \leq 1 \text{ or } x > 2\}$ or, using interval notation, $(-2, 1] \cup (2, \infty)$.

Now Work PROBLEM 33

To solve a rational inequality algebraically, we follow the same approach that we used to solve a polynomial inequality algebraically. However, we must also identify the zeros of the denominator of the rational function, because the sign of a rational function may change on either side of a vertical asymptote. Convince yourself of this by looking at Figure 47. Notice that the function values are negative for $x < -2$ and are positive for $x > -2$ (but less than 1).

EXAMPLE 4

How to Solve a Rational Inequality Algebraically

Solve the inequality $\dfrac{4x+5}{x+2} \geq 3$ algebraically, and graph the solution set.

Step-by-Step Solution

Step 1: Write the inequality so that a rational expression f is on the left side and zero is on the right side.

Rearrange the inequality so that 0 is on the right side.

$$\frac{4x+5}{x+2} \geq 3$$

$$\frac{4x+5}{x+2} - 3 \geq 0 \qquad \text{Subtract 3 from both sides of the inequality.}$$

$$\frac{4x+5}{x+2} - 3 \cdot \frac{x+2}{x+2} \geq 0 \qquad \text{Multiply 3 by } \frac{x+2}{x+2}.$$

$$\frac{4x+5-3x-6}{x+2} \geq 0 \qquad \text{Write as a single quotient.}$$

$$\frac{x-1}{x+2} \geq 0 \qquad \text{Combine like terms.}$$

Step 2: Determine the real zeros (x-intercepts of the graph) of f and the real numbers for which f is undefined.

The zero of $f(x) = \dfrac{x-1}{x+2}$ is 1. Also, f is undefined for $x = -2$.

Step 3: Use the zeros and undefined values found in Step 2 to divide the real number line into intervals.

Use the zero and undefined value to separate the real number line into three intervals:

$$(-\infty, -2) \quad (-2, 1) \quad (1, \infty)$$

Step 4: Select a number in each interval, evaluate f at the number, and determine whether f is positive or negative. If f is positive, all values of f in the interval are positive. If f is negative, all values of f in the interval are negative.

Select a test number in each interval found in Step 3 and evaluate $f(x) = \dfrac{x-1}{x+2}$ at each number to determine if $f(x)$ is positive or negative. See Table 17.

Table 17

Interval	$(-\infty, -2)$	$(-2, 1)$	$(1, \infty)$
Number chosen	-3	0	2
Value of f	$f(-3) = 4$	$f(0) = -\dfrac{1}{2}$	$f(2) = \dfrac{1}{4}$
Conclusion	Positive	Negative	Positive

Since we want to know where $f(x)$ is positive or zero, we conclude that $f(x) \geq 0$ for all numbers x for which $x < -2$ or $x \geq 1$. Notice we do not include -2 in the solution because -2 is not in the domain of f. The solution set of the inequality $\dfrac{4x+5}{x+2} \geq 3$ is $\{x \mid x < -2 \text{ or } x \geq 1\}$ or, using interval notation, $(-\infty, -2) \cup [1, \infty)$. Figure 48 shows the graph of the solution set.

Figure 48

↳

━━━━ **Now Work** PROBLEM **39**

SUMMARY Steps for Solving Polynomial and Rational Inequalities Algebraically

STEP 1: Write the inequality so that a polynomial or rational expression f is on the left side and zero is on the right side in one of the following forms:

$$f(x) > 0 \qquad f(x) \geq 0 \qquad f(x) < 0 \qquad f(x) \leq 0$$

For rational expressions, be sure that the left side is written as a single quotient and find the domain of f.

STEP 2: Determine the real numbers at which the expression f equals zero and, if the expression is rational, the real numbers at which the expression f is undefined.

STEP 3: Use the numbers found in Step 2 to separate the real number line into intervals.

STEP 4: Select a number in each interval and evaluate f at the number.

(a) If the value of f is positive, then $f(x) > 0$ for all numbers x in the interval.

(b) If the value of f is negative, then $f(x) < 0$ for all numbers x in the interval.

If the inequality is not strict (\geq or \leq), include the solutions of $f(x) = 0$ that are in the domain of f in the solution set. Be careful to exclude values of x where f is undefined.

5.4 Assess Your Understanding

'Are You Prepared?' *Answers are given at the end of these exercises. If you get a wrong answer, read the pages listed in red.*

1. Solve the inequality $3 - 4x > 5$. Graph the solution set. (pp. 123–125)

2. Solve the inequality $x^2 - 5x \leq 24$. Graph the solution set. (pp. 309–311)

Concepts and Vocabulary

3. *True or False* A test number for the interval $-2 < x < 5$ could be 4.

4. *True or False* The graph of $f(x) = \dfrac{x}{x-3}$ is above the x-axis for $x < 0$ or $x > 3$, so the solution set of the inequality $\dfrac{x}{x-3} \geq 0$ is $\{x \mid x \leq 0 \text{ or } x \geq 3\}$.

Skill Building

In Problems 5–8, use the graph of the function f to solve the inequality.

5. (a) $f(x) > 0$
(b) $f(x) \leq 0$

6. (a) $f(x) < 0$
(b) $f(x) \geq 0$

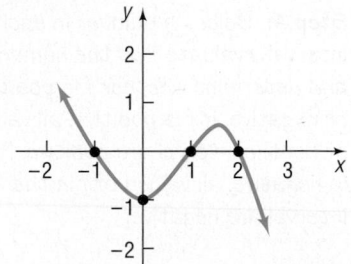

7. (a) $f(x) < 0$
(b) $f(x) \geq 0$

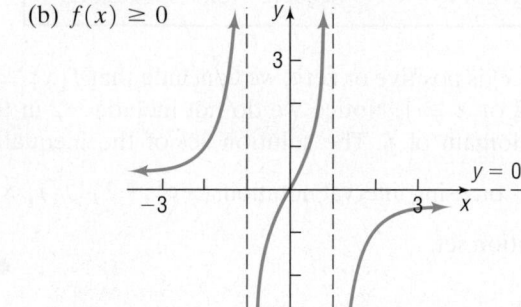

8. (a) $f(x) > 0$
(b) $f(x) \leq 0$

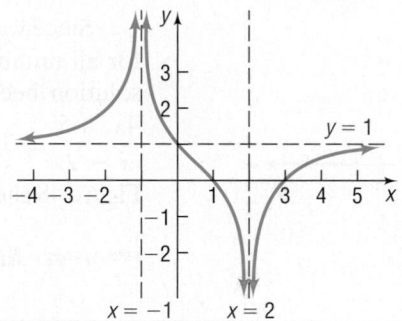

In Problems 9–14, solve the inequality by using the graph of the function.

[Hint: The graphs were drawn in Problems 69–74 of Section 5.1.]

9. Solve $f(x) < 0$, where $f(x) = x^2(x-3)$.

10. Solve $f(x) \leq 0$, where $f(x) = x(x+2)^2$.

11. Solve $f(x) \geq 0$, where $f(x) = (x+4)(x-2)^2$.

12. Solve $f(x) > 0$, where $f(x) = (x-1)(x+3)^2$.

13. Solve $f(x) \leq 0$, where $f(x) = -2(x+2)(x-2)^3$.

14. Solve $f(x) < 0$, where $f(x) = -\dfrac{1}{2}(x+4)(x-1)^3$.

In Problems 15–18, solve the inequality by using the graph of the function.

[Hint: The graphs were drawn in Problems 7–10 of Section 5.3.]

15. Solve $R(x) > 0$, where $R(x) = \dfrac{x+1}{x(x+4)}$.

16. Solve $R(x) < 0$, where $R(x) = \dfrac{x}{(x-1)(x+2)}$.

17. Solve $R(x) \leq 0$, where $R(x) = \dfrac{3x+3}{2x+4}$.

18. Solve $R(x) \geq 0$, where $R(x) = \dfrac{2x+4}{x-1}$.

In Problems 19–48, solve each inequality algebraically.

19. $(x-5)^2(x+2) < 0$

20. $(x-5)(x+2)^2 > 0$

21. $x^3 - 4x^2 > 0$

22. $x^3 + 8x^2 < 0$

23. $2x^3 > -8x^2$

24. $3x^3 < -15x^2$

25. $(x-1)(x-2)(x-3) \leq 0$

26. $(x+1)(x+2)(x+3) \leq 0$

27. $x^3 - 2x^2 - 3x > 0$

28. $x^3 + 2x^2 - 3x > 0$

29. $x^4 > x^2$

30. $x^4 < 9x^2$

31. $x^4 > 1$

32. $x^3 > 1$

33. $\dfrac{x+1}{x-1} > 0$

34. $\dfrac{x-3}{x+1} > 0$

35. $\dfrac{(x-1)(x+1)}{x} \le 0$

36. $\dfrac{(x-3)(x+2)}{x-1} \le 0$

37. $\dfrac{(x-2)^2}{x^2-1} \ge 0$

38. $\dfrac{(x+5)^2}{x^2-4} \ge 0$

39. $\dfrac{x+4}{x-2} \le 1$

40. $\dfrac{x+2}{x-4} \ge 1$

41. $\dfrac{3x-5}{x+2} \le 2$

42. $\dfrac{x-4}{2x+4} \ge 1$

43. $\dfrac{1}{x-2} < \dfrac{2}{3x-9}$

44. $\dfrac{5}{x-3} > \dfrac{3}{x+1}$

45. $\dfrac{x^2(3+x)(x+4)}{(x+5)(x-1)} \ge 0$

46. $\dfrac{x(x^2+1)(x-2)}{(x-1)(x+1)} \ge 0$

47. $\dfrac{(3-x)^3(2x+1)}{x^3-1} < 0$

48. $\dfrac{(2-x)^3(3x-2)}{x^3+1} < 0$

Mixed Practice

In Problems 49–60, solve each inequality algebraically.

49. $(x+1)(x-3)(x-5) > 0$

50. $(2x-1)(x+2)(x+5) < 0$

51. $7x - 4 \ge -2x^2$

52. $x^2 + 3x \ge 10$

53. $\dfrac{x+1}{x-3} \le 2$

54. $\dfrac{x-1}{x+2} \ge -2$

55. $3(x^2-2) < 2(x-1)^2 + x^2$

56. $(x-3)(x+2) < x^2 + 3x + 5$

57. $6x - 5 < \dfrac{6}{x}$

58. $x + \dfrac{12}{x} < 7$

59. $x^3 - 9x \le 0$

60. $x^3 - x \ge 0$

Applications and Extensions

61. For what positive numbers will the cube of a number exceed four times its square?

62. For what positive numbers will the cube of a number be less than the number?

63. What is the domain of the function $f(x) = \sqrt{x^4 - 16}$?

64. What is the domain of the function $f(x) = \sqrt{x^3 - 3x^2}$?

65. What is the domain of the function $f(x) = \sqrt{\dfrac{x-2}{x+4}}$?

66. What is the domain of the function $f(x) = \sqrt{\dfrac{x-1}{x+4}}$?

In Problems 67–70, determine where the graph of f is below the graph of g by solving the inequality $f(x) \le g(x)$. Graph f and g together.

67. $f(x) = x^4 - 1$
$g(x) = -2x^2 + 2$

68. $f(x) = x^4 - 1$
$g(x) = x - 1$

69. $f(x) = x^4 - 4$
$g(x) = 3x^2$

70. $f(x) = x^4$
$g(x) = 2 - x^2$

71. Average Cost Suppose that the daily cost C of manufacturing bicycles is given by $C(x) = 80x + 5000$. Then the average daily cost \overline{C} is given by $\overline{C}(x) = \dfrac{80x + 5000}{x}$. How many bicycles must be produced each day for the average cost to be no more than $100?

72. Average Cost See Problem 71. Suppose that the government imposes a $1000 per day tax on the bicycle manufacturer so that the daily cost C of manufacturing x bicycles is now given by $C(x) = 80x + 6000$. Now the average daily cost \overline{C} is given by $\overline{C}(x) = \dfrac{80x + 6000}{x}$. How many bicycles must be produced each day for the average cost to be no more than $100?

73. Bungee Jumping Originating on Pentecost Island in the Pacific, the practice of a person jumping from a high place harnessed to a flexible attachment was introduced to western culture in 1979 by the Oxford University Dangerous Sport Club. One important parameter to know before attempting a bungee jump is the amount the cord will stretch at the bottom of the fall. The stiffness of the cord is related to the amount of stretch by the equation

$$K = \dfrac{2W(S+L)}{S^2}$$

where W = weight of the jumper (pounds)
 K = cord's stiffness (pounds per foot)
 L = free length of the cord (feet)
 S = stretch (feet)

(a) A 150-pound person plans to jump off a ledge attached to a cord of length 42 feet. If the stiffness of the cord is no less than 16 pounds per foot, how much will the cord stretch?

(b) If safety requirements will not permit the jumper to get any closer than 3 feet to the ground, what is the minimum height required for the ledge in part (a)?

Source: American Institute of Physics, Physics News Update, No. 150, November 5, 1993.

74. Gravitational Force According to Newton's Law of universal gravitation, the attractive force F between two bodies is given by

$$F = G\frac{m_1 m_2}{r^2}$$

where $m_1, m_2 =$ the masses of the two bodies
$r =$ distance between the two bodies
$G =$ gravitational constant $= 6.6742 \times 10^{-11}$ newtons meter2 kilogram^{-2}

Suppose an object is traveling directly from Earth to the moon. The mass of Earth is 5.9742×10^{24} kilograms, the mass of the moon is 7.349×10^{22} kilograms, and the mean distance from Earth to the moon is 384,400 kilometers. For an object between Earth and the moon, how far from Earth is the force on the object due to the moon greater than the force on the object due to Earth?

Source: www.solarviews.com;en.wikipedia.org

75. Field Trip Mrs. West has decided to take her fifth grade class to a play. The manager of the theater agreed to discount the regular $40 price of the ticket by $0.20 for each ticket sold. The cost of the bus, $500, will be split equally among each of the students. How many students must attend to keep the cost per student at or below $40?

Explaining Concepts: Discussion and Writing

76. Make up an inequality that has no solution. Make up one that has exactly one solution.

77. The inequality $x^4 + 1 < -5$ has no solution. Explain why.

78. A student attempted to solve the inequality $\dfrac{x + 4}{x - 3} \le 0$ by multiplying both sides of the inequality by $x - 3$ to get

$x + 4 \le 0$. This led to a solution of $\{x \mid x \le -4\}$. Is the student correct? Explain.

79. Write a rational inequality whose solution set is $\{x \mid -3 < x \le 5\}$.

'Are You Prepared?' Answers

1. $\left\{ x \mid x < -\dfrac{1}{2} \right\}$ or $\left(-\infty, -\dfrac{1}{2} \right)$ ⟵———⟶ $-2 \quad -1\frac{1}{2} \ 0 \qquad 1$

2. $\{x \mid -3 \le x \le 8\}$ or $[-3, 8]$ ⟵———⟶ $-3 \qquad 0 \qquad\qquad 8$

5.5 The Real Zeros of a Polynomial Function

PREPARING FOR THIS SECTION *Before getting started, review the following:*

- Evaluating Functions (Section 3.1, pp. 203–206)
- Factoring Polynomials (Chapter R, Section R.5, pp. 49–55)
- Synthetic Division (Chapter R, Section R.6, pp. 58–61)
- Polynomial Division (Chapter R, Section R.4, pp. 44–47)
- Zeros of a Quadratic Function (Section 4.3, pp. 292–293)

Now Work the 'Are You Prepared?' problems on page 384.

OBJECTIVES **1** Use the Remainder and Factor Theorems (p. 375)
 2 Use the Rational Zeros Theorem to List the Potential Rational Zeros of a Polynomial Function (p. 378)
 3 Find the Real Zeros of a Polynomial Function (p. 378)
 4 Solve Polynomial Equations (p. 381)
 5 Use the Theorem for Bounds on Zeros (p. 381)
 6 Use the Intermediate Value Theorem (p. 382)

In Section 5.1, we were able to identify the real zeros of a polynomial function because either the polynomial function was in factored form or it could be easily factored. But how do we find the real zeros of a polynomial function if it is not factored or cannot be easily factored?

Recall that if r is a real zero of a polynomial function f then $f(r) = 0$, r is an x-intercept of the graph of f, $x - r$ is a factor of f, and r is a solution of the equation $f(x) = 0$. For example, if $x - 4$ is a factor of f, then 4 is a real zero of f and 4 is a solution to the equation $f(x) = 0$. For polynomial and rational functions, we have seen the importance of the real zeros for graphing. In most cases, however, the real zeros of a polynomial function are difficult to find using algebraic methods. No nice formulas like the quadratic formula are available to help us find zeros for polynomials of degree 3 or higher. Formulas do exist for solving any third- or fourth-degree polynomial equation, but they are somewhat complicated. No general formulas exist for polynomial equations of degree 5 or higher. Refer to the Historical Feature at the end of this section for more information.

1 Use the Remainder and Factor Theorems

When we divide one polynomial (the dividend) by another (the divisor), we obtain a quotient polynomial and a remainder, the remainder being either the zero polynomial or a polynomial whose degree is less than the degree of the divisor. To check our work, we verify that

$$(\text{Quotient})(\text{Divisor}) + \text{Remainder} = \text{Dividend}$$

This checking routine is the basis for a famous theorem called the **division algorithm* for polynomials,** which we now state without proof.

THEOREM

Division Algorithm for Polynomials

If $f(x)$ and $g(x)$ denote polynomial functions and if $g(x)$ is a polynomial whose degree is greater than zero, then there are unique polynomial functions $q(x)$ and $r(x)$ such that

$$\frac{f(x)}{g(x)} = q(x) + \frac{r(x)}{g(x)} \quad \text{or} \quad f(x) = q(x)g(x) + r(x) \qquad \textbf{(1)}$$

dividend quotient divisor remainder

where $r(x)$ is either the zero polynomial or a polynomial of degree less than that of $g(x)$.

In equation (1), $f(x)$ is the **dividend,** $g(x)$ is the **divisor,** $q(x)$ is the **quotient,** and $r(x)$ is the **remainder.**

If the divisor $g(x)$ is a first-degree polynomial of the form

$$g(x) = x - c \quad c \text{ a real number}$$

then the remainder $r(x)$ is either the zero polynomial or a polynomial of degree 0. As a result, for such divisors, the remainder is some number, say R, and

$$f(x) = (x - c)q(x) + R \qquad \textbf{(2)}$$

This equation is an identity in x and is true for all real numbers x. Suppose that $x = c$. Then equation (2) becomes

$$f(c) = (c - c)q(c) + R$$
$$f(c) = R$$

*A systematic process in which certain steps are repeated a finite number of times is called an **algorithm.** For example, long division is an algorithm.

Substitute $f(c)$ for R in equation (2) to obtain

$$f(x) = (x - c)q(x) + f(c) \qquad \textbf{(3)}$$

We have now proved the **Remainder Theorem.**

REMAINDER THEOREM
Let f be a polynomial function. If $f(x)$ is divided by $x - c$, then the remainder is $f(c)$.

EXAMPLE 1 **Using the Remainder Theorem**

Find the remainder if $f(x) = x^3 - 4x^2 - 5$ is divided by

(a) $x - 3$ (b) $x + 2$

Solution (a) We could use long division or synthetic division, but it is easier to use the Remainder Theorem, which says that the remainder is $f(3)$.

$$f(3) = (3)^3 - 4(3)^2 - 5 = 27 - 36 - 5 = -14$$

The remainder is -14.

(b) To find the remainder when $f(x)$ is divided by $x + 2 = x - (-2)$, evaluate $f(-2)$.

$$f(-2) = (-2)^3 - 4(-2)^2 - 5 = -8 - 16 - 5 = -29$$

The remainder is -29.

Compare the method used in Example 1(a) with the method used in Example 1 of Chapter R, Section R.6. Which method do you prefer? Give reasons.

 COMMENT A graphing utility provides another way to find the value of a function using the eVALUEate feature. Consult your manual for details. Then check the results of Example 1. ∎

An important and useful consequence of the Remainder Theorem is the **Factor Theorem.**

FACTOR THEOREM
Let f be a polynomial function. Then $x - c$ is a factor of $f(x)$ if and only if $f(c) = 0$.

The Factor Theorem actually consists of two separate statements:

1. If $f(c) = 0$, then $x - c$ is a factor of $f(x)$.
2. If $x - c$ is a factor of $f(x)$, then $f(c) = 0$.

The proof requires two parts.

Proof

1. Suppose that $f(c) = 0$. Then, by equation (3), we have

$$f(x) = (x - c)q(x)$$

for some polynomial $q(x)$. That is, $x - c$ is a factor of $f(x)$.

2. Suppose that $x - c$ is a factor of $f(x)$. Then there is a polynomial function q such that

$$f(x) = (x - c)q(x)$$

Replacing x by c, we find that

$$f(c) = (c - c)q(c) = 0 \cdot q(c) = 0$$

This completes the proof. ∎

| **EXAMPLE 2** | **Using the Factor Theorem** |

Use the Factor Theorem to determine whether the function

$$f(x) = 2x^3 - x^2 + 2x - 3$$

has the factor

(a) $x - 1$ (b) $x + 3$

Solution The Factor Theorem states that if $f(c) = 0$ then $x - c$ is a factor.

(a) Because $x - 1$ is of the form $x - c$ with $c = 1$, we find the value of $f(1)$. We choose to use substitution.

$$f(1) = 2(1)^3 - (1)^2 + 2(1) - 3 = 2 - 1 + 2 - 3 = 0$$

By the Factor Theorem, $x - 1$ is a factor of $f(x)$.

(b) To test the factor $x + 3$, we first need to write it in the form $x - c$. Since $x + 3 = x - (-3)$, we find the value of $f(-3)$. We choose to use synthetic division.

$$
\begin{array}{r|rrrr}
-3 & 2 & -1 & 2 & -3 \\
 & & -6 & 21 & -69 \\
\hline
 & 2 & -7 & 23 & -72
\end{array}
$$

Because $f(-3) = -72 \neq 0$, we conclude from the Factor Theorem that $x - (-3) = x + 3$ is not a factor of $f(x)$. ♩

Now Work PROBLEM 11

In Example 2(a), we found that $x - 1$ is a factor of f. To write f in factored form, use long division or synthetic division. Using synthetic division,

$$
\begin{array}{r|rrrr}
1 & 2 & -1 & 2 & -3 \\
 & & 2 & 1 & 3 \\
\hline
 & 2 & 1 & 3 & 0
\end{array}
$$

The quotient is $q(x) = 2x^2 + x + 3$ with a remainder of 0, as expected. We can write f in factored form as

$$f(x) = 2x^3 - x^2 + 2x - 3 = (x - 1)(2x^2 + x + 3)$$

The next theorem concerns the number of real zeros that a polynomial function may have. In counting the zeros of a polynomial, we count each zero as many times as its multiplicity.

THEOREM **Number of Real Zeros**

A polynomial function cannot have more real zeros than its degree. ⌋

Proof The proof is based on the Factor Theorem. If r is a real zero of a polynomial function f, then $f(r) = 0$ and, hence, $x - r$ is a factor of $f(x)$. Each real zero corresponds to a factor of degree 1. Because f cannot have more first-degree factors than its degree, the result follows. ∎

2 Use the Rational Zeros Theorem to List the Potential Rational Zeros of a Polynomial Function

The next result, called the **Rational Zeros Theorem,** provides information about the rational zeros of a polynomial *with integer coefficients*.

THEOREM **Rational Zeros Theorem**

Let f be a polynomial function of degree 1 or higher of the form

$$f(x) = a_n x^n + a_{n-1} x^{n-1} + \cdots + a_1 x + a_0 \quad a_n \neq 0, \quad a_0 \neq 0$$

where each coefficient is an integer. If $\dfrac{p}{q}$, in lowest terms, is a rational zero of f, then p must be a factor of a_0 and q must be a factor of a_n.

EXAMPLE 3 **Listing Potential Rational Zeros**

List the potential rational zeros of

$$f(x) = 2x^3 + 11x^2 - 7x - 6$$

Solution Because f has integer coefficients, we may use the Rational Zeros Theorem. First, list all the integers p that are factors of the constant term $a_0 = -6$ and all the integers q that are factors of the leading coefficient $a_3 = 2$.

$$p: \quad \pm 1, \pm 2, \pm 3, \pm 6 \quad \text{Factors of } -6$$
$$q: \quad \pm 1, \pm 2 \quad \text{Factors of 2}$$

> **In Words**
> For the polynomial function $f(x) = 2x^3 + 11x^2 - 7x - 6$, we know 5 is not a zero, because 5 is not in the list of potential rational zeros. However, -1 may or may not be a zero.

Now form all possible ratios $\dfrac{p}{q}$.

$$\frac{p}{q}: \quad \pm 1, \pm 2, \pm 3, \pm 6, \pm \frac{1}{2}, \pm \frac{3}{2}$$

If f has a rational zero, it will be found in this list, which contains 12 possibilities.

Now Work PROBLEM 33

Be sure that you understand what the Rational Zeros Theorem says: For a polynomial with integer coefficients, *if* there is a rational zero, it is one of those listed. It may be the case that the function does not have any rational zeros.

Long division, synthetic division, or substitution can be used to test each potential rational zero to determine whether it is indeed a zero. To make the work easier, integers are usually tested first.

3 Find the Real Zeros of a Polynomial Function

EXAMPLE 4 **How to Find the Real Zeros of a Polynomial Function**

Find the real zeros of the polynomial function $f(x) = 2x^3 + 11x^2 - 7x - 6$. Write f in factored form.

Step-by-Step Solution

Step 1: Use the degree of the polynomial to determine the maximum number of zeros.

Since f is a polynomial of degree 3, there are at most three real zeros.

Step 2: If the polynomial has integer coefficients, use the Rational Zeros Theorem to identify those rational numbers that potentially can be zeros. Use the Factor Theorem to determine if each potential rational zero is a zero. If it is, use synthetic division or long division to factor the polynomial function. Repeat Step 2 until all the zeros of the polynomial function have been identified and the polynomial function is completely factored.

List the potential rational zeros obtained in Example 3:

$$\pm 1, \pm 2, \pm 3, \pm 6, \pm \frac{1}{2}, \pm \frac{3}{2}$$

From our list of potential rational zeros, we will test 6 to determine if it is a zero of f. Because $f(6) = 780 \neq 0$, we know that 6 is not a zero of f. Now, let's test if -6 is a zero. Because $f(-6) = 0$, we know that -6 is a zero and $x - (-6) = x + 6$ is a factor of f. Use long division or synthetic division to factor f. (We will not show the division here, but you are encouraged to verify the results shown.) After dividing f by $x + 6$, the quotient is $2x^2 - x - 1$, so

$$f(x) = 2x^3 + 11x^2 - 7x - 6$$
$$= (x + 6)(2x^2 - x - 1)$$

Now any solution of the equation $2x^2 - x - 1 = 0$ will be a zero of f. We call the equation $2x^2 - x - 1 = 0$ a **depressed equation** of f. Because any solution to the equation $2x^2 - x - 1 = 0$ is a zero of f, we work with the depressed equation to find the remaining zeros of f.

The depressed equation $2x^2 - x - 1 = 0$ is a quadratic equation with discriminant $b^2 - 4ac = (-1)^2 - 4(2)(-1) = 9 > 0$. The equation has two real solutions, which can be found by factoring.

$$2x^2 - x - 1 = (2x + 1)(x - 1) = 0$$
$$2x + 1 = 0 \quad \text{or} \quad x - 1 = 0$$
$$x = -\frac{1}{2} \quad \text{or} \quad x = 1$$

The zeros of f are $-6, -\frac{1}{2}$, and 1.

We completely factor f as follows:

$$f(x) = 2x^3 + 11x^2 - 7x - 6 = (x + 6)(2x^2 - x - 1)$$
$$= (x + 6)(2x + 1)(x - 1)$$

Notice that the three zeros of f are in the list of potential rational zeros.

SUMMARY Steps for Finding the Real Zeros of a Polynomial Function

STEP 1: Use the degree of the polynomial to determine the maximum number of real zeros.

STEP 2: (a) If the polynomial has integer coefficients, use the Rational Zeros Theorem to identify those rational numbers that potentially could be zeros.

(b) Use substitution, synthetic division, or long division to test each potential rational zero. Each time that a zero (and thus a factor) is found, repeat Step 2 on the depressed equation.

In attempting to find the zeros, remember to use (if possible) the factoring techniques that you already know (special products, factoring by grouping, and so on).

EXAMPLE 5 **Finding the Real Zeros of a Polynomial Function**

Find the real zeros of $f(x) = x^5 - 5x^4 + 12x^3 - 24x^2 + 32x - 16$. Write f in factored form.

Solution **STEP 1:** There are at most five real zeros.

STEP 2: Because the leading coefficient is $a_5 = 1$, the potential rational zeros are the integers $\pm 1, \pm 2, \pm 4, \pm 8$, and ± 16, the factors of the constant term, 16.

We test the potential rational zero 1 first, using synthetic division.

$$
\begin{array}{r|rrrrr}
1) & 1 & -5 & 12 & -24 & 32 & -16 \\
 & & 1 & -4 & 8 & -16 & 16 \\
\hline
 & 1 & -4 & 8 & -16 & 16 & 0
\end{array}
$$

The remainder is $f(1) = 0$, so 1 is a zero and $x - 1$ is a factor of f. Using the entries in the bottom row of the synthetic division, we can begin to factor f.

$$f(x) = x^5 - 5x^4 + 12x^3 - 24x^2 + 32x - 16$$
$$= (x - 1)(x^4 - 4x^3 + 8x^2 - 16x + 16)$$

We now work with the first depressed equation:

$$q_1(x) = x^4 - 4x^3 + 8x^2 - 16x + 16 = 0$$

REPEAT STEP 2: The potential rational zeros of q_1 are still $\pm 1, \pm 2, \pm 4, \pm 8,$ and ± 16. We test 1 first, since it may be a repeated zero of f.

$$
\begin{array}{r|rrrr}
1) & 1 & -4 & 8 & -16 & 16 \\
 & & 1 & -3 & 5 & -11 \\
\hline
 & 1 & -3 & 5 & -11 & 5
\end{array}
$$

Since the remainder is 5, 1 is not a repeated zero. Try 2 next.

$$
\begin{array}{r|rrrr}
2) & 1 & -4 & 8 & -16 & 16 \\
 & & 2 & -4 & 8 & -16 \\
\hline
 & 1 & -2 & 4 & -8 & 0
\end{array}
$$

The remainder is $f(2) = 0$, so 2 is a zero and $x - 2$ is a factor of f. Again using the bottom row, we find

$$f(x) = x^5 - 5x^4 + 12x^3 - 24x^2 + 32x - 16$$
$$= (x - 1)(x - 2)(x^3 - 2x^2 + 4x - 8)$$

The remaining zeros satisfy the new depressed equation

$$q_2(x) = x^3 - 2x^2 + 4x - 8 = 0$$

Notice that $q_2(x)$ can be factored using grouping. (Alternatively, you could repeat Step 2 and check the potential rational zero 2.) Then

$$x^3 - 2x^2 + 4x - 8 = 0$$
$$x^2(x - 2) + 4(x - 2) = 0$$
$$(x^2 + 4)(x - 2) = 0$$
$$x^2 + 4 = 0 \quad \text{or} \quad x - 2 = 0$$
$$x = 2$$

Since $x^2 + 4 = 0$ has no real solutions, the real zeros of f are 1 and 2, with 2 being a zero of multiplicity 2. The factored form of f is

$$f(x) = x^5 - 5x^4 + 12x^3 - 24x^2 + 32x - 16$$
$$= (x - 1)(x - 2)^2(x^2 + 4)$$

♩

Now Work PROBLEM 45

4 Solve Polynomial Equations

EXAMPLE 6 **Solving a Polynomial Equation**

Find the real solutions of the equation: $x^5 - 5x^4 + 12x^3 - 24x^2 + 32x - 16 = 0$

Solution The real solutions of this equation are the real zeros of the polynomial function

$$f(x) = x^5 - 5x^4 + 12x^3 - 24x^2 + 32x - 16$$

Using the result of Example 5, the real zeros of f are 1 and 2. So, $\{1, 2\}$ is the solution set of the equation

$$x^5 - 5x^4 + 12x^3 - 24x^2 + 32x - 16 = 0$$

Now Work PROBLEM 57

In Example 5, the quadratic factor $x^2 + 4$ that appears in the factored form of f is called *irreducible*, because the polynomial $x^2 + 4$ cannot be factored over the real numbers. In general, a quadratic factor $ax^2 + bx + c$ is **irreducible** if it cannot be factored over the real numbers, that is, if it is prime over the real numbers.

Refer to Examples 4 and 5. The polynomial function of Example 4 has three real zeros, and its factored form contains three linear factors. The polynomial function of Example 5 has two distinct real zeros, and its factored form contains two distinct linear factors and one irreducible quadratic factor.

THEOREM Every polynomial function with real coefficients can be uniquely factored into a product of linear factors and/or irreducible (prime) quadratic factors.

We prove this result in Section 5.6, and, in fact, shall draw several additional conclusions about the zeros of a polynomial function. One conclusion is worth noting now. If a polynomial with real coefficients is of odd degree, it must contain at least one linear factor. (Do you see why? Consider the end behavior of polynomial functions of odd degree.) This means that it must have at least one real zero.

THEOREM A polynomial function of odd degree that has real coefficients has at least one real zero.

5 Use the Theorem for Bounds on Zeros

The search for the real zeros of a polynomial function can be reduced somewhat if *bounds* on the zeros are found. A number M is a **bound** on the zeros of a polynomial if every zero lies between $-M$ and M, inclusive. That is, M is a bound on the zeros of a polynomial f if

$$-M \leq \text{any real zero of } f \leq M$$

THEOREM **Bounds on Zeros**

Let f denote a polynomial function whose leading coefficient is 1.

$$f(x) = x^n + a_{n-1}x^{n-1} + \cdots + a_1x + a_0$$

A bound M on the real zeros of f is the smaller of the two numbers

$$\text{Max}\{1, |a_0| + |a_1| + \cdots + |a_{n-1}|\}, 1 + \text{Max}\{|a_0|, |a_1|, \cdots, |a_{n-1}|\} \quad \textbf{(4)}$$

where Max { } means "choose the largest entry in { }."

| EXAMPLE 7 | **Using the Theorem for Finding Bounds on Zeros** |

Find a bound on the real zeros of each polynomial function.

(a) $f(x) = x^5 + 3x^3 - 9x^2 + 5$ (b) $g(x) = 4x^5 - 2x^3 + 2x^2 + 1$

Solution

(a) The leading coefficient of f is 1.

$$f(x) = x^5 + 3x^3 - 9x^2 + 5 \quad a_4 = 0, a_3 = 3, a_2 = -9, a_1 = 0, a_0 = 5$$

Evaluate the two expressions in (4).

$$\text{Max}\{1, |a_0| + |a_1| + \cdots + |a_{n-1}|\} = \text{Max}\{1, |5| + |0| + |-9| + |3| + |0|\}$$
$$= \text{Max}\{1, 17\} = 17$$

$$1 + \text{Max}\{|a_0|, |a_1|, \cdots, |a_{n-1}|\} = 1 + \text{Max}\{|5|, |0|, |-9|, |3|, |0|\}$$
$$= 1 + 9 = 10$$

The smaller of the two numbers, 10, is the bound. Every real zero of f lies between -10 and 10.

(b) First write g so that it is the product of a constant times a polynomial whose leading coefficient is 1 by factoring out the leading coefficient of g, 4.

$$g(x) = 4x^5 - 2x^3 + 2x^2 + 1 = 4\left(x^5 - \frac{1}{2}x^3 + \frac{1}{2}x^2 + \frac{1}{4}\right)$$

Next evaluate the two expressions in (4) with $a_4 = 0$, $a_3 = -\frac{1}{2}$, $a_2 = \frac{1}{2}$, $a_1 = 0$, and $a_0 = \frac{1}{4}$.

$$\text{Max}\{1, |a_0| + |a_1| + \cdots + |a_{n-1}|\} = \text{Max}\left\{1, \left|\frac{1}{4}\right| + |0| + \left|\frac{1}{2}\right| + \left|-\frac{1}{2}\right| + |0|\right\}$$
$$= \text{Max}\left\{1, \frac{5}{4}\right\} = \frac{5}{4}$$

$$1 + \text{Max}\{|a_0|, |a_1|, \ldots, |a_{n-1}|\} = 1 + \text{Max}\left\{\left|\frac{1}{4}\right|, |0|, \left|\frac{1}{2}\right|, \left|-\frac{1}{2}\right|, |0|\right\}$$
$$= 1 + \frac{1}{2} = \frac{3}{2}$$

The smaller of the two numbers, $\frac{5}{4}$, is the bound. Every real zero of g lies between $-\frac{5}{4}$ and $\frac{5}{4}$.

 COMMENT The bounds on the zeros of a polynomial provide good choices for setting *Xmin* and *Xmax* of the viewing rectangle. With these choices, all the *x*-intercepts of the graph can be seen. ■

━━━━ **Now Work** PROBLEM 69

6 Use the Intermediate Value Theorem

The next result, called the **Intermediate Value Theorem,** is based on the fact that the graph of a polynomial function is continuous; that is, it contains no "holes" or "gaps." Although the proof of this result requires advanced methods in calculus, it is easy to "see" why the result is true. Look at Figure 49.

Figure 49 If $f(a) < 0$ and $f(b) > 0$, there is a zero between a and b.

THEOREM **Intermediate Value Theorem**

Let f denote a polynomial function. If $a < b$ and if $f(a)$ and $f(b)$ are of opposite sign, there is at least one real zero of f between a and b.

EXAMPLE 8

Using the Intermediate Value Theorem to Locate a Real Zero

Show that $f(x) = x^5 - x^3 - 1$ has a zero between 1 and 2.

Solution

Evaluate f at 1 and at 2.

$$f(1) = -1 \quad \text{and} \quad f(2) = 23$$

Because $f(1) < 0$ and $f(2) > 0$, it follows from the Intermediate Value Theorem that the polynomial function f has at least one zero between 1 and 2.

═══ **Now Work** PROBLEM **77**

Let's look at the polynomial f of Example 8 more closely. Based on the Rational Zeros Theorem, ± 1 are the only potential rational zeros. Since $f(1) \neq 0$, we conclude that the zero between 1 and 2 is irrational. We can use the Intermediate Value Theorem to approximate it.

Approximating the Real Zeros of a Polynomial Function

STEP 1: Find two consecutive integers a and $a + 1$ such that f has a zero between them.
STEP 2: Divide the interval $[a, a + 1]$ into 10 equal subintervals.
STEP 3: Evaluate f at each endpoint of the subintervals until the Intermediate Value Theorem applies; this subinterval then contains a zero.
STEP 4: Repeat the process starting at Step 2 until the desired accuracy is achieved.

EXAMPLE 9

Approximating a Real Zero of a Polynomial Function

$f(x) = x^5 - x^3 - 1$ has exactly one zero between 1 and 2. Approximate it correct to two decimal places.

Solution

Divide the interval $[1, 2]$ into 10 equal subintervals: $[1, 1.1]$, $[1.1, 1.2]$, $[1.2, 1.3]$, $[1.3, 1.4]$, $[1.4, 1.5]$, $[1.5, 1.6]$, $[1.6, 1.7]$, $[1.7, 1.8]$, $[1.8, 1.9]$, $[1.9, 2]$. Now find the value of f at each endpoint until the Intermediate Value Theorem applies.

$$f(x) = x^5 - x^3 - 1$$
$$f(1.0) = -1 \qquad f(1.2) = -0.23968$$
$$f(1.1) = -0.72049 \quad f(1.3) = 0.51593$$

COMMENT The TABLE feature of a graphing calculator makes the computations in the solution to Example 9 a lot easier. ■

We can stop here and conclude that the zero is between 1.2 and 1.3. Now divide the interval $[1.2, 1.3]$ into 10 equal subintervals and proceed to evaluate f at each endpoint.

$$f(1.20) = -0.23968 \qquad f(1.23) \approx -0.0455613$$
$$f(1.21) \approx -0.1778185 \quad f(1.24) \approx 0.025001$$
$$f(1.22) \approx -0.1131398$$

Figure 50

We conclude that the zero lies between 1.23 and 1.24, and so, correct to two decimal places, the zero is 1.23.

Exploration
We examine the polynomial f given in Example 9. The Theorem on Bounds of Zeros tells us that every zero is between -2 and 2. If we graph f using $-2 \leq x \leq 2$ (see Figure 50), we see that f has exactly one x-intercept. Using ZERO or ROOT, we find this zero to be 1.24 rounded to two decimal places. Correct to two decimal places, the zero is 1.23.

═══ **Now Work** PROBLEM **89**

There are many other numerical techniques for approximating the zeros of a polynomial. The one outlined in Example 9 (a variation of the *bisection method*) has the advantages that it will always work, it can be programmed rather easily on a computer, and each time it is used another decimal place of accuracy is achieved. See Problem 115 for the bisection method, which places the zero in a succession of intervals, with each new interval being half the length of the preceding one.

Historical Feature

Formulas for the solution of third- and fourth-degree polynomial equations exist, and, while not very practical, they do have an interesting history.

In the 1500s in Italy, mathematical contests were a popular pastime, and persons possessing methods for solving problems kept them secret. (Solutions that were published were already common knowledge.) Niccolo of Brescia (1499–1557), commonly referred to as Tartaglia ("the stammerer"), had the secret for solving cubic (third-degree) equations, which gave him a decided advantage in the contests. Girolamo Cardano (1501–1576) found out that Tartaglia had the secret, and, being interested in cubics, he requested it from Tartaglia. The reluctant Tartaglia hesitated for some time, but finally, swearing Cardano to secrecy with midnight oaths by candlelight, told him the secret. Cardano then published the solution in his book

Ars Magna (1545), giving Tartaglia the credit but rather compromising the secrecy. Tartaglia exploded into bitter recriminations, and each wrote pamphlets that reflected on the other's mathematics, moral character, and ancestry.

The quartic (fourth-degree) equation was solved by Cardano's student Lodovico Ferrari, and this solution also was included, with credit and this time with permission, in the *Ars Magna*.

Attempts were made to solve the fifth-degree equation in similar ways, all of which failed. In the early 1800s, P. Ruffini, Niels Abel, and Evariste Galois all found ways to show that it is not possible to solve fifth-degree equations by formula, but the proofs required the introduction of new methods. Galois's methods eventually developed into a large part of modern algebra.

Historical Problems

Problems 1–8 develop the Tartaglia–Cardano solution of the cubic equation and show why it is not altogether practical.

1. Show that the general cubic equation $y^3 + by^2 + cy + d = 0$ can be transformed into an equation of the form $x^3 + px + q = 0$ by using the substitution $y = x - \dfrac{b}{3}$.

2. In the equation $x^3 + px + q = 0$, replace x by $H + K$. Let $3HK = -p$, and show that $H^3 + K^3 = -q$.

3. Based on Problem 2, we have the two equations

$$3HK = -p \quad \text{and} \quad H^3 + K^3 = -q$$

Solve for K in $3HK = -p$ and substitute into $H^3 + K^3 = -q$. Then show that

$$H = \sqrt[3]{\frac{-q}{2} + \sqrt{\frac{q^2}{4} + \frac{p^3}{27}}}$$

[**Hint:** Look for an equation that is quadratic in form.]

4. Use the solution for H from Problem 3 and the equation $H^3 + K^3 = -q$ to show that

$$K = \sqrt[3]{\frac{-q}{2} - \sqrt{\frac{q^2}{4} + \frac{p^3}{27}}}$$

5. Use the results from Problems 2 to 4 to show that the solution of $x^3 + px + q = 0$ is

$$x = \sqrt[3]{\frac{-q}{2} + \sqrt{\frac{q^2}{4} + \frac{p^3}{27}}} + \sqrt[3]{\frac{-q}{2} - \sqrt{\frac{q^2}{4} + \frac{p^3}{27}}}$$

6. Use the result of Problem 5 to solve the equation $x^3 - 6x - 9 = 0$.

7. Use a calculator and the result of Problem 5 to solve the equation $x^3 + 3x - 14 = 0$.

8. Use the methods of this section to solve the equation $x^3 + 3x - 14 = 0$.

5.5 Assess Your Understanding

'Are You Prepared?' *Answers are given at the end of these exercises. If you get a wrong answer, read the pages listed in red.*

1. Find $f(-1)$ if $f(x) = 2x^2 - x$. (pp. 203–206)

2. Factor the expression $6x^2 + x - 2$. (pp. 49–55)

3. Find the quotient and remainder if $3x^4 - 5x^3 + 7x - 4$ is divided by $x - 3$. (pp. 44–47 or 58–61)

4. Find the zeros of $f(x) = x^2 + x - 3$. (pp. 292–293)

Concepts and Vocabulary

5. In the process of polynomial division, (Divisor)(Quotient) + _____ = _____ .

6. When a polynomial function f is divided by $x - c$, the remainder is _____ .

7. If a function f, whose domain is all real numbers, is even and if 4 is a zero of f, then _____ is also a zero.

8. *True or False* Every polynomial function of degree 3 with real coefficients has exactly three real zeros.

9. If f is a polynomial function and $x - 4$ is a factor of f, then $f(4) =$ _____ .

10. *True or False* If f is a polynomial function of degree 4 and if $f(2) = 5$, then

$$\frac{f(x)}{x - 2} = p(x) + \frac{5}{x - 2}$$

where $p(x)$ is a polynomial of degree 3.

Skill Building

In Problems 11–20, use the Remainder Theorem to find the remainder when $f(x)$ is divided by $x - c$. Then use the Factor Theorem to determine whether $x - c$ is a factor of $f(x)$.

11. $f(x) = 4x^3 - 3x^2 - 8x + 4; \ x - 2$

12. $f(x) = -4x^3 + 5x^2 + 8; \ \ x + 3$

13. $f(x) = 3x^4 - 6x^3 - 5x + 10; \ x - 2$

14. $f(x) = 4x^4 - 15x^2 - 4; \ x - 2$

15. $f(x) = 3x^6 + 82x^3 + 27; \ x + 3$

16. $f(x) = 2x^6 - 18x^4 + x^2 - 9; \ x + 3$

17. $f(x) = 4x^6 - 64x^4 + x^2 - 15; \ x + 4$

18. $f(x) = x^6 - 16x^4 + x^2 - 16; \ x + 4$

19. $f(x) = 2x^4 - x^3 + 2x - 1; \ x - \dfrac{1}{2}$

20. $f(x) = 3x^4 + x^3 - 3x + 1; \ x + \dfrac{1}{3}$

In Problems 21–32, tell the maximum number of real zeros that each polynomial function may have. Do not attempt to find the zeros.

21. $f(x) = -4x^7 + x^3 - x^2 + 2$

22. $f(x) = 5x^4 + 2x^2 - 6x - 5$

23. $f(x) = 2x^6 - 3x^2 - x + 1$

24. $f(x) = -3x^5 + 4x^4 + 2$

25. $f(x) = 3x^3 - 2x^2 + x + 2$

26. $f(x) = -x^3 - x^2 + x + 1$

27. $f(x) = -x^4 + x^2 - 1$

28. $f(x) = x^4 + 5x^3 - 2$

29. $f(x) = x^5 + x^4 + x^2 + x + 1$

30. $f(x) = x^5 - x^4 + x^3 - x^2 + x - 1$

31. $f(x) = x^6 - 1$

32. $f(x) = x^6 + 1$

In Problems 33–44, list the potential rational zeros of each polynomial function. Do not attempt to find the zeros.

33. $f(x) = 3x^4 - 3x^3 + x^2 - x + 1$

34. $f(x) = x^5 - x^4 + 2x^2 + 3$

35. $f(x) = x^5 - 6x^2 + 9x - 3$

36. $f(x) = 2x^5 - x^4 - x^2 + 1$

37. $f(x) = -4x^3 - x^2 + x + 2$

38. $f(x) = 6x^4 - x^2 + 2$

39. $f(x) = 6x^4 - x^2 + 9$

40. $f(x) = -4x^3 + x^2 + x + 6$

41. $f(x) = 2x^5 - x^3 + 2x^2 + 12$

42. $f(x) = 3x^5 - x^2 + 2x + 18$

43. $f(x) = 6x^4 + 2x^3 - x^2 + 20$

44. $f(x) = -6x^3 - x^2 + x + 10$

In Problems 45–56, use the Rational Zeros Theorem to find all the real zeros of each polynomial function. Use the zeros to factor f over the real numbers.

45. $f(x) = x^3 + 2x^2 - 5x - 6$

46. $f(x) = x^3 + 8x^2 + 11x - 20$

47. $f(x) = 2x^3 - x^2 + 2x - 1$

48. $f(x) = 2x^3 + x^2 + 2x + 1$

49. $f(x) = 2x^3 - 4x^2 - 10x + 20$

50. $f(x) = 3x^3 + 6x^2 - 15x - 30$

51. $f(x) = 2x^4 + x^3 - 7x^2 - 3x + 3$

52. $f(x) = 2x^4 - x^3 - 5x^2 + 2x + 2$

53. $f(x) = x^4 + x^3 - 3x^2 - x + 2$

54. $f(x) = x^4 - x^3 - 6x^2 + 4x + 8$

55. $f(x) = 4x^4 + 5x^3 + 9x^2 + 10x + 2$

56. $f(x) = 3x^4 + 4x^3 + 7x^2 + 8x + 2$

In Problems 57–68, solve each equation in the real number system.

57. $x^4 - x^3 + 2x^2 - 4x - 8 = 0$

58. $2x^3 + 3x^2 + 2x + 3 = 0$

59. $3x^3 + 4x^2 - 7x + 2 = 0$

60. $2x^3 - 3x^2 - 3x - 5 = 0$

61. $3x^3 - x^2 - 15x + 5 = 0$

62. $2x^3 - 11x^2 + 10x + 8 = 0$

63. $x^4 + 4x^3 + 2x^2 - x + 6 = 0$

64. $x^4 - 2x^3 + 10x^2 - 18x + 9 = 0$

65. $x^3 - \frac{2}{3}x^2 + \frac{8}{3}x + 1 = 0$

66. $x^3 + \frac{3}{2}x^2 + 3x - 2 = 0$

67. $2x^4 - 19x^3 + 57x^2 - 64x + 20 = 0$

68. $2x^4 + x^3 - 24x^2 + 20x + 16 = 0$

In Problems 69–76, find bounds on the real zeros of each polynomial function.

69. $f(x) = x^4 - 3x^2 - 4$

70. $f(x) = x^4 - 5x^2 - 36$

71. $f(x) = x^4 + x^3 - x - 1$

72. $f(x) = x^4 - x^3 + x - 1$

73. $f(x) = 3x^4 + 3x^3 - x^2 - 12x - 12$

74. $f(x) = 3x^4 - 3x^3 - 5x^2 + 27x - 36$

75. $f(x) = 4x^5 - x^4 + 2x^3 - 2x^2 + x - 1$

76. $f(x) = 4x^5 + x^4 + x^3 + x^2 - 2x - 2$

In Problems 77–82, use the Intermediate Value Theorem to show that each polynomial function has a zero in the given interval.

77. $f(x) = 8x^4 - 2x^2 + 5x - 1; [0, 1]$

78. $f(x) = x^4 + 8x^3 - x^2 + 2; [-1, 0]$

79. $f(x) = 2x^3 + 6x^2 - 8x + 2; [-5, -4]$

80. $f(x) = 3x^3 - 10x + 9; [-3, -2]$

81. $f(x) = x^5 - x^4 + 7x^3 - 7x^2 - 18x + 18; [1.4, 1.5]$

82. $f(x) = x^5 - 3x^4 - 2x^3 + 6x^2 + x + 2; [1.7, 1.8]$

In Problems 83–86, each equation has a solution r in the interval indicated. Use the method of Example 9 to approximate this solution correct to two decimal places.

83. $8x^4 - 2x^2 + 5x - 1 = 0; 0 \le r \le 1$

84. $x^4 + 8x^3 - x^2 + 2 = 0; -1 \le r \le 0$

85. $2x^3 + 6x^2 - 8x + 2 = 0; -5 \le r \le -4$

86. $3x^3 - 10x + 9 = 0; -3 \le r \le -2$

In Problems 87–90, each polynomial function has exactly one positive zero. Use the method of Example 9 to approximate the zero correct to two decimal places.

87. $f(x) = x^3 + x^2 + x - 4$

88. $f(x) = 2x^4 + x^2 - 1$

89. $f(x) = 2x^4 - 3x^3 - 4x^2 - 8$

90. $f(x) = 3x^3 - 2x^2 - 20$

Mixed Practice

In Problems 91–102, graph each polynomial function.

91. $f(x) = x^3 + 2x^2 - 5x - 6$

92. $f(x) = x^3 + 8x^2 + 11x - 20$

93. $f(x) = 2x^3 - x^2 + 2x - 1$

94. $f(x) = 2x^3 + x^2 + 2x + 1$

95. $f(x) = x^4 + x^2 - 2$

96. $f(x) = x^4 - 3x^2 - 4$

97. $f(x) = 4x^4 + 7x^2 - 2$

98. $f(x) = 4x^4 + 15x^2 - 4$

99. $f(x) = x^4 + x^3 - 3x^2 - x + 2$

100. $f(x) = x^4 - x^3 - 6x^2 + 4x + 8$

101. $f(x) = 4x^5 - 8x^4 - x + 2$

102. $f(x) = 4x^5 + 12x^4 - x - 3$

Applications and Extensions

103. Find k such that $f(x) = x^3 - kx^2 + kx + 2$ has the factor $x - 2$.

104. Find k such that $f(x) = x^4 - kx^3 + kx^2 + 1$ has the factor $x + 2$.

105. What is the remainder when $f(x) = 2x^{20} - 8x^{10} + x - 2$ is divided by $x - 1$?

106. What is the remainder when $f(x) = -3x^{17} + x^9 - x^5 + 2x$ is divided by $x + 1$?

107. Use the Factor Theorem to prove that $x - c$ is a factor of $x^n - c^n$ for any positive integer n.

108. Use the Factor Theorem to prove that $x + c$ is a factor of $x^n + c^n$ if $n \ge 1$ is an odd integer.

109. One solution of the equation $x^3 - 8x^2 + 16x - 3 = 0$ is 3. Find the sum of the remaining solutions.

110. One solution of the equation $x^3 + 5x^2 + 5x - 2 = 0$ is -2. Find the sum of the remaining solutions.

111. Geometry What is the length of the edge of a cube if, after a slice 1 inch thick is cut from one side, the volume remaining is 294 cubic inches?

112. Geometry What is the length of the edge of a cube if its volume could be doubled by an increase of 6 centimeters in one edge, an increase of 12 centimeters in a second edge, and a decrease of 4 centimeters in the third edge?

113. Let $f(x)$ be a polynomial function whose coefficients are integers. Suppose that r is a real zero of f and that the leading coefficient of f is 1. Use the Rational Zeros Theorem to show that r is either an integer or an irrational number.

114. Prove the Rational Zeros Theorem.

[**Hint:** Let $\dfrac{p}{q}$, where p and q have no common factors except 1 and -1, be a zero of the polynomial function

$$f(x) = a_n x^n + a_{n-1} x^{n-1} + \cdots + a_1 x + a_0$$

whose coefficients are all integers. Show that

$$a_n p^n + a_{n-1} p^{n-1} q + \cdots + a_1 p q^{n-1} + a_0 q^n = 0$$

Now, because p is a factor of the first n terms of this equation, p must also be a factor of the term $a_0 q^n$. Since p is not a factor of q (why?), p must be a factor of a_0. Similarly, q must be a factor of a_n.]

115. Bisection Method for Approximating Zeros of a Function f We begin with two consecutive integers, a and $a + 1$, such that $f(a)$ and $f(a + 1)$ are of opposite sign. Evaluate f at the midpoint m_1 of a and $a + 1$. If $f(m_1) = 0$, then m_1 is the zero of f, and we are finished. Otherwise, $f(m_1)$ is of opposite sign to either $f(a)$ or $f(a + 1)$. Suppose that it is $f(a)$ and $f(m_1)$ that are of opposite sign. Now evaluate f at the midpoint m_2 of a and m_1. Repeat this process until the desired degree of accuracy is obtained. Note that each iteration places the zero in an interval whose length is half that of the previous interval. Use the bisection method to approximate the zero of $f(x) = 8x^4 - 2x^2 + 5x - 1$ in the interval $[0, 1]$ correct to three decimal places.

[**Hint:** The process ends when both endpoints agree to the desired number of decimal places.]

Explaining Concepts: Discussion and Writing

116. Is $\dfrac{1}{3}$ a zero of $f(x) = 2x^3 + 3x^2 - 6x + 7$? Explain.

117. Is $\dfrac{1}{3}$ a zero of $f(x) = 4x^3 - 5x^2 - 3x + 1$? Explain.

118. Is $\dfrac{3}{5}$ a zero of $f(x) = 2x^6 - 5x^4 + x^3 - x + 1$? Explain.

119. Is $\dfrac{2}{3}$ a zero of $f(x) = x^7 + 6x^5 - x^4 + x + 2$? Explain.

'Are You Prepared?' Answers

1. 3 **2.** $(3x + 2)(2x - 1)$ **3.** Quotient: $3x^3 + 4x^2 + 12x + 43$; Remainder: 125 **4.** $\dfrac{-1 - \sqrt{13}}{2}, \dfrac{-1 + \sqrt{13}}{2}$

5.6 Complex Zeros; Fundamental Theorem of Algebra

PREPARING FOR THIS SECTION *Before getting started, review the following:*

- Complex Numbers (Section 1.3, pp. 104–109)
- Complex Solutions of a Quadratic Equation (Section 1.3, pp. 109–111)

Now Work the 'Are You Prepared?' problems on page 392.

OBJECTIVES **1** Use the Conjugate Pairs Theorem (p. 389)

2 Find a Polynomial Function with Specified Zeros (p. 390)

3 Find the Complex Zeros of a Polynomial Function (p. 391)

In Section 1.2, we found the real solutions of a quadratic equation. That is, we found the real zeros of a polynomial function of degree 2. Then, in Section 1.3 we found the complex solutions of a quadratic equation. That is, we found the complex zeros of a polynomial function of degree 2.

In Section 5.5, we found the real zeros of polynomial functions of degree 3 or higher. In this section we will find the *complex zeros* of polynomial functions of degree 3 or higher.

DEFINITION

A variable in the complex number system is referred to as a **complex variable.** A **complex polynomial function** f of degree n is a function of the form

$$f(x) = a_n x^n + a_{n-1} x^{n-1} + \cdots + a_1 x + a_0 \qquad (1)$$

where $a_n, a_{n-1}, \ldots, a_1, a_0$ are complex numbers, $a_n \neq 0$, n is a nonnegative integer, and x is a complex variable. As before, a_n is called the **leading coefficient** of f. A complex number r is called a **complex zero** of f if $f(r) = 0$.

In most of our work the coefficients in (1) will be real numbers.

We have learned that some quadratic equations have no real solutions, but that in the complex number system every quadratic equation has a solution, either real or complex. The next result, proved by Karl Friedrich Gauss (1777–1855) when he was 22 years old,* gives an extension to complex polynomial equations. In fact, this result is so important and useful that it has become known as the **Fundamental Theorem of Algebra.**

FUNDAMENTAL THEOREM OF ALGEBRA

Every complex polynomial function $f(x)$ of degree $n \geq 1$ has at least one complex zero.

We shall not prove this result, as the proof is beyond the scope of this book. However, using the Fundamental Theorem of Algebra and the Factor Theorem, we can prove the following result:

THEOREM

Every complex polynomial function $f(x)$ of degree $n \geq 1$ can be factored into n linear factors (not necessarily distinct) of the form

$$f(x) = a_n(x - r_1)(x - r_2) \cdot \cdots \cdot (x - r_n) \qquad (2)$$

where $a_n, r_1, r_2, \ldots, r_n$ are complex numbers. That is, every complex polynomial function of degree $n \geq 1$ has exactly n complex zeros, some of which may repeat.

Proof Let

$$f(x) = a_n x^n + a_{n-1} x^{n-1} + \cdots + a_1 x + a_0$$

By the Fundamental Theorem of Algebra, f has at least one zero, say r_1. Then, by the Factor Theorem, $x - r_1$ is a factor, and

$$f(x) = (x - r_1)q_1(x)$$

where $q_1(x)$ is a complex polynomial of degree $n - 1$ whose leading coefficient is a_n. Repeating this argument n times, we arrive at

$$f(x) = (x - r_1)(x - r_2) \cdot \cdots \cdot (x - r_n)q_n(x)$$

where $q_n(x)$ is a complex polynomial of degree $n - n = 0$ whose leading coefficient is a_n. That is, $q_n(x) = a_n x^0 = a_n$, and so

$$f(x) = a_n(x - r_1)(x - r_2) \cdot \cdots \cdot (x - r_n)$$

We conclude that every complex polynomial function $f(x)$ of degree $n \geq 1$ has exactly n (not necessarily distinct) zeros. ∎

*In all, Gauss gave four different proofs of this theorem, the first one in 1799 being the subject of his doctoral dissertation.

1 Use the Conjugate Pairs Theorem

We can use the Fundamental Theorem of Algebra to obtain valuable information about the complex zeros of polynomial functions whose coefficients are real numbers.

CONJUGATE PAIRS THEOREM

Let $f(x)$ be a polynomial function whose coefficients are real numbers. If $r = a + bi$ is a zero of f, the complex conjugate $\bar{r} = a - bi$ is also a zero of f.

In other words, for polynomial functions whose coefficients are real numbers, the complex zeros occur in conjugate pairs. This result should not be all that surprising since the complex solutions of a quadratic equation occurred in conjugate pairs.

Proof Let

$$f(x) = a_n x^n + a_{n-1} x^{n-1} + \cdots + a_1 x + a_0$$

where $a_n, a_{n-1}, \ldots, a_1, a_0$ are real numbers and $a_n \neq 0$. If $r = a + bi$ is a zero of f, then $f(r) = f(a + bi) = 0$, so

$$a_n r^n + a_{n-1} r^{n-1} + \cdots + a_1 r + a_0 = 0$$

Take the conjugate of both sides to get

$$\overline{a_n r^n + a_{n-1} r^{n-1} + \cdots + a_1 r + a_0} = \overline{0}$$

$$\overline{a_n r^n} + \overline{a_{n-1} r^{n-1}} + \cdots + \overline{a_1 r} + \overline{a_0} = \overline{0} \qquad \text{The conjugate of a sum equals the sum of the conjugates (see Section 1.3).}$$

$$\overline{a_n}(\bar{r})^n + \overline{a_{n-1}}(\bar{r})^{n-1} + \cdots + \overline{a_1}\,\bar{r} + \overline{a_0} = \overline{0} \qquad \text{The conjugate of a product equals the product of the conjugates.}$$

$$a_n(\bar{r})^n + a_{n-1}(\bar{r})^{n-1} + \cdots + a_1 \bar{r} + a_0 = 0 \qquad \text{The conjugate of a real number equals the real number.}$$

This last equation states that $f(\bar{r}) = 0$; that is, $\bar{r} = a - bi$ is a zero of f. ∎

The importance of this result should be clear. Once we know that, say, $3 + 4i$ is a zero of a polynomial function with real coefficients, then we know that $3 - 4i$ is also a zero. This result has an important corollary.

COROLLARY

A polynomial function f of odd degree with real coefficients has at least one real zero.

Proof Because complex zeros occur as conjugate pairs in a polynomial function with real coefficients, there will always be an even number of zeros that are not real numbers. Consequently, since f is of odd degree, one of its zeros has to be a real number. ∎

For example, the polynomial function $f(x) = x^5 - 3x^4 + 4x^3 - 5$ has at least one zero that is a real number, since f is of degree 5 (odd) and has real coefficients.

EXAMPLE 1 | **Using the Conjugate Pairs Theorem**

A polynomial function f of degree 5 whose coefficients are real numbers has the zeros $1, 5i$, and $1 + i$. Find the remaining two zeros.

Solution Since f has coefficients that are real numbers, complex zeros appear as conjugate pairs. It follows that $-5i$, the conjugate of $5i$, and $1 - i$, the conjugate of $1 + i$, are the two remaining zeros.

Now Work PROBLEM 7

2 Find a Polynomial Function with Specified Zeros

EXAMPLE 2

Finding a Polynomial Function Whose Zeros Are Given

Find a polynomial function f of degree 4 whose coefficients are real numbers that has the zeros 1, 1, and $-4 + i$.

Solution

Since $-4 + i$ is a zero, by the Conjugate Pairs Theorem, $-4 - i$ must also be a zero of f. Because of the Factor Theorem, if $f(c) = 0$, then $x - c$ is a factor of $f(x)$. So we can now write f as

$$f(x) = a(x - 1)(x - 1)[x - (-4 + i)][x - (-4 - i)]$$

where a is any real number. Then

$$
\begin{aligned}
f(x) &= a(x - 1)(x - 1)[x - (-4 + i)][x - (-4 - i)] \\
&= a(x^2 - 2x + 1)[x^2 - (-4 + i)x - (-4 - i)x + (-4 + i)(-4 - i)] \\
&= a(x^2 - 2x + 1)(x^2 + 4x - ix + 4x + ix + 16 + 4i - 4i - i^2) \\
&= a(x^2 - 2x + 1)(x^2 + 8x + 17) \\
&= a(x^4 + 8x^3 + 17x^2 - 2x^3 - 16x^2 - 34x + x^2 + 8x + 17) \\
&= a(x^4 + 6x^3 + 2x^2 - 26x + 17)
\end{aligned}
$$

Exploration

Graph the function f found in Example 2 for $a = 1$. Does the value of a affect the zeros of f? How does the value of a affect the graph of f? What information about f is sufficient to uniquely determine a?

Result A quick analysis of the polynomial function f tells us what to expect:

At most three turning points.

For large $|x|$, the graph will behave like $y = x^4$.

A repeated real zero at 1 of even multiplicity, so the graph will touch the x-axis at 1.

The only x-intercept is at 1; the y-intercept is 17.

Figure 51 shows the complete graph. (Do you see why? The graph has exactly three turning points.) The value of a causes a stretch or compression; a reflection also occurs if $a < 0$. The zeros are not affected.

If any point other than an x-intercept on the graph of f is known, then a can be determined. For example, if $(2, 3)$ is on the graph, then $f(2) = 3 = a(37)$, so $a = 3/37$. Why won't an x-intercept work?

Figure 51

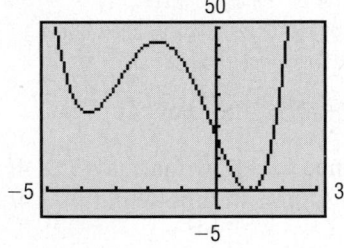

Now we can prove the theorem we conjectured in Section 5.5.

THEOREM

Every polynomial function with real coefficients can be uniquely factored over the real numbers into a product of linear factors and/or irreducible quadratic factors.

Proof Every complex polynomial function f of degree n has exactly n zeros and can be factored into a product of n linear factors. If its coefficients are real, those zeros that are complex numbers will always occur as conjugate pairs. As a result, if $r = a + bi$ is a complex zero, then so is $\bar{r} = a - bi$. Consequently, when the linear factors $x - r$ and $x - \bar{r}$ of f are multiplied, we have

$$(x - r)(x - \bar{r}) = x^2 - (r + \bar{r})x + r\bar{r} = x^2 - 2ax + a^2 + b^2$$

This second-degree polynomial has real coefficients and is irreducible (over the real numbers). Thus, the factors of f are either linear or irreducible quadratic factors. ■

─── **Now Work** PROBLEM 17

3 Find the Complex Zeros of a Polynomial Function

The steps for finding the complex zeros of a polynomial function are the same as those for finding the real zeros.

EXAMPLE 3	**Finding the Complex Zeros of a Polynomial Function**

Find the complex zeros of the polynomial function

$$f(x) = 3x^4 + 5x^3 + 25x^2 + 45x - 18$$

Write f in factored form.

Solution **STEP 1:** The degree of f is 4. So f will have four complex zeros.

STEP 2: The Rational Zeros Theorem provides information about the potential rational zeros of polynomial functions with integer coefficients. For this polynomial function (which has integer coefficients), the potential rational zeros are

$$\pm\frac{1}{3}, \pm\frac{2}{3}, \pm1, \pm2, \pm3, \pm6, \pm9, \pm18$$

Test 1 first:
$$1\overline{)3\quad 5\quad 25\quad 45\quad -18}$$
$$\quad\ 3\quad\ 8\quad 33\quad\ 78$$
$$\overline{3\quad\ 8\quad 33\quad 78\quad\ 60}\quad\text{1 is not a zero.}$$

Test -1:
$$-1\overline{)3\quad 5\quad 25\quad 45\quad -18}$$
$$-3\ -2\ -23\ -22$$
$$\overline{3\quad 2\quad 23\quad 22\ -40}\quad\text{-1 is not a zero.}$$

Test 2:
$$2\overline{)3\quad 5\quad 25\quad 45\quad -18}$$
$$\quad\ 6\quad 22\quad 94\quad 278$$
$$\overline{3\quad 11\quad 47\ 139\quad 260}\quad\text{2 is not a zero.}$$

Test -2:
$$-2\overline{)3\quad\ 5\quad 25\quad\ 45\quad -18}$$
$$\quad -6\quad\ 2\ -54\quad\ 18$$
$$\overline{3\ -1\quad 27\quad -9\quad\ \ 0}$$

Since $f(-2) = 0$, then -2 is a zero and $x + 2$ is a factor of f. The depressed equation is

$$3x^3 - x^2 + 27x - 9 = 0$$

REPEAT STEP 2: Factor the depressed equation by grouping.

$$3x^3 - x^2 + 27x - 9 = 0$$
$$x^2(3x - 1) + 9(3x - 1) = 0 \quad\text{Factor x^2 from $3x^3 - x^2$ and 9 from $27x - 9$.}$$
$$(x^2 + 9)(3x - 1) = 0 \quad\text{Factor out the common factor $3x - 1$.}$$

$$x^2 + 9 = 0 \quad\text{or}\quad 3x - 1 = 0 \quad\text{Apply the Zero-Product Property.}$$
$$x^2 = -9 \quad\text{or}\quad x = \frac{1}{3}$$
$$x = -3i, \quad x = 3i \quad\text{or}\quad x = \frac{1}{3}$$

The four complex zeros of f are $\left\{-3i, 3i, -2, \dfrac{1}{3}\right\}$.

The factored form of f is

$$f(x) = 3x^4 + 5x^3 + 25x^2 + 45x - 18$$
$$= 3(x + 3i)(x - 3i)(x + 2)\left(x - \frac{1}{3}\right)$$

Now Work PROBLEM 33

5.6 Assess Your Understanding

'Are You Prepared?' *Answers are given at the end of these exercises. If you get a wrong answer, read the pages listed in red.*

1. Find the sum and the product of the complex numbers $3 - 2i$ and $-3 + 5i$. (pp. 104–109)

2. In the complex number system, find the complex solutions of the equation $x^2 + 2x + 2 = 0$. (pp. 104–109)

Concepts and Vocabulary

3. Every polynomial function of odd degree with real coefficients will have at least _____ real zero(s).

4. If $3 + 4i$ is a zero of a polynomial function of degree 5 with real coefficients, then so is _____.

5. **True or False** A polynomial function of degree n with real coefficients has exactly n complex zeros. At most n of them are real zeros.

6. **True or False** A polynomial function of degree 4 with real coefficients could have $-3, 2 + i, 2 - i$, and $-3 + 5i$ as its zeros.

Skill Building

In Problems 7–16, information is given about a polynomial function $f(x)$ whose coefficients are real numbers. Find the remaining zeros of f.

7. Degree 3; zeros: $3, 4 - i$

8. Degree 3; zeros: $4, 3 + i$

9. Degree 4; zeros: $i, 1 + i$

10. Degree 4; zeros: $1, 2, 2 + i$

11. Degree 5; zeros: $1, i, 2i$

12. Degree 5; zeros: $0, 1, 2, i$

13. Degree 4; zeros: $i, 2, -2$

14. Degree 4; zeros: $2 - i, -i$

15. Degree 6; zeros: $2, 2 + i, -3 - i, 0$

16. Degree 6; zeros: $i, 3 - 2i, -2 + i$

In Problems 17–22, form a polynomial function $f(x)$ with real coefficients having the given degree and zeros. Answers will vary depending on the choice of the leading coefficient.

17. Degree 4; zeros: $3 + 2i; 4$, multiplicity 2

18. Degree 4; zeros: $i, 1 + 2i$

19. Degree 5; zeros: $2; -i; 1 + i$

20. Degree 6; zeros: $i, 4 - i; 2 + i$

21. Degree 4; zeros: 3, multiplicity $2; -i$

22. Degree 5; zeros: 1, multiplicity $3; 1 + i$

In Problems 23–30, use the given zero to find the remaining zeros of each function.

23. $f(x) = x^3 - 4x^2 + 4x - 16$; zero: $2i$

24. $g(x) = x^3 + 3x^2 + 25x + 75$; zero: $-5i$

25. $f(x) = 2x^4 + 5x^3 + 5x^2 + 20x - 12$; zero: $-2i$

26. $h(x) = 3x^4 + 5x^3 + 25x^2 + 45x - 18$; zero: $3i$

27. $h(x) = x^4 - 9x^3 + 21x^2 + 21x - 130$; zero: $3 - 2i$

28. $f(x) = x^4 - 7x^3 + 14x^2 - 38x - 60$; zero: $1 + 3i$

29. $h(x) = 3x^5 + 2x^4 + 15x^3 + 10x^2 - 528x - 352$; zero: $-4i$

30. $g(x) = 2x^5 - 3x^4 - 5x^3 - 15x^2 - 207x + 108$; zero: $3i$

In Problems 31–40, find the complex zeros of each polynomial function. Write f in factored form.

31. $f(x) = x^3 - 1$

32. $f(x) = x^4 - 1$

33. $f(x) = x^3 - 8x^2 + 25x - 26$

34. $f(x) = x^3 + 13x^2 + 57x + 85$

35. $f(x) = x^4 + 5x^2 + 4$

36. $f(x) = x^4 + 13x^2 + 36$

37. $f(x) = x^4 + 2x^3 + 22x^2 + 50x - 75$

38. $f(x) = x^4 + 3x^3 - 19x^2 + 27x - 252$

39. $f(x) = 3x^4 - x^3 - 9x^2 + 159x - 52$

40. $f(x) = 2x^4 + x^3 - 35x^2 - 113x + 65$

Explaining Concepts: Discussion and Writing

In Problems 41 and 42, explain why the facts given are contradictory.

41. $f(x)$ is a polynomial function of degree 3 whose coefficients are real numbers; its zeros are $4 + i, 4 - i,$ and $2 + i$.

42. $f(x)$ is a polynomial function of degree 3 whose coefficients are real numbers; its zeros are $2, i,$ and $3 + i$.

43. $f(x)$ is a polynomial function of degree 4 whose coefficients are real numbers; three of its zeros are $2, 1 + 2i,$ and $1 - 2i$. Explain why the remaining zero must be a real number.

44. $f(x)$ is a polynomial function of degree 4 whose coefficients are real numbers; two of its zeros are -3 and $4 - i$. Explain why one of the remaining zeros must be a real number. Write down one of the missing zeros.

'Are You Prepared?' Answers

1. Sum: $3i$; product: $1 + 21i$ **2.** $-1 - i, \quad -1 + i$

CHAPTER REVIEW

Things to Know

Power function (pp. 321–324)

$f(x) = x^n, \ n \geq 2$ even

Domain: all real numbers Range: nonnegative real numbers

Passes through $(-1, 1), (0, 0), (1, 1)$

Even function

Decreasing on $(-\infty, 0)$, increasing on $(0, \infty)$

$f(x) = x^n, \ n \geq 3$ odd

Domain: all real numbers Range: all real numbers

Passes through $(-1, -1), (0, 0), (1, 1)$

Odd function

Increasing on $(-\infty, \infty)$

Polynomial function (pp. 320, 329–331)

$f(x) = a_n x^n + a_{n-1} x^{n-1}$
$\qquad + \cdots + a_1 x + a_0, \ a_n \neq 0$

Domain: all real numbers

At most $n - 1$ turning points

End behavior: Behaves like $y = a_n x^n$ for large $|x|$

Real zeros of a polynomial function f (p. 325)

Real numbers for which $f(x) = 0$; the real zeros of f are the x-intercepts of the graph of f.

Rational function (pp. 343–350)

$R(x) = \dfrac{p(x)}{q(x)}$

p, q are polynomial functions and q is not the zero polynomial.

Domain: $\{x | q(x) \neq 0\}$

Vertical asymptotes: With $R(x)$ in lowest terms, if $q(r) = 0$ for some real number, then $x = r$ is a vertical asymptote.

Horizontal or oblique asymptote: See the summary on page 350.

Remainder Theorem (p. 376)

If a polynomial function $f(x)$ is divided by $x - c$, then the remainder is $f(c)$.

Factor Theorem (p. 376)

$x - c$ is a factor of a polynomial function $f(x)$ if and only if $f(c) = 0$.

Rational Zeros Theorem (p. 378)

Let f be a polynomial function of degree 1 or higher of the form

$$f(x) = a_n x^n + a_{n-1} x^{n-1} + \cdots + a_1 x + a_0 \qquad a_n \neq 0, a_0 \neq 0$$

where each coefficient is an integer. If $\dfrac{p}{q}$, in lowest terms, is a rational zero of f, then p must be a factor of a_0, and q must be a factor of a_n.

Intermediate Value Theorem (p. 382)

Let f be a polynomial function. If $a < b$ and $f(a)$ and $f(b)$ are of opposite sign, then there is at least one real zero of f between a and b.

Fundamental Theorem of Algebra (p. 383)

Every complex polynomial function $f(x)$ of degree $n \geq 1$ has at least one complex zero.

Conjugate Pairs Theorem (p. 389)

Let $f(x)$ be a polynomial whose coefficients are real numbers. If $r = a + bi$ is a zero of f, then its complex conjugate $\bar{r} = a - bi$ is also a zero of f.

Objectives

Section	You should be able to . . .	Example(s)	Review Exercises
5.1	**1** Identify polynomial functions and their degree (p. 320)	1	1–4
	2 Graph polynomial functions using transformations (p. 324)	2, 3	5–10
	3 Identify the real zeros of a polynomial function and their multiplicity (p. 325)	4–6	11–18 (a) and (b)
	4 Analyze the graph of a polynomial function (p. 332)	9–11	11–18
	5 Build cubic models from data (p. 336)	12	88
5.2	**1** Find the domain of a rational function (p. 343)	1	19–22
	2 Find the vertical asymptotes of a rational function (p. 346)	4	19–22
	3 Find the horizontal or oblique asymptote of a rational function (p. 347)	5–8	19–22
5.3	**1** Analyze the graph of a rational function (p. 353)	1–5	23–34
	2 Solve applied problems involving rational functions (p. 364)	7	87
5.4	**1** Solve polynomial inequalities (p. 368)	1, 2	35, 36
	2 Solve rational inequalities (p. 369)	3, 4	37–44
5.5	**1** Use the Remainder and Factor Theorems (p. 375)	1, 2	45–50
	2 Use the Rational Zeros Theorem to list the potential rational zeros of a polynomial function (p. 378)	3	51–58
	3 Find the real zeros of a polynomial function (p. 378)	4, 5	53–58
	4 Solve polynomial equations (p. 381)	6	59–62
	5 Use the Theorem for Bounds on Zeros (p. 381)	7	63–66
	6 Use the Intermediate Value Theorem (p. 382)	8	67–74
5.6	**1** Use the Conjugate Pairs Theorem (p. 389)	1	75–78
	2 Find a polynomial function with specified zeros (p. 390)	2	75–78
	3 Find the complex zeros of a polynomial (p. 391)	3	79–86

Review Exercises

In Problems 1–4, determine whether the function is a polynomial function, rational function, or neither. For those that are polynomial functions, state the degree. For those that are not polynomial functions, tell why not.

1. $f(x) = 4x^5 - 3x^2 + 5x - 2$

2. $f(x) = \dfrac{3x^5}{2x + 1}$

3. $f(x) = 3x^2 + 5x^{1/2} - 1$

4. $f(x) = 3$

In Problems 5–10, graph each function using transformations (shifting, compressing, stretching, and reflection). Show all the stages.

5. $f(x) = (x + 2)^3$

6. $f(x) = -x^3 + 3$

7. $f(x) = -(x - 1)^4$

8. $f(x) = (x - 1)^4 - 2$

9. $f(x) = (x - 1)^4 + 2$

10. $f(x) = (1 - x)^3$

In Problems 11–18, analyze each polynomial function by following Steps 1 through 6 on page 333.

11. $f(x) = x(x + 2)(x + 4)$

12. $f(x) = x(x - 2)(x - 4)$

13. $f(x) = (x - 2)^2(x + 4)$

14. $f(x) = (x - 2)(x + 4)^2$

15. $f(x) = -2x^3 + 4x^2$

16. $f(x) = -4x^3 + 4x$

17. $f(x) = (x - 1)^2(x + 3)(x + 1)$

18. $f(x) = (x - 4)(x + 2)^2(x - 2)$

In Problems 19–22, find the domain of each rational function. Find any horizontal, vertical, or oblique asymptotes.

19. $R(x) = \dfrac{x + 2}{x^2 - 9}$

20. $R(x) = \dfrac{x^2 + 4}{x - 2}$

21. $R(x) = \dfrac{x^2 + 3x + 2}{(x + 2)^2}$

22. $R(x) = \dfrac{x^3}{x^3 - 1}$

In Problems 23–34, discuss each rational function following the eight steps given on page 336.

23. $R(x) = \dfrac{2x - 6}{x}$
24. $R(x) = \dfrac{4 - x}{x}$
25. $H(x) = \dfrac{x + 2}{x(x - 2)}$
26. $H(x) = \dfrac{x}{x^2 - 1}$

27. $R(x) = \dfrac{x^2 + x - 6}{x^2 - x - 6}$
28. $R(x) = \dfrac{x^2 - 6x + 9}{x^2}$
29. $F(x) = \dfrac{x^3}{x^2 - 4}$
30. $F(x) = \dfrac{3x^3}{(x - 1)^2}$

31. $R(x) = \dfrac{2x^4}{(x - 1)^2}$
32. $R(x) = \dfrac{x^4}{x^2 - 9}$
33. $G(x) = \dfrac{x^2 - 4}{x^2 - x - 2}$
34. $F(x) = \dfrac{(x - 1)^2}{x^2 - 1}$

In Problems 35–44, solve each inequality. Graph the solution set.

35. $x^3 + x^2 < 4x + 4$
36. $x^3 + 4x^2 \geq x + 4$
37. $\dfrac{6}{x + 3} \geq 1$
38. $\dfrac{-2}{1 - 3x} < 1$

39. $\dfrac{2x - 6}{1 - x} < 2$
40. $\dfrac{3 - 2x}{2x + 5} \geq 2$
41. $\dfrac{(x - 2)(x - 1)}{x - 3} \geq 0$
42. $\dfrac{x + 1}{x(x - 5)} \leq 0$

43. $\dfrac{x^2 - 8x + 12}{x^2 - 16} > 0$
44. $\dfrac{x(x^2 + x - 2)}{x^2 + 9x + 20} \leq 0$

In Problems 45–48, find the remainder R when $f(x)$ is divided by $g(x)$. Is g a factor of f?

45. $f(x) = 8x^3 - 3x^2 + x + 4;\quad g(x) = x - 1$
46. $f(x) = 2x^3 + 8x^2 - 5x + 5;\quad g(x) = x - 2$

47. $f(x) = x^4 - 2x^3 + 15x - 2;\quad g(x) = x + 2$
48. $f(x) = x^4 - x^2 + 2x + 2;\quad g(x) = x + 1$

49. Find the value of $f(x) = 12x^6 - 8x^4 + 1$ at $x = 4$.

50. Find the value of $f(x) = -16x^3 + 18x^2 - x + 2$ at $x = -2$.

51. List all the potential rational zeros of $f(x) = 12x^8 - x^7 + 6x^4 - x^3 + x - 3$.

52. List all the potential rational zeros of $f(x) = -6x^5 + x^4 + 2x^3 - x + 1$.

In Problems 53–58, use the Rational Zeros Theorem to find all the real zeros of each polynomial function. Use the zeros to factor f over the real numbers.

53. $f(x) = x^3 - 3x^2 - 6x + 8$
54. $f(x) = x^3 - x^2 - 10x - 8$

55. $f(x) = 4x^3 + 4x^2 - 7x + 2$
56. $f(x) = 4x^3 - 4x^2 - 7x - 2$

57. $f(x) = x^4 - 4x^3 + 9x^2 - 20x + 20$
58. $f(x) = x^4 + 6x^3 + 11x^2 + 12x + 18$

In Problems 59–62, solve each equation in the real number system.

59. $2x^4 + 2x^3 - 11x^2 + x - 6 = 0$
60. $3x^4 + 3x^3 - 17x^2 + x - 6 = 0$

61. $2x^4 + 7x^3 + x^2 - 7x - 3 = 0$
62. $2x^4 + 7x^3 - 5x^2 - 28x - 12 = 0$

In Problems 63–66, find bounds to the real zeros of each polynomial function.

63. $f(x) = x^3 - x^2 - 4x + 2$
64. $f(x) = x^3 + x^2 - 10x - 5$

65. $f(x) = 2x^3 - 7x^2 - 10x + 35$
66. $f(x) = 3x^3 - 7x^2 - 6x + 14$

In Problems 67–70, use the Intermediate Value Theorem to show that each polynomial function has a zero in the given interval.

67. $f(x) = 3x^3 - x - 1;\quad [0, 1]$
68. $f(x) = 2x^3 - x^2 - 3;\quad [1, 2]$

69. $f(x) = 8x^4 - 4x^3 - 2x - 1;\quad [0, 1]$
70. $f(x) = 3x^4 + 4x^3 - 8x - 2;\quad [1, 2]$

In Problems 71–74, each polynomial function has exactly one positive zero. Approximate the zero correct to two decimal places.

71. $f(x) = x^3 - x - 2$

72. $f(x) = 2x^3 - x^2 - 3$

73. $f(x) = 8x^4 - 4x^3 - 2x - 1$

74. $f(x) = 3x^4 + 4x^3 - 8x - 2$

In Problems 75–78, information is given about a complex polynomial $f(x)$ whose coefficients are real numbers. Find the remaining zeros of f. Then find a polynomial function with real coefficients that has the zeros.

75. Degree 3; zeros: $4 + i, 6$

76. Degree 3; zeros: $3 + 4i, 5$

77. Degree 4; zeros: $i, 1 + i$

78. Degree 4; zeros: $1, 2, 1 + i$

In Problems 79–86, find the complex zeros of each polynomial function $f(x)$. Write f in factored form.

79. $f(x) = x^3 - 3x^2 - 6x + 8$

80. $f(x) = x^3 - x^2 - 10x - 8$

81. $f(x) = 4x^3 + 4x^2 - 7x + 2$

82. $f(x) = 4x^3 - 4x^2 - 7x - 2$

83. $f(x) = x^4 - 4x^3 + 9x^2 - 20x + 20$

84. $f(x) = x^4 + 6x^3 + 11x^2 + 12x + 18$

85. $f(x) = 2x^4 + 2x^3 - 11x^2 + x - 6$

86. $f(x) = 3x^4 + 3x^3 - 17x^2 + x - 6$

87. Making a Can A can in the shape of a right circular cylinder is required to have a volume of 250 cubic centimeters.
 (a) Express the amount A of material to make the can as a function of the radius r of the cylinder.
 (b) How much material is required if the can is of radius 3 centimeters?
 (c) How much material is required if the can is of radius 5 centimeters?
 (d) Graph $A = A(r)$. For what value of r is A smallest?

88. Model It: Poverty Rates The following data represent the percentage of families in the United States whose income is below the poverty level.

Year, t	Percent below Poverty Level, p
1990, 1	10.9
1991, 2	11.5
1992, 3	11.9
1993, 4	12.3
1994, 5	11.6
1995, 6	10.8
1996, 7	11.0
1997, 8	10.3
1998, 9	10.0
1999, 10	9.3
2000, 11	8.7
2001, 12	9.2
2002, 13	9.6
2003, 14	10.0
2004, 15	10.2

Source: U.S. Census Bureau

 (a) With a graphing utility, draw a scatter diagram of the data. Comment on the type of relation that appears to exist between the two variables.

 (b) Decide on a function of best fit to these data (linear, quadratic, or cubic), and use this function to predict the percentage of U.S. families that were below the poverty level in 2005 ($t = 16$).
 (c) Draw the function of best fit on the scatter diagram drawn in part (a).

89. Design a polynomial function with the following characteristics: degree 6; four real zeros, one of multiplicity 3; y-intercept 3; behaves like $y = -5x^6$ for large values of $|x|$. Is this polynomial unique? Compare your polynomial with those of other students. What terms will be the same as everyone else's? Add some more characteristics, such as symmetry or naming the real zeros. How does this modify the polynomial?

90. Design a rational function with the following characteristics: three real zeros, one of multiplicity 2; y-intercept 1; vertical asymptotes $x = -2$ and $x = 3$; oblique asymptote $y = 2x + 1$. Is this rational function unique? Compare yours with those of other students. What will be the same as everyone else's? Add some more characteristics, such as symmetry or naming the real zeros. How does this modify the rational function?

91. The illustration shows the graph of a polynomial function.
 (a) Is the degree of the polynomial even or odd?
 (b) Is the leading coefficient positive or negative?
 (c) Is the function even, odd, or neither?
 (d) Why is x^2 necessarily a factor of the polynomial?
 (e) What is the minimum degree of the polynomial?
 (f) Formulate five different polynomials whose graphs could look like the one shown. Compare yours to those of other students. What similarities do you see? What differences?

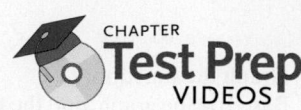

CHAPTER

Test Prep

VIDEOS

The Chapter Test Prep Videos are step-by-step test solutions available in the Video Resources DVD, in *MyMathLab* , or on this text's You Tube Channel. Flip back to the Student Resources page to see the exact web address for this text's YouTube channel.

CHAPTER TEST

1. Graph $f(x) = (x - 3)^4 - 2$ using transformations.
2. For the polynomial function $g(x) = 2x^3 + 5x^2 - 28x - 15$,
 (a) Determine the maximum number of real zeros that the function may have.
 (b) Find bounds to the zeros of the function.
 (c) List the potential rational zeros.
 (d) Determine the real zeros of g. Factor g over the reals.
 (e) Find the x- and y-intercepts of the graph of g.
 (f) Determine whether the graph crosses or touches the x-axis at each x-intercept.
 (g) Find the power function that the graph of g resembles for large values of $|x|$.
 (h) Determine the behavior of the graph of g near each x-intercept.
 (i) Put all the information together to obtain the graph of g.
3. Find the complex zeros of $f(x) = x^3 - 4x^2 + 25x - 100$.
4. Solve $3x^3 + 2x - 1 = 8x^2 - 4$ in the complex number system.

In Problems 5 and 6, find the domain of each function. Find any horizontal, vertical, or oblique asymptotes.

5. $g(x) = \dfrac{2x^2 - 14x + 24}{x^2 + 6x - 40}$

6. $r(x) = \dfrac{x^2 + 2x - 3}{x + 1}$

7. Sketch the graph of the function in Problem 6. Label all intercepts, vertical asymptotes, horizontal asymptotes, and oblique asymptotes.

In Problems 8 and 9, write a function that meets the given conditions.

8. Fourth-degree polynomial with real coefficients; zeros: $-2, 0, 3 + i$.

9. Rational function; asymptotes: $y = 2, x = 4$; domain: $\{x \mid x \neq 4, x \neq 9\}$

10. Use the Intermediate Value Theorem to show that the function $f(x) = -2x^2 - 3x + 8$ has at least one real zero on the interval $[0, 4]$.

11. Solve: $\dfrac{x + 2}{x - 3} < 2$

CUMULATIVE REVIEW

1. Find the distance between the points $P = (1, 3)$ and $Q = (-4, 2)$.
2. Solve the inequality $x^2 \geq x$ and graph the solution set.
3. Solve the inequality $x^2 - 3x < 4$ and graph the solution set.
4. Find a linear function with slope -3 that contains the point $(-1, 4)$. Graph the function.
5. Find the equation of the line parallel to the line $y = 2x + 1$ and containing the point $(3, 5)$. Express your answer in slope–intercept form and graph the line.
6. Graph the equation $y = x^3$.
7. Does the relation $\{(3, 6), (1, 3), (2, 5), (3, 8)\}$ represent a function? Why or why not?
8. Solve the equation $x^3 - 6x^2 + 8x = 0$.
9. Solve the inequality $3x + 2 \leq 5x - 1$ and graph the solution set.
10. Find the center and radius of the circle $x^2 + 4x + y^2 - 2y - 4 = 0$. Graph the circle.
11. For the equation $y = x^3 - 9x$, determine the intercepts and test for symmetry.
12. Find an equation of the line perpendicular to $3x - 2y = 7$ that contains the point $(1, 5)$.
13. Is the following the graph of a function? Why or why not?

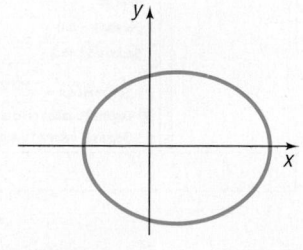

14. For the function $f(x) = x^2 + 5x - 2$, find
 (a) $f(3)$ (b) $f(-x)$
 (c) $-f(x)$ (d) $f(3x)$
 (e) $\dfrac{f(x + h) - f(x)}{h} \quad h \neq 0$

15. Answer the following questions regarding the function
 $$f(x) = \frac{x + 5}{x - 1}$$
 (a) What is the domain of f?
 (b) Is the point $(2, 6)$ on the graph of f?
 (c) If $x = 3$, what is $f(x)$? What point is on the graph of f?
 (d) If $f(x) = 9$, what is x? What point is on the graph of f?
 (e) Is f a polynomial or rational function?

16. Graph the function $f(x) = -3x + 7$.

17. Graph $f(x) = 2x^2 - 4x + 1$ by determining whether its graph opens up or down and by finding its vertex, axis of symmetry, y-intercept, and x-intercepts, if any.

18. Find the average rate of change of $f(x) = x^2 + 3x + 1$ from 1 to 2. Use this result to find the equation of the secant line containing $(1, f(1))$ and $(2, f(2))$.

19. In parts (a) to (f) on page 398, use the following graph.

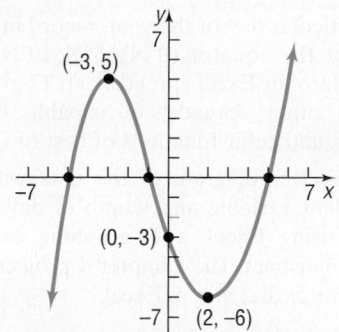

397

(a) Determine the intercepts.

(b) Based on the graph, tell whether the graph is symmetric with respect to the x-axis, the y-axis, and/or the origin.

(c) Based on the graph, tell whether the function is even, odd, or neither.

(d) List the intervals on which f is increasing. List the intervals on which f is decreasing.

(e) List the numbers, if any, at which f has a local maximum value. What are these local maxima values?

(f) List the numbers, if any, at which f has a local minimum value. What are these local minima values?

20. Determine algebraically whether the function

$$f(x) = \frac{5x}{x^2 - 9}$$

is even, odd, or neither.

21. For the function $f(x) = \begin{cases} 2x + 1 & \text{if } -3 < x < 2 \\ -3x + 4 & \text{if } x \ge 2 \end{cases}$

(a) Find the domain of f.

(b) Locate any intercepts.

(c) Graph the function.

(d) Based on the graph, find the range.

22. Graph the function $f(x) = -3(x + 1)^2 + 5$ using transformations.

23. Suppose that $f(x) = x^2 - 5x + 1$ and $g(x) = -4x - 7$.

(a) Find $f + g$ and state its domain.

(b) Find $\dfrac{f}{g}$ and state its domain.

24. Demand Equation The price p (in dollars) and the quantity x sold of a certain product obey the demand equation

$$p = -\frac{1}{10}x + 150,$$

(a) Express the revenue R as a function of x.

(b) What is the revenue if 100 units are sold?

(c) What quantity x maximizes revenue? What is the maximum revenue?

(d) What price should the company charge to maximize revenue?

CHAPTER PROJECTS

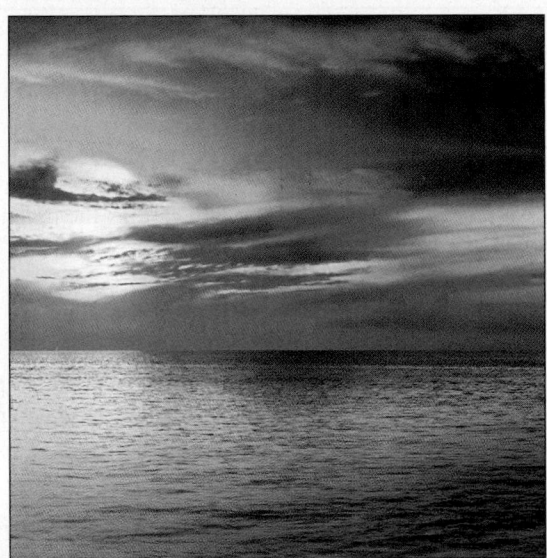

Internet-based Project

I. Length of Day Go to *http://en.wikipedia.org/wiki/Latitude* and read about latitude through the subhead "Effect of Latitude". Now go to *http://www.orchidculture.com/COD/ daylength.html#60N*.

1. For a particular day of the year, record in a table the length of day for the equator (0°N), 5°N, 10°N, . . . , 60°N. Enter the data into an Excel spreadsheet, TI-graphing calculator, or some other spreadsheet capable of finding linear, quadratic, and cubic functions of best fit.

2. Draw a scatter diagram of the data with latitude as the independent variable and length of day as the dependent variable using Excel, a TI-graphing calculator, or some other spreadsheet. The Chapter 4 project describes how to draw a scatter diagram in Excel.

3. Determine the linear function of best fit. Graph the linear function of best fit on the scatter diagram. To do this in Excel, click on any data point in the scatter diagram. Now click the Layout menu, select Trendline within the Analysis region, select More Trendline Options. Select the Linear radio button and select Display Equation on Chart. See Figure 52. Move the Trendline Options window off to the side and you will see the linear function of best fit displayed on the scatter diagram. Do you think the function accurately describes the relation between latitude and length of the day?

Figure 52

4. Determine the quadratic function of best fit. Graph the quadratic function of best fit on the scatter diagram. To do this in Excel, click on any data point in the scatter diagram. Now click the Layout menu, select Trendline within the Analysis region, select More Trendline Options. Select the Polynomial radio button with Order set to 2. Select Display Equation on Chart. Move the Trendline Options window off to the side and you will see the quadratic function of best fit displayed on the scatter diagram. Do you think the function accurately describes the relation between latitude and length of the day?

5. Determine the cubic function of best fit. Graph the cubic function of best fit on the scatter diagram. To do this in Excel, click on any data point in the scatter diagram. Now click the Layout menu, select Trendline within the Analysis region, select More Trendline Options. Select the

Polynomial radio button with Order set to 3. Select Display Equation on Chart. Move the Trendline Options window off to the side and you will see the cubic function of best fit displayed on the scatter diagram. Do you think the function accurately describes the relation between latitude and length of the day?

6. Which of the three models seems to fit the data best? Explain your reasoning.

7. Use your model to predict the hours of daylight on the day you selected for Chicago (41.85 degrees north latitude). Go to the Old Farmer's Almanac or other website (such as *http://astro.unl.edu/classaction/animations/coordsmotion/ daylighthoursexplorer.html*) to determine the hours of daylight in Chicago for the day you selected. How do the two compare?

The following project is available at the Instructor's Resource Center (IRC):

II. Theory of Equations The coefficients of a polynomial function can be found if its zeros are known, an advantage of using polynomials in modeling.

Citation: Excel © 2010 Microsoft Corporation. Used with permission from Microsoft.

6

Exponential and Logarithmic Functions

Outline

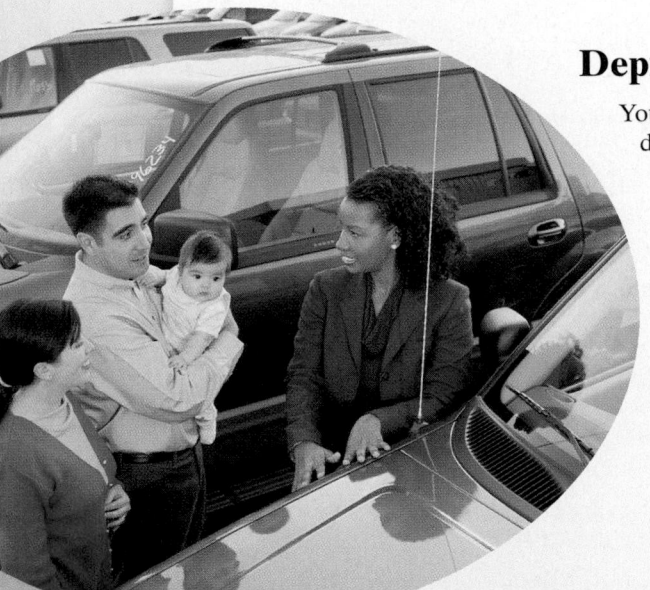

Depreciation of Cars

You are ready to buy that first new car. You know that cars lose value over time due to depreciation and that different cars have different rates of depreciation. So you will research the depreciation rates for the cars you are thinking of buying. After all, the lower the depreciation rate is, the more the car will be worth each year.

—See the Internet-based Chapter Project I—

◁ **A Look Back** Until now, our study of functions has concentrated on polynomial and rational functions. These functions belong to the class of **algebraic functions,** that is, functions that can be expressed in terms of sums, differences, products, quotients, powers, or roots of polynomials. Functions that are not algebraic are termed **transcendental** (they transcend, or go beyond, algebraic functions).

A Look Ahead ▷ In this chapter, we study two transcendental functions: the exponential function and the logarithmic function. These functions occur frequently in a wide variety of applications, such as biology, chemistry, economics, and psychology.

The chapter begins with a discussion of composite, one-to-one, and inverse functions, concepts needed to see the relationship between exponential and logarithmic functions.

6.1 Composite Functions

PREPARING FOR THIS SECTION *Before getting started, review the following:*

- Find the Value of a Function (Section 3.1, pp. 203–206)
- Domain of a Function (Section 3.1, pp. 206–208)

Now Work the **'Are You Prepared?'** problems on page 406.

OBJECTIVES **1** Form a Composite Function (p. 401)

2 Find the Domain of a Composite Function (p. 402)

1 Form a Composite Function

Figure 1

Suppose that an oil tanker is leaking oil and you want to determine the area of the circular oil patch around the ship. See Figure 1. It is determined that the oil is leaking from the tanker in such a way that the radius of the circular patch of oil around the ship is increasing at a rate of 3 feet per minute. Therefore, the radius r of the oil patch at any time t, in minutes, is given by $r(t) = 3t$. So after 20 minutes the radius of the oil patch is $r(20) = 3(20) = 60$ feet.

The area A of a circle as a function of the radius r is given by $A(r) = \pi r^2$. The area of the circular patch of oil after 20 minutes is $A(60) = \pi(60)^2 = 3600\pi$ square feet. Notice that $60 = r(20)$, so $A(60) = A(r(20))$. The argument of the function A is the output a function!

In general, we can find the area of the oil patch as a function of time t by evaluating $A(r(t))$ and obtaining $A(r(t)) = A(3t) = \pi(3t)^2 = 9\pi t^2$. The function $A(r(t))$ is a special type of function called a *composite function*.

As another example, consider the function $y = (2x + 3)^2$. If we write $y = f(u) = u^2$ and $u = g(x) = 2x + 3$, then, by a substitution process, we can obtain the original function: $y = f(u) = f(g(x)) = (2x + 3)^2$.

In general, suppose that f and g are two functions and that x is a number in the domain of g. By evaluating g at x, we get $g(x)$. If $g(x)$ is in the domain of f, then we may evaluate f at $g(x)$ and obtain the expression $f(g(x))$. The correspondence from x to $f(g(x))$ is called a *composite function* $f \circ g$.

DEFINITION

Given two functions f and g, the **composite function,** denoted by $f \circ g$ (read as "f composed with g"), is defined by

$$(f \circ g)(x) = f(g(x))$$

The domain of $f \circ g$ is the set of all numbers x in the domain of g such that $g(x)$ is in the domain of f.

Look carefully at Figure 2. Only those x's in the domain of g for which $g(x)$ is in the domain of f can be in the domain of $f \circ g$. The reason is that if $g(x)$ is not in the domain of f then $f(g(x))$ is not defined. Because of this, the domain of $f \circ g$ is a subset of the domain of g; the range of $f \circ g$ is a subset of the range of f.

Figure 2

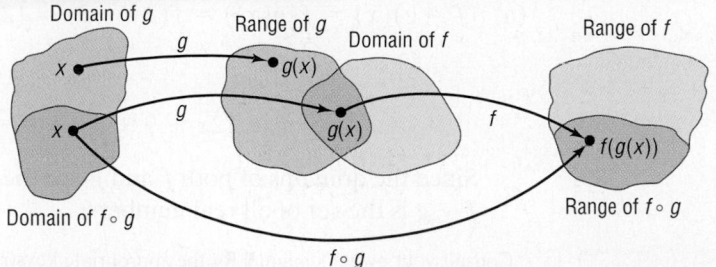

Figure 3 provides a second illustration of the definition. Here x is the input to the function g, yielding $g(x)$. Then $g(x)$ is the input to the function f, yielding $f(g(x))$. Notice that the "inside" function g in $f(g(x))$ is done first.

Figure 3

INPUT x ⟶ g $\xrightarrow{g(x)}$ f ⟶ OUTPUT $f(g(x))$

EXAMPLE 1 **Evaluating a Composite Function**

Suppose that $f(x) = 2x^2 - 3$ and $g(x) = 4x$. Find:

(a) $(f \circ g)(1)$ (b) $(g \circ f)(1)$ (c) $(f \circ f)(-2)$ (d) $(g \circ g)(-1)$

Solution (a) $(f \circ g)(1) = f(g(1)) = f(4) = 2 \cdot 4^2 - 3 = 29$

$\qquad\qquad\quad\uparrow\qquad\quad\uparrow$
$\qquad\qquad g(x) = 4x \quad f(x) = 2x^2 - 3$
$\qquad\qquad g(1) = 4$

(b) $(g \circ f)(1) = g(f(1)) = g(-1) = 4 \cdot (-1) = -4$

$\qquad\qquad\quad\uparrow\qquad\qquad\uparrow$
$\qquad\quad f(x) = 2x^2 - 3 \quad g(x) = 4x$
$\qquad\quad f(1) = -1$

(c) $(f \circ f)(-2) = f(f(-2)) = f(5) = 2 \cdot 5^2 - 3 = 47$

$\qquad\qquad\qquad\quad\uparrow$
$\qquad\quad f(-2) = 2(-2)^2 - 3 = 5$

(d) $(g \circ g)(-1) = g(g(-1)) = g(-4) = 4 \cdot (-4) = -16$

$\qquad\qquad\qquad\quad\uparrow$
$\qquad\qquad g(-1) = -4$

Figure 4

 COMMENT *Graphing calculators can be used to evaluate composite functions.* Let $Y_1 = f(x) = 2x^2 - 3$ and $Y_2 = g(x) = 4x$. Then, using a TI-84 Plus graphing calculator, $(f \circ g)(1)$ is found as shown in Figure 4. Notice that this is the result obtained in Example 1(a).* ■

Now Work PROBLEM 11

2 Find the Domain of a Composite Function

EXAMPLE 2 **Finding a Composite Function and Its Domain**

Suppose that $f(x) = x^2 + 3x - 1$ and $g(x) = 2x + 3$.

Find: (a) $f \circ g$ (b) $g \circ f$

Then find the domain of each composite function.

Solution The domain of f and the domain of g are the set of all real numbers.

(a) $(f \circ g)(x) = f(g(x)) = f(2x + 3) = (2x + 3)^2 + 3(2x + 3) - 1$

$\qquad\qquad\qquad\qquad\qquad\quad\uparrow$
$\qquad\qquad\qquad\qquad f(x) = x^2 + 3x - 1$

$\qquad\qquad = 4x^2 + 12x + 9 + 6x + 9 - 1 = 4x^2 + 18x + 17$

Since the domains of both f and g are the set of all real numbers, the domain of $f \circ g$ is the set of all real numbers.

*Consult your owner's manual for the appropriate keystrokes.

(b) $(g \circ f)(x) = g(f(x)) = g(x^2 + 3x - 1) = 2(x^2 + 3x - 1) + 3$

$$\uparrow$$
$$g(x) = 2x + 3$$

$$= 2x^2 + 6x - 2 + 3 = 2x^2 + 6x + 1$$

Since the domains of both f and g are the set of all real numbers, the domain of $g \circ f$ is the set of all real numbers.

Look back at Figure 2 on page 401. In determining the domain of the composite function $(f \circ g)(x) = f(g(x))$, keep the following two thoughts in mind about the input x.

1. Any x not in the domain of g must be excluded.
2. Any x for which $g(x)$ is not in the domain of f must be excluded.

EXAMPLE 3 **Finding the Domain of $f \circ g$**

Find the domain of $f \circ g$ if $f(x) = \dfrac{1}{x + 2}$ and $g(x) = \dfrac{4}{x - 1}$.

Solution For $(f \circ g)(x) = f(g(x))$, first note that the domain of g is $\{x | x \neq 1\}$, so exclude 1 from the domain of $f \circ g$. Next note that the domain of f is $\{x | x \neq -2\}$, which means that $g(x)$ cannot equal -2. Solve the equation $g(x) = -2$ to determine what additional value(s) of x to exclude.

$$\frac{4}{x - 1} = -2 \qquad g(x) = -2$$
$$4 = -2(x - 1)$$
$$4 = -2x + 2$$
$$2x = -2$$
$$x = -1$$

Also exclude -1 from the domain of $f \circ g$.
The domain of $f \circ g$ is $\{x | x \neq -1, x \neq 1\}$.

✔**Check:** For $x = 1$, $g(x) = \dfrac{4}{x - 1}$ is not defined, so $(f \circ g)(x) = f(g(x))$ is not defined.

For $x = -1$, $g(-1) = \dfrac{4}{-2} = -2$, and $(f \circ g)(-1) = f(g(-1)) = f(-2)$ is not defined.

━━━▶ **Now Work** PROBLEM 21

EXAMPLE 4 **Finding a Composite Function and Its Domain**

Suppose that $f(x) = \dfrac{1}{x + 2}$ and $g(x) = \dfrac{4}{x - 1}$.

Find: (a) $f \circ g$ (b) $f \circ f$

Then find the domain of each composite function.

Solution The domain of f is $\{x | x \neq -2\}$ and the domain of g is $\{x | x \neq 1\}$.

(a) $(f \circ g)(x) = f(g(x)) = f\left(\dfrac{4}{x - 1}\right) = \dfrac{1}{\dfrac{4}{x - 1} + 2} = \dfrac{x - 1}{4 + 2(x - 1)} = \dfrac{x - 1}{2x + 2} = \dfrac{x - 1}{2(x + 1)}$

$$\uparrow \qquad\qquad\qquad\qquad \uparrow$$
$$f(x) = \frac{1}{x + 2} \qquad\qquad \text{Multiply by } \frac{x - 1}{x - 1}.$$

In Example 3, we found the domain of $f \circ g$ to be $\{x | x \neq -1, x \neq 1\}$.

We could also find the domain of $f \circ g$ by first looking at the domain of g: $\{x \mid x \neq 1\}$. We exclude 1 from the domain of $f \circ g$ as a result. Then we look at $f \circ g$ and notice that x cannot equal -1, since $x = -1$ results in division by 0. So we also exclude -1 from the domain of $f \circ g$. Therefore, the domain of $f \circ g$ is $\{x \mid x \neq -1, x \neq 1\}$.

(b) $(f \circ f)(x) = f(f(x)) = f\left(\dfrac{1}{x+2}\right) = \dfrac{1}{\dfrac{1}{x+2} + 2} = \dfrac{x+2}{1+2(x+2)} = \dfrac{x+2}{2x+5}$

\uparrow $\dfrac{1}{x+2}$ \uparrow

$f(x) = \dfrac{1}{x+2}$ Multiply by $\dfrac{x+2}{x+2}$.

The domain of $f \circ f$ consists of those x in the domain of f, $\{x \mid x \neq -2\}$, for which

$$f(x) = \dfrac{1}{x+2} \neq -2 \qquad \dfrac{1}{x+2} = -2$$

$$1 = -2(x+2)$$
$$1 = -2x - 4$$
$$2x = -5$$
$$x = -\dfrac{5}{2}$$

or, equivalently,

$$x \neq -\dfrac{5}{2}$$

The domain of $f \circ f$ is $\left\{ x \,\middle|\, x \neq -\dfrac{5}{2}, x \neq -2 \right\}$.

We could also find the domain of $f \circ f$ by recognizing that -2 is not in the domain of f and so should be excluded from the domain of $f \circ f$. Then, looking at $f \circ f$, we see that x cannot equal $-\dfrac{5}{2}$. Do you see why? Therefore, the domain of $f \circ f$ is $\left\{ x \,\middle|\, x \neq -\dfrac{5}{2}, x \neq -2 \right\}$.

───**Now Work** PROBLEMS 33 AND 35

Look back at Example 2, which illustrates that, in general, $f \circ g \neq g \circ f$. Sometimes $f \circ g$ does equal $g \circ f$, as shown in the next example.

EXAMPLE 5 **Showing That Two Composite Functions Are Equal**

If $f(x) = 3x - 4$ and $g(x) = \dfrac{1}{3}(x + 4)$, show that

$$(f \circ g)(x) = (g \circ f)(x) = x$$

for every x in the domain of $f \circ g$ and $g \circ f$.

Solution $(f \circ g)(x) = f(g(x))$

$= f\left(\dfrac{x+4}{3}\right)$ $g(x) = \dfrac{1}{3}(x+4) = \dfrac{x+4}{3}$

$= 3\left(\dfrac{x+4}{3}\right) - 4$ Substitute $g(x)$ into the rule for f, $f(x) = 3x - 4$.

$= x + 4 - 4 = x$

 Seeing the Concept

Using a graphing calculator, let

$$Y_1 = f(x) = 3x - 4$$

$$Y_2 = g(x) = \frac{1}{3}(x + 4)$$

$$Y_3 = f \circ g, Y_4 = g \circ f$$

Using the viewing window $-3 \le x \le 3$, $-2 \le y \le 2$, graph only Y_3 and Y_4. What do you see? TRACE to verify that $Y_3 = Y_4$.

$$
\begin{aligned}
(g \circ f)(x) &= g(f(x)) \\
&= g(3x - 4) \qquad \text{\small } f(x) = 3x - 4 \\
&= \frac{1}{3}[(3x - 4) + 4] \qquad \text{\small Substitute } f(x) \text{ into the rule for } g, g(x) = \frac{1}{3}(x + 4). \\
&= \frac{1}{3}(3x) = x
\end{aligned}
$$

We conclude that $(f \circ g)(x) = (g \circ f)(x) = x$.

In Section 6.2, we shall see that there is an important relationship between functions f and g for which $(f \circ g)(x) = (g \circ f)(x) = x$.

Now Work PROBLEM **45**

Calculus Application

 Some techniques in calculus require that we be able to determine the components of a composite function. For example, the function $H(x) = \sqrt{x + 1}$ is the composition of the functions f and g, where $f(x) = \sqrt{x}$ and $g(x) = x + 1$, because $H(x) = (f \circ g)(x) = f(g(x)) = f(x + 1) = \sqrt{x + 1}$.

EXAMPLE 6 | **Finding the Components of a Composite Function**

Find functions f and g such that $f \circ g = H$ if $H(x) = (x^2 + 1)^{50}$.

Solution The function H takes $x^2 + 1$ and raises it to the power 50. A natural way to decompose H is to raise the function $g(x) = x^2 + 1$ to the power 50. If we let $f(x) = x^{50}$ and $g(x) = x^2 + 1$, then

Figure 5

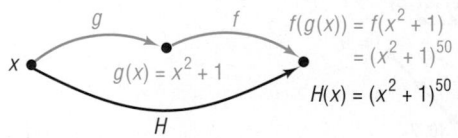

$$
\begin{aligned}
(f \circ g)(x) &= f(g(x)) \\
&= f(x^2 + 1) \\
&= (x^2 + 1)^{50} = H(x)
\end{aligned}
$$

See Figure 5.

Other functions f and g may be found for which $f \circ g = H$ in Example 6. For example, if $f(x) = x^2$ and $g(x) = (x^2 + 1)^{25}$, then

$$(f \circ g)(x) = f(g(x)) = f((x^2 + 1)^{25}) = [(x^2 + 1)^{25}]^2 = (x^2 + 1)^{50}$$

Although the functions f and g found as a solution to Example 6 are not unique, there is usually a "natural" selection for f and g that comes to mind first.

EXAMPLE 7 | **Finding the Components of a Composite Function**

Find functions f and g such that $f \circ g = H$ if $H(x) = \dfrac{1}{x + 1}$.

Solution Here H is the reciprocal of $g(x) = x + 1$. If we let $f(x) = \dfrac{1}{x}$ and $g(x) = x + 1$, we find that

$$(f \circ g)(x) = f(g(x)) = f(x + 1) = \frac{1}{x + 1} = H(x)$$

Now Work PROBLEM **53**

6.1 Assess Your Understanding

'Are You Prepared?' *Answers are given at the end of these exercises. If you get a wrong answer, read the pages listed in red.*

1. Find $f(3)$ if $f(x) = -4x^2 + 5x$. (pp. 203–206)

2. Find $f(3x)$ if $f(x) = 4 - 2x^2$. (pp. 203–206)

3. Find the domain of the function $f(x) = \dfrac{x^2 - 1}{x^2 - 25}$. (pp. 206–208)

Concepts and Vocabulary

4. Given two functions f and g, the _____ _____, denoted $f \circ g$, is defined by $f \circ g(x) =$ _____.

5. *True or False* $f(g(x)) = f(x) \cdot g(x)$.

6. *True or False* The domain of the composite function $(f \circ g)(x)$ is the same as the domain of $g(x)$.

Skill Building

In Problems 7 and 8, evaluate each expression using the values given in the table.

7.

x	−3	−2	−1	0	1	2	3
f(x)	−7	−5	−3	−1	3	5	7
g(x)	8	3	0	−1	0	3	8

(a) $(f \circ g)(1)$
(b) $(f \circ g)(-1)$
(c) $(g \circ f)(-1)$
(d) $(g \circ f)(0)$
(e) $(g \circ g)(-2)$
(f) $(f \circ f)(-1)$

8.

x	−3	−2	−1	0	1	2	3
f(x)	11	9	7	5	3	1	−1
g(x)	−8	−3	0	1	0	−3	−8

(a) $(f \circ g)(1)$
(b) $(f \circ g)(2)$
(c) $(g \circ f)(2)$
(d) $(g \circ f)(3)$
(e) $(g \circ g)(1)$
(f) $(f \circ f)(3)$

In Problems 9 and 10, evaluate each expression using the graphs of $y = f(x)$ and $y = g(x)$ shown in the figure.

9. (a) $(g \circ f)(-1)$
(b) $(g \circ f)(0)$
(c) $(f \circ g)(-1)$
(d) $(f \circ g)(4)$

10. (a) $(g \circ f)(1)$
(b) $(g \circ f)(5)$
(c) $(f \circ g)(0)$
(d) $(f \circ g)(2)$

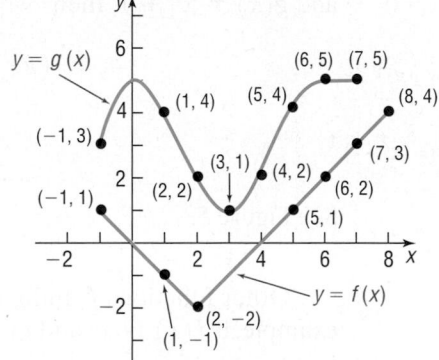

In Problems 11–20, for the given functions f and g, find:

(a) $(f \circ g)(4)$ (b) $(g \circ f)(2)$ (c) $(f \circ f)(1)$ (d) $(g \circ g)(0)$

11. $f(x) = 2x;\ \ g(x) = 3x^2 + 1$

12. $f(x) = 3x + 2;\ \ g(x) = 2x^2 - 1$

13. $f(x) = 4x^2 - 3;\ \ g(x) = 3 - \dfrac{1}{2}x^2$

14. $f(x) = 2x^2;\ \ g(x) = 1 - 3x^2$

15. $f(x) = \sqrt{x};\ \ g(x) = 2x$

16. $f(x) = \sqrt{x + 1};\ \ g(x) = 3x$

17. $f(x) = |x|;\ \ g(x) = \dfrac{1}{x^2 + 1}$

18. $f(x) = |x - 2|;\ \ g(x) = \dfrac{3}{x^2 + 2}$

19. $f(x) = \dfrac{3}{x + 1};\ \ g(x) = \sqrt[3]{x}$

20. $f(x) = x^{3/2};\ \ g(x) = \dfrac{2}{x + 1}$

In Problems 21–28, find the domain of the composite function $f \circ g$.

21. $f(x) = \dfrac{3}{x - 1};\ \ g(x) = \dfrac{2}{x}$

22. $f(x) = \dfrac{1}{x + 3};\ \ g(x) = -\dfrac{2}{x}$

23. $f(x) = \dfrac{x}{x - 1}$; $g(x) = -\dfrac{4}{x}$

24. $f(x) = \dfrac{x}{x + 3}$; $g(x) = \dfrac{2}{x}$

25. $f(x) = \sqrt{x}$; $g(x) = 2x + 3$

26. $f(x) = x - 2$; $g(x) = \sqrt{1 - x}$

27. $f(x) = x^2 + 1$; $g(x) = \sqrt{x - 1}$

28. $f(x) = x^2 + 4$; $g(x) = \sqrt{x - 2}$

In Problems 29–44, for the given functions f and g, find:

 (a) $f \circ g$ *(b)* $g \circ f$ *(c)* $f \circ f$ *(d)* $g \circ g$

State the domain of each composite function.

29. $f(x) = 2x + 3$; $g(x) = 3x$

30. $f(x) = -x$; $g(x) = 2x - 4$

31. $f(x) = 3x + 1$; $g(x) = x^2$

32. $f(x) = x + 1$; $g(x) = x^2 + 4$

33. $f(x) = x^2$; $g(x) = x^2 + 4$

34. $f(x) = x^2 + 1$; $g(x) = 2x^2 + 3$

35. $f(x) = \dfrac{3}{x - 1}$; $g(x) = \dfrac{2}{x}$

36. $f(x) = \dfrac{1}{x + 3}$; $g(x) = -\dfrac{2}{x}$

37. $f(x) = \dfrac{x}{x - 1}$; $g(x) = -\dfrac{4}{x}$

38. $f(x) = \dfrac{x}{x + 3}$; $g(x) = \dfrac{2}{x}$

39. $f(x) = \sqrt{x}$; $g(x) = 2x + 3$

40. $f(x) = \sqrt{x - 2}$; $g(x) = 1 - 2x$

41. $f(x) = x^2 + 1$; $g(x) = \sqrt{x - 1}$

42. $f(x) = x^2 + 4$; $g(x) = \sqrt{x - 2}$

43. $f(x) = \dfrac{x - 5}{x + 1}$; $g(x) = \dfrac{x + 2}{x - 3}$

44. $f(x) = \dfrac{2x - 1}{x - 2}$; $g(x) = \dfrac{x + 4}{2x - 5}$

In Problems 45–52, show that $(f \circ g)(x) = (g \circ f)(x) = x$.

45. $f(x) = 2x$; $g(x) = \dfrac{1}{2}x$

46. $f(x) = 4x$; $g(x) = \dfrac{1}{4}x$

47. $f(x) = x^3$; $g(x) = \sqrt[3]{x}$

48. $f(x) = x + 5$; $g(x) = x - 5$

49. $f(x) = 2x - 6$; $g(x) = \dfrac{1}{2}(x + 6)$

50. $f(x) = 4 - 3x$; $g(x) = \dfrac{1}{3}(4 - x)$

51. $f(x) = ax + b$; $g(x) = \dfrac{1}{a}(x - b)$ $a \neq 0$

52. $f(x) = \dfrac{1}{x}$; $g(x) = \dfrac{1}{x}$

In Problems 53–58, find functions f and g so that $f \circ g = H$.

53. $H(x) = (2x + 3)^4$

54. $H(x) = (1 + x^2)^3$

55. $H(x) = \sqrt{x^2 + 1}$

56. $H(x) = \sqrt{1 - x^2}$

57. $H(x) = |2x + 1|$

58. $H(x) = |2x^2 + 3|$

Applications and Extensions

59. If $f(x) = 2x^3 - 3x^2 + 4x - 1$ and $g(x) = 2$, find $(f \circ g)(x)$ and $(g \circ f)(x)$.

60. If $f(x) = \dfrac{x + 1}{x - 1}$, find $(f \circ f)(x)$.

61. If $f(x) = 2x^2 + 5$ and $g(x) = 3x + a$, find a so that the graph of $f \circ g$ crosses the y-axis at 23.

62. If $f(x) = 3x^2 - 7$ and $g(x) = 2x + a$, find a so that the graph of $f \circ g$ crosses the y-axis at 68.

In Problems 63 and 64, use the functions f and g to find:

 (a) $f \circ g$ *(b)* $g \circ f$

 (c) the domain of $f \circ g$ *and of* $g \circ f$

 (d) the conditions for which $f \circ g = g \circ f$

63. $f(x) = ax + b$; $g(x) = cx + d$

64. $f(x) = \dfrac{ax + b}{cx + d}$; $g(x) = mx$

65. **Surface Area of a Balloon** The surface area S (in square meters) of a hot-air balloon is given by

$$S(r) = 4\pi r^2$$

where r is the radius of the balloon (in meters). If the radius r is increasing with time t (in seconds) according to the formula $r(t) = \dfrac{2}{3}t^3, t \geq 0$, find the surface area S of the balloon as a function of the time t.

66. **Volume of a Balloon** The volume V (in cubic meters) of the hot-air balloon described in Problem 65 is given by $V(r) = \dfrac{4}{3}\pi r^3$. If the radius r is the same function of t as in Problem 65, find the volume V as a function of the time t.

67. **Automobile Production** The number N of cars produced at a certain factory in one day after t hours of operation is given by $N(t) = 100t - 5t^2, 0 \leq t \leq 10$. If the cost C

(in dollars) of producing N cars is $C(N) = 15,000 + 8000N$, find the cost C as a function of the time t of operation of the factory.

68. **Environmental Concerns** The spread of oil leaking from a tanker is in the shape of a circle. If the radius r (in feet) of the spread after t hours is $r(t) = 200\sqrt{t}$, find the area A of the oil slick as a function of the time t.

69. **Production Cost** The price p, in dollars, of a certain product and the quantity x sold obey the demand equation

$$p = -\frac{1}{4}x + 100 \quad 0 \le x \le 400$$

Suppose that the cost C, in dollars, of producing x units is

$$C = \frac{\sqrt{x}}{25} + 600$$

Assuming that all items produced are sold, find the cost C as a function of the price p.

[**Hint:** Solve for x in the demand equation and then form the composite.]

70. **Cost of a Commodity** The price p, in dollars, of a certain commodity and the quantity x sold obey the demand equation

$$p = -\frac{1}{5}x + 200 \quad 0 \le x \le 1000$$

Suppose that the cost C, in dollars, of producing x units is

$$C = \frac{\sqrt{x}}{10} + 400$$

Assuming that all items produced are sold, find the cost C as a function of the price p.

71. **Volume of a Cylinder** The volume V of a right circular cylinder of height h and radius r is $V = \pi r^2 h$. If the height is twice the radius, express the volume V as a function of r.

72. **Volume of a Cone** The volume V of a right circular cone is $V = \frac{1}{3}\pi r^2 h$. If the height is twice the radius, express the volume V as a function of r.

73. **Foreign Exchange** Traders often buy foreign currency in hope of making money when the currency's value changes. For example, on June 5, 2009, one U.S. dollar could purchase 0.7143 Euros, and one Euro could purchase 137.402 yen. Let $f(x)$ represent the number of Euros you can buy with x dollars, and let $g(x)$ represent the number of yen you can buy with x Euros.
 (a) Find a function that relates dollars to Euros.
 (b) Find a function that relates Euros to yen.
 (c) Use the results of parts (a) and (b) to find a function that relates dollars to yen. That is, find $(g \circ f)(x) = g(f(x))$.
 (d) What is $g(f(1000))$?

74. **Temperature Conversion** The function $C(F) = \frac{5}{9}(F - 32)$ converts a temperature in degrees Fahrenheit, F, to a temperature in degrees Celsius, C. The function $K(C) = C + 273$, converts a temperature in degrees Celsius to a temperature in kelvins, K.
 (a) Find a function that converts a temperature in degrees Fahrenheit to a temperature in kelvins.
 (b) Determine 80 degrees Fahrenheit in kelvins.

75. **Discounts** The manufacturer of a computer is offering two discounts on last year's model computer. The first discount is a \$200 rebate and the second discount is 20% off the regular price, p.
 (a) Write a function f that represents the sale price if only the rebate applies.
 (b) Write a function g that represents the sale price if only the 20% discount applies.
 (c) Find $f \circ g$ and $g \circ f$. What does each of these functions represent? Which combination of discounts represents a better deal for the consumer? Why?

76. If f and g are odd functions, show that the composite function $f \circ g$ is also odd.

77. If f is an odd function and g is an even function, show that the composite functions $f \circ g$ and $g \circ f$ are both even.

'Are You Prepared?' Answers

1. -21 **2.** $4 - 18x^2$ **3.** $\{x \mid x \ne -5, x \ne 5\}$

6.2 One-to-One Functions; Inverse Functions

PREPARING FOR THIS SECTION *Before getting started, review the following:*

- Functions (Section 3.1, pp. 200–208)
- Increasing/Decreasing Functions (Section 3.3, pp. 224–225)
- Rational Expressions (Chapter R, Section R.7, pp. 62–69)

Now Work the 'Are You Prepared?' problems on page 417.

OBJECTIVES 1 Determine Whether a Function Is One-to-One (p. 409)
 2 Determine the Inverse of a Function Defined by a Map or a Set of Ordered Pairs (p. 411)
 3 Obtain the Graph of the Inverse Function from the Graph of the Function (p. 413)
 4 Find the Inverse of a Function Defined by an Equation (p. 414)

1 Determine Whether a Function Is One-to-One

In Section 3.1, we presented four different ways to represent a function as (1) a map, (2) a set of ordered pairs, (3) a graph, and (4) an equation. For example, Figures 6 and 7 illustrate two different functions represented as mappings. The function in Figure 6 shows the correspondence between states and their population (in millions). The function in Figure 7 shows a correspondence between animals and life expectancy (in years).

Figure 6

Figure 7

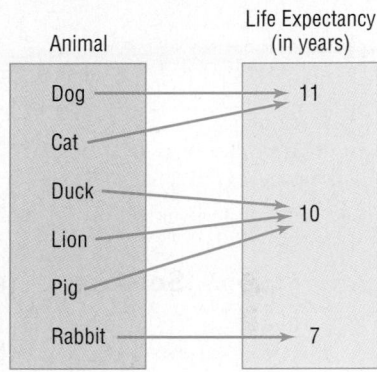

Suppose we asked a group of people to name the state that has a population of 0.8 million based on the function in Figure 6. Everyone in the group would respond South Dakota. Now, if we asked the same group of people to name the animal whose life expectancy is 11 years based on the function in Figure 7, some would respond dog, while others would respond cat. What is the difference between the functions in Figures 6 and 7? In Figure 6, we can see that no two elements in the domain correspond to the same element in the range. In Figure 7, this is not the case: two different elements in the domain correspond to the same element in the range. Functions such as the one in Figure 6 are given a special name.

DEFINITION

A function is **one-to-one** if any two different inputs in the domain correspond to two different outputs in the range. That is, if x_1 and x_2 are two different inputs of a function f, then f is one-to-one if $f(x_1) \neq f(x_2)$.

> **In Words**
> A function is not one-to-one if two different inputs correspond to the same output.

Put another way, a function f is one-to-one if no y in the range is the image of more than one x in the domain. A function is not one-to-one if two different elements in the domain correspond to the same element in the range. So the function in Figure 7 is not one-to-one because two different elements in the domain, *dog* and *cat*, both correspond to 11. Figure 8 illustrates the distinction among one-to-one functions, functions that are not one-to-one, and relations that are not functions.

Figure 8

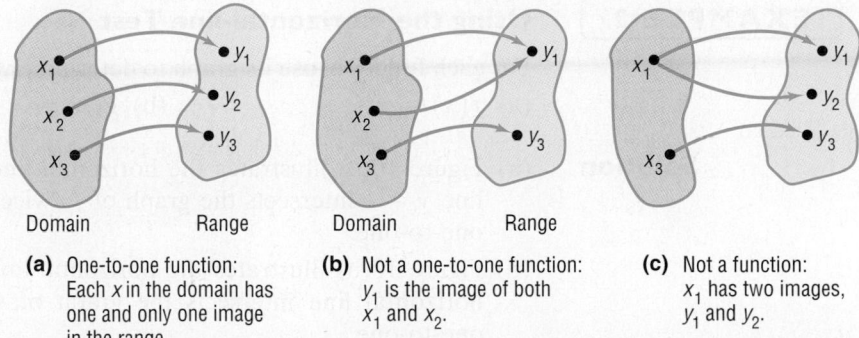

(a) One-to-one function: Each *x* in the domain has one and only one image in the range.

(b) Not a one-to-one function: y_1 is the image of both x_1 and x_2.

(c) Not a function: x_1 has two images, y_1 and y_2.

EXAMPLE 1 **Determining Whether a Function Is One-to-One**

Determine whether the following functions are one-to-one.

(a) For the following function, the domain represents the age of five males and the range represents their HDL (good) cholesterol (mg/dL).

(b) $\{(-2, 6), (-1, 3), (0, 2), (1, 5), (2, 8)\}$

Solution (a) The function is not one-to-one because there are two different inputs, 55 and 61, that correspond to the same output, 38.

(b) The function is one-to-one because there are no two distinct inputs that correspond to the same output.

■

Now Work PROBLEMS 11 AND 15

For functions defined by an equation $y = f(x)$ and for which the graph of f is known, there is a simple test, called the **horizontal-line test,** to determine whether f is one-to-one.

THEOREM **Horizontal-line Test**

If every horizontal line intersects the graph of a function f in at most one point, then f is one-to-one.

■

Figure 9
$f(x_1) = f(x_2) = h$ and $x_1 \neq x_2$; f is not a one-to-one function.

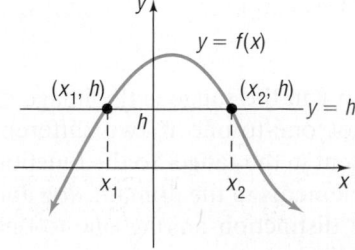

The reason that this test works can be seen in Figure 9, where the horizontal line $y = h$ intersects the graph at two distinct points, (x_1, h) and (x_2, h). Since h is the image of both x_1 and x_2 and $x_1 \neq x_2$, f is not one-to-one. Based on Figure 9, we can state the horizontal-line test in another way: If the graph of any horizontal line intersects the graph of a function f at more than one point, then f is not one-to-one.

EXAMPLE 2 **Using the Horizontal-line Test**

For each function, use its graph to determine whether the function is one-to-one.

(a) $f(x) = x^2$ (b) $g(x) = x^3$

Solution (a) Figure 10(a) illustrates the horizontal-line test for $f(x) = x^2$. The horizontal line $y = 1$ intersects the graph of f twice, at $(1, 1)$ and at $(-1, 1)$, so f is not one-to-one.

(b) Figure 10(b) illustrates the horizontal-line test for $g(x) = x^3$. Because every horizontal line intersects the graph of g exactly once, it follows that g is one-to-one.

Figure 10

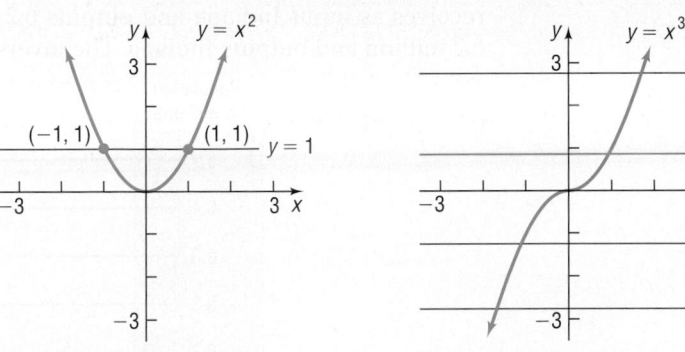

(a) A horizontal line intersects the graph twice; f is not one-to-one

(b) Every horizontal line intersects the graph exactly once; g is one-to-one

Now Work PROBLEM 19

Look more closely at the one-to-one function $g(x) = x^3$. This function is an increasing function. Because an increasing (or decreasing) function will always have different y-values for unequal x-values, it follows that a function that is increasing (or decreasing) over its domain is also a one-to-one function.

THEOREM

A function that is increasing on an interval I is a one-to-one function on I.
A function that is decreasing on an interval I is a one-to-one function on I.

2 Determine the Inverse of a Function Defined by a Map or a Set of Ordered Pairs

DEFINITION

Suppose that f is a one-to-one function. Then, to each x in the domain of f, there is exactly one y in the range (because f is a function); and to each y in the range of f, there is exactly one x in the domain (because f is one-to-one). The correspondence from the range of f back to the domain of f is called the **inverse function of f.** The symbol f^{-1} is used to denote the inverse of f.

In Words
Suppose that we have a one-to-one function f where the input 5 corresponds to the output 10. In the inverse function f^{-1}, the input 10 would correspond to the output 5.

We will discuss how to find inverses for all four representations of functions: (1) maps, (2) sets of ordered pairs, (3) graphs, and (4) equations. We begin with finding inverses of functions represented by maps or sets of ordered pairs.

EXAMPLE 3

Finding the Inverse of a Function Defined by a Map

Find the inverse of the following function. Let the domain of the function represent certain states, and let the range represent the state's population (in millions). State the domain and the range of the inverse function.

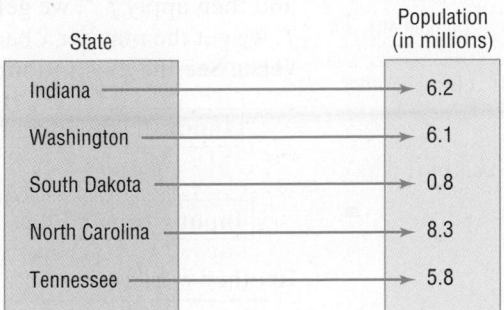

Solution

The function is one-to-one. To find the inverse function, we interchange the elements in the domain with the elements in the range. For example, the function

receives as input Indiana and outputs 6.2 million. So the inverse receives as input 6.2 million and outputs Indiana. The inverse function is shown next.

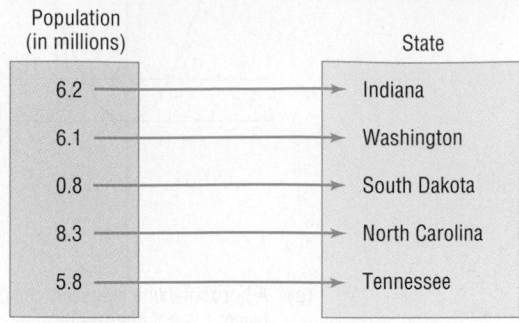

The domain of the inverse function is $\{6.2, 6.1, 0.8, 8.3, 5.8\}$. The range of the inverse function is $\{$Indiana, Washington, South Dakota, North Carolina, Tennessee$\}$.

If the function f is a set of ordered pairs (x, y), then the inverse of f, denoted f^{-1}, is the set of ordered pairs (y, x).

EXAMPLE 4

Finding the Inverse of a Function Defined by a Set of Ordered Pairs

Find the inverse of the following one-to-one function:

$$\{(-3, -27), (-2, -8), (-1, -1), (0, 0), (1, 1), (2, 8), (3, 27)\}$$

State the domain and the range of the function and its inverse.

Solution

The inverse of the given function is found by interchanging the entries in each ordered pair and so is given by

$$\{(-27, -3), (-8, -2), (-1, -1), (0, 0), (1, 1), (8, 2), (27, 3)\}$$

The domain of the function is $\{-3, -2, -1, 0, 1, 2, 3\}$. The range of the function is $\{-27, -8, -1, 0, 1, 8, 27\}$. The domain of the inverse function is $\{-27, -8, -1, 0, 1, 8, 27\}$. The range of the inverse function is $\{-3, -2, -1, 0, 1, 2, 3\}$.

\qquad **Now Work** PROBLEMS 25 AND 29

Remember, if f is a one-to-one function, it has an inverse function, f^{-1}. See Figure 11.

Based on the results of Example 4 and Figure 11, two facts are now apparent about a one-to-one function f and its inverse f^{-1}.

Figure 11

Domain of f \qquad Range of f

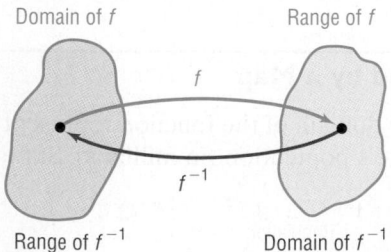

Range of f^{-1} \qquad Domain of f^{-1}

$$\boxed{\text{Domain of } f = \text{Range of } f^{-1} \qquad \text{Range of } f = \text{Domain of } f^{-1}}$$

WARNING Be careful! f^{-1} is a symbol for the inverse function of f. The -1 used in f^{-1} is not an exponent. That is, f^{-1} does not mean the reciprocal of f; $f^{-1}(x)$ is not equal to $\dfrac{1}{f(x)}$. ∎

Look again at Figure 11 to visualize the relationship. If we start with x, apply f, and then apply f^{-1}, we get x back again. If we start with x, apply f^{-1}, and then apply f, we get the number x back again. To put it simply, what f does, f^{-1} undoes, and vice versa. See the illustration that follows.

$$\boxed{\text{Input } x \text{ from domain of } f} \xrightarrow{\text{Apply } f} \boxed{f(x)} \xrightarrow{\text{Apply } f^{-1}} \boxed{f^{-1}(f(x)) = x}$$

$$\boxed{\text{Input } x \text{ from domain of } f^{-1}} \xrightarrow{\text{Apply } f^{-1}} \boxed{f^{-1}(x)} \xrightarrow{\text{Apply } f} \boxed{f(f^{-1}(x)) = x}$$

In other words,

$$\boxed{\begin{array}{ll} f^{-1}(f(x)) = x & \text{where } x \text{ is in the domain of } f \\ f(f^{-1}(x)) = x & \text{where } x \text{ is in the domain of } f^{-1} \end{array}}$$

Consider the function $f(x) = 2x$, which multiplies the argument x by 2. The inverse function f^{-1} undoes whatever f does. So the inverse function of f is $f^{-1}(x) = \dfrac{1}{2}x$, which divides the argument by 2. For example, $f(3) = 2(3) = 6$ and

Figure 12

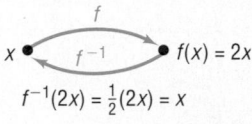

$f^{-1}(2x) = \frac{1}{2}(2x) = x$

$f^{-1}(6) = \dfrac{1}{2}(6) = 3$, so f^{-1} undoes what f did. We can verify this by showing that

$$f^{-1}(f(x)) = f^{-1}(2x) = \frac{1}{2}(2x) = x \quad \text{and} \quad f(f^{-1}(x)) = f\left(\frac{1}{2}x\right) = 2\left(\frac{1}{2}x\right) = x$$

See Figure 12.

EXAMPLE 5 **Verifying Inverse Functions**

(a) Verify that the inverse of $g(x) = x^3$ is $g^{-1}(x) = \sqrt[3]{x}$ by showing that

$$g^{-1}(g(x)) = g^{-1}(x^3) = \sqrt[3]{x^3} = x \quad \text{for all } x \text{ in the domain of } g$$
$$g(g^{-1}(x)) = g(\sqrt[3]{x}) = (\sqrt[3]{x})^3 = x \quad \text{for all } x \text{ in the domain of } g^{-1}$$

(b) Verify that the inverse of $f(x) = 2x + 3$ is $f^{-1}(x) = \dfrac{1}{2}(x - 3)$ by showing that

$$f^{-1}(f(x)) = f^{-1}(2x + 3) = \frac{1}{2}[(2x + 3) - 3] = \frac{1}{2}(2x) = x \qquad \begin{array}{l}\text{for all } x \text{ in the} \\ \text{domain of } f\end{array}$$

$$f(f^{-1}(x)) = f\left(\frac{1}{2}(x - 3)\right) = 2\left[\frac{1}{2}(x - 3)\right] + 3 = (x - 3) + 3 = x \quad \begin{array}{l}\text{for all } x \text{ in the} \\ \text{domain of } f^{-1}\end{array}$$

EXAMPLE 6 **Verifying Inverse Functions**

Verify that the inverse of $f(x) = \dfrac{1}{x - 1}$ is $f^{-1}(x) = \dfrac{1}{x} + 1$. For what values of x is $f^{-1}(f(x)) = x$? For what values of x is $f(f^{-1}(x)) = x$?

Solution The domain of f is $\{x \mid x \neq 1\}$ and the domain of f^{-1} is $\{x \mid x \neq 0\}$. Now

$$f^{-1}(f(x)) = f^{-1}\left(\frac{1}{x - 1}\right) = \frac{1}{\dfrac{1}{x - 1}} + 1 = x - 1 + 1 = x \qquad \text{provided } x \neq 1$$

$$f(f^{-1}(x)) = f\left(\frac{1}{x} + 1\right) = \frac{1}{\dfrac{1}{x} + 1 - 1} = \frac{1}{\dfrac{1}{x}} = x \qquad \text{provided } x \neq 0$$

Now Work PROBLEM 33

3 Obtain the Graph of the Inverse Function from the Graph of the Function

Figure 13

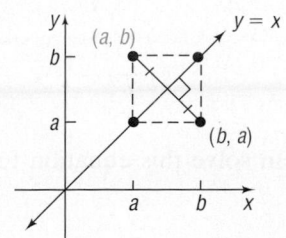

Suppose that (a, b) is a point on the graph of a one-to-one function f defined by $y = f(x)$. Then $b = f(a)$. This means that $a = f^{-1}(b)$, so (b, a) is a point on the graph of the inverse function f^{-1}. The relationship between the point (a, b) on f and the point (b, a) on f^{-1} is shown in Figure 13. The line segment with endpoints (a, b) and (b, a) is perpendicular to the line $y = x$ and is bisected by the line $y = x$. (Do you see why?) It follows that the point (b, a) on f^{-1} is the reflection about the line $y = x$ of the point (a, b) on f.

THEOREM The graph of a one-to-one function f and the graph of its inverse f^{-1} are symmetric with respect to the line $y = x$.

Figure 14 illustrates this result. Notice that, once the graph of f is known, the graph of f^{-1} may be obtained by reflecting the graph of f about the line $y = x$.

Figure 14

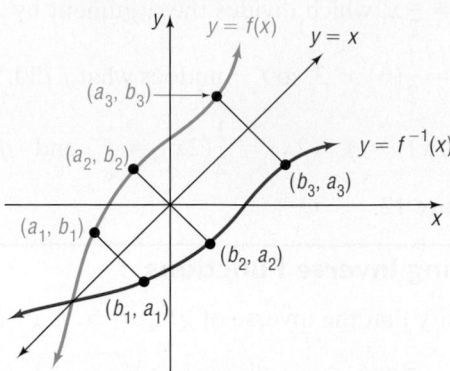

EXAMPLE 7 **Graphing the Inverse Function**

The graph in Figure 15(a) is that of a one-to-one function $y = f(x)$. Draw the graph of its inverse.

Solution Begin by adding the graph of $y = x$ to Figure 15(a). Since the points $(-2, -1), (-1, 0)$, and $(2, 1)$ are on the graph of f, the points $(-1, -2), (0, -1)$, and $(1, 2)$ must be on the graph of f^{-1}. Keeping in mind that the graph of f^{-1} is the reflection about the line $y = x$ of the graph of f, draw f^{-1}. See Figure 15(b).

Figure 15

(a)

(b)

━━━ **Now Work** PROBLEM 43

4 Find the Inverse of a Function Defined by an Equation

The fact that the graphs of a one-to-one function f and its inverse function f^{-1} are symmetric with respect to the line $y = x$ tells us more. It says that we can obtain f^{-1} by interchanging the roles of x and y in f. Look again at Figure 14. If f is defined by the equation

$$y = f(x)$$

then f^{-1} is defined by the equation

$$x = f(y)$$

The equation $x = f(y)$ defines f^{-1} *implicitly*. If we can solve this equation for y, we will have the *explicit* form of f^{-1}, that is,

$$y = f^{-1}(x)$$

Let's use this procedure to find the inverse of $f(x) = 2x + 3$. (Since f is a linear function and is increasing, we know that f is one-to-one and so has an inverse function.)

EXAMPLE 8

How to Find the Inverse Function

Find the inverse of $f(x) = 2x + 3$. Graph f and f^{-1} on the same coordinate axes.

Step-by-Step Solution

Step 1: Replace $f(x)$ with y. In $y = f(x)$, interchange the variables x and y to obtain $x = f(y)$. This equation defines the inverse function f^{-1} implicitly.

Replace $f(x)$ with y in $f(x) = 2x + 3$ and obtain $y = 2x + 3$. Now interchange the variables x and y to obtain

$$x = 2y + 3$$

This equation defines the inverse f^{-1} implicitly.

Step 2: If possible, solve the implicit equation for y in terms of x to obtain the explicit form of f^{-1}, $y = f^{-1}(x)$.

To find the explicit form of the inverse, solve $x = 2y + 3$ for y.

$$x = 2y + 3$$
$$2y + 3 = x \qquad \text{Reflexive Property; If } a = b, \text{ then } b = a.$$
$$2y = x - 3 \qquad \text{Subtract 3 from both sides.}$$
$$y = \frac{1}{2}(x - 3) \qquad \text{Divide both sides by 2.}$$

The explicit form of the inverse f^{-1} is

$$f^{-1}(x) = \frac{1}{2}(x - 3)$$

Step 3: Check the result by showing that $f^{-1}(f(x)) = x$ and $f(f^{-1}(x)) = x$.

We verified that f and f^{-1} are inverses in Example 5(b).

Figure 16

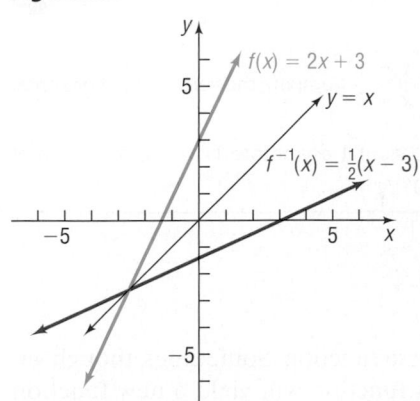

The graphs of $f(x) = 2x + 3$ and its inverse $f^{-1}(x) = \frac{1}{2}(x - 3)$ are shown in Figure 16 . Note the symmetry of the graphs with respect to the line $y = x$.

Procedure for Finding the Inverse of a One-to-One Function

STEP 1: In $y = f(x)$, interchange the variables x and y to obtain

$$x = f(y)$$

This equation defines the inverse function f^{-1} implicitly.

STEP 2: If possible, solve the implicit equation for y in terms of x to obtain the explicit form of f^{-1}:

$$y = f^{-1}(x)$$

STEP 3: Check the result by showing that

$$f^{-1}(f(x)) = x \quad \text{and} \quad f(f^{-1}(x)) = x$$

EXAMPLE 9

Finding the Inverse Function

The function

$$f(x) = \frac{2x + 1}{x - 1} \qquad x \neq 1$$

is one-to-one. Find its inverse and check the result.

Solution

STEP 1: Replace $f(x)$ with y and interchange the variables x and y in

$$y = \frac{2x + 1}{x - 1}$$

to obtain

$$x = \frac{2y + 1}{y - 1}$$

STEP 2: Solve for y.

$$x = \frac{2y + 1}{y - 1}$$

$$x(y - 1) = 2y + 1 \qquad \text{Multiply both sides by } y - 1.$$

$$xy - x = 2y + 1 \qquad \text{Apply the Distributive Property.}$$

$$xy - 2y = x + 1 \qquad \text{Subtract } 2y \text{ from both sides; add } x \text{ to both sides.}$$

$$(x - 2)y = x + 1 \qquad \text{Factor.}$$

$$y = \frac{x + 1}{x - 2} \qquad \text{Divide by } x - 2.$$

The inverse is

$$f^{-1}(x) = \frac{x + 1}{x - 2} \quad x \neq 2 \quad \text{Replace } y \text{ by } f^{-1}(x).$$

STEP 3: ✓**Check:**

$$f^{-1}(f(x)) = f^{-1}\left(\frac{2x + 1}{x - 1}\right) = \frac{\dfrac{2x + 1}{x - 1} + 1}{\dfrac{2x + 1}{x - 1} - 2} = \frac{2x + 1 + x - 1}{2x + 1 - 2(x - 1)} = \frac{3x}{3} = x \quad x \neq 1$$

$$f(f^{-1}(x)) = f\left(\frac{x + 1}{x - 2}\right) = \frac{2\left(\dfrac{x + 1}{x - 2}\right) + 1}{\dfrac{x + 1}{x - 2} - 1} = \frac{2(x + 1) + x - 2}{x + 1 - (x - 2)} = \frac{3x}{3} = x \quad x \neq 2$$

↵

Exploration

In Example 9, we found that, if $f(x) = \dfrac{2x + 1}{x - 1}$, then $f^{-1}(x) = \dfrac{x + 1}{x - 2}$. Compare the vertical and horizontal asymptotes of f and f^{-1}.

Result The vertical asymptote of f is $x = 1$, and the horizontal asymptote is $y = 2$. The vertical asymptote of f^{-1} is $x = 2$, and the horizontal asymptote is $y = 1$.

═══ **Now Work** PROBLEMS 51 AND 65

If a function is not one-to-one, it has no inverse function. Sometimes, though, an appropriate restriction on the domain of such a function will yield a new function that is one-to-one. Then the function defined on the restricted domain has an inverse function. Let's look at an example of this common practice.

EXAMPLE 10

Finding the Inverse of a Domain-restricted Function

Find the inverse of $y = f(x) = x^2$ if $x \geq 0$. Graph f and f^{-1}.

Solution

The function $y = x^2$ is not one-to-one. [Refer to Example 2(a).] However, if we restrict the domain of this function to $x \geq 0$, as indicated, we have a new function that is increasing and therefore is one-to-one. As a result, the function defined by $y = f(x) = x^2$, $x \geq 0$, has an inverse function, f^{-1}.

Follow the steps given previously to find f^{-1}.

STEP 1: In the equation $y = x^2$, $x \geq 0$, interchange the variables x and y. The result is

$$x = y^2 \qquad y \geq 0$$

This equation defines (implicitly) the inverse function.

Figure 17

STEP 2: Solve for y to get the explicit form of the inverse. Since $y \geq 0$, only one solution for y is obtained: $y = \sqrt{x}$. So $f^{-1}(x) = \sqrt{x}$.

STEP 3: ✓ **Check:** $f^{-1}(f(x)) = f^{-1}(x^2) = \sqrt{x^2} = |x| = x$ since $x \geq 0$

$$f(f^{-1}(x)) = f(\sqrt{x}) = (\sqrt{x})^2 = x$$

Figure 17 illustrates the graphs of $f(x) = x^2$, $x \geq 0$, and $f^{-1}(x) = \sqrt{x}$.

SUMMARY

1. If a function f is one-to-one, then it has an inverse function f^{-1}.
2. Domain of f = Range of f^{-1}; Range of f = Domain of f^{-1}.
3. To verify that f^{-1} is the inverse of f, show that $f^{-1}(f(x)) = x$ for every x in the domain of f and $f(f^{-1}(x)) = x$ for every x in the domain of f^{-1}.
4. The graphs of f and f^{-1} are symmetric with respect to the line $y = x$.

6.2 Assess Your Understanding

'Are You Prepared?' *Answers are given at the end of these exercises. If you get a wrong answer, read the pages listed in red.*

1. Is the set of ordered pairs $\{(1, 3), (2, 3), (-1, 2)\}$ a function? Why or why not? (pp. 200–208)

2. Where is the function $f(x) = x^2$ increasing? Where is it decreasing? (pp. 224–225)

3. What is the domain of $f(x) = \dfrac{x + 5}{x^2 + 3x - 18}$? (pp. 200–208)

4. Simplify: $\dfrac{\dfrac{1}{x} + 1}{\dfrac{1}{x^2} - 1}$ (pp. 62–69)

Concepts and Vocabulary

5. If x_1 and x_2 are two different inputs of a function f, then f is one-to-one if _____.

6. If every horizontal line intersects the graph of a function f at no more than one point, f is a(n) _____ function.

7. If f is a one-to-one function and $f(3) = 8$, then $f^{-1}(8) =$ _____.

8. If f^{-1} denotes the inverse of a function f, then the graphs of f and f^{-1} are symmetric with respect to the line _____.

9. If the domain of a one-to-one function f is $[4, \infty)$, the range of its inverse, f^{-1}, is _____.

10. **True or False** If f and g are inverse functions, the domain of f is the same as the range of g.

Skill Building

In Problems 11–18, determine whether the function is one-to-one.

11.

12.

13.

14.

15. $\{(2, 6), (-3, 6), (4, 9), (1, 10)\}$

16. $\{(-2, 5), (-1, 3), (3, 7), (4, 12)\}$

17. $\{(0, 0), (1, 1), (2, 16), (3, 81)\}$

18. $\{(1, 2), (2, 8), (3, 18), (4, 32)\}$

In Problems 19–24, the graph of a function f is given. Use the horizontal-line test to determine whether f is one-to-one.

19.

20.

21.

22.

23.

24.

In Problems 25–32, find the inverse of each one-to-one function. State the domain and the range of each inverse function.

25.

Source: *Information Please Almanac*

26.

Title	Domestic Gross (in millions)
Star Wars	$461
Star Wars: Episode One – The Phantom Menace	$431
E.T. the Extra Terrestrial	$400
Jurassic Park	$357
Forrest Gump	$330

Source: *Information Please Almanac*

27.

Source: eterm.com

28.

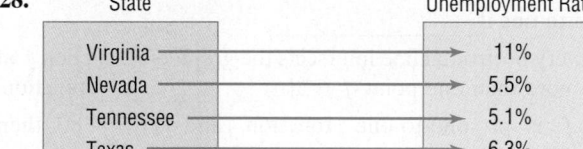

Source: *United States Statistical Abstract*

29. $\{(-3, 5), (-2, 9), (-1, 2), (0, 11), (1, -5)\}$

30. $\{(-2, 2), (-1, 6), (0, 8), (1, -3), (2, 9)\}$

31. $\{(-2, 1), (-3, 2), (-10, 0), (1, 9), (2, 4)\}$

32. $\{(-2, -8), (-1, -1), (0, 0), (1, 1), (2, 8)\}$

In Problems 33–42, verify that the functions f and g are inverses of each other by showing that $f(g(x)) = x$ and $g(f(x)) = x$. Give any values of x that need to be excluded from the domain of f and the domain of g.

33. $f(x) = 3x + 4$; $g(x) = \dfrac{1}{3}(x - 4)$

34. $f(x) = 3 - 2x$; $g(x) = -\dfrac{1}{2}(x - 3)$

35. $f(x) = 4x - 8$; $g(x) = \dfrac{x}{4} + 2$

36. $f(x) = 2x + 6$; $g(x) = \dfrac{1}{2}x - 3$

37. $f(x) = x^3 - 8$; $g(x) = \sqrt[3]{x + 8}$

38. $f(x) = (x - 2)^2, x \geq 2$; $g(x) = \sqrt{x} + 2$

39. $f(x) = \dfrac{1}{x}$; $g(x) = \dfrac{1}{x}$

40. $f(x) = x$; $g(x) = x$

41. $f(x) = \dfrac{2x + 3}{x + 4}$; $g(x) = \dfrac{4x - 3}{2 - x}$

42. $f(x) = \dfrac{x - 5}{2x + 3}$; $g(x) = \dfrac{3x + 5}{1 - 2x}$

In Problems 43–48, the graph of a one-to-one function f is given. Draw the graph of the inverse function f^{-1}. For convenience (and as a hint), the graph of y = x is also given.

43.

44.

45.

46.

47.

48.

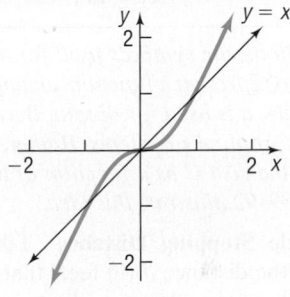

In Problems 49–60, the function f is one-to-one. Find its inverse and check your answer. Graph f, f^{-1}, and y = x on the same coordinate axes.

49. $f(x) = 3x$

50. $f(x) = -4x$

51. $f(x) = 4x + 2$

52. $f(x) = 1 - 3x$

53. $f(x) = x^3 - 1$

54. $f(x) = x^3 + 1$

55. $f(x) = x^2 + 4 \quad x \geq 0$

56. $f(x) = x^2 + 9 \quad x \geq 0$

57. $f(x) = \dfrac{4}{x}$

58. $f(x) = -\dfrac{3}{x}$

59. $f(x) = \dfrac{1}{x - 2}$

60. $f(x) = \dfrac{4}{x + 2}$

In Problems 61–72, the function f is one-to-one. Find its inverse and check your answer.

61. $f(x) = \dfrac{2}{3 + x}$

62. $f(x) = \dfrac{4}{2 - x}$

63. $f(x) = \dfrac{3x}{x + 2}$

64. $f(x) = -\dfrac{2x}{x - 1}$

65. $f(x) = \dfrac{2x}{3x - 1}$

66. $f(x) = -\dfrac{3x + 1}{x}$

67. $f(x) = \dfrac{3x + 4}{2x - 3}$

68. $f(x) = \dfrac{2x - 3}{x + 4}$

69. $f(x) = \dfrac{2x + 3}{x + 2}$

70. $f(x) = \dfrac{-3x - 4}{x - 2}$

71. $f(x) = \dfrac{x^2 - 4}{2x^2} \quad x > 0$

72. $f(x) = \dfrac{x^2 + 3}{3x^2} \quad x > 0$

Applications and Extensions

73. Use the graph of $y = f(x)$ given in Problem 43 to evaluate the following:

 (a) $f(-1)$ (b) $f(1)$ (c) $f^{-1}(1)$ (d) $f^{-1}(2)$

74. Use the graph of $y = f(x)$ given in Problem 44 to evaluate the following:

 (a) $f(2)$ (b) $f(1)$ (c) $f^{-1}(0)$ (d) $f^{-1}(-1)$

75. If $f(7) = 13$ and f is one-to-one, what is $f^{-1}(13)$?

76. If $g(-5) = 3$ and g is one-to-one, what is $g^{-1}(3)$?

77. The domain of a one-to-one function f is $[5, \infty)$, and its range is $[-2, \infty)$. State the domain and the range of f^{-1}.

78. The domain of a one-to-one function f is $[0, \infty)$, and its range is $[5, \infty)$. State the domain and the range of f^{-1}.

79. The domain of a one-to-one function g is $(-\infty, 0]$, and its range is $[0, \infty)$. State the domain and the range of g^{-1}.

80. The domain of a one-to-one function g is $[0, 15]$, and its range is $(0, 8)$. State the domain and the range of g^{-1}.

81. A function $y = f(x)$ is increasing on the interval $(0, 5)$. What conclusions can you draw about the graph of $y = f^{-1}(x)$?

82. A function $y = f(x)$ is decreasing on the interval $(0, 5)$. What conclusions can you draw about the graph of $y = f^{-1}(x)$?

83. Find the inverse of the linear function
$$f(x) = mx + b \quad m \neq 0$$

84. Find the inverse of the function
$$f(x) = \sqrt{r^2 - x^2} \quad 0 \le x \le r$$

85. A function f has an inverse function. If the graph of f lies in quadrant I, in which quadrant does the graph of f^{-1} lie?

86. A function f has an inverse function. If the graph of f lies in quadrant II, in which quadrant does the graph of f^{-1} lie?

87. The function $f(x) = |x|$ is not one-to-one. Find a suitable restriction on the domain of f so that the new function that results is one-to-one. Then find the inverse of f.

88. The function $f(x) = x^4$ is not one-to-one. Find a suitable restriction on the domain of f so that the new function that results is one-to-one. Then find the inverse of f.

In applications, the symbols used for the independent and dependent variables are often based on common usage. So, rather than using $y = f(x)$ to represent a function, an applied problem might use $C = C(q)$ to represent the cost C of manufacturing q units of a good since, in economics, q is used for output. Because of this, the inverse notation f^{-1} used in a pure mathematics problem is not used when finding inverses of applied problems. Rather, the inverse of a function such as $C = C(q)$ will be $q = q(C)$. So $C = C(q)$ is a function that represents the cost C as a function of the output q, while $q = q(C)$ is a function that represents the output q as a function of the cost C. Problems 89–92 illustrate this idea.

89. **Vehicle Stopping Distance** Taking into account reaction time, the distance d (in feet) that a car requires to come to a complete stop while traveling r miles per hour is given by the function
$$d(r) = 6.97r - 90.39$$
(a) Express the speed r at which the car is traveling as a function of the distance d required to come to a complete stop.
(b) Verify that $r = r(d)$ is the inverse of $d = d(r)$ by showing that $r(d(r)) = r$ and $d(r(d)) = d$.
(c) Predict the speed that a car was traveling if the distance required to stop was 300 feet.

90. **Height and Head Circumference** The head circumference C of a child is related to the height H of the child (both in inches) through the function
$$H(C) = 2.15C - 10.53$$
(a) Express the head circumference C as a function of height H.
(b) Verify that $C = C(H)$ is the inverse of $H = H(C)$ by showing that $H(C(H)) = H$ and $C(H(C)) = C$.
(c) Predict the head circumference of a child who is 26 inches tall.

91. **Ideal Body Weight** One model for the ideal body weight W for men (in kilograms) as a function of height h (in inches) is given by the function
$$W(h) = 50 + 2.3(h - 60)$$
(a) What is the ideal weight of a 6-foot male?
(b) Express the height h as a function of weight W.
(c) Verify that $h = h(W)$ is the inverse of $W = W(h)$ by showing that $h(W(h)) = h$ and $W(h(W)) = W$.
(d) What is the height of a male who is at his ideal weight of 80 kilograms?

[**Note:** The ideal body weight W for women (in kilograms) as a function of height h (in inches) is given by $W(h) = 45.5 + 2.3(h - 60)$.]

92. **Temperature Conversion** The function $F(C) = \dfrac{9}{5}C + 32$ converts a temperature from C degrees Celsius to F degrees Fahrenheit.

(a) Express the temperature in degrees Celsius C as a function of the temperature in degrees Fahrenheit F.
(b) Verify that $C = C(F)$ is the inverse of $F = F(C)$ by showing that $C(F(C)) = C$ and $F(C(F)) = F$.
(c) What is the temperature in degrees Celsius if it is 70 degrees Fahrenheit?

93. **Income Taxes** The function
$$T(g) = 4675 + 0.25(g - 33,950)$$
represents the 2009 federal income tax T (in dollars) due for a "single" filer whose modified adjusted gross income is g dollars, where $33,950 \le g \le 82,250$.
(a) What is the domain of the function T?
(b) Given that the tax due T is an increasing linear function of modified adjusted gross income g, find the range of the function T.
(c) Find adjusted gross income g as a function of federal income tax T. What are the domain and the range of this function?

94. **Income Taxes** The function
$$T(g) = 1670 + 0.15(g - 16,700)$$
represents the 2009 federal income tax T (in dollars) due for a "married filing jointly" filer whose modified adjusted gross income is g dollars, where $16,700 \le g \le 67,900$.
(a) What is the domain of the function T?
(b) Given that the tax due T is an increasing linear function of modified adjusted gross income g, find the range of the function T.
(c) Find adjusted gross income g as a function of federal income tax T. What are the domain and the range of this function?

95. **Gravity on Earth** If a rock falls from a height of 100 meters on Earth, the height H (in meters) after t seconds is approximately
$$H(t) = 100 - 4.9t^2$$
(a) In general, quadratic functions are not one-to-one. However, the function H is one-to-one. Why?
(b) Find the inverse of H and verify your result.
(c) How long will it take a rock to fall 80 meters?

96. Period of a Pendulum The period T (in seconds) of a simple pendulum as a function of its length l (in feet) is given by

$$T(l) = 2\pi\sqrt{\frac{l}{32.2}}$$

(a) Express the length l as a function of the period T.
(b) How long is a pendulum whose period is 3 seconds?

97. Given

$$f(x) = \frac{ax + b}{cx + d}$$

find $f^{-1}(x)$. If $c \neq 0$, under what conditions on $a, b, c,$ and d is $f = f^{-1}$?

Explaining Concepts: Discussion and Writing

98. Can a one-to-one function and its inverse be equal? What must be true about the graph of f for this to happen? Give some examples to support your conclusion.

99. Draw the graph of a one-to-one function that contains the points $(-2, -3)$, $(0, 0)$, and $(1, 5)$. Now draw the graph of its inverse. Compare your graph to those of other students. Discuss any similarities. What differences do you see?

100. Give an example of a function whose domain is the set of real numbers and that is neither increasing nor decreasing on its domain, but is one-to-one.

[**Hint:** Use a piecewise-defined function.]

101. Is every odd function one-to-one? Explain.

102. Suppose that $C(g)$ represents the cost C, in dollars, of manufacturing g cars. Explain what $C^{-1}(800,000)$ represents.

103. Explain why the horizontal-line test can be used to identify one-to-one functions from a graph.

'Are You Prepared?' Answers

1. Yes; for each input x there is one output y.

2. Increasing on $(0, \infty)$; decreasing on $(-\infty, 0)$

3. $\{x | x \neq -6, x \neq 3\}$

4. $\dfrac{x}{1 - x}, x \neq 0, x \neq -1$

6.3 Exponential Functions

PREPARING FOR THIS SECTION *Before getting started, review the following:*

- Exponents (Chapter R, Section R.2, pp. 21–24, and Section R.8, pp. 73–77)
- Graphing Techniques: Transformations (Section 3.5, pp. 244–253)
- Solving Equations (Section 1.1, pp. 82–87 and Section 1.2, pp. 92–99)

- Average Rate of Change (Section 3.3, pp. 228–230)
- Quadratic Functions (Section 4.3, pp. 288–296)
- Linear Functions (Section 4.1, pp. 272–275)
- Horizontal Asymptotes (Section 5.2, pp. 345–346)

Now Work the **'Are You Prepared?'** problems on page 432.

OBJECTIVES **1** Evaluate Exponential Functions (p. 421)
2 Graph Exponential Functions (p. 425)
3 Define the Number e (p. 428)
4 Solve Exponential Equations (p. 430)

1 Evaluate Exponential Functions

In Chapter R, Section R.8, we give a definition for raising a real number a to a rational power. Based on that discussion, we gave meaning to expressions of the form

$$a^r$$

where the base a is a positive real number and the exponent r is a rational number. But what is the meaning of a^x, where the base a is a positive real number and the exponent x is an irrational number? Although a rigorous definition requires methods discussed in calculus, the basis for the definition is easy to follow: Select a rational number r that is formed by truncating (removing) all but a finite number of digits from the irrational number x. Then it is reasonable to expect that

$$a^x \approx a^r$$

For example, take the irrational number $\pi = 3.14159\ldots$. Then an approximation to a^π is

$$a^\pi \approx a^{3.14}$$

where the digits after the hundredths position have been removed from the value for π. A better approximation would be

$$a^\pi \approx a^{3.14159}$$

where the digits after the hundred-thousandths position have been removed. Continuing in this way, we can obtain approximations to a^π to any desired degree of accuracy.

Most calculators have an $\boxed{x^y}$ key or a caret key $\boxed{\wedge}$ for working with exponents. To evaluate expressions of the form a^x, enter the base a, then press the $\boxed{x^y}$ key (or the $\boxed{\wedge}$ key), enter the exponent x, and press $\boxed{=}$ (or $\boxed{\text{ENTER}}$).

EXAMPLE 1 **Using a Calculator to Evaluate Powers of 2**

Using a calculator, evaluate:

(a) $2^{1.4}$ (b) $2^{1.41}$ (c) $2^{1.414}$ (d) $2^{1.4142}$ (e) $2^{\sqrt{2}}$

Solution (a) $2^{1.4} \approx 2.639015822$ (b) $2^{1.41} \approx 2.657371628$

(c) $2^{1.414} \approx 2.66474965$ (d) $2^{1.4142} \approx 2.665119089$

(e) $2^{\sqrt{2}} \approx 2.665144143$

Now Work PROBLEM 15

It can be shown that the familiar laws for rational exponents hold for real exponents.

THEOREM **Laws of Exponents**

If s, t, a, and b are real numbers with $a > 0$ and $b > 0$, then

$$a^s \cdot a^t = a^{s+t} \qquad (a^s)^t = a^{st} \qquad (ab)^s = a^s \cdot b^s$$

$$1^s = 1 \qquad a^{-s} = \frac{1}{a^s} = \left(\frac{1}{a}\right)^s \qquad a^0 = 1 \qquad \textbf{(1)}$$

Introduction to Exponential Growth

Suppose a function f has the following two properties:

1. The value of f doubles with every 1-unit increase in the independent variable x.

2. The value of f at $x = 0$ is 5, so $f(0) = 5$.

Table 1 shows values of the function f for $x = 0, 1, 2, 3$, and 4.

We seek an equation $y = f(x)$ that describes this function f. The key fact is that the value of f doubles for every 1-unit increase in x.

$f(0) = 5$

$f(1) = 2f(0) = 2 \cdot 5 = 5 \cdot 2^1$ *Double the value of f at 0 to get the value at 1.*

$f(2) = 2f(1) = 2(5 \cdot 2) = 5 \cdot 2^2$ *Double the value of f at 1 to get the value at 2.*

$f(3) = 2f(2) = 2(5 \cdot 2^2) = 5 \cdot 2^3$

$f(4) = 2f(3) = 2(5 \cdot 2^3) = 5 \cdot 2^4$

The pattern leads us to

$$f(x) = 2f(x - 1) = 2(5 \cdot 2^{x-1}) = 5 \cdot 2^x$$

Table 1

x	f(x)
0	5
1	10
2	20
3	40
4	80

DEFINITION

An **exponential function** is a function of the form

$$f(x) = Ca^x$$

where a is a positive real number $(a > 0)$, $a \neq 1$, and $C \neq 0$ is a real number. The domain of f is the set of all real numbers. The base a is the **growth factor,** and because $f(0) = Ca^0 = C$, we call C the **initial value.**

WARNING It is important to distinguish a power function, $g(x) = ax^n$, $n \geq 2$, an integer, from an exponential function, $f(x) = C \cdot a^x$, $a \neq 1$, $a > 0$. In a power function, the base is a variable and the exponent is a constant. In an exponential function, the base is a constant and the exponent is a variable. ∎

In the definition of an exponential function, we exclude the base $a = 1$ because this function is simply the constant function $f(x) = C \cdot 1^x = C$. We also need to exclude bases that are negative; otherwise, we would have to exclude many values of x from the domain, such as $x = \dfrac{1}{2}$ and $x = \dfrac{3}{4}$. [Recall that $(-2)^{1/2} = \sqrt{-2}$, $(-3)^{3/4} = \sqrt[4]{(-3)^3} = \sqrt[4]{-27}$, and so on, are not defined in the set of real numbers.] Finally, transformations (vertical shifts, horizontal shifts, reflections, and so on) of a function of the form $f(x) = Ca^x$ also represent exponential functions.

Some examples of exponential functions are

$$f(x) = 2^x \qquad F(x) = \left(\frac{1}{3}\right)^x + 5 \qquad G(x) = 2 \cdot 3^{x-3}$$

Notice for each function that the base of the exponential expression is a constant and the exponent contains a variable.

In the function $f(x) = 5 \cdot 2^x$, notice that the ratio of consecutive outputs is constant for 1-unit increases in the input. This ratio equals the constant 2, the base of the exponential function. In other words,

$$\frac{f(1)}{f(0)} = \frac{5 \cdot 2^1}{5} = 2 \qquad \frac{f(2)}{f(1)} = \frac{5 \cdot 2^2}{5 \cdot 2^1} = 2 \qquad \frac{f(3)}{f(2)} = \frac{5 \cdot 2^3}{5 \cdot 2^2} = 2 \quad \text{and so on}$$

This leads to the following result.

THEOREM

For an exponential function $f(x) = Ca^x$, where $a > 0$ and $a \neq 1$, if x is any real number, then

$$\frac{f(x + 1)}{f(x)} = a \quad \text{or} \quad f(x + 1) = af(x)$$

In Words

For 1-unit changes in the input x of an exponential function $f(x) = C \cdot a^x$, the ratio of consecutive outputs is the constant a.

Proof

$$\frac{f(x + 1)}{f(x)} = \frac{Ca^{x+1}}{Ca^x} = a^{x+1-x} = a^1 = a \qquad \blacksquare$$

EXAMPLE 2

Identifying Linear or Exponential Functions

Determine whether the given function is linear, exponential, or neither. For those that are linear, find a linear function that models the data. For those that are exponential, find an exponential function that models the data.

(a)

x	y
-1	5
0	2
1	-1
2	-4
3	-7

(b)

x	y
-1	32
0	16
1	8
2	4
3	2

(c)

x	y
-1	2
0	4
1	7
2	11
3	16

Solution For each function, compute the average rate of change of y with respect to x and the ratio of consecutive outputs. If the average rate of change is constant, then the function is linear. If the ratio of consecutive outputs is constant, then the function is exponential.

Table 2

x	y	Average Rate of Change	Ratio of Consecutive Outputs
-1	5		
		$\dfrac{\Delta y}{\Delta x} = \dfrac{2-5}{0-(-1)} = -3$	$\dfrac{2}{5}$
0	2		
		-3	$-\dfrac{1}{2}$
1	-1		
		-3	4
2	-4		
		-3	$\dfrac{7}{4}$
3	-7		

(a)

x	y	Average Rate of Change	Ratio of Consecutive Outputs
-1	32		
		$\dfrac{\Delta y}{\Delta x} = \dfrac{16-32}{0-(-1)} = -16$	$\dfrac{16}{32} = \dfrac{1}{2}$
0	16		
		-8	$\dfrac{8}{16} = \dfrac{1}{2}$
1	8		
		-4	$\dfrac{4}{8} = \dfrac{1}{2}$
2	4		
		-2	$\dfrac{2}{4} = \dfrac{1}{2}$
3	2		

(b)

x	y	Average Rate of Change	Ratio of Consecutive Outputs
-1	2		
		$\dfrac{\Delta y}{\Delta x} = \dfrac{4-2}{0-(-1)} = 2$	2
0	4		
		3	$\dfrac{7}{4}$
1	7		
		4	$\dfrac{11}{7}$
2	11		
		5	$\dfrac{16}{11}$
3	16		

(c)

(a) See Table 2(a). The average rate of change for every 1-unit increase in x is -3. Therefore, the function is a linear function. In a linear function the average rate of change is the slope m, so $m = -3$. The y-intercept b is the value of the function at $x = 0$, so $b = 2$. The linear function that models the data is $f(x) = mx + b = -3x + 2$.

(b) See Table 2(b). For this function, the average rate of change from -1 to 0 is -16, and the average rate of change from 0 to 1 is -8. Because the average rate of change is not constant, the function is not a linear function. The ratio of consecutive outputs for a 1-unit increase in the inputs is a constant, $\dfrac{1}{2}$. Because the ratio of consecutive outputs is constant, the function is an exponential function with growth factor $a = \dfrac{1}{2}$. The initial value of the exponential function

is $C = 16$. Therefore, the exponential function that models the data is

$$g(x) = Ca^x = 16 \cdot \left(\frac{1}{2}\right)^x.$$

(c) See Table 2(c). For this function, the average rate of change from -1 to 0 is 2, and the average rate of change from 0 to 1 is 3. Because the average rate of change is not constant, the function is not a linear function. The ratio of consecutive outputs from -1 to 0 is 2, and the ratio of consecutive outputs from 0 to 1 is $\frac{7}{4}$. Because the ratio of consecutive outputs is not a constant, the function is not an exponential function.

—**Now Work** PROBLEM 25

2 Graph Exponential Functions

If we know how to graph an exponential function of the form $f(x) = a^x$, then we could use transformations (shifting, stretching, and so on) to obtain the graph of any exponential function.

First, we graph the exponential function $f(x) = 2^x$.

| EXAMPLE 3 | **Graphing an Exponential Function** |

Graph the exponential function: $f(x) = 2^x$

Solution The domain of $f(x) = 2^x$ is the set of all real numbers. We begin by locating some points on the graph of $f(x) = 2^x$, as listed in Table 3.

Since $2^x > 0$ for all x, the range of f is $(0, \infty)$. From this, we conclude that the graph has no x-intercepts, and, in fact, the graph will lie above the x-axis for all x. As Table 3 indicates, the y-intercept is 1. Table 3 also indicates that as $x \to -\infty$ the values of $f(x) = 2^x$ get closer and closer to 0. We conclude that the x-axis ($y = 0$) is a horizontal asymptote to the graph as $x \to -\infty$. This gives us the end behavior for x large and negative.

To determine the end behavior for x large and positive, look again at Table 3. As $x \to \infty$, $f(x) = 2^x$ grows very quickly, causing the graph of $f(x) = 2^x$ to rise very rapidly. It is apparent that f is an increasing function and hence is one-to-one.

Using all this information, we plot some of the points from Table 3 and connect them with a smooth, continuous curve, as shown in Figure 18.

Table 3

x	f(x) = 2ˣ
-10	$2^{-10} \approx 0.00098$
-3	$2^{-3} = \frac{1}{8}$
-2	$2^{-2} = \frac{1}{4}$
-1	$2^{-1} = \frac{1}{2}$
0	$2^0 = 1$
1	$2^1 = 2$
2	$2^2 = 4$
3	$2^3 = 8$
10	$2^{10} = 1024$

Figure 18

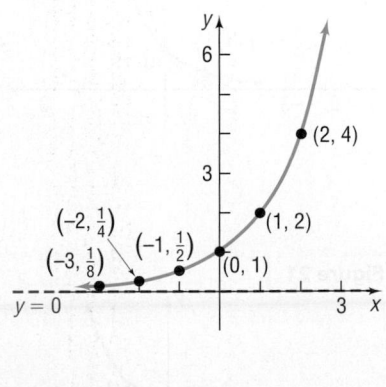

As we shall see, graphs that look like the one in Figure 18 occur very frequently in a variety of situations. For example, the graph in Figure 19 illustrates the number

of cellular telephone subscribers at the end of each year from 1985 to 2008. We might conclude from this graph that the number of cellular telephone subscribers is growing *exponentially*.

Figure 19

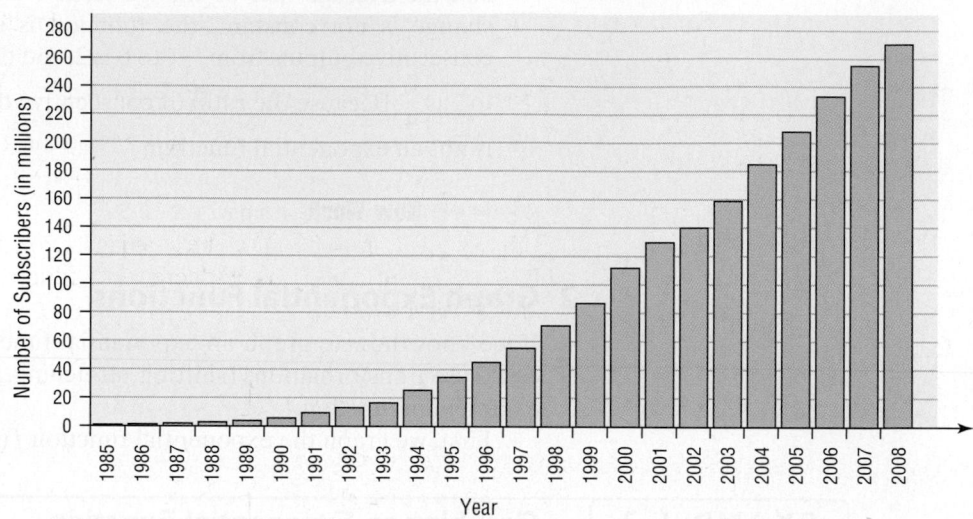

Number of Cellular Phone Subscribers at Year End

Source: ©2010 CTIA–The Wireless Association®.

We shall have more to say about situations that lead to exponential growth later in this chapter. For now, we continue to seek properties of exponential functions.

The graph of $f(x) = 2^x$ in Figure 18 is typical of all exponential functions of the form $f(x) = a^x$ with $a > 1$. Such functions are increasing functions and hence are one-to-one. Their graphs lie above the x-axis, pass through the point $(0, 1)$, and thereafter rise rapidly as $x \to \infty$. As $x \to -\infty$, the x-axis ($y = 0$) is a horizontal asymptote. There are no vertical asymptotes. Finally, the graphs are smooth and continuous with no corners or gaps.

Figure 20 illustrates the graphs of two more exponential functions whose bases are larger than 1. Notice that the larger the base, the steeper the graph is when $x > 0$, and when $x < 0$, the larger the base, the closer the graph of the equation is to the x-axis.

Figure 20

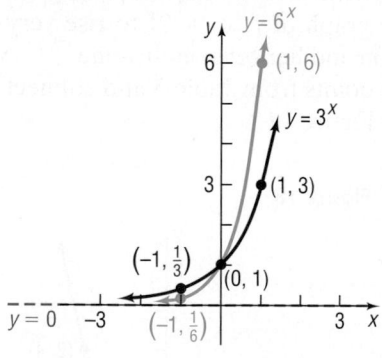

Seeing the Concept

Graph $Y_1 = 2^x$ and compare what you see to Figure 18. Clear the screen and graph $Y_1 = 3^x$ and $Y_2 = 6^x$ and compare what you see to Figure 20. Clear the screen and graph $Y_1 = 10^x$ and $Y_2 = 100^x$.

Properties of the Exponential Function $f(x) = a^x$, $a > 1$

1. The domain is the set of all real numbers or $(-\infty, \infty)$ using interval notation; the range is the set of positive real numbers or $(0, \infty)$ using interval notation.

2. There are no x-intercepts; the y-intercept is 1.

3. The x-axis ($y = 0$) is a horizontal asymptote as $x \to -\infty$ $\left[\lim\limits_{x \to -\infty} a^x = 0\right]$.

4. $f(x) = a^x$, where $a > 1$, is an increasing function and is one-to-one.

5. The graph of f contains the points $(0, 1)$, $(1, a)$, and $\left(-1, \dfrac{1}{a}\right)$.

Figure 21

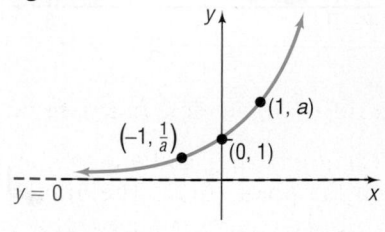

6. The graph of f is smooth and continuous, with no corners or gaps. See Figure 21.

Now consider $f(x) = a^x$ when $0 < a < 1$.

| EXAMPLE 4 | **Graphing an Exponential Function** |

Graph the exponential function: $f(x) = \left(\dfrac{1}{2}\right)^x$

Solution

The domain of $f(x) = \left(\dfrac{1}{2}\right)^x$ consists of all real numbers. As before, we locate some points on the graph by creating Table 4. Since $\left(\dfrac{1}{2}\right)^x > 0$ for all x, the range of f is the interval $(0, \infty)$. The graph lies above the x-axis and so has no x-intercepts. The y-intercept is 1. As $x \to -\infty$, $f(x) = \left(\dfrac{1}{2}\right)^x$ grows very quickly. As $x \to \infty$, the values of $f(x)$ approach 0. The x-axis ($y = 0$) is a horizontal asymptote as $x \to \infty$. It is apparent that f is a decreasing function and so is one-to-one. Figure 22 illustrates the graph.

Table 4

x	$f(x) = \left(\dfrac{1}{2}\right)^x$
-10	$\left(\dfrac{1}{2}\right)^{-10} = 1024$
-3	$\left(\dfrac{1}{2}\right)^{-3} = 8$
-2	$\left(\dfrac{1}{2}\right)^{-2} = 4$
-1	$\left(\dfrac{1}{2}\right)^{-1} = 2$
0	$\left(\dfrac{1}{2}\right)^{0} = 1$
1	$\left(\dfrac{1}{2}\right)^{1} = \dfrac{1}{2}$
2	$\left(\dfrac{1}{2}\right)^{2} = \dfrac{1}{4}$
3	$\left(\dfrac{1}{2}\right)^{3} = \dfrac{1}{8}$
10	$\left(\dfrac{1}{2}\right)^{10} \approx 0.00098$

Figure 22

We could have obtained the graph of $y = \left(\dfrac{1}{2}\right)^x$ from the graph of $y = 2^x$ using transformations. The graph of $y = \left(\dfrac{1}{2}\right)^x = 2^{-x}$ is a reflection about the y-axis of the graph of $y = 2^x$ (replace x by $-x$). See Figures 23(a) and (b).

Figure 23

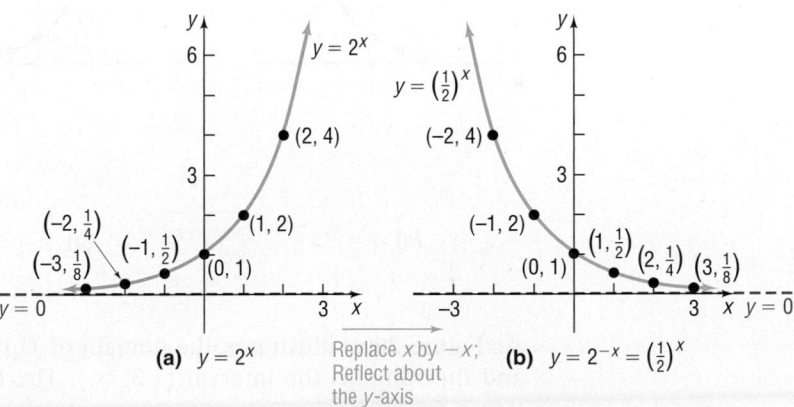

(a) $y = 2^x$ Replace x by $-x$; (b) $y = 2^{-x} = \left(\dfrac{1}{2}\right)^x$
Reflect about
the y-axis

 Seeing the Concept

Using a graphing utility, simultaneously graph:

(a) $Y_1 = 3^x, Y_2 = \left(\dfrac{1}{3}\right)^x$

(b) $Y_1 = 6^x, Y_2 = \left(\dfrac{1}{6}\right)^x$

Conclude that the graph of $Y_2 = \left(\dfrac{1}{a}\right)^x$, for $a > 0$, is the reflection about the y-axis of the graph of $Y_1 = a^x$.

The graph of $f(x) = \left(\dfrac{1}{2}\right)^x$ in Figure 22 is typical of all exponential functions of the form $f(x) = a^x$ with $0 < a < 1$. Such functions are decreasing and one-to-one. Their graphs lie above the x-axis and pass through the point $(0, 1)$. The graphs rise rapidly as $x \to -\infty$. As $x \to \infty$, the x-axis ($y = 0$) is a horizontal asymptote. There are no vertical asymptotes. Finally, the graphs are smooth and continuous, with no corners or gaps.

Figure 24

Figure 24 illustrates the graphs of two more exponential functions whose bases are between 0 and 1. Notice that the smaller base results in a graph that is steeper when $x < 0$. When $x > 0$, the graph of the equation with the smaller base is closer to the x-axis.

Properties of the Exponential Function $f(x) = a^x$, $0 < a < 1$

1. The domain is the set of all real numbers or $(-\infty, \infty)$ using interval notation; the range is the set of positive real numbers or $(0, \infty)$ using interval notation.

2. There are no x-intercepts; the y-intercept is 1.

3. The x-axis $(y = 0)$ is a horizontal asymptote as $x \to \infty$ $\left[\lim\limits_{x \to \infty} a^x = 0 \right]$.

4. $f(x) = a^x$, $0 < a < 1$, is a decreasing function and is one-to-one.

5. The graph of f contains the points $\left(-1, \dfrac{1}{a} \right)$, $(0, 1)$, and $(1, a)$.

6. The graph of f is smooth and continuous, with no corners or gaps. See Figure 25.

Figure 25

EXAMPLE 5

Graphing Exponential Functions Using Transformations

Graph $f(x) = 2^{-x} - 3$ and determine the domain, range, and horizontal asymptote of f.

Solution Begin with the graph of $y = 2^x$. Figure 26 shows the stages.

Figure 26

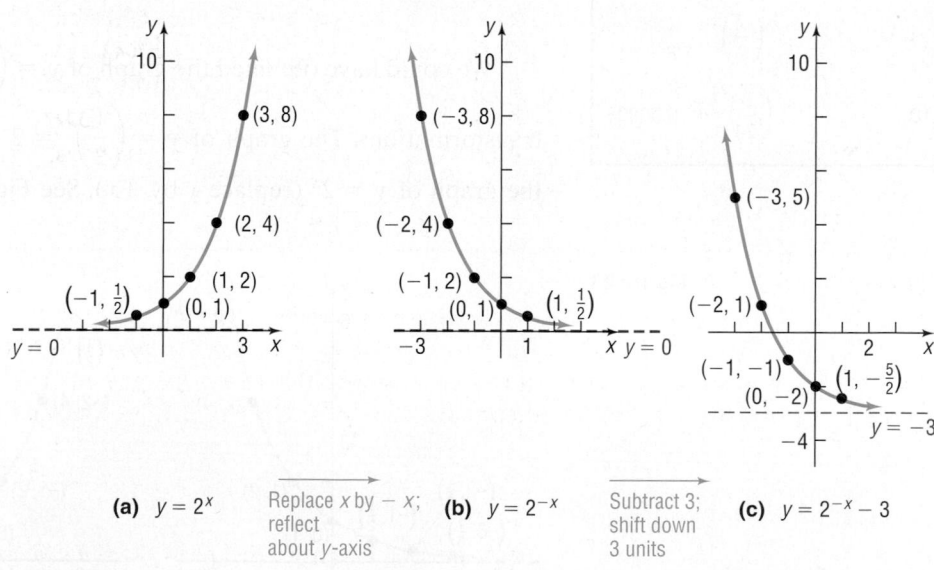

(a) $y = 2^x$ Replace x by $-x$; reflect about y-axis **(b)** $y = 2^{-x}$ Subtract 3; shift down 3 units **(c)** $y = 2^{-x} - 3$

As Figure 26(c) illustrates, the domain of $f(x) = 2^{-x} - 3$ is the interval $(-\infty, \infty)$ and the range is the interval $(-3, \infty)$. The horizontal asymptote of f is the line $y = -3$.

━━━ **Now Work** PROBLEM 41

3 Define the Number e

As we shall see shortly, many problems that occur in nature require the use of an exponential function whose base is a certain irrational number, symbolized by the letter e.

One way of arriving at this important number e is given next.

DEFINITION

The **number e** is defined as the number that the expression

$$\left(1 + \frac{1}{n}\right)^n \qquad (2)$$

approaches as $n \to \infty$. In calculus, this is expressed using limit notation as

$$e = \lim_{n \to \infty}\left(1 + \frac{1}{n}\right)^n$$

Table 5 illustrates what happens to the defining expression (2) as n takes on increasingly large values. The last number in the right column in the table is correct to nine decimal places and is the same as the entry given for e on your calculator (if expressed correctly to nine decimal places).

The exponential function $f(x) = e^x$, whose base is the number e, occurs with such frequency in applications that it is usually referred to as *the* exponential function. Indeed, most calculators have the key $\boxed{e^x}$ or $\boxed{\exp(x)}$, which may be used to evaluate the exponential function for a given value of x.*

Table 5

n	$\dfrac{1}{n}$	$1 + \dfrac{1}{n}$	$\left(1 + \dfrac{1}{n}\right)^n$
1	1	2	2
2	0.5	1.5	2.25
5	0.2	1.2	2.48832
10	0.1	1.1	2.59374246
100	0.01	1.01	2.704813829
1,000	0.001	1.001	2.716923932
10,000	0.0001	1.0001	2.718145927
100,000	0.00001	1.00001	2.718268237
1,000,000	0.000001	1.000001	2.718280469
1,000,000,000	10^{-9}	$1 + 10^{-9}$	2.718281827

Table 6

x	e^x
-2	$e^{-2} \approx 0.14$
-1	$e^{-1} \approx 0.37$
0	$e^0 \approx 1$
1	$e^1 \approx 2.72$
2	$e^2 \approx 7.39$

Figure 27
$y = e^x$

Now use your calculator to approximate e^x for $x = -2$, $x = -1$, $x = 0$, $x = 1$, and $x = 2$, as we have done to create Table 6. The graph of the exponential function $f(x) = e^x$ is given in Figure 27. Since $2 < e < 3$, the graph of $y = e^x$ lies between the graphs of $y = 2^x$ and $y = 3^x$. Do you see why? (Refer to Figures 18 and 20.)

Seeing the Concept

Graph $Y_1 = e^x$ and compare what you see to Figure 27. Use eVALUEate or TABLE to verify the points on the graph shown in Figure 27. Now graph $Y_2 = 2^x$ and $Y_3 = 3^x$ on the same screen as $Y_1 = e^x$. Notice that the graph of $Y_1 = e^x$ lies between these two graphs.

EXAMPLE 6

Graphing Exponential Functions Using Transformations

Graph $f(x) = -e^{x-3}$ and determine the domain, range, and horizontal asymptote of f.

Solution Begin with the graph of $y = e^x$. Figure 28 shows the stages.

*If your calculator does not have one of these keys, refer to your Owner's Manual.

Figure 28

(a) $y = e^x$ — Multiply by -1; Reflect about x-axis. — (b) $y = -e^x$ — Replace x by $x - 3$; Shift right 3 units. — (c) $y = -e^{x-3}$

As Figure 28(c) illustrates, the domain of $f(x) = -e^{x-3}$ is the interval $(-\infty, \infty)$, and the range is the interval $(-\infty, 0)$. The horizontal asymptote is the line $y = 0$.

Now Work PROBLEM 53

4 Solve Exponential Equations

Equations that involve terms of the form a^x, where $a > 0$ and $a \neq 1$, are referred to as **exponential equations.** Such equations can sometimes be solved by appropriately applying the Laws of Exponents and property (3):

In Words
When two exponential expressions with the same base are equal, then their exponents are equal.

$$\text{If} \quad a^u = a^v, \quad \text{then} \quad u = v. \tag{3}$$

Property (3) is a consequence of the fact that exponential functions are one-to-one. To use property (3), each side of the equality must be written with the same base.

EXAMPLE 7 | **Solving Exponential Equations**

Solve each exponential equation.

(a) $3^{x+1} = 81$ (b) $4^{2x-1} = 8^{x+3}$

Solution (a) Since $81 = 3^4$, write the equation as

$$3^{x+1} = 81 = 3^4$$

Now we have the same base, 3, on each side. Set the exponents equal to each other to obtain

$$x + 1 = 4$$
$$x = 3$$

The solution set is $\{3\}$.

(b)
$$4^{2x-1} = 8^{x+3}$$
$$(2^2)^{(2x-1)} = (2^3)^{(x+3)} \qquad 4 = 2^2; 8 = 2^3$$
$$2^{2(2x-1)} = 2^{3(x+3)} \qquad (a^r)^s = a^{rs}$$
$$2(2x - 1) = 3(x + 3) \qquad \text{If } a^u = a^v, \text{ then } u = v.$$
$$4x - 2 = 3x + 9$$
$$x = 11$$

The solution set is $\{11\}$.

Now Work PROBLEM 63

EXAMPLE 8 **Solving an Exponential Equation**

Solve: $e^{-x^2} = (e^x)^2 \cdot \dfrac{1}{e^3}$

Solution Use the Laws of Exponents first to get a single expression with the base e on the right side.

$$(e^x)^2 \cdot \frac{1}{e^3} = e^{2x} \cdot e^{-3} = e^{2x-3}$$

As a result,

$$
\begin{aligned}
e^{-x^2} &= e^{2x-3} \\
-x^2 &= 2x - 3 \qquad \text{Apply property (3).} \\
x^2 + 2x - 3 &= 0 \qquad \text{Place the quadratic equation in standard form.} \\
(x + 3)(x - 1) &= 0 \qquad \text{Factor.} \\
x = -3 \quad \text{or} \quad x &= 1 \qquad \text{Use the Zero-Product Property.}
\end{aligned}
$$

The solution set is $\{-3, 1\}$.

EXAMPLE 9 **Exponential Probability**

Between 9:00 PM and 10:00 PM cars arrive at Burger King's drive-thru at the rate of 12 cars per hour (0.2 car per minute). The following formula from statistics can be used to determine the probability that a car will arrive within t minutes of 9:00 PM.

$$F(t) = 1 - e^{-0.2t}$$

(a) Determine the probability that a car will arrive within 5 minutes of 9 PM (that is, before 9:05 PM).

(b) Determine the probability that a car will arrive within 30 minutes of 9 PM (before 9:30 PM).

(c) Graph F using your graphing utility.

(d) What value does F approach as t increases without bound in the positive direction?

Solution (a) The probability that a car will arrive within 5 minutes is found by evaluating $F(t)$ at $t = 5$.

$$F(5) = 1 - e^{-0.2(5)} \approx 0.63212$$
$$\underset{\text{Use a calculator.}}{\uparrow}$$

We conclude that there is a 63% probability that a car will arrive within 5 minutes.

(b) The probability that a car will arrive within 30 minutes is found by evaluating $F(t)$ at $t = 30$.

$$F(30) = 1 - e^{-0.2(30)} \approx 0.9975$$
$$\underset{\text{Use a calculator.}}{\uparrow}$$

Figure 29

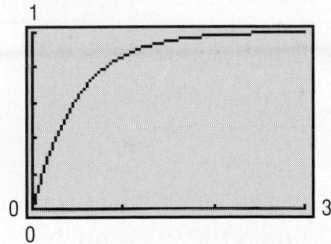

There is a 99.75% probability that a car will arrive within 30 minutes.

(c) See Figure 29 for the graph of F.

(d) As time passes, the probability that a car will arrive increases. The value that F approaches can be found by letting $t \rightarrow \infty$. Since $e^{-0.2t} = \dfrac{1}{e^{0.2t}}$, it follows that $e^{-0.2t} \rightarrow 0$ as $t \rightarrow \infty$. We conclude that F approaches 1 as t gets large. The algebraic analysis is confirmed by Figure 29.

——Now Work PROBLEM 107

SUMMARY Properties of the Exponential Function

$f(x) = a^x, \quad a > 1$ Domain: the interval $(-\infty, \infty)$; range: the interval $(0, \infty)$
 x-intercepts: none; y-intercept: 1
 Horizontal asymptote: x-axis $(y = 0)$ as $x \to -\infty$
 Increasing; one-to-one; smooth; continuous
 See Figure 21 for a typical graph.

$f(x) = a^x, \quad 0 < a < 1$ Domain: the interval $(-\infty, \infty)$; range: the interval $(0, \infty)$
 x-intercepts: none; y-intercept: 1
 Horizontal asymptote: x-axis $(y = 0)$ as $x \to \infty$
 Decreasing; one-to-one; smooth; continuous
 See Figure 25 for a typical graph.

If $a^u = a^v$, then $u = v$.

6.3 Assess Your Understanding

'Are You Prepared?' *Answers are given at the end of these exercises. If you get a wrong answer, read the pages listed in red.*

1. $4^3 =$ _____; $8^{2/3} =$ _____; $3^{-2} =$ _____. (pp. 21–24 and pp. 73–77)

2. Solve: $x^2 + 3x = 4$ (pp. 92–99)

3. **True or False** To graph $y = (x - 2)^3$, shift the graph of $y = x^3$ to the left 2 units. (pp. 244–253)

4. Find the average rate of change of $f(x) = 3x - 5$ from $x = 0$ to $x = 4$. (pp. 223–230; 272–275)

5. **True or False** The function $f(x) = \dfrac{2x}{x - 3}$ has $y = 2$ as a horizontal asymptote. (pp. 345–346)

Concepts and Vocabulary

6. A(n) _____ _____ is a function of the form $f(x) = Ca^x$, where $a > 0, a \neq 1$, and $C \neq 0$ are real numbers. The base a is the _____ _____ and C is the _____ _____.

7. For an exponential function $f(x) = Ca^x$, $\dfrac{f(x + 1)}{f(x)} =$ ___.

8. **True or False** The domain of the exponential function $f(x) = a^x$, where $a > 0$ and $a \neq 1$, is the set of all real numbers.

9. **True or False** The range of the exponential function $f(x) = a^x$, where $a > 0$ and $a \neq 1$, is the set of all real numbers.

10. **True or False** The graph of the exponential function $f(x) = a^x$, where $a > 0$ and $a \neq 1$, has no x-intercept.

11. The graph of every exponential function $f(x) = a^x$, where $a > 0$ and $a \neq 1$, passes through three points: _____, _____, and _____.

12. If the graph of the exponential function $f(x) = a^x$, where $a > 0$ and $a \neq 1$, is decreasing, then a must be less than _____.

13. If $3^x = 3^4$, then $x =$ _____.

14. **True or False** The graphs of $y = 3^x$ and $y = \left(\dfrac{1}{3}\right)^x$ are identical.

Skill Building

In Problems 15–24, approximate each number using a calculator. Express your answer rounded to three decimal places.

15. (a) $3^{2.2}$ (b) $3^{2.23}$ (c) $3^{2.236}$ (d) $3^{\sqrt{5}}$

16. (a) $5^{1.7}$ (b) $5^{1.73}$ (c) $5^{1.732}$ (d) $5^{\sqrt{3}}$

17. (a) $2^{3.14}$ (b) $2^{3.141}$ (c) $2^{3.1415}$ (d) 2^{π}

18. (a) $2^{2.7}$ (b) $2^{2.71}$ (c) $2^{2.718}$ (d) 2^{e}

19. (a) $3.1^{2.7}$ (b) $3.14^{2.71}$ (c) $3.141^{2.718}$ (d) π^{e}

20. (a) $2.7^{3.1}$ (b) $2.71^{3.14}$ (c) $2.718^{3.141}$ (d) e^{π}

21. $e^{1.2}$

22. $e^{-1.3}$

23. $e^{-0.85}$

24. $e^{2.1}$

In Problems 25–32, determine whether the given function is linear, exponential, or neither. For those that are linear functions, find a linear function that models the data; for those that are exponential, find an exponential function that models the data.

25.

x	f(x)
−1	3
0	6
1	12
2	18
3	30

26.

x	g(x)
−1	2
0	5
1	8
2	11
3	14

27.

x	H(x)
−1	$\frac{1}{4}$
0	1
1	4
2	16
3	64

28.

x	F(x)
−1	$\frac{2}{3}$
0	1
1	$\frac{3}{2}$
2	$\frac{9}{4}$
3	$\frac{27}{8}$

29.

x	f(x)
−1	$\frac{3}{2}$
0	3
1	6
2	12
3	24

30.

x	g(x)
−1	6
0	1
1	0
2	3
3	10

31.

x	H(x)
−1	2
0	4
1	6
2	8
3	10

32.

x	F(x)
−1	$\frac{1}{2}$
0	$\frac{1}{4}$
1	$\frac{1}{8}$
2	$\frac{1}{16}$
3	$\frac{1}{32}$

In Problems 33–40, the graph of an exponential function is given. Match each graph to one of the following functions.

(a) $y = 3^x$ (b) $y = 3^{-x}$ (c) $y = -3^x$ (d) $y = -3^{-x}$

(e) $y = 3^x - 1$ (f) $y = 3^{x-1}$ (g) $y = 3^{1-x}$ (h) $y = 1 - 3^x$

33.

34.

35.

36.

37.

38.

39.

40.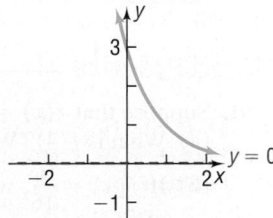

In Problems 41–52, use transformations to graph each function. Determine the domain, range, and horizontal asymptote of each function.

41. $f(x) = 2^x + 1$ **42.** $f(x) = 3^x - 2$ **43.** $f(x) = 3^{x-1}$ **44.** $f(x) = 2^{x+2}$

45. $f(x) = 3 \cdot \left(\frac{1}{2}\right)^x$ **46.** $f(x) = 4 \cdot \left(\frac{1}{3}\right)^x$ **47.** $f(x) = 3^{-x} - 2$ **48.** $f(x) = -3^x + 1$

49. $f(x) = 2 + 4^{x-1}$ **50.** $f(x) = 1 - 2^{x+3}$ **51.** $f(x) = 2 + 3^{x/2}$ **52.** $f(x) = 1 - 2^{-x/3}$

In Problems 53–60, begin with the graph of $y = e^x$ [Figure 27] and use transformations to graph each function. Determine the domain, range, and horizontal asymptote of each function.

53. $f(x) = e^{-x}$ **54.** $f(x) = -e^x$ **55.** $f(x) = e^{x+2}$ **56.** $f(x) = e^x - 1$

57. $f(x) = 5 - e^{-x}$ **58.** $f(x) = 9 - 3e^{-x}$ **59.** $f(x) = 2 - e^{-x/2}$ **60.** $f(x) = 7 - 3e^{2x}$

In Problems 61–80, solve each equation.

61. $7^x = 7^3$

62. $5^x = 5^{-6}$

63. $2^{-x} = 16$

64. $3^{-x} = 81$

65. $\left(\dfrac{1}{5}\right)^x = \dfrac{1}{25}$

66. $\left(\dfrac{1}{4}\right)^x = \dfrac{1}{64}$

67. $2^{2x-1} = 4$

68. $5^{x+3} = \dfrac{1}{5}$

69. $3^{x^3} = 9^x$

70. $4^{x^2} = 2^x$

71. $8^{-x+14} = 16^x$

72. $9^{-x+15} = 27^x$

73. $3^{x^2-7} = 27^{2x}$

74. $5^{x^2+8} = 125^{2x}$

75. $4^x \cdot 2^{x^2} = 16^2$

76. $9^{2x} \cdot 27^{x^2} = 3^{-1}$

77. $e^x = e^{3x+8}$

78. $e^{3x} = e^{2-x}$

79. $e^{x^2} = e^{3x} \cdot \dfrac{1}{e^2}$

80. $(e^4)^x \cdot e^{x^2} = e^{12}$

81. If $4^x = 7$, what does 4^{-2x} equal?

82. If $2^x = 3$, what does 4^{-x} equal?

83. If $3^{-x} = 2$, what does 3^{2x} equal?

84. If $5^{-x} = 3$, what does 5^{3x} equal?

In Problems 85–88, determine the exponential function whose graph is given.

85.

86.

87.

88.

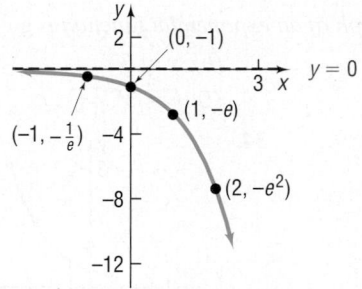

89. Find an exponential function with horizontal asymptote $y = 2$ whose graph contains the points $(0, 3)$ and $(1, 5)$.

90. Find an exponential function with horizontal asymptote $y = -3$ whose graph contains the points $(0, -2)$ and $(-2, 1)$.

Mixed Practice

91. Suppose that $f(x) = 2^x$.
 (a) What is $f(4)$? What point is on the graph of f?
 (b) If $f(x) = \dfrac{1}{16}$, what is x? What point is on the graph of f?

92. Suppose that $f(x) = 3^x$.
 (a) What is $f(4)$? What point is on the graph of f?
 (b) If $f(x) = \dfrac{1}{9}$, what is x? What point is on the graph of f?

93. Suppose that $g(x) = 4^x + 2$.
 (a) What is $g(-1)$? What point is on the graph of g?
 (b) If $g(x) = 66$, what is x? What point is on the graph of g?

94. Suppose that $g(x) = 5^x - 3$.
 (a) What is $g(-1)$? What point is on the graph of g?
 (b) If $g(x) = 122$, what is x? What point is on the graph of g?

95. Suppose that $H(x) = \left(\dfrac{1}{2}\right)^x - 4$.
 (a) What is $H(-6)$? What point is on the graph of H?
 (b) If $H(x) = 12$, what is x? What point is on the graph of H?
 (c) Find the zero of H.

96. Suppose that $F(x) = \left(\dfrac{1}{3}\right)^x - 3$.
 (a) What is $F(-5)$? What point is on the graph of F?
 (b) If $F(x) = 24$, what is x? What point is on the graph of F?
 (c) Find the zero of F.

In Problems 97–100, graph each function. Based on the graph, state the domain and the range and find any intercepts.

97. $f(x) = \begin{cases} e^{-x} & \text{if } x < 0 \\ e^x & \text{if } x \geq 0 \end{cases}$

98. $f(x) = \begin{cases} e^x & \text{if } x < 0 \\ e^{-x} & \text{if } x \geq 0 \end{cases}$

99. $f(x) = \begin{cases} -e^x & \text{if } x < 0 \\ -e^{-x} & \text{if } x \geq 0 \end{cases}$

100. $f(x) = \begin{cases} -e^{-x} & \text{if } x < 0 \\ -e^x & \text{if } x \geq 0 \end{cases}$

Applications and Extensions

101. Optics If a single pane of glass obliterates 3% of the light passing through it, the percent p of light that passes through n successive panes is given approximately by the function

$$p(n) = 100(0.97)^n$$

(a) What percent of light will pass through 10 panes?
(b) What percent of light will pass through 25 panes?

102. Atmospheric Pressure The atmospheric pressure p on a balloon or plane decreases with increasing height. This pressure, measured in millimeters of mercury, is related to the height h (in kilometers) above sea level by the function

$$p(h) = 760e^{-0.145h}$$

(a) Find the atmospheric pressure at a height of 2 kilometers (over a mile).
(b) What is it at a height of 10 kilometers (over 30,000 feet)?

103. Depreciation The price p, in dollars, of a Honda Civic DX Sedan that is x years old is modeled by

$$p(x) = 16,630(0.90)^x$$

(a) How much should a 3-year-old Civic DX Sedan cost?
(b) How much should a 9-year-old Civic DX Sedan cost?

104. Healing of Wounds The normal healing of wounds can be modeled by an exponential function. If A_0 represents the original area of the wound and if A equals the area of the wound, then the function

$$A(n) = A_0 e^{-0.35n}$$

describes the area of a wound after n days following an injury when no infection is present to retard the healing. Suppose that a wound initially had an area of 100 square millimeters.
(a) If healing is taking place, how large will the area of the wound be after 3 days?
(b) How large will it be after 10 days?

105. Drug Medication The function

$$D(h) = 5e^{-0.4h}$$

can be used to find the number of milligrams D of a certain drug that is in a patient's bloodstream h hours after the drug has been administered. How many milligrams will be present after 1 hour? After 6 hours?

106. Spreading of Rumors A model for the number N of people in a college community who have heard a certain rumor is

$$N = P(1 - e^{-0.15d})$$

where P is the total population of the community and d is the number of days that have elapsed since the rumor began. In a community of 1000 students, how many students will have heard the rumor after 3 days?

107. Exponential Probability Between 12:00 PM and 1:00 PM, cars arrive at Citibank's drive-thru at the rate of 6 cars per hour (0.1 car per minute). The following formula from probability can be used to determine the probability that a car will arrive within t minutes of 12:00 PM:

$$F(t) = 1 - e^{-0.1t}$$

(a) Determine the probability that a car will arrive within 10 minutes of 12:00 PM (that is, before 12:10 PM).
(b) Determine the probability that a car will arrive within 40 minutes of 12:00 PM (before 12:40 PM).
(c) What value does F approach as t becomes unbounded in the positive direction?
(d) Graph F using a graphing utility.
(e) Using INTERSECT, determine how many minutes are needed for the probability to reach 50%.

108. Exponential Probability Between 5:00 PM and 6:00 PM, cars arrive at Jiffy Lube at the rate of 9 cars per hour (0.15 car per minute). The following formula from probability can be used to determine the probability that a car will arrive within t minutes of 5:00 PM:

$$F(t) = 1 - e^{-0.15t}$$

(a) Determine the probability that a car will arrive within 15 minutes of 5:00 PM (that is, before 5:15 PM).
(b) Determine the probability that a car will arrive within 30 minutes of 5:00 PM (before 5:30 PM).
(c) What value does F approach as t becomes unbounded in the positive direction?
(d) Graph F using a graphing utility.
(e) Using INTERSECT, determine how many minutes are needed for the probability to reach 60%.

109. Poisson Probability Between 5:00 PM and 6:00 PM, cars arrive at McDonald's drive-thru at the rate of 20 cars per hour. The following formula from probability can be used to determine the probability that x cars will arrive between 5:00 PM and 6:00 PM.

$$P(x) = \frac{20^x e^{-20}}{x!}$$

where

$$x! = x \cdot (x - 1) \cdot (x - 2) \cdot \cdots \cdot 3 \cdot 2 \cdot 1$$

(a) Determine the probability that $x = 15$ cars will arrive between 5:00 PM and 6:00 PM.
(b) Determine the probability that $x = 20$ cars will arrive between 5:00 PM and 6:00 PM.

110. Poisson Probability People enter a line for the *Demon Roller Coaster* at the rate of 4 per minute. The following formula from probability can be used to determine the probability that x people will arrive within the next minute.

$$P(x) = \frac{4^x e^{-4}}{x!}$$

where

$$x! = x \cdot (x - 1) \cdot (x - 2) \cdot \cdots \cdot 3 \cdot 2 \cdot 1$$

(a) Determine the probability that $x = 5$ people will arrive within the next minute.

(b) Determine the probability that $x = 8$ people will arrive within the next minute.

111. Relative Humidity The relative humidity is the ratio (expressed as a percent) of the amount of water vapor in the air to the maximum amount that it can hold at a specific temperature. The relative humidity, R, is found using the following formula:

$$R = 10^{\left(\frac{4221}{T+459.4} - \frac{4221}{D+459.4} + 2\right)}$$

where T is the air temperature (in °F) and D is the dew point temperature (in °F).

(a) Determine the relative humidity if the air temperature is 50° Fahrenheit and the dew point temperature is 41° Fahrenheit.

(b) Determine the relative humidity if the air temperature is 68° Fahrenheit and the dew point temperature is 59° Fahrenheit.

(c) What is the relative humidity if the air temperature and the dew point temperature are the same?

112. Learning Curve Suppose that a student has 500 vocabulary words to learn. If the student learns 15 words after 5 minutes, the function

$$L(t) = 500(1 - e^{-0.0061t})$$

approximates the number of words L that the student will learn after t minutes.

(a) How many words will the student learn after 30 minutes?

(b) How many words will the student learn after 60 minutes?

113. Current in a *RL* Circuit The equation governing the amount of current I (in amperes) after time t (in seconds) in a single *RL* circuit consisting of a resistance R (in ohms), an inductance L (in henrys), and an electromotive force E (in volts) is

$$I = \frac{E}{R}[1 - e^{-(R/L)t}]$$

(a) If $E = 120$ volts, $R = 10$ ohms, and $L = 5$ henrys, how much current I_1 is flowing after 0.3 second? After 0.5 second? After 1 second?

(b) What is the maximum current?

(c) Graph this function $I = I_1(t)$, measuring I along the y-axis and t along the x-axis.

(d) If $E = 120$ volts, $R = 5$ ohms, and $L = 10$ henrys, how much current I_2 is flowing after 0.3 second? After 0.5 second? After 1 second?

(e) What is the maximum current?

(f) Graph the function $I = I_2(t)$ on the same coordinate axes as $I_1(t)$.

114. Current in a *RC* Circuit The equation governing the amount of current I (in amperes) after time t (in microseconds) in a single *RC* circuit consisting of a resistance R (in ohms), a capacitance C (in microfarads), and an electromotive force E (in volts) is

$$I = \frac{E}{R} e^{-t/(RC)}$$

(a) If $E = 120$ volts, $R = 2000$ ohms, and $C = 1.0$ microfarad, how much current I_1 is flowing initially ($t = 0$)? After 1000 microseconds? After 3000 microseconds?

(b) What is the maximum current?

(c) Graph the function $I = I_1(t)$, measuring I along the y-axis and t along the x-axis.

(d) If $E = 120$ volts, $R = 1000$ ohms, and $C = 2.0$ microfarads, how much current I_2 is flowing initially? After 1000 microseconds? After 3000 microseconds?

(e) What is the maximum current?

(f) Graph the function $I = I_2(t)$ on the same coordinate axes as $I_1(t)$.

115. If f is an exponential function of the form $f(x) = C \cdot a^x$ with growth factor 3 and $f(6) = 12$, what is $f(7)$?

116. Another Formula for e Use a calculator to compute the values of

$$2 + \frac{1}{2!} + \frac{1}{3!} + \cdots + \frac{1}{n!}$$

for $n = 4, 6, 8,$ and 10. Compare each result with e.

[**Hint:** $1! = 1, 2! = 2 \cdot 1, 3! = 3 \cdot 2 \cdot 1,$
$n! = n(n - 1) \cdot \cdots \cdot (3)(2)(1).]$

117. Another Formula for e Use a calculator to compute the various values of the expression. Compare the values to e.

$$2 + \cfrac{1}{1 + \cfrac{1}{2 + \cfrac{2}{3 + \cfrac{3}{4 + \cfrac{4}{\text{etc.}}}}}}$$

118. Difference Quotient If $f(x) = a^x$, show that

$$\frac{f(x + h) - f(x)}{h} = a^x \cdot \frac{a^h - 1}{h} \qquad h \neq 0$$

119. If $f(x) = a^x$, show that $f(A + B) = f(A) \cdot f(B)$.

120. If $f(x) = a^x$, show that $f(-x) = \frac{1}{f(x)}$.

121. If $f(x) = a^x$, show that $f(\alpha x) = [f(x)]^\alpha$.

Problems 122 and 123 provide definitions for two other transcendental functions.

122. The **hyperbolic sine function,** designated by sinh x, is defined as

$$\sinh x = \frac{1}{2}(e^x - e^{-x})$$

(a) Show that $f(x) = \sinh x$ is an odd function.

(b) Graph $f(x) = \sinh x$ using a graphing utility.

123. The **hyperbolic cosine function,** designated by cosh x, is defined as

$$\cosh x = \frac{1}{2}(e^x + e^{-x})$$

(a) Show that $f(x) = \cosh x$ is an even function.

(b) Graph $f(x) = \cosh x$ using a graphing utility.

(c) Refer to Problem 122. Show that, for every x,

$$(\cosh x)^2 - (\sinh x)^2 = 1$$

124. Historical Problem Pierre de Fermat (1601–1665) conjectured that the function

$$f(x) = 2^{(2^x)} + 1$$

for $x = 1, 2, 3, \ldots$, would always have a value equal to a prime number. But Leonhard Euler (1707–1783) showed that this formula fails for $x = 5$. Use a calculator to determine the prime numbers produced by f for $x = 1, 2, 3, 4$. Then show that $f(5) = 641 \times 6{,}700{,}417$, which is not prime.

Explaining Concepts: Discussion and Writing

125. The bacteria in a 4-liter container double every minute. After 60 minutes the container is full. How long did it take to fill half the container?

126. Explain in your own words what the number e is. Provide at least two applications that use this number.

127. Do you think that there is a power function that increases more rapidly than an exponential function whose base is greater than 1? Explain.

128. As the base a of an exponential function $f(x) = a^x$, where $a > 1$ increases, what happens to the behavior of its graph for $x > 0$? What happens to the behavior of its graph for $x < 0$?

129. The graphs of $y = a^{-x}$ and $y = \left(\dfrac{1}{a}\right)^x$ are identical. Why?

'Are You Prepared?' Answers

1. 64; 4; $\dfrac{1}{9}$ **2.** $\{-4, 1\}$ **3.** False **4.** 3 **5.** True

6.4 Logarithmic Functions

PREPARING FOR THIS SECTION *Before getting started, review the following:*

- Solving Inequalities (Section 1.5, pp. 119–126)
- Quadratic Inequalities (Section 4.5, pp. 309–311)
- Polynomial and Rational Inequalities (Section 5.4, pp. 368–371)
- Solve Linear Equations (Section 1.1, pp. 82–87)

Now Work the 'Are You Prepared?' problems on page 446.

OBJECTIVES 1 Change Exponential Statements to Logarithmic Statements and Logarithmic Statements to Exponential Statements (p. 438)

2 Evaluate Logarithmic Expressions (p. 438)

3 Determine the Domain of a Logarithmic Function (p. 439)

4 Graph Logarithmic Functions (p. 440)

5 Solve Logarithmic Equations (p. 444)

Recall that a one-to-one function $y = f(x)$ has an inverse function that is defined (implicitly) by the equation $x = f(y)$. In particular, the exponential function $y = f(x) = a^x$, where $a > 0$ and $a \neq 1$, is one-to-one and hence has an inverse function that is defined implicitly by the equation

$$x = a^y, \quad a > 0, \quad a \neq 1$$

This inverse function is so important that it is given a name, the *logarithmic function*.

DEFINITION

The **logarithmic function to the base a,** where $a > 0$ and $a \neq 1$, is denoted by $y = \log_a x$ (read as "y is the logarithm to the base a of x") and is defined by

$$y = \log_a x \quad \text{if and only if} \quad x = a^y$$

The domain of the logarithmic function $y = \log_a x$ is $x > 0$.

In Words
When you read $\log_a x$, think to yourself "a raised to what power gives me x."

As this definition illustrates, **a logarithm is a name for a certain exponent.** So, $\log_a x$ represents the exponent to which a must be raised to obtain x.

EXAMPLE 1 **Relating Logarithms to Exponents**

(a) If $y = \log_3 x$, then $x = 3^y$. For example, the logarithmic statement $4 = \log_3 81$ is equivalent to the exponential statement $81 = 3^4$.

(b) If $y = \log_5 x$, then $x = 5^y$. For example, $-1 = \log_5\left(\dfrac{1}{5}\right)$ is equivalent to $\dfrac{1}{5} = 5^{-1}$.

1 Change Exponential Statements to Logarithmic Statements and Logarithmic Statements to Exponential Statements

We can use the definition of a logarithm to convert from exponential form to logarithmic form, and vice versa, as the following two examples illustrate.

EXAMPLE 2 **Changing Exponential Statements to Logarithmic Statements**

Change each exponential statement to an equivalent statement involving a logarithm.

(a) $1.2^3 = m$ (b) $e^b = 9$ (c) $a^4 = 24$

Solution Use the fact that $y = \log_a x$ and $x = a^y$, where $a > 0$ and $a \neq 1$, are equivalent.

(a) If $1.2^3 = m$, then $3 = \log_{1.2} m$.
(b) If $e^b = 9$, then $b = \log_e 9$.
(c) If $a^4 = 24$, then $4 = \log_a 24$.

Now Work PROBLEM 9

EXAMPLE 3 **Changing Logarithmic Statements to Exponential Statements**

Change each logarithmic statement to an equivalent statement involving an exponent.

(a) $\log_a 4 = 5$ (b) $\log_e b = -3$ (c) $\log_3 5 = c$

Solution (a) If $\log_a 4 = 5$, then $a^5 = 4$.
(b) If $\log_e b = -3$, then $e^{-3} = b$.
(c) If $\log_3 5 = c$, then $3^c = 5$.

Now Work PROBLEM 17

2 Evaluate Logarithmic Expressions

To find the exact value of a logarithm, we write the logarithm in exponential notation using the fact that $y = \log_a x$ is equivalent to $a^y = x$ and use the fact that if $a^u = a^v$, then $u = v$.

EXAMPLE 4 **Finding the Exact Value of a Logarithmic Expression**

Find the exact value of:

(a) $\log_2 16$

(b) $\log_3 \dfrac{1}{27}$

Solution (a) To evaluate $\log_2 16$, think "2 raised to what power yields 16." So,

$$y = \log_2 16$$

$$2^y = 16 \qquad \textit{Change to exponential form.}$$

$$2^y = 2^4 \qquad 16 = 2^4$$

$$y = 4 \qquad \textit{Equate exponents.}$$

Therefore, $\log_2 16 = 4$.

(b) To evaluate $\log_3 \dfrac{1}{27}$, think "3 raised to what power yields $\dfrac{1}{27}$." So,

$$y = \log_3 \dfrac{1}{27}$$

$$3^y = \dfrac{1}{27} \qquad \textit{Change to exponential form.}$$

$$3^y = 3^{-3} \qquad \dfrac{1}{27} = \dfrac{1}{3^3} = 3^{-3}$$

$$y = -3 \qquad \textit{Equate exponents.}$$

Therefore, $\log_3 \dfrac{1}{27} = -3$.

Now Work PROBLEM 25

3 Determine the Domain of a Logarithmic Function

The logarithmic function $y = \log_a x$ has been defined as the inverse of the exponential function $y = a^x$. That is, if $f(x) = a^x$, then $f^{-1}(x) = \log_a x$. Based on the discussion given in Section 6.2 on inverse functions, for a function f and its inverse f^{-1}, we have

$$\text{Domain of } f^{-1} = \text{Range of } f \quad \text{and} \quad \text{Range of } f^{-1} = \text{Domain of } f$$

Consequently, it follows that

> Domain of the logarithmic function = Range of the exponential function = $(0, \infty)$
>
> Range of the logarithmic function = Domain of the exponential function = $(-\infty, \infty)$

In the next box, we summarize some properties of the logarithmic function:

> $y = \log_a x$ (defining equation: $x = a^y$)
>
> Domain: $0 < x < \infty$ Range: $-\infty < y < \infty$

The domain of a logarithmic function consists of the *positive* real numbers, so the argument of a logarithmic function must be greater than zero.

EXAMPLE 5 **Finding the Domain of a Logarithmic Function**

Find the domain of each logarithmic function.

(a) $F(x) = \log_2(x + 3)$ (b) $g(x) = \log_5\left(\dfrac{1 + x}{1 - x}\right)$ (c) $h(x) = \log_{1/2}|x|$

Solution (a) The domain of F consists of all x for which $x + 3 > 0$, that is, $x > -3$. Using interval notation, the domain of f is $(-3, \infty)$.

(b) The domain of g is restricted to

$$\dfrac{1 + x}{1 - x} > 0$$

Solving this inequality, we find that the domain of g consists of all x between -1 and 1, that is, $-1 < x < 1$ or, using interval notation, $(-1, 1)$.

(c) Since $|x| > 0$, provided that $x \neq 0$, the domain of h consists of all real numbers except zero or, using interval notation, $(-\infty, 0) \cup (0, \infty)$.

Now Work PROBLEMS 39 AND 45

4 Graph Logarithmic Functions

Since exponential functions and logarithmic functions are inverses of each other, the graph of the logarithmic function $y = \log_a x$ is the reflection about the line $y = x$ of the graph of the exponential function $y = a^x$, as shown in Figure 30.

Figure 30

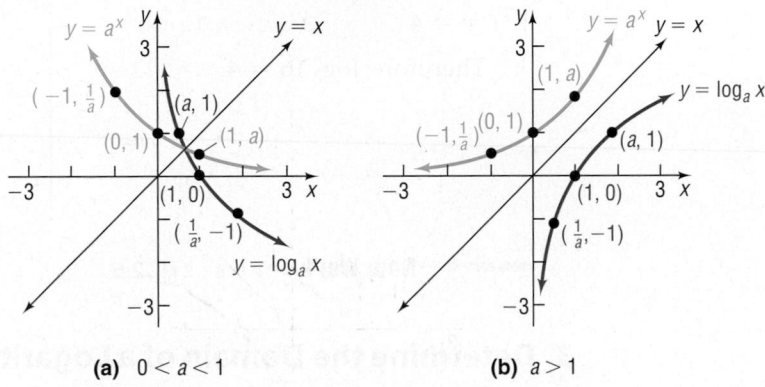

(a) $0 < a < 1$ **(b)** $a > 1$

For example, to graph $y = \log_2 x$, graph $y = 2^x$ and reflect it about the line $y = x$. See Figure 31. To graph $y = \log_{1/3} x$, graph $y = \left(\dfrac{1}{3}\right)^x$ and reflect it about the line $y = x$. See Figure 32.

Figure 31

Figure 32

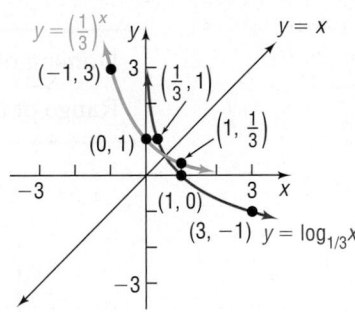

Now Work PROBLEM 59

The graphs of $y = \log_a x$ in Figures 30(a) and (b) lead to the following properties.

Properties of the Logarithmic Function $f(x) = \log_a x$

1. The domain is the set of positive real numbers or $(0, \infty)$ using interval notation; the range is the set of all real numbers or $(-\infty, \infty)$ using interval notation.

2. The x-intercept of the graph is 1. There is no y-intercept.

3. The y-axis ($x = 0$) is a vertical asymptote of the graph.

4. A logarithmic function is decreasing if $0 < a < 1$ and increasing if $a > 1$.

5. The graph of f contains the points $(1, 0)$, $(a, 1)$, and $\left(\dfrac{1}{a}, -1\right)$.

6. The graph is smooth and continuous, with no corners or gaps.

If the base of a logarithmic function is the number e, then we have the **natural logarithm function.** This function occurs so frequently in applications that it is given a special symbol, **ln** (from the Latin, *logarithmus naturalis*). That is,

<div style="border:1px solid">

$$y = \ln x \quad \text{if and only if} \quad x = e^y \tag{1}$$

</div>

In Words
$y = \log_e x$ is written $y = \ln x$

Since $y = \ln x$ and the exponential function $y = e^x$ are inverse functions, we can obtain the graph of $y = \ln x$ by reflecting the graph of $y = e^x$ about the line $y = x$. See Figure 33.

Using a calculator with an ⌊ ln ⌋ key, we can obtain other points on the graph of $f(x) = \ln x$. See Table 7.

Figure 33

Table 7

x	$\ln x$
$\frac{1}{2}$	-0.69
2	0.69
3	1.10

Seeing the Concept

Graph $Y_1 = e^x$ and $Y_2 = \ln x$ on the same square screen. Use eVALUEate to verify the points on the graph given in Figure 33. Do you see the symmetry of the two graphs with respect to the line $y = x$?

EXAMPLE 6

Graphing a Logarithmic Function and Its Inverse

(a) Find the domain of the logarithmic function $f(x) = -\ln(x - 2)$.

(b) Graph f.

(c) From the graph, determine the range and vertical asymptote of f.

(d) Find f^{-1}, the inverse of f.

(e) Find the domain and the range of f^{-1}.

(f) Graph f^{-1}.

Solution

(a) The domain of f consists of all x for which $x - 2 > 0$ or, equivalently, $x > 2$. The domain of f is $\{x \mid x > 2\}$ or $(2, \infty)$ in interval notation.

(b) To obtain the graph of $y = -\ln(x - 2)$, we begin with the graph of $y = \ln x$ and use transformations. See Figure 34.

Figure 34

(a) $y = \ln x$ Multiply by -1; reflect about x-axis (b) $y = -\ln x$ Replace x by $x - 2$; shift right 2 units. (c) $y = -\ln(x - 2)$

(c) The range of $f(x) = -\ln(x - 2)$ is the set of all real numbers. The vertical asymptote is $x = 2$. [Do you see why? The original asymptote ($x = 0$) is shifted to the right 2 units.]

(d) To find f^{-1}, begin with $y = -\ln(x - 2)$. The inverse function is defined (implicitly) by the equation

$$x = -\ln(y - 2)$$

Proceed to solve for y.

$$-x = \ln(y - 2) \quad \text{Isolate the logarithm.}$$
$$e^{-x} = y - 2 \quad \text{Change to an exponential statement.}$$
$$y = e^{-x} + 2 \quad \text{Solve for y.}$$

The inverse of f is $f^{-1}(x) = e^{-x} + 2$.

(e) The domain of f^{-1} equals the range of f, which is the set of all real numbers, from part (c). The range of f^{-1} is the domain of f, which is $(2, \infty)$ in interval notation.

(f) To graph f^{-1}, use the graph of f in Figure 34(c) and reflect it about the line $y = x$. See Figure 35. We could also graph $f^{-1}(x) = e^{-x} + 2$ using transformations.

Figure 35

Figure 36

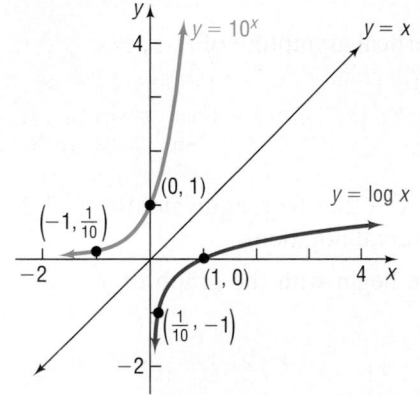

Now Work PROBLEM **71**

If the base of a logarithmic function is the number 10, then we have the **common logarithm function.** If the base a of the logarithmic function is not indicated, it is understood to be 10. That is,

$$y = \log x \quad \text{if and only if} \quad x = 10^y$$

Since $y = \log x$ and the exponential function $y = 10^x$ are inverse functions, we can obtain the graph of $y = \log x$ by reflecting the graph of $y = 10^x$ about the line $y = x$. See Figure 36.

EXAMPLE 7 **Graphing a Logarithmic Function and Its Inverse**

(a) Find the domain of the logarithmic function $f(x) = 3 \log (x - 1)$.
(b) Graph f.
(c) From the graph, determine the range and vertical asymptote of f.
(d) Find f^{-1}, the inverse of f.
(e) Find the domain and the range of f^{-1}.
(f) Graph f^{-1}.

Solution (a) The domain of f consists of all x for which $x - 1 > 0$ or, equivalently, $x > 1$. The domain of f is $\{x | x > 1\}$ or $(1, \infty)$ in interval notation.

(b) To obtain the graph of $y = 3 \log(x - 1)$, begin with the graph of $y = \log x$ and use transformations. See Figure 37.

Figure 37

Replace x by x − 1;
horizontal shift right
1 unit

Multiply by 3; vertical
stretch by a factor of 3.

(a) $y = \log x$ (b) $y = \log(x - 1)$ (c) $y = 3 \log(x - 1)$

(c) The range of $f(x) = 3 \log(x - 1)$ is the set of all real numbers. The vertical asymptote is $x = 1$.

(d) Begin with $y = 3 \log(x - 1)$. The inverse function is defined (implicitly) by the equation

$$x = 3 \log(y - 1)$$

Proceed to solve for y.

$$\frac{x}{3} = \log(y - 1) \quad \text{Isolate the logarithm.}$$

$$10^{x/3} = y - 1 \quad \text{Change to an exponential statement.}$$

$$y = 10^{x/3} + 1 \quad \text{Solve for y.}$$

The inverse of f is $f^{-1}(x) = 10^{x/3} + 1$.

(e) The domain of f^{-1} is the range of f, which is the set of all real numbers, from part (c). The range of f^{-1} is the domain of f, which is $(1, \infty)$ in interval notation.

(f) To graph f^{-1}, we use the graph of f in Figure 37(c) and reflect it about the line $y = x$. See Figure 38. We could also graph $f^{-1}(x) = 10^{x/3} + 1$ using transformations.

Figure 38

Now Work PROBLEM **79**

5 Solve Logarithmic Equations

Equations that contain logarithms are called **logarithmic equations.** Care must be taken when solving logarithmic equations algebraically. In the expression $\log_a M$, remember that a and M are positive and $a \neq 1$. Be sure to check each apparent solution in the original equation and discard any that are extraneous.

Some logarithmic equations can be solved by changing the logarithmic equation to exponential form using the fact that $y = \log_a x$ means $a^y = x$.

EXAMPLE 8

Solving Logarithmic Equations

Solve:

(a) $\log_3(4x - 7) = 2$ (b) $\log_x 64 = 2$

Solution

(a) We can obtain an exact solution by changing the logarithmic equation to exponential form.

$$\log_3(4x - 7) = 2$$
$$4x - 7 = 3^2 \quad \text{Change to exponential form using } y = \log_a x$$
$$\text{means } a^y = x.$$
$$4x - 7 = 9$$
$$4x = 16$$
$$x = 4$$

✓ **Check:** $\log_3(4x - 7) = \log_3(4 \cdot 4 - 7) = \log_3 9 = 2 \quad 3^2 = 9$

The solution set is $\{4\}$.

(b) We can obtain an exact solution by changing the logarithmic equation to exponential form.

$$\log_x 64 = 2$$
$$x^2 = 64 \quad \text{Change to exponential form.}$$
$$x = \pm\sqrt{64} = \pm 8 \quad \text{Square Root Method}$$

The base of a logarithm is always positive. As a result, we discard -8. We check the solution 8.

✓ **Check:** $\log_8 64 = 2 \quad 8^2 = 64$

The solution set is $\{8\}$.

EXAMPLE 9

Using Logarithms to Solve an Exponential Equation

Solve: $e^{2x} = 5$

Solution

We can obtain an exact solution by changing the exponential equation to logarithmic form.

$$e^{2x} = 5$$
$$\ln 5 = 2x \quad \text{Change to logarithmic form using the}$$
$$\text{fact that if } e^y = x \text{ then } y = \ln x.$$
$$x = \frac{\ln 5}{2} \quad \text{Exact solution}$$
$$\approx 0.805 \quad \text{Approximate solution}$$

The solution set is $\left\{\dfrac{\ln 5}{2}\right\}$.

Now Work PROBLEMS 87 AND 99

EXAMPLE 10 | **Alcohol and Driving**

Blood alcohol concentration (BAC) is a measure of the amount of alcohol in a person's bloodstream. A BAC of 0.04% means that a person has 4 parts alcohol per 10,000 parts blood in the body. Relative risk is defined as the likelihood of one event occurring divided by the likelihood of a second event occurring. For example, if an individual with a BAC of 0.02% is 1.4 times as likely to have a car accident as an individual that has not been drinking, the relative risk of an accident with a BAC of 0.02% is 1.4. Recent medical research suggests that the relative risk R of having an accident while driving a car can be modeled by an equation of the form

$$R = e^{kx}$$

where x is the percent of concentration of alcohol in the bloodstream and k is a constant.

(a) Research indicates that the relative risk of a person having an accident with a BAC of 0.02% is 1.4. Find the constant k in the equation.

(b) Using this value of k, what is the relative risk if the concentration is 0.17%?

(c) Using this same value of k, what BAC corresponds to a relative risk of 100?

(d) If the law asserts that anyone with a relative risk of 5 or more should not have driving privileges, at what concentration of alcohol in the bloodstream should a driver be arrested and charged with a DUI (driving under the influence)?

Solution

(a) For a concentration of alcohol in the blood of 0.02% and a relative risk of 1.4, we let $x = 0.02$ and $R = 1.4$ in the equation and solve for k.

$$R = e^{kx}$$
$$1.4 = e^{k(0.02)} \qquad R = 1.4; x = 0.02$$
$$0.02k = \ln 1.4 \qquad \text{Change to a logarithmic statement.}$$
$$k = \frac{\ln 1.4}{0.02} \approx 16.82 \quad \text{Solve for } k.$$

(b) For a concentration of 0.17%, we have $x = 0.17$. Using $k = 16.82$ in the equation, we find the relative risk R to be

$$R = e^{kx} = e^{(16.82)(0.17)} \approx 17.5$$

For a concentration of alcohol in the blood of 0.17%, the relative risk of an accident is about 17.5. That is, a person with a BAC of 0.17% is 17.5 times as likely to have a car accident as a person with no alcohol in the bloodstream.

(c) For a relative risk of 100, we have $R = 100$. Using $k = 16.82$ in the equation $R = e^{kx}$, we find the concentration x of alcohol in the blood obeys

$$100 = e^{16.82x} \qquad R = e^{kx}, R = 100; k = 16.82$$
$$16.82x = \ln 100 \qquad \text{Change to a logarithmic statement.}$$
$$x = \frac{\ln 100}{16.82} \approx 0.27 \quad \text{Solve for } x.$$

NOTE A BAC of 0.30% results in a loss of consciousness in most people. ∎

For a concentration of alcohol in the blood of 0.27%, the relative risk of an accident is 100.

(d) For a relative risk of 5, we have $R = 5$. Using $k = 16.82$ in the equation $R = e^{kx}$, we find the concentration x of alcohol in the bloodstream obeys

$$5 = e^{16.82x}$$
$$16.82x = \ln 5$$
$$x = \frac{\ln 5}{16.82} \approx 0.096$$

NOTE Most states use 0.08% or 0.10% as the blood alcohol content at which a DUI citation is given. ∎

A driver with a BAC of 0.096% or more should be arrested and charged with DUI.

SUMMARY Properties of the Logarithmic Function

$f(x) = \log_a x, \quad a > 1$

($y = \log_a x$ means $x = a^y$)

Domain: the interval $(0, \infty)$; Range: the interval $(-\infty, \infty)$

x-intercept: 1; y-intercept: none; vertical asymptote: $x = 0$ (y-axis); increasing; one-to-one

See Figure 39(a) for a typical graph.

$f(x) = \log_a x, \quad 0 < a < 1$

($y = \log_a x$ means $x = a^y$)

Domain: the interval $(0, \infty)$; Range: the interval $(-\infty, \infty)$

x-intercept: 1; y-intercept: none; vertical asymptote: $x = 0$ (y-axis); decreasing; one-to-one

See Figure 39(b) for a typical graph.

Figure 39

(a) $a > 1$ **(b)** $0 < a < 1$

6.4 Assess Your Understanding

'Are You Prepared?' *Answers are given at the end of these exercises. If you get a wrong answer, read the pages listed in red.*

1. Solve each inequality:
 (a) $3x - 7 \le 8 - 2x$ (pp. 119–126)
 (b) $x^2 - x - 6 > 0$ (pp. 309–311)

2. Solve the inequality: $\dfrac{x - 1}{x + 4} > 0$ (pp. 368–371)

3. Solve: $2x + 3 = 9$ (pp. 82–87)

Concepts and Vocabulary

4. The domain of the logarithmic function $f(x) = \log_a x$ is _____.

5. The graph of every logarithmic function $f(x) = \log_a x$, where $a > 0$ and $a \ne 1$, passes through three points: _____, _____, and _____.

6. If the graph of a logarithmic function $f(x) = \log_a x$, where $a > 0$ and $a \ne 1$, is increasing, then its base must be larger than _____.

7. ***True or False*** If $y = \log_a x$, then $y = a^x$.

8. ***True or False*** The graph of $f(x) = \log_a x$, where $a > 0$ and $a \ne 1$, has an x-intercept equal to 1 and no y-intercept.

Skill Building

In Problems 9–16, change each exponential statement to an equivalent statement involving a logarithm.

9. $9 = 3^2$

10. $16 = 4^2$

11. $a^2 = 1.6$

12. $a^3 = 2.1$

13. $2^x = 7.2$

14. $3^x = 4.6$

15. $e^x = 8$

16. $e^{2.2} = M$

In Problems 17–24, change each logarithmic statement to an equivalent statement involving an exponent.

17. $\log_2 8 = 3$

18. $\log_3\left(\dfrac{1}{9}\right) = -2$

19. $\log_a 3 = 6$

20. $\log_b 4 = 2$

21. $\log_3 2 = x$

22. $\log_2 6 = x$

23. $\ln 4 = x$

24. $\ln x = 4$

In Problems 25–36, find the exact value of each logarithm without using a calculator.

25. $\log_2 1$

26. $\log_8 8$

27. $\log_5 25$

28. $\log_3\left(\dfrac{1}{9}\right)$

29. $\log_{1/2} 16$

30. $\log_{1/3} 9$

31. $\log_{10}\sqrt{10}$

32. $\log_5 \sqrt[3]{25}$

33. $\log_{\sqrt{2}} 4$

34. $\log_{\sqrt{3}} 9$

35. $\ln\sqrt{e}$

36. $\ln e^3$

In Problems 37–48, find the domain of each function.

37. $f(x) = \ln(x - 3)$

38. $g(x) = \ln(x - 1)$

39. $F(x) = \log_2 x^2$

40. $H(x) = \log_5 x^3$

41. $f(x) = 3 - 2\log_4\left[\dfrac{x}{2} - 5\right]$

42. $g(x) = 8 + 5\ln(2x + 3)$

43. $f(x) = \ln\left(\dfrac{1}{x + 1}\right)$

44. $g(x) = \ln\left(\dfrac{1}{x - 5}\right)$

45. $g(x) = \log_5\left(\dfrac{x + 1}{x}\right)$

46. $h(x) = \log_3\left(\dfrac{x}{x - 1}\right)$

47. $f(x) = \sqrt{\ln x}$

48. $g(x) = \dfrac{1}{\ln x}$

In Problems 49–56, use a calculator to evaluate each expression. Round your answer to three decimal places.

49. $\ln\dfrac{5}{3}$

50. $\dfrac{\ln 5}{3}$

51. $\dfrac{\ln\dfrac{10}{3}}{0.04}$

52. $\dfrac{\ln\dfrac{2}{3}}{-0.1}$

53. $\dfrac{\ln 4 + \ln 2}{\log 4 + \log 2}$

54. $\dfrac{\log 15 + \log 20}{\ln 15 + \ln 20}$

55. $\dfrac{2\ln 5 + \log 50}{\log 4 - \ln 2}$

56. $\dfrac{3\log 80 - \ln 5}{\log 5 + \ln 20}$

57. Find a so that the graph of $f(x) = \log_a x$ contains the point $(2, 2)$.

58. Find a so that the graph of $f(x) = \log_a x$ contains the point $\left(\dfrac{1}{2}, -4\right)$.

In Problems 59–62, graph each function and its inverse on the same Cartesian plane.

59. $f(x) = 3^x;\ f^{-1}(x) = \log_3 x$

60. $f(x) = 4^x;\ f^{-1}(x) = \log_4 x$

61. $f(x) = \left(\dfrac{1}{2}\right)^x;\ f^{-1}(x) = \log_{\frac{1}{2}} x$

62. $f(x) = \left(\dfrac{1}{3}\right)^x;\ f^{-1}(x) = \log_{\frac{1}{3}} x$

In Problems 63–70, the graph of a logarithmic function is given. Match each graph to one of the following functions:

(a) $y = \log_3 x$

(b) $y = \log_3(-x)$

(c) $y = -\log_3 x$

(d) $y = -\log_3(-x)$

(e) $y = \log_3 x - 1$

(f) $y = \log_3(x - 1)$

(g) $y = \log_3(1 - x)$

(h) $y = 1 - \log_3 x$

63.

64.

65.

66.

67.

68.

69.

70.

In Problems 71–86, use the given function f to:

(a) Find the domain of f. (b) Graph f. (c) From the graph, determine the range and any asymptotes of f.
(d) Find f^{-1}, the inverse of f. (e) Find the domain and the range of f^{-1}. (f) Graph f^{-1}.

71. $f(x) = \ln(x + 4)$

72. $f(x) = \ln(x - 3)$

73. $f(x) = 2 + \ln x$

74. $f(x) = -\ln(-x)$

75. $f(x) = \ln(2x) - 3$ **76.** $f(x) = -2\ln(x + 1)$ **77.** $f(x) = \log(x - 4) + 2$ **78.** $f(x) = \frac{1}{2}\log x - 5$

79. $f(x) = \frac{1}{2}\log(2x)$ **80.** $f(x) = \log(-2x)$ **81.** $f(x) = 3 + \log_3(x + 2)$ **82.** $f(x) = 2 - \log_3(x + 1)$

83. $f(x) = e^{x+2} - 3$ **84.** $f(x) = 3e^x + 2$ **85.** $f(x) = 2^{x/3} + 4$ **86.** $f(x) = -3^{x+1}$

In Problems 87–110, solve each equation.

87. $\log_3 x = 2$ **88.** $\log_5 x = 3$ **89.** $\log_2(2x + 1) = 3$ **90.** $\log_3(3x - 2) = 2$

91. $\log_x 4 = 2$ **92.** $\log_x\left(\frac{1}{8}\right) = 3$ **93.** $\ln e^x = 5$ **94.** $\ln e^{-2x} = 8$

95. $\log_4 64 = x$ **96.** $\log_5 625 = x$ **97.** $\log_3 243 = 2x + 1$ **98.** $\log_6 36 = 5x + 3$

99. $e^{3x} = 10$ **100.** $e^{-2x} = \frac{1}{3}$ **101.** $e^{2x+5} = 8$ **102.** $e^{-2x+1} = 13$

103. $\log_3(x^2 + 1) = 2$ **104.** $\log_5(x^2 + x + 4) = 2$ **105.** $\log_2 8^x = -3$ **106.** $\log_3 3^x = -1$

107. $5e^{0.2x} = 7$ **108.** $8 \cdot 10^{2x-7} = 3$ **109.** $2 \cdot 10^{2-x} = 5$ **110.** $4e^{x+1} = 5$

Mixed Practice

111. Suppose that $G(x) = \log_3(2x + 1) - 2$.
 (a) What is the domain of G?
 (b) What is $G(40)$? What point is on the graph of G?
 (c) If $G(x) = 3$, what is x? What point is on the graph of G?
 (d) What is the zero of G?

112. Suppose that $F(x) = \log_2(x + 1) - 3$.
 (a) What is the domain of F?
 (b) What is $F(7)$? What point is on the graph of F?
 (c) If $F(x) = -1$, what is x? What point is on the graph of F?
 (d) What is the zero of F?

In Problems 113–116, graph each function. Based on the graph, state the domain and the range and find any intercepts.

113. $f(x) = \begin{cases} \ln(-x) & \text{if } x < 0 \\ \ln x & \text{if } x > 0 \end{cases}$ **114.** $f(x) = \begin{cases} \ln(-x) & \text{if } x \leq -1 \\ -\ln(-x) & \text{if } -1 < x < 0 \end{cases}$

115. $f(x) = \begin{cases} -\ln x & \text{if } 0 < x < 1 \\ \ln x & \text{if } x \geq 1 \end{cases}$ **116.** $f(x) = \begin{cases} \ln x & \text{if } 0 < x < 1 \\ -\ln x & \text{if } x \geq 1 \end{cases}$

Applications and Extensions

117. Chemistry The pH of a chemical solution is given by the formula

$$\text{pH} = -\log_{10}[H^+]$$

where $[H^+]$ is the concentration of hydrogen ions in moles per liter. Values of pH range from 0 (acidic) to 14 (alkaline).
 (a) What is the pH of a solution for which $[H^+]$ is 0.1?
 (b) What is the pH of a solution for which $[H^+]$ is 0.01?
 (c) What is the pH of a solution for which $[H^+]$ is 0.001?
 (d) What happens to pH as the hydrogen ion concentration decreases?
 (e) Determine the hydrogen ion concentration of an orange (pH = 3.5).
 (f) Determine the hydrogen ion concentration of human blood (pH = 7.4).

118. Diversity Index **Shannon's diversity index** is a measure of the diversity of a population. The diversity index is given by the formula

$$H = -(p_1 \log p_1 + p_2 \log p_2 + \cdots + p_n \log p_n)$$

where p_1 is the proportion of the population that is species 1, p_2 is the proportion of the population that is species 2, and so on.
 (a) According to the U.S. Census Bureau, the distribution of race in the United States in 2007 was as follows:

Race	Proportion
American Indian or Native Alaskan	0.015
Asian	0.042
Black or African American	0.129
Hispanic	0.125
Native Hawaiian or Pacific Islander	0.003
White	0.686

Source: U.S. Census Bureau

Compute the diversity index of the United States in 2007.

(b) The largest value of the diversity index is given by $H_{max} = \log(S)$, where S is the number of categories of race. Compute H_{max}.

(c) The **evenness ratio** is given by $E_H = \dfrac{H}{H_{max}}$, where $0 \le E_H \le 1$. If $E_H = 1$, there is complete evenness. Compute the evenness ratio for the United States.

(d) Obtain the distribution of race for the United States in 2010 from the Census Bureau. Compute Shannon's diversity index. Is the United States becoming more diverse? Why?

119. **Atmospheric Pressure** The atmospheric pressure p on an object decreases with increasing height. This pressure, measured in millimeters of mercury, is related to the height h (in kilometers) above sea level by the function

$$p(h) = 760e^{-0.145h}$$

(a) Find the height of an aircraft if the atmospheric pressure is 320 millimeters of mercury.

(b) Find the height of a mountain if the atmospheric pressure is 667 millimeters of mercury.

120. **Healing of Wounds** The normal healing of wounds can be modeled by an exponential function. If A_0 represents the original area of the wound and if A equals the area of the wound, then the function

$$A(n) = A_0 e^{-0.35n}$$

describes the area of a wound after n days following an injury when no infection is present to retard the healing. Suppose that a wound initially had an area of 100 square millimeters.

(a) If healing is taking place, after how many days will the wound be one-half its original size?

(b) How long before the wound is 10% of its original size?

121. **Exponential Probability** Between 12:00 PM and 1:00 PM, cars arrive at Citibank's drive-thru at the rate of 6 cars per hour (0.1 car per minute). The following formula from statistics can be used to determine the probability that a car will arrive within t minutes of 12:00 PM.

$$F(t) = 1 - e^{-0.1t}$$

(a) Determine how many minutes are needed for the probability to reach 50%.

(b) Determine how many minutes are needed for the probability to reach 80%.

(c) Is it possible for the probability to equal 100%? Explain.

122. **Exponential Probability** Between 5:00 PM and 6:00 PM, cars arrive at Jiffy Lube at the rate of 9 cars per hour (0.15 car per minute). The following formula from statistics can be used to

determine the probability that a car will arrive within t minutes of 5:00 PM.

$$F(t) = 1 - e^{-0.15t}$$

(a) Determine how many minutes are needed for the probability to reach 50%.

(b) Determine how many minutes are needed for the probability to reach 80%.

123. **Drug Medication** The formula

$$D = 5e^{-0.4h}$$

can be used to find the number of milligrams D of a certain drug that is in a patient's bloodstream h hours after the drug was administered. When the number of milligrams reaches 2, the drug is to be administered again. What is the time between injections?

124. **Spreading of Rumors** A model for the number N of people in a college community who have heard a certain rumor is

$$N = P(1 - e^{-0.15d})$$

where P is the total population of the community and d is the number of days that have elapsed since the rumor began. In a community of 1000 students, how many days will elapse before 450 students have heard the rumor?

125. **Current in a *RL* Circuit** The equation governing the amount of current I (in amperes) after time t (in seconds) in a simple *RL* circuit consisting of a resistance R (in ohms), an inductance L (in henrys), and an electromotive force E (in volts) is

$$I = \frac{E}{R}\left[1 - e^{-(R/L)t}\right]$$

If $E = 12$ volts, $R = 10$ ohms, and $L = 5$ henrys, how long does it take to obtain a current of 0.5 ampere? Of 1.0 ampere? Graph the equation.

126. **Learning Curve** Psychologists sometimes use the function

$$L(t) = A(1 - e^{-kt})$$

to measure the amount L learned at time t. The number A represents the amount to be learned, and the number k measures the rate of learning. Suppose that a student has an amount A of 200 vocabulary words to learn. A psychologist determines that the student learned 20 vocabulary words after 5 minutes.

(a) Determine the rate of learning k.

(b) Approximately how many words will the student have learned after 10 minutes?

(c) After 15 minutes?

(d) How long does it take for the student to learn 180 words?

Loudness of Sound Problems 127–130 use the following discussion: The **loudness** $L(x)$, measured in decibels (dB), of a sound of intensity x, measured in watts per square meter, is defined as $L(x) = 10 \log \dfrac{x}{I_0}$, where $I_0 = 10^{-12}$ watt per square meter is the least intense sound that a human ear can detect. Determine the loudness, in decibels, of each of the following sounds.

127. Normal conversation: intensity of $x = 10^{-7}$ watt per square meter.

128. Amplified rock music: intensity of 10^{-1} watt per square meter.

129. Heavy city traffic: intensity of $x = 10^{-3}$ watt per square meter.

130. Diesel truck traveling 40 miles per hour 50 feet away: intensity 10 times that of a passenger car traveling 50 miles per hour 50 feet away whose loudness is 70 decibels.

The Richter Scale Problems 131 and 132 use the following discussion: The **Richter scale** is one way of converting seismographic readings into numbers that provide an easy reference for measuring the magnitude M of an earthquake. All earthquakes are compared to a **zero-level earthquake** whose seismographic reading measures 0.001 millimeter at a distance of 100 kilometers from the epicenter. An earthquake whose seismographic reading measures x millimeters has **magnitude** M(x), given by

$$M(x) = \log\left(\frac{x}{x_0}\right)$$

where $x_0 = 10^{-3}$ is the reading of a zero-level earthquake the same distance from its epicenter. In Problems 131 and 132, determine the magnitude of each earthquake.

131. Magnitude of an Earthquake Mexico City in 1985: seismographic reading of 125,892 millimeters 100 kilometers from the center

132. Magnitude of an Earthquake San Francisco in 1906: seismographic reading of 50,119 millimeters 100 kilometers from the center

133. Alcohol and Driving The concentration of alcohol in a person's bloodstream is measurable. Suppose that the relative risk R of having an accident while driving a car can be modeled by an equation of the form

$$R = e^{kx}$$

where x is the percent of concentration of alcohol in the bloodstream and k is a constant.

(a) Suppose that a concentration of alcohol in the bloodstream of 0.03 percent results in a relative risk of an accident of 1.4. Find the constant k in the equation.

(b) Using this value of k, what is the relative risk if the concentration is 0.17 percent?

(c) Using the same value of k, what concentration of alcohol corresponds to a relative risk of 100?

(d) If the law asserts that anyone with a relative risk of having an accident of 5 or more should not have driving privileges, at what concentration of alcohol in the bloodstream should a driver be arrested and charged with a DUI?

(e) Compare this situation with that of Example 10. If you were a lawmaker, which situation would you support? Give your reasons.

Explaining Concepts: Discussion and Writing

134. Is there any function of the form $y = x^\alpha, 0 < \alpha < 1$, that increases more slowly than a logarithmic function whose base is greater than 1? Explain.

135. In the definition of the logarithmic function, the base a is not allowed to equal 1. Why?

136. Critical Thinking In buying a new car, one consideration might be how well the price of the car holds up over time. Different makes of cars have different depreciation rates. One way to compute a depreciation rate for a car is given here. Suppose that the current prices of a certain automobile are as shown in the table.

	Age in Years				
New	**1**	**2**	**3**	**4**	**5**
$38,000	$36,600	$32,400	$28,750	$25,400	$21,200

Use the formula New = Old(e^{Rt}) to find R, the annual depreciation rate, for a specific time t. When might be the best time to trade in the car? Consult the NADA ("blue") book and compare two like models that you are interested in. Which has the better depreciation rate?

'Are You Prepared?' Answers

1. (a) $x \leq 3$ (b) $x < -2$ or $x > 3$ **2.** $x < -4$ or $x > 1$ **3.** {3}

6.5 Properties of Logarithms

OBJECTIVES **1** Work with the Properties of Logarithms (p. 450)

2 Write a Logarithmic Expression as a Sum or Difference of Logarithms (p. 452)

3 Write a Logarithmic Expression as a Single Logarithm (p. 453)

4 Evaluate Logarithms Whose Base Is Neither 10 Nor e (p. 455)

1 Work with the Properties of Logarithms

Logarithms have some very useful properties that can be derived directly from the definition and the laws of exponents.

| EXAMPLE 1 | **Establishing Properties of Logarithms** |

(a) Show that $\log_a 1 = 0$. (b) Show that $\log_a a = 1$.

Solution (a) This fact was established when we graphed $y = \log_a x$ (see Figure 30 on page 440). To show the result algebraically, let $y = \log_a 1$. Then

$$y = \log_a 1$$
$$a^y = 1 \qquad \text{\textit{Change to an exponential statement.}}$$
$$a^y = a^0 \qquad \text{\textit{$a^0 = 1$ since $a > 0, a \neq 1$}}$$
$$y = 0 \qquad \text{\textit{Solve for y.}}$$
$$\log_a 1 = 0 \qquad \text{\textit{$y = \log_a 1$}}$$

(b) Let $y = \log_a a$. Then

$$y = \log_a a$$
$$a^y = a \qquad \text{\textit{Change to an exponential statement.}}$$
$$a^y = a^1 \qquad \text{\textit{$a = a^1$}}$$
$$y = 1 \qquad \text{\textit{Solve for y.}}$$
$$\log_a a = 1 \qquad \text{\textit{$y = \log_a a$}}$$

To summarize:

$$\log_a 1 = 0 \qquad \log_a a = 1$$

THEOREM **Properties of Logarithms**

In the properties given next, M and a are positive real numbers, $a \neq 1$, and r is any real number.

The number $\log_a M$ is the exponent to which a must be raised to obtain M. That is,

$$a^{\log_a M} = M \tag{1}$$

The logarithm to the base a of a raised to a power equals that power. That is,

$$\log_a a^r = r \tag{2}$$

The proof uses the fact that $y = a^x$ and $y = \log_a x$ are inverses.

Proof of Property (1) For inverse functions,

$$f(f^{-1}(x)) = x \quad \text{for all x in the domain of f^{-1}}$$

Using $f(x) = a^x$ and $f^{-1}(x) = \log_a x$, we find

$$f(f^{-1}(x)) = a^{\log_a x} = x \quad \text{for $x > 0$}$$

Now let $x = M$ to obtain $a^{\log_a M} = M$, where $M > 0$. ■

Proof of Property (2) For inverse functions,

$$f^{-1}(f(x)) = x \quad \text{for all x in the domain of f}$$

Using $f(x) = a^x$ and $f^{-1}(x) = \log_a x$, we find

$$f^{-1}(f(x)) = \log_a a^x = x \quad \text{for all real numbers x}$$

Now let $x = r$ to obtain $\log_a a^r = r$, where r is any real number. ■

EXAMPLE 2 **Using Properties (1) and (2)**

(a) $2^{\log_2 \pi} = \pi$ (b) $\log_{0.2} 0.2^{-\sqrt{2}} = -\sqrt{2}$ (c) $\ln e^{kt} = kt$

➤ **Now Work** PROBLEM 15

Other useful properties of logarithms are given next.

THEOREM

Properties of Logarithms

In the following properties, M, N, and a are positive real numbers, $a \neq 1$, and r is any real number.

The Log of a Product Equals the Sum of the Logs

$$\log_a(MN) = \log_a M + \log_a N \tag{3}$$

The Log of a Quotient Equals the Difference of the Logs

$$\log_a\left(\frac{M}{N}\right) = \log_a M - \log_a N \tag{4}$$

The Log of a Power Equals the Product of the Power and the Log

$$\log_a M^r = r \log_a M \tag{5}$$

$$a^x = e^{x \ln a} \tag{6}$$

We shall derive properties (3), (5), and (6) and leave the derivation of property (4) as an exercise (see Problem 109).

Proof of Property (3) Let $A = \log_a M$ and let $B = \log_a N$. These expressions are equivalent to the exponential expressions

$$a^A = M \quad \text{and} \quad a^B = N$$

Now

$$\begin{aligned}
\log_a(MN) = \log_a(a^A a^B) &= \log_a a^{A+B} && \text{Law of Exponents} \\
&= A + B && \text{Property (2) of logarithms} \\
&= \log_a M + \log_a N
\end{aligned}$$ ∎

Proof of Property (5) Let $A = \log_a M$. This expression is equivalent to

$$a^A = M$$

Now

$$\begin{aligned}
\log_a M^r = \log_a(a^A)^r &= \log_a a^{rA} && \text{Law of Exponents} \\
&= rA && \text{Property (2) of logarithms} \\
&= r \log_a M
\end{aligned}$$ ∎

Proof of Property (6) From property (1), with $a = e$, we have

$$e^{\ln M} = M$$

Now let $M = a^x$ and apply property (5).

$$e^{\ln a^x} = e^{x \ln a} = a^x$$ ∎

➤ **Now Work** PROBLEM 19

2 Write a Logarithmic Expression as a Sum or Difference of Logarithms

Logarithms can be used to transform products into sums, quotients into differences, and powers into factors. Such transformations prove useful in certain types of calculus problems.

EXAMPLE 3 **Writing a Logarithmic Expression as a Sum of Logarithms**

Write $\log_a\left(x\sqrt{x^2+1}\right)$, $x > 0$, as a sum of logarithms. Express all powers as factors.

Solution $\log_a\left(x\sqrt{x^2+1}\right) = \log_a x + \log_a\sqrt{x^2+1}$ $\log_a(M \cdot N) = \log_a M + \log_a N$

$$= \log_a x + \log_a(x^2+1)^{1/2}$$

$$= \log_a x + \frac{1}{2}\log_a(x^2+1) \quad \log_a M^r = r\log_a M$$

EXAMPLE 4 **Writing a Logarithmic Expression as a Difference of Logarithms**

Write

$$\ln\frac{x^2}{(x-1)^3} \qquad x > 1$$

as a difference of logarithms. Express all powers as factors.

Solution $$\ln\frac{x^2}{(x-1)^3} = \ln x^2 - \ln(x-1)^3 = 2\ln x - 3\ln(x-1)$$

$$\underset{\log_a\left(\frac{M}{N}\right) = \log_a M - \log_a N}{\uparrow} \qquad \underset{\log_a M^r = r\log_a M}{\uparrow}$$

EXAMPLE 5 **Writing a Logarithmic Expression as a Sum and Difference of Logarithms**

Write

$$\log_a\frac{\sqrt{x^2+1}}{x^3(x+1)^4} \qquad x > 0$$

as a sum and difference of logarithms. Express all powers as factors.

Solution $\log_a\frac{\sqrt{x^2+1}}{x^3(x+1)^4} = \log_a\sqrt{x^2+1} - \log_a[x^3(x+1)^4]$ Property (4)

WARNING In using properties (3) through (5), be careful about the values that the variable may assume. For example, the domain of the variable for $\log_a x$ is $x > 0$ and for $\log_a(x-1)$ it is $x > 1$. If we add these functions, the domain is $x > 1$. That is, the equality

$\log_a x + \log_a(x-1) = \log_a[x(x-1)]$

is true only for $x > 1$. ■

$$= \log_a\sqrt{x^2+1} - [\log_a x^3 + \log_a(x+1)^4] \quad \text{Property (3)}$$

$$= \log_a(x^2+1)^{1/2} - \log_a x^3 - \log_a(x+1)^4$$

$$= \frac{1}{2}\log_a(x^2+1) - 3\log_a x - 4\log_a(x+1) \quad \text{Property (5)}$$

Now Work PROBLEM 51

3 Write a Logarithmic Expression as a Single Logarithm

Another use of properties (3) through (5) is to write sums and/or differences of logarithms with the same base as a single logarithm. This skill will be needed to solve certain logarithmic equations discussed in the next section.

EXAMPLE 6 **Writing Expressions as a Single Logarithm**

Write each of the following as a single logarithm.

(a) $\log_a 7 + 4\log_a 3$ (b) $\frac{2}{3}\ln 8 - \ln(5^2 - 1)$

(c) $\log_a x + \log_a 9 + \log_a(x^2+1) - \log_a 5$

Solution (a) $\log_a 7 + 4\log_a 3 = \log_a 7 + \log_a 3^4 \quad {\scriptstyle r\log_a M = \log_a M^r}$

$$= \log_a 7 + \log_a 81$$

$$= \log_a(7 \cdot 81) \qquad {\scriptstyle \log_a M + \log_a N = \log_a(M\cdot N)}$$

$$= \log_a 567$$

(b) $\dfrac{2}{3}\ln 8 - \ln(5^2 - 1) = \ln 8^{2/3} - \ln(25 - 1) \quad {\scriptstyle r\log_a M = \log_a M^r}$

$$= \ln 4 - \ln 24 \qquad {\scriptstyle 8^{2/3} = (\sqrt[3]{8})^2 = 2^2 = 4}$$

$$= \ln\left(\frac{4}{24}\right) \qquad {\scriptstyle \log_a M - \log_a N = \log_a\left(\frac{M}{N}\right)}$$

$$= \ln\left(\frac{1}{6}\right)$$

$$= \ln 1 - \ln 6$$

$$= -\ln 6 \qquad {\scriptstyle \ln 1 = 0}$$

(c) $\log_a x + \log_a 9 + \log_a(x^2 + 1) - \log_a 5 = \log_a(9x) + \log_a(x^2 + 1) - \log_a 5$

$$= \log_a[9x(x^2 + 1)] - \log_a 5$$

$$= \log_a\left[\frac{9x(x^2 + 1)}{5}\right]$$

WARNING A common error made by some students is to express the logarithm of a sum as the sum of logarithms.

$$\log_a(M + N) \quad \text{is not equal to} \quad \log_a M + \log_a N$$

Correct statement $\log_a(MN) = \log_a M + \log_a N$ Property (3)

Another common error is to express the difference of logarithms as the quotient of logarithms.

$$\log_a M - \log_a N \quad \text{is not equal to} \quad \frac{\log_a M}{\log_a N}$$

Correct statement $\log_a M - \log_a N = \log_a\left(\frac{M}{N}\right)$ Property (4)

A third common error is to express a logarithm raised to a power as the product of the power times the logarithm.

$$(\log_a M)^r \quad \text{is not equal to} \quad r\log_a M$$

Correct statement $\log_a M^r = r\log_a M$ Property (5) ∎

═══ **Now Work** PROBLEM **57**

Two other properties of logarithms that we need to know are consequences of the fact that the logarithmic function $y = \log_a x$ is a one-to-one function.

THEOREM **Properties of Logarithms**

In the following properties, M, N, and a are positive real numbers, $a \neq 1$.

> If $M = N$, then $\log_a M = \log_a N$. (7)
>
> If $\log_a M = \log_a N$, then $M = N$. (8)

When property (7) is used, we start with the equation $M = N$ and say "take the logarithm of both sides" to obtain $\log_a M = \log_a N$.

Properties (7) and (8) are useful for solving *exponential and logarithmic equations*, a topic discussed in the next section.

4 Evaluate Logarithms Whose Base Is Neither 10 Nor e

Logarithms to the base 10, common logarithms, were used to facilitate arithmetic computations before the widespread use of calculators. (See the Historical Feature at the end of this section.) Natural logarithms, that is, logarithms whose base is the number e, remain very important because they arise frequently in the study of natural phenomena.

Common logarithms are usually abbreviated by writing **log,** with the base understood to be 10, just as natural logarithms are abbreviated by **ln,** with the base understood to be e.

Most calculators have both $\boxed{\log}$ and $\boxed{\ln}$ keys to calculate the common logarithm and natural logarithm of a number. Let's look at an example to see how to approximate logarithms having a base other than 10 or e.

EXAMPLE 7 **Approximating a Logarithm Whose Base Is Neither 10 Nor e**

Approximate $\log_2 7$. Round the answer to four decimal places.

Solution Remember, $\log_2 7$ means "2 raised to what exponent equals 7." If we let $y = \log_2 7$, then $2^y = 7$. Because $2^2 = 4$ and $2^3 = 8$, we expect $\log_2 7$ to be between 2 and 3.

$$2^y = 7$$
$$\ln 2^y = \ln 7 \qquad \text{Property (7)}$$
$$y \ln 2 = \ln 7 \qquad \text{Property (5)}$$
$$y = \frac{\ln 7}{\ln 2} \qquad \text{Exact value}$$
$$y \approx 2.8074 \qquad \text{Approximate value rounded to four decimal places}$$

Example 7 shows how to approximate a logarithm whose base is 2 by changing to logarithms involving the base e. In general, we use the **Change-of-Base Formula.**

THEOREM **Change-of-Base Formula**

If $a \neq 1$, $b \neq 1$, and M are positive real numbers, then

$$\log_a M = \frac{\log_b M}{\log_b a} \qquad \qquad (9)$$

Proof We derive this formula as follows: Let $y = \log_a M$. Then

$$a^y = M$$
$$\log_b a^y = \log_b M \qquad \text{Property (7)}$$
$$y \log_b a = \log_b M \qquad \text{Property (5)}$$
$$y = \frac{\log_b M}{\log_b a} \qquad \text{Solve for } y.$$
$$\log_a M = \frac{\log_b M}{\log_b a} \qquad y = \log_a M \qquad \blacksquare$$

Since calculators have keys only for $\boxed{\log}$ and $\boxed{\ln}$, in practice, the Change-of-Base Formula uses either $b = 10$ or $b = e$. That is,

$$\log_a M = \frac{\log M}{\log a} \quad \text{and} \quad \log_a M = \frac{\ln M}{\ln a} \qquad (10)$$

> [**EXAMPLE 8**] **Using the Change-of-Base Formula**

Approximate:

(a) $\log_5 89$ \hspace{3cm} (b) $\log_{\sqrt{2}} \sqrt{5}$

Round answers to four decimal places.

Solution (a) $\log_5 89 = \dfrac{\log 89}{\log 5} \approx \dfrac{1.949390007}{0.6989700043}$
≈ 2.7889

or

$\log_5 89 = \dfrac{\ln 89}{\ln 5} \approx \dfrac{4.48863637}{1.609437912}$
≈ 2.7889

(b) $\log_{\sqrt{2}} \sqrt{5} = \dfrac{\log \sqrt{5}}{\log \sqrt{2}} = \dfrac{\frac{1}{2}\log 5}{\frac{1}{2}\log 2}$

$= \dfrac{\log 5}{\log 2} \approx 2.3219$

or

$\log_{\sqrt{2}} \sqrt{5} = \dfrac{\ln \sqrt{5}}{\ln \sqrt{2}} = \dfrac{\frac{1}{2}\ln 5}{\frac{1}{2}\ln 2}$

$= \dfrac{\ln 5}{\ln 2} \approx 2.3219$

➡ **Now Work** PROBLEMS **23** AND **71**

COMMENT To graph logarithmic functions when the base is different from *e* or 10 requires the Change-of-Base Formula. For example, to graph $y = \log_2 x$, we would instead graph $y = \dfrac{\ln x}{\ln 2}$. Try it. ∎

➡ **Now Work** PROBLEM **79**

SUMMARY Properties of Logarithms

In the list that follows, a, b, M, N, and r are real numbers. Also, $a > 0$, $a \neq 1$, $b > 0$, $b \neq 1$, $M > 0$, and $N > 0$.

Definition $y = \log_a x$ means $x = a^y$

Properties of logarithms $\log_a 1 = 0; \quad \log_a a = 1$ \hspace{2cm} $\log_a M^r = r \log_a M$

$a^{\log_a M} = M; \quad \log_a a^r = r$ \hspace{2cm} $a^x = e^{x \ln a}$

$\log_a(MN) = \log_a M + \log_a N$ \hspace{1.5cm} If $M = N$, then $\log_a M = \log_a N$.

$\log_a\left(\dfrac{M}{N}\right) = \log_a M - \log_a N$ \hspace{1cm} If $\log_a M = \log_a N$, then $M = N$.

Change-of-Base Formula $\log_a M = \dfrac{\log_b M}{\log_b a}$

Historical Feature

John Napier (1550–1617)

Logarithms were invented about 1590 by John Napier (1550–1617) and Joost Bürgi (1552–1632), working independently. Napier, whose work had the greater influence, was a Scottish lord, a secretive man whose neighbors were inclined to believe him to be in league with the devil. His approach to logarithms was very different from ours; it was based on the relationship between arithmetic and geometric sequences, discussed in a later chapter, and not on the inverse function relationship of logarithms to exponential functions (described in Section 6.4).

Napier's tables, published in 1614, listed what would now be called *natural logarithms* of sines and were rather difficult to use. A London professor, Henry Briggs, became interested in the tables and visited Napier. In their conversations, they developed the idea of common logarithms, which were published in 1617. Their importance for calculation was immediately recognized, and by 1650 they were being printed as far away as China. They remained an important calculation tool until the advent of the inexpensive handheld calculator about 1972, which has decreased their calculational, but not their theoretical, importance.

A side effect of the invention of logarithms was the popularization of the decimal system of notation for real numbers.

6.5 Assess Your Understanding

Concepts and Vocabulary

1. $\log_a 1 =$ _____

2. $\log_a a =$ _____

3. $a^{\log_a M} =$ _____

4. $\log_a a^r =$ _____

5. $\log_a(MN) =$ _____ + _____

6. $\log_a\left(\dfrac{M}{N}\right) =$ _____ − _____

7. $\log_a M^r =$ _____

8. If $\log_a x = \log_a 6$, then $x =$ _____.

9. If $\log_8 M = \dfrac{\log_5 7}{\log_5 8}$, then $M =$ _____.

10. *True or False* $\ln(x + 3) - \ln(2x) = \dfrac{\ln(x + 3)}{\ln(2x)}$

11. *True or False* $\log_2(3x^4) = 4\log_2(3x)$

12. *True or False* $\dfrac{\ln 8}{\ln 4} = 2$

Skill Building

In Problems 13–28, use properties of logarithms to find the exact value of each expression. Do not use a calculator.

13. $\log_3 3^{71}$

14. $\log_2 2^{-13}$

15. $\ln e^{-4}$

16. $\ln e^{\sqrt{2}}$

17. $2^{\log_2 7}$

18. $e^{\ln 8}$

19. $\log_8 2 + \log_8 4$

20. $\log_6 9 + \log_6 4$

21. $\log_6 18 - \log_6 3$

22. $\log_8 16 - \log_8 2$

23. $\log_2 6 \cdot \log_6 8$

24. $\log_3 8 \cdot \log_8 9$

25. $3^{\log_3 5 - \log_3 4}$

26. $5^{\log_5 6 + \log_5 7}$

27. $e^{\log_{e^2} 16}$

28. $e^{\log_{e^2} 9}$

In Problems 29–36, suppose that $\ln 2 = a$ and $\ln 3 = b$. Use properties of logarithms to write each logarithm in terms of a and b.

29. $\ln 6$

30. $\ln\dfrac{2}{3}$

31. $\ln 1.5$

32. $\ln 0.5$

33. $\ln 8$

34. $\ln 27$

35. $\ln \sqrt[5]{6}$

36. $\ln \sqrt[4]{\dfrac{2}{3}}$

In Problems 37–56, write each expression as a sum and/or difference of logarithms. Express powers as factors.

37. $\log_5(25x)$

38. $\log_3 \dfrac{x}{9}$

39. $\log_2 z^3$

40. $\log_7 x^5$

41. $\ln(ex)$

42. $\ln \dfrac{e}{x}$

43. $\ln \dfrac{x}{e^x}$

44. $\ln(xe^x)$

45. $\log_a(u^2 v^3)$ $\quad u > 0, v > 0$

46. $\log_2\left(\dfrac{a}{b^2}\right)$ $\quad a > 0, b > 0$

47. $\ln\left(x^2\sqrt{1 - x}\right)$ $\quad 0 < x < 1$

48. $\ln\left(x\sqrt{1 + x^2}\right)$ $\quad x > 0$

49. $\log_2\left(\dfrac{x^3}{x - 3}\right)$ $\quad x > 3$

50. $\log_5\left(\dfrac{\sqrt[3]{x^2 + 1}}{x^2 - 1}\right)$ $\quad x > 1$

51. $\log\left[\dfrac{x(x + 2)}{(x + 3)^2}\right]$ $\quad x > 0$

52. $\log\left[\dfrac{x^3\sqrt{x + 1}}{(x - 2)^2}\right]$ $\quad x > 2$

53. $\ln\left[\dfrac{x^2 - x - 2}{(x + 4)^2}\right]^{1/3}$ $\quad x > 2$

54. $\ln\left[\dfrac{(x - 4)^2}{x^2 - 1}\right]^{2/3}$ $\quad x > 4$

55. $\ln \dfrac{5x\sqrt{1 + 3x}}{(x - 4)^3}$ $\quad x > 4$

56. $\ln\left[\dfrac{5x^2\sqrt[3]{1 - x}}{4(x + 1)^2}\right]$ $\quad 0 < x < 1$

In Problems 57–70, write each expression as a single logarithm.

57. $3\log_5 u + 4\log_5 v$

58. $2\log_3 u - \log_3 v$

59. $\log_3 \sqrt{x} - \log_3 x^3$

60. $\log_2\left(\dfrac{1}{x}\right) + \log_2\left(\dfrac{1}{x^2}\right)$

61. $\log_4(x^2 - 1) - 5\log_4(x + 1)$

62. $\log(x^2 + 3x + 2) - 2\log(x + 1)$

63. $\ln\left(\dfrac{x}{x - 1}\right) + \ln\left(\dfrac{x + 1}{x}\right) - \ln(x^2 - 1)$

64. $\log\left(\dfrac{x^2 + 2x - 3}{x^2 - 4}\right) - \log\left(\dfrac{x^2 + 7x + 6}{x + 2}\right)$

65. $8\log_2\sqrt{3x - 2} - \log_2\left(\dfrac{4}{x}\right) + \log_2 4$

66. $21 \log_3 \sqrt[3]{x} + \log_3(9x^2) - \log_3 9$ **67.** $2 \log_a(5x^3) - \frac{1}{2} \log_a(2x + 3)$ **68.** $\frac{1}{3} \log(x^3 + 1) + \frac{1}{2}\log(x^2 + 1)$

69. $2 \log_2(x + 1) - \log_2(x + 3) - \log_2(x - 1)$ **70.** $3 \log_5(3x + 1) - 2 \log_5(2x - 1) - \log_5 x$

In Problems 71–78, use the Change-of-Base Formula and a calculator to evaluate each logarithm. Round your answer to three decimal places.

71. $\log_3 21$ **72.** $\log_5 18$ **73.** $\log_{1/3} 71$ **74.** $\log_{1/2} 15$

75. $\log_{\sqrt{2}} 7$ **76.** $\log_{\sqrt{5}} 8$ **77.** $\log_\pi e$ **78.** $\log_\pi \sqrt{2}$

In Problems 79–84, graph each function using a graphing utility and the Change-of-Base Formula.

79. $y = \log_4 x$ **80.** $y = \log_5 x$ **81.** $y = \log_2(x + 2)$

82. $y = \log_4(x - 3)$ **83.** $y = \log_{x-1}(x + 1)$ **84.** $y = \log_{x+2}(x - 2)$

Mixed Practice

85. If $f(x) = \ln x$, $g(x) = e^x$, and $h(x) = x^2$, find:
 (a) $(f \circ g)(x)$. What is the domain of $f \circ g$?
 (b) $(g \circ f)(x)$. What is the domain of $g \circ f$?
 (c) $(f \circ g)(5)$
 (d) $(f \circ h)(x)$. What is the domain of $f \circ h$?
 (e) $(f \circ h)(e)$

86. If $f(x) = \log_2 x$, $g(x) = 2^x$, and $h(x) = 4x$, find:
 (a) $(f \circ g)(x)$. What is the domain of $f \circ g$?
 (b) $(g \circ f)(x)$. What is the domain of $g \circ f$?
 (c) $(f \circ g)(3)$
 (d) $(f \circ h)(x)$. What is the domain of $f \circ h$?
 (e) $(f \circ h)(8)$

Applications and Extensions

In Problems 87–96, express y as a function of x. The constant C is a positive number.

87. $\ln y = \ln x + \ln C$ **88.** $\ln y = \ln(x + C)$

89. $\ln y = \ln x + \ln(x + 1) + \ln C$ **90.** $\ln y = 2 \ln x - \ln(x + 1) + \ln C$

91. $\ln y = 3x + \ln C$ **92.** $\ln y = -2x + \ln C$

93. $\ln(y - 3) = -4x + \ln C$ **94.** $\ln(y + 4) = 5x + \ln C$

95. $3 \ln y = \frac{1}{2} \ln(2x + 1) - \frac{1}{3} \ln(x + 4) + \ln C$ **96.** $2 \ln y = -\frac{1}{2} \ln x + \frac{1}{3} \ln(x^2 + 1) + \ln C$

97. Find the value of $\log_2 3 \cdot \log_3 4 \cdot \log_4 5 \cdot \log_5 6 \cdot \log_6 7 \cdot \log_7 8$. **98.** Find the value of $\log_2 4 \cdot \log_4 6 \cdot \log_6 8$.

99. Find the value of $\log_2 3 \cdot \log_3 4 \cdot \cdots \cdot \log_n(n + 1) \cdot \log_{n+1} 2$. **100.** Find the value of $\log_2 2 \cdot \log_2 4 \cdot \cdots \cdot \log_2 2^n$.

101. Show that $\log_a\left(x + \sqrt{x^2 - 1}\right) + \log_a\left(x - \sqrt{x^2 - 1}\right) = 0$.

102. Show that $\log_a\left(\sqrt{x} + \sqrt{x - 1}\right) + \log_a\left(\sqrt{x} - \sqrt{x - 1}\right) = 0$.

103. Show that $\ln(1 + e^{2x}) = 2x + \ln(1 + e^{-2x})$.

104. Difference Quotient If $f(x) = \log_a x$, show that $\dfrac{f(x + h) - f(x)}{h} = \log_a\left(1 + \dfrac{h}{x}\right)^{1/h}$, $h \neq 0$.

105. If $f(x) = \log_a x$, show that $-f(x) = \log_{1/a} x$. **106.** If $f(x) = \log_a x$, show that $f(AB) = f(A) + f(B)$.

107. If $f(x) = \log_a x$, show that $f\left(\dfrac{1}{x}\right) = -f(x)$. **108.** If $f(x) = \log_a x$, show that $f(x^\alpha) = \alpha f(x)$.

109. Show that $\log_a\left(\dfrac{M}{N}\right) = \log_a M - \log_a N$, where a, M, and N are positive real numbers and $a \neq 1$.

110. Show that $\log_a\left(\dfrac{1}{N}\right) = -\log_a N$, where a and N are positive real numbers and $a \neq 1$.

Explaining Concepts: Discussion and Writing

111. Graph $Y_1 = \log(x^2)$ and $Y_2 = 2 \log(x)$ using a graphing utility. Are they equivalent? What might account for any differences in the two functions?

112. Write an example that illustrates why $(\log_a x)^r \neq r \log_a x$.

113. Write an example that illustrates why $\log_2(x + y) \neq \log_2 x + \log_2 y$.

114. Does $3^{\log_3(-5)} = -5$? Why or why not?

6.6 Logarithmic and Exponential Equations

PREPARING FOR THIS SECTION *Before getting started, review the following:*

- Solving Equations Using a Graphing Utility (Appendix, Section 4, pp. A8–A10)
- Solving Quadratic Equations (Section 1.2, pp. 92–99)
- Solving Equations Quadratic in Form (Section 1.4, pp. 114–116)

➤ Now Work the *'Are You Prepared?'* problems on page 463.

OBJECTIVES 1 Solve Logarithmic Equations (p. 459)

2 Solve Exponential Equations (p. 461)

3 Solve Logarithmic and Exponential Equations Using a Graphing Utility (p. 462)

1 Solve Logarithmic Equations

In Section 6.4 we solved logarithmic equations by changing a logarithmic expression to an exponential expression. That is, we used the definition of a logarithm:

$$y = \log_a x \quad \text{is equivalent to} \quad x = a^y \qquad a > 0, a \neq 1$$

For example, to solve the equation $\log_2(1 - 2x) = 3$, we write the logarithmic equation as an equivalent exponential equation $1 - 2x = 2^3$ and solve for x.

$$\log_2(1 - 2x) = 3$$
$$1 - 2x = 2^3 \qquad \text{Change to an exponential statement.}$$
$$-2x = 7 \qquad \text{Simplify.}$$
$$x = -\frac{7}{2} \qquad \text{Solve.}$$

You should check this solution for yourself.

For most logarithmic equations, some manipulation of the equation (usually using properties of logarithms) is required to obtain a solution. Also, to avoid extraneous solutions with logarithmic equations, we determine the domain of the variable first.

We begin with an example of a logarithmic equation that requires using the fact that a logarithmic function is a one-to-one function:

If $\log_a M = \log_a N$, then $M = N$ $\qquad M, N,$ and a are positive and $a \neq 1$.

EXAMPLE 1 | **Solving a Logarithmic Equation**

Solve: $2 \log_5 x = \log_5 9$

Solution The domain of the variable in this equation is $x > 0$. Because each logarithm is to the same base, 5, we can obtain an exact solution as follows:

$$2 \log_5 x = \log_5 9$$
$$\log_5 x^2 = \log_5 9 \qquad\qquad r \log_a M = \log_a M^r$$
$$x^2 = 9 \qquad\qquad \text{If } \log_a M = \log_a N, \text{ then } M = N.$$
$$x = 3 \quad \text{or} \quad x = -3$$

Recall that the domain of the variable is $x > 0$. Therefore, -3 is extraneous and we discard it.

✓**Check:** $2 \log_5 3 \overset{?}{=} \log_5 9$

$\log_5 3^2 \overset{?}{=} \log_5 9 \qquad r \log_a M = \log_a M^r$

$\log_5 9 = \log_5 9$

The solution set is $\{3\}$.

━━━━ **Now Work** PROBLEM 13

Often we need to use one or more properties of logarithms to rewrite the equation as a single logarithm. In the next example we employ the log of a product property to solve a logarithmic equation.

EXAMPLE 2 **Solving a Logarithmic Equation**

Solve: $\log_5(x + 6) + \log_5(x + 2) = 1$

Solution The domain of the variable requires that $x + 6 > 0$ and $x + 2 > 0$, so $x > -6$ and $x > -2$. This means any solution must satisfy $x > -2$. To obtain an exact solution, we need to express the left side as a single logarithm. Then we will change the equation to an equivalent exponential equation.

$\log_5(x + 6) + \log_5(x + 2) = 1$

$\log_5[(x + 6)(x + 2)] = 1 \qquad \log_a M + \log_a N = \log_a(MN)$

$(x + 6)(x + 2) = 5^1 = 5 \quad$ Change to an exponential statement.

$x^2 + 8x + 12 = 5 \qquad$ Simplify.

$x^2 + 8x + 7 = 0 \qquad$ Place the quadratic equation in standard form.

$(x + 7)(x + 1) = 0 \qquad$ Factor.

$x = -7 \quad \text{or} \quad x = -1 \qquad$ Zero-Product Property

WARNING A negative solution is not automatically extraneous. You must determine whether the potential solution causes the argument of any logarithmic expression in the equation to be negative. ∎

Only $x = -1$ satisfies the restriction that $x > -2$, so $x = -7$ is extraneous. The solution set is $\{-1\}$, which you should check.

━━━━ **Now Work** PROBLEM 21

EXAMPLE 3 **Solving a Logarithmic Equation**

Solve: $\ln x = \ln(x + 6) - \ln(x - 4)$

Solution The domain of the variable requires that $x > 0$, $x + 6 > 0$, and $x - 4 > 0$. As a result, the domain of the variable here is $x > 4$. We begin the solution using the log of a difference property.

$\ln x = \ln(x + 6) - \ln(x - 4)$

$\ln x = \ln\left(\dfrac{x + 6}{x - 4}\right) \qquad \ln M - \ln N = \ln\left(\dfrac{M}{N}\right)$

$x = \dfrac{x + 6}{x - 4} \qquad$ If $\ln M = \ln N$, then $M = N$.

$x(x - 4) = x + 6 \qquad$ Multiply both sides by $x - 4$.

$x^2 - 4x = x + 6 \qquad$ Simplify.

$x^2 - 5x - 6 = 0 \qquad$ Place the quadratic equation in standard form.

$(x - 6)(x + 1) = 0 \qquad$ Factor.

$x = 6 \quad \text{or} \quad x = -1 \qquad$ Zero-Product Property

Since the domain of the variable is $x > 4$, we discard -1 as extraneous. The solution set is $\{6\}$, which you should check.

WARNING In using properties of logarithms to solve logarithmic equations, avoid using the property $\log_a x^r = r \log_a x$, when r is even. The reason can be seen in this example:

Solve: $\log_3 x^2 = 4$

Solution: The domain of the variable x is all real numbers except 0.

(a) $\log_3 x^2 = 4$

$x^2 = 3^4 = 81$ *Change to exponential form.*

$x = -9$ or $x = 9$

(b) $\log_3 x^2 = 4$ $\log_a x^r = r \log_a x$

$2 \log_3 x = 4$ *Domain of variable is $x > 0$.*

$\log_3 x = 2$

$x = 9$

Both -9 and 9 are solutions of $\log_3 x^2 = 4$ (as you can verify). The solution in part (b) does not find the solution -9 because the domain of the variable was further restricted due to the application of the property $\log_a x^r = r \log_a x$. ■

━━━**Now Work** PROBLEM 31

2 Solve Exponential Equations

In Sections 6.3 and 6.4, we solved exponential equations algebraically by expressing each side of the equation using the same base. That is, we used the one-to-one property of the exponential function:

$$\text{If } a^u = a^v, \quad \text{then } u = v \qquad a > 0, a \neq 1$$

For example, to solve the exponential equation $4^{2x+1} = 16$, notice that $16 = 4^2$ and apply the property above to obtain $2x + 1 = 2$, from which we find $x = \dfrac{1}{2}$.

For most exponential equations, we cannot express each side of the equation using the same base. In such cases, algebraic techniques can sometimes be used to obtain exact solutions.

EXAMPLE 4

Solving Exponential Equations

Solve: (a) $2^x = 5$ (b) $8 \cdot 3^x = 5$

Solution

(a) Since 5 cannot be written as an integer power of 2 ($2^2 = 4$ and $2^3 = 8$), write the exponential equation as the equivalent logarithmic equation.

$$2^x = 5$$

$$x = \log_2 5 = \frac{\ln 5}{\ln 2}$$

↑

Change-of-Base Formula (10), Section 6.5

Alternatively, we can solve the equation $2^x = 5$ by taking the natural logarithm (or common logarithm) of each side. Taking the natural logarithm,

$$2^x = 5$$

$$\ln 2^x = \ln 5 \qquad \textit{If } M = N, \textit{ then } \ln M = \ln N.$$

$$x \ln 2 = \ln 5 \qquad \ln M^r = r \ln M$$

$$x = \frac{\ln 5}{\ln 2} \qquad \textit{Exact solution}$$

$$\approx 2.322 \qquad \textit{Approximate solution}$$

The solution set is $\left\{ \dfrac{\ln 5}{\ln 2} \right\}$.

(b) $8 \cdot 3^x = 5$

$$3^x = \frac{5}{8} \qquad \textit{Solve for } 3^x.$$

$$x = \log_3\left(\frac{5}{8}\right) = \frac{\ln\left(\frac{5}{8}\right)}{\ln 3} \qquad \textit{Exact solution}$$

$$\approx -0.428 \qquad \textit{Approximate solution}$$

The solution set is $\left\{\dfrac{\ln\left(\dfrac{5}{8}\right)}{\ln 3}\right\}$.

Now Work PROBLEM 35

EXAMPLE 5 **Solving an Exponential Equation**

Solve: $5^{x-2} = 3^{3x+2}$

Solution Because the bases are different, we first apply property (7), Section 6.5 (take the natural logarithm of each side), and then use a property of logarithms. The result is an equation in x that we can solve.

$$5^{x-2} = 3^{3x+2}$$
$$\ln 5^{x-2} = \ln 3^{3x+2} \qquad \textit{If } M = N, \ln M = \ln N.$$
$$(x-2)\ln 5 = (3x+2)\ln 3 \qquad \textit{ln } M^r = r \ln M$$
$$(\ln 5)x - 2\ln 5 = (3\ln 3)x + 2\ln 3 \qquad \textit{Distribute.}$$
$$(\ln 5)x - (3\ln 3)x = 2\ln 3 + 2\ln 5 \qquad \textit{Place terms involving x on the left.}$$
$$(\ln 5 - 3\ln 3)x = 2(\ln 3 + \ln 5) \qquad \textit{Factor.}$$
$$x = \frac{2(\ln 3 + \ln 5)}{\ln 5 - 3\ln 3} \qquad \textit{Exact solution}$$
$$\approx -3.212 \qquad \textit{Approximate solution}$$

The solution set is $\left\{\dfrac{2(\ln 3 + \ln 5)}{\ln 5 - 3\ln 3}\right\}$.

Now Work PROBLEM 45

EXAMPLE 6 **Solving an Exponential Equation That Is Quadratic in Form**

Solve: $4^x - 2^x - 12 = 0$

Solution We note that $4^x = (2^2)^x = 2^{(2x)} = (2^x)^2$, so the equation is quadratic in form, and we can rewrite it as

$$(2^x)^2 - 2^x - 12 = 0 \quad \textit{Let } u = 2^x; \textit{ then } u^2 - u - 12 = 0.$$

Now we can factor as usual.

$$(2^x - 4)(2^x + 3) = 0 \qquad (u - 4)(u + 3) = 0$$
$$2^x - 4 = 0 \quad \text{or} \quad 2^x + 3 = 0 \qquad u - 4 = 0 \quad \text{or} \quad u + 3 = 0$$
$$2^x = 4 \qquad\qquad 2^x = -3 \qquad u = 2^x = 4 \qquad u = 2^x = -3$$

The equation on the left has the solution $x = 2$, since $2^x = 4 = 2^2$; the equation on the right has no solution, since $2^x > 0$ for all x. The only solution is 2. The solution set is $\{2\}$.

Now Work PROBLEM 53

3 Solve Logarithmic and Exponential Equations Using a Graphing Utility

The algebraic techniques introduced in this section to obtain exact solutions apply only to certain types of logarithmic and exponential equations. Solutions for other types are usually studied in calculus, using numerical methods. For such types, we can use a graphing utility to approximate the solution.

EXAMPLE 7 **Solving Equations Using a Graphing Utility**

Solve: $x + e^x = 2$

Express the solution(s) rounded to two decimal places.

Solution The solution is found by graphing $Y_1 = x + e^x$ and $Y_2 = 2$. Since Y_1 is an increasing function (do you know why?), there is only one point of intersection for Y_1 and Y_2. Figure 40 shows the graphs of Y_1 and Y_2. Using the INTERSECT command, the solution is 0.44 rounded to two decimal places.

Figure 40

Now Work PROBLEM 63

6.6 Assess Your Understanding

'Are You Prepared?' *Answers are given at the end of these exercises. If you get a wrong answer, read the pages listed in red.*

1. Solve $x^2 - 7x - 30 = 0$. (pp. 92–99)

2. Solve $(x + 3)^2 - 4(x + 3) + 3 = 0$. (pp. 114–116)

3. Approximate the solution(s) to $x^3 = x^2 - 5$ using a graphing utility. (pp. A8–A10)

4. Approximate the solution(s) to $x^3 - 2x + 2 = 0$ using a graphing utility. (pp. A8–A10)

Skill Building

In Problems 5–32, solve each logarithmic equation. Express irrational solutions in exact form and as a decimal rounded to three decimal places.

5. $\log_4 x = 2$

6. $\log(x + 6) = 1$

7. $\log_2(5x) = 4$

8. $\log_3(3x - 1) = 2$

9. $\log_4(x + 2) = \log_4 8$

10. $\log_5(2x + 3) = \log_5 3$

11. $\dfrac{1}{2}\log_3 x = 2\log_3 2$

12. $-2\log_4 x = \log_4 9$

13. $3\log_2 x = -\log_2 27$

14. $2\log_5 x = 3\log_5 4$

15. $3\log_2(x - 1) + \log_2 4 = 5$

16. $2\log_3(x + 4) - \log_3 9 = 2$

17. $\log x + \log(x + 15) = 2$

18. $\log x + \log(x - 21) = 2$

19. $\log(2x + 1) = 1 + \log(x - 2)$

20. $\log(2x) - \log(x - 3) = 1$

21. $\log_2(x + 7) + \log_2(x + 8) = 1$

22. $\log_6(x + 4) + \log_6(x + 3) = 1$

23. $\log_8(x + 6) = 1 - \log_8(x + 4)$

24. $\log_5(x + 3) = 1 - \log_5(x - 1)$

25. $\ln x + \ln(x + 2) = 4$

26. $\ln(x + 1) - \ln x = 2$

27. $\log_3(x + 1) + \log_3(x + 4) = 2$

28. $\log_2(x + 1) + \log_2(x + 7) = 3$

29. $\log_{1/3}(x^2 + x) - \log_{1/3}(x^2 - x) = -1$

30. $\log_4(x^2 - 9) - \log_4(x + 3) = 3$

31. $\log_a(x - 1) - \log_a(x + 6) = \log_a(x - 2) - \log_a(x + 3)$

32. $\log_a x + \log_a(x - 2) = \log_a(x + 4)$

In Problems 33–60, solve each exponential equation. Express irrational solutions in exact form and as a decimal rounded to three decimal places.

33. $2^{x-5} = 8$

34. $5^{-x} = 25$

35. $2^x = 10$

36. $3^x = 14$

37. $8^{-x} = 1.2$

38. $2^{-x} = 1.5$

39. $5(2^{3x}) = 8$

40. $0.3(4^{0.2x}) = 0.2$

41. $3^{1-2x} = 4^x$

42. $2^{x+1} = 5^{1-2x}$

43. $\left(\dfrac{3}{5}\right)^x = 7^{1-x}$

44. $\left(\dfrac{4}{3}\right)^{1-x} = 5^x$

45. $1.2^x = (0.5)^{-x}$

46. $0.3^{1+x} = 1.7^{2x-1}$

47. $\pi^{1-x} = e^x$

48. $e^{x+3} = \pi^x$

49. $2^{2x} + 2^x - 12 = 0$ **50.** $3^{2x} + 3^x - 2 = 0$ **51.** $3^{2x} + 3^{x+1} - 4 = 0$ **52.** $2^{2x} + 2^{x+2} - 12 = 0$

53. $16^x + 4^{x+1} - 3 = 0$ **54.** $9^x - 3^{x+1} + 1 = 0$ **55.** $25^x - 8 \cdot 5^x = -16$ **56.** $36^x - 6 \cdot 6^x = -9$

57. $3 \cdot 4^x + 4 \cdot 2^x + 8 = 0$ **58.** $2 \cdot 49^x + 11 \cdot 7^x + 5 = 0$ **59.** $4^x - 10 \cdot 4^{-x} = 3$ **60.** $3^x - 14 \cdot 3^{-x} = 5$

In Problems 61–74, use a graphing utility to solve each equation. Express your answer rounded to two decimal places.

61. $\log_5(x + 1) - \log_4(x - 2) = 1$ **62.** $\log_2(x - 1) - \log_6(x + 2) = 2$

63. $e^x = -x$ **64.** $e^{2x} = x + 2$ **65.** $e^x = x^2$ **66.** $e^x = x^3$

67. $\ln x = -x$ **68.** $\ln(2x) = -x + 2$ **69.** $\ln x = x^3 - 1$ **70.** $\ln x = -x^2$

71. $e^x + \ln x = 4$ **72.** $e^x - \ln x = 4$ **73.** $e^{-x} = \ln x$ **74.** $e^{-x} = -\ln x$

Mixed Practice

In Problems 75–86, solve each equation. Express irrational solutions in exact form and as a decimal rounded to three decimal places.

75. $\log_2(x + 1) - \log_4 x = 1$
[**Hint:** Change $\log_4 x$ to base 2.]

76. $\log_2(3x + 2) - \log_4 x = 3$

77. $\log_{16} x + \log_4 x + \log_2 x = 7$

78. $\log_9 x + 3 \log_3 x = 14$

79. $\left(\sqrt[3]{2}\right)^{2-x} = 2^{x^2}$

80. $\log_2 x^{\log_2 x} = 4$

81. $\dfrac{e^x + e^{-x}}{2} = 1$
[**Hint:** Multiply each side by e^x.]

82. $\dfrac{e^x + e^{-x}}{2} = 3$

83. $\dfrac{e^x - e^{-x}}{2} = 2$

84. $\dfrac{e^x - e^{-x}}{2} = -2$

85. $\log_5 x + \log_3 x = 1$
[**Hint:** Use the Change-of-Base Formula.]

86. $\log_2 x + \log_6 x = 3$

87. $f(x) = \log_2(x + 3)$ and $g(x) = \log_2(3x + 1)$.
 (a) Solve $f(x) = 3$. What point is on the graph of f?
 (b) Solve $g(x) = 4$. What point is on the graph of g?
 (c) Solve $f(x) = g(x)$. Do the graphs of f and g intersect? If so, where?
 (d) Solve $(f + g)(x) = 7$.
 (e) Solve $(f - g)(x) = 2$.

88. $f(x) = \log_3(x + 5)$ and $g(x) = \log_3(x - 1)$.
 (a) Solve $f(x) = 2$. What point is on the graph of f?
 (b) Solve $g(x) = 3$. What point is on the graph of g?
 (c) Solve $f(x) = g(x)$. Do the graphs of f and g intersect? If so, where?
 (d) Solve $(f + g)(x) = 3$.
 (e) Solve $(f - g)(x) = 2$.

89. (a) If $f(x) = 3^{x+1}$ and $g(x) = 2^{x+2}$, graph f and g on the same Cartesian plane.
 (b) Find the point(s) of intersection of the graphs of f and g by solving $f(x) = g(x)$. Round answers to three decimal places. Label any intersection points on the graph drawn in part (a).
 (c) Based on the graph, solve $f(x) > g(x)$.

90. (a) If $f(x) = 5^{x-1}$ and $g(x) = 2^{x+1}$, graph f and g on the same Cartesian plane.
 (b) Find the point(s) of intersection of the graphs of f and g by solving $f(x) = g(x)$. Label any intersection points on the graph drawn in part (a).
 (c) Based on the graph, solve $f(x) > g(x)$.

91. (a) Graph $f(x) = 3^x$ and $g(x) = 10$ on the same Cartesian plane.

 (b) Shade the region bounded by the y-axis, $f(x) = 3^x$, and $g(x) = 10$ on the graph drawn in part (a).
 (c) Solve $f(x) = g(x)$ and label the point of intersection on the graph drawn in part (a).

92. (a) Graph $f(x) = 2^x$ and $g(x) = 12$ on the same Cartesian plane.
 (b) Shade the region bounded by the y-axis, $f(x) = 2^x$, and $g(x) = 12$ on the graph drawn in part (a).
 (c) Solve $f(x) = g(x)$ and label the point of intersection on the graph drawn in part (a).

93. (a) Graph $f(x) = 2^{x+1}$ and $g(x) = 2^{-x+2}$ on the same Cartesian plane.
 (b) Shade the region bounded by the y-axis, $f(x) = 2^{x+1}$, and $g(x) = 2^{-x+2}$ on the graph draw in part (a).
 (c) Solve $f(x) = g(x)$ and label the point of intersection on the graph drawn in part (a).

94. (a) Graph $f(x) = 3^{-x+1}$ and $g(x) = 3^{x-2}$ on the same Cartesian plane.
 (b) Shade the region bounded by the y-axis, $f(x) = 3^{-x+1}$, and $g(x) = 3^{x-2}$ on the graph draw in part (a).
 (c) Solve $f(x) = g(x)$ and label the point of intersection on the graph drawn in part (a).

95. (a) Graph $f(x) = 2^x - 4$.
 (b) Find the zero of f.
 (c) Based on the graph, solve $f(x) < 0$.

96. (a) Graph $g(x) = 3^x - 9$.
 (b) Find the zero of g.
 (c) Based on the graph, solve $g(x) > 0$.

Applications and Extensions

97. A Population Model The resident population of the United States in 2008 was 304 million people and was growing at a rate of 0.9% per year. Assuming that this growth rate continues, the model $P(t) = 304(1.009)^{t-2008}$ represents the population P (in millions of people) in year t.
(a) According to this model, when will the population of the United States be 354 million people?
(b) According to this model, when will the population of the United States be 416 million people?
Source: Statistical Abstract of the United States, 125th ed., 2009.

98. A Population Model The population of the world in 2009 was 6.78 billion people and was growing at a rate of 1.14% per year. Assuming that this growth rate continues, the model $P(t) = 6.78(1.0114)^{t-2009}$ represents the population P (in billions of people) in year t.
(a) According to this model, when will the population of the world be 8.7 billion people?

(b) According to this model, when will the population of the world be 14 billion people?
Source: U.S. Census Bureau.

99. Depreciation The value V of a Chevy Cobalt that is t years old can be modeled by $V(t) = 16,500(0.82)^t$.
(a) According to the model, when will the car be worth $9000?
(b) According to the model, when will the car be worth $4000?
(c) According to the model, when will the car be worth $2000?
Source: Kelley Blue Book

100. Depreciation The value V of a Honda Civic DX that is t years old can be modeled by $V(t) = 16,775(0.905)^t$.
(a) According to the model, when will the car be worth $15,000?
(b) According to the model, when will the car be worth $8000?
(c) According to the model, when will the car be worth $4000?
Source: Kelley Blue Book

Explaining Concepts: Discussion and Writing

101. Fill in reasons for each step in the following two solutions.
Solve: $\log_3(x-1)^2 = 2$

Solution A	**Solution B**
$\log_3(x-1)^2 = 2$	$\log_3(x-1)^2 = 2$
$(x-1)^2 = 3^2 = 9$ ____	$2\log_3(x-1) = 2$ ____
$(x-1) = \pm 3$ ____	$\log_3(x-1) = 1$ ____
$x - 1 = -3$ or $x - 1 = 3$ ____	$x - 1 = 3^1 = 3$ ____
$x = -2$ or $x = 4$ ____	$x = 4$ ____

Both solutions given in Solution A check. Explain what caused the solution $x = -2$ to be lost in Solution B.

'Are You Prepared?' Answers

1. $\{-3, 10\}$ **2.** $\{-2, 0\}$ **3.** $\{-1.43\}$ **4.** $\{-1.77\}$

6.7 Financial Models

PREPARING FOR THIS SECTION *Before getting started, review the following:*
- Simple Interest (Section 1.7, p. 136)

Now Work the **'Are You Prepared?'** problems on page 472.

OBJECTIVES 1 Determine the Future Value of a Lump Sum of Money (p. 466)
2 Calculate Effective Rates of Return (p. 469)
3 Determine the Present Value of a Lump Sum of Money (p. 470)
4 Determine the Rate of Interest or Time Required to Double a Lump Sum of Money (p. 471)

1 Determine the Future Value of a Lump Sum of Money

Interest is money paid for the use of money. The total amount borrowed (whether by an individual from a bank in the form of a loan or by a bank from an individual in the form of a savings account) is called the **principal.** The **rate of interest,** expressed as a percent, is the amount charged for the use of the principal for a given period of time, usually on a yearly (that is, per annum) basis.

THEOREM

Simple Interest Formula

If a principal of P dollars is borrowed for a period of t years at a per annum interest rate r, expressed as a decimal, the interest I charged is

$$I = Prt \tag{1}$$

Interest charged according to formula (1) is called **simple interest.**

In working with problems involving interest, we define the term **payment period** as follows:

Annually:	Once per year	**Monthly:**	12 times per year
Semiannually:	Twice per year	**Daily:**	365 times per year*
Quarterly:	Four times per year		

When the interest due at the end of a payment period is added to the principal so that the interest computed at the end of the next payment period is based on this new principal amount (old principal + interest), the interest is said to have been **compounded. Compound interest** is interest paid on the principal and previously earned interest.

EXAMPLE 1

Computing Compound Interest

A credit union pays interest of 8% per annum compounded quarterly on a certain savings plan. If $1000 is deposited in such a plan and the interest is left to accumulate, how much is in the account after 1 year?

Solution

We use the simple interest formula, $I = Prt$. The principal P is $1000 and the rate of interest is 8% = 0.08. After the first quarter of a year, the time t is $\frac{1}{4}$ year, so the interest earned is

$$I = Prt = (\$1000)(0.08)\left(\frac{1}{4}\right) = \$20$$

* Most banks use a 360-day "year." Why do you think they do?

The new principal is $P + I = \$1000 + \$20 = \$1020$. At the end of the second quarter, the interest on this principal is

$$I = (\$1020)(0.08)\left(\frac{1}{4}\right) = \$20.40$$

At the end of the third quarter, the interest on the new principal of $\$1020 + \$20.40 = \$1040.40$ is

$$I = (\$1040.40)(0.08)\left(\frac{1}{4}\right) = \$20.81$$

Finally, after the fourth quarter, the interest is

$$I = (\$1061.21)(0.08)\left(\frac{1}{4}\right) = \$21.22$$

After 1 year the account contains $\$1061.21 + \$21.22 = \$1082.43$.

The pattern of the calculations performed in Example 1 leads to a general formula for compound interest. To fix our ideas, let P represent the principal to be invested at a per annum interest rate r that is compounded n times per year, so the time of each compounding period is $\frac{1}{n}$ years. (For computing purposes, r is expressed as a decimal.) The interest earned after each compounding period is given by formula (1).

$$\text{Interest} = \text{principal} \times \text{rate} \times \text{time} = P \cdot r \cdot \frac{1}{n} = P \cdot \left(\frac{r}{n}\right)$$

The amount A after one compounding period is

$$A = P + P \cdot \left(\frac{r}{n}\right) = P \cdot \left(1 + \frac{r}{n}\right)$$

After two compounding periods, the amount A, based on the new principal $P \cdot \left(1 + \frac{r}{n}\right)$, is

$$A = \underbrace{P \cdot \left(1 + \frac{r}{n}\right)}_{\substack{\text{New} \\ \text{principal}}} + \underbrace{P \cdot \left(1 + \frac{r}{n}\right)\left(\frac{r}{n}\right)}_{\substack{\text{Interest on} \\ \text{new principal}}} = \overset{\uparrow}{P} \cdot \left(1 + \frac{r}{n}\right)\left(1 + \frac{r}{n}\right) = P \cdot \left(1 + \frac{r}{n}\right)^2$$

$$\text{Factor out } P \cdot \left(1 + \frac{r}{n}\right).$$

After three compounding periods, the amount A is

$$A = P \cdot \left(1 + \frac{r}{n}\right)^2 + P \cdot \left(1 + \frac{r}{n}\right)^2\left(\frac{r}{n}\right) = P \cdot \left(1 + \frac{r}{n}\right)^2 \cdot \left(1 + \frac{r}{n}\right) = P \cdot \left(1 + \frac{r}{n}\right)^3$$

Continuing this way, after n compounding periods (1 year), the amount A is

$$A = P \cdot \left(1 + \frac{r}{n}\right)^n$$

Because t years will contain $n \cdot t$ compounding periods, after t years we have

$$A = P \cdot \left(1 + \frac{r}{n}\right)^{nt}$$

THEOREM **Compound Interest Formula**

The amount A after t years due to a principal P invested at an annual interest rate r compounded n times per year is

$$A = P \cdot \left(1 + \frac{r}{n}\right)^{nt} \tag{2}$$

Exploration

To see the effects of compounding interest monthly on an initial deposit of $1, graph $Y_1 = \left(1 + \dfrac{r}{12}\right)^{12x}$ with $r = 0.06$ and $r = 0.12$ for $0 \leq x \leq 30$. What is the future value of $1 in 30 years when the interest rate per annum is $r = 0.06$ (6%)? What is the future value of $1 in 30 years when the interest rate per annum is $r = 0.12$ (12%)? Does doubling the interest rate double the future value?

For example, to rework Example 1, use $P = \$1000$, $r = 0.08$, $n = 4$ (quarterly compounding), and $t = 1$ year to obtain

$$A = P \cdot \left(1 + \frac{r}{n}\right)^{nt} = 1000\left(1 + \frac{0.08}{4}\right)^{4 \cdot 1} = \$1082.43$$

In equation (2), the amount A is typically referred to as the **future value** of the account, while P is called the **present value.**

───────**Now Work** PROBLEM 7

EXAMPLE 2

Comparing Investments Using Different Compounding Periods

Investing $1000 at an annual rate of 10% compounded annually, semiannually, quarterly, monthly, and daily will yield the following amounts after 1 year:

Annual compounding ($n = 1$): $\qquad A = P \cdot (1 + r)$

$$= (\$1000)(1 + 0.10) = \$1100.00$$

Semiannual compounding ($n = 2$): $\qquad A = P \cdot \left(1 + \dfrac{r}{2}\right)^2$

$$= (\$1000)(1 + 0.05)^2 = \$1102.50$$

Quarterly compounding ($n = 4$): $\qquad A = P \cdot \left(1 + \dfrac{r}{4}\right)^4$

$$= (\$1000)(1 + 0.025)^4 = \$1103.81$$

Monthly compounding ($n = 12$): $\qquad A = P \cdot \left(1 + \dfrac{r}{12}\right)^{12}$

$$= (\$1000)\left(1 + \frac{0.10}{12}\right)^{12} = \$1104.71$$

Daily compounding ($n = 365$): $\qquad A = P \cdot \left(1 + \dfrac{r}{365}\right)^{365}$

$$= (\$1000)\left(1 + \frac{0.10}{365}\right)^{365} = \$1105.16$$

From Example 2, we can see that the effect of compounding more frequently is that the amount after 1 year is higher: $1000 compounded 4 times a year at 10% results in $1103.81, $1000 compounded 12 times a year at 10% results in $1104.71, and $1000 compounded 365 times a year at 10% results in $1105.16. This leads to the following question: What would happen to the amount after 1 year if the number of times that the interest is compounded were increased without bound?

Let's find the answer. Suppose that P is the principal, r is the per annum interest rate, and n is the number of times that the interest is compounded each year. The amount after 1 year is

$$A = P \cdot \left(1 + \frac{r}{n}\right)^n$$

Rewrite this expression as follows:

$$A = P \cdot \left(1 + \frac{r}{n}\right)^n = P \cdot \left(1 + \frac{1}{\dfrac{n}{r}}\right)^n = P \cdot \left[\left(1 + \frac{1}{\dfrac{n}{r}}\right)^{n/r}\right]^r = P \cdot \left[\left(1 + \frac{1}{h}\right)^h\right]^r \quad \textbf{(3)}$$

$$\underset{\substack{\uparrow \\ h = \frac{n}{r}}}{}$$

Now suppose that the number n of times that the interest is compounded per year gets larger and larger; that is, suppose that $n \to \infty$. Then $h = \dfrac{n}{r} \to \infty$, and the expression in brackets in equation (3) equals e. That is, $A \to Pe^r$.

Table 8 compares $\left(1 + \dfrac{r}{n}\right)^n$, for large values of n, to e^r for $r = 0.05$, $r = 0.10$, $r = 0.15$, and $r = 1$. The larger that n gets, the closer $\left(1 + \dfrac{r}{n}\right)^n$ gets to e^r. No matter how frequent the compounding, the amount after 1 year has the definite ceiling Pe^r.

Table 8

	$\left(1 + \frac{r}{n}\right)^n$			
	$n = 100$	$n = 1000$	$n = 10{,}000$	e^r
$r = 0.05$	1.0512580	1.0512698	1.051271	1.0512711
$r = 0.10$	1.1051157	1.1051654	1.1051704	1.1051709
$r = 0.15$	1.1617037	1.1618212	1.1618329	1.1618342
$r = 1$	2.7048138	2.7169239	2.7181459	2.7182818

When interest is compounded so that the amount after 1 year is Pe^r, we say that the interest is **compounded continuously.**

THEOREM

Continuous Compounding

The amount A after t years due to a principal P invested at an annual interest rate r compounded continuously is

$$A = Pe^{rt} \qquad (4)$$

EXAMPLE 3

Using Continuous Compounding

The amount A that results from investing a principal P of $1000 at an annual rate r of 10% compounded continuously for a time t of 1 year is

$$A = \$1000e^{0.10} = (\$1000)(1.10517) = \$1105.17$$

──── **Now Work** PROBLEM 13

2 Calculate Effective Rates of Return

Suppose that you have $1000 and a bank offers to pay you 3% annual interest on a savings account with interest compounded monthly. What annual interest rate do you need to earn to have the same amount at the end of the year if the interest is compounded annually (once per year)? To answer this question, first determine the value of the $1000 in the account that earns 3% compounded monthly.

$$A = \$1000\left(1 + \frac{0.03}{12}\right)^{12} \quad \text{Use } A = P\left(1 + \frac{r}{n}\right)^n \text{ with } P = \$1000, r = 0.03, n = 12.$$

$$= \$1030.42$$

So the interest earned is $30.42. Using $I = Prt$ with $t = 1$, $I = \$30.42$, and $P = \$1000$, we find the annual simple interest rate is $0.03042 = 3.042\%$. This interest rate is known as the *effective rate of interest.*

The **effective rate of interest** is the equivalent annual simple interest rate that would yield the same amount as compounding n times per year, or continuously, after 1 year.

THEOREM

Effective Rate of Interest

The effective rate of interest r_e of an investment earning an annual interest rate r is given by

Compounding n times per year: $r_e = \left(1 + \dfrac{r}{n}\right)^n - 1$

Continuous compounding: $\qquad r_e = e^r - 1$

EXAMPLE 4

Computing the Effective Rate of Interest—Which Is the Best Deal?

Suppose you want to open a money market account. You visit three banks to determine their money market rates. Bank A offers you 6% annual interest compounded daily and Bank B offers you 6.02% compounded quarterly. Bank C offers 5.98% compounded continuously. Determine which bank is offering the best deal.

Solution

The bank that offers the best deal is the one with the highest effective interest rate.

Bank A	Bank B	Bank C
$r_e = \left(1 + \dfrac{0.06}{365}\right)^{365} - 1$	$r_e = \left(1 + \dfrac{0.0602}{4}\right)^4 - 1$	$r_e = e^{0.0598} - 1$
$\approx 1.06183 - 1$	$\approx 1.06157 - 1$	$\approx 1.06162 - 1$
$= 0.06183$	$= 0.06157$	$= 0.06162$
$= 6.183\%$	$= 6.157\%$	$= 6.162\%$

Since the effective rate of interest is highest for Bank A, Bank A is offering the best deal.

═══ **Now Work** PROBLEM 23

3 Determine the Present Value of a Lump Sum of Money

When people in finance speak of the "time value of money," they are usually referring to the *present value* of money. The **present value** of A dollars to be received at a future date is the principal that you would need to invest now so that it will grow to A dollars in the specified time period. The present value of money to be received at a future date is always less than the amount to be received, since the amount to be received will equal the present value (money invested now) *plus* the interest accrued over the time period.

We use the compound interest formula (2) to get a formula for present value. If P is the present value of A dollars to be received after t years at a per annum interest rate r compounded n times per year, then, by formula (2),

$$A = P \cdot \left(1 + \frac{r}{n}\right)^{nt}$$

To solve for P, divide both sides by $\left(1 + \dfrac{r}{n}\right)^{nt}$. The result is

$$\frac{A}{\left(1 + \dfrac{r}{n}\right)^{nt}} = P \quad \text{or} \quad P = A \cdot \left(1 + \frac{r}{n}\right)^{-nt}$$

THEOREM **Present Value Formulas**

The present value P of A dollars to be received after t years, assuming a per annum interest rate r compounded n times per year, is

$$P = A \cdot \left(1 + \frac{r}{n}\right)^{-nt} \tag{5}$$

If the interest is compounded continuously,

$$P = Ae^{-rt} \tag{6}$$

To derive (6), solve formula (4) for P.

EXAMPLE 5 **Computing the Value of a Zero-coupon Bond**

A zero-coupon (noninterest-bearing) bond can be redeemed in 10 years for $1000. How much should you be willing to pay for it now if you want a return of
(a) 8% compounded monthly? (b) 7% compounded continuously?

Solution (a) We are seeking the present value of $1000. Use formula (5) with $A = \$1000$, $n = 12$, $r = 0.08$, and $t = 10$.

$$P = A \cdot \left(1 + \frac{r}{n}\right)^{-nt} = \$1000\left(1 + \frac{0.08}{12}\right)^{-12(10)} = \$450.52$$

For a return of 8% compounded monthly, you should pay $450.52 for the bond.
(b) Here use formula (6) with $A = \$1000$, $r = 0.07$, and $t = 10$.

$$P = Ae^{-rt} = \$1000e^{-(0.07)(10)} = \$496.59$$

For a return of 7% compounded continuously, you should pay $496.59 for the bond.

 Now Work PROBLEM 15

4 Determine the Rate of Interest or Time Required to Double a Lump Sum of Money

EXAMPLE 6 **Rate of Interest Required to Double an Investment**

What annual rate of interest compounded annually should you seek if you want to double your investment in 5 years?

Solution If P is the principal and we want P to double, the amount A will be $2P$. We use the compound interest formula with $n = 1$ and $t = 5$ to find r.

$$A = P \cdot \left(1 + \frac{r}{n}\right)^{nt}$$

$$2P = P \cdot (1 + r)^5 \qquad \text{\textcolor{gray}{$A = 2P, n = 1, t = 5$}}$$

$$2 = (1 + r)^5 \qquad \text{\textcolor{gray}{Divide both sides by } P.}$$

$$1 + r = \sqrt[5]{2} \qquad \text{\textcolor{gray}{Take the fifth root of each side.}}$$

$$r = \sqrt[5]{2} - 1 \approx 1.148698 - 1 = 0.148698$$

The annual rate of interest needed to double the principal in 5 years is 14.87%.

Now Work PROBLEM 31

> **EXAMPLE 7** **Time Required to Double or Triple an Investment**

(a) How long will it take for an investment to double in value if it earns 5% compounded continuously?

(b) How long will it take to triple at this rate?

Solution (a) If P is the initial investment and we want P to double, the amount A will be $2P$. We use formula (4) for continuously compounded interest with $r = 0.05$. Then

$$A = Pe^{rt}$$
$$2P = Pe^{0.05t} \qquad A = 2P, r = 0.05$$
$$2 = e^{0.05t} \qquad \text{Cancel the P's.}$$
$$0.05t = \ln 2 \qquad \text{Rewrite as a logarithm.}$$
$$t = \frac{\ln 2}{0.05} \approx 13.86 \qquad \text{Solve for t.}$$

It will take about 14 years to double the investment.

(b) To triple the investment, we set $A = 3P$ in formula (4).

$$A = Pe^{rt}$$
$$3P = Pe^{0.05t} \qquad A = 3P, r = 0.05$$
$$3 = e^{0.05t} \qquad \text{Cancel the P's.}$$
$$0.05t = \ln 3 \qquad \text{Rewrite as a logarithm.}$$
$$t = \frac{\ln 3}{0.05} \approx 21.97 \qquad \text{Solve for t.}$$

It will take about 22 years to triple the investment.

──**Now Work** PROBLEM 35

6.7 Assess Your Understanding

'Are You Prepared?' *Answers are given at the end of these exercises. If you get a wrong answer, read the pages listed in red.*

1. What is the interest due if $500 is borrowed for 6 months at a simple interest rate of 6% per annum? (p. 136)

2. If you borrow $5000 and, after 9 months, pay off the loan in the amount of $5500, what per annum rate of interest was charged? (p. 136)

Concepts and Vocabulary

3. The total amount borrowed (whether by an individual from a bank in the form of a loan or by a bank from an individual in the form of a savings account) is called the _____.

4. If a principal of P dollars is borrowed for a period of t years at a per annum interest rate r, expressed as a decimal, the interest I charged is _____ = _____. Interest charged according to this formula is called _____ _____.

5. In working problems involving interest, if the payment period of the interest is quarterly, then interest is paid _____ times per year.

6. The _____ _____ _____ _____ is the equivalent annual simple interest rate that would yield the same amount as compounding n times per year, or continuously, after 1 year.

Skill Building

In Problems 7–14, find the amount that results from each investment.

7. $100 invested at 4% compounded quarterly after a period of 2 years

8. $50 invested at 6% compounded monthly after a period of 3 years

9. $500 invested at 8% compounded quarterly after a period of $2\frac{1}{2}$ years

10. $300 invested at 12% compounded monthly after a period of $1\frac{1}{2}$ years

11. $600 invested at 5% compounded daily after a period of 3 years

12. $700 invested at 6% compounded daily after a period of 2 years

13. $1000 invested at 11% compounded continuously after a period of 2 years

14. $400 invested at 7% compounded continuously after a period of 3 years

In Problems 15–22, find the principal needed now to get each amount; that is, find the present value.

15. To get $100 after 2 years at 6% compounded monthly

16. To get $75 after 3 years at 8% compounded quarterly

17. To get $1000 after $2\frac{1}{2}$ years at 6% compounded daily

18. To get $800 after $3\frac{1}{2}$ years at 7% compounded monthly

19. To get $600 after 2 years at 4% compounded quarterly

20. To get $300 after 4 years at 3% compounded daily

21. To get $80 after $3\frac{1}{4}$ years at 9% compounded continuously

22. To get $800 after $2\frac{1}{2}$ years at 8% compounded continuously

In Problems 23–26, find the effective rate of interest.

23. For 5% compounded quarterly

24. For 6% compounded monthly

25. For 5% compounded continuously

26. For 6% compounded continuously

In Problems 27–30, determine the rate that represents the better deal.

27. 6% compounded quarterly or $6\frac{1}{4}$% compounded annually

28. 9% compounded quarterly or $9\frac{1}{4}$% compounded annually

29. 9% compounded monthly or 8.8% compounded daily

30. 8% compounded semiannually or 7.9% compounded daily

31. What rate of interest compounded annually is required to double an investment in 3 years?

32. What rate of interest compounded annually is required to double an investment in 6 years?

33. What rate of interest compounded annually is required to triple an investment in 5 years?

34. What rate of interest compounded annually is required to triple an investment in 10 years?

35. (a) How long does it take for an investment to double in value if it is invested at 8% compounded monthly?
 (b) How long does it take if the interest is compounded continuously?

36. (a) How long does it take for an investment to triple in value if it is invested at 6% compounded monthly?
 (b) How long does it take if the interest is compounded continuously?

37. What rate of interest compounded quarterly will yield an effective interest rate of 7%?

38. What rate of interest compounded continuously will yield an effective interest rate of 6%?

Applications and Extensions

39. **Time Required to Reach a Goal** If Tanisha has $100 to invest at 8% per annum compounded monthly, how long will it be before she has $150? If the compounding is continuous, how long will it be?

40. **Time Required to Reach a Goal** If Angela has $100 to invest at 10% per annum compounded monthly, how long will it be before she has $175? If the compounding is continuous, how long will it be?

41. **Time Required to Reach a Goal** How many years will it take for an initial investment of $10,000 to grow to $25,000? Assume a rate of interest of 6% compounded continuously.

42. **Time Required to Reach a Goal** How many years will it take for an initial investment of $25,000 to grow to $80,000? Assume a rate of interest of 7% compounded continuously.

43. **Price Appreciation of Homes** What will a $90,000 condominium cost 5 years from now if the price appreciation for condos over that period averages 3% compounded annually?

44. **Credit Card Interest** A department store charges 1.25% per month on the unpaid balance for customers with charge accounts (interest is compounded monthly). A customer charges $200 and does not pay her bill for 6 months. What is the bill at that time?

45. **Saving for a Car** Jerome will be buying a used car for $15,000 in 3 years. How much money should he ask his parents for now so that, if he invests it at 5% compounded continuously, he will have enough to buy the car?

46. **Paying off a Loan** John requires $3000 in 6 months to pay off a loan that has no prepayment privileges. If he has the $3000 now, how much of it should he save in an account paying 3% compounded monthly so that in 6 months he will have exactly $3000?

47. **Return on a Stock** George contemplates the purchase of 100 shares of a stock selling for $15 per share. The stock pays no dividends. The history of the stock indicates that it should grow at an annual rate of 15% per year.

How much should the 100 shares of stock be worth in 5 years?

48. Return on an Investment A business purchased for $650,000 in 2005 is sold in 2008 for $850,000. What is the annual rate of return for this investment?

49. Comparing Savings Plans Jim places $1000 in a bank account that pays 5.6% compounded continuously. After 1 year, will he have enough money to buy a computer system that costs $1060? If another bank will pay Jim 5.9% compounded monthly, is this a better deal?

50. Savings Plans On January 1, Kim places $1000 in a certificate of deposit that pays 6.8% compounded continuously and matures in 3 months. Then Kim places the $1000 and the interest in a passbook account that pays 5.25% compounded monthly. How much does Kim have in the passbook account on May 1?

51. Comparing IRA Investments Will invests $2000 in his IRA in a bond trust that pays 9% interest compounded semiannually. His friend Henry invests $2000 in his IRA in a certificate of deposit that pays $8\frac{1}{2}$% compounded continuously. Who has more money after 20 years, Will or Henry?

52. Comparing Two Alternatives Suppose that April has access to an investment that will pay 10% interest compounded continuously. Which is better: to be given $1000 now so that she can take advantage of this investment opportunity or to be given $1325 after 3 years?

53. College Costs The average annual cost of college at 4-year private colleges was $25,143 in the 2008–2009 academic year. This was a 5.9% increase from the previous year.
Source: The College Board

(a) If the cost of college increases by 5.9% each year, what will be the average cost of college at a 4-year private college for the 2028–2029 academic year?

(b) College savings plans, such as a 529 plan, allow individuals to put money aside now to help pay for college later. If one such plan offers a rate of 4% compounded continuously, how much should be put in a college savings plan in 2010 to pay for 1 year of the cost of college at a 4-year private college for an incoming freshman in 2028?

54. Analyzing Interest Rates on a Mortgage Colleen and Bill have just purchased a house for $650,000, with the seller holding a second mortgage of $100,000. They promise to pay the seller $100,000 plus all accrued interest 5 years from now. The seller offers them three interest options on the second mortgage:

(a) Simple interest at 12% per annum

(b) $11\frac{1}{2}$% interest compounded monthly

(c) $11\frac{1}{4}$% interest compounded continuously

Which option is best; that is, which results in the least interest on the loan?

55. 2009 Federal Stimulus Package In February 2009, President Obama signed into law a $787 billion federal stimulus package. At that time, 20-year Series EE bonds had a fixed rate of 1.3% compounded semiannually. If the federal government financed the stimulus through EE bonds, how much would it have to pay back in 2029? How much interest was paid to finance the stimulus?
Source: U.S. Treasury Department

56. Per Capita Federal Debt In 2008, the federal debt was about $10 trillion. In 2008, the U.S. population was about 304 million. Assuming that the federal debt is increasing about 7.8% per year and the U.S. population is increasing about 0.9% per year, determine the per capita debt (total debt divided by population) in 2020.

Inflation Problems 57–62 require the following discussion. *Inflation* is a term used to describe the erosion of the purchasing power of money. For example, if the annual inflation rate is 3%, then $1000 worth of purchasing power now will have only $970 worth of purchasing power in 1 year because 3% of the original $1000 (0.03 × 1000 = 30) has been eroded due to inflation. In general, if the rate of inflation averages r per annum over n years, the amount A that $P will purchase after n years is

$$A = P \cdot (1 - r)^n$$

where r is expressed as a decimal.

57. Inflation If the inflation rate averages 3%, how much will $1000 purchase in 2 years?

58. Inflation If the inflation rate averages 2%, how much will $1000 purchase in 3 years?

59. Inflation If the amount that $1000 will purchase is only $950 after 2 years, what was the average inflation rate?

60. Inflation If the amount that $1000 will purchase is only $930 after 2 years, what was the average inflation rate?

61. Inflation If the average inflation rate is 2%, how long is it until purchasing power is cut in half?

62. Inflation If the average inflation rate is 4%, how long is it until purchasing power is cut in half?

Problems 63–66 involve zero-coupon bonds. A *zero-coupon bond* is a bond that is sold now at a discount and will pay its face value at the time when it matures; no interest payments are made.

63. Zero-Coupon Bonds A zero-coupon bond can be redeemed in 20 years for $10,000. How much should you be willing to pay for it now if you want a return of:
(a) 10% compounded monthly?
(b) 10% compounded continuously?

64. Zero-Coupon Bonds A child's grandparents are considering buying a $40,000 face-value, zero-coupon bond at birth so that she will have enough money for her college education 17 years later. If they want a rate of return of 8% compounded annually, what should they pay for the bond?

65. Zero-Coupon Bonds How much should a $10,000 face-value, zero-coupon bond, maturing in 10 years, be sold for now if its rate of return is to be 8% compounded annually?

67. Time to Double or Triple an Investment The formula

$$t = \frac{\ln m}{n \ln\left(1 + \dfrac{r}{n}\right)}$$

can be used to find the number of years t required to multiply an investment m times when r is the per annum interest rate compounded n times a year.
(a) How many years will it take to double the value of an IRA that compounds annually at the rate of 12%?
(b) How many years will it take to triple the value of a savings account that compounds quarterly at an annual rate of 6%?
(c) Give a derivation of this formula.

66. Zero-Coupon Bonds If Pat pays $12,485.52 for a $25,000 face-value, zero-coupon bond that matures in 8 years, what is his annual rate of return?

68. Time to Reach an Investment Goal The formula

$$t = \frac{\ln A - \ln P}{r}$$

can be used to find the number of years t required for an investment P to grow to a value A when compounded continuously at an annual rate r.
(a) How long will it take to increase an initial investment of $1000 to $8000 at an annual rate of 10%?
(b) What annual rate is required to increase the value of a $2000 IRA to $30,000 in 35 years?
(c) Give a derivation of this formula.

Problems 69–72 require the following discussion. The **Consumer Price Index (CPI)** *indicates the relative change in price over time for a fixed basket of goods and services. It is a cost of living index that helps measure the effect of inflation on the cost of goods and services. The CPI uses the base period 1982–1984 for comparison (the CPI for this period is 100). The CPI for January 2006 was 198.3. This means that $100 in the period 1982–1984 had the same purchasing power as $198.30 in January 2006. In general, if the rate of inflation averages r per annum over n years, then the CPI index after n years is*

$$\text{CPI} = \text{CPI}_0\left(1 + \frac{r}{100}\right)^n$$

where CPI$_0$ is the CPI index at the beginning of the n-year period.
Source: U.S. Bureau of Labor Statistics

69. Consumer Price Index
(a) The CPI was 163.0 for 1998 and 215.3 for 2008. Assuming that annual inflation remained constant for this time period, determine the average annual inflation rate.
(b) Using the inflation rate from part (a), in what year will the CPI reach 300?

70. Consumer Price Index If the current CPI is 234.2 and the average annual inflation rate is 2.8%, what will be the CPI in 5 years?

71. Consumer Price Index If the average annual inflation rate is 3.1%, how long will it take for the CPI index to double? (A doubling of the CPI index means purchasing power is cut in half.)

72. Consumer Price Index The base period for the CPI changed in 1998. Under the previous weight and item structure, the CPI for 1995 was 456.5. If the average annual inflation rate was 5.57%, what year was used as the base period for the CPI?

Explaining Concepts: Discussion and Writing

73. Explain in your own words what the term *compound interest* means. What does *continuous compounding* mean?

74. Explain in your own words the meaning of *present value*.

75. Critical Thinking You have just contracted to buy a house and will seek financing in the amount of $100,000. You go to several banks. Bank 1 will lend you $100,000 at the rate of 8.75% amortized over 30 years with a loan origination fee of 1.75%. Bank 2 will lend you $100,000 at the rate of 8.375% amortized over 15 years with a loan origination fee of 1.5%. Bank 3 will lend you $100,000 at the rate of 9.125% amortized over 30 years with no loan origination fee. Bank 4 will lend you $100,000 at the rate of 8.625% amortized over 15 years with no loan origination fee. Which loan would you take? Why? Be sure to have sound reasons for your choice.

Use the information in the table to assist you. If the amount of the monthly payment does not matter to you, which loan would you take? Again, have sound reasons for your choice. Compare your final decision with others in the class. Discuss.

	Monthly Payment	Loan Origination Fee
Bank 1	$786.70	$1,750.00
Bank 2	$977.42	$1,500.00
Bank 3	$813.63	$0.00
Bank 4	$992.08	$0.00

'Are You Prepared?' Answers

1. $15

2. $13\frac{1}{3}\%$

6.8 Exponential Growth and Decay Models; Newton's Law; Logistic Growth and Decay Models

OBJECTIVES **1** Find Equations of Populations That Obey the Law of Uninhibited Growth (p. 476)

2 Find Equations of Populations That Obey the Law of Decay (p. 478)

3 Use Newton's Law of Cooling (p. 479)

4 Use Logistic Models (p. 481)

1 Find Equations of Populations That Obey the Law of Uninhibited Growth

Many natural phenomena have been found to follow the law that an amount A varies with time t according to the function

Figure 41

(a) $A(t) = A_0 e^{kt}, k > 0$
Exponential growth

(b) $A(t) = A_0 e^{kt}, k < 0$
Exponential decay

$$A(t) = A_0 e^{kt} \qquad \textbf{(1)}$$

Here A_0 is the original amount ($t = 0$) and $k \neq 0$ is a constant.

If $k > 0$, then equation (1) states that the amount A is increasing over time; if $k < 0$, the amount A is decreasing over time. In either case, when an amount A varies over time according to equation (1), it is said to follow the **exponential law** or the **law of uninhibited growth** ($k > 0$) **or decay** ($k < 0$). See Figure 41.

For example, we saw in Section 6.7 that continuously compounded interest follows the law of uninhibited growth. In this section we shall look at some additional phenomena that follow the exponential law.

Cell division is the growth process of many living organisms, such as amoebas, plants, and human skin cells. Based on an ideal situation in which no cells die and no by-products are produced, the number of cells present at a given time follows the law of uninhibited growth. Actually, however, after enough time has passed, growth at an exponential rate will cease due to the influence of factors such as lack of living space and dwindling food supply. The law of uninhibited growth accurately models only the early stages of the cell division process.

The cell division process begins with a culture containing N_0 cells. Each cell in the culture grows for a certain period of time and then divides into two identical cells. We assume that the time needed for each cell to divide in two is constant and does not change as the number of cells increases. These new cells then grow, and eventually each divides in two, and so on.

Uninhibited Growth of Cells

A model that gives the number N of cells in a culture after a time t has passed (in the early stages of growth) is

$$N(t) = N_0 e^{kt} \qquad k > 0 \qquad \textbf{(2)}$$

where N_0 is the initial number of cells and k is a positive constant that represents the growth rate of the cells.

In using formula (2) to model the growth of cells, we are using a function that yields positive real numbers, even though we are counting the number of cells, which must be an integer. This is a common practice in many applications.

EXAMPLE 1 **Bacterial Growth**

A colony of bacteria that grows according to the law of uninhibited growth is modeled by the function $N(t) = 100e^{0.045t}$, where N is measured in grams and t is measured in days.

(a) Determine the initial amount of bacteria.

(b) What is the growth rate of the bacteria?

(c) What is the population after 5 days?

(d) How long will it take for the population to reach 140 grams?

(e) What is the doubling time for the population?

Solution (a) The initial amount of bacteria, N_0, is obtained when $t = 0$, so

$$N_0 = N(0) = 100e^{0.045(0)} = 100 \text{ grams}$$

(b) Compare $N(t) = 100e^{0.045t}$ to $N(t) = N_0e^{kt}$. The value of k, 0.045, indicates a growth rate of 4.5%.

(c) The population after 5 days is $N(5) = 100e^{0.045(5)} \approx 125.2$ grams.

(d) To find how long it takes for the population to reach 140 grams, solve the equation $N(t) = 140$.

$$100e^{0.045t} = 140$$
$$e^{0.045t} = 1.4 \qquad \text{\small Divide both sides of the equation by 100.}$$
$$0.045t = \ln 1.4 \qquad \text{\small Rewrite as a logarithm.}$$
$$t = \frac{\ln 1.4}{0.045} \qquad \text{\small Divide both sides of the equation by 0.045.}$$
$$\approx 7.5 \text{ days}$$

(e) The population doubles when $N(t) = 200$ grams, so we find the doubling time by solving the equation $200 = 100e^{0.045t}$ for t.

$$200 = 100e^{0.045t}$$
$$2 = e^{0.045t} \qquad \text{\small Divide both sides of the equation by 100.}$$
$$\ln 2 = 0.045t \qquad \text{\small Rewrite as a logarithm.}$$
$$t = \frac{\ln 2}{0.045} \qquad \text{\small Divide both sides of the equation by 0.045.}$$
$$\approx 15.4 \text{ days}$$

The population doubles approximately every 15.4 days.

──────── **Now Work** PROBLEM 1

EXAMPLE 2 **Bacterial Growth**

A colony of bacteria increases according to the law of uninhibited growth.

(a) If N is the number of cells and t is the time in hours, express N as a function of t.

(b) If the number of bacteria doubles in 3 hours, find the function that gives the number of cells in the culture.

(c) How long will it take for the size of the colony to triple?

(d) How long will it take for the population to double a second time (that is, increase four times)?

Solution (a) Using formula (2), the number N of cells at time t is

$$N(t) = N_0e^{kt}$$

where N_0 is the initial number of bacteria present and k is a positive number.

(b) We seek the number k. The number of cells doubles in 3 hours, so

$$N(3) = 2N_0$$

But $N(3) = N_0 e^{k(3)}$, so

$$N_0 e^{k(3)} = 2N_0$$
$$e^{3k} = 2 \qquad \text{Divide both sides by } N_0.$$
$$3k = \ln 2 \qquad \text{Write the exponential equation as a logarithm.}$$
$$k = \frac{1}{3} \ln 2 \approx 0.23105$$

The function that models this growth process is therefore

$$N(t) = N_0 e^{0.23105t}$$

(c) The time t needed for the size of the colony to triple requires that $N = 3N_0$. Substitute $3N_0$ for N to get

$$3N_0 = N_0 e^{0.23105t}$$
$$3 = e^{0.23105t}$$
$$0.23105t = \ln 3$$
$$t = \frac{\ln 3}{0.23105} \approx 4.755 \text{ hours}$$

It will take about 4.755 hours or 4 hours, 45 minutes for the size of the colony to triple.

(d) If a population doubles in 3 hours, it will double a second time in 3 more hours, for a total time of 6 hours. ⌐

2 Find Equations of Populations That Obey the Law of Decay

Radioactive materials follow the law of uninhibited decay.

Uninhibited Radioactive Decay

The amount A of a radioactive material present at time t is given by

$$A(t) = A_0 e^{kt} \qquad k < 0 \qquad \text{(3)}$$

where A_0 is the original amount of radioactive material and k is a negative number that represents the rate of decay.

All radioactive substances have a specific **half-life,** which is the time required for half of the radioactive substance to decay. In **carbon dating,** we use the fact that all living organisms contain two kinds of carbon, carbon 12 (a stable carbon) and carbon 14 (a radioactive carbon with a half-life of 5600 years). While an organism is living, the ratio of carbon 12 to carbon 14 is constant. But when an organism dies, the original amount of carbon 12 present remains unchanged, whereas the amount of carbon 14 begins to decrease. This change in the amount of carbon 14 present relative to the amount of carbon 12 present makes it possible to calculate when the organism died.

| EXAMPLE 3 | **Estimating the Age of Ancient Tools** |

Traces of burned wood along with ancient stone tools in an archeological dig in Chile were found to contain approximately 1.67% of the original amount of carbon 14. If the half-life of carbon 14 is 5600 years, approximately when was the tree cut and burned?

Solution Using formula (3), the amount A of carbon 14 present at time t is

$$A(t) = A_0 e^{kt}$$

where A_0 is the original amount of carbon 14 present and k is a negative number. We first seek the number k. To find it, we use the fact that after 5600 years half of the original amount of carbon 14 remains, so $A(5600) = \dfrac{1}{2} A_0$. Then

$$\frac{1}{2} A_0 = A_0 e^{k(5600)}$$

$$\frac{1}{2} = e^{5600k} \qquad\qquad \text{Divide both sides of the equation by } A_0.$$

$$5600k = \ln \frac{1}{2} \qquad\qquad \text{Rewrite as a logarithm.}$$

$$k = \frac{1}{5600} \ln \frac{1}{2} \approx -0.000124$$

Formula (3) therefore becomes

$$A(t) = A_0 e^{-0.000124t}$$

If the amount A of carbon 14 now present is 1.67% of the original amount, it follows that

$$0.0167 A_0 = A_0 e^{-0.000124t}$$

$$0.0167 = e^{-0.000124t} \qquad\qquad \text{Divide both sides of the equation by } A_0.$$

$$-0.000124t = \ln 0.0167 \qquad\qquad \text{Rewrite as a logarithm.}$$

$$t = \frac{\ln 0.0167}{-0.000124} \approx 33{,}003 \text{ years}$$

The tree was cut and burned about 33,003 years ago. Some archeologists use this conclusion to argue that humans lived in the Americas 33,000 years ago, much earlier than is generally accepted.

Now Work PROBLEM 3

3 Use Newton's Law of Cooling

Newton's Law of Cooling* states that the temperature of a heated object decreases exponentially over time toward the temperature of the surrounding medium.

> **Newton's Law of Cooling**
>
> The temperature u of a heated object at a given time t can be modeled by the following function:
>
> $$u(t) = T + (u_0 - T)e^{kt} \qquad k < 0 \tag{4}$$
>
> where T is the constant temperature of the surrounding medium, u_0 is the initial temperature of the heated object, and k is a negative constant.

EXAMPLE 4 **Using Newton's Law of Cooling**

An object is heated to 100°C (degrees Celsius) and is then allowed to cool in a room whose air temperature is 30°C.

(a) If the temperature of the object is 80°C after 5 minutes, when will its temperature be 50°C?

(b) Determine the elapsed time before the temperature of the object is 35°C.

(c) What do you notice about the temperature as time passes?

* Named after Sir Isaac Newton (1643–1727), one of the cofounders of calculus.

Solution (a) Using formula (4) with $T = 30$ and $u_0 = 100$, the temperature $u(t)$ (in degrees Celsius) of the object at time t (in minutes) is

$$u(t) = 30 + (100 - 30)e^{kt} = 30 + 70e^{kt}$$

where k is a negative constant. To find k, use the fact that $u = 80$ when $t = 5$. Then

$$u(t) = 30 + 70e^{kt}$$
$$80 = 30 + 70e^{k(5)} \qquad u(5) = 80$$
$$50 = 70e^{5k} \qquad \text{Simplify.}$$
$$e^{5k} = \frac{50}{70} \qquad \text{Solve for } e^{5k}.$$
$$5k = \ln\frac{5}{7} \qquad \text{Take ln of both sides.}$$
$$k = \frac{1}{5}\ln\frac{5}{7} \approx -0.0673 \quad \text{Solve for } k.$$

Formula (4) therefore becomes

$$u(t) = 30 + 70e^{-0.0673t} \qquad\qquad\qquad\qquad (5)$$

We want to find t when $u = 50°C$, so

$$50 = 30 + 70e^{-0.0673t}$$
$$20 = 70e^{-0.0673t} \qquad \text{Simplify.}$$
$$e^{-0.0673t} = \frac{20}{70}$$
$$-0.0673t = \ln\frac{2}{7} \qquad \text{Take ln of both sides.}$$
$$t = \frac{\ln\dfrac{2}{7}}{-0.0673} \approx 18.6 \text{ minutes} \quad \text{Solve for } t.$$

The temperature of the object will be 50°C after about 18.6 minutes or 18 minutes, 36 seconds.

(b) If $u = 35°C$, then, based on equation (5), we have

$$35 = 30 + 70e^{-0.0673t}$$
$$5 = 70e^{-0.0673t} \qquad \text{Simplify.}$$
$$e^{-0.0673t} = \frac{5}{70}$$
$$-0.0673t = \ln\frac{5}{70} \qquad \text{Take ln of both sides.}$$
$$t = \frac{\ln\dfrac{5}{70}}{-0.0673} \approx 39.2 \text{ minutes} \quad \text{Solve for } t.$$

The object will reach a temperature of 35°C after about 39.2 minutes.

(c) Look at equation (5). As t increases, the exponent $-0.0673t$ becomes unbounded in the negative direction. As a result, the value of $e^{-0.0673t}$ approaches zero so the value of u, the temperature of the object, approaches 30°C, the air temperature of the room. ♩

Now Work PROBLEM 13

4 Use Logistic Models

The exponential growth model $A(t) = A_0 e^{kt}$, $k > 0$, assumes uninhibited growth, meaning that the value of the function grows without limit. Recall that we stated that cell division could be modeled using this function, assuming that no cells die and no by-products are produced. However, cell division eventually is limited by factors such as living space and food supply. The **logistic model,** given next, can describe situations where the growth or decay of the dependent variable is limited.

Logistic Model

In a logistic model, the population P after time t is given by the function

$$P(t) = \frac{c}{1 + ae^{-bt}} \qquad (6)$$

where a, b, and c are constants with $a > 0$ and $c > 0$. The model is a growth model if $b > 0$; the model is a decay model if $b < 0$.

The number c is called the **carrying capacity** (for growth models) because the value $P(t)$ approaches c as t approaches infinity; that is, $\lim\limits_{t \to \infty} P(t) = c$. The number $|b|$ is the growth rate for $b > 0$ and the decay rate for $b < 0$. Figure 42(a) shows the graph of a typical logistic growth function, and Figure 42(b) shows the graph of a typical logistic decay function.

Figure 42

(a) (b)

Based on the figures, we have the following properties of logistic growth functions.

Properties of the Logistic Model, Equation (6)

1. The domain is the set of all real numbers. The range is the interval $(0, c)$, where c is the carrying capacity.

2. There are no x-intercepts; the y-intercept is $P(0)$.

3. There are two horizontal asymptotes: $y = 0$ and $y = c$.

4. $P(t)$ is an increasing function if $b > 0$ and a decreasing function if $b < 0$.

5. There is an **inflection point** where $P(t)$ equals $\dfrac{1}{2}$ of the carrying capacity.

 The inflection point is the point on the graph where the graph changes from being curved upward to curved downward for growth functions and the point where the graph changes from being curved downward to curved upward for decay functions.

6. The graph is smooth and continuous, with no corners or gaps.

EXAMPLE 5 Fruit Fly Population

Fruit flies are placed in a half-pint milk bottle with a banana (for food) and yeast plants (for food and to provide a stimulus to lay eggs). Suppose that the fruit fly population after t days is given by

$$P(t) = \frac{230}{1 + 56.5e^{-0.37t}}$$

(a) State the carrying capacity and the growth rate.

(b) Determine the initial population.

(c) What is the population after 5 days?

(d) How long does it take for the population to reach 180?

(e) Use a graphing utility to determine how long it takes for the population to reach one-half of the carrying capacity by graphing $Y_1 = P(t)$ and $Y_2 = 115$ and using INTERSECT.

Solution (a) As $t \to \infty$, $e^{-0.37t} \to 0$ and $P(t) \to \dfrac{230}{1}$. The carrying capacity of the half-pint bottle is 230 fruit flies. The growth rate is $|b| = |0.37| = 37\%$ per day.

(b) To find the initial number of fruit flies in the half-pint bottle, evaluate $P(0)$.

$$\begin{aligned} P(0) &= \frac{230}{1 + 56.5e^{-0.37(0)}} \\ &= \frac{230}{1 + 56.5} \\ &= 4 \end{aligned}$$

So, initially, there were 4 fruit flies in the half-pint bottle.

(c) To find the number of fruit flies in the half-pint bottle after 5 days, evaluate $P(5)$.

$$P(5) = \frac{230}{1 + 56.5e^{-0.37(5)}} \approx 23 \text{ fruit flies}$$

After 5 days, there are approximately 23 fruit flies in the bottle.

(d) To determine when the population of fruit flies will be 180, solve the equation $P(t) = 180$.

$$\frac{230}{1 + 56.5e^{-0.37t}} = 180$$

$$230 = 180(1 + 56.5e^{-0.37t})$$

$$1.2778 = 1 + 56.5e^{-0.37t} \qquad \text{Divide both sides by 180.}$$

$$0.2778 = 56.5e^{-0.37t} \qquad \text{Subtract 1 from both sides.}$$

$$0.0049 = e^{-0.37t} \qquad \text{Divide both sides by 56.5.}$$

$$\ln(0.0049) = -0.37t \qquad \text{Rewrite as a logarithmic expression.}$$

$$t \approx 14.4 \text{ days} \qquad \text{Divide both sides by } -0.37.$$

Figure 43

It will take approximately 14.4 days (14 days, 10 hours) for the population to reach 180 fruit flies.

(e) One-half of the carrying capacity is 115 fruit flies. We solve $P(t) = 115$ by graphing $Y_1 = \dfrac{230}{1 + 56.5e^{-0.37t}}$ and $Y_2 = 115$ and using INTERSECT. See Figure 43. The population will reach one-half of the carrying capacity in about 10.9 days (10 days, 22 hours).

 Look back at Figure 43. Notice the point where the graph reaches 115 fruit flies (one-half of the carrying capacity): the graph changes from being curved upward to being curved downward. Using the language of calculus, we say the graph changes from increasing at an increasing rate to increasing at a decreasing rate. For any logistic growth function, when the population reaches one-half the carrying capacity, the population growth starts to slow down.

Now Work PROBLEM 23

Exploration

On the same viewing rectangle, graph

$$Y_1 = \frac{500}{1 + 24e^{-0.03t}} \quad \text{and} \quad Y_2 = \frac{500}{1 + 24e^{-0.08t}}$$

What effect does the growth rate $|b|$ have on the logistic growth function?

EXAMPLE 6 | **Wood Products**

The EFISCEN wood product model classifies wood products according to their life-span. There are four classifications: short (1 year), medium short (4 years), medium long (16 years), and long (50 years). Based on data obtained from the European Forest Institute, the percentage of remaining wood products after t years for wood products with long life-spans (such as those used in the building industry) is given by

$$P(t) = \frac{100.3952}{1 + 0.0316e^{0.0581t}}$$

(a) What is the decay rate?

(b) What is the percentage of remaining wood products after 10 years?

(c) How long does it take for the percentage of remaining wood products to reach 50%?

(d) Explain why the numerator given in the model is reasonable.

Solution

(a) The decay rate is $|b| = |-0.0581| = 5.81\%$.

(b) Evaluate $P(10)$.

$$P(10) = \frac{100.3952}{1 + 0.0316e^{0.0581(10)}} \approx 95.0$$

So 95% of long-life-span wood products remain after 10 years.

(c) Solve the equation $P(t) = 50$.

$$\frac{100.3952}{1 + 0.0316e^{0.0581t}} = 50$$

$$100.3952 = 50(1 + 0.0316e^{0.0581t})$$

$$2.0079 = 1 + 0.0316e^{0.0581t} \qquad \text{Divide both sides by 50.}$$

$$1.0079 = 0.0316e^{0.0581t} \qquad \text{Subtract 1 from both sides.}$$

$$31.8956 = e^{0.0581t} \qquad \text{Divide both sides by 0.0316.}$$

$$\ln(31.8956) = 0.0581t \qquad \text{Rewrite as a logarithmic expression.}$$

$$t \approx 59.6 \text{ years} \qquad \text{Divide both sides by 0.0581.}$$

It will take approximately 59.6 years for the percentage of long-life-span wood products remaining to reach 50%.

(d) The numerator of 100.3952 is reasonable because the maximum percentage of wood products remaining that is possible is 100%.

6.8 Assess Your Understanding

Applications and Extensions

1. **Growth of an Insect Population** The size P of a certain insect population at time t (in days) obeys the function $P(t) = 500e^{0.02t}$.
 (a) Determine the number of insects at $t = 0$ days.
 (b) What is the growth rate of the insect population?
 (c) What is the population after 10 days?
 (d) When will the insect population reach 800?
 (e) When will the insect population double?

2. **Growth of Bacteria** The number N of bacteria present in a culture at time t (in hours) obeys the law of uninhibited growth $N(t) = 1000e^{0.01t}$.
 (a) Determine the number of bacteria at $t = 0$ hours.
 (b) What is the growth rate of the bacteria?
 (c) What is the population after 4 hours?
 (d) When will the number of bacteria reach 1700?
 (e) When will the number of bacteria double?

3. **Radioactive Decay** Strontium 90 is a radioactive material that decays according to the function $A(t) = A_0 e^{-0.0244t}$, where A_0 is the initial amount present and A is the amount present at time t (in years). Assume that a scientist has a sample of 500 grams of strontium 90.
 (a) What is the decay rate of strontium 90?
 (b) How much strontium 90 is left after 10 years?
 (c) When will 400 grams of strontium 90 be left?
 (d) What is the half-life of strontium 90?

4. **Radioactive Decay** Iodine 131 is a radioactive material that decays according to the function $A(t) = A_0 e^{-0.087t}$, where A_0 is the initial amount present and A is the amount present at time t (in days). Assume that a scientist has a sample of 100 grams of iodine 131.
 (a) What is the decay rate of iodine 131?
 (b) How much iodine 131 is left after 9 days?
 (c) When will 70 grams of iodine 131 be left?
 (d) What is the half-life of iodine 131?

5. **Growth of a Colony of Mosquitoes** The population of a colony of mosquitoes obeys the law of uninhibited growth.
 (a) If N is the population of the colony and t is the time in days, express N as a function of t.
 (b) If there are 1000 mosquitoes initially and there are 1800 after 1 day, what is the size of the colony after 3 days?
 (c) How long is it until there are 10,000 mosquitoes?

6. **Bacterial Growth** A culture of bacteria obeys the law of uninhibited growth.
 (a) If N is the number of bacteria in the culture and t is the time in hours, express N as a function of t.
 (b) If 500 bacteria are present initially and there are 800 after 1 hour, how many will be present in the culture after 5 hours?
 (c) How long is it until there are 20,000 bacteria?

7. **Population Growth** The population of a southern city follows the exponential law.
 (a) If N is the population of the city and t is the time in years, express N as a function of t.

 (b) If the population doubled in size over an 18-month period and the current population is 10,000, what will the population be 2 years from now?

8. **Population Decline** The population of a midwestern city follows the exponential law.
 (a) If N is the population of the city and t is the time in years, express N as a function of t.
 (b) If the population decreased from 900,000 to 800,000 from 2008 to 2010, what will the population be in 2012?

9. **Radioactive Decay** The half-life of radium is 1690 years. If 10 grams is present now, how much will be present in 50 years?

10. **Radioactive Decay** The half-life of radioactive potassium is 1.3 billion years. If 10 grams is present now, how much will be present in 100 years? In 1000 years?

11. **Estimating the Age of a Tree** A piece of charcoal is found to contain 30% of the carbon 14 that it originally had. When did the tree die from which the charcoal came? Use 5600 years as the half-life of carbon 14.

12. **Estimating the Age of a Fossil** A fossilized leaf contains 70% of its normal amount of carbon 14. How old is the fossil?

13. **Cooling Time of a Pizza Pan** A pizza pan is removed at 5:00 PM from an oven whose temperature is fixed at 450°F into a room that is a constant 70°F. After 5 minutes, the pan is 300°F.
 (a) At what time is the temperature of the pan 135°F?
 (b) Determine the time that needs to elapse before the pan is 160°F.
 (c) What do you notice about the temperature as time passes?

14. **Newton's Law of Cooling** A thermometer reading 72°F is placed in a refrigerator where the temperature is a constant 38°F.
 (a) If the thermometer reads 60°F after 2 minutes, what will it read after 7 minutes?
 (b) How long will it take before the thermometer reads 39°F?
 (c) Determine the time needed to elapse before the thermometer reads 45°F.
 (d) What do you notice about the temperature as time passes?

15. **Newton's Law of Heating** A thermometer reading 8°C is brought into a room with a constant temperature of 35°C. If the thermometer reads 15°C after 3 minutes, what will it read after being in the room for 5 minutes? For 10 minutes?

 [**Hint:** You need to construct a formula similar to equation (4).]

16. **Warming Time of a Beer Stein** A beer stein has a temperature of 28°F. It is placed in a room with a constant temperature of 70°F. After 10 minutes, the temperature of the stein has risen to 35°F. What will the temperature of the stein be after 30 minutes? How long will it take the stein to reach a temperature of 45°F? (See the hint given for Problem 15.)

17. **Decomposition of Chlorine in a Pool** Under certain water conditions, the free chlorine (hypochlorous acid, HOCl) in a swimming pool decomposes according to the law of uninhibited decay. After shocking his pool, Ben tested the water and found the amount of free chlorine to be 2.5 parts per million (ppm). Twenty-four hours later, Ben tested the water again and found the amount of free chlorine to be 2.2 ppm. What will be the reading after 3 days (that is, 72 hours)? When the chlorine level reaches 1.0 ppm, Ben must shock the pool again. How long can Ben go before he must shock the pool again?

18. **Decomposition of Dinitrogen Pentoxide** At 45°C, dinitrogen pentoxide (N_2O_5) decomposes into nitrous dioxide (NO_2) and oxygen (O_2) according to the law of uninhibited decay. An initial amount of 0.25 M of dinitrogen pentoxide decomposes to 0.15 M in 17 minutes. How much dinitrogen pentoxide will remain after 30 minutes? How long will it take until 0.01 M of dinitrogen pentoxide remains?

19. **Decomposition of Sucrose** Reacting with water in an acidic solution at 35°C, sucrose ($C_{12}H_{22}O_{11}$) decomposes into glucose ($C_6H_{12}O_6$) and fructose ($C_6H_{12}O_6$)* according to the law of uninhibited decay. An initial amount of 0.40 M of sucrose decomposes to 0.36 M in 30 minutes. How much sucrose will remain after 2 hours? How long will it take until 0.10 M of sucrose remains?

20. **Decomposition of Salt in Water** Salt (NaCl) decomposes in water into sodium (Na^+) and chloride (Cl^-) ions according to the law of uninhibited decay. If the initial amount of salt is 25 kilograms and, after 10 hours, 15 kilograms of salt is left, how much salt is left after 1 day? How long does it take until $\frac{1}{2}$ kilogram of salt is left?

21. **Radioactivity from Chernobyl** After the release of radioactive material into the atmosphere from a nuclear power plant at Chernobyl (Ukraine) in 1986, the hay in Austria was contaminated by iodine 131 (half-life 8 days). If it is safe to feed the hay to cows when 10% of the iodine 131 remains, how long did the farmers need to wait to use this hay?

22. **Pig Roasts** The hotel Bora-Bora is having a pig roast. At noon, the chef put the pig in a large earthen oven. The pig's original temperature was 75°F. At 2:00 PM the chef checked the pig's temperature and was upset because it had reached only 100°F. If the oven's temperature remains a constant

325°F, at what time may the hotel serve its guests, assuming that pork is done when it reaches 175°F?

23. **Population of a Bacteria Culture** The logistic growth model

$$P(t) = \frac{1000}{1 + 32.33e^{-0.439t}}$$

 represents the population (in grams) of a bacterium after t hours.
 (a) Determine the carrying capacity of the environment.
 (b) What is the growth rate of the bacteria?
 (c) Determine the initial population size.
 (d) What is the population after 9 hours?
 (e) When will the population be 700 grams?
 (f) How long does it take for the population to reach one-half the carrying capacity?

24. **Population of an Endangered Species** Often environmentalists capture an endangered species and transport the species to a controlled environment where the species can produce offspring and regenerate its population. Suppose that six American bald eagles are captured, transported to Montana, and set free. Based on experience, the environmentalists expect the population to grow according to the model

$$P(t) = \frac{500}{1 + 83.33e^{-0.162t}}$$

 where t is measured in years.

 (a) Determine the carrying capacity of the environment.
 (b) What is the growth rate of the bald eagle?
 (c) What is the population after 3 years?
 (d) When will the population be 300 eagles?
 (e) How long does it take for the population to reach one-half of the carrying capacity?

25. **The *Challenger* Disaster** After the *Challenger* disaster in 1986, a study was made of the 23 launches that preceded the fatal flight. A mathematical model was developed involving the relationship between the Fahrenheit temperature x around the O-rings and the number y of eroded or leaky primary O-rings. The model stated that

$$y = \frac{6}{1 + e^{-(5.085 - 0.1156x)}}$$

 where the number 6 indicates the 6 primary O-rings on the spacecraft.

* Author's Note: Surprisingly, the chemical formulas for glucose and fructose are the same: This is not a typo.

(a) What is the predicted number of eroded or leaky primary O-rings at a temperature of 100°F?

(b) What is the predicted number of eroded or leaky primary O-rings at a temperature of 60°F?

(c) What is the predicted number of eroded or leaky primary O-rings at a temperature of 30°F?

(d) Graph the equation using a graphing utility. At what temperature is the predicted number of eroded or leaky O-rings 1? 3? 5?

Source: Linda Tappin, *"Analyzing Data Relating to the* Challenger *Disaster,"* Mathematics Teacher, *Vol. 87, No. 6, September 1994, pp. 423–426.*

6.9　Building Exponential, Logarithmic, and Logistic Models from Data

PREPARING FOR THIS SECTION　*Before getting started, review the following:*

- Building Linear Models from Data (Section 4.2, pp. 282–285)
- Building Cubic Models from Data (Section 5.1, pp. 336–337)
- Building Quadratic Models from Data (Section 4.4, pp. 300–305)

OBJECTIVES　**1** Build an Exponential Model from Data　(p. 487)

　　　　　　　2 Build a Logarithmic Model from Data　(p. 488)

　　　　　　　3 Build a Logistic Model from Data　(p. 489)

In Section 4.2, we discussed how to find the linear function of best fit $(y = ax + b)$, in Section 4.4, we discussed how to find the quadratic function of best fit $(y = ax^2 + bx + c)$, and in Section 5.1, we discussed how to find the cubic function of best fit $(y = ax^3 + bx^2 + cx + d)$.

In this section we discuss how to use a graphing utility to find equations of best fit that describe the relation between two variables when the relation is thought to be exponential $(y = ab^x)$, logarithmic $(y = a + b \ln x)$, or logistic $\left(y = \dfrac{c}{1 + ae^{-bx}}\right)$. As before, we draw a scatter diagram of the data to help to determine the appropriate model to use.

Figure 44 shows scatter diagrams that will typically be observed for the three models. Below each scatter diagram are any restrictions on the values of the parameters.

Figure 44

$y = ab^x, a > 0, b > 1$	$y = ab^x, 0 < b < 1, a > 0$	$y = a + b \ln x, a > 0, b < 0$	$y = a + b \ln x, a > 0, b > 0$	$y = \dfrac{c}{1 + ae^{-bx}}, a > 0, b > 0, c > 0$
Exponential	Exponential	Logarithmic	Logarithmic	Logistic

Most graphing utilities have REGression options that fit data to a specific type of curve. Once the data have been entered and a scatter diagram obtained, the type of curve that you want to fit to the data is selected. Then that REGression option is used to obtain the curve of *best fit* of the type selected.

The correlation coefficient r will appear only if the model can be written as a linear expression. As it turns out, r will appear for the linear, power, exponential, and logarithmic models, since these models can be written as a linear expression. Remember, the closer $|r|$ is to 1, the better the fit.

1 Build an Exponential Model from Data

We saw in Section 6.7 that the future value of money behaves exponentially, and we saw in Section 6.8 that growth and decay models also behave exponentially. The next example shows how data can lead to an exponential model.

EXAMPLE 1

Fitting an Exponential Function to Data

Kathleen is interested in finding a function that explains the growth of cell phone usage in the United States. She gathers data on the number (in millions) of U.S. cell phone subscribers from 1985 through 2008. The data are shown in Table 9.

(a) Using a graphing utility, draw a scatter diagram with year as the independent variable.
(b) Using a graphing utility, build an exponential model from the data.
(c) Express the function found in part (b) in the form $A = A_0 e^{kt}$.
(d) Graph the exponential function found in part (b) or (c) on the scatter diagram.
(e) Using the solution to part (b) or (c), predict the number of U.S. cell phone subscribers in 2009.
(f) Interpret the value of k found in part (c).

Solution

(a) Enter the data into the graphing utility, letting 1 represent 1985, 2 represent 1986, and so on. We obtain the scatter diagram shown in Figure 45.

(b) A graphing utility fits the data in Figure 45 to an exponential function of the form $y = ab^x$ using the EXPonential REGression option. From Figure 46 we find that $y = ab^x = 0.86498(1.31855)^x$. Notice that $|r|$ is close to 1, indicating a good fit.

Table 9

Year, x	Number of Subscribers (in millions), y
1985 (x = 1)	0.34
1986 (x = 2)	0.68
1987 (x = 3)	1.23
1988 (x = 4)	2.07
1989 (x = 5)	3.51
1990 (x = 6)	5.28
1991 (x = 7)	7.56
1992 (x = 8)	11.03
1993 (x = 9)	16.01
1994 (x = 10)	24.13
1995 (x = 11)	33.76
1996 (x = 12)	44.04
1997 (x = 13)	55.31
1998 (x = 14)	69.21
1999 (x = 15)	86.05
2000 (x = 16)	109.48
2001 (x = 17)	128.37
2002 (x = 18)	140.77
2003 (x = 19)	158.72
2004 (x = 20)	182.14
2005 (x = 21)	207.90
2006 (x = 22)	233.00
2007 (x = 23)	255.40
2008 (x = 24)	270.33

Source: ©2009 CTIA–The Wireless Association®. All Rights Reserved.

Figure 45

275

0 |_____ 25
0

Figure 46

```
ExpReg
y=a*b^x
a=.8649775385
b=1.318554023
r²=.9310544659
r=.9649116363
```

(c) To express $y = ab^x$ in the form $A = A_0 e^{kt}$, where $x = t$ and $y = A$, proceed as follows:

$$ab^x = A_0 e^{kt} \quad x = t$$

When $x = t = 0$, we find that $a = A_0$. This leads to

$$a = A_0 \qquad b^x = e^{kt}$$
$$b^x = (e^k)^t$$
$$b = e^k \qquad x = t$$

Since $y = ab^x = 0.86498(1.31855)^x$, we find that $a = 0.86498$ and $b = 1.31855$.

$$a = A_0 = 0.86498 \quad \text{and} \quad b = e^k = 1.31855$$

We want to find k, so we rewrite $e^k = 1.31855$ as a logarithm and obtain

$$k = \ln(1.31855) \approx 0.2765$$

As a result, $A = A_0 e^{kt} = 0.86498 e^{0.2765t}$.

(d) See Figure 47 for the graph of the exponential function of best fit.

(e) Let $t = 25$ (end of 2009) in the function found in part (c). The predicted number (in millions) of cell phone subscribers in the United States in 2009 is

$$A_0 e^{kt} = 0.86498 e^{0.2765(25)} \approx 869$$

This prediction (869 million) far exceeds what the U.S. population was in 2009 (currently the U.S. population is about 304 million). See the answer in part (f).

(f) The value of $k = 0.2765$ represents the growth rate of the number of cell phone subscribers in the United States. Over the period 1985 through 2008, the number of cell phone subscribers grew at an annual rate of 27.65% compounded continuously. This growth rate is not sustainable as we learned in part (e). In Problem 10 you are asked to build a better model from these data.

Figure 47

Now Work PROBLEM 1

2 Build a Logarithmic Model from Data

Many relations between variables do not follow an exponential model; instead, the independent variable is related to the dependent variable using a logarithmic model.

EXAMPLE 2

Fitting a Logarithmic Function to Data

Jodi, a meteorologist, is interested in finding a function that explains the relation between the height of a weather balloon (in kilometers) and the atmospheric pressure (measured in millimeters of mercury) on the balloon. She collects the data shown in Table 10.

(a) Using a graphing utility, draw a scatter diagram of the data with atmospheric pressure as the independent variable.

(b) It is known that the relation between atmospheric pressure and height follows a logarithmic model. Using a graphing utility, build a logarithmic model from the data.

(c) Draw the logarithmic function found in part (b) on the scatter diagram.

(d) Use the function found in part (b) to predict the height of the weather balloon if the atmospheric pressure is 560 millimeters of mercury.

Table 10

Atmospheric Pressure, p	Height, h
760	0
740	0.184
725	0.328
700	0.565
650	1.079
630	1.291
600	1.634
580	1.862
550	2.235

Solution

(a) After entering the data into the graphing utility, we obtain the scatter diagram shown in Figure 48.

(b) A graphing utility fits the data in Figure 48 to a logarithmic function of the form $y = a + b \ln x$ by using the LOGarithm REGression option. See Figure 49. The logarithmic model from the data is

$$h(p) = 45.7863 - 6.9025 \ln p$$

where h is the height of the weather balloon and p is the atmospheric pressure. Notice that $|r|$ is close to 1, indicating a good fit.

(c) Figure 50 shows the graph of $h(p) = 45.7863 - 6.9025 \ln p$ on the scatter diagram.

Figure 48

Figure 49

Figure 50

(d) Using the function found in part (b), Jodi predicts the height of the weather balloon when the atmospheric pressure is 560 to be

$$h(560) = 45.7863 - 6.9025 \ln 560$$
$$\approx 2.108 \text{ kilometers}$$

Now Work PROBLEM 5

 ## 3 Build a Logistic Model from Data

Logistic growth models can be used to model situations for which the value of the dependent variable is limited. Many real-world situations conform to this scenario. For example, the population of the human race is limited by the availability of natural resources such as food and shelter. When the value of the dependent variable is limited, a logistic growth model is often appropriate.

EXAMPLE 3 **Fitting a Logistic Function to Data**

The data in Table 11 represent the amount of yeast biomass in a culture after t hours.

Table 11

Time (in hours)	Yeast Biomass	Time (in hours)	Yeast Biomass
0	9.6	10	513.3
1	18.3	11	559.7
2	29.0	12	594.8
3	47.2	13	629.4
4	71.1	14	640.8
5	119.1	15	651.1
6	174.6	16	655.9
7	257.3	17	659.6
8	350.7	18	661.8
9	441.0		

Source: Tor Carlson (Über Geschwindigkeit und Grösse der Hefevermehrung in Würze, Biochemische Zeitschrift, Bd. 57, pp. 313–334, 1913)

(a) Using a graphing utility, draw a scatter diagram of the data with time as the independent variable.

(b) Using a graphing utility, build a logistic model from the data.

(c) Using a graphing utility, graph the function found in part (b) on the scatter diagram.

(d) What is the predicted carrying capacity of the culture?

(e) Use the function found in part (b) to predict the population of the culture at $t = 19$ hours.

Solution

(a) See Figure 51 for a scatter diagram of the data.

(b) A graphing utility fits a logistic growth model of the form $y = \dfrac{c}{1 + ae^{-bx}}$ by using the LOGISTIC regression option. See Figure 52. The logistic model from the data is

$$y = \frac{663.0}{1 + 71.6e^{-0.5470x}}$$

where y is the amount of yeast biomass in the culture and x is the time.

(c) See Figure 53 for the graph of the logistic model.

Figure 51

Figure 52

Figure 53

(d) Based on the logistic growth model found in part (b), the carrying capacity of the culture is 663.

(e) Using the logistic growth model found in part (b), the predicted amount of yeast biomass at $t = 19$ hours is

$$y = \frac{663.0}{1 + 71.6e^{-0.5470(19)}} \approx 661.5$$

─ **Now Work** PROBLEM 7

6.9 Assess Your Understanding

Applications and Extensions

1. Biology A strain of E-coli Beu 397-recA441 is placed into a nutrient broth at 30° Celsius and allowed to grow. The following data are collected. Theory states that the number of bacteria in the petri dish will initially grow according to the law of uninhibited growth. The population is measured using an optical device in which the amount of light that passes through the petri dish is measured.

Time (hours), x	Population, y
0	0.09
2.5	0.18
3.5	0.26
4.5	0.35
6	0.50

Source: Dr. Polly Lavery, Joliet Junior College

(a) Draw a scatter diagram treating time as the independent variable.

(b) Using a graphing utility, build an exponential model from the data.

(c) Express the function found in part (b) in the form $N(t) = N_0 e^{kt}$.

(d) Graph the exponential function found in part (b) or (c) on the scatter diagram.

(e) Use the exponential function from part (b) or (c) to predict the population at $x = 7$ hours.

(f) Use the exponential function from part (b) or (c) to predict when the population will reach 0.75.

2. Biology A strain of E-coli SC18del-recA718 is placed into a nutrient broth at 30° Celsius and allowed to grow. The data on the following page are collected. Theory states that the number of bacteria in the petri dish will initially grow according to the law of uninhibited growth. The population

is measured using an optical device in which the amount of light that passes through the petri dish is measured.

Time (hours), x	Population, y
2.5	0.175
3.5	0.38
4.5	0.63
4.75	0.76
5.25	1.20

Source: Dr. Polly Lavery, Joliet Junior College

(a) Draw a scatter diagram treating time as the independent variable.
(b) Using a graphing utility, build an exponential model from the data.
(c) Express the function found in part (b) in the form $N(t) = N_0 e^{kt}$.
(d) Graph the exponential function found in part (b) or (c) on the scatter diagram.
(e) Use the exponential function from part (b) or (c) to predict the population at $x = 6$ hours.
(f) Use the exponential function from part (b) or (c) to predict when the population will reach 2.1.

3. Chemistry A chemist has a 100-gram sample of a radioactive material. He records the amount of radioactive material every week for 7 weeks and obtains the following data:

Week	Weight (in Grams)
0	100.0
1	88.3
2	75.9
3	69.4
4	59.1
5	51.8
6	45.5

(a) Using a graphing utility, draw a scatter diagram with week as the independent variable.
(b) Using a graphing utility, build an exponential model from the data.
(c) Express the function found in part (b) in the form $A(t) = A_0 e^{kt}$.
(d) Graph the exponential function found in part (b) or (c) on the scatter diagram.
(e) From the result found in part (b), determine the half-life of the radioactive material.
(f) How much radioactive material will be left after 50 weeks?
(g) When will there be 20 grams of radioactive material?

4. Cigarette Exports The following data represent the number of cigarettes (in billions) exported from the United States by year.

Year	Cigarette Exports (in billions of pieces)
1995	231.1
1998	201.3
1999	151.4
2000	147.9
2001	133.9
2002	127.4
2003	121.5
2004	118.7

Source: Statistical Abstract of the United States, 2006

(a) Let t = the number of years since 1995. Using a graphing utility, draw a scatter diagram of the data using t as the independent variable and number of cigarettes as the dependent variable.
(b) Using a graphing utility, build an exponential model from the data.
(c) Express the function found in part (b) in the form $A(t) = A_0 e^{kt}$.
(d) Graph the exponential function found in part (b) or (c) on the scatter diagram.
(e) Use the exponential function from part (b) or (c) to predict the number of cigarettes that will be exported from the United States in 2010.
(f) Use the exponential function from part (b) or (c) to predict when the number of cigarettes exported from the United States will decrease to 50 billion.

5. Economics and Marketing The following data represent the price and quantity demanded in 2009 for Dell personal computers.

Price ($/Computer)	Quantity Demanded
2300	152
2000	159
1700	164
1500	171
1300	176
1200	180
1000	189

(a) Using a graphing utility, draw a scatter diagram of the data with price as the dependent variable.
(b) Using a graphing utility, build a logarithmic model from the data.
(c) Using a graphing utility, draw the logarithmic function found in part (b) on the scatter diagram.
(d) Use the function found in part (b) to predict the number of Dell personal computers that will be demanded if the price is $1650.

6. Economics and Marketing The following data represent the price and quantity supplied in 2009 for Dell personal computers.

Price ($/Computer)	Quantity Supplied
2300	180
2000	173
1700	160
1500	150
1300	137
1200	130
1000	113

(a) Using a graphing utility, draw a scatter diagram of the data with price as the dependent variable.
(b) Using a graphing utility, build a logarithmic model from the data.
(c) Using a graphing utility, draw the logarithmic function found in part (b) on the scatter diagram.
(d) Use the function found in part (b) to predict the number of Dell personal computers that will be supplied if the price is $1650.

7. **Population Model** The following data represent the population of the United States. An ecologist is interested in building a model that describes the population of the United States.

Year	Population
1900	76,212,168
1910	92,228,496
1920	106,021,537
1930	123,202,624
1940	132,164,569
1950	151,325,798
1960	179,323,175
1970	203,302,031
1980	226,542,203
1990	248,709,873
2000	281,421,906

Source: U.S. Census Bureau

(a) Using a graphing utility, draw a scatter diagram of the data using years since 1900 as the independent variable and population as the dependent variable.
(b) Using a graphing utility, build a logistic model from the data.
(c) Using a graphing utility, draw the function found in part (b) on the scatter diagram.
(d) Based on the function found in part (b), what is the carrying capacity of the United States?
(e) Use the function found in part (b) to predict the population of the United States in 2004.
(f) When will the United States population be 300,000,000?
(g) Compare actual U.S. Census figures to the predictions found in parts (e) and (f). Discuss any differences.

8. **Population Model** The following data represent the world population. An ecologist is interested in building a model that describes the world population.

Year	Population (in Billions)
2001	6.17
2002	6.25
2003	6.32
2004	6.40
2005	6.48
2006	6.55
2007	6.63
2008	6.71
2009	6.79

Source: U.S. Census Bureau

(a) Using a graphing utility, draw a scatter diagram of the data using years since 2000 as the independent variable and population as the dependent variable.
(b) Using a graphing utility, build a logistic model from the data.
(c) Using a graphing utility, draw the function found in part (b) on the scatter diagram.
(d) Based on the function found in part (b), what is the carrying capacity of the world?
(e) Use the function found in part (b) to predict the population of the world in 2015.
(f) When will world population be 10 billion?

9. **Cable Subscribers** The following data represent the number of basic cable TV subscribers in the United States. A market researcher believes that external factors, such as satellite TV, have affected the growth of cable subscribers. She is interested in building a model that can be used to describe the number of cable TV subscribers in the United States.

Year	Subscribers (1,000)
1975 ($t = 5$)	9,800
1980 ($t = 10$)	17,500
1985 ($t = 15$)	35,440
1990 ($t = 20$)	50,520
1992 ($t = 22$)	54,300
1994 ($t = 24$)	58,373
1996 ($t = 26$)	62,300
1998 ($t = 28$)	64,650
2000 ($t = 30$)	66,250
2002 ($t = 32$)	66,472
2004 ($t = 34$)	65,727
2006 ($t = 36$)	65,319

Source: Statistical Abstract of the United States, 2009

(a) Using a graphing utility, draw a scatter diagram of the data using the number of years after 1970, t, as the independent variable and number of subscribers as the dependent variable.
(b) Using a graphing utility, build a logistic model from the data.

(c) Using a graphing utility, draw the function found in part (b) on the scatter diagram.

(d) Based on the model found in part (b), what is the maximum number of cable TV subscribers in the United States?

(e) Use the model found in part (b) to predict the number of cable TV subscribers in the United States in 2015.

10. Cell Phone Users Refer to the data in Table 9.

(a) Using a graphing utility, build a logistic model from the data.

(b) Graph the logistic function found in part (b) on a scatter diagram of the data.

(c) What is the predicted carrying capacity of U.S. cell phone subscribers?

(d) Use the model found in part (b) to predict the number of U.S. cell phone subscribers at the end of 2009.

(e) Compare the answer to part (d) above with the answer to Example 1, part (e). How do you explain the different predictions?

Mixed Practice

11. Age versus Total Cholesterol The following data represent the age and average total cholesterol for adult males at various ages.

Age	Total Cholesterol
27	189
40	205
50	215
60	210
70	210
80	194

(a) Using a graphing utility, draw a scatter diagram of the data using age, x, as the independent variable and total cholesterol, y, as the dependent variable.

(b) Based on the scatter diagram drawn in part (a), decide on a model (linear, quadratic, cubic, exponential, logarithmic, or logistic) that you think best describes the relation between age and total cholesterol. Be sure to justify your choice of model.

(c) Using a graphing utility, find the model of best fit.

(d) Using a graphing utility, draw the model of best fit on the scatter diagram drawn in part (a).

(e) Use your model to predict the total cholesterol of a 35-year-old male.

12. Income versus Crime Rate The following data represent crime rate against individuals (crimes per 1000 households) and their income in the United States in 2006.

Income	Crime Rate
$5000	217.3
11,250	195.7
20,000	183.1
30,000	179.4
42,500	166.2
62,500	166.8
85,000	162.0

Source: Statistical Abstract of the United States, 2009

(a) Using a graphing utility, draw a scatter diagram of the data using income, x, as the independent variable and crime rate, y, as the dependent variable.

(b) Based on the scatter diagram drawn in part (a), decide on a model (linear, quadratic, cubic, exponential, logarithmic, or logistic) that you think best describes the relation between income and crime rate. Be sure to justify your choice of model.

(c) Using a graphing utility, find the model of best fit.

(d) Using a graphing utility, draw the model of best fit on the scatter diagram drawn in part (a).

(e) Use your model to predict the crime rate of a household whose income is $55,000.

13. Depreciation of a Chevrolet Impala The following data represent the asking price and age of a Chevrolet Impala SS.

Age	Asking Price
1	$27,417
1	26,750
2	22,995
2	23,195
3	17,999
4	16,995
4	16,490

Source: cars.com

(a) Using a graphing utility, draw a scatter diagram of the data using age, x, as the independent variable and asking price, y, as the dependent variable.

(b) Based on the scatter diagram drawn in part (a), decide on a model (linear, quadratic, cubic, exponential, logarithmic, or logistic) that you think best describes the relation between age and asking price. Be sure to justify your choice of model.

(c) Using a graphing utility, find the model of best fit.

(d) Using a graphing utility, draw the model of best fit on the scatter diagram drawn in part (a).

(e) Use your model to predict the asking price of a Chevrolet Impala SS that is 5 years old.

CHAPTER REVIEW

Things to Know

Composite function (p. 401)

$(f \circ g)(x) = f(g(x))$ The domain of $f \circ g$ is the set of all numbers x in the domain of g for which $g(x)$ is in the domain of f.

One-to-one function f (p. 409)

A function for which any two different inputs in the domain correspond to two different outputs in the range
For any choice of elements x_1, x_2 in the domain of f, if $x_1 \neq x_2$, then $f(x_1) \neq f(x_2)$.

Horizontal-line test (p. 410)

If every horizontal line intersects the graph of a function f in at most one point, f is one-to-one.

Inverse function f^{-1} of f (pp. 411–414)

Domain of f = range of f^{-1}; range of f = domain of f^{-1}
$f^{-1}(f(x)) = x$ for all x in the domain of f
$f(f^{-1}(x)) = x$ for all x in the domain of f^{-1}
The graphs of f and f^{-1} are symmetric with respect to the line $y = x$.

Properties of the exponential function (pp. 423, 426, 428)

$f(x) = Ca^x$, $a > 1, C > 0$

Domain: the interval $(-\infty, \infty)$

Range: the interval $(0, \infty)$

x-intercepts: none; y-intercept: C

Horizontal asymptote: x-axis ($y = 0$) as $x \to -\infty$

Increasing; one-to-one; smooth; continuous

See Figure 21 for a typical graph.

$f(x) = Ca^x$, $0 < a < 1, C > 0$

Domain: the interval $(-\infty, \infty)$

Range: the interval $(0, \infty)$

x-intercepts: none; y-intercept: C

Horizontal asymptote: x-axis ($y = 0$) as $x \to \infty$

Decreasing; one-to-one; smooth; continuous

See Figure 25 for a typical graph.

Number e (p. 429)

Value approached by the expression $\left(1 + \dfrac{1}{n}\right)^n$ as $n \to \infty$; that is, $\lim\limits_{n \to \infty} \left(1 + \dfrac{1}{n}\right)^n = e$.

Property of exponents (p. 430)

If $a^u = a^v$, then $u = v$.

Properties of the logarithmic function (pp. 438–440)

$f(x) = \log_a x$, $a > 1$
($y = \log_a x$ means $x = a^y$)

Domain: the interval $(0, \infty)$

Range: the interval $(-\infty, \infty)$

x-intercept: 1; y-intercept: none

Vertical asymptote: $x = 0$ (y-axis)

Increasing; one-to-one; smooth; continuous

See Figure 39(a) for a typical graph.

$f(x) = \log_a x$, $0 < a < 1$
($y = \log_a x$ means $x = a^y$)

Domain: the interval $(0, \infty)$

Range: the interval $(-\infty, \infty)$

x-intercept: 1; y-intercept: none

Vertical asymptote: $x = 0$ (y-axis)

Decreasing; one-to-one; smooth; continuous

See Figure 39(b) for a typical graph.

Natural logarithm (p. 441)

$y = \ln x$ means $x = e^y$.

Properties of logarithms (pp. 451–452, 454)

$\log_a 1 = 0$ $\log_a a = 1$ $a^{\log_a M} = M$ $\log_a a^r = r$

$\log_a(MN) = \log_a M + \log_a N$ $\log_a\left(\dfrac{M}{N}\right) = \log_a M - \log_a N$

$\log_a M^r = r \log_a M$

If $M = N$, then $\log_a M = \log_a N$. $a^x = e^{x \ln a}$

If $\log_a M = \log_a N$, then $M = N$.

Formulas

Change-of-Base Formula (p. 455)	$\log_a M = \dfrac{\log_b M}{\log_b a}$
Compound Interest Formula (p. 467)	$A = P \cdot \left(1 + \dfrac{r}{n}\right)^{nt}$
Continuous compounding (p. 469)	$A = Pe^{rt}$
Effective rate of interest (p. 470)	Compounding n times per year: $r_e = \left(1 + \dfrac{r}{n}\right)^n - 1$
	Continuous compounding: $r_e = e^r - 1$
Present Value Formulas (p. 471)	$P = A \cdot \left(1 + \dfrac{r}{n}\right)^{-nt}$ or $P = Ae^{-rt}$
Growth and decay (p. 476, 478)	$A(t) = A_0 e^{kt}$
Newton's Law of Cooling (p. 479)	$u(t) = T + (u_0 - T)e^{kt}$ $k < 0$
Logistic model (p. 481)	$P(t) = \dfrac{c}{1 + ae^{-bt}}$

Objectives

Section		You should be able to ...	Example(s)	Review Exercises
6.1	1	Form a composite function (p. 401)	1, 2, 4, 5	1–12
	2	Find the domain of a composite function (p. 402)	2–2	7–12
6.2	1	Determine whether a function is one-to-one (p. 409)	1, 2	13(a), 14(a), 15, 16
	2	Determine the inverse of a function defined by a map or a set of ordered pairs (p. 411)	3, 4	13(b), 14(b)
	3	Obtain the graph of the inverse function from the graph of the function (p. 413)	7	15, 16
	4	Find the inverse of a function defined by an equation (p. 414)	8, 9, 10	17–22
6.3	1	Evaluate exponential functions (p. 421)	1	23(a), (c), 24(a), (c), 87(a)
	2	Graph exponential functions (p. 425)	3–6	55–60
	3	Define the number e (p. 428)	pg. 429	59, 60
	4	Solve exponential equations (p. 430)	7, 8	63–66, 71, 72, 74–76
6.4	1	Change exponential statements to logarithmic statements and logarithmic statements to exponential statements (p. 438)	2, 3	25–28
	2	Evaluate logarithmic expressions (p. 438)	4	23(b), (d), 24(b), (d), 33, 34, 83(b), 84(b), 85, 86, 88(a), 89
	3	Determine the domain of a logarithmic function (p. 439)	5	29–32, 61(a), 62(a)
	4	Graph logarithmic functions (p. 440)	6, 7	61, 62, 83(a), 84(a)
	5	Solve logarithmic equations (p. 444)	8, 9	67, 68, 73, 83(c), 84(c), 88(b)
6.5	1	Work with the properties of logarithms (p. 450)	1, 2	35–38
	2	Write a logarithmic expression as a sum or difference of logarithms (p. 452)	3–5	39–44
	3	Write a logarithmic expression as a single logarithm (p. 453)	6	45–50
	4	Evaluate logarithms whose base is neither 10 nor e (p. 455)	7, 8	51, 52
6.6	1	Solve logarithmic equations (p. 459)	1–3	67, 68, 77, 78
	2	Solve exponential equations (p. 461)	4–6	63–66, 69–72, 74–76, 79–82
	3	Solve logarithmic and exponential equations using a graphing utility (p. 462)	7	69–82
6.7	1	Determine the future value of a lump sum of money (p. 466)	1–3	90, 92, 97
	2	Calculate effective rates of return (p. 469)	4	90
	3	Determine the present value of a lump sum of money (p. 470)	5	91
	4	Determine the rate of interest or time required to double a lump sum of money (p. 471)	6, 7	90

Section		You should be able to ...	Example(s)	Review Exercises
6.8	1	Find equations of populations that obey the law of uninhibited growth (p. 476)	1, 2	95
	2	Find equations of populations that obey the law of decay (p. 478)	3	93, 96
	3	Use Newton's Law of Cooling (p. 479)	4	94
	4	Use logistic models (p. 481)	5, 6	98
6.9	1	Build an exponential model from data (p. 487)	1	99
	2	Build a logarithmic model from data (p. 488)	2	100
	3	Build a logistic model from data (p. 489)	3	101

Review Exercises

In Problems 1–6, for the given functions f and g find:

(a) $(f \circ g)(2)$ (b) $(g \circ f)(-2)$ (c) $(f \circ f)(4)$ (d) $(g \circ g)(-1)$

1. $f(x) = 3x - 5$; $g(x) = 1 - 2x^2$ **2.** $f(x) = 4 - x$; $g(x) = 1 + x^2$ **3.** $f(x) = \sqrt{x + 2}$; $g(x) = 2x^2 + 1$

4. $f(x) = 1 - 3x^2$; $g(x) = \sqrt{4 - x}$ **5.** $f(x) = e^x$; $g(x) = 3x - 2$ **6.** $f(x) = \dfrac{2}{1 + 2x^2}$; $g(x) = 3x$

In Problems 7–12, find $f \circ g$, $g \circ f$, $f \circ f$, and $g \circ g$ for each pair of functions. State the domain of each composite function.

7. $f(x) = 2 - x$; $g(x) = 3x + 1$ **8.** $f(x) = 2x - 1$; $g(x) = 2x + 1$ **9.** $f(x) = 3x^2 + x + 1$; $g(x) = |3x|$

10. $f(x) = \sqrt{3x}$; $g(x) = 1 + x + x^2$ **11.** $f(x) = \dfrac{x + 1}{x - 1}$; $g(x) = \dfrac{1}{x}$ **12.** $f(x) = \sqrt{x - 3}$; $g(x) = \dfrac{3}{x}$

In Problems 13 and 14, (a) verify that the function is one-to-one, and (b) find the inverse of the given function.

13. $\{(1, 2), (3, 5), (5, 8), (6, 10)\}$ **14.** $\{(-1, 4), (0, 2), (1, 5), (3, 7)\}$

In Problems 15 and 16, state why the graph of the function is one-to-one. Then draw the graph of the inverse function f^{-1}. For convenience (and as a hint), the graph of $y = x$ is also given.

15.

16.

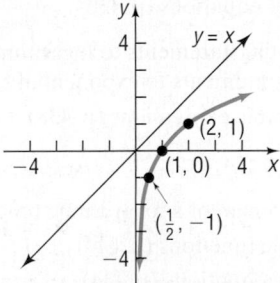

In Problems 17–22, the function f is one-to-one. Find the inverse of each function and check your answer.

17. $f(x) = \dfrac{2x + 3}{5x - 2}$ **18.** $f(x) = \dfrac{2 - x}{3 + x}$ **19.** $f(x) = \dfrac{1}{x - 1}$

20. $f(x) = \sqrt{x - 2}$ **21.** $f(x) = \dfrac{3}{x^{1/3}}$ **22.** $f(x) = x^{1/3} + 1$

In Problems 23 and 24, $f(x) = 3^x$ and $g(x) = \log_3 x$.

23. Evaluate: (a) $f(4)$ (b) $g(9)$ (c) $f(-2)$ (d) $g\left(\dfrac{1}{27}\right)$

24. Evaluate: (a) $f(1)$ (b) $g(81)$ (c) $f(-4)$ (d) $g\left(\dfrac{1}{243}\right)$

In Problems 25 and 26, convert each exponential statement to an equivalent statement involving a logarithm. In Problems 27 and 28, convert each logarithmic statement to an equivalent statement involving an exponent.

25. $5^2 = z$ **26.** $a^5 = m$ **27.** $\log_5 u = 13$ **28.** $\log_a 4 = 3$

In Problems 29–32, find the domain of each logarithmic function.

29. $f(x) = \log(3x - 2)$ **30.** $F(x) = \log_5(2x + 1)$ **31.** $H(x) = \log_2(x^2 - 3x + 2)$ **32.** $F(x) = \ln(x^2 - 9)$

In Problems 33–38, evaluate each expression. Do not use a calculator.

33. $\log_2\left(\dfrac{1}{8}\right)$ **34.** $\log_3 81$ **35.** $\ln e^{\sqrt{2}}$ **36.** $e^{\ln 0.1}$ **37.** $2^{\log_2 0.4}$ **38.** $\log_2 2^{\sqrt{3}}$

In Problems 39–44, write each expression as the sum and/or difference of logarithms. Express powers as factors.

39. $\log_3\left(\dfrac{uv^2}{w}\right), \quad u > 0, v > 0, w > 0$ **40.** $\log_2\left(a^2 \sqrt{b}\right)^4, \quad a > 0, b > 0$ **41.** $\log\left(x^2 \sqrt{x^3 + 1}\right), \quad x > 0$

42. $\log_5\left(\dfrac{x^2 + 2x + 1}{x^2}\right), \quad x > 0$ **43.** $\ln\left(\dfrac{x\sqrt[3]{x^2 + 1}}{x - 3}\right), \quad x > 3$ **44.** $\ln\left(\dfrac{2x + 3}{x^2 - 3x + 2}\right)^2, \quad x > 2$

In Problems 45–50, write each expression as a single logarithm.

45. $3 \log_4 x^2 + \dfrac{1}{2} \log_4 \sqrt{x}$

46. $-2 \log_3\left(\dfrac{1}{x}\right) + \dfrac{1}{3} \log_3 \sqrt{x}$

47. $\ln\left(\dfrac{x - 1}{x}\right) + \ln\left(\dfrac{x}{x + 1}\right) - \ln(x^2 - 1)$

48. $\log(x^2 - 9) - \log(x^2 + 7x + 12)$

49. $2 \log 2 + 3 \log x - \dfrac{1}{2}[\log(x + 3) + \log(x - 2)]$

50. $\dfrac{1}{2} \ln(x^2 + 1) - 4 \ln\dfrac{1}{2} - \dfrac{1}{2}[\ln(x - 4) + \ln x]$

In Problems 51 and 52, use the Change-of-Base Formula and a calculator to evaluate each logarithm. Round your answer to three decimal places.

51. $\log_4 19$

52. $\log_2 21$

In Problems 53 and 54, graph each function using a graphing utility and the Change-of-Base Formula.

53. $y = \log_3 x$

54. $y = \log_7 x$

In Problems 55–62, use the given function f to:
 (a) Find the domain of f. (b) Graph f. (c) From the graph, determine the range and any asymptotes of f.
 (d) Find f^{-1}, the inverse of f. (e) Find the domain and the range of f^{-1}. (f) Graph f^{-1}.

55. $f(x) = 2^{x-3}$ **56.** $f(x) = -2^x + 3$ **57.** $f(x) = \dfrac{1}{2}(3^{-x})$ **58.** $f(x) = 1 + 3^{-x}$

59. $f(x) = 1 - e^{-x}$ **60.** $f(x) = 3e^{x-2}$ **61.** $f(x) = \dfrac{1}{2} \ln(x + 3)$ **62.** $f(x) = 3 + \ln(2x)$

In Problems 63–82, solve each equation. Express irrational solutions in exact form and as a decimal rounded to 3 decimal places.

63. $4^{1-2x} = 2$ **64.** $8^{6+3x} = 4$ **65.** $3^{x^2+x} = \sqrt{3}$ **66.** $4^{x-x^2} = \dfrac{1}{2}$

67. $\log_x 64 = -3$ **68.** $\log_{\sqrt{2}} x = -6$ **69.** $5^x = 3^{x+2}$ **70.** $5^{x+2} = 7^{x-2}$

71. $9^{2x} = 27^{3x-4}$ **72.** $25^{2x} = 5^{x^2-12}$ **73.** $\log_3 \sqrt{x - 2} = 2$ **74.** $2^{x+1} \cdot 8^{-x} = 4$

75. $8 = 4^{x^2} \cdot 2^{5x}$ **76.** $2^x \cdot 5 = 10^x$ **77.** $\log_6(x + 3) + \log_6(x + 4) = 1$

78. $\log(7x - 12) = 2 \log x$ **79.** $e^{1-x} = 5$ **80.** $e^{1-2x} = 4$

81. $9^x + 4 \cdot 3^x - 3 = 0$ **82.** $4^x - 14 \cdot 4^{-x} = 5$

83. Suppose that $f(x) = \log_2(x - 2) + 1$.
 (a) Graph f.
 (b) What is $f(6)$? What point is on the graph of f?
 (c) Solve $f(x) = 4$. What point is on the graph of f?
 (d) Based on the graph drawn in part (a), solve $f(x) > 0$.
 (e) Find $f^{-1}(x)$. Graph f^{-1} on the same Cartesian plane as f.

84. Suppose that $f(x) = \log_3(x + 1) - 4$.
 (a) Graph f.
 (b) What is $f(8)$? What point is on the graph of f?
 (c) Solve $f(x) = -3$. What point is on the graph of f?
 (d) Based on the graph drawn in part (a), solve $f(x) < 0$.
 (e) Find $f^{-1}(x)$. Graph f^{-1} on the same Cartesian plane as f.

In Problems 85 and 86, use the following result: If x is the atmospheric pressure (measured in millimeters of mercury), then the formula for the altitude h(x) (measured in meters above sea level) is

$$h(x) = (30T + 8000) \log\left(\frac{P_0}{x}\right)$$

where T is the temperature (in degrees Celsius) and P_0 is the atmospheric pressure at sea level, which is approximately 760 millimeters of mercury.

85. Finding the Altitude of an Airplane At what height is a Piper Cub whose instruments record an outside temperature of 0°C and a barometric pressure of 300 millimeters of mercury?

86. Finding the Height of a Mountain How high is a mountain if instruments placed on its peak record a temperature of 5°C and a barometric pressure of 500 millimeters of mercury?

87. Amplifying Sound An amplifier's power output P (in watts) is related to its decibel voltage gain d by the formula

$$P = 25e^{0.1d}$$

(a) Find the power output for a decibel voltage gain of 4 decibels.

(b) For a power output of 50 watts, what is the decibel voltage gain?

88. Limiting Magnitude of a Telescope A telescope is limited in its usefulness by the brightness of the star that it is aimed at and by the diameter of its lens. One measure of a star's brightness is its *magnitude;* the dimmer the star, the larger its magnitude. A formula for the limiting magnitude L of a telescope, that is, the magnitude of the dimmest star that it can be used to view, is given by

$$L = 9 + 5.1 \log d$$

where d is the diameter (in inches) of the lens.

(a) What is the limiting magnitude of a 3.5-inch telescope?

(b) What diameter is required to view a star of magnitude 14?

89. Salvage Value The number of years n for a piece of machinery to depreciate to a known salvage value can be found using the formula

$$n = \frac{\log s - \log i}{\log(1 - d)}$$

where s is the salvage value of the machinery, i is its initial value, and d is the annual rate of depreciation.

(a) How many years will it take for a piece of machinery to decline in value from $90,000 to $10,000 if the annual rate of depreciation is 0.20 (20%)?

(b) How many years will it take for a piece of machinery to lose half of its value if the annual rate of depreciation is 15%?

90. Funding a College Education A child's grandparents purchase a $10,000 bond fund that matures in 18 years to be used for her college education. The bond fund pays 4% interest compounded semiannually. How much will the bond fund be worth at maturity? What is the effective rate of interest? How long will it take the bond to double in value under these terms?

91. Funding a College Education A child's grandparents wish to purchase a bond that matures in 18 years to be used for her college education. The bond pays 4% interest

compounded semiannually. How much should they pay so that the bond will be worth $85,000 at maturity?

92. Funding an IRA First Colonial Bankshares Corporation advertised the following IRA investment plans.

Target IRA Plans

For each $5000 Maturity Value Desired	
Deposit:	At a Term of:
$620.17	20 Years
$1045.02	15 Years
$1760.92	10 Years
$2967.26	5 Years

(a) Assuming continuous compounding, what annual rate of interest did they offer?

(b) First Colonial Bankshares claims that $4000 invested today will have a value of over $32,000 in 20 years. Use the answer found in part (a) to find the actual value of $4000 in 20 years. Assume continuous compounding.

93. Estimating the Date That a Prehistoric Man Died The bones of a prehistoric man found in the desert of New Mexico contain approximately 5% of the original amount of carbon 14. If the half-life of carbon 14 is 5600 years, approximately how long ago did the man die?

94. Temperature of a Skillet A skillet is removed from an oven whose temperature is 450°F and placed in a room whose temperature is 70°F. After 5 minutes, the temperature of the skillet is 400°F. How long will it be until its temperature is 150°F?

95. World Population The annual growth rate of the world's population in 2005 was $k = 1.15\% = 0.0115$. The population of the world in 2005 was 6,451,058,790. Letting $t = 0$ represent 2005, use the uninhibited growth model to predict the world's population in the year 2015.

Source: U.S. Census Bureau

96. Radioactive Decay The half-life of radioactive cobalt is 5.27 years. If 100 grams of radioactive cobalt is present now, how much will be present in 20 years? In 40 years?

97. Federal Deficit In fiscal year 2005, the federal deficit was $319 billion. At that time, 10-year treasury notes were paying 4.25% interest per annum. If the federal government financed this deficit through 10-year notes, how much would it have to pay back in 2015?

Source: U.S. Treasury Department

98. Logistic Growth The logistic growth model

$$P(t) = \frac{0.8}{1 + 1.67e^{-0.16t}}$$

represents the proportion of new cars with a global positioning system (GPS). Let $t = 0$ represent 2006, $t = 1$ represent 2007, and so on.
(a) What proportion of new cars in 2006 had a GPS?
(b) Determine the maximum proportion of new cars that have a GPS.
(c) Using a graphing utility, graph $P = P(t)$.
(d) When will 75% of new cars have a GPS?

99. CBL Experiment The following data were collected by placing a temperature probe in a portable heater, removing the probe, and then recording temperature over time.

Time (sec.)	Temperature (°F)
0	165.07
1	164.77
2	163.99
3	163.22
4	162.82
5	161.96
6	161.20
7	160.45
8	159.35
9	158.61
10	157.89
11	156.83
12	156.11
13	155.08
14	154.40
15	153.72

According to Newton's Law of Cooling, these data should follow an exponential model.
(a) Using a graphing utility, draw a scatter diagram for the data.
(b) Using a graphing utility, build an exponential model from the data.
(c) Graph the exponential function found in part (b) on the scatter diagram.
(d) Predict how long it will take for the probe to reach a temperature of 110°F.

100. Wind Chill Factor The following data represent the wind speed (mph) and wind chill factor at an air temperature of 15°F.
(a) Using a graphing utility, draw a scatter diagram with wind speed as the independent variable.
(b) Using a graphing utility, build a logarithmic model from the data.
(c) Using a graphing utility, draw the logarithmic function found in part (b) on the scatter diagram.

(d) Use the function found in part (b) to predict the wind chill factor if the air temperature is 15°F and the wind speed is 23 mph.

Wind Speed (mph)	Wind Chill Factor (°F)
5	7
10	3
15	0
20	−2
25	−4
30	−5
35	−7

Source: U.S. National Weather Service

101. Spreading of a Disease Jack and Diane live in a small town of 50 people. Unfortunately, both Jack and Diane have a cold. Those who come in contact with someone who has this cold will themselves catch the cold. The following data represent the number of people in the small town who have caught the cold after t days.

Days, t	Number of People with Cold, C
0	2
1	4
2	8
3	14
4	22
5	30
6	37
7	42
8	44

(a) Using a graphing utility, draw a scatter diagram of the data. Comment on the type of relation that appears to exist between the days and number of people with a cold.
(b) Using a graphing utility, build a logistic model from the data.
(c) Graph the function found in part (b) on the scatter diagram.
(d) According to the function found in part (b), what is the maximum number of people who will catch the cold? In reality, what is the maximum number of people who could catch the cold?
(e) Sometime between the second and third day, 10 people in the town had a cold. According to the model found in part (b), when did 10 people have a cold?
(f) How long will it take for 46 people to catch the cold?

CHAPTER TEST

CHAPTER
Test Prep
VIDEOS

The Chapter Test Prep Videos are step-by-step test solutions available in the Video Resources DVD, in **MyMathLab**, or on this text's **YouTube** Channel. Flip back to the Student Resources page to see the exact web address for this text's YouTube channel.

1. Given $f(x) = \dfrac{x + 2}{x - 2}$ and $g(x) = 2x + 5$, find:
 (a) $f \circ g$ and state its domain
 (b) $(g \circ f)(-2)$
 (c) $(f \circ g)(-2)$

2. Determine whether the function is one-to-one.
 (a) $y = 4x^2 + 3$
 (b) $y = \sqrt{x + 3} - 5$

3. Find the inverse of $f(x) = \dfrac{2}{3x - 5}$ and check your answer. State the domain and the range of f and f^{-1}.

4. If the point $(3, -5)$ is on the graph of a one-to-one function f, what point must be on the graph of f^{-1}?

In Problems 5–7, solve each equation.

5. $3^x = 243$ 6. $\log_b 16 = 2$

7. $\log_5 x = 4$

In Problems 8–11, use a calculator to evaluate each expression. Round your answer to three decimal places.

8. $e^3 + 2$ 9. $\log 20$

10. $\log_3 21$ 11. $\ln 133$

In Problems 12 and 13, use the given function f to:
 (a) Find the domain of f.
 (b) Graph f.
 (c) From the graph, determine the range and any asymptotes of f.
 (d) Find f^{-1}, the inverse of f.
 (e) Find the domain and the range of f^{-1}.
 (f) Graph f^{-1}.

12. $f(x) = 4^{x+1} - 2$

13. $f(x) = 1 - \log_5(x - 2)$

In Problems 14–19, solve each equation.

14. $5^{x+2} = 125$ 15. $\log(x + 9) = 2$

16. $8 - 2e^{-x} = 4$ 17. $\log(x^2 + 3) = \log(x + 6)$

18. $7^{x+3} = e^x$ 19. $\log_2(x - 4) + \log_2(x + 4) = 3$

20. Write $\log_2\left(\dfrac{4x^3}{x^2 - 3x - 18}\right)$ as the sum and/or difference of logarithms. Express powers as factors.

21. A 50-mg sample of a radioactive substance decays to 34 mg after 30 days. How long will it take for there to be 2 mg remaining?

22. (a) If $1000 is invested at 5% compounded monthly, how much is there after 8 months?
 (b) If you want to have $1000 in 9 months, how much do you need to place in a savings account now that pays 5% compounded quarterly?
 (c) How long does it take to double your money if you can invest it at 6% compounded annually?

23. The decibel level, D, of sound is given by the equation $D = 10 \log\left(\dfrac{I}{I_0}\right)$, where I is the intensity of the sound and $I_0 = 10^{-12}$ watt per square meter.
 (a) If the shout of a single person measures 80 decibels, how loud will the sound be if two people shout at the same time? That is, how loud would the sound be if the intensity doubled?
 (b) The pain threshold for sound is 125 decibels. If the Athens Olympic Stadium 2004 (Olympiako Stadio Athinas 'Spyros Louis') can seat 74,400 people, how many people in the crowd need to shout at the same time for the resulting sound level to meet or exceed the pain threshold? (Ignore any possible sound dampening.)

CUMULATIVE REVIEW

1. Is the following graph the graph of a function? If it is, is the function one-to-one?

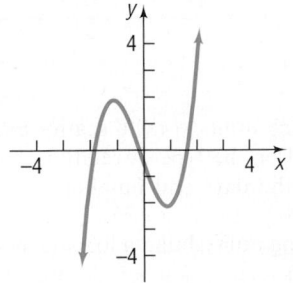

2. For the function $f(x) = 2x^2 - 3x + 1$, find the following:
 (a) $f(3)$ (b) $f(-x)$ (c) $f(x + h)$

3. Determine which of the following points are on the graph of $x^2 + y^2 = 1$.
 (a) $\left(\dfrac{1}{2}, \dfrac{1}{2}\right)$ (b) $\left(\dfrac{1}{2}, \dfrac{\sqrt{3}}{2}\right)$

4. Solve the equation $3(x - 2) = 4(x + 5)$.

5. Graph the line $2x - 4y = 16$.

6. (a) Graph the quadratic function $f(x) = -x^2 + 2x - 3$ by determining whether its graph opens up or down and by finding its vertex, axis of symmetry, y-intercept, and x-intercept(s), if any.
 (b) Solve $f(x) \leq 0$.

7. Determine the quadratic function whose graph is given in the figure.

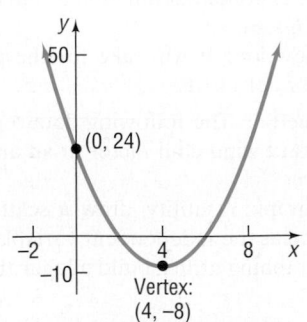

8. Graph $f(x) = 3(x + 1)^3 - 2$ using transformations.

9. Given that $f(x) = x^2 + 2$ and $g(x) = \dfrac{2}{x - 3}$, find $f(g(x))$ and state its domain. What is $f(g(5))$?

10. For the polynomial function $f(x) = 4x^3 + 9x^2 - 30x - 8$:
 (a) Find the real zeros of f.
 (b) Determine the intercepts of the graph of f.
 (c) Use a graphing utility to approximate the local maxima and local minima.
 (d) Draw a complete graph of f. Be sure to label the intercepts and turning points.

11. For the function $g(x) = 3^x + 2$:
 (a) Graph g using transformations. State the domain, range, and horizontal asymptote of g.
 (b) Determine the inverse of g. State the domain, range, and vertical asymptote of g^{-1}.
 (c) On the same graph as g, graph g^{-1}.

12. Solve the equation $4^{x-3} = 8^{2x}$.

13. Solve the equation: $\log_3(x + 1) + \log_3(2x - 3) = \log_9 9$

14. Suppose that $f(x) = \log_3(x + 2)$. Solve:
 (a) $f(x) = 0$ (b) $f(x) > 0$
 (c) $f(x) = 3$

15. **Data Analysis** The following data represent the percent of all drivers by age that have been stopped by the police for any reason within the past year. The median age represents the midpoint of the upper and lower limit for the age range.

Age Range	Median Age, x	Percentage Stopped, y
16–19	17.5	18.2
20–29	24.5	16.8
30–39	34.5	11.3
40–49	44.5	9.4
50–59	54.5	7.7
≥60	69.5	3.8

(a) Using your graphing utility, draw a scatter diagram of the data treating median age, x, as the independent variable.
(b) Determine a model that you feel best describes the relation between median age and percentage stopped. You may choose from among linear, quadratic, cubic, exponential, logarithmic, or logistic models.
(c) Provide a justification for the model that you selected in part (b).

CHAPTER PROJECTS

(((•))) **Internet-based Project**
I. **Depreciation of Cars** Kelley Blue Book is an official guide that provides the current retail price of cars. You can access the Kelley Blue Book at your library or online at *www.kbb.com*.

1. Identify three cars that you are considering purchasing and find the Kelley Blue Book value of the cars for 0 (brand new), 1, 2, 3, 4, and 5 years of age. Online, the value of the car can be found by selecting Used Cars, then Used Car Values. Enter the year, make, and model of the car you are selecting. To be consistent, we will

assume the cars will be driven 12,000 miles per year, so a 1-year-old car will have 12,000 miles, a 2-year-old car will have 24,000 miles, and so on. Choose the same options for each year, and finally determine the suggested retail price for cars that are in Excellent, Good, and Fair shape. So, you should have a total of 16 observations (one for a brand new car, 3 for a 1-year-old car, 3 for a 2-year-old car, and so on).

2. Draw a scatter diagram of the data with age as the independent variable and value as the dependent variable using Excel, a TI-graphing calculator, or some other spreadsheet. The Chapter 4 project describes how to draw a scatter diagram in Excel.

3. Determine the exponential function of best fit. Graph the exponential function of best fit on the scatter diagram. To do this in Excel, click on any data point in the scatter diagram. Now click the Layout menu, select Trendline within the Analysis region, select More Trendline Options. Select the Exponential radio button and select Display Equation on Chart. See Figure 54. Move the Trendline Options window off to the side and you will see the exponential function of best fit displayed on the scatter diagram. Do you think the function accurately describes the relation between age of the car and suggested retail price?

Figure 54

4. The exponential function of best fit is of the form $y = Ce^{rx}$ where y is the suggested retail value of the car and x is the age of the car (in years). What does the value of C represent? What does the value of r represent? What is the depreciation rate for each car that you are considering?

5. Write a report detailing which car you would purchase based on the depreciation rate you found for each car.

The following projects are available on the Instructor's Resource Center (IRC):

II. Hot Coffee A fast-food restaurant wants a special container to hold coffee. The restaurant wishes the container to quickly cool the coffee from 200° to 130°F and keep the liquid between 110° and 130°F as long as possible. The restaurant has three containers to select from. Which one should be purchased?

III. Project at Motorola *Thermal Fatigue of Solder Connections* Product reliability is a major concern of a manufacturer. Here a logarithmic transformation is used to simplify the analysis of a cell phone's ability to withstand temperature change.

Citation: Excel © 2010 Microsoft Corporation. Used with permission from Microsoft.

Analytic Geometry

7

Outline

The Orbit of the Hale-Bopp Comet

The orbits of the Hale-Bopp Comet and Earth can be modeled using *ellipses*, the subject of Section 7.3. The Internet-based Project at the end of this chapter explores the possibility of the Hale-Bopp Comet colliding with Earth.

 —See the Internet-based Chapter Project I—

◁ **A Look Back** In Chapter 2, we introduced rectangular coordinates and showed how geometry problems can be solved algebraically. We defined a circle geometrically and then used the distance formula and rectangular coordinates to obtain an equation for a circle.

A Look Ahead ▷ In this chapter, we give geometric definitions for the conics and use the distance formula and rectangular coordinates to obtain their equations.

Historically, Apollonius (200 BC) was among the first to study *conics* and discover some of their interesting properties. Today, conics are still studied because of their many uses. *Paraboloids of revolution* (parabolas rotated about their axes of symmetry) are used as signal collectors (the satellite dishes used with radar and dish TV, for example), as solar energy collectors, and as reflectors (telescopes, light projection, and so on). The planets circle the Sun in approximately *elliptical* orbits. Elliptical surfaces can be used to reflect signals such as light and sound from one place to another. A third conic, the *hyperbola,* can be used to determine the location of lightning strikes.

The Greeks used the methods of Euclidean geometry to study conics. However, we shall use the more powerful methods of analytic geometry, which uses both algebra and geometry, for our study of conics.

7.1 Conics

OBJECTIVE **1** Know the Names of the Conics (p. 504)

1 Know the Names of the Conics

The word *conic* derives from the word *cone,* which is a geometric figure that can be constructed in the following way: Let *a* and *g* be two distinct lines that intersect at a point *V*. Keep the line *a* fixed. Now rotate the line *g* about *a* while maintaining the same angle between *a* and *g*. The collection of points swept out (generated) by the line *g* is called a **(right circular) cone.** See Figure 1. The fixed line *a* is called the **axis** of the cone; the point *V* is its **vertex;** the lines that pass through *V* and make the same angle with *a* as *g* are **generators** of the cone. Each generator is a line that lies entirely on the cone. The cone consists of two parts, called **nappes,** that intersect at the vertex.

Figure 1

Axis, *a*

Generators

Vertex, *V*

g

Conics, an abbreviation for **conic sections,** are curves that result from the intersection of a right circular cone and a plane. The conics we shall study arise when the plane does not contain the vertex, as shown in Figure 2. These conics are **circles** when the plane is perpendicular to the axis of the cone and intersects each generator; **ellipses** when the plane is tilted slightly so that it intersects each generator, but intersects only one nappe of the cone; **parabolas** when the plane is tilted farther so that it is parallel to one (and only one) generator and intersects only one nappe of the cone; and **hyperbolas** when the plane intersects both nappes.

If the plane does contain the vertex, the intersection of the plane and the cone is a point, a line, or a pair of intersecting lines. These are usually called **degenerate conics.**

Figure 2

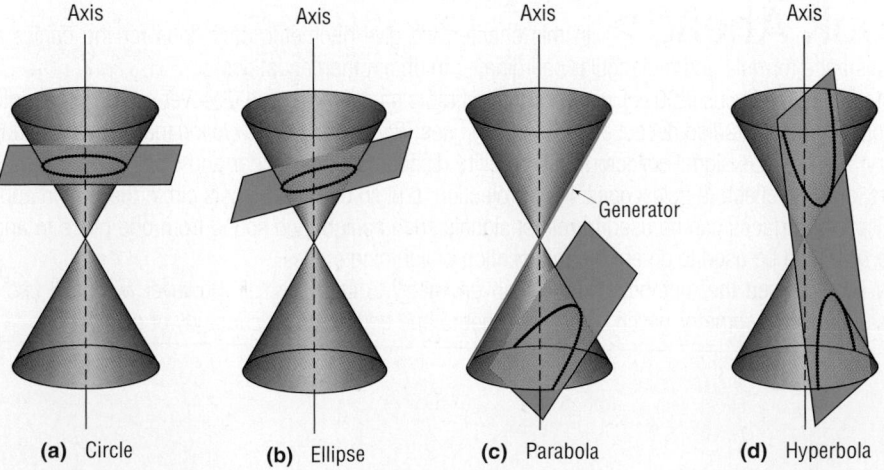

Axis Axis Axis Axis

Generator

(a) Circle **(b)** Ellipse **(c)** Parabola **(d)** Hyperbola

Conic sections are used in modeling many different applications. For example parabolas are used in describing satellite dishes and telescopes (see Figures 14 and 15 on page 510). Ellipses are used to model the orbits of planets and whispering chambers (see pages 520–521). And hyperbolas are used to locate lightning strikes and model nuclear cooling towers (see Problems 76 and 77 in Section 7.4).

7.2 The Parabola

PREPARING FOR THIS SECTION *Before getting started, review the following:*

- Distance Formula (Section 2.1, p. 151)
- Symmetry (Section 2.2, pp. 160–162)
- Square Root Method (Section 1.2, pp. 94–95)
- Completing the Square (Chapter R, Review, Section R.5, p. 56)
- Graphing Techniques: Transformations (Section 3.5, pp. 244–253)

Now Work the 'Are You Prepared?' problems on page 511.

OBJECTIVES **1** Analyze Parabolas with Vertex at the Origin (p. 505)

 2 Analyze Parabolas with Vertex at (h, k) (p. 508)

 3 Solve Applied Problems Involving Parabolas (p. 510)

We stated earlier (Section 4.3) that the graph of a quadratic function is a parabola. In this section, we give a geometric definition of a parabola and use it to obtain an equation.

DEFINITION

A **parabola** is the collection of all points P in the plane that are the same distance d from a fixed point F as they are from a fixed line D. The point F is called the **focus** of the parabola, and the line D is its **directrix.** As a result, a parabola is the set of points P for which

$$d(F, P) = d(P, D) \qquad \textbf{(1)}$$

Figure 3

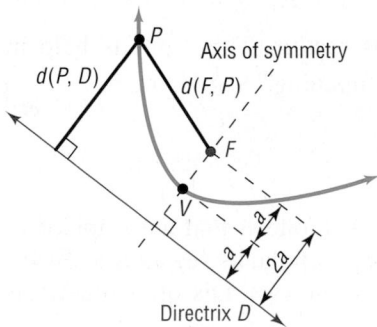

Figure 3 shows a parabola (in blue). The line through the focus F and perpendicular to the directrix D is called the **axis of symmetry** of the parabola. The point of intersection of the parabola with its axis of symmetry is called the **vertex V.**

Because the vertex V lies on the parabola, it must satisfy equation (1): $d(F, V) = d(V, D)$. The vertex is midway between the focus and the directrix. We shall let a equal the distance $d(F, V)$ from F to V. Now we are ready to derive an equation for a parabola. To do this, we use a rectangular system of coordinates positioned so that the vertex V, focus F, and directrix D of the parabola are conveniently located.

1 Analyze Parabolas with Vertex at the Origin

If we choose to locate the vertex V at the origin $(0, 0)$, we can conveniently position the focus F on either the x-axis or the y-axis. First, consider the case where the focus F is on the positive x-axis, as shown in Figure 4. Because the distance from F to V is a, the coordinates of F will be $(a, 0)$ with $a > 0$. Similarly, because the distance from V to the directrix D is also a and, because D must be perpendicular to the x-axis (since the x-axis is the axis of symmetry), the equation of the directrix D must be $x = -a$.

Figure 4

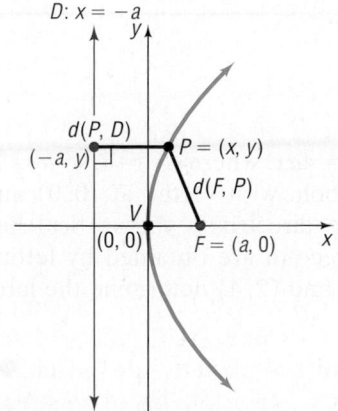

Now, if $P = (x, y)$ is any point on the parabola, P must obey equation (1):

$$d(F, P) = d(P, D)$$

So we have

$$\sqrt{(x - a)^2 + (y - 0)^2} = |x + a| \qquad \text{Use the Distance Formula.}$$

$$(x - a)^2 + y^2 = (x + a)^2 \qquad \text{Square both sides.}$$

$$x^2 - 2ax + a^2 + y^2 = x^2 + 2ax + a^2 \qquad \text{Remove parentheses.}$$

$$y^2 = 4ax \qquad \text{Simplify.}$$

THEOREM

Equation of a Parabola: Vertex at (0, 0), Focus at (a, 0), a > 0

The equation of a parabola with vertex at $(0, 0)$, focus at $(a, 0)$, and directrix $x = -a$, $a > 0$, is

$$y^2 = 4ax \tag{2}$$

Recall that a is the distance from the vertex to the focus of a parabola. When graphing the parabola $y^2 = 4ax$ it is helpful to determine the "opening" by finding the points that lie directly above or below the focus $(a, 0)$. This is done by letting $x = a$ in $y^2 = 4ax$, so $y^2 = 4a(a) = 4a^2$, or $y = \pm 2a$. The line segment joining the two points, $(a, 2a)$ and $(a, -2a)$, is called the **latus rectum;** its length is $4a$.

EXAMPLE 1

Finding the Equation of a Parabola and Graphing It

Find an equation of the parabola with vertex at $(0, 0)$ and focus at $(3, 0)$. Graph the equation.

Solution

Figure 5

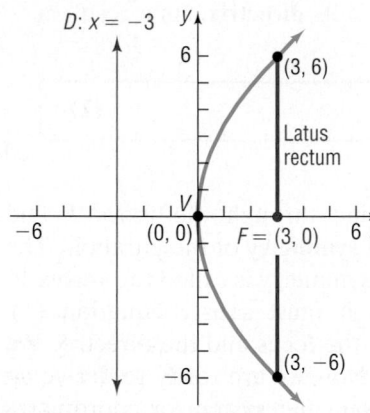

The distance from the vertex $(0, 0)$ to the focus $(3, 0)$ is $a = 3$. Based on equation (2), the equation of this parabola is

$$y^2 = 4ax$$
$$y^2 = 12x \quad a = 3$$

To graph this parabola, we find the two points that determine the latus rectum by letting $x = 3$. Then

$$y^2 = 12x = 12(3) = 36$$
$$y = \pm 6 \qquad \text{Solve for } y.$$

The points $(3, 6)$ and $(3, -6)$ determine the latus rectum. These points help in graphing the parabola because they determine the "opening." See Figure 5.

Now Work PROBLEM 19

 COMMENT To graph the parabola $y^2 = 12x$ discussed in Example 1, we need to graph the two functions $Y_1 = \sqrt{12x}$ and $Y_2 = -\sqrt{12x}$. Do this and compare what you see with Figure 5. ∎

By reversing the steps used to obtain equation (2), it follows that the graph of an equation of the form of equation (2), $y^2 = 4ax$, is a parabola; its vertex is at $(0, 0)$, its focus is at $(a, 0)$, its directrix is the line $x = -a$, and its axis of symmetry is the x-axis.

For the remainder of this section, the direction **"Analyze the equation"** will mean to find the vertex, focus, and directrix of the parabola and graph it.

EXAMPLE 2

Analyzing the Equation of a Parabola

Analyze the equation: $y^2 = 8x$

Figure 6

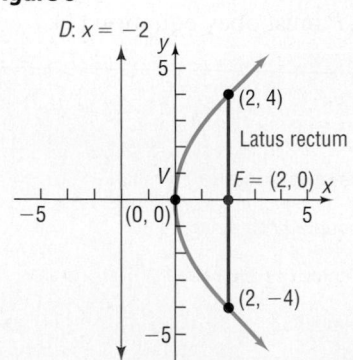

Solution The equation $y^2 = 8x$ is of the form $y^2 = 4ax$, where $4a = 8$, so $a = 2$. Consequently, the graph of the equation is a parabola with vertex at $(0, 0)$ and focus on the positive x-axis at $(a, 0) = (2, 0)$. The directrix is the vertical line $x = -2$. The two points that determine the latus rectum are obtained by letting $x = 2$. Then $y^2 = 16$, so $y = \pm 4$. The points $(2, -4)$ and $(2, 4)$ determine the latus rectum. See Figure 6 for the graph.

Recall that we obtained equation (2) after placing the focus on the positive x-axis. If the focus is placed on the negative x-axis, positive y-axis, or negative y-axis, a different form of the equation for the parabola results. The four forms of the equation of a parabola with vertex at $(0, 0)$ and focus on a coordinate axis a distance a from $(0, 0)$ are given in Table 1, and their graphs are given in Figure 7. Notice that each graph is symmetric with respect to its axis of symmetry.

Table 1

Equations of a Parabola: Vertex at $(0, 0)$; Focus on an Axis; $a > 0$				
Vertex	**Focus**	**Directrix**	**Equation**	**Description**
$(0, 0)$	$(a, 0)$	$x = -a$	$y^2 = 4ax$	Axis of symmetry is the x-axis, opens right
$(0, 0)$	$(-a, 0)$	$x = a$	$y^2 = -4ax$	Axis of symmetry is the x-axis, opens left
$(0, 0)$	$(0, a)$	$y = -a$	$x^2 = 4ay$	Axis of symmetry is the y-axis, opens up
$(0, 0)$	$(0, -a)$	$y = a$	$x^2 = -4ay$	Axis of symmetry is the y-axis, opens down

Figure 7

(a) $y^2 = 4ax$ (b) $y^2 = -4ax$ (c) $x^2 = 4ay$ (d) $x^2 = -4ay$

EXAMPLE 3

Analyzing the Equation of a Parabola

Analyze the equation: $x^2 = -12y$

Figure 8

Solution The equation $x^2 = -12y$ is of the form $x^2 = -4ay$, with $a = 3$. Consequently, the graph of the equation is a parabola with vertex at $(0, 0)$, focus at $(0, -3)$, and directrix the line $y = 3$. The parabola opens down, and its axis of symmetry is the y-axis. To obtain the points defining the latus rectum, let $y = -3$. Then $x^2 = 36$, so $x = \pm 6$. The points $(-6, -3)$ and $(6, -3)$ determine the latus rectum. See Figure 8 for the graph.

Now Work PROBLEM 39

EXAMPLE 4

Finding the Equation of a Parabola

Find the equation of the parabola with focus at $(0, 4)$ and directrix the line $y = -4$. Graph the equation.

Figure 9

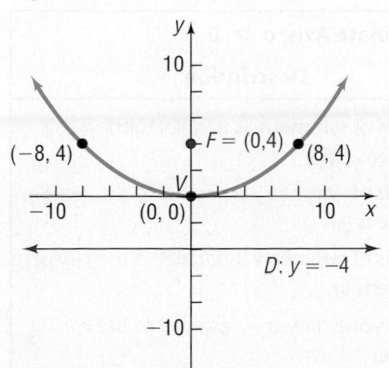

Solution A parabola whose focus is at $(0, 4)$ and whose directrix is the horizontal line $y = -4$ will have its vertex at $(0, 0)$. (Do you see why? The vertex is midway between the focus and the directrix.) Since the focus is on the positive y-axis at $(0, 4)$, the equation of this parabola is of the form $x^2 = 4ay$, with $a = 4$; that is,

$$x^2 = 4ay = 4(4)y = 16y$$
$$\uparrow$$
$$a = 4$$

The points $(8, 4)$ and $(-8, 4)$ determine the latus rectum. Figure 9 shows the graph of $x^2 = 16y$.

| EXAMPLE 5 | **Finding the Equation of a Parabola** |

Find the equation of a parabola with vertex at $(0, 0)$ if its axis of symmetry is the x-axis and its graph contains the point $\left(-\dfrac{1}{2}, 2\right)$. Find its focus and directrix, and graph the equation.

Solution The vertex is at the origin, the axis of symmetry is the x-axis, and the graph contains a point in the second quadrant, so the parabola opens to the left. We see from Table 1 that the form of the equation is

$$y^2 = -4ax$$

Because the point $\left(-\dfrac{1}{2}, 2\right)$ is on the parabola, the coordinates $x = -\dfrac{1}{2}$, $y = 2$ must satisfy $y^2 = -4ax$. Substituting $x = -\dfrac{1}{2}$ and $y = 2$ into this equation, we find

$$4 = -4a\left(-\dfrac{1}{2}\right) \qquad y^2 = -4ax; \; x = -\dfrac{1}{2}, y = 2$$

$$a = 2$$

The equation of the parabola is

$$y^2 = -4(2)x = -8x$$

The focus is at $(-2, 0)$ and the directrix is the line $x = 2$. Let $x = -2$. Then $y^2 = 16$, so $y = \pm 4$. The points $(-2, 4)$ and $(-2, -4)$ determine the latus rectum. See Figure 10.

Figure 10

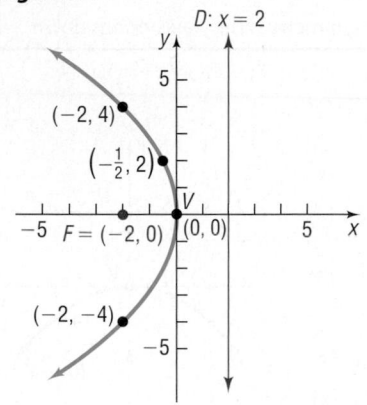

D: $x = 2$

$(-2, 4)$

$\left(-\dfrac{1}{2}, 2\right)$

V $(0, 0)$

$F = (-2, 0)$

$(-2, -4)$

═══ **Now Work** PROBLEM 27

2 Analyze Parabolas with Vertex at (h, k)

If a parabola with vertex at the origin and axis of symmetry along a coordinate axis is shifted horizontally h units and then vertically k units, the result is a parabola with vertex at (h, k) and axis of symmetry parallel to a coordinate axis. The equations of such parabolas have the same forms as those in Table 1, but with x replaced by $x - h$ (the horizontal shift) and y replaced by $y - k$ (the vertical shift). Table 2 gives the forms of the equations of such parabolas. Figures 11(a)–(d) on page 509, illustrate the graphs for $h > 0$, $k > 0$.

> **NOTE** It is not recommended that Table 2 be memorized. Rather use the ideas of transformations (shift horizontally h units, vertically k units) along with the fact that a represents the distance from the vertex to the focus to determine the various components of a parabola. It is also helpful to remember that parabolas of the form "$x^2 = $ " will open up or down, while parabolas of the form "$y^2 = $ " will open left or right. ∎

Table 2

Equations of a Parabola: Vertex at (h, k); Axis of Symmetry Parallel to a Coordinate Axis; $a > 0$				
Vertex	**Focus**	**Directrix**	**Equation**	**Description**
(h, k)	$(h + a, k)$	$x = h - a$	$(y - k)^2 = 4a(x - h)$	Axis of symmetry is parallel to the x-axis, opens right
(h, k)	$(h - a, k)$	$x = h + a$	$(y - k)^2 = -4a(x - h)$	Axis of symmetry is parallel to the x-axis, opens left
(h, k)	$(h, k + a)$	$y = k - a$	$(x - h)^2 = 4a(y - k)$	Axis of symmetry is parallel to the y-axis, opens up
(h, k)	$(h, k - a)$	$y = k + a$	$(x - h)^2 = -4a(y - k)$	Axis of symmetry is parallel to the y-axis, opens down

Figure 11

(a) $(y - k)^2 = 4a(x - h)$

(b) $(y - k)^2 = -4a(x - h)$

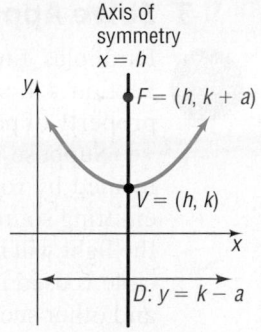

(c) $(x - h)^2 = 4a(y - k)$

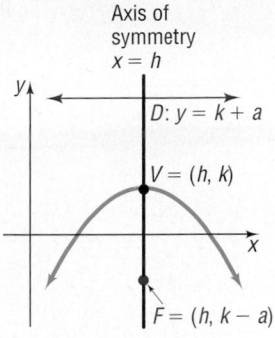

(d) $(x - h)^2 = -4a(y - k)$

EXAMPLE 6 **Finding the Equation of a Parabola, Vertex Not at the Origin**

Find an equation of the parabola with vertex at $(-2, 3)$ and focus at $(0, 3)$. Graph the equation.

Solution

Figure 12

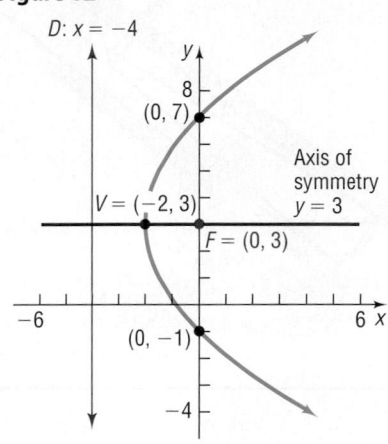

The vertex $(-2, 3)$ and focus $(0, 3)$ both lie on the horizontal line $y = 3$ (the axis of symmetry). The distance a from the vertex $(-2, 3)$ to the focus $(0, 3)$ is $a = 2$. Also, because the focus lies to the right of the vertex, the parabola opens to the right. Consequently, the form of the equation is

$$(y - k)^2 = 4a(x - h)$$

where $(h, k) = (-2, 3)$ and $a = 2$. Therefore, the equation is

$$(y - 3)^2 = 4 \cdot 2[x - (-2)]$$
$$(y - 3)^2 = 8(x + 2)$$

To find the points that define the latus rectum, let $x = 0$, so that $(y - 3)^2 = 16$. Then $y - 3 = \pm 4$, so $y = -1$ or $y = 7$. The points $(0, -1)$ and $(0, 7)$ determine the latus rectum; the line $x = -4$ is the directrix. See Figure 12.

Now Work PROBLEM **29**

Polynomial equations define parabolas whenever they involve two variables that are quadratic in one variable and linear in the other.

EXAMPLE 7 **Analyzing the Equation of a Parabola**

Analyze the equation: $x^2 + 4x - 4y = 0$

Solution

Figure 13

To analyze the equation $x^2 + 4x - 4y = 0$, complete the square involving the variable x.

$$x^2 + 4x - 4y = 0$$
$$x^2 + 4x = 4y \qquad \text{Isolate the terms involving } x \text{ on the left side.}$$
$$x^2 + 4x + 4 = 4y + 4 \qquad \text{Complete the square on the left side.}$$
$$(x + 2)^2 = 4(y + 1) \qquad \text{Factor.}$$

This equation is of the form $(x - h)^2 = 4a(y - k)$, with $h = -2$, $k = -1$, and $a = 1$. The graph is a parabola with vertex at $(h, k) = (-2, -1)$ that opens up. The focus is at $(-2, 0)$, and the directrix is the line $y = -2$. See Figure 13.

Now Work PROBLEM **47**

3 Solve Applied Problems Involving Parabolas

Parabolas find their way into many applications. For example, as discussed in Section 4.4, suspension bridges have cables in the shape of a parabola. Another property of parabolas that is used in applications is their reflecting property.

Suppose that a mirror is shaped like a **paraboloid of revolution,** a surface formed by rotating a parabola about its axis of symmetry. If a light (or any other emitting source) is placed at the focus of the parabola, all the rays emanating from the light will reflect off the mirror in lines parallel to the axis of symmetry. This principle is used in the design of searchlights, flashlights, certain automobile headlights, and other such devices. See Figure 14.

Conversely, suppose that rays of light (or other signals) emanate from a distant source so that they are essentially parallel. When these rays strike the surface of a parabolic mirror whose axis of symmetry is parallel to these rays, they are reflected to a single point at the focus. This principle is used in the design of some solar energy devices, satellite dishes, and the mirrors used in some types of telescopes. See Figure 15.

Figure 14
Searchlight

Figure 15
Telescope

EXAMPLE 8

Satellite Dish

A satellite dish is shaped like a paraboloid of revolution. The signals that emanate from a satellite strike the surface of the dish and are reflected to a single point, where the receiver is located. If the dish is 8 feet across at its opening and 3 feet deep at its center, at what position should the receiver be placed? That is, where is the focus?

Solution

Figure 16(a) shows the satellite dish. Draw the parabola used to form the dish on a rectangular coordinate system so that the vertex of the parabola is at the origin and its focus is on the positive *y*-axis. See Figure 16(b).

Figure 16

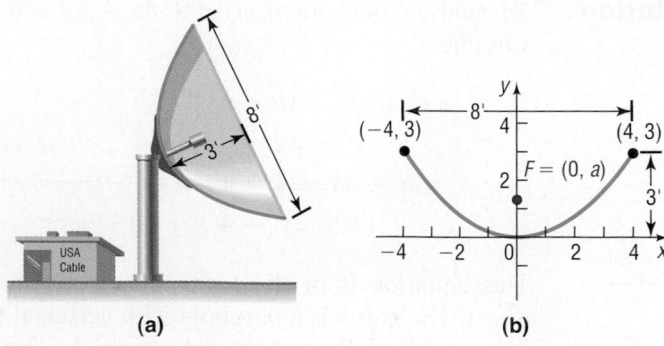

(a) (b)

The form of the equation of the parabola is

$$x^2 = 4ay$$

and its focus is at $(0, a)$. Since $(4, 3)$ is a point on the graph, we have

$$4^2 = 4a(3) \quad \text{\textcolor{gray}{$x^2 = 4ay; x = 4, y = 3$}}$$

$$a = \frac{4}{3} \quad \text{\textcolor{gray}{Solve for a.}}$$

The receiver should be located $1\frac{1}{3}$ feet (1 foot, 4 inches) from the base of the dish, along its axis of symmetry.

Now Work PROBLEM 63

7.2 Assess Your Understanding

'Are You Prepared?' *Answers are given at the end of these exercises. If you get a wrong answer, read the pages listed in red.*

1. The formula for the distance d from $P_1 = (x_1, y_1)$ to $P_2 = (x_2, y_2)$ is $d =$ _____. (p. 151)

2. To complete the square of $x^2 - 4x$, add _____. (p. 56)

3. Use the Square Root Method to find the real solutions of $(x + 4)^2 = 9$. (pp. 94–95)

4. The point that is symmetric with respect to the x-axis to the point $(-2, 5)$ is _____. (pp. 160–162)

5. To graph $y = (x - 3)^2 + 1$, shift the graph of $y = x^2$ to the right _____ units and then _____ 1 unit. (pp. 244–253)

Concepts and Vocabulary

6. A(n) _____ is the collection of all points in the plane such that the distance from each point to a fixed point equals its distance to a fixed line.

Answer Problems 7–10 using the figure.

7. If $a > 0$, the equation of the parabola is of the form
 (a) $(y - k)^2 = 4a(x - h)$
 (b) $(y - k)^2 = -4a(x - h)$
 (c) $(x - h)^2 = 4a(y - k)$
 (d) $(x - h)^2 = -4a(y - k)$

8. The coordinates of the vertex are _____.

9. If $a = 4$, then the coordinates of the focus are _____.

10. If $a = 4$, then the equation of the directrix is _____.

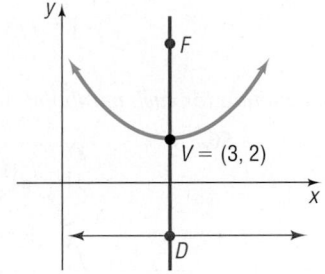

Skill Building

In Problems 11–18, the graph of a parabola is given. Match each graph to its equation.

(a) $y^2 = 4x$ (c) $y^2 = -4x$ (e) $(y - 1)^2 = 4(x - 1)$ (g) $(y - 1)^2 = -4(x - 1)$
(b) $x^2 = 4y$ (d) $x^2 = -4y$ (f) $(x + 1)^2 = 4(y + 1)$ (h) $(x + 1)^2 = -4(y + 1)$

11.

12.

13.

14.

15.

16.

17.

18.

In Problems 19–36, find the equation of the parabola described. Find the two points that define the latus rectum, and graph the equation.

19. Focus at $(4, 0)$; vertex at $(0, 0)$

20. Focus at $(0, 2)$; vertex at $(0, 0)$

21. Focus at $(0, -3)$; vertex at $(0, 0)$

22. Focus at $(-4, 0)$; vertex at $(0, 0)$

23. Focus at $(-2, 0)$; directrix the line $x = 2$

24. Focus at $(0, -1)$; directrix the line $y = 1$

25. Directrix the line $y = -\dfrac{1}{2}$; vertex at $(0, 0)$

26. Directrix the line $x = -\dfrac{1}{2}$; vertex at $(0, 0)$

27. Vertex at $(0, 0)$; axis of symmetry the y-axis; containing the point $(2, 3)$

28. Vertex at $(0, 0)$; axis of symmetry the x-axis; containing the point $(2, 3)$

29. Vertex at $(2, -3)$; focus at $(2, -5)$

30. Vertex at $(4, -2)$; focus at $(6, -2)$

31. Vertex at $(-1, -2)$; focus at $(0, -2)$

32. Vertex at $(3, 0)$; focus at $(3, -2)$

33. Focus at $(-3, 4)$; directrix the line $y = 2$

34. Focus at $(2, 4)$; directrix the line $x = -4$

35. Focus at $(-3, -2)$; directrix the line $x = 1$

36. Focus at $(-4, 4)$; directrix the line $y = -2$

In Problems 37–54, find the vertex, focus, and directrix of each parabola. Graph the equation.

37. $x^2 = 4y$

38. $y^2 = 8x$

39. $y^2 = -16x$

40. $x^2 = -4y$

41. $(y - 2)^2 = 8(x + 1)$

42. $(x + 4)^2 = 16(y + 2)$

43. $(x - 3)^2 = -(y + 1)$

44. $(y + 1)^2 = -4(x - 2)$

45. $(y + 3)^2 = 8(x - 2)$

46. $(x - 2)^2 = 4(y - 3)$

47. $y^2 - 4y + 4x + 4 = 0$

48. $x^2 + 6x - 4y + 1 = 0$

49. $x^2 + 8x = 4y - 8$

50. $y^2 - 2y = 8x - 1$

51. $y^2 + 2y - x = 0$

52. $x^2 - 4x = 2y$

53. $x^2 - 4x = y + 4$

54. $y^2 + 12y = -x + 1$

In Problems 55–62, write an equation for each parabola.

55.

56.

57.

58.

59.

60.

61.

62.
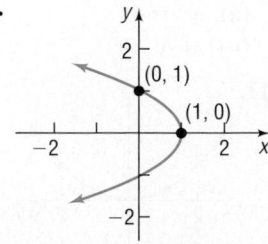

Applications and Extensions

63. Satellite Dish A satellite dish is shaped like a paraboloid of revolution. The signals that emanate from a satellite strike the surface of the dish and are reflected to a single point, where the receiver is located. If the dish is 10 feet across at its opening and 4 feet deep at its center, at what position should the receiver be placed?

64. Constructing a TV Dish A cable TV receiving dish is in the shape of a paraboloid of revolution. Find the location of the receiver, which is placed at the focus, if the dish is 6 feet across at its opening and 2 feet deep.

65. Constructing a Flashlight The reflector of a flashlight is in the shape of a paraboloid of revolution. Its diameter is 4 inches

and its depth is 1 inch. How far from the vertex should the light bulb be placed so that the rays will be reflected parallel to the axis?

66. **Constructing a Headlight** A sealed-beam headlight is in the shape of a paraboloid of revolution. The bulb, which is placed at the focus, is 1 inch from the vertex. If the depth is to be 2 inches, what is the diameter of the headlight at its opening?

67. **Suspension Bridge** The cables of a suspension bridge are in the shape of a parabola, as shown in the figure. The towers supporting the cable are 600 feet apart and 80 feet high. If the cables touch the road surface midway between the towers, what is the height of the cable from the road at a point 150 feet from the center of the bridge?

68. **Suspension Bridge** The cables of a suspension bridge are in the shape of a parabola. The towers supporting the cable are 400 feet apart and 100 feet high. If the cables are at a height of 10 feet midway between the towers, what is the height of the cable at a point 50 feet from the center of the bridge?

69. **Searchlight** A searchlight is shaped like a paraboloid of revolution. If the light source is located 2 feet from the base along the axis of symmetry and the opening is 5 feet across, how deep should the searchlight be?

70. **Searchlight** A searchlight is shaped like a paraboloid of revolution. If the light source is located 2 feet from the base along the axis of symmetry and the depth of the searchlight is 4 feet, what should the width of the opening be?

71. **Solar Heat** A mirror is shaped like a paraboloid of revolution and will be used to concentrate the rays of the sun at its focus, creating a heat source. See the figure. If the mirror is 20 feet across at its opening and is 6 feet deep, where will the heat source be concentrated?

72. **Reflecting Telescope** A reflecting telescope contains a mirror shaped like a paraboloid of revolution. If the mirror is 4 inches across at its opening and is 3 inches deep, where will the collected light be concentrated?

73. **Parabolic Arch Bridge** A bridge is built in the shape of a parabolic arch. The bridge has a span of 120 feet and a maximum height of 25 feet. See the illustration. Choose a suitable

rectangular coordinate system and find the height of the arch at distances of 10, 30, and 50 feet from the center.

74. **Parabolic Arch Bridge** A bridge is to be built in the shape of a parabolic arch and is to have a span of 100 feet. The height of the arch a distance of 40 feet from the center is to be 10 feet. Find the height of the arch at its center.

75. **Gateway Arch** The Gateway Arch in St. Louis is often mistaken to be parabolic in shape. In fact, it is a *catenary*, which has a more complicated formula than a parabola. The Arch is 625 feet high and 598 feet wide at its base.
 (a) Find the equation of a parabola with the same dimensions. Let x equal the horizontal distance from the center of the arc.
 (b) The table below gives the height of the Arch at various widths; find the corresponding heights for the parabola found in (a).

Width (ft)	Height (ft)
567	100
478	312.5
308	525

 (c) Do the data support the notion that the Arch is in the shape of a parabola?
 Source: Wikipedia, the free encyclopedia

76. Show that an equation of the form
$$Ax^2 + Ey = 0, \qquad A \neq 0, E \neq 0$$
is the equation of a parabola with vertex at $(0, 0)$ and axis of symmetry the y-axis. Find its focus and directrix.

77. Show that an equation of the form
$$Cy^2 + Dx = 0, \qquad C \neq 0, D \neq 0$$
is the equation of a parabola with vertex at $(0, 0)$ and axis of symmetry the x-axis. Find its focus and directrix.

78. Show that the graph of an equation of the form
$$Ax^2 + Dx + Ey + F = 0, \qquad A \neq 0$$
 (a) Is a parabola if $E \neq 0$.
 (b) Is a vertical line if $E = 0$ and $D^2 - 4AF = 0$.
 (c) Is two vertical lines if $E = 0$ and $D^2 - 4AF > 0$.
 (d) Contains no points if $E = 0$ and $D^2 - 4AF < 0$.

79. Show that the graph of an equation of the form
$$Cy^2 + Dx + Ey + F = 0, \qquad C \neq 0$$
 (a) Is a parabola if $D \neq 0$.
 (b) Is a horizontal line if $D = 0$ and $E^2 - 4CF = 0$.
 (c) Is two horizontal lines if $D = 0$ and $E^2 - 4CF > 0$.
 (d) Contains no points if $D = 0$ and $E^2 - 4CF < 0$.

'Are You Prepared?' Answers

1. $\sqrt{(x_2 - x_1)^2 + (y_2 - y_1)^2}$ **2.** 4 **3.** $x + 4 = \pm 3; \{-7, -1\}$ **4.** $(-2, -5)$ **5.** 3; up

7.3 The Ellipse

PREPARING FOR THIS SECTION *Before getting started, review the following:*

- Distance Formula (Section 2.1, p. 151)
- Completing the Square (Chapter R, Review, Section R.5, p. 56)
- Intercepts (Section 2.2, pp. 159–160)
- Symmetry (Section 2.2, pp. 160–162)
- Circles (Section 2.4, pp. 182–185)
- Graphing Techniques: Transformations (Section 3.5, pp. 244–253)

Now Work the *'Are You Prepared?'* problems on page 521.

OBJECTIVES 1 Analyze Ellipses with Center at the Origin (p. 514)
 2 Analyze Ellipses with Center at (h, k) (p. 518)
 3 Solve Applied Problems Involving Ellipses (p. 520)

DEFINITION

An **ellipse** is the collection of all points in the plane, the sum of whose distances from two fixed points, called the **foci,** is a constant.

Figure 17

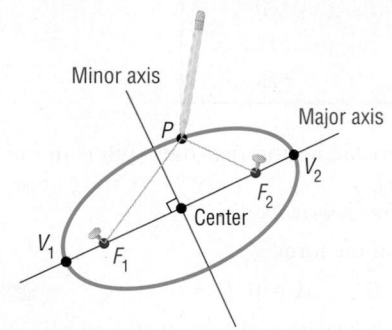

The definition contains within it a physical means for drawing an ellipse. Find a piece of string (the length of this string is the constant referred to in the definition). Then take two thumbtacks (the foci) and stick them into a piece of cardboard so that the distance between them is less than the length of the string. Now attach the ends of the string to the thumbtacks and, using the point of a pencil, pull the string taut. See Figure 17. Keeping the string taut, rotate the pencil around the two thumbtacks. The pencil traces out an ellipse, as shown in Figure 17.

In Figure 17, the foci are labeled F_1 and F_2. The line containing the foci is called the **major axis.** The midpoint of the line segment joining the foci is the **center** of the ellipse. The line through the center and perpendicular to the major axis is the **minor axis.**

The two points of intersection of the ellipse and the major axis are the **vertices,** V_1 and V_2, of the ellipse. The distance from one vertex to the other is the **length of the major axis.** The ellipse is symmetric with respect to its major axis, with respect to its minor axis, and with respect to its center.

1 Analyze Ellipses with Center at the Origin

With these ideas in mind, we are ready to find the equation of an ellipse in a rectangular coordinate system. First, place the center of the ellipse at the origin. Second, position the ellipse so that its major axis coincides with a coordinate axis, say the x-axis, as shown in Figure 18. If c is the distance from the center to a focus, one focus will be at $F_1 = (-c, 0)$ and the other at $F_2 = (c, 0)$. As we shall see, it is

Figure 18

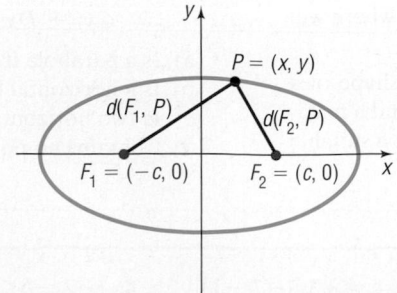

convenient to let $2a$ denote the constant distance referred to in the definition. Then, if $P = (x, y)$ is any point on the ellipse, we have

$$d(F_1, P) + d(F_2, P) = 2a$$ Sum of the distances from P to the foci equals a constant, $2a$.

$$\sqrt{(x + c)^2 + y^2} + \sqrt{(x - c)^2 + y^2} = 2a$$ Use the Distance Formula.

$$\sqrt{(x + c)^2 + y^2} = 2a - \sqrt{(x - c)^2 + y^2}$$ Isolate one radical.

$$(x + c)^2 + y^2 = 4a^2 - 4a\sqrt{(x - c)^2 + y^2} + (x - c)^2 + y^2$$ Square both sides.

$$x^2 + 2cx + c^2 + y^2 = 4a^2 - 4a\sqrt{(x - c)^2 + y^2} + x^2 - 2cx + c^2 + y^2$$ Remove parentheses.

$$4cx - 4a^2 = -4a\sqrt{(x - c)^2 + y^2}$$ Simplify; isolate the radical.

$$cx - a^2 = -a\sqrt{(x - c)^2 + y^2}$$ Divide each side by 4.

$$(cx - a^2)^2 = a^2[(x - c)^2 + y^2]$$ Square both sides again.

$$c^2x^2 - 2a^2cx + a^4 = a^2(x^2 - 2cx + c^2 + y^2)$$ Remove parentheses.

$$(c^2 - a^2)x^2 - a^2y^2 = a^2c^2 - a^4$$ Rearrange the terms.

$$(a^2 - c^2)x^2 + a^2y^2 = a^2(a^2 - c^2)$$ Multiply each side by -1; factor a^2 on the right side. **(1)**

To obtain points on the ellipse off the x-axis, it must be that $a > c$. To see why, look again at Figure 18. Then

$$d(F_1, P) + d(F_2, P) > d(F_1, F_2)$$ The sum of the lengths of two sides of a triangle is greater than the length of the third side.

$$2a > 2c$$ $d(F_1, P) + d(F_2, P) = 2a,\ d(F_1, F_2) = 2c$

$$a > c$$

Since $a > c > 0$, we also have $a^2 > c^2$, so $a^2 - c^2 > 0$. Let $b^2 = a^2 - c^2$, $b > 0$. Then $a > b$ and equation (1) can be written as

$$b^2x^2 + a^2y^2 = a^2b^2$$

$$\frac{x^2}{a^2} + \frac{y^2}{b^2} = 1$$ Divide each side by a^2b^2.

As you can verify, the graph of this equation has symmetry with respect to the x-axis, y-axis, and origin.

Because the major axis is the x-axis, we find the vertices of this ellipse by letting $y = 0$. The vertices satisfy the equation $\frac{x^2}{a^2} = 1$, the solutions of which are $x = \pm a$. Consequently, the vertices of this ellipse are $V_1 = (-a, 0)$ and $V_2 = (a, 0)$. The y-intercepts of the ellipse, found by letting $x = 0$, have coordinates $(0, -b)$ and $(0, b)$. These four intercepts, $(a, 0)$, $(-a, 0)$, $(0, b)$, and $(0, -b)$, are used to graph the ellipse.

THEOREM **Equation of an Ellipse: Center at (0, 0); Major Axis along the x-Axis**

An equation of the ellipse with center at $(0, 0)$, foci at $(-c, 0)$ and $(c, 0)$, and vertices at $(-a, 0)$ and $(a, 0)$ is

$$\frac{x^2}{a^2} + \frac{y^2}{b^2} = 1, \quad \text{where } a > b > 0 \text{ and } b^2 = a^2 - c^2 \quad \textbf{(2)}$$

The major axis is the x-axis. See Figure 19.

Figure 19

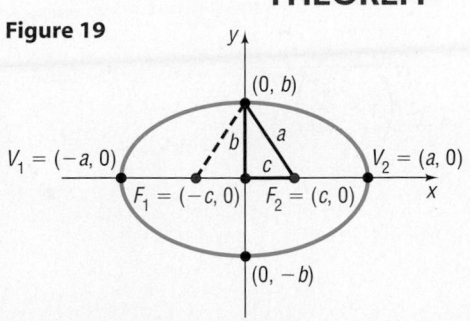

Notice in Figure 19 the right triangle formed by the points $(0, 0)$, $(c, 0)$, and $(0, b)$. Because $b^2 = a^2 - c^2$ (or $b^2 + c^2 = a^2$), the distance from the focus at $(c, 0)$ to the point $(0, b)$ is a.

This can be seen another way. Look at the two right triangles in Figure 19. They are congruent. Do you see why? Because the sum of the distances from the foci to a point on the ellipse is $2a$, it follows that the distance from $(c, 0)$ to $(0, b)$ is a.

EXAMPLE 1 — Finding an Equation of an Ellipse

Find an equation of the ellipse with center at the origin, one focus at $(3, 0)$, and a vertex at $(-4, 0)$. Graph the equation.

Solution

Figure 20

$$\frac{x^2}{16} + \frac{y^2}{7} = 1$$

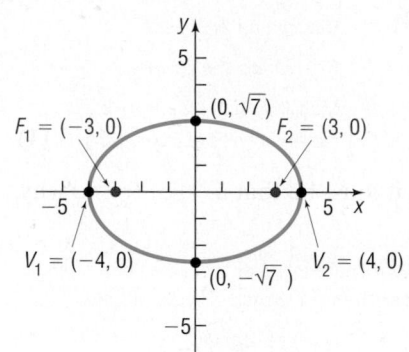

The ellipse has its center at the origin and, since the given focus and vertex lie on the x-axis, the major axis is the x-axis. The distance from the center, $(0, 0)$, to one of the foci, $(3, 0)$, is $c = 3$. The distance from the center, $(0, 0)$, to one of the vertices, $(-4, 0)$, is $a = 4$. From equation (2), it follows that

$$b^2 = a^2 - c^2 = 16 - 9 = 7$$

so an equation of the ellipse is

$$\frac{x^2}{16} + \frac{y^2}{7} = 1$$

Figure 20 shows the graph.

In Figure 20, the intercepts of the equation are used to graph the ellipse. Following this practice will make it easier for you to obtain an accurate graph of an ellipse when graphing.

COMMENT The intercepts of the ellipse also provide information about how to set the viewing rectangle for graphing an ellipse. To graph the ellipse

$$\frac{x^2}{16} + \frac{y^2}{7} = 1$$

discussed in Example 1, set the viewing rectangle using a square screen that includes the intercepts, perhaps $-4.5 \leq x \leq 4.5$, $-3 \leq y \leq 3$. Then proceed to solve the equation for y:

$$\frac{x^2}{16} + \frac{y^2}{7} = 1$$

$$\frac{y^2}{7} = 1 - \frac{x^2}{16} \qquad \text{Subtract } \frac{x^2}{16} \text{ from each side.}$$

$$y^2 = 7\left(1 - \frac{x^2}{16}\right) \qquad \text{Multiply both sides by 7.}$$

$$y = \pm\sqrt{7\left(1 - \frac{x^2}{16}\right)} \qquad \text{Take the square root of each side.}$$

Now graph the two functions

$$Y_1 = \sqrt{7\left(1 - \frac{x^2}{16}\right)} \quad \text{and} \quad Y_2 = -\sqrt{7\left(1 - \frac{x^2}{16}\right)}$$

Figure 21

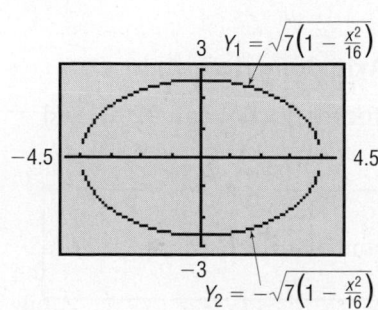

Figure 21 shows the result.

Now Work PROBLEM 27

An equation of the form of equation (2), with $a^2 > b^2$, is the equation of an ellipse with center at the origin, foci on the x-axis at $(-c, 0)$ and $(c, 0)$, where $c^2 = a^2 - b^2$, and major axis along the x-axis.

For the remainder of this section, the direction **"Analyze the equation"** will mean to find the center, major axis, foci, and vertices of the ellipse and graph it.

EXAMPLE 2　　**Analyzing the Equation of an Ellipse**

Analyze the equation: $\dfrac{x^2}{25} + \dfrac{y^2}{9} = 1$

Solution　The given equation is of the form of equation (2), with $a^2 = 25$ and $b^2 = 9$. The equation is that of an ellipse with center $(0, 0)$ and major axis along the x-axis. The vertices are at $(\pm a, 0) = (\pm 5, 0)$. Because $b^2 = a^2 - c^2$, we find that

$$c^2 = a^2 - b^2 = 25 - 9 = 16$$

The foci are at $(\pm c, 0) = (\pm 4, 0)$. Figure 22 shows the graph.

Figure 22

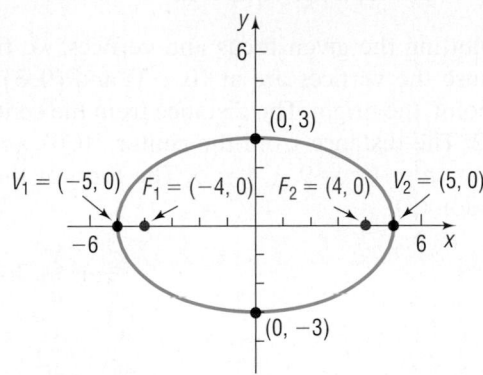

$V_1 = (-5, 0)$ $F_1 = (-4, 0)$ $F_2 = (4, 0)$ $V_2 = (5, 0)$
$(0, 3)$
$(0, -3)$

—**Now Work** PROBLEM 17

If the major axis of an ellipse with center at $(0, 0)$ lies on the y-axis, the foci are at $(0, -c)$ and $(0, c)$. Using the same steps as before, the definition of an ellipse leads to the following result:

THEOREM　　**Equation of an Ellipse: Center at (0, 0); Major Axis along the y-Axis**

An equation of the ellipse with center at $(0, 0)$, foci at $(0, -c)$ and $(0, c)$, and vertices at $(0, -a)$ and $(0, a)$ is

$$\boxed{\dfrac{x^2}{b^2} + \dfrac{y^2}{a^2} = 1 \qquad \text{where } a > b > 0 \text{ and } b^2 = a^2 - c^2} \qquad \text{(3)}$$

The major axis is the y-axis.

Figure 23

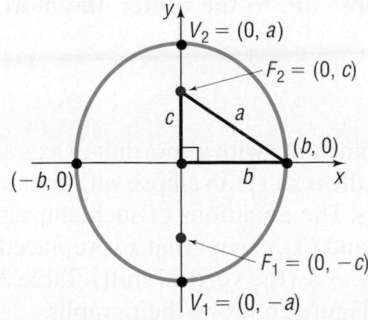

$V_2 = (0, a)$
$F_2 = (0, c)$
$(b, 0)$
$(-b, 0)$
$F_1 = (0, -c)$
$V_1 = (0, -a)$

Figure 23 illustrates the graph of such an ellipse. Again, notice the right triangle formed by the points at $(0, 0)$, $(b, 0)$, and $(0, c)$, so that $a^2 = b^2 + c^2$ (or $b^2 = a^2 - c^2$).

Look closely at equations (2) and (3). Although they may look alike, there is a difference! In equation (2), the larger number, a^2, is in the denominator of the x^2-term, so the major axis of the ellipse is along the x-axis. In equation (3), the larger number, a^2, is in the denominator of the y^2-term, so the major axis is along the y-axis.

EXAMPLE 3 **Analyzing the Equation of an Ellipse**

Analyze the equation: $9x^2 + y^2 = 9$

Figure 24

Solution To put the equation in proper form, divide each side by 9.

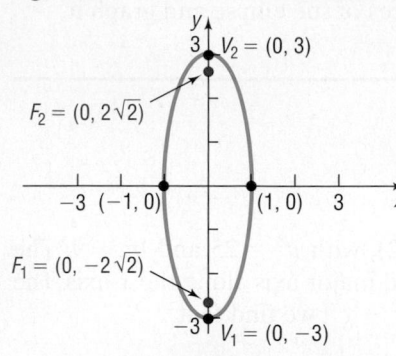

$$x^2 + \frac{y^2}{9} = 1$$

The larger denominator, 9, is in the y^2-term so, based on equation (3), this is the equation of an ellipse with center at the origin and major axis along the y-axis. Also, we conclude that $a^2 = 9$, $b^2 = 1$, and $c^2 = a^2 - b^2 = 9 - 1 = 8$. The vertices are at $(0, \pm a) = (0, \pm 3)$, and the foci are at $(0, \pm c) = (0, \pm 2\sqrt{2})$. Figure 24 shows the graph.

Now Work PROBLEM 21

EXAMPLE 4 **Finding an Equation of an Ellipse**

Find an equation of the ellipse having one focus at $(0, 2)$ and vertices at $(0, -3)$ and $(0, 3)$. Graph the equation.

Figure 25

Solution By plotting the given focus and vertices, we find that the major axis is the y-axis. Because the vertices are at $(0, -3)$ and $(0, 3)$, the center of this ellipse is at their midpoint, the origin. The distance from the center, $(0, 0)$, to the given focus, $(0, 2)$, is $c = 2$. The distance from the center, $(0, 0)$, to one of the vertices, $(0, 3)$, is $a = 3$. So $b^2 = a^2 - c^2 = 9 - 4 = 5$. The form of the equation of this ellipse is given by equation (3).

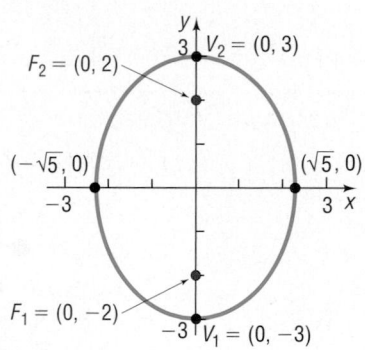

$$\frac{x^2}{b^2} + \frac{y^2}{a^2} = 1$$

$$\frac{x^2}{5} + \frac{y^2}{9} = 1$$

Figure 25 shows the graph.

Now Work PROBLEM 29

The circle may be considered a special kind of ellipse. To see why, let $a = b$ in equation (2) or (3). Then

$$\frac{x^2}{a^2} + \frac{y^2}{a^2} = 1$$

$$x^2 + y^2 = a^2$$

This is the equation of a circle with center at the origin and radius a. The value of c is

$$c^2 = a^2 - b^2 = 0$$
$$\uparrow$$
$$a = b$$

We conclude that the closer the two foci of an ellipse are to the center, the more the ellipse will look like a circle.

2 Analyze Ellipses with Center at (h, k)

If an ellipse with center at the origin and major axis coinciding with a coordinate axis is shifted horizontally h units and then vertically k units, the result is an ellipse with center at (h, k) and major axis parallel to a coordinate axis. The equations of such ellipses have the same forms as those given in equations (2) and (3), except that x is replaced by $x - h$ (the horizontal shift) and y is replaced by $y - k$ (the vertical shift). Table 3 gives the forms of the equations of such ellipses, and Figure 26 shows their graphs.

Table 3

NOTE It is not recommended that Table 3 be memorized. Rather, use the ideas of transformations (shift horizontally h units, vertically k units) along with the fact that a represents the distance from the center to the vertices, c represents the distance from the center to the foci, and $b^2 = a^2 - c^2$ (or $c^2 = a^2 - b^2$). ∎

Equations of an Ellipse: Center at (h, k); Major Axis Parallel to a Coordinate Axis

Center	Major Axis	Foci	Vertices	Equation
(h, k)	Parallel to the x-axis	$(h + c, k)$ $(h - c, k)$	$(h + a, k)$ $(h - a, k)$	$\dfrac{(x - h)^2}{a^2} + \dfrac{(y - k)^2}{b^2} = 1,$ $a > b > 0$ and $b^2 = a^2 - c^2$
(h, k)	Parallel to the y-axis	$(h, k + c)$ $(h, k - c)$	$(h, k + a)$ $(h, k - a)$	$\dfrac{(x - h)^2}{b^2} + \dfrac{(y - k)^2}{a^2} = 1,$ $a > b > 0$ and $b^2 = a^2 - c^2$

Figure 26

(a) $\dfrac{(x - h)^2}{a^2} + \dfrac{(y - k)^2}{b^2} = 1$

(b) $\dfrac{(x - h)^2}{b^2} + \dfrac{(y - k)^2}{a^2} = 1$

EXAMPLE 5

Finding an Equation of an Ellipse, Center Not at the Origin

Find an equation for the ellipse with center at $(2, -3)$, one focus at $(3, -3)$, and one vertex at $(5, -3)$. Graph the equation.

Solution

The center is at $(h, k) = (2, -3)$, so $h = 2$ and $k = -3$. If we plot the center, focus, and vertex, we notice that the points all lie on the line $y = -3$, so the major axis is parallel to the x-axis. The distance from the center $(2, -3)$ to a focus $(3, -3)$ is $c = 1$; the distance from the center $(2, -3)$ to a vertex $(5, -3)$ is $a = 3$. Then $b^2 = a^2 - c^2 = 9 - 1 = 8$. The form of the equation is

$$\frac{(x - h)^2}{a^2} + \frac{(y - k)^2}{b^2} = 1 \quad \text{where } h = 2, k = -3, a = 3, b = 2\sqrt{2}$$

$$\frac{(x - 2)^2}{9} + \frac{(y + 3)^2}{8} = 1$$

To graph the equation, use the center $(h, k) = (2, -3)$ to locate the vertices. The major axis is parallel to the x-axis, so the vertices are $a = 3$ units left and right of the center $(2, -3)$. Therefore, the vertices are

$$V_1 = (2 - 3, -3) = (-1, -3) \quad \text{and} \quad V_2 = (2 + 3, -3) = (5, -3)$$

Since $c = 1$ and the major axis is parallel to the x-axis, the foci are 1 unit left and right of the center. Therefore, the foci are

$$F_1 = (2 - 1, -3) = (1, -3) \quad \text{and} \quad F_2 = (2 + 1, -3) = (3, -3)$$

Finally, use the value of $b = 2\sqrt{2}$ to find the two points above and below the center.

$$\left(2, -3 - 2\sqrt{2}\right) \quad \text{and} \quad \left(2, -3 + 2\sqrt{2}\right)$$

Figure 27 shows the graph.

Figure 27

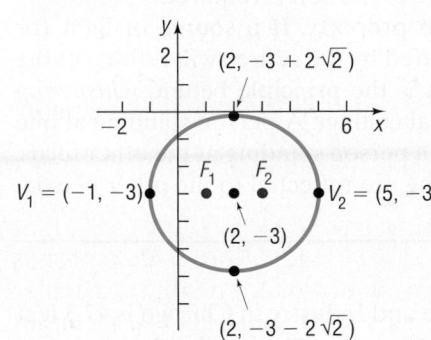

Now Work PROBLEM 55

EXAMPLE 6

Analyzing the Equation of an Ellipse

Analyze the equation: $4x^2 + y^2 - 8x + 4y + 4 = 0$

Solution

Proceed to complete the squares in x and in y.

$$4x^2 + y^2 - 8x + 4y + 4 = 0$$
$$4x^2 - 8x + y^2 + 4y = -4 \qquad \text{Group like variables; place the constant on the right side.}$$
$$4(x^2 - 2x) + (y^2 + 4y) = -4 \qquad \text{Factor out 4 from the first two terms.}$$
$$4(x^2 - 2x + 1) + (y^2 + 4y + 4) = -4 + 4 + 4 \qquad \text{Complete each square.}$$
$$4(x - 1)^2 + (y + 2)^2 = 4 \qquad \text{Factor.}$$
$$(x - 1)^2 + \frac{(y + 2)^2}{4} = 1 \qquad \text{Divide each side by 4.}$$

Figure 28

This is the equation of an ellipse with center at $(1, -2)$ and major axis parallel to the y-axis. Since $a^2 = 4$ and $b^2 = 1$, we have $c^2 = a^2 - b^2 = 4 - 1 = 3$. The vertices are at $(h, k \pm a) = (1, -2 \pm 2)$ or $(1, -4)$ and $(1, 0)$. The foci are at $(h, k \pm c) = (1, -2 \pm \sqrt{3})$ or $(1, -2 - \sqrt{3})$ and $(1, -2 + \sqrt{3})$. Figure 28 shows the graph.

━━━ Now Work PROBLEM 47

3 Solve Applied Problems Involving Ellipses

Ellipses are found in many applications in science and engineering. For example, the orbits of the planets around the Sun are elliptical, with the Sun's position at a focus. See Figure 29.

Figure 29

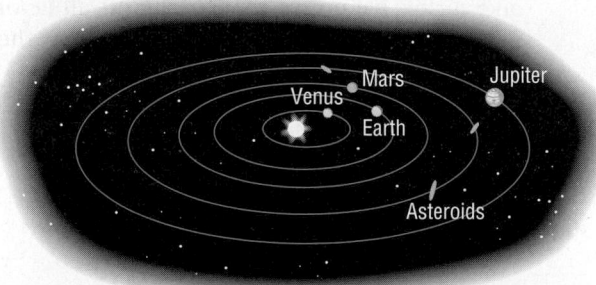

Stone and concrete bridges are often shaped as semielliptical arches. Elliptical gears are used in machinery when a variable rate of motion is required.

Ellipses also have an interesting reflection property. If a source of light (or sound) is placed at one focus, the waves transmitted by the source will reflect off the ellipse and concentrate at the other focus. This is the principle behind *whispering galleries*, which are rooms designed with elliptical ceilings. A person standing at one focus of the ellipse can whisper and be heard by a person standing at the other focus, because all the sound waves that reach the ceiling are reflected to the other person.

EXAMPLE 7

A Whispering Gallery

The whispering gallery in the Museum of Science and Industry in Chicago is 47.3 feet long. The distance from the center of the room to the foci is 20.3 feet. Find an equation that describes the shape of the room. How high is the room at its center?

Source: Chicago Museum of Science and Industry Web site; www.msichicago.org

Solution Set up a rectangular coordinate system so that the center of the ellipse is at the origin and the major axis is along the x-axis. The equation of the ellipse is

$$\frac{x^2}{a^2} + \frac{y^2}{b^2} = 1$$

Since the length of the room is 47.3 feet, the distance from the center of the room to each vertex (the end of the room) will be $\dfrac{47.3}{2} = 23.65$ feet; so $a = 23.65$ feet. The distance from the center of the room to each focus is $c = 20.3$ feet. See Figure 30.

Since $b^2 = a^2 - c^2$, we find $b^2 = 23.65^2 - 20.3^2 = 147.2325$. An equation that describes the shape of the room is given by

$$\frac{x^2}{23.65^2} + \frac{y^2}{147.2325} = 1$$

Figure 30

The height of the room at its center is $b = \sqrt{147.2325} \approx 12.1$ feet.

─ **Now Work** PROBLEM **71**

7.3 Assess Your Understanding

'Are You Prepared?' *Answers are given at the end of these exercises. If you get a wrong answer, read the pages listed in red.*

1. The distance d from $P_1 = (2, -5)$ to $P_2 = (4, -2)$ is $d =$ ___. (p. 151)

2. To complete the square of $x^2 - 3x$, add ___. (p. 56)

3. Find the intercepts of the equation $y^2 = 16 - 4x^2$. (pp. 159–160)

4. The point that is symmetric with respect to the y-axis to the point $(-2, 5)$ is ___. (pp. 160–162)

5. To graph $y = (x + 1)^2 - 4$, shift the graph of $y = x^2$ to the (left/right) ___ unit(s) and then (up/down) ___ unit(s). (pp. 244–253)

6. The standard equation of a circle with center at $(2, -3)$ and radius 1 is ___. (pp. 182–185)

Concepts and Vocabulary

7. A(n) ___ is the collection of all points in the plane the sum of whose distances from two fixed points is a constant.

8. For an ellipse, the foci lie on a line called the ___ axis.

9. For the ellipse $\dfrac{x^2}{4} + \dfrac{y^2}{25} = 1$, the vertices are the points ___ and ___.

10. For the ellipse $\dfrac{x^2}{25} + \dfrac{y^2}{9} = 1$, the value of a is ___, the value of b is ___, and the major axis is the ___-axis.

11. If the center of an ellipse is $(2, -3)$, the major axis is parallel to the x-axis, and the distance from the center of the ellipse to its vertices is $a = 4$ units, then the coordinates of the vertices are ___ and ___.

12. If the foci of an ellipse are $(-4, 4)$ and $(6, 4)$, then the coordinates of the center of the ellipse are ___.

Skill Building

In Problems 13–16, the graph of an ellipse is given. Match each graph to its equation.

(A) $\dfrac{x^2}{4} + y^2 = 1$ (B) $x^2 + \dfrac{y^2}{4} = 1$ (C) $\dfrac{x^2}{16} + \dfrac{y^2}{4} = 1$ (D) $\dfrac{x^2}{4} + \dfrac{y^2}{16} = 1$

13.

14.

15.

16.

In Problems 17–26, find the vertices and foci of each ellipse. Graph each equation.

17. $\dfrac{x^2}{25} + \dfrac{y^2}{4} = 1$ **18.** $\dfrac{x^2}{9} + \dfrac{y^2}{4} = 1$ **19.** $\dfrac{x^2}{9} + \dfrac{y^2}{25} = 1$ **20.** $x^2 + \dfrac{y^2}{16} = 1$

21. $4x^2 + y^2 = 16$ **22.** $x^2 + 9y^2 = 18$ **23.** $4y^2 + x^2 = 8$ **24.** $4y^2 + 9x^2 = 36$

25. $x^2 + y^2 = 16$ **26.** $x^2 + y^2 = 4$

In Problems 27–38, find an equation for each ellipse. Graph the equation.

27. Center at $(0, 0)$; focus at $(3, 0)$; vertex at $(5, 0)$

28. Center at $(0, 0)$; focus at $(-1, 0)$; vertex at $(3, 0)$

29. Center at $(0, 0)$; focus at $(0, -4)$; vertex at $(0, 5)$

30. Center at $(0, 0)$; focus at $(0, 1)$; vertex at $(0, -2)$

31. Foci at $(\pm 2, 0)$; length of the major axis is 6

32. Foci at $(0, \pm 2)$; length of the major axis is 8

33. Focus at $(-4, 0)$; vertices at $(\pm 5, 0)$

34. Focus at $(0, -4)$; vertices at $(0, \pm 8)$

35. Foci at $(0, \pm 3)$; x-intercepts are ± 2

36. Vertices at $(\pm 4, 0)$; y-intercepts are ± 1

37. Center at $(0, 0)$; vertex at $(0, 4)$; $b = 1$

38. Vertices at $(\pm 5, 0)$; $c = 2$

In Problems 39–42, write an equation for each ellipse.

39. **40.** **41.** **42.**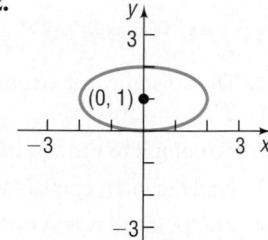

In Problems 43–54, analyze each equation; that is, find the center, foci, and vertices of each ellipse. Graph each equation.

43. $\dfrac{(x-3)^2}{4} + \dfrac{(y+1)^2}{9} = 1$ **44.** $\dfrac{(x+4)^2}{9} + \dfrac{(y+2)^2}{4} = 1$ **45.** $(x+5)^2 + 4(y-4)^2 = 16$

46. $9(x-3)^2 + (y+2)^2 = 18$ **47.** $x^2 + 4x + 4y^2 - 8y + 4 = 0$ **48.** $x^2 + 3y^2 - 12y + 9 = 0$

49. $2x^2 + 3y^2 - 8x + 6y + 5 = 0$ **50.** $4x^2 + 3y^2 + 8x - 6y = 5$ **51.** $9x^2 + 4y^2 - 18x + 16y - 11 = 0$

52. $x^2 + 9y^2 + 6x - 18y + 9 = 0$ **53.** $4x^2 + y^2 + 4y = 0$ **54.** $9x^2 + y^2 - 18x = 0$

In Problems 55–64, find an equation for each ellipse. Graph the equation.

55. Center at $(2, -2)$; vertex at $(7, -2)$; focus at $(4, -2)$

56. Center at $(-3, 1)$; vertex at $(-3, 3)$; focus at $(-3, 0)$

57. Vertices at $(4, 3)$ and $(4, 9)$; focus at $(4, 8)$

58. Foci at $(1, 2)$ and $(-3, 2)$; vertex at $(-4, 2)$

59. Foci at $(5, 1)$ and $(-1, 1)$; length of the major axis is 8

60. Vertices at $(2, 5)$ and $(2, -1)$; $c = 2$

61. Center at $(1, 2)$; focus at $(4, 2)$; contains the point $(1, 3)$

62. Center at $(1, 2)$; focus at $(1, 4)$; contains the point $(2, 2)$

63. Center at $(1, 2)$; vertex at $(4, 2)$; contains the point $(1, 5)$

64. Center at $(1, 2)$; vertex at $(1, 4)$; contains the point $(1 + \sqrt{3}, 3)$

In Problems 65–68, graph each function. Be sure to label all the intercepts.
[**Hint:** *Notice that each function is half an ellipse.*]

65. $f(x) = \sqrt{16 - 4x^2}$ **66.** $f(x) = \sqrt{9 - 9x^2}$ **67.** $f(x) = -\sqrt{64 - 16x^2}$ **68.** $f(x) = -\sqrt{4 - 4x^2}$

Applications and Extensions

69. Semielliptical Arch Bridge An arch in the shape of the upper half of an ellipse is used to support a bridge that is to span a river 20 meters wide. The center of the arch is 6 meters above the center of the river. See the figure. Write an equation for the ellipse in which the *x*-axis coincides with the water level and the *y*-axis passes through the center of the arch.

70. Semielliptical Arch Bridge The arch of a bridge is a semiellipse with a horizontal major axis. The span is 30 feet, and the top of the arch is 10 feet above the major axis. The roadway is horizontal and is 2 feet above the top of the arch. Find the vertical distance from the roadway to the arch at 5-foot intervals along the roadway.

71. Whispering Gallery A hall 100 feet in length is to be designed as a whispering gallery. If the foci are located 25 feet from the center, how high will the ceiling be at the center?

72. Whispering Gallery Jim, standing at one focus of a whispering gallery, is 6 feet from the nearest wall. His friend is standing at the other focus, 100 feet away. What is the length of this whispering gallery? How high is its elliptical ceiling at the center?

73. Semielliptical Arch Bridge A bridge is built in the shape of a semielliptical arch. The bridge has a span of 120 feet and a maximum height of 25 feet. Choose a suitable rectangular coordinate system and find the height of the arch at distances of 10, 30, and 50 feet from the center.

74. Semielliptical Arch Bridge A bridge is to be built in the shape of a semielliptical arch and is to have a span of 100 feet. The height of the arch, at a distance of 40 feet from the center, is to be 10 feet. Find the height of the arch at its center.

75. Racetrack Design Consult the figure. A racetrack is in the shape of an ellipse, 100 feet long and 50 feet wide. What is the width 10 feet from a vertex?

76. Semielliptical Arch Bridge An arch for a bridge over a highway is in the form of half an ellipse. The top of the arch is 20 feet above the ground level (the major axis). The highway has four lanes, each 12 feet wide; a center safety strip 8 feet wide; and two side strips, each 4 feet wide. What should the span of the bridge be (the length of its major axis) if the height 28 feet from the center is to be 13 feet?

77. Installing a Vent Pipe A homeowner is putting in a fireplace that has a 4-inch-radius vent pipe. He needs to cut an elliptical hole in his roof to accommodate the pipe. If the pitch of his roof is $\dfrac{5}{4}$, (a rise of 5, run of 4) what are the dimensions of the hole?

Source: www.doe.virginia.gov

78. Volume of a Football A football is in the shape of a **prolate spheroid,** which is simply a solid obtained by rotating an ellipse $\left(\dfrac{x^2}{a^2} + \dfrac{y^2}{b^2} = 1\right)$ about its major axis. An inflated NFL football averages 11.125 inches in length and 28.25 inches in center circumference. If the volume of a prolate spheroid is $\dfrac{4}{3}\pi a b^2$, how much air does the football contain? (Neglect material thickness).

Source: www.answerbag.com

*In Problems 79–82, use the fact that the orbit of a planet about the Sun is an ellipse, with the Sun at one focus. The **aphelion** of a planet is its greatest distance from the Sun, and the **perihelion** is its shortest distance. The **mean distance** of a planet from the Sun is the length of the semimajor axis of the elliptical orbit. See the illustration.*

79. Earth The mean distance of Earth from the Sun is 93 million miles. If the aphelion of Earth is 94.5 million miles, what is the perihelion? Write an equation for the orbit of Earth around the Sun.

80. Mars The mean distance of Mars from the Sun is 142 million miles. If the perihelion of Mars is 128.5 million miles, what is the aphelion? Write an equation for the orbit of Mars about the Sun.

81. Jupiter The aphelion of Jupiter is 507 million miles. If the distance from the center of its elliptical orbit to the Sun is 23.2 million miles, what is the perihelion? What is the mean distance? Write an equation for the orbit of Jupiter around the Sun.

82. Pluto The perihelion of Pluto is 4551 million miles, and the distance from the center of its elliptical orbit to the Sun is 897.5 million miles. Find the aphelion of Pluto. What is the mean distance of Pluto from the Sun? Write an equation for the orbit of Pluto about the Sun.

83. Show that an equation of the form

$$Ax^2 + Cy^2 + F = 0, \qquad A \neq 0, C \neq 0, F \neq 0$$

where *A* and *C* are of the same sign and *F* is of opposite sign,
(a) Is the equation of an ellipse with center at $(0, 0)$ if $A \neq C$.
(b) Is the equation of a circle with center $(0, 0)$ if $A = C$.

84. Show that the graph of an equation of the form

$$Ax^2 + Cy^2 + Dx + Ey + F = 0, \qquad A \neq 0, C \neq 0$$

where A and C are of the same sign,

(a) Is an ellipse if $\dfrac{D^2}{4A} + \dfrac{E^2}{4C} - F$ is the same sign as A.

(b) Is a point if $\dfrac{D^2}{4A} + \dfrac{E^2}{4C} - F = 0$.

(c) Contains no points if $\dfrac{D^2}{4A} + \dfrac{E^2}{4C} - F$ is of opposite sign to A.

Explaining Concepts: Discussion and Writing

85. The **eccentricity** e of an ellipse is defined as the number $\dfrac{c}{a}$, where a is the distance of a vertex from the center and c is the distance of a focus from the center. Because $a > c$, it follows that $e < 1$. Write a brief paragraph about the general shape of each of the following ellipses. Be sure to justify your conclusions.

(a) Eccentricity close to 0 (b) Eccentricity = 0.5 (c) Eccentricity close to 1

'Are You Prepared?' Answers

1. $\sqrt{13}$ **2.** $\dfrac{9}{4}$ **3.** $(-2, 0), (2, 0), (0, -4), (0, 4)$ **4.** $(2, 5)$ **5.** left; 1; down: 4 **6.** $(x - 2)^2 + (y + 3)^2 = 1$

7.4 The Hyperbola

PREPARING FOR THIS SECTION *Before getting started, review the following:*

- Distance Formula (Section 2.1, p. 151)
- Completing the Square (Chapter R, Review, Section R.5, p. 56)
- Intercepts (Section 2.2, pp. 159–160)
- Symmetry (Section 2.2, pp. 160–162)
- Asymptotes (Section 5.2, pp. 345–348)
- Graphing Techniques: Transformations (Section 3.5, pp. 244–253)
- Square Root Method (Section 1.2, pp. 94–95)

Now Work the 'Are You Prepared?' problems on page 534.

OBJECTIVES **1** Analyze Hyperbolas with Center at the Origin (p. 524)
 2 Find the Asymptotes of a Hyperbola (p. 529)
 3 Analyze Hyperbolas with Center at *(h, k)* (p. 531)
 4 Solve Applied Problems Involving Hyperbolas (p. 532)

DEFINITION

A **hyperbola** is the collection of all points in the plane, the difference of whose distances from two fixed points, called the **foci,** is a constant.

Figure 31

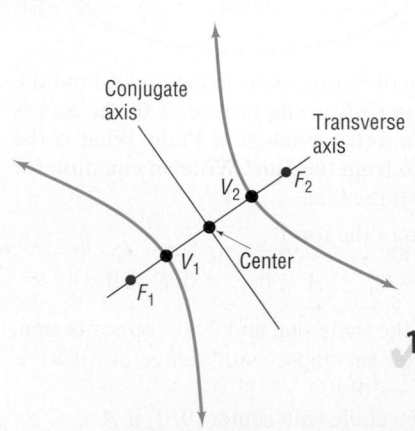

Figure 31 illustrates a hyperbola with foci F_1 and F_2. The line containing the foci is called the **transverse axis.** The midpoint of the line segment joining the foci is the **center** of the hyperbola. The line through the center and perpendicular to the transverse axis is the **conjugate axis.** The hyperbola consists of two separate curves, called **branches,** that are symmetric with respect to the transverse axis, conjugate axis, and center. The two points of intersection of the hyperbola and the transverse axis are the **vertices,** V_1 and V_2, of the hyperbola.

1 Analyze Hyperbolas with Center at the Origin

With these ideas in mind, we are now ready to find the equation of a hyperbola in the rectangular coordinate system. First, place the center at the origin. Next,

Figure 32

$d(F_1, P) - d(F_2, P) = \pm 2a$

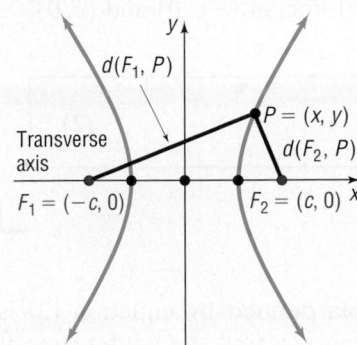

position the hyperbola so that its transverse axis coincides with a coordinate axis. Suppose that the transverse axis coincides with the x-axis, as shown in Figure 32.

If c is the distance from the center to a focus, one focus will be at $F_1 = (-c, 0)$ and the other at $F_2 = (c, 0)$. Now we let the constant difference of the distances from any point $P = (x, y)$ on the hyperbola to the foci F_1 and F_2 be denoted by $\pm 2a$. (If P is on the right branch, the $+$ sign is used; if P is on the left branch, the $-$ sign is used.) The coordinates of P must satisfy the equation

$$d(F_1, P) - d(F_2, P) = \pm 2a \qquad \text{Difference of the distances from } P \text{ to the foci equals } \pm 2a.$$

$$\sqrt{(x + c)^2 + y^2} - \sqrt{(x - c)^2 + y^2} = \pm 2a \qquad \text{Use the Distance Formula.}$$

$$\sqrt{(x + c)^2 + y^2} = \pm 2a + \sqrt{(x - c)^2 + y^2} \qquad \text{Isolate one radical.}$$

$$(x + c)^2 + y^2 = 4a^2 \pm 4a\sqrt{(x - c)^2 + y^2} \qquad \text{Square both sides.}$$
$$+ (x - c)^2 + y^2$$

Next we remove the parentheses.

$$x^2 + 2cx + c^2 + y^2 = 4a^2 \pm 4a\sqrt{(x - c)^2 + y^2} + x^2 - 2cx + c^2 + y^2$$

$$4cx - 4a^2 = \pm 4a\sqrt{(x - c)^2 + y^2} \qquad \text{Simplify; isolate the radical.}$$

$$cx - a^2 = \pm a\sqrt{(x - c)^2 + y^2} \qquad \text{Divide each side by 4.}$$

$$(cx - a^2)^2 = a^2[(x - c)^2 + y^2] \qquad \text{Square both sides.}$$

$$c^2x^2 - 2ca^2x + a^4 = a^2(x^2 - 2cx + c^2 + y^2) \qquad \text{Simplify.}$$

$$c^2x^2 + a^4 = a^2x^2 + a^2c^2 + a^2y^2 \qquad \text{Remove parentheses and simplify.}$$

$$(c^2 - a^2)x^2 - a^2y^2 = a^2c^2 - a^4 \qquad \text{Rearrange terms.}$$

$$(c^2 - a^2)x^2 - a^2y^2 = a^2(c^2 - a^2) \qquad \text{Factor } a^2 \text{ on the right side.} \quad \textbf{(1)}$$

To obtain points on the hyperbola off the x-axis, it must be that $a < c$. To see why, look again at Figure 32.

$$d(F_1, P) < d(F_2, P) + d(F_1, F_2) \qquad \text{Use triangle } F_1PF_2.$$

$$d(F_1, P) - d(F_2, P) < d(F_1, F_2)$$

$$2a < 2c \qquad \begin{array}{l} P \text{ is on the right branch, so} \\ d(F_1, P) - d(F_2, P) = 2a; \\ d(F_1, F_2) = 2c. \end{array}$$

$$a < c$$

Since $a < c$, we also have $a^2 < c^2$, so $c^2 - a^2 > 0$. Let $b^2 = c^2 - a^2$, $b > 0$. Then equation (1) can be written as

$$b^2x^2 - a^2y^2 = a^2b^2$$

$$\frac{x^2}{a^2} - \frac{y^2}{b^2} = 1 \qquad \text{Divide each side by } a^2b^2.$$

To find the vertices of the hyperbola defined by this equation, let $y = 0$. The vertices satisfy the equation $\dfrac{x^2}{a^2} = 1$, the solutions of which are $x = \pm a$. Consequently, the vertices of the hyperbola are $V_1 = (-a, 0)$ and $V_2 = (a, 0)$. Notice that the distance from the center $(0, 0)$ to either vertex is a.

THEOREM

Equation of a Hyperbola: Center at (0, 0); Transverse Axis along the x-Axis

An equation of the hyperbola with center at $(0, 0)$, foci at $(-c, 0)$ and $(c, 0)$, and vertices at $(-a, 0)$ and $(a, 0)$ is

$$\frac{x^2}{a^2} - \frac{y^2}{b^2} = 1, \quad \text{where } b^2 = c^2 - a^2 \tag{2}$$

The transverse axis is the x-axis.

Figure 33

$\frac{x^2}{a^2} - \frac{y^2}{b^2} = 1, \quad b^2 = c^2 - a^2$

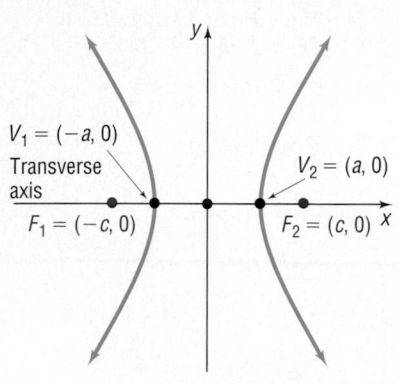

See Figure 33. As you can verify, the hyperbola defined by equation (2) is symmetric with respect to the x-axis, y-axis, and origin. To find the y-intercepts, if any, let $x = 0$ in equation (2). This results in the equation $\frac{y^2}{b^2} = -1$, which has no real solution, so the hyperbola defined by equation (2) has no y-intercepts. In fact, since $\frac{x^2}{a^2} - 1 = \frac{y^2}{b^2} \geq 0$, it follows that $\frac{x^2}{a^2} \geq 1$. There are no points on the graph for $-a < x < a$.

EXAMPLE 1

Finding and Graphing an Equation of a Hyperbola

Find an equation of the hyperbola with center at the origin, one focus at $(3, 0)$, and one vertex at $(-2, 0)$. Graph the equation.

Solution

The hyperbola has its center at the origin. Plot the center, focus, and vertex. Since they all lie on the x-axis, the transverse axis coincides with the x-axis. One focus is at $(c, 0) = (3, 0)$, so $c = 3$. One vertex is at $(-a, 0) = (-2, 0)$, so $a = 2$. From equation (2), it follows that $b^2 = c^2 - a^2 = 9 - 4 = 5$, so an equation of the hyperbola is

$$\frac{x^2}{4} - \frac{y^2}{5} = 1$$

To graph a hyperbola, it is helpful to locate and plot other points on the graph. For example, to find the points above and below the foci, we let $x = \pm 3$. Then

$$\frac{x^2}{4} - \frac{y^2}{5} = 1$$

$$\frac{(\pm 3)^2}{4} - \frac{y^2}{5} = 1 \qquad x = \pm 3$$

$$\frac{9}{4} - \frac{y^2}{5} = 1$$

$$\frac{y^2}{5} = \frac{5}{4}$$

$$y^2 = \frac{25}{4}$$

$$y = \pm \frac{5}{2}$$

Figure 34

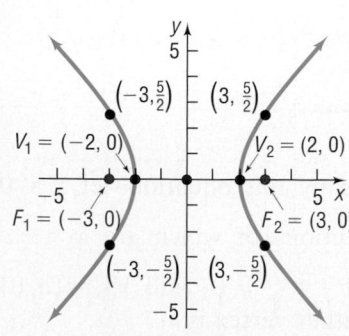

The points above and below the foci are $\left(\pm 3, \frac{5}{2}\right)$ and $\left(\pm 3, -\frac{5}{2}\right)$. These points determine the "opening" of the hyperbola. See Figure 34.

COMMENT To graph the hyperbola $\dfrac{x^2}{4} - \dfrac{y^2}{5} = 1$ discussed in Example 1, we need to graph the two functions $Y_1 = \sqrt{5}\sqrt{\dfrac{x^2}{4} - 1}$ and $Y_2 = -\sqrt{5}\sqrt{\dfrac{x^2}{4} - 1}$. Do this and compare what you see with Figure 34. ■

Now Work PROBLEM 19

An equation of the form of equation (2) is the equation of a hyperbola with center at the origin, foci on the x-axis at $(-c, 0)$ and $(c, 0)$, where $c^2 = a^2 + b^2$, and transverse axis along the x-axis.

For the next two examples, the direction **"Analyze the equation"** will mean to find the center, transverse axis, vertices, and foci of the hyperbola and graph it.

EXAMPLE 2 ### Analyzing the Equation of a Hyperbola

Analyze the equation: $\dfrac{x^2}{16} - \dfrac{y^2}{4} = 1$

Solution The given equation is of the form of equation (2), with $a^2 = 16$ and $b^2 = 4$. The graph of the equation is a hyperbola with center at $(0, 0)$ and transverse axis along the x-axis. Also, we know that $c^2 = a^2 + b^2 = 16 + 4 = 20$. The vertices are at $(\pm a, 0) = (\pm 4, 0)$, and the foci are at $(\pm c, 0) = (\pm 2\sqrt{5}, 0)$.

To locate the points on the graph above and below the foci, we let $x = \pm 2\sqrt{5}$. Then

$$\frac{x^2}{16} - \frac{y^2}{4} = 1$$

$$\frac{\left(\pm 2\sqrt{5}\right)^2}{16} - \frac{y^2}{4} = 1 \qquad x = \pm 2\sqrt{5}$$

$$\frac{20}{16} - \frac{y^2}{4} = 1$$

$$\frac{5}{4} - \frac{y^2}{4} = 1$$

$$\frac{y^2}{4} = \frac{1}{4}$$

$$y = \pm 1$$

The points above and below the foci are $(\pm 2\sqrt{5}, 1)$ and $(\pm 2\sqrt{5}, -1)$. See Figure 35. ♩

Figure 35

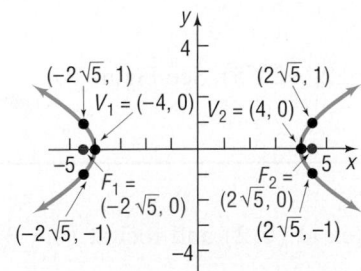

THEOREM ### Equation of a Hyperbola: Center at (0, 0); Transverse Axis along the y-Axis

An equation of the hyperbola with center at $(0, 0)$, foci at $(0, -c)$ and $(0, c)$, and vertices at $(0, -a)$ and $(0, a)$ is

$$\boxed{\dfrac{y^2}{a^2} - \dfrac{x^2}{b^2} = 1, \qquad \text{where } b^2 = c^2 - a^2} \tag{3}$$

The transverse axis is the y-axis.

Figure 36

$$\frac{y^2}{a^2} - \frac{x^2}{b^2} = 1, b^2 = c^2 - a^2$$

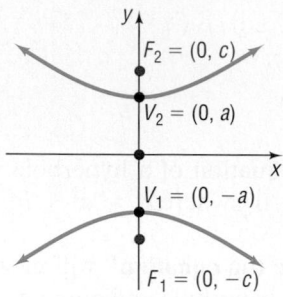

Figure 36 shows the graph of a typical hyperbola defined by equation (3).

An equation of the form of equation (2), $\frac{x^2}{a^2} - \frac{y^2}{b^2} = 1$, is the equation of a hyperbola with center at the origin, foci on the x-axis at $(-c, 0)$ and $(c, 0)$, where $c^2 = a^2 + b^2$, and transverse axis along the x-axis.

An equation of the form of equation (3), $\frac{y^2}{a^2} - \frac{x^2}{b^2} = 1$, is the equation of a hyperbola with center at the origin, foci on the y-axis at $(0, -c)$ and $(0, c)$, where $c^2 = a^2 + b^2$, and transverse axis along the y-axis.

Notice the difference in the forms of equations (2) and (3). When the y^2-term is subtracted from the x^2-term, the transverse axis is along the x-axis. When the x^2-term is subtracted from the y^2-term, the transverse axis is along the y-axis.

EXAMPLE 3

Analyzing the Equation of a Hyperbola

Analyze the equation: $y^2 - 4x^2 = 4$

Solution

To put the equation in proper form, divide each side by 4:

$$\frac{y^2}{4} - x^2 = 1$$

Since the x^2-term is subtracted from the y^2-term, the equation is that of a hyperbola with center at the origin and transverse axis along the y-axis. Also, comparing the above equation to equation (3), we find $a^2 = 4$, $b^2 = 1$, and $c^2 = a^2 + b^2 = 5$. The vertices are at $(0, \pm a) = (0, \pm 2)$, and the foci are at $(0, \pm c) = (0, \pm\sqrt{5})$.

To locate other points on the graph, let $x = \pm 2$. Then

Figure 37

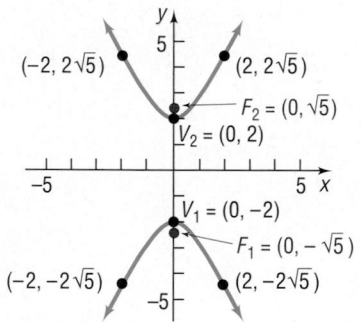

$$y^2 - 4x^2 = 4$$
$$y^2 - 4(\pm 2)^2 = 4 \qquad x = \pm 2$$
$$y^2 - 16 = 4$$
$$y^2 = 20$$
$$y = \pm 2\sqrt{5}$$

Four other points on the graph are $(\pm 2, 2\sqrt{5})$ and $(\pm 2, -2\sqrt{5})$. See Figure 37.

EXAMPLE 4

Finding an Equation of a Hyperbola

Find an equation of the hyperbola having one vertex at $(0, 2)$ and foci at $(0, -3)$ and $(0, 3)$. Graph the equation.

Solution

Since the foci are at $(0, -3)$ and $(0, 3)$, the center of the hyperbola, which is at their midpoint, is the origin. Also, the transverse axis is along the y-axis. The given information also reveals that $c = 3$, $a = 2$, and $b^2 = c^2 - a^2 = 9 - 4 = 5$. The form of the equation of the hyperbola is given by equation (3):

Figure 38

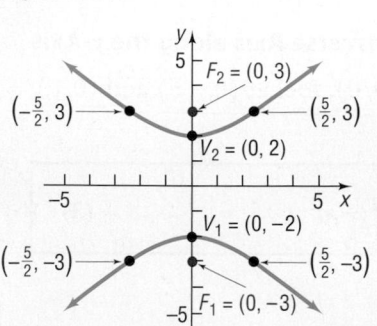

$$\frac{y^2}{a^2} - \frac{x^2}{b^2} = 1$$
$$\frac{y^2}{4} - \frac{x^2}{5} = 1$$

Let $y = \pm 3$ to obtain points on the graph on either side of each focus. See Figure 38.

Now Work PROBLEM **21**

Look at the equations of the hyperbolas in Examples 2 and 4. For the hyperbola in Example 2, $a^2 = 16$ and $b^2 = 4$, so $a > b$; for the hyperbola in Example 4, $a^2 = 4$ and $b^2 = 5$, so $a < b$. We conclude that, for hyperbolas, there are no requirements involving the relative sizes of a and b. Contrast this situation to the case of an ellipse, in which the relative sizes of a and b dictate which axis is the major axis. Hyperbolas have another feature to distinguish them from ellipses and parabolas: Hyperbolas have asymptotes.

2 Find the Asymptotes of a Hyperbola

Recall from Section 5.2 that a horizontal or oblique asymptote of a graph is a line with the property that the distance from the line to points on the graph approaches 0 as $x \to -\infty$ or as $x \to \infty$. Asymptotes provide information about the end behavior of the graph of a hyperbola.

THEOREM

Asymptotes of a Hyperbola

The hyperbola $\dfrac{x^2}{a^2} - \dfrac{y^2}{b^2} = 1$ has the two oblique asymptotes

$$y = \frac{b}{a}x \quad \text{and} \quad y = -\frac{b}{a}x \tag{4}$$

Proof We begin by solving for y in the equation of the hyperbola.

$$\frac{x^2}{a^2} - \frac{y^2}{b^2} = 1$$

$$\frac{y^2}{b^2} = \frac{x^2}{a^2} - 1$$

$$y^2 = b^2\left(\frac{x^2}{a^2} - 1\right)$$

Since $x \neq 0$, we can rearrange the right side in the form

$$y^2 = \frac{b^2 x^2}{a^2}\left(1 - \frac{a^2}{x^2}\right)$$

$$y = \pm\frac{bx}{a}\sqrt{1 - \frac{a^2}{x^2}}$$

Now, as $x \to -\infty$ or as $x \to \infty$, the term $\dfrac{a^2}{x^2}$ approaches 0, so the expression under the radical approaches 1. So, as $x \to -\infty$ or as $x \to \infty$, the value of y approaches $\pm\dfrac{bx}{a}$; that is, the graph of the hyperbola approaches the lines

$$y = -\frac{b}{a}x \quad \text{and} \quad y = \frac{b}{a}x$$

These lines are oblique asymptotes of the hyperbola. ∎

The asymptotes of a hyperbola are not part of the hyperbola, but they do serve as a guide for graphing a hyperbola. For example, suppose that we want to graph the equation

$$\frac{x^2}{a^2} - \frac{y^2}{b^2} = 1$$

Figure 39

$$\frac{x^2}{a^2} - \frac{y^2}{b^2} = 1$$

Begin by plotting the vertices $(-a, 0)$ and $(a, 0)$. Then plot the points $(0, -b)$ and $(0, b)$ and use these four points to construct a rectangle, as shown in Figure 39. The diagonals of this rectangle have slopes $\frac{b}{a}$ and $-\frac{b}{a}$, and their extensions are the asymptotes $y = \frac{b}{a}x$ and $y = -\frac{b}{a}x$ of the hyperbola. If we graph the asymptotes, we can use them to establish the "opening" of the hyperbola and avoid plotting other points.

THEOREM

Asymptotes of a Hyperbola

The hyperbola $\dfrac{y^2}{a^2} - \dfrac{x^2}{b^2} = 1$ has the two oblique asymptotes

$$y = \frac{a}{b}x \quad \text{and} \quad y = -\frac{a}{b}x \tag{5}$$

You are asked to prove this result in Problem 84.

For the remainder of this section, the direction **"Analyze the equation"** will mean to find the center, transverse axis, vertices, foci, and asymptotes of the hyperbola and graph it.

EXAMPLE 5

Analyzing the Equation of a Hyperbola

Analyze the equation: $\dfrac{y^2}{4} - x^2 = 1$

Figure 40

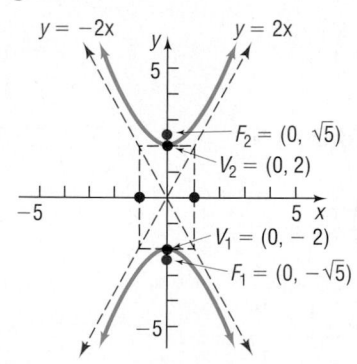

Solution Since the x^2-term is subtracted from the y^2-term, the equation is of the form of equation (3) and is a hyperbola with center at the origin and transverse axis along the y-axis. Also, comparing this equation to equation (3), we find that $a^2 = 4$, $b^2 = 1$, and $c^2 = a^2 + b^2 = 5$. The vertices are at $(0, \pm a) = (0, \pm 2)$, and the foci are at $(0, \pm c) = (0, \pm\sqrt{5})$. Using equation (5) with $a = 2$ and $b = 1$, the asymptotes are the lines $y = \dfrac{a}{b}x = 2x$ and $y = -\dfrac{a}{b}x = -2x$. Form the rectangle containing the points $(0, \pm a) = (0, \pm 2)$ and $(\pm b, 0) = (\pm 1, 0)$. The extensions of the diagonals of this rectangle are the asymptotes. Now graph the rectangle, the asymptotes, and the hyperbola. See Figure 40.

EXAMPLE 6

Analyzing the Equation of a Hyperbola

Analyze the equation: $9x^2 - 4y^2 = 36$

Solution Divide each side of the equation by 36 to put the equation in proper form.

$$\frac{x^2}{4} - \frac{y^2}{9} = 1$$

The center of the hyperbola is the origin. Since the x^2-term is first in the equation, the transverse axis is along the x-axis and the vertices and foci will lie on the x-axis. Using equation (2), we find $a^2 = 4$, $b^2 = 9$, and $c^2 = a^2 + b^2 = 13$. The vertices are $a = 2$ units left and right of the center at $(\pm a, 0) = (\pm 2, 0)$, the foci are $c = \sqrt{13}$

Figure 41

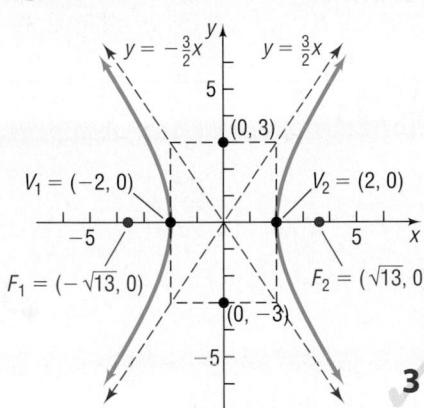

units left and right of the center at $(\pm c, 0) = (\pm\sqrt{13}, 0)$, and the asymptotes have the equations

$$y = \frac{b}{a}x = \frac{3}{2}x \quad \text{and} \quad y = -\frac{b}{a}x = -\frac{3}{2}x$$

To graph the hyperbola, form the rectangle containing the points $(\pm a, 0)$ and $(0, \pm b)$, that is, $(-2, 0)$, $(2, 0)$, $(0, -3)$, and $(0, 3)$. The extensions of the diagonals of this rectangle are the asymptotes. See Figure 41 for the graph.

Now Work PROBLEM 31

3 Analyze Hyperbolas with Center at (h, k)

If a hyperbola with center at the origin and transverse axis coinciding with a coordinate axis is shifted horizontally h units and then vertically k units, the result is a hyperbola with center at (h, k) and transverse axis parallel to a coordinate axis. The equations of such hyperbolas have the same forms as those given in equations (2) and (3), except that x is replaced by $x - h$ (the horizontal shift) and y is replaced by $y - k$ (the vertical shift). Table 4 gives the forms of the equations of such hyperbolas. See Figure 42 for typical graphs.

Table 4

Equations of a Hyperbola: Center at (h, k); Transverse Axis Parallel to a Coordinate Axis					
Center	Transverse Axis	Foci	Vertices	Equation	Asymptotes
(h, k)	Parallel to the x-axis	$(h \pm c, k)$	$(h \pm a, k)$	$\dfrac{(x - h)^2}{a^2} - \dfrac{(y - k)^2}{b^2} = 1$, $b^2 = c^2 - a^2$	$y - k = \pm\dfrac{b}{a}(x - h)$
(h, k)	Parallel to the y-axis	$(h, k \pm c)$	$(h, k \pm a)$	$\dfrac{(y - k)^2}{a^2} - \dfrac{(x - h)^2}{b^2} = 1$, $b^2 = c^2 - a^2$	$y - k = \pm\dfrac{a}{b}(x - h)$

Figure 42

NOTE It is not recommended that Table 4 be memorized. Rather use the ideas of transformations (shift horizontally h units, vertically k units) along with the fact that a represents the distance from the center to the vertices, c represents the distance from the center to the foci, and $b^2 = c^2 - a^2$ (or $c^2 = a^2 + b^2$). ■

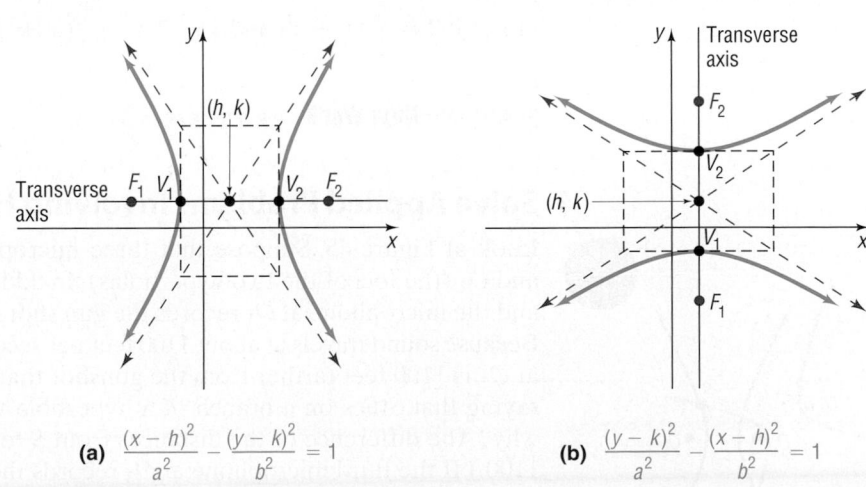

(a) $\dfrac{(x - h)^2}{a^2} - \dfrac{(y - k)^2}{b^2} = 1$

(b) $\dfrac{(y - k)^2}{a^2} - \dfrac{(x - h)^2}{b^2} = 1$

EXAMPLE 7 | Finding an Equation of a Hyperbola, Center Not at the Origin

Find an equation for the hyperbola with center at $(1, -2)$, one focus at $(4, -2)$, and one vertex at $(3, -2)$. Graph the equation.

Solution The center is at $(h, k) = (1, -2)$, so $h = 1$ and $k = -2$. Since the center, focus, and vertex all lie on the line $y = -2$, the transverse axis is parallel to the x-axis. The distance from the center $(1, -2)$ to the focus $(4, -2)$ is $c = 3$; the distance from

Figure 43

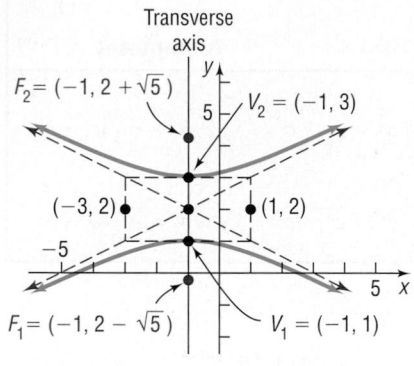

the center $(1, -2)$ to the vertex $(3, -2)$ is $a = 2$. Then $b^2 = c^2 - a^2 = 9 - 4 = 5$. The equation is

$$\frac{(x - h)^2}{a^2} - \frac{(y - k)^2}{b^2} = 1$$

$$\frac{(x - 1)^2}{4} - \frac{(y + 2)^2}{5} = 1$$

See Figure 43.

━━━━ **Now Work** PROBLEM 41

EXAMPLE 8 | **Analyzing the Equation of a Hyperbola**

Analyze the equation: $-x^2 + 4y^2 - 2x - 16y + 11 = 0$

Solution | Complete the squares in x and in y.

$$-x^2 + 4y^2 - 2x - 16y + 11 = 0$$

$$-(x^2 + 2x) + 4(y^2 - 4y) = -11 \qquad \text{Group terms.}$$

$$-(x^2 + 2x + 1) + 4(y^2 - 4y + 4) = -11 - 1 + 16 \qquad \text{Complete each square.}$$

$$-(x + 1)^2 + 4(y - 2)^2 = 4 \qquad \text{Factor.}$$

$$(y - 2)^2 - \frac{(x + 1)^2}{4} = 1 \qquad \text{Divide each side by 4.}$$

Figure 44

This is the equation of a hyperbola with center at $(-1, 2)$ and transverse axis parallel to the y-axis. Also, $a^2 = 1$ and $b^2 = 4$, so $c^2 = a^2 + b^2 = 5$. Since the transverse axis is parallel to the y-axis, the vertices and foci are located a and c units above and below the center, respectively. The vertices are at $(h, k \pm a) = (-1, 2 \pm 1)$, or $(-1, 1)$ and $(-1, 3)$. The foci are at $(h, k \pm c) = (-1, 2 \pm \sqrt{5})$. The asymptotes are $y - 2 = \frac{1}{2}(x + 1)$ and $y - 2 = -\frac{1}{2}(x + 1)$. Figure 44 shows the graph.

━━━━ **Now Work** PROBLEM 55

4 Solve Applied Problems Involving Hyperbolas

Figure 45

Look at Figure 45. Suppose that three microphones are located at points O_1, O_2, and O_3 (the foci of the two hyperbolas). In addition, suppose that a gun is fired at S and the microphone at O_1 records the gun shot 1 second after the microphone at O_2. Because sound travels at about 1100 feet per second, we conclude that the microphone at O_1 is 1100 feet farther from the gunshot than O_2. We can model this situation by saying that S lies on a branch of a hyperbola with foci at O_1 and O_2. (Do you see why? The difference of the distances from S to O_1 and from S to O_2 is the constant 1100.) If the third microphone at O_3 records the gunshot 2 seconds after O_1, then S will lie on a branch of a second hyperbola with foci at O_1 and O_3. In this case, the constant difference will be 2200. The intersection of the two hyperbolas will identify the location of S.

EXAMPLE 9 | **Lightning Strikes**

Suppose that two people standing 1 mile apart both see a flash of lightning. After a period of time, the person standing at point A hears the thunder. One second later, the person standing at point B hears the thunder. If the person at B is due west of

the person at A and the lightning strike is known to occur due north of the person standing at point A, where did the lightning strike occur?

Solution See Figure 46 in which the ordered pair (x, y) represents the location of the lightning strike. We know that sound travels at 1100 feet per second, so the person at point A is 1100 feet closer to the lightning strike than the person at point B. Since the difference of the distance from (x, y) to B and the distance from (x, y) to A is the constant 1100, the point (x, y) lies on a hyperbola whose foci are at A and B.

Figure 46

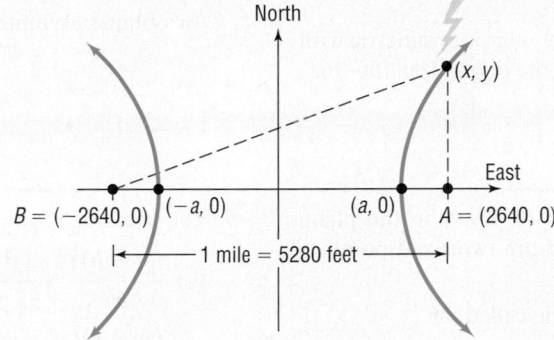

An equation of the hyperbola is

$$\frac{x^2}{a^2} - \frac{y^2}{b^2} = 1$$

where $2a = 1100$, so $a = 550$.

Because the distance between the two people is 1 mile (5280 feet) and each person is at a focus of the hyperbola, we have

$$2c = 5280$$

$$c = \frac{5280}{2} = 2640$$

Since $b^2 = c^2 - a^2 = 2640^2 - 550^2 = 6{,}667{,}100$, the equation of the hyperbola that describes the location of the lightning strike is

$$\frac{x^2}{550^2} - \frac{y^2}{6{,}667{,}100} = 1$$

Refer to Figure 46. Since the lightning strike occurred due north of the individual at the point $A = (2640, 0)$, we let $x = 2640$ and solve the resulting equation.

$$\frac{2640^2}{550^2} - \frac{y^2}{6{,}667{,}100} = 1$$

$$-\frac{y^2}{6{,}667{,}100} = -22.04 \qquad \text{Subtract } \frac{2640^2}{550^2} \text{ from both sides.}$$

$$y^2 = 146{,}942{,}884 \qquad \text{Multiply both sides by } -6{,}667{,}100.$$

$$y = 12{,}122 \qquad \begin{array}{l}y > 0 \text{ since the lightning strike} \\ \text{occurred in quadrant I.}\end{array}$$

The lightning strike occurred 12,122 feet north of the person standing at point A.

✓**Check:** The difference between the distance from $(2640, 12{,}122)$ to the person at the point $B = (-2640, 0)$ and the distance from $(2640, 12{,}122)$ to the person at the point $A = (2640, 0)$ should be 1100. Using the distance formula, we find the difference in the distances is

$$\sqrt{[2640 - (-2640)]^2 + (12{,}122 - 0)^2} - \sqrt{(2640 - 2640)^2 + (12{,}122 - 0)^2} = 1100$$

as required. ♪

Now Work PROBLEM 7 5

7.4 Assess Your Understanding

'Are You Prepared?' *Answers are given at the end of these exercises. If you get a wrong answer, read the pages listed in red.*

1. The distance d from $P_1 = (3, -4)$ to $P_2 = (-2, 1)$ is $d =$ _____. (p. 151)

2. To complete the square of $x^2 + 5x$, add ____. (p. 56)

3. Find the intercepts of the equation $y^2 = 9 + 4x^2$. (pp. 159–160)

4. **True or False** The equation $y^2 = 9 + x^2$ is symmetric with respect to the x-axis, the y-axis, and the origin. (pp. 160–162)

5. To graph $y = (x - 5)^3 - 4$, shift the graph of $y = x^3$ to the (left/right) ____ unit(s) and then (up/down) ____ unit(s). (pp. 249–253)

6. Find the vertical asymptotes, if any, and the horizontal or oblique asymptote, if any, of $y = \dfrac{x^2 - 9}{x^2 - 4}$. (pp. 345–348)

Concepts and Vocabulary

7. A(n) _____ is the collection of points in the plane the difference of whose distances from two fixed points is a constant.

8. For a hyperbola, the foci lie on a line called the _____ _____.

Answer Problems 9–11 using the figure.

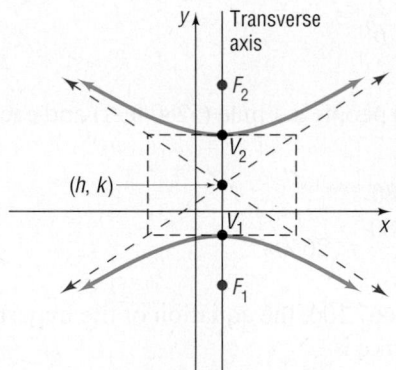

9. The equation of the hyperbola is of the form

 (a) $\dfrac{(x - h)^2}{a^2} - \dfrac{(y - k)^2}{b^2} = 1$

 (b) $\dfrac{(y - k)^2}{a^2} - \dfrac{(x - h)^2}{b^2} = 1$

10. If the center of the hyperbola is $(2, 1)$ and $a = 3$, then the coordinates of the vertices are _____ and _____.

11. If the center of the hyperbola is $(2, 1)$ and $c = 5$, then the coordinates of the foci are _____ and _____.

12. In a hyperbola, if $a = 3$ and $c = 5$, then $b =$ _____.

13. For the hyperbola $\dfrac{x^2}{4} - \dfrac{y^2}{9} = 1$, the value of a is _____, the value of b is _____, and the transverse axis is the _____-axis.

14. For the hyperbola $\dfrac{y^2}{16} - \dfrac{x^2}{81} = 1$, the asymptotes are _____ and _____.

Skill Building

In Problems 15–18, the graph of a hyperbola is given. Match each graph to its equation.

(A) $\dfrac{x^2}{4} - y^2 = 1$ (B) $x^2 - \dfrac{y^2}{4} = 1$ (C) $\dfrac{y^2}{4} - x^2 = 1$ (D) $y^2 - \dfrac{x^2}{4} = 1$

15.

16.

17.

18.
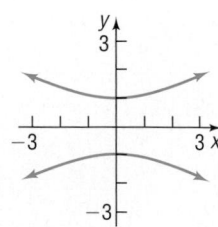

In Problems 19–28, find an equation for the hyperbola described. Graph the equation.

19. Center at $(0, 0)$; focus at $(3, 0)$; vertex at $(1, 0)$

20. Center at $(0, 0)$; focus at $(0, 5)$; vertex at $(0, 3)$

21. Center at $(0, 0)$; focus at $(0, -6)$; vertex at $(0, 4)$

22. Center at $(0, 0)$; focus at $(-3, 0)$; vertex at $(2, 0)$

23. Foci at $(-5, 0)$ and $(5, 0)$; vertex at $(3, 0)$

24. Focus at $(0, 6)$; vertices at $(0, -2)$ and $(0, 2)$

25. Vertices at $(0, -6)$ and $(0, 6)$; asymptote the line $y = 2x$

26. Vertices at $(-4, 0)$ and $(4, 0)$; asymptote the line $y = 2x$

27. Foci at $(-4, 0)$ and $(4, 0)$; asymptote the line $y = -x$

28. Foci at $(0, -2)$ and $(0, 2)$; asymptote the line $y = -x$

In Problems 29–36, find the center, transverse axis, vertices, foci, and asymptotes. Graph each equation.

29. $\dfrac{x^2}{25} - \dfrac{y^2}{9} = 1$

30. $\dfrac{y^2}{16} - \dfrac{x^2}{4} = 1$

31. $4x^2 - y^2 = 16$

32. $4y^2 - x^2 = 16$

33. $y^2 - 9x^2 = 9$

34. $x^2 - y^2 = 4$

35. $y^2 - x^2 = 25$

36. $2x^2 - y^2 = 4$

In Problems 37–40, write an equation for each hyperbola.

37.

38.

39.

40.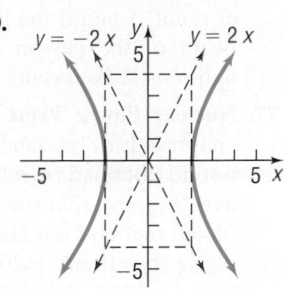

In Problems 41–48, find an equation for the hyperbola described. Graph the equation.

41. Center at $(4, -1)$; focus at $(7, -1)$; vertex at $(6, -1)$

42. Center at $(-3, 1)$; focus at $(-3, 6)$; vertex at $(-3, 4)$

43. Center at $(-3, -4)$; focus at $(-3, -8)$; vertex at $(-3, -2)$

44. Center at $(1, 4)$; focus at $(-2, 4)$; vertex at $(0, 4)$

45. Foci at $(3, 7)$ and $(7, 7)$; vertex at $(6, 7)$

46. Focus at $(-4, 0)$ vertices at $(-4, 4)$ and $(-4, 2)$

47. Vertices at $(-1, -1)$ and $(3, -1)$; asymptote the line $y + 1 = \dfrac{3}{2}(x - 1)$

48. Vertices at $(1, -3)$ and $(1, 1)$; asymptote the line $y + 1 = \dfrac{3}{2}(x - 1)$

In Problems 49–62, find the center, transverse axis, vertices, foci, and asymptotes. Graph each equation.

49. $\dfrac{(x - 2)^2}{4} - \dfrac{(y + 3)^2}{9} = 1$

50. $\dfrac{(y + 3)^2}{4} - \dfrac{(x - 2)^2}{9} = 1$

51. $(y - 2)^2 - 4(x + 2)^2 = 4$

52. $(x + 4)^2 - 9(y - 3)^2 = 9$

53. $(x + 1)^2 - (y + 2)^2 = 4$

54. $(y - 3)^2 - (x + 2)^2 = 4$

55. $x^2 - y^2 - 2x - 2y - 1 = 0$

56. $y^2 - x^2 - 4y + 4x - 1 = 0$

57. $y^2 - 4x^2 - 4y - 8x - 4 = 0$

58. $2x^2 - y^2 + 4x + 4y - 4 = 0$

59. $4x^2 - y^2 - 24x - 4y + 16 = 0$

60. $2y^2 - x^2 + 2x + 8y + 3 = 0$

61. $y^2 - 4x^2 - 16x - 2y - 19 = 0$

62. $x^2 - 3y^2 + 8x - 6y + 4 = 0$

In Problems 63–66, graph each function. Be sure to label any intercepts.
[**Hint:** *Notice that each function is half a hyperbola.*]

63. $f(x) = \sqrt{16 + 4x^2}$

64. $f(x) = -\sqrt{9 + 9x^2}$

65. $f(x) = -\sqrt{-25 + x^2}$

66. $f(x) = \sqrt{-1 + x^2}$

Mixed Practice

In Problems 67–74, analyze each conic.

67. $\dfrac{(x - 3)^2}{4} - \dfrac{y^2}{25} = 1$

68. $\dfrac{(y + 2)^2}{16} - \dfrac{(x - 2)^2}{4} = 1$

69. $x^2 = 16(y - 3)$

70. $y^2 = -12(x + 1)$

71. $25x^2 + 9y^2 - 250x + 400 = 0$

72. $x^2 + 36y^2 - 2x + 288y + 541 = 0$

73. $x^2 - 6x - 8y - 31 = 0$

74. $9x^2 - y^2 - 18x - 8y - 88 = 0$

Applications and Extensions

75. Fireworks Display Suppose that two people standing 2 miles apart both see the burst from a fireworks display. After a period of time, the first person standing at point A hears the burst. One second later, the second person standing at point B hears the burst. If the person at point B is due west of the person at point A and if the display is known to occur due

north of the person at point A, where did the fireworks display occur?

76. Lightning Strikes Suppose that two people standing 1 mile apart both see a flash of lightning. After a period of time, the first person standing at point A hears the thunder. Two seconds later, the second person standing at point B hears the thunder. If the person at point B is due west of the person at point A and if the lightning strike is known to occur due north of the person standing at point A, where did the lightning strike occur?

77. Nuclear Power Plant Some nuclear power plants utilize "natural draft" cooling towers in the shape of a **hyperboloid,** a solid obtained by rotating a hyperbola about its conjugate axis. Suppose that such a cooling tower has a base diameter of 400 feet and the diameter at its narrowest point, 360 feet above the ground, is 200 feet. If the diameter at the top of the tower is 300 feet, how tall is the tower?

Source: Bay Area Air Quality Management District

78. An Explosion Two recording devices are set 2400 feet apart, with the device at point A to the west of the device at point B. At a point between the devices, 300 feet from point B, a small amount of explosive is detonated. The recording devices record the time until the sound reaches each. How far directly north of point B should a second explosion be done so that the measured time difference recorded by the devices is the same as that for the first detonation?

79. Rutherford's Experiment In May 1911, Ernest Rutherford published a paper in *Philosophical Magazine*. In this article, he described the motion of alpha particles as they are shot at a piece of gold foil 0.00004 cm thick. Before conducting this experiment, Rutherford expected that the alpha particles would shoot through the foil just as a bullet would shoot through snow. Instead, a small fraction of the alpha particles bounced off the foil. This led to the conclusion that the nucleus of an atom is dense, while the remainder of the atom is sparse. Only the density of the nucleus could cause the alpha particles to deviate from their path. The figure shows a diagram from Rutherford's paper that indicates that the deflected alpha particles follow the path of one branch of a hyperbola.

(a) Find an equation of the asymptotes under this scenario.
(b) If the vertex of the path of the alpha particles is 10 cm from the center of the hyperbola, find a model that describes the path of the particle.

80. Hyperbolic Mirrors Hyperbolas have interesting reflective properties that make them useful for lenses and mirrors. For example, if a ray of light strikes a convex hyperbolic mirror on a line that would (theoretically) pass through its rear focus, it is reflected through the front focus. This property, and that of the parabola, were used to develop the *Cassegrain* telescope in 1672. The focus of the parabolic mirror and the rear focus of the hyperbolic mirror are the same point. The rays are collected by the parabolic mirror, reflected toward the (common) focus, and thus are reflected by the hyperbolic mirror through the opening to its front focus, where the eyepiece is located. If the equation of the hyperbola is $\dfrac{y^2}{9} - \dfrac{x^2}{16} = 1$ and the focal length (distance from the vertex to the focus) of the parabola is 6, find the equation of the parabola.

Source: www.enchantedlearning.com

81. The **eccentricity** e of a hyperbola is defined as the number $\dfrac{c}{a}$, where a is the distance of a vertex from the center and c is the distance of a focus from the center. Because $c > a$, it follows that $e > 1$. Describe the general shape of a hyperbola whose eccentricity is close to 1. What is the shape if e is very large?

82. A hyperbola for which $a = b$ is called an **equilateral hyperbola.** Find the eccentricity e of an equilateral hyperbola.

[**Note:** The eccentricity of a hyperbola is defined in Problem 81.]

83. Two hyperbolas that have the same set of asymptotes are called **conjugate.** Show that the hyperbolas

$$\frac{x^2}{4} - y^2 = 1 \quad \text{and} \quad y^2 - \frac{x^2}{4} = 1$$

are conjugate. Graph each hyperbola on the same set of coordinate axes.

84. Prove that the hyperbola

$$\frac{y^2}{a^2} - \frac{x^2}{b^2} = 1$$

has the two oblique asymptotes

$$y = \frac{a}{b}x \quad \text{and} \quad y = -\frac{a}{b}x$$

85. Show that the graph of an equation of the form

$$Ax^2 + Cy^2 + F = 0 \qquad A \neq 0, C \neq 0, F \neq 0$$

where A and C are of opposite sign, is a hyperbola with center at $(0, 0)$.

86. Show that the graph of an equation of the form

$$Ax^2 + Cy^2 + Dx + Ey + F = 0 \qquad A \neq 0, C \neq 0$$

where A and C are of opposite sign,

(a) is a hyperbola if $\dfrac{D^2}{4A} + \dfrac{E^2}{4C} - F \neq 0$.

(b) is two intersecting lines if $\dfrac{D^2}{4A} + \dfrac{E^2}{4C} - F = 0$.

'Are You Prepared?' Answers

1. $5\sqrt{2}$ **2.** $\dfrac{25}{4}$ **3.** $(0, -3), (0, 3)$ **4.** True **5.** right; 5; down; 4 **6.** Vertical: $x = -2, x = 2$; horizontal: $y = 1$

CHAPTER REVIEW

Things to Know

Equations

Parabola (pp. 505–511)	See Tables 1 and 2 (pp. 507 and 508).
Ellipse (pp. 514–521)	See Table 3 (p. 519).
Hyperbola (pp. 524–533)	See Table 4 (p. 531).

Definitions

Parabola (p. 505)	Set of points P in the plane for which $d(F, P) = d(P, D)$, where F is the focus and D is the directrix
Ellipse (p. 514)	Set of points P in the plane, the sum of whose distances from two fixed points (the foci) is a constant
Hyperbola (p. 524)	Set of points P in the plane, the difference of whose distances from two fixed points (the foci) is a constant

Objectives

Section		You should be able to . . .	Example(s)	Review Exercises
7.1	1	Know the names of the conics (p. 504)		1–32
7.2	1	Analyze parabolas with vertex at the origin (p. 505)	1–5	1, 2, 21, 24
	2	Analyze parabolas with vertex at (h, k) (p. 508)	6, 7	7, 11, 12, 17, 18, 27, 30
	3	Solve applied problems involving parabolas (p. 510)	8	41, 42
7.3	1	Analyze ellipses with center at the origin (p. 514)	1–4	5, 6, 10, 22, 25, 39
	2	Analyze ellipses with center at (h, k) (p. 518)	5, 6	14–16, 19, 28, 31
	3	Solve applied problems involving ellipses (p. 520)	7	43, 44
7.4	1	Analyze hyperbolas with center at the origin (p. 524)	1–4	3, 4, 8, 9, 23, 26, 40
	2	Find the asymptotes of a hyperbola (p. 529)	5, 6	3, 4, 8, 9
	3	Analyze hyperbolas with center at (h, k) (p. 531)	7, 8	13, 20, 29, 32–36
	4	Solve applied problems involving hyperbolas (p. 532)	9	45

Review Exercises

In Problems 1–20, identify each equation. If it is a parabola, give its vertex, focus, and directrix; if it is an ellipse, give its center, vertices, and foci; if it is a hyperbola, give its center, vertices, foci, and asymptotes.

1. $y^2 = -16x$

2. $16x^2 = y$

3. $\dfrac{x^2}{25} - y^2 = 1$

4. $\dfrac{y^2}{25} - x^2 = 1$

5. $\dfrac{y^2}{25} + \dfrac{x^2}{16} = 1$

6. $\dfrac{x^2}{9} + \dfrac{y^2}{16} = 1$

7. $x^2 + 4y = 4$

8. $3y^2 - x^2 = 9$

9. $4x^2 - y^2 = 8$

10. $9x^2 + 4y^2 = 36$

11. $x^2 - 4x = 2y$

12. $2y^2 - 4y = x - 2$

13. $y^2 - 4y - 4x^2 + 8x = 4$

14. $4x^2 + y^2 + 8x - 4y + 4 = 0$

15. $4x^2 + 9y^2 - 16x - 18y = 11$

16. $4x^2 + 9y^2 - 16x + 18y = 11$

17. $4x^2 - 16x + 16y + 32 = 0$

18. $4y^2 + 3x - 16y + 19 = 0$

19. $9x^2 + 4y^2 - 18x + 8y = 23$

20. $x^2 - y^2 - 2x - 2y = 1$

In Problems 21–36, find an equation of the conic described. Graph the equation.

21. Parabola; focus at $(-2, 0)$; directrix the line $x = 2$

22. Ellipse; center at $(0, 0)$; focus at $(0, 3)$; vertex at $(0, 5)$

23. Hyperbola; center at $(0, 0)$; focus at $(0, 4)$; vertex at $(0, -2)$

24. Parabola; vertex at $(0, 0)$; directrix the line $y = -3$

25. Ellipse; foci at $(-3, 0)$ and $(3, 0)$; vertex at $(4, 0)$

26. Hyperbola; vertices at $(-2, 0)$ and $(2, 0)$; focus at $(4, 0)$

27. Parabola; vertex at $(2, -3)$; focus at $(2, -4)$

28. Ellipse; center at $(-1, 2)$; focus at $(0, 2)$; vertex at $(2, 2)$

29. Hyperbola; center at $(-2, -3)$; focus at $(-4, -3)$; vertex at $(-3, -3)$

30. Parabola; focus at $(3, 6)$; directrix the line $y = 8$

31. Ellipse; foci at $(-4, 2)$ and $(-4, 8)$; vertex at $(-4, 10)$

32. Hyperbola; vertices at $(-3, 3)$ and $(5, 3)$; focus at $(7, 3)$

33. Center at $(-1, 2)$; $a = 3$; $c = 4$; transverse axis parallel to the x-axis

34. Center at $(4, -2)$; $a = 1$; $c = 4$; transverse axis parallel to the y-axis

35. Vertices at $(0, 1)$ and $(6, 1)$; asymptote the line $3y + 2x = 9$

36. Vertices at $(4, 0)$ and $(4, 4)$; asymptote the line $y + 2x = 10$

37. Find an equation of the hyperbola whose foci are the vertices of the ellipse $4x^2 + 9y^2 = 36$ and whose vertices are the foci of this ellipse.

38. Find an equation of the ellipse whose foci are the vertices of the hyperbola $x^2 - 4y^2 = 16$ and whose vertices are the foci of this hyperbola.

39. Describe the collection of points in a plane so that the distance from each point to the point $(3, 0)$ is three-fourths of its distance from the line $x = \dfrac{16}{3}$.

40. Describe the collection of points in a plane so that the distance from each point to the point $(5, 0)$ is five-fourths of its distance from the line $x = \dfrac{16}{5}$.

41. Searchlight A searchlight is shaped like a paraboloid of revolution. If a light source is located 1 foot from the vertex along the axis of symmetry and the opening is 2 feet across, how deep should the mirror be in order to reflect the light rays parallel to the axis of symmetry?

42. Parabolic Arch Bridge A bridge is built in the shape of a parabolic arch. The bridge has a span of 60 feet and a maximum height of 20 feet. Find the height of the arch at distances of 5, 10, and 20 feet from the center.

43. Semielliptical Arch Bridge A bridge is built in the shape of a semielliptical arch. The bridge has a span of 60 feet and

a maximum height of 20 feet. Find the height of the arch at distances of 5, 10, and 20 feet from the center.

44. Whispering Gallery The figure below shows the specifications for an elliptical ceiling in a hall designed to be a whispering gallery. Where are the foci located in the hall?

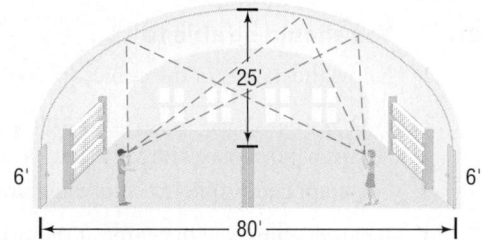

45. Calibrating Instruments In a test of their recording devices, a team of seismologists positioned two of the devices 2000 feet apart, with the device at point A to the west of the device at point B. At a point between the devices and 200 feet from point B, a small amount of explosive was detonated and a note made of the time at which the sound reached each device. A second explosion is to be carried out at a point directly north of point B. How far north should the site of the second explosion be chosen so that the measured time difference recorded by the devices for the second detonation is the same as that recorded for the first detonation?

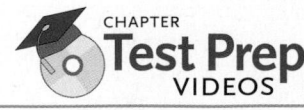

CHAPTER TEST

The Chapter Test Prep Videos are step-by-step test solutions available in the Video Resources DVD, in *MyMathLab* , or on this text's You Tube Channel. Flip back to the Student Resources page to see the exact web address for this text's YouTube channel.

In Problems 1–3, identify each equation. If it is a parabola, give its vertex, focus, and directrix; if an ellipse, give its center, vertices, and foci; if a hyperbola, give its center, vertices, foci, and asymptotes.

1. $\dfrac{(x + 1)^2}{4} - \dfrac{y^2}{9} = 1$

2. $8y = (x - 1)^2 - 4$

3. $2x^2 + 3y^2 + 4x - 6y = 13$

In Problems 4–6, find an equation of the conic described; graph the equation.

4. Parabola: focus $(-1, 4.5)$, vertex $(-1, 3)$

5. Ellipse: center $(0, 0)$, vertex $(0, -4)$, focus $(0, 3)$

6. Hyperbola: center $(2, 2)$, vertex $(2, 4)$, contains the point $\left(2 + \sqrt{10}, 5\right)$

7. A parabolic reflector (paraboloid of revolution) is used by TV crews at football games to pick up the referee's announcements, quarterback signals, and so on. A microphone is placed at the focus of the parabola. If a certain reflector is 4 feet wide and 1.5 feet deep, where should the microphone be placed?

CUMULATIVE REVIEW

1. For $f(x) = -3x^2 + 5x - 2$, find

$$\frac{f(x + h) - f(x)}{h} \quad h \neq 0$$

2. In the complex number system, solve the equation

$$9x^4 + 33x^3 - 71x^2 - 57x - 10 = 0$$

3. For what numbers x is $6 - x \geq x^2$?

4. (a) Find the domain and range of $y = 3^x + 2$.
 (b) Find the inverse of $y = 3^x + 2$ and state its domain and range.

5. $f(x) = \log_4(x - 2)$
 (a) Solve $f(x) = 2$.
 (b) Solve $f(x) \leq 2$.

6. Find an equation for each of the following graphs:

(a) Line:

(b) Circle:

(c) Ellipse:

(d) Parabola:

(e) Hyperbola:

(f) Exponential:

CHAPTER PROJECTS

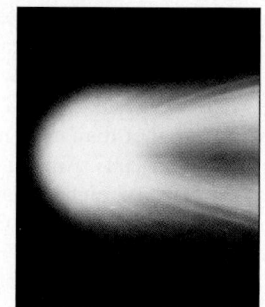

Internet-based Project

I. **The Hale-Bopp Comet** The orbits of planets and some comets about the Sun are ellipses, with the Sun at one focus. The **aphelion** of a planet is its greatest distance from the Sun

and the **perihelion** is its shortest distance. The **mean distance** of a planet from the Sun is the length of the semimajor axis of the elliptical orbit. See the illustration.

1. Research the history of the Hale-Bopp on the Internet. In particular, determine the aphelion and perihelion. Often these values are given in terms of astronomical units. What is an astronomical unit? What is it equivalent to in miles? In kilometers? What is the orbital period of the Hale-Bopp Comet? When will it next be visible from Earth? How close does it come to Earth?

2. Find a model for the orbit of the Hale-Bopp Comet around the Sun. Use the x-axis as the major axis.

3. The Hale-Bopp Comet has an orbit that is roughly perpendicular to that of Earth. Find a model for the orbit of Earth using the y-axis as the major axis.

4. Use a graphing utility or some other graphing technology to graph the paths of the orbits. Based on the graphs, do the paths of the orbits intersect? Does this mean the Hale-Bopp Comet will collide with Earth?

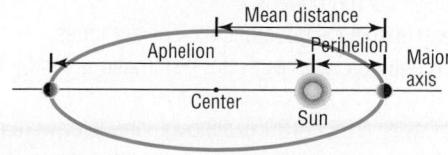

The following projects can be found at the Instructor's Resource Center (IRC):

II. **The Orbits of Neptune and Pluto** Pluto and Neptune travel around the Sun in elliptical orbits. Pluto, at times, comes closer to the Sun than Neptune, the outermost planet. This project examines and analyzes the two orbits.

III. **Project at Motorola** *Distorted Deployable Space Reflector Antennas* An engineer designs an antenna that will deploy in space to collect sunlight.

IV. **Constructing a Bridge over the East River** The size of ships using a river and fluctuations in water height due to tides or flooding must be considered when designing a bridge that will cross a major waterway.

8 Systems of Equations and Inequalities

Outline

Economic Outcomes
Annual Earnings of Young Adults

For both males and females, earnings increase with education: full-time workers with at least a bachelor's degree have higher median earnings than those with less education. For example, in 2003, male college graduates earned 93 percent more than male high school completers. Females with a bachelor's or higher degree earned 91 percent more than female high school completers. Males and females who dropped out of high school earned 37 and 39 percent less, respectively, than male and female high school completers.

The median earnings of young adults who have at least a bachelor's degree declined in the 1970s relative to their counterparts who were high school completers, before increasing between 1980 and 2003. Males with a bachelor's degree or higher had earnings 19 percent higher than male high school completers in 1980 and had earnings 93 percent higher in 2003. Among females, those with at least a bachelor's degree had earnings 34 percent higher than female high school completers in 1980, compared with earnings 91 percent higher in 2003.

 —See the Internet-based Chapter Project I—

◁ A Look Back In Chapters 1, 4, 5, and 6, we solved various kinds of equations and inequalities involving a single variable.

A Look Ahead ▷ In this chapter we take up the problem of solving equations and inequalities containing two or more variables. There are various ways to solve such problems.

The *method of substitution* for solving equations in several unknowns goes back to ancient times.

The *method of elimination*, although it had existed for centuries, was put into systematic order by Karl Friedrich Gauss (1777–1855) and by Camille Jordan (1838–1922).

The theory of *matrices* was developed in 1857 by Arthur Cayley (1821–1895), although only later were matrices used as we use them in this chapter. Matrices have become a very flexible instrument, useful in almost all areas of mathematics.

The method of *determinants* was invented by Takakazu Seki Kôwa (1642–1708) in 1683 in Japan and by Gottfried Wilhelm von Leibniz (1646–1716) in 1693 in Germany. *Cramer's Rule* is named after Gabriel Cramer (1704–1752) of Switzerland, who popularized the use of determinants for solving linear systems.

Section 8.5, on *partial fraction decomposition*, provides an application of systems of equations. This particular application is one that is used in integral calculus.

Section 8.8 introduces *linear programming*, a modern application of linear inequalities. This topic is particularly useful for students interested in operations research.

8.1 Systems of Linear Equations: Substitution and Elimination

PREPARING FOR THIS SECTION *Before getting started, review the following:*

- Linear Equations (Section 1.1, pp. 82–87)
- Lines (Section 2.3, pp. 167–177)

Now Work the '**Are You Prepared?**' problems on page 552.

OBJECTIVES
1 Solve Systems of Equations by Substitution (p. 543)
2 Solve Systems of Equations by Elimination (p. 545)
3 Identify Inconsistent Systems of Equations Containing Two Variables (p. 546)
4 Express the Solution of a System of Dependent Equations Containing Two Variables (p. 547)
5 Solve Systems of Three Equations Containing Three Variables (p. 548)
6 Identify Inconsistent Systems of Equations Containing Three Variables (p. 550)
7 Express the Solution of a System of Dependent Equations Containing Three Variables (p. 550)

EXAMPLE 1 **Movie Theater Ticket Sales**

A movie theater sells tickets for $8.00 each, with seniors receiving a discount of $2.00. One evening the theater took in $3580 in revenue. If x represents the number of tickets sold at $8.00 and y the number of tickets sold at the discounted price of $6.00, write an equation that relates these variables.

Solution Each nondiscounted ticket brings in $8.00, so x tickets will bring in $8x$ dollars. Similarly, y discounted tickets bring in $6y$ dollars. Since the total brought in is $3580, we must have

$$8x + 6y = 3580$$

In Example 1, suppose that we also know that 525 tickets were sold that evening. Then we have another equation relating the variables x and y:

$$x + y = 525$$

The two equations

$$8x + 6y = 3580$$
$$x + y = 525$$

form a *system* of equations.

In general, a **system of equations** is a collection of two or more equations, each containing one or more variables. Example 2 gives some illustrations of systems of equations.

EXAMPLE 2 **Examples of Systems of Equations**

(a) $\begin{cases} 2x + y = 5 & (1) \\ -4x + 6y = -2 & (2) \end{cases}$ *Two equations containing two variables, x and y*

(b) $\begin{cases} x + y^2 = 5 & (1) \\ 2x + y = 4 & (2) \end{cases}$ *Two equations containing two variables, x and y*

(c) $\begin{cases} x + y + z = 6 & \text{(1)} \\ 3x - 2y + 4z = 9 & \text{(2)} \\ x - y - z = 0 & \text{(3)} \end{cases}$ Three equations containing three variables, x, y, and z

(d) $\begin{cases} x + y + z = 5 & \text{(1)} \\ x - y \quad\;\;\; = 2 & \text{(2)} \end{cases}$ Two equations containing three variables, x, y, and z

(e) $\begin{cases} x + y + z = 6 & \text{(1)} \\ 2x \quad\;\; + 2z = 4 & \text{(2)} \\ \quad\;\; y + z = 2 & \text{(3)} \\ x \quad\qquad\;\; = 4 & \text{(4)} \end{cases}$ Four equations containing three variables, x, y, and z

We use a brace, as shown, to remind us that we are dealing with a system of equations. We also will find it convenient to number each equation in the system.

A **solution** of a system of equations consists of values for the variables that are solutions of each equation of the system. To **solve** a system of equations means to find all solutions of the system.

For example, $x = 2, y = 1$ is a solution of the system in Example 2(a), because

$$\begin{cases} 2x + y = 5 & \text{(1)} \\ -4x + 6y = -2 & \text{(2)} \end{cases} \qquad \begin{cases} 2(2) + 1 = 4 + 1 = 5 \\ -4(2) + 6(1) = -8 + 6 = -2 \end{cases}$$

We may also write this solution as the ordered pair $(2, 1)$.

A solution of the system in Example 2(b) is $x = 1, y = 2$, because

$$\begin{cases} x + y^2 = 5 & \text{(1)} \\ 2x + y = 4 & \text{(2)} \end{cases} \qquad \begin{cases} 1 + 2^2 = 1 + 4 = 5 \\ 2(1) + 2 = 2 + 2 = 4 \end{cases}$$

Another solution of the system in Example 2(b) is $x = \dfrac{11}{4}, y = -\dfrac{3}{2}$, which you can check for yourself.

A solution of the system in Example 2(c) is $x = 3, y = 2, z = 1$, because

$$\begin{cases} x + y + z = 6 & \text{(1)} \\ 3x - 2y + 4z = 9 & \text{(2)} \\ x - y - z = 0 & \text{(3)} \end{cases} \qquad \begin{cases} 3 + 2 + 1 = 6 \\ 3(3) - 2(2) + 4(1) = 9 - 4 + 4 = 9 \\ 3 - 2 - 1 = 0 \end{cases}$$

We may also write this solution as the ordered triplet $(3, 2, 1)$.

Note that $x = 3, y = 3, z = 0$ is not a solution of the system in Example 2(c).

$$\begin{cases} x + y + z = 6 & \text{(1)} \\ 3x - 2y + 4z = 9 & \text{(2)} \\ x - y - z = 0 & \text{(3)} \end{cases} \qquad \begin{cases} 3 + 3 + 0 = 6 \\ 3(3) - 2(3) + 4(0) = 3 \neq 9 \\ 3 - 3 - 0 = 0 \end{cases}$$

Although $x = 3, y = 3$, and $z = 0$ satisfy equations (1) and (3), they do not satisfy equation (2). Any solution of the system must satisfy *each* equation of the system.

Now Work PROBLEM 9

When a system of equations has at least one solution, it is said to be **consistent**. If a system of equations has no solution, it is called **inconsistent**.

An equation in n variables is said to be **linear** if it is equivalent to an equation of the form

$$a_1 x_1 + a_2 x_2 + \cdots + a_n x_n = b$$

where x_1, x_2, \dots, x_n are n distinct variables, a_1, a_2, \dots, a_n, b are constants, and at least one of the a's is not 0.

Some examples of linear equations are

$$2x + 3y = 2 \qquad 5x - 2y + 3z = 10 \qquad 8x + 8y - 2z + 5w = 0$$

If each equation in a system of equations is linear, we have a **system of linear equations.** The systems in Examples 2(a), (c), (d), and (e) are linear, whereas the system in Example 2(b) is nonlinear. In this chapter we shall solve linear systems in Sections 8.1 to 8.3. We discuss nonlinear systems in Section 8.6.

We begin by discussing a system of two linear equations containing two variables. We can view the problem of solving such a system as a geometry problem. The graph of each equation in such a system is a line. So a system of two linear equations containing two variables represents a pair of lines. The lines either (1) intersect or (2) are parallel or (3) are **coincident** (that is, identical).

1. If the lines intersect, the system of equations has one solution, given by the point of intersection. The system is **consistent** and the equations are **independent.** See Figure 1(a).

2. If the lines are parallel, the system of equations has no solution, because the lines never intersect. The system is **inconsistent.** See Figure 1(b).

3. If the lines are coincident (the lines lie on top of each other), the system of equations has infinitely many solutions, represented by the totality of points on the line. The system is **consistent** and the equations are **dependent.** See Figure 1(c).

Figure 1

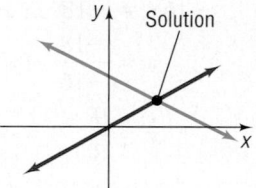

(a) Intersecting lines; system has one solution

(b) Parallel lines; system has no solution

(c) Coincident lines; system has infinitely many solutions

EXAMPLE 3

Graphing a System of Linear Equations

Graph the system:
$$\begin{cases} 2x + y = 5 & (1) \\ -4x + 6y = 12 & (2) \end{cases}$$

Solution

Equation (1) in slope–intercept form is $y = -2x + 5$, which has slope -2 and y-intercept 5. Equation (2) in slope–intercept form is $y = \dfrac{2}{3}x + 2$, which has slope $\dfrac{2}{3}$ and y-intercept 2. Figure 2 shows their graphs.

Figure 2

From the graph in Figure 2, we see that the lines intersect, so the system given in Example 3 is consistent. We can also use the graph as a means of approximating the solution. For this system, the solution appears to be close to the point $(1, 3)$. The actual solution, which you should verify, is $\left(\dfrac{9}{8}, \dfrac{11}{4}\right)$.

Seeing the Concept

Graph the lines $2x + y = 5$ $(Y_1 = -2x + 5)$ and $-4x + 6y = 12$ $\left(Y_2 = \dfrac{2}{3}x + 2\right)$ and compare what you see with Figure 2. Use INTERSECT to verify that the point of intersection is $(1.125, 2.75)$.

1 Solve Systems of Equations by Substitution

Most of the time we must use algebraic methods to obtain exact solutions. A number of methods are available to us for solving systems of linear equations algebraically. In this section, we introduce two methods: *substitution* and *elimination*. We illustrate the **method of substitution** by solving the system given in Example 3.

| **EXAMPLE 4** | **How to Solve a System of Linear Equations by Substitution** |

Solve: $\begin{cases} 2x + y = 5 & (1) \\ -4x + 6y = 12 & (2) \end{cases}$

Step-by-Step Solution

Step 1: Pick one of the equations and solve for one of the variables in terms of the remaining variable(s).

Solve equation (1) for y.

$$2x + y = 5 \qquad \text{Equation (1)}$$
$$y = -2x + 5 \qquad \text{Subtract 2x from each side of (1).}$$

Step 2: Substitute the result into the remaining equation(s).

Substitute $-2x + 5$ for y in equation (2). The result is an equation containing just the variable x, which we can solve.

$$-4x + 6y = 12 \qquad \text{Equation (2)}$$
$$-4x + 6(-2x + 5) = 12 \qquad \text{Substitute } -2x + 5 \text{ for y in (2).}$$

Step 3: If one equation in one variable results, solve this equation. Otherwise, repeat Steps 1 and 2 until a single equation with one variable remains.

$$-4x - 12x + 30 = 12 \qquad \text{Distribute.}$$
$$-16x + 30 = 12 \qquad \text{Combine like terms.}$$
$$-16x = -18 \qquad \text{Subtract 30 from both sides.}$$
$$x = \frac{-18}{-16} \qquad \text{Divide both sides by } -16.$$
$$x = \frac{9}{8} \qquad \text{Simplify.}$$

Step 4: Find the values of the remaining variables by back-substitution.

Because we know that $x = \frac{9}{8}$, we can find the value of y by **back-substitution,** that is, by substituting $\frac{9}{8}$ for x in one of the original equations. Equation (1) seems easier to work with, so we will back-substitute into equation (1).

$$2x + y = 5 \qquad \text{Equation (1)}$$
$$2\left(\frac{9}{8}\right) + y = 5 \qquad \text{Substitute } x = \frac{9}{8} \text{ into equation (1).}$$
$$\frac{9}{4} + y = 5 \qquad \text{Simplify.}$$
$$y = 5 - \frac{9}{4} \qquad \text{Subtract } \frac{9}{4} \text{ from both sides.}$$
$$y = \frac{11}{4} \qquad 5 - \frac{9}{4} = \frac{20}{4} - \frac{9}{4} = \frac{11}{4}$$

Step 5: Check the solution found.

We have $x = \frac{9}{8}$ and $y = \frac{11}{4}$. We verify that both equations are satisfied (true) for these values.

$$\begin{cases} 2x + y = 5: & 2\left(\frac{9}{8}\right) + \frac{11}{4} = \frac{9}{4} + \frac{11}{4} = \frac{20}{4} = 5 \\ -4x + 6y = 12: & -4\left(\frac{9}{8}\right) + 6\left(\frac{11}{4}\right) = -\frac{9}{2} + \frac{33}{2} = \frac{24}{2} = 12 \end{cases}$$

The solution of the system is $x = \frac{9}{8} = 1.125$ and $y = \frac{11}{4} = 2.75$. We can also write the solution as the ordered pair $\left(\frac{9}{8}, \frac{11}{4}\right)$.

Now Use Substitution to Work PROBLEM 19

2 Solve Systems of Equations by Elimination

A second method for solving a system of linear equations is the *method of elimination*. This method is usually preferred over substitution if substitution leads to fractions or if the system contains more than two variables. Elimination also provides the necessary motivation for solving systems using matrices (the subject of Section 8.2).

The idea behind the **method of elimination** is to replace the original system of equations by an equivalent system so that adding two of the equations eliminates a variable. The rules for obtaining equivalent equations are the same as those studied earlier. However, we may also interchange any two equations of the system and/or replace any equation in the system by the sum (or difference) of that equation and a nonzero multiple of any other equation in the system.

> **In Words**
> When using elimination, we want to get the coefficients of one of the variables to be negatives of one another.

Rules for Obtaining an Equivalent System of Equations

1. Interchange any two equations of the system.
2. Multiply (or divide) each side of an equation by the same nonzero constant.
3. Replace any equation in the system by the sum (or difference) of that equation and a nonzero multiple of any other equation in the system.

An example will give you the idea. As you work through the example, pay particular attention to the pattern being followed.

EXAMPLE 5

How to Solve a System of Linear Equations by Elimination

Solve: $\begin{cases} 2x + 3y = 1 & (1) \\ -x + y = -3 & (2) \end{cases}$

Step-by-Step Solution

Step 1: Multiply both sides of one or both equations by a nonzero constant so that the coefficients of one of the variables are additive inverses.

Multiply both sides of equation (2) by 2 so that the coefficients of x in the two equations are additive inverses.

$$\begin{cases} 2x + 3y = 1 & (1) \\ -x + y = -3 & (2) \end{cases}$$

$$\begin{cases} 2x + 3y = 1 & (1) \\ 2(-x + y) = 2(-3) & (2) \quad \text{Multiply by 2.} \end{cases}$$

$$\begin{cases} 2x + 3y = 1 & (1) \\ -2x + 2y = -6 & (2) \end{cases}$$

Step 2: Add the equations to eliminate the variable. Solve the resulting equation for the remaining unknown.

$$\begin{cases} 2x + 3y = 1 & (1) \\ \underline{-2x + 2y = -6} & (2) \\ 5y = -5 & \text{Add equations (1) and (2).} \\ y = -1 & \text{Divide both sides by 5.} \end{cases}$$

Step 3: Back-substitute the value of the variable found in Step 2 into one of the *original equations* to find the value of the remaining variable.

Back-substitute $y = -1$ into equation (1) and solve for x.

$$\begin{aligned} 2x + 3y &= 1 & \text{Equation (1)} \\ 2x + 3(-1) &= 1 & \text{Substitute } y = -1 \text{ into equation (1).} \\ 2x - 3 &= 1 & \text{Simplify.} \\ 2x &= 4 & \text{Add 3 to both sides.} \\ x &= 2 & \text{Divide both sides by 2.} \end{aligned}$$

Step 4: Check the solution found.

We leave the check to you.

The solution of the system is $x = 2$ and $y = -1$. We can also write the solution as the ordered pair $(2, -1)$.

━━━━━ **Now Use Elimination to Work** PROBLEM 19

EXAMPLE 6 **Movie Theater Ticket Sales**

A movie theater sells tickets for $8.00 each, with seniors receiving a discount of $2.00. One evening the theater sold 525 tickets and took in $3580 in revenue. How many of each type of ticket were sold?

Solution If x represents the number of tickets sold at $8.00 and y the number of tickets sold at the discounted price of $6.00, then the given information results in the system of equations

$$\begin{cases} 8x + 6y = 3580 & (1) \\ x + y = 525 & (2) \end{cases}$$

We use the method of elimination. First, multiply the second equation by -6, and then add the equations.

$$\begin{cases} 8x + 6y = 3580 \\ -6x - 6y = -3150 \end{cases}$$
$$\overline{\qquad 2x = 430} \qquad \text{Add the equations.}$$
$$x = 215$$

Since $x + y = 525$, then $y = 525 - x = 525 - 215 = 310$. So 215 nondiscounted tickets and 310 senior discount tickets were sold.

3 Identify Inconsistent Systems of Equations Containing Two Variables

The previous examples dealt with consistent systems of equations that had a single solution. The next two examples deal with two other possibilities that may occur, the first being a system that has no solution.

EXAMPLE 7 **An Inconsistent System of Linear Equations**

Solve: $\begin{cases} 2x + y = 5 & (1) \\ 4x + 2y = 8 & (2) \end{cases}$

Solution We choose to use the method of substitution and solve equation (1) for y.

$$2x + y = 5 \qquad (1)$$
$$y = -2x + 5 \qquad \text{Subtract } 2x \text{ from each side.}$$

Now substitute $-2x + 5$ for y in equation (2) and solve for x.

$$4x + 2y = 8 \qquad (2)$$
$$4x + 2(-2x + 5) = 8 \qquad \text{Substitute } y = -2x + 5 \text{ in (2).}$$
$$4x - 4x + 10 = 8 \qquad \text{Remove parentheses.}$$
$$0 = -2 \qquad \text{Subtract 10 from both sides.}$$

This statement is false. We conclude that the system has no solution and is therefore inconsistent.

Figure 3

Figure 3 illustrates the pair of lines whose equations form the system in Example 7. Notice that the graphs of the two equations are lines, each with slope -2;

one has a *y*-intercept of 5, the other a *y*-intercept of 4. The lines are parallel and have no point of intersection. This geometric statement is equivalent to the algebraic statement that the system has no solution.

 Seeing the Concept

Graph the lines $2x + y = 5$ ($Y_1 = -2x + 5$) and $4x + 2y = 8$ ($Y_2 = -2x + 4$) and compare what you see with Figure 3. How can you be sure that the lines are parallel?

4 Express the Solution of a System of Dependent Equations Containing Two Variables

EXAMPLE 8

Solving a System of Dependent Equations

Solve: $\begin{cases} 2x + y = 4 & (1) \\ -6x - 3y = -12 & (2) \end{cases}$

Solution

We choose to use the method of elimination.

$$\begin{cases} 2x + y = 4 & (1) \\ -6x - 3y = -12 & (2) \end{cases}$$

$$\begin{cases} 6x + 3y = 12 & (1) \quad \textit{Multiply each side of equation (1) by 3.} \\ -6x - 3y = -12 & (2) \end{cases}$$

$$\overline{\qquad 0 = 0 \qquad} \quad \textit{Add equations (1) and (2).}$$

The statement $0 = 0$ is true. This means the equation $6x + 3y = 12$ is equivalent to $-6x - 3y = -12$. Therefore, the original system is equivalent to a system containing one equation, so the equations are dependent. This means that any values of *x* and *y* that satisfy $6x + 3y = 12$ or, equivalently, $2x + y = 4$ are solutions. For example, $x = 2$, $y = 0$; $x = 0$, $y = 4$; $x = -2$, $y = 8$; $x = 4$, $y = -4$; and so on, are solutions. There are, in fact, infinitely many values of *x* and *y* for which $2x + y = 4$, so the original system has infinitely many solutions. We will write the solution of the original system either as

$$y = -2x + 4 \quad \text{where } x \text{ can be any real number}$$

or as

$$x = -\frac{1}{2}y + 2 \quad \text{where } y \text{ can be any real number.}$$

We can also express the solution as $\{(x, y)\mid y = -2x + 4, x \text{ is any real number}\}$ or as $\left\{(x, y)\left|x = -\dfrac{1}{2}y + 2, y \text{ is any real number}\right.\right\}$.

Figure 4

(-1, 6)
$2x + y = 4$
$-6x - 3y = -12$
(0, 4)
$\left(\frac{1}{2}, 3\right)$
-4
(2, 0)
4 *x*
-2
(3, -2)

Figure 4 illustrates the situation presented in Example 8. Notice that the graphs of the two equations are lines, each with slope -2 and each with *y*-intercept 4. The lines are coincident. Notice also that equation (2) in the original system is -3 times equation (1), indicating that the two equations are dependent.

For the system in Example 8, we can write down some of the infinite number of solutions by assigning values to *x* and then finding $y = -2x + 4$.

If $x = -1$, then $y = -2(-1) + 4 = 6$.

If $x = 0$, then $y = 4$.

If $x = 2$, then $y = 0$.

The ordered pairs $(-1, 6)$, $(0, 4)$, and $(2, 0)$ are three of the points on the line in Figure 4.

> **Seeing the Concept**
>
> Graph the lines $2x + y = 4$ ($Y_1 = -2x + 4$) and $-6x - 3y = -12$ ($Y_2 = -2x + 4$) and compare what you see with Figure 4. How can you be sure that the lines are coincident?

———**Now Work** PROBLEMS 25 AND 29

5 Solve Systems of Three Equations Containing Three Variables

Just as with a system of two linear equations containing two variables, a system of three linear equations containing three variables has either (1) exactly one solution (a consistent system with independent equations), or (2) no solution (an inconsistent system), or (3) infinitely many solutions (a consistent system with dependent equations).

We can view the problem of solving a system of three linear equations containing three variables as a geometry problem. The graph of each equation in such a system is a plane in space. A system of three linear equations containing three variables represents three planes in space. Figure 5 illustrates some of the possibilities.

Figure 5

(a) Consistent system; one solution

(b) Consistent system; infinite number of solutions

(c) Inconsistent system; no solution

Recall that a **solution** to a system of equations consists of values for the variables that are solutions of each equation of the system. For example, $x = 3$, $y = -1$, $z = -5$ or, using an ordered triplet, $(3, -1, -5)$ is a solution to the system of equations

$$\begin{cases} x + y + z = -3 & (1) \\ 2x - 3y + 6z = -21 & (2) \\ -3x + 5y = -14 & (3) \end{cases} \qquad \begin{aligned} & 3 + (-1) + (-5) = -3 \\ & 2(3) - 3(-1) + 6(-5) = 6 + 3 - 30 = -21 \\ & -3(3) + 5(-1) = -9 - 5 = -14 \end{aligned}$$

because these values of the variables are solutions of each equation.

Typically, when solving a system of three linear equations containing three variables, we use the method of elimination. Recall that the idea behind the method of elimination is to form equivalent equations so that adding two of the equations eliminates a variable.

EXAMPLE 9

Solving a System of Three Linear Equations with Three Variables

Use the method of elimination to solve the system of equations.

$$\begin{cases} x + y - z = -1 & (1) \\ 4x - 3y + 2z = 16 & (2) \\ 2x - 2y - 3z = 5 & (3) \end{cases}$$

Solution

For a system of three equations, we attempt to eliminate one variable at a time, using pairs of equations until an equation with a single variable remains. Our strategy for solving this system will be to use equation (1) to eliminate the variable x from

equations (2) and (3). We can then treat the new equations (2) and (3) as a system with two unknowns. Alternatively, we could use equation (1) to eliminate either y or z from equations (2) and (3). Try one of these approaches for yourself.

Begin by multiplying each side of equation (1) by -4 and adding the result to equation (2). (Do you see why? The coefficients of x are now negatives of one another.) We also multiply equation (1) by -2 and add the result to equation (3). Notice that these two procedures result in the elimination of the variable x from equations (2) and (3).

$$
\begin{aligned}
x + y - z &= -1 \quad (1) \quad \text{Multiply by } -4. \\
4x - 3y + 2z &= 16 \quad (2)
\end{aligned}
\qquad
\begin{aligned}
-4x - 4y + 4z &= 4 \quad (1) \\
4x - 3y + 2z &= 16 \quad (2) \\
\hline
-7y + 6z &= 20 \quad \text{Add.}
\end{aligned}
$$

$$
\begin{aligned}
x + y - z &= -1 \quad (1) \quad \text{Multiply by } -2. \\
2x - 2y - 3z &= 5 \quad (3)
\end{aligned}
\qquad
\begin{aligned}
-2x - 2y + 2z &= 2 \quad (1) \\
2x - 2y - 3z &= 5 \quad (3) \\
\hline
-4y - z &= 7 \quad \text{Add.}
\end{aligned}
$$

$$
\begin{cases}
x + y - z = -1 & (1) \\
-7y + 6z = 20 & (2) \\
-4y - z = 7 & (3)
\end{cases}
$$

Now concentrate on the new equations (2) and (3), treating them as a system of two equations containing two variables. It is easiest to eliminate z. Multiply each side of equation (3) by 6 and add equations (2) and (3). The result is the new equation (3).

$$
\begin{aligned}
-7y + 6z &= 20 \quad (2) \\
-4y - z &= 7 \quad (3) \quad \text{Multiply by 6.}
\end{aligned}
\qquad
\begin{aligned}
-7y + 6z &= 20 \quad (2) \\
-24y - 6z &= 42 \quad (3) \\
\hline
-31y &= 62 \quad \text{Add.}
\end{aligned}
\qquad
\begin{cases}
x + y - z = -1 & (1) \\
-7y + 6z = 20 & (2) \\
-31y = 62 & (3)
\end{cases}
$$

Now solve equation (3) for y by dividing both sides of the equation by -31.

$$
\begin{cases}
x + y - z = -1 & (1) \\
-7y + 6z = 20 & (2) \\
y = -2 & (3)
\end{cases}
$$

Back-substitute $y = -2$ in equation (2) and solve for z.

$$
\begin{aligned}
-7y + 6z &= 20 && (2) \\
-7(-2) + 6z &= 20 && \text{Substitute } y = -2 \text{ in (2).} \\
6z &= 6 && \text{Subtract 14 from both sides of the equation.} \\
z &= 1 && \text{Divide both sides of the equation by 6.}
\end{aligned}
$$

Finally, back-substitute $y = -2$ and $z = 1$ in equation (1) and solve for x.

$$
\begin{aligned}
x + y - z &= -1 && (1) \\
x + (-2) - 1 &= -1 && \text{Substitute } y = -2 \text{ and } z = 1 \text{ in (1).} \\
x - 3 &= -1 && \text{Simplify.} \\
x &= 2 && \text{Add 3 to both sides.}
\end{aligned}
$$

The solution of the original system is $x = 2$, $y = -2$, $z = 1$ or, using an ordered triplet, $(2, -2, 1)$. You should check this solution.

Look back over the solution given in Example 9. Note the pattern of removing one of the variables from two of the equations, followed by solving this system of two equations and two unknowns. Although which variables to remove is your choice, the methodology remains the same for all systems.

═══ **Now Work** PROBLEM 43

6 Identify Inconsistent Systems of Equations Containing Three Variables

EXAMPLE 10

Identify an Inconsistent System of Linear Equations

Solve:
$$\begin{cases} 2x + y - z = -2 & (1) \\ x + 2y - z = -9 & (2) \\ x - 4y + z = 1 & (3) \end{cases}$$

Solution Our strategy is the same as in Example 9. However, in this system, it seems easiest to eliminate the variable z first. Do you see why?

Multiply each side of equation (1) by -1 and add the result to equation (2). Also, add equations (2) and (3).

$$\begin{array}{l} 2x + y - z = -2 \quad (1) \\ x + 2y - z = -9 \quad (2) \end{array} \quad \text{Multiply by } -1. \quad \begin{array}{l} -2x - y + z = 2 \quad (1) \\ x + 2y - z = -9 \quad (2) \\ \hline -x + y = -7 \quad \text{Add.} \end{array}$$

$$\begin{array}{l} x + 2y - z = -9 \quad (2) \\ x - 4y + z = 1 \quad (3) \\ \hline 2x - 2y = -8 \quad \text{Add.} \end{array}$$

$$\begin{cases} x + 2y - z = -9 & (1) \\ -x + y = -7 & (2) \\ 2x - 2y = -8 & (3) \end{cases}$$

Now concentrate on the new equations (2) and (3), treating them as a system of two equations containing two variables. Multiply each side of equation (2) by 2 and add the result to equation (3).

$$\begin{array}{l} -x + y = -7 \quad (2) \\ 2x - 2y = -8 \quad (3) \end{array} \quad \text{Multiply by 2.} \quad \begin{array}{l} -2x + 2y = -14 \quad (2) \\ 2x - 2y = -8 \quad (3) \\ \hline 0 = -22 \quad \text{Add.} \end{array} \quad \begin{cases} x + 2y - z = -9 & (1) \\ -x + y = -7 & (2) \\ 0 = -22 & (3) \end{cases}$$

Equation (3) has no solution, so the system is inconsistent.

7 Express the Solution of a System of Dependent Equations Containing Three Variables

EXAMPLE 11

Solving a System of Dependent Equations

Solve:
$$\begin{cases} x - 2y - z = 8 & (1) \\ 2x - 3y + z = 23 & (2) \\ 4x - 5y + 5z = 53 & (3) \end{cases}$$

Solution Our plan is to eliminate x from equations (2) and (3). Multiply each side of equation (1) by -2 and add the result to equation (2). Also, multiply each side of equation (1) by -4 and add the result to equation (3).

$$\begin{array}{l} x - 2y - z = 8 \quad (1) \\ 2x - 3y + z = 23 \quad (2) \end{array} \quad \text{Multiply by } -2. \quad \begin{array}{l} -2x + 4y + 2z = -16 \quad (1) \\ 2x - 3y + z = 23 \quad (2) \\ \hline y + 3z = 7 \quad \text{Add.} \end{array}$$

$$\begin{array}{l} x - 2y - z = 8 \quad (1) \\ 4x - 5y + 5z = 53 \quad (3) \end{array} \quad \text{Multiply by } -4. \quad \begin{array}{l} -4x + 8y + 4z = -32 \quad (1) \\ 4x - 5y + 5z = 53 \quad (3) \\ \hline 3y + 9z = 21 \quad \text{Add.} \end{array}$$

$$\begin{cases} x - 2y - z = 8 & (1) \\ y + 3z = 7 & (2) \\ 3y + 9z = 21 & (3) \end{cases}$$

Treat equations (2) and (3) as a system of two equations containing two variables, and eliminate the variable y by multiplying both sides of equation (2) by -3 and adding the result to equation (3).

$$
\begin{array}{l}
\begin{aligned}
y + 3z &= 7 \quad \text{(2)} \;\; \text{Multiply by } -3. \\
3y + 9z &= 21 \quad \text{(3)}
\end{aligned}
\qquad
\begin{aligned}
-3y - 9z &= -21 \\
\underline{3y + 9z} &= \underline{\;\;21} \quad \text{Add.} \\
0 &= 0
\end{aligned}
\longrightarrow
\left\{
\begin{aligned}
x - 2y - z &= 8 \quad \text{(1)} \\
y + 3z &= 7 \quad \text{(2)} \\
0 &= 0 \quad \text{(3)}
\end{aligned}
\right.
\end{array}
$$

The original system is equivalent to a system containing two equations, so the equations are dependent and the system has infinitely many solutions. If we solve equation (2) for y, we can express y in terms of z as $y = -3z + 7$. Substitute this expression into equation (1) to determine x in terms of z.

$$
\begin{aligned}
x - 2y - z &= 8 & \text{(1)} \\
x - 2(-3z + 7) - z &= 8 & \text{Substitute } y = -3z + 7 \text{ in (1).} \\
x + 6z - 14 - z &= 8 & \text{Remove parentheses.} \\
x + 5z &= 22 & \text{Combine like terms.} \\
x &= -5z + 22 & \text{Solve for } x.
\end{aligned}
$$

We will write the solution to the system as

$$
\begin{cases}
x = -5z + 22 \\
y = -3z + 7
\end{cases}
\quad \text{where } z \text{ can be any real number.}
$$

This way of writing the solution makes it easier to find specific solutions of the system. To find specific solutions, choose any value of z and use the equations $x = -5z + 22$ and $y = -3z + 7$ to determine x and y. For example, if $z = 0$, then $x = 22$ and $y = 7$, and if $z = 1$, then $x = 17$ and $y = 4$.

Using ordered triplets, the solution is

$$
\{(x, y, z) \mid x = -5z + 22, y = -3z + 7, z \text{ is any real number}\}
$$

Now Work PROBLEM 45

Two distinct points in the Cartesian plane determine a unique line. Given three noncollinear points, we can find the (unique) quadratic function whose graph contains these three points.

EXAMPLE 12 **Curve Fitting**

Find real numbers a, b, and c so that the graph of the quadratic function $y = ax^2 + bx + c$ contains the points $(-1, -4)$, $(1, 6)$, and $(3, 0)$.

Solution We require that the three points satisfy the equation $y = ax^2 + bx + c$.

For the point $(-1, -4)$ we have: $\quad -4 = a(-1)^2 + b(-1) + c \quad -4 = a - b + c$
For the point $(1, 6)$ we have: $\qquad 6 = a(1)^2 + b(1) + c \qquad 6 = a + b + c$
For the point $(3, 0)$ we have: $\qquad 0 = a(3)^2 + b(3) + c \qquad 0 = 9a + 3b + c$

We wish to determine a, b, and c so that each equation is satisfied. That is, we want to solve the following system of three equations containing three variables:

$$
\begin{cases}
a - b + c = -4 & \text{(1)} \\
a + b + c = 6 & \text{(2)} \\
9a + 3b + c = 0 & \text{(3)}
\end{cases}
$$

Figure 6

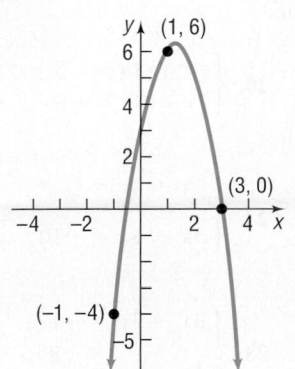

Solving this system of equations, we obtain $a = -2$, $b = 5$, and $c = 3$. So the quadratic function whose graph contains the points $(-1, -4)$, $(1, 6)$, and $(3, 0)$ is

$$
y = -2x^2 + 5x + 3 \quad y = ax^2 + bx + c, \quad a = -2, b = 5, c = 3
$$

Figure 6 shows the graph of the function along with the three points.

Now Work PROBLEM 69

8.1 Assess Your Understanding

'Are You Prepared?' *Answers are given at the end of these exercises. If you get a wrong answer, read the pages listed in red.*

1. Solve the equation: $3x + 4 = 8 - x$. (pp. 82–87)

2. (a) Graph the line: $3x + 4y = 12$.
(b) What is the slope of a line parallel to this line? (pp. 167–177)

Concepts and Vocabulary

3. If a system of equations has no solution, it is said to be _____.

4. If a system of equations has one solution, the system is _____ and the equations are _____.

5. If the solution to a system of two linear equations containing two unknowns is $x = 3$, $y = -2$, then the lines intersect at the point _____.

6. If the lines that make up a system of two linear equations are coincident, then the system is _____ and the equations are _____.

Skill Building

In Problems 7–16, verify that the values of the variables listed are solutions of the system of equations.

7. $\begin{cases} 2x - y = 5 \\ 5x + 2y = 8 \end{cases}$
$x = 2, y = -1; (2, -1)$

8. $\begin{cases} 3x + 2y = 2 \\ x - 7y = -30 \end{cases}$
$x = -2, y = 4; (-2, 4)$

9. $\begin{cases} 3x - 4y = 4 \\ \dfrac{1}{2}x - 3y = -\dfrac{1}{2} \end{cases}$
$x = 2, y = \dfrac{1}{2}; \left(2, \dfrac{1}{2}\right)$

10. $\begin{cases} 2x + \dfrac{1}{2}y = 0 \\ 3x - 4y = -\dfrac{19}{2} \end{cases}$
$x = -\dfrac{1}{2}, y = 2; \left(-\dfrac{1}{2}, 2\right)$

11. $\begin{cases} x - y = 3 \\ \dfrac{1}{2}x + y = 3 \end{cases}$
$x = 4, y = 1; (4, 1)$

12. $\begin{cases} x - y = 3 \\ -3x + y = 1 \end{cases}$
$x = -2, y = -5; (-2, -5)$

13. $\begin{cases} 3x + 3y + 2z = 4 \\ x - y - z = 0 \\ 2y - 3z = -8 \end{cases}$
$x = 1, y = -1, z = 2;$
$(1, -1, 2)$

14. $\begin{cases} 4x - z = 7 \\ 8x + 5y - z = 0 \\ -x - y + 5z = 6 \end{cases}$
$x = 2, y = -3, z = 1;$
$(2, -3, 1)$

15. $\begin{cases} 3x + 3y + 2z = 4 \\ x - 3y + z = 10 \\ 5x - 2y - 3z = 8 \end{cases}$
$x = 2, y = -2, z = 2; (2, -2, 2)$

16. $\begin{cases} 4x - 5z = 6 \\ 5y - z = -17 \\ -x - 6y + 5z = 24 \end{cases}$
$x = 4, y = -3, z = 2; (4, -3, 2)$

In Problems 17–54, solve each system of equations. If the system has no solution, say that it is inconsistent.

17. $\begin{cases} x + y = 8 \\ x - y = 4 \end{cases}$

18. $\begin{cases} x + 2y = -7 \\ x + y = -3 \end{cases}$

19. $\begin{cases} 5x - y = 21 \\ 2x + 3y = -12 \end{cases}$

20. $\begin{cases} x + 3y = 5 \\ 2x - 3y = -8 \end{cases}$

21. $\begin{cases} 3x = 24 \\ x + 2y = 0 \end{cases}$

22. $\begin{cases} 4x + 5y = -3 \\ -2y = -8 \end{cases}$

23. $\begin{cases} 3x - 6y = 2 \\ 5x + 4y = 1 \end{cases}$

24. $\begin{cases} 2x + 4y = \dfrac{2}{3} \\ 3x - 5y = -10 \end{cases}$

25. $\begin{cases} 2x + y = 1 \\ 4x + 2y = 3 \end{cases}$

26. $\begin{cases} x - y = 5 \\ -3x + 3y = 2 \end{cases}$

27. $\begin{cases} 2x - y = 0 \\ 4x + 2y = 12 \end{cases}$

28. $\begin{cases} 3x + 3y = -1 \\ 4x + y = \dfrac{8}{3} \end{cases}$

29. $\begin{cases} x + 2y = 4 \\ 2x + 4y = 8 \end{cases}$

30. $\begin{cases} 3x - y = 7 \\ 9x - 3y = 21 \end{cases}$

31. $\begin{cases} 2x - 3y = -1 \\ 10x + y = 11 \end{cases}$

32. $\begin{cases} 3x - 2y = 0 \\ 5x + 10y = 4 \end{cases}$

33. $\begin{cases} 2x + 3y = 6 \\ x - y = \dfrac{1}{2} \end{cases}$

34. $\begin{cases} \dfrac{1}{2}x + y = -2 \\ x - 2y = 8 \end{cases}$

35. $\begin{cases} \dfrac{1}{2}x + \dfrac{1}{3}y = 3 \\ \dfrac{1}{4}x - \dfrac{2}{3}y = -1 \end{cases}$

36. $\begin{cases} \dfrac{1}{3}x - \dfrac{3}{2}y = -5 \\ \dfrac{3}{4}x + \dfrac{1}{3}y = 11 \end{cases}$

37. $\begin{cases} 3x - 5y = 3 \\ 15x + 5y = 21 \end{cases}$

38. $\begin{cases} 2x - y = -1 \\ x + \dfrac{1}{2}y = \dfrac{3}{2} \end{cases}$

39. $\begin{cases} \dfrac{1}{x} + \dfrac{1}{y} = 8 \\ \dfrac{3}{x} - \dfrac{5}{y} = 0 \end{cases}$

40. $\begin{cases} \dfrac{4}{x} - \dfrac{3}{y} = 0 \\ \dfrac{6}{x} + \dfrac{3}{2y} = 2 \end{cases}$

$$\left[\textbf{Hint:} \text{ Let } u = \frac{1}{x} \text{ and } v = \frac{1}{y}, \text{ and solve for } u \text{ and } v. \text{ Then } x = \frac{1}{u} \right.$$
$$\left. \text{ and } y = \frac{1}{v}. \right]$$

41. $\begin{cases} x - y = 6 \\ 2x - 3z = 16 \\ 2y + z = 4 \end{cases}$

42. $\begin{cases} 2x + y = -4 \\ -2y + 4z = 0 \\ 3x - 2z = -11 \end{cases}$

43. $\begin{cases} x - 2y + 3z = 7 \\ 2x + y + z = 4 \\ -3x + 2y - 2z = -10 \end{cases}$

44. $\begin{cases} 2x + y - 3z = 0 \\ -2x + 2y + z = -7 \\ 3x - 4y - 3z = 7 \end{cases}$

45. $\begin{cases} x - y - z = 1 \\ 2x + 3y + z = 2 \\ 3x + 2y = 0 \end{cases}$

46. $\begin{cases} 2x - 3y - z = 0 \\ -x + 2y + z = 5 \\ 3x - 4y - z = 1 \end{cases}$

47. $\begin{cases} x - y - z = 1 \\ -x + 2y - 3z = -4 \\ 3x - 2y - 7z = 0 \end{cases}$

48. $\begin{cases} 2x - 3y - z = 0 \\ 3x + 2y + 2z = 2 \\ x + 5y + 3z = 2 \end{cases}$

49. $\begin{cases} 2x - 2y + 3z = 6 \\ 4x - 3y + 2z = 0 \\ -2x + 3y - 7z = 1 \end{cases}$

50. $\begin{cases} 3x - 2y + 2z = 6 \\ 7x - 3y + 2z = -1 \\ 2x - 3y + 4z = 0 \end{cases}$

51. $\begin{cases} x + y - z = 6 \\ 3x - 2y + z = -5 \\ x + 3y - 2z = 14 \end{cases}$

52. $\begin{cases} x - y + z = -4 \\ 2x - 3y + 4z = -15 \\ 5x + y - 2z = 12 \end{cases}$

53. $\begin{cases} x + 2y - z = -3 \\ 2x - 4y + z = -7 \\ -2x + 2y - 3z = 4 \end{cases}$

54. $\begin{cases} x + 4y - 3z = -8 \\ 3x - y + 3z = 12 \\ x + y + 6z = 1 \end{cases}$

Applications and Extensions

55. The perimeter of a rectangular floor is 90 feet. Find the dimensions of the floor if the length is twice the width.

56. The length of fence required to enclose a rectangular field is 3000 meters. What are the dimensions of the field if it is known that the difference between its length and width is 50 meters?

57. Orbital Launches In 2005 there was a total of 55 commercial and noncommercial orbital launches worldwide. In addition, the number of noncommercial orbital launches was one more than twice the number of commercial orbital launches. Determine the number of commercial and noncommercial orbital launches in 2005.

Source: Federal Aviation Administration

58. Movie Theater Tickets A movie theater charges $9.00 for adults and $7.00 for senior citizens. On a day when 325 people paid an admission, the total receipts were $2495. How many who paid were adults? How many were seniors?

59. Mixing Nuts A store sells cashews for $5.00 per pound and peanuts for $1.50 per pound. The manager decides to mix 30 pounds of peanuts with some cashews and sell the mixture for $3.00 per pound. How many pounds of cashews should be mixed with the peanuts so that the mixture will produce the same revenue as would selling the nuts separately?

60. Financial Planning A recently retired couple needs $12,000 per year to supplement their Social Security. They

have $150,000 to invest to obtain this income. They have decided on two investment options: AA bonds yielding 10% per annum and a Bank Certificate yielding 5%.

(a) How much should be invested in each to realize exactly $12,000?

(b) If, after 2 years, the couple requires $14,000 per year in income, how should they reallocate their investment to achieve the new amount?

61. Computing Wind Speed With a tail wind, a small Piper aircraft can fly 600 miles in 3 hours. Against this same wind, the Piper can fly the same distance in 4 hours. Find the average wind speed and the average airspeed of the Piper.

62. Computing Wind Speed The average airspeed of a single-engine aircraft is 150 miles per hour. If the aircraft flew the same distance in 2 hours with the wind as it flew in 3 hours against the wind, what was the wind speed?

63. Restaurant Management A restaurant manager wants to purchase 200 sets of dishes. One design costs $25 per set, while another costs $45 per set. If she only has $7400 to spend, how many of each design should be ordered?

64. Cost of Fast Food One group of people purchased 10 hot dogs and 5 soft drinks at a cost of $35.00. A second bought 7 hot dogs and 4 soft drinks at a cost of $25.25. What is the cost of a single hot dog? A single soft drink?

65. Computing a Refund The grocery store we use does not mark prices on its goods. My wife went to this store, bought three 1-pound packages of bacon and two cartons of eggs, and paid a total of $13.45. Not knowing that she went to the store, I also went to the same store, purchased two 1-pound packages of bacon and three cartons of eggs, and paid a total of $11.45. Now we want to return two 1-pound packages of bacon and two cartons of eggs. How much will be refunded?

66. Finding the Current of a Stream Pamela requires 3 hours to swim 15 miles downstream on the Illinois River. The

return trip upstream takes 5 hours. Find Pamela's average speed in still water. How fast is the current? (Assume that Pamela's speed is the same in each direction.)

67. Pharmacy A doctor's prescription calls for a daily intake containing 40 milligrams (mg) of vitamin C and 30 mg of vitamin D. Your pharmacy stocks two liquids that can be used: one contains 20% vitamin C and 30% vitamin D, the other 40% vitamin C and 20% vitamin D. How many milligrams of each compound should be mixed to fill the prescription?

68. Pharmacy A doctor's prescription calls for the creation of pills that contain 12 units of vitamin B_{12} and 12 units of vitamin E. Your pharmacy stocks two powders that can be used to make these pills: one contains 20% vitamin B_{12} and 30% vitamin E, the other 40% vitamin B_{12} and 20% vitamin E. How many units of each powder should be mixed in each pill?

69. Curve Fitting Find real numbers a, b, and c so that the graph of the function $y = ax^2 + bx + c$ contains the points $(-1, 4)$, $(2, 3)$, and $(0, 1)$.

70. Curve Fitting Find real numbers a, b, and c so that the graph of the function $y = ax^2 + bx + c$ contains the points $(-1, -2)$, $(1, -4)$, and $(2, 4)$.

71. IS–LM Model in Economics In economics, the IS curve is a linear equation that represents all combinations of income Y and interest rates r that maintain an equilibrium in the market for goods in the economy. The LM curve is a linear equation that represents all combinations of income Y and interest rates r that maintain an equilibrium in the market for money in the economy. In an economy, suppose that the equilibrium level of income (in millions of dollars) and interest rates satisfy the system of equations

$$\begin{cases} 0.06Y - 5000r = 240 \\ 0.06Y + 6000r = 900 \end{cases}$$

Find the equilibrium level of income and interest rates.

72. IS–LM Model in Economics In economics, the IS curve is a linear equation that represents all combinations of income Y and interest rates r that maintain an equilibrium in the market for goods in the economy. The LM curve is a linear equation that represents all combinations of income Y and interest rates r that maintain an equilibrium in the market for money in the economy. In an economy, suppose that the equilibrium level of income (in millions of dollars) and interest rates satisfy the system of equations

$$\begin{cases} 0.05Y - 1000r = 10 \\ 0.05Y + 800r = 100 \end{cases}$$

Find the equilibrium level of income and interest rates.

73. Electricity: Kirchhoff's Rules An application of Kirchhoff's Rules to the circuit shown on page 555 results in the following system of equations:

$$\begin{cases} I_2 = I_1 + I_3 \\ 5 - 3I_1 - 5I_2 = 0 \\ 10 - 5I_2 - 7I_3 = 0 \end{cases}$$

Find the currents I_1, I_2, and I_3.

Source: Physics for Scientists & Engineers, 3rd ed., by Serway. © 1990 Brooks/Cole, a division of Thomson Learning.

74. Electricity: Kirchhoff's Rules An application of Kirchhoff's Rules to the circuit shown results in the following system of equations:

$$\begin{cases} I_3 = I_1 + I_2 \\ 8 = 4I_3 + 6I_2 \\ 8I_1 = 4 + 6I_2 \end{cases}$$

Find the currents I_1, I_2, and I_3.

Source: Physics for Scientists & Engineers, 3rd ed., by Serway. © 1990 Brooks/Cole, a division of Thomson Learning.

75. Theater Revenues A Broadway theater has 500 seats, divided into orchestra, main, and balcony seating. Orchestra seats sell for $50, main seats for $35, and balcony seats for $25. If all the seats are sold, the gross revenue to the theater is $17,100. If all the main and balcony seats are sold, but only half the orchestra seats are sold, the gross revenue is $14,600. How many are there of each kind of seat?

76. Theater Revenues A movie theater charges $8.00 for adults, $4.50 for children, and $6.00 for senior citizens. One day the theater sold 405 tickets and collected $2320 in receipts. Twice as many children's tickets were sold as adult tickets. How many adults, children, and senior citizens went to the theater that day?

77. Nutrition A dietitian wishes a patient to have a meal that has 66 grams (g) of protein, 94.5 g of carbohydrates, and

910 milligrams (mg) of calcium. The hospital food service tells the dietitian that the dinner for today is chicken, corn, and 2% milk. Each serving of chicken has 30 g of protein, 35 g of carbohydrates, and 200 mg of calcium. Each serving of corn has 3 g of protein, 16 g of carbohydrates, and 10 mg of calcium. Each glass of 2% milk has 9 g of protein, 13 g of carbohydrates, and 300 mg of calcium. How many servings of each food should the dietitian provide for the patient?

78. Investments Kelly has $20,000 to invest. As her financial planner, you recommend that she diversify into three investments: Treasury bills that yield 5% simple interest, Treasury bonds that yield 7% simple interest, and corporate bonds that yield 10% simple interest. Kelly wishes to earn $1390 per year in income. Also, Kelly wants her investment in Treasury bills to be $3000 more than her investment in corporate bonds. How much money should Kelly place in each investment?

79. Prices of Fast Food One group of customers bought 8 deluxe hamburgers, 6 orders of large fries, and 6 large colas for $26.10. A second group ordered 10 deluxe hamburgers, 6 large fries, and 8 large colas and paid $31.60. Is there sufficient information to determine the price of each food item? If not, construct a table showing the various possibilities. Assume that the hamburgers cost between $1.75 and $2.25, the fries between $0.75 and $1.00, and the colas between $0.60 and $0.90.

80. Prices of Fast Food Use the information given in Problem 79. Suppose that a third group purchased 3 deluxe hamburgers, 2 large fries, and 4 large colas for $10.95. Now is there sufficient information to determine the price of each food item? If so, determine each price.

81. Painting a House Three painters, Beth, Bill, and Edie, working together, can paint the exterior of a home in 10 hours (hr). Bill and Edie together have painted a similar house in 15 hr. One day, all three worked on this same kind of house for 4 hr, after which Edie left. Beth and Bill required 8 more hr to finish. Assuming no gain or loss in efficiency, how long should it take each person to complete such a job alone?

Explaining Concepts: Discussion and Writing

82. Make up a system of three linear equations containing three variables that has:
 (a) No solution
 (b) Exactly one solution
 (c) Infinitely many solutions
 Give the three systems to a friend to solve and critique.

83. Write a brief paragraph outlining your strategy for solving a system of two linear equations containing two variables.

84. Do you prefer the method of substitution or the method of elimination for solving a system of two linear equations containing two variables? Give reasons.

'Are You Prepared?' Answers

1. $\{1\}$ 2. (a)

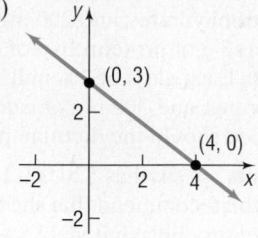

(b) $-\dfrac{3}{4}$

8.2 Systems of Linear Equations: Matrices

OBJECTIVES **1** Write the Augmented Matrix of a System of Linear Equations (p. 557)
 2 Write the System of Equations from the Augmented Matrix (p. 557)
 3 Perform Row Operations on a Matrix (p. 558)
 4 Solve a System of Linear Equations Using Matrices (p. 559)

The systematic approach of the method of elimination for solving a system of linear equations provides another method of solution that involves a simplified notation. Consider the following system of linear equations:

$$\begin{cases} x + 4y = 14 \\ 3x - 2y = 0 \end{cases}$$

If we choose not to write the symbols used for the variables, we can represent this system as

$$\left[\begin{array}{cc|c} 1 & 4 & 14 \\ 3 & -2 & 0 \end{array} \right]$$

where it is understood that the first column represents the coefficients of the variable x, the second column the coefficients of y, and the third column the constants on the right side of the equal signs. The vertical line serves as a reminder of the equal signs. The large square brackets are used to denote a *matrix* in algebra.

DEFINITION A **matrix** is defined as a rectangular array of numbers,

$$\begin{array}{c} \\ \text{Row 1} \\ \text{Row 2} \\ \\ \text{Row } i \\ \\ \text{Row } m \end{array} \overset{\begin{array}{ccccc} \text{Column 1} & \text{Column 2} & & \text{Column } j & & \text{Column } n \end{array}}{\left[\begin{array}{ccccccc} a_{11} & a_{12} & \cdots & a_{1j} & \cdots & a_{1n} \\ a_{21} & a_{22} & \cdots & a_{2j} & \cdots & a_{2n} \\ \vdots & \vdots & & \vdots & & \vdots \\ a_{i1} & a_{i2} & \cdots & a_{ij} & \cdots & a_{in} \\ \vdots & \vdots & & \vdots & & \vdots \\ a_{m1} & a_{m2} & \cdots & a_{mj} & \cdots & a_{mn} \end{array} \right]} \qquad \textbf{(1)}$$

Each number a_{ij} of the matrix has two indexes: the **row index** i and the **column index** j. The matrix shown in display (1) has m rows and n columns. The numbers a_{ij} are usually referred to as the **entries** of the matrix. For example, a_{23} refers to the entry in the second row, third column.

1 Write the Augmented Matrix of a System of Linear Equations

Now we will use matrix notation to represent a system of linear equations. The matrix used to represent a system of linear equations is called an **augmented matrix.** In writing the augmented matrix of a system, the variables of each equation must be on the left side of the equal sign and the constants on the right side. A variable that does not appear in an equation has a coefficient of 0.

EXAMPLE 1

Writing the Augmented Matrix of a System of Linear Equations

Write the augmented matrix of each system of equations.

(a) $\begin{cases} 3x - 4y = -6 & (1) \\ 2x - 3y = -5 & (2) \end{cases}$
(b) $\begin{cases} 2x - y + z = 0 & (1) \\ x + z - 1 = 0 & (2) \\ x + 2y - 8 = 0 & (3) \end{cases}$

Solution
(a) The augmented matrix is

$$\begin{bmatrix} 3 & -4 & | & -6 \\ 2 & -3 & | & -5 \end{bmatrix}$$

(b) Care must be taken that the system be written so that the coefficients of all variables are present (if any variable is missing, its coefficient is 0). Also, all constants must be to the right of the equal sign. We need to rearrange the given system as follows:

$$\begin{cases} 2x - y + z = 0 & (1) \\ x + z - 1 = 0 & (2) \\ x + 2y - 8 = 0 & (3) \end{cases}$$

$$\begin{cases} 2x - y + z = 0 & (1) \\ x + 0 \cdot y + z = 1 & (2) \\ x + 2y + 0 \cdot z = 8 & (3) \end{cases}$$

The augmented matrix is

$$\begin{bmatrix} 2 & -1 & 1 & | & 0 \\ 1 & 0 & 1 & | & 1 \\ 1 & 2 & 0 & | & 8 \end{bmatrix}$$

If we do not include the constants to the right of the equal sign, that is, to the right of the vertical bar in the augmented matrix of a system of equations, the resulting matrix is called the **coefficient matrix** of the system. For the systems discussed in Example 1, the coefficient matrices are

$$\begin{bmatrix} 3 & -4 \\ 2 & -3 \end{bmatrix} \quad \text{and} \quad \begin{bmatrix} 2 & -1 & 1 \\ 1 & 0 & 1 \\ 1 & 2 & 0 \end{bmatrix}$$

Now Work PROBLEM 7

2 Write the System of Equations from the Augmented Matrix

EXAMPLE 2

Writing the System of Linear Equations from the Augmented Matrix

Write the system of linear equations corresponding to each augmented matrix.

(a) $\begin{bmatrix} 5 & 2 & | & 13 \\ -3 & 1 & | & -10 \end{bmatrix}$
(b) $\begin{bmatrix} 3 & -1 & -1 & | & 7 \\ 2 & 0 & 2 & | & 8 \\ 0 & 1 & 1 & | & 0 \end{bmatrix}$

Solution

(a) The matrix has two rows and so represents a system of two equations. The two columns to the left of the vertical bar indicate that the system has two variables. If x and y are used to denote these variables, the system of equations is

$$\begin{cases} 5x + 2y = 13 & (1) \\ -3x + y = -10 & (2) \end{cases}$$

(b) Since the augmented matrix has three rows, it represents a system of three equations. Since there are three columns to the left of the vertical bar, the system contains three variables. If $x, y,$ and z are the three variables, the system of equations is

$$\begin{cases} 3x - y - z = 7 & (1) \\ 2x \quad\quad + 2z = 8 & (2) \\ y + z = 0 & (3) \end{cases}$$

♩

3 Perform Row Operations on a Matrix

Row operations on a matrix are used to solve systems of equations when the system is written as an augmented matrix. There are three basic row operations.

> ### Row Operations
>
> 1. Interchange any two rows.
> 2. Replace a row by a nonzero multiple of that row.
> 3. Replace a row by the sum of that row and a constant nonzero multiple of some other row.

These three row operations correspond to the three rules given earlier for obtaining an equivalent system of equations. When a row operation is performed on a matrix, the resulting matrix represents a system of equations equivalent to the system represented by the original matrix.

For example, consider the augmented matrix

$$\begin{bmatrix} 1 & 2 & | & 3 \\ 4 & -1 & | & 2 \end{bmatrix}$$

Suppose that we want to apply a row operation to this matrix that results in a matrix whose entry in row 2, column 1 is a 0. The row operation to use is

Multiply each entry in row 1 by -4 and add the result to the corresponding entries in row 2.　　　**(2)**

If we use R_2 to represent the new entries in row 2 and r_1 and r_2 to represent the original entries in rows 1 and 2, respectively, we can represent the row operation in statement (2) by

$$R_2 = -4r_1 + r_2$$

Then

$$\begin{bmatrix} 1 & 2 & | & 3 \\ 4 & -1 & | & 2 \end{bmatrix} \rightarrow \begin{bmatrix} 1 & 2 & | & 3 \\ -4(1)+4 & -4(2)+(-1) & | & -4(3)+2 \end{bmatrix} = \begin{bmatrix} 1 & 2 & | & 3 \\ 0 & -9 & | & -10 \end{bmatrix}$$

$$\uparrow$$
$$R_2 = -4r_1 + r_2$$

As desired, we now have the entry 0 in row 2, column 1.

EXAMPLE 3

Applying a Row Operation to an Augmented Matrix

Apply the row operation $R_2 = -3r_1 + r_2$ to the augmented matrix

$$\begin{bmatrix} 1 & -2 & | & 2 \\ 3 & -5 & | & 9 \end{bmatrix}$$

Solution

The row operation $R_2 = -3r_1 + r_2$ tells us that the entries in row 2 are to be replaced by the entries obtained after multiplying each entry in row 1 by -3 and adding the result to the corresponding entries in row 2.

$$\begin{bmatrix} 1 & -2 & | & 2 \\ 3 & -5 & | & 9 \end{bmatrix} \underset{R_2 = -3r_1 + r_2}{\rightarrow} \begin{bmatrix} 1 & -2 & | & 2 \\ -3(1)+3 & (-3)(-2)+(-5) & | & -3(2)+9 \end{bmatrix} = \begin{bmatrix} 1 & -2 & | & 2 \\ 0 & 1 & | & 3 \end{bmatrix}$$

Now Work PROBLEM 17

EXAMPLE 4

Finding a Particular Row Operation

Find a row operation that will result in the augmented matrix

$$\begin{bmatrix} 1 & -2 & | & 2 \\ 0 & 1 & | & 3 \end{bmatrix}$$

having a 0 in row 1, column 2.

Solution

We want a 0 in row 1, column 2. Because there is a 1 in row 2, column 2, this result can be accomplished by multiplying row 2 by 2 and adding the result to row 1. That is, we apply the row operation $R_1 = 2r_2 + r_1$.

$$\begin{bmatrix} 1 & -2 & | & 2 \\ 0 & 1 & | & 3 \end{bmatrix} \underset{R_1 = 2r_2 + r_1}{\rightarrow} \begin{bmatrix} 2(0)+1 & 2(1)+(-2) & | & 2(3)+2 \\ 0 & 1 & | & 3 \end{bmatrix} = \begin{bmatrix} 1 & 0 & | & 8 \\ 0 & 1 & | & 3 \end{bmatrix}$$

A word about the notation introduced here. A row operation such as $R_1 = 2r_2 + r_1$ changes the entries in row 1. Note also that for this type of row operation we change the entries in a given row by multiplying the entries in some other row by an appropriate nonzero number and adding the results to the original entries of the row to be changed.

4 Solve a System of Linear Equations Using Matrices

To solve a system of linear equations using matrices, we use row operations on the augmented matrix of the system to obtain a matrix that is in *row echelon form*.

DEFINITION

A matrix is in **row echelon form** when the following conditions are met:

1. The entry in row 1, column 1 is a 1, and only 0's appear below it.

2. The first nonzero entry in each row after the first row is a 1, only 0's appear below it, and the 1 appears to the right of the first nonzero entry in any row above.

3. Any rows that contain all 0's to the left of the vertical bar appear at the bottom.

For example, for a system of three equations containing three variables, x, y, and z, with a unique solution, the augmented matrix is in row echelon form if it is of the form

$$\left[\begin{array}{ccc|c} 1 & a & b & d \\ 0 & 1 & c & e \\ 0 & 0 & 1 & f \end{array}\right]$$

where a, b, c, d, e, and f are real numbers. The last row of this augmented matrix states that $z = f$. We can then determine the value of y using back-substitution with $z = f$, since row 2 represents the equation $y + cz = e$. Finally, x is determined using back-substitution again.

Two advantages of solving a system of equations by writing the augmented matrix in row echelon form are the following:

1. The process is algorithmic; that is, it consists of repetitive steps that can be programmed on a computer.

2. The process works on any system of linear equations, no matter how many equations or variables are present.

The next example shows how to write a matrix in row echelon form.

EXAMPLE 5 **How to Solve a System of Linear Equations Using Matrices**

Solve: $\begin{cases} 2x + 2y = 6 & (1) \\ x + y + z = 1 & (2) \\ 3x + 4y - z = 13 & (3) \end{cases}$

Step-by-Step Solution

Step 1: Write the augmented matrix that represents the system.

Write the augmented matrix of the system.

$$\left[\begin{array}{ccc|c} 2 & 2 & 0 & 6 \\ 1 & 1 & 1 & 1 \\ 3 & 4 & -1 & 13 \end{array}\right]$$

Step 2: Perform row operations that result in the entry in row 1, column 1 becoming 1.

To get a 1 in row 1, column 1, interchange rows 1 and 2. [Note that this is equivalent to interchanging equations (1) and (2) of the system.]

$$\left[\begin{array}{ccc|c} 1 & 1 & 1 & 1 \\ 2 & 2 & 0 & 6 \\ 3 & 4 & -1 & 13 \end{array}\right]$$

Step 3: Perform row operations that leave the entry in row 1, column 1 a 1, while causing the entries in column 1 below row 1 to become 0's.

Next, we want a 0 in row 2, column 1 and a 0 in row 3, column 1. Use the row operations $R_2 = -2r_1 + r_2$ and $R_3 = -3r_1 + r_3$ to accomplish this. Notice that row 1 is unchanged using these row operations. Also, do you see that performing these row operations simultaneously is the same as doing one followed by the other?

$$\left[\begin{array}{ccc|c} 1 & 1 & 1 & 1 \\ 2 & 2 & 0 & 6 \\ 3 & 4 & -1 & 13 \end{array}\right] \rightarrow \left[\begin{array}{ccc|c} 1 & 1 & 1 & 1 \\ 0 & 0 & -2 & 4 \\ 0 & 1 & -4 & 10 \end{array}\right]$$

$$R_2 = -2r_1 + r_2$$
$$R_3 = -3r_1 + r_3$$

Step 4: Perform row operations that result in the entry in row 2, column 2 becoming 1 with 0's below it.

We want the entry in row 2, column 2 to be 1. We also want to have a 0 below the 1 in row 2, column 2. Interchanging rows 2 and 3 will accomplish both goals.

$$\left[\begin{array}{ccc|c} 1 & 1 & 1 & 1 \\ 0 & 0 & -2 & 4 \\ 0 & 1 & -4 & 10 \end{array}\right] \rightarrow \left[\begin{array}{ccc|c} 1 & 1 & 1 & 1 \\ 0 & 1 & -4 & 10 \\ 0 & 0 & -2 & 4 \end{array}\right]$$

Step 5: Repeat Step 4, placing a 1 in row 3, column 3.

To obtain a 1 in row 3, column 3, we use the row operation $R_3 = -\dfrac{1}{2}r_3$. The result is

$$\begin{bmatrix} 1 & 1 & 1 & | & 1 \\ 0 & 1 & -4 & | & 10 \\ 0 & 0 & -2 & | & 4 \end{bmatrix} \rightarrow \begin{bmatrix} 1 & 1 & 1 & | & 1 \\ 0 & 1 & -4 & | & 10 \\ 0 & 0 & 1 & | & -2 \end{bmatrix}$$

$$\uparrow$$
$$R_3 = -\frac{1}{2}r_3$$

Step 6: The matrix on the right in Step 5 is the row echelon form of the augmented matrix. Use back-substitution to solve the original system.

The third row of the augmented matrix represents the equation $z = -2$. Using $z = -2$, back-substitute into the equation $y - 4z = 10$ (from the second row) and obtain

$$y - 4z = 10$$
$$y - 4(-2) = 10 \quad z = -2$$
$$y = 2 \quad \text{Solve for } y.$$

Finally, back-substitute $y = 2$ and $z = -2$ into the equation $x + y + z = 1$ (from the first row) and obtain

$$x + y + z = 1$$
$$x + 2 + (-2) = 1 \quad y = 2, z = -2$$
$$x = 1 \quad \text{Solve for } x.$$

The solution of the system is $x = 1$, $y = 2$, $z = -2$ or, using an ordered triplet, $(1, 2, -2)$.

In Words

To obtain an augmented matrix in row echelon form:
- Add rows, exchange rows, or multiply a row by a nonzero constant.
- Work from top to bottom and left to right.
- Get 1's in the main diagonal with 0's below the 1's.
- Once the entry in row 1, column 1 is 1 with 0's below it, we do not use row 1 in our row operations. Once the entries in row 1, column 1 and row 2, column 2 are 1 with 0's below, we do not use rows 1 or 2 in our row operations (and so on).

Matrix Method for Solving a System of Linear Equations (Row Echelon Form)

STEP 1: Write the augmented matrix that represents the system.

STEP 2: Perform row operations that place the entry 1 in row 1, column 1.

STEP 3: Perform row operations that leave the entry 1 in row 1, column 1 unchanged, while causing 0's to appear below it in column 1.

STEP 4: Perform row operations that place the entry 1 in row 2, column 2, but leave the entries in columns to the left unchanged. If it is impossible to place a 1 in row 2, column 2, proceed to place a 1 in row 2, column 3. Once a 1 is in place, perform row operations to place 0's below it. (Place any rows that contain only 0's on the left side of the vertical bar, at the bottom of the matrix.)

STEP 5: Now repeat Step 4, placing a 1 in the next row, but one column to the right. Continue until the bottom row or the vertical bar is reached.

STEP 6: The matrix that results is the row echelon form of the augmented matrix. Analyze the system of equations corresponding to it to solve the original system.

| **EXAMPLE 6** | **Solving a System of Linear Equations Using Matrices (Row Echelon Form)** |

Solve:
$$\begin{cases} x - y + z = 8 & (1) \\ 2x + 3y - z = -2 & (2) \\ 3x - 2y - 9z = 9 & (3) \end{cases}$$

Solution

STEP 1: The augmented matrix of the system is

$$\begin{bmatrix} 1 & -1 & 1 & | & 8 \\ 2 & 3 & -1 & | & -2 \\ 3 & -2 & -9 & | & 9 \end{bmatrix}$$

STEP 2: Because the entry 1 is already present in row 1, column 1, we can go to step 3.

STEP 3: Perform the row operations $R_2 = -2r_1 + r_2$ and $R_3 = -3r_1 + r_3$. Each of these leaves the entry 1 in row 1, column 1 unchanged, while causing 0's to appear under it.

$$\begin{bmatrix} 1 & -1 & 1 & | & 8 \\ 2 & 3 & -1 & | & -2 \\ 3 & -2 & -9 & | & 9 \end{bmatrix} \rightarrow \begin{bmatrix} 1 & -1 & 1 & | & 8 \\ 0 & 5 & -3 & | & -18 \\ 0 & 1 & -12 & | & -15 \end{bmatrix}$$

$$R_2 = -2r_1 + r_2$$
$$R_3 = -3r_1 + r_3$$

STEP 4: The easiest way to obtain the entry 1 in row 2, column 2 without altering column 1 is to interchange rows 2 and 3 (another way would be to multiply row 2 by $\frac{1}{5}$, but this introduces fractions).

$$\begin{bmatrix} 1 & -1 & 1 & | & 8 \\ 0 & 1 & -12 & | & -15 \\ 0 & 5 & -3 & | & -18 \end{bmatrix}$$

To get a 0 under the 1 in row 2, column 2, perform the row operation $R_3 = -5r_2 + r_3$.

$$\begin{bmatrix} 1 & -1 & 1 & | & 8 \\ 0 & 1 & -12 & | & -15 \\ 0 & 5 & -3 & | & -18 \end{bmatrix} \rightarrow \begin{bmatrix} 1 & -1 & 1 & | & 8 \\ 0 & 1 & -12 & | & -15 \\ 0 & 0 & 57 & | & 57 \end{bmatrix}$$

$$R_3 = -5r_2 + r_3$$

STEP 5: Continuing, we obtain a 1 in row 3, column 3 by using $R_3 = \frac{1}{57}r_3$.

$$\begin{bmatrix} 1 & -1 & 1 & | & 8 \\ 0 & 1 & -12 & | & -15 \\ 0 & 0 & 57 & | & 57 \end{bmatrix} \rightarrow \begin{bmatrix} 1 & -1 & 1 & | & 8 \\ 0 & 1 & -12 & | & -15 \\ 0 & 0 & 1 & | & 1 \end{bmatrix}$$

$$R_3 = \frac{1}{57}r_3$$

STEP 6: The matrix on the right is the row echelon form of the augmented matrix. The system of equations represented by the matrix in row echelon form is

$$\begin{cases} x - y + z = 8 & (1) \\ y - 12z = -15 & (2) \\ z = 1 & (3) \end{cases}$$

Using $z = 1$, we back-substitute to get

$$\begin{cases} x - y + 1 = 8 & (1) \\ y - 12(1) = -15 & (2) \end{cases} \xrightarrow{\text{Simplify.}} \begin{cases} x - y = 7 & (1) \\ y = -3 & (2) \end{cases}$$

We get $y = -3$ and, back-substituting into $x - y = 7$, we find that $x = 4$. The solution of the system is $x = 4$, $y = -3$, $z = 1$ or, using an ordered triplet, $(4, -3, 1)$.

Sometimes it is advantageous to write a matrix in **reduced row echelon form.** In this form, row operations are used to obtain entries that are 0 above (as well as below) the leading 1 in a row. For example, the row echelon form obtained in the solution to Example 6 is

$$\begin{bmatrix} 1 & -1 & 1 & | & 8 \\ 0 & 1 & -12 & | & -15 \\ 0 & 0 & 1 & | & 1 \end{bmatrix}$$

To write this matrix in reduced row echelon form, we proceed as follows:

$$\begin{bmatrix} 1 & -1 & 1 & | & 8 \\ 0 & 1 & -12 & | & -15 \\ 0 & 0 & 1 & | & 1 \end{bmatrix} \rightarrow \begin{bmatrix} 1 & 0 & -11 & | & -7 \\ 0 & 1 & -12 & | & -15 \\ 0 & 0 & 1 & | & 1 \end{bmatrix} \rightarrow \begin{bmatrix} 1 & 0 & 0 & | & 4 \\ 0 & 1 & 0 & | & -3 \\ 0 & 0 & 1 & | & 1 \end{bmatrix}$$
$$\qquad\qquad\qquad\qquad\qquad\quad R_1 = r_2 + r_1 \qquad\qquad\quad R_1 = 11r_3 + r_1$$
$$\qquad\qquad\qquad\qquad\qquad\qquad\qquad\qquad\qquad\qquad\qquad R_2 = 12r_3 + r_2$$

The matrix is now written in reduced row echelon form. The advantage of writing the matrix in this form is that the solution to the system, $x = 4$, $y = -3$, $z = 1$, is readily found, without the need to back-substitute. Another advantage will be seen in Section 8.4, where the inverse of a matrix is discussed. The methodology used to write a matrix in reduced row echelon form is called **Gauss-Jordan elimination.**

━━▶**Now Work** PROBLEMS 37 AND 47

The matrix method for solving a system of linear equations also identifies systems that have infinitely many solutions and systems that are inconsistent.

EXAMPLE 7 **Solving a Dependent System of Linear Equations Using Matrices**

Solve:
$$\begin{cases} 6x - y - z = 4 & (1) \\ -12x + 2y + 2z = -8 & (2) \\ 5x + y - z = 3 & (3) \end{cases}$$

Solution Start with the augmented matrix of the system and proceed to obtain a 1 in row 1, column 1 with 0's below.

$$\begin{bmatrix} 6 & -1 & -1 & | & 4 \\ -12 & 2 & 2 & | & -8 \\ 5 & 1 & -1 & | & 3 \end{bmatrix} \rightarrow \begin{bmatrix} 1 & -2 & 0 & | & 1 \\ -12 & 2 & 2 & | & -8 \\ 5 & 1 & -1 & | & 3 \end{bmatrix} \rightarrow \begin{bmatrix} 1 & -2 & 0 & | & 1 \\ 0 & -22 & 2 & | & 4 \\ 0 & 11 & -1 & | & -2 \end{bmatrix}$$
$$\qquad\qquad\qquad\qquad\qquad\quad R_1 = -1r_3 + r_1 \qquad\qquad R_2 = 12r_1 + r_2$$
$$\qquad\qquad\qquad\qquad\qquad\qquad\qquad\qquad\qquad\qquad\qquad R_3 = -5r_1 + r_3$$

Obtaining a 1 in row 2, column 2 without altering column 1 can be accomplished by $R_2 = -\dfrac{1}{22}r_2$ or by $R_3 = \dfrac{1}{11}r_3$ and interchanging rows 2 and 3 or by $R_2 = \dfrac{23}{11}r_3 + r_2$. We shall use the first of these.

$$\begin{bmatrix} 1 & -2 & 0 & | & 1 \\ 0 & -22 & 2 & | & 4 \\ 0 & 11 & -1 & | & -2 \end{bmatrix} \rightarrow \begin{bmatrix} 1 & -2 & 0 & | & 1 \\ 0 & 1 & -\dfrac{1}{11} & | & -\dfrac{2}{11} \\ 0 & 11 & -1 & | & -2 \end{bmatrix} \rightarrow \begin{bmatrix} 1 & -2 & 0 & | & 1 \\ 0 & 1 & -\dfrac{1}{11} & | & -\dfrac{2}{11} \\ 0 & 0 & 0 & | & 0 \end{bmatrix}$$
$$\qquad\qquad\qquad\qquad\qquad R_2 = -\dfrac{1}{22}r_2 \qquad\qquad\qquad\qquad R_3 = -11r_2 + r_3$$

This matrix is in row echelon form. Because the bottom row consists entirely of 0's, the system actually consists of only two equations.

$$\begin{cases} x - 2y = 1 \quad (1) \\ y - \dfrac{1}{11}z = -\dfrac{2}{11} \quad (2) \end{cases}$$

To make it easier to write down some of the solutions, we express both x and y in terms of z.

From the second equation, $y = \dfrac{1}{11}z - \dfrac{2}{11}$. Now back-substitute this solution for y into the first equation to get

$$x = 2y + 1 = 2\left(\frac{1}{11}z - \frac{2}{11}\right) + 1 = \frac{2}{11}z + \frac{7}{11}$$

The original system is equivalent to the system

$$\begin{cases} x = \dfrac{2}{11}z + \dfrac{7}{11} \quad (1) \\ y = \dfrac{1}{11}z - \dfrac{2}{11} \quad (2) \end{cases} \qquad \text{where } z \text{ can be any real number.}$$

Let's look at the situation. The original system of three equations is equivalent to a system containing two equations. This means that any values of x, y, z that satisfy both

$$x = \frac{2}{11}z + \frac{7}{11} \quad \text{and} \quad y = \frac{1}{11}z - \frac{2}{11}$$

will be solutions. For example, $z = 0$, $x = \dfrac{7}{11}$, $y = -\dfrac{2}{11}$; $z = 1$, $x = \dfrac{9}{11}$, $y = -\dfrac{1}{11}$; and $z = -1$, $x = \dfrac{5}{11}$, $y = -\dfrac{3}{11}$ are some of the solutions of the original system.

There are, in fact, infinitely many values of x, y, and z for which the two equations are satisfied. That is, the original system has infinitely many solutions. We will write the solution of the original system as

$$\begin{cases} x = \dfrac{2}{11}z + \dfrac{7}{11} \\ y = \dfrac{1}{11}z - \dfrac{2}{11} \end{cases} \qquad \text{where } z \text{ can be any real number}$$

or, using ordered triplets, as

$$\left\{ (x, y, z) \,\middle|\, x = \frac{2}{11}z + \frac{7}{11}, \; y = \frac{1}{11}z - \frac{2}{11}, \, z \text{ any real number} \right\}.$$

We can also find the solution by writing the augmented matrix in reduced row echelon form. Starting with the row echelon form, we have

$$\begin{bmatrix} 1 & -2 & 0 & \bigm| & 1 \\ 0 & 1 & -\dfrac{1}{11} & \bigm| & -\dfrac{2}{11} \\ 0 & 0 & 0 & \bigm| & 0 \end{bmatrix} \rightarrow \begin{bmatrix} 1 & 0 & -\dfrac{2}{11} & \bigm| & \dfrac{7}{11} \\ 0 & 1 & -\dfrac{1}{11} & \bigm| & -\dfrac{2}{11} \\ 0 & 0 & 0 & \bigm| & 0 \end{bmatrix}$$

$$\uparrow \\ R_1 = 2r_2 + r_1$$

The matrix on the right is in reduced row echelon form. The corresponding system of equations is

$$\begin{cases} x - \dfrac{2}{11}z = \dfrac{7}{11} & (1) \\[2mm] y - \dfrac{1}{11}z = -\dfrac{2}{11} & (2) \end{cases}$$ where z can be any real number

or, equivalently,

$$\begin{cases} x = \dfrac{2}{11}z + \dfrac{7}{11} \\[2mm] y = \dfrac{1}{11}z - \dfrac{2}{11} \end{cases}$$ where z can be any real number

━━━━━ **Now Work** PROBLEM 53

EXAMPLE 8 **Solving an Inconsistent System of Linear Equations Using Matrices**

Solve: $\begin{cases} x + y + z = 6 \\ 2x - y - z = 3 \\ x + 2y + 2z = 0 \end{cases}$

Solution Begin with the augmented matrix.

$$\begin{bmatrix} 1 & 1 & 1 & | & 6 \\ 2 & -1 & -1 & | & 3 \\ 1 & 2 & 2 & | & 0 \end{bmatrix} \rightarrow \begin{bmatrix} 1 & 1 & 1 & | & 6 \\ 0 & -3 & -3 & | & -9 \\ 0 & 1 & 1 & | & -6 \end{bmatrix} \rightarrow \begin{bmatrix} 1 & 1 & 1 & | & 6 \\ 0 & 1 & 1 & | & -6 \\ 0 & -3 & -3 & | & -9 \end{bmatrix} \rightarrow \begin{bmatrix} 1 & 1 & 1 & | & 6 \\ 0 & 1 & 1 & | & -6 \\ 0 & 0 & 0 & | & -27 \end{bmatrix}$$

$R_2 = -2r_1 + r_2$ Interchange rows 2 and 3. $R_3 = 3r_2 + r_3$
$R_3 = -1r_1 + r_3$

This matrix is in row echelon form. The bottom row is equivalent to the equation

$$0x + 0y + 0z = -27$$

which has no solution. The original system is inconsistent.

━━━━━ **Now Work** PROBLEM 27

The matrix method is especially effective for systems of equations for which the number of equations and the number of variables are unequal. Here, too, such a system is either inconsistent or consistent. If it is consistent, it will have either exactly one solution or infinitely many solutions.

EXAMPLE 9 **Solving a System of Linear Equations Using Matrices**

Solve: $\begin{cases} x - 2y + z = 0 & (1) \\ 2x + 2y - 3z = -3 & (2) \\ y - z = -1 & (3) \\ -x + 4y + 2z = 13 & (4) \end{cases}$

Solution Begin with the augmented matrix.

$$
\begin{bmatrix}
1 & -2 & 1 & | & 0 \\
2 & 2 & -3 & | & -3 \\
0 & 1 & -1 & | & -1 \\
-1 & 4 & 2 & | & 13
\end{bmatrix}
\rightarrow
\begin{bmatrix}
1 & -2 & 1 & | & 0 \\
0 & 6 & -5 & | & -3 \\
0 & 1 & -1 & | & -1 \\
0 & 2 & 3 & | & 13
\end{bmatrix}
\rightarrow
\begin{bmatrix}
1 & -2 & 1 & | & 0 \\
0 & 1 & -1 & | & -1 \\
0 & 6 & -5 & | & -3 \\
0 & 2 & 3 & | & 13
\end{bmatrix}
$$

$R_2 = -2r_1 + r_2$ Interchange rows 2 and 3.
$R_4 = r_1 + r_4$

$$
\rightarrow
\begin{bmatrix}
1 & -2 & 1 & | & 0 \\
0 & 1 & -1 & | & -1 \\
0 & 0 & 1 & | & 3 \\
0 & 0 & 5 & | & 15
\end{bmatrix}
\rightarrow
\begin{bmatrix}
1 & -2 & 1 & | & 0 \\
0 & 1 & -1 & | & -1 \\
0 & 0 & 1 & | & 3 \\
0 & 0 & 0 & | & 0
\end{bmatrix}
$$

$R_3 = -6r_2 + r_3$ $R_4 = -5r_3 + r_4$
$R_4 = -2r_2 + r_4$

We could stop here, since the matrix is in row echelon form, and back-substitute $z = 3$ to find x and y. Or we can continue and obtain the reduced row echelon form.

$$
\begin{bmatrix}
1 & -2 & 1 & | & 0 \\
0 & 1 & -1 & | & -1 \\
0 & 0 & 1 & | & 3 \\
0 & 0 & 0 & | & 0
\end{bmatrix}
\rightarrow
\begin{bmatrix}
1 & 0 & -1 & | & -2 \\
0 & 1 & -1 & | & -1 \\
0 & 0 & 1 & | & 3 \\
0 & 0 & 0 & | & 0
\end{bmatrix}
\rightarrow
\begin{bmatrix}
1 & 0 & 0 & | & 1 \\
0 & 1 & 0 & | & 2 \\
0 & 0 & 1 & | & 3 \\
0 & 0 & 0 & | & 0
\end{bmatrix}
$$

$R_1 = 2r_2 + r_1$ $R_1 = r_3 + r_1$
 $R_2 = r_3 + r_2$

The matrix is now in reduced row echelon form, and we can see that the solution is $x = 1$, $y = 2$, $z = 3$ or, using an ordered triplet, $(1, 2, 3)$.

Now Work PROBLEM 69

EXAMPLE 10 | **Financial Planning**

Adam and Michelle require an additional $25,000 in annual income (beyond their pension benefits). They are rather risk averse and have narrowed their investment choices down to Treasury notes that yield 3%, Treasury bonds that yield 5%, or corporate bonds that yield 6%. If they have $600,000 to invest and want the amount invested in Treasury notes to equal the total amount invested in Treasury bonds and corporate bonds, how much should be placed in each investment?

Solution Let n, b, and c represent the amounts invested in Treasury notes, Treasury bonds, and corporate bonds, respectively. There is a total of $600,000 to invest, which means that the sum of the amounts invested in Treasury notes, Treasury bonds, and corporate bonds should equal $600,000. The first equation is

$$n + b + c = 600,000 \quad (1)$$

If $100,000 was invested in Treasury notes, the income would be $0.03(\$100,000) = \3000. In general, if n dollars was invested in Treasury notes, the income would be $0.03n$. Since the total income is to be $25,000, the second equation is

$$0.03n + 0.05b + 0.06c = 25,000 \quad (2)$$

The amount invested in Treasury notes should equal the amount invested in Treasury bonds and corporate bonds, so the third equation is

$$n = b + c \quad \text{or} \quad n - b - c = 0 \quad (3)$$

We have the following system of equations:

$$\begin{cases} n + b + c = 600{,}000 & (1) \\ 0.03n + 0.05b + 0.06c = 25{,}000 & (2) \\ n - b - c = 0 & (3) \end{cases}$$

Begin with the augmented matrix and proceed as follows:

$$\left[\begin{array}{ccc|c} 1 & 1 & 1 & 600{,}000 \\ 0.03 & 0.05 & 0.06 & 25{,}000 \\ 1 & -1 & -1 & 0 \end{array}\right] \rightarrow \left[\begin{array}{ccc|c} 1 & 1 & 1 & 600{,}000 \\ 0 & 0.02 & 0.03 & 7000 \\ 0 & -2 & -2 & -600{,}000 \end{array}\right]$$

$$R_2 = -0.03r_1 + r_2$$
$$R_3 = -r_1 + r_2$$

$$\rightarrow \left[\begin{array}{ccc|c} 1 & 1 & 1 & 600{,}000 \\ 0 & 1 & 1.5 & 350{,}000 \\ 0 & -2 & -2 & -600{,}000 \end{array}\right] \rightarrow \left[\begin{array}{ccc|c} 1 & 1 & 1 & 600{,}000 \\ 0 & 1 & 1.5 & 350{,}000 \\ 0 & 0 & 1 & 100{,}000 \end{array}\right]$$

$$R_2 = \frac{1}{0.02}r_2 \qquad\qquad R_3 = 2r_2 + r_3$$

The matrix is now in row echelon form. The final matrix represents the system

$$\begin{cases} n + b + c = 600{,}000 & (1) \\ b + 1.5c = 350{,}000 & (2) \\ c = 100{,}000 & (3) \end{cases}$$

COMMENT Most graphing utilities have the capability to put an augmented matrix into row echelon form (ref) and also reduced row echelon form (rref). See the Appendix, Section 7, for a discussion. ∎

From equation (3), we determine that Adam and Michelle should invest $100,000 in corporate bonds. Back-substitute $100,000 into equation (2) to find that $b = 200{,}000$, so Adam and Michelle should invest $200,000 in Treasury bonds. Back-substitute these values into equation (1) and find that $n = \$300{,}000$, so $300,000 should be invested in Treasury notes.

8.2 Assess Your Understanding

Concepts and Vocabulary

1. An m by n rectangular array of numbers is called a(n) _____.

2. The matrix used to represent a system of linear equations is called a(n) _____ matrix.

3. The notation a_{35} refers to the entry in the _____ row and _____ column of a matrix.

4. *True or False* The matrix $\left[\begin{array}{cc|c} 1 & 3 & -2 \\ 0 & 1 & 5 \\ 0 & 0 & 0 \end{array}\right]$ is in row echelon form.

Skill Building

In Problems 5–16, write the augmented matrix of the given system of equations.

5. $\begin{cases} x - 5y = 5 \\ 4x + 3y = 6 \end{cases}$

6. $\begin{cases} 3x + 4y = 7 \\ 4x - 2y = 5 \end{cases}$

7. $\begin{cases} 2x + 3y - 6 = 0 \\ 4x - 6y + 2 = 0 \end{cases}$

8. $\begin{cases} 9x - y = 0 \\ 3x - y - 4 = 0 \end{cases}$

9. $\begin{cases} 0.01x - 0.03y = 0.06 \\ 0.13x + 0.10y = 0.20 \end{cases}$

10. $\begin{cases} \dfrac{4}{3}x - \dfrac{3}{2}y = \dfrac{3}{4} \\ -\dfrac{1}{4}x + \dfrac{1}{3}y = \dfrac{2}{3} \end{cases}$

11. $\begin{cases} x - y + z = 10 \\ 3x + 3y = 5 \\ x + y + 2z = 2 \end{cases}$

12. $\begin{cases} 5x - y - z = 0 \\ x + y = 5 \\ 2x - 3z = 2 \end{cases}$

13. $\begin{cases} x + y - z = 2 \\ 3x - 2y = 2 \\ 5x + 3y - z = 1 \end{cases}$

14. $\begin{cases} 2x + 3y - 4z = 0 \\ x - 5z + 2 = 0 \\ x + 2y - 3z = -2 \end{cases}$

15. $\begin{cases} x - y - z = 10 \\ 2x + y + 2z = -1 \\ -3x + 4y = 5 \\ 4x - 5y + z = 0 \end{cases}$

16. $\begin{cases} x - y + 2z - w = 5 \\ x + 3y - 4z + 2w = 2 \\ 3x - y - 5z - w = -1 \end{cases}$

In Problems 17–24, write the system of equations corresponding to each augmented matrix. Then perform the indicated row operation(s) on the given augmented matrix.

17. $\begin{bmatrix} 1 & -3 & | & -2 \\ 2 & -5 & | & 5 \end{bmatrix}$ $R_2 = -2r_1 + r_2$

18. $\begin{bmatrix} 1 & -3 & | & -3 \\ 2 & -5 & | & -4 \end{bmatrix}$ $R_2 = -2r_1 + r_2$

19. $\begin{bmatrix} 1 & -3 & 4 & | & 3 \\ 3 & -5 & 6 & | & 6 \\ -5 & 3 & 4 & | & 6 \end{bmatrix}$ $\begin{array}{l} R_2 = -3r_1 + r_2 \\ R_3 = 5r_1 + r_3 \end{array}$

20. $\begin{bmatrix} 1 & -3 & 3 & | & -5 \\ -4 & -5 & -3 & | & -5 \\ -3 & -2 & 4 & | & 6 \end{bmatrix}$ $\begin{array}{l} R_2 = 4r_1 + r_2 \\ R_3 = 3r_1 + r_3 \end{array}$

21. $\begin{bmatrix} 1 & -3 & 2 & | & -6 \\ 2 & -5 & 3 & | & -4 \\ -3 & -6 & 4 & | & 6 \end{bmatrix}$ $\begin{array}{l} R_2 = -2r_1 + r_2 \\ R_3 = 3r_1 + r_3 \end{array}$

22. $\begin{bmatrix} 1 & -3 & -4 & | & -6 \\ 6 & -5 & 6 & | & -6 \\ -1 & 1 & 4 & | & 6 \end{bmatrix}$ $\begin{array}{l} R_2 = -6r_1 + r_2 \\ R_3 = r_1 + r_3 \end{array}$

23. $\begin{bmatrix} 5 & -3 & 1 & | & -2 \\ 2 & -5 & 6 & | & -2 \\ -4 & 1 & 4 & | & 6 \end{bmatrix}$ $\begin{array}{l} R_1 = -2r_2 + r_1 \\ R_3 = 2r_2 + r_3 \end{array}$

24. $\begin{bmatrix} 4 & -3 & -1 & | & 2 \\ 3 & -5 & 2 & | & 6 \\ -3 & -6 & 4 & | & 6 \end{bmatrix}$ $\begin{array}{l} R_1 = -r_2 + r_1 \\ R_3 = r_2 + r_3 \end{array}$

In Problems 25–36, the reduced row echelon form of a system of linear equations is given. Write the system of equations corresponding to the given matrix. Use x, y; or x, y, z; or x_1, x_2, x_3, x_4 as variables. Determine whether the system is consistent or inconsistent. If it is consistent, give the solution.

25. $\begin{bmatrix} 1 & 0 & | & 5 \\ 0 & 1 & | & -1 \end{bmatrix}$

26. $\begin{bmatrix} 1 & 0 & | & -4 \\ 0 & 1 & | & 0 \end{bmatrix}$

27. $\begin{bmatrix} 1 & 0 & 0 & | & 1 \\ 0 & 1 & 0 & | & 2 \\ 0 & 0 & 0 & | & 3 \end{bmatrix}$

28. $\begin{bmatrix} 1 & 0 & 0 & | & 0 \\ 0 & 1 & 0 & | & 0 \\ 0 & 0 & 0 & | & 2 \end{bmatrix}$

29. $\begin{bmatrix} 1 & 0 & 2 & | & -1 \\ 0 & 1 & -4 & | & -2 \\ 0 & 0 & 0 & | & 0 \end{bmatrix}$

30. $\begin{bmatrix} 1 & 0 & 4 & | & 4 \\ 0 & 1 & 3 & | & 2 \\ 0 & 0 & 0 & | & 0 \end{bmatrix}$

31. $\begin{bmatrix} 1 & 0 & 0 & 0 & | & 1 \\ 0 & 1 & 0 & 1 & | & 2 \\ 0 & 0 & 1 & 2 & | & 3 \end{bmatrix}$

32. $\begin{bmatrix} 1 & 0 & 0 & 0 & | & 1 \\ 0 & 1 & 0 & 2 & | & 2 \\ 0 & 0 & 1 & 3 & | & 0 \end{bmatrix}$

33. $\begin{bmatrix} 1 & 0 & 0 & 4 & | & 2 \\ 0 & 1 & 1 & 3 & | & 3 \\ 0 & 0 & 0 & 0 & | & 0 \end{bmatrix}$

34. $\begin{bmatrix} 1 & 0 & 0 & 0 & | & 1 \\ 0 & 1 & 0 & 0 & | & 2 \\ 0 & 0 & 1 & 2 & | & 3 \end{bmatrix}$

35. $\begin{bmatrix} 1 & 0 & 0 & 1 & | & -2 \\ 0 & 1 & 0 & 2 & | & 2 \\ 0 & 0 & 1 & -1 & | & 0 \\ 0 & 0 & 0 & 0 & | & 0 \end{bmatrix}$

36. $\begin{bmatrix} 1 & 0 & 0 & 0 & | & 1 \\ 0 & 1 & 0 & 0 & | & 2 \\ 0 & 0 & 1 & 0 & | & 3 \\ 0 & 0 & 0 & 1 & | & 0 \end{bmatrix}$

In Problems 37–72, solve each system of equations using matrices (row operations). If the system has no solution, say that it is inconsistent.

37. $\begin{cases} x + y = 8 \\ x - y = 4 \end{cases}$

38. $\begin{cases} x + 2y = 5 \\ x + y = 3 \end{cases}$

39. $\begin{cases} 2x - 4y = -2 \\ 3x + 2y = 3 \end{cases}$

40. $\begin{cases} 3x + 3y = 3 \\ 4x + 2y = \dfrac{8}{3} \end{cases}$

41. $\begin{cases} x + 2y = 4 \\ 2x + 4y = 8 \end{cases}$

42. $\begin{cases} 3x - y = 7 \\ 9x - 3y = 21 \end{cases}$

43. $\begin{cases} 2x + 3y = 6 \\ x - y = \dfrac{1}{2} \end{cases}$

44. $\begin{cases} \dfrac{1}{2}x + y = -2 \\ x - 2y = 8 \end{cases}$

45. $\begin{cases} 3x - 5y = 3 \\ 15x + 5y = 21 \end{cases}$

46. $\begin{cases} 2x - y = -1 \\ x + \dfrac{1}{2}y = \dfrac{3}{2} \end{cases}$

47. $\begin{cases} x - y = 6 \\ 2x - 3z = 16 \\ 2y + z = 4 \end{cases}$

48. $\begin{cases} 2x + y = -4 \\ -2y + 4z = 0 \\ 3x - 2z = -11 \end{cases}$

49. $\begin{cases} x - 2y + 3z = 7 \\ 2x + y + z = 4 \\ -3x + 2y - 2z = -10 \end{cases}$

50. $\begin{cases} 2x + y - 3z = 0 \\ -2x + 2y + z = -7 \\ 3x - 4y - 3z = 7 \end{cases}$

51. $\begin{cases} 2x - 2y - 2z = 2 \\ 2x + 3y + z = 2 \\ 3x + 2y = 0 \end{cases}$

52. $\begin{cases} 2x - 3y - z = 0 \\ -x + 2y + z = 5 \\ 3x - 4y - z = 1 \end{cases}$

53. $\begin{cases} -x + y + z = -1 \\ -x + 2y - 3z = -4 \\ 3x - 2y - 7z = 0 \end{cases}$

54. $\begin{cases} 2x - 3y - z = 0 \\ 3x + 2y + 2z = 2 \\ x + 5y + 3z = 2 \end{cases}$

55. $\begin{cases} 2x - 2y + 3z = 6 \\ 4x - 3y + 2z = 0 \\ -2x + 3y - 7z = 1 \end{cases}$

56. $\begin{cases} 3x - 2y + 2z = 6 \\ 7x - 3y + 2z = -1 \\ 2x - 3y + 4z = 0 \end{cases}$

57. $\begin{cases} x + y - z = 6 \\ 3x - 2y + z = -5 \\ x + 3y - 2z = 14 \end{cases}$

58. $\begin{cases} x - y + z = -4 \\ 2x - 3y + 4z = -15 \\ 5x + y - 2z = 12 \end{cases}$

59. $\begin{cases} x + 2y - z = -3 \\ 2x - 4y + z = -7 \\ -2x + 2y - 3z = 4 \end{cases}$

60. $\begin{cases} x + 4y - 3z = -8 \\ 3x - y + 3z = 12 \\ x + y + 6z = 1 \end{cases}$

61. $\begin{cases} 3x + y - z = \dfrac{2}{3} \\ 2x - y + z = 1 \\ 4x + 2y = \dfrac{8}{3} \end{cases}$

62. $\begin{cases} x + y = 1 \\ 2x - y + z = 1 \\ x + 2y + z = \dfrac{8}{3} \end{cases}$

63. $\begin{cases} x + y + z + w = 4 \\ 2x - y + z = 0 \\ 3x + 2y + z - w = 6 \\ x - 2y - 2z + 2w = -1 \end{cases}$

64. $\begin{cases} x + y + z + w = 4 \\ -x + 2y + z = 0 \\ 2x + 3y + z - w = 6 \\ -2x + y - 2z + 2w = -1 \end{cases}$

65. $\begin{cases} x + 2y + z = 1 \\ 2x - y + 2z = 2 \\ 3x + y + 3z = 3 \end{cases}$

66. $\begin{cases} x + 2y - z = 3 \\ 2x - y + 2z = 6 \\ x - 3y + 3z = 4 \end{cases}$

67. $\begin{cases} x - y + z = 5 \\ 3x + 2y - 2z = 0 \end{cases}$

68. $\begin{cases} 2x + y - z = 4 \\ -x + y + 3z = 1 \end{cases}$

69. $\begin{cases} 2x + 3y - z = 3 \\ x - y - z = 0 \\ -x + y + z = 0 \\ x + y + 3z = 5 \end{cases}$

70. $\begin{cases} x - 3y + z = 1 \\ 2x - y - 4z = 0 \\ x - 3y + 2z = 1 \\ x - 2y = 5 \end{cases}$

71. $\begin{cases} 4x + y + z - w = 4 \\ x - y + 2z + 3w = 3 \end{cases}$

72. $\begin{cases} -4x + y - 5 \\ 2x - y + z - w = 5 \\ z + w = 4 \end{cases}$

Applications and Extensions

73. Curve Fitting Find the function $y = ax^2 + bx + c$ whose graph contains the points $(1, 2)$, $(-2, -7)$, and $(2, -3)$.

74. Curve Fitting Find the function $y = ax^2 + bx + c$ whose graph contains the points $(1, -1)$, $(3, -1)$, and $(-2, 14)$.

75. Curve Fitting Find the function $f(x) = ax^3 + bx^2 + cx + d$ for which $f(-3) = -112, f(-1) = -2, f(1) = 4$, and $f(2) = 13$.

76. Curve Fitting Find the function $f(x) = ax^3 + bx^2 + cx + d$ for which $f(-2) = -10, f(-1) = 3, f(1) = 5$, and $f(3) = 15$.

77. Nutrition A dietitian at Palos Community Hospital wants a patient to have a meal that has 78 grams (g) of protein, 59 g of carbohydrates, and 75 milligrams (mg) of vitamin A. The hospital food service tells the dietitian that the dinner for today is salmon steak, baked eggs, and acorn squash. Each serving of salmon steak has 30 g of protein, 20 g of carbohydrates, and 2 mg of vitamin A. Each serving of baked eggs contains 15 g of protein, 2 g of carbohydrates, and 20 mg of vitamin A. Each serving of acorn squash contains 3 g of protein, 25 g of carbohydrates, and 32 mg of vitamin A. How many servings of each food should the dietitian provide for the patient?

78. Nutrition A dietitian at General Hospital wants a patient to have a meal that has 47 grams (g) of protein, 58 g

of carbohydrates, and 630 milligrams (mg) of calcium. The hospital food service tells the dietitian that the dinner for today is pork chops, corn on the cob, and 2% milk. Each serving of pork chops has 23 g of protein, 0 g of carbohydrates, and 10 mg of calcium. Each serving of corn on the cob contains 3 g of protein, 16 g of carbohydrates, and 10 mg of calcium. Each glass of 2% milk contains 9 g of protein, 13 g of carbohydrates, and 300 mg of calcium. How many servings of each food should the dietitian provide for the patient?

79. Financial Planning Carletta has $10,000 to invest. As her financial consultant, you recommend that she invest in Treasury bills that yield 6%, Treasury bonds that yield 7%, and corporate bonds that yield 8%. Carletta wants to have an annual income of $680, and the amount invested in corporate bonds must be half that invested in Treasury bills. Find the amount in each investment.

80. Landscaping A landscape company is hired to plant trees in three new subdivisions. The company charges the developer for each tree planted, an hourly rate to plant the trees, and a fixed delivery charge. In one subdivision it took 166 labor hours to plant 250 trees for a cost of $7520. In a second subdivision it took 124 labor hours to plant 200 trees for a cost of $5945. In the final subdivision it took 200 labor hours to plant 300 trees for a cost of $8985. Determine the cost for each tree, the hourly labor charge, and the fixed delivery charge.

Sources: www.bx.org

81. Production To manufacture an automobile requires painting, drying, and polishing. Epsilon Motor Company produces three types of cars: the Delta, the Beta, and the Sigma. Each Delta requires 10 hours (hr) for painting, 3 hr for drying, and 2 hr for polishing. A Beta requires 16 hr for painting, 5 hr for drying, and 3 hr for polishing, and a Sigma requires 8 hr for painting, 2 hr for drying, and 1 hr for polishing. If the company has 240 hr for painting, 69 hr for drying, and 41 hr for polishing per month, how many of each type of car are produced?

82. Production A Florida juice company completes the preparation of its products by sterilizing, filling, and labeling bottles. Each case of orange juice requires 9 minutes (min) for sterilizing, 6 min for filling, and 1 min for labeling. Each case of grapefruit juice requires 10 min for sterilizing, 4 min for filling, and 2 min for labeling. Each case of tomato juice requires 12 min for sterilizing, 4 min for filling, and 1 min for labeling. If the company runs the sterilizing machine for 398 min, the filling machine for 164 min, and the labeling machine for 58 min, how many cases of each type of juice are prepared?

83. Electricity: Kirchhoff's Rules An application of Kirchhoff's Rules to the circuit shown results in the following system of equations:

$$\begin{cases} -4 + 8 - 2I_2 = 0 \\ 8 = 5I_4 + I_1 \\ 4 = 3I_3 + I_1 \\ I_3 + I_4 = I_1 \end{cases}$$

Find the currents $I_1, I_2, I_3,$ and I_4.

Source: Based on Raymond Serway. Physics, 3rd ed. (Philadelphia: Saunders, 1990), Prob. 34. p. 790.

84. Electricity: Kirchhoff's Rules An application of Kirchhoff's Rules to the circuit shown results in the following system of equations:

$$\begin{cases} I_1 = I_3 + I_2 \\ 24 - 6I_1 - 3I_3 = 0 \\ 12 + 24 - 6I_1 - 6I_2 = 0 \end{cases}$$

Find the currents $I_1, I_2,$ and I_3.

Source: Ibid., Prob. 38, p. 791.

85. Financial Planning Three retired couples each require an additional annual income of $2000 per year. As their financial consultant, you recommend that they invest some money in Treasury bills that yield 7%, some money in corporate bonds that yield 9%, and some money in junk bonds that yield 11%. Prepare a table for each couple showing the various ways that their goals can be achieved:
(a) If the first couple has $20,000 to invest.
(b) If the second couple has $25,000 to invest.
(c) If the third couple has $30,000 to invest.
(d) What advice would you give each couple regarding the amount to invest and the choices available?
[**Hint:** Higher yields generally carry more risk.]

86. Financial Planning A young couple has $25,000 to invest. As their financial consultant, you recommend that they invest some money in Treasury bills that yield 7%, some money in corporate bonds that yield 9%, and some money in junk bonds that yield 11%. Prepare a table showing the various ways that this couple can achieve the following goals:
(a) $1500 per year in income
(b) $2000 per year in income
(c) $2500 per year in income
(d) What advice would you give this couple regarding the income that they require and the choices available?
[**Hint:** Higher yields generally carry more risk.]

87. Pharmacy A doctor's prescription calls for a daily intake of a supplement containing 40 milligrams (mg) of vitamin C and 30 mg of vitamin D. Your pharmacy stocks three supplements that can be used: one contains 20% vitamin C and 30% vitamin D; a second, 40% vitamin C and 20% vitamin D; and a third, 30% vitamin C and 50% vitamin D. Create a table showing the possible combinations that could be used to fill the prescription.

88. Pharmacy A doctor's prescription calls for the creation of pills that contain 12 units of vitamin B_{12} and 12 units of vitamin E. Your pharmacy stocks three powders that can be used to make these pills: one contains 20% vitamin B_{12} and 30% vitamin E; a second, 40% vitamin B_{12} and 20% vitamin E; and a third, 30% vitamin B_{12} and 40% vitamin E. Create a table showing the possible combinations of each powder that could be mixed in each pill.

Explaining Concepts: Discussion and Writing

89. Write a brief paragraph or two that outline your strategy for solving a system of linear equations using matrices.

90. When solving a system of linear equations using matrices, do you prefer to place the augmented matrix in row echelon form or in reduced row echelon form? Give reasons for your choice.

91. Make up a system of three linear equations containing three variables that has:
(a) No solution
(b) Exactly one solution
(c) Infinitely many solutions
Give the three systems to a friend to solve and critique.

8.3 Systems of Linear Equations: Determinants

OBJECTIVES **1** Evaluate 2 by 2 Determinants (p. 571)

2 Use Cramer's Rule to Solve a System of Two Equations Containing Two Variables (p. 572)

3 Evaluate 3 by 3 Determinants (p. 574)

4 Use Cramer's Rule to Solve a System of Three Equations Containing Three Variables (p. 576)

5 Know Properties of Determinants (p. 577)

In the preceding section, we described a method of using matrices to solve a system of linear equations. This section deals with yet another method for solving systems of linear equations; however, it can be used only when the number of equations equals the number of variables. Although the method will work for any system (provided that the number of equations equals the number of variables), it is most often used for systems of two equations containing two variables or three equations containing three variables. This method, called *Cramer's Rule*, is based on the concept of a *determinant*.

1 Evaluate 2 by 2 Determinants

DEFINITION

If $a, b, c,$ and d are four real numbers, the symbol

$$D = \begin{vmatrix} a & b \\ c & d \end{vmatrix}$$

is called a **2 by 2 determinant.** Its value is the number $ad - bc$; that is,

$$D = \begin{vmatrix} a & b \\ c & d \end{vmatrix} = ad - bc \qquad \textbf{(1)}$$

The following device may be helpful for remembering the value of a 2 by 2 determinant:

$$\begin{vmatrix} a & b \\ c & d \end{vmatrix} = ad - bc$$

Minus

EXAMPLE 1 **Evaluating a 2 × 2 Determinant**

Evaluate: $\begin{vmatrix} 3 & -2 \\ 6 & 1 \end{vmatrix}$

Solution

$$\begin{vmatrix} 3 & -2 \\ 6 & 1 \end{vmatrix} = (3)(1) - (6)(-2)$$
$$= 3 - (-12)$$
$$= 15$$

Now Work PROBLEM 7

2 Use Cramer's Rule to Solve a System of Two Equations Containing Two Variables

Let's see the role that a 2 by 2 determinant plays in the solution of a system of two equations containing two variables. Consider the system

$$\begin{cases} ax + by = s & (1) \\ cx + dy = t & (2) \end{cases} \quad \textbf{(2)}$$

We use the method of elimination to solve this system.

Provided $d \neq 0$ and $b \neq 0$, this system is equivalent to the system

$$\begin{cases} adx + bdy = sd & (1) \quad \text{Multiply by } d. \\ bcx + bdy = tb & (2) \quad \text{Multiply by } b. \end{cases}$$

Subtract the second equation from the first equation and obtain

$$\begin{cases} (ad - bc)x + 0 \cdot y = sd - tb & (1) \\ bcx + bdy = tb & (2) \end{cases}$$

Now the first equation can be rewritten using determinant notation.

$$\begin{vmatrix} a & b \\ c & d \end{vmatrix} x = \begin{vmatrix} s & b \\ t & d \end{vmatrix}$$

If $D = \begin{vmatrix} a & b \\ c & d \end{vmatrix} = ad - bc \neq 0$, we can solve for x to get

$$x = \frac{\begin{vmatrix} s & b \\ t & d \end{vmatrix}}{\begin{vmatrix} a & b \\ c & d \end{vmatrix}} = \frac{\begin{vmatrix} s & b \\ t & d \end{vmatrix}}{D} \quad \textbf{(3)}$$

Return now to the original system (2). Provided that $a \neq 0$ and $c \neq 0$, the system is equivalent to

$$\begin{cases} acx + bcy = cs & (1) \quad \text{Multiply by } c. \\ acx + ady = at & (2) \quad \text{Multiply by } a. \end{cases}$$

Subtract the first equation from the second equation and obtain

$$\begin{cases} acx + bcy = cs & (1) \\ 0 \cdot x + (ad - bc)y = at - cs & (2) \end{cases}$$

The second equation can now be rewritten using determinant notation.

$$\begin{vmatrix} a & b \\ c & d \end{vmatrix} y = \begin{vmatrix} a & s \\ c & t \end{vmatrix}$$

If $D = \begin{vmatrix} a & b \\ c & d \end{vmatrix} = ad - bc \neq 0$, we can solve for y to get

$$y = \frac{\begin{vmatrix} a & s \\ c & t \end{vmatrix}}{\begin{vmatrix} a & b \\ c & d \end{vmatrix}} = \frac{\begin{vmatrix} a & s \\ c & t \end{vmatrix}}{D} \quad \textbf{(4)}$$

Equations (3) and (4) lead us to the following result, called **Cramer's Rule.**

THEOREM

Cramer's Rule for Two Equations Containing Two Variables

The solution to the system of equations

$$\begin{cases} ax + by = s & (1) \\ cx + dy = t & (2) \end{cases} \qquad (5)$$

is given by

$$x = \frac{\begin{vmatrix} s & b \\ t & d \end{vmatrix}}{\begin{vmatrix} a & b \\ c & d \end{vmatrix}} \qquad y = \frac{\begin{vmatrix} a & s \\ c & t \end{vmatrix}}{\begin{vmatrix} a & b \\ c & d \end{vmatrix}} \qquad (6)$$

provided that

$$D = \begin{vmatrix} a & b \\ c & d \end{vmatrix} = ad - bc \neq 0$$

In the derivation given for Cramer's Rule, we assumed that none of the numbers a, b, c, and d was 0. In Problem 61 you will be asked to complete the proof under the less stringent condition that $D = ad - bc \neq 0$.

Now look carefully at the pattern in Cramer's Rule. The denominator in the solution (6) is the determinant of the coefficients of the variables.

$$\begin{cases} ax + by = s \\ cx + dy = t \end{cases} \qquad D = \begin{vmatrix} a & b \\ c & d \end{vmatrix}$$

In the solution for x, the numerator is the determinant, denoted by D_x, formed by replacing the entries in the first column (the coefficients of x) of D by the constants on the right side of the equal sign.

$$D_x = \begin{vmatrix} s & b \\ t & d \end{vmatrix}$$

In the solution for y, the numerator is the determinant, denoted by D_y, formed by replacing the entries in the second column (the coefficients of y) of D by the constants on the right side of the equal sign.

$$D_y = \begin{vmatrix} a & s \\ c & t \end{vmatrix}$$

Cramer's Rule then states that, if $D \neq 0$,

$$x = \frac{D_x}{D} \qquad y = \frac{D_y}{D} \qquad (7)$$

EXAMPLE 2

Solving a System of Linear Equations Using Determinants

Use Cramer's Rule, if applicable, to solve the system

$$\begin{cases} 3x - 2y = 4 & (1) \\ 6x + y = 13 & (2) \end{cases}$$

Solution

The determinant D of the coefficients of the variables is

$$D = \begin{vmatrix} 3 & -2 \\ 6 & 1 \end{vmatrix} = (3)(1) - (6)(-2) = 15$$

Because $D \neq 0$, Cramer's Rule (7) can be used.

$$x = \frac{D_x}{D} = \frac{\begin{vmatrix} 4 & -2 \\ 13 & 1 \end{vmatrix}}{15} \qquad y = \frac{D_y}{D} = \frac{\begin{vmatrix} 3 & 4 \\ 6 & 13 \end{vmatrix}}{15}$$

$$= \frac{(4)(1) - (13)(-2)}{15} \qquad = \frac{(3)(13) - (6)(4)}{15}$$

$$= \frac{30}{15} \qquad = \frac{15}{15}$$

$$= 2 \qquad = 1$$

The solution is $x = 2$, $y = 1$ or, using an ordered pair, $(2, 1)$.

In attempting to use Cramer's Rule, if the determinant D of the coefficients of the variables is found to equal 0 (so that Cramer's Rule is not applicable), then the system is either inconsistent or has infinitely many solutions.

Now Work PROBLEM 15

3 Evaluate 3 by 3 Determinants

To use Cramer's Rule to solve a system of three equations containing three variables, we need to define a 3 by 3 determinant.

A **3 by 3 determinant** is symbolized by

$$\begin{vmatrix} a_{11} & a_{12} & a_{13} \\ a_{21} & a_{22} & a_{23} \\ a_{31} & a_{32} & a_{33} \end{vmatrix} \tag{8}$$

in which a_{11}, a_{12}, \ldots, are real numbers.

As with matrices, we use a double subscript to identify an entry by indicating its row and column numbers. For example, the entry a_{23} is in row 2, column 3.

The value of a 3 by 3 determinant may be defined in terms of 2 by 2 determinants by the following formula:

$$\begin{vmatrix} a_{11} & a_{12} & a_{13} \\ a_{21} & a_{22} & a_{23} \\ a_{31} & a_{32} & a_{33} \end{vmatrix} = a_{11}\begin{vmatrix} a_{22} & a_{23} \\ a_{32} & a_{33} \end{vmatrix} \overset{\text{Minus}}{-} a_{12}\begin{vmatrix} a_{21} & a_{23} \\ a_{31} & a_{33} \end{vmatrix} \overset{\text{Plus}}{+} a_{13}\begin{vmatrix} a_{21} & a_{22} \\ a_{31} & a_{32} \end{vmatrix} \tag{9}$$

$$\begin{array}{ccc}
\uparrow & \uparrow & \uparrow \\
\text{2 by 2} & \text{2 by 2} & \text{2 by 2} \\
\text{determinant} & \text{determinant} & \text{determinant} \\
\text{left after} & \text{left after} & \text{left after} \\
\text{removing the row} & \text{removing the row} & \text{removing the row} \\
\text{and column} & \text{and column} & \text{and column} \\
\text{containing } a_{11} & \text{containing } a_{12} & \text{containing } a_{13}
\end{array}$$

The 2 by 2 determinants shown in formula (9) are called **minors** of the 3 by 3 determinant. For an n by n determinant, the **minor** M_{ij} of entry a_{ij} is the determinant resulting from removing the ith row and jth column.

EXAMPLE 3

Finding Minors of a 3 by 3 Determinant

For the determinant $A = \begin{vmatrix} 2 & -1 & 3 \\ -2 & 5 & 1 \\ 0 & 6 & -9 \end{vmatrix}$, find: (a) M_{12} (b) M_{23}

Solution (a) M_{12} is the determinant that results from removing the first row and second column from A.

$$A = \begin{vmatrix} 2 & -1 & 3 \\ -2 & 5 & 1 \\ 0 & 6 & -9 \end{vmatrix} \qquad M_{12} = \begin{vmatrix} -2 & 1 \\ 0 & -9 \end{vmatrix} = (-2)(-9) - (0)(1) = 18$$

(b) M_{23} is the determinant that results from removing the second row and third column from A.

$$A = \begin{vmatrix} 2 & -1 & 3 \\ -2 & 5 & 1 \\ 0 & 6 & -9 \end{vmatrix} \qquad M_{23} = \begin{vmatrix} 2 & -1 \\ 0 & 6 \end{vmatrix} = (2)(6) - (0)(-1) = 12$$

Referring back to formula (9), we see that each element a_{ij} in the first row of the determinant is multiplied by its minor, but sometimes this term is added and other times subtracted. To determine whether to add or subtract a term, we must consider the *cofactor*.

DEFINITION For an n by n determinant A, the **cofactor** of entry a_{ij}, denoted by A_{ij}, is given by

$$\boxed{A_{ij} = (-1)^{i+j} M_{ij}}$$

where M_{ij} is the minor of entry a_{ij}.

The exponent of $(-1)^{i+j}$ is the sum of the row and column of the entry a_{ij}, so if $i + j$ is even, $(-1)^{i+j}$ will equal 1, and if $i + j$ is odd, $(-1)^{i+j}$ will equal -1.

To find the value of a determinant, multiply each entry in any row or column by its cofactor and sum the results. This process is referred to as **expanding across a row or column.** For example, the value of the 3 by 3 determinant in formula (9) was found by expanding across row 1.

If we choose to expand down column 2, we obtain

$$\begin{vmatrix} a_{11} & a_{12} & a_{13} \\ a_{21} & a_{22} & a_{23} \\ a_{31} & a_{32} & a_{33} \end{vmatrix} = (-1)^{1+2} a_{12} \begin{vmatrix} a_{21} & a_{23} \\ a_{31} & a_{33} \end{vmatrix} + (-1)^{2+2} a_{22} \begin{vmatrix} a_{11} & a_{13} \\ a_{31} & a_{33} \end{vmatrix} + (-1)^{3+2} a_{32} \begin{vmatrix} a_{11} & a_{13} \\ a_{21} & a_{23} \end{vmatrix}$$

Expand down column 2.

If we choose to expand across row 3, we obtain

$$\begin{vmatrix} a_{11} & a_{12} & a_{13} \\ a_{21} & a_{22} & a_{23} \\ a_{31} & a_{32} & a_{33} \end{vmatrix} = (-1)^{3+1} a_{31} \begin{vmatrix} a_{12} & a_{13} \\ a_{22} & a_{23} \end{vmatrix} + (-1)^{3+2} a_{32} \begin{vmatrix} a_{11} & a_{13} \\ a_{21} & a_{23} \end{vmatrix} + (-1)^{3+3} a_{33} \begin{vmatrix} a_{11} & a_{12} \\ a_{21} & a_{22} \end{vmatrix}$$

Expand across row 3.

It can be shown that the value of a determinant does not depend on the choice of the row or column used in the expansion. However, expanding across a row or column that has an entry equal to 0 reduces the amount of work needed to compute the value of the determinant.

EXAMPLE 4 **Evaluating a 3 × 3 Determinant**

Find the value of the 3 by 3 determinant: $\begin{vmatrix} 3 & 0 & -1 \\ 4 & 6 & 2 \\ 8 & -2 & 3 \end{vmatrix}$

Solution Because of the 0 in row 1, column 2, it is easiest to expand across row 1 or down column 2. We choose to expand across row 1.

$$\begin{vmatrix} 3 & 0 & -1 \\ 4 & 6 & 2 \\ 8 & -2 & 3 \end{vmatrix} = (-1)^{1+1} \cdot 3 \cdot \begin{vmatrix} 6 & 2 \\ -2 & 3 \end{vmatrix} + (-1)^{1+2} \cdot 0 \cdot \begin{vmatrix} 4 & 2 \\ 8 & 3 \end{vmatrix} + (-1)^{1+3} \cdot (-1) \cdot \begin{vmatrix} 4 & 6 \\ 8 & -2 \end{vmatrix}$$

$$= 3(18 - (-4)) - 0 + (-1)(-8 - 48)$$
$$= 3(22) + (-1)(-56)$$
$$= 66 + 56 = 122$$

───Now Work PROBLEM 11

4 Use Cramer's Rule to Solve a System of Three Equations Containing Three Variables

Consider the following system of three equations containing three variables.

$$\begin{cases} a_{11}x + a_{12}y + a_{13}z = c_1 \\ a_{21}x + a_{22}y + a_{23}z = c_2 \\ a_{31}x + a_{32}y + a_{33}z = c_3 \end{cases} \tag{10}$$

It can be shown that if the determinant D of the coefficients of the variables is not 0, that is, if

$$D = \begin{vmatrix} a_{11} & a_{12} & a_{13} \\ a_{21} & a_{22} & a_{23} \\ a_{31} & a_{32} & a_{33} \end{vmatrix} \neq 0$$

the unique solution of system (10) is given by

THEOREM **Cramer's Rule for Three Equations Containing Three Variables**

$$x = \frac{D_x}{D} \qquad y = \frac{D_y}{D} \qquad z = \frac{D_z}{D}$$

where

$$D_x = \begin{vmatrix} c_1 & a_{12} & a_{13} \\ c_2 & a_{22} & a_{23} \\ c_3 & a_{32} & a_{33} \end{vmatrix} \qquad D_y = \begin{vmatrix} a_{11} & c_1 & a_{13} \\ a_{21} & c_2 & a_{23} \\ a_{31} & c_3 & a_{33} \end{vmatrix} \qquad D_z = \begin{vmatrix} a_{11} & a_{12} & c_1 \\ a_{21} & a_{22} & c_2 \\ a_{31} & a_{32} & c_3 \end{vmatrix}$$

Do you see the similarity of this pattern and the pattern observed earlier for a system of two equations containing two variables?

EXAMPLE 5 **Using Cramer's Rule**

Use Cramer's Rule, if applicable, to solve the following system:

$$\begin{cases} 2x + y - z = 3 & (1) \\ -x + 2y + 4z = -3 & (2) \\ x - 2y - 3z = 4 & (3) \end{cases}$$

Solution The value of the determinant D of the coefficients of the variables is

$$D = \begin{vmatrix} 2 & 1 & -1 \\ -1 & 2 & 4 \\ 1 & -2 & -3 \end{vmatrix} = (-1)^{1+1} \cdot 2 \cdot \begin{vmatrix} 2 & 4 \\ -2 & -3 \end{vmatrix} + (-1)^{1+2} \cdot 1 \cdot \begin{vmatrix} -1 & 4 \\ 1 & -3 \end{vmatrix} + (-1)^{1+3}(-1) \begin{vmatrix} -1 & 2 \\ 1 & -2 \end{vmatrix}$$

$$= 2(2) - 1(-1) + (-1)(0)$$
$$= 4 + 1 = 5$$

Because $D \neq 0$, we proceed to find the values of D_x, D_y, and D_z. To find D_x, we replace the coefficients of x in D with the constants and then evaluate the determinant.

$$D_x = \begin{vmatrix} 3 & 1 & -1 \\ -3 & 2 & 4 \\ 4 & -2 & -3 \end{vmatrix} = (-1)^{1+1} \cdot 3 \cdot \begin{vmatrix} 2 & 4 \\ -2 & -3 \end{vmatrix} + (-1)^{1+2} \cdot 1 \cdot \begin{vmatrix} -3 & 4 \\ 4 & -3 \end{vmatrix} + (-1)^{1+3}(-1) \begin{vmatrix} -3 & 2 \\ 4 & -2 \end{vmatrix}$$

$$= 3(2) - 1(-7) + (-1)(-2) = 15$$

$$D_y = \begin{vmatrix} 2 & 3 & -1 \\ -1 & -3 & 4 \\ 1 & 4 & -3 \end{vmatrix} = (-1)^{1+1} \cdot 2 \cdot \begin{vmatrix} -3 & 4 \\ 4 & -3 \end{vmatrix} + (-1)^{1+2} \cdot 3 \cdot \begin{vmatrix} -1 & 4 \\ 1 & -3 \end{vmatrix} + (-1)^{1+3}(-1) \begin{vmatrix} -1 & -3 \\ 1 & 4 \end{vmatrix}$$

$$= 2(-7) - 3(-1) + (-1)(-1) = -10$$

$$D_z = \begin{vmatrix} 2 & 1 & 3 \\ -1 & 2 & -3 \\ 1 & -2 & 4 \end{vmatrix} = (-1)^{1+1} \cdot 2 \cdot \begin{vmatrix} 2 & -3 \\ -2 & 4 \end{vmatrix} + (-1)^{1+2} \cdot 1 \cdot \begin{vmatrix} -1 & -3 \\ 1 & 4 \end{vmatrix} + (-1)^{1+3} \cdot 3 \cdot \begin{vmatrix} -1 & 2 \\ 1 & -2 \end{vmatrix}$$

$$= 2(2) - 1(-1) + 3(0) = 5$$

As a result,

$$x = \frac{D_x}{D} = \frac{15}{5} = 3 \qquad y = \frac{D_y}{D} = \frac{-10}{5} = -2 \qquad z = \frac{D_z}{D} = \frac{5}{5} = 1$$

The solution is $x = 3, y = -2, z = 1$ or, using an ordered triplet, $(3, -2, 1)$.

We already know that Cramer's Rule does not apply when the determinant of the coefficients on the variables, D, is 0. But can we learn anything about the system other than it is not a consistent and independent system if $D = 0$? The answer is yes!

Cramer's Rule with Inconsistent or Dependent Systems

- If $D = 0$ and at least one of the determinants D_x, D_y, or D_z is different from 0, then the system is inconsistent and the solution set is \varnothing or $\{\ \}$.
- If $D = 0$ and all the determinants D_x, D_y, and D_z equal 0, then the system is consistent and dependent so that there are infinitely many solutions. The system must be solved using row reduction techniques.

Now Work PROBLEM 33

5 Know Properties of Determinants

Determinants have several properties that are sometimes helpful for obtaining their value. We list some of them here.

THEOREM The value of a determinant changes sign if any two rows (or any two columns) are interchanged.

(11)

Proof for 2 by 2 Determinants

$$\begin{vmatrix} a & b \\ c & d \end{vmatrix} = ad - bc \quad \text{and} \quad \begin{vmatrix} c & d \\ a & b \end{vmatrix} = bc - ad = -(ad - bc) \quad \blacksquare$$

EXAMPLE 6 **Demonstrating Theorem (11)**

$$\begin{vmatrix} 3 & 4 \\ 1 & 2 \end{vmatrix} = 6 - 4 = 2 \qquad \begin{vmatrix} 1 & 2 \\ 3 & 4 \end{vmatrix} = 4 - 6 = -2$$

THEOREM If all the entries in any row (or any column) equal 0, the value of the determinant is 0.

(12)

Proof Expand across the row (or down the column) containing the 0's. ∎

THEOREM If any two rows (or any two columns) of a determinant have corresponding entries that are equal, the value of the determinant is 0. **(13)**

You are asked to prove this result for a 3 by 3 determinant in which the entries in column 1 equal the entries in column 3 in Problem 64.

EXAMPLE 7 **Demonstrating Theorem (13)**

$$\begin{vmatrix} 1 & 2 & 3 \\ 1 & 2 & 3 \\ 4 & 5 & 6 \end{vmatrix} = (-1)^{1+1} \cdot 1 \cdot \begin{vmatrix} 2 & 3 \\ 5 & 6 \end{vmatrix} + (-1)^{1+2} \cdot 2 \cdot \begin{vmatrix} 1 & 3 \\ 4 & 6 \end{vmatrix} + (-1)^{1+3} \cdot 3 \cdot \begin{vmatrix} 1 & 2 \\ 4 & 5 \end{vmatrix}$$

$$= 1(-3) - 2(-6) + 3(-3) = -3 + 12 - 9 = 0$$

THEOREM If any row (or any column) of a determinant is multiplied by a nonzero number k, the value of the determinant is also changed by a factor of k. **(14)**

You are asked to prove this result for a 3 by 3 determinant using row 2 in Problem 63.

EXAMPLE 8 **Demonstrating Theorem (14)**

$$\begin{vmatrix} 1 & 2 \\ 4 & 6 \end{vmatrix} = 6 - 8 = -2$$

$$\begin{vmatrix} k & 2k \\ 4 & 6 \end{vmatrix} = 6k - 8k = -2k = k(-2) = k \begin{vmatrix} 1 & 2 \\ 4 & 6 \end{vmatrix}$$

THEOREM If the entries of any row (or any column) of a determinant are multiplied by a nonzero number k and the result is added to the corresponding entries of another row (or column), the value of the determinant remains unchanged. **(15)**

In Problem 65, you are asked to prove this result for a 3 by 3 determinant using rows 1 and 2.

EXAMPLE 9 **Demonstrating Theorem (15)**

$$\begin{vmatrix} 3 & 4 \\ 5 & 2 \end{vmatrix} = -14 \qquad \begin{vmatrix} 3 & 4 \\ 5 & 2 \end{vmatrix} \xrightarrow{\uparrow} \begin{vmatrix} -7 & 0 \\ 5 & 2 \end{vmatrix} = -14$$

Multiply row 2 by −2 and add to row 1.

8.3 Assess Your Understanding

Concepts and Vocabulary

1. $D = \begin{vmatrix} a & b \\ c & d \end{vmatrix} = $ _____ .

2. Using Cramer's Rule, the value of x that satisfies the system of equations $\begin{cases} 2x + 3y = 5 \\ x - 4y = -3 \end{cases}$ is $x = \dfrac{}{\begin{vmatrix} 2 & 3 \\ 1 & -4 \end{vmatrix}}$.

3. *True or False* A determinant can never equal 0.

4. *True or False* When using Cramer's Rule, if $D = 0$, then the system of linear equations is inconsistent.

5. *True or False* The value of a determinant remains unchanged if any two rows or any two columns are interchanged.

6. *True or False* If any row (or any column) of a determinant is multiplied by a nonzero number k, the value of the determinant remains unchanged.

Skill Building

In Problems 7–14, find the value of each determinant.

7. $\begin{vmatrix} 6 & 4 \\ -1 & 3 \end{vmatrix}$

8. $\begin{vmatrix} 8 & -3 \\ 4 & 2 \end{vmatrix}$

9. $\begin{vmatrix} -3 & -1 \\ 4 & 2 \end{vmatrix}$

10. $\begin{vmatrix} -4 & 2 \\ -5 & 3 \end{vmatrix}$

11. $\begin{vmatrix} 3 & 4 & 2 \\ 1 & -1 & 5 \\ 1 & 2 & -2 \end{vmatrix}$

12. $\begin{vmatrix} 1 & 3 & -2 \\ 6 & 1 & -5 \\ 8 & 2 & 3 \end{vmatrix}$

13. $\begin{vmatrix} 4 & -1 & 2 \\ 6 & -1 & 0 \\ 1 & -3 & 4 \end{vmatrix}$

14. $\begin{vmatrix} 3 & -9 & 4 \\ 1 & 4 & 0 \\ 8 & -3 & 1 \end{vmatrix}$

In Problems 15–42, solve each system of equations using Cramer's Rule if it is applicable. If Cramer's Rule is not applicable, say so.

15. $\begin{cases} x + y = 8 \\ x - y = 4 \end{cases}$

16. $\begin{cases} x + 2y = 5 \\ x - y = 3 \end{cases}$

17. $\begin{cases} 5x - y = 13 \\ 2x + 3y = 12 \end{cases}$

18. $\begin{cases} x + 3y = 5 \\ 2x - 3y = -8 \end{cases}$

19. $\begin{cases} 3x = 24 \\ x + 2y = 0 \end{cases}$

20. $\begin{cases} 4x + 5y = -3 \\ -2y = -4 \end{cases}$

21. $\begin{cases} 3x - 6y = 24 \\ 5x + 4y = 12 \end{cases}$

22. $\begin{cases} 2x + 4y = 16 \\ 3x - 5y = -9 \end{cases}$

23. $\begin{cases} 3x - 2y = 4 \\ 6x - 4y = 0 \end{cases}$

24. $\begin{cases} -x + 2y = 5 \\ 4x - 8y = 6 \end{cases}$

25. $\begin{cases} 2x - 4y = -2 \\ 3x + 2y = 3 \end{cases}$

26. $\begin{cases} 3x + 3y = 3 \\ 4x + 2y = \dfrac{8}{3} \end{cases}$

27. $\begin{cases} 2x - 3y = -1 \\ 10x + 10y = 5 \end{cases}$

28. $\begin{cases} 3x - 2y = 0 \\ 5x + 10y = 4 \end{cases}$

29. $\begin{cases} 2x + 3y = 6 \\ x - y = \dfrac{1}{2} \end{cases}$

30. $\begin{cases} \dfrac{1}{2}x + y = -2 \\ x - 2y = 8 \end{cases}$

31. $\begin{cases} 3x - 5y = 3 \\ 15x + 5y = 21 \end{cases}$

32. $\begin{cases} 2x - y = -1 \\ x + \dfrac{1}{2}y = \dfrac{3}{2} \end{cases}$

33. $\begin{cases} x + y - z = 6 \\ 3x - 2y + z = -5 \\ x + 3y - 2z = 14 \end{cases}$

34. $\begin{cases} x - y + z = -4 \\ 2x - 3y + 4z = -15 \\ 5x + y - 2z = 12 \end{cases}$

35. $\begin{cases} x + 2y - z = -3 \\ 2x - 4y + z = -7 \\ -2x + 2y - 3z = 4 \end{cases}$

36. $\begin{cases} x + 4y - 3z = -8 \\ 3x - y + 3z = 12 \\ x + y + 6z = 1 \end{cases}$

37. $\begin{cases} x - 2y + 3z = 1 \\ 3x + y - 2z = 0 \\ 2x - 4y + 6z = 2 \end{cases}$

38. $\begin{cases} x - y + 2z = 5 \\ 3x + 2y = 4 \\ -2x + 2y - 4z = -10 \end{cases}$

39. $\begin{cases} x + 2y - z = 0 \\ 2x - 4y + z = 0 \\ -2x + 2y - 3z = 0 \end{cases}$

40. $\begin{cases} x + 4y - 3z = 0 \\ 3x - y + 3z = 0 \\ x + y + 6z = 0 \end{cases}$

41. $\begin{cases} x - 2y + 3z = 0 \\ 3x + y - 2z = 0 \\ 2x - 4y + 6z = 0 \end{cases}$

42. $\begin{cases} x - y + 2z = 0 \\ 3x + 2y = 0 \\ -2x + 2y - 4z = 0 \end{cases}$

In Problems 43–50, use properties of determinants to find the value of each determinant if it is known that

$$\begin{vmatrix} x & y & z \\ u & v & w \\ 1 & 2 & 3 \end{vmatrix} = 4$$

43. $\begin{vmatrix} 1 & 2 & 3 \\ u & v & w \\ x & y & z \end{vmatrix}$

44. $\begin{vmatrix} x & y & z \\ u & v & w \\ 2 & 4 & 6 \end{vmatrix}$

45. $\begin{vmatrix} x & y & z \\ -3 & -6 & -9 \\ u & v & w \end{vmatrix}$

46. $\begin{vmatrix} 1 & 2 & 3 \\ x - u & y - v & z - w \\ u & v & w \end{vmatrix}$

47. $\begin{vmatrix} 1 & 2 & 3 \\ x - 3 & y - 6 & z - 9 \\ 2u & 2v & 2w \end{vmatrix}$

48. $\begin{vmatrix} x & y & z - x \\ u & v & w - u \\ 1 & 2 & 2 \end{vmatrix}$

49. $\begin{vmatrix} 1 & 2 & 3 \\ 2x & 2y & 2z \\ u - 1 & v - 2 & w - 3 \end{vmatrix}$

50. $\begin{vmatrix} x + 3 & y + 6 & z + 9 \\ 3u - 1 & 3v - 2 & 3w - 3 \\ 1 & 2 & 3 \end{vmatrix}$

Mixed Practice

In Problems 51–56, solve for x.

51. $\begin{vmatrix} x & x \\ 4 & 3 \end{vmatrix} = 5$

52. $\begin{vmatrix} x & 1 \\ 3 & x \end{vmatrix} = -2$

53. $\begin{vmatrix} x & 1 & 1 \\ 4 & 3 & 2 \\ -1 & 2 & 5 \end{vmatrix} = 2$

54. $\begin{vmatrix} 3 & 2 & 4 \\ 1 & x & 5 \\ 0 & 1 & -2 \end{vmatrix} = 0$

55. $\begin{vmatrix} x & 2 & 3 \\ 1 & x & 0 \\ 6 & 1 & -2 \end{vmatrix} = 7$

56. $\begin{vmatrix} x & 1 & 2 \\ 1 & x & 3 \\ 0 & 1 & 2 \end{vmatrix} = -4x$

Applications and Extensions

57. Geometry: Equation of a Line An equation of the line containing the two points (x_1, y_1) and (x_2, y_2) may be expressed as the determinant

$$\begin{vmatrix} x & y & 1 \\ x_1 & y_1 & 1 \\ x_2 & y_2 & 1 \end{vmatrix} = 0$$

Prove this result by expanding the determinant and comparing the result to the two-point form of the equation of a line.

58. Geometry: Collinear Points Using the result obtained in Problem 57, show that three distinct points (x_1, y_1), (x_2, y_2), and (x_3, y_3) are collinear (lie on the same line) if and only if

$$\begin{vmatrix} x_1 & y_1 & 1 \\ x_2 & y_2 & 1 \\ x_3 & y_3 & 1 \end{vmatrix} = 0$$

59. Geometry: Area of a Triangle A triangle has vertices (x_1, y_1), (x_2, y_2), and (x_3, y_3). The area of the triangle is given by the absolute value of D, where $D = \frac{1}{2} \begin{vmatrix} x_1 & x_2 & x_3 \\ y_1 & y_2 & y_3 \\ 1 & 1 & 1 \end{vmatrix}$.

Use this formula to find the area of a triangle with vertices $(2, 3), (5, 2)$, and $(6, 5)$.

60. Show that $\begin{vmatrix} x^2 & x & 1 \\ y^2 & y & 1 \\ z^2 & z & 1 \end{vmatrix} = (y - z)(x - y)(x - z)$.

61. Complete the proof of Cramer's Rule for two equations containing two variables.

[**Hint:** In system (5), page 573, if $a = 0$, then $b \neq 0$ and $c \neq 0$, since $D = -bc \neq 0$. Now show that equation (6) provides a solution of the system when $a = 0$. Then three cases remain: $b = 0, c = 0$, and $d = 0$.]

62. Interchange columns 1 and 3 of a 3 by 3 determinant. Show that the value of the new determinant is -1 times the value of the original determinant.

63. Multiply each entry in row 2 of a 3 by 3 determinant by the number $k, k \neq 0$. Show that the value of the new determinant is k times the value of the original determinant.

64. Prove that a 3 by 3 determinant in which the entries in column 1 equal those in column 3 has the value 0.

65. Prove that, if row 2 of a 3 by 3 determinant is multiplied by $k, k \neq 0$, and the result is added to the entries in row 1, there is no change in the value of the determinant.

8.4 Matrix Algebra

OBJECTIVES **1** Find the Sum and Difference of Two Matrices (p. 582)
2 Find Scalar Multiples of a Matrix (p. 583)
3 Find the Product of Two Matrices (p. 584)
4 Find the Inverse of a Matrix (p. 589)
5 Solve a System of Linear Equations Using an Inverse Matrix (p. 593)

In Section 8.2, we defined a matrix as a rectangular array of real numbers and used an augmented matrix to represent a system of linear equations. There is, however, a branch of mathematics, called **linear algebra,** that deals with matrices in such a way that an algebra of matrices is permitted. In this section, we provide a survey of how this **matrix algebra** is developed.

Before getting started, we restate the definition of a matrix.

DEFINITION

A **matrix** is defined as a rectangular array of numbers:

$$
\begin{array}{c}
\text{Row 1} \\
\text{Row 2} \\
\vdots \\
\text{Row } i \\
\vdots \\
\text{Row } m
\end{array}
\begin{bmatrix}
a_{11} & a_{12} & \cdots & a_{1j} & \cdots & a_{1n} \\
a_{21} & a_{22} & \cdots & a_{2j} & \cdots & a_{2n} \\
\vdots & \vdots & & \vdots & & \vdots \\
a_{i1} & a_{i2} & \cdots & a_{ij} & \cdots & a_{in} \\
\vdots & \vdots & & \vdots & & \vdots \\
a_{m1} & a_{m2} & \cdots & a_{mj} & \cdots & a_{mn}
\end{bmatrix}
$$

Column 1, Column 2, Column j, Column n

Each number a_{ij} of the matrix has two indexes: the **row index** i and the **column index** j. The matrix shown here has m rows and n columns. The numbers a_{ij} are usually referred to as the **entries** of the matrix. For example, a_{23} refers to the entry in the second row, third column.

EXAMPLE 1

Arranging Data in a Matrix

In a survey of 900 people, the following information was obtained:

200 males	Thought federal defense spending was too high
150 males	Thought federal defense spending was too low
45 males	Had no opinion
315 females	Thought federal defense spending was too high
125 females	Thought federal defense spending was too low
65 females	Had no opinion

We can arrange these data in a rectangular array as follows:

	Too High	Too Low	No Opinion
Male	200	150	45
Female	315	125	65

or as the matrix

$$
\begin{bmatrix}
200 & 150 & 45 \\
315 & 125 & 65
\end{bmatrix}
$$

This matrix has two rows (representing male and female) and three columns (representing "too high," "too low," and "no opinion").

The matrix we developed in Example 1 has 2 rows and 3 columns. In general, a matrix with m rows and n columns is called an **m by n matrix.** The matrix we developed in Example 1 is a 2 by 3 matrix and contains $2 \cdot 3 = 6$ entries. An m by n matrix will contain $m \cdot n$ entries.

If an m by n matrix has the same number of rows as columns, that is, if $m = n$, then the matrix is referred to as a **square matrix.**

EXAMPLE 2

Examples of Matrices

(a) $\begin{bmatrix} 5 & 0 \\ -6 & 1 \end{bmatrix}$ A 2 by 2 square matrix

(b) $\begin{bmatrix} 1 & 0 & 3 \end{bmatrix}$ A 1 by 3 matrix

(c) $\begin{bmatrix} 6 & -2 & 4 \\ 4 & 3 & 5 \\ 8 & 0 & 1 \end{bmatrix}$ A 3 by 3 square matrix

1 Find the Sum and Difference of Two Matrices

We begin our discussion of matrix algebra by first defining what is meant by two matrices being equal and then defining the operations of addition and subtraction. It is important to note that these definitions require each matrix to have the same number of rows *and* the same number of columns as a condition for equality and for addition and subtraction.

We usually represent matrices by capital letters, such as A, B, and C.

DEFINITION
Two matrices A and B are said to be **equal,** written as

$$A = B$$

provided that A and B have the same number of rows and the same number of columns and each entry a_{ij} in A is equal to the corresponding entry b_{ij} in B.

For example,

$$\begin{bmatrix} 2 & 1 \\ 0.5 & -1 \end{bmatrix} = \begin{bmatrix} \sqrt{4} & 1 \\ \frac{1}{2} & -1 \end{bmatrix} \quad \text{and} \quad \begin{bmatrix} 3 & 2 & 1 \\ 0 & 1 & -2 \end{bmatrix} = \begin{bmatrix} \sqrt{9} & \sqrt{4} & 1 \\ 0 & 1 & \sqrt[3]{-8} \end{bmatrix}$$

$$\begin{bmatrix} 4 & 1 \\ 6 & 1 \end{bmatrix} \neq \begin{bmatrix} 4 & 0 \\ 6 & 1 \end{bmatrix}$$ Because the entries in row 1, column 2 are not equal

$$\begin{bmatrix} 4 & 1 & 2 \\ 6 & 1 & 2 \end{bmatrix} \neq \begin{bmatrix} 4 & 1 & 2 & 3 \\ 6 & 1 & 2 & 4 \end{bmatrix}$$ Because the matrix on the left has 3 columns and the matrix on the right has 4 columns

Suppose that A and B represent two m by n matrices. We define their **sum** $A + B$ to be the m by n matrix formed by adding the corresponding entries a_{ij} of A and b_{ij} of B. The **difference $A - B$** is defined as the m by n matrix formed by subtracting the entries b_{ij} in B from the corresponding entries a_{ij} in A. Addition and subtraction of matrices are allowed only for matrices having the same number m of rows and the same number n of columns. For example, a 2 by 3 matrix and a 2 by 4 matrix cannot be added or subtracted.

EXAMPLE 3

Adding and Subtracting Matrices

Suppose that

$$A = \begin{bmatrix} 2 & 4 & 8 & -3 \\ 0 & 1 & 2 & 3 \end{bmatrix} \quad \text{and} \quad B = \begin{bmatrix} -3 & 4 & 0 & 1 \\ 6 & 8 & 2 & 0 \end{bmatrix}$$

Find: (a) $A + B$ (b) $A - B$

Solution
(a) $A + B = \begin{bmatrix} 2 & 4 & 8 & -3 \\ 0 & 1 & 2 & 3 \end{bmatrix} + \begin{bmatrix} -3 & 4 & 0 & 1 \\ 6 & 8 & 2 & 0 \end{bmatrix}$

$$= \begin{bmatrix} 2 + (-3) & 4 + 4 & 8 + 0 & -3 + 1 \\ 0 + 6 & 1 + 8 & 2 + 2 & 3 + 0 \end{bmatrix}$$ Add corresponding entries.

$$= \begin{bmatrix} -1 & 8 & 8 & -2 \\ 6 & 9 & 4 & 3 \end{bmatrix}$$

(b) $A - B = \begin{bmatrix} 2 & 4 & 8 & -3 \\ 0 & 1 & 2 & 3 \end{bmatrix} - \begin{bmatrix} -3 & 4 & 0 & 1 \\ 6 & 8 & 2 & 0 \end{bmatrix}$

$$= \begin{bmatrix} 2 - (-3) & 4 - 4 & 8 - 0 & -3 - 1 \\ 0 - 6 & 1 - 8 & 2 - 2 & 3 - 0 \end{bmatrix}$$ Subtract corresponding entries.

$$= \begin{bmatrix} 5 & 0 & 8 & -4 \\ -6 & -7 & 0 & 3 \end{bmatrix}$$

Figure 7

Seeing the Concept

Graphing utilities can make the sometimes tedious process of matrix algebra easy. In fact, most graphing calculators can handle matrices as large as 9 by 9, some even larger ones. Enter the matrices into a graphing utility. Name them [A] and [B]. Figure 7 shows the results of adding and subtracting [A] and [B].

Now Work PROBLEM 9

Many of the algebraic properties of sums of real numbers are also true for sums of matrices. Suppose that A, B, and C are m by n matrices. Then matrix addition is **commutative.** That is,

Commutative Property of Matrix Addition

$$A + B = B + A$$

Matrix addition is also **associative.** That is,

Associative Property of Matrix Addition

$$(A + B) + C = A + (B + C)$$

Although we shall not prove these results, the proofs, as the following example illustrates, are based on the commutative and associative properties for real numbers.

EXAMPLE 4 **Demonstrating the Commutative Property**

$$\begin{bmatrix} 2 & 3 & -1 \\ 4 & 0 & 7 \end{bmatrix} + \begin{bmatrix} -1 & 2 & 1 \\ 5 & -3 & 4 \end{bmatrix} = \begin{bmatrix} 2 + (-1) & 3 + 2 & -1 + 1 \\ 4 + 5 & 0 + (-3) & 7 + 4 \end{bmatrix}$$

$$= \begin{bmatrix} -1 + 2 & 2 + 3 & 1 + (-1) \\ 5 + 4 & -3 + 0 & 4 + 7 \end{bmatrix}$$

$$= \begin{bmatrix} -1 & 2 & 1 \\ 5 & -3 & 4 \end{bmatrix} + \begin{bmatrix} 2 & 3 & -1 \\ 4 & 0 & 7 \end{bmatrix}$$

A matrix whose entries are all equal to 0 is called a **zero matrix.** Each of the following matrices is a zero matrix.

$$\begin{bmatrix} 0 & 0 \\ 0 & 0 \end{bmatrix}$$ 2 by 2 square zero matrix $$\begin{bmatrix} 0 & 0 & 0 \\ 0 & 0 & 0 \end{bmatrix}$$ 2 by 3 zero matrix $$\begin{bmatrix} 0 & 0 & 0 \end{bmatrix}$$ 1 by 3 zero matrix

Zero matrices have properties similar to the real number 0. If A is an m by n matrix and 0 is the m by n zero matrix, then

$$A + 0 = 0 + A = A$$

In other words, the zero matrix is the additive identity in matrix algebra.

2 Find Scalar Multiples of a Matrix

We can also multiply a matrix by a real number. If k is a real number and A is an m by n matrix, the matrix kA is the m by n matrix formed by multiplying each entry a_{ij} in A by k. The number k is sometimes referred to as a **scalar,** and the matrix kA is called a **scalar multiple** of A.

EXAMPLE 5 **Operations Using Matrices**

Suppose that

$$A = \begin{bmatrix} 3 & 1 & 5 \\ -2 & 0 & 6 \end{bmatrix} \qquad B = \begin{bmatrix} 4 & 1 & 0 \\ 8 & 1 & -3 \end{bmatrix} \qquad C = \begin{bmatrix} 9 & 0 \\ -3 & 6 \end{bmatrix}$$

Find: (a) $4A$ (b) $\dfrac{1}{3}C$ (c) $3A - 2B$

Solution (a) $4A = 4\begin{bmatrix} 3 & 1 & 5 \\ -2 & 0 & 6 \end{bmatrix} = \begin{bmatrix} 4 \cdot 3 & 4 \cdot 1 & 4 \cdot 5 \\ 4(-2) & 4 \cdot 0 & 4 \cdot 6 \end{bmatrix} = \begin{bmatrix} 12 & 4 & 20 \\ -8 & 0 & 24 \end{bmatrix}$

(b) $\dfrac{1}{3}C = \dfrac{1}{3}\begin{bmatrix} 9 & 0 \\ -3 & 6 \end{bmatrix} = \begin{bmatrix} \dfrac{1}{3} \cdot 9 & \dfrac{1}{3} \cdot 0 \\ \dfrac{1}{3}(-3) & \dfrac{1}{3} \cdot 6 \end{bmatrix} = \begin{bmatrix} 3 & 0 \\ -1 & 2 \end{bmatrix}$

(c) $3A - 2B = 3\begin{bmatrix} 3 & 1 & 5 \\ -2 & 0 & 6 \end{bmatrix} - 2\begin{bmatrix} 4 & 1 & 0 \\ 8 & 1 & -3 \end{bmatrix}$

$\qquad\qquad\qquad = \begin{bmatrix} 3 \cdot 3 & 3 \cdot 1 & 3 \cdot 5 \\ 3(-2) & 3 \cdot 0 & 3 \cdot 6 \end{bmatrix} - \begin{bmatrix} 2 \cdot 4 & 2 \cdot 1 & 2 \cdot 0 \\ 2 \cdot 8 & 2 \cdot 1 & 2(-3) \end{bmatrix}$

$\qquad\qquad\qquad = \begin{bmatrix} 9 & 3 & 15 \\ -6 & 0 & 18 \end{bmatrix} - \begin{bmatrix} 8 & 2 & 0 \\ 16 & 2 & -6 \end{bmatrix}$

$\qquad\qquad\qquad = \begin{bmatrix} 9 - 8 & 3 - 2 & 15 - 0 \\ -6 - 16 & 0 - 2 & 18 - (-6) \end{bmatrix}$

$\qquad\qquad\qquad = \begin{bmatrix} 1 & 1 & 15 \\ -22 & -2 & 24 \end{bmatrix}$

✓ **Check:** Enter the matrices $[A]$, $[B]$, and $[C]$ into a graphing utility. Then find $4A, \dfrac{1}{3}C$, and $3A - 2B$.

━ **Now Work** PROBLEM 13

We list next some of the algebraic properties of scalar multiplication. Let h and k be real numbers, and let A and B be m by n matrices. Then

Properties of Scalar Multiplication

$$k(hA) = (kh)A$$
$$(k + h)A = kA + hA$$
$$k(A + B) = kA + kB$$

3 Find the Product of Two Matrices

Unlike the straightforward definition for adding two matrices, the definition for multiplying two matrices is not what we might expect. In preparation for this definition, we need the following definitions:

DEFINITION

A **row vector** R is a 1 by n matrix

$$R = \begin{bmatrix} r_1 & r_2 & \cdots & r_n \end{bmatrix}$$

A **column vector** C is an n by 1 matrix

$$C = \begin{bmatrix} c_1 \\ c_2 \\ \vdots \\ c_n \end{bmatrix}$$

The **product** RC of R times C is defined as the number

$$RC = \begin{bmatrix} r_1 & r_2 \cdots r_n \end{bmatrix} \begin{bmatrix} c_1 \\ c_2 \\ \vdots \\ c_n \end{bmatrix} = r_1 c_1 + r_2 c_2 + \cdots + r_n c_n$$

Notice that a row vector and a column vector can be multiplied only if they contain the same number of entries.

EXAMPLE 6 | **The Product of a Row Vector and a Column Vector**

If $R = \begin{bmatrix} 3 & -5 & 2 \end{bmatrix}$ and $C = \begin{bmatrix} 3 \\ 4 \\ -5 \end{bmatrix}$, then

$$RC = \begin{bmatrix} 3 & -5 & 2 \end{bmatrix} \begin{bmatrix} 3 \\ 4 \\ -5 \end{bmatrix} = 3 \cdot 3 + (-5)4 + 2(-5) = 9 - 20 - 10 = -21$$

EXAMPLE 7 | **Using Matrices to Compute Revenue**

A clothing store sells men's shirts for $40, silk ties for $20, and wool suits for $400. Last month, the store had sales consisting of 100 shirts, 200 ties, and 50 suits. What was the total revenue due to these sales?

Solution

Set up a row vector R to represent the prices of each item and a column vector C to represent the corresponding number of items sold. Then

$$\begin{array}{cc} \text{Prices} & \text{Number} \\ \text{Shirts Ties Suits} & \text{sold} \end{array}$$

$$R = \begin{bmatrix} 40 & 20 & 400 \end{bmatrix} \qquad C = \begin{bmatrix} 100 \\ 200 \\ 50 \end{bmatrix} \begin{matrix} \text{Shirts} \\ \text{Ties} \\ \text{Suits} \end{matrix}$$

The total revenue obtained is the product RC. That is,

$$RC = \begin{bmatrix} 40 & 20 & 400 \end{bmatrix} \begin{bmatrix} 100 \\ 200 \\ 50 \end{bmatrix}$$

$$= \underbrace{40 \cdot 100}_{\text{Shirt revenue}} + \underbrace{20 \cdot 200}_{\text{Tie revenue}} + \underbrace{400 \cdot 50}_{\text{Suit revenue}} = \underbrace{\$28,000}_{\text{Total revenue}}$$

The definition for multiplying two matrices is based on the definition of a row vector times a column vector.

DEFINITION

Let A denote an m by r matrix and let B denote an r by n matrix. The **product** AB is defined as the m by n matrix whose entry in row i, column j is the product of the ith row of A and the jth column of B.

The definition of the product AB of two matrices A and B, in this order, requires that the number of columns of A equal the number of rows of B; otherwise, no product is defined.

In Words

To find the product AB, the number of columns in A must equal the number of rows in B.

An example will help to clarify the definition.

EXAMPLE 8 **Multiplying Two Matrices**

Find the product AB if

$$A = \begin{bmatrix} 2 & 4 & -1 \\ 5 & 8 & 0 \end{bmatrix} \quad \text{and} \quad B = \begin{bmatrix} 2 & 5 & 1 & 4 \\ 4 & 8 & 0 & 6 \\ -3 & 1 & -2 & -1 \end{bmatrix}$$

Solution First, notice that A is 2 by 3 and B is 3 by 4, so the product AB is defined and will be a 2 by 4 matrix.

Suppose we want the entry in row 2, column 3 of AB. To find it, we find the product of the row vector from row 2 of A and the column vector from column 3 of B.

$$\underset{\text{Row 2 of } A}{[5 \quad 8 \quad 0]} \overset{\text{Column 3 of } B}{\begin{bmatrix} 1 \\ 0 \\ -2 \end{bmatrix}} = 5 \cdot 1 + 8 \cdot 0 + 0(-2) = 5$$

So far, we have

$$AB = \begin{bmatrix} \underline{} & \underline{} & \overset{\text{Column 3}}{\underset{5}{\downarrow}} & \underline{} \end{bmatrix} \quad \leftarrow \text{Row 2}$$

Now, to find the entry in row 1, column 4 of AB, we find the product of row 1 of A and column 4 of B.

$$\underset{\text{Row 1 of } A}{[2 \quad 4 \quad -1]} \overset{\text{Column 4 of } B}{\begin{bmatrix} 4 \\ 6 \\ -1 \end{bmatrix}} = 2 \cdot 4 + 4 \cdot 6 + (-1)(-1) = 33$$

Continuing in this fashion, we find AB.

$$AB = \begin{bmatrix} 2 & 4 & -1 \\ 5 & 8 & 0 \end{bmatrix} \begin{bmatrix} 2 & 5 & 1 & 4 \\ 4 & 8 & 0 & 6 \\ -3 & 1 & -2 & -1 \end{bmatrix}$$

$$= \begin{bmatrix} \text{Row 1 of } A & \text{Row 1 of } A & \text{Row 1 of } A & \text{Row 1 of } A \\ \text{times} & \text{times} & \text{times} & \text{times} \\ \text{column 1 of } B & \text{column 2 of } B & \text{column 3 of } B & \text{column 4 of } B \\ \\ \text{Row 2 of } A & \text{Row 2 of } A & \text{Row 2 of } A & \text{Row 2 of } A \\ \text{timcs} & \text{times} & \text{times} & \text{times} \\ \text{column 1 of } B & \text{column 2 of } B & \text{column 3 of } B & \text{column 4 of } B \end{bmatrix}$$

$$= \begin{bmatrix} 2\cdot 2 + 4\cdot 4 + (-1)(-3) & 2\cdot 5 + 4\cdot 8 + (-1)1 & 2\cdot 1 + 4\cdot 0 + (-1)(-2) & 33(\text{from earlier}) \\ 5\cdot 2 + 8\cdot 4 + 0(-3) & 5\cdot 5 + 8\cdot 8 + 0\cdot 1 & 5(\text{from earlier}) & 5\cdot 4 + 8\cdot 6 + 0(-1) \end{bmatrix}$$

$$= \begin{bmatrix} 23 & 41 & 4 & 33 \\ 42 & 89 & 5 & 68 \end{bmatrix}$$

✓**Check:** Enter the matrices $[A]$ and $[B]$. Then find AB. (See what happens if you try to find BA.)

Now Work PROBLEM 25

Notice that for the matrices given in Example 8 the product BA is not defined, because B is 3 by 4 and A is 2 by 3.

Another result that can occur when multiplying two matrices is illustrated in the next example.

EXAMPLE 9 **Multiplying Two Matrices**

If

$$A = \begin{bmatrix} 2 & 1 & 3 \\ 1 & -1 & 0 \end{bmatrix} \quad \text{and} \quad B = \begin{bmatrix} 1 & 0 \\ 2 & 1 \\ 3 & 2 \end{bmatrix}$$

find: (a) AB (b) BA

Solution (a) $AB = \begin{bmatrix} 2 & 1 & 3 \\ 1 & -1 & 0 \end{bmatrix} \begin{bmatrix} 1 & 0 \\ 2 & 1 \\ 3 & 2 \end{bmatrix} = \begin{bmatrix} 13 & 7 \\ -1 & -1 \end{bmatrix}$

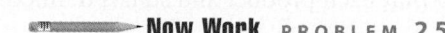

 2 by 3 3 by 2 2 by 2

(b) $BA = \begin{bmatrix} 1 & 0 \\ 2 & 1 \\ 3 & 2 \end{bmatrix} \begin{bmatrix} 2 & 1 & 3 \\ 1 & -1 & 0 \end{bmatrix} = \begin{bmatrix} 2 & 1 & 3 \\ 5 & 1 & 6 \\ 8 & 1 & 9 \end{bmatrix}$

 3 by 2 2 by 3 3 by 3

Notice in Example 9 that AB is 2 by 2 and BA is 3 by 3. It is possible for both AB and BA to be defined, yet be unequal. In fact, even if A and B are both n by n matrices so that AB and BA are each defined and n by n, AB and BA will usually be unequal.

EXAMPLE 10 **Multiplying Two Square Matrices**

If

$$A = \begin{bmatrix} 2 & 1 \\ 0 & 4 \end{bmatrix} \quad \text{and} \quad B = \begin{bmatrix} -3 & 1 \\ 1 & 2 \end{bmatrix}$$

find: (a) AB (b) BA

Solution (a) $AB = \begin{bmatrix} 2 & 1 \\ 0 & 4 \end{bmatrix} \begin{bmatrix} -3 & 1 \\ 1 & 2 \end{bmatrix} = \begin{bmatrix} -5 & 4 \\ 4 & 8 \end{bmatrix}$

(b) $BA = \begin{bmatrix} -3 & 1 \\ 1 & 2 \end{bmatrix} \begin{bmatrix} 2 & 1 \\ 0 & 4 \end{bmatrix} = \begin{bmatrix} -6 & 1 \\ 2 & 9 \end{bmatrix}$

The preceding examples demonstrate that an important property of real numbers, the commutative property of multiplication, is not shared by matrices. In general:

THEOREM Matrix multiplication is not commutative.

━━━━━━ **Now Work** PROBLEMS 15 AND 17

Next we give two of the properties of real numbers that are shared by matrices. Assuming that each product and sum is defined, we have the following:

Associative Property of Matrix Multiplication

$$A(BC) = (AB)C$$

Distributive Property

$$A(B + C) = AB + AC$$

For an n by n square matrix, the entries located in row i, column i, $1 \le i \le n$, are called the **diagonal entries** or **the main diagonal.** The n by n square matrix whose diagonal entries are 1's, while all other entries are 0's, is called the **identity matrix I_n.** For example,

$$I_2 = \begin{bmatrix} 1 & 0 \\ 0 & 1 \end{bmatrix} \qquad I_3 = \begin{bmatrix} 1 & 0 & 0 \\ 0 & 1 & 0 \\ 0 & 0 & 1 \end{bmatrix}$$

and so on.

EXAMPLE 11 **Multiplication with an Identity Matrix**

Let

$$A = \begin{bmatrix} -1 & 2 & 0 \\ 0 & 1 & 3 \end{bmatrix} \quad \text{and} \quad B = \begin{bmatrix} 3 & 2 \\ 4 & 6 \\ 5 & 2 \end{bmatrix}$$

Find: (a) AI_3 (b) I_2A (c) BI_2

Solution

(a) $AI_3 = \begin{bmatrix} -1 & 2 & 0 \\ 0 & 1 & 3 \end{bmatrix} \begin{bmatrix} 1 & 0 & 0 \\ 0 & 1 & 0 \\ 0 & 0 & 1 \end{bmatrix} = \begin{bmatrix} -1 & 2 & 0 \\ 0 & 1 & 3 \end{bmatrix} = A$

(b) $I_2 A = \begin{bmatrix} 1 & 0 \\ 0 & 1 \end{bmatrix} \begin{bmatrix} -1 & 2 & 0 \\ 0 & 1 & 3 \end{bmatrix} = \begin{bmatrix} -1 & 2 & 0 \\ 0 & 1 & 3 \end{bmatrix} = A$

(c) $BI_2 = \begin{bmatrix} 3 & 2 \\ 4 & 6 \\ 5 & 2 \end{bmatrix} \begin{bmatrix} 1 & 0 \\ 0 & 1 \end{bmatrix} = \begin{bmatrix} 3 & 2 \\ 4 & 6 \\ 5 & 2 \end{bmatrix} = B$

Example 11 demonstrates the following property:

Identity Property

If A is an m by n matrix, then

$$I_m A = A \quad \text{and} \quad AI_n = A$$

If A is an n by n square matrix,

$$AI_n = I_n A = A$$

An identity matrix has properties similar to those of the real number 1. In other words, the identity matrix is a multiplicative identity in matrix algebra.

4 Find the Inverse of a Matrix

DEFINITION

Let A be a square n by n matrix. If there exists an n by n matrix A^{-1}, read "A inverse," for which

$$AA^{-1} = A^{-1}A = I_n$$

then A^{-1} is called the **inverse** of the matrix A.

NOTE If the determinant of A is zero, A is singular. (Refer to Section 8.3.) ∎

As we shall soon see, not every square matrix has an inverse. When a matrix A does have an inverse A^{-1}, then A is said to be **nonsingular.** If a matrix A has no inverse, it is called **singular.**

EXAMPLE 12 | **Multiplying a Matrix by Its Inverse**

Show that the inverse of

$$A = \begin{bmatrix} 3 & 1 \\ 2 & 1 \end{bmatrix} \quad \text{is} \quad A^{-1} = \begin{bmatrix} 1 & -1 \\ -2 & 3 \end{bmatrix}$$

Solution

We need to show that $AA^{-1} = A^{-1}A = I_2$.

$$AA^{-1} = \begin{bmatrix} 3 & 1 \\ 2 & 1 \end{bmatrix} \begin{bmatrix} 1 & -1 \\ -2 & 3 \end{bmatrix} = \begin{bmatrix} 1 & 0 \\ 0 & 1 \end{bmatrix} = I_2$$

$$A^{-1}A = \begin{bmatrix} 1 & -1 \\ -2 & 3 \end{bmatrix} \begin{bmatrix} 3 & 1 \\ 2 & 1 \end{bmatrix} = \begin{bmatrix} 1 & 0 \\ 0 & 1 \end{bmatrix} = I_2$$

We now show one way to find the inverse of

$$A = \begin{bmatrix} 3 & 1 \\ 2 & 1 \end{bmatrix}$$

Suppose that A^{-1} is given by

$$A^{-1} = \begin{bmatrix} x & y \\ z & w \end{bmatrix} \qquad \textbf{(1)}$$

where x, y, z, and w are four variables. Based on the definition of an inverse, if A has an inverse,

$$AA^{-1} = I_2$$

$$\begin{bmatrix} 3 & 1 \\ 2 & 1 \end{bmatrix} \begin{bmatrix} x & y \\ z & w \end{bmatrix} = \begin{bmatrix} 1 & 0 \\ 0 & 1 \end{bmatrix}$$

$$\begin{bmatrix} 3x + z & 3y + w \\ 2x + z & 2y + w \end{bmatrix} = \begin{bmatrix} 1 & 0 \\ 0 & 1 \end{bmatrix}$$

Because corresponding entries must be equal, it follows that this matrix equation is equivalent to two systems of linear equations.

$$\begin{cases} 3x + z = 1 \\ 2x + z = 0 \end{cases} \qquad \begin{cases} 3y + w = 0 \\ 2y + w = 1 \end{cases}$$

The augmented matrix of each system is

$$\begin{bmatrix} 3 & 1 & | & 1 \\ 2 & 1 & | & 0 \end{bmatrix} \qquad \begin{bmatrix} 3 & 1 & | & 0 \\ 2 & 1 & | & 1 \end{bmatrix} \qquad \textbf{(2)}$$

The usual procedure would be to transform each augmented matrix into reduced row echelon form. Notice, though, that the left sides of the augmented matrices are equal, so the same row operations (see Section 8.2) can be used to reduce each one. We find it more efficient to combine the two augmented matrices (2) into a single matrix, as shown next, and then transform it into reduced row echelon form.

$$\begin{bmatrix} 3 & 1 & | & 1 & 0 \\ 2 & 1 & | & 0 & 1 \end{bmatrix}$$

We attempt to transform the left side into an identity matrix.

$$\begin{bmatrix} 3 & 1 & | & 1 & 0 \\ 2 & 1 & | & 0 & 1 \end{bmatrix} \underset{\underset{R_1 = -1r_2 + r_1}{\uparrow}}{\rightarrow} \begin{bmatrix} 1 & 0 & | & 1 & -1 \\ 2 & 1 & | & 0 & 1 \end{bmatrix}$$

$$\underset{\underset{R_2 = -2r_1 + r_2}{\uparrow}}{\rightarrow} \begin{bmatrix} 1 & 0 & | & 1 & -1 \\ 0 & 1 & | & -2 & 3 \end{bmatrix} \qquad \textbf{(3)}$$

Matrix (3) is in reduced row echelon form.

Now reverse the earlier step of combining the two augmented matrices in (2) and write the single matrix (3) as two augmented matrices.

$$\begin{bmatrix} 1 & 0 & | & 1 \\ 0 & 1 & | & -2 \end{bmatrix} \quad \text{and} \quad \begin{bmatrix} 1 & 0 & | & -1 \\ 0 & 1 & | & 3 \end{bmatrix}$$

We conclude from these matrices that $x = 1$, $z = -2$, and $y = -1$, $w = 3$. Substituting these values into matrix (1), we find that

$$A^{-1} = \begin{bmatrix} 1 & -1 \\ -2 & 3 \end{bmatrix}$$

Notice in display (3) that the 2 by 2 matrix to the right of the vertical bar is, in fact, the inverse of A. Also notice that the identity matrix I_2 is the matrix that appears to the left of the vertical bar. These observations and the procedures followed to get display (3) will work in general.

Procedure for Finding the Inverse of a Nonsingular Matrix*

To find the inverse of an n by n nonsingular matrix A, proceed as follows:

STEP 1: Form the matrix $[A|I_n]$.
STEP 2: Transform the matrix $[A|I_n]$ into reduced row echelon form.
STEP 3: The reduced row echelon form of $[A|I_n]$ will contain the identity matrix I_n on the left of the vertical bar; the n by n matrix on the right of the vertical bar is the inverse of A.

In Words

If A is nonsingular, begin with the matrix $[A|I_n]$ and, after transforming it into reduced row echelon form, you end up with the matrix $[I_n|A^{-1}]$.

EXAMPLE 13 **Finding the Inverse of a Matrix**

The matrix

$$A = \begin{bmatrix} 1 & 1 & 0 \\ -1 & 3 & 4 \\ 0 & 4 & 3 \end{bmatrix}$$

is nonsingular. Find its inverse.

Solution First, form the matrix

$$[A|I_3] = \left[\begin{array}{ccc|ccc} 1 & 1 & 0 & 1 & 0 & 0 \\ -1 & 3 & 4 & 0 & 1 & 0 \\ 0 & 4 & 3 & 0 & 0 & 1 \end{array}\right]$$

Next, use row operations to transform $[A|I_3]$ into reduced row echelon form.

$$\left[\begin{array}{ccc|ccc} 1 & 1 & 0 & 1 & 0 & 0 \\ -1 & 3 & 4 & 0 & 1 & 0 \\ 0 & 4 & 3 & 0 & 0 & 1 \end{array}\right] \rightarrow \left[\begin{array}{ccc|ccc} 1 & 1 & 0 & 1 & 0 & 0 \\ 0 & 4 & 4 & 1 & 1 & 0 \\ 0 & 4 & 3 & 0 & 0 & 1 \end{array}\right] \rightarrow \left[\begin{array}{ccc|ccc} 1 & 1 & 0 & 1 & 0 & 0 \\ 0 & 1 & 1 & \frac{1}{4} & \frac{1}{4} & 0 \\ 0 & 4 & 3 & 0 & 0 & 1 \end{array}\right]$$

$$R_2 = r_1 + r_2 \qquad\qquad R_2 = \frac{1}{4}r_2$$

$$\rightarrow \left[\begin{array}{ccc|ccc} 1 & 0 & -1 & \frac{3}{4} & -\frac{1}{4} & 0 \\ 0 & 1 & 1 & \frac{1}{4} & \frac{1}{4} & 0 \\ 0 & 0 & -1 & -1 & -1 & 1 \end{array}\right] \rightarrow \left[\begin{array}{ccc|ccc} 1 & 0 & -1 & \frac{3}{4} & -\frac{1}{4} & 0 \\ 0 & 1 & 1 & \frac{1}{4} & \frac{1}{4} & 0 \\ 0 & 0 & 1 & 1 & 1 & -1 \end{array}\right]$$

$$\begin{array}{l} R_1 = -1r_2 + r_1 \\ R_3 = -4r_2 + r_3 \end{array} \qquad\qquad R_3 = -1r_3$$

$$\rightarrow \left[\begin{array}{ccc|ccc} 1 & 0 & 0 & \frac{7}{4} & \frac{3}{4} & -1 \\ 0 & 1 & 0 & -\frac{3}{4} & -\frac{3}{4} & 1 \\ 0 & 0 & 1 & 1 & 1 & -1 \end{array}\right]$$

$$\begin{array}{l} R_1 = r_3 + r_1 \\ R_2 = -1r_3 + r_2 \end{array}$$

* For 2×2 matrices there is a simple formula that can be used. See Problem 89.

The matrix $[A|I_3]$ is now in reduced row echelon form, and the identity matrix I_3 is on the left of the vertical bar. The inverse of A is

$$A^{-1} = \begin{bmatrix} \dfrac{7}{4} & \dfrac{3}{4} & -1 \\[2mm] -\dfrac{3}{4} & -\dfrac{3}{4} & 1 \\[2mm] 1 & 1 & -1 \end{bmatrix}$$

Figure 8

You can (and should) verify that this is the correct inverse by showing that

$$AA^{-1} = A^{-1}A = I_3$$

 ✓**Check:** Enter the matrix A into a graphing utility. Figure 8 shows A^{-1}.

Now Work PROBLEM 33

If transforming the matrix $[A|I_n]$ into reduced row echelon form does not result in the identity matrix I_n to the left of the vertical bar, A is singular and has no inverse.

EXAMPLE 14

Showing That a Matrix Has No Inverse

Show that the matrix $A = \begin{bmatrix} 4 & 6 \\ 2 & 3 \end{bmatrix}$ has no inverse.

Solution　Proceeding as in Example 13, form the matrix

$$[A|I_2] = \begin{bmatrix} 4 & 6 & | & 1 & 0 \\ 2 & 3 & | & 0 & 1 \end{bmatrix}$$

Then use row operations to transform $[A|I_2]$ into reduced row echelon form.

$$[A|I_2] = \begin{bmatrix} 4 & 6 & | & 1 & 0 \\ 2 & 3 & | & 0 & 1 \end{bmatrix} \underset{\underset{R_1 = \frac{1}{4}r_1}{\uparrow}}{\rightarrow} \begin{bmatrix} 1 & \dfrac{3}{2} & | & \dfrac{1}{4} & 0 \\[2mm] 2 & 3 & | & 0 & 1 \end{bmatrix} \underset{\underset{R_2 = -2r_1 + r_2}{\uparrow}}{\rightarrow} \begin{bmatrix} 1 & \dfrac{3}{2} & | & \dfrac{1}{4} & 0 \\[2mm] 0 & 0 & | & -\dfrac{1}{2} & 1 \end{bmatrix}$$

The matrix $[A|I_2]$ is sufficiently reduced for us to see that the identity matrix cannot appear to the left of the vertical bar. We conclude that A is singular and so has no inverse.

It can be shown that if the determinant of a matrix is 0 then the matrix is singular. The determinant of matrix A from Example 14 is

$$\begin{vmatrix} 4 & 6 \\ 2 & 3 \end{vmatrix} = 4 \cdot 3 - 6 \cdot 2 = 0$$

 ✓**Check:** Enter the matrix A. Try to find its inverse. What happens?

Now Work PROBLEM 61

5 Solve a System of Linear Equations Using an Inverse Matrix

Inverse matrices can be used to solve systems of equations in which the number of equations is the same as the number of variables.

EXAMPLE 15

Using the Inverse Matrix to Solve a System of Linear Equations

Solve the system of equations: $\begin{cases} x + y & = 3 \\ -x + 3y + 4z = -3 \\ 4y + 3z = 2 \end{cases}$

Solution If we let

$$A = \begin{bmatrix} 1 & 1 & 0 \\ -1 & 3 & 4 \\ 0 & 4 & 3 \end{bmatrix} \qquad X = \begin{bmatrix} x \\ y \\ z \end{bmatrix} \qquad B = \begin{bmatrix} 3 \\ -3 \\ 2 \end{bmatrix}$$

the original system of equations can be written compactly as the matrix equation

$$AX = B \tag{4}$$

From Example 13, the matrix A has the inverse A^{-1}. Multiply each side of equation (4) by A^{-1}.

$$
\begin{aligned}
AX &= B \\
A^{-1}(AX) &= A^{-1}B & &\text{Multiply both sides by } A^{-1}. \\
(A^{-1}A)X &= A^{-1}B & &\text{Associative Property of multiplication} \\
I_3X &= A^{-1}B & &\text{Definition of an inverse matrix} \\
X &= A^{-1}B & &\text{Property of the identity matrix} \tag{5}
\end{aligned}
$$

Now use (5) to find $X = \begin{bmatrix} x \\ y \\ z \end{bmatrix}$.

$$X = \begin{bmatrix} x \\ y \\ z \end{bmatrix} = A^{-1}B = \underset{\underset{\text{Example 13}}{\uparrow}}{\begin{bmatrix} \frac{7}{4} & \frac{3}{4} & -1 \\ -\frac{3}{4} & -\frac{3}{4} & 1 \\ 1 & 1 & -1 \end{bmatrix}} \begin{bmatrix} 3 \\ -3 \\ 2 \end{bmatrix} = \begin{bmatrix} 1 \\ 2 \\ -2 \end{bmatrix}$$

The solution is $x = 1$, $y = 2$, $z = -2$ or, using an ordered triplet, $(1, 2, -2)$.

The method used in Example 15 to solve a system of equations is particularly useful when it is necessary to solve several systems of equations in which the constants appearing to the right of the equal signs change, while the coefficients of the variables on the left side remain the same. See Problems 41–60 for some illustrations. Be careful; this method can only be used if the inverse exists. If it does not exist, row reduction must be used since the system is either inconsistent or dependent.

Now Work PROBLEM 45

Historical Feature

Arthur Cayley
(1821–1895)

Matrices were invented in 1857 by Arthur Cayley (1821–1895) as a way of efficiently computing the result of substituting one linear system into another (see Historical Problem 3). The resulting system had incredible richness, in the sense that a wide variety of mathematical systems could be mimicked by the matrices. Cayley and his friend James J. Sylvester (1814–1897) spent much of the rest of their lives elaborating the theory. The torch was then passed to Georg Frobenius (1849–1917), whose deep investigations established a central place for matrices in modern mathematics. In 1924, rather to the surprise of physicists, it was found that matrices (with complex numbers in them) were exactly the right tool for describing the behavior of atomic systems. Today, matrices are used in a wide variety of applications.

Historical Problems

1. **Matrices and Complex Numbers** Frobenius emphasized in his research how matrices could be used to mimic other mathematical systems. Here, we mimic the behavior of complex numbers using matrices. Mathematicians call such a relationship an *isomorphism*.

$$\text{Complex number} \longleftrightarrow \text{Matrix}$$

$$a + bi \longleftrightarrow \begin{bmatrix} a & b \\ -b & a \end{bmatrix}$$

Note that the complex number can be read off the top line of the matrix. Thus,

$$2 + 3i \longleftrightarrow \begin{bmatrix} 2 & 3 \\ -3 & 2 \end{bmatrix} \quad \text{and} \quad \begin{bmatrix} 4 & -2 \\ 2 & 4 \end{bmatrix} \longleftrightarrow 4 - 2i$$

(a) Find the matrices corresponding to $2 - 5i$ and $1 + 3i$.

(b) Multiply the two matrices.

(c) Find the corresponding complex number for the matrix found in part (b).

(d) Multiply $2 - 5i$ and $1 + 3i$. The result should be the same as that found in part (c).

The process also works for addition and subtraction. Try it for yourself.

2. Compute $(a + bi)(a - bi)$ using matrices. Interpret the result.

3. **Cayley's Definition of Matrix Multiplication** Cayley invented matrix multiplication to simplify the following problem:

$$\begin{cases} u = ar + bs \\ v = cr + ds \end{cases} \quad \begin{cases} x = ku + lv \\ y = mu + nv \end{cases}$$

(a) Find x and y in terms of r and s by substituting u and v from the first system of equations into the second system of equations.

(b) Use the result of part (a) to find the 2 by 2 matrix A in

$$\begin{bmatrix} x \\ y \end{bmatrix} = A \begin{bmatrix} r \\ s \end{bmatrix}$$

(c) Now look at the following way to do it. Write the equations in matrix form.

$$\begin{bmatrix} u \\ v \end{bmatrix} = \begin{bmatrix} a & b \\ c & d \end{bmatrix} \begin{bmatrix} r \\ s \end{bmatrix} \quad \begin{bmatrix} x \\ y \end{bmatrix} = \begin{bmatrix} k & l \\ m & n \end{bmatrix} \begin{bmatrix} u \\ v \end{bmatrix}$$

So

$$\begin{bmatrix} x \\ y \end{bmatrix} = \begin{bmatrix} k & l \\ m & n \end{bmatrix} \begin{bmatrix} a & b \\ c & d \end{bmatrix} \begin{bmatrix} r \\ s \end{bmatrix}$$

Do you see how Cayley defined matrix multiplication?

8.4 Assess Your Understanding

Concepts and Vocabulary

1. A matrix that has the same number of rows as columns is called a(n) _____ matrix.

2. *True or False* Matrix addition is commutative.

3. To find the product AB of two matrices A and B, the number of _____ in matrix A must equal the number of _____ in matrix B.

4. *True or False* Matrix multiplication is commutative.

5. Suppose that A is a square n by n matrix that is nonsingular. The matrix B such that $AB = BA = I_n$ is called the _____ of the matrix A.

6. If a matrix A has no inverse, it is called _____.

7. *True or False* The identity matrix has properties similar to those of the real number 1.

8. If $AX = B$ represents a matrix equation where A is a nonsingular matrix, then we can solve the equation using $X = $ _____.

Skill Building

In Problems 9–24, use the following matrices to evaluate the given expression.

$$A = \begin{bmatrix} 0 & 3 & -5 \\ 1 & 2 & 6 \end{bmatrix} \quad B = \begin{bmatrix} 4 & 1 & 0 \\ -2 & 3 & -2 \end{bmatrix} \quad C = \begin{bmatrix} 4 & 1 \\ 6 & 2 \\ -2 & 3 \end{bmatrix}$$

9. $A + B$

10. $A - B$

11. $4A$

12. $-3B$

13. $3A - 2B$

14. $2A + 4B$

15. AC

16. BC

17. CA **18.** CB **19.** $C(A + B)$ **20.** $(A + B)C$

21. $AC - 3I_2$ **22.** $CA + 5I_3$ **23.** $CA - CB$ **24.** $AC + BC$

In Problems 25–30, find the product.

25. $\begin{bmatrix} 2 & -2 \\ 1 & 0 \end{bmatrix} \begin{bmatrix} 2 & 1 & 4 & 6 \\ 3 & -1 & 3 & 2 \end{bmatrix}$

26. $\begin{bmatrix} 4 & 1 \\ 2 & 1 \end{bmatrix} \begin{bmatrix} -6 & 6 & 1 & 0 \\ 2 & 5 & 4 & -1 \end{bmatrix}$

27. $\begin{bmatrix} 1 & 2 & 3 \\ 0 & -1 & 4 \end{bmatrix} \begin{bmatrix} 1 & 2 \\ -1 & 0 \\ 2 & 4 \end{bmatrix}$

28. $\begin{bmatrix} 1 & -1 \\ -3 & 2 \\ 0 & 5 \end{bmatrix} \begin{bmatrix} 2 & 8 & -1 \\ 3 & 6 & 0 \end{bmatrix}$

29. $\begin{bmatrix} 1 & 0 & 1 \\ 2 & 4 & 1 \\ 3 & 6 & 1 \end{bmatrix} \begin{bmatrix} 1 & 3 \\ 6 & 2 \\ 8 & -1 \end{bmatrix}$

30. $\begin{bmatrix} 4 & -2 & 3 \\ 0 & 1 & 2 \\ -1 & 0 & 1 \end{bmatrix} \begin{bmatrix} 2 & 6 \\ 1 & -1 \\ 0 & 2 \end{bmatrix}$

In Problems 31–40, each matrix is nonsingular. Find the inverse of each matrix.

31. $\begin{bmatrix} 2 & 1 \\ 1 & 1 \end{bmatrix}$

32. $\begin{bmatrix} 3 & -1 \\ -2 & 1 \end{bmatrix}$

33. $\begin{bmatrix} 6 & 5 \\ 2 & 2 \end{bmatrix}$

34. $\begin{bmatrix} -4 & 1 \\ 6 & -2 \end{bmatrix}$

35. $\begin{bmatrix} 2 & 1 \\ a & a \end{bmatrix}$ $a \neq 0$

36. $\begin{bmatrix} b & 3 \\ b & 2 \end{bmatrix}$ $b \neq 0$

37. $\begin{bmatrix} 1 & -1 & 1 \\ 0 & -2 & 1 \\ -2 & -3 & 0 \end{bmatrix}$

38. $\begin{bmatrix} 1 & 0 & 2 \\ -1 & 2 & 3 \\ 1 & -1 & 0 \end{bmatrix}$

39. $\begin{bmatrix} 1 & 1 & 1 \\ 3 & 2 & -1 \\ 3 & 1 & 2 \end{bmatrix}$

40. $\begin{bmatrix} 3 & 3 & 1 \\ 1 & 2 & 1 \\ 2 & -1 & 1 \end{bmatrix}$

In Problems 41–60, use the inverses found in Problems 31–40 to solve each system of equations.

41. $\begin{cases} 2x + y = 8 \\ x + y = 5 \end{cases}$

42. $\begin{cases} 3x - y = 8 \\ -2x + y = 4 \end{cases}$

43. $\begin{cases} 2x + y = 0 \\ x + y = 5 \end{cases}$

44. $\begin{cases} 3x - y = 4 \\ -2x + y = 5 \end{cases}$

45. $\begin{cases} 6x + 5y = 7 \\ 2x + 2y = 2 \end{cases}$

46. $\begin{cases} -4x + y = 0 \\ 6x - 2y = 14 \end{cases}$

47. $\begin{cases} 6x + 5y = 13 \\ 2x + 2y = 5 \end{cases}$

48. $\begin{cases} -4x + y = 5 \\ 6x - 2y = -9 \end{cases}$

49. $\begin{cases} 2x + y = -3 \\ ax + ay = -a \end{cases}$ $a \neq 0$

50. $\begin{cases} bx + 3y = 2b + 3 \\ bx + 2y = 2b + 2 \end{cases}$ $b \neq 0$

51. $\begin{cases} 2x + y = \dfrac{7}{a} \\ ax + ay = 5 \end{cases}$ $a \neq 0$

52. $\begin{cases} bx + 3y = 14 \\ bx + 2y = 10 \end{cases}$ $b \neq 0$

53. $\begin{cases} x - y + z = 0 \\ -2y + z = -1 \\ -2x - 3y = -5 \end{cases}$

54. $\begin{cases} x + 2z = 6 \\ -x + 2y + 3z = -5 \\ x - y = 6 \end{cases}$

55. $\begin{cases} x - y + z = 2 \\ -2y + z = 2 \\ -2x - 3y = \dfrac{1}{2} \end{cases}$

56. $\begin{cases} x + 2z = 2 \\ -x + 2y + 3z = -\dfrac{3}{2} \\ x - y = 2 \end{cases}$

57. $\begin{cases} x + y + z = 9 \\ 3x + 2y - z = 8 \\ 3x + y + 2z = 1 \end{cases}$

58. $\begin{cases} 3x + 3y + z = 8 \\ x + 2y + z = 5 \\ 2x - y + z = 4 \end{cases}$

59. $\begin{cases} x + y + z = 2 \\ 3x + 2y - z = \dfrac{7}{3} \\ 3x + y + 2z = \dfrac{10}{3} \end{cases}$

60. $\begin{cases} 3x + 3y + z = 1 \\ x + 2y + z = 0 \\ 2x - y + z = 4 \end{cases}$

In Problems 61–66, show that each matrix has no inverse.

61. $\begin{bmatrix} 4 & 2 \\ 2 & 1 \end{bmatrix}$

62. $\begin{bmatrix} -3 & \dfrac{1}{2} \\ 6 & -1 \end{bmatrix}$

63. $\begin{bmatrix} 15 & 3 \\ 10 & 2 \end{bmatrix}$

64. $\begin{bmatrix} -3 & 0 \\ 4 & 0 \end{bmatrix}$

65. $\begin{bmatrix} -3 & 1 & -1 \\ 1 & -4 & -7 \\ 1 & 2 & 5 \end{bmatrix}$

66. $\begin{bmatrix} 1 & 1 & -3 \\ 2 & -4 & 1 \\ -5 & 7 & 1 \end{bmatrix}$

In Problems 67–70, use a graphing utility to find the inverse, if it exists, of each matrix. Round answers to two decimal places.

67. $\begin{bmatrix} 25 & 61 & -12 \\ 18 & -2 & 4 \\ 8 & 35 & 21 \end{bmatrix}$

68. $\begin{bmatrix} 18 & -3 & 4 \\ 6 & -20 & 14 \\ 10 & 25 & -15 \end{bmatrix}$

69. $\begin{bmatrix} 44 & 21 & 18 & 6 \\ -2 & 10 & 15 & 5 \\ 21 & 12 & -12 & 4 \\ -8 & -16 & 4 & 9 \end{bmatrix}$

70. $\begin{bmatrix} 16 & 22 & -3 & 5 \\ 21 & -17 & 4 & 8 \\ 2 & 8 & 27 & 20 \\ 5 & 15 & -3 & -10 \end{bmatrix}$

In Problems 71–74, use the idea behind Example 15 with a graphing utility to solve the following systems of equations. Round answers to two decimal places.

71. $\begin{cases} 25x + 61y - 12z = 10 \\ 18x - 12y + 7y = -9 \\ 3x + 4y - z = 12 \end{cases}$

72. $\begin{cases} 25x + 61y - 12z = 15 \\ 18x - 12y + 7z = -3 \\ 3x + 4y - z = 12 \end{cases}$

73. $\begin{cases} 25x + 61y - 12z = 21 \\ 18x - 12y + 7z = 7 \\ 3x + 4y - z = -2 \end{cases}$

74. $\begin{cases} 25x + 61y - 12z = 25 \\ 18x - 12y + 7z = 10 \\ 3x + 4y - z = -4 \end{cases}$

Mixed Practice

In Problems 75–82, algebraically solve each system of equations using any method you wish.

75. $\begin{cases} 2x + 3y = 11 \\ 5x + 7y = 24 \end{cases}$

76. $\begin{cases} 2x + 8y = -8 \\ x + 7y = -13 \end{cases}$

77. $\begin{cases} x - 2y + 4z = 2 \\ -3x + 5y - 2z = 17 \\ 4x - 3y = -22 \end{cases}$

78. $\begin{cases} 2x + 3y - z = -2 \\ 4x + 3z = 6 \\ 6y - 2z = 2 \end{cases}$

79. $\begin{cases} 5x - y + 4z = 2 \\ -x + 5y - 4z = 3 \\ 7x + 13y - 4z = 17 \end{cases}$

80. $\begin{cases} 3x + 2y - z = 2 \\ 2x + y + 6z = -7 \\ 2x + 2y - 14z = 17 \end{cases}$

81. $\begin{cases} 2x - 3y + z = 4 \\ -3x + 2y - z = -3 \\ -5y + z = 6 \end{cases}$

82. $\begin{cases} -4x + 3y + 2z = 6 \\ 3x + y - z = -2 \\ x + 9y + z = 6 \end{cases}$

Applications and Extensions

83. College Tuition Nikki and Joe take classes at a community college, LCCC, and a local university, SIUE. The number of credit hours taken and the cost per credit hour (2009–2010 academic year, tuition only) are as follows:

	LCCC	SIUE
Nikki	6	9
Joe	3	12

Cost per Credit Hour	
LCCC	$80.00
SIUE	$277.80

(a) Write a matrix A for the credit hours taken by each student and a matrix B for the cost per credit hour.

(b) Compute AB and interpret the results.

Sources: www.lc.edu, www.siue.edu

84. School Loan Interest Jamal and Stephanie each have school loans issued from the same two banks. The amounts borrowed and the monthly interest rates are given next (interest is compounded monthly):

	Lender 1	Lender 2
Jamal	$4000	$3000
Stephanie	$2500	$3800

Monthly Interest Rate	
Lender 1	0.011 (1.1%)
Lender 2	0.006 (0.6%)

(a) Write a matrix A for the amounts borrowed by each student and a matrix B for the monthly interest rates.

(b) Compute AB and interpret the results.

(c) Let $C = \begin{bmatrix} 1 \\ 1 \end{bmatrix}$. Compute $A(C + B)$ and interpret the results.

85. Computing the Cost of Production The Acme Steel Company is a producer of stainless steel and aluminum containers. On a certain day, the following stainless steel containers were manufactured: 500 with 10-gallon (gal) capacity, 350 with 5-gal capacity, and 400 with 1-gal capacity. On the same day, the following aluminum containers were manufactured: 700 with 10-gal capacity, 500 with 5-gal capacity, and 850 with 1-gal capacity.

(a) Find a 2 by 3 matrix representing these data. Find a 3 by 2 matrix to represent the same data.

(b) If the amount of material used in the 10-gal containers is 15 pounds (lb), the amount used in the 5-gal containers is 8 lb, and the amount used in the 1-gal containers is 3 lb, find a 3 by 1 matrix representing the amount of material used.

(c) Multiply the 2 by 3 matrix found in part (a) and the 3 by 1 matrix found in part (b) to get a 2 by 1 matrix showing the day's usage of material.

(d) If stainless steel costs Acme $0.10 lb and aluminum costs $0.05 lb, find a 1 by 2 matrix representing cost.

(e) Multiply the matrices found in parts (c) and (d) to determine the total cost of the day's production.

86. Computing Profit Rizza Ford has two locations, one in the city and the other in the suburbs. In January, the city location sold 400 subcompacts, 250 intermediate-size cars, and 50 SUVs; in February, it sold 350 subcompacts, 100 intermediates, and 30 SUVs. At the suburban location in January, 450 subcompacts, 200 intermediates, and 140 SUVs were sold. In February, the suburban location sold 350 subcompacts, 300 intermediates, and 100 SUVs.

(a) Find 2 by 3 matrices that summarize the sales data for each location for January and February (one matrix for each month).

(b) Use matrix addition to obtain total sales for the 2-month period.

(c) The profit on each kind of car is $100 per subcompact, $150 per intermediate, and $200 per SUV. Find a 3 by 1 matrix representing this profit.

(d) Multiply the matrices found in parts (b) and (c) to get a 2 by 1 matrix showing the profit at each location.

87. Cryptography One method of encryption is to use a matrix to encrypt the message and then use the corresponding inverse matrix to decode the message. The encrypted matrix, E, is obtained by multiplying the message matrix, M, by a key matrix, K. The original message can be retrieved by multiplying the encrypted matrix by the inverse of the key matrix. That is, $E = M \cdot K$ and $M = E \cdot K^{-1}$.

(a) Given the key matrix $K = \begin{bmatrix} 2 & 1 & 1 \\ 1 & 1 & 0 \\ 1 & 1 & 1 \end{bmatrix}$, find its inverse, K^{-1}. [**Note:** This key matrix is known as the Q_2^3 Fibonacci encryption matrix.]

(b) Use your result from part (a) to decode the encrypted

$$\text{matrix } E = \begin{bmatrix} 47 & 34 & 33 \\ 44 & 36 & 27 \\ 47 & 41 & 20 \end{bmatrix}.$$

(c) Each entry in your result for part (b) represents the position of a letter in the English alphabet ($A = 1, B = 2, C = 3,$ and so on). What is the original message?

Source: goldenmuseum.com

88. Economic Mobility The relative income of a child (low, medium, or high) generally depends on the relative income of the child's parents. The matrix P, given by

$$\begin{array}{c} \qquad\qquad \text{Parent's Income} \\ \qquad\quad \text{L} \quad\ \text{M} \quad\ \text{H} \\ P = \begin{bmatrix} 0.4 & 0.2 & 0.1 \\ 0.5 & 0.6 & 0.5 \\ 0.1 & 0.2 & 0.4 \end{bmatrix} \begin{matrix} \text{L} \\ \text{M} \\ \text{H} \end{matrix} \ \text{Child's income} \end{array}$$

is called a *left stochastic transition matrix*. For example, the entry $P_{21} = 0.5$ means that 50% of the children of low relative income parents will transition to the medium level of income. The diagonal entry P_{ii} represents the percent of children who remain in the same income level as their parents. Assuming that the transition matrix is valid from one generation to the next, compute and interpret P^2.

Source: Understanding Mobility in America, April 2006

89. Consider the 2 by 2 square matrix

$$A = \begin{bmatrix} a & b \\ c & d \end{bmatrix}$$

If $D = ad - bc \neq 0$, show that A is nonsingular and that

$$A^{-1} = \frac{1}{D}\begin{bmatrix} d & -b \\ -c & a \end{bmatrix}$$

Explaining Concepts: Discussion and Writing

90. Create a situation different from any found in the text that can be represented by a matrix.

91. Explain why the number of columns in matrix A must equal the number of rows in matrix B when finding the product AB.

92. If $a, b,$ and $c \neq 0$ are real numbers with $ac = bc$, then $a = b$. Does this same property hold for matrices? In other words, if $A, B,$ and C, are matrices and $AC = BC$, must $A = B$?

93. What is the solution of the system of equations $AX = \varnothing$, if A^{-1} exists? Discuss the solution of $AX = \varnothing$ if A^{-1} does not exist.

8.5 Partial Fraction Decomposition

PREPARING FOR THIS SECTION *Before getting started, review the following:*

- Identity (Section 1.1, p. 82)
- Proper and Improper Rational Functions (Section 5.2, p. 347)

- Factoring Polynomials (Chapter R, Review, Section R.5, pp. 49–55)
- Complex Zeros; Fundamental Theorem of Algebra (Section 5.6, p. 388)

Now Work the 'Are You Prepared?' problems on page 604.

OBJECTIVES **1** Decompose $\dfrac{P}{Q}$, Where Q Has Only Nonrepeated Linear Factors (p. 598)

2 Decompose $\dfrac{P}{Q}$, Where Q Has Repeated Linear Factors (p. 600)

3 Decompose $\dfrac{P}{Q}$, Where Q Has a Nonrepeated Irreducible Quadratic Factor (p. 602)

4 Decompose $\dfrac{P}{Q}$, Where Q Has a Repeated Irreducible Quadratic Factor (p. 603)

Consider the problem of adding two rational expressions:

$$\frac{3}{x + 4} \quad \text{and} \quad \frac{2}{x - 3}$$

The result is

$$\frac{3}{x + 4} + \frac{2}{x - 3} = \frac{3(x - 3) + 2(x + 4)}{(x + 4)(x - 3)} = \frac{5x - 1}{x^2 + x - 12}$$

The reverse procedure, starting with the rational expression $\dfrac{5x - 1}{x^2 + x - 12}$ and writing it as the sum (or difference) of the two simpler fractions $\dfrac{3}{x + 4}$ and $\dfrac{2}{x - 3}$, is referred to as **partial fraction decomposition,** and the two simpler fractions are called **partial fractions.** Decomposing a rational expression into a sum of partial fractions is important in solving certain types of calculus problems. This section presents a systematic way to decompose rational expressions.

We begin by recalling that a rational expression is the ratio of two polynomials, say, P and $Q \neq 0$. We assume that P and Q have no common factors. Recall also that a rational expression $\dfrac{P}{Q}$ is called **proper** if the degree of the polynomial in the numerator is less than the degree of the polynomial in the denominator. Otherwise, the rational expression is termed **improper.**

Because any improper rational expression can be reduced by long division to a mixed form consisting of the sum of a polynomial and a proper rational expression, we shall restrict the discussion that follows to proper rational expressions.

The partial fraction decomposition of the rational expression $\dfrac{P}{Q}$ depends on the factors of the denominator Q. Recall from Section 5.6 that any polynomial whose coefficients are real numbers can be factored (over the real numbers) into products of linear and/or irreducible quadratic factors. This means that the denominator Q of the rational expression $\dfrac{P}{Q}$ will contain only factors of one or both of the following types:

1. *Linear factors* of the form $x - a$, where a is a real number.
2. *Irreducible quadratic factors* of the form $ax^2 + bx + c$, where a, b, and c are real numbers, $a \neq 0$, and $b^2 - 4ac < 0$ (which guarantees that $ax^2 + bx + c$ cannot be written as the product of two linear factors with real coefficients).

As it turns out, there are four cases to be examined. We begin with the case for which Q has only nonrepeated linear factors.

1 Decompose $\dfrac{P}{Q}$, Where Q Has Only Nonrepeated Linear Factors

Case 1: Q has only nonrepeated linear factors.

Under the assumption that Q has only nonrepeated linear factors, the polynomial Q has the form

$$Q(x) = (x - a_1)(x - a_2) \cdots \cdots (x - a_n)$$

where no two of the numbers a_1, a_2, \ldots, a_n are equal. In this case, the partial fraction decomposition of $\dfrac{P}{Q}$ is of the form

$$\frac{P(x)}{Q(x)} = \frac{A_1}{x - a_1} + \frac{A_2}{x - a_2} + \cdots + \frac{A_n}{x - a_n} \tag{1}$$

where the numbers A_1, A_2, \ldots, A_n are to be determined.

We show how to find these numbers in the example that follows.

EXAMPLE 1 **Nonrepeated Linear Factors**

Write the partial fraction decomposition of $\dfrac{x}{x^2 - 5x + 6}$.

Solution First, factor the denominator,

$$x^2 - 5x + 6 = (x - 2)(x - 3)$$

and conclude that the denominator contains only nonrepeated linear factors. Then decompose the rational expression according to equation (1):

$$\frac{x}{x^2 - 5x + 6} = \frac{A}{x - 2} + \frac{B}{x - 3} \qquad (2)$$

where A and B are to be determined. To find A and B, clear the fractions by multiplying each side by $(x - 2)(x - 3) = x^2 - 5x + 6$. The result is

$$x = A(x - 3) + B(x - 2) \qquad (3)$$

or

$$x = (A + B)x + (-3A - 2B)$$

This equation is an identity in x. Equate the coefficients of like powers of x to get

$$\begin{cases} 1 = A + B & \text{Equate coefficients of } x: 1x = (A + B)x. \\ 0 = -3A - 2B & \text{Equate the constants: } 0 = -3A - 2B. \end{cases}$$

This system of two equations containing two variables, A and B, can be solved using whatever method you wish. Solving it, we get

$$A = -2 \qquad B = 3$$

From equation (2), the partial fraction decomposition is

$$\frac{x}{x^2 - 5x + 6} = \frac{-2}{x - 2} + \frac{3}{x - 3}$$

✓**Check:** The decomposition can be checked by adding the rational expressions.

$$\frac{-2}{x - 2} + \frac{3}{x - 3} = \frac{-2(x - 3) + 3(x - 2)}{(x - 2)(x - 3)} = \frac{x}{(x - 2)(x - 3)}$$

$$= \frac{x}{x^2 - 5x + 6}$$

♩

The numbers to be found in the partial fraction decomposition can sometimes be found more readily by using suitable choices for x (which may include complex numbers) in the identity obtained after fractions have been cleared. In Example 1, the identity after clearing fractions is equation (3):

$$x = A(x - 3) + B(x - 2)$$

If we let $x = 2$ in this expression, the term containing B drops out, leaving $2 = A(-1)$, or $A = -2$. Similarly, if we let $x = 3$, the term containing A drops out, leaving $3 = B$. As before, $A = -2$ and $B = 3$.

Now Work PROBLEM 13

2 Decompose $\dfrac{P}{Q}$, Where Q Has Repeated Linear Factors

Case 2: Q has repeated linear factors.

If the polynomial Q has a repeated linear factor, say $(x - a)^n$, $n \geq 2$ an integer, then, in the partial fraction decomposition of $\dfrac{P}{Q}$, we allow for the terms

$$\frac{A_1}{x - a} + \frac{A_2}{(x - a)^2} + \cdots + \frac{A_n}{(x - a)^n}$$

where the numbers A_1, A_2, \ldots, A_n are to be determined.

EXAMPLE 2	**Repeated Linear Factors**

Write the partial fraction decomposition of $\dfrac{x + 2}{x^3 - 2x^2 + x}$.

Solution First, factor the denominator,

$$x^3 - 2x^2 + x = x(x^2 - 2x + 1) = x(x - 1)^2$$

and find that the denominator has the nonrepeated linear factor x and the repeated linear factor $(x - 1)^2$. By Case 1, we must allow for the term $\dfrac{A}{x}$ in the decomposition; and by Case 2 we must allow for the terms $\dfrac{B}{x - 1} + \dfrac{C}{(x - 1)^2}$ in the decomposition. We write

$$\frac{x + 2}{x^3 - 2x^2 + x} = \frac{A}{x} + \frac{B}{x - 1} + \frac{C}{(x - 1)^2} \tag{4}$$

Again, clear fractions by multiplying each side by $x^3 - 2x^2 + x = x(x - 1)^2$. The result is the identity

$$x + 2 = A(x - 1)^2 + Bx(x - 1) + Cx \tag{5}$$

If we let $x = 0$ in this expression, the terms containing B and C drop out, leaving $2 = A(-1)^2$, or $A = 2$. Similarly, if we let $x = 1$, the terms containing A and B drop out, leaving $3 = C$. Then equation (5) becomes

$$x + 2 = 2(x - 1)^2 + Bx(x - 1) + 3x$$

Now let $x = 2$ (any choice other than 0 or 1 will work as well). The result is

$$4 = 2(1)^2 + B(2)(1) + 3(2)$$
$$4 = 2 + 2B + 6$$
$$2B = -4$$
$$B = -2$$

We have $A = 2$, $B = -2$, and $C = 3$.

From equation (4), the partial fraction decomposition is

$$\frac{x + 2}{x^3 - 2x^2 + x} = \frac{2}{x} + \frac{-2}{x - 1} + \frac{3}{(x - 1)^2}$$

EXAMPLE 3 **Repeated Linear Factors**

Write the partial fraction decomposition of $\dfrac{x^3 - 8}{x^2(x - 1)^3}$.

Solution The denominator contains the repeated linear factors x^2 and $(x - 1)^3$. The partial fraction decomposition takes the form

$$\frac{x^3 - 8}{x^2(x - 1)^3} = \frac{A}{x} + \frac{B}{x^2} + \frac{C}{x - 1} + \frac{D}{(x - 1)^2} + \frac{E}{(x - 1)^3} \qquad \text{(6)}$$

As before, clear fractions and obtain the identity

$$x^3 - 8 = Ax(x - 1)^3 + B(x - 1)^3 + Cx^2(x - 1)^2 + Dx^2(x - 1) + Ex^2 \quad \text{(7)}$$

Let $x = 0$. (Do you see why this choice was made?) Then

$$-8 = B(-1)$$
$$B = 8$$

Let $x = 1$ in equation (7). Then

$$-7 = E$$

Use $B = 8$ and $E = -7$ in equation (7) and collect like terms.

$$x^3 - 8 = Ax(x - 1)^3 + 8(x - 1)^3$$
$$+ Cx^2(x - 1)^2 + Dx^2(x - 1) - 7x^2$$
$$x^3 - 8 - 8(x^3 - 3x^2 + 3x - 1) + 7x^2 = Ax(x - 1)^3$$
$$+ Cx^2(x - 1)^2 + Dx^2(x - 1)$$
$$-7x^3 + 31x^2 - 24x = x(x - 1)[A(x - 1)^2 + Cx(x - 1) + Dx]$$
$$x(x - 1)(-7x + 24) = x(x - 1)[A(x - 1)^2 + Cx(x - 1) + Dx]$$
$$-7x + 24 = A(x - 1)^2 + Cx(x - 1) + Dx \qquad \text{(8)}$$

Now work with equation (8). Let $x = 0$. Then

$$24 = A$$

Let $x = 1$ in equation (8). Then

$$17 = D$$

Use $A = 24$ and $D = 17$ in equation (8).

$$-7x + 24 = 24(x - 1)^2 + Cx(x - 1) + 17x$$

Let $x = 2$ and simplify.

$$-14 + 24 = 24 + C(2) + 34$$
$$-48 = 2C$$
$$-24 = C$$

We now know all the numbers $A, B, C, D,$ and E, so, from equation (6), we have the decomposition

$$\frac{x^3 - 8}{x^2(x - 1)^3} = \frac{24}{x} + \frac{8}{x^2} + \frac{-24}{x - 1} + \frac{17}{(x - 1)^2} + \frac{-7}{(x - 1)^3}$$

♩

══════ **Now Work Example 3** by solving the system of five equations containing **five variables that the expansion of equation (7) leads to.**

══════ **Now Work** PROBLEM 19

The final two cases involve irreducible quadratic factors. A quadratic factor is irreducible if it cannot be factored into linear factors with real coefficients. A quadratic expression $ax^2 + bx + c$ is irreducible whenever $b^2 - 4ac < 0$. For example, $x^2 + x + 1$ and $x^2 + 4$ are irreducible.

3 Decompose $\dfrac{P}{Q}$, Where Q Has a Nonrepeated Irreducible Quadratic Factor

Case 3: Q contains a nonrepeated irreducible quadratic factor.

If Q contains a nonrepeated irreducible quadratic factor of the form $ax^2 + bx + c$, then, in the partial fraction decomposition of $\dfrac{P}{Q}$, allow for the term

$$\frac{Ax + B}{ax^2 + bx + c}$$

where the numbers A and B are to be determined.

EXAMPLE 4 **Nonrepeated Irreducible Quadratic Factor**

Write the partial fraction decomposition of $\dfrac{3x - 5}{x^3 - 1}$.

Solution Factor the denominator,

$$x^3 - 1 = (x - 1)(x^2 + x + 1)$$

and find that it has a nonrepeated linear factor $x - 1$ and a nonrepeated irreducible quadratic factor $x^2 + x + 1$. Allow for the term $\dfrac{A}{x - 1}$ by Case 1, and allow for the term $\dfrac{Bx + C}{x^2 + x + 1}$ by Case 3. We write

$$\frac{3x - 5}{x^3 - 1} = \frac{A}{x - 1} + \frac{Bx + C}{x^2 + x + 1} \qquad \textbf{(9)}$$

Clear fractions by multiplying each side of equation (9) by $x^3 - 1 = (x - 1)(x^2 + x + 1)$ to obtain the identity

$$3x - 5 = A(x^2 + x + 1) + (Bx + C)(x - 1) \qquad \textbf{(10)}$$

Expand the identity in (10) to obtain

$$3x - 5 = (A + B)x^2 + (A - B + C)x + (A - C)$$

This identity leads to the system of equations

$$\begin{cases} A + B && = & 0 & \text{(1)} \\ A - B + C & = & 3 & \text{(2)} \\ A && - C & = & -5 & \text{(3)} \end{cases}$$

The solution of this system is $A = -\dfrac{2}{3}, B = \dfrac{2}{3}, C = \dfrac{13}{3}$. Then, from equation (9), we see that

$$\frac{3x - 5}{x^3 - 1} = \frac{-\dfrac{2}{3}}{x - 1} + \frac{\dfrac{2}{3}x + \dfrac{13}{3}}{x^2 + x + 1}$$

 Now Work Example 4 using equation (10) and assigning values to x.

Now Work PROBLEM 21

4 Decompose $\dfrac{P}{Q}$, Where Q Has a Repeated Irreducible Quadratic Factor

Case 4: Q contains a repeated irreducible quadratic factor.

If the polynomial Q contains a repeated irreducible quadratic factor $(ax^2 + bx + c)^n, n \geq 2, n$ an integer, then, in the partial fraction decomposition of $\dfrac{P}{Q}$, allow for the terms

$$\frac{A_1 x + B_1}{ax^2 + bx + c} + \frac{A_2 x + B_2}{(ax^2 + bx + c)^2} + \cdots + \frac{A_n x + B_n}{(ax^2 + bx + c)^n}$$

where the numbers $A_1, B_1, A_2, B_2, \ldots, A_n, B_n$ are to be determined.

EXAMPLE 5

Repeated Irreducible Quadratic Factor

Write the partial fraction decomposition of $\dfrac{x^3 + x^2}{(x^2 + 4)^2}$.

Solution

The denominator contains the repeated irreducible quadratic factor $(x^2 + 4)^2$, so we write

$$\frac{x^3 + x^2}{(x^2 + 4)^2} = \frac{Ax + B}{x^2 + 4} + \frac{Cx + D}{(x^2 + 4)^2} \qquad \textbf{(11)}$$

Clear fractions to obtain

$$x^3 + x^2 = (Ax + B)(x^2 + 4) + Cx + D$$

Collecting like terms yields the identity

$$x^3 + x^2 = Ax^3 + Bx^2 + (4A + C)x + 4B + D$$

Equating coefficients, we arrive at the system

$$\begin{cases} A = 1 \\ B = 1 \\ 4A + C = 0 \\ 4B + D = 0 \end{cases}$$

The solution is $A = 1, B = 1, C = -4, D = -4$. From equation (11),

$$\frac{x^3 + x^2}{(x^2 + 4)^2} = \frac{x + 1}{x^2 + 4} + \frac{-4x - 4}{(x^2 + 4)^2}$$

Now Work PROBLEM 35

8.5 Assess Your Understanding

'Are You Prepared?' *Answers are given at the end of these exercises. If you get a wrong answer, read the pages listed in red.*

1. **True or False** The equation $(x - 1)^2 - 1 = x(x - 2)$ is an example of an identity. (p. 82)

2. **True or False** The rational expression $\dfrac{5x^2 - 1}{x^3 + 1}$ is proper. (p. 347)

3. Factor completely: $3x^4 + 6x^3 + 3x^2$ (pp. 48–55)

4. **True or False** Every polynomial with real numbers as coefficients can be factored into products of linear and/or irreducible quadratic factors. (p. 388)

Skill Building

In Problems 5–12, tell whether the given rational expression is proper or improper. If improper, rewrite it as the sum of a polynomial and a proper rational expression.

5. $\dfrac{x}{x^2 - 1}$

6. $\dfrac{5x + 2}{x^3 - 1}$

7. $\dfrac{x^2 + 5}{x^2 - 4}$

8. $\dfrac{3x^2 - 2}{x^2 - 1}$

9. $\dfrac{5x^3 + 2x - 1}{x^2 - 4}$

10. $\dfrac{3x^4 + x^2 - 2}{x^3 + 8}$

11. $\dfrac{x(x - 1)}{(x + 4)(x - 3)}$

12. $\dfrac{2x(x^2 + 4)}{x^2 + 1}$

In Problems 13–46, write the partial fraction decomposition of each rational expression.

13. $\dfrac{4}{x(x - 1)}$

14. $\dfrac{3x}{(x + 2)(x - 1)}$

15. $\dfrac{1}{x(x^2 + 1)}$

16. $\dfrac{1}{(x + 1)(x^2 + 4)}$

17. $\dfrac{x}{(x - 1)(x - 2)}$

18. $\dfrac{3x}{(x + 2)(x - 4)}$

19. $\dfrac{x^2}{(x - 1)^2(x + 1)}$

20. $\dfrac{x + 1}{x^2(x - 2)}$

21. $\dfrac{1}{x^3 - 8}$

22. $\dfrac{2x + 4}{x^3 - 1}$

23. $\dfrac{x^2}{(x - 1)^2(x + 1)^2}$

24. $\dfrac{x + 1}{x^2(x - 2)^2}$

25. $\dfrac{x - 3}{(x + 2)(x + 1)^2}$

26. $\dfrac{x^2 + x}{(x + 2)(x - 1)^2}$

27. $\dfrac{x + 4}{x^2(x^2 + 4)}$

28. $\dfrac{10x^2 + 2x}{(x - 1)^2(x^2 + 2)}$

29. $\dfrac{x^2 + 2x + 3}{(x + 1)(x^2 + 2x + 4)}$

30. $\dfrac{x^2 - 11x - 18}{x(x^2 + 3x + 3)}$

31. $\dfrac{x}{(3x - 2)(2x + 1)}$

32. $\dfrac{1}{(2x + 3)(4x - 1)}$

33. $\dfrac{x}{x^2 + 2x - 3}$

34. $\dfrac{x^2 - x - 8}{(x + 1)(x^2 + 5x + 6)}$

35. $\dfrac{x^2 + 2x + 3}{(x^2 + 4)^2}$

36. $\dfrac{x^3 + 1}{(x^2 + 16)^2}$

37. $\dfrac{7x + 3}{x^3 - 2x^2 - 3x}$

38. $\dfrac{x^3 + 1}{x^5 - x^4}$

39. $\dfrac{x^2}{x^3 - 4x^2 + 5x - 2}$

40. $\dfrac{x^2 + 1}{x^3 + x^2 - 5x + 3}$

41. $\dfrac{x^3}{(x^2 + 16)^3}$

42. $\dfrac{x^2}{(x^2 + 4)^3}$

43. $\dfrac{4}{2x^2 - 5x - 3}$

44. $\dfrac{4x}{2x^2 + 3x - 2}$

45. $\dfrac{2x + 3}{x^4 - 9x^2}$

46. $\dfrac{x^2 + 9}{x^4 - 2x^2 - 8}$

'Are You Prepared?' Answers

1. True 2. True 3. $3x^2(x + 1)^2$ 4. True

8.6 Systems of Nonlinear Equations

PREPARING FOR THIS SECTION *Before getting started, review the following:*

- Lines (Section 2.3, pp. 167–177)
- Circles (Section 2.4, pp. 182–185)
- Parabolas (Section 7.2, pp. 505–509)
- Ellipses (Section 7.3, pp. 514–520)
- Hyperbolas (Section 7.4, pp. 524–532)

Now Work the *'Are You Prepared?'* problems on page 610.

OBJECTIVES **1** Solve a System of Nonlinear Equations Using Substitution (p. 605)
2 Solve a System of Nonlinear Equations Using Elimination (p. 606)

1 Solve a System of Nonlinear Equations Using Substitution

In Section 8.1, we observed that the solution to a system of linear equations could be found geometrically by determining the point(s) of intersection (if any) of the equations in the system. Similarly, when solving systems of nonlinear equations, the solution(s) also represents the point(s) of intersection (if any) of the graphs of the equations.

There is no general methodology for solving a system of nonlinear equations. At times substitution is best; other times, elimination is best; and sometimes neither of these methods works. Experience and a certain degree of imagination are your allies here.

Before we begin, two comments are in order.

1. If the system contains two variables and if the equations in the system are easy to graph, then graph them. By graphing each equation in the system, you can get an idea of how many solutions a system has and approximately where they are located.

2. Extraneous solutions can creep in when solving nonlinear systems, so it is imperative that all apparent solutions be checked.

EXAMPLE 1 **Solving a System of Nonlinear Equations Using Substitution**

Solve the following system of equations:

$$\begin{cases} 3x - y = -2 & (1) \\ 2x^2 - y = 0 & (2) \end{cases}$$

Solution

First, notice that the system contains two variables and that we know how to graph each equation. Equation (1) is the line $y = 3x + 2$ and equation (2) is the parabola $y = 2x^2$. See Figure 9. The system apparently has two solutions.

Figure 9

To use substitution to solve the system, we choose to solve equation (1) for y.

$$3x - y = -2 \qquad \text{Equation (1)}$$
$$y = 3x + 2$$

Substitute this expression for y in equation (2). The result is an equation containing just the variable x, which we can then solve.

$$2x^2 - y = 0 \qquad \text{Equation (2)}$$
$$2x^2 - (3x + 2) = 0 \qquad \text{Substitute } 3x + 2 \text{ for } y.$$
$$2x^2 - 3x - 2 = 0 \qquad \text{Remove parentheses.}$$
$$(2x + 1)(x - 2) = 0 \qquad \text{Factor.}$$
$$2x + 1 = 0 \quad \text{or} \quad x - 2 = 0 \qquad \text{Apply the Zero-Product Property.}$$
$$x = -\frac{1}{2} \quad \text{or} \quad x = 2$$

Using these values for x in $y = 3x + 2$, we find

$$y = 3\left(-\frac{1}{2}\right) + 2 = \frac{1}{2} \quad \text{or} \quad y = 3(2) + 2 = 8$$

The apparent solutions are $x = -\frac{1}{2}$, $y = \frac{1}{2}$ and $x = 2$, $y = 8$.

Check: For $x = -\frac{1}{2}$, $y = \frac{1}{2}$,

$$\begin{cases} 3\left(-\frac{1}{2}\right) - \frac{1}{2} = -\frac{3}{2} - \frac{1}{2} = -2 & (1) \\ 2\left(-\frac{1}{2}\right)^2 - \frac{1}{2} = 2\left(\frac{1}{4}\right) - \frac{1}{2} = 0 & (2) \end{cases}$$

For $x = 2$, $y = 8$,

$$\begin{cases} 3(2) - 8 = 6 - 8 = -2 & (1) \\ 2(2)^2 - 8 = 2(4) - 8 = 0 & (2) \end{cases}$$

Each solution checks. Now we know that the graphs in Figure 9 intersect at the points $\left(-\frac{1}{2}, \frac{1}{2}\right)$ and $(2, 8)$.

 Check: Graph $3x - y = -2$ ($Y_1 = 3x + 2$) and $2x^2 - y = 0$ ($Y_2 = 2x^2$) and compare what you see with Figure 9. Use INTERSECT (twice) to find the two points of intersection.

Now Work PROBLEM 15 USING SUBSTITUTION

2 Solve a System of Nonlinear Equations Using Elimination

| EXAMPLE 2 | **Solving a System of Nonlinear Equations Using Elimination** |

Solve: $\begin{cases} x^2 + y^2 = 13 & (1) \\ x^2 - y = 7 & (2) \end{cases}$

Solution Equation (1) is a circle and equation (2) is the parabola $y = x^2 - 7$. We graph each equation, as shown in Figure 10. Based on the graph, we expect four solutions. By subtracting equation (2) from equation (1), the variable x can be eliminated.

Figure 10

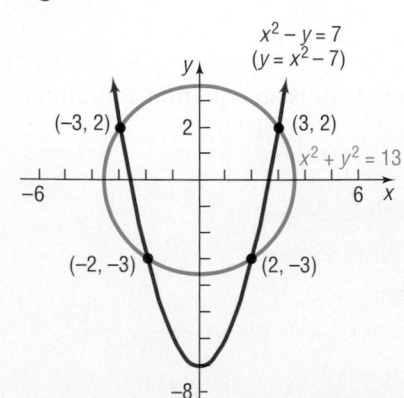

$$\begin{cases} x^2 + y^2 = 13 \\ \underline{x^2 - y = 7} \end{cases} \quad \text{Subtract.}$$
$$y^2 + y = 6$$

This quadratic equation in y can be solved by factoring.

$$y^2 + y - 6 = 0$$
$$(y + 3)(y - 2) = 0$$
$$y = -3 \quad \text{or} \quad y = 2$$

Use these values for y in equation (2) to find x.

If $y = 2$, then $x^2 = y + 7 = 9$, so $x = 3$ or -3.
If $y = -3$, then $x^2 = y + 7 = 4$, so $x = 2$ or -2.

We have four solutions: $x = 3, y = 2$; $x = -3, y = 2$; $x = 2, y = -3$; and $x = -2, y = -3$.

You should verify that, in fact, these four solutions also satisfy equation (1), so all four are solutions of the system. The four points, $(3, 2)$, $(-3, 2)$, $(2, -3)$, and $(-2, -3)$, are the points of intersection of the graphs. Look again at Figure 10.

 ✓**Check:** Graph $x^2 + y^2 = 13$ and $x^2 - y = 7$. (Remember that to graph $x^2 + y^2 = 13$ requires two functions: $Y_1 = \sqrt{13 - x^2}$ and $Y_2 = -\sqrt{13 - x^2}$.) Use INTERSECT to find the points of intersection.

 Now Work PROBLEM **13** USING ELIMINATION

| EXAMPLE 3 | **Solving a System of Nonlinear Equations** |

Solve: $\begin{cases} x^2 - y^2 = 4 & (1) \\ y = x^2 & (2) \end{cases}$

Figure 11

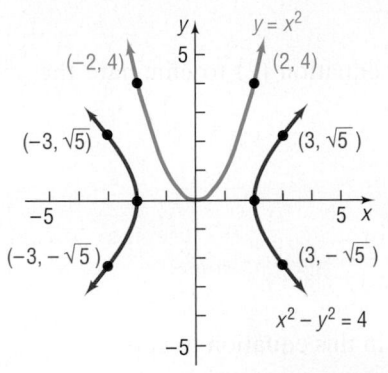

Solution Equation (1) is a hyperbola and equation (2) is a parabola. See Figure 11. It appears the system has no solution. We verify this using substitution. Replace x^2 by y in equation (1). The result is

$$x^2 - y^2 = 4 \quad \text{Equation (1)}$$
$$y - y^2 = 4 \quad y = x^2$$
$$y^2 - y + 4 = 0 \quad \text{Place in standard form.}$$

This is a quadratic equation. Its discriminant is $(-1)^2 - 4 \cdot 1 \cdot 4 = 1 - 16 = -15 < 0$. The equation has no real solutions, so the system is inconsistent. The graphs of these two equations do not intersect.

| EXAMPLE 4 | **Solving a System of Nonlinear Equations Using Elimination** |

Solve: $\begin{cases} x^2 + x + y^2 - 3y + 2 = 0 & (1) \\ x + 1 + \dfrac{y^2 - y}{x} = 0 & (2) \end{cases}$

Solution First, multiply equation (2) by x to eliminate the fraction. The result is an equivalent system because x cannot be 0. [Look at equation (2) to see why.]

$$\begin{cases} x^2 + x + y^2 - 3y + 2 = 0 & (1) \\ x^2 + x + y^2 - y = 0 & (2) \quad x \neq 0 \end{cases}$$

Now subtract equation (2) from equation (1) to eliminate x. The result is

$$-2y + 2 = 0$$
$$y = 1 \quad \text{Solve for } y.$$

To find x, we back-substitute $y = 1$ in equation (1):

$$x^2 + x + y^2 - 3y + 2 = 0 \qquad \text{Equation (1)}$$
$$x^2 + x + 1 - 3 + 2 = 0 \qquad \text{Substitute 1 for } y \text{ in (1).}$$
$$x^2 + x = 0 \qquad \text{Simplify.}$$
$$x(x + 1) = 0 \qquad \text{Factor.}$$
$$x = 0 \quad \text{or} \quad x = -1 \qquad \text{Apply the Zero-Product Property.}$$

Because x cannot be 0, the value $x = 0$ is extraneous, and we discard it.

✔**Check:** Check $x = -1$, $y = 1$:

$$\begin{cases} (-1)^2 + (-1) + 1^2 - 3(1) + 2 = 1 - 1 + 1 - 3 + 2 = 0 & (1) \\ -1 + 1 + \dfrac{1^2 - 1}{-1} = 0 + \dfrac{0}{-1} = 0 & (2) \end{cases}$$

The solution is $x = -1$, $y = 1$. The point of intersection of the graphs of the equations is $(-1, 1)$.

In Problem 55 you are asked to graph the equations given in Example 4. Be sure to show holes in the graph of equation (2) for $x = 0$.

Now Work PROBLEMS 29 AND 49

EXAMPLE 5 **Solving a System of Nonlinear Equations**

Solve: $\begin{cases} 3xy - 2y^2 = -2 & (1) \\ 9x^2 + 4y^2 = 10 & (2) \end{cases}$

Solution Multiply equation (1) by 2 and add the result to equation (2) to eliminate the y^2 terms.

$$\begin{cases} 6xy - 4y^2 = -4 & (1) \\ 9x^2 + 4y^2 = 10 & (2) \end{cases}$$
$$\overline{\hspace{2cm}}$$
$$9x^2 + 6xy = 6 \qquad \text{Add.}$$
$$3x^2 + 2xy = 2 \qquad \text{Divide each side by 3.}$$

Since $x \neq 0$ (do you see why?), we can solve for y in this equation to get

$$y = \frac{2 - 3x^2}{2x} \qquad x \neq 0 \tag{3}$$

Now substitute for y in equation (2) of the system.

$$9x^2 + 4y^2 = 10 \qquad \text{Equation (2)}$$

$$9x^2 + 4\left(\frac{2 - 3x^2}{2x}\right)^2 = 10 \qquad \text{Substitute } y = \frac{2 - 3x^2}{2x} \text{ in (2).}$$

$$9x^2 + \frac{4 - 12x^2 + 9x^4}{x^2} = 10 \qquad \text{Expand and simplify.}$$

$$9x^4 + 4 - 12x^2 + 9x^4 = 10x^2 \qquad \text{Multiply both sides by } x^2.$$

$$18x^4 - 22x^2 + 4 = 0 \qquad \text{Subtract } 10x^2 \text{ from both sides.}$$

$$9x^4 - 11x^2 + 2 = 0 \qquad \text{Divide both sides by 2.}$$

This quadratic equation (in x^2) can be factored:

$$(9x^2 - 2)(x^2 - 1) = 0$$

$$9x^2 - 2 = 0 \qquad \text{or} \quad x^2 - 1 = 0$$

$$x^2 = \frac{2}{9} \qquad \text{or} \qquad x^2 = 1$$

$$x = \pm\sqrt{\frac{2}{9}} = \pm\frac{\sqrt{2}}{3} \qquad \text{or} \qquad x = \pm 1$$

To find y, use equation (3):

$$\text{If } x = \frac{\sqrt{2}}{3}: \quad y = \frac{2 - 3x^2}{2x} = \frac{2 - \dfrac{2}{3}}{2\left(\dfrac{\sqrt{2}}{3}\right)} = \frac{4}{2\sqrt{2}} = \sqrt{2}$$

$$\text{If } x = -\frac{\sqrt{2}}{3}: \quad y = \frac{2 - 3x^2}{2x} = \frac{2 - \dfrac{2}{3}}{2\left(-\dfrac{\sqrt{2}}{3}\right)} = \frac{4}{-2\sqrt{2}} = -\sqrt{2}$$

$$\text{If } x = 1: \quad y = \frac{2 - 3x^2}{2x} = \frac{2 - 3(1)^2}{2} = -\frac{1}{2}$$

$$\text{If } x = -1: \quad y = \frac{2 - 3x^2}{2x} = \frac{2 - 3(-1)^2}{-2} = \frac{1}{2}$$

The system has four solutions: $\left(\dfrac{\sqrt{2}}{3}, \sqrt{2}\right), \left(-\dfrac{\sqrt{2}}{3}, -\sqrt{2}\right), \left(1, -\dfrac{1}{2}\right), \left(-1, \dfrac{1}{2}\right).$
Check them for yourself.

Now Work PROBLEM 47

The next example illustrates an imaginative solution to a system of nonlinear equations.

EXAMPLE 6

Running a Long Distance Race

3 miles ⟷ 1 mile

In a 50-mile race, the winner crosses the finish line 1 mile ahead of the second-place runner and 4 miles ahead of the third-place runner. Assuming that each runner maintains a constant speed throughout the race, by how many miles does the second-place runner beat the third-place runner?

Solution

Let v_1, v_2, and v_3 denote the speeds of the first-, second-, and third-place runners, respectively. Let t_1 and t_2 denote the times (in hours) required for the first-place runner and second-place runner to finish the race. Then we have the system of equations

$$\begin{cases} 50 = v_1 t_1 & \text{(1)} \quad \text{First-place runner goes 50 miles in } t_1 \text{ hours.} \\ 49 = v_2 t_1 & \text{(2)} \quad \text{Second-place runner goes 49 miles in } t_1 \text{ hours.} \\ 46 = v_3 t_1 & \text{(3)} \quad \text{Third-place runner goes 46 miles in } t_1 \text{ hours.} \\ 50 = v_2 t_2 & \text{(4)} \quad \text{Second-place runner goes 50 miles in } t_2 \text{ hours.} \end{cases}$$

We seek the distance d of the third-place runner from the finish at time t_2. At time t_2, the third-place runner has gone a distance of $v_3 t_2$ miles, so the distance d remaining is $50 - v_3 t_2$. Now

$$d = 50 - v_3 t_2$$

$$= 50 - v_3\left(t_1 \cdot \frac{t_2}{t_1}\right)$$

$$= 50 - (v_3 t_1) \cdot \frac{t_2}{t_1}$$

$$= 50 - 46 \cdot \dfrac{\dfrac{50}{v_2}}{\dfrac{50}{v_1}}$$

$$\begin{cases} \text{From (3), } v_3 t_1 = 46 \\ \text{From (4), } t_2 = \dfrac{50}{v_2} \\ \text{From (1), } t_1 = \dfrac{50}{v_1} \end{cases}$$

$$= 50 - 46 \cdot \dfrac{v_1}{v_2}$$

$$= 50 - 46 \cdot \dfrac{50}{49} \qquad \text{From the quotient of (1) and (2).}$$

$$\approx 3.06 \text{ miles}$$

Historical Feature

In the beginning of this section, we said that imagination and experience are important in solving systems of nonlinear equations. Indeed, these kinds of problems lead into some of the deepest and most difficult parts of modern mathematics. Look again at the graphs in Examples 1 and 2 of this section (Figures 9 and 10). We see that Example 1 has two solutions, and Example 2 has four solutions. We might conjecture that the number of solutions is equal to the product of the degrees of the equations involved. This conjecture was indeed

made by Étienne Bézout (1730–1783), but working out the details took about 150 years. It turns out that, to arrive at the correct number of intersections, we must count not only the complex number intersections, but also those intersections that, in a certain sense, lie at infinity. For example, a parabola and a line lying on the axis of the parabola intersect at the vertex and at infinity. This topic is part of the study of algebraic geometry.

Historical Problem

A papyrus dating back to 1950 BC contains the following problem: "A given surface area of 100 units of area shall be represented as the sum of two squares whose sides are to each other as $1:\dfrac{3}{4}$."

Solve for the sides by solving the system of equations

$$\begin{cases} x^2 + y^2 = 100 \\ x = \dfrac{3}{4}y \end{cases}$$

8.6 Assess Your Understanding

'Are You Prepared?' *Answers are given at the end of these exercises. If you get a wrong answer, read the pages listed in red.*

1. Graph the equation: $y = 3x + 2$ (pp. 167–177)
2. Graph the equation: $y + 4 = x^2$ (pp. 505–509)
3. Graph the equation: $y^2 = x^2 - 1$ (pp. 524–532)
4. Graph the equation: $x^2 + 4y^2 = 4$ (pp. 514–520)

Skill Building

In Problems 5–24, graph each equation of the system. Then solve the system to find the points of intersection.

5. $\begin{cases} y = x^2 + 1 \\ y = x + 1 \end{cases}$

6. $\begin{cases} y = x^2 + 1 \\ y = 4x + 1 \end{cases}$

7. $\begin{cases} y = \sqrt{36 - x^2} \\ y = 8 - x \end{cases}$

8. $\begin{cases} y = \sqrt{4 - x^2} \\ y = 2x + 4 \end{cases}$

9. $\begin{cases} y = \sqrt{x} \\ y = 2 - x \end{cases}$

10. $\begin{cases} y = \sqrt{x} \\ y = 6 - x \end{cases}$

11. $\begin{cases} x = 2y \\ x = y^2 - 2y \end{cases}$

12. $\begin{cases} y = x - 1 \\ y = x^2 - 6x + 9 \end{cases}$

13. $\begin{cases} x^2 + y^2 = 4 \\ x^2 + 2x + y^2 = 0 \end{cases}$

14. $\begin{cases} x^2 + y^2 = 8 \\ x^2 + y^2 + 4y = 0 \end{cases}$

15. $\begin{cases} y = 3x - 5 \\ x^2 + y^2 = 5 \end{cases}$

16. $\begin{cases} x^2 + y^2 = 10 \\ y = x + 2 \end{cases}$

17. $\begin{cases} x^2 + y^2 = 4 \\ y^2 - x = 4 \end{cases}$

18. $\begin{cases} x^2 + y^2 = 16 \\ x^2 - 2y = 8 \end{cases}$

19. $\begin{cases} xy = 4 \\ x^2 + y^2 = 8 \end{cases}$

20. $\begin{cases} x^2 = y \\ xy = 1 \end{cases}$

21. $\begin{cases} x^2 + y^2 = 4 \\ y = x^2 - 9 \end{cases}$

22. $\begin{cases} xy = 1 \\ y = 2x + 1 \end{cases}$

23. $\begin{cases} y = x^2 - 4 \\ y = 6x - 13 \end{cases}$

24. $\begin{cases} x^2 + y^2 = 10 \\ xy = 3 \end{cases}$

In Problems 25–54, solve each system. Use any method you wish.

25. $\begin{cases} 2x^2 + y^2 = 18 \\ xy = 4 \end{cases}$

26. $\begin{cases} x^2 - y^2 = 21 \\ x + y = 7 \end{cases}$

27. $\begin{cases} y = 2x + 1 \\ 2x^2 + y^2 = 1 \end{cases}$

28. $\begin{cases} x^2 - 4y^2 = 16 \\ 2y - x = 2 \end{cases}$

29. $\begin{cases} x + y + 1 = 0 \\ x^2 + y^2 + 6y - x = -5 \end{cases}$

30. $\begin{cases} 2x^2 - xy + y^2 = 8 \\ xy = 4 \end{cases}$

31. $\begin{cases} 4x^2 - 3xy + 9y^2 = 15 \\ 2x + 3y = 5 \end{cases}$

32. $\begin{cases} 2y^2 - 3xy + 6y + 2x + 4 = 0 \\ 2x - 3y + 4 = 0 \end{cases}$

33. $\begin{cases} x^2 - 4y^2 + 7 = 0 \\ 3x^2 + y^2 = 31 \end{cases}$

34. $\begin{cases} 3x^2 - 2y^2 + 5 = 0 \\ 2x^2 - y^2 + 2 = 0 \end{cases}$

35. $\begin{cases} 7x^2 - 3y^2 + 5 = 0 \\ 3x^2 + 5y^2 = 12 \end{cases}$

36. $\begin{cases} x^2 - 3y^2 + 1 = 0 \\ 2x^2 - 7y^2 + 5 = 0 \end{cases}$

37. $\begin{cases} x^2 + 2xy = 10 \\ 3x^2 - xy = 2 \end{cases}$

38. $\begin{cases} 5xy + 13y^2 + 36 = 0 \\ xy + 7y^2 = 6 \end{cases}$

39. $\begin{cases} 2x^2 + y^2 = 2 \\ x^2 - 2y^2 + 8 = 0 \end{cases}$

40. $\begin{cases} y^2 - x^2 + 4 = 0 \\ 2x^2 + 3y^2 = 6 \end{cases}$

41. $\begin{cases} x^2 + 2y^2 = 16 \\ 4x^2 - y^2 = 24 \end{cases}$

42. $\begin{cases} 4x^2 + 3y^2 = 4 \\ 2x^2 - 6y^2 = -3 \end{cases}$

43. $\begin{cases} \dfrac{5}{x^2} - \dfrac{2}{y^2} + 3 = 0 \\ \dfrac{3}{x^2} + \dfrac{1}{y^2} = 7 \end{cases}$

44. $\begin{cases} \dfrac{2}{x^2} - \dfrac{3}{y^2} + 1 = 0 \\ \dfrac{6}{x^2} - \dfrac{7}{y^2} + 2 = 0 \end{cases}$

45. $\begin{cases} \dfrac{1}{x^4} + \dfrac{6}{y^4} = 6 \\ \dfrac{2}{x^4} - \dfrac{2}{y^4} = 19 \end{cases}$

46. $\begin{cases} \dfrac{1}{x^4} - \dfrac{1}{y^4} = 1 \\ \dfrac{1}{x^4} + \dfrac{1}{y^4} = 4 \end{cases}$

47. $\begin{cases} x^2 - 3xy + 2y^2 = 0 \\ x^2 + xy = 6 \end{cases}$

48. $\begin{cases} x^2 - xy - 2y^2 = 0 \\ xy + x + 6 = 0 \end{cases}$

49. $\begin{cases} y^2 + y + x^2 - x - 2 = 0 \\ y + 1 + \dfrac{x - 2}{y} = 0 \end{cases}$

50. $\begin{cases} x^3 - 2x^2 + y^2 + 3y - 4 = 0 \\ x - 2 + \dfrac{y^2 - y}{x^2} = 0 \end{cases}$

51. $\begin{cases} \log_x y = 3 \\ \log_x(4y) = 5 \end{cases}$

52. $\begin{cases} \log_x(2y) = 3 \\ \log_x(4y) = 2 \end{cases}$

53. $\begin{cases} \ln x = 4 \ln y \\ \log_3 x = 2 + 2 \log_3 y \end{cases}$

54. $\begin{cases} \ln x = 5 \ln y \\ \log_2 x = 3 + 2 \log_2 y \end{cases}$

55. Graph the equations given in Example 4.

56. Graph the equations given in Problem 49.

In Problems 57–64, use a graphing utility to solve each system of equations. Express the solution(s) rounded to two decimal places.

57. $\begin{cases} y = x^{2/3} \\ y = e^{-x} \end{cases}$

58. $\begin{cases} y = x^{3/2} \\ y = e^{-x} \end{cases}$

59. $\begin{cases} x^2 + y^3 = 2 \\ x^3 y = 4 \end{cases}$

60. $\begin{cases} x^3 + y^2 = 2 \\ x^2 y = 4 \end{cases}$

61. $\begin{cases} x^4 + y^4 = 12 \\ xy^2 = 2 \end{cases}$

62. $\begin{cases} x^4 + y^4 = 6 \\ xy = 1 \end{cases}$

63. $\begin{cases} xy = 2 \\ y = \ln x \end{cases}$

64. $\begin{cases} x^2 + y^2 = 4 \\ y = \ln x \end{cases}$

Mixed Practice

In Problems 65–70, graph each equation and find the point(s) of intersection, if any.

65. The line $x + 2y = 0$ and the circle $(x - 1)^2 + (y - 1)^2 = 5$

66. The line $x + 2y + 6 = 0$ and the circle $(x + 1)^2 + (y + 1)^2 = 5$

67. The circle $(x - 1)^2 + (y + 2)^2 = 4$ and the parabola $y^2 + 4y - x + 1 = 0$

68. The circle $(x + 2)^2 + (y - 1)^2 = 4$ and the parabola $y^2 - 2y - x - 5 = 0$

69. $y = \dfrac{4}{x - 3}$ and the circle $x^2 - 6x + y^2 + 1 = 0$

70. $y = \dfrac{4}{x + 2}$ and the circle $x^2 + 4x + y^2 - 4 = 0$

Applications and Extensions

71. The difference of two numbers is 2 and the sum of their squares is 10. Find the numbers.

72. The sum of two numbers is 7 and the difference of their squares is 21. Find the numbers.

73. The product of two numbers is 4 and the sum of their squares is 8. Find the numbers.

74. The product of two numbers is 10 and the difference of their squares is 21. Find the numbers.

75. The difference of two numbers is the same as their product, and the sum of their reciprocals is 5. Find the numbers.

76. The sum of two numbers is the same as their product, and the difference of their reciprocals is 3. Find the numbers.

77. The ratio of a to b is $\dfrac{2}{3}$. The sum of a and b is 10. What is the ratio of $a + b$ to $b - a$?

78. The ratio of a to b is $4:3$. The sum of a and b is 14. What is the ratio of $a - b$ to $a + b$?

79. Geometry The perimeter of a rectangle is 16 inches and its area is 15 square inches. What are its dimensions?

80. Geometry An area of 52 square feet is to be enclosed by two squares whose sides are in the ratio of $2:3$. Find the sides of the squares.

81. Geometry Two circles have circumferences that add up to 12π centimeters and areas that add up to 20π square centimeters. Find the radius of each circle.

82. Geometry The altitude of an isosceles triangle drawn to its base is 3 centimeters, and its perimeter is 18 centimeters. Find the length of its base.

83. The Tortoise and the Hare In a 21-meter race between a tortoise and a hare, the tortoise leaves 9 minutes before the hare. The hare, by running at an average speed of 0.5 meter per hour faster than the tortoise, crosses the finish line 3 minutes before the tortoise. What are the average speeds of the tortoise and the hare?

84. Running a Race In a 1-mile race, the winner crosses the finish line 10 feet ahead of the second-place runner and 20 feet ahead of the third-place runner. Assuming that each runner maintains a constant speed throughout the race, by how many feet does the second-place runner beat the third-place runner?

85. Constructing a Box A rectangular piece of cardboard, whose area is 216 square centimeters, is made into an open box by cutting a 2-centimeter square from each corner and turning up the sides. See the figure. If the box is to have a volume of 224 cubic centimeters, what size cardboard should you start with?

86. Constructing a Cylindrical Tube A rectangular piece of cardboard, whose area is 216 square centimeters, is made into a cylindrical tube by joining together two sides of the rectangle. See the figure. If the tube is to have a volume of 224 cubic centimeters, what size cardboard should you start with?

87. Fencing A farmer has 300 feet of fence available to enclose a 4500-square-foot region in the shape of adjoining squares, with sides of length x and y. See the figure. Find x and y.

88. Bending Wire A wire 60 feet long is cut into two pieces. Is it possible to bend one piece into the shape of a square and the other into the shape of a circle so that the total area enclosed by the two pieces is 100 square feet? If this is possible, find the length of the side of the square and the radius of the circle.

89. Geometry Find formulas for the length l and width w of a rectangle in terms of its area A and perimeter P.

90. Geometry Find formulas for the base b and one of the equal sides l of an isosceles triangle in terms of its altitude h and perimeter P.

91. Descartes's Method of Equal Roots Descartes's method for finding tangents depends on the idea that, for many graphs, the tangent line at a given point is the *unique* line that intersects the graph at that point only. We will apply his method to find an equation of the tangent line to the parabola $y = x^2$ at the point $(2, 4)$. See the figure.

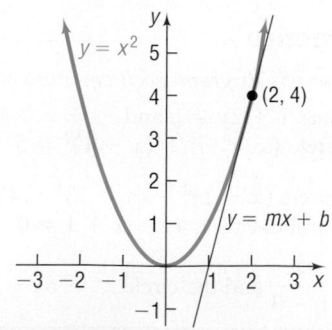

First, we know that the equation of the tangent line must be in the form $y = mx + b$. Using the fact that the point $(2, 4)$ is on the line, we can solve for b in terms of m and get the equation $y = mx + (4 - 2m)$. Now we want $(2, 4)$ to be the *unique* solution to the system

$$\begin{cases} y = x^2 \\ y = mx + 4 - 2m \end{cases}$$

From this system, we get $x^2 - mx + (2m - 4) = 0$. By using the quadratic formula, we get

$$x = \frac{m \pm \sqrt{m^2 - 4(2m - 4)}}{2}$$

To obtain a unique solution for x, the two roots must be equal; in other words, the discriminant $m^2 - 4(2m - 4)$ must be 0. Complete the work to get m, and write an equation of the tangent line.

In Problems 92–98, use Descartes's method from Problem 91 to find the equation of the line tangent to each graph at the given point.

92. $x^2 + y^2 = 10$; at $(1, 3)$

93. $y = x^2 + 2$; at $(1, 3)$

94. $x^2 + y = 5$; at $(-2, 1)$

95. $2x^2 + 3y^2 = 14$; at $(1, 2)$

96. $3x^2 + y^2 = 7$; at $(-1, 2)$

97. $x^2 - y^2 = 3$; at $(2, 1)$

98. $2y^2 - x^2 = 14$; at $(2, 3)$

99. If r_1 and r_2 are two solutions of a quadratic equation $ax^2 + bx + c = 0$, it can be shown that

$$r_1 + r_2 = -\frac{b}{a} \quad \text{and} \quad r_1 r_2 = \frac{c}{a}$$

Solve this system of equations for r_1 and r_2.

Explaining Concepts: Discussion and Writing

100. A circle and a line intersect at most twice. A circle and a parabola intersect at most four times. Deduce that a circle and the graph of a polynomial of degree 3 intersect at most six times. What do you conjecture about a polynomial of degree 4? What about a polynomial of degree n? Can you explain your conclusions using an algebraic argument?

101. Suppose that you are the manager of a sheet metal shop. A customer asks you to manufacture 10,000 boxes, each box being open on top. The boxes are required to have a square base and a 9-cubic-foot capacity. You construct the boxes by cutting out a square from each corner of a square piece of sheet metal and folding along the edges.

(a) What are the dimensions of the square to be cut if the area of the square piece of sheet metal is 100 square feet?

(b) Could you make the box using a smaller piece of sheet metal? Make a list of the dimensions of the box for various pieces of sheet metal.

'Are You Prepared?' Answers

1.

2.

3.

4.

8.7 Systems of Inequalities

PREPARING FOR THIS SECTION *Before getting started, review the following:*

- Solving Linear Inequalities (Section 1.5, pp. 123–126)
- Lines (Section 2.3, pp. 167–177)
- Circles (Section 2.4, pp. 182–185)
- Graphing Techniques: Transformations (Section 3.5, pp. 244–253)

Now Work the *'Are You Prepared?'* problems on page 619.

OBJECTIVES **1** Graph an Inequality (p. 614)
 2 Graph a System of Inequalities (p. 616)

In Section 1.5, we discussed inequalities in one variable. In this section, we discuss inequalities in two variables.

EXAMPLE 1 **Examples of Inequalities in Two Variables**

(a) $3x + y \leq 6$ (b) $x^2 + y^2 < 4$ (c) $y^2 > x$

1 Graph an Inequality

An inequality in two variables x and y is **satisfied** by an ordered pair (a, b) if, when x is replaced by a and y by b, a true statement results. The **graph of an inequality in two variables** x and y consists of all points (x, y) whose coordinates satisfy the inequality.

EXAMPLE 2 **Graphing an Inequality**

Graph the linear inequality: $3x + y \leq 6$

Solution We begin by graphing the equation

$$3x + y = 6$$

formed by replacing (for now) the \leq symbol with an $=$ sign. The graph of the equation is a line. See Figure 12(a). This line is part of the graph of the inequality that we seek because the inequality is nonstrict, so we draw the line as a solid line. (Do you see why? We are seeking points for which $3x + y$ is less than *or equal to* 6.)

Figure 12

(a) $3x + y = 6$

(b) Graph of $3x + y \leq 6$

Now test a few randomly selected points to see whether they belong to the graph of the inequality.

	$3x + y \leq 6$	Conclusion
$(4, -1)$	$3(4) + (-1) = 11 > 6$	Does not belong to the graph
$(5, 5)$	$3(5) + 5 = 20 > 6$	Does not belong to the graph
$(-1, 2)$	$3(-1) + 2 = -1 \leq 6$	Belongs to the graph
$(-2, -2)$	$3(-2) + (-2) = -8 \leq 6$	Belongs to the graph

Look again at Figure 12(a). Notice that the two points that belong to the graph both lie on the same side of the line, and the two points that do not belong to the graph lie on the opposite side. As it turns out, all the points that satisfy the inequality will lie on one side of the line or on the line itself. All the points that do not satisfy the inequality will lie on the other side. The graph we seek consists of all points that lie on the line or on the same side of the line as $(-1, 2)$ and $(-2, -2)$ and is shown as the shaded region in Figure 12(b).

─────── **Now Work** PROBLEM 15

The graph of any inequality in two variables may be obtained in a like way. The steps to follow are given next.

Steps for Graphing an Inequality

STEP 1: Replace the inequality symbol by an equal sign and graph the resulting equation. If the inequality is strict, use dashes; if it is nonstrict, use a solid mark. This graph separates the xy-plane into two or more regions.

STEP 2: In each region, select a test point P.
 (a) If the coordinates of P satisfy the inequality, so do all the points in that region. Indicate this by shading the region.
 (b) If the coordinates of P do not satisfy the inequality, none of the points in that region do.

NOTE The strict inequalities are $<$ and $>$. The nonstrict inequalities are \leq and \geq. ∎

EXAMPLE 3

Graphing an Inequality

Graph: $x^2 + y^2 \leq 4$

Figure 13

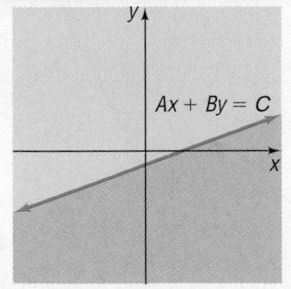

Solution

STEP 1: Graph the equation $x^2 + y^2 = 4$, a circle of radius 2, center at the origin. A solid circle will be used because the inequality is not strict.

STEP 2: Use two test points, one inside the circle, the other outside.

Inside $(0, 0)$: $x^2 + y^2 = 0^2 + 0^2 = 0 \leq 4$ Belongs to the graph

Outside $(4, 0)$: $x^2 + y^2 = 4^2 + 0^2 = 16 > 4$ Does not belong to the graph

All the points inside and on the circle satisfy the inequality. See Figure 13.

─────── **Now Work** PROBLEM 17

Linear Inequalities

A linear inequality is an inequality in one of the forms

$$Ax + By < C \qquad Ax + By > C \qquad Ax + By \leq C \qquad Ax + By \geq C$$

where A and B are not both zero.

Figure 14

The graph of the corresponding equation of a linear inequality is a line that separates the xy-plane into two regions, called **half-planes.** See Figure 14.

As shown, $Ax + By = C$ is the equation of the boundary line, and it divides the plane into two half-planes: one for which $Ax + By < C$ and the other for which $Ax + By > C$. Because of this, for linear inequalities, only one test point is required.

EXAMPLE 4 **Graphing Linear Inequalities**

Graph: (a) $y < 2$ (b) $y \geq 2x$

Solution (a) Points on the horizontal line $y = 2$ are not part of the graph of the inequality, so we show the graph as a dashed line. Since $(0, 0)$ satisfies the inequality, the graph consists of the half-plane below the line $y = 2$. See Figure 15.

(b) Points on the line $y = 2x$ are part of the graph of the inequality, so we show the graph as a solid line. Using $(3, 0)$ as a test point, we find it does not satisfy the inequality $[0 < 2 \cdot 3]$. Points in the half-plane on the opposite side of $(3, 0)$ satisfy the inequality. See Figure 16.

Figure 15

Figure 16

COMMENT A graphing utility can be used to graph inequalities. To see how, read Section 6 in the Appendix. ∎

━━ **Now Work** PROBLEM **13**

2 Graph a System of Inequalities

The **graph of a system of inequalities** in two variables x and y is the set of all points (x, y) that simultaneously satisfy *each* inequality in the system. The graph of a system of inequalities can be obtained by graphing each inequality individually and then determining where, if at all, they intersect.

EXAMPLE 5 **Graphing a System of Linear Inequalities**

Graph the system: $\begin{cases} x + y \geq 2 \\ 2x - y \leq 4 \end{cases}$

Solution Begin by graphing the lines $x + y = 2$ and $2x - y = 4$ using a solid line since both inequalities are nonstrict. Use the test point $(0, 0)$ on each inequality. For example, $(0, 0)$ does not satisfy $x + y \geq 2$, so we shade above the line $x + y = 2$. See Figure 17(a). Also, $(0, 0)$ does satisfy $2x - y \leq 4$, so we shade above the line $2x - y = 4$. See Figure 17(b). The intersection of the shaded regions (in purple) gives us the result presented in Figure 17(c).

Figure 17

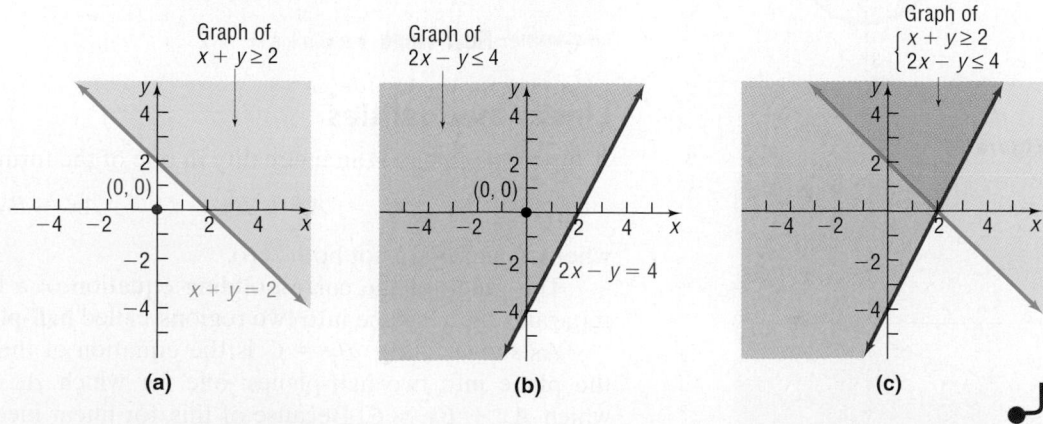

(a) (b) (c)

━━ **Now Work** PROBLEM **23**

EXAMPLE 6

Graphing a System of Linear Inequalities

Graph the system: $\begin{cases} x + y \leq 2 \\ x + y \geq 0 \end{cases}$

Solution

See Figure 18. The overlapping purple-shaded region between the two boundary lines is the graph of the system.

Figure 18

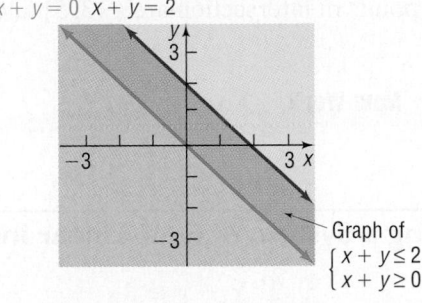

Graph of
$\begin{cases} x + y \leq 2 \\ x + y \geq 0 \end{cases}$

─── **Now Work** PROBLEM 29

EXAMPLE 7

Graphing a System of Linear Inequalities

Graph the systems:

(a) $\begin{cases} 2x - y \geq 0 \\ 2x - y \geq 2 \end{cases}$

(b) $\begin{cases} x + 2y \leq 2 \\ x + 2y \geq 6 \end{cases}$

Solution

(a) See Figure 19. The overlapping purple-shaded region is the graph of the system. Note that the graph of the system is identical to the graph of the single inequality $2x - y \geq 2$.

(b) See Figure 20. Because no overlapping region results, there are no points in the xy-plane that simultaneously satisfy each inequality. The system has no solution.

Figure 19

Graph of
$\begin{cases} 2x - y \geq 0 \\ 2x - y \geq 2 \end{cases}$

Figure 20

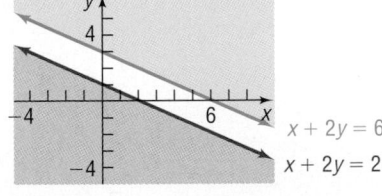

EXAMPLE 8

Graphing a System of Nonlinear Inequalities

Graph the region below the graph of $x + y = 2$ and above the graph of $y = x^2 - 4$ by graphing the system:

$$\begin{cases} y \geq x^2 - 4 \\ x + y \leq 2 \end{cases}$$

Label all points of intersection.

Solution

Figure 21 on the following page shows the graph of the region above the graph of the parabola $y = x^2 - 4$ and below the graph of the line $x + y = 2$. The points of intersection are found by solving the system of equations

$$\begin{cases} y = x^2 - 4 \\ x + y = 2 \end{cases}$$

Figure 21

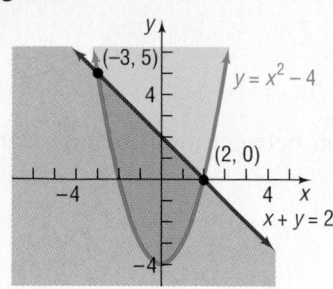

Using substitution, we find

$$x + (x^2 - 4) = 2$$
$$x^2 + x - 6 = 0$$
$$(x + 3)(x - 2) = 0$$
$$x = -3 \quad \text{or} \quad x = 2$$

The two points of intersection are $(-3, 5)$ and $(2, 0)$.

Now Work PROBLEM 37

EXAMPLE 9

Graphing a System of Four Linear Inequalities

Figure 22

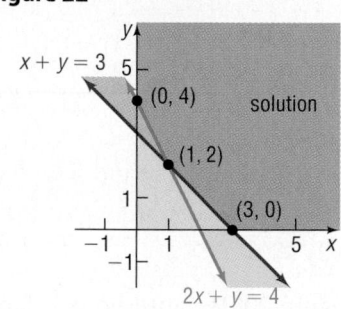

Graph the system: $\begin{cases} x + y \geq 3 \\ 2x + y \geq 4 \\ x \geq 0 \\ y \geq 0 \end{cases}$

Solution See Figure 22. The two inequalities $x \geq 0$ and $y \geq 0$ require the graph of the system to be in quadrant I, which is shaded light gray. We concentrate on the remaining two inequalities. The intersection of the graphs of these two inequalities and quadrant I is shown in dark purple.

EXAMPLE 10

Financial Planning

A retired couple can invest up to $25,000. As their financial adviser, you recommend that they place at least $15,000 in Treasury bills yielding 2% and at most $5000 in corporate bonds yielding 3%.

(a) Using x to denote the amount of money invested in Treasury bills and y the amount invested in corporate bonds, write a system of linear inequalities that describes the possible amounts of each investment. We shall assume that x and y are in thousands of dollars.

(b) Graph the system.

Solution

(a) The system of linear inequalities is

$$\begin{cases} x \geq 0 & \text{\textit{x} and \textit{y} are nonnegative variables since they represent} \\ y \geq 0 & \text{money invested in thousands of dollars.} \\ x + y \leq 25 & \text{The total of the two investments, \textit{x} + \textit{y}, cannot} \\ & \text{exceed \$25,000.} \\ x \geq 15 & \text{At least \$15,000 in Treasury bills} \\ y \leq 5 & \text{At most \$5000 in corporate bonds} \end{cases}$$

Figure 23

(b) See the shaded region in Figure 23. Note that the inequalities $x \geq 0$ and $y \geq 0$ require that the graph of the system be in quadrant I.

The graph of the system of linear inequalities in Figure 23 is said to be **bounded,** because it can be contained within some circle of sufficiently large radius. A graph that cannot be contained in any circle is said to be **unbounded.** For example, the graph of the system of linear inequalities in Figure 22 is unbounded, since it extends indefinitely in the positive x and positive y directions.

Notice in Figures 22 and 23 that those points belonging to the graph that are also points of intersection of boundary lines have been plotted. Such points are referred to as **vertices** or **corner points** of the graph. The system graphed in Figure 22 has three corner points: $(0, 4)$, $(1, 2)$, and $(3, 0)$. The system graphed in Figure 23 has four corner points: $(15, 0)$, $(25, 0)$, $(20, 5)$, and $(15, 5)$.

These ideas will be used in the next section in developing a method for solving linear programming problems, an important application of linear inequalities.

Now Work PROBLEM 45

8.7 Assess Your Understanding

'Are You Prepared?' *Answers are given at the end of these exercises. If you get a wrong answer, read the pages listed in* red.

1. Solve the inequality: $3x + 4 < 8 - x$ (pp. 123–126)
2. Graph the equation: $3x - 2y = 6$ (pp. 167–177)
3. Graph the equation: $x^2 + y^2 = 9$ (pp. 182–185)
4. Graph the equation: $y = x^2 + 4$ (pp. 244–253)

5. **True or False** The lines $2x + y = 4$ and $4x + 2y = 0$ are parallel. (pp. 167–177)
6. The graph of $y = (x - 2)^2$ may be obtained by shifting the graph of _____ to the (left/right) a distance of _____ units. (pp. 244–253)

Concepts and Vocabulary

7. When graphing an inequality in two variables, if the inequality is strict use _____; if the inequality is nonstrict use a _____ mark.
8. The graph of the corresponding equation of a linear inequality is a line that separates the xy-plane into two regions. The two regions are called _____.

9. **True or False** The graph of a system of inequalities must have an overlapping region.
10. If a graph of a system of linear inequalities cannot be contained in any circle, then it is said to be _____.

Skill Building

In Problems 11–22, graph each inequality.

11. $x \geq 0$

12. $y \geq 0$

13. $x \geq 4$

14. $y \leq 2$

15. $2x + y \geq 6$

16. $3x + 2y \leq 6$

17. $x^2 + y^2 > 1$

18. $x^2 + y^2 \leq 9$

19. $y \leq x^2 - 1$

20. $y > x^2 + 2$

21. $xy \geq 4$

22. $xy \leq 1$

In Problems 23–34, graph each system of linear inequalities.

23. $\begin{cases} x + y \leq 2 \\ 2x + y \geq 4 \end{cases}$

24. $\begin{cases} 3x - y \geq 6 \\ x + 2y \leq 2 \end{cases}$

25. $\begin{cases} 2x - y \leq 4 \\ 3x + 2y \geq -6 \end{cases}$

26. $\begin{cases} 4x - 5y \leq 0 \\ 2x - y \geq 2 \end{cases}$

27. $\begin{cases} 2x - 3y \leq 0 \\ 3x + 2y \leq 6 \end{cases}$

28. $\begin{cases} 4x - y \geq 2 \\ x + 2y \geq 2 \end{cases}$

29. $\begin{cases} x - 2y \leq 6 \\ 2x - 4y \geq 0 \end{cases}$

30. $\begin{cases} x + 4y \leq 8 \\ x + 4y \geq 4 \end{cases}$

31. $\begin{cases} 2x + y \geq -2 \\ 2x + y \geq 2 \end{cases}$

32. $\begin{cases} x - 4y \leq 4 \\ x - 4y \geq 0 \end{cases}$

33. $\begin{cases} 2x + 3y \geq 6 \\ 2x + 3y \leq 0 \end{cases}$

34. $\begin{cases} 2x + y \geq 0 \\ 2x + y \geq 2 \end{cases}$

In Problems 35–42, graph each system of inequalities.

35. $\begin{cases} x^2 + y^2 \leq 9 \\ x + y \geq 3 \end{cases}$

36. $\begin{cases} x^2 + y^2 \geq 9 \\ x + y \leq 3 \end{cases}$

37. $\begin{cases} y \geq x^2 - 4 \\ y \leq x - 2 \end{cases}$

38. $\begin{cases} y^2 \leq x \\ y \geq x \end{cases}$

39. $\begin{cases} x^2 + y^2 \leq 16 \\ y \geq x^2 - 4 \end{cases}$

40. $\begin{cases} x^2 + y^2 \leq 25 \\ y \leq x^2 - 5 \end{cases}$

41. $\begin{cases} xy \geq 4 \\ y \geq x^2 + 1 \end{cases}$

42. $\begin{cases} y + x^2 \leq 1 \\ y \geq x^2 - 1 \end{cases}$

In Problems 43–52, graph each system of linear inequalities. Tell whether the graph is bounded or unbounded, and label the corner points.

43. $\begin{cases} x \geq 0 \\ y \geq 0 \\ 2x + y \leq 6 \\ x + 2y \leq 6 \end{cases}$

44. $\begin{cases} x \geq 0 \\ y \geq 0 \\ x + y \geq 4 \\ 2x + 3y \geq 6 \end{cases}$

45. $\begin{cases} x \geq 0 \\ y \geq 0 \\ x + y \geq 2 \\ 2x + y \geq 4 \end{cases}$

46. $\begin{cases} x \geq 0 \\ y \geq 0 \\ 3x + y \leq 6 \\ 2x + y \leq 2 \end{cases}$

47. $\begin{cases} x \geq 0 \\ y \geq 0 \\ x + y \geq 2 \\ 2x + 3y \leq 12 \\ 3x + y \leq 12 \end{cases}$

48. $\begin{cases} x \geq 0 \\ y \geq 0 \\ x + y \geq 2 \\ x + y \leq 8 \\ 2x + y \geq 10 \end{cases}$

49. $\begin{cases} x \geq 0 \\ y \geq 0 \\ x + y \geq 2 \\ x + y \leq 8 \\ 2x + y \leq 10 \end{cases}$

50. $\begin{cases} x \geq 0 \\ y \geq 0 \\ x + y \geq 2 \\ x + y \leq 8 \\ x + 2y \geq 1 \end{cases}$

51. $\begin{cases} x \geq 0 \\ y \geq 0 \\ x + 2y \geq 1 \\ x + 2y \leq 10 \end{cases}$

52. $\begin{cases} x \geq 0 \\ y \geq 0 \\ x + 2y \geq 1 \\ x + 2y \leq 10 \\ x + y \geq 2 \\ x + y \leq 8 \end{cases}$

In Problems 53–56, write a system of linear inequalities for the given graph.

53.

54.

55.

56.

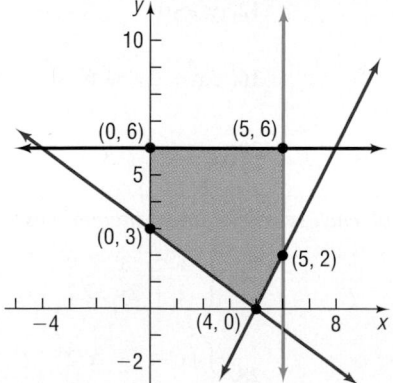

Applications and Extensions

57. **Financial Planning** A retired couple has up to $50,000 to invest. As their financial adviser, you recommend that they place at least $35,000 in Treasury bills yielding 1% and at most $10,000 in corporate bonds yielding 3%.
 (a) Using x to denote the amount of money invested in Treasury bills and y the amount invested in corporate bonds, write a system of linear inequalities that describes the possible amounts of each investment.
 (b) Graph the system and label the corner points.

58. **Manufacturing Trucks** Mike's Toy Truck Company manufactures two models of toy trucks, a standard model and a deluxe model. Each standard model requires 2 hours (hr) for painting and 3 hr for detail work; each deluxe model requires 3 hr for painting and 4 hr for detail work. Two painters and three detail workers are employed by the company, and each works 40 hr per week.
 (a) Using x to denote the number of standard-model trucks and y to denote the number of deluxe-model trucks,

write a system of linear inequalities that describes the possible number of each model of truck that can be manufactured in a week.

(b) Graph the system and label the corner points.

59. Blending Coffee Bill's Coffee House, a store that specializes in coffee, has available 75 pounds (lb) of *A* grade coffee and 120 lb of *B* grade coffee. These will be blended into 1-lb packages as follows: An economy blend that contains 4 ounces (oz) of *A* grade coffee and 12 oz of *B* grade coffee and a superior blend that contains 8 oz of *A* grade coffee and 8 oz of *B* grade coffee.

(a) Using *x* to denote the number of packages of the economy blend and *y* to denote the number of packages of the

superior blend, write a system of linear inequalities that describes the possible number of packages of each kind of blend.

(b) Graph the system and label the corner points.

60. Mixed Nuts Nola's Nuts, a store that specializes in selling nuts, has available 90 pounds (lb) of cashews and 120 lb of peanuts. These are to be mixed in 12-ounce (oz) packages as follows: a lower-priced package containing 8 oz of peanuts and 4 oz of cashews and a quality package containing 6 oz of peanuts and 6 oz of cashews.

(a) Use *x* to denote the number of lower-priced packages and use *y* to denote the number of quality packages. Write a system of linear inequalities that describes the possible number of each kind of package.

(b) Graph the system and label the corner points.

61. Transporting Goods A small truck can carry no more than 1600 pounds (lb) of cargo nor more than 150 cubic (ft³) of cargo. A printer weighs 20 lb and occupies 3 ft³ of space. A microwave oven weighs 30 lb and occupies 2 ft³ of space.

(a) Using *x* to represent the number of microwave ovens and *y* to represent the number of printers, write a system of linear inequalities that describes the number of ovens and printers that can be hauled by the truck.

(b) Graph the system and label the corner points.

'Are You Prepared?' Answers

1. $\{x \mid x < 1\}$ or $(-\infty, 1)$

2.

3.

4.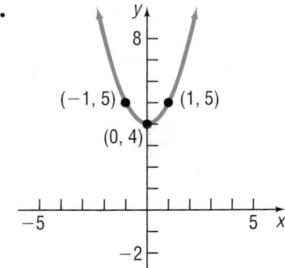

5. True

6. $y = x^2$; right; 2

8.8 Linear Programming

OBJECTIVES **1** Set up a Linear Programming Problem (p. 622)
2 Solve a Linear Programming Problem (p. 622)

Historically, linear programming evolved as a technique for solving problems involving resource allocation of goods and materials for the U.S. Air Force during World War II. Today, linear programming techniques are used to solve a wide variety of problems, such as optimizing airline scheduling and establishing telephone lines. Although most practical linear programming problems involve systems of several hundred linear inequalities containing several hundred variables, we will limit our discussion to problems containing only two variables, because we can solve such problems using graphing techniques.*

* The **simplex method** is a way to solve linear programming problems involving many inequalities and variables. This method was developed by George Dantzig in 1946 and is particularly well suited for computerization. In 1984, Narendra Karmarkar of Bell Laboratories discovered a way of solving large linear programming problems that improves on the simplex method.

1 Set up a Linear Programming Problem

We begin by returning to Example 10 of the previous section.

EXAMPLE 1 **Financial Planning**

A retired couple has up to $25,000 to invest. As their financial adviser, you recommend that they place at least $15,000 in Treasury bills yielding 2% and at most $5000 in corporate bonds yielding 3%. Develop a model that can be used to determine how much money should be placed in each investment so that income is maximized.

Solution The problem is typical of a *linear programming problem*. The problem requires that a certain linear expression, the income, be maximized. If I represents income, x the amount invested in Treasury bills at 2%, and y the amount invested in corporate bonds at 3%, then

$$I = 0.02x + 0.03y$$

We shall assume, as before, that I, x, and y are in thousands of dollars.

The linear expression $I = 0.02x + 0.03y$ is called the **objective function.** Further, the problem requires that the maximum income be achieved under certain conditions or **constraints,** each of which is a linear inequality involving the variables. (See Example 10 in Section 8.7.) The linear programming problem may be modeled as

$$\text{Maximize} \quad I = 0.02x + 0.03y$$

subject to the conditions that

$$\begin{cases} x \geq 0, & y \geq 0 \\ & x + y \leq 25 \\ & x \geq 15 \\ & y \leq 5 \end{cases}$$

In general, every linear programming problem has two components:

1. A linear objective function that is to be maximized or minimized

2. A collection of linear inequalities that must be satisfied simultaneously

DEFINITION A **linear programming problem** in two variables x and y consists of maximizing (or minimizing) a linear objective function

$$z = Ax + By \qquad A \text{ and } B \text{ are real numbers, not both } 0$$

subject to certain conditions, or constraints, expressible as linear inequalities in x and y.

2 Solve a Linear Programming Problem

To maximize (or minimize) the quantity $z = Ax + By$, we need to identify points (x, y) that make the expression for z the largest (or smallest) possible. But not all points (x, y) are eligible; only those that also satisfy each linear inequality (constraint) can be used. We refer to each point (x, y) that satisfies the system of linear inequalities (the constraints) as a **feasible point.** In a linear programming problem, we seek the feasible point(s) that maximizes (or minimizes) the objective function.

Look again at the linear programming problem in Example 1.

EXAMPLE 2 **Analyzing a Linear Programming Problem**

Consider the linear programming problem

$$\text{Maximize} \qquad I = 0.02x + 0.03y$$

subject to the conditions that

$$\begin{cases} x \geq 0 \\ y \geq 0 \\ x + y \leq 25 \\ x \geq 15 \\ y \leq 5 \end{cases}$$

Graph the constraints. Then graph the objective function for $I = 0, 0.3, 0.45, 0.55$, and 0.6.

Solution Figure 24 shows the graph of the constraints. We superimpose on this graph the graph of the objective function for the given values of I.

For $I = 0$, the objective function is the line $0 = 0.02x + 0.03y$.
For $I = 0.3$, the objective function is the line $0.3 = 0.02x + 0.03y$.
For $I = 0.45$, the objective function is the line $0.45 = 0.02x + 0.03y$.
For $I = 0.55$, the objective function is the line $0.55 = 0.02x + 0.03y$.
For $I = 0.6$, the objective function is the line $0.6 = 0.02x + 0.03y$.

Figure 24

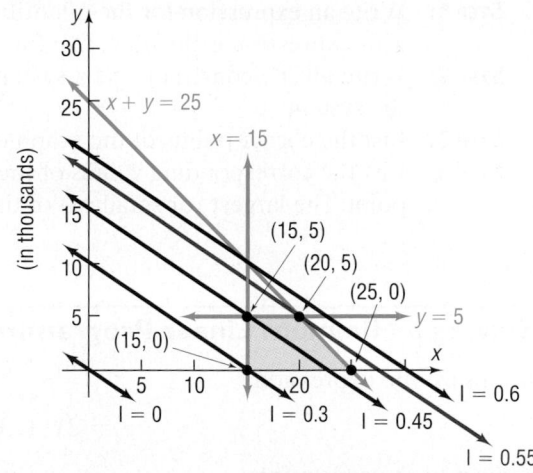

DEFINITION A **solution** to a linear programming problem consists of a feasible point that maximizes (or minimizes) the objective function, together with the corresponding value of the objective function.

One condition for a linear programming problem in two variables to have a solution is that the graph of the feasible points be bounded. (Refer to page 618.)

If none of the feasible points maximizes (or minimizes) the objective function or if there are no feasible points, the linear programming problem has no solution.

Consider the linear programming problem stated in Example 2, and look again at Figure 24. The feasible points are the points that lie in the shaded region. For example, $(20, 3)$ is a feasible point, as are $(15, 5)$, $(20, 5)$, $(18, 4)$, and so on. To find the solution of the problem requires that we find a feasible point (x, y) that makes $I = 0.02x + 0.03y$ as large as possible. Notice that, as I increases in value from

$I = 0$ to $I = 0.3$ to $I = 0.45$ to $I = 0.55$ to $I = 0.6$, we obtain a collection of parallel lines. Further, notice that the largest value of I that can be obtained using feasible points is $I = 0.55$, which corresponds to the line $0.55 = 0.02x + 0.03y$. Any larger value of I results in a line that does not pass through any feasible points. Finally, notice that the feasible point that yields $I = 0.55$ is the point $(20, 5)$, a corner point. These observations form the basis of the following result, which we state without proof.

THEOREM

Location of the Solution of a Linear Programming Problem

If a linear programming problem has a solution, it is located at a corner point of the graph of the feasible points.

If a linear programming problem has multiple solutions, at least one of them is located at a corner point of the graph of the feasible points.

In either case, the corresponding value of the objective function is unique.

We shall not consider here linear programming problems that have no solution. As a result, we can outline the procedure for solving a linear programming problem as follows:

Procedure for Solving a Linear Programming Problem

STEP 1: Write an expression for the quantity to be maximized (or minimized). This expression is the objective function.

STEP 2: Write all the constraints as a system of linear inequalities and graph the system.

STEP 3: List the corner points of the graph of the feasible points.

STEP 4: List the corresponding values of the objective function at each corner point. The largest (or smallest) of these is the solution.

EXAMPLE 3

Solving a Minimum Linear Programming Problem

Minimize the expression

$$z = 2x + 3y$$

subject to the constraints

$$y \leq 5 \qquad x \leq 6 \qquad x + y \geq 2 \qquad x \geq 0 \qquad y \geq 0$$

Solution **STEP 1:** The objective function is $z = 2x + 3y$.

STEP 2: We seek the smallest value of z that can occur if x and y are solutions of the system of linear inequalities

$$\begin{cases} y \leq 5 \\ x \leq 6 \\ x + y \geq 2 \\ x \geq 0 \\ y \geq 0 \end{cases}$$

Figure 25

STEP 3: The graph of this system (the set of feasible points) is shown as the shaded region in Figure 25. We have also plotted the corner points.

STEP 4: Table 1 lists the corner points and the corresponding values of the objective function. From the table, we can see that the minimum value of z is 4, and it occurs at the point $(2, 0)$.

Table 1

Corner Point (x, y)	Value of the Objective Function $z = 2x + 3y$
$(0, 2)$	$z = 2(0) + 3(2) = 6$
$(0, 5)$	$z = 2(0) + 3(5) = 15$
$(6, 5)$	$z = 2(6) + 3(5) = 27$
$(6, 0)$	$z = 2(6) + 3(0) = 12$
$(2, 0)$	$z = 2(2) + 3(0) = 4$

─────Now Work PROBLEMS 5 AND 11

EXAMPLE 4 **Maximizing Profit**

At the end of every month, after filling orders for its regular customers, a coffee company has some pure Colombian coffee and some special-blend coffee remaining. The practice of the company has been to package a mixture of the two coffees into 1-pound (lb) packages as follows: a low-grade mixture containing 4 ounces (oz) of Colombian coffee and 12 oz of special-blend coffee and a high-grade mixture containing 8 oz of Colombian and 8 oz of special-blend coffee. A profit of $0.30 per package is made on the low-grade mixture, whereas a profit of $0.40 per package is made on the high-grade mixture. This month, 120 lb of special-blend coffee and 100 lb of pure Colombian coffee remain. How many packages of each mixture should be prepared to achieve a maximum profit? Assume that all packages prepared can be sold.

Solution **STEP 1:** We begin by assigning symbols for the two variables.

$$x = \text{Number of packages of the low-grade mixture}$$

$$y = \text{Number of packages of the high-grade mixture}$$

If P denotes the profit, then

$$P = \$0.30x + \$0.40y \quad \textit{Objective function}$$

STEP 2: We seek to maximize P subject to certain constraints on x and y. Because x and y represent numbers of packages, the only meaningful values for x and y are nonnegative integers. So we have the two constraints

$$x \geq 0 \quad y \geq 0 \quad \textit{Nonnegative constraints}$$

We also have only so much of each type of coffee available. For example, the total amount of Colombian coffee used in the two mixtures cannot exceed 100 lb, or 1600 oz. Because we use 4 oz in each low-grade package and 8 oz in each high-grade package, we are led to the constraint

$$4x + 8y \leq 1600 \quad \textit{Colombian coffee constraint}$$

Similarly, the supply of 120 lb, or 1920 oz, special-blend coffee leads to the constraint

$$12x + 8y \leq 1920 \quad \textit{Special-blend coffee constraint}$$

The linear programming problem may be stated as

$$\text{Maximize} \quad P = 0.3x + 0.4y$$

subject to the constraints

$$x \geq 0 \quad y \geq 0 \quad 4x + 8y \leq 1600 \quad 12x + 8y \leq 1920$$

Figure 26

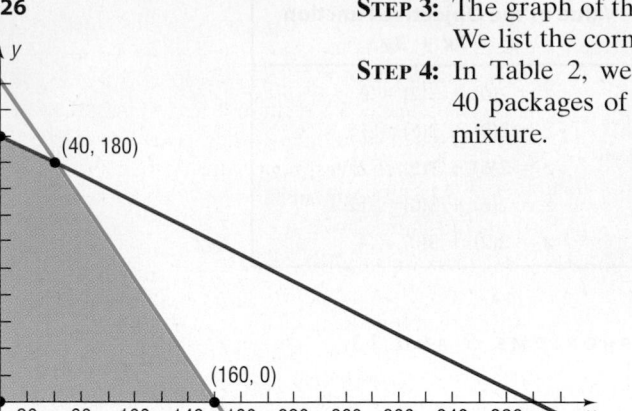

$12x + 8y = 1920$

$4x + 8y = 1600$

STEP 3: The graph of the constraints (the feasible points) is illustrated in Figure 26. We list the corner points and evaluate the objective function at each.

STEP 4: In Table 2, we can see that the maximum profit, $84, is achieved with 40 packages of the low-grade mixture and 180 packages of the high-grade mixture.

Table 2

Corner Point (x, y)	Value of Profit $P = 0.3x + 0.4y$
(0, 0)	$P = 0$
(0, 200)	$P = 0.3(0) + 0.4(200) = \80
(40, 180)	$P = 0.3(40) + 0.4(180) = \84
(160, 0)	$P = 0.3(160) + 0.4(0) = \48

───── **Now Work** PROBLEM 19

8.8 Assess Your Understanding

Concepts and Vocabulary

1. A linear programming problem requires that a linear expression, called the _____ _____, be maximized or minimized.

2. *True or False* If a linear programming problem has a solution, it is located at a corner point of the graph of the feasible points.

Skill Building

In Problems 3–8, find the maximum and minimum value of the given objective function of a linear programming problem. The figure illustrates the graph of the feasible points.

3. $z = x + y$

4. $z = 2x + 3y$

5. $z = x + 10y$

6. $z = 10x + y$

7. $z = 5x + 7y$

8. $z = 7x + 5y$

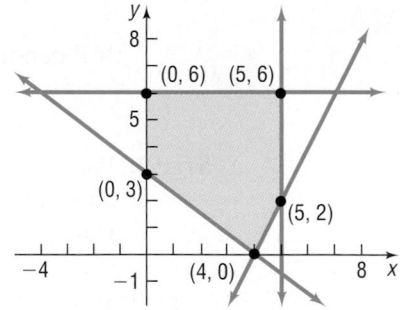

In Problems 9–18, solve each linear programming problem.

9. Maximize $z = 2x + y$ subject to $x \geq 0$, $y \geq 0$, $x + y \leq 6$, $x + y \geq 1$

10. Maximize $z = x + 3y$ subject to $x \geq 0$, $y \geq 0$, $x + y \geq 3$, $x \leq 5$, $y \leq 7$

11. Minimize $z = 2x + 5y$ subject to $x \geq 0$, $y \geq 0$, $x + y \geq 2$, $x \leq 5$, $y \leq 3$

12. Minimize $z = 3x + 4y$ subject to $x \geq 0$, $y \geq 0$, $2x + 3y \geq 6$, $x + y \leq 8$

13. Maximize $z = 3x + 5y$ subject to $x \geq 0$, $y \geq 0$, $x + y \geq 2$, $2x + 3y \leq 12$, $3x + 2y \leq 12$

14. Maximize $z = 5x + 3y$ subject to $x \geq 0$, $y \geq 0$, $x + y \geq 2$, $x + y \leq 8$, $2x + y \leq 10$

15. Minimize $z = 5x + 4y$ subject to $x \geq 0$, $y \geq 0$, $x + y \geq 2$, $2x + 3y \leq 12$, $3x + y \leq 12$

16. Minimize $z = 2x + 3y$ subject to $x \geq 0$, $y \geq 0$, $x + y \geq 3$, $x + y \leq 9$, $x + 3y \geq 6$

17. Maximize $z = 5x + 2y$ subject to $x \geq 0$, $y \geq 0$, $x + y \leq 10$, $2x + y \geq 10$, $x + 2y \geq 10$

18. Maximize $z = 2x + 4y$ subject to $x \geq 0$, $y \geq 0$, $2x + y \geq 4$, $x + y \leq 9$

Applications and Extensions

19. Maximizing Profit A manufacturer of skis produces two types: downhill and cross-country. Use the following table to determine how many of each kind of ski should be produced to achieve a maximum profit. What is the maximum profit? What would the maximum profit be if the time available for manufacturing is increased to 48 hours?

	Downhill	Cross-country	Time Available
Manufacturing time per ski	2 hours	1 hour	40 hours
Finishing time per ski	1 hour	1 hour	32 hours
Profit per ski	$70	$50	

20. Farm Management A farmer has 70 acres of land available for planting either soybeans or wheat. The cost of preparing the soil, the workdays required, and the expected profit per acre planted for each type of crop are given in the following table:

	Soybeans	Wheat
Preparation cost per acre	$60	$30
Workdays required per acre	3	4
Profit per acre	$180	$100

The farmer cannot spend more than $1800 in preparation costs nor use more than a total of 120 workdays. How many acres of each crop should be planted to maximize the profit? What is the maximum profit? What is the maximum profit if the farmer is willing to spend no more than $2400 on preparation?

21. Banquet Seating A banquet hall offers two types of tables for rent: 6-person rectangular tables at a cost of $28 each and 10-person round tables at a cost of $52 each. Kathleen would like to rent the hall for a wedding banquet and needs tables for 250 people. The room can have a maximum of 35 tables and the hall only has 15 rectangular tables available. How many of each type of table should be rented to minimize cost and what is the minimum cost?

Source: facilities.princeton.edu

22. Spring Break The student activities department of a community college plans to rent buses and vans for a spring-break trip. Each bus has 40 regular seats and 1 handicapped seat; each van has 8 regular seats and 3 handicapped seats. The rental cost is $350 for each van and $975 for each bus. If 320 regular and 36 handicapped seats are required for the trip, how many vehicles of each type should be rented to minimize cost?

Source: www.busrates.com

23. Return on Investment An investment broker is instructed by her client to invest up to $20,000, some in a junk bond yielding 9% per annum and some in Treasury bills yielding 7% per annum. The client wants to invest at least $8000 in T-bills and no more than $12,000 in the junk bond.

(a) How much should the broker recommend that the client place in each investment to maximize income if the client insists that the amount invested in T-bills must equal or exceed the amount placed in junk bonds?

(b) How much should the broker recommend that the client place in each investment to maximize income if the client insists that the amount invested in T-bills must not exceed the amount placed in junk bonds?

24. Production Scheduling In a factory, machine 1 produces 8-inch (in.) pliers at the rate of 60 units per hour (hr) and 6-in. pliers at the rate of 70 units/hr. Machine 2 produces 8-in. pliers at the rate of 40 units/hr and 6-in. pliers at the rate of 20 units/hr. It costs $50/hr to operate machine 1, and machine 2 costs $30/hr to operate. The production schedule requires that at least 240 units of 8-in. pliers and at least 140 units of 6-in. pliers be produced during each 10-hr day. Which combination of machines will cost the least money to operate?

25. Managing a Meat Market A meat market combines ground beef and ground pork in a single package for meat loaf. The ground beef is 75% lean (75% beef, 25% fat) and costs the market $0.75 per pound (lb). The ground pork is 60% lean and costs the market $0.45/lb. The meat loaf must be at least 70% lean. If the market wants to use at least 50 lb of its available pork, but no more than 200 lb of its available ground beef, how much ground beef should be mixed with ground pork so that the cost is minimized?

26. Ice Cream The Mom and Pop Ice Cream Company makes two kinds of chocolate ice cream: regular and premium. The properties of 1 gallon (gal) of each type are shown in the table:

	Regular	Premium
Flavoring	24 oz	20 oz
Milk-fat products	12 oz	20 oz
Shipping weight	5 lbs	6 lbs
Profit	$0.75	$0.90

In addition, current commitments require the company to make at least 1 gal of premium for every 4 gal of regular. Each day, the company has available 725 pounds (lb) of flavoring and 425 lb of milk-fat products. If the company can ship no more than 3000 lb of product per day, how many gallons of each type should be produced daily to maximize profit?

Source: www.scitoys.com/ingredients/ice_cream.html

27. Maximizing Profit on Ice Skates A factory manufactures two kinds of ice skates: racing skates and figure skates. The racing skates require 6 work-hours in the fabrication department, whereas the figure skates require 4 work-hours there. The racing skates require 1 work-hour in the finishing

department, whereas the figure skates require 2 work-hours there. The fabricating department has available at most 120 work-hours per day, and the finishing department has no more than 40 work-hours per day available. If the profit on each racing skate is $10 and the profit on each figure skate is $12, how many of each should be manufactured each day to maximize profit? (Assume that all skates made are sold.)

28. Financial Planning A retired couple has up to $50,000 to place in fixed-income securities. Their financial adviser suggests two securities to them: one is an AAA bond that yields 8% per annum; the other is a certificate of deposit (CD) that yields 4%. After careful consideration of the alternatives, the couple decides to place at most $20,000 in the AAA bond and at least $15,000 in the CD. They also instruct the financial adviser to place at least as much in the CD as in the AAA bond. How should the financial adviser proceed to maximize the return on their investment?

29. Product Design An entrepreneur is having a design group produce at least six samples of a new kind of fastener that he wants to market. It costs $9.00 to produce each metal fastener and $4.00 to produce each plastic fastener. He wants to have at least two of each version of the fastener and needs to have all the samples 24 hours (hr) from now. It takes 4 hr to produce each metal sample and 2 hr to produce each plastic sample. To minimize the cost of the samples, how many of each kind should the entrepreneur order? What will be the cost of the samples?

30. Animal Nutrition Kevin's dog Amadeus likes two kinds of canned dog food. Gourmet Dog costs 40 cents a can and has 20 units of a vitamin complex; the calorie content is 75 calories. Chow Hound costs 32 cents a can and has 35 units of vitamins and 50 calories. Kevin likes Amadeus to have at least 1175 units of vitamins a month and at least 2375 calories during the same time period. Kevin has space to store only 60 cans of dog food at a time. How much of each kind of dog food should Kevin buy each month to minimize his cost?

31. Airline Revenue An airline has two classes of service: first class and coach. Management's experience has been that each aircraft should have at least 8 but no more than 16 first-class seats and at least 80 but not more than 120 coach seats.

(a) If management decides that the ratio of first class to coach seats should never exceed 1:12, with how many of each type of seat should an aircraft be configured to maximize revenue?

(b) If management decides that the ratio of first class to coach seats should never exceed 1:8, with how many of each type of seat should an aircraft be configured to maximize revenue?

(c) If you were management, what would you do?

[**Hint:** Assume that the airline charges $C for a coach seat and $F for a first-class seat; $C > 0, F > C$.]

Explaining Concepts: Discussion and Writing

32. Explain in your own words what a linear programming problem is and how it can be solved.

CHAPTER REVIEW

Things to Know

Systems of equations (pp. 541–543)

Systems with no solutions are inconsistent. Systems with a solution are consistent.

Consistent systems of linear equations have either a unique solution (independent) or an infinite number of solutions (dependent).

Matrix (p. 556)
Rectangular array of numbers, called entries

Augmented matrix (p. 557)

Row operations (p. 558)

Row echelon form (p. 559)

Determinants and Cramer's Rule (pp. 571, 573, 574–575, and 576)

Matrix (p. 581)

m by n matrix (p. 581) — Matrix with m rows and n columns

Identity matrix I_n (p. 588) — An n by n square matrix whose diagonal entries are 1's, while all other entries are 0's

Inverse of a matrix (p. 589) — A^{-1} is the inverse of A if $AA^{-1} = A^{-1}A = I_n$.

Nonsingular matrix (p. 589) — A square matrix that has an inverse

Linear programming problem (p. 622)

Maximize (or minimize) a linear objective function, $z = Ax + By$, subject to certain conditions, or constraints, expressible as linear inequalities in x and y. A feasible point (x, y) is a point that satisfies the constraints (linear inequalities) of a linear programming problem.

Location of solution (p. 624)

If a linear programming problem has a solution, it is located at a corner point of the graph of the feasible points. If a linear programming problem has multiple solutions, at least one of them is located at a corner point of the graph of the feasible points. In either case, the corresponding value of the objective function is unique.

Objectives

Section		You should be able to . . .	Example(s)	Review Exercises
8.1	1	Solve systems of equations by substitution (p. 543)	4	1–14, 101, 102, 105–107
	2	Solve systems of equations by elimination (p. 545)	5, 6	1–14, 101, 102, 105–107
	3	Identify inconsistent systems of equations containing two variables (p. 546)	7	9, 10, 13, 98
	4	Express the solution of a system of dependent equations containing two variables (p. 547)	8	14, 97
	5	Solve systems of three equations containing three variables (p. 548)	9	15–18, 99, 100, 103
	6	Identify inconsistent systems of equations containing three variables (p. 550)	10	18
	7	Express the solution of a system of dependent equations containing three variables (p. 550)	11	17
8.2	1	Write the augmented matrix of a system of linear equations (p. 557)	1	35–44
	2	Write the system of equations from the augmented matrix (p. 557)	2	19, 20
	3	Perform row operations on a matrix (p. 558)	3, 4	35–44
	4	Solve a system of linear equations using matrices (p. 559)	5–10	35–44
8.3	1	Evaluate 2 by 2 determinants (p. 571)	1	45, 46
	2	Use Cramer's Rule to solve a system of two equations containing two variables (p. 572)	2	51–54
	3	Evaluate 3 by 3 determinants (p. 574)	4	47–50
	4	Use Cramer's Rule to solve a system of three equations containing three variables (p. 576)	5	55, 56
	5	Know properties of determinants (p. 577)	6–9	57, 58
8.4	1	Find the sum and difference of two matrices (p. 582)	3, 4	21, 22
	2	Find scalar multiples of a matrix (p. 583)	5	23, 24
	3	Find the product of two matrices (p. 584)	6–11	25–28
	4	Find the inverse of a matrix (p. 589)	12–14	29–34
	5	Solve a system of linear equations using an inverse matrix (p. 593)	15	35–44
8.5	1	Decompose $\dfrac{P}{Q}$, where Q has only nonrepeated linear factors (p. 598)	1	59, 60
	2	Decompose $\dfrac{P}{Q}$, where Q has repeated linear factors (p. 600)	2, 3	61, 62
	3	Decompose $\dfrac{P}{Q}$, where Q has a nonrepeated irreducible quadratic factor (p. 602)	4	63, 64, 67, 68
	4	Decompose $\dfrac{P}{Q}$, where Q has a repeated irreducible quadratic factor (p. 603)	5	65, 66
8.6	1	Solve a system of nonlinear equations using substitution (p. 605)	1, 3	69–78
	2	Solve a system of nonlinear equations using elimination (p. 606)	2, 4	69–78
8.7	1	Graph an inequality (p. 614)	2–4	79–82
	2	Graph a system of inequalities (p. 616)	5–10	83–92, 104
8.8	1	Set up a linear programming problem (p. 622)	1	108, 109
	2	Solve a linear programming problem (p. 622)	2–4	93–96, 108, 109

Review Exercises

In Problems 1–18, solve each system of equations using the method of substitution or the method of elimination. If the system has no solution, say that it is inconsistent.

1. $\begin{cases} 2x - y = 5 \\ 5x + 2y = 8 \end{cases}$

2. $\begin{cases} 2x + 3y = 2 \\ 7x - y = 3 \end{cases}$

3. $\begin{cases} 3x - 4y = 4 \\ x - 3y = \dfrac{1}{2} \end{cases}$

4. $\begin{cases} 2x + y = 0 \\ 5x - 4y = -\dfrac{13}{2} \end{cases}$

5. $\begin{cases} x - 2y - 4 = 0 \\ 3x + 2y - 4 = 0 \end{cases}$

6. $\begin{cases} x - 3y + 5 = 0 \\ 2x + 3y - 5 = 0 \end{cases}$

7. $\begin{cases} y = 2x - 5 \\ x = 3y + 4 \end{cases}$

8. $\begin{cases} x = 5y + 2 \\ y = 5x + 2 \end{cases}$

9. $\begin{cases} x - 3y + 4 = 0 \\ \dfrac{1}{2}x - \dfrac{3}{2}y + \dfrac{4}{3} = 0 \end{cases}$

10. $\begin{cases} x + \dfrac{1}{4}y = 2 \\ y + 4x + 2 = 0 \end{cases}$

11. $\begin{cases} 2x + 3y - 13 = 0 \\ 3x - 2y = 0 \end{cases}$

12. $\begin{cases} 4x + 5y = 21 \\ 5x + 6y = 42 \end{cases}$

13. $\begin{cases} 3x - 2y = 8 \\ x - \dfrac{2}{3}y = 12 \end{cases}$

14. $\begin{cases} 2x + 5y = 10 \\ 4x + 10y = 20 \end{cases}$

15. $\begin{cases} x + 2y - z = 6 \\ 2x - y + 3z = -13 \\ 3x - 2y + 3z = -16 \end{cases}$

16. $\begin{cases} x + 5y - z = 2 \\ 2x + y + z = 7 \\ x - y + 2z = 11 \end{cases}$

17. $\begin{cases} 2x - 4y + z = -15 \\ x + 2y - 4z = 27 \\ 5x - 6y - 2z = -3 \end{cases}$

18. $\begin{cases} x - 4y + 3z = 15 \\ -3x + y - 5z = -5 \\ -7x - 5y - 9z = 10 \end{cases}$

In Problems 19 and 20, write the system of equations corresponding to the given augmented matrix.

19. $\left[\begin{array}{cc|c} 3 & 2 & 8 \\ 1 & 4 & -1 \end{array}\right]$

20. $\left[\begin{array}{ccc|c} 1 & 2 & 5 & -2 \\ 5 & 0 & -3 & 8 \\ 2 & -1 & 0 & 0 \end{array}\right]$

In Problems 21–28, use the following matrices to compute each expression.

$$A = \begin{bmatrix} 1 & 0 \\ 2 & 4 \\ -1 & 2 \end{bmatrix} \qquad B = \begin{bmatrix} 4 & -3 & 0 \\ 1 & 1 & -2 \end{bmatrix} \qquad C = \begin{bmatrix} 3 & -4 \\ 1 & 5 \\ 5 & 2 \end{bmatrix}$$

21. $A + C$

22. $A - C$

23. $6A$

24. $-4B$

25. AB

26. BA

27. CB

28. BC

In Problems 29–34, find the inverse, if there is one, of each matrix. If there is not an inverse, say that the matrix is singular.

29. $\begin{bmatrix} 4 & 6 \\ 1 & 3 \end{bmatrix}$

30. $\begin{bmatrix} -3 & 2 \\ 1 & -2 \end{bmatrix}$

31. $\begin{bmatrix} 1 & 3 & 3 \\ 1 & 2 & 1 \\ 1 & -1 & 2 \end{bmatrix}$

32. $\begin{bmatrix} 3 & 1 & 2 \\ 3 & 2 & -1 \\ 1 & 1 & 1 \end{bmatrix}$

33. $\begin{bmatrix} 4 & -8 \\ -1 & 2 \end{bmatrix}$

34. $\begin{bmatrix} -3 & 1 \\ -6 & 2 \end{bmatrix}$

In Problems 35–44, solve each system of equations using matrices. If the system has no solution, say that it is inconsistent.

35. $\begin{cases} 3x - 2y = 1 \\ 10x + 10y = 5 \end{cases}$

36. $\begin{cases} 3x + 2y = 6 \\ x - y = -\dfrac{1}{2} \end{cases}$

37. $\begin{cases} 5x - 6y - 3z = 6 \\ 4x - 7y - 2z = -3 \\ 3x + y - 7z = 1 \end{cases}$

38. $\begin{cases} 2x + y + z = 5 \\ 4x - y - 3z = 1 \\ 8x + y - z = 5 \end{cases}$

39. $\begin{cases} x - 2z = 1 \\ 2x + 3y = -3 \\ 4x - 3y - 4z = 3 \end{cases}$

40. $\begin{cases} x + 2y - z = 2 \\ 2x - 2y + z = -1 \\ 6x + 4y + 3z = 5 \end{cases}$

41. $\begin{cases} x - y + z = 0 \\ x - y - 5z - 6 = 0 \\ 2x - 2y + z - 1 = 0 \end{cases}$

42. $\begin{cases} 4x - 3y + 5z = 0 \\ 2x + 4y - 3z = 0 \\ 6x + 2y + z = 0 \end{cases}$

43. $\begin{cases} x - y - z - t = 1 \\ 2x + y - z + 2t = 3 \\ x - 2y - 2z - 3t = 0 \\ 3x - 4y + z + 5t = -3 \end{cases}$

44. $\begin{cases} x - 3y + 3z - t = 4 \\ x + 2y - z = -3 \\ x + 3z + 2t = 3 \\ x + y + 5z = 6 \end{cases}$

In Problems 45–50, find the value of each determinant.

45. $\begin{vmatrix} 3 & 4 \\ 1 & 3 \end{vmatrix}$

46. $\begin{vmatrix} -4 & 0 \\ 1 & 3 \end{vmatrix}$

47. $\begin{vmatrix} 1 & 4 & 0 \\ -1 & 2 & 6 \\ 4 & 1 & 3 \end{vmatrix}$

48. $\begin{vmatrix} 2 & 3 & 10 \\ 0 & 1 & 5 \\ -1 & 2 & 3 \end{vmatrix}$

49. $\begin{vmatrix} 2 & 1 & -3 \\ 5 & 0 & 1 \\ 2 & 6 & 0 \end{vmatrix}$

50. $\begin{vmatrix} -2 & 1 & 0 \\ 1 & 2 & 3 \\ -1 & 4 & 2 \end{vmatrix}$

In Problems 51–56, use Cramer's Rule, if applicable, to solve each system.

51. $\begin{cases} x - 2y = 4 \\ 3x + 2y = 4 \end{cases}$

52. $\begin{cases} x - 3y = -5 \\ 2x + 3y = 5 \end{cases}$

53. $\begin{cases} 2x + 3y - 13 = 0 \\ 3x - 2y = 0 \end{cases}$

54. $\begin{cases} 3x - 4y - 12 = 0 \\ 5x + 2y + 6 = 0 \end{cases}$

55. $\begin{cases} x + 2y - z = 6 \\ 2x - y + 3z = -13 \\ 3x - 2y + 3z = -16 \end{cases}$

56. $\begin{cases} x - y + z = 8 \\ 2x + 3y - z = -2 \\ 3x - y - 9z = 9 \end{cases}$

In Problems 57 and 58, use properties of determinants to find the value of each determinant if it is known that $\begin{vmatrix} x & y \\ a & b \end{vmatrix} = 8.$

57. $\begin{vmatrix} 2x & y \\ 2a & b \end{vmatrix}$

58. $\begin{vmatrix} y & x \\ b & a \end{vmatrix}$

In Problems 59–68, write the partial fraction decomposition of each rational expression.

59. $\dfrac{6}{x(x - 4)}$

60. $\dfrac{x}{(x + 2)(x - 3)}$

61. $\dfrac{x - 4}{x^2(x - 1)}$

62. $\dfrac{2x - 6}{(x - 2)^2(x - 1)}$

63. $\dfrac{x}{(x^2 + 9)(x + 1)}$

64. $\dfrac{3x}{(x - 2)(x^2 + 1)}$

65. $\dfrac{x^3}{(x^2 + 4)^2}$

66. $\dfrac{x^3 + 1}{(x^2 + 16)^2}$

67. $\dfrac{x^2}{(x^2 + 1)(x^2 - 1)}$

68. $\dfrac{4}{(x^2 + 4)(x^2 - 1)}$

In Problems 69–78, solve each system of equations.

69. $\begin{cases} 2x + y + 3 = 0 \\ x^2 + y^2 = 5 \end{cases}$

70. $\begin{cases} x^2 + y^2 = 16 \\ 2x - y^2 = -8 \end{cases}$

71. $\begin{cases} 2xy + y^2 = 10 \\ 3y^2 - xy = 2 \end{cases}$

72. $\begin{cases} 3x^2 - y^2 = 1 \\ 7x^2 - 2y^2 - 5 = 0 \end{cases}$

73. $\begin{cases} x^2 + y^2 = 6y \\ x^2 = 3y \end{cases}$

74. $\begin{cases} 2x^2 + y^2 = 9 \\ x^2 + y^2 = 9 \end{cases}$

75. $\begin{cases} 3x^2 + 4xy + 5y^2 = 8 \\ x^2 + 3xy + 2y^2 = 0 \end{cases}$

76. $\begin{cases} 3x^2 + 2xy - 2y^2 = 6 \\ xy - 2y^2 + 4 = 0 \end{cases}$

77. $\begin{cases} x^2 - 3x + y^2 + y = -2 \\ \dfrac{x^2 - x}{y} + y + 1 = 0 \end{cases}$

78. $\begin{cases} x^2 + x + y^2 = y + 2 \\ x + 1 = \dfrac{2 - y}{x} \end{cases}$

In Problems 79–82 graph each inequality.

79. $3x + 4y \le 12$

80. $2x - 3y \ge 6$

81. $y \le x^2$

82. $x \ge y^2$

In Problems 83–88, graph each system of inequalities. Tell whether the graph is bounded or unbounded, and label the corner points.

83. $\begin{cases} -2x + y \le 2 \\ x + y \ge 2 \end{cases}$

84. $\begin{cases} x - 2y \le 6 \\ 2x + y \ge 2 \end{cases}$

85. $\begin{cases} x \ge 0 \\ y \ge 0 \\ x + y \le 4 \\ 2x + 3y \le 6 \end{cases}$

86. $\begin{cases} x \geq 0 \\ y \geq 0 \\ 3x + y \geq 6 \\ 2x + y \geq 2 \end{cases}$
87. $\begin{cases} x \geq 0 \\ y \geq 0 \\ 2x + y \leq 8 \\ x + 2y \geq 2 \end{cases}$
88. $\begin{cases} x \geq 0 \\ y \geq 0 \\ 3x + y \leq 9 \\ 2x + 3y \geq 6 \end{cases}$

In Problems 89–92, graph each system of inequalities.

89. $\begin{cases} x^2 + y^2 \leq 16 \\ x + y \geq 2 \end{cases}$
90. $\begin{cases} y^2 \leq x - 1 \\ x - y \leq 3 \end{cases}$
91. $\begin{cases} y \leq x^2 \\ xy \leq 4 \end{cases}$
92. $\begin{cases} x^2 + y^2 \geq 1 \\ x^2 + y^2 \leq 4 \end{cases}$

In Problems 93–96, solve each linear programming problem.

93. Maximize $z = 3x + 4y$ subject to $x \geq 0, y \geq 0, 3x + 2y \geq 6, x + y \leq 8$

94. Maximize $z = 2x + 4y$ subject to $x \geq 0, y \geq 0, x + y \leq 6, x \geq 2$

95. Minimize $z = 3x + 5y$ subject to $x \geq 0, y \geq 0, x + y \geq 1, 3x + 2y \leq 12, x + 3y \leq 12$

96. Minimize $z = 3x + y$ subject to $x \geq 0, y \geq 0, x \leq 8, y \leq 6, 2x + y \geq 4$

97. Find A so that the system of equations has infinitely many solutions.

$$\begin{cases} 2x + 5y = 5 \\ 4x + 10y = A \end{cases}$$

98. Find A so that the system in Problem 97 is inconsistent.

99. Curve Fitting Find the quadratic function $y = ax^2 + bx + c$ that passes through the three points $(0, 1), (1, 0)$, and $(-2, 1)$.

100. Curve Fitting Find the general equation of the circle that passes through the three points $(0, 1), (1, 0)$, and $(-2, 1)$.

[Hint: The general equation of a circle is $x^2 + y^2 + Dx + Ey + F = 0$.]

101. Blending Coffee A coffee distributor is blending a new coffee that will cost $6.90 per pound. It will consist of a blend of $6.00 per pound coffee and $9.00 per pound coffee. What amounts of each type of coffee should be mixed to achieve the desired blend?

[Hint: Assume that the weight of the blended coffee is 100 pounds.]

$6.00/lb $6.90/lb $9.00/lb

102. Farming A 1000-acre farm in Illinois is used to grow corn and soybeans. The cost per acre for raising corn is $65, and the cost per acre for soybeans is $45. If $54,325 has been budgeted for costs and all the acreage is to be used, how many acres should be allocated for each crop?

103. Cookie Orders A cookie company makes three kinds of cookies, oatmeal raisin, chocolate chip, and shortbread, packaged in small, medium, and large boxes. The small box contains 1 dozen oatmeal raisin and 1 dozen chocolate chip; the medium box has 2 dozen oatmeal raisin, 1 dozen chocolate chip, and 1 dozen shortbread; the large box contains 2 dozen oatmeal raisin, 2 dozen chocolate chip, and 3 dozen shortbread. If you require exactly 15 dozen oatmeal raisin, 10 dozen chocolate chip, and 11 dozen shortbread, how many of each size box should you buy?

104. Mixed Nuts A store that specializes in selling nuts has available 72 pounds (lb) of cashews and 120 lb of peanuts. These are to be mixed in 12-ounce (oz) packages as follows: a lower-priced package containing 8 oz of peanuts and 4 oz of cashews and a quality package containing 6 oz of peanuts and 6 oz of cashews.

(a) Use x to denote the number of lower-priced packages and use y to denote the number of quality packages. Write a system of linear inequalities that describes the possible number of each kind of package.

(b) Graph the system and label the corner points.

105. Determining the Speed of the Current of the Aguarico River On a recent trip to the Cuyabeno Wildlife Reserve in the Amazon region of Ecuador, Mike took a 100-kilometer trip by speedboat down the Aguarico River from Chiritza to the Flotel Orellana. As Mike watched the Amazon unfold, he wondered how fast the speedboat was going and how fast the current of the white-water Aguarico River was. Mike timed the trip downstream at 2.5 hours and the return trip at 3 hours. What were the two speeds?

106. Finding the Speed of the Jet Stream On a flight between Midway Airport in Chicago and Ft. Lauderdale, Florida, a Boeing 737 jet maintains an airspeed of 475 miles per hour. If the trip from Chicago to Ft. Lauderdale takes 2 hours, 30 minutes and the return flight takes 2 hours, 50 minutes, what is the speed of the jet stream? (Assume that the speed of the jet stream remains constant at the various altitudes of the plane and that the plane flies with the jet stream one way and against it the other way.)

107. Constant Rate Jobs If Bruce and Bryce work together for 1 hour and 20 minutes, they will finish a certain job. If Bryce and Marty work together for 1 hour and 36 minutes, the same job can be finished. If Marty and Bruce work together, they can complete this job in 2 hours and 40 minutes. How long will it take each of them working alone to finish the job?

108. Maximizing Profit on Figurines A factory manufactures two kinds of ceramic figurines: a dancing girl and a mermaid. Each requires three processes: molding, painting, and glazing. The daily labor available for molding is no more than 90 work-hours, labor available for painting does not exceed 120 work-hours, and labor available for glazing is no more than 60 work-hours. The dancing girl requires 3 work-hours for molding, 6 work-hours for painting, and 2 work-hours for glazing. The mermaid requires 3 work-hours for molding, 4 work-hours for painting, and 3 work-hours for glazing. If the profit on each figurine is $25 for dancing girls and $30 for mermaids, how many of each should be produced each day to maximize profit? If management decides to produce the number of each figurine that maximizes profit, determine which of these processes has work-hours assigned to it that are not used.

109. Minimizing Production Cost A factory produces gasoline engines and diesel engines. Each week the factory is obligated to deliver at least 20 gasoline engines and at least 15 diesel engines. Due to physical limitations, however, the factory cannot make more than 60 gasoline engines nor more than 40 diesel engines in any given week. Finally, to prevent layoffs, a total of at least 50 engines must be produced. If gasoline engines cost $450 each to produce and diesel engines cost $550 each to produce, how many of each should be produced per week to minimize the cost? What is the excess capacity of the factory; that is, how many of each kind of engine is being produced in excess of the number that the factory is obligated to deliver?

110. Describe four ways of solving a system of three linear equations containing three variables. Which method do you prefer? Why?

CHAPTER TEST

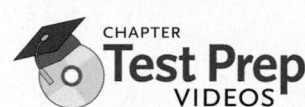

The Chapter Test Prep Videos are step-by-step test solutions available in the Video Resources DVD, in **MyMathLab**, or on this text's You**Tube**™ Channel. Flip back to the Student Resources page to see the exact web address for this text's YouTube channel.

In Problems 1–4, solve each system of equations using the method of substitution or the method of elimination. If the system has no solution, say that it is inconsistent.

1. $\begin{cases} -2x + y = -7 \\ 4x + 3y = 9 \end{cases}$

2. $\begin{cases} \dfrac{1}{3}x - 2y = 1 \\ 5x - 30y = 18 \end{cases}$

3. $\begin{cases} x - y + 2z = 5 \\ 3x + 4y - z = -2 \\ 5x + 2y + 3z = 8 \end{cases}$

4. $\begin{cases} 3x + 2y - 8z = -3 \\ -x - \dfrac{2}{3}y + z = 1 \\ 6x - 3y + 15z = 8 \end{cases}$

5. Write the augmented matrix corresponding to the system of equations: $\begin{cases} 4x - 5y + z = 0 \\ -2x - y + 6 = -19 \\ x + 5y - 5z = 10 \end{cases}$

6. Write the system of equations corresponding to the augmented matrix: $\begin{bmatrix} 3 & 2 & 4 & | & -6 \\ 1 & 0 & 8 & | & 2 \\ -2 & 1 & 3 & | & -11 \end{bmatrix}$

In Problems 7–10, use the given matrices to compute each expression.

$$A = \begin{bmatrix} 1 & -1 \\ 0 & -4 \\ 3 & 2 \end{bmatrix} \quad B = \begin{bmatrix} 1 & -2 & 5 \\ 0 & 3 & 1 \end{bmatrix} \quad C = \begin{bmatrix} 4 & 6 \\ 1 & -3 \\ -1 & 8 \end{bmatrix}$$

7. $2A + C$

8. $A - 3C$

9. CB

10. BA

In Problems 11 and 12, find the inverse of each nonsingular matrix.

11. $A = \begin{bmatrix} 3 & 2 \\ 5 & 4 \end{bmatrix}$

12. $B = \begin{bmatrix} 1 & -1 & 1 \\ 2 & 5 & -1 \\ 2 & 3 & 0 \end{bmatrix}$

In Problems 13–16, solve each system of equations using matrices. If the system has no solution, say that it is inconsistent.

13. $\begin{cases} 6x + 3y = 12 \\ 2x - y = -2 \end{cases}$

14. $\begin{cases} x + \dfrac{1}{4}y = 7 \\ 8x + 2y = 56 \end{cases}$

15. $\begin{cases} x + 2y + 4z = -3 \\ 2x + 7y + 15z = -12 \\ 4x + 7y + 13z = -10 \end{cases}$

16. $\begin{cases} 2x + 2y - 3z = 5 \\ x - y + 2z = 8 \\ 3x + 5y - 8z = -2 \end{cases}$

In Problems 17 and 18, find the value of each determinant.

17. $\begin{vmatrix} -2 & 5 \\ 3 & 7 \end{vmatrix}$

18. $\begin{vmatrix} 2 & -4 & 6 \\ 1 & 4 & 0 \\ -1 & 2 & -4 \end{vmatrix}$

In Problems 19 and 20, use Cramer's Rule, if possible, to solve each system.

19. $\begin{cases} 4x + 3y = -23 \\ 3x - 5y = 19 \end{cases}$

20. $\begin{cases} 4x - 3y + 2z = 15 \\ -2x + y - 3z = -15 \\ 5x - 5y + 2z = 18 \end{cases}$

In Problems 21 and 22, solve each system of equations.

21. $\begin{cases} 3x^2 + y^2 = 12 \\ \qquad y^2 = 9x \end{cases}$

22. $\begin{cases} 2y^2 - 3x^2 = 5 \\ \quad y - x = 1 \end{cases}$

23. Graph the system of inequalities: $\begin{cases} x^2 + y^2 \le 100 \\ 4x - 3y \ge 0 \end{cases}$

In Problems 24 and 25, write the partial fraction decomposition of each rational expression.

24. $\dfrac{3x + 7}{(x + 3)^2}$

25. $\dfrac{4x^2 - 3}{x(x^2 + 3)^2}$

26. Graph the system of inequalities. Tell whether the graph is bounded or unbounded, and label all corner points.

$$\begin{cases} x \ge 0 \\ y \ge 0 \\ x + 2y \ge 8 \\ 2x - 3y \ge 2 \end{cases}$$

27. Maximize $z = 5x + 8y$ subject to $x \ge 0$, $2x + y \le 8$, and $x - 3y \le -3$.

28. Megan went clothes shopping and bought 2 pairs of flare jeans, 2 camisoles, and 4 T-shirts for $90.00. At the same store, Paige bought one pair of flare jeans and 3 T-shirts for $42.50, while Kara bought 1 pair of flare jeans, 3 camisoles, and 2 T-shirts for $62.00. Determine the price of each clothing item.

CUMULATIVE REVIEW

In Problems 1–6, solve each equation.

1. $2x^2 - x = 0$

2. $\sqrt{3x + 1} = 4$

3. $2x^3 - 3x^2 - 8x - 3 = 0$

4. $3^x = 9^{x+1}$

5. $\log_3(x - 1) + \log_3(2x + 1) = 2$

6. $3^x = e$

7. Determine whether the function $g(x) = \dfrac{2x^3}{x^4 + 1}$ is even, odd, or neither. Is the graph of g symmetric with respect to the x-axis, y-axis, or origin?

8. Find the center and radius of the circle $x^2 + y^2 - 2x + 4y - 11 = 0$. Graph the circle.

9. Graph $f(x) = 3^{x-2} + 1$ using transformations. What is the domain, range, and horizontal asymptote of f?

10. The function $f(x) = \dfrac{5}{x + 2}$ is one-to-one. Find f^{-1}. Find the domain and the range of f and the domain and the range of f^{-1}.

11. Graph each equation.
(a) $y = 3x + 6$
(b) $x^2 + y^2 = 4$
(c) $y = x^3$
(d) $y = \dfrac{1}{x}$
(e) $y = \sqrt{x}$
(f) $y = e^x$
(g) $y = \ln x$
(h) $2x^2 + 5y^2 = 1$
(i) $x^2 - 3y^2 = 1$
(j) $x^2 - 2x - 4y + 1 = 0$

12. $f(x) = x^3 - 3x + 5$
(a) Using a graphing utility, graph f and approximate the zero(s) of f.
(b) Using a graphing utility, approximate the local maxima and local minima.
(c) Determine the intervals on which f is increasing.

CHAPTER PROJECTS

Internet-based Project

I. Markov Chains A **Markov chain** (or process) is one in which future outcomes are determined by a current state. Future outcomes are based on probabilities. The probability of moving to a certain state depends only on the state previously occupied and does not vary with time. An example of a Markov chain is the maximum education achieved by children based on the highest education attained by their parents, where the states are (1) earned college degree, (2) high school diploma only, (3) elementary school only. If p_{ij} is the probability of moving from state i to state j, the **transition matrix** is the $m \times m$ matrix

$$P = \begin{bmatrix} p_{11} & p_{12} & \cdots & p_{1m} \\ \vdots & \vdots & & \vdots \\ p_{m1} & p_{m2} & \cdots & p_{mm} \end{bmatrix}$$

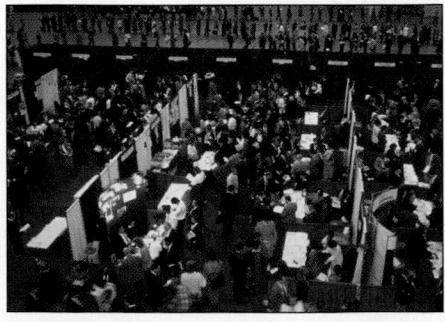

The table represents the probabilities of the highest educational level of children based on the highest educational level of their parents. For example, the table shows that the probability p_{21} is 40% that parents with a high-school

education (row 2) will have children with a college education (column 1).

Highest Educational Level of Parents	Maximum Education That Children Achieve		
	College	High School	Elementary
College	80%	18%	2%
High school	40%	50%	10%
Elementary	20%	60%	20%

1. Convert the percentages to decimals.
2. What is the transition matrix?
3. Sum across the rows. What do you notice? Why do you think that you obtained this result?
4. If P is the transition matrix of a Markov chain, the (i, j)th entry of P^n (nth power of P) gives the probability of passing from state i to state j in n stages. What is the probability that a grandchild of a college graduate is a college graduate?
5. What is the probability that the grandchild of a high school graduate finishes college?
6. The row vector $v^{(0)} = \begin{bmatrix} 0.288 & 0.569 & 0.143 \end{bmatrix}$ represents the proportion of the U.S. population 25 years or older that has college, high school, and elementary school, respectively, as the highest educational level in 2007.* In a Markov chain the probability distribution $v^{(k)}$ after k stages is $v^{(k)} = v^{(0)}P^k$, where P^k is the kth power of the transition matrix. What will be the distribution of highest educational attainment of the grandchildren of the current population?
7. Calculate P^3, P^4, P^5, \ldots. Continue until the matrix does not change. This is called the long-run distribution. What is the long-run distribution of highest educational attainment of the population?

__Source:__ U.S. Census Bureau.

The following projects are available at the Instructor's Resource Center (IRC):

II. Project at Motorola: *Error Control Coding* The high-powered engineering needed to assure that wireless communications are transmitted correctly is analyzed using matrices to control coding errors.

III. Using Matrices to Find the Line of Best Fit Have you wondered how our calculators get a line of best fit? See how to find the line by solving a matrix equation.

IV. CBL Experiment Simulate two people walking toward each other at a constant rate. Then solve the resulting system of equations to determine when and where they will meet.

9
Sequences; Induction; the Binomial Theorem

Outline

World Population Prospects

In July 2009, the world population reached 6.8 billion, 313 million more than in 2005 or a gain of 78 million persons annually. Assuming that fertility levels continue to decline, the world population is expected to reach 9.1 billion in 2050 and to be increasing by about 33 million persons annually at that time.

Future population growth is highly dependent on the path that future fertility takes. Fertility is projected to decline from 2.56 children per woman in 2005–2010 to 2.02 children per woman in 2045–2050. If fertility were to remain about half a child above the levels projected, world population would reach 10.5 billion by 2050. A fertility path half a child below the levels projected would lead to a population of 8 billion by mid-century. Consequently, population growth until 2050 is inevitable even if the decline of fertility accelerates.

In the more developed regions, fertility has increased slightly in recent years so that its estimated level in 2005–2010, 1.64 children per woman according to the *2008 Revision,* is higher than the one reported in the *2006 Revision* (1.60 children per woman). As a result of the slightly higher projected fertility and a sustained net immigration averaging 2.4 million annually, the population of the more developed regions is expected to increase slightly from 1.23 billion in 2009 to 1.28 billion in 2050.

The population of the 49 least developed countries is still the fastest growing in the world, at 2.3 percent per year. Although its rate of increase is expected to moderate significantly over the next decades, the population of the least developed countries is projected to double, passing from 0.84 billion in 2009 to 1.7 billion in 2050. Growth in the rest of the developing world is also projected to be robust, though less rapid, with its population rising from 4.8 billion to 6.2 billion between 2009 and 2050 according to the medium variant.

Although the population of all countries is expected to age over the foreseeable future, the population will remain relatively young in countries where fertility is still high, many of which are experiencing very rapid population growth. High population growth rates prevail in many developing countries, most of which are least developed. Between 2010 and 2050, the populations of 31 countries, the majority of which are least developed, will double or more. Among them, the populations of Afghanistan, Burkina Faso, Niger, Somalia, Timor-Leste, and Uganda are projected to increase by 150 percent or more.

Source: Adapted with permission from Millennium Development Goals Report 2008, pp. IX–X. © United Nations, 2009.

—See the Internet-based Chapter Project I—

◁ A Look Back, A Look Ahead ▷

This chapter may be divided into three independent parts: Sections 9.1–9.3, Section 9.4, and Section 9.5.

In Chapter 3, we defined a function and its domain, which was usually some set of real numbers. In Sections 9.1–9.3, we discuss a sequence, which is a function whose domain is the set of positive integers.

Throughout this text, where it seemed appropriate, we have given proofs of many of the results. In Section 9.4, a technique for proving theorems involving natural numbers is discussed.

In Chapter R, Review, Section R.4, there are formulas for expanding $(x + a)^2$ and $(x + a)^3$. In Section 9.5, we discuss the Binomial Theorem, a formula for the expansion of $(x + a)^n$, where n is any positive integer.

The topics introduced in this chapter are covered in more detail in courses titled *Discrete Mathematics.* Applications of these topics can be found in the fields of computer science, engineering, business and economics, the social sciences, and the physical and biological sciences.

9.1 Sequences

PREPARING FOR THIS SECTION *Before getting started, review the following concept:*
• Functions (Section 3.1, pp. 200–206)

Now Work the 'Are You Prepared?' problems on page 643.

OBJECTIVES 1 Write the First Several Terms of a Sequence (p. 637)
2 Write the Terms of a Sequence Defined by a Recursive Formula (p. 640)
3 Use Summation Notation (p. 641)
4 Find the Sum of a Sequence (p. 642)

When you hear the word *sequence* as in "a sequence of events," you likely think of something that happens first, then second, and so on. In mathematics, the word *sequence* also deals with outcomes that are first, second, and so on.

DEFINITION A **sequence** is a function whose domain is the set of positive integers.

So in a sequence the inputs are $1, 2, 3, \ldots$. Because a sequence is a function, it will have a graph. In Figure 1(a), we show the graph of the function $f(x) = \dfrac{1}{x}, x > 0$. If all the points on this graph were removed except those whose x-coordinates are positive integers, that is, if all points were removed except $(1, 1), \left(2, \dfrac{1}{2}\right), \left(3, \dfrac{1}{3}\right)$, and so on, the remaining points would be the graph of the sequence $f(n) = \dfrac{1}{n}$, as shown in Figure 1(b). Notice that we use n to represent the independent variable in a sequence. This serves to remind us that n is a positive integer.

Figure 1

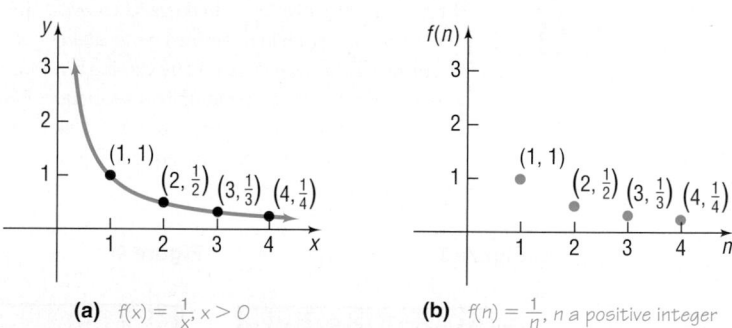

(a) $f(x) = \frac{1}{x}, x > 0$ **(b)** $f(n) = \frac{1}{n}, n$ a positive integer

1 Write the First Several Terms of a Sequence

A sequence is usually represented by listing its values in order. For example, the sequence whose graph is given in Figure 1(b) might be represented as

$$f(1), f(2), f(3), f(4), \ldots \quad \text{or} \quad 1, \frac{1}{2}, \frac{1}{3}, \frac{1}{4}, \ldots$$

The list never ends, as the ellipsis indicates. The numbers in this ordered list are called the **terms** of the sequence.

In dealing with sequences, we usually use subscripted letters, such as a_1, to represent the first term, a_2 for the second term, a_3 for the third term, and so on. For the sequence $f(n) = \dfrac{1}{n}$, we write

$$\underbrace{a_1 = f(1) = 1,}_{\text{first term}} \quad \underbrace{a_2 = f(2) = \frac{1}{2},}_{\text{second term}} \quad \underbrace{a_3 = f(3) = \frac{1}{3},}_{\text{third term}} \quad \underbrace{a_4 = f(4) = \frac{1}{4}, \ldots}_{\text{fourth term}} \quad \underbrace{a_n = f(n) = \frac{1}{n}, \ldots}_{n^{\text{th}} \text{ term}}$$

In other words, we usually do not use the traditional function notation $f(n)$ for sequences. For this particular sequence, we have a rule for the nth term, which is $a_n = \dfrac{1}{n}$, so it is easy to find any term of the sequence.

When a formula for the nth term (sometimes called the **general term**) of a sequence is known, rather than write out the terms of the sequence, we usually represent the entire sequence by placing braces around the formula for the nth term.

For example, the sequence whose nth term is $b_n = \left(\dfrac{1}{2}\right)^n$ may be represented as

$$\{b_n\} = \left\{\left(\dfrac{1}{2}\right)^n\right\}$$

or by

$$b_1 = \dfrac{1}{2}, \quad b_2 = \dfrac{1}{4}, \quad b_3 = \dfrac{1}{8}, \dots, \quad b_n = \left(\dfrac{1}{2}\right)^n, \dots$$

EXAMPLE 1 Writing the First Several Terms of a Sequence

Write down the first six terms of the following sequence and graph it.

$$\{a_n\} = \left\{\dfrac{n-1}{n}\right\}$$

Figure 2

Solution The first six terms of the sequence are

$$a_1 = 0, \quad a_2 = \dfrac{1}{2}, \quad a_3 = \dfrac{2}{3}, \quad a_4 = \dfrac{3}{4}, \quad a_5 = \dfrac{4}{5}, \quad a_6 = \dfrac{5}{6}$$

See Figure 2 for the graph. ⌐

COMMENT Graphing utilities can be used to write the terms of a sequence and graph them. Figure 3 shows the sequence given in Example 1 generated on a TI-84 Plus graphing calculator. We can see the first few terms of the sequence on the viewing window. You need to press the right arrow key to scroll right to see the remaining terms of the sequence. Figure 4 shows a graph of the sequence. Notice that the first term of the sequence is not visible since it lies on the x-axis. TRACEing the graph will allow you to see the terms of the sequence. The TABLE feature can also be used to generate the terms of the sequence. See Table 1.

Figure 3 **Figure 4** Table 1

━━ **Now Work** PROBLEM 17

EXAMPLE 2 Writing the First Several Terms of a Sequence

Write down the first six terms of the following sequence and graph it.

$$\{b_n\} = \left\{(-1)^{n+1}\left(\dfrac{2}{n}\right)\right\}$$

Figure 5

Solution

The first six terms of the sequence are

$$b_1 = 2, \quad b_2 = -1, \quad b_3 = \frac{2}{3}, \quad b_4 = -\frac{1}{2}, \quad b_5 = \frac{2}{5}, \quad b_6 = -\frac{1}{3}$$

See Figure 5 for the graph.

Notice in the sequence $\{b_n\}$ in Example 2 that the signs of the terms **alternate.** This occurs when we use factors such as $(-1)^{n+1}$, which equals 1 if n is odd and -1 if n is even, or $(-1)^n$, which equals -1 if n is odd and 1 if n is even.

EXAMPLE 3

Writing the First Several Terms of a Sequence

Write down the first six terms of the following sequence and graph it.

$$\{c_n\} = \left\{ \begin{array}{ll} n & \text{if } n \text{ is even} \\ \dfrac{1}{n} & \text{if } n \text{ is odd} \end{array} \right\}$$

Figure 6

Solution

The first six terms of the sequence are

$$c_1 = 1, \quad c_2 = 2, \quad c_3 = \frac{1}{3}, \quad c_4 = 4, \quad c_5 = \frac{1}{5}, \quad c_6 = 6$$

See Figure 6 for the graph.

Now Work PROBLEM 19

Sometimes a sequence is indicated by an observed pattern in the first few terms that makes it possible to infer the makeup of the nth term. In the example that follows, a sufficient number of terms of the sequence is given so that a natural choice for the nth term is suggested.

EXAMPLE 4

Determining a Sequence from a Pattern

(a) $e, \dfrac{e^2}{2}, \dfrac{e^3}{3}, \dfrac{e^4}{4}, \ldots$ $a_n = \dfrac{e^n}{n}$

(b) $1, \dfrac{1}{3}, \dfrac{1}{9}, \dfrac{1}{27}, \ldots$ $b_n = \dfrac{1}{3^{n-1}}$

(c) $1, 3, 5, 7, \ldots$ $c_n = 2n - 1$

(d) $1, 4, 9, 16, 25, \ldots$ $d_n = n^2$

(e) $1, -\dfrac{1}{2}, \dfrac{1}{3}, -\dfrac{1}{4}, \dfrac{1}{5}, \ldots$ $e_n = (-1)^{n-1}\left(\dfrac{1}{n}\right)$

Now Work PROBLEM 27

The Factorial Symbol

Some sequences in mathematics involve a special product called a *factorial*.

DEFINITION

If $n \geq 0$ is an integer, the **factorial symbol** $n!$ is defined as follows:

$$0! = 1 \qquad 1! = 1$$
$$n! = n(n - 1) \cdots \cdot 3 \cdot 2 \cdot 1 \qquad \text{if } n \geq 2$$

Table 2

n	$n!$
0	1
1	1
2	2
3	6
4	24
5	120
6	720

For example, $2! = 2 \cdot 1 = 2$, $3! = 3 \cdot 2 \cdot 1 = 6$, $4! = 4 \cdot 3 \cdot 2 \cdot 1 = 24$, and so on. Table 2 lists the values of $n!$ for $0 \leq n \leq 6$.

Because

$$n! = \underbrace{n(n-1)(n-2) \cdot \ldots \cdot 3 \cdot 2 \cdot 1}_{(n-1)!}$$

we can use the formula

$$\boxed{n! = n(n-1)!}$$

to find successive factorials. For example, because $6! = 720$, we have

$$7! = 7 \cdot 6! = 7(720) = 5040$$

and

$$8! = 8 \cdot 7! = 8(5040) = 40{,}320$$

Now Work PROBLEM 11

Exploration

Your calculator has a factorial key. Use it to see how fast factorials increase in value. Find the value of 69!. What happens when you try to find 70!? In fact, 70! is larger than 10^{100} (a googol).

2 Write the Terms of a Sequence Defined by a Recursive Formula

A second way of defining a sequence is to assign a value to the first (or the first few) term(s) and specify the nth term by a formula or equation that involves one or more of the terms preceding it. Sequences defined this way are said to be defined **recursively,** and the rule or formula is called a **recursive formula.**

EXAMPLE 5

Writing the Terms of a Recursively Defined Sequence

Write down the first five terms of the following recursively defined sequence.

$$s_1 = 1, \qquad s_n = n s_{n-1}$$

Solution

The first term is given as $s_1 = 1$. To get the second term, we use $n = 2$ in the formula $s_n = n s_{n-1}$ to get $s_2 = 2s_1 = 2 \cdot 1 = 2$. To get the third term, we use $n = 3$ in the formula to get $s_3 = 3s_2 = 3 \cdot 2 = 6$. To get a new term requires that we know the value of the preceding term. The first five terms are

$$s_1 = 1$$
$$s_2 = 2 \cdot 1 = 2$$
$$s_3 = 3 \cdot 2 = 6$$
$$s_4 = 4 \cdot 6 = 24$$
$$s_5 = 5 \cdot 24 = 120$$

Do you recognize this sequence? $s_n = n!$

EXAMPLE 6

Writing the Terms of a Recursively Defined Sequence

Write down the first five terms of the following recursively defined sequence.

$$u_1 = 1, \qquad u_2 = 1, \qquad u_n = u_{n-2} + u_{n-1}$$

Solution

We are given the first two terms. To get the third term requires that we know both of the previous two terms. That is,

$$u_1 = 1$$
$$u_2 = 1$$
$$u_3 = u_1 + u_2 = 1 + 1 = 2$$
$$u_4 = u_2 + u_3 = 1 + 2 = 3$$
$$u_5 = u_3 + u_4 = 2 + 3 = 5$$

The sequence defined in Example 6 is called the **Fibonacci sequence,** and the terms of this sequence are called **Fibonacci numbers.** These numbers appear in a wide variety of applications (see Problems 85–88).

Now Work PROBLEMS 35 AND 43

3 Use Summation Notation

It is often important to find the sum of the first n terms of a sequence $\{a_n\}$, that is,

$$a_1 + a_2 + a_3 + \cdots + a_n$$

Rather than write down all these terms, we introduce a more concise way to express the sum, called **summation notation.** Using summation notation, we write the sum as

$$a_1 + a_2 + a_3 + \cdots + a_n = \sum_{k=1}^{n} a_k$$

The symbol Σ (the Greek letter sigma, which is an S in our alphabet) is simply an instruction to sum, or add up, the terms. The integer k is called the **index** of the sum; it tells you where to start the sum and where to end it. The expression

$$\sum_{k=1}^{n} a_k$$

is an instruction to add the terms a_k of the sequence $\{a_n\}$ starting with $k = 1$ and ending with $k = n$. We read the expression as "the sum of a_k from $k = 1$ to $k = n$."

EXAMPLE 7

Expanding Summation Notation

Write out each sum.

(a) $\displaystyle\sum_{k=1}^{n} \frac{1}{k}$ (b) $\displaystyle\sum_{k=1}^{n} k!$

Solution (a) $\displaystyle\sum_{k=1}^{n} \frac{1}{k} = 1 + \frac{1}{2} + \frac{1}{3} + \cdots + \frac{1}{n}$ (b) $\displaystyle\sum_{k=1}^{n} k! = 1! + 2! + \cdots + n!$

Now Work PROBLEM 51

EXAMPLE 8

Writing a Sum in Summation Notation

Express each sum using summation notation.

(a) $1^2 + 2^2 + 3^2 + \cdots + 9^2$ (b) $1 + \dfrac{1}{2} + \dfrac{1}{4} + \dfrac{1}{8} + \cdots + \dfrac{1}{2^{n-1}}$

Solution (a) The sum $1^2 + 2^2 + 3^2 + \cdots + 9^2$ has 9 terms, each of the form k^2, and starts at $k = 1$ and ends at $k = 9$:

$$1^2 + 2^2 + 3^2 + \cdots + 9^2 = \sum_{k=1}^{9} k^2$$

(b) The sum

$$1 + \frac{1}{2} + \frac{1}{4} + \frac{1}{8} + \cdots + \frac{1}{2^{n-1}}$$

has n terms, each of the form $\dfrac{1}{2^{k-1}}$, and starts at $k = 1$ and ends at $k = n$:

$$1 + \frac{1}{2} + \frac{1}{4} + \frac{1}{8} + \cdots + \frac{1}{2^{n-1}} = \sum_{k=1}^{n} \frac{1}{2^{k-1}}$$

The index of summation need not always begin at 1 or end at n; for example, we could have expressed the sum in Example 8(b) as

$$\sum_{k=0}^{n-1}\frac{1}{2^k} = 1 + \frac{1}{2} + \frac{1}{4} + \cdots + \frac{1}{2^{n-1}}$$

Letters other than k may be used as the index. For example,

$$\sum_{j=1}^{n} j! \quad \text{and} \quad \sum_{i=1}^{n} i!$$

each represent the same sum as the one given in Example 7(b).

Now Work PROBLEM 61

4 Find the Sum of a Sequence

Next we list some properties of sequences using summation notation. These properties are useful for adding the terms of a sequence.

THEOREM **Properties of Sequences**

If $\{a_n\}$ and $\{b_n\}$ are two sequences and c is a real number, then:

$$\sum_{k=1}^{n} (ca_k) = ca_1 + ca_2 + \cdots + ca_n = c(a_1 + a_2 + \cdots + a_n) = c\sum_{k=1}^{n} a_k \quad \textbf{(1)}$$

$$\sum_{k=1}^{n} (a_k + b_k) = \sum_{k=1}^{n} a_k + \sum_{k=1}^{n} b_k \quad \textbf{(2)}$$

$$\sum_{k=1}^{n} (a_k - b_k) = \sum_{k=1}^{n} a_k - \sum_{k=1}^{n} b_k \quad \textbf{(3)}$$

$$\sum_{k=j+1}^{n} a_k = \sum_{k=1}^{n} a_k - \sum_{k=1}^{j} a_k \quad \text{where } 0 < j < n \quad \textbf{(4)}$$

The proof of property (1) follows from the distributive property of real numbers. The proofs of properties 2 and 3 are based on the commutative and associative properties of real numbers. Property (4) states that the sum from $j + 1$ to n equals the sum from 1 to n minus the sum from 1 to j. It can be helpful to employ this property when the index of summation begins at a number larger than 1.

THEOREM **Formulas for Sums of Sequences**

$$\sum_{k=1}^{n} c = \underbrace{c + c + \cdots + c}_{n \text{ terms}} = cn \quad c \text{ is a real number} \quad \textbf{(5)}$$

$$\sum_{k=1}^{n} k = 1 + 2 + 3 + \cdots + n = \frac{n(n + 1)}{2} \quad \textbf{(6)}$$

$$\sum_{k=1}^{n} k^2 = 1^2 + 2^2 + 3^2 + \cdots + n^2 = \frac{n(n + 1)(2n + 1)}{6} \quad \textbf{(7)}$$

$$\sum_{k=1}^{n} k^3 = 1^3 + 2^3 + 3^3 + \cdots + n^3 = \left[\frac{n(n + 1)}{2}\right]^2 \quad \textbf{(8)}$$

The proof of formula (5) follows from the definition of summation notation. You are asked to prove formula (6) in Problem 92. The proofs of formulas (7) and (8) require mathematical induction, which is discussed in Section 9.4.

Notice the difference between formulas (5) and (6). In (5), the constant c is being summed from 1 to n, while in (6) the index of summation k is being summed from 1 to n.

EXAMPLE 9

Finding the Sum of a Sequence

Find the sum of each sequence.

(a) $\displaystyle\sum_{k=1}^{5} (3k)$ (b) $\displaystyle\sum_{k=1}^{10} (k^3 + 1)$

(c) $\displaystyle\sum_{k=1}^{24} (k^2 - 7k + 2)$ (d) $\displaystyle\sum_{k=6}^{20} (4k^2)$

Solution

(a) $\displaystyle\sum_{k=1}^{5} (3k) = 3\sum_{k=1}^{5} k$ Property (1)

$\displaystyle = 3\left(\frac{5(5+1)}{2}\right)$ Formula (6)

$= 3(15)$

$= 45$

(b) $\displaystyle\sum_{k=1}^{10} (k^3 + 1) = \sum_{k=1}^{10} k^3 + \sum_{k=1}^{10} 1$ Property (2)

$\displaystyle = \left(\frac{10(10+1)}{2}\right)^2 + 1(10)$ Formulas (8) and (5)

$= 3025 + 10$

$= 3035$

(c) $\displaystyle\sum_{k=1}^{24} (k^2 - 7k + 2) = \sum_{k=1}^{24} k^2 - \sum_{k=1}^{24} (7k) + \sum_{k=1}^{24} 2$ Properties (2) and (3)

$\displaystyle = \sum_{k=1}^{24} k^2 - 7\sum_{k=1}^{24} k + \sum_{k=1}^{24} 2$ Property (1)

$\displaystyle = \frac{24(24+1)(2\cdot 24+1)}{6} - 7\left(\frac{24(24+1)}{2}\right) + 2(24)$ Formulas (7), (6), (5)

$= 4900 - 2100 + 48$

$= 2848$

(d) Notice that the index of summation starts at 6. We use property (4) as follows:

$$\sum_{k=6}^{20} (4k^2) = 4\sum_{k=6}^{20} k^2 = 4\left[\sum_{k=1}^{20} k^2 - \sum_{k=1}^{5} k^2\right] = 4\left[\frac{20(21)(41)}{6} - \frac{5(6)(11)}{6}\right]$$

 Property (1) Property (4) Formula (7)

$$= 4[2870 - 55] = 11{,}260$$

♩

Now Work PROBLEM 73

9.1 Assess Your Understanding

'Are You Prepared?' *Answers are given at the end of these exercises. If you get a wrong answer, read the pages listed in red.*

1. For the function $f(x) = \dfrac{x-1}{x}$, find $f(2)$ and $f(3)$. (pp. 200–206)

2. *True or False* A function is a relation between two sets D and R so that each element x in the first set D is related to exactly one element y in the second set R. (pp. 200–206)

Concepts and Vocabulary

3. A(n) _____ is a function whose domain is the set of positive integers.

4. *True or False* The notation a_5 represents the fifth term of a sequence.

5. If $n \geq 0$ is an integer, then $n! = $ _____ when $n \geq 2$.

6. The sequence $a_1 = 5$, $a_n = 3a_{n-1}$ is an example of a _____ sequence.

7. The notation $a_1 + a_2 + a_3 + \cdots + a_n = \sum_{k=1}^{n} a_k$ is an example of _____ notation.

8. *True or False* $\sum_{k=1}^{n} k = 1 + 2 + 3 + \cdots + n = \dfrac{n(n+1)}{2}$

Skill Building

In Problems 9–14, evaluate each factorial expression.

9. $10!$

10. $9!$

11. $\dfrac{9!}{6!}$

12. $\dfrac{12!}{10!}$

13. $\dfrac{3! \, 7!}{4!}$

14. $\dfrac{5! \, 8!}{3!}$

In Problems 15–26, write down the first five terms of each sequence.

15. $\{s_n\} = \{n\}$

16. $\{s_n\} = \{n^2 + 1\}$

17. $\{a_n\} = \left\{\dfrac{n}{n+2}\right\}$

18. $\{b_n\} = \left\{\dfrac{2n+1}{2n}\right\}$

19. $\{c_n\} = \{(-1)^{n+1} n^2\}$

20. $\{d_n\} = \left\{(-1)^{n-1}\left(\dfrac{n}{2n-1}\right)\right\}$

21. $\{s_n\} = \left\{\dfrac{2^n}{3^n + 1}\right\}$

22. $\{s_n\} = \left\{\left(\dfrac{4}{3}\right)^n\right\}$

23. $\{t_n\} = \left\{\dfrac{(-1)^n}{(n+1)(n+2)}\right\}$

24. $\{a_n\} = \left\{\dfrac{3^n}{n}\right\}$

25. $\{b_n\} = \left\{\dfrac{n}{e^n}\right\}$

26. $\{c_n\} = \left\{\dfrac{n^2}{2^n}\right\}$

In Problems 27–34, the given pattern continues. Write down the nth term of a sequence $\{a_n\}$ suggested by the pattern.

27. $\dfrac{1}{2}, \dfrac{2}{3}, \dfrac{3}{4}, \dfrac{4}{5}, \cdots$

28. $\dfrac{1}{1 \cdot 2}, \dfrac{1}{2 \cdot 3}, \dfrac{1}{3 \cdot 4}, \dfrac{1}{4 \cdot 5}, \cdots$

29. $1, \dfrac{1}{2}, \dfrac{1}{4}, \dfrac{1}{8}, \cdots$

30. $\dfrac{2}{3}, \dfrac{4}{9}, \dfrac{8}{27}, \dfrac{16}{81}, \cdots$

31. $1, -1, 1, -1, 1, -1, \ldots$

32. $1, \dfrac{1}{2}, 3, \dfrac{1}{4}, 5, \dfrac{1}{6}, 7, \dfrac{1}{8}, \ldots$

33. $1, -2, 3, -4, 5, -6, \ldots$

34. $2, -4, 6, -8, 10, \ldots$

In Problems 35–48, a sequence is defined recursively. Write down the first five terms.

35. $a_1 = 2$; $a_n = 3 + a_{n-1}$

36. $a_1 = 3$; $a_n = 4 - a_{n-1}$

37. $a_1 = -2$; $a_n = n + a_{n-1}$

38. $a_1 = 1$; $a_n = n - a_{n-1}$

39. $a_1 = 5$; $a_n = 2a_{n-1}$

40. $a_1 = 2$; $a_n = -a_{n-1}$

41. $a_1 = 3$; $a_n = \dfrac{a_{n-1}}{n}$

42. $a_1 = -2$; $a_n = n + 3a_{n-1}$

43. $a_1 = 1$; $a_2 = 2$; $a_n = a_{n-1} \cdot a_{n-2}$

44. $a_1 = -1$; $a_2 = 1$; $a_n = a_{n-2} + na_{n-1}$

45. $a_1 = A$; $a_n = a_{n-1} + d$

46. $a_1 = A$; $a_n = ra_{n-1}$, $r \neq 0$

47. $a_1 = \sqrt{2}$; $a_n = \sqrt{2 + a_{n-1}}$

48. $a_1 = \sqrt{2}$; $a_n = \sqrt{\dfrac{a_{n-1}}{2}}$

In Problems 49–58, write out each sum.

49. $\sum_{k=1}^{n} (k + 2)$

50. $\sum_{k=1}^{n} (2k + 1)$

51. $\sum_{k=1}^{n} \dfrac{k^2}{2}$

52. $\sum_{k=1}^{n} (k + 1)^2$

53. $\sum_{k=0}^{n} \dfrac{1}{3^k}$

54. $\sum_{k=0}^{n} \left(\dfrac{3}{2}\right)^k$

55. $\sum_{k=0}^{n-1} \dfrac{1}{3^{k+1}}$

56. $\sum_{k=0}^{n-1} (2k + 1)$

57. $\sum_{k=2}^{n} (-1)^k \ln k$

58. $\sum_{k=3}^{n} (-1)^{k+1} 2^k$

In Problems 59–68, express each sum using summation notation.

59. $1 + 2 + 3 + \cdots + 20$

60. $1^3 + 2^3 + 3^3 + \cdots + 8^3$

61. $\dfrac{1}{2} + \dfrac{2}{3} + \dfrac{3}{4} + \cdots + \dfrac{13}{13 + 1}$

62. $1 + 3 + 5 + 7 + \cdots + [2(12) - 1]$

63. $1 - \dfrac{1}{3} + \dfrac{1}{9} - \dfrac{1}{27} + \cdots + (-1)^6 \left(\dfrac{1}{3^6}\right)$

64. $\dfrac{2}{3} - \dfrac{4}{9} + \dfrac{8}{27} - \cdots + (-1)^{12} \left(\dfrac{2}{3}\right)^{11}$

65. $3 + \dfrac{3^2}{2} + \dfrac{3^3}{3} + \cdots + \dfrac{3^n}{n}$

66. $\dfrac{1}{e} + \dfrac{2}{e^2} + \dfrac{3}{e^3} + \cdots + \dfrac{n}{e^n}$

67. $a + (a + d) + (a + 2d) + \cdots + (a + nd)$

68. $a + ar + ar^2 + \cdots + ar^{n-1}$

In Problems 69–80, find the sum of each sequence.

69. $\displaystyle\sum_{k=1}^{40} 5$

70. $\displaystyle\sum_{k=1}^{50} 8$

71. $\displaystyle\sum_{k=1}^{40} k$

72. $\displaystyle\sum_{k=1}^{24} (-k)$

73. $\displaystyle\sum_{k=1}^{20} (5k + 3)$

74. $\displaystyle\sum_{k=1}^{26} (3k - 7)$

75. $\displaystyle\sum_{k=1}^{16} (k^2 + 4)$

76. $\displaystyle\sum_{k=0}^{14} (k^2 - 4)$

77. $\displaystyle\sum_{k=10}^{60} (2k)$

78. $\displaystyle\sum_{k=8}^{40} (-3k)$

79. $\displaystyle\sum_{k=5}^{20} k^3$

80. $\displaystyle\sum_{k=4}^{24} k^3$

Applications and Extensions

81. Credit Card Debt John has a balance of $3000 on his Discover card that charges 1% interest per month on any unpaid balance. John can afford to pay $100 toward the balance each month. His balance each month after making a $100 payment is given by the recursively defined sequence

$$B_0 = \$3000 \qquad B_n = 1.01B_{n-1} - 100$$

Determine John's balance after making the first payment. That is, determine B_1.

82. Trout Population A pond currently has 2000 trout in it. A fish hatchery decides to add an additional 20 trout each month. In addition, it is known that the trout population is growing 3% per month. The size of the population after n months is given by the recursively defined sequence

$$p_0 = 2000 \qquad p_n = 1.03p_{n-1} + 20$$

How many trout are in the pond after two months? That is, what is p_2?

83. Car Loans Phil bought a car by taking out a loan for $18,500 at 0.5% interest per month. Phil's normal monthly payment is $434.47 per month, but he decides that he can afford to pay $100 extra toward the balance each month. His balance each month is given by the recursively defined sequence

$$B_0 = \$18,\!500 \qquad B_n = 1.005B_{n-1} - 534.47$$

Determine Phil's balance after making the first payment. That is, determine B_1.

84. Environmental Control The Environmental Protection Agency (EPA) determines that Maple Lake has 250 tons of pollutant as a result of industrial waste and that 10% of the pollutant present is neutralized by solar oxidation every year. The EPA imposes new pollution control laws that result in 15 tons of new pollutant entering the lake each year. The amount of pollutant in the lake after n years is given by the recursively defined sequence

$$p_0 = 250 \qquad p_n = 0.9p_{n-1} + 15$$

Determine the amount of pollutant in the lake after 2 years. That is, determine p_2.

85. Growth of a Rabbit Colony A colony of rabbits begins with one pair of mature rabbits, which will produce a pair of offspring (one male, one female) each month. Assume that all rabbits mature in 1 month and produce a pair of offspring (one

male, one female) after 2 months. If no rabbits ever die, how many pairs of mature rabbits are there after 7 months?

[**Hint:** A Fibonacci sequence models this colony. Do you see why?]

1 mature pair	
1 mature pair	
2 mature pairs	
3 mature pairs	

86. Fibonacci Sequence Let

$$u_n = \frac{\left(1 + \sqrt{5}\right)^n - \left(1 - \sqrt{5}\right)^n}{2^n\sqrt{5}}$$

define the nth term of a sequence.
(a) Show that $u_1 = 1$ and $u_2 = 1$.
(b) Show that $u_{n+2} = u_{n+1} + u_n$.
(c) Draw the conclusion that $\{u_n\}$ is a Fibonacci sequence.

87. Pascal's Triangle Divide the triangular array shown (called Pascal's triangle) using diagonal lines as indicated. Find the sum of the numbers in each diagonal row. Do you recognize this sequence?

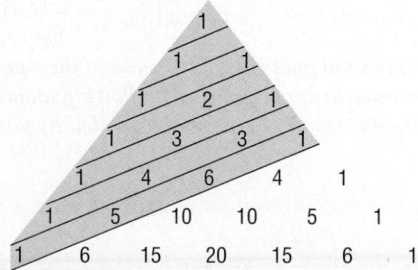

88. Fibonacci Sequence Use the result of Problem 86 to do the following problems:
(a) Write the first 11 terms of the Fibonacci sequence.
(b) Write the first 10 terms of the ratio $\dfrac{u_{n+1}}{u_n}$.
(c) As n gets large, what number does the ratio approach? This number is referred to as the **golden ratio**. Rectangles whose sides are in this ratio were considered pleasing to

the eye by the Greeks. For example, the façade of the Parthenon was constructed using the golden ratio.

(d) Write down the first 10 terms of the ratio $\dfrac{u_n}{u_{n+1}}$.

(e) As n gets large, what number does the ratio approach? This number is referred to as the **conjugate golden ratio.** This ratio is believed to have been used in the construction of the Great Pyramid in Egypt. The ratio equals the sum of the areas of the four face triangles divided by the total surface area of the Great Pyramid.

 89. Approximating $f(x) = e^x$ In calculus, it can be shown that

$$f(x) = e^x = \sum_{k=0}^{\infty} \frac{x^k}{k!}$$

We can approximate the value of $f(x) = e^x$ for any x using the following sum

$$f(x) = e^x \approx \sum_{k=0}^{n} \frac{x^k}{k!}$$

for some n.
(a) Approximate $f(1.3)$ with $n = 4$
(b) Approximate $f(1.3)$ with $n = 7$.
(c) Use a calculator to approximate $f(1.3)$.
(d) Using trial and error along with a graphing utility's SEQuence mode, determine the value of n required to approximate $f(1.3)$ correct to eight decimal places.

 90. Approximating $f(x) = e^x$ Refer to Problem 89.
(a) Approximate $f(-2.4)$ with $n = 3$.
(b) Approximate $f(-2.4)$ with $n = 6$.
(c) Use a calculator to approximate $f(-2.4)$.
(d) Using trial and error along with a graphing utility's SEQuence mode, determine the value of n required to approximate $f(-2.4)$ correct to eight decimal places.

91. Bode's Law In 1772, Johann Bode published the following formula for predicting the mean distances, in astronomical units (AU), of the planets from the sun:

$$a_1 = 0.4 \qquad a_n = 0.4 + 0.3 \cdot 2^{n-2}, \ n \ge 2$$

where n is the number of the planet from the sun.

(a) Determine the first eight terms of this sequence.
(b) At the time of Bode's publication, the known planets were Mercury (0.39 AU), Venus (0.72 AU), Earth (1 AU), Mars (1.52 AU), Jupiter (5.20 AU), and Saturn (9.54 AU). How do the actual distances compare to the terms of the sequence?
(c) The planet Uranus was discovered in 1781 and the asteroid Ceres was discovered in 1801. The mean orbital distances from the sun to Uranus and Ceres* are 19.2 AU and 2.77 AU, respectively. How well do these values fit within the sequence?
(d) Determine the ninth and tenth terms of Bode's sequence.
(e) The planets Neptune and Pluto* were discovered in 1846 and 1930, respectively. Their mean orbital distances from the sun are 30.07 AU and 39.44 AU, respectively. How do these actual distances compare to the terms of the sequence?
(f) On July 29, 2005, NASA announced the discovery of a dwarf planet* ($n = 11$), which has been named Eris. Use Bode's Law to predict the mean orbital distance of Eris from the sun. Its actual mean distance is not yet known, but Eris is currently about 97 astronomical units from the sun.

Source: NASA.

92. Show that

$$1 + 2 + \cdots + (n - 1) + n = \frac{n(n + 1)}{2}$$

[Hint: Let
$$S = 1 + 2 + \cdots + (n - 1) + n$$
$$S = n + (n - 1) + (n - 2) + \cdots + 1$$

Add these equations. Then

$$2S = \underbrace{[1 + n] + [2 + (n - 1)] + \cdots + [n + 1]}_{n \text{ terms in brackets}}$$

Now complete the derivation.**]**

* Ceres, Haumea, Makemake, Pluto, and Eris are referred to as dwarf planets.

Computing Square Roots *A method for approximating \sqrt{p} can be traced back to the Babylonians. The formula is given by the recursively defined sequence*

$$a_0 = k \qquad a_n = \frac{1}{2}\left(a_{n-1} + \frac{p}{a_{n-1}}\right)$$

where k is an initial guess as to the value of the square root. Use this recursive formula to approximate the following square roots by finding a_5. Compare this result to the value provided by your calculator.

93. $\sqrt{5}$

94. $\sqrt{8}$

95. $\sqrt{21}$

96. $\sqrt{89}$

97. Triangular Numbers A **triangular number** is a term of the sequence

$$u_1 = 1, \quad u_{n+1} = u_n + (n + 1)$$

Write down the first seven triangular numbers.

98. For the sequence given in Problem 97, show that

$$u_{n+1} = \frac{(n + 1)(n + 2)}{2}.$$

99. For the sequence given in Problem 97, show that

$$u_{n+1} + u_n = (n + 1)^2.$$

Explaining Concepts: Discussion and Writing

100. Investigate various applications that lead to a Fibonacci sequence, such as art, architecture, or financial markets. Write an essay on these applications.

101. Write a paragraph that explains why the numbers found in Problem 97 are called triangular.

'Are You Prepared?' Answers

1. $f(2) = \dfrac{1}{2}; f(3) = \dfrac{2}{3}$ **2.** True

9.2 Arithmetic Sequences

OBJECTIVES **1** Determine If a Sequence Is Arithmetic (p. 647)
 2 Find a Formula for an Arithmetic Sequence (p. 648)
 3 Find the Sum of an Arithmetic Sequence (p. 649)

1 Determine If a Sequence Is Arithmetic

When the difference between successive terms of a sequence is always the same number, the sequence is called **arithmetic.**

DEFINITION

An **arithmetic sequence*** may be defined recursively as $a_1 = a$, $a_n - a_{n-1} = d$, or as

$$a_1 = a, \qquad a_n = a_{n-1} + d \qquad \qquad (1)$$

where $a_1 = a$ and d are real numbers. The number a is the first term, and the number d is called the **common difference.**

The terms of an arithmetic sequence with first term a_1 and common difference d follow the pattern

$$a_1, \quad a_1 + d, \quad a_1 + 2d, \quad a_1 + 3d, \ldots$$

EXAMPLE 1

Determining If a Sequence Is Arithmetic

The sequence

$$4, \quad 6, \quad 8, \quad 10, \ldots$$

is arithmetic since the difference of successive terms is 2. The first term is $a_1 = 4$, and the common difference is $d = 2$.

EXAMPLE 2

Determining If a Sequence Is Arithmetic

Show that the following sequence is arithmetic. Find the first term and the common difference.

$$\{s_n\} = \{3n + 5\}$$

Solution

The first term is $s_1 = 3 \cdot 1 + 5 = 8$. The nth term and the $(n - 1)$st term of the sequence $\{s_n\}$ are

$$s_n = 3n + 5 \quad \text{and} \quad s_{n-1} = 3(n - 1) + 5 = 3n + 2$$

Their difference d is

$$d = s_n - s_{n-1} = (3n + 5) - (3n + 2) = 5 - 2 = 3$$

Since the difference of any two successive terms is the constant 3, the sequence $\{s_n\}$ is arithmetic and the common difference is 3.

EXAMPLE 3

Determining If a Sequence Is Arithmetic

Show that the sequence $\{t_n\} = \{4 - n\}$ is arithmetic. Find the first term and the common difference.

* Sometimes called an **arithmetic progression.**

Solution The first term is $t_1 = 4 - 1 = 3$. The nth term and the $(n - 1)$st term are

$$t_n = 4 - n \quad \text{and} \quad t_{n-1} = 4 - (n - 1) = 5 - n$$

Their difference d is

$$d = t_n - t_{n-1} = (4 - n) - (5 - n) = 4 - 5 = -1$$

Since the difference of any two successive terms is the constant -1, $\{t_n\}$ is an arithmetic sequence whose common difference is -1.

Now Work PROBLEM 7

2 Find a Formula for an Arithmetic Sequence

Suppose that a is the first term of an arithmetic sequence whose common difference is d. We seek a formula for the nth term, a_n. To see the pattern, we write down the first few terms.

$$a_1 = a$$
$$a_2 = a_1 + d = a_1 + 1 \cdot d$$
$$a_3 = a_2 + d = (a_1 + d) + d = a_1 + 2 \cdot d$$
$$a_4 = a_3 + d = (a_1 + 2 \cdot d) + d = a_1 + 3 \cdot d$$
$$a_5 = a_4 + d = (a_1 + 3 \cdot d) + d = a_1 + 4 \cdot d$$
$$\vdots$$
$$a_n = a_{n-1} + d = [a_1 + (n - 2)d] + d = a_1 + (n - 1)d$$

We are led to the following result:

THEOREM **nth Term of an Arithmetic Sequence**

For an arithmetic sequence $\{a_n\}$ whose first term is a_1 and whose common difference is d, the nth term is determined by the formula

$$\boxed{a_n = a_1 + (n - 1)d} \qquad (2)$$

EXAMPLE 4 **Finding a Particular Term of an Arithmetic Sequence**

Find the forty-first term of the arithmetic sequence: $2, 6, 10, 14, 18, \ldots$

Solution The first term of this arithmetic sequence is $a_1 = 2$, and the common difference is $d = 4$. By formula (2), the nth term is

$$a_n = 2 + (n - 1)4 \qquad a_n = a_1 + (n - 1)d; a_1 = 2, d = 4$$

The forty-first term is

$$a_{41} = 2 + (41 - 1) \cdot 4 = 162$$

EXAMPLE 5 **Finding a Recursive Formula for an Arithmetic Sequence**

The eighth term of an arithmetic sequence is 75, and the twentieth term is 39.

(a) Find the first term and the common difference.

(b) Give a recursive formula for the sequence.

(c) What is the nth term of the sequence?

Solution (a) By formula (2), we know that $a_n = a_1 + (n - 1)d$. As a result,

$$\begin{cases} a_8 = a_1 + 7d = 75 \\ a_{20} = a_1 + 19d = 39 \end{cases}$$

This is a system of two linear equations containing two variables, a_1 and d, which we can solve by elimination. Subtracting the second equation from the first, we get

$$-12d = 36$$
$$d = -3$$

With $d = -3$, we use $a_1 + 7d = 75$ and find that $a_1 = 75 - 7d = 75 - 7(-3) = 96$. The first term is $a_1 = 96$, and the common difference is $d = -3$.

(b) Using formula (1), a recursive formula for this sequence is

$$a_1 = 96, \qquad a_n = a_{n-1} - 3$$

(c) Using formula (2), a formula for the nth term of the sequence $\{a_n\}$ is

$$a_n = a_1 + (n - 1)d = 96 + (n - 1)(-3) = 99 - 3n$$

Exploration

Graph the recursive formula from Example 5, $a_1 = 96$, $a_n = a_{n-1} - 3$, using a graphing utility. Conclude that the graph of the recursive formula behaves like the graph of a linear function. How is d, the common difference, related to m, the slope of a line?

Now Work PROBLEMS **23** AND **29**

3 Find the Sum of an Arithmetic Sequence

The next result gives two formulas for finding the sum of the first n terms of an arithmetic sequence.

THEOREM

Sum of the First n Terms of an Arithmetic Sequence

Let $\{a_n\}$ be an arithmetic sequence with first term a_1 and common difference d. The sum S_n of the first n terms of $\{a_n\}$ may be found in two ways:

$$S_n = a_1 + a_2 + a_3 + \cdots + a_n$$

$$= \sum_{k=1}^{n} [a_1 + (k - 1)d] = \frac{n}{2}[2a_1 + (n - 1)d] \tag{3}$$

$$= \frac{n}{2}(a_1 + a_n) \tag{4}$$

Proof

$$S_n = a_1 + a_2 + a_3 + \cdots + a_n \qquad \text{Sum of first } n \text{ terms}$$

$$= a_1 + (a_1 + d) + (a_1 + 2d) + \cdots + [a_1 + (n - 1)d] \qquad \text{Formula (2)}$$

$$= \underbrace{(a_1 + a_1 + \cdots + a_1)}_{n \text{ terms}} + [d + 2d + \cdots + (n - 1)d] \qquad \text{Rearrange terms.}$$

$$= na_1 + d[1 + 2 + \cdots + (n - 1)]$$

$$= na_1 + d\left[\frac{(n - 1)n}{2}\right] \qquad \text{Formula 6, Section 9.1}$$

$$= na_1 + \frac{n}{2}(n - 1)d$$

$$= \frac{n}{2}[2a_1 + (n - 1)d] \qquad \text{Factor out } \frac{n}{2}; \text{ this is Formula (3).}$$

$$= \frac{n}{2}[a_1 + a_1 + (n - 1)d]$$

$$= \frac{n}{2}(a_1 + a_n) \qquad \text{Use Formula (2); this is Formula (4).}$$

∎

There are two ways to find the sum of the first n terms of an arithmetic sequence. Notice that formula (3) involves the first term and common difference, whereas formula (4) involves the first term and the nth term. Use whichever form is easier.

EXAMPLE 6	**Finding the Sum of an Arithmetic Sequence**

Find the sum S_n of the first n terms of the sequence $\{a_n\} = \{3n + 5\}$; that is, find

$$8 + 11 + 14 + \cdots + (3n + 5) = \sum_{k=1}^{n} (3k + 5)$$

Solution The sequence $\{a_n\} = \{3n + 5\}$ is an arithmetic sequence with first term $a_1 = 8$ and nth term $a_n = 3n + 5$. To find the sum S_n, use formula (4).

$$S_n = \sum_{k=1}^{n} (3k + 5) = \frac{n}{2}[8 + (3n + 5)] = \frac{n}{2}(3n + 13)$$

$$\uparrow$$
$$S_n = \frac{n}{2}(a_1 + a_n)$$

━━━━━**Now Work** PROBLEM 37

EXAMPLE 7	**Finding the Sum of an Arithmetic Sequence**

Find the sum: $60 + 64 + 68 + 72 + \cdots + 120$

Solution This is the sum S_n of an arithmetic sequence $\{a_n\}$ whose first term is $a_1 = 60$ and whose common difference is $d = 4$. The nth term is $a_n = 120$. Use formula (2) to find n.

$$
\begin{aligned}
a_n &= a_1 + (n - 1)d &&\text{Formula (2)} \\
120 &= 60 + (n - 1) \cdot 4 &&a_n = 120, a_1 = 60, d = 4 \\
60 &= 4(n - 1) &&\text{Simplify.} \\
15 &= n - 1 &&\text{Simplify.} \\
n &= 16 &&\text{Solve for } n.
\end{aligned}
$$

Now use formula (4) to find the sum S_{16}.

$$60 + 64 + 68 + \cdots + 120 = S_{16} = \frac{16}{2}(60 + 120) = 1440$$

$$\uparrow$$
$$S_n = \frac{n}{2}(a_1 + a_n)$$

━━━━━**Now Work** PROBLEM 41

EXAMPLE 8	**Creating a Floor Design**

A ceramic tile floor is designed in the shape of a trapezoid 20 feet wide at the base and 10 feet wide at the top. See Figure 7. The tiles, 12 inches by 12 inches, are to be placed so that each successive row contains one less tile than the preceding row. How many tiles will be required?

Figure 7

Solution The bottom row requires 20 tiles and the top row, 10 tiles. Since each successive row requires one less tile, the total number of tiles required is

$$S = 20 + 19 + 18 + \cdots + 11 + 10$$

This is the sum of an arithmetic sequence; the common difference is -1. The number of terms to be added is $n = 11$, with the first term $a_1 = 20$ and the last term $a_{11} = 10$. The sum S is

$$S = \frac{n}{2}(a_1 + a_{11}) = \frac{11}{2}(20 + 10) = 165$$

In all, 165 tiles will be required.

9.2 Assess Your Understanding

Concepts and Vocabulary

1. In a(n) _____ sequence, the difference between successive terms is a constant.

2. *True or False* For an arithmetic sequence $\{a_n\}$ whose first term is a_1 and whose common difference is d, the nth term is determined by the formula $a_n = a_1 + nd$.

3. If the fifth term of an arithmetic sequence is 12 and the common difference is 5, then the sixth term of the sequence is _____.

4. *True or False* The sum S_n of the first n terms of an arithmetic sequence $\{a_n\}$ whose first term is a_1 can be found using the formula $S_n = \frac{n}{2}(a_1 + a_n)$.

Skill Building

In Problems 5–14, show that each sequence is arithmetic. Find the common difference and write out the first four terms.

5. $\{s_n\} = \{n + 4\}$ **6.** $\{s_n\} = \{n - 5\}$ **7.** $\{a_n\} = \{2n - 5\}$ **8.** $\{b_n\} = \{3n + 1\}$ **9.** $\{c_n\} = \{6 - 2n\}$

10. $\{a_n\} = \{4 - 2n\}$ **11.** $\{t_n\} = \left\{\frac{1}{2} - \frac{1}{3}n\right\}$ **12.** $\{t_n\} = \left\{\frac{2}{3} + \frac{n}{4}\right\}$ **13.** $\{s_n\} = \{\ln 3^n\}$ **14.** $\{s_n\} = \{e^{\ln n}\}$

In Problems 15–22, find the nth term of the arithmetic sequence $\{a_n\}$ whose initial term a and common difference d are given. What is the fifty-first term?

15. $a_1 = 2$; $d = 3$ **16.** $a_1 = -2$; $d = 4$ **17.** $a_1 = 5$; $d = -3$ **18.** $a_1 = 6$; $d = -2$

19. $a_1 = 0$; $d = \frac{1}{2}$ **20.** $a_1 = 1$; $d = -\frac{1}{3}$ **21.** $a_1 = \sqrt{2}$; $d = \sqrt{2}$ **22.** $a_1 = 0$; $d = \pi$

In Problems 23–28, find the indicated term in each arithmetic sequence.

23. 100th term of $2, 4, 6, \ldots$ **24.** 80th term of $-1, 1, 3, \ldots$ **25.** 90th term of $1, -2, -5, \ldots$

26. 80th term of $5, 0, -5, \ldots$ **27.** 80th term of $2, \frac{5}{2}, 3, \frac{7}{2}, \ldots$ **28.** 70th term of $2\sqrt{5}, 4\sqrt{5}, 6\sqrt{5}, \ldots$

In Problems 29–36, find the first term and the common difference of the arithmetic sequence described. Give a recursive formula for the sequence. Find a formula for the nth term.

29. 8th term is 8; 20th term is 44 **30.** 4th term is 3; 20th term is 35 **31.** 9th term is -5; 15th term is 31

32. 8th term is 4; 18th term is -96 **33.** 15th term is 0; 40th term is -50 **34.** 5th term is -2; 13th term is 30

35. 14th term is -1; 18th term is -9 **36.** 12th term is 4; 18th term is 28

In Problems 37–54, find each sum.

37. $1 + 3 + 5 + \cdots + (2n - 1)$ **38.** $2 + 4 + 6 + \cdots + 2n$ **39.** $7 + 12 + 17 + \cdots + (2 + 5n)$

40. $-1 + 3 + 7 + \cdots + (4n - 5)$ **41.** $2 + 4 + 6 + \cdots + 70$ **42.** $1 + 3 + 5 + \cdots + 59$

43. $5 + 9 + 13 + \cdots + 49$ **44.** $2 + 5 + 8 + \cdots + 41$ **45.** $73 + 78 + 83 + 88 + \cdots + 558$

46. $7 + 1 - 5 - 11 - \cdots - 299$ **47.** $4 + 4.5 + 5 + 5.5 + \cdots + 100$ **48.** $8 + 8\frac{1}{4} + 8\frac{1}{2} + 8\frac{3}{4} + 9 + \cdots + 50$

49. $\displaystyle\sum_{n=1}^{80}(2n-5)$

50. $\displaystyle\sum_{n=1}^{90}(3-2n)$

51. $\displaystyle\sum_{n=1}^{100}\left(6-\frac{1}{2}n\right)$

52. $\displaystyle\sum_{n=1}^{80}\left(\frac{1}{3}n+\frac{1}{2}\right)$

53. The sum of the first 120 terms of the sequence 14, 16, 18, 20,

54. The sum of the first 46 terms of the sequence 2, −1, −4, −7,

Applications and Extensions

55. Find x so that $x+3$, $2x+1$, and $5x+2$ are consecutive terms of an arithmetic sequence.

56. Find x so that $2x$, $3x+2$, and $5x+3$ are consecutive terms of an arithmetic sequence.

57. How many terms must be added in an arithmetic sequence whose first term is 11 and whose common difference is 3 to obtain a sum of 1092?

58. How many terms must be added in an arithmetic sequence whose first term is 78 and whose common difference is −4 to obtain a sum of 702?

59. Drury Lane Theater The Drury Lane Theater has 25 seats in the first row and 30 rows in all. Each successive row contains one additional seat. How many seats are in the theater?

60. Football Stadium The corner section of a football stadium has 15 seats in the first row and 40 rows in all. Each successive row contains two additional seats. How many seats are in this section?

61. Creating a Mosaic A mosaic is designed in the shape of an equilateral triangle, 20 feet on each side. Each tile in the mosaic is in the shape of an equilateral triangle, 12 inches to a side. The tiles are to alternate in color as shown in the illustration. How many tiles of each color will be required?

62. Constructing a Brick Staircase A brick staircase has a total of 30 steps. The bottom step requires 100 bricks. Each successive step requires two less bricks than the prior step.
(a) How many bricks are required for the top step?
(b) How many bricks are required to build the staircase?

63. Cooling Air As a parcel of air rises (for example, as it is pushed over a mountain), it cools at the *dry adiabatic lapse rate* of 5.5°F per 1000 feet until it reaches its dew point. If the ground temperature is 67°F, write a formula for the sequence of temperatures, $\{T_n\}$, of a parcel of air that has risen n thousand feet. What is the temperature of a parcel of air if it has risen 5000 feet?

Source: National Aeronautics and Space Administration

64. Citrus Ladders Ladders used by fruit pickers are typically tapered with a wide bottom for stability and a narrow top for ease of picking. If the bottom rung of such a ladder is 49 inches wide and the top rung is 24 inches wide, how many rungs does the ladder have if each rung is 2.5 inches shorter than the one below it? How much material would be needed to make the rungs for the ladder described?

Source: www.stokesladders.com

65. Seats in an Amphitheater An outdoor amphitheater has 35 seats in the first row, 37 in the second row, 39 in the third row, and so on. There are 27 rows altogether. How many can the amphitheater seat?

66. Stadium Construction How many rows are in the corner section of a stadium containing 2040 seats if the first row has 10 seats and each successive row has 4 additional seats?

67. Salary If you received a job offer with a starting salary of $35,000 per year and a guaranteed raise of $1400 per year, how many years will it take before your aggregate salary is $280,000?

[**Hint:** Your aggregate salary after 2 years is $35,000 + ($35,000 + $1400).]

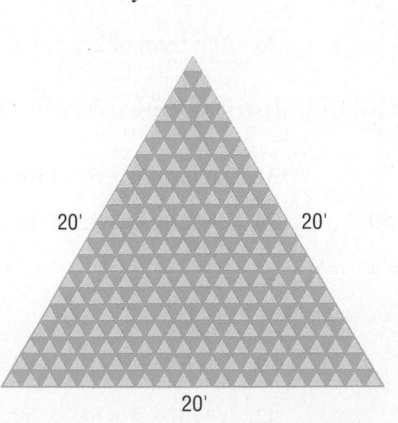

20' 20'

20'

Explaining Concepts: Discussion and Writing

68. Make up an arithmetic sequence. Give it to a friend and ask for its twentieth term.

69. Describe the similarities and differences between arithmetic sequences and linear functions.

9.3 Geometric Sequences; Geometric Series

PREPARING FOR THIS SECTION *Before getting started, review the following:*

- Compound Interest (Section 6.7, pp. 466–472)

Now Work the '**Are You Prepared?**' problems on page 661.

OBJECTIVES 1 Determine If a Sequence Is Geometric (p. 653)
 2 Find a Formula for a Geometric Sequence (p. 654)
 3 Find the Sum of a Geometric Sequence (p. 655)
 4 Determine Whether a Geometric Series Converges or Diverges (p. 656)
 5 Solve Annuity Problems (p. 659)

1 Determine If a Sequence Is Geometric

When the ratio of successive terms of a sequence is always the same nonzero number, the sequence is called **geometric**.

DEFINITION

A **geometric sequence*** may be defined recursively as $a_1 = a$, $\dfrac{a_n}{a_{n-1}} = r$, or as

$$a_1 = a, \qquad a_n = ra_{n-1} \tag{1}$$

where $a_1 = a$ and $r \neq 0$ are real numbers. The number a_1 is the first term, and the nonzero number r is called the **common ratio.**

The terms of a geometric sequence with first term a_1 and common ratio r follow the pattern

$$a_1, \quad a_1r, \quad a_1r^2, \quad a_1r^3, \dots$$

EXAMPLE 1 **Determining If a Sequence Is Geometric**

The sequence

$$2, \ 6, \ 18, \ 54, \ 162, \dots$$

is geometric since the ratio of successive terms is 3; $\left(\dfrac{6}{2} = \dfrac{18}{6} = \dfrac{54}{18} = \dots = 3 \right)$. The first term is $a_1 = 2$, and the common ratio is 3.

EXAMPLE 2 **Determining If a Sequence Is Geometric**

Show that the following sequence is geometric.

$$\{s_n\} = 2^{-n}$$

Find the first term and the common ratio.

Solution The first term is $s_1 = 2^{-1} = \dfrac{1}{2}$. The nth term and the $(n-1)$st term of the sequence $\{s_n\}$ are

$$s_n = 2^{-n} \quad \text{and} \quad s_{n-1} = 2^{-(n-1)}$$

Their ratio is

$$\frac{s_n}{s_{n-1}} = \frac{2^{-n}}{2^{-(n-1)}} = 2^{-n+(n-1)} = 2^{-1} = \frac{1}{2}$$

* Sometimes called a **geometric progression.**

Because the ratio of successive terms is the nonzero constant $\frac{1}{2}$, the sequence $\{s_n\}$ is geometric with common ratio $\frac{1}{2}$.

EXAMPLE 3

Determining If a Sequence Is Geometric

Show that the following sequence is geometric.

$$\{t_n\} = \{3 \cdot 4^n\}$$

Find the first term and the common ratio.

Solution The first term is $t_1 = 3 \cdot 4^1 = 12$. The nth term and the $(n-1)$st term are

$$t_n = 3 \cdot 4^n \quad \text{and} \quad t_{n-1} = 3 \cdot 4^{n-1}$$

Their ratio is

$$\frac{t_n}{t_{n-1}} = \frac{3 \cdot 4^n}{3 \cdot 4^{n-1}} = 4^{n-(n-1)} = 4$$

The sequence, $\{t_n\}$, is a geometric sequence with common ratio 4.

━━━━━━► **Now Work** PROBLEM 11

2 Find a Formula for a Geometric Sequence

Suppose that a_1 is the first term of a geometric sequence with common ratio $r \neq 0$. We seek a formula for the nth term a_n. To see the pattern, we write down the first few terms:

$$a_1 = a_1 \cdot 1 = a_1 r^0$$
$$a_2 = r a_1 = a_1 r^1$$
$$a_3 = r a_2 = r(a_1 r) = a_1 r^2$$
$$a_4 = r a_3 = r(a_1 r^2) = a_1 r^3$$
$$a_5 = r a_4 = r(a_1 r^3) = a_1 r^4$$
$$\vdots$$
$$a_n = r a_{n-1} = r(a_1 r^{n-2}) = a_1 r^{n-1}$$

We are led to the following result:

THEOREM

nth Term of a Geometric Sequence

For a geometric sequence $\{a_n\}$ whose first term is a_1 and whose common ratio is r, the nth term is determined by the formula

$$a_n = a_1 r^{n-1} \qquad r \neq 0 \tag{2}$$

EXAMPLE 4

Finding a Particular Term of a Geometric Sequence

(a) Find the nth term of the geometric sequence: $10, 9, \dfrac{81}{10}, \dfrac{729}{100} \cdots$

(b) Find the ninth term of this sequence.

(c) Find a recursive formula for this sequence.

Solution

(a) The first term of this geometric sequence is $a_1 = 10$ and the common ratio is $\dfrac{9}{10}$. (Use $\dfrac{9}{10}$ or $\dfrac{\frac{81}{10}}{9} = \dfrac{9}{10}$ or any two successive terms.) Then, by formula (2), the nth term is

$$a_n = 10\left(\frac{9}{10}\right)^{n-1} \qquad a_n = a_1 r^{n-1};\ a_1 = 10,\ r = \frac{9}{10}$$

(b) The ninth term is

$$a_9 = 10\left(\frac{9}{10}\right)^{9-1} = 10\left(\frac{9}{10}\right)^8 \approx 4.3046721$$

(c) The first term in the sequence is 10 and the common ratio is $r = \dfrac{9}{10}$. Using formula (1), the recursive formula is $a_1 = 10,\ a_n = \dfrac{9}{10}a_{n-1}$.

The Exploration sidebar text.

Exploration

Use a graphing utility to find the ninth term of the sequence given in Example 4. Use it to find the twentieth and fiftieth terms. Now use a graphing utility to graph the recursive formula found in Example 4(c). Conclude that the graph of the recursive formula behaves like the graph of an exponential function. How is r, the common ratio, related to a, the base of the exponential function $y = a^x$?

Now Work PROBLEMS 19, 27, AND 35

3 Find the Sum of a Geometric Sequence

THEOREM

Sum of the First n Terms of a Geometric Sequence

Let $\{a_n\}$ be a geometric sequence with first term a_1 and common ratio r, where $r \neq 0, r \neq 1$. The sum S_n of the first n terms of $\{a_n\}$ is

$$S_n = a_1 + a_1 r + a_1 r^2 + \cdots + a_1 r^{n-1} = \sum_{k=1}^{n} a_1 r^{k-1}$$

$$= a_1 \cdot \frac{1 - r^n}{1 - r} \qquad r \neq 0, 1 \tag{3}$$

Proof The sum S_n of the first n terms of $\{a_n\} = \{a_1 r^{n-1}\}$ is

$$S_n = a_1 + a_1 r + \cdots + a_1 r^{n-1} \tag{4}$$

Multiply each side by r to obtain

$$r S_n = a_1 r + a_1 r^2 + \cdots + a_1 r^n \tag{5}$$

Now, subtract (5) from (4). The result is

$$S_n - r S_n = a_1 - a_1 r^n$$
$$(1 - r)S_n = a_1(1 - r^n)$$

Since $r \neq 1$, we can solve for S_n.

$$S_n = a_1 \cdot \frac{1 - r^n}{1 - r}$$

∎

EXAMPLE 5

Finding the Sum of the First n Terms of a Geometric Sequence

Find the sum S_n of the first n terms of the sequence $\left\{\left(\dfrac{1}{2}\right)^n\right\}$; that is, find

$$\frac{1}{2} + \frac{1}{4} + \frac{1}{8} + \cdots + \left(\frac{1}{2}\right)^n = \sum_{k=1}^{n} \frac{1}{2}\left(\frac{1}{2}\right)^{k-1}$$

Solution The sequence $\left\{\left(\frac{1}{2}\right)^n\right\}$ is a geometric sequence with $a_1 = \frac{1}{2}$ and $r = \frac{1}{2}$. Use formula (3) to get

$$S_n = \sum_{k=1}^{n} \frac{1}{2}\left(\frac{1}{2}\right)^{k-1} = \frac{1}{2} + \frac{1}{4} + \frac{1}{8} + \cdots + \left(\frac{1}{2}\right)^n$$

$$= \frac{1}{2}\left[\frac{1 - \left(\frac{1}{2}\right)^n}{1 - \frac{1}{2}}\right] \qquad \text{Formula (3); } a_1 = \frac{1}{2}, r = \frac{1}{2}$$

$$= \frac{1}{2}\left[\frac{1 - \left(\frac{1}{2}\right)^n}{\frac{1}{2}}\right]$$

$$= 1 - \left(\frac{1}{2}\right)^n$$

━━━ **Now Work** PROBLEM 41

EXAMPLE 6 **Using a Graphing Utility to Find the Sum of a Geometric Sequence**

Use a graphing utility to find the sum of the first 15 terms of the sequence $\left\{\left(\frac{1}{3}\right)^n\right\}$; that is, find

$$\frac{1}{3} + \frac{1}{9} + \frac{1}{27} + \cdots + \left(\frac{1}{3}\right)^{15} = \sum_{k=1}^{15} \frac{1}{3}\left(\frac{1}{3}\right)^{k-1}$$

Figure 8

Solution Figure 8 shows the result obtained using a TI-84 Plus graphing calculator. The sum of the first 15 terms of the sequence $\left\{\left(\frac{1}{3}\right)^n\right\}$ is approximately 0.4999999652.

━━━ **Now Work** PROBLEM 47

4 Determine Whether a Geometric Series Converges or Diverges

DEFINITION An infinite sum of the form

$$a_1 + a_1 r + a_1 r^2 + \cdots + a_1 r^{n-1} + \cdots$$

with first term a_1 and common ratio r, is called an **infinite geometric series** and is denoted by

$$\sum_{k=1}^{\infty} a_1 r^{k-1}$$

Based on formula (3), the sum S_n of the first n terms of a geometric series is

$$S_n = a_1 \cdot \frac{1 - r^n}{1 - r} = \frac{a_1}{1 - r} - \frac{a_1 r^n}{1 - r} \tag{6}$$

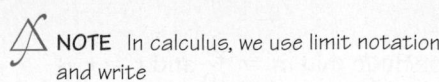

NOTE In calculus, we use limit notation and write

$$L = \lim_{n \to \infty} S_n = \lim_{n \to \infty} \sum_{k=1}^{n} a_1 r^{k-1} = \sum_{k=1}^{\infty} a_1 r^{k-1}$$ ∎

If this finite sum S_n approaches a number L as $n \to \infty$, we say the infinite geometric series $\sum_{k=1}^{\infty} a_1 r^{k-1}$ **converges.** We call L the **sum of the infinite geometric series** and we write

$$L = \sum_{k=1}^{\infty} a_1 r^{k-1}$$

If a series does not converge, it is called a **divergent series.**

THEOREM

Convergence of an Infinite Geometric Series

If $|r| < 1$, the infinite geometric series $\sum_{k=1}^{\infty} a_1 r^{k-1}$ converges. Its sum is

$$\sum_{k=1}^{\infty} a_1 r^{k-1} = \frac{a_1}{1-r} \qquad (7)$$

Intuitive Proof Since $|r| < 1$, it follows that $|r^n|$ approaches 0 as $n \to \infty$. Then, based on formula (6), the term $\dfrac{a_1 r^n}{1-r}$ approaches 0, so the sum S_n approaches $\dfrac{a_1}{1-r}$ as $n \to \infty$. ∎

EXAMPLE 7

Determining Whether a Geometric Series Converges or Diverges

Determine if the geometric series

$$\sum_{k=1}^{\infty} 2 \left(\frac{2}{3} \right)^{k-1} = 2 + \frac{4}{3} + \frac{8}{9} + \cdots$$

converges or diverges. If it converges, find its sum.

Solution Comparing $\sum_{k=1}^{\infty} 2 \left(\dfrac{2}{3} \right)^{k-1}$ to $\sum_{k=1}^{\infty} a_1 r^{k-1}$, the first term is $a_1 = 2$ and the common ratio is $r = \dfrac{2}{3}$. Since $|r| < 1$, the series converges. Use formula (7) to find its sum:

$$\sum_{k=1}^{\infty} 2 \left(\frac{2}{3} \right)^{k-1} = 2 + \frac{4}{3} + \frac{8}{9} + \cdots = \frac{2}{1 - \dfrac{2}{3}} = 6$$

━━━━ **Now Work** PROBLEM 53

EXAMPLE 8

Repeating Decimals

Show that the repeating decimal $0.999\ldots$ equals 1.

Solution The decimal $0.999\ldots = 0.9 + 0.09 + 0.009 + \cdots = \dfrac{9}{10} + \dfrac{9}{100} + \dfrac{9}{1000} + \cdots$ is an

infinite geometric series. We will write it in the form $\sum_{k=1}^{\infty} a_1 r^{k-1}$ so that we can use formula (7).

$$0.999\ldots = \frac{9}{10} + \frac{9}{100} + \frac{9}{1000} + \cdots = \sum_{k=1}^{\infty} \frac{9}{10^k} = \sum_{k=1}^{\infty} \frac{9}{10 \cdot 10^{k-1}} = \sum_{k=1}^{\infty} \frac{9}{10} \left(\frac{1}{10} \right)^{k-1}$$

Now we can compare this series to $\sum_{k=1}^{\infty} a_1 r^{k-1}$ and conclude that $a_1 = \dfrac{9}{10}$ and $r = \dfrac{1}{10}$. Since $|r| < 1$, the series converges and its sum is

$$0.999\ldots = \dfrac{\dfrac{9}{10}}{1 - \dfrac{1}{10}} = \dfrac{\dfrac{9}{10}}{\dfrac{9}{10}} = 1$$

The repeating decimal $0.999\ldots$ equals 1.

EXAMPLE 9

Pendulum Swings

Initially, a pendulum swings through an arc of 18 inches. See Figure 9. On each successive swing, the length of the arc is 0.98 of the previous length.

(a) What is the length of the arc of the 10th swing?
(b) On which swing is the length of the arc first less than 12 inches?
(c) After 15 swings, what total distance will the pendulum have swung?
(d) When it stops, what total distance will the pendulum have swung?

Figure 9

Solution

(a) The length of the first swing is 18 inches.
The length of the second swing is 0.98(18) inches.
The length of the third swing is $0.98(0.98)(18) = 0.98^2(18)$ inches.
The length of the arc of the 10th swing is

$$(0.98)^9(18) \approx 15.007 \text{ inches}$$

(b) The length of the arc of the nth swing is $(0.98)^{n-1}(18)$. For this to be exactly 12 inches requires that

$$(0.98)^{n-1}(18) = 12$$

$$(0.98)^{n-1} = \dfrac{12}{18} = \dfrac{2}{3} \qquad \text{Divide both sides by 18.}$$

$$n - 1 = \log_{0.98}\left(\dfrac{2}{3}\right) \qquad \text{Express as a logarithm.}$$

$$n = 1 + \dfrac{\ln\left(\dfrac{2}{3}\right)}{\ln 0.98} \approx 1 + 20.07 = 21.07 \qquad \text{Solve for } n; \text{ use the Change of Base Formula.}$$

The length of the arc of the pendulum exceeds 12 inches on the 21st swing and is first less than 12 inches on the 22nd swing.

(c) After 15 swings, the pendulum will have swung the following total distance L:

$$L = 18 + \underset{\text{1st}}{18} + \underset{\text{2nd}}{0.98(18)} + \underset{\text{3rd}}{(0.98)^2(18)} + \underset{\text{4th}}{(0.98)^3(18)} + \cdots + \underset{\text{15th}}{(0.98)^{14}(18)}$$

This is the sum of a geometric sequence. The common ratio is 0.98; the first term is 18. The sum has 15 terms, so

$$L = 18 \cdot \dfrac{1 - 0.98^{15}}{1 - 0.98} \approx 18(13.07) \approx 235.3 \text{ inches}$$

The pendulum will have swung through approximately 235.3 inches after 15 swings.

(d) When the pendulum stops, it will have swung the following total distance T:

$$T = 18 + 0.98(18) + (0.98)^2(18) + (0.98)^3(18) + \cdots$$

This is the sum of an infinite geometric series. The common ratio is $r = 0.98$; the first term is $a_1 = 18$. Since $|r| < 1$, the series converges. Its sum is

$$T = \dfrac{a_1}{1 - r} = \dfrac{18}{1 - 0.98} = 900$$

The pendulum will have swung a total of 900 inches when it finally stops.

Now Work PROBLEM **87**

5 Solve Annuity Problems

In Section 6.7 we developed the compound interest formula that gives the future value when a fixed amount of money is deposited in an account that pays interest compounded periodically. Often, though, money is invested in small amounts at periodic intervals. An **annuity** is a sequence of equal periodic deposits. The periodic deposits may be made annually, quarterly, monthly, or daily.

When deposits are made at the same time that the interest is credited, the annuity is called **ordinary.** We will only deal with ordinary annuities here. The **amount of an annuity** is the sum of all deposits made plus all interest paid.

Suppose that the interest rate that an account earns is i percent per payment period (expressed as a decimal). For example, if an account pays 12% compounded monthly (12 times a year), then $i = \dfrac{0.12}{12} = 0.01$. If an account pays 8% compounded quarterly (4 times a year), then $i = \dfrac{0.08}{4} = 0.02$.

To develop a formula for the amount of an annuity, suppose that $\$P$ is deposited each payment period for n payment periods in an account that earns i percent per payment period. When the last deposit is made at the nth payment period, the first deposit of $\$P$ has earned interest compounded for $n - 1$ payment periods, the second deposit of $\$P$ has earned interest compounded for $n - 2$ payment periods, and so on. Table 3 shows the value of each deposit after n deposits have been made.

Table 3

Deposit	1	2	3	...	$n - 1$	n
Amount	$P(1 + i)^{n-1}$	$P(1 + i)^{n-2}$	$P(1 + i)^{n-3}$...	$P(1 + i)$	P

The amount A of the annuity is the sum of the amounts shown in Table 3; that is,

$$A = P \cdot (1 + i)^{n-1} + P \cdot (1 + i)^{n-2} + \cdots + P \cdot (1 + i) + P$$

$$= P[1 + (1 + i) + \cdots + (1 + i)^{n-1}]$$

The expression in brackets is the sum of a geometric sequence with n terms and a common ratio of $(1 + i)$. As a result,

$$A = P[1 + (1 + i) + \cdots + (1 + i)^{n-2} + (1 + i)^{n-1}]$$

$$= P\frac{1 - (1 + i)^n}{1 - (1 + i)} = P\frac{1 - (1 + i)^n}{-i} = P\frac{(1 + i)^n - 1}{i}$$

We have established the following result:

THEOREM

Amount of an Annuity

Suppose that P is the deposit in dollars made at the end of each payment period for an annuity paying i percent interest per payment period. The amount A of the annuity after n deposits is

$$A = P\frac{(1 + i)^n - 1}{i} \tag{8}$$

NOTE In using formula (8), remember that when the nth deposit is made the first deposit has earned interest for $n - 1$ compounding periods. ∎

EXAMPLE 10

Determining the Amount of an Annuity

To save for retirement, Brett decides to place $4000 into an Individual Retirement Account (IRA) each year for the next 30 years. What will the value of the IRA be when Brett makes his 30th deposit? Assume that the rate of return of the IRA is 10% per annum compounded annually. (This is the historical rate of return in the stock market.)

Solution This is an ordinary annuity with $n = 30$ annual deposits of $P = \$4000$. The rate of interest per payment period is $i = \dfrac{0.10}{1} = 0.10$. The amount A of the annuity after 30 deposits is

$$A = \$4000 \,\frac{(1 + 0.10)^{30} - 1}{0.10} \approx \$4000(164.494023) \approx \$657,976.09$$

EXAMPLE 11 **Determining the Amount of an Annuity**

To save for her daughter's college education, Ms. Miranda decides to put \$50 aside every month in a credit union account paying 10% interest compounded monthly. She begins this savings program when her daughter is 3 years old. How much will she have saved by the time she makes the 180th deposit? How old is her daughter at this time?

Solution This is an annuity with $P = \$50$, $n = 180$, and $i = \dfrac{0.10}{12}$. The amount A saved is

$$A = \$50 \,\frac{\left(1 + \dfrac{0.10}{12}\right)^{180} - 1}{\dfrac{0.10}{12}} \approx \$50(414.47035) \approx \$20,723.52$$

Since there are 12 deposits per year, when the 180th deposit is made $\dfrac{180}{12} = 15$ years have passed and Ms. Miranda's daughter is 18 years old.

Now Work PROBLEM 91

Historical Feature

Fibonacci

Sequences are among the oldest objects of mathematical investigation, having been studied for over 3500 years. After the initial steps, however, little progress was made until about 1600.

Arithmetic and geometric sequences appear in the Rhind papyrus, a mathematical text containing 85 problems copied around 1650 BC by the Egyptian scribe Ahmes from an earlier work (see Historical Problem 1). Fibonacci (AD 1220) wrote about problems similar to those found in the Rhind papyrus, leading one to suspect that Fibonacci may have had material available that is now lost. This material would have been in the non-Euclidean Greek tradition of Heron (about AD 75) and Diophantus (about AD 250).

The Rhind papyrus indicates that the Egyptians knew how to add up the terms of an arithmetic or geometric sequence, as did the Babylonians. The rule for summing up a geometric sequence is found in Euclid's *Elements* (Book IX, 35, 36), where, like all Euclid's algebra, it is presented in a geometric form.

Investigations of other kinds of sequences began in the 1500s, when algebra became sufficiently developed to handle the more complicated problems. The development of calculus in the 1600s added a powerful new tool, especially for finding the sum of an infinite series, and the subject continues to flourish today.

Historical Problem

1. *Arithmetic sequence problem from the Rhind papyrus (statement modified slightly for clarity)* One hundred loaves of bread are to be divided among five people so that the amounts that they receive form an arithmetic sequence. The first two together receive one-seventh of what the last three receive. How many loaves does each receive?

 [*Partial answer:* First person receives $1\dfrac{2}{3}$ loaves.]

9.3 Assess Your Understanding

'Are You Prepared?' *Answers are given at the end of these exercises. If you get a wrong answer, read the pages listed in red.*

1. If \$1000 is invested at 4% per annum compounded semiannually, how much is in the account after 2 years? (pp. 466–472)

2. How much do you need to invest now at 5% per annum compounded monthly so that in 1 year you will have \$10,000? (pp. 466–472)

Concepts and Vocabulary

3. In a(n) _____ sequence the ratio of successive terms is a constant.

4. If $|r| < 1$, the sum of the geometric series $\sum_{k=1}^{\infty} ar^{k-1}$ is _____.

5. If a series does not converge, it is called a _____ _____.

6. **True or False** A geometric sequence may be defined recursively.

7. **True or False** In a geometric sequence the common ratio is always a positive number.

8. **True or False** For a geometric sequence with first term a_1 and common ratio r, where $r \neq 0$, $r \neq 1$, the sum of the first n terms is $S_n = a_1 \cdot \dfrac{1 - r^n}{1 - r}$.

Skill Building

In Problems 9–18, show that each sequence is geometric. Then find the common ratio and write out the first four terms.

9. $\{s_n\} = \{3^n\}$

10. $\{s_n\} = \{(-5)^n\}$

11. $\{a_n\} = \left\{ -3\left(\dfrac{1}{2}\right)^n \right\}$

12. $\{b_n\} = \left\{ \left(\dfrac{5}{2}\right)^n \right\}$

13. $\{c_n\} = \left\{ \dfrac{2^{n-1}}{4} \right\}$

14. $\{d_n\} = \left\{ \dfrac{3^n}{9} \right\}$

15. $\{e_n\} = \{2^{n/3}\}$

16. $\{f_n\} = \{3^{2n}\}$

17. $\{t_n\} = \left\{ \dfrac{3^{n-1}}{2^n} \right\}$

18. $\{u_n\} = \left\{ \dfrac{2^n}{3^{n-1}} \right\}$

In Problems 19–26, find the fifth term and the nth term of the geometric sequence whose initial term a_1 and common ratio r are given.

19. $a_1 = 2; \quad r = 3$

20. $a_1 = -2; \quad r = 4$

21. $a_1 = 5; \quad r = -1$

22. $a_1 = 6; \quad r = -2$

23. $a_1 = 0; \quad r = \dfrac{1}{2}$

24. $a_1 = 1; \quad r = -\dfrac{1}{3}$

25. $a_1 = \sqrt{2}; \quad r = \sqrt{2}$

26. $a_1 = 0; \quad r = \dfrac{1}{\pi}$

In Problems 27–32, find the indicated term of each geometric sequence.

27. 7th term of $1, \dfrac{1}{2}, \dfrac{1}{4}, \ldots$

28. 8th term of $1, 3, 9, \ldots$

29. 9th term of $1, -1, 1, \ldots$

30. 10th term of $-1, 2, -4, \ldots$

31. 8th term of $0.4, 0.04, 0.004, \ldots$

32. 7th term of $0.1, 1.0, 10.0, \ldots$

In Problems 33–40, find the nth term a_n of each geometric sequence. When given, r is the common ratio.

33. $7, 14, 28, 56, \ldots$

34. $5, 10, 20, 40, \ldots$

35. $-3, 1, -\dfrac{1}{3}, \dfrac{1}{9}, \ldots$

36. $4, 1, \dfrac{1}{4}, \dfrac{1}{16}, \ldots$

37. $a_6 = 243; \quad r = -3$

38. $a_2 = 7; \quad r = \dfrac{1}{3}$

39. $a_2 = 7; \quad a_4 = 1575$

40. $a_3 = \dfrac{1}{3}; \quad a_6 = \dfrac{1}{81}$

In Problems 41–46, find each sum.

41. $\dfrac{1}{4} + \dfrac{2}{4} + \dfrac{2^2}{4} + \dfrac{2^3}{4} + \cdots + \dfrac{2^{n-1}}{4}$

42. $\dfrac{3}{9} + \dfrac{3^2}{9} + \dfrac{3^3}{9} + \cdots + \dfrac{3^n}{9}$

43. $\sum_{k=1}^{n} \left(\dfrac{2}{3}\right)^k$

44. $\sum_{k=1}^{n} 4 \cdot 3^{k-1}$

45. $-1 - 2 - 4 - 8 - \cdots - (2^{n-1})$

46. $2 + \dfrac{6}{5} + \dfrac{18}{25} + \cdots + 2\left(\dfrac{3}{5}\right)^{n-1}$

For Problems 47–52, use a graphing utility to find the sum of each geometric sequence.

47. $\dfrac{1}{4} + \dfrac{2}{4} + \dfrac{2^2}{4} + \dfrac{2^3}{4} + \cdots + \dfrac{2^{14}}{4}$

48. $\dfrac{3}{9} + \dfrac{3^2}{9} + \dfrac{3^3}{9} + \cdots + \dfrac{3^{15}}{9}$

49. $\sum_{n=1}^{15} \left(\dfrac{2}{3}\right)^n$

50. $\sum_{n=1}^{15} 4 \cdot 3^{n-1}$

51. $-1 - 2 - 4 - 8 - \cdots - 2^{14}$

52. $2 + \dfrac{6}{5} + \dfrac{18}{25} + \cdots + 2\left(\dfrac{3}{5}\right)^{15}$

In Problems 53–68, determine whether each infinite geometric series converges or diverges. If it converges, find its sum.

53. $1 + \dfrac{1}{3} + \dfrac{1}{9} + \cdots$

54. $2 + \dfrac{4}{3} + \dfrac{8}{9} + \cdots$

55. $8 + 4 + 2 + \cdots$

56. $6 + 2 + \dfrac{2}{3} + \cdots$

57. $2 - \dfrac{1}{2} + \dfrac{1}{8} - \dfrac{1}{32} + \cdots$

58. $1 - \dfrac{3}{4} + \dfrac{9}{16} - \dfrac{27}{64} + \cdots$

59. $8 + 12 + 18 + 27 + \cdots$

60. $9 + 12 + 16 + \dfrac{64}{3} + \cdots$

61. $\displaystyle\sum_{k=1}^{\infty} 5\left(\dfrac{1}{4}\right)^{k-1}$

62. $\displaystyle\sum_{k=1}^{\infty} 8\left(\dfrac{1}{3}\right)^{k-1}$

63. $\displaystyle\sum_{k=1}^{\infty} \dfrac{1}{2}\, 3^{k-1}$

64. $\displaystyle\sum_{k=1}^{\infty} 3\left(\dfrac{3}{2}\right)^{k-1}$

65. $\displaystyle\sum_{k=1}^{\infty} 6\left(-\dfrac{2}{3}\right)^{k-1}$

66. $\displaystyle\sum_{k=1}^{\infty} 4\left(-\dfrac{1}{2}\right)^{k-1}$

67. $\displaystyle\sum_{k=1}^{\infty} 3\left(\dfrac{2}{3}\right)^{k}$

68. $\displaystyle\sum_{k=1}^{\infty} 2\left(\dfrac{3}{4}\right)^{k}$

Mixed Practice

In Problems 69–82, determine whether the given sequence is arithmetic, geometric, or neither. If the sequence is arithmetic, find the common difference; if it is geometric, find the common ratio. If the sequence is arithmetic or geometric, find the sum of the first 50 terms.

69. $\{n + 2\}$

70. $\{2n - 5\}$

71. $\{4n^2\}$

72. $\{5n^2 + 1\}$

73. $\left\{3 - \dfrac{2}{3}n\right\}$

74. $\left\{8 - \dfrac{3}{4}n\right\}$

75. $1, 3, 6, 10, \ldots$

76. $2, 4, 6, 8, \ldots$

77. $\left\{\left(\dfrac{2}{3}\right)^n\right\}$

78. $\left\{\left(\dfrac{5}{4}\right)^n\right\}$

79. $-1, 2, -4, 8, \ldots$

80. $1, 1, 2, 3, 5, 8, \ldots$

81. $\{3^{n/2}\}$

82. $\{(-1)^n\}$

Applications and Extensions

83. Find x so that x, $x + 2$, and $x + 3$ are consecutive terms of a geometric sequence.

84. Find x so that $x - 1$, x, and $x + 2$ are consecutive terms of a geometric sequence.

85. Salary Increases If you have been hired at an annual salary of $18,000 and expect to receive annual increases of 5%, what will your salary be when you begin your fifth year?

86. Equipment Depreciation A new piece of equipment cost a company $15,000. Each year, for tax purposes, the company depreciates the value by 15%. What value should the company give the equipment after 5 years?

87. Pendulum Swings Initially, a pendulum swings through an arc of 2 feet. On each successive swing, the length of the arc is 0.9 of the previous length.
 (a) What is the length of the arc of the 10th swing?
 (b) On which swing is the length of the arc first less than 1 foot?
 (c) After 15 swings, what total length will the pendulum have swung?
 (d) When it stops, what total length will the pendulum have swung?

88. Bouncing Balls A ball is dropped from a height of 30 feet. Each time it strikes the ground, it bounces up to 0.8 of the previous height.

 (a) What height will the ball bounce up to after it strikes the ground for the third time?
 (b) How high will it bounce after it strikes the ground for the nth time?
 (c) How many times does the ball need to strike the ground before its bounce is less than 6 inches?
 (d) What total distance does the ball travel before it stops bouncing?

89. Retirement Christine contributes $100 each month to her 401(k). What will be the value of Christine's 401(k) after the 360th deposit (30 years) if the per annum rate of return is assumed to be 12% compounded monthly?

90. Saving for a Home Jolene wants to purchase a new home. Suppose that she invests $400 per month into a mutual fund. If the per annum rate of return of the mutual fund is assumed to be 10% compounded monthly, how much will Jolene have for a down payment after the 36th deposit (3 years)?

91. Tax Sheltered Annuity Don contributes $500 at the end of each quarter to a tax-sheltered annuity (TSA). What will the value of the TSA be after the 80th deposit (20 years) if the per annum rate of return is assumed to be 8% compounded quarterly?

92. Retirement Ray contributes $1000 to an Individual Retirement Account (IRA) semiannually. What will the value of the IRA be when Ray makes his 30th deposit (after 15 years) if the per annum rate of return is assumed to be 10% compounded semiannually?

93. Sinking Fund Scott and Alice want to purchase a vacation home in 10 years and need $50,000 for a down payment. How much should they place in a savings account each month if the per annum rate of return is assumed to be 6% compounded monthly?

94. Sinking Fund For a child born in 1996, the cost of a 4-year college education at a public university is projected to be $150,000. Assuming an 8% per annum rate of return compounded monthly, how much must be contributed to a

college fund every month to have $150,000 in 18 years when the child begins college?

95. Grains of Wheat on a Chess Board In an old fable, a commoner who had saved the king's life was told he could ask the king for any just reward. Being a shrewd man, the commoner said, "A simple wish, sire. Place one grain of wheat on the first square of a chessboard, two grains on the second square, four grains on the third square, continuing until you have filled the board. This is all I seek." Compute the total number of grains needed to do this to see why the request, seemingly simple, could not be granted. (A chessboard consists of $8 \times 8 = 64$ squares.)

96. Look at the figure. What fraction of the square is eventually shaded if the indicated shading process continues indefinitely?

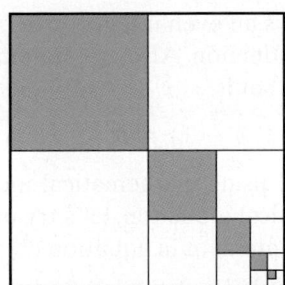

97. Multiplier Suppose that, throughout the U.S. economy, individuals spend 90% of every additional dollar that they earn. Economists would say that an individual's **marginal propensity to consume** is 0.90. For example, if Jane earns an

additional dollar, she will spend $0.9(1) = \$0.90$ of it. The individual that earns $0.90 (from Jane) will spend 90% of it or $0.81. This process of spending continues and results in an infinite geometric series as follows:

$$1, 0.90, 0.90^2, 0.90^3, 0.90^4, \ldots$$

The sum of this infinite geometric series is called the **multiplier.** What is the multiplier if individuals spend 90% of every additional dollar that they earn?

98. Multiplier Refer to Problem 97. Suppose that the marginal propensity to consume throughout the U.S. economy is 0.95. What is the multiplier for the U.S. economy?

99. Stock Price One method of pricing a stock is to discount the stream of future dividends of the stock. Suppose that a stock pays $\$P$ per year in dividends and, historically, the dividend has been increased $i\%$ per year. If you desire an annual rate of return of $r\%$, this method of pricing a stock states that the price that you should pay is the present value of an infinite stream of payments:

$$\text{Price} = P + P\frac{1+i}{1+r} + P\left(\frac{1+i}{1+r}\right)^2 + P\left(\frac{1+i}{1+r}\right)^3 + \cdots$$

The price of the stock is the sum of an infinite geometric series. Suppose that a stock pays an annual dividend of $4.00 and, historically, the dividend has been increased 3% per year. You desire an annual rate of return of 9%. What is the most you should pay for the stock?

100. Stock Price Refer to Problem 99. Suppose that a stock pays an annual dividend of $2.50 and, historically, the dividend has increased 4% per year. You desire an annual rate of return of 11%. What is the most that you should pay for the stock?

101. A Rich Man's Promise A rich man promises to give you $1000 on September 1, 2010. Each day thereafter he will give you $\dfrac{9}{10}$ of what he gave you the previous day. What is the first date on which the amount you receive is less than 1¢? How much have you received when this happens?

Explaining Concepts: Discussion and Writing

102. Critical Thinking You are interviewing for a job and receive two offers:

 A: $20,000 to start, with guaranteed annual increases of 6% for the first 5 years

 B: $22,000 to start, with guaranteed annual increases of 3% for the first 5 years

Which offer is better if your goal is to be making as much as possible after 5 years? Which is better if your goal is to make as much money as possible over the contract (5 years)?

103. Critical Thinking Which of the following choices, A or B, results in more money?

 A: To receive $1000 on day 1, $999 on day 2, $998 on day 3, with the process to end after 1000 days

 B: To receive $1 on day 1, $2 on day 2, $4 on day 3, for 19 days

104. Critical Thinking You have just signed a 7-year professional football league contract with a beginning salary of

$2,000,000 per year. Management gives you the following options with regard to your salary over the 7 years.

1. A bonus of $100,000 each year
2. An annual increase of 4.5% per year beginning after 1 year
3. An annual increase of $95,000 per year beginning after 1 year

Which option provides the most money over the 7-year period? Which the least? Which would you choose? Why?

105. Critical Thinking Suppose you were offered a job in which you would work 8 hours per day for 5 workdays per week for 1 month at hard manual labor. Your pay the first day would be 1 penny. On the second day your pay would be two pennies; the third day 4 pennies. Your pay would double on each successive workday. There are 22 workdays in the month. There will be no sick days. If you miss a day of work, there is no pay or pay increase. How much would you get

paid if you work all 22 days? How much do you get paid for the 22nd workday? What risks do you run if you take this job offer? Would you take the job?

106. Can a sequence be both arithmetic and geometric? Give reasons for your answer.

107. Make up a geometric sequence. Give it to a friend and ask for its 20th term.

108. Make up two infinite geometric series, one that has a sum and one that does not. Give them to a friend and ask for the sum of each series.

109. Describe the similarities and differences between geometric sequences and exponential functions.

'Are You Prepared?' Answers

1. $1082.43 **2.** $9513.28

9.4 Mathematical Induction

> OBJECTIVE **1** Prove Statements Using Mathematical Induction (p. 664)

1 Prove Statements Using Mathematical Induction

Mathematical induction is a method for proving that statements involving natural numbers are true for all natural numbers.*

For example, the statement "$2n$ is always an even integer" can be proved for all natural numbers n by using mathematical induction. Also, the statement "the sum of the first n positive odd integers equals n^2," that is,

$$1 + 3 + 5 + \cdots + (2n - 1) = n^2 \qquad (1)$$

can be proved for all natural numbers n by using mathematical induction.

Before stating the method of mathematical induction, let's try to gain a sense of the power of the method. We shall use the statement in equation (1) for this purpose by restating it for various values of $n = 1, 2, 3, \ldots$.

$n = 1$	The sum of the first positive odd integer is 1^2; $1 = 1^2$.
$n = 2$	The sum of the first 2 positive odd integers is 2^2; $1 + 3 = 4 = 2^2$.
$n = 3$	The sum of the first 3 positive odd integers is 3^2; $1 + 3 + 5 = 9 = 3^2$.
$n = 4$	The sum of the first 4 positive odd integers is 4^2; $1 + 3 + 5 + 7 = 16 = 4^2$.

Although from this pattern we might conjecture that statement (1) is true for any choice of n, can we really be sure that it does not fail for some choice of n? The method of proof by mathematical induction will, in fact, prove that the statement is true for all n.

THEOREM **The Principle of Mathematical Induction**

Suppose that the following two conditions are satisfied with regard to a statement about natural numbers:

> CONDITION I: The statement is true for the natural number 1.
>
> CONDITION II: If the statement is true for some natural number k, it is also true for the next natural number $k + 1$.

Then the statement is true for all natural numbers.

* Recall that the natural numbers are the numbers $1, 2, 3, 4, \ldots$. In other words, the terms *natural numbers* and *positive integers* are synonymous.

Figure 10

We shall not prove this principle. However, we can provide a physical interpretation that will help us to see why the principle works. Think of a collection of natural numbers obeying a statement as a collection of infinitely many dominoes. See Figure 10.

Now, suppose that we are told two facts:

1. The first domino is pushed over.

2. If one domino falls over, say the kth domino, so will the next one, the $(k + 1)$st domino.

Is it safe to conclude that *all* the dominoes fall over? The answer is yes, because if the first one falls (Condition I), the second one does also (by Condition II); and if the second one falls, so does the third (by Condition II); and so on.

EXAMPLE 1　**Using Mathematical Induction**

Show that the following statement is true for all natural numbers n.

$$1 + 3 + 5 + \cdots + (2n - 1) = n^2 \qquad (2)$$

Solution　We need to show first that statement (2) holds for $n = 1$. Because $1 = 1^2$, statement (2) is true for $n = 1$. Condition I holds.

Next, we need to show that Condition II holds. From statement (2), we assume that

$$1 + 3 + \cdots + (2k - 1) = k^2 \qquad (3)$$

is true for some natural number k.

We wish to show that, based on equation (3), statement (2) holds for $k + 1$. We look at the sum of the first $k + 1$ positive odd integers to determine whether this sum equals $(k + 1)^2$.

$$1 + 3 + \cdots + (2k - 1) + [2(k + 1)-1] = \underbrace{[1 + 3 + \cdots + (2k - 1)]}_{= \, k^2 \text{ by equation (3)}} + (2k + 1)$$

$$= k^2 + (2k + 1)$$
$$= k^2 + 2k + 1 = (k + 1)^2$$

Conditions I and II are satisfied; by the Principle of Mathematical Induction, statement (2) is true for all natural numbers n.

EXAMPLE 2　**Using Mathematical Induction**

Show that the following statement is true for all natural numbers n.

$$2^n > n$$

Solution　First, we show that the statement $2^n > n$ holds when $n = 1$. Because $2^1 = 2 > 1$, the inequality is true for $n = 1$. Condition I holds.

Next, we assume, for some natural number k, that $2^k > k$. We wish to show that the formula holds for $k + 1$; that is, we wish to show that $2^{k+1} > k + 1$. Now

$$2^{k+1} = 2 \cdot 2^k > 2 \cdot k = k + k \geq k + 1$$
$$\qquad\qquad \uparrow \qquad\qquad\quad \uparrow$$
$$\qquad\quad \text{We know that} \qquad k \geq 1$$
$$\qquad\qquad 2^k > k.$$

If $2^k > k$, then $2^{k+1} > k + 1$, so Condition II of the Principle of Mathematical Induction is satisfied. The statement $2^n > n$ is true for all natural numbers n.

EXAMPLE 3 **Using Mathematical Induction**

Show that the following formula is true for all natural numbers n.

$$1 + 2 + 3 + \cdots + n = \frac{n(n + 1)}{2} \qquad \textbf{(4)}$$

Solution First, we show that formula (4) is true when $n = 1$. Because

$$\frac{1(1 + 1)}{2} = \frac{1(2)}{2} = 1$$

Condition I of the Principle of Mathematical Induction holds.

Next, we assume that formula (4) holds for some k, and we determine whether the formula then holds for $k + 1$. We assume that

$$1 + 2 + 3 + \cdots + k = \frac{k(k + 1)}{2} \quad \text{for some } k \qquad \textbf{(5)}$$

Now we need to show that

$$1 + 2 + 3 + \cdots + k + (k + 1) = \frac{(k + 1)[(k + 1) + 1]}{2} = \frac{(k + 1)(k + 2)}{2}$$

We do this as follows:

$$1 + 2 + 3 + \cdots + k + (k + 1) = \underbrace{[1 + 2 + 3 + \cdots + k]}_{= \frac{k(k+1)}{2} \quad \text{by equation (5)}} + (k + 1)$$

$$= \frac{k(k + 1)}{2} + (k + 1)$$

$$= \frac{k^2 + k + 2k + 2}{2}$$

$$= \frac{k^2 + 3k + 2}{2} = \frac{(k + 1)(k + 2)}{2}$$

Condition II also holds. As a result, formula (4) is true for all natural numbers n.

Now Work PROBLEM 1

EXAMPLE 4 **Using Mathematical Induction**

Show that $3^n - 1$ is divisible by 2 for all natural numbers n.

Solution First, we show that the statement is true when $n = 1$. Because $3^1 - 1 = 3 - 1 = 2$ is divisible by 2, the statement is true when $n = 1$. Condition I is satisfied.

Next, we assume that the statement holds for some k, and we determine whether the statement then holds for $k + 1$. We assume that $3^k - 1$ is divisible by 2 for some k. We need to show that $3^{k+1} - 1$ is divisible by 2. Now

$$3^{k+1} - 1 = 3^{k+1} - 3^k + 3^k - 1 \qquad \text{Subtract and add } 3^k.$$
$$= 3^k(3 - 1) + (3^k - 1) = 3^k \cdot 2 + (3^k - 1)$$

Because $3^k \cdot 2$ is divisible by 2 and $3^k - 1$ is divisible by 2, it follows that $3^k \cdot 2 + (3^k - 1) = 3^{k+1} - 1$ is divisible by 2. Condition II is also satisfied. As a result, the statement "$3^n - 1$ is divisible by 2" is true for all natural numbers n.

WARNING The conclusion that a statement involving natural numbers is true for all natural numbers is made only after *both* Conditions I and II of the Principle of Mathematical Induction have been satisfied. Problem 28 demonstrates a statement for which only Condition I holds, but the statement is *not* true for all natural numbers. Problem 29 demonstrates a statement for which only Condition II holds, but the statement is *not* true for any natural number. ■

9.4 Assess Your Understanding

Skill Building

In Problems 1–22, use the Principle of Mathematical Induction to show that the given statement is true for all natural numbers n.

1. $2 + 4 + 6 + \cdots + 2n = n(n + 1)$

2. $1 + 5 + 9 + \cdots + (4n - 3) = n(2n - 1)$

3. $3 + 4 + 5 + \cdots + (n + 2) = \dfrac{1}{2}n(n + 5)$

4. $3 + 5 + 7 + \cdots + (2n + 1) = n(n + 2)$

5. $2 + 5 + 8 + \cdots + (3n - 1) = \dfrac{1}{2}n(3n + 1)$

6. $1 + 4 + 7 + \cdots + (3n - 2) = \dfrac{1}{2}n(3n - 1)$

7. $1 + 2 + 2^2 + \cdots + 2^{n-1} = 2^n - 1$

8. $1 + 3 + 3^2 + \cdots + 3^{n-1} = \dfrac{1}{2}(3^n - 1)$

9. $1 + 4 + 4^2 + \cdots + 4^{n-1} = \dfrac{1}{3}(4^n - 1)$

10. $1 + 5 + 5^2 + \cdots + 5^{n-1} = \dfrac{1}{4}(5^n - 1)$

11. $\dfrac{1}{1 \cdot 2} + \dfrac{1}{2 \cdot 3} + \dfrac{1}{3 \cdot 4} + \cdots + \dfrac{1}{n(n + 1)} = \dfrac{n}{n + 1}$

12. $\dfrac{1}{1 \cdot 3} + \dfrac{1}{3 \cdot 5} + \dfrac{1}{5 \cdot 7} + \cdots + \dfrac{1}{(2n - 1)(2n + 1)} = \dfrac{n}{2n + 1}$

13. $1^2 + 2^2 + 3^2 + \cdots + n^2 = \dfrac{1}{6}n(n + 1)(2n + 1)$

14. $1^3 + 2^3 + 3^3 + \cdots + n^3 = \dfrac{1}{4}n^2(n + 1)^2$

15. $4 + 3 + 2 + \cdots + (5 - n) = \dfrac{1}{2}n(9 - n)$

16. $-2 - 3 - 4 - \cdots - (n + 1) = -\dfrac{1}{2}n(n + 3)$

17. $1 \cdot 2 + 2 \cdot 3 + 3 \cdot 4 + \cdots + n(n + 1) = \dfrac{1}{3}n(n + 1)(n + 2)$

18. $1 \cdot 2 + 3 \cdot 4 + 5 \cdot 6 + \cdots + (2n - 1)(2n) = \dfrac{1}{3}n(n + 1)(4n - 1)$

19. $n^2 + n$ is divisible by 2.

20. $n^3 + 2n$ is divisible by 3.

21. $n^2 - n + 2$ is divisible by 2.

22. $n(n + 1)(n + 2)$ is divisible by 6.

Applications and Extensions

In Problems 23–27, prove each statement.

23. If $x > 1$, then $x^n > 1$.

24. If $0 < x < 1$, then $0 < x^n < 1$.

25. $a - b$ is a factor of $a^n - b^n$.
 [**Hint:** $a^{k+1} - b^{k+1} = a(a^k - b^k) + b^k(a - b)$]

26. $a + b$ is a factor of $a^{2n+1} + b^{2n+1}$.

27. $(1 + a)^n \geq 1 + na$, for $a > 0$

28. Show that the statement "$n^2 - n + 41$ is a prime number" is true for $n = 1$, but is not true for $n = 41$.

29. Show that the formula
$$2 + 4 + 6 + \cdots + 2n = n^2 + n + 2$$
obeys Condition II of the Principle of Mathematical Induction. That is, show that if the formula is true for some k it is also true for $k + 1$. Then show that the formula is false for $n = 1$ (or for any other choice of n).

30. Use mathematical induction to prove that if $r \neq 1$ then
$$a + ar + ar^2 + \cdots + ar^{n-1} = a\dfrac{1 - r^n}{1 - r}$$

31. Use mathematical induction to prove that
$$a + (a + d) + (a + 2d)$$
$$+ \cdots + [a + (n - 1)d] = na + d\dfrac{n(n - 1)}{2}$$

32. Extended Principle of Mathematical Induction The Extended Principle of Mathematical Induction states that if Conditions I and II hold, that is,
 (I) A statement is true for a natural number j.
 (II) If the statement is true for some natural number $k \geq j$, then it is also true for the next natural number $k + 1$.

then the statement is true for all natural numbers $\geq j$. Use the Extended Principle of Mathematical Induction to show that the number of diagonals in a convex polygon of n sides is $\dfrac{1}{2}n(n - 3)$.
 [**Hint:** Begin by showing that the result is true when $n = 4$ (Condition I).]

33. Geometry Use the Extended Principle of Mathematical Induction to show that the sum of the interior angles of a convex polygon of n sides equals $(n - 2) \cdot 180°$.

Explaining Concepts: Discussion and Writing

34. How would you explain the Principle of Mathematical Induction to a friend?

9.5 The Binomial Theorem

OBJECTIVES 1 Evaluate $\binom{n}{j}$ (p. 668)

2 Use the Binomial Theorem (p. 670)

Formulas have been given for expanding $(x + a)^n$ for $n = 2$ and $n = 3$. The *Binomial Theorem** is a formula for the expansion of $(x + a)^n$ for any positive integer n. If $n = 1, 2, 3,$ and 4, the expansion of $(x + a)^n$ is straightforward.

$$(x + a)^1 = x + a$$

Two terms, beginning with x^1 and ending with a^1

$$(x + a)^2 = x^2 + 2ax + a^2$$

Three terms, beginning with x^2 and ending with a^2

$$(x + a)^3 = x^3 + 3ax^2 + 3a^2x + a^3$$

Four terms, beginning with x^3 and ending with a^3

$$(x + a)^4 = x^4 + 4ax^3 + 6a^2x^2 + 4a^3x + a^4$$

Five terms, beginning with x^4 and ending with a^4

Notice that each expansion of $(x + a)^n$ begins with x^n and ends with a^n. As you read from left to right, the powers of x are decreasing by 1, while the powers of a are increasing by 1. Also, the number of terms equals $n + 1$. Notice, too, that the degree of each monomial in the expansion equals n. For example, in the expansion of $(x + a)^3$, each monomial $(x^3, 3ax^2, 3a^2x, a^3)$ is of degree 3. As a result, we might conjecture that the expansion of $(x + a)^n$ would look like this:

$$(x + a)^n = x^n + \underline{} ax^{n-1} + \underline{} a^2x^{n-2} + \cdots + \underline{} a^{n-1}x + a^n$$

where the blanks are numbers to be found. This is, in fact, the case, as we shall see shortly.

Before we can fill in the blanks, we need to introduce the symbol $\binom{n}{j}$.

1 Evaluate $\binom{n}{j}$

COMMENT On a graphing calculator, the symbol $\binom{n}{j}$ may be denoted by the key \boxed{nCr}. ∎

We define the symbol $\binom{n}{j}$, read "n taken j at a time," as follows:

DEFINITION

If j and n are integers with $0 \leq j \leq n$, the symbol $\binom{n}{j}$ is defined as

$$\binom{n}{j} = \frac{n!}{j!(n - j)!} \tag{1}$$

EXAMPLE 1

Evaluating $\binom{n}{j}$

Find:

(a) $\binom{3}{1}$ (b) $\binom{4}{2}$ (c) $\binom{8}{7}$ (d) $\binom{65}{15}$

*The name *binomial* is derived from the fact that $x + a$ is a binomial; that is, it contains two terms.

Solution (a) $\dbinom{3}{1} = \dfrac{3!}{1!(3-1)!} = \dfrac{3!}{1!\,2!} = \dfrac{3 \cdot 2 \cdot 1}{1(2 \cdot 1)} = \dfrac{6}{2} = 3$

(b) $\dbinom{4}{2} = \dfrac{4!}{2!(4-2)!} = \dfrac{4!}{2!\,2!} = \dfrac{4 \cdot 3 \cdot 2 \cdot 1}{(2 \cdot 1)(2 \cdot 1)} = \dfrac{24}{4} = 6$

(c) $\dbinom{8}{7} = \dfrac{8!}{7!(8-7)!} = \dfrac{8!}{7!\,1!} = \dfrac{8 \cdot \cancel{7!}}{\cancel{7!} \cdot 1!} = \dfrac{8}{1} = 8$

$8! = 8 \cdot 7!$

Figure 11

65 nCr 15
 2.073746998ᴇ14

(d) Figure 11 shows the solution using a TI-84 Plus graphing calculator. So

$$\binom{65}{15} \approx 2.073746998 \times 10^{14}$$

Now Work PROBLEM 5

Four useful formulas involving the symbol $\dbinom{n}{j}$ are

$$\binom{n}{0} = 1 \qquad \binom{n}{1} = n \qquad \binom{n}{n-1} = n \qquad \binom{n}{n} = 1$$

Proof $\dbinom{n}{0} = \dfrac{n!}{0!(n-0)!} = \dfrac{n!}{0!\,n!} = \dfrac{1}{1} = 1$

$\dbinom{n}{1} = \dfrac{n!}{1!(n-1)!} = \dfrac{n!}{(n-1)!} = \dfrac{n(n-1)!}{(n-1)!} = n$

You are asked to prove the remaining two formulas in Problem 45. ∎

Suppose that we arrange the values of the symbol $\dbinom{n}{j}$ in a triangular display, as shown next and in Figure 12.

$$\binom{0}{0}$$

$$\binom{1}{0} \quad \binom{1}{1}$$

$$\binom{2}{0} \quad \binom{2}{1} \quad \binom{2}{2}$$

$$\binom{3}{0} \quad \binom{3}{1} \quad \binom{3}{2} \quad \binom{3}{3}$$

$$\binom{4}{0} \quad \binom{4}{1} \quad \binom{4}{2} \quad \binom{4}{3} \quad \binom{4}{4}$$

$$\binom{5}{0} \quad \binom{5}{1} \quad \binom{5}{2} \quad \binom{5}{3} \quad \binom{5}{4} \quad \binom{5}{5}$$

Figure 12
Pascal's triangle

This display is called the **Pascal triangle,** named after Blaise Pascal (1623–1662), a French mathematician.

The Pascal triangle has 1's down the sides. To get any other entry, add the two nearest entries in the row above it. The shaded triangles in Figure 12 illustrate this feature of the Pascal triangle. Based on this feature, the row corresponding to $n = 6$ is found as follows:

$$
\begin{array}{c}
n = 5 \rightarrow \quad 1 \quad 5 \quad 10 \quad 10 \quad 5 \quad 1 \\[4pt]
n = 6 \rightarrow \quad 1 \quad 6 \quad 15 \quad 20 \quad 15 \quad 6 \quad 1
\end{array}
$$

Later we shall prove that this addition always works (see the theorem on page 672).

Although the Pascal triangle provides an interesting and organized display of the symbol $\binom{n}{j}$, in practice it is not all that helpful. For example, if you wanted to know the value of $\binom{12}{5}$, you would need to produce 13 rows of the triangle before seeing the answer. It is much faster to use definition (1).

2 Use the Binomial Theorem

THEOREM

Binomial Theorem

Let x and a be real numbers. For any positive integer n, we have

$$
\begin{aligned}
(x + a)^n &= \binom{n}{0}x^n + \binom{n}{1}ax^{n-1} + \cdots + \binom{n}{j}a^j x^{n-j} + \cdots + \binom{n}{n}a^n \\
&= \sum_{j=0}^{n} \binom{n}{j}x^{n-j}a^j \tag{2}
\end{aligned}
$$

Now you know why we needed to introduce the symbol $\binom{n}{j}$; these symbols are the numerical coefficients that appear in the expansion of $(x + a)^n$. Because of this, the symbol $\binom{n}{j}$ is called a **binomial coefficient.**

EXAMPLE 2

Expanding a Binomial

Use the Binomial Theorem to expand $(x + 2)^5$.

Solution In the Binomial Theorem, let $a = 2$ and $n = 5$. Then

$$
(x + 2)^5 = \binom{5}{0}x^5 + \binom{5}{1}2x^4 + \binom{5}{2}2^2x^3 + \binom{5}{3}2^3x^2 + \binom{5}{4}2^4x + \binom{5}{5}2^5
$$

↑
Use equation (2).

$$
= 1 \cdot x^5 + 5 \cdot 2x^4 + 10 \cdot 4x^3 + 10 \cdot 8x^2 + 5 \cdot 16x + 1 \cdot 32
$$

↑
Use row $n = 5$ of the Pascal triangle or formula (1) for $\binom{n}{j}$.

$$
= x^5 + 10x^4 + 40x^3 + 80x^2 + 80x + 32
$$

EXAMPLE 3

Expanding a Binomial

Expand $(2y - 3)^4$ using the Binomial Theorem.

Solution

First, rewrite the expression $(2y - 3)^4$ as $[2y + (-3)]^4$. Now use the Binomial Theorem with $n = 4$, $x = 2y$, and $a = -3$.

$$[2y + (-3)]^4 = \binom{4}{0}(2y)^4 + \binom{4}{1}(-3)(2y)^3 + \binom{4}{2}(-3)^2(2y)^2$$

$$+ \binom{4}{3}(-3)^3(2y) + \binom{4}{4}(-3)^4$$

$$= 1 \cdot 16y^4 + 4(-3)8y^3 + 6 \cdot 9 \cdot 4y^2 + 4(-27)2y + 1 \cdot 81$$

↑

Use row $n = 4$ of the Pascal triangle or formula (1) for $\binom{n}{j}$.

$$= 16y^4 - 96y^3 + 216y^2 - 216y + 81$$

In this expansion, note that the signs alternate due to the fact that $a = -3 < 0$.

───**Now Work** PROBLEM 21

EXAMPLE 4

Finding a Particular Coefficient in a Binomial Expansion

Find the coefficient of y^8 in the expansion of $(2y + 3)^{10}$.

Solution

Write out the expansion using the Binomial Theorem.

$$(2y + 3)^{10} = \binom{10}{0}(2y)^{10} + \binom{10}{1}(2y)^9(3)^1 + \binom{10}{2}(2y)^8(3)^2 + \binom{10}{3}(2y)^7(3)^3$$

$$+ \binom{10}{4}(2y)^6(3)^4 + \cdots + \binom{10}{9}(2y)(3)^9 + \binom{10}{10}(3)^{10}$$

From the third term in the expansion, the coefficient of y^8 is

$$\binom{10}{2}(2)^8(3)^2 = \frac{10!}{2! \, 8!} \cdot 2^8 \cdot 9 = \frac{10 \cdot 9 \cdot 8!}{2 \cdot 8!} \cdot 2^8 \cdot 9 = 103,680$$

As this solution demonstrates, we can use the Binomial Theorem to find a particular term in an expansion without writing the entire expansion.

Based on the expansion of $(x + a)^n$, the term containing x^j is

$$\boxed{\binom{n}{n - j}a^{n-j}x^j} \qquad \text{(3)}$$

For example, we can solve Example 4 by using formula (3) with $n = 10$, $a = 3$, $x = 2y$, and $j = 8$. Then the term containing y^8 is

$$\binom{10}{10 - 8}3^{10-8}(2y)^8 = \binom{10}{2} \cdot 3^2 \cdot 2^8 \cdot y^8 = \frac{10!}{2! \, 8!} \cdot 9 \cdot 2^8 y^8$$

$$= \frac{10 \cdot 9 \cdot 8!}{2 \cdot 8!} \cdot 9 \cdot 2^8 y^8 = 103,680y^8$$

EXAMPLE 5 **Finding a Particular Term in a Binomial Expansion**

Find the sixth term in the expansion of $(x + 2)^9$.

Solution A Expand using the Binomial Theorem until the sixth term is reached.

$$(x + 2)^9 = \binom{9}{0}x^9 + \binom{9}{1}x^8 \cdot 2 + \binom{9}{2}x^7 \cdot 2^2 + \binom{9}{3}x^6 \cdot 2^3 + \binom{9}{4}x^5 \cdot 2^4$$
$$+ \binom{9}{5}x^4 \cdot 2^5 + \cdots$$

The sixth term is

$$\binom{9}{5}x^4 \cdot 2^5 = \frac{9!}{5! \, 4!} \cdot x^4 \cdot 32 = 4032x^4$$

Solution B The sixth term in the expansion of $(x + 2)^9$, which has 10 terms total, contains x^4. (Do you see why?) By formula (3), the sixth term is

$$\binom{9}{9 - 4}2^{9-4}x^4 = \binom{9}{5}2^5x^4 = \frac{9!}{5! \, 4!} \cdot 32x^4 = 4032x^4$$

━━━**Now Work** PROBLEMS 29 AND 35

Next we show that the *triangular addition* feature of the Pascal triangle illustrated in Figure 12 always works.

THEOREM If n and j are integers with $1 \le j \le n$, then

$$\binom{n}{j - 1} + \binom{n}{j} = \binom{n + 1}{j} \tag{4}$$

Proof

$$\binom{n}{j - 1} + \binom{n}{j} = \frac{n!}{(j - 1)![n - (j - 1)]!} + \frac{n!}{j!(n - j)!}$$

$$= \frac{n!}{(j - 1)!(n - j + 1)!} + \frac{n!}{j!(n - j)!}$$

$$= \frac{jn!}{j(j - 1)!(n - j + 1)!} + \frac{(n - j + 1)n!}{j!(n - j + 1)(n - j)!} \qquad \text{Multiply the first term by } \frac{j}{j} \text{ and}$$
$$\text{the second term by } \frac{n - j + 1}{n - j + 1}.$$

$$= \frac{jn!}{j!(n - j + 1)!} + \frac{(n - j + 1)n!}{j!(n - j + 1)!} \qquad \text{Now the denominators are}$$
$$\text{equal.}$$

$$= \frac{jn! + (n - j + 1)n!}{j!(n - j + 1)!}$$

$$= \frac{n!(j + n - j + 1)}{j!(n - j + 1)!}$$

$$= \frac{n!(n + 1)}{j!(n - j + 1)!} = \frac{(n + 1)!}{j![(n + 1) - j]!} = \binom{n + 1}{j}$$

■

Historical Feature

*Omar Khayyám
(1050–1123)*

The case $n = 2$ of the Binomial Theorem, $(a + b)^2$, was known to Euclid in 300 BC, but the general law seems to have been discovered by the Persian mathematician and astronomer Omar Khayyám (1050–1123), who is also well known as the author of the *Rubáiyát*, a collection of four-line poems making observations on the human condition. Omar Khayyám did not state the Binomial Theorem explicitly, but he claimed to have a method for extracting third, fourth, fifth roots, and so on. A little study shows that one must know the Binomial Theorem to create such a method.

The heart of the Binomial Theorem is the formula for the numerical coefficients, and, as we saw, they can be written in a symmetric triangular form. The Pascal triangle appears first in the books of Yang Hui (about 1270) and Chu Shih-chieh (1303). Pascal's name is attached to the triangle because of the many applications he made of it, especially to counting and probability. In establishing these results, he was one of the earliest users of mathematical induction.

Many people worked on the proof of the Binomial Theorem, which was finally completed for all n (including complex numbers) by Niels Abel (1802–1829).

9.5 Assess Your Understanding

Concepts and Vocabulary

1. _____ _____ is a triangular display of the binomial coefficients.

2. $\dbinom{n}{0} =$ ___ and $\dbinom{n}{1} =$ ___ .

3. *True or False* $\dbinom{n}{j} = \dfrac{j!}{(n-j)!\,n!}$

4. The _____ _____ can be used to expand expressions like $(2x + 3)^6$.

Skill Building

In Problems 5–16, evaluate each expression.

5. $\dbinom{5}{3}$

6. $\dbinom{7}{3}$

7. $\dbinom{7}{5}$

8. $\dbinom{9}{7}$

9. $\dbinom{50}{49}$

10. $\dbinom{100}{98}$

11. $\dbinom{1000}{1000}$

12. $\dbinom{1000}{0}$

13. $\dbinom{55}{23}$

14. $\dbinom{60}{20}$

15. $\dbinom{47}{25}$

16. $\dbinom{37}{19}$

In Problems 17–28, expand each expression using the Binomial Theorem.

17. $(x + 1)^5$

18. $(x - 1)^5$

19. $(x - 2)^6$

20. $(x + 3)^5$

21. $(3x + 1)^4$

22. $(2x + 3)^5$

23. $(x^2 + y^2)^5$

24. $(x^2 - y^2)^6$

25. $\left(\sqrt{x} + \sqrt{2}\right)^6$

26. $\left(\sqrt{x} - \sqrt{3}\right)^4$

27. $(ax + by)^5$

28. $(ax - by)^4$

In Problems 29–42, use the Binomial Theorem to find the indicated coefficient or term.

29. The coefficient of x^6 in the expansion of $(x + 3)^{10}$

30. The coefficient of x^3 in the expansion of $(x - 3)^{10}$

31. The coefficient of x^7 in the expansion of $(2x - 1)^{12}$

32. The coefficient of x^3 in the expansion of $(2x + 1)^{12}$

33. The coefficient of x^7 in the expansion of $(2x + 3)^9$

34. The coefficient of x^2 in the expansion of $(2x - 3)^9$

35. The fifth term in the expansion of $(x + 3)^7$

36. The third term in the expansion of $(x - 3)^7$

37. The third term in the expansion of $(3x - 2)^9$

38. The sixth term in the expansion of $(3x + 2)^8$

39. The coefficient of x^0 in the expansion of $\left(x^2 + \dfrac{1}{x}\right)^{12}$

40. The coefficient of x^0 in the expansion of $\left(x - \dfrac{1}{x^2}\right)^9$

41. The coefficient of x^4 in the expansion of $\left(x - \dfrac{2}{\sqrt{x}}\right)^{10}$

42. The coefficient of x^2 in the expansion of $\left(\sqrt{x} + \dfrac{3}{\sqrt{x}}\right)^8$

Applications and Extensions

43. Use the Binomial Theorem to find the numerical value of $(1.001)^5$ correct to five decimal places.
[**Hint:** $(1.001)^5 = (1 + 10^{-3})^5$]

44. Use the Binomial Theorem to find the numerical value of $(0.998)^6$ correct to five decimal places.

45. Show that $\dbinom{n}{n-1} = n$ and $\dbinom{n}{n} = 1$.

46. Show that if n and j are integers with $0 \le j \le n$ then

$$\binom{n}{j} = \binom{n}{n-j}$$

Conclude that the Pascal triangle is symmetric with respect to a vertical line drawn from the topmost entry.

47. If n is a positive integer, show that

$$\binom{n}{0} + \binom{n}{1} + \cdots + \binom{n}{n} = 2^n$$

[**Hint:** $2^n = (1 + 1)^n$; now use the Binomial Theorem.]

48. If n is a positive integer, show that

$$\binom{n}{0} - \binom{n}{1} + \binom{n}{2} - \cdots + (-1)^n \binom{n}{n} = 0$$

49. $\dbinom{5}{0}\left(\dfrac{1}{4}\right)^5 + \dbinom{5}{1}\left(\dfrac{1}{4}\right)^4\left(\dfrac{3}{4}\right) + \dbinom{5}{2}\left(\dfrac{1}{4}\right)^3\left(\dfrac{3}{4}\right)^2$

$+ \dbinom{5}{3}\left(\dfrac{1}{4}\right)^2\left(\dfrac{3}{4}\right)^3 + \dbinom{5}{4}\left(\dfrac{1}{4}\right)\left(\dfrac{3}{4}\right)^4 + \dbinom{5}{5}\left(\dfrac{3}{4}\right)^5 = ?$

50. Stirling's Formula An approximation for $n!$, when n is large, is given by

$$n! \approx \sqrt{2n\pi}\left(\frac{n}{e}\right)^n\left(1 + \frac{1}{12n - 1}\right)$$

Calculate 12!, 20!, and 25! on your calculator. Then use Stirling's formula to approximate 12!, 20!, and 25!.

CHAPTER REVIEW

Things to Know

Sequence (p. 637)	A function whose domain is the set of positive integers
Factorials (p. 639)	$0! = 1, 1! = 1, n! = n(n-1) \cdot \cdots \cdot 3 \cdot 2 \cdot 1$ if $n \ge 2$ is an integer
Arithmetic sequence (pp. 647 and 648)	$a_1 = a, \ a_n = a_{n-1} + d$, where $a_1 = a = $ first term, $d = $ common difference $a_n = a_1 + (n-1)d$
Sum of the first n terms of an arithmetic sequence (p. 649)	$S_n = \dfrac{n}{2}[2a_1 + (n-1)d] = \dfrac{n}{2}(a_1 + a_n)$
Geometric sequence (pp. 653 and 654)	$a_1 = a, \ a_n = ra_{n-1}$, where $a_1 = a = $ first term, $r = $ common ratio $a_n = a_1 r^{n-1}, \ r \ne 0$
Sum of the first n terms of a geometric sequence (p. 655)	$S_n = a_1 \dfrac{1 - r^n}{1 - r}, \ r \ne 0, 1$
Infinite geometric series (p. 656)	$a_1 + a_1 r + \cdots + a_1 r^{n-1} + \cdots = \displaystyle\sum_{k=1}^{\infty} a_1 r^{k-1}$
Sum of a convergent infinite geometric series (p. 657)	If $\lvert r \rvert < 1, \ \displaystyle\sum_{k=1}^{\infty} a_1 r^{k-1} = \dfrac{a_1}{1 - r}$
Principle of Mathematical Induction (p. 664)	If the following two conditions are satisfied, Condition I: The statement is true for the natural number 1. Condition II: If the statement is true for some natural number k, it is also true for $k + 1$. then the statement is true for all natural numbers.
Binomial coefficient (p. 668)	$\dbinom{n}{j} = \dfrac{n!}{j!(n-j)!}$
Pascal's triangle (p. 669)	See Figure 12.
Binomial Theorem (p. 670)	$(x + a)^n = \dbinom{n}{0}x^n + \dbinom{n}{1}ax^{n-1} + \cdots + \dbinom{n}{j}a^j x^{n-j} + \cdots + \dbinom{n}{n}a^n = \displaystyle\sum_{j=0}^{n}\dbinom{n}{j}x^{n-j}a^j$

Objectives

Section		You should be able to . . .	Example(s)	Review Exercises
9.1	1	Write the first several terms of a sequence (p. 637)	1–4	1–4
	2	Write the terms of a sequence defined by a recursive formula (p. 640)	5, 6	5–8
	3	Use summation notation (p. 641)	7, 8	9–12
	4	Find the sum of a sequence (p. 642)	9	25, 26
9.2	1	Determine if a sequence is arithmetic (p. 647)	1–3	13–24
	2	Find a formula for an arithmetic sequence (p. 648)	4, 5	31, 32, 35, 37–40, 65
	3	Find the sum of an arithmetic sequence (p. 649)	6–8	13, 14, 19, 20, 27, 28, 65, 66
9.3	1	Determine if a sequence is geometric (p. 653)	1–3	13–24
	2	Find a formula for a geometric sequence (p. 654)	4	21, 22, 33, 34, 36, 67(a)–(c), 70
	3	Find the sum of a geometric sequence (p. 655)	5, 6	17, 18, 21, 22, 29, 30
	4	Determine whether a geometric series converges or diverges (p. 656)	7–9	41–48, 67(d)
	5	Solve annuity problems (p. 659)	10, 11	68, 69
9.4	1	Prove statements using mathematical induction (p. 664)	1–4	49–54
9.5	1	Evaluate $\binom{n}{j}$ (p. 668)	1	55, 56
	2	Use the Binomial Theorem (p. 670)	2–5	57–64

Review Exercises

In Problems 1–8, write down the first five terms of each sequence.

1. $\{a_n\} = \left\{(-1)^n\left(\dfrac{n+3}{n+2}\right)\right\}$ **2.** $\{b_n\} = \{(-1)^{n+1}(2n+3)\}$ **3.** $\{c_n\} = \left\{\dfrac{2^n}{n^2}\right\}$ **4.** $\{d_n\} = \left\{\dfrac{e^n}{n}\right\}$

5. $a_1 = 3;\quad a_n = \dfrac{2}{3}a_{n-1}$ **6.** $a_1 = 4;\quad a_n = -\dfrac{1}{4}a_{n-1}$ **7.** $a_1 = 2;\quad a_n = 2 - a_{n-1}$ **8.** $a_1 = -3;\quad a_n = 4 + a_{n-1}$

In Problems 9 and 10, write out each sum.

9. $\displaystyle\sum_{k=1}^{4}(4k+2)$ **10.** $\displaystyle\sum_{k=1}^{3}(3-k^2)$

In Problems 11 and 12, express each sum using summation notation.

11. $1 - \dfrac{1}{2} + \dfrac{1}{3} - \dfrac{1}{4} + \cdots + \dfrac{1}{13}$ **12.** $2 + \dfrac{2^2}{3} + \dfrac{2^3}{3^2} + \cdots + \dfrac{2^{n+1}}{3^n}$

In Problems 13–24, determine whether the given sequence is arithmetic, geometric, or neither. If the sequence is arithmetic, find the common difference and the sum of the first n terms. If the sequence is geometric, find the common ratio and the sum of the first n terms.

13. $\{a_n\} = \{n+5\}$ **14.** $\{b_n\} = \{4n+3\}$ **15.** $\{c_n\} = \{2n^3\}$ **16.** $\{d_n\} = \{2n^2 - 1\}$

17. $\{s_n\} = \{2^{3n}\}$ **18.** $\{u_n\} = \{3^{2n}\}$ **19.** $0, 4, 8, 12, \ldots$ **20.** $1, -3, -7, -11, \ldots$

21. $3, \dfrac{3}{2}, \dfrac{3}{4}, \dfrac{3}{8}, \dfrac{3}{16}, \ldots$ **22.** $5, -\dfrac{5}{3}, \dfrac{5}{9}, -\dfrac{5}{27}, \dfrac{5}{81}, \ldots$ **23.** $\dfrac{2}{3}, \dfrac{3}{4}, \dfrac{4}{5}, \dfrac{5}{6}, \ldots$ **24.** $\dfrac{3}{2}, \dfrac{5}{4}, \dfrac{7}{6}, \dfrac{9}{8}, \dfrac{11}{10}, \ldots$

In Problems 25–30, find each sum.

25. $\displaystyle\sum_{k=1}^{50}(3k)$ **26.** $\displaystyle\sum_{k=1}^{30}k^2$ **27.** $\displaystyle\sum_{k=1}^{30}(3k-9)$

28. $\displaystyle\sum_{k=1}^{40}(-2k+8)$ **29.** $\displaystyle\sum_{k=1}^{7}\left(\dfrac{1}{3}\right)^k$ **30.** $\displaystyle\sum_{k=1}^{10}(-2)^k$

In Problems 31–36, find the indicated term in each sequence. [**Hint:** Find the general term first.]

31. 9th term of $3, 7, 11, 15, \ldots$

32. 8th term of $1, -1, -3, -5, \ldots$

33. 11th term of $1, \dfrac{1}{10}, \dfrac{1}{100}, \ldots$

34. 11th term of $1, 2, 4, 8, \ldots$

35. 9th term of $\sqrt{2}, 2\sqrt{2}, 3\sqrt{2}, \ldots$

36. 9th term of $\sqrt{2}, 2, 2^{3/2}, \ldots$

In Problems 37–40, find a general formula for each arithmetic sequence.

37. 7th term is 31; 20th term is 96

38. 8th term is -20; 17th term is -47

39. 10th term is 0; 18th term is 8

40. 12th term is 30; 22nd term is 50

In Problems 41–48, determine whether each infinite geometric series converges or diverges. If it converges, find its sum.

41. $3 + 1 + \dfrac{1}{3} + \dfrac{1}{9} + \cdots$

42. $2 + 1 + \dfrac{1}{2} + \dfrac{1}{4} + \cdots$

43. $2 - 1 + \dfrac{1}{2} - \dfrac{1}{4} + \cdots$

44. $6 - 4 + \dfrac{8}{3} - \dfrac{16}{9} + \cdots$

45. $\dfrac{1}{2} + \dfrac{3}{4} + \dfrac{9}{8} + \cdots$

46. $\displaystyle\sum_{k=1}^{\infty} 5 \cdot \left(-\dfrac{5}{4}\right)^{k-1}$

47. $\displaystyle\sum_{k=1}^{\infty} 4\left(\dfrac{1}{2}\right)^{k-1}$

48. $\displaystyle\sum_{k=1}^{\infty} 3\left(-\dfrac{3}{4}\right)^{k-1}$

In Problems 49–54, use the Principle of Mathematical Induction to show that the given statement is true for all natural numbers.

49. $3 + 6 + 9 + \cdots + 3n = \dfrac{3n}{2}(n + 1)$

50. $2 + 6 + 10 + \cdots + (4n - 2) = 2n^2$

51. $2 + 6 + 18 + \cdots + 2 \cdot 3^{n-1} = 3^n - 1$

52. $3 + 6 + 12 + \cdots + 3 \cdot 2^{n-1} = 3(2^n - 1)$

53. $1^2 + 4^2 + 7^2 + \cdots + (3n - 2)^2 = \dfrac{1}{2}n(6n^2 - 3n - 1)$

54. $1 \cdot 3 + 2 \cdot 4 + 3 \cdot 5 + \cdots + n(n + 2) = \dfrac{n}{6}(n + 1)(2n + 7)$

In Problems 55 and 56, evaluate each binomial coefficient.

55. $\dbinom{5}{2}$

56. $\dbinom{8}{6}$

In Problems 57–60, expand each expression using the Binomial Theorem.

57. $(x + 2)^5$

58. $(x - 3)^4$

59. $(2x + 3)^5$

60. $(3x - 4)^4$

61. Find the coefficient of x^7 in the expansion of $(x + 2)^9$.

62. Find the coefficient of x^3 in the expansion of $(x - 3)^8$.

63. Find the coefficient of x^2 in the expansion of $(2x + 1)^7$.

64. Find the coefficient of x^6 in the expansion of $(2x + 1)^8$.

65. Constructing a Brick Staircase A brick staircase has a total of 25 steps. The bottom step requires 80 bricks. Each successive step requires three less bricks than the prior step.
 (a) How many bricks are required for the top step?
 (b) How many bricks are required to build the staircase?

66. Creating a Floor Design A mosaic tile floor is designed in the shape of a trapezoid 30 feet wide at the base and 15 feet wide at the top. The tiles, 12 inches by 12 inches, are to be placed so that each successive row contains one less tile than the row below. How many tiles will be required?

67. Bouncing Balls A ball is dropped from a height of 20 feet. Each time it strikes the ground, it bounces up to three-quarters of the previous height.
 (a) What height will the ball bounce up to after it strikes the ground for the third time?

 (b) How high will it bounce after it strikes the ground for the *n*th time?
 (c) How many times does the ball need to strike the ground before its bounce is less than 6 inches?
 (d) What total distance does the ball travel before it stops bouncing?

68. Retirement Planning Chris gets paid once a month and contributes $200 each pay period into his 401(k). If Chris plans on retiring in 20 years, what will be the value of his 401(k) if the per annum rate of return of the 401(k) is 10% compounded monthly?

69. Retirement Planning Jacky contributes $500 every quarter to an IRA. If Jacky plans on retiring in 30 years, what will be the value of the IRA if the per annum rate of return of the IRA is 8% compounded quarterly?

70. Salary Increases Your friend has just been hired at an annual salary of $20,000. If she expects to receive annual increases of 4%, what will be her salary as she begins her fifth year?

CHAPTER TEST

The Chapter Test Prep Videos are step-by-step test solutions available in the Video Resources DVD, in **MyMathLab**, or on this text's You Tube Channel. Flip back to the Student Resources page to see the exact web address for this text's YouTube channel.

In Problems 1 and 2, write down the first five terms of each sequence.

1. $\{s_n\} = \left\{\dfrac{n^2 - 1}{n + 8}\right\}$

2. $a_1 = 4, a_n = 3a_{n-1} + 2$

5. Write the following sum using summation notation.

$$-\frac{2}{5} + \frac{3}{6} - \frac{4}{7} + \cdots + \frac{11}{14}$$

In Problems 3 and 4, write out each sum. Evaluate each sum.

3. $\displaystyle\sum_{k=1}^{3} (-1)^{k+1}\left(\frac{k + 1}{k^2}\right)$

4. $\displaystyle\sum_{k=1}^{4}\left[\left(\frac{2}{3}\right)^k - k\right]$

In Problems 6–11, determine whether the given sequence is arithmetic, geometric, or neither. If the sequence is arithmetic, find the common difference and the sum of the first n terms. If the sequence is geometric, find the common ratio and the sum of the first n terms.

6. $6, 12, 36, 144, \ldots$

7. $\left\{-\dfrac{1}{2} \cdot 4^n\right\}$

8. $-2, -10, -18, -26, \ldots$

9. $\left\{-\dfrac{n}{2} + 7\right\}$

10. $25, 10, 4, \dfrac{8}{5}, \ldots$

11. $\left\{\dfrac{2n - 3}{2n + 1}\right\}$

12. Determine whether the infinite geometric series

$$256 - 64 + 16 - 4 + \cdots$$

converges or diverges. If it converges, find its sum.

13. Expand $(3m + 2)^5$ using the Binomial Theorem.

14. Use the Principle of Mathematical Induction to show that the given statement is true for all natural numbers.

$$\left(1 + \frac{1}{1}\right)\left(1 + \frac{1}{2}\right)\left(1 + \frac{1}{3}\right)\cdots\left(1 + \frac{1}{n}\right) = n + 1$$

15. A new car sold for $31,000. If the vehicle loses 15% of its value each year, how much will it be worth after 10 years?

16. A weightlifter begins his routine by benching 100 pounds and increases the weight by 30 pounds for each set. If he does 10 repetitions in each set, what is the total weight lifted after 5 sets?

CUMULATIVE REVIEW

1. Find all the solutions, real and complex, of the equation $|x^2| = 9$

2. (a) Graph the circle $x^2 + y^2 = 100$ and the parabola $y = 3x^2$.

 (b) Solve the system of equations: $\begin{cases} x^2 + y^2 = 100 \\ \quad\quad y = 3x^2 \end{cases}$

 (c) Where do the circle and the parabola intersect?

3. Solve the equation $2e^x = 5$.

4. Find an equation of the line with slope 5 and x-intercept 2.

5. Find the standard equation of the circle whose center is the point $(-1, 2)$ if $(3, 5)$ is a point on the circle.

6. $f(x) = \dfrac{3x}{x - 2}$, $g(x) = 2x + 1$

 Find:

 (a) $(f \circ g)(2)$ (b) $(g \circ f)(4)$

 (c) $(f \circ g)(x)$ (d) The domain of $(f \circ g)(x)$

 (e) $(g \circ f)(x)$ (f) The domain of $(g \circ f)(x)$

 (g) The function g^{-1} and its domain

 (h) The function f^{-1} and its domain

7. Find the equation of an ellipse with center at the origin, a focus at $(0, 3)$, and a vertex at $(0, 4)$.

8. Find the equation of a parabola with vertex at $(-1, 2)$ and focus at $(-1, 3)$.

CHAPTER PROJECTS

Internet-based Project

I. Population Growth The size of the population of the United States essentially depends on its current population, the birth and death rates of the population, and immigration. Let b represent the birth rate of the U.S. population and d represent its death rate. Then $r = b - d$ represents the growth rate of the population, where r varies from year to year. The U.S. population after n years can be modeled using the recursive function

$$p_n = (1 + r)p_{n-1} + I$$

where I represents net immigration into the United States.

1. Using data from the CIA World Factbook at *www.cia.gov /cia/publications/factbook/index.html*, determine the birth and death rates for all races for the most recent year that data are available. Birth rates and death rates are given as

the number of live births per 1000 population. Each must be computed as the number of births (deaths) per individual. For example, in 2009, the birth rate was 13.82 per 1000 and the death rate was 8.38 per 1,000, so $b = \dfrac{13.82}{1000} = 0.01382$, while $d = \dfrac{8.38}{1000} = 0.00838$.

 Next, using data from the Immigration and Naturalization Service *www.fedstats.gov*, determine the net immigration to the United States for the same year used to obtain b and d in Problem 1.

2. Determine the value of r, the growth rate of the population.

3. Find a recursive formula for the population of the United States.

4. Use the recursive formula to predict the population of the United States in the following year. In other words, if data are available up to the year 2009, predict the U.S. population in 2010.

5. Compare your prediction to actual data.

6. Repeat Problems 1–5 for Uganda using the CIA World Factbook (in 2009, the birth rate was 47.84 per 1000 and the death rate was 12.09 per 1000).

7. Do your results for the United States (a developed country) and Uganda (a developing country) seem in line with the article in the chapter opener? Explain.

8. Do you think the recursive formula found in Problem 3 will be useful in predicting future populations? Why or why not?

The following projects are available at the Instructor's Resource Center (IRC):

II. Project at Motorola *Digital Wireless Communication* Cell phones take speech and change it into digital code using only zeros and ones. See how the code length can be modeled using a mathematical sequence.

III. Economics Economists use the current price of a good and a recursive model to predict future consumer demand and to determine future production.

IV. Standardized Tests Many tests of intelligence, aptitude, and achievement contain questions asking for the terms of a mathematical sequence.

Counting and Probability

<div align="right">

10

</div>

Outline

Deal or No Deal

By LYNN ELBER, AP Television Writer—LOS ANGELES—The promise of an easy million bucks, a stage crowded with sexy models and the smoothly calibrated charm of host **Howie Mandel** made "Deal or No Deal" an unexpected hit in television's December dead zone. Based on a series that debuted in Holland in 2002 and became an international hit, "Deal or No Deal" is about luck and playing the odds. Contestants are faced with 26 briefcases held by 26 models, each case with a hidden value ranging from a penny to the top prize that will escalate by week's end to $3 million. As the game progresses and cases are eliminated, a contestant weighs the chance of snaring a big prize against lesser but still tempting offers made by the show's "bank," represented by an anonymous, silhouetted figure.

Source: Adapted from Lynn Elber, "'Deal or No Deal' Back with Bigger Prizes," Associated Press, February 24, 2006. © 2006 Associated Press.

—See Chapter Project I—

◁ A Look Back We introduced sets in Chapter R, Review, and have been using them to represent solutions of equations and inequalities and to represent the domain and range of functions.

A Look Ahead ▷ Here we discuss methods for counting the number of elements in a set and the role of sets in probability.

10.1 Counting

PREPARING FOR THIS SECTION *Before getting started, review the following:*

• Sets (Chapter R, Review, Section R.1, pp. 2–3)

Now Work the '**Are You Prepared?**' problems on page 684.

OBJECTIVES **1** Find All the Subsets of a Set (p. 680)
 2 Count the Number of Elements in a Set (p. 680)
 3 Solve Counting Problems Using the Multiplication Principle (p. 682)

Counting plays a major role in many diverse areas, such as probability, statistics, and computer science; counting techniques are a part of a branch of mathematics called **combinatorics.**

1 Find All the Subsets of a Set

We begin by reviewing the ways that two sets can be compared.

If two sets A and B have precisely the same elements, we say that A and B are **equal** and write $A = B$.

If each element of a set A is also an element of a set B, we say that A is a **subset** of B and write $A \subseteq B$.

If $A \subseteq B$ and $A \neq B$, we say that A is a **proper subset** of B and write $A \subset B$.

If $A \subseteq B$, every element in set A is also in set B, but B may or may not have additional elements. If $A \subset B$, every element in A is also in B, and B has at least one element not found in A.

Finally, we agree that the empty set, \varnothing, is a subset of every set; that is,

$$\varnothing \subseteq A \qquad \text{for any set } A$$

EXAMPLE 1 | **Finding All the Subsets of a Set**

Write down all the subsets of the set $\{a, b, c\}$.

Solution | To organize the work, write down all the subsets with no elements, then those with one element, then those with two elements, and finally those with three elements. These will give us all the subsets. Do you see why?

0 Elements	1 Element	2 Elements	3 Elements
\varnothing	$\{a\}, \{b\}, \{c\}$	$\{a, b\}, \{b, c\}, \{a, c\}$	$\{a, b, c\}$

↵

Now Work PROBLEM 9

2 Count the Number of Elements in a Set

As you count the number of students in a classroom or the number of pennies in your pocket, what you are really doing is matching, on a one-to-one basis, each object to be counted with the set of counting numbers, $1, 2, 3, \ldots, n$, for some number n. If a set A matched up in this fashion with the set $\{1, 2, \ldots, 25\}$, you would conclude that there are 25 elements in the set A. We use the notation $n(A) = 25$ to indicate that there are 25 elements in the set A.

Because the empty set has no elements, we write

$$n(\varnothing) = 0$$

If the number of elements in a set is a nonnegative integer, we say that the set is **finite.** Otherwise, it is **infinite.** We shall concern ourselves only with finite sets.

Look again at Example 1. A set with 3 elements has $2^3 = 8$ subsets. This result can be generalized.

In Words
We use the notation $n(A)$ to mean the number of elements in set A.

> If A is a set with n elements, A has 2^n subsets.

For example, the set $\{a, b, c, d, e\}$ has $2^5 = 32$ subsets.

EXAMPLE 2 **Analyzing Survey Data**

In a survey of 100 college students, 35 were registered in College Algebra, 52 were registered in Computer Science I, and 18 were registered in both courses.

(a) How many students were registered in College Algebra or Computer Science I?

(b) How many were registered in neither course?

Solution (a) First, let A = set of students in College Algebra

B = set of students in Computer Science I

Figure 1

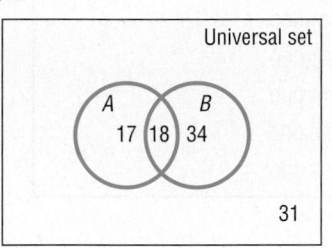

Then the given information tells us that

$$n(A) = 35 \qquad n(B) = 52 \qquad n(A \cap B) = 18$$

Refer to Figure 1. Since $n(A \cap B) = 18$, we know that the common part of the circles representing set A and set B has 18 elements. In addition, we know that the remaining portion of the circle representing set A will have $35 - 18 = 17$ elements. Similarly, we know that the remaining portion of the circle representing set B has $52 - 18 = 34$ elements. We conclude that $17 + 18 + 34 = 69$ students were registered in College Algebra or Computer Science I.

(b) Since 100 students were surveyed, it follows that $100 - 69 = 31$ were registered in neither course.

🔁

Now Work PROBLEMS 17 AND 27

The solution to Example 2 contains the basis for a general counting formula. If we count the elements in each of two sets A and B, we necessarily count twice any elements that are in both A and B, that is, those elements in $A \cap B$. To count correctly the elements that are in A or B, that is, to find $n(A \cup B)$, we need to subtract those in $A \cap B$ from $n(A) + n(B)$.

THEOREM **Counting Formula**

If A and B are finite sets,

$$n(A \cup B) = n(A) + n(B) - n(A \cap B) \qquad \textbf{(1)}$$

Refer back to Example 2. Using (1), we have

$$n(A \cup B) = n(A) + n(B) - n(A \cap B)$$
$$= 35 + 52 - 18$$
$$= 69$$

There are 69 students registered in College Algebra or Computer Science I.

A special case of the counting formula (1) occurs if A and B have no elements in common. In this case, $A \cap B = \varnothing$, so $n(A \cap B) = 0$.

THEOREM **Addition Principle of Counting**

If two sets A and B have no elements in common, that is,

$$\text{if } A \cap B = \varnothing, \text{ then } n(A \cup B) = n(A) + n(B) \qquad \textbf{(2)}$$

We can generalize formula (2).

THEOREM **General Addition Principle of Counting**

If, for n sets A_1, A_2, \ldots, A_n, no two have elements in common,

$$n(A_1 \cup A_2 \cup \cdots \cup A_n) = n(A_1) + n(A_2) + \cdots + n(A_n) \qquad (3)$$

EXAMPLE 3 **Counting**

Table 1 lists the level of education for all United States residents 25 years of age or older in 2007.

Table 1

Level of Education	Number of U.S. Residents at Least 25 Years Old
Not a high school graduate	27,787,000
High school graduate	61,404,000
Some college, but no degree	32,451,000
Associate's degree	16,711,000
Bachelor's degree	36,726,000
Advanced degree	19,237,000

Source: Statistical Abstract of the United States, 2009

(a) How many U.S. residents 25 years of age or older had an associate's degree or bachelor's degree?

(b) How many U.S. residents 25 years of age or older had an associate's degree, a bachelor's degree, or an advanced degree?

Solution Let A represent the set of associate's degree holders, B represent the set of bachelor's degree holders, and C represent the set of advanced degree holders. No two of the sets A, B, or C have elements in common (while the holder of an advanced degree certainly also holds a bachelor's degree, the individual would be part of the set for which the highest degree has been conferred). Then

$$n(A) = 16,711,000 \quad n(B) = 36,726,000 \quad n(C) = 19,237,000$$

(a) Using formula (2),

$$n(A \cup B) = n(A) + n(B) = 16,711,000 + 36,726,000 = 53,437,000$$

There were 53,437,000 U.S. residents 25 years of age or older who had an associate's degree or bachelor's degree.

(b) Using formula (3),

$$\begin{aligned} n(A \cup B \cup C) &= n(A) + n(B) + n(C) \\ &= 16,711,000 + 36,726,000 + 19,237,000 \\ &= 72,674,000 \end{aligned}$$

There were 72,674,000 U.S. residents 25 years of age or older who had an associate's degree, bachelor's degree, or advanced degree.

 Now Work PROBLEM 31

3 Solve Counting Problems Using the Multiplication Principle

EXAMPLE 4 **Counting the Number of Possible Meals**

The fixed-price dinner at Mabenka Restaurant provides the following choices:

Appetizer: soup or salad
Entrée: baked chicken, broiled beef patty, baby beef liver, or roast beef au jus
Dessert: ice cream or cheese cake

How many different meals can be ordered?

Solution Ordering such a meal requires three separate decisions:

Choose an Appetizer	**Choose an Entree**	**Choose a Dessert**
2 choices	4 choices	2 choices

Look at the **tree diagram** in Figure 2. We see that, for each choice of appetizer, there are 4 choices of entrees. And for each of these $2 \cdot 4 = 8$ choices, there are 2 choices for dessert. A total of

$$2 \cdot 4 \cdot 2 = 16$$

different meals can be ordered.

Figure 2

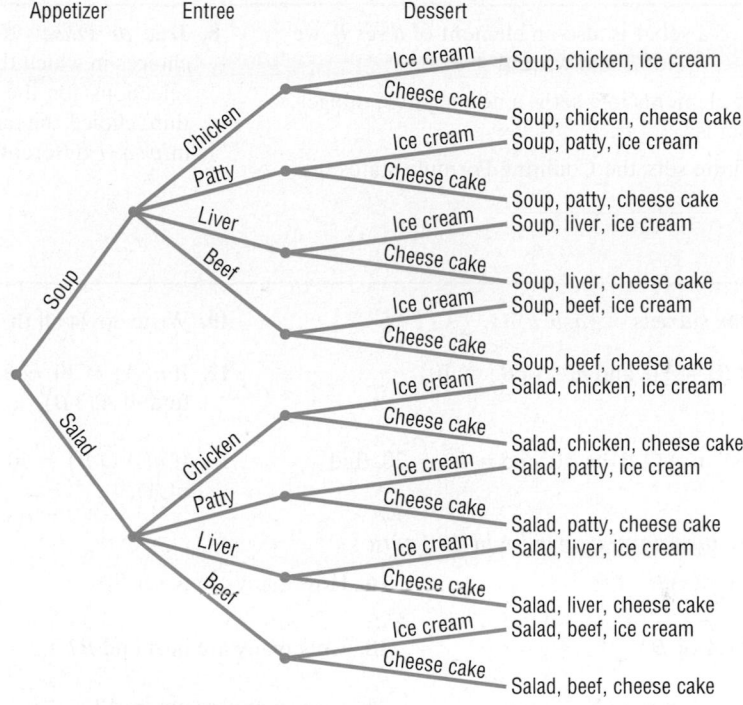

Example 4 demonstrates a general principle of counting.

THEOREM **Multiplication Principle of Counting**

If a task consists of a sequence of choices in which there are p selections for the first choice, q selections for the second choice, r selections for the third choice, and so on, the task of making these selections can be done in

$$p \cdot q \cdot r \cdot \cdots$$

different ways.

EXAMPLE 5 **Forming Codes**

How many two-symbol code words can be formed if the first symbol is an uppercase letter and the second symbol is a digit?

Solution It sometimes helps to begin by listing some of the possibilities. The code consists of an uppercase letter followed by a digit, so some possibilities are A1, A2, B3, X0, and so on. The task consists of making two selections: the first selection requires choosing an uppercase letter (26 choices) and the second task requires choosing a digit (10 choices). By the Multiplication Principle, there are

$$26 \cdot 10 = 260$$

different code words of the type described.

Now Work PROBLEM 23

10.1 Assess Your Understanding

1. The _____ of A and B consists of all elements in either A or B or both. (pp. 2–3)

2. The _____ of A with B consists of all elements in both A and B. (pp. 2–3)

3. *True or False* The intersection of two sets is always a subset of their union. (pp. 2–3)

4. *True or False* If A is a set, the complement of A is the set of all the elements in the universal set that are not in A. (pp. 2–3)

Concepts and Vocabulary

5. If each element of a set A is also an element of a set B, we say that A is a _____ of B and write A ____ B.

6. If the number of elements in a set is a nonnegative integer, we say that the set is _____.

7. If A and B are finite sets, the Counting Formula states that $n(A \cup B) =$ _____.

8. *True or False* If a task consists of a sequence of three choices in which there are p selections for the first choice, q selections for the second choice, and r selections for the third choice, the task of making these selections can be done in $p \cdot q \cdot r$ different ways.

Skill Building

9. Write down all the subsets of $\{a, b, c, d\}$.

10. Write down all the subsets of $\{a, b, c, d, e\}$.

11. If $n(A) = 15, n(B) = 20$, and $n(A \cap B) = 10$, find $n(A \cup B)$.

12. If $n(A) = 30, n(B) = 40$, and $n(A \cup B) = 45$, find $n(A \cap B)$.

13. If $n(A \cup B) = 50, n(A \cap B) = 10$, and $n(B) = 20$, find $n(A)$.

14. If $n(A \cup B) = 60, n(A \cap B) = 40$, and $n(A) = n(B)$, find $n(A)$.

In Problems 15–22, use the information given in the figure.

15. How many are in set A?

16. How many are in set B?

17. How many are in A or B?

18. How many are in A and B?

19. How many are in A but not C?

20. How many are not in A?

21. How many are in A and B and C?

22. How many are in A or B or C?

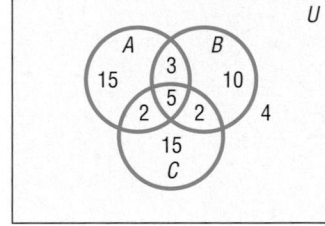

Applications and Extensions

23. Shirts and Ties A man has 5 shirts and 3 ties. How many different shirt and tie arrangements can he wear?

24. Blouses and Skirts A woman has 5 blouses and 8 skirts. How many different outfits can she wear?

25. Four-digit Numbers How many four-digit numbers can be formed using the digits 0, 1, 2, 3, 4, 5, 6, 7, 8, and 9 if the first digit cannot be 0? Repeated digits are allowed.

26. Five-digit Numbers How many five-digit numbers can be formed using the digits 0, 1, 2, 3, 4, 5, 6, 7, 8, and 9 if the first digit cannot be 0 or 1? Repeated digits are allowed.

27. Analyzing Survey Data In a consumer survey of 500 people, 200 indicated that they would be buying a major appliance within the next month, 150 indicated that they would buy a car, and 25 said that they would purchase both a major appliance and a car. How many will purchase neither? How many will purchase only a car?

28. Analyzing Survey Data In a student survey, 200 indicated that they would attend Summer Session I and 150 indicated Summer Session II. If 75 students plan to attend both summer sessions and 275 indicated that they would attend neither session, how many students participated in the survey?

29. Analyzing Survey Data In a survey of 100 investors in the stock market,

50 owned shares in IBM
40 owned shares in AT&T
45 owned shares in GE
20 owned shares in both IBM and GE
15 owned shares in both AT&T and GE
20 owned shares in both IBM and AT&T
5 owned shares in all three

(a) How many of the investors surveyed did not have shares in any of the three companies?
(b) How many owned just IBM shares?
(c) How many owned just GE shares?
(d) How many owned neither IBM nor GE?
(e) How many owned either IBM or AT&T but no GE?

30. Classifying Blood Types Human blood is classified as either Rh+ or Rh−. Blood is also classified by type: A, if it contains an A antigen but not a B antigen; B, if it contains a B antigen

but not an A antigen; AB, if it contains both A and B antigens; and O, if it contains neither antigen. Draw a Venn diagram illustrating the various blood types. Based on this classification, how many different kinds of blood are there?

31. Demographics The following data represent the marital status of males 18 years old and older in 2007.

Marital Status	Number (in thousands)
Married	63,318
Widowed	2,723
Divorced	9,200
Never married	29,608

Source: Current Population Survey

(a) Determine the number of males 18 years old and older who are widowed or divorced.

(b) Determine the number of males 18 years old and older who are married, widowed, or divorced.

32. Demographics The following data represent the marital status of females 18 years old and older in 2007.

Marital Status	Number (in thousands)
Married	63,971
Widowed	11,105
Divorced	12,932
Never married	24,318

Source: Current Population Survey

(a) Determine the number of females 18 years old and older who are widowed or divorced.

(b) Determine the number of females 18 years old and older who are married, widowed, or divorced.

33. Stock Portfolios As a financial planner, you are asked to select one stock each from the following groups: 8 DOW stocks, 15 NASDAQ stocks, and 4 global stocks. How many different portfolios are possible?

Explaining Concepts: Discussion and Writing

34. Make up a problem different from any found in the text that requires the addition principle of counting to solve. Give it to a friend to solve and critique.

35. Investigate the notion of counting as it relates to infinite sets. Write an essay on your findings.

'Are You Prepared?' Answers

1. union **2.** intersection **3.** True **4.** True

10.2 Permutations and Combinations

PREPARING FOR THIS SECTION *Before getting started, review the following:*

• Factorial (Section 9.1, p. 639)

Now Work the 'Are You Prepared?' problems on page 691.

OBJECTIVES **1** Solve Counting Problems Using Permutations Involving *n* Distinct Objects (p. 685)

2 Solve Counting Problems Using Combinations (p. 688)

3 Solve Counting Problems Using Permutations Involving *n* Nondistinct Objects (p. 690)

1 Solve Counting Problems Using Permutations Involving *n* Distinct Objects

DEFINITION A **permutation** is an ordered arrangement of *r* objects chosen from *n* objects.

We discuss three types of permutations:

1. The *n* objects are distinct (different), and repetition is allowed in the selection of *r* of them. [Distinct, with repetition]

2. The n objects are distinct (different), and repetition is not allowed in the selection of r of them, where $r \leq n$. [Distinct, without repetition]

3. The n objects are not distinct, and we use all of them in the arrangement. [Not distinct]

We take up the first two types here and deal with the third type at the end of this section.

The first type of permutation (n distinct objects, repetition allowed) is handled using the Multiplication Principle.

EXAMPLE 1 | **Counting Airport Codes [Permutation: Distinct, with Repetition]**

The International Airline Transportation Association (IATA) assigns three-letter codes to represent airport locations. For example, the airport code for Ft. Lauderdale, Florida, is FLL. Notice that repetition is allowed in forming this code. How many airport codes are possible?

Solution We are choosing 3 letters from 26 letters and arranging them in order. In the ordered arrangement a letter may be repeated. This is an example of a permutation with repetition in which 3 objects are chosen from 26 distinct objects.

The task of counting the number of such arrangements consists of making three selections. Each selection requires choosing a letter of the alphabet (26 choices). By the Multiplication Principle, there are

$$26 \cdot 26 \cdot 26 = 26^3 = 17{,}576$$

possible airport codes.

The solution given to Example 1 can be generalized.

THEOREM | **Permutations: Distinct Objects with Repetition**

The number of ordered arrangements of r objects chosen from n objects, in which the n objects are distinct and repetition is allowed, is n^r.

Now Work PROBLEM 33

Now let's consider permutations in which the objects are distinct and repetition is not allowed.

EXAMPLE 2 | **Forming Codes [Permutation: Distinct, without Repetition]**

Suppose that we wish to establish a three-letter code using any of the 26 uppercase letters of the alphabet, but we require that no letter be used more than once. How many different three-letter codes are there?

Solution Some of the possibilities are ABC, ABD, ABZ, ACB, CBA, and so on. The task consists of making three selections. The first selection requires choosing from 26 letters. Because no letter can be used more than once, the second selection requires choosing from 25 letters. The third selection requires choosing from 24 letters. (Do you see why?) By the Multiplication Principle, there are

$$26 \cdot 25 \cdot 24 = 15{,}600$$

different three-letter codes with no letter repeated.

For the second type of permutation, we introduce the following notation.

The notation $P(n, r)$ represents the number of ordered arrangements of r objects chosen from n distinct objects, where $r \leq n$ and repetition is not allowed.

For example, the question posed in Example 2 asks for the number of ways that the 26 letters of the alphabet can be arranged in order using three nonrepeated letters. The answer is

$$P(26, 3) = 26 \cdot 25 \cdot 24 = 15,600$$

EXAMPLE 3

Lining People Up

In how many ways can 5 people be lined up?

Solution

The 5 people are distinct. Once a person is in line, that person will not be repeated elsewhere in the line; and, in lining people up, order is important. We have a permutation of 5 objects taken 5 at a time. We can line up 5 people in

$$P(5, 5) = \underbrace{5 \cdot 4 \cdot 3 \cdot 2 \cdot 1}_{5 \text{ factors}} = 120 \text{ ways}$$

══════ Now Work PROBLEM 35

To arrive at a formula for $P(n, r)$, note that the task of obtaining an ordered arrangement of n objects in which only $r \leq n$ of them are used, without repeating any of them, requires making r selections. For the first selection, there are n choices; for the second selection, there are $n - 1$ choices; for the third selection, there are $n - 2$ choices; ...; for the rth selection, there are $n - (r - 1)$ choices. By the Multiplication Principle, we have

$$
\begin{array}{cccc}
\text{1st} & \text{2nd} & \text{3rd} & \text{rth} \\
\end{array}
$$
$$P(n, r) = n \cdot (n - 1) \cdot (n - 2) \cdot \cdots \cdot [n - (r - 1)]$$
$$= n \cdot (n - 1) \cdot (n - 2) \cdot \cdots \cdot (n - r + 1)$$

This formula for $P(n, r)$ can be compactly written using factorial notation.*

$$P(n, r) = n \cdot (n - 1) \cdot (n - 2) \cdot \cdots \cdot (n - r + 1)$$
$$= n \cdot (n - 1) \cdot (n - 2) \cdot \cdots \cdot (n - r + 1) \cdot \frac{(n - r) \cdot \cdots \cdot 3 \cdot 2 \cdot 1}{(n - r) \cdot \cdots \cdot 3 \cdot 2 \cdot 1} = \frac{n!}{(n - r)!}$$

THEOREM

Permutations of r Objects Chosen from n Distinct Objects without Repetition

The number of arrangements of n objects using $r \leq n$ of them, in which

1. the n objects are distinct,
2. once an object is used it cannot be repeated, and
3. order is important,

is given by the formula

$$P(n, r) = \frac{n!}{(n - r)!} \tag{1}$$

*Recall that $0! = 1$, $1! = 1$, $2! = 2 \cdot 1, \ldots, n! = n(n - 1) \cdot \cdots \cdot 3 \cdot 2 \cdot 1$.

| EXAMPLE 4 | **Computing Permutations** |

Evaluate: (a) $P(7, 3)$ (b) $P(6, 1)$ (c) $P(52, 5)$

Solution We work parts (a) and (b) in two ways.

(a) $P(7, 3) = \underbrace{7 \cdot 6 \cdot 5}_{3 \text{ factors}} = 210$

or

$$P(7, 3) = \frac{7!}{(7 - 3)!} = \frac{7!}{4!} = \frac{7 \cdot 6 \cdot 5 \cdot 4!}{4!} = 210$$

(b) $P(6, 1) = \underbrace{6}_{1 \text{ factor}} = 6$

or

$$P(6, 1) = \frac{6!}{(6 - 1)!} = \frac{6!}{5!} = \frac{6 \cdot 5!}{5!} = 6$$

Figure 3

```
52 nPr 5
        311875200
```

(c) Figure 3 shows the solution using a TI-84 Plus graphing calculator. So

$$P(52, 5) = 311,875,200$$

──── **Now Work** PROBLEM 7

| EXAMPLE 5 | **The Birthday Problem** |

All we know about Shannon, Patrick, and Ryan is that they have different birthdays. If we listed all the possible ways this could occur, how many would there be? Assume that there are 365 days in a year.

Solution This is an example of a permutation in which 3 birthdays are selected from a possible 365 days, and no birthday may repeat itself. The number of ways that this can occur is

$$P(365, 3) = \frac{365!}{(365 - 3)!} = \frac{365 \cdot 364 \cdot 363 \cdot 362!}{362!} = 365 \cdot 364 \cdot 363 = 48,228,180$$

There are 48,228,180 ways in a group of three people for each to have a different birthday.

──── **Now Work** PROBLEM 47

2 Solve Counting Problems Using Combinations

In a permutation, order is important. For example, the arrangements ABC, CAB, BAC, ... are considered different arrangements of the letters A, B, and C. In many situations, though, order is unimportant. For example, in the card game of poker, the order in which the cards are received does not matter; it is the *combination* of the cards that matters.

DEFINITION A **combination** is an arrangement, without regard to order, of r objects selected from n distinct objects without repetition, where $r \leq n$. The notation $C(n, r)$ represents the number of combinations of n distinct objects using r of them.

| EXAMPLE 6 | **Listing Combinations** |

List all the combinations of the 4 objects a, b, c, d taken 2 at a time. What is $C(4, 2)$?

Solution One combination of a, b, c, d taken 2 at a time is

$$ab$$

We exclude *ba* from the list because order is not important in a combination (this means that we do not distinguish *ab* from *ba*). The list of all combinations of *a, b, c, d* taken 2 at a time is

$$ab, \quad ac, \quad ad, \quad bc, \quad bd, \quad cd$$

so

$$C(4, 2) = 6$$

We can find a formula for $C(n, r)$ by noting that the only difference between a permutation of type 2 (distinct, without repetition) and a combination is that we disregard order in combinations. To determine $C(n, r)$, we need only eliminate from the formula for $P(n, r)$ the number of permutations that are simply rearrangements of a given set of *r* objects. This can be determined from the formula for $P(n, r)$ by calculating $P(r, r) = r!$. So, if we divide $P(n, r)$ by $r!$, we will have the desired formula for $C(n, r)$:

$$C(n, r) = \frac{P(n, r)}{r!} = \frac{\dfrac{n!}{(n - r)!}}{r!} = \frac{n!}{(n - r)!r!}$$

$$\underset{\text{Use formula (1).}}{\uparrow}$$

We have proved the following result:

THEOREM **Number of Combinations of *n* Distinct Objects Taken *r* at a Time**

The number of arrangements of *n* objects using $r \leq n$ of them, in which

1. the *n* objects are distinct,
2. once an object is used, it cannot be repeated, and
3. order is not important,

is given by the formula

$$C(n, r) = \frac{n!}{(n - r)!r!} \tag{2}$$

Based on formula (2), we discover that the symbol $C(n, r)$ and the symbol $\begin{pmatrix} n \\ r \end{pmatrix}$ for the binomial coefficients are, in fact, the same. Pascal's triangle (see Section 9.5) can be used to find the value of $C(n, r)$. However, because it is more practical and convenient, we will use formula (2) instead.

EXAMPLE 7 **Using Formula (2)**

Use formula (2) to find the value of each expression.

(a) $C(3, 1)$ (b) $C(6, 3)$ (c) $C(n, n)$ (d) $C(n, 0)$ (e) $C(52, 5)$

Solution (a) $C(3, 1) = \dfrac{3!}{(3 - 1)!1!} = \dfrac{3!}{2!1!} = \dfrac{3 \cdot 2 \cdot 1}{2 \cdot 1 \cdot 1} = 3$

(b) $C(6, 3) = \dfrac{6!}{(6 - 3)!3!} = \dfrac{6 \cdot 5 \cdot 4 \cdot 3!}{3! \cdot 3!} = \dfrac{6 \cdot 5 \cdot 4}{6} = 20$

(c) $C(n, n) = \dfrac{n!}{(n - n)!n!} = \dfrac{n!}{0!n!} = \dfrac{1}{1} = 1$

Figure 4

(d) $C(n, 0) = \dfrac{n!}{(n - 0)!0!} = \dfrac{n!}{n!0!} = \dfrac{1}{1} = 1$

(e) Figure 4 shows the solution using a TI-84 Plus graphing calculator. So

$$C(52, 5) = 2{,}598{,}960$$

⟍

Now Work PROBLEM 15

EXAMPLE 8	**Forming Committees**

How many different committees of 3 people can be formed from a pool of 7 people?

Solution The 7 people are distinct. More important, though, is the observation that the order of being selected for a committee is not significant. The problem asks for the number of combinations of 7 objects taken 3 at a time.

$$C(7, 3) = \dfrac{7!}{4!3!} = \dfrac{7 \cdot 6 \cdot 5 \cdot 4!}{4!3!} = \dfrac{7 \cdot 6 \cdot 5}{6} = 35$$

Thirty-five different committees can be formed.

⟍

EXAMPLE 9	**Forming Committees**

In how many ways can a committee consisting of 2 faculty members and 3 students be formed if 6 faculty members and 10 students are eligible to serve on the committee?

Solution The problem can be separated into two parts: the number of ways that the faculty members can be chosen, $C(6, 2)$, and the number of ways that the student members can be chosen, $C(10, 3)$. By the Multiplication Principle, the committee can be formed in

$$C(6, 2) \cdot C(10, 3) = \dfrac{6!}{4!2!} \cdot \dfrac{10!}{7!3!} = \dfrac{6 \cdot 5 \cdot 4!}{4!2!} \cdot \dfrac{10 \cdot 9 \cdot 8 \cdot 7!}{7!3!}$$

$$= \dfrac{30}{2} \cdot \dfrac{720}{6} = 1800 \text{ ways}$$

⟍

Now Work PROBLEM 49

3 Solve Counting Problems Using Permutations Involving *n* Nondistinct Objects

EXAMPLE 10	**Forming Different Words**

How many different words (real or imaginary) can be formed using all the letters in the word REARRANGE?

Solution Each word formed will have 9 letters: 3 R's, 2 A's, 2 E's, 1 N, and 1 G. To construct each word, we need to fill in 9 positions with the 9 letters:

$$\overline{1}\ \overline{2}\ \overline{3}\ \overline{4}\ \overline{5}\ \overline{6}\ \overline{7}\ \overline{8}\ \overline{9}$$

The process of forming a word consists of five tasks:

Task 1: Choose the positions for the 3 R's.
Task 2: Choose the positions for the 2 A's.
Task 3: Choose the positions for the 2 E's.

Task 4: Choose the position for the 1 N.
Task 5: Choose the position for the 1 G.

Task 1 can be done in $C(9, 3)$ ways. There then remain 6 positions to be filled, so Task 2 can be done in $C(6, 2)$ ways. There remain 4 positions to be filled, so Task 3 can be done in $C(4, 2)$ ways. There remain 2 positions to be filled, so Task 4 can be done in $C(2, 1)$ ways. The last position can be filled in $C(1, 1)$ way. Using the Multiplication Principle, the number of possible words that can be formed is

$$C(9, 3) \cdot C(6, 2) \cdot C(4, 2) \cdot C(2, 1) \cdot C(1, 1) = \frac{9!}{3! \cdot 6!} \cdot \frac{6!}{2! \cdot 4!} \cdot \frac{4!}{2! \cdot 2!} \cdot \frac{2!}{1! \cdot 1!} \cdot \frac{1!}{0! \cdot 1!}$$

$$= \frac{9!}{3! \cdot 2! \cdot 2! \cdot 1! \cdot 1!} = 15{,}120$$

15,120 possible words can be formed.

The form of the expression before the answer to Example 10 is suggestive of a general result. Had the letters in REARRANGE each been different, there would have been $P(9, 9) = 9!$ possible words formed. This is the numerator of the answer. The presence of 3 R's, 2 A's, and 2 E's reduces the number of different words, as the entries in the denominator illustrate. We are led to the following result:

THEOREM

Permutations Involving n Objects That Are Not Distinct

The number of permutations of n objects of which n_1 are of one kind, n_2 are of a second kind, . . . , and n_k are of a kth kind is given by

$$\frac{n!}{n_1! \cdot n_2! \cdot \cdots \cdot n_k!} \tag{3}$$

where $n = n_1 + n_2 + \cdots + n_k$.

EXAMPLE 11

Arranging Flags

How many different vertical arrangements are there of 8 flags if 4 are white, 3 are blue, and 1 is red?

Solution

We seek the number of permutations of 8 objects, of which 4 are of one kind, 3 are of a second kind, and 1 is of a third kind. Using formula (3), we find that there are

$$\frac{8!}{4! \cdot 3! \cdot 1!} = \frac{8 \cdot 7 \cdot 6 \cdot 5 \cdot 4!}{4! \cdot 3! \cdot 1!} = 280 \text{ different arrangements}$$

━━━**Now Work** PROBLEM 51

10.2 Assess Your Understanding

'Are You Prepared?' *Answers are given at the end of these exercises. If you get a wrong answer, read the pages listed in red.*

1. $0! = $ _____ ; $1! = $ _____ . (p. 639)

2. *True or False* $n! = \dfrac{(n + 1)!}{n}$. (p. 639)

Concepts and Vocabulary

3. A(n) _____ is an ordered arrangement of r objects chosen from n objects.

4. A(n) _____ is an arrangement of r objects chosen from n distinct objects, without repetition and without regard to order.

5. $P(n, r) = $ _____ .

6. $C(n, r) = $ _____ .

Skill Building

In Problems 7–14, find the value of each permutation.

7. $P(6, 2)$

8. $P(7, 2)$

9. $P(4, 4)$

10. $P(8, 8)$

11. $P(7, 0)$

12. $P(9, 0)$

13. $P(8, 4)$

14. $P(8, 3)$

In Problems 15–22, use formula (2) to find the value of each combination.

15. $C(8, 2)$

16. $C(8, 6)$

17. $C(7, 4)$

18. $C(6, 2)$

19. $C(15, 15)$

20. $C(18, 1)$

21. $C(26, 13)$

22. $C(18, 9)$

Applications and Extensions

23. List all the ordered arrangements of 5 objects a, b, c, d, and e choosing 3 at a time without repetition. What is $P(5, 3)$?

24. List all the ordered arrangements of 5 objects a, b, c, d, and e choosing 2 at a time without repetition. What is $P(5, 2)$?

25. List all the ordered arrangements of 4 objects $1, 2, 3$, and 4 choosing 3 at a time without repetition. What is $P(4, 3)$?

26. List all the ordered arrangements of 6 objects $1, 2, 3, 4, 5,$ and 6 choosing 3 at a time without repetition. What is $P(6, 3)$?

27. List all the combinations of 5 objects a, b, c, d, and e taken 3 at a time. What is $C(5, 3)$?

28. List all the combinations of 5 objects a, b, c, d, and e taken 2 at a time. What is $C(5, 2)$?

29. List all the combinations of 4 objects $1, 2, 3$, and 4 taken 3 at a time. What is $C(4, 3)$?

30. List all the combinations of 6 objects $1, 2, 3, 4, 5,$ and 6 taken 3 at a time. What is $C(6, 3)$?

31. Forming Codes How many two-letter codes can be formed using the letters $A, B, C,$ and D? Repeated letters are allowed.

32. Forming Codes How many two-letter codes can be formed using the letters $A, B, C, D,$ and E? Repeated letters are allowed.

33. Forming Numbers How many three-digit numbers can be formed using the digits 0 and 1? Repeated digits are allowed.

34. Forming Numbers How many three-digit numbers can be formed using the digits $0, 1, 2, 3, 4, 5, 6, 7, 8,$ and 9? Repeated digits are allowed.

35. Lining People Up In how many ways can 4 people be lined up?

36. Stacking Boxes In how many ways can 5 different boxes be stacked?

37. Forming Codes How many different three-letter codes are there if only the letters $A, B, C, D,$ and E can be used and no letter can be used more than once?

38. Forming Codes How many different four-letter codes are there if only the letters $A, B, C, D, E,$ and F can be used and no letter can be used more than once?

39. Stocks on the NYSE Companies whose stocks are listed on the New York Stock Exchange (NYSE) have their company name represented by either 1, 2, or 3 letters (repetition of letters is allowed). What is the maximum number of companies that can be listed on the NYSE?

40. Stocks on the NASDAQ Companies whose stocks are listed on the NASDAQ stock exchange have their company name represented by either 4 or 5 letters (repetition of letters is allowed). What is the maximum number of companies that can be listed on the NASDAQ?

41. Establishing Committees In how many ways can a committee of 4 students be formed from a pool of 7 students?

42. Establishing Committees In how many ways can a committee of 3 professors be formed from a department having 8 professors?

43. Possible Answers on a True/False Test How many arrangements of answers are possible for a true/false test with 10 questions?

44. Possible Answers on a Multiple-choice Test How many arrangements of answers are possible in a multiple-choice test with 5 questions, each of which has 4 possible answers?

45. Arranging Books Five different mathematics books are to be arranged on a student's desk. How many arrangements are possible?

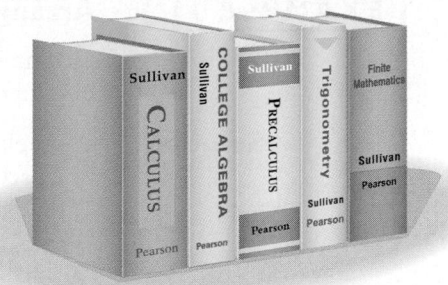

46. Forming License Plate Numbers How many different license plate numbers can be made using 2 letters followed by 4 digits selected from the digits 0 through 9, if
(a) letters and digits may be repeated?
(b) letters may be repeated, but digits may not be repeated?
(c) neither letters nor digits may be repeated?

47. Birthday Problem In how many ways can 2 people each have different birthdays? Assume that there are 365 days in a year.

48. Birthday Problem In how many ways can 5 people each have different birthdays? Assume that there are 365 days in a year.

49. Forming a Committee A student dance committee is to be formed consisting of 2 boys and 3 girls. If the membership is to be chosen from 4 boys and 8 girls, how many different committees are possible?

50. Forming a Committee The student relations committee of a college consists of 2 administrators, 3 faculty members, and 5 students. Four administrators, 8 faculty members, and 20 students are eligible to serve. How many different committees are possible?

51. Forming Words How many different 9-letter words (real or imaginary) can be formed from the letters in the word ECONOMICS?

52. Forming Words How many different 11-letter words (real or imaginary) can be formed from the letters in the word MATHEMATICS?

53. Selecting Objects An urn contains 7 white balls and 3 red balls. Three balls are selected. In how many ways can the 3 balls be drawn from the total of 10 balls:
(a) If 2 balls are white and 1 is red?
(b) If all 3 balls are white?
(c) If all 3 balls are red?

54. Selecting Objects An urn contains 15 red balls and 10 white balls. Five balls are selected. In how many ways can the 5 balls be drawn from the total of 25 balls:
(a) If all 5 balls are red?
(b) If 3 balls are red and 2 are white?
(c) If at least 4 are red balls?

55. Senate Committees The U.S. Senate has 100 members. Suppose that it is desired to place each senator on exactly 1 of 7 possible committees. The first committee has 22 members, the second has 13, the third has 10, the fourth has 5, the fifth has 16, and the sixth and seventh have 17 apiece. In how many ways can these committees be formed?

56. Football Teams A defensive football squad consists of 25 players. Of these, 10 are linemen, 10 are linebackers, and 5 are safeties. How many different teams of 5 linemen, 3 linebackers, and 3 safeties can be formed?

57. Baseball In the American Baseball League, a designated hitter may be used. How many batting orders is it possible for a manager to use? (There are 9 regular players on a team.)

58. Baseball In the National Baseball League, the pitcher usually bats ninth. If this is the case, how many batting orders is it possible for a manager to use?

59. Baseball Teams A baseball team has 15 members. Four of the players are pitchers, and the remaining 11 members can play any position. How many different teams of 9 players can be formed?

60. World Series In the World Series the American League team (*A*) and the National League team (*N*) play until one team wins four games. If the sequence of winners is designated by letters (for example, *NAAAA* means that the National League team won the first game and the American League won the next four), how many different sequences are possible?

61. Basketball Teams A basketball team has 6 players who play guard (2 of 5 starting positions). How many different teams are possible, assuming that the remaining 3 positions are filled and it is not possible to distinguish a left guard from a right guard?

62. Basketball Teams On a basketball team of 12 players, 2 only play center, 3 only play guard, and the rest play forward (5 players on a team: 2 forwards, 2 guards, and 1 center). How many different teams are possible, assuming that it is not possible to distinguish left and right guards and left and right forwards?

63. Combination Locks A combination lock displays 50 numbers. To open it, you turn to a number, then rotate clockwise to a second number, and then counterclockwise to the third number.
(a) How many different lock combinations are there?
(b) Comment on the description of such a lock as a *combination* lock.

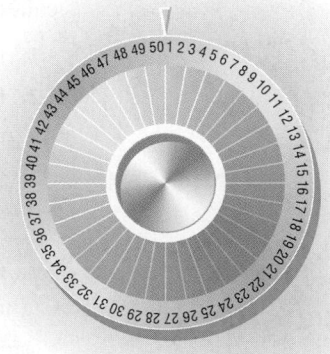

Explaining Concepts: Discussion and Writing

64. Create a problem different from any found in the text that requires a permutation to solve. Give it to a friend to solve and critique.

65. Create a problem different from any found in the text that requires a combination to solve. Give it to a friend to solve and critique.

66. Explain the difference between a permutation and a combination. Give an example to illustrate your explanation.

'Are You Prepared?' Answers

1. 1; 1 **2.** False

10.3 Probability

> **OBJECTIVES** 1 Construct Probability Models (p. 694)
> 2 Compute Probabilities of Equally Likely Outcomes (p. 696)
> 3 Find Probabilities of the Union of Two Events (p. 698)
> 4 Use the Complement Rule to Find Probabilities (p. 699)

Probability is an area of mathematics that deals with experiments that yield random results, yet admit a certain regularity. Such experiments do not always produce the same result or outcome, so the result of any one observation is not predictable. However, the results of the experiment over a long period do produce regular patterns that enable us to predict with remarkable accuracy.

EXAMPLE 1 **Tossing a Fair Coin**

In tossing a fair coin, we know that the outcome is either a head or a tail. On any particular throw, we cannot predict what will happen, but, if we toss the coin many times, we observe that the number of times that a head comes up is approximately equal to the number of times that we get a tail. It seems reasonable, therefore, to assign a probability of $\frac{1}{2}$ that a head comes up and a probability of $\frac{1}{2}$ that a tail comes up.

1 Construct Probability Models

The discussion in Example 1 constitutes the construction of a **probability model** for the experiment of tossing a fair coin once. A probability model has two components: a sample space and an assignment of probabilities. A **sample space** S is a set whose elements represent all the possibilities that can occur as a result of the experiment. Each element of S is called an **outcome.** To each outcome, we assign a number, called the **probability** of that outcome, which has two properties:

1. The probability assigned to each outcome is nonnegative.
2. The sum of all the probabilities equals 1.

DEFINITION

A **probability model** with the sample space

$$S = \{e_1, e_2, \ldots, e_n\}$$

where e_1, e_2, \ldots, e_n are the possible outcomes and $P(e_1), P(e_2), \ldots, P(e_n)$ are the respective probabilities of these outcomes, requires that

$$P(e_1) \geq 0, P(e_2) \geq 0, \ldots, P(e_n) \geq 0 \tag{1}$$

$$\sum_{i=1}^{n} P(e_i) = P(e_1) + P(e_2) + \cdots + P(e_n) = 1 \tag{2}$$

EXAMPLE 2 **Determining Probability Models**

In a bag of M&Ms,™ the candies are colored red, green, blue, brown, yellow, and orange. A candy is drawn from the bag and the color is recorded. The sample space of this experiment is {red, green, blue, brown, yellow, orange}. Determine which of the following are probability models.

(a)

Outcome	Probability
red	0.3
green	0.15
blue	0
brown	0.15
yellow	0.2
orange	0.2

(b)

Outcome	Probability
red	0.1
green	0.1
blue	0.1
brown	0.4
yellow	0.2
orange	0.3

(c)

Outcome	Probability
red	0.3
green	−0.3
blue	0.2
brown	0.4
yellow	0.2
orange	0.2

(d)

Outcome	Probability
red	0
green	0
blue	0
brown	0
yellow	1
orange	0

Solution

(a) This model is a probability model since all the outcomes have probabilities that are nonnegative and the sum of the probabilities is 1.

(b) This model is not a probability model because the sum of the probabilities is not 1.

(c) This model is not a probability model because $P(\text{green})$ is less than 0. Recall, all probabilities must be nonnegative.

(d) This model is a probability model because all the outcomes have probabilities that are nonnegative, and the sum of the probabilities is 1. Notice that $P(\text{yellow}) = 1$, meaning that this outcome will occur with 100% certainty each time that the experiment is repeated. This means that the bag of M&Ms™ has only yellow candies.

══════Now Work PROBLEM 7

EXAMPLE 3 **Constructing a Probability Model**

An experiment consists of rolling a fair die once. A die is a cube with each face having either 1, 2, 3, 4, 5, or 6 dots on it. See Figure 5. Construct a probability model for this experiment.

Figure 5

Solution A sample space S consists of all the possibilities that can occur. Because rolling the die will result in one of six faces showing, the sample space S consists of

$$S = \{1, 2, 3, 4, 5, 6\}$$

Because the die is fair, one face is no more likely to occur than another. As a result, our assignment of probabilities is

$$P(1) = \frac{1}{6} \qquad P(2) = \frac{1}{6}$$

$$P(3) = \frac{1}{6} \qquad P(4) = \frac{1}{6}$$

$$P(5) = \frac{1}{6} \qquad P(6) = \frac{1}{6}$$

Now suppose that a die is loaded (weighted) so that the probability assignments are

$$P(1) = 0 \quad P(2) = 0 \quad P(3) = \frac{1}{3} \quad P(4) = \frac{2}{3} \quad P(5) = 0 \quad P(6) = 0$$

This assignment would be made if the die were loaded so that only a 3 or 4 could occur and the 4 is twice as likely as the 3 to occur. This assignment is consistent with the definition, since each assignment is nonnegative and the sum of all the probability assignments equals 1.

──── **Now Work** PROBLEM 23

EXAMPLE 4 **Constructing a Probability Model**

An experiment consists of tossing a coin. The coin is weighted so that heads (H) is three times as likely to occur as tails (T). Construct a probability model for this experiment.

Solution The sample space S is $S = \{H, T\}$. If x denotes the probability that a tail occurs,

$$P(T) = x \quad \text{and} \quad P(H) = 3x$$

Since the sum of the probabilities of the possible outcomes must equal 1, we have

$$P(T) + P(H) = x + 3x = 1$$
$$4x = 1$$
$$x = \frac{1}{4}$$

Assign the probabilities

$$P(T) = \frac{1}{4} \quad P(H) = \frac{3}{4}$$

──── **Now Work** PROBLEM 27

In working with probability models, the term **event** is used to describe a set of possible outcomes of the experiment. An event E is some subset of the sample space S. The **probability of an event** E, $E \neq \varnothing$, denoted by $P(E)$, is defined as the sum of the probabilities of the outcomes in E. We can also think of the probability of an event E as the likelihood that the event E occurs. If $E = \varnothing$, then $P(E) = 0$; if $E = S$, then $P(E) = P(S) = 1$.

2 Compute Probabilities of Equally Likely Outcomes

When the same probability is assigned to each outcome of the sample space, the experiment is said to have **equally likely outcomes.**

THEOREM **Probability for Equally Likely Outcomes**

If an experiment has n equally likely outcomes and if the number of ways that an event E can occur is m, then the probability of E is

$$P(E) = \frac{\text{Number of ways that } E \text{ can occur}}{\text{Number of possible outcomes}} = \frac{m}{n} \qquad (3)$$

If S is the sample space of this experiment,

$$P(E) = \frac{n(E)}{n(S)} \qquad (4)$$

EXAMPLE 5

Calculating Probabilities of Events Involving Equally Likely Outcomes

Calculate the probability that in a 3-child family there are 2 boys and 1 girl. Assume equally likely outcomes.

Solution

Begin by constructing a tree diagram to help in listing the possible outcomes of the experiment. See Figure 6, where B stands for boy and G for girl. The sample space S of this experiment is

$$S = \{BBB, BBG, BGB, BGG, GBB, GBG, GGB, GGG\}$$

so $n(S) = 8$.

Figure 6

1st child 2nd child 3rd child

We wish to know the probability of the event E: "having two boys and one girl." From Figure 6, we conclude that $E = \{BBG, BGB, GBB\}$, so $n(E) = 3$. Since the outcomes are equally likely, the probability of E is

$$P(E) = \frac{n(E)}{n(S)} = \frac{3}{8}$$

Now Work PROBLEM 37

So far, we have calculated probabilities of single events. Now we compute probabilities of multiple events, called **compound probabilities.**

EXAMPLE 6

Computing Compound Probabilities

Consider the experiment of rolling a single fair die. Let E represent the event "roll an odd number," and let F represent the event "roll a 1 or 2."

(a) Write the event E and F. What is $n(E \cap F)$?
(b) Write the event E or F. What is $n(E \cup F)$?
(c) Compute $P(E)$. Compute $P(F)$.
(d) Compute $P(E \cap F)$.
(e) Compute $P(E \cup F)$.

Solution

The sample space S of the experiment is $\{1, 2, 3, 4, 5, 6\}$, so $n(S) = 6$. Since the die is fair, the outcomes are equally likely. The event E: "roll an odd number" is $\{1, 3, 5\}$, and the event F: "roll a 1 or 2" is $\{1, 2\}$, so $n(E) = 3$ and $n(F) = 2$.

(a) The word *and* in probability means the intersection of two events. The event E and F is

$$E \cap F = \{1, 3, 5\} \cap \{1, 2\} = \{1\} \qquad n(E \cap F) = 1$$

(b) The word *or* in probability means the union of the two events. The event E or F is

$$E \cup F = \{1, 3, 5\} \cup \{1, 2\} = \{1, 2, 3, 5\} \qquad n(E \cup F) = 4$$

(c) We use formula (4). Then

$$P(E) = \frac{n(E)}{n(S)} = \frac{3}{6} = \frac{1}{2} \qquad P(F) = \frac{n(F)}{n(S)} = \frac{2}{6} = \frac{1}{3}$$

(d) $P(E \cap F) = \dfrac{n(E \cap F)}{n(S)} = \dfrac{1}{6}$

(e) $P(E \cup F) = \dfrac{n(E \cup F)}{n(S)} = \dfrac{4}{6} = \dfrac{2}{3}$

3 Find Probabilities of the Union of Two Events

The next formula can be used to find the probability of the union of two events.

THEOREM

For any two events E and F,

$$P(E \cup F) = P(E) + P(F) - P(E \cap F) \qquad (5)$$

This result is a consequence of the Counting Formula discussed earlier in Section 10.1.

For example, we can use formula (5) to find $P(E \cup F)$ in Example 6(e). Then

$$P(E \cup F) = P(E) + P(F) - P(E \cap F) = \frac{1}{2} + \frac{1}{3} - \frac{1}{6} = \frac{3}{6} + \frac{2}{6} - \frac{1}{6} = \frac{4}{6} = \frac{2}{3}$$

as before.

EXAMPLE 7

Computing Probabilities of the Union of Two Events

If $P(E) = 0.2$, $P(F) = 0.3$, and $P(E \cap F) = 0.1$, find the probability of E or F. That is, find $P(E \cup F)$.

Solution

Use formula (5).

$$\text{Probability of } E \text{ or } F = P(E \cup F) = P(E) + P(F) - P(E \cap F)$$
$$= 0.2 + 0.3 - 0.1 = 0.4$$

A Venn diagram can sometimes be used to obtain probabilities. To construct a Venn diagram representing the information in Example 7, we draw two sets E and F. We begin with the fact that $P(E \cap F) = 0.1$. See Figure 7(a). Then, since $P(E) = 0.2$ and $P(F) = 0.3$, we fill in E with $0.2 - 0.1 = 0.1$ and F with $0.3 - 0.1 = 0.2$. See Figure 7(b). Since $P(S) = 1$, we complete the diagram by inserting $1 - (0.1 + 0.1 + 0.2) = 0.6$ outside the circles. See Figure 7(c). Now it is easy to see, for example, that the probability of F, but not E, is 0.2. Also, the probability of neither E nor F is 0.6.

Figure 7

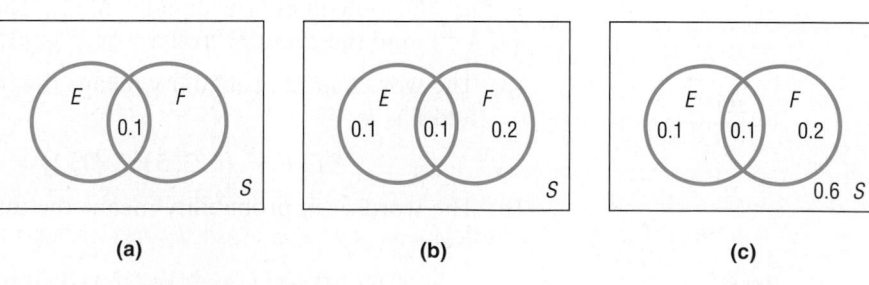

 (a) (b) (c)

Now Work PROBLEM 45

If events E and F are disjoint so that $E \cap F = \varnothing$, we say they are **mutually exclusive.** In this case, $P(E \cap F) = 0$, and formula (5) takes the following form:

THEOREM

Mutually Exclusive Events

If E and F are mutually exclusive events,

$$P(E \cup F) = P(E) + P(F) \qquad (6)$$

EXAMPLE 8 Computing Probabilities of the Union of Two Mutually Exclusive Events

If $P(E) = 0.4$ and $P(F) = 0.25$, and E and F are mutually exclusive, find $P(E \cup F)$.

Solution Since E and F are mutually exclusive, use formula (6).

$$P(E \cup F) = P(E) + P(F) = 0.4 + 0.25 = 0.65$$

━━━━━━Now Work PROBLEM 47

4 Use the Complement Rule to Find Probabilities

Recall, if A is a set, the complement of A, denoted \overline{A}, is the set of all elements in the universal set U not in A. We similarly define the complement of an event.

DEFINITION **Complement of an Event**

Let S denote the sample space of an experiment, and let E denote an event. The **complement of E,** denoted \overline{E}, is the set of all outcomes in the sample space S that are not outcomes in the event E.

The complement of an event E, that is, \overline{E}, in a sample space S has the following two properties:

$$E \cap \overline{E} = \emptyset \qquad E \cup \overline{E} = S$$

Since E and \overline{E} are mutually exclusive, it follows from (6) that

$$P(E \cup \overline{E}) = P(S) = 1 \qquad P(E) + P(\overline{E}) = 1 \qquad P(\overline{E}) = 1 - P(E)$$

We have the following result:

THEOREM **Computing Probabilities of Complementary Events**

If E represents any event and \overline{E} represents the complement of E, then

$$P(\overline{E}) = 1 - P(E) \qquad\qquad (7)$$

EXAMPLE 9 Computing Probabilities Using Complements

On the local news the weather reporter stated that the probability of rain tomorrow is 40%. What is the probability that it will not rain?

Solution The complement of the event "rain" is "no rain."

$$P(\text{no rain}) = 1 - P(\text{rain}) = 1 - 0.4 = 0.6$$

There is a 60% chance of no rain tomorrow.

━━━━━━Now Work PROBLEM 51

EXAMPLE 10 Birthday Problem

What is the probability that in a group of 10 people at least 2 people have the same birthday? Assume that there are 365 days in a year.

Solution We assume that a person is as likely to be born on one day as another, so we have equally likely outcomes.

We first determine the number of outcomes in the sample space S. There are 365 possibilities for each person's birthday. Since there are 10 people in the group, there are 365^{10} possibilities for the birthdays. [For one person in the group, there are 365 days on which his or her birthday can fall; for two people, there are $(365)(365) = 365^2$ pairs of days; and, in general, using the Multiplication Principle, for n people there are 365^n possibilities.] So

$$n(S) = 365^{10}$$

We wish to find the probability of the event E: "at least two people have the same birthday." It is difficult to count the elements in this set; it is much easier to count the elements of the complementary event \overline{E}: "no two people have the same birthday."

We find $n(\overline{E})$ as follows: Choose one person at random. There are 365 possibilities for his or her birthday. Choose a second person. There are 364 possibilities for this birthday, if no two people are to have the same birthday. Choose a third person. There are 363 possibilities left for this birthday. Finally, we arrive at the tenth person. There are 356 possibilities left for this birthday. By the Multiplication Principle, the total number of possibilities is

$$n(\overline{E}) = 365 \cdot 364 \cdot 363 \cdot \ \cdots \ \cdot 356$$

The probability of the event \overline{E} is

$$P(\overline{E}) = \frac{n(\overline{E})}{n(S)} = \frac{365 \cdot 364 \cdot 363 \cdot \ \cdots \ \cdot 356}{365^{10}} \approx 0.883$$

The probability of two or more people in a group of 10 people having the same birthday is then

$$P(E) = 1 - P(\overline{E}) \approx 1 - 0.883 = 0.117$$

The birthday problem can be solved for any group size. The following table gives the probabilities for two or more people having the same birthday for various group sizes. Notice that the probability is greater than $\frac{1}{2}$ for any group of 23 or more people.

	Number of People																
	5	**10**	**15**	**20**	**21**	**22**	**23**	**24**	**25**	**30**	**40**	**50**	**60**	**70**	**80**	**90**	
Probability That Two or More Have the Same Birthday	0.027	0.117	0.253	0.411	0.444	0.476	0.507	0.538	0.569	0.706	0.891	0.970	0.994	0.99916	0.99991	0.99999	

Now Work PROBLEM **71**

Historical Feature

Blaise Pascal (1623–1662)

Set theory, counting, and probability first took form as a systematic theory in an exchange of letters (1654) between Pierre de Fermat (1601–1665) and Blaise Pascal (1623–1662). They discussed the problem of how to divide the stakes in a game that is interrupted before completion, knowing how many points each player needs to win. Fermat solved the problem by listing all possibilities and counting the favorable ones, whereas Pascal made use of the triangle that now bears his name. As mentioned in the text, the entries in

Pascal's triangle are equivalent to $C(n, r)$. This recognition of the role of $C(n, r)$ in counting is the foundation of all further developments.

The first book on probability, the work of Christiaan Huygens (1629–1695), appeared in 1657. In it, the notion of mathematical expectation is explored. This allows the calculation of the profit or loss that a gambler might expect, knowing the probabilities involved in the game (see the Historical Problem that follows).

Although Girolamo Cardano (1501–1576) wrote a treatise on probability, it was not published until 1663 in Cardano's collected works, and this was too late to have any effect on the early development of the theory.

In 1713, the posthumously published *Ars Conjectandi* of Jakob Bernoulli (1654–1705) gave the theory the form it would have until 1900. Recently, both combinatorics (counting) and probability have undergone rapid development due to the use of computers.

A final comment about notation. The notations $C(n, r)$ and $P(n, r)$ are variants of a form of notation developed in England after 1830. The notation $\binom{n}{r}$ for $C(n, r)$ goes back to Leonhard Euler (1707–1783), but is now losing ground because it has no clearly related symbolism of the same type for permutations. The set symbols \cup and \cap were introduced by Giuseppe Peano (1858-1932) in 1888 in a slightly different context. The inclusion symbol \subset was introduced by E. Schroeder (1841–1902) about 1890. The treatment of set theory in the text is due to George Boole (1815-1864), who wrote $A + B$ for $A \cup B$ and AB for $A \cap B$ (statisticians still use AB for $A \cap B$).

Historical Problem

1. *The Problem Discussed by Fermat and Pascal* A game between two equally skilled players, *A* and *B*, is interrupted when *A* needs 2 points to win and *B* needs 3 points. In what proportion would the stakes be divided?

 (a) *Fermat's solution* List all possible outcomes that can occur as a result of four more plays. The probabilities for *A* to win and *B* to win then determine how the stakes should be divided.

 (b) *Pascal's solution* Use combinations to determine the number of ways that the 2 points needed for *A* to win could occur in four plays. Then use combinations to determine the number of ways that the 3 points needed for *B* to win could occur. This is trickier than it looks, since *A* can win with 2 points in either two plays, three plays, or four plays. Compute the probabilities and compare with the results in part (a).

10.3 Assess Your Understanding

Concepts and Vocabulary

1. When the same probability is assigned to each outcome of a sample space, the experiment is said to have _____ _____ outcomes.

2. The _____ of an event *E* is the set of all outcomes in the sample space *S* that are not outcomes in the event *E*.

3. *True or False* The probability of an event can never equal 0.

4. *True or False* In a probability model, the sum of all probabilities is 1.

Skill Building

5. In a probability model, which of the following numbers could be the probability of an outcome?

$$0 \quad 0.01 \quad 0.35 \quad -0.4 \quad 1 \quad 1.4$$

6. In a probability model, which of the following numbers could be the probability of an outcome?

$$1.5 \quad \frac{1}{2} \quad \frac{3}{4} \quad \frac{2}{3} \quad 0 \quad -\frac{1}{4}$$

7. Determine whether the following is a probability model.

Outcome	Probability
1	0.2
2	0.3
3	0.1
4	0.4

8. Determine whether the following is a probability model.

Outcome	Probability
Steve	0.4
Bob	0.3
Faye	0.1
Patricia	0.2

9. Determine whether the following is a probability model.

Outcome	Probability
Linda	0.3
Jean	0.2
Grant	0.1
Jim	0.3

10. Determine whether the following is a probability model.

Outcome	Probability
Erica	0.3
Joanne	0.2
Laura	0.1
Donna	0.5
Angela	−0.1

In Problems 11–16, construct a probability model for each experiment.

11. Tossing a fair coin twice

12. Tossing two fair coins once

13. Tossing two fair coins, then a fair die

14. Tossing a fair coin, a fair die, and then a fair coin

15. Tossing three fair coins once

16. Tossing one fair coin three times

In Problems 17–22, use the following spinners to construct a probability model for each experiment.

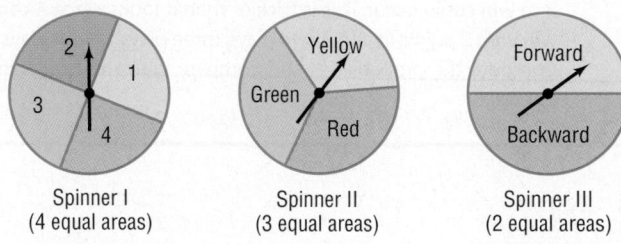

Spinner I
(4 equal areas)

Spinner II
(3 equal areas)

Spinner III
(2 equal areas)

17. Spin spinner I, then spinner II. What is the probability of getting a 2 or a 4, followed by Red?

18. Spin spinner III, then spinner II. What is the probability of getting Forward, followed by Yellow or Green?

19. Spin spinner I, then II, then III. What is the probability of getting a 1, followed by Red or Green, followed by Backward?

20. Spin spinner II, then I, then III. What is the probability of getting Yellow, followed by a 2 or a 4, followed by Forward?

21. Spin spinner I twice, then spinner II. What is the probability of getting a 2, followed by a 2 or a 4, followed by Red or Green?

22. Spin spinner III, then spinner I twice. What is the probability of getting Forward, followed by a 1 or a 3, followed by a 2 or a 4?

In Problems 23–26, consider the experiment of tossing a coin twice. The table lists six possible assignments of probabilities for this experiment. Using this table, answer the following questions.

| Assignments | Sample Space | | | |
	HH	HT	TH	TT
A	$\frac{1}{4}$	$\frac{1}{4}$	$\frac{1}{4}$	$\frac{1}{4}$
B	0	0	0	1
C	$\frac{3}{16}$	$\frac{5}{16}$	$\frac{5}{16}$	$\frac{3}{16}$
D	$\frac{1}{2}$	$\frac{1}{2}$	$-\frac{1}{2}$	$\frac{1}{2}$
E	$\frac{1}{4}$	$\frac{1}{4}$	$\frac{1}{4}$	$\frac{1}{8}$
F	$\frac{1}{9}$	$\frac{2}{9}$	$\frac{2}{9}$	$\frac{4}{9}$

23. Which of the assignments of probabilities is(are) consistent with the definition of a probability model?

24. Which of the assignments of probabilities should be used if the coin is known to be fair?

25. Which of the assignments of probabilities should be used if the coin is known to always come up tails?

26. Which of the assignments of probabilities should be used if tails is twice as likely as heads to occur?

27. **Assigning Probabilities** A coin is weighted so that heads is four times as likely as tails to occur. What probability should we assign to heads? to tails?

28. **Assigning Probabilities** A coin is weighted so that tails is twice as likely as heads to occur. What probability should we assign to heads? to tails?

29. **Assigning Probabilities** A die is weighted so that an odd-numbered face is twice as likely to occur as an even-numbered face. What probability should we assign to each face?

30. **Assigning Probabilities** A die is weighted so that a six cannot appear. The other faces occur with the same probability. What probability should we assign to each face?

For Problems 31–34, the sample space is $S = \{1, 2, 3, 4, 5, 6, 7, 8, 9, 10\}$. Suppose that the outcomes are equally likely.

31. Compute the probability of the event $E = \{1, 2, 3\}$.

32. Compute the probability of the event $F = \{3, 5, 9, 10\}$.

33. Compute the probability of the event E: "an even number."

34. Compute the probability of the event F: "an odd number."

For Problems 35 and 36, an urn contains 5 white marbles, 10 green marbles, 8 yellow marbles, and 7 black marbles.

35. If one marble is selected, determine the probability that it is white.

36. If one marble is selected, determine the probability that it is black.

In Problems 37–40, assume equally likely outcomes.

37. Determine the probability of having 3 boys in a 3-child family.

38. Determine the probability of having 3 girls in a 3-child family.

39. Determine the probability of having 1 girl and 3 boys in a 4-child family.

40. Determine the probability of having 2 girls and 2 boys in a 4-child family.

For Problems 41–44, two fair dice are rolled.

41. Determine the probability that the sum of the two dice is 7.

42. Determine the probability that the sum of the two dice is 11.

43. Determine the probability that the sum of the two dice is 3.

44. Determine the probability that the sum of the two dice is 12.

In Problems 45–48, find the probability of the indicated event if P(A) = 0.25 and P(B) = 0.45.

45. $P(A \cup B)$ if $P(A \cap B) = 0.15$

46. $P(A \cap B)$ if $P(A \cup B) = 0.6$

47. $P(A \cup B)$ if A, B are mutually exclusive

48. $P(A \cap B)$ if A, B are mutually exclusive

49. If $P(A) = 0.60$, $P(A \cup B) = 0.85$, and $P(A \cap B) = 0.05$, find $P(B)$.

50. If $P(B) = 0.30$, $P(A \cup B) = 0.65$, and $P(A \cap B) = 0.15$, find $P(A)$.

51. Automobile Theft According to the Insurance Information Institute, in 2004 there was a 13% probability that an automobile theft in the United States would be cleared by arrests. If an automobile theft case from 2004 is randomly selected, what is the probability that it was not cleared by an arrest?

52. Pet Ownership According to the American Pet Products Manufacturers Association's *2005-2006 National Pet Owners Survey,* there is a 63% probability that a U.S. household owns a pet. If a U.S. household is randomly selected, what is the probability that it does not own a pet?

53. Cat Ownership According to the American Pet Products Manufacturers Association's *2005–2006 National Pet Owners Survey,* in 2004 there was a 34% probability that a U.S. pet owner owned a cat. If a U.S. pet owner is randomly selected, what is the probability that he or she does not own a cat?

54. Doctorate Degrees According to the National Science Foundation, in 2004 there was a 13.7% probability that a doctoral degree awarded at a U.S. university was awarded in engineering. If a 2004 U.S. doctoral recipient is randomly selected, what is the probability that his or her degree was not in engineering?

55. Online Gambling According to a Harris poll (January 12–17, 2006), 5% of U.S. adults admitted to having spent money gambling online. If a U.S. adult is selected at random, what is the probability that he or she has never spent any money gambling online?

56. Girl Scout Cookies According to the Girl Scouts of America, in March 2006, 9% of all Girl Scout cookies sold are shortbread/trefoils. If a box of Girl Scout cookies is selected at random, what is the probability that it is not shortbread/trefoils?

For Problems 57–60, a golf ball is selected at random from a container. If the container has 9 white balls, 8 green balls, and 3 orange balls, find the probability of each event.

57. The golf ball is white or green.

58. The golf ball is white or orange.

59. The golf ball is not white.

60. The golf ball is not green.

61. On the "Price is Right" there is a game in which a bag is filled with 3 strike chips and 5 numbers. Let's say that the numbers in the bag are 0, 1, 3, 6, and 9. What is the probability of selecting a strike chip or the number 1?

62. Another game on the "Price is Right" requires the contestant to spin a wheel with numbers 5, 10, 15, 20, . . . , 100. What is the probability that the contestant spins 100 or 30?

Problems 63–66 are based on a consumer survey of annual incomes in 100 households. The following table gives the data.

Income	$0–9999	$10,000–19,999	$20,000–29,999	$30,000–39,999	$40,000 or more
Number of households	5	35	30	20	10

63. What is the probability that a household has an annual income of $30,000 or more?

64. What is the probability that a household has an annual income between $10,000 and $29,999, inclusive?

65. What is the probability that a household has an annual income of less than $20,000?

66. What is the probability that a household has an annual income of $20,000 or more?

67. Surveys In a survey about the number of TV sets in a house, the following probability table was constructed:

Number of TV sets	0	1	2	3	4 or more
Probability	0.05	0.24	0.33	0.21	0.17

Find the probability of a house having:
(a) 1 or 2 TV sets
(b) 1 or more TV sets
(c) 3 or fewer TV sets
(d) 3 or more TV sets

(e) Fewer than 2 TV sets
(f) Fewer than 1 TV set
(g) 1, 2, or 3 TV sets
(h) 2 or more TV sets

68. Checkout Lines Through observation, it has been determined that the probability for a given number of people waiting in line at the "5 items or less" checkout register of a supermarket is as follows:

Number waiting in line	0	1	2	3	4 or more
Probability	0.10	0.15	0.20	0.24	0.31

Find the probability of:
(a) At most 2 people in line
(b) At least 2 people in line
(c) At least 1 person in line

69. In a certain College Algebra class, there are 18 freshmen and 15 sophomores. Of the 18 freshmen, 10 are male, and of the 15 sophomores, 8 are male. Find the probability that a randomly selected student is:

(a) A freshman or female
(b) A sophomore or male

70. The faculty of the mathematics department at Joliet Junior College is composed of 4 females and 9 males. Of the 4 females, 2 are under age 40, and of the males 3 are under age 40. Find the probability that a randomly selected faculty member is:

(a) Female or under age 40

(b) Male or over age 40

71. Birthday Problem What is the probability that at least 2 people have the same birthday in a group of 12 people? Assume that there are 365 days in a year.

72. Birthday Problem What is the probability that at least 2 people have the same birthday in a group of 35 people? Assume that there are 365 days in a year.

73. Winning a Lottery In a certain lottery, there are ten balls, numbered 1, 2, 3, 4, 5, 6, 7, 8, 9, 10. Of these, five are drawn in order. If you pick five numbers that match those drawn in the correct order, you win $1,000,000. What is the probability of winning such a lottery?

CHAPTER REVIEW

Things to Know

Counting formula (p. 681)	$n(A \cup B) = n(A) + n(B) - n(A \cap B)$
Addition Principle of Counting (p. 681)	If $A \cap B = \varnothing$, then $n(A \cup B) = n(A) + n(B)$.
Multiplication Principle of Counting (p. 683)	If a task consists of a sequence of choices in which there are p selections for the first choice, q selections for the second choice, and so on, the task of making these selections can be done in $p \cdot q \cdot \cdots$ different ways.
Permutation (p. 685)	An ordered arrangement of r objects chosen from n objects
Number of permutations: Distinct, with repetition (p. 686)	n^r The n objects are distinct (different), and repetition is allowed in the selection of r of them.
Number of permutations: Distinct, without repetition (p. 687)	$P(n, r) = n(n - 1) \cdot \cdots \cdot [n - (r - 1)] = \dfrac{n!}{(n - r)!}$ The n objects are distinct (different), and repetition is not allowed in the selection of r of them, where $r \leq n$.
Combinations (p. 688)	An arrangement, without regard to order, of r objects selected from n distinct objects, where $r \leq n$
Number of combinations (p. 689)	$C(n, r) = \dfrac{P(n, r)}{r!} = \dfrac{n!}{(n - r)!r!}$
Number of permutations: Not distinct (p. 691)	$\dfrac{n!}{n_1! n_2! \cdots n_k!}$ The number of permutations of n objects of which n_1 are of one kind, n_2 are of a second kind, ..., and n_k are of a kth kind, where $n = n_1 + n_2 + \cdots + n_k$
Sample space (p. 694)	Set whose elements represent the possible outcomes that can occur as a result of an experiment
Probability (p. 694)	A nonnegative number assigned to each outcome of a sample space; the sum of all the probabilities of the outcomes equals 1.
Probability for equally likely outcomes (p. 696)	$P(E) = \dfrac{n(E)}{n(S)}$ The same probability is assigned to each outcome.
Probability of the union of two events (p. 698)	$P(E \cup F) = P(E) + P(F) - P(E \cap F)$
Probability of the complement of an event (p. 699)	$P(\overline{E}) = 1 - P(E)$

Objectives

Section	You should be able to . . .	Example(s)	Review Exercises
10.1	**1** Find all the subsets of a set (p. 680)	1	1, 2
	2 Count the number of elements in a set (p. 680)	2, 3	3–10
	3 Solve counting problems using the Multiplication Principle (p. 682)	4, 5	15–18
10.2	**1** Solve counting problems using permutations involving n distinct objects (p. 685)	1–5	11, 12, 19, 20, 24–28, 33(a)
	2 Solve counting problems using combinations (p. 688)	6–9	13, 14, 21–23, 31, 32
	3 Solve counting problems using permutations involving n nondistinct objects (p. 690)	10, 11	29, 30
10.3	**1** Construct probability models (p. 694)	2–4	33(b)
	2 Compute probabilities of equally likely outcomes (p. 696)	5, 6	33(b), 34(a), 35(a), 36–39
	3 Find probabilities of the union of two events (p. 698)	7, 8	40
	4 Use the Complement Rule to find probabilities (p. 699)	9, 10	33(c), 34(b), 35(b), 36

Review Exercises

1. Write down all the subsets of the set {Dave, Joanne, Erica}.

2. Write down all the subsets of the set {Green, Blue, Red}.

3. If $n(A) = 8$, $n(B) = 12$, and $n(A \cap B) = 3$, find $n(A \cup B)$.

4. If $n(A) = 12$, $n(A \cup B) = 30$, and $n(A \cap B) = 6$, find $n(B)$.

In Problems 5–10, use the information supplied in the figure.

5. How many are in A?

6. How many are in A or B?

7. How many are in A and C?

8. How many are not in B?

9. How many are in neither A nor C?

10. How many are in B but not in C?

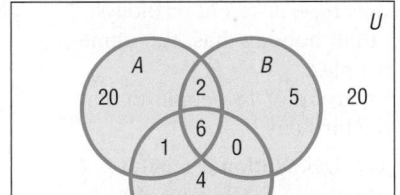

In Problems 11–14, compute the given expression.

11. $P(8, 3)$ **12.** $P(7, 3)$ **13.** $C(8, 3)$ **14.** $C(7, 3)$

15. Stocking a Store A clothing store sells pure wool and polyester-wool suits. Each suit comes in 3 colors and 10 sizes. How many suits are required for a complete assortment?

16. Wiring In connecting a certain electrical device, 5 wires are to be connected to 5 different terminals. How many different wirings are possible if 1 wire is connected to each terminal?

17. Baseball On a given day, the American Baseball League schedules 7 games. How many different outcomes are possible, assuming that each game is played to completion?

18. Baseball On a given day, the National Baseball League schedules 6 games. How many different outcomes are possible, assuming that each game is played to completion?

19. Choosing Seats If 4 people enter a bus having 9 vacant seats, in how many ways can they be seated?

20. Arranging Letters How many different arrangements are there of the letters in the word ROSE?

21. Choosing a Team In how many ways can a squad of 4 relay runners be chosen from a track team of 8 runners?

22. Writing a Test A professor has 10 similar problems to put on a test that has 3 problems. How many different tests can she design?

23. Baseball In how many ways can 2 teams from 14 teams in the American League be chosen without regard to which team is at home?

24. Arranging Books on a Shelf There are 5 different French books and 5 different Spanish books. How many ways are there to arrange them on a shelf if:
(a) Books of the same language must be grouped together, French on the left, Spanish on the right?
(b) French and Spanish books must alternate in the grouping, beginning with a French book?

25. Telephone Numbers Using the digits 0, 1, 2, . . . , 9, how many 7-digit numbers can be formed if the first digit cannot be 0 or 9 and if the last digit is greater than or equal to 2 and less than or equal to 3? Repeated digits are allowed.

26. Home Choices A contractor constructs homes with 5 different choices of exterior finish, 3 different roof arrangements, and 4 different window designs. How many different types of homes can be built?

27. License Plate Possibilities A license plate consists of 1 letter, excluding O and I, followed by a 4-digit number that cannot have a 0 in the lead position. How many different plates are possible?

28. Binary Codes Using the digits 0 and 1, how many different numbers consisting of 8 digits can be formed?

29. Forming Different Words How many different words, real or imaginary, can be formed using all the letters in the word MISSING?

30. Arranging Flags How many different vertical arrangements are there of 10 flags if 4 are white, 3 are blue, 2 are green, and 1 is red?

31. Forming Committees A group of 9 people is going to be formed into committees of 4, 3, and 2 people. How many committees can be formed if:
(a) A person can serve on any number of committees?
(b) No person can serve on more than one committee?

32. Forming Committees A group consists of 5 men and 8 women. A committee of 4 is to be formed from this group, and policy dictates that at least 1 woman be on this committee.
(a) How many committees can be formed that contain exactly 1 man?
(b) How many committees can be formed that contain exactly 2 women?
(c) How many committees can be formed that contain at least 1 man?

33. Birthday Problem For this problem, assume that a year has 365 days.
(a) How many ways can 18 people have different birthdays?
(b) What is the probability that nobody has the same birthday in a group of 18 people?
(c) What is the probability in a group of 18 people that at least 2 people have the same birthday?

34. Death Rates According to the U.S. National Center for Health Statistics, 29% of all deaths in 2001 were due to heart disease.
(a) What is the probability that a randomly selected death in 2001 was due to heart disease?
(b) What is the probability that a randomly selected death in 2001 was not due to heart disease?

35. Unemployment According to the U.S. Bureau of Labor Statistics, 5.8% of the U.S. labor force was unemployed in 2002.
(a) What is the probability that a randomly selected member of the U.S. labor force was unemployed in 2002?

(b) What is the probability that a randomly selected member of the U.S. labor force was not unemployed in 2002?

36. From a box containing three 40-watt bulbs, six 60-watt bulbs, and eleven 75-watt bulbs, a bulb is drawn at random. What is the probability that the bulb is 40 watts? What is the probability that it is not a 75-watt bulb?

37. You have four $1 bills, three $5 bills, and two $10 bills in your wallet. If you pick a bill at random, what is the probability that it will be a $1 bill?

38. Each letter in the word ROSE is written on an index card and the cards are shuffled. When the cards are dealt, what is the probability that they spell the word ROSE?

39. Each of the numbers, 1, 2, . . . , 100 is written on an index card and the cards are shuffled. If a card is selected at random, what is the probability that the number on the card is divisible by 5? What is the probability that the card selected is either a 1 or names a prime number?

40. At the Milex tune-up and brake repair shop, the manager has found that a car will require a tune-up with a probability of 0.6, a brake job with a probability of 0.1, and both with a probability of 0.02.
(a) What is the probability that a car requires either a tune-up or a brake job?
(b) What is the probability that a car requires a tune-up but not a brake job?
(c) What is the probability that a car requires neither a tune-up nor a brake job?

CHAPTER TEST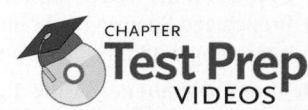

The Chapter Test Prep Videos are step-by-step test solutions available in the Video Resources DVD, in *MyMathLab*, or on this text's You[Tube] Channel. Flip back to the Student Resources page to see the exact web address for this text's YouTube channel.

In Problems 1–4, a survey of 70 college freshmen asked whether students planned to take biology, chemistry, or physics during their first year. Use the diagram to answer each question.

1. How many of the surveyed students plan to take physics during their first year?

2. How many of the surveyed students do not plan to take biology, chemistry, or physics during their first year?

3. How many of the surveyed students plan to take only biology and chemistry during their first year?

4. How many of the surveyed students plan to take physics or chemistry during their first year?

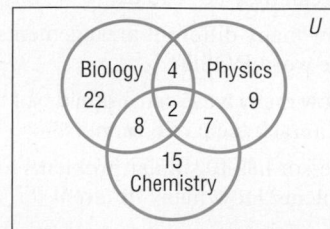

In Problems 5–7, compute the value of the given expression.

5. $7!$ **6.** $P(10, 6)$ **7.** $C(11, 5)$

8. M&M's® offers customers the opportunity to create their own color mix of candy. There are 21 colors to choose from, and customers are allowed to select up to 6 different colors. How many different color mixes are possible, assuming that no color is selected more than once and 6 different colors are chosen?

9. How many distinct 8-letter words (real or imaginary) can be formed from the letters in the word REDEEMED?

10. In horse racing, an exacta bet requires the bettor to pick the first two horses in the exact order. If there are 8 horses in a race, in how many ways could you make an exacta bet?

11. On February 20, 2004, the Ohio Bureau of Motor Vehicles unveiled the state's new license plate format. The plate consists of three letters (A–Z) followed by 4 digits (0–9). Assume that all letters and digits may be used except that the third letter cannot be O, I, or Z. If repetitions are allowed, how many different plates are possible?

12. Kiersten applies for admission to the University of Southern California (USC) and Florida State University (FSU). She estimates that she has a 60% chance of being admitted to USC, a 70% chance of being admitted to FSU, and a 35% chance of being admitted to both universities.
(a) What is the probability that she will be admitted to either USC or FSU?
(b) What is the probability that she will not be admitted to FSU?

13. A cooler contains 8 bottles of Pepsi, 5 bottles of Coke, 4 bottles of Mountain Dew, and 3 bottles of IBC.
(a) What is the probability that a bottle chosen at random is Coke?
(b) What is the probability that a bottle chosen at random is either Pepsi or IBC?

14. A study on the age distribution of students at a community college gave the following table:

Age	17 and under	18–20	21–24	25–34	35–64	65 and over
Probability	0.03	???	0.23	0.29	0.25	0.01

What must be the probability that a randomly selected student at the college is between 18 and 20 years old?

15. Powerball is a multistate lottery in which 5 white balls from a drum with 53 balls and 1 red ball from a drum with 42 red balls are selected. For a $1 ticket, players get one chance at winning the jackpot by matching all 6 numbers. What is the probability of selecting the winning numbers on a $1 play?

16. If you roll a die five times, what is the probability that you obtain exactly 2 fours?

CUMULATIVE REVIEW

1. Solve: $3x^2 - 2x = -1$

2. Graph $f(x) = x^2 + 4x - 5$ by determining whether the graph opens up or down and by finding the vertex, axis of symmetry, and intercepts.

3. Graph $f(x) = 2(x + 1)^2 - 4$ using transformations.

4. Solve: $|x - 4| \leq 0.01$

5. Find the complex zeros of
$$f(x) = 5x^4 - 9x^3 - 7x^2 - 31x - 6$$

6. Graph $g(x) = 3^{x-1} + 5$ using transformations. Determine the domain, the range, and horizontal asymptote of g.

7. What is the exact value of $\log_3 9$?

8. Solve: $\log_2(3x - 2) + \log_2 x = 4$

9. Solve the system: $\begin{cases} x - 2y + z = 15 \\ 3x + y - 3z = -8 \\ -2x + 4y - z = -27 \end{cases}$

10. What is the 33rd term in the sequence $-3, 1, 5, 9, \ldots$? What is the sum of the first 20 terms?

CHAPTER PROJECTS

I. The Monty Hall Game The *Monty Hall Game,* based on a segment from the game show *Let's Make a Deal,* is a classic probability problem that continues to stir debate. A more recent game show, *Deal or No Deal* (see the chapter opening vignette) has often been compared to the classic *Monty Hall Game.*

1. Research the *Monty Hall Game* and *Deal or No Deal.*

2. In the *Monty Hall Game,* what is the probability that a contestant wins if she does not switch? What is the probability of winning if she does switch? Perform a simulation to estimate the probabilities. Do the values agree with your research?

3. Suppose the game *Deal or No Deal* is played with only three suitcases. Explain why this game is not the same as the *Monty Hall Game*.

4. Suppose the game *Deal or No Deal* is played with 26 suitcases and the contestant is not allowed to switch at the end. What is the probability that the contestant will win the grand prize?

5. Suppose the game *Deal or No Deal* is played with 26 suitcases and the contestant is allowed to switch at the end. Perform a simulation to estimate the probability that the contestant will win the grand prize if he does not switch at the end.

6. Repeat Problem 5, but assume that the contestant will always switch at the end.

The following projects are available at the Instructor's Resource Center (IRC):

II. Project at Motorola *Probability of Error in Digital Wireless Communications* Transmission errors in digital communications can often be detected by adding an extra digit of code to each transmitted signal. Investigate the probability of identifying an erroneous code using this simple coding method.

III. Surveys Polling (or taking a survey) is big business in the United States. Take and analyze a survey; then consider why different pollsters might get different results.

IV. Law of Large Numbers The probability that an event occurs, such as a head in a coin toss, is the proportion of heads you expect in the long run. A simulation is used to show that as a coin is flipped more and more times, the proportion of heads gets close to 0.5.

V. Simulation Electronic simulation of an experiment is often an economical way to investigate a theoretical probability. Develop a theory without leaving your desk.

Appendix
Graphing Utilities

Outline

1 The Viewing Rectangle

Figure 1
$y = 2x$

Figure 2

All graphing utilities, that is, all graphing calculators and all computer software graphing packages, graph equations by plotting points on a screen. The screen itself actually consists of small rectangles, called **pixels.** The more pixels the screen has, the better the resolution. Most graphing calculators have 2048 pixels per square inch; most computer screens have 4096 to 8192 pixels per square inch. When a point to be plotted lies inside a pixel, the pixel is turned on (lights up). The graph of an equation is a collection of pixels. Figure 1 shows how the graph of $y = 2x$ looks on a TI-84 Plus graphing calculator.

The screen of a graphing utility will display the coordinate axes of a rectangular coordinate system. However, you must set the scale on each axis. You must also include the smallest and largest values of x and y that you want included in the graph. This is called **setting the viewing rectangle** or **viewing window.** Figure 2 illustrates a typical viewing window.

To select the viewing window, we must give values to the following expressions:

Xmin: the smallest value of x
Xmax: the largest value of x
Xscl: the number of units per tick mark on the x-axis
Ymin: the smallest value of y
Ymax: the largest value of y
Yscl: the number of units per tick mark on the y-axis

Figure 3 illustrates these settings and their relation to the Cartesian coordinate system.

Figure 3

A1

If the scale used on each axis is known, we can determine the minimum and maximum values of x and y shown on the screen by counting the tick marks. Look again at Figure 2. For a scale of 1 on each axis, the minimum and maximum values of x are -10 and 10, respectively; the minimum and maximum values of y are also -10 and 10. If the scale is 2 on each axis, then the minimum and maximum values of x are -20 and 20, respectively; and the minimum and maximum values of y are -20 and 20, respectively.

Conversely, if we know the minimum and maximum values of x and y, we can determine the scales being used by counting the tick marks displayed. We shall follow the practice of showing the minimum and maximum values of x and y in our illustrations so that you will know how the viewing window was set. See Figure 4.

Figure 4

means

$$X\text{min} = -3 \qquad Y\text{min} = -4$$
$$X\text{max} = 3 \qquad Y\text{max} = 4$$
$$X\text{scl} = 1 \qquad Y\text{scl} = 2$$

EXAMPLE 1 **Finding the Coordinates of a Point Shown on a Graphing Utility Screen**

Find the coordinates of the point shown in Figure 5. Assume that the coordinates are integers.

Figure 5

Solution First we note that the viewing window used in Figure 5 is

$$X\text{min} = -3 \qquad Y\text{min} = -4$$
$$X\text{max} = 3 \qquad Y\text{max} = 4$$
$$X\text{scl} = 1 \qquad Y\text{scl} = 2$$

The point shown is 2 tick units to the left on the horizontal axis (scale = 1) and 1 tick up on the vertical axis (scale = 2). The coordinates of the point shown are $(-2, 2)$.

1 Exercises

In Problems 1–4, determine the coordinates of the points shown. Tell in which quadrant each point lies. Assume that the coordinates are integers.

1.

2.

3.

4.

In Problems 5–10, determine the viewing window used.

5.

6.

7.

8.

9.

10.

In Problems 11–16, select a setting so that each of the given points will lie within the viewing rectangle.

11. $(-10, 5)$, $(3, -2)$, $(4, -1)$ **12.** $(5, 0)$, $(6, 8)$, $(-2, -3)$ **13.** $(40, 20)$, $(-20, -80)$, $(10, 40)$

14. $(-80, 60)$, $(20, -30)$, $(-20, -40)$ **15.** $(0, 0)$, $(100, 5)$, $(5, 150)$ **16.** $(0, -1)$, $(100, 50)$, $(-10, 30)$

2 Using a Graphing Utility to Graph Equations

From Examples 2 and 3 in Chapter 2, Section 2.2, we see that a graph can be obtained by plotting points in a rectangular coordinate system and connecting them. Graphing utilities perform these same steps when graphing an equation. For example, the TI-84 Plus determines 95 evenly spaced input values,* starting at Xmin and ending at Xmax, uses the equation to determine the output values, plots these points on the screen, and finally (if in the connected mode) draws a line between consecutive points.

To graph an equation in two variables x and y using a graphing utility requires that the equation be written in the form $y = \{expression\ in\ x\}$. If the original equation is not in this form, replace it by equivalent equations until the form $y = \{expression\ in\ x\}$ is obtained.

Steps for Graphing an Equation Using a Graphing Utility

STEP 1: Solve the equation for y in terms of x.

STEP 2: Get into the graphing mode of your graphing utility. The screen will usually display $Y_1 = $, prompting you to enter the expression involving x that you found in Step 1. (Consult your manual for the correct way to enter the expression; for example, $y = x^2$ might be entered as $x^{\wedge}2$ or as $x*x$ or as $x\ x^Y\ 2$).

STEP 3: Select the viewing window. Without prior knowledge about the behavior of the graph of the equation, it is common to select the **standard viewing window**** initially. The viewing window is then adjusted based on the graph that appears. In this text the standard viewing window is

$$X\text{min} = -10 \qquad Y\text{min} = -10$$
$$X\text{max} = 10 \qquad Y\text{max} = 10$$
$$X\text{scl} = 1 \qquad Y\text{scl} = 1$$

STEP 4: Graph.

STEP 5: Adjust the viewing window until a complete graph is obtained.

EXAMPLE 1 **Graphing an Equation on a Graphing Utility**

Graph the equation: $6x^2 + 3y = 36$

Solution **STEP 1:** Solve for y in terms of x.

$$6x^2 + 3y = 36$$
$$3y = -6x^2 + 36 \quad \text{Subtract } 6x^2 \text{ from both sides of the equation.}$$
$$y = -2x^2 + 12 \quad \text{Divide both sides of the equation by 3 and simplify.}$$

*These input values depend on the values of Xmin and Xmax. For example, if Xmin $= -10$ and Xmax $= 10$, then the first input value will be -10 and the next input value will be $-10 + \dfrac{10 - (-10)}{94} = -9.7872$, and so on.

**Some graphing utilities have a ZOOM-STANDARD feature that automatically sets the viewing window to the standard viewing window and graphs the equation.

Step 2: From the $Y_1 =$ screen, enter the expression $-2x^2 + 12$ after the prompt.

Step 3: Set the viewing window to the standard viewing window.

Step 4: Graph. The screen should look like Figure 6.

Step 5: The graph of $y = -2x^2 + 12$ is not complete. The value of Ymax must be increased so that the top portion of the graph is visible. After increasing the value of Ymax to 12, we obtain the graph in Figure 7. The graph is now complete.

Figure 6

Figure 7

Figure 8

Look again at Figure 7. Although a complete graph is shown, the graph might be improved by adjusting the values of Xmin and Xmax. Figure 8 shows the graph of $y = -2x^2 + 12$ using Xmin $= -4$ and Xmax $= 4$. Do you think this is a better choice for the viewing window?

EXAMPLE 2 **Creating a Table and Graphing an Equation**

Create a table and graph the equation: $y = x^3$

Solution Most graphing utilities have the capability of creating a table of values for an equation. (Check your manual to see if your graphing utility has this capability.) Table 1 illustrates a table of values for $y = x^3$ on a TI-84 Plus. See Figure 9 for the graph.

Table 1

Figure 9

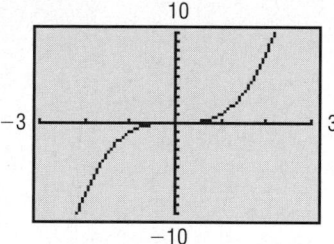

2 Exercises

In Problems 1–16, graph each equation using the following viewing windows:

(a) Xmin $= -5$
 Xmax $= 5$
 Xscl $= 1$
 Ymin $= -4$
 Ymax $= 4$
 Yscl $= 1$

(b) Xmin $= -10$
 Xmax $= 10$
 Xscl $= 2$
 Ymin $= -8$
 Ymax $= 8$
 Yscl $= 2$

1. $y = x + 2$ **2.** $y = x - 2$ **3.** $y = -x + 2$ **4.** $y = -x - 2$

5. $y = 2x + 2$ **6.** $y = 2x - 2$ **7.** $y = -2x + 2$ **8.** $y = -2x - 2$

9. $y = x^2 + 2$ **10.** $y = x^2 - 2$ **11.** $y = -x^2 + 2$ **12.** $y = -x^2 - 2$

13. $3x + 2y = 6$ **14.** $3x - 2y = 6$ **15.** $-3x + 2y = 6$ **16.** $-3x - 2y = 6$

17–32. *For each of the above equations, create a table, $-3 \leq x \leq 3$, and list points on the graph.*

3 Using a Graphing Utility to Locate Intercepts and Check for Symmetry

Value and Zero (or Root)

Most graphing utilities have an eVALUEate feature that, given a value of x, determines the value of y for an equation. We can use this feature to evaluate an equation at $x = 0$ to determine the y-intercept. Most graphing utilities also have a ZERO (or ROOT) feature that can be used to determine the x-intercept(s) of an equation.

EXAMPLE 1 **Finding Intercepts Using a Graphing Utility**

Use a graphing utility to find the intercepts of the equation $y = x^3 - 8$.

Solution Figure 10(a) shows the graph of $y = x^3 - 8$.

Figure 10

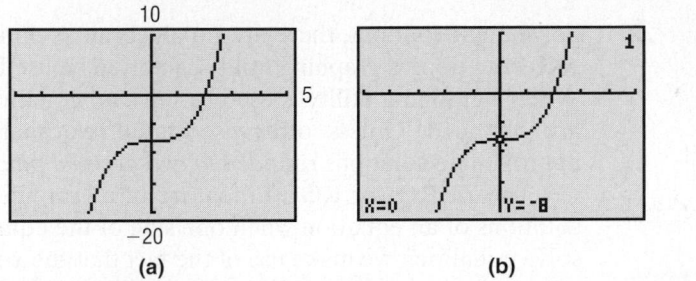

(a) (b) (c)

The eVALUEate feature of a TI-84 Plus graphing calculator accepts as input a value of x and determines the value of y. If we let $x = 0$, we find that the y-intercept is -8. See Figure 10(b).

The ZERO feature of a TI-84 Plus is used to find the x-intercept(s). See Figure 10(c). The x-intercept is 2.

EXAMPLE 2 **Graphing the Equation $y = \dfrac{1}{x}$**

Graph the equation $y = \dfrac{1}{x}$. Based on the graph, infer information about intercepts and symmetry.

Solution Figure 11 illustrates the graph. We infer from the graph that there are no intercepts; we may also infer that symmetry with respect to the origin is a possibility. The TABLE feature on a graphing utility can provide further evidence of symmetry with respect to the origin. Using a TABLE, we observe that for any ordered pair (x, y) the ordered pair $(-x, -y)$ is also a point on the graph.

Figure 11

3 Exercises

In Problems 1–6, use ZERO (or ROOT) to approximate the smaller of the two x-intercepts of each equation. Express the answer rounded to two decimal places.

1. $y = x^2 + 4x + 2$

2. $y = x^2 + 4x - 3$

3. $y = 2x^2 + 4x + 1$

4. $y = 3x^2 + 5x + 1$

5. $y = 2x^2 - 3x - 1$

6. $y = 2x^2 - 4x - 1$

*In Problems 7–12, use ZERO (or ROOT) to approximate the **positive** x-intercepts of each equation. Express each answer rounded to two decimal places.*

7. $y = x^3 + 3.2x^2 - 16.83x - 5.31$

8. $y = x^3 + 3.2x^2 - 7.25x - 6.3$

9. $y = x^4 - 1.4x^3 - 33.71x^2 + 23.94x + 292.41$

10. $y = x^4 + 1.2x^3 - 7.46x^2 - 4.692x + 15.2881$

11. $y = x^3 + 19.5x^2 - 1021x + 1000.5$

12. $y = x^3 + 14.2x^2 - 4.8x - 12.4$

4 Using a Graphing Utility to Solve Equations

For many equations, there are no algebraic techniques that lead to a solution. For such equations, a graphing utility can often be used to investigate possible solutions. When a graphing utility is used to solve an equation, usually *approximate* solutions are obtained. Unless otherwise stated, we shall follow the practice of giving approximate solutions *rounded to two decimal places*.

The ZERO (or ROOT) feature of a graphing utility can be used to find the solutions of an equation when one side of the equation is 0. In using this feature to solve equations, we make use of the fact that the x-intercepts (or zeros) of the graph of an equation are found by letting $y = 0$ and solving the equation for x. Solving an equation for x when one side of the equation is 0 is equivalent to finding where the graph of the corresponding equation crosses or touches the x-axis.

EXAMPLE 1 **Using ZERO (or ROOT) to Approximate Solutions of an Equation**

Find the solution(s) of the equation $x^2 - 6x + 7 = 0$. Round answers to two decimal places.

Solution The solutions of the equation $x^2 - 6x + 7 = 0$ are the same as the x-intercepts of the graph of $Y_1 = x^2 - 6x + 7$. We begin by graphing the equation. See Figure 12(a).

From the graph there appear to be two x-intercepts (solutions to the equation): one between 1 and 2, the other between 4 and 5.

Figure 12

(a) (b) (c)

Using the ZERO (or ROOT) feature of our graphing utility, we determine that the x-intercepts, and so the solutions to the equation, are $x = 1.59$ and $x = 4.41$, rounded to two decimal places. See Figures 12(b) and (c). ↵

A second method for solving equations using a graphing utility involves the INTERSECT feature of the graphing utility. This feature is used most effectively when one side of the equation is not 0.

EXAMPLE 2	**Using INTERSECT to Approximate Solutions of an Equation**

Find the solution(s) to the equation $3(x - 2) = 5(x - 1)$.

Solution Begin by graphing each side of the equation as follows: graph $Y_1 = 3(x - 2)$ and $Y_2 = 5(x - 1)$. See Figure 13(a).

Figure 13

(a) (b)

At the point of intersection of the graphs, the value of the y-coordinate is the same. We conclude that the x-coordinate of the point of intersection represents the solution to the equation. Do you see why? The INTERSECT feature on a graphing utility determines the point of intersection of the graphs. Using this feature, we find that the graphs intersect at $(-0.5, -7.5)$. See Figure 13(b). The solution of the equation is therefore $x = -0.5$.

SUMMARY

The steps to follow for approximating solutions of equations are given next.

Steps for Approximating Solutions of Equations Using ZERO (or ROOT)

STEP 1: Write the equation in the form $\{expression\ in\ x\} = 0$.

STEP 2: Graph $Y_1 = \{expression\ in\ x\}$.

Be sure that the graph is complete. That is, be sure that all the intercepts are shown on the screen.

STEP 3: Use ZERO (or ROOT) to determine each x-intercept of the graph.

Steps for Approximating Solutions of Equations Using INTERSECT

STEP 1: Graph $Y_1 = \{expression\ in\ x\ on\ the\ left\ side\ of\ the\ equation\}$.

Graph $Y_2 = \{expression\ in\ x\ on\ the\ right\ side\ of\ the\ equation\}$.

STEP 2: Use INTERSECT to determine each x-coordinate of the point(s) of intersection, if any.

Be sure that the graphs are complete. That is, be sure that all the points of intersection are shown on the screen.

EXAMPLE 3	**Solving a Radical Equation**

Find the real solutions of the equation $\sqrt[3]{2x - 4} - 2 = 0$.

Solution Figure 14 shows the graph of the equation $Y_1 = \sqrt[3]{2x - 4} - 2$. From the graph, we see one x-intercept near 6. Using ZERO (or ROOT), we find that the x-intercept is 6. The only solution is $x = 6$.

Figure 14

5 Square Screens

Figure 15

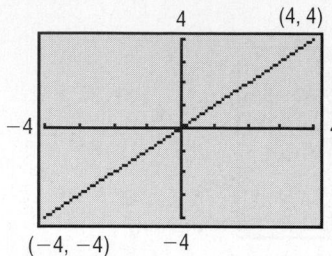

Most graphing utilities have a rectangular screen. Because of this, using the same settings for both x and y will result in a distorted view. For example, Figure 15 shows the graph of the line $y = x$ connecting the points $(-4, -4)$ and $(4, 4)$.

We expect the line to bisect the first and third quadrants, but it doesn't. We need to adjust the selections for Xmin, Xmax, Ymin, and Ymax so that a **square screen** results. On most graphing utilities, this is accomplished by setting the ratio of x to y at $3:2$.* For example, if

$$X\text{min} = -6 \qquad Y\text{min} = -4$$
$$X\text{max} = 6 \qquad Y\text{max} = 4$$

then the ratio of x to y is

$$\frac{X\text{max} - X\text{min}}{Y\text{max} - Y\text{min}} = \frac{6 - (-6)}{4 - (-4)} = \frac{12}{8} = \frac{3}{2}$$

for a ratio of $3:2$, resulting in a square screen.

EXAMPLE 1 **Examples of Viewing Rectangles That Result in Square Screens**

Figure 16

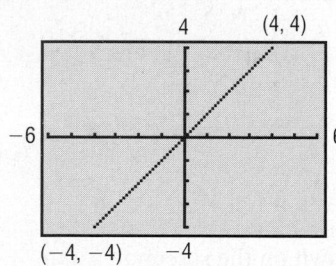

(a) Xmin = -3 (b) Xmin = -6 (c) Xmin = -12
 Xmax = 3 Xmax = 6 Xmax = 12
 Xscl = 1 Xscl = 1 Xscl = 1
 Ymin = -2 Ymin = -4 Ymin = -8
 Ymax = 2 Ymax = 4 Ymax = 8
 Yscl = 1 Yscl = 1 Yscl = 2

Figure 16 shows the graph of the line $y = x$ on a square screen using the viewing rectangle given in part (b). Notice that the line now bisects the first and third quadrants. Compare this illustration to Figure 15.

5 Exercises

In Problems 1–8, determine which of the given viewing rectangles result in a square screen.

1. Xmin = -3
 Xmax = 3
 Xscl = 2
 Ymin = -2
 Ymax = 2
 Yscl = 2

2. Xmin = -5
 Xmax = 5
 Xscl = 1
 Ymin = -4
 Ymax = 4
 Yscl = 1

3. Xmin = 0
 Xmax = 9
 Xscl = 3
 Ymin = -2
 Ymax = 4
 Yscl = 2

4. Xmin = -6
 Xmax = 6
 Xscl = 1
 Ymin = -4
 Ymax = 4
 Yscl = 2

5. Xmin = -6
 Xmax = 6
 Xscl = 1
 Ymin = -2
 Ymax = 2
 Yscl = 0.5

6. Xmin = -6
 Xmax = 6
 Xscl = 2
 Ymin = -4
 Ymax = 4
 Yscl = 1

7. Xmin = 0
 Xmax = 9
 Xscl = 1
 Ymin = -2
 Ymax = 4
 Yscl = 1

8. Xmin = -6
 Xmax = 6
 Xscl = 2
 Ymin = -4
 Ymax = 4
 Yscl = 2

9. If Xmin = -4, Xmax = 8, and Xscl = 1, how should Ymin, Ymax, and Yscl be selected so that the viewing rectangle contains the point $(4, 8)$ and the screen is square?

10. If Xmin = -6, Xmax = 12, and Xscl = 2, how should Ymin, Ymax, and Yscl be selected so that the viewing rectangle contains the point $(4, 8)$ and the screen is square?

* Some graphing utilities have a built-in function that automatically squares the screen. For example, the TI-84 has a ZSquare function that does this. Some graphing utilities require a ratio other than $3:2$ to square the screen. For example, the HP 48G requires the ratio of x to y to be $2:1$ for a square screen. Consult your manual.

6 Using a Graphing Utility to Graph Inequalities

EXAMPLE 1	**Graphing an Inequality Using a Graphing Utility**

Use a graphing utility to graph: $3x + y - 6 \le 0$

Solution We begin by graphing the equation $3x + y - 6 = 0$ $(Y_1 = -3x + 6)$. See Figure 17.

As with graphing by hand, we need to test points selected from each region and determine whether they satisfy the inequality. To test the point $(-1, 2)$, for example, enter $3(-1) + 2 - 6 \le 0$. See Figure 18(a). The 1 that appears indicates that the statement entered (the inequality) is true. When the point $(5, 5)$ is tested, a 0 appears, indicating that the statement entered is false. Thus, $(-1, 2)$ is a part of the graph of the inequality and $(5, 5)$ is not. Figure 18(b) shows the graph of the inequality on a TI-84 Plus.*

Figure 17

$Y_1 = -3x + 6$

Figure 18

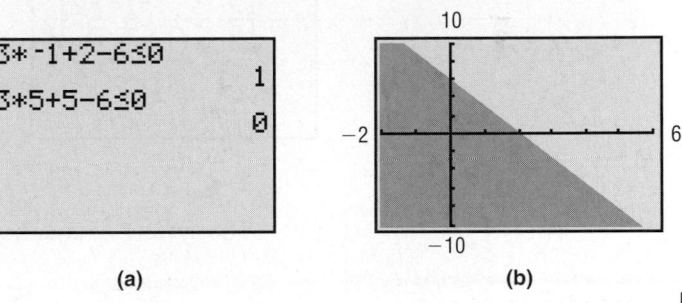

(a) (b)

Steps for Graphing an Inequality Using a Graphing Utility

STEP 1: Replace the inequality symbol by an equal sign, solve the equation for y, and graph the equation.

STEP 2: In each region, select a test point P and determine if the coordinates of P satisfy the inequality.

(a) If the test point satisfies the inequality, then so do all the points in the region. Indicate this by using the graphing utility to shade the region.

(b) If the coordinates of P do not satisfy the inequality, then none of the points in that region do.

7 Using a Graphing Utility to Solve Systems of Linear Equations

Most graphing utilities have the capability to put the augmented matrix of a system of linear equations in row echelon form. The next example, Example 6 from Section 8.2, demonstrates this feature using a TI-84 Plus graphing calculator.

* Consult your owner's manual for shading techniques.

| EXAMPLE 1 | **Solving a System of Linear Equations Using a Graphing Utility** |

Solve: $\begin{cases} x - y + z = 8 & (1) \\ 2x + 3y - z = -2 & (2) \\ 3x - 2y - 9z = 9 & (3) \end{cases}$

Solution The augmented matrix of the system is

$$\begin{bmatrix} 1 & -1 & 1 & | & 8 \\ 2 & 3 & -1 & | & -2 \\ 3 & -2 & -9 & | & 9 \end{bmatrix}$$

Enter this matrix into a graphing utility and name it A. See Figure 19(a). Using the REF (row echelon form) command on matrix A, we obtain the results shown in Figure 19(b). Since the entire matrix does not fit on the screen, you need to scroll right to see the rest of it. See Figure 19(c).

Figure 19

(a) (b) (c)

The system of equations represented by the matrix in row echelon form is

$$\begin{bmatrix} 1 & -\frac{2}{3} & -3 & | & 3 \\ 0 & 1 & \frac{15}{13} & | & -\frac{24}{13} \\ 0 & 0 & 1 & | & 1 \end{bmatrix} \qquad \begin{cases} x - \frac{2}{3}y - 3z = 3 & (1) \\ y + \frac{15}{13}z = -\frac{24}{13} & (2) \\ z = 1 & (3) \end{cases}$$

Using $z = 1$, back-substitute to get

$$\begin{cases} x - \frac{2}{3}y - 3(1) = 3 & (1) \\ y + \frac{15}{13}(1) = -\frac{24}{13} & (2) \end{cases} \xrightarrow{\text{Simplify.}} \begin{cases} x - \frac{2}{3}y = 6 & (1) \\ y = \frac{-39}{13} = -3 & (2) \end{cases}$$

Solving the second equation for y, we find that $y = -3$. Back-substituting $y = -3$ into $x - \frac{2}{3}y = 6$, we find that $x = 4$. The solution of the system is $x = 4$, $y = -3$, $z = 1$. ♩

Notice that the row echelon form of the augmented matrix using the graphing utility differs from the row echelon form in Chapter 8 (p. 562), yet both matrices provide the same solution! This is because the two solutions used different row operations to obtain the row echelon form. In all likelihood, the two solutions parted ways in Step 4 of the algebraic solution, where we avoided introducing fractions by interchanging rows 2 and 3.

Figure 20

Most graphing utilities also have the ability to put a matrix in reduced row echelon form. Figure 20 shows the reduced row echelon form of the augmented matrix from Example 1 using the RREF command on a TI-84 Plus graphing calculator. Using this command, we see that the solution of the system is $x = 4$, $y = -3$, $z = 1$.

Answers

CHAPTER R Review

R.1 Assess Your Understanding *(page 15)*

1. rational **2.** 31 **3.** Distributive **4.** $5(x + 3) = 6$ **5.** T **6.** F **7.** F **8.** T **9.** $\{1, 2, 3, 4, 5, 6, 7, 8, 9\}$ **11.** $\{4\}$ **13.** $\{1, 3, 4, 6\}$

15. $\{0, 2, 6, 7, 8\}$ **17.** $\{0, 1, 2, 3, 5, 6, 7, 8, 9\}$ **19.** $\{0, 1, 2, 3, 5, 6, 7, 8, 9\}$ **21. (a)** $\{2, 5\}$ **(b)** $\{-6, 2, 5\}$ **(c)** $\left\{-6, \dfrac{1}{2}, -1.333\cdots, 2, 5\right\}$ **(d)** $\{\pi\}$

(e) $\left\{-6, \dfrac{1}{2}, -1.333..., \pi, 2, 5\right\}$ **23. (a)** $\{1\}$ **(b)** $\{0, 1\}$ **(c)** $\left\{0, 1, \dfrac{1}{2}, \dfrac{1}{3}, \dfrac{1}{4}\right\}$ **(d)** None **(e)** $\left\{0, 1, \dfrac{1}{2}, \dfrac{1}{3}, \dfrac{1}{4}\right\}$ **25. (a)** None **(b)** None **(c)** None

(d) $\left\{\sqrt{2}, \pi, \sqrt{2} + 1, \pi + \dfrac{1}{2}\right\}$ **(e)** $\left\{\sqrt{2}, \pi, \sqrt{2} + 1, \pi + \dfrac{1}{2}\right\}$ **27. (a)** 18.953 **(b)** 18.952 **29. (a)** 28.653 **(b)** 28.653 **31. (a)** 0.063 **(b)** 0.062

33. (a) 9.999 **(b)** 9.998 **35. (a)** 0.429 **(b)** 0.428 **37. (a)** 34.733 **(b)** 34.733 **39.** $3 + 2 = 5$ **41.** $x + 2 = 3 \cdot 4$ **43.** $3y = 1 + 2$

45. $x - 2 = 6$ **47.** $\dfrac{x}{2} = 6$ **49.** 7 **51.** 6 **53.** 1 **55.** $\dfrac{13}{3}$ **57.** -11 **59.** 11 **61.** -4 **63.** 1 **65.** 6 **67.** $\dfrac{2}{7}$ **69.** $\dfrac{4}{45}$ **71.** $\dfrac{23}{20}$ **73.** $\dfrac{79}{30}$ **75.** $\dfrac{13}{36}$

77. $-\dfrac{16}{45}$ **79.** $\dfrac{1}{60}$ **81.** $\dfrac{15}{22}$ **83.** 1 **85.** $\dfrac{15}{8}$ **87.** $6x + 24$ **89.** $x^2 - 4x$ **91.** $\dfrac{3}{2}x - 1$ **93.** $x^2 + 6x + 8$ **95.** $x^2 - x - 2$ **97.** $x^2 - 10x + 16$

99. $2x + 3x = (2 + 3)x = 5x$ **101.** $2(3 \cdot 4) = 2 \cdot 12 = 24; (2 \cdot 3) \cdot (2 \cdot 4) = 6 \cdot 8 = 48$ **103.** No; $2 - 3 \neq 3 - 2$ **105.** No; $\dfrac{2}{3} \neq \dfrac{3}{2}$

107. Symmetric Property **109.** No; no

R.2 Assess Your Understanding *(page 26)*

1. variable **2.** origin **3.** strict **4.** base; exponent or power **5.** 1.2345678×10^3 **6.** T **7.** T **8.** F **9.** F **10.** F

11. **13.** $>$ **15.** $>$ **17.** $>$ **19.** $=$ **21.** $<$ **23.** $x > 0$ **25.** $x < 2$ **27.** $x \leq 1$

29. **31.** **33.** 1 **35.** 2 **37.** 6 **39.** 4 **41.** -28 **43.** $\dfrac{4}{5}$ **45.** 0 **47.** 1 **49.** 5 **51.** 1

53. 22 **55.** 2 **57.** $x = 0$ **59.** $x = 3$ **61.** None **63.** $x = 0, x = 1, x = -1$ **65.** $\{x | x \neq 5\}$ **67.** $\{x | x \neq -4\}$ **69.** 0°C **71.** 25°C **73.** 16 **75.** $\dfrac{1}{16}$

77. $\dfrac{1}{9}$ **79.** 9 **81.** 5 **83.** 4 **85.** $64x^6$ **87.** $\dfrac{x^4}{y^2}$ **89.** $\dfrac{x}{y}$ **91.** $-\dfrac{8x^3z}{9y}$ **93.** $\dfrac{16x^2}{9y^2}$ **95.** -4 **97.** 5 **99.** 4 **101.** 2 **103.** $\sqrt{5}$ **105.** $\dfrac{1}{2}$ **107.** $10; 0$

109. 81 **111.** 304,006.671 **113.** 0.004 **115.** 481.890 **117.** 0.000 **119.** 4.542×10^2 **121.** 1.3×10^{-2} **123.** 3.2155×10^4 **125.** 4.23×10^{-4}

127. 61,500 **129.** 0.001214 **131.** 110,000,000 **133.** 0.081 **135.** $A = lw$ **137.** $C = \pi d$ **139.** $A = \dfrac{\sqrt{3}}{4}x^2$ **141.** $V = \dfrac{4}{3}\pi r^3$ **143.** $V = x^3$

145. (a) \$6000 **(b)** \$8000 **147.** $|x - 4| \geq 6$ **149. (a)** $2 \leq 5$ **(b)** $6 > 5$ **151. (a)** Yes **(b)** No **153.** 400,000,000 m **155.** 0.0000005 m

157. 5×10^{-4} in. **159.** 5.865696×10^{12} mi **161.** No; $\dfrac{1}{3}$ is larger; 0.000333... **163.** No

R.3 Assess Your Understanding *(page 36)*

1. right; hypotenuse **2.** $A = \dfrac{1}{2}bh$ **3.** $C = 2\pi r$ **4.** similar **5.** T **6.** T **7.** F **8.** T **9.** T **10.** F **11.** 13 **13.** 26 **15.** 25

17. Right triangle; 5 **19.** Not a right triangle **21.** Right triangle; 25 **23.** Not a right triangle **25.** 8 in.2 **27.** 4 in.2 **29.** $A = 25\pi$ m^2; $C = 10\pi$ m

31. $V = 224$ ft^3; $S = 232$ ft^2 **33.** $V = \dfrac{256}{3}\pi$ cm^3; $S = 64\pi$ cm^2 **35.** $V = 648\pi$ in.3; $S = 306\pi$ in.2 **37.** π square units **39.** 2π square units

41. $x = 4$ units; $A = 90°$; $B = 60°$; $C = 30°$ **43.** $x = 67.5$ units; $A = 60°$; $B = 95°$; $C = 25°$ **45.** About 16.8 ft **47.** 64 ft^2

49. $24 + 2\pi \approx 30.28$ ft^2; $16 + 2\pi \approx 22.28$ ft **51.** 160 paces **53.** About 5.477 mi **55.** From 100 ft: 12.2 mi; From 150 ft: 15.0 mi

R.4 Assess Your Understanding *(page 47)*

1. 4; 3 **2.** $x^4 - 16$ **3.** $x^3 - 8$ **4.** F **5.** T **6.** F **7.** Monomial; variable: x; coefficient: 2; degree: 3 **9.** Not a monomial; the exponent of the variable is not a nonnegative integer **11.** Monomial; variables: x, y; coefficient: -2; degree: 3 **13.** Not a monomial; the exponent of one of the variables is not a nonnegative integer **15.** Not a monomial; it has more than one term **17.** Yes; 2 **19.** Yes; 0 **21.** No; the exponent of the variable of one of the terms is not a nonnegative integer **23.** Yes; 3 **25.** No; the polynomial of the denominator has a degree greater than 0 **27.** $x^2 + 7x + 2$
29. $x^3 - 4x^2 + 9x + 7$ **31.** $6x^5 + 5x^4 + 3x^2 + x$ **33.** $7x^2 - x - 7$ **35.** $-2x^3 + 18x^2 - 18$ **37.** $2x^2 - 4x + 6$ **39.** $15y^2 - 27y + 30$
41. $x^3 + x^2 - 4x$ **43.** $-8x^5 - 10x^2$ **45.** $x^3 + 3x^2 - 2x - 4$ **47.** $x^2 + 6x + 8$ **49.** $2x^2 + 9x + 10$ **51.** $x^2 - 2x - 8$ **53.** $x^2 - 5x + 6$
55. $2x^2 - x - 6$ **57.** $-2x^2 + 11x - 12$ **59.** $2x^2 + 8x + 8$ **61.** $x^2 - xy - 2y^2$ **63.** $-6x^2 - 13xy - 6y^2$ **65.** $x^2 - 49$ **67.** $4x^2 - 9$
69. $x^2 + 8x + 16$ **71.** $x^2 - 8x + 16$ **73.** $9x^2 - 16$ **75.** $4x^2 - 12x + 9$ **77.** $x^2 - y^2$ **79.** $9x^2 - y^2$ **81.** $x^2 + 2xy + y^2$ **83.** $x^2 - 4xy + 4y^2$
85. $x^3 - 6x^2 + 12x - 8$ **87.** $8x^3 + 12x^2 + 6x + 1$ **89.** $4x^2 - 11x + 23$; remainder -45 **91.** $4x - 3$; remainder $x + 1$

93. $5x^2 - 13$; remainder $x + 27$ **95.** $2x^2$; remainder $-x^2 + x + 1$ **97.** $x^2 - 2x + \frac{1}{2}$; remainder $\frac{5}{2}x + \frac{1}{2}$ **99.** $-4x^2 - 3x - 3$; remainder -7
101. $x^2 - x - 1$; remainder $2x + 2$ **103.** $x^2 + ax + a^2$; remainder 0

R.5 Assess Your Understanding (page 57)

1. $3x(x - 2)(x + 2)$ **2.** Prime **3.** T **4.** F **5.** $3(x + 2)$ **7.** $a(x^2 + 1)$ **9.** $x(x^2 + x + 1)$ **11.** $2x(x - 1)$ **13.** $3xy(x - 2y + 4)$
15. $(x + 1)(x - 1)$ **17.** $(2x + 1)(2x - 1)$ **19.** $(x + 4)(x - 4)$ **21.** $(5x + 2)(5x - 2)$ **23.** $(x + 1)^2$ **25.** $(x + 2)^2$ **27.** $(x - 5)^2$
29. $(2x + 1)^2$ **31.** $(4x + 1)^2$ **33.** $(x - 3)(x^2 + 3x + 9)$ **35.** $(x + 3)(x^2 - 3x + 9)$ **37.** $(2x + 3)(4x^2 - 6x + 9)$ **39.** $(x + 2)(x + 3)$
41. $(x + 6)(x + 1)$ **43.** $(x + 5)(x + 2)$ **45.** $(x - 8)(x - 2)$ **47.** $(x - 8)(x + 1)$ **49.** $(x + 8)(x - 1)$ **51.** $(x + 2)(2x + 3)$
53. $(x - 2)(2x + 1)$ **55.** $(2x + 3)(3x + 2)$ **57.** $(3x + 1)(x + 1)$ **59.** $(z + 1)(2z + 3)$ **61.** $(x + 2)(3x - 4)$ **63.** $(x - 2)(3x + 4)$
65. $(x + 4)(3x + 2)$ **67.** $(x + 4)(3x - 2)$ **69.** $25; (x + 5)^2$ **71.** $9; (y - 3)^2$ **73.** $\frac{1}{16}; \left(x - \frac{1}{4}\right)^2$ **75.** $(x + 6)(x - 6)$ **77.** $2(1 + 2x)(1 - 2x)$
79. $(x + 1)(x + 10)$ **81.** $(x - 7)(x - 3)$ **83.** $4(x^2 - 2x + 8)$ **85.** Prime **87.** $-(x - 5)(x + 3)$ **89.** $3(x + 2)(x - 6)$ **91.** $y^2(y + 5)(y + 6)$
93. $(2x + 3)^2$ **95.** $2(3x + 1)(x + 1)$ **97.** $(x - 3)(x + 3)(x^2 + 9)$ **99.** $(x - 1)^2(x^2 + x + 1)^2$ **101.** $x^5(x - 1)(x + 1)$ **103.** $(4x + 3)^2$
105. $-(4x - 5)(4x + 1)$ **107.** $(2y - 5)(2y - 3)$ **109.** $-(3x - 1)(3x + 1)(x^2 + 1)$ **111.** $(x + 3)(x - 6)$ **113.** $(x + 2)(x - 3)$
115. $(3x - 5)(9x^2 - 3x + 7)$ **117.** $(x + 5)(3x + 11)$ **119.** $(x - 1)(x + 1)(x + 2)$ **121.** $(x - 1)(x + 1)(x^2 - x + 1)$
123. $2(3x + 4)(9x + 13)$ **125.** $2x(3x + 5)$ **127.** $5(x + 3)(x - 2)^2(x + 1)$ **129.** $3(4x - 3)(4x - 1)$ **131.** $6(3x - 5)(2x + 1)^2(5x - 4)$
133. The possibilities are $(x \pm 1)(x \pm 4) = x^2 \pm 5x + 4$ or $(x \pm 2)(x \pm 2) = x^2 \pm 4x + 4$, none of which equals $x^2 + 4$.

R.6 Assess Your Understanding (page 61)

1. quotient; divisor; remainder **2.** $-3\overline{)2\ 0\ -5\ 1}$ **3.** T **4.** T **5.** $x^2 + x + 4$; remainder 12 **7.** $3x^2 + 11x + 32$; remainder 99
9. $x^4 - 3x^3 + 5x^2 - 15x + 46$; remainder -138 **11.** $4x^5 + 4x^4 + x^3 + x^2 + 2x + 2$; remainder 7 **13.** $0.1x^2 - 0.11x + 0.321$; remainder -0.3531
15. $x^4 + x^3 + x^2 + x + 1$; remainder 0 **17.** No **19.** Yes **21.** Yes **23.** No **25.** Yes **27.** -9

R.7 Assess Your Understanding (page 70)

1. lowest terms **2.** least common multiple **3.** T **4.** F **5.** $\dfrac{3}{x - 3}$ **7.** $\dfrac{x}{3}$ **9.** $\dfrac{4x}{2x - 1}$ **11.** $\dfrac{y + 5}{2(y + 1)}$ **13.** $\dfrac{x + 5}{x - 1}$ **15.** $-(x + 7)$ **17.** $\dfrac{3}{5x(x - 2)}$
19. $\dfrac{2x(x^2 + 4x + 16)}{x + 4}$ **21.** $\dfrac{8}{3x}$ **23.** $\dfrac{x - 3}{x + 7}$ **25.** $\dfrac{4x}{(x - 2)(x - 3)}$ **27.** $\dfrac{4}{5(x - 1)}$ **29.** $-\dfrac{(x - 4)^2}{4x}$ **31.** $\dfrac{(x + 3)^2}{(x - 3)^2}$ **33.** $\dfrac{(x - 4)(x + 3)}{(x - 1)(2x + 1)}$
35. $\dfrac{x + 5}{2}$ **37.** $\dfrac{(x - 2)(x + 2)}{2x - 3}$ **39.** $\dfrac{3x - 2}{x - 3}$ **41.** $\dfrac{x + 9}{2x - 1}$ **43.** $\dfrac{4 - x}{x - 2}$ **45.** $\dfrac{2(x + 5)}{(x - 1)(x + 2)}$ **47.** $\dfrac{3x^2 - 2x - 3}{(x + 1)(x - 1)}$ **49.** $\dfrac{-(11x + 2)}{(x + 2)(x - 2)}$
51. $\dfrac{2(x^2 - 2)}{x(x - 2)(x + 2)}$ **53.** $(x - 2)(x + 2)(x + 1)$ **55.** $x(x - 1)(x + 1)$ **57.** $x^3(2x - 1)^2$ **59.** $x(x - 1)^2(x + 1)(x^2 + x + 1)$
61. $\dfrac{5x}{(x - 6)(x - 1)(x + 4)}$ **63.** $\dfrac{2(2x^2 + 5x - 2)}{(x - 2)(x + 2)(x + 3)}$ **65.** $\dfrac{5x + 1}{(x - 1)^2(x + 1)^2}$ **67.** $\dfrac{-x^2 + 3x + 13}{(x - 2)(x + 1)(x + 4)}$ **69.** $\dfrac{x^3 - 2x^2 + 4x + 3}{x^2(x + 1)(x - 1)}$
71. $\dfrac{-1}{x(x + h)}$ **73.** $\dfrac{x + 1}{x - 1}$ **75.** $\dfrac{(x - 1)(x + 1)}{2x(2x + 1)}$ **77.** $\dfrac{2(5x - 1)}{(x - 2)(x + 1)^2}$ **79.** $\dfrac{-2x(x^2 - 2)}{(x + 2)(x^2 - x - 3)}$ **81.** $\dfrac{-1}{x - 1}$ **83.** $\dfrac{3x - 1}{2x + 1}$ **85.** $\dfrac{19}{(3x - 5)^2}$
87. $\dfrac{(x + 1)(x - 1)}{(x^2 + 1)^2}$ **89.** $\dfrac{x(3x + 2)}{(3x + 1)^2}$ **91.** $-\dfrac{(x + 3)(3x - 1)}{(x^2 + 1)^2}$ **93.** $f = \dfrac{R_1 \cdot R_2}{(n - 1)(R_1 + R_2)}; \dfrac{2}{15}$ m

R.8 Assess Your Understanding (page 78)

3. index **4.** T **5.** cube root **6.** F **7.** 3 **9.** -2 **11.** $2\sqrt{2}$ **13.** $-2x\sqrt[3]{x}$ **15.** x^3y^2 **17.** x^2y **19.** $6\sqrt{x}$ **21.** $6x\sqrt{x}$ **23.** $15\sqrt[3]{3}$
25. $12\sqrt{3}$ **27.** $7\sqrt{2}$ **29.** $\sqrt{2}$ **31.** $2\sqrt{3}$ **33.** $-\sqrt[3]{2}$ **35.** $x - 2\sqrt{x} + 1$ **37.** $(2x - 1)\sqrt[3]{2x}$ **39.** $(2x - 15)\sqrt{2x}$ **41.** $-(x + 5y)\sqrt[3]{2xy}$
43. $\dfrac{\sqrt{2}}{2}$ **45.** $-\dfrac{\sqrt{15}}{5}$ **47.** $\dfrac{(5 + \sqrt{2})\sqrt{3}}{23}$ **49.** $\dfrac{8\sqrt{5} - 19}{41}$ **51.** $\dfrac{5\sqrt[3]{4}}{2}$ **53.** $\dfrac{2x + h - 2\sqrt{x^2 + xh}}{h}$ **55.** 4 **57.** -3 **59.** 64 **61.** $\dfrac{1}{27}$ **63.** $\dfrac{27\sqrt{2}}{32}$
65. $\dfrac{27\sqrt{2}}{32}$ **67.** $x^{7/12}$ **69.** xy^2 **71.** $x^{2/3}y$ **73.** $\dfrac{8x^{5/4}}{y^{3/4}}$ **75.** $\dfrac{3x + 2}{(1 + x)^{1/2}}$ **77.** $\dfrac{x(3x^2 + 2)}{(x^2 + 1)^{1/2}}$ **79.** $\dfrac{22x + 5}{10\sqrt{x} - 5\sqrt{4x + 3}}$ **81.** $\dfrac{2 + x}{2(1 + x)^{3/2}}$ **83.** $\dfrac{4 - x}{(x + 4)^{3/2}}$
85. $\dfrac{1}{x^2(x^2 - 1)^{1/2}}$ **87.** $\dfrac{1 - 3x^2}{2\sqrt{x}(1 + x^2)^2}$ **89.** $\dfrac{1}{2}(5x + 2)(x + 1)^{1/2}$ **91.** $2x^{1/2}(3x - 4)(x + 1)$ **93.** $(x^2 + 4)^{1/3}(11x^2 + 12)$
95. $(3x + 5)^{1/3}(2x + 3)^{1/2}(17x + 27)$ **97.** $\dfrac{3(x + 2)}{2x^{1/2}}$ **99.** 1.41 **101.** 1.59 **103.** 4.89 **105.** 2.15 **107.** **(a)** 15,660.4 gal **(b)** 390.7 gal
109. $2\sqrt{2\pi} \approx 8.89$ sec **111.** $\dfrac{\pi\sqrt{3}}{6} \approx 0.91$ sec

CHAPTER 1 Equations and Inequalities

1.1 Assess Your Understanding (page 90)

4. F **5.** identity **6.** linear; first-degree **7.** F **8.** T **9.** $\{3\}$ **11.** $\{-5\}$ **13.** $\left\{\dfrac{3}{2}\right\}$ **15.** $\left\{\dfrac{5}{4}\right\}$ **17.** $\{-2\}$ **19.** $\{3\}$ **21.** $\{-1\}$ **23.** $\{-2\}$ **25.** $\{-18\}$
27. $\{-4\}$ **29.** $\left\{-\dfrac{3}{4}\right\}$ **31.** $\{-20\}$ **33.** $\{2\}$ **35.** $\{0.5\}$ **37.** $\left\{\dfrac{29}{10}\right\}$ **39.** $\{2\}$ **41.** $\{8\}$ **43.** $\{2\}$ **45.** $\{-1\}$ **47.** $\{3\}$ **49.** No solution **51.** No solution

53. $\{-6\}$ **55.** $\{34\}$ **57.** $\left\{-\dfrac{20}{39}\right\}$ **59.** $\{-1\}$ **61.** $\left\{-\dfrac{11}{6}\right\}$ **63.** $\{-6\}$ **65.** $\{5.91\}$ **67.** $\{0.41\}$ **69.** $x = \dfrac{b+c}{a}$ **71.** $x = \dfrac{abc}{a+b}$ **73.** $x = a^2$

75. $a = 3$ **77.** $R = \dfrac{R_1 R_2}{R_1 + R_2}$ **79.** $R = \dfrac{mv^2}{F}$ **81.** $r = \dfrac{S-a}{S}$ **83.** \$11,500 will be invested in bonds and \$8500 in CDs. **85.** The regular hourly rate is \$8.50. **87.** Brooke needs a score of 85. **89.** The original price was \$500,000; purchasing the model saves \$75,000. **91.** The bookstore paid \$68.15 for the book. **93.** There were 2187 adults. **95.** The length is 19 ft; the width is 11 ft. **97.** Judy pays \$10.80 and Tom, \$7.20.

Historical Problems *(page 100)*

1. The area of each shaded square is 9, so the larger square will have area $85 + 4(9) = 121$. The area of the larger square is also given by the expression $(x+6)^2$ so $(x+6)^2 = 121$. Taking the positive square root of each side, $x + 6 = 11$ or $x = 5$.
2. Let $z = -6$, so $z^2 + 12z - 85 = -121$. We get the equation $u^2 - 121 = 0$ or $u^2 = 121$. Thus $u = \pm 11$, so $x = \pm 11 - 6$. $x = -17$ or $x = 5$.

3.
$$\left(x + \frac{b}{2a}\right)^2 = \left(\frac{\sqrt{b^2 - 4ac}}{2a}\right)^2$$
$$\left(x + \frac{b}{2a}\right)^2 - \left(\frac{\sqrt{b^2 - 4ac}}{2a}\right)^2 = 0$$
$$\left(x + \frac{b}{2a} - \frac{\sqrt{b^2 - 4ac}}{2a}\right)\left(x + \frac{b}{2a} + \frac{\sqrt{b^2 - 4ac}}{2a}\right) = 0$$
$$\left(x + \frac{b - \sqrt{b^2 - 4ac}}{2a}\right)\left(x + \frac{b + \sqrt{b^2 - 4ac}}{2a}\right) = 0$$
$$x = \frac{-b + \sqrt{b^2 - 4ac}}{2a} \quad \text{or} \quad x = \frac{-b - \sqrt{b^2 - 4ac}}{2a}$$

1.2 Assess Your Understanding *(page 101)*

6. discriminant; negative **7.** F **8.** F **9.** $\{0, 9\}$ **11.** $\{-5, 5\}$ **13.** $\{-3, 2\}$ **15.** $\left\{-\dfrac{1}{2}, 3\right\}$ **17.** $\{-4, 4\}$ **19.** $\{2, 6\}$ **21.** $\left\{\dfrac{3}{2}\right\}$ **23.** $\left\{-\dfrac{2}{3}, \dfrac{3}{2}\right\}$

25. $\left\{-\dfrac{2}{3}, \dfrac{3}{2}\right\}$ **27.** $\left\{-\dfrac{3}{4}, 2\right\}$ **29.** $\{-5, 5\}$ **31.** $\{-1, 3\}$ **33.** $\{-3, 0\}$ **35.** $\{-7, 3\}$ **37.** $\left\{-\dfrac{1}{4}, \dfrac{3}{4}\right\}$ **39.** $\left\{\dfrac{-1 - \sqrt{7}}{6}, \dfrac{-1 + \sqrt{7}}{6}\right\}$

41. $\{2 - \sqrt{2}, 2 + \sqrt{2}\}$ **43.** $\{2 - \sqrt{5}, 2 + \sqrt{5}\}$ **45.** $\left\{1, \dfrac{3}{2}\right\}$ **47.** No real solution **49.** $\left\{\dfrac{-1 - \sqrt{5}}{4}, \dfrac{-1 + \sqrt{5}}{4}\right\}$ **51.** $\left\{0, \dfrac{9}{4}\right\}$ **53.** $\left\{\dfrac{1}{3}\right\}$

55. $\left\{-\dfrac{2}{3}, 1\right\}$ **57.** $\left\{\dfrac{3 - \sqrt{29}}{10}, \dfrac{3 + \sqrt{29}}{10}\right\}$ **59.** $\left\{\dfrac{-2 - \sqrt{10}}{2}, \dfrac{-2 + \sqrt{10}}{2}\right\}$ **61.** $\left\{\dfrac{1 - \sqrt{33}}{8}, \dfrac{1 + \sqrt{33}}{8}\right\}$ **63.** $\left\{\dfrac{9 - \sqrt{73}}{2}, \dfrac{9 + \sqrt{73}}{2}\right\}$

65. $\{0.63, 3.47\}$ **67.** $\{-2.80, 1.07\}$ **69.** $\{-0.85, 1.17\}$ **71.** No real solution **73.** Repeated real solution **75.** Two unequal real solutions

77. $\{-\sqrt{5}, \sqrt{5}\}$ **79.** $\left\{\dfrac{1}{4}\right\}$ **81.** $\left\{-\dfrac{3}{5}, \dfrac{5}{2}\right\}$ **83.** $\left\{-\dfrac{1}{2}, \dfrac{2}{3}\right\}$ **85.** $\left\{\dfrac{-\sqrt{2} + 2}{2}, \dfrac{-\sqrt{2} - 2}{2}\right\}$ **87.** $\left\{\dfrac{-1 - \sqrt{17}}{2}, \dfrac{-1 + \sqrt{17}}{2}\right\}$ **89.** $\{5\}$

91. 2; 5 meters, 12 meters, 13 meters; 20 meters, 21 meters; 29 meters **93.** The dimensions are 11 ft by 13 ft. **95.** The dimensions are 5 m by 8 m.
97. The dimensions should be 4 ft by 4 ft. **99. (a)** The ball strikes the ground after 6 sec. **(b)** The ball passes the top of the building on its way down after 5 sec. **101.** The dimensions should be 11.55 cm by 6.55 cm by 3 cm. **103.** The border will be 2.71 ft wide.
105. The border will be 2.56 ft wide. **107.** The screen of a 37-inch TV in 4:3 format has an area of 657.12 square inches; the screen of a 37-inch TV in 16:9 format has an area of 584.97 square inches. The traditional TV has a larger screen. **109.** 36 consecutive integers must be added.

111. $\dfrac{-b + \sqrt{b^2 - 4ac}}{2a} + \dfrac{-b - \sqrt{b^2 - 4ac}}{2a} = \dfrac{-2b}{2a} = -\dfrac{b}{a}$ **113.** $k = \dfrac{1}{2}$ or $k = -\dfrac{1}{2}$

115. $ax^2 + bx + c = 0, x = \dfrac{-b \pm \sqrt{b^2 - 4ac}}{2a}; ax^2 - bx + c = 0, x = \dfrac{b \pm \sqrt{(-b)^2 - 4ac}}{2a} = \dfrac{b \pm \sqrt{b^2 - 4ac}}{2a} = -\dfrac{-b \pm \sqrt{b^2 - 4ac}}{2a}$ **117. (b)**

1.3 Assess Your Understanding *(page 111)*

4. real; imaginary; imaginary unit **5.** $\{-2i, 2i\}$ **6.** F **7.** T **8.** F **9.** $8 + 5i$ **11.** $-7 + 6i$ **13.** $-6 - 11i$ **15.** $6 - 18i$ **17.** $6 + 4i$ **19.** $10 - 5i$

21. 37 **23.** $\dfrac{6}{5} + \dfrac{8}{5}i$ **25.** $1 - 2i$ **27.** $\dfrac{5}{2} - \dfrac{7}{2}i$ **29.** $-\dfrac{1}{2} + \dfrac{\sqrt{3}}{2}i$ **31.** $2i$ **33.** $-i$ **35.** i **37.** -6 **39.** $-10i$ **41.** $-2 + 2i$ **43.** 0 **45.** 0 **47.** $2i$

49. $5i$ **51.** $5i$ **53.** $\{-2i, 2i\}$ **55.** $\{-4, 4\}$ **57.** $\{3 - 2i, 3 + 2i\}$ **59.** $\{3 - i, 3 + i\}$ **61.** $\left\{\dfrac{1}{4} - \dfrac{1}{4}i, \dfrac{1}{4} + \dfrac{1}{4}i\right\}$ **63.** $\left\{\dfrac{1}{5} - \dfrac{2}{5}i, \dfrac{1}{5} + \dfrac{2}{5}i\right\}$

65. $\left\{-\dfrac{1}{2} - \dfrac{\sqrt{3}}{2}i, -\dfrac{1}{2} + \dfrac{\sqrt{3}}{2}i\right\}$ **67.** $\{2, -1 - \sqrt{3}i, -1 + \sqrt{3}i\}$ **69.** $\{-2, 2, -2i, 2i\}$ **71.** $\{-3i, -2i, 2i, 3i\}$ **73.** Two complex solutions that are conjugates of each other **75.** Two unequal real solutions **77.** A repeated real solution **79.** $2 - 3i$ **81.** 6 **83.** 25 **85.** $2 + 3i$ ohms
87. $z + \bar{z} = (a + bi) + (a - bi) = 2a; z - \bar{z} = (a + bi) - (a - bi) = 2bi$
89. $\overline{z + w} = \overline{(a + bi) + (c + di)} = \overline{(a + c) + (b + d)i} = (a + c) - (b + d)i = (a - bi) + (c - di) = \bar{z} + \bar{w}$

1.4 Assess Your Understanding *(page 117)*

4. extraneous **5.** quadratic in form **6.** T **7.** $\{1\}$ **9.** No real solution **11.** $\{-13\}$ **13.** $\{4\}$ **15.** $\{-1\}$ **17.** $\{0, 64\}$ **19.** $\{3\}$ **21.** $\{2\}$ **23.** $\left\{-\dfrac{8}{5}\right\}$

25. $\{8\}$ **27.** $\{-1, 3\}$ **29.** $\{1, 5\}$ **31.** $\{1\}$ **33.** $\{5\}$ **35.** $\{2\}$ **37.** $\{-4, 4\}$ **39.** $\{0, 3\}$ **41.** $\{-2, -1, 1, 2\}$ **43.** $\{-1, 1\}$ **45.** $\{-2, 1\}$ **47.** $\{-6, -5\}$

49. $\left\{-\dfrac{1}{3}\right\}$ **51.** $\left\{-\dfrac{3}{2}, 2\right\}$ **53.** $\left\{0, \dfrac{1}{16}\right\}$ **55.** $\{16\}$ **57.** $\{1\}$ **59.** $\left\{\left(\dfrac{9 - \sqrt{17}}{8}\right)^4, \left(\dfrac{9 + \sqrt{17}}{8}\right)^4\right\}$ **61.** $\{\sqrt{2}, \sqrt{3}\}$ **63.** $\{-4, 1\}$ **65.** $\left\{-2, -\dfrac{1}{2}\right\}$

67. $\left\{-\dfrac{3}{2}, \dfrac{1}{3}\right\}$ **69.** $\left\{-\dfrac{1}{8}, 27\right\}$ **71.** $\left\{-2, -\dfrac{4}{5}\right\}$ **73.** $\{-3, 0, 3\}$ **75.** $\left\{0, \dfrac{3}{4}\right\}$ **77.** $\{-5, 0, 4\}$ **79.** $\{-1, 1\}$ **81.** $\{-2, 2, 3\}$ **83.** $\left\{-2, \dfrac{1}{2}, 2\right\}$

85. $\left\{\dfrac{2}{5}\right\}$ **87.** $\left\{0, \dfrac{5}{2}, 3\right\}$ **89.** $\{0.34, 11.66\}$ **91.** $\{-1.03, 1.03\}$ **93.** $\{-1.85, 0.17\}$ **95.** $\left\{\dfrac{3}{2}, 5\right\}$ **97.** The depth of the well is 229.94 ft. **99.** 220.7 ft

103. -1 is extraneous.

1.5 Assess Your Understanding (page 127)

3. negative **4.** closed interval **5.** Multiplication Properties **6.** T **7.** T **8.** T **9.** F **10.** T **11.** $[0, 2]$; $0 \le x \le 2$ **13.** $[2, \infty)$; $x \ge 2$
15. $[0, 3)$; $0 \le x < 3$ **17.** (a) $6 < 8$ (b) $-2 < 0$ (c) $9 < 15$ (d) $-6 > -10$ **19.** (a) $7 > 0$ (b) $-1 > -8$ (c) $12 > -9$
(d) $-8 < 6$ **21.** (a) $2x + 4 < 5$ (b) $2x - 4 < -3$ (c) $6x + 3 < 6$ (d) $-4x - 2 > -4$

23. $[0, 4]$

25. $[4, 6)$

27. $[4, \infty)$

29. $(-\infty, -4)$

31. $2 \le x \le 5$

33. $-3 < x < -2$

35. $x \ge 4$

37. $x < -3$

39. $<$ **41.** $>$ **43.** \ge **45.** $<$ **47.** \le **49.** $>$ **51.** \ge
53. $\{x \mid x < 4\}$ or $(-\infty, 4)$ **55.** $\{x \mid x \ge -1\}$ or $[-1, \infty)$ **57.** $\{x \mid x > 3\}$ or $(3, \infty)$ **59.** $\{x \mid x \ge 2\}$ or $[2, \infty)$

61. $\{x \mid x > -7\}$ or $(-7, \infty)$ **63.** $\left\{x \mid x \le \dfrac{2}{3}\right\}$ or $\left(-\infty, \dfrac{2}{3}\right]$ **65.** $\{x \mid x < -20\}$ or $(-\infty, -20)$ **67.** $\left\{x \mid x \ge \dfrac{4}{3}\right\}$ or $\left[\dfrac{4}{3}, \infty\right)$

69. $\{x \mid 3 \le x \le 5\}$ or $[3, 5]$ **71.** $\left\{x \mid \dfrac{2}{3} \le x \le 3\right\}$ or $\left[\dfrac{2}{3}, 3\right]$ **73.** $\left\{x \mid -\dfrac{11}{2} < x < \dfrac{1}{2}\right\}$ or $\left(-\dfrac{11}{2}, \dfrac{1}{2}\right)$ **75.** $\{x \mid -6 < x < 0\}$ or $(-6, 0)$

77. $\{x \mid x < -5\}$ or $(-\infty, -5)$ **79.** $\{x \mid x \ge -1\}$ or $[-1, \infty)$ **81.** $\left\{x \mid \dfrac{1}{2} \le x < \dfrac{5}{4}\right\}$ or $\left[\dfrac{1}{2}, \dfrac{5}{4}\right)$ **83.** $\left\{x \mid x < -\dfrac{1}{2}\right\}$ or $\left(-\infty, -\dfrac{1}{2}\right)$

85. $\left\{x \mid x > \dfrac{10}{3}\right\}$ or $\left(\dfrac{10}{3}, \infty\right)$ **87.** $\{x \mid x > 3\}$ or $(3, \infty)$

89. $a = 3, b = 5$ **91.** $a = -12, b = -8$ **93.** $a = 3, b = 11$
95. $a = \dfrac{1}{4}, b = 1$ **97.** $a = 4, b = 16$ **99.** $\{x \mid x \ge -2\}$ **101.** $21 < \text{Age} < 30$

103. (a) Male ≥ 76.6 years (b) Female ≥ 81.03 years (c) A female can
expect to live 4.43 years longer.

105. The agent's commission ranges from \$45,000 to \$95,000, inclusive. As a percent of selling price, the commission ranges from 5% to 8.6%, inclusive.
107. The amount withheld varies from \$84.10 to \$134.10, inclusive. **109.** The usage varied from 675 kW · hr to 2500 kW · hr, inclusive.
111. The dealer's cost varies from \$15,254.24 to \$16,071.43, inclusive.
113. (a) You need at least a 74 on the fifth test. (b) You need at least a 77 on the fifth test.
115. $\dfrac{a + b}{2} - a = \dfrac{a + b - 2a}{2} = \dfrac{b - a}{2} > 0$; therefore, $a < \dfrac{a + b}{2}$. $b - \dfrac{a + b}{2} = \dfrac{2b - a - b}{2} = \dfrac{b - a}{2} > 0$; therefore, $b > \dfrac{a + b}{2}$.
117. $(\sqrt{ab})^2 - a^2 = ab - a^2 = a(b - a) > 0$; thus $(\sqrt{ab})^2 > a^2$ and $\sqrt{ab} > a$.
 $b^2 - (\sqrt{ab})^2 = b^2 - ab = b(b - a) > 0$; thus $b^2 > (\sqrt{ab})^2$ and $b > \sqrt{ab}$.
119. $h - a = \dfrac{2ab}{a + b} - a = \dfrac{ab - a^2}{a + b} = \dfrac{a(b - a)}{a + b} > 0$; thus $h > a$. $b - h = b - \dfrac{2ab}{a + b} = \dfrac{b^2 - ab}{a + b} = \dfrac{b(b - a)}{a + b} > 0$; thus $h < b$.

121. Since $0 < a < b$, then $a - b < 0$ and $\dfrac{a - b}{ab} < 0$. So $\dfrac{a}{ab} - \dfrac{b}{ab} < 0$, or $\dfrac{1}{b} - \dfrac{1}{a} < 0$. Therefore, $\dfrac{1}{b} < \dfrac{1}{a}$. And $0 < \dfrac{1}{b}$ because $b > 0$.

1.6 Assess Your Understanding (page 132)

3. $\{-5, 5\}$ **4.** $\{x \mid -5 < x < 5\}$ **5.** T **6.** T **7.** $\{-3, 3\}$ **9.** $\{-4, 1\}$ **11.** $\left\{-1, \dfrac{3}{2}\right\}$ **13.** $\{-4, 4\}$ **15.** $\{2\}$ **17.** $\left\{-\dfrac{27}{2}, \dfrac{27}{2}\right\}$ **19.** $\left\{-\dfrac{36}{5}, \dfrac{24}{5}\right\}$

21. No solution **23.** $\left\{-\dfrac{1}{2}, \dfrac{1}{2}\right\}$ **25.** $\{-3, 3\}$ **27.** $\{-1, 3\}$ **29.** $\{-2, -1, 0, 1\}$ **31.** $\left\{\dfrac{8}{7}, 4\right\}$ **33.** $\left\{-\dfrac{1}{2}, 0\right\}$

35. $\{x \mid -4 < x < 4\}$; $(-4, 4)$ **37.** $\{x \mid x < -4 \text{ or } x > 4\}$; $(-\infty, -4) \cup (4, \infty)$ **39.** $\{x \mid 1 < x < 3\}$; $(1, 3)$ **41.** $\left\{t \mid -\dfrac{2}{3} \le t \le 2\right\}$; $\left[-\dfrac{2}{3}, 2\right]$

43. $\left\{x \mid x \leq \dfrac{1}{2} \text{ or } x \geq \dfrac{5}{2}\right\}$; $\left(-\infty, \dfrac{1}{2}\right] \cup \left[\dfrac{5}{2}, \infty\right)$ **45.** $\left\{x \mid -1 < x < \dfrac{3}{2}\right\}$; $\left(-1, \dfrac{3}{2}\right)$ **47.** $\{x \mid x < -1 \text{ or } x > 2\}$; $(-\infty, -1)$ or $(2, \infty)$

49. No solution

51. $\left\{x \mid x < -\dfrac{3}{2} \text{ or } x > \dfrac{3}{2}\right\}$; $\left(-\infty, -\dfrac{3}{2}\right) \cup \left(\dfrac{3}{2}, \infty\right)$ **53.** $\{x \mid -1 \leq x \leq 2\}$; $[-1, 2]$

55. No solution **57.** All real numbers; $(-\infty, \infty)$ **59.** $\left\{x \mid -\dfrac{9}{4} < x < \dfrac{3}{4}\right\}$; $\left(-\dfrac{9}{4}, \dfrac{3}{4}\right)$ **61.** All real numbers; $(-\infty, \infty)$

63. $|x - 98.6| \geq 1.5$; $x \leq 97.1$ or $x \geq 100.1$ **65.** $|x - 13.4| < 1.35$; between 12.05 and 14.75 books per year are read. **67.** $|x - 3| < \dfrac{1}{2}$; $\dfrac{5}{2} < x < \dfrac{7}{2}$

69. $|x + 3| > 2$; $x < -5$ or $x > -1$ **71.** $a = 2, b = 8$ **73.** $a = -15, b = -7$ **75.** $a = -1, b = -\dfrac{1}{15}$ **77.** $b - a = (\sqrt{b} - \sqrt{a})(\sqrt{b} + \sqrt{a})$.

Since $\sqrt{b} - \sqrt{a} > 0$, $\sqrt{a} > 0$, $\sqrt{b} > 0$, then $b - a > 0$, so $a < b$. **79.** $(a + b)^2 = a^2 + 2ab + b^2 \leq |a|^2 + 2|a||b| + |b|^2 = (|a| + |b|)^2$;

thus, $|a + b| \leq |a| + |b|$. **81.** $x^2 - a < 0$; $(x - \sqrt{a})(x + \sqrt{a}) < 0$; therefore, $-\sqrt{a} < x < \sqrt{a}$. **83.** $\{x \mid -1 < x < 1\}$ **85.** $\{x \mid x \leq -3 \text{ or } x \geq 3\}$

87. $\{x \mid -4 \leq x \leq 4\}$ **89.** $\{x \mid x < -2 \text{ or } x > 2\}$ **91.** $\{-1, 5\}$

1.7 Assess Your Understanding (page 140)

1. mathematical modeling **2.** interest **3.** uniform motion **4.** F **5.** T **6.** $100 - x$ **7.** $A = \pi r^2$; r = radius, A = area
9. $A = s^2$; A = area, s = length of a side **11.** $F = ma$; F = force, m = mass, a = acceleration **13.** $W = Fd$; W = work, F = force, d = distance
15. $C = 150x$; C = total variable cost, x = number of dishwashers **17.** Invest \$31,250 in bonds and \$18,750 in CDs.
19. \$11,600 was loaned out at 8%. **21.** Mix 75 lb of Earl Grey tea with 25 lb of Orange Pekoe tea. **23.** Mix 160 lb of cashews with the almonds.
25. The speed of the current is 2.286 mi/hr. **27.** The speed of the current is 5 mi/hr. **29.** Karen walked at 4.05 ft/sec. **31.** A doubles tennis court
is 78 feet long and 36 feet wide. **33.** Working together, it takes 12 min. **35. (a)** The dimensions are 10 ft by 5 ft. **(b)** The area is 50 sq ft. **(c)** The
dimensions would be 7.5 ft by 7.5 ft. **(d)** The area would be 56.25 sq ft. **37.** The defensive back catches up to the tight end at the tight end's 45-yd
line. **39.** Add $\dfrac{2}{3}$ gal of water. **41.** Evaporate 10.67 oz of water. **43.** 40 g of 12-karat gold should be mixed with 20 g of pure gold. **45.** Mike passes
Dan $\dfrac{1}{3}$ mile from the start, 2 min from the time Mike started to run. **47.** Start the auxiliary pump at 9:45 AM. **49.** The tub will fill in 1 hr.
51. Run: 12 miles; bicycle: 75 miles **53.** Lewis would beat Burke by 16.75 m. **55.** Set the original price at \$40. At 50% off, there will be no profit.
59. The tail wind was 91.47 knots.

Review Exercises (page 145)

1. $\{-18\}$ **3.** $\{6\}$ **5.** $\left\{\dfrac{1}{5}\right\}$ **7.** $\{6\}$ **9.** No real solution **11.** $\left\{\dfrac{11}{8}\right\}$ **13.** $\left\{-2, \dfrac{3}{2}\right\}$ **15.** $\left\{\dfrac{1 - \sqrt{13}}{4}, \dfrac{1 + \sqrt{13}}{4}\right\}$ **17.** $\{-3, 3\}$ **19.** No real solution

21. $\{-2, -1, 1, 2\}$ **23.** $\{2\}$ **25.** $\left\{\dfrac{13}{2}\right\}$ **27.** $\left\{\dfrac{\sqrt{5}}{2}\right\}$ **29.** $\left\{\dfrac{9}{4}\right\}$ **31.** $\left\{-1, \dfrac{1}{2}\right\}$ **33.** $\left\{\dfrac{m}{1 - n}, \dfrac{m}{1 + n}\right\}$ **35.** $\left\{-\dfrac{9b}{5a}, \dfrac{2b}{a}\right\}$ **37.** $\left\{-\dfrac{9}{5}\right\}$ **39.** $\{-5, 2\}$

41. $\left\{-\dfrac{5}{3}, 3\right\}$ **43.** $\left\{0, \dfrac{3}{2}\right\}$ **45.** $\left\{-\dfrac{5}{2}, -2, 2\right\}$

47. $\{x \mid x \geq 14\}$; $[14, \infty)$ **49.** $\left\{x \mid -\dfrac{31}{2} \leq x \leq \dfrac{33}{2}\right\}$; $\left[-\dfrac{31}{2}, \dfrac{33}{2}\right]$ **51.** $\{x \mid -23 < x < -7\}$; $(-23, -7)$ **53.** $\left\{x \mid -\dfrac{3}{2} < x < -\dfrac{7}{6}\right\}$; $\left(-\dfrac{3}{2}, -\dfrac{7}{6}\right)$

55. $\{x \mid x \leq -2 \text{ or } x \geq 7\}$; $(-\infty, -2] \cup [7, \infty)$ **57.** $\left\{x \mid 0 \leq x \leq \dfrac{4}{3}\right\}$; $\left[0, \dfrac{4}{3}\right]$ **59.** $\left\{x \mid x < -1 \text{ or } x > \dfrac{7}{3}\right\}$; $\left(-\infty, -1\right) \cup \left(\dfrac{7}{3}, \infty\right)$

61. $4 + 7i$ **63.** $-3 + 2i$ **65.** $\dfrac{9}{10} - \dfrac{3}{10}i$ **67.** -1 **69.** $-46 + 9i$ **71.** $\left\{-\dfrac{1}{2} - \dfrac{\sqrt{3}}{2}i, -\dfrac{1}{2} + \dfrac{\sqrt{3}}{2}i\right\}$ **73.** $\left\{\dfrac{-1 - \sqrt{17}}{4}, \dfrac{-1 + \sqrt{17}}{4}\right\}$

75. $\left\{\dfrac{1}{2} - \dfrac{\sqrt{11}}{2}i, \dfrac{1}{2} + \dfrac{\sqrt{11}}{2}i\right\}$ **77.** $\left\{\dfrac{1}{2} - \dfrac{\sqrt{23}}{2}i, \dfrac{1}{2} + \dfrac{\sqrt{23}}{2}i\right\}$ **79.** $p = 2l + 2w$ **81.** The interest is \$630. **83.** The storm is 3300 ft away.
85. The search plane can go as far as 616 miles. **87.** The helicopter will reach the life raft in a little less than 1 hr 35 min.
89. The Metra commuter averages 30 mi/hr; the Amtrak averages 80 mi/hr. **91.** It takes Clarissa 10 days by herself. **93.** Add 256 oz of water.
95. 5 cm and 12 cm **97.** Mix 90 cm³ of 15% HCl with the 60 cm³ of 40% of HCl to obtain 150 cm³ of 25% HCl. **99.** It will take the smaller pump 2 hr.
101. Scott receives \$400,000, Alice receives \$300,000, and Tricia receives \$200,000. **103.** It would take the older copier 180 min or 3 hr.
105. The freight train is 190.67 feet long.

Chapter Test *(page 147)*

1. $\left\{\dfrac{5}{2}\right\}$ 2. $\{-2, 3\}$ 3. $\{-2, 2\}$ 4. $\left\{\dfrac{9}{2}\right\}$ 5. $\{0, 3\}$ 6. $\left\{-2, -\dfrac{2}{3}, 2\right\}$ 7. No real solution

8. $\left[-\dfrac{2}{3}, \dfrac{16}{3}\right]$

9. $\left(-4, \dfrac{4}{3}\right)$

10. $(-\infty, -1] \cup [6, \infty)$

11. $-\dfrac{3}{5} - \dfrac{1}{5}i$ 12. $\left\{\dfrac{1}{2} - i, \dfrac{1}{2} + i\right\}$

13. Add $6\dfrac{2}{3}$ lb of \$8/lb coffee to get $26\dfrac{2}{3}$ lb of \$5/lb coffee.

CHAPTER 2 Graphs

2.1 Assess Your Understanding *(page 154)*

7. *x*-coordinate or abscissa; *y*-coordinate or ordinate 8. quadrants 9. midpoint 10. F 11. F 12. T

13. (a) Quadrant II (b) *x*-axis 15. The points will be on a 17. $\sqrt{5}$ 29. $d(A, B) = \sqrt{13}$
(c) Quadrant III (d) Quadrant I vertical line that is 2 units 19. $\sqrt{10}$ $d(B, C) = \sqrt{13}$
(e) *y*-axis (f) Quadrant IV to the right of the *y*-axis. 21. $2\sqrt{17}$ $d(A, C) = \sqrt{26}$
23. $\sqrt{85}$ $(\sqrt{13})^2 + (\sqrt{13})^2 = (\sqrt{26})^2$
25. $\sqrt{53}$ Area $= \dfrac{13}{2}$ square units
27. $\sqrt{a^2 + b^2}$

31. $d(A, B) = \sqrt{130}$ 33. $d(A, B) = 4$ 35. $(4, 0)$ 37. $\left(\dfrac{3}{2}, 1\right)$
$d(B, C) = \sqrt{26}$ $d(B, C) = \sqrt{41}$
$d(A, C) = 2\sqrt{26}$ $d(A, C) = 5$ 39. $(5, -1)$ 41. $\left(\dfrac{a}{2}, \dfrac{b}{2}\right)$
$(\sqrt{26})^2 + (2\sqrt{26})^2 = (\sqrt{130})^2$ $4^2 + 5^2 = (\sqrt{41})^2$
Area $= 26$ square units Area $= 10$ square units 43. $(5, 3)$ 45. $(3, -13), (3, 11)$

47. $(4 + 3\sqrt{3}, 0); (4 - 3\sqrt{3}, 0)$ 49. $(1, 2)$ 51. $\sqrt{17}; 2\sqrt{5}; \sqrt{29}$ 53. $\left(\dfrac{s}{2}, \dfrac{s}{2}\right)$ 55. $d(P_1, P_2) = 6; d(P_2, P_3) = 4; d(P_1, P_3) = 2\sqrt{13}$; right triangle

57. $d(P_1, P_2) = 2\sqrt{17}; d(P_2, P_3) = \sqrt{34}; d(P_1, P_3) = \sqrt{34}$; isosceles right triangle 59. $90\sqrt{2} \approx 127.28$ ft 61. (a) $(90, 0), (90, 90), (0, 90)$
(b) $5\sqrt{2161} \approx 232.43$ ft (c) $30\sqrt{149} \approx 366.20$ ft 63. $d = 50t$ mi 65. (a) $(2.65, 1.6)$ (b) Approximately 1.285 units

2.2 Assess Your Understanding *(page 164)*

3. intercepts 4. $y = 0$ 5. *y*-axis 6. 4 7. $(-3, 4)$ 8. T 9. F 10. F 11. $(0, 0)$ is on the graph. 13. $(0, 3)$ is on the graph.
15. $(0, 2)$ and $(\sqrt{2}, \sqrt{2})$ are on the graph.

17. $(-2, 0), (0, 2)$ 19. $(-4, 0), (0, 8)$ 21. $(-1, 0), (1, 0), (0, -1)$ 23. $(-2, 0), (2, 0), (0, 4)$ 25. $(3, 0), (0, 2)$

27. $(-2, 0), (2, 0), (0, 9)$ 29. 31. 33. 35.

37. 39. (a) $(-1, 0), (1, 0)$ 41. (a) $\left(-\dfrac{\pi}{2}, 0\right), (0, 1), \left(\dfrac{\pi}{2}, 0\right)$ 43. (a) $(0, 0)$
(b) Symmetric with respect to the (b) Symmetric with respect to
x-axis, the *y*-axis, and the origin (b) Symmetric with respect to the *y*-axis the *x*-axis
45. (a) $(-2, 0), (0, 0), (2, 0)$ 47. (a) $(x, 0), -2 \le x \le 1$ 49. (a) No intercepts
(b) Symmetric with respect to the origin (b) No symmetry (b) Symmetric with respect to the origin

51. 53. 55. $(-4, 0), (0, -2), (0, 2)$; symmetric with respect to the *x*-axis 57. $(0, 0)$; symmetric with
respect to the origin 59. $(0, 9), (3, 0), (-3, 0)$; symmetric with respect to the *y*-axis
61. $(-2, 0), (2, 0), (0, -3), (0, 3)$; symmetric with respect to the *x*-axis, *y*-axis, and origin
63. $(0, -27), (3, 0)$; no symmetry 65. $(0, -4), (4, 0), (-1, 0)$; no symmetry 67. $(0, 0)$;
symmetric with respect to the origin 69. $(0, 0)$; symmetric with respect to the origin

71. **73.**

75. $b = 13$ **77.** $a = -4$ or $a = 1$ **79.** $(-1, -2)$ **81.** 4 **83.** (a) $(0, 0), (2, 0), (0, 1), (0, -1)$
(b) x-axis symmetry **85.** (a) $y = \sqrt{x^2}$ and $y = |x|$ have the same graph. (b) $\sqrt{x^2} = |x|$
(c) $x \geq 0$ for $y = (\sqrt{x})^2$, while x can be any real number for $y = x$.
(d) $y \geq 0$ for $y = \sqrt{x^2}$

2.3 Assess Your Understanding (page 178)

1. undefined; 0 **2.** 3; 2 **3.** $y = b$; y-intercept **4.** T **5.** F **6.** T **7.** $m_1 = m_2$; y-intercepts; $m_1 m_2 = -1$ **8.** 2 **9.** $-\dfrac{1}{2}$ **10.** False

11. (a) Slope $= \dfrac{1}{2}$ **13.** (a) Slope $= -\dfrac{1}{3}$ **15.** Slope $= -\dfrac{3}{2}$ **17.** Slope $= -\dfrac{1}{2}$ **19.** Slope $= 0$ **21.** Slope undefined

(b) If x increases by 2 units, y will increase by 1 unit. (b) If x increases by 3 units, y will decrease by 1 unit.

23. **25.** **27.** **29.**

31. $(2, 6); (3, 10); (4, 14)$
33. $(4, -7); (6, -10); (8, -13)$
35. $(-1, -5); (0, -7); (1, -9)$
37. $x - 2y = 0$ or $y = \dfrac{1}{2}x$

39. $x + y = 2$ or $y = -x + 2$ **41.** $2x - y = 3$ or $y = 2x - 3$ **43.** $x + 2y = 5$ or $y = -\dfrac{1}{2}x + \dfrac{5}{2}$ **45.** $3x - y = -9$ or $y = 3x + 9$

47. $2x + 3y = -1$ or $y = -\dfrac{2}{3}x - \dfrac{1}{3}$ **49.** $x - 2y = -5$ or $y = \dfrac{1}{2}x + \dfrac{5}{2}$ **51.** $3x + y = 3$ or $y = -3x + 3$ **53.** $x - 2y = 2$ or $y = \dfrac{1}{2}x - 1$

55. $x = 2$; no slope–intercept form **57.** $y = 2$ **59.** $2x - y = -4$ or $y = 2x + 4$ **61.** $2x - y = 0$ or $y = 2x$ **63.** $x = 4$; no slope–intercept form

65. $2x + y = 0$ or $y = -2x$ **67.** $x - 2y = -3$ or $y = \dfrac{1}{2}x + \dfrac{3}{2}$ **69.** $y = 4$

71. Slope $= 2$; y-intercept $= 3$ **73.** Slope $= 2$; y-intercept $= -2$ **75.** Slope $= \dfrac{1}{2}$; y-intercept $= 2$ **77.** Slope $= -\dfrac{1}{2}$; y-intercept $= 2$

 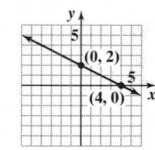

79. Slope $= \dfrac{2}{3}$; y-intercept $= -2$ **81.** Slope $= -1$; y-intercept $= 1$ **83.** Slope undefined; no y-intercept **85.** Slope $= 0$; y-intercept $= 5$

 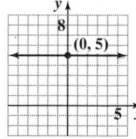

87. Slope $= 1$; y-intercept $= 0$ **89.** Slope $= \dfrac{3}{2}$; y-intercept $= 0$ **91.** (a) x-intercept: 3; y-intercept: 2 **93.** (a) x-intercept: -10; y-intercept: 8
(b) (b)

95. (a) x-intercept: 3; y-intercept: $\dfrac{21}{2}$ **97.** (a) x-intercept: 2; y-intercept: 3 **99.** (a) x-intercept: 5; y-intercept: -2 **101.** $y = 0$
(b) (b) (b) **103.** Parallel
105. Neither
107. $x - y = -2$ or $y = x + 2$
109. $x + 3y = 3$ or $y = -\dfrac{1}{3}x + 1$

111. $P_1 = (-2, 5), P_2 = (1, 3), m_1 = -\dfrac{2}{3}; P_2 = (1, 3), P_3 = (-1, 0), m_2 = \dfrac{3}{2};$ because $m_1 m_2 = -1$, the lines are perpendicular and the
points $(-2, 5), (1, 3),$ and $(-1, 0)$ are the vertices of a right triangle; thus, the points $P_1, P_2,$ and P_3 are the vertices of a right triangle.

113. $P_1 = (-1, 0), P_2 = (2, 3), m = 1; P_3 = (1, -2), P_4 = (4, 1), m = 1; P_1 = (-1, 0), P_3 = (1, -2), m = -1; P_2 = (2, 3), P_4 = (4, 1), m = -1;$
opposite sides are parallel, and adjacent sides are perpendicular; the points are the vertices of a rectangle.

115. $C = 0.20x + 29$; $51.00; $75.00 **117.** $C = 0.15x + 1289$

119. (a) $C = 0.0944x + 10.55, 0 \le x \le 600$ **121.** $^\circ C = \dfrac{5}{9}(^\circ F - 32)$; approximately 21.1°C **123. (a)** $y = -\dfrac{2}{25}x + 30$ **(b)** x-intercept: 375;

(b)

The ramp meets the floor 375 in. (31.25 ft) from the base of the platform. **(c)** The ramp does not meet design requirements. It has a run of 31.25 ft long. **(d)** The only slope possible for the ramp to comply with the requirement is for it to drop 1 in. for every 12-in. run.

125. (a) $A = \dfrac{1}{5}x + 20,000$ **(b)** $80,000 **(c)** Each additional box sold requires an additional $0.20 in advertising. **127.** All have the same slope, 2; the lines are parallel.

(c) $29.43 **(d)** $57.75

(e) Each additional kW-hr used adds $0.0944 to the bill.

129. (b),(c),(e),(g) **131.** (c) **137.** No; no

139. They are the same line. **141.** Yes, if the y-intercept is 0.

2.4 Assess Your Understanding *(page 185)*

3. F **4.** radius **5.** T **6.** F **7.** Center $(2, 1)$; radius $= 2$; $(x - 2)^2 + (y - 1)^2 = 4$ **9.** Center $\left(\dfrac{5}{2}, 2\right)$; radius $= \dfrac{3}{2}$; $\left(x - \dfrac{5}{2}\right)^2 + (y - 2)^2 = \dfrac{9}{4}$

11. $x^2 + y^2 = 4$; **13.** $x^2 + (y - 2)^2 = 4$; **15.** $(x - 4)^2 + (y + 3)^2 = 25$; **17.** $(x + 2)^2 + (y - 1)^2 = 16$;
$x^2 + y^2 - 4 = 0$ $x^2 + y^2 - 4y = 0$ $x^2 + y^2 - 8x + 6y = 0$ $x^2 + y^2 + 4x - 2y - 11 = 0$

19. $\left(x - \dfrac{1}{2}\right)^2 + y^2 = \dfrac{1}{4}$; **21. (a)** $(h, k) = (0, 0)$; $r = 2$ **23. (a)** $(h, k) = (3, 0)$; $r = 2$ **25. (a)** $(h, k) = (1, 2)$; $r = 3$
$x^2 + y^2 - x = 0$ **(b)** **(b)** **(b)**

 (c) $(\pm 2, 0)$; $(0, \pm 2)$ **(c)** $(1, 0)$; $(5, 0)$ **(c)** $(1 \pm \sqrt{5}, 0)$; $(0, 2 \pm 2\sqrt{2})$

27. (a) $(h, k) = (-2, 2)$; $r = 3$ **29. (a)** $(h, k) = \left(\dfrac{1}{2}, -1\right)$; $r = \dfrac{1}{2}$ **31. (a)** $(h, k) = (3, -2)$; $r = 5$ **33. (a)** $(h, k) = (-2, 0)$; $r = 2$

(b) **(b)** **(b)** **(b)**

 (c) $(-2 \pm \sqrt{5}, 0)$; $(0, 2 \pm \sqrt{5})$ **(c)** $(3 \pm \sqrt{21}, 0)$; $(0, -6)$, $(0, 2)$ **(c)** $(0, 0)$, $(-4, 0)$

 (c) $(0, -1)$

35. $x^2 + y^2 = 13$ **37.** $(x - 2)^2 + (y - 3)^2 = 9$ **39.** $(x + 1)^2 + (y - 3)^2 = 5$ **41.** $(x + 1)^2 + (y - 3)^2 = 1$ **43.** (c) **45.** (b) **47.** 18 units2

49. $x^2 + (y - 139)^2 = 15,625$ **51.** $x^2 + y^2 + 2x + 4y - 4168.16 = 0$ **53.** $\sqrt{2}x + 4y - 9\sqrt{2} = 0$ **55.** $(1, 0)$ **57.** $y = 2$ **59.** (b), (c), (e), (g)

2.5 Assess Your Understanding *(page 191)*

1. $y = kx$ **2.** F **3.** $y = \dfrac{1}{5}x$ **5.** $A = \pi x^2$ **7.** $F = \dfrac{250}{d^2}$ **9.** $z = \dfrac{1}{5}(x^2 + y^2)$ **11.** $M = \dfrac{9d^2}{2\sqrt{x}}$ **13.** $T^2 = \dfrac{8a^3}{d^2}$ **15.** $V = \dfrac{4\pi}{3}r^3$ **17.** $A = \dfrac{1}{2}bh$

19. $F = 6.67 \times 10^{-11}\left(\dfrac{mM}{d^2}\right)$ **21.** $p = 0.00649B$; $941.05 **23.** 144 ft; 2 sec **25.** 2.25 **27.** $R = 3.95g$; $41.48 **29. (a)** $D = \dfrac{429}{p}$ **(b)** 143 bags

31. 450 cm^3 **33.** 124.76 lb **35.** $V = \pi r^2 h$ **37.** 54.86 lb **39.** $\sqrt[3]{6} \approx 1.82$ in. **41.** 2812.5 joules **43.** 384 psi

Review Exercises *(page 195)*

1. (a) $2\sqrt{5}$ **(b)** $(2, 1)$ **(c)** $\dfrac{1}{2}$ **(d)** For each run of 2, there is a rise of 1. **3. (a)** 5 **(b)** $\left(-\dfrac{1}{2}, 1\right)$ **(c)** $-\dfrac{4}{3}$ **(d)** For each run of 3, there is a rise of -4.

5. (a) 12 **(b)** $(4, 2)$ **(c)** undefined **(d)** no change in x

7.

 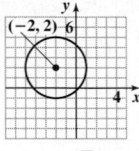

9. $(0, 0)$; symmetric with respect to the x-axis **11.** $(\pm 4, 0)$, $(0, \pm 2)$; symmetric with respect to the x-axis, y-axis, and origin
13. $(0, 1)$; symmetric with respect to the y-axis **15.** $(0, 0)$, $(-1, 0)$, $(0, -2)$; no symmetry
17. $(x + 2)^2 + (y - 3)^2 = 16$ **19.** $(x + 1)^2 + (y + 2)^2 = 1$

21. Center $(0, 1)$; radius $= 2$

Intercepts: $(-\sqrt{3}, 0), (\sqrt{3}, 0),$
$(0, -1), (0, 3)$

23. Center $(1, -2)$; radius $= 3$

Intercepts: $(1 - \sqrt{5}, 0), (1 + \sqrt{5}, 0),$
$(0, -2 - 2\sqrt{2}), (0, -2 + 2\sqrt{2})$

25. Center $(1, -2)$; radius $= \sqrt{5}$

Intercepts: $(0, 0), (2, 0), (0, -4)$

27. $2x + y = 5$ or $y = -2x + 5$

29. $x = -3$; no slope–intercept form

31. $x + 5y = -10$ or $y = -\dfrac{1}{5}x - 2$

33. $2x - 3y = -19$ or $y = \dfrac{2}{3}x + \dfrac{19}{3}$

35. $x - y = 7$ or $y = x - 7$

37. Slope $= \dfrac{4}{5}$; y-intercept $= 4$ **39.** Slope $= \dfrac{3}{2}$; y-intercept $= \dfrac{1}{2}$ **41.** Intercepts: $(6, 0), (0, -4)$ **43.** Intercepts: $(4, 0), (0, 6)$ **45.**

47.

49. (a) $d(A, B) = 2\sqrt{5}$; $d(B, C) = \sqrt{145}$; $d(A, C) = 5\sqrt{5}$; $[d(A, B)]^2 + [d(A, C)]^2 = (2\sqrt{5})^2 + (5\sqrt{5})^2 =$
$20 + 125 = 145 = [d(B, C)]^2$ (b) Slope from A to B is -2; slope from A to C is $\dfrac{1}{2}$. Since $(-2)\left(\dfrac{1}{2}\right) = -1$,
the lines are perpendicular. **51.** Slope from A to B is -1; slope from A to C is -1. **53.** $R = 3.59g$; $\$40.21$
55. $a \approx 36$ million miles

Chapter Test *(page 197)*

1. $d = 2\sqrt{13}$ **2.** $(2, 1)$ **3.** (a) $m = -\dfrac{2}{3}$ (b) For every 3-unit change in x, y will change by -2 units.

4.

5.

6. Intercepts:
$(-3, 0), (3, 0), (0, 9)$;
symmetric with
respect to the y-axis

7. $y = -2x + 2$

8. $x^2 + y^2 - 8x + 6y = 0$

9. Center: $(-2, 1)$;
radius: 3

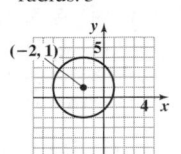

10. Parallel line: $y = -\dfrac{2}{3}x - \dfrac{1}{3}$; perpendicular line: $y = \dfrac{3}{2}x + 3$ **11.** 14.69 ohms

Cumulative Review *(page 197)*

1. $\left\{\dfrac{5}{3}\right\}$ **2.** $\{-3, 4\}$ **3.** $\left\{-\dfrac{1}{2}, 3\right\}$ **4.** $\{1 - \sqrt{3}, 1 + \sqrt{3}\}$ **5.** No real solution **6.** $\{4\}$ **7.** $\{1, 3\}$ **8.** $\{-2 - 2\sqrt{2}, -2 + 2\sqrt{2}\}$ **9.** $\{-3i, 3i\}$

10. $\{1 - 2i, 1 + 2i\}$ **11.** $\{x | x \le 5\}$ or $(\infty, 5]$; **12.** $\{x | -5 < x < 1\}$ or $(-5, 1)$;

13. $\{x | 1 \le x \le 3\}$ or $[1, 3]$; **14.** $\{x | x < -5 \text{ or } x > 1\}$ or $(\infty, -5) \cup (1, \infty)$;

15. $5\sqrt{2}; \left(\dfrac{3}{2}, \dfrac{1}{2}\right)$ **16.** (a), (b) **17.**

18. $y = -2x + 2$ **19.** $y = -\dfrac{1}{2}x + \dfrac{13}{2}$;

20.

CHAPTER 3 Functions and Their Graphs

3.1 Assess Your Understanding *(page 210)*

5. independent; dependent **6.** range **7.** $[0, 5]$ **8.** \ne; f; g **9.** $(g - f)(x)$ **10.** F **11.** T **12.** T **13** F **14.** F
15. Function; Domain: {Elvis, Colleen, Kaleigh, Marissa}; Range: {January 8, March 15, September 17} **17.** Not a function **19.** Not a function
21. Function; Domain: $\{1, 2, 3, 4\}$, Range: $\{3\}$ **23.** Not a function **25.** Function; Domain: $\{-2, -1, 0, 1\}$; Range: $\{0, 1, 4\}$ **27.** Function **29.** Function
31. Not a function **33.** Not a function **35.** Function **37.** Not a function **39.** (a) -4 (b) 1 (c) -3 (d) $3x^2 - 2x - 4$ (e) $-3x^2 - 2x + 4$
(f) $3x^2 + 8x + 1$ (g) $12x^2 + 4x - 4$ (h) $3x^2 + 6xh + 3h^2 + 2x + 2h - 4$ **41.** (a) 0 (b) $\dfrac{1}{2}$ (c) $-\dfrac{1}{2}$ (d) $\dfrac{-x}{x^2 + 1}$ (e) $\dfrac{-x}{x^2 + 1}$
(f) $\dfrac{x + 1}{x^2 + 2x + 2}$ (g) $\dfrac{2x}{4x^2 + 1}$ (h) $\dfrac{x + h}{x^2 + 2xh + h^2 + 1}$ **43.** (a) 4 (b) 5 (c) 5 (d) $|x| + 4$ (e) $-|x| - 4$ (f) $|x + 1| + 4$ (g) $2|x| + 4$
(h) $|x + h| + 4$ **45.** (a) $-\dfrac{1}{5}$ (b) $-\dfrac{3}{2}$ (c) $\dfrac{1}{8}$ (d) $\dfrac{2x - 1}{3x + 5}$ (e) $\dfrac{-2x - 1}{3x - 5}$ (f) $\dfrac{2x + 3}{3x - 2}$ (g) $\dfrac{4x + 1}{6x - 5}$ (h) $\dfrac{2x + 2h + 1}{3x + 3h - 5}$ **47.** All real numbers
49. All real numbers **51.** $\{x | x \ne -4, x \ne 4\}$ **53.** $\{x | x \ne 0\}$ **55.** $\{x | x \ge 4\}$ **57.** $\{x | x > 9\}$ **59.** $\{x | x > 1\}$ **61.** $\{t | t \ge 4, t \ne 7\}$

63. (a) $(f + g)(x) = 5x + 1$; All real numbers **(b)** $(f - g)(x) = x + 7$; All real numbers **(c)** $(f \cdot g)(x) = 6x^2 - x - 12$; All real numbers

(d) $\left(\dfrac{f}{g}\right)(x) = \dfrac{3x + 4}{2x - 3}$; $\left\{x \middle| x \neq \dfrac{3}{2}\right\}$ **(e)** 16 **(f)** 11 **(g)** 10 **(h)** -7 **65. (a)** $(f + g)(x) = 2x^2 + x - 1$; All real numbers

(b) $(f - g)(x) = -2x^2 + x - 1$; All real numbers **(c)** $(f \cdot g)(x) = 2x^3 - 2x^2$; All real numbers **(d)** $\left(\dfrac{f}{g}\right)(x) = \dfrac{x - 1}{2x^2}$; $\{x | x \neq 0\}$

(e) 20 **(f)** -29 **(g)** 8 **(h)** 0 **67. (a)** $(f + g)(x) = \sqrt{x} + 3x - 5$; $\{x | x \geq 0\}$ **(b)** $(f - g)(x) = \sqrt{x} - 3x + 5$; $\{x | x \geq 0\}$

(c) $(f \cdot g)(x) = 3x\sqrt{x} - 5\sqrt{x}$; $\{x | x \geq 0\}$ **(d)** $\left(\dfrac{f}{g}\right)(x) = \dfrac{\sqrt{x}}{3x - 5}$; $\left\{x \middle| x \geq 0, x \neq \dfrac{5}{3}\right\}$ **(e)** $\sqrt{3} + 4$ **(f)** -5 **(g)** $\sqrt{2}$ **(h)** $-\dfrac{1}{2}$

69. (a) $(f + g)(x) = 1 + \dfrac{2}{x}$; $\{x | x \neq 0\}$ **(b)** $(f - g)(x) = 1$; $\{x | x \neq 0\}$ **(c)** $(f \cdot g)(x) = \dfrac{1}{x} + \dfrac{1}{x^2}$; $\{x | x \neq 0\}$ **(d)** $\left(\dfrac{f}{g}\right)(x) = x + 1$; $\{x | x \neq 0\}$

(e) $\dfrac{5}{3}$ **(f)** 1 **(g)** $\dfrac{3}{4}$ **(h)** 2 **71. (a)** $(f + g)(x) = \dfrac{6x + 3}{3x - 2}$; $\left\{x \middle| x \neq \dfrac{2}{3}\right\}$ **(b)** $(f - g)(x) = \dfrac{-2x + 3}{3x - 2}$; $\left\{x \middle| x \neq \dfrac{2}{3}\right\}$

(c) $(f \cdot g)(x) = \dfrac{8x^2 + 12x}{(3x - 2)^2}$; $\left\{x \middle| x \neq \dfrac{2}{3}\right\}$ **(d)** $\left(\dfrac{f}{g}\right)(x) = \dfrac{2x + 3}{4x}$; $\left\{x \middle| x \neq 0, x \neq \dfrac{2}{3}\right\}$ **(e)** 3 **(f)** $-\dfrac{1}{2}$ **(g)** $\dfrac{7}{2}$ **(h)** $\dfrac{5}{4}$ **73.** $g(x) = 5 - \dfrac{7}{2}x$

75. 4 **77.** $2x + h - 1$ **79.** $\dfrac{-(2x + h)}{x^2(x + h)^2}$ **81.** $\dfrac{1}{\sqrt{x + h} + \sqrt{x}}$ **83.** $A = -\dfrac{7}{2}$ **85.** $A = -4$ **87.** $A = 8$; undefined at $x = 3$ **89.** $A(x) = \dfrac{1}{2}x^2$

91. $G(x) = 10x$ **93. (a)** P is the dependent variable; a is the independent variable. **(b)** $P(20) = 197.34$ million; In 2005, there were 197.34 million people 20 years of age or older. **(c)** $P(0) = 290.580$ million; In 2005, there were 290.580 million people. **95. (a)** 15.1 m, 14.071 m, 12.944 m, 11.719 m

(b) 1.01 sec, 1.43 sec, 1.75 sec **(c)** 2.02 sec **97. (a)** \$222 **(b)** \$225 **(c)** \$220 **(d)** \$230 **99.** $R(x) = \dfrac{L(x)}{P(x)}$ **101.** $H(x) = P(x) \cdot I(x)$

103. (a) $P(x) = -0.05x^3 + 0.8x^2 + 155x - 500$ **(b)** $P(15) = \$1836.25$ **(c)** When 15 hundred cellphones are sold, the profit is \$1836.25.

105. Only $h(x) = 2x$

3.2 Assess Your Understanding *(page 218)*

3. vertical **4.** 5; -3 **5.** $a = -2$ **6.** F **7.** F **8.** T **9. (a)** $f(0) = 3$; $f(-6) = -3$ **(b)** $f(6) = 0$; $f(11) = 1$ **(c)** Positive **(d)** Negative

(e) $-3, 6$, and 10 **(f)** $-3 < x < 6$; $10 < x \leq 11$ **(g)** $\{x | -6 \leq x \leq 11\}$ **(h)** $\{y | -3 \leq y \leq 4\}$ **(i)** $-3, 6, 10$ **(j)** 3 **(k)** 3 times **(l)** Once

(m) 0, 4 **(n)** $-5, 8$ **11.** Not a function **13.** Function **(a)** Domain: $\{x | -\pi \leq x \leq \pi\}$; Range: $\{y | -1 \leq y \leq 1\}$ **(b)** $\left(-\dfrac{\pi}{2}, 0\right), \left(\dfrac{\pi}{2}, 0\right), (0, 1)$

(c) y-axis **15.** Not a function **17.** Function **(a)** Domain: $\{x | x > 0\}$; Range: all real numbers **(b)** $(1, 0)$ **(c)** None **19.** Function **(a)** Domain: all real numbers; Range: $\{y | y \leq 2\}$ **(b)** $(-3, 0), (3, 0), (0, 2)$ **(c)** y-axis **21.** Function **(a)** Domain: all real numbers; Range:

$\{y | y \geq -3\}$ **(b)** $(1, 0), (3, 0), (0, 9)$ **(c)** None **23. (a)** Yes **(b)** $f(-2) = 9$; $(-2, 9)$ **(c)** $0, \dfrac{1}{2}$; $(0, -1), \left(\dfrac{1}{2}, -1\right)$ **(d)** All real numbers

(e) $-\dfrac{1}{2}, 1$ **(f)** -1 **25. (a)** No **(b)** $f(4) = -3$; $(4, -3)$ **(c)** 14; $(14, 2)$ **(d)** $\{x | x \neq 6\}$ **(e)** -2 **(f)** $-\dfrac{1}{3}$ **27. (a)** Yes

(b) $f(2) = \dfrac{8}{17}$; $\left(2, \dfrac{8}{17}\right)$ **(c)** $-1, 1$; $(-1, 1), (1, 1)$ **(d)** All real numbers **(e)** 0 **(f)** 0

29. (a) Approximately 10.4 ft high
(b) Approximately 9.9 ft high
(c)

(d) The ball will not go through the hoop; $h(15) \approx 8.4$ ft.
If $v = 30$ ft/sec, $h(15) = 10$ ft.

31. (a) About 81.07 ft **(b)** About 129.59 ft **(c)** About 26.63 ft
(d) About 528.13 ft
(e)

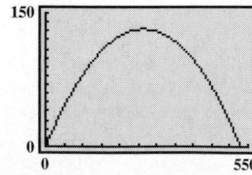

(f) About 115.07 ft and 413.05 ft **(g)** 275 ft; maximum height shown in the table is 131.8 ft **(h)** 264 ft

33. (a)

(b)

(c) 600 mi/hr

35. (a) 3 **(b)** -2 **(c)** -1 **(d)** 1 **(e)** 2 **(f)** $-\dfrac{1}{3}$

37. The x-intercepts can number anywhere from 0 to infinitely many. There is at most one y-intercept.

39. (a) III **(b)** IV **(c)** I **(d)** V **(e)** II

41.

43. (a) 2 hr elapsed during which Kevin was between 0 and 3 mi from home **(b)** 0.5 hr elapsed during which Kevin was 3 mi from home **(c)** 0.3 hr elapsed during which Kevin was between 0 and 3 mi from home **(d)** 0.2 hr elapsed during which Kevin was 0 mi from home **(e)** 0.9 hr elapsed during which Kevin was between 0 and 2.8 mi from home **(f)** 0.3 hr elapsed during which Kevin was 2.8 mi from home **(g)** 1.1 hr elapsed during which Kevin was between 0 and 2.8 mi from home **(h)** 3 mi **(i)** 2 times **45.** No points whose x-coordinate is 5 or whose y-coordinate is 0 can be on the graph.

3.3 Assess Your Understanding (page 230)

6. increasing **7.** even; odd **8.** T **9.** T **10.** F **11.** Yes **13.** No **15.** $(-8, -2); (0, 2); (5, \infty)$ **17.** Yes; 10 **19.** $-2, 2; 6, 10$
21. (a) $(-2, 0), (0, 3), (2, 0)$ **(b)** Domain: $\{x | -4 \le x \le 4\}$ or $[-4, 4]$; Range: $\{y | 0 \le y \le 3\}$ or $[0, 3]$ **(c)** Increasing on $(-2, 0)$ and $(2, 4)$;
Decreasing on $(-4, -2)$ and $(0, 2)$ **(d)** Even **23. (a)** $(0, 1)$ **(b)** Domain: all real numbers; Range: $\{y | y > 0\}$ or $(0, \infty)$ **(c)** Increasing on
$(-\infty, \infty)$ **(d)** Neither **25. (a)** $(-\pi, 0), (0, 0), (\pi, 0)$ **(b)** Domain: $\{x | -\pi \le x \le \pi\}$ or $[-\pi, \pi]$; Range: $\{y | -1 \le y \le 1\}$ or $[-1, 1]$
(c) Increasing on $\left(-\dfrac{\pi}{2}, \dfrac{\pi}{2}\right)$; Decreasing on $\left(-\pi, -\dfrac{\pi}{2}\right)$ and $\left(\dfrac{\pi}{2}, \pi\right)$ **(d)** Odd **27. (a)** $\left(0, \dfrac{1}{2}\right), \left(\dfrac{1}{3}, 0\right), \left(\dfrac{5}{2}, 0\right)$ **(b)** Domain: $\{x | -3 \le x \le 3\}$ or
$[-3, 3]$; Range: $\{y | -1 \le y \le 2\}$ or $[-1, 2]$ **(c)** Increasing on $(2, 3)$; Decreasing on $(-1, 1)$; Constant on $(-3, -1)$ and $(1, 2)$ **(d)** Neither
29. (a) $0; 3$ **(b)** $-2, 2; 0, 0$ **31. (a)** $\dfrac{\pi}{2}; 1$ **(b)** $-\dfrac{\pi}{2}; -1$ **33.** Odd **35.** Even **37.** Odd **39.** Neither **41.** Even **43.** Odd **45.** Absolute
maximum: $f(1) = 4$; absolute minimum: $f(5) = 1$ **47.** Absolute maximum: $f(3) = 4$; absolute minimum: $f(1) = 1$ **49.** Absolute maximum:
none; absolute minimum: $f(0) = 0$ **51.** Absolute maximum: none; absolute minimum: none

53.
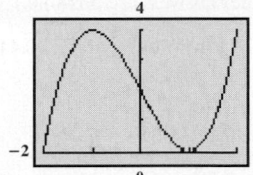

Increasing: $(-2, -1), (1, 2)$
Decreasing: $(-1, 1)$
Local maximum: $(-1, 4)$
Local minimum: $(1, 0)$

55.

Increasing: $(-2, -0.77), (0.77, 2)$
Decreasing: $(-0.77, 0.77)$
Local maximum: $(-0.77, 0.19)$
Local minimum: $(0.77, -0.19)$

57.
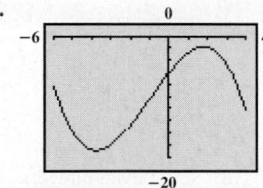

Increasing: $(-3.77, 1.77)$
Decreasing: $(-6, -3.77), (1.77, 4)$
Local maximum: $(-1.77, -1.91)$
Local minimum: $(-3.77, -18.89)$

59.

Increasing: $(-1.87, 0), (0.97, 2)$
Decreasing: $(-3, -1.87), (0, 0.97)$
Local maximum: $(0, 3)$
Local minima: $(-1.87, 0.95), (0.97, 2.65)$

61. (a) -4 **(b)** -8 **(c)** -10 **63. (a)** 17 **(b)** -1 **(c)** 11 **65. (a)** 5 **(b)** $y = 5x - 2$ **67. (a)** -1 **(b)** $y = -x$ **69. (a)** 4 **(b)** $y = 4x - 8$
71. (a) Odd **(b)** Local maximum value: 54 at $x = -3$ **73. (a)** Even **(b)** Local maximum value: 24 at $x = -2$ **(c)** 47.4 sq. units

75. (a)

(b) 10 riding lawn mowers/hr
(c) \$239/mower

77. (a) On average, the population is increasing at a rate of 0.036 g/hr from 0 to 2.5 hr.
(b) On average, from 4.5 to 6 hr, the population is increasing at a rate of 0.1 g/hr. **(c)** The average rate of change is increasing over time.

79. (a) 1 **(b)** 0.5 **(c)** 0.1 **(d)** 0.01
(e) 0.001
(f)
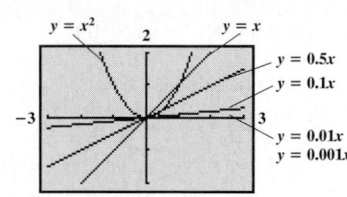

(g) They are getting closer to the tangent line at $(0, 0)$.
(h) They are getting closer to 0.

81. (a) 2
(b) 2; 2; 2; 2
(c) $y = 2x + 5$
(d)
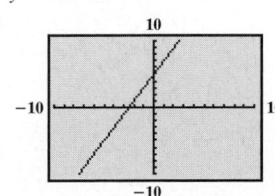

83. (a) $2x + h + 2$
(b) 4.5; 4.1; 4.01; 4
(c) $y = 4.01x - 1.01$
(d)
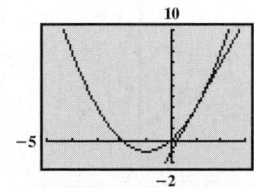

85. (a) $4x + 2h - 3$
(b) 2; 1.2; 1.02; 1
(c) $y = 1.02x - 1.02$
(d)
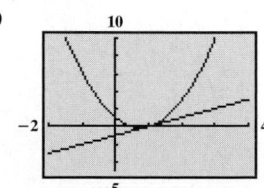

87. (a) $-\dfrac{1}{(x + h)x}$
(b) $-\dfrac{2}{3}; -\dfrac{10}{11}; -\dfrac{100}{101}; -1$
(c) $y = -\dfrac{100}{101}x + \dfrac{201}{101}$
(d)
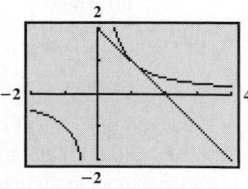

91. At most one **93.** Yes; the function $f(x) = 0$ is both even and odd. **95.** Not necessarily. It just means $f(5) > f(2)$.

3.4 Assess Your Understanding (page 241)

4. $(-\infty, 0)$ **5.** piccewise-defined **6.** T **7.** F **8.** F **9.** C **11.** E **13.** B **15.** F

17.

19.

21.

23.

25. (a) 4 **(b)** 2 **(c)** 5 **27. (a)** -4 **(b)** -2 **(c)** 0 **(d)** 25

29. (a) All real numbers
(b) $(0, 1)$
(c)

(d) $\{y | y \neq 0\}; (-\infty, 0) \cup (0, \infty)$
(e) Discontinuous at $x = 0$

31. (a) All real numbers
(b) $(0, 3)$
(c)

(d) $\{y | y \geq 1\}; [1, \infty)$
(e) Continuous

33. (a) $\{x | x \geq -2\}; [-2, \infty)$
(b) $(0, 3), (2, 0)$
(c)

(d) $\{y | y < 4, y = 5\}; (-\infty, 4) \cup \{5\}$
(e) Discontinuous at $x = 1$

35. (a) All real numbers
(b) $(-1, 0), (0, 0)$
(c)

(d) All real numbers
(e) Discontinuous at $x = 0$

37. (a) $\{x | x \geq -2, x \neq 0\}; [-2, 0) \cup (0, \infty)$
(b) No intercepts
(c)

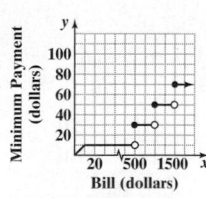

(d) $\{y | y > 0\}; (0, \infty)$
(e) Discontinuous at $x = 0$

39. (a) All real numbers
(b) $(x, 0)$ for $0 \leq x < 1$
(c)

(d) Set of even integers
(e) Discontinuous at $\{x | x \text{ is an integer}\}$

41. $f(x) = \begin{cases} -x & \text{if } -1 \leq x \leq 0 \\ \frac{1}{2}x & \text{if } 0 < x \leq 2 \end{cases}$ (Other answers are possible.)

43. $f(x) = \begin{cases} -x & \text{if } x \leq 0 \\ -x + 2 & \text{if } 0 < x \leq 2 \end{cases}$ (Other answers are possible.)

45. (a) 2 **(b)** 3 **(c)** -4 **47. (a)** \$39.99 **(b)** \$46.74 **(c)** \$40.44
49. (a) \$52.45 **(b)** \$277.36
(c) $C(x) = \begin{cases} 0.73006x + 15.95 & \text{if } 0 \leq x \leq 50 \\ 0.4998x + 27.463 & \text{if } x > 50 \end{cases}$
(d)

51. For schedule X: $f(x) = \begin{cases} 0.10x \text{ if} & 0 < x \leq 8350 \\ 835 + 0.15(x - 8350) \text{ if} & 8350 < x \leq 33,950 \\ 4675 + 0.25(x - 33,950) \text{ if} & 33,950 < x \leq 82,250 \\ 16,750 + 0.28(x - 82,250) \text{ if} & 82,250 < x \leq 171,550 \\ 41,754 + 0.33(x - 171,550) \text{ if} & 171,550 < x \leq 372,950 \\ 108,216 + 0.35(x - 372,950) \text{ if} & x > 372,950 \end{cases}$

53. (a)

(b) $C(x) = 10 + 0.4x$ **(c)** $C(x) = 70 + 0.25x$

55. $f(x) = \begin{cases} x & \text{if } 0 \leq x < 10 \\ 10 & \text{if } 10 \leq x < 500 \\ 30 & \text{if } 500 \leq x < 1000 \\ 50 & \text{if } 1000 \leq x < 1500 \\ 70 & \text{if } x \geq 1500 \end{cases}$

57. (a) 10°C **(b)** 4°C
(c) -3°C **(d)** -4°C
(e) The wind chill is equal to the air temperature.
(f) At wind speed greater than 20 m/s, the wind chill factor depends only on the air temperature.

59. $C(x) = \begin{cases} 1.17 & 0 < x \leq 1 \\ 1.34 & 1 < x \leq 2 \\ 1.51 & 2 < x \leq 3 \\ 1.68 & 3 < x \leq 4 \\ 1.85 & 4 < x \leq 5 \\ 2.02 & 5 < x \leq 6 \\ 2.19 & 6 < x \leq 7 \\ 2.36 & 7 < x \leq 8 \\ 2.53 & 8 < x \leq 9 \\ 2.70 & 9 < x \leq 10 \\ 2.87 & 10 < x \leq 11 \\ 3.04 & 11 < x \leq 12 \\ 3.21 & 12 < x \leq 13 \end{cases}$

61. Each graph is that of $y = x^2$, but shifted horizontally. If $y = (x - k)^2, k > 0$, the shift is right k units; if $y = (x + k)^2, k > 0$, the shift is left k units.
63. The graph of $y = -f(x)$ is the reflection about the x-axis of the graph of $y = f(x)$. **65.** Yes. The graph of $y = (x - 1)^3 + 2$ is the graph of $y = x^3$ shifted right 1 unit and up 2 units. **67.** They all have the same general shape. All three go through the points $(-1, -1), (0, 0)$, and $(1, 1)$. As the exponent increases, the steepness of the curve increases (except near $x = 0$).

3.5 Assess Your Understanding (page 253)

1. horizontal; right **2.** y **3.** vertical; up **4.** T **5.** F **6.** T **7.** B **9.** H **11.** I **13.** L **15.** F **17.** G **19.** $y = (x - 4)^3$ **21.** $y = x^3 + 4$
23. $y = -x^3$ **25.** $y = 4x^3$ **27.** $y = -(\sqrt{-x} + 2)$ **29.** $y = -\sqrt{x + 3} + 2$ **31.** (c) **33.** (c) **35. (a)** -7 and 1 **(b)** -3 and 5 **(c)** -5 and 3
(d) -3 and 5 **37. (a)** $(-3, 3)$ **(b)** $(4, 10)$ **(c)** Decreasing on $(-1, 5)$ **(d)** Decreasing on $(-5, 1)$

39.

Domain: $(-\infty, \infty)$;
Range: $[-1, \infty)$

41.

Domain: $(-\infty, \infty)$;
Range: $(-\infty, \infty)$

43.

Domain: $[2, \infty)$;
Range: $[0, \infty)$

45.

Domain: $(-\infty, \infty)$;
Range: $(-\infty, \infty)$

47.

Domain: $[0, \infty)$;
Range: $[0, \infty)$

49.

Domain: $(-\infty, \infty)$;
Range: $(-\infty, \infty)$

51.
Domain: $(-\infty, \infty)$;
Range: $[-3, \infty)$

53.
Domain: $[2, \infty)$;
Range: $[1, \infty)$

55.
Domain: $(-\infty, 0]$;
Range: $[-2, \infty)$

57.
Domain: $(-\infty, \infty)$;
Range: $(-\infty, \infty)$

59.
Domain: $(-\infty, \infty)$;
Range: $[0, \infty)$

61.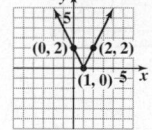
Domain: $(-\infty, \infty)$;
Range: $\{y \mid y \text{ is an even integer}\}$

63. (a) $F(x) = f(x) + 3$ **(b)** $G(x) = f(x + 2)$ **(c)** $P(x) = -f(x)$

(d) $H(x) = f(x + 1) - 2$ **(e)** $Q(x) = \dfrac{1}{2}f(x)$ **(f)** $g(x) = f(-x)$ **(g)** $h(x) = f(2x)$ **65. (a)** $F(x) = f(x) + 3$

(b) $G(x) = f(x + 2)$ **(c)** $P(x) = -f(x)$ **(d)** $H(x) = f(x + 1) - 2$ **(e)** $Q(x) = \dfrac{1}{2}f(x)$ **(f)** $g(x) = f(-x)$

(g) $h(x) = f(2x)$ **67.** $f(x) = (x + 1)^2 - 1$ **69.** $f(x) = (x - 4)^2 - 15$ **71.** $f(x) = 2(x - 3)^2 + 1$ **73.** $f(x) = -3(x + 2)^2 - 5$

75.

77. (a) 72°F; 65°F
(b) The temperature decreases by 2° to 70°F during the day and 63°F overnight.

(c) The time at which the temperature adjusts between the daytime and overnight settings is moved to 1 hr sooner. It begins warming up at 5:00 AM instead of 6:00 AM, and it begins cooling down at 8:00 PM instead of 9:00 PM.

79.

81. (a)

(b) 10% tax
(c) Y_1 is the graph of $p(x)$ shifted down vertically 10,000 units. Y_2 is the graph of $p(x)$ vertically compressed by a factor of 0.9.
(d) 10% tax

83. (a) **(b)**

85. (a) $(-4, 2)$ **(b)** $(1, -12)$ **(c)** $(-4, 5)$

87. The graph of $y = f(x) - 2$ is the graph of $y = f(x)$ shifted down 2 units. The graph of $y = f(x - 2)$ is the graph of $y = f(x)$ shifted right 2 units.

3.6 Assess Your Understanding *(page 260)*

1. (a) $d(x) = \sqrt{x^4 - 15x^2 + 64}$
 (b) $d(0) = 8$ **(c)** $d(1) = \sqrt{50} \approx 7.07$
 (d)

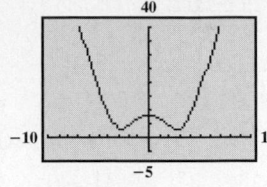

 (e) d is smallest when $x \approx -2.74$ or $x \approx 2.74$

3. (a) $d(x) = \sqrt{x^2 - x + 1}$
 (b)

 (c) d is smallest when $x = 0.5$.

5. $A(x) = \dfrac{1}{2}x^4$ **7. (a)** $A(x) = x(16 - x^2)$
 (b) Domain: $\{x | 0 < x < 4\}$
 (c) The area is largest when $x \approx 2.31$.

9. (a) $A(x) = 4x\sqrt{4 - x^2}$
 (c) A is largest when $x \approx 1.41$.

(b) $p(x) = 4x + 4\sqrt{4 - x^2}$
 (d) p is largest when $x \approx 1.41$.

11. (a) $A(x) = x^2 + \dfrac{25 - 20x + 4x^2}{\pi}$
 (b) Domain: $\{x | 0 < x < 2.5\}$
 (c) A is smallest when $x \approx 1.40$ m.

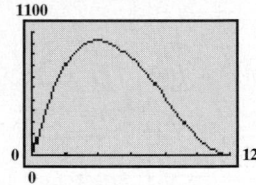

13. (a) $C(x) = x$ **(b)** $A(x) = \dfrac{x^2}{4\pi}$ **15. (a)** $A(r) = 2r^2$ **(b)** $p(r) = 6r$ **17.** $A(x) = \left(\dfrac{\pi}{3} - \dfrac{\sqrt{3}}{4}\right)x^2$

19. (a) $d(t) = \sqrt{2500t^2 - 360t + 13}$ **21.** $V(r) = \dfrac{\pi H(R - r)r^2}{R}$ **23. (a)** $T(x) = \dfrac{12 - x}{5} + \dfrac{\sqrt{x^2 + 4}}{3}$ **25. (a)** $V(x) = x(24 - 2x)^2$
 (b) d is smallest when $t \approx 0.07$ hr. **(b)** $\{x | 0 \le x \le 12\}$ **(b)** 972 in.3 **(c)** 160 in.3
 (c) 3.09 hr **(d)** V is largest when $x = 4$.
 (d) 3.55 hr

Review Exercises *(page 265)*

1. Function; domain $\{-1, 2, 4\}$, range $\{0, 3\}$ **3. (a)** 2 **(b)** -2 **(c)** $-\dfrac{3x}{x^2 - 1}$ **(d)** $-\dfrac{3x}{x^2 - 1}$ **(e)** $\dfrac{3(x - 2)}{x^2 - 4x + 3}$ **(f)** $\dfrac{6x}{4x^2 - 1}$ **5. (a)** 0 **(b)** 0

(c) $\sqrt{x^2 - 4}$ **(d)** $-\sqrt{x^2 - 4}$ **(e)** $\sqrt{x^2 - 4x}$ **(f)** $2\sqrt{x^2 - 1}$ **7. (a)** 0 **(b)** 0 **(c)** $\dfrac{x^2 - 4}{x^2}$ **(d)** $-\dfrac{x^2 - 4}{x^2}$ **(e)** $\dfrac{x(x - 4)}{(x - 2)^2}$ **(f)** $\dfrac{x^2 - 1}{x^2}$

9. $\{x | x \neq -3, x \neq 3\}$ **11.** $\{x | x \le 2\}$ **13.** $\{x | x > 0\}$ **15.** $\{x | x \neq -3, x \neq 1\}$

17. $(f + g)(x) = 2x + 3$; Domain: all real numbers
 $(f - g)(x) = -4x + 1$; Domain: all real numbers
 $(f \cdot g)(x) = -3x^2 + 5x + 2$; Domain: all real numbers
 $\left(\dfrac{f}{g}\right)(x) = \dfrac{2 - x}{3x + 1}$; Domain: $\left\{x | x \neq -\dfrac{1}{3}\right\}$

19. $(f + g)(x) = 3x^2 + 4x + 1$; Domain: all real numbers
 $(f - g)(x) = 3x^2 - 2x + 1$; Domain: all real numbers
 $(f \cdot g)(x) = 9x^3 + 3x^2 + 3x$; Domain: all real numbers
 $\left(\dfrac{f}{g}\right)(x) = \dfrac{3x^2 + x + 1}{3x}$; Domain: $\{x | x \neq 0\}$

21. $(f + g)(x) = \dfrac{x^2 + 2x - 1}{x(x - 1)}$; Domain: $\{x | x \neq 0, x \neq 1\}$

 $(f - g)(x) = \dfrac{x^2 + 1}{x(x - 1)}$; Domain: $\{x | x \neq 0, x \neq 1\}$

 $(f \cdot g)(x) = \dfrac{x + 1}{x(x - 1)}$; Domain: $\{x | x \neq 0, x \neq 1\}$

 $\left(\dfrac{f}{g}\right)(x) = \dfrac{x(x + 1)}{x - 1}$; Domain: $\{x | x \neq 0, x \neq 1\}$

23. $-4x + 1 - 2h$ **25. (a)** Domain: $\{x | -4 \le x \le 3\}$; Range: $\{y | -3 \le y \le 3\}$
 (b) $(0, 0)$ **(c)** -1 **(d)** -4 **(e)** $\{x | 0 < x \le 3\}$
 (f)

 (g)

 (h)

27. (a) Domain: $\{x | -4 \le x \le 4\}$ or $[-4, 4]$
 Range: $\{y | -3 \le y \le 1\}$ or $[-3, 1]$
 (b) Increasing on $(-4, -1)$ and $(3, 4)$; Decreasing on $(-1, 3)$
 (c) Local maximum value is 1 and occurs at $x = -1$.
 Local minimum value is -3 and occurs at $x = 3$.

 (d) Absolute maximum: $f(-1) = 1$
 Absolute minimum: $f(3) = -3$
 (e) No symmetry **(f)** Neither
 (g) x-intercepts: $-2, 0, 4$; y-intercept: 0
29. Odd **31.** Even **33.** Neither **35.** Odd

37.

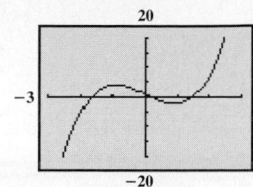

Local maximum value: 4.04 at $x = -0.91$
Local minimum value: -2.04 at $x = 0.91$
Increasing: $(-3, -0.91); (0.91, 3)$
Decreasing: $(-0.91, 0.91)$

39.

Local maximum value: 1.53 at $x = 0.41$
Local minima values: -0.54 at $x = -0.34$
and -3.56 at $x = 1.80$
Increasing: $(-0.34, 0.41); (1.80, 3)$
Decreasing: $(-2, -0.34); (0.41, 1.80)$

41. (a) 23 **(b)** 7 **(c)** 47 **43.** -5 **45.** -17 **47.** No **49.** Yes

51.

53.

55.

Intercepts: $(-4, 0), (4, 0), (0, -4)$
Domain: all real numbers
Range: $\{y | y \geq -4\}$ or $[-4, \infty)$

57.

Intercept: $(0, 0)$
Domain: all real numbers
Range: $\{y | y \leq 0\}$ or $(-\infty, 0]$

59.

Intercept: $(1, 0)$
Domain: $\{x | x \geq 1\}$ or $[1, \infty)$
Range: $\{y | y \geq 0\}$ or $[0, \infty)$

61.

Intercepts: $(0, 1), (1, 0)$
Domain: $\{x | x \leq 1\}$ or $(-\infty, 1]$
Range: $\{y | y \geq 0\}$ or $[0, \infty)$

63.

Intercept: $(0, 3)$
Domain: all real numbers
Range: $\{y | y \geq 2\}$ or $[2, \infty)$

65.

Intercepts: $(0, -2)$,
$\left(1 - \dfrac{\sqrt[3]{9}}{3}, 0\right)$ or about $(0.3, 0)$
Domain: all real numbers
Range: all real numbers

67. (a) $\{x | x > -2\}$ or $(-2, \infty)$
(b) $(0, 0)$
(c)

(d) $\{y | y > -6\}$ or $(-6, \infty)$
(e) Discontinuous at $x = 1$

69. (a) $\{x | x \geq -4\}$ or $[-4, \infty)$
(b) $(0, 1)$
(c)

(d) $\{y | -4 \leq y < 0 \text{ or } y > 0\}$
or $[-4, 0) \cup (0, \infty)$
(e) Discontinuous at $x = 0$

71. $A = 11$ **73. (a)** $A(x) = (8.5 - 2x)(11 - 2x)$
(b) $0 \leq x < 4.25, 0 < A \leq 93.5$
(c) $A(1) = 58.5$ in.2, $A(1.2) = 52.46$ in.2, $A(1.5) = 44$ in.2
(d)

75. (a) $A(x) = 10x - x^3$
(b) The largest area that can be enclosed by the
rectangle is approximately 12.17 square units.

Chapter Test *(page 268)*

1. (a) Function; domain: $\{2, 4, 6, 8\}$; range: $\{5, 6, 7, 8\}$ **(b)** Not a function **(c)** Not a function **(d)** Function; domain; all real numbers;
range: $\{y | y \geq 2\}$ **2.** Domain: $\left\{x \middle| x \leq \dfrac{4}{5}\right\}$; $f(-1) = 3$ **3.** Domain: $\{x | x \neq -2\}$; $g(-1) = 1$ **4.** Domain: $\{x | x \neq -9, x \neq 4\}$; $h(-1) = \dfrac{1}{8}$

5. (a) Domain: $\{x | -5 \leq x \leq 5\}$; range: $\{y | -3 \leq y \leq 3\}$ **(b)** $(0, 2), (-2, 0)$, and $(2, 0)$ **(c)** $f(1) = 3$ **(d)** $x = -5$ and $x = 3$
(e) $\{x | -5 \leq x < -2 \text{ or } 2 < x \leq 5\}$ or $[-5, -2) \cup (2, 5]$ **6.** Local maxima values: $f(-0.85) \approx -0.86$; $f(2.35) \approx 15.55$; local minima values: $f(0) = -2$;
the function is increasing on the intervals $(-5, -0.85)$ and $(0, 2.35)$ and decreasing on the intervals $(-0.85, 0)$ and $(2.35, 5)$.

7. (a)

(b) $(0, -4), (4, 0)$
(c) $g(-5) = -9$
(d) $g(2) = -2$

8. 19 **9. (a)** $(f - g)(x) = 2x^2 - 3x + 3$
(b) $(f \cdot g)(x) = 6x^3 - 4x^2 + 3x - 2$
(c) $f(x + h) - f(x) = 4xh + 2h^2$

10. (a)

(b)

11. (a) 8.67% occurring in 1997 $(x \approx 5)$ **(b)** The model predicts that the interest rate will be -10.343%. This is not reasonable.

12. (a) $V(x) = \dfrac{x^2}{8} - \dfrac{5x}{4} + \dfrac{\pi x^2}{64}$ **(b)** 1297.61 ft^3

Cumulative Review (page 269)

1. $\{6\}$ **2.** $\left\{0, \dfrac{1}{3}\right\}$ **3.** $\{-1, 9\}$ **4.** $\left\{\dfrac{1}{3}, \dfrac{1}{2}\right\}$ **5.** $\left\{-\dfrac{7}{2}, \dfrac{1}{2}\right\}$ **6.** $\left\{\dfrac{1}{2}\right\}$

7. $\left\{x \middle| x < -\dfrac{4}{3}\right\}; \left(-\infty, -\dfrac{4}{3}\right)$ **8.** $\{x \mid 1 < x < 4\}; (1, 4)$ **9.** $\left\{x \middle| x \le -2 \text{ or } x \ge \dfrac{3}{2}\right\}; (-\infty, -2] \cup \left[\dfrac{3}{2}, \infty\right)$

10. (a) distance: $\sqrt{29}$ **(b)** midpoint: $\left(\dfrac{1}{2}, -4\right)$ **(c)** slope: $-\dfrac{2}{5}$

11. **12.** **13.** **14.** **15.** Intercepts: $(0, -3), (-2, 0), (2, 0)$; symmetry with respect to the y-axis

16. $y = \dfrac{1}{2}x + 5$

17. **18.** **19.**

CHAPTER 4 Linear and Quadratic Functions

4.1 Assess Your Understanding (page 278)

7. slope; y-intercept **8.** -4; 3 **9.** positive **10.** T **11.** F **12.** F

13. (a) $m = 2$; $b = 3$ **15. (a)** $m = -3$; $b = 4$ **17. (a)** $m = \dfrac{1}{4}$; $b = -3$ **19. (a)** $m = 0$; $b = 4$ **21.** Linear; -3 **23.** Nonlinear

(b) **(b)** **(b)** **(b)** **25.** Nonlinear **27.** Linear; 0

(c) 2 **(d)** Increasing **(c)** -3 **(d)** Decreasing **(c)** $\dfrac{1}{4}$ **(d)** Increasing **(c)** 0 **(d)** Constant

29. (a) $\dfrac{1}{4}$ **(b)** $\left\{x \middle| x > \dfrac{1}{4}\right\}$ or $\left(\dfrac{1}{4}, \infty\right)$ **31. (a)** 40 **(b)** 88 **(c)** -40 **(d)** $\{x \mid x > 40\}$ or $(40, \infty)$ **(e)** $\{x \mid x \le 88\}$ or $(-\infty, 88]$

(c) 1 **(d)** $\{x \mid x \le 1\}$ or $(-\infty, 1]$ **(f)** $\{x \mid -40 < x < 88\}$ or $(-40, 88)$ **33. (a)** -4 **(b)** $\{x \mid x < -4\}$ or $(-\infty, -4)$

(e) **35. (a)** -6 **(b)** $\{x \mid -6 \le x < 5\}$ or $[-6, 5)$ **37. (a)** \$45 **(b)** 180 mi **(c)** 260 mi

(d) $\{x \mid x \ge 0\}$ or $[0, \infty)$ **39. (a)** \$16; 600 T-shirts **(b)** $\$0 \le p < \16 **(c)** The price will increase.

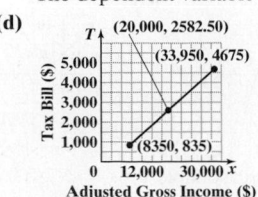

41. (a) $\{x \mid 8350 \le x \le 33{,}950\}$ or $[8350, 33{,}950]$ **43. (a)** $x = 5000$ **45. (a)** $V(x) = -1000x + 3000$ **47. (a)** $C(x) = 90x + 1800$

(b) \$2582.50 **(b)** $x > 5000$ **(b)** $\{x \mid 0 \le x \le 3\}$ or $[0, 3]$ **(b)**

(c) The independent variable is adjusted gross income, x. **(c)**

The dependent variable is the tax bill, T.

(d)

 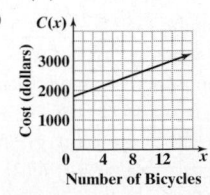

(c) \$3060

(d) \$1000 **(d)** 22 bicycles

(e) After 1 year

(e) \$27,500

49. (a) $C(x) = 0.07x + 29$ **(b)** $36.70; $45.10

51. (a)

(e)

53. (d), (e) **55.** $b = 0$; yes, $f(x) = b$

(b) Since each input (price) corresponds to a single output (quantity demanded), we know that quantity demanded is a function of price. Also, because the average rate of change is a constant -0.4 24″ LCD monitor per dollar, the function is linear.

(c) $q(p) = -0.4p + 160$

(d) $\{p \mid 0 \le p \le 400\}$ or $[0, 400]$

(f) If price increases by $1, quantity demanded of 24″ LCD monitors decreases by 0.4 monitor.

(g) q-intercept: When the price is $0, 160 24″ LCD monitors will be demanded p-intercept: There will be 0 24″ LCD monitors demanded when the price is $400.

4.2 Assess Your Understanding *(page 285)*

3. scatter diagram **4.** T **5.** Linear relation, $m > 0$ **7.** Linear relation, $m < 0$ **9.** Nonlinear relation

11. (a)

(c)

(e)

(b) Answers vary. Using $(4, 6)$ and $(8, 14)$, $y = 2x - 2$.

(d) $y = 2.0357x - 2.3571$

13. (a)

(c)

(e)

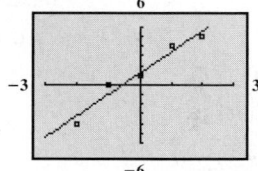

(b) Answers will vary. Using $(-2, -4)$ and $(2, 5)$, $y = \dfrac{9}{4}x + \dfrac{1}{2}$.

(d) $y = 2.2x + 1.2$

15. (a)

(c)

(e)

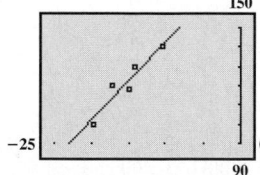

(b) Answers will vary. Using $(-20, 100)$ and $(-10, 140)$, $y = 4x + 180$.

(d) $y = 3.8613x + 180.2920$

17. (a)

(c) Answers will vary. Using the points $(39.52, 210)$ and $(66.45, 280)$, $y = 2.599x + 107.288$.

(d)

(e) 269 calories

(f) If the weight of a candy bar is increased by 1 gram, the number of calories will increase by 2.599, on average.

(b) Linear with positive slope

19. (a) The independent variable is the number of hours spent playing video games and cumulative grade-point average is the dependent variable because we are using number of hours playing video games to predict (or explain) cumulative grade-point average.

(b)

(c) $G(h) = -0.0942h + 3.2763$

(d) If the number of hours playing video games in a week increases by 1 hour, the cumulative grade-point average decreases 0.09, on average.

(e) 2.52

(f) Approximately 9.3 hours

21. (a) No

(b)
Demand (pairs of jeans sold per day)
Price (dollars/pair)

(c) $D = -1.3355p + 86.1974$

(d) If the price increases \$1, the quantity sold per day decreases by about 1.34 pairs of jeans, on average.

(e) $D(p) = -1.3355p + 86.1974$

(f) $\{p|0 < p \le 64\}$

(g) About 49 pairs

23.
Incidence Rate (per 1000)
Age of Mother

No, the data do not follow a linear pattern.

25. No linear relation **27.** 34.8 hours; A student whose GPA is 0 spends 34.8 hours each week playing video games.; $G(0) = 3.28$; The average GPA of a student who does not play video games is 3.28.

4.3 Assess Your Understanding (page 297)

5. parabola **6.** axis or axis of symmetry **7.** $-\dfrac{b}{2a}$ **8.** T **9.** T **10.** T **11.** C **13.** F **15.** G **17.** H

19.

21.

23. $f(x) = (x + 2)^2 - 2$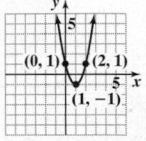

25. $f(x) = 2(x - 1)^2 - 1$

27. $f(x) = -(x + 1)^2 + 1$

29. $f(x) = \dfrac{1}{2}(x + 1)^2 - \dfrac{3}{2}$

31. (a)
$x = -1$

(b) Domain: $(-\infty, \infty)$
Range: $[-1, \infty)$

(c) Decreasing: $(-\infty, -1)$
Increasing: $(-1, \infty)$

33. (a)
$x = -3$

(b) Domain: $(-\infty, \infty)$
Range: $(-\infty, 9]$

(c) Increasing: $(-\infty, -3)$
Decreasing: $(-3, \infty)$

35. (a)
$x = -1$

(b) Domain: $(-\infty, \infty)$
Range: $[-9, \infty)$

(c) Decreasing: $(-\infty, -1)$
Increasing: $(-1, \infty)$

37. (a)
$x = -1$

(b) Domain: $(-\infty, \infty)$
Range: $[0, \infty)$

(c) Decreasing: $(-\infty, -1)$
Increasing: $(-1, \infty)$

39. (a)
$x = \dfrac{1}{4}$

(b) Domain: $(-\infty, \infty)$
Range: $\left[\dfrac{15}{8}, \infty\right)$

(c) Decreasing: $\left(-\infty, \dfrac{1}{4}\right)$
Increasing: $\left(\dfrac{1}{4}, \infty\right)$

41. (a)
$x = \dfrac{1}{2}$

(b) Domain: $(-\infty, \infty)$
Range: $\left(-\infty, -\dfrac{5}{2}\right]$

(c) Increasing: $\left(-\infty, \dfrac{1}{2}\right)$
Decreasing: $\left(\dfrac{1}{2}, \infty\right)$

43. (a)
$x = -1$

(b) Domain: $(-\infty, \infty)$
Range: $[-1, \infty)$

(c) Decreasing: $(-\infty, -1)$
Increasing: $(-1, \infty)$

45. (a)
$x = -\dfrac{3}{4}$

(b) Domain: $(-\infty, \infty)$
Range: $\left(-\infty, \dfrac{17}{4}\right]$

(c) Increasing: $\left(-\infty, -\dfrac{3}{4}\right)$
Decreasing: $\left(-\dfrac{3}{4}, \infty\right)$

47. $f(x) = (x + 1)^2 - 2 = x^2 + 2x - 1$ **49.** $f(x) = -(x + 3)^2 + 5 = -x^2 - 6x - 4$ **51.** $f(x) = 2(x - 1)^2 - 3 = 2x^2 - 4x - 1$

53. Minimum value; -18 **55.** Minimum value; -21 **57.** Maximum value; 21 **59.** Maximum value; 13 **61.** $a = 6, b = 0, c = 2$

63. (a), (c), (d)

(b) $\{-1, 3\}$

65. (a), (c), (d)

(b) $\{-1, 3\}$

67. (a), (c), (d)

(b) $\{-1, 2\}$

69. (a) $a = 1: f(x) = (x + 3)(x - 1) = x^2 + 2x - 3$
$a = 2: f(x) = 2(x + 3)(x - 1) = 2x^2 + 4x - 6$
$a = -2: f(x) = -2(x + 3)(x - 1) = -2x^2 - 4x + 6$
$a = 5: f(x) = 5(x + 3)(x - 1) = 5x^2 + 10x - 15$

(b) The value of a does not affect the x-intercepts, but it changes the y-intercept by a factor of a.

(c) The value of a does not affect the axis of symmetry. It is $x = -1$ for all values of a.

(d) The value of a does not affect the x-coordinate of the vertex. However, the y-coordinate of the vertex is multiplied by a.

(e) The mean of the x-intercepts is the x-coordinate of the vertex.

71. (a) $(-2, -25)$
 (b) $-7, 3$
 (c) $-4, 0; (-4, -21), (0, -21)$
 (d)

73. $(2, 2)$ **75.** $500; $1,000,000 **77. (a)** 70,000 mp3 players **(b)** $2500 **79. (a)** 187 or 188 watches; $7031.20
(b) $P(x) = -0.2x^2 + 43x - 1750$ **(c)** 107 or 108 watches; $561.20 **81. (a)** 171 ft **(b)** 49 mph
(c) Reaction time **83.** $f(x) = 2(x + 4)(x - 2)$ **85.** If x is even, then ax^2 and bx are even and $ax^2 + bx$ is even, which means that $ax^2 + bx + c$ is odd. If x is odd, then ax^2 and bx are odd and $ax^2 + bx$ is even, which means that $ax^2 + bx + c$ is odd. In either case, $f(x)$ is odd.
87.

89. $b^2 - 4ac < 0$ **91.** No

4.4 Assess Your Understanding *(page 305)*

3. (a) $R(x) = -\dfrac{1}{6}x^2 + 100x$ **(b)** $\{x | 0 \le x \le 600\}$ **(c)** $13,333.33 **(d)** $300; $15,000 **(e)** $50 **5. (a)** $R(x) = -\dfrac{1}{5}x^2 + 20x$ **(b)** $255

(c) 50; $500 **(d)** $10 **(e)** Between $8 and $12 **7. (a)** $A(w) = -w^2 + 200w$ **(b)** A is largest when $w = 100$ yd. **(c)** 10,000 yd^2 **9.** 2,000,000 m^2

11. (a) $\dfrac{625}{16} \approx 39$ ft **(b)** $\dfrac{7025}{32} \approx 219.5$ ft **(c)** About 170 ft **13.** 18.75 m **15. (a)** 3 in. **(b)** Between 2 in. and 4 in.

(d)

(f) When the height is 100 ft, the projectile is about 135.7 ft from the cliff.
17. $\dfrac{750}{\pi} \approx 238.73$ m by 375 m
19. $x = \dfrac{a}{2}$ **21.** $\dfrac{38}{3}$ **23.** $\dfrac{248}{3}$

25. (a)

The data appear to follow a quadratic relation with $a < 0$.
(b) $I(x) = -45.466x^2 + 4314.374x - 55,961.675$ **(c)** About 47.4 years of age **(d)** Approximately $46,388 **(e)**

27. (a)

The data appear to be linearly related with positive slope.
(b) $R(x) = 0.836x + 1032.273$ **(c)** $1743

29. (a)

The data appear to follow a quadratic relation with $a > 0$.
(b) $P(a) = 0.015a^2 - 1.332a + 39.823$ **(c)** 11.6%

4.5 Assess Your Understanding *(page 312)*

3. (a) $\{x | x < -2 \text{ or } x > 2\}; (-\infty, -2) \text{ or } (2, \infty)$ **(b)** $\{x | -2 \le x \le 2\}; [-2, 2]$
5. (a) $\{x | -2 \le x \le 1\}; [-2, 1]$ **(b)** $\{x | x < -2 \text{ or } x > 1\}; (-\infty, -2) \text{ or } (1, \infty)$
7. $\{x | -2 < x < 5\}; (-2, 5)$ **9.** $\{x | x < 0 \text{ or } x > 4\}; (-\infty, 0) \text{ or } (4, \infty)$ **11.** $\{x | -3 < x < 3\}; (-3, 3)$

13. $\{x | x < -4 \text{ or } x > 3\}; (-\infty, -4) \text{ or } (3, \infty)$ **15.** $\left\{x \mid -\dfrac{1}{2} < x < 3\right\}; \left(-\dfrac{1}{2}, 3\right)$ **17.** No real solution **19.** No real solution

21. $\left\{x \mid x < -\dfrac{2}{3} \text{ or } x > \dfrac{3}{2}\right\}; \left(-\infty, -\dfrac{2}{3}\right) \text{ or } \left(\dfrac{3}{2}, \infty\right)$ **23.** $\{x | x \le -4 \text{ or } x \ge 4\}; (-\infty, -4] \text{ or } [4, \infty)$ **25. (a)** $\{-1, 1\}$ **(b)** $\{-1\}$ **(c)** $\{-1, 4\}$
(d) $\{x | x < -1 \text{ or } x > 1\}; (-\infty, -1) \text{ or } (1, \infty)$ **(e)** $\{x | x \le -1\} \text{ or } (-\infty, -1]$ **(f)** $\{x | x < -1 \text{ or } x > 4\}; (-\infty, -1) \text{ or } (4, \infty)$
(g) $\left\{x \mid x \le -\sqrt{2} \text{ or } x \ge \sqrt{2}\right\}; \left(-\infty, -\sqrt{2}\right] \text{ or } \left[\sqrt{2}, \infty\right)$

27. (a) $\{-1, 1\}$ **(b)** $\left\{\dfrac{1}{4}\right\}$ **(c)** $\{-4, 0\}$ **(d)** $\{x | -1 < x < 1\}; (-1, 1)$ **(e)** $\left\{x \mid x \le -\dfrac{1}{4}\right\} \text{ or } \left(-\infty, -\dfrac{1}{4}\right]$ **(f)** $\{x | -4 < x < 0\}; (-4, 0)$ **(g)** $\{0\}$

29. (a) $\{-2, 2\}$ **(b)** $\{-2, 2\}$ **(c)** $\{-2, 2\}$ **(d)** $\{x | x < -2 \text{ or } x > 2\}; (-\infty, -2) \text{ or } (2, \infty)$ **(e)** $\{x | x \le -2 \text{ or } x \ge 2\}; (-\infty, -2] \text{ or } [2, \infty)$
(f) $\{x | x < -2 \text{ or } x > 2\}; (-\infty, -2) \text{ or } (2, \infty)$ **(g)** $\left\{x \mid x \le -\sqrt{5} \text{ or } x \ge \sqrt{5}\right\}; \left(-\infty, -\sqrt{5}\right] \text{ or } \left[\sqrt{5}, \infty\right)$

31. (a) $\{-1, 2\}$ **(b)** $\{-2, 1\}$ **(c)** $\{0\}$ **(d)** $\{x | x < -1 \text{ or } x > 2\}; (-\infty, -1) \text{ or } (2, \infty)$ **(e)** $\{x | -2 \le x \le 1\}; [-2, 1]$ **(f)** $\{x | x < 0\}; (-\infty, 0)$

(g) $\left\{x \mid x \le \dfrac{1 - \sqrt{13}}{2} \text{ or } x \ge \dfrac{1 + \sqrt{13}}{2}\right\}; \left(-\infty, \dfrac{1 - \sqrt{13}}{2}\right] \text{ or } \left[\dfrac{1 + \sqrt{13}}{2}, \infty\right)$ **33. (a)** 5 sec **(b)** The ball is more than 96 ft above the ground
for time t between 2 and 3 sec, $2 < t < 3$. **35. (a)** $0, $1000 **(b)** The revenue is more than $800,000 for prices between $276.39 and $723.61,
$276.39 < p < $723.61. **37. (a)** $\{c \mid 0.112 < c < 81.907\}; (0.112, 81.907)$ **(b)** It is possible to hit a target 75 km away if $c = 0.651$ or $c = 1.536$.

Review Exercises *(page 314)*

1. (a) $m = 2; b = -5$ **(b)** 2 **3. (a)** $m = \dfrac{4}{5}; b = -6$ **(b)** $\dfrac{4}{5}$ **5. (a)** $m = 0; b = 4$ **(b)** 0 **7.** Linear; Slope: 5

(c)

(c)

(c)

(d) Increasing

(d) Increasing

(d) Constant

9.

11.

13.

15. (a)

(b) Domain: $(-\infty, \infty)$
Range: $[2, \infty)$

(c) Decreasing: $(-\infty, 2)$
Increasing: $(2, \infty)$

17. (a)

19. (a)

21. (a)

23. (a)

(b) Domain: $(-\infty, \infty)$

Range: $[-16, \infty)$

(c) Decreasing: $(-\infty, 0)$

Increasing: $(0, \infty)$

(b) Domain: $(-\infty, \infty)$

Range: $(-\infty, 1]$

(c) Increasing: $\left(-\infty, \dfrac{1}{2}\right)$

Decreasing: $\left(\dfrac{1}{2}, \infty\right)$

(b) Domain: $(-\infty, \infty)$

Range: $\left[\dfrac{1}{2}, \infty\right)$

(c) Decreasing: $\left(-\infty, -\dfrac{1}{3}\right)$

Increasing: $\left(-\dfrac{1}{3}, \infty\right)$

(b) Domain: $(-\infty, \infty)$

Range: $\left[-\dfrac{7}{3}, \infty\right)$

(c) Decreasing: $\left(-\infty, -\dfrac{2}{3}\right)$

Increasing: $\left(-\dfrac{2}{3}, \infty\right)$

25. Minimum value; 1 **27.** Maximum value; 12 **29.** Maximum value; 16 **31.** $\{x \mid -8 < x < 2\}$; $(-8, -2)$

33. $\left\{x \mid x \le -\dfrac{1}{3} \text{ or } x \ge 5\right\}$; $\left(-\infty, -\dfrac{1}{3}\right]$ or $[5, \infty)$ **35.** $y = x^2 + 2x + 3$

37. (a) Company A: $C(x) = 0.06x + 7$; Company B: $C(x) = 0.08x$ **(b)** 350 min **(c)** $0 \le x < 350$

39. (a) $R(x) = -\dfrac{1}{10}x^2 + 150x$ **(b)** \$14,000 **(c)** 750; \$56,250 **(d)** \$75 **41.** 4,166,666.7 m^2 **43. (a)** 63 clubs **(b)** \$151.90 **45.** 3.6 ft

47. (a) Quadratic, $a < 0$ **(b)** About \$26.5 thousand

(c) \$6408 thousand

(e)

Chapter Test *(page 316)*

1. (a) Slope: -4; y-intercept: 3 **2.** $\left(-\dfrac{4}{3}, 0\right)$, $(2, 0)$, $(0, -8)$ **4.** $\{-1, 3\}$ **5.**

(b) -4 **(c)** Decreasing

(d)

3. $\left(\dfrac{2 - \sqrt{6}}{2}, 0\right)$, $\left(\dfrac{2 + \sqrt{6}}{2}, 0\right)$, $(0, 1)$

6. (a) Opens up

(b) $(2, -8)$

(c) $x = 2$

(d) x-intercepts: $\dfrac{6 - 2\sqrt{6}}{3}$, $\dfrac{6 + 2\sqrt{6}}{3}$;

y-intercept: 4

(e)

7. Maximum value; 21 **8.** $\{x \mid x \le 4 \text{ or } x \ge 6\}$; $(-\infty, 4]$ or $[6, \infty)$

9. (a) $C(m) = 0.15m + 129.50$ **(b)** \$258.50 **(c)** 562 miles

Cumulative Review *(page 317)*

1. $5\sqrt{2}; \left(\frac{3}{2}, \frac{1}{2}\right)$ **2.** $(-2, -1)$ and $(2, 3)$ are on the graph.

3. $\left\{x \mid x \geq -\frac{3}{5}\right\}$ or $\left[-\frac{3}{5}, \infty\right)$ **4.** $y = -2x + 2$ **5.** $y = -\frac{1}{2}x + \frac{13}{2}$ **6.** $(x - 2)^2 + (y + 4)^2 = 25$

7. Yes **8. (a)** -3 **(b)** $x^2 - 4x - 2$ **(c)** $x^2 + 4x + 1$ **(d)** $-x^2 + 4x - 1$ **(e)** $x^2 - 3$ **(f)** $2x + h - 4$ **9.** $\left\{z \mid z \neq \frac{7}{6}\right\}$

10. Yes **11. (a)** No **(b)** $-1; (-2, -1)$ is on the graph. **(c)** $-8; (-8, 2)$ is on the graph. **12.** Neither **13.** Local maximum value is 5.30 and occurs at $x = -1.29$. Local minimum value is -3.30 and occurs at $x = 1.29$. Increasing: $(-4, -1.29)$ and $(1.29, 4)$; Decreasing: $(-1.29, 1.29)$

14. (a) -4 **(b)** $\{x \mid x > -4\}$ or $(-4, \infty)$

15. (a) Domain: $\{x \mid -4 \leq x \leq 4\}$; Range: $\{y \mid -1 \leq y \leq 3\}$ **(b)** $(-1, 0), (0, -1), (1, 0)$ **(c)** y-axis **(d)** 1 **(e)** -4 and 4 **(f)** $\{x \mid -1 < x < 1\}$

(g)

(h)

(i)

(j) Even **(k)** $(0, 4)$

CHAPTER 5 Polynomial and Rational Functions

5.1 Assess Your Understanding *(page 337)*

7. smooth; continuous **8.** touches **9.** $(-1, 1); (0, 0); (1, 1)$ **10.** r is a real zero of f; r is an x-intercept of the graph of f; $x - r$ is a factor of f.
11. turning points **12.** $y = 3x^4$ **13.** $\infty; -\infty$ **14.** As x increases in the positive direction, $f(x)$ decreases without bound. **15.** Yes; degree 3

17. Yes; degree 2 **19.** No; x is raised to the -1 power. **21.** No; x is raised to the $\frac{3}{2}$ power. **23.** Yes; degree 4 **25.** Yes; degree 4

27.

29.

31.

33.

35.

37.

39.

41. $f(x) = x^3 - 3x^2 - x + 3$ for $a = 1$ **43.** $f(x) = x^3 - x^2 - 12x$ for $a = 1$
45. $f(x) = x^4 - 15x^2 + 10x + 24$ for $a = 1$ **47.** $f(x) = x^3 - 5x^2 + 3x + 9$ for $a = 1$
49. (a) 7, multiplicity 1; -3, multiplicity 2 **(b)** Graph touches the x-axis at -3 and crosses it at 7. **(c)** Near -3: $f(x) \approx -30(x + 3)^2$; Near 7: $f(x) \approx 300(x - 7)$
(d) 2 **(e)** $y = 3x^3$

51. (a) 2, multiplicity 3 **(b)** Graph crosses the x-axis at 2. **(c)** Near 2: $f(x) \approx 20(x - 2)^3$ **(d)** 4 **(e)** $y = 4x^5$ **53. (a)** $-\frac{1}{2}$, multiplicity 2; -4, multiplicity 3 **(b)** Graph touches the x-axis at $-\frac{1}{2}$ and crosses at -4 **(c)** Near $-\frac{1}{2}$: $f(x) \approx -85.75\left(x + \frac{1}{2}\right)^2$; Near -4: $f(x) \approx -24.5(x + 4)^3$

(d) 4 **(e)** $y = -2x^5$ **55. (a)** 5, multiplicity 3; -4, multiplicity 2 **(b)** Graph touches the x-axis at -4 and crosses it at 5.
(c) Near -4: $f(x) \approx -729(x + 4)^2$; Near 5: $f(x) \approx 81(x - 5)^3$ **(d)** 4 **(e)** $y = x^5$ **57. (a)** No real zeros **(b)** Graph neither crosses nor touches the x-axis. **(c)** No real zeros **(d)** 5 **(e)** $y = 3x^6$ **59. (a)** 0, multiplicity 2; $-\sqrt{2}, \sqrt{2}$, multiplicity 1 **(b)** Graph touches the x-axis at 0 and crosses at $-\sqrt{2}$ and $\sqrt{2}$. **(c)** Near $-\sqrt{2}$: $f(x) \approx 11.31(x + \sqrt{2})$; Near 0: $f(x) \approx 4x^2$; Near $\sqrt{2}$: $f(x) \approx -11.31(x - \sqrt{2})$ **(d)** 3 **(e)** $y = -2x^4$
61. Could be; zeros: $-1, 1, 2$; Least degree is 3. **63.** Cannot be the graph of a polynomial; gap at $x = -1$ **65.** $f(x) = x(x - 1)(x - 2)$

67. $f(x) = -\frac{1}{2}(x + 1)(x - 1)^2(x - 2)$

69. Step 1: $y = x^3$
 Step 2: x-intercepts: $0, 3$; y-intercept: 0
 Step 3: 0: multiplicity 2, touches; 3: multiplicity 1, crosses
 Step 4: At most 2 turning points
 Step 5: Near 0: $f(x) \approx -3x^2$; Near 3: $f(x) \approx 9(x - 3)$
 Step 6:

71. Step 1: $y = x^3$
 Step 2: x-intercepts: $-4, 2$; y-intercept: 16
 Step 3: -4: multiplicity 1, crosses; 2: multiplicity 2, touches
 Step 4: At most 2 turning points
 Step 5: Near -4: $f(x) \approx 36(x + 4)$; Near 2: $f(x) \approx 6(x - 2)^2$
 Step 6:

73. Step 1: $y = -2x^4$
Step 2: x-intercepts: $-2, 2$; y-intercept: 32
Step 3: -2: multiplicity 1, crosses; 2: multiplicity 3, crosses
Step 4: At most 3 turning points
Step 5: Near -2: $f(x) \approx 128(x + 2)$; Near 2: $f(x) \approx -8(x - 2)^3$
Step 6:

75. Step 1: $y = x^3$
Step 2: x-intercepts: $-4, -1, 2$; y-intercept: -8
Step 3: $-4, -1, 2$: multiplicity 1, crosses
Step 4: At most 2 turning points
Step 5: Near -4: $f(x) \approx 18(x + 4)$; Near -1: $f(x) \approx -9(x + 1)$; Near 2: $f(x) \approx 18(x - 2)$
Step 6:

77. Step 1: $y = x^4$
Step 2: x-intercepts: $-2, 0, 2$; y-intercept: 0
Step 3: $-2, 2$, multiplicity 1, crosses; 0, multiplicity 2, touches
Step 4: At most 3 turning points
Step 5: Near -2: $f(x) \approx -16(x + 2)$; Near 0: $f(x) \approx -4x^2$; Near 2: $f(x) \approx 16(x - 2)$
Step 6:

79. Step 1: $y = x^4$
Step 2: x-intercepts: $-1, 2$; y-intercept: 4
Step 3: $-1, 2$: multiplicity 2, touches
Step 4: At most 3 turning points
Step 5: Near -1: $f(x) \approx 9(x + 1)^2$; Near 2: $f(x) \approx 9(x - 2)^2$
Step 6:

81. Step 1: $y = x^4$
Step 2: x-intercepts: $-1, 0, 3$; y-intercept: 0
Step 3: $-1, 3$: multiplicity 1, crosses; 0: multiplicity 2, touches
Step 4: At most 3 turning points
Step 5: Near -1: $f(x) \approx -4(x + 1)$; Near 0: $f(x) \approx -3x^2$; Near 3: $f(x) \approx 36(x - 3)$
Step 6:

83. Step 1: $y = x^4$
Step 2: x-intercepts: $-2, 4$; y-intercept: 64
Step 3: $-2, 4$, multiplicity 2, touches
Step 4: At most 3 turning points
Step 5: Near -2: $f(x) \approx 36(x + 2)^2$; Near 4: $f(x) \approx 36(x - 4)^2$
Step 6:

85. Step 1: $y = x^5$
Step 2: x-intercepts: $0, 2$; y-intercept: 0
Step 3: 2, multiplicity 1, crosses; 0, multiplicity 2, touches
Step 4: At most 4 turning points
Step 5: Near 0: $f(x) \approx -6x^2$; Near 2: $f(x) \approx 28(x - 2)$
Step 6:

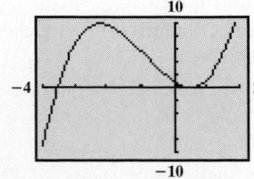

87. Step 1: $y = x^3$
Step 2:

Step 3: x-intercepts: $-1.26, -0.20, 1.26$
y-intercept: -0.31752
Step 4:

Step 5: $(-0.80, 0.57)$; $(0.66, -0.99)$
Step 6:

Step 7: Domain: $(-\infty, \infty)$;
Range: $(-\infty, \infty)$
Step 8: Increasing on $(-\infty, -0.80)$
and $(0.66, \infty)$
Decreasing on $(-0.80, 0.66)$

89. Step 1: $y = x^3$
Step 2:

Step 3: x-intercepts: $-3.56, 0.50$
y-intercept: 0.89

Step 4:

Step 5: $(-2.21, 9.91)$; $(0.50, 0)$

Step 6:

Step 7: Domain: $(-\infty, \infty)$;
Range: $(-\infty, \infty)$
Step 8: Increasing on $(-\infty, -2.21)$
and $(0.50, \infty)$
Decreasing on $(-2.21, 0.50)$

91. Step 1: $y = x^4$

Step 2:

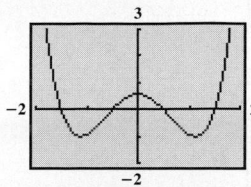

Step 3: x-intercepts: $-1.5, -0.5, 0.5, 1.5$

y-intercept: 0.5625

Step 4:

Step 5: $(-1.12, -1); (1.12, -1), (0, 0.56)$

Step 6:

Step 7: Domain: $(-\infty, \infty)$;

Range: $[-1, \infty)$

Step 8: Increasing on $(-1.12, 0)$ and $(1.12, \infty)$

Decreasing on $(-\infty, -1.12)$

and $(0, 1.12)$

93. Step 1: $y = 2x^4$

Step 2:

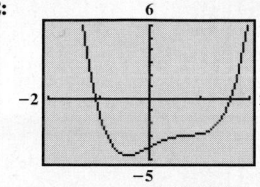

Step 3: x-intercepts: $-1.07, 1.62$;

y-intercept: -4

Step 4:

Step 5: $(-0.42, -4.64)$

Step 6:

Step 7: Domain: $(-\infty, \infty)$;

Range: $[-4.64, \infty)$

Step 8: Increasing on $(-0.42, \infty)$

Decreasing on $(-\infty, -0.42)$

95. $f(x) = -x(x + 2)(x - 2)$

Step 1: $y = -x^3$

Step 2: x-intercepts: $-2, 0, 2$; y-intercept: 0

Step 3: $-2, 0, 2$: multiplicity 1, crosses

Step 4: At most 2 turning points

Step 5: Near -2: $f(x) \approx -8(x + 2)$;

Near 0: $f(x) \approx 4x$;

Near 2: $f(x) \approx -8(x - 2)$

Step 6:

97. $f(x) = x(x + 4)(x - 3)$

Step 1: $y = x^3$

Step 2: x-intercepts: $-4, 0, 3$; y-intercept: 0

Step 3: $-4, 0, 3$: multiplicity 1, crosses

Step 4: At most 2 turning points

Step 5: Near -4: $f(x) \approx 28(x + 4)$;

Near 0: $f(x) \approx -12x$;

Near 3: $f(x) \approx 21(x - 3)$

Step 6:

99. $f(x) = 2x(x + 6)(x - 2)(x + 2)$

Step 1: $y = 2x^4$

Step 2: x-intercepts: $-6, -2, 0, 2$;

y-intercept: 0

Step 3: $-6, -2, 0, 2$: multiplicity 1, crosses

Step 4: At most 3 turning points

Step 5: Near -6: $f(x) \approx -384(x + 6)$;

Near -2: $f(x) \approx 64(x + 2)$;

Near 0: $f(x) \approx -48x$;

Near 2: $f(x) \approx 128(x - 2)$

Step 6:

101. $f(x) = -x^2(x + 1)^2(x - 1)$

Step 1: $y = -x^5$

Step 2: x-intercepts: $-1, 0, 1$;

y-intercept: 0

Step 3: 1: multiplicity 1, crosses;

$-1, 0$: multiplicity 2, touches

Step 4: At most 4 turning points

Step 5: Near -1: $f(x) \approx 2(x + 1)^2$;

Near 0: $f(x) \approx x^2$;

Near 1: $-4(x - 1)$

Step 6:

103. $f(x) = 3(x + 3)(x - 1)(x - 4)$

105. $f(x) = -2(x + 5)^2(x - 2)(x - 4)$

107. (a) $-3, 2$ **(b)** $-6, -1$

109. (a)

The relation appears to be cubic.

(b) $H(x) = 0.1591x^3 - 2.3203x^2 + 9.3301x - 2.2143$

(c) About 6 major hurricanes

(d)

(e) Approximately 10 major hurricanes

111. (a)

The relation appears to be cubic.

(b) Average rate of change: 2.27°/h

(c) Average rate of change: 0°/h

(d) $T(x) = -0.0103x^3 + 0.3174x^2 - 1.3742x + 45.3929$; The predicted temperature at 5 PM is 63.2°F.

(e)

(f) The y-intercept, 45.4°, is the predicted temperature at midnight.

113. (a)

(b)

(c)

(d) As more terms are added, the values of the polynomial function get closer to the values of f. The approximations near 0 are better than those near -1 or 1.

119. (a)–(d)

5.2 Assess Your Understanding (page 350)

5. F **6.** horizontal asymptote **7.** vertical asymptote **8.** proper **9.** T **10.** F **11.** $y = 0$ **12.** T **13.** All real numbers except 3; $\{x \mid x \neq 3\}$

15. All real numbers except 2 and -4; $\{x \mid x \neq 2, x \neq -4\}$ **17.** All real numbers except $-\frac{1}{2}$ and 3; $\left\{x \mid x \neq -\frac{1}{2}, x \neq 3\right\}$

19. All real numbers except 2; $\{x \mid x \neq 2\}$ **21.** All real numbers **23.** All real numbers except -3 and 3; $\{x \mid x \neq -3, x \neq 3\}$

25. (a) Domain: $\{x \mid x \neq 2\}$; Range: $\{y \mid y \neq 1\}$ **(b)** $(0,0)$ **(c)** $y = 1$ **(d)** $x = 2$ **(e)** None

27. (a) Domain: $\{x \mid x \neq 0\}$; Range: all real numbers **(b)** $(-1,0),(1,0)$ **(c)** None **(d)** $x = 0$ **(e)** $y = 2x$

29. (a) Domain: $\{x \mid x \neq -2, x \neq 2\}$; Range: $\{y \mid y \leq 0, y > 1\}$ **(b)** $(0,0)$ **(c)** $y = 1$ **(d)** $x = -2, x = 2$ **(e)** None

31. **33.** **35.** **37.** **39.** **41.**

43. Vertical asymptote: $x = -4$; horizontal asymptote: $y = 3$ **45.** Vertical asymptote: $x = 3$; oblique asymptote: $y = x + 5$

47. Vertical asymptotes: $x = 1, x = -1$; horizontal asymptote: $y = 0$ **49.** Vertical asymptote: $x = -\frac{1}{3}$; horizontal asymptote: $y = \frac{2}{3}$

51. Vertical asymptote: none; oblique asymptote: $y = 2x - 1$ **53.** Vertical asymptote: $x = 0$; no horizontal or oblique asymptote

55. (a) 9.8208 m/sec^2 **(b)** 9.8195 m/sec^2 **(c)** 9.7936 m/sec^2 **(d)** h-axis **(e)** \varnothing

57. (a) **(b)** Horizontal: $R_{\text{tot}} = 10$; as the resistance of R_2 increases without bound, the total resistance approaches 10 ohms, the resistance R_1. **(c)** $R_1 \approx 103.5$ ohms

5.3 Assess Your Understanding (page 365)

2. in lowest terms **3.** vertical **4.** True **5.** True **6. (a)** $\{x \mid x \neq 2\}$ **(b)** 0 **7. 1.** Domain: $\{x \mid x \neq 0, x \neq -4\}$ **2.** R is in lowest terms
3. no y-intercept; x-intercept: -1 **4.** R is in lowest terms; vertical asymptotes: $x = 0, x = -4$ **5.** Horizontal asymptote: $y = 0$, intersected at $(-1,0)$
6.

Interval	$(-\infty, -4)$	$(-4, -1)$	$(-1, 0)$	$(0, \infty)$
Number Chosen	-5	-2	$-\frac{1}{2}$	1
Value of R	$R(-5) = -\frac{4}{5}$	$R(-2) = \frac{1}{4}$	$R\left(-\frac{1}{2}\right) = -\frac{2}{7}$	$R(1) = \frac{2}{5}$
Location of Graph	Below x-axis	Above x-axis	Below x-axis	Above x-axis
Point on Graph	$\left(-5, -\frac{4}{5}\right)$	$\left(-2, \frac{1}{4}\right)$	$\left(-\frac{1}{2}, -\frac{2}{7}\right)$	$\left(1, \frac{2}{5}\right)$

7/8.

9. 1. $R(x) = \dfrac{3(x + 1)}{2(x + 2)}$; domain: $\{x \mid x \neq -2\}$ **2.** R is in lowest terms **3.** y-intercept: $\dfrac{3}{4}$; x-intercept: -1

4. R is in lowest terms; vertical asymptote: $x = -2$ **5.** Horizontal asymptote: $y = \dfrac{3}{2}$, not intersected

6.

Interval	$(-\infty, -2)$	$(-2, -1)$	$(-1, \infty)$
Number Chosen	-3	$-\dfrac{3}{2}$	0
Value of R	$R(-3) = 3$	$R\left(-\dfrac{3}{2}\right) = -\dfrac{3}{2}$	$R(0) = \dfrac{3}{4}$
Location of Graph	Above x-axis	Below x-axis	Above x-axis
Point on Graph	$(-3, 3)$	$\left(-\dfrac{3}{2}, -\dfrac{3}{2}\right)$	$\left(0, \dfrac{3}{4}\right)$

7/8.

11. 1. $R(x) = \dfrac{3}{(x + 2)(x - 2)}$; domain: $\{x \mid x \neq -2, x \neq 2\}$ **2.** R is in lowest terms **3.** y-intercept: $-\dfrac{3}{4}$; no x-intercept

4. R is in lowest terms; vertical asymptotes: $x = 2, x = -2$ **5.** Horizontal asymptote: $y = 0$, not intersected

6.

Interval	$(-\infty, -2)$	$(-2, 2)$	$(2, \infty)$
Number Chosen	-3	0	3
Value of R	$R(-3) = \dfrac{3}{5}$	$R(0) = -\dfrac{3}{4}$	$R(3) = \dfrac{3}{5}$
Location of Graph	Above x-axis	Below x-axis	Above x-axis
Point on Graph	$\left(-3, \dfrac{3}{5}\right)$	$\left(0, -\dfrac{3}{4}\right)$	$\left(3, \dfrac{3}{5}\right)$

7/8.

13. 1. $P(x) = \dfrac{(x^2 + x + 1)(x^2 - x + 1)}{(x + 1)(x - 1)}$; domain: $\{x \mid x \neq -1, x \neq 1\}$ **2.** P is in lowest terms **3.** y-intercept: -1; no x-intercept

4. P is in lowest terms; vertical asymptotes: $x = -1, x = 1$ **5.** No horizontal or oblique asymptote

6.

Interval	$(-\infty, -1)$	$(-1, 1)$	$(1, \infty)$
Number Chosen	-2	0	2
Value of P	$P(-2) = 7$	$P(0) = -1$	$P(2) = 7$
Location of Graph	Above x-axis	Below x-axis	Above x-axis
Point on Graph	$(-2, 7)$	$(0, -1)$	$(2, 7)$

7/8.

15. 1. $H(x) = \dfrac{(x - 1)(x^2 + x + 1)}{(x + 3)(x - 3)}$; domain: $\{x \mid x \neq -3, x \neq 3\}$ **2.** H is in lowest terms **3.** y-intercept: $\dfrac{1}{9}$; x-intercept: 1

4. H is in lowest terms; vertical asymptotes: $x = 3, x = -3$ **5.** Oblique asymptote: $y = x$, intersected at $\left(\dfrac{1}{9}, \dfrac{1}{9}\right)$

6.

Interval	$(-\infty, -3)$	$(-3, 1)$	$(1, 3)$	$(3, \infty)$
Number Chosen	-4	0	2	4
Value of H	$H(-4) \approx -9.3$	$H(0) = \dfrac{1}{9}$	$H(2) = -1.4$	$H(4) = 9$
Location of Graph	Below x-axis	Above x-axis	Below x-axis	Above x-axis
Point on Graph	$(-4, -9.3)$	$\left(0, \dfrac{1}{9}\right)$	$(2, -1.4)$	$(4, 9)$

7/8.

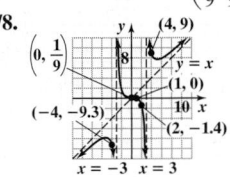

17. 1. $R(x) = \dfrac{x^2}{(x + 3)(x - 2)}$; domain: $\{x \neq -3, x \neq 2\}$ **2.** R is in lowest terms **3.** y-intercept: 0; x-intercept: 0

4. R is in lowest terms; vertical asymptotes: $x = 2, x = -3$ **5.** Horizontal asymptote: $y = 1$, intersected at $(6, 1)$

6.

Interval	$(-\infty, -3)$	$(-3, 0)$	$(0, 2)$	$(2, \infty)$
Number Chosen	-6	-1	1	3
Value of R	$R(-6) = 1.5$	$R(-1) = -\dfrac{1}{6}$	$R(1) = -0.25$	$R(3) = 1.5$
Location of Graph	Above x-axis	Below x-axis	Below x-axis	Above x-axis
Point on Graph	$(-6, 1.5)$	$\left(-1, -\dfrac{1}{6}\right)$	$(1, -0.25)$	$(3, 1.5)$

7/8.

19. 1. $G(x) = \dfrac{x}{(x + 2)(x - 2)}$; domain: $\{x \mid x \neq -2, x \neq 2\}$ **2.** G is in lowest terms **3.** y-intercept: 0; x-intercept: 0

4. G is in lowest terms; vertical asymptotes: $x = -2, x = 2$ **5.** Horizontal asymptote: $y = 0$, intersected at $(0, 0)$

6.

Interval	$(-\infty, -2)$	$(-2, 0)$	$(0, 2)$	$(2, \infty)$
Number Chosen	-3	-1	1	3
Value of G	$G(-3) = -\dfrac{3}{5}$	$G(-1) = \dfrac{1}{3}$	$G(1) = -\dfrac{1}{3}$	$G(3) = \dfrac{3}{5}$
Location of Graph	Below x-axis	Above x-axis	Below x-axis	Above x-axis
Point on Graph	$\left(-3, -\dfrac{3}{5}\right)$	$\left(-1, \dfrac{1}{3}\right)$	$\left(1, -\dfrac{1}{3}\right)$	$\left(3, \dfrac{3}{5}\right)$

7/8.

21. **1.** $R(x) = \dfrac{3}{(x-1)(x+2)(x-2)}$; domain: $\{x \mid x \neq 1, x \neq -2, x \neq 2\}$ **2.** R is in lowest terms **3.** y-intercept: $\dfrac{3}{4}$; no x-intercept

4. R is in lowest terms; vertical asymptotes: $x = -2, x = 1, x = 2$ **5.** Horizontal asymptote: $y = 0$, not intersected

6.

Interval	$(-\infty, -2)$	$(-2, 1)$	$(1, 2)$	$(2, \infty)$
Number Chosen	-3	0	1.5	3
Value of R	$R(-3) = -\frac{3}{20}$	$R(0) = \frac{3}{4}$	$R(1.5) = -\frac{24}{7}$	$R(3) = \frac{3}{10}$
Location of Graph	Below x-axis	Above x-axis	Below x-axis	Above x-axis
Point on Graph	$\left(-3, -\frac{3}{20}\right)$	$\left(0, \frac{3}{4}\right)$	$\left(1.5, -\frac{24}{7}\right)$	$\left(3, \frac{3}{10}\right)$

7/8.

23. **1.** $H(x) = \dfrac{(x+1)(x-1)}{(x^2+4)(x+2)(x-2)}$; domain: $\{x \mid x \neq -2, x \neq 2\}$ **2.** H is in lowest terms **3.** y-intercept: $\dfrac{1}{16}$; x-intercepts: $-1, 1$

4. H is in lowest terms; vertical asymptotes: $x = -2, x = 2$ **5.** Horizontal asymptote: $y = 0$, intersected at $(-1, 0)$ and $(1, 0)$

6.

Interval	$(-\infty, -2)$	$(-2, -1)$	$(-1, 1)$	$(1, 2)$	$(2, \infty)$
Number Chosen	-3	-1.5	0	1.5	3
Value of H	$H(-3) \approx 0.12$	$H(-1.5) \approx -0.11$	$H(0) = \frac{1}{16}$	$H(1.5) \approx -0.11$	$H(3) \approx 0.12$
Location of Graph	Above x-axis	Below x-axis	Above x-axis	Below x-axis	Above x-axis
Point on Graph	$(-3, 0.12)$	$(-1.5, -0.11)$	$\left(0, \frac{1}{16}\right)$	$(1.5, -0.11)$	$(3, 0.12)$

7/8.

25. **1.** $F(x) = \dfrac{(x+1)(x-4)}{x+2}$; domain: $\{x \mid x \neq -2\}$ **2.** F is in lowest terms **3.** y-intercept: -2; x-intercepts: $-1, 4$

4. F is in lowest terms; vertical asymptote: $x = -2$ **5.** Oblique asymptote: $y = x - 5$, not intersected

6.

Interval	$(-\infty, -2)$	$(-2, -1)$	$(-1, 4)$	$(4, \infty)$
Number Chosen	-3	-1.5	0	5
Value of F	$F(-3) = -14$	$F(-1.5) = 5.5$	$F(0) = -2$	$F(5) \approx 0.86$
Location of Graph	Below x-axis	Above x-axis	Below x-axis	Above x-axis
Point on Graph	$(-3, -14)$	$(-1.5, 5.5)$	$(0, -2)$	$(5, 0.86)$

7/8.

27. **1.** $R(x) = \dfrac{(x+4)(x-3)}{x-4}$; domain: $\{x \mid x \neq 4\}$ **2.** R is in lowest terms **3.** y-intercept: 3; x-intercepts: $-4, 3$

4. R is in lowest terms; vertical asymptote: $x = 4$ **5.** Oblique asymptote: $y = x + 5$, not intersected

6.

Interval	$(-\infty, -4)$	$(-4, 3)$	$(3, 4)$	$(4, \infty)$
Number Chosen	-5	0	3.5	5
Value of R	$R(-5) = -\frac{8}{9}$	$R(0) = 3$	$R(3.5) = -7.5$	$R(5) = 18$
Location of Graph	Below x-axis	Above x-axis	Below x-axis	Above x-axis
Point on Graph	$\left(-5, -\frac{8}{9}\right)$	$(0, 3)$	$(3.5, -7.5)$	$(5, 18)$

7/8.

29. **1.** $F(x) = \dfrac{(x+4)(x-3)}{x+2}$; domain: $\{x \mid x \neq -2\}$ **2.** F is in lowest terms **3.** y-intercept: -6; x-intercepts: $-4, 3$

4. F is in lowest terms; vertical asymptote: $x = -2$ **5.** Oblique asymptote: $y = x - 1$, not intersected

6.

Interval	$(-\infty, -4)$	$(-4, -2)$	$(-2, 3)$	$(3, \infty)$
Number Chosen	-5	-3	0	4
Value of F	$F(-5) = -\frac{8}{3}$	$F(-3) = 6$	$F(0) = -6$	$F(4) = \frac{4}{3}$
Location of Graph	Below x-axis	Above x-axis	Below x-axis	Above x-axis
Point on Graph	$\left(-5, -\frac{8}{3}\right)$	$(-3, 6)$	$(0, -6)$	$\left(4, \frac{4}{3}\right)$

7/8.

31. **1.** Domain: $\{x \mid x \neq -3\}$ **2.** R is in lowest terms **3.** y-intercept: 0; x-intercepts: $0, 1$ **4.** Vertical asymptote: $x = -3$

5. Horizontal asymptote: $y = 1$, not intersected

6.

Interval	$(-\infty, -3)$	$(-3, 0)$	$(0, 1)$	$(1, \infty)$
Number Chosen	-4	-1	$\frac{1}{2}$	2
Value of R	$R(-4) = 100$	$R(-1) = -0.5$	$R\left(\frac{1}{2}\right) \approx 0.003$	$R(2) = 0.016$
Location of Graph	Above x-axis	Below x-axis	Above x-axis	Above x-axis
Point on Graph	$(-4, 100)$	$(-1, -0.5)$	$\left(\frac{1}{2}, 0.003\right)$	$(2, 0.016)$

7/8. See enlarged view at right.

33. 1. $R(x) = \dfrac{(x + 4)(x - 3)}{(x + 2)(x - 3)}$; domain: $\{x | x \neq -2, x \neq 3\}$ **2.** In lowest terms, $R(x) = \dfrac{x + 4}{x + 2}$ **3.** y-intercept: 2; x-intercept: -4

4. Vertical asymptote: $x = -2$; hole at $\left(3, \dfrac{7}{5}\right)$ **5.** Horizontal asymptote: $y = 1$, not intersected

6.

Interval	$(-\infty, -4)$	$(-4, -2)$	$(-2, 3)$	$(3, \infty)$
Number Chosen	-5	-3	0	4
Value of R	$R(-5) = \frac{1}{3}$	$R(-3) = -1$	$R(0) = 2$	$R(4) = \frac{4}{3}$
Location of Graph	Above x-axis	Below x-axis	Above x-axis	Above x-axis
Point on Graph	$\left(-5, \frac{1}{3}\right)$	$(-3, -1)$	$(0, 2)$	$\left(4, \frac{4}{3}\right)$

7/8.

35. 1. $R(x) = \dfrac{(3x + 1)(2x - 3)}{(x - 2)(2x - 3)}$; domain: $\left\{x \middle| x \neq \dfrac{3}{2}, x \neq 2\right\}$ **2.** In lowest terms, $R(x) = \dfrac{3x + 1}{x - 2}$ **3.** y-intercept: $-\dfrac{1}{2}$; x-intercept: $-\dfrac{1}{3}$

4. Vertical asymptote: $x = 2$; hole at $\left(\dfrac{3}{2}, -11\right)$ **5.** Horizontal asymptote: $y = 3$, not intersected

6.

Interval	$\left(-\infty, -\frac{1}{3}\right)$	$\left(-\frac{1}{3}, \frac{3}{2}\right)$	$\left(\frac{3}{2}, 2\right)$	$(2, \infty)$
Number Chosen	-1	0	1.7	6
Value of R	$R(-1) = \frac{2}{3}$	$R(0) = -\frac{1}{2}$	$R(1.7) \approx -20.3$	$R(6) = 4.75$
Location of Graph	Above x-axis	Below x-axis	Below x-axis	Above x-axis
Point on Graph	$\left(-1, \frac{2}{3}\right)$	$\left(0, -\frac{1}{2}\right)$	$(1.7, -20.3)$	$(6, 4.75)$

7/8.

37. 1. $R(x) = \dfrac{(x + 3)(x + 2)}{x + 3}$; domain: $\{x | x \neq -3\}$ **2.** In lowest terms, $R(x) = x + 2$ **3.** y-intercept: 2; x-intercept: -2

4. Vertical asymptote: none; hole at $(-3, -1)$ **5.** Oblique asymptote: $y = x + 2$ intersected at all points except $x = -3$

6.

Interval	$(-\infty, -3)$	$(-3, -2)$	$(-2, \infty)$
Number Chosen	-4	$-\frac{5}{2}$	0
Value of R	$R(-4) = -2$	$R\left(-\frac{5}{2}\right) = -\frac{1}{2}$	$R(0) = 2$
Location of Graph	Below x-axis	Below x-axis	Above x-axis
Point on Graph	$(-4, -2)$	$\left(-\frac{5}{2}, -\frac{1}{2}\right)$	$(0, 2)$

7/8.

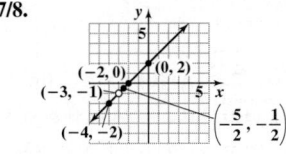

39. 1. $f(x) = \dfrac{x^2 + 1}{x}$; domain: $\{x | x \neq 0\}$ **2.** f is in lowest terms **3.** no y-intercept; no x-intercepts

4. f is in lowest terms; vertical asymptote: $x = 0$ **5.** Oblique asymptote: $y = x$, not intersected

6.

Interval	$(-\infty, 0)$	$(0, \infty)$
Number Chosen	-1	1
Value of f	$f(-1) = -2$	$f(1) = 2$
Location of Graph	Below x-axis	Above x-axis
Point on Graph	$(-1, -2)$	$(1, 2)$

7/8.

41. 1. $f(x) = \dfrac{x^3 + 1}{x} = \dfrac{(x + 1)(x^2 - x + 1)}{x}$; domain: $\{x | x \neq 0\}$ **2.** f is in lowest terms **3.** no y-intercept; x-intercept: -1

4. f is in lowest terms; vertical asymptote: $x = 0$ **5.** No horizontal or oblique asymptote

6.

Interval	$(-\infty, -1)$	$(-1, 0)$	$(0, \infty)$
Number Chosen	-2	$-\frac{1}{2}$	1
Value of f	$f(-2) = \frac{7}{2}$	$f\left(-\frac{1}{2}\right) = -\frac{7}{4}$	$f(1) = 2$
Location of Graph	Above x-axis	Below x-axis	Above x-axis
Point on Graph	$\left(-2, \frac{7}{2}\right)$	$\left(-\frac{1}{2}, -\frac{7}{4}\right)$	$(1, 2)$

7/8.

43. 1. $f(x) = \dfrac{x^4 + 1}{x^3}$; domain: $\{x | x \neq 0\}$ **2.** f is in lowest terms **3.** no y-intercept; no x-intercepts

4. f is in lowest terms; vertical asymptote: $x = 0$ **5.** Oblique asymptote: $y = x$, not intersected

6.

Interval	$(-\infty, 0)$	$(0, \infty)$
Number Chosen	-1	1
Value of f	$f(-1) = -2$	$f(1) = 2$
Location of Graph	Below x-axis	Above x-axis
Point on Graph	$(-1, -2)$	$(1, 2)$

7/8.

45. One possibility: $R(x) = \dfrac{x^2}{x^2 - 4}$ **47.** One possibility: $R(x) = \dfrac{(x - 1)(x - 3)\left(x^2 + \dfrac{4}{3}\right)}{(x + 1)^2(x - 2)^2}$

49. (a) t-axis; $C(t) \to 0$

(b)

(c) 0.71 h after injection

51. (a) $C(x) = 16x + \dfrac{5000}{x} + 100$

(b) $x > 0$

(c)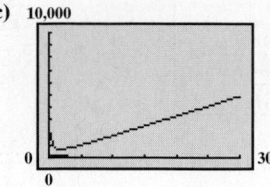

(d) Approximately 17.7 ft by 56.6 ft (longer side parallel to river)

53. (a) $S(x) = 2x^2 + \dfrac{40{,}000}{x}$

(b)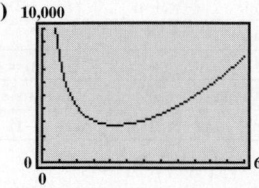

(c) 2784.95 in.2

(d) 21.54 in. \times 21.54 in. \times 21.54 in.

(e) To minimize the cost of materials needed for construction

55. (a) $C(r) = 12\pi r^2 + \dfrac{4000}{r}$

(b)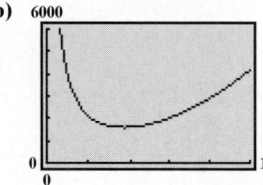

The cost is smallest when $r = 3.76$ cm.

57. No. Each function is a quotient of polynomials, but it is not written in lowest terms. Each function is undefined for $x = 1$; each graph has a hole at $x = 1$.

63. If there is a common factor between the numerator and the denominator, and the factor yields a real zero, then the graph will have a hole.

5.4 Assess Your Understanding *(page 371)*

3. T **4.** F **5. (a)** $\{x \mid 0 < x < 1 \text{ or } x > 2\}$; $(0, 1) \cup (2, \infty)$ **(b)** $\{x \mid x \le 0 \text{ or } 1 \le x \le 2\}$; $(-\infty, 0] \cup [1, 2]$

7. (a) $\{x \mid -1 < x < 0 \text{ or } x > 1\}$; $(-1, 0) \cup (1, \infty)$ **(b)** $\{x \mid x < -1 \text{ or } 0 \le x < 1\}$; $(-\infty, -1) \cup [0, 1)$

9. $\{x \mid x < 0 \text{ or } 0 < x < 3\}$; $(-\infty, 0) \cup (0, 3)$ **11.** $\{x \mid x \ge -4\}$; $[-4, \infty)$ **13.** $\{x \mid x \le -2 \text{ or } x \ge 2\}$; $(-\infty, -2] \cup [2, \infty)$

15. $\{x \mid -4 < x < -1 \text{ or } x > 0\}$; $(-4, -1) \cup (0, \infty)$ **17.** $\{x \mid -2 < x \le -1\}$; $(-2, -1]$ **19.** $\{x \mid x < -2\}$; $(-\infty, -2)$ **21.** $\{x \mid x > 4\}$; $(4, \infty)$

23. $\{x \mid -4 < x < 0 \text{ or } x > 0\}$; $(-4, 0) \cup (0, \infty)$ **25.** $\{x \mid x \le 1 \text{ or } 2 \le x \le 3\}$; $(-\infty, 1] \cup [2, 3]$ **27.** $\{x \mid -1 < x < 0 \text{ or } x > 3\}$; $(-1, 0) \cup (3, \infty)$

29. $\{x \mid x < -1 \text{ or } x > 1\}$; $(-\infty, -1) \cup (1, \infty)$ **31.** $\{x \mid x < -1 \text{ or } x > 1\}$; $(-\infty, -1) \cup (1, \infty)$ **33.** $\{x \mid x < -1 \text{ or } x > 1\}$; $(-\infty, -1) \cup (1, \infty)$

35. $\{x \mid x \le -1 \text{ or } 0 < x \le 1\}$; $(-\infty, -1] \cup (0, 1]$ **37.** $\{x \mid x < -1 \text{ or } x > 1\}$; $(-\infty, -1) \cup (1, \infty)$ **39.** $\{x \mid x < 2\}$; $(-\infty, 2)$

41. $\{x \mid -2 < x \le 9\}$; $(-2, 9]$ **43.** $\{x \mid x < 2 \text{ or } 3 < x < 5\}$; $(-\infty, 2) \cup (3, 5)$ **45.** $\{x \mid x < -5 \text{ or } -4 \le x \le -3 \text{ or } x = 0 \text{ or } x > 1\}$;

$(-\infty, -5) \cup [-4, -3] \cup \{0\} \cup (1, \infty)$ **47.** $\left\{x \mid -\dfrac{1}{2} < x < 1 \text{ or } x > 3\right\}$; $\left(-\dfrac{1}{2}, 1\right) \cup (3, \infty)$ **49.** $\{x \mid -1 < x < 3 \text{ or } x > 5\}$; $(-1, 3) \cup (5, \infty)$

51. $\left\{x \mid x \le -4 \text{ or } x \ge \dfrac{1}{2}\right\}$; $(-\infty, -4] \cup \left[\dfrac{1}{2}, \infty\right)$ **53.** $\{x \mid x < 3 \text{ or } x \ge 7\}$; $(-\infty, 3) \cup [7, \infty)$ **55.** $\{x \mid x < 2\}$; $(-\infty, 2)$

57. $\left\{x \mid x < -\dfrac{2}{3} \text{ or } 0 < x < \dfrac{3}{2}\right\}$; $\left(-\infty, -\dfrac{2}{3}\right) \cup \left(0, \dfrac{3}{2}\right)$ **59.** $\{x \mid x \le -3 \text{ or } 0 \le x \le 3\}$; $(-\infty, -3] \cup [0, 3]$ **61.** $\{x \mid x > 4\}$; $(4, \infty)$

63. $\{x \mid x \le -2 \text{ or } x \ge 2\}$; $(-\infty, -2] \cup [2, \infty)$ **65.** $\{x \mid x < -4 \text{ or } x \ge 2\}$; $(-\infty, -4) \cup [2, \infty)$

67.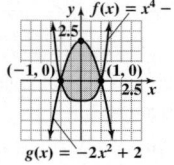

$f(x) \le g(x)$ if $-1 \le x \le 1$

69.

$f(x) \le g(x)$ if $-2 \le x \le 2$

71. Produce at least 250 bicycles

73. (a) The stretch is less than 39 ft.

(b) The ledge should be at least 84 ft above the ground for a 150-lb jumper.

75. At least 50 students must attend.

Historical Problems *(page 384)*

1.
$$\left(x - \frac{b}{3}\right)^3 + b\left(x - \frac{b}{3}\right)^2 + c\left(x - \frac{b}{3}\right) + d = 0$$

$$x^3 - bx^2 + \frac{b^2 x}{3} - \frac{b^3}{27} + bx^2 - \frac{2b^2 x}{3} + \frac{b^3}{9} + cx - \frac{bc}{3} + d = 0$$

$$x^3 + \left(c - \frac{b^2}{3}\right)x + \left(\frac{2b^3}{27} - \frac{bc}{3} + d\right) = 0$$

Let $p = c - \dfrac{b^2}{3}$ and $q = \dfrac{2b^3}{27} - \dfrac{bc}{3} + d$. Then $x^3 + px + q = 0$.

2.
$$(H + K)^3 + p(H + K) + q = 0$$

$$H^3 + 3H^2 K + 3HK^2 + K^3 + pH + pK + q = 0$$

Let $3HK = -p$.

$$H^3 - pH - pK + K^3 + pH + pK + q = 0, \quad H^3 + K^3 = -q$$

3.
$$3HK = -p$$
$$K = -\frac{p}{3H}$$
$$H^3 + \left(-\frac{p}{3H}\right)^3 = -q$$
$$H^3 - \frac{p^3}{27H^3} = -q$$
$$27H^6 - p^3 = -27qH^3$$
$$27H^6 + 27qH^3 - p^3 = 0$$
$$H^3 = \frac{-27q \pm \sqrt{(27q)^2 - 4(27)(-p^3)}}{2 \cdot 27}$$
$$H^3 = \frac{-q}{2} \pm \sqrt{\frac{27^2 q^2}{2^2(27^2)} + \frac{4(27)p^3}{2^2(27^2)}}$$
$$H^3 = \frac{-q}{2} \pm \sqrt{\frac{q^2}{4} + \frac{p^3}{27}}$$
$$H = \sqrt[3]{\frac{-q}{2} + \sqrt{\frac{q^2}{4} + \frac{p^3}{27}}}$$

Choose the positive root for now.

4.
$$H^3 + K^3 = -q$$
$$K^3 = -q - H^3$$
$$K^3 = -q - \left[\frac{-q}{2} + \sqrt{\frac{q^2}{4} + \frac{p^3}{27}}\right]$$
$$K^3 = \frac{-q}{2} - \sqrt{\frac{q^2}{4} + \frac{p^3}{27}}$$
$$K = \sqrt[3]{\frac{-q}{2} - \sqrt{\frac{q^2}{4} + \frac{p^3}{27}}}$$

5. $x = H + K$
$$x = \sqrt[3]{\frac{-q}{2} + \sqrt{\frac{q^2}{4} + \frac{p^3}{27}}} + \sqrt[3]{\frac{-q}{2} - \sqrt{\frac{q^2}{4} + \frac{p^3}{27}}}$$

(Note that had we used the negative root in 3 the result would be the same.)

6. $x = 3$ **7.** $x = 2$ **8.** $x = 2$

5.5 Assess Your Understanding *(page 384)*

5. Remainder; dividend **6.** $f(c)$ **7.** -4 **8.** F **9.** 0 **10.** T **11.** $R = f(2) = 8$; no **13.** $R = f(2) = 0$; yes **15.** $R = f(-3) = 0$; yes

17. $R = f(-4) = 1$; no **19.** $R = f\left(\frac{1}{2}\right) = 0$; yes **21.** 7 **23.** 6 **25.** 3 **27.** 4 **29.** 5 **31.** 6 **33.** $\pm 1, \pm\frac{1}{3}$ **35.** $\pm 1, \pm 3$

37. $\pm 1, \pm 2, \pm\frac{1}{4}, \pm\frac{1}{2}$ **39.** $\pm 1, \pm 3, \pm 9, \pm\frac{1}{2}, \pm\frac{1}{3}, \pm\frac{1}{6}, \pm\frac{3}{2}, \pm\frac{9}{2}$ **41.** $\pm 1, \pm 2, \pm 3, \pm 4, \pm 6, \pm 12, \pm\frac{1}{2}, \pm\frac{3}{2}$

43. $\pm 1, \pm 2, \pm 4, \pm 5, \pm 10, \pm 20, \pm\frac{1}{2}, \pm\frac{5}{2}, \pm\frac{1}{3}, \pm\frac{2}{3}, \pm\frac{4}{3}, \pm\frac{5}{3}, \pm\frac{10}{3}, \pm\frac{20}{3}, \pm\frac{1}{6}, \pm\frac{5}{6}$ **45.** $-3, -1, 2; f(x) = (x + 3)(x + 1)(x - 2)$

47. $\frac{1}{2}; f(x) = 2\left(x - \frac{1}{2}\right)(x^2 + 1)$ **49.** $2, \sqrt{5}, -\sqrt{5}; f(x) = 2(x - 2)(x - \sqrt{5})(x + \sqrt{5})$

51. $-1, \frac{1}{2}, \sqrt{3}, -\sqrt{3}; f(x) = 2(x + 1)\left(x - \frac{1}{2}\right)(x - \sqrt{3})(x + \sqrt{3})$ **53.** 1, multiplicity 2; $-2, -1; f(x) = (x + 2)(x + 1)(x - 1)^2$

55. $-1, -\frac{1}{4}; f(x) = 4(x + 1)\left(x + \frac{1}{4}\right)(x^2 + 2)$ **57.** $\{-1, 2\}$ **59.** $\left\{\frac{2}{3}, -1 + \sqrt{2}, -1 - \sqrt{2}\right\}$ **61.** $\left\{\frac{1}{3}, \sqrt{5}, -\sqrt{5}\right\}$ **63.** $\{-3, -2\}$ **65.** $\left\{-\frac{1}{3}\right\}$

67. $\left\{\frac{1}{2}, 2, 5\right\}$ **69.** 5 **71.** 2 **73.** 5 **75.** $\frac{3}{2}$ **77.** $f(0) = -1; f(1) = 10$ **79.** $f(-5) = -58; f(-4) = 2$ **81.** $f(1.4) = -0.17536; f(1.5) = 1.40625$

83. 0.21 **85.** -4.04 **87.** 1.15 **89.** 2.53

91.

93.

95.

97.

99.

101.

103. $k = 5$ **105.** -7 **107.** If $f(x) = x^n - c^n$, then $f(c) = c^n - c^n = 0$, so $x - c$ is a factor of f. **109.** 5 **111.** 7 in. **113.** All the potential rational zeros are integers, so r is either an integer or is not a rational zero (and is therefore irrational).

115. 0.215 **117.** No; by the Rational Zeros Theorem, $\frac{1}{3}$ is not a potential rational zero.

119. No; by the Rational Zeros Theorem, $\frac{2}{3}$ is not a potential rational zero.

5.6 Assess Your Understanding *(page 392)*

3. one **4.** $3 - 4i$ **5.** T **6.** F **7.** $4 + i$ **9.** $-i, 1 - i$ **11.** $-i, -2i$ **13.** $-i$ **15.** $2 - i, -3 + i$ **17.** $f(x) = x^4 - 14x^3 + 77x^2 - 200x + 208; a = 1$

19. $f(x) = x^5 - 4x^4 + 7x^3 - 8x^2 + 6x - 4; a = 1$ **21.** $f(x) = x^4 - 6x^3 + 10x^2 - 6x + 9; a = 1$ **23.** $-2i, 4$ **25.** $2i, -3, \frac{1}{2}$

27. $3 + 2i, -2, 5$ **29.** $4i, -\sqrt{11}, \sqrt{11}, -\frac{2}{3}$ **31.** $1, -\frac{1}{2} - \frac{\sqrt{3}}{2}i, -\frac{1}{2} + \frac{\sqrt{3}}{2}i; f(x) = (x - 1)\left(x + \frac{1}{2} + \frac{\sqrt{3}}{2}i\right)\left(x + \frac{1}{2} - \frac{\sqrt{3}}{2}i\right)$

33. $2, 3 - 2i, 3 + 2i; f(x) = (x - 2)(x - 3 + 2i)(x - 3 - 2i)$ **35.** $-i, i, -2i, 2i; f(x) = (x + i)(x - i)(x + 2i)(x - 2i)$

37. $-5i, 5i, -3, 1; f(x) = (x + 5i)(x - 5i)(x + 3)(x - 1)$ **39.** $-4, \frac{1}{3}, 2 - 3i, 2 + 3i; f(x) = 3(x + 4)\left(x - \frac{1}{3}\right)(x - 2 + 3i)(x - 2 - 3i)$

41. Zeros that are complex numbers must occur in conjugate pairs; or a polynomial with real coefficients of odd degree must have at least one real zero.

43. If the remaining zero were a complex number, its conjugate would also be a zero, creating a polynomial of degree 5.

Review Exercises (page 394)

1. Polynomial of degree 5 **3.** Neither

5.

7.

9.

11. 1. $y = x^3$
2. x-intercepts: $-4, -2, 0$; y-intercept: 0
3. $-4, -2, 0$, multiplicity 1, crosses
4. 2
5. Near -4: $f(x) \approx 8(x + 4)$
Near -2: $f(x) \approx -4(x + 2)$
Near 0: $f(x) \approx 8x$
6.

13. 1. $y = x^3$
2. x-intercepts: $-4, 2$; y-intercept: 16
3. -4, multiplicity 1, crosses; 2, multiplicity 2, touches
4. 2
5. Near -4: $f(x) \approx 36(x + 4)$
Near 2: $f(x) \approx 6(x - 2)^2$
6.

15. 1. $y = -2x^3$
2. $f(x) = -2x^2(x - 2)$
x-intercepts: $0, 2$; y-intercept: 0
3. 0, multiplicity 2, touches; 2, multiplicity 1, crosses
4. 2
5. Near 0: $f(x) \approx 4x^2$
Near 2: $f(x) \approx -8(x - 2)$
6.

17. 1. $y = x^4$
2. x-intercepts: $-3, -1, 1$; y-intercept: 3
3. $-3, -1$, multiplicity 1, crosses; 1, multiplicity 2, touches
4. 3
5. Near -3: $f(x) \approx -32(x + 3)$
Near -1: $f(x) \approx 8(x + 1)$; Near 1: $f(x) \approx 8(x - 1)^2$

6.

19. Domain: $\{x | x \neq -3, x \neq 3\}$: horizontal asymptote: $y = 0$; vertical asymptotes: $x = -3, x = 3$

21. Domain: $\{x | x \neq -2\}$; horizontal asymptote: $y = 1$; vertical asymptote: $x = -2$

23. 1. $R(x) = \dfrac{2(x - 3)}{x}$; domain: $\{x | x \neq 0\}$ **2.** R is in lowest terms **3.** no y-intercept; x-intercept: 3
4. R is in lowest terms; vertical asymptote: $x = 0$ **5.** Horizontal asymptote: $y = 2$; not intersected

6.

Interval	$(-\infty, 0)$	$(0, 3)$	$(3, \infty)$
Number Chosen	-2	1	4
Value of R	$R(-2) = 5$	$R(1) = -4$	$R(4) = \frac{1}{2}$
Location of Graph	Above x-axis	Below x-axis	Above x-axis
Point on Graph	$(-2, 5)$	$(1, -4)$	$\left(4, \frac{1}{2}\right)$

7/8.

25. 1. Domain: $\{x | x \neq 0, x \neq 2\}$ **2.** H is in lowest terms **3.** no y-intercept; x-intercept: -2 **4.** H is in lowest terms;
vertical asymptote: $x = 0, x = 2$ **5.** Horizontal asymptote: $y = 0$; intersected at $(-2, 0)$

6.

Interval	$(-\infty, -2)$	$(-2, 0)$	$(0, 2)$	$(2, \infty)$
Number Chosen	-3	-1	1	3
Value of H	$H(-3) = -\frac{1}{15}$	$H(-1) = \frac{1}{3}$	$H(1) = -3$	$H(3) = \frac{5}{3}$
Location of Graph	Below x-axis	Above x-axis	Below x-axis	Above x-axis
Point on Graph	$\left(-3, -\frac{1}{15}\right)$	$\left(-1, \frac{1}{3}\right)$	$(1, -3)$	$\left(3, \frac{5}{3}\right)$

7/8.

27. 1. $R(x) = \dfrac{(x + 3)(x - 2)}{(x - 3)(x + 2)}$; domain: $\{x | x \neq -2, x \neq 3\}$ **2.** R is in lowest terms **3.** y-intercept: 1; x-intercepts: $-3, 2$
4. R is in lowest terms; vertical asymptotes: $x = -2, x = 3$ **5.** Horizontal asymptote: $y = 1$; intersected at $(0, 1)$

6.

Interval	$(-\infty, -3)$	$(-3, -2)$	$(-2, 2)$	$(2, 3)$	$(3, \infty)$
Number Chosen	-4	$-\frac{5}{2}$	0	$\frac{5}{2}$	4
Value of R	$R(-4) = \frac{3}{7}$	$R\left(-\frac{5}{2}\right) = -\frac{9}{11}$	$R(0) = 1$	$R\left(\frac{5}{2}\right) = -\frac{11}{9}$	$R(4) = \frac{7}{3}$
Location of Graph	Above x-axis	Below x-axis	Above x-axis	Below x-axis	Above x-axis
Point on Graph	$\left(-4, \frac{3}{7}\right)$	$\left(-\frac{5}{2}, -\frac{9}{11}\right)$	$(0, 1)$	$\left(\frac{5}{2}, -\frac{11}{9}\right)$	$\left(4, \frac{7}{3}\right)$

7/8.

29. 1. $F(x) = \dfrac{x^3}{(x + 2)(x - 2)}$; domain: $\{x \mid x \neq -2, x \neq 2\}$ **2.** F is in lowest terms **3.** y-intercept: 0; x-intercept: 0
4. F is in lowest terms; vertical asymptotes: $x = -2, x = 2$ **5.** Oblique asymptote: $y = x$; intersected at $(0, 0)$
6.

Interval	$(-\infty, -2)$	$(-2, 0)$	$(0, 2)$	$(2, \infty)$
Number Chosen	-3	-1	1	3
Value of F	$F(-3) = -\frac{27}{5}$	$F(-1) = \frac{1}{3}$	$F(1) = -\frac{1}{3}$	$F(3) = \frac{27}{5}$
Location of Graph	Below x-axis	Above x-axis	Below x-axis	Above x-axis
Point on Graph	$\left(-3, -\frac{27}{5}\right)$	$\left(-1, \frac{1}{3}\right)$	$\left(1, -\frac{1}{3}\right)$	$\left(3, \frac{27}{5}\right)$

7/8.

31. 1. Domain: $\{x \mid x \neq 1\}$ **2.** R is in lowest terms **3.** y-intercept: 0; x-intercept: 0 **4.** R is in lowest terms; vertical asymptote: $x = 1$
5. No oblique or horizontal asymptote
6.

Interval	$(-\infty, 0)$	$(0, 1)$	$(1, \infty)$
Number Chosen	-2	$\frac{1}{2}$	2
Value of R	$R(-2) = \frac{32}{9}$	$R\left(\frac{1}{2}\right) = \frac{1}{2}$	$R(2) = 32$
Location of Graph	Above x-axis	Above x-axis	Above x-axis
Point on Graph	$\left(-2, \frac{32}{9}\right)$	$\left(\frac{1}{2}, \frac{1}{2}\right)$	$(2, 32)$

7/8.

33. 1. $G(x) = \dfrac{(x + 2)(x - 2)}{(x + 1)(x - 2)}$; domain: $\{x \mid x \neq -1, x \neq 2\}$ **2.** In lowest terms, $G(x) = \dfrac{x + 2}{x + 1}$ **3.** y-intercept: 2; x-intercept: -2
4. Vertical asymptote: $x = -1$; hole at $\left(2, \dfrac{4}{3}\right)$ **5.** Horizontal asymptote: $y = 1$, not intersected
6.

Interval	$(-\infty, -2)$	$(-2, -1)$	$(-1, 2)$	$(2, \infty)$
Number Chosen	-3	$-\frac{3}{2}$	0	3
Value of G	$G(-3) = \frac{1}{2}$	$G\left(-\frac{3}{2}\right) = -1$	$G(0) = 2$	$G(3) = \frac{5}{4}$
Location of Graph	Above x-axis	Below x-axis	Above x-axis	Above x-axis
Point on Graph	$\left(-3, \frac{1}{2}\right)$	$\left(-\frac{3}{2}, -1\right)$	$(0, 2)$	$\left(3, \frac{5}{4}\right)$

7/8.

35. $\{x \mid x < -2 \text{ or } -1 < x < 2\}$; $(-\infty, -2) \cup (-1, 2)$

37. $\{x \mid -3 < x \leq 3\}$; $(-3, 3]$

39. $\{x \mid x < 1 \text{ or } x > 2\}$; $(-\infty, 1) \cup (2, \infty)$

41. $\{x \mid 1 \leq x \leq 2 \text{ or } x > 3\}$; $[1, 2] \cup (3, \infty)$

43. $\{x \mid x < -4 \text{ or } 2 < x < 4 \text{ or } x > 6\}$;
$(-\infty, -4) \cup (2, 4) \cup (6, \infty)$

45. $R = 10$; g is not a factor of f. **47.** $R = 0$; g is a factor of f. **49.** $f(4) = 47{,}105$ **51.** $\pm 1, \pm 3, \pm\frac{1}{2}, \pm\frac{3}{2}, \pm\frac{1}{3}, \pm\frac{1}{4}, \pm\frac{3}{4}, \pm\frac{1}{6}, \pm\frac{1}{12}$

53. $-2, 1, 4$; $f(x) = (x + 2)(x - 1)(x - 4)$ **55.** $\frac{1}{2}$, multiplicity 2; -2; $f(x) = 4\left(x - \frac{1}{2}\right)^2(x + 2)$ **57.** 2, multiplicity 2; $f(x) = (x - 2)^2(x^2 + 5)$

59. $\{-3, 2\}$ **61.** $\left\{-3, -1, -\frac{1}{2}, 1\right\}$ **63.** 5 **65.** $\frac{37}{2}$ **67.** $f(0) = -1$; $f(1) = 1$ **69.** $f(0) = -1$; $f(1) = 1$ **71.** 1.52 **73.** 0.93

75. $4 - i$; $f(x) = x^3 - 14x^2 + 65x - 102$ **77.** $-i, 1 - i$; $f(x) = x^4 - 2x^3 + 3x^2 - 2x + 2$ **79.** $-2, 1, 4$; $f(x) = (x + 2)(x - 1)(x - 4)$

81. $-2, \frac{1}{2}$ (multiplicity 2); $f(x) = 4(x + 2)\left(x - \frac{1}{2}\right)^2$ **83.** 2 (multiplicity 2), $-\sqrt{5}i, \sqrt{5}i$; $f(x) = (x + \sqrt{5}i)(x - \sqrt{5}i)(x - 2)^2$

85. $-3, 2, -\dfrac{\sqrt{2}}{2}i, \dfrac{\sqrt{2}}{2}i$; $f(x) = 2(x + 3)(x - 2)\left(x + \dfrac{\sqrt{2}}{2}i\right)\left(x - \dfrac{\sqrt{2}}{2}i\right)$

87. (a) $A(r) = 2\pi r^2 + \dfrac{500}{r}$

(b) 223.22 cm^2 **(c)** 257.08 cm^2

(d)

A is smallest when $r \approx 3.41$ cm.

91. (a) Even
(b) Positive
(c) Even
(d) The graph touches the x-axis at $x = 0$, but does not cross it there.
(e) 8

Chapter Test (page 397)

1.

2. (a) 3

(b) Every zero of g lies between -15 and 15.

(c) $\frac{p}{q}: \pm\frac{1}{2}, \pm1, \pm\frac{3}{2}, \pm\frac{5}{2}, \pm3, \pm5, \pm\frac{15}{2}, \pm15$

(d) $-5, -\frac{1}{2}, 3; g(x) = (x + 5)(2x + 1)(x - 3)$

(e) y-intercept: -15; x-intercepts: $-5, -\frac{1}{2}, 3$

(f) Crosses at $-5, -\frac{1}{2}, 3$

(g) $y = 2x^3$

(h) Near -5: $g(x) \approx 72(x + 5)$

Near $-\frac{1}{2}$: $g(x) \approx -\frac{63}{4}(2x + 1)$

Near 3: $g(x) \approx 56(x - 3)$

(i)

3. $4, -5i, 5i$ **4.** $\left\{1, \dfrac{5 - \sqrt{61}}{6}, \dfrac{5 + \sqrt{61}}{6}\right\}$ **5.** Domain: $\{x | x \neq -10, x \neq 4\}$; asymptotes: $x = -10, y = 2$

6. Domain: $\{x | x \neq -1\}$; asymptotes: $x = -1, y = x + 1$

7.

8. Answers may vary. One possibility is
$f(x) = x^4 - 4x^3 - 2x^2 + 20x$.

9. Answers may vary. One possibility is $r(x) = \dfrac{2(x - 9)(x - 1)}{(x - 4)(x - 9)}$.

10. $f(0) = 8; f(4) = -36$

Since $f(0) = 8 > 0$ and $f(4) = -36 < 0$, the Intermediate Value Theorem guarantees that there is at least one real zero between 0 and 4.

11. $\{x | x < 3 \text{ or } x > 8\}; (-\infty, 3) \cup (8, \infty)$

Cumulative Review (page 397)

1. $\sqrt{26}$ **2.** $\{x | x \leq 0 \text{ or } x \geq 1\};$ $(-\infty, 0] \text{ or } [1, \infty)$ **3.** $\{x | -1 < x < 4\}; (-1, 4)$ **4.** $f(x) = -3x + 1$ **5.** $y = 2x - 1$ **6.**

7. Not a function; 3 has two images. **8.** $\{0, 2, 4\}$ **9.** $\left\{x \,\middle|\, x \geq \dfrac{3}{2}\right\}; \left[\dfrac{3}{2}, \infty\right)$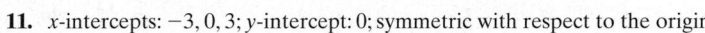

10. Center: $(-2, 1)$; radius: 3

11. x-intercepts: $-3, 0, 3$; y-intercept: 0; symmetric with respect to the origin

12. $y = -\dfrac{2}{3}x + \dfrac{17}{3}$

13. Not a function; it fails the Vertical Line Test.

14. (a) 22 **(b)** $x^2 - 5x - 2$ **(c)** $-x^2 - 5x + 2$ **(d)** $9x^2 + 15x - 2$ **(e)** $2x + h + 5$

15. (a) $\{x | x \neq 1\}$ **(b)** No; $(2, 7)$ is on the graph. **(c)** 4; $(3, 4)$ is on the graph. **(d)** $\dfrac{7}{4}; \left(\dfrac{7}{4}, 9\right)$ is on the graph.

(e) Rational

16.

17.

18. $6; y = 6x - 1$

19. (a) x-intercepts: $-5, -1, 5$; y-intercept: -3

(b) No symmetry

(c) Neither

(d) Increasing: $(-\infty, -3)$ and $(2, \infty)$; decreasing: $(-3, 2)$

(e) Local maximum value is 5 and occurs at $x = -3$.

(f) Local minimum value is -6 and occurs at $x = 2$.

20. Odd

21. (a) Domain: $\{x | x > -3\}$ or $(-3, \infty)$

(b) x-intercept: $-\dfrac{1}{2}$; y-intercept: 1

(c)

(d) Range: $\{y | y < 5\}$ or $(-\infty, 5)$

22.

23. (a) $(f + g)(x) = x^2 - 9x - 6$; domain: all real numbers

(b) $\left(\dfrac{f}{g}\right)(x) = \dfrac{x^2 - 5x + 1}{-4x - 7}$; domain: $\left\{x \,\middle|\, x \neq -\dfrac{7}{4}\right\}$

24. (a) $R(x) = -\dfrac{1}{10}x^2 + 150x$

(b) \$14,000

(c) 750, \$56,250

(d) \$75

CHAPTER 6 Exponential and Logarithmic Functions

6.1 Assess Your Understanding (page 406)

4. composite function; $f(g(x))$ **5.** F **6.** F **7. (a)** -1 **(b)** -1 **(c)** 8 **(d)** 0 **(e)** 8 **(f)** -7 **9. (a)** 4 **(b)** 5 **(c)** -1 **(d)** -2 **11. (a)** 98

(b) 49 **(c)** 4 **(d)** 4 **13. (a)** 97 **(b)** $-\dfrac{163}{2}$ **(c)** 1 **(d)** $-\dfrac{3}{2}$ **15. (a)** $2\sqrt{2}$ **(b)** $2\sqrt{2}$ **(c)** 1 **(d)** 0 **17. (a)** $\dfrac{1}{17}$ **(b)** $\dfrac{1}{5}$ **(c)** 1 **(d)** $\dfrac{1}{2}$

19. (a) $\dfrac{3}{\sqrt[3]{4}+1}$ **(b)** 1 **(c)** $\dfrac{6}{5}$ **(d)** 0 **21.** $\{x|x \neq 0,\ x \neq 2\}$ **23.** $\{x|x \neq -4,\ x \neq 0\}$ **25.** $\left\{x \,\middle|\, x \geq -\dfrac{3}{2}\right\}$ **27.** $\{x|x \geq 1\}$

29. (a) $(f \circ g)(x) = 6x + 3$; all real numbers **31. (a)** $(f \circ g)(x) = 3x^2 + 1$; all real numbers **33. (a)** $(f \circ g)(x) = x^4 + 8x^2 + 16$; all
(b) $(g \circ f)(x) = 6x + 9$; all real numbers **(b)** $(g \circ f)(x) = 9x^2 + 6x + 1$; all real numbers
(c) $(f \circ f)(x) = 4x + 9$; all real numbers numbers **(b)** $(g \circ f)(x) = x^4 + 4$; all real numbers
(d) $(g \circ g)(x) = 9x$; all real numbers **(c)** $(f \circ f)(x) = 9x + 4$; all real numbers **(c)** $(f \circ f)(x) = x^4$; all real numbers
 (d) $(g \circ g)(x) = x^4$; all real numbers **(d)** $(g \circ g)(x) = x^4 + 8x^2 + 20$; all real
 numbers

35. (a) $(f \circ g)(x) = \dfrac{3x}{2 - x}$; $\{x|x \neq 0,\ x \neq 2\}$ **(b)** $(g \circ f)(x) = \dfrac{2(x - 1)}{3}$; $\{x|x \neq 1\}$ **(c)** $(f \circ f)(x) = \dfrac{3(x - 1)}{4 - x}$; $\{x|x \neq 1,\ x \neq 4\}$

(d) $(g \circ g)(x) = x$; $\{x|x \neq 0\}$ **37. (a)** $(f \circ g)(x) = \dfrac{4}{4 + x}$; $\{x|x \neq -4,\ x \neq 0\}$ **(b)** $(g \circ f)(x) = \dfrac{-4(x - 1)}{x}$; $\{x|x \neq 0,\ x \neq 1\}$

(c) $(f \circ f)(x) = x$; $\{x|x \neq 1\}$ **(d)** $(g \circ g)(x) = x$; $\{x|x \neq 0\}$ **39. (a)** $(f \circ g)(x) = \sqrt{2x + 3}$; $\left\{x \,\middle|\, x \geq -\dfrac{3}{2}\right\}$

(b) $(g \circ f)(x) = 2\sqrt{x} + 3$; $\{x|x \geq 0\}$ **(c)** $(f \circ f)(x) = \sqrt[4]{x}$; $\{x|x \geq 0\}$ **(d)** $(g \circ g)(x) = 4x + 9$; all real numbers

41. (a) $(f \circ g)(x) = x$; $\{x|x \geq 1\}$ **(b)** $(g \circ f)(x) = |x|$; all real numbers **(c)** $(f \circ f)(x) = x^4 + 2x^2 + 2$; all real numbers

(d) $(g \circ g)(x) = \sqrt{\sqrt{x - 1} - 1}$; $\{x|x \geq 2\}$ **43. (a)** $(f \circ g)(x) = -\dfrac{4x - 17}{2x - 1}$; $\left\{x \,\middle|\, x \neq 3;\ x \neq \dfrac{1}{2}\right\}$

(b) $(g \circ f)(x) = -\dfrac{3x - 3}{2x + 8}$; $\{x|x \neq -4;\ x \neq -1\}$ **(c)** $(f \circ f)(x) = -\dfrac{2x + 5}{x - 2}$; $\{x|x \neq -1;\ x \neq 2\}$

(d) $(g \circ g)(x) = -\dfrac{3x - 4}{2x - 11}$; $\left\{x \,\middle|\, x \neq \dfrac{11}{2};\ x \neq 3\right\}$ **45.** $(f \circ g)(x) = f(g(x)) = f\!\left(\dfrac{1}{2}x\right) = 2\!\left(\dfrac{1}{2}x\right) = x$; $(g \circ f)(x) = g(f(x)) = g(2x) = \dfrac{1}{2}(2x) = x$

47. $(f \circ g)(x) = f(g(x)) = f(\sqrt[3]{x}) = (\sqrt[3]{x})^3 = x$; $(g \circ f)(x) = g(f(x)) = g(x^3) = \sqrt[3]{x^3} = x$

49. $(f \circ g)(x) = f(g(x)) = f\!\left(\dfrac{1}{2}(x + 6)\right) = 2\!\left[\dfrac{1}{2}(x + 6)\right] - 6 = x + 6 - 6 = x$;

 $(g \circ f)(x) = g(f(x)) = g(2x - 6) = \dfrac{1}{2}(2x - 6 + 6) = \dfrac{1}{2}(2x) = x$

51. $(f \circ g)(x) = f(g(x)) = f\!\left(\dfrac{1}{a}(x - b)\right) = a\!\left[\dfrac{1}{a}(x - b)\right] + b = x$; $(g \circ f)(x) = g(f(x)) = g(ax + b) = \dfrac{1}{a}(ax + b - b) = x$

53. $f(x) = x^4$; $g(x) = 2x + 3$ (Other answers are possible.) **55.** $f(x) = \sqrt{x}$; $g(x) = x^2 + 1$ (Other answers are possible.)

57. $f(x) = |x|$; $g(x) = 2x + 1$ (Other answers are possible.) **59.** $(f \circ g)(x) = 11$; $(g \circ f)(x) = 2$ **61.** $-3, 3$

63. (a) $(f \circ g)(x) = acx + ad + b$ **(b)** $(g \circ f)(x) = acx + bc + d$ **(c)** The domains of both $f \circ g$ and $g \circ f$ are all real numbers.

(d) $f \circ g = g \circ f$ when $ad + b = bc + d$ **65.** $S(t) = \dfrac{16}{9}\pi t^6$ **67.** $C(t) = 15{,}000 + 800{,}000t - 40{,}000t^2$

69. $C(p) = \dfrac{2\sqrt{100 - p}}{25} + 600, 0 \leq p \leq 100$ **71.** $V(r) = 2\pi r^3$ **73. (a)** $f(x) = 0.7143x$ **(b)** $g(x) = 137.402x$

(c) $g(f(x)) = g(0.7143x) = 98.1462486x$ **(d)** 98,146.2486 yen **75. (a)** $f(p) = p - 200$ **(b)** $g(p) = 0.8p$

(c) $(f \circ g)(p) = 0.8p - 200$; $(g \circ f)(p) = 0.8p - 160$; The 20% discount followed by the \$200 rebate is the better deal.

77. f is an odd function, so $f(-x) = -f(x)$. g is an even function, so $g(-x) = g(x)$. Then $(f \circ g)(-x) = f(g(-x)) = f(g(x)) = (f \circ g)(x)$. So $f \circ g$
 is even. Also, $(g \circ f)(-x) = g(f(-x)) = g(-f(x)) = g(f(x)) = (g \circ f)(x)$, so $g \circ f$ is even.

6.2 Assess Your Understanding *(page 417)*

5. $f(x_1) \neq f(x_2)$ **6.** one-to-one **7.** 3 **8.** $y = x$ **9.** $[4, \infty)$ **10.** T **11.** one-to-one **13.** not one-to-one **15.** not one-to-one **17.** one-to-one
19. one-to-one **21.** not one-to-one **23.** one-to-one

25.

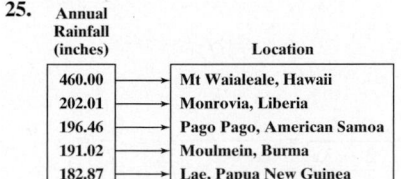

Annual Rainfall (inches)	Location
460.00	Mt Waialeale, Hawaii
202.01	Monrovia, Liberia
196.46	Pago Pago, American Samoa
191.02	Moulmein, Burma
182.87	Lae, Papua New Guinea

Domain: $\{460.00, 202.01, 196.46, 191.02, 182.87\}$
Range: $\{$Mt Waialeale, Monrovia, Pago Pago, Moulmein, Lae$\}$

27.

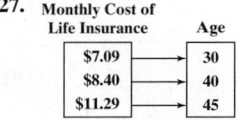

Monthly Cost of Life Insurance	Age
\$7.09	30
\$8.40	40
\$11.29	45

Domain: $\{\$7.09, \$8.40, \$11.29\}$
Range: $\{30, 40, 45\}$

29. $\{(5, -3), (9, -2), (2, -1), (11, 0), (-5, 1)\}$ **31.** $\{(1, -2), (2, -3), (0, -10), (9, 1), (4, 2)\}$
Domain: $\{5, 9, 2, 11, -5\}$ Domain: $\{1, 2, 0, 9, 4\}$
Range: $\{-3, -2, -1, 0, 1\}$ Range: $\{-2, -3, -10, 1, 2\}$

33. $f(g(x)) = f\!\left(\dfrac{1}{3}(x - 4)\right) = 3\!\left[\dfrac{1}{3}(x - 4)\right] + 4 = (x - 4) + 4 = x$ **35.** $f(g(x)) = f\!\left(\dfrac{x}{4} + 2\right) = 4\!\left[\dfrac{x}{4} + 2\right] - 8 = (x + 8) - 8 = x$

 $g(f(x)) = g(3x + 4) = \dfrac{1}{3}[(3x + 4) - 4] = \dfrac{1}{3}(3x) = x$ $g(f(x)) = g(4x - 8) = \dfrac{4x - 8}{4} + 2 = (x - 2) + 2 = x$

37. $f(g(x)) = f(\sqrt[3]{x + 8}) = (\sqrt[3]{x + 8})^3 - 8 = (x + 8) - 8 = x$

$g(f(x)) = g(x^3 - 8) = \sqrt[3]{(x^3 - 8) + 8} = \sqrt[3]{x^3} = x$

39. $f(g(x)) = f\left(\dfrac{1}{x}\right) = \dfrac{1}{\left(\dfrac{1}{x}\right)} = x; \; x \neq 0, \; g(f(x)) = g\left(\dfrac{1}{x}\right) = \dfrac{1}{\left(\dfrac{1}{x}\right)} = x, \; x \neq 0$

41. $f(g(x)) = f\left(\dfrac{4x - 3}{2 - x}\right) = \dfrac{2\left(\dfrac{4x - 3}{2 - x}\right) + 3}{\dfrac{4x - 3}{2 - x} + 4}$

$= \dfrac{2(4x - 3) + 3(2 - x)}{4x - 3 + 4(2 - x)} = \dfrac{5x}{5} = x, \; x \neq 2$

$g(f(x)) = g\left(\dfrac{2x + 3}{x + 4}\right) = \dfrac{4\left(\dfrac{2x + 3}{x + 4}\right) - 3}{2 - \dfrac{2x + 3}{x + 4}}$

$= \dfrac{4(2x + 3) - 3(x + 4)}{2(x + 4) - (2x + 3)} = \dfrac{5x}{5} = x, \; x \neq -4$

43.

45.

47.

49. $f^{-1}(x) = \dfrac{1}{3}x$

$f(f^{-1}(x)) = f\left(\dfrac{1}{3}x\right) = 3\left(\dfrac{1}{3}x\right) = x$

$f^{-1}(f(x)) = f^{-1}(3x) = \dfrac{1}{3}(3x) = x$

51. $f^{-1}(x) = \dfrac{x}{4} - \dfrac{1}{2}$

$f(f^{-1}(x)) = f\left(\dfrac{x}{4} - \dfrac{1}{2}\right) = 4\left(\dfrac{x}{4} - \dfrac{1}{2}\right) + 2$

$= (x - 2) + 2 = x$

$f^{-1}(f(x)) = f^{-1}(4x + 2) = \dfrac{4x + 2}{4} - \dfrac{1}{2}$

$= \left(x + \dfrac{1}{2}\right) - \dfrac{1}{2} = x$

53. $f^{-1}(x) = \sqrt[3]{x + 1}$

$f(f^{-1}(x)) = f(\sqrt[3]{x + 1})$

$= (\sqrt[3]{x + 1})^3 - 1 = x$

$f^{-1}(f(x)) = f^{-1}(x^3 - 1)$

$= \sqrt[3]{(x^3 - 1) + 1} = x$

55. $f^{-1}(x) = \sqrt{x - 4}, \; x \geq 4$

$f(f^{-1}(x)) = f(\sqrt{x - 4}) = (\sqrt{x - 4})^2 + 4 = x$

$f^{-1}(f(x)) = f^{-1}(x^2 + 4) = \sqrt{(x^2 + 4) - 4} = \sqrt{x^2} = x, \; x \geq 0$

57. $f^{-1}(x) = \dfrac{4}{x}$

$f(f^{-1}(x)) = f\left(\dfrac{4}{x}\right) = \dfrac{4}{\left(\dfrac{4}{x}\right)} = x$

$f^{-1}(f(x)) = f^{-1}\left(\dfrac{4}{x}\right) = \dfrac{4}{\left(\dfrac{4}{x}\right)} = x$

59. $f^{-1}(x) = \dfrac{2x + 1}{x}$

$f(f^{-1}(x)) = f\left(\dfrac{2x + 1}{x}\right) = \dfrac{1}{\dfrac{2x + 1}{x} - 2} = \dfrac{x}{(2x + 1) - 2x} = x$

$f^{-1}(f(x)) = f^{-1}\left(\dfrac{1}{x - 2}\right) = \dfrac{2\left(\dfrac{1}{x - 2}\right) + 1}{\dfrac{1}{x - 2}} = \dfrac{2 + (x - 2)}{1} = x$

61. $f^{-1}(x) = \dfrac{2 - 3x}{x}$

$f(f^{-1}(x)) = f\left(\dfrac{2 - 3x}{x}\right) = \dfrac{2}{3 + \dfrac{2 - 3x}{x}} = \dfrac{2x}{3x + 2 - 3x} = \dfrac{2x}{2} = x$

$f^{-1}(f(x)) = f^{-1}\left(\dfrac{2}{3 + x}\right) = \dfrac{2 - 3\left(\dfrac{2}{3 + x}\right)}{\dfrac{2}{3 + x}} = \dfrac{2(3 + x) - 3 \cdot 2}{2} = \dfrac{2x}{2} = x$

63. $f^{-1}(x) = \dfrac{-2x}{x-3}$

$$f(f^{-1}(x)) = f\left(\dfrac{-2x}{x-3}\right) = \dfrac{3\left(\dfrac{-2x}{x-3}\right)}{\dfrac{-2x}{x-3}+2}$$

$$= \dfrac{3(-2x)}{-2x+2(x-3)} = \dfrac{-6x}{-6} = x$$

$$f^{-1}(f(x)) = f^{-1}\left(\dfrac{3x}{x+2}\right) = \dfrac{-2\left(\dfrac{3x}{x+2}\right)}{\dfrac{3x}{x+2}-3}$$

$$= \dfrac{-2(3x)}{3x-3(x+2)} = \dfrac{-6x}{-6} = x$$

65. $f^{-1}(x) = \dfrac{x}{3x-2}$

$$f(f^{-1}(x)) = f\left(\dfrac{x}{3x-2}\right) = \dfrac{2\left(\dfrac{x}{3x-2}\right)}{3\left(\dfrac{x}{3x-2}\right)-1}$$

$$= \dfrac{2x}{3x-(3x-2)} = \dfrac{2x}{2} = x$$

$$f^{-1}(f(x)) = f^{-1}\left(\dfrac{2x}{3x-1}\right) = \dfrac{\dfrac{2x}{3x-1}}{3\left(\dfrac{2x}{3x-1}\right)-2}$$

$$= \dfrac{2x}{6x-2(3x-1)} = \dfrac{2x}{2} = x$$

67. $f^{-1}(x) = \dfrac{3x+4}{2x-3}$

$$f(f^{-1}(x)) = f\left(\dfrac{3x+4}{2x-3}\right) = \dfrac{3\left(\dfrac{3x+4}{2x-3}\right)+4}{2\left(\dfrac{3x+4}{2x-3}\right)-3}$$

$$= \dfrac{3(3x+4)+4(2x-3)}{2(3x+4)-3(2x-3)}$$

$$= \dfrac{17x}{17} = x$$

$$f^{-1}(f(x)) = f^{-1}\left(\dfrac{3x+4}{2x-3}\right) = \dfrac{3\left(\dfrac{3x+4}{2x-3}\right)+4}{2\left(\dfrac{3x+4}{2x-3}\right)-3}$$

$$= \dfrac{3(3x+4)+4(2x-3)}{2(3x+4)-3(2x-3)}$$

$$= \dfrac{17x}{17} = x$$

69. $f^{-1}(x) = \dfrac{-2x+3}{x-2}$

$$f(f^{-1}(x)) = f\left(\dfrac{-2x+3}{x-2}\right) = \dfrac{2\left(\dfrac{-2x+3}{x-2}\right)+3}{\dfrac{-2x+3}{x-2}+2}$$

$$= \dfrac{2(-2x+3)+3(x-2)}{-2x+3+2(x-2)} = \dfrac{-x}{-1} = x$$

$$f^{-1}(f(x)) = f^{-1}\left(\dfrac{2x+3}{x+2}\right) = \dfrac{-2\left(\dfrac{2x+3}{x+2}\right)+3}{\dfrac{2x+3}{x+2}-2}$$

$$= \dfrac{-2(2x+3)+3(x+2)}{2x+3-2(x+2)} = \dfrac{-x}{-1} = x$$

71. $f^{-1}(x) = \dfrac{2}{\sqrt{1-2x}}$

$$f(f^{-1}(x)) = f\left(\dfrac{2}{\sqrt{1-2x}}\right) = \dfrac{\dfrac{4}{1-2x}-4}{2\cdot\dfrac{4}{1-2x}} = \dfrac{4-4(1-2x)}{2\cdot 4}$$

$$= \dfrac{8x}{8} = x$$

$$f^{-1}(f(x)) = f^{-1}\left(\dfrac{x^2-4}{2x^2}\right) = \dfrac{2}{\sqrt{1-2\left(\dfrac{x^2-4}{2x^2}\right)}} = \dfrac{2}{\sqrt{\dfrac{4}{x^2}}} = \sqrt{x^2}$$

$$= x, \text{ since } x > 0$$

73. (a) 0 (b) 2 (c) 0 (d) 1
75. 7 **77.** Domain of f^{-1}: $[-2, \infty)$; range of f^{-1}: $[5, \infty)$
79. Domain of g^{-1}: $[0, \infty)$; range of g^{-1}: $(-\infty, 0]$
81. Increasing on the interval $(f(0), f(5))$
83. $f^{-1}(x) = \dfrac{1}{m}(x-b), m \neq 0$ **85.** Quadrant I
87. Possible answer: $f(x) = |x|, x \geq 0$, is one-to-one; $f^{-1}(x) = x, x \geq 0$
89. (a) $r(d) = \dfrac{d+90.39}{6.97}$

(b) $r(d(r)) = \dfrac{6.97r - 90.39 + 90.39}{6.97} = \dfrac{6.97r}{6.97} = r$

$$d(r(d)) = 6.97\left(\dfrac{d+90.39}{6.97}\right) - 90.39 = d + 90.39 - 90.39 = d$$

(c) 56 miles per hour

91. (a) 77.6 kg

(b) $h(W) = \dfrac{W-50}{2.3} + 60 = \dfrac{W+88}{2.3}$

(c) $h(W(h)) = \dfrac{50 + 2.3(h-60) + 88}{2.3} = \dfrac{2.3h}{2.3} = h$

$$W(h(W)) = 50 + 2.3\left(\dfrac{W+88}{2.3} - 60\right)$$

$$= 50 + W + 88 - 138 = W$$

(d) 73 inches

93. (a) $\{g \mid 33,950 \leq g \leq 82,250\}$

(b) $\{T \mid 4675 \leq T \leq 16,750\}$

(c) $g(T) = \dfrac{T - 4675}{0.25} + 33,950$

Domain: $\{T \mid 4675 \leq T \leq 16,750\}$
Range: $\{g \mid 33,950 \leq g \leq 82,250\}$

97. $f^{-1}(x) = \dfrac{-dx+b}{cx-a}; f = f^{-1}$ if $a = -d$ **101.** No

95. (a) t represents time, so $t \geq 0$.

(b) $t(H) = \sqrt{\dfrac{H-100}{-4.9}} = \sqrt{\dfrac{100-H}{4.9}}$

(c) 2.02 seconds

6.3 Assess Your Understanding (page 432)

6. Exponential function; growth factor; initial value **7.** a **8.** T **9.** F **10.** T **11.** $\left(-1, \dfrac{1}{a}\right); (0, 1); (1, a)$ **12.** 1 **13.** 4 **14.** F

15. (a) 11.212 (b) 11.587 (c) 11.664 (d) 11.665 **17.** (a) 8.815 (b) 8.821 (c) 8.824 (d) 8.825 **19.** (a) 21.217 (b) 22.217 (c) 22.440
(d) 22.459 **21.** 3.320 **23.** 0.427 **25.** Neither **27.** Exponential; $H(x) = 4^x$ **29.** Exponential; $f(x) = 3(2^x)$ **31.** Linear; $H(x) = 2x + 4$
33. B **35.** D **37.** A **39.** E

41.
Domain: All real numbers
Range: $\{y|y > 1\}$ or $(1, \infty)$
Horizontal asymptote: $y = 1$

43.
Domain: All real numbers
Range: $\{y|y > 0\}$ or $(0, \infty)$
Horizontal asymptote: $y = 0$

45.
Domain: All real numbers
Range: $\{y|y > 0\}$ or $(0, \infty)$
Horizontal asymptote: $y = 0$

47.
Domain: All real numbers
Range: $\{y|y > -2\}$ or $(-2, \infty)$
Horizontal asymptote: $y = -2$

49.
Domain: All real numbers
Range: $\{y|y > 2\}$ or $(2, \infty)$
Horizontal asymptote: $y = 2$

51.
Domain: All real numbers
Range: $\{y|y > 2\}$ or $(2, \infty)$
Horizontal asymptote: $y = 2$

53.
Domain: All real numbers
Range: $\{y|y > 0\}$ or $(0, \infty)$
Horizontal asymptote: $y = 0$

55.
Domain: All real numbers
Range: $\{y|y > 0\}$ or $(0, \infty)$
Horizontal asymptote: $y = 0$

57.
Domain: All real numbers
Range: $\{y|y < 5\}$ or $(-\infty, 5)$
Horizontal asymptote: $y = 5$

59.
Domain: All real numbers
Range: $\{y|y < 2\}$ or $(-\infty, 2)$
Horizontal asymptote: $y = 2$

61. $\{3\}$ **63.** $\{-4\}$ **65.** $\{2\}$ **67.** $\left\{\dfrac{3}{2}\right\}$ **69.** $\{-\sqrt{2}, 0, \sqrt{2}\}$ **71.** $\{6\}$

73. $\{-1, 7\}$ **75.** $\{-4, 2\}$ **77.** $\{-4\}$ **79.** $\{1, 2\}$ **81.** $\dfrac{1}{49}$ **83.** $\dfrac{1}{4}$

85. $f(x) = 3^x$ **87.** $f(x) = -6^x$ **89.** $f(x) = 3^x + 2$

91. (a) $16; (4, 16)$ (b) $-4; \left(-4, \dfrac{1}{16}\right)$ **93.** (a) $\dfrac{9}{4}; \left(-1, \dfrac{9}{4}\right)$
(b) $3; (3, 66)$ **95.** (a) $60; (-6, 60)$ (b) $-4; (-4, 12)$ (c) -2

97.
Domain: $(-\infty, \infty)$
Range: $[1, \infty)$
Intercept: $(0, 1)$

99.
Domain: $(-\infty, \infty)$
Range: $[-1, 0)$
Intercept: $(0, -1)$

101. (a) 74% (b) 47%
103. (a) $\$12,123$ (b) $\$6443$
105. 3.35 mg; 0.45 mg

107. (a) 0.632 (b) 0.982 (c) 1
(d)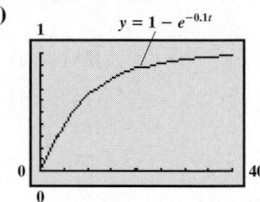
(e) About 7 min

109. (a) 0.0516 (b) 0.0888 **111.** (a) 70.95% (b) 72.62% (c) 100% **115.** 36
113. (a) 5.41 amp, 7.59 amp, 10.38 amp (b) 12 amp
(d) 3.34 amp, 5.31 amp, 9.44 amp (e) 24 amp
(c), (f)

117.

Final Denominator	Value of Expression	Compare Value to $e \approx$ **2.718281828**
$1 + 1$	2.5	$2.5 < e$
$2 + 2$	2.8	$2.8 > e$
$3 + 3$	2.7	$2.7 < e$
$4 + 4$	2.721649485	$2.721649485 > e$
$5 + 5$	2.717770035	$2.717770035 < e$
$6 + 6$	2.718348855	$2.718348855 > e$

119. $f(A + B) = a^{A+B} = a^A \cdot a^B = f(A) \cdot f(B)$ **121.** $f(\alpha x) = a^{\alpha x} = (a^x)^\alpha = [f(x)]^\alpha$

123. (a) $f(-x) = \dfrac{1}{2}(e^{-x} + e^{-(-x)})$
$= \dfrac{1}{2}(e^{-x} + e^x)$
$= \dfrac{1}{2}(e^x + e^{-x})$
$= f(x)$

(b)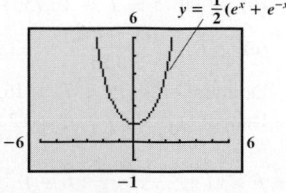
$y = \dfrac{1}{2}(e^x + e^{-x})$

(c) $(\cosh x)^2 - (\sinh x)^2$
$= \left[\dfrac{1}{2}(e^x + e^{-x})\right]^2 - \left[\dfrac{1}{2}(e^x - e^{-x})\right]^2$
$= \dfrac{1}{4}[e^{2x} + 2 + e^{-2x} - e^{2x} + 2 - e^{-2x}]$
$= \dfrac{1}{4}(4) = 1$

125. 59 minutes

6.4 Assess Your Understanding (page 446)

4. $\{x|x > 0\}$ or $(0, \infty)$ **5.** $\left(\dfrac{1}{a}, -1\right), (1, 0), (a, 1)$ **6.** 1 **7.** F **8.** T **9.** $2 = \log_3 9$ **11.** $2 = \log_a 1.6$ **13.** $x = \log_2 7.2$ **15.** $x = \ln 8$

17. $2^3 = 8$ **19.** $a^6 = 3$ **21.** $3^x = 2$ **23.** $e^x = 4$ **25.** 0 **27.** 2 **29.** -4 **31.** $\dfrac{1}{2}$ **33.** 4 **35.** $\dfrac{1}{2}$ **37.** $\{x|x > 3\}; (3, \infty)$

39. All real numbers except 0; $\{x \mid x \neq 0\}$ **41.** $\{x \mid x > 10\}$; $(10, \infty)$ **43.** $\{x \mid x > -1\}$; $(-1, \infty)$ **45.** $\{x \mid x < -1 \text{ or } x > 0\}$; $(-\infty, -1) \cup (0, \infty)$
47. $\{x \mid x \geq 1\}$; $[1, \infty)$ **49.** 0.511 **51.** 30.099 **53.** 2.303 **55.** -53.991 **57.** $\sqrt{2}$

59.

61.

63. B **65.** D **67.** A **69.** E

71. (a) Domain: $(-4, \infty)$
(b)

(c) Range: $(-\infty, \infty)$
Vertical asymptote: $x = -4$
(d) $f^{-1}(x) = e^x - 4$
(e) Domain of f^{-1}: $(-\infty, \infty)$
Range of f^{-1}: $(-4, \infty)$
(f)

73. (a) Domain: $(0, \infty)$
(b)

(c) Range: $(-\infty, \infty)$
Vertical asymptote: $x = 0$
(d) $f^{-1}(x) = e^{x-2}$
(e) Domain of f^{-1}: $(-\infty, \infty)$
Range of f^{-1}: $(0, \infty)$
(f)

75. (a) Domain: $(0, \infty)$
(b)

(c) Range: $(-\infty, \infty)$
Vertical asymptote: $x = 0$
(d) $f^{-1}(x) = \dfrac{1}{2} e^{x+3}$
(e) Domain of f^{-1}: $(-\infty, \infty)$
Range of f^{-1}: $(0, \infty)$
(f)

77. (a) Domain: $(4, \infty)$
(b)

(c) Range: $(-\infty, \infty)$
Vertical asymptote: $x = 4$
(d) $f^{-1}(x) = 10^{x-2} + 4$
(e) Domain of f^{-1}: $(-\infty, \infty)$
Range of f^{-1}: $(4, \infty)$
(f)

79. (a) Domain: $(0, \infty)$
(b)

(c) Range: $(-\infty, \infty)$
Vertical asymptote: $x = 0$
(d) $f^{-1}(x) = \dfrac{1}{2} \cdot 10^{2x}$
(e) Domain of f^{-1}: $(-\infty, \infty)$
Range of f^{-1}: $(0, \infty)$
(f)

81. (a) Domain: $(-2, \infty)$
(b)

(c) Range: $(-\infty, \infty)$
Vertical asymptote: $x = -2$
(d) $f^{-1}(x) = 3^{x-3} - 2$
(e) Domain of f^{-1}: $(-\infty, \infty)$
Range of f^{-1}: $(-2, \infty)$
(f)

83. (a) Domain: $(-\infty, \infty)$
(b)

(c) Range: $(-3, \infty)$
Horizontal asymptote: $y = -3$
(d) $f^{-1}(x) = \ln(x + 3) - 2$
(e) Domain of f^{-1}: $(-3, \infty)$
Range of f^{-1}: $(-\infty, \infty)$
(f)

85. (a) Domain: $(-\infty, \infty)$
(b)

(c) Range: $(4, \infty)$
Horizontal asymptote: $y = 4$
(d) $f^{-1}(x) = 3 \log_2 (x - 4)$
(e) Domain of f^{-1}: $(4, \infty)$
Range of f^{-1}: $(-\infty, \infty)$
(f)

87. $\{9\}$ **89.** $\left\{\dfrac{7}{2}\right\}$ **91.** $\{2\}$ **93.** $\{5\}$

95. $\{3\}$ **97.** $\{2\}$ **99.** $\left\{\dfrac{\ln 10}{3}\right\}$

101. $\left\{\dfrac{\ln 8 - 5}{2}\right\}$ **103.** $\{-2\sqrt{2}, 2\sqrt{2}\}$

105. $\{-1\}$ **107.** $\left\{5 \ln \dfrac{7}{5}\right\}$

109. $\left\{2 - \log \dfrac{5}{2}\right\}$

111. (a) $\left\{x \mid x > -\dfrac{1}{2}\right\}$; $\left(-\dfrac{1}{2}, \infty\right)$
(b) $2; (40, 2)$ **(c)** $121; (121, 3)$ **(d)** 4

113.

Domain: $\{x \mid x \neq 0\}$
Range: $(-\infty, \infty)$
Intercepts: $(-1, 0), (1, 0)$

115.

Domain: $\{x \mid x > 0\}$
Range: $\{y \mid y \geq 0\}$
Intercept: $(1, 0)$

117. (a) 1 **(b)** 2 **(c)** 3
(d) It increases. **(e)** 0.000316
(f) 3.981×10^{-8}
119. (a) 5.97 km **(b)** 0.90 km
121. (a) 6.93 min **(b)** 16.09 min
123. $h \approx 2.29$, so the time between injections is about 2 h, 17 min.

125. 0.2695 s
0.8959 s

127. 50 decibels (dB) **129.** 90 dB **131.** 8.1 **133. (a)** $k \approx 11.216$ **(b)** 6.73 **(c)** 0.41% **(d)** 0.14%
135. Because $y = \log_1 x$ means $1^y = 1 = x$, which cannot be true for $x \neq 1$

6.5 Assess Your Understanding (page 457)

1. 0 **2.** 1 **3.** M **4.** r **5.** $\log_a M$; $\log_a N$ **6.** $\log_a M$; $\log_a N$ **7.** $r \log_a M$ **8.** 6 **9.** 7 **10.** F **11.** F **12.** F **13.** 71 **15.** -4 **17.** 7 **19.** 1
21. 1 **23.** 3 **25.** $\dfrac{5}{4}$ **27.** 4 **29.** $a + b$ **31.** $b - a$ **33.** $3a$ **35.** $\dfrac{1}{5}(a + b)$ **37.** $2 + \log_5 x$ **39.** $3 \log_2 z$ **41.** $1 + \ln x$ **43.** $\ln x - x$

45. $2 \log_a u + 3 \log_a v$ **47.** $2 \ln x + \dfrac{1}{2}\ln(1 - x)$ **49.** $3 \log_2 x - \log_2(x - 3)$ **51.** $\log x + \log(x + 2) - 2 \log(x + 3)$

53. $\dfrac{1}{3}\ln(x - 2) + \dfrac{1}{3}\ln(x + 1) - \dfrac{2}{3}\ln(x + 4)$ **55.** $\ln 5 + \ln x + \dfrac{1}{2}\ln(1 + 3x) - 3\ln(x - 4)$ **57.** $\log_5 u^3 v^4$ **59.** $\log_3\left(\dfrac{1}{x^{5/2}}\right)$ **61.** $\log_4\left[\dfrac{x - 1}{(x + 1)^4}\right]$

63. $-2 \ln(x - 1)$ **65.** $\log_2[x(3x - 2)^4]$ **67.** $\log_a\left(\dfrac{25x^6}{\sqrt{2x + 3}}\right)$ **69.** $\log_2\left[\dfrac{(x + 1)^2}{(x + 3)(x - 1)}\right]$ **71.** 2.771 **73.** -3.880 **75.** 5.615 **77.** 0.874

79. $y = \dfrac{\log x}{\log 4}$

81. $y = \dfrac{\log(x + 2)}{\log 2}$

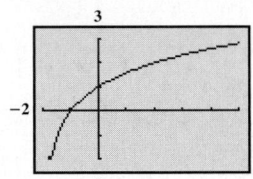

83. $y = \dfrac{\log(x + 1)}{\log(x - 1)}$

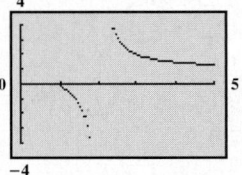

85. (a) $(f \circ g)(x) = x$; $\{x \mid x$ is any real number$\}$ or $(-\infty, \infty)$
(b) $(g \circ f)(x) = x$; $\{x \mid x > 0\}$ or $(0, \infty)$ **(c)** 5
(d) $(f \circ h)(x) = \ln x^2$; $\{x \mid x \neq 0\}$ or $(-\infty, 0) \cup (0, \infty)$ **(e)** 2

87. $y = Cx$ **89.** $y = Cx(x + 1)$ **91.** $y = Ce^{3x}$ **93.** $y = Ce^{-4x} + 3$

95. $y = \dfrac{\sqrt[3]{C}(2x + 1)^{1/6}}{(x + 4)^{1/9}}$ **97.** 3 **99.** 1

101. $\log_a\left(x + \sqrt{x^2 - 1}\right) + \log_a\left(x - \sqrt{x^2 - 1}\right) = \log_a\left[\left(x + \sqrt{x^2 - 1}\right)\left(x - \sqrt{x^2 - 1}\right)\right] = \log_a[x^2 - (x^2 - 1)] = \log_a 1 = 0$
103. $\ln(1 + e^{2x}) = \ln[e^{2x}(e^{-2x} + 1)] = \ln e^{2x} + \ln(e^{-2x} + 1) = 2x + \ln(1 + e^{-2x})$
105. $y = f(x) = \log_a x$; $a^y = x$ implies $a^y = \left(\dfrac{1}{a}\right)^{-y} = x$, so $-y = \log_{1/a} x = -f(x)$.

107. $f(x) = \log_a x$; $f\left(\dfrac{1}{x}\right) = \log_a\dfrac{1}{x} = \log_a 1 - \log_a x = -f(x)$

109. $\log_a \dfrac{M}{N} = \log_a(M \cdot N^{-1}) = \log_a M + \log_a N^{-1} = \log_a M - \log_a N$, since $a^{\log_a N^{-1}} = N^{-1}$ implies $a^{-\log_a N^{-1}} = N$; i.e., $\log_a N = -\log_a N^{-1}$.

6.6 Assess Your Understanding (page 463)

5. $\{16\}$ **7.** $\left\{\dfrac{16}{5}\right\}$ **9.** $\{6\}$ **11.** $\{16\}$ **13.** $\left\{\dfrac{1}{3}\right\}$ **15.** $\{3\}$ **17.** $\{5\}$ **19.** $\left\{\dfrac{21}{8}\right\}$ **21.** $\{-6\}$ **23.** $\{-2\}$ **25.** $\left\{-1 + \sqrt{1 + e^4}\right\} \approx \{6.456\}$

27. $\left\{\dfrac{-5 + 3\sqrt{5}}{2}\right\} \approx \{0.854\}$ **29.** $\{2\}$ **31.** $\left\{\dfrac{9}{2}\right\}$ **33.** $\{8\}$ **35.** $\{\log_2 10\} = \left\{\dfrac{\ln 10}{\ln 2}\right\} \approx \{3.322\}$ **37.** $\{-\log_8 1.2\} = \left\{-\dfrac{\ln 1.2}{\ln 8}\right\} \approx \{-0.088\}$

39. $\left\{\dfrac{1}{3}\log_2\dfrac{8}{5}\right\} = \left\{\dfrac{\ln\dfrac{8}{5}}{3 \ln 2}\right\} \approx \{0.226\}$ **41.** $\left\{\dfrac{\ln 3}{2 \ln 3 + \ln 4}\right\} \approx \{0.307\}$ **43.** $\left\{\dfrac{\ln 7}{\ln 0.6 + \ln 7}\right\} \approx \{1.356\}$ **45.** $\{0\}$ **47.** $\left\{\dfrac{\ln \pi}{1 + \ln \pi}\right\} \approx \{0.534\}$

49. $\left\{\dfrac{\ln 3}{\ln 2}\right\} \approx \{1.585\}$ **51.** $\{0\}$ **53.** $\left\{\log_4\left(-2 + \sqrt{7}\right)\right\} \approx \{-0.315\}$ **55.** $\{\log_5 4\} \approx \{0.861\}$ **57.** No real solution **59.** $\{\log_4 5\} \approx \{1.161\}$

61. $\{2.79\}$ **63.** $\{-0.57\}$ **65.** $\{-0.70\}$ **67.** $\{0.57\}$ **69.** $\{0.39, 1.00\}$ **71.** $\{1.32\}$ **73.** $\{1.31\}$ **75.** $\{1\}$ **77.** $\{16\}$ **79.** $\left\{-1, \dfrac{2}{3}\right\}$ **81.** $\{0\}$

83. $\left\{\ln\left(2 + \sqrt{5}\right)\right\} \approx \{1.444\}$ **85.** $\left\{\dfrac{\ln 5 \cdot \ln 3}{\ln 15}\right\} \approx \{1.921\}$ **87. (a)** $\{5\}; (5, 3)$ **(b)** $\{5\}; (5, 4)$ **(c)** $\{1\}$; yes, at $(1, 2)$ **(d)** $\{5\}$ **(e)** $\left\{-\dfrac{1}{11}\right\}$

89. (a), (b)

(c) $\{x \mid x > 0.710\}$ or $(0.710, \infty)$

91. (a), (b), (c)

93. (a), (b), (c)

95. (a)

(b) 2

(c) $\{x \mid x < 2\}$ or $(-\infty, 2)$

97. (a) 2025 **(b)** 2043

99. (a) After 3.1 yr

(b) After 7.1 yr

(c) After 10.6 yr

6.7 Assess Your Understanding (page 472)

3. principal **4.** *I; Prt;* simple interest **5.** 4 **6.** effective rate of interest **7.** $108.29 **9.** $609.50 **11.** $697.09 **13.** $1246.08 **15.** $88.72

17. $860.72 **19.** $554.09 **21.** $59.71 **23.** 5.095% **25.** 5.127% **27.** $6\dfrac{1}{4}$% compounded annually **29.** 9% compounded monthly **31.** 25.992%

33. 24.573% **35. (a)** About 8.69 yr **(b)** About 8.66 yr **37.** 6.823% **39.** 5.09 yr; 5.07 yr **41.** 15.27 yr or 15 yr, 3 mo **43.** $104,335

45. $12,910.62 **47.** About $30.17 per share or $3017 **49.** Not quite. Jim will have $1057.60. The second bank gives a better deal, since Jim will have $1060.62 after 1 yr. **51.** Will has $11,632.73; Henry has $10,947.89. **53. (a)** $79,129 **(b)** $38,516 **55.** About $1019 billion; about $232 billion

57. $940.90 **59.** 2.53% **61.** 34.31 yr **63. (a)** $1364.62 **(b)** $1353.35 **65.** $4631.93

67. (a) 6.12 yr **(c)** $mP = P\left(1 + \dfrac{r}{n}\right)^{nt}$

(b) 18.45 yr

$$m = \left(1 + \dfrac{r}{n}\right)^{nt}$$

$$\ln m = \ln\left(1 + \dfrac{r}{n}\right)^{nt} = nt \ln\left(1 + \dfrac{r}{n}\right)$$

$$t = \dfrac{\ln m}{n \ln\left(1 + \dfrac{r}{n}\right)}$$

69. (a) 2.82% **(b)** In 2020 or after 12 yr **71.** 22.7 yr

6.8 Assess Your Understanding (page 484)

1. (a) 500 insects **(b)** $0.02 = 2\%$ per day **(c)** About 611 insects **(d)** After about 23.5 days **(e)** After about 34.7 days

3. (a) $-0.0244 = -2.44\%$ per year **(b)** About 391.7 g **(c)** After about 9.1 yr **(d)** 28.4 yr **5. (a)** $N(t) = N_0 e^{kt}$ **(b)** 5832 **(c)** 3.9 days

7. (a) $N(t) = N_0 e^{kt}$ **(b)** 25,198 **9.** 9.797 g **11.** 9727 yr ago **13. (a)** 5:18 PM **(b)** About 14.3 min **(c)** The temperature of the pan approaches 70°F. **15.** 18.63°C; 25.1°C **17.** 1.7 ppm; 7.17 days or 172 hr **19.** 0.26 M; 6.58 hr or 395 min **21.** 26.6 days **23. (a)** 1000 g **(b)** 43.9% **(c)** 30 g

(d) 616.6 g **(e)** After 9.85 h **(f)** About 7.9 h **25. (a)** 9.23×10^{-3}, or about 0 **(b)** 0.81, or about 1 **(c)** 5.01, or about 5$^+$

(d) 57.91°, 43.99°, 30.07°

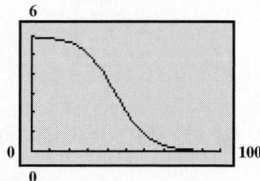

6.9 Assess Your Understanding (page 490)

1. (a)

(d)

3. (a)

(d)

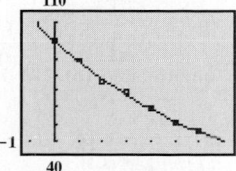

(b) $y = 0.0903(1.3384)^x$ **(e)** 0.69 **(b)** $y = 100.326(0.8769)^x$ **(e)** 5.3 weeks **(f)** 0.14 g

(c) $N(t) = 0.0903 e^{0.2915t}$ **(f)** After about 7.26 hr **(c)** $A(t) = 100.326 e^{-0.1314t}$ **(g)** After about 12.3 weeks

5. (a)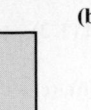

(b) $y = 32{,}741.02 - 6070.96 \ln x$ **(c)**

(d) Approximately 168 computers

7. (a)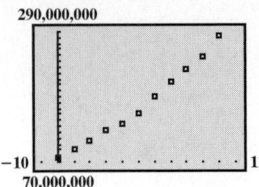

(b) $y = \dfrac{799{,}475{,}916.5}{1 + 9.1968e^{-0.0160x}}$

(c)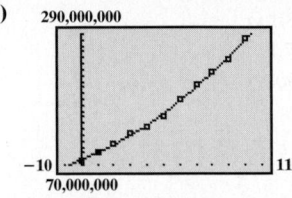

(d) 799,475,917
(e) Approximately 291,599,733
(f) 2006

9. (a)

(c)

(b) $y = \dfrac{67{,}856.6}{1 + 19.844e^{-0.2029x}}$

(d) About 67,856,600 subscribers
(e) About 67,711,000 subscribers

11. (a)

(b) Quadratic with $a < 0$ because of the "upside down U-shape" of the data
(c) $y = -0.0311x^2 + 3.4444x + 118.2493$

(d)

(e) 201

13. (a)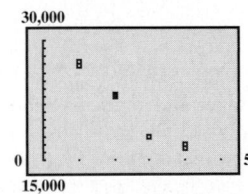

(b) Exponential because depreciation of a car is described by exponential models in the theory of finance.
(c) $y = 31{,}808.51(0.8474)^x$

(d)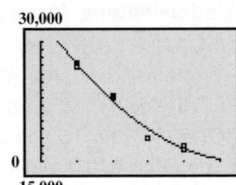

(e) $13,899

Review Exercises *(page 496)*

1. (a) -26 **(b)** -241 **(c)** 16 **(d)** -1 **3. (a)** $\sqrt{11}$ **(b)** 1 **(c)** $\sqrt{\sqrt{6}+2}$ **(d)** 19 **5. (a)** e^4 **(b)** $3e^{-2} - 2$ **(c)** e^{e^4} **(d)** -17

7. $(f \circ g)(x) = 1 - 3x$, all real numbers; $(g \circ f)(x) = 7 - 3x$, all real numbers; $(f \circ f)(x) = x$, all real numbers; $(g \circ g)(x) = 9x + 4$, all real numbers

9. $(f \circ g)(x) = 27x^2 + 3|x| + 1$, all real numbers; $(g \circ f)(x) = 3|3x^2 + x + 1|$, all real numbers; $(f \circ f)(x) = 27x^4 + 18x^3 + 24x^2 + 7x + 5$, all real numbers; $(g \circ g)(x) = 9|x|$, all real numbers

11. $(f \circ g)(x) = \dfrac{1+x}{1-x}$, $\{x | x \neq 0, x \neq 1\}$; $(g \circ f)(x) = \dfrac{x-1}{x+1}$, $\{x | x \neq -1, x \neq 1\}$; $(f \circ f)(x) = x$, $\{x | x \neq 1\}$; $(g \circ g)(x) = x$, $\{x | x \neq 0\}$

13. (a) one-to-one **(b)** $\{(2,1), (5,3), (8,5), (10,6)\}$

15.

17. $f^{-1}(x) = \dfrac{2x + 3}{5x - 2}$

$$f(f^{-1}(x)) = \frac{2\left(\dfrac{2x + 3}{5x - 2}\right) + 3}{5\left(\dfrac{2x + 3}{5x - 2}\right) - 2} = x$$

$$f^{-1}(f(x)) = \frac{2\left(\dfrac{2x + 3}{5x - 2}\right) + 3}{5\left(\dfrac{2x + 3}{5x - 2}\right) - 2} = x$$

Domain of f = range of f^{-1} = all real numbers except $\dfrac{2}{5}$

Range of f = domain of f^{-1} = all real numbers except $\dfrac{2}{5}$

19. $f^{-1}(x) = \dfrac{x + 1}{x}$

$$f(f^{-1}(x)) = \frac{1}{\dfrac{x + 1}{x} - 1} = x$$

$$f^{-1}(f(x)) = \frac{\dfrac{1}{x - 1} + 1}{\dfrac{1}{x - 1}} = x$$

Domain of f = range of f^{-1} = all real numbers except 1
Range of f = domain of f^{-1} = all real numbers except 0

21. $f^{-1}(x) = \dfrac{27}{x^3}$

$$f(f^{-1}(x)) = \frac{3}{\left(\dfrac{27}{x^3}\right)^{1/3}} = x$$

$$f^{-1}(f(x)) = \frac{27}{\left(\dfrac{3}{x^{1/3}}\right)^3} = x$$

Domain of f = range of f^{-1} = all real numbers except 0
Range of f = domain of f^{-1} = all real numbers except 0

23. (a) 81 **(b)** 2 **(c)** $\dfrac{1}{9}$ **(d)** -3 **25.** $\log_5 z = 2$ **27.** $5^{13} = u$

29. $\left\{x \middle| x > \dfrac{2}{3}\right\}; \left(\dfrac{2}{3}, \infty\right)$ **31.** $\{x | x < 1 \text{ or } x > 2\}; (-\infty, 1) \cup (2, \infty)$

33. -3 **35.** $\sqrt{2}$ **37.** 0.4 **39.** $\log_3 u + 2\log_3 v - \log_3 w$

41. $2\log x + \dfrac{1}{2}\log(x^3 + 1)$ **43.** $\ln x + \dfrac{1}{3}\ln(x^2 + 1) - \ln(x - 3)$ **45.** $\dfrac{25}{4}\log_4 x$

47. $-2\ln(x + 1)$ **49.** $\log\left(\dfrac{4x^3}{[(x + 3)(x - 2)]^{1/2}}\right)$ **51.** 2.124

53.
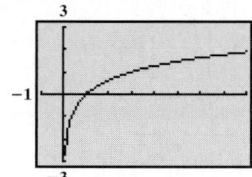

55. (a) Domain of f: $(-\infty, \infty)$
(b)
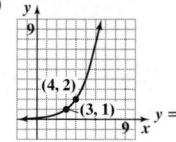
(c) Range of f: $(0, \infty)$
Horizontal asymptote: $y = 0$
(d) $f^{-1}(x) = 3 + \log_2 x$
(e) Domain of f^{-1}: $(0, \infty)$
Range of f^{-1}: $(-\infty, \infty)$
(f)
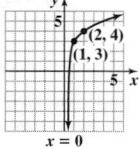

57. (a) Domain of f: $(-\infty, \infty)$
(b)

(c) Range of f: $(0, \infty)$
Horizontal asymptote: $y = 0$
(d) $f^{-1}(x) = -\log_3(2x)$
(e) Domain of f^{-1}: $(0, \infty)$
Range of f^{-1}: $(-\infty, \infty)$
(f)

59. (a) Domain of f: $(-\infty, \infty)$
(b)

(c) Range of f: $(-\infty, 1)$
Horizontal asymptote: $y = 1$
(d) $f^{-1}(x) = -\ln(1 - x)$
(e) Domain of f^{-1}: $(-\infty, 1)$
Range of f^{-1}: $(-\infty, \infty)$
(f)

61. (a) Domain of f: $(-3, \infty)$
(b)
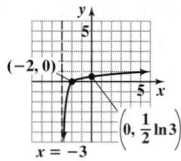
(c) Range of f: $(-\infty, \infty)$
Vertical asymptote: $x = -3$
(d) $f^{-1}(x) = e^{2x} - 3$
(e) Domain of f^{-1}: $(-\infty, \infty)$
Range of f^{-1}: $(-3, \infty)$
(f)

63. $\left\{\dfrac{1}{4}\right\}$ **65.** $\left\{\dfrac{-1 - \sqrt{3}}{2}, \dfrac{-1 + \sqrt{3}}{2}\right\} \approx \{-1.366, 0.366\}$ **67.** $\left\{\dfrac{1}{4}\right\}$ **69.** $\left\{\dfrac{2\ln 3}{\ln 5 - \ln 3}\right\} \approx \{4.301\}$ **71.** $\left\{\dfrac{12}{5}\right\}$ **73.** $\{83\}$ **75.** $\left\{\dfrac{1}{2}, -3\right\}$

77. $\{-1\}$ **79.** $\{1 - \ln 5\} \approx \{-0.609\}$ **81.** $\left\{\log_3\left(-2 + \sqrt{7}\right)\right\} = \left\{\dfrac{\ln\left(-2 + \sqrt{7}\right)}{\ln 3}\right\} \approx \{-0.398\}$

83. (a), (e)

(b) 3; (6, 3) **(c)** 10; (10, 4)
(d) $\left\{x \middle| x > \dfrac{5}{2}\right\}$ or $\left(\dfrac{5}{2}, \infty\right)$
(e) $f^{-1}(x) = 2^{x-1} + 2$

85. 3229.5 m **87. (a)** 37.3 W **(b)** 6.9 dB
89. (a) 9.85 yr **(b)** 4.27 yr **91.** \$41,668.97 **93.** 24,203 yr ago
95. 7,237,271,501 **97.** \$483.67 billion

99. (a)

(c)

101. (a)

(c)

(b) $y = 165.73(0.9951)^x$ **(d)** Approximately 83 s

(b) $C = \dfrac{46.93}{1 + 21.273e^{-0.7306t}}$

(d) About 47 people; 50 people
(e) 2.4 days; during the tenth hour of day 3
(f) 9.5 days

Chapter Test *(page 500)*

1. (a) $f \circ g = \dfrac{2x + 7}{2x + 3}$; domain: $\left\{x \mid x \neq -\dfrac{3}{2}\right\}$ **(b)** $(g \circ f)(-2) = 5$ **(c)** $(f \circ g)(-2) = -3$

2. (a) The function is not one-to-one. **(b)** The function is one-to-one.

3. $f^{-1}(x) = \dfrac{2 + 5x}{3x}$; domain of $f = \left\{x \mid x \neq \dfrac{5}{3}\right\}$, range of $f = \{y \mid y \neq 0\}$; domain of $f^{-1} = \{x \mid x \neq 0\}$; range of $f^{-1} = \left\{y \mid y \neq \dfrac{5}{3}\right\}$

4. The point $(-5, 3)$ must be on the graph of f^{-1}. **5.** $x = 5$ **6.** $b = 4$ **7.** $x = 625$ **8.** $e^3 + 2 \approx 22.086$ **9.** $\log 20 \approx 1.301$

10. $\log_3 21 = \dfrac{\ln 21}{\ln 3} \approx 2.771$ **11.** $\ln 133 \approx 4.890$

12. (a) Domain of f: $\{x \mid -\infty < x < \infty\}$ or $(-\infty, \infty)$

(b)

(e) Domain of f^{-1}: $\{x \mid x > -2\}$ or $(-2, \infty)$
Range of f^{-1}: $\{y \mid -\infty < y < \infty\}$ or $(-\infty, \infty)$

(f)

(c) Range of f: $\{y \mid y > -2\}$ or $(-2, \infty)$;
Horizontal asymptote: $y = -2$

(d) $f^{-1}(x) = \log_4 (x + 2) - 1$

13. (a) Domain of f: $\{x \mid x > 2\}$ or $(2, \infty)$

(b)

(c) Range of f: $\{y \mid -\infty < y < \infty\}$ or $(-\infty, \infty)$; vertical asymptote: $x = 2$

(d) $f^{-1}(x) = 5^{1-x} + 2$

(e) Domain of f^{-1}: $\{x \mid -\infty < x < \infty\}$ or $(-\infty, \infty)$
Range of f^{-1}: $\{y \mid y > 2\}$ or $(2, \infty)$

(f)

14. $\{1\}$ **15.** $\{91\}$ **16.** $\{-\ln 2\} \approx \{-0.693\}$ **17.** $\left\{\dfrac{1 - \sqrt{13}}{2}, \dfrac{1 + \sqrt{13}}{2}\right\} \approx \{-1.303, 2.303\}$ **18.** $\left\{\dfrac{3 \ln 7}{1 - \ln 7}\right\} \approx \{-6.172\}$

19. $\{2\sqrt{6}\} \approx \{4.899\}$ **20.** $2 + 3 \log_2 x - \log_2(x - 6) - \log_2(x + 3)$ **21.** About 250.39 days **22. (a)** \$1033.82 **(b)** \$963.42 **(c)** 11.9 yr

23. (a) About 83 dB **(b)** The pain threshold will be exceeded if 31,623 people shout at the same time.

Cumulative Review *(page 500)*

1. Yes; no **2. (a)** 10 **(b)** $2x^2 + 3x + 1$ **(c)** $2x^2 + 4xh + 2h^2 - 3x - 3h + 1$ **3.** $\left(\dfrac{1}{2}, \dfrac{\sqrt{3}}{2}\right)$ is on the graph. **4.** $\{-26\}$

5.

6. (a)

(b) $\{x \mid -\infty < x < \infty\}$

7. $f(x) = 2(x - 4)^2 - 8 = 2x^2 - 16x + 24$

8.

9. $f(g(x)) = \dfrac{4}{(x - 3)^2} + 2$; domain: $\{x \mid x \neq 3\}$; 3

10. (a) Zeros: $-4, -\dfrac{1}{4}, 2$

(b) x-intercepts: $-4, -\dfrac{1}{4}, 2$; y-intercept: -8

(c) Local maximum value of 60.75 occurs at $x = -2.5$.
Local minimum value of -25 occurs at $x = 1$.

(d)

11. (a), (c)

Domain $g = $ range $g^{-1} = (-\infty, \infty)$
Range $g = $ domain $g^{-1} = (2, \infty)$

(b) $g^{-1}(x) = \log_3(x - 2)$

12. $\left\{-\dfrac{3}{2}\right\}$ **13.** $\{2\}$ **14. (a)** $\{-1\}$ **(b)** $\{x\,|\,x > -1\}$ or $(-1, \infty)$ **(c)** $\{25\}$

15. (a) **(b)** Logarithmic; $y = 49.293 - 10.563 \ln x$ **(c)** Highest value of $|r|$

CHAPTER 7 Analytic Geometry

7.2 Assess Your Understanding *(page 511)*

6. parabola **7.** (c) **8.** $(3, 2)$ **9.** $(3, 6)$ **10.** $y = -2$ **11.** B **13.** E **15.** H **17.** C

19. $y^2 = 16x$

21. $x^2 = -12y$

23. $y^2 = -8x$

25. $x^2 = 2y$

27. $x^2 = \dfrac{4}{3}y$

29. $(x - 2)^2 = -8(y + 3)$

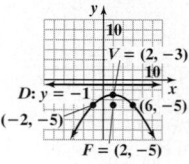

31. $(y + 2)^2 = 4(x + 1)$

33. $(x + 3)^2 = 4(y - 3)$

35. $(y + 2)^2 = -8(x + 1)$

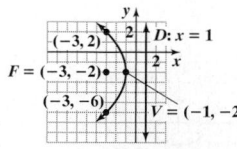

37. Vertex: $(0, 0)$; focus: $(0, 1)$; directrix: $y = -1$

39. Vertex: $(0, 0)$; focus: $(-4, 0)$; directrix: $x = 4$

41. Vertex: $(-1, 2)$; focus: $(1, 2)$; directrix: $x = -3$

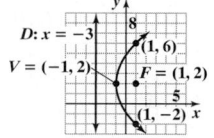

43. Vertex: $(3, -1)$; focus: $\left(3, -\dfrac{5}{4}\right)$; directrix: $y = -\dfrac{3}{4}$

45. Vertex: $(2, -3)$; focus: $(4, -3)$; directrix: $x = 0$

47. Vertex: $(0, 2)$; focus: $(-1, 2)$; directrix: $x = 1$

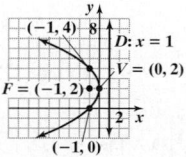

49. Vertex: $(-4, -2)$; focus: $(-4, -1)$; directrix: $y = -3$

51. Vertex: $(-1, -1)$; focus: $\left(-\dfrac{3}{4}, -1\right)$; directrix: $x = -\dfrac{5}{4}$

53. Vertex: $(2, -8)$; focus: $\left(2, -\dfrac{31}{4}\right)$; directrix: $y = -\dfrac{33}{4}$

55. $(y - 1)^2 = x$ **57.** $(y - 1)^2 = -(x - 2)$

59. $x^2 = 4(y - 1)$ **61.** $y^2 = \dfrac{1}{2}(x + 2)$

63. 1.5625 ft from the base of the dish, along the axis of symmetry

65. 1 in. from the vertex, along the axis of symmetry

67. 20 ft **69.** 0.78125 ft

71. 4.17 ft from the base, along the axis of symmetry

73. 24.31 ft, 18.75 ft, 7.64 ft

75. (a) $y = -\dfrac{625}{(299)^2}x^2 + 625$

(b) 567 ft: 63.12 ft; 478 ft: 225.67 ft; 308 ft: 459.2 ft **(c)** No

77. $Cy^2 + Dx = 0, C \neq 0, D \neq 0$ This is the equation of a parabola with vertex at $(0, 0)$ and axis of symmetry the x-axis.

$$Cy^2 = -Dx$$ The focus is $\left(-\dfrac{D}{4C}, 0\right)$; the directrix is the line $x = \dfrac{D}{4C}$. The parabola opens to the right if

$$y^2 = -\dfrac{D}{C}x$$ $-\dfrac{D}{C} > 0$ and to the left if $-\dfrac{D}{C} < 0$.

79. $Cy^2 + Dx + Ey + F = 0, C \neq 0$

$$Cy^2 + Ey = -Dx - F$$

$$y^2 + \dfrac{E}{C}y = -\dfrac{D}{C}x - \dfrac{F}{C}$$

$$\left(y + \dfrac{E}{2C}\right)^2 = -\dfrac{D}{C}x - \dfrac{F}{C} + \dfrac{E^2}{4C^2}$$

$$\left(y + \dfrac{E}{2C}\right)^2 = -\dfrac{D}{C}x + \dfrac{E^2 - 4CF}{4C^2}$$

(a) If $D \neq 0$, then the equation may be written as

$$\left(y + \dfrac{E}{2C}\right)^2 = -\dfrac{D}{C}\left(x - \dfrac{E^2 - 4CF}{4CD}\right).$$

This is the equation of a parabola with vertex at $\left(\dfrac{E^2 - 4CF}{4CD}, -\dfrac{E}{2C}\right)$

and axis of symmetry parallel to the x-axis.

(b)–(d) If $D = 0$, the graph of the equation contains no points if $E^2 - 4CF < 0$, is a single horizontal line if $E^2 - 4CF = 0$, and is two horizontal lines if $E^2 - 4CF > 0$.

7.3 Assess Your Understanding *(page 521)*

7. ellipse **8.** major **9.** $(0, -5); (0, 5)$ **10.** $5; 3; x$ **11.** $(-2, -3); (6, -3)$ **12.** $(1, 4)$ **13.** C **15.** B

17. Vertices: $(-5, 0), (5, 0)$
Foci: $(-\sqrt{21}, 0), (\sqrt{21}, 0)$

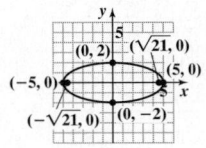

19. Vertices: $(0, -5), (0, 5)$
Foci: $(0, -4), (0, 4)$

21. $\dfrac{x^2}{4} + \dfrac{y^2}{16} = 1$
Vertices: $(0, -4), (0, 4)$
Foci: $(0, -2\sqrt{3}), (0, 2\sqrt{3})$

23. $\dfrac{x^2}{8} + \dfrac{y^2}{2} = 1$
Vertices: $\left(-2\sqrt{2}, 0\right), \left(2\sqrt{2}, 0\right)$
Foci: $\left(-\sqrt{6}, 0\right), \left(\sqrt{6}, 0\right)$

25. $\dfrac{x^2}{16} + \dfrac{y^2}{16} = 1$
Vertices: $(-4, 0), (4, 0),$
$(0, -4), (0, 4);$ Focus: $(0, 0)$

27. $\dfrac{x^2}{25} + \dfrac{y^2}{16} = 1$

29. $\dfrac{x^2}{9} + \dfrac{y^2}{25} = 1$

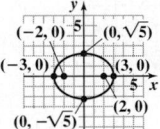

31. $\dfrac{x^2}{9} + \dfrac{y^2}{5} = 1$

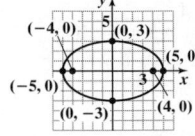

33. $\dfrac{x^2}{25} + \dfrac{y^2}{9} = 1$

35. $\dfrac{x^2}{4} + \dfrac{y^2}{13} = 1$

37. $x^2 + \dfrac{y^2}{16} = 1$

39. $\dfrac{(x + 1)^2}{4} + (y - 1)^2 = 1$

41. $(x - 1)^2 + \dfrac{y^2}{4} = 1$

43. Center: $(3, -1)$; vertices: $(3, -4), (3, 2)$;
foci: $\left(3, -1 - \sqrt{5}\right), \left(3, -1 + \sqrt{5}\right)$

45. $\dfrac{(x + 5)^2}{16} + \dfrac{(y - 4)^2}{4} = 1$
Center: $(-5, 4)$; vertices: $(-9, 4), (-1, 4)$;
foci: $\left(-5 - 2\sqrt{3}, 4\right), \left(-5 + 2\sqrt{3}, 4\right)$

47. $\dfrac{(x + 2)^2}{4} + (y - 1)^2 = 1$
Center: $(-2, 1)$; vertices: $(-4, 1), (0, 1)$;
foci: $\left(-2 - \sqrt{3}, 1\right), \left(-2 + \sqrt{3}, 1\right)$

49. $\dfrac{(x-2)^2}{3}+\dfrac{(y+1)^2}{2}=1$

Center: $(2,-1)$; vertices: $\left(2-\sqrt{3},-1\right)$;

$\left(2+\sqrt{3},-1\right)$; foci: $(1,-1),(3,-1)$

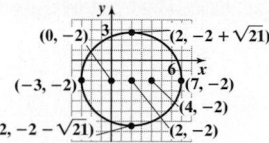

51. $\dfrac{(x-1)^2}{4}+\dfrac{(y+2)^2}{9}=1$

Center: $(1,-2)$; vertices: $(1,-5),(1,1)$;

foci: $\left(1,-2-\sqrt{5}\right),\left(1,-2+\sqrt{5}\right)$

53. $x^2+\dfrac{(y+2)^2}{4}=1$

Center: $(0,-2)$; vertices: $(0,-4),(0,0)$;

foci: $\left(0,-2-\sqrt{3}\right),\left(0,-2+\sqrt{3}\right)$

55. $\dfrac{(x-2)^2}{25}+\dfrac{(y+2)^2}{21}=1$

57. $\dfrac{(x-4)^2}{5}+\dfrac{(y-6)^2}{9}=1$

59. $\dfrac{(x-2)^2}{16}+\dfrac{(y-1)^2}{7}=1$

61. $\dfrac{(x-1)^2}{10}+(y-2)^2=1$

63. $\dfrac{(x-1)^2}{9}+\dfrac{(y-2)^2}{9}=1$

65.

67.

69. $\dfrac{x^2}{100}+\dfrac{y^2}{36}=1$ **71.** 43.3 ft **73.** 24.65 ft, 21.65 ft, 13.82 ft **75.** 30 ft **77.** The elliptical hole will have a

major axis of length $2\sqrt{41}$ in. and a minor axis of length 8 in. **79.** 91.5 million mi; $\dfrac{x^2}{(93)^2}+\dfrac{y^2}{8646.75}=1$

81. Perihelion: 460.6 million mi; mean distance: 483.8 million mi; $\dfrac{x^2}{(483.8)^2}+\dfrac{y^2}{233,524.2}=1$

83. (a) $Ax^2+Cy^2+F=0$ 	 If A and C are of the same sign and F is of opposite sign, then the equation takes the form

$Ax^2+Cy^2=-F$ 	 $\dfrac{x^2}{\left(-\dfrac{F}{A}\right)}+\dfrac{y^2}{\left(-\dfrac{F}{C}\right)}=1$, where $-\dfrac{F}{A}$ and $-\dfrac{F}{C}$ are positive. This is the equation of an ellipse with center at $(0,0)$.

(b) If $A=C$, the equation may be written as $x^2+y^2=-\dfrac{F}{A}$.

This is the equation of a circle with center at $(0,0)$ and radius equal to $\sqrt{-\dfrac{F}{A}}$.

7.4 Assess Your Understanding *(page 534)*

7. hyperbola **8.** transverse axis **9.** b

10. $(2,4);(2,-2)$ **11.** $(2,6);(2,-4)$

12. 4 **13.** $2;3;x$

14. $y=-\dfrac{4}{9}x;\ y=\dfrac{4}{9}x$ **15.** B **17.** A

19. $x^2-\dfrac{y^2}{8}=1$

21. $\dfrac{y^2}{16}-\dfrac{x^2}{20}=1$

23. $\dfrac{x^2}{9}-\dfrac{y^2}{16}=1$

25. $\dfrac{y^2}{36}-\dfrac{x^2}{9}=1$

27. $\dfrac{x^2}{8}-\dfrac{y^2}{8}=1$

29. $\dfrac{x^2}{25} - \dfrac{y^2}{9} = 1$

Center: $(0, 0)$

Transverse axis: x-axis

Vertices: $(-5, 0)$, $(5, 0)$

Foci: $\left(-\sqrt{34}, 0\right)$, $\left(\sqrt{34}, 0\right)$

Asymptotes: $y = \pm\dfrac{3}{5}x$

31. $\dfrac{x^2}{4} - \dfrac{y^2}{16} = 1$

Center: $(0, 0)$

Transverse axis: x-axis

Vertices: $(-2, 0)$, $(2, 0)$

Foci: $\left(-2\sqrt{5}, 0\right)$, $\left(2\sqrt{5}, 0\right)$

Asymptotes: $y = \pm 2x$

33. $\dfrac{y^2}{9} - x^2 = 1$

Center: $(0, 0)$

Transverse axis: y-axis

Vertices: $(0, -3)$, $(0, 3)$

Foci: $\left(0, -\sqrt{10}\right)$, $\left(0, \sqrt{10}\right)$

Asymptotes: $y = \pm 3x$

35. $\dfrac{y^2}{25} - \dfrac{x^2}{25} = 1$

Center: $(0, 0)$

Transverse axis: y-axis

Vertices: $(0, -5)$, $(0, 5)$

Foci: $\left(0, -5\sqrt{2}\right)$, $\left(0, 5\sqrt{2}\right)$

Asymptotes: $y = \pm x$

37. $x^2 - y^2 = 1$

39. $\dfrac{y^2}{36} - \dfrac{x^2}{9} = 1$

41. $\dfrac{(x-4)^2}{4} - \dfrac{(y+1)^2}{5} = 1$

43. $\dfrac{(y+4)^2}{4} - \dfrac{(x+3)^2}{12} = 1$

45. $(x-5)^2 - \dfrac{(y-7)^2}{3} = 1$

47. $\dfrac{(x-1)^2}{4} - \dfrac{(y+1)^2}{9} = 1$

49. $\dfrac{(x-2)^2}{4} - \dfrac{(y+3)^2}{9} = 1$

Center: $(2, -3)$

Transverse axis: parallel to x-axis

Vertices: $(0, -3)$, $(4, -3)$

Foci: $\left(2 - \sqrt{13}, -3\right)$, $\left(2 + \sqrt{13}, -3\right)$

Asymptotes: $y + 3 = \pm\dfrac{3}{2}(x - 2)$

51. $\dfrac{(y-2)^2}{4} - (x+2)^2 = 1$

Center: $(-2, 2)$

Transverse axis: parallel to y-axis

Vertices: $(-2, 0)$, $(-2, 4)$

Foci: $\left(-2, 2 - \sqrt{5}\right)$, $\left(-2, 2 + \sqrt{5}\right)$

Asymptotes: $y - 2 = \pm 2(x + 2)$

53. $\dfrac{(x+1)^2}{4} - \dfrac{(y+2)^2}{4} = 1$

Center: $(-1, -2)$

Transverse axis: parallel to x-axis

Vertices: $(-3, -2)$, $(1, -2)$

Foci: $\left(-1 - 2\sqrt{2}, -2\right)$, $\left(-1 + 2\sqrt{2}, -2\right)$

Asymptotes: $y + 2 = \pm(x + 1)$

55. $(x-1)^2 - (y+1)^2 = 1$

Center: $(1, -1)$

Transverse axis: parallel to x-axis

Vertices: $(0, -1)$, $(2, -1)$

Foci: $\left(1 - \sqrt{2}, -1\right)$, $\left(1 + \sqrt{2}, -1\right)$

Asymptotes: $y + 1 = \pm(x - 1)$

57. $\dfrac{(y-2)^2}{4} - (x+1)^2 = 1$

Center: $(-1, 2)$
Transverse axis: parallel to y-axis
Vertices: $(-1, 0)$, $(-1, 4)$
Foci: $\left(-1, 2 - \sqrt{5}\right)$, $\left(-1, 2 + \sqrt{5}\right)$
Asymptotes: $y - 2 = \pm 2(x + 1)$

59. $\dfrac{(x-3)^2}{4} - \dfrac{(y+2)^2}{16} = 1$

Center: $(3, -2)$
Transverse axis: parallel to x-axis
Vertices: $(1, -2)$, $(5, -2)$
Foci: $\left(3 - 2\sqrt{5}, -2\right)$, $\left(3 + 2\sqrt{5}, -2\right)$
Asymptotes: $y + 2 = \pm 2(x - 3)$

61. $\dfrac{(y-1)^2}{4} - (x+2)^2 = 1$

Center: $(-2, 1)$
Transverse axis: parallel to y-axis
Vertices: $(-2, -1)$, $(-2, 3)$
Foci: $\left(-2, 1 - \sqrt{5}\right)$, $\left(-2, 1 + \sqrt{5}\right)$
Asymptotes: $y - 1 = \pm 2(x + 2)$

63.

65.

67. Center: $(3, 0)$
Transverse axis: parallel to x-axis
Vertices: $(1, 0)$, $(5, 0)$
Foci: $(3 - \sqrt{29}, 0)$, $(3 + \sqrt{29}, 0)$
Asymptotes: $y = \pm\dfrac{5}{2}(x - 3)$

69. Vertex: $(0, 3)$; focus: $(0, 7)$;
directrix: $y = -1$

71. $\dfrac{(x-5)^2}{9} + \dfrac{y^2}{25} = 1$

Center: $(5, 0)$; vertices: $(5, 5)$, $(5, -5)$;
foci: $(5, -4)$, $(5, 4)$

73. $(x - 3)^2 = 8(y + 5)$
Vertex: $(3, -5)$; focus: $(3, -3)$;
directrix: $y = -7$

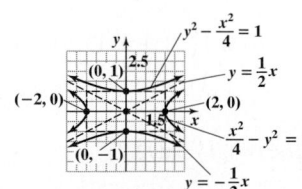

75. The fireworks display is 50,138 ft north of the person at point A. **77.** The tower is 592.4 ft tall. **79. (a)** $y = \pm x$ **(b)** $\dfrac{x^2}{100} - \dfrac{y^2}{100} = 1, \; x \geq 0$

81. If the eccentricity is close to one, the "opening" of the hyperbola is very small. As e increases, the opening gets bigger.

83. $\dfrac{x^2}{4} - y^2 = 1$; asymptotes $y = \pm\dfrac{1}{2}x$

$y^2 - \dfrac{x^2}{4} = 1$; asymptotes $y = \pm\dfrac{1}{2}x$

85. $Ax^2 + Cy^2 + F = 0$

$Ax^2 + Cy^2 = -F$

If A and C are of opposite sign and $F \neq 0$, this equation may be written as $\dfrac{x^2}{\left(-\dfrac{F}{A}\right)} + \dfrac{y^2}{\left(-\dfrac{F}{C}\right)} = 1$,

where $-\dfrac{F}{A}$ and $-\dfrac{F}{C}$ are opposite in sign. This is the equation of a hyperbola with center $(0, 0)$.
The transverse axis is the x-axis if $-\dfrac{F}{A} > 0$; the transverse axis is the y-axis if $-\dfrac{F}{A} < 0$.

Review Exercises *(page 537)*

1. Parabola; vertex $(0, 0)$, focus $(-4, 0)$, directrix $x = 4$ **3.** Hyperbola; center $(0, 0)$, vertices $(5, 0)$ and $(-5, 0)$, foci $\left(\sqrt{26}, 0\right)$ and $\left(-\sqrt{26}, 0\right)$,

asymptotes $y = \dfrac{1}{5}x$ and $y = -\dfrac{1}{5}x$ **5.** Ellipse; center $(0, 0)$, vertices $(0, 5)$ and $(0, -5)$, foci $(0, 3)$ and $(0, -3)$ **7.** $x^2 = -4(y - 1)$: Parabola;

vertex $(0, 1)$, focus $(0, 0)$, directrix $y = 2$ **9.** $\dfrac{x^2}{2} - \dfrac{y^2}{8} = 1$: Hyperbola; center $(0, 0)$, vertices $\left(\sqrt{2}, 0\right)$ and $\left(-\sqrt{2}, 0\right)$, foci $\left(\sqrt{10}, 0\right)$ and $\left(-\sqrt{10}, 0\right)$,

asymptotes $y = 2x$ and $y = -2x$ **11.** $(x - 2)^2 = 2(y + 2)$: Parabola; vertex $(2, -2)$, focus $\left(2, -\dfrac{3}{2}\right)$, directrix $y = -\dfrac{5}{2}$

13. $\dfrac{(y-2)^2}{4} - (x-1)^2 = 1$: Hyperbola; center $(1,2)$, vertices $(1,4)$ and $(1,0)$, foci $\left(1, 2+\sqrt{5}\right)$ and $\left(1, 2-\sqrt{5}\right)$, asymptotes $y-2 = \pm 2(x-1)$

15. $\dfrac{(x-2)^2}{9} + \dfrac{(y-1)^2}{4} = 1$: Ellipse; center $(2,1)$, vertices $(5,1)$ and $(-1,1)$, foci $\left(2+\sqrt{5},1\right)$ and $\left(2-\sqrt{5},1\right)$

17. $(x-2)^2 = -4(y+1)$: Parabola; vertex $(2,-1)$, focus $(2,-2)$, directrix $y = 0$

19. $\dfrac{(x-1)^2}{4} + \dfrac{(y+1)^2}{9} = 1$: Ellipse; center $(1,-1)$, vertices $(1,2)$ and $(1,-4)$, foci $\left(1, -1+\sqrt{5}\right)$ and $\left(1, -1-\sqrt{5}\right)$

21. $y^2 = -8x$

23. $\dfrac{y^2}{4} - \dfrac{x^2}{12} = 1$

25. $\dfrac{x^2}{16} + \dfrac{y^2}{7} = 1$

27. $(x-2)^2 = -4(y+3)$

29. $(x+2)^2 - \dfrac{(y+3)^2}{3} = 1$

31. $\dfrac{(x+4)^2}{16} + \dfrac{(y-5)^2}{25} = 1$
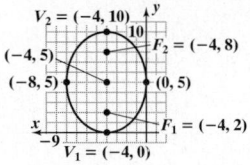

33. $\dfrac{(x+1)^2}{9} - \dfrac{(y-2)^2}{7} = 1$

35. $\dfrac{(x-3)^2}{9} - \dfrac{(y-1)^2}{4} = 1$
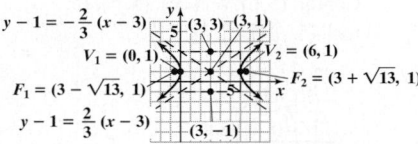

37. $\dfrac{x^2}{5} - \dfrac{y^2}{4} = 1$ **39.** The ellipse $\dfrac{x^2}{16} + \dfrac{y^2}{7} = 1$ **41.** $\dfrac{1}{4}$ ft or 3 in. **43.** 19.72 ft, 18.86 ft, 14.91 ft **45.** 450 ft

Chapter Test *(page 538)*

1. Hyperbola; center: $(-1,0)$; vertices: $(-3,0)$ and $(1,0)$; foci: $\left(-1-\sqrt{13},0\right)$ and $\left(-1+\sqrt{13},0\right)$; asymptotes: $y = -\dfrac{3}{2}(x+1)$ and $y = \dfrac{3}{2}(x+1)$

2. Parabola; vertex: $\left(1, -\dfrac{1}{2}\right)$; focus: $\left(1, \dfrac{3}{2}\right)$; directrix: $y = -\dfrac{5}{2}$

3. Ellipse; center: $(-1,1)$; foci: $\left(-1-\sqrt{3},1\right)$ and $\left(-1+\sqrt{3},1\right)$; vertices: $(-4,1)$ and $(2,1)$

4. $(x+1)^2 = 6(y-3)$
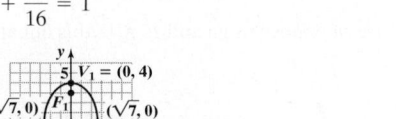

5. $\dfrac{x^2}{7} + \dfrac{y^2}{16} = 1$

6. $\dfrac{(y-2)^2}{4} - \dfrac{(x-2)^2}{8} = 1$

7. The microphone should be located $\dfrac{2}{3}$ ft from the base of the reflector, along its axis of symmetry.

Cumulative Review *(page 539)*

1. $-6x + 5 - 3h$ **2.** $\left\{-5, -\dfrac{1}{3}, 2\right\}$ **3.** $\{x \mid -3 \le x \le 2\}$ or $[-3, 2]$ **4.** (a) Domain: $(-\infty, \infty)$; range: $(2, \infty)$

(b) $y = \log_3(x-2)$; domain: $(2, \infty)$; range: $(-\infty, \infty)$ **5.** (a) $\{18\}$ (b) $(2, 18]$ **6.** (a) $y = 2x - 2$ (b) $(x-2)^2 + y^2 = 4$ (c) $\dfrac{x^2}{9} + \dfrac{y^2}{4} = 1$

(d) $y = 2(x-1)^2$ (e) $y^2 - \dfrac{x^2}{3} = 1$ (f) $y = 4^x$

CHAPTER 8 Systems of Equations and Inequalities

8.1 Assess Your Understanding (page 552)

3. inconsistent **4.** consistent; independent **5.** $(3, -2)$ **6.** consistent; dependent

7. $\begin{cases} 2(2) - (-1) = 5 \\ 5(2) + 2(-1) = 8 \end{cases}$ **9.** $\begin{cases} 3(2) - 4\left(\dfrac{1}{2}\right) = 4 \\ \dfrac{1}{2}(2) - 3\left(\dfrac{1}{2}\right) = -\dfrac{1}{2} \end{cases}$ **11.** $\begin{cases} 4 - 1 = 3 \\ \dfrac{1}{2}(4) + 1 = 3 \end{cases}$ **13.** $\begin{cases} 3(1) + 3(-1) + 2(2) = 4 \\ 1 - (-1) - 2 = 0 \\ 2(-1) - 3(2) = -8 \end{cases}$ **15.** $\begin{cases} 3(2) + 3(-2) + 2(2) = 4 \\ 2 - 3(-2) + 2 = 10 \\ 5(2) - 2(-2) - 3(2) = 8 \end{cases}$

17. $x = 6, y = 2; (6, 2)$ **19.** $x = 3, y = -6; (3, -6)$ **21.** $x = 8, y = -4; (8, -4)$ **23.** $x = \dfrac{1}{3}, y = -\dfrac{1}{6}; \left(\dfrac{1}{3}, -\dfrac{1}{6}\right)$ **25.** Inconsistent

27. $x = \dfrac{3}{2}, y = 3; \left(\dfrac{3}{2}, 3\right)$ **29.** $\{(x, y)|x = 4 - 2y, y$ is any real number$\}$ or $\left\{(x, y)\left|y = \dfrac{4 - x}{2}, x$ is any real number$\right\}\right.$ **31.** $x = 1, y = 1; (1, 1)$

33. $x = \dfrac{3}{2}, y = 1; \left(\dfrac{3}{2}, 1\right)$ **35.** $x = 4, y = 3; (4, 3)$ **37.** $x = \dfrac{4}{3}, y = \dfrac{1}{5}; \left(\dfrac{4}{3}, \dfrac{1}{5}\right)$ **39.** $x = \dfrac{1}{5}, y = \dfrac{1}{3}; \left(\dfrac{1}{5}, \dfrac{1}{3}\right)$ **41.** $x = 8, y = 2, z = 0; (8, 2, 0)$

43. $x = 2, y = -1, z = 1; (2, -1, 1)$ **45.** Inconsistent **47.** $\{(x, y, z)|x = 5z - 2, y = 4z - 3; z$ is any real number$\}$ **49.** Inconsistent

51. $x = 1, y = 3, z = -2; (1, 3, -2)$ **53.** $x = -3, y = \dfrac{1}{2}, z = 1; \left(-3, \dfrac{1}{2}, 1\right)$ **55.** Length 30 ft; width 15 ft

57. There were 18 commercial launches and 37 noncommercial launches in 2005. **59.** 22.5 lb

61. Average wind speed 25 mph; average airspeed 175 mph **63.** 80 \$25 sets and 120 \$45 sets **65.** \$9.96

67. Mix 50 mg of first compound with 75 mg of second. **69.** $a = \dfrac{4}{3}, b = -\dfrac{5}{3}, c = 1$ **71.** $Y = 9000, r = 0.06$ **73.** $I_1 = \dfrac{10}{71}, I_2 = \dfrac{65}{71}, I_3 = \dfrac{55}{71}$

75. 100 orchestra, 210 main, and 190 balcony seats **77.** 1.5 chicken, 1 corn, 2 milk

79. If $x =$ price of hamburgers, $y =$ price of fries, and $z =$ price of colas, then

$x = 2.75 - z, y = \dfrac{41}{60} + \dfrac{1}{3}z, \$0.60 \le z \le \$0.90.$

There is not sufficient information:

x	\$2.13	\$2.01	\$1.86
y	\$0.89	\$0.93	\$0.98
z	\$0.62	\$0.74	\$0.89

81. It will take Beth 30 hr, Bill 24 hr, and Edie 40 hr.

8.2 Assess Your Understanding (page 567)

1. matrix **2.** augmented **3.** third; fifth **4.** T

5. $\begin{bmatrix} 1 & -5 & | & 5 \\ 4 & 3 & | & 6 \end{bmatrix}$ **7.** $\begin{bmatrix} 2 & 3 & | & 6 \\ 4 & -6 & | & -2 \end{bmatrix}$ **9.** $\begin{bmatrix} 0.01 & -0.03 & | & 0.06 \\ 0.13 & 0.10 & | & 0.20 \end{bmatrix}$ **11.** $\begin{bmatrix} 1 & -1 & 1 & | & 10 \\ 3 & 3 & 0 & | & 5 \\ 1 & 1 & 2 & | & 2 \end{bmatrix}$ **13.** $\begin{bmatrix} 1 & 1 & -1 & | & 2 \\ 3 & -2 & 0 & | & 2 \\ 5 & 3 & -1 & | & 1 \end{bmatrix}$ **15.** $\begin{bmatrix} 1 & -1 & -1 & | & 10 \\ 2 & 1 & 2 & | & -1 \\ -3 & 4 & 0 & | & 5 \\ 4 & -5 & 1 & | & 0 \end{bmatrix}$

17. $\begin{cases} x - 3y = -2 & (1) \\ 2x - 5y = 5 & (2) \end{cases}$ $\begin{bmatrix} 1 & -3 & | & -2 \\ 0 & 1 & | & 9 \end{bmatrix}$ **19.** $\begin{cases} x - 3y + 4z = 3 & (1) \\ 3x - 5y + 6z = 6 & (2); \\ -5x + 3y + 4z = 6 & (3) \end{cases}$ $\begin{bmatrix} 1 & -3 & 4 & | & 3 \\ 0 & 4 & -6 & | & -3 \\ 0 & -12 & 24 & | & 21 \end{bmatrix}$

21. $\begin{cases} x - 3y + 2z = -6 & (1) \\ 2x - 5y + 3z = -4 & (2); \\ -3x - 6y + 4z = 6 & (3) \end{cases}$ $\begin{bmatrix} 1 & -3 & 2 & | & -6 \\ 0 & 1 & -1 & | & 8 \\ 0 & -15 & 10 & | & -12 \end{bmatrix}$ **23.** $\begin{cases} 5x - 3y + z = -2 & (1) \\ 2x - 5y + 6z = -2 & (2); \\ -4x + y + 4z = 6 & (3) \end{cases}$ $\begin{bmatrix} 1 & 7 & -11 & | & 2 \\ 2 & -5 & 6 & | & -2 \\ 0 & -9 & 16 & | & 2 \end{bmatrix}$

25. $\begin{cases} x = 5 \\ y = -1 \end{cases}$
Consistent; $x = 5, y = -1$ or $(5, -1)$

27. $\begin{cases} x = 1 \\ y = 2 \\ 0 = 3 \end{cases}$
Inconsistent

29. $\begin{cases} x + 2z = -1 \\ y - 4z = -2 \\ 0 = 0 \end{cases}$
Consistent:
$\begin{cases} x = -1 - 2z \\ y = -2 + 4z \\ z \text{ is any real number or} \end{cases}$
$\{(x, y, z) \mid x = -1 - 2z,$
$y = -2 + 4z, z \text{ is any real}$
number$\}$

31. $\begin{cases} x_1 = 1 \\ x_2 + x_4 = 2 \\ x_3 + 2x_4 = 3 \end{cases}$
Consistent:
$\begin{cases} x_1 = 1, x_2 = 2 - x_4 \\ x_3 = 3 - 2x_4 \\ x_4 \text{ is any real number or} \end{cases}$
$\{(x_1, x_2, x_3, x_4) \mid x_1 = 1,$
$x_2 = 2 - x_4, x_3 = 3 - 2x_4,$
$x_4 \text{ is any real number}\}$

33. $\begin{cases} x_1 + 4x_4 = 2 \\ x_2 + x_3 + 3x_4 = 3 \\ 0 = 0 \end{cases}$
Consistent:
$\begin{cases} x_1 = 2 - 4x_4 \\ x_2 = 3 - x_3 - 3x_4 \\ x_3, x_4 \text{ are any real numbers or} \end{cases}$
$\{(x_1, x_2, x_3, x_4) \mid x_1 = 2 - 4x_4,$
$x_2 = 3 - x_3 - 3x_4, x_3, x_4 \text{ are}$
any real numbers$\}$

35. $\begin{cases} x_1 + x_4 = -2 \\ x_2 + 2x_4 = 2 \\ x_3 - x_4 = 0 \\ 0 = 0 \end{cases}$
Consistent:
$\begin{cases} x_1 = -2 - x_4 \\ x_2 = 2 - 2x_4 \\ x_3 = x_4 \\ x_4 \text{ is any real number or} \end{cases}$
$\{(x_1, x_2, x_3, x_4) \mid x_1 = -2 - x_4,$
$x_2 = 2 - 2x_4, x_3 = x_4, x_4 \text{ is any}$
real number$\}$

37. $x = 6, y = 2; (6, 2)$ **39.** $x = \dfrac{1}{2}, y = \dfrac{3}{4}; \left(\dfrac{1}{2}, \dfrac{3}{4}\right)$ **41.** $x = 4 - 2y, y$ is any real number; $\{(x, y) \mid x = 4 - 2y, y$ is any real number$\}$

43. $x = \dfrac{3}{2}, y = 1; \left(\dfrac{3}{2}, 1\right)$ **45.** $x = \dfrac{4}{3}, y = \dfrac{1}{5}; \left(\dfrac{4}{3}, \dfrac{1}{5}\right)$ **47.** $x = 8, y = 2, z = 0; (8, 2, 0)$ **49.** $x = 2, y = -1, z = 1; (2, -1, 1)$

51. Inconsistent **53.** $x = 5z - 2, y = 4z - 3$, where z is any real number; $\{(x, y, z) \mid x = 5z - 2, y = 4z - 3, z$ is any real number$\}$

55. Inconsistent **57.** $x = 1, y = 3, z = -2; (1, 3, -2)$ **59.** $x = -3, y = \dfrac{1}{2}, z = 1; \left(-3, \dfrac{1}{2}, 1\right)$ **61.** $x = \dfrac{1}{3}, y = \dfrac{2}{3}, z = 1; \left(\dfrac{1}{3}, \dfrac{2}{3}, 1\right)$

63. $x = 1, y = 2, z = 0, w = 1; (1, 2, 0, 1)$ **65.** $y = 0, z = 1 - x, x$ is any real number; $\{(x, y, z) \mid y = 0, z = 1 - x, x$ is any real number$\}$

67. $x = 2, y = z - 3, z$ is any real number; $\{(x, y, z) \mid x = 2, y = z - 3, z$ is any real number$\}$ **69.** $x = \dfrac{13}{9}, y = \dfrac{7}{18}, z = \dfrac{19}{18}; \left(\dfrac{13}{9}, \dfrac{7}{18}, \dfrac{19}{18}\right)$

71. $x = \dfrac{7}{5} - \dfrac{3}{5}z - \dfrac{2}{5}w, y = -\dfrac{8}{5} + \dfrac{7}{5}z + \dfrac{13}{5}w$, where z and w are any real numbers; $\left\{(x, y, z, w) \,\middle|\, x = \dfrac{7}{5} - \dfrac{3}{5}z - \dfrac{2}{5}w, y = -\dfrac{8}{5} + \dfrac{7}{5}z + \dfrac{13}{5}w,\right.$

z and w are any real numbers $\Big\}$ **73.** $y = -2x^2 + x + 3$ **75.** $f(x) = 3x^3 - 4x^2 + 5$ **77.** 1.5 salmon steak, 2 baked eggs, 1 acorn squash

79. \$4000 in Treasury bills, \$4000 in Treasury bonds, \$2000 in corporate bonds **81.** 8 Deltas, 5 Betas, 10 Sigmas **83.** $I_1 = \dfrac{44}{23}, I_2 = 2, I_3 = \dfrac{16}{23}, I_4 = \dfrac{28}{23}$

85. (a)

7%	9%	11%
0	10,000	10,000
1000	8000	11,000
2000	6000	12,000
3000	4000	13,000
4000	2000	14,000
5000	0	15,000

(b)

7%	9%	11%
12,500	12,500	0
14,500	8500	2000
16,500	4500	4000
18,750	0	6250

(c) All the money invested at 7% provides \$2100, more than what is required.

87.

First Supplement	Second Supplement	Third Supplement
50 mg	75 mg	0 mg
36 mg	76 mg	8 mg
22 mg	77 mg	16 mg
8 mg	78 mg	24 mg

8.3 Assess Your Understanding *(page 578)*

1. $ad - bc$ **2.** $\begin{vmatrix} 5 & 3 \\ -3 & -4 \end{vmatrix}$ **3.** F **4.** F **5.** F **6.** F **7.** 22 **9.** -2 **11.** 10 **13.** -26 **15.** $x = 6, y = 2; (6, 2)$ **17.** $x = 3, y = 2; (3, 2)$

19. $x = 8, y = -4; (8, -4)$ **21.** $x = 4, y = -2; (4, -2)$ **23.** Not applicable **25.** $x = \dfrac{1}{2}, y = \dfrac{3}{4}; \left(\dfrac{1}{2}, \dfrac{3}{4}\right)$ **27.** $x = \dfrac{1}{10}, y = \dfrac{2}{5}; \left(\dfrac{1}{10}, \dfrac{2}{5}\right)$

29. $x = \dfrac{3}{2}, y = 1; \left(\dfrac{3}{2}, 1\right)$ **31.** $x = \dfrac{4}{3}, y = \dfrac{1}{5}; \left(\dfrac{4}{3}, \dfrac{1}{5}\right)$ **33.** $x = 1, y = 3, z = -2; (1, 3, -2)$ **35.** $x = -3, y = \dfrac{1}{2}, z = 1; \left(-3, \dfrac{1}{2}, 1\right)$

37. Not applicable **39.** $x = 0, y = 0, z = 0; (0, 0, 0)$ **41.** Not applicable **43.** -4 **45.** 12 **47.** 8 **49.** 8 **51.** -5 **53.** $\dfrac{13}{11}$ **55.** 0 or -9

57. $(y_1 - y_2)x - (x_1 - x_2)y + (x_1y_2 - x_2y_1) = 0$ **59.** The triangle has an area of 5 square units.

$$(y_1 - y_2)x + (x_2 - x_1)y = x_2y_1 - x_1y_2$$
$$(x_2 - x_1)y - (x_2 - x_1)y_1 = (y_2 - y_1)x + x_2y_1 - x_1y_2 - (x_2 - x_1)y_1$$
$$(x_2 - x_1)(y - y_1) = (y_2 - y_1)x - (y_2 - y_1)x_1$$
$$y - y_1 = \dfrac{y_2 - y_1}{x_2 - x_1}(x - x_1)$$

61. If $a = 0$, we have

$$by = s$$
$$cx + dy = t$$

Thus, $y = \dfrac{s}{b}$ and

$$x = \dfrac{t - dy}{c} = \dfrac{tb - ds}{bc}$$

Using Cramer's Rule, we get

$$x = \dfrac{sd - tb}{-bc} = \dfrac{tb - sd}{bc}$$
$$y = \dfrac{-sc}{-bc} = \dfrac{s}{b}$$

If $b = 0$, we have

$$ax = s$$
$$cx + dy = t$$

Since $D = ad \neq 0$, then
$a \neq 0$ and $d \neq 0$.

Thus, $x = \dfrac{s}{a}$ and

$$y = \dfrac{t - cx}{d} = \dfrac{ta - cs}{ad}$$

Using Cramer's Rule, we get

$$x = \dfrac{sd}{ad} = \dfrac{s}{a}$$
$$y = \dfrac{ta - cs}{ad}$$

If $c = 0$, we have

$$ax + by = s$$
$$dy = t$$

Since $D = ad \neq 0$, then
$a \neq 0$ and $d \neq 0$.

Thus, $y = \dfrac{t}{d}$ and

$$x = \dfrac{s - by}{a} = \dfrac{sd - bt}{ad}$$

Using Cramer's Rule, we get

$$x = \dfrac{sd - bt}{ad}$$
$$y = \dfrac{at}{ad} = \dfrac{t}{d}$$

If $d = 0$, we have

$$ax + by = s$$
$$cx = t$$

Since $D = -bc \neq 0$, then
$b \neq 0$ and $c \neq 0$.

Thus, $x = \dfrac{t}{c}$ and

$$y = \dfrac{s - ax}{b} = \dfrac{sc - at}{bc}$$

Using Cramer's Rule, we get

$$x = \dfrac{-tb}{-bc} = \dfrac{t}{c}$$
$$y = \dfrac{at - sc}{-bc} = \dfrac{sc - at}{bc}$$

63. $\begin{vmatrix} a_{11} & a_{12} & a_{13} \\ ka_{21} & ka_{22} & ka_{23} \\ a_{31} & a_{32} & a_{33} \end{vmatrix} = -ka_{21}(a_{12}a_{33} - a_{32}a_{13}) + ka_{22}(a_{11}a_{33} - a_{31}a_{13}) - ka_{23}(a_{11}a_{32} - a_{31}a_{12})$

$$= k[-a_{21}(a_{12}a_{33} - a_{32}a_{13}) + a_{22}(a_{11}a_{33} - a_{31}a_{13}) - a_{23}(a_{11}a_{32} - a_{31}a_{12})] = k\begin{vmatrix} a_{11} & a_{12} & a_{13} \\ a_{21} & a_{22} & a_{23} \\ a_{31} & a_{32} & a_{33} \end{vmatrix}$$

65. $\begin{vmatrix} a_{11} + ka_{21} & a_{12} + ka_{22} & a_{13} + ka_{23} \\ a_{21} & a_{22} & a_{23} \\ a_{31} & a_{32} & a_{33} \end{vmatrix} = (a_{11} + ka_{21})(a_{22}a_{33} - a_{32}a_{23}) - (a_{12} + ka_{22})(a_{21}a_{33} - a_{31}a_{23}) + (a_{13} + ka_{23})(a_{21}a_{32} - a_{31}a_{22})$

$$= a_{11}a_{22}a_{33} - a_{11}a_{32}a_{23} + \overline{ka_{21}a_{22}a_{33}} - \overline{ka_{21}a_{32}a_{23}} - a_{12}a_{21}a_{33} + a_{12}a_{31}a_{23}$$
$$- \overline{ka_{22}a_{21}a_{33}} + \overline{ka_{22}a_{31}a_{23}} + a_{13}a_{21}a_{32} - a_{13}a_{31}a_{22} + \overline{ka_{23}a_{21}a_{32}} - \overline{ka_{23}a_{31}a_{22}}$$
$$= a_{11}a_{22}a_{33} - a_{11}a_{32}a_{23} - a_{12}a_{21}a_{33} + a_{12}a_{31}a_{23} + a_{13}a_{21}a_{32} - a_{13}a_{31}a_{22}$$
$$= a_{11}(a_{22}a_{33} - a_{32}a_{23}) - a_{12}(a_{21}a_{33} - a_{31}a_{23}) + a_{13}(a_{21}a_{32} - a_{31}a_{22})$$
$$= \begin{vmatrix} a_{11} & a_{12} & a_{13} \\ a_{21} & a_{22} & a_{23} \\ a_{31} & a_{32} & a_{33} \end{vmatrix}$$

Historical Problems *(page 594)*

1. (a) $2 - 5i \longleftrightarrow \begin{bmatrix} 2 & -5 \\ 5 & 2 \end{bmatrix}, 1 + 3i \longleftrightarrow \begin{bmatrix} 1 & 3 \\ -3 & 1 \end{bmatrix}$ **(b)** $\begin{bmatrix} 2 & -5 \\ 5 & 2 \end{bmatrix}\begin{bmatrix} 1 & 3 \\ -3 & 1 \end{bmatrix} = \begin{bmatrix} 17 & 1 \\ -1 & 17 \end{bmatrix}$ **(c)** $17 + i$ **(d)** $17 + i$

2. $\begin{bmatrix} a & b \\ -b & a \end{bmatrix}\begin{bmatrix} a & -b \\ b & a \end{bmatrix} = \begin{bmatrix} a^2 + b^2 & 0 \\ 0 & b^2 + a^2 \end{bmatrix}$; the product is a real number.

3. (a) $x = k(ar + bs) + l(cr + ds) = r(ka + lc) + s(kb + ld)$ **(b)** $A = \begin{bmatrix} ka + lc & kb + ld \\ ma + nc & mb + nd \end{bmatrix}$
$\quad\quad y = m(ar + bs) + n(cr + ds) = r(ma + nc) + s(mb + nd)$

8.4 Assess Your Understanding *(page 594)*

1. square **2.** T **3.** columns; rows **4.** F **5.** inverse **6.** singular **7.** T **8.** $A^{-1}B$

9. $\begin{bmatrix} 4 & 4 & -5 \\ -1 & 5 & 4 \end{bmatrix}$ **11.** $\begin{bmatrix} 0 & 12 & -20 \\ 4 & 8 & 24 \end{bmatrix}$ **13.** $\begin{bmatrix} -8 & 7 & -15 \\ 7 & 0 & 22 \end{bmatrix}$ **15.** $\begin{bmatrix} 28 & -9 \\ 4 & 23 \end{bmatrix}$ **17.** $\begin{bmatrix} 1 & 14 & -14 \\ 2 & 22 & -18 \\ 3 & 0 & 28 \end{bmatrix}$ **19.** $\begin{bmatrix} 15 & 21 & -16 \\ 22 & 34 & -22 \\ -11 & 7 & 22 \end{bmatrix}$ **21.** $\begin{bmatrix} 25 & -9 \\ 4 & 20 \end{bmatrix}$

23. $\begin{bmatrix} -13 & 7 & -12 \\ -18 & 10 & -14 \\ 17 & -7 & 34 \end{bmatrix}$ **25.** $\begin{bmatrix} -2 & 4 & 2 & 8 \\ 2 & 1 & 4 & 6 \end{bmatrix}$ **27.** $\begin{bmatrix} 5 & 14 \\ 9 & 16 \end{bmatrix}$ **29.** $\begin{bmatrix} 9 & 2 \\ 34 & 13 \\ 47 & 20 \end{bmatrix}$ **31.** $\begin{bmatrix} 1 & -1 \\ -1 & 2 \end{bmatrix}$ **33.** $\begin{bmatrix} 1 & -\dfrac{5}{2} \\ -1 & 3 \end{bmatrix}$ **35.** $\begin{bmatrix} 1 & -\dfrac{1}{a} \\ -1 & \dfrac{2}{a} \end{bmatrix}$

37. $\begin{bmatrix} 3 & -3 & 1 \\ -2 & 2 & -1 \\ -4 & 5 & -2 \end{bmatrix}$ **39.** $\begin{bmatrix} -\dfrac{5}{7} & \dfrac{1}{7} & \dfrac{3}{7} \\ \dfrac{9}{7} & \dfrac{1}{7} & -\dfrac{4}{7} \\ \dfrac{3}{7} & -\dfrac{2}{7} & \dfrac{1}{7} \end{bmatrix}$ **41.** $x = 3, y = 2; (3, 2)$ **43.** $x = -5, y = 10; (-5, 10)$ **45.** $x = 2, y = -1; (2, -1)$

47. $x = \dfrac{1}{2}, y = 2; \left(\dfrac{1}{2}, 2\right)$ **49.** $x = -2, y = 1; (-2, 1)$ **51.** $x = \dfrac{2}{a}, y = \dfrac{3}{a}; \left(\dfrac{2}{a}, \dfrac{3}{a}\right)$ **53.** $x = -2, y = 3, z = 5; (-2, 3, 5)$

55. $x = \dfrac{1}{2}, y = -\dfrac{1}{2}, z = 1; \left(\dfrac{1}{2}, -\dfrac{1}{2}, 1\right)$ **57.** $x = -\dfrac{34}{7}, y = \dfrac{85}{7}, z = \dfrac{12}{7}; \left(-\dfrac{34}{7}, \dfrac{85}{7}, \dfrac{12}{7}\right)$ **59.** $x = \dfrac{1}{3}, y = 1, z = \dfrac{2}{3}; \left(\dfrac{1}{3}, 1, \dfrac{2}{3}\right)$

61. $\left[\begin{array}{cc|cc} 4 & 2 & 1 & 0 \\ 2 & 1 & 0 & 1 \end{array}\right] \rightarrow \left[\begin{array}{cc|cc} 1 & \dfrac{1}{2} & \dfrac{1}{4} & 0 \\ 2 & 1 & 0 & 1 \end{array}\right] \rightarrow \left[\begin{array}{cc|cc} 1 & \dfrac{1}{2} & \dfrac{1}{4} & 0 \\ 0 & 0 & -\dfrac{1}{2} & 1 \end{array}\right]$ **63.** $\left[\begin{array}{cc|cc} 15 & 3 & 1 & 0 \\ 10 & 2 & 0 & 1 \end{array}\right] \rightarrow \left[\begin{array}{cc|cc} 1 & \dfrac{1}{5} & \dfrac{1}{15} & 0 \\ 10 & 2 & 0 & 1 \end{array}\right] \rightarrow \left[\begin{array}{cc|cc} 1 & \dfrac{1}{5} & \dfrac{1}{15} & 0 \\ 0 & 0 & -\dfrac{2}{3} & 1 \end{array}\right]$

65. $\left[\begin{array}{ccc|ccc} -3 & 1 & -1 & 1 & 0 & 0 \\ 1 & -4 & -7 & 0 & 1 & 0 \\ 1 & 2 & 5 & 0 & 0 & 1 \end{array}\right] \rightarrow \left[\begin{array}{ccc|ccc} 1 & 2 & 5 & 0 & 0 & 1 \\ 1 & -4 & -7 & 0 & 1 & 0 \\ -3 & 1 & -1 & 1 & 0 & 0 \end{array}\right] \rightarrow \left[\begin{array}{ccc|ccc} 1 & 2 & 5 & 0 & 0 & 1 \\ 0 & -6 & -12 & 0 & 1 & -1 \\ 0 & 7 & 14 & 1 & 0 & 3 \end{array}\right]$

$$\rightarrow \left[\begin{array}{ccc|ccc} 1 & 2 & 5 & 0 & 0 & 1 \\ 0 & 1 & 2 & 0 & -\dfrac{1}{6} & \dfrac{1}{6} \\ 0 & 1 & 2 & \dfrac{1}{7} & 0 & \dfrac{3}{7} \end{array}\right] \rightarrow \left[\begin{array}{ccc|ccc} 1 & 2 & 5 & 0 & 0 & 1 \\ 0 & 1 & 2 & 0 & -\dfrac{1}{6} & \dfrac{1}{6} \\ 0 & 0 & 0 & \dfrac{1}{7} & \dfrac{1}{6} & \dfrac{11}{42} \end{array}\right]$$

67. $\begin{bmatrix} 0.01 & 0.05 & -0.01 \\ 0.01 & -0.02 & 0.01 \\ -0.02 & 0.01 & 0.03 \end{bmatrix}$ **69.** $\begin{bmatrix} 0.02 & -0.04 & -0.01 & 0.01 \\ -0.02 & 0.05 & 0.03 & -0.03 \\ 0.02 & 0.01 & -0.04 & 0.00 \\ -0.02 & 0.06 & 0.07 & 0.06 \end{bmatrix}$ **71.** $x = 4.57, y = -6.44, z = -24.07$ or $(4.57, -6.44, -24.07)$

73. $x = -1.19, y = 2.46, z = 8.27$ or $(-1.19, 2.46, 8.27)$ **75.** $x = -5, y = 7; (-5, 7)$ **77.** $x = -4, y = 2, z = \dfrac{5}{2}; \left(-4, 2, \dfrac{5}{2}\right)$

79. Inconsistent; \varnothing **81.** $x = -\dfrac{1}{5}z + \dfrac{1}{5}, y = \dfrac{1}{5}z - \dfrac{6}{5}$, where z is any real number; $\left\{ (x, y, z) \,\middle|\, x = -\dfrac{1}{5}z + \dfrac{1}{5}, y = \dfrac{1}{5}z - \dfrac{6}{5}, z \text{ is any real number} \right\}$

83. (a) $A = \begin{bmatrix} 6 & 9 \\ 3 & 12 \end{bmatrix}; B = \begin{bmatrix} 80.00 \\ 277.80 \end{bmatrix}$ **(b)** $AB = \begin{bmatrix} 2980.20 \\ 3573.60 \end{bmatrix}$; Nikki's total tuition is \$2980.20, and Joe's total tuition is \$3573.60.

85. (a) $\begin{bmatrix} 500 & 350 & 400 \\ 700 & 500 & 850 \end{bmatrix}; \begin{bmatrix} 500 & 700 \\ 350 & 500 \\ 400 & 850 \end{bmatrix}$ **(b)** $\begin{bmatrix} 15 \\ 8 \\ 3 \end{bmatrix}$ **(c)** $\begin{bmatrix} 11,500 \\ 17,050 \end{bmatrix}$ **(d)** $[0.10 \quad 0.05]$ **(e)** \$2002.50

87. (a) $K^{-1} = \begin{bmatrix} 1 & 0 & -1 \\ -1 & 1 & 1 \\ 0 & -1 & 1 \end{bmatrix}$ **(b)** $M = \begin{bmatrix} 13 & 1 & 20 \\ 8 & 9 & 19 \\ 6 & 21 & 14 \end{bmatrix}$ **(c)** Math is fun.

89. If $D = ad - bc \neq 0$, then $a \neq 0$ and $d \neq 0$, or $b \neq 0$ and $c \neq 0$. Assuming the former,

$$\begin{bmatrix} a & b & | & 1 & 0 \\ c & d & | & 0 & 1 \end{bmatrix} \rightarrow \begin{bmatrix} 1 & \dfrac{b}{a} & | & \dfrac{1}{a} & 0 \\ c & d & | & 0 & 1 \end{bmatrix} \rightarrow \begin{bmatrix} 1 & \dfrac{b}{a} & | & \dfrac{1}{a} & 0 \\ 0 & \dfrac{D}{a} & | & -\dfrac{c}{a} & 1 \end{bmatrix} \rightarrow \begin{bmatrix} 1 & \dfrac{b}{a} & | & \dfrac{1}{a} & 0 \\ 0 & 1 & | & -\dfrac{c}{D} & \dfrac{a}{D} \end{bmatrix} \rightarrow \begin{bmatrix} 1 & 0 & | & \dfrac{d}{D} & -\dfrac{b}{D} \\ 0 & 1 & | & -\dfrac{c}{D} & \dfrac{a}{D} \end{bmatrix}$$

$$R_1 = \dfrac{1}{a}r_1 \qquad R_2 = -cr_1 + r_2 \qquad R_2 = \dfrac{a}{D}r_2 \qquad R_1 = -\dfrac{b}{a}r_2 + r_1$$

8.5 Assess Your Understanding (page 604)

5. Proper **7.** Improper; $1 + \dfrac{9}{x^2 - 4}$ **9.** Improper; $5x + \dfrac{22x - 1}{x^2 - 4}$ **11.** Improper; $1 + \dfrac{-2(x - 6)}{(x + 4)(x - 3)}$ **13.** $\dfrac{-4}{x} + \dfrac{4}{x - 1}$ **15.** $\dfrac{1}{x} + \dfrac{-x}{x^2 + 1}$

17. $\dfrac{-1}{x - 1} + \dfrac{2}{x - 2}$ **19.** $\dfrac{\frac{1}{4}}{x + 1} + \dfrac{\frac{3}{4}}{x - 1} + \dfrac{\frac{1}{2}}{(x - 1)^2}$ **21.** $\dfrac{\frac{1}{12}}{x - 2} + \dfrac{-\frac{1}{12}(x + 4)}{x^2 + 2x + 4}$ **23.** $\dfrac{\frac{1}{4}}{x - 1} + \dfrac{\frac{1}{4}}{(x - 1)^2} + \dfrac{-\frac{1}{4}}{x + 1} + \dfrac{\frac{1}{4}}{(x + 1)^2}$

25. $\dfrac{-5}{x + 2} + \dfrac{5}{x + 1} + \dfrac{-4}{(x + 1)^2}$ **27.** $\dfrac{\frac{1}{4}}{x} + \dfrac{1}{x^2} + \dfrac{-\frac{1}{4}(x + 4)}{x^2 + 4}$ **29.** $\dfrac{\frac{2}{3}}{x + 1} + \dfrac{\frac{1}{3}(x + 1)}{x^2 + 2x + 4}$ **31.** $\dfrac{\frac{2}{7}}{3x - 2} + \dfrac{\frac{1}{7}}{2x + 1}$ **33.** $\dfrac{\frac{3}{4}}{x + 3} + \dfrac{\frac{1}{4}}{x - 1}$

35. $\dfrac{1}{x^2 + 4} + \dfrac{2x - 1}{(x^2 + 4)^2}$ **37.** $\dfrac{-1}{x} + \dfrac{2}{x - 3} + \dfrac{-1}{x + 1}$ **39.** $\dfrac{4}{x - 2} + \dfrac{-3}{x - 1} + \dfrac{-1}{(x - 1)^2}$ **41.** $\dfrac{x}{(x^2 + 16)^2} + \dfrac{-16x}{(x^2 + 16)^3}$

43. $\dfrac{-\frac{8}{7}}{2x + 1} + \dfrac{\frac{4}{7}}{x - 3}$ **45.** $\dfrac{-\frac{2}{9}}{x} + \dfrac{-\frac{1}{3}}{x^2} + \dfrac{\frac{1}{6}}{x - 3} + \dfrac{\frac{1}{18}}{x + 3}$

Historical Problem (page 610)

$x = 6$ units, $y = 8$ units

8.6 Assess Your Understanding (page 610)

5.

7.

9.

11.

13.

15.

17.

19.

21. No points of intersection

$y = x^2 - 9$
$x^2 + y^2 = 4$

23.

$y = x^2 - 4$ $y = 6x - 13$
$(3, 5)$

25. $x = 1, y = 4; x = -1, y = -4; x = 2\sqrt{2}, y = \sqrt{2}; x = -2\sqrt{2}, y = -\sqrt{2}$ or $(1, 4), (-1, -4), (2\sqrt{2}, \sqrt{2}), (-2\sqrt{2}, -\sqrt{2})$

27. $x = 0, y = 1; x = -\frac{2}{3}, y = -\frac{1}{3}$ or $(0,1), \left(-\frac{2}{3}, -\frac{1}{3}\right)$ **29.** $x = 0, y = -1; x = \frac{5}{2}, y = -\frac{7}{2}$ or $(0, -1), \left(\frac{5}{2}, -\frac{7}{2}\right)$

31. $x = 2, y = \frac{1}{3}; x = \frac{1}{2}, y = \frac{4}{3}$ or $\left(2, \frac{1}{3}\right), \left(\frac{1}{2}, \frac{4}{3}\right)$ **33.** $x = 3, y = 2; x = 3, y = -2; x = -3, y = 2; x = -3, y = -2$ or

$(3, 2), (3, -2), (-3, 2), (-3, -2)$ **35.** $x = \frac{1}{2}, y = \frac{3}{2}; x = \frac{1}{2}, y = -\frac{3}{2}; x = -\frac{1}{2}, y = \frac{3}{2}; x = -\frac{1}{2}, y = -\frac{3}{2}$ or $\left(\frac{1}{2}, \frac{3}{2}\right), \left(\frac{1}{2}, -\frac{3}{2}\right), \left(-\frac{1}{2}, \frac{3}{2}\right), \left(-\frac{1}{2}, -\frac{3}{2}\right)$

37. $x = \sqrt{2}, y = 2\sqrt{2}; x = -\sqrt{2}, y = -2\sqrt{2}$ or $(\sqrt{2}, 2\sqrt{2}), (-\sqrt{2}, -2\sqrt{2})$ **39.** No real solution exists.

41. $x = \frac{8}{3}, y = \frac{2\sqrt{10}}{3}; x = -\frac{8}{3}, y = \frac{2\sqrt{10}}{3}; x = \frac{8}{3}, y = -\frac{2\sqrt{10}}{3}; x = -\frac{8}{3}, y = -\frac{2\sqrt{10}}{3}$ or $\left(\frac{8}{3}, \frac{2\sqrt{10}}{3}\right), \left(-\frac{8}{3}, \frac{2\sqrt{10}}{3}\right), \left(\frac{8}{3}, -\frac{2\sqrt{10}}{3}\right), \left(-\frac{8}{3}, -\frac{2\sqrt{10}}{3}\right)$

43. $x = 1, y = \frac{1}{2}; x = -1, y = \frac{1}{2}; x = 1, y = -\frac{1}{2}; x = -1, y = -\frac{1}{2}$ or $\left(1, \frac{1}{2}\right), \left(-1, \frac{1}{2}\right), \left(1, -\frac{1}{2}\right), \left(-1, -\frac{1}{2}\right)$ **45.** No real solution exists.

47. $x = \sqrt{3}, y = \sqrt{3}; x = -\sqrt{3}, y = -\sqrt{3}; x = 2, y = 1; x = -2, y = -1$ or $(\sqrt{3}, \sqrt{3}), (-\sqrt{3}, -\sqrt{3}), (2, 1), (-2, -1)$

49. $x = 0, y = -2; x = 0, y = 1; x = 2, y = -1$ or $(0, -2), (0, 1), (2, -1)$ **51.** $x = 2, y = 8$ or $(2, 8)$ **53.** $x = 81, y = 3$ or $(81, 3)$

55.

$x^2 + x + y^2 - 3y + 2 = 0$
$(-1, 1)$
$x + 1 + \frac{y^2 - y}{x} = 0$

57. $x = 0.48, y = 0.62$ **59.** $x = -1.65, y = -0.89$

61. $x = 0.58, y = 1.86; x = 1.81, y = 1.05; x = 0.58, y = -1.86; x = 1.81, y = -1.05$

63. $x = 2.35, y = 0.85$

65. $\left(-\frac{6}{5}, \frac{3}{5}\right)$

$(x - 1)^2 + (y - 1)^2 = 5$
$x + 2y = 0$
$(2, -1)$

67.

$(1, 0)$
$(0, \sqrt{3} - 2)$
$y^2 + 4y - x + 1 = 0$
$(x - 1)^2 + (y + 2)^2 = 4$
$(0, -\sqrt{3} - 2)$ $(1, -4)$

69.

$y = \frac{4}{x - 3}$
$(5, 2)$
$x^2 - 6x + y^2 + 1 = 0$
$(1, -2)$

71. 3 and 1; -1 and -3 **73.** 2 and 2; -2 and -2 **75.** $\frac{1}{2}$ and $\frac{1}{3}$ **77.** 5 **79.** 5 in. by 3 in. **81.** 2 cm and 4 cm **83.** tortoise: 7 m/hr, hare: $7\frac{1}{2}$ m/hr

85. 12 cm by 18 cm **87.** $x = 60$ ft; $y = 30$ ft **89.** $l = \frac{P + \sqrt{P^2 - 16A}}{4}; w = \frac{P - \sqrt{P^2 - 16A}}{4}$ **91.** $y = 4x - 4$ **93.** $y = 2x + 1$

95. $y = -\frac{1}{3}x + \frac{7}{3}$ **97.** $y = 2x - 3$ **99.** $r_1 = \frac{-b + \sqrt{b^2 - 4ac}}{2a}; r_2 = \frac{-b - \sqrt{b^2 - 4ac}}{2a}$ **101.** (a) 4.274 ft by 4.274 ft or 0.093 ft by 0.093 ft

8.7 Assess Your Understanding (page 619)

7. dashes; solid **8.** half-planes **9.** F **10.** unbounded

11.

$x \geq 0$

13.

$x \geq 4$

15.

$2x + y \geq 6$

17.

$x^2 + y^2 > 1$

19.

$y \leq x^2 - 1$

21.

$xy \geq 4$

23.

$x + y = 2$ $2x + y = 4$

25.

$3x + 2y = -6$ $2x - y = 4$

27.

$2x - 3y = 0$ $3x + 2y = 6$

29.

$2x - 4y = 0$
$x - 2y = 6$

31.

$2x + y = -2$ $2x + y = 2$

33. No solution

$2x + 3y = 6$
$2x + 3y = 0$

35.

37.

39.

41.

43. Bounded

45. Unbounded

47. Bounded

49. Bounded

51. Bounded

53. $\begin{cases} x \le 4 \\ x + y \le 6 \\ x \ge 0 \\ y \ge 0 \end{cases}$

55. $\begin{cases} x \le 20 \\ y \ge 15 \\ x + y \le 50 \\ x - y \le 0 \\ x \ge 0 \end{cases}$

57. (a) $\begin{cases} x + y \le 50{,}000 \\ x \ge 35{,}000 \\ y \le 10{,}000 \\ y \ge 0 \end{cases}$

59. (a) $\begin{cases} x \ge 0 \\ y \ge 0 \\ x + 2y \le 300 \\ 3x + 2y \le 480 \end{cases}$

61. (a) $\begin{cases} 3x + 2y \le 160 \\ 2x + 3y \le 150 \\ x \ge 0 \\ y \ge 0 \end{cases}$

(b)

(b)

(b)

8.8 Assess Your Understanding (page 626)

1. objective function **2.** T **3.** Maximum value is 11; minimum value is 3. **5.** Maximum value is 65; minimum value is 4. **7.** Maximum value is 67; minimum value is 20. **9.** The maximum value of z is 12, and it occurs at the point $(6, 0)$. **11.** The minimum value of z is 4, and it occurs at the point $(2, 0)$. **13.** The maximum value of z is 20, and it occurs at the point $(0, 4)$. **15.** The minimum value of z is 8, and it occurs at the point $(0, 2)$. **17.** The maximum value of z is 50, and it occurs at the point $(10, 0)$. **19.** Produce 8 downhill and 24 cross-country; $1760; $1920 which is the profit when producing 16 downhill and 16 cross-country. **21.** Rent 15 rectangular tables and 16 round tables for a minimum cost of $1252. **23. (a)** $10,000 in a junk bond and $10,000 in Treasury bills **(b)** $12,000 in a junk bond and $8000 in Treasury bills **25.** 100 lb of ground beef should be mixed with 50 lb of pork. **27.** Manufacture 10 racing skates and 15 figure skates. **29.** Order 2 metal samples and 4 plastic samples; $34 **31. (a)** Configure with 10 first class seats and 120 coach seats. **(b)** Configure with 15 first class seats and 120 coach seats.

Review Exercises (page 629)

1. $x = 2, y = -1$ or $(2, -1)$ **3.** $x = 2, y = \dfrac{1}{2}$ or $\left(2, \dfrac{1}{2}\right)$ **5.** $x = 2, y = -1$ or $(2, -1)$ **7.** $x = \dfrac{11}{5}, y = -\dfrac{3}{5}$ or $\left(\dfrac{11}{5}, -\dfrac{3}{5}\right)$ **9.** Inconsistent

11. $x = 2, y = 3$ or $(2, 3)$ **13.** Inconsistent **15.** $x = -1, y = 2, z = -3$ or $(-1, 2, -3)$

17. $x = \dfrac{7}{4}z + \dfrac{39}{4}, y = \dfrac{9}{8}z + \dfrac{69}{8}$, where z is any real number or $\left\{ (x, y, z) \,\middle|\, x = \dfrac{7}{4}z + \dfrac{39}{4}, y = \dfrac{9}{8}z + \dfrac{69}{8}, z \text{ is any real number} \right\}$

19. $\begin{cases} 3x + 2y = 8 \\ x + 4y = -1 \end{cases}$ **21.** $\begin{bmatrix} 4 & -4 \\ 3 & 9 \\ 4 & 4 \end{bmatrix}$ **23.** $\begin{bmatrix} 6 & 0 \\ 12 & 24 \\ -6 & 12 \end{bmatrix}$ **25.** $\begin{bmatrix} 4 & -3 & 0 \\ 12 & -2 & -8 \\ -2 & 5 & -4 \end{bmatrix}$ **27.** $\begin{bmatrix} 8 & -13 & 8 \\ 9 & 2 & -10 \\ 22 & -13 & -4 \end{bmatrix}$ **29.** $\begin{bmatrix} \dfrac{1}{2} & -1 \\ -\dfrac{1}{6} & \dfrac{2}{3} \end{bmatrix}$

31. $\begin{bmatrix} -\dfrac{5}{7} & \dfrac{9}{7} & \dfrac{3}{7} \\ \dfrac{1}{7} & \dfrac{1}{7} & -\dfrac{2}{7} \\ \dfrac{3}{7} & -\dfrac{4}{7} & \dfrac{1}{7} \end{bmatrix}$ **33.** Singular **35.** $x = \dfrac{2}{5}, y = \dfrac{1}{10}$ or $\left(\dfrac{2}{5}, \dfrac{1}{10}\right)$ **37.** $x = 9, y = \dfrac{13}{3}, z = \dfrac{13}{3}$ or $\left(9, \dfrac{13}{3}, \dfrac{13}{3}\right)$

39. $x = -\dfrac{1}{2}, y = -\dfrac{2}{3}, z = -\dfrac{3}{4}$, or $\left(-\dfrac{1}{2}, -\dfrac{2}{3}, -\dfrac{3}{4}\right)$

41. $z = -1, x = y + 1$, where y is any real number or $\{(x, y, z) \mid x = y + 1, z = -1, y \text{ is any real number}\}$

43. $x = 4, y = 2, z = 3, t = -2$ or $(4, 2, 3, -2)$ **45.** 5 **47.** 108 **49.** -100 **51.** $x = 2, y = -1$ or $(2, -1)$ **53.** $x = 2, y = 3$ or $(2, 3)$

55. $x = -1, y = 2, z = -3$ or $(-1, 2, -3)$ **57.** 16 **59.** $\dfrac{-\frac{3}{2}}{x} + \dfrac{\frac{3}{2}}{x - 4}$ **61.** $\dfrac{-3}{x - 1} + \dfrac{3}{x} + \dfrac{4}{x^2}$ **63.** $\dfrac{-\frac{1}{10}}{x + 1} + \dfrac{\frac{1}{10}x + \frac{9}{10}}{x^2 + 9}$

65. $\dfrac{x}{x^2 + 4} + \dfrac{-4x}{(x^2 + 4)^2}$ **67.** $\dfrac{\frac{1}{2}}{x^2 + 1} + \dfrac{\frac{1}{4}}{x - 1} + \dfrac{-\frac{1}{4}}{x + 1}$ **69.** $x = -\dfrac{2}{5}, y = -\dfrac{11}{5}; x = -2, y = 1$ or $\left(-\dfrac{2}{5}, -\dfrac{11}{5}\right), (-2, 1)$

71. $x = 2\sqrt{2}, y = \sqrt{2}; x = -2\sqrt{2}, y = -\sqrt{2}$ or $(2\sqrt{2}, \sqrt{2}), (-2\sqrt{2}, -\sqrt{2})$ **73.** $x = 0, y = 0; x = -3, y = 3; x = 3, y = 3$ or $(0, 0), (-3, 3), (3, 3)$ **75.** $x = \sqrt{2}, y = -\sqrt{2}; x = -\sqrt{2}, y = \sqrt{2}; x = \dfrac{4}{3}\sqrt{2}, y = -\dfrac{2}{3}\sqrt{2}; x = -\dfrac{4}{3}\sqrt{2}, y = \dfrac{2}{3}\sqrt{2}$ or $(\sqrt{2}, -\sqrt{2}), (-\sqrt{2}, \sqrt{2}), \left(\dfrac{4}{3}\sqrt{2}, -\dfrac{2}{3}\sqrt{2}\right), \left(-\dfrac{4}{3}\sqrt{2}, \dfrac{2}{3}\sqrt{2}\right)$ **77.** $x = 1, y = -1$ or $(1, -1)$

79.

81.

83. Unbounded

85. Bounded

87. Bounded

89.

91.

93. The maximum value is 32 when $x = 0$ and $y = 8$. **95.** The minimum value is 3 when $x = 1$ and $y = 0$. **97.** 10

99. $y = -\dfrac{1}{3}x^2 - \dfrac{2}{3}x + 1$ **101.** Mix 70 lb of \$6.00 coffee and 30 lb of \$9.00 coffee. **103.** Buy 1 small, 5 medium, and 2 large.

105. Speedboat: 36.67 km/hr; Aguarico River: 3.33 km/hr **107.** Bruce: 4 hr; Bryce: 2 hr; Marty: 8 hr

109. Produce 35 gasoline engines and 15 diesel engines; the factory is producing an excess of 15 gasoline engines and 0 diesel engines.

Chapter Test (page 633)

1. $x = 3, y = -1$ or $(3, -1)$ **2.** Inconsistent **3.** $x = -z + \dfrac{18}{7}, y = z - \dfrac{17}{7}$, where z is any real number or

$\left\{(x, y, z) \mid x = -z + \dfrac{18}{7}, y = z - \dfrac{17}{7}, z \text{ is any real number}\right\}$ **4.** $x = \dfrac{1}{3}, y = -2, z = 0$ or $\left(\dfrac{1}{3}, -2, 0\right)$ **5.** $\begin{bmatrix} 4 & -5 & 1 & | & 0 \\ -2 & -1 & 0 & | & -25 \\ 1 & 5 & -5 & | & 10 \end{bmatrix}$

6. $\begin{cases} 3x + 2y + 4z = -6 \\ 1x + 0y + 8z = 2 \\ -2x + 1y + 3z = -11 \end{cases}$ or $\begin{cases} 3x + 2y + 4z = -6 \\ x + 8z = 2 \\ -2x + y + 3z = -11 \end{cases}$ **7.** $\begin{bmatrix} 6 & 4 \\ 1 & -11 \\ 5 & 12 \end{bmatrix}$ **8.** $\begin{bmatrix} -11 & -19 \\ -3 & 5 \\ 6 & -22 \end{bmatrix}$ **9.** $\begin{bmatrix} 4 & 10 & 26 \\ 1 & -11 & 2 \\ -1 & 26 & 3 \end{bmatrix}$

10. $\begin{bmatrix} 16 & 17 \\ 3 & -10 \end{bmatrix}$ **11.** $\begin{bmatrix} 2 & -1 \\ -\dfrac{5}{2} & \dfrac{3}{2} \end{bmatrix}$ **12.** $\begin{bmatrix} 3 & 3 & -4 \\ -2 & -2 & 3 \\ -4 & -5 & 7 \end{bmatrix}$ **13.** $x = \dfrac{1}{2}, y = 3$ or $\left(\dfrac{1}{2}, 3\right)$ **14.** $x = -\dfrac{1}{4}y + 7$, where y is any real number or

$\left\{(x, y) \mid x = -\dfrac{1}{4}y + 7, y \text{ is any real number}\right\}$ **15.** $x = 1, y = -2, z = 0$ or $(1, -2, 0)$ **16.** Inconsistent **17.** -29 **18.** -12

19. $x = -2, y = -5$ or $(-2, -5)$ **20.** $x = 1, y = -1, z = 4$ or $(1, -1, 4)$ **21.** $(1, -3)$ and $(1, 3)$ **22.** $(3, 4)$ and $(1, 2)$

23.

24. $\dfrac{3}{x + 3} + \dfrac{-2}{(x + 3)^2}$

25. $\dfrac{-\frac{1}{3}}{x} + \dfrac{\frac{1}{3}x}{(x^2 + 3)} + \dfrac{5x}{(x^2 + 3)^2}$

26. Unbounded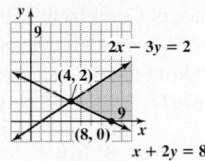

27. The maximum value of z is 64, and it occurs at the point $(0, 8)$. **28.** Flare jeans cost \$24.50, camisoles cost \$8.50, and T-shirts cost \$6.00.

Cumulative Review *(page 634)*

1. $\left\{0, \dfrac{1}{2}\right\}$ **2.** $\{5\}$ **3.** $\left\{-1, -\dfrac{1}{2}, 3\right\}$ **4.** $\{-2\}$ **5.** $\left\{\dfrac{5}{2}\right\}$ **6.** $\left\{\dfrac{1}{\ln 3}\right\}$ **7.** Odd; symmetric with respect to the origin

8. Center: $(1, -2)$; radius $= 4$

9. Domain: all real numbers
Range: $\{y \mid y > 1\}$
Horizontal asymptote: $y = 1$

10. $f^{-1}(x) = \dfrac{5}{x} - 2$

Domain of f: $\{x \mid x \neq -2\}$
Range of f: $\{y \mid y \neq 0\}$
Domain of f^{-1}: $\{x \mid x \neq 0\}$
Range of f^{-1}: $\{y \mid y \neq -2\}$

11. (a)

(b)

(c)

(d)

(e)

(f)

(g)

(h)

(i)

(j)

12. (a)

$; -2.28$

(b) Local maximum of 7 at $x = -1$;
local minimum of 3 at $x = 1$

(c) $(-\infty, -1), (1, \infty)$

CHAPTER 9 Sequences; Induction; the Binomial Theorem

9.1 Assess Your Understanding *(page 643)*

3. sequence **4.** True **5.** $n(n-1) \cdots 3 \cdot 2 \cdot 1$ **6.** recursive **7.** summation **8.** True **9.** 3,628,800 **11.** 504 **13.** 1260 **15.** $s_1 = 1, s_2 = 2, s_3 = 3,$
$s_4 = 4, s_5 = 5$ **17.** $a_1 = \dfrac{1}{3}, a_2 = \dfrac{1}{2}, a_3 = \dfrac{3}{5}, a_4 = \dfrac{2}{3}, a_5 = \dfrac{5}{7}$ **19.** $c_1 = 1, c_2 = -4, c_3 = 9, c_4 = -16, c_5 = 25$ **21.** $s_1 = \dfrac{1}{2}, s_2 = \dfrac{2}{5}, s_3 = \dfrac{2}{7}, s_4 = \dfrac{8}{41}, s_5 = \dfrac{8}{61}$

23. $t_1 = -\dfrac{1}{6}, t_2 = \dfrac{1}{12}, t_3 = -\dfrac{1}{20}, t_4 = \dfrac{1}{30}, t_5 = -\dfrac{1}{42}$ **25.** $b_1 = \dfrac{1}{e}, b_2 = \dfrac{2}{e^2}, b_3 = \dfrac{3}{e^3}, b_4 = \dfrac{4}{e^4}, b_5 = \dfrac{5}{e^5}$ **27.** $a_n = \dfrac{n}{n+1}$ **29.** $a_n = \dfrac{1}{2^{n-1}}$

31. $a_n = (-1)^{n+1}$ **33.** $a_n = (-1)^{n+1}n$ **35.** $a_1 = 2, a_2 = 5, a_3 = 8, a_4 = 11, a_5 = 14$ **37.** $a_1 = -2, a_2 = 0, a_3 = 3, a_4 = 7, a_5 = 12$

39. $a_1 = 5, a_2 = 10, a_3 = 20, a_4 = 40, a_5 = 80$ **41.** $a_1 = 3, a_2 = \dfrac{3}{2}, a_3 = \dfrac{1}{2}, a_4 = \dfrac{1}{8}, a_5 = \dfrac{1}{40}$ **43.** $a_1 = 1, a_2 = 2, a_3 = 2, a_4 = 4, a_5 = 8$

45. $a_1 = A, a_2 = A + d, a_3 = A + 2d, a_4 = A + 3d, a_5 = A + 4d$

47. $a_1 = \sqrt{2}, a_2 = \sqrt{2 + \sqrt{2}}, a_3 = \sqrt{2 + \sqrt{2 + \sqrt{2}}},$ **49.** $3 + 4 + \cdots + (n+2)$ **51.** $\dfrac{1}{2} + 2 + \dfrac{9}{2} + \cdots + \dfrac{n^2}{2}$ **53.** $1 + \dfrac{1}{3} + \dfrac{1}{9} + \cdots + \dfrac{1}{3^n}$

$a_4 = \sqrt{2 + \sqrt{2 + \sqrt{2 + \sqrt{2}}}},$

55. $\dfrac{1}{3} + \dfrac{1}{9} + \cdots + \dfrac{1}{3^n}$ **57.** $\ln 2 - \ln 3 + \ln 4 - \cdots + (-1)^n \ln n$ **59.** $\displaystyle\sum_{k=1}^{20} k$

$a_5 = \sqrt{2 + \sqrt{2 + \sqrt{2 + \sqrt{2 + \sqrt{2}}}}}$

61. $\displaystyle\sum_{k=1}^{13} \dfrac{k}{k+1}$ **63.** $\displaystyle\sum_{k=0}^{6} (-1)^k \left(\dfrac{1}{3^k}\right)$ **65.** $\displaystyle\sum_{k=1}^{n} \dfrac{3^k}{k}$ **67.** $\displaystyle\sum_{k=0}^{n} (a + kd)$ or $\displaystyle\sum_{k=1}^{n+1} [a + (k-1)d]$ **69.** 200 **71.** 820 **73.** 1110 **75.** 1560 **77.** 3570

79. 44,000 **81.** \$2930 **83.** \$18,058.03 **85.** 21 pairs **87.** Fibonacci sequence **89. (a)** 3.630170833 **(b)** 3.669060828 **(c)** 3.669296668 **(d)** 12

91. (a) $a_1 = 0.4; a_2 = 0.7; a_3 = 1; a_4 = 1.6; a_5 = 2.8; a_6 = 5.2; a_7 = 10; a_8 = 19.6$

(b) Except for term 5, which has no match, Bode's formula provides excellent approximations for the mean distances of the planets from the Sun.

(c) The mean distance of Ceres from the Sun is approximated by $a_5 = 2.8$ and that of Uranus is $a_8 = 19.6$.

(d) $a_9 = 38.8; a_{10} = 77.2$ **(e)** Pluto's distance is approximated by a_9, but no term approximates Neptune's mean distance from the Sun.

(f) According to Bode's Law, the mean orbital distance of Eris will be 154 AU from the Sun.

93. $a_0 = 2; a_5 = 2.236067977; 2.236067977$ **95.** $a_0 = 4; a_5 = 4.582575695; 4.582575695$ **97.** 1, 3, 6, 10, 15, 21, 28

99. $u_n = 1 + 2 + 3 + \cdots + n = \displaystyle\sum_{k=1}^{n} k = \dfrac{n(n+1)}{2}$, and from Problem 98, $u_{n+1} = \dfrac{(n+1)(n+2)}{2}$.

Thus, $u_{n+1} + u_n = \dfrac{(n+1)(n+2)}{2} + \dfrac{n(n+1)}{2} = \dfrac{(n+1)[(n+2) + n]}{2} = (n+1)^2.$

9.2 Assess Your Understanding *(page 651)*

1. arithmetic **2.** F **3.** 17 **4.** T **5.** $s_n - s_{n-1} = (n + 4) - [(n - 1) + 4] = n + 4 - (n + 3) = n + 4 - n - 3 = 1$, a constant;
$d = 1; s_1 = 5, s_2 = 6, s_3 = 7, s_4 = 8$

7. $a_n - a_{n-1} = (2n - 5) - [2(n - 1) - 5] = 2n - 5 - (2n - 2 - 5) = 2n - 5 - (2n - 7) = 2n - 5 - 2n + 7 = 2$, a constant;
$d = 2; a_1 = -3, a_2 = -1, a_3 = 1, a_4 = 3$

9. $c_n - c_{n-1} = (6 - 2n) - [6 - 2(n - 1)] = 6 - 2n - (6 - 2n + 2) = 6 - 2n - (8 - 2n) = 6 - 2n - 8 + 2n = -2$, a constant;
$d = -2; c_1 = 4, c_2 = 2, c_3 = 0, c_4 = -2$

11. $t_n - t_{n-1} = \left(\frac{1}{2} - \frac{1}{3}n\right) - \left[\frac{1}{2} - \frac{1}{3}(n - 1)\right] = \frac{1}{2} - \frac{1}{3}n - \left(\frac{1}{2} - \frac{1}{3}n + \frac{1}{3}\right) = \frac{1}{2} - \frac{1}{3}n - \left(\frac{5}{6} - \frac{1}{3}n\right) = \frac{1}{2} - \frac{1}{3}n - \frac{5}{6} + \frac{1}{3}n = -\frac{1}{3}$, a constant;
$d = -\frac{1}{3}; t_1 = \frac{1}{6}, t_2 = -\frac{1}{6}, t_3 = -\frac{1}{2}, t_4 = -\frac{5}{6}$

13. $s_n - s_{n-1} = \ln 3^n - \ln 3^{n-1} = n \ln 3 - (n - 1) \ln 3 = n \ln 3 - (n \ln 3 - \ln 3) = n \ln 3 - n \ln 3 + \ln 3 = \ln 3$, a constant;
$d = \ln 3; s_1 = \ln 3, s_2 = 2 \ln 3, s_3 = 3 \ln 3, s_4 = 4 \ln 3$

15. $a_n = 3n - 1; a_{51} = 152$ **17.** $a_n = 8 - 3n; a_{51} = -145$ **19.** $a_n = \frac{1}{2}(n - 1); a_{51} = 25$ **21.** $a_n = \sqrt{2}n; a_{51} = 51\sqrt{2}$ **23.** 200 **25.** -266 **27.** $\frac{83}{2}$

29. $a_1 = -13; d = 3; a_n = a_{n-1} + 3; a_n = -16 + 3n$ **31.** $a_1 = -53; d = 6; a_n = a_{n-1} + 6; a_n = -59 + 6n$

33. $a_1 = 28; d = -2; a_n = a_{n-1} - 2; a_n = 30 - 2n$ **35.** $a_1 = 25; d = -2; a_n = a_{n-1} - 2; a_n = 27 - 2n$ **37.** n^2 **39.** $\frac{n}{2}(9 + 5n)$ **41.** 1260

43. 324 **45.** 30,919 **47.** 10,036 **49.** 6080 **51.** -1925 **53.** 15,960 **55.** $-\frac{3}{2}$ **57.** 24 terms **59.** 1185 seats

61. 210 beige and 190 blue **63.** $\{T_n\} = \{-5.5n + 67\}; T_5 = 39.5°F$ **65.** The amphitheater has 1647 seats. **67.** 8 yr

Historical Problem *(page 660)*

1. $1\frac{2}{3}$ loaves, $10\frac{5}{6}$ loaves, 20 loaves, $29\frac{1}{6}$ loaves, $38\frac{1}{3}$ loaves

9.3 Assess Your Understanding *(page 661)*

3. geometric **4.** $\frac{a}{1 - r}$ **5.** divergent series **6.** T **7.** F **8.** T **9.** $r = 3; s_1 = 3, s_2 = 9, s_3 = 27, s_4 = 81$

11. $r = \frac{1}{2}; a_1 = -\frac{3}{2}, a_2 = -\frac{3}{4}, a_3 = -\frac{3}{8}, a_4 = -\frac{3}{16}$ **13.** $r = 2; c_1 = \frac{1}{4}, c_2 = \frac{1}{2}, c_3 = 1, c_4 = 2$ **15.** $r = 2^{1/3}; e_1 = 2^{1/3}, e_2 = 2^{2/3}, e_3 = 2, e_4 = 2^{4/3}$

17. $r = \frac{3}{2}; t_1 = \frac{1}{2}, t_2 = \frac{3}{4}, t_3 = \frac{9}{8}, t_4 = \frac{27}{16}$ **19.** $a_5 = 162; a_n = 2 \cdot 3^{n-1}$ **21.** $a_5 = 5; a_n = 5 \cdot (-1)^{n-1}$ **23.** $a_5 = 0; a_n = 0$

25. $a_5 = 4\sqrt{2}; a_n = (\sqrt{2})^n$ **27.** $a_7 = \frac{1}{64}$ **29.** $a_9 = 1$ **31.** $a_8 = 0.00000004$ **33.** $a_n = 7 \cdot 2^{n-1}$ **35.** $a_n = -3 \cdot \left(-\frac{1}{3}\right)^{n-1} = \left(-\frac{1}{3}\right)^{n-2}$

37. $a_n = -(-3)^{n-1}$ **39.** $a_n = \frac{7}{15}(15)^{n-1} = 7 \cdot 15^{n-2}$ **41.** $-\frac{1}{4}(1 - 2^n)$ **43.** $2\left[1 - \left(\frac{2}{3}\right)^n\right]$ **45.** $1 - 2^n$

47.

49.

51.

53. Converges; $\frac{3}{2}$ **55.** Converges; 16 **57.** Converges; $\frac{8}{5}$ **59.** Diverges **61.** Converges; $\frac{20}{3}$ **63.** Diverges **65.** Converges; $\frac{18}{5}$ **67.** Converges; 6

69. Arithmetic; $d = 1$; 1375 **71.** Neither **73.** Arithmetic; $d = -\frac{2}{3}$; -700 **75.** Neither **77.** Geometric; $r = \frac{2}{3}; 2\left[1 - \left(\frac{2}{3}\right)^{50}\right]$

79. Geometric; $r = -2; -\frac{1}{3}[1 - (-2)^{50}]$ **81.** Geometric; $r = 3^{1/2}; -\frac{\sqrt{3}}{2}(1 + \sqrt{3})(1 - 3^{25})$ **83.** -4 **85.** \$21,879.11 **87. (a)** 0.775 ft **(b)** 8th
(c) 15.88 ft **(d)** 20 ft **89.** \$349,496.41 **91.** \$96,885.98 **93.** \$305.10 **95.** 1.845×10^{19} **97.** 10 **99.** \$72.67 per share **101.** December 20, 2010;
\$9999.92 **103.** Option B results in more money (\$524,287 versus \$500,500). **105.** Total pay: \$41,943.03; pay on day 22: \$20,971.52

9.4 Assess Your Understanding *(page 667)*

1. (I) $n = 1: 2(1) = 2$ and $1(1 + 1) = 2$
 (II) If $2 + 4 + 6 + \cdots + 2k = k(k + 1)$, then $2 + 4 + 6 + \cdots + 2k + 2(k + 1) = (2 + 4 + 6 + \cdots + 2k) + 2(k + 1)$
 $= k(k + 1) + 2(k + 1) = k^2 + 3k + 2 = (k + 1)(k + 2) = (k + 1)[(k + 1) + 1]$.

3. (I) $n = 1: 1 + 2 = 3$ and $\frac{1}{2}(1)(1 + 5) = \frac{1}{2}(6) = 3$

 (II) If $3 + 4 + 5 + \cdots + (k + 2) = \frac{1}{2}k(k + 5)$, then $3 + 4 + 5 + \cdots + (k + 2) + [(k + 1) + 2]$

 $= [3 + 4 + 5 + \cdots + (k + 2)] + (k + 3) = \frac{1}{2}k(k + 5) + k + 3 = \frac{1}{2}(k^2 + 7k + 6) = \frac{1}{2}(k + 1)(k + 6) = \frac{1}{2}(k + 1)[(k + 1) + 5]$.

5. (I) $n = 1$: $3(1) - 1 = 2$ and $\frac{1}{2}(1)[3(1) + 1] = \frac{1}{2}(4) = 2$

(II) If $2 + 5 + 8 + \cdots + (3k - 1) = \frac{1}{2}k(3k + 1)$, then $2 + 5 + 8 + \cdots + (3k - 1) + [3(k + 1) - 1]$

$= [2 + 5 + 8 + \cdots + (3k - 1)] + (3k + 2) = \frac{1}{2}k(3k + 1) + (3k + 2) = \frac{1}{2}(3k^2 + 7k + 4) = \frac{1}{2}(k + 1)(3k + 4)$

$= \frac{1}{2}(k + 1)[3(k + 1) + 1]$.

7. (I) $n = 1$: $2^{1-1} = 1$ and $2^1 - 1 = 1$

(II) If $1 + 2 + 2^2 + \cdots + 2^{k-1} = 2^k - 1$, then $1 + 2 + 2^2 + \cdots + 2^{k-1} + 2^{(k+1)-1} = (1 + 2 + 2^2 + \cdots + 2^{k-1}) + 2^k$
$= 2^k - 1 + 2^k = 2(2^k) - 1 = 2^{k+1} - 1$.

9. (I) $n = 1$: $4^{1-1} = 1$ and $\frac{1}{3}(4^1 - 1) = \frac{1}{3}(3) = 1$

(II) If $1 + 4 + 4^2 + \cdots + 4^{k-1} = \frac{1}{3}(4^k - 1)$, then $1 + 4 + 4^2 + \cdots + 4^{k-1} + 4^{(k+1)-1} = (1 + 4 + 4^2 + \cdots + 4^{k-1}) + 4^k$

$= \frac{1}{3}(4^k - 1) + 4^k = \frac{1}{3}[4^k - 1 + 3(4^k)] = \frac{1}{3}[4(4^k) - 1] = \frac{1}{3}(4^{k+1} - 1)$.

11. (I) $n = 1$: $\frac{1}{1 \cdot 2} = \frac{1}{2}$ and $\frac{1}{1 + 1} = \frac{1}{2}$

(II) If $\frac{1}{1 \cdot 2} + \frac{1}{2 \cdot 3} + \frac{1}{3 \cdot 4} + \cdots + \frac{1}{k(k + 1)} = \frac{k}{k + 1}$, then $\frac{1}{1 \cdot 2} + \frac{1}{2 \cdot 3} + \frac{1}{3 \cdot 4} + \cdots + \frac{1}{k(k + 1)} + \frac{1}{(k + 1)[(k + 1) + 1]}$

$= \left[\frac{1}{1 \cdot 2} + \frac{1}{2 \cdot 3} + \frac{1}{3 \cdot 4} + \cdots + \frac{1}{k(k + 1)}\right] + \frac{1}{(k + 1)(k + 2)} = \frac{k}{k + 1} + \frac{1}{(k + 1)(k + 2)} = \frac{k(k + 2) + 1}{(k + 1)(k + 2)}$

$= \frac{k^2 + 2k + 1}{(k + 1)(k + 2)} = \frac{(k + 1)^2}{(k + 1)(k + 2)} = \frac{k + 1}{k + 2} = \frac{k + 1}{(k + 1) + 1}$.

13. (I) $n = 1$: $1^2 = 1$ and $\frac{1}{6} \cdot 1 \cdot 2 \cdot 3 = 1$

(II) If $1^2 + 2^2 + 3^2 + \cdots + k^2 = \frac{1}{6}k(k + 1)(2k + 1)$, then $1^2 + 2^2 + 3^2 + \cdots + k^2 + (k + 1)^2$

$= (1^2 + 2^2 + 3^2 + \cdots + k^2) + (k + 1)^2 = \frac{1}{6}k(k + 1)(2k + 1) + (k + 1)^2 = \frac{1}{6}(2k^3 + 9k^2 + 13k + 6)$

$= \frac{1}{6}(k + 1)(k + 2)(2k + 3) = \frac{1}{6}(k + 1)[(k + 1) + 1][2(k + 1) + 1]$.

15. (I) $n = 1$: $5 - 1 = 4$ and $\frac{1}{2}(1)(9 - 1) = \frac{1}{2} \cdot 8 = 4$

(II) If $4 + 3 + 2 + \cdots + (5 - k) = \frac{1}{2}k(9 - k)$, then $4 + 3 + 2 + \cdots + (5 - k) + [5 - (k + 1)]$

$= [4 + 3 + 2 + \cdots + (5 - k)] + 4 - k = \frac{1}{2}k(9 - k) + 4 - k = \frac{1}{2}(9k - k^2 + 8 - 2k) = \frac{1}{2}(-k^2 + 7k + 8)$

$= \frac{1}{2}(k + 1)(8 - k) = \frac{1}{2}(k + 1)[9 - (k + 1)]$.

17. (I) $n = 1$: $1 \cdot (1 + 1) = 2$ and $\frac{1}{3} \cdot 1 \cdot 2 \cdot 3 = 2$

(II) If $1 \cdot 2 + 2 \cdot 3 + 3 \cdot 4 + \cdots + k(k + 1) = \frac{1}{3}k(k + 1)(k + 2)$, then $1 \cdot 2 + 2 \cdot 3 + 3 \cdot 4 + \cdots + k(k + 1)$

$+ (k + 1)[(k + 1) + 1] = [1 \cdot 2 + 2 \cdot 3 + 3 \cdot 4 + \cdots + k(k + 1)] + (k + 1)(k + 2)$

$= \frac{1}{3}k(k + 1)(k + 2) + \frac{1}{3} \cdot 3(k + 1)(k + 2) = \frac{1}{3}(k + 1)(k + 2)(k + 3) = \frac{1}{3}(k + 1)[(k + 1) + 1][(k + 1) + 2]$.

19. (I) $n = 1$: $1^2 + 1 = 2$, which is divisible by 2.

(II) If $k^2 + k$ is divisible by 2, then $(k + 1)^2 + (k + 1) = k^2 + 2k + 1 + k + 1 = (k^2 + k) + 2k + 2$. Since $k^2 + k$ is divisible by 2 and $2k + 2$ is divisible by 2, $(k + 1)^2 + (k + 1)$ is divisible by 2.

21. (I) $n = 1$: $1^2 - 1 + 2 = 2$, which is divisible by 2.

(II) If $k^2 - k + 2$ is divisible by 2, then $(k + 1)^2 - (k + 1) + 2 = k^2 + 2k + 1 - k - 1 + 2 = (k^2 - k + 2) + 2k$. Since $k^2 - k + 2$ is divisible by 2 and $2k$ is divisible by 2, $(k + 1)^2 - (k + 1) + 2$ is divisible by 2.

23. (I) $n = 1$: If $x > 1$, then $x^1 = x > 1$.

(II) Assume, for an arbitrary natural number k, that if $x > 1$ then $x^k > 1$. Multiply both sides of the inequality $x^k > 1$ by x. If $x > 1$, then $x^{k+1} > x > 1$.

25. (I) $n = 1$: $a - b$ is a factor of $a^1 - b^1 = a - b$.

(II) If $a - b$ is a factor of $a^k - b^k$, then $a^{k+1} - b^{k+1} = a(a^k - b^k) + b^k(a - b)$.
Since $a - b$ is a factor of $a^k - b^k$ and $a - b$ is a factor of $a - b$, then $a - b$ is a factor of $a^{k+1} - b^{k+1}$.

27. (I) $n = 1$: $(1 + a)^1 = 1 + a \geq 1 + 1 \cdot a$

(II) Assume that there is an integer k for which the inequality holds. So $(1 + a)^k \geq 1 + ka$. We need to show that $(1 + a)^{k+1} \geq 1 + (k + 1)a$.
$(1 + a)^{k+1} = (1 + a)^k (1 + a) \geq (1 + ka)(1 + a) = 1 + ka^2 + a + ka = 1 + (k + 1)a + ka^2 \geq 1 + (k + 1)a.$

29. If $2 + 4 + 6 + \cdots + 2k = k^2 + k + 2$, then $2 + 4 + 6 + \cdots + 2k + 2(k + 1)$
$= (2 + 4 + 6 + \cdots + 2k) + 2k + 2 = k^2 + k + 2 + 2k + 2 = k^2 + 3k + 4 = (k^2 + 2k + 1) + (k + 1) + 2$
$= (k + 1)^2 + (k + 1) + 2.$
But $2 \cdot 1 = 2$ and $1^2 + 1 + 2 = 4$. The fact is that $2 + 4 + 6 + \cdots + 2n = n^2 + n$, not $n^2 + n + 2$ (Problem 1).

31. (I) $n = 1$: $[a + (1 - 1)d] = a$ and $1 \cdot a + d\dfrac{1 \cdot (1 - 1)}{2} = a.$

(II) If $a + (a + d) + (a + 2d) + \cdots + [a + (k - 1)d] = ka + d\dfrac{k(k - 1)}{2}$, then

$a + (a + d) + (a + 2d) + \cdots + [a + (k - 1)d] + [a + ((k + 1) - 1)d] = ka + d\dfrac{k(k - 1)}{2} + a + kd$

$= (k + 1)a + d\dfrac{k(k - 1) + 2k}{2} = (k + 1)a + d\dfrac{(k + 1)(k)}{2} = (k + 1)a + d\dfrac{(k + 1)[(k + 1) - 1]}{2}.$

33. (I) $n = 3$: The sum of the angles of a triangle is $(3 - 2) \cdot 180° = 180°$.

(II) Assume for some $k \geq 3$ that the sum of the angles of a convex polygon of k sides is $(k - 2) \cdot 180°$. A convex polygon of $k + 1$ sides consists of a convex polygon of k sides plus a triangle (see the illustration). The sum of the angles is
$(k - 2) \cdot 180° + 180° = (k - 1) \cdot 180° = [(k + 1) - 2] \cdot 180°.$

k sides

$k + 1$ sides

9.5 Assess Your Understanding *(page 673)*

1. Pascal's triangle **2.** $1; n$ **3.** F **4.** Binomial Theorem **5.** 10 **7.** 21 **9.** 50 **11.** 1 **13.** $\approx 1.8664 \times 10^{15}$ **15.** $\approx 1.4834 \times 10^{13}$
17. $x^5 + 5x^4 + 10x^3 + 10x^2 + 5x + 1$ **19.** $x^6 - 12x^5 + 60x^4 - 160x^3 + 240x^2 - 192x + 64$ **21.** $81x^4 + 108x^3 + 54x^2 + 12x + 1$
23. $x^{10} + 5x^8y^2 + 10x^6y^4 + 10x^4y^6 + 5x^2y^8 + y^{10}$ **25.** $x^3 + 6\sqrt{2}x^{5/2} + 30x^2 + 40\sqrt{2}x^{3/2} + 60x + 24\sqrt{2}x^{1/2} + 8$
27. $a^5x^5 + 5a^4bx^4y + 10a^3b^2x^3y^2 + 10a^2b^3x^2y^3 + 5ab^4xy^4 + b^5y^5$ **29.** 17,010 **31.** $-101,376$ **33.** 41,472 **35.** $2835x^3$
37. $314,928x^7$ **39.** 495 **41.** 3360 **43.** 1.00501

45. $\dbinom{n}{n - 1} = \dfrac{n!}{(n - 1)! \, [n - (n - 1)]!} = \dfrac{n!}{(n - 1)! \, 1!} = \dfrac{n \cdot (n - 1)!}{(n - 1)!} = n; \dbinom{n}{n} = \dfrac{n!}{n! \, (n - n)!} = \dfrac{n!}{n! 0!} = \dfrac{n!}{n!} = 1$

47. $2^n = (1 + 1)^n = \dbinom{n}{0}1^n + \dbinom{n}{1}(1)^{n-1}(1) + \cdots + \dbinom{n}{n}1^n = \dbinom{n}{0} + \dbinom{n}{1} + \cdots + \dbinom{n}{n}$ **49.** 1

Review Exercises *(page 675)*

1. $a_1 = -\dfrac{4}{3}, a_2 = \dfrac{5}{4}, a_3 = -\dfrac{6}{5}, a_4 = \dfrac{7}{6}, a_5 = -\dfrac{8}{7}$ **3.** $c_1 = 2, c_2 = 1, c_3 = \dfrac{8}{9}, c_4 = 1, c_5 = \dfrac{32}{25}$ **5.** $a_1 = 3, a_2 = 2, a_3 = \dfrac{4}{3}, a_4 = \dfrac{8}{9}, a_5 = \dfrac{16}{27}$

7. $a_1 = 2, a_2 = 0, a_3 = 2, a_4 = 0, a_5 = 2$ **9.** $6 + 10 + 14 + 18 = 48$ **11.** $\displaystyle\sum_{k=1}^{13} (-1)^{k+1}\dfrac{1}{k}$ **13.** Arithmetic; $d = 1; S_n = \dfrac{n}{2}(n + 11)$ **15.** Neither

17. Geometric; $r = 8; S_n = \dfrac{8}{7}(8^n - 1)$ **19.** Arithmetic; $d = 4; S_n = 2n(n - 1)$ **21.** Geometric; $r = \dfrac{1}{2}; S_n = 6\left[1 - \left(\dfrac{1}{2}\right)^n\right]$ **23.** Neither

25. 3825 **27.** 1125 **29.** $\dfrac{1093}{2187} \approx 0.49977$ **31.** 35 **33.** $\dfrac{1}{10^{10}}$ **35.** $9\sqrt{2}$ **37.** $\{a_n\} = \{5n - 4\}$ **39.** $\{a_n\} = \{n - 10\}$ **41.** Converges; $\dfrac{9}{2}$

43. Converges; $\dfrac{4}{3}$ **45.** Diverges **47.** Converges; 8

49. (I) $n = 1$: $3 \cdot 1 = 3$ and $\dfrac{3 \cdot 1}{2}(1 + 1) = 3$

(II) If $3 + 6 + 9 + \cdots + 3k = \dfrac{3k}{2}(k + 1)$, then $3 + 6 + 9 + \cdots + 3k + 3(k + 1) = (3 + 6 + 9 + \cdots + 3k) + (3k + 3)$

$= \dfrac{3k}{2}(k + 1) + (3k + 3) = \dfrac{3k^2}{2} + \dfrac{3k}{2} + \dfrac{6k}{2} + \dfrac{6}{2} = \dfrac{3}{2}(k^2 + 3k + 2) = \dfrac{3}{2}(k + 1)(k + 2) = \dfrac{3(k + 1)}{2}[(k + 1) + 1].$

51. (I) $n = 1$: $2 \cdot 3^{1-1} = 2$ and $3^1 - 1 = 2$

(II) If $2 + 6 + 18 + \cdots + 2 \cdot 3^{k-1} = 3^k - 1$, then $2 + 6 + 18 + \cdots + 2 \cdot 3^{k-1} + 2 \cdot 3^{(k+1)-1} = (2 + 6 + 18 + \cdots + 2 \cdot 3^{k-1}) + 2 \cdot 3^k$
$= 3^k - 1 + 2 \cdot 3^k = 3 \cdot 3^k - 1 = 3^{k+1} - 1.$

53. (I) $n = 1$: $(3 \cdot 1 - 2)^2 = 1$ and $\dfrac{1}{2} \cdot 1 \cdot [6(1)^2 - 3(1) - 1] = 1$

(II) If $1^2 + 4^2 + 7^2 + \cdots + (3k - 2)^2 = \dfrac{1}{2}k(6k^2 - 3k - 1)$, then $1^2 + 4^2 + 7^2 + \cdots + (3k - 2)^2 + [3(k + 1) - 2]^2$

$= [1^2 + 4^2 + 7^2 + \cdots + (3k - 2)^2] + (3k + 1)^2 = \dfrac{1}{2}k(6k^2 - 3k - 1) + (3k + 1)^2 = \dfrac{1}{2}(6k^3 - 3k^2 - k) + (9k^2 + 6k + 1)$

$= \dfrac{1}{2}(6k^3 + 15k^2 + 11k + 2) = \dfrac{1}{2}(k + 1)(6k^2 + 9k + 2) = \dfrac{1}{2}(k + 1)[6(k + 1)^2 - 3(k + 1) - 1].$

55. 10 **57.** $x^5 + 10x^4 + 40x^3 + 80x^2 + 80x + 32$ **59.** $32x^5 + 240x^4 + 720x^3 + 1080x^2 + 810x + 243$ **61.** 144 **63.** 84

65. (a) 8 bricks **(b)** 1100 bricks **67. (a)** $20\left(\dfrac{3}{4}\right)^3 = \dfrac{135}{16}$ ft **(b)** $20\left(\dfrac{3}{4}\right)^n$ ft **(c)** 13 times **(d)** 140 ft **69.** $244,129.08

Chapter Test *(page 677)*

1. $0, \dfrac{3}{10}, \dfrac{8}{11}, \dfrac{5}{4}, \dfrac{24}{13}$ **2.** $4, 14, 44, 134, 404$ **3.** $2 - \dfrac{3}{4} + \dfrac{4}{9} = \dfrac{61}{36}$ **4.** $-\dfrac{1}{3} - \dfrac{14}{9} - \dfrac{73}{27} - \dfrac{308}{81} = -\dfrac{680}{81}$ **5.** $\displaystyle\sum_{k=1}^{10}(-1)^k\left(\dfrac{k+1}{k+4}\right)$ **6.** Neither

7. Geometric; $r = 4$; $S_n = \dfrac{2}{3}(1 - 4^n)$ **8.** Arithmetic: $d = -8$; $S_n = n(2 - 4n)$ **9.** Arithmetic; $d = -\dfrac{1}{2}$; $S_n = \dfrac{n}{4}(27 - n)$

10. Geometric; $r = \dfrac{2}{5}$; $S_n = \dfrac{125}{3}\left[1 - \left(\dfrac{2}{5}\right)^n\right]$ **11.** Neither **12.** Converges; $\dfrac{1024}{5}$ **13.** $243m^5 + 810m^4 + 1080m^3 + 720m^2 + 240m + 32$

14. First we show that the statement holds for $n = 1$. $\left(1 + \dfrac{1}{1}\right) = 1 + 1 = 2$. The equality is true for $n = 1$, so Condition I holds. Next we assume that

$$\left(1 + \dfrac{1}{1}\right)\left(1 + \dfrac{1}{2}\right)\left(1 + \dfrac{1}{3}\right)\cdots\left(1 + \dfrac{1}{n}\right) = n + 1 \text{ is true for some } k, \text{ and we determine whether the formula then holds for } k + 1. \text{ We assume that}$$

$$\left(1 + \dfrac{1}{1}\right)\left(1 + \dfrac{1}{2}\right)\left(1 + \dfrac{1}{3}\right)\cdots\left(1 + \dfrac{1}{k}\right) = k + 1. \text{ Now we need to show that } \left(1 + \dfrac{1}{1}\right)\left(1 + \dfrac{1}{2}\right)\left(1 + \dfrac{1}{3}\right)\cdots\left(1 + \dfrac{1}{k}\right)\left(1 + \dfrac{1}{k+1}\right)$$

$= (k + 1) + 1 = k + 2$. We do this as follows:

$$\left(1 + \dfrac{1}{1}\right)\left(1 + \dfrac{1}{2}\right)\left(1 + \dfrac{1}{3}\right)\cdots\left(1 + \dfrac{1}{k}\right)\left(1 + \dfrac{1}{k+1}\right) = \left[\left(1 + \dfrac{1}{1}\right)\left(1 + \dfrac{1}{2}\right)\left(1 + \dfrac{1}{3}\right)\cdots\left(1 + \dfrac{1}{k}\right)\right]\left(1 + \dfrac{1}{k+1}\right)$$

$= (k + 1)\left(1 + \dfrac{1}{k+1}\right)$ (induction assumption) $= (k + 1)\cdot 1 + (k + 1)\cdot\dfrac{1}{k+1} = k + 1 + 1 = k + 2$

Condition II also holds. Thus, the formula holds true for all natural numbers.

15. After 10 years, the Durango will be worth $6103.11. **16.** The weightlifter will have lifted a total of 8000 pounds after 5 sets.

Cumulative Review *(page 677)*

1. $\{-3, 3, -3i, 3i\}$ **2. (a)**

(b) $\left\{\left(\sqrt{\dfrac{-1 + \sqrt{3601}}{18}}, \dfrac{-1 + \sqrt{3601}}{6}\right), \left(-\sqrt{\dfrac{-1 + \sqrt{3601}}{18}}, \dfrac{-1 + \sqrt{3601}}{6}\right)\right\}$

(c) The circle and the parabola intersect at

$\left(\sqrt{\dfrac{-1 + \sqrt{3601}}{18}}, \dfrac{-1 + \sqrt{3601}}{6}\right), \left(-\sqrt{\dfrac{-1 + \sqrt{3601}}{18}}, \dfrac{-1 + \sqrt{3601}}{6}\right)$.

3. $\left\{\ln\left(\dfrac{5}{2}\right)\right\}$ **4.** $y = 5x - 10$ **5.** $(x + 1)^2 + (y - 2)^2 = 25$ **6. (a)** 5 **(b)** 13 **(c)** $\dfrac{6x + 3}{2x - 1}$ **(d)** $\left\{x \middle| x \neq \dfrac{1}{2}\right\}$ **(e)** $\dfrac{7x - 2}{x - 2}$ **(f)** $\{x | x \neq 2\}$

(g) $g^{-1}(x) = \dfrac{1}{2}(x - 1)$; all reals **(h)** $f^{-1}(x) = \dfrac{2x}{x - 3}$; $\{x | x \neq 3\}$ **7.** $\dfrac{x^2}{7} + \dfrac{y^2}{16} = 1$ **8.** $(x + 1)^2 = 4(y - 2)$

CHAPTER 10 Counting and Probability

10.1 Assess Your Understanding *(page 684)*

5. subset; \subseteq **6.** finite **7.** $n(A) + n(B) - n(A \cap B)$ **8.** T **9.** \varnothing, $\{a\}$, $\{b\}$, $\{c\}$, $\{d\}$, $\{a, b\}$, $\{a, c\}$, $\{a, d\}$, $\{b, c\}$, $\{b, d\}$, $\{c, d\}$, $\{a, b, c\}$, $\{b, c, d\}$, $\{a, c, d\}$, $\{a, b, d\}$, $\{a, b, c, d\}$ **11.** 25 **13.** 40 **15.** 25 **17.** 37 **19.** 18 **21.** 5 **23.** 15 different arrangements **25.** 9000 numbers **27.** 175; 125
29. (a) 15 **(b)** 15 **(c)** 15 **(d)** 25 **(e)** 40 **31. (a)** 11,923 thousand **(b)** 75,241 thousand **33.** 480 portfolios

10.2 Assess Your Understanding *(page 691)*

3. permutation **4.** combination **5.** $\dfrac{n!}{(n - r)!}$ **6.** $\dfrac{n!}{(n - r)!r!}$ **7.** 30 **9.** 24 **11.** 1 **13.** 1680 **15.** 28 **17.** 35 **19.** 1 **21.** 10,400,600

23. $\{abc, abd, abe, acb, acd, ace, adb, adc, ade, aeb, aec, aed, bac, bad, bae, bca, bcd, bce, bda, bdc, bde, bea, bec, bed, cab, cad, cae, cba, cbd, cbe, cda,$
$cdb, cde, cea, ceb, ced, dab, dac, dae, dba, dbc, dbe, dca, dcb, dce, dea, deb, dec, eab, eac, ead, eba, ebc, ebd, eca, ecb, ecd, eda, edb, edc\}$; 60
25. $\{123, 124, 132, 134, 142, 143, 213, 214, 231, 234, 241, 243, 312, 314, 321, 324, 341, 342, 412, 413, 421, 423, 431, 432\}$; 24
27. $\{abc, abd, abe, acd, ace, ade, bcd, bce, bde, cde\}$; 10 **29.** $\{123, 124, 134, 234\}$; 4 **31.** 16 **33.** 8 **35.** 24 **37.** 60 **39.** 18,278 **41.** 35
43. 1024 **45.** 120 **47.** 132,860 **49.** 336 **51.** 90,720 **53. (a)** 63 **(b)** 35 **(c)** 1 **55.** 1.157×10^{76} **57.** 362,880 **59.** 660 **61.** 15
63. (a) 125,000; 117,600 **(b)** A better name for a *combination* lock would be a *permutation* lock because the order of the numbers matters.

Historical Problem *(page 701)*

1. (a) $\{AAAA, AAAB, AABA, AABB, ABAA, ABAB, ABBA, ABBB, BAAA, BAAB, BABA, BABB, BBAA, BBAB, BBBA, BBBB\}$

(b) $P(A \text{ wins}) = \dfrac{C(4, 2) + C(4, 3) + C(4, 4)}{2^4} = \dfrac{6 + 4 + 1}{16} = \dfrac{11}{16}$; $P(B \text{ wins}) = \dfrac{C(4, 3) + C(4, 4)}{2^4} = \dfrac{4 + 1}{16} = \dfrac{5}{16}$

10.3 Assess Your Understanding (page 701)

1. equally likely **2.** complement **3.** F **4.** T **5.** 0, 0.01, 0.35, 1 **7.** Probability model **9.** Not a probability model

11. $S = \{HH, HT, TH, TT\}$; $P(HH) = \frac{1}{4}$, $P(HT) = \frac{1}{4}$, $P(TH) = \frac{1}{4}$, $P(TT) = \frac{1}{4}$ **13.** $S = \{HH1, HH2, HH3, HH4, HH5, HH6, HT1, HT2, HT3,$
HT4, HT5, HT6, TH1, TH2, TH3, TH4, TH5, TH6, TT1, TT2, TT3, TT4, TT5, TT6$\}$; each outcome has the probability of $\frac{1}{24}$.

15. $S = \{HHH, HHT, HTH, HTT, THH, THT, TTH, TTT\}$; each outcome has the probability of $\frac{1}{8}$.

17. $S = \{1\ Yellow, 1\ Red, 1\ Green, 2\ Yellow, 2\ Red, 2\ Green, 3\ Yellow, 3\ Red, 3\ Green, 4\ Yellow, 4\ Red, 4\ Green\}$; each outcome has the probability
of $\frac{1}{12}$; thus, $P(2\ Red) + P(4\ Red) = \frac{1}{12} + \frac{1}{12} = \frac{1}{6}$.

19. $S = \{1\ Yellow\ Forward, 1\ Yellow\ Backward, 1\ Red\ Forward, 1\ Red\ Backward, 1\ Green\ Forward, 1\ Green\ Backward, 2\ Yellow\ Forward, 2\ Yellow$
Backward, 2 Red Forward, 2 Red Backward, 2 Green Forward, 2 Green Backward, 3 Yellow Forward, 3 Yellow Backward, 3 Red Forward, 3 Red
Backward, 3 Green Forward, 3 Green Backward, 4 Yellow Forward, 4 Yellow Backward, 4 Red Forward, 4 Red Backward, 4 Green Forward, 4 Green
Backward$\}$; each outcome has the probability of $\frac{1}{24}$; thus, $P(1\ Red\ Backward) + P(1\ Green\ Backward) = \frac{1}{24} + \frac{1}{24} = \frac{1}{12}$.

21. $S = \{11\ Red, 11\ Yellow, 11\ Green, 12\ Red, 12\ Yellow, 12\ Green, 13\ Red, 13\ Yellow, 13\ Green, 14\ Red, 14\ Yellow, 14\ Green, 21\ Red, 21\ Yellow, 21\ Green,$
22 Red, 22 Yellow, 22 Green, 23 Red, 23 Yellow, 23 Green, 24 Red, 24 Yellow, 24 Green, 31 Red, 31 Yellow, 31 Green, 32 Red, 32 Yellow, 32 Green, 33 Red,
33 Yellow, 33 Green, 34 Red, 34 Yellow, 34 Green, 41 Red, 41 Yellow, 41 Green, 42 Red, 42 Yellow, 42 Green, 43 Red, 43 Yellow, 43 Green, 44 Red,
44 Yellow, 44 Green$\}$; each outcome has the probability of $\frac{1}{48}$; thus, $E = \{22\ Red, 22\ Green, 24\ Red, 24\ Green\}$; $P(E) = \frac{n(E)}{n(S)} = \frac{4}{48} = \frac{1}{12}$.

23. A, B, C, F **25.** B **27.** $P(H) = \frac{4}{5}$; $P(T) = \frac{1}{5}$ **29.** $P(1) = P(3) = P(5) = \frac{2}{9}$; $P(2) = P(4) = P(6) = \frac{1}{9}$ **31.** $\frac{3}{10}$ **33.** $\frac{1}{2}$ **35.** $\frac{1}{6}$ **37.** $\frac{1}{8}$

39. $\frac{1}{4}$ **41.** $\frac{1}{6}$ **43.** $\frac{1}{18}$ **45.** 0.55 **47.** 0.70 **49.** 0.30 **51.** 0.87 **53.** 0.66 **55.** 0.95 **57.** $\frac{17}{20}$ **59.** $\frac{11}{20}$ **61.** $\frac{1}{2}$ **63.** $\frac{3}{10}$ **65.** $\frac{2}{5}$

67. (a) 0.57 **(b)** 0.95 **(c)** 0.83 **(d)** 0.38 **(e)** 0.29 **(f)** 0.05 **(g)** 0.78 **(h)** 0.71 **69. (a)** $\frac{25}{33}$ **(b)** $\frac{25}{33}$ **71.** 0.167 **73.** 0.000033069

Review Exercises (page 705)

1. \varnothing, {Dave}, {Joanne}, {Erica}, {Dave, Joanne}, {Dave, Erica}, {Joanne, Erica}, {Dave, Joanne, Erica}
3. 17 **5.** 29 **7.** 7 **9.** 25 **11.** 336 **13.** 56 **15.** 60 **17.** 128 **19.** 3024 **21.** 1680 **23.** 91 **25.** 1,600,000 **27.** 216,000 **29.** 1260
31. (a) 381,024 **(b)** 1260 **33. (a)** $8.634628387 \times 10^{45}$ **(b)** 0.6531 **(c)** 0.3469 **35. (a)** 0.058 **(b)** 0.942 **37.** $\frac{4}{9}$ **39.** 0.2; 0.26

Chapter Test (page 706)

1. 22 **2.** 3 **3.** 8 **4.** 45 **5.** 5040 **6.** 151,200 **7.** 462 **8.** There are 54,264 ways to choose 6 different colors from the 21 available colors.
9. There are 840 distinct arrangements of the letters in the word REDEEMED. **10.** There are 56 different exacta bets for an 8-horse race.
11. There are 155,480,000 possible license plates using the new format. **12. (a)** 0.95 **(b)** 0.30 **13. (a)** 0.25 **(b)** 0.55 **14.** 0.19
15. $P(\text{win on \$1 play}) = \frac{1}{120,526,770} \approx 0.0000000083$ **16.** $P(\text{exactly 2 fours}) = \frac{1250}{7776} \approx 0.1608$

Cumulative Review (page 707)

1. $\left\{\frac{1}{3} - \frac{\sqrt{2}}{3}i, \frac{1}{3} + \frac{\sqrt{2}}{3}i\right\}$ **2.** **3.** **4.** $\{x \mid 3.99 \leq x \leq 4.01\}$ or $[3.99, 4.01]$

5. $\left\{-\frac{1}{2} + \frac{\sqrt{7}}{2}i, -\frac{1}{2} - \frac{\sqrt{7}}{2}i, -\frac{1}{5}, 3\right\}$ **6.** **7.** 2 **8.** $\left\{\frac{8}{3}\right\}$ **9.** $x = 2$, $y = -5$, $z = 3$ **10.** 125; 700

Domain: all real numbers
Range: $\{y \mid y > 5\}$
Horizontal asymptote: $y = 5$

APPENDIX Graphing Utilities

1 Exercises (page A2)

1. $(-1, 4)$; II **3.** $(3, 1)$; I **5.** $X\text{min} = -6$, $X\text{max} = 6$, $X\text{scl} = 2$, $Y\text{min} = -4$, $Y\text{max} = 4$, $Y\text{scl} = 2$
7. $X\text{min} = -6$, $X\text{max} = 6$, $X\text{scl} = 2$, $Y\text{min} = -1$, $Y\text{max} = 3$, $Y\text{scl} = 1$ **9.** $X\text{min} = 3$, $X\text{max} = 9$, $X\text{scl} = 1$, $Y\text{min} = 2$, $Y\text{max} = 10$, $Y\text{scl} = 2$
11. $X\text{min} = -11$, $X\text{max} = 5$, $X\text{scl} = 1$, $Y\text{min} = -3$, $Y\text{max} = 6$, $Y\text{scl} = 1$
13. $X\text{min} = -30$, $X\text{max} = 50$, $X\text{scl} = 10$, $Y\text{min} = -90$, $Y\text{max} = 50$, $Y\text{scl} = 10$
15. $X\text{min} = -10$, $X\text{max} = 110$, $X\text{scl} = 10$, $Y\text{min} = -10$, $Y\text{max} = 160$, $Y\text{scl} = 10$

2 Exercises *(page A4)*

1. (a) **(b)** **3. (a)** **(b)**

5. (a) **(b)** **7. (a)** **(b)**

9. (a) **(b)** **11. (a)** **(b)**

13. (a) **(b)** **15. (a)** **(b)**

17. $Y_1 = X+2$ **19.** $Y_1 = -X+2$ **21.** $Y_1 = 2X+2$ **23.** $Y_1 = -2X+2$

25. $Y_1 = X^2+2$ **27.** $Y_1 = -X^2+2$ **29.** $Y_1 = -(3/2)X+3$ **31.** $Y_1 = (3/2)X+3$

3 Exercises *(page A6)*

1. -3.41 **3.** -1.71 **5.** -0.28 **7.** 3.00 **9.** 4.50 **11.** $1.00, 23.00$

5 Exercises *(page A8)*

1. Yes **3.** Yes **5.** No **7.** Yes **9.** Answers may vary. A possible answer is $Y\min = 4$, $Y\max = 12$, and $Y\mathrm{scl} = 1$.

Index

ENCYCLOPEDIA OF AESTHETICS

ENCYCLOPEDIA OF

AESTHETICS

MICHAEL KELLY

Editor in Chief

Volume 2

OXFORD UNIVERSITY PRESS

New York 1998 Oxford

OXFORD UNIVERSITY PRESS

Oxford New York
Athens Auckland Bangkok Bogotá
Buenos Aires Calcutta Cape Town Chennai
Dar es Salaam Delhi Florence Hong Kong Istanbul
Karachi Kuala Lumpur Madrid Melbourne Mexico City
Mumbai Nairobi Paris São Paulo Singapore
Taipei Tokyo Toronto Warsaw

and associated companies in
Berlin Ibadan

Library of Congress Cataloging-in-Publication Data
Encyclopedia of aesthetics / editor in chief, Michael Kelly.
p. cm
Includes bibliographical references and index.
1. Aesthetics—Encyclopedias. I. Kelly, Michael, 1953–.
BH56.E53 1998 111'.85'03—dc21 98-18741 CIP
ISBN 0-19-511307-1 (set)
ISBN 0-19-512646-7 (vol. 2)

Printing (last digit): 9 8 7 6 5 4 3 2 1

Printed in the United States of America
on acid-free paper

D

CONTINUED

DERRIDA, JACQUES. [*To treat the thought of contemporary French philosopher Jacques Derrida, this entry consists of four essays:*

Survey of Thought
Derrida and Deconstruction
Derrida and Literature
Derrida and Kant

The first essay is on Derrida's philosophy in general and its relevance to aesthetics. The second essay traces the origins and development of "deconstruction," a term introduced by Derrida and exemplified in many of his works. The third essay concerns his theory of literature, the main art form he discusses when he analyzes matters aesthetic. The fourth essay discusses his interpretation of Kant, a major aesthetic theorist in dialogue with whom Derrida articulated his critique of aesthetics. See also Metaphor, *article on* Derrida and de Man on Metaphor; Postcolonialism; Postmodernism, *historical and conceptual overview article;* Text; *and* Truth.]

Survey of Thought

Jacques Derrida was born in 1930 in El-Biar, Algeria, and is now affiliated primarily with the École des Hautes Études en Sciences Sociales in Paris. Derrida began emerging in the late 1960s as an increasingly influential figure across a wide range of fields, including philosophy, literary theory, architecture and its theory, and art and its theory.

Time and Thesis. In 1980 Derrida chose to present a selection from his published texts as his doctoral thesis. The brief remarks with which he prefaced his defense, later published as "The Time of a Thesis: Punctuations," took the form of a perhaps surprisingly straightforward narrative of the changing shape of both his thesis project and his relationship to it. The twenty-five years of work that resulted in his thesis break down in this narrative into several quasi periods, which might be distinguished as follows:

1. The time of the first proposed thesis, duly registered and actively pursued by Derrida under the title "The Ideality of the Literary Object": The primary philosophical reference inscribed in this version of the thesis is to Edmund Husserl, and its most visible outcome is Derrida's first publication, a critical introduction to Husserl's *Origin of Geometry*.

2. Ten years later, in 1967, Derrida recast the thesis; it is now to be an interpretation of G. W. F. Hegel's semiology.

The most prominent artifact of this version of the thesis is the essay that opens *Margins of Philosophy*, "The Pit and the Pyramid," but the project may also seem more generally visible in the sustained explications with Hegel that run from the early essay on Georges Bataille in *Writing and Difference* up to and perhaps through 1975's *Glas*—Derrida's most sustained exercise of what one might call experimental writing.

3. This period also overlaps with another, defined less by the project than by a shifted relation to the very idea of submitting a thesis. Derrida links this variously to the events of 1968 (opening a period of his neglect of the thesis) and the university reforms of the early seventies (opening a period of more deliberate refusal of institutional advancement).

4. The moment, presumably in early 1980, at which he decided, for reasons he finds obscure at the time of his defense, to go ahead with it after all, and to do so on the basis of a selection from his writings.

His remarks, as they unfold, include various observations about what from one or another period has been included in or excluded from the thesis. To the extent that one can make out any criteria informing these choices, they seem to be what one might call "stylistic" or "formal." In any case, the decision to present a thesis does not, evidently, entail on Derrida's part any attempt to characterize it as a thesis or as the completion of any particular project. Telling the time or times of the thesis has apparently come to stand in for the thesis "itself." Or perhaps one will rather say that "the thesis" has become simply the fact of its writing.

All of this amounts, of course, to but one moment of retrospection among others in the course of a career, and there is no reason to take it as definitive or even particularly accurate. But it is, I suggest, interesting in a number of respects.

It carries, for example, a certain amount of information about Derrida's location in a particular French inter- and intragenerational intellectual space. Engagement with—translation of or commentary on—one of a restricted number of German thinkers, Husserl above all, is characteristic of French philosophical beginnings from sometime in the thirties up to around 1960. Derrida is squarely in—and toward the end of—this line. At the same time, a certain aggressive indifference toward Jean-Paul Sartre and/or Maurice Merleau-Ponty invoked by Derrida is a characteristic shared by certain of the thinkers who come to prominence

1

in the late sixties or early seventies, like Derrida and Michel Foucault, and not of others, for example, Jean-François Lyotard or Gilles Deleuze, both of whom are plausibly described as following from Merleau-Ponty in particular. The desire to do philosophy as a form of making peace with Hegel is likewise broadly characteristic of French intellectual life from the early thirties until perhaps the middle or late seventies, when other names—perhaps most notably that of Immanuel Kant but also those of such others as Baruch Spinoza and Henri Bergson—make a bid for attention.

While "structuralism," Ferdinand de Saussure, and so on do find a certain place in this narrative, it is a distinctly limited one—a matter, more or less, of a stimulating atmosphere within which Derrida worked even as the work itself found its direction elsewhere (here he mentions Martin Heidegger, Emmanuel Lévinas, and Maurice Blanchot). One ought probably see this quiet refusal to characterize himself as a theorist or philosopher of language alongside the way in which Derrida, speaking of his 1957 self, poses the question of literature and its "ideality" as *directly* a question of writing. In the United States Derrida has been received overwhelmingly as a "poststructuralist," which is to say, as someone for whom Saussure's work constitutes a kind of minimal condition of entry. I will return to some of the costs of this view.

That early thesis title—"The Ideality of Literature"—is what perhaps rings most strangely to the contemporary ear, accustomed as it is to hearing "materiality" as the continuous bass of so much contemporary theoretical writing. But, as Derrida notes, it matters to hear the implicit claim of the title within the context of the phenomenological orientation toward rigor and exact science, an implicit description of the ideality of its objects that precludes any ideality of literature, so that an argument for such ideality amounts to a reconstruing of ideality altogether (an argument very much like the one Derrida deploys against John Searle much later). What the title might be said to mark, from the very beginning of Derrida's career, is the ways in which he will refuse simple oppositions of "materialism" and "idealism" in favor of opening out the pressures inherent in such imaginations.

There are also some transitions worth noticing in Derrida's narrative. Perhaps the most interesting is the one that begins with the taking of a certain distance from the structuralist atmosphere in favor of something represented by the names Heidegger, Lévinas, and Blanchot, and glossed by Derrida first of all in terms of their solitudes, their separateness from schools or groups—even as they also are, in their very solitude, thinkers "to whom, strangely enough, one may consider oneself most close" ("The Time of a Thesis: Punctuations," p. 41). The passage effected here from "schools" to "solitude" seems to serve in Derrida's remarks as the ground for questions about the university and *univer-sitas*, which open into accounts of his work with the Groupe de recherches sur l'enseignment philosophique—all of which serves in its turn to anchor the political and institutional emphasis Derrida weaves through his story. What one wants to note here is how this is, from the beginning, a matter of the possibilities for community—an outcome that becomes fully explicit in Derrida's work of the 1980s and 1990s as well as in various exchanges, most notably between Jean-Luc Nancy and Blanchot, about this topic. This is an orientation that will sit oddly with the emergent American orientation to identity over roughly the same stretch of time.

Finally, there is the whole point of the narrative in which a project—aimed first at the ideality of literature and then at Hegel's semiology—finds itself volatilized in favor of the simple fact of its punctuation, the rhythm of a work coming to stand in for its shape or what one might imagine as its content. Derrida does of course manage to sketch in along the way various pictures of his activity:

> During the years that followed, from about 1963 to 1968, I tried to work out . . . what was in no way meant to be a system but rather a sort of strategic device, opening onto its own abyss, an unclosed, unenclosable, not wholly formalizable ensemble of rules for reading, interpretation, and writing. This type of device may have enabled me to detect not only in the history of philosophy and in the related socio-historical totality, but also in what are alleged to be sciences and in so-called post-philosophical discourses that figure among the most modern (in linguistics, in anthropology, in psychoanalysis), to detect in these an evaluation of writing, or, to tell the truth, rather a devaluation of writing whose insistent, repetitive, even obscurely compulsive, character was the sign of a whole set of long-standing constraints.
>
> ("The Time of a Thesis: Punctuations," p. 40)

The "strategic device" Derrida speaks of here is an attention to that moment or system of moments in a text through which it appears at once to name itself as a text and to disavow or repress that status. Derrida claims, for example, that the term *supplément* as a name for writing and culture and masturbation does this work in Rousseau's writing. To read a text in a way that takes such moments as central is, in effect, to open a seam between what it means and what it does (different texts will of course characterize this seam differently), thus opening out the network by which its thought is placed at odds with itself by the very condition of its expression. This can be seen as an inheritance and transformation of Heidegger's interpretive practices insofar as they engage a certain violence on the ground of a "forgottenness of Being" taken to be integral to the construction of a philosophical text. Derrida's frequent formulations in terms of a certain "repression of writing" clearly invite a Freudian gloss as well—and as with the Freudian cure, the object appears to be less undoing repression, which remains an integral condition of consciousness as such, than a sort of making peace with it that will make a crucial difference to one's inhabitation of one's self or world. One might speak

equally here in terms of an attempt to do justice to a text or object that cannot be elevated above the play of debt and excess that is its actual—wordly, cultural, and temporal—condition.

A number of Derrida's earliest essays set themselves up, unsurprisingly, as corrections or revisions or complications of more standard readings (in some cases, they will appear to increase the incoherence or contradictions of the text on the face of that reading; in other cases, like that of Rousseau, they may seem more nearly to find the actual thread along which a text perceived as contradictory actually hangs together). In taking this form, they follow a familiar enough model for conducting a reading, even if the reading itself outraces the usual criteria for such activity. Increasingly, however, Derrida's writings take on a structure that inverts that typical of the interpretive or academic essay: Beginning in something that feels like the middle, they work their way to a sort of conclusion inseparable from a (re)beginning that unfolds once more into the dispersion of the middle. These formal considerations are perhaps more helpful than his more celebrated neologistic play (a part of his Heideggerian inheritance) in thinking about Derrida's philosophical location and stakes: an essay shaped in this way is responsive to—can be said to derive from—those regions of Hegel's writing, like his aesthetics, where the apparent subject is in fact not sustained but rather is undone by the movement of "its" dialectic. One might also compare Derrida, as a writer of philosophy, to Nietzsche trying to write an account of the birth of tragedy that does not betray the division and contradiction that is its origin and the very shape of the unity that will also be its betrayal. An essay unplotted in Derrida's way intends no delivery of itself from time, and counts perhaps above all as a rhythmic articulation of it.

"The Time of a Thesis: Punctuations" is thus unusual in doing something Derrida's writing rarely does: It lets a straightforward narrative, almost completely unburdened of neologisms and other forms of linguistic play, and written in ordinary enough language, do the work of what he calls "writing"—a work one might then describe as a simultaneous doing and undoing of the propositions or theses it deploys, or as a rendering of them responsible to the time not of their enunciation (something like their momentary present) but of their inscription (an activity that both sustains itself in and eludes any simple capture by such a present). Picking up Derrida this way makes it clear that he inherits a particular vision of philosophical work as essentially argumentative and logical. The particular shape of this inheritance is the one given by the passage from Kant's claim to transcendental argument through its speculative Hegelian transformation to Heidegger's meditations on time; in this passage argument is increasingly understood to be responsible to—and so our understanding of it transformable by—the conditions of its realization or expression. In this context, Derrida's appeal to writing is neither simply "metaphorical" nor arbitrary but absolutely integral to a view of philosophy as such. That this view of philosophy may seem in his work to be approaching a certain exhaustion; that one might choose rather to take up philosophy as something other than argument (as, for example, the work of grammatical attention fostered by Ludwig Wittgenstein or the invention of concepts urged by Derrida's near contemporary Deleuze) would be one way of specifying Derrida's philosophic stakes.

As one examines this pattern of proximities and competitions, two things may seem worth remarking in particular. The first is that the Derridean, Wittgensteinian, and Deleuzian visions of philosophy all place a central value on the encounter with literature as writing (rather than, say, meaning or form or any number of other such candidates). The second, which bears sharply on the reception of Derrida's work in the United States, is that not all of them communicate in the same way with ideas or images of "theory."

Lucidity. Derrida's work can be strongly characterized as a reception of Heidegger or of Sigmund Freud, or indeed of both. The first of these will make prominent, among many other things, the ways in which Derrida's interest in literature extends and transforms Heidegger's attention to poetry (where the transformation may be emblematically caught in the contrast between Heidegger's interest in Friedrich Hölderlin's words and Derrida's in Stéphane Mallarmé's blanks and intervals); his interest in writing similarly extends and revises Heidegger's interest in Saying; and his interest in reading renews and deepens the stance toward the past of philosophy that Heidegger variously poses as "destruction," "retrieve," "the step back," and so on.

If one's attention then shifts to the second, Freudian characterization—a characterization that would need to be fleshed out through an account of the particular version of Freud advanced by the French psychoanalyst Jacques Lacan—what will be most striking is an apparent retreat from Heidegger's claims to a certain kind of radically philosophical propriety (say, all the elements in Heidegger that conspire to present him as moving more deeply than ever into the actual heart of philosophy, its self-concealing center) in favor of an engagement with a "science" one can only imagine Heidegger dismissing as "merely ontic." This means, at a minimum, that much of what Heidegger wants to attribute to something like (on perhaps the most generous reading) the pure, logical structure of Being will find itself redistributed through terms like *desire* and *repression* that cannot be grounded that way. One way of taking this would be to say that for Derrida philosophy finds itself no longer simply or deeply within itself but ineluctably outside itself and in the world (and it would be important to recognize that this is, precisely, something Heidegger meant to say, something both at stake and lost within his elaboration of *Da-sein*).

Taking the two receptions together brings to focus a deeply shared interest in what is in thought or in us as its or our other side—as structurally and necessarily "unthought" (Heidegger) or "unconscious" (Freud). This orientation toward a fundamental acknowledgment of alterity is perhaps the defining feature of post-Hegelian thought, especially in France, organizing the new place and construal it gives to such key terms as *repetition* and *difference*. It is tempting—and the temptation is often given into in the literature—to imagine this as addressing a difficulty in self-relation, but it is better taken as a consequence of understanding identity as itself relational. This is, for example, how the central place assigned to difference and opposition in defining the core elements of language made Saussure's linguistics an important resource for extending and transforming the Heideggerian view of language. [*See* Freud; *and* Heidegger.]

Imaginations of such dualities as thought and unthought or conscious and unconscious often fill themselves out through appeals to a certain imagery of light and dark. This tendency, perhaps because of the influence of Johann Wolfgang von Goethe, seems particularly marked in the tradition of post-Kantian German and French philosophy, and the particular turn it takes in Derrida's work offers an interesting view onto his central concerns. One might begin by looking to Hegel's wariness of philosophy's collapse into "a night in which all cows are black"—a figure that seems to work in his writing to name both the threat of skepticism, of indeterminate or absolute negation, and the countervailing threat of a claim to philosophical absoluteness devoid of any articulation whatsoever. The writing of Hegel's Absolute appears against this background as the measured passage of the world through all its possibilities of light and darkness so that in the end we can claim to have seen what is completely, in its conditions both of visibility and invisibility. In Hegel's wake, the philosophical complementarity of light and dark falls out increasingly otherwise, and the world is imagined, in Heidegger, for example, as a lighting or coming into light that is necessarily also horizoned by darkness and shot through with shadow, neither of which can be overcome except through some more or less epochal transformation that will itself be equally a constellation of light and darkness, the relation of which to the earlier moment will not be recoverable, dialectically or otherwise. These figures thus render a certain otherness as a permanent concomitant of what is; philosophy's job is not to overcome this—it cannot—but to acknowledge it, to render what is unthought palpable as a dimension of thought.

Derrida has often cited the importance of his early encounter with the work of the critic and novelist Maurice Blanchot, himself strongly influenced by Heidegger. This reference is relevant here because of the emblematic significance of Blanchot's turns—in both his fictions and his criticism—on this motif. Blanchot's insistence is, in effect,

that the real problem is that of light—say, lucidity—itself. [*See* Blanchot.] Heidegger's writing in this area had typically insisted on a structure of "double concealment": one concealment the shadow and mutual blockage of things that comes inevitably as a consequence of one light rather than another, and the other—much harder to phrase—simply the fact of this world (rather than some other; rather than none at all) and so is simply continuous with its availability. It is as if Heidegger wants "light" to figure Being, and yet the light he invokes cannot but appear as a being, thus betraying the thing (or perhaps the nothing) it means to figure: lucidity betrays itself and does so necessarily—just as one might imagine that the psychoanalytic Unconscious is not some dark thing hidden behind or beneath consciousness but something more like its invisible lining, the integral cost of its existence. These thoughts are well caught in Blanchot's figure of "the madness of the day"—a figure that offers a powerful image of where Derrida's stakes are—between, on the one hand, the threat of a sort of blank empiricism easily convertible into a radical skepticism and, on the other, the appearance of an empty self-reflexive recession that hands itself finally over to a more palpably ironic variety of skepticism. These more or less symmetrical possibilities have figured prominently in both appropriations of and attacks on Derrida's work; and both miss the terrain he means to claim under the name of a "critique of presence." Both turn it toward something it will be more natural to call a "critique of representation."

Whether Derrida's ground is indeed there to claim or not has no deeper test or more independent proof than the strength of his readings. And these readings take, necessarily, the form of writing, an insistence on the exposure of thought and speech to the visible—an insistence, then, on their problematic lucidity.

"Theory." Perceptions of Derrida's work have for some time now been considerably, some might say fatally, complicated by his entry into the U.S. literary academy beginning in the early 1970s. This is perhaps reason enough for the more or less "pictorial" turn I have given the account so far.

The difficulties here are multiple. If one begins from the attacks on Derrida, the charge of a certain "political" failure or refusal will certainly be among the most prominent. As suggested here, this is in no small part a simple missing of the actual political drift of Derrida's work, which is toward questions of friendship, community, and the like (questions of solidarity), rather than toward the questions of identity and power that came to the fore in American cultural and academic politics during the 1970s and 1980s. But it also rests on a vision of the relation between action and reason or knowledge (of, in particular, the former depending on and following from the latter) that seems essentially foreign to Derrida (as well as to his much more overtly activist contemporary Michel Foucault). It is at this level that the polit-

ical critique of Derrida shares in the broader terms of Derrida's initial reception as a "theorist."

The term *theory,* and the recent verb forms built from it— *to do theory* and, somewhat later, *to theorize*—are perhaps the single most enduring legacy of American criticism's encounter with late-twentieth-century French thought. This development has one foundation in a particular moment of that thought, associated with the journal of the late sixties *Tel Quel* and the presentation of an epitome of its collective work under the title *Théorie d'ensemble* (Group theory). It seems fair enough to say that at this moment there was a marked sense of a continuity between, say, the mathematical work growing out of group theory and published under the name of Nicolas Bourbaki, a certain strain of avant-garde writing, and various new departures in philosophy, literary criticism, and intellectual history. But this was clearly a vanishing moment, the various authors represented in *Théorie d'ensemble* quickly separating out into a range of positions neither easily assimilable to one another nor subsumable under any common ambition to "theory."

But there is another foundation as well: Something called "literary theory" or "theory of literature" was, by the mid- to late 1960s very much on the agenda of American literary study as the name for what was needed to move its practices beyond either a vaguely historical attention or the more sharply formulated, and very sharply limited, strictures of the New Criticism. The New Critics had, of course, formulated their own theoretical foundations, very much in the face of a perceived challenge from the sciences and the philosophical support then offered to them by various positivisms, but these foundations seemed to preclude any large general statements—about the nature, unity, or inner articulation of some domain called "literature," or about relations between things literary and things nonliterary—of the sort literary critics found themselves wanting to make. In this light, such magisterial works as Northrop Frye's *Anatomy of Criticism,* Wayne Booth's *Rhetoric of Fiction,* or Kenneth Burke's various forays into what he distinguished as rhetoric, grammar, and logic were important symptoms of a thirst for "theory," as was the turn by a number of younger critics, some of whom later took up Derrida's work, toward the phenomenological criticism of the "Geneva School" (Georges Poulet, Jean-Pierre Richard, Jean Starobinski, and others). These various earlier appeals for "theory" were, in effect, rendered nugatory by the turn toward French structuralism and poststructuralism that dominated the late sixties and the seventies. Some of this material was theoretical in a very straightforward sense; much of it, including Derrida's work, was not—or at least entertained a complex and ambiguous stance toward the very idea of "theory." The blurring over of such differences was enabled to a high degree by the sense that what was at issue across the quite varied work of such figures as Claude Lévi-Strauss, Roland Barthes, Lacan, Foucault, Derrida, and others, was a more or less systematic development of the implications of Saussure's linguistics—a sense reinforced in Derrida's case by the privileged position assigned to both Saussure and Lévi-Strauss in opening up the issues of his major early work, *Of Grammatology* (itself primarily a reading of Rousseau).

The picture of Derrida as advancing a theory of language or of literary language, characterized to a high degree by a belief in the arbitrariness of the linguistic sign and so also in some radical indeterminacy of meaning alongside a notion of the literary as in some sense radically self-reflexive, has had and continues to have a considerable hold on current views of his work. As indicated by, for example, the proposed first title for Derrida's thesis work, by his continued interest in such sustained and properly theoretical enterprises as Lacan's explorations and revisions of psychoanalysis, and by his general interest in the limits of argumentative philosophy, this view cannot be simply wrong. But it is to a high degree partial and misleading. [*See* Lacan.]

The most compelling and influential appropriations of Derrida to literary study have centered on Romanticism, a tendency sharply devalued by the New Criticism. Romanticism was thought to be rescuable on the condition that one read in it a writerly awareness and problematic—a play of irony and allegory—outracing both the (especially English) Romantic appeal to a real depth of symbolic meaning and the modernist counterclaim to a poetry of explicit artifice and strong conventionality. These readings inclined strongly—especially in the hands of Paul de Man, Derrida's most prominent American interpreter—to a view of "literary language" as enacting a temporal play of mystification and demystification (thus the privilege of Romanticism over its presumably demystified modernist sequel) that was a key at once to the dynamics of literary history and to the general rhetorical undoing of the subject by the prior fact of language.

De Man's work culminates in a collection of posthumously published essays entitled, by the book's editor with some support from de Man's notes, *Aesthetic Ideology.* That de Man's itinerary should lead to an attempt to undo the stability of "the aesthetic" or to "void" its categories or to refuse the possibility of a specifically aesthetic judgment (all formulations that figure along his way) is perhaps hardly surprising given his explicit impulse to read Romanticism against its apparent grain. One can nonetheless raise the question of how far this does or does not reflect the actual drift of Derrida's writing in this area.

Aesthetic, Anti-aesthetic. It is clear enough, I hope, that Derrida's work makes an almost continuous claim on the attention of aestheticians, but its most explicit intersections with the established texts of aesthetics are probably to

be found in his readings of Kant's *Critique of Judgment* and Heidegger's "Origin of the Work of Art." Both readings are long, complex, and rendered considerably more difficult by the pronounced formal strategies (the Kant texts are punctuated by a series of breaks or cutouts; the Heidegger reading unfolds as a conversation among an unspecified number of voices) that clearly mean to short-circuit any attempt to establish an authoritative line through them.

Nonetheless, it seems possible to venture a few remarks about their general bearing. In both cases, Derrida's interest seems to be less in the adequacy or inadequacy of the proffered account of art or the aesthetic than in the relation between that account and its limits. In the Kant essays this means focusing to a high degree on things Kant distinguishes as "parergonal"—related to but not actually proper to the work, such as frames for painting, color for sculpture, or colonnades in architecture; in the Heidegger "polylogue," essentially similar questions are played out on the rather different axis of the pertinence of one or another interpretation of the shoes in a van Gogh painting to the painting itself. One can say that in each case the question of limits unfolds into a nest of questions about what attaches itself to or detaches itself from the work of art, or about what the work attaches itself to and detaches itself from, or, finally, about the various rhythms these movements set up. These questions have a fairly clear Kantian ground, not simply as concerns with limits but as concerns phrasable as internal to Kant's account of judgment: The beautiful, one can say, is for Kant that which detaches itself from the world completely even as it attaches itself to nothing other than itself, while the sublime is the (not quite reciprocal) discovery in us of a self-attachment proper to our detachment from the world. The work of art is itself neither sublime nor beautiful, but seems to become visible only against the ground of their articulation and distinction. This is, perhaps, the sense of Derrida's interest in those structures by which the artwork simultaneously marks and undoes its self-containment—whether Kant's parergonal frame or the laces of van Gogh's Heideggerian boots. To the extent that one is justified in imagining "aesthetics" itself as one of those structures, it will figure as an object of neither affirmation nor critique but of reading. And this would be to say that the "anti-aesthetic" vector in Derrida, as in Heidegger, is one that aims less to display the shortcomings of the aesthetic in the face of the claims of knowledge or practice than to show the artwork as itself doing work both essential to and uncapturable within philosophy. But this is simply to say that the critical impulse that one might want to call "anti-aesthetic" in Derrida is consonant with a very high valorization of the work of art and is of a piece with what he has meant at various times by "writing" and by "the ideality of literature." The object of critique here is the philosophical effort to contain and regulate the claims of art, and the affirmative effort is to display philosophy as necessarily caught up and exposed beyond itself.

BIBLIOGRAPHY

Works by Derrida

Edmund Husserl's "Origin of Geometry": An Introduction (1962). Translated by John P. Leavey, Jr., edited by David B. Allison. Reprint, Lincoln, Nebr., 1989.

Of Grammatology (1967). Translated by Gayatri Chakravorty Spivak. Baltimore, 1976.

Speech and Phenomena and Other Essays on Husserl's Theory of Signs (1967). Translated by David B. Allison. Evanston, Ill., 1973.

Writing and Difference (1967). Translated by Alan Bass. Chicago, 1978.

Dissemination (1972). Translated by Barbara Johnson. Chicago, 1981.

Margins of Philosophy (1972). Translated by Alan Bass. Chicago, 1982.

Positions (1972). Translated by Alan Bass. Chicago, 1981.

Glas (1974). Translated by John P. Leavey, Jr., and Richard Rand. Lincoln, Nebr., 1986.

The Archaeology of the Frivolous: Reading Condillac (1976). Translated by John P. Leavey, Jr. Pittsburgh, 1980.

Spurs: Nietzsche's Styles (1978). Translated by Barbara Harlow. Chicago, 1979.

The Truth in Painting (1978). Translated by Geoff Bennington and Ian McLeod. Chicago, 1987.

The Post Card: From Socrates to Freud and Beyond (1980). Translated by Alan Bass. Chicago, 1987.

The Ear of the Other: Otobiography, Transference, Translation (1982). Edited by Christie V. McDonald, translated by Peggy Kamuf. New York, 1985.

Signéponge/Signsponge (1983). Translated by Richard Rand. New York, 1984.

"The Time of a Thesis: Punctuations." Translated by Kathleen McLaughlin. In *Philosophy in France Today,* edited by Alan Montefiore, pp. 34–50. Cambridge and New York, 1983.

Mémoires: For Paul de Man (1986). Translated by Cecile Lindsay, Jonathan Culler, and Eduardo Cadava. New York, 1986.

Cinders (1987). Translated and edited by Ned Luckacher. Lincoln, Nebr., 1992.

Limited Inc. (1988). Translated by Samuel Weber. Evanston, Ill., 1988.

Of Spirit: Heidegger and the Question (1989). Translated and edited by Geoffrey Bennington and Rachel Bowlby. Chicago, 1989.

Memoirs of the Blind: The Self-Portrait and Other Ruins (1990). Translated by Pascale-Anne Brault and Michael B. Naas. Chicago, 1993.

Given Time I: Counterfeit Money (1991). Translated by Peggy Kamuf. Chicago, 1992.

The Other Heading: Reflections on Today's Europe (1991). Translated by Pascale-Anne Brault and Michael B. Nass. Bloomington, Ill., 1992.

The Gift of Death (1992). Translated by David Wills. Chicago, 1995.

Points . . . Interviews, 1974–1994. Edited by Elisabeth Weber, translated by Peggy Kamuf et al. Stanford, Calif., 1995.

Aporias (1993). Translated by Thomas Dutoit. Stanford, Calif., 1993.

On the Name (1993). Edited by Thomas Dutoit, translated by David Wood, John P. Leavey, Jr., and Ian McLeod. Stanford, Calif., 1995.

Specters of Marx: The State of the Debt, the Work of Mourning, and the New International (1993). Translated by Peggy Kamuf. New York, 1994.

Archive Fever: A Freudian Impression (1995). Translated by Eric Prenowitz. Chicago, 1996.

Other Sources

Bennington, Geoffrey, and Jacques Derrida. *Jacques Derrida.* Translated by Geoffrey Bennington. Chicago, 1993.

STEPHEN MELVILLE

Derrida and Deconstruction

The term *deconstruction* was coined by the French philosopher Jacques Derrida (b. 1930) between the two appearances of the material that, in its second version, became part of *De la grammatologie* (1967). In its first appearance, in 1965–1966, as a series of reviews in the French journal *Critique*, it contained the term *destruction*. The word owed something to Martin Heidegger, especially to the projected second part of his *Kant and the Problem of Metaphysics* (1929), which was to have been titled *The Fundamental Characteristics of a Phenomenological Destruction of the History of Ontology under the Guidance of the Problematic of Temporality*. The naming of "deconstruction," then, is, among other things, something like a definitive modification of a Heideggerian program. It should be remembered that Heidegger was a strong reader of Friedrich Nietzsche, in whose work also "destruction" played a special part. [*See* Heidegger.]

This essay interprets deconstruction specifically in Jacques Derrida's work and focuses on the issue of "the setting to work on deconstruction."

Deconstruction, as it emerged in Derrida's early writings, examined how texts of philosophy, when they established definitions as starting points, did not attend to the fact that all such gestures involved setting each defined item off from all that it was not. It was possible, Derrida said, to show that the elaboration of a definition as a theme or an argument was a pushing away of these antonyms. Such demonstrations involved tracking the rhetorical maneuvers performed by words such as *supplément* in Jean-Jacques Rousseau *(Grammatology)* or *pharmakos* and *hama* in Plato and Aristotle ("Plato's Pharmacy," 1968, in *Dissemination; "Ousia* and *grammè*: Note on a Note from *Being and Time*," 1968, in *Margins*). What these maneuvers seemed to conceal was the track of the first *différance* (a word coined by Derrida)—the setting off described above—as well as its continuation—the pushing away, also described above. This track, of a previous differentiation and a continuous deferment, is called "trace."

The structuralists had emphasized language, or rather sign systems, as explanatory models of the last instance. In *Of Grammatology,* Derrida submitted that Ferdinand de Saussure had not been able to admit in his work the implications of his insight that the origin of the possibility of language was the capacity to articulate differences among linguistic and verbal units rather than some internalized knowledge or reservoir of chunks of language. In "Speech and Phenomena" (1967, in *Speech and Phenomena*), Derrida argued that Edmund Husserl's phenomenological notion of "the living present" entailed the subject's death, since it implied a present extending before and after any given subject's livingness or life. In "Différance" (1968, in *Margins of Philosophy*), an important theoretical intervention presented before the Société française de philosophie, he named this inevitability of the differentiation (setting off) from, and deferment (pushing away) of the trace or track of all that is not what is being defined or posited, as différance. It was a "necessary but impossible" move (a formula to become useful for deconstruction), because, in being named, différance has already submitted to its own law, as outlined.

This irreducible work of the trace not only produces an unrestricted economy of same and other, rather than a relatively restricted dialectic of negation and sublation, in all philosophical oppositions. It also places our selfhood (ipseity) in a relationship of différance with what can only be "named" radical alterity (and thus necessarily effaced). This rich essay suggests certain rules of thumb for the deconstructive philosopher.

In "Signature Event Context" (1977, in *Margins*), Derrida suggested that J. L. Austin, in founding speech act theory, which investigates language as not merely statement but act, acknowledged the role of force in signification. He could not, however, admit the consequences of his irreducibly "locutionary" perception of language: that truth telling is also a performative convention, producing an effect, not limited to the transference of a semantic content. Each effective situation alters the truth iterated. "Speech" shares the structure we commonly call "writing," which is given over to the openness of use in unmarked and heterogeneous situations.

Derrida's kinship with other philosophers such as Immanuel Kant, Georg Friedrich Hegel, Søren Kierkegaard, Nietzsche, Sigmund Freud, Husserl, Walter Benjamin, and Emmanuel Lévinas, to name only a few, has been established. But it cannot be denied that, all through these philosophical essays and his other early work, the Heideggerian theme of the priority of the question (of Being, to all ontological investigation, as indicated in the projected title of the Kantbook 2, for example) is never absent. It is therefore significant that, in "Violence and Metaphysics: An Essay on the Thought of Emmanuel Lévinas" (1964, in *Writing and Difference*), Derrida embraced Lévinas's critique of Heidegger, even as he subjected it to a dismantling similar to the ones already mentioned.

Such critical intimacy—rather than the usual critical distance—is a mark of affirmative deconstruction.

In "The Ends of Man" (1968/72, in *Margins*), Derrida once again laid out his own project by distinguishing it from Heidegger's. That essay was perhaps Derrida's first articulation of the argument, continued in *Of Spirit* (1987) and beyond, that after the famous turn or *Kehre* of the thirties, Heidegger betrayed his insistence that, at the start of all investigative questioning was a prior question that could not be adequately answered. It is to be noted that Derrida's important essay is open-ended, signaling toward an indefinite future.

At the conference entitled "The Ends of Man" held in 1982 at Cerisy-la-Salle, Derrida described a movement in

his own work as well. It was a turn from "guarding the question"—insisting on the priority of an unanswerable question, the question of différance—to a "call to the wholly other"—that which must be differed-deferred so that we can posit ourselves, as it were. As was evident in the above discussion of radical alterity in "Différance," a similar double program figured in his work from the start. The movement now announced by Derrida, understood as an other-directed swerve away from mere philosophical correctness, alerts us to a greater emphasis on ethics and its relationship to the political.

An early text prefiguring the turn is "Declarations of Independence" (1976/82). Here Derrida, borrowing terminology from Austinian speech act theory, argues that the constitutional subject is produced by the performative of a declaration of independence, which must necessarily state itself as already given, in a constative statement of national identity. (For the important distinction between "performative" and "constative," see J. L. Austin, *How to Do Things with Words*. This text illuminates Derrida's many incursions into the question of philosophical nationalisms, and indeed his readings of all acts of institution.)

"Force of Law: The 'Mystical Foundation of Authority'" (1989) can be identified as the central statement of Derrida's ethical turn: from "guarding the question" to a "call to the wholly other [radical alterity]." If we consider *Given Time* (1991), *The Gift of Death* (1992), and *Aporias* (1993) with it, we will see some major ideas in play.

The earlier work—broadly grasped as the necessary yet impossible argument from différance—insisted that all institutions of origin concealed the splitting off from something other than the origin, in order for the origin to be instituted. This was a making indeterminate of any answer to questions of origin, as to what it was from which the supposedly original thing or thought, in description or definition, was being differentiated. It is this question, instituted at the origin, that had to be guarded or kept as a task in the first phase of deconstruction.

If Derrida's own ad hoc periodization is to be credited, the second phase is more "affirmative," a word he used in the mid-seventies. The affirmative call or appeal to the wholly other presumably addressed whatever may be prior to the trace of the other-than-origin instituting the origin, most often through the new concept-metaphor of "the experience of the impossible." If radical alterity was earlier conceived of as a methodologically necessary presupposition that is effaced in being named, now the category of presupposition is deliberately blurred and made more vulnerable as "experience."

Now such imponderables as justice and ethics can be seen as "experiences of the impossible": experiences of radical alterity. As such, they are undeconstructible, for to open them to deconstruction is to open them to the law of différance. Decisions based on such experiences involve aporias, or nonpassages. Aporias are distinguished from logical categories such as dilemmas or paradoxes, as experience is from presupposition. Aporias are known in the experience of being passed through, although they are nonpassages; they are thus disclosed in effacement, thus experience of the impossible. Formalization is achieved by passing through or "solving" aporias, treating them as practical logical problems. In the second phase of deconstruction, then, formalizations can therefore be seen as a halfway house toward the open end of a "setting to work." (The last theme had been broached in a 1983 text: "The Principle of Reason: The University in the Eyes of its Pupils," invoking not only a Heideggerian text of the same title, but also the older philosopher's famous Rectorate address of 1933.)

"Law is not justice, [although] it is just that there be law," says "Force of Law" (notice that the connective has to be supplied; Derrida philosophizes interactively—the reader provides connections in order to make the text work—because he uses the rhetorical dimension of language).

Justice cannot pass in a direct line to law; that line is a nonpassage, an aporia. Yet justice is disclosed in law, even as its own effacement. This is the peculiar nature of the deconstructive embrace. Ethics as "the experience of the impossible" and politics as the calculus of action are also in a deconstructive embrace. The space of being is the gift of time (so to speak): We fall into time, we begin to "be," unanticipatably. To call it a gift is to solve that aporia by thinking of some other (one) that "gives" time. Thus life is lived as the call of the wholly other, which must necessarily be answered (in its forgetting, of course, assuming there had been a gift in the first place in the subject's unanticipatable insertion into temporality) by a responsibility bound by accountable reason. Ethics as experience of the impossible—therefore incalculable—is lived as the possible calculus that covers the range between self-interest and responsibility that includes the politico-legal. Justice and law, ethics and politics, gift and responsibility are structureless structures because the first item of each pair is neither available nor unavailable. It is in view of justice and ethics as undeconstructible, as experiences of the impossible, that legal and political decisions must be made, empirically scrupulous but philosophically errant. (Even this opposition, of course, is not tenable to the last degree.) Here is a summary, made in view of the possibility that to summarize is to efface necessary discontinuities: The calculus of the second item in each pair such as the ones named above is imperative for responsible action, always in view of this peculiarity. These pairs are not interchangeable, but move on an unconcatenated chain of displacements. In each case, the "and" in the pair opens up the task entailed by what Derrida had formalized in "The Supplement of Copula: Philosophy Before Linguistics" (1971): that the copula "and" is a "supplement"—that slippery ("undecidable") word that he had first tracked in Rousseau—covering an indefinite variety of relationships,

since the supplement both supplies a lack and adds an excess. As "Principle of Reason" and "*Mochlos;* or, The Conflict of the Faculties" (1980/1984, in *Logomachia*) argue, if responsible action is fully formulated or justified within the system of the calculus, it cannot retain its accountability to the trace of the other. It must open itself to being judged by a setting to work that cannot be defined from within the system. One instantiation of this is to be found in the discussion of messianism in *Specters of Marx* (1993).

Is there a relation of reinscription between this "setting to work" and the rather carefully defined *"ins Werke setzen"*—setting or positing in the work—to be found not only in *Being and Time,* but most particularly elaborated in the later Heidegger's "Origin of the Work of Art" (1935/1950/1960), a piece discussed by Derrida in *The Truth in Painting* (1978)? In this brief compass, suffice it to say that whereas in Heidegger every conflict of "worlding" upon resistant ground is posited *in* the lineaments of the work of art as work, for Derrida what the word *work* marks is outside and discontinuous with the formulations of philosophy as an end in itself, with a logical systematicity that is mere calculus. This idea of a work outside or beside the disciplinary work of philosophy is laid out by Derrida, thematically as well as rhetorically, in parts of essays entitled exergue or parergon. In "Of Grammatology as Positive Science," Derrida had repeatedly stated that grammatology could not be a positive science because the philosopher could not or would not "venture up to the perilous necessity" of facing unanswerable questions at the origin. It is as if the mature philosopher now acknowledges the peril and steps out of "the shelter" he had invoked in the earlier text. At the origin now is the necessary experience of the impossible, which is lived as a calculus without guarantee.

Literature—more specifically poetry—remains a figure that provides an experience of the impossible, as suggested in the discussion of Paul Celan's work in *Schibboleth* (1986). Derrida's earlier discussions of Stéphane Mallarmé ("The Double Session," 1970, in *Disseminations*), Francis Ponge (*Signsponge,* 1975), and Maurice Blanchot (in *Parages,* 1986), circulate and cluster around this position. His intuitions about the visual arts are not inconsonant, but less assured. *Truth in Painting* (1978), focusing on "Peasant Boots" by Vincent van Gogh, the picture that Heidegger considers in the essay mentioned above, asks the simple question: what would be "idiom" (a subsystemic production of meaning) in art, a signifying logic that could lead to a calculable meaning system, which would then provide a springboard for any investigation of "truth" in painting?

It is a curious fact that many so-called ethnophilosophies (such as the Dao, Zen, Sunyavada, the philosophy of Nāgārjuna, varieties of Sufism, and the like) show affinities with parts of deconstruction. This may relate to their critique of the intending subject. Insofar as they transcendentalize extrasubjective authority, they are not quite "the same

thing" as deconstruction. But insofar as they locate agency in the radically other (commonly called "fatalism"), the exorbitancy of the sphere of work in the ethical as figured by Derrida has something like a relationship with them. Thus, although Derrida himself is at best cautious about resemblances between his own system and any "theologies" ("Onto-Theology of National-Humanism: Prolegomena to a Hypothesis," 1992), deconstruction after the turn, in its "setting-to-work" mode, may be of interest for many marginalized cultural systems as a development from within the aftermath of the Kantian Enlightenment, whereby their own calculuses, dominant in reaction, have become as compromised (especially gender compromised) and stagnant as anything perceived by Heidegger in the Kantian line itself. Of course, the possibility of these connections remains dubious as long as the "setting-to-work" mode remains caught within the descriptive and/or formalizing practices of the academic or disciplinary calculus. As long as the othering of deconstructive philosophy remains confined to discourses at least accessible to related academic disciplines (such as literature, architecture, theology, or feminism), it gives rise to restricted but useful debates.

Currently, the most critical and dynamic enclave of marginalized cultural systems is in counterglobalist or alternative-development activism (just as the financialization of the globe is the most robust vanguard of the Enlightenment). In this area, the "setting-to-work" mode of deconstruction breaks hesitantly into an active resistance to the inexorable calculus of globalization, where "democratization" is often a description of the political restructuring entailed by the transformation of state capitalisms and their colonies to tributary economies of rationalized financialization; or it may be engaged in displacing the binary opposition between economic growth and well-being by proposing alternatives to "development." These efforts do not, of course, produce a sustained formalized theory that is recognizably deconstructive. This is the risk of a deconstruction without reserve.

The aporia of exemplarity is most keenly felt here. The subjects and collectivities that produce the examples are in an aporetic bind with those who, far from and often ignorant of their field of work—globalization and development—yet produce the systematic formalizations. The situation may be described by way of the definition of irony (akin to our general sense of allegory) given by the deconstructionist literary critic Paul de Man: permanent parabasis or sustained interruption from a source relating "otherwise" (*allegorein* = speaking otherwise) to the continuous unfolding of the main system of meaning—both the formalization of deconstruction and, on another level of abstraction, the logic of global development. Further, if the splitting off of socialism from capitalism is perceived as grounded in the prior economy between self-preservation and the call of the other, this setting-to-work of deconstruc-

tion without reserve, quite unlike the failures of establishing an alternative system, may be described as a constant pushing away—a differing and a deferral—of the *capital*-ist harnessing of the *social* productivity of capital.

The structureless structure described above, where an item of a pair is both available and unavailable in an experience of the impossible, can be aesthetically figured in various ways. In the novel *Beloved* (1987), Toni Morrison places the "Africa" that is the prehistory of Afro-America or New World Africa—to be strictly distinguished from the named contemporary continent—in the undeconstructible experience of the impossible. As this call of the other is lived in the calculus of an Afro-America conscious of its rights, *Beloved* figures this disclosure, in effacement, as a maternal sacrifice, "not to be passed on." History requires it on the impossible passage, and does not stay the mother's hand. The central character kills her child to save it from the white world. The ring of the covenant—the brand on her own nameless slave-mother's breast—does not ensure continuity. Historiality is not changed into genealogy.

Two matters should be mentioned in conclusion. First, Derrida's own position, as a Franco-Maghrebian—being of Algerian-Jewish extraction he has described himself this way—tends toward global struggles in a general rather than a deconstructive way: in the call for an economically aware human rights vision, advanced in *Specters*. His more elaborated arguments are drawn from migrancy: the double responsibility of the New Europe (*The Other Heading*, 1991); a critique of "ontopology"—something like (multi)cultural identitarianism—"an axiomatics linking indissociably the ontological value of present-being [*on*] to its *situation*, to the stable and presentable determination of a locality, the *topos* of a territory, native soil, city, body in general" (*Specters*, p. 82); and figures of the absolute *arrivant* (the undeconstructible figure of alterity that is lived as any calculable diaspora). When he refers to his early years in Algeria ("Circumfessions," 1991, in *Jacques Derrida*), Derrida is not speaking of a country that has undergone a recent national liberation, and he is therefore not "postcolonial" in any precise sense. And second, the scholarship on Derrida's ethical turn and his relationship to Heidegger as well as on postcolonialism and deconstruction, when in the rare case it risks setting itself to work by breaking its frame, is still not identical with the setting to work of deconstruction altogether outside the formalizing calculus of the academic institution.

BIBLIOGRAPHY

Works by Derrida

Of Grammatology (1967). Translated by Gayatri Chakravorty Spivak. Baltimore, 1976. See "Of Grammatology as a Positive Science," pp. 74–93.
Speech and Phenomena and Other Essays on Husserl's Theory of Signs (1967). Translated and edited by David B. Allison. Evanston, Ill., 1973.

"Violence and Metaphysics." In *Writing and Difference* (1967), translated by Alan Bass, pp. 79–153. Chicago, 1978.
"Différance," "*Ousiu* and Grammè: Note on a Note from *Being and Time*," "The Ends of Man," "The Supplement of Coppula: Philosophy before Linguistics," "Signature Event Context." In *Margins of Philosophy* (1972), translated by Alan Bass, pp. 1–27, 29–67, 109–136, 175–205, 307–330. Chicago, 1982.
"Plato's Pharmacy" and "The Double Session." In *Dissemination* (1972), translated by Barbara Johnson, pp. 61–171 and 172–286. Chicago, 1981.
"Declarations of Independence" (1976/1982). Translated by Tom Keenan and Tom Pepper. *New Political Science* 15 (Summer 1986): 7–15.
The Truth in Painting (1978). Translated by Geoff Bennington and Ian McLeod. Chicago, 1987.
"*Mochlos*; or, The Conflict of the Faculties" (1980/1984). In *Logomachia: The Conflict of the Faculties*, edited by Richard Rand, pp. 1–34. Lincoln, Nebr., 1992.
"The Principle of Reason: The University in the Eyes of its Pupils." *Diacritics* 13.3 (Fall 1983): 3–21.
Signéponge/Signsponge (1983). Translated by Richard Rand. New York, 1984.
Parages. Paris, 1986.
"Shibboleth: For Paul Celan" (1986). Partial translation in *Acts of Literature*, edited by Derek Attridge, pp. 370–413. New York and London, 1992.
Of Spirit: Heidegger and the Question (1987). Translated by Geoffrey Bennington and Rachel Bowlby. Chicago, 1989.
"Force of Law: The 'Mystical Foundation of Authority'" (1989). In *Deconstruction and the Possibility of Justice*, edited by David Gray Carlson et al., pp. 3–67. New York and London, 1992.
Memoirs of the Blind: The Self-Portrait and Other Ruins (1990). Translated by Pascale-Anne Brault and Michael Naas. Chicago, 1993.
"Circumfessions" (1991). In *Jacques Derrida*, by Geoff Bennington and Jacques Derrida, pp. 3–315. Chicago, 1993.
Given Time: I, Counterfeit Money (1991). Translated by Peggy Kamuf. Chicago, 1992.
The Other Heading: Reflections on Today's Europe (1991). Translated by Pascale-Anne Brault and Michael B. Naas. Bloomington, Ind., 1992.
The Gift of Death (1992). Translated by David Wills. Chicago, 1995.
"Onto-Theology of a National-Humanism: Prolegomena to a Hypothesis." *Oxford Literary Review* 14.1–2 (1992).
Aporias. Translated by Thomas Dutoit. Stanford, Calif., 1993.
Specters of Marx: The State of the Debt, the Work of Mourning, and the New International (1993). Translated by Peggy Kamuf. New York and London, 1994.

Primary Texts

Austin, J. L. *How to Do Things with Words*. Cambridge, Mass., 1962.
Heidegger, Martin. "The Origin of the Work of Art" (1935). In *Basic Writings: From Being and Time, 1927, to the Task of Thinking, 1964*, rev. exp. ed., edited by David Farrell Krell, pp. 137–212. San Francisco, 1993.
Lévinas, Emmanuel. *Totality and Infinity: An Essay on Exteriority*. Translated by Alphonso Lingis. Dordrecht and Boston, 1991.
Morrison, Toni. *Beloved*. New York, 1987.

Other Sources

Balibar, Étienne. "Violence et politique—quelques questions." In *Le passage des frontières*, pp. 203–213. Paris, 1994.
Bernasconi, Robert, and David Wood, eds. *The Provocation of Levinas: Rethinking the Other*. London and New York, 1988.
Butler, Judith, and Joan W. Scott, eds. *Feminists Theorize the Political*. New York and London, 1992.

Cornell, Drucilla. *Beyond Accommodation: Ethical Feminism, Deconstruction, and the Law.* New York and London, 1991.

Gasché, Rodolphe. *The Tain of the Mirror: Derrida and the Philosophy of Reflection.* Cambridge, Mass., 1986.

Heidegger, Martin. *Kant and the Problem of Metaphysics* (1929). 4th enl. ed. Translated by Richard Taft. Bloomington, Ind., 1990.

Keenan, Thomas W. *Fables of Responsibility.* Stanford, Calif., 1997.

Lacoue-Labarthe, Philippe. *Heidegger, Art, and Politics: The Fiction of the Political.* Translated by Chris Turner. Oxford and Cambridge, Mass., 1990.

Laclau, Ernesto, and Chantal Mouffe. *Hegemony and Socialist Strategy: Towards a Radical Democratic Politics.* Translated by Winston Moore and Paul Cammack. London, 1985.

Nancy, Jean-Luc. *The Inoperative Community.* Edited by Peter Connor, translated by Peter Connor et al. Minneapolis, 1991.

Sallis, John, ed. *Deconstruction and Philosophy: The Texts of Jacques Derrida.* Chicago, 1987.

Spivak, Gayatri Chakravorty. "Responsibility." *boundary 2* 21.3 (Fall 1994): 19–64.

Spivak, Gayatri Chakravorty. "Supplementing Marxism." In *Whither Marxism? Global Crises in International Perspective,* edited by Bernd Magnus and Stephen Cullenberg, pp. 118–119, New York and London, 1995.

Weber, Samuel. *Institution and Interpretation.* Minneapolis, 1987.

Wood, David, ed. *Derrida: A Critical Reader.* Oxford and Cambridge, Mass., 1992.

GAYATRI CHAKRAVORTY SPIVAK

Derrida and Literature

The topic of literature has had without question a privileged place in the work of Jacques Derrida. This is not surprising given that his thinking took its point of departure from the attempted exclusion of writing whereby philosophy sought to ground itself in the truth of presence. In his earliest books and essays (for example, *Speech and Phenomena, Of Grammatology,* "Plato's Pharmacy," "Signature Event Context"), Derrida argues both that the banishment of writing from the domain of serious philosophical reflection has determined Western philosophy's most consistent project, giving it its identity across historical differences, and that the failure of this project has been no less necessary to the very possibility of philosophy.

In these essays, the term *writing* is exploded out of the conceptual limits within which it has been confined by philosophical tradition, from Plato and Aristotle to Edmund Husserl, Ferdinand de Saussure, and J. L. Austin. As in Aristotle's famous formulation, "Spoken words are the symbols of mental experience and written words are the symbols of spoken words" *(De Interpretatione),* the conventional concept of writing invariably links three features: Writing is secondary to speech, it is therefore at one further remove from the interiority or self-presence of "mental experience," and it is formed through imitation or representation of spoken sounds. Each of these features is predetermined by the supposed priority of an ideal self-presence, a pure identity without difference, and an absolute "interiority" uncontaminated by representation or symbolization in

the outside. It is this presumed priority, the hallmark of metaphysics, that is submitted to Derrida's deconstructive displacement whereby the secondary figure of writing comes to designate the originary possibility of all signification, whether spoken, written, or otherwise proffered in the world. The displacement turns on an analysis of *iterability,* that is, of the sign's essential constitution as a repeatable trace that can bear all sorts of intentions in all sorts of contexts. But iterability also exposes the trace necessarily and in an essential way to all these empirical differences, therefore to chance and to everything that in metaphysical discourse has been called accident (as distinct from essence). Derrida's analysis of iterability overturns metaphysical priorities, for it amounts to thinking an originary repetition the operation of which is not guaranteed by any prior ideal self-presence of meaning; on the contrary, ideality would itself be an effect of iterability, a construction after the fact, rather than the realm of pure identity of meaning to itself from which signifying practices, in all their impurity and exteriority, derive in a secondary moment. This analysis has crucial consequences all across the board, for it locates an irreducible instability in the very possibility of retrieving meaning from its essential drift through the accidents of repetition. For Derrida, however, there is no question of deploring or still less correcting this instability, because, as the condition of what he has called deconstruction, it is what allows instituted structures of meaning—which are always ultimately or initially political in that they are overlaid on differences of force—to be opened from within, to be transformed, to be historical.

Within this general iterability, "literature" would name not only a recent and historically determined institution but a space of inscription in the margins of both philosophical and historical "truth." Inscribed there is neither what *must* be the case (ontological essence or transcendental law) nor what *was* the case (empirical accident). And yet, in its practice of what Derrida has called "reference without referent" ("The Double Session"), the literary or poetic text is never simply separable or clearly distinct from the process of these truthful discourses. That is, literary or poetic inscription touches on the truths of history or philosophy but from a certain outside or margin, which is best figured not as a localizable space or discursive domain, but rather as a supplementary folding of the representational, referential potential itself: referential*ity,* which, like iterabil*ity,* names but the possibility of repetition. As a practice of reference or referring, literature unfolds in the absence of the anchoring referent on which both philosophical and historical truth processes depend. This is why a philosophy of "literature" (or poetry), as inaugurated by Plato's *Republic,* and a history of literature, as inaugurated perhaps by G. W. F. Hegel and the Romantics, have each, for different but related reasons, been unable to fix the "truth" or essence of literature. Instead, literature will always have designated a "space" (iter-

ability, referentiality) in excess of whatever truthful referent can be assigned to it. This does not amount to saying, as many critics of deconstructive literary thought have misconstrued, that literature is simply self-referential, a system closed on itself without relation to, for example, historical context. In fact, it is saying almost precisely the opposite, since the condition of reference without referent is one that must remain radically open to all contexts within which a text can come to be read. Such "openness" is even the condition of a text's readability beyond the time and place of its composition and is indeed but another way of figuring the referential power of a text. This power, therefore, is not to be accounted for by projections of an immutable form, as attempted by Kantian formal aesthetics and by the various schools of formalist literary criticism that it inspired, including the American New Critics. Once again, historicist criticism goes wide of the mark when it construes deconstructive thinking about literature as a new formalism or a new New Criticism. One could even say, in a certain narrowly historical sense, that the path of this thinking was initially opened by a deconstruction of the concept of literary form, for one of Derrida's earliest published essays (1963) and his first to engage in print with an explicit literary thematic is a long review of a book of literary criticism titled *Form and Signification*, a title Derrida appropriates after having made a pointed substitution: "Force and Signification." It is a question of force rather than form because the referentiality at play in a literary text ("play" being the closest term we have for an essence of what is called literature) must remain always a potential without ever settling into the outlines of a determined referent, which could then be "contained" by some form. This force is that of a fold that can never be finally unfolded or flattened out by an interpretation (deconstruction is also, therefore, not a new hermeneutics), because referentiality *itself* can never be just itself, a referent. It always engages a reference to the not-itself, to that which is not itself present, and as such, it "is" "itself" nothing that can be made present.

What we call literature, then, is not a thing, but a force of play (rather than a play of force: literature has no power with which to impose its laws) on the margins of ontological discourses and their institutions. This description dictates a significant difference between deconstructive thinking about literature (as well as the other arts) and traditional, post-Kantian aesthetics or "philosophy of art." To the extent that the latter is mobilized by the philosopher's classic question, "What is . . .?" it seeks to subsume the literary or the poetic to a concept. Preeminently, the concept is that of representable truth (Hegel's *Aesthetics* is the crowning example), which, because "representation" is already determined philosophically, serves less to distinguish the literary than to subordinate it to the enterprise of philosophy. The concern of deconstructive thought is rather to bring out the manifold ways in which the question "What is . . .?" gets dislocated, thrown out of joint when it takes as its "object" literature. For this reason one may say that this thinking is less *about* literature and more an attempt to think *with* literature, in the sense both of accompanying it and of taking its procedures as the "tools" of thought. Such a "thinking-with" has a precedent perhaps in Heidegger's "double-speech" *(Zweisprache)* between the thinker and the poet, although Derrida has also marked his distance from Heidegger's engagement with poetry, which he sees as still too tempted to gather everything together under the *logos* (see "*Geschlecht* II: Heidegger's Hand"). It would not even be going too far to say that the force of literature, that which one must try to think with, is what Derrida has always meant by deconstruction. This would explain, for example, the differences one may remark between that part of his writings devoted to explicitly literary works from the part engaged primarily with texts of the philosophical tradition (see *Glas*, which opposes on facing pages a reading of Hegel to a reading of Jean Genet). With the latter, Derrida always proceeds by a kind of sifting or triage, as he is intent to show above all how the philosophical text is constructed out of different and frequently contrary gestures that can cross each other up. It is this crossing pattern that allows the text to become unwoven when pressure is applied at certain points. Although this pressure can appear to be applied *against* the text, in fact, it is working only against a dominant logic that would suppress whatever differences cannot be made to fit under its rule. With the literary work, however, Derrida rarely if ever makes it a point to read against the apparent dynamics of the text, which does not mean at all that his approach to the work is merely that of an admirer who seeks to share his aesthetic pleasure. Instead, he approaches the literary text as something that, more insistently but also more passively than philosophical discourse, calls for response. Without denying that the latter may well include the response normally called "pleasure" (if one can presume against all odds that there could ever be a norm of pleasure), Derrida has shown that at stake in this response, in an altogether exemplary way, are the indispensable conditions of the ethical relation to otherness and others.

To understand how literature could assume this exemplary role in deconstructive thought, one must consider again the general theory of writing or the trace as it has been elaborated there. As the possibility of meaning, iterability does not depend on any specific intention to mean, which is why the same words or phrases can accommodate an infinite or at least indefinite number of different intentions. It is also, however, that which allows my intention to have meaning for another even when I am absent, for example, when someone reads what I have written years later, perhaps even after my death (a description could also apply to transcendental ideality, which is the reduction—or absence—of every empirical particular). Derrida has argued at length that such absence of the "author" is not an accident that

may or may not befall writing; it is not, in other words, on the order of that which may simply be tolerated by the act of meaning, but is rather a constitutive condition and therefore a requirement of any meaning act. Even when I am "present," my speech act or my address to another is conditioned by my possible absence. Every finite act of address thus anticipates the death of the addressor, who in this sense must already be "dead" in order to address anything at all. Writing, in the conventional meaning of the term, clearly illustrates this "already-dead" condition more patently than so-called living speech, which is why Derrida has always insisted on the testimonial nature of all written inscription.

Given this testimonial structure, to whom or, perhaps, to what does one respond in responding to a literary or poetic text? And why does this response engage a responsibility that, if taken seriously, would place the various acts of engagement with literature closer to an ethics than an aesthetics? Briefly put: The responsibility cannot be primarily or finally to *return* the text's meaning to its author by *reproducing* it. The circle or circulation of meaning, in which "understanding" is figured as such a return to the source, is necessarily interrupted from the very outset of the address. Or rather, there was never a circle in the first place that could be interrupted, since, when I write, it is only with my possible and therefore necessary absence that my intention to mean could begin to "circulate" meaningfully in the world. Derrida has therefore preferred to speak neither of signification nor of "circulation" (which remains the dominant figure in most communication theory), but of "dissemination" where the "dis-" has also a privative sense, which yields something like "not-semination." Not semination, not the sowing of the seeds whose product of sense is owed to the father of the book, but a dispersal that can never return: literature dis-seminates itself, and always at great risk of accident, contamination, even total loss.

Nevertheless, despite this dispersal, our relation to a text remains one of response and responsibility, if not of debt to be repaid. To whom or to what, then, can one be responsible if not to the work's "author"? This is essentially the question that every reading poses to itself as its *task*. And the answer will never be the same, which is why deconstructive thought does not propose a general theory of literature. This task (and, again, it is not at all incompatible with pleasure) is incumbent on whoever receives the text as destined *for* finitude, for that condition that is inscribed on every act of address as its constitutive possibility. The literary text is the exemplary text of finitude, which does not mean that it is by nature any more perishable than, for instance, the philosophical text, perhaps even on the contrary. What it does mean is that the absence necessarily inscribed there is a finite, *singular* one, and it is finally to this finite singularity that one responds and for which one takes responsibility. In this manner the relation to the literary text—but the same could be said of the relation to any singular work, which is

to say to all works of art as well as to all acts of address that bear, implicitly or explicitly, a signature—gives us to *think* and to *enact*, to think actively, the relation to the other, to the infinite otherness of the other's singularity, that which can signify itself, repeat itself, only in the mode of what is already absent because no longer singular and never present or presentable as such. The other's absolute singularity is the distinctly nonphilosophical, nongeneralizable, and therefore nontrue "truth" of literature.

Derrida's thinking *with* literature, then, is not a generalizable theory, but this is so for reasons that can be generalized. Instead, we could say that it offers itself as an example, one that is less to be followed than dispersed and dis-seminated among all the singular readers of these singular texts.

[*See also* Literature, *article on* Literary Aesthetics.]

BIBLIOGRAPHY

Works by Derrida

Speech and Phenomena and Other Essays on Husserl's Theory of Signs (1967). Edited and translated by David B. Allison. Evanston, Ill., 1973.

Of Grammatology (1967). Translated by Gayatri Chakravorty Spivak. Baltimore, 1976.

"Force and Signification." In *Writing and Difference* (1967), translated by Alan Bass, pp. 3–30, Chicago, 1978.

"Plato's Pharmacy." In *Dissemination* (1972), translated by Barbara Johnson, pp. 61–171. Chicago, 1981.

"The Double Session." In *Dissemination* (1972), translated by Barbara Johnson, pp. 173–286. Chicago, 1981.

"Signature Event Context." In *Margins of Philosophy* (1972), translated by Alan Bass, pp. 307–330. Chicago, 1982.

Glas (1974). Translated by John P. Leavey, Jr., and Richard Rand. Lincoln, Nebr., 1986.

"*Geschlecht* II: Heidegger's Hand." Translated by John P. Leavey, Jr. In *Deconstruction and Philosophy: The Texts of Jacques Derrida,* edited by John Sallis, pp. 161–196. Chicago, 1987.

PEGGY KAMUF

Derrida and Kant

Jacques Derrida's readings of Immanuel Kant are largely confined to two major texts—the chapter titled "The Parergon" in *The Truth in Painting* and the essay "Economimesis." Both treat the analytic of aesthetic judgment as set out in Kant's *Critique of Judgment*. Derrida's arguments may also be considered in a broader context. The philosophy of Kant in general and the problem of judgment in particular have been major concerns in French philosophy of the 1980s and 1990s.

Derrida's deconstruction of Kant's analytic of aesthetic judgment proposes three issues of primary concern: first, the autonomy of the aesthetic as an interiorized, subjective activity as opposed to a social and collective one; second, the value and self-identity of autonomous art as free of monetary value; and finally, the way in which Kant's conception of the ends and activities of art strategically obscure

the inability of Enlightenment philosophy to bridge or to re- solve distinctions between mind and nature, subject and ob- ject. Derrida examines these issues through three figures: the parergon, economimesis, and exemplorality.

In *The Truth in Painting*, the chapter on the parergon traces how the domain of aesthetic inquiry emerges in Kant's philosophy; that is, how the aesthetic attempts to de- fine itself, to mark off its borders, and to give itself distinct activities and ends. For Kant parerga include all of those things "attached" to the work of art yet that are not part of its intrinsic form or meaning: the frame of a painting, the colonnades of palaces, or drapery on statues. They are or- namental—an adjunct or supplement to the intrinsic beauty of the artwork. Parerga border the work but are not part of it, they resemble the work without being identical to it, and they belong to the work while being subsidiary to it. As such, the question of the parergon initiates a series of divi- sions that define for Kant the proper object of the pure judgment of taste through a process of increasing interior- ization.

The parergon involves, both literally and conceptually, the idea of the frame and therefore the ontological particularity of the aesthetic. But the paradox of Kant's analysis is that his solution to the specificity of aesthetic judgments *creates* the dilemma it was designed to resolve. The insistence on enframing—defining, on one hand, the self-identity of art and, on the other, the specificity of aesthetic judgments—is what in fact *produces* the divisions between object and sub- ject, inside and outside, mind and nature, that the third *Cri- tique* claims to resolve in completing Kant's transcendental system. While enclosing and protecting an interior, the frame also produces an outside with which it must communicate. If the third *Critique* is to complete its teleological movement, this externality must also be enframed—a process creating a new outside, a new necessity for enframement, and so on ad infinitum.

Derrida also emphasizes that the goal of the third *Critique* is to identify art as a middle term bridging an opposition between mind and nature, internal and external phenom- ena, the inside and the outside, and so on. Kant poses two separate worlds that are absolutely divided through the following concepts: object/subject, nature/mind, external/ internal, outside/inside, sensible/supersensible, natural con- cept or understanding/concept of freedom or reason. This divide, however, is not to be bridged by pure reason, that is, determinate concepts, since this would render aesthetic and scientific judgments as equivalent. Rather, the aesthetic judgment must bridge these discontinuous worlds through, on one hand, the discursive structure of the example and, on the other, a logic of semblance without identity originat- ing in analogies referring to the model of speech.

Judgment must also be universal. This appeal to univer- sality informs Kant's famous emphasis on the disinterested- ness of aesthetic judgments that is defined by freedom and by a noncognitive pleasure: the *Wohlgefallen* proper to the object of pure taste. Freedom, as the realm of the concept of the supersensible, is especially important in Kant's attempt to unify his philosophical system. This indicates, first of all, that aesthetic judgments are "free" from all contingent demands or extrinsic motives, especially economic ones. There must be an absolute lack of interest in the object's ex- istence, otherwise the critic cannot operate with perfect freedom. If the critic invests in the object, as it were, his or her judgment cannot transcend its subjective origins and pretend to universal communicability.

Nevertheless, in the analytic of judgment the question of value is paramount. Derrida expresses this idea in coining the term *economimesis*. This is not simply a question of con- joining the aesthetic (mimesis as a process of imitation in relation to nature) and the economic to demonstrate the al- legiance of art to ideology as well as its reliance on capital. Rather, value is produced both in the self-identity of art as formulated by the logic of the parergon, and in the hierar- chies that adjudicate the relative value of different media and practices in relation to an identification of freedom and reason with speech.

The concept of mimesis has a curious role, then, in the third *Critique*. It is governed not by a logic of semblance or imitation but rather by a logic of analogy. Kant's theory of mimesis asserts the superiority of beauty in nature and de- rives the beautiful in art from its relation to nature. This is not a logic of the copy, but rather a rhetoric of production and reproduction. In finding common ground between art, nature, and genius, mimesis requires a logic of equivalent activities, not one of mirrors. Thus free and mercenary art, artist and craftsperson, are distinguished as the difference between a reproductive imagination and a productive imag- ination that is originary, spontaneous, and playful. In Der- rida's view, the value of play in Kant defines a form of pro- ductivity that is purer, freer, and more human, as opposed to labor, which is ends-directed, unpleasant, and exchanged against a salary. Reproductive imagination is therefore a vulgar realism—reproduction in the form of likeness. In contrast, productive imagination—regardless of whether it applies to acts of creating or of judging aesthetic objects—is characterized by a paradoxical freedom that is the imagina- tion's "free conformity to law."

Resemblance limits the freedom of the imagination, if for no other reason than it may function as an aim, purpose, or end. And the more semblance between sign and referent, the more extreme are these limitations. On the other hand, without an underlying "lawfulness" there would be no ground for uniting understanding, moral judgments, and judgments of taste, and no language with which to commu- nicate them. By a process of analogy this sense of lawfulness without a law and purposiveness without purpose, whose original territory is that of nature, informs and "naturalizes" every reference in the third *Critique* to representation, signi-

fication, or language. Nature limits what is most "wildly free" in the production of art through its (silent and nonconceptual) "dictation" of rules to genius. In like manner, the forms of analogy derived from mimesis efface all the oppositions—mind/nature, subject/object, supersensible/sensible—that divide Kant's philosophical system. This process of naturalization also obscures the role of economy in artistic production. The aesthetic implies an ideal freedom for humankind, by analogy like nature yet entirely separate from it, conceived as a "nonexchangeable productivity." Through mimesis, then, art does not imitate nature in the sense of reproducing its visible signs. Art does not reproduce nature; it must *produce like* nature, that is, in perfect freedom.

The modality of aesthetic judgment is tied to the forms of speech. Derrida writes: "We are familiar with the example: I stand before a palace. Someone asks me whether I think it is beautiful, or rather whether I can say 'this is beautiful.' It is a question of judgment, of a judgment of universal validity, and everything must therefore be in the form of statements, questions, and answers. Even though the aesthetic affect is irreducible, judgment demands that I say 'this is beautiful' or 'this is not beautiful'" ("The Parergon," p. 11). Judgment formulates itself as statements, questions, and answers. It is a kind of dialogue, but of what sort? Across a series of divisions—between interlocutors engaged in "aesthetic" conversation, between the subject (spectator) and the object (palace), and within the philosophical subject internally divided in its faculties—a filigree of words is woven. Within the space of the statement universal communication must occur freely between spatially detached and isolated parts. Everything eventually returns to the power of *logos* to breathe life into judgment and harmonize the faculties.

At this point Derrida reemphasizes how a divine teleology, in fact, a process of ontotheological naturalization, underwrites the logic of economimesis, securing the identification of human action with divine action. This is already apparent in the hierarchies of value and identity established by the opposition of free to mercenary art. The artist produces in his or her activity a divine subjectivity. Economimesis secures the figure of Genius as the exemplar of a divine agency in art where the artist creates—without concepts as a pure and free productivity of the imagination—in a fashion analogous to the way God produces divine works in nature. For Kant, fine art is the art of genius and genius is a gift of nature, an endowment of its productive freedom. And what nature gives to genius, genius gives to art in the form of "nonconceptual rules." In so doing, genius "capitalizes freedom but in the same gesture naturalizes the whole of *economimesis*" ("Economimesis," p. 10). At the origin of all analogy, then, is the word of God; in the third *Critique* everything returns to logos as origin. And for this reason, Derrida argues that the "origin of analogy, that from which analogy proceeds and towards which it returns, is the logos, reason and

word, the source as mouth and as an outlet [*embouchure*]" ("Economimesis," p. 13). This is a discourse of/on the aesthetic that is governed ultimately by the logic of logocentrism.

The same divine teleology that ranks and orders artistic labor and subjectivity also organizes a hierarchy within the fine arts. In this manner, Kant's privileging of oral examples, the "exemplorality," of the *Critique of Judgment,* underwrites the crucial function of mimesis in Kant's attempt to resolve the dilemmas of subject and object formulated in his philosophical system. Kant's portrayal of "pure" judgments of taste relies on a rejection of empirical sensation and a withdrawal of the physical body. Curious, then, how the centrality of the mouth figures in the *Critique of Judgment.* Above all in the section "On the Division of the Fine Arts," it organizes a hierarchy among the arts, and in the terms of aesthetic value (taste or disgust), by defining them with respect to the expressive organization of the human body. For Derrida, the figured circle of the mouth, and the circularity of immaculate commerce in spoken communication, organizes a parergonal logic of the subject in Kant. Just as the "frame" of painting had both to protect the intrinsic purity of art and open up commerce with the outside, the mouth establishes a privileged border between the interiority of the subject and an outside that be must represented and communicated to others, whose purest form of expression is speech.

Therefore, in the third *Critique,* poetry is the highest art because "it is the art which imitates the least, and which therefore resembles most closely divine productivity. It produces more by liberating the imagination; it is more playful because the forms of external sensible nature no longer serve to limit it" ("Economimesis," p. 17). By the same token, poetic genius is the highest form of aesthetic subjectivity because, in its analogous relation to the divine *logos,* it is the freest and confers the most liberty on the imagination of individuals. In Kant's view, by freeing us from the limits of external, sensual nature poetry binds linguistic presentation to the fullness of thought, rendering the presence of ideas to thought, in a way that no other art can.

Thus Derrida rightly insists that Kant derives a theory of value from the arbitrariness of the vocal signifier, that is, from its difference with respect to external sensible nature. The difference, immateriality, and interiority of the vocal signifier aligns it with the realm of freedom. The nature of this freedom is marked in every case by a profound interiorization, a retreat from the external signs of nature into a purely subjective autonomy whose measure is the autoaffective structure of logocentrism. The circle of orality passes again through three otherwise autonomous realms: those of nature (God), art (poetry), and philosophy (judgment). The nonconceptual pleasure inherent in judgments of pure taste is thus associated with the play of freedom as "a lawfulness without a law, and a subjective harmony of the

imagination without an objective harmony" (Kant, *Critique of Judgment*, p. 92) in a movement of idealizing interiorization. Everything recedes—from the extrinsic, the empirical, and the corporeal—into the subjective, the internal, and the spiritual.

For this reason, Derrida states that the *Wohlgefallen*, the pleasure-affect proper to art in the Kantian sense, takes the form of an autoaffection, an interiorized and self-authenticating dialogue. In *Of Grammatology*, the logocentric circle of autoaffection is critiqued as a self-producing and self-authenticating movement that identifies reason and fullness of being with the temporality of speech. Thinking, at least the pure thought of philosophy, is represented as hearing-oneself-speak, a formula Derrida reprises in relation to Kant's *Critique of Judgment*. The comparative authenticity and veracity of poetic speech—its capacity for mimesis without semblance, the indissociable relation between the mouth and the ear, the irreplaceability of hearing, the association of speech with interiority, with concepts, and with internal sense—demonstrate that the position of logos in Kant's system is not one analogy among others. The linguistic signifier is that which "regulates all analogy," writes Derrida, "and which itself is not analogical, since it forms the ground of analogy, the *logos* of analogy towards which everything flows back but which itself remains without system, outside of the system that it orients as its ends and its origin, its embouchure and its source" ("Economimesis," p. 19).

Just as there could not be beauty in nature if there were not, by analogy, a poetry of nature, a discourse could not emerge from the beautiful if the beautiful were not itself discursive. This is why the orality of poetry has the purest affinity with that of aesthetic judgment. Not only because they are the most purely internal and autoaffecting, but because art and judgment share the same frame, that is, the circle of the mouth. Judgment must speak or state the beautiful, even if the beautiful eludes it conceptually, in order to supplement beauty's nonconceptual lack and return it to the space of philosophy. The autoaffective circle that produces the judgment of pure taste informs Kant's understanding of how God figures his order in nature, how the gift of "natural" creativity is "dictated" to genius, how genius bestows the gift of form on poetic language, and in turn how a judgment of pure taste is engendered by contemplation of the beautiful forms of poetry or of nature. As parerga, there is an essential relation here between the circle of the mouth in relation to exemplorality, on one hand, and the frame and the signature, on the other. Just as the inscription of the signature assures an external authorizing presence within the purportedly pure aesthetic interiority delimited by the frame, the figure of the mouth and the circularity between speech and hearing assure a passageway between mind and nature, the inside and the outside, subject and object, where heteroaffection and autoaf-

fection fly into and out of one another gliding on the wings of speech.

The parergon is therefore a metaphysical logic of "controlled indeterminacy" or of a ceaseless vibration between inside and outside, the intrinsic and extrinsic, subject and object, the reflective and the determinant, the singular and the universal, the conceptual and the nonconceptual, mind and nature. In short, the ontological question of "what is," a question that is meant to define the integral being of art and of aesthetic subjectivity, paradoxically appeals to, and is infected with, the outside in the very asking of the question. This indeterminacy is, in fact, the ontological uncertainty of the very idea of the aesthetic. The frame is summoned to give an ontological presence and shape to a space that otherwise threatens to dissolve in aporia; the circle is there to give form to what is otherwise an absent center, and to provide a concept for an otherwise conceptless blank space. This is another way of saying that the aesthetic is an imaginary concept, but in the psychoanalytic rather than Kantian manner. Feeding a regressive fantasy of presence and autonomy, it detaches the work from the field of history by resolutely occluding any social meaning, including the economic and the political, in the figure of econo-mimesis.

While representing the drive for enframing and enclosure that informs the ontological imperative of the aesthetic, the parergon simultaneously presents its empty center, in fact, the absence of a center as ontological lack. In this manner, Derrida's genealogical critique demonstrates the breadth and complexity of what must be deconstructed in the idea of the aesthetic. What is most important is understanding how philosophy has produced the problem of the self-definition of the arts, as well as the autonomy of fine art and of aesthetic meaning, as a response to the very indeterminacy or undecidability of all ontological questions.

[*See also* Kant.]

BIBLIOGRAPHY

Works by Derrida

De la grammatologie. Paris, 1967. Translated by Gayatri Chakravorty Spivak as *Of Grammatology* (Baltimore, 1976).

"Economimesis." In *Mimesis des articulations*, edited by Sylviane Agacinski et al., pp. 57–93. Paris, 1975. Translated by R. Klein in *Diacritics* 11 (1981): 3–25.

La vérité en peinture. Paris, 1978. Translated by Geoff Bennington and Ian McLeod as *The Truth in Painting* (Chicago, 1987).

"The Parergon." Translated by Craig Owens. *October* 9 (Summer 1979): 3–40.

Derrida, Jacques, et al. *La faculté de juger*. Paris, 1985.

Other Sources

Caygill, Howard. *Art of Judgement*. Oxford and Cambridge, Mass., 1989.

Eagleton, Terry. *The Ideology of the Aesthetic*. Oxford and Cambridge, Mass., 1990.

Kant, Immanuel. *Kritik der Urteilskraft* (1790). Frankfurt am Main, 1974. Translated by Werner S. Pluhar as *Critique of Judgment* (Indianapolis, 1987).

Rodowick, D. N. "Impure Mimesis; or, The Ends of the Aesthetic." In *Deconstruction and the Visual Arts: Art, Media, Architecture*, edited by Peter Brunette and David Wills, pp. 96–117. Cambridge and New York, 1993.

D. N. RODOWICK

DESIGN. Design as a concept necessarily occupies an ambiguous position in the analysis of the aesthetic, and exemplifies many of the current debates about postmodernism and valuation in philosophical aesthetics. Its definition alone is the subject of controversy, but may be characterized by some combination of the following:

1. the possibility of a separation of the maker of an artifact from someone who is responsible for its "blueprint";
2. the location of decisions about what is to be produced in the hands of the person who commissions the artifact, usually on the basis of a brief;
3. the possibility of multiple "runs" of the object for which the designer constructs a model;
4. a tight relationship between this modeling and the economic function of the object in question.

According to such a definition, the ontology of the designed object is determined in a process where the role of "creator" is fragmented by the production process, where the latter is inseparable from economic calculations and where the artifact's communicative capacity is premised on some feature or features of the potential user. A definition based on these elements implies that design is to be seen as a segment of the division of labor in society. Moreover, the details of this definition imply that the design process is specific to societies with a high degree of the division of labor, in other words, modern societies. Alternative approaches to the definition of design are based on the distinction between those elements of an object that derive from its function, and those elements that may be pleasing to somebody independently of the function in question. Certainly, objects produced in societies with a low degree of the division of labor, and which are therefore not amenable to definition as "designed objects" in the first sense outlined above, may be characterized by the latter distinction (Gombrich, 1984). By the same token, objects that may now, in our society, be considered "designed objects" may or may not have been considered such in the past, since this judgment is part of the wider classification system to be found in different societies.

Whichever of these definitions is used, the designed artifact must be considered inseparable from its place in a social network. The place of objects within the constitution of social networks has been the subject of divergent analyses. It may be seen as the result of desire in the form of consumer demand and thus subject to economic and historical explanations. It may be analyzed in terms of the satisfaction of universal needs, whether ethical and material or aesthetic. In the school of thought associated with Jean Baudrillard, it is seen as a sui generis process in which human-object relations are fundamentally changed, introducing a new phase of human history.

In short, the ontology of the designed artifact is subject to a logic of localization: Here, the significance of objects—in other words, their meanings for users, the forms of pleasure they give, and the criteria of appreciation to which they are subjected—is established in some context of production, distribution, and use by some social group; these contexts are often summarized in the term *function*. As a result, designed artifacts are difficult to judge according to the criteria of any universalizing aesthetics; noteworthy attempts to do so are to be found in the analysis of architecture and in programmatic attempts to reconcile traditional forms of aesthetic judgment with the needs of modern industry. Correlatively, the relationship of designed artifacts to the realm of the aesthetic is more readily compatible with theories in which localization is an integral dimension. By the same token, any aesthetics rooted in the empirical data of the design process is likely to present problems of conceptual compatibility with an aesthetics grounded in the existence of a canon of "great works" or "masterpieces."

Clearly, many if not all objects traditionally classified as "art" objects in fact share some of the characteristics here attributed to the designed object. Patronage has been commonplace in the history of art forms, and its economic and political mechanisms are usually held to influence the production of art objects. Many art forms have involved the distinction of different roles: architect and builder, composer or author and performer, for instance. All artistic production involves some economic calculations, since all production involves the allocation of finite resources such as time and raw materials. Much art is difficult to separate from its social function—in Europe the relationship between music, ceremony, and dance, for example, is widely held to have been very close until the Romantic era. To this extent, the distinction between art and design is eroded by historical analysis. However, it is often considered that in objects currently classified as "art" these aspects of their existence are relatively less important than the dimension that is relevant to aesthetic appreciation, whereas in design this disintrication is commonly felt to be more difficult.

Aesthetic theories that reject the relevance of such localization claim competence to handle designed artifacts by operating a rigorous separation between the dimensions of the object that derive from the realm of the aesthetic and the dimensions that derive from its social, economic, or other function, often under the headings of "beauty" and "utility," respectively. Since it is the social and economic func-

tions of objects that are responsible for their localization, such an aesthetic is able to make claims to universality only insofar as it denies the relevance of function. This process is given an institutional basis in galleries and museums, where artifacts are commonly separated in the act of classification as either "art" or "design." The tension between these divergent appreciations of artifacts may be seen in the development of the museum. Here, artifacts are necessarily removed from their original context and offered to public scrutiny in ways that may implicitly emphasize their formal value and deemphasize their functionality; or that may—as in the case of some design museums and anthropological collections—attempt to reinstitute functionality through an explanation of their original context.

Thus, the distinction of the aesthetic realm involves an insistence on the superordinate pertinence of a single dimension of artifacts. In contradistinction, in theories that start from the design elements of artifacts, the opposite emphasis obtains: here the element of aesthetic appreciation and evaluation is grounded in an account of the social meanings attached to objects.

More generally, such accounts point to the centrality of the separation of art and design in a historical process in which the separation is a constitutive element of the ontology of both categories of artifact. This separation can be examined through a study of the etymologies of the two terms.

Art originally meant a "skill," and in pre-Renaissance European thought the term was commonly applied to an area of activity for which the usual modern label would be "manual" or "mechanical" crafts; an exception occurs when the qualifier "liberal" is used, indicating an art characteristically practiced by free men or citizens. Thus the "arts" at this time were divided into the "liberal arts," which did not produce money, and the "mechanical arts," which did, a division that corresponds approximately to the subsequent distinction between "intellectual" and "manual" labor. In the Renaissance the term *art* began to acquire some of its modern range of meanings, namely, the product of a highly specialized and high-status intellectual activity, and also that activity itself. At the same time the activity referred to by the term *design* began also to acquire a high status: the artist's design or *disegno* (drawing, plan, or blueprint) was the conceptual or spiritual basis for the final material artifact. The separation of the designer from the actual fabricator of the product was not initially, therefore, a separation that led to the distinction between universalizing aesthetic activity and a localizing manufacturing activity. To the contrary, the ability of the artist to grasp the essential conceptual framework needed to carry out a beautiful work of art through *disegno* was readily comprehensible by reference to a syncretic philosophical tradition—derived ultimately from Greek antiquity—in which the

realm of the ideal is reached through the realization of conceptual activity. The change in the meaning of "design" to one that emphasizes the local and the particular demands of a specific set of consumers without any strong link to universalizing features took place over several centuries, from the Renaissance to the mid-twentieth century. Two crucial conditions for this change of meaning were the increasingly rapid development of industrialized manufacturing and the concomitant emergence of a consumer (or "wasteful") society. In a society typified by these two forms of activity economic growth is central, and it is widely held that the phenomenon of fashion is intrinsic to this process: insofar as changes in public taste can be subjected to a predictable rhythm, fashion acts as an incentive to a rapid turnover of consumer items.

This use of the term *design* has only recently received strong support from postmodern theory and practice. Until the "postmodern moment," the dominant trend in art history and art theory in this respect had been to postulate variations on the notion of "good design" as an activity based on principles of organizing matter and thought that—while they might certainly be articulated differently in different cultural contexts—were nonetheless universally valid. This attempt to promote design, even in manufacturing, as an activity of this nature, can be traced through a variety of genealogies. Noteworthy landmarks in this terrain are: William Hogarth (1697–1764) and Sir Joshua Reynolds (1723–1792); John Ruskin (1819–1900) and Augustus Welby Pugin (1812–1852); the Arts and Crafts movement (c.1885–1914) (Naylor, 1971) and the Rational Dress movement (c.1850–1900); de Stijl (1917–1932) (Banham, 1962), the Bauhaus, and Constructivism (c.1913–1933).

Reynolds and Hogarth attempted to find universalizing principles of abstract design that could be used in both applied and fine art. Their project was always in opposition to the local and ephemeral varieties of fashion during the very period when the modern phenomenon of fashion was itself emerging. Ruskin and Pugin argued that art, craft, and design were central to the good society and pointed to the Middle Ages as the prime example of such integration. The Arts and Crafts and Rational Dress movements attempted a fusion of universal aesthetic principles with a similarly universal conception of rational utility. For the practitioners and the theoreticians of the Bauhaus and the Constructivist movement, modern industry was an extension of the human species' rational control over its environment. At the same time, they believed in a language of forms that could be discovered or articulated through industrial production. Thus, by combining aesthetic judgment and the practices of industrial production, they believed that it would be possible to satisfy simultaneously both the ethical and the aesthetic demands of the good society. It can be seen that all these attempts to define "good design" involved the devel-

opment of norms that were implicitly if not explicitly grounded in principles of universal validity.

This tradition of thought has been influential in several key areas. The state and corporate interest groups have founded or promoted design museums and schools of architecture and design, as well as organizing national trade fairs and exhibitions, some of which have been closely associated with design movements, for example, the Great Exhibition in London of 1851 or the Paris "Arts Décoratifs" exposition of 1925. Influential practitioners and arbiters of taste have accepted this tradition of thought. From Josiah Wedgwood onward, entrepreneurs who not only manufactured and distributed but also designed consumer products both incorporated ideas from this tradition in their work as well as entered into the public debates on the nature of "good design." These debates were further pursued and elaborated in the critical and academic literature that was devoted to the history and nature of the designed object and the design process during the same period of European industrialization.

The public interest in these arguments has been premised on their potential as a strategic tool in production decisions. Discussions of what constituted "good design" had an impact on design practice. They were based on principles whose universality was established in neoclassical debates: truth and beauty defined in terms of rationality and permanence. These subsequently became slogans in the modern period such as form follows function and truth to materials, slogans through which generations of designers have been encouraged to appraise their own and each others' work in terms that transcend the economic function of the objects in question, even if such judgments have rarely been able to sustain a protracted rejection of economic considerations. Since the principles were capable of multiple applications, discussions based on them could accommodate considerable stylistic shifts—for example, the move from Art Nouveau to the Bauhaus.

Within twentieth-century debates on these topics, the relationship between "good design" and fashion has been particularly problematic. Here fashion has stood for the local and the ephemeral at their most intense, whereas "good design" has been seen as the incarnation of neoclassical principles and their modern derivatives. The problematic relationship has resulted from two fundamental tensions in the theory and practice of design aesthetics. On the one hand, the normative judgment of aesthetic success was always in fact a temporary judgment because of the desire for innovation under the aegis of progress. On the other hand, industrial practice in every branch of industry has resulted in the thorough imbrication of design and fashion, since the economic role of the designer includes, centrally, aesthetic incitements to increased turnover. This tendency is especially pronounced in what economists term "mature markets" where the ready availability of relevant raw materials, production techniques, and labor skills has reduced the price of the commodities in question to a point where they are no longer rare luxuries but staple items in the average consumer budget.

As a result, the defense of the hypostasized "universal design aesthetic" has been obliged to make concessions to arguments based on the observation of historical variations. The most sophisticated recent articulation of the defense against localization has developed a conception of an innate patterning capacity of the human species that finds its expression in varied cultural forms all of which have the same fundamental root. At the same time, the postmodern movement in design practice has rejected the norms of "good design" as articulated in earlier debates: slogans such as form follows function, and "truth to materials," which were taken as fundamental axioms of good practice, have been rejected in favor of playful eclecticism and colorful ornamentation for its own sake. In combination with a deliberate erosion of the traditional distinction between canonical and popular art forms, this has produced a series of styles in relation to which critics have found it difficult to establish or apply universalizing norms. In such a context, the definition of "fashion" as something distinct from "art" or "good design" on the basis of its localization and ephemerality has become a tentative enterprise.

The debates that have been inspired by recent shifts in design practice turn on the same central problem that has been a focal point in much recent argument in philosophical aesthetics, namely, the extent to which regimes of valuation are entirely local and the extent to which they may transcend such limitations. An early resolution to these arguments is not expected.

[*See also* Bauhaus; Constructivism; *and* Morris.]

BIBLIOGRAPHY

Appadurai, Arjun, ed. *The Social Life of Things: Commodities in Cultural Perspective.* Cambridge and New York, 1986.

Banham, Reyner. *Theory and Design in the First Machine Age.* London, 1960.

Baudrillard, Jean. *Symbolic Exchange and Death.* Translated by Iain Hamilton Grant. London, 1993.

Bourdieu, Pierre. *Distinction: A Social Critique of the Judgement of Taste.* Translated by Richard Nice. Cambridge, Mass., 1984.

Campbell, Colin. *The Romantic Ethic and the Spirit of Modern Consumerism.* Oxford and New York, 1987.

Clifford, James. *The Predicament of Culture: Twentieth-Century Ethnography, Literature, and Art.* Cambridge, Mass., 1988.

Curtius, Ernst Robert. *European Literature and the Latin Middle Ages.* Translated by Willard R. Trask. Reprint, London, 1979.

Douglas, Mary, and Baron Isherwood. *The World of Goods.* New York, 1979.

Doyal, Len, and Ian Gough. *A Theory of Human Need.* London, 1991.

Forty, Adrian. *Objects of Desire: Design and Society, 1750–1980.* London, 1986.

Gombrich, E. H. *The Sense of Order: A Study in the Psychology of Decorative Art.* 2d ed. Oxford, 1984.

Lodder, Christina. *Russian Constructivism: From Fine Art into Design, 1913–1933.* New Haven, 1983.

Mckendrick, Neil, John Brewer, and J. H. Plumb. *The Birth of a Consumer Society: The Commercialisation of Eighteenth-Century England.* London, 1982.

Naylor, Gillian. *The Arts and Crafts Movement: A Study of Its Sources, Ideals, and Influence on Design Theory.* London, 1971.

Newton, Stella Mary. *Health, Art and Reason: Dress Reformers of the Nineteenth Century.* London, 1974.

Palmer, Jerry, and Mo Dodson, eds. *Design and Aesthetics.* London and New York, 1995.

Pevsner, Nikolaus. *Pioneers of Modern Design: From William Morris to Walter Gropius.* London and New York, 1991.

Scruton, Roger. *The Aesthetics of Architecture.* London, 1979.

Smith, Barbara Herrnstein. *Contingencies of Value: Alternative Perspectives for Critical Theory.* Cambridge, Mass., 1988.

Whitford, Frank. *Bauhaus.* London, 1984.

Williams, Raymond. *Culture and Society, 1780–1950.* London, 1958.

Williams, Raymond. *Keywords.* Rev. exp. ed. London, 1983.

Wollheim, Richard. *Art and Its Objects.* 2d ed. with six supplementary essays. Cambridge and New York, 1980.

JERRY PALMER and MO DODSON

DEWEY, JOHN. [*To treat the thought of the twentieth-century American philosopher John Dewey, this entry consists of three essays:*

Survey of Thought
Theory of Expression
Experience and the Organic Unity of Artworks

The first essay is an overview of Dewey's pragmatism, which is followed by essays on three major concepts in his aesthetic theory: expression, experience, and organic unity. For related discussion, see Education, Aesthetic.]

Survey of Thought

John Dewey (1859–1952) was an American philosopher, social critic, and cofounder, with Charles Sanders Peirce and William James, of the pragmatist tradition in philosophy. Over a career spanning seven decades Dewey made major contributions to a wide range of discussions within and without professional philosophy, ranging from metaphysics, logic, and scientific method to social psychology, politics, education, ethics, and aesthetics, as represented in his influential book *Art as Experience* (1934). After an overview of his philosophical contributions we will consider some principal themes of his aesthetic theory.

Overview of Dewey's Philosophy. Dewey's philosophical career spanned seven decades and appointments at the University of Michigan (1884–1894), the University of Chicago (1894–1904), and Columbia University, from which he retired in 1930. The lectures published as *Art as Experience* were delivered at Harvard University in 1931. By that time Dewey had developed the mature philosophical outlook, popularly known as Deweyan pragmatism, which

he variously referred to as the "experimental philosophy" and "instrumentalism." This outlook is broadly characterizable as empirical (though not empiricist in the traditional sense), emergent naturalist, historicist, and reconstructive. The last of these terms reflects Dewey's conviction that philosophy is, and should view itself, not as a purely intellectual inquiry but as a contributor to the self-critique and institutional transformation of its larger society. For Dewey this meant specifically democratic society, which he viewed as the most promising modern setting for the development of institutions that could foster the developmental potential of human experience. This pragmatic vision of philosophy's function is mirrored in Dewey's account of science, which viewed the truth (or warranted assertability) of a statement or hypothesis not in terms of an idealized correspondence between thought and the world but rather in terms of whether the acceptance of that statement or hypothesis ultimately leads to experiences that promote the practical aims of the inquirer or inquiring community.

Intelligence. Many elements of Dewey's mature pragmatism are evident in his early work, which developed rapidly toward a synthesis of two dominant influences. One was Hegelian metaphysical idealism, which emphasized the developmental and socially imbedded character of all thought and experience. The other was Darwinian evolutionary theory, which offered a naturalistic corrective to the idealizing extremes of Hegelianism and a framework congenial to scientific articulation of the developmental theme. With this background, Dewey gradually came to view the relationship between intelligence (a term he preferred to *mind*) and nature as broadly dynamic and holistic in character. Nature is not simply a blind mechanism, ordered along austerely mathematizable lines, onto which a detached mind projects subjective ideas of meaning and purpose. Rather, it is the other way around: before we understand nature in mechanistic and causal terms we must first experience it as a whole replete with an array of objectively real qualities including not only perceptual properties but also values. This pattern of realist argument decisively rejects a distinction central to the major tradition in modern philosophy that runs from classical empiricism and Kantianism through the twentieth-century movement of logical positivism: the distinction between subjective and objective (or mental and physical) orders of existence. In a similar spirit Dewey rejected what he called the "spectator" theory of mind favored by the empiricist/Kantian tradition in favor of a vision of intelligence as a highly developed participant in the causal order of nature. Intelligence, he maintained, is in all of its individual and social forms experimental and creative; in addition, it is disposed to employ symbolic and material resources to bring about increasing organization or unity in its environment. Given Dewey's continual attention to science throughout his writings, many readers have supposed that he saw science to epitomize this last role among

modern cultural practices. But his scientist leanings were always in the end held in check by a deeper romantic tendency to view art as epitomizing the activity of intelligence in the modern world. Dewey accordingly remarked in *Experience and Nature* (and repeated again in *Art as Experience*) that "Art, the mode of activity that is charged with meanings capable of immediately enjoyed possession, is the complete culmination of nature, and . . . science is properly a handmaiden that conducts events to this happy issue."

Nowhere is Dewey's departure from the Kantian/empiricist tradition more apparent and dramatic than in his critical account of what modern philosophy has come to know as "the mind." We are best off, he insists, dropping the idea of *a* mind as a distinct mental substance or transcendental seat of consciousness. It is better to speak instead of physical events interacting at a variety of organizational levels, and to think of what emerges from such interactions as corresponding to what we usually mean when we speak, for example, of perception, emotion, volition, and aesthetic experience. These forms of the activity of intelligence occur in the context of what Dewey terms "situations." A situation is an interaction between intelligence (an individual or a community) and an environment; the situation's further components include the desires and behavior of the intelligent agent(s) involved and the impinging forces from the environment. These components can stand to each other in varying degrees of unity or disunity, relative to the purposes of intelligence, and are registered by intelligence with varying degrees of explicitness. So described, situations may be relatively simple—such as scratching one's foot—or more complex—such as producing a film or writing a novel. But all situations, regardless of their degree of complexity and regardless of whether they emphasize sensuous or intellectual forms of activity, possess a distinctive "feel" and a pervasive quality of unity or disunity, prior to further description or interpretation by the agent(s) constituting them. This fact, for Dewey, lends an aesthetic dimension to all experiences regardless of whether or not they occur within settings institutionally recognized as aesthetic.

Although the account of the relationship between intelligence and the world just sketched would not fully emerge until Dewey's later writings, it was anticipated by his earlier views of human behavior. In an influential article of 1896, "The Reflex Arc Concept in Psychology," Dewey examined a paradigmatic case of behavior: a child reaching to touch a burning candle. Contrary to the then prevailing orthodoxies of stimulus-response psychology, it is not the case, he argued, that in such an instance the seeing of the candle and the act of reaching toward it are distinct sensorimotor events, devoid of intrinsic connection to one another. What really happens is that the seeing and the grasping exhibit a reciprocal relationship that is continually conditioned and, if these events are repeated, reconditioned by the context. The initial act of seeing informs the direction

and attack of the grasping, whose (in this case painful) outcome confers a new meaning upon the seeing; that in turn leads to a modification in further reaching, and so on. In this way, behavior and environment are continuous, just as, within behavior, are mental and physical processes. All in the final analysis belong to a larger dynamic complex possessing an immediately felt unity and discursively describable meanings that evolve as action proceeds. This account is echoed throughout discussions in *Art as Experience* of how an artist goes through a creative process and how a spectator or audience member actively constitutes meaning from an artwork. It is reflected also, for example, in Dewey's hypothesis of "growth" as a fundamental end of action in the *Ethics* and in his account in *Logic: The Theory of Inquiry* of how inquiry within and without science (including, in an extended sense, in art) is generated by intelligence's drive toward the resolution or unification of "problematic situations."

Critique of dualism. These subjects cannot be further addressed here, but some additional words deserve to be said about their relation to a topic that increasingly occupied Dewey and that informs *Art as Experience*'s argument in various ways: dualism in philosophy and its reconstruction. This theme figures prominently in his naturalized version of the kind of sweeping interpretation of the history of Western philosophy pioneered by G. W. F. Hegel. In Hegel's version various distinctions central to earlier philosophical thought, such as those between the individual and the state, between moral autonomy and heteronomy, and between mind and body, were mistakenly granted absolute validity (i.e., treated dualistically). This, Hegel argued, was due to the failure of philosophers to understand how each such distinction expresses part of a larger absolute unity whose development requires a different method of reflection to be grasped. [*See* Hegel.] In Dewey's version, developed in books like *The Quest for Certainty,* a Hegel-style account of the historical emergence of dualisms remains; the notion of an absolute unifying perspective is dropped; and new emphasis is placed on how the impulse to harden distinctions into dualisms is shaped equally by institutional pressures and the struggle for biological survival. Within this context Dewey was relentlessly critical of what he saw as modern philosophy's archaic habit of picturing experience as situated between two worlds: nature construed as a mechanistic realm of causality and contingency, and a supersensible realm of mind or spirit, as supposedly reflected in our conceptions of moral, aesthetic, and religious values.

Means and ends. The last-mentioned legacy of dualism—its effect on the understanding of values—Dewey found particularly problematic. Within modern philosophy, he observed, values are commonly understood as falling into two irreducible and hierarchical categories: final or intrinsic values, which derive their superior normative force from their reference to a reality outside the contingencies of history;

and instrumental values, whose implication in the desiderative processes of practical reasoning links them to the causality of nature and so confers on them a supposedly second-class status. Within this hierarchy, intellectual or spiritual goods and activities—and cultural practices centering on their preservation and production—are associated with final values and thereby accorded metaphysical pride of place over satisfactions and practices oriented toward instrumentality and the body. This results in overgeneralized distinctions between "higher" and "lower" cultural practices and ideals of universal cultural development that, while conferring legitimacy on certain institutional arrangements (notably, as Dewey would emphasize in *Art as Experience,* those of the modern art world), fail to be sustained by investigations of a more empirical and anthropological nature. Dewey traces the legacy of this modern Western outlook from classical philosophy through Judeo-Christian theology and on into modern secular philosophies of value that maintain strong versions of the final/instrumental dualism, such as Kantianism.

Dewey struggled throughout his long career to formulate an adequate pragmatic account of valuation. The details of this project cannot be pursued here, but two central points in his critique of traditional approaches to valuation are particularly relevant to his aesthetic theory. First, he maintains that the dualistic approach to picturing human values traditionally severs value from valuation, and so neglects how the things and situations we value are part of the dynamic structure of experience. The very word *value* is itself the source of much confusion in this regard, since it can give the impression that *a* value is a metaphysically reified entity abiding outside the order of natural events. On Dewey's analysis, when we are warranted in judging an object good or bad or an action right or wrong, we are making a highly contextualized claim about a quality of a situation that arises from the process of interaction between intelligence—in this case ourselves—and our cultural and natural environment. Value, in the end, just *is* valuation. To the charge that this naturalization of value slides toward an untenable relativism, Dewey's basic response is that just as there is no contradiction in speaking of other qualities in the world of our experience as emergent from the interaction between intelligence and environment but also *real,* so is there no contradiction in speaking of values in these terms.

Second, Dewey argues that the philosophical practice of erecting a dualism of two incommensurable forms of value, intrinsic and instrumental, and of supposing that this dualism adequately captures the way valued objects and states of affairs are related to each other as means and ends, reflects a basic confusion about the relationship between means and ends and their representation in practical reasoning. To illustrate this point, suppose someone sets out to bake a cake and discovers that it requires hard-to-obtain ingredients and difficult-to-master skills. In an obvious sense

the obtaining of the ingredients and the exercising of the skills are means to the end of the existence of the cake. One might at first try to argue that the cake, construed as an end, has a fixed intrinsic value independent of whatever needs to be done to bring this end about. Thus it might seem irrelevant to the cake's value whether it was produced by an individual taking great care in the process or by, say, an assembly line in a factory found to engage in unfair employment practices. On such an analysis, if one were presented with two perceptually indistinguishable cakes possessing these different histories, one would have no particular reason to value, and choose, one rather than the other. But this way of thinking obscures how the cakes in these two scenarios are in an important sense two very different cakes calling for different evaluations. For rather than being an autonomous object of enjoyment, each cake is an organic part of a larger process in which no part is fully intelligible as a means or an end in abstraction from the whole. (This analysis is analogous, it will be noted, to familiar arguments for why "perfect" forgeries of artworks and their originals call for different evaluations.) In both cake-producing scenarios, Dewey would argue, means and end are reciprocally related: not only is the value of the means conditioned by the end result, but the value of the end is also conditioned by the nature of the means. This way of thinking about values has ramifications for a wide spectrum of human problems in areas like ethics, science, politics, and religion. And it has particular relevance, as we will see, to how Dewey came to understand the values of the objects and situations we describe as artistic and aesthetic.

Aesthetic Theory. Despite occasional digressions on aesthetic topics in earlier writings, Dewey produced no systematic discussion of these matters until *Art as Experience.* Owing in part to his dominant influence within American philosophy during the early decades of the twentieth century, *Art as Experience* became the most influential work in philosophical aesthetics of that century by an American writer. Readers unacquainted with the background sketched above may, upon casually perusing the book, get the impression that in its essentials *Art as Experience* was another treatise on the "philosophy of fine art" in the mainstream tradition of such idealist-flavored books of the day as Benedetto Croce's *Aesthetics; or, The Science of General Linguistic* and R. G. Collingwood's *Principles of Art.* In fact, *Art as Experience* represented a radical assault on the very idea of aesthetics as understood by most of its practitioners throughout the nineteenth and early twentieth centuries, given that idea's ties to the metaphysics of idealism. Dewey's opposition to this tradition is apparent on the first page, where he complains about theories that relegate art to a "separate realm" and declares that the primary task of a philosophy of art is "to restore continuity between the refined and intensified forms of experience that are works of art and the everyday events, doings, and sufferings that are universally recognized

to constitute experience." Everyday experience includes the stylized rituals and folkways of a specific culture, informed by more basic patterns of prizing, desiring, avoiding, hoping, and fearing arising from our biological constitution as "live creatures."

The above remark about continuity reflects Dewey's conviction that traditional aesthetic theory—as it develops from German idealism through the emerging varieties of formalism in the academic art writing of the twentieth century—is guided by an untenable dualism between art and life that is symptomatic of a deeper impulse to locate the sources of the most important human values in a realm outside nature. This dualism shows itself in a range of more specific distinctions commonly employed within traditional aesthetics: for example, between the form and content of an artwork; between aesthetic value as a form of intrinsic value, and other forms of instrumental value; and between fine art, on the one hand, and craft and popular art, on the other. Dewey's point is not that such distinctions are inherently dualistic. Rather, it is that they have been employed by traditional aesthetics in such a way as to lead modern culture to understand art, somewhat paradoxically, as a compartment of activity that derives its specially valued status within the scheme of human practices because of its irrelevance to the activities and purposes occupying most people most of the time.

The larger reconstructive question *Art as Experience* poses and sets out to answer thus amounts to this: How can the subject matter of aesthetics be redescribed in a way that lacks this air of paradox? Needless to say, Dewey will insist that this subject matter be describable as part of nature. In addition, he emphasizes its social character. Here *Art as Experience* shows notable affinities with much writing on aesthetics in the Marxist tradition, but with an important difference. Dewey, like pragmatists generally, rejects the idea of historical inevitability, along with classical Marxism's commitment to the explanatory reduction of cultural facts to economic facts. (Thus in substance and method *Art as Experience* may be read as occupying a middle position between the antiempirical extremes of classical idealist aesthetics, and the reductively sociological approaches characteristic of early Marxist aesthetics.) It might seem to go without saying that the subject matter of a pragmatic aesthetic theory should also be describable as artistic. But for Dewey, whether this is so will depend on how one defines "artistic." Although much of *Art as Experience*'s argument deals with issues and examples drawn directly from the fine arts, Dewey's considered definition of "art" is by no means orthodox, for it rejects two conceptual connections that had been regarded as axiomatic by most philosophical aestheticians since Hegel. One is the treatment of "art" as equivalent to "fine art"; the other is the assumption that the subject matter of aesthetics is, properly speaking, fine art, as contrasted with other kinds of cultural production. To un-

derstand the account of art Dewey offers instead, more needs to be said first about *Art as Experience*'s core notion of experience.

Experience. We have seen how Dewey views all experience as emerging out of an ongoing interaction between intelligence and environment. In *Art as Experience* he introduces the further theme that experience is developmental in nature. Situations in which we engage in an activity mainly for the sake of some end or ends external to the activity itself (for example, going to the market or tying a shoe) exhibit what Dewey calls experience's "instrumental" phase of development. Activities we engage in primarily for their own sakes (for example, playing a game or eating ice cream) exhibit experience's "final" phase. We live our experiences, that is, either as means, as ends, or—and this represents a further developmental possibility—as both. Our most developed experiences exhibit the characteristics of both the instrumental and the final phases. And they also possess another quality: They are *consummatory*. The consummatory phase of experience is characterized by a felt sense that in the immediacy of the present moment one's prior efforts are brought to fruition. One refers to such experiences with phrases like "*that* trip I took" or "*that* conversation we had." Consummatory experiences surpass those that are only instrumental or final in terms of their degree of felt unity, a quality that pervades not only the subject's state of mind but the entire situation in which the experience occurs.

Dewey regards experience in its consummatory mode as being, in a literal, axiological, and phenomenological sense, life at its fullest. His discussion of this subject evokes the romantic image of an individual life as a narrative whose "chapters" are punctuated by that individual's history of striving to achieve objectives in the world, along with the failures and consummatory successes attending these efforts. The idea of consummatory experience further extends the idea of *growth* as the most basic human end that Dewey set out earlier in the *Ethics*. When one has an experience of a consummatory character, one has a sense that life has achieved enough narrative closure to release energies for the reassessment of past purposes and the pursuit of new ones. Some consummatory experiences, in addition, are characterized by an intensified awareness of the qualities of an object. Under such conditions, the experience and the object itself are both *aesthetic*: "An object is peculiarly and dominantly aesthetic . . . when the factors that determine anything which can be called an experience are lifted high above the threshold of perception and are made manifest for their own sake." That experiences of this kind can occur in a vast range of human situations including (but not only in) the fine arts Dewey regards as a central datum for aesthetic theory.

Art. So what, then, is art? Art occurs when a consummatory experience with a marked degree of aesthetic quality

leads, through the agency of the subject, to the production of further such experiences, not merely for the subject, but for others as well. Art is thus literally a form of experience:

> Art is a quality of doing and of what is done. . . . Since it adheres to the manner and content of doing, it is adjectival in nature. When we say that tennis-playing, singing, acting, and a multitude of other activities are arts, we engage in an elliptical way of saying that there is art in the conduct of these activities, and that this art so qualifies what is done and made as to induce activities in those who perceive them in which there is also art. The *product* of art—temple, painting, statue, poem—is not the *work* of art. The work takes place when a human being cooperates with a product so that the outcome is an experience that is enjoyed because of its liberating and ordered properties.
>
> (*Art as Experience,* chapter 10)

This definition implies (1) that a work of art is an interaction between an individual—the artist—and some means of production, intellectual or material; (2) that this interaction is the consummation of a specific history of activity on the artist's part, funded by the qualitatively diverse experiences constituting that history; (3) that these qualities are expressed in the art product; and (4) that through the art product, a spectator or audience can have an experience that is, to a significant extent, a "reconstruction" of the artist's experience.

Dewey sees fine arts like painting, poetry, sculpture, music, and architecture as providing central examples of this experientially oriented definition of art within modern societies. But he does not mean the definition to apply *uniquely* to the fine arts. His argument here shows partial affinities with those of others, such as Friedrich Nietzsche and various postmodern theorists, who propound radically aestheticizing accounts of human affairs in which traditional distinctions between life and art are collapsed. But Dewey in the end endeavors to strike a balance between the competing impulses of a radical aestheticism and the more compartmentalized approach of traditional aesthetics. To the question of whether there is any area of intelligent activity in our lives that could *not*, on his definition, count as art, his answer seems to be: no, *under the right conditions.* In his version of the aestheticist thesis, most forms of productive experience possess the *potential* to be art. But this is not to say that under foreseeable historical conditions, most productive activities will actually realize that potential, or, more radically still, that an entire life might itself literally become, as Nietzsche held, a work of art. [*See* Nietzsche.] Although there are significant resonances between Dewey's and Nietzsche's views on other subjects (notably in their accounts of the cultural sources of human values), a central motive underlying Dewey's own aestheticism is a very un-Nietzschean social meliorism of a sort more common among certain Marxist writers. Like such writers, Dewey saw that many characteristic problems of modern civilization arise from a general lack of integration between means and ends (or in Dewey's idiom, between the instrumental and final phases of experience) in daily life. Most forms of work manifestly fail to afford settings for the realization of experience's consummatory potential. This situation is worsened by the institutionalization of absolutized distinctions between "higher" and "lower" cultural pursuits. For these can function to degrade activities falling on the lower end of this hierarchy where intelligence is often exercised in highly developed ways, for example, with certain kinds of skilled labor and popular culture. Traditional aesthetic theories all too frequently also contribute to this situation by exalting "fine" over "useful" forms of production without acknowledging how the fine/useful distinction, far from being a logically necessary feature of thought, arose, like various other distinctions of modern philosophy, under specific historical conditions and must be evaluated accordingly. From such considerations, Dewey draws the conclusion that although the practices we know as the fine arts may be superior to other areas of "useful" production as sources of art in experience, this superiority is not intrinsic to those practices but is rather a property of a larger set of contingent social relationships that calls for criticism at a variety of levels.

Form and expression. Two other important applications of Dewey's reconstructive approach remain to be mentioned here: his treatments of form and expression. In his account of form, Dewey agrees with conventional formalist writers of his generation like Clive Bell or Roger Fry to the extent of holding that in experiencing an object aesthetically we are naturally drawn to certain of its features more than others (its "formal" features); that in so doing we abstract perceptually from those other features; and that we undergo and value the experience that results for its own sake. Yet this does not mean that the features of an object that facilitate such experiences possess their significance independent of the larger fund of meanings and interests that we bring to the object, and that we are continually revising (or "reconstructing") as our lives move forward. Artistic form in this sense is significant because of the way it functions both as an end and as a means within our experience, and draws on the same larger patterns of shifting and developing energies that constitute all natural and cultural processes.

Dewey gives a similarly reconstructive treatment to the phenomenon of artistic expression. Like form in the arts, expression does not spring from self-enclosed artistic or aesthetic origins. Rather, it is the culmination of a general process that occurs throughout the spectrum of intelligent behavior in all cultures. A simple gush of tears or a cry of joy is not by itself expressive. Expression occurs only when an initial emotionally charged response to a situation is experienced as meaningful and is then further refined into a communicative gesture guided by an intention to transform

the initial situation. Expression in the fine arts exhibits this basic structure in conjunction with more specific types of communicative intention directed at the production of aesthetic response, and takes on further characteristics depending on the type of artistic medium employed.

An instrumental aesthetic. To conclude, it may be remarked that Dewey's aesthetic theory is an "instrumental" theory in two related senses that reflect the larger tendencies of his pragmatism. First, he holds that the fine arts themselves are instrumental to a variety of further ends. Like the sciences, the arts contribute to our continual quest for reflective connections among the meanings in our day-to-day lived experience. One thinks here, for example, of how various stylistic innovations in the fine arts of Dewey's generation—such as Cubism in the visual arts, atonality in music, and stream-of-consciousness narrative in literature—served as a site of new expressive "experiments" in the representation of space and time, reflecting some of the same currents of cultural experience as did emerging developments in science such as relativity theory. Dewey also holds, with a long tradition of philosophers and critics reaching back to Aristotle, that the arts can enrich our ethical lives by conveying insights into experience that are not captured by conventional moral codes and that resist communication in ordinary discursive speech. A tragedy by Euripides, a novel by Dostoevsky, or a ballad by Billie Holiday, for example, may reveal more to us about ourselves than does any set of ethical or religious precepts. To this extent, "works of art are the only media of complete communication between man and man that can occur in a world full of gulfs and walls that limit community of experience."

A second and more general sense in which Dewey's aesthetic theory is instrumental emerges from considering what it means to call aesthetics a form of *inquiry* in his sense of that term. A basic theme of his pragmatism, again, is that intelligence is inextricably involved in the world it examines. An inquiring individual or group is motivated by the desire to resolve a disharmony or tension between its purposes and the pressures of a changing world. The disharmonies Dewey, in the end, came to see as most in need of resolution in aesthetics are the discontinuities between philosophies of art that overcharacterize the autonomy of their subject and a world in which art, in the narrow and broader senses of the term, is connected in increasingly complex ways to larger social and cultural structures. To this extent, the greatest problem for aesthetics is aesthetics itself as traditionally understood. For Dewey, and for pragmatists in aesthetics generally, this suggests that a mark of a viable aesthetic theory will be its instrumentality toward the criticism of that legacy.

[*See also* Historicism, *article on* Historicism and Philosophy; *and* Peirce.]

BIBLIOGRAPHY

Works by Dewey

Reconstruction in Philosophy. New York, 1920.
Experience and Nature. Chicago, 1925.
The Quest for Certainty: A Study of the Relation of Knowledge and Action. New York, 1929.
Art as Experience. New York, 1934.
Theory of Valuation (1939). In *John Dewey: The Later Works, 1925–1953,* edited by Jo Ann Boydston, vol. 13, *1938–1939,* pp. 196–202. Carbondale, Ill., 1981–1990.

Other Sources

Alexander, Thomas M. *John Dewey's Theory of Art, Experience, and Nature: The Horizons of Feeling.* Albany, N.Y., 1987.
Buettner, Stewart. "John Dewey and the Visual Arts in America." *Journal of Aesthetics and Art Criticism* 33.4 (Summer 1975): 383–391.
Haskins, Casey. "Dewey's *Art as Experience:* The Tension between Aesthetics and Aestheticism." *Transactions of the Charles S. Peirce Society* 28.2 (Spring 1992): 217–245.
Pepper, Stephen C. "Some Questions on Dewey's Aesthetics." In *The Philosophy of John Dewey,* edited by Paul Arthur Schilpp, pp. 369–390. New York, 1939. For Dewey's response, see pp. 549–554. The volume contains other critical essays on Dewey's philosophy, along with Dewey's responses; also, a bibliography of all of Dewey's writings up to 1937.
Romanell, Patrick. "Comment on Dewey's and Croce's Aesthetics" (1948). In *John Dewey: The Later Works, 1925–1953,* edited by Jo Ann Boydston, vol. 16, *1948–1952,* pp. 463–467. Carbondale, Ill., 1981–1990. Dewey's response is in the same volume, pp. 395–398.
Seiple, David, and Casey Haskins, eds. *Democracy and the Aesthetics of Intelligence: Essays in Deweyan Pragmatism.* Albany, N.Y., 1998.
Shusterman, Richard. *Pragmatist Aesthetics: Living Beauty, Rethinking Art.* Oxford and Cambridge, Mass., 1992.
Tiles, J. E. *Dewey.* London and New York, 1988.
Westbrook, Robert B. *John Dewey and American Democracy.* Ithaca, N.Y., 1991. Extensive bibliography.

CASEY HASKINS

Theory of Expression

John Dewey's theory of expression is not only central to his aesthetic theory, it is also commonly misunderstood. The expression theory is primarily associated with the views of two Idealists, Benedetto Croce and his follower, R. G. Collingwood, and it would be helpful to begin by summarizing these better-known views. While Dewey, Croce, and Collingwood were all deeply indebted to Romanticism and Hegelian Idealism, Dewey eventually rejected the metaphysics of idealism for what he called "cultural (or empirical) naturalism." Thus his theory of expression is profoundly different from theirs. (*Instrumentalism,* a term often used for Dewey's thought, designates only his theory of inquiry.) Idealism regards self-consciousness as the criterion for determining ultimate reality. G. W. F. Hegel provided an important basis for the expression theory by holding that Spirit comes to be conscious of itself by expressing itself (that is, by making its subjectivity objective) in ever higher

and more inclusive stages: from matter to life, from life to consciousness, from consciousness to self-consciousness, a process that develops from sensation on up to art, religion, and philosophy. For Hegel, art thus is expressive of Spirit's self-knowledge at any given cultural and historical epoch.

While Hegel's objective version of the expression theory has affected interpretations of art history, Croce's more individualistic and subjective version has been more influential in theories of artistic creation. Croce's famous claim is that "intuition = expression"; that is, the work of art is essentially a mental, imaginative synthesis, fusing impressions into a unified whole. [*See* Croce.] Expression is achieved internally for the artist whether any physical artwork is produced or not. Collingwood agrees with this, emphasizing that it is only through the process of organizing the image that the artist comes to know what it was he or she desired to express. Expression is a subjective process of self-articulation in the creator. He also discusses expression as essentially expression of emotion, adding that there are no emotions prior to their expression. Finally, Collingwood believes that for the audience to understand the work, it must feel the same emotion. [*See* Collingwood.]

Numerous criticisms of these views have been raised, especially by O. K. Bouwsma, John Hospers, and Monroe Beardsley. Insofar as their criticisms have been taken by many analytic philosophers as holding for all versions of the expression theory, including Dewey's, it is important to understand how Dewey's position is significantly different from the Croce-Collingwood theory and does not succumb to its problems. Beardsley, for example, questions whether we need to grasp the creator's intentions to understand a work of art. Dewey also offered serious objections to the subjective version of the expression theory. He questions whether art ought to be the narcissistic activity of self-exploration and self-expression Croce and Collingwood think it is and asks whether we can create works of art apart from actively working with a physical medium. Instead of rejecting the expression theory, Dewey put forward a more promising version that does not limit expression to the internal or creative process of the artist; it is social and ongoing, not fixed or final at any moment in its history. Dewey likewise denies that expression is of emotion; rather, it is of emotionalized meanings.

Dewey's theory of expression stems from his naturalism. Unlike many naturalists, Dewey is not a reductionist, materialist, or positivist. Nature for him is a locus of dynamic interactions and processes in which structure and creativity are manifest. More complex forms of organic order gradually develop, presenting novel, irreducible features. Today we might term this position *ecological emergentism*. Dewey thinks in terms of fields and processes as temporally unfolding events, repudiating the traditional fundamental dichotomies of mind and matter, subjective and objective, cognition and feeling. Meaning develops from the human organism as it interacts with its environment and from our existence as social, communicating beings. The living body, acting, responding, and learning, inhabits the world in a preverbal but expressive way. We also develop expression through the interplay of gesture, communication, and the acquisition of language and culture as we grow.

Although Dewey's extended treatment of expression is found in *Art as Experience,* his more general theory of meaning as communication is best presented in *Experience and Nature.* We do not begin as autonomous individuals, rational minds, or centers of pure subjectivity. We begin as infants who are neither self-conscious nor independent. We do not have "thoughts" distinguished from "emotions," and we do not "express" ourselves. Although infants are preverbal, the world is not meaningless for them. But a crying infant does not "express" anger or hunger if it is merely reacting to a feeling. The infant may be hungry or angry, but there is no explicit use of gestures as symbols to communicate with another person to remedy the situation. We gradually learn a complex language of gesture because others are trying to communicate with us. When we begin to use a gesture like crying to attract attention to be fed as a symbol with a common meaning the process of communication has begun. We learn to talk to others because they talk to us, thereby making possible an internal, expressive dialogue that becomes our "self-consciousness." The self, for Dewey, is constituted by the ways we can create meaningful actions and responses to the world. The self is its "arts."

Communication is broader and more basic than verbal language. Attempts to treat it as a syntactic system or rule-governed "game" miss the fact that language gets meaning from the pragmatics of preverbal interaction. For Dewey, the *process* of communicating with others is more fundamental for meaning than the formal system used, just as for him the process of learning is more significant for intelligence than knowing. Communication is a dialogue in which sympathetic imagination must be used. It occurs when one individual becomes capable of putting herself into the position of another so that she can see herself from the other's point of view, and use this as a way of guiding her own gestures so they elicit the desired response from the other. When the other can do this as well, the process can go forward. The "meaning" at any given time may be more or less indeterminate, but through the process become more determinate in surprising, creative ways. As we grow up, the self develops insofar as we become expressive individuals. This also means learning to be receptive, to know how to listen to others. The self emerges both in terms of imaginatively extending our experience to the social standpoint and of developing a unique and creative response to it. This last point is very important: The symbols of a culture lose their expressive ability unless they are continually transformed. Dewey thus sees communication as a tense balance between the pull toward established social practices and individual

novelty. Creative expression lies in the successful combination of the two.

Art, for Dewey, emerges from the process of life. Our delight in colors or sounds or words or shapes leads us to play with them and organize them in a variety of ways. By nature, human beings enjoy organizing experience to explore its creative potential. In short, we enjoy making and finding meaning in the world and in each other. Art—not science or language—taps most deeply into this fundamental impulse. At the beginning of *Art as Experience,* Dewey warns us that if we begin to reflect on art beginning with the objects found in museums and the experiences associated with them, we will be led to develop theories that separate art from life, that speak of art in terms of "pure forms" evoking equally pure "aesthetic emotions." Instead, the fine arts should be seen as highly developed and profoundly deepened attempts to seek the kind of embodied expressive meaning that we all seek in our lives. To sever the continuity between art and life turns art into the sterile narcissism Croce and Collingwood thought it naturally was and leaves the rest of life ill-formed, crude, and shallow.

All human activity begins, says Dewey, as an "impulsion," a general drive toward meaning that "calls the whole self into play" (Dewey, 1987, p. 64). Because we must act in a world where we have needs and conditions change, we must develop organized responses whereby action may be directed toward fulfilling ends. The environment calls forth the need to have a "self" and makes us aware that there is a "world." Action must become thoughtful; memory and hope dilate the present moment so that histories and possibilities are seen. We comprehend things in terms of their potentialities, not just in terms of brute identity or "isness." Because there is drama in life, there is emotion in experience. But it is not merely discharged; it must direct action to explore the world, finding means or media and organizing them. By discovering and remaking the world, experience takes on order and meaning itself. "Only where material is employed as media is there expression and art" (Dewey, 1987, p. 69). For Dewey, emotion thus plays a role in expression. But not only must it be objectified in a medium; as it organizes experience into a whole, it becomes transformed. Expression is expression of meaning, whether verbal or not.

Expression is also a process in time. All works of art take time to be experienced or understood. Their rhythmic temporality builds up the ordered play of form and possibility that defines "*an* experience"—the movement of experience toward its meaningful fulfillment or consummatory closure. As the inchoate emotion becomes linked with the medium, directed by its possibilities and selecting among them, it becomes ever more organized and meaningful. Emotion fuses with intelligence, so that there is no sharp distinction between the medium and the form. The form is the continuity of the process of aesthetic experience generated by the medium itself. To the extent that emotion, form, medium, and subject matter are all integrated, there is expression in the interactive relation between the artwork and its audience.

Thus "expression" for Dewey does not refer uniquely to the creative act, and so is not primarily limited to an act of "self-expression" in the artist. The audience takes an active role, too, in responding to the work. The work of art is an ongoing process in which the object interacts in the experience of others and becomes generative of meaning. Nor are these experiences themselves discrete and isolated, but rather are part of an ongoing history. Culture is the process of learning from the experience of the past to shape the experience of the future. Indeed, if we focus on what transpires in the best of learning experiences, we can grasp better Dewey's understanding of the relation of art and its audiences. Art challenges us, forces us to use our senses and our minds in new ways, and gives us a new world and a new self in the end. In learning it is pointless to ask whether the experience is individual *or* social: It is both, requiring individual creativity and openness to the other.

The expressive nature of Shakespeare's *Tempest* is not the sum total of subjective responses. We cannot verbalize its "meaning," though critics try to help us understand it through discussion. There is no one nameable idea, much less emotion, it "expresses." We can never determine what Shakespeare's "mental intuition" may have been. When we encounter the work and respond to it, we have a profound, meaningful experience. The work has transformed us, given us something "rich and strange," and we have brought it to new life. Audiences must become creatively involved with the artwork for it to become expressive, just as actors or musicians must. We may see the play in relation to other events in terms of which it is also expressive; for example, it expresses Shakespeare's "art" or "Elizabethan drama" or even "the Renaissance." The reception—the variety of interpretations, debates, and artistic renditions—is not external to the work; the work lives and develops in them. The work does not stand in a strictly active-passive relationship to the audience (including the performers) any more than it did to the creator. We only perceive the work of art if we use our senses, imagination, and intelligence. Yet there is something in the object that calls forth these abilities and directs them.

A work of art also expresses because it produces an experience that gives a concrete "local habitation" to meaning. Sophisticated codes, linguistic, cultural, or perceptual, may be involved (as in Dante's *Commedia*). But the experience generated by the interaction of the work and its audience is palpable and vital, rich in play, ambiguity, and surprise, evoking fluid moods that cannot be fixed and named. Art deserving of the name opens up the dynamic, tensive relationship between the "live creature" and the "world," awakening the full, sensitive intelligence of the one and the sig-

nificant mystery of the other. Wherever the potentiality of experience to take on a sense of profound meaning that awakens us to ourselves and the world is consummated, there is art. It is perhaps because of this that Dewey observed that "experience in the form of art . . . solves more problems which have troubled philosophers and resolves more hard and fast dualisms than any other theme of thought" (Dewey, 1981, p. 293).

[*See also* Expression Theory of Art.]

BIBLIOGRAPHY

Works by Dewey

Experience and Nature. In *John Dewey: The Later Works, 1925–1953,* vol. 1, 1925, edited by Jo Ann Boydston. Carbondale, Ill., 1981, see chaps. 5 and 9.
Art as Experience, 1934, In *John Dewey: The Later Works, 1925–1953,* vol. 10, edited by Jo Ann Boydston. Carbondale, Ill., 1987.

Other Sources

Alexander, Thomas M. *John Dewey's Theory of Art, Experience, and Nature: The Horizons of Feeling.* Albany, N.Y., 1987.
Kestenbaum, Victor. *The Phenomenological Sense of John Dewey.* Atlantic Highlands, N.J., 1977.
Meyer, Leonard B. *Emotion and Meaning in Music.* Chicago, 1956.
Shusterman, Richard. *Pragmatist Aesthetics: Living Beauty, Rethinking Art.* Oxford and Cambridge, Mass., 1992.
Zeltner, Philip M. *John Dewey's Aesthetic Philosophy.* Amsterdam, 1975.

THOMAS ALEXANDER

Experience and the Organic Unity of Artworks

John Dewey's aesthetics can be naturally viewed from two rather distinct angles—the one emphasizing the experiential process leading to consummation, the other accentuating the immediacy and intensity of discrete moments within that process. His entire theory of art owes much of its interest, as well as its applicability beyond the experience of art proper, to the wedding of these two theoretical elements (the aspects of instrumentality and intrinsicality, respectively), and this gets articulated in Dewey's doctrine of organic unity.

Traditionally, the doctrine of organic unity in art, or "aesthetic organicism," refers to a certain kind of formal interrelatedness that is alleged to hold among elements within an artwork, where these parts maintain such close affinity with each other that the nature of the whole cannot resist an alteration in any of its essential components.

Without further specifying the nature of the "wholes" in question, however, this qualification alone is obviously barren. Viewed as a mere aggregate, any collection of items is of course altered when one of its elements changes. What traditional exponents of the theory have maintained instead, from the Greeks to the New Critics, is that the aesthetic whole is more than just the sum of its parts. Aristotle held in

the *Poetics* that a well-crafted drama presents "a single piece of action" having a "beginning, middle, and end." And Formalists like Clive Bell insisted that aesthetic experience capture the "significant" form of the artwork in an immediate apprehension of an irreducible quality of the work as a harmonious whole. Dewey's own view, as stated in *Art as Experience* and elsewhere, remains a classical one in just this sense.

But Dewey's aesthetic theory is not, like the Formalists', a view about the inherent qualities of the physical artifact, and for that reason this aspect of Dewey shares interesting affinities with certain themes current today (e.g., Arthur Danto's notion of "artistic identification"). For Dewey, the physical artifact itself is not in fact the "work of art" at all, and use of familiar expressions such as "the art object" can be seriously misleading. This is not perhaps as odd as it may at first seem (even given the fact that Dewey's view implies that a new work of art is created in the perception of each spectator) because speaking about "an object" of inquiry does not guarantee any concrete physical reference, anymore than recounting the story of Ponce de León commits us to the existence of the Fountain of Youth. The object of inquiry is the label we give to the intentional focus of procedures we undertake in relation to a situation we want to know more about. Similarly, Dewey wants to say that "being a work of art" is a relational property, and invites us to regard art "as experience" (as his famous title suggests). The artwork is the artifact *working*—working "to make our experiences as complete, as full, as total in their vitality" as possible ("The Philosophy of the Arts," p. 364). The physical artifact is an occasion for a certain sort of interaction between percipient and object, and the "pervasive quality" of that experiential process is the mark of organic unity.

This begins to suggest how Dewey seeks to combine the notions of instrinsic and instrumentality, and also how indebted Dewey is to the aesthetics of Romanticism. Since the advent of Romanticism in the late eighteenth century, the point of designating aesthetic form as "organic" has typically been to mark a distinction from the "mechanical" form of physical causation. Aesthetic perception no less than mundane physical causation is a process, but (Dewey insists) it differs from the latter in its dependence on the "fundedness" of the percipient's sensibilities—by which Dewey means that the informed spectator approaches an art object with a repertoire of emotionally imbued responses that treat the object as one especially arresting solution to a set of aesthetic problems. The perception of the unity of a work is really the culmination of a creative process by the spectator that corresponds, generically if not in specific historical detail, to the actual genesis of such an artifact—a process that makes use of both the apprehension of the instrinsic qualities of the medium and an intuitive sense for the promising directions that the artist's efforts can take.

Dewey's own language here, however, is not always straightforward. He declares, for example, that "we lay hold of the full import of a work of art only as we go through in our own vital processes the processes the artist went through in producing the work" (*Art as Experience,* p. 328). This might seem to commit Dewey to a strong and perhaps suspect theory of art as communication, but this could easily be misleading. What is involved here for the viewer is a "communication" of a rather odd sort—not the reception of information, nor any direct intuition of the historical artist's state of mind. What is really crucial for Dewey is the notion that intelligence, as it applies to aesthetic perception, has certain constant features that are modeled after artistic creativity itself and that make up the general contours of the process of aesthetic appreciation. This involves most prominently a "regulation by an underlying and pervasive quality" ("Qualitative Thought," p. 246). Here we have one of the driving ideas behind not only Dewey's aesthetics but his entire philosophical enterprise as well. "Control by a single pervasive quality" for Dewey implies both a sensitivity to the emotionally charged immediacy of the medium and an intuition for the unfolding artistic solution expressed in the finished artifact—in other words, both intrinsicality and instrumentality. At this point Dewey's view is teleological, naturalistic, and romantic: the work of art is said to "grow" (for the perceiver as well as for the original artist) as an "end-in-view," out of the natural organic process of experience.

Now what enables this process to occur, according to Dewey, is the fact that experience itself, when conditions are right, exhibits organic unity: "the *form* of the whole" is then "present in every member" (*Art as Experience,* p. 62). Experiential processes of this sort are "arts"—activities composed of phases that are internally related by a single pervasive quality, which can be discerned early on by an immediate "total seizure, an inclusive qualitative whole not yet articulated" (*Art as Experience,* p. 195). The salient feature of such a qualitative whole Dewey calls a "consummation"—"a summing up and a fulfillment" of its prior phases (*Art as Experience,* p. 177). What would otherwise be a disconnected series of acts is here transfigured into a cumulative process, and this for Dewey is an entirely biological occurrence, a function of appropriate environmental cues on the one hand and settled human habits on the other. These latter (which Dewey designates as "intelligence") get formed over similar previous occasions, in which a person's "doings and undergoings" join an action and its experienced consequences into a "fused" relationship that remains embodied in the dispositional makeup of the agent. Dewey would say that aesthetic experience is "expressive" of those settled dispositions.

It is apparent here how a full understanding of Dewey's view of organic unity introduces topics that, traditionally, have not been considered especially relevant to aesthetics proper. But recent commentators have come to regard this as a major source of interest: Virtually the whole of Dewey's mature work, and much of what he did prior, is grounded on his wider views about the aesthetics of experience. For this Dewey has stretched not just the concept of "aesthetics" but also "experience" far beyond its familiar connotations. "Experience" for Dewey is not simply a phenomenological category. The term is a placeholder for the complex of operations and apprehensions that occur in what he calls "the interactive situation"—which is his way of describing the functional relationship between the human organism and its environment. All perceived qualities themselves emerge from this kind of complex, but "pervasive qualities" in particular emerge only when the human organism has learned to sustain a certain range of responses (emotional as well as cognitive) throughout the experiential process. In that case, and in the presence of appropriate environmental conditions, the result is a particular "system"—an assembly of parts connected together so that a particular end (consummation) is attained. Dewey's familiar model here is biological (as in the case of the respiratory system of the human body, comprising the lungs and their associated vessels and airways). But he extends the reach of such a system beyond the boundaries of the organism proper, to include not only external elements that temporally coexist with the present state of the organism, but also future conditions that are created out of tensions internal to that present state. Dewey's substantial, and controversial, claim at this point seems to be that what emerges from this interaction—and what constitutes the entire interactive situation as an organic unity—is more than simply the "pervasive quality" itself: it is also a telic element within the system that orients it toward consummatory "equilibrium," for which the agent's immediate apprehension of intrinsic pervasive quality functions instrumentally as an orienting clue for conduct.

So for Dewey, what has organic unity in such cases is not a physical artifact but a succession of interrelated events, and a physical art object may serve as an occasion not only for consummatory satisfaction but for a reinforcement of a spectator's capacity for intelligent conduct in any field of practical inquiry. This (Dewey insists) is a central aspect of human life that Formalists missed by cordoning off "fine art" from other human products, and the fundamental impulse behind Dewey's aesthetics is to "restore the continuity" between art and life that prevalent modernist theories had been ignoring. The encompassing breadth of the term *aesthetic* in Dewey is an indication of this. Life has real human value, Dewey would say, insofar as it exhibits the organic unity of experience.

All this places Dewey in an interesting relationship to recent developments in aesthetic theory. If we identify "modernism" with Clement Greenberg's view of a "self-critical tendency" that focuses exclusively on the intrinsic properties of the medium, Dewey is not a modernist because he

does not hold that preoccupation with material properties, apart from other practical concerns, is even possible. But if we identify "postmodernism" with the Derridean rejection of any fixed frame for the artwork that could definitively set it apart from "outside" elements, then Dewey is no postmodernist either because he holds that experience itself does naturally exhibit structures of organic unity that define interactive situations. Here, as Richard Shusterman has pointed out (*Pragmatist Aesthetics*, pp. 71ff.), the contention between Dewey and postmodernism is not over the notion of organic unity per se but rather over the priority one should ascribe to the experience of aesthetic richness. Dewey takes such experience to be of supreme value for human life, and this fact he believes should determine our critical interests in art. His theory of organic unity is intended to provide a basic naturalistic framework for those critical interests.

BIBLIOGRAPHY

Works by Dewey

Human Nature and Conduct. In *John Dewey: The Middle Works, 1899–1924,* vol. 14, *1922,* edited by Jo Ann Boydston, pp. 13–118. Carbondale, Ill., 1983.

Experience and Nature. In *John Dewey: The Later Works, 1925–1953,* vol. 1, *1925,* edited by Jo Ann Boydston. Carbondale, Ill., 1981. See "Experience, Nature, and Art," pp. 266–294.

"Qualitative Thought." In *John Dewey: The Later Works, 1925–1953,* vol. 5, *1929–1930,* edited by Jo Ann Boydston, pp. 243–262. Carbondale, Ill., 1984.

Art as Experience. In *John Dewey: The Later Works, 1925–1953,* vol. 10, *1934,* edited by Jo Ann Boydston. Carbondale, Ill., 1987.

"The Philosophy of the Arts." In *John Dewey: The Later Works, 1925–1953,* vol. 13, *1938–1939,* edited by Jo Ann Boydston, pp. 357–368. Carbondale, Ill., 1988.

Other Sources

Aristotle. *Poetics.* In *The Basic Works of Aristotle,* edited by Richard McKeon, chap. 7. New York, 1941.

Barnes, Albert C. "The Aesthetic Values of Painting." In *The Art in Painting,* 2d ed., pp. 34–70. New York, 1928.

Bell, Clive. *Art* (1914). New York, 1958.

Danto, Arthur C. "The Artworld." *Journal of Philosophy* 61.19 (October 15, 1964): 571–584.

Seiple, David I. "John Dewey and the Aesthetics of Moral Intelligence." Ph.D. diss., Columbia University, 1993.

Shusterman, Richard. *Pragmatist Aesthetics: Living Beauty, Rethinking Art.* Oxford and Cambridge, Mass., 1992. See pp. 62–83.

DAVID I. SEIPLE

DICKIE, GEORGE (b. 1926), contemporary American philosopher specializing in aesthetics. Since the early 1960s Dickie has made numerous important contributions to the philosophy of art. Among the most influential of his contributions are his attacks on key aspects of widely held aesthetic theories and his creation and critical development of the *institutional theory of art.* His critique of aesthetic theory addresses a number of theses about what is involved in people's experiencing something's aesthetic qualities (and associated theses about aesthetic objects), whereas his Institutional Theory provides an account of the concept of art that locates art's essence within a special category of social practices attributed to a social group Dickie calls *the artworld.*

A widely held view among aesthetic theorists is that someone must in some way invoke a special mode of perception in himself or herself in order to experience something's aesthetic qualities (or in order to experience something as an aesthetic object). Invoking this special mode of perception is commonly equated with adopting a special attitude toward what is being experienced, a disinterested attitude, for example. Speaking generally, Dickie shows that experiencing aesthetic qualities cannot require adopting a special attitude by providing counterexamples to the various attempts philosophers have made to show that there is a distinct kind of experience (properly classified as aesthetic experience) that people must have in order to experience something's aesthetic qualities, and that having this kind of experience requires adopting a special attitude. [*See* Attitude, *article on* Aesthetic Attitude.]

Early on in his attack on aesthetic attitude theorists, Dickie argued against the view that experiencing something's aesthetic qualities required attending to it disinterestedly. He did this by providing examples to show that the difference between people who are experiencing something's aesthetic qualities and people who are experiencing the same object without being aware of its aesthetic qualities merely is a function of which characteristics of the thing each person is paying attention to, regardless of the interests motivating his or her attention. Since the difference in what is experienced is explained by what is being attended to, not the mode of attention, it is not necessary to introduce notions like disinterested attention or other special modes of perception (identified in terms of the perceiver's interests, purposes, or motives) in order to understand the experience of something's aesthetic qualities. [*See* Qualities, Aesthetic.]

Acknowledging Dickie's criticism of aesthetic theories that made disinterested attention the special attitude necessary for experiencing aesthetic qualities, aesthetic attitude theorists replied that the relevant distinction was between aesthetic and nonaesthetic perception, where these two modes of perception are distinguished by reference to three things: the perceiver's motives, purposes, and interests, the qualities of the thing attended to, and attention. [*See* Disinterestedness.] Dickie pointed out that since attention is common to both aesthetic and nonaesthetic perception, and since nonaesthetic qualities can be the object of aesthetic perception, the perceiver's motives, purposes, and interests are what must distinguish aesthetic from nonaesthetic perception. He then provided examples to show that someone

whose motives, purposes, and interests concern only something's nonaesthetic qualities can nonetheless notice its aesthetic qualities, and that someone concerned only to discover something's aesthetic qualities nonetheless can fail to do so.

In regard to those versions of aesthetic theory that attempt to characterize aesthetic perception in terms of someone's perceiving without regard for practical ends (as distinct from someone's perceiving in an altogether disinterested way), Dickie noted that qualities that aesthetic attitude theorists agree are not aesthetic, such as being square, are perceivable without regard to practical ends. To this he added that someone with a practical end in mind (Dickie's example is someone intending to write a critical article) can perceive the qualities aesthetic attitude theorists agree are aesthetic. Dickie concluded that the notion of a special kind of aesthetic perception or aesthetic attitude does not work in aesthetic theory. Taken together, Dickie's arguments seriously undermine attempts by various aesthetic theorists to hold that someone's experiencing something's aesthetic properties requires his or her adopting a special sort of psychological state.

Not limiting himself to revealing flaws in the theories of others, Dickie developed his own theory about the nature of art. He credits Paul Ziff and Morris Weitz with correctly rejecting the traditional method that seeks to discover art's essence in some function common to all works of art. Ziff and Weitz (separately) argued that the attempt to define art in this way invariably fails, and on this basis concluded that art lacks an essence. Dickie agrees that the traditional method is flawed, but he does not conclude from this that art has no essence (that there can be no necessary and sufficient conditions for something's being art). Instead, he maintains that the traditional method fails because in each case it concentrates on a single, simple relation (such as the relation between an artist and her emotions). He says that art's essence can be discovered if we adopt an approach that "places the work of art within a multi-placed network of much greater complexity than anything envisaged by earlier theories." [See Essentialism.] The network of relations Dickie has in mind is presumed to be instantiated by "the artworld." Dickie's institutional theory of art is his account of the network of social practices that he believes make works of art possible.

Dickie's institutional theory has two major versions (referred to here as the *standard theory* and the *new theory*), the first of which underwent numerous revisions from the late 1960s until Dickie's publication of his new theory in *The Art Circle* in 1984. Both versions hold that something is a work of art because of its position in an institutional context, but differ significantly in what they say about this context. Dickie summarizes the numerous versions of the standard theory thus:

> A work of art in the classificatory sense is (1) an artifact (2) a set of the aspects of which has had conferred upon it the status of candidate for appreciation by some person or persons acting on behalf of a certain social institution (the Artworld).
>
> *(The Art Circle)*

The standard theory is criticized in a variety of ways, many of which Dickie says attack a view he did not hold, but which he agrees his presentation of his position caused others to think he held. Dickie's artifactuality condition is attacked by citing counterexamples thought by their proponents to be both obvious and decisive. The manufactured urinal Marcel Duchamp titled *The Fountain* by R. Mutt is a much cited counterexample, despite Duchamp's insistence that *The Fountain* was not art and could not accomplish his aim—challenging elitist aesthetics—if it were art. On some occasions Dickie attempts to avoid the problem created by the appropriation of natural or manufactured objects by artists through maintaining that an act of conferring artifactuality occurs at the same time as our conferring on something the status of a candidate for appreciation. In his new theory he maintains that something becomes an artifact upon being located in an artworld context ("*Fountain* is the urinal used as it is, as an artistic medium").

Another frequent objection to the standard theory is that it is circular because Dickie cannot avoid using the concepts of art, the artist, and the artworld in presenting the theory that is supposed to define these concepts. This objection is motivated by Dickie's circular statement of his view in conjunction with the assumption that, given the sort of view it is, it cannot be stated without circularity. When Dickie presents his new theory he agrees that circularity is unavoidable and attempts to explain why circularity does not present a problem for his statement of the new theory (this topic will be resumed later).

The standard theory also is widely criticized for assuming that the social context required for art is far more formally structured than is the actual segment of human society involved in the arts. Critics claim that if the standard theory were true, then for art to exist there would have to be individuals whom a highly structured organization had officially appointed to confer art status, much as a jury is officially appointed in a court to establish someone's guilt or innocence. In reply, Dickie insists that his intent was to characterize the social institution required for art as something informally structured, while agreeing that his own choice of examples, in conjunction with his use of expressions like "acting on behalf of" and "conferred upon," led his readers to think otherwise.

Dickie's use of "acting on behalf of" and "conferred upon" also results in the further misunderstanding that something has to be accepted by the artworld (or by someone acting on the artworld's behalf) in order to be art. This "acceptance" condition currently is so entrenched that

many philosophers view it as an essential element of any institutional theory of art. This leads some philosophers to accept as a definitive objection to all institutional theories that it is obvious that something can be art prior to any social group's having the opportunity to accept it. If this objection to versions of institutional theories that contain an acceptance condition is sound, this fact is irrelevant to Dickie's understanding of his standard theory. Because the flawed but nonetheless widespread characterizations of the standard theory face strong objections, Dickie's critics fail to consider seriously the significance complex social factors might have for something's being a work of art.

Dickie's *new institutional theory* holds that something is art because the intentions with which it is created relate it in an appropriate way to an appropriate social group (a group that is part of a social system composed of such groups). Understanding which intentions are appropriate requires understanding the social practices that define the type of group Dickie calls "the artworld." Dickie's presentation of his new theory proceeds on the assumption that this information cannot be conveyed to readers without circularity, but Dickie does not see this as a problem. In defense of his circular presentation he asserts that this circularity reflects the inflected nature of art, by which he means the interdependency on one another of a set of practices, both for their individual identities and for their place in the network of practices constitutive of what Dickie calls "the artworld." Because of this interdependency, the concepts of art, artist, and artworld cannot be understood independently of one another. Dickie also is aware that circular presentations are uninformative for people who are not already in the know, and maintains that what he is doing when he presents his theory is clarifying something his readers already know. In case there are people who want to understand the new theory who are not already informed, it is important to note that since an account can be given of the social factors definitive of "the artworld" without using the concepts *art, artist,* or *artworld,* the "uninformativeness-to-the-uninitiated" of Dickie's circular account of this set of inflected concepts can be overcome.

Understanding Dickie's new institutional theory requires understanding five statements that, taken at face value, reveal the structure but not the full content of the new theory:

> (I) An artist is a person who participates with understanding in the making of a work of art. (II) A work of art is an artifact of a kind created to be presented to an Artworld public. (III) A public is a set of persons the members of which are prepared in some degree to understand an object which is presented to them. (IV) The Artworld is the totality of all Artworld systems. (V) An Artworld system is a framework for the presentation of a work of art by an artist to an Artworld public.

These five statements are distillations of Dickie's understanding of the practices that define the social system he believes makes art possible. Understood in conjunction with (II)–(V), the significant information presented in statement (I) is that by "work of art" is meant something that can be created only by someone possessing certain knowledge and who (in part) creates as he or she does because of what he or she knows. Dickie states that "accidents, fortuitous and not, may occur *within* the creative process, but the overall activity is not accidental." Statement (II) indicates that the required knowledge makes possible the intentions with which something must be created to be of a kind created to be presented to the appropriate group (what here will be referred to as group G). Statement (II) also indicates that by a "work of art" is meant anything created with the appropriate intentions, whether it ever actually is presented, or even was made with the intention that it actually be presented to group G. Statement (III) indicates that membership in group G requires (among other things) being prepared to understand artifacts that are intentionally created (by people applying certain knowledge of the practices that define group G) to be of a kind created to be presented to members of group G. Statement (IV) indicates that group G is a subgroup linked with other groups that together make up a social system, and (V) that the practices definitive of this social system fix the intentions with which something must be created in order to be of a kind created for presentation to group G.

The role of the members of the system responsible for creating the artifacts other members of the system are (by definition) prepared to understand in appropriate ways is using creative techniques to produce artifacts of a kind to be presented to the group (group G) as a whole. Knowing when creative activities fulfill this role and when they do not requires knowing the practices that give the group its identity and in terms of which the notion "being of a kind created to be presented to the group" gets its content. That is, it is the practices of the group as such that determine which activities constitute artmaking, and which of the members of the group are artists. What are these practices?

Dickie's answer is (1) an ability to perform the role of determining whether what is presented is or is not art, and (2) a wide variety of abilities and sensitivities that enable one to perform certain roles involving perceiving and understanding the particular kind of art with which one is presented. As stated, both conditions require that the reader already understand what art is. However, Dickie does offer examples of the relevant practices that define the roles, the totality of which define the artworld. Had he introduced these examples independently of using the notions whose content ultimately depends on the practices his examples explicate—notions such as "the artworld"—and had he offered the examples in greater number (for, as he says, "a list of such conventions would be very long"), his description of these interrelated practices would present a noncircular answer to the question of which practices define something

as a "work of art." Such a list would be tedious to produce. But it would also be very informative for the uninitiated if Dickie's basic insight that art exists only in relation to an interrelated complex of social practices is correct.

Here are samples (taken from *The Art Circle*) of practices (conventions) that would occur on Dickie's list for a given group within the system: the "non-participation by the spectator" convention; the convention directing people's attention when in a gallery or theater; the convention of hanging paintings with the unworked surface to the wall. To this brief list can be added, for the sake of illustration, the practice of holding the person who created the artifact responsible for its perceived merits or lack of same; the practice of treating the artifact as if it were valuable and as if it should be taken seriously prior to establishing whether or not it deserves such treatment; the practice of assuming a right to critique the merits of the artifact; the practice of picking some artifacts for presentation and rejecting others and of assuming that doing this well requires training and experience; the practice of taking an interest in whether an artifact was created by someone with any training or not, and if so, what sort of training; the practice of seeking to preserve artifacts in the state they were in when first completed, insofar as it is possible, and with the exception of artifacts intentionally made to be temporary (in this case, the preservation practice applies to their photographed or videotaped traces); and so on. To eliminate overlap with practices that it is unreasonable to consider artworld practices and that are structurally similar to those definitive of the artworld, this list has to be extended considerably, and the interrelated character of the listed practices has to be made clear. In principle, either at some point this exercise will suffice to individuate practices definitive of the artworld from practices definitive of types of groups whose types are inconsistent with being the artworld (in addition to whatever else they are), or the failure to accomplish this will count against the notion that "the artworld" refers to a distinct kind of social group. That a group might have mirrored in its practices those practices definitive of the artworld, if any, in the absence of any use of terms like "art," does not count against Dickie's view, provided that the principle by which the group was initially identified is not incompatible with its also being the artworld. If Dickie is correct, the instantiation of parallel practices by a group initially identified as of some kind other than the artworld would bring an artworld into existence regardless of what the group called the artifacts that were the focus of its practices. This is so because the link between being a work of art and these practices is not itself a matter of convention; this complex relation is the essence of art.

The approach outlined above leaves unanswered how to decide which practices should be on Dickie's list and which should not, and thus which social groups make up the totality of the artworld system. Dickie's answer is that these practices or groups have nothing in common that makes them rather than others the practices (and groups) that make art possible: this fact is arbitrary. According to Dickie, the reason this arbitrariness does not render the notion of art absurd is that in picking items for the list of interrelated practices and in describing how the practices are interrelated we are making clear something we already "in some sense know." Dickie believes that any other answer is a return to the method of understanding art's essence criticized by Ziff and Weitz.

In 1988 Dickie published *Evaluating Art*. The work presents Dickie's *instrumentalist theory of art evaluation* by way of critiquing (primarily) the views of Paul Ziff, Monroe Beardsley, Frank Sibley, Nelson Goodman, Nicholas Wolterstorff, David Hume, Bruce Vermazen, and J. O. Urmson. Dickie's modification of what remains from each of these philosopher's views after Dickie's critique becomes part of what Dickie calls his "compromise" theory of artistic evaluation. Dickie's argument for his compromise theory is that the views of the philosophers he discusses are the best of what is available on evaluating art, and that it is reasonable to accept a theory combining those aspects of these views that withstand critique and that can, with modification, be united into one view. Dickie's instrumentalist theory of art evaluation has no necessary connection with his institutional theory of art, which Dickie says is proper, since his institutional theory is purely classificatory, in that it explains why a work of art is a work of art, not why a work of art does or does not have value.

Dickie holds that a work of art is valuable when it is an instrumental source of a valuable experience. In general, some aspects of works of art are instrumentally valuable because they can produce valuable experiences without referring to anything outside the experience of the work of art. Other aspects of a work of art are instrumentally valuable because they can produce valuable experiences in which these aspects are experienced as standing in relation to things outside the immediate experience of the work. An experience is valuable either because the experience or an element of it is intrinsically valued by the person having it or because the experience or an element of it is of a representational feature of the work of art that refers to a valued thing. A work of art's aesthetic and cognitive features are an instrumental source of a valuable experience.

Dickie distinguishes universal weak principles of evaluation that ground judgments that something has or lacks value from nonuniversal strong principles that ground judgments that something is good or bad in specific respects and that ground specific comparative judgments. Working from his critique of Beardsley, Dickie holds that since (at least two) aesthetic characteristics are standard features of works of art, there are at least two necessity principles on a work of art's being valuable: nothing is a valuable work of art unless it is unified, and nothing is a valuable

work of art unless it is complex. Critiquing Sibley leads Dickie to conclude that the presence of numerous aesthetic properties in a work, including but not limited to unity and complexity, satisfies weak universal "positive sufficiency" principles, since such properties are always valuable. Unity and complexity also satisfy weak universal "necessity" principles, since the absence of these properties is always disvaluable. Nothing can be a valuable work of art unless it is unified and complex. Some properties, such as garishness, are always disvaluable.

On the cognitive value front, Dickie's critique of Goodman leads him to conclude that art can have three kinds of positive and negative cognitive value: imitative, supportive, and referent-centered. Something has imitative value if someone derives satisfaction from noticing that the work or something about it is true of the world. When noticing that something of a work is not true of the world, the work has cognitive disvalue. When a work's referential features are responsible for its having valuable or disvaluable aesthetic properties, the work has supportive cognitive value or disvalue. A work has referential cognitive value or disvalue by virtue of referring to something that has value or disvalue.

With regard to the strong principles that ground judgments that a particular work of art is good or bad in specific respects and that ground specific comparative judgments between works of art, Dickie maintains (following his critique of Francis Hutcheson and David Hume) that there cannot be a single standard of taste covering all cases. Since the value of a work of art, on Dickie's instrumentalist view, derives from its capacity to be the source of a valuable experience, the denial of a single standard of taste means that there is no single kind of experience that universally grounds the value of works of art.

In 1996 Dickie published *A Century of Taste*, in which he defends a Humean theory of taste. Dickie's aim is to utilize eighteenth-century theories of taste as a basis for his theoretical account of the pleasure or displeasure we take in certain experiences and the properties we ascribe to the objects of such experiences (beautiful, sublime). Dickie examines the views of Hutcheson, Alexander Gerard, Archibald Alison, Immanuel Kant, and Hume. He argues that Hutcheson initiated and Hume " 'all but' perfected" the theory of taste, whereas the associationist theories of Gérard and Alison lead to dead ends, and Kant's work continues to lead philosophers away from an adequate theory of the phenomena that theories of taste attempt to explicate.

In his critique of Kant, Dickie stresses the need to understand Kant's theory of taste in light of Kant's teleology. For Kant, the goal of fine art is the realization of beauty. Beauty (teleologically understood) is an object's form of purposiveness (purposive forms invigorate and entertain our mental powers). Dickie objects that this view is too broad, and that it does not allow some things to be more beautiful than others (a problem also faced by Hutcheson's view). Dickie also

objects that Kant's view rejects color as an aspect of the experience of beauty, which Dickie says is by itself sufficient to show that Kant's theory is "badly wrong."

Dickie opens his critique of Hume by noting that "the superiority of Hume's theory in *Of the Standard of Taste* is exhibited as much by what he does not attempt to do as by what he does do." Hume shows (by means of a "disproportionate pair" argument) that there are differences in artistic value. He then seeks the principles that underlie judgments of taste. Hume's theory is superior to Hutcheson's, since it allows that something might have beauty-making characteristics and still not be beautiful. Dickie criticizes Hume for confusing the task of discovering principles of taste with the task of discovering how to convince cognitively insensitive persons that objects can have subtle merits that insensitive people cannot detect. As a result, Hume's search for principles of taste becomes a description of the characteristics of a good critic. Dickie says that Hume's discussion of a good critic fails to address the problem that the people who lack good taste cannot recognize the people who do. Nonetheless, Dickie does think that Hume gives us the "key" to a limited resolution of this problem. Hume observes that there is evidence that one person is a better critic than another when the one person can get the other to notice an aspect of beauty the other has failed to notice on his or her own.

Dickie notes that Hume does not succeed in presenting a method of recognizing good critics that will work for everyone, but does succeed in giving the criteria of a good critic. Dickie says that, despite the fact that Hume does not succeed in telling us how to discover good critics, he does give an excellent account of how to discover the standard of taste and principles of taste. The standard of taste is conceived of by Hume as the joint verdicts of good critics about the kinds of merits and defects art can have, these joint verdicts constituting principles of taste. Dickie says that "the standard of taste for Hume's theory is the complete set of principles of taste," by which Dickie means the joint verdict of everyone.

Dickie concludes *A Century of Taste* by arguing that Hume's theory of taste is "far superior" to the other theories he has discussed. Hume's theory is superior to Hutcheson's and Kant's in that Hutcheson and Kant focus only on one characteristic as beauty-making (excluding color, for example). Further, Kant's theory cannot admit of degrees of beauty and so cannot accommodate the fact that there is a threshold for being beautiful (having a form of purposiveness does not admit of degrees). For Hume, there are many merits and demerits that contribute to or subtract from something's beauty. Dickie's counterexamples (some natural systems are not beautiful) to Kant's notion that the form of purposiveness is a beauty-making characteristic present a further, serious problem for Kant's theory of taste. Gerard's and Alison's use of the notion of the associa-

tion and coalescence of ideas, according to Dickie, "hopelessly dilutes" notions of taste. Dickie says that Kant's arguments for a faculty of taste are "inconclusive," and that he sees no way to establish the existence of a faculty of taste as understood by Hutcheson, Gerard, or Kant. Hume, on the other hand, establishes the objectivity of taste and gives us a means for discovering the principles of taste, without having to introduce the notion of a faculty of taste.

[*See also* Art World; Definition of Art; Evaluation, *article on* Aesthetic Evaluation; *and* Institutional Theory of Art.]

BIBLIOGRAPHY

Books by Dickie

Aesthetics: An Introduction. Indianapolis, 1971. A revised edition titled *Introduction to Aesthetics: An Analytic Approach* is forthcoming from Oxford University Press.
Art and the Aesthetic: An Institutional Analysis. Ithaca, N.Y., 1974.
Aesthetics: A Critical Anthology. Coedited by Richard Sclafani. New York, 1977; 2d ed., New York, 1989.
The Art Circle: A Theory of Art. New York, 1984.
Evaluating Art. Philadelphia, 1988.
The Century of Taste: The Philosophical Odyssey of Taste in the Eighteenth Century. New York and Oxford, 1996.

Selected Articles by Dickie

"The Myth of the Aesthetic Attitude." *American Philosophical Quarterly* 1.1 (January 1964): 56–65.
"Beardsley's Phantom Aesthetic Experience." *Journal of Philosophy* 62.5 (March 4, 1965): 129–136.
"Defining Art." *American Philosophical Quarterly* 6.3 (July 1969): 253–256.
"Taste and Attitude: The Origin of the Aesthetic." *Theoria* 39.1–3 (1973): 153–170.
"Art and the Romantic Artist." In *Essays on Aesthetics: Perspectives on the Work of Monroe C. Beardsley,* edited by John Fisher, pp. 99–108. Philadelphia, 1983.
"Instrumental Inference." *Journal of Aesthetics and Art Criticism* 42.2 (Winter 1983): 151–154.
"Stolnitz's Attitude: Taste and Perception." *Journal of Aesthetics and Art Criticism* 43.2 (Winter 1984): 195–203.
"Evaluating Art." *British Journal of Aesthetics* 25.1 (Winter 1985): 3–16.
"Beardsley, Sibley, and the Critical Principles." *Journal of Aesthetics and Art Criticism* 46.2 (Winter 1987): 229–237.

Other Sources

Beardsley, Monroe. "Is Art Essentially Institutional?" In *Culture and Art,* edited by Lars Aagaard-Mogensen, pp. 194–209. Atlantic Highlands, N.J., 1976.
Cohen, Ted. "The Possibility of Art: Remarks on a Proposal by Dickie." *Philosophical Review* 82.1 (January 1973): 69–82.
Davies, Stephen. *Definitions of Art.* Ithaca, N.Y., 1991.
Stolnitz, Jerome. *Aesthetics and Philosophy of Art Criticism.* Boston, 1960.
Weitz, Morris. "The Role of Theory in Aesthetics." *Journal of Aesthetics and Art Criticism* 15.1 (September 1956): 27–35.
Wieand, Jeffery. "Can There Be an Institutional Theory of Art?" *Journal of Aesthetics and Art Criticism* 39.4 (Summer 1981): 409–417.
Yanal, Robert, ed. *Institutions of Art: Reconsiderations of George Dickie's Philosophy.* University Park, Pa., 1994.
Ziff, Paul. "The Task of Defining a Work of Art." *Philosophical Review* 62.1 (January 1953): 58–78.

GEORGE W. S. BAILEY

DIDEROT, DENIS. [*To appreciate fully the relevance of Diderot to aesthetics and art criticism as well as to the origins of the encyclopedia, this entry comprises two essays:*
Survey of Thought
Diderot and the Salon
The first covers the full breadth of his work, whereas the second focuses specifically on Diderot's writings on the salon art exhibitions in eighteenth-century France. For a related discussion, see Theater.]

Survey of Thought

In 1746, Denis Diderot (1713–1784) published his first work, the *Pensées philosophiques* (Philosophic Thoughts), a small book that had been printed secretly, appeared anonymously, was promptly condemned by the Parlement de Paris, the highest court in France, and attracted widespread interest and controversy. Already in this short essay, Diderot's own intellectual and aesthetic stance is unmistakable. He opens with a spirited defense and rehabilitation of the passions, generally denounced by both Christian apologists and lay moralists as dangerous obstacles to spiritual perfection and emotional well-being. Diderot, for his part, proclaims that nothing great in the sciences, arts, or even morality can be accomplished without the driving force of the passions, without which man inevitably sinks to a state of apathy and mediocrity.

Diderot's keen interest in the psychology, ethics, and aesthetics of an individual deprived of one or two senses is evidenced by two works: his *Lettre sur les aveugles à l'usage de ceux qui voient* (Letter on the Blind for the Use of Those Who See), published in 1749, which would raise an uproar and cause the arrest and three-month imprisonment of Diderot at Vincennes, a medieval castle six miles east of Paris, and the less controversial *Lettre sur les sourds et muets à l'usage de ceux qui entendent et parlent* (Letter on the Deaf and Mute, for the Use of Those Who Hear and Speak), published in 1751. While the *Letter on the Blind* owed a great deal to Lockean sensationalism and empiricism and contained imaginative scientific conjectures that already foreshadowed Darwin's theory of the evolution and survival of the species, it did not deal with aesthetics. On the other hand, the *Letter on the Deaf and Mute* focuses almost exclusively on aesthetic and linguistic questions. It presents a convincing refutation of the Abbé Charles Batteux's rather simplistic concept of art as imitation of nature set forth in the widely read *Beaux-Arts réduits à un même principe* (The Arts Reduced to a Common Principle), published in 1746.

Before Gotthold Ephraim Lessing, whose influential treatise *Laocoön* was to appear in 1766, Diderot had become aware that the notion of the sisterhood of the arts, inherited from the Renaissance, required qualification, for each medium depends on a different set of conventions and symbols. At this time Diderot was especially concerned with the

possibilities and limitations of the language of the poet, as distinct from that of the prose writer. Vigorously rejecting the descriptive role generally assigned to poetry in the eighteenth century, Diderot emphasized its suggestive, symbolic function and its reliance on a subtle intermingling of "hieroglyphs," which surpass the explicit meaning of words and represent the most complex manipulation of language and its associative potentialities.

Despite its rambling form and occasional abstruseness, the *Letter on the Deaf and Mute* has been acknowledged as an important contribution to modern understanding of the psychological, aesthetic, and epistemological nature of human speech and verbal communication, with keen insights into such nonverbal means of expression as gesture and body movement, a preoccupation that reappears in Diderot's theories of drama and fiction. This work also clearly shows Diderot's growing ability to deal with aesthetic problems that cut across traditional boundaries of knowledge.

Less than a year after the appearance of the *Letter on the Deaf and Mute,* Diderot's extensive article "Beau" was published as part of the second volume of the *Encyclopédie;* it encompasses a masterful survey of major aesthetic doctrines as well as a personal theoretical definition of beauty. After successively summarizing and evaluating the ideas of such predecessors as Plato, Saint Augustine, Francis Hutcheson, Anthony Ashley Cooper, earl of Shaftesbury, Christian von Wolff, Jean-Pierre Crousaz, and Père André, Diderot proposes his own notion of beauty. It has the merit of avoiding metaphysical abstractions and of resting on the human capacity for perception of relationships, thereby firmly linking his aesthetics to his materialistic, empirical philosophy, and at the same time avoiding the excesses of Lockean sensationalism. Despite the diversity in our aesthetic judgments, which he readily acknowledged, Diderot posited the existence of an objective, universal law enabling us to perceive beauty in a harmonious relationship of the parts to the whole, in unity within multiplicity. While acknowledging cultural and individual vagaries in our ideas of beauty, Diderot lays strong emphasis on the universal nature of our aesthetic sensibility, which derives from our need for harmony and order and our awareness of the underlying analogies that link our sensory perceptions. Among the first to hail the importance and originality of Diderot's essay on beauty was Immanuel Kant, who, as early as 1759, recommended it warmly to his disciple Johann Georg Hamann.

Before long, Diderot would turn from theoretical speculations to more practical and specific considerations on the arts. In 1759 he agreed to contribute critical reviews of the biennial exhibits of paintings and sculptures at the Louvre to the *Correspondance littéraire,* edited by his friend Friedrich Melchior Grimm, a German literary journalist and publicist settled in Paris. The *Correspondance littéraire* was circulated in manuscript form for the benefit of foreign subscribers of note, Catherine the Great and Frederick the Great among others, who wished to keep abreast of the cultural, political, artistic, and social news and activities in the French capital.

Diderot's *Salons,* the product of this commitment as contributor to his friend's newsletter, ranks among his most delightful and personal works. It shows his growing critical expertise and familiarity with the widely divergent pictorial styles and genres popular in his time as well as his thorough knowledge, not only of such diverse talents of his own time as François Boucher, Jean-Honoré Fragonard, Jean-Baptiste Greuze, Joseph Vernet, Quentin de La Tour, Jean-Baptiste-Siméon Chardin, and the Van Loos, but also of many great artists of earlier periods he undertook to study tirelessly by visiting public and private collections.

Putting to brilliant use his stylistic virtuosity and his awesome abilities as a polemicist, Diderot launched with enthusiasm into art criticism and gave it new status as a literary genre worthy of the best efforts of a *philosophe.* That Diderot greatly relished his role as art critic is borne out by the zest, humor, and piquancy of his descriptions and comments, and by the considerable length of some of his *Salons,* especially those of 1765 and 1767. The tone throughout is engagingly informal, the style ranges from playful bawdiness to highly lyrical flights, and the judgments are blunt but hardly ever capricious and arbitrary. From Diderot's detailed and lively descriptions of the numerous paintings and sculptures he reviewed it is obvious that he welcomed the challenge of transposing line, color, and form into words, since the far-off subscribers to the *Correspondance littéraire* could not see these works of art for themselves.

If Diderot could be less than generous in recognizing the qualities of such rococo artists as Boucher and Fragonard, and, on the other hand, at times embarrassingly maudlin in extolling Greuze's sentimental and moralistic compositions, his reactions and comments must be understood in the context of his aesthetic and ethical ideals and models, which exemplified a serious, uplifting, and socially useful art. To be sure, his approach to painting could be considered too "literary," as well as too strongly oriented toward subject and content, especially according to our present-day standards and criteria of taste in these matters. But painting in general, and especially historical painting, then held as the noblest and most demanding genre, was closely associated in the minds of most artists with literature, history, mythology, and religious iconography.

But lest we judge Diderot's art criticism unfairly, we should keep in mind that he could look at a painting on its own terms, without superimposing on it his own ideas or extrapictorial considerations. That he could appreciate purely painterly values is amply demonstrated by his sensitive comments on Chardin's unpretentious still lifes, which fascinated him endlessly. His expertise in technical matters,

acquired through his personal friendship with many artists and numerous visits to their studios over several decades, is repeatedly demonstrated in his *Salons*. He offered an overview of his experience as art critic and theorist in his *Essais sur la peinture* (Notes on Painting), written in 1766 and first published in 1796, which elicited the enthusiasm of Friedrich von Schiller and Johann Wolfgang von Goethe (the latter published in 1798 a German translation with lengthy commentaries), as well as in the *Pensées détachées sur la peinture* (Random Thoughts on Painting), written between 1774 and 1781 and first published in 1798.

As an art critic, Diderot is undoubtedly at his best when he gives free play to his aesthetic sensibilities without allowing ideological preoccupations to interfere unduly with his direct response to a painting. His enthusiastic response to Vernet's landscapes and seascapes, to Chardin's still lifes and genre paintings, and to La Tour's portraits bespeaks his keen delight in analyzing the process by which, through the magic of color and line, an artist renders and transfigures into transcendent form the features of a particular individual, a segment of natural scenery, or a few utensils and some food on a kitchen table. His enthusiasm knew no bounds when he found himself in the presence of originality of vision reinforced by mastery of execution. What Diderot searched for among contemporary painters was an art at once grandoise, high-minded, forceful, and morally uplifting, as well as intimate and immediate in its depiction of human situations, conflicts, and emotions. Diderot yearned for an art that could be spiritually ennobling, yet closely related to life and its everyday concerns. Hence, his basic distrust of such rococo masters of erotic pleasures as Boucher and Fragonard and his eagerness to praise Greuze and Chardin for their virtuosity in representing the dignity that can be found in domestic, quotidian scenes.

If Diderot harked back to the great masters of the past, notably to Nicholas Poussin, Peter Paul Rubens, and Rembrandt, it was because they could combine the sublime and the ordinary, the noble and the simple, by capturing the passions and drives that impel humanity in revealing moments of conflict or introspection. Diderot dreamed of a French art revivified by a forceful, passionate, civic, and virile ideal. No wonder, therefore, that in his last *Salon*, that of 1781, he warmly greeted a young newcomer by the name of Jacques-Louis David, whose historical composition, *Blind Belisarius Asking Alms*, showing the famed Roman general reduced to begging, Diderot singled out for praise because it so effectively and convincingly exemplified sublimity and pathos.

An enduring preoccupation of Diderot closely linked with his aesthetics in general and his dramatic theories in particular is his insistent emphasis on the notion of virtue, already evident in his 1745 translation of Shaftesbury's *Inquiry concerning Virtue and Merit*, as a prerequisite not only for personal happiness but also for greatness in the arts. [*See* Cooper.*] Even more than when dealing with painting, Diderot was determined to endow the theater with great moral seriousness. The new drama he envisioned would fulfill several important functions: It would supersede the church and pulpit as a vehicle for proselytizing the masses; it would be promoted as a form of entertainment worthy of men and women of refined taste; it would serve to propagandize the philosophy of the Enlightenment; and it would be invested with the respectability necessary to justify its already widespread appeal and to counteract its vulnerable position vis-à-vis ecclesiastical disapproval. Following the example of Voltaire, but more daringly than his illustrious predecessor, he would make his mark as a major playwright and dramatic reformer and innovator.

Diderot sought a complete renewal of the drama and therefore found a great deal to learn from comedies of character exemplified by Terence and Molière. He saw especially useful models in the contemporary English theater, notably in George Lillo's realistic yet dramatically effective treatment of middle-class characters in his *London Merchant* and in Edward Moore's *Gamester*. In his eagerness to present a comprehensive program encompassing both theory and practice, Diderot wrote two plays in quick succession, *Le fils naturel* (The Illegitimate Son, 1757) and *Le père de famille* (A Father and His Family, 1758). Simultaneously, he propounded his dramatic theories in two essays that accompanied the plays, the first, *Entretiens sur le "Fils naturel"* (Conversations on "The Illegitimate Son"), in the form of three dialogues between the author and Dorval, the hero of the play, and the second, the more formally structured *Discours sur la poésie dramatique* (Discourse on Dramatic Poetry).

Both in his plays and his essays Diderot endeavored to create a new type of drama that, unlike classical tragedy, deals with real-life situations and dilemmas and unexceptional individuals. Central to Diderot's theory is the social milieu, especially the family and profession of the hero, as a primary motivating force in shaping his attitudes, prejudices, and beliefs. Henceforth, character would be defined within a specific sociological context. He furthermore stressed the importance of such nonverbal modes of dramatic expression and communication as *tableaux vivants* (living pictures), silent pauses, gestures, mimicry, and body language. He urged greater verisimilitude in the staging of a play and less shouting, singsonging, declaiming, and posturing on the part of the actors. In the dialogue *Paradoxe sur le comédien* (Paradox on Acting or The Actor's Paradox), written over a period of approximately ten years, from 1769 to 1778, but not published until 1830, Diderot put forth the thesis that untrammeled emotionalism in matters of artistic creation and interpretation is a potential obstacle to greatness, and extolled cool-headed judgment and discrimination, so necessary in the mastery of techniques and of what it takes to give a concept its ultimate concrete form, whether

it be on the stage, on the canvas, or on the blank page. In so doing, Diderot strongly qualified his long-held belief that originality and genius require unrestrained enthusiasm and the uninhibited expression of strong passions.

In 1760 Diderot switched from the theater to fiction as a genre eminently suited for achieving both moral seriousness and a high degree of verisimilitude. The failure of the theater to fulfill Diderot's expectations and the revelation of Samuel Richardson prompted him to reexamine the novel, an officially decried yet immensely popular genre, its aesthetic potentialities and moral import. As he eloquently proclaimed in his *Eloge de Richardson* (Eulogy of Richardson), written in 1761 and first published in the *Journal étranger* of 1762, fiction nccd not be a frivolous, inconsequential pastime primarily intended for bored, idle, and restless women. What the *Eloge* makes clear is Diderot's admiration for the uniquely effective and forceful combination of realism and moralism that made such novels as *Pamela* and especially *Clarissa Harlowe* so compelling as well as edifying. *La religieuse* (The Nun), written in 1760 and first published in 1796, is Diderot's first full-fledged novel. It shows that Diderot had learned a great deal from Richardson in the way a novelist can make extensive use of small touches of everyday life, as well as of characters drawn from the middle class in order to create the illusion of reality and appeal to the interest and the sympathy of the reader. In epistolary form, it offers the sad recital of the misfortunes of Suzanne, a young woman who is forced to take religious vows as an atonement for being the offspring of her mother's adulterous liaison. *La religieuse* introduces some daring new themes that foreshadow the darker side of Romanticism through its episodes of sadism and lesbianism and through its pervasive mood of penumbral melancholy and anxiety.

In 1761 Diderot wrote the initial draft of a dialogue, *Le neveu de Rameau* (Rameau's Nephew), periodically revising and expanding it for nearly twenty years. Because of the subversive nature of the work and its many ferociously satirical allusions to influential living contemporaries, Diderot wisely decided to refrain from publishing it in his lifetime and to place his faith in posterity, keeping silent about it even with his most intimate friends. After his death in 1784, a copy of the manuscript somehow reached Schiller, who recognized its exceptional significance and hastened to show it to Goethe. The latter immediately set out to translate it, and thus it was that in 1805 what many view as Diderot's masterpiece first appeared in Germany. The first French edition in 1821 was merely a retranslation of the German version. A holograph copy of the manuscript was fortuitously discovered in 1890 among the bookstalls of the Quai Voltaire, eventually ending up at the Pierpont Morgan Library.

Le neveu de Rameau is entirely in the form of a dialogue between two historical characters, Diderot himself and Jean-François Rameau, the parasitic, ne'er-do-well nephew of the famous composer. However, MOI (or me)-Diderot represents a slyly caricatured portrait of a rather self-satisfied, stuffy *philosophe*, while LUI (or him)-Rameau is endowed with far greater imaginative powers and self-awareness than the mediocre, lackluster individual we know from contemporary accounts and documents. One of the most dramatically effective features of *Le neveu de Rameau* derives from the sharp contrast between the relative conservatism and smugness of MOI and the volatility of mood and thought exhibited by the cynically outspoken LUI. Bitter resentment and jealousy, smoldering hatred and vindictiveness, helpless frustration and provocative self-mockery, desperate yearning for success and wealth, genuine enthusiasm for the arts and especially for music, stunning virtuosity in the execution of impromptu pantomimes, thought-provoking insights into the nature of talent versus genius are flashed in rapid succession before an amused and bemused MOI.

As the contemptuously discarded artist barely subsisting on the fringes of an indifferent society he in turn despises, LUI sporadically attains true insight, not only into the injustices and absurdities he sees everywhere, but also into his own shortcomings and lack of true creative genius. In these fleeting but revealing moments, he attains a tragic dimension. But these flashes of pitiless lucidity are soon replaced by more typical sarcastic recriminations and salacious, gossipy anecdotes. As an alienated outsider, the Nephew is the ideal observer and satirist, and it is through his comic mimicry that Diderot holds up to ridicule a whole range of social and political abuses and corrupt practices, emphasizing the pervasiveness of money with special force and in a way that foreshadows Honoré de Balzac. Through ferocious caricatures of the rich and powerful of the day and their fawning satellites we are presented with striking examples of how highly vaunted moral values are subverted and invoked solely in order to deceive the gullible and simpleminded.

In an even more experimental vcin is *Jacques le fataliste* (written from 1765 to 1784, and first published in 1796), a highly personal adaptation of the picaresque genre. Written at a time when Diderot had read with delight Laurence Sterne's *Tristram Shandy*, it quite deliberately sets out to parody fictional conventions and techniques, and has therefore frequently been characterized as an early antinovel. Loosely based on a device going back to *Don Quixote*, the multilayered narrative, in lively dialogue, revolves around a master-servant relationship while both are journeying on horseback. But this narrative framework and the plot involving the adventures of Jacques, the servant, and his nameless master, frequently interrupted by the author's asides and interventions, are primarily meant to thwart the reader's expectations, largely derived from novelistic traditions and clichés.

Unlike Cervantes, Diderot gives the servant rather than the master the leading role, for it is through Jacques's rambling, digressive, and oft-interrupted tale that the ethical

dilemma of determinism versus free choice, which had greatly preoccupied Diderot ever since his endorsement of scientific materialism, is repeatedly tested. Jacques's incessant protestations of fatalism may suggest a passive, pessimistic philosophy of life, but his actions belie his simplistic credence in a foreordained world and are motivated by impulses that he is unable to explain in rational terms. It is evident that Diderot's intent is to treat the complex problem of determinism in a satirical, humorous mode, but more importantly to reaffirm his humanistic faith in the individual's ability to overcome, at least to a certain extent, the blind forces that are supposed to shape his destiny. *Jacques le fataliste* is an uncannily modern work of fiction with a special appeal for twentieth-century critics, theorists, and novelists. It has been singled out for generous praise by André Gide in his *Journal*, and by Milan Kundera, who wrote a play based on it titled *Jacques and His Master: An Homage to Diderot in Three Acts* (1971).

Diderot also wrote several *contes* or tales that exemplify his artistic mastery of the short story as a means to develop a single central theme, usually of a moral nature, through the behavior and actions of single-minded, willful characters who find themselves in conflict with the accepted codes and conventions of society, notably *Entretien d'un père avec ses enfants* (Conversation of a Father with His Children; written in 1770, first published in 1773), *Les deux amis de Bourbonne* (The Two Friends from Bourbonne; written in 1770, first published in 1773), *Ceci n'est pas un conte* (*This is Not a Story;* written in 1772, first published in 1798), and *Madame de La Carlière*, also known as *Sur l'inconséquence du jugement public* (On the Inconsistency of Public Opinion; written in 1772, first published in 1798). From Diderot's choice of topics it is evident that he is particularly preoccupied with themes of exceptional devotion, courage, and self-sacrifice, or egotism, obstinacy, and treachery, and with the ambiguities, contradictions, misunderstandings, ironies, and cruelties that so often characterize relations between men and women, especially when their passions are aroused.

Diderot is a naturally gifted storyteller who knows how to extract from his own experience and from real-life situations completely believable characters, at once easily recognizable yet uncommon in the intensity of their emotional involvement or moral and psychological predicament. With a keen eye for the telling detail, gesture, or word, he excels at creating scenes of stark dramatic power through outward simplicity and even banality. Thus he created a new kind of short story that strives to produce the illusion of reality but at the same time sets into sharp relief the characteristics, details, and idiosyncrasies of exceptionally strong-willed individuals pitted against the laws and taboos of society.

One of Diderot's major goals was to formulate a new relation between culture and society, as well as between the creative individual and society. His writings on literature and the arts make clear that his aesthetics were inseparable from his ethical and political ideas. Unlike Rousseau, however, Diderot looked on the role of the arts in society, even a corrupt one as he undoubtedly viewed the France of Louis XV and Louis XVI, as an immensely positive force. [*See* Rousseau.] All his life, he approached the artist's creative process and procedures and the interrelation between art and society with the liveliest interest and with a keen personal sense of sympathetic, enthusiastic involvement. Whereas Rousseau viewed the arts with suspicion if not outright hostility, Diderot found in them an unending source of delight and spiritual enrichment and renewal. Whereas Rousseau perceived a profound cleavage between art and moral values, Diderot tirelessly sought to reconcile the respective exigencies of the aesthetically pleasing and the socially useful. He did not look on the arts as above or separate from life, and from the outset he acknowledged the essential roles of passion and enthusiasm in the creative process and of subjective sensibility in aesthetic appreciation and pleasure. The sublime, in particular, had great appeal for him, especially in conjunction with Edmund Burke's *Philosophical Enquiry into the Origin of Our Ideas of the Sublime and Beautiful* (1757), which reinforced his rejection of neoclassicism in favor of a new aesthetic stressing the overpowering, nostalgic, and melancholy rather than pleasing emotions and experiences.

BIBLIOGRAPHY

Editions of Works by Diderot

Contes. Edited by Herbert Dieckmann. London, 1963.

Contes et entretiens. Edited by Lucette Perol. Paris, 1977.

Essais sur la peinture: Salons de 1759, 1761, 1763. Edited by Gita May and Jacques Chouillet. Paris, 1984.

Lettre sur les aveugles. Edited by Robert Niklaus. Geneva, 1951.

Lettre sur les sourds et muets. Edited by Paul Hugo Meyer. Diderot Studies 7. Geneva, 1965.

Le neveu de Rameau. Edited by Jean Fabre. Geneva, 1950.

Quatre contes. Edited by Jacques Proust. Geneva, 1964.

La religieuse. Edited by Roland Desné. Paris, 1968.

Salon de 1765. Edited by Else Marie Bukdahl and Annette Lorenceau. Paris, 1984.

Salons. 4 vols. Edited by Jean Seznec and Jean Adhémar. Oxford, 1957–1967. Reissued and revised by Jean Seznec (Oxford, 1975–1979).

Salons, III: Ruines et paysages: Salons de 1767. Edited by Else Marie Bukdahl, Michel Delon, and Annette Lorenceau. Paris, 1995.

Salons, IV: Heros et martyrs: Salons de 1769, 1771, 1775, 1781; Pensées detachées sur la peinture. Edited by Else Marie Bukdahl, Michel Delon, Didier Kahn, Annette Lorenceau, and Gita May. Paris, 1995.

Collected Works

Oeuvres. Edited by André Billy. Paris, 1951.

Oeuvres complètes. 20 vols. Edited by Jules Assézat and Maurice Tourneux. Paris, 1875–1877.

Oeuvres complètes. 15 vols. Edited by Roger Lewinter. Paris, 1969–1973.

Oeuvres complètes. Edited by Herbert Dieckmann, Jean Fabre, Jacques Proust, Jean Varloot, et al. Paris, 1975– . 34 volumes when completed (22 thus far published).

Oeuvres esthétiques. Edited by Paul Vernière. Paris, 1963.

Oeuvres politiques. Edited by Paul Vernière. Paris, 1963.

Translations

Rameau's Nephew and Other Works. Translated by Jacques Barzun and Ralph H. Bowen. Garden City, N.Y., 1956.

The Paradox of Acting. Translated by Walter Herries Pollock. New York, 1957.

Jacques the Fatalist and His Master. Translated by John Robert Loy. New York, 1959.

Memoirs of a Nun. Translated by Francis Birrell. London, 1959.

Diderot on Art, vol. 1, *The Salon of 1765 and Notes on Painting;* vol. 2, *The Salon of 1767.* Translated and edited by John Goodman. New Haven, 1995.

Critical Studies

Barzun, Jacques. "Why Diderot?" In *Varieties of Literary Experience,* edited by Stanley Burnshaw. New York, 1962.

Belaval, Yvon. *L'Esthétique sans paradoxe de Diderot.* Paris, 1950.

Bukdahl, Else Marie. *Diderot, critique d'art.* 2 vols. Copenhagen, 1980–1982.

Chouillet, Jacques. *La Formation des idées esthétiques de Diderot, 1745–1763.* Paris, 1973.

Crocker, Lester. *Two Diderot Studies: Ethics and Esthetics.* Baltimore, 1952.

Diderot et l'art de Boucher à David: Les salons, 1759–1781. Paris, 1984. Catalogue of an exhibition curated by Marie-Cathérine Sahut and Nathalie Volle.

Dieckmann, Herbert. *Cinq Leçons sur Diderot.* Geneva, 1959.

Fellows, Otis. *Diderot.* Upd. ed. Boston, 1989.

Fredman, Alice. *Diderot and Sterne.* New York, 1955.

Fried, Michael. *Absorption and Theatricality: Painting and Beholder in the Age of Diderot.* Berkeley, 1980.

Furbank, P. N. *Diderot: A Critical Biography.* New York, 1992.

Josephs, Herbert. *Diderot's Dialogue of Gesture and Language: "Le Neveu de Rameau."* Columbus, Ohio, 1969.

Kavanagh, Thomas M. *The Vacant Mirror: A Study of Mimesis through Diderot's "Jacques le fataliste."* Studies on Voltaire and the Eighteenth Century, 104. Oxford, 1973.

May, Georges. *Diderot et "La religieuse."* New Haven, 1954.

May, Gita. *Diderot et Baudelaire, critiques d'art.* Geneva, 1957.

Mehlman, Jeffrey. *Cataract: A Study in Diderot.* Middletown, Conn., 1979.

Seznec, Jean. *Essais sur Diderot et l'antiquité.* Oxford, 1957.

Sherman, Carol. *Diderot and the Art of Dialogue.* Geneva, 1976.

Undank, Jack. *Diderot: Inside, Outside, and In-Between.* Madison, Wis., 1979.

Wilson, Arthur M. *Diderot.* New York and Oxford, 1972.

GITA MAY

Diderot and the Salon

Since the nineteenth century, Denis Diderot's *Salons* have been regarded as the inaugural works of modern art criticism, and the reputation is deserved. The greatest of these texts, occasioned by the official Louvre exhibitions of works by members of the Royal Academy in 1765 and 1767, rank with Charles Baudelaire as the most compelling reads in the Western art-critical tradition, and they were influential from the moment they became widely accessible in the Naigeon edition of Diderot's works (1798). Their influence on nineteenth-century criticism, theoretical writing, and even artistic production is vastly underestimated, and their engaging lack of system makes them especially valuable today, for they present us with the spectacle of a brilliant intellect grappling with the central issues of visual representation and signification at the moment of transition between the classic and romantic approaches to these problems, between the regimes dominated by, respectively, theories of mimesis and of subjective expression.

Since Charles-Augustin Sainte-Beuve, the richness of these texts has been obscured by a reductive view of Diderot's art criticism as "literary." In this context, the term is meant to evoke his penchant for reading the dramatic content of paintings, and for using them as jumping-off points for his own narrative invention. It is easy to understand how this notion gained currency, for there is no denying the important role in these works of self-contained stories, from witty anecdotes and smutty jokes to ingeniously conceived set pieces such as the salacious dialogue with the subject of Jean-Baptiste Greuze's *Young Girl Crying over Her Dead Bird* (in which Diderot reveals that she mourns the loss of something far more precious than her canary) or the dream fantasia on Jean-Honoré Fragonard's *Corésus and Callirhoé* (in which he recounts the entire sequence of events climaxing in the depicted moment of self-sacrifice prompted by unrequited love, a vision described in terms evocative of Plato's parable of the cave). Indeed, at several points in *1767* he says to Grimm, the close friend to whom the *Salons* were addressed, "Let's continue to tell stories." But the tendency to focus on isolated passages—in the English-speaking world, a by-product of the lack of integral translations prior to 1995—has bred misconceptions, making it difficult to appreciate the critical subtlety, historical importance, and full literary interest of these texts. After completing *1765,* Diderot wrote his beloved Sophie Volland that he thought it the best thing he had yet written, and *1767,* though less poised and resolved than its predecessor, is perhaps greater still. The last four Salons—*1769, 1771, 1775,* and *1781*—contain many memorable passages, but they had to be pried out of him and show signs of creative exhaustion. No wonder: He had broken the mold in *1767,* which overflows with ruminative excurses, provocative insights, and playful *racontage.* Thus there is some irony in his famous paean in the *Salon of 1781* to Jacques-Louis David's *Belisarius,* paraphrasing Jean Racine ("Every day I see it and every day I think I'm seeing it for the first time"), for the enterprise of art criticism had long since gone stale for him.

When in 1759 Diderot became an art critic for the *Correspondance littéraire,* the exclusive manuscript newssheet for which he wrote the *Salons,* he followed the same procedure he used in preparing the craft entries for the *Encyclopédie:* He sought the tutelage of practitioners themselves—Jean-Baptiste-Siméon Chardin, the sculptors Jean-Baptiste Lemoyne and Étienne-Maurice Falconet, and other artists—concerning technical questions, listening attentively to their

DIDEROT: Diderot and the Salon. Gabriel Jacques Saint-Aubin, *The Salon of 1765* (1765), water-color; Musée du Louvre, Paris. (Courtesy of Giraudon/Art Resource, New York.)

remarks about theoretical issues as well, for instance, the nature of imitation (in the opening pages of *1767* he discusses the "metaphysics" of mimesis with Falconet), the character of genius (Diderot implies it is analogous to criminality, anticipating the modern notion of the transgressive artist), and the desire for posthumous fame as a spur to artistic ambition (the subject of a fascinating but rather wooly epistolary debate between Diderot and Falconet). He learned studio lingo from Charles-Nicolas Cochin (1715–1790), the influential printmaker, aesthetic propagandist, and art-bureaucratic gadfly whose many publications downplaying subject matter in favor of formal values made him the intellectual heir of Roger de Piles. [*See* Cochin.] Newly enlightened by this education, which prompted him to ponder a host of aesthetic questions ("On my own I've reflected on what I've seen and heard, and on artistic terms such as *unity, variety, contrast, symmetry, disposition, composition, character,* and *expression,* formerly so comfortable on my lips but so indistinct in my mind," *1767*), Diderot set about recasting the project of art criticism in light of the changing realities of French society and the Parisian art world.

At the beginning of the *Salon of 1763* he profiles the ideal Salon reviewer: He must have "taste of every sort, a heart sensitive to all charms, a soul susceptible to an infinity of different enthusiasms, a variety of style that answers to the variety of brushes." He adhered to this commodious program, thinking himself into many points of view and ven-

triloquizing the remarks of a broad range of observers (dialogue is almost as prominent in his art criticism as it is in *Rameau's Nephew, Jacques the Fatalist,* and *D'Alembert's Dream*). As a result, the reader is often convinced that he or she has access not only to Diderot's own thought processes but to something like the full range of contemporary response to the visual arts.

This open conception of criticism has made it possible for the influence of these texts to operate in seemingly contradictory ways. They have functioned, for example, as a bridge between a predominantly literary critical agenda and a more formalist one. Diderot's delight in fabulation is countered by his gift for description and his acute awareness of how style and subject matter interact. The *Salons* contain the longest and most technically sophisticated evocations of works of art penned up to that time in any European language. They considerably upped the stakes in the matter of translating the visual into the verbal (a problematic explicitly investigated in *1767*), heightening awareness of medium and technique, as opposed to iconography, and underscoring the limitations of the word when confronted with the concrete image. But the same texts proved important for the Realist project as formulated by Gustave Courbet's friends Champfleury and Duranty, whose *La nouvelle peinture* (1876) cites Diderot's *Essais sur la peinture* (Notes on Painting, a kind of technical-theoretical appendix to the *Salon of 1767*) on the inevitable imprinting of lived

experience on the body, an idea now indelibly associated with the work of Edgar Degas. These two readings—one Formalist, the other Realist—are at odds with one another, yet both find justification in Diderot, and their very incompatibility points up the proleptic potency and intellectual breadth of his texts.

Insofar as the *Salons* explore the many contradictions generated by the emergence of the public sphere in eighteenth-century French society, they also lend themselves to a Habermasian reading. Perhaps the most piquant of these contradictions is exemplified by the curious circumstances in which these texts first appeared. The *Correspondance littéraire* numbered at most fifteen subscribers, all of them members of European ruling families. The withering frankness with which Diderot articulated his views about the failings of many artists he discussed was made possible by the texts's virtual inaccessibility during his lifetime, for the rules of journalistic decorum precluded such pointed commentary in the public discourse of 1760s France. Yet his judgments were based on solid convictions about the important role of art as the symbolic expression of communal values and aims fostered by his encounters with Charles Du Bos, Montesquieu, Johann Joachim Winckelmann, and Jean-Jacques Rousseau, views that prepare the way for a conception of liberal democracy most prominently linked with the early work of Jürgen Habermas.

Similarly subject to feint and parry were Diderot's ideas about the hierarchy of genres, an important element in the aesthetic ideology of the old regime since its codification by the French Academy in the second half of the seventeenth century. He pronounced the distinction between history painting and genre painting—rooted in the aristocratic humanist culture of the Renaissance and fundamental to the organization's intensely status-conscious agenda—to be "intelligent" but reformulated it, suggesting that human interest and technical accomplishment should be the decisive considerations, not the social class of their figures or the canonical status of their literary source: "I submit that Greuze's treatment of a father reading to his family, an ungrateful son, and a marriage contract, that Vernet's marines, which offer all kinds of scenes and incidents, are for me no less history paintings than are Poussin's *Sacraments*, Le Brun's *Family of Darius*, or Van Loo's *Susanna" (Notes on Painting)*. Arguments along these lines, which proliferate in the *Salons*, tend to blur distinctions between the eighteenth-century equivalents of high and low culture in ways consistent with the emergence of a vocal proto-egalitarian "public." He called for a "modern" history painting that would take recent events of compelling public interest as its subject matter, a notion that would reverberate throughout the nineteenth century. One might go further: It is not entirely frivolous to see Diderot as a precursor, through Baudelaire and other intermediaries, of today's cultural studies agenda,

which is premised on a romanticized conception of critique implicit in Habermas's thesis.

And then there is the postmodern reading of Diderot, the one that currently prevails among literary theorists. It proves an apt platform from which to explore the art criticism, which is rife with deconstructive strategies. Tone can shift in these pages from the high-flown to the down-and-dirty with dizzying speed, and Diderot takes pleasure in driving home the instability of subject-object relations in that ambiguous space that might be called the scene of looking-at-painting. The richly detailed excursion through several spectacular landscapes subsequently revealed to be paintings by Joseph Vernet *(1767)* is only the most celebrated example of his tactics of purposive disorientation and genre violation, which also figure prominently in his fictional works. Indeed, the literary conceits in the *Salons* can be understood properly only in the context of Diderot's determination to investigate epistemological problems as they bear on visual perception and the psychology of human interaction.

Diderot's awareness of the erotic component of visual experience, intensified by a sustained encounter with Edmund Burke's *Philosophical Enquiry* in 1765–1767, was exceptional among eighteenth-century writers on art (as opposed to artists themselves). As a result, interesting relations can be mapped between his criticism and recent work on gender, sexuality, and reception. Many of his notorious ex post facto reworkings of paintings and drawings take the form of sexual projections (the fantasy scenario prompted by Greuze's *Young Girl Crying over Her Dead Bird* being only the most flagrant example of many), and there are hints he sensed that sexual stimulation and the aesthetic sense were more thoroughly imbricated than most of his French contemporaries were ready to admit, at least in print. In *1765*, for example, some racy works by P.-A. Baudouin (Boucher's son-in-law) occasion a pun on the word *prunelle* that turns on an analogy between the pupils of the eyes and male testicles. Such passages suggest Diderot was groping toward views implicit in Burke and recently developed within a Lacanian framework in *The Judgment of Paris* by Hubert Damisch (1993/1996), which posits the sex/beauty relation as the dirty little secret of Western aesthetics, a fundamental truth that many of its canonical texts—most crucially Kant's *Critique of Judgment*—obscure more or less deliberately. Diderot probed this issue time and again, preparing the way for more fully theorized explorations along these lines in the post-Freudian era. Indeed, his "line of liaison," the swath traced over a painted surface by the perceiving eye as it follows pathways laid out for it by the artist *(1767*, discussions of J.-M. Vien and G.-F. Doyen), marks a crucial step in the prehistory of the subjectifying "gaze," that obsessive preoccupation of post-Lacanian and feminist cultural criticism. Consequently, Diderot should be

allotted a place of honor next to Lessing as one of the primary progenitors of the critical enterprise as we have come to understand it under the modern scopic regime.

[*See also* Criticism, *article on* Art Criticism.]

BIBLIOGRAPHY

Work by Diderot

Diderot on Art, vol. 1, *The Salon of 1765 and Notes on Painting;* vol. 2, *The Salon of 1767.* Translated and edited by John Goodman. New Haven, 1995.

Other Sources

Bryson, Norman. *Word and Image: French Painting of the Ancien Régime.* Cambridge and New York, 1981. See chap. 6, "Diderot and the Word," pp. 154–178, and chap. 7, "Diderot and the Image," pp. 179–203.
Diderot et l'art de Boucher à David: Les salons, 1759–1781. Paris, 1984. Catalog of an exhibition curated by Marie-Cathérine Sahut and Nathalie Volle. See especially the essay by Jean Starobinski, "Diderot dans l'espace des peintres," pp. 21–40.
Lévi-Strauss, Claude. *Regarder Écouter Lire.* Paris, 1993.

JOHN GOODMAN

DIFFERENCE (*différance*). *See* Deleuze; Derrida; Feminism; Irigaray; Lyotard; *and* Politics and Aesthetics, *article on* Difference and Culture.

DIFFICULTY, AESTHETICS OF. [*This entry traces the development of the aesthetics of difficulty (namely, writing or creating art that seems deliberately hard to understand or interpret) from the beginning of modernity up through contemporary art and art theory. For related discussion, see* Ineffability; *and* Romanticism.]

G. W. F. Hegel opens the *Phenomenology of Spirit* by warning the reader against using his preface as a shortcut. By objecting to the notion of an abridged, more intelligible version of his work, Hegel was placing difficulty as the cornerstone of the philosophical inquiry; philosophy was neither to provide solutions, nor to reconcile the reader with an increasingly fragmented world. Far from making reconciliation (*Versöhnung*) the goal of his philosophy, he advanced negativity—the disclosing of the contradictory nature of experience—as one of his aims. Thought, writes Hegel, is presented "in a negative relation to the figure to which it goes forth and with which it wrestles, all the time trying to work itself into [einzubilden] it." The act of reading must go through this continuous wrestling with meaning.

German critics like Wolfgang Iser, Hans Robert Jauss, or Jurij Striedter, affiliated with the so-called Constance school and influenced by Edmund Husserl, Roman Ingarden, Hans-Georg Gadamer, and speech-act theorists have analyzed at great length the connection between difficulty and the way we perceive and concretize works of literature.

Inspired by Russian Formalism, in particular Viktor Shklovsky's conception of dehabitualization (*ostranenie*), and Gadamer's idea of horizon of expectations (*Erwartungshorizont*) in *Truth and Method,* they explored how works appeal to their readers by breaking and creating expectations. Works of art go through a "difficult" reception phase when they usher in new tropes, new literary devices into the canon. This was the case when Gustave Flaubert revolutionized the history of the novel by introducing the *style indirect libre* (free indirect discourse) in *Madame Bovary.* Until the device is absorbed into everyday discourse, it stands between the reader and the proper concretization of the artwork. Iser and Jauss, the *chefs de file* of what is now known as reader-response criticism, often chose these very points of indeterminacy (*Unbestimmtheitstellen*) to understand the importance of difficulty for the art process. They distinguish the texts that elicit a quasi-pragmatic reading (based on immediate consumption, and therefore doomed to be absorbed into the language of clichés) from those that lengthen our perception, thereby securing a more strenuous, but far more lasting, encounter with the work's content. What became clear in their findings was the appeal derived from what is denied to the reader. An aesthetics of negativity is far more suggestive than one that plays on primary levels of identification between reader and work. If the public were denied the much awaited catharsis at the end of *Oedipus,* then the impact of the work might be even more devastating.

The Marxist poet and playwright Bertolt Brecht discusses these very issues in his essays on the theater. His self-proclaimed anti-Aristotelian plays pivot around the spectator's productive frustration and alienation. The famous *V-Effekt* (*Verfremdungseffekt*, principle of alienation) originated in his belief that only through displacement and estrangement can the spectator be genuinely affected by the play or the poem. When art is based on empathy, on identification, the catharsis it elicits can only be short-lived. Like Theodor W. Adorno, Brecht has the ambition to turn art into a lasting, scarring experience. Resolution and closure are to be shunned; expectations must be thwarted. In the performances of his plays, he often had women perform the roles of men, older actors play younger characters, or even switched around the actors' roles minutes before the performance, to induce a halting, strange delivery in his actors. All these anomalies ended up shaking the viewers' expectations, making it impossible for them to settle into a reassuring and solacing relationship to art. The principle of difficulty preserves the unity of the work of art. It thwarts any attempt on the part of the reader or viewer to make the work his or her own; first and foremost, it must remain irreducible to itself.

Franz Kafka's stories illustrate such irreducibility by depriving readers of a key that would direct their understand-

ing and steer their interpretation one way or the other. With this relentless play of meanings, stories like "Josephine the Singer" or "Before the Law" turn reading back onto itself, putting readers through a series of taxing, precarious exegeses that write themselves anew with each reading. This leaves readers not simply with a sense of the cryptic, the impregnable, but produces difficulty, the way a religious parable would, when it leaves the final word to disciples. Rather than providing explanations, completing spots of indeterminacy, Kafka plays on these gaps, stretching them out, so that his readers can never achieve a certain intimacy with the text. The narrator's ultimate move is to pull back, thereby prolonging the reader's inquiry. The truth can only be sought, never found; texts that place such a premium on their public's confusion are making the point that art is not about finding but about searching. Finding is coextensive with appropriation, and appropriation is what ends up freezing the aesthetic experience into a commonplace, hackneyed view of the world.

This resistance to the public's desires can be traced back to the distinction made by Immanuel Kant and Alexander Gottlieb Baumgarten between disinterested and teleological art. It is the former that interests us here, as it can never be confused with any clear or practical application; it is difficult in its very essence. According to Kant, the Beautiful always requires reflection, and is therefore inevitably fraught with obscurity. In his masterful work on allegory, Angus Fletcher reminds us that this equation of art and obscurity, aesthetics and variability, stems from Aristotle's *Nicomachean Ethics:* "Deliberation is concerned with things that happen in a certain way for the most part, but in which the event is obscure, and with things in which it is indeterminate. We call in others to aid us in deliberation on important questions, distrusting ourselves as not being equal to deciding." Fletcher points out that this "might be the case with allegorical obscurity," although "allegory teems with cases where the particular meaning of the work is not immediately obvious, where the intention of the work is to shroud abstract ideas under a cover of dimly understandable imagery, as in the mystical or doctrinal passages of *The Divine Comedy . . .* This enigmatic art thrives because its 'difficult ornament' arouses the reader's curiosity. It therefore deliberately tries to be obscure" (Fletcher, 1964, pp. 310–311).

Paul de Man offered his own interpretation of allegory by establishing its relationship to difficulty. He rebukes the Romantic privileging of symbol over allegory ("The Rhetoric of Temporality") by defying the assumption that the symbol heals the gap between subject and object, human being and nature. Allegory, de Man argues, does just the reverse; it renounces the nostalgic desire to harmonize the self with an illusory identification with the nonself. It is the wish to maintain this painful discontinuity that is at the core of the allegorical project, making it not about healing the otherness of the world, but about emphasizing the difficult and unresolved quality of experience. Like irony, allegory undermines at will the readability of a text by refusing to reconcile sign and meaning, signifier and signified. This antiessentialism denies the existence, to quote Marcuse, of the "one true Being" who is spared "the constantly changing multiplicity of appearances."

For the Yale school (de Man, Harold Bloom, Geoffrey Hartman) Romanticism's self-conscious mode of writing opened up a rich field of interpretation where difficulty and nonidentity became conditions of the artistic experience. This contiguity may explain the prime place of the fragment in an aesthetics of difficulty. As the writings of Roland Barthes, Gilles Deleuze, or Jacques Lacan testify, the fragmentary encourages the beholder to relinquish the idea of psychological coherence. Contrary to expectations, what is incoherent, difficult to assimilate, arrests dramatically the reader's attention; not surprisingly, the act of putting fragments together might even convince the reader that art is about reconstruction, not assimilation. In this scenario, works of art are not about mirroring one's identity, but about having it altered. This transformation reveals itself best in spaces of indeterminacy: the arduous passages of a poem, the clouded lines of a portrait, or the garbled explanation of an analysand. The tremendous changes that have occurred in the history of interpretation—in domains as different as music and psychoanalysis—have had a lot to do with the incorporation of the difficult, the unassimilable, in the process of understanding. If allegory insists on a perverse elusiveness, it is in order to preserve an autonomous aesthetic realm that will ultimately increase the public's mental involvement. To thinkers like Adorno, Herbert Marcuse, or Peter Bürger, so-called difficult art, art that embraces negativity, is desirable because it provides a refuge against commodification. These writers naturally gravitated toward authors who, like Samuel Beckett, Peter Handke, or the Surrealists, refused to turn their works into pat life lessons. Bürger's analysis of commodification versus artistic autonomy *(Theory of the Avant-Garde)* successfully situates the function of subversion and estrangement in the history of literature. He traces some of the symptoms of the increasingly widespread desire to be difficult (not dissimilar to what Friedrich Nietzsche called *Muth zur Langweiligkeit,* the courage to be boring) to the antagonism between artists and the so-called *littérature facile.* Bürger scrutinizes the type of poetic obscurity rooted in linguistic hermeticism: why do writers try to conceal, rather than disclose their message? Linking William Empson's *Seven Types of Ambiguity* to the Brechtian alienation effect, he concludes that this obscurity "would . . . aim at creating a treasure trove of new meanings within the poverty of common language." He sees in poetic language "a game of multiple, diverse, and opposing meanings, the most authentic effect [of which] would be ambiguity." In other words, it is ambiguity (which translates into the reader's experiencing difficulties) that

guarantees that the text will preserve its shock value, thus ensuring a lasting impact. Bürger demonstrates how difficult works—Friedrich Schlegel's *Lucinde,* for example—successfully stymied the overbearing needs of patrons, publishing houses, and the public by showing up the same deception Charles Baudelaire denounced in the opening lines of *Les fleurs du mal:* "Hypocrite lecteur! Mon semblable, mon frère." Baudelaire mocks the hypocritical reader for hiding the fact that his or her only interest in poetry stems from the desire to be amused.

Some of the most provocative discussions about the benefits of difficulty are to be found in the writings of the first generation of German Romantics. It was the so-called Jena brotherhood that began embarking most actively on a campaign against the equation of art and recreation, targeting most pointedly the reader as the prime suspect in the practice of indulging in literature for private purposes. Schlegel, one of Romanticism's main theorists, advocated artistic difficulty ("On Incomprehensibility") as a buffer against the public's demand for increasingly spicy and intense stimulations. One of Schlegel's aphorisms reads: "In order to write well about something, one shouldn't be interested in it any longer." The same would apply to the act of reading. Schlegel worried that the public only read what could apply directly to its situation; readers prized what was most shocking, so "when the interesting has become expected, men will demand more intense stimulations . . . until in the end the shocking itself is expected and only boredom remains." Difficulty is recognized as a protective agent, a strategy designed to retard the commodification and the trivialization of art. It achieved this by moving from the external to the internal, a move that Hegel traced to Romanticism, the first school to postulate the spirit's own inwardness, a spirit he contrasted with "the flesh, the external reality, and the world," all posited as negative *(Nichtiges).* To the German Romantics, and this is a current that will dominate the artistic production of authors ranging from Edgar Allan Poe, Baudelaire, Stéphane Mallarmé, and Nietzsche, to Marcel Proust and James Joyce, aesthetics evolves in the realm of negativity. It stands antithetically to the world, severing itself firmly from the realm of ethics.

The Neoplatonic ring of some of Novalis's or Baudelaire's writings derives from a confidence that the *ici-bas,* the prosaic, is a mere physical incarnation of the Idea, of the essence. For such authors, the role of the artist is to translate into common speech the language of the unsayable, to find the *correspondances* with the beyond. For Swedenborg, one of the most influential voices for the Symbolist poets, there "is somewhere in the world a mysterious book in which the eternal laws of Beauty are written. Only artists may decipher its meaning." Art, in other words, is the mere physical manifestation of another text, far more important, but almost impossible to decipher. The task of the poet-translator is to overcome this difficulty, and force open the text through a learned, rigorous act of contemplation. In an unprecedented move to increase and complicate the reader's participation, authors like Schlegel, Novalis, and Charles Nodier made reading the odd meeting point between faith and puzzle-solving. Following the lead of some of their innovative predecessors (Miguel de Cervantes's *Don Quixote,* Laurence Sterne's *Tristram Shandy,* Denis Diderot's *Jacques le fataliste*), they used unreliability as one of the principles of their paradoxical philosophies of reading, forcing their public to enter into a complicated maze of significations and enigmas. Art, in other words, played the role of an initiation in which difficulty was one of the ritualized components. Romantic texts often ridicule the reader's desire for predigested art (E. T. W. Hoffmann's *Der goldene Topf,* Ludwig Tieck's *Franz Sternbalds Wanderungen,* Nikolay Gogol's *The Portrait*), giving center stage to the philistine who calls art whatever will entertain him, mocking artists who are not willing to make themselves immediately understood. The writer, the painter, or the musician who is willing to make art against the audience's wishes (Hoffmann's *Kreisleriana*) has a religious status. The vocabulary of martyrdom, asceticism, or *Weltflucht* (escape from the world) is recurrent in these works, suggesting a very important link between creation and self-sacrifice. Difficulty, in turn, derives from the contemplative effort demanded from the public. German Romanticism's grim depictions of the emerging world of museums heralded the loss of the sacred. The hordes of spectators, rushing from painting to painting, were missing the ceremonial, near-devotional relationship to art. According to Novalis, Wilhelm Wackenroder, or Hoffmann, it was this slow process of elucidation that would make art a transfiguring experience. Without difficulty, without the arduous passage from text to translation, from obscurity to revelation, the prosaic would never be turned into the sacred. George Santayana's influential *The Sense of Beauty* also concluded that "there must always remain a penumbra and fringe of suggestion if the most explicit representation is to communicate a truth. When there is real profundity . . . there will accordingly be felt inadequacy of expression, and an appeal to the observer to piece out our imperfections with his thoughts" (Santayana, 1988).

The separation between the beautiful and the useful is particularly salient in works that hold to the principle of art for art's sake. If art is severed from the useful, if it is free from any moral judgment or self-interest, then it will focus primarily on the formal properties of the object, precisely what cannot be readily appropriated. It will emphasize the role of the artist as an artisan who must fashion his or her project so precisely that each isolated part (sentence, brushstroke, musical phrase) is perfect unto itself, self-governing, distinct from the spectator's needs. A work that holds such finality (Kant's *Zweckmässigkeit ohne Zweck*) relinquishes a relationship of empathy and identification with the reader, signifying only its own immanence. Any blatantly self-

contained composition naturally obstructs the audience's emotional gate of entry. If Arnold Schoenberg's atonal music was considered so difficult by his contemporaries, it was because it did not allow the audience to project onto it pre-existing ideas of harmony or melody. He and Anton Webern exemplified the terror brought about by the fear of the new. Their music, notes Adorno in a famous dithyramb against jazz and the regression of listening, "gives form to that anxiety, that terror, that insight into the catastrophic situation which others merely evade by regressing" (1978). Adorno rages against the power of the banal, the artistic turned culinary. The distinction Franz Joseph Haydn had made between the "gallant" and the "learned" was increasingly reflected in the social preference for light against serious, that is, easy versus difficult, music. What Adorno calls the vulgarization of taste is the substitution of the multilevel unity of the whole work for isolated popular passages. He argues that the dynamics of such compositions are so predetermined that they do away with any form of tension; as opposed to works that require effort, deconcentrated listening makes the perception of a whole impossible. The listener, in other words, is selecting whatever he or she has already understood, rather than perceiving art as a formative voyage into the difficult and the unknown.

In *Le plaisir du texte*, Barthes also singles out the reader who cannot tolerate difficulty and is solely concerned with the outcome of the story. By shunning what cannot be immediately integrated into his or her worldview, the reader misses the intrinsic qualities of the work, precisely what gives it its distinctness. Barthes argues that such consumeristic methods would render completely opaque the modernist text or the French *nouveau roman*. The entire meaning of such works derives from the cracks of language, from tensions between form and content—indeed, from whatever falls between the lines. If you practice speed-reading on the modernist text, warns Barthes, then it will be as good as lost; if you are looking for action on the level of plot, then action will be denied since whatever occurs on the linguistic level does not show itself on the level of discourse. It is precisely the difficult, boring, often painstakingly descriptive passages that hide the meaning of the work.

Difficulty arises when the very elements that would have secured a smooth and familiar entrance into the work of art are elided. When form takes precedence over content (whether plot, theme, or melody), the public is forced to absorb the particularities of the *inventio* before being able to apprehend its meaning. It is this permutation that forces the reader to consider the intrinsic values of antimimetic strategies, the divergence of figurative vs. abstract. Works like Beckett's *Endgame* or John Fowles's *French Lieutenant's Woman* deliver their meaning only once the reader has grasped their underlying formal principle; they experiment with the transaction between reader and text, rather than attempt to satisfy curiosity.

The difficult text denies itself the comforts of mimesis. From Flaubert to Joyce, its drive against conventional storytelling is grounded in the desire to make difficulty an integral part of the artistic experience. While a snapshot imparts the more obvious qualities of phenomena, it actually ends up freezing the object, lulling the reader into a false sense of epistemological security. An elusive, nonfigurative type of depiction, on the other hand, forces an ongoing, deeply speculative relationship with the object. Romanticism turned away from mimesis because it privileged the external laws of phenomena, rather than its internal, intrinsic principles. Some of its great figures (Baudelaire, Poe) clamored against the emergence of realism (Honoré Balzac or Charles Dickens) because its emissaries couched their prose in the *style coulant*, an easy, flowing style that trickled down effortlessly into the popular imagination. As a reaction to this, writers like Flaubert, Joris-Karl Huysmans, Adalbert Stifter, Wilhelm Raabe, or Herman Melville took to its limits the notion of realism, making their descriptions so meticulous, so lengthy, that they managed to occupy a quasi-autonomous realm, much like Homer's catalog of ships in the *Iliad*. The disjunction between style and content, plot and technique, was meant to sever the ties between a pleasure-seeking reader and the deep significance of the work.

Critics have often compared the aims of Romanticism to those of abstract art. Both erect their aesthetics around what cannot be recognized immediately, around that which requires translation through an effort of reconstruction. Even though the nonfigurative, nonmimetic, frustrates concretization, it ends up having a more steadfast hold on the subject. The link between abstraction, iconoclasm, and an aesthetics of difficulty brings out the effectiveness of works of art that practice concealment, not disclosure. The unrepresentability of God in the Old Testament, for example, and the direct link between faith and unrepresentability (God is a *deus absconditus*, a hidden god) suggest that it is precisely that which has been denied representation (God's image) that will be incorruptible, never appropriated, and therefore endure the longest. In other words, the desire not to represent, the effort to make revelation a difficult process, guarantees ultimately a dialogical, highly charged relationship to art. Fredric Jameson describes modern artists who practice an aesthetics of difficulty as a breed that relentlessly insists "on the need for modern art and thought to be difficult, to guard their truth and freshness from the austere demands they make on the powers of concentration of their participants, by their refusal of all habitual response in their attempt to reawaken numb thinking and deadened perception to a raw, wholly unfamiliar real world."

BIBLIOGRAPHY

Adorno, Theodor W. *Aesthetic Theory* (1970). Edited by Gretel Adorno and Rolf Tiedemann, translated by C. Lenhardt. London and Boston, 1984.

Adorno, Theodor W. "On the Fetish-Character of Music and the Regression of Listening." In *The Essential Frankfurt School Reader*, edited by Andrew Arato and Eike Gebhardt. New York, 1978.

Amossy, Ruth, and Elisheva Rosen. *Le discours du cliché*. Paris, 1982.

Arato, Andrew, and Eike Gebhardt, eds. *The Essential Frankfurt School Reader*. New York, 1978.

Barthes, Roland. *Le plaisir du texte*. Paris, 1973.

Brooks, Peter. *Reading for the Plot: Design and Intention in Narrative*. New York, 1984.

Bürger, Peter. *Theory of the Avant-Garde*. Translated by Michael Shaw. Minneapolis, 1984.

Cheetham, Mark A. *The Rhetoric of Purity: Essentialist Theory and the Advent of Abstract Painting*. Cambridge and New York, 1991.

Compagnon, Antoine. *La seconde main ou le travail de la citation*. Paris, 1979.

de Man, Paul. "The Rhetoric of Temporality." In *Blindness and Insight: Essays in the Rhetoric of Contemporary Criticism*, 2d rev. ed. Minneapolis: 1983.

Erlich, Victor. *Russian Formalism: History, Doctrine*. 2d rev. ed. The Hague, 1965.

Fletcher, Angus. *Allegory: The Theory of a Symbolic Mode*. Ithaca, N.Y., 1964.

Gadamer, Hans-Georg. *Truth and Method*. 2d rev. ed. Translation revised by Joel Weinsheimer and Donald G. Marshall. New York, 1989.

Holub, Robert C. *Reception Theory: An Introduction*. London and New York, 1984.

Ingarden, Roman. *The Cognition of the Literary Work of Art*. Translated by Ruth Ann Crowley and Kenneth R. Olson. Evanston, Ill., 1973.

Iser, Wolfgang. *The Implied Reader: Patterns of Communication in Prose Fiction from Bunyan to Beckett*. Baltimore, 1974.

Iser, Wolfgang. *The Act of Reading: A Theory of Aesthetic Response*. Baltimore, 1978.

Jameson, Fredric. "T. W. Adorno: or, Historical Tropes." In *Marxism and Form: Twentieth-Century Dialectical Theories of Literature*. Princeton, 1971.

Jauss, Hans Robert. *Aesthetic Experience and Literary Hermeneutics*. Translated by Michael Shaw. Minneapolis, 1982.

Jauss, Hans Robert. *Toward an Aesthetic of Reception*. Translated by Timothy Bahti. Minneapolis, 1982.

Lemon, Lee T., and Marion J. Reis, eds. *Russian Formalist Criticism: Four Essays*. Lincoln, Nebr., 1965.

Marcuse, Herbert. "The Concept of Essence." In *Negations: Essays in Critical Theory*, translated by Jeremy J. Shapiro. Reprint, London, 1988.

Santayana, George. *The Sense of Beauty: Being the Outlines of Aesthetic Theory* (1896). Cambridge, Mass., 1988.

Schlegel, Friedrich. *Friedrich Schlegel's Lucinde and the Fragments*. Translated by Peter Firchow. Minneapolis, 1971.

Schulte-Sasse, Jochen, ed. *Die Kritik an der Trivialliteratur seit der Aufklärung*. Munich, 1971.

Shklovsky, Viktor. "Art as Device." In *Russian Formalist Criticism: Four Essays*, translated and edited by Lee T. Lemon and Marion J. Reis. Lincoln, Nebr., 1965.

Simpson, David, ed. *German Aesthetic and Literary Criticism: Kant, Fichte, Schelling, Schopenhauer, Hegel*. Cambridge and New York, 1984.

Stierle, K. H. "The Reading of Fictional Texts." In *The Reader in the Text: Essays on Audience and Interpretation*, edited by Susan R. Suleiman and Inge Crosman. Princeton, 1980.

Striedter, Jurij, ed. *Texte der russischen Formalisten*, vol. 1. Munich, 1969.

van Zuylen, Marina. *Difficulty as an Aesthetic Principle: Realism and Unreadability in Stifter, Melville, and Flaubert*. Tübingen, 1994.

Weinrich, Harald, ed. *Positionen der Negativität*. Poetik und Hermeneutik, vol. 6. Munich, 1975.

MARINA VAN ZUYLEN

DIGITAL MEDIA. [*To explain the meaning and relevance of digital media, this entry comprises two essays:*
　　　　An Overview
　　　　The Postphotographic Image
The first essay explains the history and present status of digital media, while the second essay explores the aesthetics of images produced by these media. For further discussion, see Computer Art; Cyberspace; Hypertext; Medium; Multimedia; *and* Virtual Reality.]

An Overview

Media are the substances out of which artists have crafted their creations since ancient times. But these substrates of communication are more than mere raw materials. A medium also has a culturally determined instrumentality that defines how and where it is used. Highly evolved social practices supply the structures and contexts for understanding the formed matter that constitutes the objects of cultural expression. Until the late twentieth century, the institutions that established the primary channels of communication harbored at their core the physical objects, processes, and events that ground what we now call analog media. These media are based on the concrete transcription of form from one physical object or event to another. This transference involves the imprinting of shapes into the medium by hand or with specialized tools designed to mold it through physical forces. Paint is applied to surfaces with brushes and palette knives, sounds are made into music with reeds and bows, plastic materials are modeled into sculptures with fingers and chisels, and light is exposed onto photographic film with cameras and lenses. In each case, the shape of the end result bears a physical analogy to the shape of its source because some type of causal link engraves the medium through physical contact. Once put in place, these vulnerable embodiments are protected by galleries, museums, concert halls, and a host of other institutions dedicated to their preservation.

With the proliferation of computers a new kind of medium has emerged. Digital media store information in numbers and symbols rather than in objects and events. They are based on abstract rather than concrete forms of expression that portend a different role for media in culture. Their substrates range from paper cards to magnetic disks to silicon circuits, but they use these compounds as structured registers for logging data rather than as supple surfaces for stamping impressions. They not only use materials differently, they also support different kinds of activities and require different kinds of institutions for organizing their use. One might summarize the contrast between the two fundamental types of media by saying that analog media store and preserve traces of events, while digital media host and compute symbols for numbers. This simple polarity harbors a variety of salient dissimilarities that dra-

matically distinguish analog media from their digital successors.

An analog medium typically sets its stage in an arena primed to receive impressions. Painting serves as a useful paradigm. The canvas is gessoed into a uniform tabula rasa ready to accept imprints of the artist's gestures. Form and matter are melded to embody an expression in an object. The fundamental purpose of an analog medium is to preserve and present specific perceptual displays. It does this by securing the transcription of form from one physical substance to another. The arced movement of a handheld brush is deposited into a curved stroke of paint as one smudges the other. The result of the process is a relatively resilient self-contained object that carries the artistic vision across space and time. Until recently, becoming an artist involved honing the crafts required to forge matter into precious objects or shape its flow into special events.

Although they are capable of simulating the analog model, digital media have a very different purpose. Instead of crystallizing into freestanding products for display, they remain changeable and hidden components of complex integrated processes. Digital media are one of many elements in networked computational systems that facilitate an ongoing ebb and flow of numerical information rendered in a variety of forms of input and output. The emphasis is on promoting a process versus protecting a product. Instead of shaping materials by directing physical forces, the computer artist can generate something with self-motivated vitality by articulating mathematical formulas. This cognitive aspect of computer creation sometimes takes place explicitly through the writing of programs; but as interfaces become more elaborate, intelligent, and friendly, the creative activity is becoming more implicit and intuitive.

Analog media are primed for imprint, but digital ones are structured to symbolize. One is focused on concrete preservation and presentation, the other on abstract storage and manipulation. As a result, rather than being prepared into uniform substrates, digital media are preformatted with rigid structures to enable them to store digits. The difference is clearly demonstrated by comparing a painting or a photograph with a punched card or a floppy disk. A canvas is delicately smoothed into an unstructured continuum ready to experience subtle movements of the artist's brush. Similarly, film is carefully coated with an even layer of chemicals and kept in complete darkness until the moment of exposure to the outside world. The same attentiveness to unblemished uniformity as a preparation for bearing marks is also apparent in modern electronic-based analog media, such as the phosphor coatings on television tubes and the iron oxide layers on magnetic tape. The goal of any analog medium is to maximize sensitivity. A punched card, on the other hand, is crudely organized into a highly structured array of elements. Its vertical columns and horizontal rows constitute predetermined positions where a most unsubtle

hole may appear. The underlying numerical system has the now standard binary base: "0" is symbolized by the absence of a hole and "1" by its presence. Different symbols are punched into the card by various combinations of these holes. Punched cards are rarely used anymore, having been superseded by faster hardware and better user interfaces. But the same general framework is employed in all digital media, including the floppy disks, optical disks, digital magnetic tapes, and electronic circuits that are widely used today. Floppy disks, for example, are divided into circular tracks and wedge-shaped sectors before receiving any data. Any digital medium needs to be formatted in advance as an array of receptacles, each one of which can be put into one of two radically different physical states as a way of symbolizing one of two numbers (0 or 1). The aim is to inscribe symbols unambiguously rather than to respond sensitively to subtle perturbations.

The process of digital recording (also known as digitizing) is one of mensuration rather than maculation; information gets converted from physical events into abstract numbers rather than transcribed onto one material substance by another. The substrate that holds information in a digital medium is therefore completely extrinsic to it, whereas an analog medium's substrate is an intrinsic part of what is embodied in it. For example, when sounds are recorded onto an analog magnetic tape, they become transcribed, not into some abstract representation, but rather into concrete physical properties of magnetic fields in iron oxide that bear an analogous relationship to the sounds that caused them. As a result, these recordings have properties endemic to the medium of tape. Hiss is one of the most noticeable byproducts because of its often annoying interference with the sounds that are not associated with tape (i.e., the music that was recorded). The same is true of all analog media. Paintings have a characteristic appearance inherited from the look of paint, a fugue sounds different on a harpsichord than it does when played on an organ, and movies on television do not look quite the same as they do on the big screen. These essential material properties of analog media are sometimes heightened and sometimes diminished, depending on the artist's goals. But digital information bears no telltale traces of the magnetism, electricity, or cardboard that might happen to host its abstract symbols.

The purpose of analog media is to flaunt their essential perceptual qualities. We see, hear, and touch their messages directly. If they contain any measurable quantities, these are tied to a physical unit. A color might be identified as a particular wavelength of light, or a sound as a frequency of vibrating air. The size of a sculpture is measured in meters, and its weight in kilograms. But the bits of digital information stored in a digital medium are divorced from any physical unit. They are pure naked numbers that function outside any intimate alliances with material objects. When a digit is placed in a digital medium, it is stored not as a phys-

ical unit of electricity or magnetism, but rather as a pure numerical entity logged in a ledger. This is why digital media can be integral components of computing systems. Since their quantities are stored abstractly, they can be readily subjected to mathematical processes. This is done by composing commands that are equally abstract sequences of symbols, not by prodding chunks of physical substances with manual tools.

Their abstractness makes digital media both more temporary and more permanent than analog media. Computers can be quickly responsive to new input and change their contents millions of times per second. Yet information recorded as numbers is less subject to the ravages of time and mischief than material objects. It is more resilient since data can sustain a certain amount of deterioration without changing. But the dependency of software on particular hardware configurations also makes it difficult to maintain its operability in the face of rapid changes in computer architecture.

The lack of intinsic physicality gives digital media a variety of unique powers, but also imposes special limitations on them. Since numerical information is readily computed, its creative possibilities are infused with all the heady resources of mathematics that are circumscribed by logical rather than physical limitations. Strange new universes are opened to us in which heretofore impossible visions and vistas become navigable and palpable. These capabilities are reaping profound harvests, not only in the arts, but in nearly every field of human endeavor across the cultural landscape. Most of the sciences have not only benefited from computer-based "scientific visualization," but have also been reinvigorated by their ability to view and manipulate data in a virtual environment. But these same ethereal powers also make digital media incapable of functioning as freestanding perceptual objects. They require interfaced computers to have any utility. Machine-readable symbols must be translated into human-perceivable form for them to have any efficacy, and this often makes digital media parasitic on their analog ancestors. But it also links them in an interactive cycle that offers new cultural possibilities in the unlimited repetitions of conversion from events to numbers and back again that establish the embrace between computation and human action.

Not all the traditional arts fit the analog paradigm perfectly, but most conform to some aspects of it. Literature and music may seem to be the most glaring exceptions. Books often contain pictures, but they are also full of words made up of letters from a symbolic alphabet. Similarly, musical notation is written in signs rather than in sounds. In both these cases, however, the symbols involved are representations of auditory events. They are not numerical and hence are not computable.

Performances also may not seem to match this canonical paradigm until they are recorded. But today more people

experience recorded than live performances, and one of the most extensive uses of analog media is sound and video recordings on magnetic tape.

Furthermore, for both literature and music, there is an original physical manuscript penned by the artist's hand that serves as the standard for judging the accuracy of subsequent instances of the work. Herein lies another important difference. Digital media have no original. The information is immediately spirited away into electronic circuits and magnetic disks that are inscrutable by the human eye and ear. It moves freely from one digital medium to another, and by the time it is finally "saved" in what is likely to be a tentative version after a session at the machine, it has to be reinscribed and masticated numerous times. There is no authoritative original.

Particularly since the Renaissance, artistic expression has been explicitly channeled through media. The concept of aesthetics was invented in the eighteenth century, and its affirmation of perception as the focus of art consolidated the taxonomy of the arts into analog genres. This approach reached its apex in the twentieth century as modernism became obsessed with the self-critical process of ferreting out the essence of each medium. By that time, institutions for artistic creation, presentation, education, and preservation had become well-entrenched bastions of media, consolidating their hegemony into separate realms. Paintings hang on museum walls, concerts are played in performance halls, libraries house books, and television broadcasts programs to specialized receivers in homes.

The mimetic process of transcription has been efficiently automated and refined in the industrial technologies of photography, sound recording, printing, and electromagnetic broadcast, which gave rise to the analog media that epitomize communication in the twentieth century. Newspapers, magazines, radio, and television are sometimes called "mass media" or simply "the media" since their unprecedented reach crosses all geographical, political, and cultural barriers to include simply everyone. The goal of analog transcription has always been to disseminate crafted percepts, and the mass media finally and fully realized that goal by making it possible for a single individual to speak to everyone simultaneously. The key factor in consolidating this hegemony is the transcriptive recursion realized through mechanical analog devices. By taking advantage of assembly line methods, they perfected ways of making repeated imprints of words, sounds, and images so that these cultural expressions could cascade from their source to the population as a whole. A scene is photographed onto a negative, which is transcribed onto a positive, then onto a printing plate, and finally onto the pages of a book. In a similar fashion, a video signal is repeatedly transcribed from light to electronic circuitry to radio waves and then broadcast to multiple receivers, where it is profusely imprinted back again into electronics and ultimately into light emanating

from the television tube. Everyone becomes a member of Marshall McLuhan's "global village" by huddling around this glow and basking in its synchronized emanations.

Analog recursion is physical transcription in the service of dissemination. Digital media, however, are not recursive in this way. Once a physical magnitude is converted to a number, it cannot be converted again—except from a number back into an event to complete the interactive cycle. A measurement cannot be measured. Computers frequently use recursion, but it is of a logical rather than a physical nature, and it serves the purpose of focused individual discourse rather than anonymous dispersal. As a result, the model for connectivity in the digital age is based not on being a receiver in a hierarchical broadcast system, but rather on being a participant in a network. Digital functionality does not emanate from points of distribution, such as museums, bookstores, and television stations. It is ubiquitously distributed through the Internet and even resides in our appliances. The so-called cyberspace where computed activity takes place cannot be localized since whatever space it contains is nothing more than references to an abstract coordinate system.

Institutions of information are focused on virtual databases and fluid interactive exchanges as opposed to physical objects and frozen passive presentations. The industrial revolution at its core was all about the uniform automation of product manufacture. The information revolution is about individuating do-it-yourself interactive processes.

Digital media diminish the importance of media. This may seem odd, since the computer has given rise to an enormous market for so-called multimedia products that appear to proliferate media by mingling them promiscuously. But appearances can be deceiving. Conceptual art began the retreat from perceptually based analog media by demonstrating that you could make art with naked ideas as well as clothed objects. But the computer carries this a significant step further by basing itself in substrates whose sole purpose is to symbolize numbers. Multimedia do not proliferate media, but rather represent a retreat from media by consolidating everything into a common denominator of bits. Multimedia is a misnomer since it is neither "multi" nor "media." The digital nature of information makes it easy to move from one medium to another. A numerical file is quickly read from a magnetic disk into electronic memory, where it is manipulated and then displayed as a video image of light on a cathode ray tube. After further manipulations back and forth between light and electricity, a changed file is written back out to the magnetic disk. This fluidity is anathema to analog media that struggle for permanence, stability, and fidelity; but it is the essence of digital media that emphasize process over product.

BIBLIOGRAPHY

Binkley, Timothy. "Digital Dilemmas." *Leonardo* supp. vol. 23 (Fall 1990): 13–19.

Heim, Michael. *The Metaphysics of Virtual Reality*. New York and Oxford, 1994.
Margolis, Joseph. "Film as Fine Art." *Millennium* 14/15 (Fall/Winter 1984–1985).
McLuhan, Marshall. *Understanding Media: The Extensions of Man*. New York, 1964.
Mitchell, William J. *The Reconfigured Eye: Visual Truth in the Post-photographic Era*. Cambridge, Mass., 1992.
Negroponte, Nicholas. *Being Digital*. New York, 1995.

TIMOTHY BINKLEY

The Postphotographic Image

Pliny the Elder tells us in his *Natural History* that the art of painting was born when a Corinthian maiden traced the shadow, cast by the sun, of her departing lover's profile. At this mythic moment, the three basic components of all subsequent imaging practices were defined: the sun's rays provided a way of *projecting* information about a three-dimensional scene onto a two-dimensional surface; the artist's drawing instrument made possible the *recording* of marks on this surface; and the artist employed some strategy for *filtering* the incoming graphic information to select and record that which was important to her—the outline of the profile, presumably. Subsequently, a variety of technical means—perspective construction procedures and devices, the camera obscura, the photographic process, and now digital imaging techniques—have been employed, in different combinations, to mechanize these tasks. Each new stage of mechanization has changed the ways that images are produced and consumed, their social and cultural roles, and their aesthetic status.

In the closing decades of the twentieth century—a century and a half after the invention of photography, and more than five centuries after Filippo Brunelleschi's famous demonstration of linear perspective—an alliance of the digital camera, the personal computer, computer graphics software, and worldwide computer networking has given this old story a radical new turn. A clear way to identify many of the properties and roles of the postphotographic, digital image is by considering the associated changes in practices of projecting, recording, filtering, and receiving graphic information.

Projecting. The points of interest in any scene can be specified by coordinate triples x, y, z, while the points on a picture plane can be specified by coordinate pairs x, y, so the task of creating a projected image of the scene on the picture plane is one of following some rule or procedure to convert x, y, z triples to x, y pairs. The simplest way to do this, of course, is just to throw away the z (depth) coordinates; this yields a life-size parallel projection—as traced out by Pliny's Corinthian maiden, for example.

This elementary technique has the disadvantage that it completely destroys the depth information contained in the original x, y, z coordinates, so artists have often been moti-

DIGITAL MEDIA: Post-Photographic Image. *Parallel Projections of a Building.* (From Mitchell [1992], p. 127; courtesy of William J. Mitchell.)

vated to employ slightly more complicated rules to express that depth information in some way. One possibility is to add some fraction of the z-coordinate to the x-coordinate and the y-coordinate, so that coordinate triples, x, y, z are converted to coordinate pairs $(x + az)$, $(y + bz)$, where a and b are constants. This had the effect of translating points across the picture plane in proportion to their depth back in the scene, as in the oblique projections traditionally employed by Chinese and Japanese painters, and in the vertical oblique projections sometimes used in architectural drawings.

Another obvious approach is to divide x-coordinates and y-coordinates by z-coordinates, so that coordinate triples, x, y, z are converted to coordinate pairs (x/cz), (y/cz), where c is a constant scale factor. Thus shapes diminish in size proportionally to their depth back in the scene. This is the effect produced by linear perspective constructions, by the pinhole camera obscura, and approximately (with some lens distortions) by the photographic camera. Images produced in this way have widely been accepted as truthful and

accurate, since they are generated by consistent, mechanical application of a simple, well-understood rule.

Digital cameras simply replace the film plane by a CCD sensor array, so they also record perspective images. However, digital recording devices of various kinds (magnetic resonance imaging systems for medical use, laser surveying equipment, arrangements employing multiple cameras, and the like) can also be employed to capture three-dimensional rather than two-dimensional data about a scene. Subsequently, the recorded x, y, z coordinates comprising the three-dimensional digital model can be processed by computer to generate two-dimensional images.

This processing may apply rules to generate traditional parallel or perspective projections, but the computer graphics software need not simply function as a "virtual camera" in this way. In fact, the mathematical function employed to convert x, y, z coordinates to x, y coordinates may be completely arbitrary. The consequence of this is profound: where traditional photography bound perspective projection and visual recording technology tightly together, mak-

ing perspective seem synonymous with visual truth, three-dimensional computer graphics completely separates the two. In this context, traditional perspective projection loses its privileged status; it appears as one among infinitely many projection possibilities—any one of which might be employed, in a particular setting, for some particular technical or artistic purpose.

Recording. In drawing and painting, visual information is recorded through the action of the artist's hand. Clearly, this a subjective and error-prone operation. In the photographic process, by contrast, light intensities are recorded automatically, and with high levels of accuracy and consistency, through the direct effect of the light on photographic emulsion. Thus, there is far less room for recording error or for introducing an artist's subjective choices.

At first glance, digital cameras and similar devices seem closely analogous to traditional photographic cameras in this respect; they just employ electronic instead of chemical processes to record light intensities. But beyond this, there is a fundamental difference: they yield digital rather than analog data. Whereas photographic film records essentially continuous spatial and tonal variation within a scene, a digital camera subdivides the picture plane into a finite array of pixels, and records the intensity of each pixel at some finite level of resolution. The result of this digital recording process is an array of integers that can be interpreted by display and printing devices as tonal values of colors. The higher the spatial and tonal resolution of this integer array, the more computer memory it consumes, but the more closely it approximates the continuous variation of a photographic image.

The most obvious consequence of this distinction is that digital images cannot be enlarged indefinitely. Enlargement does not reveal anything new (as photographic enlargement may), but simply increases pixel size. The eventual result of increasing enlargement is that pixels become more and more prominent, the illusion of continuous variation is lost, and in the end the image breaks down completely and becomes unreadable.

A less obvious consequence is that the traditional distinction between original and reproduction disappears. There is no negative—just a file of digital data. And, since it is digital, this file can be replicated absolutely exactly. There is never any generational loss, as, for example, in photographs of photographs. There is no variation of the kind found among different prints from the same negative or etched plate. Large number of copies can be made rapidly and inexpensively, and quickly distributed through computer networks. In traditional media, quality differences, aura, and rarity value often distinguish an original artwork from reproductions, fakes, and copies, but this sort of distinction is almost completely eliminated in the digital world.

Furthermore, digital images—unlike photographic ones—can quickly and easily be transformed and combined. Visual elements can be inserted, deleted, recolored, reshaped, and rearranged to construct fictional scenes from factual fragments. This is accomplished by using paint or image-processing software to substitute or alter pixel values, then redisplaying or reprinting the image. Insertions and changes can be made seamlessly, since digital images are divided into discrete pixels anyway, so as to be very difficult to detect. And anyone who has access to image files, a personal computer, and the necessary software has easy opportunity to perform such manipulations. Thus, though digital images may look exactly like photographs (and, indeed, may even be scanned versions of photographs) we must remember that they have been produced and transmitted in a very different way—a way that allows many chances for surreptitious manipulation, and provides far less ground for confidence in the presumption that they are unmediated results of instantaneous, automatic recording processes.

Finally, digital images may appear to be photographic, but may in fact be completely synthetic. Images of this type are created by applying photorealistic rendering algorithms to digital models of three-dimensional scenes. These digital models may be electronically captured from real objects or scenes, they may result from manual measurements and surveys, or they may simply be constructed by artists. The rendering algorithms perform perspective projection operations, and calculate surface shading from information about surface properties and light sources. The resulting images may stand alone, or—as is common with movie special effects—they may be blended electronically with captured images of real scenes to produce fictional hybrids.

Clearly then, digital imaging technology has opened up new possibilities for creating realistic fictions within the graphic conventions that have been established and popularized by photography. Because it is often hard to trace the provenance of a digital image or to detect whether alterations have been made to it, there is relatively little difficulty in passing off such fictions as truthful records. As photojournalists and the courts have been quick to recognize, this has begun seriously to undermine the widespread presumption—created by more than a century and a half of photographic practice—that a realistically shaded perspective image is probably reliable evidence of what actually happened, at some particular place, at some precise moment.

Filtering. With any graphic medium, information about a scene must be expressed in terms of the particular repertoire of marks that the medium offers—pencil marks, brush strokes, mosaic tesserae, and so on. The artist must choose where and how to make marks to record the scene, and this choice process may be a complex and subtle one. In making a drawing of a scene, for example, a draftsperson usually locates lines at the sharp intensity discontinuities occurring along the edges of objects; and painters often approximate fine detail with broad brush strokes. These are processes of

filtering visual information, in particular ways, so that some of it is retained and emphasized while the remainder is discarded.

In the photographic process, filtering is accomplished automatically, subject to general choices made by the photographer. Lens filters shift color and tonal ranges. Aperture and focus settings determine whether images (and parts of images) will be sharp or fuzzy. Exposure settings alter the overall tone of an image, and may cause areas to lose detail through over- or underexposure. Choices of film stock, paper, and darkroom operations may vary tonal range, contrast, and granularity of the final image. Tonality and levels of detail may be varied selectively by burning and dodging. Occasionally, more exotic procedures such as solarization, or printing through patterned glass, are used to filter images in unusual ways.

In digital imaging, a first level of filtering is automatically performed by the recording device, when visual information is approximated to the pixel grid; in effect, the scene is expressed in terms of square pixels colored from a finite palette. Subsequently, further filtering in a wide variety of ways may be accomplished by processing pixel values. Thus digital images may quickly be lightened, darkened, varied in their contrast, sharpened, made fuzzier, and color-shifted. Less obvious digital filtering procedures may be applied to emphasize or suppress fine detail, to transform tonal images into line drawings, and to simulate effects of traditional graphic media such as pencil, charcoal, watercolor, and mosaic. Filters may be applied singly or in combination, and to an image as a whole or to selected parts.

Now, filtering operations in drawing, painting, and photography are normally executed and controlled by the image producer, and the consumer does not participate in this. But with digital imaging, filtering capabilities may be available anywhere in the image production, distribution, and reception chain—in recording devices such as digital cameras, in the graphic workstations of editors and art directors, and in the browsing and viewing software employed by consumers. So the traditional division of labor among producers, distributors, and consumers of images is fundamentally changed: the originator of a digital image—unlike a painter or photographer—cannot control or be held responsible for the filtering and final appearance of that image at the point of reception.

Viewing. Where digital imaging systems capture and distribute three-dimensional models rather than two-dimensional images, an even more startling new condition emerges. Here, not only choices about filtering but also directorial choices of viewpoint, framing, and shading devolve from the originator to the consumer. So, for example, an architect's client may not be limited to seeing a few perspective renderings of a project from predetermined viewpoints, but may employ interactive viewing software to "walk" through it at will. VRML and other formats for pre-

senting three-dimensional "worlds" on the World Wide Web allow viewers to control camera pans and zooms. And in emerging object-based television systems, camera angles and lighting can potentially be controlled at the reception rather than the production end of a transmission; the viewer thus takes over traditional directorial functions.

In the digital world, then, viewers of images are far from passive consumers of prepackaged information. They may play very active roles in constructing particular visual interpretations of the graphic information that they receive. Since they cannot safely assume that this information has been produced by means of highly standardized, well-understood processes—as in the usual practices of photography—they are well advised to be particularly vigilant and critical in their uses of it. Distinctions among visual facts, fictions, and falsehoods can still be drawn, but not so straightforwardly or confidently as in the past.

Conclusions. As the term *digital photography* suggests, digital imaging is, in some respects, a straightforward extension of the discourse of photography and of the even older traditions of perspective draftsmanship. But closer, more critical examination reveals that digital images possess some fundamentally new and different technical properties. These new properties are generating new image production and distribution practices. And these practices, in turn, are radically changing the ways in which images are used, understood, and valued.

[*See also* Photography, *article on* Photography and Digital Technology.]

BIBLIOGRAPHY

Abrams, Janet. "Little Photoshop of Horrors: The Ethics of Manipulating Journalistic Imagery." *Print* 49.6 (November/December 1995): 24–45, 159–164.

Bove, V. Michael, Jr. "Beyond Images." *Convergence: The Journal of Research into New Media Technologies* 2.2 (Autumn 1996): 30–45.

Bove, V. Michael, Jr. "Multimedia Based on Object Models: Some Whys and Hows." *IBM Systems Journal* 35.3–4 (1996): 337–348.

Glassner, Andrew S. *Principles of Digital Image Synthesis*. San Francisco, 1995.

Gortler, Steven J., Radek Grzeszczuk, Richard Szeliski, and Michael F. Cohen. "The Lumigraph." In *Computer Graphics Proceedings: Annual Conference Series*, pp. 43–54. New York, 1996.

Greenberg, Donald P. "Computers and Architecture." *Scientific American* 264.2 (1991): 104–109.

Grundberg, Andy. "Photography in the Age of Electronic Simulation." In *Crisis of the Real: Writings on Photography, 1974–1989*. New York, 1990.

Holzmann, Gerard J. *Beyond Photography: The Digital Darkroom*. Englewood Cliffs, N.J., 1988.

Mitchell, William J. *The Reconfigured Eye: Visual Truth in the Post-Photographic Era*. Cambridge, Mass., 1992.

Mitchell, William J. "When Is Seeing Believing?" *Scientific American* (February 1994): 44–49.

Mitchell, William J., and Malcolm McCullough. *Digital Design Media*. 2d ed. New York, 1995.

Ojeda, Oscar Riera, and Lucas H. Guerra. *Hyper-Realistic: Computer Generated Architectural Renderings*. Rockport, Mass., 1996.

Ritchin, Fred. *In Our Own Image: The Coming Revolution in Photography.* New York, 1990.

WILLIAM J. MITCHELL

DILTHEY, WILHELM (1833–1911) German philosopher. All the arts, but especially poetry and music, play a key role in Wilhelm Dilthey's philosophy of life. His main works on aesthetics are "The Imagination of the Poet" of 1887 and "The Three Epochs of Modern Aesthetics and Its Present Task" of 1892, but reflections on the arts are scattered throughout his theoretical writing on hermeneutics and the human sciences. In conjunction with his historical writings, Dilthey also produced a considerable amount of literary criticism. The most important literary essays are the four on Gotthold Ephraim Lessing, Novalis, Johann Wolfgang von Goethe, and Friedrich Hölderlin that were collected in *Das Erlebnis und die Dichtung* (Lived Experience and Poetry; 1906). Dilthey used literary figures and their works to help focus larger cultural and social movements. His success in doing so inspired many similar efforts at cultural literary history, including *Soul and Form* (1911) by György Lukács and *Freiheit und Form* (Freedom and Form; 1916) by Ernst Cassirer. Moreover, Dilthey's hermeneutical approach to the arts influenced Martin Heidegger and Hans-Georg Gadamer.

Dilthey opens "The Imagination of the Poet," also known as his *Poetics,* with the observation that modern artists no longer exhibit the traditional aesthetic concerns for ideal beauty and the unities of pure form. A naturalistic art is being produced that apprehends this world as it is and exposes the power relations behind the multiplicity of its contents. Dilthey's own proposal can be located somewhere in between traditional ideals and naturalism; namely, he thinks that the arts should reinforce our sense of reality while also adding coherence by a process of heightening and deepening it. Artists should reflect the attitudes and thoughts of their age, and yet they must also shape all this into a new unity by formulating a vision or worldview. In the Renaissance the visual arts were still capable of presenting worldviews, but now reality has become so complex that only literature can fully do so. Poetry still has the power to make life visible through the imagination.

The poetic imagination comes to stand for the telos of all the arts, and its play with possibilities must fulfill three main functions. First, it must exclude from a situation—whether actual or possible—whatever distracts us from noting its most important components. Such exclusion teaches us to concentrate our attention on certain features of things and to abstract from others. Second, it must heighten or intensify those components of experience that have been preserved. This aspect of the poetic imagination serves to enliven our ordinary experience *(Erfahrung)* of reality and transforms it into lived experience *(Erlebnis).* The third and most important function of the imagination is to complete lived experience. Whereas exclusion and intensification introduce quantitative changes that can affect the quality of our experience, Dilthey conceives completion as a fundamental structural change whereby some central component of experience is unfolded to incorporate other aspects of experience that had been pressed into the background. This completed nucleus comes then to symbolically represent the storehouse of experience. Poetry serves to articulate the meaning of our lived experience by creating certain points of impression that typify life more generally. These points of impression organize the artist's apperception of reality to give it a distinctive stylistic unity.

"The Imagination of the Poet" is representative of the stage in Dilthey's thought when he looked to the psyche of the artist to account for the way certain conscious and focused images can symbolize many other experiences. Although all people acquire a fund of lived experience that serves as a kind of background nexus guiding their behavior and influencing their attitudes toward life, poets are distinctive in being more aware of this acquired psychic nexus and in being able to articulate its overall structure. What remains relatively unconscious in most people is raised to consciousness by poets. The fact that reality is filtered through the acquired psychic nexus of individual artists means that the significance of a work of art is never conceptually universalizable like a Hegelian idea, but possesses the more restricted generality of typicality.

Whereas Georg Wilhelm Friedrich Hegel saw art as the first sphere of absolute spirit, embodying rational ideas that are then imaginatively elaborated in religion and conceptually refined in philosophy, Dilthey preferred artistic formulations of worldviews to religious and philosophical formulations. The reason for this preference is that the arts stay within the confines of life. Religious worldviews tend to appeal to invisible forces transcending life and philosophical formulations appeal to something that can ground life. Only artistic worldviews respect Dilthey's conviction that it is impossible to go behind life.

The broadest appropriate framework for understanding the typical meaning of literature is a historical epoch. Dilthey recognized the importance of history as a basis for human understanding even when he had great hopes that a new descriptive psychology could provide a neutral foundation for the human sciences. Such a psychology was not supposed to sever psychological states from their surroundings. One of the best examples of such psychological description is to be found in the section of the *Poetics* analyzing six different spheres of feeling and their aesthetic relevance. Some of these spheres of feeling are pleasurable, but according to Dilthey we are drawn to art less by the pleasures it can provide than by the way it can move us to expand ourselves. What we seek from poetry, for instance, is "powerful stimuli, even when mixed with strong pains."

As Dilthey moved away from his view of psychology as the first of the human sciences, he came to relate understanding more directly to objective historical frameworks than to the subjective psychic nexus of the author or reader. Although Dilthey had long been interested in hermeneutics because of his work on Friedrich Schleiermacher, the hermeneutical perspective did not fully shape his philosophical standpoint until his 1900 essay "The Rise of Hermeneutics." There Dilthey formulates the position that the interpretation of human objectifications is just as essential for self-understanding as it obviously is for the understanding of others. Our own experience does not constitute self-understanding until we come to terms with their expressions.

Schleiermacher had divided hermeneutics into two main parts: the grammatical, which stressed the indispensability of language for thought, and the psychological, which attempted to explain the unity of a text by an authorial seminal decision. Dilthey places less stress on the grammatical aspect of interpretation because he expands the scope of hermeneutics beyond spoken and written discourse in order to encompass any human objectification, including the visual arts, public monuments, and historical deeds. Moreover, he modifies the psychological part of hermeneutics by rejecting any explanatory schema such as an authorial seminal decision. He questions the idea that the unity of a work must already preexist in its conception. The overall unity of a work is the result of a formative process with several possible points of origin, some of them external to the author. Hermeneutics is the art of interweaving the various genetic strands that make it possible to understand the meaning of a work of art contextually.

If Dilthey's psychological *Poetics* was already implicitly hermeneutical in showing how works of art help us to integrate what is felt in experience with what is thought and willed, his later reflections on poetry are hermeneutical in a much more impersonal way. Thus in *The Formation of the Historical World in Relation to the Human Sciences* of 1910, Dilthey writes: "The structure of a drama consists in a peculiar relation of material, poetic mood, motif, plot, and means of representation. Each of these moments fulfills a function in the structure of the work. . . . Thus the object which poetics or the history of literature deals with is at first completely distinct from psychic processes in the poet or his readers." The qualification "at first" indicates that Dilthey is not ruling out the possibility that psychology can at some point contribute to his hermeneutic approach to the arts. There is clearly a tension between a hermeneutic approach to the arts, which proceeds from the outside to the inside, and a psychological approach, which moves from the inside out. But since the meaning of what is inner shifts, the two approaches need not be exclusive. When we begin by interpreting a drama like *Faust* as an objective whole with its own inherent structure to give its parts meaning, there

may come a point where either its ambiguities or its peculiar richness lead us to also recognize it as a human product. The object before us is an objectification that also may have an inner = subjective reference. But if the psychological understanding of Goethe is to add to the hermeneutic process of interpreting his *Faust*, then it must be motivated by the latter. Goethe is not to be understood for his own sake, but as the vehicle for a larger process. Accordingly, there is no clear way of demarcating the poet's lived experience from its poetic expression because the former was already steeped in a world of literary conventions and expectations. In a 1911 discussion of musical understanding Dilthey makes a similar point: "There is no duality of lived experience and music, no double world, no carry-over from the one into the other. Genius involves simply living in the tonal sphere, as though this sphere alone existed."

Many of Dilthey's late aesthetic reflections refer to music. He comes to a new appreciation of the musicality of Hölderlin's lyric poetry for expressing the nonobjective power of moods that relate us to life in invisible ways. Although unable to produce the historically focused meaning of literature, especially of drama, music can provide us with a general sense of life. Heidegger would elaborate the way in which such moods attune us to the world. The highest mode of music for Dilthey is instrumental music. It has no determinate object, but has life itself as its object.

The essay "The Three Epochs of Modern Aesthetics and Its Present Task" begins with an account of the rationalist aesthetics of the seventeenth century, which is criticized for approaching the arts primarily from the normative perspective of the critic. The second epoch is represented by the psychological aesthetics found mainly in eighteenth-century Britain, which analyzes the aesthetic impressions of the spectator. Dilthey's criticism of this second approach is that it reduces the work of art to an aggregate of impressions so that style that pertains to its overall effect remains unaccounted for. Dilthey characterizes the third epoch as a constellation of four factors that came together at the beginning of the nineteenth century. The first of these is transcendental philosophy and its conception of the role of the mind in actively shaping our worldview. This contributed greatly to the appreciation of artistic creativity and genius. The second factor involves a more systematic concern with the relation between natural beauty and artistic beauty. Here Dilthey refers to Friedrich Schelling's metaphysical claim that what is created unconsciously in nature can be imitated and made conscious by the artist. A third aspect involves the attempts by Friedrich Schiller, Friedrich Schlegel, and Hegel to distinguish the main epochs of art in human history as a sequence of possible attitudes to reality. A fourth factor encompasses inquiries into the way the specific medium and means of representation chosen by the artist influence the work of art. Although these four aspects of the third epoch of modern aesthetics constitute great ad-

vances in the understanding of the arts, Dilthey criticizes all but the last as being too speculative. They are rooted in an idealism that can no longer speak to the naturalistic artists of his day. The present task of aesthetics is to bring general philosophical theories into contact with reality. Transcendental and metaphysical theses must be tested by empirical descriptions; constructions of developmental patterns common to all the arts must be reexamined in light of the different formal and material conditions of the various arts. We may be able to delineate a series of epochs within the history of aesthetics and within specific arts, but not for a history of art in general.

As Dilthey's aesthetics became increasingly historical, his hermeneutics turned more impersonal. A great work of art is not *true* in any descriptive sense, but "*truthful* in itself." The arts establish a sphere in which life can be disclosed at a depth not open to observation, reflection, and theory. Here we see intimations of the ontological hermeneutics of Heidegger and Gadamer, who go on to demand *truth* of art.

Heidegger expands Dilthey's insight that aesthetic understanding must draw on our participation in historical life into the thesis that artistic understanding is ontological. This makes art less humanistic, for it not only sets up a historical world shaped by human action, but also sets forth an ahistorical earth that tends to elude our grasp. There is no longer anything playful about our engagement with art, as the cultural tradition from Immanuel Kant, Schiller, and Dilthey might suggest. Instead, Heidegger stresses the seriousness of art understood as truth setting itself to work. Accordingly, he self-consciously gives up Dilthey's word *Erlebnis*, which stands for a lived experience in which meaning is possessed, and reinstates *Erfahrung* to indicate that art involves an existential struggle to wrest truth from concealment—a struggle for which ordinary experience already gives us a preunderstanding.

Gadamer in turn reappropriates the humanistic, historical, and cultural concerns of Dilthey in his account of the arts. The *Erfahrung* of art is neither an individual possession nor an existential struggle, but a communal project based on tradition. Aesthetic play is reinstated, however, not as a private imaginative process, but as the play of a public performance that encompasses an interplay among its participants.

Gadamer's critique of Dilthey's hermeneutics as still too method oriented in the traditional sense is in effect answered by Heidegger, who argued that Dilthey's real concern was much more radical. Increasingly, Dilthey replaced methodical psychological descriptions of individual aesthetic experiences with less formal, but more disclosive, anthropological reflections on aesthetic experience in general. Similarly, the 1907–1908 notes for a revised *Poetics* show Dilthey conceiving the poetic imagination not in terms of psychological images but in terms of publicly accessible lin-guistic imagery. Ultimately, Dilthey's poetic imagination becomes a mode of linguistic imagination *(Sprachphantasie)*.

[*See also* Hermeneutics; *and* Poetics.]

BIBLIOGRAPHY

Works by Dilthey

Das Erlebnis und die Dichtung: Lessing, Goethe, Novalis, Hölderlin (1906). 13th ed. Stuttgart, 1957.
Die grosse Phantasiedichtung und andere Studien zur vergleichenden Literaturgeschichte. Edited by Herman Nohl. Göttingen, 1954.
Gesammelte Schriften. Stuttgart, 1957.
Selected Works, vol. 5, *Poetry and Experience.* Edited by Rudolf A. Makkreel and Frithjof Rodi. Princeton, N.J., 1985. Contains: (1) "The Imagination of the Poet: Elements for a Poetics" (1887), pp. 29–173; (2) "The Three Epochs of Modern Aesthetics and Its Present Task" (1892), pp. 175–222; (3) "Fragments for a Poetics" (1907–1908), pp. 223–231; (4) "Goethe and the Poetic Imagination" (1910), pp. 235–302; and (5) "Friedrich Hölderlin" (1910), pp. 303–383.
Von deutscher Dichtung und Musik: Aus den Studien zur Geschichte des deutschen Geistes. 2d ed. Edited by Herman Nohl and Georg Misch. Stuttgart, 1957.

Other Sources

Makkreel, Rudolf A. *Dilthey: Philosopher of the Human Studies.* 2d ed. Princeton, N.J., 1992. See especially chaps. 2–5, 9, 10.
Müller-Vollmer, Kurt. *Towards a Phenomenological Theory of Literature: A Study of Wilhelm Dilthey's "Poetik."* The Hague, 1963.
Rodi, Frithjof. *Morphologie und Hermeneutik: Zur Methode von Diltheys Ästhetik.* Stuttgart, 1969.

RUDOLF MAKKREEL

DISCOURSE. Discourse in art refers to the use of words and texts by visual artists. Words enter art in various forms: as visual image, verbal construct, essay, or spoken text. For the purposes of the following discussion, "discourse" is synonymous with language and text.

The twentieth century has witnessed an eruption of language into the field of the visual arts (Owens, 1979, p. 126). This eruption began early in the century, exploded on the scene in Conceptual Art, and is still evident today in performance modes and multimedia installations. Indeed, the pervasiveness and the variety of "language works" produced during the past few decades are remarkable. Language works encompass paintings with words as dominant images (Edward Rusha), words painted on walls (Lawrence Weiner, Robert Barry, Mel Bochner), words printed on sheets of vinyl (Edwin Schlossberg), visual signs contextualized by theoretical essays (Daniel Buren), essays on aesthetics presented as art (the Art and Language Group), wall texts that are stories (Jean Le Gac), collage or sculptural objects with fictional texts (Alexis Smith, Edward Kienholz), performed texts (Guy de Cointet, Joseph Beuys, Jacki Apple), discussion art (Ian Wilson), books (Lawrence Weiner, Dan Graham), card files (Robert Morris), post-

cards with words (On Kawara), puns conveyed through body art (Bruce Nauman, Chris Burden), thesaurus works (Joseph Kosuth, Vito Acconci), sculptures with texts (Alice Aycock), rooms filled with painted words (Allan Kaprow), installations designed like books (Alison Knowles's *The Big Book*) or like novels to walk through (Acconci's *Plot*), scripted events (George Brecht), the exchange of spoken or written secrets (Vito Acconci, Douglas Huebler), videos with spoken narratives (Lynda Benglis), multimedia installations with texts (Jonathan Borofsky and Marcel Broodthaers), photographs with words (Barbara Kruger), neon signs (Jenny Holzer), and more. Numerous art movements generated within the past two decades are also defined by their use of language: conceptual art, narrative art, discussion art, and performance art all include programs for using language as a primary material. Since the 1960s, language has become as important to many artists as the brushstroke has remained to others.

What is most distinctive in the art of this time is the way in which the visual debate concerning art's possibilities, limits, and boundaries becomes a verbal one. Words are used not only in the production of art (Joseph Kosuth, Weiner, Barry, and Huebler), but also in the production of theories that become inseparable from it. Traditionally, artists create objects and leave critics free to formulate the conceptual implications of their art. "As such," Kosuth says, "the traditional artist functions like the 'valet's assistant' to his marksman master (the critic)" (Kosuth, 1973, p. 100). The Art and Language Group, Kosuth, Buren, Robert Smithson, and Joseph Beuys are just some of the artists who appropriate the critic's role and create critical discourse as an essential part of art; they generate the ideas through which the artistic objects and actions are viewed, or present theoretical discourse as the substance of art. Each artist manipulates the relation of theory and art in a slightly different way. Members of the Art and Language group write essays that are presented as "visual" art. Smithson intertwines art criticism and art forms, sustaining a shifting balance through a constant interplay of verbal and visual components. Theory becomes performance in Beuys's *How to Explain Pictures to a Dead Hare*, in which the artist tours a gallery talking to a dead hare about art. The piece is a poignant commentary on the necessity and the futility of talking about art. Kosuth places priority on theory: his visual "forms of presentation" are secondary manifestations of his ideas; nevertheless, they constitute procedures essential to his art.

The prevalent use of language in contemporary visual art is just one manifestation of the whole orientation of postmodern culture toward language. "The sixties," writes Julia Kristeva, "witnessed a theoretical ebullience that could roughly be summarized as leading to the discovery of the determinative role of *language* in all the human sciences" (*Desire in Language*, p. vii). Structuralism in general "turned to language as a model for understanding all cultural phenomena" (Steiner, 1982, p. 51). As Roland Barthes puts it in *Elements of Semiology*, "The world of signifieds is none other than that of language. . . . We shall therefore postulate that there exists a general category language/speech, which embraces all the systems of signs; since there are no better ones, we shall keep the terms language and speech, even when they are applied to communications whose substance is not verbal" (Barthes, 1967, pp. 10, 25). W. J. T. Mitchell observes that semiotics in general failed to provide an adequate theory of the image, even when individual theorists (Umberto Eco, for example) did not subscribe entirely to Barthes's "linguistic imperialism" (Mitchell, 1986, p. 56). Viewed in this context, the prioritization of language in conceptual art, where the texts themselves have become the art, was not an aberration of art history so much as the fullest expression of the intellectual thought of its time. It is not a coincidence that one of the most important texts on the postmodern style was titled *The Language of Post-Modern Architecture* and that in it Charles Jencks asked us to attend to architectural "syntax" and "semantics" (Jencks, 1981, p. 39).

A shift has occurred since the sixties to modify this view of language, on the part of both theorists and artists. Language and image are now frequently presented without priority, and there is a concern for the interplay of sign systems rather than any hierarchy between them. Smithson, Laurie Anderson, and other artists like Borofsky set image and text in constantly changing relations to one another; these artists would concur with Mitchell's observation that the real task—not only of theory but of artistic production—"is not to renounce this dialogue" between word and image (Mitchell, 1986, p. 46).

The use of language within art is hardly new. Early fusions of the verbal and the visual occur in Egyptian hieroglyphs, Chinese ideograms, and Mayan glyph writing. Illuminated manuscripts, seventeenth-century emblem books, pattern poetry (from George Herbert to the present), and William Blake's mixed art are among some of the diverse and "curious" combinations of media and arts that predate our time. The most immediate and important context for a discussion of contemporary "language works," however, is modern art, in works by such artists as Juan Gris, Stuart Davis, Gino Severini, and Carlo Carrà, that also includes words.

Georges Braque's *The Portuguese* (1911) was the first Cubist collage to include lettering, and it paved the way for numerous canvases with words used as elements in the composition. As Douglas Cooper suggests, the fragments of verbal material in a Cubist painting serve a double function: they operate as mimetic elements, heightening the representational aspect of the work, while at the same time foregrounding the two-dimensionality of the canvas (1994). Richard Kostelanetz proposes, however, that Cubist paint-

ings represent a complementary art, like opera, where even at best a libretto can viably be separated from the music rather than an intermedium where one element is entwined with the other (1993).

Juan Gris's *Still Life with a Poem* (1915) is an example of a "complementary art"; the poem by Pierre Reverdy affixed to its border in no way totalizes the painting, nor does the painting resolve any of the poem's ambiguities. The relation of poem and painting in no way confuses the two, in no way contaminates the autonomous integrity of the separate artifacts. Duchamp's *The Bride Stripped Bare by Her Bachelors, Even* (1915–1923), however, constitutes an "intermedium" in which the verbal and the visual are intertwined. *The Green Box* includes Duchamp's fragmented narratives, drawings, and anecdotes for *The Bride Stripped Bare by Her Bachelors, Even.* When the images are seen "through" this text and the text, as it were, seen "through" the glass, *The Large Glass* as a whole is suddenly animated, lively, intriguing, and even, in a strange way, beautiful. Suddenly the mechanical images are comically humanized and implicated in a complex love story, an erotic encounter with an indecisive conclusion. *The Large Glass* is a hybrid work, neither purely visual nor purely verbal, neither "pictorial" nor "literary" in any conventional sense of the terms: "My research," Duchamp said, "was in the direction of finding some way of expressing myself without being a painter, without being a writer, without taking one of those labels" (quoted in Schwarz, 1996, 7). *The Large Glass* effects a unique interdependence or interconnection between verbal and visual elements. It is a perfect example of what Kostelanetz would call an "intermedial" art.

Duchamp influenced an entire generation of artists in the Conceptual Art movement, in which the art was essentially linguistic. Artists such as Kosuth, Weiner, Huebler, and Barry produced works that were predominately linguistic. Kosuth, in particular, produced an art influenced by language philosophy and poised dramatically between philosophy and art. His dictionary definitions and his book, *Art Investigations and "Problematics" since 1965*, present art as an exploration of linguistic meaning (the meaning of "art").

Robert Smithson is another interdisciplinary artist whose essays constitute an essential component of his art. *The Spiral Jetty* (a Barthean "text") cuts across three works: earthwork, essay, and film. In *The Spiral Jetty* (1970), Smithson presents linguistic data as solid objects. "Language tended to inform my structures," he says. "I guess if there was any kind of notation, it was a kind of linguistic notation" (Smithson, 1979, p. 154). Language may, in fact, be one of the many axes around which the verbal and visual material of *The Spiral Jetty* accumulates:

> *Helix:* A spiral in many dimensions, in which the Jetty "could be considered one layer."

> *Helicopter:* "From the Greeks, *helix, helikos* meaning spiral" (p. 113).
> *Helicline:* A curving ramp that ascends gradually. *The Spiral Jetty* curves and "ascends" ("scala . . . a means of ascent") (p. 112).
> *Heliogram:* Signaling by flashing sun's rays from a mirror. The water serves as a "vast thermal mirror" reflecting sunlight.
> *Heliotypes:* Printed reproductions. Smithson's work in general is presented as a series of duplications and replicas for lost and absent originals.
> *Helios:* The sun. The film begins with a shot of the sun and fictively ends in "sunstroke." The narration includes a definition of sunstroke from *Black's Medical Dictionary.*

Thus not only does Smithson use language to describe *The Spiral Jetty;* he also uses the jetty to inscribe words and ideas into matter. Not only does he shape and mold language like "material to put together" (p. 104), treating language as a thing, he also treats objects linguistically, to produce a kind of "writing beyond the book" (Ulmer, 1985, pp. 15–16, 242).

Another artist to use language as an essential component of her art is the performance artist Laurie Anderson. Beginning as a kind of Conceptual artist, Anderson turned to performance art because of a fascination with the spoken word:

> Originally I was a sculptor. During the Seventies I did large photographs with texts. At that point I thought it's silly to have all these giant words on the wall when they have very different kinds of meanings. When you get a letter from somebody, you get a feeling of what's going on with them. But if you get a phone call and they say the same things, you get so much more information from the voice and the way they pause. So in looking at my pieces, I thought, why flatten the words out like this. Just say them. So I started giving these performances which were really just spoken stories. (quoted in Smith, 1983, p. 61)

Anderson's work includes language as one component of a multifarious art, but it is also about language, how we think about it and use it. One way, in fact, to come to terms with the relation of Anderson's art to that of her modern predecessors is to note her ambivalent attitudes toward language and the complexity of her linguistic uses. Anderson's art includes all of the linguistic pessimism traditionally associated with modernism (Sheppard, 1978). Assaulted as vigorously as in Dadaism or in Futurism, words frequently collapse into pure noise, garbled signals, high-pitched squeaks, and electronic sounds. Anderson's innovative technological devices—the tape bow violin, for instance—or the "Door and Wall Songs" actively transform spoken phrases into dissociated sound in a way that parallels the disintegrating effect of Duchamp's *Anémic cinéma* upon language. "I'm very

suspicious of language," Anderson said in one interview; "it's so slippery, so slippery to seize."

At the same time, a pervasive desire for communication can be traced throughout Anderson's art. While her work describes semantic and semiotic breakdowns and contains numerous parables of failed communication, it also expresses the artist's own desire to communicate. "I feel a personal responsibility to communicate," Anderson says, "not to make art that's so coded that people can't receive it." Thus much of her language is just plain, straightforward talk, everyday colloquial speech turned to the task of storytelling. "I think of the work that I do as the most ancient in the world: storytelling," she says (quoted in Jarmusch, 1984, p. 130), and her stories are told in the vernacular languages that her audiences speak. The key to her American vision, it might be said, is language and a vocabulary rooted in ordinary everyday speech. Neither Duchamp nor Filippo Marinetti would ever have said, as Anderson has, that the most important thing is that people learn to talk to each other. Nor would they have used the slangy, straightforward, even prosaic language with which she constructs her narrative vignettes.

Numerous other artists use language as a constitutive feature of their art—Kruger, Holzer, Borofsky—to mention a few notable examples. All of this experimentation with discourse in art makes one thing clear: never before have the "visual arts" been so verbal. Artists are producing truly interdisciplinary art, art balancing precariously between the disciplines: philosophy, photography, art, fiction, poetry, art criticism, and music mix in new combinations and novel forms. This interdisciplinarity and this explosion of the verbal into the field of the visual are the hallmarks of postmodernism in the arts.

[*See also* Conceptual Art, *overview article;* Ecphrasis; Imagery; Ineffability; Postmodernism, *overview article;* Text; Theories of Art; *and* Ut Pictura Poesis.]

BIBLIOGRAPHY

Anderson, Laurie. *United States.* New York, 1984.
Barthes, Roland. *Elements of Semiology.* Translated by Annette Lavers and Colin Smith. New York, 1967.
Collins, James. "Narrative." In *Narrative Art: An Exhibition of Works by David Askevol, Didier Bay, Bill Beckley, Robert Cumming, Peter Hutchinson, Jean Le Gac, and Roger Welch.* Brussels, 1974. Exhibition catalog, Palais des Beaux-Arts.
Cooper, Douglas. *Cubist Epoch.* London, 1994.
Duchamp, Marcel. *The Bride Stripped Bare by Her Bachelors, Even.* Translated by George Heard Hamilton, edited by Richard Hamilton. London, 1960.
Flynt, Henry. "Concept Art." In *An Anthology of Chance Operations.* Edited by La Monte Young and Jackson MacLow (1963). New York, 1979.
Freeman, Judi. "Layers of Meaning: The Multiple Readings of Dada and Surrealist Word-Images." In *The Dada and Surrealist Word-Image,* edited by Judi Freeman, pp. 13–56. Cambridge, Mass., 1989. Los Angeles County Museum of Art catalog.
Goldberg, RoseLee. *Performance: Live Art, 1909 to the Present.* New York, 1979.
Hutcheon, Linda. *The Politics of Postmodernism.* London and New York, 1989.
Jarmusch, Ann. Review. *Art News* 83 (May 1984): 128–31.
Jencks, Charles A. *The Language of Post-Modern Architecture.* 3d ed. New York, 1981.
Kahnweiler, Daniel-Henry. *Juan Gris: His Life and Work.* Translated by Douglas Cooper. New York, 1947.
Kostelanetz, Richard. *The Dictionary of the Avant Gardes.* Pennington, N.J. 1993.
Kosuth, Joseph. *Art Investigations and "Problematics" since 1965.* Lucerne, 1973.
Kristeva, Julia. *Desire in Language: A Semiotic Approach to Literature and Art.* Edited by Leon S. Roudiez, translated by Thomas Gora, Alice Jardine, and Leon S. Roudiez. New York, 1980.
Lippard, Lucy R., ed. *Six Years: The Dematerialization of the Art Object from 1966 to 1972.* London, 1973.
Meyer, Ursula, ed. *Conceptual Art.* New York, 1972.
Mitchell, W. J. T. *Iconology: Image, Text, Ideology.* Chicago, 1986.
Owens, Craig. "Earthwords." *October* 10 (Fall 1979): 121–30.
Schwarz, Arturo, ed. *The Complete Works of Marcel Duchamp.* New York, 1969; 3d rev. exp. ed., 1996.
Sheppard, Richard. "The Crises of Lanaguage." In *Modernism, 1890–1930.* Edited by Malcolm Bradbury and James McFarlane, pp. 323–326. New York, 1978.
Smith, Philip. "A Laurie Anderson Story." Interview. *Arts Magazine* 57 (January 1983): 60–61.
Smithson, Robert. *The Writings of Robert Smithson: Essays with Illustrations.* Edited by Nancy Holt. New York, 1979.
Steiner, Wendy. *The Colors of Rhetoric: Problems in the Relation between Modern Literature and Painting.* Chicago, 1982.
Ulmer, Gregory L. *Applied Grammatology: Post(e) Pedagogy from Jacques Derrida to Joseph Beuys.* Baltimore, 1985.

JESSICA PRINZ

DISINTERESTEDNESS. [*This entry explains the meaning and history of a central notion (namely, the exclusion of nonaesthetic interests—related to desire or politics, for example—from judgments of art) in the history of aesthetics and, though largely in critical terms, in contemporary aesthetics as well.*]

When philosophers have attempted to characterize aesthetic experience they have frequently looked for a special kind of attitude or way of attending to the object on the part of the perceiver. Aesthetic attitude theorists, as they have come to be called, often propose a stance of *disinterestedness* on the part of the subject with regard to the aesthetic object. Broadly speaking, this means that when attending to an object's beauty, the perceiving subject's relation to the object is different in important respects from other sorts of cognitive or affective relations. More specifically, it means that the perceiver is somehow engaged with the object in a way that separates and for a time disregards nonaesthetic concerns.

Details of the account of this detachment vary from theorist to theorist, as does the role that disinterestedness plays in their theories. The most recent defense of the notion of

aesthetic disinterest, that of Jerome Stolnitz, closely identifies the concept with aesthetic attitude theory, but it cannot simply be assumed that all philosophers who make use of a notion of disinterest are committed to the existence of a purely aesthetic *attitude*. George Dickie has argued most vehemently against making this assumption, especially for the British theorists of taste in the eighteenth century (Dickie, 1964; 1974, pp. 53–77; 1984). He argues that these philosophers use "disinterestedness" as an adjective describing a type of motivation for attending to an object, not the attention itself. But whether disinterestedness is a property of attitudes, motivations, or states of pleasure, the view that disinterest in some form is a necessary condition of aesthetic experience is a common thread running through much Western aesthetic theory since the eighteenth century. In what follows some of the most important of these disinterest-based theories will be discussed, beginning with a brief account of disinterestedness as a concept prior to the rise of aesthetic theory in the eighteenth century.

Historical Background. A precursor concept, and perhaps also a direct influence on theories of disinterested contemplation of art, was the notion of the disinterested love of God found in Neoplatonic Christian philosophy and also in Pietism. Anthony Ashley Cooper, the third earl of Shaftesbury (1671–1713), speaks of it, and it is affirmed in eighteenth-century Germany in the work of Karl Philipp Moritz. Both wrote important works that transferred their religious notion of disinterestedness to aesthetics (Woodmansee, 1994, pp. 19ff.; Stolnitz, 1961, pp. 132ff.). Significant accounts of religious disinterest can be found much earlier, however. John Scotus Erigena, for instance, spoke of the "wise man's" way of viewing a beautiful vase:

[He] simply refers its beauty to the glory of God alone. He is not seized with the temptation of avidity, no poison of greed infects the intention of his pure mind, and he is not sullied by any lust.
(Quoted in Tartarkiewicz, 1970, vol. I, p. 103)

Notable, too, is the account of disinterestedness in the writings of Meister Eckhart (1260–1328). Eckhart uses the term *Abgeschiedenheit* to refer to a particular state of virtue that involves as complete a detachment as possible from earthly objects in order to be open to God, and to approximate as far as possible a contemplation of the object "from its own point of view" rather than from the particular experience and desires of the individual (Eckhart, 1941, p. 83). Eckhart himself cites Ibn Sīnā (Avicenna), whose philosophy is influenced by both Platonic and Aristotelian elements, in elaborating his own view that detachment is a necessary condition of "seeing" God (p. 84).

In general, the notion of a detached and isolated attention to a "higher" object of contemplation found in these philosophers can be traced to Plato's views on the contemplation of the forms and the state of mind necessary for that sort of intellectual activity. Yet the Platonic notion of disinterest encountered in these early works is decidedly intellectual, if not completely nonaesthetic. The combination of a conception of disinterested contemplation with the view that *sense* objects are worthy in their own right of disinterested attention is required before a notion of aesthetic, that is, sense-based, disinterest can emerge. This synthesis appears to have occurred first within the Aristotelian tradition, where Aristotle's notion of a common sense is brought to bear on accounts of disinterest. Aristotle had proposed that human beings have a general, inner "sense" or ability for discriminating objects of the five outward senses. For Aristotle, the common sense is the human capacity for making basic perceptual judgments and thus makes coherent sense experience possible. When the detachment formerly reserved for contemplating the forms or God is turned on the physical world through this "organ" of inner sense, that is, through the common sense, then *aesthetic* disinterest is possible. Just such a synthesis of the religious disinterest and Aristotelian common sense may be found in the philosophies of Muḥammad al-Fārābī and Thomas Aquinas, both of whom make reference to a kind of disinterested perceptual contemplation (Summers, 1987, pp. 63ff., 225ff.).

The concepts of detached contemplation and a common sense for perceiving physical beauty were both theorized and even synthesized into a notion of aesthetic disinterestedness long before a sustained and systematic account of the latter was developed. The concept of a disinterested "sense" or ability for distinguishing beautiful perceptual forms in nature was not a central part of aesthetic theory for the simple reason that aesthetics as a separate discipline did not itself arise until the early eighteenth century in Europe. When it did develop, however, disinterestedness was among the central tenets of the most important of these early theories.

Particularly in Britain and Germany, the concept of disinterestedness played an important role in the new field of aesthetic theory. Historians of aesthetics have recently begun to explain the significance of the introduction of a concept of aesthetic disinterest in terms of the economic and social conditions of art during the period in which it was introduced. Thus Martha Woodmansee, in her discussion of the notion of disinterestedness and the autonomy of art in the works of Moritz in Germany, points to the proliferation of art meant to pander to a new reading public's desire for entertainment as a factor shaping his theory of disinterest. The new popular art, on this account, threatened neoclassical art and the artists who produced it. The neoclassical definition of good ("fine") art developed by French theorists such as Abbé Batteux required the *idealized* imitation of nature, and artists working from this paradigm saw pleasing mass audiences neither as desirable in itself nor likely to happen. But theories such as Moses Mendelssohn's and Gotthold Ephraim Lessing's in Germany emphasized the importance of art's producing pleasure and sympathetic re-

action in the audience. These rival theories served to legitimate the burgeoning industry in popular romances, plays, and so on, and threatened to create a public unfit or unwilling to appreciate more formal and demanding arts. Thus, on this account it was necessary for authors to promote in the reading public a notion of high culture, in which the art object is seen as a self-contained and independent piece that could only be appreciated by a more refined audience, that is, an audience capable of putting aside its particular feelings and interests in order to contemplate its object in a purely disinterested fashion. Elizabeth A. Bohls gives an account of similar developments in England, linking the emphasis on disinterestedness to the legitimation of upper-class power and interests in the name of disinterested civic virtue and refined taste. As eighteenth-century English culture grew more specialized, heterogeneous, and potentially volatile, Bohls suggests that "an aesthetic paradigm of perception thus becomes a cultural means toward the ultimately political end of homogeneity and solidarity among England's governing classes" (Bohls, 1993).

Theories of Aesthetic Disinterest. It is in the work of the third earl of Shaftesbury that the notion of a purely aesthetic disinterest first is fleshed out in the context of aesthetics. Under the influence of the Neoplatonist school at Cambridge, his account unites both the religious/Platonic notion of detached contemplation, a doctrine of a common sense for the perception of beautiful forms in nature, and an affirmation of neoclassical views of idealization in art. Disinterest for Shaftesbury is a condition of a "rational and refined contemplation of beauty" that distinguishes itself from other "enjoyments" inasmuch as it involves no desire for possession or consumption of the object (Cooper, 1964, vol. 2, pp. 127–128). Particular interests destroy this intellectual perception. Seeing beauty requires disinterest lest we make the absurd mistake of "seeking the enjoyment elsewhere than in the subject loved" (vol. 2, p. 246). In other words, a true judgment of beauty requires a disinterested contemplation of the object for its own sake. It should be noted that for Shaftesbury, the disinterested enjoyment of nature is still conceived as highly intellectual. It is by means of an "inward eye" for the eternal "absconded" beautiful forms behind the particular natural objects, as well as a sense for the virtuous character ("form" as opposed to "deformity") of actions, that disinterested enjoyment is possible (vol. 2, p. 137).

Philosophers in the British tradition who followed Shaftesbury continued to make use of the idea of disinterestedness in various ways. One of the most notable developments in these philosophers was a move away from Shaftesbury's notion of a common sense or *sensus communis* as a social sense or feeling for community, toward a more purely perceptual model. Thus Francis Hutcheson (1694–1746) made use of the notion of an inner sense for beauty that was much closer to the actual senses (Hutcheson, 1973, p. 34).

He also added the further prohibition against allowing interest in obtaining knowledge of the object to interfere with one's appreciation of it in a judgment of beauty (p. 84). Joseph Addison and Edmund Burke also hold that spectators' disinterestedness is crucial to making their judgments aesthetic, but without postulating a special common sense or "inward eye" for beauty at all. This is an important development, because in referring the experience of aesthetic properties solely to the five senses, they in effect break the last link between aesthetic and religious experience.

Burke, like Hutcheson, severs ties between experiences of beauty and knowledge, but goes even further in arguing that morality, too, should not be "confounded" with beauty (Burke, 1958, pp. 112, 107). Later in the eighteenth century, Archibald Alison (1757–1839) is more explicitly concerned with the state of the subject necessary for aesthetic appreciation, and suggests more than do his predecessors that this state is at least partially constitutive of the aesthetic object. In this respect he comes closer to Immanuel Kant's and Arthur Schopenhauer's views, and to being an aesthetic attitude theorist. Alison argues that aesthetic contemplation requires subjective involvement in the form of an active exercise of the imagination and takes the negative account of disinterest to new extremes. In a tone reminiscent of Eckhart's view that "disinterest comes close to zero" (Eckhart, 1941, p. 83), and oddly prefiguring Schopenhauer, Alison argues that objects of beauty are most accessible to the "vacant and unemployed" mind (Alison, 1817, pp. 10–11).

In Germany in the latter half of the eighteenth century, Moritz was influential in developing an account of disinterestedness in conjunction with an account of the self-contained nature of the artwork. He describes the nature of the object as independent and isolated from ordinary experience, capable of causing in us a "forgetfulness" that he identifies with disinterested pleasure. Although his focus is on the nature of the object, he also emphasizes the importance of the proper state of mind that contemplation of art objects produces.

It was Kant, however, who was primarily responsible for developing the notion in a systematic way and incorporating it within the larger context of an account of reflective judgment. Like Moritz, Kant gave an account in the *Critique of Judgment* of the properties that the aesthetic object must possess in order to facilitate aesthetic reflection. But his primary focus in describing what he called the "pure judgment of taste" was to state the conditions under which a judgment that an object is beautiful could justifiably demand the agreement of others. A crucial step in Kant's argument is his claim that in order for an aesthetic judgment to have validity it must be free of any interest, where interest is defined as "the liking *(Wohlgefallen)* we connect with the presentation of the object's existence" (Kant, 1987, p. 204). The demand for detachment that requires ignoring or abstracting from our cognitive, moral, and private interests in

its extreme form is a disinterest in the very existence of the object. As with Alison, or Eckhart, this radical detachment should not be taken as apathy. For Kant it is rather a focusing of attention on the purely formal, outward presentation that the object affords. He argues that in judging properly about an object we must be concerned only with the formal appearance of purposiveness in the design and composition of the object (p. 225). In order to maintain this sort of relationship to the presentation, no interests may be present: "All interest ruins a judgment of taste and deprives it of its impartiality" (p. 223). Impartiality is important because it is the only basis for the universality of a judgment based on feeling (liking) of subjects, and proving this universality is one of Kant's primary goals. Yet aesthetic reflective judgment, for Kant, "cannot be other than subjective," that is, determined by the state of the subject rather than the object (p. 203). For Kant this determination is the result of the free play of the subject's imagination and understanding, where no concept of the object—of its use, nature, or goodness—is predicated on the resulting judgment. Disinterested pleasure, Kant says, is the consciousness of this free play and a condition of its perpetuation (p. 222). Since the experience, that is, the pleasure felt, is disinterested, and because we can assume that all individuals have the capacity for feeling such a pleasure (they have a *sensus communis aestheticus*), Kant argues, we may be assured that our pure judgments of taste may justifiably claim the agreement of others.

After Kant the concept continues to play an important role in aesthetics of the Romantic period. It may be found, for example, in Friedrich Schiller and Samuel Taylor Coleridge, and is brought to new theoretical heights in the philosophy of Schopenhauer. Schopenhauer's account relies on a particularly strong version of disinterestedness in which the genuine aesthetic attitude requires the perceiver to lose him- or herself completely in the aesthetic object, to be a "disinterested spectator" to such an extent that the perceiver's own existence is transcended, at least temporarily. On this view the aesthetic object is wholly the result of the subject's disinterested contemplative activity.

Schopenhauer's aesthetics represents a kind of watershed in the development of the concept of disinterest. Borrowing Kant's ban on all relation to desire in aesthetic reflection, Schopenhauer defines disinterest as a state in which aesthetic objects "stand in no relation to our will" (Schopenhauer, 1957, vol. 2, p. 369). This absolutely will-less state differs from Kant's, however, in its demand on the subject to transcend him- or herself. Although Schopenhauer's Platonist metaphysics is in part responsible for the demands of his aesthetics, it is interesting to note that the tendency of his view is to focus on the description of the aesthetic experience rather than on the justification of its claim to universality. As David E. W. Fenner points out, this new focus on the nature of the experience itself approximates the approach of twentieth-century attitude theorists to a far

greater degree than did the aesthetics of disinterest in the British theorists or even in Kant (Fenner, 1996, pp. 53ff.).

At the beginning of the twentieth century, the theory of disinterest became the centerpiece of several theorists' approaches. E. F. Carritt, H. Münsterberg, Ethel Puffer, and Edward Bullough are representatives of this tradition. Of these, Bullough has had the greatest impact. His account of the aesthetic attitude in terms of "psychical distance" is primarily a negative one, in which the "practical side of things" is "inhibited." Like Schopenhauer, Bullough emphasizes the rare nature of the resulting experience. Unlike Schopenhauer, he does not insist on a complete but temporary transcendence of the self. Psychical distancing is for him a matter of fine-tuning our attitude to produce "the utmost decrease of Distance without its disappearance." What he means by this is that the personal, in the sense of one's own deeply felt life experiences, is crucial to our appreciation of many works of art. At the same time aesthetic appreciation is jeopardized by immediate, strong personal emotions or desires. The ideal situation is one in which the subject is able to set aside these immediate personal reactions while maintaining a personal predisposition in favor of the object's qualities. Thus, unlike the "disinterest" theorists before him, Bullough argues that one can have too much distance, as well as not enough. Distance, unlike disinterest for most of the earlier theorists, comes in degrees. Too much distance would mean the complete loss of the intensely felt connection to the work of art that Bullough argues is definitive of the aesthetic experience.

By far the staunchest defender of disinterestedness recently has been Jerome Stolnitz, whose aesthetic attitude theory incorporates the concept of disinterest as a central necessary condition of the aesthetic attitude. Like Bullough, he is concerned to explain what exactly it is that makes an experience distinctively aesthetic, but Stolnitz (1961) argues that the concept of disinterestedness contained in his definition of the aesthetic attitude has its roots in the eighteenth-century British theories of beauty and art. He characterizes it as a way of perceiving the object in which

> we do not look at the object out of concern for any ulterior purpose which it may serve. We are not trying to use or manipulate the object. There is no purpose governing the experience other than the purpose of just *having* the experience. Our interest comes to rest upon the object alone. (Stolnitz, 1960, p. 35)

This definition makes clear that for Stolnitz, just as for others who use the concept, "disinterest" does not mean apathy or lack of interest toward the object, but rather refers to a special kind of interest in the object "for its own sake." Interests that are excluded on his account include preoccupation with possessing or owning the object, cognitive interests (of the sort a scientist might take in the physical nature of the object, or that a sociologist might take in the cultural significance of it), or even, Stolnitz says, the interest that an

art critic might take in passing judgment on the value of the object: "For the aesthetic attitude, things are not to be classified or studied or judged. They are in themselves pleasant or exciting to look at" (p. 35). Here Stolnitz echoes Kant's claim that "a judgment of beauty [cannot] be determined by an *a priori* proof, in accordance with determinate rules," and that

> if someone reads me his poem, or takes me to a play that in the end I simply cannot find to my taste, then let him adduce *Batteaux* or *Lessing* to prove that his poem is beautiful, or still older and more famous critics of taste with all the rules they have laid down; . . . I shall stop my ears, shall refuse to listen to reasons and arguments, and shall sooner assume that those rules of the critics are false . . . than allow my judgment to be determined by *a priori* bases of proof. (Kant, 1987, p. 284)

This affinity with Kant (and Schopenhauer) separates Stolnitz from theorists of taste like the earl of Shaftesbury for whom aesthetic experience cannot be isolated from aesthetic evaluation. To this extent at least, he breaks with the British tradition.

This short survey cannot do justice to the differences and disagreements that may be found among the various accounts of aesthetic disinterest. It can, however, serve to show what general feature these approaches have in common, namely, the view that aesthetic experience involves an important feature—attitude, motivation, pleasure or "enjoyment"—that is removed from the particularities of individual concerns. The disinterested aspect of the aesthetic is believed to provide the subject with a brief withdrawal from the mire of personal and historical particulars into a more all-encompassing realm.

Criticisms of Aesthetic Disinterest. This influential concept has not surprisingly had many critics. One straightforward objection is simply the denial that disinterestedness is a necessary condition of aesthetic experience. That is, the argument goes, it may be the case that at least on some occasions, having a particular kind of interest is important to true appreciation of an aesthetic object. An example might be the case of Duchamp's *In Advance of a Broken Arm* where it is important to know the function of the object, a snow shovel, in order to appreciate the work (Fenner, 1996).

A second sort of argument against disinterestedness as an attitude is based on the view that there is no special aesthetic attitude in the first place. Dickie argues that what happens when we view or listen to an aesthetic work is simply focused attention. Adducing a special state of consciousness beyond attention, he argues, amounts to an unnecessary multiplying of explanatory entities. Similarly, he argues that the postulating of an attitude of disinterest is unnecessary and confusing. To be disinterested, he argues, means (for theorists like Stolnitz) to fail to be interested, but failing to be interested can only mean being distracted. To be distracted is just a form of inattention, so once again it

looks as if all the attitude theorist is saying is that it is necessary to attend to the object when viewing it aesthetically. This move is similar to that of empiricists like Addison and Burke, who, as we saw, denied the need to postulate a common sense for beauty and other aesthetic properties. Dickie also takes issue with the claim that criticism and appreciation are different sorts of activities, arguing that the careful, sensitive viewing of the appreciator is not distinct from the critics viewing an object with certain standards in mind (Dickie, 1964).

A third type of criticism is broader, and is typically put forth by those who, in Jacques Taminiaux's formulation, incline toward the "death of art" pole in modern aesthetics (Taminiaux, 1993, pp. 55ff.). This approach argues that no aesthetic concept is universal or transhistorical, nor can there be such a thing as *the* aesthetic experience. Aesthetic experience, where it exists at all, must be viewed with an eye toward political, social, and economic factors that shape the very notion of aesthetics in a culture at a time. Thus, Pierre Bourdieu sees the disinterested attitude as an artifact of a "visceral disgust at vulgarity" and the intellectual's need to set up and internalize a distinction between the pleasures of "mere" sense and taste:

> "Empirical" interest enters into the composition of the most disinterested pleasures of pure taste, because the principle of the pleasure derived from these refined games for refined players lies, in the last analysis, in the denied experience of a social relationship of membership and exclusion. (Bourdieu, 1984, p. 499)

The denial of "lower" pleasures, Bourdieu argues, affirms the superiority of those whose taste is disinterested. Thus the theories that espouse the concept "fulfil a social function of legitimating social difference" (p. 7).

Feminist criticism has argued along similar lines, adding that those excluded by this concept are not only the "vulgar" lower classes, but also those typically associated with the particular, with sensuousness and the body, namely, women (Bohls, 1993). Moreover, the insistence on ignoring the instrumental value of the work has the tendency of elevating nonfunctional works like painting and sculpture over functional works like quilting and handcrafts of various sorts that are typically the domain of women and "mere" craftsmen. It follows that those who create these useful objects must not be artists, so that many feminists would agree that "women are not well served by the 'disinterestedness thesis'" (Hein, 1993, p. 12). It should be noted, however, that recently some feminists have decided to take "a second look" at disinterestedness on the grounds that it may have potential value in formulating accounts of political solidarity among diverse groups (Bohls, 1993) and for open-minded and sensitive advocacy (Hein, forthcoming).

The number and vehemence of these criticisms attest to the significance of the concept of aesthetic disinterest that they attack. Supporters and detractors all agree that the no-

tion has had enormous influence on contemporary views of art, artists, and aesthetic appreciation, and will continue to have influence for a long time to come.

[*See also* Art for Art's Sake; Attitude, *article on* Aesthetic Attitude; Autonomy, *historical and conceptual overview article;* Japanese Aesthetics, *historical overview article;* Nietzsche, *article on* Nietzsche, Schopenhauer, and Disinterestedness; *and* Taste.]

BIBLIOGRAPHY

Addison, Joseph, and Richard Steele. *Addison and Steele: Selections from "The Tatler" and "The Spectator."* Edited by Robert Allen. New York, 1957.

al-Farabi. *Alfarabi's Philosophy of Plato and Aristotle.* Rev. ed. Translated by Muhsin Mahdi. Ithaca, N.Y., 1969.

Alison, Archibald. *Essays on the Nature and Principles of Taste.* 5th ed. Edinburgh, 1817.

Aquinas, Thomas. *Sententia libri ethicorum.* In *Opera omnia,* vol. 47. Rome, 1969.

Bohls, Elizabeth A. "Disinterestedness and the Denial of the Particular: Locke, Adam Smith and the Subject of Aesthetics." In *Eighteenth-Century Aesthetics and the Reconstruction of Art,* edited by Paul Mattick, Jr., pp. 16–51. Cambridge and New York, 1993.

Bordieu, Pierre. *Distinction: A Social Critique of the Judgment of Taste.* Translated by Richard Nice. Cambridge, 1984.

Burke, Edmund. *A Philosophical Enquiry into the Origin of Our Ideas of the Sublime and Beautiful.* Edited by James T. Boulton. New York, 1958.

Cooper, Anthony Ashley, Earl of Shaftesbury. "The Moralists." In *Characteristics of Men, Manners, Opinions, Times,* edited by John M. Robertson, pt. 3, sec. 2. Indianapolis, 1964.

Dickie, George. "The Myth of the Aesthetic Attitude." *American Philosophical Quarterly* 1.1 (January 1964): 56–65.

Dickie, George. *Art and the Aesthetic: An Institutional Analysis.* Ithaca, N.Y., 1974.

Dickie, George. "Stolnitz's Attitude: Taste and Perception." *Journal of Aesthetics and Art Criticism* 43.2 (Winter 1984): 195–203.

Eckhart, Meister. *Meister Eckhardt: A Modern Translation.* Translated by Raymond Bernard Blakney. New York, 1941.

Fenner, David E. W. *The Aesthetic Attitude.* Atlantic Highlands, N.J., 1996.

Guyer, Paul. "The Dialectic of Disinterestedness: I. Eighteenth-Century Aesthetics" and "II. Kant and Schiller on Interest in Disinterest." In *Kant and the Experience of Freedom: Essays on Aesthetics and Morality.* Cambridge and New York, 1993.

Hein, Hilde. "Refining Feminist Theory: Lessons from Aesthetics." In *Aesthetics in Feminist Perspective,* edited by Hilde Hein and C. Korsmeyer, pp. 3–18. Bloomington, Ind., 1993.

Hein, Hilde. "The Internalization of Disinterestedness." In *Advocacy and the Classroom,* edited by Patricia Meyer Spack. New York, forthcoming.

Hutcheson, Francis. *An Inquiry concerning Beauty, Order, Harmony, Design.* Edited by Peter Kivy. The Hague, 1973.

Kant, Immanuel. *First Introduction to the Critique of Judgment.* Translated by James Haden. Indianapolis, 1965.

Kant, Immanuel. *Critique of Judgment.* Translated by Werner S. Pluhar. Indianapolis, 1987.

Moritz, Karl Philipp. "Versuch einer Vereinigung aller schönen Künste und wissenschaften unter dem Begriff des in sich selbst Vollendeten." In *Schriften zur Ästhetik und Poetik,* edited by Hans Joachim Schrimpf. Tübingen, 1962.

Rogerson, Kenneth. "Dickie's Disinterest." *Philosophia* (1987).

Schopenhauer, Arthur. *The World as Will and Idea.* 3 vols. 2d ed. Translated by R. B. Haldane and J. Kemp. London, 1957.

Stolnitz, Jerome. *Aesthetics and Philosophy of Art Criticism: A Critical Introduction.* Boston, 1960.

Stolnitz, Jerome. "On the Origins of 'Aesthetic Disinterest'." *Journal of Aesthetics and Art Criticism* 20.2 (Winter 1961): 131–143.

Stolnitz, Jerome. " 'The Aesthetic Attitude' in the Rise of Modern Aesthetics." *Journal of Aesthetics and Art Criticism* 36.4 (Summer 1978): 409–423.

Summers, David. *The Judgment of Sense: Renaissance Naturalism and the Rise of Aesthetics.* Cambridge and New York, 1987.

Taminiaux, Jacques. "Between the Aesthetic Attitude and the Death of Art." In *Poetics, Speculation, and Judgment: The Shadow of the Work of Art from Kant to Phenomenology,* translated and edited by Michael Gendre. Albany, N.Y., 1993.

Tartarkiewicz, Wladislaw. *History of Aesthetics.* 3 vols. Edited by C. Barnett. The Hague, 1970–1974.

Vivas, Eliseo. "Contextualism Reconsidered." *Journal of Aesthetics and Art Criticism* 18 (1959).

Woodmansee, Martha. *The Author, Art, and the Market: Rereading the History of Aesthetics.* New York, 1994.

Zangwill, Nick. "UnKantian Notions of Disinterest." *British Journal of Aesthetics* 32.2 (April 1992): 149–152.

JANE KNELLER

DOCUMENTARY FILM. *See* Film.

DRAWING. Drawing, the skilled activity, and drawings, the products of that activity, have been tremendously important to the history of painting, sculpture, architecture, and the decorative arts, though little attention has been paid to them by philosophers of art. One factor contributing to this lack of attention is that drawings have historically been (largely, but by no means exclusively) preparatory studies for other projects, rather than completed works of art in themselves. Even as completed artworks, they do not carry the prestige of paintings, music, or sculpture. Philosophical interest in drawing, however, *is* implicated by interests in painting (and other arts, such as sculpture) as a skilled activity where understanding the generative process is essential to understanding the nature and value of the product, that is, the painting or picture itself. Philosophical interest in drawing and drawing practices is also implicated by interests in the production of art as part of a larger social and cultural history, where social status, economic conditions, and educational practices are all part of that history. Workshops and academies where drawing was taught reflected economic, class, and gender divisions prevalent in society at large, and these in turn influence what values we attribute to drawing.

It is not difficult to figure out why, art historically, drawings are so important. They are a record of what the artist knew and could do. Drawings have been created to try out

different ways of composing a painting, to practice rendering a figure or scene, and to perfect the details of a picture, such as the angle of the foot, hand, or head. Full-scale "cartoons" served as "ground plans" for wall and ceiling murals. Paintings may replicate, expand upon, or diverge from their underdrawings. Drawings provided designs to be copied for monuments, candlesticks, and elaborate cornices. They accompanied text, and provided the basis for multiple reproduction processes. They were produced in reverse for engraving and for tapestries. They were architectural plans and designs for facades; they served to work out architectural ideas, and as part of contracts for what was to be engraved or built.

Distinctively philosophical problems also arise around drawing itself, however. How does drawing differ from just making marks, or from writing (especially when the writing is pictographic)? What is different about the psyche of someone who knows how to draw and someone who does not? What is the relation between knowing how to draw and knowing how to produce art? Are the methods of depiction in drawing more or less realistic than those used in painting? How do the values drawings are recognized as having reflect changes in the material conditions and technology of drawing? How does the changing role of drawings as experimental or preparatory studies affect the way we view the arts? Each of these issues will be addressed in turn through the course of this essay.

Consider first how drawing is different from just making marks. Scratches on a rock or piece of bone present problems for the anthropologist. Did accidental abrasion of one rock against another produce them? Did human beings produce them intentionally, just as marks, the way a child might hack at a surface? Are they not merely marks, but lines that are *drawn*, as lines, so that a consciousness of their linearity informs the gesture that produces those incisions on the surface of rock or stone? Do they in addition represent something, for example, quantities of things? Or are they pictures, for example, of grasses or grain, of poles or walls? These interpretative decisions depend on what mental properties an interpreter is willing to attribute to whoever made those marks. These mental properties are important because, insofar as a line is drawn, they control or inform the motor behavior of that person. When a line is *drawn*—not merely when a mark is made—it is an exercise of a kind of skill or ability. As a skill or ability, a kind of psychological control is exercised over that gesture, over the making of the line. The greater the skill, the greater the control of relevant characteristics of the line. If the thickness of the line is significant, how one's movement and gesture increases or decreases its thickness makes a difference to what movement one makes. And the same is true of its length, curvature or rectilinearity, and smoothness or jaggedness.

DRAWING. Honoré Daumier, *Confidence;* Victoria & Albert Museum, London. (Photograph copyright Board of Trustees of the Victoria & Albert Museum, courtesy of Victoria & Albert Museum, London/Art Resource, New York; used by permission.)

Drawing ability, like many other abilities, has a feature we can call "extendability": if one knows how to draw, one can draw things fairly well that one has never drawn before. Knowledge of a language is also extendable: if one knows how to speak English, one can generate novel utterances, and one is not confined to repeating, parrotlike, utterances already learned. Psychologically, what accounts for the extendability of drawing ability? One of the most popular models to explain the extendability of linguistic ability—the compositional model—is clearly not adequate to explain the extendability of drawing ability. On the compositional model, novel performances are explained by substituting new units (for language, they are semantic units; for drawing, they would be graphic units, each of which need not have representational significance in itself) according to grammatical rules. But learning to draw is not a matter of learning a set of graphic units—a set of lines and blobs of different curvatures and shapes—along with rules about how to combine them. *Part* of what distinguishes great drawers such as Henri Matisse, Rembrandt van Rijn, and Il Guercino (Giovanni Barbieri) is the quality of the lines they produce. The quality of these lines may in turn be due to the

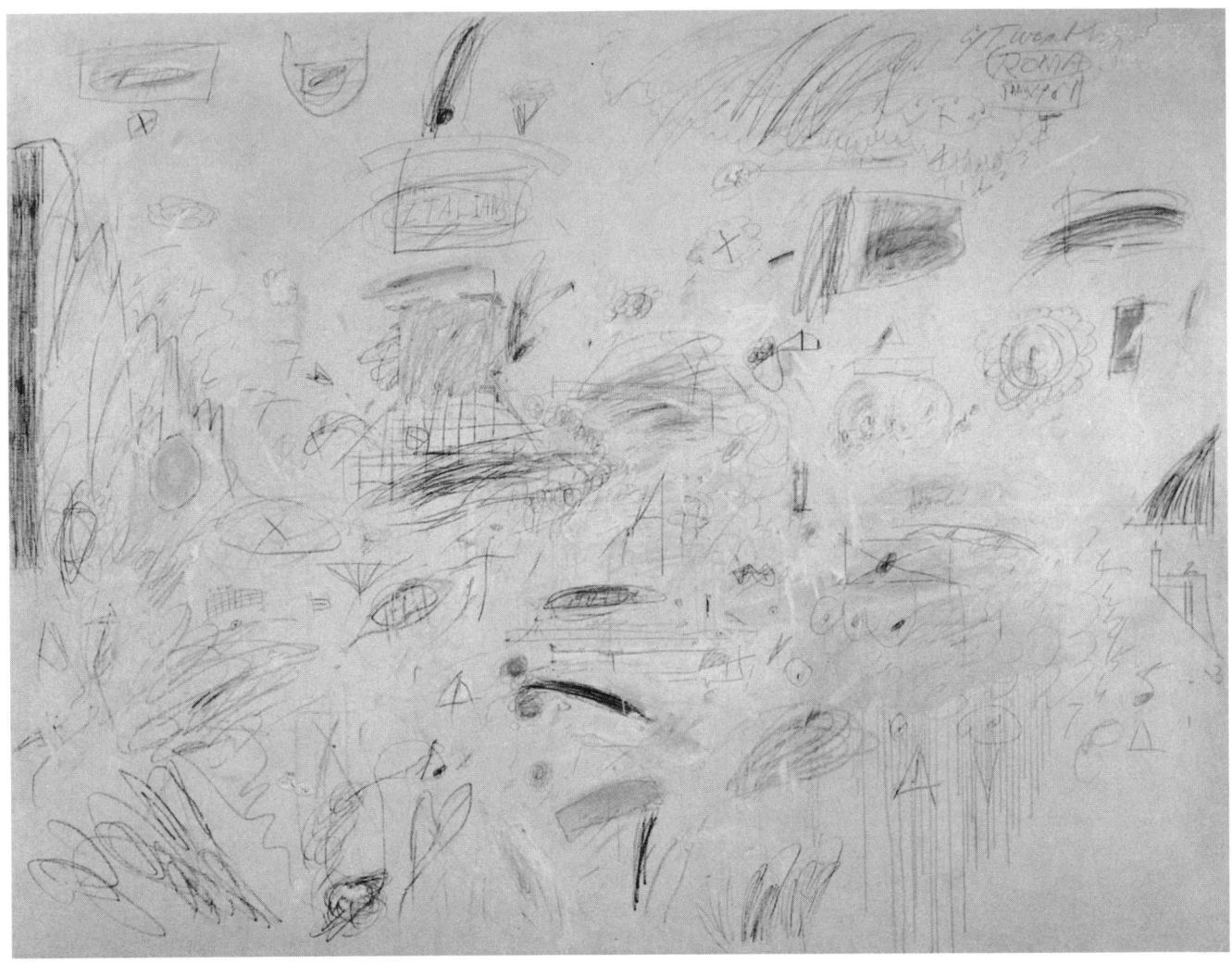

DRAWING. Cy Twombly, *The Italians* (1961), oil, pencil, and crayon on canvas, 6′ 6 5/8″ × 8′ 6 1/4″ (199.5 × 259.6 cm); Museum of Modern Art, New York; Blanchette Rockefeller Fund. (Photograph copyright 1998 by the Museum of Modern Art; used by permission.)

overall conception of the depiction, one that uses economy of line to suggest space, agitated and congested lines to suggest movement, and so on.

Part of the folklore of drawing addresses the question, what is different about the psyche of someone who can draw and those who cannot? The folklore says that learning to draw involves learning to see. It has good foundation: in learning to draw, one comes to see things in graphically useful ways. A good drawing instructor can teach you how to "conceptualize" what you see in a way that enables you to control the movements of your arm and hand in ways that produce artistically desirable lines and images. One has to recruit certain patterns of visual stimuli not (merely?) as one usually does, to identify what one sees, but to control the way one moves one's arm. These perceptual skills help account for the ex-

tendability of drawing ability, and their solidification into a distinctive style helps to elevate drawing into an art.

Another popular saying in the drawing classroom is that there are no lines in nature. There *are* no lines in nature, but there are edges. We generally recognize edges by abrupt changes in one expanse of color or brightness to another; outline drawings instead use lines to represent edges. Lines may also be used to represent contours within the body of the object, such as noses or mouths within a face. But objects do not come with black or gray lines surrounding them and highlighting their internal contours. Lines are, in this sense, an unrealistic means of depiction. Thus, a drawing is less realistic to the extent that lines—rather than an abrupt change in an area of color or brightness—are used to represent the edges of objects.

The affordability and availability of paper since the mid-fifteenth century created a paradigm shift for both the nature of drawing practices and the value we find in drawings. What are some examples of pre-Renaissance drawings? Petroglyphs—pictures incised in rock—were drawn by Native Americans as long ago as 1000 BCE; for what reasons, we do not know. Greek vases were decorated with designs and pictures of mythological subjects, both incised and painted. During the Middle Ages, working drawings for manuscript illuminations were done on prepared wooden or wax tablets. Pictures were generally composed by studying other pictures, rather than by studying "nature," or what they were pictures *of*. In the 1230s, Villard de Honnecourt composed a pattern book that served as a source of ideas for projects undertaken by craftsmen in his workshop. The late Gothic model book, derived in concept from the pattern book, served as a set of pictures to be copied, and widespread replication of images contributed to the "international" style of late Gothic art. These "books" were produced on vellum or parchment, and sometimes wooden tablets, to enhance durability. Paper as we know it did not appear in the West until the twelfth century, and remained costly until demand increased precipitously after the invention of movable type in the mid-fifteenth century. The availability of relatively low-cost paper made experimentation on paper possible, and also made it possible to preserve an artist's experimental drawings and sketches, unlike those previously produced on wax or prepared wood. The relative permanence and availability of paper, initiated during the Renaissance, opened the door to appreciating drawings as providing insight into the workings of the artist's mind, with accompanying values of spontaneity, intimacy, and expressiveness.

In the early modern academies from the seventeenth century on, copying from other drawings, from sculpture, and "from life" marked stages in the education of a painter or sculptor. Drawing is still a major part of the foundation of the education of visual artists of all types, including painters, sculptors, and printmakers. Its diminishing importance for computer graphics and graphic design is symptomatic of a major change taking place in the visual arts.

Drawings in the late twentieth century have emerged in a major way as forms of art in their own right. Paradigms for drawing are also changing in ways that reflect trends within the art world at large. First, many artworks challenge traditional boundaries among media, and the same is true of drawings. Drawing media are often mixed within a single work—graphite, paint, lipstick, rabbit's blood—and prints, bits of newspaper, magazines, photographs, and photocopies may be added. Second, there is a self-consciousness about the nature of art and what is involved in the creation of art. Questions about the nature of art have become part of its subject matter. One question is whether drawing as *art*

requires skill in the traditional sense, as taught in the academies. Drawings raise this question by incorporating writing, popular and formulaic imagery, and crude drawing that is reminiscent of graffiti or children's drawings, and other references to "outsiders"—the mentally disturbed, folk artists—anyone without formal artistic education. Someone other than the acknowledged artist may actually make the drawings (just as some artists whose work is photographs do not actually snap the shutter), a phenomenon in accord with R. G. Collingwood's idea that art consists of the clarification of thought and feeling, and not the craft of embodying the thought in external form. We may be on the verge of another paradigm shift in drawing, reflecting a somewhat altered view of its nature as a skilled activity, how it is used to generate art, and the values of those artworks.

BIBLIOGRAPHY

Ames-Lewis, Francis. *Drawing in Early Renaissance Italy.* New Haven, 1981.

Feagin, Susan L. "Pictorial Representation and the Act of Drawing." *American Philosophical Quarterly* 24.2 (April 1987): 161–170.

Gombrich, E. H. *Art and Illusion: A Study in the Psychology of Pictorial Representation.* 2d rev. ed. New York, 1961; reprint, Princeton, N.J., 1969.

Monnier, Geneviève, and Bernice Rose. *Drawing: History of an Art.* Translations by Barbara Bray. Geneva and New York, 1979.

Rose, Barbara. *Drawing Now.* New York, 1976.

Wollheim, Richard. "On Drawing an Object." In *On Art and the Mind: Essays and Lectures,* chap. 1. London, 1973; reprint, Cambridge, Mass., 1974.

Wollheim, Richard. "What the Artist Does." In *Painting as an Art,* chap. 4. Princeton, N.J., 1987.

SUSAN L. FEAGIN

DRESS. *See* Fashion.

DRYDEN, JOHN (1631–1700), English author of the Restoration, equally renowned for drama, poetry, criticism, and translation. John Dryden was born in Aldwinkle, Northamptonshire, and raised in a Puritan household. He was educated first at the famed Westminster School in London, and then at Trinity College, Cambridge. He began his writing career as a dramatist during the Restoration, quickly reaching such prominence that in 1668, with the death of William Davenant, he became the obvious choice for poet laureate, and in 1670, was appointed Historiographer Royal as well, thereby becoming the official voice of the nation. As laureate, he turned increasingly to poetry, displaying an extraordinary range of genres and topics: he wrote elegies, odes, dedicatory poems, songs, panegyrics, satires, and fables, as well as exploratory verse on such subjects as science, religion, philosophy, politics, and, of course, poetry itself. He not only introduced the genre of verse satire,

carried into the next generation most notably by Alexander Pope, but also accomplished so much within the heroic couplet that he helped make it the dominant English verse medium for the next century. Because of his conversion to Catholicism in 1685, Dryden lost his court positions in 1689 when William III came to the throne. Dryden now had to make a major adjustment, from official spokesman of the nation to political outcast, a well-known Catholic and Jacobite sympathizer distrusted but nervously tolerated by a Protestant regime. His writing in no way diminished, however. If anything, the richness and the scope of his poetry expanded in his last twelve years, when he continued writing occasional verse, but also returned to the stage and embarked on an ambitious program of translations, both strategies undertaken in part to earn an income now that he no longer received a pension.

Throughout his career, Dryden wrote critical essays, which tended to be occasional rather than theoretical, typically written to justify something he was undertaking, or to defend himself from attacks. He never wrote about the perception of beauty, the quality of the sublime, or the nature of taste, except to deplore the English for not having any. Ever the practical author, he was more interested in the production than in the reception of art, more interested in learning how to suit his work to the tastes of the times than in determining how taste operated within the human psyche. But if he seldom addressed aesthetic issues directly, he was nonetheless engaged in all the aesthetic controversies of the day and, as a major author, reshaped the cultural landscape by the sheer force of his reputation and the vitality of his poetry.

Dryden wrote no essays on music, although he wrote masques, opera librettos, and two songs for St. Cecilia's Day. All of these display an awareness of the affective powers of music, to raise the passions, infuse pity, induce pleasure, but he made no original contribution, instead relying on traditional associations between musical modes and emotional responses. He knew even less about painting, though he is often cited as the first English critic to explore the topic of the sister arts in "A Parallel of Poetry and Painting," an essay he wrote to accompany his 1695 translation of Charles Dufresnoy's *De arte graphica*. That so eminent an author treated this topic makes it worth noting, but the essay itself, a piece of hackwork churned out in twelve days, is scarcely a significant contribution. Two of Dryden's poems deal specifically with painting, one the 1685 Ode to Anne Killigrew, "Excellent in the Two Sister-Arts of Poësie, and Painting," the other his 1694 poem "To Sir Godfrey Kneller," addressed to the leading portrait painter of the day. Despite the ostensible subject, it becomes apparent that Dryden is interested in Killigrew and Kneller not so much as painters but as fellow artists, whose successes are measured against Dryden's own accomplishments.

Dryden's primary concern, and the central topic of most of his criticism, was literature. In discussing drama, he argued for the superiority of rhyme to blank verse, disputed the Aristotelian truism that drama was mimetic, and questioned the utility of the classical unities of time, place, and action, claiming that they could "raise perfection higher where it is," but that they were "not sufficient to give it where it is not" (actually, it was Dryden who first used the phrase "the three unities" in English). These views were not part of a systematic analysis but rather a response to the specific circumstances in the English playhouse. Since he was attempting to revive drama after the eighteen-year Puritan interregnum during which the playhouses were closed, he was understandably reluctant to compete against the glorious authors of the Elizabethan and Jacobean stage. Instead, he wanted to move English drama in a different direction by introducing a new genre, the heroic drama, which he described as an epic in miniature. For heroic drama, rhyme and heightened imitation, though "unnatural," were more appropriate than blank verse, because the characters were heroes, not ordinary men, and heroes could not be expected to speak or act as common people do. As for Dryden's attitude toward the unities, this too was a response to "the experience of our English stage," which demonstrated that dutifully unified French plays, when acted in London, fared poorly, whereas the incorrect and unclassical William Shakespeare invariably succeeded. As an indication of Dryden's pragmatism, one might add that he eventually abandoned rhyme in drama as well as heightened imitation, simply because heroic drama ceased to be popular with English audiences, and it was time to move on to something else.

In his attempts to restore the theater, Dryden found himself not merely defending the English stage, promoting Shakespeare for his variety and copiousness over the French playwrights who were confined by the unities, but also, in effect, inventing a national literary tradition. In his essays on drama and satire, and in his many poems on poetry, Dryden took pains to name his literary heroes, who included Geoffrey Chaucer, Shakespeare, Edmund Spenser, and John Milton, arranging them in a literary succession, which both implied a progress and allowed him to place himself in the lineage. In his final collection of poems, *Fables* (1700), he begins by invoking Homer and Virgil, claims that Chaucer was to England what they were to Greece and Rome, and then steps in as Chaucer's successor, thereby further enriching his heritage by basing it on classical literature, and then assuming the mantle as the next national poet in a noble line.

As he prepared his collections of translations, Dryden had the opportunity to theorize about the nature of translations, dividing them into three modes: the doggedly literal "metaphrase" at one end, the freely inventive "imitation" at

the other, with "paraphrase," the method he claimed to adopt, holding the middle ground. In fact, Dryden's versions of Juvenal, Ovid, Persius, Homer, and Virgil are actually imitations, with frequent omissions, additions, expansions, and rearrangements of material. However liberal they may be, it has been argued that they capture the spirit of the classical authors far better than "accurate" translations that remain true instead to the surface details of the originals. By "Englishing" the classical authors, so that we get, for example, not Virgil but Dryden's Virgil (he translated all of Virgil), Dryden quite deliberately attempts to subsume the classical within the native literary tradition he was attempting to assemble. His translations succeeded so well, and demonstrated what could be done in the idiom, that the genre of the "imitation" became prominent in the next generation, particularly in the works of Jonathan Swift, John Gay, Pope, whose most popular work was his translation of Homer, and Samuel Johnson, whose most famous poems, "London" (1738) and "The Vanity of Human Wishes" (1749), are both imitations of Juvenal.

Dryden seldom commented on the beauty of poetry; instead, he focused on the power of poetry, consistently suggesting that because poets reshape reality, they have profound influence on the world of human affairs, although he recognized that in the "real world" that influence would seldom be recognized. In his early political poetry, such as *Absalom and Achitophel* (1681), when he was speaking as poet laureate, he not only criticized Charles II but also assumed the role of prophet, predicting how political events, which were then unsettled, would and should turn out. Even late in his career, when he had lost the laureateship and was an opposition figure rather than a public spokesman, Dryden continued to comment critically on affairs of state, arguing against the ascension to the throne of a nonlineal monarch, William III, and then criticizing many of William's policies, from his reluctance to support the arts to his desire to maintain a standing army in time of peace. Whatever the subject, painting, music, poetics, Dryden tended to use the occasion to make social and political commentary. *Alexander's Feast* (1697), for instance, supposedly an ode to music, is also a variation on the theme of the artist's power, for it depicts the poet-orator-musician Timotheus manipulating the usurping monarch, Alexander the Great, in ways Dryden liked to think he was managing his own monarch, the usurping William III. Similarly, Dryden claims that he offers his translation of the *Aeneis* (1697) as "good Counsel" to his king, following Virgil, who, in directing his epic to the emperor Augustus, "dextr'ously . . . mannag'd both the Prince and People." While we often associate claims for the power of art and the moral responsibility of artists with the Romantic sensibility, epitomized by Percy Bysshe Shelley's proclamation in his *Defence of Poetry* (1821) that "Poets are the unacknowledged legislators of the World," this attitude begins with Dryden and is apparent in his practice rather than in his theories.

[*See also* Literature, *article on* Literary Aesthetics; Poetics; *and* Theater.]

BIBLIOGRAPHY

Works by Dryden

The Letters of John Dryden: With Letters Addressed to Him. Edited by Charles E. Ward. Durham, N.C., 1942.
The Works of John Dryden. 20 vols. Edited by Edward Niles Hooker, H. T. Swedenberg, Jr., et al. Berkeley, 1956–.

Other Sources

Corse, Taylor. *Dryden's* Aeneid: *The English Virgil.* Newark, Del., 1991.
Harth, Phillip. *Contexts of Dryden's Thought.* Chicago, 1968.
Harth, Phillip. *Pen for a Party: Dryden's Tory Propaganda in Its Contexts.* Princeton, 1993.
Hopkins, David. *John Dryden.* Cambridge and New York, 1986.
Hume, Robert D. *Dryden's Criticism.* Ithaca, N.Y., 1970.
Miner, Earl. *Dryden's Poetry.* Bloomington, Ind., 1967.
Pechter, Edward. *Dryden's Classical Theory of Literature.* Cambridge and New York, 1975.
Reverand, Cedric D., II. *Dryden's Final Poetic Mode: The Fables.* Philadelphia, 1988.
Roper, Alan. *Dryden's Poetic Kingdoms.* London, 1965.
Schilling, Bernard Nicholas. *Dryden and the Conservative Myth: A Reading of Absalom and Achitophel.* New Haven, 1961.
Winn, James Anderson. *John Dryden and His World.* New Haven, 1987.
Zwicker, Steven N. *Politics and Language in Dryden's Poetry: The Arts of Disguise.* Princeton, N.J., 1984.

CEDRIC D. REVERAND II

DU BOS, JEAN-BAPTISTE

DU BOS, JEAN-BAPTISTE (1670–1742), early eighteenth-century French art theorist. Earlier in his life Du Bos was a diplomat and historian, as well as a traveler with knowledge of intellectual life in England, the Netherlands, and Italy. He published the first version of his *Réflexions critiques sur la poésie et la peinture* in 1719; but the book is best read in the enlarged and rearranged later editions, the fourth (1740) being the last Du Bos supervised. It is divided into three parts, of which the third is devoted to the archaeology of classical drama, not addressed here. The first part, formally a discussion of the relative beauties of painting and poetry, is also an informal flow of ideas set off by differences between the arts, often with finely detailed instances; the second part is an anthropology of art. "It is not a methodical book; but the author thinks and makes the reader think" (Voltaire). He is commonly associated with the new empiricist psychology of John Locke, although too much can be made of this.

It is the function of the arts to feed the soul's hunger for activity at times when attention to both external perception and internal reflection is disordered, the painful condition of ennui (Du Bos, 1740, part I, p. 1). The most immediate medium for this is an appeal to the passions—these being

what we are destructively led into by our need for psychic activity—as surrogates for passions: thus strong subject matters (pp. 2–9). But painting differs from poetry in many ways. It engages our attention through manner more than matter of representation, its *mécanique* being more difficult than that of poetry and able to retain attention even with banal matter (pp. 10–11, 24). But its subject matter is defining: it cannot represent complex thoughts in its actors, and only generic passions; but it is inherently more specific about individual character and attributes like age and temperament than would be tolerable in poetry. It must represent one instant. Unlike poetry it can only represent matter already known to the beholder (p. 13); yet its power derives partly from using not the artificial signs of language but natural signs (including gesture) whose energy does not depend on the beholder's culture (p. 40). Taste, a sixth sense as it were, is determined by individual physical organization, different persons being most affected by different stimulations, so it is not discussible; anything like the debate between color and design in painting is absurd (p. 49).

Part 2 is a sustained discussion of three interrelated topics: *génie*, its constitution and nurture (especially pp. 1–11); historical, moral, and climatic determinants of *génie*, with regard to periods and regions (pp. 12–20); and the mechanisms of reception and "the public" (pp. 21–39). The public has less authority and effect in evaluating painting than poetry because painting is the more technical, and also because, unlike the case of poetry and language, the public has no active experience of creation in its medium (pp. 22, 27).

The book, much reprinted, was translated into English in 1748, but Du Bos's ideas were particularly fruitful in Germany. He was important to Gotthold Ephraim Lessing, and Johann Georg Sulzer valued him, strangely, for making the first attempt to base the theory of art on a general principle. In the next century some of his ideas were still crucial for the aesthetic of Jena Romanticism, though they were not explicitly acknowledged, presumably because no longer directly derived.

[*See also* French Aesthetics, *article on* Eighteenth-Century French Aesthetics; *and* Origins of Aesthetics.]

BIBLIOGRAPHY

Work by Du Bos

Réflexions critiques sur la poésie et sur la peinture. 4th rev. corr. ed. Paris, 1740.

Other Sources

Chouillet, Jacques. *L'Esthétique des lumières.* Paris, 1974. See pp. 37–47.
Lombard, Alfred. *L'Abbé Du Bos: Un initiateur de la pensée moderne.* Paris, 1913.
Muntaneo, B. "Survivances antiques: L'abbé Du Bos, esthéticien de la persuasion passionelle." *Revue de Littérature comparée* 30 (1956): 318–350.

MICHAEL BAXANDALL

DUCHAMP, MARCEL. [*To explain the work and influence of the twentieth-century artist Marcel Duchamp, this entry comprises two essays:*

Survey of Art
En-Gendering of the Artistic Subject

The first essay clarifies Duchamp's art from the 1910s to the 1960s and its critical impact on aesthetics. The second essay analyzes the continued significance of Duchamp's aesthetics, focusing on a specific topic: Duchamp's figure of Rrose Sélavy and gender/sexual identity. For related discussion, see Avant-Garde; Conceptual Art; Kant, *article on* Kant, Duchamp, and Judgment; *and* Modernism.]

Survey of Art

The contributions to contemporary art and to the aesthetic principles that inform it made by Marcel Duchamp (1887–1968) are universally recognized. His primary achievement involved undermining our basic ideas about what it means to do a "work" of art. (He essentially abrogated the physical part of the process and replaced it with having an idea.) Interestingly enough, he managed to undermine the notion of the "work" of art while he was engaged in a very labor-intensive project, *The Bride Stripped Bare by Her Bachelors, Even* (a complex work known as *The Large Glass*). Duchamp was deeply involved with *The Large Glass,* directly for some ten years between around 1913 and 1923, and indirectly for almost his entire career.

Between 1910 and 1911, Duchamp moved through several turn-of-the-century artistic styles including Post-Impressionism, Symbolism, Fauvism, and Cubism. By late 1911, he had become interested in depicting movement in static form, something he referred to as "elementary parallelism." In early 1912, he painted *Nude Descending a Staircase No. 2,* the most famous of his "cubo-futurist" pictures, which made his reputation, first by being rejected by the supposedly avant-garde members of the Cubist section of the Salon des Indépendants in Paris later that year, and then by causing a *succès de scandale* at the Armory Show in New York in early 1913. During the first few months of 1912, Duchamp continued to explore "elementary parallelism," a pursuit inspired by his interests in chronophotography and geometry, particularly the notion of the fourth dimension.

During the summer of 1912, after he completed two of his best pictures, *The Passage from the Virgin to the Bride* and *The Bride,* Duchamp came to feel that he could no longer aspire to his goals by using traditional means. At this juncture, he abandoned painting on canvas and began an examination of art at more basic levels. In the fall of 1912, and continuing for the next several years, he produced notes and preliminary studies for *The Large Glass* in which he investigated his philosophical concerns and elaborated his growing preoccupation not just with four-dimensional but with n-dimensional geometry. In the process, he began his

lifelong project of questioning art by disclosing the conventionality of making aesthetic choices (much as one of his primary mathematical sources, Henri Poincaré, had discussed the conventional nature of choosing one geometry over another).

Along with *The Large Glass*, Duchamp's most effective means of realizing his project were the "readymades," the first of which date from 1913. These important "works" (one might almost call them "devices") were not really works, and they were not "made" (at least not by the artist); they were simply chosen by him from the world of things that already existed. From an aesthetic point of view, the artist discovered a basic indistinctness between what is made and what is ready-made. An ordinary store-bought article—say, a bicycle wheel, a bottle rack, a urinal, a comb, or a hat rack—could play the same role as a handcrafted work of art. This general ability of almost any object to simulate the specific aspects of art informs Duchamp's most important achievements. Thus, the readymades are perhaps Duchamp's central works, particularly when they are taken to be closely related to *The Large Glass*. In this more encompassing view, Duchamp himself thought they were his most important contributions. For him, they had the right kind of theoretical edge. By choosing readymades (and by incorporating ready-made categories such as projective geometry in *The Large Glass*), he suggested that the qualities of objects were less significant than their definitions: how an object looked mattered less than how it was looked at.

For Duchamp, the dynamics of how objects were "looked at" found expression in the world of art, and his activities were more an exercise in redefining art than in examining art's larger social context. But this observation aside, his art (or his nonart) has had important impacts on the practical politics of art. Especially since the 1960s, many artistic approaches ranging from Pop art and New Realism to Conceptual art and performance art, from Minimalism to Picture/Theory art, have benefited from his example. All of these directions interrogate art in Duchampian terms—terms that recognize the conditional nature of choice and its involvement with power.

Many of Duchamp's major works were created (or chosen) during the heyday of Dada, a movement fundamentally affected by conflict. At that time, the "Great War" was destroying Western Europe using devastating new technology—machine guns, airplanes, mechanized armor. Duchamp's response to the war was to leave Paris for New York. Soon after he arrived in 1915, he began choosing a series of particularly American readymades (the snow shovel, the Underwood typewriter cover, and most infamously, the Mott Works urinal submitted as a "Fountain" to the New York Independents Exhibition in 1917 and signed "R. Mutt").

In New York, Duchamp also began the actual construction of *The Large Glass*. As the title implies, this work was executed on large sheets of glass using unusual techniques and innovative materials, including lead wire affixed to the glass with varnish, mirror coating, and dust accumulated over many months. *The Large Glass* was essentially an elaboration of the mass of notes and calculations that Duchamp had been accumulating since late 1912. It was not only a new kind of painting but also an ancillary domain for the readymades (which are discussed at length in the notes). Both *The Large Glass* and the readymades involved irony and paradox, both hinged on language games, and both played with such notions as authenticity and originality.

The readymades and *The Large Glass* allowed Duchamp to make the general point that good reasons exist for no longer believing old stories about truth and beauty. As the Great War seemed to indicate, such stories were complicit with the ruination of the world. Duchamp used the readymades and *The Large Glass* to speak about the aesthetic conditions of his particular historical moment and to make long-standing observations about the art process in general (insofar as it involved truth and beauty). He felt that such matters were conventional, if not wholly arbitrary. The fate of the readymades in particular (the way they could behave as art objects or play havoc with art exhibitions) seemed to bear out his presuppositions.

Duchamp once said that his readymades "seemed trivial, but they were not," and they have been surprisingly effective. Their ability to lie in wait, as it were, ready to demolish this or that aesthetic theory, has made them consequential for any number of subsequent artistic developments. That a bicycle wheel, a bottle rack, a urinal, a comb, or a hat rack could become an aesthetic object raised questions about how art should be defined. Duchamp was skeptical about definitions of art, and he oppugned our ability to do the things that have traditionally been considered essential in establishing such definitions—things like determining rules for assessing beauty.

Duchamp's overall oeuvre relied on multiply contingent principles, ready-made terms, and n-dimensional geometries. As members of an interconnected body of works, individual pieces could be opened up and folded back upon themselves in an almost infinite number of ways—ways that were both Euclidean and non-Euclidean. Duchamp's interest in mathematics, in particular, amounted to a radical technique for casting doubt on meaning. Geometry gave him a way of dealing with complex spaces (and by implication, with complex intellectual domains, with, say, the way art works). Manipulating conceptual objects such as circles, spheres, and hyperspheres—or squares, cubes, and hypercubes—by speculating about shifting his field of operations into the fourth dimension, Duchamp could give extradimensionality to his readymades and other works. At least at the conceptual level, geometry allowed these objects to be flattened into projective planes and expanded outward along the infinite directions of n-dimensional continua.

That the readymades and many of the principal iconographic elements of *The Large Glass* such as the "chocolate grinder" were taken from store windows (and thus readymade in the commodity sense of the term) allowed the imagery of *The Large Glass,* which was itself a kind of shattered window, to work on various levels. It could be ordered and then reordered around Duchamp's basic doubts about art. The process was also *n*-dimensionally circular: he could come back around to query the same issues again and again. This returning was apparently among his reasons for recasting the two-dimensional, abstract imagery of *The Large Glass* in the three-dimensional, realistic imagery of his last piece, which he worked on for the last twenty years of his life, *Given: 1st the Waterfall, 2nd the Illuminating Gas.* Like the readymades and *The Large Glass,* this work is all about how something is "looked at," and also like them, it turns the responsibility for content and meaning back on the viewer.

Duchamp often said that all the artist needed was an idea, and what distinguished him from many of his fellow artists working in Dada and Surrealism, the movements with which he can most readily be identified, is that he practiced systematic doubt (even though he was suspicious of systems). He refuted certain artistic assumptions and showed that basic principles widely believed to be true were defeasible. By way of demonstration, he stood back and watched his readymades become works of art. Since aesthetic first principles would not hold through time (shifting even to the point where non-art objects could become art objects), he established that judgments of taste were far from infractible—indeed, they could be broken apart all too easily.

Duchamp came to admire the cracks in *The Large Glass* much as he liked the breaks in the rules of the game. By means of his particular game of chance, he dismantled the pretensions of Modernist art (for example, the way formalist reductionism and essentialism were taken for granted). He was uninterested in neo-Romantic notions of meaning bubbling forth from the inner wellsprings of creativity. He was more impressed by an artist coming up with an original idea, with something new to say, and he advocated the kinds of "lucubrations" that led to innovation. Duchamp's famous statements about the importance of concept, such as "I'm interested in putting painting once again at the service of the mind," are intended to shift the viewer's attention away from the purely "retinal" aspects of art, as he called it. Duchamp believed that how you characterized a thing had a great deal to do with how you specified it. A bottle rack when named "hardware" and used to dry bottles was an ordinary object, but that same bottle rack named a "readymade" and placed within the context of art became an aesthetic object, a work of art.

The implications for aesthetics in this simple state of affairs have been far-reaching. If choosing anything is all you really need to do to make a work of art, then what role do skill and the ability to recognize beauty play in producing and evaluating artworks? The problem seems to be more a matter of value than status. Even though we may be willing to give Duchamp and his chosen objects the benefit of the doubt and concede that they are art, we may be more hesitant to accept them as good (or beautiful) art. But do not both categories (status and value) hinge on our naming them such and such a thing possessing such and such characteristics? The workings of this "namism," as Duchamp once called it, were central to his project.

Duchamp was not so much interested in revealing the conventional nature of deciding on categories—whether or not something was art or nonart—as he was in disclosing the arbitrary nature of making aesthetic decisions in the first place. In other words, Duchamp was among the first artists to understand the political nature of making value judgments. He knew that such judgments were grounded in power relationships. He adopted his particular anti-aesthetic practices, which challenged traditional notions of truth and beauty, largely because he knew very well how such notions could be appropriated for uses he despised. Recognition of truth and appreciation of beauty presuppose a certain sensitivity, even an "infinite acuteness," on the part of the beholder; the viewer needs to possess sophisticated eyes. Conveniently, the rich and the powerful have such visual systems and can perceive the finer things in life. Duchamp questioned the handy ways in which special objects (art) were separated from ordinary objects (nonart). He asked his questions primarily by replacing craft objects (which were handmade) with manufactured objects (which were readymade).

Duchamp's actions reversed the earlier criticism of manufacturing articulated by such critics as William Morris. The term *manufacture* literally means "to make by hand," but it was used as a scathing epithet by Morris to designate the shoddiest outputs of machine production. These were contrasted to the artlike, handcrafted objects that Morris admired. Duchamp, although ostensibly quite different from Morris, raised a similar set of issues. Duchamp removed the handcrafted aspect of the work of art by replacing it with a machine-made object produced on an assembly line. He put the "art" operation in the mind rather than in the hand. What this accomplished was to replace the demeaning physical effort required to produce the original object—the bottle rack or the urinal—with an act of intellectual insight. The play, the pleasure, now resided in the thinking rather than in the physical making.

Duchamp was not necessarily making fun of high art by replacing it with low objects. His project was more complex. At the same time he was choosing readymades, he was producing fine, difficult-to-make, handcrafted objects. All of these works, the readymades and *The Large Glass,* were created as challenges to conventional aesthetics by conflating

issues of "work" with issues of conceptualization. Duchamp used contradictory practices: he adopted machine iconography, he drew in lead wire, he appreciated the cracks put in the *Glass* by its accidental breakage, and so on. All of these unusual techniques and attitudes tend to block seeing the art as beautiful (although by now it looks really good).

The Large Glass, like the readymades, is not something to look at directly, as one might, say, a Renaissance picture. Seeing *The Large Glass* also involves an unusual way of looking at things: it involves a glance out of the corner of the eye, and its appearance is filtered through a mass of notes that involve the viewer with any number of difficult conceptual operations, not the least of which is deciding what art is.

Despite the easy ways in which the readymades can be overlooked as inconsequential, and the easy ways in which *The Large Glass* can be disregarded as incomprehensible, such works continue to engage the imaginations of artists and scholars. Duchamp wanted to demonstrate the conventional nature of defining art, and his small gestures have gone a good distance toward accomplishing his goals. Indeed, the fact that so much can be encased in small spaces—both physically and intellectually—has fascinated artists and critics ever since choosing a bottle rack as a work of art or submitting a urinal to an art exhibition created a furor. These moves in Duchamp's game collapsed the conventional rules of art and suggested that the intellectual structures supporting art were weaker than had been supposed.

Duchamp was a game player par excellence. Not only did he play chess at the master level, he also approached almost everything he did in terms of its effectiveness as strategy. In this sense, he was intellectual in his outlook, even though he carried out several of his projects (such as *The Large Glass*) with diligence and exquisite skill. Perhaps his consummate gamesmanship combined with his physical and mental adroitness is what most fascinates his followers. There are not that many readymades, perhaps sixteen or seventeen (even though some counts that include virtually everything that Duchamp touched go as high as fifty). Among the most important readymades are the bicycle wheel, the bottle rack, the landscape chromo, the snow shovel, the hat rack, the coat rack, the comb, and the typewriter cover. Several were lost, the ventilator cover being the most famous. A number were hybrids: the bird cage filled with marble cubes; the ball of twine that produces a "secret" noise; the strips cut from bathing caps and stretched from the four corners of the room. Some were conceptual and exist only in the notes: "Buy a pair of ice tongs as a readymade." "Find an inscription for the Woolworth Building as a readymade." Some were intended to scandalize the public: the urinal, *Fountain,* and the mustache and goatee on the Mona Lisa, *L.H.O.O.Q.,* are perhaps the most notorious members of this category.

All of these readymades were simple, but none was simplistic. They involved making judgments about the bound-ary conditions of art. Duchamp uncovered the difficulty of deciding where to locate the demarcation. A bicycle wheel chosen ready-made and mounted on a stool can be taken simultaneously as a substantial challenge to accepted artistic practice or dismissed as an inconsequential joke. These kinds of antinomies are not easily reconciled. Contingency fascinated Duchamp, and he provided questions, not answers. He challenged the surety of modernism, and his methods exemplify the crosscurrent of avant-garde resistance that brushes against the grain of modernism from the very beginning. Duchamp represents the destructive, or perhaps better, he anticipates the deconstructive branch of modernism, a postmodernist *avant la lettre.*

Duchamp's "namism" has become part of a larger contemporary cultural discourse. By naming an object a readymade, Duchamp radically altered the way we deal with it. Friedrich Nietzsche must have had something similar in mind when he pointed out that "every idea originates through equating the unequal." Duchamp equates the unequal: art and nonart. His new ideas challenge the art object's uniqueness and the belief that the art object's value lies in its history. What he makes clear is that whatever value there may be in the object, it is placed there by choice (and thereby through chance operations), and aesthetics then becomes a matter of assigning value, rather than discovering it.

BIBLIOGRAPHY

Works by Duchamp

Dialogues with Marcel Duchamp. Interview by Pierre Cabanne. Translated by Ron Padgett. New York, 1971.
Salt Seller: The Writings of Marcel Duchamp (Marchand du Sel). Edited by Michel Sanouillet and Elmer Peterson. New York and Oxford, 1973.

Other Sources

Adcock, Craig. *Marcel Duchamp's Notes from the Large Glass: An N-Dimensional Analysis.* Ann Arbor, 1983.
Adcock, Craig. "Conventionalism in Henri Poincaré and Marcel Duchamp." *Art Journal* 44 (Fall 1984): 249–258.
Adcock, Craig. "Marcel Duchamp's Approach to New York: 'Find an Inscription for the Woolworth Building as a Ready-Made'." *Dada/Surrealism* 14 (1985): 52–65.
Adcock, Craig. "Duchamp's Eroticism: A Mathematical Analysis." *Dada/Surrealism* 16 (1988): 149–167.
Adcock, Craig. "Duchamp's Way: Twisting Our Memory of the Past 'for the Fun of It'." In *The Definitively Unfinished Marcel Duchamp,* edited by Thierry de Duve, pp. 311–349. Cambridge, Mass., 1991.
Adcock, Craig. "Marcel Duchamp's Gap Music: Operations in the Space between Art and Noise." In *Wireless Imagination: Sound, Radio, and the Avant-Garde,* edited by Douglas Kahn and Gregory Whitehead, pp. 105–138. Cambridge, Mass., 1992.
de Duve, Thierry. *Pictorial Nominalism: On Marcel Duchamp's Passage from Painting to the Readymade.* Translated by Dana Polan with the author. Minneapolis, 1991.
de Duve, Thierry, ed. *The Definitively Unfinished Marcel Duchamp.* Cambridge, Mass., 1991.
de Duve, Thierry. *Kant after Duchamp.* Cambridge, Mass., 1996.
d'Harnoncourt, Anne, and Kynaston McShine, eds. *Marcel Duchamp.* New York, 1973.

Jones, Amelia. *Postmodernism and the En-Gendering of Marcel Duchamp.* Cambridge and New York, 1994.

Judovitz, Dalia. *Unpacking Duchamp: Art in Transit.* Berkeley, 1995.

Kuenzli, Rudolf E., and Francis M. Naumann, eds. *Marcel Duchamp: Artist of the Century.* Cambridge, Mass., 1989.

Lebel, Robert. *Marcel Duchamp.* Translated by George Heard Hamilton. New York, 1959. With chapters by Marcel Duchamp, André Breton, and H. P. Roché.

Paz, Octavio. *Marcel Duchamp: Appearance Stripped Bare.* Translated by Rachel Phillips and Donald Gardner. New York, 1978.

Schwarz, Arturo, ed. *The Complete Works of Marcel Duchamp.* New York, 1969; 3d rev. ed., New York, 1997.

Seigel, Jerrold. *The Private Worlds of Marcel Duchamp: Desire, Liberation, and the Self in Modern Culture.* Berkeley, 1995.

CRAIG ADCOCK

En-Gendering of the Artistic Subject

In 1920 or 1921 Marcel Duchamp donned bourgeois women's clothing (at least two different outfits) and posed coyly for Man Ray, his friend and New York Dada collaborator, to memorialize himself in a series of photographs as "Rrose Sélavy." The figure of Rrose Sélavy (in French, a homophone for "Eros, c'est la vie," or "Eros, that's life") offers up an alternative to the strident machismo of the male modernist, epitomized in Western art history by the bluff womanizing and macho persona of the prodigiously prolific Pablo Picasso or the strained athleticism of the action painter Jackson Pollock. Rrose can be seen in retrospect as an early interrogation of the complex structures by which the artistic subject is *en-gendered* (constituted within sexual difference) in Western culture.

Duchamp has engendered the artist as an open question, a site of desire, transference, and ambivalence (the latter of which is defined psychoanalytically as the simultaneous existence of contradictory tendencies). Through Marcel as Rrose (displayed via the photographic vision of "Man") we reach *eros (that's life):* the eroticism that Sigmund Freud recognized as being constitutive of all interpretation (as an intersubjective process). Is eros feminine? Can the artist be a woman? The artist, stated Duchamp in 1949, "is only the mother [of the work], in other words." Conditioned through interpretation (which, as Freud knew, is a form of love), the artist-as-subject is revealed to us in the feminine: as an object of interpretative desire. And yet, as procreative, maternal origin of the work, she authorizes herself as its ideological core. This mother, Rrose/eros, is a man.

There s/he is, in two different versions: (1) delicate face, dark lips, and aquiline nose, surrounded by a lush fur collar (the accoutrement of upper-class feminine display); eyes softly, demurely gazing toward us from under the saucy brim of a women's bowler; the graceful hands (which may have been those of an-*other*, Francis Picabia's lover, Germaine Everling) are held decorously up to the throat; the hair appears to be "marcelled"; (2) the pearl choker is a jarring contrast to the masculine facial contours, which are here emphasized by harsh white light (revealing the slightly pitted skin of one who shaves); the hair is overtly feminine in its aggressive, looping curls, which sit uncomfortably on an excessive gathered collar. Number one: this is "really" a woman—this is Claudette Colbert (a shade on the demure side). Number two: this is "really" a man (Jack Lemmon in bad drag in *Some Like It Hot*).

What does it mean for a male artist to represent himself and nominate himself as a woman? Not only an image and a named identity, Rrose Sélavy also, crucially, enunciated readymades, letters, and other authorial activities, signing them or authenticating them by inscribing them with "her" name (a nomination that nonetheless takes its value through that of the *master,* Marcel). See, for example, the last frames of Duchamp's film (in collaboration with Marc Allégret and Man Ray), *Anémic Cinéma* (1926), signed "Rrose Sélavy" and thumbprinted; Rrose Sélavy's "Baggage Tags" (c.1921); her signature on the sketch *Témoins oculistes* (1920); her well-known intervention (as "Le président") in the mass-produced lithograph *Obligations pour la roulette de Monte Carlo* (1924) of Marcel (noted simply as "Un administrateur"), and so on.

What Rrose speaks to us about is the way in which artistic meaning is formulated in relation to the image of the artist (the artist's *body* as an instantiation of her/his intentional force as a subject within art-historical systems of value). Art interpretation, like Freudian analysis, takes its insights from the (admittedly often veiled) body of the (artistic) subject, on which are inscribed her/his symptoms: the body is a surface on which is written "subjectivity." Case in point: Pollock's toiling body (ejaculating paint over the stretched expanse of canvas on the floor) informs our experience of his work (by means of the textual and photographic fabulations of Harold Rosenberg, Clement Greenberg, Hans Namuth). Whether acknowledged or not, art history has embedded that body into the skeins of paint that are the action paintings. They are traces (indexes) of Pollock's *having been there.*

But Marcel, who is known for having removed precisely just such traces, gives himself "directly" to us in Rrose: "Here *I* am." But who, after all, is "I"? The "I," in normal speech, is the person who speaks; the "you" the person to whom the "I" speaks. The lines are clearly drawn—until the "you" speaks back. Subjectivity, Rrose says, is contingent (the *I* on the *you* who is both the *I* and yet also *other*). The asymmetrical split between the "I" and the "you" (after all, it is Marcel who initially speaks Rrose) retraces the split within the subject (his ambivalence marked as his masculine/feminine "self"). "Call it a little game between 'I' and 'me'," he asserted in a 1962 interview with Katherine Kuh.

Who, again, is the "I" spoken by Marcel? With Rrose, s/he is a name and a picture (of a woman no less), a photograph that is, like the signature itself, an indexical trace of the subject's "presence." But what is the meaning of a "presence"

that is so clearly fabricated as object/other (and thus, implicitly, as absence or lack)? Being a "woman," constructed through highly sexualized period codes of desirability drawn from a commercially produced vocabulary (the then brand-new rhetoric of the print advertisement as well as, perhaps, the Parisian *chanteuse*), the nonexistent yet authoring "subject" Rrose Sélavy exaggerates the seductive appeal of the artist (as well as that of the art object—the images giving us both at once). The figure of Rrose Sélavy, a "woman" but also a "man," an object in the portrait (and, arguably, of Duchamp's machinations) but also a subject who signs (and sings?), works to unhinge the masculine-feminine, analyst-analysand, subject-object polarities that ground Western interpretative systems.

Rrose dislocates these oppositional positions because she is not secured as (straight) woman or as "straight man" or as "gay man in drag," and so on, and because, as noted, she opens the gap between the "I" and the "me" as well as that between the "I" and the "you." In marking the incoherence of subjectivity itself, she illustrates the dialectic of desire that propels interpretation. She solicits interpretative desire through standardized codes of female desirability, yet speaks these through the recognizable voice of Marcel. Compelled by her image, we think we love her for herself, but, she insists with resignation, it is Marcel we desire to "know" through her. In this way, Rrose Sélavy rides the invisible line of *inframince*: that incongruent asymmetry of and/or. Rrose *and* Marcel—Rrose *or* Marcel: both apply. I *and/or* you; I *and/or* me; feminine *and/or* masculine; self *and/or* other; artist *and/or* interpreter. Rrose marks the split within the self as well as that between self and other. The oppositional logic of gender is dispersed.

Duchamp was not by any means alone among the early-twentieth-century French and U.S. (male-dominated) avant-gardes in adopting this seemingly bizarre cross-dressing strategy. Among these very men who have had exclusive claim to the masculine privilege of artistic authority in Western narratives of cultural history a number of avant-gardists, irritated by the drab constraints of normative bourgeois male dress and attitude (figures such as Oscar Wilde, Duchamp, and Charlie Chaplin [in *A Woman*, 1915]), adopted women's clothing and gestures to various ends. From Wilde's queer politics, to Duchamp's ambivalence, to Chaplin's masculinity-confirming parodic drag, these performative gender strategies exacerbated the radical shift in Western culture away from Victorianism (with its prescriptive notions of sexuality and gender identity), throwing into question the knowability of a person's sexual self.

The typical male modernist from the nineteenth century on had thus negotiated the banality of bourgeois masculinity by taking on the sartorial and attitudinal equipment of its antipodes: hence the artist-laborer (Pollock, Paul Cézanne) and the artist-dandy (Duchamp, Éduoard Manet, Andy Warhol). Rrose is, historically, an extension of these negoti-ations, skewing Marcel firmly away from the platitudes associated with bourgeois masculinity and into the mysterious, ambiguously gendered realm of the dandy. The effusion of gender play from the late teens into the early 1920s, however, turned out to have been but a brief respite from the gender police: not until the 1960s (viz., Jack Smith, Warhol) would overt drag or oblique gender confusion emerge again as a key strategy for male artists.

In her role-playing, Rrose is not just about the polarity of gender, however. Concocted by two heterosexually identified male modernists alone in a dark room, scheming together with a pile of women's clothes and a camera, Rrose is decidedly queer. It is perhaps through her queerness that she can best be seen as disturbing the seemingly self-evident, corporeally defined boundaries of conventional gender. While we assume Marcel—not to mention Man—was anatomically male, and that Rrose is thus "really" an anatomically determined "man" wearing women's clothing, what this means is not altogether clear. The very assumption that we are dealing with a definitively determinable heterosexual masculinity here (which would be the case were Duchamp to be understood as a conventional Western artist: heterosexual, white, male) is challenged.

Openly soliciting our spectatorial desire, Rrose opens painful questions about identity. What if I were an anatomically male cross-dresser who identified with the woman in Rrose/Marcel? What if I were a gay male viewer turned off by Marcel's overt femininity? Or turned on by it? What would Rrose mean to me if I were a straight, white, female art historian invested on some level in retaining a fixable function "Duchamp" on which I could pin my expertise? (Undoubtedly, I would position Rrose as a clear example of Marcel's radical approach to artistic authorship.) When looked at through a poststructuralist lens of interpretative relations, the strategic formulation Rrose Sélavy—picture, name, signature—seems to point productively to the dislocation of authorship in postmodern art and poststructuralist theory: the reemergence of gender/sex crisis in the post-1960 period.

Within this framework Rrose leads us, then, to the more overtly queer politics of the figure of Warhol and beyond. From the late 1950s onward, with the explosion of rights movements (civil, women's, gay and lesbian) and the economic transformations of late capitalism, all of which put in question normative conceptions of subjectivity, performativity (both literal, in body art and performance, and figurative, in the Warholian self-construction on the public stage) became central to artistic subjectivity. Warhol and Europeans such as Yves Klein and Urs Lüthi refined the performativity of the artist (as an open-ended function of spectatorial desire)—the performativity that had been so auspiciously introduced by Duchamp and some of his avant-garde colleagues.

Through dress, statements, and the wide-scale development of an entire alternative universe of media-generated

art/celebrity production (the Factory, *Interview* magazine), Warhol expanded Duchamp's modest one-shot foray into cross-gendering into an industry. Whereas Duchamp was, to the end, a product of bourgeois capitalism (and one of its best critics), Warhol was an avatar of late capitalist techno-culture. With his white wigs, battle scars, and studiously maintained attitude of bland indifference, Warhol created a media empire to sustain the performance of "Warhol." Rrose, with her sweet smile and touching technological naïveté (photography, with its refinement of Enlightenment dreams of scientificity and knowledge-through-vision, being the language of the nineteenth century), seems quaint and somehow "authentic" by comparison.

As suggested, the dislocation of modernism's normative subject (its insistence on a singular, universal "I" mirrored by the monotheistic, implicitly white male "God" of Judeo-Christianity) has been the most marked during periods of economic and social upheaval: in the late 1910s and 1920s, during and just after the destruction of old Europe in World War I and as the "New Woman" arose to challenge Victorian stereotypes of femininity; and in the 1960s and following, with the consolidation of Black Power, the women's movement, and the gay and lesbian rights movements and their merciless hacking away at the pedestal that has long sustained the erection of straight white male privilege. By the 1990s, even the consensual racism within the art world (the racism that had forever condoned and supported the mythical figure of the [white male] "genius") has been irrevocably undermined. While the market may still yearn for the faces of white men to secure its economic bases, women artists and artists of color have become highly visible contenders in the redefinition of artistic authorship.

In this negotiation of artistic authority, a Duchampian or Warholian performativity is still paramount. Through performance, as Rrose has taught us, the artist opens up the circuits by which her/his identity is codified, her/his work given meaning within the structures of art criticism and art history. By crossing the codes of gender, sexuality, race, class, and so on, the artist relinquishes forever the authority granted to the mythically secure genius; at the same time, however, the artist actively performing her/himself in relation to the audience/you (soon to be interpreter/I) sustains a new kind of power—but one that is insistently marked as *provisional*. Through this marking, it is acknowledged that meaning is an ongoing negotiation (with the I/you forever battling it out, as *and/or*). Rrose's relevance for the 1990s becomes clear.

Crossing over gender, sex, and race lines, Lyle Ashton Harris, for example, speaks himself, variously, as a naked black man (torso shot: muscular chest) in white face; a naked black man, frontal view (no question about his anatomical sex), bouffant wig and feminine ballerina pose; dressed in "mammy" kerchief, child in arms, next to a large, imposing black "man" (the artist Renée Cox in drag). Harris is who we want him to be and more.

More than seventy-five years ago, by performing himself as Rrose Sélavy Marcel Duchamp opened up a range of enunciative possibilities that would ultimately (through readings such as mine here) challenge the deepest assumptions of art-historical interpretation. While inevitably (again, as here), Rrose has been continually returned to the figure of Marcel, in the moment before such a return is secured s/he offers the potential of shattering the complacency with which the interpreter asserts her/his "knowledge" in relation to the work of art (which implies the body/self of the artist). While Greenberg could veil his investment in the body of Pollock to secure his readings as disinterested "fact," the scholar approaching Duchamp's work encounters this body so insistently—and in such unpredictable (feminine) form—that such a veiling is thwarted, if only momentarily. As an authorial function marked ambivalently (both female and/or male), Rrose, then—like Harris—*eroticizes* the interpretative process, opening it to the thrilling uncertainties of human desire.

[*See also* Feminism; *and* Sexuality.]

BIBLIOGRAPHY

Works by Duchamp

Statement at the Western Roundtable on Modern Art, 1949. San Francisco Museum of Modern Art Records. Owned by the San Francisco Art Institute. Microfilmed by the Archives of American Art/Smithsonian Institution, roll 2432, frame 350.
Marcel Duchamp, Notes. Translated and edited by Paul Matisse. Boston, 1983.

Other Sources

Ades, Dawn. "Duchamp's Masquerades." In *The Portrait in Photography,* edited by Graham Clarke, pp. 94–114. London, 1992.
Desnos, Robert. "Rrose Sélavy," *Littérature* 7 (December 1, 1922): 15.
de Duve, Thierry. *Au Nom de l'art: Pour une archéologie de la modernité.* Paris, 1989.
de Duve, Thierry. *Resonances du readymade: Duchamp entre avant-garde et tradition.* Nîmes, 1989.
de Duve, Thierry, ed. *The Definitively Unfinished Marcel Duchamp.* Cambridge, Mass., 1991.
Freud, Sigmund. "The Dynamics of Transference" (1912). Translated by Joan Rivière. In *Therapy and Technique,* edited by Philip Rieff, pp. 105–115. New York, 1963.
Irigaray, Luce. "The Blind Spot of an Old Dream of Symmetry." In *Speculum of the Other Woman,* translated by Gillian C. Gill, pp. 11–129. Ithaca, N.Y., 1985.
Jones, Amelia. *Postmodernism and the En-Gendering of Marcel Duchamp.* Cambridge and New York, 1994.
Jones, Amelia. "The Ambivalence of Male Masquerade: Marcel Duchamp as Rrose Sélavy." In *The Body Imaged: The Human Form and Visual Culture since the Renaissance,* edited by Kathleen Adler and Marcia Pointon, pp. 21–31. Cambridge and New York, 1993.
Judd, Donald. "Marcel Duchamp and/or Rrose Sélavy" (1965). Reprinted in *Marcel Duchamp in Perspective,* edited by Joseph Masheck, p. 121. Englewood Cliffs, N.J., 1975.

Kuh, Katherine. "Marcel Duchamp." In *The Artist's Voice: Talks with Seventeen Artists,* pp. 81–93. New York, 1962.

McShine, Kynaston. "La Vie en Rose." In *Marcel Duchamp,* edited by Anne d'Harnoncourt and Kynaston McShine, pp. 125–134. New York, 1973.

O'Hara, Frank. "Homage to Rrose Sélavy." In *The Collected Poems of Frank O'Hara,* edited by Donald Allen, p. 10. New York, 1971; reprint, Berkeley, 1995.

Roth, Moira. "Marcel Duchamp in America: A Self Ready-Made," *Arts Magazine* 51.9 (May 1977): 92–96.

Rrose Sélavy à Paris le 28 octobre 1941. Montreal, 1984.

Schwarz, Arturo, ed. *The Complete Works of Marcel Duchamp.* New York, 1969; 3d rev. ed., New York, 1997.

Stauffer, Serge. *Marcel Duchamp: Die Schriften.* Zurich, 1981.

AMELIA JONES

DUFRENNE, MIKEL (b. 1910), French phenomenologist in the field of aesthetics during the second half of the twentieth century. Dufrenne's *Phenomenology of Aesthetic Experience* is the most insightful, cogent, and systematic study of aesthetic experience from a phenomenological point of view, and is written with the firm belief that the study of aesthetic experience is the highest road to an adequate understanding of art, aesthetic judgment, and the role of art in human life. For the *aesthetic character,* which is the ultimate goal of artistic creation and aesthetic perception, becomes actual in this event. The aesthetic experience reveals the aspect that makes an object a work of art, on the one hand, and occasions aesthetic enjoyment, on the other. This short article can be viewed as an introduction to the phenomenological analysis of aesthetic experience in general and Dufrenne's philosophy of art in particular. The discussion begins with the following question: What is the basis and structure of the aesthetic experience?

Aesthetic experience consists of three basic elements: (1) the artwork, (2) the act that grasps the work *as art,* that is, as an aesthetic object, in aesthetic perception, and (3) the unity of the object and the subject. Broadly speaking, aesthetic experience is the event in which one *responds* to and *constitutes* the aesthetic dimension of the artwork as a sensuous object.

Aesthetic Object. Dufrenne makes a distinction between "artwork" and "aesthetic object." The artwork is the object the artist "makes" during the process of artistic creation, for example, the painting we see hanging on the wall when we visit a museum. It is the "thing" that exists "independently of the act which aims at it" in aesthetic perception. The aesthetic object is the artwork insofar as it is perceived. The artwork is, in other words, the structural foundation of the aesthetic experience and the aesthetic object. Dufrenne is anxious to provide an objective basis for the aesthetic as such: the aesthetic object derives its structure and content from the artwork, and the artwork derives its truth—its *artistic* character—from the aesthetic object

(see Dufrenne, 1973, pp. 117, 232ff.). This character is revealed, however, through the "sensuous presence" of the artwork; it is what allows us to grasp the artwork as an aesthetic object. The formal, perceptible properties of the work constitute the sensuous substance of the aesthetic object: "The aesthetic object is, above all, the irresistible and magnificent presence of the sensuous. What is a melody, if not a stream of sound which washes over me?" (p. 86). The aesthetic object is essentially a *perceived object.*

But we should distinguish "aesthetic" from "brute" sensuousness. The sensuousness that belongs to the aesthetic object differs from the one that belongs to the artwork as an object of ordinary perception. Aesthetic sensuousness is the significant form that the artist embodies in his or her medium in the creative act. This embodiment is exactly what distinguishes an artwork from other types of objects. It is also what gives the work its own autonomy, in virtue of which it possesses the capacity to lay a certain claim on its creator and perceiver alike: "The work therefore has the initiative. What it expects of the spectator corresponds to what it has planned for him. And that forbids any subjectivism. Far from the work's existing in us, we exist in the work. To be a witness is to be prohibited from adding anything to the work, for the work imposes itself upon the spectator as imperiously as it does upon the performer" (p. 59). Insofar as it is perceived, then, the aesthetic object is objectively grounded in the sensuous structure of the artwork. This is why in an aesthetic experience a person submits to what the work has to offer.

In addition to aesthetic sensuousness, the aesthetic object is composed of what Dufrenne calls "sense," or meaning. The aesthetic object is a meaningful object, and this meaning is neither subjective nor ideal in character but immanent in the formal structure of the aesthetic object, "being its very organization" (p. 12). Meaning is the principle of the integrity of the aesthetic object; it unifies the sensuous by means of the spatial and temporal schemata that constitute the basic structure of the object. The aesthetic object is a spatiotemporal formation. Dufrenne argues at length that every work of art has its own space and time (pp. 241ff.). In aesthetic perception we experience the space and time of the artwork not as distinct dimensions, or realities, but as a dynamic unity: space and time together constitute the basic structure of the aesthetic experience. Here in this experience space is essentially temporal, and time is essentially spatial. Hence it would be a mistake to make a sharp and final distinction between the spatial and temporal arts.

This feature enables Dufrenne to refer to the aesthetic object as a "quasi-subject" (pp. 248, 190), that is, as an object made of spatiotemporal relationships. These relationships anchor the "world" of meaning potential in the artwork; they determine the unity and the general mood, or affective atmosphere, in terms of which we apprehend the

content of the aesthetic object. As a quasi subject, the aesthetic object possesses its own power of expressiveness; it is indeed the source of this expressiveness. In Dufrenne's words, "The aesthetic object is luminous through its very opacity—not by receiving an alien light by which a world is outlined, but by making its own light spring from itself in the act of expression" (p. 146). And the vehicle of this expression is the significant form that the artist conceives in the process of artistic creation. The aesthetic object is immanent in this form; it "does not have form. It is form" (p. 229). This is why in the aesthetic experience we are not drawn to anything outside the artwork; on the contrary, we feel attracted to it, mainly because the aesthetic meaning that is its very essence dwells within the form as a world of meaning (cf. pp. 121ff., 326ff.).

Aesthetic Perception. As a world, that is, as a complex system of aesthetic qualities and relations, the aesthetic object exists solely for the perceiver, and without the perceiver it would be nothing but an ordinary or utilitarian object: "The aesthetic object is essentially perceived. For its epiphany to take place, sometimes its performance, and always a witness or the public, is required. It manifests the sensuous in its glory" (p. 223). It would be appropriate, then, to say that the artwork *becomes* an aesthetic object. Accordingly, it (aesthetic object) is a task. The artwork acquires its aesthetic identity, it becomes complete and generates its inner life, by means of the consciousness that perceives it.

Aesthetic perception for Dufrenne is an active, constructive activity; it is the framework within which the aesthetic experience takes place. The highest point of this activity is the discovery of the meaning potential in the aesthetic object: "To perceive is not to register appearances passively—appearances which are meaningless in themselves. To perceive is to know—that is, to discover—a meaning within or beyond appearances which they offer only to the one who knows how to decipher them" (p. 335). But how do we move from the aesthetic object to the meaning that is immanent in it? Dufrenne holds that the logical structure of this movement consists of three stages: presence, representation and imagination, and reflection and feeling. First, "meaning is a demand to which I respond with my body" (p. 336). Hence the process of aesthetic perception begins in presence; a person must respond to the aesthetic object with the whole of the body: "The aesthetic object is above all the apotheosis of the sensuous, and all its meaning is given in the sensuous. Hence the latter must be amenable to the body" (p. 339). But this response is only an introduction to a higher dimension of perception, for the meaning that is immanent in the aesthetic object appeals not only to the senses or emotions but also to our higher faculties. The aesthetic object, as G. F. W. Hegel insisted, is addressed to the mind. Second, the aesthetic object "must be perceived at a proper distance and not simply lived in the proximity of presence" (p. 358); it should, in other words, be perceived as a formed object, as it is in itself. This phase of perceptibility is made possible by the imagination: "The imagination somehow creates the liaison between mind and body" (p. 345), but it does not play a vital role in the construction of the aesthetic object, nor does it make a contribution to its content, mainly because the aesthetic object presents itself to perception as a structured world. Third, grasping and enjoying the meaning of the aesthetic object takes place at the level of reflection and feeling. Dufrenne distinguishes two moments of reflection, critical and sympathetic. The former enables us to capture the form and content of the aesthetic object and to see them for what they are, whereas the latter draws us to its very heart and allows us to transcend the merely sensuous being of the object into its spiritual being, namely, meaning (see p. 393). In this moment of aesthetic perception the aesthetic object ceases to be a thing; it opens itself to us as a *world of meaning*. Aesthetic meaning, however, comes to life in the medium of feeling: "Aesthetic experience culminates in feeling as the reading of expression" (p. 437). This feeling should not, for Dufrenne, be confused with emotion; it is the mental power by which the mind discovers and apprehends the essence of the aesthetic object. It has its own depth. My ability to read what is expressed in the world of the aesthetic object, or put differently, my ability to penetrate its depth, depends on the sort of feeling in terms of which I perceive the object: "We experience the depth of things in the image of human depth" (p. 398). The depth of the aesthetic object, however, consists not merely in its complexity or distance but in its power of expression: "The depth must be sought in the object's power to express, through which it is the analogue (as a result of being the proxy) of a subjectivity" (p. 411). Thus the aesthetic experience reveals the depth of the aesthetic object, but it also reveals the nature of the act (the creator) that produced it.

Unity of Subject and Object. The aesthetic experience is the medium in which the autonomy of the aesthetic object and the autonomy of the perceiving subject are transcended and reconciled, for the aesthetic object is merely a thing apart from the consciousness that perceives it: the aesthetic character of the artwork becomes actual in the hands of the feeling that penetrates the depth of its being. But this sort of penetration is not subjective or relative, for what the subject feels is determined by the already articulated structure of the aesthetic object. It is this feature that led Dufrenne to characterize the aesthetic object as a "quasi subject" that can place a certain demand on us. The kind of form that the aesthetic object reveals sets the tone and direction of the substance of aesthetic perception: "The fulness which the aesthetic object possesses is always that of an appearance, that is, of the sensuous which is the act common to the sensing being and to what is sensed" (p. 225). This sensuousness, which should not be confused with the unity of the ordinary properties of the artwork, is the framework that

unifies the aesthetic object with the perceiving consciousness. It represents what the artwork has become and what the subject has apprehended in aesthetic perception. Accordingly, it is the basis of the communion that takes place between the subject and the object, on the one hand, and the subject and the creator, on the other. Dufrenne discusses in some detail the conditions under this sort of communion takes place.

The unity between subject and object, however, produces a deeper kind of unity, one between human being and the world. Thus the aesthetic experience makes a transition from the phenomenological to the ontological. This transition is possible only because the aesthetic object reveals the truth of being: "Aesthetic experience illuminates the real because reality is like the obverse side of being—a being of which man is the witness. Consequently, art bespeaks the real because both art and reality are subordinate to being" (p. 539). Accordingly, the truthfulness of the aesthetic object consists not in copying reality but in illuminating some aspects of its dimensions; thus it is true "because it induces us to complete the constitutive movement of a truth" (p. 540). The truth that art reveals, though, is not the truth of science but rather of being. This truth is antecedent to both aesthetic object and artist. This is one more reason why the aesthetic object can demand that we submit to it, and this is why subjects can read the world expressed in the aesthetic object only if they are themselves that expression; "that is, expression is read by the subject who is that expression. Expression is present to this subject before being taken in from the real. Expression is the truth which is given before the real. In sum, expression is the world as meaning—a world which is given before the objects in it" (p. 541). This world is the ontological foundation that unifies art and human being.

[*See also* Phenomenology.]

BIBLIOGRAPHY

Works by Dufrenne

Language and Philosophy. Translated by Henry B. Veatch. Bloomington, Ind., 1963.
The Notion of the A Priori. Translated by Edward S. Casey. Evanston, Ill., 1966.
The Phenomenology of Aesthetic Experience. Translated by Edward S. Casey. Evanston, Ill., 1973.
La Poétique. Paris, 1963.

Other Sources

Beardsley, Monore C. *Aesthetics from Classical Greece to the Present.* New York, 1966.
Coleman, Francis. "A Phenomenology of Aesthetics Reasoning." *Journal of Aesthetics and Art Criticism* 25.2 (Winter 1966): 197–203.
Heidegger, Martin. *Poetry, Language, Thought.* Translated by Albert Hofstadter. New York, 1971.
Ingarden, Roman. *The Cognition of the Literary Work of Art.* Translated by Ruth Ann Crowley and Kenneth R. Olson. Evanston, Ill., 1973.
Ingarden, Roman. *The Literary Work of Art.* Translated by George G. Grabowicz. Evanston, Ill., 1973.
Merleau-Ponty, Maurice. *Phenomenology of Perception.* Translated by Colin Smith. New York, 1962.
Spiegelberg, Herbert. *The Phenomenological Movement: A Historical Introduction.* 2d ed. The Hague, 1965.

MICHAEL MITIAS

DURAND, JEAN-NICOLAS-LOUIS (1760–1834), French architect, architectural theorist, and educator. Jean-Nicolas-Louis Durand might seem to be an architect of the most straightforward academic kind, but he was extreme in his outlook. He sought to design all buildings on the basis of a grid and to render them, as far as possible, symmetrical. He is sometimes upheld as a rationalist, but his beliefs were sustained by aesthetic criteria rather than reason.

He was born in Paris, the son of a cobbler; he worked briefly for a sculptor, but at the age of fifteen was working under the architect Pierre Panseron (1736–1787), professor of drawing at the École de génie militaire, and giving lessons of his own. The following year he entered the office of the celebrated Étienne-Louis Boullée (1728–1799), who enrolled him in the Académie d'architecture, a pupil of Julien-David Leroy (1724–1803) and Jean-Rodolphe Perronet (1708–1794). Durand competed four times for the Grand Prix but was awarded only second prizes, in 1779 and 1780. How long he remained with Boullée is uncertain, but it was no doubt on Boullée's recommendation that Durand was awarded his first commission, a house at 65, rue Faubourg-Poissonière, Paris, for Boullée's contractor, Pierre-Armand Lathuile, a project that began in 1788. This was the most ambitious of Durand's architectural undertakings; he built little more and nothing of note—the Maison Lermina, at Chessy, to the east of Paris, in 1802, and three houses at Thiais, to the south, where he lived, in 1807, 1820, and 1825. His buildings were all unpretentious illustrations of his theories, small in size and scale. But not so his competition designs.

With the suppression of the academies after the Revolution the government thought to provide a stimulus to the arts by initiating a series of competitions; in 1794 sixteen were concerned with architecture. Durand submitted no fewer than fifteen entries together with Jean-Thomas Thibault (1757–1826), whom he had met in Boullée's office and with whom he is thought to have traveled to Italy in 1788. Twelve of their entries won prizes. The public monuments proposed were all of the large-scale kind, commemorative columns, triumphal arches, dedicatory temples, and other buildings of public import. All are startling geometric compositions. The most remarkable were a Temple à l'Égalité, surrounded by square columns with human heads in place of capitals, suggesting a knowledge of the temple of Dendrah in Egypt, and a Temple Décadaire, a domed assembly hall entered through four attached porticoes, that harks back to Boullée's design for a monument to Newton.

It was on the strength of these designs that, in 1795, Durand was made professor of architecture at the newly established École polytechnique. Thereafter he taught primarily cartographers, surveyors, and engineers. His course was no more than a part—the seventh part—of a year-long introductory course offered to students who went on from the École polytechnique to more specialized engineering schools. Only a fraction of the 6,000-odd students instructed by Durand in his lifetime went on to study architecture, though some were of note: Achille Leclère, Jules de Joly, François Mazois, Émile Gilbert, Charles Rohault de Fleury, and Léonce Reynaud. The most notable of Durand's many German pupils was Leo von Klenze. But Durand's influence on nineteenth-century architecture was extended largely through the medium of books. His *Recueil et parallèle des édifices de tout genre, anciens et modernes,* issued between 1799 and 1801, was probably the most widely consulted picture book of the nineteenth century and served, almost, as a bible. When Charles Garnier (1825–1898), the future architect of the Paris Opéra, entered his *atelier* as a student of the École des beaux-arts in 1842, he was required to swear "Haine à Viollet-le-Duc," on the "Grand Durand," as it was by then known. It remained a basic reference even into the twentieth century, appearing, surprisingly, in the office of Auguste Perret (1874–1954).

Durand's second great work was the *Précis des leçons d'architecture données à l'École polytechnique,* designed as a teaching manual. The various volumes and editions faithfully reflect the change in emphasis of Durand's instruction over successive years. The first volume of the *Précis,* published in 1802, illustrated the details and component parts of architecture and described the method of their composition. The second volume (1805) contained thirty plates illustrating individual building types, corresponding to the thirty lessons Durand gave on the subject. The second volume was not much in demand and did not sell out until 1817. The first volume, however, was reprinted in 1809 and was then entirely reworked, with some of the text transferred to footnotes and new plates added, all to reflect the revised curriculum of 1811. This was published in 1813 as the *Nouveau précis,* and students were required to buy it. It was reprinted in 1819 and 1823 under the original title *Précis.* But with a growing emphasis on the rapid formulation of architectural ideas, something even more clear-cut was demanded. In 1821 Durand issued the *Partie graphique des cours d'architecture faits à l'École royale polytechnique,* yet another variant of the *Précis,* with a highly condensed text. By then, however, building types were beginning to be appreciated anew, and the initial version of the second volume was reissued in 1823 and 1825. The last Paris edition of both volumes appeared in 1840, six years after Durand's death. The final edition was published two years later, in Liège.

Durand was an arrogant and dogmatic teacher. His *Recueil* seemed to broaden the base of architectural study to include all historical styles, and most contemporary ones as well, but though he acceded to J.-G. Legrand's suggestion that his *Histoire générale de l'architecture* should be issued together with the *Recueil,* Durand was not interested in history or the styles as such. A contemporary of naturalists such as Auguste-François-César Saint-Hilaire, Antoine de Jussieu, and Georges Cuvier, he aimed to classify buildings according to type, regardless of their stylistic diversity. Hundreds of buildings are illustrated, all drawn to a common scale and grouped together on pages with regard only to their particular function.

This method of comparing buildings thus has its origin in three plates appended to Jacques Tarade's survey of St. Peter's in Rome, published in 1715, in which St. Peter's, Notre-Dame in Paris, and the cathedral in Strasbourg are investigated together. The method was taken up, around 1750, by J.-A. Meissonnier, who prepared two plates for a proposed history of architecture, the one showing no fewer than twenty-nine elevations, a plan and a section, all drawn to the same scale, of a range of antique temples, Chinese pagodas, the church of St. Hagia Sophia in Constantinople, and Gothic, Renaissance, baroque, and contemporary churches; the second plate, though it begins with Notre-Dame, comprises secular buildings, all French. The method was to be taken up more than once during the eighteenth century—notably by Gabriel Pierre Martin Dumont (1763 and 1764), J.-D. Leroy (1764 and 1770), Victor Louis (1782), and Antoine François Peyre (1795)—but the scale of Durand's undertaking was without precedent. His forerunners offered no more than one or two plates at a time; Durand provided sixty or more, illustrating temples, mosques and pagodas, churches, markets, public buildings, theaters. Other plates were devoted to columns and individual architectural features. He showed scant respect for the individual nature of the works he illustrated, and without compunction altered a number of plates to accord with his simple standards of symmetry and monumentality.

Although Durand is invariably associated with the concept of building types it is important to note that he never once employed the term; the notion was discussed at length by his contemporary, Antoine-Chrysostome Quatremère de Quincy in his dictionary of architecture, but he was concerned rather with the conceptual notion, in the manner of Plato's forms, than with the planning arrangements that occupied Durand.

The *Précis des leçons d'architecture* was even more ruthless in its outlook. In it Durand condemned Vitruvius and shrugged off the orders. He proposed an architecture based on the principles of convenience and economy, but one that derived its effects from symmetry and geometrical simplicity, which happened to be his aesthetic preferences. The nature of architecture, he held, was inherent in its efficient planning and construction; its character therefore was to be derived from the expression of these functions, the presen-

tation of a visible argument to the spectator. The column was thus not to be regarded as a decorative feature, but was to act always as a structural support, and to appear as such; the pilaster was to be employed only to buttress an angle or a junction; and the architrave used always as a horizontal beam or tie. These structural features were to be made of materials manifestly stronger and more durable than the rest of the building. Such was "natural" decoration, although Durand also permitted such devices as statues, inscriptions, and even growing plants to provide some incident and pattern in architecture.

Durand developed his themes with implacable logic, but his logic was naive and not free from sophistry. He argued that regularity and absolute symmetry resulted in an economy of means and were therefore to be upheld as ideals. Similarly, the most regular of figures, the circle—or as a more practical alternative, the square—were to be preferred to all other shapes because they enclosed the greatest area within the shortest perimeter. His arguments take no account of the real problems, the practical requirements of organization and planning.

The disingenuousness of Durand's vaunted logic is most apparent in his method of "composing." He described first the separate elements of architecture, which he saw to be columns, piers, walls, and even such "negative" elements as windows and doorways, and he analyzed them with reference to the various materials of which they might be composed to form an architectural whole. The basis of architectural unity, he conceived, lay in straightforward geometrical relationships. Rooms were always to lie on common axes; windows and doors were to be placed opposite one another and equally spaced—though this general rule was complicated by the fact that they were to be spaced also in accord with the load they were bearing. Durand proposed, therefore, the use in all plans of the grid, derived from the subdivision of a square: "La première chose à faire pour acquerir de la facilité à composer, c'est de s'excercer beaucoup sur les différentes divisions du carré" (1802, vol. 1, p. 91). The individual elements were then to be imposed and arrayed on this grid. Not surprisingly, many of his students' sketches were done on graph paper, a technique common already in the eighteenth century. Modular planning is, of course, inherent to the use of the orders, but it was only in the eighteenth century that grid planning was to be illus-

trated in textbooks, in England, in those of Robert Morris (1734), and in Italy in those of Bernardo Vittone (1760 and 1766). No one, however, had exploited this technique as radically and obsessively as Durand.

The hundreds of buildings that he composed in accord with this method and with which he illustrated the *Précis* incline thus to be nothing more than neat and regular geometrical patterns. As planning arrangements they are altogether unsatisfactory, and as three-dimensional compositions they fail entirely. Durand composed in two distinct stages; first, horizontally, to obtain the plan, and then vertically, to arrive at sections and elevations. The same grid served for both planes. Clearly, not only did he have no interest in problems of planning, he had equally little feeling for the rhythms of architectural forms and space. His compositions are repetitive, dull, and static.

Yet for all their lack of vital stimulus, for all their dull and mechanical properties, Durand's designs are undeniably impressive and indicative of a strong controlling system, and it is not difficult to understand why they were to assume such importance in the uncertain years of the early nineteenth century. Even then they were not thought of as masterpieces of architecture, but rather as expressions of formal ideas that could be taken up and rapidly mastered by anyone. "Le bon goût" could be acquired in 200-odd hours. Durand's books largely conditioned the nature of the lackluster architecture that was to be built throughout Europe—continental Europe, that is—during the early decades of the nineteenth century. In Germany, where the *Précis* appeared in translation in 1831, he was especially favored.

[*See also* Architecture.]

BIBLIOGRAPHY

Works by Durand

Recueil et parallèle des édifices de tout genre, anciens et modernes. Paris, 1799–1801.
Précis des leçons d'architecture données à l'École polytechnique. 2 vols. Paris, 1802, 1805.

Other Source

Szambien, Werner. *Jean-Nicolas-Louis Durand, 1760–1834: De l'imitation à la norme.* Paris, 1984.

ROBIN MIDDLETON

E

ECO, UMBERTO (b. 1932), Italian aesthetician, novelist, and critic of art. Eco is best known academically for his effort to promote semiotics, the study of signs, as the central humanistic discipline. Eco has exerted a powerful influence on Italian culture since the 1960s, when his views on the "openness" of modern art and the importance of scrutinizing and interpreting mass culture helped shape the vocabulary of contemporary Italian criticism.

Because Eco, unlike most scholars, operates as an outsized figure in the everyday culture of his country, making art as well as analyzing it, he poses a singular challenge to interpreters of his aesthetics. Early in his career he produced cultural programs for RAI, the Italian television network, and served as a longtime adviser to, and editor for, the Italian publisher Bompiani. Most notably, he has become, over the past two decades, the internationally renowned author of three novels: *The Name of the Rose* (1980), *Foucault's Pendulum* (1988), and *The Island of the Day Before* (1995). Those complex fictions suggest views about central aesthetic topics, such as interpretation, that jibe with the arguments in Eco's voluminous scholarly and critical work, as well as in his journalism for publications such as *L'Espresso* and *Corriere della Sera*. The relevance of Eco's fiction and semiotics to his aesthetics is discussed below.

Eco was born in the Piedmont town of Alessandria. He began his scholarly career at the University of Turin as a student of the leading Italian aesthetician of the time, the existentialist Luigi Pareyson. Pareyson's ideas constitute the roots of Eco's aesthetics. Eco has developed and extended Pareyson's anti-Crocean views, but never abandoned them.

For Pareyson, the key human activity is the creation of organic forms in all areas of life, including politics, science, and philosophy. A work of art is a special type of form in that it reifies the personality of the artist, embodying within it the marks of the artist's "concrete experiences, his interior life, his unique spirituality, his personal reaction to the world in which he lives, his thoughts, customs, feelings, beliefs, aspirations." That incarnation of the artist's personality in a material work of art is its style, which can be better deduced the more one understands the formation of the work. Pareyson's most famous articulation of this core view, which can lead to misinterpretation in English, is that art is *il formare per il formare* ("forming for the sake of forming"). Unlike "art for art's sake," however, Pareyson's view, which

Eco adopted wholeheartedly, is that such "formation" encompasses all the philosophical and historical experience that an artist brings to his project and material. It similarly includes the struggle between the artist's intention and his sometimes recalcitrant material, which culminates in a finished work. In this struggle, always a process of trial and error, the artist is guided by some concept of the intended work. Proper aesthetic contemplation and interpretation of the work require understanding the dynamic history of its development, as well as trying imaginatively to place oneself in the artist's shoes at the beginning of the process.

By advancing such an existentialist view, Pareyson rejected the vision of art that had dominated Italian aesthetics for decades: Benedetto Croce's view that art is an intuition of feeling whose expression is essentially indistinguishable from it, a mere external epiphenomenon and thus not a form of knowledge. That approach suggested that the physical properties of art, and their etiology, were barely worthy of attention. Pareyson opposed such a position completely, and Eco has been a faithful follower in that respect, indefatigably analyzing works of art as complex historical subjects. In commenting on Pareyson in *La definizione dell'arte* (1968), Eco endorsed his teacher's views, agreeing that the "content" of every artistic form is, so to speak, the artist, because the artist leaves a "concrete and personalized trace" in his work. Moreover, Eco asserts against Croce, "art understood as form cannot but have a physical existence." For Eco, too, style is a "characteristic way of forming" an artwork. In both Pareyson's and Eco's view, such a work allows for an infinity of interpretations, no one of which is "definitive or exclusive," though some will show greater "fidelity to the work" than others. [*See* Pareyson.]

Eco's popularization, refinement, and expansion of these positions did not come until after he commenced his career with a focus on medieval aesthetics. He received his doctorate in 1954, writing his dissertation under Pareyson on Aquinas. It became his first publication, *Il problema estetico in San Tommaso* (1956), later revised and republished as *Il problema estetico in Tommaso d'Aquino* (1970), and finally translated into English as *The Aesthetics of Thomas Aquinas* (1988). In 1959, Eco also contributed a chapter on the development of medieval aesthetics to a four-volume history of aesthetics—it appeared in English translation as *Art and Beauty in the Middle Ages* (1986).

In his study of Aquinas, Eco argues that his subject's system "included a coherent aesthetic theory," particularly in regard to the ontology of beauty, which Aquinas considered transcendental, created by God, and subject to the three formal criteria of proportion, integrity, and clarity. When writing his dissertation, Eco recalled later, he philosophized in a "spirit of adherence" to Aquinas's religious world that subsequently lapsed. Eco's chief scholarly contribution to Aquinean aesthetics was, he contended, "to pull together scattered remarks that appeared here and there in his immense opus" and synthesize them, because Aquinas "never wrote *ex professo* about aesthetics." Eco regards his most original contribution to Thomistic scholarship to be his argument that the experience of beauty, to Aquinas, occurs not in the so-called first operation of the mind, known in the scholarly tradition as *simplex apprehensio,* but in the second, the *compositio et divisio,* or judgment. [*See* Aquinas.] In his introduction to the English edition, Eco states that he considers his conclusions in the book "still valid." Moreover, he acknowledges that his nostalgia for the order and unified theory of the medieval world lingers in his own taste for methodological rigor. He moved on from medieval aesthetics to contemporary aesthetics, he wrote, because he was "looking for a unified approach to all those phenomena that go from Picasso to Donald Duck, from Beethoven to the Beatles." Eco's extensive attention to the work of James Joyce reflects his own assessment that Joyce experienced a similar rebellion against the systematicity of Catholicism: Eco describes *Ulysses* as a "reverse *summa.*"

Eco's first move toward contemporary aesthetics and his own rebellion came in the book that made him famous in Italy, *Opera aperta* (The Open Work, 1962), which has gone through several editions—it is generally seen as his chief contribution to aesthetics as a whole. Like Susan Sontag's *Against Interpretation* a few years later in the United States, it stirred intellectuals to debate about the relations of high and low art, modern and traditional art, quality and kitsch. Eco argued that modern composers (Karlheinz Stockhausen, Henri Pousseur), writers (Stéphane Mallarmé, Joyce), and artists (Alexander Calder), unlike traditional artists, create *opere in movimento* (works in motion) or "open works"—works that require readers, hearers, or viewers to participate in the creation of their meaning, and sometimes even their structure. Art, as a practice, spurs acts of communication and productive discourse. By violating traditional, conventional methods and rules, Eco argues, modern artists create the ambiguity that is the sine qua non for openness. A work like *Finnegans Wake,* for example, amounts to a field of possible aesthetic dreams, to be determined, in Eco's approach, by the dreamer.

The "open work," according to Eco, is nonetheless "an epistemological metaphor" that suggests the uncertainty of modern life while reflecting the artist's persona. The phrase is true to Eco's intent, which is to see the open work as a form of crystallized knowledge that impacts others. In that way, the modern open work is socially if not partisanly political, seeking to shake things up. *Opera aperta,* in fact, exercised the impact of a manifesto. It functioned as such for the influential Italian avant-garde movement, Gruppo 63, formed shortly after its publication to foster aesthetic attacks by artists on an archaic social system. The book presaged the great attention in literary theory of the subsequent twenty-five years to pluralism in interpretation and the interactive role of readers and texts.

In the wake of that declaration of independence, Eco moved on to display another enduring hallmark as an aesthetician: fascination with mass and pop culture. He first exhibited this interest in *Apocalittici e integrati* (1964), where he analyzed television shows and rock music, and apotheosized comic strips such as *Peanuts* and then fashionable pop fiction (Ian Fleming's James Bond novels). In the book, Eco distinguishes "apocalyptic" intellectuals—those who fear mass media adulterates culture and ought to be ignored—and "integrated" intellectuals who do not fear mass culture, but rather seek to understand it, distinguish among its products, and decode its ideology. Eco not only prefers the latter group and considers himself a member, but sees "integrated" types as necessary to pierce through the hegemonizing influences of traditional culture, particularly its simplification of life's complexity.

Here, too, Eco's work should be seen as contemporary with the American questioning of high and low cultural distinctions expressed by Sontag in the 1960s, and in advance of similar efforts produced later by intellectuals such as Andrew Ross in the Cultural Studies movement. Eco's maverick openness to mass art, then largely identified with postwar American cultural imperialism, and his homages to pop cult figures like Sam Spade, helped take the colonialist edge off that attitude, providing a model to Italian and other intellectuals of how to appreciate bourgeois culture cautiously. Eco's original Thomistic tendencies fuel an almost palpable sense of schoolboy naughtiness when he expounds on Disney and the like.

Just as Eco began to look seriously at popular culture, however, he grew influenced by structuralism. That school of thought helped shape the third prominent characteristic of Eco as an aesthetician: his proclivity for methodological rather than metaphysical structuralism. In *La struttura assente* (1968), Eco attacked Claude Lévi-Strauss for promoting the latter by presuming we can identify ontologically objective features and laws of the world. He recommended instead his "methodological" alternative that sees structures as revisable procedures. Eco argued in that book that culture is a system of communication and every act of culture or cultural event is a system of signs. To interpret that system of signs, he went on to contend in *A Theory of Semiotics*

(1976) and *Semiotics and the Philosophy of Language* (1984), that we draw upon an "encyclopedia," the set of cultural and historical information we have available to us.

Eco's theory of semiotics, though applicable to far more than the world of art, is hardly irrelevant to his aesthetics. Believing that signs and sign systems are everywhere, he set out in *A Theory of Semiotics* to articulate their architecture. Eco's paramount tenet is that signs must be understood contextually, in relation to all other signs, and not as referents to objects in the external world—he calls that condition "unlimited semiosis." Understanding signs in their contexts requires the "abductive" methods of the detective, who tests inferences and imagines explanatory hypotheses that make signs intelligible. Interpreters must confront and understand the "aesthetic idiolects" of works of art, their personal and private languages, by evenhanded research in the history of ideas. Semiotics thus should not be frozen into a discipline, but rather seen as a call to look at and clarify the world under the aspect of communication. Art, in turn, is matter "rendered semiotically interesting." It is important to Eco that semiotics be understood not as a "science," but as a method of investigation that displays a "scientific attitude."

Eco's analytical bent can make his approach to aesthetics seem wholly clinical, full of willingness to "murder to dissect" the work of art. But he denies that in *La definizione dell'arte*, one of his lesser-known books outside Italy. In it, Eco concedes that his powerful concerns with the "poetics" or theoretical intentions of a work of art threaten to give short shrift to aesthetic evaluation of art—to criticism. In part, he says, the fault is that of modern art, where the "question of poetics" has become "more important than the creation of the work itself." If that is so, "aesthetic pleasure has gradually changed from the emotional and intuitive reaction it once was to a much more intellectual sort of appreciation." It may be that the "intuitive spark and emotional shiver that were once thought to accompany all aesthetic revelation are today not only dated but also limited to a particular historical period."

Nonetheless, we should not despair, Eco counsels, because throughout the centuries, the idea of art has undergone numerous changes. We should, rather, adopt the dialectic confidence and lightheartedness of the eminent Italian literary critic Francesco De Sanctis in the face of changing notions of art. One prod to that modest enjoyment of changing art is, Eco insists, that the work of art is always more than the sum of its dissectable parts. "Is it true, really true," he asks, "that the only thing we enjoy in our reading of *Finnegans Wake* is the poetics that sustains it, and that the concrete expression of the linguistic event leaves us entirely indifferent?" No. "To appreciate a work as a perceptible form means to react to the physical stimuli of the object, not just intellectually but also—so to speak—physically."

Eco thus asserts that contemporary works of art "can be evaluated critically in terms of success and failure only if we take these terms in their most organic sense. . . . the work fulfills its fullest aesthetic value only insofar as the formed product adds something to the formal model." Aesthetic valuation can therefore be brought "back within the current aesthetic horizon," regardless of the high-theoretical intentions of modern art, because "the work is more than its own poetics." That is why "every time we reread *Ulysses*, we understand things that the mere enunciation of its poetics could not have told us."

In his extensive writings on aesthetic issues, Eco weighs in on a wide variety of contested concepts. In his well-known essay "The Structure of Bad Taste," he argues that "kitsch" requires not just the production of particular effects, but a manifest intent in the work, through "palming itself off as art," to cue an audience that an aesthetic experience awaits. He believes that the best works of fiction are complex and difficult, like those of Marcel Proust and Joyce, and that readers want to be "involved in a mutual act of seduction." Throughout his work, the advocacy of humor as an explosive weapon against dogmatism, pomposity, and overconfident models of reality is repeatedly signaled.

Eco's later books, such as *The Limits of Interpretation* (1990), *Interpretation and Overinterpretation* (1992), and *Misreadings* (1993), confirm that interpretation remains the aesthetic subject that most occupies him. In them, he seeks to guard against misinterpretation of his own work as an invitation to interpret works of art in any way the viewer, listener, or reader wishes. While Eco, like Pareyson, rejects the notion that the creator of a work of art can adjudicate its meaning or constrain its possible implications by citing his intentions, he believes the work of art itself behooves an ideal reader to understand its evidence in sensible ways. An ideal reader is as open as the work itself, ready to interpret it in ways that make sense.

Eco's aesthetic views have been criticized from a number of angles. One American commentator, David Robey, remarks on Eco's "broad, synthesizing generalizations," saying they tend to stress the similarities between concepts and phenomena at the expense of the differences. Richard Rorty, in his contribution to *Interpretation and Overinterpretation,* suggests that Eco remains too loyal to the notion that a text has a "nature" that the interpreter should get right. Ernesto Paolozzi, in his *I problemi dell'estetica italiana dal secondo dopoguerra al 1985* (1985), criticizes Eco for recycling Pareyson without his master's metaphysics, and with insufficient rigor. Admittedly pro-Crocean, Paolozzi concludes after summarizing Eco's nonfiction corpus that it "has scant philosophical value."

That is a charge Eco originally accepted in regard to his novels, which he claimed should not be construed as philosophy set to literary music, no matter how much or little they

chime with his critical work. In a preface to a collection of essays on *The Name of the Rose,* he wrote, "I did not (consciously) try to put in my novel the theories I had developed in my scholarly writings" (as quoted in Inge, 1988). By the time of *Foucault's Pendulum,* however, critics had convinced him that his fiction reflected his philosophy in clear-cut ways. *Foucault's Pendulum,* for instance, offers three Milanese editors who concoct a complicated conspiracy theory based on threadbare evidence about secret religious societies, exploiting the idea that "everything might be mysteriously related to everything else." It plainly illustrates Eco's philosophical opposition to runaway interpretation not grounded in evidence; the book ferociously mocks neurotic overinterpretation as an evil that leads to tragedy in both public and private life.

Similarly, Roberto della Griva, the main character of *The Island of the Day Before,* is described as a man with a talent for absorbing contradictory truths "as if he were a sponge," and the hero of *The Name of the Rose,* William of Baskerville, shares Eco's pragmatist sense (derived from his hero Charles Sanders Peirce) of the tentativeness of all hypotheses. The novel itself, according to Eco's *Postscript* (a short book reflecting on it), is an example of "an open work," that key notion in Eco's aesthetic universe. In *Postscript,* Eco also calls the novel—as genre—"a machine for generating interpretations."

In an autobiographical essay published in 1989, Eco conceded that his fiction is also "a philosophical activity, because there are questions that can't be treated in the assertive mode of an essay, but must be, so to speak, dramatized in all their ambiguity and contradictions." In another article, he elaborates: "The distinction is this. There are texts you write in order to demonstrate something you exactly believe . . . And there are other texts in which you don't try to demonstrate something, but exactly because something is very contradictory and complex, you stage the contradiction. You show it. That's the difference between a novel and a philosophical essay." A sensible approach to Eco's aesthetics, therefore, dictates giving primacy to his critical and philosophical corpus while scrutinizing the fictions for aids to interpretation.

[*See also* Interpretation; Semiotics, *article on* Semiotics as a Theory of Art; *and* Text.]

BIBLIOGRAPHY

Works by Eco

Il problema estetico in Tommaso d'Aquino (1956). Rev. ed. Milan, 1970. Translated by Hugh Bredin as *The Aesthetics of Thomas Aquinas* (Cambridge, Mass., 1988).
"Sviluppo dell'estetica medievale." In *Momenti e problemi di storia dell'estetica.* vol. 1, pp. 115–229. Milan, 1959. Translated by Hugh Bredin as *Art and Beauty in the Middle Ages* (New Haven, 1986).
Opera aperta (1962). Rev. ed. Milan, 1967. Translated by Anna Cancogni as *The Open Work* (Cambridge, Mass., 1989).

Apocalittici e integrati (1964). Milan, 1977. Partial translation with other texts edited by Robert Lumley, *Apocalypse Postponed* (Bloomington, Ind., 1994).
La definizione dell'arte. Milan, 1968.
La struttura assente (1968). Milan, 1983. Rewritten as *A Theory of Semiotics* (Bloomington, Ind., 1976).
The Role of the Reader: Explorations in the Semiotics of Texts. Bloomington, Ind., 1979.
Il nome della rosa. Milan, 1980. Translated by William Weaver as *The Name of the Rose* (New York, 1983).
Postscript to The Name of the Rose. Translated by William Weaver. New York, 1984.
Semiotica e filosofia del linguaggio. Turin, 1984. English translation, *Semiotics and the Philosophy of Language* (Bloomington, Ind., 1984).
Il pendolo di Foucault. Milan, 1988. Translated by William Weaver as *Foucault's Pendulum* (New York, 1989).
I limiti dell'interpretazione. Milan, 1990. Translated as *The Limits of Interpretation* (Bloomington, Ind., 1990).
Interpretation and Overinterpretation. With Richard Rorty, Jonathan Culler, and Christine Brooke-Rose. Edited by Stefan Collini. Cambridge, 1992.
Misreadings. Translated by William Weaver. New York, 1993. Selected translations of early pieces published as *Diario Minimo.*
La ricerca della lingua perfetta nella culture europea. Bari, 1993. Translated by James Fentress as *The Search for the Perfect Language* (Oxford and Cambridge, Mass., 1995).
L'isola del giorno prima. Milan, 1994. Translated by William Weaver as *The Island of the Day Before* (New York, 1995).
Six Walks in the Fictional Woods. Cambridge, Mass., 1994.

Other Sources

Capozzi, Rocco, ed. *Reading Eco: An Anthology.* Bloomington, Ind., 1997.
Coletti, Theresa. *Naming the Rose: Eco, Medieval Signs and Modern Theory.* Ithaca, N.Y., 1989.
de Lauretis, Teresa. *Eco.* Florence, 1981.
Ganeri, Margherita. *Il "Caso" Eco.* Palermo, 1991.
Inge, W. Thomas ed. *Naming the Rose: Essays on Eco's "The Name of the Rose."* Jackson, Miss., 1988.
Paolozzi, Ernesto. *I problemi dell'estetica italiana dal secondo dopoguerra al 1985.* Naples, 1985.
Pareyson, Luigi. *Estetica: Teoria della formativita.* Rev. ed. Milan, 1974.

CARLIN ROMANO

ECPHRASIS. The term *ecphrasis* (or, in the more precise transliteration, *ekphrasis*—both are now commonly used) derives from the Greek *ekphradzein*—literally, "to speak out, tell, or describe fully"—is used today in its narrower sense to mean the literary description of a work of art, usually employing considerable rhetorical art and, in a broader sense, to designate entire poems or passages from longer works of fiction, poetry, or expository essays that address in various ways particular works of painting, sculpture, graphic art, or perhaps even architecture. It should be noted, however, that Jean H. Hagstrum, in his important study of the relations between art and literature, reserves the term *iconic* for most poems and prose passages in this mode, and *ecphrastic* for those that purport to give "voice and language to the otherwise mute art object" (Hagstrum,

1958). In Hellenistic times *ekphrasis* was also sometimes used to characterize particularly vivid description of any kind, including actual scenes as well.

For ancient and modern literature, it may be useful to distinguish between notional and actual ecphrasis (see Hollander, 1995). The first comprises descriptions of purely fictive works of art, brought into being only in and by means of the literary work itself. The second refers to descriptions or other poetic or narrative confrontations with actual particular works of art that preexist the literary text. (In a way, this mode of ecphrasis might be considered a reciprocal of illustration of a preexistent literary text.) A subcategory of actual ecphrasis might be set up to include descriptions or invocations in verse and prose of objects that the writer had indeed confronted but that have since been lost, leaving the present-day reader with no way to compare the work addressed with the literary text, as is of course also the case with virtual works generated by the texts themselves. The preserved ecphrastic description of an actual but lost object from classical or Hellenistic antiquity, by the way, plays an important role in Renaissance art, when paintings are based on surviving ecphrases, such as the description by Lucian of an allegory of Calumny by the celebrated painter Apelles (this is discussed in detail by Cast, 1981).

In some other instances the production of the image and of the text are more ambiguously involved, as in the case of the engraved plates of some of William Blake's works, where the question of what is illustration and what is addressed to a picture is sometimes hard to determine. Similarly, in the case of Chinese art of the Tang and other periods, a painting may illustrate a poem previously composed, or, alternatively, such a poem may be addressed to the preexistent painting, but in both cases the text will appear often beautifully calligraphed on the painting itself.

Notional ecphrases precede actual ones in Western literary history. The description of Achilles' shield in the *Iliad*, and of that of Heracles in a shorter poem of Hesiod's that is roughly contemporary with it is followed by ecphrastic passages in Theocritus, Virgil, Ovid, prose romances of Achilles Tatius and Longus (whose *Daphnis and Chloe* recounts a narrative that is offered up, from the outset, to explain the meaning of a notional picture described at the story's very opening), essays and passages in Lucian and the *Eikones* of Philostratus, a collection of short ecphrastic set pieces describing paintings purportedly seen in a house the author had visited (one or two scholars have tried to make a case for there having been actual but lost paintings here). In later literature, Dante Alighieri's *Purgatorio* (cantos 10 and 12), Geoffrey Chaucer's *The House of Fame* and *The Knight's Tale*, Ludovico Ariosto's *Orlando Furioso*, William Shakespeare's *The Rape of Lucrece*, and book 3 of Edmund Spenser's *The Faerie Queene*, all contain elaborate and significant ecphrastic moments, in which the reading or interpretation of a painted or carved image by a fic-

tional character is frequently a crucial matter for the narrative.

Actual ecphrases are to be found in Hellenistic literature, as in many epigrams in book 16 of the Greek Anthology, where short poems invoking actual, notional, and actual but lost works of art are all gathered together. This body of verse is reflected in the early-seventeenth-century collection of short ecphrastic poems comprising Giambattista Marino's *La Galeria* (1620); they number among them actual and notional ecphrases; but most of the specified objects in the actual ones are now lost. In the nineteenth century, ecphrastic poetry shifts its emphasis somewhat from praise of, or gloss on, a pictorial or sculptural image, to a contemplation of the experience of gazing at it, of encountering or apprehending it. Poets such as Charles Baudelaire and Dante Gabriel Rossetti, in particular, wrote a considerable number of ecphrastic poems, and the novel from the early nineteenth century on presents an array of occasions for ecphrastic description. In the twentieth century, Rainer Maria Rilke wrote a good many actual and notional ecphrastic poems, and in English and American poetry after World War II, the example of W. H. Auden's celebrated "Musée des Beaux-Arts" with its discussion of Pieter Brueghel's *Fall of Icarus* has led to what has almost become a contemporary conventional mode.

In fiction like Honoré de Balzac's *Le chef-d'oeuvre inconnu* (The Unknown Masterpiece) and works of Johann Wolfgang von Goethe, Edgar Allan Poe, Nathaniel Hawthorne, Nikolay Gogol, Henry James, Émile Zola, and Oscar Wilde that specifically concern art and artists, notional ecphrases—most particularly, of portraits—are central to the work's descriptive concerns; but in these and other novels one frequently encounters interesting juxtapositions of actual and notional images (at the end of Herman Melville's *Pierre*, for example, and in Hawthorne's *The Marble Faun*). In general, however, ecphrastic moments in fiction often function rather differently in narratological structures than do descriptions of place and scene.

Another mode of ecphrasis occurs in nonfictional essays and books devoted to the study and consideration of art. In the twentieth century these assumed the nonliterary and purportedly technical stance of an academic—and sometimes even a scientistically framed—discourse; but prior to the development of modern art-historical practices, writing about art remains much more part of literature generally. From Pliny's discussion of painting and sculpture, and Giorgio Vasari's ecphrases of works by the various subjects of his mid-sixteenth-century *Lives of the Painters, Sculptors, and Architects,* on into the earlier twentieth century, ecphrases are central to discussions of works of art, even when, at later times, printed reproductions of the art in question are more easily available to accompany the text. But even in the case of expository writing about art, the partiality, rhetorical pointing, and ad hoc interpretative pro-

gram of such descriptions can be fully as constructed and ulterior as that of a poem or fictional episode.

A Renaissance and baroque ecphrastic tradition that should be distinguished here is that of the emblem. Emblems, *imprese*, or devices (as they were variously called) were symbolic images represented in woodcuts or engravings, each captioned by a motto or tag identifying its moral or other general significance. Below each image there were lines of verse—originally in Latin in the early sixteenth century, and then in vernaculars thereafter—that spelled out the relation between image and idea in specific and detailed ways, allegorizing the emblematic object or situation depicted. As a special case of ecphrasis, emblem verses read the picture above them only for this significance: thus, for example, an image of a visored helmet with a plume and with bees flying about it with the motto *Ex bello pax* (Peace from War) is explained anecdotally as a warrior's helmet that, after a war ended, he presented to a swarm of bees who built their hive in it and produced honeycomb and honey. (This might be considered a version of the famous biblical passage, in Isaiah 2:3, about beating swords into plowshares and spears into pruning hooks.) But in any case, the particular pictorial presentation of the emblematic image—details of setting, of rendering, of choice of style of object (in the case above, whether the helmet was visored or a more hat-like morion; what the plume might signify, etc.)—is totally ignored. Emblematic interpretations of images embody an extreme type of reductive ecphrastic reading that attends only to the object depicted, rather than any aspect of its mode of representation.

It might be said that some of the basic rhetorical paradigms developed in notional ecphrasis still operate in actual ones (notions of centrality and peripherality in describing an image; ways in which texts construct virtual scanning processes ad hoc only to their own occasions; the devices by which temporality, in a text's unfolding over time, is figuratively made to deal with spatiality; and so on). Then, too, the impulse to produce poetic vividness that would stand in for an absent image is in some ways not too far from constructing a sort of verbal screen between a viewer's eye and the picture or sculpture in question. W. J. T. Mitchell suggests, in a somewhat more deconstructive manner, that notional and actual are really to be identified, both being fictional and both falsely invoking a presence. And yet an interesting question for notional ecphrasis turns on the extent to which such fictional descriptions manifestly or implicitly invoke pictorial or sculptural conventions of the author's own—or of an earlier—time; in a sense all notional ecphrases are constructed from particular visual material present for, and invoked by, actual ones.

Recent theorizing has also introduced into discussions of ecphrasis not only the subject-object relation to discursive text and mute image, but analogous problems of self and other and, particularly in the case of lyric poetry, a matter of the quasi-erotic and hopeless desire of language to attain the condition of image. Questions also always remain to be raised about the unspecified and unacknowledged partiality of ecphrastic representation: for example, what realms of pictorial phenomena are neglected or ignored by the literary description. What might in a literary work give the representation in language of a work of pictorial art its poetic or narrative power might lie just in its selectivity. To this degree, too, actual ecphrases might be said to partake of the character of notional ones, in that out of selected and highlighted elements of the described object, a new but notional one is constructed. But other questions of selectivity or contrived prominence seem more historically determined. For example, the transparency or even invisibility of the medium (considered at any level, such brush stroke, color, modeling, conventions of drawing) remains almost universal until the twentieth century, after some fifty years at least of modern French painting had foregrounded such matters for art itself, and made poets aware of them.

A traditional basis for varying interpretations of the relation between text and image stems from the telling phrase *ut pictura poesis* (as a painting so a poem) from Horace's *Ars poetica;* although in its original context the term likens poem and picture with respect to matters of scale, it grew later to invite more general analogies. Just as important is the transmission to the Renaissance and after of the notion (attributed by Plutarch to Simonides of Ceos) that painting is mute poetry and poetry a speaking painting. From this there easily arises the idea that ecphrasis gives voice to a silent picture, or at the very least speaks *for* it, as well as *to* it in some mode of unacknowledged personification. To the extent that prose fiction primarily *describes* works of art, which poems can address, question, openly or covertly allude to elements in particular pictures or sculptures, it might be observed that novels propound "iconic" (according to Hagstrum's distinction mentioned above) and poems, "ecphrastic" language.

The central question of the spatial and the temporal as distinguishing visual and verbal arts, respectively, was authoritatively raised by Gotthold Ephraim Lessing in his celebrated essay on the Hellenistic *Laocoön* published in 1766. It has remained an important issue for subsequent aesthetic discussions, shifting interest away from the older issue of ut pictura poesis. The implications of Lessing's writing for modern and contemporary literary critical theory have been most recently explored in great detail and ramification by Murray Krieger.

[*See also* Discourse; Imagery; Ineffability; *and* Ut Pictura Poesis.]

BIBLIOGRAPHY

Alpers, Svetlana Leontief. "*Ekphrasis* and Aesthetic Attitudes in Vasari's *Lives.*" *Journal of the Warburg and Courtauld Institutes* 23 (1960): 190–215.

Bath, Michael. *Speaking Pictures: English Emblem Books and Renaissance Culture.* London and New York, 1993.

Baxandall, Michael. *Giotto and the Orators: Humanist Observers of Painting in Italy and the Discovery of Pictorial Composition, 1350–1450.* Oxford, 1971.

Becker, Andrew Sprague. *The Shield of Achilles and the Poetics of Ekphrasis.* Lanham, Md., 1995.

Braider, Christopher. *Refiguring the Real: Picture and Modernity in Word and Image, 1400–1700.* Princeton, 1993.

Cast, David. *The Calumny of Apelles: A Study in the Humanist Tradition.* New Haven, 1981.

Cook, Albert. *Figural Choice in Poetry and Art.* Hanover, N.H., 1985.

Fowler, Alastair. *Kinds of Literature: An Introduction to the Theory of Genres and Modes.* Cambridge, Mass., 1982.

Gross, Kenneth. *The Dream of the Moving Statue.* Ithaca, N.Y., 1992.

Hagstrum, Jean H. *The Sister Arts: The Tradition of Literary Pictorialism and English Poetry from Dryden to Gray.* Chicago, 1958.

Harlan, E. "The Description of Paintings as a Literary Device and Its Application in Achilles Tatius." Ph.D. diss., Columbia University, 1958.

Heffernan, James A. W. *Museum of Words: The Poetics of Ekphrasis from Homer to Ashbery.* Chicago, 1993.

Hollander, John. "The Poetics of *Ekphrasis.*" Word and Image 4.1 (January/March 1988): 209–219.

Hollander, John. *The Gazer's Spirit: Poems Speaking to Silent Works of Art.* Chicago, 1995.

Krieger, Murray. *Ekphrasis: The Illusion of the Natural Sign.* Baltimore, 1992.

Lessing, Gotthold Ephraim. *Laocoön: An Essay on the Limits of Painting and Poetry* (1766). Translated by Edward Allen McCormick. Baltimore, 1984.

Mitchell, W. J. T. *Picture Theory: Essays on Verbal and Visual Representation.* Chicago, 1994.

Philostratus. *Imagines.* Translated by Arthur Fairbanks. Cambridge, Mass., 1931.

Rischin, Abigail. "Speaking Looks: Varieties of Ekphrastic Experience in Nineteenth-Century Literature." Ph.D. diss., Yale University, 1995.

Scott, David. *Pictorialist Poetics: Poetry and the Visual Arts in Nineteenth-Century France.* Cambridge and New York, 1988.

JOHN HOLLANDER

EDUCATION, AESTHETIC.

[*To clarify the multiple meanings of the concept of aesthetic education in the history of aesthetics, this entry comprises two essays:*

History of Aesthetic Education
Contemporary Aesthetic Education

The first essay explains the history of the philosophical literature on aesthetic education and examines contemporary research on the role of the arts and aesthetics in education. Against the background of Friedrich Schiller's notion of aesthetic education, the second essay explains the contemporary understanding of this philosophical notion. For further discussion, see Children's Art; *and* Schiller.]

History of Aesthetic Education

Two sets of issues are commonly subsumed under the rubric of aesthetic education, as the expression is used in philosophical and pedagogical literature. The first comprises a host of questions about roles to be played by the arts (chiefly the "fine arts") in the education of various populations of students, about relations between these roles and other components of academic training, and about instructive dimensions of aesthetic experience that extend outside the classroom. The second more narrowly concerns the question of what role, if any, philosophical aesthetics itself should play in the education of both arts students and others. The first is an ancient concern, traceable in the West to Plato's prescription for the training regimens of prospective leaders in the ideally just state. The second is a modern concern, inspired by the post-Kantian elevation of aesthetics into a full-fledged discipline within philosophy and motivated by the development within this new discipline of intellectual tools designed to render both the artistic enterprise and extra-artistic aesthetic experience more intelligible. While these two concerns address subjects of importance to people of all ages (in that aesthetic education, like most other forms of education, is a lifelong process), they have generally been understood to apply most immediately to the schooling of young people, especially in their precollegiate years. Recent debate in the United States concerning aesthetic education has, in conformity with this tendency, become increasingly absorbed with the manner in which aesthetics is to be integrated into the nationwide debate about arts education reform within kindergarten through grade 12 schools. Although this preoccupation with early learning may have narrowed somewhat the focus of professional inquiry into aesthetic education, it has fostered unusual public interest in the subject, fueled by burgeoning curiosity about its content on the part of art teachers and administrators and by popular expectation that this is one area where philosophical theory might have a salubrious effect on the development of young minds.

Philosophical Views of the Role of the Arts in Education. Western philosophers from Plato to John Dewey have extolled instruction in the arts as a vital component of educational programs preparing young people for rich and satisfying adult lives. The dominant rationale for this advocacy has never been that arts instruction's job is to create a layer of cultural veneer, a refined "finish" to be impressed on the more practical, and therefore more important, elements of learning (as is so often, and so regrettably, supposed today). Rather, the fundamental premise has always been that understanding the arts constitutes understanding some of the basic principles uniting the life of the state with that of the individual. Nor have philosophers concerned with this issue ever supposed that a smattering of training in various arts or a nodding acquaintance with canonical masterpieces would suffice in preparing students for enlightened adulthood. Rather, they have insisted that artistic activity and artistic products must be apprehended and interpreted in a context of theory to make their force felt.

Just as students of science or business or history must be brought to comprehend the fundamental principles of these disciplines if they are to understand their potential impact on social endeavor, so students of the arts must engage their subject with their minds as much as with their senses.

Plato appears to have been the first Western philosopher to have insisted that education in the arts is essential in the intellectual preparation of leaders. His well-known condemnation of various artistic activities in *The Republic,* book 10, in no way denigrated the importance he attaches to the same activities (in books 2, 3, and 7) as components of the educational program for eventual political leaders. As Plato saw it, the arts, by highlighting the difference between what is and what seems to be within the world of sensory experience, have the power to liberate young minds, preparing them to contemplate ideas that lie beyond that world. Moreover, whatever else the arts might contribute to a learner's knowledge of the culture whose character and values they exhibit, these arts were seen to present a model of the ultimate art, the art of living well, an art that, in the case of the philosophically adept statesman, amounts to ruling well.

Later Platonists refined and amplified this message, which eventually became the intellectual platform for Renaissance thinking about the social function of the arts. In the Florentine "New Academy," Marsilio Ficino presented a full curriculum of studies organized around Plato's conception of the unity of philosophical and artistic theory with public practice. The chief objective of the academic enterprise, as Ficino and other Renaissance thinkers conceived it, was to condition young minds for the responsibilities of statesmanship by inculcating in them those noble dispositions that, founded on (philosophically accessible) ideal harmonies, might inspire the cultivation of happy and harmonious forms of social interaction. And music, art, literature, and drama were seen as the most legible exemplars of the harmonics in question. It was out of this effort that the academic curriculum known as the "liberal arts" arose. This educational program was deemed liberal in that it held out the prospect of liberating its subscribers from conventional and local prejudices in order to partake of the pleasures and advantages of a wider citizenship.

Although Enlightenment philosophers repudiated many aspects of the entrenched liberal education doctrines they inherited, they usually endorsed the doctrine of respect for instruction in the arts on which those doctrines were founded. Arts education was still regarded as liberating, inspiring, and conducive to both philosophical wisdom and social success. This conviction is evident even in the writings of that most ardent of critics of the establishment practices and the heritage of civilization itself, Jean-Jacques Rousseau. No matter how many damning indictments Rousseau was willing to direct at the artistic practices and pretensions of his day, he remained adamant in reserving a prominent place for the arts in the cultivation of children's talents. His extraordinarily unorthodox educational manual, *Émile,* warned against the obtrusiveness of standard instruction in the cultivation of original aesthetic talent in young people, and it proposed that all measures of student progress be tied to incremental successes, rather than to some abstract, teacher-imposed ideal. *Émile* also insisted that arts education complements and enhances training in the sciences, mathematics, logic, and so on, by infusing them with the love of life and vicissitude that the arts inspire. As Rousseau saw it, the arts, no less than the sciences, supply a foundation on which young people may develop capacities for living happy, fulfilling lives untouched by the corrupting influence of modern civilization.

Immanuel Kant, as is well known, read *Émile* and was profoundly affected by it. Even though he wrote nothing resembling a treatise on education for young people, there can be no denying that his ideas about aesthetic experience and its connection with general intelligence have had a monumental effect on subsequent educational theory. Three Kantian ideas have seemed particularly salient to later generations of art educators. First, knowledge in all areas grows through stages. What is appropriate moral advice for the two-year-old is a far cry from what is appropriate for the fifty-two-year-old. Second, these stages reach a point of transition where youth becomes mature. A civilized adult, as Kant sees it, must accept the autonomy and responsibility that come from taking up the task of acting on universally applicable principles. Third, aesthetic awareness is both mysteriously subjective and decisively communal in that it depends on individual responses (of the imagination and understanding), on the one hand, and implies universal assent to aesthetic judgments, on the other. It was left to neo-Kantian thinkers to bring these three ideas together in laying a platform for modern aesthetic education theory. One way of describing this platform is to say that it understands the acquisition of aesthetic knowledge as incremental, reaching a stage ultimately when the subjective sensibilities of the percipient agent are connected to those of others through the demands of socialization. At that stage, the glue that (suggestively, if imperfectly) holds together all the subjective parts into an objective whole is the seductive notion that somehow (and in the appropriate ways) we are all alike under the skin.

However much the post-Kantian world has disputed the prospects of convergent aesthetic judgment and the demarcation of a line of aesthetic maturity, it has not abandoned the fundamental ideas of growth and staged advancement developed in Kant's theory. These ideas were, of course, recast by the next generation of philosophers and art educators, largely under the influence of G. W. F. Hegel and subsequent Hegelians, who were disposed to see all intellectual progress as connected with the ultimate fulfillment of spiritual goals—goals whose attainment would, ironically, render the educational role of the arts nugatory. While turn-of-the-

century American thinkers were willing to accept many aspects of the Hegelian worldview, they generally disavowed its cynical estimate of art's ultimate place in the progress of civilization. It was Dewey's work that finally succeeded in welding the spirit of Kantian and Hegelian thought to the practical concerns of American art educators. Dewey's fundamental idea was that art and aesthetic experience are of fundamental importance in the education of the young just because they are the most salient articulations we have of the natural, continual development of human life. Dewey was, of course, a consummate observer of children. He noticed that young children draw no distinction between work and play, and suggested that when the two come together productively in later life, we are inclined to call that happy amalgamation art. He supposed that similar qualities of fusion impart value to all of life's experiences, and that value is most intense when those experiences are patterned in such a way as to draw infancy, youth, and maturity together into an intelligible whole.

Later American thinkers who have reflected on the place the arts should occupy in the general educational scheme have all, in one way or another, sought to work their way out of Dewey's shadow by drawing on evidence accumulated in the new sciences (particularly psychology and sociology) and theories of cognition and value developed by modern philosophers. The recent history of these developments is well documented in Ralph A. Smith's companion essay here and need not be rehearsed. [*See the article on* Contemporary Aesthetic Education, *below.*] Among important major latter-day amendments to Dewey's syncretic views are two that particularly reflect the changed educational environment of post-mid-century America. The first is the observation that increased awareness of idiosyncratic differences in individuals militates against any easy connections between personal and social meaning the arts might hitherto have been supposed to suggest. The second is the observation that increasing cultural diversity demands a broadening, or even a fracturing, of the account we make of the society that reflects art's effects. Current controversies and revisionisms notwithstanding, the consensus of philosophers of all periods who have considered the question is remarkably strong. Simply put, it is that the arts occupy a place of great importance in the educational process precisely because they, more than any other topics of study, transcend limitations of time, place, and personality to reveal what human beings and their societies are and may be.

The Role of Aesthetics in Education. From the eighteenth to the mid-twentieth century, little attention was paid to the idea that philosophical aesthetics might be made part of standard courses of training in the arts, let alone part of the arts component of general education for young people. This was due to the fact that, throughout this period, aesthetics evolved as a relatively abstruse discipline within value theory, communicated in writings produced by adults for adults, and transmitted mainly between professional philosophers. It must have seemed to most practitioners of this esoteric craft that there was as little to be gained by trying to teach aesthetics to artists (let alone would-be artists) as there was to be gained by teaching philosophical theories of time to watchmakers. In the later twentieth century, however, three important developments have cooperated to make the prospects of introducing aesthetics into programs of arts education more plausible: (1) a series of concerted efforts initiated by arts educators (with the support of major foundations) aimed at improving the overall quality of American arts curricula by incorporating within them theoretical and critical subject matter to complement the traditional studio, or productive, activities; (2) a small, but important, movement within philosophy to introduce philosophical subject matter to school children; and (3) the creation of new meeting grounds, both in professional journals and in the meetings of professional associations, for interdisciplinary communication involving aestheticians, teachers, educational theorists, school administrators, psychologists, and others.

Efforts to reform arts education in America accelerated sharply in the 1980s, when various pilot programs conclusively demonstrated that aesthetics, as well as art history and criticism, could be usefully integrated into classroom arts instruction, yielding favorable results. At the same time, as part of interdisciplinary "thematic organization" initiatives, the enriched arts programs were shown to be highly suitable for integration with other disciplines throughout the precollegiate curriculum. Leaders in this movement included "Discipline-Based Art Education," a program developed by the Getty Center for Education in the Arts; "Project Zero," launched by Howard Gardner and the Harvard Graduate School of Education; "Arts Propel," developed in Pennsylvania and subsequently implemented successfully throughout the eastern seaboard; the Kettering Project at Stanford University; and several regional programs, such as the California Arts Project, the Tennessee Arts Academy, and the Aesthetic Education Project at the Central Midwestern Regional Education Laboratory. These various programs broadly share the perception that aesthetics becomes a vital component of enriched arts education as it is deployed to explore meanings and values throughout all facets of aesthetic experience, rather than positioning itself as competing with other arts disciplines for student interest. Indeed, these programs have generally succeeded in showing that, far from requiring special efforts at importation, aesthetics is nearly impossible to keep out of any reflective discussion of artistic process, creativity, appreciation, perception, and related topics, even where these are presented at very early stages of education. Illustrations of the effective integration of aesthetics with other elements in arts education are provided in the Getty Center's *Curriculum Sampler,* a volume created and field-tested by school teachers at all levels of precollegiate education.

Although the idea that school children can profitably be taught philosophy is hardly novel (and, again, has its roots in Plato), it is only very recently that this idea has been systematically explored and implemented. Since 1969, the leading force in this effort has been the Institute for the Advancement of Philosophy for Children, founded by Matthew Lipman at Montclair State College. This institute has been instrumental in the production of teaching manuals, teacher training programs, and strategic plans for stimulating young peoples' native philosophical inquisitiveness. Although aesthetics has not been a major focus of the institute's pedagogical program, several of its students have been arts instructors, and they have reported considerable success in using its techniques to introduce philosophy of art to school children. Apart from the work of Lipman and his colleagues, the efforts of a growing number of professional philosophers to engage children in philosophical discussion have proven very productive. Particularly impressive results have been reported by Gareth Matthews in *Philosophy and the Young Child* (1980) and *Dialogues with Children* (1984). Following the lead of these practical experiments, a number of aestheticians (among them Marcia Muelder Eaton, Michael J. Parsons, Cynthia Rostankowski, Robert Sitton, and Marilyn Stewart) have successfully taken aesthetics into the schoolroom, finding, on the whole, that doing so has substantially enlivened and deepened the students' engagement with artworks and artistic practices. Representative approaches and results are surveyed in Michael J. Parsons, *How We Understand Art* (1987); Howard Gardner, *Art Education and Human Development* (1990); Bennett Reimer and Ralph A. Smith, eds., *The Arts, Education, and Aesthetic Knowing* (1992); and Michael J. Parsons and H. Gene Blocker, *Aesthetics and Education* (1993).

It is likely that separate efforts by art educators and philosophers to make aesthetics part of arts instruction for young people would have foundered, or dissipated their energies in competition over pedagogical domain, were it not for the emergence of several forums where advocates of this process from various disciplines could exchange ideas in print or in person. The most important print forum of this kind is the *Journal of Aesthetic Education*, founded in 1966. Since its inception, this publication has been strikingly ecumenical in its editorial policy, welcoming participation from a wide array of professionals concerned with aesthetics and art education, and promoting new reflection on the prospects of instruction in aesthetics for the young. The Getty Center for Education in the Arts has undertaken an ambitious publication program that, while predominantly concerned with the center's own interdisciplinary approach to arts education ("Discipline-Based Art Education"), has given attention to a broad array of serious work on educational reform and interdisciplinarity. The Getty Center has also sponsored an important series of volumes, Disciplines in Art Education: Contexts of Understanding, under the general editorship of Ralph A. Smith, in which prospects for enrichment in arts education are explored by pairs of authors from differing disciplinary backgrounds. An encouraging indication of cooperation among parties engaged in various dimensions of aesthetic education is the effort forged between the American Society for Aesthetics, the *Journal of Aesthetic Education,* and the National Art Education Association, whose first publication was *Aesthetics for Young People* (Moore, ed., 1995). In addition to the opportunities for communication provided in these various published forums, scholars concerned with these issues have at their disposal a widening array of institutional meeting places. An increasing amount of meeting time at conferences of the various national and regional arts associations has been devoted to these issues, and in the late 1990s the American Society for Aesthetics created two standing committees to deal with issues of aesthetic education.

Beyond Schools: Aesthetics and Education for the Good Life. Not all of aesthetic education takes place in the schoolroom, of course. Perhaps, on balance, less does than doesn't. It is thus incumbent on those who undertake to examine the topic thoroughly to consider features of aesthetic learning that occur through various extracurricular means throughout life. Here again, it is difficult to overestimate the importance of Dewey's contribution to the modern perspective on such issues. His notion of experience as a vivid, consummatory interaction between a whole individual and that individual's environment can easily be recognized as the seedbed of many modern-day accounts that stress the importance of lived experience in supplying the arts with meaning and value. Monroe C. Beardsley and the many aestheticians who were influenced by him sustained and extended this thought in various ways, always stressing the felt gratification that arises out of experiencing a thing (or an episode) completely, taking full stock of its regional qualities as well as its formal unity. The resulting perspective pertains as much to the aesthetic apprehension of grandeur in mountain landscapes as it does to the artful composition of painted landscapes, as much to the art of living as to the life of art. This development has suggested to a number of aestheticians the close affinity of aesthetic value to other forms of value from which it is normally thought to stand apart, such as environmental integrity, psychological well-being, and certain moral concerns. In the view of these theorists, aesthetic education takes on great importance in that it cannot be detached from the learning of fundamental lessons in good living as these are drawn together from various directions.

No doubt the leading present-day exponent of such a view is Marcia Muelder Eaton. In *Aesthetics and the Good Life* (1989), she argues at length that aesthetic responses to objects, events, and other people are partially constitutive of what it means to lead a good and rational life. The patterns one discerns through the intelligent aesthetic appreciation

of all sorts of phenomena (natural as well as artistic) invite translation into the moral sphere as models of the patterning through which we organize the details of life into meaningful wholes. Some of life's patterns are drawn from culture and traditions; some are constructed through the creative application of learned instincts for wholeness and coherence. But, however they are composed, it is through the invocation of these patterns that life is aestheticized and those features of experience that are worthy of attention and reflection become objects of delight.

If Eaton's view, or something like it, is right, aesthetic education is destined to gravitate to the center of importance both in and out of schools, for the clear implication of this position is that instruction in aesthetics is always, at the same time, instruction in life skills conducive to living a rich and satisfying life. This is perhaps nothing more than the modern-dress version of Plato's observation that the ultimate art (in a sense that fully conflates moral and aesthetic senses of the term) is the art of living well.

BIBLIOGRAPHY

Alexander, Kay, and Michael Day, eds. *Discipline-based Art Education: A Curriculum Sampler.* Santa Monica, Calif., 1991.

Beardsley, Monroe C. *The Aesthetic Point of View: Selected Essays.* Edited by Michael J. Wreen and Donald M. Callen. Ithaca, N.Y., 1982.

Dewey, John. *Art as Experience.* New York, 1934.

Eaton, Marcia Muelder. *Aesthetics and the Good Life.* Rutherford, N.J., 1989.

Gardner, Howard. *Art Education and Human Development.* Los Angeles, 1990.

Lankford, E. Louis. *Aesthetics: Issues and Inquiry.* Reston, Va., 1992.

Matthews, Gareth B. *Philosophy and the Young Child.* Cambridge, Mass., 1980.

Matthews, Gareth B. *Dialogues with Children.* Cambridge, Mass., 1984.

Moore, Ronald, ed. *Aesthetics for Young People.* Reston, Va., 1995.

Parsons, Michael J. *How We Understand Art: A Cognitive Developmental Account of Aesthetic Experience.* Cambridge and New York, 1987.

Parsons, Michael J., and H. Gene Blocker. *Aesthetics and Education.* Urbana, 1993.

Plato. *Republic.* See 2.376–379; 3.392C–403; 7.530C–531; 10.

Reimer, Bennett, and Ralph A. Smith, eds. *The Arts, Education, and Aesthetic Knowing.* Chicago, 1992.

Rousseau, Jean-Jacques. *Émile: or, On Education.* Translated by Allan Bloom. New York, 1979.

RONALD MOORE

Contemporary Aesthetic Education

To place modern and contemporary philosophical writing about aesthetic education in historical context, it is instructive to understand it as reflecting two senses of the term: aesthetic education as education in the arts, and, in a broader sense, as the aesthetic dimension of all human experience. Both senses of the term are discernible in the writings of Friedrich Schiller (1759–1805) whose reflections on aesthetic education stand at the threshold of the modern era. Accordingly, after summarizing Schiller's basic idea, this article will refer to representative philosophies of art, education, and arts education that illustrate one or both senses of the term. Since the first sense of the term, aesthetic education as education in the arts, is the most prominent in the literature, writings exemplifying the second sense receive slight mention. Only a rough chronological order of writings is indicated, that is, earlier and later twentieth-century thinking on the topic.

In *On the Aesthetic Education of Man in a Series of Letters* (1793–1795), Schiller, a dramatic poet living during a time of revolutionary upheaval, analyzed the dilemma of necessity and freedom in terms of three basic human drives: the sensuous impulse *(Stofftrieb)* and the formal impulse *(Formtrieb)* whose divergent aims are synthesized by the play impulse *(Spieltrieb)* into living form *(Lebensform).* The play impulse is Schiller's version of Immanuel Kant's harmony of the imagination and the understanding, and the task of cultivating such harmony fell to aesthetic education, the goal of which was nothing less than the realization of human freedom, rationality, and moral judgment. Monroe C. Beardsley remarked that no one has ever staked out a higher claim for aesthetic education, and later writers on the subject have had more modest expectations.

It is convenient to begin a brief review of contemporary thinking about aesthetic education with the writings of Harry S. Broudy, whose article "Some Duties of an Educational Aesthetics" (1951) stated the problems any philosophy of aesthetic education must address: accounting for the marginal status of arts education in the schools, explaining the aesthetic mode of experience, articulating the nature of critical standards, and justifying aesthetic education. In a later essay, *Enlightened Cherishing* (1972), Broudy expressed the belief that aesthetic experience serves individuals well in their quest for a good and worthwhile life and held that participation in such a quest presupposes the capacity to derive satisfaction from works of serious art that express the meaning of the more complex and subtle forms of human experience. Aesthetic education thus has the task of providing a context for learning the skills of artistic expression and aesthetic impression. Broudy further took aesthetic education to be fundamentally important because the perception of works of art contributes to a rich imagic store that energizes both the perception of works of art and the interpretation of a large range of other phenomena. He envisioned a society in which the large majority of persons express a love of art that is justified by knowledge. Broudy's ideas influenced a number of major research and development ventures, including those of CEMREL (Madeja and Onuska, 1977), a regional educational laboratory, and the Getty Center for Education in the Arts (1993).

Also writing at mid-century, Thomas Munro, in *Art Education: Its Philosophy and Psychology* (1956), stated that the aims of aesthetic education were not only the fostering of the artistic and aesthetic strains of human experience and

the transmitting of the cultural heritage of great art, but also preparing youth for citizenship and a vocation. Munro's ideas were implemented in the educational programs of the Cleveland Museum of Art but did not substantially influence subsequent thinking about arts education.

Among prominent writers has been Maxine Greene, whose "Aesthetic Literacy and General Education" (1981) encapsulates the content of numerous articles and essays. Greene attempts to define the unique pleasure people seem to seek from works of art but are often prevented from fully realizing by their inadequate knowledge and lack of skill. She defines aesthetic literacy as the capacity to actualize a range of benefits artworks are capable of conferring, for example, a greater perceptual awareness of the qualities and meanings of works of art, an expanded imagination, an enhanced appreciation of ordinary life and natural phenomena, and an enlarged sense of personal freedom. These values may also help ameliorate the stringent technological imperatives of modern life and deflate perennial stereotypes. Greene stresses that since aesthetic literacy implies an understanding of art and mastery of requisite interpretative skills, to cultivate it teachers of art should have a grasp of the philosophy of art and be able to engage works critically. But, above all, teachers must be sensitive to the subjectivity of individual experience. Competency and sensitivity in the aesthetic domain will, she believes, enable persons to live more vivid and intense lives. Greene has also provided philosophical guidance for the education institute of the Lincoln Center for the Performing Arts and its satellite programs.

As an example of a philosopher of art who occasionally addressed educational questions, Beardsley, in essays reprinted in *The Aesthetic Point of View* (1982) extrapolates from his instrumental theory of aesthetic value to questions about educational aims and policy. In considering national policy objectives, he formulates a concept of aesthetic welfare that consists of the sum of all aesthetic experiences in the society at a given moment or over a stretch of time. The realization of aesthetic welfare is dependent on the existence of aesthetic wealth, which is the totality of all aesthetically worthwhile objects as potential sources of welfare. In a democratic society there must also be aesthetic justice, which is the opportunity of all to participate in the life of culture; aesthetic capacity, defined as the ability to benefit from aesthetic experiences; and aesthetic auxiliaries, who are teachers, cultural service workers, and others concerned with providing access to aesthetic value. In his discussion of the justification question, Beardsley suggests that each of the arts may have its own inherent values; for example, sympathetic understanding of others is likely to be more characteristic of literature than of music.

Toward the end of the eighties, in *An Aesthetics for Art Educators* (1989), E. F. Kaelin also philosophically argues that art should be prized for the sake of the benefits that accrue when persons accept the demands of an aesthetic situation. Such a situation requires openness to the aesthetically expressive surfaces of artworks and a willingness to permit imagination and response to be controlled by their counters. In addition to reaping the rewards of intensified and clarified experiences, individuals also receive help in choosing their futures. They may not only create new worlds of aesthetic value but also avail themselves of the possibilities of experience imagined by others. Kaelin further claims important social outcomes for aesthetic education. It can contribute to the effective operation of the institution of art (the art world) whose function is to secure the place of aesthetic value in society, and, because of its insistence that students open themselves to and interact with the unique features of artworks, it cultivates in them an appreciation of free communication that stands in contrast to the imposition of ideological frameworks. Aesthetic education therefore can contribute to the maintenance of a society that values human communication, tolerance, judgment, and freedom.

In providing a humanities justification of arts education in a number of works, for example, *The Sense of Art* (1989) and "Toward Percipience: A Humanities Curriculum for Arts Education" (1992), Ralph A. Smith, an educational theorist, holds that inasmuch as any well-developed sense of art presupposes a knowledge not only of artistic creation and performance but also of aesthetics, art history, and art criticism, teachers of the arts should strive to achieve a degree of knowledge in each of these disciplines. Smith takes works of art to be artistic statements that are distinctive for their dramatic forms of communication, history of accomplishment, and exemplification of the civilizing impulse to impose significant form on unshaped material. Works of art are particularly valuable for their capacity to yield the constitutive and revelatory values of aesthetic experience. *Reflective percipience* is the term used to designate the capacity to traverse the world of art with intelligence and sensitivity.

In *Essays in Aesthetic Education* (1990), David Swanger discusses the relations of art, ideology, and society in a democratic society. He understands ideology as a set of beliefs that stubbornly resists change and thinks that art's inherently radical and destabilizing power enables it effectively to exert pressure on the status quo, not by didactic instruction or moralizing, but through its freshness, creativity, and innovativeness. By involving students in artistic activities under the guidance of a practicing teacher-artist, aesthetic education can encourage them to seek and appreciate aesthetic values in other areas of life. As a secondary goal, Swanger suggests aesthetic education can move a materialistic consumer society toward a more conservation-conscious way of living he calls *pro*sumption. He cautions that an aesthetic education that overemphasizes cognitive studies at the expense of creative activities is less likely to achieve the objectives he envisions.

The idea of music education as aesthetic education has received its most systematic treatment from Bennett Reimer, whose justification of aesthetic education features a notion of aesthetic cognition as sui generis. In "What Knowledge Is of Most Worth in Arts Education?" (1992), Reimer states that developing the capacity of aesthetic knowing in its creating, performing, and responding aspects constitutes the goal of aesthetic education. Aesthetic cognition deserves a place among the ways of knowing because it provides access to works of art that are exemplary expressions of subjective reality. Believing that the structures and forms of artworks capture the internal dynamics of feeling in ways other modes of knowing cannot, Reimer identifies four kinds of knowing: two that constitute the essence of aesthetic knowing (knowing of or within and knowing how) and two auxiliary kinds (knowing that and knowing why). Music education is most effective when it encourages and improves musical reaction, enhances musical creativity, increases substantive understanding of music, and, ideally, disposes the young to value music.

Michael J. Parsons and H. Gene Blocker in *Aesthetics and Education* (1993) combine educational and philosophical interests in explaining how aesthetics can contribute to the goals of understanding and appreciating art, especially within the context of discipline-based art education, an approach to teaching art identified with the Getty Center for Education in the Arts, which stipulates that content for teaching art should be drawn from the four disciplines of art making, art history, art criticism, and aesthetics. Aesthetics helps art teachers by clarifying aesthetic concepts and conundrums and leads students to develop their own views about the nature, meaning, and value of art—in short, a theory of art. In discussing multiculturalism, modernism, and postmodernism, the authors reject extreme views in favor of a balanced understanding that appropriates the most useful ideas associated with each. The authors defend both the aesthetic as a separate category and criteria of excellence in art. Further points the authors emphasize are the importance of the arts for understanding the life of the emotions and the need for teachers to take into account stages of aesthetic development. The collaboration of a philosopher and an educational theorist is a feature of a series of five volumes titled "Disciplines in Art Education: Contexts of Understanding" that brings together specialists in the disciplines and education (University of Illinois Press, 1991–1997).

The foregoing sample of representative writings about aesthetic education, which reflects both Anglo-American and Continental traditions of philosophizing, centers primarily on the teaching of the arts, with occasional side glances at the aesthetic qualities of ordinary life and nature. Another literature, which reveals the strong influence of Schiller, Herbert Read (1956), and John Dewey (1934), understands the nature of aesthetic education more broadly.

For example, in *Philosophy of Education* (1967), Donald Arnstine assimilates learning in general to aesthetic experience; in *The Enlightened Eye* (1991), Elliot W. Eisner finds important similarities between connoisseurship and art criticism and educational criticism; and in *Learning By All Means: Lessons from the Arts* (1992), V. A. Howard argues that aesthetic education is basically the education of the sensibilities, particularly with regard to the refinement of imagination and expression, whatever the subject or context. In an earlier volume of invited essays titled *Aesthetic Concepts and Education* (Smith, ed., 1970), authors writing from different philosophical perspectives examine a number of aesthetic concepts (medium, creativity, imagination, style) for the light they might throw on educational concepts and settings.

In conclusion, the larger body of writings about aesthetic education tends to assume the development of aesthetic literacy in one form or another to be the general goal of aesthetic education. Yet the concepts of the aesthetic and aesthetic experience have in recent years been subjected to scrutiny, and questions have been raised about the coherence or usefulness of the idea of experience having a special aesthetic character. It is doubtful, however, if any viable concept of aesthetic education can get along without some coherent notion of aesthetic experience. That the concept is still on the agenda of aesthetics is evident in *Possibility of the Aesthetic Experience* (Mitias, ed., 1986) and in Marcia Muelder Eaton's argument in her *Aesthetics and the Good Life* (1989). But whatever the fate of the concept of aesthetic experience, a number of features of the contemporary theoretical landscape bode well for both aesthetics and aesthetic education, namely, the founding of the *Journal of Aesthetic Education* (1966), which has been hospitable to philosophical discussions of art, aesthetics, and aesthetic education; the increasing volume of writings on the subject, for example, *Aesthetics and Arts Education* (Smith and Simpson, eds., 1991) and *Aesthetics for Young People* (Moore, ed., 1995); and the creation of a committee on education within the American Society for Aesthetics.

BIBLIOGRAPHY

Arnstine, Donald. *Philosophy of Education*. New York, 1967.

Beardsley, Monroe C. "The Aesthetic Problem of Justification" (1966) and "Aesthetic Welfare, Aesthetic Justice, and Educational Policy" (1973). In *The Aesthetic Point of View: Selected Essays*, edited by Michael J. Wreen and Donald M. Callen, pp. 65–76, 111–124. Ithaca, N.Y., 1982.

Broudy, Harry S. "Some Duties of an Educational Aesthetics" (1951). In *Aesthetics and Problems of Education*, edited by Ralph A. Smith, pp. 77–106. Urbana, 1971.

Broudy, Harry S. *Enlightened Cherishing: An Essay on Aesthetic Education*. Urbana, 1972.

Dewey, John. *Art as Experience*. New York, 1934.

Eaton, Marcia Muelder. *Aesthetics and the Good Life*. Rutherford, N.J., 1989.

Eisner, Elliot W. *The Enlightened Eye*. New York, 1991.

Getty Center for Education in the Arts. *Improving Visual Arts Education: Final Report on the Los Angeles Getty Institute for Education on the Visual Arts (1982–1989)*. Santa Monica, Calif., 1993. Getty Center renamed Getty Education Institute for the Arts in 1996.

Greene, Maxine. "Aesthetic Literacy and General Education." In *Philosophy of Education*, edited by Jonas F. Soltis, pp. 115–141. Chicago, 1981.

Howard, V. A. *Learning By All Means: Lessons from the Arts*. New York, 1992.

Journal of Aesthetic Education. Urbana, Ill., 1966–.

Kaelin, E. F. *An Aesthetics for Art Educators*. New York, 1989.

Madeja, Stanley S., and Sheila Onuska. *Through the Arts to the Aesthetic: The CEMREL Aesthetic Education Curriculum*. St. Louis, 1977.

Mitias, Michael H., ed. *Possibility of the Aesthetic Experience*. Dordrecht and Boston, 1986.

Moore, Ronald, ed. *Aesthetics for Young People*. Reston, Va., 1995.

Munro, Thomas. "The Aims of Aesthetic Education." In *Art Education: Its Philosophy and Psychology*. New York, 1956.

Parsons, Michael J., and H. Gene Blocker. *Aesthetics and Education*. Urbana, 1993.

Read, Herbert. *Education through Art*. 3d ed. New York, 1956.

Redfern, H. B. *Questions in Aesthetic Education*. London, 1986.

Reimer, Bennett. "What Knowledge Is of Most Worth in Arts Education?" In *The Arts, Education, and Aesthetic Knowing*, edited by Bennett Reimer and Ralph A. Smith, pp. 20–50. Chicago, 1992.

Schiller, Friedrich. *On the Aesthetic Education of Man in a Series of Letters (1793–1795)* Edited and translated by Elizabeth M. Wilkinson and L. A. Willoughby. Oxford, 1967.

Smith, Ralph A., ed. *Aesthetic Concepts and Education*. Urbana, 1970.

Smith, Ralph A. *The Sense of Art: A Study in Aesthetic Education*. New York and London, 1989.

Smith, Ralph A., and Alan Simpson, eds. *Aesthetics and Arts Education*. Urbana, 1991.

Swanger, David. *Essays in Aesthetic Education*. San Francisco, 1990.

RALPH SMITH

EISENSTEIN, SERGEI MIKHAILOVICH (1898–1948), Russian theater and film director. Sergei Eisenstein was born in Riga (then part of the Russian Empire) into the family of a civil engineer and architect. In 1915, Eisenstein joined the Institute of Civil Engineering in Petrograd; in 1918, he volunteered for the Red Army, where he took part in army theater activities. Demobilized in 1920, Eisenstein studied directing at Vsevolod Meyerhold's theater workshop and continued theater work for the Proletkult theater in Moscow. In the three years following 1921, Eisenstein became well known as a left-wing theater director and designer, but in 1924 he left the stage for film. Four films produced during the 1920 made Eisenstein an international celebrity. Between 1930 and 1932, he toured Europe and America, but none of Eisenstein's film projects of this time were completed. When he returned to Moscow in 1932, he became active in teaching film production, while his own production projects met with resistance on the part of party authorities. Two historical epics released in 1939 and 1945 helped Eisenstein regain official recognition, but part 2 of his last film was shelved for political reasons and was only released in 1958, ten years after Eisenstein's death.

Stored in six volumes of selected writings and seven films, Eisenstein's aesthetic legacy is best understood when situated in terms of the relationship between its theory and practice. This relationship was a two-way street: while much of Eisenstein's theory emerged as a result of post analyses of his own films, the latter always contained sequences testing theoretical ideas formulated in manifestos (Eisenstein in the twenties) or in the process of teaching (Eisenstein in the thirties and forties). At the same time, his aesthetics was the opposite of an applied theory: even when triggered by a minor issue of film form, his articles often wound up discussing universal "laws" of art and culture. Eisenstein enjoyed writing and teaching, and, as Jay Leyda testifies, at difficult moments in life (for example, around 1936) he frankly stated that he preferred research to making films. Modeled on Leonardo da Vinci, Eisenstein's image hovers between that of an artist and that of a thinker, and fully makes sense only with both identities in view.

Eisenstein's vision of an artistic career was predicated on change. Even before he embarked on his first feature, Eisenstein speculated on the nature of artists' biographies, subdividing them into "evolutionary" and "revolutionary" types: the former was based on the unfolding and expanding of the same set of artistic principles (Eisenstein's example: Lev Kuleshov), whereas the latter had a trajectory of permanent self-denial (Meyerhold). While Eisenstein the filmmaker clearly adhered to the second type (in his entire work, no two films are alike), as a theorist he appears more consistent. Beneath the stream of changing subjects and references one discerns a few aesthetic principles Eisenstein never tired of reasserting.

Form as Process. One such recurring idea was that form in art should be conceived as a process rather than as a mold. He used this distinction to account both for the artist's creative work (for example, according to Eisenstein, the art of drawing, like dance, is that of movement, not space) and, more important, for the way art works (or should work) on the viewer.

In 1923 Eisenstein published an essay titled "The Montage of Attractions," in which he argued that the task of the (theater) director is to process the spectator through a preplanned series of "shocks" regardless of whether the play performed really occasioned them. Some sequences in *Strike* (1925) were designed to sound this theory by way of downplaying the film's narrative coherence (= form) and favoring "the montage of film attractions" (= process). The further vicissitudes of this tenet took Eisenstein the thinker to his famous theory of "intellectual cinema," according to which editing operates through the collision (process) rather than linkage (form) of neighboring shots. In *October* (1928), Eisenstein tested this principle hoping to generate "abstract concepts" by way of juxtaposing artifacts found in the Winter Palace (icons, Fabergé eggs, Rodin's sculpture, Napoleon statuettes, a clockwork peacock) and histor-

ical figures of 1917. His next step was to propose that "montage as form replicates the process of human thinking," which took Eisenstein the filmmaker to a few stream-of-consciousness sequences in (unrealized) films of the thirties. During the last decade of his life, Eisenstein developed his "Pathos and Ecstasy" theory, which defined how, exactly, the process (called "dialectics") of art form works to achieve a specific type of emotional involvement (called "pathos") necessary to jolt the spectator "out of the stasis" of everyday experience. The state of "ecstasy" thus achieved deletes the boundary between the "self" and "others." This theory Eisenstein would typically illustrate with the "Odessa steps" sequence from his earlier *Battleship Potemkin* (1925), in which the images of the descending firing squad and those of the massacred crowd interact to create the implacable "montage rhythm" due to which the "jumping sculpture" of the stone lion that crowns the sequence is perceived not as a joke but as a high point of tragedy.

Conflict. For Eisenstein, collision was not simply a method of montage but the principle of art in general. In the twenties, he found conflict everywhere. Unlike most of modernist acting reformers of the twenties (Meyerhold, Kuleshov, Nikolai Forreger), Eisenstein's idea of expressivity foregrounded conflict rather than accord within the moving body (gravity vs. effort, volition vs. inertia, etc.). As musical analogy gained weight in Eisenstein's later aesthetics, he would instead use the term "counterpoint" to stress that conflict was a constituent of dynamic unity. Art separates formal elements from their everyday carriers (he called this operation "de-anecdotization")—this idea was key to Eisenstein's theories of sound and color in cinema. For the artist, an element becomes instrumental only when separated from its real-life connections. The creaking of boots will work emotionally when we do not see the boots on the screen. Otherwise black-and-white, *Ivan the Terrible* (1944/46) suddenly acquires color in its last sequence—similarly, with no apparent motivation, a character's face turns blue at the high point of tragedy. Separated from its source, sound becomes a matter of artistic choice; when color is not predicated by the object, but is chosen by the filmmaker, true color cinema is born.

Organic Mimicry. Another credo that runs through Eisenstein's career is that, subliminally, the beholder responds to art with his or her entire body and, if this micro-muscular response is cued correctly (= the purpose of art), the right emotion is born.

Borrowed from Meyerhold's "biomechanics" (Meyerhold had, in turn, derived it from behaviorism and experimental psychology), and initially manifested in Eisenstein's conception of actors' "expressive movement" being motorically imitated by all members of the audience, this tenet was soon extended to a theory about how formal properties of film style immediately translated into the viewers' "organic response" to, say, the "structure of pathos." As Eisenstein assimilated post-Freudian psychoanalysis (triggered in 1918 by Sigmund Freud's work on Leonardo, an interest in psychoanalysis is felt in his entire body of work), the concept of organic mimicry regained some of its initial anthropomorphic implications. The intensely claustrophobic set design for *Ivan the Terrible* owes much to Otto Rank's "birth trauma" theory, not only in its use of labyrinth-like corridors, but also in Eisenstein's endeavor to employ various audiovisual means (conelike passages, rounded corners, low-key lighting, "spasmodic" musical accents) in order to help the viewer relive his or her own "prenatal experience" believed to dwell in deeper layers of one's unconscious memory.

Polyphony. Unlike most other artists and theorists gathered around LEF (Left Front of Arts), who would insist on exclusive use of "correct" material or techniques (Dziga Vertov, Osip Brik), in his first manifesto, "The Montage of Attractions" (1923), Eisenstein stated the primacy of impact over devices: because the basic material of theater is the audience itself, the correct means are those that lead best to the desired effect. Late in the twenties, Eisenstein's aesthetic pluralism showed up in his proposed typology of montage: metric, rhythmic, tonal, and overtonal, the last one being superior to the first three because, rather than operate through a single dominant, it works on the viewer through multiple "stimulants" of the shot. The principle of polystimulant impact was reinforced by the Japanese Kabuki theater during whose visit to Moscow in 1928 Eisenstein was struck by what he termed a "monistic ensemble": a technique of acting and staging based on a precisely doled-out impact by means of several sensory channels (e.g., acoustic and visual) in relay fashion rather than simultaneously. This experience proved handy when Eisenstein found himself faced with the problem of sound. He envisioned sound cinema as a promising medium for the stream-of-consciousness technique that he saw as a monistic ensemble of senses (images without sound, sounds without images, words alone, music alone, etc.). A firm believer in synaesthesia, Eisenstein argued that sounds and images could thus interact in "vertical montage" creating a "polyphony" of multisensory impacts, a theory he tested in *Alexander Nevsky* (1938) and developed into an audiovisual polyphony in *Ivan the Terrible*, one of the most ambitious ventures in all of film history.

The Basic Dilemma. In the mid-twenties, influenced by Aleksandr Veselovsky's "historical poetics" and James Frazer's studies in mythology, Eisenstein acquired his interest in "primitive" cultures first manifested in *Old and New* (1927–1929). Although set in the Soviet countryside, it reflected Eisenstein's vision of primeval mentality ("the old") residing in the midst of "the new" (modern civilization). In the remaining twenty years of his career, this juxtaposition was transformed into a basic aesthetic credo that Eisenstein

claimed he also experienced as the artist's moral dilemma. This new theory dates back to the early thirties, when Eisenstein's interest in "primitive" cultures merged with his concept of film form as replicating the syntax of human thought and that of "inner speech." Drawing from works of Soviet psychologists Lev Vygotsky and Aleksander Luria, Eisenstein envisaged "inner speech" as asyntactic and image-oriented (the characteristics that also allowed Eisenstein to identify "inner speech" with cinematic discourse) and posited its origins in the "prelinguistic" (ergo: emotional, undifferentiating) stage of a child's development. Always convinced that phylogenetic traits provided clues to ontogeny, Eisenstein joined "inner speech," infantile speech, and cinematic discourse into one cluster with a similar vision of the infancy of human culture mainly derived from his reading of Lucien Lévi-Bruhl and the Soviet linguist Nikolai Marr. Marr's ideas of the language of gestures preceding (and being traceable in) verbal languages and Lévi-Bruhl's concept of "prelogical" thinking prevalent in early cultures yet also latent in modern mentality led Eisenstein to posit a hidden presence of "sensuous thought" within the predominantly rational modern culture. In this area, art (and, first of all, the art of cinema) operates by way of immobilizing "logical" (rational, differentiating) mechanisms of thought and reviving "prelogical," irrational, and undifferentiated perceptions. Much of this philosophy resulted from Eisenstein's fascination with Mexico, which he portrayed in his unfinished film *Que viva Mexico!* (1931–1932) as a country where modern and ancient layers of culture coexist (a fascination reminiscent of *Old and New*). In a more premeditated form, these preoccupations went into his next project *Bezhin Meadow* (1935–1937), which was officially discontinued for packing a politically engaged narrative with mythological connotations. Around that time Eisenstein started thinking of the "sensuous thought" theory as the artist's basic dilemma (which he used to call *Grundproblem* using this German term also as an interim title for a half-finished aesthetic treatise). The dilemma was one between the artist's conscious self striving for "progress" and building a better life along with the whole (socialist) society, and the nature of art, which makes the artist wake up "regressive forces" dormant in us whether by way of recreating our "prenatal experience" (weightlessness, protean ever-changing forms, embryonic androgyny) or by cuing them through the very structure of the work ("pathos and ecstasy").

[*See also* Film, *article on* Film Theory; *and* Russian Aesthetics.]

BIBLIOGRAPHY

Works by Eisenstein

Drawings. Moscow, 1961.
Eisenstein 2: A Premature Celebration of Eisenstein's Centenary. Edited by Jay Leyda, translated by Alan Y. Upchurch et al. Calcutta, 1985.
Eisenstein at Work. Edited by Jay Leyda and Zina Voynow. New York, 1982.
Eisenstein on Disney. Edited by Jay Leyda, translated by Alan Y. Upchurch. Calcutta, 1986.
Film Essays and a Lecture. Edited and translated by Jay Leyda. London, 1968; reprint, Princeton, N.J., 1982.
Film Form: Essays in Film Theory. Edited and translated by Jay Leyda. New York, 1949.
The Film Sense. Edited and translated by Jay Leyda. New York, 1942.
Nonindifferent Nature: Film and the Structure of Things. Translated by Herbert Marshall. Cambridge and New York, 1987.
Notes of a Film Director. Edited by R. Yurenev, translated by X. Danko. Moscow, 1958; reprint, New York, 1970.
The Psychology of Composition. Edited and translated by Alan Y. Upchurch. Calcutta, 1987.
On the Composition of the Short Fiction Scenario. Translated by Alan Y. Upchurch. Calcutta, 1984.
Que viva Mexico! London, 1951; new rev. ed., London, 1972.
Selected Works. 4 vols. London, 1988–1996. Volumes to date: vol. 1, *Writings, 1922–1934,* edited and translated by Richard Taylor; vol. 2, *Towards a Theory of Montage,* edited by Richard Taylor and Michael Glenny, translated by Michael Glenny; vol. 3, *Writings, 1934–1947,* edited by Richard Taylor, translated by William Powell; and vol. 4, *Beyond the Stars: The Memoirs of Sergei Eisenstein,* edited by Richard Taylor, translated by William Powell.

YURI TSIVIAN

EKPHRASIS. *See* Ecphrasis.

ELIOT, THOMAS STEARNS. *See* Modernism, *article on* Modern Literature; *and* New Criticism.

EMERSON, RALPH WALDO (1803–1881), American essayist, poet, and sage who is almost always denied the title of "philosopher." From Margaret Fuller, Henry David Thoreau, and Emily Dickinson to John Dewey, Charles Ives, Robert Frost, and Wallace Stevens, the topics and tropes of Emerson's work have provided at least some of the main courses at the feast of American letters. Emerson's effect on Friedrich Nietzsche was so pervasive—down to the actual wording of sentences—that it is evidently hard to bear in mind the general fact of this influence. Oscar Wilde and Marcel Proust also bear the marks of their reading of Emerson. A later generation of American philosophers such as Dewey, George Santayana, and William James praised Emerson but did not take him up philosophically.

In a recent series of essays, Stanley Cavell (1990) has presented the fluctuations of Emerson's reputation as a function of his repression as a philosopher. He has suggested further that the very constitution of contemporary American philosophy, along with a significant portion of our literary culture, depends on the repression of the ways of thinking that Emerson enacts and represents. Here Cavell points

to the banishing of Emersonian perfectionism as an option for moral philosophy and to the neglect of reading as a path and a category of philosophical method. What both versions of this repression point to is Emerson's vision of a self that must be read in order to be surpassed and must be, to some extent, surpassed in order to be read.

Emerson's work is commonly said to emerge at the confluence of British Romanticism and German Idealism. But this does not explain his capacity for transforming the scope and power of these movements into the terms of a nascent American culture. From the Romantics, he took the sense of the sufficiency of our everyday life as the source and topic of the arts, especially of poetry. From Idealism, he inherited an apprehension of the intimate correlation of the self and its world. From both, he took the sense that the powers of freedom are slumbering in the ordinary condition of a human being and that we need to awaken those powers in order to exercise them. Emerson characterizes the Romantic sense of our present lack of interest in our overfamiliar lives not merely as a mood of ennui but as a world of dullness. Both the mood and the world will be hard to leave behind, but leaving them behind is a condition of our awakening, hence of taking what Emerson calls a new step.

Like Johann Gottlieb Fichte and Friedrich von Schelling, Emerson was concerned to combat the false dichotomies of atheism and traditional theology, along with the political alternation of materialism and superstition in our explanations and justifications of human conduct. But the despotism of the *ancien régimes* and the various priestly ideologies that supported them were only shadows compared to the despotism of custom and passivity in the daily life of Emerson's neighbors. Like his near contemporary John Stuart Mill, Emerson responds as much to the anticipated success of the democratic masses as to the threat of a reaction against that success. He follows out the consequences of two further Romantic perceptions: (1) the dominating weight of the customary and the habitual in ordinary human life; and (2) the pressure of cities and crowds on the human need for stimulus and the human capacity for the cultivation of individual feeling and responsiveness.

Emerson's inheritance of the positive and negative perceptions of Romanticism help us to locate his aesthetics as part of a program for the renovation of human experience. Emerson does not test the results of reason against the given conditions of our experience: he tests the impoverishment of our experience against the possibilities and freedoms that our experience might still contain. Since we have stunted our own capacity for human experience, we must subject this capacity for experience to a critique from the ground up. Emerson's inheritance of philosophy and his announcement of a revisionary aesthetic are therefore not separate from his critique of religion and politics. And his approaches to art and literature are not separate from their capacity to liberate and fortify the voices of self-trust and resistance to inner and outer stagnation.

Emerson thinks of human beings not only as stunted in their growth but as shrinking from those possibilities that still remain to them. His recurrent and deepening perception of this situation is the starting point for his entire project as a writer. His dominant note of "self-reliance" is invariably misunderstood as a will to strengthen the self as it stands and to reinforce its current capacities. It is better understood as a refusal to rely on the current constitution of the self, and hence as a rejection of that constitution. A major part of what distinguishes Emerson's demand for our "aversion" to conformity is his knowledge of our thirst for it. He sounds the depth of the human craving for passivity and the wish for a secrecy that negates or perverts the possibilities of our self-expression.

The imitativeness and the passivity of the mind are, for Emerson, among its most primordial features. Imitation is the source of what he calls the "capital error of the infant Man," which, understandably enough, "seeks to be great by following the great, and hopes to derive advantages from another" ("Address" to the Divinity School, 1957 p. 103). When we try to alter anything fundamental about the self, we will come to see ourselves as for the most part exactly what we have learned to be from the example and expectations of others. At such a moment, to take a step toward self-reliance will seem like taking a step from quicksand into emptiness. As we stand, the self is quicksand, swallowing us up in our very efforts to extricate ourselves ("Experience," 1957, p. 259). But the self we catch a glimpse of is so far empty: the possibility of moving onward is an occasion for anxiety and even terror, not for self-congratulation. The self's capacity to make a new statement or affirmation out of its present condition will always come upon us "like an abyss of skepticism" ("Circles," p. 170).

In all these projects of human awakening, the arts will have a special role to play. The arts are capable of a sublimity so intimate that we feel we ought to be able to have produced it for ourselves. They teach us to abide by the permanent possibility of the birth of new powers in us or of achieving a new exercise of the powers we now abuse. In what initially takes the form of a paradox, the greatness of art presents itself to Emerson as a kind of alienation or theft of the voice in our experience. Art awakens our self-reliance by appropriating it to its own ends. "In every work of genius, we recognize our own rejected thoughts; they come back to us with a certain alienated majesty" ("Self-Reliance," p. 147).

Emerson's strategy for a new aesthetics of reading is anything but genteel. The sublime of art is partly an effect of the artist's capacity to anticipate and appropriate our thoughts and to present them in a form that extorts a kind of recognition from us. Then the countersublime of the reception of the work lies in the readers' ability to steal back

their own thoughts. If part of the therapy of art is learning to overcome the repression by which we rejected our own thoughts, the process of (re-) accepting our own thoughts is never a neutral one. The title to a thought or a feeling, and hence the means of expressing them, is always a matter of a struggle with the artist. Of course, we are also struggling with own tendency to make an idol of the person or voice of the artist. To adapt a remark of T. S. Eliot's, great readers (and great viewers and great listeners) do not borrow, they steal.

Less violently, Emerson's idea of self-reliance asks at least for an aesthetics of reading or viewing the work of art, which acknowledges the dangerous intermixing of our passive capacity for imitation and appreciation with our active capacities for independence and creation. If we are to learn from the work of art the knowledge of regions of self and world that we have not entered, we must learn to separate the active and passive faculties within ourselves. Such an aesthetics must give a central place to both gratitude and suspiciousness as genuine—though partial—responses to the individual works of art. The work of art puts us back in the infantile condition of error and dependency. But whereas politics and religion as they stand are inclined to further capitalize on the fundamental condition of human dependency, art gives us a chance to rectify that condition.

Emerson's aesthetics can therefore make the formal integrity of the work seem secondary to its capacity to foment our self-discoveries. We are receptive to the best portions of a book only in our best hours of reading. And in those hours, we are less likely to care about the parts of the work that do not speak directly to our renewed capacity for reception. This is the danger that accompanies the promise of an approach to art that is not so much antiformalist as outright iconoclastic or antinomian: the very specificity of the work can be, and even ought to be, obliterated by the voices and powers that it awakens in the right reader. (It is a danger that Nietzsche picked up from Emerson, and it is part of what he cultivates under the heading of "the Dionysian.") Emerson's response to this sense of danger to the work includes the reminder that the moment of reception will also shatter the individuality of the reader or beholder. But this will be small consolation to someone who wishes to extract a more usable aesthetic from Emerson's work.

Within this radical tendency to subordinate all experience to the experience of the conditions of self-liberation, Emerson's positive aesthetics turns out, on closer examination, to be deeply involved with the question of form. Emerson's form is not related to the *imitatio* of classical or neoclassical theory, but neither is it quite equivalent to Romantic theories of expressive form. Emerson does not define form as the provider of pattern or shape to an independently specifiable content. His idea of form can be glimpsed in his demand for the reconciliation of form and power, however uneasy or momentary the achievement. Form is always the counterpart to power: form allows the poet the power to make a more timely appearance within the events of the ordinary world. It allows this power to make a demand on those who are present, allowing the affected bystanders to become a genuine audience. Form is thus the means by which the deliverance that power potentially contains will actually manage to get delivered. Hence, for Emerson, we cannot know what the form of a work is until after it has begun to work on us.

Naturally, whatever else Emerson wants the forms of art and other human action to provide for us, he wants them to provide passages of reversal, routes by which we learn the sources of power within ourselves from the power that seems to begin outside of us. The power in question is already tied to form, and form is already dependent on the soul ("The Poet," 1957, pp. 222–223). The usurpation of the mere formalist occurs when form becomes detached from the presence of an active soul. Formalists, in Emerson's view, are just as likely to be found in the pulpit, the legislature, or the judicial bench, as in the academy of the arts.

In one direction, this ties Emerson's account of form to what the mind can learn to recognize as its own region or its own habitat. In the reverse direction, however, it is precisely certain forms that the mind learns how to use and to live with. In a central statement of his poetics Emerson says this: "For it is not meters, but a meter-making argument that makes a poem,—a thought so passionate and alive that like the spirit of a plant or an animal it has an architecture of its own, and adorns nature with a new thing. The thought and the form are equal in the order of time, but in the order of genesis the thought is prior to the form. The poet has a new thought, he has a whole new experience to unfold; he will tell us how it was with him, and all men will be the richer in his fortune" (p. 225).

That form is secondary to the originality of the thought does not tell us that the form is unimportant but tells us the kind of importance it has. The very idea that you could measure the power of a thought or an experience by the meter it learns to make is more than a little alien to us. The idea that a thought that is sufficiently vital will need a kind of architecture in which to live suggests that the role of form has to do with the inhabitation of a thought and perhaps also its intelligibility.

Some practical lessons of Emerson's reading and his doctrine of form and power can be seen in his relation to Romanticism. Emerson fastened on the Romantic accomplishments in literature, especially on the reclamation of ordinary life as a scene of epic resolution. The Romantic poets were already teaching us not to require what we can no longer have: the epics of wrath, of warriors struggling to return home, or of fallen angels. Each of those dramas occurs every day, in a person's most immediate surroundings if she knows how to make them visible or audible. "The

poorest experience is rich enough for all the purposes of expressing thought. . . . We are far from having exhausted the significance of the few symbols we use" (p. 229).

Beyond this insight, Emerson accepts the poverty of our experience, as a condition of our owning it again. Homer's epics were epics of destiny, by means of which we learn to read the poet's search for the freedom of his expression, itself an allegory of the soul's coming home to itself. Romanticisim purifies the allegory of the poet's path to the work, wherein the poet becomes the hero of human expression. The poet's path to his or her own secrets is a path that is henceforth open to all (p. 239).

Expressing an experience is not for Emerson just a matter of aligning some conventional symbols with the experiences that they are more or less meant to express. Discovering a new form is not a matter of devising new symbols or other new externalities with which our inner lives can be associated and the inner lives of other can be provoked. The experience itself must be interrogated for the thought that it contains or even conceals. The role of forms and symbols is at least as much to unlock the sense of the experience to its possessor as to proclaim it outwardly. For Emerson, there is no point in just proclaiming one's capacity for raw experience, for what looks like raw experience already conceals the secret thoughts of the one who is supposedly having that experience. Most of us maintain a certain secrecy about the very conditions under which we have our experiences. For instance, we grow secretive about our despair of ever possessing our experience and protective about our nostalgia for the experiences that we use to measure our present disappointment with experience.

Eventually, secrecy itself becomes one of the conditions under which experience occurs to human beings. The poet becomes representative of the human problem, because we are already "study[ing] to utter our painful secret" (p. 223). The pleasures of artistic form are not merely window dressing or compensation. They are signs that someone has found a way to make the painful secret yield pleasure in the utterance of it. Expression is always more than a passage outward to an audience. It is a passage from a region of the mind that the mind itself has kept covert to another part of the mind, which is now ready to read and receive its own confession.

Form opens up the work of art to itself, because it imitiates or reproduces the path the artist took to his or her own work (p. 239). As a consequence, form can open the work to an audience. Form is not merely a matter of finish or polish. Form makes possible a new act of vision, and this possibility of action is the source of the work's effectiveness with us: "The new prospect is power" ("Circles," p. 170). Power is not the cause of our taking the new step: it flows from the perspective the new step affords us. The step must be taken in weakness and anxiety, and the confidence of self-reliance is usually very far from confident in any conventional sense.

Art belongs to the moments in which we forget ourselves, allowing ourselves to be surprised out of our propriety. Emerson looked for an American poetry that was not synchonized with the present state of American culture and the present abuses of political power. Thus, he was perhaps readier to recognize the accomplishment of Walt Whitman. He saw him not as the celebrator of national expansion but as the confessor of our national secrets, the poet whose genius for openness would shame us out of our national shames.

[*See also* Romanticism.]

BIBLIOGRAPHY

Works by Emerson

Emerson's many individual works appeared in various editions. All of the following were first published in Boston, in the year provided.

Nature (1836).
"The American Scholar" (1837).
The *Address* to the Divinity School (1838).
Essays. First (1841) and Second (1844) Series (see especially "Art" and "The Poet").
Representative Men (1850). Includes essays on Plato, Goethe, Shakespeare and Montaigne.
"The Fugitive Slave Law" (1851), "John Brown" (1859), and "Woman" (1855).
The Conduct of Life (1860). See especially "Fate" and "Power."
Letters and Social Aims (1875). See especially "Poetry and Imagination," "The Comic," "Perisan Poetry," and "Quotation and Originality."
"The Sovereignty of Ethics" (1878).

Collections of Emerson's Writings

"The Character of Socrates" (1820) and "The Present State of Ethical Philosophy" (1821). In Edward Everett Hale, *Ralph Waldo Emerson.* Reprint, Boston, 1977. College Essays of Emerson.
The Collected Works of Ralph Waldo Emerson. 5 vols. Edited by Robert E. Spiller and Alfred Fergusson et al. Cambridge, Mass., 1971–1994.
The Complete Works of Ralph Waldo Emerson. 12 vols. Edited by Edward Waldo Emerson, Centenary Edition. Boston, 1903–1904.
Selections from Ralph Waldo Emerson: An Organic Anthology. Edited by Stephen E. Whicher. Boston, 1957. All references are to this edition.

Other Sources

Bloom, Harold. "Emerson and Whitman: The American Sublime." In *Poetry and Repression: Revisionism from Blake to Stevens.* New Haven, 1976.
Bloom, Harold. *Agon: Towards a Theory of Revisionism.* New York and Oxford, 1982.
Cavell, Stanley. *In Quest of the Ordinary: Lines of Skepticism and Romanticism.* Chicago, 1988.
Cavell, Stanley. *This New Yet Unapproachable America: Lectures after Emerson after Wittgenstein.* Chicago, 1989.
Cavell, Stanley. *Conditions Handsome and Unhandsome: The Constitution of Emersonian Perfectionism.* Chicago, 1990.
Ellison, Julie. *Emerson's Romantic Style.* Princeton, N.J., 1984.
Goodman, Russel B. *American Philosophy and the Romantic Tradition.* Cambridge and New York, 1991.
Gould, Timothy. "Ralph Waldo Emerson." In *Encyclopedia of Ethics,* edited by Lawrence C. Becker and Charlotte B. Becker, pp. 229–301. New York, 1992.

Matthiesson, F. O. *American Renaissance: Art and Expression in the Age of Emerson and Whitman.* New York and Oxford, 1941.

Miller, Perry, ed. *The Transcendentalists: An Anthology.* Cambridge, Mass., 1950.

Packer, Barbara L. *Emerson's Fall: A New Interpretation of the Major Essays.* New York, 1982.

Packer, Barbara L. *The Transcendentalists.* In *The Cambridge History of American Literature,* vol. 2, pp. 329–604, general editor, Sacvan Bercovitch. Cambridge and New York, 1995.

Poirier, Richard. *The Renewal of Literature: Emersonian Reflections.* New York, 1987.

Zwarg, Christina. *Feminist Conversations: Fuller, Emerson, and the Play of Reading.* Ithaca, N.Y., 1995.

TIMOTHY GOULD

EMOTIONS. [*To explain the complex history of the role of the emotions in art, this entry comprises three essays:*

An Overview
Emotions and Literature
Emotions and Music

The first is an overview of the topic of emotions in the history of aesthetics, that is, how emotions are expressed, evoked, or otherwise (re)present(ed) in art. Then there is one essay each on how this topic has been understood in connection with literature and music, two art forms in which the topic of emotions has been a major artistic and philosophical concern. For related discussion, see Expression Theory of Art; Katharsis; Literature, *article on* Literary Aesthetics; Music, *overview article;* Pleasure; *and* Romanticism.]

An Overview

The history of emotions in aesthetics is bound up with the fate of emotions in philosophy. The history of the treatment of emotions in philosophy, however, is a history of abuse, often tempered with fear and awe. Insofar as (what we now call) aesthetics has been bound up with the expression of emotion, it has been sometimes villainized, often marginalized, irregularly celebrated, and sometimes even worshiped. Plato and Aristotle vigorously debated the place of the arts in the good life and the good society, and much of that debate had to do with the role and effects of emotions in the arts. Judaism, Christianity, and Islam witnessed many disagreements within and between themselves concerning the place of music and "graven images" in their respective religions, at least in part because of the emotional effects of art in the context of the larger emotion of faith and what is variously understood as spirituality. The history of modern (Western) philosophy—at least as it is usually taught and included on syllabi today—gives relatively scant recognition to either the arts or aesthetics, first of all because they do not fit into the largely epistemological and metaphysical program that has gripped so much of philosophy since the seventeenth century but also because philosophy has so

keenly allied itself with reason and science that there is little room for the appreciation of art and the emotions art expresses and evokes. This picture is challenged, of course, by many philosophers, especially in Germany. Immanuel Kant praised the arts and the production of beauty despite the fact that he was so famously awed by "the starry skies above and the moral law within." G. W. F. Hegel recognized the arts as one of the penultimate expressions of cosmic "Spirit"; Arthur Schopenhauer recommended the contemplation of beauty as a temporary respite from the ravages of the Will, and Friedrich Nietzsche celebrated art (at least in his early philosophy) as the highest form of human activity. Nevertheless, for much of the twentieth century, especially in England and the United States, both aesthetics and emotions were all but ignored. Today, however, both aesthetics and the study of emotions are flourishing in philosophy.

Insofar as philosophy has grown up as the pursuit of reason, the emotions have always lurked in the background, as a threat to reason and a danger to philosophy and philosophers. One of the most enduring models of reason and emotion has been the metaphor of master and slave, reason firmly in control and the dangerous impulses of emotion safely suppressed, channeled or, ideally, in harmony with reason. Plato and later Horace compared the emotions to a spirited horse. The arts, accordingly, have often been viewed in the context of this metaphor, as a dangerous provocation and stimulus to the emotions, as a relatively safe and socially acceptable channel for expressing emotion, as emotional support and accompaniment for the right way defined by reason. The same philosopher or philosophy, however, might warn against one art or one set of emotions and encourage another. The different arts have often been deemed of different value because of the emotions they tend to express or evoke, and different emotions have been variously valued on the basis of their different roles in the philosophy, social life, and religion of a people. To mention just two outstanding examples, Plato condemns flute music as conducive to licentiousness, but he defends and encourages drama that supports the emotions of good citizenship. Nietzsche praises those arts (especially music) that evoke the grand passions he summarizes in the phrase "will to power," but he condemns literature that encourages such oversentimental and self-demeaning emotions as pity and resentment. The history of emotions in aesthetics is rarely so simpleminded as to defend or condemn all of the arts or all of the emotions without distinction.

The idea that emotions are more primitive, less intelligent, more bestial, less dependable, and more dangerous, thus to be controlled by reason, led to both the marginalization and a certain view of the arts in terms of how much *control* the artist exercised (what Nietzsche later referred to as the "Apollonian"). Those arts that tended to overflow their bounds and provoke excessive emotion were condemned (though many of the early Greeks and, later, Nietz-

sche celebrated them as "Dionysian"). Insofar as emotions were akin to reason, they were clearly inferior ("confused perception" or "distorted judgment"). Thus it was a mark of his considerable iconoclasm that the eighteenth-century Scottish skeptic David Hume famously declared that "reason is, and ought to be, the slave of the passions." In aesthetics, Hume rejected the suggestion that there could be rational disputing of matters of taste, falling back on the conservative position that what "men of good taste" had long judged excellent would serve as a guide to what deserves our admiration and praise.

In an early dialogue, *Symposium,* Plato discusses *eros* as the love of Beauty, thus suggesting that certain grand emotions are essential to wisdom, but the arts play no role in the progression from vulgar desire to philosophical enlightenment. Aristotle, on the other hand, suggests that both art and the emotions play an essential role. In his *Rhetoric,* he defines emotion "as that which leads one's condition to become so transformed that his judgment is affected, and which is accompanied by pleasure and pain." Aristotle discussed certain emotions at length, notably anger, which he describes in remarkably modern terms. He defines anger as "a distressed desire for conspicuous vengeance in return for a conspicuous and unjustifiable contempt of one's person or friends." It is worth noting that Aristotle had little to say about "feeling," presumably not because the Greeks were anaesthetic but rather because what we (inconsistently) call "affect" and inner sensation generally held little interest for them and played no significant role in their language or their psychological discourse. Nevertheless, in his *Poetics,* Aristotle is quite clear about the healthy role of shared fear and pity in the theater, and the purpose of tragedy is to evoke and release such emotions (catharsis). Perhaps the most important single point to make about Aristotle's view of emotion, therefore, is that his analyses make sense only in the context of a broader *ethical* concern. Anger is of interest to him because it is a natural reaction to offense and a moral force, which can be cultivated and provoked by reason, poetry and rhetoric. So, too, fear and pity.

In Roman times, ethics and emotion conjoin in the philosophy of the Stoics. But whereas Aristotle took emotion to be essential to the good life, the Stoics analyzed emotions as conceptual errors, conducive to misery. The virtuous life depended on freeing oneself from such worldly values as health, wealth, and honor. Accordingly, Stoicism defended the arts only insofar as they encourage virtue. Their purpose was not catharsis, nor to evoke emotion. The Roman Stoic Seneca's version of *Oedipus* (one of his nine tragedies) can be contrasted with the better-known version by Sophocles, as Aristotle understands that play. Whereas Aristotle's Sophocles confronts us full-face with fear and pity, Seneca instructs us to endure tragedy. In general (for the movement lasted almost five hundred years, through two successive cultures), the Stoics insisted on the presumptuousness of moral judgment in anger, the intolerable vulnerability of love, the self-absorption of security in fear, and the inevitability of suffering. The alternative, "psychic indifference" or *apatheia,* could be found in reason and by getting straight about the pointlessness of emotional attachments and involvement. Thus the appreciation of beauty was subsumed under the more general category of virtue, and poetry, in particular, was appreciated for its beauty only insofar as it enhanced virtue.

Throughout the Middle Ages, the study of emotion was again typically attached to ethics, and it was central to Christian psychology and the theories of human nature in terms of which the medievals understood themselves. There were elaborate, quasi-medical studies of the effects of the various "humors" (gall, spleen, choler and blood) on emotional temperament, but there were also rich studies of the effects of the emotions on faith and moral behavior. Many emotions were linked with self-absorbed desires, and so the Christian preoccupation with sin led to elaborate analyses of those emotions identified as sins (notably, greed, lust, anger, envy, and pride). It might be noted that the highest virtues, such as love, hope, and faith, were not classified as emotions as such but were elevated to higher status. Medieval debates about the role of emotions in the arts (and the role of the arts in society) thus turned on the question of sinfulness, what encouraged sin and what encouraged faith and piety. Christian art and music flourished as the church defined the proper depiction of religious topics and appropriate musical forms. Medieval Judaism proscribed "graven images" as blasphemous, and Islam prohibited music. (Chanting the Qur'ān was not considered music.) These arts, accordingly, were not easily developed, and the emotions they tended to inspire were mixed with fear and revulsion.

In modern philosophy, with its emphasis on epistemology and logic, both aesthetics and emotion have tended to be pushed aside. René Descartes, for example, was a scientist and a mathematician by temperament, awed by "the natural light of reason." Nevertheless, he paid more attention than most philosophers to the nature of emotions. Descartes defended a view in which the mind and body "meet" in a small gland at the base of the brain, and the latter affects the former by means of the agitation of "animal spirits"—minute particles of blood—which brings about the emotions and their physical effects in the body. Nevertheless, the passions (wonder, love, hatred, desire, joy, and sadness) are not merely agitations of the body but important perceptions of the soul. Although Descartes did not write about beauty, his inclusion of wonder and joy as basic emotions suggests its inclusion, and his strict rationalism would provide a guide for aesthetics for centuries to come.

Hume, in carrying out the Enlightenment directives of reason to challenge, debate, and question, came to examine the role and capacities of reason itself. What motivates us to

right (and wrong) behavior, Hume insisted, were our passions, and rather than being relegated to the margins of ethics and philosophy the "sentiments," in particular, deserved central respect and consideration. The sentiments became particularly important in aesthetics, where the moral sentiment theorists (Anthony Ashley Cooper, earl of Shaftesbury, Francis Hutcheson, and Hume) defended an analysis of beauty and the sublime in terms of immediate "perception," "impressions," or "taste." This raised the question of how differing tastes, like differing sentiments, could be disputed. Hume gave the emotions great importance, but he also concurred in the traditional view that the emotions, as impressions, are not rational. In order for aesthetics to progress, one might argue, the conception of emotions needed to be enhanced.

Such an enhancement can be found, for instance, in the aesthetic theories of Edmund Burke. He explained the feeling of beauty as love without desire and the sense of the sublime as including an experience of controlled or nonthreatening horror. Like Descartes, he speculated on the physiology of such emotions, but more important for our purposes here is his attempt to develop an enriched conception of the aesthetic emotions, an effort which was also ongoing in Germany. Kant reinforced the distinction between reason and what he called "the inclinations" (emotions, moods, and desires) and dismissed the latter (including the moral sentiments) as inessential to morals. Nevertheless, in an early essay he proclaims compassion to be "beautiful," and in his *Critique of Judgment* he attempts a full-blown aesthetic theory, at least partially in terms of "feelings." Such feelings are most pronounced in our experience of the sublime, but in his analysis of the experience of beauty Kant can be read as attempting a deep structural analysis of our aesthetic emotions. Moreover, it was Kant who celebrated the importance of shared ("intersubjective") feeling in the appreciation of beauty and the awe with which we try to comprehend the wonder of God's creation, and it was Kant, a quarter century before Hegel (who is credited with it) who insisted that "nothing great is ever done without passion."

In Hegel's aesthetics, the attempt to develop an enhanced conception of emotion in the arts reaches its natural culmination as, in a sense, the distinct concepts of emotion and feeling virtually drop out of view. Feelings and the merely sensuous are integrated into a larger conception of spirituality, and emotions are expressions of "the Idea," a holistic vision that is the conclusion of all of Hegel's "dialectical" works. Art develops along with human consciousness in distinctive stages, from symbolic through classical and culminating in the deeply emotional and self-reflective romantic art of his own day. One might with some benefit compare the music of Ludwig van Beethoven, especially, with the developmental counterpoint of Hegel's major work, *The Phenomenology of Spirit* (1807). One should not think of "the

Idea" as a particularly "intellectual" conception. Rather, it is Hegel's attempt to synthesize the most spiritual emotions with reason.

Hegel's efforts were soon compromised, first by the German Romantics, who considered art to be a distinctive expression of emotion and then by the Formalists, for example, Eduard Hanslick with regard to music. In the case of the Romantics, the emotions were not trivialized but bloated and rendered ineffable. Inspired by the English Romantics (William Blake, William Wordsworth, Percy Bysshe Shelley, Samuel Taylor Coleridge, Thomas Carlyle), who emphasized the freedom but also the control of the imagination, the Germans (Friedrich von and August Schlegel, Novalis) celebrated the unbounded idea of the world as art and the work of art as a symbolic, organic whole. In its basic form, the Romantic view resembled Hegel's vision, but Hegel, who remained a rationalist, rejected the Romantics. So did Johann Wolfgang von Goethe, the leading *Dichter* of Germany, who declared Romanticism "sickly" for its uncontrolled excess of emotion. In terms of giving the emotions their due in aesthetics, the Romantics perhaps showed us that there can certainly be too much of a good thing.

Hanslick dismissed the claim that emotions are expressed in music, thus setting up one of the central debates in current-day aesthetics. Hanslick had nothing but abuse for those good bourgeois citizens who attended concerts in order to indulge their sentimentality. Indeed, it is about this time that the notion of sentimentality itself became a term of abuse, minimally suggesting an excess or deformity of emotion but, arguably, indicating a deep suspicion of emotions in the arts. But Hanslick did not deny that music arouses emotions. Indeed, he thought that there were good grounds for emotional responses to music, since the dynamics of music resembles the dynamic properties of human feelings. While the beauty of music was not a consequence of the emotions experienced while listening, musical beauty was nevertheless a function of "tonally moving forms" appreciated by the intellect.

The connection between emotions and music is, indeed, particularly baffling. On the one hand, since Plato it has been thought that music is the clearest expression of emotion, a belief endorsed in modern times by Schopenhauer and Nietzsche. But, on the other hand, the source of that connection has proved notoriously hard to pin down. In the representational arts, it is often and obviously the subject matter that evokes emotion. When pure color or shape evokes emotions this can often be explained by reference to associations or resemblances (red as blood, the familiar shapes of female anatomy). But, except for music with texts, "program music," and music associated with particular circumstances or natural sounds (e.g., a wedding march, bird calls), music does not seem to lend itself to such explanation. A more technical question, answered in one form by Hanslick, is whether the emotions are *in* the music (rather

than merely associated *with* the music). One might mention, in this regard, rhythm, melody, harmony, counterpoint, interaction, complexity, key, amplitude, pitch, and instrumental timbre. One should also distinguish between music as an *expression* of emotion, music as *provoking* emotion, and music as *representing* or *imitating* emotion (apart from circumstances of composition and performance). What we should also note is how impoverished the range of such emotions (e.g., sadness, joy, and cheerfulness) is compared to the richness of the music itself.

Schopenhauer and Nietzsche were two philosophers for whom aesthetics took center stage. But for Nietzsche, passion was essential. Reason was a source of deep suspicion. Schopenhauer distrusted the passions and defended only a modest role for reason, but for him, the world was revealed as Will, which manifested itself in our passions. These passions had no ultimate purpose, however, and so Schopenhauer famously emerged the great pessimist, for whom life was essentially pointless and all willing a "burden." Art, and music in particular, was both an expression of the Will and a temporary escape from willing, and thus took on its central importance. Nietzsche rejected Schopenhauer's extreme pessimism, but nevertheless he anticipated the global skepticism and conceptual chaos of the twentieth century, and like Sigmund Freud (who admired him), he described the darker, more instinctual human passions. The arts expressed these passions, and it was in our artistic expressions of suffering that we could come to see ourselves, like the early Greeks, as "beautiful." But this was not to say that all passions are wise; some, Nietzsche declared, "drag us down with their stupidity," and others, notably the "slave morality" emotion of resentment, are devious and clever but to a disastrous end, the "leveling" of the virtuous passions and the defense of mediocrity.

Following Nietzsche, Martin Heidegger gave emotions and art a central place in human existence. Heidegger's aesthetic views are by no means clear, but certain aspects are evident. First, Heidegger argues (in *Being and Time*) that mood is one of the basic "existential" categories through which we understand our "being-in-the-world." He also argues that it is our preverbal practices, not our pursuit of knowledge and articulate philosophy, through which our world is "disclosed." In his "Origin of the Work of Art" (1935) he argues that works of art (at least, great works of art) not only express the vision of their creators but, more important, shape the world. In a very different vein, the French existentialist Jean-Paul Sartre developed a phenomenological analysis of emotions as "magical transformations of the world"—willful stratagems for coping with difficulty. But Sartre's later views commited him to the role of "the engaged intellectual," in which he would give up his conception of emotions as "escape behavior" and view them instead as willful commitments to change society through his art.

In contemporary aesthetics in the English-speaking world, the role of emotion is still a matter of considerable debate. Much of this debate turns on the nature of emotion, which, as this brief history suggests, is no simple matter. How we conceive of emotion depends not only on science but also on ethics, one's conception of human nature and the good life. And to this short list we can add one's conception of the arts and their role in the good life. Insofar as emotions are conceived as primitive, unintelligent reactions or forces, straining for release, then aesthetics will reflect the satisfactions and dangers of such catharsis. On the other hand, insofar as one's conceptions of the emotions become more complex and sophisticated, aesthetics will become more complex and sophisticated as well.

BIBLIOGRAPHY

Aristotle. *Nicomachean Ethics* and *Rhetoric*. In *The Basic Works of Aristotle*, edited by Richard McKeon. New York, 1941.
Carroll, Noël. *The Philosophy of Horror*. New York and London, 1991.
Descartes, René. *Passions of the Soul*. Translated by Stephen Voss. Indianapolis, 1989.
de Sousa, Ronald. *The Rationality of Emotion*. Cambridge, Mass., 1989.
Heidegger, Martin. *Being and Time*. Translated by John Macquarrie and Edward Robinson. New York, 1962.
Higgins, Kathleen Marie. *The Music of Our Lives*. Philadelphia, 1991.
Hume, David. *A Treatise of Human Nature*. 2d ed. Edited by L. A. Selby-Bigge, revised by P. H. Nidditch. Oxford, 1978.
James, William. *What Is an Emotion?* New York, 1890.
Neu, Jerome. *Emotion, Thought and Therapy*. Berkeley, 1977.
Nietzsche, Friedrich. *On the Genealogy of Morals* (with *Ecce Homo*). Translated by Walter Kaufmann. New York, 1967.
Plato. *The Republic*. Translated by G. M. A. Grube. Indianapolis, 1974.
Plato. *Symposium*. Translated by Alexander Nehamas and Paul Woodruff. Indianapolis, 1989.
Sartre, Jean-Paul. *The Emotions: Outline of a Theory*. Translated by Bernard Frechtman. New York, 1948.
Seneca. "On Anger." In *Moral and Political Essays*, edited and translated by John M. Cooper and J. F. Procope. Cambridge and New York, 1995.
Solomon, Robert C. *The Passions*. Garden City, N.Y., 1976.
Solomon, Robert C. "In Defense of Sentimentality." *Philosophy and Literature* 14.4 (Fall 1990).

ROBERT SOLOMON and KATHLEEN HIGGINS

Emotions and Literature

The written word has enormous power to move its readers in various ways. Poems can make us feel lighthearted or melancholy; short stories can induce sadness, surprise, and delight. Novels produce a plethora of feelings and emotions, including those where we empathize and sympathize with various characters, or where we come to hate them, or wish a character would act differently. Plays can be read to oneself and not simply viewed in performance, and as Aristotle pointed out, a tragedy can induce pity and fear just in the reading without viewing the acting or spectacle. Not only fiction but also nonfiction—history, biography, news and magazine reports, travel writing, essays on culture, and

opinion columns—can be emotionally moving; we may be horrified at events that have occurred, admire a courageous person, shocked by the actions of public officials, awestruck by descriptions of natural beauty, angry with the opinions expressed.

Although hardly anything seems more commonplace than such responses, they raise a whole host of philosophical questions, some of which show how central philosophy of mind is to doing aesthetics. For example, what is an emotion, anyway? How do various emotions differ from one another? How do emotions differ from other affective states, such as feelings, moods, and desires? Other questions focus on the relationships between responses to literature and responses to actual events. Are our responses to literature really emotions, that is, emotions in the same sense as emotions that arise when we actually experience an event in the course of ordinary life? Are the explanations of what produces responses to literature, especially fiction, different from the explanations of responses to real-life situations? Other questions focus more directly on aesthetic issues in particular. How is it possible to judge the appropriateness or inappropriateness of a response to literature? What kinds of assessments can be made of the relevance, appropriateness, or reasonableness of a response? Why is it important to respond to literature at all?

Currently, the most popular theory of emotions is a cognitive theory that holds that emotions have cognitive components as one of their defining features. A cognitive component is a belief, or in some theories it may be an "unasserted thought," that is, a thought that is entertained but not believed. (Aristotle is credited with the first cognitive theory of emotion; Roger Scruton is responsible for the term *unasserted thought*.) Thus, the emotions fear, pity, anger, and pride require having a particular sort of belief (respectively, that one is in danger; that someone is suffering unjustly; that one has been done an injustice; and that one is responsible for bringing about something good in a way that required special skill, courage, or effort). Cognitive theories also typically require that one have certain sorts of values or desires (for one's self-preservation, that people not suffer unjustly, etc.). Feelings, on the other hand, include such states as tension, uneasiness, lethargy, or well-being, and are generally not taken to have cognitive components. They also need not be linked with a particular external object (think of a generalized sense of uneasiness, contentment, or well-being). Our vocabulary for real-life feelings, much less responses to literature, is woefully inadequate, metaphorical, and derivative of sensations of bodily states (light-headedness, tingling), which may also tell us something about the nature of these states. Moods—being cheerful, bored, belligerent—are mental states that color one's actions and reactions during a particular period of time. Reading also produces desires, everything from sexual

desire to the desire that a fictional character act or not act a given way (e.g., a desire that Othello not kill Desdemona). Some desires are also cognitive in that they require a belief that what one desires would be good. It may be only because of that belief that one has the desire (such as a desire that I do the dishes). But some desires are desires to do things, desires to act, whether or not one has a belief that doing such a thing would be good (such as desiring to take a walk in the woods). In such cases, beliefs that doing what one desires to do would be good may arise because one desires it, rather than that one desires it because one believes it is good. Furthermore, we sometimes have irrational desires, desires for things we know will harm us or that we believe are not good. Plato believed that it was possible for fiction to give rise to such desires.

What is the relationship between affective responses to literature and responses to real-life situations? Responses to literature can often (though not always) be described using the same vocabulary we use for identifying emotions and affective responses to events in actual life. It would be rash, however, to assume that responses to literature are exactly the same as responses to real-life events simply because they go by the same names. Many philosophers have asserted that our responses to art, and hence literature viewed as art, are sui generis and entirely different from our responses to life. One consideration in support of this view is that our responses to literature are rarely accompanied by actions or desires to act. We do not try to console a grieving character, and we do not throw things or yell and shout at a character who makes us angry. Since affective responses to literature are not typically connected with desires and actions in the way responses to actual events are, there is something fundamentally different about our affective responses to literature. In addition, very often we know that what we are responding to is fictional (so we do not believe that, for example, a disaster happened, a person was betrayed, an orphan was rescued, and so on). This knowledge that what we are reading is fictional can account for the absence of action. Any theory about the nature of responses to fiction will have to account both for our use of ordinary emotion vocabulary to describe them, and for the different roles of belief and action in responses to real-life situations and to fiction.

If one is reading nonfiction, it is plausible to explain responses in terms of one's existing beliefs and desires or values. Suppose I read in a history book about life and death in Auschwitz during World War II; I am overcome with horror. Why? I believe what I read in the book, and it grossly offends my most basic sense of decency about how people ought to be treated. Thus, I am moved because reading produces (or calls to mind) a *belief*, which, in this case, conflicts with one or more of my *values or desires*. Horror is an emotion whose identity depends on my having certain beliefs

and desires. I would not be horrified if I were a Nazi sympathizer and thought there was nothing wrong with, say, killing Jews just because they are Jewish.

But what if I am reading a work of fiction, and I do not believe those events actually occurred? In an influential article, Colin Radford challenged philosophers to explain how it was both possible and desirable for us to have emotional responses to fictional literature. Either such responses are irrational (and hence undesirable) because they depend on evidentially unsupported beliefs (after all, we are reading fiction), or we do not have the unsupported belief and the response is not an emotion as normally construed (because it does not have the relevant cognitive component). One reply to Radford is that the beliefs involved are of a more general sort. That is, I (justifiably) believe that events of the *sort* described have occurred, or that they are likely to occur, even if I do not believe in the existence of the particular people and situations described. Thus, I may respond as a result of my justified general beliefs about what the world is like.

Another reply to Radford questions his assumption that belief in the truth of what one reads is necessary for having an emotional response. It is alleged instead that the cognitive component of an emotion might be an unasserted thought rather than a belief, and hence one is not subject to the charge that the emotion is irrational because the belief is irrational. Simply *thinking of* things, imagining they happened, may be sufficient to generate an emotional response: simply imagining people being subjected to certain tortures can generate horror, even if one does not believe anyone was actually treated that way. But this cannot be the whole story. One reason Radford and others claim emotions require beliefs is that the belief *explains* why you get emotionally worked up in a way an unasserted thought does not. If you *believe* your friend was on the plane that crashed, you are distraught; when you find out she was not on the plane (you then believe she was not on the plane), you are no longer distraught but relieved. Furthermore, a lot of unasserted thoughts pass into our heads. Why are some emotionally affecting and not others? Something else besides simply having the thought (and the relevant desires or values) is necessary to explain why reading that generates such thoughts can be emotionally affecting. Finally, if you continue to imagine that your friend was on the plane when you know it was not so, and this imagining makes you distraught, it *still* seems you are being irrational. Why get so worked up over something when you know it is not true? One important factor that can influence the emotive power of unasserted thoughts or imaginings is the words that generate the thoughts or imaginings, such as diction, syntax, rhythm of the writing, metaphor and other imagery, narrative voice. (Is the story told in the first person? From the standpoint of an omniscient observer or an unreliable nar-

rator? Are multiple points of view combined so that we do not know what to believe about what is true in the story?) Responses to poetry obviously require sensitivities to these features, but responses to prose, when considered as an art form whose medium is words, will also depend on them. The way a story is told leads the reader to read *with* feeling.

Moreover, the problem is actually broader than Radford proposes, for we respond to fiction not only with emotions but also with feelings, moods, and desires. Feelings and moods are not generally held to have cognitive components in the form of beliefs or unasserted thoughts (the jury is still out about desires). Thus, it is unclear how beliefs or unasserted thoughts could help to explain nonemotion affective responses, such as feelings and moods, that do not have cognitive components. Indeed, how *does* one explain such responses?

An explanation of the emotional effectiveness of fictional literature, precisely because it is fiction, will need to develop some account of how imagination is involved. Kendall Walton is one of the few who have developed a systematic account of imagination, arguing that reading and responding to fiction are an extension of ordinary games of make-believe, such as those that children play. He holds that in reading a work of literature we treat it as a prop in a game of make-believe, and it becomes true in the game that I believe Anna Karenina suffers unjustly, and hence true in the game of make-believe that I pity Anna. On Walton's view, just as it may be true in the game of make-believe that I am riding a horse when it is not true of me (simpliciter) that I am riding a horse, so also it may be true in the game of make-believe that I pity Anna even though it is not true of me (simpliciter) that I pity Anna.

It is also possible for a work—history as well as fiction—to glorify what we actually take to be abhorrent, or to denigrate what we value, leading us to respond in ways that are contrary to what we really value. This can happen because we identify or empathize with a character who sees things from a different point of view, or simply because we get "caught up" in a story and "forget" our usual values. These phenomena also cry out for a theory of imagination. Plato worried that poetry and drama could lead us both to believe things that were not reasonable or true (simply because we get seduced by the persuasive power of the words), and to change our values or desires. Thus, readers may respond without the relevant beliefs and without the relevant desires.

It certainly seems that anyone who cheers on Iago and laughs when Desdemona dies is responding inappropriately to William Shakespeare's play. What makes these cheers and laughter inappropriate? Little work has been done explaining how to make any such assessments of our responses, but clearly they are of the greatest import. For if one laughs when Desdemona dies, one has missed a very important part of what is valuable about Shakespeare's play: it is *sup-*

posed to move us to sorrow and pity, and to tears. If we don't experience those feelings, we haven't gotten the value out of the work.

Assessments of responses as appropriate or inappropriate are intimately connected with questions about why it is important to respond to literature at all. So what if Shakespeare leaves you cold, if you don't find Oscar Wilde amusing, if Marcel Proust does not make you reflective and pensive, if *The Death of Ivan Ilych* does not fill you with feelings of poignancy and compassion, if William Wordsworth's poetry does not instill a desire to explore nature, if "Kubla Khan" does not fill you with a sense of awe and mystery, and so on and so on? Is anything lost? Without such responses, one may wonder why bother reading literature at all. The value and importance of literature are wrapped up with our responses to it, and a failure to respond is (to that extent) a failure to appreciate the work. And this helps us understand why it is so important to have not just any old response to a work, but to have appropriate responses (conceding that several responses may be equally appropriate, revealing different qualities or aspects of a work).

But rather than settling the issue, this solution merely raises another question. Surely affective responses, when appropriate, are part of what it is to appreciate a work, but why is it important to appreciate a work? Put another way, why is it good to have things around such as works of literature that are supposed to (and presumably do) move us in the sorts of ways discussed earlier? A very old idea is that the function of the arts is to delight and instruct. (And/or: so that one point of having such responses is simply that we find it delightful or pleasurable. If you separate them, you can look at it this way. But, below, you can see the delight coming through the instruction.) Suppose we say that the function of art is to please or delight us. Then it would appear that we should respond with the emotions and feelings that provide the most delight or pleasure, which removes any objective standard for the responses (i.e., a standard that resides in the properties of the object, the literary work, to which we respond), and accepts the preferences of each reader. On the other hand, some ways of approaching literature may be more likely, in the long run, once they are mastered, to produce richer and more fulfilling pleasures than simply reacting or responding in any way one wishes. If so, the approach (or approaches) with the richest potential overall in the long run could provide a basis for judgments of the appropriateness of responses.

Delight, however, might not be separate from instruction. Aristotle pointed out that learning can be pleasurable. How can emotions and other affective responses instruct? Martha Nussbaum has argued persuasively that literature teaches us to "see," where seeing involves not merely a cognitive or sensory dimension, but is broadly experiential and affective as well. Our ethical sensitivities, for example, are affective as much as cognitive, and influence how we experience the world and how we relate to other people. Others have argued that we can learn in a very straightforward way what it is like to be a certain sort of person, or in a certain sort of situation—how such a person would feel, and how he or she would experience things. It can thus enhance our capacity for empathy and understanding of other people in general. Susan L. Feagin argues that developing the cognitive and affective skills required to appreciate literary works of art expands the capacities of the human mind, increasing its strength and flexibility. It is important on Feagin's account that one genuinely appreciate a work, rather than merely accept whatever affective responses one happens to have, since only then will the powers of one's mind be enhanced, as opposed to simply further entrenching one's usual modes of experience and understanding.

Finally, there is one quite different unpleasant possibility to be explored. Jerome Shaffer (1968) has defended the view that emotions are irrational and indefensible. He agrees that here and there an emotion might be beneficial because it is pleasant or has some practical advantage, but his point is that, as a group, there is no good reason to have them and we would be better off without them. This view tends to make people very excited and angry (adding more evidence for or against Shaffer's thesis?). In contrast to Shaffer, the neurologist Antonio R. Damasio has argued that a capacity for emotion is necessary for practical judgments and actions based on them. Emotional commitments are also involved in our ethical lives—from the feelings that bind friends and family, to pangs of remorse and indignation over injustice. Is the importance of emotions to practical judgments and actions and our ethical lives reflected at all in our responses to literature? If so, how; and if not, why not?

[*See also* Fiction.]

BIBLIOGRAPHY

Aristotle. *The Poetics.* Translated by James Hutton. New York, 1982.

Damasio, Antonio R. *Descartes' Error: Emotion, Reason, and the Human Brain.* New York, 1994.

Feagin, Susan L. *Reading with Feeling: The Aesthetics of Appreciation.* Ithaca, N.Y., 1996.

Lamarque, Peter. "How Can We Fear and Pity Fictions?" *British Journal of Aesthetics* 21 (1981): 291–304.

Lyons, William. *Emotion.* Cambridge and New York, 1980.

Neill, Alex. "Fiction and the Emotions." *American Philosophical Quarterly* 30 (1993): 1–13.

Novitz, David. *Knowledge, Fiction, and Imagination.* Philadelphia, 1987.

Nussbaum, Martha. "'Finely Aware and Richly Responsible': Moral Attention and the Moral Task of Literature." *Journal of Philosophy* 82.10 (October 1985): 516–529.

Plato. *Republic.* Books 3 and 10.

Radford, Colin. "How Can We Be Moved by the Fate of Anna Karenina?" *Proceedings of the Aristotelian Society* supp. vol. 49 (1975): 67–80.

Ryle, Gilbert. *The Concept of Mind.* New York, 1949. See chapter 4, "Emotion."

Scruton, Roger. *Art and Imagination: A Study in the Philosophy of Mind.* London, 1974.

Shaffer, Jerome A. *Philosophy of Mind.* Englewood Cliffs, N.J., 1968.
Walton, Kendall L. *Mimesis as Make-Believe: On the Foundations of the Representational Arts.* Cambridge, Mass., 1990.

<div align="right">SUSAN L. FEAGIN</div>

Emotions and Music

That there is a relation of some kind between music and emotion has always been acknowledged, even by those who doubt that much of aesthetic moment hangs on the fact. The problem, however, is to account for it. A satisfactory account would provide answers to at least three questions: How is the relation between music and emotion grounded? How is it experienced? What value, if any, does the experience of it have, and why?

This way of posing the questions derives indirectly from Eduard Hanslick's *On the Musically Beautiful,* published in 1854, the book that first established the problem as a philosophically interesting one. Hanslick was not the first to write provocatively on music and emotion: the Greeks had attributed the relation between the two to the distinctive characters of the various modes or scales; eighteenth-century thinkers had conceived of music as exciting or imitating those "animal spirits" in whose movements René Descartes had located the origin of emotion; and Arthur Schopenhauer had attested more gloriously than anybody to the sheer depth that a passionate engagement with music can have. But it was Hanslick who sharpened the problem and gave it the focus it still retains—by denying that the value of music has anything to do with emotion at all. "Only on the basis of a number of ideas and judgments," he argues, "can our state of mind congeal into this or that specific feeling." Feeling, so construed, "can only be precisely set forth in concepts" (for instance, "Love cannot be thought without the representation of a beloved person"), and these concepts "lie beyond the scope of music." Music is capable of "reproducing" only "the fluctuations of our inner activity," and these "can be similar with different feelings." Therefore, music cannot express or represent particular feelings; and this means that the relation of music to emotion cannot function as "an aesthetic principle." The beauty and the value of music, Hanslick concludes, "is a specifically musical kind of beauty," which "consists simply and solely of tones and their artistic combination." Thus Hanslick inaugurates musical formalism (in theory, at least: in his criticism he is unexpectedly fond of describing music in emotional terms).

The formalist view that the value of music is rooted exclusively in "tones and their artistic combination" encourages the claim that the relation between music and emotion, though real, is merely *causal,* and hence of little philosophical or aesthetic interest. This claim comes in two main forms. One involves the postulation of an innate or natural psychological mechanism, such that certain musical stimuli automatically produce an emotional response in the listener. A comparatively sophisticated version of this claim has been developed by Leonard B. Meyer. But even his account renders the connection between music and emotion philosophically boring—in much the way that the connection between the smell of food and salivation is philosophically boring. The other form of the claim, of which no sophisticated version is even possible, involves the postulation of an associational mechanism, such that the listener feels an emotion in response to music because he or she associates that music with something else (a love affair, perhaps, or a period of deep gloom). This, too, removes all philosophical and aesthetic interest from the picture. Just as the bell that set Ivan Pavlov's dogs drooling might as well have been an air-raid siren, so music, on this account, might as well be wallpaper, so long as someone has come to associate it with something. Nothing of moment swings here on the relation between music, specifically, and emotion. Hence this account, like the previous one, leaves no room for the antiformalist thought that music might be valuable at least partly in virtue of its relation to emotion.

It is, of course, perfectly possible that these causal accounts contain an element of truth—indeed, they surely do. But neither (nor both together) can be exhaustive. For even if we are sometimes triggered into feeling something by an innate or associational mechanism, we can nonetheless *recognize* an emotional quality in a piece of music without ourselves feeling anything ("What a jolly tune!"). And this fact shows, first, that the power of music to *arouse* emotion is not the only link between music and emotion; and, second, that the philosophically interesting questions cannot be removed by the mere postulation of mechanism.

Observations of this kind have prompted a critical return to Hanslick, and a variety of efforts to avoid his conclusion that the relation between music and emotion cannot function as "an aesthetic principle." Opposite means to this end are well illustrated by two mid-twentieth-century writers. Susanne K. Langer takes issue with Hanslick's claim that emotion "can only be precisely set forth in concepts," and argues instead that emotion can be "set forth" nonconceptually in "presentational symbols"—such symbols being found in pieces of music. Deryck Cooke, by contrast, appears to deny that the conceptual resources required to "set forth" emotion "lie beyond the scope of music": his claim is that music really functions as a "language" of emotion, in which various musical phrases stand, like words, for various kinds of feelings. Neither of these accounts, however, commands much support today. Among their many difficulties, Langer's "presentational symbols" consistently defy explication, while Cooke's linguistic turn is both naive about language and mistaken about the degree to which any stable emotional connotation can be assigned to the same musical phrase in different musical contexts. Hanslick's doctrine of the "specifically musical kind of beauty" can be expected to

withstand criticism from either of these positions, however artfully refined.

In one sense, though, Langer's approach has proved the more durable. If no one now takes the idea of a "presentational symbol" very seriously, it is nonetheless widely agreed that Langer was right to question the claim, central to Hanslick's formalism, that the conceptual inarticulacy of music necessarily debars it from "setting forth" particular feelings. Indeed, skepticism about this claim is clearly discernible in the work of Donald Ferguson, whose book *Music as Metaphor* introduces a number of themes that have become central to the philosophy of music and emotion. Concepts are not everything, Ferguson suggests, for the movement of music "may suggest distinctive characteristics of physical or nervous energy", and these, in conjunction with "the intrinsic motor impulses of consonant or dissonant harmony," may present "an apparently corporeal musical mass whose weight and volume may be adjusted (as, for instance, no dancer's body could be adjusted) to the portrayal" of the "vital impulse to motion activating a sentient being." In effect, Ferguson denies Hanslick's claim that "the fluctuations of inner activity," which music *is* capable of "reproducing," are always "similar with different feelings." Rather, in Ferguson's view, those motions (or "fluctuations") may be distinctive, and so may allow music to "set forth" particular emotions. Much of the remainder of Ferguson's book is devoted, appropriately, to the presentation of evidence in support of this claim, through close musical analyses that remain models of their kind.

Ferguson's position can be construed in either or both of two ways: as asserting a link of some kind between music and the outward manifestations of emotion, that is, expressive human behavior; or as asserting a link of some kind between music and the inward dimension of the experience of emotion itself. These two constructions mark the limits within which most subsequent philosophical investigation has been conducted. Peter Kivy has inclined strongly to the former limit. In a series of books and articles he has revived and reinterpreted the eighteenth-century idea that the relation between music and emotion is grounded in the resemblance of the "contours" of musical movement to the "contours" of characteristic pieces of expressive behavior, including speech—so that "if the criteria of human expression are public, objective, immune from philosophical skepticism, so too are the criteria of expressiveness in music." In Kivy's view, to say of a piece of music that it has the property of "sadness" may be to say something that is literally true, in just the way that it may be literally true to say that the piece has the property of being in E minor. Thus Kivy agrees with Hanslick that the beauty of music is a "specifically musical kind of beauty," but he does not accept that talk of the "specifically musical" is confined "simply and solely" to talk of "tones and their artistic combination."

Rather, the expressiveness of a piece of music is every bit as much a musical feature of it as its tonality is.

This conclusion has won widespread acceptance, even if Kivy's warrant for drawing it is open to dispute. But perhaps the main problem with his account is that it cannot explain why someone might value a piece of music *because* of its expressive properties. To say that Mozart's Requiem would be a different piece of music without its terror-stricken quality is one thing; but to say that Mozart's work is valuable at least partly *in virtue of* its terror-stricken quality is quite another. Kivy's emphasis on the purely cognitive aspects of the listening experience—an experience in which the listener recognizes resemblances between musical and behavioral "contours"—renders him mute on this question. It is not surprising, then, that others have inclined to the alternative view that the crucial link is not so much between music and the outward manifestations of emotion as between music and the inward dimension of the experience of emotion itself. Malcolm Budd, for instance, impressed by Carroll C. Pratt's dictum that "music *sounds* the way moods *feel*," has argued that an imaginative engagement with music can enable "the listener to experience imaginatively (or really) the inner nature of emotional states in a peculiarly vivid, satisfying and poignant form"—and in a form, moreover, whose value is inseparable from the value of the music so experienced.

It seems likely, however, that the truth lies somewhere between these views. For while Kivy's externalist position is hard-pressed to explain why it *matters* that music is expressive, the internalist position is equally at a loss to explain how the imaginative experience of music might be constrained in such a way as to allow the link between music and emotion to function as "an aesthetic principle." A hint here can be gained from Jerrold Levinson. He suggests that "recognizing emotion in music and experiencing emotion from music may not be as separable in principle as one might have liked." If this is correct (and it is certainly persuasive), it is perhaps possible to answer the questions posed at the outset in the following way. How is the relation between music and emotion grounded? In the resemblance between various features of music and the outward (and maybe also the inward) features of emotion. How is it experienced? Imaginatively, so that the listener grasps the expressive quality of a piece of music in coming "to experience imaginatively (or really)" the nature of the emotion expressed. What value does that experience have, and why? It has the value of granting the listener access to states of mind not otherwise available—for the experience described is conceivable *only* as a mode of experiencing particular pieces of music.

These answers may fail to convince, of course. If they do, it may be because Hanslick was right all along. But few contemporary philosophers of music appear ready to accept

that conclusion. Whatever their disagreements, most do seem to be persuaded that the relation between music and emotion is both philosophically interesting and aesthetically important; and that, surely, is just as it should be.

[*See also* Hanslick; *and* Langer.]

BIBLIOGRAPHY

Budd, Malcolm. *Music and the Emotions: The Philosophical Theories.* London and Boston, 1985.

Budd, Malcolm. *Values of Art: Pictures, Poetry, and Music.* London and New York, 1995.

Cooke, Deryck. *The Language of Music.* New York and Oxford, 1959.

Davies, Stephen. *Musical Meaning and Expression.* Ithaca, N.Y., 1994.

Ferguson, Donald. *Music as Metaphor: The Elements of Expression.* Minneapolis, 1960.

Hanslick, Eduard. *On the Musically Beautiful: A Contribution towards the Revision of the Aesthetics of Music.* Translated and edited by Geoffrey Payzant. Indianapolis, 1986.

Kivy, Peter. *The Corded Shell: Reflections on Musical Expression.* Princeton, N.J. 1980. Expanded and reissued as *Sound Sentiment: An Essay on the Musical Emotions* (Philadelphia, 1989).

Kivy, Peter. *Music Alone: Philosophical Reflections on the Purely Musical Experience.* Ithaca, N.Y., 1990.

Langer, Susanne K. *Philosophy in a New Key: A Study in the Symbolism of Reason, Rite, and Art.* Cambridge, Mass., 1942.

le Huray, Peter, and James Day, eds. *Music and Aesthetics in the Eighteenth and Early-Nineteenth Centuries.* Cambridge and New York, 1981.

Levinson, Jerrold. *Music, Art, and Metaphysics: Essays in Philosophical Aesthetics.* Ithaca, N.Y., 1990.

Levinson, Jerrold. *The Pleasures of Aesthetics: Philosophical Essays.* Ithaca, N.Y., 1996.

Meyer, Leonard B. *Emotion and Meaning in Music.* Chicago, 1956.

Pratt, Carroll C. *The Meaning of Music: A Study in Psychological Aesthetics.* New York, 1931.

Ridley, Aaron. *Music, Value, and the Passions.* Ithaca, N.Y., 1995.

Scruton, Roger. *The Aesthetic Understanding: Essays in the Philosophy of Art and Culture.* London and New York, 1983.

Walton, Kendall. "What Is Abstract about the Art of Music?" *Journal of Aesthetics and Art Criticism* 46.3 (Spring 1988): 351–364.

AARON RIDLEY

EMPIRICAL AESTHETICS. The study of beauty and of the arts using the theories and methods of the scientific behavioral sciences, especially psychology, is generally known as empirical aesthetics, a field that began with the publication of Gustav Theodor Fechner's *Vorschule der Ästhetik* (1876). Fechner, who fifteen years earlier had established the entire discipline of experimental psychology, aimed to base aesthetics on an empirical as opposed to a speculative or deductive approach. As such, he opposed the philosophical approach, which he viewed as coming "von oben." That is, philosophical aesthetics attempts to deduce aesthetic principles from abstract generalities. Fechner argued for an induction approach "von unten"—that is, gathering basic data empirically and attempting to devise general principles from them.

Fechner proposed three methods that are still the basic ones used in empirical aesthetics. With the method of choice, people are asked to indicate their preference or liking for objects either by choosing one over another (the method of paired comparisons) or by indicating, as on a rating scale, their degree of preference for a single object (the method of single stimuli). With the method of production, people are asked to create a maximally pleasing object. Unless one is dealing with trained artists, this method only works for the study of very simple stimuli. The method of use involves examination of everyday objects, the notion being that pleasing proportions, colors, and so on, should be most used. For example, if golden section rectangles are pleasing, then they should be found in rectangular objects such as tabletops, writing paper, and picture frames. Fechner, incidentally, did find evidence for preference for golden section rectangles with the method of choice but not with the method of use.

Fechner did not propose a general theory but did lay down a number of aesthetic principles, many of which have stood the test of time. Some of the more important of these are as follows (the evidence in general comes from later research rather than from Fechner's own studies):

1. *The principle of the aesthetic threshold.* There is some threshold (in terms of duration or intensity) that must be exceeded before a stimulus induces feelings of pleasure or displeasure. By way of comparison, for the dark-adapted eye, the threshold for determining the color of a light is much higher than the threshold for detecting whether or not a light is present. Experiments with reaction times suggest that the aesthetic threshold is very close to sensory or perceptual thresholds. That is, virtually as soon as we have perceived a stimulus, we have also determined our preference for it.

2. *The principle of repetition.* Several repetitions may be necessary before we derive maximal pleasure from a stimulus. This is the case for complex stimuli such as classical music. Fechner's principle of fatigue is that too many exposures will lead to a decrease in preference. Massed repetitions of any stimulus will lead to habituation or a decrease in pleasure induced by the stimulus. The simpler the stimulus, the more rapidly habituation occurs. Fechner did not foresee what is called the mere exposure effect (Zajonc, 1980). Distributed as opposed to massed repetitions of virtually any stimulus lead to increases in preference. The more often something has been repeated in this way, the better it is liked.

3. *The principle of association.* Preference is due not only to stimulus characteristics but also to associations. This is the implicit theory of aesthetics of the person in the street. People believe that they like, say, different colors because of things the colors remind them of, when, in fact, their preferences are more parsimoniously explained by factors such as

saturation or lightness. Although not negligible, associations are generally not as important in determining preference as is commonly believed.

4. *The principle of the aesthetic mean.* When stimuli vary along one or more dimensions, we tend to prefer the average stimulus. This is definitely false for some dimensions—for example, loudness—but there is some evidence that that is the case for other dimensions (see the discussion of prototypicality below).

5. *The principle of secondary pleasure and displeasure.* A person's own mood and mental and physical conditions can contribute to preference for aesthetic objects. If we include extreme cases, this is so obvious that essentially no research has been done on the question.

6. *The principle of expression of pleasure and displeasure.* The mood of the environment is transferred to a person and thereby to the object being judged.

7. *The principle of the unified connection of the manifold.* Maximal variety and maximal unity lead to maximal pleasure. This is simply Fechner's version of the widely held principle of unity in variety. For example, a row of columns is hypothetically more pleasing than a single column. Taken in this sense, research supports the principle. A circle made up of smaller circles is more pleasing than a circle made up of squares, which is in turn more pleasing than a circle made up of heterogeneous elements.

8. *The principles of aesthetic contrast and aesthetic sequence.* A stimulus elicits more pleasure if it occurs with or follows a less pleasurable stimulus. Conversely, a stimulus elicits less pleasure if it occurs with or follows a more pleasurable stimulus. Fechner stressed that the stimuli must be quite similar. Later research shows that they must be even more similar than Fechner thought. Such hedonic contrast effects are found for colors that are shades of the same hue but are absent in the case of different hues. Thus, a beautiful shade of red will make a dull shade of red less pleasing but will have no effect at all on shades of blue, green, and so on.

9. *The principle of aesthetic help.* If a variety of noncontradictory stimuli are gathered together, preference for the resultant ensemble is greater than the sum of preference for the component stimuli. A dozen roses, for example, are more beautiful than the preference for a single rose multiplied by twelve. There is not much research on this principle, but I mention it because it is often mistakenly thought that psychological aestheticians believe in a simple sort of hedonic summation whereby we could determine, say, preferences for a painting by simply adding up preferences for the component colors and forms.

10. *The principle of economical application of means.* The least effort applied so as to achieve the maximum effect is maximally pleasurable. Thus, a sketch easily capturing a subject should be more pleasing than a laboriously drawn one. This is probably true if there is a way for people to determine accurately how much effort was involved in producing an artifact.

The structuralist or introspectionist psychologists such as Wilhelm Wundt and Oswald Külpe saw empirical aesthetics as an important subdivision of psychology. Early work on empirical aesthetics, mainly concerned with simple stimuli such as shapes and colors, is carefully surveyed by John Gilbert Beebe-Center (1932). In 1933, the mathematician George D. Birkhoff published *Aesthetic Measure*. In this book, Birkhoff proposed explicit measures of order and complexity for visual forms, music, and poetry. He proposed that the beauty of a stimulus is a function of amount of order divided by amount of complexity. In empirical tests of this contention, Hans Eysenck (1941) showed that Birkhoff's formula describes preference judgments quite poorly. Eysenck proposed that the correct formula is that preference is a function of order multiplied by complexity. That is, people show maximal preference for stimuli that show both maximal order and maximal complexity. As he points out, this is consistent with the philosophical notion of unity in variety. With the rise of behaviorism, interest in empirical aesthetics essentially collapsed among experimental psychologists. Work on the topic, though not usually of an experimental nature, did continue among Gestalt psychologists and their followers such as Rudolf Arnheim (1954) and Sir Ernst Gombrich (1956).

With the collapse of the behaviorist paradigm in the 1960s came a renewed interest in psychological aesthetics. Although he thought of himself as a behaviorist, Daniel E. Berlyne—with his studies demonstrating that curiosity, preference for novelty, and so on, are basic rather than secondary needs even in lower animals—had done as much as anyone to destroy the paradigm. Berlyne's *Aesthetics and Psychobiology* (1971) ushered in the modern era of empirical aesthetics. Berlyne proposed an elegant and general theory of aesthetic preference. The theory is based on the well-established fact that organisms prefer a moderate level of overall arousal or tension, whereas both high and low levels of arousal are unpleasurable. Any stimulus has what Berlyne called arousal potential—the capability to induce arousal. The arousal potential of a stimulus comes from psychophysical properties (properties of the stimulus such as loudness or brightness), ecological properties (meaning or signal value), and collative properties (which derive from comparing or collating different parts of a stimulus, e.g., incongruity, or from comparing the stimulus with expectations, e.g., novelty, surprise). Preference for a stimulus is hypothetically a function of its arousal potential *plus* the arousal potential of surrounding environmental stimuli. Because of habituation, the arousal potential of a stimulus does not remain constant. Habituation causes a stimulus to lose an increment of arousal potential with each repetition. In the case of aesthetic stimuli, Berlyne assumed that colla-

tive properties are the most important determinants of preference.

Berlyne and others amassed a good bit of evidence supportive of his theory. For both logical and empirical reasons, however, the theory has had to be considerably revised (Martindale, 1988). The most important problem with the original theory is that it draws one into an isohedonic trap: according to the theory, any stimuli that elicit equal amounts of arousal must be equally pleasing. This cannot be the case, however. If we determine how much arousal a beautiful painting elicits, we can produce exactly the same amount of arousal with, say, a mild electric shock. Clearly the painting and the electric shock will not be equally pleasing. An empirical problem is that Berlyne vastly overrated the importance of collative variables. In general, meaning has been found to be a far more important determinant of aesthetic preference than are collative variables.

The current consensus among cognitive psychologists is that natural categories are defined not by lists of features but by prototypes. The prototype may be thought of as a best example or "average" of the category. Given the dominance of meaning in determining aesthetic preference, a good deal of recent work has focused on the relationship between prototypicality and preference. It has been known for a century that prototypes formed by taking composite photographs of a group of faces are notably more attractive than faces of individual members of the group. For purely perceptual categories such as colors, there is a strong positive relationship between prototypicality and preference (Martindale, 1988). For more conceptual categories, the relationship is often U-shaped, with preference being shown for both prototypical and exotic or extremely atypical category members.

Berlyne (1974) also introduced important new methods into the field. Earlier studies in empirical aesthetics tended to deal with very simple stimuli. A set of paintings, say, differs along thousands of dimensions, so it was not immediately obvious how to go about studying them. Berlyne introduced techniques such as multidimensional scaling that make the study of complex stimuli tractable. Multidimensional scaling takes a matrix of similarity judgments—obtained by asking people to rate the similarity of a set of objects—and tells us how many dimensions are needed to represent the similarities. A perfect representation of the similarities of 100 objects would require a 99-dimensional space. Multidimensional scaling discards dimensions that are of negligible importance so that we may arrive at the, say, two or three dimensions to which people are really attending. As such, it is a valuable tool for approaching taxonomic questions and for determining what people notice and what they ignore in aesthetic stimuli.

New methods such as computerized content analysis have recently allowed large-scale studies of historical trends in the content of art and literature as well as studies of whether there are stylistic or contentual markers of artistic merit or quality (e.g., Martindale, 1990; Simonton, 1984). Another recent trend is the burgeoning of interest in the psychology of music (e.g., Deutsch, 1982). Although studies of music have tended to focus on perceptual rather than strictly aesthetic questions, they may be viewed as belonging to the broadly defined area of empirical aesthetics.

In 1965, Berlyne—along with Robert Francès, Albert Wellek, and Carmelo Genovese—founded the International Association of Empirical Aesthetics. The biannual congresses of the association have, since that time, provided an international forum for researchers on all aspects of empirical aesthetics as well as allied disciplines such as the psychology of creativity, semiotics, and the sociology of art. The main general journal in the field is *Empirical Studies of the Arts,* the successor to *Scientific Aesthetics,* which Berlyne founded shortly before his death in 1976. Specialized journals include *Music Perception, Psychology of Music,* and *Visual Arts Research.* Christian G. Allesch (1987) provides a comprehensive bibliography of research in the area. Beebe-Center (1932) reviews early work. Reviews of later work may be found in Berlyne (1971), Hans Kreitler and Shulamith Kreitler (1972), Martindale (1988), and Ellen Winner (1982).

[*See also* Arnheim; Gombrich; *and* Psychology of Art.]

BIBLIOGRAPHY

Allesch, Christian G. *Geschichte der psychologischen Ästhetik: Untersuchungen zur historischen Entwicklung eines psychologischen Verstandnisses ästhetischer Phänomene.* Göttingen, 1987.
Arnheim, Rudolf. *Art and Visual Perception: A Psychology of the Creative Eye* (1954). Exp. rev. ed. Berkeley, 1974.
Beebe-Center, John Gilbert. *The Psychology of Pleasantness and Unpleasantness.* New York, 1932.
Berlyne, Daniel E. *Aesthetics and Psychobiology.* New York, 1971.
Berlyne, Daniel E., ed. *Studies in the New Experimental Aesthetics: Steps toward an Objective Psychology of Aesthetic Appreciation.* Washington, D.C., 1974.
Birkhoff, George D. *Aesthetic Measure.* Cambridge, Mass., 1933.
Deutsch, Diana, ed. *The Psychology of Music.* New York, 1982.
Eysenck, Hans J. "The Empirical Determination of an Aesthetic Formula." *Psychological Review* 48 (1941): 83–92.
Fechner, Gustav Theodor. *Vorschule der Ästhetik.* Leipzig, 1876.
Gombrich, E. H. *Art and Illusion: A Study in the Psychology of Pictorial Representation* (1960). 2d rev. ed. New York, 1961; reprint, Princeton, N.J., 1969.
Kreitler, Hans, and Shulamith Kreitler. *Psychology of the Arts.* Durham, N.C., 1972.
Martindale, Colin. "Aesthetics, Psychobiology, and Cognition." In *The Foundations of Aesthetics, Art, and Art Education,* edited by Frank H. Furley and Ronald W. Nepend, pp. 7–42. New York, 1988.
Martindale, Colin. *The Clockwork Muse: The Predictability of Artistic Change.* New York, 1990.
Simonton, Dean Keith. *Genius, Creativity, and Leadership.* Cambridge, Mass., 1984.
Winner, Ellen. *Invented Worlds: The Psychology of the Arts.* Cambridge, Mass., 1982.

Zajonc, Robert B. "Feeling and Thinking: Preferences Need No Inferences." *American Psychologist* 35 (1980): 151–175.

COLIN MARTINDALE

END OF ART. *See* Baudrillard; Danto; Heidegger; *and* Hegel.

ENGLISH AESTHETICS. *See* Addison; Alison; Arnold; Bell; Berkeley; Blair; Bullough; Burke; Coleridge; Collingwood; Cooper; Dryden; Gerard; Home; Hume; Hutcheson; Knight; Morris; Pater; Price; Priestley; Reid; Reynolds; Ruskin; Sidney; Smith; Stokes; Wilde; Williams; Wollheim; *and* Wordsworth.

ENTERTAINMENT. *See* Comics; Cultural Studies; Popular Culture; *and* Television.

ENVIRONMENTAL AESTHETICS. The natural world has long held aesthetic attraction, but environmental aesthetics is only now emerging as a discipline in its own right, with distinctive concepts, issues, and theories. In the last two decades, scholars have begun in earnest to develop the field from scattered beginnings that predate the current environmental movement. Environmental aesthetics does not stand apart from other kinds of research. It draws from philosophy, anthropology, psychology, literary theory and criticism, cultural geography, architecture, and environmental design, as well as from the arts. It is also related to philosophy, in particular, for ontology, ethics, and the theory of art. Moreover, environmental aesthetics has implications as well for governmental policy and social practice. Like environment itself, the field of environmental aesthetics extends broadly and on many levels. This article will explore this field in several directions: historical, conceptual, experiential, and practical.

History. The history of environmental aesthetics before the twentieth century is the history of the aesthetics of nature. An aesthetic interest in nature has probably been present since our hominid ancestors evolved, although physical evidence goes back only some thirty thousand years to the cave paintings and engravings in southwestern Europe. Curiously enough, these early manifestations of art are, in fact, environmental, for the caves were not picture galleries but total environments that even today require active participation. Preliterate societies characteristically exhibit an aesthetic sensibility that is thoroughly integrated into their religious beliefs and everyday activities, giving us good reason to think that an aesthetic aspect infused people's attitudes and practices toward the natural world throughout prehistory.

Since the early history of Western civilization, not to mention the longer duration of Egyptian and Asian cultures, people have found aesthetic value in nature. Twenty-four hundred years ago, Aristotle recognized the beauty and order of nature, while a century later Chrysippus, the Stoic, claimed that the peacock's tail is proof that beauty is a value in nature. An aesthetic sensibility infuses the didactic nature poetry of Virgil and the natural philosophy of Lucretius, while the New Testament writers extolled the lilies of the field, the charm in the bending of a stalk of grain, and the gaping of ripe figs. Nature has long been a source and inspiration for aesthetic appreciation.

Attention to the aesthetic dimensions of landscape developed gradually in Europe during succeeding centuries. As the slow but cumulative process of clearing the forests left great reaches of open land, the practice increased of planting trees deliberately, and people began to discover a sense of pleasure in woodlands. This culminated in the eighteenth century, when attitudes shifted definitively away from regarding wilderness as savage and threatening and toward viewing the forest as a domain of beauty. Indeed, the idea of the picturesque became popular in England at this time to express aesthetic delight in the charm of a rural landscape. The most influential picturesque theorists, William Gilpin, Richard Payne Knight, and Uvedale Price, were in close agreement in rejecting regularity of design and systematic order in favor of irregularity, variation, wildness, change, and decay. The picturesque, moreover, typifies the eighteenth-century aesthetic of gentlemanly contemplative observation. Aided by viewing the landscape through the frame of a Claude glass, a small viewer of convex black or colored glass, whose curved surface reflects the landscape in miniature and reduced color, and framed as if it were a painting, attention was exclusively visual. Attention was directed to surface appearances only, ignoring, for example, the condition of the picturesque rural poor. Yet the interest in the picturesque both aestheticized nature and naturalized art, implying that these have a common mode of appreciation.

Much of the direct aesthetic attention to nature, though, lay in using it as a model for art. The imitation of nature dominated art from classical times through the eighteenth century, and has continued in attenuated form into our own. Gardens often reversed this relationship, applying artistic principles in refashioning nature. The designs of medieval and Renaissance gardens and the formal garden traditions in Italy and France reflect the same careful fashioning as the arts, and, in fact, until the eighteenth century gardens were often included among the fine arts. Other traditions in garden design, such as the Chinese and Japanese, select elements from the larger landscape, often incorporating distant views. In England during the late eighteenth and the nineteenth centuries, the naturalistic landscape designs of William Kent, Humphrey Repton, and Lancelot "Capa-

bility" Brown deliberately shaped the beauty recognized in uncultivated nature and, as was the practice in Asian gardens, incorporated distant views as "borrowed landscapes."

This aesthetic interest in nature signified a shift in attitude toward the environment, a change that is still taking place. The great Kantian synthesis that reconciled the human realm with the natural one was grounded in subjective purposiveness, since purpose could be found in nature and humans both, and it used the power of imagination to direct our response to nature. Yet it was the imagination of poets and essayists more than of philosophers that pursued the interest in natural beauty that began during those centuries. Some nineteenth-century philosophers and writers, such as G. W. F. Hegel, Gustave Flaubert, and Charles Baudelaire, tilted the balance between the natural and the human to find the ideal in art rather than nature. Art and nature achieved greater equality among the Romantics and inspired William Wordsworth to particular eloquence. John Ruskin's tendency to equalize the relation of the human world with the natural was a precursor of certain contemporary views. For him nature is not only the source of visual beauty but retains its original aesthetic value, and the great artist is the one who is able to convey its reality. This connection becomes yet more explicit in Friedrich Wilhelm von Schelling, who united in aesthetic activity both the ideal world of art and the real world of objects.

The theorists and practitioners of the eighteenth and nineteenth centuries attempted to regulate natural beauty by the constraints of design, order, harmony, and, for Immanuel Kant, by purposiveness. During this same period, however, the idea of the sublime emerged as the correlative of beauty, designating aesthetic experience that surpasses the controls that maintain art within rational bounds. As far back as the third century, Longinus had recognized that the sublime possesses the capacity to overpower the constraints of reason and order. He observed that it induces ecstasy in the appreciator, who prances and rears like a horse, fancying himself the creator of the literary work being recited. The sublime, in fact, disrupts the balance of nature and art to make nature preeminent. During the eighteenth century, the sublime assumed a central place in aesthetic discussion. It is not literature, however, but the natural world that became its exemplar, for it is unbridled nature that most strikingly exceeds the constraints of order and limit. Early in that century, Joseph Addison cited one source of aesthetic pleasure in the great, the vast, and the grand, such as a boundless desert, mountain ranges, and a wide expanse of water. Kant, late in the same century, retained Edmund Burke's association of the sublime in literature with the emotion of terror and its power over the imagination, but ascribed sublimity to nature only, not to art. It is the size or might of nature that excites this response, as in the innumerable stars in the heavens or the sands of the seashore, and in contemplating the force of a storm at sea or a torrent of falling water. While this led Kant back to human reason and the moral dignity of human beings, it recognized the capacity of nature, unbounded in its size and power, to surpass the human. Nature, indeed, excites aesthetic admiration in its very excesses.

Over the past two centuries the aesthetic appeal of environment has broadened still further to incorporate architectural and interior design, on the one hand, and to include the city and the commercial and industrial landscape, on the other. The normative range of aesthetics has also been extended to allow for negative values: a significant part of the criticism by the environmental movement is aesthetic in character. Efforts have also been made to place environmental aesthetics on a scientific basis by environmental psychologists, cultural geographers, and others.

Nature as an inspiration for art, the aesthetics of nature, ideas of the picturesque and the sublime, and environmental aesthetics are all different, yet they can be seen as stages in a single evolutionary process. The recognition has grown, too, not only of the power, magnitude, and importance of nature, but also of its fragility. By the last part of the twentieth century, this increased awareness has combined with a heightened recognition of our interdependence with nature to encourage the reinterpretation of nature as environment. Such a modification of the sense of nature has also stimulated efforts to fashion ideas that reflect its aesthetic dimension and to understand the human place in an aesthetics of environment. Replacing the concept of nature with that of environment is more than a shift in terminological style: it represents a stage in the transformation of our understanding.

Terminology and Scope. Perhaps here more than in most fields, the use of familiar terms in environmental aesthetics is tendentious. Words like *nature, landscape,* and *environment* carry with them long associations with conventional beliefs and philosophical thought that make it difficult for environmental aesthetics to develop as a free and independent field of study. Social influences on our understanding of the natural world are powerful, while ethnic traditions incorporate a wide range of beliefs and practices concerning environment. When theories of beauty and aesthetic value are directed to environment, they face a multitude of meanings and practices that are even more varied and diffuse than with art. In fact, the scope of environmental aesthetics is still more difficult to circumscribe than that of art.

Environment is a word that has acquired many meanings. Taken narrowly in its etymological sense, it denotes the region that surrounds something (Fr. *en* in + *viron* circuit). Conceived most broadly, it is sometimes confused with *ecology,* which refers to the complex set of relations that bind an organism to its environment, or with *ecosystem,* which signifies those relations taken as a functional, interactive system of organisms and their environment. Most meanings of en-

vironment lie within this range and have a highly significant implication: they retain to some degree the assumption of an object and its surroundings or a self and its setting, bound together in varying degrees of intimacy but ultimately distinct and separate. This dualistic presumption has been challenged increasingly by scientific and philosophical developments in the twentieth century. Relativity physics, quantum mechanics, and operationalism, together with the increasing influence of philosophical pragmatism, phenomenology, and hermeneutics all deny the principal presupposition of our conventional terminology that organism and environment are discrete entities that can be located and defined separately. These more recent developments have encouraged attempts to understand environment in ways that incorporate ideas of inclusion and continuity. The biological concept of an ecosystem is one such proposal. Another, still more general one, is to consider environment as a unified field incorporating a complex order of animate and inanimate objects bound together in spatial and causal transactions, and whose fluid boundaries respond to geographical conditions, human activity, and other such influences.

Landscape raises similar difficulties. Geographers tend to define it in visual ways as the space of the earth's surface that extends from the viewer's eye to the horizon. A similar understanding is reflected by the common idea of a natural landscape as an expanse of scenery seen in a single view, and in the genre of landscape painting as representing such a view. Yet with landscape, too, we can identify a range of alternatives, from the panoramic landscape that easily turns an environment into a visual object, to the participatory landscape that incorporates the appreciator perceptually and relinquishes any sense of separateness. As with environment, the issue of separation or continuity with the perceiver remains problematic.

These alternatives suggest that the heart of the problem lies in different conceptions of *nature,* the most inclusive of these terms. Perhaps we can recognize the changing meanings and implications of the alternatives more clearly here. The familiar notion of nature as everything outside the human sphere places the natural realm separate and apart. Until the early decades of this century, natural science adopted and refined this model, considering nature as an objective system of spatiotemporal objects and events that can be designated in specific and absolute terms. Related to this conception is the familiar idea of nature as that part of the world unchanged by human action, a view codified in the conventional distinction between the natural and the artificial.

As a consequence of increasingly influential developments in philosophy and theoretical physics, on the one hand, and of pressing ecological issues and the conservation movement, on the other, people have begun to rethink the customary understanding of nature. Challenges have been mounted against the many practical and legal barriers that circumscribe the scope of what is regarded as part of the natural world and that attempt to restrict its effects. The difficulty, perhaps the impossibility, of locating regions on the earth that have not been affected in some significant way by human activity, and the recognition that human actions have transformed the planet, have led to the realization that the distinction between the natural and the artificial no longer holds. In the light of what we now know about the far-reaching effects of human actions—on the atmosphere, on the oceans and seas, on the polar ice caps, on climatic patterns and geographical features—it is no longer plausible to think of nature, in any significant sense, as separate from humans. Nor, conversely, can we insulate human life from the reciprocal effects of these changes. We are all bound up in one great natural system, an ecosystem of universal proportions in which no part is immune from the events and changes in the others. The natural world is, then, incorrigibly artificial and, in the largest sense, includes human beings and human works. We can only conclude that nature has become all-embracing, either in Baruch Spinoza's sense of a total order or in Martin Heidegger's sense of existential habitation, of dwelling poetically. This brings the issue back to aesthetics.

Environmental aesthetics encompasses these divergent ideas, and its various meanings reflect the disciplinary interests and goals of different investigators. Environmental psychologists, urban and regional planners, and other behavioral scientists commonly associate environmental aesthetics with the visual beauty of landscapes. They attempt to measure it quantitatively through studies of preferential selection and behavior with the goal of formulating guidelines for design decisions and for governmental environmental policies. *Aesthetics* here is usually taken as what is visually pleasing. Factors such as coherence, complexity, legibility, mystery, and attractiveness are identified for purposes of empirical studies, which determine and measure their value from the preferences of experimental subjects. Others, including philosophers and some social scientists, consider the quantitative bias of such empirical research to be restricted and even flawed by being conceptually naive, perceptually undiscriminating, and heavily assumptive. Some choose a qualitative orientation and identify environmental aesthetics with the beauty of objects or scenes as apprehended by a skilled viewer. Those who adopt a phenomenological approach emphasize the activity of perception, the formative contribution of the perceiver in aesthetic experience of environment, and the fundamental reciprocity of perceiver and environment.

In its largest sense, environmental aesthetics denotes the appreciative engagement of humans as part of a total environmental complex, where the intrinsic experience of sensory qualities and immediate meanings predominates. The experience of environment as an inclusive perceptual system includes such factors as space, mass, volume, time,

movement, color, light, smell, sound, tactility, kinesthesia, pattern, order, and meaning. Environmental experience here is not exclusively visual but actively involves all the sensory modalities synaesthetically, engaging the participant in intense awareness. Moreover, a normative dimension suffuses the perceptual range, and this underlies positive or negative value judgments of an environment. Environmental aesthetics thus becomes the study of environmental experience in the immediate and intrinsic value of its perceptual and cognitive dimensions.

In practice, we always inhabit a particular environment whose boundary is the horizon of our perceptual field. Yet distinguishing among individual environments may be more troublesome than helpful, for their boundaries may overlap or be defined differently, and the environments themselves may merge with one another. Moreover, there are different types of environment and these have different emphases.

Architecture, not ordinarily thought of as a form of environment, is coming to be regarded more as the design of the built environment rather than of isolated physical structures. Architecture shapes both interior and exterior spaces. It crafts volumes, surfaces, and patterns of movement for various purposes—domestic, industrial, governmental, celebratory. Moreover, architectural structures occupy sites that are contiguous with other environmental configurations and are often part of an urban area. The aesthetics of the architectural environment therefore merges with that of landscape architecture, as its concerns move beyond the physical boundaries of a structure to embrace its connections to its site. Architectural aesthetics also coalesces with urban design by including relationships and groupings of multiple structures and patterns of human activity.

Landscape aesthetics concerns larger domains, often defined visually, as we have seen, yet not necessarily so, as we begin to understand the aesthetic inhabitation of a landscape. On one end of the spectrum it may include landscape architecture, from foundation planting and landscaping to the design of gardens and parks as perceptual wholes. On the other end of its range, it may reach to the perceptual horizon, even extending to a geographic region perceived cumulatively as a whole, because of similar or complementary landforms and vegetation or by unifying human activity. Landscape aesthetics, understood most generally, may be thought of as synonymous with the aesthetics of environment or with the aesthetics of nature.

Urban aesthetics focuses on a special landscape, the built environment, shaped almost entirely by human direction for human purposes. We do not, however, have to oppose the city aesthetically to the countryside or to wilderness, a common tendency. The city is rather a particular environment, made from materials obtained or derived from the natural world and embodying the same perceptual elements as other environments, but designed and controlled by human agency. Moreover, although the city is a distinctively human environment, it is nevertheless an integral part of the geography of its region, from which it usually has no distinct boundaries and with which it has a reciprocal relation.

Urban aesthetics deals with the same perceptual factors that are part of all environmental experience. And as the preeminent cultural environment, the city's social and historical dimensions are inseparable from its sensory ones. Aesthetic value here, then, is more than a matter of urban beauty; it encompasses the perceptual experience of meanings, traditions, familiarity, and contrast, as well. Further, urban aesthetics must also include a consideration of negative aesthetic values: the obstruction of perceptual interest by noise pollution, air pollution, strident signage, utility lines, littered streets, dull, trite, or oppressive building designs, and the destruction of traditional neighborhoods. Indeed, an aesthetic critique should be a key factor in evaluating a city's character and its success. To incorporate aesthetic considerations into urban planning is to place the city in the service of the values and goals that we associate with the full meaning of civilization.

Environmental Appreciation. The aesthetic appreciation of environment draws people to many different activities: visiting gardens and parks, hiking, camping, bicycling, canoeing, sailing, and even gliding and flying. For some it is the main appeal of certain sports, such as golf. Yet while environment is increasingly appreciated for its aesthetic character, there is active debate on the explanation and theory of that appreciation.

Certain complexities make it even more difficult to develop an aesthetics of environment than of art. For one thing, environment involves perceptual categories that are wider and more numerous than those usually recognized in the arts. No single sense dominates the situation; rather, all the modes of sensibility are involved. Sight, touch, hearing, smell, and its correlate, taste, are all active in environmental experience. So, too, is the proprioceptive awareness of internal muscular and visceral sensations and the kinesthetic sense of movement. Cognitive factors often are present in environmental appreciation to a degree found only occasionally in art. An awareness of geological and land use history, for example, is very much part of traversing an old footpath, viewing an eroded hillside, or climbing the rough heights of a morainal landscape.

Second, in part because of the range and activity of environmental perception, it is more difficult and often impossible to adopt the model of appreciation traditionally assigned to the arts. In fact, it is necessary to question the suitability for environment of traditional ideas of aesthetic appreciation. The usual account stresses a contemplative attitude, a receptive attention toward a special object separated from the appreciator and taken in isolation. This seems to lend itself to some environmental occasions, such as viewing a for-

mal garden or enjoying the vista from a scenic overlook. Yet, much environmental experience requires more active participation, such as strolling through a garden, hiking along a mountain trail, paddling a canoe down a flowing stream, or driving through a scenic countryside. Even when an environment does not demand physical engagement, part of its appeal lies in the magnetic forces that seem to emanate from it. One can feel the invitation of an entryway or the pull of a serpentine garden path. Even when standing still, the embrace of a sunset can draw one into an intimate relation. Such experiences make it difficult to accept the usual account of appreciation as disinterested contemplation, and this has led to theoretical accounts, such as pragmatic and phenomenological ones, that emphasize the active qualities of environmental experience. Some, indeed, have argued for an aesthetic engagement that rejects the traditional separation of viewer and object of appreciation in favor of their total absorption in environment. Not only do the theoretical demands of environmental aesthetics require a radical rethinking of traditional aesthetic theory; they raise doubts about the appropriateness of that tradition for the arts themselves.

Redefining aesthetic appreciation requires expanding other traditional aesthetic concepts when they are applied to environment. Beauty, for example, no longer concerns the formal perfection of a prized object but becomes the pervasive aesthetic value of an environmental situation. That value, moreover, is measured less by formal traits than by perceptual immediacy and intensity in enhancing the intimate bond of person and place. The sublime comes into its own as an aesthetic category here. It designates experience that is not so much in contrast to beauty as an aesthetic force that comes from the sense of being part of a perceptual matrix of overwhelming magnitude or power. Creation, often important for the theory of the arts, is transmuted into an awareness and awe of natural processes coupled with the formative contribution of an active, participating perceiver.

Implications and Applications. Theoretical considerations and practical purposes are inseparable in environmental aesthetics, and properly so. As the landscape of daily life and the landscape of appreciation are indivisible in practice, so too are the theory of environmental aesthetics and its applications. The importance of the aesthetic dimension of environment becomes increasingly compelling as industrialization and its products irretrievably affect the land surfaces, the seas, and the atmosphere of our planetary environment.

Environmental aesthetics can make a distinctive contribution to many areas of environmental discussion, although its influence so far has been limited. Empirical studies of landscape preference have been used in formulating policy for parks, forests, and wilderness preserves, but theoretical and critical debate on environmental appreciation, though

highly relevant, has until now had little effect. Critical aesthetic studies of specific urban and rural environments have implications for planning, where decisions affect the quality of our experience of city and countryside. Debate on issues such as wilderness preservation, safeguarding endangered species, and conservation tends to center on ethical issues, but some have argued that the underlying foundation of such ethical concerns lies in aesthetics. Architecture, landscape architecture, and urban design include aesthetic values as an integral part of the discipline. Zoning and other land use practices may embody aesthetic considerations explicitly, as in the desire to maintain scenic views, but always implicitly, since such practices cannot help but affect the quality of environmental experience. In clarifying concepts such as environmental aesthetic value, scenic beauty, and aesthetic degradation, environmental aesthetics can contribute to formulating clearer and sounder criteria for judgment. In contributing to methods for assessing aesthetic value, it can help raise the quality and sophistication of standards. And in developing a better understanding of environmental appreciation, aesthetics can extend the scope of aesthetic appreciation. Whether in agriculture or forestry, town planning or landscape design, environmental aesthetics can contribute to increasing the awareness of the aesthetic dimension of environment in the education of professionals in these fields and in the practice of these and similar environmental professions. Most generally, it is important to argue for the inclusion of aesthetic value, along with economic, biological, and ethical values, as an essential factor in all environmental decision making.

The scope of the aesthetics of environment therefore extends far beyond the conventional limits of the work of art as an aesthetic object created for contemplative appreciation. Any environment in which an aesthetic aspect is significant possesses aesthetic value. Any human context in which the aesthetic dimension predominates is an aesthetic environment: policies or actions that affect its aesthetic value act to that extent, then, as arts of environment. In some contexts, such as architecture and landscape design, a sense of the aesthetic has long been cultivated; in others, such as city and regional planning, it has yet to be even recognized. While there is a sensitivity to the aesthetic character of interior design, architecture, and landscape architecture, this has had little theoretical articulation. Yet by focusing on our immediate, active environment, these design fields challenge the object-oriented aesthetic of traditional theory. And by integrating an aesthetic function into their practical one, they embody the denial of the convention that separates beauty and practice into separate, uncommunicating camps.

Recognizing aesthetic values in the broader environment is more recent. As the natural landscape is overtaken by human uses and engulfed by urban sprawl, the threat to these values has become increasingly grave and their importance

more commanding. As a result, landscape assessment has developed in recent years along with the environmental movement. It attempts to identify landscapes whose beauty is a value for society as a whole, and therefore whose preservation lies in the public interest, although as we have already noted, the quantitative measures commonly adopted are partial, at best. The aesthetic dimension of urban design and regional planning has not yet achieved clear recognition in the United States, although in some other countries it is debated intensely. In political forums and in planning circles, practical and economic considerations are virtually unchallenged as the sole determinants of design decisions, although a literature on urban aesthetics is beginning to appear. Another theoretical consequence of environmental philosophy is that all these areas exhibit the intimate relation of the aesthetic domain with the ethical. The influence of environmental conditions on health, contentment, human fulfillment, and happiness is powerful and pervasive, and the aesthetic character of those conditions is a major factor in that influence. Aesthetic values are also entering into discussions about the future of wilderness areas and other environmental resources, where economic interests involving short-term private gain and local economic benefits are pitted against long-term public values.

A number of scholars have contributed to the development of the field of environmental aesthetics. The early essays of R. W. Hepburn had a wide influence in stimulating interest in the aesthetics of nature. Allen Carlson's compelling arguments for a cognitivist approach in the aesthetic appreciation of nature, and especially for the role of scientific knowledge, have evoked considerable debate. Yrjö Sepänmaa was one of the first to develop a systematic view of environmental aesthetics that includes works of art and environments other than nature. Arnold Berleant has joined environment with the arts in a contextual account that leads to a reconsideration of traditional aesthetic theory on the basis of environmental experience. These scholars and others have helped shape the direction of philosophical discussion on the aesthetics of environment.

Recent work in environmental aesthetics has pushed inquiry in further directions. One of these is the aesthetics of gardens, considered historically, cross-culturally, and in relation to the traditional arts and crafts. Another is comparative aesthetics, particularly the comparison and interrelation of Western with Japanese and other Asian traditions. Sensitive studies of the aesthetic character of specific environmental features, such as snow, rain, individual colors, and light and shadow, have been highly revealing and theoretically significant. The discussion of issues includes such topics as the comparison of nature to art, the relation between the built environment and the natural one, the history and character of landscape appreciation, and the aesthetic criticism of specific environments. These areas and issues will undoubtedly continue to multiply as environ-mental aesthetics achieves greater visibility and a clearer identity.

The aesthetics of environment is, then, a rich and varied field, and its significance is just as broad. It extends the range of traditional aesthetics to encompass the many different settings in which we participate, from rooms and buildings to streets and neighborhoods, from villages and cities to the countryside and the wilderness, from the presence of an individual to the active engagement of a social group. In all these, environment must not be construed as our material surroundings alone but rather as the socio-physical context in which we participate. The salience of different factors in an environment varies with the individual case. Sometimes physical features, such as mountains, rivers, or a great tree or rock, have a dominating influence and give the situation its prevailing tone. Sometimes the social aspect predominates, as in a classroom, a party, or a sporting event. Whatever the situation, environment is always inclusive and it encompasses a multiplicity of physical, social, and perceptual features.

The significance of environmental aesthetics is equally broad. Its study demonstrates the need to restructure aesthetic theory to accommodate the varied circumstances and applications outside the arts in which the aesthetic emerges as an important factor. For an adequate theory should be able to account equally well for the presence and character of aesthetic experience in all its occurrences, whether in art or in other domains of human culture. This includes acknowledging the presence of an aesthetic factor, not only in the landscape, but in environments of all sorts, including human situations and social relationships. Aesthetic value is not only omnipresent in environment but universal in its scope.

[*See also* Landscape; Nature; *and* Picturesque.]

BIBLIOGRAPHY

Berleant, Arnold. "Architecture as Environmental Design." In *Art and Engagement*. Philadelphia, 1991.

Berleant, Arnold. *The Aesthetics of Environment*. Philadelphia, 1992.

Berleant, Arnold. *Living in the Landscape: Toward an Aesthetics of Environment*. Lawrence, Kans., 1997.

Berleant, Arnold, and Allen Carlson, eds. Special Issue on Environmental Aesthetics, *Journal of Aesthetics and Art Criticism* 56 (1998).

Bourassa, Steven C. *The Aesthetics of Landscape*. London and New York, 1991.

Carlson, Allen. "On the Possibility of Quantifying Scenic Beauty." *Landscape Planning* 4 (1977): 131–172.

Carlson, Allen. "Appreciation and the Natural Environment." *Journal of Aesthetics and Art Criticism* 37.3 (Spring 1979): 267–275.

Carlson, Allen, and Barry Sadler, eds. *Environmental Aesthetics: Essays in Interpretation* Victoria, B.C., 1982.

Hepburn, R. W. "Nature in the Light of Art." In *Philosophy and the Arts*, edited by Godfrey Vesey, pp. 242 258. London, 1973.

Hepburn, R. W. "Contemporary Aesthetics and the Neglect of Natural Beauty." In *"Wonder" and Other Essays*, pp. 9–35. Edinburgh, 1984.

Kemal, Salim, and Ivan Gaskell, eds. *Landscape, Natural Beauty, and the Arts*. Cambridge and New York, 1993.

Miller, Mara. *The Garden as an Art.* Albany, N.Y., 1993.

Nasar, Jack L., ed. *Environmental Aesthetics: Theory, Research, and Applications.* Cambridge and New York, 1988.

Seamon, David, ed. *Dwelling, Seeing, and Designing: Toward a Phenomenological Ecology.* Albany, N.Y., 1993.

Seamon, David, and Robert Mugerauer, eds. *Dwelling, Place and Environment: Towards a Phenomenology of Person and World* (1985). Reprint, New York, 1989.

Sepänmaa, Yrjö. *The Beauty of Environment: A General Model for Environmental Aesthetics.* 2d ed. Denton, Tex., 1993.

ARNOLD BERLEANT

EPICURUS. *See* Hellenistic Aesthetics.

EPISTEMOLOGY AND AESTHETICS. [*This entry explores how certain philosophical issues raised in epistemology (that is, the theory of knowledge) are also relevant to aesthetics.*]

Claims to knowledge feature prominently in the way that we talk about art. Critics often claim to know something about the works of which they speak and write that, for instance, Ivan Turgenev's *A Month in the Country* is a tragicomedy, or that a production of it is elegant or good or clumsy or inept. They also claim to know that certain emotional responses to works of art are mistaken or unwarranted or that others are appropriate. It is fitting, some think, to despise John Willoughby for his inconstancy in Jane Austen's *Sense and Sensibility,* and to admire Elinor Dashwood for her self-control in the face of Lucy Steele's taunting behavior. More obviously, perhaps, people also claim to learn from art, and it is widely maintained of insightful works that they engender beliefs about the world in which we live or that they can even afford knowledge of that world. Thus, for instance, many of the critical reviews of Tom Stoppard's *Arcadia* claim that the play teaches us something about the pretensions of academics and the difficulty of understanding past lives.

There are, thus, at least three different types of knowledge claims about the arts. These are distinguished by their objects. The first concerns what we know (or believe) *about the work* and whatever imaginary worlds that it posits; the second is concerned with what we know (or believe) to be an *appropriate emotional response* to the work; and the third is concerned with what art enables us to know or understand *about the actual world* in which we live.

All three claims to knowledge have been the object of scholarly discussion within aesthetics. The first concerns the validity and objectivity of criticism. Can interpretations and evaluations be true and be known to be true? Or is their truth always relative to a certain culture or period within a culture? Is the "knowledge" that a work means such and such always knowledge relative to a set of cultural myths? The second concerns belief that works of art can properly

be understood only if we have appropriate emotional responses to them; and the problem, of course, is to know which emotional responses are appropriate, which inappropriate to a particular work. The third concerns the widespread idea that some works of art convey understanding of, and insight into, human affairs, and can bring people to have new beliefs about and knowledge of the world in which they live.

The focus here is with the third of these claims, for it obviously challenges a good deal of mainstream epistemology. After all, much of the art that we regard as instructive, insightful, or informative is also fictional. It is art bred of a fanciful or creative imagination—an imagination that does not purport to represent or describe the world as it actually is. An empiricist epistemology cannot easily explain how it is possible to learn about actual situations from such works. The worlds that they delineate do not exist, and so cannot furnish us either directly or indirectly with the experience that empiricism requires for genuine knowledge. Nor will a rationalist epistemology do service. Both rationalists and empiricists are scathing of the idea that the creative or fanciful imagination can properly be invoked in an explanation of knowledge. Rationalists, such as René Descartes and Nicolas de Malebranche, dismissed "the blundering constructions of the imagination" as a source of knowledge (Descartes, 1931, p. 7). Empiricists, such as Francis Bacon, Thomas Hobbes, John Locke, David Hume, and, eventually, the neorationalist Immanuel Kant, all rejected fantasy (the creative imagination) as a source of knowledge. According to Locke, for instance, "if the fancy be allowed the place of judgement at first in sport, it afterwards comes to usurp it" (Locke, 1961, vol. 1, book 2, sect. 33, pp. 336–337). Kant is similarly dismissive in the *Anthropologie* of "fantasy," which he sees as an "unruly" imagination incapable of affording knowledge.

Whatever their achievements, the epistemological theories of the Enlightenment clearly encouraged the unfortunate but widely held belief that only empirical science can furnish us with useful knowledge about the world. Perhaps as a result, Anglo-American philosophers have continued for most of this century to speak of the growth of knowledge only within the context of scientific inquiry. This is strange, especially in light of the fact that many people claim to learn, and to learn a lot, from art. According to Hans-Georg Gadamer, an emphasis on empirical science as the only legitimate source of knowledge tends to cut people off from other sources and fountainheads—in this case, artistic sources—of knowledge. If, however, one thinks that at least some art is by its nature informative or insightful, we at once assume that art is not always an end in itself, is not purely autonomous, but can properly have a didactic function. This, of course, is controversial, but it is not a controversy to which attention can be given here.

Of more importance is the fact that any explanation that we offer of how people learn from art invariably assumes

some or other theory of knowledge. What we need to do is to determine what sort of epistemology would best enable us to explain how people acquire not just beliefs, but understanding, insight, and knowledge from art. Unfortunately, and with only a few exceptions, Anglo-American aestheticians have not engaged robustly with this question. One of the earlier exceptions was John Hospers, who claimed that we acquire knowledge about the world from art by "a kind of transference" that is based on analogical or inductive inferences (Hospers, 1946). In so arguing, Hospers subscribed to an account of how people may come to learn and know. What he did not do is explore his assumed epistemology in order to see whether we really can learn anything in this way. There are some reasons for thinking that we cannot, for we can only learn what we do not already know or believe, and it is arguable that induction of the sort referred to by Hospers involves the application in new situations of what we already know or believe. When, for instance, one infers inductively that the trees in the garden will shed their leaves during the coming fall, one merely applies what one already knows about these trees to the next season, and do not appear to learn anything new in so doing. In just the same way, when the resemblances between the imaginary world of a novel and the actual world are so obvious that inductive inferences about the actual world are straightforwardly drawn from the novel, it is at least arguable that such inferences occur only because we *already* believe or know that what we have learned of fictional situations applies to their actual counterparts. If we already know or believe this, then we do not learn anything new by inferring inductively from the one to the other. Some other and better account of how we learn from works of art is required.

Since contemporary epistemology has said so little about art as a source of knowledge (Bender, 1993), and since rationalist and empiricist epistemologies clearly are not equal to the task, it is helpful to look to the Romantic movement, which, it is well known, protested vigorously against the passive empiricist accounts of knowledge encouraged by Locke and Hume. Johann Fichte, Friedrich Wilhelm von Schelling, G. W. F. Hegel, and Arthur Schopenhauer in Germany, like Samuel Taylor Coleridge and William Wordsworth in England, took from Kant the idea that our concepts or our ideas help fix the nature of reality. Unlike Kant, however, they believed that the imagination is free to create its own concepts or ideas, so that (on the Romantic view) there is no fixed, unchanging core of concepts, given a priori, in terms of which to experience the world. Since, in addition, the Romantics discarded Kant's distinction between the phenomenal and noumenal world, they contended that our imaginatively produced concepts do not just determine our experiences, but help produce the world as it really is.

A Romantic epistemology thus emphasizes the role of the imagination, not just in our experience of the world, but in the creation of the world itself. It is thus closely allied to Romantic idealism—a position that is notoriously difficult to defend. This notwithstanding, any epistemology that emphasizes the active and creative role of the imagination in the genesis and growth of knowledge seems well placed not just to explain how we come to know the world about us, but also to explain how we learn from art.

There are very good reasons for thinking that the creative imagination plays a crucial role in the genesis and growth of all knowledge. When, for instance, we run out of knowledge or established belief with which to solve a problem, we are forced to conjecture, to guess, to imagine creatively. Our guesses or imaginings are made to do service, and, if successful, will eventually pass as knowledge. The newborn infant is required, on pain of extinction, to make sense of its environment but can do so neither by bringing objects under empirical concepts (since it has none), nor by induction or deduction (since there is nothing to induce or deduce from). Even if we allow that the infant is to some extent "hardwired," and even if it is helped in some measure by its parents, it will still need to guess (or imagine creatively and tentatively) in order eventually to make sense of the many different objects and events of its environment (Novitz, 1987, chap. 2).

Let us suppose, then, that something like a Romantic epistemology is correct. Can it really help us to explain how people learn from art? It seems that it can. Works of art, most especially fictional literature, invite viewers to imagine derivatively what the artist has imagined creatively. They give readers new mental sets, new categories, concepts, and possibilities in terms of which to construe their own experience of the world. In this way, they bring readers to experience the world differently. For instance, in reading a novel readers will often imagine what it is like to be in a specific situation; this will alert them, feelingly, to the demands that the situation makes on those who are caught up in it, and brings them to experience "from the inside" the pitfalls and the advantages of the situation—whatever it may be. In light of this, readers inevitably formulate certain hypotheses and beliefs about what it feels like to be in such situations, and to the extent that these are borne out in the readers' experience, they may pass as knowledge—as something, moreover, that has been learned from the work (Novitz, 1987, chap. 6).

Imaginative projection of this sort enables a reader to learn in ways that are importantly different from the dispassionate, more or less passive, observation encouraged by the empirical sciences. Bernard Harrison agrees, and argues that such learning is "dangerous" since it requires "personal change as a condition of coming to possess it" (Harrison, 1991, p. 3). In arguing along these lines, Harrison explicitly rejects the paradigm of knowledge handed down to us by Locke and other Enlightenment philosophers "which includes the ideal of a unified science whose basic deliver-

ances would be expressed in an austere, pre-theoretical language of bare empirical description" (Harrison, 1991, p. 9).

Since the imagination can work variously with very different effects, a Romantic epistemology helps us to avoid a common error that besets a good deal of the contemporary literature on this topic—the mistake, that is, of thinking that there must be a single way in which we learn from art. There are many different things that we learn from art—so different, in fact, that it seems most unlikely we will learn them in a single way. Not only do we claim to acquire factual beliefs and propositional knowledge from fiction, we also claim to acquire cognitive, conceptual, intellectual, and strategic skills, values, and an empathic understanding of what it is like to be caught up in certain situations. To suggest that there could be just one way in which we learn such a diverse array of things is as silly as the claim that we learn to play ice hockey in the way that we learn algebra.

The versatility of the creative imagination, plus the role accorded it by the Romantics in the acquisition and growth of knowledge, enables us to explain the different ways in which we learn from art. It is through our capacity to play imaginatively by combining or suspending at will various ideas, concepts, or images that we acquire different ways of construing our experience and that we discover new possibilities, many of which are insightful, instructive, helpful, and informative. It is a way, Gadamer has argued, of enlarging our understanding by getting us tentatively to suspend the categories or "prejudices" in terms of which we see and understand. In the end, through a process that he calls the fusion of horizons, we come to understand differently (Gadamer, 1989).

The epistemology advocated by Gadamer is effectively a Romantic epistemology. Whether such an epistemology is ultimately tenable, and whether it can properly be divorced from the antirealism that has traditionally accompanied it, are important issues that need further attention.

[*See also* Alienation, Aesthetic; Fiction; Imagination; Interpretation; Literature, *article on* Literature and Cognition; *and* Perception.]

BIBLIOGRAPHY

Bender, John W. "Art as a Source of Knowledge." In *Contemporary Philosophy of Art: Readings in Analytic Aesthetics,* edited by John W. Bender and H. Gene Blocker, pp. 593–607. Englewood Cliffs, N.J., 1993.
Descartes, René. *The Philosophical Works of Descartes.* 2 vols. Translated by Elizabeth S. Haldane and G. R. T. Ross. Cambridge, 1931.
Gadamer, Hans-Georg. *Truth and Method.* 2d rev. ed. Translation revised by Joel Weinsheimer and Donald G. Marshall. New York, 1989.
Harrison, Bernard. *Inconvenient Fictions: Literature and the Limits of Theory.* New Haven, 1991.
Hospers, John. *Meaning and Truth in the Arts.* Chapel Hill, N.C., 1946.
Locke, John. *Essay concerning Human Understanding.* 2 vols. Edited by John W. Yolton. London, 1961.
Novitz, David. *Knowledge, Fiction, and Imagination.* Philadelphia, 1987.
Walsh, Dorothy. *Literature and Knowledge.* Middletown, Conn., 1969.

DAVID NOVITZ

ESSENTIALISM. [*This entry comprises two essays that explain essentialism in aesthetics, that is, the attempt to uncover the "nature," "necessary and sufficient conditions," or "definition" of art:*

Conceptual Overview
Anti-Essentialism

While the first essay clarifies what essentialism is in philosophy and aesthetics, the second examines the critiques of essentialism in twentieth-century analytic and Continental aesthetics. This critique is one of the principal conceptual connections between these otherwise distinct, often competing, seemingly incompatible aesthetic traditions. For further discussion, see Anti-Art; Definition of Art; Historicism; Ontology of Art; *and* Relativism.]

Conceptual Overview

By *essentialism* is understood "belief in essence or essences." The distinction here between singular and plural forms of *essence* is not philosophically important, except possibly for Baruch Spinoza. The term essentialism ranges in meaning, depending on how *essence*, which is used variably by different philosophers, is understood. Thus, Sir Karl Popper in his *Open Society and Its Enemies* understands by essentialism Plato's theory of forms, since he identifies Plato's Forms with essences (vol. 2, pp. 5–6), but it is possible to be an essentialist without making this identification. Plato's thought is essentialist, but not all essentialism is Platonist. For example, *essentialism* has been used recently in aesthetics to mean the essential properties, if any, that a work of art must have. (See, for instance, Currie, 1990.)

The doctrine of essences belongs to metaphysics. Its bearing on aesthetics is quite difficult to chart, since contemporary aesthetics has been hostile to it, particularly Anglo-American aesthetics, the temper of which, at any rate, was until recently resolutely anti-essentialist. (See Diffey, 1973, and Åhlberg, 1995.) Anti-essentialism is also a powerful current in contemporary cultural and gender studies. But since the search for an adequate definition of art has been a dominant concern of much twentieth-century Anglo-American aesthetics and read back by this tradition into the history of aesthetics, where it is also seen as a major concern, an excursion into these other branches of philosophy such as metaphysics is important for understanding what is at stake in the question of the definition of art.

The most important writers on essence are Plato, Aristotle, John Locke, George Santayana, and Robin George Collingwood. (Saul Kripke's *Naming and Necessity* on essential properties is also important, but as a contribution to the theory of names and not, or has not yet been shown to be, important to aesthetics.)

The account of essence given by Aristotle, particularly in his *Metaphysics,* is the most important source in philosophy

for the topic. Essence is one of the ways in which substance is defined, is one of the marks by which we determine substance, and, according to Aristotle, "Substance alone is definable." Aristotle offers four meanings for *substance*, only two of which are pertinent here: (1) simple bodies that "are not predicated of a subject while everything else is predicated of them" and (4) "the essence, whose formula is a definition"; hence "'substance' has two senses: (a) The ultimate substratum which cannot be further predicated of anything else, and (b) that which is individual and separable, that is, the shape or form." "Substance, in the truest and primary and most definite sense of the word, is that which is neither predicable of a subject nor present in a subject; for instance, the individual man or horse."

"The essence of a thing is that which it is said to be *per se* . . . The formula of a thing's essence is that which defines the term but does not contain it." Definition "is the formula of the essence, and that essence belongs only to substances or to them alone in the proper, primary, and unqualified sense." "A 'definition' is a phrase signifying a thing's essence." "[D]efinition and essence in the primary and simple sense belong to substances" *(Selections)*.

The part that Aristotle's view of essence in the sense labeled (4) above—"essence, whose formula is a definition"—plays in his aesthetics and, through his influence, in the subsequent histories of literature, criticism (for example, seventeenth-century genre criticism, especially in France), and aesthetics, is through Aristotle's method of definition *(per genus et differentia)*. To apply this method is to identify the kind to which the thing being defined belongs, and to distinguish that from the other species also belonging to that kind. By this method Aristotle develops the theory of tragedy that he sets out in his *Poetics*.

The most obvious candidate in aesthetics for the status of substance in Aristotle's first sense—simple bodies that "are not predicated of a subject while everything else is predicated of them"—is the work of art. This is one reason why the notions of substance and essence are important to understanding the question "What is art?" It is over the question whether art is or has an essence that debate has most fiercely raged in aesthetics in recent years. (See, for example, Bell, 1914, for one account of the essence of, at any rate, visual art.) Traditionally beauty was favored as a substance and was therefore held to have an essence. This assumption is evident in the Dialogues of Plato, where Socrates, for example, acts on the belief that we ought to be able to define or state the essence of beauty. Begging the question against Plato, let us deny *beauty* to be the name of a substance term, and hold that the adjectival *beautiful* is the primary form philosophically, and is thus that which applies to *substances*, for example, works of art, without being a substance term itself. Conversely, if *work of art* is a substance term, it is absurd to ask what the term *work of art* applies to, thus treating

it as if it were a predicate term, which philosophers, when considering the question of what is art, have recently been inclined to do. (See, for example, William Kennick's argument in his justly influential essay, "Does Traditional Aesthetics Rest on a Mistake?")

A distinction is sometimes drawn between real and nominal definitions, for example, by Richard Robinson in *Definition*: "Real definition is concerned with things in general; but nominal definition is concerned only with a peculiar sort of thing, namely, symbols" (Robinson, 1954, p. 16).

An early version of this distinction is found in John Locke's *Essay concerning Human Understanding.* Locke mentions "the several significations of the word essence" but takes *real essence* most usefully to mean the internal constitution of a thing, that which belongs to it in virtue of its nature as that thing. By the *nominal essence* of a thing Locke means how we speak about it, what its characteristics or properties are that determine its correct name. Thus the "real essence" of gold is the structure, chemical or molecular, that gold has, whereas its "nominal essence" is those characteristics, such as its color, that lead us to say a ring made of gold is gold (Locke, 1975, book 3, chap. 3, especially sections 15–18). Like most things in philosophy this distinction is controversial, including the question whether it can be maintained.

The essence of something is generally taken to be matter for real definition, not nominal. On Aristotle's interpretation, "if '*x* is *yz*' is a significant and true real definition of *x*, then *x* is a thing and *yz* is the essence of that thing" (Robinson, p. 154). Robinson objects that Aristotle's account in his *Metaphysics* of "essence" is obscure, and in fact a useless circle, since, "A definition states essence, and essence is what you find in a definition" (p. 154). Robinson continues, however: "there has been from Plato onwards a very strong and persistent tendency to believe in essence. In particular, it has been a tendency to believe in the distinction between essence and property" (p. 154). What motivates this distinction is the idea that we can separate in thought a thing from its characteristics. To take the example of the work of art used above: on the one hand, we can think of the work of art, and on the other, we can think of what its properties or characteristics are as experienced in our perception of the work. Of course, since, as we noted at the outset, it seems open to us to ask which properties of a thing are essential to it and which are not, and given that essence belongs only to substance, the ambition to draw a cogent distinction between substance and its properties seems thwarted. But this is to give up too soon and to reckon without the work of Collingwood.

In his *Essay on Philosophical Method*, Collingwood distinguishes between definitions in the exact sciences, such as mathematics, where a person "possessing a definition knows the essence of the concept perfectly" and a kind

of concept where "the later part of the exposition . . . served to qualify or explain the earlier" (Collingwood, 1933, pp. 94–95). In the former case the essence is distinct from the properties that, flowing logically "from the essence, are stated in theorems." But in the latter kind of case, which Collingwood thinks is exemplified by philosophical concepts, "no line could be drawn between definition and theorems; the entire exposition [of the concept] would be a statement at once of its essence, and of its properties regarded as the elements constituting that essence" (p. 95). In the exposition, Collingwood continues, "there would be no one phrase or sentence which could be taken out of its context and called the definition." On Collingwood's view, a philosophical concept such as justice is appropriately the subject of an extended and reasoned exposition of the kind just described, such that no one formula can be picked out as the definition of justice (pp. 95–96). The concept of art can be considered in the light of the model Collingwood offers for philosophical concepts, which is not unreasonable since Collingwood in any case regarded art as a philosophical concept. Art, along with certain other things such as religion and history, "is somehow a universal and necessary characteristic of things" (Collingwood, 1965, p. 122). If we follow this approach, the charge sometimes leveled against Collingwood that he commits the so-called essentialist fallacy, by implying that a work of art can be defined in terms of a formulaic statement to the effect that its essence consists, say, in its expressing the emotions of the artist who created it, falls wide of the mark. The *whole* of *The Principles of Art*, the primary work in which Collingwood sets out his theory of art, has to be read as constituting Collingwood's definition of art or account of its essence.

Popper, on the other hand, is scathing about essentialism: "the degree to which the various sciences have been able to make any progress depended on the degree to which they have been able to get rid" of Aristotle's *"essentialist method of Definitions"* (vol. 2, p. 9). Popper criticizes Aristotle's belief that "we obtain all knowledge ultimately by an intuitive grasp of the essence of things" (p. 10). Knowledge is not "an encyclopaedia containing the intuitive definitions of all essences" but rather consists of empirical hypotheses that so far have withstood rigorous attempts to refute them (pp. 12–13). But analysis of the concepts of aesthetics, whether or not such analysis is held to require the positing of essences, is better thought of as increasing or promoting understanding rather than knowledge.

Popper, in making these charges, has Plato in his sights, too. In this hostility he is joined by Arnold Hauser, who maintains, in an early expression of the bias against essentialism now so ubiquitous in cultural studies, that Plato's "theory of Ideas is the classical philosophic expression of conservatism, the pattern of all subsequent reactionary idealism" (Hauser, 1951, vol. 1, p. 88).

A doctrine of essence plays an important part in the philosophy of George Santayana. Santayana was himself an idealist in philosophy and is often thought of as a conservative philosopher by those for whom these labels matter. He developed his views on essence *after* the publication of his book *The Sense of Beauty,* the work he is best known for in aesthetics. Timothy L. S. Sprigge, in his book on Santayana's philosophy (see especially chap. 4, "The Doctrine of Essence"), notes that "one of the main tenets of Santayana's so-called *doctrine of essence* is that nothing is immediately given except essence, and that when I am conscious of any matter of fact some immediately given essence is acting as a symbol of it" (Sprigge, 1974, p. 65). By "the given" Santayana understands that which we are conscious of immediately, "that is, one's consciousness of it must not be mediated by one's consciousness of something else" and also that which we are conscious of indubitably, "in the sense that one's consciousness cannot distort its character" (p. 66).

The idea of pure intuitions given in experience is controversial. Much recent philosophy has resisted the idea that we can cogently conceive of anything given as it were as the ground floor of experience. The interest for aesthetics of Santayana's essences, if they are possible in the face of recent philosophical disbelief, is that the route to them through skepticism—that is, essences are what remain in consciousness when everything that is mediate and doubtful is winnowed out—is not "the sole path to such pure intuition" (Sprigge, p. 66). Essences may also be reached "in certain aesthetic or mystical experiences, when I merely, so to speak, 'drink in' the way things appear to me without any concern as to how things really are" (p. 66). Sprigge gives as examples "the brief moment that I give myself up wholly to the beautiful smell of a rose, or to the beautiful pattern of a sunset" (p. 66).

Santayana holds that the beautiful is itself an essence; that is, "What is found to be beautiful . . . is a given essence, not the physical thing which presents it" (Sprigge, pp. 86–87). Because it is "the sense of beauty which most typically leads one to contemplate the given essences as the eternal objects which essentially they are," Sprigge finds in Santayana's views an echo of Schopenhauer's theory of aesthetic experience (p. 88) and points out that this constitutes a development in Santayana's theory of beauty "somewhat divergent" from what is offered in *The Sense of Beauty* (p. 87).

In spite of a contemporary philosophical climate, then, not friendly to belief in essences, essences are less easily dispensed with than philosophers have assumed. For one thing, the more thoroughly the notion that art has an essence is repudiated, the more difficult it becomes to explain exactly how there can be the philosophy of art as distinct from the history or sociology of art. If, as has been re-

cently the case, the terms *aesthetics* and *philosophy of art* have been used more or less synonymously, and given the prominence in aesthetics of the question, what is art, the theory of essences is important for two reasons: first, definition is impossible to understand philosophically without reference to it; second, accounts of essence offer a means of explaining how "art" is a philosophical concept and why there can thus be a philosophy of art. Arthur C. Danto quite admirably takes this last question seriously. Moreover, he recently proclaimed himself to be an essentialist: "Indeed, I would have taken the entire burden of my . . . work . . ., *The Transfiguration of the Commonplace*, to have been to underwrite essentialism in the philosophy of art" (Danto, 1996, p. 284). Thus, there are now encouraging signs of a greater willingness to allow essentialism than was the case in the heyday of the linguistic analysis movement in aesthetics.

[*See also* Aristotle; Collingwood; Plato; *and* Santayana.]

BIBLIOGRAPHY

Åhlberg, Lars-Olof. "The Nature and Limits of Analytic Aesthetics." *British Journal of Aesthetics* 33.1 (1993): 5–16.

Åhlberg, Lars-Olof. "Analytic Aesthetics and Anti-Essentialism: A Reply to Richard Shusterman." *British Journal of Aesthetics* 35.4 (1995): 387–389.

Aristotle. *Selections.* Edited by W. D. Ross. New York, 1927.

Aristotle. *Metaphysics.* Translated and edited by John Warrington. London, 1956.

Bell, Clive. *Art.* London, 1914.

Collingwood, R. G. *An Essay on Philosophical Method.* Oxford, 1933.

Collingwood, R. G. *The Principles of Art.* Oxford, 1938.

Collingwood, R. G. *Essays in the Philosophy of History.* Edited by William Debbins. Austin, 1965.

Currie, Gregory. "Supervenience, Essentialism and Aesthetic Properties." *Philosophical Studies* 58.3 (1990): 243–257.

Danto, Arthur C. "Art, Essence, History, and Beauty: A Reply to Carrier, a Response to Higgins." *Journal of Aesthetics and Art Criticism* 54.3 (1996): 284–287.

Diffey, T. J. "Essentialism and the Definition of 'Art'." *British Journal of Aesthetics* 13.2 (1973): 103–120. Reprinted in *The Republic of Art and Other Essays* (New York, 1991).

Diffey, T. J. "Aesthetics and Philosophical Method." In *Philosophy, History and Civilization: Interdisciplinary Perspectives on R. G. Collingwood,* edited by David Boucher, James Connelly, and Tariq Modood, pp. 62–78. Cardiff, 1995. Also printed in *The Republic of Art and Other Essays* (New York, 1991).

Hauser, Arnold. *The Social History of Art: From Prehistoric Times to the Middle Ages.* 2 vols. Translated in collaboration with the author by Stanley Goodman. London, 1951.

Kennick, William E. "Does Traditional Aesthetics Rest on a Mistake?" *Mind* 67 (1958): 317–334. Reprinted in *Collected Papers on Aesthetics,* edited by Cyril Barrett (Oxford, 1965).

Kripke, Saul A. *Naming and Necessity.* Cambridge, Mass., 1980.

Locke, John. *An Essay concerning Human Understanding* (1689). Edited by Peter H. Nidditch. Oxford, 1975.

Popper, K. R. *The Open Society and Its Enemies,* vol. 2, *The High Tide of Prophecy: Hegel, Marx, and the Aftermath.* 5th rev. ed. London, 1966.

Robinson, Richard. *Definition.* Oxford, 1954.

Shusterman, Richard. "On Analysing Analytic Aesthetics." *British Journal of Aesthetics* 34.4 (1994): 389–394.

Sprigge, Timothy L. S. *Santayana: An Examination of His Philosophy.* London and Boston, 1974.

T. J. DIFFEY

Anti-Essentialism

From the late nineteenth century to the 1950s one of the main foci of aesthetic inquiry was the attempt to develop definitions of art and such related concepts as visual art, music, tragedy, beauty, and metaphor. Clive Bell (1958) famously stated that either all works of visual art have some common quality or when we speak of "work of art" we speak nonsense. DeWitt H. Parker (1939) argued more generally that the assumption underlying every philosophy of art is the existence of some common nature present in all the arts. This search for a common quality or nature of art was generally taken to be a form of essentialism.

When analytic philosophers came to aesthetics in the 1950s they saw it as dominated by the essentialism of G. W. F. Hegel and such idealist followers as Benedetto Croce (1922) and Robin George Collingwood (1938). It was in this context that John A. Passmore (1954) attacked aestheticians for pretentiously saying nothing and for trying to *retain* mystery rather than dispel it. Aesthetics, he maintained, has presented us with empty and accommodating formulas based on an attempt to impose "a spurious unity" on a conflicted field. Most aesthetics books are dull because they fail to bring sharpness to their subject matters, and because they bring mechanical methods to something that is, by nature, spontaneous. They are "wooly" because they try to theorize on a too general level.

During this time Ludwig Wittgenstein's *Philosophical Investigations* (1958) had a profound influence on analytic aesthetics. Although Wittgenstein was not totally opposed to the existence of essences he did object to the assumption that there is an essence that corresponds to every abstract term or concept. Instead, he asked us to "look and see" whether there really was something in common between all entities called x, where x stands for any general term. He observed that there are certain terms, for example, *game,* for which there are no necessary and sufficient conditions. When we look into the concept of game we find only a "complicated network of similarities." Wittgenstein called these similarities "family resemblances," since they are like the resemblances we might find between members of a family. Philosophers were quick to apply this idea to the concept of art. [*See* Wittgenstein.]

Several aestheticians in the late 1940s and 1950s took an anti-essentialist line. W. B. Gallie called the assumption that to define something we must know its essential nature the "essentialist fallacy." Drawing on Wittgenstein's family re-

semblance analogy, he held that the referents of abstract words such as *art* do not necessarily have some one thing in common. He also insisted that essentialism leads to errors in criticism. Gallie later (1956) rejected some of his earlier views, in particular that "art" is a family resemblance concept and that we do not need a concept of art in criticism. He came to see the concept of art as essentially complex and essentially contested, much like the concept of democracy. Anticipating more recent writers, Gallie argued that analysis should be pursued in a historically sensitive way.

Paul Ziff (1953) stressed that there are different senses of *work of art*. He argued that there are no clear-cut cases of works of art in the sense that there are clear-cut cases of tables. Taking Nicolas Poussin's *The Rape of the Sabine Women* as a relatively clear case, Ziff constructed a set of characteristics that he considered sufficient conditions for something to be art in the sense that *The Rape of the Sabine Women* is art. (None was a necessary condition.) Anything that satisfies these conditions (for instance, *The Rape of the Sabine Women* itself) is a "characteristic case," and anything that is sufficiently similar to a characteristic case (meeting some subset of the sufficient conditions) is a work of art in *that sense* of *work of art*. Sufficient similarity is judged on the basis of our conception of the function of art. Thus a definition of art simply describes one use of *work of art* considered reasonable in light of social consequences and the functions of a work of art in our society.

Perhaps the best known of the anti-essentialists was Morris Weitz. Weitz (1956) believed that all of the great theories of art attempted to define art by stating its necessary and sufficient properties, and that each of these theories failed because of counterexamples, circularity, theoretical vagueness, and untestability. Instead of asking after the essence of art, philosophers should analyze the logic of the concept of art: what the term *art* does in the language. We then discover that art is not definable in terms of necessary and sufficient properties or conditions. It is not even necessary for a work of art to be an artifact. Art, rather, is a family resemblance concept in Wittgenstein's sense. There are paradigms of art, and networks of similarities between works of art, but there is no exhaustive set of conditions for correctly applying the concept. New conditions and cases can always be envisaged, since art is an "open" concept and we can always decide to extend it. Art by its nature is expansive, and a closed definition of art would make creativity impossible. This does not mean, however, that aesthetic theory is worthless: traditional definitions should simply be seen as encouraging us to attend to previously neglected qualities in works of art. (See also Weitz 1977.)

William Kennick (1958) similarly asserted that the assumption that there is some common nature to art or some set of necessary and sufficient conditions for something being a work of art is a mistake. What makes the question "What is art?" difficult is nothing mysterious about works of art but that the word *art* has a complex logic. Kennick insisted that we know how to separate works of art from other things because we know the correct English use of *art*. If we sent a man into a warehouse full of diverse objects and asked him to pick out all of the works of art, he could do it reasonably well. But if we asked him to pick out all of the objects that fit the criteria of one of the famous definitions of art he would fail, since phrases like *significant form* (which Clive Bell used to define art) are even more obscure than *art*. We know what art is because we know how to use the term *art* correctly, but we cannot produce a simple or even a complex formula to define it.

Maurice Mandelbaum (1965) was the most influential critic of anti-essentialism. He questioned the family resemblance analogy used by Gallie and Weitz, noting that although fortune-telling bears a resemblance to solitaire, it is not then a form of playing cards; that a literal family resemblance requires an underlying genetic connection between members of the family; and that, similarly, although there may not be any directly exhibited properties that distinguish art objects from other things, this does not preclude nonexhibited (perhaps intentional and relational) properties doing so.

When Arthur Danto (1964) introduced the concept of the artworld and George Dickie (1969) the institutional theory of art where art is defined in terms of nonexhibited properties, it seemed that essentialism was back in business. Danto is quite explicit about his essentialism, claiming that "art is always the same" and that there are conditions necessary and sufficient for something to be an artwork (1996). He couples his essentialism with historicism: what is a work of art at one time cannot be at another, and the essence of art is brought to consciousness through history. [*See* Danto.] Although Dickie does not declare himself an essentialist he does think he can define the concept of art *we* have, where the "we" seems to include at least present-day Americans and maybe Westerners since the eighteenth century.

These theories, in the original versions and in the current updates, are still influenced by anti-essentialism, however. Danto has criticized the way that essentialism in art criticism has inhibited art practice, although he thinks his own theory, with its Hegelian "death of art" thesis, *encourages* art practice through encouraging pluralism (1992). And, although Dickie has provided a series of progressively refined definitions of art in terms of necessary and sufficient conditions, he is skeptical that there is any Platonic essence of art that can be discovered by philosophical reflection or intuition (1984). Also, unlike earlier essentialists, Dickie deliberately limits himself to a classificatory definition of art, leaving out all evaluative components. [*See* Dickie.]

The banner of essentialism in the traditional functionalist mode continued to be carried by Harold Osborne (1955) and Monroe Beardsley (1981) during the height of the anti-essentialist movement and afterward. Both of these authors

attempted to define art in terms of its ability to provide valuable aesthetic experience.

Nelson Goodman (1978) seems anti-essentialist insofar as he seeks to replace the question "What is art?" with "When is art?" Yet, his answer to the latter question pushes him closer to essentialism. For Goodman, something is art when it functions *as* art, that is, functions as a symbol in a certain way. This would imply that "being a symbol" is a necessary condition of a work of art: indeed, art *must* either represent, express, or exemplify. Goodman refuses to give a sufficient condition for art, but does give five "symptoms of the aesthetic," none of which is necessary, and suggests that "nontransparency" (the tendency of properties to focus our attention on the symbol rather than on what is symbolized) might set works of art off from other symbols. [*See* Goodman.]

Going Wittgenstein one better, Benjamin R. Tilghman (1984) maintains that it is not even coherent to *look* for essences. To ask what two or more things have in common outside of all context is just another form of what Wittgenstein called "language gone on holiday." Thus, asking what *all* games or works of art have in common is unintelligible: even if they are all "unified" they are so in different ways. More recently (1989), Tilghman has claimed that there cannot be theories of human practices such as art since there is nothing philosophically significant hidden behind those practices. Aesthetic theories are caught up in the confusions of philosophical theory in general.

Noël Carroll (1988, 1993) wants to reorient the "What is art?" question. Art for him is a cultural practice governed by reasons internal to that practice. A cultural practice tends to reproduce itself in ways that sustain continuity with the tradition to which it belongs. Identification of new objects as works of art is related to this process. Carroll believes we have several rational strategies for justifying inclusion of such objects. These include repetition, amplification, and even repudiation of works already included in the tradition. We identify works of art not by applying formal definitions but by telling stories about their place in a historic narrative. (See Thomas Leddy [1993] for a criticism.)

Richard Shusterman (1992, 1993), who has proposed a pragmatist aesthetics, is concerned that attempts to define art put it in a box, disenfranchise it, neutralize it, and further detach it from the rest of life. Shusterman's anti-essentialism is directed not only against traditional theories but also against Danto's historicist essentialism. Danto simply collapses the philosophy of art into a representation of art history's representation of art, which, in turn, is inevitably a history of high art. Pragmatist aesthetics, by contrast, takes an active role in reshaping art, particularly in the direction of overcoming the distinction between high and low art and between art and life.

Anita Silvers (1989) levels a parallel attack against Danto, arguing that his grounding of aesthetics, interpretation, and art itself in seemingly purely descriptive art history fails because art history cannot escape evaluative aesthetic criteria. Such criteria are relevant, for example, in the selection of objects for historical study as canonical. Art historians revise history in light of changed evaluations. Thus, insofar as it attempts to gain stability through disassociation of evaluation from theory, essentialist aesthetic theory seems doomed once again.

Jerrold Levinson, somewhat like Dickie, seeks to provide a definition of art that captures what the concept of art is at the present time: what it *now* means for an object created at any time to count as art at that time (1990a, b). Levinson thus seeks to describe "our concept of art," by which he means the one used in enlightened and informed contemporary discourse about art in the West. Levinson is an anti-essentialist insofar as he does not posit an unchanging essence of art. He does seem, however, to be an essentialist about the *concept* of art since he provides a definition in terms of (not strictly) necessary and sufficient conditions. He argues, against Wittgensteinians, that cultural concepts can have "extractable, fairly serviceable, essences" and that we can try to tease out the "most central core of meaning" for a concept like art. Moreover, he thinks his definition captures an essential aspect of artmaking through *all* periods, at least for the past three thousand years (1996): it reveals the "minimal essence" of art that has only come to light with the advent of avant-garde art in the early twentieth century. A short version of his definition is as follows: "an artwork is a thing . . . that has been seriously intended for regard-as-a-work-of-art—i.e., regard in any *way preexisting artworks are or were correctly regarded*" (1990b, pp. 38–39).

Anti-essentialism has also played a role in Continental aesthetics during the second half of the twentieth century, particularly in poststructuralism. Poststructuralism began as a reaction against the essentialism of structuralism (i.e., the belief that there are stable underlying structures that explain human behavior) but was quickly extended to an attack on the entire tradition of Western metaphysics as essentialist. [*See* Poststructuralism.] The search for essences was seen by Jacques Derrida (1976) as an example of the fallacious "metaphysics of presence." This was tied to his critique of "logocentrism": the belief that words represent meanings present in the speaker's mind. Meaning, rather, is always and necessarily in the process of being deferred. This would of course imply that one could never come up with a definition of art in terms of necessary and sufficient conditions. If one believes that meaning is never determinate or stable, then one is not going to waste energy on trying to arrive at stable definitions of key terms in aesthetics, or even in trying to refute attempts to do so. Deconstructionists would not even *begin* this task, since, they would say, it assumes the existence of a "transcendental signified." [*See* Derrida.] Jean-François Lyotard's (1984) critique of "grand

narratives" or "metanarratives" (such as Christianity and Marxism) is also anti-essentialist. For example, it would rule out Danto's Hegelian historicist version of essentialism, and no doubt Levinson's as well.

The dialectic between essentialism and anti-essentialism has played a central role in contemporary analytic and postanalytic aesthetics. Although very few would hold to a Platonic/Aristotelian essentialism that takes essences to be real entities that are also unchanging, the view that important aesthetic concepts such as art can be defined in terms of necessary and sufficient conditions relativized in some way to history, or in terms of something more modest such as necessary or sufficient conditions alone, is still attractive to many. Nonetheless, this *conceptual* essentialism has been under consistent attack by anti-essentialists, most of whom were strongly influenced by Wittgenstein, and some more recently by poststructuralist anti-essentialism. Institutional definitions like Dickie's, and historical definitions like Levinson's, have been resistant to such attacks largely because they share with anti-essentialism a skepticism concerning the traditional functionalist approach to defining art. (For a recent discussion, see Davies, 1991.) Such definitions are usually self-consciously circular, however (that is, the term *art* appears in the definiens, as in Levinson's definition given above), and this is problematic for some.

BIBLIOGRAPHY

Beardsley, Monroe. *Aesthetics: Problems in the Philosophy of Criticism* (1958). Indianapolis, 1981.
Bell, Clive. *Art* (1914). New York, 1958.
Carroll, Noël. "Art, Practice, and Narrative." *Monist* 71.2 (April 1988): 140–156.
Carroll, Noël. "Historical Narratives and the Philosophy of Art." *Journal of Aesthetics and Art Criticism* 51.3 (Summer 1993): 313–326.
Collingwood, R. G. *The Principles of Art.* Oxford, 1938.
Croce, Benedetto. *Aesthetic.* Translated by Douglas Ainslee (1909). 2d ed. New York, 1922.
Danto, Arthur. "The Artworld." *Journal of Philosophy* 61.19 (15 October 1964): 571–584.
Danto, Arthur C. "Learning to Live with Pluralism." In *Beyond the Brillo Box,* pp. 222–223. New York, 1992.
Danto, Arthur C. "From Aesthetics to Art Criticism and Back." *Journal of Aesthetics and Art Criticism* 54.2 (Spring 1996): 105–115.
Davies, Stephen. *Definitions of Art.* Ithaca, N.Y., 1991.
Derrida, Jacques. *Of Grammatology.* Translated by Gayatri Chakravorty Spivak. Baltimore, 1976.
Dickie, George. "Defining Art." *American Philosophical Quarterly* 6.3 (July 1969): 253–256.
Dickie, George. *The Art Circle: A Theory of Art.* New York, 1984.
Elton, William, ed. *Aesthetics and Language.* Oxford, 1954.
Gallie, W. B. "The Function of Philosophical Aesthetics." In *Aesthetics and Language,* edited by William Elton, pp. 13–35. Oxford, 1954.
Gallie, W. B. "Art as an Essentially Contested Concept." *Philosophical Quarterly* 6.23 (April 1956): 97–114.
Goodman, Nelson. "When Is Art?" In *Ways of Worldmaking,* pp. 57–70. Indianapolis, 1978.
Kennick, William E. "Does Traditional Aesthetics Rest on a Mistake?" *Mind* 67 (July 1958): 317–334.
Leddy, Thomas. "The Socratic Quest in Art and Philosophy." *Journal of Aesthetics and Art Criticism* 51.3 (Summer 1993): 399–410.
Levinson, Jerrold. "Defining Art Historically." In *Music, Art, and Metaphysics: Essays in Philosophical Aesthetics,* pp. 3–25. Ithaca, N.Y., 1990a.
Levinson, Jerrold. "Refining Art Historically." In *Music, Art, and Metaphysics: Essays in Philosophical Aesthetics,* pp. 37–59. Ithaca, N.Y., 1990b.
Levinson, Jerrold. "Extending Art Historically." In *The Pleasures of Aesthetics,* pp. 150–171. Ithaca, N.Y., 1996.
Lyotard, Jean-François. *The Postmodern Condition: A Report on Knowledge.* Translated by Geoff Bennington and Brian Massumi. Minneapolis, 1984.
Mandelbaum, Maurice. "Family Resemblances and Generalization concerning the Art." *American Philosophical Quarterly* 2.3 (July 1965): 219–228.
Osborne, Harold. *Aesthetics and Criticism.* New York, 1955.
Parker, DeWitt H. "The Nature of Art." *Revue internationale de philosophie* 1.4 (July 1939): 684–702.
Passmore, John A. "The Dreariness of Aesthetics." In *Aesthetics and Language,* edited by William Elton, pp. 36–55. Oxford, 1954.
Shusterman, Richard. *Pragmatist Aesthetics: Living Beauty, Rethinking Art.* Oxford and Cambridge, Mass., 1992.
Shusterman, Richard. "Art in a Box." In *Danto and His Critics,* edited by Mark Rollins, pp. 161–174. Oxford and Cambridge, Mass., 1993.
Silvers, Anita. "Once upon a Time in the Artworld." *Aesthetics: A Critical Anthology,* 2d ed., edited by George Dickie, Richard Sclafani, and Ronald Roblin, pp. 183–195. New York, 1989.
Tilghman, B. R. *But Is It Art? The Value of Art and the Temptation of Theory.* Oxford and New York, 1984.
Tilghman, B. R. "Reflections on Aesthetic Theory." In *Aesthetics: A Critical Anthology,* 2d ed., edited by George Dickie, Richard Sclafani, and Ronald Roblin, pp. 160–170. New York, 1989.
Weitz, Morris. "The Role of Theory in Aesthetics." *Journal of Aesthetics and Art Criticism* 15.1 (September 1956): 27–35.
Weitz, Morris. *The Opening Mind: A Philosophical Study of Humanistic Concepts.* Chicago, 1977.
Wittgenstein, Ludwig. *Philosophical Investigations* (1953). 3d ed. Translated by G. E. M. Anscombe. New York, 1958.
Ziff, Paul. "The Task of Defining a Work of Art." *Philosophical Review* 62.1 (January 1953): 58–78.

THOMAS LEDDY

EVALUATION. [*To analyze the different notions of evaluation, this entry comprises two essays:*

Aesthetic Evaluation

Cultural Evaluations

The first essay concerns the treatment of evaluation in the history of aesthetics. The second essay discusses the same topic but from a different philosophical perspective and in relation to cultural theory in general. These two essays combined represent only a sampling of the wide spectrum of views on how art and culture can be evaluated critically and aesthetically. For further discussion, see Appreciation; Criticism; Interpretation; Taste; *and* Value.]

Aesthetic Evaluation

Theorizing about the evaluation of art did not have a very high profile in philosophy until the eighteenth century.

From ancient times until the eighteenth century, it was held that either beauty or representation or both were what made art valuable, although this view was more an assumption than a conclusion that was argued for. Beauty and representation were thought of by the earlier philosophers in one way or another as *objective* properties. With the advent of the theory of taste at the beginning of the eighteenth century, theorizing about the evaluation of art took on a *subjective* cast, making the reactions of persons (subjects) the central focus. During the eighteenth century, a host of philosophers—some of the first rank—worked out theories of taste.

Theorizing in the taste mode was a reaction against the objectivist theory of beauty and against the view that beauty (however conceived) is the only valuable property of art. The various theories of taste subsume and subordinate beauty, sublimity, and other aesthetic characteristics as well as representation under the primary category of taste in different ways. In general, theories of taste claim that works of art (and aspects of nature) are valuable because they have specific properties that are instrumental in producing disinterested pleasure. In focusing on specific properties of objects, these theories retain a foothold in the objective world within a basically subjective theory.

Francis Hutcheson, an early, influential theorist of taste (1725), claimed that the experience of beauty consists of the perception of uniformity in variety (the object of taste) triggering the sense of beauty (a faculty of taste) and producing a feeling of disinterested pleasure. Hutcheson identified beauty with this pleasure. On Hutcheson's view, the pleasure taken in uniformity in variety is disinterested because the perception of it triggers pleasure immediately and automatically; the sense of beauty responds only to uniformity in variety and cannot be influenced by any envisaged self-interested consequences. Most of the objects with which Hutcheson illustrated his theory were natural objects, but he did apply the theory to art. For example, he claimed that artistic representation involves a kind of uniformity; a representation and the object it represents resemble one another and as a pair constitute a unified whole. Concerning beauty, Hutcheson is a formalist; he claimed that only uniformity in variety—a kind of form—is beauty-making. Hutcheson tried to avoid relativism—the view that one person could be blameless pleased and another person blameless not pleased by uniformity in variety—by holding that basic human nature is pleased by uniformity in variety. He explained actual disagreements as deriving from physical defect such as poor vision or from the perverting result of the association of ideas. Hutcheson also suggested that there is a sense of sublimity as well as a sense of beauty, but he did not attempt an account of it.

According to Edmund Burke's theory of taste (1757), experiences of beauty consist in the triggering of a feeling of pleasure by the perception of certain properties—smallness, smoothness, gently curving lines, and the like. Burke also claimed that there is an experience of sublimity separate from that of beauty and in a sense opposed to it; this experience consists in the perception of largeness, obscurity, jaggedness, and the like, triggering a feeling of *relative* pleasure, that is, pleasure that results from the removal of the threat of pain. A significant difference between Hutcheson's and Burke's theories is that Hutcheson specified *one* complex property—uniformity in variety—as the experiential thing that triggers the pleasure of beauty, whereas Burke specified a short *list* of properties as beauty-making (and another short list as sublimity-making). Burke also denied that there is a specific faculty of taste.

David Hume's short essay, "Of the Standard of Taste" (1757), presents the eighteenth century's most sophisticated and successful theory of taste. Hume, mentioning only works of art, cited a long but clearly not exhaustive list of "beauties" that trigger pleasure and "blemishes" that trigger pain; today many of the "beauties" and "blemishes" he mentioned would be called aesthetic qualities. He cited, among others, uniformity, variety, clarity of expression, and luster of color as "beauties." He also specified exactness of imitation and the imitation of valuable characteristics as sources of merit in art. Hume thought that all persons would agree about the value of the "aesthetic" and the other valuable properties he cited, but he did admit a mild relativism in that there can be blameless disagreement among persons about the preference for one of the "beauties" to another at a particular time. Unlike Hutcheson, he did not claim that there is a specific faculty of taste. Also, Hume, unlike many theorists of taste, specified moral point of view as a source of merit and blame in artworks.

As the eighteenth century drew to a close, Immanuel Kant presented in the *Critique of Judgment* (1790) what many regard as the culminating theory of taste. In fact, his theory of beauty is in some ways a throwback to Hutcheson's. Like Hutcheson, with regard to beauty, Kant is a formalist; that is, he claimed that the object of taste (beauty) is a kind of form. Kant, unlike Hutcheson, tried to characterize beauty in terms of teleology, claiming objects of taste are forms of *purpose*. The form of a natural object is God's purpose, and the form of an artwork is an artist's purpose. Kant theorized that the faculty of taste, which produces the disinterested pleasure of taste, is the cognitive faculties functioning in a disengaged way. The pleasure of taste is disinterested, on Kant's theory, because the disengaged cognitive faculties are conceived of as isolating the object of attention (a *form* of purpose) from other objects of consciousness and, hence, from any connection to any interest or purpose. Because of the limits of their formalisms, neither Hutcheson's nor Kant's theory has any way to account for the beauty of color or many of the other valuable qualities present in art and nature. Also, many natural objects ("forms of purpose") are not objects of beauty. Kant, like Burke, devel-

oped a conception of sublimity that is in a sense the opposite of his conception of beauty. In giving the cognitive faculties a central role in taste, Kant set the stage for the development of the *aesthetic attitude* theories that displaced theories of taste.

Theorizing in the aesthetic attitude mode begins with Arthur Schopenhauer (1819, 1969), who claimed that there are two kinds of cognition or consciousness: (1) ordinary consciousness, dominated by the will, in which the world is experienced as causally related objects used to gratify the will; and (2) disinterested aesthetic consciousness, which has broken free from the will and experiences objects as causally relationless instances of Platonic Forms. Schopenhauer conceived of aesthetic consciousness as the experience of an object for its own sake and as unrelated to other things. His conception greatly influenced the development of nineteenth- and twentieth-century aesthetics. This kind of theorizing focuses almost entirely on the *description* of the aesthetic attitude (aesthetic consciousness, psychical distance, etc.) as a disinterested phenomenon and has little to say about the evaluation of art. While aesthetic attitude theorizing dominated aesthetics, little work was done on the evaluation of art.

Monroe Beardsley's 1958 book, *Aesthetics*, revived theorizing about art evaluation. Beardsley claimed that *the* function of artworks is to produce aesthetic experience and that artworks are to be evaluated according to their capacity to produce such experience. Following the aesthetic attitude theorists to a degree, Beardsley claimed that aesthetic experience is detached from the remainder of experience and that referential aspects of art (which relate art to the world outside aesthetic experience) are irrelevant to its evaluation as art. On his view, only *aesthetic* characteristics are relevant for the evaluation of art.

Beardsley also thought every aesthetic quality is subsumable under unity, intensity, or complexity, and he saw these three as *the* criteria of artistic value. There are, on his theory, three corresponding evaluative principles with the three criteria as their subjects: for example, "Unity in an artwork is always valuable (for producing aesthetic experience)." Beardsley's three principles, however, even with complete information about an artwork's unity, intensity, and complexity, can never entail specific, narrowly focused evaluations such as "This artwork is *good*" but only unspecific evaluations such as "This artwork is *valuable to some degree.*"

Although he did not formulate them, on Beardsley's theory, principles that have aesthetic experience as their subject are also possible: for example, "Aesthetic experiences of a fairly great magnitude are always good." This particular principle, together with the information that an artwork can produce an aesthetic experience of fairly great magnitude, entails that the artwork can produce a *good* aesthetic experience. If an artwork can produce a *good* aesthetic experience,

then the artwork itself is instrumentally *good*—a specific, narrowly focused evaluation.

Beardsley tried to avoid relativism by holding that art's value depends only on its capacity to produce aesthetic experience and not merely on persons' being pleased by its features. (If art's value depends simply on persons' being pleased, some might be pleased and some displeased by the same work, and relativism would be the case.) Also, according to his theory, the value of every artwork can be compared with that of every other work because artworks' values are determined by their capacity to produce one kind of thing, namely, aesthetic experience.

Beardsley's conception of aesthetic experience as detached from the remainder of experience is flawed because it makes it impossible for him to explain the cognitive value that some artworks have.

In his 1968 book, *Languages of Art*, Nelson Goodman denied that the possession of aesthetic qualities makes art valuable and claimed that art's value is *solely* a function of the symbolizing or referring of its cognitive aspects. The crucial cases for Goodman's theory are nonobjective paintings and instrumental music, which do not appear to symbolize in any obvious way. He claimed that such works do symbolize and are valuable because their dominant characteristics symbolize by exemplifying themselves; thus, he held that these characteristics are valuable because they are *samples* (exemplifying symbols) of themselves, like paint chips at a paint store.

Is Goodman right? Consider a nonobjective painting that is uniformly a brilliant red. Does it function like a paint chip? Nonobjective paintings as they are typically experienced do not exemplify. Contrary to Goodman's theory, then, such artworks do not typically symbolize in this way. Such valuable nonobjective paintings are not valuable because they exemplify.

Consider a second argument against Goodman's theory. The brilliant red of a sunset is a paradigm of aesthetic value, and it does not typically exemplify anything. Assume the nonobjective painting to be the same shade of red as an aesthetically valuable sunset; such a painting will be valuable in the same nonreferential way that the sunset's color is. So, even if the painting were valuable because it is a sample, it is also valuable, contrary to Goodman's theory, in a nonreferential way.

Goodman was right that the cognitive content of art can have artistic value, but he neglected aesthetic qualities. Beardsley was right that the possession of aesthetic qualities can have artistic value, but he neglected cognitive value in art.

In 1983, Frank Sibley criticized Beardsley's account of evaluational criteria. He noted that Beardsley held that the presence or increase of any one of his criteria—unity, intensity, complexity—must always contribute positively to an artwork's overall value. For Beardsley, a critical principle

must have the form: "X in an artwork is always valuable," with "valuable" being understood to mean "contributes positively to overall value." Sibley argued that Beardsley's requirement cannot be maintained because adverse *interactions* among his three criteria are possible; for example, a work's complexity may cause a work to be disunified and its overall value to be less than it would be if it were less complex.

Sibley also maintained that many aesthetic properties are not subsumable under unity, intensity, and complexity. There are *positive* aesthetic properties besides unity, intensity, and complexity—for example, gracefulness and delicacy. He also claimed that there are many negative aesthetic properties—garishness, insipidness, and the like.

On Sibley's view, there will be as many critical principles as there are aesthetic properties—positive and negative. Since the aesthetic properties of artworks may interact with their other aesthetic or nonaesthetic properties to reduce or enhance overall value, critical principles must be formulated in a way that reflects this. Principles must have the form: "X in an artwork, in isolation from the other properties of the work, is always valuable." For example, "Gracefulness in an artwork, in isolation from the other properties of the work, is always valuable." The application of such principles to artworks always requires qualifications about interactions among the various properties of artworks. Sibley's principles can help entail only unspecific evaluations of artworks such as, "This artwork is valuable to some degree."

For Sibley, artworks are instrumentally valuable because they can produce valuable experiences, but he did not claim, as Beardsley did, that artworks produce a single kind of valuable experience. On his view, it is possible that persons can derive different value experiences from an artwork, and relativism is not ruled out.

Beardsley and Goodman attempted to evaluate artworks on the basis of *one* complex property, namely, the capacity of artworks to produce a particular kind of valuable experience. This would make the value of artworks the function of a single property and make the value of every artwork comparable to that of every other. If the value of every work could be compared to the value of every other work, then all existing works could be envisaged as ranked in a hierarchical value matrix. We could then assign *specific* values to artworks, saying that those works at the top of the envisaged matrix are *excellent* works, those in the middle are *good* works, those at the bottom are *bad* works, and so on. Unfortunately, Beardsley's and Goodman's conclusions are not satisfactory.

In 1975, Bruce Vermazen argued persuasively against theories such as Beardsley's and Goodman's; he claimed that artworks' values are a function of many independently valuable properties. If two artworks have different independently valuable properties, it will be impossible to compare their values. If two artworks each have the same independently valuable property and only that one valuable property, then their values can be compared and ranked. Vermazen then showed that if two artworks have the same independently valuable properties but have two or more such properties, it will be possible to compare and rank the values of the two artworks if the values of the properties are of a certain sort but that it will not be *generally* possible to compare and rank the values of all such pairs of artworks.

Even if the values of works cannot be generally compared and ranked, Vermazen did show that the values of some works can be compared to those of others. His conclusions can be applied and extended in the following way. For any given work, there *may* be other existing works to which its value can be compared, and for any given work, nonexisting (possible) works can be imagined to which its value can be compared. Thus, for any given work, a set consisting of existing (if there are any existing comparable works) and possible works or a set of only possible works (if there are not any existing comparable works) can be imagined as a hierarchical value matrix, and the value of the given work can be compared to the values of the other members of the imagined matrix. The place of the given work within the imagined matrix determines the *specific* evaluation to be assigned to it—a rank at the top means *excellent*, and so on.

Relativism is still possible on Vermazen's view because it does not guarantee that all persons will agree about the independently valuable characteristics of works of art.

[*See also* Beardsley; Burke; Hume; Hutcheson; Kant; *and* Schopenhauer.]

BIBLIOGRAPHY

Beardsley, Monroe C. *Aesthetics: Problems in the Philosophy of Criticism* (1958). 2d ed., with postscripts. Indianapolis, 1981.

Beardsley, Monroe C. "On the Generality of Critical Reasons." *Journal of Philosophy* 59.18 (1962): 477–486.

Beardsley, Monroe C. "In Defense of Aesthetic Value." *Proceedings and Addresses of the American Philosophical Association* 52.6 (1979): 723–749.

Burke, Edmund. *A Philosophical Enquiry into the Origin of Our Ideas of the Sublime and Beautiful* (1757). 2d ed. Edited by James T. Boulton. Notre Dame, Ind., 1968.

Dickie, George. *Evaluating Art.* Philadelphia, 1988.

Goodman, Nelson. *Languages of Art: An Approach to a Theory of Symbols* (1968). 2d. ed. Indianapolis, 1976.

Hume, David. "Of the Standard of Taste" (1757). In *Essays: Moral, Political, and Literary,* rev. ed., edited by Eugene F. Miller, pp. 226–249. Indianapolis, 1987.

Hutcheson, Francis. *An Inquiry concerning Beauty, Order, Harmony, Design* (1725). Edited by Peter Kivy. The Hague, 1973.

Kant, Immanuel. *Critique of Judgment* (1790). Translated by Werner S. Pluhar. Indianapolis, 1987.

Schopenhauer, Arthur. *The World as Will and Representation,* 2 vols. Translated by E. F. J. Payne. Reprint, New York, 1969.

Sibley, Frank. "General Criteria and Reasons in Aesthetics." In *Essays on Aesthetics: Perspectives on the Work of Monroe Beardsley,* edited by John Fisher, pp. 3–20. Philadelphia, 1983.

Vermazen, Bruce. "Comparing Evaluations of Works of Art." *Journal of Aesthetics and Art Criticism* 34.1 (Fall 1975): 7–14.

Wolterstorff, Nicholas. *Art in Action: Toward a Christian Aesthetic.* Grand Rapids, Mich., 1980.

GEORGE DICKIE

Cultural Evaluation

Evaluation can be understood generally as any or all of the more or less complex processes of sampling, discriminating, classifying, comparing, assessing, and selecting that constitute the ongoing activities of responsive creatures in their interactions with their environments. So understood, evaluation (or evaluative behavior) is not confined to human beings and not necessarily deliberative, conscious, rational, verbal, overt, or intentionally communicative—though it can, of course, be all these as well.

With respect to art and other cultural productions, the term *evaluation* can be understood to embrace a quite extensive array of value-(re)marking practices, covert as well as overt, informal as well as formal, unarticulated as well as verbal, and institutional as well as individual, including the following:

1. the innumerable acts of trial and revision, approval and rejection—not necessarily discrete, verbalized, or clearly sequential—that make up the process of artistic creation or cultural production itself;
2. the acts of ongoing discrimination and covert or implicit preference performed by anyone who engages with a cultural artifact, whether to read, view, or listen to it, or to display, produce, print, perform, quote, cite, translate, purchase, preserve, parody, or imitate it;
3. the more explicit and specifically verbal but still relatively casual judgments framed, debated, and negotiated in informal social contexts by and among those for whom the artwork or cultural production figures in some (not necessarily conventionally aesthetic) way; and
4. the more or less specialized activities of professional and institutionally authorized evaluators (academic, curatorial, journalistic, and so forth): not only their explicit ratings, rankings, reviews, and appraisals, but also the anthologies they compile, the catalogs they compose, the prizes they award, the conferences they convene, the classroom readings they assign, and the less professionally expert judgments they endorse or reject.

With respect to specifically verbal evaluations, preoccupation with their logical or epistemic status compared with that of other types of proposition or utterance, such as presumptively factual statements or objective descriptions (cf., e.g., Hirsch, 1976), has perpetuated a number of questionable dichotomies: personal versus impersonal judgments, expressive versus descriptive utterances, and so forth. It has also obscured both the wide variety of evaluative utterances, aesthetic and other, and what could be seen as the continuities between verbal judgments and other types of verbal utterance or communicative behavior more generally.

Verbal judgments are not altogether discontinuous from other forms of social communication or from the evaluative behavior of other creatures. Animals, for example, including birds and insects, may signal the quality of a territory or food supply to other members of their groups by some form of specialized overt behavior. And, in all human societies, people solicit from one another both indications of personal experiences of value and observations or estimates of the value of objects more generally. Other people's ("mere") expressions of their personal likes and dislikes, though often assigned inferior epistemic status, may nevertheless be interesting and informative to us under a variety of familiar conditions: if, for example, we ourselves (as artists, say, or manufacturers) have produced the objects in question, or if we have an independently motivated interest in the current states or general structure of tastes and preferences of those particular people (as their parents, for example, or potential associates). Moreover, with regard to artworks as to virtually anything else, other people's reports of their personal responses and preferences may serve us as measures of the social validity of our own tastes or as indexes of the general structure of preferences of various otherwise interesting groups of people (for example, Southerners, lawyers, cosmopolites, the younger generation).

Some of the operations and limits of aesthetic or cultural value judgments are instructively illustrated by the assessments of professional or commercial evaluators: film and book reviewers, commissioned art connoisseurs, writers of restaurant and travel guides, and so forth. The ratings and recommendations issued by such critics can be seen as highly specialized and more or less commoditized versions of the estimates and predictions of value exchanged more generally and informally among associates in any society. Professional evaluators, economically and otherwise constrained to increase the likelihood that their assessments of a particular object will be found appropriate by (certain) other people, typically sample the object (novel, film, restaurant) themselves, operating as stand-ins for or, in effect, metonymic representatives of their particular clients or audiences (wealthy art patrons, subscribers to the *New York Review of Books,* film buffs, purchasers of *The Michelin Guide to Italy*). For the same reasons, such evaluators are typically attentive to the particular tastes, needs, interests, and expectations of those clients or audiences. An assessment (of a novel, film, restaurant, or anything else) that reflected only the evaluator's superficial sampling, confined or uninformed comparisons, idiosyncratic responses, or mistaken assumptions concerning his or her clients' tastes and interests would, of course, be of only limited value to those clients and relatively worthless to most other people as well. Recurrent attempts to specify the qualifications of ideal

critics and to define the conditions for the rightful claim of judgments to objective validity can be seen as efforts to eliminate or, more plausibly, to minimize just such unhappy, but perhaps always to some extent inevitable, possibilities.

It is sometimes claimed, especially in efforts to quash arguments for aesthetic or cultural relativism, that an objective and universal standard of taste can be established by reference to certain fundamental and invariant human characteristics (e.g., innate and uniform psychophysiological structures or dispositions, or ubiquitous and universally shared experiences) and certified by the convergent judgments of duly qualified critics. Such claims, however, can be seen as both empirically and logically dubious: empirically insofar as the posited invariant human characteristics are recognized (or conceded) to admit of many exceptions and variations or to require many implausible conditions to display themselves in their duly grounding form; and logically inasmuch as the qualifications attributed to such critics (e.g., innate powers of discrimination, due experience, cultivated taste, freedom from bias) either presuppose the very norms their convergent judgments are alleged to prove or beg the question of the possibility of an objective determination of tastes. (For examples of some of these claims and problems, see Hume, 1965, pp. 8–10; Richards, 1960, pp. 202–205; and the discussion of both in Smith, 1988, pp. 36–38, 55–64.)

To be sure, insofar as the responses and critical judgments of individual people reflect similar specific histories of acculturation, similar specific experiences, and similar communicative contexts, they are likely to coincide and to be mutually predictive (for a sociological analysis, see Bourdieu, 1984). Also, because the structures and dispositions of human beings are often relatively similar and because some of the other conditions for what we call aesthetic experience may be relatively widespread or recurrent, the responses and judgments of otherwise diversely acculturated people may be relatively convergent. It must be added, however, that people's experiences of the value of objects appear always to be to some extent subject-variable and sensitive to contingent circumstances, that their experiences of the value of cultural productions seem to be exceptionally complex and heterogeneous (see Thompson, 1979), and that the operations of all physiological structures and psychological dispositions interact with and are affected by numerous other contingent and variable factors (see Oyama, 1985). Consequently, the fact that experiences and judgments of the value of certain cultural productions are sometimes stable across different eras or common across different cultures cannot be taken as evidence of either the absolute objectivity of those values or the basic or potential universality of all such experiences and judgments. Nor, for the same reasons, does the coincidence of the judgments of academic specialists or other critical authorities underwrite the claims of objective validity or absolute normative propriety that may be made for their evaluations.

According to the argument developed by Kant in the "Analytic of the Beautiful" (1951, sections 1–22), a judgment of taste—specifically, the discrimination/statement that some object is beautiful—could rightfully claim objective (in the sense of universal intersubjective) validity if the sensation it expressed were free of concepts, desires, practical interests, and other elements or considerations peculiar to the individual subject having the sensation and making the judgment. These conditions might be satisfied, Kant indicates, if the judgment were elicited by a subject's reflection on his or her sensation of the maximally harmonious operation of certain of his or her mental faculties as excited by and responding to pure form, and if we could assume that the operations of the relevant mental faculties were uniform for all subjects.

There is considerable question, however, as to whether these strenuous requirements can be fulfilled. For one thing, it is questionable whether personally individuating features can be eliminated from aesthetic experiences and judgments any more than from any other experiences or judgments. Second, it is doubtful that the formation of a specifically verbal judgment, such as a remark to the effect that some object is beautiful, could be freed—or, put another way, dissociated—from the specific social conditions in which the judgment occurs and to which it must be responsive in order to have communicative force at all. Such conditions would include the particular social and institutional context in which the judgment is produced (a gallery salesroom, for example, as distinct from a college classroom or the art review columns of a daily newspaper), the particular social and institutional relations that obtain between the evaluator and his or her audience (the relationship between an art teacher and an art student, for example, as distinct from that between a painter and a rival artist, or an official state censor and the citizens of that state), and the particular interests, assumptions, and expectations that motivate and constrain the entire communicative transaction in which the judgment figures.

Given the logic of Kant's argument, then, it is doubtful that an aesthetic judgment rightfully claiming objective validity can actually occur—as he seems, for different reasons, to acknowledge (sections 19–22). Whether, as he goes on to suggest (section 22), the possibility of such a claim remains a regulative ideal for such judgments is itself a debatable issue. It could be argued, for example, that unanimity, consensus, or what Kant refers to as "the confluence of any one man's feeling with that of any other" is by no means a self-evident good. On the contrary, it could be maintained that diversity and even conflicts of feelings and judgments in regard to, among other things, what we call aesthetically pleasing objects are not only inevitable but also in some ways communally and individually desirable: as sources, for

example, of social enrichment and as stimulants to increased perceptual responsiveness and renewed artistic creativity. (For related discussion, see Smith, 1988, pp. 64–72; for other views of Kant's arguments in the *Critique of Judgment*, see Guyer, 1979, and Cohen and Guyer, eds., 1982.)

In discussions of the logical or epistemic status of value judgments, distinctions are sometimes drawn between two putatively discrete and opposed types of evaluative statement. One type, exemplified by "I like it" or "It pleases me," is said to claim only personal validity; the other type, exemplified by "It is beautiful" or "It is good," is said to claim objective validity. These distinctions and contrasts, however, can be seen as both dubious and distracting, depending on questionable ideas about the operations of discourse, doing little to explain the differential communicative effects of different evaluative statements, and obscuring the sorts of gradient differences among utterances, evaluative and other, that may enable us to relate them to each other in more illuminating ways.

The differences between what appear to be "(merely) personal" and "(genuinely) impersonal" value judgments, for example, could be analyzed and described, not in terms of their dichotomous inherent claims—redeemable or not—to objective validity, but in relation to various, often quite subtle, differences in the relevant beliefs and assumptions of particular evaluators articulating such judgments under specific conditions. Thus, it may be observed that we tend, as evaluators, to frame our overt assessments in relatively unqualified or "impersonal" terms (e.g., "a beautiful performance of the Brahms Opus 116, one of the best ever recorded") to the extent that we believe (1) that our relevant experiences, interests, and tastes coincide relatively closely with those of our presumptive audience, (2) that the range of conditions under which our assessment would be applicable is relatively broad and stable, and (3) that the set of people for whom our assessment would be appropriable is relatively extensive. Conversely, it seems that the narrower or more unstable we suspect the conditions of applicability of our assessment to be, the more confined and distinctive the set of people we think would find it appropriable; and the more divergent we believe our relevant experiences, interests, and tastes may be from those of our immediate audience, the more likely that we will qualify our judgment accordingly, phrasing it in increasingly conditional or guarded terms ("Not bad . . . unless you've heard Glenn Gould do it . . . and you have a taste for Gould") or, perhaps, as a "(mere) personal preference" ("Well, anyway, *I* enjoyed it"). Different verbal idioms have different relevant conventions, of course, and other differences of phrasing may be introduced by the styles of individual speakers or the constraints of particular communicative contexts. The tendencies just described, however, appear quite common in most of the idioms relevant to these discussions.

Contrary, then, to much so-called linguistic intuition (MacIntyre, 1984, p. 9) and to Kant's apparent supposition (section 7), a failure to specify any limiting conditions for the validity of a value judgment does not in itself imply a claim of its unconditional—objective, transcendent, or universally intersubjective—validity. Strong claims of that kind may, of course, be intended or framed explicitly by particular critics, but, it seems, most unqualified judgments of value—including ostensibly impersonal judgments of what we call aesthetic or literary value ("It's totally gorgeous"; "A really first-rate poem"; "Great film—you must see it"; "Pure bathos"; and so forth)—operate not as presumptively valid certifications of unconditional value but as statements of the speaker's belief that the object in question will have the kind and degree of value indicated (a) for some more or less specific, though perhaps only implicitly defined, set of people (b) with certain more or less specific, though perhaps only implicitly defined, tastes, interests, and expectations, (c) who are conceived as experiencing the object under some more or less specific, though perhaps only implicitly defined, range of conditions. If we do not, in framing our judgments, make these specifications explicit, it is commonly because we assume that they are taken for granted along similar enough lines by our audiences or are more or less readily inferable from the verbal and circumstantial context of the judgment itself. Our beliefs and assumptions in all these regards may, of course, be only tenuously grounded or quite erroneous. But such possibilities seem to be an ineradicable condition of all discourse and do not in themselves compromise the validity, in the sense of relative communicative effectiveness and appropriability, of our value judgments—at least not any more than that of any other type of utterance. Accordingly, the epistemic value of overt evaluations seems generally to be on a par with that of all other utterances, including what we call purely factual or descriptive statements.

[*See also* Value.]

BIBLIOGRAPHY

Bourdieu, Pierre. *Distinction: A Social Critique of the Judgement of Taste* (1979). Translated by Richard Nice. Cambridge, Mass., 1984.

Cohen, Ted, and Paul Guyer, eds. *Essays in Kant's Aesthetics.* Chicago, 1982.

Dewey, John. *Theory of Valuation.* Chicago, 1939.

Frye, Northrop. *Anatomy of Criticism: Four Essays.* Princeton, N.J., 1957.

Guillory, John. *Cultural Capital: The Problem of Literary Canon Formation.* Chicago, 1993.

Guyer, Paul. *Kant and the Claims of Taste.* Cambridge, Mass., 1979.

Hirsch, E. D., Jr. "Evaluation as Knowledge." In *The Aims of Interpretation,* pp. 95–109. Chicago, 1976.

Hume, David. "Of the Standard of Taste" (1757). In *Of the Standard of Taste, and Other Essays,* edited by John W. Lenz, pp. 3–28. Indianapolis, 1965.

Kant, Immanuel. *Critique of Judgment* (1790). Translated by J. H. Bernard. New York, 1951.

MacIntyre, Alaisdair. *After Virtue: A Study in Moral Theory.* 2d ed. Notre Dame, Ind., 1984.

Oyama, Susan. *The Ontogeny of Information: Developmental Systems and Evolution.* Cambridge and New York, 1985.

Richards, I. A. *Principles of Literary Criticism* (1924). London, 1960.

Smith, Barbara Herrnstein. *Contingencies of Value: Alternative Perspectives for Critical Theory.* Cambridge, Mass., 1988.

Thompson, Michael. *Rubbish Theory: The Creation and Destruction of Value.* New York and Oxford, 1979.

BARBARA HERRNSTEIN SMITH

EXISTENTIALISM. *See* Heidegger; Kierkegaard; Merleau-Ponty; *and* Sartre.

EXPERIENCE, AESTHETIC. *See* Appreciation; Aristotle; Attitude, *article on* Aesthetic Attitude; Dewey; Dufrenne; Emerson; Emotions; Gadamer, *survey article*; Hypertext; Imagery; Japanese Aesthetics, *historical overview article;* Mendelssohn; Perception; Pleasure; Qualities; *and* Rasa.

EXPRESSIONISM. As a historical artistic phenomenon, Expressionism is most frequently identified with the progressive, antinaturalistic art and literature of Germany and Austria during the years 1905 to 1920. Beyond this broad chronological and geographic identification, however, on such fundamental questions as who the Expressionist artists and writers were, or what stylistic and other criteria characterize an Expressionist work, there is little agreement. While such imprecision is not unusual in the definition of modern art movements, usually it appears in establishing the extent of a movement and its peripheries—whether it is appropriate to identify a Czech artist working in Prague with the Paris-centered movement of Cubism, for example—and seldom are essential defining qualities so disputed. Indeed, because of its imprecision, the very concept of Expressionism and its usability in systematic historical and critical studies have been called into question repeatedly. Nonetheless, the concept refuses to disappear, and a review of its most widely accepted components and associated aesthetic theories or concepts is therefore useful.

In art history and literary criticism today, German Expressionism is linked initially to the artists groups Die Brücke, founded in Dresden in 1905 by Ernst Ludwig Kirchner, Erich Heckel, Karl Schmidt-Rottluff, and Fritz Bleyl, and Der Blaue Reiter, founded in Munich in 1911 by Vassily Kandinsky, Franz Marc, Gabriele Münter, and August Macke. This core is then extended to include the periodicals *Der Sturm* and *Die Aktion*—which sponsored exhibitions, and published and defended works by the artists of the two groups—and to other artists associated with them, as well as to writers whose poems, plays, novellas, and critical or theoretical essays were published in them. From this expanded core, since these artists and writers also joined or founded other groups, periodicals, and exhibition societies, the identification of German Expressionist works and artists is additively extended to incorporate others joining or appearing in these ventures. The height of German Expressionist activity and its greatest extension then appears during World War I, the November Revolution of 1918, and the initial years of the Weimar Republic, and 1920 is generally accepted as marking the end of Expressionism as a movement. Works by the various artists and authors associated with Expressionism, but produced after 1920, generally continue to be identified as Expressionist, however, as do other isolated individual pieces, and the concept of Expressionism is also extended to film, music, and politics, as well as to other countries. According to this prolonged chronology, the coming to power of Adolf Hitler in 1933, his anti-modern art politics, and the Degenerate Art Exhibition in Munich in 1937 definitively terminate the Expressionist movement.

It is an awkward and imprecise historical construct. Unlike Futurism, Dada, or Surrealism, *Expressionism* was a name very few artists themselves employed to identify their work or a movement, and it was not—as Impressionism, Cubism, or Fauvism were—a name invented by critics but then accepted by many of the artists and their apologists. Moreover, ironically, it was Nazi opposition to, and propaganda against, modernist art, music, and literature that lent the term *Expressionism* its seemingly unshakable chronological and conceptual ambiguity while also extending its meaning into further indefinite realms of ideology and weltanschauung.

Although expression is a fundamental concept in art theory and aesthetics since antiquity, and the terms *expressionist* and *expressionism* appear sporadically in English, French, and German literary and art criticism since the mid-nineteenth century, not until around 1910 do these terms become frequently employed. Initially they were adopted in London by Roger Fry and critics supporting his first Postimpressionist exhibition in 1910–1911, as simple terminological indicators of a perceived stylistic antithesis to Impressionism in the works of the exhibition's artists, who ranged from Vincent van Gogh, Paul Gauguin, and Paul Cézanne to Henri Matisse and Pablo Picasso.

This linkage to artists active in France also characterized the first significant documented German usage, in the catalog of the Twenty-second Berlin Secession exhibition in April 1911 that listed "a number of works by younger French artists, Expressionists. The exhibition shall present an overview of the movement that has been called Expressionism." Rooted in Latin, just as the widely accepted term *Impressionism* was, the word *Expressionism* dramatically proclaimed its position of opposition and partial *Aufhebung.* It was quickly applied prolifically by German art critics as an

EXPRESSIONISM. Ernst Ludwig Kirchner, *Five Tarts (Fünf Kokotten)* (1914), woodcut on blotting paper, plate: 19 × 14 1/2 in. (48.5 × 37 cm), sheet: 21 1/2 × 15 1/4 in. (52 × 39 cm); National Gallery of Art, Washington; Ruth and Jacob Kainen Collection. (Photograph by Dean Beasom; courtesy Board of Trustees, National Gallery of Art.)

overarching characterization of various modes of imported modernism—Cubism, Fauvism, Futurism—but also of the work of Die Brücke and Der Blaue Reiter artists in Germany, and more sporadically by literary critics. Similarly, gallery owners, especially Herward Walden of Berlin's Der Sturm gallery, and the organizers of the first extensive survey of Modernist art, the Sonderbund exhibition in Cologne in 1912, numbered Germany among the nations contributing to the newly identified "most recent art movement," Expressionism. After the first book-length treatments of Expressionism appeared in 1914 and 1916, by the Dresden art critic Paul Fechter and the Viennese writer Hermann Bahr, respectively, both using the term *Expressionismus* as the book title, during the course of World War I the identification of Expressionism as a manifestation of a particularly Germanic and German manner of artistic creation became widely accepted. In 1916, the art historian Wilhelm Worringer thus summarized and concluded

Clearly indicated in the [artistic] evolution prior to the war is a unified . . . drive toward an enrichment in the expressive vo-

cabulary of art . . . [that is] the turnaround from Impressionism to Expressionism. . . . The new program announced in [the term] *Expressionism* could perhaps be formulated in France in its full qualities, but could not find a true evolutionary, historically fruitful resonance there. Rather, the necessary resonance can only be in the land whose entire artistic evolution . . . always struggled for heightened expressive power in artistic form such as defied pure sensual perception, that is, Germany. The impetuous-expressionistic in contrast to a quiet purified observation of nature has always lain within the inner being of German art.

Worringer is the theorist and historian who remains most readily linked to Expressionism. His extensively disseminated dissertation and book, *Abstraktion und Einfühlung* (Abstraction and Empathy; 1908), formulated an aesthetic that rejected the imitation of nature as art's primary goal and likewise rejected the widely accepted perception of artistic evolution as being from nonillusionistic to illusionistic. "Our studies are grounded in the premise that the artwork appears as independent organism equal to nature and in its deepest, innermost essence has no relationship to nature, insofar as one understands nature to be the visible surface of things." Following principles enunciated by the Viennese art historian Alois Riegl, Worringer argued that styles differed according to what a nation or period willed its art to do, the *Kunstwollen*, not because of varying degrees of skill in illusionistic rendering. Antithetical styles thus are manifested historically and geographically: abstraction, which rejects the emulation of natural forms and accents inorganic configurations, and naturalism, which reproduces the appearances of life so as to permit the viewer to engage in a process of empathy or self-identification with the artwork. The two represent different psychological artistic needs, abstraction corresponding to an agoraphobic anxiety about the chaotic and confusing appearance of the external world and seeking to supplant this with the calm derived from an abstract order, and empathy corresponding to a pantheistic desire to achieve happiness through the mysterious powers of organic forms in which one could recognize one's own organism, whereby art becomes objectified self-enjoyment. While the drive to abstraction is found largely in the early phases of civilizations and among *Naturvölker* but also in Asia, and there is an evolution toward an empathetic or naturalist art in the West, the rationalism deriving from the study of nature reaches a degree where it too seeks the abstract "thing-in-itself," but as a product of study based in Arthur Schopenhauer's recognition of the illusion of all reality and not instinctively, as was the case earlier. Ironically, the primitive and the modern thereby approach each other, although the individualism of modernity makes the earlier artistic productivity rooted in a sense of community impossible.

Worringer, however, also set up an intermediary between abstraction and empathy as he considered the artistic pro-

duction of northern Europe. In its attitudes toward nature failing to share the beneficence that characterizes Mediterranean antiquity and the Renaissance, Nordic humanity shares in the alienation of abstraction even as it manifests the empathetic desires of the West, Worringer argued.

> The inner desires of life and expression of these disharmonic peoples do not take the near at hand path to the organic because the quiet harmonic movement of the organic does not state enough for them, because their disharmony cannot express itself with the means of the organic, but rather they needed the intensification of resistance, they needed the disquieting pathos that is manifested when the anorganic is brought to life.

A hybrid of abstraction and empathetic emulation, employing exaggeration and distortion of natural forms by submitting them to an abstracting process to achieve utmost pathos and expressive force, the art of northern Europe reaches its greatest manifestation in the Gothic. Worringer concluded his study by bemoaning the loss of this last "style" and its great values, defeated by the Italian Renaissance's organic naturalism, and devoted his next major study, *Formprobleme der Gotik* (Problems of Form in the Gothic; 1911), to an extensive exploration of these characteristics attributed to Gothic art and architecture.

The applicability of Worringer's concepts to contemporary art was recognized immediately. "The way Worringer describes Gothic creations is how the works of the Expressionists and Cubists look," wrote one reviewer. Offering both historical precedent and aesthetic analysis and justification for an expressive antinaturalistic art, Worringer's view of the Gothic found immediate application in the critical approaches to contemporary Postimpressionist art. Worringer himself applied his concepts to modern art in his polemical essay of 1911, published in *Der Sturm*, "Zur Entwicklungsgeschichte der modernen Kunst" (Concerning the Evolutionary History of Modern Art), and his ideas likewise served as the basis for Paul Ferdinand Schmidt's similarly early defining essay, "Über die Expressionisten" (Concerning the Expressionists; 1911). Linked to medieval Gothic precedent, Expressionism was defined as a modern antimaterialism in art, a manifestation of a new spirituality in which the anxieties and hopes of modern humanity were given constructive form and a new communal "modern style" was taking shape, with Germany serving a leading role in bringing it to fulfillment.

Worringer's model shaped the critical discourse on Modernist art in Germany during the following years. His concept of the Gothic and its continuous, frequently unrecognized antinaturalist, "spiritual" tradition in German art was extended in Bahr's and Fechter's books, and largely adopted even by the socialist (although inconsistently materialist) art critic Wilhelm Hausenstein, who saw a "new religiosity" in Expressionist art that he equated with socialism (*Die bildende Kunst der Gegenwart,* 1914). Such characterizations

were most dramatically and succinctly summarized by the poet Kasimir Edschmid's polemical argument that the German Expressionist artist "does not see, he has visions; he does not depict, he experiences; he does not reproduce, he creates forms . . . Objects are significant only insofar as they make it possible for the artist's deeply searching hand to grasp the existence behind them" (*Expressionismus in der Dichtung,* 1919).

During the last years of World War I and the initial ones of the Weimar Republic, numerous manifestos and critical studies sought to explicate further the inherent qualities of an Expressionism that became the quasi-official artistic vocabulary of the early Weimar Republic. By 1920, however, Expressionism as an art movement was being declared dead or incapable of addressing the fundamental issues of the day, "a ghostly play with empty gestures," as Worringer now proclaimed (1921). As artists turned increasingly either to geometrically ordered abstraction or to the sharp objectivity of *Neue Sachlichkeit* (New Sobriety), Expressionism failed to gain new adherents and ceased to be a vital component of German artistic and aesthetic conviction.

There are significant problems with this standard prioritizing of Worringer and the critical tradition of Expressionism his writings fostered as I have recounted it. Most apparent is that artists and artworks have no role in this narrative and that, indeed, this coherent, unified Expressionism is fundamentally a construct or, as Charles W. Haxthausen has recently argued, even an "illusion" without historical basis outside the writings of the art critics who invented it. By accenting unity and continuity, moreover, this narrative ignores the diversity and multiplicity of critical and aesthetic voices that sought to confront and comprehend modernism in Germany during the 1905–1920 "Era of Expressionism."

The most notable missing critical voices in this narrative, only now beginning to be rediscovered and analyzed, are those of Adolf Behne, Carl Einstein, Richard Hamann, Wilhelm Hausenstein, Max Raphael, and Herwarth Walden. Significantly, while they begin their aesthetic concerns in different intellectual and ideological traditions, they later share an adherence, informal and with varying rigor, to socialism and to Marxist materialism. Most, however, did not overtly profess their adherence until after World War I, by which time Hausenstein already begins to reject it. Their approaches thus vary significantly, and their views of Marxist values are both individually shaped and accepted at varying times, but nonetheless as a group they can be distinguished from Worringer and his followers in that they diminish the importance of national and racial identity or psychology, and instead prioritize the defining consciousness of a modern capitalist, democratizing society as determinant of the subjectivity and individualism apparent in the newest manifestations of the arts. Similarly, it is in what may be considered as a direct materialist confrontation with particular artworks and artists, not in principles of national

EXPRESSIONISM. Kathe Kollwitz, *Death Embracing a Woman* (1921), etching; Kunsthalle, Bremen. (Copyright 1998 by the Artists Rights Society, New York/VG Bild-Kunst, Bonn; photograph courtesy of Foto Marburg/Art Resource, New York; used by permission.)

psychological identity, that they both ground and seek evidence for the aesthetics of Expressionism they enunciate.

Despite unquestionable, fundamental differences among the critics and their aesthetic formulations that should not be ignored or forgotten, several further shared factors of a common aesthetic can be recognized, many of them interrelated and overlapping. The most significant of these might be identified as the individuality and subjectivity of the artist, an accent on the art object as independent material or spiritual entity rather than an imitation of nature, the fragmentation of forms and an additive or serializing practice of composition rather than one of harmonization and unity, and a perception of the new art as predictive of a future unified style in a similarly transformed cohesive postcapitalist society. To these characteristics should be added, especially in the visual arts, a frequent recognition not unlike Worringer's, although now based in principles of societal organization rather than national psychology, of kinship be-

tween the new art and the so-called primitive art of pre-Renaissance Europe, of folk and native artists as well as children, and of the indigenous peoples of Africa, Oceanea, the Americas, and parts of Asia.

Significant here in contrast to the dominant theories of Worringer is the rejection of a nationalistic grounding and argumentation, and indeed these critics sought much more to identify the qualities of modernist art generally than to establish a unique, separable German form of it. Rather than a German Expressionist aesthetic, it can be argued, their writings and theories identify and grant positive value to the aesthetic of an international modernist avant-garde around and immediately after 1910. It seems appropriate, however, to retain for them their particular geographic and historical specificity and to recognize their acknowledged dependence precisely on the works and, often, personal interaction with artists in Germany during the "Era of Expressionism." Indeed, if the confusion and ambiguity that

remain inherent in historically identifying a German Expressionist art movement and aesthetic, as noted at the beginning of this essay, are to be overcome, it may well be in the ongoing study and analysis of these and other critical voices alternative to Worringer and in the further matching of their observations and principles to their contemporary artistic milieu and historical situation.

[*See also* Abstract Expressionism; Expression Theory of Art; *and* Worringer.]

BIBLIOGRAPHY

Behne, Adolf. *Zur neuen Kunst.* Sturmbücher, vol. 7. Berlin, 1915.

Behne, Adolf. *Die Wiederkehr der Kunst.* Leipzig, 1919.

Brinkmann, Richard. *Expressionismus: Internationale Forschungen zu einem internationalen Phänomen.* Stuttgart, 1980.

Edschmid, Kasimir. *Über den Expressionismus in der Literatur und die neue Dichtung.* Tribüne der Kunst und Zeit, vol. 1. Berlin, 1919.

Einstein, Carl. *Gesammelte Werke.* Edited by Ernst Nef. Wiesbaden, 1962.

Gordon, Donald. "On the Origin of the Word 'Expressionism'." *Journal of the Warburg and Courtauld Institutes* 29 (1966): 368–385.

Hamann, Richard. *Krieg, Kunst und Gegenwart: Aufsätze.* Marburg, 1914.

Hausenstein, Wilhelm. *Der nackte Mensch in der Kunst aller Zeiten und Völker.* Munich, 1913.

Hausenstein, Wilhelm. "Versuch einer Soziologie der bildenden Kunst." *Archiv für Sozialwissenschaft und Sozialpolitik* 36 (1913): 760–792.

Hausenstein, Wilhelm. *Die bildende Kunst der Gegenwart.* Stuttgart, 1914.

Hausenstein, Wilhelm. *Die Kunst in diesem Augenblick.* Munich, 1920.

Haxthausen, Charles. "A Critical Illusion: 'Expressionism' in the Writings of Wilhelm Hausenstein." In *The Ideological Crisis of Expressionism: The Literary and Artistic German War Colony in Belgium, 1914–1918,* edited by Rainer Rumold and O. K. Werckmeister. Columbia, S.C., 1990.

Heinrichs, Hans-Jürgen, ed. *Wir lassen uns die Welt nicht zerbrechen: Max Raphaels Werk in der Diskussion.* Frankfurt am Main, 1989.

Long, Rose-Carol Washton, ed. *German Expressionism: Documents from the End of the Wilhelmine Empire to the Rise of National Socialism.* New York, 1993; reprint, Berkeley, 1995.

Mannheim, Ron. "'Expressionismus': Zur Entstehung eines kunsthistorischen Stil- und Periodenbegriffs." *Zeitschrift für Kunstgeschichte* 49.1 (1986): 73–91.

Perkins, Geoffrey. *Contemporary Theory of Expressionism.* Britische und irische Studien zur deutschen Sprache und Literatur, vol. 1. Bern, 1974.

Raphael, Max. *Das schöpferische Auge; oder, Die Geburt des Expressionismus: Die frühen Schriften, 1910–1913.* Edited by Patrick Healy and Hans-Jürgen Heinrichs with Ron Mannheim. Vienna, 1993.

Raphael, Max. *Von Monet zu Picasso: Grundzüge einer Ästhetik und Entwicklung der modernen Malerei.* Munich, 1913. New edition, edited by Klaus Binder (Frankfurt am Main, 1983).

Walden, Herwarth. *Gesammelte Schriften,* vol. 1, *Kunstkritiker und Kunstmaler.* Berlin, 1916.

Walden, Herwarth. *Einblick in Kunst: Expressionismus, Futurismus, Kubismus.* Berlin, 1917.

Walden, Herwarth. *Die neue Malerei.* Berlin, 1919.

Weinstein, Joan, "Wilhelm Hausenstein, the Leftist Promotion of Expressionism, and the First World War." In *The Ideological Crisis of Expressionism: The Literary and Artistic German War Colony in Belgium, 1914–1918,* edited by Rainer Rumold and O. K. Werckmeister, pp. 193–217. Columbia, S.C., 1990.

Werenskiold, Marit. *The Concept of Expressionism: Origin and Metamorphoses.* Translated by Ronald Walford. Oslo, 1984.

Worringer, Wilhelm. *Abstraktion und Einfühlung: Ein Beitrag zur Stilpsychologie.* Munich, 1908. Translated as *Abstraction and Empathy: A Contribution to the Psychology of Style* by Michael Bullock (New York, 1953, 4th ed., New York, 1980).

Worringer, Wilhelm. "Entwicklungsgeschichtliches zur modernen Kunst." In *Im Kampf um die Kunst: Die Antwort auf den "Protest deutscher Künstler,"* pp. 92–99. Munich, 1911. Reprinted as "Zur Entwicklungsgeschichte der modernen Malerei," *Der Sturm* 75 (1911): 597–598. Partial translation as "The Historical Development of Modern Art" in *German Expressionism: Documents from the End of the Wilhelmine Empire to the Rise of National Socialism,* edited by Rose-Carol Washton Long (New York, 1993; reprint, Berkeley, 1995), pp. 9–13.

Worringer, Wilhelm. *Formprobleme der Gotik.* Munich, 1911. Translated as *Form Problems of the Gothic.* (New York, 1920). Also translated as *Form in Gothic,* edited by Herbert Read (London, 1927; rev. ed., New York, 1964).

Worringer, Wilhelm. "Künstlerische Zukunftsfragen." *Kunst und Künstler* 14 (1916): 259–264.

Worringer, Wilhelm. *Künstlerische Zeitfragen.* Munich, 1921. Reprinted in *Fragen und Gegenfragen: Schriften zum Kunstproblem* (Munich, 1956), pp. 106–129.

REINHOLD HELLER

EXPRESSION THEORY OF ART. Proponents of the expression theory of art typically claim that art should be viewed as the expression of the mind of the artist—of his or her emotions, feelings, thoughts, and so on. In comparison with the much more ancient conception of art as imitation or representation of the world, expression theories of art are of relatively recent provenance. The question of how expression theories are different from imitation theories, however, is itself a significant one. An imitation theory might after all claim that art is sometimes the representation of emotion or other mental states, and even of the artist's mental states in particular. However the matter is decided, the shift from imitation theories to expression theories involves at least an important shift in emphasis. While imitation theories tend to emphasize the relation between the artwork and what it imitates, expression theories have traditionally stressed the relation between the artwork and the artist, with a corresponding emphasis on authenticity of feeling and expression rather than faithfulness of representation.

The concept of expression has, however, been used in a variety of other ways to elucidate the nature of art, and one of these has been especially prominent. This is the notion that whether or not artworks should be seen as expressions of the artist, they are at any rate themselves *expressive* of emotions or feelings; for instance, a piece of music may strike us as melancholy or a painting as solemn. Expression theories characteristically seek to explain art's expressiveness as deriving from the fact that the artist expressed his or her emotions in making it; the emotion an artwork ex-

presses is the emotion expressed in it by the artist. But expressiveness and expression have often been regarded as independent, and other explanations of expressiveness have been sought. Whether they are in fact independent is an open question, but the two uses have at any rate given rise to two bodies of literature that at least to some degree are concerned with distinct issues and have distinct aims. We should thus consider these two notions—art as expression and art as expressive—separately, bearing in mind the question of whether they can ultimately be seen as wholly independent.

Traditional Expression Theories. What are usually thought of as the classic philosophical expression theories of art—those of Benedetto Croce and Robin George Collingwood—emerged only around the turn of the twentieth century. But while there may be some temptation, in light of this fact, to view expression theories as arising in direct response to the development, at around the same time, of styles of art that did not fit so easily into the model of art as the imitation of nature, such a view would be at best an oversimplification. In various ways, art has been viewed as the expression of the artist's mind since at least the late eighteenth century, and this notion was indeed a key feature in the development of Romanticism. William Wordsworth's "Poetry is the spontaneous overflow of powerful feelings" is but one well-known example. (For the development of metaphors of art as expression during this period, see M. H. Abrams's classic study *The Mirror and the Lamp* [1953].)

It is important to bear in mind that the notion of art as expression that took hold in the Romantic period tended to be based on a quite particular conception of expression. Expression, on this view, is not merely one human activity among others, but a mode of knowledge distinct from, and prior to, conceptual knowledge. In the theories of Croce's *Aesthetic* (1992) and Collingwood's *Principles of Art* (1938), this distinction becomes explicitly thematized. For Croce and Collingwood, the key point is not just that through the creation of art the artist expresses his or her feelings, but that in this process of expression the inchoate feelings that impelled the artistic act come to be clarified and better understood—and in a way not possible by means of concepts. Art is in fact here so strongly identified with expression that the notion of the "aesthetic" becomes radically broadened: not only is all art expression, but all expression is aesthetic. The accounts given by Croce and Collingwood are in their essentials quite similar, Collingwood having developed his views very much under the influence of Croce. The discussion to follow will focus mainly on Croce's presentation of the ideas.

Croce speaks of the form of knowledge embodied by art not only as expression but as "intuition." (In a similar vein, Collingwood speaks of "imagination.") Intuition, for Croce, is that fundamental activity whereby passively received impressions (sensations, feelings, emotions, or impulses) are given form and unity by the human spirit. It is because in Croce's view intuition is active that he also calls it expression: in each act of forming an intuition the human spirit expresses its particular state at that time. To express a mental state in this sense is not merely, to use Collingwood's term, to betray it, as a blush may betray embarrassment. Expression is a conscious activity in which, by giving form and articulation to our impressions, we come to know them, and hence liberate ourselves from them. Intuition expression is thus not merely the province of artists. The everyday world of each person is, indeed, made up of a series of "little" expressions—artists are merely those with a greater aptitude for expression. This move, undercutting the dichotomy between art and everyday life, is characteristic of many expression theories; such a dichotomy, says Croce, threatens to turn art into an "aristocratic club or specialist activity."

In the sense in which Croce and Collingwood use the term, however, *expression* does not refer to a physical activity or the creation of a physical artifact. Both ascribe to an idealist ontology of artworks: the real artwork is "internal," a mental entity. Since for Croce expression is identical with the act of intuition, he distinguishes sharply between the expression of mental states and their *externalization* in a painting, text, or sequence of sounds. Any artwork, as Collingwood puts it, is already complete and perfect as it exists in the artist's head. So, too, there can be no *technique* for making the real, internal artwork, though there may be techniques of externalization.

The idealist ontology of art to which these versions of the expression theory ascribe has provided fuel for a good deal of criticism of them. It has often been pointed out that the position has some very counterintuitive consequences: on this view, the physical object hanging on the wall that we call Édouard Manet's *Le déjeuner sur l'herbe* is not an artwork; and on this account, artworks come in and out of existence when we think of them and stop thinking of them, and cease to exist entirely if not thought of at all. It is true that Collingwood concedes that it is often essential, in order for a painter to express a given thing, that he or she have the experience of actually putting paint on canvas in a certain way. But if this is granted, it becomes questionable whether what the painter imagines can really be defined as independently of the physical canvas as the idealist theory claims. (See Wollheim, 1980.)

There are, however, versions of the expression theory of art that do not rely on such a questionable aesthetic ontology. The expression account developed by the novelist Leo Tolstoy in his 1898 essay *What Is Art?* (Tolstoy, 1930), for instance, avoids such commitments. Similarly, the influential expression theory proposed by John Dewey in *Art as Experience* (1934) eschews an idealist ontology of art in favor of a more naturalistic one. Like Croce and Colling-

wood, Dewey stresses the continuity between art and every-day experience; indeed, for him art is the "clarified and intensified" development of features present in certain kinds of vital everyday experiences. These are experiences that are complete, harmonious, and unified in that, guided by some impulse, they run their course naturally and continuously toward their fulfillment—as when, for instance, we finish our work successfully or solve a problem. In such experiences lie the roots of the artist's concern with unity and harmony. Any such experience begins with an "emotional impulsion" that is initially inchoate; and again like Croce and Collingwood, Dewey views the process of artistic expression as the artist's effort to bring such an emotion to consciousness. Expression cannot be mere unthinking discharge—for Dewey it essentially involves the intention to communicate. But precisely for this reason, it is also essential on his view that it be embodied in an external medium that makes it communicable. Thus, while for Dewey an artwork cannot simply be identified with a physical object, neither can it be defined as independently of physical objects as it can for Croce and Collingwood.

Expression theories of art have, however, been widely criticized on grounds other than ontological ones. One focus of these critiques has been on the expression theory as a definition of art. In order to count as art, for these theories, an entity must satisfy several conditions: (1) the artist must at the time of creating the work have felt some emotion, (2) whose nature was unclear to her or him, and (3) intentionally created that entity in such a way that its form makes manifest that emotion, (4) thereby clarifying the emotion to him- or herself and perhaps others. The main difficulty with this account has sometimes been said to be that it leaves out some of the things we call art: it is doubtful whether the Parthenon, for instance, or a finely wrought altar, would meet these conditions. But expression theorists would just admit this point—what they are offering, after all, is a definition of "real" art, as opposed to those things that are merely *called* art but fail to satisfy the above conditions. Collingwood, for example, banishes to this latter category craft, "amusement art," and art that serves a practical purpose (such as religious icons).

Other difficulties have been raised for the theory, however, that are more immediately damaging. First, it is only rarely possible to know whether in fact a given artwork was produced by such a procedure; so at any rate, the notion that art is expression could not function as a criterion for deciding whether something is a work of art. But more important, it must be asked whether even if we did somehow discover that the creator of a work had not at the time he created it felt any emotion that we take the work to embody, would we really reject our former opinion that it is a work of art? Igor Stravinsky, for instance, wrote his Symphony in C at the time of the death of his wife and his mother, but he did not view the music as reflecting his grief (Davies, 1994).

Is it any less art for this reason? The expression theory seems to make the question of whether something is art too dependent on the actual psychology of the artist.

Criticism of the expression theory as a definition of art has also come from other quarters. It has sometimes been suggested that while the notion that art is personal expression may have been a useful one at the beginning of the twentieth century, as artists shifted away from naturalistic representation, the succession of artistic movements in this century has increasingly revealed its inadequacy as a definition of art (Danto, 1986). It is hard to see the differences of style between Fauvism, Cubism, and Dada as simply a matter of differences in the way the artists' emotions are expressed. The "aesthetic of expression" has, indeed, sometimes been associated with modernism itself and, by some critics, has been said to be on the wane with the development of postmodernism (Jameson, 1985). This kind of point is, however, perhaps best seen as a critique not of the notion that artists may convey aspects of themselves through their art, but of a particular ideal of art as self-expression, one in which the internal state can be identified independently of its external expression.

The expression theory has been criticized, further, not only as a definition of art but for its account of art's expressiveness (see Tormey, 1971; Wollheim, 1980; Davies, 1994). On this account, in order for a work to have a given expressive quality, the artist must have felt a corresponding emotion and made that emotion manifest in the work. But if Bach's Suites for Solo Cello strike us as melancholy, we would not, it seems, withdraw our interpretation solely on the basis of discovering that Bach had not actually felt melancholy at the time of composing them, or even if we learned he had been very happy. This is not to say that facts about the biography of the artist may not influence our interpretation. What it does suggest is that art's expressiveness does not derive solely from the intention of the artist to convey his or her actually felt emotions. This is supported by the observation that we commonly view not only artworks but natural objects as expressive—willow trees as sad, for instance. There may, of course, be ways in which the traditional expression theory's account of expressiveness could be modified to withstand this sort of objection, and we will return to this possibility later. But points such as these have led many theorists in the past few decades to turn their attention toward giving accounts of expressiveness itself, independently of the artist's emotions.

Expressiveness without Expression. Few of those interested in explaining expressiveness have viewed it as the basis for a *definition* of art. Expressiveness is certainly one element of the significance of all of the arts—a musical melody may be lonely, a painting full of gaiety, and a dance mournful. It is characteristic of the experience of expressiveness that we take the emotional quality to inhere in the form of the artwork itself, and the question is then, as it has

sometimes been put, how a feeling or emotion can get into an object.

The account offered by the traditional expression theory is a causal one: to say that an artwork is sad is to say that its creator was sad and made that emotional state manifest in the work. But in light of the shortcomings of that view, another kind of causal account—often called the arousal theory—has sometimes been offered as more plausible. Perhaps to say that an artwork is sad is to say that it has the power, at least under the right circumstances, to make its *audience* sad, this sadness then being projected onto the object. But the arousal theory faces difficulties of its own. The central problem is that it seems we often see an expressive quality in an artwork without feeling the corresponding emotion at all; we might see that a piece of music is intended to be solemn while feeling rather detached, or even jubilant, ourselves. Even if art may sometimes rouse emotions in us, then, it would seem difficult to explain art's expressiveness in terms of those emotions. (See Bouwsma, 1970, and for some recent qualified versions of the arousal view, Davies, 1994.)

The point is, in short, that our experience of the sadness of an artwork seems often to be not a matter of *feeling* any kind of sadness but rather of *recognizing* sadness in the work. In light of this, a number of theorists have developed accounts that seek to explain art's expressiveness, not causally, but in terms of its ability to *symbolize* emotion. One prominent example of such an approach is the view of Suzanne Langer (1942). In order to function symbolically, for Langer, a thing must be similar in structure to what it symbolizes, but in her view art is a symbolic medium of a different kind from language. Unlike language, art lacks a vocabulary and syntax, and hence it is ideally suited to symbolize what language cannot, namely, the detail and complexity of the world of our feelings, in which she includes everything from emotions to states of excitement and repose. More particularly, what art can symbolize is not the feelings themselves but the temporal process of feeling, that is, patterns of motion and rest, tension and release that may be shared by different feelings. Thus, the expressiveness of music derives from the similarities between its formal properties and these changing patterns of feeling; visual art synthesizes such processes into an atemporal form.

Langer's account has often been thought seriously limited, however, by its use of a model of emotion according to which even the most complex emotions are to be conceived as *felt* bodily changes in the organism (see, e.g., Budd, 1985). This model has been widely criticized in recent philosophy, for most emotions clearly have a cognitive component that relates them intrinsically to circumstances in the world, but which the feeling model leaves out. It does not seem possible to differentiate many emotions—for instance, fear and anger—solely on the basis of bodily feelings that might accompany the emotions. Fear is in part identified by

something like the belief that one is in danger, and anger by the belief that a wrong has been done. Correspondingly, Langer misrepresents art's capacity to tell us about the emotions: if joyful music could only tell us about the bodily feelings that might accompany joy rather than about joy in all of its aspects, we would not have nearly the interest in it that we do.

Perhaps the most influential approach to expressiveness at present is what might be called the "similarity theory," which pursues a rather different tack. Rather than explaining expressiveness causally or as a symbolization of emotion, such an approach seeks to demystify the question of how an emotion can "get into an object" by finding analogues of this phenomenon outside the realm of art. Two of the most significant examples of this approach are the views of Peter Kivy (1980) and Stephen Davies (1994), both of which concern expressiveness in music. What such theories point to is our natural tendency to view the human body and its behavior—the face, voice, gait, gestures, carriage, and so on—as themselves bearers of expressive qualities. A sad face, drooping posture, and sad-sounding voice, of course, often tell us that the person exhibiting them is sad; and many philosophers would indeed go further and say that there is a *conceptual* connection between the behavior and the mental state: part of what it *means* to be sad is to be disposed to "sad behavior" of this kind. Nonetheless, on any particular occasion such "sad behavior" does not necessarily indicate that the person in question is sad—he or she might be acting, or just have a permanently sad-looking face. Many human features and elements of behavior can, then, present an emotional appearance, be expressive of emotion, without actually being expressions of the agent's emotion on that occasion. We sometimes regard the sadness as belonging to the appearances themselves—and this is just what we do in the case of music or painting as well.

For Kivy and Davies, then, artistic expressiveness derives primarily from the expressiveness of behavior. To hear a piece of music as sad is to perceive a similarity between its formal features and some of the behavioral expressions of sadness, as funeral marches mimic the slow, heavy movements and speech of sad people. On this view, the range of emotions that music can express will be restricted to those that have characteristic bodily expressions—music could not, presumably, express envy or nostalgia. While musical expressiveness in this sense may be partially structured by convention, it is not simply a matter of convention. Kivy and Davies do allow, however, for another, more explicitly conventional source of expressiveness, as, for instance, minor keys have come to be associated with grief.

Similarity theories of this kind undoubtedly uncover an important element that commonly underlies our sense of a musical work's expressiveness. But it does not seem they can offer an exhaustive account of what determines expressiveness in all of the arts, or even in music itself. The bright

yellows, blues, and reds of Henri Matisse's late paintings and cutouts, for instance, are often joyous, but it would be hard to find any similarity between these colors and joyful behavior. So too Charles Baudelaire once described the painting of Eugène Delacroix as "melancholy" in part because of his simple, abundant, and harmonious use of color. Baudelaire's writings suggest that what connected this use of color with melancholy for him was his association of color, as opposed to line, with the poetic and the sensual. What these examples suggest is that the connections between an artwork and an emotion that underlie expressiveness may be more various than the similarity theory allows, and may include complex chains of association and belief. If this is right, it may be precipitous to restrict a priori the emotions that music or any artwork can express to those that have characteristic bodily expressions.

A broader worry about the similarity theory has sometimes been raised as well. Emotive descriptions of art like "that music is sad" have been compared to what Ludwig Wittgenstein spoke of as language having a "secondary sense" (Wittgenstein, 1953). If—to take two of Wittgenstein's examples—someone wants to say that Wednesday is "fat" and Tuesday "lean," rather than vice versa, or that the vowel *e* is "yellow," *fat* and *yellow* here have a secondary sense. Such usage is idiosyncratic, but its central features are shared by many other uses of language, including emotive descriptions of art. *Fat* in this use is "secondary," for Wittgenstein, in that it presupposes a prior understanding of the "primary" application of the word to people; we could not explain what *fat* means here by pointing to Wednesday as an example, but only by reference to its ordinary meaning. For this reason, although the secondary use of the word occurs in a quite different context from its primary use, it cannot be regarded as having a new, independent meaning. But neither, Wittgenstein suggests, does the fact that the secondary use differs from the primary one mean that *fat* is the wrong word to use in this case; for no other word would, by the speaker's own lights, adequately convey what he wants to say. [*See* Wittgenstein.]

Different conclusions have been drawn from the comparison between emotive descriptions of art and secondary uses (see, e.g., Scruton, 1974; Tilghman, 1984; and Davies, 1994). One suggestion—though a disputed one—has been that if the comparison is apt, the use of *sad* to describe music does not have a new meaning independent of its primary one; and thus the features we use to support that application cannot be independent criteria of a kind analogous to the criteria that govern the literal meanings of words. The ultimate ground for the application can only be a subjective inclination to use *that* word in the new situation. If this is right, then pointing to the similarities or other connections between an artwork and sadness, while it may be illuminating in many ways, cannot serve to *justify* the description of that work as sad—nor can it serve, as at least Kivy intends,

as a response to the skeptic who doubts the legitimacy of such descriptions.

Back to the Expression Theory? Can the notion of art's expressiveness really be conceived of as wholly independent of the notion that art expresses the artist's psychological states? While few today would accept any version of the expression theory as a *definition* of art, several recent philosophers have in fact suggested that some modified form of the expression theory might be required in order to give an adequate account of art's expressiveness. For despite the shortcomings of the idea that art's expressiveness derives from emotions felt by the artist, it has been pointed out that we take a much deeper interest in the expressiveness of art than in the expressiveness of natural objects, and an interest of a different kind as well. And as Richard Wollheim, for one, suggests, the obvious thought is that this difference derives from the role of the artist in creating the artwork (Wollheim, 1993). Wollheim acknowledges our tendency to see even natural objects as bearing emotional qualities as one source of expressiveness. But the expressiveness of an artwork is also controlled and refined by the artist's intentional activity, and Wollheim's conclusion is thus that what an artwork expresses is in all cases the emotion the artist intended to express in making it. Wollheim's account makes several crucial modifications in the traditional expression theory: first, the artist need not actually have felt the emotion expressed at the time of creation, but only at least have had access to it in memory; and second, the sense of "intention" in which the artist's intention determines what the work expresses should be construed quite broadly, so as to include not only what was deliberately willed but all of the thoughts and emotions that caused and guided the creative act.

While it is plausible to think that there is some connection between an artwork's expressive qualities and the artist's mind, it might be questioned whether Wollheim's modifications still preserve too much of the traditional expression theory. While the boundary between feeling an emotion and merely thinking of it is not a sharp one, there seems no reason to suppose that what an artwork expresses must be an emotion actually felt by the artist at some time in the past, rather than merely the artist's *conception* of the emotion in question. Further, there seems no reason to say that an artwork can never have an expressive quality its maker did not intend it to have; poorly made films, for instance, often have feeling tones far from what was intended. A more moderate form of expression theory is one proposed by Stephen Davies (1994). Music may, Davies allows, sometimes express emotions that do not derive from the composer's intentions. But ordinarily, part of what is involved in the composer's crafting of materials is an attempt to exert intentional control over the expressive content of the product; and thus usually what the music expresses is some emotion of which the composer had at least some conception, even if

he or she never actually felt it, and which he or she intended the work to express. If this were not normally the case, we would have no reason to care about the emotions in art in the way that we do.

[*See also* Abstract Expressionism; Collingwood; Croce; Dewey; Emotions; Expressionism; Gurney; Langer; Theories of Art; *and* Tolstoy.]

BIBLIOGRAPHY

Abrams, M. H. *The Mirror and the Lamp: Romantic Theory and the Critical Tradition.* New York and Oxford, 1953.

Bouwsma, O. K. "The Expression Theory of Art" (1954). In *Artistic Expression,* edited by John Hospers. New York, 1970.

Budd, Malcolm. *Music and the Emotions: The Philosophical Theories.* London and Boston, 1985.

Collingwood, R. G. *The Principles of Art.* Oxford, 1938.

Croce, Benedetto. *The Aesthetic as the Science of Expression and of the Linguistic in General* (1902). Translated by Colin Lyas. Cambridge and New York, 1992.

Danto, Arthur C. *The Philosophical Disenfranchisement of Art.* New York, 1986.

Davies, Stephen. *Musical Meaning and Expression.* Ithaca, N.Y., 1994.

Dewey, John. *Art as Experience.* New York, 1934.

Goodman, Nelson. *Languages of Art: An Approach to a Theory of Symbols.* 2d ed. Indianapolis, 1976.

Hanslick, Eduard. *The Beautiful in Music.* Translated by Gustav Cohen. Indianapolis, 1957.

Hospers, John. "The Croce-Collingwood Theory of Art." In *Artistic Expression,* edited by John Hospers. New York, 1971.

Jameson, Fredric. "Postmodernism and Consumer Society." In *Postmodern Culture,* edited by Hal Foster. London, 1985.

Kivy, Peter. *The Corded Shell: Reflections on Musical Expression.* Princeton, N.J., 1980.

Langer, Suzanne. *Philosophy in a New Key: A Study in the Symbolism of Reason, Rite, and Art.* Cambridge, Mass., 1942.

Santayana, George. *The Sense of Beauty: Being the Outlines of Aesthetic Theory* (1896). In *The Works of George Santayana,* edited by William G. Holzberger and Herman J. Saatkamp, Jr. Vol. 2. Cambridge, Mass., 1988.

Scruton, Roger. *Art and Imagination: A Study in the Philosophy of Mind.* London, 1974.

Sircello, Guy. *Mind and Art: An Essay on the Varieties of Expression.* Princeton, N.J., 1972.

Sparshott, Francis. *The Theory of the Arts.* Princeton, N.J., 1982.

Tilghman, B. R. *But Is It Art? The Value of Art and the Temptation of Theory.* Oxford and New York, 1984.

Tolstoy, Leo. *What Is Art? and Essays on Art.* Translated by Aylmer Maude. Oxford, 1930.

Tormey, Alan. *The Concept of Expression: A Study in Philosophical Psychology and Aesthetics.* Princeton, N.J., 1971.

Wittgenstein, Ludwig. *Philosophical Investigations.* Translated by G. E. M. Anscombe. Oxford, 1953.

Wollheim, Richard. *Art and Its Objects.* 2d ed. with six supplementary essays. Cambridge and New York, 1980.

Wollheim, Richard. "Correspondence, Projective Properties, and Expression in the Arts." In *The Mind and Its Depths.* Cambridge, Mass., 1993.

JOHN SPACKMAN

F

FĀRĀBĪ, MUHAMMAD AL-, tenth-century Arab philosopher. The little information available about al-Fārābī's life comes from biographers writing during the tenth and thirteenth centuries. He was probably born in 870 CE in Fārāb, in Turkistan, and in adult life studied logic in Baghdad under Abū Bishr Mattā, one of the translators of Aristotle's works into Arabic. The list of philosophical writings ascribed by medieval biographers to al-Fārābī includes more than a hundred works, with the largest part devoted to logic and the philosophy of language. In addition, he wrote on metaphysics, psychology, political philosophy, and the philosophy of religion. The concern with logic provides the context for al-Fārābī's reflections on philosophical poetics, in which he maintains that the validity of poetic language depends on its logical status.

Earlier critics within the Arab Islamic tradition relied on the exegetical and normative method of Quranic and philological sciences. These sciences justified the validity of particular examples of poetry by finding an original accepted model of poetic excellence and showing that the example followed accepted usage in that it bore significant analogies (*qiyās*) with the accepted model and generated consensus (*ijmaʿ*) through the responses of its audience.

Al-Fārābī sought an alternative account of poetic validity. His principal concern was to explore the validity of poetic language in terms of Aristotelian logic in the form in which the latter entered the Arab tradition. Later Greek commentators organized Aristotle's writings into a single text, *The Organon*, that was made up of eight Aristotelian treatises. Demonstrative arguments were of principal concern in this text, which therefore began with the *Analytics,* the *Categories,* and *On Interpretation*. These examined the nature of demonstrative arguments and assertions and the scope and power of logical and cognitive necessity. Logic made up the rules for thinking of scientific truths. The last four books, including the *Topics,* the *Sophists,* the *Rhetoric,* and the *Poetics,* examined the validity of modes of thought that had a progressively less secure grasp of truth; but rhetoric and poetic statements were still part of reasoning. This organization of the treatises was present in the Syriac translations of the original Greek text and, through translation from Syriac into Arabic, become the context for the Arabic study of the Stagrite's works. As a consequence of this context and transmission of texts, the *Poetics* gained a logical coloring in Arabic writing that it did not clearly have in the original. And since al-Fārābī considered poetics a part of reasoning, he looked for its justification in logic and sought to develop a conception of a poetic syllogism.

An initial step in developing this logical justification of poetry was to show the inadequacy of the grammatical method. Al-Fārābī argues that the methodology of the Islamic sciences can only succeed because it relies on logic. Although its exponents refuse to countenance this possibility, preferring to see grammar as the principal mode of justification, al-Fārābī explains that discovering accepted usage, examining analogies, and establishing consensus are secure only when they satisfy the demands of logical thought.

Having established the need for logic, al-Fārābī considers the logic nature of poetic discourses in a number of texts, including the "Canons of Poetry," *Kitāb al-shirʿ,* his commentary on Aristotle's *Poetics,* and the "Catalogue of the Sciences." In the "Canons of Poetry," he justifies the validity of poetic language by analyzing a distinctive "imaginative syllogistic proof by example" lying at its basis (al-Fārābī, 1975, p. 115). In this al-Fārābī relies on Aristotle's definition in *Prior Analytics* of a syllogism as a "discourse in which, certain things being posited, something other than what is posited follows of necessity from their being so" (Aristotle, 1989, 1.24b16).

The imaginative nature of poetic discourse is crucial to its distinction from other syllogisms. In the "Catalogue of the Sciences" al-Fārābī explains that poetic discourse brings to mind an image or imagined representation that lacks truth value yet, at a prerational but ratiocinative level, still has an effect on us. Something happens to us "through the imaginative creation [representation, *takhyīl*] which takes place in our souls" (al-Fārābī, 1975, p. 118). In the "Canons of Poetry" he contends that we "create an illusion" to a "circumstance or characteristic . . . of the object one speaks about" (p. 116). We associate things in imagination that do not themselves have this association, for example, the span of a day and the span of a life, thereby constituting poetic similes and comparisons that give objects and events new meanings.

In the "Canons" al-Fārābī refers these constructions of imagination to the form of proof by example, which is a subset of arguments by analogy. Poetic similes are like ex-

amples that work by associating two objects that resemble each other in some respect, say, because both possess a property, P, and extrapolating that since they possess property P in common, both must also possess another property, Q. This extrapolation has the form of a syllogism insofar as acceptance of the first association generates a disposition to accept the second. Here imagination has the power to construct the association of representations that are present in examples and poetic similes and comparisons. It supplies the middle or enabling premise that may be universal but only imagined or that associates elements so closely that they carry as much force of conviction for subjects as universal premises do. The association between a day and a life seems so right that we infer for the life characteristics that we would normally associate only with the day, perhaps thinking of old age as the evening of a life, in which the restfulness felt at the end of a day spent in hard work is the relevant similarity. Other associations are also possible here, too, but the important thing is that for the comparison to be meaningful it must have rules, and al-Fārābī explains the latter as having this form of an argument by example.

The inference secures our acquiescence as if it were a legitimate and warranted syllogism, because once we find plausible and concede the initial representation of day and life we also accept the resulting association of evening and old age. The warrant for this plausibility is our ready acceptance of the imaginative construct. And al-Fārābī suggests that we assess poets by their insight in making optimally remote but maximally convincing comparisons in their discourse.

There are problems with al-Fārābī's theory. First, the acceptance of the imaginative construct remains subjective and arbitrary. It is merely psychological, being dependent on how easily representations generate a conviction that may vary from subject to subject. It cannot then so much secure agreement as obtain it only in those contingent cases where subjects' psychology coincides. Second, examples neither yield nor rely on propositions that are generally applicable elsewhere; yet, if poetic discourse is constituted by arguments by example, and if the premises of these arguments cannot be justified in the standard way by deriving them from other generalizations or categories, then all we can do is simply accept or reject the premises, where the psychological basis of such acceptance or rejection remains unjustified. But this arbitrariness of poetic syllogisms renders arbitrary the poetic discourse it sustains, and denies it legitimacy. Legitimacy implies an expectation we can have of how any subject must respond; but, because their success depends ultimately on a contingent and variable psychological conviction, arguments by example do not satisfy such expectations. Third, arguably al-Fārābī does not clearly explain the relation between aesthetic value and the logical form of the poetic syllogism. He mentions pleasure as a part of this value but gives little indication of how it gains validity.

These weakness in al-Fārābī's theory find some resolution in the work of Ibn Sīnā. Relying on Aristotle's conception of demonstrative syllogisms, Ibn Sīnā argues for the formal role of pleasure in constituting poetic syllogisms and for their moral value.

[*See also* Arab Aesthetics; Aristotle; Ibn Rushd; Ibn Sīnā; Islamic Aesthetics; *and* Poetics.]

BIBLIOGRAPHY

Works by al-Fārābī

"Catalogue of the Sciences." In *Arabic Poetics in the Golden Age.* Edited and translated by Vicente Cantarino. Leiden, 1975.
"Canons of Poetry." In *Arabic Poetics in the Golden Age*, edited and translated by Vicente Cantarino. Leiden, 1975.

Other Sources

Aristotle. *Prior Analytics.* Translated by Robin Smith. Indianapolis, 1989.
Matar, Nabil I. "Al Farabi on Imagination: With a Translation of 'Treatise on Poetry'." *College Literature* 23 (1996): 100–111.

SALIM KEMAL

FASHION. [*To explore the history and aesthetics of fashion, including clothes in general, this entry comprises four essays:*

> Dress in the World
> La Haute Couture
> Fashion as Art
> Fashion and Philosophy

The first essay is an overview of the aesthetics of dress from the beginning of civilization to the present in cultures around the world. The second essay is a survey of the distinct Western tradition of la haute couture, or "high" fashion, which began only in the nineteenth century. The third essay is a discussion of whether fashion is an art. The fourth essay is an analysis both of why philosophers, even aestheticians, have said little about clothing and fashion and of what they have said when they have discussed them.] *For related discussion, see* Cultural Studies; *and* Popular Culture.

Dress in the World

If we follow biblical legend, we find that clothes were the first artifacts. Adam and Eve began by making aprons of leaves when they saw they were naked; only later came useful tools when they saw they were helpless. It has long been understood that getting dressed is fundamentally an imaginative rather than a utilitarian undertaking in human life, and that it proves itself a necessary condition for all other behavior. It is the outward form of an enabling self-recognition. Human beings feel they need to be completed by some form of applied physical change, some adornment, expansion, distor-

tion, or mutilation, before they can proceed with any other imaginative or practical acts, even before putting on strictly protective gear. They seem to need to put on their humanity first, in some defining style for their group and place. Any utility clothing has is thus primarily psychological or aesthetic and only secondarily practical.

It is therefore possible to speak of the art of dress, if by that we mean the art of getting dressed, and not the art of making garments or ornaments. The living body is essential to this art, and the ephemeral nature of it is therefore assured, no matter how long certain elements in it may endure. Bracelets and breastplates or certain combinations of woven stuffs may last for centuries, but they are not, properly speaking, works of art in the art of dress; they are works of applied art, or of craft, that were intended only to contribute to the art of dress. The true work of art is never the garment, but the clothed person on a given occasion.

This distinction can sometimes get blurred. In ancient Japanese civilization, for example, certain garments of the type now generally called "kimono" in the West, although then each shape had its own Japanese name, were designed to stand alone as works of art when they are not being worn. They were not cut and fitted and embellished to follow and enhance the shape of the human body, but composed the way a box, vase, or screen was—that is, conceived and ornamented with respect to the integrity only of their own shape, not to echo any use to which they might be put. They might be vessels, but they did not need to hold anything, or even look as if they should.

The distinction actually remained, however, because each really was a garment, not an object that is permanently separate from a human body. A separate object, which certainly can be a work of art, might be called a body in itself, inanimate but vital, a kind of presence; but any garment is fundamentally different, even if it is spread out flat and displayed for itself. It might be worn rarely, and perhaps by a sequence of persons in different generations rather than by just one, but it is in fact made only to be worn. When it is put on, the shape, line, and surface embellishment all change their character when they form a composition with the living person; and only then does the complete work of art come into existence. The separate integrity of the garment proves to be a beautiful illusion, dispelled when it is brought to its true life.

The intrinsic beauty of clothing has been most compelling in civilizations that developed a mode of art based on refining and perfecting an original and carefully stabilized formal scheme, for both representation and ornament. Artists would create new examples based on old models, gradually absorbing new technical or imaginative discoveries into the prevailing scheme without letting them subvert or reform it. Such art characterizes many nonmodern civilizations, old and new, of which a great example is ancient Egypt, where styles in art, artifacts, and building were re-fined over thousands of years with considerable variation in detail and hardly any basic formal change.

In such a culture, dress has a similar slow evolution in its essential forms along with a great refinement of texture and embellishment in the components, the result being simple garments made either of complex fabrics or of simple fabrics with complex ornamentation: North African djellabas, Mexican serapes, Chinese robes, Persian coats, ancient Greek and Roman tunics. Any carefully wrought example would have the kind of abiding elegance created by harmony among stably integrated forms and refined surface textures. A deeply satisfying aesthetic appeal, an evolved, familiar beauty would reside in the garment itself.

Modernity arose in late medieval Europe to propose for art the model of individuality, gradually coming to measure collective aesthetic accomplishments according to the trajectory of a single human life—a birth and a rise, a flourishing and a zenith, a decline and a death. Art was conceived in terms of individual experience, with the movement of artistic style in art resembling the motion of a single consciousness undergoing continual action and reaction, both internal and external, making unsteady efforts to reconcile opposing inner and outer forces as it goes through life. Viewed this way, art is in a perpetual revolution, which moves onward at varying speeds spurred by a state of conflict between old and new, present and future.

It was not surprising that modern European civilization, committed to the sense of dialectical forward movement and of personal history, should invent fashion in dress in the early Renaissance. Fashion went along with other artistic effects that aim to intensify the sense of time's passage by showing one moment of it, such as perspective, chiaroscuro, and the cast shadow. The beauty of elegant Western clothing thereafter became contingent on a progressive movement of taste, feeling, and fantasy, and the idea of what is new perpetually trumping what is old. To match this new scheme founded on revolution and individuality, complex forms for clothes were invented that related to the forms of the individual body, always with special reference to its gender. These forms were realized by complicated cutting and seaming, sustained by padding and stiffening, and then permitted to keep shifting in shape, line, and surface character. Shapes once invented might swell to extremes and then collapse and reform, as fantasy demanded different ways to look, doubtless spurred by ideology, politics, economics, and morality, but also by the overriding force of aesthetic boredom and lust for internal change.

The ideal look of the clothed figure as a mobile living unit kept shifting, too, in shape and in general flavor. Loose gave way to stiff, fleshy to slender, careless to orderly, crisp to vague; and later on, everything shifted back, but differently. Conscious revivals regularly occurred, with nostalgic or ironic returns to long-outmoded sartorial themes. Beauty in

fashion-bound clothing thus became truly dependent on complete bodily effect, since the aim of its design was to create a vivid satisfying form out of *this* clad body at *this* moment in history, not to make something itself unchangingly beautiful that would absorb and adapt any wearer's body into its perfection. Historic fashionable clothes exhibited on mannequins consequently have somewhat the look of corpses, as Mughal coats or kimonos spread out on stands never do, and as church vestments also never do.

Fashion is bound to the waywardness of fantasy, which can pretend to situations that do not exist. It makes suggestions and projections at odds with the truth; it creates states of peace or turbulence that politics cannot immediately manage. It therefore reflects its times a little obliquely, or sometimes presciently, suggesting the desire for political or cultural movements before they are put into effect. One strong fantasy in fashion is that of reverting to a world without it; and one of the recurrent fashions is that of anti-fashion, with a return to nonmodern simplicities or complexities. Reformers of the past have longed for classical draperies; more recently, varying forms of ethnic or folk clothing have had an occasional vogue in fashionable dress.

What anthropologists have called the folk order, by contrast to the history of fashion, is the development and history of regional dress all over the world. This kind of clothing has clear links to antiquity and seems founded on similar principles as that of all nonmodern civilizations of the highest sophistication, old and new. The draperies now worn by Ghanaians are not unlike the draped clothing of ancient Greece and Egypt; the same is true of India and Nepal or of Middle Eastern countries. Until the present century, European folk dress had a greater relation to fashion than did regional clothing outside Europe, since it was much more exposed to its influence than was possible elsewhere, with the exception of Japan in the nineteenth century.

After Japan had been isolated from all other civilizations for three hundred years, Europeans and European fashion arrived in the 1860s and overnight made the Japanese feel like quaint peasants instead of normal people. An instantaneous movement to adapt local dress to fashionable Western standards was begun and continued with great intensity. In three generations, the movement had virtually destroyed not only the variety and richness of indigenous Japanese clothing but also its multiform traditions. Japan adopted Western dress except for the geisha, the stage, the religious, and wholly rural or very old-fashioned people, most of them women. This success was measured by the equal intensity with which, some years after World War II, the extinct costume of prosperous merchants from a particular class, period, and region was deliberately revived for women, codified anew, christened "kimono," and called "traditional Japanese costume," just as if it had always been there, and as if it were the only example of its kind.

Western dress had all along been especially important for Japanese men, active in international as well as local enterprises, whereas women had kept longer to the old ways; and this is a pattern of behavior that can still be seen elsewhere, now that Western pants and shirts have gone everywhere in the world. In countries such as Nepal, men have switched to them, and local tailors make them for local customers, whereas women have not switched to the equivalent dresses and skirts of the West.

During the hundreds of years when European folk dress flourished, fashionable elements might get adopted in a particular region or village in Switzerland or Italy and then immediately become fixed into the traditional, nonfashionable scheme. Bits of elegant eighteenth- or nineteenth-century modes from specific fashionable moments would appear in peasant bodices, waistcoats, or hairstyles, and then stay there for the next century or two, modified and decorated but not changing when fashion did, and soon looking traditional themselves. In parts of eastern Europe and Spain, farther from the centers of fashion and closer to Turkish and Moorish influence, this effect seems to have been less strong. But whether near to or far from the influence of fashion, the folk art of dress from all these cultures, especially for women, has had a similar style of multilayered

FASHION: Dress in the World. *Japanese Kimono (kosode) with Pattern of Stream and Wild Iris.* (Courtesy of the Japan Society.)

shapelessness overlaid with differing and detailed embellishment. This broadly apparent folk style has a great impersonality, a generality freighted with all the collective authority of ancient custom and none of the personalized fluency and perversity of fashion.

In western Europe, male folk dress has generally had much more erotic appeal, in that it tends to show off bodily shape and to include dashing accessories such as one-shouldered capes. Since Greek antiquity, it has been the case that Western traditional schemes of costume emphasize the modesty of women and the activity of men, usually by tending to obscure the physical shape of the former and reveal that of the latter. In eastern Europe and much of the Eastern Hemisphere, as well as in the extreme North, both are obscured, but differently; in the East, the legs of women may be divided and visible and mens' not, whereas the opposite is true in the West. Female modesty, however, with a great change observed in its expression on marriage, is fairly widespread, often involving covering the hair after marriage but not before.

In fashionable Western clothing, sexuality has had a greater female role, matching the erotic element in male dress, and gradually establishing a separate scheme for erotic expression. At first women followed old ways, as usual, keeping to long skirts and robes, but they added corseting and low necks, exposed arms, and finally bare legs; exposure became a female fashionable device. Also following old ways, men tended to have free hair and to wear hats, while women tended to tame their hair in braids and buns, subject to fashion change, and to wear hoods, caps, and veils, also of perpetually changing shapes. But fashionable male bodily covering changed much more radically than female, usually to allow for articulating the whole body differently, but without exposing its surface. Both sexes, however, have tended to switch back and forth between each other's motifs to create disturbing fashionable interest, in a way impossible for traditional clothing. Molded torsos and exposed chests and limbs and curled long hair have been considered suitable for both married and unmarried women during the history of fashion; meanwhile, similar devices, but without exposure—shaped doublet, tight hose, and curled hair—were eventually given up by men, who sought masculine appeal in shorter, neater hair and a flexible envelope for the body—trousers and jacket—that did not tightly outline its shape or reveal any skin.

Women have lately turned around and adopted those male ways of looking attractive, leaving the way open for men to go back to tights and doublets—approximations occur in ski clothing and skating gear—or indeed to take up skirts, if men should again find them virile in the kilt tradition. Long loose hair and jewelry have already been repossessed by men; and in general, international ordinary fashion has lately produced an androgynous condition, especially in the realm of sports and leisure wear, that distinguishes it profoundly from nonfashionable, nonmodern schemes of dress with their fixed male and female division.

It is tempting to use the word *timeless* in referring to much traditional regional clothing, since time has had so much less importance to it than place. Stylistic change has occurred chiefly through the external influence of other traditions in a way quite different from fashionable change, which is generated from within, perpetually reforming its notable expressions with opposing ones. Fashion-based work clothing and casual dress, however, have lately tended to imitate traditional schemes in seeking a certain basic stability: the fundamental designs of zipper jackets, sport shirts, work shirts, and all forms of casual trousers have stabilized, so that marginal variation has more importance than basic change, just as in the case of folk costume; except it has become homogeneous and regionally undifferentiated. The same might be said about the modern fashion tradition of classic male tailoring.

Fashion affects these clothes differently from the original fashionable ideal of extreme change, which tends now to prevail chiefly for those women's clothes specially created to be conspicuously fashionable. Women who wear fashionable garments may now also wear the traditional forms of leisure clothing; so "peasant" clothes and fashionable clothes may now apparently be worn by the same person. Both kinds are nevertheless products of the fashion system, which now simply incorporates the aesthetic of timeless tradition as well as that of perpetual revolution.

In many nations of the Southern Hemisphere, there is an entire scheme of human dress that is founded not on garments but on decoration, some of which is incised on the body in the form of elaborate tattoo or patterns of scarification, or of such effects as tooth filing and the stretching of lips, necks, and earlobes, and some of which is created by permanent body dyeing and impermanent body painting. Applied ornaments often accompany such arrangements: metal, ceramic, shells, gourds, feathers, pebbles, animal bones and teeth, fur, plant fibers, or modern synthetic versions of all these, as well as woven fabric. Such ornaments may be attached fairly permanently, inserted through holes or slits in parts of the body, or worn and removed the way garments are. Some indeed are garments, of a greater formal simplicity than those in the Northern Hemisphere, since they are wrapped and not fitted.

There is obviously a great distance between these forms of dress and those of either the regional tradition or the international fashion scheme of the Northern Hemisphere, which have interacted for so long using garments as a chief medium. Among the widely differing southern schemes, however, northern objects have arrived: plastic sunglasses and fountain pens or metal rings from cola cans may form part of a feather headdress or a shell necklace, and T-shirts or baseball caps can now be seen in ensembles that include bones and body painting. Similarly, body piercing and tattooing have entered the fashion scheme of the North.

Masks form a large part of this general mode of dress that does not conceal the body but uses it as a canvas. A headdress that hides the face with another while the body is visible obviously alters the human appearance as much as changing one set of garments for another does when the face remains the same, perhaps more. Habits can clearly have similar social or ritual aims and widely different aesthetic means; and, of course, they may have widely different aesthetic aims altogether. North and South both use cosmetic devices for the face and hair, and these may be the most universal habits of all. But the great universal fact is that human beings are dressed animals. Some applied imaginative change is required by all civilizations for unfinished human creatures before they feel at home in their skin, in the world, and in the cosmos.

[*See also* Anthropology and Aesthetics; *and* Comparative Aesthetics.]

BIBLIOGRAPHY

Boucher, François. *20,000 Years of Fashion: The History of Costume and Personal Adornment.* Exp. ed. New York, 1987.
Brain, Robert. *The Decorated Body.* New York, 1979.
Broby-Johansen, R. *Body and Clothes: An Illustrated History of Costume.* Translated by Erik I. Friis and Karen Rush. New York, 1968.
Dalby, Liza Crihfield. *Kimono: Fashioning Culture.* New Haven, 1993.
Davenport, Millia. *The Book of Costume.* 2 vols. New York, 1948.
Deslandres, Yvonne. *Le costume: Image de l'homme.* Paris, 1976.
Fairservis, Walter A., Jr. *Costumes of the East.* New York, 1971. Published in association with the American Museum of Natural History.
Flügel, J. C. *The Psychology of Clothes.* London, 1930.
Hiler, Hilaire. *From Nudity to Raiment: An Introduction to the Study of Costume.* London, 1929.
Kern, Stephen. *Anatomy and Destiny: A Cultural History of the Human Body.* Indianapolis, 1975.
König, René. *A La Mode: On the Social Psychology of Fashion.* Translated by F. Bradley. New York, 1973.
Oakes, Alma, and Margot Hamilton Hill. *Rural Costume: Its Origin and Development in Western Europe and the British Isles.* London, 1970.
Polhemus, Ted, and Lynn Procter. *Fashion and Anti-Fashion: An Anthropology of Clothing and Adornment.* London, 1978.
Roach, Mary Ellen, and Joanne Bubolz Eicher, eds. *Dress, Adornment, and the Social Order.* New York, 1965.
Roach, Mary Ellen, and Joanne B. Eicher. *The Visible Self: Perspectives on Dress.* Englewood Cliffs, N.J., 1973.
Rudofsky, Bernard. *Are Clothes Modern? An Essay on Contemporary Apparel.* Chicago, 1947.
Rudofsky, Bernard. *The Unfashionable Human Body.* Garden City, N.Y., 1971.
Snowden, James. *The Folk Dress of Europe.* New York, 1979.
Tilke, Max. *Kostümschnitte und Gewandformen.* Tübingen, 1945.

ANNE HOLLANDER

La Haute Couture

Dress design, considered as a serious profession for both sexes, is a modern Western phenomenon dating from the second half of the nineteenth century. Its early ascendance was contemporary with the rise of the realistic novel and the rise of feminism, and like them it follows on shifts in Western views about the cultural relation between men and women. The idea that originality in the design of female clothing—by contrast with the artistic design of stage and festival costume—might be a serious and profitable endeavor arrived very late in clothing's long history.

Since the Middle Ages, and even before, male tailors made clothing to order for both men and women, as cobblers made shoes and glovers made gloves. They followed stylistic trends without necessarily seeking originality of design except at the request of a client. Basic design thus developed in an evolutionary manner, with new individual details—worked out by artisan and client together—spurring innovation at irregular intervals.

No famous tailors' names survive from the period of Albrecht Dürer and Leonardo da Vinci, for example, although those artists were known to have designed costumes. Judging from European portraiture, however, male and female elegant dress in the early and late Renaissance, as well as in the Baroque period, was rich and interesting. But garments were identified with their wearers, not their makers, and nobody got credit for designing them. The success or failure of the effect feminine clothes produced was assimilated into the effect produced by the woman herself, as if her own qualities had generated her clothing and precipitated it onto her surface.

Credit for design seems to have been given first to Rose Bertin, the *modiste* of Marie Antoinette and other noble and royal ladies of late-eighteenth-century France. By that time, male tailors had been joined by female dressmakers, who specialized in making female clothing; but a further self-consciousness about the character of feminine elegance had prompted the rise of the *modiste*. Such a practitioner made feminine accessories and trimmings that created charm with superficial piquant effects. Meanwhile, the couturiere, the dressmaker who cut out, fitted, and constructed the basic dress, was not responsible for its ultimate flavor and produced a fairly standard product. Bertin the *modiste* had great renown in her day, as much for the novelty of her métier as for the prominence of her clients.

Even greater self-consciousness about appearance resulted in the renown of Leroy, the couturier of the empress Josephine and the designer of a great deal of imperial pageantry at the Napoleonic court. Legitimizing the regime was one of Leroy's tasks, just as it was that of the painter Jacques-Louis David; and they both produced many sartorial variations on ancient Roman themes for contemporary public personages, whether for actual wear or for fantasy paintings and fancy dress occasions.

But Leroy and his predecessor Bertin were each working in a rather theatrical style within an extremely small courtly world, and their fame spread mainly to other courts. Real clothes for most people, rich or poor, still had no designers that anyone knew of. Male tailors and female dressmakers

on all levels were still artisans, privately engaged by private clients. Credit for their masterpieces or failures was still generally attributed only to the taste of clients, with allowances for their rank or pocketbook. Poor women who could not afford good dressmakers were blessed if they had a talent for sewing and could make their own clothes with professional flair. Homemade garments were often visibly unprofessional and awkward, although all women were expected to know how to sew, to make undergarments, hem and mend household linen, and make children's clothes.

In the face of the increasingly more showy and complicated female fashions of the nineteenth century (by contrast to the increasing sobriety of tailored male clothing) good dressmakers, while not exactly thought of as designers, could nevertheless give clients ready professional advice about style, fabrics, and trim, about composing outfits of different garments and about choosing accessories. Materials for construction and trimming and all accessories were purchased separately by the client. Fashion plates and fashion journalism, in existence since Bertin's time, gave suggestions in fashion magazines; but it was important that an elegant woman's resources for making herself look elegant were her own business. Dressmakers' fame, sometimes serious, was nevertheless limited to a small circle of present and potential customers who might exchange information, but it was not a public matter. The idea was still strong that the woman's own taste was the only thing reflected in every aesthetic detail of her costume. Her income, of course, was reflected in its materials.

Charles Frederick Worth (1825–1895), an Englishman who worked for a Parisian dry-goods shop before opening his innovative dressmaking establishment in 1858, is acknowledged to be the founder of haute couture. This was a new French system whereby the design, construction, and embellishment of a woman's garment and its accessories would be undertaken as a single creation by a named artist. Such an artist would thus practice not just "couture," which is simply ordinary dressmaking, but "haute couture," a sort of visionary creative art like modern costume design. It meant giving a woman's clothed appearance a vivid shape and a distinct style by cutting, fitting, arranging, and combining carefully chosen materials to give her a unified flavor that would make a unique impression on a given occasion. The work of *modiste* and couturier was thus blended, along with that of purveyor of dry goods—Worth provided all the materials in his one establishment, and clearly pitched his own contribution as the work of an artist. An artist's fame soon followed.

It was of importance to Worth's success that he was a man, and working at a period when great male novelists and poets were creating strong female types, characters with distinct sexual and personal charms—Leo Tolstoy's Anna Karenina and Gustave Flaubert's Emma Bovary, Émile Zola's Nana and Alexandre Dumas's consumptive Mar-

guerite among many others in Europe, England, and America. The last two examples were indeed founded on real counterparts among the *grandes horizontales,* the fabulous courtesans whose caprices could ruin the fabulously rich merchants and financiers of the booming century, or eat up the ancient patrimony of a dazzled young nobleman.

Male painters were filling the salons with images of similarly compelling women; masculine imagination was brilliantly at work engendering images and models of female sexual power and force of character. Meanwhile three generations of women had already been dressing themselves up to actualize the new Romantic fantasies shared by both sexes, while men had become the more and more discreetly clad doers and thinkers, the takers and makers of ideas, the originators of fantasy. It was no wonder that a male magician of complete feminine appearance, calling himself an artist, should succeed so well. It did not hurt that the new empress Eugénie, a great beauty incidentally committed to promoting the French silk industry, became his chief client and patron.

The true secret of Worth's success, however, was his method of working. Although he did design and make unique dresses to individual order, he also began the custom of designing a "collection," a whole set of brilliant outfits designed for an ideal lady, destined for whoever might wish to lend herself to his genius and be created through it. He thus took the individual creative emphasis away from the client and kept it for himself, as designers have done ever since. Since Worth, the haute couture client has been given the credit not for choosing the dress but for choosing the designer.

At the Paris Exposition of 1900, haute couture was publicly represented for the first time as a French profession. Outfits from the collections of Worth's son and successor and of his newly arisen competitors, some of them women, were put on display along with the names of clients who had ordered them. At that epoch the whole matter of the couturier's clientele became public. The lasting connection between publicity and haute couture was formed and became respectable. Thereafter, the press was permitted to attend couture showings, to which formerly only private clients had been privately invited. Who actually patronized which designer became news, along with what each designer created for merely possible customers to choose from.

Haute couture was exclusive only by reason of its high cost. Since, in fact, anyone with the money could go to the same dressmaker as the empress, it was obvious that the couture could best flourish in a republican and commercial atmosphere, a long historical distance from Marie Antoinette or indeed Napoléon I. After the fall of the Second Empire, the successive French republics saw the triumph of haute couture as an internationally important French commercial enterprise, with strict rules eventually enforced by its own Chambre syndicale.

Those without means to patronize the real thing found approximations and adaptations and downright copies of French couture clothing in other countries, all promoted as French, and ordinary French dressmakers were raised on the aesthetic scale by the prestige of haute couture. The so-called tyranny of couture was engendered by the manifest superiority of France in fostering and promoting dress design as a French institution. Americans, who were increasingly the ones with the most money to spend, believed in and promoted the same idea, and materially helped the rest of the world to believe it, too. The couturiers of Rome and Vienna, for example, could not match the fame of Parisian ones, although immensely gifted designers were at work in those and other non-French cities. Their names do not resonate in the present as French ones still do, but their works survive in costume collections, eloquent material witnesses to the general rise of dress design as a field for serious aesthetic effort during the last years of the nineteenth century and the first of the twentieth.

The most resonant name in Paris just before World War I was that of Paul Poiret, who helped refashion the old, tight-waisted, and full-skirted image of clothed feminine beauty into a sleek, abstract shape. Poiret also helped link art and fashion together in the realm of modern design, a new aesthetic concept achieving material expression and commercial power at the same period. The beginning of this modern century saw the decorative arts and architecture already receiving renewed attention as spheres for the exercise of original talent, with a prestige comparable to the talent of painters and sculptors. Soon haute couture could join the decorative arts with its own roster of distinguished original designers, of whom Poiret was the first to be aesthetically alert and enterprising in the modern way. He caused his designs, for example, to be interpreted in images by well-known modern graphic artists who emphasized their stylized and abstract quality, rather than by lowly fashion illustrators; and he arranged to have his fabrics designed by painters such as Raoul Dufy.

After World War I, fashion for women began to modernize more abruptly, and Poiret's distinctive, leisurely elegance gave way to brisker modes as fashionable women went to work, took to more active sports, and wished to appear immature rather than ripe. Gabrielle "Coco" Chanel is now the most famous *couturière* associated with the period between the wars, partly because her vision of modern feminine style proved so prophetic as to allow her to renew her career well after World War II, following nearly twenty years of obscure retirement. Others shared the limelight with her during the first period, however, among them Patou, whose neat sportswear rivaled Chanel's; Alix, whose draped dresses still remain unsurpassed; Vionnet, who excelled in the bias cut and a caressing fit; Molyneux; Lelong; and Elsa Schiaparelli. These were all distinguished designers whose gifts sustained the prestige of haute couture in the modern world. Clients by this time included movie stars, whose images in French couture clothing further enhanced that prestige among a mass audience.

During and just after World War II, while French haute couture could not continue its accustomed output nor maintain its world prestige, American designers, already numerous, gifted, and known for their vivid informality, began to become famous. Adrian, who had designed for many movie stars, and Mainbocher, who dressed the rich on both sides of the Atlantic, were the first in a group that later included Claire McCardell and Norman Norell, and eventually Charles James and Geoffrey Beene (who is still at work) in an identifiably American version of haute couture. The standard was always as high as the French, except perhaps for the hand embroidery and trimmings that remain a French specialty, whereas the flavor had a definition and forthrightness distinct from the French emphasis on elegance, nuance, or delicate perversity.

French couture recovered after the war, most famously by means of Christian Dior's neoromantic vision, christened the "New Look." Dior revived the long, full skirt and tight waist of earlier centuries, and engaged in a considerable reversal of the fluid ease characteristic of all modern fashion between the wars. Dior's new version of old-fashioned formal elegance, which was shared during the 1940s and 1950s by other French designers such as Balenciaga and Givenchy, was opposed by the renewed work of Chanel, who went back to her prewar simple suits in soft fabrics and school-girl styles. Her modern look eventually triumphed a second time, after exaggerations of formal clarity had resulted in abstract, rigid hair and stiff little dresses by the first half of the 1960s, exemplified in the work of André Courrèges and Pierre Cardin.

By this time a thriving ready-to-wear women's clothing industry, launched in the United States even before World War I, existed all over the industrialized world. The garment manufacturing business was staffed by its own clever designers, most of them no longer looking to French haute couture for ideas about dressing the multitudes of women and girls for whom modifications of advanced Parisian elegance were irrelevant. Also by this time Italy had produced a new brand of haute couture, as vital and dashing as the American version but much more sensuous, along with dramatic designs for active sportswear. Milan became (and remains) an important fashion capital, and French haute couture was embattled in new ways, challenged from beyond national borders, from the lower ranks, and from the street.

The countercultural revolution of the late 1960s arose to attack many institutions and customs, costume among the most noticeable. Elegance altogether came under suspicion, and ordinary fashion broke up into many separate streams, some of them flowing in the direction of agricultural work clothes, blue jeans, army and navy surplus garments, and the stock of cheap thrift shops, others toward ethnic effects

borrowed from Africa and the Far East, India and Native American cultures, both North and South, still others toward the old-fashioned flowered dresses and shawls of rural communities, and even others toward elements of earlier twentieth-century elegance retrievable from old movies. More recently, the effects of menace and depravity have been ascendant in urban fashion, with a great deal of black leather and triumphantly synthetic fabric. The garment business has adjusted to all this and produced clothing expressly manufactured to imitate or suggest such politicized and otherwise costume-like effects; and eventually designers at the couture level also tried it. Pressure on haute couture over a twenty-five year period has indeed produced a tense and insecure condition in its aesthetic character, so that its productions have tended more and more to seem like theatrical and sensational experimentation, instead of original ideas that vary and play against a stable ideal of elegance.

The feminist movement that had begun as one of the revolutions of the late 1960s eventually, however, produced a highly paid phalanx of working women, many of whom were in need of elegance as a professional advantage and could afford refined examples. The ready-to-wear industry thereafter began to annex the couturiers, who designed collections specifically to be mass-produced to a high standard, sold at high prices, and to bear the designer's name. This development has increased and enlivened the scope of haute couture design, but it has also helped sustain its originally rarefied and exclusive life, of which the immediate end has consistently but inaccurately been predicted for three decades.

New haute couture designers may still create luxurious distinctive works for a custom clientele, and can even hope to thrive in an increasingly inegalitarian world. Such a hope is only possible, however, because the increased marketing of mass-produced garments no longer opposes haute couture, many of whose designers are now responsible for mass-produced style at various levels of price, but has been largely combined with it. Haute couture has, moreover, long been sustained by a system that promotes famous names in fashion through commercially profitable links with other costly goods licensed to bear the same names, perfume being only the chief one.

More recently, haute couture has come to profit from the general international consolidation of the whole luxury goods business along with the availability of its stock on the public market. The only threat to the future life of haute couture may indeed be the lack of disciplined training available for the skills it requires. Sartorial imagination may not now be lacking, but it will tend to wither later if there are no extant techniques for realizing its ideas—no deft-handed cutters, drapers, fitters, finishers, embroiderers. New methods for exclusive manufacture may yet arise, but they will need time to develop.

Chanel's modern vision of the self-sufficient woman, whose youth was eternal because her sexuality was always in confident use, and put on display with a comfortable elegance analogous to that of modern male clothing, has been realized by many other designers in the haute couture tradition. The most famous of these has been Yves St. Laurent, who incorporated trousers, gradually adopted by all women during the last third of this century, into high feminine fashion with no loss of feminine sexual overtones. Plain transvestism, especially women dressing in men's clothes, has also been a persistent theme in late-twentieth-century avant-garde and bohemian fashion, following the unobtrusive plebeian version constantly expressed in the androgyny of casual sportswear. Thus, crew cuts and tuxedos may be seen on some women, and so may long hair with jewelry and velvet tunics on some men, but zippered nylon jackets and cotton work pants are seen on everybody.

Serious masculine fashion, however, long eclipsed by the salient feminine fantasies that began with the work of eighteenth-century *modistes*, is forming a new field for haute couture. Certain current designers have begun to take advantage of the recently blurred distinction between appropriate male and female clothing. Jean-Paul Gaultier and others, including the Englishwoman Vivienne Westwood, are creating extraordinary masculine ensembles based on the traditionally feminine principles of display and sensuality. To do this they are using mixtures of male and female convention for men's clothing, just as designers for women have done during the whole of this century, but as designers for men have not.

The old order of things certainly persists in the world at large, with only women wearing "creations" consisting of artfully draped skirts surmounted by artfully cut and decorated tops, or of interesting ensembles in which cleverly modified male motifs combine with standard female elements. Most men who accompany such women are still confidently wearing the tailored suits created by classically minded tailoring establishments, set off by classic shirts, ties, shoes and hair, with no feminine suggestions whatsoever.

Male heads of state and powerful men in business and the professions are still appearing in these generally becoming modern costumes, at least on the job and in the media, so suits and ties show no real signs of extinguishing. It therefore remains to be seen whether highly ornamental and suggestive costumes, once worn by such powerful heads of state as Henry VIII of England, for example, but abandoned by modern leaders for more than two centuries, will ever be adopted by powerful men in the contemporary world. If powerful men ever again wish to embody brilliant personal display and variable sexual fantasy, it may well happen. But certainly not immediately.

BIBLIOGRAPHY

Barthes, Roland. *The Fashion System.* Translated by Richard Howard and Matthew Ward. New York, 1983.

Bell, Quentin. *On Human Finery* (1947). Rev. ed. New York, 1976.

Coleman, Elizabeth Ann. *The Genius of Charles James.* Brooklyn, 1982. Exhibition catalog, the Brooklyn Museum.

Davis, Fred. *Fashion, Culture, and Identity.* Chicago, 1992.

de Marly, Diana. *The History of Haute Couture, 1850–1950.* London, 1980.

Gibbs-Smith, Charles H. *The Fashionable Lady in the Nineteenth Century.* London, 1960.

Hollander, Anne. *Seeing through Clothes.* New York, 1978; reprint, Berkeley, 1993.

Hollander, Anne. *Sex and Suits: The Evolution of Modern Dress.* New York, 1994.

Kidwell, Claudia B., and Margaret C. Christman. *Suiting Everyone: The Democratization of Clothing in America.* Washington, D.C., 1974. Exhibition catalog, the Smithsonian Institution.

Langlade, Emile. *Rose Bertin: The Creator of Fashion at the Court of Marie-Antoinette.* Translated by Angelo S. Rappoport. London, 1913.

Latour, Anny. *Kings of Fashion.* Translated by Mervyn Savill. London, 1958.

Lee, Sarah Tomerlin, ed. *American Fashion: The Life and Lines of Adrian, Mainbocher, McCardell, Norell, Trigere.* New York, 1975.

Leymarie, Jean. *Chanel.* New York, 1987.

Lipovetsky, Gilles. *The Empire of Fashion: Dressing Modern Democracy.* Translated by Catherine Porter. Princeton, N.J., 1994.

Lynam, Ruth, ed. *Paris Fashion: The Great Designers and Their Creations.* London, 1972.

Milbank, Caroline Rennolds. *Couture: The Great Designers.* New York, 1985.

Poiret, Paul. *En habillant l'époque* (1930). Paris, 1974.

Pritchard, Mrs. Eric. *The Cult of Chiffon.* London, 1902.

Ribeiro, Aileen. *Dress in Eighteenth Century Europe, 1715–1789.* London, 1984.

St. Laurent, Yves, Diana Vreeland, Pierre Berge, and René Huyghe. *Yves St. Laurent.* New York, 1983. Exhibition catalog, the Metropolitan Museum of Art.

Squire, Geoffrey. *Dress, Art, and Society, 1560–1970.* New York, 1974.

Steele, Valerie. *Paris Fashion: A Cultural History.* New York and Oxford, 1988.

White, Palmer. *Elsa Schiaparelli: Empress of Paris Fashion.* New York, 1986.

Wilson, Elizabeth. *Adorned in Dreams: Fashion and Modernity.* Berkeley, 1985.

ANNE HOLLANDER

Fashion as Art

It is not absolute, but it is ubiquitous. Apparel might be its name; fashion is its modern configuration; costume is its consolidated expression; garments are its elements; decoration, identity, and socialization are its variable objectives. Vanity is its epithet, often richly deserved. It is a singular art, but an art directed toward acculturation. Denis Diderot posited clothing as a major encyclopedic entity; Charles Baudelaire observed and analyzed fashion with enthusiasm as a token of the modern; Roland Barthes cast his glance, largely secondhand, onto fashion representations, but with consummate, if not consuming, curiosity. Gilles Lipovetsky in *The Empire of Fashion* offered a serviceable definition of fashion as "the permanent theater of ephemeral metamorphoses" (Lipovetsky, 1994, p. 47).

Fashion is pervasive and persuasive. A design art consummated by the consumer, fashion is chiefly identified with the second half of the nineteenth century and the twentieth century. Like modern art, it arose from new conditions of spectatorship and is a profoundly socialized and socializing form. The invention of the sewing machine, the emergence of the House of Worth in the nineteenth century, and the twentieth-century proliferation of designer identities even in ready-to-wear, department stores and the availability of ready-to-wear, the boulevards and places of public spectacle have set the course of modern fashion. Its race is demonstrably swift, a pursuit of volatile consumer desire, an apparent delight in obsolescence and ephemerality of style, and in accompaniment with many arts that have been chiefly propagated by fashion and its cognate consumer yearnings, including posters, advertising, fashion illustration and photography, display windows, and store techniques and environment from architecture to point-of-purchase ambience. Fashion breeds and promulgates a visual aura that is an inescapable condition of modern life.

In its ubiquity, fashion is difficult to treat as an isolated phenomenon or as a discrete art or design. Its industry and related design encircle and enlarge its role in an inextricable embrace. Further, the style in dress is often commingled with "lifestyle"; Oscar Wilde conflated the two in declaring in a famous aphorism that one should either wear a work of art or *be* a work of art.

Further, fashion's matrix is one of the fundamental human needs. Clothing is a necessity as proved by anthropology, if sometimes exaggerated by claims of modesty. Correspondingly, fashion never leaves the body. By definition, it is an anthropocentric art whose form is partly predetermined or fixed to the options of the human body. Thus, it lacks an option of pure abstraction, and it returns again and again to the body form for which fashion is frame and exoexpression. Inasmuch as modern fashion is often thought of as pertaining chiefly to women, its link to the body may seem specific to the female body. If the fine arts were limited to life rendering of the human model, would the fine arts be fine?

Finery is not so fine. But it is undeniably engaged. Fashion may address the aesthetic needs and expression of the individual, but it also places the individual into a social context. In traditional societies, clothing marks rank and status with surety as a primary function of dress. Modern fashion indexes indeterminate statuses and allows, through production, significant involvement of personal choice, but nonetheless still serves a role in nonverbal social discourse.

One of the most vexing issues of considering fashion is its oscillation between the expressions of self and society, neither supreme in modern clothing's equivocation between an aesthetic objective and a cultural condition.

Society, that ever-present and palpable partner to fashion, affords many problems for the aesthetic witness and evaluation. In one example on 21 May 1997, President William Clinton decried "heroin chic," or as he stated it, "the allure of heroin among young people," the appearance of slender, drugged, young models, often with macabre, stoned stares in contemporary fashion advertising and editorial photography. Clinton's remarks depended on a front-page essay on the same topic by fashion reporter Amy M. Spindler appearing in the *New York Times* on 20 May 1997. Remorse and rebuttal ricocheted through the media in subsequent days, touching not on the design and artifacts of fashion, but on its influence, its capacity to cast charisma and glamour, and its social implication. Fern Mallis, executive director of the Council of Fashion Designers of America, responded (quoted in the *New York Times,* 22 May 1997) in part and quite reasonably, "the fashion industry should not be the easy target to blame for society's woes." Interestingly enough, Mallis was taken to task subsequently by editorial writer A. M. Rosenthal in the *New York Times* (23 May 1997) for that exculpating sentiment. Can the industry or the aesthetic of fashion be held responsible for "society's woes," as characterized in a maudlin, Dickensian way by Mallis? The puny, but graphic, contretemps suggests the appeal of fashion as the easy social scapegoat, the public art that is begging to be chastised as fine arts had been nearly a decade earlier in the case of Robert Mapplethorpe's photographs. Moreover, at the crux of the "heroin chic" debate was the elisions among fashion, advertising, and media imagery. To such moralizing about fashion, there is a ready aesthetic response. In Lipovetsky's words, "We have to move beyond moralistic vituperation against fashion. Its irrationality and its apparent wastefulness aside, fashion contributes to a more rational edification of society because it socializes human beings to change and prepares them for perpetual recycling" (p. 149). But presidents seldom read or rely on philosophers.

Fashion is not easily or even responsibly segregated from the social fabric if we understand its motives. A classic case of aesthetic interpretation and social value is the consideration of Christian Dior. Aestheticians, notably Remy Saisselin in a shocking encomium "From Baudelaire to Christian Dior: The Poetics of Fashion" (*Journal of Aesthetics and Art Criticism,* September 1959), have notably defended or deferred to Dior's signal place in mid-century design. For others, however, Dior represented the oppressive return of the corseted silhouette, boned bodice, and archetypal femininity. He became thereby for many feminist critics of the 1970s and 1980s a bête noir of 1950s' gender divergence and stereotyping. One's view of Dior, of his commercial and socially propelled art, must be holistic, including both the social and the aesthetic dimensions as they overlap and complement. To vilify Dior for a purported misogyny or domestication of women is, of course, to ignore the positive consumer acclaim for the design, willingly bought and copied by many. To see his soul as purely "poetic" in projecting an ideal woman is to slight the social terms of an aesthetic that changed the way women and men lived.

Fashion's insinuation into the social is also evident in gender ramifications. If modern fashion has traditionally been taken to obtain chiefly to women, evolving concepts of gender alter the purview and concept of fashion. Indisputably, fashion pertains to men and masculinity. In a review in *Artforum* (November 1996), the critic Bruce Hainley chafed at the relative indifference of an exhibition at the Metropolitan Museum of Art to menswear and men's bodies, noting, "When porn star Jeff Stryker dons leather chaps and nothing else . . ., his butt and cock bared for all to admire, it seems a moment fashion (or at least its historians) cannot ignore. Fashion thrives on that *between*—one skin meeting another." Hainley's argument not only includes men but also insists on the body. Cultural studies of the 1980s and 1990s have, in particular, sought to increase the discourse on body articulation and gender as pertinent to fashion. Hainley implies the further proposition that not only gender but also sexual orientation plays a role in fashion witness as the gay man will see Stryker's exposure and attire differently perhaps than the heterosexual man.

Of late, fashion's claim to originality is in dispute. Yves St. Laurent sued Ralph Lauren in 1994 for infringement of the design concept for a man's tuxedo modified to a woman's dress. This legal dispute lays bare fashion's relation to an absolute art. Fashion has long borrowed design concept in a manner ruder and crasser than the other visual arts in which appropriation remained until the 1980s a genteel, if constant, tradition. Copying and modified copying may be inevitable in an aesthetic expression that is limited to the ergonomics of the body, the constraints of culture (e.g., erogenous zones, somatic support, work and exercise needs), and a relatively restricted medium (textiles and cognate covers for the body). A dispute about the man's tuxedo accommodated to a woman would have to begin with a popular or ad hoc idea in Marlene Dietrich's image in menswear, understand that clothing recycling exists in all times and places and was specifically recommended in a 1943 *Harper's Bazaar* feature that showed how to retrofit an unused man's tuxedo into a new outfit for a woman, and assimilate the context of Coco Chanel in couture, Claire McCardell in sportswear, and popular culture in every aspect borrowing from men's apparel and prerogatives to serve and empower women in the second two thirds of the twentieth century. A design copyright or primacy of creation would, under these circum-

stances, be genuinely difficult to establish, as would any principles of wholly convincing "originality." Fashion often feigns an avant-garde style in runway shows and deliberately exploits a rhetoric of "originals" and "designer" creativity that surrounds familiar design and styling differentiation, but that never yearns for the wholly new or transgressive. Dress that was genuinely socially subversive would be at best socially outlandish or outsider-producing in an aesthetic realm that seeks its social place, its consumer satisfaction, and its acceptance of the individual into the group. Twentieth-century designers from Elsa Schiaparelli to Franco Moschino and Gianni Versace have practiced only the ironic *épater le bourgeois* etiquettes of fashion sedition.

Given its deep involvement in social workings, can fashion be considered a work of art? Knowing that all works of art incubate in the social fabric as much as in inspiration alone, fashion is not solitary in being culturally engaged and thereby vitiated from a status as a pure art. More than most, though, fashion has been an avoided topic of consideration, perhaps largely because of its primary association with women and thereby denial to power and prestige. Similarly, a "Puritan ethic" may obtain in skirting, as it were, an art that so unabashedly pertains to the body and never can be entirely removed from the body. By dint of fashion's tendency to soft form (corresponding to the body's flexure and mobility), it is not in most instances a freestanding three-dimensional art (that is, sculpture, perhaps) unless it is inhabited. The phantom and the figure/physique that must occupy clothing for its coherence are a frightening prospect to some moralism and to some who want their art without either ghosts or viscera.

Commerce seems to discredit fashion. Ironically, an art that seeks grace is founded for some in an inherent disgrace. Fashion is energetic and willful in commerce. In particular, contemporary fashion is so generative and joyous in its commercial partnership that it can seem the most mercenary of arts. Ironically, fashion's prodigious linkages to commerce sponsor technological accomplishment and innovation, often applying most advanced technology (that nylon applied first to synthetic stockings popularly known as "nylons" in 1939–1940 is an example of fashion's appeal to science for modern life), stimulate imaginative forms of marketing and advertising, and represent and are present in modern life in a manner predicted by Baudelaire. Knowing that all art needs the looking glass of patronage and consumption, fashion is perhaps not a minor art for its recognized exchanges with business.

Indeed, its overt commerce may be positive in opposition to the covert trafficking of other arts. No one would argue that Paloma Picasso is as important as Pablo Picasso or that a garment is a transcendent, unmitigated masterpiece in the manner of the fine arts. But a dress by Vionnet may be as fully and compellingly an expression of the Cubist and Fu-

turist desire for reviewing the world, revising an optic, and renewing a design and art viability apposite to modern times. The Vionnet dress is not self-sufficient only in the limited sense in which one can probably not see it without a mannequin or other armature, but one can certainly see art and history configured therein. The eschewal of corset, fluid semitransparency, and adaptation of devices of lingerie may be socially implicit and art form–specific variations on Cubism and Futurism, but these dressmaking and body-proximate situations may not vitiate the modern objective; instead, they impose the modern objective more fully and intimately into the lives of modern people.

Is fashion merely recipient to the boulevardier's glance, transmuted in some instances by erotic desire and social strategy and stratification, or is fashion a converging gaze, absorbing cultural and aesthetic force into a visual deliberation? The gaze of fashion makes no ephemeral garment last, nor for that matter does art's serious scrutiny. Apparel, so intimately connected with the body, may seem more like the mirror image of the individual of beauty, a fleeting appearance inevitably in transit with time. But have we not realized that even art as object endures and is intended to last, aesthetics change? An art often less durable, generally more inclined to variation, is perhaps not a lesser art in its grasp of time and society. Indeed, fashion touches on what we long to be personally and socially; reciprocally, it touches us.

Finery and artifice, neither fine nor art. Fashion has been and ever will be dependent on the body; it has been accused of a narcissism surpassing human vanity. Clothing remains a need and a social constituent; a beauty associated with the human body likewise abides; it may be expressed in fashion.

[*See also* Definition of Art.]

BIBLIOGRAPHY

Barthes, Roland. *The Fashion System.* Translated by Matthew Ward and Richard Howard. New York, 1983.

Beaton, Cecil. *The Glass of Fashion.* Garden City, N.Y., 1954; reprint, London, 1989.

Benstock, Shari, and Suzanne Ferriss, eds. *On Fashion.* New Brunswick, N.J., 1994.

Breward, Christopher. *The Culture of Fashion: A New History of Fashionable Dress.* Manchester and New York, 1995.

Flugel, J. C. *The Psychology of Clothes.* London, 1930.

Hollander, Anne. *Seeing through Clothes.* New York, 1978; reprint, Berkeley, 1993.

Langner, Lawrence. *The Importance of Wearing Clothes.* Rev. upd. ed. Los Angeles, 1991.

Lipovetsky, Gilles. *The Empire of Fashion: Dressing Modern Democracy.* Translated by Catherine Porter. Princeton, N.J., 1994.

Polhemus, Ted. *Street Style: From Sidewalk to Catwalk.* London and New York, 1994.

Ribeiro, Aileen. *Dress and Morality.* New York, 1986.

Roche, Daniel. *The Culture of Clothing: Dress and Fashion in the Ancien Régime.* Cambridge and New York, 1994.

Rudofsky, Bernard. *Are Clothes Modern?* Chicago, 1947.

Wilson, Elizabeth. *Adorned in Dreams: Fashion and Modernity.* Berkeley, 1985.

RICHARD MARTIN

Fashion and Philosophy

In the main, philosophy likes to pretend inattention to fashion. Ralph Waldo Emerson is representative of this attitude when he derides "the consideration of personal appearance of face and form and manners, with all which the wise have simply nothing to do. No fashion is the best fashion" (Emerson, 1972, vol. 3). Of course even the wise must *have* some personal appearance, must present themselves in *some* fashion or other. Why should those whose business is *thought* be advised *not* to give thought to these matters?

One reason to scorn attention to fashions in clothing and personal appearance is evidently drawn from what could literally be called the superficiality of these concerns. Clothing, cosmetics, hair styles—all are situated at what might be thought of as the surface of the human being. Philosophy often understands itself to probe beyond the surface, to be interested in what *is*, as opposed to what merely *seems*, to be the case. Philosophy recurrently seeks the essential, the underlying substance, the fundamental. No particular fashion is essential to human beings. Clothing is added; it is put on the body and may be changed at will, and nothing is essential to fashion except distinctions of and changes in appearance.

There are, to be sure, a few philosophers who have paused at the sight of clothing and seen in fashion some fundamental elements of identity, if not essence. William James, for example, in prising apart the divisions he calls "the constituents of the self," and then peeling away the various layers of what he calls "the material self," finds clothes right next to the body, at "the innermost part" of each of us, more intimately a part of our identity, he says, than our family, our work, and our possessions:

> We so appropriate our clothes and identify ourselves with them that there are few of us who, if asked to choose between having a beautiful body clad in raiment perpetually shabby and unclean, and having an ugly and blemished form always spotlessly attired, would not hesitate a moment before making a decisive reply. (James, 1890, vol. 1)

(If there are more of us now, than in James's time, who would not hesitate a moment, that may only show that fashion has changed.)

The prospect of fashion's contribution to personal identity can be, in any case, even for those who grant it, a ground for further disdain. Some feminist philosophy has critically explored this territory, protesting not just the superficiality of the identity given by appearance but also what

are seen as the gendered hardships of fashion's classifications. Simone de Beauvoir complains that by dressing up, "the woman who is deprived of *doing* anything feels that she expresses what she *is*." In reality, however,

> [t]he purpose of the fashions to which she is enslaved is not to reveal her as an independent individual, but rather to cut her off from her transcendence in order to offer her as prey to male desires. (*Beauvoir,* 1952, book 2, part 5, chap. 8)

The connection between fashion and desire may in fact explain the peculiar character of much of the philosophical antipathy to fashion. Attention to fashionable dress is inseparable from attention to the body. From Plato to René Descartes, and to their successors, the interests and inclinations of the human body have been treated as embarrassments and seen as dispensable, inessential to a person's true identity and highest aspirations. Fashion, on the other hand, cannot dispense with the body. Fashion achieves its aspirations in being worn, and the body collaborates with costume to produce the effects of fashionable dress.

The relation between fashion and the body may even be among the reasons that philosophical aesthetics has tended to neglect the study of costume and personal appearance. Aesthetics, after all, could afford to be interested in surfaces and a beautiful appearance, but fashion is not much studied even here. The impediment may be connected to the unavoidable body. Compare the materials and activities of fashion with other arts: metalsmithing and stone carving can produce sculpture, in forms limited only by the artist's imagination; but if stones and metal are to become jewelry, there must be some thought of necks or wrists, ears or ankles. Tapestries and fiber constructions seem to offer unbounded possibilities for art, compared to the fabric and stitching that must work with, within the possibilities of, the human form. There is, of course, portraiture within the possibilities of painting. Human forms as well as purely geometrical shapes can be chipped out of marble, and dance is inherently embodied. But there is in all these other arts a kind of abstraction from the actual human being who is model or vehicle, or there is an appropriation of the person's appearance and contours to serve ends intrinsic to the art. The relation between fashionable dress and the body on which it is worn is both tighter and more ambivalent. Clothing conceals and reveals particular bodies; it always depends for its sense on the thought of actual human beings, whose embodiment is *their* mode of being in the world, who have their *own* ends and who have their own idiosyncratic ways of moving through, and presenting *themselves* to, being seen in, the world.

This is a point important to the critics of fashion. Henry David Thoreau, for example, who famously commands us to "beware of all enterprises that require new clothes, and not rather a new wearer of clothes," and who asserts that "it

cannot be maintained that dressing has in this or any country risen to the dignity of an art," says that "[a]ll costume off a man is pitiful or grotesque. It is only the serious eye peering from and the sincere life passed within it, which restrain laughter and consecrate the costume of any people" (Thoreau, 1966). But even the historically significant champions of fashion concede the point. Charles Baudelaire, for example, says that "if one wants to appreciate them properly, fashions should never be considered as dead things. . . . Rather they should be thought of as vitalized and animated by the beautiful women who wore them" (Baudelaire, 1964). Thoreau's remark, of course, also echoes the general philosophical complaint about fashion's superficiality. Baudelaire, on the other hand, refigures the worth of both the superficial and the transitory, and he is willing to count as art both dress and the dressing table. But even as he makes a brief for what he calls "accidents" of beauty, for the artificial overlay that he understands fashionable costume and cosmetics to be, he seems to disclose some acceptance of an evaluational scheme not so different from the one operating among fashion's critics.

Baudelaire wants black eyeliner and bright rouge to "represent life," not to enhance or draw attention to the actual vitality of the particular living being on whom they are placed. Face powder is "to create an abstract unity in the color and texture of the skin," to "[approximate] the human being to the statue, that is to something superior and divine." The wayward human body, blemished and full of peculiarities, is evidently still the problem to be overcome, even in Baudelaire's aesthetics. He praises fashion and cosmetics precisely because he sees them as an attempt to triumph over our ordinary bodies. "External finery," in his view, is "one of the signs of the primitive nobility of the human soul": "Every fashion is . . . a new . . . effort in the direction of Beauty, some kind of approximation to an ideal for which the restless human mind feels a constant, titillating hunger." Fashion is thus defended as an interest of the mind, a spiritual ideal, something that reveals the noble soul—not something that indexes the base and vulgar body.

Seeing Baudelaire occupy this argumentative position can be a shock, and placing him there may run against a common interpretation of his historical force. His sarcasm about the value of "infallible Mother Nature" is always understood as directed to his defense of art and artifice, but this defense has a moral framework. Baudelaire derides Nature as the voice of self-interest, the venue of "patricide and cannibilism and a thousand other abominations." Crime is natural, he says, while "[g]ood is always the product of some art." Nature needs reformation, even "deformation," and not just in moral matters but also in the realm of Beauty. The reformer, the redeemer, is, in both spheres, Reason. With Reason set against—above—Nature, however, the familiar dichotomies and hierarchies fall into place. Reason goes with the soul, the immaterial, the ideal;

and Nature includes the body, the material, the terrestrial. Baudelaire may praise the textures and appeal of "iridescent fabrics" and "artificial forms," but it is because the attractions of these embellishments, these adornments that redeem the body from its natural state, "give proof . . . of the immateriality of the soul." It may not be surprising that a philosophical tradition so impressed by varieties of dualism should generally try to disparage or take no interest in the wrappings of the body—the latter itself is often seen as a dispensable wrapping of the mind; but it is remarkable that even one who loves fashion seems to find it difficult to accept and straightforwardly celebrate the body.

In the notes collected under the title *Culture and Value*, Ludwig Wittgenstein remarks on "a beautiful garment that is transformed (coagulates, as it were) into worms and serpents if its wearer looks smugly at himself in the mirror" (1980). One might see, here again, the condemnation of a self-satisfaction that is confused about the self, that rests only on the most superficial trappings. But this fantasy of maggots and snakes, the suggestion of clotted fluids, hints vividly at a range of additional worries that may work to repress the philosophical embrace of fashion. There is a whiff of both sexuality and death in this strange image, and alarm about the powers of transformation that may be connected to enchanting clothes.

Sexuality is itself an underanalyzed topic in Western philosophy, and the tradition's avoidance of this subject has provoked some telling criticism from feminist philosophers. It is plain, however, that if philosophy has historically preferred to ignore sexual interests and activities, and issues of gender, then fashion—virtually always rendered and clearly interested in new presentations of the body—will also be ignored. Feminist philosophy, interested in every inflection of gender, should feel freer to give thought to fashion, and, indeed, a philosopher such as Beauvoir offers extended observations to support her claim that fashion is a system organized by male desire. Some clothing and makeup, she says, advertise erotic appeal—and adopting these fashions may count as "dressing like a streetwalker"—but more modest dress and cosmetics still signal acceptance of one's gender role, and even opting out is not an option, as eccentricity attracts attention. The curious determinism of this account may strike both men and women as a bit depressing, but the ineluctability of sexuality might, at least on some accounts, and in a less embattled context, be assumed to have a playful side. [*See* Sexuality.]

More sustained dejection may ensue, however, from focusing on another fact connected with our embodiment, namely, our mortality. If one is buried in a beautiful garment, worms will indeed grow out of the clothes—but they will also grow out of one's flesh. Fashion may somehow, subtly, remind us of this. Fashions not only cover our changing, always aging bodies, they also underscore our transitoriness by changing themselves. They become old

FASHION: Fashion and Philosophy. I. D. de St.-Jean, *Suit Worn with a Sword* (1685), French fashion print. (Courtesy of Anne Hollander.)

and tired; they die. The philosophical tradition attracted to permanence—eternal forms, stable foundations for knowledge, lasting truths—may seem to have a principled antagonism to fashion, and in refusing to look at fashion it may understand itself to be keeping faith with higher or deeper concerns. This faith, in turn, is often tied to another: that the decline and the death of the body are unimportant because the human being's essence is mental or spiritual. But even these soulful philosophers can see that we are not now angels. And even their opponents can see that we differ from mere brutes—we wear clothes. Our suspension between these more carefree alternatives may be evident in the anxiety of our peculiar capacity for self-awareness, the self-awareness that is materially portrayed in the image—in Wittgenstein, in the horrible image—of the looking glass.

Part of the point of a mirror, of course, is to see how one might appear to others, but acknowledging this interest can also injure the philosopher's self-conception. A philosopher may aim to be a seer, but an interest in fashion depends on a responsiveness to the fact that one is *seen*. Admitting that one can be an object to others—a thing seen, known, evaluated—may be particularly difficult for one who aspires to be the thinker, the knower, the evaluator. Discomfort

with an interest in fashion may thus attest to philosophy's willful neglect of the possibilities of human passivity, and it can also signal a desire to escape the force of thoughtless social norms. When Thoreau's request for a garment of a certain sort receives the response, "They do not make them so now," he says he turns this "oracular sentence" over in his mind,

> emphasizing to myself each word separately that I may come at the meaning of it, that I may find out by what degree of consanguinity *They* are related to *me*, and what authority they may have in an affair which affects me so nearly.

He of course insists on his own style, condemning his "tailoress" with the philosophical question, "Of what use this measuring of me if she does not measure my character, but only the breadth of my shoulders, as it were a peg to hang the coat on?"

The most significant exceptions to philosophy's studied indifference to fashion have also taken the form of condemnation. Some contemporary applications of moral philosophy have objected to the use of animals in the service of fashion—animal pelts in garments and their living tissues in the testing of cosmetics. And the conditions of textile and apparel production and distribution in the modern world have sometimes worried political philosophers. But these concerns are not clearly aesthetic, and they are not really about *fashion*. Philosophy's deepest problems with fashion lie elsewhere.

Just as a full account of a personal phobia may disclose not only the individual's life history but also his or her underlying needs and desires, so an account of philosophy's fear and hatred of fashion may reveal not only some neglected history of philosophy but also some of the subject's buried motivations. There might be diagnostic value in surveying traditional philosophy's antagonism to fashion, not because of the profundity of philosophy's past questions about fashion, but because of something like the reverse: because attention to the struggle of philosophy against fashion can suggest a deep interrogation of philosophy.

Some of the lineaments of this sort of inquiry can already be sketched. Philosophy has derided fashion for its superficiality. But why has philosophy coped so poorly with surface detail and with the transient, sought so recurrently to ignore or deny these phenomena in a quest for noumena, for absolutes, for timeless essences? Fashion, clothing, cannot be understood apart from the human body, and fashion is virtually always gendered. But how has philosophy been shaped, perhaps deformed, by its squeamishness about sex and its attempts to place itself outside, perhaps above, distinctions of gender? Hypotheses about the character and the potency of philosophy's unmet needs may be constructed from a study of its peculiar evasions and oversights, and reflection on philosophy's surly and inapt comments on clothing and fashion appears to provide abundant data.

Additional evidence that philosophy has had problems here—and hence that there are here interesting philosophical problems yet to be explored—can be discerned in the disciplines, the intellectual constraints, adopted by the thinkers who wish to present themselves as unashamed in their attraction to fashion. As already suggested, Baudelaire wants to show in his praise of cosmetics and changing fashion that he is not sacrificing adherence to a traditional philosophical faith—in ideal Beauty, in the distinction between mind and body and the assumption that the former is superior to the latter. And Roland Barthes's sustained analysis of "the fashion system"—an analysis that proceeds from an avowed preference for the universal to the particular, the general to the specific—involves willfully forgoing, in the name of these philosophical allegiances, any attention whatsoever to real fashion, to actual clothes. He is interested, he says, in "Fashion in general," in "Fashion . . ."re-constitute[d] as a formal system" (Barthes, 1983); he is interested, one might say, in disembodied fashion.

Barthes examines not clothes or the wearing of fashions, and not even pictures of clothes, but *descriptions* of clothes, textual copy from a year's worth of fashion magazines. He uses these "statements of Fashion" as the linguist or philosopher of language would use examples of grammar, to get at what he says is a general *structure,* as opposed to the material particulars, of fashion. The "mental model" of fashion—fashion so construed—is, Barthes claims, formal logic. Thus, he turns his back on "the history of Fashion," on the "substance of Fashion," on all that might be called "contingent Fashion." Fashion description has as its "primary function," Barthes claims, "the immobilization of perception." While even the still photographs in fashion magazines contain endless possibilities for perception and interpretation, the words, Barthes says, "determine a single certainty."

However dubious one might find that claim, however in need of unmasking or diagnostic explanation, Barthes himself acknowledges that his choice to work with descriptions alone, his quest for structure and system, his employment, in short, of what he calls "the taxonomical imagination" is "both psychoanalyzable and subject to historical criticism." While it seems likely that the prose of fashion description, together with its function, varies from one magazine to another, one general editor to another, as well as from one era and one cultural context to another, so that the effort to uncover a general structure on the model of formal logic is quixotic, the fantasy of escape from contingency that is represented by this undertaking has abundant historical precedent. The desire to arrest the change inherent in fashion, to avoid the complexities of thought and emotion that might be aroused by the contemplation of clothing "vitalized and animated," as Baudelaire says, by the beautiful women wearing it—in sum, the obsessive flight from contingency—

reaches a kind of apotheosis in Barthes's claim to study fashion in terms of immobilized perception, in terms of descriptions that "determine a single certainty."

Barthes quotes with approval G. W. F. Hegel's claim that "it is clothing which gives the body's attitude its relief, and for this reason it must be considered as an advantage, in the sense that it protects us from the direct view of what, as sentience, is devoid of signification"; but Barthes asserts that the body signified by the fashion garment is "no one's body," "a pure form." Philosophers with a more robust sense of the contribution to fashion's significance made by real bodies might try to read differently the character of that significance. (Indeed, Barthes's view that "the cover girl's body possesses no attribute . . . [and so] by a sort of tautology . . . refers to the garment itself," an impression perhaps derived from his sampling French fashion magazines of 1958–1959, is directly confounded by the 1980s and 1990s phenomenon of "supermodels," cover girls such as Cindy Crawford, Claudia Schiffer, and Naomi Campbell, who became internationally famous for their own physical attributes and whose personal appearance began to matter more than the garments they wore, whose celebrity began to shape and then to obscure even couture.) In addition, the Hegelian reciprocities of self and other to which Barthes refers need to be more accurately untangled. In what ways, and through what means, can clothing and fashion achieve signification—on the one hand, for the wearer and on the other, to the observer? What could the dynamics of fashionable dress, with its peculiar blends of self-consciousness and social attentiveness, reveal about the metaphysics of personhood?

In a time when opposition to dualism and essentialism is not seen as heretical, when gender sensitivities cast new light on old theories, when vigorous debate focuses on the import of issues of perspective for a variety of epistemological and metaphysical questions, the conditions would seem favorable for fresh philosophical investigations of the nature and effects on fashion. Thoreau's claim that "our whole life is startlingly moral"—"There is never an instant's truce between virtue and vice"—seems to resonate for those who worry about the use of furs in fashion, those who object to animal testing of cosmetics, those committed to boycotting clothing lines produced in sweatshops or with child labor. For some, in at least some moods, a parallel claim could be asserted, that our whole life is startlingly aesthetic; and this assertion would bring parallel imperatives for thought, an additional range of responsibilities for action. For those with even the most modest economic prosperity, enough to allow some choice of clothing, the aesthetic dimension seems always available. Philosophy is called for to chart this dimension, to make sense of the ever-changing but constantly present phenomena of fashion.

[*See also* Barthes; *and* Baudelaire.]

BIBLIOGRAPHY

Barthes, Roland. *The Fashion System.* Translated by Matthew Ward and Richard Howard. New York, 1983.

Baudelaire, Charles. *The Painter of Modern Life and Other Essays.* Translated and edited by Jonathan Mayne. London, 1964.

Beauvoir, Simone de. *The Second Sex.* Translated by H. M. Parshley. New York, 1952.

Bell, Quentin. *On Human Finery.* London, 1947.

Emerson, Ralph Waldo. "Comedy." In *The Early Lectures,* vol. 3, edited by Robert Spiller and Wallace Williams, pp. 121–137. Cambridge, Mass., 1972.

Hanson, Karen. "Dressing Down Dressing Up." *Hypatia* 5.2 (Spring 1990): 107–212. Reprinted in *Aesthetics in Feminist Perspective,* edited by Hilde Hein and Carolyn Korsmeyer (Bloomington, Ind., 1993), pp. 229–241.

Hollander, Anne. *Seeing through Clothes.* New York, 1978; reprint, Berkeley, 1993.

Hegel, Georg Wilhelm Friedrich. *Aesthetics: Lectures on Fine Art.* Translated by T. M. Knox. Oxford, 1975. Barthes's quotation of Hegel is from *Esthétique* (Paris, 1944).

James, William. *The Principles of Psychology,* vol. 1. New York, 1890.

Steele, Valerie. *Paris Fashion: A Cultural History.* New York and Oxford, 1988.

Thoreau, Henry David. *Walden and Civil Disobedience.* Edited by Owen Thomas. New York, 1966.

Veblen, Thorstein. *The Theory of the Leisure Class: An Economic Study of Institutions.* New York, 1899.

Wilson, Elizabeth. *Adorned in Dreams: Fashion and Modernity.* Berkeley, 1985.

Wittgenstein, Ludwig. *Culture and Value.* Edited by G. H. Von Wright, translated by Peter Winch. Chicago, 1980.

KAREN HANSON

FEMINISM. [*To explain what feminism is and how it has transformed aesthetics, this entry comprises four essays:*

An Overview
Feminism and Tradition
Critique of Feminist Aesthetics
Feminist Art History

The first essay is an analysis of the role and development of feminism in aesthetic theory. The second is a discussion of the notion(s) of tradition in aesthetics long before feminists had a voice, and it offers a critical account of tradition from the perspective of feminism. The third is a critique of the notion of feminist aesthetics. The last essay explores the history of feminism in the fields of art history and, to a less extent, art itself. For further discussion, see Autonomy, *article on* Autonomy and Its Feminist Critics; Canon, *article on* The Canon in Aesthetics; Contemporary Art, *article on* Images and Desire; Duchamp, *article on* En-Gendering of the Artistic Subject; Film, *article on* Feminist Film Theory; Genius, *article on* Genius and Feminism; Irigaray; Kant, *article on* Feminism and Kantian Aesthetics; Kristeva; Lesbian Aesthetics; Music, *article on* Music and Feminism; Performance Art, *article on* Feminist Performance Art; Stein; *and* Sublime, *article on* Feminine Sublime.]

An Overview

In the art world of the last quarter of the twentieth century, women have thrived as never before. So testifies a legend inscribed on the wall of the gallery displaying the Women's Museum of Art's permanent collection in Washington, D.C.:

> At a certain point during the 1970s it suddenly occurred to me that half of the interesting new artists in America were women— nothing like it had ever happened in the history of art. Were these women so phenomenally gifted that their talent simply swept away the obstacles? Or had changes in society and in cultural attitudes made possible a kind of acceptance that these women could not have won twenty years earlier?
> —Calvin Tomkins, *Righting the Balance* (1989)

Whether women themselves have extended their talents, or society has advanced its treatment of them by improving women's access to how culture is made and changed, their ascent in the art world during the past quarter century is well heralded and documented. Their flourishing goes far beyond there being more women crafting more esteemed objects of aesthetic appreciation in more media than formerly.

What marks this as a time of remarkable progress for women is that they have achieved an unprecedented level of leadership as they write about and manage art in ways that compel respect for works made by women. Previously, the few women who wrote criticism introduced women's concerns infrequently. They did not see themselves eliding female cultural politics into their personal responses to art.

But by 1994 Marcelle Marini could write:

> Women in twentieth-century Western societies participate in cultural production to an unprecedented extent . . . For the first time in history the women's movement achieved a true cultural dimension, and women's demands in the cultural arena achieved a true social impact.

To demonstrate how visible women have become in the world of art and culture, Marini cites the somewhat extravagant reaction exhibited by the author of "The 80 Women Who Rule the World of Letters," an article in the very conservative *Figaro littéraire:* "Not only as successful novelists, biographers, historians, and academics but also as editors, women have taken remarkable revenge in the world of letters." Both admiring of, and taken aback by, the growing influence of women, the author reads their success as showing how they have transcended their difference from men: "They have given striking proof that literary genius has no sex any more than it has race or nationality."

Cultural Politics. This comment demonstrates failure to comprehend the current feminist program, for assimilation is precisely not the conclusion the women's movement has striven to reach. The prize on which the collective eyes

of progressive women are fixed is the recognition of an artistic identity that pertains to their sex or gender. Feminists seek to prevail over the heretofore unchallenged authority the art world affords to practices preferred by men. To advance recognition of both women's personal and their group identities in the arts and culture, they have devised a variety of well-publicized strategies that promote women's artistic preferences and practices.

This was not always the case. Women's consciousness of how to relate their (group's) art to their (group) identity usually is portrayed as having risen through several stages. For centuries previously and in the present century through the early 1970s, women's attempts at artistic self-definition focused mainly on the politics of their subordination by men, their lack of equality, and their material disadvantage. Mary Wollstonecraft's *A Vindication of the Rights of Women* (1792), Virginia Woolf's *A Room of One's Own* (1929), and Simone de Beauvoir's *The Second Sex* (1952) stand among the progenitors of this first stage.

In writing about themselves during the early development of feminism, women explored their social confinement. Thus, their artistic biographies of this period often are vehicles of social criticism. In this stage, art writing about women in the popular journals and in the art-related scholarly disciplines typically is political, undertaken with the goal of dissolving the differences that obstruct women's artistic endeavors from being assimilated to men's. Linda Nochlin's immensely influential essay "Why Have There Been No Great Women Artists?" (1971; reprint, 1988) exemplifies the thought characteristic of this stage. The answer, she urges, lies not in women's lack of artistic talent compared to men, but in their comparative lack of encouragement and in the absence of opportunities to acquire the knowledge and skills to produce art like men.

A Women's Tradition. Subsequently, strategies shifted. For on closer inspection it appeared that the suppression of women artists had not been absolute. On the walls of the Women's Museum, the permanent collection displays work by women whose art made them respected and secure in their own times: Sofonisba Anguissola, Artemisia Gentileschi, Judith Leyster, Élisabeth Vigée-Lebrun, Angelica Kauffmann, Rosa Bonheur, Käthe Kollwitz. Likewise, Dr. Johnson celebrated Fanny Burney's novels far beyond those of her male contemporaries, and Henry James despaired of ever matching the success enjoyed by Edith Wharton. Yet, in subsequent scholarship—in the retrospective art-historical and art-critical treatment of them—successful women art makers are made out to be of less interest than their male counterparts. However well these and other women may have been integrated into the art world of their own time, there is something systematic about the way their reputations fade as they are distanced by the history of their art.

And this seems to be the case even where their history is not so distant. The French volume *Notre siècle* cites 180 men, but only eight women, as significant contributors to twentieth-century art, literature, and philosophy (Marini, 1994). Unlike work by (some) men, it has been almost unheard of for art made by women to grow in stature rather than diminish with age.

In the second stage through which feminist reflection about art ascended, the urgent goal thus became to combat this discounting of the durable significance of female art makers. The strategy was to identify and revalue the work of female artists whom history had obscured. Thus, in the middle 1970s, feminists turned to an approach sometimes termed "gynocriticism," namely, the serious analysis of women's artistic traditions and the fashioning of a woman's canon.

Feminists refocused the histories of the arts to give more prominence to work made by women. Much as feminist ethics of roughly the same period elevated the caretaking roles traditionally occupied by women to a moral status equivalent to the contractual relationships men customarily assume, gynocriticism elevated the status of the (kind of) work customarily associated with women. One of many indications that their scholarship has realized the desired change is the growing permeability of the line between art and craft, a movement that has carried domestic objects like quilts onto museum walls. It has successfully challenged such distinctions as that between high and low art, as well as the artistic ranking that diminishes genres associated with women—novels, miniatures, still lifes, portraits.

Aesthetics and Philosophy of Art. As feminist revaluations of women's art advanced in sophistication, however, it began to seem vital to determine whether the historical suppression of art by women and, concomitantly, their prospective liberation stem primarily from their art's internal features or, instead, from what is beyond the work itself, namely, the social status of its maker. Reflecting this division in conceptualizing what is central to art, feminist treatments of art diverged in their next stage. Two somewhat different directions or emphases, both of which had been implicit in the earlier scholarship on women's art, emerged.

Gynesis argued that a distinctive female identity lies in the distinctive features of liberated women's art (and sometimes in men's art as well). *Gender theory* argued that women's fundamental artistic identity is constituted by their roles in cultural practices rather than by the expressive forms of their art. While both seek to identify criteria or principles that are indigenous to women collectively, the former approach naturalizes the artistic identity of women while the latter makes female identity an artifact of social organization.

Moreover, gender theory leads us to a feminist philosophy of art, while gynesis provides material for a feminist aesthetic. In grasping the difference between philosophy of art and aesthetics, it is helpful to refer back to some of the sources that set the agenda for philosophical aesthetics in

the latter part of the century. Stuart Hampshire observes: "Nothing but holding an object still in attention, by itself and for its own sake, would count as having an aesthetic interest in it. A great part . . . is to place a frame upon the object and upon its parts and features" (Hampshire, 1954, pp. 166–167). And John Passmore writes in the same vein: "sometimes the aesthetician substitutes for aesthetics something quite different, for example, the theory of art . . . art (good and bad) . . . considered as a social institution" (Passmore, 1954).

Much more recently, Peggy Brand (1995) identifies a focus on formal properties of the work, disinterestedness and detachment, and pleasure as three traditional indicators of the aesthetic and argues that only the first is necessary to a feminist aesthetic. Aesthetics, then, focuses on properties internal to the work that are supposed to make it what it is and to constitute its value. Philosophy of art, on the other hand, addresses properties beyond the work that relate to it in ways that compose the conditions of the work's significance and that constitute the source of its (good or bad) value.

Gynesis. Gynesis claims that certain properties of works of art are the natural expressions and representations of women. Emerging from and reversing the structuralist preoccupation with the coherence of symbolic expressions, French feminist writers such as Hélène Cixous (1976, 1986) and Luce Irigaray (see Whitford, 1991) famously urge that women write with their bodies, employing such techniques as fragmenting gaps and disruptions, heteroglossic voices and multiplexic diction, cumulative or repetitive as opposed to linear thought, metaphors of fluidity and flow, a passion for process, use of the continuous present, and a decentering of point of view that precludes privileging the male viewpoint. This natural language of subjectivity, they affirm, has been depreciated by the equally nature-driven masculine obsession with pursuing objectivity through disengagement and distancing.

In an analogous vein, Renée Lorraine (1993; see also Cox, 1991) points out that in the Romantic era the distinction between masculine and feminine music is already strongly drawn. Feminine music was thought to be lyrical and legato, with delicate instrumentation and small intervals, inconclusive or implicative. She suggests that the musical analogue to writing with the body would disrupt linearity, avoid definitive closures, deconstruct the dialectical juxtaposition and resolution of definitive closures, and have a flexible, cyclical form.

In the visual arts, a female sensibility analogous to the female imagination expressed in women's distinctive writings is proposed. According to Joanna Frueh (1988), this sensibility privileges visceral responses to materials, obsessive or repetitive imagery, and completed rather than empty rounds. Female engagement with art advances feeling over intellect by embracing rather than abstracting from subject.

San MacColl (1993) exemplifies how this sort of feminist aesthetic operates. To show how Georgia O'Keeffe's painting is woman's work, she cites how the artist Judy Chicago developed a style that expressed her womanliness. In her autobiography, Chicago writes:

> I used the flower as the symbol of femininity, as O'Keeffe had done. But in my images the petals of the flower are parting to reveal an inviting but undefined space, the space beyond the confines of our own femininity. (Chicago, 1975)

MacColl comments:

> The significance of the body comes out in two important ways in O'Keeffe's work. First, it is integrated into, and fundamental to the nature of, her subjectivity—"she paints with her body" . . . Second, this body which pervades her subjectivity is sexually specific. . . . In O'Keeffe's work we have an expression of how female subjectivity is sexed through our experience of the body.

Gynesis accepts that women may suppress or ignore female modes of expression but warns that they do so at the cost of diluting their creative vitality. Thus, Cixous acknowledges, just a few women so far have freed themselves sufficiently to write with their bodies. To liberate women's art, gynesis contends, women must substitute the symbolic expressions and representations of their own nature for those favored by the patriarchal canon. Women's identity is positioned somatically. Women experience from this place and should create from this place as well.

Carolyn Korsmeyer (1990) points out that, for much of women's history, touch, taste, and smell have been thought to prevail over hearing and sight in being exercised and relied on in women's everyday lives. But for Korsmeyer this is a cultural rather than a biological fact: touch, taste, and smell are more closely identified with corporeal being than are hearing and sight, the so-called intellectual senses central to executing the social roles customarily assigned to men. Suppose sight and hearing typically are given primacy over touch, taste, and smell as the aesthetic senses because the former pair of senses are associated with masculinity. For gynesis, this suggests that touch, taste, and smell rather than hearing and sight should have primacy in feminist art, and that properties perceived through the former three senses rather than the latter two should be the concern of feminist aesthetics. Korsmeyer (1998) herself, however, questions whether such "gendering" of the senses exaggerates and distorts how they operate.

Gender theory. From the point of view of gender theory, gynesis is tainted with the sin of essentialism. Gender theory denies that there is a female nature—psychological, phenomenological, biological, or otherwise—that is independent of the accidents of race, class, and custom. Consequently, gender theory urges, gynesis's claiming to have identified universal expressions of femaleness inevitably

marginalizes some women. About French feminists' practice of gynesis, Elaine Showalter writes: "These . . . terms echo classical European stereotypes . . . they are constructed by suppressing the history of black and lesbian feminist theory" (Showalter, 1990).

Contrary to gynesis, gender theory portrays how women are embodied as the effect rather than the source of discursive and other symbol-making practices. The body is a construct rather than a natural constant. Sex-related properties become relevant to art making and art writing only because they are made so by social structures. Gender theory, which owes more than a little to the work of Michel Foucault, considers the properties found in women's art to be expressions not of their nature but of the social roles they occupy. Women's and men's art equally are products of whatever practices distinguish female and male lives in the historical and cultural contexts in which the works emerge.

Against gynesis, gender theory challenges both that women have foundational female experiences and that they have sovereign experiences of art. As for the former, women's experience is not uniform but differs in respect to a wide range of personal and social characteristics, so proposals purporting to describe female nature are likely to be so abstracted from the identifying differences and the lived experiences of actual women as to become mere stereotypes. As for the latter, contrary to what Hampshire recommends for aesthetics, gender theory finds it implausible to try "to place a frame upon the object" in an attempt to isolate aesthetic experience from social context.

What the frame shows is not independent of who holds it in place. Elaine Showalter (1979) points out that it is a regular practice in both making and writing about art to conjecture an ideal audience, and that it also is customary to idealize that audience as primarily male. If women's experience, beliefs, and values diverge from those of the males of the traditional imagined audience, the frame they put on the object will bring unexpected features into prominence.

As Jenefer Robinson and Stephanie Ross note: "Men and women are likely, because of their diverse experiences, to have different assumptions and perspectives, and so too . . . feminists and non-feminists. Indeed, Kolodny's feminist reading of *The Yellow Wallpaper* only became possible through developments in contemporary thought, in particular, the growth of feminism. . . . What is salient to most women of the 1890s and what is salient to contemporary feminist women is very different" (Robinson and Ross, 1993).

Gender theory presumes neither that the practices in which women characteristically engage in any social context are preferable for them, nor that they are not so. Rather, gender theory compares the works and ways of women and men to determine which result from destructive imbalances of power between the genders, and which are beneficial or at least benign. In contrast to gynesis, which assumes that male-devised culture must disadvantage women by forcing them into unnatural modes of expression at which they do not succeed very well, gender theory explores prevailing practices to discover whether they disadvantage women. The frequent answer is that they do so because they are instruments to help males retain power over females. In this regard, gender theory approaches and weighs the differences between genders somewhat as Marxism assesses the differences between economic classes.

To illustrate, the novelist Marilyn French proposes a functional revolutionary account of gendered art. She formulates two fundamental principles, both of which relativize artworks to what is beyond their boundaries, as marking an artwork as "feminist":

> First, it approaches reality from a feminist perspective; second it endorses female experience. The feminist perspective is partly a reversal of patriarchal views . . . feminist art focuses on people as whole . . . part of a community, connected to others . . . and—on the broadest level—to nature in both positive and negative aspects. The second principle is equally complicated. To endorse women's experience, feminist art must present it honestly, wholly. This is difficult because literature, like all art, is made up of conventions which are particularly marked in the area of gender. . . . This means that either considerable time—decades or even centuries—must elapse before readers can concentrate on what is being shown, or the work will be forgotten without this ever happening. . . . Feminist work often focuses on groups, community, people as part of a context, and helps to remind us of a reality alternative to the Western tradition of individualistic, alienated man, lonely in a hostile, aggressive world.
>
> (French, 1993)

Linking Women's Art to Women's Worlds. Feminist writing about art characteristically, albeit somewhat confusingly, draws on both gynesis and gender theory for themes that lie together uneasily and not completely effectively. As Ismay Barwell explains:

> Actual experience does not simply "reflect" economic and institutional conditions, nor does the . . . structure of a work of art simply "mirror" the psychological structure of the individual or individuals who produce it. . . . These mistakes point to a need for an adequate theory of how sexual differentiation is produced under variable conditions and how this sexual difference is related to features of paintings, sculpture, films, and literature.
>
> (Barwell, 1993)

Adherents of both approaches are inclined to privilege woman as a site from which to discern the failings of phallocentrism and patriarchy, and neither approach easily escapes the difficulty that to so conceive of woman is once more to define women in terms of how they are not men. Both therefore need to consider how to link the gender-identified internal properties of an artwork to the social construction of gendered identities in the world beyond the work in some way that escapes continuing contamination by controlling male figurations.

Gynesis looks at the properties of a work that identify it as women's art as naturally linked to the artist's sex or sexuality; gender theory as conventionally so. Gynesis sees women's art as individuals' expressing and representing their personally experienced femaleness. Gender theory assimilates the individual to the identity of the group, substituting collective for personal artistic agency.

Indeed, in the spirit of both the postmodern and the Marxist writers who inspire it, gender theory is inclined to dismiss art writing that exalts singular artistic exploits rather than reads off artifacts, or their contexts, as texts. In this vein, Foucault is as insistent as Monroe Beardsley and William Wimsatt, the analytic authors of "The Intentional Fallacy," in rejecting the practice of referring to artwork artists. In the context of feminist applications of gender theory, the thrust of such proposals is that, by discarding the idea that art originates in singular achievement, we liberate it from male individualism and elitism, and thus from the art-historical approaches that have given us a male-dominated past.

But the literary critic Henry Louis Gates, Jr., voices concern about whether the suppression of self might not devitalize reform:

> Precisely when we . . . obtain the complex wherewithal to define our . . . subjectivity in the republic of . . . letters, our theoretical colleagues declare that there ain't no such thing as a subject. . . . those of us in feminist criticism or African-American criticism who are engaged in the necessary work of . . . reformation confront . . . those who are allies on other fronts . . . over this matter of the death of the subject and our own discursive subjectivity . . . (for) . . . self-identification proves a condition for agency, for social change. (Gates, 1992)

To acknowledge the importance of the hero(in)ic artistic subject to the attainment of feminist goals, and, as well, to mediate between gynesis and gender theory, Anita Silvers (1990) extends suggestions in the work of Richard Wollheim (1987), Kendall Walton (1979), and Jenefer Robinson (1989). All these propose, in one way or another, that we perceive at least some of the internal properties of artworks as purposeful, particularly those that express and signify. To facilitate their cohering, we construct an idea of a singular agency, an artistic agent, male or female, whose hand and voice remain present in the work. Thus, we properly describe a work's gendered character in the language of feeling and intention, not in terms of a work's formal properties.

Nor must the art-historical narratives of gendered "apparent" artistic agents thoroughly correspond to, nor need they do more than plausibly reflect, the known facts of the life of the actual maker of the work. The real woman artist's life is the occasion, while the constructed female figure of the apparent artist is the vehicle, for an artwork's having import of the sort feminists celebrate. Our philosophy of art accounts for the significance of the real artist's contribu-

tion, while our aesthetics directs our regard of the latter's role. The female voices and hands of gynesis's subjects center the themes of feminist art writing, but they are *gender-theoretical* constructs rather than the natural extensions of female bodies.

Silvers observes: "It is the mediating figures of the rare hero(ines) of the story of art that transform past into future by making art not for their peers alone but for posterity." This explains why it was so stimulating for feminist scholarship, at a stage prior to the appearance of both gynesis and gender theory, to rediscover and revalue historical women art makers.

Analytic Aesthetics. Despite how dynamically feminism flourished through the 1980s, in the arts and in criticism and scholarship about the arts and in many fields of philosophy as well, it was not until 1990 that a noticeable body of feminist work in philosophical aesthetics appeared. In this year, the feminist philosophy journal *Hypatia* published a special issue on feminist aesthetics (5.2, Summer 1990, edited by Hilde Hein and Carolyn Korsmeyer), as did the analytically inclined *Journal of Aesthetics and Art Criticism* (48.4, Fall 1990, edited by Peggy Brand and Carolyn Korsmeyer). But why before 1990 was there virtually no feminist philosophical aesthetics?

Brand and Korsmeyer (1995) believe feminism challenges two presumptions on which traditional aesthetics relies so very greatly as to hardly ever call them into question. Implicit in their diagnosis, then, is the suggestion that feminism may seem incompatible with traditional aesthetics. The first traditional theme Brand and Korsmeyer identify is the persuasion that art has universal value. Mary Devereaux (1995) agrees. As she puts it, traditional aesthetics supposes that art paradigmatically speaks for and to everyone, and that philosophy's job is to understand how art transcends differences among members of its universal audience. In contrast, feminism supposes that art does not speak for everyone and that it is philosophy's job to understand how art's customary forms privilege some kinds of speakers over others. As we have seen, this is the import of gynesis.

Second, Brand and Korsmeyer say, traditional aesthetics takes the value of art to be intrinsic, isolated from practical concerns and independent of other sorts of value. Brand (1995) characterizes traditional aesthetics as ignoring the messages of artworks except occasionally to blame them for politicizing a piece of art and consequently diminishing its aesthetic value. But the very rarity and strangeness of female art cannot help but carry the message that males monopolize the art world. Hilde Hein concurs (1995), saying that feminists "face the challenge of recasting . . . experiences *as they are undergone by women,* so as to reveal an aspect of them that has been ignored. In doing so, they expose . . . the politics . . . of traditional art." The influence of gender theory is evident here.

Is this explanation fully convincing? No, because it remains unclear why feminism's refusal to view aesthetic value as universal should alarm contemporary philosophical aestheticians. For it has been with challenging precisely this preanalytic tradition that contemporary analytic aesthetics has occupied itself for nearly half a century. That is, from midway through the twentieth century, analytic philosophers have argued that art has no set of necessary and sufficient conditions, no fixed definition, no universal essence. Far from being foundationalist, they acknowledged the low degree of closure in regard to standards of aesthetic interpretation and judgment. They applauded the openness of our concept of art; indeed, they considered this conceptual analysis to be a key accomplishment of their methodology.

Moreover, they contrasted aesthetics with philosophical fields whose subject matter seemed much more fixed, namely, the discourses addressed by ethics, epistemology, and philosophy of science. Extending Brand and Korsmeyer's diagnosis, one would predict that the greater magnitude of feminism's challenge to the fixity of those fields would result in its achieving even less of a profile in them than in aesthetics. Yet these philosophical fields have proven most receptive to feminist thought.

As for the second traditional assumption—namely, that the aesthetic is autonomous and detached—by the mid-1960s, Maurice Mandelbaum (1989; 1965) and Arthur Danto (1964) both were recommending that art must be understood relationally and as amenable to a historicized definition. From this point, the default position for aesthetics has been to deny—certainly not to affirm—that art has a separate and timeless essence. It thus seems just as unlikely to have been feminism's challenge to the presumption of aesthetic detachment as to have been feminism's denial of aesthetic universalism that impeded its surfacing in aesthetics.

Analytic aestheticians have paid serious attention to a variety of proposals that attach art to the social institutions of the past and present. Indeed, during the last quarter century, various institutional theories of art—broadly, those which hold that identifying what objects are art requires attending to convention rather than natural kinds—have monopolized conversation in the field. Yet comparatively little attention has been expended to account for how such collectively conventional responses emerge from our very personal experiences with art. Here, then, is a disconnection between analytic aesthetics and feminism, for in contrast feminism addresses this connection with such intensity that it famously elides the personal with the political.

There is yet another disconnection. In contemporary aesthetics, feminism initially found no invitation comparable to the preexisting intellectual environment it encountered when it turned to traditional criticism of the arts. Prevailing theory in the latter fields of scholarship readily encouraged the introduction of concepts amenable to inflection by sex or gender. So on the Continent as well as in English-speaking literary and art scholarship, gynesis introduced itself into discussions already framed by Sigmund Freud's focus on the differences arising from sex. Likewise, gender theory found a fertile field cultivated by both Karl Marx's and Foucault's focus on differences arising from social structure.

On the other hand, analytic aesthetics could not similarly incorporate female inflections simply by adapting its prevailing approaches to another kind of difference. In this regard, bell hooks's warning in *Art on My Mind: Visual Politics* (1995) to feminists holds much wisdom for philosophical aesthetics as well: "Despite feminist thinking and practice . . . no longer bound by sexist, racist, or class constraints that tell us we cannot . . . create . . . compelling work, we remain bound by limitations on our imagination." Yet, even now those boundaries are being stretched by the magnitude of feminist activity in the fields analysis targets. A way in which considerations of sex or gender can connect up to the prevailing themes in aesthetics now presents itself, thanks to the robust development of feminism in other fields of arts scholarship.

As Showalter (1990) observes: "Although feminism as a political ideology can be traced back at least to the seventeenth century, feminist literary criticism is a very recent invention." During the last half century philosophical aesthetics has seen itself as a meta-activity, devoted to understanding how discourses about art operate; yet only recently have sex and gender been discerned as a subject and an operating principle of these discourses. It has been less than a quarter century since feminist scholarship began to reform how we talk about the arts. Now, as these changes have taken hold, we begin to see them acknowledged and explored in philosophy.

BIBLIOGRAPHY

Barwell, Ismay. "Feminist Perspectives and Narrative Points of View." In *Aesthetics in Feminist Perspective*, edited by Hilde Hein and Carolyn Korsmeyer, pp. 93–104. Bloomington, Ind., 1993.

Battersby, Christine. *Gender and Genius: Towards a Feminist Aesthetics.* London, 1989.

Beauvoir, Simone de. *The Second Sex.* Translated by H. M. Parshley. New York, 1952.

Brand, Peggy. "Revising the Aesthetic-Nonaesthetic Distinction: The Aesthetic Value of Activist Art." In *Feminism and Tradition in Aesthetics,* edited by Peggy Zeglin Brand and Carolyn Korsmeyer, pp. 245–272. University Park, Pa., 1995.

Brand, Peggy Zeglin, and Carolyn Korsmeyer. "Introduction: Aesthetics and Its Traditions." In *Feminism and Tradition in Aesthetics,* edited by Peggy Zeglin Brand and Carolyn Korsmeyer, pp. 1–22. University Park, Pa., 1995.

Chadwick, Whitney. *Women, Art, and Society.* London, 1990.

Chicago, Judy. *Through the Flower: My Struggle as a Woman Artist.* Garden City, N.Y., 1975.

Cixous, Hélène. "The Laugh of the Medusa." *Signs* 1.4 (1976): 875–894.

Cixous, Hélène, and Catherine Clément. *The Newly Born Woman.* Translated by Betsy Wing. Minneapolis, 1986.

Cox, Renée. "Recovering Jouissance: Feminist Aesthetics and Music." In *Women and Music: A History,* edited by Karin Pendle. Bloomington, Ind., 1991.

Danto, Arthur. "The Artworld." *Journal of Philosophy* 61.19 (15 October, 1964): 571–584.

Devereaux, Mary. "Oppressive Texts, Resisting Readers, and the Gendered Spectator: The 'New' Aesthetics." In *Feminism and Tradition in Aesthetics,* edited by Peggy Zeglin Brand and Carolyn Korsmeyer, pp. 121–141. University Park, Pa., 1995.

Ecker, Gisela, ed. *Feminist Aesthetics.* Translated by Harriet Anderson. Boston, 1986.

French, Marilyn. "Is There a Feminist Aesthetics?" In *Aesthetics in Feminist Perspective,* edited by Hilde Hein and Carolyn Korsmeyer, pp. 68–76. Bloomington, Ind., 1993.

Frueh, Joanna. "Towards a Feminist Theory of Art." In *Feminist Art Criticism: An Anthology,* edited by Arlene Raven, Cassandra L. Langer, and Joanna Frueh, pp. 153–165. Ann Arbor, 1988.

Fuss, Diana. "'Essentially Speaking': Luce Irigaray's Language of Essence." *Hypatia* 3.3 (1989): 62–80.

Gates, Henry Louis. "The Master's Pieces: On Canon Formation and the Afro-American Tradition." *South Atlantic Quarterly* 89.1 (1990).

Hampshire, Stuart. "Logic and Appreciation." In *Aesthetics and Language,* edited by William Elton, pp. 161–169. Oxford, 1954.

Hein, Hilde. "The Role of Feminist Aesthetics in Feminist Theory." In *Feminism and Tradition in Aesthetics,* edited by Peggy Zeglin Brand and Carolyn Korsmeyer, pp. 446–463. University Park, Pa., 1995.

hooks, bell. *Art on My Mind: Visual Politics.* New York, 1995.

hooks, bell. "The Oppositional Gaze: Black Female Spectators." In *Feminism and Tradition in Aesthetics,* edited by Peggy Zeglin Brand and Carolyn Korsmeyer, pp. 142–159. University Park, Pa., 1995.

Jehlen, Myra. "Gender." In *Critical Terms for Literary Study,* edited by Frank Lentricchia and Thomas McLaughlin, pp. 263–273. Chicago, 1990.

Johnson, Barbara. "Writing." In *Critical Terms for Literary Study,* edited by Frank Lentricchia and Thomas McLaughlin, pp. 39–49. Chicago, 1990.

Korsmeyer, Carolyn. "Gender Bias in Aesthetics." *APA Newsletter* 89.2 (1990): 45–47.

Korsmeyer, Carolyn. "Perceptions, Pleasures, Arts: Considering Aesthetics." In *Philosophy in a Feminist Voice: Critiques and Reconstructions,* edited by Janet A. Kourany. Princeton, N.J., 1998.

Lorraine, Renée. "A Gynecentric Aesthetics." In *Aesthetics in Feminist Perspective,* edited by Hilde Hein and Carolyn Korsmeyer, pp. 35–52. Bloomington, Ind., 1993.

MacColl, San. "A Woman on Paper." In *Aesthetics in Feminist Perspective,* edited by Hilde Hein and Carolyn Korsmeyer, pp. 150–168. Bloomington, Ind., 1993.

Mandelbaum, Maurice. "Family Resemblance and Generalizations concerning the Arts." In *Aesthetics: A Critical Anthology,* 2d ed., edited by George Dickie, R. J. Sclafani, and Ronald Roblin, pp. 138–151. New York, 1989.

Marini, Marcelle. "The Creators of Culture in France." In *A History of Women in the West,* edited by Françoise Thebaud, pp. 297–323. Cambridge, Mass., 1994.

Moore, Catriona, ed. *Dissonance: Feminism and the Arts, 1970–1990.* St. Leonards, Australia, 1994.

Mulvey, Laura. *Visual and Other Pleasures.* Bloomington, Ind., 1989.

Nochlin, Linda. *Women, Art, and Power and Other Essays.* New York, 1988.

Passmore, John. "The Dreariness of Aesthetics." In *Aesthetics and Language,* edited by William Elton, pp. 36–55. Oxford, 1954.

Robinson, Jenefer. "Style and Personality in Literary Work." In *Aesthetics: A Critical Anthology,* 2d ed., edited by George Dickie, R. J. Sclafani, and Ronald Roblin, pp. 455–468. New York, 1989.

Robinson, Jenefer, and Stephanie Ross. "Women, Morality and Fiction." In *Aesthetics in Feminist Perspective,* edited by Hilde Hein and Carolyn Korsmeyer, pp. 105–118. Bloomington, Ind., 1993.

Showalter, Elaine. "Towards a Feminist Poetics." In *Women Writing and Writing about Women,* edited by Mary Jacobus, pp. 22–41. New York, 1979.

Showalter, Elaine, ed. *The New Feminist Criticism: Essays on Women, Literature, and Theory.* New York, 1985.

Showalter, Elaine. "A Criticism of Our Own: Autonomy and Assimilation in Afro-American and Feminist Literary Theory." In *The Future of Literary Theory,* edited by Ralph Cohen. New York and London, 1989.

Showalter, Elaine. "Feminism and Literature." In *Literary Theory Today,* edited by Peter Collier and Helga Geyer-Ryan, pp. 179–202. Ithaca, N.Y., 1990.

Shrage, Laurie. "Feminist Film Aesthetics: A Contextual Approach." In *Aesthetics in Feminist Perspective,* edited by Hilde Hein and Carolyn Korsmeyer, pp. 139–149. Bloomington, Ind., 1993.

Silvers, Anita. "Has Her(oine's) Time Now Come?" *Journal of Aesthetics and Art Criticism* 48.4 (Fall 1990): 365–379.

Silvers, Anita. "Pure Historicism and the Heritage of Hero(in)es: Who Grows in Phillis Wheatley's Garden?" *Journal of Aesthetics and Art Criticism* 51.3 (Summer 1993): 475–482.

Smith, Valerie. "Gender Theory and Afro-American Literary Theory and Criticism." In *Speaking of Gender,* edited by Elaine Showalter, pp. 56–70. New York and London, 1989.

Walton, Kendall. "Style and the Products and Processes of Art." In *The Concept of Style,* edited by Berel Lang, pp. 46–66. Philadelphia, 1979.

Whitford, Margaret. *Luce Irigaray: Philosophy in the Feminine.* London and New York, 1991.

Wollheim, Richard. *Painting as an Art.* Princeton, N.J., 1987.

Wollenstonecraft, Mary. *A Vindication of the Rights of Women.* London, 1792.

Woolf, Virginia. *A Room of One's Own.* New York, 1929.

ANITA SILVERS

Feminism and Tradition

Feminism has come to the history of aesthetics only recently in spite of advances made by feminist scholars in related fields such as literary theory, art history, art criticism, and film studies since the early 1970s. Basic questions that define the aesthetic "tradition" have historically been answered in ways that discouraged the influx of feminist thinking from outside the established parameters of debate. Traditional aesthetics—proclaiming objectivity and universality—sought to eliminate personalized or idiosyncratic responses from the range of effects proper to the experiencing of works of nature and art. Individualized (e.g., gendered) reactions yielding a multiplicity of diverse experiences and value judgments were not tolerated, and uniformity of taste was idealized and encouraged. Explorations into the nature of art, theorizing about perception, and establishing criteria of beauty presupposed a particular type of person whose judgment was educated, discriminating, and universalizable. In contrast, recent proponents of feminism have challenged these entrenched theoretical and philosophical traditions not only by objecting to the criteria utilized in the ongoing dialectic but also by bringing attention to its source. As characterized

by Katy Deepwell, this person is predominantly upper class, white, and male, otherwise known as the privileged, white "man of taste."

Before looking at the rise of feminism per se, it will be helpful to ask what sense can be made of the term *tradition*. One all-encompassing sense of tradition is that of a philosophical backdrop to the emergence of the activism of 1960s' feminist artists and the accompanying theorizing. Also, separate segments can be delineated within tradition's long duration. Another approach is in terms of the many traditions that have emerged within feminism itself as it has evolved in its first twenty-five years. Let us look at these in turn.

First, the most familiar form—known as the Western tradition—is the most inclusive and captures the longest theoretical continuity. It extends back to classical antiquity with the writings of Plato and Aristotle and flourishes in the eighteenth century with theories of taste, considered the beginning of the field of aesthetics: Francis Hutcheson, Anthony Ashley Cooper, the Earl of Shaftesbury, David Hume, and Immanuel Kant. This continuum is refined and expanded in the nineteenth century with G. W. F. Hegel, Arthur Schopenhauer, and Friedrich Nietzsche. As radical new art seriously affected early-twentieth-century philosophical thought, focus shifted to the proliferating dimensions of the aesthetic: Clive Bell, Edward Bullough, John Dewey, Benedetto Croce, and R. G. Collingwood. Under the influence of Ludwig Wittgenstein, far-reaching objections arose to idealism and essentialism in art by authors Morris Weitz, John Passmore, W. B. Gallie, and Monroe Beardsley. The task of defining was reintroduced, but in a manner different than before; the Artworld and its institutionalized practices came to play a crucial role in the theories of Arthur C. Danto and George Dickie. Steeped in a self-consciousness of the art world context and its unique history, issues of representation, interpretation, and aesthetic value resurfaced as well in the work of Nelson Goodman, Joseph Margolis, and Richard Wollheim.

Within the vast scope of this two-thousand-year-old tradition, it is commonly held that the value of art transcends cultural differences and is a source of timeless and everlasting value. This presumption underlies the status attributed to so-called masterpieces of art—works created by "genius"—which are seen as universally valuable. Such an experience of a work of art lifts a person of taste out of her or his menial concerns and into a lofty state of mind shared by others. Also basic to this tradition is the belief that aesthetic contemplation and appreciation are due to the work's intrinsic value. A work of art derives its significance and enduring value from elements internal to the work, independent of extraneous ties.

Other, more narrow senses of *tradition* within philosophy have also arisen. For instance, in the 1940s and 1950s, analytic philosophers developed their own idiosyncratic sense of the term in which it came to mean the preanalytic writings by essentialist, Idealist authors like Arthur Schopenhauer prior to Bertrand Russell and G. E. Moore. In one of the clearest statements of this phase, William E. Kennick stipulated that "traditional aesthetics" was the philosophical discipline that concerned itself with questions of beauty, aesthetic experience, creativity, aesthetic judgment, taste, and the role of criticism.

Another sense of "tradition" arose during the period just after World War II: a number of writings in analytic aesthetics coincided with the shift of the art world from Paris to New York, and the rise of Abstract Expressionism and its attendant critical discourse. Not all philosophers, however, found this newest "tradition" to be the remedy for all critical and philosophical ills. What was new eventually became old and more recently—in the 1980s and 1990s—philosophers have come to critique the writings of their predecessors from the 1950s through the mid-1980s. At times, the tradition included not only the body of analytic writings that dominated during this period but also the preanalytic—essentialist and Idealist phase—thereby combining two recent shorter traditions into one, more continuous one. This exercise of "counting" traditions is admittedly less important than noting their presence as a stimulus to philosophical dialectic, and it is worth noting that advances in philosophical thinking often utilized the rhetoric of "the tradition" to symbolize the past.

Just as there is a multiplicity of historical notions of "tradition" within the history of aesthetics, there is variety within feminism in its emergence from those traditions and its confrontation with existing philosophical practices. Steeped in the political roots of the women's movement of the 1960s, feminist theorizing is multifaceted. There is no monolithic feminist approach to traditions in aesthetics, just as there is no one feminist approach to political, ethical, or legal issues. Whatever sense of unity is apparent, it rests on a foundation of numerous feminisms. There is ample room for variation in the shared goal of addressing inequalities toward women's creativity within the art-world and art-historical tradition: some feminist artists would do away entirely with the notion of artistic value, whereas others seek to be valued within entrenched systems of institutional recognition. The practical application of these views results in artists choosing to show their works in women-only cooperative galleries instead of competing to gain entry into mainstream art-world venues. By and large, however, feminism challenges oppressive historical and contemporary art-world practices that serve to undermine the import of women's contributions. Current statistics show that women's artworks constitute under 10 percent of the permanent holdings of major American museums, in spite of women comprising over 50 percent of art school graduates and self-declared artists. The number of works by artists of color in these collections is even lower. Thus, the beginnings

of feminism within aesthetics can be traced back directly to women authors, artists, art historians, and critics clearly outside any philosophical tradition.

In the late 1960s and early 1970s, feminist scholars issued a challenge to art historians and literary critics that ironically shared a certain orientation with analytic aesthetics, especially a skepticism about the definition and autonomy of "art," the purported universal appeal of art, and the enormous impact of art-critical language and the success of critics like Harold Rosenberg, Leo Steinberg, and Clement Greenberg in bringing about unparalleled success for male artists within the art world. Linda Nochlin's groundbreaking essay "Why Have There Been No Great Women Artists?" (1971) launched an entire movement centered on women's involvement in the arts. In her review of the systematic exclusion of female artists from teaching studios and other realms of art instruction that persisted for centuries, Nochlin suggested that the historical notion of "fine art" continued to overlook creative output by women.

From then on, feminist scholarship unearthed women writers and artists of the past who were well known, amply commissioned, and self-supporting in their day but subsequently omitted from the canonical histories of art. Exhibitions of works by women initiated a rediscovery that brought to light works that had been attributed to male artists, lost in archives, and obscured by the preponderance of male-created "masterpieces" dominating the limelight. As a result, the entire tradition of the male-defined canon was jeopardized, leading feminists in the direction of more theoretical and abstract pursuits. They began to question the underlying assumptions of so-called universal standards of taste and their resulting value judgments.

A new, revisionist art history developed that emphasized the work of art within its sociohistorical context rather than as an example of a particular genre, style, or artistic trend. Definitions of "art" were rejected as limiting and oppressive. Attention was brought to the privileging of high art over low, fine art over craft, men's art over women's. Early on, enthusiasm ran high that something like a female nature was discernible and that a woman's art expressing a feminine sensibility or a female aesthetic could be easily delineated. It was eventually replaced by a strain of anti-essentialism that focused instead on gender differences and suggested that all art forms are open to multiple interpretations, all meanings are subject dependent. The entire foundation of interpretation and evaluation came under attack as feminists, in rejecting innate meaning, also rejected conventional modes of determining a work's intrinsic value.

The first collection of feminist art-historical essays, *Feminism and Art History: Questioning the Litany* (1982), edited by Norma Broude and Mary D. Garrard, sought to distinguish itself from catalogs and monographs by examining Western art history and the extent to which the discipline had been distorted in every major period by sexual bias. Similarly, Rozsika Parker and Griselda Pollock's pioneering work, *Old Mistresses: Women, Art, and Ideology* (1981), introduced strategies to subvert and collapse stereotypes about women's art by analyzing women's historical and ideological position within the world of art production. Hilary Robinson's anthology, *Visibly Female: Feminism and Art Today,* introduced Americans to British art and criticism in 1987. In Germany, Gisela Ecker's 1986 collection of essays, *Feminist Aesthetics,* sought to demarcate important terminological differences between *feminine* and *feminist.* French feminists coming from poststructuralist and psychoanalytic backgrounds rejected hierarchical binary oppositions typical of patriarchal thinking and sought to develop unique linguistic expressions of the *jouissance* of women's writing and thought; such authors included Julia Kristeva, Luce Irigaray, and Hélène Cixous.

In retrospect, feminist theorists such as Toril Moi (1985), Rita Felski (1989), and Janet Wolff (1990), note at least two clearly demarcated "traditions" established in less than twenty years: Anglo-American and French. Feminist theorizing continues to move rapidly beyond these two approaches as it expands to encompass global concerns among women while respecting differences. Although this task has been made increasingly more complex by issues of class, ethnicity, and sexual orientation, feminist inquiry has truly become a site of interdisciplinary intersection: of historical reclamation, linguistic analysis, sociological explanation, and philosophical questioning of the most basic, underlying assumptions of the traditions of a variety of fields.

Perhaps not surprisingly, the rhetoric of these inquiries is carefully formulated and subtle. French feminists have boasted of a unique form of women's writing—*l'écriture féminine;* they have subsequently been criticized for their elitism and lack of concern for the praxis of women's everyday hardships. Some feminist film theorists have sought to identify one single *feminine aesthetic* that constitutes a unified or coherent tradition while others such as Ismay Barwell concede only individual feminine *aesthetics.* There have been calls for a "nondominative aesthetic" by Josephine Donovan and "a positive female aesthetic" by Christine Battersby. But only some theorists, for example, Battersby, Barwell, Marilyn French, and Laurie Shrage, are comfortable extending these calls to a notion of a "feminist aesthetic" (see Hein and Korsmeyer, eds., 1993, for Barwell, Donovan, French, and Schrage). Finally, even the name and constitution of the entire enterprise are under suspicion; some feminists (e.g., Battersby and French) define their activity in terms of "feminist aesthetics" while others (e.g., Brand and Korsmeyer) prefer less ideologically loaded labels, such as "feminist theory," "gendered aesthetics," "aesthetics in feminist perspective," and "feminism in aesthetics." Rita Felski has gone so far as to argue against a

single feminist aesthetics, exploring what it means to move "beyond" feminist aesthetics:

> "the question of a feminist aesthetics, defined as a theory which would subordinate all aesthetic categories to the interests of feminist ideology, reveals itself to be something of a nonissue, a chimera which feminist critics have needlessly pursued."
>
> (Felski, 1989)

Speculations about the future of feminism and philosophical aesthetics have questioned their compatibility. In our supposedly postfeminist and postanalytic phase of aesthetics at the end of the twentieth century, suggestions for the future vary considerably. Joanne Waugh encourages feminists to break more severely with past aesthetic traditions; Joseph Margolis advises reconciling the two approaches while admitting others into the fold. Felski (1995) recommends transitioning to a "paraesthetics" that captures feminism's resistance to definition. As Hilde Hein (1995) notes, "Feminist theory is still in its infancy, and feminist aesthetic theory is only beginning to find itself." Her recommendation provides a vision of future theorizing that results in change that is both internal and external to aesthetics. She sees feminist aesthetics fulfilling a role *within* feminist theory by which women's experiences—and their expression through art—become fundamental for all theoretical speculation. She also envisions feminist aesthetics at the core of feminist theory gravitating away from "the margins" and toward the center of traditional Western philosophy. (See Brand and Korsmeyer, 1995, for Waugh, Margolis, Felski, and Hein.)

The majority of philosophers are slowly coming to realize the importance of women's contributions to the arts, in spite of encouragement from leaders in the field. Consider these remarks by Arthur Danto, art critic for the *Nation*: "were we to select the most innovative artists of the particular period, most of them would probably be women . . ." (1989); "the best performance artists are for some reason women . . ." (1987). In addition, Danto urges philosophers who seek solace in old-fashioned traditions to see that "feminism is neither a restricted topic nor solely of interest to a restricted audience" (see Brand and Korsmeyer, 1995).

The announcement of a postfeminist phase in aesthetics appears to have been somewhat premature.

BIBLIOGRAPHY

Battersby, Christine. *Gender and Genius: Towards a Feminist Aesthetics.* Bloomington, Ind., 1989.

Brand, Peggy Zeglin, and Carolyn Korsmeyer, eds. *Feminism and Tradition in Aesthetics.* University Park, Pa., 1995. Contains essays by Rita Felski, Hilde Hein, Joseph Margolis, Joanne Waugh, and a preface by Arthur Danto.

Broude, Norma, and Mary D. Garrard, eds. *Feminism and Art History: Questioning the Litany.* New York, 1982.

Danto, Arthur C. "Cindy Sherman." *Nation* (15 August 1987): 134–137.

Danto, Arthur C. "Women Artists, 1970–85." *Nation* (25 December 1989): 794–798.

Deepwell, Katy, ed. *New Feminist Art Criticism: Critical Strategies.* Manchester and New York, 1995.

Ecker, Gisela, ed. *Feminist Aesthetics.* Translated by Harriet Anderson. Boston, 1986.

Felski, Rita. *Beyond Feminist Aesthetics: Feminist Literature and Social Change.* Cambridge, Mass., 1989.

Hein, Hilde, and Carolyn Korsmeyer, eds. *Aesthetics in Feminist Perspective.* Bloomington, Ind., 1993. Contains essays by Ismay Barwell, Josephine Donovan, Marilyn French, and Laurie Shrage.

Korsmeyer, Carolyn. "Perceptions, Pleasures, Arts: Considering Aesthetics." In *Philosophy in a Feminist Voice: Critiques and Recollections,* edited by Janet A. Kourany. Princeton, N.J., 1998.

Moi, Toril. *Sexual/Textual Politics: Feminist Literary Theory.* London and New York, 1985.

Nochlin, Linda. "Why Have There Been No Great Women Artists?" In *Women, Art, and Power and Other Essays,* pp. 145–178. New York, 1988.

Parker, Rozsika, and Griselda Pollock. *Old Mistresses: Women, Art, and Ideology.* New York, 1981.

Robinson, Hilary. *Visibly Female: Feminism and Art: An Anthology.* London, 1987.

Wolff, Janet. *Feminine Sentences: Essays on Women and Culture.* Berkeley, 1990.

PEGGY ZEGLIN BRAND

Critique of Feminist Aesthetics

"Feminist aesthetics" does not seem a particularly useful concept for contemporary feminist theorists. This is not only because of the by now well-rehearsed problems of gender essentialism (the difficulties arising out of any attempt to formulate a global definition of women's language, women's art, etc.), but also because it typically involves a reductive approach to questions of aesthetics. The history, theory, and practice of art is entangled with questions of gender politics in multifarious, if often contradictory ways, but the nature of these entanglements is not usefully illuminated by arguments for a feminist aesthetic. Such arguments (e.g., Ecker, 1985; Battersby, 1989) usually rely on two assumptions: that existing art and aesthetic theory is an unequivocal expression of patriarchal ideology, and that feminism needs to develop its own autonomous criteria for the production, interpretation, and appreciation of art. Both claims are problematic in certain key respects.

The gender politics of traditional aesthetics can be addressed at a variety of levels: through analysis of particular artworks, through a critique of aesthetic theory and philosophy of art, and by developing feminist histories and sociologies of the art institution. Important work has been done at all three levels by feminist scholars who have ably documented the pervasive gender ideologies underpinning Western conceptions of art. In their influential book *Old Mistresses: Women, Art, and Ideology* (1981), Rozsika Parker and Griselda Pollock began to chart the histories of female artists in relation to the various institutional structures and

aesthetic theories that have often devalued or trivialized female creativity. Artistic genius has invariably been deemed a quintessentially masculine phenomenon. At the same time, the authors warn against a tragic feminist narrative that assumes women's inevitable and unilateral absence from the history of art, as well as an essentialist aesthetic that valorizes female creativity as radically and mysteriously other. Rather, they go on to argue, certain female artists achieved varying degrees of prominence and critical acknowledgment in different historical periods, even as their own creation and understanding of art were necessarily enmeshed within the dominant ideologies of both art and gender prevalent within a particular era. Women's art, while it may differ in certain, though not always foreseeable, ways from that of men, cannot be understood outside prevailing discourses of the aesthetic.

If women have not been completely "outside" the history of art, it is also true that art and aesthetics cannot be seen as the simple expression of a preexisting, monolithic maleness. Within the modern era, for example, artists have often assumed a dissonant or adversarial relationship to dominant social norms and values. From the late nineteenth century onward, this antagonism expressed itself not simply at the level of content, but in the turn toward experimental form and the creation of polyvocal, self-reflexive artworks that explicitly refused norms of utility and function associated with bourgeois masculinity. As a result, literary theorists such as Julia Kristeva (1984) have famously valorized the experimental art of a male avant-garde tradition as an exemplary site of resistance to a patriarchal symbolic order; the "revolution in poetic language" epitomized by writers such as Stéphane Mallarmé is identified with the eruption of a polysemic textuality associated with the feminine in its subversion of phallocentric structures of logic and identity.

Kristeva's argument is in many respects a problematic one for feminism, both in its attribution of an inherently gendered nature to particular forms of language (logic as masculine, experimental writing as feminine) and in its overestimation of the radical effects of artistic experimentation (Felski, 1989). Nevertheless, it is surely important to acknowledge the relative autonomy of modern art as a particular and differentiated set of practices with its own distinctive traditions, logics, procedures, and modes of representation. Of course, these practices do not exist outside of broader social structures and gender ideologies; thus male avant-garde artists of the fin de siècle who allied themselves with a notion of transgressive femininity often legitimated their own transvestism through a vilification of the natural embodiment of "real women." On the other hand, many of the aesthetically self-conscious, experimental techniques of representation developed by such artists have been taken over and used in new and provocative ways by contemporary feminist artists to challenge assumptions about gender. In this sense, women's art cannot be understood outside ex-

isting traditions of aesthetic vocabularies, techniques, and modes of representation, even as particular women will draw on, contest, or enrich that tradition in diverse ways.

In this regard, arguments for the autonomous nature of women's art have conventionally resorted to some generalized notion of distinctive female experience to substantiate their claims. Such appeals, however, rely on a host of problematic assumptions: first, that one can identify a universally shared kernel of femininity, a notion that has been powerfully challenged by those women whose lives, practices, and modes of self-understanding have been excluded from the norm of white, middle-class, heterosexual femininity governing mainstream feminism; second, that art can be assessed in terms of its accuracy in rendering that experience, a form of mimeticism that has been systematically undermined by recent theories of representation as well as by much of the practice of modern art; and third, that the goal of feminist aesthetics should be the expression of a distinctive femaleness, a belief questioned by those feminist artists and critics who see female identity as a problem rather than a solution, a mythology to be deconstructed rather than an underlying truth to be recovered and embraced.

This in turn raises the question of the relationship between female and feminist aesthetics, terms that are often confusingly elided in critical discussion. Unless one subscribes to the view that *any* form of female artistic expression is in and of itself oppositional, it becomes necessary to develop some account of the relationship between particular forms of art and the goals of a political ideology such as feminism. Attempts to construct such a relationship have tended to fall on either side of a form/content divide. Either the feminism of a text is defined in terms of its identifiable ideological content (in which case the whole question of the aesthetic as a distinctive mode of representation is effectively elided) or else a feminist aesthetic is equated with a particular set of formal techniques. Thus, at certain moments in the history of feminist criticism, realism has been identified as the most authentically political form for women, whereas modernist art has been deemed to serve patriarchal interests. At other moments, exactly the opposite has been true, as the impact of French feminist notions of *l'écriture féminine* and the more recent feminist interest in performance, transvestism, and parody has led to an almost exclusive focus on experimental, self-conscious artworks by women.

Attributing an inherent political meaning to particular artistic forms, however, is ahistorical, reductive, and unproductive for feminism (Felski, 1994). This is not to deny the related, more modest point, namely, that particular styles of representation have acquired distinctively gendered meanings within specific historical and cultural contexts. Thus, for example, literary realism and naturalism were often codified as masculine in the nineteenth century, particularly in

France, because of their commitment to a hard-nosed verisimilitude and their disdain for a "feminine" domain of melodrama and the sentimental. Yet in recent years women writers have appropriated and revitalized realist forms and genres to explore issues of female subjectivity, thereby in turn confounding totalizing models of literary periodization that seek to chart a unilinear, sequential, historical path from realism to modernism to postmodernism. Indeed, the variety of artistic forms and styles currently being used by female writers and artists to explore feminist concerns tends to destabilize any attempt to develop a single feminist "textual politics."

How, then, should feminist critics address the question of the relationship between art and politics? Not by attempting to create a single, all-encompassing aesthetic theory grounded in an abstract notion of the female psyche or of feminine textuality, but rather through careful analysis of the historically and culturally specific relationships between conceptions of the aesthetic, ideologies of gender, and women's cultural practices, relationships that are in turn informed by broader discursive and social relations. The 1970s and early 1980s, for example, saw an enormous outpouring of women's writing dealing with questions of female identity and the search for self, writings that often downplayed "literariness" in favor of a confessional and often didactic realism. These texts resulted from and contributed to a particular conception of the female self that was largely coterminous with the exigencies of second-wave feminism as an emerging social movement preoccupied with questions of identity politics. As the influence of feminism has disseminated into the academy and the domain of professional art practice, some feminists have developed critical vocabularies and representational strategies that challenge and subvert commonsense notions of female identity, including those harbored within mainstream feminism itself. Many of the most prominent contemporary feminist artists, writers, and theorists—figures as different as Barbara Kruger, Nancy Spero, Jeanette Winterson, Kathy Acker, Judith Butler, and Donna Haraway—share a common investment in aligning themselves with innovative forms of representation that will challenge, deconstruct, or even explode the category of "woman." Their intellectual and aesthetic frameworks and reference points are shaped as much by their familiarity with recent developments in poststructuralist theory and postmodernist art practice as by gender affiliation.

In other words, the relationship between "feminism" and "aesthetics" is necessarily imbricated not just by questions of class, race, and sexuality, but perhaps more pertinently by educational background, cultural capital, and professional status. There can be no single "feminist aesthetic" because women belong to different, socially stratified cultures characterized by different aesthetic preferences and vocabularies of interpretation. Moreover, the tradition of feminist aesthetics, with its typical focus on works of high art or on an idealized tradition of authentic female folk art (quilting, embroidery), has been singularly reluctant to deal with the texts of the mass media that constitute the primary cultural reference point for the majority of women. Recent work in feminist cultural studies, by contrast, has sought to rehabilitate such conventionally feminine popular forms as sentimental novels, soap operas, and romance fiction. While usefully enlarging our understanding of the semiotic richness of such forms and the complex processes through which they are interpreted and rendered meaningful by female audiences, such analyses often idealize the popular as a site of resistance in order to dismiss high culture as inherently elitist and patriarchal. Yet, this is to deny the critical potential of any feminist intervention in the realm of high art and to subscribe to a nostalgic populism that underscores rather than effaces the feminist critic's own relatively privileged position. Some of the most interesting recent feminist scholarship seeks to move beyond such high/popular binaries, for example, Lynda Nead's (1992) exploration of the complex intersections, affinities, and tensions between representations of the female body in the discourses of high art criticism, legal theory, and pornography.

The assumption that a particular politics will generate a particular aesthetics, once a mainstay of Marxist criticism, has become an increasingly untenable position. To refuse the notion of a feminist aesthetic is neither to negate feminism nor to abandon aesthetics, but merely to suggest that their conjuncture needs to be theorized differently.

BIBLIOGRAPHY

Battersby, Christine. *Gender and Genius: Towards a Feminist Aesthetics.* Bloomington, Ind., 1989.

Ecker, Gisela, ed. *Feminist Aesthetics.* London, 1985.

Felski, Rita. *Beyond Feminist Aesthetics: Feminist Literature and Social Change.* Cambridge, Mass., 1989.

Felski, Rita. "Why Feminism Doesn't Need an Aesthetic (and Why It Can't Ignore Aesthetics)." In *Feminism and Tradition in Aesthetics,* edited by Peggy Zeglin Brand and Carolyn Korsmeyer, pp. 431–445. University Park, Pa., 1995.

Felski, Rita. *The Gender of Modernity.* Cambridge, Mass., 1995.

Kristeva, Julia. *Revolution in Poetic Language.* Translated by Margaret Waller. New York, 1984.

Nead, Lynda. *The Female Nude: Art, Obscenity and Sexuality.* London and New York, 1992.

Parker, Rozsika, and Griselda Pollock. *Old Mistresses: Women, Art, and Ideology.* New York, 1981.

Shusterman, Richard. *Pragmatist Aesthetics: Living Beauty, Rethinking Art.* Oxford and Cambridge, Mass., 1992.

RITA FELSKI

Feminist Art History

Although art history has always been a discourse enriched by its necessary extensions into other fields, until recently, and often without regard for the roots in philosophy and so-

cial history that informed the texts of the founders, a tenacious commitment to a particular ideology has dominated most areas of research. With few exceptions, scholarship has explored the content of the object, its periodicity, and the particular signifying marks of its maker. The desire behind these practices is to read art objects as transparent documents of their time and as cultural signifiers of artistic genius viewed for their essential aesthetic attributes. Still at play are concepts rooted in Giorgio Vasari's *Lives of the Artists* written in the sixteenth century, the source for the analogy between the artist and God in respect to creation.

While an overview of the field uncovers many furrows, the primary exceptions to interpretations of the artist's style and the object's meaning reside in the work of those art historians who consider that made objects should be interrogated as the material products of particular social activities, or as textlike instances of the construction of cultural signs. The designation of certain things as "art" is now understood to be that kind of construction emerging in the discourse of the eighteenth century in response to the urge to separate spheres of knowledge from those considered to be about emotional responses. Critique of the discipline is primarily found in the writings of art historians who are informed by Marxism, semiotics, or postmodern critical theory, in rare cases blending aspects of these positions. Svetlana Alpers accomplished that in her seminal work *The Art of Describing* (1983). She exposed the Italic-centricity at the very core of art history and its history: narrative and text-based rather than visual, as she proposed was the case for Dutch art of the seventeenth century. Recently, the texts of Walter Benjamin have provided art historians with a model for the exploration of visual culture and visual studies transforming the discipline through that extension to other practices implied in its origins. The political philosopher and social theorist Susan Buck-Morss, in *The Dialectics of Seeing: Walter Benjamin and the Arcades Project* (1989), provides provocative evidence of this kind of enrichment. Rosalind Krauss develops an interesting critique of the "new" art history from the vantage point of semiotics in "Welcome to the Cultural Revolution" in *October* (Summer 1996), one of the leading journals of contemporary neo-Marxist and postmodern critique in the domain of the visual. The very notion of a field of study like "visual culture," however interdisciplinary the claims made for it, is already under scrutiny not least of all as a result of the new technologies that integrate sound, text, and image. Nonetheless, at this moment the radical crossing of the boundaries between separate disciplinary studies is beginning to recast feminist art history as a relation between theory and practice in the domain of the social and political. This direction is seeded by engagement with the work of contemporary feminist artists and critics in which issues of gender, sexuality, race, and class are centered. First, some reflections on the history of feminist art history.

A comprehensive overview of feminist art history up to 1987, "The Feminist Critique of Art History," appeared in the *Art Bulletin* of that year. The authors of the review, Thalia Gouma-Peterson and Patricia Mathews, found it necessary to discuss not only art history and theory but the production of art as well. It was immediately apparent that attempts to keep discourses apart, particularly in relation to a topic like feminism, were exceedingly difficult, if at all desirable. The necessity of constructing a new model for the discipline of art history as a result of crossing all these boundaries was noted, but not yet explored. It was also evident that the groundbreaking work related to feminist art activities was not then centered in art history, but in postmodern art activities and in French critical theory in its psychoanalytic form.

It is important to read this review in the light of what was considered feminist art history prior to 1987. The article gives the founding date for a feminist inquiry into art history as 1971, when Linda Nochlin published "Why Have There Been No Great Women Artists?" Nochlin has since become the prime exponent of "first-generation feminism." Gouma-Peterson and Mathews structure their survey of feminist art, art history, and critique in two parts based on a generational division. Their first section discusses the work of the first group of American feminists, who, like Nochlin, "were concerned with issues pertaining to the nature, evaluation, and status of female artistic production" (p. 329), while British feminists about the same time "arising from a Marxist ideology . . . have been politically active since the beginning of the movement" (p. 331) and emphasize radical action by women in a socialist, political platform. This was followed by a second generation in which English and Euro-American feminists were inspired by postmodern (for the most part French) critical theory (founded by men like Claude Lévi-Strauss, Roland Barthes, Jacques Lacan, Jacques Derrida, and Michel Foucault). Lisa Tickner, for example, defines the shift in emphasis in Britain from equal rights struggles in the sexual division of labor and a cultural feminism founded on the revaluation of an existing biological or social femininity to recognition of the processes of sexual "differentiation," the instability of gender positions, and the hopelessness of excavating a free or original femininity beneath the layers of patriarchal oppression (Tickner, 1988).

This direction was described by Gouma-Peterson and Mathews in the last sentence of their article as a world view that "goes beyond attention to women's issues to embrace a totally new consideration of the production and evaluation of art and the role of the artist" (p. 357). Tickner also divided her inquiry into national categories, rather than relying on a generation-based analysis; however, she considered only North American, British, and French writers (1988).

As noted above, the primary task undertaken by the first generation of feminist art historians in the 1970s and early

1980s was one of recuperation. A few women artists from the Renaissance on were added to the established male canon of recognized genius. An interesting consequence of this recuperative process for women artists was its recovery of many heretofore excluded male artists from the "museum basement." The few women who made it were still a minuscule part of the male-dominated list. Gouma-Peterson and Mathews question the very practice of recuperation, adding that the consequences of continued work in that vein is self-defeating for two reasons: because the preexisting structure remains unquestioned and, "since many of the same women artists have been repeatedly discussed, feminist art history has come dangerously close to creating its own canon of white female artists (predominantly painters), a canon that is almost as restrictive and exclusionary as its male counterpart" (p. 327). This canonization applies even when dealing with artists of color. A leading example is the sanctification of the Euro-Mexican artist Frida Kahlo through the production of books, articles, exhibitions, and memorabilia to the exclusion, until recently, of almost all other women artists from Latin America.

It was in Rozsika Parker and Griselda Pollock's *Old Mistresses: Women, Art, and Ideology* (1981) that Gouma-Peterson and Mathews located second-generation feminism with its emphasis on issues of gender formation and ideology and the difference in women's lives and perspectives in relation to the production of art. First-generation feminist art historians like Linda Nochlin were trained in accordance with the canonical texts by the German originators of modern art-historical methodology, or by their students. Nochlin's work on Gustave Courbet followed that tradition, which was rooted in philosophy and social history, but extended it to emphasize the political nature of his painting while also providing "misreadings" from a feminist perspective. The second generation turned to semiotics, film criticism, and French psychoanalytic theory in order to raise questions about imaging gender. The first generation were still practitioners of the Panofskyan enterprise in which style, iconography, and iconology dominated the mode of inquiry, and applied this to the woman artist and her art. The second generation, in erasing the boundaries between word and image, appropriated the discourse of the text. Less interested in notions about particular artists, they emphasized systems of representation (see, e.g., Krauss, 1985; Solomon-Godeau, 1991). Some of this work was indebted to the construction of the *feminine* in writings by French feminist theoreticians such as Julia Kristeva, Luce Irigaray, and Hélène Cixous, students of the texts of Sigmund Freud and Lacan. Cixous herself used that approach in an article written for a catalog of Nancy Spero's mixed-media work on the victimization and empowerment of women. Although these texts seemed to "universalize" representation through the absence of specificity and difference in women's experiences within the same as well as diverse cultures, they provided the agenda and the discourse for the present attention to these issues. For example, Pollock (1988) examines the reception of works of art by the female viewer as a central issue; that aspect of difference was not an essential concern in the 1980s.

Feminist art historians, as their male counterparts, usually work on artists and issues that relate to the past; the fifteenth through the seventeenth centuries and the nineteenth century are major sources for feminist scholarship. Mary D. Garrard's *Artemisia Gentileschi: The Image of the Female Hero in Italian Baroque Art* (1989) and Kathleen Adler and Tamar Garb's *Berthe Morisot* (1987) focus on the subjects and the manner of painting them that they consider to be particular qualities of the female artist. Eight of the twenty-nine contributions to Norma Broude and Barbara Buhler Lynes (eds.), *The Expanding Discourse: Feminism and Art History*, are on Renaissance and Baroque topics, and nine discuss nineteenth-century issues; and of the remaining twelve essays, only four concern contemporary artists or subjects. Two of these focus on women artists identified with modern art, often the wives of the modernist superstars, as in Anne Wagner's article on Lee Krasner and Barbara Buhler Lynes's on Georgia O'Keeffe.

A recent collection edited by Richard Kendall and Griselda Pollock, *Dealing with Degas: Representations of Women and the Politics of Vision* (1992), addresses feminist issues in relation to that nineteenth-century artist. There has been a continuing interest over the past twenty years in Edgar Degas's pictures of women as representations that interfere with the conventions established in the Renaissance. Degas has been singled out as the favorite painter of that period to praise for his daring images of women, or to castigate for his misogyny. Although some of the essays introduce interesting issues, the form of presentation—exhibition, symposium, publication of papers, and finally the requisite reviews that critique the papers—reinforces the usual system of featuring a master artist in a major exhibition.

As previously indicated, some of the important writings that have influenced revisionist views of art history can be found in neo-Marxist analysis of material culture indebted to the works of Benjamin and the Frankfurt School. The early writings of Meyer Schapiro are important, as are texts by Arnold Hauser, Albert Boime, and Robert L. Herbert, and the more radical writings of T. J. Clark and O. K. Werckmeister. For feminist art history, in addition to the aforementioned women art historians, Eleanor Tufts, Susan Rubin Suleiman, Patricia Mainardi, Moira Roth, Alessandra Comini, Natalie Kampen, Carol Armstrong, and Carol Duncan are among those whose writings are of particular importance (see Gouma-Peterson and Mathews for bibliography). Carol Duncan's "The MoMA'S Hot Mamas" (1993) decodes the placement of paintings in New York's

Museum of Modern Art that not only creates the story of "Modern Art," but demonstrates that the arrangement is a spatial narrative written over the bodies of women.

Outside of the discipline of art history itself, the writings of the critic and novelist John Berger, the sociologist Janet Wolff, and the film theoretician Laura Mulvey have been particularly influential. Several provocative writings prior to the 1990s have examined representations of women as the object of the male gaze, both related to and different from the interrogation of the gaze in film theory that takes Freud and Lacan as models for the discourse. These texts explore the role of women in the social class constructions of the nineteenth century. Griselda Pollock's work is again important in isolating examples of the victimization of women through representation and, in the case of Mary Cassatt, their modes of redress. Issues of Victorian misogyny and the bourgeois family structure, prostitution, and the emerging impact of public spaces for women occasioned, according to Bram Dijkstra, "the evolution of images of women from the nuns, madonnas and invalids of the mid-Victorian period to the vampires and man-eaters of the 1890s." This inquiry has more recently been directed to the twentieth century, and Maria Tatar's *Lustmord: Sexual Murder in Weimar Germany* (1995) is an important example. The focus has also turned toward the history and theory of contemporary media and popular culture. The film historian Anne Friedberg's *Window Shopping* (1993) is an example of a recent work inspired by Benjamin's Arcades Project. She locates contemporary modes of being, instanced by movement through shopping malls and multiplex cinemas, to the manner of viewing goods under the nineteenth-century arcades and in department stores, the latter designed for the female consumer. Susan Buck-Morss's *Dialectics of Seeing* grounds postmodern culture in those "ruins" of nineteenth-century material culture.

The emergence in the 1970s of performative and conceptually based art activities and of collaborative work in alternative spaces, or in the "streets" like the performances facilitated by Adrian Piper on the East Coast and Suzanne Lacy on the West Coast were often intrinsically theoretical. Similarly, artist-critics like Lucy Lippard, Martha Rossler, Adrian Piper, Coco Fusco, and Trinh T. Minh-ha have merged theory and practice in their studio work as well as in their critical writings. At the "cutting edge" of contemporary practices, electronic imaging, that connection is crucial, as is the role of the artist redefined as collaborator-facilitator. Art practices have also been responsive to the dialectic that is centered in the connections between image and text. As feminist art historians engage these contemporary practices and discourses they have of necessity pursued diverse areas of inquiry: sociology, anthropology, literary theory, and cultural history are prominent examples. This mode of inquiry has also influenced the direction of

studies about the past. An interesting illustration of this is provided by the British feminist art historian Annie E. Coombes in *Reinventing Africa: Museums, Material Culture and Popular Imagination in late Victorian and Edwardian England* (1994), in which she engages issues of colonialism through the interrogation of museum practices. For the most part, feminists in the United States have had an uneasy relationship with Marxism and until recently have turned to semiotics and French deconstruction and psychoanalytic theory. Craig Owens points out a crucial problem for feminists who are also historical materialists:

> Marxism privileges the characteristically masculine activity of production as the *definitively human* activity (Marx: men "begin to distinguish themselves from animals as soon as they begin to produce their means of substance"); women, historically consigned to the spheres of nonproductive or reproductive labor, are thereby situated outside the society of male producers, in a state of nature. . . . What is at issue, however, is not simply the oppressiveness of Marxist Discourse, but its totalizing ambitions, its claim to account for every form of social experience. But this claim is characteristic of all theoretical discourse, which is one reason women frequently condemn it as phallocratic.
>
> (Owens, 1983)

While the notions of work, production, and ideology are suspect, among feminists there seems to be little initiative at this time, mirroring the general culture, to explore a collective politics centered in issues about gender, sexuality, agency, community, collaboration, economic and cultural diversity, and racial exploitation. Some of these issues were central to first-generation feminist artists, and their formation of collaborative galleries created a system that supported the work of many women artists of renown, as well as often dedicating their exhibitions to political causes. With the exception of some artists and theoreticians of color and some in the gay and lesbian communities, the struggle for economic and cultural empowerment has been redirected to issues of personal history, bypassing even emended versions of the class struggle. The very definition of class has become the source of theoretical struggle that takes place in the academy.

Karen-Edis Barzman has admonished feminist art historians for not being essentially political or oppositional as are feminists working in other domains. She notes that most still practice their discipline in order to include women artists who have been systematically excluded from the canon. "Yet these gestures constitute an immanent critique of art history at best, leaving the field's most problematic assumptions and practices intact" (Barzman, 1994, p. 327). Her suggestions for the redirection of the enterprise start with the observation that "perhaps it is time to stop worrying about female artists and their objects, at least temporarily, in order to cast a critical eye on art history itself. . . . The question is: Why do feminists still spend so much time

talking about artists and their objects?" (p. 327). This query is a more radical formulation of the description of the task than in Gouma-Peterson and Mathews's view, although Barzman is in general accord with their prescription for the future. This direction has been labeled "third-generation feminism." Although issues that are of particular concern to women must still be studied, and neglected women artists given appropriate acknowledgment—the recent focus on the "cutting-edge" collages of Hannah Höch are a case in point—feminism must also take its place within the general concern for "the predicament of culture" that James Clifford has presented (1988).

As Barzman suggests, a work like the one on Degas, supposedly interventionist, follows traditional art-historical practices in relation to meaning "that must be uncovered and *correctly* interpreted, or . . . the operations of ideology in the dominant cultures' 'misreading' of this meaning must be exposed in feminist 'masterreadings,' presented rhetorically as though beyond refute" (p. 327). Her suggestion for the future of art history is to move from the centrality of object and producer to viewers' acknowledging their difference and the provisional nature of our constructed narratives. She presents with a radical reading of postmodernism in which "we *produce* meaning—*we* produce meaning—and the meaning we produce is partial, contingent, cannot be universalized" (p. 335).

Postmodern inquiry has been successful in providing a critique of traditional art-historical practices. The quest for difference in the production of meaning is particularly applicable to issues that are crucial for the work of women from diverse cultures and outside of the frame of "high art." Almost all the feminist art history writing mentioned here concerns art and artists whose heritage is Anglo-European. The concerns raised by women from other countries has been summarized by the Marxist literary theorist Gayatri Chakravorty Spivak: "I am suggesting, then, that a *deliberate* application of the doctrines of French High 'Feminism' to a different situation of political specificity might misfire" (Spivak, 1981, p. 141). Herself a product of these doctrines and in particular of Lacanian theory, Spivak questions the application of French feminism in the works of Cixous, Kristeva, and Irigaray, who posit an essential *feminine,* to the condition of women of all classes all over the world. Although Spivak has used essentialism, it has been in the service of "political analysis and mobilization" (Butler, 1993, p. 281). A critique of essentialism has also been a central concern for Lucy Lippard, particularly articulated in *Mixed Blessings: New Art in a Multicultural America* (1990), and in works of women of color such as the theoretician bell hooks and the performance artist and critic Coco Fusco, the latter of whom writes:

> At this historical moment, then, the postmodern fascination with the exchange of cultural property and with completely deraci-
> nated identity can seem for many people of color less like emancipation and more like intensified alienation. Instead, for many, the times demand what Gayatri Spivak has called "strategic essentialism," that is, a critical position that validates identity as politically necessary but not as ahistorical or unchangeable.
>
> (Fusco, 1995 27)

Can those aspects of postmodern practices pertinent to social and political activities remain effective at the same time that women from diverse cultures are beginning to form their individual modes of expression and postmodern discourse critiques the notion of the author-artist as the producer of a text-object? Many of these women, and in particular women of color, use traditional mediums and symbolic systems out of necessity as well as choice in order to situate themselves as subjects and agents empowered in their own cultures and in the ones to which they have immigrated (see Lippard, 1990, for the work of the late Ana Mendieta, Faith Rhinggold, Howardena Pindell, and Amalia Mesa-Bains, among others). Many women (e.g., Arbie Williams; see Leon, 1993) also introduce ways of producing art through skills often (erroneously) labeled "women's work," for example, weaving, quilting, and sewing. As Barzman insists, the answer may lie in understanding that the postmodern enterprise is complex; that it is anti ideology; that it exalts difference; and that it engages activity in the social and political domains.

> If traditional [art] history writing has been in a sense a process of collecting, it has also been a process of marginalizing, omitting. Still, it speaks *at* us, if not *to* us, with the authority of all discourses that seek to demonstrate cause-and-effect. . . . the new historicism is not necessarily a History of the Victors. It disrupts the option of the "major" event. It is anti-hierarchical and questions the narratives of chronology. It is cross-disciplinary: its most productive tools of analysis are originating in feminist literary-critical studies and in their rereading of psychoanalytic texts; in poststructural, sociolinguistic examinations of ideology construction and its operation through the political, cultural and social; and out of a recognition of the power of the image, its centrality in ideological formations and its usefulness in analyzing change and reformation.
>
> (Kruger and Mariani, eds., 1989, pp. ix–x)

Art history, feminist or otherwise, is an offspring of the project of modernity; postmodernity is now in the process of seeding its progeny. As the idea of "art" was a creation of modernism its time is also past, not because art is dead in the Dantoesque sense, having become philosophy or theory, but because creative activities are practiced in multifarious ways, by persons from diverse backgrounds and with many different agendas. Some feminist art historians have played a role in forming that emerging paradigm; like the death of the parents in many species it has done so at the cost of normative art history's probable extinction.

BIBLIOGRAPHY

Adler, Kathleen, and Tamar Garb. *Berthe Morisot.* Oxford, 1987.

Alpers, Svetlana. *The Art of Describing: Dutch Art in the Seventeenth Century.* Chicago, 1983.

Barzman, Karen-Edis. "Beyond the Canon: Feminist, Postmodernism, and the History of Art." *Journal of Aesthetics and Art Criticism* 52.3 (Summer 1994): 327–339.

Benjamin, Walter. *Illuminations.* Edited by Hannah Arendt, translated by Harry Zohn. New York, 1969.

Berger, John. *Ways of Seeing.* Reprint, New York, 1977.

Boime, Albert. *The Academy and French Painting in the Nineteenth Century* (1971). New Haven, 1986.

Brodsky, Joyce. "Only the End of Art." In *The Death of Art,* edited by Berel Lang. New York, 1984.

Broude, Norma, and Mary D. Garrard, eds. *The Expanding Discourse: Feminism and Art History.* New York, 1992.

Buck-Morss, Susan. *The Dialectics of Seeing: Walter Benjamin and the Arcades Project.* Cambridge, Mass., 1989.

Butler, Judith. *Bodies That Matter: On the Discursive Limits of "Sex."* New York and London, 1993.

Clark, T. J. *The Painting of Modern Life: Paris in the Art of Manet and His Followers.* New York, 1985.

Clifford, James. *The Predicament of Culture: Twentieth-Century Ethnography, Literature, and Art.* Cambridge, Mass., 1988.

Coombes, Annie E. *Reinventing Africa: Museums, Material Culture, and Popular Imagination in Late Victorian and Edwardian England.* New Haven, 1994.

Crimp, Douglas. *On the Museum's Ruins.* Cambridge, Mass., 1993.

Danto, Arthur C., "The End of Art." In *The Death of Art,* edited by Berel Lang. New York, 1984.

Dijkstra, Bram. *Idols of Perversity: Fantasies of Feminine Evil in Fin de Siècle Culture.* New York and Oxford, 1986.

Duncan, Carol. "The MoMA's Hot Mamas." In *The Aesthetics of Power: Essays in Critical Art History,* pp. 183–207. Cambridge and New York, 1993.

Friedberg, Anne. *Window Shopping: Cinema and the Postmodern.* Berkeley, 1993.

Fusco, Coco. *English Is Broken Here: Notes on Cultural Fusion in the Americas.* New York, 1995.

Garrard, Mary D. *Artemisia Gentileschi: The Image of the Female Hero in Italian Baroque Art.* Princeton, N.J., 1989.

Gouma-Peterson, Thalia, and Patricia Mathews. "The Feminist Critique of Art History." *Art Bulletin* 69.3 (September 1987): 326–357.

Hauser, Arnold. *The Social History of Art.* 2 vols. Translated in collaboration with the author by Stanley Goodman. New York, 1951.

Herbert, Robert L. *Impressionism: Art, Leisure, and Parisian Society.* New Haven, 1988.

hooks, bell. *Black Looks: Race and Representation.* Boston, 1992.

Kendall, Richard, and Griselda Pollock, eds. *Dealing with Degas: Representations of Women and the Politics of Vision.* London, 1992.

Krauss, Rosalind E. *The Originality of the Avant-Garde and Other Modernist Myths.* Cambridge, Mass., 1985.

Kruger, Barbara, and Phil Mariani, eds. *Remaking History.* Seattle, 1989.

Leon, Eli. *Arbie Williams Transforms the Britches Quilt.* Santa Cruz, Calif., 1993.

Lippard, Lucy R. *Mixed Blessings: New Art in a Multicultural America.* New York, 1990.

Lippard, Lucy R. *The Pink Glass Swan: Selected Essays on Feminist Art.* New York, 1995.

Lynes, Barbara Buhler. "Georgia O'Keeffe and Feminism." In *The Expanding Discourse: Feminism and Art History,* edited by Norma Broude and Mary D. Garrard, pp. 437–448. New York, 1992.

Moi, Tori. *Sexual/Textual Politics: Feminist Literary Theory.* London and New York, 1985.

Moi, Tori. *French Feminist Thought: A Reader.* Oxford and New York, 1987.

Mulvey, Laura. "Visual Pleasure and Narrative Cinema." *Screen* 16 (Autumn 1975): 6–18.

Nochlin, Linda. "Why Are There No Great Women Artists?" In *Women, Art, and Power and Other Essays.* New York, 1988.

Owens, Craig. "The Discourses of Others: Feminists and Postmodernism." In *The Anti-Aesthetic: Essays on Postmodern Culture,* edited by Hal Foster, pp. 57–82. Port Townsend, Wash., 1983.

Parker, Rozsika, and Griselda Pollock. *Old Mistresses: Women, Art, and Ideology.* New York, 1981.

Pollock, Griselda. *Vision and Difference: Feminism, Feminity, and the Histories of Art.* London and New York, 1988.

Schapiro, Meyer. *Romanesque Art.* New York, 1977.

Solomon-Godeau, Abigail. *Photography at the Dock: Essays on Photographic History, Institutions, and Practices.* Minneapolis, 1991.

Spivak, Gayatri Chakravorty. "French Feminism in an International Frame." In *Other Worlds: Essays in Cultural Politics.* New York and London, 1987.

Tatar, Maria. *Lustmord: Sexual Murder in Weimar Germany.* Princeton, N.J., 1995.

Tickner, Lisa. *The Spectacle of Women: Imagery of the Suffrage Campaign 1907–1914.* Chicago, 1988.

Werckmeister, O. K. *Citadel Culture.* Chicago, 1991.

Wolff, Janet. *Feminine Sentences: Essays on Women and Culture.* Berkeley, 1990.

Wolff, Janet. *The Social Production of Art.* 2d ed. New York, 1993.

JOYCE BRODSKY

FICTION. [*To explore the concept of fiction in the history of aesthetics, this entry comprises three essays:*

Historical and Conceptual Overview
Contemporary Literary Account of the Fictive
Epistemology of Fiction

The first essay explains the various philosophical issues (realism, truth, etc.) that arise in the history of the analysis of literary fictions. The second essay presents a contemporary account of the understanding of fiction within literary theory. The final essay returns to some of the epistemological (theory of knowledge) issues that fictions provoke (specifically, whether we can acquire knowledge from fictions). For further discussion, see Epistemology and Aesthetics; Imagination; Literature; Narrative; Play; Realism, *article on* Realism and Aesthetics; Representation, *overview article; and* Truth.]

Historical and Conceptual Overview

Whether in philosophy of law, philosophy of science, or metaphysics and epistemology, fictions continue to prompt much philosophical reflection today. Discussion remains most lively, however, in the philosophy of art. Still, contemporary reflection on fictions has yet to take full critical account of the long history of philosophical discussion of fictions.

The etymological sense of the English word *fiction* comes through the French from the Latin past-participle root,

fict-, of the transitive verb, *fingere*. Although this word's range of reference is wide—to fashion, form, conceive, contrive, or feign—the basic sense here is that of an activity rather than that of a product.

By contrast, contemporary uses of *fiction* generally stress the thing as well as the action. Thus, a fiction can be either an imaginatively invented narrative, statement, idea, or the action of imaginatively inventing such things. More specifically, contemporary uses of "fiction" usually refer either to literary matters (novels or stories), conventionally accepted pretenses in, for example, social situations (polite fictions), or to some elements of legal reasoning (legal fictions). The latter uses include suggestions of something false, untrue, or not genuine that are nonetheless conventionally acceptable. Recently, some literary scholars and philosophers have used the word *fictionalization* to particularize both the process and the product of constructive versions of actual events and entities. Before returning briefly to the problems with fictions today, we need to look beyond our contemporary preoccupations to several of the very different ways fictions have been considered problematic in the history of philosophy.

European reflection on fictions begins mainly in the Homeric representations of the effects of bardic singing on heroic characters. Philosophically, Greek reflection focuses on forms of argument, especially Aristotle's key notion of *eks hypotheseos* (for instance, in *Metaphysics* 1004a13), and Stoic reflection on both incorporeal nonexistent things (Seneca on "giants" and "centaurs" in *Letters* 58.13–15) and incorporeal subsistent things (Galen on the notion of *huphistasthai* in *On Medical Method* 10, 155). Roman rhetoric, especially Quintilian's *Institutio oratorio,* explores the vague notion of *hypothesis* in the context of arguments that posit something as true that, if it were true, would either resolve or be useful in resolving a disputed issue. Similarly, Roman jurisprudence, especially Gaius's *Institutio juris civilis,* develops the notion of fictions as hypotheses that, in the context of particular legal situations, can allow some otherwise insoluble matters to be adjudicated equitably.

Regardless of late medieval discussions of, for example, chimeras in William of Ockham (*Expositio in librum Perihermenias Aristotelis,* Prooemium no. 7) or fictional entities in Francisco Suarez (*Disputatio metaphysicae* 31.2.2), the key notion of fictions as hypotheses is subjected to reexamination in the early modern period. Isaac Newton distinguishes in the domain of physics between hypotheses and fictions. And in mathematics Gottfried Wilhelm Leibniz introduces a further distinction between positive fictions like imaginary numbers that contribute to the solution of mathematical problems and negative fictions like formless first matter that have no connection with the nature of things. In the eighteenth century these distinctions are still active in Alexander Gottlieb Baumgarten's aesthetics, in still clearer form in the

work of Baumgarten's successor, Georg Friedrich Meier, and, later, in the scattered reflections of Jeremy Bentham.

The major modern transformations and systematizations of the classical reflection on fictions however are twofold. Besides the radical work of Newton and Leibniz, Immanuel Kant introduces a second fundamental rearticulation. Already in the *Critique of Pure Reason* (B 799ff.), Kant is at work on specifying the concepts of reason as "heuristic fictions" or, in the more familiar phrase, as "regulative principles" that are required for the systematic articulation of the uses of the understanding and of the unity of reason itself. In his later work, however, in the *Critique of Judgment* and especially in the *Opus postumum,* Kant tries to rethink his earlier reflection in terms of a typology of ideas and ideals of reason.

This late work in turn stimulates further developments in neo-Kantian circles. Rudolf Hermann Lotze, for example, in his 1843 *Logik,* while returning to the Newtonian preoccupation with distinguishing hypotheses from fictions construes fictions as "assumptions" that are made in full awareness of their not holding in the actual world. Much later, in his 1902 work, *Über Annahmen,* Alexius Meinong will extensively develop this notion of "assumptions." Roughly around the same time, from 1902 to 1906 in his correspondence with Anton Marty, Franz Brentano will explore in great detail the "nonreal" status of mental objects including fictions. In his attempts to systematize neo-Kantian reflection on fictions under the Kantian notion of "as if," Hans Vaihinger in his 1923 book, *The Philosophy of As If,* will take "fiction" as what some would call today a "term of art," that is, a subjective formulation that unlike hypotheses is an element of several kinds of methodical investigation rather than a formulation of the probable results of an investigation. This stress on the practical and instrumental value of fictions was also to be pursued independently in the domains of psychology and education by Giorgio Marchesini in Italy.

With these historical contexts in view we need now to look at some elements in contemporary reflections on fiction. If we start with the distinction between fictions as products and fictions as processes, a first set of contemporary philosophical issues appears. As a product a fiction is some kind of thing. But of precisely what kind? One reply is that a fiction, say the character Emma Bovary, is a thing that does not exist in the actual spatiotemporal world. Are we then to suppose, however, that, besides the existing entities comprising the actual world, other nonexisting entities comprise other worlds? But such suppositions today remain controversial.

Perhaps a fiction is a thing that does not exist but could exist. But, instead of a proliferation of nonexistent, possible entities, we now have a problem with accounting for those fictions, say square circles, that are impossible, that is, enti-

ties with contradictory properties. Are fictions then to be construed as, say, quarks are, as theoretical entities? But many fictions appear to be concrete particulars, whereas theoretical entities are abstract. Perhaps a fiction is an ontological type only of which some actual entities are tokens. But, as we have noted, many fictions lack nonfictional counterparts. Now, just as there is no problem with having an idea of something that does not exist, there may be no problem with postulating types that have no existing tokens. But when we are specifically concerned with fictions as types, it cannot be assumed without careful argument that there are fictional types without fictional tokens.

Philosophers of course have devised strategies to deal with the simple objections here. But these strategies have given rise to more complicated objections. The result is that saying what fictions as products are remains quite difficult. Similar difficulties arise in articulating what fictions are as processes. To see this general difficulty one needs only to examine just why almost all answers to questions such as "What is the opposite of the fictional?" or "How do we quantify over fictions?" remain deeply problematic.

One approach to resolving such difficulties has been to focus on sentences that include fictions as nouns with a view toward more precisely establishing two things, the referent of such expressions and the truth or falsity of assertions that incorporate such expressions. In the first case, the problem has been to distinguish between two kinds of objects. On the one hand, the referent of a fictional expression might be taken as a nonactual object of some sort that subsists, that might exist (i.e., that is not impossible), but will never exist in the sense of being a spatiotemporal segment of the one actual world. Emma Bovary will never exist. On the other hand, the referent might be taken as a possible object of some sort, that is, an object that subsists, that might exist, and that may in fact sometimes become part of the one actual world. The peculiar mental state of Emma's quite particular boredom with Charles may one day be exactly the kind of boredom Louise Colet will come to experience with Flaubert. But, however thoroughly developed, neither alternative so far has found consensus.

In the second case, the truth or falsity of sentences incorporating fictional expressions, similar debates continue. The critical issues here seem to turn on whether using such expressions commits one to asserting the existence of their problematic referents in some strong sense. Could one use such expressions without necessarily affirming the existence of an entity to which the subject of a sentence apparently refers? If yes, then exactly what would the appropriate theory of reference look like in detail? If no, then what other distinctions than the shaky one here between weak and strong senses are required to deal effectively with the putative truth or falsity of such ordinary utterances as "Just like Emma, Diana was monumentally bored with Charles."

When we look back over both historical and contemporary discussions about fictions, including some explored in the works cited in the bibliography below, we may observe how some central aspects of the fictional require further reflection. These aspects include questions about fictions as sources of knowledge and rational belief, as capable of generating genuine emotional involvement, as heuristic resources, and as capable of exercising a formative influence on the rationalization of certain actions. It remains to be seen whether philosophers today will continue to pursue these larger concerns with the still problematic uses, nature, kinds, and roles of fictions.

BIBLIOGRAPHY

Chisholm, Roderick M. "Fictitious Objects." In *A Realistic Theory of Categories: An Essay on Ontology*, pp. 121–123. Cambridge and New York, 1996.

Currie, Gregory. *The Nature of Fiction*. Cambridge and New York, 1990.

Iser, Wolfgang. *The Fictive and the Imaginary: Charting Literary Anthropology*. Baltimore, 1993.

Lamarque, Peter, and Stein Haugom Olsen. *Truth, Fiction, and Literature: A Philosophical Perspective*. Oxford, 1994.

Loetzsch, F. "Fiktion." In *Historisches Wörterbuch der Philosophie*, edited by Joachim Ritter et al., vol. 2, pp. 951–953. Basel, 1972.

McCormick, Peter J. *Fictions, Philosophies, and the Problems of Poetics*. Ithaca, N.Y., 1988.

Parsons, Terence. *Nonexistent Objects*. New Haven, 1980.

Proudfoot, Diana. "Fictional Entities." In *A Companion to Aesthetics*, edited by David Cooper, pp. 152–155. Oxford and Cambridge, Mass., 1992.

Walton, Kendall L. *Mimesis as Make-Believe: On the Foundations of the Representational Arts*. Cambridge, Mass., 1990.

Wilkenson, Jennifer. "How Not to Do Things with Fictional Discourse." *South African Journal of Philosophy* 14 (1995): 162–167.

Woods, John. *The Logic of Fiction: A Philosophical Sounding of Deviant Logic*. The Hague, 1974.

PETER MCCORMICK

Contemporary Literary Account of the Fictive

Most people associate the term *fiction* with the storytelling branch of literature, but in its other guise it is what Samuel Johnson called "a falsehood; a lye." The equivocalness of the word is revealing, for each meaning sheds light on the other. Both meanings entail similar processes, which we might term "overstepping" what is: the lie oversteps the truth, and the literary work oversteps the real world that it incorporates. It is therefore not surprising that literary fictions were so often branded as lies, since they talk of what does not exist, even though they present its nonreality as if it *did* exist.

Plato's complaint that poets lie met its first strong opposition in the Renaissance, when Sir Philip Sidney rejoined that "the Poet . . . nothing affirmeth, and therefore never lieth," since poets do not talk of what is, but of what ought to

be, and this form of overstepping is quite different from lying. Fiction and fictionalizing entail a duality: the liar must conceal the truth, but the truth is potentially present in the mask disguising it. In literary fictions, existing worlds are overstepped, and although they are individually still recognizable, they are set in a context that defamiliarizes them. Thus, both lie and literature contain two worlds: the lie incorporates the truth and the purpose for which it must be concealed; literary fictions incorporate an identifiable reality that is subjected to an unforeseeable refashioning.

Even if nowadays literary fictions are no longer charged with lying, they are still stigmatized as being unreal, regardless of the vital role fictions play in our everyday lives. In *Ways of Worldmaking* (Indianapolis, 1978), Nelson Goodman shows that we do not live in one reality but in many, and each of these realities is the result of a processing that can never be traced back to "something stolid underneath." There is no single underlying world, but instead we create new worlds out of old ones in a process that Goodman describes as making "fact from fiction." Fictions, then, are not the unreal side of reality, let alone the opposite of reality, which our "tacit knowledge" still takes them to be; they are, rather, conditions that enable the production of worlds whose reality, in turn, is not to be doubted.

This may be one of the reasons why we cannot talk of fiction as such, for it can only be described by way of its functions, that is, the manifestations of its use and the products resulting from it. This is evident even to cursory observation: in epistemology we find fictions as presuppositions; in science they are hypotheses; fictions provide the foundations for world pictures; and the assumptions that guide our actions are fictions as well. In each of these cases, fiction has a different task to perform: with epistemological positing, it is a premise; with the hypothesis, it is a test; with world pictures, it is a dogma whose fictional nature must remain concealed if the foundation is not to be impaired; and with our actions, it is anticipation. Since fictions have such manifold applications, we might well ask what they appear to be like, and what they reveal in literature.

The literary text is a mixture of reality and fiction, and as such it brings about an interaction between the given and the imagined. Because this interaction produces far more than just a contrast between the two, we might do better to discard the old opposition of fiction and reality altogether, and to replace this duality with a triad: the real, the fictive, and the imaginary. Undoubtedly, the text is permeated by a vast range of identifiable items selected from social and other extratextual realities. The mere importation into the text, however, of such realities—even though they are not being represented in the text for their own sake—does not ipso facto make them fictive. Instead, the text's apparent reproduction of items within the fictional text brings to light purposes, attitudes, and experiences that are decidedly *not*

part of the reality reproduced. Hence, they appear in the text as a product of a fictionalizing act.

Because this act of fictionalizing cannot be deduced from the reality repeated in the text, it clearly brings into play an imaginary quality that does not belong to the reality reproduced in the text but cannot be disentangled from it. Thus the fictionalizing act converts the reality reproduced into a sign, simultaneously casting the imaginary as a form that allows us to conceive what the sign points to. As this transformation is brought about by the fictionalizing act, the basic quality of this act begins to emerge: it is a crossing of boundaries. Two distinct processes are set in motion by the act of fictionalizing. Reproduced reality is made to point to a "reality" beyond itself, while the imaginary is lured into form. In each case there is a crossing of boundaries: The determinacy of reality is exceeded at the same time that the diffuseness of the imaginary is called into form. The fictionalizing act leads the real to the imaginary and the imaginary to the real; it thus conditions the extent to which a given world is to be transcoded, a nongiven world is to be conceived, and the reshuffled worlds are to be made accessible to the reader's experience.

As the product of an author, the literary text evidences a particular attitude through which the author directs himself or herself to the world. As this attitude does not exist in the world to which the author refers, it can take on a form only by being literally inserted into the real world. The insertion takes place not by plain mimesis of existing structures but by a process of restructuring them. Every literary text inevitably contains a selection from a variety of social, historical, cultural, and literary systems that exist as referential fields outside the text. This selection is itself a stepping beyond boundaries, in that the elements selected are lifted out of the systems in which they fulfill their specific function. This applies both to cultural norms and to literary allusions, which are incorporated into every new literary text in such a way that the structure and the semantics of the systems concerned are decomposed. The elements that are now incorporated into the text are not in themselves fictive, but their selection is an act of fictionalizing.

A complement to the act of selection is the act of combination. The different elements that are combined within the text range from words and their meanings through encapsulated extratextual items to the patterns in which characters and actions are organized. Combination, too, is an act of fictionalizing, with the same basic mode of operation: the crossing of boundaries.

On the lexical level this is to be seen, for instance, with neologisms such as James Joyce's coining of the term *benefiction*, which combines the words *benefaction, benediction,* and *fiction.* Here lexical meanings are used to derestrict semantic limitations. The lexical meaning of a particular word is faded out and a new meaning faded in, without the

loss of the original meaning. This establishes a figure-and-ground relationship allowing both the separation of the individual elements and a continuous switching of the perspective between them. In accordance with whichever reference forms the foreground or background, the semantic weighting will be shifted.

A second level of relating is to be seen in the organization of specific semantic enclosures within the text. These give rise to intratextual fields of reference themselves brought about by the relationship between and among external items encapsulated in the text. Generally, these fields of reference provide an occasion for the hero to step over their boundaries. Such boundary crossing is a subject-creating event. It is a "revolutionary element" because it breaks down "accepted classifications."

Fictions also play vital roles in the activities of cognition and behavior, as in the founding of institutions, societies, and world pictures. Unlike such nonliterary fictions, the literary text reveals its own fictionality. Because of this, its function must be radically different from that of related activities that mask their fictional nature. The masking, of course, need not necessarily occur with the intention to deceive; it occurs because the fiction is meant to provide an explanation, or even a foundation, and would not do so if its fictive nature were to be exposed. The concealment of fictionality endows an explanation with an *appearance* of reality, which is vital, because fiction—as explanation—functions as the constitutive basis of this reality.

When a fiction signals its own fictionality, it necessitates an attitude different from that adopted toward fictions hiding their fictionality. The incorporated "real" world is, so to speak, placed in brackets to indicate that it is not something given but is merely to be understood *as if* it were given. In the self-disclosure of its fictionality, an important feature of the fictional text comes to the fore: it turns the whole of the world organized in the text into an "as-if" construction. In the light of this qualification (implicitly accepted the moment we embark on our reading), it is clear that we must and do suspend all natural attitudes adopted toward the "real" world once we are confronted with the represented world. This is not present in the text for its own sake, nor is its function exhausted merely by its denoting reality. Just as the incorporated "real" world is bracketed off, so too are our natural attitudes.

Whenever bracketing occurs, a purpose makes itself felt that can never be a property of the world represented, not least because the represented world is built out of a selection of items from the world outside. In this overarching purpose the pragmatic function of the fictional work is adumbrated, for fictions are inextricably tied to their use. The reality represented in the text is not meant to represent reality; it is a pointer to something that is not given, although its function is to make that something conceivable.

Thus the purpose of the self-disclosing fiction comes to light. If the world represented is not meant to denote a given world, and hence is turned into an analogue, it may serve two different purposes at once. The reaction provoked by the represented world could be directed toward conceiving what it is meant to "figure forth." The analogue, however, could simultaneously direct the reaction to the empirical world from which the textual world was drawn, allowing this very world to be perceived from a vantage point that has never been part of it. In this case the reverse side of things will come into view. The duality of the analogue will never exclude either of the two possibilities; in fact, they appear to interpenetrate, making conceivable what would otherwise remain hidden.

All the acts of fictionalizing that can be distinguished within the fictional text are acts of boundary crossing. Selection transgresses the limits of extratextual systems as well as the boundaries of the text itself by pointing to the referential fields that link the text to what is beyond the page. Combination transgresses the semantic enclosures established by the text, ranging from the derestriction of lexical meanings to the hero's infringement of strictly enforced borderlines. Finally, the "as-if" construction discloses the fictionality of fiction, thus transgressing the represented world set up by the acts of selection and combination. It brackets off this world, thereby indicating that it is to be used for an unverbalized, though overarching, purpose.

The acts of fictionalizing can be clearly distinguished by the different gestalt each of them brings about: selection results in intentionality; combination results in relatedness; and self-disclosure leads to bracketing. All these cases are "facts from fiction," as none of them is inherent in what it refers to, nor identical to the imaginary.

Furthermore, the various acts of fictionalizing carry with them whatever has been outstripped, and the resultant doubleness might therefore be defined as the simultaneity of the mutually exclusive. This formula allows for describing the structure of the fictional component of literature. It gives rise to a dynamic oscillation resulting in a constant interpenetration of things that are set off from one another without ever losing their difference. The tension ensuing from the attempt to resolve this ineradicable difference creates an aesthetic potential that, as a source of meaning, can never be substituted by anything else. This does not imply that the fictional component of literature is the actual work of art. What it does imply is that the fictional component makes the work of art possible.

BIBLIOGRAPHY

Anderegg, Johannes. *Fiktion und Kommunikation: Ein Beitrag zur Theorie der Prosa.* Göttingen, 1973.

Assmann, Aleida. *Die Legitimität der Fiktion: Ein Beitrag zur Geschichte der literarischen Kommunikation.* Munich, 1980.

Cebik, L. B. *Fictional Narrative and Truth: An Epistemic Analysis.* Lanham, Md., 1984.

Costa Lima, Luíz. *The Dark Side of Reason: Fictionality and Power.* Translated by Paulo Henriques Britto. Stanford, Calif., 1992.

Henrich, Dieter, and Wolfgang Iser, eds. *Funktionen des Fiktiven.* Munich, 1983.

Iser, Wolfgang. *The Fictive and the Imaginary: Charting Literary Anthropology.* Baltimore, 1993.

Kasics, Kaspar. *Literatur und Fiktion: Zur Theorie und Geschichte der literarischen Kommunikation.* Heidelberg, 1990.

Keller, Ulrich. *Fiktionalität als literaturwissenschaftliche Kategorie.* Heidelberg, 1980.

Kermode, Frank. *The Sense of an Ending: Studies in the Theory of Fiction.* New York and Oxford, 1967.

Martinez-Bonati, Félix. *Fictive Discourse and the Structures of Literature: A Phenomenological Approach.* Translated by Philip W. Silver with the author. Ithaca, N.Y., 1981.

Merrell, Floyd. *Pararealities: The Nature of Our Fictions and How We Know Them.* Amsterdam and Philadelphia, 1983.

Pavel, Thomas G. *Fictional Worlds.* Cambridge, Mass., 1986.

Prado, C. G. *Making Believe: Philosophical Reflections on Fiction.* Westport, Conn., 1984.

Riffaterre, Michael. *Fictional Truth.* Baltimore, 1990.

Roberts, Thomas J. *When Is Something Fiction?* Carbondale, Ill., 1972.

Rockwell, Joan. *Fact from Fiction.* London, 1974.

Schlaffer, Heinz. *Poesie und Wissen.* Frankfurt am Main, 1990.

Searle, John. "The Logical Status of Fictional Discourse." *New Literary History* 6 (1975): 319–332.

WOLFGANG ISER

Epistemology of Fiction

Since the time of David Hume, it has been thought obvious that fictional narrative evokes a different response in the reader than historical narrative. Hume says that we read history with a cognitive attitude of belief, an attitude attributable to the fact that persons and events uncreated and uninfluenced by our wills are the sources of the reading experience. [*See* Hume.]

Hume's claim that fiction construed as such does not induce beliefs might seem to settle the case against the possibility of acquiring knowledge from fiction. For if fictions do not even produce beliefs, they assuredly cannot, in standard theories, produce the justified or otherwise warranted true belief that is knowledge. Even if fiction does produce some beliefs, it might be argued, these beliefs concern only what fictionally happens to fictional characters in fictional worlds, and these events have no determinate relationship to what happens in our world. There are methods by means of which one finds out what our world is like, methods involving controlled observation, experimentation, and statistical analysis. Novel reading is not another such method.

Nevertheless, it is to literary fiction—and more recently to film and television—that people turn in order to express and communicate their views about life and the world and in order to learn. In eighteenth-century Germany, the novel gradually displaced didactic poetry, the sermon, and the advice-giving magazines after 1740, and lonely young readers everywhere have always devoured novels in the hope of finding insight into the problems of how to love and how to leave, how to determine the limits of parental authority, and how to choose a modus vivendi. Have they thereby refuted Hume, or do they fail to grasp that what they are reading is only fiction? As Paul Veyne has noted, the ancients did not distinguish systematically between history and fable: the tradition was there and it was truth, that was all, and contemporary consumers of books, films, fiction, and daytime serials may appear to exhibit the same sublime indifference (1988). Indeed, the real world as we experience it is permeated by fictions of our own authorship—by scenes of fantasy, hope, and fear that reside in an indeterminate region between what we expect to happen, really, and what we hope or fear may or will happen. Our own privately concocted fictions and quasi fictions appear to be indispensable to our thinking and planning and so to our cognitive functioning. As scientists perform thought experiments to generate knowledge of the physical world, so ordinary people might seem to use their capacities to envision and imagine to generate knowledge of the human world.

Although all this may be conceded, it is still far from clear that a stranger's imaginings can give one anything worthy of the name of knowledge. The question whether literary art in the public realm can be a source of knowledge has been discussed since Plato's time. Three main positions have been taken: (1) Fictional works make a negative contribution to our knowledge of the world, including awareness of moral truths, arousing beliefs and expectations that are inaccurate and distorted; (2) fictional works are able to make a positive contribution to our moral, psychological, and historical knowledge of manners, mores, and political events; and (3) fictional works are a source of ideas and exert a powerful influence on behavior, but they cannot be supposed to provide knowledge in any important sense.

Plato, Augustine, and Jean-Jacques Rousseau may be counted among the proponents of the view that the relationship of literature to knowledge, specifically to moral knowledge, is a negative one. "Poetry maims the thought," Plato decides in book 10 of *The Republic*. Fictions are thrice removed from reality as copies of copies of Forms, and they "keep company with that part of the soul that is not the best." By this Plato means that a preoccupation with fictional entities supplants the search for true knowledge, and that attention to fictional persons—the Homeric heroes, with their displays of unmanly weakness and emotionality—provokes imitation of inappropriate models. [*See* Plato.] Augustine, who wanted to live in the truth, found that the theater provoked only tears and lust, and Rousseau argues that the illusions of the stage wrongly glamorize the situations and actions of monstrous women like Phaedra and Medea, and that moralistic plays in which the virtuous are rewarded and the wicked punished fail to make an impression on the audience: "[A]s a result of showing them that we want to in-

struct them, we no longer instruct them" (Rousseau, 1968). The early novel, especially the "French" novel with its encouragement to *galanterie* and to an active social role for women, was attended by warnings and reproaches from clergymen. Indeed, Immanuel Kant contrasted the effeminacy and amorality of a novel-reading society with sublime and bibliophobic ideals of virtue. Sigmund Freud extends the Plato-Rousseau analysis by arguing that fictional works are public versions of private fantasies of violence and eroticism. They deliver libidinal gratification to the reader within the framework of a moralistic story line that is unfolded with technical and stylistic expertise. The framework mitigates the guilt and disgust that would otherwise attend the sharing of forbidden fantasies. Readers believe themselves edified—but that is not why they read.

Freud's theory is perhaps hard pressed to account for contemporary popular fiction, which is highly explicit in its depictions of violence and eroticism. This literature, however, tends to be loaded with factual and pseudofactual information about the financial and glamour industries, politics, spying, medicine, weapons, boats, and human geography, suggesting that aesthetics and didacticism continue to play their mitigating role. Latter-day Platonists insist nevertheless that consumption of popular fiction can lead to personal moral degradation, to misogyny or misanthropy, and to a loss of sympathy and humanity. Romance novels as well have been attacked as engendering delusive representations and expectations in women that prevent them from adjusting to reality. Such fictions, it is argued, do not provide a temporary respite from reality, but actually substitute for it with detrimental consequences, precisely as the older critics charged.

Since the rise of literary criticism, however, generations of critics have dedicated themselves to defending the value of fiction as a source of knowledge. Their orientation may be *representational,* in case they regard fiction as containing philosophical and moral truths of propositional form, which the reader by analysis and reflection can find in the text. Or it may be *practical,* in case they regard fiction as producing not knowledge *that* but knowledge of *how* or knowledge of what certain experiences are *like.* Praxis-oriented critics argue that fiction has a morally and psychologically ennobling or enabling effect on the reader through its creation of certain states of awareness, receptivity, and consciousness of alternatives.

Both representation- and praxis-based accounts suggest that reading a novel is rather like sitting next to a voluble stranger on a long train trip: although you take in only one person's admittedly partial and unverified view of the world, you go away edified, or so it seems. But what have you in fact learned? The futility of worldly ambition, the vanity of love, the inevitability of conflict between parents and children, the possibility of reversals of fortune, the volatility of secrets, the treachery of men, and the costs of naïveté are common themes of novels. But recognizing that such and such is the theme of a novel is not the same as recognizing that the corresponding proposition is true. Fortunately, closer arguments for a positive connection between literature and knowledge are available. The following discussion provides some examples.

1. Fiction writers are typically close observers of human affairs who themselves possess a high level of psychological and moral awareness. Thus, knowledge is worked into the text and retrievable from it, even in the absence of a specific didactic intention.

2. Fiction acquaints us with the first-person experience and thought patterns of those who are radically other to us: members of the opposite sex, of other cultures, of different ages and stations.

3. Recognizing, as Aristotle said, is a kind of learning attended by pleasure (*Poetics* 1448b). Recognizing fictional types—the shallow suitor, the vain wife, the miser, the unscrupulous politician—is a cognitive activity and one that equips readers to identify unsavory and attractive characteristics in themselves and others.

4. Reading a fictional work or otherwise engaging with one mobilizes the reader's empathetic reactions. The strengthening of these reactions develops the response of sympathy, which is the basis of moral feeling.

5. The dialectical structure of fictional works in which opposing values clash in a way finally decided by the outcome encourages the reader's development of inner deliberative resources, including the important moral ability to see both sides of a question. Readers moreover tend to discuss and argue about books and characters with other readers, and this discussion produces moral and psychological insight.

Argument one is not altogether convincing. While fiction writers must have a greater-than-average interest in human affairs, they cannot be assumed to be good observers simply because they have adopted writing as a profession. That they are accurate transcribers of reality rather than remakers of a believable artificial reality may reasonably be doubted. Even in those aspects of life in which it is easiest to accept their authority (descriptions of landscape, costume, public reactions to historical events) because they have less motive to rewrite the world, we should prefer the historical findings of the trained and fully equipped expert.

Argument two is similarly unconvincing. As long as fictional works do not acquaint us with the interior thoughts of actual members of the classes named, but with thoughts imaginatively ascribed to members of those classes by an author, they do not provide knowledge of the actual world. Novelistic sequences are plausible not, in the statistical sense, probable: in no case do novelistic sequences reflect rules of cause and effect. Crime may in fact pay, ambition may satisfy, secrets may stay buried, passionate love may endure.

Argument three appears at first to be a fully acceptable argument. When a child, after numerous trials, is able to

recognize animals like lions and giraffes in a picture book, we believe that it has learned something, something it would have been hard to learn in the real world. So why can one not learn from encountering and recognizing character types in books? There are two problems with the analogy. First, characters in works of fiction are not drawn, labeled, and subject to editorial verification as the animals are in children's books. Second, according to social psychologists, these supposed human "species" do not really exist in nature, in which people's behavior is demonstrably more determined by context than by dispositions and drives. Our sorting of people into types—our stereotyping—may help to make our moral and practical reasoning simple and easy, but it does not produce knowledge of the world.

Argument four is convincing only to the extent that one accepts sympathy as the basis of moral feeling and the heart as a reliable guide to action and rejects Kantian or utilitarian formulations. That literature of a certain complex, probing, psychological type increases sensitivity and understanding has been eloquently argued. It is a factual question, however, whether people who consume more literature or who are more involved with and interested in the fiction they do consume are more sensitive or sympathetic than readers of nonfiction or nonreaders.

Argument five, that literature assists us in envisioning alternative courses of action, provides no assurance that the outcome of the deliberation and conversation engaged in by habitual readers of fiction is more likely to be morally or practically correct than the unspontaneous action of the unreflective moral rigorist. "Thought experiments" are not appropriate when the type of knowledge we are seeking is a posteriori, not a priori.

With the failure of these arguments, it may be seem that the following conclusion is warranted. Literature is a source of vivid ideation and profound cognitive and emotional satisfaction. It induces imitative behavior and a tendency to self-scrutiny, both of which may, but need not, be morally beneficial. As well, it generates sophisticated moral discourse through the discussion of fictional characters, the proliferation of interpretations. Yet none of this implies that fiction readers as a class, or even certain kinds of fiction readers, acquire moral and psychological knowledge, or authoritative historical knowledge, and it certainly does not imply, as is sometimes claimed, that literature supplies a kind of moral-psychological knowledge that cannot be acquired through systematic instruction or general life experience. Readers, it may be noted, are accommodating creatures who willingly embrace the relevance to their own condition of stories about life in a London slum or in the world of cocktail-drinking, heel-clicking diplomats and top generals, and about hobbits and bunnies as well.

It may be reasonably argued, however, that the anticognitivist's conclusion is too harsh. It does not take into account the manner in which imaginative representations are controlled by cultural beliefs and assumptions or the importance of the "pang of recognition" in psychological and moral life. Relativists may deny that "moral knowledge" refers to anything beyond the mores of a given culture, and it might be pointed out that even if the stereotyping that literature arguably facilitates is, in the terms of empirical psychology, irrational, and to be distinguished from scientific concept formation, learning about the world implies learning about the stereotypes others form. Thus, coherentist conceptions of truth tend to favor the notion that literature is a source of knowledge. Novelists cannot help but betray in their work what their society or their local subculture believes and what norms it lives by, and this mirroring, however unintentional, instills a kind of knowledge in readers. Even in literature that appears to be profoundly critical of social norms and expectations or that presents a utopian alternative, there is revelation and mirroring. That fiction may be a powerful force in cultural indoctrination suggests that there is a cognitive reciprocity between writers and readers. If literature was not intended to and did not teach us, was not adapted to us in this respect, we would not be able to understand it at all.

Second, fiction constantly presents readers with situations in which outward actions disguise true plans, intentions, and feelings, and it frequently offers an implied commentary on the manifest and the hidden, the exercise of power, and the secret resistance or inner chaos that pressure outwardly applied may produce. The novelist, the dramatist, and the filmmaker are able to present the first-person experience of the protagonist as ludicrously or poignantly out of synchrony with the objective third-person view of an event, merely by exercising technical control over point of view. They thereby reveal the first-person perspective as the limited one it really is. The "pang of recognition" typically attaches to fictional situations that bring home this fact—which is a fact—to us.

Modern critics from Charles Baudelaire to Theodor W. Adorno have tended to reject all suggestions of a direct didactic role for fiction, and literary modernism has experimented with forms designed to mock or subvert the reader's hope and expectation of finding a message in literature. Typically, social and geographic locations are not identified, psychologically charged situations are presented in a confusing and disengaged way, there is no recognizable plot based on cause-and-effect sequences, mysteries may remain permanently unexplained, and the outcome offers no satisfaction of the reader's moral and emotional wishes. Postmodern literature has, however, returned to the elaborately plotted novel, to psychologism, to social criticism, and the high moral register. It rejects the ideal of objectivity and the search for a truth behind appearances that informed both classical and modernist fiction. The abandonment of these ideals, together with its powerful sense of cultural engagement, will assuredly alter the nature of debates

about the epistemology of fiction. The ubiquity of copies, models, and simulacra in contemporary life and the recognition of constant leakage between the fictional and real worlds promise to influence future discussion in interesting ways.

BIBLIOGRAPHY

Freud, Sigmund. "The Relation of the Poet to Day-dreaming." In *On Creativity and the Unconscious: Papers on the Psychology of Arts, Literature, Love, Religion,* edited by Benjamin Nelson. New York, 1958.

Hunter, J. P. "The Young, the Ignorant, and the Idle: Some Notes on Readers and the Beginnings of the English Novel." In *Anticipations of the Enlightenment in England, France, and Germany,* edited by Alan Charles Kors and Paul J. Korshin, pp. 259–282. Philadelphia, 1987.

Iser, Wolfgang. *The Implied Reader: Patterns of Communication in Prose Fiction from Bunyan to Beckett.* Baltimore, 1974.

Kermode, Frank. *The Sense of an Ending: Studies in the Theory of Fiction.* New York and Oxford, 1967.

Nussbaum, Martha. *Love's Knowledge: Essays on Philosophy and Literature.* New York and Oxford, 1990.

Putnam, Hilary. "Literature, Science, and Reflection." *New Literary History* 2 (1976): 483–491.

Rousseau, Jean-Jacques. "Letter to D'Alembert on the Theatre." In *Politics and the Arts: Letter to M. D'Alembert on the Theatre,* translated by Allan Bloom. Glencoe, Ill., 1960; reprint, Ithaca, N.Y., 1968.

Veyne, Paul. *Did the Greeks Believe in their Myths? An Essay on the Constitutive Imagination.* Translated by Paula Wissing. Chicago, 1988.

CATHERINE WILSON

FILM. [*To explore the aesthetics of film, this entry comprises seven essays:*

> Motion Pictures
> Visual Literacy
> Film and Documentary
> Film Noir
> Film Theory
> Feminist Film Theory
> Film and Ideology

After an overview essay on the history and current status of film aesthetics, there is a discussion of the issue of visual literacy, namely, whether we must already know certain (visual) conventions to understand film. The third and fourth essays are about two genres of film—documentary and film noir—that are important here because of the aesthetic issues they raise relative to other art forms as well as to film. The fifth essay explains how various types of theory have been incorporated into film studies. The sixth essay examines the role of feminists in the production and reception of film and in the recent development of film theory. The last essay analyzes how the ideology of film has been studied and why film, particularly because of its narrative structure, is often analyzed in terms of the concept of ideology. For related discussions, see Arnheim; Bazin; Cavell, *article on* Cavell and Film; Cultural Studies; Deleuze; Eisenstein; Lacan, *article on* Visual and Literary Arts; Photography; Television; *and* Video.]

Motion Pictures

The term *motion pictures* (also *film, cinema, moving pictures,* or *movies*) generally refers to the projection of a succession of still photographic images (called "frames"), mounted on a flexible, translucent celluloid strip, at such a speed that they yield the impression of movement to the normal viewer. Ironically, much of the technology of motion pictures was originally developed so that movement could be arrested, broken down, and analyzed in "frozen moments." But, instead, this technology was eventually deployed in order to reanimate movement, often for purposes of entertainment.

The motion picture apparatus can be used for many purposes—from bank surveillance to ballistics tests. Many of these uses have nothing to do with art. It is as art, however, particularly as popular art, that the motion pictures have come to prominence in the twentieth century. Yet, this outcome could have hardly been predicted during the 1890s, when motion pictures first appeared (the first, *screened* motion pictures became popular in 1895).

The primary reason that the motion picture seemed to be an unlikely candidate for art status was that one of its main constituents was photography. At the turn of the century, the belief held sway that photography was merely the slavish reproduction of reality. Unlike painting, to which it was frequently compared, photography was conceived to be a purely mechanical process of recording that provided no opportunity for the exercise of creativity or expression on the part of the artist. It was simply an automatic copying process, and, as such, it could not be art.

This prejudice against the artistic potentials of photography was inherited by the motion picture. It could not be art because it was nothing but the mechanical recording of reality. Or sometimes it was said that photography added nothing to the normal perception of what there is. Furthermore, when early motion pictures turned to theater for their subject matter, a variation on the charge of slavish reproduction—of reality or of normal perception—arose in a new form. Motion pictures were said to be involved in nothing more than the mechanical recording of plays. Motion pictures, that is, were not artworks in their own right. They were at best only recordings of other artworks, such as plays.

Motion pictures arrived on the scene at just that point in the history of aesthetics when imitation theories of art were in retreat. But since motion pictures, because of their photographic provenance, were popularly construed to be essentially mimetic, the prospect that they might aspire to art status appeared to many to be extremely dubious. By the early twentieth century, it was no longer enough for a "genuine" artwork such as a painting merely to reproduce perceptual reality. Imitation was not enough to warrant art status. Thus, the proponents of motion pictures, filmmakers and film theoreticians alike, took as one of their first orders

of business to show that motion pictures were not merely copies of anything—neither of reality nor of stage plays.

Filmmakers of the silent period (which extended through the late 1920s) developed a language of film, especially in terms of editing, but also comprising sophisticated developments in lighting techniques, special effects, camera movement, and so on, that diverged significantly from the standard conventions of theatrical representation, thereby effectively refuting the worry that film was nothing but "canned" theater; at the same time, these devices were also evidence that filmmakers had the power to reconstitute reality creatively and expressively, thus demonstrating the misconception involved in supposing that motion pictures were nothing but slavish copies of reality.

Early film theorists took note of this stylistic evolution and used it as a basis of arguments bent on showing that film could be art. As early as 1916, Hugo Münsterberg, a member of the Harvard philosophy department, published a treatise titled *The Photoplay: A Psychological Study* in which he argued that in virtue of techniques such as the close-up, parallel editing, flashbacks, and flash-forwards, motion pictures diverged from both the mere mechanical recording of both reality and theater in ways that could enable motion pictures to satisfy the presiding criteria for art status.

Throughout the twenties, a great deal of emphasis was placed on the theoretical importance of film editing, since editing was the dimension of cinematic articulation that most clearly challenged the prejudice that film was nothing but the mechanical reproduction of either reality or theater. Through editing, the filmmaker had the capacity to rearrange the spatial/temporal continuum of "reality" in ways that also typically diverged from conventions of theatrical representation.

The editing dimension of cinema, called "montage," became the special focus of theoretical concern for Soviet filmmaker-theorists during the 1920s, including Sergei Eisenstein, V. I. Pudovkin, Lev Kuleshov, and Dziga Vertov. For them, montage was the essence of cinema. That is, on their view, the photographic element of film was less important than editing; cinematic expression was dependent on editing. The photographic image acquired its significance through the way in which it was embedded in shot chains.

The theoretical relevance of this approach for the film-as-art debate should be clear. If some charged that film could not be art because the motion picture is nothing but a photographic medium, others rejected this on the grounds that, in fact, film was essentially a matter of montage, not photography. Furthermore, montage clearly gave filmmakers the capacity to reshape reality; it was not committed simply to copying reality or the "normal" perception of reality as critics (wrongly) supposed was all of which photography was capable.

The essentialism of the montage approach, moreover, was supposed to have implications for artistic practice. On this view, the essence of the medium was montage, and, therefore, artistic success, all things being equal, depended on the exploitation of montage. Truly cinematic motion pictures would be exercises in montage. Films that did not exploit the possibilities of montage were apt to go badly. Success and failure hinged on respecting the nature or essence of the medium, which for montagists, was identified with editing.

By at least the late 1930s, the debate for the artistic status of motion pictures had been won. On the theoretical side, this was attributable to arguments that stressed the ways in which the film medium was not a matter of slavishly recording anything—whether reality or "normal" perception or theater—of which the montage arguments were perhaps the weightiest. But with the coming of sound, the nature of the debate shifted.

If, for obvious polemical reasons, silent film theoreticians emphasized the importance of the editing component of film over its photographic element, beginning in the 1930s, film theorists, notably André Bazin, began to reconsider the photographic provenance of motion pictures. This reevaluation took place in the context of the development of new styles of cinematic realism in the work of Jean Renoir, Orson Welles, William Wyler, and, later, the Italian Neorealists. Theoretically, the realist view of the nature of cinema is still strong at least as late as 1960 when Siegfried Kracauer published his masterwork, *Theory of Film: The Redemption of Physical Reality*.

According to realist theorists like Bazin and Kracauer, photography is the essence of the medium of motion pictures. In terms of the argument about whether film can be art, given its photographic provenance, Bazin and Kracauer respond that it is a new kind of art. Moreover, given its origins in photography, cinema has a stylistic commitment to realism. Films that emphasize the realist tendencies of the photographic essence of the cinema are apt to succeed; films that attempt to go against the grain of the essence of the medium are apt to fail. Like the montagists, the realists believe that the motion picture medium has an essence and that this essence entails what is and is not cinematic. The difference between them, of course, resides in what they identify as the essence of cinema: is it editing or photography?

The great debate over the nature of film was staged in terms of editing versus photography. It was a debate that was intimately connected to the stylistic rivalry between montage and realism. By at least the 1970s, however, the debate began to seem hollow. It was difficult to understand how one might go about adjudicating, in a principled way, the contesting claims about whether the essence of the motion picture was editing or cinematography. Both seemed to be material to the creation of typical motion pictures. Any-

FILM: Motion Pictures. *Citizen Kane,* Orson Welles film (1941), film still; Museum of Modern Art, New York. (Print courtesy of the Museum of Modern Art/Film Stills Archive.)

way, why suppose that film could be reduced to one essential feature? Moreover, suspiciously enough, the ontological claims of montagists and realists often seemed to amount to nothing more than the special pleadings of the proponents of one style over another. Yet, neither the montage style nor realism appeared to be woven into the fabric of cinema in some ultimately privileged way for the simple reason that masterpieces in either style could be produced in film without any noticeable metaphysical repercussions.

Nevertheless, though motion pictures do not appear to possess a single essential feature of the sort championed by montagists and realists, they do have a series of definable, necessary conditions.

Photographic realists, like Bazin and, more recently, Kendall Walton, maintain that there is a peculiar relation of identity that persists between the photographic (and cine-matographic) image and its referent. They believe that, in some sense, the photographic (and cinematic) image is transparent: when one looks at a photo of Franklin Roosevelt, one is somehow seeing FDR, rather than an imitation of FDR. That is, in some sense, seeing through photographs is virtually the same as seeing directly. Seeing through photographs is on a par with the direct perception of reality. Whereas a telescope enables us to see directly objects at a distance in space, a photographic (and a cinematographic) image enables us to see objects temporally distant from us.

This does not seem correct, however. Seeing through photographic (and cinematographic) images is not analogous to direct perception. In direct perception, one is able to orient one's body in the direction of whatever one is looking at. In nature, so to speak, one sees a tree and, all things being equal, one can find one's way to it. But with films and

photographs, this is not possible. One sees the image of a tree in *The Yearling*, but has no idea of how to get from one's seat to that place in the world where the tree is (or was) on the basis of the image. The arrays that one sees in films and photos are detached displays with respect to one's body. They are not part of the space the body can navigate. One cannot relate one's body to them in the way one can relate it to the visual arrays that one perceives directly. One can move one's body toward what one sees directly, but that kind of bodily information about the referent of a film image is not available in the screening room. Seeing film images is not just like seeing reality. It is a species of alienated perception because something is a motion picture image only if it is a detached display.

Nevertheless, even if being a detached display supplies us with a necessary condition for being a cinematic image, it does not provide us with the conceptual wherewithal to differentiate motion picture images from things like paintings. Pictures are also detached displays. One cannot point one's body in the direction of Arcola on the basis of *Napoleon at Arcola*.

In order to draw the distinction between film and paintings, it is useful to consider the way in which film is often referred to in ordinary language, namely, in terms of "motion pictures" or "moving pictures." In both cases, the capacity for movement figures in our labels. Motion pictures move or, at least, are capable of moving, and this distinguishes them from still pictures, such as paintings, slides, and photographs.

Of course, there are motion pictures that are composed exclusively of still images, such as *One Second in Montreal* by Michael Snow. When viewing such a film for the first time, however, it is intelligible to wonder whether or not there might be some movement in the imagery, since movement is an essential capacity of film. On the other hand, to anticipate the possibility of movement in a painting or a slide is an absurdity. Thus, the possibility of movement in the imagery is a necessary condition of motion pictures. One might even say: that's why they're called *motion* pictures. So, x is a motion picture only if x is a detached display and only if x belongs to the class of things wherein the impression of movement is possible.

Even if, however, the possibility of movement supplies us with a way to differentiate between motion pictures and *still* pictures (such as paintings), it does not distinguish motion pictures from theatrical spectacles, since movement is also a possibility on the stage. What, then, differentiates motion pictures from live theatrical performances?

Both theatrical performances and film performances (i.e., film showings) are tokens of a type. The performance of *The Taming of the Shrew* at the repertory theater is a token of William Shakespeare's play, while the screening of *Kids* at the local multiplex is a token of Larry Clark's film. Nevertheless, the route from the film type to the token performance in cinema differs from the route from play type to the stage in drama.

The play by Shakespeare is a recipe to be filled out by an interpretation—or a series of interpretations—by actors, directors, costumers, makeup artists, and so on. Such interpretations, which are themselves types, supply the bridge between the play type by Shakespeare and its token performance instances. On the other hand, in order to screen a token performance of a film type, such as Clark's *Kids*, one needs a template, such as a filmstrip, which is itself a token.

One should not confuse the original negative or master of the film with the film type. For it can be destroyed, as can any of the other prints of the film, whereas the film itself is said to still exist. The original print of F. W. Murnau's *Nosferatu* was destroyed in the early twenties, but this seminal vampire masterpiece will continue to exist as long as at least one template of it does, whether in the form of a film print, a videocassette, or a computer program. So, x is a motion picture image (1) only if x is a detached display, (2) only if x is a member of the class of things wherein the impression of movement is possible, and (3) only if a token performance of x is generated by means of a template that is itself a token.

These are not the only necessary features of the motion picture image, nor are these conditions jointly sufficient for determining whether a candidate counts as a motion picture; however, they are adequate to the question that has most concerned film theorists: namely, how is one to differentiate the film image from the so-called normal perception of profilmic reality (i.e., the scene in front of the camera), from painting and still pictures, and from theater imagery? According to the preceding hypothesis concerning the necessary features of the moving picture: film differs from normal perception inasmuch as the perception it affords is alienated, disjunct from the spatial continuum our bodies inhabit in that the image is a detached display; film differs from still pictures because it possesses, generically, the possibility of movement; and film differs from theater, since token theater performances are generated by interpretations that are types, whereas token motion picture performances are generated by means of templates that are themselves tokens of types (e.g., my videocassette is a token-template of the film type *Clueless* through which I can generate token performances of *Clueless* in my living room).

BIBLIOGRAPHY

Arnheim, Rudolf. *Film as Art*. Berkeley, 1957.

Bazin, André. *What Is Cinema?* vol. 1. Translated and edited by Hugh Gray. Berkeley, 1967.

Carroll, Noël. *Philosophical Problems of Classical Film Theory*. Princeton, N.J., 1988.

Carroll, Noël. "Defining the Moving Image." In *Theorizing the Moving Image*, pp. 49–74. Cambridge and New York, 1996.

Danto, Arthur C. "Moving Pictures." *Quarterly Review of Film Studies* 4.1 (Winter 1979): 1–21.

Eisenstein, Sergei. *Film Form*. Edited and translated by Jay Leyda. (New York, 1949).

Ingarden, Roman. "On the Borderline between Literature and Painting." In *Ontology of the Work of Art: The Musical Work, the Picture, the Architectual Work, the Film,* translated by Raymond Meyer and John T. Goldthwait. Athens, Ohio, 1989.

Kracauer, Siegfried. *Theory of Film: The Redemption of Physical Reality*. New York and Oxford, 1960.

Pudovkin, Vsevolod. *Film Technique and Film Acting*. Translated by Ivor Montagu. New York, 1960.

Scharf, Aaron. *Art and Photography*. Harmondsworth, 1975.

Sparshott, F. E. "Vision and Dream in the Cinema." *Philosophic Exchange* (Summer 1971).

Walton, Kendall. "Transparent Pictures: On the Nature of Photographic Realism." *Critical Inquiry* 11.2 (December 1984): 246–277.

NoËl Carroll

Visual Literacy

People who write about movies have traditionally referred to the conventions of cinematic representation—such things as low-angle shots, fade-outs, or flashbacks—as the "language" of film. Does the ability to understand this language require previous experience? Or, to put this question differently, would a "naive" viewer, someone who had never seen a movie before, be able to make any sense of his or her first encounter with this medium? The term *visual literacy,* popular among media scholars, reflects the widely held belief that the comprehension of cinematic conventions is indeed an acquired skill, comparable to fluency in reading or writing. In contemporary film scholarship, this belief is based largely on an extrapolation from the writings of E. H. Gombrich and others on the cross-cultural variability of pictorial conventions. This body of literature is commonly assumed to have shown that any perceived similarity between pictures and the things they represent is simply the result of viewers' unwitting assimilation of the representational standards of a particular culture or historical period. Consequently, it is argued, the ability to connect a picture to its intended referent must depend on prior familiarity with the conventions employed in that picture. As far as film is concerned, this argument has occasionally been supported by stories about misinterpretations reportedly experienced by early-twentieth-century filmgoers or other inexperienced viewers.

The assumption that film interpretation requires a "literacy" derived from previous experience with film may be the dominant view in academic cinema studies. In the writings of film scholars and critics, however, there is also a contrary position that appears to have made some gains in popularity in recent years. The central ingredient of this contrary position is the argument that the conventions of visual media can be analogues of informational cues that people use in making sense of unmediated, real-world experience. In the presence of an analogy of this sort, viewers may be able to sense the intended meaning of a convention by drawing on their real-world perceptual skills rather than on any specific knowledge gained from previous encounters with movies or other visual media. According to this view, then, the concept of a medium-specific literacy may be a misleading metaphor for describing the cognitive skills that are brought into play in the interpretation of picture-based media.

The skeptical view of visual literacy is sometimes premised on the belief that analogy to real-world informational cues is the characteristic principle of all pictorial representation. The theoretical support and the empirical evidence for this belief vary considerably, however, depending on what aspect of visual media one is talking about. What is true of individual images, whether in a still picture or in the context of a movie, is not necessarily true of the juxtaposition of images in an edited sequence. Moreover, representational principles governing an image's "literal" content—the objects and events depicted in it—may not carry over to such formal features as compositional style, camera angle, or editing rhythm.

Single Images. The relationship between pictorial conventions and the information available in our real visual surroundings has been clarified considerably by recent research on how the mind deals with visual data. As the late David Marr and others have pointed out, the perception of pictures is affected critically by the so-called modularity of the mental operations involved in vision. In this context, modularity refers to the mind's ability to derive meaning from a particular element of a visual scene without necessarily taking account of the other elements. For example, the identification of an object's structure and location in space can be achieved independently of the sensing of color—as is demonstrated by cases of pinpoint brain injury affecting the color-processing "module." As a result of the modularity of real-world vision, a broad array of pictorial styles that could be described as artificial and unrealistic may nonetheless be capable of giving us the same kinds of visual information that our minds make use of when we look at real objects and events. Color photographs lack the depth cues associated with binocular vision and motion parallax, but they can convey a sense of depth through occlusion and texture gradients; black-and-white photographs and movies lack information about color, but they can provide detailed accounts of shape; animated cartoons lack many details of shape, but their depictions of objects' structural features can apparently match the informational cues used in real-world object identification.

The mind's ability to derive meaning from the correspondence between individual pictorial elements and real-world informational cues, even when other cues are missing, is arguably the essence of analogical representation. In other words, the efficacy of analogy as a substitute for a specifically pictorial literacy may be said to depend on the modularity of real-world vision. This conception of picture per-

ception is in accord with a substantial body of experimental research, including the pioneering work of Julian E. Hochberg and, more recently, John M. Kennedy's studies on the ability of congenitally blind people to identify pictured objects on the basis of raised outline drawings. A full interpretation of the empirical evidence on picture perception must, however, await a more complete understanding of exactly how the various real-world visual cues operate.

Editing: Single Location/Time Frame. The presence of editing in a movie creates a radical discontinuity in the flow of visual information displayed to the viewer. The relationship between this discontinuity and anything that the viewer might experience in real-world vision will depend crucially on whether the editing occurs within a single location and time frame. When the camera stays within the confines of one time and place, what the viewer typically sees on the screen is a succession of partial views of that place and the action occurring within it. In order to derive a sense of coherent objects and events in a continuous space, the viewer must perform the mental task of stitching these partial views together into one whole. Strictly speaking, there is no real-world visual experience that matches precisely the kind of visual sequence with which a viewer is confronted in this kind of situation, unless the camera is positioned in a single spot and the successive views are achieved only through panning or tilting interrupted by cuts. Since the latter possibility is exceedingly rare, one could conclude in principle that analogy to real-world vision does not seem to furnish an adequate basis for the interpretation of editing of this general type.

This conclusion is not supported, however, by the findings of research on inexperienced viewers' interpretations of narrative editing. These findings come from an ongoing investigation by Renée Hobbs and others. Working in a Kenyan village whose inhabitants had had no previous exposure to movies, Hobbs and her colleagues tested the villagers' comprehension of videotaped stories based on the local culture and produced in two different versions, one with editing, the other without. The unedited versions of the videos were shot in one long take in which camera movement was used to follow a continuous action. In the edited versions, the same action was interrupted by conventional, Hollywood-style cutting. The villagers' comprehension of both versions was good, and, more to the point, there were no significant differences in level of comprehension between the two versions. It is conceivable, then, that editing in a single location and time frame is actually an adequate analogue of real-world informational cues, despite its superficial divergence from the possibilities of real-world perception. It is certainly true that in the process of real-world vision our minds have to create a coherent whole out of information collected by the eyes through a process of jumping around from one momentary view to another. Still, these momentary views are all produced from a single vantage point, whereas the typical edited

sequence in a movie is not. Furthermore, the brain's piecing together of successive real-world views is not yet understood well enough to provide a theoretical explanation for any assumption of an analogical connection between that process and movie editing. These considerations leave open the alternative conclusion that the empirical findings in this area are due to some other factor, including the possibility that contextual knowledge related to the movies' familiar subject matter allows viewers to override any difficulties encountered in making sense of the editing.

Editing Across Locations or Time Frames. When an editing transition breaks the continuity of time or space—as is the case in such situations as scene changes, flashbacks, montage sequences, or most kinds of nonnarrative editing—there can be no question of an analogy to real-world visual experience. Instead, film theorists have sometimes argued that transitions of this sort replicate the conceptual juxtapositions that occur in other modes of communication, most notably language. Such affinities between movies and other modes undoubtedly do exist, although the direction of influence is often unclear. There is also some reason to believe that familiarity with verbal narrative conventions can take the place of visual literacy as a basis for the comprehension of scene-change editing. Hobbs's research in Kenya investigated viewers' interpretations of a movie with space-time transitions. Despite the fact that one of the transitions was a flashback, these inexperienced viewers were able to give fluent accounts of the story presented in the movie. As with the data discussed earlier, however, it is possible that these findings reflect viewers' abilities to use contextual information as a guide for their interpretations, as opposed to decoding the meaning of individual editing transitions by analogy to specific verbal devices.

In general, then, the commonly held belief that film interpretation requires a specifically cinematic literacy does not appear to mesh with available research findings, and this seems true not only of individual images but also of editing. On the other hand, the theoretical basis for these findings is by no means secure. While the argument for an analogy to real-world perception is relatively clear as far as single images are concerned, it is less so in the case of editing within a single location and time frame, and irrelevant in the case of other kinds of editing. It should be emphasized, though, that these observations apply mainly to that aspect of interpretation that is concerned with identifying the "literal" content of a movie: the characters, actions, and places depicted in it. Arguments against the need for visual literacy have been much less concerned with the connotations of a movie's stylistic characteristics, perhaps because it is taken for granted that analogical connections to real-world experience cannot occur in the case of such features. Yet, evidence indicative of such connections is not hard to come by. For example, in an experiment on the gender connotations of visual form, Aletha C. Huston and others found that

young children were able to assign gender to advertisements purely on the basis of such stylistic features as editing rhythm, equating staccato editing with masculinity and smooth editing with femininity. Unless the children's performance was simply a result of learned associations between these editing styles and gender connotations, these findings suggest that the children may have had an implicit sense of an analogical connection between the editing and traditional images of gender characteristics. Such a connection could also be seen as a canonical example of perceptual modularity: in these children's interpretations the temporal quality of action or behavior appears to have carried meaning in and of itself, independently of the entities embodying that quality. It should not be difficult to imagine other stylistic features—shapes, colors, camera movements, and the like—that could function similarly for viewers, whether with regard to gender or in other areas of meaning.

[*See also* Imagery; *and* Psychology of Art.]

BIBLIOGRAPHY

Anderson, Joseph D. "A Cognitive Approach to Continuity." *Post Script: Essays in Film and the Humanities* 13.1 (Fall 1993): 61–66.

Carroll, Noël. *Theorizing the Moving Image.* Cambridge and New York, 1996.

Gombrich, E. H. *Art and Illusion: A Study in the Psychology of Pictorial Representation* (1960). 2d rev. ed. Reprint, Princeton, N.J. 1969.

Goodman, Nelson. *Languages of Art: An Approach to a Theory of Symbols,* (1968). 2d ed. Indianapolis, 1976.

Grodal, Torben Kragh. *Cognition, Emotion, and Visual Fiction: Theory and Typology of Affective Patterns and Genres in Film and Television.* Copenhagen, 1994.

Hobbs, Renée, Richard Frost, Arthur Davis, and John Stauffer. "How First-Time Viewers Comprehend Editing Conventions." *Journal of Communication* 38.4 (Autumn 1988): 50–60.

Hochberg, Julian E., and Virginia Brooks. "The Perception of Motion Pictures." In *Handbook of Perception,* edited by Edward C. Carterette and Morton P. Friedman, vol. 10, pp. 259–304. New York, 1978.

Huston, Aletha C., Douglas Greer, John C. Wright, Renate Welch, and Rhonda Ross. "Children's Comprehension of Televised Formal Features with Masculine and Feminine Connotations." *Developmental Psychology* 20.4 (1984): 707–716.

Kennedy, John M. *Drawing and the Blind: Pictures to Touch.* New Haven, 1993.

Marr, David. *Vision: A Computational Investigation into the Human Representation and Processing of Visual Information.* San Francisco, 1982.

Messaris, Paul. *Visual "Literacy": Image, Mind, and Reality.* Boulder, Colo., 1994.

Prince, Stephen. "The Discourse of Pictures: Iconicity and Film Study." *Film Quarterly* 47.1 (Fall 1993): 16–28.

Spiegel, Alan. *Fiction and the Camera Eye: Visual Consciousness in Film and the Modern Novel.* Charlottesville, Va., 1976.

PAUL MESSARIS

Film and Documentary

The documentary film or video is often distinguished from its poorer relation—the information or instructional film—by its fusion of "dramatic" or "creative" elements with the presentation of information. This creative or dramatic requirement has traditionally served to distinguish the documentary from the nonfiction film, the latter designation more broadly covering all films not fictional. Thus, Errol Morris's *The Thin Blue Line* (1988), a documentary, is elevated above the U.S. Army instructional film, *How to Make Your Bed,* in part as a result of differences in artistic merit.

When British filmmaker and producer John Grierson called documentary "the creative treatment of actuality," he initiated an ongoing discussion of its purposes and forms. In fact, the two poles his succinct phrase suggests—first, presenting information and recording actual scenes and sounds ("treatment of actuality"), and second, rhetoric and aesthetics ("creative treatment")—often exist in uneasy alliance. The simultaneous need to record/reveal/inform *and* to create has long been central to discussions of documentary film.

Popular notions of the documentary downplay the poetic role of the filmmaker. One schema for documentary, derived from television journalism, considers it to be objective. Here "objective" refers to requirements for balance, fairness, and restraint in representation. According to this characterization, prototypical documentaries are journalistic and include, for example, episodes of *CBS Reports, 48 Hours, 60 Minutes,* or *Frontline.* Critics of journalistic documentary find its claim of objectivity to be deceptive; objectivity, they say, is impossible, since no filmmaker can escape perspective or point of view. For these critics, objectivity in television journalism is a conventional practice that masks subtle biases and perspectives. Such criticisms fail, however, to consider *perspective-relative,* as opposed to absolute, objectivity. Although every documentary has a perspective or point of view, one film may still be more or less objective than another, depending on its treatment of a subject. All else being equal, the documentary portraying alternative points of view A, B, and C on a controversial subject is more objective than one portraying only A, though neither documentary may be objective in an absolute sense. Neither does the presence of point of view cancel requirements for good evidence. Recognizing that all documentaries are perspective-relative does not make the practice of journalistic objectivity necessarily illicit, nor does it prove its impossibility. It does make suspect all claims for absolute objectivity, however.

Considerations of objectivity extend to print journalism and are not specific to the documentary. The use of moving photographic images and recorded sounds, however, is unique to documentary film and video. A second popular conception of the documentary derives from this, characterizing it as the filmic (and sonic) recording, ostensibly veracious, of an event or other subject. This definition stems from direct cinema, or cinema verité, a style of documentary made possible by the development of lightweight, portable filmmaking equipment in the late fifties. With

portable cameras, lightweight sound recorders, and fast film stocks, a crew of two could film and record synchronous sound with an ease previously unknown. This led to an ethos of "capturing reality," where the task of the filmmaker was to present a transparent record of the events unfolding before the camera, and to allow film spectators more freedom to interpret an ambiguous representation of indeterminate reality. Direct cinema filmmakers influence the profilmic event as little as possible. Direct cinema also avoids stylistic techniques—voice-over narration and nondiegetic music, for example—thought to explain or interpret material for the spectator. While some practitioners of direct cinema claimed transparency for their films, more subtle filmmakers—for example, Fred Wiseman—have always recognized that the method does not eliminate subjectivity and mediation. Thus, direct cinema has taught us the power of the camera to record and to reveal, but also its tendency to subtly manipulate the spectator's perception of events through editing and visual means.

A third way to think of the documentary—what might be called a traditional rhetorical approach—builds on distinctions between fiction and nonfiction discourse. Put simply, where fictional discourse presents states of affairs as fictions, nonfiction discourse asserts that its represented states of affairs actually occur(red). A fictional discourse may assert broad truths, for example, about the follies of war or the vagaries of love, but its particulars are presented as fictions. Alternatively, the nonfiction discourse may imply the broad truths found in fiction, but also asserts its particulars to be true.

This is clear in documentaries making use of voice-over narration, since the voice-over narrator makes direct linguistic assertions. With regard to photographic images in documentary, direct assertion becomes an implicit claim that the images and sounds presented give accurate visual and sonic information. In addition, stylistic and structural elements imply a point of view toward what is presented through image and sound. Whether the assertions and implications of documentary are made through voice-over narration, images, structure, style, or point of view, they all are asserted or implied to be true of the actual world.

This rhetorical approach has an advantage over those that conceive of documentaries either as objective or as imitative recordings; it can better account for creative expression in documentaries. If the documentary is thought to be objective and/or a mere recording, then any expressive device becomes alien to its purpose. But when we find the nonfictional elements of documentary to arise in the intent of its assertions and implications rather than in their means of presentation (through objectivity or recording), then they clearly may be embodied within the same expressive presentation as the fiction film, and still make assertive claims. At the most basic level, then, the nonfiction film/video implies neither objectivity nor the simple imitation of the profilmic scene; it is nonfiction in its assertion and implication that its states of affairs actually occur or occurred.

An alternative to the view outlined above is broadly "poststructuralist" or "postmodernist." The poststructuralist finds the very practice of documentary to be inherently deceptive in its illusionism and feigned transparence. The poststructuralist foregrounds the documentary use of motion photography, together with realist film techniques, as means of implying truth, transparence, and reality, where there exist only assertion, mediation, and fabrication. By means of moving photographs, imbued with an aura of scientificity, the documentary authenticates its claims. The alluring realism of the image, together with devices such as continuity editing, also work to mask the constructed, mediated nature of documentary texts. Through various "strategies of authentication," the documentary implies spontaneity, transparency, and lack of mediation. These various deceptions are historically determined and "built into" the genre and the medium. To combat these rhetorical effects, poststructuralists advocate reflexive practices that foreground the documentary as a constructed text.

Where the traditional rhetorical approach sees persuasion as a normative function of discourse, the poststructuralist characterizes documentary rhetoric as inherently deceptive. In the traditional approach, the teacher of visual rhetoric would attempt to educate the documentary spectator, since sophisticated viewers must possess a degree of self-consciousness about their role as interpreters. The poststructuralist, alternatively, makes education secondary to a call for radical changes in the forms and practices of realist documentary. While both approaches recognize possible deceptions in documentary representation, their characterizations of the nature of, extent of, and appropriate responses to these deceptions differ markedly.

Most agree that a poetics of documentary film has implications for a politics and ethics of representation. Creative decisions have subtle effects on the representation of reality and on the spectator's perception of the pictured events. Documentary filmmaking inevitably involves creative choices, both aesthetic and rhetorical, that manipulate mood, perspective, and spectator identification. Through editing, for example, the filmmaker not only presents selected visual information but also creates point of view, a sense of energy or its lack (through pacing), feelings of balance or discomfort, and other perceptual effects. Other relevant techniques are variations in sound (music, speech, noise), framing, angle, camera movement, and lighting.

Rhetorical purpose also becomes significant in a filmmaker's structural choices. For example, consider Michael Moore's *Roger and Me* (1989), a film about General Motors' layoffs of workers and the economic and social devastation of Flint, Michigan. The narrative structure of the film—the means by which it orders and presents historical

events—follows the conventional dramatic structure of the classic Hollywood film, featuring preliminary exposition, a protagonist with clear-cut goals, an antagonist, decisive turning points or reversals, and even a loose three-act structure. Yet to create such a drama, Moore misrepresented the chronology of historical events. For example, while the layoffs in Flint actually occurred incrementally over dozens of years, they are represented as occurring in 1987, in a single catastrophic blow. Flint's response to rising unemployment was to attempt to attract tourists to the area. Although the resultant Hyatt Regency Hotel, Water Street Pavilion, and AutoWorld (an amusement park) had all opened and in two cases failed before 1987, Moore represents them as attempts to counter the effects of the 1987 layoffs. Whether we see these manipulations as normative for documentary filmmaking, and justified by artistic license, or as avoidable manipulations that deceive the audience, depends on our view of the nature and function of this film and of the documentary in general.

Although all documentaries have a poetic dimension, in the sense that they fashion their materials of representation, some films have foregrounded aesthetic concerns more than others. Any account of documentary aesthetics must account for these "art" or "poetic" documentaries. Art documentaries include early films such as the "city symphonies"—*Berlin: Symphony of the City* (*Berlin: Sinfonie der Grosstadt*, 1927) and *Only the Hours* (*Rien que les heures*, 1926)—that represent the complexities of great cities through associational editing and visual effects. Dziga Vertov, in *Man with a Movie Camera* (1929), explores Soviet life through virtuosic editing, split-screen framing, superimpositions, and other expressive techniques. Dutch filmmaker Joris Ivens later found fault in his own early "esthetic approach," but his early films, *The Bridge* (*De Brug*, 1928) and *Rain* (*Regen*, 1929), emphasize the formal and sensual elements of their subjects. Jean Mitry's *Pacific 231* (1944) uses accelerating patterns of editing to represent the building momentum of a locomotive, as it slowly moves from a stationary position to a whirl of powerful motion. One might also include the films of Robert Flaherty among those of the art documentary. Although his films of Eskimos (*Nanook of the North*, 1922) or Aran Islanders (*Man of Aran*, 1934), for example, have an anthropological purpose, his practice of engaging locals as "actors," staging scenes of native practices long abandoned, and the repetition of great themes (humanity versus nature) suggest the artist at work.

Recent filmmakers carry on the tradition of art documentary, with films experimenting in form and function. Michele Citron's *Daughter Rite* (1978) poses as a documentary exploration of mother-daughter relationships, but reveals its artifice in its manipulations of Super-8 home movies, and in the end credits, as we see the names of the actresses who play the two sisters. Ross McElwee's self-absorbed film diaries, *Sherman's March* (1985) and *Time In-*

definite (1994), represent the filmmaker's life in innovative ways, exploring themes of angst, love, family, and mortality. Recent filmmakers who have extended documentary form are Chris Marker and Errol Morris. Marker's *Letter from Siberia* (1957), *Far from Vietnam* (1967), and *Sans Soleil* (1982), for example, all are unique experiments in documentary style, structure, and point of view. Morris's films— *Gates of Heaven* (1978), *Vernon, Florida* (1981), *The Thin Blue Line* (1988), and *A Brief History of Time* (1992)—are classical in their unity, harmony, and clarity. Yet, certain of Morris's practices (e.g., the building of sets for interviews and the use of staged reenactments) reveal his penchant for pushing the boundaries of documentary technique.

What these art documentaries clearly show is the inadequacy of conceptions of the documentary as either merely objective, or as a photographic imitation of reality. At best, these characterize specific, limited types of documentary filmmaking. An aesthetics of documentary must recognize the diversity of films and purposes. Moreover, for all documentaries, considerations of rhetoric *and* poetics—in both structure and style—are central.

BIBLIOGRAPHY

Barnouw, Erik. *Documentary: A History of the Nonfiction Film.* 2d rev. ed. New York and Oxford, 1993.

Barsam, Richard. *Nonfiction Film: A Critical History.* Rev. exp. ed. Bloomington, Ind., 1992. Extensive bibliography.

Carroll, Noël. "From Real to Reel: Entangled in the Nonfiction Film." *Philosophic Exchange* 14 (1983): 5–45.

Nichols, Bill. *Representing Reality: Issues and Concepts in Documentary.* Bloomington, Ind., 1991.

Plantinga, Carl. "The Mirror Framed: A Case for Expression in the Documentary." *Wide Angle* 13.2 (Summer 1991): 40–53.

Plantinga, Carl. "Motion Pictures and the Rhetoric of Nonfiction Film: Two Approaches." In *Post-Theory: Reconstructing Film Studies,* edited by David Bordwell and Noël Carroll, pp. 307–324. Madison, Wis., 1996.

Renov, Michael, ed. *Theorizing Documentary.* New York and London, 1993.

Rosenthal, Alan, ed. *New Challenges for Documentary.* Berkeley, 1988.

CARL PLANTINGA

Film Noir

Film noir remains one of the best-known, yet least understood, categories of the narrative sound film. Critics and scholars continue to debate whether it designates a coherent genre, a historical period in film history, or a consistent cinematic aesthetic. Regardless of their diverse answers there can be little doubt that to a greater degree than other formulaic approaches to filmmaking (such as the science fiction or horror film), film noir connotes a set of cultural styles and attitudes that have reached beyond cinema and permeate fashion, advertising, television, and varied products of the culture industry. Combining elements of the crime, detective, gangster, melodrama, and thriller genres,

film noir has become a principal catalyst for the recrudescence of these narrative types in current filmmaking and functions increasingly as a relay between the contemporary film industry and an idealized understanding of the Hollywood cinema of the 1940s and 1950s.

According to standard histories, film noir consists in a body of American films—estimations of its size vary from several dozen titles to more than five hundred—produced between 1940 and the final years of the 1950s. French critics, unable to view American films during the Occupation, were struck in the postwar moment by the despairing mood in 1944 releases such as *Double Indemnity* (Billy Wilder) and *Murder, My Sweet* (Edward Dmytryk) and popularized the phrase in a series of articles published in 1946. Early examples of film noir include *The Stranger on the Third Floor* (Boris Ingster, 1940), *The Maltese Falcon* (John Huston, 1941), *Phantom Lady* (Robert Siodmak, 1944), *Detour* (Edgar G. Ulmer, 1945), *Scarlet Street* (Fritz Lang, 1946), *The Big Sleep* (Howard Hawks, 1946), and *Gilda* (Charles Vidor, 1946), although the innovative narrative techniques employed in *Citizen Kane* (Orson Welles, 1941) suggest it as a key progenitor of the cycle. Later contributions include *Out of the Past* (Jacques Tourneur, 1947), *T-Men* (Anthony Mann, 1948), *The Asphalt Jungle* (John Huston, 1950), *In a Lonely Place* (Nicholas Ray, 1950), and *The Big Combo* (Joseph H. Lewis, 1955).

The earliest films noir were frequently adaptations of crime novels written in the 1920s and 1930s by Dashiell Hammett, W. R. Burnett, James M. Cain, Cornell Woolrich, and Raymond Chandler, whose reworking by the studio system and screenplay writers after World War II introduced a range of topical allusions into their stories. Social historians interpret films noir as an expression of postwar anxiety and cynicism accompanying the reentry of veterans into domestic life and the relocation of women from the workplace to home and family. Allegorical readings of the film cycle emphasize its treatment of the political assaults of McCarthyism, cold-war fears of nuclear apocalypse, and perceived threats to the stability of the American way of life.

Typical character types such as the femme fatale, the innocent victim, and the loser trapped in a deadening urban routine are frequently viewed as an inheritance from the cinema of Weimar Germany (itself deeply marked by the traumas of World War I) introduced into American film by German-speaking exile directors (Fritz Lang, Robert Siodmak, Edgar G. Ulmer, Billy Wilder). Film noir also drew heavily on gangster films of the 1930s and the cycle of melodramas produced in the 1940s, but displays more a sharply drawn social portraiture made possible by the loosening of postwar censorship restrictions.

One of the earliest beneficiaries of advances in optics, lighting technology, and film chemistry that introduced deep-focus cinematography in the early 1940s, films noir typically appropriate composition in-depth and highly stylized mise-en-scène in their visual repertoire. Elaborate camera movement and filming in long takes (rather than the traditional alternation between establishing, medium, and close-up shots) appear frequently even in low-budget B productions, and established the cycle as a favorite venue for some of Hollywood's most gifted cinematographers and technicians, many of whom (John Alton, Rudolph Maté) were European émigrés.

The earliest films noir were generally photographed on studio sets and depicted a darkened and claustrophobic city populated by indifferent strangers. This transposition of the iconography of French and German cinema (if not representations of the metropolis associated with the naturalist novel and photographic modernism) to an American context remains among the film cycle's most decisive contributions to the modernist lexicon of narrative cinema. Diagonal shadows cast by venetian blinds on the wall of an office or a blinking neon sign visible through the window of a cheap hotel room are but two of the many visual conventions associated with the film cycle and its thematics of urban decay, danger, and the instability of vision itself.

After 1948, the improvement of lightweight cameras and portable lighting devices facilitated location photography and led to the use of actual urban settings (*The Naked City*, Jules Dassin, 1948; *Call Northside 777*, Henry Hathaway, 1948; *Force of Evil*, Abraham Polonsky, 1948), an instance of semidocumentary filmmaking that evolved perhaps in response to the influence of Italian Neorealism. Films noir produced after 1949 frequently relinquish the "centripetal" metropolis of railroad stations, neighborhood bars, and pedestrian promenades in favor of the spatially dispersed "centrifugal" settings of the suburbs, highways, and the rural countryside (*Gun Crazy*, Joseph H. Lewis, 1950).

By the close of the 1950s, the self-conscious and frequently parodic deployment of earlier film noir conventions in *Kiss Me Deadly* (Robert Aldrich, 1955) and *Touch of Evil* (Orson Welles, 1958) suggest the cycle's imminent exhaustion. The disappearance of the double bill and the B movie, growing competition from television, and the enforced inactivity of many seasoned writers and directors occasioned by blacklisting also contributed to its increasing desiccation. Since the late 1960s the resurgence of film noir tendencies in *Point Blank* (John Boorman, 1967), *Chinatown* (Roman Polanski, 1974), *Body Heat* (Lawrence Kasdan, 1981), *Bladerunner* (Ridley Scott, 1982), *Red Rock West* (John Dahl, 1994), *Pulp Fiction* (Quentin Tarantino, 1994), and *Devil in a Blue Dress* (Carl Franklin, 1995), as well as in many lesser-known "cable noir" television productions, has led some historians to proclaim its rejuvenation or the rebirth of a postmodern "neo-noir" cycle.

Yet the intellectual coherence of this account—like most definitions of film noir—suffers on close scrutiny. Unlike the classical Hollywood film genres of the western or the musical, film noir never functioned as an industrial category

that determined how films were produced and marketed. Classified by the studio system during the 1940s and 1950s within the "melodrama," "thriller," "mystery," or "suspense" genres, films one might today recognize as film noir were plentiful during the postwar era, although the term itself was not employed as a merchandising technique to attract audiences before the 1970s.

The syncretic character of film noir and its status as an amalgam of elements from other literary, artistic, and film types (detective fiction, painting and photography of the city, gangster and crime cinema) hinder the identification of a dominant narrative pattern. The prominence of mood and atmosphere in films noir, often formally excessive and unrelated to advancing the narrative, suggests the limitations of any purely narratological definition. Attempts to establish a typology of essential plot structures and character types yield abstract formulas such as that developed by James Damico:

> Either because he is fated to do so by chance, or because he has been hired for a job specifically associated with her, a man whose experience of life has left him sanguine and often bitter meets a not-innocent woman of similar outlook to whom he is sexually and fatally attracted. Through this attraction, either because the woman induces him to it or because it is the natural result of their relationship, the man comes to cheat, attempt to murder, or actually murder a second man to whom the woman is unhappily or unwillingly attached (generally he is her husband or lover), an act which often leads to the woman's betrayal of the protagonist, but which in any event brings about the sometimes metaphoric, but usually literal destruction of the woman, the man to whom she is attached, and frequently the protagonist himself. (Damico, 1978)

Perhaps the greatest value of this structural model is the attention it directs toward recurrent narrative features in the film cycle, such as the instability of relationships between protagonists and the destructive consequences accompanying the consummation of desire, that form apposite starting points for the development of feminist, psychoanalytic, and sociocultural approaches to its films.

Considered as definitions of film noir as a genre, these ideal types tend to be simultaneously too narrow and too inclusive, excluding some appropriate films while including others that would be better understood utilizing different categories. Designations of film noir as a specific tone and mood, such as Paul Schrader's, emphasize the prevalence of "low-key" lighting that creates areas of high contrast and deep shadows, scenes filmed to create the impression of night, oblique and vertical compositions, and a predilection for nocturnally illuminated rainy city streets. First-person camera, flashback devices, and voice-over commentary express a sense of predetermined fate and hopelessness. The temporal and visual indeterminacy present in many films noir, a slippage of narrative agency and stability, constitute widely imitated trademark features of the film cycle.

Yet the predilection for low-key lighting in many Hollywood films of the 1940s, as well as the prominence of the mise-en-scène and visual style associated with film noir in much silent cinema (*Sylvester*, Lupu Pick, 1923; *The Docks of New York*, Josef von Sternberg, 1928), suggests the cycle can no more be defined satisfactorily by reference to these features. The presence of flashback narration in both silent and sound films unrelated to the film noir no longer proves helpful in formulating a cogent theory of its narrative specificity.

To abandon the quest for an essential definition of film noir that has engaged (and sometimes obsessed) film critics and historians for the past half century is not to pronounce the term ineffable but to recognize instead, as James Naremore suggests, that film noir also "belongs to the history of ideas" (Naremore, 1966). Rather than designating a definable period in film history, a coherent genre, or a single visual or narrative style, the idea of film noir is best comprehended as a reading practice generated by different film-critical discourses that have attempted to grapple with a complex aesthetic object.

As a point of convergence between film history and the contemporary situation of the film critic, the idea of film noir has proved attractive to students of the cinema as a means of conjuring recurrent (and hence temporally imprecise) tensions in the experience of modernity. Representations in film noir of violence, fear, sexuality, racial and ethnic difference, political repression, technological domination, urbanization, and the search for meaning in secular industrial society continually capture the attention of subsequent generations of critics who consistently respond to its films as an occasion for reflection on the ongoing traumas of life in the modern world.

Contrary to widespread belief, the notion of film noir circulated widely in French film culture prior to 1946. Charles O'Brien notes that by 1939 it assumed a pejorative character and designated the allegedly immoral or socially scandalous character of cinematic poetic realism *Quai des brumes* (Marcel Carné, 1938) and naturalism *La bête humaine* (Jean Renoir, 1939). Yet the attempts to distance postwar French culture from that of the late 1930s led critics to reserve the term for American films produced just before the end of the war. This disavowal of the earlier role occupied by film noir in French film culture suggests its fundamental connection to the social transformations of the postwar period.

Critics such as Nino Frank, Henri-François Rey, and Jean-Pierre Chartier praised the abandonment in film noir of the formulaic narratives of the detective genre and delineation of characters endowed with moral and emotional complexity. They noted the despairing image of American society, and posited a link between the anxiety evident in film narratives and the postwar popularization of psychoanalysis. As these French intellectuals grappled with the so-

FILM: Film Noir. *Murder My Sweet,* Edward Dmytryk film (1944), film still; Museum of Modern Art, New York. (Print courtesy of the Museum of Modern Art/Film Stills Archive.)

cial and political aftermath of the Occupation, they approached the meditations on agency and responsibility implicit in many films noir—as well as in the nascent philosophy of existentialism—with passionate interest. Trading on the French phrase for the Gothic novel, the *roman noir,* and the *Série noire* crime fiction paperbacks launched after the war by Surrealist Marcel Duhamel, film noir became a privileged site for a wide range of French political, literary, and cinematic preoccupations.

The publication in 1955 of *Panorama du film noir américain* by Raymond Borde and Étienne Chaumeton, the first full-length book devoted to the cycle, further alludes to its modernist pedigree. Praising the oneiricism and treatment of death in film noir, the authors identified the "disappearance of psychological bearings or guideposts" and the delivery of irrational shocks *(insolite)* as evidence of its affinities with Surrealism. Written several years before the emergence of the French New Wave, their influential book underscores the nonconformist and antigeneric elements of film noir later manifested in the early films of Jean-Luc Godard and François Truffaut.

English-language treatments of film noir in the 1960s and 1970s primarily emphasized directorial style, frequently by arguing for continuity between a filmmaker's European and American career. As the older generation of émigrés became less active, film noir inflected the 1970s work of prominent younger directors, including Robert Altman,

Francis Ford Coppola, Martin Scorsese, and Paul Schrader. Writing in a period of political unrest and disillusionment, film noir critic-directors such as Schrader discerned the continued relevance of the cycle as a vehicle for social commentary.

Feminist film noir critics in the 1970s analyzed the film cycle's representations of women as expressions of the misogyny prevalent in patriarchal society and explored the value of psychoanalysis to the interpretation of its corpus. They also directed renewed attention to the cultural construction of the femme fatale and the careers of Barbara Stanwyck, Gloria Grahame, Rita Hayworth, and other female stars identified with their roles in films noir.

By the 1980s, film noir criticism overlapped increasingly with cultural studies and admitted a growing range of methodologies and sociocultural concerns into the study of its films. Increasing acknowledgment of the films noir produced in Spain, Mexico, and Latin America (some dating back to the 1940s), as well as the influence of film noir on postwar national cinemas in Japan, Hong Kong, Germany, Italy, and Scandinavia also increased awareness of its international status.

Today, when elements of film noir surface in mass cultural products of every sort, when the paintings of Edward Hopper are presented as a source of its visual style, or alternately the high modernism of Jackson Pollock is studied in relation to the film cycle, earlier efforts to construct a co-

herent genealogy appear increasingly futile. Yet, a certain measure of imprecision is perhaps inevitable in any attempt to grapple with a cultural phenomenon as ubiquitous and entangled in the upheavals of the twentieth century as film noir.

BIBLIOGRAPHY

Alton, John. *Painting with Light.* New York, 1949; reprint, Berkeley, 1995.

Borde, Raymond, and Étienne Chaumeton. *Panorama du film noir américain.* Paris, 1955.

Cameron, Ian, ed. *The Book of Film Noir.* New York, 1993.

Chartier, Jean-Pierre. "Les américains aussi font des films 'noirs'." *Revue du Cinema* 2 (September 1946).

Copjec, Joan, ed. *Shades of Noir.* London and New York, 1993.

Damico, James. "Film Noir: A Modest Proposal." *Film Reader* 3 (February 1978).

Dimendberg, Edward. *Film Noir and the Spaces of Modernity.* Cambridge, Mass., forthcoming.

Durgnat, Raymond, "The Family Tree of Film Noir." *Cinema* (U.K.) 6–7 (1970).

Frank, Nino. "Un nouveau genre 'policier'." *L'écran français* 61 (1946).

Higham, Charles, and Joel Greenberg. *Hollywood in the Forties.* New York, 1968.

Kaplan, E. Anne, ed. *Women in Film Noir.* London, 1978.

Leja, Michael. *Reframing Abstract Expressionism: Subjectivity and Painting in the 1940s.* New Haven, 1993.

Naremore, James. "American Film Noir: The Origins of an Idea." *Film Quarterly* (Winter 1996).

O'Brien, Charles. "Film Noir and Film Culture during the Avant-Guerre." *Iris* 21 (1996).

Palmer, R. Barton, ed. *Perspectives on Film Noir.* New York, 1996.

Place, J. A., and L. S. Peterson. "Some Visual Motifs in Film Noir." *Film Comment* 10.1 (1974).

Polan, Dana. *Power and Paranoia: History, Narrative, and the American Cinema, 1940–1950.* New York, 1986.

Rey, Henri-François, "Demonstration par l'absurde: Les films noir." *L'écran français* 157 (June 1948).

Schrader, Paul. "Notes on Film Noir." *Film Comment* 8.1 (1972).

Silver, Alain, Elizabeth Ward et al., eds. *Film Noir: An Encyclopedic Reference to the American Style.* 3d. rev. ed. Woodstock, N.Y., 1992.

EDWARD DIMENDBERG

Film Theory

Film theories offer systematic, general explanations of aspects of the nature, functions, and effects of cinema. They may conceive of cinema as a medium, independent of particular uses; as a social or institutional practice; or as a set of films to be described or analyzed. Film theorists have been journalists, practicing filmmakers, and scholars in a range of academic disciplines. On the whole, film theory developed to only a small extent in the social sciences, largely by psychologists studying perception and sociologists interested in the effects of mass media generally. Most film theory has been resolutely humanistic: speculative, linked to the critical study of films and filmmakers, and aesthetic in orientation.

History. For a medium so young—the earliest films date from the mid-1890s—cinema has hosted a rich array of theories. The era of silent film, which ended around 1930, disclosed areas of inquiry that are still pursued today.

The first ambitious efforts to understand cinema emerged during the 1910s, when producers were beginning to lure the middle class to what had been considered a disreputable working-class entertainment. Intellectuals, smitten by film's visual beauty and emotional power, began to argue that cinema was a distinct art. From the start, this entailed distinguishing film from theater. Hugo Münsterberg wrote perhaps the most substantial early book of film theory in 1915, *The Photoplay: A Psychological Study,* proposing that cinema differentiated itself from theater by mimicking activities of eye and mind, as when a sudden close-up simulates a focusing of attention on a detail. At the same period, the film's creator was being set forth as an artist in her or his own right. French journalists like Louis Delluc celebrated American directors like Cecil B. DeMille, while German masters of drama and prose fiction proposed to make *Autorenfilme,* works that had the range and subtlety of literary works. The 1910s initiated the effort of intellectuals to legitimize this popular pastime as what has been called "the seventh art."

Over the next decade an extraordinarily diverse and imaginative body of work developed conceptual foundations for film theory to follow. Writers began to distinguish particular techniques of the medium, categorizing types of cutting, lighting, and camera movements. They identified a canon of milestones, from the trick films of Georges Méliès through the works of Edwin S. Porter (*The Great Train Robbery,* 1903) and D. W. Griffith (particularly *The Birth of a Nation,* 1915) to the works of German masters like F. W. Murnau (*Nosferatu,* 1922; *The Last Laugh,* 1924) and Fritz Lang (*Siegfried's Death,* 1924), French directors such as Abel Gance (*La roue,* 1923), and especially Soviet filmmakers. Dziga Vertov (*The Man with a Movie Camera,* 1929), Lev Kuleshov (*By the Law,* 1926), Vsevolod Pudovkin (*Mother,* 1926), and Sergei Eisenstein (*Battleship Potemkin,* 1925) became famous as film theorists themselves. Seeking the principles by which film exercised its extraordinary influence over audiences, the Russians based their aesthetic largely on montage. For them, montage was not merely a synonym for editing but a conception of filmic construction whereby the juxtaposition of shots governed the spectator's emotional and intellectual response to the filmed material. This idea proved central both to film aestheticians and to cultural critics such as Walter Benjamin, who treated film as a symptom of the jolting, distracting ambience of the modern city.

The ambitions of the silent film's enthusiasts were synthesized in Rudolf Arnheim's *Film as Art (Film als Kunst,* 1932). By then, however, the silent film had been killed by the "talkie." Many aestheticians had defended the silent cinema as a highly stylized, form-based medium; but these arguments rang hollow in the face of synchronized sound and

the more realistic approaches to acting, staging, cutting, and camera work that emerged.

Well into the 1940s, writers struggled to construct viable theories of the sound cinema. Eisenstein advocated an operatic conception, whereby music and image would be welded into an overpowering "synthetic spectacle." More influential proved a strain of realistic theory identified with the Parisian critic André Bazin. Deliberately repudiating silent-film tendencies, Bazin claimed that cinema was different from painting and theater in that its raison d'être was mechanical reproduction. The film gloried in its ability to record reality, and efforts to turn it into painting-in-motion misunderstood the medium's unique properties. Everywhere Bazin saw confirmation of cinema's realistic vocation—in the recent masterpieces of Italian Neorealism (*Paisan*, 1946; *The Bicycle Thief*, 1947), in the growing intelligence with which plays and novels were adapted, and in the experiments with deep-focus cinematography and long-take filming pursued by Orson Welles (*Citizen Kane*, 1941) and William Wyler (*The Little Foxes*, 1941). Bazin believed that the richest films would be those that pursued an "aesthetic of reality," relinquishing artifice and convention to some degree in favor of film's inclination to record and reveal phenomenal reality.

World War II had revealed that cinema had enormous power as a vehicle for propaganda and mass indoctrination. After the war, some film theorists began to search for a scientific basis for understanding the medium's social and psychological effects. One path was pursued by social thinkers and experimentalists interested in film's appeal to stereotypes and social conventions. This tendency often led to virulent criticism, most famously in Theodor W. Adorno and Max Horkheimer's denunciation of "the culture industry" in *Dialectic of Enlightenment* (1947). A more neutral line of inquiry was pursued by the French-based "filmology" movement, which combined experimental research into perception, cognition, and emotional responses to film with broader speculation on the social functions of the mass media.

Within the humanities, another quasi-scientific tendency took firmer root. Under the influence of French structuralists of the 1950s, there emerged a semiology of cinema. Drawing on models derived from Ferdinand de Saussure and other Continental linguists, semiologists sought to understand how representations functioned as signs—coded units that could combine in rule-governed ways. The most famous film semiologist, Christian Metz, treated both the film image and methods of combining images as convention-bound, fairly arbitrary signifying procedures. Metz argued, for instance, that there was a large-scale code systematically ordering the representation of time and space in mainstream narrative film. Other semiologists studied film techniques as codes for conveying denotative and connotative meanings. Between 1965 and 1975 semiology was believed to have put film theory on a rigorous scientific basis. [*See* Semiotics.]

During the same years, political upheavals around the peace movement, radical social critique, Third World liberation movements, and feminist activism had an impact on art theory generally. Film theorists responded on several fronts. First, there emerged a politically critical semiology—one that saw the task of understanding film as one of demystifying those sign systems that maintained capitalist economic relations. Second, an appeal to psychoanalytic explanations for political domination and submission, circulating in Paris during the riots of May 1968, became strongly entrenched in the humanities. Film theorists proposed that cinema engaged with the viewer's unconscious in ways that reinforced mechanisms creating stable social identity. Third, feminist theorists began to argue that orthodox cinema's representational processes were gender governed. The mainstream movie, asserted Laura Mulvey in *"Visual and Other Pleasures*" (1989), tacitly addressed a male viewer, objectifying women and negating such unconscious processes as castration anxiety in order to sustain pleasure.

These trends were all generally known as theories of cinematic "subjectivity," and during the 1970s and 1980s many syntheses emerged. One, often called "apparatus theory," relied on Lacanian psychoanalysis to posit that the very situation of watching the screen encourages the spectator to identify with the image, in the way that the child recognizes a mirror reflection as being both another and the child itself. This identification was the basis for social subjectivity. Another tendency combined feminism's political critique of feminism with a modified psychoanalysis; sexist though Sigmund Freud or Jacques Lacan might be, their theories offered tools to reveal how patriarchy made women docile subjects. Notable variants of these arguments were played out in the pages of the British journal *Screen*.

Semiology had minimized aesthetics as a domain of inquiry, putting artistic and nonartistic signs on the same footing. Subjectivity theories went further, seeing the very category of art as suspiciously elitist. The emphasis fell on the social and ideological functions of cinema. As a result, the favored objects of study were mainstream entertainments from Hollywood (often genre pieces previously considered marginal) and radical avant-garde works that challenged dominant ideology at all levels—political stance, form and style, and psychic response (perhaps by cultivating "unpleasure"). The new film theory set itself against the canons of the silent and mature sound cinema: writers now studied Max Ophuls's melodramas and Alfred Hitchcock's thrillers alongside the austere political modernism of Jean-Luc Godard (*Vent d'est*, 1969) and Jean-Marie Straub and Danièle Huillet (*Moses and Aaron*, 1974).

Subjectivity theory remains an active trend, but during the late 1980s its influence slackened. Controversial from the start, it was vulnerable to the objection that in taking cinema as a powerful machine for producing spectator-subjects, it granted too little active control to the viewer.

Some writers proposed that some versions of cultural theory retained the advances of semiology—particularly, the capacity to treat representation as socially constructed and used—while also granting that spectators appropriated films in a wide variety of ways, some running distinctly counter to the dictates of the "text." Instead of a viewer "positioned" by the dynamic of the unconscious, cultural theorists described a viewer who is conscious, bringing to a film those experiences that form racial and gender-based identity.

One variant of this view, allied with the broader phenomenon of postmodernism, treated the "active subject" as the result of the present historical moment, when signs had lost their referents and codes had lost their force. Another variant, allying itself with the movement known as Cultural Studies, sought to understand how subcultures appropriated media artifacts. Ranging from speculative theory to ethnomethodology, Cultural Studies work treated films, like all media representations, as operating within open-ended processes of socially negotiated identity.

If cultural theory formed the main challenge to theories of subjectivity, it was not the only one. Some literary scholars applied deconstructionist theory to cinema. Other writers proposed that questions about meaning and spectatorial effect posed by psychoanalytic theories could be better answered by theories based in cognition. The "cognitive revolution" in the social sciences caught up with film theory in the work of both experimentalists and speculative theorists who sought to understand films as designed to channel perceptual-cognitive activity. Analytic philosophers of art began to examine cinema along lines that return to issues of aesthetic function and value. There have also been efforts to revive silent-film concerns with the artistic principles of film form and style, notably in a self-conscious "historical poetics" of cinema. By the time movies celebrated their hundredth anniversary in 1995, a variety of competing theories about them had taken the floor.

Cinematic specificity. Since the earliest years, theorists have reflected on the ontology of film. Presuming that film is a photographic medium, what features differentiate it from the empirical reality it records? Are there significant properties that distinguish film from other media, such as painting or theater?

As long as cinema was conceived as a medium reproducing reality, it could hardly harbor artistic possibilities. Silent-era theorists eagerly sought "the cinematic" in hopes that it would ground an aesthetic, a means of showing that cinema deserved consideration as one of the fine arts. Moreover, specifically cinematic features would provide the basis of film style, those resources that creative artists would tap.

Most generally, 1910s and 1920s theorists argued that cinema becomes artistic exactly through breaking from passively recording what happened in front of the lens. Manipulations of camera angle and composition, along with choices about how to assemble shots, transformed what was filmed into something meaningful, expressive, and artistically significant. Arnheim's *Film as Art* was a far-ranging statement of this view. More specifically, theorists sought a cinematic essence that could ground aesthetic preferences. One candidate for a cinematic essence was that of "light in motion"; another was *photogénie,* a mystical affinity of the lens for certain aspects of reality. The most popular candidate for a cinematic essence was the Soviet version of constructive montage. The point was that only cinema could create an imaginary time and space through the assembly of moving images. Films that exploited editing to generate feelings and ideas were fulfilling the medium's potential; those that did not fell back into "canned theater."

This view was challenged by Bazin and his contemporaries. They claimed that unlike traditional arts, cinema had a privileged relation to phenomenal reality because it was photographic. Silent-era theorists erred in treating this as a fault to be overcome; cinema's distinctive aesthetic could arise from its ontological link to the world, its capacity to record concrete events in our spatiotemporal world. Thus not only the masterworks of Jean Renoir, Welles, and Roberto Rossellini, but also documentaries could display the medium's essence to artistic advantage.

Since Bazin, film theorists have often declared the search for a cinematic essence a fruitless task. The subjectivity theorists of the 1970s denounced such ontological quests for ignoring the ultimately social and material determinations shaping cinema. Significantly, however, the semiologist Metz continued to pursue the idea of cinematic specificity: first as a "heterogeneous specificity" consisting of distinct materials and codes, later as an "imaginary signifier," an image that solicits the spectator to oscillate between belief and knowledge. Most recently, analytic philosophers of film have returned to ontological questions, such as those involving the nature of "illusory movement" in film.

Film style and form. In searching for a cinematic essence, silent-era theorists raised other pertinent questions. What are the resources (unique or not) of the film medium? What techniques are available for achieving artistic effects?

The theorists of the 1920s defined many of the techniques fundamental to the medium: camera angles (high or low), camera distances (long, medium, or close-up shots), camera movements, and many possibilities of editing. The theorists' ontological concerns and their prescriptive bent did not prevent them from launching systematic inquiry into film style. The Russians articulated how shot design and editing could be systematically integrated; Eisenstein's essays of the late 1920s and early 1930s offer many suggestions in this regard. Bazin built on the Soviets' insights by showing that the most favored technical choices, such as abstract montage, remained only one cluster of options. He pioneered the realization that staging in depth, focusing on several significant planes of action at once, and allowing the

camera to record action uninterrupted all offered expressive alternatives to montage.

Like paintings, plays, and novels, films are produced and consumed as wholes; yet film theory lacks a vigorous tradition of inquiry into large-scale form. How shall we understand films as integral formal constructs?

The Russian Formalist literary theorists, who also contributed to the theory of style, offered some suggestions in this regard. So too did Eisenstein, when he proposed conceiving of films as spectacles based on audiovisual "polyphony." Bazin had a tacit theory of form, treating traditional storytelling as a pattern that could be elided or opened up for the sake of suggesting a wider reality. On the whole, however, systematic inquiry into films as formal wholes had to await the 1960s. Under the influence of serial theories of musical composition, Noël Burch suggested studying films as through-composed "structures" based on stylistic parameters. Film scholars also borrowed theories of narrative form from anthropology and literary theory. Marie-Claire Ropars Wuilleumier, Raymond Bellour, and Thierry Kuntzel, while largely confining themselves to "textual analysis" of particular works, offered avenues for treating films as large-scale wholes. In the 1980s, Anglo-American writers such as Seymour Chatman have proposed theories of cinematic storytelling. David Bordwell and Kristin Thompson have outlined principles of narrative and nonnarrative form in cinema.

Filmic comprehension. Films are understood, often by many people. How can we explain that process? Silent-era theorists were not preoccupied with this, nor were realists like Bazin: they presupposed some basic psychology of comprehension. With the rise of semiology, however, such presuppositions seemed less obvious.

Film had been considered a "language" in a loose sense since its invention, but the semiologists of the 1960s sought to give the rough analogy real theoretical heft. First, they took as a methodological precept the idea that film could be studied as if it were a language. The researcher could break it into minimal units and postulate rules for substitution and combination that would yield a coherent hierarchy of alternatives. But many semiologists mounted a more substantive analogy between film and language. They held that both the individual image and the ordering of images in montage were grasped through a process akin to reading written language. Umberto Eco proposed a nested set of codes that allowed spectators to build up images of objects and agents out of smaller graphic units. Metz set out eight categories of shot combinations as a paradigm of options for denoting narrative space and time. The presumption was that spectators learned to decipher shots and sequences much as a native speaker learned to understand natural language. This idea chimed with the "social constructivist" emphasis of semiology.

Both "film language" tendencies were seldom explicitly repudiated, but they gradually fell from favor as structuralist theory came to seem too "scientist." A more popular model was that of film as "text" or "discourse," analogies based not on linguistic grammars but on literary rhetoric and hermeneutics. Insofar as subjectivity theory and cultural theory are concerned with comprehension, these analogies have come to the fore, emphasizing interpretation rather than problems of denotation. By contrast, the recent cognitivist trend has focused centrally on comprehension. A cognitive model typically emphasizes the role of spectators in picking up cues proffered by the film and in building inferences, guided by perceptual predispositions and conceptual structures. Several theorists have proposed models of how spectators follow narratives or process film style, and some researchers have sought to test these and kindred theories.

Social and political functions of cinema. Silent-era theorists were quite interested in the cultural role that cinema played, and some left-wing thinkers sought ways in which film's mass popularity could be tapped for progressive ends. Vertov believed that a montage-based "Kino-Eye" could demystify reactionary ideology and reveal the necessity of building socialism. Benjamin saw in cinema's kinetic rhythm both a reiteration of the sensory barrage of urban experience and a robust protest against a dying bourgeois culture.

Subjectivity theory, arising in the wake of failed rebellions of the late 1960s, often sought to understand how semiotic systems solicited audiences into surrendering their critical capacities. Borrowing from Bertolt Brecht as well as from Louis Althusser's theory of ideology, theorists held that cinema recruited people as "subjects," not only by the content of the world depicted on screen, but also through the very tools and techniques of the medium. Thus Jean-Louis Comolli and others argued that cinema was a direct descendant of pictorial systems tied to the rise of bourgeois interests; the camera lens, reproducing Renaissance perspective, also reproduced the effect of a centered, all-knowing subject at the point of perfect visibility. Similarly, the "invisible style" of mainstream cinema, with its smooth cuts and unobtrusive camera movements, created the sense of an omniscient witness, an illusory mastery of a fictional world.

More deeply, cinema asked spectators to find "themselves" in this representation, by channeling unconscious drives into more socially acceptable forms. The pleasure of film viewing was a matter of literally subjecting oneself: the spectator submitted to an ideological programming that offered a satisfaction that in the final analysis served the social status quo. Feminist writers urged that mainstream cinema obliged women to find pleasure in an alienated identity, in scenes in which the woman's body was on display or in which the woman becomes the object of pursuit or mystery.

Only by breaking with cinema's traditional means—its spatial systems, its ideology of invisible technique, the hollow satisfaction of its narrative formulas—could filmmakers offer audiences the kind of distance that made genuine critique possible. These alternatives would sometimes be, by the standards of mainstream entertainment, unpleasant (boring, offensive), but they would avoid the "subject positions" perpetuated by tradition.

It did not take long for feminist theorists to object that this rather ascetic and pessimistic conception of film's social effects left a narrow band of choice for the female spectator. Within psychoanalytic circles, some feminists began to search for sources of "specifically feminine" pleasure, even in mainstream Hollywood cinema. Other theorists minimized the unconscious dimension altogether. One reason for the emergence of culturalist theories of film has been their promise of restoring freedom to the spectator.

Many writers have accordingly moved toward more "decentered" accounts of the psychosocial functions of cinema. Current feminist, gay and lesbian, and minority conceptions of cinematic pleasure emphasize the variety of appeals that any given film can solicit. A film, on this account, can be appropriated as both oppressive and liberatory, depending partly on the negotiations of its audience with their own sense of identity and community. This conception of "negotiated reading," however, has been articulated chiefly through interpretations of particular films. It has not yet yielded a theory of cinema's psychological and social functions as coherent and general as the subjectivity accounts that it contests. The social and political functions of cinema, like other basic questions, remain fair game for the young discipline of film theory.

Silent-era theorists proffered general accounts, but these accounts remained centered in film and films; perhaps the very concept of a cinematic essence kept theorists close to the ground. With the rise of semiology and subjectivity theory, propositions about cinema became part of a much larger weltanschauung, a totalizing theory of signification, social relations, and psychic identity. Cinema became one more signifying system, one more machine of subject positioning and social control. Cultural theory offers a similarly broad, if somewhat more flexible, framework. To these "grand theories," developments since the 1980s offer an alternative. Middle-level problems (of comprehension, or narrative structure) invite a plurality of comparatively modest theories focused on particular questions.

BIBLIOGRAPHY

Bazin, André. *What Is Cinema?* 2 vols. Translated and edited by Hugh Gray. Berkeley, 1967, 1971.
Bordwell, David, and Noël Carroll, eds. *Post-Theory: Reconstructing Film Studies.* Madison, Wis., 1996.
Bordwell, David, and Kristin Thompson. *Film Art: An Introduction.* 5th ed. New York, 1996.
Branigan, Edward. *Narrative Comprehension and Film.* London and New York, 1992.
Carroll, Noël. *Philosophical Problems of Classical Film Theory.* Princeton, N.J. 1988.
Carroll, Noël. *Mystifying Movies: Fads and Fallacies of Contemporary Film Theory.* New York, 1988.
Chatman, Seymour. *Story and Discourse: Narrative Structure in Fiction and Film.* Ithaca, N.Y., 1978.
Currie, Gregory. *Image and Mind: Film, Philosophy, and Cognitive Science.* Cambridge and New York, 1995.
Eikhenbaum, B. M., ed. *The Poetics of Cinema* (1927). Edited by Richard Taylor. *Russian Poetics in Translation* 9 (1982): 1–126.
Eisenstein, S. M. *Selected Writings; 1922–34.* Edited and translated by Richard Taylor. Bloomington, Ind., 1988.
Eisenstein, S. M. *Towards a Theory of Montage.* Edited by Richard Taylor and Michael Glenny, translated by Michael Glenny. Bloomington, Ind., 1992.
Kuleshov, Lev. *Kuleshov on Film.* Edited and translated by Ronald Levaco. Berkeley, 1974.
Lowry, Edward. *The Filmology Movement and Film Study in France.* Ann Arbor, 1985.
Metz, Christian. *Film Language: A Semiotics of the Cinema* (1968). Translated by Michael Taylor. New York and Oxford, 1974.
Metz, Christian. *The Imaginary Signifier: Psychoanalysis and the Cinema.* Translated by Celia Britton et al. Bloomington, Ind., 1982.
Mulvey, Laura. *Visual and Other Pleasures.* Bloomington, Ind., 1989.
Nichols, Bill, ed. *Movies and Methods.* 2 vols. Berkeley, 1976, 1985.
Rosen, Philip, ed. *Narrative, Apparatus, Ideology: A Film Theory Reader.* New York, 1986.
Tan, Ed S. *Emotion and the Structure of Narrative Film: Film as an Emotion Machine.* Hillsdale, N.J., 1996.

DAVID BORDWELL

Feminist Film Theory

Feminist film theory has emerged in the past twenty years to become a large and flourishing field. Its dominant approach, exemplified by such journals as *Screen* and *Camera Obscura,* involves a theoretical combination of semiotics, Althusserian Marxism, and Lacanian psychoanalysis. On this view, human subjects are formed through complex significatory processes, including cinema; traditional Hollywood cinema's "classic realist texts" are purveyors of bourgeois ideology. Added to this theory by Laura Mulvey's now-classic essay, "Visual Pleasure and Narrative Cinema" (1975), was the feminist claim that men and women are differentially positioned by cinema: men as subjects identifying with agents who drive the film's narrative forward, women as objects for masculine desire and fetishistic gazing. Mulvey's essay is heavily invested in theory. It is cited as "the founding document of feminist film theory" (Modleski, 1988), as providing "the theoretical grounds for the rejection of Hollywood and its pleasures" (Penley, 1988), and even as setting out feminist film theory's "axioms" (Silverman, quoted in Byars, 1991). Mulvey assumed a general picture of cinema as a symbolic medium that, like other aspects of mass culture, forms spectators as bourgeois sub-

jects. She used Lacanian psychoanalysis to ground her account of gendered subjectivity, desire, and visual pleasure.

Mulvey allowed little possibility of resistance or critical spectatorship, and recognized no variations in structure or effect of realist cinema. Unsurprisingly, her view has been much criticized and further refined, as writers (including Mulvey herself) have noted issues raised by differences among women, phenomena like male masochism, or genres that function in distinctive ways, such as comedy, melodrama, and horror. Still, writers in feminist film theory commonly assume Mulvey's basic parameters and take some version of psychoanalytic theory as a desideratum. Key issues are often seen only in terms of some refinement or qualification of psychoanalytic theory. Thus, Barbara Creed's book *The Monstrous-Feminine: Film, Feminism, Psychoanalysis*, argues that the fact that women in horror films are often not victims but monsters "necessitates a rereading of key aspects of Freudian theory, particularly his theory of the Oedipus complex and castration crisis" (Creed, 1993). Creed turns instead to Julia Kristeva's theory of the abject and the maternal. Far less often, Mulvey's critics have adopted more sharply different theoretical bases such as cultural studies, identity politics, deconstruction, or the philosophy of Michel Foucault.

The resulting "theory" in feminist film theory is peculiar. What justification does a specifically feminist theory have for adopting the patriarchal theory of psychoanalysis? Why is theory needed at all; what is it a theory of or about? What are its data; do they supply evidence in a noncircular way? How is theory related to feminist action and social change? What is the relevant theory of feminism itself? Theory has usually been more problematic in feminism. Feminist philosophers question patriarchal theories and urge the need to link theory with practice. Jane Flax in "Can Women Do Theory?" describes patriarchal theory as "territorial" or "entrepreneurial"—something used to prop up forms of dominance (Flax, 1979). In the face of theoretical structures that are abstract, hostile, unintelligible, and disempowering, she says, women understandably panic. Similarly, feminist philosophers like Karen Hanson question why writers in film studies have assumed science as a paradigm of theory (Hanson, 1995). In doing so, they set up film *theory* as distinct from and superior or even foundational to film *criticism*. Theory stands somehow over and above the more primitive "data": it is ideal, abstract, permanent, austere, universal, and true. Allegedly, science/theory has the virtues of being unifying, coherent, well grounded, and explanatory. But film theorists naively invoke concepts that are quite contested, such as explanation, justification, and systematicity. Nor is there operational agreement within the discipline for what counts as evidence, testing, or confirmation of a theory. This differs sharply from feminism's more usual emphasis on the experiential.

As Hilde Hein notes, "Some feminists advocate a new definition of theory that decenters, displaces, and foregrounds the inessential and that does not flee from experience but 'muses at its edges'" (Hein, 1993). Early feminists took consciousness-raising as a model for feminist theory because it is necessarily both experiential and transformative. Feminist film theory, however, is often universalizing, and it makes use of complex language and alienating categories that deny women's experiences as active spectators enjoying films or reading them critically. Bell hooks argues, for example, that "mainstream feminist film criticism in no way acknowledges black female spectatorship. It does not even consider the possibility that women can construct an oppositional gaze via an understanding and awareness of the politics of race and racism (hooks, 1995)."

In sum, many feminists would criticize feminist film theory as overabstract, totalizing, jargon prone, and nonexperiential. The dominant psychoanalytic focus has created a narrow framework for the analysis of subjects, pleasure, and desire, while alternative feminist accounts are not considered. For example, Kaja Silverman notes in *Male Subjectivity at the Margins* that "the implicit starting point for virtually every formulation this book will propose is the assumption that lack of being is the irreducible condition of subjectivity" (Silverman, 1992). This approach is odd. Silverman posits a theory of the subject without saying why and without considering alternatives. Similarly, Mulvey posits a theory of desire and pleasure rooted in Jacques Lacan's theory of the self and desire, a view stemming from a particular and highly contested philosophical tradition. To begin from a certain theory of the self or pleasure in interpreting film commits one in advance to categorizations that create the evidence that allegedly confirms them.

An alternative approach would ask how films depict the self and pleasure, or whether viewers can find gaps and ruptures in a film's depictions. Feminist philosophers present alternative views about the construction of women as subjects of knowledge, vision, or pleasure. Carolyn Korsmeyer argues that feminist aesthetics offers a picture of emotional response to artworks different from the traditional one and from that employed in mainstream feminist film theory (Korsmeyer, 1993). Traditional aesthetics is problematic because it sought only a supposedly "disinterested" pleasure. Psychoanalytic theories of the emotions are also problematic: in treating emotions as a matter of the unconscious, they ignore key questions about description, introspection, and moral recommendation. Feminist aesthetics by contrast holds that "perception and appreciation not only entail some particular social standpoint, but are also formed out of the responsive dynamic operating within an embodied viewer" (Korsmeyer).

Philosophers of film can use philosophical defenses of the rationality of emotions to offer new, nonpsychoanalytic

studies of film pleasures. An emotional response to a film (or other artwork) can be rational, permitting positive claims about the viewer: that she is active not passive, cognizing not simply reacting, and potentially critical not simply absorbing ideological effects. Laurie Shrage argues that concentration on film texts has led to a universalizing of psychological subjects and an overemphasis on readers' or viewers' passivity (Shrage, 1993). Shrage proposes a contextual approach that recognizes considerable variation among an audience's "cinematic habits." Similarly, Flo Leibowitz uses a rhetorical and cognitive approach, rather than a psychoanalytic one, to discuss processes of audience identification with certain forms of melodrama, which she thinks may be a form of rational reflection (Leibowitz, 1995). Noël Carroll supposes that emotions are complex learned forms of behavior acquired from certain "paradigm scenarios," and that film, among other sources, can offer such scenarios (Carroll, 1995). Thus, sexist films like *Fatal Attraction* present a purportedly valid but problematic paradigm scenario about the omnivorously sexual career woman, and about the "appropriate" level of male emotional response to such women.

Some feminist philosophers think that what is needed is not so much one feminist film *theory* as a number of strategies of feminist critical *readings* of films. Hanson mentions Stanley Cavell as an example of someone who offers deeply theoretical and philosophical readings by exploring films individually and attentively. Although Cavell's writings on film are idiosyncratic and not necessarily feminist, they offer a springboard for philosophical critiques of assumptions about subjectivity and pleasure still dominant in psychoanalytic feminist film theory (Cavell, 1981, 1987). For example, Naomi Scheman notes that Cavell offers broad and varied notions of the gaze and visual pleasure, enabling one to criticize Mulvey and reject her position that women are subjects or viewers only "in drag" (Scheman, 1995). But Scheman also criticizes Cavell for reading films to uncover not a *feminist* but only a *feminine* gaze, one "conscripted" by a masculinist world. Scheman seeks a more promising form of female subjectivity in film, and cites Foucault's point that dominant modes of specularity are quite complex and do "not define women as exclusively as either the seers or the seen."

Significant alternative feminist theories might also inform feminist film theory. Liberal, socialist, and postmodern feminism all suggest new questions and frameworks. Liberal feminism has emphasized traditionally female attributes in constructing alternatives to standard ethics, such as maternal ethics, the ethics of care, or lesbian ethics. These models may romanticize women, but they offer more complex accounts of the social nature of the self than the Althusserian-Lacanian ones, since the self is essentially configured in relation to others (mothers, children, sisters, friends, lovers). Alternatively, work by socialist feminists or feminists in cultural studies foregrounds the linked oppressions of gender,

ethnicity, race, and class. Ann Ferguson, in *Blood at the Root: Motherhood, Sexuality, and Male Dominance*, describes women's traditional unpaid forms of labor as a form of "sex-affective production" that has been exploited by men (Ferguson, 1989). Thomas E. Wartenberg draws on this kind of notion in offering a critical reading of *White Palace*, a Hollywood film that seems to attend to factors of class and ethnicity (Wartenberg, 1995). Wartenberg shows, however, that the film relies on a stereotyped representation of the older, divorced woman, romanticizes the working class, and oversimplifies the nature of class divisions.

Postmodern feminists also present alternatives to a Lacanian-Althusserian theory of the self or subject, since they question standard notions of human identity based on categories of bodily integrity, race, ethnicity, class standpoint, or even gender. Identity is fractured by complex intersecting social technologies. Hooks points to Julie Dash's film *Daughters of the Dust* as exemplifying a specifically black feminist gaze. Other black writers similarly point to the complicated ways in which a specifically black female identity is represented in films. The postmodern approach to contemporary fractured identities takes filmic signifiers to be recirculated and utilized in a larger system of mass media and popular culture. This sort of approach is used by Tania Modleski in her essay "Three Men and Baby M" to link a popular film with contemporary social and legal issues (Modleski, 1991). Modleski criticizes the allegedly comical inability of the three male adoptive parents in *Three Men and a Baby* to deal with infantile bodily functions. This comedy climaxes in a scene she describes as shockingly voyeuristic, a close and lingering examination of the (female) infant's genitalia. But Modleski does not link this sort of voyeurism to an *individual* viewer's castration anxiety or threat to subjectivity. Instead, she criticizes the film's heroic, simplistic resolution as revealing a current *social* anxiety about changing gender roles.

Alternatives to psychoanalytic feminist film theory raise new questions about the representation of women in films because of their different accounts of the self, agency, identity, and the cultural surroundings of the subject. They reflect more textured and nuanced views about the self's complexity and emphasize the film viewer's potential to construct critical readings. In so doing, they offer more scope for feminist social change than a view that maintains that we are, in effect, products of the texts around us.

[*See also* Feminism.]

BIBLIOGRAPHY

Byars, Jackie. *All That Hollywood Allows: Re-reading Gender In 1950s Melodrama.* Chapel Hill, N.C., 1991.

Carroll, Noël. "The Images of Women in Film: A Defense of a Paradigm." *Journal of Aesthetics and Art Criticism* 48.4 (Fall 1990): 349–360. Reprinted in *Feminism and Tradition in Aesthetics,* edited by

Peggy Zeglin Brand and Carolyn Korsmeyer (University Park, Pa., 1995), pp. 371–391.

Cavell, Stanley. *Pursuits of Happiness: The Hollywood Comedy of Remarriage.* Cambridge, Mass., 1981.

Cavell, Stanley. "Psychoanalysis and Cinema: The Melodrama of the Unknown Woman." In *Images in Our Souls: Cavell, Psychoanalysis, and Cinema,* edited by Joseph H. Smith and William Kerrigan, pp. 11–43. Psychiatry and the Humanities, vol. 10. Baltimore, 1987.

Creed, Barbara. *The Monstrous-Feminine: Film, Feminism, Psychoanalysis.* London and New York, 1993.

Ferguson, Ann. *Blood at the Root: Motherhood, Sexuality, and Male Dominance.* London, 1989.

Flax, Jane. "Can Women Do Theory?" *Quest* 5 (1979): 20–26. Reprinted in *Women and Values,* edited by Marilyn Pearsall (Belmont, Calif., 1986) pp. 2–7.

Hanson, Karen. "Provocations and Justifications of Film." In *Philosophy and Film,* edited by Cynthia A. Freeland and Thomas E. Wartenberg, pp. 33–48. New York and London, 1995.

Hein, Hilde. "The Role of Feminist Aesthetics in Feminist Theory." *Journal of Aesthetics and Art Criticism* 48.4 (Fall 1990): 281–291. Reprinted in *Feminism and Tradition in Aesthetics,* edited by Peggy Zeglin Brand and Carolyn Korsmeyer (University Park, Pa., 1995), pp. 446–459.

hooks, bell. "The Oppositional Gaze: Black Female Spectators." In *Black Looks: Race and Representation,* pp. 115–131. Boston, 1992. Reprinted in *Feminism and Tradition in Aesthetics,* edited by Peggy Zeglin Brand and Carolyn Korsmeyer (University Park, Pa., 1995), pp. 142–159.

Korsmeyer, Carolyn. "Pleasure: Reflections on Aesthetics and Feminism." *Journal of Aesthetics and Art Criticism* 51.2 (Spring 1993): 199–206.

Leibowitz, F. "Apt Feelings, or Why Women's Films Aren't Trivial." In *Post-Theory,* edited by David Bordwell and Noël Carroll, pp. 219–229. Madison, Wis., 1996.

Modleski, Tania. *The Women Who Knew Too Much: Hitchcock and Feminist Theory.* New York and London, 1988.

Modleski, Tania. *Feminism without Women: Culture and Criticism in a "Postfeminist" Age.* New York and London, 1991.

Mulvey, Laura. "Visual Pleasure and Narrative Cinema." *Screen* 16.3 (Autumn 1975): 6–18. Reprinted in *Visual and Other Pleasures* (Bloomington, Ind., 1989), pp. 14–26.

Penley, Constance, ed. *Feminism and Film Theory.* New York and London, 1988.

Scheman, Naomi. "Missing Mothers/Desiring Daughters: Framing the Sight of Women." *Critical Inquiry* 15.1 (Autumn 1988): 62–89. Reprinted, abridged, in *Philosophy and Film,* edited by Cynthia A. Freeland and Thomas E. Wartenberg (New York and London, 1995), pp. 89–108.

Silverman, Kaja. *Male Subjectivity at the Margins.* New York and London, 1992.

Wartenberg, Thomas E. "An Unlikely Couple: The Significance of Difference in *White Palace.*" In *Philosophy and Film,* edited by Cynthia A. Freeland and Thomas E. Wartenberg, pp. 161–179. New York and London, 1995.

CYNTHIA FREELAND

Film and Ideology

Critics and scholars may study film for its ideological content when they wish to examine the relationship between movies and society and, specifically, how films represent social and political realities. Ideologies are systems and patterns of thought characteristic of a particular society or social subgroup. These patterns of thought stipulate particular judgments and assumptions about the nature of right and wrong, good and evil, justice and injustice, law and social order, and human nature and behavior. Because a society contains a multitude of ideologies, films may take up a variety of positions with respect to the issues that are dramatized in a given film's narrative and images. Furthermore, narrative film is an excellent vehicle for ideology because narrative is about process. A film's narrative traces and explains a series of changes in a situation or state of affairs and, by doing so, can embody an ideological *perspective.* For this reason, and because films are so obviously marked socially by the period in which they are made, ideological analysis has been one of the most common ways of studying film.

Scholars examining the ways film models social and political topics may take either an inclusive or a restrictive approach to issues of ideology. Scholars taking an inclusive approach tend to see all films, irrespective of the conditions of their production or their designation as entertainments or as more serious "art" films, as being inevitably ideological. Thus, Tony Scott's *Top Gun* (1986) and Ingmar Bergman's *Persona* (1966), very dissimilar films, may each be regarded as works that give voice to various ideologies, though with differing inflections and degrees of subtlety.

In addition, for the inclusive position, scholars see ideology as coextensive with all film production, and they discern it within the very mechanisms and institutions of the cinema itself. By creating realistic images and by effacing the conditions of that production, the camera and projector are said to subject viewers to "the ideology of the visible," by which viewers come to accept as natural and given the ideologically charged narratives they consume. For the inclusive position, ideology is everywhere present in society and cinema, and the critic's task is to expose its workings while acknowledging the impossibility of ever getting to a place beyond ideology.

Several problems beset the inclusive approach. By stressing the notion that the camera and projector, as mechanisms, embody an ideological operation that substitutes visual illusions masquerading as "reality" in place of an unveiling of social forces and conditions, this approach necessarily remains suspicious of *all* cinema images, viewing them as deceptions foisted on viewers who remain dupes of the illusions created on-screen. Accordingly, the inclusive approach minimizes the rational bases on which filmic understanding rests, namely, the ways that camera work, editing, and narrative construction may cue viewers to apply real-world perceptual and social skills to the medium as a way of making sense of film images and stories. As understood by the inclusive approach, viewers tend to remain passive and worked on by the film's images and narrative, engulfed by spectacle and the ideologies it embodies.

While the inclusive approach has achieved considerable acceptance among film scholars, those taking a more restrictive approach do so based on the conviction that a critical and social construct like "ideology" loses its explanatory usefulness when it becomes coextensive with all the phenomena of cinema that might be studied. Accordingly, for the restrictive approach the task becomes to see in what ways certain films are ideological. Typically, a limited definition of ideology is applied to constrain the range of films and cinematic styles that are said to be ideological. For example, a restrictive critic might elect to study as ideological productions only those films—like Oliver Stone's *Platoon* (1986), which deals with the Vietnam War, or John Milius's *Red Dawn* (1984), which portrays a fictional Soviet invasion of America—that emphasize their sociopolitical discourse in an immediately denotative way. A restrictive critic might also view ideology in film as being a matter, primarily, of a given film's content and of the ways that content may be inflected by visual and narrative structure, rather than regarding ideology (as inclusive scholars do) as being an inherent property of certain visual and narrative styles. By limiting the concept of ideology in these ways, restrictive approaches can avoid portraying viewers as passive dupes and can view them as more active and critical interpreters of a film's social messages.

Whether employing an inclusive or a restrictive definition, the ideological critic studies the ways films portray social and political issues and give voice to one or more social ideologies. Accordingly, the critic may begin by describing certain social trends or habits of thought and then demonstrating how these are represented in given bodies of film. For example, an ideological critic could point to the close correspondence between the anti-Soviet political rhetoric and policies of the Reagan era and a myriad of films produced in those years demonizing the Soviet Union: *Red Dawn* (1984), *Rocky IV* (1985), *Invasion USA* (1985), *Top Gun* (1986), *Rambo: First Blood, Part II* (1985), and *Rambo III* (1988). The job of ideological criticism, here, would be to explicate the political discourse of these films and relate it to the public political rhetoric and policies of the administration.

Sociopolitical ideologies may exist on several levels in a film. In the case of these anti-Soviet productions, the political discourse exists on a first-order level—it is immediate, explicit, and denotative. *Rambo* (1985), for example, deals with an American super warrior (Sylvester Stallone) who returns to Vietnam to rescue American MIAs and score a symbolic military victory against the Vietnamese. Its narrative directly addresses lingering national anxieties (circa the mid-1980s) about the role of the United States as a world superpower and its defeat in Southeast Asia. As such, the film's images and narrative are ideological in an immediate, explicit, first-order way.

More intriguing, in some ways, are those films whose ideological content is of a second-order nature, recessive and implicit within narrative and visual design. The social values and portraits of society in such films tend to be more subtle and indirect. An ideological critic, for example, could show how *Back to the Future* (1985) and *Field of Dreams* (1989) are strongly correlated with the nostalgic myth of a return to the past represented by Reagan-era political culture, particularly that folklore of small towns and close-knit communities that underlay its vision of the idealized national past. Likewise, an ideological critic could explore the highly charged design of Adrian Lyne's *Fatal Attraction* (1987), especially its polarized female characters—the good wife and mother and the evil, single, career woman—to illuminate the social values and attitudes at work there.

Whether films are ideological in a first- or a second-order way, they may adopt any of three basic and distinct perspectives on the sociopolitical material they handle. First, obviously, films may be supportive of established social values. *Top Gun*, for example, demonstrates, through its hero, the grit and resolve needed to become a U.S. Navy pilot. The military is a basic social institution, and the film's glowing portrait of military heroism places the film in a position of clear support for the institution and its values.

The second distinct ideological perspective a film may have is a genuinely critical view of established social values. Within the Hollywood system, such films are rare because hard-edged social criticism is hard to market to a large audience. It is potentially more profitable to reinforce existing ideologies than to challenge them in fundamental ways. Nevertheless, in the late 1960s, films like Dennis Hopper's *Easy Rider* (1969) and Sam Peckinpah's *The Wild Bunch* (1969) offered critical views of established society and clearly sided with rebel heroes.

If a critic wishes to argue that a film either supports or criticizes established social values, the critic also needs to identify the social groups to whom those values correspond. Because society contains multiple communities and multiple, often contending and conflicting ideologies, films may be supportive of values that have currency within one community or subgroup but are disdained or rejected by others. *Longtime Companion* (1990), for example, dealing with the spread of AIDS in the gay community in the 1980s, was controversial because its affectionate, supportive portrayal of gay life clashed with the values of subgroups convinced that gay sexuality is wrong. Descriptions of the ideological position of a film, then, must specify three things: the constellation of social values within the film, the film's attitude toward them, and the social groups to whom these values belong.

Because major studio productions are marketed to a large, heterogeneous audience, ideological ambiguity, the third potential perspective a film may adopt on its sociopolitical material, is actually far more common than straight criticism or support of established social values in major Hollywood productions. By putting, ideologically, a little of

this and a little of that into the film, studios create a sufficiently ambiguous mix that can, ideally, attract as many potential, and ideologically diverse, viewers as possible. The acidic anticapitalist social satire of Paul Verhoeven's *Robocop* (1987), for example, is mitigated by its wholly conventional reliance on action and violent spectacle. In a similar way, the anticorporate satire of Verhoeven's *Total Recall* (1990) is undercut by the film's heavy use of brand-name product placements. Complete ideological coherence is a rarity in American film, largely because of the nation's own tradition of bipartisan and centrist politics and because of the marketing pressures faced by studios seeking distribution for multimillion dollar productions.

Ideological criticism seeks to explicate the first- and second-order levels of ideology present in a given film or body of films, as well as to ascertain whether the ideological perspectives that are present are supportive or critical of given sets of social values or are more ambiguous. The major weakness of ideological criticism occurs when a critic collapses levels of meaning by moving too quickly from the dense pattern of images and the particular and concrete details of a film's characters and story situations to a more abstract social message said to be present in those details. Sophisticated ideological analysis does not reduce a film to a blanket, generalized "message" but maintains clear separation between levels of analysis and respect for the extraordinary density of local and concrete details in the film's surface structure. Good ideological criticism clarifies the multidimensional aspect of film content and keeps us vigilant against the sometimes egregious distortions of social issues that occur in film productions.

[*See also* Ideology.]

BIBLIOGRAPHY

Abercrombie, Nicholas, Stephen Hill, and Bryan S. Turner. *The Dominant Ideology Thesis.* London and Boston, 1980.

Baudry, Jean-Louis. "Ideological Effects of the Basic Cinematographic Apparatus." In *Narrative, Apparatus, Ideology: A Film Theory Reader,* edited by Philip Rosen, pp. 286–318. New York, 1986.

Carroll, Noël. *Mystifying Movies: Fads and Fallacies in Contemporary Film Theory.* New York, 1988.

Comolli, Jean-Luc, and Jean Narboni. "Cinema/Ideology/Criticism." In *Movies and Methods,* vol. 1, edited by Bill Nichols, pp. 22–30. Berkeley, 1976.

Curran, James, Michael Gurevitch, and Janet Woollacott. "The Study of the Media: Theoretical Approaches." In *Culture, Society and the Media,* edited by Michael Gurevitch, Tony Bennett, James Curran, and Janet Woollacott. London and New York, 1982.

Grossberg, Lawrence. "Strategies of Marxist Cultural Interpretation." *Critical Studies in Mass Communication* 1 (December 1984): 392–421.

Hall, Stuart. "Encoding/Decoding." In *Culture, Media, Language: Working Papers in Cultural Studies, 1972–1979,* edited by Stuart Hall et al. London, 1980.

Nichols, Bill. *Ideology and the Image: Social Representation in the Cinema and Other Media.* Bloomington, Ind., 1981.

Polan, Dana. *Power and Paranoia: History, Narrative, and the American Cinema, 1940–1950.* New York, 1986.

Prince, Stephen. *Visions of Empire: Political Imagery in Contemporary American Film.* New York, 1992.

Rosen, Phillip, ed. *Narrative, Apparatus, Ideology: A Film Theory Reader.* New York, 1986.

STEPHEN PRINCE

FIRST AMENDMENT AND ART. *See* Law and Art, *article on* Government-Funded Art and the First Amendment.

FOLK ART. The separation of art and craft in the eighteenth century clearly left what later came to be called folk art on the side of craft since fine art was a class concept, defined not only by its exclusion of utility and mere entertainment but also by the exclusion of popular genres. Johann Gottfried Herder and the early Romantics discovered aesthetic value in the arts of the people or folk *(das Volk),* primarily in traditional ballads and tales that were considered to embody the uncorrupted genius of each nation. In the nineteenth century, as industrialization and urbanization shrank the remaining areas of traditional culture, collecting and studying the traditional arts of rural folk became a focus of scholarly interest and national pride and generated societies for the study of "folklore" by the end of the century. In the twentieth century the general idea of folk art has played an important role in the work of a number of art critics and cultural historians and has led to the institutionalization of the academic discipline of folklore. Folk art has also come to be recognized by the art world as an area of collection, exhibition, and criticism with its own museums and galleries. Although the idea of folk art has not been central to traditional philosophical aesthetics, an examination of the debates about its meaning can illuminate the failure of twentieth-century attempts to define art.

Until the 1970s many cultural critics, art historians, and sociologists used the concepts of folk art and popular art as foils for defining fine art (Hauser, 1985). In these tripartite models folk art was often presented as the expression of an authentic culture that was fast disappearing before the onslaught of modern "mass culture." In this view the arts of the unspoiled folk who once made and enjoyed their own music, stories, and artifacts are contrasted with the popular arts supposedly manufactured by a distant and exploitative culture industry for mass consumption.

The academic discipline of folklore has combined the approaches of literary history (and more recently, art history) with those of anthropology and sociology to study folk communities. Traditionally, the folklore profession has defined folk art as works or performances done in a traditional

style, for consumption in the local community, by members of a socially homogeneous group that is usually poor and relatively isolated from the influences of both fine art and mass culture. Until recently, folklorists still gave the bulk of their attention to folktales, ballads, and other literary and musical genres. When folklorists look at folk cultures more broadly, they tend to divide study along the lines of the canonical divisions of the fine arts—folk literature, folk music, folk dance, folk art (painting, sculpture, decorative arts). As in almost any discipline, definitions of what constitutes the "folk" and what constitutes "art" or "music" have been a matter of dispute. Some folklorists, for example, would distinguish works of folk "art" (paintings, nonutilitarian decoration or carvings) from folk "craft" (chairs, quilts, churns), but others claim that the art/craft distinction is inapplicable to folk societies.

A somewhat different conception of folk art has prevailed in the art world of critics, dealers, collectors, and museum curators. In Europe at the turn of the century a number of modernist artists turned to the arts of the "uncivilized," finding inspiration and models in the works of peasants, primitive peoples, children, and the mentally ill. Part of the American echo of this trend was the "discovery" and exhibition of folk paintings, weather vanes, and other decorated and carved objects by a group of modernist painters in the 1920s. The Newark Museum and the Museum of Modern Art in New York took up the cause in the early 1930s, and soon folk art was being collected, exhibited, and written about as part of the art world. By the 1960s some of its advocates claimed parity with fine art (Bishop, ed., 1982).

Beginning in the 1970s the collecting and exhibition of folk art began to accelerate. A popular version of the art-world idea of folk art inspired this enthusiasm; in this view anything homemade, quaint, or crude has been called either folk art or "naive," "outsider," "grassroots," "vernacular" art. The causes of this American penchant are varied and probably include the 1976 bicentennial, the art world's acceptance of almost any kind of artifact or performance as art, the feminist critique of excluding traditional domestic arts such as quilt making from the fine arts. By the mid-1980s isolated rural carvers, weavers, or chair makers and self-taught urban painters or sculptors were besieged by collectors, dealers, and curators. As the popularity of folk art accelerated, folklorists worried about the erosion of tradition, skeptics decried financial exploitation, and the concept of folk art itself became highly contested.

Folklorists, for example, have attacked the art-world idea of folk art for focusing on the isolated "work of art" rather than the communal context that produced it. Other folklorists have shown that many of the so-called folk portraits that form the core of many private or museum collections were mostly done by itinerant middle-class artists with for-

mal training. Some folklorists have concluded that the modern concept of art is simply not appropriate to genuine folk communities and have recommended the term *material culture* instead (Ames, 1977). Others have gone to the opposite extreme and claim that given the prestige of the term *art*, the concept of art should be expanded to include all folk objects, performances, and practices (Glassie, 1989). Others have suggested not only expanding the concept of art but expanding the definition of *folk* to include the everyday lives of contemporary urban and suburban people (Jones, 1987). One prominent cultural historian has argued that what we call popular culture should simply be regarded as "the folk art of industrial society" (Levine, 1992).

Whereas earlier in the twentieth century the idea of folk art formed a more or less stable placeholder in a conceptual set that included fine art and popular art, it has now become part of the debate about whether the idea of art has any defining boundaries. Although the meaning of both *folk* and *art* are highly contested, the term *folk art* itself does not seem likely to vanish given the heavy intellectual, institutional, and financial investment in it. Unlike the term *primitive art*, which is rapidly disappearing because of the deprecatory connotations of *primitive*, the word *folk* still conveys the positive connotations of honesty, simplicity, naturalness, and links with feelings of ethnic or national pride. The debates about the meaning of *folk art*, however, would appear to bolster the view of some philosophers and art historians that the basic concepts of aesthetics, including that of fine art itself, are Euro-American historical constructions rather than human universals.

[*See also* Craft; Fashion; Outsider Art; Popular Culture; *and* Tribal Art.]

BIBLIOGRAPHY

Ames, Kenneth L. *Beyond Necessity: Art in the Folk Tradition.* Winterthur, Del., 1977.
Bishop, Robert, ed. *American Folk Art: Expressions of a New Spirit.* New York, 1982.
Glassie, Henry. *The Spirit of Folk Art.* New York, 1989.
Hauser, Arnold. *The Philosophy of Art History.* Reprint, Evanston, Ill., 1985.
Jones, Michael Owen. *Exploring Folk Art: Twenty Years of Thought on Craft, Work, and Aesthetics.* Ann Arbor, 1987.
Levine, Lawrence. "The Folklore of Industrial Society: Popular Culture and Its Audiences." *American Historical Review* 97 (1992): 1369–1399.
Metcalf, Eugene W. "Black Folk Art and the Politics of Art." In *Art, Ideology, and Politics,* edited by Judith H. Balfe and Margaret Jane Wyzomirski. New York, 1985.
Oettinger, Marion, Jr. *The Folk Art of Latin America: Visiones del Pueblo.* New York, 1992.
Shapiro, Gary. "High Art, Folk Art, and Other Social Distinctions." In *The Folk: Identity, Landscapes, and Lores,* edited by Robert J. Smith and Jerry Stannard. Lawrence, Kans., 1989.

LARRY SHINER

FOOD. Foodstuffs have long been made into objects that are claimed to be visual works of art: the nineteenth-century pastry cook Marie-Antoine Carême, who was famous for his elaborate sculptures made out of sugar, claimed that confectionery was the principal branch of architecture. But such creations do not raise any aesthetic questions peculiar to food (unless the knowledge that they are edible affects the way we look at them, as Marienne L. Quinet suggests). What follows therefore assumes that food aesthetics concerns the possibility of aesthetic reactions to, and works of art based primarily on, the qualities *peculiar* to food and drink: taste and smell.

There is a tendency to dismiss the possibility of an art based on food on the grounds of sharp differences between the putative art and accepted ones. But standard arts themselves differ sharply. Painting is visual, music aural; architecture is spatial, music temporal, dancing both; sculptors create locatable objects, poets create abstract things. The strategy of this essay will be to show that general points that can be made about the arts can often be made about some food creations and that although an art of food would differ from some standard art forms, it would resemble others.

Aesthetic Reactions. It seems to be agreed that at least there can be aesthetic reactions to food. But the notion of an aesthetic reaction is obscure. In simple cases it can be characterized as a noninstrumental, vivid pleasure in (or pain at) something as perceived by the senses. The term *noninstrumental* is used to distinguish being pleased with something because it is useful from aesthetic pleasure in the thing in itself, and *vivid* to distinguish vague pleasures from the more focused pleasure that we think of as an aesthetic reaction. Aesthetic reactions may be simple (enjoying the beauty of the blue sky) or more complex (enjoying the pattern made by differently colored fields in a landscape). They can be sought by deliberately paying attention to the intrinsic qualities of an object, or borne in on someone who initially looks at an object instrumentally or pays no attention to it at all.

It is not clear how this kind of analysis fits abstract things such as poetry. But it fits food rather well. When we get a vivid pleasure from the taste and smell of some food, not because (for example) we think it will sell but for its own sake, this is an aesthetic pleasure like that felt by the admirer of the sky or the landscape, and it can arise whether the eater is a gourmet looking for just this experience or an ordinary eater who is thinking at first of something else but whose attention is caught by something special in the taste.

Aesthetic pleasure is sometimes spoken of as *disinterested*, rather than noninstrumental. But that term, or rather the German word usually translated by it in English, is also used in a special sense by Immanuel Kant to distinguish pleasure in the arts, which he calls disinterested, from pleasure in the agreeable, which he calls interested. Since he classifies the pleasure of food as interested, his distinction is relevant to the question whether food can be the basis of an art. A disinterested pleasure for Kant is one that neither arises from nor produces desire. He thinks that pleasure in the agreeable *produces* desire; for example, if we enjoy the first chocolate we want another one. But this does not always happen; there are also sated pleasures, such as enjoying a last chocolate and *not* wanting another one. More important, we should ask whether the tendency to produce desire really distinguishes pleasure in the agreeable from pleasure in the arts. If a work of art gives us pleasure, must we not want to go on contemplating it, or to look at or listen to similar things in the future? Admittedly, when a play is over we feel that our pleasure is somehow complete, but this is like the sense in which our pleasure at the end of a perfect meal is complete.

Aesthetic reactions often include a judgmental component: observers not only enjoy the sights or sounds, but also think that their pleasure is appropriate to the quality of the object and that others ought to share their reactions. They may try to give reasons for their judgment; but as these typically consist of pointing out different features that please them in the object, they may have no reasons to give for admiring a *simple* object, like the blueness of the sky or the taste of a wild raspberry.

Some think that whereas the notion of judgments that claim objectivity makes sense in other aesthetic spheres, it cannot apply to food, where people simply like some things and dislike others. Sometimes writers contrast a single taste with a complex work of art, as Marcia Mueder Eaton does when she asks whether our reactions to *War and Peace* are like our reactions to the taste of peanut butter. But this difference may be more apparent than real. A fairer comparison would be between a short poem and a complex dish of food. In both these cases we find people arguing the merits of the object in question and hoping to convince others that their view of it is right; and in both cases we would expect to find considerable agreement between discerning people.

David Hume, in his essay "Of the Standard of Taste" (*taste* in what he calls the "metaphorical" sense), writes of "true and just" standards of taste in aesthetic matters. Perhaps aesthetic judgments are not as objective as Hume thinks, but it may be maintained that there can be judgments of taste in the metaphorical sense about tastes in the literal sense, and that such judgments may be no less objective than those in other aesthetic spheres.

Works of Art and Food. If there can be aesthetic reactions to food, can there be works of art based on taste and smell? This is a complex question, because what counts as a work of art is not a simple matter. For one thing, the phrase "work of art" can be used either in a neutral, classifying sense or in an honorific sense. It might indeed be claimed that whereas food is sometimes a work of art in the classifying sense, the honorific status of "work of art" cannot be bestowed on anything consisting primarily of tastes and smells.

We can define work of art in the classifying sense of the term by means of the notion of aesthetic reaction I have sketched above: a work of art is created by a human being and intended by its maker to be studied primarily for the sake of the aesthetic reactions it evokes. The word "primarily" allows useful objects to count as works of art if usefulness is not their main purpose. According to this definition, some foods (by no means all, of course) can be classed as works of art: they are made not mainly for nutritive value or even casual pleasure but to be studied for the sake of those reactions that count as aesthetic.

This definition does not cover objects, such as cave paintings and cathedrals, that were not necessarily intended for aesthetic contemplation by their makers but that come to be treated in that way. We therefore need a second clause in the definition, whereby something becomes a work of art if it is studied for the sake of aesthetic reactions by critics or public. Dishes at an unpretentious restaurant become works of art if a food critic urges readers to sample them and look for various features. So does a traditional recipe recommended by a food writer for its subtle balance of flavors or its sure handling of ingredients.

The claim that recipes as well as actual dishes of food are possible works of art can be supported by appealing to parallels in other arts: musical compositions or plays are works of art that are abstract blueprints realized in performance, rather as a recipe is realized by being cooked. Admittedly, in drama or music we do not call the performances as well as the blueprint *works* of art, as we might do with food, but we do think of them as artistic creations.

So far the candidates for food works of art in the classifying sense are recipes and dishes. But dishes can form other works of art such as courses and meals, just as works of art such as statues can form parts of larger works such as cathedrals. (Sometimes, indeed, we can see the food aspect of a meal as only one aspect of a multimedia art form in which the diners are not only audience but also actors in a partly improvised, partly organized "happening": a dinner party.)

Art and Craft. It is sometimes claimed that cookery is not an art but a craft. One reason for this claim, the usefulness of the products of cookery, is not a barrier to art, according to the definition here of a work of art. Arnold Whittick suggests other differences between arts and crafts: an artist is original and creative, whereas an artisan follows instructions, conforms to conventions, applies techniques, and so on. According to these criteria most artists are also artisans. The inventor of a recipe usually works within conventions, and is in that respect an artisan, but if he or she intends the dish to be scrutinized aesthetically he or she also counts as an artist in the classifying sense. An assistant who follows a chef's exact instructions is an artisan rather than an artist. But a cook following a recipe may be an interpretative artist. She or he needs technique and so is an artisan as well. But a cook may also make decisions about interpretation that call for aesthetic imagination and raise questions analogous to those about interpreting music, such as whether to use different ingredients from those specified—perhaps ingredients that were not current when the original recipe was created—which might make a better though inauthentic version.

Many recipes do not have an individual creator: they have developed through many hands. But it does not follow that the eventual recipe is not a work of art; the question in such cases is whether the recipe and its interpretations are *treated* as works of art.

Foods as Works of Art in the Evaluative Sense. Can some food constitute a work of art in the evaluative sense? In other words, can it deserve or repay the kind of scrutiny that is characteristically given to works of art, and is often in fact given to food?

The usefulness of food need not prevent it from being worth aesthetic scrutiny, but treating useful things as art is sometimes criticized as precious. Whether this criticism is apt, however, depends on the degree to which an object repays aesthetic scrutiny. Where food does do so, treating it as art is not in itself precious; but attention may become precious if it is out of proportion to the food's aesthetic importance.

The bodily nature of the way we appreciate food, and of its eventual fate, might seem unworthy of the dignity of art, but only to an extremist who seeks to decry anything bodily. If these considerations do have importance it is because they underlie other difficulties about food as an art.

The main type of argument against the possibility of food works of art in the evaluative sense appeals to the supposed limitations of our capacities to discern tastes and smells, or to the comparative simplicity of tastes and smells themselves. (The distinction between limitations in our capacities and limitations in the phenomena is dubious; it is safer to say that some objections are presented in terms of the perceiver, others in terms of what is perceived.) Arguments about the limitations in our capacities typically maintain that eye and ear can make finer discriminations, and have greater powers of recuperation, than nose and palate; and that it is more difficult to remember tastes and smells than sights or sounds. On discrimination, one might reply that all senses can be cultivated, and that if food were more widely regarded as an art more people would cultivate their powers of taste and smell. *Power of recuperation* refers to the ability to switch a faculty quickly and cleanly from one object to another. Admittedly, tastes sometimes linger and spoil the next taste. But there are analogous problems in painting and music. In all these spheres the experienced observer can allow for recuperation. Memory for food, like discrimination, may be amenable to training. But in any case it seems more relevant to criticism than to appreciation in an art form such as food, which need not depend on remembering what has gone before.

One argument about the limitations of tastes themselves is that of Monroe C. Beardsley, who suggests that tastes cannot be arranged in "systematic, repeatable, regular combinations" because they do not form sequences. [*See article on* Beardsley.] D. W. Prall makes similar claims. This argument is unconvincing on several grounds: tastes do seem to form sequences (from sweet to sour, for example); a work of art does not seem to require regular combinations, though that is characteristic of some arts, such as music; food does seem to allow for regular combinations (though, admittedly, these are partly created by the rhythm of the diner's eating).

A second argument, by Edmund Gurney, suggests that form is not possible in tastes, as it is in a chord in music, and that form is necessary for beauty. [*See article on* Gurney.] This argument too is puzzling: it seems that tastes can be combined in ways rather analogous to chords in music, with elements varying along several dimensions at once. Tastes are also thought not to admit of balance and climax, which works of art may be thought to need. But this claim also is unconvincing. Certainly, cooks think in terms of these ideas when planning meals.

Writers rejecting the possibility of food works of art tend to hold that unless food creations are complex they do not deserve to be considered works of art at all: for example, Beardsley asks why there are no "taste symphonies and smell sonatas." But symphonies and sonatas are particularly complex works of art: one can concede that food creations will be comparatively simple but still maintain that some deserve to be treated as works of art.

A more important type of objection concerns the difficulty of sharing experiences of food. Whereas each observer of a painting looks at the same thing, each eater of a dish tastes a different thing: his or her own set of mouthfuls. It might be claimed that this fact limits the possibility of public debate about food and of recommending fine food to others—activities that are characteristic of the arts. But the difficulty is only theoretical when the dish is reasonably homogeneous or when realizations of a recipe are very similar, and such debate and recommendation often take place.

A related objection concerns the transience of food: how can something that is destroyed in the very act of appreciating it deserve to be called a work of art? One possible answer is that dishes of food are like transient performances, which are not thought of as works of art but which are nevertheless artistic creations that can be discussed and commended. Moreover, food can be like the repeated performances of a play, rather than the single performances of music: a food critic can recommend a restaurant's bouillabaisse to readers rather as a theater critic recommends a particular actor's performance of a role.

There is, however, an important distinction between food creations and other performances. Performances of drama, music, and dance can now be recorded, allowing succeeding generations to contemplate and compare them and enabling some performances to attain classic or definitive status by this process. Food creations cannot be recorded in this way, not because preservation techniques are not adequate, but because what is preserved is not a representation but the thing itself: the food will be eaten before sufficient numbers of people can taste it. There cannot therefore be classic realizations of a dish in the way that there can be classic performances of a play or piece of music. Recipes, which as we saw earlier are analogous to a composition rather than to particular performances of it, are more promising candidates for permanence; there can be classic recipes that have stood the test of time.

Despite the similarities between food art and other arts, we tend not to think that even classic recipes have the stature that a classic piece of music or of literature does, or that a fine example of a classic dish is comparable to a great musical performance. In other words, even if food can sometimes be art in the evaluative sense of the term, we tend to think of it as a *minor* art.

One reason for this attitude is the impermanence of food, which has just been discussed. But a more important reason concerns food's lack of *meaning*. Food is rich in some kinds of meaning: the food we eat, and the ways in which we eat it, have great historical and sociological significance and often various symbolic meanings too. But there are other kinds of meaning that food lacks. It cannot represent the world, as painting and literature can, and it cannot be said to embody or convey truths as they can also be said to do. In these respects food resembles music. But food cannot express emotion either, as music often does: there is no language of food, as there is said to be a language of music.

What these limitations signify depends on one's view of art. To a formalist none of them constitutes reasons why food is any less an art, as they all concern nonaesthetic features. But formalists and nonformalists can agree that factors that depend on representation or on expression of emotion are among the reasons why we *value* art, and so the absence of them is a reason to downgrade the art of food. More important still is the fact that although good food can please, excite, or exhilarate us, it does not seem capable of moving us. (This is not the same as saying that food does not *express* emotion: we can be moved by the beauty of something austere that itself expresses no emotion at all.) The capacity to move us seems a very important element in what we value about the arts. Lacking this, the art of food must remain a minor art form.

Practical Implications. We hold that the state should subsidize the arts and that we should teach children about them. Do these requirements apply to food art, since it is only a minor art? The state need not subsidize food art by supporting creative chefs, since they will survive, minor artists or not, without state support. It might be thought that children need to be taught only about major arts. But

this is not necessarily so. Good food is a more universally accessible art form than many major ones. It is also one way of making elegance part of ordinary life, and it therefore has an aesthetic role that the major arts cannot play.

[*See also* Craft; Pleasure; *and* Taste.]

BIBLIOGRAPHY

Beardsley, Monroe C. *Aesthetics: Problems in the Philosophy of Criticism* (1958). 2d ed. Indianapolis, 1981.

Eaton, Marcia Muelder. *Basic Issues in Aesthetics.* Belmont, Calif., 1988.

Fisher, M. F. K. *The Art of Eating.* New York, 1954.

Gurney, Edmund. *The Power of Sound.* London, 1880.

Heldke, Lisa M. "Foodmaking as a Thoughtful Practice." In *Cooking, Eating, Thinking: Transformative Philosophies of Food,* edited by Deane M. Curtin and Lisa M. Heldke, pp. 203–229. Bloomington, Ind., 1992.

Hume, David. "Of the Standard of Taste." In *David Hume: Selected Essays,* edited by Stephen Copley and Andrew Edgar, pp. 133–154. Oxford and New York, 1993.

Kant, Immanuel. *Critique of Judgement.* Translated by James Creed Meredith. Oxford, 1952.

Prall, D. W. "The Elements of Aesthetic Surface in General." In *The Problems of Aesthetics,* edited by Eliseo Vivas and Murray Krieger, pp. 182–193. New York, 1953.

Quinet, Marienne L. "Food as Art: The Problem of Function." *British Journal of Aesthetics* 21.2 (Spring 1981): 159–171.

Telfer, Elizabeth. *Food for Thought: Philosophy and Food.* London and New York, 1996.

Urmson, J. O. "What Makes a Situation Aesthetic?" *Proceedings of the Aristotelian Society* suppl. vol. 31 (1957): 75–92.

Whittick, Arnold. "Towards Precise Distinctions of Art and Craft." *British Journal of Aesthetics* 24.1 (Winter 1984): 47–52.

ELIZABETH TELFER

FORGERY. A forgery is a work of art whose history of production is intentionally misrepresented to a buyer or audience, usually for financial gain. Fraudulent intention is necessary for a work to be a forgery; this distinguishes forgeries from honest copies and merely mistaken attributions. A forger normally draws, paints, or sculpts a work in the style of a famous artist and tries to sell the result, often in collusion with an unscrupulous dealer, as coming from the hand of the famous artist. Forgers seldom try to execute exact copies of existing authentic works, as such fakes are impossible to sell to informed buyers.

Forgery has vexed critics and philosophers as an aesthetic issue. The market value of any work of art will normally plummet if it can be shown to be faked, and museums will relegate to their basements paintings that are revealed to be forgeries, even though they may have delighted generations of museum goers. Is this reasonable? If a work of art remains the same visual object after the revelation that it is a forgery, why should it be rejected aesthetically? Some theorists, notably Arthur Koestler and Alfred Lessing, have argued that only confusion and snobbery could be behind such rejection. The aesthetic value of art derives from the pleasure of immediate auditory or visual experience, they claim, and it therefore can make no aesthetic difference if a work is revealed to be a forgery. This position can be developed for (1) a forgery that is an exact copy of an existing work, (2) a forgery that is a new work in his or her style of and attributed to another artist, and can be extended even to (3) an honest copy of an existing work or an honestly presented new work in another artist's style.

The celebrated case of the Dutch forger Han van Meegeren (1889–1947) is a locus classicus of modern forgery. As van Meegeren's promising artistic career faltered and declined in the 1920s, he began to see himself as a victim of critics, dealers, and academics. Partly from resentment and partly from greed, he began by forging a *Laughing Cavalier* in the style of Franz Hals. Later, he moved on to the work of Jan Vermeer, whose paintings were of considerable rarity and market value (fewer than forty authentic Vermeers had survived into the twentieth century). His efforts culminated in *Christ and the Disciples at Emmaeus,* a pseudo-Vermeer he completed in 1937. To produce this painting he undertook an exacting study of formulas for seventeenth-century paints and experimented with volatile flower oils to produce a pigment surface that had both the hardness of old paint and at the same time made it possible to create *craquelure,* the system of cracking found on the surface of old paintings. When *Emmaeus* was first put on display in 1937, the eminent historian Abraham Bredius pronounced it perhaps Vermeer's greatest masterpiece. Van Meegeren continued to forge Vermeers until he was arrested soon after the war for having sold a Dutch national treasure to the enemy: one of his fakes had been bought by Hermann Göring. Confessing his forgery, van Meegeren was treated as something of a folk hero and dealt with lightly by the court. He died in prison before the end of his one-year sentence.

The Koestler and Lessing position would demand that van Meegeren's pseudo-Vermeers should continue to hang, although correctly labeled, on museums walls: since no one could tell the difference between them and real Vermeers before the confession, it follows that there can be no aesthetic difference between them.

Other aestheticians have opposed this position. In his influential treatment of the problem, Nelson Goodman has argued that just because we cannot tell today the difference between an original and an apparently indiscernible copy, it does not follow that we will not be able to see a difference in the future. For Goodman, far too much is made of the notion "perfect," indistinguishable fakes and copies: though indistinguishable to an uneducated child, he says, two paintings might well be easy to tell apart one day when that child has become a museum director. This fact about the development of connoisseurship leads to two more considerations for Goodman. First, if one is informed that a work is a forgery, the sheer possibility that one might be able to see a difference in the future justifiably makes a difference

to one's present viewing. Second, knowledge that a work is forged "assigns the present looking a role as training toward . . . perceptual discrimination" (Goodman, 1976). By trying to detect the subtle qualities that distinguish an original from a fake, one actually learns to see such differences. These factors entail that in the mind of the serious art aficionado, an enormous gulf properly separates an original even from an apparently indiscernible forgery.

Goodman's argument applies both to identical twin copies and to works forged with the intent to include them in an existing oeuvre as newly discovered works, and his stress on the importance of the informed eye is nicely supported by van Meegeren's forgeries. What had looked to many like a perfectly acceptable Vermeer in 1937, *Christ and the Disciples at Emmaeus*, was in 1947 less plausible, while by the end of the century it seems almost laughable that the painting was once considered to be a Vermeer. In retrospect, *Emmaeus* looks out of place for a seventeenth-century painting: its faces are both sentimental and oddly photographic; one even bears a resemblance to Greta Garbo. Nevertheless, it is perhaps the most acceptable of van Meegeren's Vermeer fakes. Once *Emmaeus* was included in the class of authentic Vermeers, van Meegeren's next effort did not need to strain as hard to look like a Vermeer: it had to resemble the new understanding of Vermeer's oeuvre, which was modified by the inclusion of *Emmaeus*. So as each new van Meegeren Vermeer distorted the historical understanding of Vermeer by adding yet another fake to the class of accepted Vermeers, the forger found he could get away with his fraud merely by adopting the style of his previous forgeries, rather than following the original precedent-class of authentic paintings. His final pseudo-Vermeers thus more resembled twentieth-century German Expressionist paintings than seventeenth-century Vermeers. When van Meegeren confessed, it also became easy for anyone to spot the paintings by van Meegeren. (Still, it should be noted that there were doubts about van Meegeren's Vermeers even at the beginning: the Dutch agent for the New York dealer Duveen went to the first showing of *Emmaeus* and wired his employer that it was a "rotten fake.")

In another attempt to counter the aesthetic empiricism of Lessing and Koestler, Leonard Meyer has argued that our culturally determined response to art requires that it be seen not merely in terms of immediate experience: we cannot help but see a work as a historical artifact. Hence, a work's status as original or forged is inevitably tied to how we see and understand it. This is because works of art are solutions to historically determined stylistic and technical problems, he says, and their solutions earn respect only if they occur in a particular historic context: if we know that a work is forged, he says, "our whole set of attitudes and resulting responses is necessarily altered" (Meyer, in Dutton, 1983). Meyer claims that it is pointless to say that it is all in the mind, since we can no more escape from our cultural con-

ditioning than we can breathe in a vacuum. (In fact, Meyer turns back on itself Koestler's charge of snobbery: "to continue to admire a proved forgery," as though its provenance were an irrelevancy, "is a pretentious form of inverted snobbery.") Along the same line, Denis Dutton has argued that failure and achievement are intrinsic to the practice of art, and that forgeries constitute a misrepresentation of achievement because they misrepresent their origins (Dutton, 1983). The van Meegeren Vermeers might well be excellent van Meegerens, but they are solutions to the problem of how to forge works by a seventeenth-century artist; Vermeer himself created work in his own time that answered to a completely different set of problems. In this sense, the forger's achievement, though considerable, cannot be understood in the same way as the achievement of the original artist. Both Meyer and Dutton reject aesthetic empiricism by insisting that an aesthetic response to a work takes into account the whole human context of its creation, and is not merely a reaction to its sensuous surface.

Michael Wreen has pointed out, on the other hand, that such positions as Meyer's and Dutton's cannot distinguish between the aesthetic value of an original work of art and that of a forgery produced in the same historic context— say, by a contemporary forger of Pablo Picasso—because both the original artist and the forger faced the same technical problems (Wreen, in Dutton, 1983). In this respect, the forger's task might be imagined to be even more difficult than the original artist's, because the forger must overcome historic technical problems and at the same time produce a passable imitation of another artist's work. Jack W. Meiland has also criticized what he considers the overemphasis on originality, which he calls a "derivative," rather than primary, aesthetic value (Meiland, in Dutton, 1983). Being an original artwork—perhaps a work that initiates a tradition or a genre—is valuable only in the sense that later works in the established tradition themselves have what Meiland calls primary aesthetic value. The foundation of aesthetic value is therefore still in the basic appearances of works of art, and not in their special historical circumstances. Strikingly original artworks, like the first lightbulb or waterwheel, may well be fascinating historical curios, but their aesthetic interest, if they possess any, still stems from their aesthetic appearances.

That forgery as a form of fraud is immoral is accepted by all aestheticians who have dealt with the problem. What remains in dispute is the extent to which the moral question should be allowed to affect the aesthetic response. In a provocative approach to this question, Francis Sparshott says, "In seeking to appreciate a work we rely on its promise of a human significance and loyally entrust ourselves to that promise" (Sparshott, in Dutton, 1983). What the art forger "exploits and betrays is just the self-giving on which all human relationship depends." The analogy Sparshott presents is that of making passionate love in the dark to someone

who, alas, turns out to be the wrong person. The aesthetician who claims that it "makes no difference" whether a painting one enjoys is forged or not is, for Sparshott, rather like the proponent of free sex for all; legitimate aesthetic experience, he says, like sexual experience, "depends on imaginative construction and association, for only an imaginatively funded vision detects and responds to the meaningful structure of a picture or a musical piece. And, because all social and personal bonds are reciprocal but directly known from one end only, all the relationships we know or think we know ourselves to be living in are as fragile and subject to illusion as the art lover's confidence in the authenticity of a work and the integrity of its artist."

Sparshott's position suggests that the appreciation of art is transactional, a communicative event requiring trust and good faith. Meyer's insistence that the artwork be seen in its historical context also construes aesthetic experience as requiring a properly understood relation between artistic creator and audience. If so, the future might see greater confusion in a proliferation of fakes, copies, and altered performances. For example, digital techniques now make possible not only the easy altering of news photos and other visual material, they also make such changes possible in the field of audio recording to correct a pianist's mistakes in a performance, or even to improve a singer's pitch. Whether such doctoring will in the future be considered normal procedure or a form of cheating remains to be seen. Copying technologies for painting and sculpture are not yet as technically impressive as they are in the digital recording of sound, but that may change in the next generation. Whether this will alter our view of copies and create new possibilities for faking remains to be seen.

[See also Appropriation.]

BIBLIOGRAPHY

Arnau, Frank [pseud. for Heinrich Schmitt]. *The Art of the Faker: Three Thousand Years of Deception.* Translated by J. Maxwell Brownjohn. Boston, 1961.

Bredius, Abraham. "A New Vermeer." *Burlington Magazine* 71 (November 1937): 210–211.

Cebik, L. B. *Nonaesthetic Issues in the Philosophy of Art: Art as a Social Realm.* Lewiston, N.Y., 1995.

Coremans, P. B. *Van Meegeren's Faked Vermeers and de Hooghs.* Translated by A. Hardy and C. Hutt. London, 1949.

Dutton, Denis, ed. *The Forger's Art: Forgery and the Philosophy of Art.* Berkeley, 1983.

Goodman, Nelson. *Languages of Art: An Approach to a Theory of Symbols.* 2d ed. Indianapolis, 1976.

Grafton, Anthony. *Forgers and Critics: Creativity and Duplicity in Western Scholarship.* Princeton, N.J., 1990.

Hebborn, Eric. *Drawn to Trouble: The Forging of an Artist; An Autobiography.* Edinburgh, 1991. Also published under the title *Master Faker.*

Hebborn, Eric. *The Art Forger's Handbook.* London, 1997.

Hoving, Thomas. *False Impressions: The Hunt for Big-Time Art Fakes.* New York, 1996.

Jones, Mark, et al., eds. *Fake? The Art of Detection.* Berkeley, 1990.

Kilbracken, John Raymond Godley. *Van Meegeren: Master Forger.* New York, 1967.

Koestler, Arthur. *The Act of Creation.* New York, 1964.

Koobatian, James, ed. *Faking It: An International Bibliography of Art and Literary Forgeries, 1949–1986.* Washington, D.C., 1987.

Kurz, Otto. *Fakes: A Handbook for Collectors and Students.* New Haven, 1948.

Van Bemmelen, J. M., et al., eds. *Aspects of Art Forgery.* The Hague, 1962.

DENIS DUTTON

FORM. *See* Aristotle; *and* Formalism.

FORMALISM. [*To explore the multiple meanings of formalism, this entry comprises three essays:*

> Conceptual and Historical Overview
> Formalism in Analytic Aesthetics
> Formalism in Art History

The first essay is an overview of the conceptual history of formalism and its different stages and meanings, especially in the context of modern art. The second essay explains the critique of formalism by two analytic philosophers, Arnold Isenberg and Richard Wollheim, especially the latter; this critique serves to distinguish different types of formalism. The third essay examines how formalism has been understood in art history since the eighteenth century, especially in terms of the contrast between formalism and historicism. For further discussion, see Hanslick; Prague School; Russian Aesthetics, article on Russian Formalism; and Structuralism.]

Conceptual and Historical Overview

Form, as a term in aesthetics, refers to the perceptual elements of an artwork and to the relationships holding between them. *Formalism* names the aesthetic doctrine in which these related (formal) elements are said to be the primary locus of aesthetic value, a value that is independent of such other characteristics of an artwork as meaning, reference, or utility.

Although formalism has a long theoretical history, its emergence as an aesthetic ideology, one that provides an evaluative measure for both artistic styles and individual works, occurred during the advent of modernism in the late nineteenth and early twentieth centuries. Even in its modern guise, however, formalism is not a monolithic doctrine, but has a number of versions. The differences between these versions can be understood through their relative emphasis on one or the other of the following theses: the first is that all artworks are to be primarily valued *for* their formal properties; and the second is that the most valuable artworks are those whose primary emphasis is *on* their formal properties.

An important philosophical source for the first thesis can be found in the aesthetics of Immanuel Kant, specifically in

his requirements for the appreciation, or judgment, of beauty. Beauty, as it functions in Kant's overall system, is located in both art and nature, but the conditions for judging that something is beautiful are consistent between each realm. The requirements in his system of particular concern for formalist theory are two: that beautiful things must be "purposive in form," and that their appreciation must be "disinterested." These requirements apply, respectively, to the thing perceived and to the nature of perception. Purposiveness is a manifestation of order and rationality that does not point to any specific purpose. Here, Kant argues that no practical identity or stipulated task can be assigned to an entity that is beautiful, although that entity, be it artwork or mountain meadow, may in a nonaesthetic context have practical purposes of various kinds. Additionally, the requirement of purposiveness stipulates that there are no rules through which the sense of formal order and internal cohesiveness that constitutes beauty can be documented and didactically transferred, for otherwise beauty would be an empirical rather than an aesthetic property.

The other of Kant's requirements that emphasizes form in the judgment of beauty is disinterestness. This requirement stipulates that in the perception—the pure aesthetic appreciation—of beauty we must forgo all interests we might otherwise have in an entity's physical makeup or in the various ways it may be used. Such interests are appropriate to our practical concerns, but they interfere with our aesthetic capacities. Within the realm of art, Kant's restriction on "interest" extends to the symbolic characteristics of artworks. Although paintings of his time were all representational, he stipulates that appreciation discount the specific identities, origins, or social import of the subjects represented. In music, likewise, the ceremonial, liturgical, and dramatic functions must remain secondary to the pure appreciation of tone and harmonic sequence. For if appreciation were to include interest in these practical considerations, attention would be diverted from a work's internal purposiveness—its purely formal existence—into its variable life as a commodity. In this latter context, the appreciation of beauty is undermined.

Although the two requirements listed above are but a part of Kant's aesthetic doctrine, they illuminate his contribution to formalist theory, namely, that as a characteristic of objects and as subject of perception, form is a sufficient demarcator of the aesthetic. The historical effect of this thesis on later theories worked to minimize other factors such as expression, verisimilitude, or narrative in interpretations of art. With the advent of modern art, the emphasis on form had developed into a dominant criterion of aesthetic value.

It is a historically short although contentious passage from identifying form as the dominant aesthetic value to valorizing artworks that emphasize form to the exclusion of other symbolic characteristics. The difference that separates these is between what *has always* been aesthetically impor-

tant in art and that which artworks *only presently* have been able to offer as their main concern. This passage introduces the other version of formalist theory that emerges in the modernism of the late nineteenth century, the version that couples "form" with the notion of "progress" in art.

Presenting formalism as a historical achievement rather than an enduring aesthetic value identifies the thesis that historical change in artistic style is both value laden and goal directed. The major source of this thesis can be found in the aesthetics of G. F. W. Hegel. "Progress," understood as both value and goal, is a teleological notion that characterizes Hegel's central philosophical principle, the "evolution of spirit." Here, progressive change in artistic style functions as one important documentation of spirit's manifestation in culture. To so interpret change as progress requires evidence, and this is found in what Hegel calls "dematerialization," the historical tendency within all forms of culture to reduce dependency on their material component. This tendency is evidenced in both physical and conceptual forms, and thus the negatives of materiality are characterized as bulk, opaqueness, and inertia, as well as categorical rigidity and persistent duality. The contrasting positive values, evidently, are those such as fluency, transparency, intersection, and the anticipation of ongoing syntheses. Two points should be noted here: Hegel's process-centered theory can be taken as a critique of the architectonic nature or "rigidity" of Kantian philosophy, and his displacement of value from nature to history, his emphasis on artistic rather than on natural beauty, is one part of that critique. Given his doctrinal concern with synthesis and the reconciliation of opposites, however, Hegel does not tolerate a "form-content" distinction. Art is a comprehensive symbol, an adequate presentation of a culture's "ethos," its historically achieved stage of "spirit's progress." In this context, it would be meaningless to, for example, separate form from what is represented—to valorize one aspect of an artwork over another. Nevertheless, Hegel conceives of the different arts as themselves being susceptible to historical progress. He begins his narrative with the lowest, most material, art of ancient architecture, and he concludes art's philosophical task with the art of discourse, the poetry of his own day. Greek art occupies the middle point of this historical progression, and it is here, particularly in classical sculpture, that Hegel locates a perfect—if ephemeral—balance between matter and spirit, one that can be seen as forming the "ideal" of art qua art. But the more internalized demands of later historical periods—those of the Christian West—shift aesthetic dominance to the more allusive arts of, successively, painting, music, and poetry. It is in poetry, which shares the medium of discursive language with philosophy, that Hegel locates the culmination of art's historical progression: the linguistic form is the end point of the dematerialization of art. This notion of "progress in art"—that symbol systems are historically superseded by other more

adequate ones—is offered here as an enabling concept for the preoccupation with formal value in modern art; it is the Hegelian antagonism between "spirit" and "matter" that presages the modernist move to locate artistic content in an artwork's form.

Although the arts of modernism can be seen as doctrinally much influenced by the Hegelian notion of progress, they do not subscribe to the scale of value on which he weighs each of the various arts. The normative consequences of Hegel's historical hierarchy of art forms—the notion that certain arts are inherently inferior to others—was unacceptable in the egalitarian climate of the early twentieth century. Here is where the Kantian construal of form offers a cogent emendation: form is not only to be emphasized, it can also be purified of its obscurants, its historical atavisms and limitations. This was regarded as equally true for the various arts; each is free to devise unique ways to move beyond the stylistic and utilitarian limitations of its traditional past. Such divestiture, and the creativity that ensues, would then give all the arts immunity from the Hegelian critique of historical limitation.

In this effort, the Kantian value of "pure form" becomes the modern value of "radical form." Within the antiacademic climate of modern art, the normative signposts for all the arts are novelty and invention, and these serve to conjoin Kant's standard of formal purity with Hegel's standard of progress in art. Change becomes an overt and self-conscious value, and the rejection of traditional forms becomes a programmatic ideological imperative. Some of the relevant stylistic transformations of this period are the move from representation to abstraction in the visual arts, from harmony to polytonality in music, and from linear narrative to recursiveness in literature.

Twentieth-century versions of formalist theory tend to center on particular art forms. Here, I locate the contrasting senses of formalism described above in two such theories, both of which are concerned with the visual arts. The first of these theories, which I take to have strong Kantian roots, is the "significant form" hypothesis of Clive Bell. Bell argues for a causal link between a work's formal relationships and an "aesthetic emotion" that such relationships purportedly arouse in the viewer. Thus, the having of this special emotion and the value—or "significance"—of the particular form are, reciprocally, evidence for each other. While this theory has been much criticized for its apparent circularity, it does articulate the importance of form as a value in modern aesthetics. Additionally, through its relegation of mimetic or narrative factors to a secondary status, the theory draws attention to the modern movements (e.g., abstract and nonobjective art) that emphasize form to the exclusion of these other factors. Nevertheless, "significant form" is not strictly a modernist theory for, within it, the discrimination of formal value, such as would separate good from bad art, is applicable to all art, historical as well as modern. Consequently, for Bell, the history of art does not so much exhibit continuing progress as it does peaks and valleys in which the value of form is either revealed or obscured. It is this historical catholicity that puts Bell more in the Kantian than the Hegelian tradition. The normative accommodation between aesthetic and nonaesthetic is maintained in his theory by distinguishing between levels of appreciative capacity and artistic talent such that formal value emerges as the dominant concern only at the higher levels in each. Because such distinctions can be made in all periods of art, primitive, classical, non-Western as well as modern, this separation of aesthetic value from historical position echoes the universality requirement of the Kantian categorical schema. Bell also leans on Kant when, in contrast to avant-garde ideology, he points to social institutions of "competent judges" as being the pedagogical sources of the sophistication necessary to discriminate levels of formal significance in artworks. A critical application of Bell's theory can be found in Roger Fry's essay on Paul Cézanne, and a similar approach in Eduard Hanslick's writings on music.

Bell's theory can be contrasted with other, more teleological, versions in which normative distinctions are not so much categorical as historical. In such theories the emphasis shifts from the discrimination of valuable works within historical styles to the valorization of modern styles for their emphases on particular formal characteristics that are both novel and historically prescient. These theories place formalist values in a context of progressive historical achievement, and, accordingly, they bring a Hegelian dynamic to the Kantian construct of pure form.

Much of early modernist work can be characterized as reductive or eliminative, the target of these efforts being the pictorial conventions of older—academic—art. These conventions were variously perceived as creatively inhibiting, socially decadent, and aesthetically impure; and the path to freedom lay in the discovery of the "essential nature" of artistic form. The move in Impressionist painting toward a pure perception of nature, the Cubist intersection of pictorial foreground and background, and Constructivist attempts to develop an abstract pictorial language, all can be seen as efforts at specifying what is essential in form and at eliminating all that is inessential. These efforts have a critical dimension as well, for they were also directed against the modes of appreciation identified with academic pictorial traditions, and such criticism then expanded to include conservative social institutions. With this juncture of the aesthetic and social, the concept of art exhibits a radical Hegelianism, for the process of its internal—formal—purification is coupled with an insistence on commensurate changes in appreciation, in the sensibilities and values of its audience.

The critic Clement Greenberg wrote extensively on the "new American painting" of the postwar years, 1945–1960,

an era in which, he believed, the historical center of artistic achievement had moved from Europe to America. As an influential spokesman for, as well as critic of, the new art, Greenberg articulated the principles of its claim to historical prominence. Here, as does Bell, he places the primary value on form, but it is a value achieved not by looking through a work's representational and illusionistic aspects to its formal structure, but by eliminating these aspects in favor of the ones that belong to a painting's internal aesthetic identity. As Greenberg puts it, "The irreducibility of pictorial art consists in but two constitutive conventions or norms: flatness and the delimitation of flatness" (Greenberg, 1961).

This notion of delimited flatness had a considerable influence on the stylistic developments of that period. Traced through such artists as Jackson Pollock and Morris Louis, it champions an art that rejects not only representational subject matter but also such pictorial conventions as atmospheric light and perspectival illusion. *Flatness*, in Greenberg's usage, is actually more a theoretical than a material designator, for its achievement is seen as a culmination of painting's historical efforts to reject all encumbrances to its pure internal exposition and thereby to attain the autonomy of self-reference. It must be said here that Greenberg's identification of *flatness*, or such cognate terms as *frontality*, as aesthetic values requires a commitment to a Hegelian historicism wherein a construal of form that, on other readings, could be taken as merely surface embellishment, is theoretically charged and presented as pictorial essence.

Greenberg's extension of historicist formalism into the social arena, into matters of appreciation and collective sensibility, is to be found in his distinction between "high art" and "novelty art." Here the tension is not so much between recidivist and radical formal practices as between authentic and inauthentic perceptions of the new. In the American climate, modernism achieved a social success that contrasts with the indifference and hostility surrounding the European avant-garde. Whereas, in Europe, the institutional world of art—academies, museums, historians, and critics—was regarded as unfriendly to modernism, in the United States, by 1955, such institutions came to function as popularizers and purveyors of the new art to the larger society. In this receptive climate, both creative and appreciative practices flourished and variations on successful modernist themes proliferated. As a consequence, the distinction between truly advanced art and its imitations was obscured, and much artistic production became a packaging of modernist images for popular consumption—a new academic modernism. By acknowledging this collusion in late modernism between popularity and progress, between the middle class and the avant-garde, Greenberg abandoned his earlier belief that an advanced art—a purely formal abstraction—could lift popular taste to a higher level

and thus achieve an aesthetic emancipation paralleling the social and economic achievements of the time. Consequently, Greenberg was led to a critical position that does not rest solely on formal criteria. New movements such as "Pop" and "Minimal" art challenged his earlier teleology: neither pictorial abstractness nor flatness could guarantee excellence—this could only be ascertained by a sophisticated critical taste. But this ideological shift also brought Greenberg's position closer to the one associated earlier here with Bell: the criteria for aesthetic value—the identification of what is formally significant—remain the province of competent judges; they are not to be independently located in the characteristics of an ascendant historical style. With this realization, the importance of formalism as a modernist ideology diminishes—indeed, the very notion of a dominant historical style comes to be seen as irrelevant—and the era of diverse practices we now know as postmodernism begins.

In the new climate of postmodernism—the art of our own times—the concept of form loses its polemical overtones. It no longer functions as an imperative to support a historicostylistic strategy, as in Greenberg's call for an art of "pure form." Neither is it an admonition to discern what is truly important in art, as in Bell's evocation of "significant form." Rather, attention to form has become normatively neutral; it is now simply a discriminatory tool that gives us a way of distinguishing between the plethora of present artistic styles and motifs through the formal characteristics that are specific to each—and perhaps through the historical allegiances that these characteristics reveal. In this pluralistic context, the new location of aesthetic value becomes an interesting philosophical issue.

[*See also* Abstraction; Bell; Greenberg; *and* Kant.]

BIBLIOGRAPHY

Adorno, Theodor W. *Philosophy of Modern Music*. Translated by Anne G. Mitchell and Wesley V. Blomster. New York, 1973.

Bell, Clive. "Art as Significant Form." In *Aesthetics: A Critical Anthology*, edited by George Dickie and R. J. Sclafani. New York, 1977.

Fried, Michael. "Art and Objecthood." In *Minimal Art: A Critical Anthology*, edited by Gregory Battcock. New York, 1968.

Fry, Roger. *Transformations: Critical and Speculative Essays on Art* (1926). Garden City, N.Y., 1956.

Gabo, Naum, and Antoine Pevsner. "Constructivist (Realistic) Manifesto." In *Theories of Modern Art*, edited by Herschel B. Chipp. Berkeley, 1968.

Gass, William. *On Being Blue: A Philosophical Inquiry*. Boston, 1975.

Greenberg, Clement. *Art and Culture: Critical Essays*. Boston, 1961.

Hanslick, Eduard. *On the Musically Beautiful: A Contribution towards the Revision of the Aesthetics of Music*. Edited and translated by Geoffrey Payzant. Indianapolis, 1986.

Hegel, G. W. F. *Aesthetics: Lectures on Fine Art*. 2 vols. Translated by T. M. Knox. Oxford, 1975.

Kant, Immanuel. *Critique of Judgment*. Translated by J. H. Bernard. New York, 1951.

Krukowski, Lucian. *Aesthetic Legacies*. Philadelphia, 1992.

Mondrian, Piet. *Plastic Art and Pure Plastic Art, 1937, and Other Essays, 1941–1943.* 2d ed. New York, 1947.

Zuckerkandl, Victor. *Sound and Symbol: Music and the External World.* 2 vols. Translated by Willard R. Trask and Norbert Guterman. Princeton, N.J. 1956–1973.

LUCIAN KRUKOWSKI

Formalism in Analytic Aesthetics

The prevailing response within analytic philosophy to formalist theories in the visual arts is that formalist theorists and critics have failed to offer a notion of form that can support a general thesis about aesthetic judgment or aesthetic regard. Here, as elsewhere, analytic philosophers have sought out a claim—in this case, a general claim about the form of a work of art—that is susceptible to their particular methods of criticism and conceptual analysis. Such a claim may well appear in late formalist theories, particularly in Clive Bell and Roger Fry, whose aesthetic theories share some of the philosophical and, more specifically, psychological preoccupations of midcentury analytic philosophy. But the philosophical criticism of such theories has failed to account for—and therefore has obscured—other interests and motivations that have been at work in earlier formalist criticism and theory, specifically, in the theories and accounts of form that we find in Leon Battista Alberti, Roger de Piles, Gotthold Ephraim Lessing, Immanuel Kant, and others.

The most trenchant critiques of formalist aesthetic theory by Anglo-American analytic philosophers appear in two well-known essays by Arnold Isenberg, "Formalism" and "Perception, Meaning, and the Subject Matter of Art," and in a recent essay by Richard Wollheim, "On Formalism and Its Kinds." Isenberg and Wollheim both present formalism as a general thesis or theory about aesthetic appreciation. They take the formalist to maintain that what we appreciate or judge when we appreciate or judge a work of art aesthetically (that is, *as* a work of art) are just its formal attributes, roughly, the composition or organization of the work as opposed to its representational content; and they take the formalist to maintain this generally, for works from all periods of history, of all types. Thus, both Isenberg and Wollheim single out a formalism that is itself philosophically minded (or at any rate psychologically minded), a thesis, explicit or implicit, that the only or the essential element of aesthetic value in a work of art is its form. In their critiques, both ask if a general account can be given of the notion of form needed here, that is, the notion of a significant organization or composition of a work *distinct from* subject or content. Both find attempts to provide such a general account to fail or to be futile and, with them, the formalist thesis that what we aesthetically appreciate and judge in a work of art is just (essentially) its form (as opposed to its subject or content).

In his critique, Wollheim, following Isenberg, distinguishes two types of formalism, analytic and normative, the first of which (analytic or "descriptive" formalism) he characterizes as a theory about how paintings essentially are or, again, a theory "that [paintings] necessarily . . . are composed in a certain way, and that this composition is what we need to take account of, and all that we need to take account of, in coming to understand them" (Wollheim, 1995). Wollheim directs his criticisms solely against this first type of formalism, presumably because he takes it to define the notion of form that is needed in the second type (normative or "prescriptive" formalism). If the first type of formalism fails, the second must as well.

The notion that form is the essence of the work of art, that everything else about the work is an accretion on particular forms, does not of itself, however, give a very specific idea of what is to count as form. The idea can be made more specific if instead of thinking of form as the essence of the work or of our understanding of the work (notions that are perhaps unclear), we think of our understanding of the work as requiring a conception of the essence of form: namely, something like line, shape, or color. On this view (discussed in Isenberg's essay "Perception, Meaning, and the Subject Matter of Art"), the form of a work is to be described by describing its abstract organization (its lines, shapes, colors, and their combination). Anything else that we might consider part of the meaning of the work—for example, what any part of the work represents—is a matter of association. Isenberg dismisses this view in a few sentences: "Representation introduces an entirely new set of forms. It contributes new designs. It does not intrude itself, with an effect of disturbance, into the midst of a set of abstract forms, which somehow hang on as pure line and color and shape." Wollheim, for his part, allows that a formalist may be interested in more complex notions of form than the one supposed here, and, consequently, his refutation of what he calls analytic formalism is more complex. Nevertheless, he conceives the formalist, as Isenberg does, to be trying to separate the form of a work from its subject matter.

The formalist view that Isenberg dismisses derives from a particular conception of perception, according to which we can be said to see lines, shapes, and colors apart from anything they might represent or mean. The theory of perception here is psychological, that is, broadly associational and empirical (as opposed to transcendental). It is the sort of theory that came to occupy Anglo-American analytic philosophers after the turn of the century. Such theories—sense-data theories, gestalt theories, and the like—were a significant preoccupation of or provided the framing assumptions of Anglo-American aesthetic theory even through the 1960s and 1970s. (A particular consequence of this is that questions of representation, of how paintings and other works of art represent things, dominated analytic aes-

thetic theory through this period.) Both Isenberg and Wollheim have approached issues in aesthetics with views shaped by or in response to the sorts of notions of perception roughly indicated here. Thus, while both reject a formalism based in these notions, for example, the formalism of Roger Fry (cited by Isenberg as one of the two "leading formalists" and by Wollheim as his example of an analytic and a normative formalist), they both share Fry's expectation or hope that we can give a general psychological account of aesthetic appreciation. It is easy to see how the expectation or hope of such an account might reinforce Fry's idea that formalism (as a theory of aesthetic appreciation) is a theory about our regard for works of art generally, that it might be applied to the appreciation of art in the past and in other cultures as well as in our present culture. By the same token, Isenberg's and Wollheim's expectations and hopes for a general psychological account of aesthetic appreciation may lead them to consider only formalisms like Fry's, formalisms that presume to offer an account of past, as well as current, art.

In a number of respects, Isenberg's and Wollheim's philosophical and psychological preoccupations obscure the range of motivations that have entered and informed formalism (meaning now not the formalism of Bell and Fry, but what Isenberg terms the "old debate of the aestheticians" about form, joined "by . . . Goethe, Kant, Reynolds, Burke, perhaps even Michelangelo"). An instance here is Wollheim's very emphasis on analytic ("descriptive") formalism, his taking it as the object of his attack. Wollheim remarks in *Art and Its Objects* that "the thesis that . . . [the] aesthetic content [of paintings] consists exclusively in flat coloured surfaces and their juxtapositions . . . is how a great deal of 'formalist' aesthetics is arrived at" (Wollheim, 1980). But this is not what we find if we look to traditional formalist theories. For example, neither Alberti nor Kant is an analytic ("descriptive") formalist in Wollheim's sense, nor, for that matter, a normative ("prescriptive") formalist (neither thinks that form is the *only* thing that we need to attend to in order to understand what a painting is nor the *only* thing that we need to attend to in order to judge it). I am not even sure that Fry comes to his theory of form in the way that Wollheim's account suggests, by asking a question about the essence or essential content of painting. Nonetheless, with Alberti, Kant, and Fry, notions of form are integral both to their conceptions of what a painting is or does or means and to their conceptions of how we judge painting.

Isenberg's "formalism" and Wollheim's "analytic formalism" are philosophical constructions, something that a philosopher *makes* of our interests in form or something that a formalist critic, in a philosophical moment, might make of his own or her own interests in form. They also are *late* constructions, deriving from a modern philosophical

interest in empirical or psychological theories of perception. The notion of formalism that is criticized by Isenberg and Wollheim arises at a moment when formalists like Fry (and, subsequently, his philosophical critics) perceive form in a very limited fashion, in the first place, as a pictorial feature of works of visual art and, second, as a feature that might enter a psychological account of our response to pictures (Fry) or of their meaning (as Wollheim prefers to see the matter).

There are other formalisms, arising at other moments in history, in which the appeal to form is not part of a psychological theory and, indeed, encompasses features of works that are not solely or purely pictorial. Such accounts often single out formal features not by an appeal to general facts about perception or sight, but by appeal to rhetorical or literary notions or to conceptions of the ways that language, that is, structures of words, represent or communicate significant aspects of experience. These rhetorical or literary or language-centered origins of formalism in the visual arts slip *between* Wollheim's various distinctions, for example, between what he calls analytic and normative formalism.

Notions of form that derive from literary theory or rhetoric or the theory of language can be found in Alberti and Kant and in other critics and theorists of the visual arts, including Piles and Lessing. It is tempting to speak here of something like a tradition of formalism, although it remains to be seen if a history can be traced *through* these various critics and theorists, that is, if they can be seen to have developed their accounts and theories of form in response to each other. At this time we have histories only of individual formal theories, *histories* of formalism rather than *a* history. Notable examples are Michael Baxandall, *Giotto and the Orators: Humanist Observers of Painting in Italy and the Discovery of Pictorial Composition, 1350–1450* (1971); Michael Fried, *Absorption and Theatricality: Painting and Beholder in the Age of Diderot* (1980); Thomas Puttfarken, *Roger de Piles' Theory of Art* (1985); David Summers, *The Judgment of Sense: Renaissance Naturalism and the Rise of Aesthetics* (1987); and John White, *The Birth and Rebirth of Pictorial Space* (1987).

The question of how literary or rhetorical or linguistic notions of form came to be applied to works of visual art involves matters of the training and education of critics, the nature of the critical models (whether in literature or in the visual arts) most prominent or respected at a given moment, and the appropriateness of literary-critical models to the particular concerns of the visual arts at a given moment. Of course, the specifics here will differ from one moment in the history of the visual arts to another. It is instructive to consider some examples, to see how traditional formalist expressions may differ from those of the early twentieth century.

Baxandall traces the notion of *compositio* in book 2 of Alberti's *On Painting* to a contemporary humanist conception of the structure of the sentence:

> *Compositio* was a technical concept every schoolboy in a humanist school had been taught to apply to language. It did not mean what we mean by literary composition, but rather the putting together of the single evolved sentence or period, this being done within the framework of a four-level hierarchy of elements: words go to make up phrases, phrases to make clauses, clauses to make sentences.

Alberti invokes this notion of *compositio,* a notion in contemporary rhetoric, to describe the proper ordering and subordination of elements of a painting to one another—of the parts of the depicted human body to each another, of individual persons in a painting to each other, and so on. This account supplements his initial account of painting, in book 1 of his treatise, in terms of notions of the visual pyramid and what he calls the intersection. A painting, Alberti claims in book 1, represents things as they would appear on a plane that intersects the visual pyramid or cone of vision. Here, a geometric theory applied to vision (rather than a rhetorical notion, applied to what we might call the subjects of a painting) specifies how objects are to be proportioned and placed in the space of the painting.

Both Alberti's theory of vision (his perspective theory) and his notion of *compositio* are totalizing. Both claim to describe what regulates the proper proportions and ordering of objects in the space of a painting, the first by defining the nature of this space (as White has remarked, by freeing the notion of space from the "tyranny" of the individual object), the second by specifying the proper relations of each element and object in the painting to every other. Both notions contribute to a conception of the unity of painting, a unity that in Alberti encompasses considerations not only of spatial arrangement and proportion, but of conformity of parts (in the depiction of the body), grace (in the depiction of bodies and actions), and dignity (in the depiction both of particular figures, human or divine, and in their combination in what Alberti calls an *historia*).

It might be objected that Alberti's theory of vision and his notion of *compositio* regard different features of the unity of painting. His perspective theory dictates how the objects represented in a painting, whatever they are, must be proportioned to one another on the canvas to reflect actual relations in space; it mandates a consistency of proportion, a regularity in relations of size. His notion of *compositio,* on the other hand, governs the *choice* of subject or object matter (for example, how many human or divine figures might be included in a *historia*). But this way of putting the difference between Alberti's notions reduces the range of his notion of *compositio*. He applies this not just to the choice of subject matter, but to questions of how each figure in a painting is posed and how the various figures are disposed in relation to one another—matters that are as pertinent to what we might call the form or the composition of a painting as the requirements of Alberti's perspective theory. Alberti's perspective theory and his notion of *compositio* together convey a demand that pictorial space be unified throughout—in depth and in every detail, much as our experience of the natural world is unified throughout, that is, continuous and consistent in every one of its parts. On Alberti's notions, these unities are not merely spatial and temporal, but, as we might put it, causal and moral.

But, it might be asked, even if Alberti's notion of *compositio* is (also) a formal notion, why call it a rhetorical or literary notion, at least in its particular application to painting? In origin the notion of *compositio* may be a notion of rhetoric, but surely (the objection here might go) it is at best an *analogy* for the organization of a painting. (After all, Alberti does not describe the relations among the parts of a painting as the *same* as the relations among the parts of sentences, that is, as grammatical relations.) But again this oversimplifies Alberti's notion. As Baxandall points out, the more specific notions that Alberti uses in describing whether or not a painted *historia* will have *compositio*—notions like variety *(varietas),* abundance *(copia),* restraint *(gravis),* modesty *(moderata),* dignity *(dignitas),* disconnectedness *(dissolutus)*—are themselves rhetorical notions that are used in the criticism of prose style. Alberti is not just making a general analogy between the structure of a painting and the structure of prose; he is singling out specific features of painting by reference to their rhetorical counterparts. In doing so, Baxandall argues, Alberti means also to draw the attention of his contemporaries to parallels between the *compositio* or the *dissolutus* of specific contemporary painters—Baxandall cites Andrea Mantegna and Piero della Francesca, on the one hand, and Antonio Pisanello, on the other—and the *compositio* or *dissolutus* in the prose style of specific contemporary writers.

Another, later formalism that is not purely pictorial (and has at best a distant connection to a psychology of vision) is Kant's account of the form of a beautiful object. In section 14 of the *Critique of Judgment,* Kant remarks: "In painting, sculpture, and in all the formative arts . . . the *delineation* is the essential thing; and here it is not what gratifies in sensation but what pleases by means of its form that is fundamental for taste." Kant appeals to notions of form with temporal arts such as dance or music as well:

> Every form of the objects of sense . . . is either *figure* or *play.* In the latter case it is either play of figures (in space, viz. pantomime and dancing) or the mere play of sensations (in time) . . . the *delineation* in the first case and the composition in the second constitute the proper object of the pure judgment of taste.

These remarks about the form of works of art, their delineation or composition, appear in a section titled "Elucida-

tion by Means of Examples" that follows Kant's discussion of what he calls "purposiveness without purpose," a notion that he explains in terms of the "free play of the representative powers [of imagination and understanding]." The notion of free play here is crucial to Kant's account, in sections 8–13 of the *Critique*, of how we can make judgments of the beautiful, that is, how we can express our pleasure in contemplating a beautiful object in a judgment that, as Kant puts it, "claim[s] the assent of everyone." Kant's argument, in brief, is that in contemplating a beautiful object we feel the free play of our imagination and understanding, our two powers of thought, *as* a pleasure—indeed, as Kant notes, our pleasure just *is* one of "*linger[ing]* over the contemplation of the beautiful," lingering as if at play—and this pleasure, which depends on nothing other than our powers or capacity to think and to communicate with others, provides the ground of our judgment of beauty, our demand that others agree with us. This is familiar, if not everywhere easy, territory in Kant's *Critique*. Two further observations are needed about later sections of the *Critique*:

1. Kant appeals to the form of a work of art to explain what we are regarding or what we are dwelling on when we judge something to be beautiful. The notion of form here enters an account of our capacity to make what we nowadays call critical judgments, including critical judgments about works of art. Later in the *Critique*, however, in sections 44–50, when Kant turns to the question of how works of art are possible (how can an artist have the purpose of making a beautiful object if no purpose can be in view in making a judgment of the beautiful?), he argues that it is not enough that an object *have* a form that satisfies taste—it is not enough that the object be judged beautiful—for it to be a work of art. What more is needed for an object to be a work of art, he claims, is that it present an aesthetical idea.

2. Kant's account in section 49 of the *Critique* of what he calls an aesthetical idea closely parallels his account of the free play of imagination and understanding in the judgment of the beautiful. He defines an aesthetical idea as a "representation of the imagination which occasions much thought without . . . any definite thought, i.e. any *concept,* being capable of being adequate to it." He also says that an aesthetical idea gives rise to a "feeling . . . which quickens the cognitive faculties" and evidences "the happy relation [between these faculties]," a relation he describes as an "unsought undesigned subjective purposiveness in the free accordance of the imagination with the legality of the understanding." Just as a beautiful object's form engages the mind by prompting a free play of the imagination and understanding, so an aesthetical idea "enlivens" or "animates" the mind by prompting a free play of images and thought.

Although Kant uses the term *form* in the *Critique of Judgment* in connection with spatial or temporal delineation or composition, his account of an aesthetical idea defines another notion of formal coherence: namely, a play of images and thoughts under the aegis of an intellectual idea:

> The poet ventures to realize to sense, rational ideas of invisible beings, the kingdom of the blessed, hell, eternity, creation, etc.; or even if he deals with things of which there are examples in experience—e.g. death, envy and all vices, also love, fame, and the like—he tries, by means of imagination, . . . to go beyond the limits of experience and to present them to sense with a completeness of which there is no example in nature.

The terms in which Kant defines an aesthetical idea ("a representation of the imagination . . . which occasions in itself more thought than can ever be comprehended in a definite concept and which consequently aesthetically enlarges the concept itself in an unbounded fashion") precisely describe what Stanley Cavell has called "the burgeoning meaning" in poetic metaphor, the literary trope that for Samuel Taylor Coleridge and William Wordsworth stood for the powers of poetry altogether. This suggests a general affinity between Kant's notion of an aesthetical idea and the literary or rhetorical concept of metaphor (an affinity that goes beyond the more specific claim that poetic metaphors in particular are expressions of aesthetical ideas) and, further, an affinity between Kant's formalism and specific literary ambitions at the turn of the century.

We might say that what makes something a work of art for Kant is that it displays free beauties of imagination and thought, rather than free beauties of line ("delineation") or of spatial or temporal arrangement ("composition"). In the judgment of beauty, the free play of our cognitive faculties arises in reflection *on* particular formal (spatial or temporal) attributes; in the work of art, the free play of our cognitive faculties that is prompted by an aesthetical idea gives rise to an awareness of formal coherence. What matter or content may enter the formal order of a work of art is not limited by our capacities of sense (which constitute the orders of space and time), but may be as extensive as the capacity of our understanding altogether (which is constituted by the categories of our thought in general). The formal order of the work of art is an order not just of spatial and temporal attributes, but of the lines of thought, of emotion, and of human relation depicted in a painting.

Kant's notions of form and of an aesthetical idea may not be taken immediately from rhetoric, but his explanations of their significance involve considerations of the possibility and the limits of communication and of thought—in other words, considerations of the possibility of language (and, as noted, of poetry).

Formalism, as we find it in Alberti and Kant and in other critics and theorists from the Renaissance to the nineteenth century, has been motivated by a complex of considerations: by the art of the moment; by the formalist critic's or theorist's ambitions for painting (which may derive from

painting, but also, as in Alberti's notion of *compositio* or seventeenth-century appeals to the Aristotelian unities, from literary theory); by the critic's or theorist's conceptions of our capacities of sight or perception; but also, both deeply and throughout the tradition of formalism, by conceptions of the way language works, of how it shapes sense. The history of formalism is important to our understanding of what painters were doing at particular moments, of how they might have conceived what a painting is; it also is important to our understanding of how painters' formal choices at a given moment—what they perceived to be formal choices—engaged significant concerns of their culture, concerns not only with the nature of seeing and of viewing, but with the nature of language and of thought. Isenberg's and Wollheim's discussions of formalism, not so much in what they argue as in what they leave out of their argument, obscure the range and the depth of these considerations. They also obscure the fact that formalism has a history, that there is no single theory of the form of a painting. Roger Fry's is only one theory of formalism and not even the last in its line.

In opposition to what he calls analytic formalism, Wollheim proposes an account of pictorial meaning in terms of what he calls "representational content." This account, too, has its roots in a particular historical moment, roots that include the way in which it makes particular appeal to psychological notions, especially, Wollheim's notion of our perceptual capacity of "seeing in," our specialized and culturally evolved ability to see things in particular sorts of visual objects, most notably, works of art. Still, the roots of Wollheim's account are not entirely or purely psychological, but are based in a conception of language as well.

In *Art and Its Objects*, where Wollheim first offers an account of our capacity of "seeing in," he traces its similarities to and its differences from our ability to see an object *as* something, that is, under a particular term or description or, as we might put it, under a particular concept. Invoking Ludwig Wittgenstein, Wollheim argues that "seeing as" is a perceptual capacity in which a concept "permeates or mixes into . . . perception: the concept does not stand outside the perception." Whatever this means, Wollheim would presumably say it as well of our seeing something *in* an object, where we also describe what we see by a word or a phrase. "Seeing in" is a notion, like Kant's notion of an aesthetical idea, that depends on a particular conception of the way language words the world; as such, it is as much a notion of how we *conceive* the world as a notion of how we see it. In the end, Wollheim cannot avoid this leading idea of the formalists: that in looking at a painting, we do not just see something, we take thought about the world, an activity informed and formed by our ability to speak (to word our world) altogether.

[*See also* Alberti; Greenberg; Kant; *and* Wollheim.]

BIBLIOGRAPHY

Alberti, Leon Battista. *On Painting.* Translated by Cecil Grayson. London, 1991.

Baxandall, Michael. *Giotto and the Orators: Humanist Observers of Painting in Italy and the Discovery of Pictorial Composition, 1350–1450.* Oxford, 1971.

Cavell, Stanley. "Aesthetic Problems of Modern Philosophy." In *Must We Mean What We Say? A Book of Essays.* New York, 1969; reprint, Cambridge and New York, 1976.

Fried, Michael. *Absorption and Theatricality: Painting and Beholder in the Age of Diderot.* Berkeley, 1980.

Isenberg, Arnold. "Formalism" and "Perception, Meaning, and the Subject Matter of Art." In *Aesthetics and the Theory of Criticism: Selected Essays of Arnold Isenberg,* edited by William Callaghan et al., pp. 22–35 and 36–52. Chicago, 1973.

Kant, Immanuel. *Critique of Judgment.* Translated by J. H. Bernard. New York, 1951.

Puttfarken, Thomas. *Roger de Piles' Theory of Art.* New Haven, 1985.

Summers, David. *The Judgment of Sense: Renaissance Naturalism and the Rise of Aesthetics.* Cambridge and New York, 1987.

White, John. *The Birth and Rebirth of Pictorial Space.* 3d ed. Cambridge, Mass., 1987.

Wollheim, Richard. *Art and Its Objects.* 2d ed. with six supplementary essays. Cambridge and New York, 1980.

Wollheim, Richard. "On Formalism and Its Kinds." In *On Formalism and Its Kinds/Sobre el formalisme i els seus tipus,* pp. 7–47. Barcelona, 1995.

NORTON BATKIN

Formalism in Art History

In *Formalism and Its Kinds* (1995) Richard Wollheim distinguishes between "two kinds of formalism"—manifest and latent formalism. Manifest formalism attempts literally to diagram the essential form of a pictorial or sculptural configuration. It has been embodied most notoriously in such studies as Earle Loran's analysis of Paul Cézanne's compositions. Latent formalism often conceives works of art to have a linguistic or at least language-like identity. It attempts to identify the system within and through which a pictorial or sculptural configuration has been organized, form being the result or precipitate of the system. Latent formalism has been embodied in a range of art-critical and art-historical writings in recent years. But it can also be traced in the work of art historians much earlier in this century.

In the discipline of art history, there are deep connections between formalism, both manifest and latent, and historicism, not least their common origin in the writing of J. J. Winckelmann in the 1760s. Unfortunately, however, in a good deal of contemporary writing formalism and historicism are said to be antithetical; they have been polarized as distinctive methodological and theoretical options. (Here it will not be possible to consider an equally debilitating confusion between formalism and stylistic analysis. Briefly put, whereas it is possible for formalism to consider the morphology of a unique artifact or an artifact taken entirely on

its own, stylistic analysis is essentially and inherently comparative—a statement about patterns of similarity and dissimilarity in groups of artifacts.) Thus, it is necessary briefly to recall the interconnections of formalist and historicist projects. In art history, formalism can be said to be the attempt to write art history—that is, to identify the historicity of art—without being historicist. The basis of the formalist claim to produce not only a penetrating description of pictorial or sculptural configuration but also a historical explanation of it is the putative, though highly tendentious, objectivity of formalist analysis. This matter will be explored briefly here.

The manifest formalist is exposed to a central difficulty identified by Wollheim. To the extent that in analysis manifest formalism does not just copy a painting exactly, reproducing it for us as the "formal analysis" of the painting, it fails to analyze what the form of the painting essentially is. This is not because pictorial form is essentially autographic (to use Nelson Goodman's term for describing the irreducibly expressive role of every visible feature of the configuration) and hence could only be analyzed formally—or for its form—by identifying its modulation so minutely that we would do best simply to reproduce it. (It is interesting to note, however, that Roger Fry objected on just these grounds to the manifest formalism with which his art-critical and art-historical project had threatened to become identified by the early 1930s.) According to Wollheim, manifest formalism—its attempt to diagram essential form—fails because no graphic or discursive specification of the form of a painting can simultaneously represent its two-dimensional layout and what three-dimensionally is or might be seen in it, even though these two fields are mutually determining. What we see in the surface patterns inflects how we see those patterns or, more exactly, what we see them to be, and vice versa. manifest formalism seemingly cannot capture what Wollheim calls the "representationality" of a painting—namely, the fact that visual perception is presented in the painting, as it is anywhere in the world, with apparent if sometimes materially fictive edges and occlusions that certainly sustain (and suitably arranged can reliably provoke) the perception of depth, whether or not this fiction has an objective correlate in actual distance between apparently overlapping or occluded surfaces. Representationality in this sense intersects with graphic format, the painted planar array of apparent edges and occlusions, in such a way that the essential form of the array (what the manifest formalist hopes to describe or diagram) can never be experienced or seen as such. At best the array constructs an apparent specific, if unlabeled, "thingliness," a perception or impression of this-in-front-of-thatness-so-distant-from-over-thereness.

For the art historian, if not for the philosopher, what is troublesome about manifest formalism is not so much that it produces the wrong analysis of what the painting essentially is formally. The "wrong analysis" of form is simply a given perceiver's limited, partial, or idiosyncratic assumption of a particular representationality on the basis of which essential form has then been specified and might be diagrammed. Such "wrong analysis" could be historically significant precisely because it could tell the historian what a given perceiver took to be essential form upon making the assumption of a certain representationality. But in fact what manifest formalism analyzes as the painting's form is ineluctably the potentially unreplicable result of the formalist's own unique construction, based on his or her own modes of response and interpretation, of the representationality of the array. It is the formalist's finding of a particular this-in-front-of-thatness for him. Hence, one cannot easily credit it as a discovery of the painting's essential form for the painter or indeed for any viewers other than the formalist him- or herself. Thus, manifest formalism appears to be preliminary to historical or so-called contextual analysis. Without further confirmation or documentation, the art historian cannot credit the manifest formalist's analytic discovery because it is falsely advertised as the discovery of essential form as it supposedly objectively is—that is, as if a given particular representationality of the array were universally shared or shareable and hence can be taken for granted in a diagram or description of the array.

By contrast, the latent formalist seems to secure a more adequate or more scientific analysis by trading on the putative objectivity—the sharedness or universality—of the structural-generative elements, conceived (for instance) as the basic units of an extrasubjective "language." Here, however, Wollheim urges that the latent formalist cannot actually show there to be any extrasubjective order—any social criteria of well-formedness or grammar of pictorial or sculptural configuration. In fact, for the latent formalist the problem of representationality appears in the next register down, namely, in the seeing-in that organizes not the manifest structure of a finished configuration but rather the latent system of its generation: despite the claims of certain semioticians, the elements of the so-called language cannot be arbitrarily related one to the other by mere difference but must have meaning in virtue of some primal distribution of references, of representationalities.

In sum, both manifest and latent formalism are troublesome to the extent that they advertise essential form to be objective—what is the case about the painting as if its representationality, being the same for all perceivers in either its two-dimensional structure (for the manifest formalist) or its generative system (for the latent formalist), can be taken for granted. Yet, the commonest complaint against formalism has been that it is "subjective"—as if formalism, were it to become more and more and finally completely objective, as it seemingly aspires to do, would shade into something like a fully historical, an anthropologically and psychologically exact, account of how painters and viewers other than our-

selves painted and viewed their pictures or at least of what they thought they had configured and depicted there. The success of formalism as critical description suggests that it might also be successful in offering history—an anthropology and psychology of others' perception of the works they have made and viewed. The very existence of a critical language and its ability to get one inaugurally to see a representationality of or in the work that can henceforth be there for one are the initial conditions if not the sufficient guarantees of its potential universalizability or possible universality— that is, of its hypothetically historical validity.

For just this reason, strong formalist writing has not typically thought itself to require the supplement—the confirmation or documentation—provided by any history in the extrinsic sense, any so-called context that independently documents the psychological situation (described above as the assumption of a representationality) that formal analysis both presumes and reconstructs. The more urgent a formalism, the more vividly it seems to open onto the artist's process of constructing form in relation to an assumed or desired representationality. We feel translated from the critic's to the artist's perception of form (in both the work of art and in the world that work supposedly represents). Consider, for example, Fry writing on Vincent van Gogh's *Yellow House at Arles*:

> Nowhere else did he express so fully the feeling of ecstatic wonder, the intoxicated delight with which he greeted the radiance of Provençal light and colour. Here he shrank at nothing in order to find a pictorial phrase in which to sing of the violence of the sun's rays upon red-tiled roofs and white stone walls. To do this he has saturated and loaded his sky, making it, not indeed the conventional cobalt which is popularly attributed to Mediterranean climes, but something more intense, more dramatic and almost menacing. . . . Here all is stated with the crudity of first impressions but also with all their importunate intensity—and they are the first impressions of a nature that vibrated instantly and completely to certain dramatic appeals in the appearances of nature. . . . The artist's interest was entirely held by the dramatic conflict of houses and sky, and the rest has been little more than an introduction to that theme. (Fry, 1926)

In Wollheim's terms, Fry practices both a manifest and a latent formalism. On the one hand, he attends to color values and contrasts—for example, red, white, and yellow dividing houses and sky—and, on the other, he recalls the "phrases" and conventions in which this story had been and could be told. Clearly, we encounter a patently subjective formalism; Fry allows himself normative judgment—an absolute if apparently principled personal determination of what is artistically valuable or pictorially successful in van Gogh—and poetical-rhetorical cadencing that are unapologetically his own. But his subjectivism is almost turning into what would seem to be diametrically opposite to it, namely, a plausible historical psychologism—an analysis not of Fry's but of van Gogh's seeing-in, the jump or transfer from Fry's to van

Gogh's "sensibility." This is achieved partly through the assertive confidence of Fry's criticism and his selection of an example in which (for instance) the yellows are so distinctly and especially yellow that they must as it were be so noticeably yellow for practically anybody. And it is partly achieved through a highly characteristic feature of Fry's formalism, namely, the way in which he does not conduct much formal analysis at all, or, at least, only briefly conducts business as either a manifest or a latent formalist in Wollheim's sense.

To be sure, Fry makes gestures at enumerating painterly elements and forces in their array and at imagining a pictorial language or language-like source for and organization of the whole. But he is less interested in characterizing pictorial or even painterly form than in identifying a global artistic expressivity, what lies *behind* the painting's essential form in the painter's character and his culture in the strict sense. Fry wants to address what we might call the form of the form, that is, the underlying and determining intentional significance and structure of the essential form of the painting or what art history in the formalist tradition derived from Alois Riegl has quite precisely called the *Kunstwollen* (will-to-form). Successfully to identify the *Kunstwollen* is to have effected the translation from the formalist's assumption of the representationality of the configuration to the original maker's and viewers' assumption. That is to say, if we know what original makers and viewers intended to represent, our formal analysis—otherwise subjective—is objectively constrained to describe the manifest formal structures or latent formal systems psychologically and historically compatible with just that representationality. If there is a problem here, it is not theoretical but instead methodological. How do we know what form makers intended to represent—or, in slightly different models, what they did in fact or were taken by contemporary viewers to represent—except by way of an analysis of the form they made?

Manifest and latent formalism—or "small-*f*" formalism— attempt to establish the scientificity or universality of criticism and hence the objectivity of form and hence the potential historicality of formalist identifications. (The theoretical mediations stretching between criticism and historicism are such that the two fields can be polarized, even though the formalist ambition is to lead us from the one to the other.) "Big-*F*" formalism, however, attempts *directly* to specify the historically determining intentionalities of form. This is the formalism not so much of the English art-critical school circa 1910, exemplified by Fry, as it is the formalism of the Vienna school circa 1900, manifested in Riegl's *Spätrömische Kunstindustrie* and Emanuel Loewy's *Rendering of Nature in Early Greek Art*, both of 1901, and their influential descendants in non-Western and pre-Greek art studies and in the *Strukturforschung* of Heinrich Schaefer, Friedrich Matz, Guido von Kaschnitz-Weinberg, and others. Important American works such as Meyer Schapiro's studies of Romanesque tympana and George Kubler's

Shape of Time should also be connected with this tradition. (Its relations to critical idealism deserve extended treatment [see Podro, 1982] but cannot be pursued here.)

According to Vienna school formalism, the objectivity—that is, the social communicability—of the formal procedures of the Egyptians, the archaic Greeks, and the late Romans or early Byzantines was not then and is not now historically in doubt. But we have little or no subjective access to these objective traditions of meaning. Applied to them, manifest and latent formalism, as such or unqualified, seem irrelevant, despite Fry's persistent and admirable efforts to come to terms with Islamic, classical Chinese, and other non-Western traditions. In contrast, the *Strukturforscher* or big-*F* formalists produced their most lasting works precisely as historical psychology. As such, their histories are open to the objections of historical sociology. Today's so-called social historians of art tend to dismiss big-*F* as well as small-*f* formalism as just so much subjectivity and psychologization. But within the spectrum of formalisms we should note the difference between subjectivism and psychologism, represented by small-*f* and big-*F* or English and Vienna school formalisms, and notice, further, that big-*F* formalism probably escapes some of Wollheim's strictures on formalism.

The Vienna school and the *Strukturforscher* elaborated little if any manifest or latent formalism in Wollheim's sense. The matter turns on the method by which big-*F* formalism identifies the determining historical intentionalities of art. Its approach is not empathetic but archaeological. Whereas a small-f formalist like Fry is like a person caught in a psychological transference between himself and an other, a big-*F* formalist like Riegl or Loewy is like a psychoanalyst interpreting his transference with an other to have an objective etiology in both himself and the other: in tracking form making through time and over numerous instances, the archaeological distribution of what is proper to his own seeing-in and what is proper to the other sorts itself out. (Not surprisingly, big-*F* formalism offers explicit theories of typology and seriation and of evolution and mutation, as in Riegl's *Stilfragen*. These establish not only the directionality or progress of form making in history. They also demonstrate that form has a history of deposition and diffusion—a historical fact grasped not through empathetic response but through an archaeology of style, in which measurement or other formalizations of stylistic criticism in principle can show that a work of art possesses a certain formal identity even if the formalist does not initially see it.) Whereas the material of small-*f* formalism derives from the art gallery or art show, then, for big-*F* formalism it derives from the museum of archaeology, ethnography, or natural history. Whereas the instrument of small-*f* formalism is the unaided if "trained" eye of the connoisseur, the instrument of big-*F* formalism could well be calipers or a color chart.

It might well be impossible to overcome the Manifest Formalist's quandary, in which analysis must try to grasp what Wollheim calls the "meeting point" of the surface array *on* and the three-dimensionalities *in* the pictorial or sculptural configuration. But whereas English school formalism continually tries to reduce the representationality of the configuration, Vienna school formalism explicitly concentrates on what it sometimes calls plastic structure, the virtual *relief* of the pictorial extension. Virtual relief can be understood as the "meeting point" between patterning, governed by various interests such as ornamentation or heraldry, and projection systems for representing three-dimensional objects. The latter are understood to be governed algorithmically, but the original rules are not at all like those of Western perspective configuration (they include the projection systems considered by Schaefer for Egyptian and Loewy for early Greek art). Fry does concern himself vigorously with the nonperspective orders constructed by van Gogh. But since he does not treat them as based on an algorithm, an extrasubjective rulebook for which all kinds of evidence (for instance, in the archaeology of the production process) might survive, he must perforce treat them as the manifestation of individual sensibility to be grasped empathetically—one of the reasons his subjectivism proceeds toward psychologism without becoming historical or anthropological. By contrast, the Vienna school formalist offers an independent account of the cultural interests in pattern and in projection whose meeting point is plastic structure; Riegl, Loewy, and Schaefer all founded their analysis in part on technical study of modes of manufacture in which various metrological, proportional, and projection systems historically had been developed, applied, and refined. There may well be methodological circularity here. We might know about the dynamics of communication and ideology only from the resulting plastic structure of art itself: in reading big-*F* formalism or the *Strukturforscher* one has the weird sense of a complete social history being told through a sheerly formal one. But this can fairly be taken as the formalist species of history anyway: for formalism, art history is the tautology not so much of the social history of form but of the formal history of society.

In this final regard, we can again note the difference between English and Vienna school formalism. At points Fry will awkwardly punctuate his formal criticism to offer essentially paranoid claims about the social determination of art. In his remarkable essay "Plastic Colour," for example, he finds that in the eighteenth century

> refined and aristocratic society managed to impose its demands on the art of the day and dictated a reversion to more frankly decorative painting, painting which would enter freely into the ensemble of the salon. A certain range of local colours was almost prescribed, and . . . degradations of local colour might not go so far as to lead to heavy tones of a dull or disagreeable quality in themselves.

Such heavy-handedness leads to nothing but banal functionalism—at best, a social history of form in which the cognitive specificity of pictorial or sculptural configuration has been reduced to nil. By contrast, Vienna school formalism sticks with an internal teleology: modes of plastic structure bespeak the evolution of modes of consciousness in necessarily social embodiment. To the dismay of the self-appointed sociologist of art, such formalism need never once refer to the supposed uses and situations of art as an account of art's formal constitution—and not because it ignores the social dimension but precisely because it assumes from the start that the social dimension is known to us historically only through and as the formal development of plastic structure. It embraces the methodological circularity of formalism as the necessary consequence of the thesis that form making in art—or perhaps art making as such—has or is a historically and cognitively specific mode of consciousness.

Nonetheless, if the psychologism of Vienna school formalism is more latent a formalism than manifest formalism, it remains a manifest psychologism. Supposedly, the essential mode of thought of a culture is unmediatedly manifest in form, understood as plastic structure; thought *is* plastic structure. For this reason, Vienna school formalism cannot escape Wollheim's deepest stricture on formalism. Concepts of the plastic structure, virtual relief, apparent restlessness, and so on, of the pictorial or sculptural configuration do avoid the manifest formalist's hypostatization of the picture plane. Instead of flattening and splaying the virtual sculpture that is painting, Vienna school formalism treats this representationality—this plastic significance of a system of apparent edges and occlusions—as formally manifest, a thought about the world annealed in the act, or, more exactly, in the whole arc of its directedness. For the Vienna school formalist and manifest psychologist, plastic structure or virtual relief has no sub- or understructure or formal skeleton around which it has been pasted and plastered and which formal analysis somehow should try to retrieve; plastic structure or virtual relief *is* the representationality of the work in sum and in substance. But this approach assumes that the representationality remains constant throughout its life history—in other words, that the seeing-as from which it derives and the seeing-in that it provokes has the virtual unity of the plastic structure or even the material extension itself. This ignores the psychological essence of representationality as a continuous history of reorganization under continuously changing perceptual, affective, and social conditions. In other words, plastic structure can only be the latest, or, better, the materially annealed, reversion or replication of a *history* of representationality.

The manifest psychologism of formalism, then, would seem to require a *latent* psychologism—an inquiry into the past, or reorganized, and the possible, or immanent, significances of the plastic structure or virtual relief, the arrested and annealed intentionality. Here it is not possible to consider the tensions between, and possible historical relations of, the Vienna school manifest psychologism of the art historians and the Vienna school latent psychologism of psychoanalysis. (We could, for example, notice their common partial origins in materialist psychophysics and intentionalist analytics of the later nineteenth century.) Instead, we can simply take Wollheim at his word that his manifest and latent formalism have no psychoanalytic overtones. But it is possible to say that they do, or could, and that this recognition enables us quite properly to consider the ways and means of the various formalisms.

[*See also* Bell; Historicism, *article on* Historicism in Art History; *and* Riegl.]

BIBLIOGRAPHY

Davis, Whitney. "Style and History in Art History." In *Replications: Archaeology, Art History, Psychoanalysis*, pp. 171–198. University Park, Pa., 1996.

Fry, Roger. *Transformations: Critical and Speculative Essays on Art.* London, 1926.

Fry, Roger. *Last Lectures.* Cambridge, 1939.

Goodman, Nelson. *Languages of Art: An Approach to a Theory of Symbols.* 2d ed. Indianapolis, 1976.

Kaschnitz-Weinberg, Guido von. *Ausgewählte Schriften.* 3 vols. Edited by Helga von Heintze. Berlin, 1965.

Kubler, George. *The Shape of Time: Remarks on the History of Things.* New Haven, 1962.

Loewy, Emanuel. *The Rendering of Nature in Early Greek Art* (1900–1901). Translated by John Fothergill. London, 1907.

Loewy, Emanuel. *Die griechische Plastik.* 2 vols. Leipzig, 1911.

Loran, Earle. *Cézanne's Composition: Analysis of His Form with Diagrams and Photographs of His Motifs.* 3d ed. Berkeley, 1963.

Matz, Friedrich. *Geschichte der griechischen Kunst,* vol. 1. Frankfurt am Main, 1950.

Matz, Friedrich. *The Art of Crete and Early Greece: The Prelude to Greek Art.* Translated by Ann E. Keep. London and New York, 1962.

Podro, Michael. *The Critical Historians of Art.* New Haven, 1982.

Riegl, Alois. *Problems of Style: Foundations for a History of Ornament* (1893). Translated by Evelyn Kain. Princeton, N.J., 1992.

Riegl, Alois. *Late Roman Art Industry* (1901). Translated by Rolf Winkes. Rome, 1985.

Riegl, Alois. *Gesammelte Aufsätze.* Augsburg, 1929.

Riegl, Alois. *Historische Grammatik der bildenden Künste.* Edited by Karl M. Swoboda and Otto Pacht. Graz, 1966.

Schaefer, Heinrich. *Principles of Egyptian Art* (1919). 4th ed. Edited by Emma Brunner-Traut, translated by John Baines. Oxford, 1974.

Schapiro, Meyer. *Late Antique, Early Christian, and Medieval Art: Selected Papers.* New York, 1979.

Winckelmann, J. J. *History of Ancient Art* (1764). 4 vols. Translated by G. H. Lodge. Boston, 1884–1888.

Wollheim, Richard. *On Formalism and Its Kinds/Sobre el formalisme i els seus tipus.* Barcelona, 1995.

WHITNEY DAVIS

FOUCAULT, MICHEL (1926–1984), French philosopher. Michel Foucault was born in Poitiers, France. The son of a physician, Foucault pursued a career in psychology,

serving an internship at St. Anne's mental hospital in Paris. It was in this context that he began to question the relation of knowledge and power and to develop his critique of "normalizing practices," such as medicine. After receiving both the Licence de Philosophie and the Licence de Psychologie from the Sorbonne, Foucault was awarded the Diplome de Psycho-Pathologie from the University of Paris in 1952. From 1970 until his death he held a chair at the Collège de France, where he was appointed professor of history and systems of thought. His major works include *Madness and Civilization:A History of Insanity in the Age of Reason* (1961), *The Birth of the Clinic* (1963), *The Order of Things* (1966), *Discipline and Punish:The Birth of the Prison* (1975), and *The History of Sexuality,* 4 volumes (1976–1984).

Foucault has often been classified as a sociologist or historian, rather than a philosopher; however, the problems he formulates and strategies he pursues clearly fall within the horizon of epistemology. Foucault's project is to analyze shapes or figures of knowledge in relation to the complex material forces that determine them. His object is the "discursive formation" or "apparatus": the array of concepts and rules for producing statements that defines a given system of truth, as well as the authorities, institutions, practices, technologies, and artifacts in which the system of truth is embodied and by which it is maintained. Foucault's critical question is not "What is knowledge?" or "What is truth?", but "What are the structures and procedures by which a given body of knowledge is produced and circulated?" Truth, Foucault argues, is not discovered but produced, according to regular and identifiable procedures that determine in any given historical situation what it is possible to say, who is authorized to speak, what can become an object of scientific inquiry, and how knowledge is to be tested, accumulated, and dispersed.

In studying various "regimes of truth" Foucault develops the Nietzschean idea that all truth is produced according to determinable conditions, that all truth is historically specific and intertwined with contingencies of force and power. Foucault puts into question the classical epistemological commitments of modern philosophy: to the universality of truth, to the necessity of regular procedures that dehistoricize and disindividuate the scientific knower, to the essential separability of the faculty of knowledge and the faculty of will, to the conception of knowledge in terms of representation and self-representation. His analyses, however, concentrate primarily on the "human sciences," because his aim is not only analytic but also strategic. His writings, he remarks, have always had a "political" purpose: to disrupt the operation of "normalizing practices."

Through the "human sciences"—those that study the labor, life, and language of actual human beings—philosophical concepts attain a normative status and are translated into real effects on the living bodies of actual individuals. These sciences—medicine, economics, biology, penology, psychia-

try, and the science of sexuality, as well as linguistics—set norms and standards, make measurements and diagnoses, impose regulations on the basis of statistical generalizations or projections, and establish mechanisms for dividing the normal from the abnormal so as to appropriately regulate each. Through a subversive writing Foucault strategically questions how the general images by which these sciences measure the particular are established and administered in their truth. By planting questions just where these concepts and images meet a living body, Foucault aims to undermine the regulatory power of the bureaucracy of normalcy, while reformulating the basic concepts of epistemology.

Early on in his career Foucault focused on the artist as a romantic figure who has escaped the forces of normalization, whose madness and eccentricity provide a critical purchase on the homogenizing power of bureaucracy and convention. In the almost ecstatic conclusion to *Madness and Civilization,* Foucault argues that from the literary works of the Marquis de Sade and the paintings of Francisco Goya "the Western world received the possibility of transcending its reason in violence, and of recovering tragic experience beyond the promises of dialectic." The apotheosis of truth and triumph of bureaucracy celebrated in Hegelian philosophy finds itself put in question by the work of art. Joining the "nothingness of unreason" with the "power to annihilate," Sade and Goya precipitate an intertwining of madness and the work of art: "since them, unreason, has belonged to whatever is decisive, for the modern world, in any work of art: that is, what ever any work of art contains that is both murderous and constraining" (Foucault, 1965). Friedrich Nietzsche, Vincent van Gogh, and Antonin Artaud are "able to receive" Goya's transmission of the "barely audible voices of classical unreason . . . giving them for the first time an expression, a *droit de cite,* and a hold on Western culture." Not only are the works of these artists invulnerable to that "gigantic moral imprisonment" through which the modern sciences of psychiatry and public health rationalize and normalize madness, they also oblige the world of the human sciences to "justify" itself. "[A] work of art opens a void, a moment of silence, a question without answer, provokes a breach without reconciliation where the world is forced to question itself." It is as if the positive and intrinsic voice of madness, so thoroughly silenced by the carefully elaborated diagnostic categories and treatments of psychiatry, erupted through the power of artistic genius to interrogate rational beings and the bureaucracy of normalcy.

In his 1963 essay on the writer Raymond Roussel, Foucault argues that the language of such a literary artist strategically interrupts the logic of identity that governs Western metaphysics. This "logic of the Same" reproduces each individual as the same through the general categories of science. Language serves merely to represent the variety of individuals in the truth of their general form. The language of

a literary artist like Roussel, Foucault argues, "insert[s] itself in the minute separations between the imitation and what it imitates, to bring out the flaws and duplicate that imitation to its greatest extent" (Foucault, 1986). Roussel's language is a "thin blade that slits the identity of things, showing them as hopelessly double and self-divided even as they are repeated [as so many instances of some form]." Language has the power to "wound" objects, and Foucault often describes the application of the discourse of the human sciences to the living bodies of actual individuals as a kind of repressive violence that produces the homogeneity of a normalized subject. Literary language strategically interrupts and interrogates this language of the Same.

This valorization of literature, however, depends on a conception of power as essentially repressive. Foucault becomes critical of this conception after *Discipline and Punish*, for he comes to understand power as fundamentally productive, and repression as only one type of power rather than its essence. The error of identifying power with repression is linked to a second and more egregious one. In celebrating the wild, nonnormalized voice of the artist, Foucault had posited behind the objectivities constituted in the scientific discourse of psychiatry, a pristine, natural truth of madness, and he accorded the work of art the power to reveal this original truth. Not only did he assume the existence of a positive objectivity not itself produced by discursive practices, he also invested art with the power to release this truth by severing the link between knowledge and power. In *Discipline and Punish*, however, Foucault locates the general concept of power in the notions of technique and strategy, rather than repression. All power is productive, and any objectivity is the effect or product of some system of truth or "apparatus" in which discursive and nondiscursive realities, power and truth are complicated. "We should admit that power produces knowledge that power and knowledge directly imply one another . . . it is not the activity of a subject of knowledge that produces a corpus of knowledge, useful or resistant to power, but power-knowledge . . . that determines the possible forms of knowledge" (Foucault, 1977). Although Foucault continues to be fascinated with the power of a work of art to escape forces of conventionalization and normalization, he comes to see that, like every other objectivity, it is an effect of some system of power-knowledge.

In fact, this understanding of the work of art as an effect of power-knowledge is suggested even in Foucault's early work. Nearly every work begins and develops from an emblematic image: the *stultifera navis* or ship of fools in *Madness and Civilization*, the scene of the dissecting table in *The Birth of the Clinic*, or the *panopticon* in *Discipline and Punish*. In each case the image provides a summative expression of the "apparatus" or system of power-knowledge that produced it. In *The Order of Things* (*Les Mots et les choses*) this emblematic image is a painting by Diego Rodríguez de

Silva y Velázquez. Foucault finds in Velázquez's *Las Meninas* a demonstration of the logic of representation that governs the classical period (1650–1789). The famous pictures records the scene of Velázquez painting the picture of the king and queen of Spain, who themselves appear only as reflected in a distant mirror. In an eloquent and detailed evocation of the painting, Foucault reveals how we as spectators are placed in the position of the model, at once the object being represented on the painter's canvas and the subject gazing on the painter's act of representation. The painting reveals the differences and discontinuities of the act of representation, the object of the representation, and the representation itself, calling into question that adherence and identity claimed by the logic of representation for the image and its object. Moreover, in representing representation itself, the work exposes the condition of production for knowledge as representation and reveals that the coincidence of author, model, and spectator on which representation depends, never occurs. Precisely because it represents its own production according to the "apparatus" of representation, this work of art offers Foucault a critical analytical purchase on representation itself.

Similarly, in the essay "What Is an Author?" (1969) Foucault questions the convention of referring a text to the author's creativity, profundity, inspiration, or intention, as if the work were the effect of certain psychological acts or states of the author. Foucault argues that the "author" is not a subject, but a "function of discourse," an effect of the discursive formation or apparatus that determines who can speak, what can be said, and the relation of what is said to real bodies. Foucault points out that the "author function" is not universal, insofar as literary history provides many examples of texts that were accepted and circulated without concern about the identity of their author. The identity of the author is an issue specific to those discourses in which validity is tied to property and possession. An analysis of the "author function" reveals the system of authority by which the discourse is regulated and circulated. The discursive formation or apparatus prescribes the possible positions from which a subject can speak, as well as the mechanisms and criteria that divide the authorized and unauthorized, those who may speak from whose who must remain silent. A discursive analysis proceeds, then, not by asking who the real author is and what he or she has revealed of his or her most profound self, but rather where this text comes from, how it is circulated, and who controls it. The texts of Nietzsche or Gustave Flaubert do not escape this analysis, anymore than do those of psychiatry, penology, or ethics.

A certain epistemic privilege continues to adhere to art and literature despite these critiques of repression and authority. In an essay published in 1986 "Maurice Blanchot: The Thought from Outside," Foucault continues to attribute to literature a transgressive power. In it language escapes from the "dynasty of representation" and achieves a

"breakthrough." The Marquis de Sade and Johann Christian Friedrich Hölderlin, Artaud and Stéphane Mallarmé, for example, "rend" the "interiority of our philosophical reflection and the positivity of our knowledge" and offer a language "from which the subject is excluded" (Foucault, 1986). Foucault suggests that "the being of language only appears for itself with the disappearance of the subject." As long as language is subordinated to the mechanism of representation, to the subject's task of representing herself to herself in order to establish her own self-identity, it will continue to be dominated by the Hegelian question of meaning, of saying what is meant. Writers such as Artaud or Blanchot, however, do not ask what language *means*, but rather what to write so as to reveal what it *is:* a scene of alienation and dissimulation. Literature is "language getting as far away from itself as possible." It is one pathway to what is outside the interiority or self-identity of the speaking subject, and therefore of the normalizing practices focused on that subject. The speaking subject's transparency to herself, her ability to represent herself to herself without remainder, disappears in the appearance of the dense, well-marked body of language.

> From the moment discourse ceases to follow the slope of interiorizing thought and, addressing the very being of language, returns thought to the outside[,] it becomes a meticulous narration of experiences, encounters, and improbable signs—language about the outside of all language, speech about the invisible side of words. And it becomes attentiveness to what in language already exists, has already been said, imprinted, manifested.

The writing of Blanchot or Mallarmé demonstrates that language is neither time, nor eternity, nor man, but the always undone form of the outside. Rather than returning to itself in order to confirm the circuit of its own identity, this language opens on to an outside that is irreducibly other. Speaking of and from and toward this outside, the subject disappears and the writing appears as an effect of this exteriority. Thus, Foucault celebrates Blanchot because he is "not hidden by his texts, but absent from their existence and absent by virtue of the marvelous force of their existence." [See Blanchot.] Similarly, Foucault remarks of himself that he "writes in order to have no face," in order to replace subjective intention with discursive effect. Foucault continues to valorize literature despite the critiques of repression and authority, because he no longer locates its value in a nonnormalized, liberated genius. As an apparatus or discursive formation literature is distinguished by its strategic effects and its liminal placement. It evokes the adherence of language to what is not language, and marks the site of transaction between concepts and bodies. In doing so it contributes to the possibility of contesting this site. Thus, literature remains associated for Foucault with the violence, not of repression, but of transgression.

In his last work, the four-volume *History of Sexuality,* Foucault turns his attention to the "aesthetics of existence," that is, to the ethical systems of the Greek and Roman Stoics that had as their aim making life *beautiful.* These philosophies developed "arts of existence," techniques and practices for cultivating and shaping the material of the self. Although these works contain little discussion of actual works of art or of "aesthetics" in the sense of a theory of art, they do explore this analogy between the ethical subject and the author of the work of art. Both are matters of technique: effects of an apparatus, a system of concepts and material practices that produce both objects and our knowledge of them.

[*See also* French Aesthetics, *article on* Contemporary Painting Theory; Postmodernism; *and* Poststructuralism.]

BIBLIOGRAPHY

Works by Foucault

The Birth of the Clinic: An Archeology of Medical Perception. Translated by A. M. Sheridan Smith. New York, 1973.

Death and the Labyrinth: The World of Raymond Roussel. Translated by Charles Ruas. Garden City, N.Y., 1986.

Discipline and Punish: The Birth of the Prison. Translated by Alan Sheridan. New York, 1977.

The Foucault Reader. Edited by Paul Rabinow. New York, 1984.

The History of Sexuality, vol. 1, *An Introduction.* Translated by Robert Hurley. New York, 1978.

The History of Sexuality, vol. 2, *The Use of Pleasure.* Translated by Robert Hurley. New York, 1985.

The History of Sexuality, vol. 3, *The Care of the Self.* Translated by Robert Hurley. New York, 1986.

Language, Counter-Memory, Practice: Selected Essays and Interviews. Edited by Donald F. Bouchard, translated by Donald F. Bouchard and Sherry Simon. Ithaca, N.Y., 1977.

Madness and Civilization: A History of Insanity in the Age of Reason. Translated by Richard Howard. New York, 1965.

"Maurice Blanchot: The Thought from Outside." In *Foucault/Blanchot,* translated by Jeffrey Mehlman and Brian Massumi, pp. 7–60. New York, 1986.

The Order of Things: An Archaeology of the Human Sciences. New York, 1970.

Politics, Philosophy, Culture: Interviews and Other Writings, 1977–1984. Edited by Lawrence D. Kritzman, translated by Alan Sheridan et al. New York and London, 1988.

Power/Knowledge: Selected Interviews and Other Writings, 1972–1977. Edited by Colin Gordon. New York, 1980.

This Is Not a Pipe. Translated by James Harkness. Berkeley, 1983.

Other Sources

Blanchot, Maurice. "Michel Foucault as I Imagine Him." In *Foucault/Blanchot,* translated by Jeffrey Mehlman and Brian Massumi. New York, 1986.

Deleuze, Gilles. *Foucault.* Translated and edited by Sean Hand. Minneapolis, 1988.

Gutting, Gary, ed. *The Cambridge Companion to Foucault.* Cambridge and New York, 1994.

Hoy, David Couzens, ed. *Foucault: A Critical Reader.* Oxford and New York, 1986.

MARY RAWLINSON

FRANKFURT SCHOOL. *See* Adorno; Benjamin; Habermas; *and* Marcuse.

FRENCH AESTHETICS. [*To cover the earliest and most recent stages of French aesthetics, this entry comprises two essays:*
 Eighteenth-Century French Aesthetics
 Contemporary Painting Theory
The first essay explores the period when aesthetics was just emerging and French theorists were playing major roles. The second essay analyzes the tradition(s) of post-1945 French painting theory, which has had a considerable influence on contemporary (Continental) aesthetics in connection with all the arts, not just painting. For related discussion, see Bachelard; Bataille; Baudelaire; Bergson; Blanchot; Boulez; Breton; Delacroix; Duchamp; Dufrenne; Foucault; Irigaray; Le Corbusier; Mallarmé; Maritain; Poussin; Sartre; Situationist Aesthetics; *and* Taine.]

Eighteenth-Century French Aesthetics

It is generally acknowledged that modern aesthetics began in the eighteenth century. In France, a sharp questioning of the classical ideal of imitating the ancients and following the rules was provoked by the *querelle des Anciens et des Modernes,* a dispute that needs no retelling here but one that deserves to be mentioned, for it dealt with practically all the problems, issues, and controversies that marked postclassical and Enlightenment French aesthetics. The relative merits of ancient versus modern works of literature and art were vigorously debated (as, for instance, in the proper translation of Homer). The limitations of "good" taste and accepted models of beauty were perceived, and aesthetic appreciation relying on intuitive emotion and immediate, spontaneous perception was underscored. Metaphysics of beauty and taste were challenged by an experiential aesthetic of subjective feeling and pleasure. Individual originality, genius, and creativity were increasingly valorized at the expense of tradition, rules, and academic models. The classical ideals of orderliness, dignity, grandeur, harmony, and reason receded in favor of a growing awareness of individual originality and inventiveness. The complex relations of the artist to society and the nature and modalities of creativity were repeatedly pondered. The criteria of artistic judgment and appreciation were broadened as well as sharpened. Finally, aesthetic theory itself came into its own as a field of inquiry deserving of the same status as the traditional branches of philosophy.

In 1750 Alexander Gottlieb Baumgarten, a fairly obscure German philosopher, coined the term *aesthetics* in his treatise *Aesthetica.* By this term he designated the science of knowledge, based on sensory perceptions, whose aim is beauty and the psychological response to it, in contrast to logic, whose aim is truth. He also assigned it an independent and equal status as the branch of philosophy concerned with the nature of art, its creative sources, and its forms and effects. Thus, to Baumgarten belongs the credit not only of creating the term *aesthetics* but also of promoting the "philosophy of the beautiful" to a rank on a par with other categories of philosophy. Beyond these contributions, however, he had little impact on the subsequent development of aesthetics, mainly because his views remained rooted in transcendental, abstract, and scholastic metaphysics.

French thinkers, for their part, soon came to prefer dealing with the new interest in beauty and aesthetic pleasure through an empirical approach, which owed a great deal to the English, notably John Locke, the earl of Shaftesbury, and David Hume. One cannot overlook the fact that French eighteenth-century aesthetics was practiced more by critics, amateurs, connoisseurs, antiquarians, and journalists than by philosophers. Even when a *philosophe* like Denis Diderot became deeply involved in art criticism and aesthetic theory, he deliberately avoided systematizing his thoughts in order to give full play to his rich, complex, and at times contradictory ideas on the subject. French Enlightenment thinkers were generally weary of, even hostile to, philosophical systems, which they associated with the kind of dogmatic scholasticism they were doing their very best to discredit in their writings.

Following in the footsteps of Locke, theorists and critics of the eighteenth century generally subscribed to the notion that all knowledge is derived from the senses and therefore found no difficulty in grounding aesthetics in the empirical world of experiences and sensations, placing it at the center of their theorizing and relegating to the periphery classical and academic aprioristic canons and models of beauty and artistic excellence.

Enlightenment aesthetics frequently stressed the social usefulness of art. Unlike the seventeenth-century *moralistes,* who were content to observe and point out with ironical and pessimistic detachment the inequities and absurdities of the human condition, the eighteenth-century *philosophes* were embarked on an ambitious and often perilous project of spreading knowledge far and wide and of denouncing entrenched privileges, abuses of power, religious intolerance and dogmatism, and social and political inequities in the hopes of changing the order of things. That the artist should not be satisfied with providing pleasing and decorative works for the benefit of the powerful and wealthy leisure classes was a tenet shared by such leading eighteenth-century thinkers as Diderot and Jean-Jacques Rousseau. Although this view of art in its historical, social, and political context does not represent the whole range of eighteenth-century aesthetics, it constitutes its core and dynamic force.

Before dealing with the major figures of eighteenth-century French aesthetics, we need to review several influential fore-

runners who will help clarify the gradual transition that took place from the seventeenth to the eighteenth century, from the classical preoccupation with formal beauty to the emergence of a new aesthetic, founded on the notion of the interrelatedness of art and moral values.

Roger de Piles, a diplomat by profession, was a prolific and authoritative amateur critic whose writings, notably *Dialogue sur le coloris* (1673), *Abrégé de la vie des peintres* (1699), and *Cours de peinture par principes* (1708), were promptly translated into English and went through several editions in the eighteenth century. No connoisseur was more instrumental in establishing rational as well as empirically based "principles" of art, devoid of dogmatism and academicism. In the late seventeenth century *querelle des Poussinistes et des Rubénistes,* which parallels the *querelle des Anciens et des Modernes* and opposed such powerful defenders of classicism as Charles Le Brun, who staunchly upheld the supremacy of line in the visual arts, against the rebellious advocates of color, Piles played a crucial role in championing the cause of those artists who did not conform to the academic canon of formal perfection and in ensuring the eventual triumph of the colorists, thereby making possible the emergency of the Rococo style.

Piles worked tirelessly to broaden the artistic taste and judgment of his contemporaries. Widely traveled, he was thoroughly familiar with various national schools of painting, and as a keen admirer of Dutch and Flemish artists, especially Rembrandt van Rijn and Peter Paul Rubens, he called attention to the merits of these artists and boldly questioned the absolute supremacy accorded the Italians of the High Renaissance. For Piles coloristic values constitute the very essence of painting, whereas for Le Brun color is only ornamental and always subordinated to line.

What sets Piles apart from his contemporaries is his wariness of preestablished codes and rules, his keen appreciation of individual genius, of originality, spontaneity, and verve, even when they manifest themselves at the expense of "good" taste and "correctness." For Piles there is no absolute ideal of beauty and perfection, as embodied in Raphael's serene Madonnas or Nicolas Poussin's erudite historical paintings and noble landscapes. He is one of the first to acknowledge the irreducible quality of an authentically original *style,* the unmistakable mark of a truly creative mind. What is also notable about Piles is that he took great pains to hone a discourse on art that is at once intelligible and rigorous, that retains the essential features of the elegant and clear style of a cultured man of letters while effectively incorporating technical terms borrowed directly from the painter's vocabulary, an innovative practice that will not be lost on eighteenth-century critics and aestheticians, notably Diderot in his *Salons* and theoretical essays on the visual arts.

The abbé Jean-Baptiste Du Bos was a historian, diplomat, and member of the Académie française, whose *Réflexions critiques sur la poésie et sur la peinture,* first published in 1719, was widely known and read throughout Europe. In France such major figures as Voltaire, Diderot, and Rousseau borrowed freely from him. Du Bos's ideas were grounded in the humanistic doctrines of art, yet managed to be innovative and seminal, for he was among the first to conceive of art in psychological and emotional, rather than didactic or rhetorical, terms. He had, furthermore, read and pondered Locke and Joseph Addison, and their empiricism deeply influenced his own thinking. Focusing on Horace's famous simile "Ut pictura poesis" (as is painting, so is poetry), he was the first to establish important distinctions between the two fields, and in so doing anticipated Diderot and Gotthold Ephraim Lessing in their own disquisitions on the subject, especially in his insights into how crucial it is for the artist to choose the most emotionally charged moment for a composition. He objected to allegory because of its obscurity and lack of verisimilitude, a criticism that would also be picked up by Diderot and Lessing, and he pointed out that it is the impression a work of art makes on the beholder that matters, not whether it has followed the rules. He stressed the crucial role of emotion and enthusiasm as the driving force of both the creator and critic. The great painter, dramatist, or poet possesses a capacity for profound and intense human experience. A great work of art therefore has the power to stir our deepest emotions. The pleasure we experience in viewing representations of powerful emotions, whether on the stage or on canvas, derives from this instant recognition on our part. His observation that this pleasure is keenest when the subject is awesome or dreadful would have a strong impact on Diderot in France and on Edmund Burke in England, in their elaboration of an aesthetic of the sublime.

The abbé Charles Batteux wrote a treatise titled *Les beaux-arts réduits à un même principe* (1746; reprinted in 1747 and 1773), which defined art in terms of imitation of nature. Mimesis can, of course, be traced to Aristotle, and when conveniently put in terms of a pleasingly reductive thesis, as set forth by Batteux, it appealed to the contemporary public. Its major defect, however, was underscored by Diderot in his article "Beau" for the *Encyclopédie* (1751), wherein he pointedly asked just what Batteux meant by the imitation of *la belle nature.*

Antoine Houdar de La Motte was a poet and critic who took part in the *querelle des Anciens et des Modernes,* sided with the moderns, and tirelessly championed the notion that progress operates not only in the sciences but in the arts as well. Modern translators of Homer should therefore not hesitate to "improve" on him. La Motte's own poems lacked in lyricism and differed from prose only in form, and in his various essays and *Discours* he almost rationalized poetry out of existence. His extreme "modern" position and outright rejection of such "irrational" qualities as emotion, enthusiasm, and imagination made him the leader of those

who undermined poetry in the name of a Cartesian doctrine that denied its very essence. He deserves to be mentioned here, if only because his views, as we shall see, would be directly challenged by Voltaire and Diderot.

Charles-Louis de Secondat, Baron de la Brède et de Montesquieu, is now remembered for his *Lettres persanes* (1721) and *Esprit des lois* (1748). That he also wrote on aesthetics is frequently overlooked. Montesquieu did not attempt to set forth a systematic theory of the arts, but his thoughts on the subject reflect an endlessly inquiring mind, immense erudition without pedantry, sharp powers of observation, and a sensitivity finely attuned to the arts. Unlike Voltaire and Diderot, who never managed to make the "grand tour" of Italy, he stayed there for nearly a year, from August 1728 until July 1729, and all his life he looked back with nostalgic fondness on this happy experience, which deeply marked his own aesthetic preferences and ideals. He was among the first in France to write about Michelangelo with unreserved admiration. His penetrating and lively comments range widely and deal with poetry, painting, sculpture, and architecture, as well as with such topics as beauty, taste, and genius.

As a political theorist and philosophe, Montesquieu was consistently intent on situating his aesthetic reflections in the context of history, geography, and social institutions, for he believed in the interrelatedness of all phenomena, whether natural or human, as well as in the essential kinship of all areas of knowledge. His concern with aesthetics is evidenced not only in numerous passages of his famous works, but also in his more personal notebooks and travel journals, notably in his *Voyages en Europe* (published in 1894–1896), in his *Encyclopédie* article, "Essai sur le goût" (published in volume 7, in 1755, under the title "Goût"), and in an essay, *De la manière gothique* (written in 1734 and published in 1896).

Montesquieu's cast of mind tended to be impartially analytic, scientific, and historical. He did not take a direct part in the *querelle des Anciens et des Modernes,* preferring to maintain his independence and the freedom to praise or criticize according to his own judgment. As one who, like other leading members of his generation, had repudiated religious dogmas and unquestioned universal truths, he harbored a profound wariness and skepticism toward preestablished canons. Neither did he trust any manifestations of emotionalism uncontrolled by reason such as rash enthusiasm or unbridled fancifulness. Yet, he was not averse to the demands of the heart in aesthetic matters, and he had a fine sense for psychological tensions and nuances, for those undefinably pleasing and moving qualities in a work of art that critics, at a loss to explain, referred to as the *je ne sais quoi.* On the whole, he viewed the arts as intimately bound to historical forces and social structures and individual genius as somehow managing to break free from these ties as well as from the accepted rules. Montesquieu's somewhat nostalgic

ideal of simplicity, grandeur, and dignity, his lifelong love of Italy, owe less to the preceding age of Louis XIV than to a personal vision of the arts that foreshadows the neoclassical revival of the late Enlightenment and revolutionary era.

François-Marie Arouet de Voltaire the critic, connoisseur, and aesthetician has, as is the case with Montesquieu, been all too frequently eclipsed by the philosophe. Yet, his deep and abiding interest in subjects pertaining to aesthetics is evidenced in *Le temple du goût* (1731–1733), in the *Lettres philosophiques* (1734), in *Le siècle de Louis XIV* (1751), in a number of articles of his *Dictionnaire philosophique* and *Questions sur l'encyclopédie* (1764–1772), in the prefaces to his plays written over his lifetime, and in many passing comments of his voluminous correspondence.

Voltaire's ideas, preferences, and prejudices in aesthetic matters are highly representative of an eminently cultured man of the generation of the regency who all his life remained convinced that French arts and letters had achieved their apogee under the reign of Louis XIV. Lest we confuse the forward-looking philosophe, who promptly rejected René Descartes in favor of Locke, tirelessly fought for progressive and humanitarian causes, and fearlessly excoriated the political and religious abuses of the old regime, with the conservative and at times dogmatic, shortsighted, and intolerant critic and aesthetic arbiter of the eighteenth century, we must keep in mind that Voltaire's sensibility had been formed by the "grand goût" of the classical age. He was too intelligent not to realize that changes and experimentations are unavoidable and even necessary in the arts in order to meet the ideological and emotional needs of a new age. He himself did not hesitate to tinker with revered classical models in his own tragedies and comedies: he introduced the concept of modern history in such works as *Le siècle de Louis XIV* and the *Essai sur les mœurs et l'esprit des nations,* and in *Candide* and other philosophical tales he transformed the traditional novella into a highly personal genre, perfectly suited to his polemical, satirical, and propagandist requirements. But his quibbling criticism of William Shakespeare, even though he acknowledged the latter's genius, and the less than generous attitude with which he greeted the innovations of the younger Enlightenment generation (notably Diderot and Rousseau) clearly show the limitations of his aesthetic judgment and sensibility. A purist where the French language is concerned, he fiercely derided the liberties Rousseau took with it in his novel *La nouvelle Héloïse,* and in his correspondence with Madame du Deffand he revealed his strong reservations about Diderot's *drames bourgeois.*

Even more than Montesquieu, Voltaire was suspicious of enthusiasm, which he identified with religious zealotry rather than with artistic inspiration. He had no special admiration for individual originality or for genius, which he viewed as excessively self-absorbed and disdainful of the requirements of urbanity and "good" taste (hence his harsh-

ness toward Shakespeare), for he inclined toward the notion that transcendent masterpieces are not the creations of solitary genius but rather the product of a culture that has reached its acme. As for "good" taste, it manifests itself in a few connoisseurs, those rare and privileged souls capable of, and willing to cultivate, true discernment and insight. A lover of poetry, which he himself delighted in writing in its various forms (epic, didactic, dramatic, comic, burlesque, epigrammatic), he vigorously defended it against those who, like La Motte, questioned its very existence in the name of clear, rational communication of ideas. He insisted on composing his tragedies in classical alexandrines, expressed some of his deepest philosophical and moral concerns and beliefs in poetic form, and frequently dashed off light verses and witty epigrams.

Voltaire mistrusted abstract, metaphysical systems, and it is obvious that he greatly enjoyed making fun of abstruse theories. Voltaire the skeptic and pragmatist rarely failed to bring forth specific examples to deflate and ridicule metaphysical and universal aesthetics in order to underscore matter-of-factly that beauty is in the eyes of the beholder. Thus, he begins his article "Beau" in the *Dictionnaire philosophique* by pointing out in his best satirical style that for a frog beauty is incarnated in his lady frog with her two big, protruding eyes, her flat and broad mouth, and her yellow belly.

Voltaire did not elaborate an aesthetic doctrine, but he cherished the cultural achievements of humankind during its periods of greatness such as the age of Louis XIV, and he was convinced that the positive and civilizing values that these achievements exemplified had to be nurtured and protected if they were to endure. Voltaire believed that society always teeters on the brink of barbarism and that only perpetual vigilance on the part of the more enlightened can prevent this catastrophe from happening.

This is essentially an aesthetic of elitism, aristocracy, and privilege. Hence Voltaire's outrage when an upstart like Rousseau launched a frontal attack in the explosive *Discours sur les sciences et les arts* (1750) by eloquently asserting that, far from fostering political and ethical progress and serving the cause of the betterment of the human condition, those who had brought about these superb accomplishments in the sciences, letters, and arts had been the all-too-willing and submissive servants of cruel and ruthless tyrants and oppressors.

Throughout his embattled life and career Jean-Jacques Rousseau endeavored to formulate a new relationship between culture and society. Having identified the entire world of art as he knew it as the denatured product of an unjust and corrupt society, Rousseau became its most outspoken and radical opponent and critic. His writings on literature, art, and music make clear that his aesthetics were inseparable from his ethical and political ideas and that for him the beautiful and the good are one and the same thing.

Both the *Discours sur les sciences et les arts* (1750) and the *Discours sur l'origine de l'inégalité* (1755) reflect his disapproval of the role of the poet and artist in society. Reflections on music, fiction, poetry, and the theater are intimately intertwined in almost all his works, and especially in his polemical essay, the *Lettre à d'Alembert sur les spectacles* (1758), and in his encyclopedic novel, *Julie, ou La nouvelle Héloïse* (1761).

For Rousseau the development of the arts, the progressive refinement in manners, mores, and standards of beauty and taste, and the impressive advances and achievements in architecture, theater, opera, literature, and painting had not been matched by political and ethical progress and only testified to an ever-widening rift between nature and culture as well as to the increasing degeneration of social and moral values and of the human condition in general. Art has become mere entertainment and escapism, and in order to please the public, the artist, whether he be poet, painter, or playwright, does not hesitate to flatter the passions and prejudices of his contemporaries, to conform to prevailing and popular models and fashions. Where morality is concerned he is a docile follower rather than a leader.

Rousseau's *Lettre à d'Alembert sur les spectacles* formulates some of the main arguments of his aesthetic thinking. He was acutely aware that theatricality plays a necessary role at all levels of personal, social, and political interaction. In the *Lettre* he seeks to transform the very notion of theatricality in order to bring it into greater conformity with his idea of a new society. His attack on the contemporary theater constitutes a passionate plea for a new kind, a far broader forum that would involve not just professional actors and actresses, but all citizens united in great patriotic spectacles.

Rousseau's basic thesis, using Molière's *Le misanthrope* (1666) as a paradigmatic case, is that even the greatest playwrights tend to legitimate the status quo, because they depend on pleasing the public for their very survival. And it is precisely because of his admiration for Molière's genius that he launches a ferocious attack on his representation of the honest and sincere Alceste as a ridiculous, laughable character, and of Philinte, Alceste's self-satisfied and all-too-accommodating friend, as a spokesman of wisdom and sound judgment. In *Émile, ou traité de l'éducation* (1762), Rousseau would again denounce the same kind of cynical and worldly wisdom, couched in pleasing persuasive verses, this time in Jean de La Fontaine's fables, so inappropriately given to innocent children for memorization. Poetry and the theater in particular, and the arts in general, are not places where one should look for the truth.

In Rousseau's eyes the unadulterated beauties of nature would always be far superior to the feeble and imperfect imitations produced by human beings. No painted landscape, even by a supremely talented artist, can compete with a real sunset. In his diatribes against luxury or in the disdainful way in which he dismissed the whole question of "good"

taste, Rousseau deliberately challenged and contradicted Voltaire. He also seized every opportunity to contrast the artificiality and ostentatiousness of the contemporary art world, with its hedonistic sophistication and its enervating dependence on the salon and the boudoir, with the simple and healthy pleasure that can be derived from the outdoors. In so doing, he powerfully, if unwittingly, helped to create a new aesthetic that would sweep over Europe and herald the Romantic cult of nature and fascination with rustic country life.

Rousseau also introduced a new relation of the writer to the reader, especially in his novel and in his confessional writings. Taking us into his confidence, he laid bare a whole range of intimate thoughts and emotions and some of the most private aspects of his life and experience. His valorization of subjective feeling and intuition as superior to the rational faculties in his bold and at times painful quest for inner truth set the tone for our modern preoccupation with uninhibited self-revelation and untrammeled self-expression. All the affective values that Montesquieu, Voltaire, and others had considered with cautious diffidence were now given absolute primacy. Those states of the soul due to the modulations of one's sensibility and imagination are prominently featured in Rousseau's novel and *Confessions,* and especially in the *Rêveries du promeneur solitaire.*

Unlike Rousseau, Denis Diderot looked on the role of art in society, even a corrupt one, as a positive force liberating the most creative human energies and impulses. From the outset, Diderot approached the artist's creative process and procedures and the interrelation between art and society with keen interest and a personal sense of sympathetic, enthusiastic involvement. Whereas Rousseau viewed the arts with puritanical suspicion, if not outright moralistic hostility, Diderot found in them an unending source of delight and spiritual enrichment. Whereas Rousseau perceived a profound and ineradicable cleavage between art and moral values, Diderot sought to reconcile the respective exigencies of the aesthetically pleasing and the socially useful.

All his life Diderot remained concerned with aesthetic questions of a theoretical nature, as well as with art and its function in society. Before Gotthold Ephraim Lessing, whose influential treatise *Laocoön* appeared in 1766, he had become aware that the notion of the sisterhood of the arts, inherited from the Renaissance, required qualifications, as his *Lettre sur les sourds et muets* (1751) clearly attests. Diderot was especially concerned with the possibilities and limitations of the language of the poet, as distinct from that of the prose writer, and emphasized its suggestive, symbolic function.

In 1751 Diderot's article "Beau" appeared in the second volume of the *Encyclopédie.* It was widely admired, and as early as 1759 Immanuel Kant recommended it to his friends and disciples. The article presents a masterful survey of major aesthetic doctrines of the time as well as a theoretical definition of beauty that avoids metaphysical abstractions and rejected the notion of an internal, God-given sense of beauty. Diderot opted instead for the Lockean idea that the human capacity for perception of relationships is at the core of our sense of the beautiful. Although "Beau" represents one of Diderot's earliest statements on art, it firmly links his aesthetics to his materialistic, empirical philosophy. Diderot acknowledges diversity in individual aesthetic judgments but also proclaims the existence of an objective, universal law enabling us to perceive beauty in a harmonious relationship of the parts to the whole: the unity within multiplicity. Recognizing cultural and individual vagaries in our ideas of beauty, he emphasizes the universal nature of aesthetic sensibility, which derives from a basic human need for harmony and order. Before long, Diderot turned from theoretical to more practical and specific considerations pertaining to drama, the novel, as well as the visual arts.

The last decades of the eighteenth century in France witnessed the emergence of such categories as the sublime, the picturesque, the Gothic, and the fantastic, exemplifying models of perception and symbolization superseding the rational and beautiful, and stressing the overpowering, nostalgic, and melancholy rather than pleasing emotions and experiences. The sublime, in particular, which had been a rhetorical concept in France since the seventeenth century, acquired a new importance through a growing awareness of an aesthetic of awe and terror: those expressive potentialities offered by frighteningly vast and overwhelming scenes in nature, as well as by extreme human situations of emotional tension, violence, and drama.

That sublimity could be a more significant aesthetic category than beauty was demonstrated in Edmund Burke's influential *Philosophical Enquiry into the Origin of Our Ideas of the Sublime and Beautiful* (1757), which was promptly translated into French and whose impact is noticeable in Diderot's own writings, especially in his *Salon* of 1767. Basing his analysis on Locke's pleasure-pain principle and linking it to the human passions related to self-preservation, Burke defines qualities of beauty as merely pleasing ones, and those of the sublime as related to such intense emotions as fright and astonishment. Natural scenes connoting unleashed and untamed power, immensity, mystery, and dread—such as towering, snow-clad mountains, steep gorges and dizzying precipices, roaring torrents, violent storms, beclouded and darkened skies, or heavy, turbulent seas—are likely to induce sublimity. All these features were taken into consideration when Diderot commented on the seascapes by Claude-Joseph Vernet or on landscapes with ancient ruins by Aubert Robert. That this new fascination denoted a certain blurring of nature and art should not surprise us, for eighteenth-century aesthetics never established a clear distinction between the two realms.

The appearance of new aesthetic categories stressing strong emotions and effects rather than pleasing ones even-

tually spelled the demise of the teasingly playful Rococo. Indeed, the last decades of the eighteenth century were marked by a revival of the Greco-Roman ideal, which was extolled, among others, by the antiquarian André Caylus, the engraver and connoisseur Charles-Nicolas Cochin, the critic La Fond de Saint-Yenne, and even more than occasionally by Diderot, especially when he became irritated with the slyly erotic extravagances of a François Boucher or a Jean-Honoré Fragonard. Johann Winckelmann's widely read and admired *Geschichte der Kunst des Altertums* (History of Ancient Art; 1764), glorifying a simple, dignified, virile aesthetic, was excerpted in French journals right after its publication, and two years later translated and published in full under the title *L'histoire de l'art*. The classical grandeur of the preceding century, generally identified with the austere compositions of Poussin and Eustache Le Sueur, would also frequently be evoked by Diderot and other critics in stern admonitions to contemporary painters. Joseph-Marie Vien, Jacques-Louis David's mentor, earnestly endeavored to incorporate Greek motifs in his sleekly classicizing compositions.

Even before the Revolution a renewed ethical preoccupation with civic and patriotic virtues, which of course owed a great deal to Rousseau's inflamed rhetoric, fostered this trend. After 1789 what had been a Parisian fashion became a national aesthetic ideology, grandiosely enacted by David in his heroic compositions and masterfully orchestrated public feasts and spectacles.

The eighteenth century did not look on the arts as above or separate from the difficult business of living in a complex, modern society, and it produced a creative, at times explosive, confrontation between the old and the new that paved the way for the Romantic revolution. From the outset the century acknowledged, sometimes reluctantly, the essential roles of passion and enthusiasm in the creative process and of subjective sensibility in aesthetic appreciation and pleasure. At the same time, Enlightenment writers and thinkers rarely omitted the historical, social, and moral context when dealing with aesthetics. That eighteenth-century aesthetics demanded and fully expected this relation of art to life, morality, and politics is its most distinctive feature and its greatest legacy to us.

[*See also* Alembert; Boileau-Despreaux; Cochin; Condillac; Cousin; Diderot; Du Bos; Piles; Rousseau; Sensibilité; *and* Voltaire.]

BIBLIOGRAPHY

Baumgarten, Alexander Gottlieb. *Aesthetica.* Hildesheim, 1961.

Bryson, Norman. *Word and Image: French Painting of the Ancient Regime.* Cambridge and New York, 1981.

Burke, Edmund. *A Philosophical Enquiry into the Origin of Our Ideas of the Sublime and Beautiful.* Edited by J. T. Boulton. New York, 1958.

Cassirer, Ernst. *The Philosophy of the Enlightenment.* Translated by Fritz C. A. Koelln and James P. Pettegrove. Princeton, N.J., 1951.

Chouillet, Jacques. *L'esthétique des lumières.* Paris, 1974.

Crow, Thomas. *Painters and Public Life in Eighteenth-Century Paris.* New Haven, 1985.

Dacier, Anne. *Madame Dacier's Remarks upon Mr. Pope's Account of Homer, Prefixed to His Translation of the Iliad.* Translated by Mr. Parnell. London, 1724.

Du Bos, Jean-Baptiste (abbé). *Réflexions critiques sur la poésie et sur la peinture.* Edited by Dominique Desirat. Paris, 1993.

Ehrard, Jean. *Montesquieu critique d'art.* Paris, 1965.

Fayolle, Roger. *La critique littéraire.* 2d ed. Paris, 1966.

Folkierski, Wladyslaw. *Entre le classicisme et le romantisme: Étude sur l'esthétique et les esthéticiens du XVIIIe siècle.* Paris, 1925.

Fontaine, André. *Les doctrines d'art en France: Peintres, amateurs, critiques, de Poussin à Diderot.* Paris, 1909.

Gilman, Margaret. *The Idea of Poetry in France from Houdar de La Motte to Baudelaire.* Cambridge, Mass., 1958.

Lessing, Gotthold Ephraim. *Laocoön: An Essay on the Limits of Painting and Poetry.* Translated by Ellen Frothingham. New York, 1957.

Marshall, David. *The Surprising Effects of Sympathy: Marivaux, Diderot, Rousseau, and Mary Shelley.* Chicago, 1988.

May, Gita. "Diderot and Burke: A Study in Aesthetic Affinity." *Publications of the Modern Language Association of America* 75 (1960): 527–539.

May, Gita. "Diderot et Roger de Piles." *Publications of the Modern Language Association of America* 85 (1970): 444–455.

May, Gita. "Beauty in Context." In *A New History of French Literature,* edited by Denis Hollier. Cambridge, Mass., 1989.

May, Gita. "Rousseau's *Letter to d'Alembert on the Theatre* and Revolutionary Aesthetics." In *Jean-Jacques Rousseau and the Revolution,* edited by J. Roy, pp. 199–207. Ottawa, Ontario, 1991.

Montesquieu, Charles-Louis de Secondat, baron de la Brède et de. *Oeuvres complètes.* 2 vols. Edited by Roger Caillois. Paris, 1949–1951.

Mortier, Roland. *L'originalité: Une nouvelle catégorie esthétique au siècle des lumières.* Geneva, 1982.

Navès, Raymond. *Le goût de Voltaire.* Paris, 1938.

Perrault, Charles. *Parallèle des anciens et des modernes en ce qui regarde les arts et les sciences.* Paris, 1962; reprint, Geneva, 1971.

Robinson, Philip E. J. *Jean-Jacques Rousseau's Doctrine of the Arts.* Bern and New York, 1984.

Rousseau, Jean-Jacques. *Oeuvres complètes.* 4 vols. Edited by Bernard Gagnebin and Marcel Raymond. Paris, 1959–1964.

Rousseau, Jean-Jacques. *The First and Second Discourses.* Translated by Roger D. and Judith R. Masters. New York, 1964.

Rousseau, Jean-Jacques. *La nouvelle Héloïse; Julie, or, the New Eloise.* Translated and abridged by Judith H. McDowell. University Park, Pa., 1968.

Rousseau, Jean-Jacques. *Politics and the Arts: Letter to M. d'Alembert on the Theatre.* Translated by Allan Bloom. Reprint, Ithaca, N.Y., 1968.

Rousseau, Jean-Jacques. *Emile: or, On Education.* Translated by Allan Bloom. New York, 1979.

Rousseau, Jean-Jacques. *Reveries of the Solitary Walker.* Translated by Peter France. Harmondsworth, 1979.

Rousseau, Jean-Jacques. *A Discourse on Inequality.* Translated by Maurice Cranston. Harmondsworth, 1984.

Rousseau, Jean-Jacques. *The Confessions.* Translated by J. M. Cohen (1953). Reprint. Harmondsworth, 1988.

Saisselin, Remy G. *Taste in Eighteenth-Century France.* Syracuse, N.Y., 1965.

Teyssèdre, Bernard. *L'histoire de l'art vue du grand siècle: Recherches sur "L'abrégé de la vie des peintres," par Roger de Piles et ses sources (1699).* Paris, 1964.

Voltaire, François-Marie Arouet de. *Oeuvres complètes.* 52 vols. Edited by L. Molland. Paris, 1877–1885.

Voltaire, François-Marie Arouet de. *Dictionnaire philosophique.* Paris, 1954.

Voltaire, François-Marie Arouet de. *Lettres philosophiques.* Paris, 1964.

Voltaire, François-Marie Arouet de. *The Complete Works of Voltaire.* Edited by W. H. Barber and Ulla Kolving (edition in progress). Oxford, 1968–.

Williams, David. *Voltaire: Literary Critic.* Studies on Voltaire and the Eighteenth Century, 48. Geneva, 1966.

GITA MAY

Contemporary Painting Theory

One peculiar feature of the Anglo-American reception of French thought since about 1970 is the view that the variety of thinkers and tendencies involved reduces everything to language. One crucial place to test such a reading is with regard to a set of texts devoted to painting and the visual arts, for the latter would seem to be situated at or beyond the boundaries of language, a place that Julia Kristeva calls the semiotic. The alleged reductionism of the French is usually construed as the claim that language is a seamless whole in which all meanings are defined in terms of one another. But it is characteristic of French poststructuralist thinkers to deny precisely this view, and to emphasize the fissures and fractures in linguistic systems, their susceptibility to psychological, social, or institutional power, and their tendency to generate inconsistencies and aporias. So even if the alleged reduction were to occur, it would be to a language that was already understood in what might be called a "materialistic" fashion. On the other hand, if language is construed more narrowly and conventionally as verbal, then the leading tendency among poststructuralist thinkers would seem to be the desire to distinguish between the linguistic and the visual while at the same time tracking and articulating the structure and play of their incursions into and intersections with one another. In this respect Jacques Derrida speaks for these theorists when he suggests (in *The Truth in Painting*) that we should question the traditional philosophical hierarchy or system of the arts, according to which it is the arts of language to which the others aspire and which complete their mission (the traditional view is found to be especially strong in G. W. F. Hegel and Martin Heidegger). Given this approach, it becomes an important task for poststructuralist thinkers to be self-critically vigilant about the position of the speaker or writer who addresses painting.

Maurice Merleau-Ponty. Merleau-Ponty, the most immediate French target of poststructuralist critiques of phenomenology, develops a parallel between the work of the phenomenologist and the painter (he differs here from his contemporary Jean-Paul Sartre, for whom the artist produces images that are capable of phenomenological analysis; for Sartre it is the writer of philosophy or prose who can disclose the truth of consciousness). At first this parallel seems to respect the differences between the visual and the discursive, with painter and philosopher working in quite distinct and equally significant media. In "Cézanne's Doubt" Merleau-Ponty sees Paul Cézanne's project not in terms of the finished products of his work (indeed, Cézanne denied that most of his paintings were finished, and refused to sign them), but as a continually renewed effort to work back to the primordial roots of perception. Just as Edmund Husserl found the articulation of the levels of intentionality that structure all experience an endless task, so Cézanne, on this view, kept going back beyond the easy abstractions of convention and tradition in order to disclose the nature of vision. For the painter this entailed a questioning of the residues of quattrocento perspectivism, according to which the world is seen by a monocular, immobile gaze that transcends the field it dominates. It also entailed Cézanne's parting ways with the impressionists, rejecting their dissolution of everything into light; he insisted on the obstinacy of the object. Implicit in Cézanne's stylistic development is a refusal to divorce sight from touch or from the temporal experience in which vision is always more than a momentary impression; it is the latter that should be considered an abstraction, rather than considering the object as a mere construct.

In this portrait of Cézanne Merleau-Ponty relies rather heavily on the painter's reported conversation and the literary testimonies of his contemporaries. Because Cézanne was reported to have said that he submitted himself to the *pater omnipotens,* Merleau-Ponty infers that he was "oriented to an infinite Logos" (Merleau-Ponty, 1993). In order to make Cézanne's paintings speak, the philosopher has recourse to the painter's speech, so that language seems to be the court of last resort. Especially since Merleau-Ponty's use of the written sources is rather uncritical, there seems to be an implicit priority given to the linguistic that produces unacknowledged consequences for the analysis of painting.

This apparent priority of the linguistic takes a more explicit form in "Indirect Language and the Voices of Silence," an essay inspired by André Malraux's *Voices of Silence.* As the latter title indicates, the question is whether and how one can speak for the visual, which seems to have no voice of its own. Merleau-Ponty sees Malraux as submitting visual art to the monstrous order of Hegelian history, according to which it is later works, as understood and articulated by the critic, that put us in a position to assess earlier ones. The museum, with or without walls, and its associated practices and discourses are said to overwhelm the style or "inner schema" that constitutes the artist's "life itself." So far Merleau-Ponty contrasts a primordial phenomenology of the artist's activity with a discourse that can only be secondary and external to it. The extreme expression of such a position is found in the late "Eye and Mind," where it is said the very first painting went to the farthest reaches of the future, implying that each painting involves an "advent" of meaning that can never be exhausted. "Indirect Language" begins by appealing to Ferdinand de Saussure's diacritical account of language to suggest that all meaning lies in the

interstices; in Merleau-Ponty's analysis this applies to visual art as well as to language in the narrower sense, so that the latter would have no priority over the former. Yet the essay ends with a kind of reformed Hegelianism, setting out a series of distinctions between painting and writing that hinge on painting's failure to achieve presence: "The Spirit of Painting appears only in the Museum, because it is a spirit external to itself. . . . Man does not paint paintings, but he speaks about speech, and the spirit of language wants to depend upon nothing but itself." Art seems to become more truthful, more fully present, as it approaches the condition of language. Literature of any time is said to live "entirely in the present" whereas paintings date much more easily. What emerges from these dubious contrasts is a traditional commitment that undermines Merleau-Ponty's desire to provide a phenomenological recognition of the autonomy of painting. Merleau-Ponty's late and unfinished *The Visible and the Invisible* is tantalizingly suggestive; it introduces the idea of a hyperreflection *(sur-réflexion)* that seems to hover on a boundary between the perceptual and the linguistic and that

> would set itself the task of . . . reflecting on the transcendence of the world as transcendence, speaking of it not according to the law of the word-meanings inherent in the given language, but with a perhaps difficult effort that uses the signification of words to express, beyond themselves, our mute contact with the things, when they are not yet said. (Merleau-Ponty, 1968)

Although Merleau-Ponty was not able to develop this himself with respect to articulating the relation between language and painting, it could be taken as a motto for much of the succeeding work in this intellectual tradition. *The Visible and the Invisible* is also notable for the thesis that the relation between human vision and the world is chiasmic; that is, our seeing is possible only insofar as the world is a kind of extended "flesh" that in some sense sees us. This thesis, reminiscent to some degree of Friedrich von Schelling's absolute idealism and philosophy of nature, was taken up by Jacques Lacan, who suggested that there is "a pre-existence of the gaze—I see only from one point, but in my existence I am looked at from all sides" (Lacan, 1978, p. 72). This gaze is associated with castration anxiety; it is an "evil eye" that discloses the inadequacies of the viewer. For Lacan our modes of dealing with the world are either in the imaginary mode, based on our recognition and misrecognition of ourselves in the mirror stage, or in the symbolic or linguistic register, in which language seems to form a relatively seamless set of signifiers by which we glide from one meaning to another. The visual can function as a disruption or interruption of the symbolic, making us aware of the limitations of language; and within the visual the phenomenon of distortion, or specifically anamorphosis, which Lacan analyzes in Hans Holbein's painting *The Ambassadors,* can serve to disturb the illusory integrity of the imaginary. In Holbein's painting the anamorphic skull stares back at us, challenging the assumed autonomy of our gaze and marking the inevitability of loss and death. Effects like this are "stains" on the imaginary wholeness of the visual field. Despite his famous insistence that "the unconscious is structured like a language," then, Lacan arrives at a position similar in some respects to that of Jean-François Lyotard, who accuses him of a form of linguistic reductionism (at least in Lacan's alleged omission of the visual dimension of the dream); given their common debt to Merleau-Ponty, this partial overlap is not as surprising as it might initially appear.

Michel Foucault. Foucault's *The Order of Things: An Archaeology of the Human Sciences* begins with a celebrated discussion of Diego Rodríguez de Silva y Velázquez's *Las Meninas,* which inserts the painting into history (or reads it archaeologically, to preserve the author's language) by suggesting that the position that the painting allocates to its artist, model, and spectator is one that is occupied in oscillating fashion by each, but that these roles must fail to coincide. This is taken to be an indication that the epoch of which the painting is typical does not yet have a conception of human beings as self-knowing subjects, one that would be the ground of all other knowledge. While some have criticized the essay for carelessness in understanding Velázquez's use of perspective (and these criticisms seem rather hasty), we might attend to another dimension of the text in which it explores the relation between painting and language. The "starting point for speech," Foucault writes, is the incompatibility of language and the visible. *Las Meninas* can be read either as a simple group portrait or as a highly self-referential work involving a complex reflection on the conditions of pictorial representation; similarly, Foucault's essay can be read either as a commentary on the painting (the level at which questions about matters such as perspective would arise) or as an artful meditation on the relations between painting and language. Consider the question of who is marking the failed coincidence of artist, model, and spectator or the absence of a human being; Foucault insinuates a voice in the first person plural, a "we" that manages a virtual incorporation of its readers. If "we do not know who we are" insofar as we occupy the position of the spectator/artist/model in front of the visible painting, no such uncertainty infects us in our role as the collective subject who observes and records such uncertainties or ambiguities. If the figures depicted in or implied by the painting are frozen in time, "we" are involved in a process of discovery that unfolds through Foucault's narrative. Much later in *The Order of Things,* when Foucault is explaining the rise and disappearance of the concept of the human being, he refers once more to the discussion of *Las Meninas,* pointing out that it was precisely that concept ("enslaved sovereign, observed subject") that was lacking in the painting. He broaches the possibility of imagining a different version of the work in which the various roles sketched earlier coin-

cide, and in a highly impersonal way recalls that earlier "one imagined" the figures in restless oscillation. Now the collective subject has disappeared. In each case Foucault's rhetoric is appropriate. In the opening essay the linguistic "we" marks a distinction between a speech that can assume concepts not open to its object, while in the later reference to the painting, since it is now the concept of the human that is being articulated and questioned, a different voice, one no longer rooted in the "we" takes over. Indeed, something like this shift had already occurred between the two parts of the opening essay, for the second part begins by explicitly noting the different registers of the linguistic and the visual; neither can be reduced to the other. Yet even this must be placed within a historical (or archaeological) context. This becomes clear in Foucault's essay on René Magritte, *This Is Not a Pipe*, in which he argues that painting has overcome the code, traditional at least since the fourteenth century, that prohibited interchanges between the linguistic and the visual. There he complicates the genre of ekphrasis, the verbal description of a visual work, by writing of a painting that, on his reading, has already entered into the linguistic realm.

Jacques Derrida. Perhaps with a nod to Merleau-Ponty, Derrida begins *The Truth in Painting* by citing a line from Cézanne: "I owe you the truth in painting and I will tell it to you" (Derrida, 1987). Derrida's commentary on this phrase, given directly in an introductory essay and indirectly throughout the book, takes the form of a series of meditations on how it is possible to speak, truthfully or adequately, of that which is silent. Derrida expects us to be familiar with what he refers to as "the canonical difficulties of description in discourse on art," which have to do, presumably, with the difference between visual and verbal media. The pitfall in speaking or writing about visual art seems to consist in forcing its translation into language and overlooking its specific character as painting. This danger is analyzed in the book's longest essay, offering an analysis of the difference between the philosopher Martin Heidegger and the art historian Meyer Schapiro with regard to the understanding of a Vincent van Gogh painting of old shoes. Heidegger had invoked the painting in "The Origin of the Work of Art (1936)," both to demonstrate something about the nature of equipment (such as shoes) and to show how art could be truthful. For Heidegger, the shoes disclose the world and earth of the peasant woman to whom he imagines that they belong; and insofar as art discloses something about equipment, which supposes both a world of meanings and a resisting element (earth), with which the world is in tension, it is truthful. Heidegger describes this illumination by saying "this painting has spoken," a phrase that Derrida takes quite seriously as indicating the point made more explicitly later in Heidegger's essay that the art of language somehow embraces and surpasses the other arts. Schapiro, on the other hand, had argued on biographical grounds that

these were the shoes of van Gogh, an artist and a man of the city. In a complex dialogue (or polylogue), Derrida deploys a number of voices who offer an analysis of the "correspondence" between the two interpreters of the painting. The dialogue suggests that the two are as one in their assumption of the project of appropriating the painting by means of language. To say that the shoes belong to the peasant woman (or artist) is to say something like "they belong to me, the interpreter." The linguistic project proceeds by finding familiar categories of meaning: a pair of shoes, an absent owner. But the several voices of the dialogue—pluralized so as to avoid the illusion of a single magisterial speaker—ask whether the shoes are really a pair (they could be two left shoes) and whether they must be thought of in relation to an owner (why do they stand there empty and very prominently unlaced?). As one voice suggests, the shoes are, in some sense, on the other side of language: "they concern us/look at us, mouth agape, that is, mute, making or letting us chatter on, dumbstruck before those who make them speak . . . and who in reality are made to speak by them." This suggestion has a psychoanalytic flavor: we are compelled to speak of painting, we may have an obsessive desire to do so, but if we fail to acknowledge and thematize that desire, then we become its plaything and are able to speak only of ourselves. Therapy would itself be verbal, although conducted without a magisterial speaker, and would consist in accepting the detachment of the work (for which the detached state of the shoes is a metaphor). Derrida is also wary of one tempting but false exit from the verbal/visual complex: this would be to fetishize the very silence of the work in such a way as to give it a kind of aphoristic authority, to endow it with a virtual and unquestionable discourse (see especially his interview with Peter Brunette and David Wills in *Deconstruction and the Visual Arts*). Strategies other than the dialogue are possible to avoid the two extremes. In *Memoirs of the Blind: The Self-Portrait and Other Ruins*, Derrida, responding to an unnamed interlocutor, explores the possibility of assuming explicitly one's own speaking position in all its fallibility and with all its limits. He attempts to show that the one who draws or paints does not see the subject (e.g., the model) in the very act or moment of marking the paper or canvas; whence the interest of the iconography of blindness and its analogues in the works Derrida selected for the exhibition that is the occasion of his essay. Writing also involves a certain blindness, a plunging headlong into an as yet undefined area. The blind person's cane, the artist's brush, the writer's fingers on the keyboard are analogues of one another. There is no all-seeing gaze for the same reason that one does not know what will be written until one writes it. The conditions of possibility of these activities become even more prominent in self-portraiture, whether visual or verbal. In this particular essay Derrida explores his own opticality, including dream visions, a paralysis of the eye that occurred while writing, and the tears that,

he suggests, are the eye's most distinctive and truly human activity. The traditional ecphrasis, with its assumption of a magisterial viewer and speaker, is replaced by a model of a fractured self for whom these activities fail to coincide with each other and are split within themselves. The banal fact that one cannot write and see at the same time has become a way of rendering the genre of ecphrasis problematic, so that the writer becomes a "ruin," like the self-portrait. (It is perhaps needless to add that many of the analogies piled on top of one another here may appear deeply questionable.)

Roland Barthes. Barthes's pioneering and wide-ranging work in literary and cultural semiotics might suggest that he applies a specifically linguistic model to all forms of art and image making. The analyses collected in his relatively early *Mythologies,* including those of such visual products as magazine advertisements and covers, photographic exhibitions, and films, tend to support this view. He also wrote of the photograph, however, that it is "a message without a code" (Barthes, 1981), suggesting that it cannot yield a specific meaning without contextual clues; and he went on to say that the *image* (a general term for any visual artifact, which can be profitably compared to Lacan's idea of the picture) is essentially polysemous and that language is needed to limit its indefinitely proliferating possible meanings. In an essay on the artist Cy Twombly, Barthes investigates the reciprocity of drawing and written language, pointing out that Twombly's drawing is a kind of trace or inscription, a mark of his having been there; at the same time the titles of his pieces, while not describing in any simple way what they might be thought to represent, provoke a certain direction of thought that leads us to see their strictly visual content in a more specific way than we would otherwise (for example, in terms of a "Mediteranean atmosphere").

Barthes's most sustained exploration of the relation between image and language is *Camera Lucida,* a book about photography that involves a meditation on death, genealogy, and modernity. Artfully designed, the text is punctuated by a series of photographs which illustrate and amplify the essay that ostensibly comments on them. Seeming to take up the project of phenomenology (dedicating the book to Sartre's *The Imagination*), Barthes first suggests a binary distinction for the analysis of photographs. Every photograph has its *studium,* a topos (e.g., a Russian street scene, a body in the road during wartime); but if it is a photograph of some real interest it will also have its *punctum,* some unexpected, unpredictable detail or mood that troubles, disturbs, or excites us (e.g., a strange cap worn by a Russian boy, a sheet carried by a woman at the death scene). The *punctum* can be compared to the role of the stain or anamorphosis in Lacan's theory (to which Barthes often alludes), insofar as it interrupts a quasi-linguistic continuum. One of Barthes's most striking gestures in *Camera Lucida* is his *not* reproducing the photograph of his mother as a five-year-old child on which the later argument of the book

turns. This absent image intensifies the realization that all photographs have a distinctive temporality: they are marks of something that has happened earlier, signifying "that-has-been," or the tense of the simple past. This mere fact of being there in the mode of being related to what was can be compared with Heidegger's conception of the artwork as simply giving or manifesting itself and with Derrida's attempt to honor the remainder in the visual work, even defending it against Heidegger's lapses; but we should also note the contrast between the present tense of the "it gives" or "it shows itself" and the past of "that-has-been." Beyond that, Barthes suggests, this photograph is emblematic of this: insofar as photography is marked by pastness it is akin to death and our concern with it can be a mode of mourning. If the awareness of death is largely repressed in contemporary society, the photograph, especially in its stark black-and-white form, may be the way in which this repressed element returns. Photography can be tamed, Barthes says, by the addition of color, which restores the illusion of life, or by film, which does something similar through motion and sound. The nonappearance of the mother's photograph might be taken at first to indicate that Barthes believes that his writing can provide its equivalent, as in the classical poetic and rhetorical genre of ecphrasis. But he tells us that the true reason is that the photograph is too personal, and too closely tied to that which ultimately ("nondialectically," he says) individualizes him, his death, to have a general meaning; in other words, it is precisely because the picture escapes the linguistic dimension to an extreme degree that its absence says something about the ineluctable inadequacy of language in the face of the photographic image. *Camera Lucida* can also be read as a translation of Lacanian concepts to photography, in which, for example, Barthes's analysis of death would correspond to Lacan's "real."

Julia Kristeva. Kristeva's early intellectual career owes much to the work of Barthes in particular, as well as to the general context of French structuralist and poststructuralist linguistics, criticism, and philosophy. Some affinities and contrasts with Barthes emerge in her major statement concerning visual art, "Motherhood according to Giovanni Bellini," for there she is concerned, like her mentor, with the way in which the maternal either escapes representation and language or interrupts the flow of the symbolic (perhaps it should be noted that this essay of 1975 predates *Camera Lucida* by several years). Painting occupies a role at the edge of the linguistic or symbolic in Kristeva's analysis. She is concerned, much in the manner of Lacan, with how human beings become "speaking animals." Kristeva argues that our initiation into what she calls the "socio-symbolic contract" always involves loss or sacrifice of a pre-Oedipal bond with the mother; once situated within the symbolic, we are caught between two dangers, to which art and literature can sometimes offer a resistance. We might accept the

symbolic without question, as a system for the exchange of meaning without affect, in which case we become nothing but functionaries of the social norms; or, we might attempt (as some radical feminists do, she suggests) to flee the symbolic altogether for the sake of a fantasized return to the prelinguistic maternal. Yet the symbolic is never all-powerful; it may be interrupted by what Kristeva calls the semiotic, that is, drives, impulses, and feelings that come from the maternal, presymbolic level. Art and literature can present and frame this interplay of the semiotic and symbolic; in the case of painting it is color and space that can break through and question or modify the linguistic and social norms. Although painting, we might say, is from the first nonlinguistic, the symbolic plays a role in it insofar as painting has recognizable themes, an iconological vocabulary, and fits into specific social, religious, or philosophical traditions that are typically articulated by means of language. The return of the semiotic in art, then, is both a way of escaping from the tight constraints of the socially and linguistically constructed self and a way of reconstituting that self in a less rigid form. As Kristeva says of color, in "Giotto's Joy," "it is through color—colors—that the subject escapes its alienation within a code (representational, ideological, symbolic, and so forth) that it, as conscious subject, accepts" (Kristeva, 1988, p. 37). The contrast with Barthes, who valorizes the contrast black and white, is striking (and while it is true that Barthes does this in the rather special case of photography, it is also true that he has very little to say about color in his other writings on the visual).

In her essay on Giovanni Bellini, Kristeva turns to the theme of the maternal as such. On the basis of rather sketchy biographical information, she supposes that Bellini lost or was in some way abandoned by his mother, and that his work can be read as a search for the lost maternal in its semiotic manifestations of color and space. On one level, she is offering a parallel and contrast to Sigmund Freud's psychoanalytic account of Leonardo da Vinci; but whereas Leonardo (for Sigmund Freud) was dealing with too many mothers (his birth mother and his stepmother), Bellini was attempting to recapture the semiotic equivalent of archaic, prelinguistic memory or desire. Another contrast has to do with painterly and religious traditions, which form part of the symbolic context for both; this is the opposition between the figurative style of Florentine painting (Leonardo) and the colorist mode of the Venetians (Bellini), as well as the stronger presence of Eastern Orthodox conceptions of the Virgin Mary that, on Kristeva's account, reinforce the artistic contrast insofar as Byzantine Christianity is less concerned with representing the mother by clearly delineated figures. Unlike Freud, in his writings on visual art, Kristeva is relatively unconcerned with iconographic or iconological meaning in Bellini's work; rather, she attempts to evoke the ways in which, through a complex series of stages constituting a long career, he fulfills the role of the

artist to let the traces of the semiotic break through the limits of the symbolic: "At the intersection of sign and rhythm, of representation and light, of the symbolic and the semiotic, the artist speaks from a place where she [the mother] is not. He delineates what, in her, is a body rejoicing [*jouissant*]" (Kristeva, 1980, p. 242). For Kristeva, to imagine that painting would be reducible to language or the symbolic could only be a symptom of the hypertrophy of rationalism and idealism; whereas to think that it could be an avenue for ecstatic flight into the purely semiotic would be a form of regression or madness inconsistent with our destiny as "speaking animals."

Jean-François Lyotard. At least since *Discours, figure* (1971), Lyotard insisted that language is not a self-sufficient system; it is, he always maintained, constantly disrupted, interrupted, and contested by something else that alternately is called presence, figure, or event. If painting cannot be reduced to the word, this provides no foundation for a cult of silence around art. Even silence, Lyotard argues in *The Differend: Phrases in Dispute* (1988) is a marked position within the world in which we speak. There is no escape from phrasing or sentence making. Merleau-Ponty's presence in this early work of Lyotard's is strongly marked, and perhaps one can see *Discours, figure* as a carrying out of the hyperreflexion of which the former spoke. In *Que peindre? Adami, Arakawa, Buren*, Lyotard examines the differend, or unresolvable dispute, between those who believe that there is nothing in painting that cannot be verbalized and those who claim that there is an irreducible level of visual presence in art. Appropriately enough, the exploration proceeds by a series of dialogues that eventually focuses on the problematic work of the three contemporary artists named in the book's subtitle. Lyotard sets up the issue in something like the following way. "A" claims that whatever is seen in art can be described linguistically; rhetorical ecphrasis, art history, or criticism can evoke images of real or imaginary works. If painting is meaningful, then its meaning must be linguistically accessible. If someone were to claim that something was missing from such linguistic performances, then it should be possible to say what it is. But as soon as such naming occurs, we are on the way to a verbal articulation. But someone else, "B," maintains that reference is not reduction, that the ability to speak about painting does not imply that there is nothing in the painting that is not translatable into language. B might be suspicious of any commentary that led us to short-circuit our looking. The rush to commentary obscures the fact that painting's point may be precisely to contest the hegemony of the linguistic. A replies that by valorizing the silence of painting, B actually makes it speak after a fashion; the painting is understood as making a statement, albeit a self-referential one about its own powers, status, effects, and integrity. If the work excludes the profane speech of the rhetor or critic, then its eloquent silence becomes authoritarian. Lyotard defines the work of

philosophy as bearing witness to the differend, that is, to such seemingly unresolvable disputes as the above. Each advocate points to a wrong that cannot be recognized in the other's perspective (for Lyotard such a perspective is a "regime of phrases" and such a wrong is a "damage"). Lyotard sets himself the task of rephrasing such oppositions. The first dialogue of *Que peindre?*, titled "Presence," centers on whether there is an irreducible dimension of presence in painting. The background of this discussion is the critique of the metaphysics (and, by implication, any aesthetics) of presence by Heidegger, Derrida, and others. "You" (a character with whom the reader may at first be tempted to identify) argues that there is no unmediated perception, especially in art, that everything we apprehend is surrounded by a halo of thought and language; and You adds in Hegelian fashion that art itself has become increasingly aware of this and has recently taken the inescapability of mediation as its theme. Yet the other participant in the discussion, "Him" (positioned initially as object rather than subject), asks that we pause for something like a phenomenological reflection here. Him attempts to evoke a sense of the event, the fact that something has taken place when we look at a painting that has interrupted the flow of discourse, external or internal. Another name for the event is presence. Him argues that the becoming-linguistic of recent art is not the sign of a Hegelian self-knowledge in which art realizes its true nature by dispensing with the myth of presence; it is rather simply a redistribution of social roles in which the work of commentary is now assumed by artists without waiting for the contributions of critics and historians. Color is perhaps the most striking case of discursive interruption in painting. The possibility of speaking of color (by, for example, articulating a theory of color symbolism or considering the price and esteem of pigments in fifteenth-century Florence) does not alter the fact that it is presented to us. Being receptive to such presentation demands a kind of nonaction, which Him evokes with echoes of Heidegger's letting-be and explicit reference to the Zen thought of Dōgen. Like these thinkers, Him (who seems increasingly to speak for Lyotard) recognizes that speaking of the nondiscursive is paradoxical in the way that Hegel claimed the attempt to utter the immediate (in "Sense-Certainty") is paradoxical. One way of evading the paradox is to write or speak in a way that opens itself up to interruptions, a form Him calls "the painting of presence." The consequence would be that an artfully constructed ecphrasis could show by its phrasing how art interrupts language. This seems to be the project of the ensuing dialogues in *Que peindre?* that explore variations on these themes in the work of three artists and in the phrasings to which they give rise. Similar strategies are at work in many of Lyotard's other writings, notably in about half of the essays collected in *The Inhuman: Reflections on Time*.

[*See also* Barthes; Baudrillard; Deleuze; Derrida; Foucault; Lacan; Lyotard; Ricoeur; *and* Sartre.]

BIBLIOGRAPHY

Barthes, Roland. *Image, Music, Text*. Edited and translated by Stephen Heath. New York, 1977.

Barthes, Roland. *Camera Lucida: Reflections of Photography*. Translated by Richard Howard. New York, 1981.

Barthes, Roland. "The Wisdom of Art." Translated by Annette Lavers. In *Calligram: Essays in New Art History from France*. edited by Norman Bryson. Cambridge and New York, 1988.

Derrida, Jacques. *The Truth in Painting*. Translated by Geoff Bennington and Ian McLeod. Chicago, 1987.

Derrida, Jacques. *Memoirs of the Blind: The Self-Portrait and Other Ruins*. Translated by Pascal-Anne Brault and Michael Naas. Chicago, 1993.

Derrida, Jacques. "The Spatial Arts: An Interview with Jacques Derrida." In *Deconstruction and the Visual Arts*, edited by Peter Brunette and David Wills pp. 9–32. Cambridge and New York, 1994.

Foucault, Michel. *The Order of Things: An Archaeology of the Human Sciences*. New York, 1970.

Foucault, Michel. *This Is Not a Pipe*. Translated and edited by James Harkness. Berkeley, 1983.

Jay, Martin. *Downcast Eyes: The Denigration of Vision in Twentieth-Century French Thought*. Berkeley, 1993.

Kristeva, Julia. "Motherhood according to Giovanni Bellini." In *Desire in Language: A Semiotic Approach to Literature and Art*, edited by Leon S. Roudiez, translated by Thomas Gora et al. New York, 1980.

Kristeva, Julia. "Giotto's Joy." In *Calligram: Essays in New Art History from France*, edited by Norman Bryson. Cambridge and New York, 1988.

Lacan, Jacques. *The Four Fundamental Concepts of Psychoanalysis*. Edited by Jacques-Alain Miller, translated by Alan Sheridan. New York, 1978.

Levin, David Michael, ed. *Modernity and the Hegemony of Vision*. Berkeley, 1993.

Lyotard, Jean-François. *Discours, figure*. Paris, 1971.

Lyotard, Jean-François. *Que peindre? Adami, Arakawa, Buren*. 2 vols. Paris, 1987.

Lyotard, Jean-François. *The Inhuman: Reflections on Time*. Translated by Geoffrey Bennington and Rachel Bowlby. Stanford, Calif., 1991.

Merleau-Ponty, Maurice. *The Visible and the Invisible*. Edited by Claude Lefort, translated by Alphonso Lingis. Evanston, Ill., 1968.

Merleau-Ponty, Maurice. "Cézanne's Doubt" and "Indirect Language and the Voices of Silence." In *The Merleau-Ponty Aesthetics Reader: Philosophy and Painting*, edited by Galen Johnson. Evanston, Ill., 1993.

GARY SHAPIRO

FREUD, SIGMUND. [*This entry comprises four essays on the founder of psychoanalysis:*

> Survey of Thought
> Pathography
> Freud and Literature
> Freud and Music

The first essay is a survey of Freud's psychoanalytic theory and his views on art and aesthetics, with particular attention paid to the visual arts. The second essay focuses on a key issue in psychoanalytic approaches to aesthetics: "pathography" (study of the artist's life with an emphasis on suffering, illness, or feeling). The last two essays focus on art forms that, in addition to the visual arts, have been studied by psychoanalysts: literature and

music. For Freud's views on tragedy, see Tragedy. *For related discussion, see* Irigaray; Klein; Kristeva; Lacan; Psychology of Art; Sexuality; *and* Winnicott.]

Survey of Thought

Sigmund Freud (1856–1939), physician and founder of psychoanalytic psychology, was born in the Moravian town of Freiberg. He lived his productive years in Vienna, from which he fled in 1938 upon the annexation of Austria by Nazi Germany. His last year was spent in London.

Freud's initial research was that of a medical doctor studying neurology. Only after working with Jean-Martin Charcot in Paris (1885–1886), and Josef Breuer in Vienna (1876–1882) on the problems of hysteria, did Freud independently set out on his research that led to the development of what he came to call "psychoanalysis." That theory of the mind and method of the treatment of neurotic disorders were systematically set forth in Freud's extensive writings in which the evolution of his theory and ideas may be traced. Many of his writings deal with problems that are central to the philosophy of art and aesthetics. Although Freud's contribution to the philosophy of art remained unsystematic and often suggestive rather than fully developed, his thinking made a fundamental change in the way the arts are understood, interpreted, and responded to. Thus his theory takes its place alongside revolutionary philosophies of art in the Western tradition: Plato, Aristotle, Augustine, Immanuel Kant, G. W. F. Hegel.

From his earliest writings Freud maintained an interest in philosophy, anthropology, archaeology, history, and literature, as well as the arts of painting and sculpture, and his interpretations of individual works of art and the interpretative theory upon which the analyses are based made a significant contribution to the body of writings in aesthetics that were contemporary with his work. Freud was aware of many of the aesthetic doctrines of the nineteenth century, and it is into that tradition that his own work can best be set. To read psychoanalytic essays on art and literature is to recognize a strong affinity between Freud's psychoanalytic interpretations and those of Hegel's lectures on the fine arts (1823, 1826, 1828), and Friedrich Nietzsche's speculations on the origins of tragedy in *The Birth of Tragedy* (1872). In addition, Freud was greatly impressed by the exciting archaeological uncoverings of the classical world taking place in his lifetime. These continuities with past philosophical aesthetics were supplemented by Freud's own wide reading in the literature of the West from Greek tragedy, to Henrik Ibsen, to the popular fiction of his own day, and by his close study of individual artists such as Leonardo da Vinci, Michelangelo, and above all, the works of William Shakespeare. From his earliest studies of neuroses and dreams, the medical doctor believed that his clinical observations could be applied outside the clinic to the arts, to culture,

and to the history of past civilizations. Thus, the theory that he developed to treat psychological problems in particular, was almost immediately extended to humanistic inquiry in general.

Formulation of the Theory. Freud regarded *The Interpretation of Dreams* (1900), the first published work in which he put forth a theory of mind, as his most important work, the only one he kept up-to-date throughout his life, and the one he considered to contain his most important discoveries. Central to all Freud's thinking is the concept of the unconscious and its fundamental role in all aspects of mental life. All of Freud's later writings pursue the investigations into the unconscious that are begun in what he called his "Dream Book."

The contents of the unconscious include thoughts, feelings, wishes, impulses that are, in psychoanalytic language, repressed, that is, unavailable to and kept out of consciousness because they are unacceptable to the conscious personality of the individual. That they are so regarded is a consequence of early familial, social, communal, personal, and moral pressures to restrain that which it is believed would cause loss of pleasure and put the individual in danger of losing love, approval, and acceptance from important figures in her or his life. That which is repressed, however, pushes for expression because it represents drives, impulses, many of a sexual nature, basic to physical and psychological gratification. According to this conception of the dynamic unconscious, the forces of repressed material, acting against the forces of repression that would keep it out of consciousness, underlie the formation of neurotic symptoms, dreams, slips of the tongue, and many other modes of psychic expression, including aspects of works of art. These expressions are considered in psychoanalytic language to be "compromise formations," that is, a compromise between the forces of repression and the impulses in the unconscious pressing for expression or discharge. Unconscious conflicts characterize all aspects of the mental life of human beings, and no mode of expression is entirely free of them. The history of an individual psychic life, a history of the products of creative individuals such as artists, may be conceived in terms of compromise formations that the psychoanalytic method can interpret, that is, whose unconscious meanings can be discovered. Much of Freud's writing about art and artists proposes interpretations of the lives of artists and their works in these terms.

Unconscious impulses and wishes have their origin in childhood. Freud's conception of infantile sexuality, the development of the sexual impulse from the earliest years, culminates in what Freud came to call the "Oedipus complex," a universal stage in human development. Indeed, that so many works of art deal with the Oedipus theme encouraged Freud to see their themes and plots and character studies as confirming evidence for his discoveries in the clinical setting. By Oedipus complex, a designation taken from Sopho-

cles' drama, *Oedipus Tyrannus,* Freud, who regarded this as one of his major discoveries, referred to the loving and hostile wishes that the child entertained toward the parents, that is, loving wishes toward the parent of the opposite sex, and hostile rivalrous wishes toward the parent of the same sex. Freud's theories of development were worked out primarily in terms of the male child. Critics of his theory have maintained that his conception of female development was then fitted into this scheme.

The Oedipus complex has its foundation in universal unconscious fantasies that characterize successive stages of infantile sexuality and the developmental stages that form a universal history of human growing up. Many of Freud's essays on art seek to uncover and to interpret the unconscious fantasy life of the characters portrayed in literature. Although Freud realized that literary characters cannot be psychoanalyzed as can living patients in the consulting room, he believed that psychoanalysis was able to interpret the conflict between repression and the unconscious forces pressing for discharge through interpretation of symbol formations along with other material that derived from the underlying dynamics of the unconscious. A basic concept in psychoanalytic theory of art is that of psychoanalytic interpretation, its methods, and its validation through comparisons between dream and art, and artworks in their representations of central themes and ideas that make up the content of psychoanalytic theory. Psychoanalytic interpretation of symbolic representations in art yields unconscious meaning.

The psychoanalytic method of interpretation rests on a distinction between *manifest content* and *latent content.* The distinction was introduced in dream interpretation and proves useful to understanding all psychic products; it has been particularly fruitful in interpreting works of art. Manifest content refers to the dream as the dreamer experiences it, as it would be in the telling of it. Latent content refers to the repressed, unconscious, initially unavailable aspect of the dream, referred to also as the "dream thoughts." Freud calls the process by which latent content is disguised and transformed into the manifest content the "dream work." By the method of free association, psychoanalytic technique addresses itself to means of moving from manifest content to latent content, to demonstrating their interrelationships, and interpreting the symbols of the dream, many of which Freud considered to be universal. Freud's "Dream Book" contains many examples of dream interpretation as well as several interpretations of works of art. In all the examples, the interpretative purpose is to reveal the unconscious meaning of the psychic product, summarized by Freud in his famous dictum, "A dream is a (disguised) fulfillment of a (supressed or repressed) wish" (Freud, 1953–1974, vol. 4, p. 160).

The concept of the unconscious in its dynamic properties is also delineated by means of the distinction between "primary process" functioning and "secondary process" functioning. Both refer to mental processes and correspond to the distinction Freud made between "the pleasure principle" and "the reality principle." Secondary process functioning is the use of reason and judgment to solve problems, to examine reality, to be able to manage one's life in the natural surroundings within which one must negotiate one's needs and purposes. In contrast, primary process functioning is the mode of thought in which psychic energy flows freely and facilitates symbolic transformations through the processes of "condensation" and "displacement."

Condensation refers to one of the ways in which unconscious processes function: one idea or image is capable of representing several chains of thought and overlap of images at a point of intersection so that one representation can condense within itself several meanings. Displacement refers to another of the ways in which unconscious mental processes move thought content around. In displacement a meaning or an emotional intensity can be detached from one idea and endow another idea with that meaning and affect. Thus, an emotional charge can be displaced, can move about, can attach itself to ideas and images that may at first seem far removed from the source of affect.

Condensation and displacement underlie all symbolic modes of expression to be found in dreams, works of art, jokes, and many other expressions. Freud's central interest in such symbolic presences is to be found in dream interpretation and in the interpretation of works of art and of literary characters. Where symbolism occurs, unconscious ideas are apt to be figuratively represented by the symbol, and an interpretation of the specific case must be undertaken without assuming that the latent content can be immediately or consistently determined. Freud did believe, however, that there are constancies of symbolic representations (i.e., universal symbols), which become apparent through repeated demonstrations in the clinical setting, so that the presence of a symbol in such instances immediately reveals its latent content.

The Work and the Artist. Many of Freud's essays on aesthetic issues deal with the ways in which a work and the creator of the work interrelate on psychological grounds. In an early essay, "The Relation of the Poet to Day-Dreaming" (1908), written soon after *The Interpretation of Dreams,* Freud asks the question, "How do artists come by their material?" And then, "Why is it we in the audience respond so deeply to that which the artist presents to us?" As simple as this essay appears to be, it sets down some of Freud's basic insights on the aesthetic response, psychoanalytically considered.

By posing a question that occurs to the many who are not artists, Freud sounds a theme that appears often in his aesthetic writings: the possibility of coming to understand the creative work of the artist by discovering "in ourselves or in people like ourselves an activity which is in some way akin

to creative writing" (vol. 9, p. 143). The essay claims we can if we see the continuity between the child at play, who creates an imaginative world of his or her own, and the creative artist at work, who creates a world of fantasy in words. As Freud points out, "The opposite of play is not what is serious but what is real" (ibid., p. 144). The adult has given up play and in its place obtains pleasure in fantasy, that is, daydreams. The artist engages in a fantasy motivated by an unfulfilled wish that is translated into art, and in its disguised expression and aesthetic form, it is then found agreeable by the audience, since all participate in basic fantasy life, essentially the expression of a wish. The content of most artistic fantasy is the self in its erotic and ambitious adventures, and much that would be offensive if stated openly is made acceptable through an aesthetic means, the product of a mysterious artistic gift, which Freud believed lies outside psychoanalytic understanding. But what psychoanalysis can interpret is the content of the fantasies artists create for our pleasure.

When Freud turned to Shakespeare's *King Lear* and *The Merchant of Venice* ("The Theme of the Three Caskets," 1913), he explored several levels of cultural expression that underlie the two dramas that he interprets: myth, fairy tale, and dream imagery are woven into the plots of both. Psychoanalytic interpretation begins by tracing a recurring theme that appears in all of those cultural manifestations, "a man's choice between three women" (vol. 12, p. 292). Freud asks what is represented by the three women as they appear in *King Lear*, Lear's three daughters, and as they appear symbolically in the three caskets of *The Merchant of Venice*. The early argument of Freud's essay fastens on a mysterious narrative observation: one of the three women is silent. Moving from fairy tale and folktale to dreams, Freud notices first that the fact of silence is recurrent, and then, if that appeared in a dream representation of a person, the interpretation would be that the silent person represented death: muteness is death. The second observation is that in dream symbolism a casket or box represents a woman. Putting these observations together, Freud begins his interpretation by explaining the symbolic meaning in the character portrayals of Cordelia in *King Lear* and of Portia in *The Merchant of Venice*.

In both dramas a man chooses one of three women, and the one chosen is the most attractive, the most desirable, and silent. How can it be, then, that the choice represents a man choosing death? Shakespeare's plays hide a truth that the psychoanalytic method can ferret out: the human wish to survive death transforms the inevitability of dying into that which is chosen as the most beautiful and sexually desirable. Both Bassanio and Lear in their choices convey to the audience that which lives in all of us as an unconscious need and wish, that we may escape the inevitability of death. At the conclusion of this essay Freud formulates a psychoanalytic identification in which the *is* of symbolism occurs

in its special psychoanalytic sense: "Cordelia *is* death." The embodiment of opposites in one figure, the goddess of love and the goddess of death, wishfully transforms man's ultimate destiny. The love-death goddess, Freud's concluding identification of the three women, symbolizes the

> three inevitable relationships that a man has with a woman—the woman who bears him, the woman who is his mate, and the woman who destroys him; or that they are the three forms taken by the mother in the course of a man's life—the mother herself, the beloved one who is chosen after her pattern, and lastly Mother Earth who receives him once more. (vol. 12, p. 301)

Freud surveyed the grounds for this assertion throughout a history of mythic repetitions, and from the immediacy of dream analysis in the consulting room. Thus, the power of the psychoanalytic method is shown by the ability to draw together cultural themes from a variety of human creative acts and to reveal their deeper, unconscious, meanings.

Freud has left the creative artist out of his account in the readings of these two literary works of art. In his writings, however, he asks if the work of art can throw light on the life of the artist, and conversely, if knowledge of the artist's life might enlighten elements in the artist's work. Freud referred to this method as "pathography," that is, an association of artistic character structure with the aesthetic individuality of a particular work, and in which each serves as an explanatory hypothesis for the other. Freud's most important effort to realize this kind of inquiry occurs in his book *Leonardo da Vinci and a Memory of His Childhood* (1910).

As always with Freud's inquiries into art, the study begins with a puzzle or question whose solution or answer is not easily found. In the case of Leonardo da Vinci, the question is posed because of the strange nature of the artist's life as reported by historians such as Giorgio Vasari. Leonardo was not understood by his contemporaries, and for all his genius he produced relatively little painting, though he was prolific in experiment and engineering. An inability to complete projects seems to have been characteristic of the artist. Thus Freud notes: "According to Vasari he spent four years in painting the portrait of Mona Lisa . . . without being able to bring it to final completion" (vol. 11, p. 67). And he continues that although Leonardo "declined to eat meat, and condemned war and bloodshed," this did not "stop him from devising the cruellest offensive weapons and from entering the service of Cesare Borgia as chief military engineer." Freud then turns to the psychosexual development of the artist: "Leonardo represented the cool repudiation of sexuality—a thing that would scarcely be expected of an artist and a portrayer of feminine beauty." His contemporaries noted his unusual frigidity (ibid., p. 69).

Psychoanalytic method in its effort to see through to the underlying dynamics of Leonardo assembles a portrait of his actions as they are known to us from contemporary reports, and then adds the one bit of private recollection left

us by the artist himself: "I recall in one of my very earliest memories that while I was in my cradle a vulture came down to me, and opened my mouth with its tail, and struck me many times with its tail against my lips" (ibid., p. 82). Having built a background of the life, Freud turns the argument in two directions, first to the psychosexual development of children as formulated by his theory, and second to the creative work of the artist. He is then in a position to fill in the picture of this extraordinary man and his fascinating works.

Recognizing that his readers may "find all pathography unpalatable," Freud offers his interpretation of Leonardo's life and work in their interconnectedness, with the reservations of the self-conscious inquirer:

> But even if the historical material at our disposal were very abundant, and if the psychical mechanisms could be dealt with with the greatest assurance, there are two important points at which psychoanalytic inquiry would not be able to make us understand how inevitable it was that the person concerned should have turned out in the way it did and in no other way. In Leonardo's case we have had to maintain the view that the accident of his illegitimate birth and the excessive tenderness of his mother had the most decisive influence on the formation of his character and on his later fortune, since the sexual repression which set in after this phase of childhood caused him to sublimate his libido into the urge to know, and established his sexual inactivity for the whole of his later life. . . . It seems at any rate as if only a man who had had Leonardo's childhood experiences could have painted the Mona Lisa and the St. Anne, have secured so melancholy a fate for his works and have embarked on such an astonishing career as a natural scientist, as if the key to all his achievements and misfortunes lay hidden in his childhood phantasy of the vulture. (ibid., pp. 135, 136)

Although critics have noted that Freud's essay on Leonardo suffers from mistakes due to inadequate translation of original documents, and despite severe criticism from art historians directed against its inferential leaps, the central thought has proved fruitful for psychoanalytic pathography in biographical studies. The essay, as Richard Wollheim has stated, "is an attempt to exhibit—not, of course, to prove, but, like the clinical case histories, to exhibit—the dependence of the adult capacities and proclivities on the infantile and in particular on infantile sexuality" (Wollheim, 1970, p. 214). Another pathographic essay, Freud's reflections on Fyodor Dostoyevsky, is presented in "Dostoyevsky and Parricide" (1928).

The second major study of the material arts, quite different in its focus from the Leonardo study, is the essay Freud wrote in 1914, "The Moses of Michelangelo." Here Freud concentrates on one work of art to determine its meaning, and he does not attempt to connect the *Moses* with Michelangelo's life in the sense of psychosexual development, but rather concentrates on the ability of a great artist to convey his intention to the perceiver. To discover *intention* and to make it available to the reader of his essay, Freud develops his *interpretation* of the work; psychoanalytic observation, because it attends to the smallest details of expression, is a method well suited to this end. Through a close study of the statue, Freud concludes that intense inspection reveals the meaning of Moses' posture:

> What we see before us is not the inception of a violent action but the remains of a movement that has already taken place. In his first transport of fury, Moses desired to act, to spring up and take vengeance and forget the Tables [of the law, held under the arm]; but he has overcome the temptation, and he will now remain seated and still in his frozen wrath and in his pain mingled with contempt. (vol. 13, p. 229)

Freud also attempts, however, to connect his interpretation to a possible characteristic of the sculptor, for he sees in both the figure of Moses and the character of Michelangelo a passion to create under restraint: in both we observe "a concrete expression of the highest mental achievement that is possible in a man, that of struggling successfully against an inward passion for the sake of a cause to which he has devoted himself" (ibid., p. 233).

Freud's most elaborate study of a single work of art is his *Delusion and Dreams in Jensen's Gradiva* (1907), a detailed study a popular novel of the day. The novel yields especially well to psychoanalytic inquiry because it contains dreams of the hero, which Freud interprets in terms of the character's career and reported fantasy life. Through Freud's interpretation, the novel yields a fuller, more complex, and more interesting character than appears from an initial reading. The analysis of *Gradiva* set a pattern for later psychoanalytic studies, since many works of literary art have characters endowed by the writer with dreams and fantasies, and contain psychosexual descriptions such that, consciously or not, the artist opens the work to a kind of interpretation the psychoanalytic method has developed.

Freud's penetrating character studies—Oedipus and Hamlet in *The Interpretation of Dreams*, Cordelia and Portia in "The Theme of the Three Caskets," and Richard III, Lady Macbeth, and Ibsen's Rebecca Gamvik in "Some Character Types Met with in Psychoanalytic Work" (1916)—have become paradigmatic in their influence on later studies. Many briefer studies are scattered through Freud's writings; one that discusses dramatic character in terms of the early psychoanalytic theory of aesthetic experience is an essay published posthumously "Psychopathic Characters on the Stage" (1905; published in 1942; vol. 7, pp. 303–310).

The psychoanalytic affinity to art and the psychoanalytic power of interpretation are also exemplified throughout the case histories that Freud wrote in his work as a clinician. They all exhibit a novelistic intensity and a structure that is truly aesthetic. In recognition of his literary as well as clinical powers, Freud was awarded the Goethe Prize in 1930 by

the city of Frankfurt. Indeed, as a writer Freud achieved artistic wholeness and stylistic elegance, so that his clinical cases and his use of his theory in the interpretation of art together make a whole of unusual literary interest. To the aesthetician and to the philosopher of art these continuities are well worth observing. The most striking case histories are those of "Dora" (1905), "Little Hans" (1909), "The Rat Man" (1909), and "The Wolf Man" (1909). In comparing them to some of Freud's studies of literary characters one sees affinities in method and structure that render the case studies truly literary in the best aesthetic sense, and the character studies truly interpretative psychoanalytically so that a manifest representation is seen through to its latent depths of dynamic action.

Although all of Freud's writings on language, on dream, and on art display a sensitivity to the problems of aesthetics, many of his inquiries, written over the years, undertake to analyze specific issues that he regarded as central to an understanding of the arts and their complex structures. In this regard should be mentioned *Jokes and Their Relation to the Unconscious* (1905), the first attempt to apply the psychoanalytic mode of thought to the problems of aesthetics. Jokes can be considered miniature narratives that employ all of the psychic processes outlined above in the introduction to Freud's theory. Condensation, displacement, and symbolic transformations of reality and of fantasy occur in jokes, and the book is to be recommended as a set of cultural experiments—as if the reader is taken to Freud's Vienna, where the test cases were to be found. In that respect they complement the "Dream Book," taking us out of the consulting room and into the coffeehouses of the city.

Years later, Freud explored a specific aesthetic category and the experience its examples afford the perceiver in "The Uncanny" (1919), and it too leads the reader into a whole cultural world, for the story under scrutiny is E. T. W. Hoffmann's "The Sand Man" of 1815. The problem generated by that tale is why we have a specific set of feelings "related to what is frightening—to what arouses dread and horror" (vol. 17, p. 219) when we read of certain events that carry us to the edges of everyday experience. The characters and actions in Hoffmann's tale involve a mysterious and threatening man who terrifies children by taking their eyes, a dreaded physical injury. Freud asks why stories of this kind give rise to an experience of the "uncanny." His method is first to study carefully all the ways in which the term *uncanny* (*unheimlich* in German) appears in the many languages that possess comparable terms for such experiences. He discovers that among its different shades of meaning the word *heimlich* exhibits one that is identical to its opposite, *unheimlich*. That which is "homey" also contains within itself a possible opposite, that which is "unhomey," or as we would say in English, "uncanny." Freud proposes that the experience of the uncanny occurs when one event condenses the familiar and the unfamiliar that

stand in these instances in a particular dynamic relation. The theme of the Sand Man depicts a man who enters houses at bedtime to tear out children's eyes. In this case, the repressed childhood fantasy of castration is merged with the strange figure of the Sand Man. The frightening experience will include "something repressed which *recurs*. This class of frightening things would then constitute the uncanny . . . this uncanny is nothing new or alien, but something which is familiar and old-established in the mind and which has become alienated from it only through the process of repression" (ibid., p. 241). And Freud concludes: "an uncanny experience occurs either when infantile complexes which have been repressed are once more revived by some impression, or when primitive beliefs which have been surmounted seem once more to be confirmed" (ibid., p. 249). This leads to a generalization about the relationship of art to life: "a great deal that is not uncanny in fiction would be so if it happened in real life; and . . . there are many more means of creating uncanny effects in fiction than there are in real life" (ibid.).

In this analysis Freud maintains that "every affect belonging to an emotional impulse, whatever its kind, is transformed, if it is repressed, into anxiety" (ibid., p. 241). This conception, in its affinity to Freud's theory that repressed sexuality is transformed into anxiety, reflects his first topographic model of the mind. Most of Freud's writings on art were done before the elaboration of the structural theory, set forth in *The Ego and the Id* (1923), and his revised theory of anxiety set forth in *Inhibitions, Symptoms, and Anxiety* (1926). The focus of Freud's theoretical writings, which he termed "metapsychology," increasingly concerned the agent of repression, the ego and its functions, as compared with the repressed content, drives, and impulses now corresponding to the id. It is important to note that in the structural theory the ego and its functions are not synonymous with consciousness. In the revised theory of anxiety, anxiety is conceived in terms of ego functions, serving as a signal of internal (i.e., psychic) danger, thus setting into motion the defensive activities of the ego.

In Freud's thinking, theory is never divorced from clinical observation. Psychoanalytic theory draws the arts close to the clinic, for psychoanalysis learns from the arts, especially literature, and the discoveries of the consulting room in turn help to interpret works of art. Freud in his work made it clear that "there is an intimate connection between all mental happenings—a fact which guarantees that a psychological discovery even in a remote field will be of an unpredictable value in other fields" (vol. 8, p. 15).

Freud's influence on the interpretation of art and literature has been both pervasive and profound. Later writers have also attempted to extend the application of psychoanalytic principles to the problems of creativity and formal aspects of works of art, including music, areas Freud explicitly excluded from his consideration. In the case of the

creative talent of the artist and the purely formal aspects of works of art, Freud stated that psychoanalysis had nothing to say. The exclusion of music he accounted for by a disclaimer that he himself lacked the capacity to respond to music and therefore was not qualified to analyze it.

Freud and Hegel. If we consider Freud's method of interpretation in the context of nineteenth-century philosophy of art, it becomes evident that there is an important continuity from Hegel's way of interpreting cultural products, which include works of art, and the psychoanalytic readings that today are to be found in art history, literary criticism, and theories of culture. For Hegel, his philosophy is his method of interpretation, just as for Freud his psychoanalytic theory is his method of interpretation. But there are important differences in the ways they think about art. Hegel has been said to write "art history without names," for from Hegel's point of view, the evolution of the arts throughout history exhibits a necessity of style and content that transcends any individual consciousness.

Although there are identifiable artists in Hegel's lectures on the fine arts, their work is never regarded as the product of an individual consciousness. The work is regarded, however, as possessing a *manifest* and a *latent* content, though to say that is to apply Freud's language, not Hegel's. The manifest in Hegel's interpretation is what the object is in itself, for example, Sophocles' *Antigone*. But the inner meaning of the drama in Hegel's analysis is a hidden latent content lying undiscovered until awakened and aroused by Hegel's philosophy as a necessary modern interpretation that the work, as it were, was waiting to reveal. It could not so reveal its hidden inner meaning before the time of philosophy's ascendancy as the level of historical consciousness realized in modernity—indeed, in Hegel's own philosophy of art. [*See* Hegel.]

If we set Freud's method of interpretation alongside of Hegel's, we shall see that it might be relevant to refer to the psychoanalytic method as "names without art history," for considered psychoanalytically, a work of art is always the product of an individual consciousness, however much it is seen to participate in the style and content of its own time. Freud never neglects historical references, but his interest is in the individual mind that produced the work, and in the work itself as exhibiting both a manifest and a latent content. The method embraces the ways in which psychoanalytic theory approaches, interprets, sets forth in explicit directives, as if creating a map, the means whereby hidden unconscious content is brought forward so that it may be explicitly regarded. A work of art once subjected to a psychoanalytic interpretation becomes suddenly much fuller and denser, and possibly a means of entry into the mind of the artist who created the work.

Thus, whereas Hegel seeks the grandest historical sweeps of cultures from the earliest times to the present, Freud focuses on the work of individual artists and the discoveries to be made when an individual work of art is interpreted psychoanalytically. Such differences of emphasis imply that where aesthetic issues are debated, the psychoanalytic method is limited in the specific works of art it can successfully and insightfully interpret. If the method is then to be applied to the artist and the ways in which the art of an individual reveals that person's inner psychic life, there is a need for biographical evidence full enough to justify a move from the work to the creative mind. Such a move is not undertaken in a Hegelian interpretation even where biographical evidence might be available. It is to the advantage of both the historical and the psychoanalytic to combine their methods in studying works of art.

[*See also* Imagination.]

BIBLIOGRAPHY

Works by Freud

The Standard Edition of the Complete Psychological Works of Sigmund Freud. 24 vols. Edited by James Strachey. London, 1953–1974. All references are to this edition; for an index to all of Freud's discussions of works of art, see vol. 24, pp. 187–194.

Other Sources

Alford, C. Fred. *The Psychoanalytic Theory of Greek Tragedy.* New Haven, 1992.

Berman, Emanuel, ed. *Essential Papers on Literature and Psychoanalysis.* New York, 1993.

Feder, Stuart. *Charles Ives, "My Father's Song": A Psychoanalytic Biography.* New Haven, 1992.

Feder, Stuart, Richard L. Karmel, and George H. Pollock, eds. *Psychoanalytic Explorations in Music: Second Series.* Madison, Conn., 1993.

Gay, Peter. *Freud: A Life for Our Time.* New York, 1988.

Gombrich, E. H. "Freud's Aesthetics." *Encounter* 26.1 (January 1966).

Holland, Norman N. *Psychoanalysis and Shakespeare.* New York, 1966.

Kris, Ernst. *Psychoanalytic Explorations in Art.* New York, 1952.

Kuhns, Richard. *Psychoanalytic Theory of Art: A Philosophy of Art on Developmental Principles.* New York, 1983.

Kuhns, Richard. *Tragedy: Contradiction and Repression.* Chicago, 1991.

Laplanche, J., and J.-B. Pontalis. *The Language of Psycho-Analysis.* Translated by Donald Nicholson-Smith. New York, 1973.

MacGregor, John M. *The Discovery of the Art of the Insane.* Princeton, N.J., 1989.

Ricoeur, Paul. *Freud and Philosophy: An Essay on Interpretation.* Translated by Denis Savage. New Haven, 1970.

Rose, Gilbert J. *The Power of Form: A Psychoanalytic Approach to Aesthetic Form.* New York, 1980.

Rudnytsky, Peter L. *Freud and Oedipus.* New York, 1987.

Skura, Meredith Anne. *The Literary Use of the Psychoanalytic Process.* New Haven, 1981.

Spector, Jack J. *The Aesthetics of Freud: A Study in Psychoanalysis and Art.* New York, 1972.

Spitz, Ellen Handler. *Art and Psyche: A Study in Psychoanalysis and Aesthetics.* New Haven, 1985.

Spitz, Ellen Handler. *Image and Insight: Essays in Psychoanalysis and the Arts.* New York, 1991.

Steinberg, Leo. *The Sexuality of Christ in Renaissance Art and in Modern Oblivion.* New York, 1983; 2d rev. exp. ed., Chicago, 1996.

Tennenhouse, Leonard, ed. *The Practice of Psychoanalytic Criticism.* Detroit, 1976.

Trosman, Harry. *Freud and the Imaginative World.* Hillsdale, N.J., 1985.

Wollheim, Richard. "Freud and the Understanding of Art." *British Journal of Aesthetics* 10.3 (1970): 211–224.
Wollheim, Richard. *Sigmund Freud.* New York, 1971; 2d ed., London, 1991.

MARGARET P. KUHNS and RICHARD KUHNS

Pathography

Sigmund Freud inaugurated the term *pathography* when he published his highly influential and controversial monograph *Leonardo da Vinci and a Memory of His Childhood* in 1910. This paradigm monograph was criticized from the time of its publication (see Editor's Note, p. 60), and has been the cause célèbre of many heated debates between art historians and psychoanalysts throughout the twentieth century, the most famous exchange being that between Meyer Schapiro (1956) and Kurt Eissler (1961). It continues to stimulate an ongoing flow of articles in the literature.

In the closing chapter of *Leonardo,* Freud avers: "We must stake out in a quite general way the limits which are set to what psycho-analysis can achieve in the field of biography" (Freud, 1957, p. 134). Although he says *biography* at this point, the word he uses earlier on to describe his psychoanalytic study of Leonardo is translated as *pathography,* a term that has come to distinguish Freud's work from more routine treatments of "the life of the artist." The term pathography implies writing about suffering, illness, or feeling, with important connotations of an empathic response on the part of the author for his or her subject. Pathography is thus a less inclusive term than biography and suggests that, upon reading, one should expect to find descriptions and explanations of *selected* aspects of a life, those dealing in particular with (mental) disease, with intrapsychic conflict, its symptoms, and their etiology.

Interestingly, Freud, in chapter 6 of *Leonardo,* defines his aims by exclusion. He tells us what pathography cannot do and what psychoanalysis cannot explain. For example, he says that "pathography does not in the least aim at making the great man's achievements intelligible" (p. 130), and that "we should be glad to give an account of the way in which artistic activity derives from the primal instincts of the mind if it were not just here that our capacities fail us" (p. 132). Despite these demurrals, however, Freud has already by this time developed a coherent interpretative account that links a narrative of the artist's life to selected paintings. During the course of developing that narrative, he seems to have kept a few key questions in mind. I propose to reflect on these questions as a means of clarifying the pathographic project as Freud conceived and practiced it.

Among the questions that can be extrapolated from Freud's narrative are the following: (1) Why did Leonardo work so slowly and have the habit of leaving work unfinished? (2) Why did he, master draftsman and painter that he was, abandon his art at a certain point and turn ever more intensely to his scientific researches? (3) What was the genesis of his alleged homosexuality? (4) How can one explain the idiosyncrasies of some of his paintings, especially the enigmatic smile of the Mona Lisa and the choice of theme and composition in the *Virgin and Child with Saint Anne?*

Reflecting on these questions and on the text per se, we may justifiably infer that Freud began with fascination, admiration, and an identification with the artist, then proceeded rapidly to judgments that certain aspects of Leonardo's behavior were maladaptive and thus suitable for psychoanalytic explanation. On a model such as this, the artist becomes equivalent to a patient. The text to be interpreted consists then in just those aspects of the person (artist) that are judged pathological by the author as well as those works of art that can be seen as symptomatic of the (presumptive) unconscious conflicts underlying other aspects of so-called aberrant behavior. Assumed here are the absolute reliability of the initial diagnosis by the author (there can be no presenting complaint on the part of a deceased artist and no negotiations); the confidence on the part of the interpreter to move directly from manifest to latent content in the absence of a live transference, free associations, and so forth; and the equation of works of art with ordinary symptoms or dreams, as if they were transparencies through which primary process could be read, rather than unusually complex compromise formations in which specific unconscious meanings can be glimpsed only darkly and after their subjection to a series of filters that (minimally) establish perimeters for interpretation.

In the case of Leonardo, there are many arguments against these assumptions in the specific form they take in Freud's narrative. Setting aside the methodological problems (inexactitudes) noted above, it is not at all clear, for instance, that Leonardo's working habits need be attributed to "inhibitions in (his) sexual life" (p. 131). It is conceivable that a mind of genius may generate ideas in such profusion that completion loses its ordinary value under the pressure of competing images. Painters who work slowly are legion. Further, the question as to when a painting is finished is surely one of the most delicate of artistic problems. Who decides when a painting is completed? Moreover, when Freud explains Leonardo's turning from "artistic creation" to more scientific researches by the statement "his infantile past had gained control over him" (p. 133), a value judgment is implied that one sort of activity would have been better for Leonardo than another. This seems presumptuous. Would Freud have taken kindly to the thought of posthumous ex cathedra judgments about the order, scope, and form of his own creative activities? Thus, the model of artist qua patient runs into serious snags. If it is to operate, it seems to require a system of controls in the absence of dialogue, since the "patient" cannot answer back.

The artist-as-patient paradigm of pathography raises a number of additional questions: (1) the relative motivating

force of unconscious conflict versus other factors (such as ego functioning—memory, judgment, etc.) in artistic creativity; (2) the persistence of conflict, that is, whether the artist's dilemmas continue or change throughout his or her life span, resistant to repeated attempts at resolution and "working through" by means of externalization in works of art; (3) the value of subjecting to pathography an artist who seems relatively free of painful conflict and whose productivity may stem from less defensive, more playful types of mental functioning; (4) the question, granted the artist's unconscious conflicts, whence arise the ego strengths to master these conflicts and use them in the service of producing highly valued works of art, rather than being destroyed by them through impulse or a crippling repression; (5) the analyst's problem of choosing among several plausible alternative interpretations of data when the ultimate goal is not therapeutic but merely explanatory, or—another way of putting it—how does the nature of psychoanalytic interpretation change when there is no therapeutic goal; (6) in the absence of confirmatory evidence in the form of the analysis of live transference and countertransference, how much projection, how much "reading in" can slip by undetected; (7) does an artist-as-patient narrative return us, de post facto, to the artist's oeuvre with a heightened sense of meaning; have we been empowered in succeeding moments with the works of art to experience them more richly, or does this sort of interpretation leave the (art) world as it finds it, turning solipsistically back upon itself rather than illuminating the objects it purports to interpret?

The last question, central for aesthetics, may seem at first glance unfair. After all, in pathography, it is not the works that constitute the principal text but the artist's psyche, particularly its troubled aspects. Yet, it seems clear that our interest in an artist (particularly one deceased) depends on our interest in his or her works; hence, if we are not returned to them with greater insight, what, may we reasonably ask, has the psychoanalytic interpreter accomplished? Of course, there may be other agendas, apart from object or text, such as the development of theory, or the attempt to organize material around a central theme (such as the artist's loss of a parent in childhood; see Spitz, 1985, 1994). But ultimately, just as all good interpretation must return us to the object, so the justification for doing pathography must be based likewise on a return to that object.

Only if, after reading Freud's *Leonardo,* we actually reexperience, say, the Mona Lisa in a new way, if we come to perceive associations among the artist's various works or notice inconsistencies that had not been there for us before, if our own responses to certain works seem clearer or more plausible in the light of the psychoanalytic interpretations offered, then, it seems reasonable to assert that Freud's pathographic analysis possesses interpretative value and aesthetic relevance.

Because of the preeminence of the case history in psychoanalysis and the clinical project per se, which *is* the close examination of a life, pathography (whether so labeled or not) has ever been and continues to be the most popular mode of introducing psychoanalytic ideas into the realm of the aesthetic. For many of the reasons given above, however, the practice is often poorly done. Freud himself, in the closing pages of *Leonardo,* made a special effort to highlight the twin factors of freedom and chance in order to avoid accusations of having written an overdeterministic book. To hold these delicate variables in mind along with notions of historical and political context is to keep the project of pathography open and thus, it is hoped, to reap its considerable rewards without falling prey to its snares.

BIBLIOGRAPHY

Works by Freud

Freud, Sigmund. *Leonardo da Vinci and a Memory of His Childhood (1910).* In *The Standard Edition of the Complete Psychological Works of Sigmund Freud,* edited by James Strachey, vol. 11, pp. 63–137. London, 1957.

Other Sources

Eissler, Kurt R. *Leonardo da Vinci: Psychoanalytic Notes on the Enigma.* New York, 1961.
Liebert, Robert S. *Michelangelo: A Psychoanalytic Study of His Life and Images.* New Haven, 1983.
Schapiro, Meyer. "Leonardo and Freud: An Art Historical Study." *Journal of the History of Ideas* 17 (1956): 303–336.
Spitz, Ellen Handler. *Art and Psyche: A Study in Psychoanalysis and Aesthetics.* New Haven, 1985.
Spitz, Ellen Handler. *Museums of the Mind: Magritte's Labyrinth and Other Essays in the Arts.* New Haven, 1994.

ELLEN HANDLER SPITZ

Freud and Literature

With his ideas of the unconscious mind, and guilt as aggression turned against the self, Sigmund Freud gave new dimensions to the vision of human beings as distinguished from other creatures by nothing so much as the capacity for self-alienation, a vision that has haunted modern literature from Blaise Pascal to Franz Kafka and Samuel Beckett. Not only as world picture, but also as clinical practice, as psychological theory, and as an idea about the creative process, psychoanalysis has from the beginning been congenial to the literary mind.

Kenneth Burke remarks that the analyst's attention to a patient's free associations is a useful model for the critic:

> One obviously cannot invite an author, especially a dead author, to oblige him by telling what the author thinks of when the critic isolates some detail or other for improvisation. But the critic can note the context of imagery and ideas in which an image takes place. . . . The motive of the work is equated with the structure of relationships within the work itself.	(Burke, 1963, p. 229)

As the critic pays close attention to the text, so the clinician listens for the rhythm of what is said; for changes in mood and tone; for what is muttered *sotto voce* or at the margin of the text or explicitly or implicitly denied; for the way in which individual utterances both shape and are inflected by the ongoing story.

Freud's central psychological theses have struck many literary critics as providing a special key to questions in their field. To oversimplify, the key was found in the interrelated notions of repression and defense, latent and manifest content, primary and secondary process, all of which are best illustrated in Freud's theory as elaborated in *The Interpretation of Dreams* (1900). This will be called here the "classical view."

Freud held that every dream has a meaning in the form of an early, unconscious, repressed, sexual wish, a wish that motivates the dream and that it represents as fulfilled, but in a form unrecognizable to the conscious mind. The dream has a meaning in two senses. First, as wish fulfilling, it expresses something like an intention. The "intention" belongs, however, to a prerational stratum of the mind, as does the unconscious "dream work" that translates the wish and its attendant thoughts into the "manifest content" of the dream. Second, because the wishes are disguised, the dream has a symbolic or representative dimension.

Hallucinatory wish fulfillment, as when a thirsty sleeper dreams of drinking from a cool stream, is an essential aspect of what Freud calls primary process mental functioning, in distinction from the secondary mental processes that are characterized by attunement to reality, and the capacity to think about how best to achieve what one wants rather than discharging an impulse immediately through action. Primary process wishing is born when the infant experiences a need, and on the basis of a remembered experience of satisfaction, "hallucinates" the object for which it longs. On Freud's view, such a wish-fulfilling image is the mental act of fantasizing in its earliest and most primitive form. (Fantasizing is similar to daydreaming, except that fantasy is unconscious; and rather than being aware of themselves as construing the world in a certain way, fantasizers take their fantasies as simply the way things are.)

Freud sees primary process as "primary" in several senses: it is chronologically first in the developmental order; it characterizes the thought processes of so-called primitive peoples; and it is the thinking to which we "regress" under the pressure of unconscious conflicts, in dreaming, or on other occasions when there is no immediate need for more rational thought processes. "Primary process" does not represent a stage, however, in which the individual or the culture outgrows, nor is it necessarily inferior to more rational processes; on the contrary, it is a permanent stratum of the human mind, ready to be tapped by creative processes as well as by psychopathology. Primary process is prepropositional and imagistic, free of the constraints of internal consistency and accord with external reality to which rational thought is subject, characterized by condensation (the compacting of many ideas and images into one) and by displacement (the centering of the manifest content away from what is truly important or meaningful at the unconscious level). For example, a dream may portray as alive a person who the dreamer knows to be dead; free association may reveal further that this person represents several different people (condensation); and that he or she is of central importance in the dream only at its manifest level: the important unconscious issues that motivated the dream lie elsewhere (displacement).

Even more than images, words, Freud saw, are naturally the nodal points for numerous ideas, thus easily lending themselves both to slippages of meaning through which one thing comes to stand for another, and to disguise. While displacement is essentially defensive in function, condensation serves the purpose of disguise but has other uses as well. Many critics, including William Empson, Kenneth Burke, and Lionel Trilling (as well as Jacques Lacan), have understood condensation as another name for metaphor. In this context Trilling remarks that Freud's greatest contribution was the idea that poetry is indigenous to the human mind (1953).

Besides primary process, Freud had other ideas about language that figure in his theory of dreams. For one, he posited an archaic form of communication in which certain words have antithetical meanings (an idea that appealed to Empson, but one that has been discredited by both linguists and philosophers). Freud believed that if we suppose that dreams regress to this archaic form, we can account for the fact that dreams "show a particular preference for combining contraries into a unity or for representing them as one and the same things" (Freud, 1957, p. 155). For another, Freud thought that a child's words are born from sensory images or "thing-presentations," and that words retain this sensual power even when rational thinking comes into its own.

Finally, as "secondary elaboration," dream work bridges the gap between primary and secondary, more or less rational thought processes, so providing the dream with a narratable shape, and further disguising the unconscious, repressed wishes that motivate it. Some psychoanalytic writers have explored the relation between secondary elaboration and the formal elements of the work of art (to be discussed below).

The concept of fantasy as hallucinatory wish fulfillment, described earlier, is less complex than the one Freud develops in other works where the wish, or cluster of wishes, in question is essentially many-layered and conflicted. In his early note on "the architecture of hysteria" Freud writes: "The aim [of the symptoms, memories, psychical structures] seems to be to arrive back at the primal scenes. In a

few cases this is achieved directly, but in others only by a roundabout path, via fantasies. For fantasies are psychical facades constructed in order to bar the way to these memories" (1910, p. 248). The fantasy (or the symptomatic acting out) will typically gratify a wish in a way that is simultaneously self-punitive; or it will be stimulated by conflicts from different periods in the person's life. In the fantasy that a child is being beaten (Freud, 1919), the child represents a sibling rival for the father's love; then, because of the guilt she came to feel for her incestuous wishes, the fantasizer herself is the child being beaten. At the same time the fantasy may represent the act of intercourse as understood by a child and colored by the feeling of guilt.

Many readers of literature have found this concept of fantasy as conflicted wish fulfillment particularly illuminating. For example, in the devastating final scene of William Shakespeare's play, Lear has his favorite daughter, Cordelia, at last to himself; it is he, not her husband, the King of France, who carries her in his arms. But this very image is of a Lear terribly punished, for Cordelia is dead, and dead through forces that he himself, in the impotent rage of his old age, had unleashed (see Adelman, 1992).

Applied to literature, the classical view suggests that the gratifications of both writer and reader derive from the repressed unconscious. The function of literature is to give expression to this unconscious, perhaps modifying it in the process. (The latter, an idea that comes into its own in the hands of writers other than Freud, will be discussed below in the paragraphs on play.) Symbols disguise primitive thinking, form provides a kind of diversionary pleasure, akin to secondary elaboration, a filter through which the work can achieve its goal of gratifying desire at a deeper, more conflicted level. (Simon O. Lesser's *Fiction and the Unconscious,* Norman Holland's *The Dynamics of Literary Response,* and Anton Ehrenzweig's *The Hidden Order of Art: A Study in the Psychology of Artistic Imagination* are sophisticated treatments of such a view of form.) The function of criticism is to make the unconscious conscious, by locating the fantasies and the means by which they are simultaneously hidden and satisfied. Freudian theory thus aligns itself with the twentieth-century critical move away from Matthew Arnold's position that at its best poetry embodies timeless truths, toward I. A. Richards's idea that poetry is a use of language whose value is the affording of certain kinds of experience, or the evoking of emotive rather than cognitive "attitudes." Since Plato literary theory has been concerned with the place of truth in literature and with the effects of literature on its audience. Freud's view of the mind is clearly relevant to these questions, as well as to others equally venerable about the meaning of a text and its interpretation, the relation between content and form, and the author's intentions.

Freud wrote most of his essays on art prior to the development of the structural model of the mind that he laid out in 1923 in *The Ego and the Id.* Here he suggests that we need a more complex idea of mental functioning than one that simply pits conscious against unconscious, for what is interesting about the unconscious mind is not its status merely as a body of unacknowledged and hidden thoughts, but of thoughts that are repressed. By whom, or by what? Freud now asks. Whereas earlier he had been content to answer by saying, as it were, "The conscious agent," over the years he increasingly became aware of resistance, the fact that the patient evidently resists the approach of painful, unconscious thoughts, and resists without conscious intent to do so. The repressing agent must then itself be unconscious. Repression, together with the fact of unconscious guilt, which presumably also issues from the self as agent, moves Freud to one of his most innovative but also most problematic concepts. He takes from René Descartes and the German tradition of self-philosophers that followed him—among them, Johann Gottlieb Fichte and Edmund Husserl—the concept of *das Ich,* "the I," translated in *The Standard Edition* as "ego"—the I that thinks and speaks, and that in thinking, necessarily knows of its own existence as the subject of thought, and transforms it into the self who does not know itself, who will not and cannot speak.

In a further departure from his philosophical fathers, Freud describes this self as formed out of its identifications with other persons. In "Mourning and Melancholia" (1914), he had memorably written that "the shadow of the object falls upon the ego"—where object means loved person, a person who in melancholia is in some way lost to the subject. As a way of keeping the object, or of maintaining control over it, the fantasizer imagines the object as inside its own body or mind. In *The Ego and the Id* these "objects" are not merely shadows on the ego—they *are* the ego. Freud writes: "It may be that . . . identification is the sole condition under which the ego can give up its objects . . . the ego is a precipitate of abandoned object-choices and . . . contains the history of those object-choices" (p. 29). What is known as psychoanalytic object relations theory (in particular, the work of Melanie Klein, D. W. Winnicott, Hannah Segal) elaborates the views on identification that Freud develops in "Mourning and Melancholia" and *The Ego and the Id.* [*See* Klein; *and* Winnicott.] Object relations theory envisions the mind as constituted in essential ways by the child's interactions, real and fantasied, with other persons. It envisions the mind as an inner theater in which various characters play out their proper or improper roles. Some of the best psychoanalytic readers have been deeply influenced by object relations theory.

Without denying the importance of Drives, *The Ego and the Id* also gives greater importance to the external world than Freud had previously. It implies that in important ways the mind is oriented from the beginning to reality, that it is endowed with an energy irreducible to libido and with capacities that are not derivative from conflict. Unconscious

mental processes are now seen to be not only defensive but also synthetic, engaged in forging a more integrated self and a more adaptive relation to reality. Expressing themselves most clearly in a new concept of play, these ideas encourage attitudes toward the creative process, the work of art, and aesthetic experience, that differ from the classical view.

In "Creative Writers and Day-Dreaming" (1908 [1907]), Freud places play on a continuum of illusion along with daydream and unconscious fantasy. But the concept of play returns in another guise in "Beyond the Pleasure Principle" (1920). There Freud muses on the meaning of his eighteen-month-old grandson's playing with a spool of thread. Again and again the child casts the spool away from him with a syllable that sounds like "fort" (gone) and reels it back with joyful "da" (there). Freud links the game to the child's attempt to master the painful experience of the mother's departure. Playing with the reel, Freud thinks, allows the child to repeat an experience in which he was painfully passive, the one left, making it tolerable by reversing the active and passive roles. But in light of the importance Freud gives to the processes of separation and individuation in such works as "Mourning and Melancholia," one might rather interpret the child's activity as an attempt to take in his discovery of interpersonal reciprocity. Freud goes on in "Beyond the Pleasure Principle" to postulate a universal compulsion to repeat a traumatic experience, the expression at the clinical level of what he calls the death instinct (an impulse not merely to keep stimuli constant but to abolish all stimuli) as opposed to Eros, or the life instinct. Most psychoanalysts have ignored this particular version of the dual instinct theory; but many think that the repetition compulsion is a genuine phenomenon, to be understood as the person's attempt to gain control over a traumatic past, or finally to make it come out right.

The Kleinian analyst Hannah Segal contrasts the play in artistic creativity with the play of nightdreaming, daydreaming, and neurotic fantasizing. The second sort is essentially private; it involves a wish-fulfilling denial of external and internal realities. The first sort is a rapprochement with one's unconscious that results in an imaginative solution in the real and public world. Creative play requires not only the talent to make works that engage the aesthetic sensitivities of others, but also a rare capacity for tolerating and working through conflict. Developing a similar theory, Ernst Kris calls artistic creativity "regression in the service of the ego" (Kris, 1952). Such views of the creative process concern not only the artist but also his or her audience, since it is assumed that the psychological processes involved in creation are to some extent replayed in the responsive reader.

In a highly influential elaboration of the idea of play D. W. Winnicott proposed that it marks out an ontological space that is neither subjective nor objective as we are used to thinking of these concepts, but transitional between the two.

His idea was that the concept of reality is constructed for any child through her interactions with some loved physical object (which might even be a tune) in the public world, interactions that are mediated by her communications with a loved other person who is responding both to her and to that same object. Whereas Freud's "drive" model of mental functioning would imply that a text can work for reader as well as author because of their shared biology, Winnicott's idea suggests rather that "meaning" is always a communal enterprise.

Psychoanalysis and literary critical theory have undergone considerable changes in the past fifty years, partly in response to the same larger cultural movements. Melanie Klein, feminism, and recent infant research have drawn classical Freudian theory away from its preoccupation with Oedipal triangulation and fathers, toward pre-Oedipal development and the dyadic relation between infant and mother. Influenced, like literary theory, by postmodern skepticisms about truth, objectivity, and meaning, psychoanalysis has taken up the theme—hinted at also by Freud—that the therapeutic task is largely constructive rather than reconstructive; that instead of attempting to reconstruct the past as it was experienced by the patient, psychoanalyst and patient are together figuring a new "narrative" of the patient's life, better than the old narrative in allowing him to act in more spontaneous and less self-destructive ways. On this view, clinical interpretation is often not the unearthing of ideas already present though unconscious in the patient's mind but is rather a joint creation.

These views about the nature of the psychoanalytic process have another source as well in a contemporary anti-Cartesian rejection of the model of the knower—the analyst—as occupying a privileged position in which he or she merely observes, does not interact with, the object being studied. As early as his postscript to the case of Dora (1905 [1901]), Freud was alerted to the phenomenon of "transference," the patient's interpreting her interactions with the analyst in terms of earlier, traumatic relationships. Increasingly, psychoanalytic theory has emphasized the importance not only of transference but also of countertransference; it holds that rather than providing a neutral screen on which the patient projects his unconscious fantasies, the analyst too draws on her own unconscious in responding to the patient, and these responses can be a useful clue to what is going on between them. These ideas about the essentially intersubjective, creative nature of the analytic relationship have a literary analogue in contemporary "reader-response" theory.

Taking a suggestion from Terry Eagleton (1983), one might describe the history of literary theory in the past one hundred years or so as marking three phases of preoccupation: with *the author* (the nineteenth century and Romanticism); with *the text* (the Russian Formalism that emerged before the Revolution and continued through the twenties,

and a little later, the New Critics, encompassing the work of I. A. Richards, William Empson, John Crowe Ransom, W. K. Wimsatt, Cleanth Brooks, Allen Tate, and R. P. Blackmur); and with *the reader* (Roman Ingarden, Wolfgang Iser, Kenneth Burke, Stanley Fish, and, among psychoanalytic critics, Norman Holland).

Psychoanalysis has had something to say to critics of all three persuasions. To the first it offered its view of motivation and the creative process. Harold Bloom's influential theory that poets write under "the anxiety of influence" might be seen as a late flower of Romanticism's concern with the author, though in practice Bloom is focusing rather on the relation between a body of work and the tradition from which it begins to depart. He sees the poet's work as a kind of Oedipal displacement in which the poet attempts to absorb, by revolting against and transforming it, the work of the earlier poets whom he admires and by whom he has been influenced.

In a sharp departure from nineteenth-century criticism, the New Critics claimed that criticism should be guided primarily by close attention to the text: authorial intention, extratextual exploration of historical influences, and the effect of a work on its audience are all irrelevant to interpretation. What matters are simply the workings of literary language itself: imagery, metaphor, diction, rhythm, and rhyme. To some of these critics, Freud's ideas about condensation and displacement, primary process and "primitive" language, dream work in general, the nature of form and its relation to content, were seen as an interesting extension of their own methods.

Although often thought of as a postmodern development, attention to the reader appears as early as the 1930s in some of the writings of I. A. Richards. Richards held that the critic locates the form of the text in discovering how it structures a reader's satisfactions and frustrations, a view also developed by Burke. In *Act of Reading: A Theory of Aesthetic Response* (1978) and other writings, Iser made a now standard distinction between the text, or what is written, and the work, or what is constituted by the interaction between reader and text. Whereas psychoanalytic interpretations that focus on either author or text have to make assumptions about the author's unconscious, reader-response critics need be insightful only about their own.

The best of the current psychoanalytic literary critics combine the traditional skills of the critic—detailed attention to the text and the culture that produced it—with the virtues of the good psychoanalyst: self-knowledge, sensitivity to language that expresses conflict, and a readiness to perceive fantasies concerning sex and gender, embodiment, separation, loss, transience, death, identity, and power, themes that have always been the matter of literature.

[*See also* Literature.]

BIBLIOGRAPHY

Works by Freud

The Interpretation of Dreams (1900). In *The Standard Edition of the Complete Psychological Works of Sigmund Freud,* edited by James Strachey, vols. 4–5. London, 1953.

"Fragment of an Analysis of a Case of Hysteria" (1905 [1901]). In *The Standard Edition of the Complete Psychological Works of Sigmund Freud,* edited by James Strachey, vol. 7, pp. 3–122. London, 1953.

"Delusions and Dreams in Jensen's *Gradiva* (1907 [1906]). In *The Standard Edition of the Complete Psychological Works of Sigmund Freud,* edited by James Strachey, vol. 9, pp. 3–95. London, 1959.

"Creative Writers and Day-Dreaming" (1908 [1907]). In *The Standard Edition of the Complete Psychological Works of Sigmund Freud,* edited by James Strachey, vol. 9, pp. 141–153. London, 1959.

"Leonardo da Vinci and a Memory of His Childhood" (1910). In *The Standard Edition of the Complete Psychological Works of Sigmund Freud,* edited by James Strachey, vol. 11, pp. 59–137. London, 1957.

"The Antithetical Meaning of Primal Words" (1910). In *The Standard Edition of the Complete Psychological Works of Sigmund Freud,* edited by James Strachey, vol. 11, pp. 153–161. London, 1957.

"The Theme of the Three Caskets" (1913). In *The Standard Edition of the Complete Psychological Works of Sigmund Freud,* edited by James Strachey, vol. 12, pp. 289–301. London, 1958.

"A Child Is Being Beaten: A Contribution to the Study of Sexual Perversions" (1919). In *The Standard Edition of the Complete Psychological Works of Sigmund Freud,* edited by James Strachey, vol. 17, pp. 175–204. London, 1955.

"Beyond the Pleasure Principle" (1920). In *The Standard Edition of the Complete Psychological Works of Sigmund Freud,* edited by James Strachey, vol. 18, pp. 3–64. London, 1955.

The Ego and the Id (1923). In *The Standard Edition of the Complete Psychological Works of Sigmund Freud,* edited by James Strachey, vol. 19, pp. 3–66. London, 1961.

"Dostoyevsky and Parricide" (1928 [1927]). In *The Standard Edition of the Complete Psychological Works of Sigmund Freud,* edited by James Strachey, vol. 21, pp. 175–196. London, 1961.

Other Sources

Adelman, Janet. *Suffocating Mothers: Fantasies of Maternal Origin in Shakespeare's Plays.* New York and London, 1992.

Bonaparte, Marie. *The Life and Works of Edgar Allan Poe: A Psychoanalytic Interpretation.* Translated by John Rodker. London, 1949; reprint, New York, 1971.

Burke, Kenneth. "Freud and the Analysis of Poetry." In *Art and Psychoanalysis,* edited by William Phillips, pp. 412–439. New York, 1963.

Crews, Frederick, ed. *Psychoanalysis and Literary Process.* Cambridge, Mass., 1970.

Eagleton, Terry. *Literary Theory: An Introduction.* Minneapolis, 1983.

Ehrenzweig, Anton. *The Hidden Order of Art: A Study in the Psychology of Artistic Imagination.* Berkeley, 1967.

Eissler, Kurt R. *Discourse on Hamlet and Hamlet: A Psychoanalytic Inquiry.* New York, 1970.

Greenacre, Phyllis. *Swift and Carroll: A Psychoanalytic Study of Two Lives.* New York, 1955.

Hartman, Geoffrey H., ed. *Psychoanalysis and the Question of the Text.* Selected papers from the English Institute, 1976–1977. Baltimore, 1978.

Hoffman, Frederick. *Freudianism and the Literary Mind.* 2d ed. Baton Rouge, La., 1957.

Holland, Norman N. *The Dynamics of Literary Response.* New York, 1968; reprint, New York, 1989.

Jones, Ernest. *Hamlet and Oedipus.* New York, 1949.

Kahn, Coppelia. *Man's Estate: Masculine Identity in Shakespeare*. Berkeley, 1981.

Kaplan, Morton, and Robert Kloss. *The Unspoken Motive: A Guide to Psychoanalytic Literary Criticism*. New York, 1973.

Kris, Ernst. *Psychoanalytic Explorations in Art*. New York, 1952.

Lesser, Simon O. *Fiction and the Unconscious*. Chicago, 1957.

Marcus, Steven. *Freud and the Culture of Psychoanalysis*. Boston, 1984; reprint, New York, 1987.

Phillips, William, ed. *Art and Psychoanalysis*. New York, 1963.

Rudnytsky, Peter L., ed. *Transitional Objects and Potential Spaces: Literary Uses of D. W. Winnicott*. New York, 1993.

Ruitenbeek, Hendrik, ed. *Psychoanalysis and Literature*. New York, 1964.

Schwartz, Murray M., and Coppelia Kahn, eds. *Representing Shakespeare: New Psychoanalytic Essays*. Baltimore, 1980.

Schwartz, Murray M., and David Wilbern. "Literature and Psychology." In *Interrelations of Literature*, edited by Jean-Pierre Barricelli and Joseph Gibaldi, pp. 205–224. New York, 1982.

Skura, Meredith Anne. *The Literary Use of the Psychoanalytic Process*. New Haven, 1981.

Trilling, Lionel. "Freud and Literature." In *The Liberal Imagination*, pp. 44–64. Reprint, Garden City, N.Y., 1953.

Winnicott, D. W. *Playing and Reality*. New York, 1971; reprint, London and New York, 1989.

Wollheim, Richard. "Freud and the Understanding of Art." In *The Cambridge Companion to Freud*, edited by Jerome Neu, pp. 249–266. Cambridge and New York, 1991.

MARCIA CAVELL

Freud and Music

Although the nonclinical psychoanalytic endeavor that came to be called applied psychoanalysis was initiated by Sigmund Freud with writings on Leonardo da Vinci and Fyodor Dostoyevsky among others, Freud himself never wrote specifically about music, and his writings contain few references even of a metaphorical nature. Indeed, he appears to have been least sensitive to music than any of the arts, an anomaly in late nineteenth-century Vienna. Until relatively recently, there has been a comparable lack of interest in this area among his followers. Even those most involved with art such as Hans Sachs and Otto Rank failed to deal with music in any specific manner. The reasons for this may lie less in any negative influence on the part of the first psychoanalyst than within the issue of musical "literacy" itself, that is, the difficulty of bridging the gap between psychoanalytic concept and method and their application in the context of a nonverbal and apparently nonobjective medium. After all, clinical analysis was dubbed "the talking cure," and music, whatever its communicative power, does not (at least in its "pure" form) "talk."

There is a second and more complex reason for the lack of development of what might seem to be the promising area of psychoanalysis and music and that is Freud's own *visual* bias and its tangible issue in the profound scientific and cultural influence of his seminal work *The Interpretation of Dreams*. It is regrettable that this bias has discouraged rather than inspired the exploration of the psychological ramifications of the auditory sphere. For there is much in it that may be applicable to music, particularly in Freud's discussion of the dream work itself and the psychology of the dream processes. Few among his followers have pursued these possibilities, most notably Otto Isakower, who wrote trenchantly of the "exceptional role of the auditory sphere."

Nevertheless, it was inevitable that there would be some interest among the psychoanalytic pioneers of turn-of-the-century Vienna. Indeed, one member of Freud's early circle was the music historian and critic Max Graf, who undertook the task of applying the psychoanalysis of his time to investigate the psychology of great musicians and the process of composing music, a mission yet to be accomplished. The field of psychoanalysis and music has been a latecomer to applied analysis, and it was not until 1990, with the publication of the anthology *Psychoanalytic Explorations in Music*, that a field of inquiry was even formulated. This field became further established with the more broadly interdisciplinary sequel (1993), which contains original contributions of a mutually enriching variety from psychoanalysts and musicians alike (here, musicologists and music historians).

If, as *The New Grove Dictionary of Music and Musicians* states, "musical aesthetics has been dominated by a single theme: the nature and import of that powerful yet indeterminate emotional impact that music has," then music and psychoanalysis are drawn together in a community of interest in the nature of affects. Nor is this the only area in which interests of music and psychoanalysis converge. A second important "theme" the two disciplines share lies in the nature of meaning, the hermeneutic dimension. Thus, each in its own way is concerned with the means of representation and the nature of interpretation. The interpretation of latent meanings encoded in mental life and its derivatives lies at the core of psychoanalysis. Music, on the other hand, in the past fifty years has had an uneasy relationship with the interpretation of the musical "text" (so to speak) as opposed to interpretation as audible performance, the result of the positivistic trend in musicology.

Finally, if one considers the relationship each discipline has with the field of biography, an important area of common interest emerges. Despite the fact that there is no entry for "Biography" in either the new or old *Grove*, musicology has necessarily had to come to terms with the life courses of musicians and, pari passu, the subject's early life, family background, psychological development, influences and motivations, not to mention possible sources of creativity—all issues that are central to psychoanalytic study. The excesses of the much maligned field of psychobiography, so called, constitute more of a cultural phenomenon than one that can properly be laid at the doorstep of professional psychoanalysis, since much of it has been practiced by journalists and other writers lacking specific and appropriate training. Frequently, the reductionism seen here, which is

justifiably deplored, is more the result of parlor psychology than responsible psychoanalytic practice. Thus, interdisciplinary "literacy" is as much an issue from the point of view of analysis as it is from that of music as noted above. The analytic biographer must be sensitive to the fact that any single interpretation may in itself tend toward the reductionistic unless put into a broader interpretive context and adhere strictly to the principles of overdeterminism and multiple function in mental life.

Against the background of the above discussion it may be useful to consider some examples of (1) contributions relating music and affect, (2) psychoanalytic interpretation of musical works, and (3) psychoanalytic studies in musical biography. The three areas are not always distinct and often impinge on one another in any given study. Brief reference will also be made to (4) developmental studies as well as a final, theoretical category, (5) analytic views on the psychology of music, a topic that would require development beyond the bounds of this brief review.

In "On Affect and Musical Motion," David Epstein, a music theorist, has provided a framework for an interdisciplinary consideration of affect. Long interested in biological correlates of music, Epstein discusses with the psychoanalyst Gilbert J. Rose the emotional response of the "receptive perceiver" resulting from the "enhanced rhythms of tension and release." "In On Form and Feeling in Music," Rose extends his psychoanalytic work in the aesthetics of art to music employing "the metaphor of art as *transmuter:* art facilitates bridging between and transforming perceptual stimuli . . . and affective meanings." Form is emphasized in the "homology between aesthetic structure and perceptual response—analogous to the attunement between patient and therapist in the safe holding environment." Peter Ostwald, an analytically oriented psychiatrist, has explored how the representation of specific affects associated with "painful memories of [Robert] Schumann's childhood" served to master a later "agonizing separation." Similarly, Stuart Feder has explored an array of affects whose representation in a song was associated with Gustav Mahler's fear of death and the mastery of anxiety. The psychoanalyst Morton F. Reiser, in a consideration of Richard Wagner's use of the leitmotiv, suggests that in music there may be . . . an extensive vocabulary of emotion—perhaps even a grammar and wonders whether communication may be effected by a counterpart physiological vocabulary and grammar. Leitmotivs may comprise "an extended nodal memory network," and, Reiser writes, "the complexities of neurotransmission and neuromodulation are sufficient . . . to lend a sense of face validity to such a notion." In his far-reaching "Reflections on the Communication of Affect and Idea through Music," the music historian Leo Treitler similarly explores the kinds of resonances music has with the collective experiences that contribute to the makeup of our conscious and unconscious selves. He further emphasizes

the significance of form in his consideration of the Adagio of Wolfgang Amadeus Mozart's Piano Concerto no. 23 (K. 488) and Béla Bartók's "From the Diary of a Fly" (from *Mikrokosmos*), for "it is not only the communication of the movement's affective progression but also our apprehension of its structure that depends on its resonance with our inner selves."

Among the several composers who have engaged the attention of psychoanalytic writers Mahler has a special place owing perhaps to the single four-hour psychoanalytic "session" he had with Freud in 1910. Perhaps, too, because, as is the case in the music of several composers, there is an element that implies the extramusical in the very surface of Mahler's music that invites biographical inquiry. Mahler himself acknowledged the *conscious* autobiographical impulse, writing early in his career, "My whole life is contained in my two symphonies. . . . To anyone who knows how to listen my whole life will become clear." Elements that may be *unconscious* constitute the challenge of responsible psychoanalytic study, which in any case, Mahler has consistently inspired since the writings of Theodor Reik. The latter devoted an entire section of *The Haunting Melody* (1953) to a consideration of the creative context of the composition of Mahler's Symphony no. 2 in C Minor. (Reik also included Freud's own scant notes on his meeting with Mahler sent to Reik by Freud in the form of a letter.)

W. E. Mooney noted that Mahler's conscious preoccupation with problems of life and death is reflected in some way in almost every composition. George H. Pollock has explored the multiple losses in Mahler's life and the role of musical composition in the process of mourning. Robert Still's paper "Gustav Mahler and Psychoanalysis" was in part a response to the publication of the first volume of Donald Mitchell's Mahler biography as an example of "work by experts in their own field, who have employed or drawn attention to psychoanalytical techniques." This interdisciplinary enthusiasm was short-lived as Mitchell's later volumes proved less analytically oriented.

Feder, in a retrospective study of the events leading up to Mahler's final illness and death, has delineated multiple psychological factors involved. Dean Collins, too, discussed the events of Mahler's last summer prior to his final illness. In a later paper Feder explored a major medical event earlier in Mahler's life in which Mahler had, in his own view, narrowly escaped death. The sequelae of this traumatic experience in the life of Mahler included stylistic changes in the music itself that are explored in the analysis of his lied "Um Mitternacht." The analysis of musical materials as data for psychological interpretation is a technique developed further in another study of Mahler's Symphony no. 2, which leads Feder to some different (although not necessarily contradictory) views from those of Reik regarding the creative roots of that work. Recently, the psychotherapeutic session within Freud has once again aroused interest with Emanuel

Garcia's review of the event and Feder's speculations regarding the nature of the therapeutic process and outcome.

There are two full-scale biographies of composers that merit comment from a specific point of view: the potential organizing power of psychoanalytic concepts. In his *Beethoven*, Maynard Solomon finds a unifying theme in Freud's concept of "the family romance." Applied to the life of Beethoven, this underlying (and presumably unconscious) fantasy of royal ancestry cogently draws together many diverse threads of the biography. Solomon, a music scholar who is psychoanalytically oriented, does not pursue possible ramifications of the psychoanalytic concept of family romance in the music itself in this particular work. Conversely, in articles where music is analyzed in detail (for example, Beethoven's Ninth), he prefers to interpret in terms of the symbolic, mythical, and philosophical, eschewing psychoanalytic concepts. Nevertheless, his discussion here of the representation of affects, if not specifically analytically oriented, enriches a psychoanalytic approach to the music.

In Feder's specifically subtitled "psychoanalytic biography" of Charles Ives, several organizing concepts contribute to an analytic view of a complex life. Chief among these is the incomplete mourning for an idealized but ambivalently loved father. The "intrapsychic collaboration" that resulted was the act of composition, and the music thus created is imbued with musical quotation of the tunes of his father's times. Considerable attention is paid to early life influences in the context of the mental development of the composer. In particular, the composer's "early auditory environment" is stressed. (With regard to developmental studies in general, the contributions of Pinchas Noy from the point of view of ego psychology should be noted.)

A third full-scale biography is worthy of mention both in its own right and by way of contrast to the above. The significant organizing element in Ostwald's biography of Schumann is the subject's mental illness, bipolar disease (or manic-depressive disorder). While dealing with the subject's mental life as in the above instances, its purview is necessarily psychiatric as opposed to psychoanalytic with emphasis on psychopathology. Ostwald has also contributed to the literature on Mahler, here in contrast, from the point of view of health rather than pathology.

Several authors have attempted to deal with the psychology of music from a theoretical point of view, Heinz Kohut addressing himself to the psychological functions of music in mental life and, along with the musicologist Siegmund Levarie, the nature of musical pleasure as well. This work on the psychoanalytic interpretation of the musical process was later continued by Martin L. Nass, who extended his studies into the areas of hearing and inspiration in musical composition as well as studies of the creative process of composition based on extended interviews with contemporary composers.

For an overview of the literature in the field of psychoanalysis and music the reader is further referred to the two volumes of *Psychoanalytic Explorations in Music*.

[*See also* Music, *overview article*.]

BIBLIOGRAPHY

Cheshire, Neil M. "The Empire of the Ear: Freud's Problem with Music." *International Journal of Psychoanalysis* 77 (1996): 1127–1167.

Diaz de Chumaceiro, Cora. "Richard Wagner's Life and Music: What Freud Knew." In *Psychoanalytic Explorations in Music: Second Series*, edited by Stuart Feder, Richard L. Karmel, and George H. Pollock, pp. 249–278, Madison, Conn., 1993.

Feder, Stuart. *Charles Ives: "My Father's Song": A Psychoanalytic Biography.* New Haven, 1992.

Feder, Stuart, Richard L. Karmel, and George H. Pollock. *Psychoanalytic Explorations in Music.* Madison, Conn., 1990.

Feder, Stuart, Richard L. Karmel, and George H. Pollock. *Psychoanalytic Explorations in Music: Second Series.* Madison, Conn., 1993.

Ostwald, Peter. *Schumann: The Inner Voices of a Musical Genius.* Boston, 1985.

Solomon, Maynard. *Beethoven.* New York, 1977.

STUART FEDER

FRY, ROGER ELIOT. *See* Bell; Bloomsbury Group; *and* Formalism.

FUTURISM. Although the term *avant-garde* seems to have first been used in its modern sense in 1845 by the Fourierist painter Gabriel-Désiré Laverdant, it began to acquire its full radical connotations with the publication of Filippo Tommaso Marinetti's *I manifesti del futurismo* first appearing in French as *Le futurisme* in the pages of *Le Figaro* on 26 February 1909. It would appear immediately thereafter in Italian in Marinetti's literary magazine *Poesìa*, which had been founded four years before—a publication that in typical Futurist fashion would cease to exist once it became fully devoted to the Futurist cause. In 1913, it was replaced as the official Futurist organ by *Lacerba*, a journal founded by the Florentine intellectuals Ardengo Soffici and Giovanni Papini.

Departing from a traditional approach to the vicissitudes of culture, the Futurist strategy, as formulated by Marinetti, was one of unremitting militant attack against all forms of bourgeois culture. This provocative modus operandi was to be exemplified by Marinetti at the very outset, as we may judge from the spirited character of his initial manifesto set forth with typical irreverence and aplomb:

> We had been awake all night my friends and I, under the mosque shaped chandeliers which starry like our souls were lit by the inner radiance of an electric heart. For hours we had trampled opulent oriental carpets; soiled quantities of paper with frantic thought . . . We all started up, at the sound of a tram rumbling past, ablaze with multicoloured lights, like a village in festival dress that the flooded Po tears from its banks and sweeps

through the gorges and rapids, down to the sea. Afterwards the silence grew deeper and we heard only the muttered devotions of the old canal and the creaking of the arthritic, ivy-bearded palaces . . . Suddenly we heard the roar of starving cars . . . Let us go, I cried, let us depart. Mythology and Mystic Idealism are defeated at last. We are in at the birth of the centaurs, we shall see the first angels fly. We must rattle the doors of life, test the hinges and the bolts. Let us go. There on earth is the first dawn of history and there is nothing to match the red sword of the sun, slashing for the first time through the shadows of a thousand years. (Marinetti, 1989)

There followed the eleven points of the founding manifesto, the first four of which extolled the virtues of temerity, energy, and audacity, while asserting the supreme magnificence of mechanical speed in the now famous passage that declared a racing car to be more beautiful than the Winged Victory of Samothrace. Points five to nine went on to idealize the driver of such a vehicle as being integral with the trajectories of the universe and to celebrate other diverse virtues, such as patriotism and the glorification of war; point ten called for the destruction of academic institutions of every kind, while point eleven characterized the emerging metropolis as the very crucible of dynamism and revolution:

> We will sing of the stirring of great crowds—workers, pleasure seekers, rioters and the confused sea of colour and sound as revolution sweeps through a modern metropolis. We will sing of the midnight fervour of arsenals and shipyards with electric moons; insatiable stations swallowing the smoking serpents of their trains; factories hung from the clouds by the twisted threads of their smoke; bridges flashing like knives in the sun, giant gymnasts that leap over rivers; adventurous steamers that scent the horizon, deep chested locomotives that paw the ground with their wheels, like stallions harnessed with steel tubing; the easy flight of aircraft, their propellers beating the wind like banners, with a sound like the applause of a mighty crowd.

Apart from its debt to Georges Sorel's *Refléxions sur la violence* of 1908 and to the *aeropoesia* of the Italian nationalist poet Gabriele D'Annuzio, this evocative passage paid evident homage to the triumph of industrialization and to the neotechnical innovations of the first decade of the century, above all, aviation and electrical power. Marinetti lost no time in translating the essential message of the movement into different cultural fields, acting as a prime mover behind the *Manifesto of Futurist Painters* (dated 11 February 1910) that was first proclaimed in the following month from the stage of the Teatro Chiarella in Turin by the proto-Futurist painter Umberto Boccioni. This initial manifesto, signed by Boccioni and four other Futurist painters, Carlo Carrà, Luigi Russolo, Giacoma Balla, and Gino Severini, set forth the following points as guidelines for their future activity; namely, to destroy the cult of the past, to invalidate imitation of all kinds, to elevate originality, and to despise all crit-

icism, together with such classic concepts as "harmony" and "good taste." *The Technical Manifesto of Futurist Painting* that followed on 11 April of the same year replaced such traditional values with the cult of mechanical dynamism and with the dematerialization of bodies through light, two ideologically charged modes of beholding that would be directly incorporated in a great deal of Futurist painting. On the one hand, then, is a style of painting that, while formally indebted to Analytic Cubism, was greatly influenced (like Marcel Duchamp's *Nude Descending a Staircase* of 1912) by the chronophotographic analysis of animal movement as systematically pursued by the French physiologist E.-J. Marey, and on the other hand, is a form of atomistic light painting, derived from the color-divisionist principles of Pointillism and Neo-Impressionism. Color division is patently the method adopted in Boccioni's early city paintings on Sorelian themes such as his *Riot in the Gallery* (1911) or *The City Rises* (1910), and the same applies to other canvases painted at approximately the same time: Severini's *Obsessive Dancer* (1911), Russolo's *Memories of a Night* (1911), or Balla's *Window in Düsseldorf* (1912). At the same time Balla's *Leash in Motion, Girl Running on a Balcony, Rhythms of a Bow*, and *Swallows*, painted between 1911 and 1913, are all more obviously "chronophotographic" in their representation of aspiration. A fuller, more symbolic condensation of this dynamic movement will wait on Boccioni's entry into sculpture and will come into being with the *Technical Manifesto of Futurist Sculpture*, dated 11 April 1912. Decrying all notions of realistic-figurative representation and any kind of sublime thematic, Boccioni went on to refute the use of traditional sculptural material such as marble and bronze and to substitute such honorific homogeneity with synthetic plastic works compounded out of everyday materials such as glass, wood, cardboard, iron, cement, hair, leather, cloth, mirrors, and electric light and to exploit such heterogeneity in order to evoke the dynamic kineticism of the modern world. Despite this extreme avant-gardist stance Boccioni's didactically dynamic works dating from this period are cast in bronze; that is to say, his *Bottle Evolving in Space* (1912) and his *Unique Forms of Continuity in Space* (1913). Notwithstanding the substantial material employed, these works would exemplify Boccioni's "spacetime" intentions as he declared them in his preface to an exhibition of his works in Paris during the summer of 1913:

> All these convictions impel me to search in sculpture not pure form, but *pure plastic rhythm;* not the construction of bodies, but the *construction of the action of bodies.* Thus I have as my ideal not a pyramidal architecture (static state) but a spiral architecture (dynamism) . . . In order to present a body in movement, I take care not to give its trajectory, that is, its passage from one state of repose to another, instead I force myself to determine the unique form that expresses its *continuity in space.*
>
> (Boccioni, as quoted in Apollonio et al.)

As one might expect, the Futurist aesthetic emerged much more readily in kinetic cultural forms such as music, as in Francesco Balilla Pratella's *Manifesto of Futurist Musicians* (1910) with its stand against the idea of "well-made" music, or in Anton Giulio Bragaglia's *Futurist Photo-Dynamism* (1911), which pretended to go beyond both chrono- and cinematic photography. As opposed to the sequential breakdown of movement in all kinds of photographically kinetic processes, photodynamism hoped to be able to amplify the aesthetic value of a flying figure, to achieve what Bragaglia termed the transcendental photography of movement. Given these preoccupations it is understandable that he would eventually end up as a filmmaker, although he was not a signatory of the September 1916 manifesto of *The Futurist Cinema* published, as usual and at Marinetti's insistence, on the auspicious eleventh day of the month. The year 1916 would see the completion of Bragaglia's film *Il perfido canto*, which given the paucity of the surviving archival material seems to have become confused over time with stills from his second film, *Thaïs*. *Il perfido canto* was followed by Amaldo Ginna's *Vita futurista*, first shown early in 1917. This film contained a split-screen sequence in which the sleeping style of a Futurist is contrasted with that of a *passéist*, the former sleeping in a sitting position with his hat on, while the latter lay in bed in the conventional posture.

Between 1912 and 1916 one manifesto would follow another in fairly rapid succession, including Bruno Corra's *Abstract Cinema—Chromatic Music* (1912), in which, in contrast to Marinetti's bombastic approach, Corra gave a measured, somewhat sobering account of his attempt to devise a dynamic form of Futurist light display, or what he called a chromatic symphony; that is, a spectacular abstract performance in ever-changing patterns of saturated color that necessitated the invention of what he called a chromatic picmo. This led almost at once to abstract films made of colored gelatin and to the projection of such films in rooms painted entirely white. Corra's somewhat thwarted aspirations in this regard were largely fulfilled by Balla's abstract setting for an actorless performance of Igor Stravinsky's *Feu d'artifice* in 1917, wherein suspended, brightly colored geometric sculptures by the artist were successively illuminated on a black stage in accordance with the unfolding narrative of the music.

Corra's chromatic manifesto was immediately followed by Russolo's *The Art of Noises*, dated 11 March 1913 and published in July in the form of a letter to the Futurist musician Pratella. While acknowledging the romantic grandeur of Richard Wagner, along with the emergence of the dissonant chord and the discovery of atonal music, Russolo nonetheless posited an art of pure noise, evoking in so doing the continual industrial clatter of the modern world. To this end he devised a battery of mechanical noisemakers—*intonarumori*, he called them—capable of rendering, in se-

quence, six different kinds of noise, including rumbles, whistles, whispers, screeches, percussion produced with different kinds of material, and finally human and animal voices, ranging from shouts and screams to wheezes and sobs. A year later Pratella vainly attempted to integrate these *intonarumori* into a repertoire of orchestral music played with traditional instruments. In the interim, Marinetti fleshed out the full range of the Futurist cultural project by turning his attention to concrete poetry and to the tradition of the variety stage, which had already provided a ready-made forum for Futurist agitational events. The first move in this direction came with his slogan "Words in Freedom," which formed the essential substance of his *Technical Manifesto of Futurist Literature* (1912) and the manifesto *Destruction of Syntax—Imagination without Strings—Words in Freedom* (1913). In *Destruction of Syntax* Marinetti instructs the reader in the virtues of onomatopoetic text and typography:

> Onomatopoeia that vivifies lyricism with crude and brutal elements was used in poetry (from Aristophanes to Pascoli) more or less timidly. We Futurists initiate the constant, audacious use of onomatopoeia. This should not be systematic . . . we abolish all stylistic bonds, all the bright buckles with which traditional poets link images together in their prosody. Instead we employ very brief anonymous mathematical and musical symbols and we put between parentheses indications such as (fast) (faster) (slower) (two beat time) to control the speed of the style. These parentheses can even cut into a word or an onomatopoeic harmony. . . . My revolution is aimed at the so-called typographical harmony of the page, which is contrary to the flux and reflux, the leaps and bursts of style that run through the page. On the same page therefore, we will use three or four colours of ink, or even twenty different typefaces if necessary. . . . Our lyric intoxications should freely deform, refresh words, cutting them short, stretching them out, reinforcing the center or the extremities, augmenting or diminishing the number of vowels and consonants.

He reworks this same theme in a 1914 manifesto titled *Geometric and Mechanical Splendor and the Numerical Sensibility*, in which he argues for onomatopoeia on the grounds of its descriptive economy.

> In my "contraband of War" in *Zang Tumb, Tumb*, the strident onomatopoeia *siiiii* gives the whistle of a tow boat on the Meuse River and is followed by the veiled onomatopoeia *ffiiiii, ffiiiii*, echo from the opposite bank. The two onomatopoeias saved me from needing to describe the width of the river, which is defined by the contrast between the two consonants *s* and *f*.

All of this was to be the literary graphic equivalent of that which the variety theater had already spontaneously offered with its traditionally irreverent interaction between actor and audience—the theater that Marinetti would characterize as "the Futurist marvelous," combining in one art form caricature, irony, absurdity, cynicism, wit, repartee, a display of imbecility bordering on insanity, light, sound, noise,

language, and, last but not least, "a cumulus of events" unfolding at great speed. As he put it in his manifesto *The Variety Theatre* (1913):

> Futurism wants to transform the Variety Theatre into a theatre of amazement, record-setting and body-madness . . . Oblige the *chanteuses* to dye their décolletage, their arms and especially their hair, in all the colours hitherto neglected as a means of seduction. Green hair, violet arms, blue décolletage, orange chignon, etc. Interrupt a song and continue with a revolutionary speech. Spew out a *romanza* of insults and profanity etc. . . . Systematically prostitute for example all the Greek, French and Italian tragedies, condensed and comically mixed up . . . Play a Beethoven symphony backwards, beginning with the last note—Boil all of Shakespeare down into a single act . . . Soap the floorboards to cause amusing tumbles at the most tragic moments.

This outrageous prescription was to be more or less repeated with patriotic, saber-rattling overtones in the 1915 manifesto *The Futurist Synthetic Theatre,* signed by Marinetti, Corra, and Emilio Stimelli.

For all the avant-garde artworks that were achieved in its name, the theater would endure as the ultimate vehicle of Futurism, as the one form that by definition was broad enough to accommodate the volatile nature of the Futurist sensibility in all its fullness. This in itself accounts for the way in which the later manifestos are devoted mainly to the theater and also for the way in which the more prominent younger artists to join the Futurist movement would make their names designing for the stage—above all, the architect-artists Enrico Prampolini and Fortunato Depero, the former publishing his *Futurist Scenography* in 1915 and the latter writing his *Notes on the Theatre* in the following year. Where Prampolini wished to reshape the received form of the theater into a comprehensive synthesis of action and setting, so that the latter was as much part of the drama as the former ("Instead of the illuminated stage, let's create the illuminating stages: luminous expression that will irradiate the colors demanded by the theatrical action with all its emotional power"), Depero demanded mechanically mobile scenery related to his kinetic sculpture of the years 1914–1915. These complementary approaches to the avant-garde stage sustained the Futurist sensibility after the demise of the initial period that came with the untimely deaths of Boccioni and Antonio Sant' Elia and with the end of World War I. Prampolini embraced almost at once Depero's concept of moving scenery, although this idea had been in the offing ever since Gordon Craig's independent project of 1908, which had depicted a stage set that was capable of continual transformation. Like László Moholy-Nagy's *Mechanized Eccentric,* first published in 1924 in the fourth *Bauhausbücher,* Prampolini's Magnetic Theater of virtually same date consisted of an actorless mechanical mobile stage lit by chromatic light. Depero and Prampolini went on to experiment with puppetlike robots and actor-machines, that is, actors performing in an artificially mechanized fashion. This literal move toward "geometrical and mechanical splendor," as Marinetti had called it in his manifesto of 1914, established the intellectual and aesthetic ground within which a Futurist architecture could be postulated. In the 1914 manifesto Marinetti had written: "Nothing in the world is more beautiful than a great humming power-station, holding back the hydraulic pressures of a whole mountain range and the electric power for a whole landscape, synthesized in control panels bristling with levers and gleaming commutators." This pristine vision of neotechnical splendor was paralleled by the young Italian architect Sant' Elia's power station designs of the same date. These projects were on display as part of the Futurist Città Nuova exhibition of 1914. The *Manifesto of Futurist Architecture* that accompanied this exhibition was apparently published in two versions, one known as the *Messàggio* and written by Ugo Nebbia, from ideas expounded by the architect, and one reworked by Marinetti and published in the journal *Lacerba* in August 1914. Apart from the interjection of the word *Futurist* on every possible occasion, Marinetti's version of the *Messàggio* hardly differed from the original. Insisting that a truly modern architecture must be based on technoscience and on the conditions of modern metropolitan life, the *Messàggio* argued that

> Modern structural materials and our scientific concepts do not lend themselves to the disciplines of historical styles. . . . We no longer feel ourselves to be the men of the cathedrals and ancient moot halls, but men of the Grand Hotels, railway stations, giant roads, colossal harbors, covered markets, glittering arcades, reconstruction areas and Salutary slum clearance. We must invent and re-build *ex-novo* our modern city like an immense and tumultuous shipyard, active, mobile and everywhere dynamic, and the modern building like a gigantic machine. . . .

More than any other manifesto of comparable date this text established the criteria for an absolutely functionalist modern architecture, one that was just as removed from historicism as from any kind of proto-Dadaist, avant-gardist trope that is often detectable as an impulse within a great deal of Futurist work.

As Umbro Apollonio has written, "Futurism cannot be understood and evaluated according to the usual forms of art criticism; it is art and it is also non-art; it is something more and something less than art." As far as the supra-artistic aspect was concerned it extended well beyond the themes usually covered by avant-garde art, as we may judge from Valentine de Saint-Point's *Futurist Manifesto of Lust* of 1913 in which she wrote, "Lust is for the body what an ideal is for the spirit—the magnificent Chimera that one ever clutches at but never captures, and which the young and the avid, intoxicated with vision, pursue without rest." This militant, anticlerical claim on the part of a feminist advocate of free love against all forms of conventional morality was typ-

ical of the period and was a perennial polemic within the radical political spectrum of Futurist activity. And yet at the same time we need to note that Marinetti's attitude toward feminism and the ill-fated tradition of *amóre* changed continually throughout his career. What did not change, however, was his unremitting opposition to both Catholicism and parliamentarianism. It was the former, in fact, that finally bought him into open conflict with Mussolini in 1921, effectively leading to the marginalization of Futurism within the overview of a triumphant Fascism that had made its peace with the Vatican. Just prior to World War I, Marinetti's sympathetic attitude toward the liberation of women made him a champion of the suffragist movement particularly in his militant tract *War, Sole Hygiene of the World* of 1915, in which, while advocating Italy's entry into the conflict, he saw universal franchise as the convenient agency by which parliamentarianism would be spontaneously destroyed. Parliamentary democracy for Marinetti was a totally corrupt and hypocritical system of governance, although what he would have liked to replace it with remained unclear. In the event, Marinetti became as alienated from the international left as from the Fascist right wing and the liberal center.

As James Joll has written in *Three Intellectuals in Politics: Blum, Rathenau and Marinetti,*

> The connection between Futurist ideas and methods and those of Fascism . . . [is] not the main reason[s] why the Futurist movement is of historical interest. . . . Futurism was only one of the paths that led to Fascism; and, if it had been the only one, Fascism would not have had to be taken so seriously. The Fascist movement was a success because it combined the systematic use of terrorism as a political weapon with an ability to appeal to the desperate and disillusioned elements of the organized working class, as well as the disappointed nationalism of middle class intellectuals and ex-officers. (Joll, 1960)

While Marinetti remained a supporter of Mussolini all his life he was only interested in intervening politically in moments of national crisis. Otherwise, he insisted that the movement restrict itself to bombast and rhetoric, that is, to a play that in the last analysis was cultural rather than political.

Politics was not the only sideline into which Futurism ventured with impunity, and in this regard it seems to have issued a manifesto on every conceivable subject from dancing to aeronautics, clothing, furniture, and food. This last came into focus with Marinetti's *La cucina futurista* (The Futurist Cookbook) published in 1932, in which he argued against the stultifying Italian diet of pasta. Even in this field,

wider political considerations were always in the wings. Thus, with his coauthor, the Futurist painter Luigi Fillia, he wrote: "It is not by chance this work is published during a world economic crisis, which has clearly inspired a dangerous, depressing panic, though its future direction remains unclear. We propose as an antidote to this panic a Futurist way of cooking, that is, optimism at the table" (Marinetti, 1986).

Indebted to a broad intellectual spectrum inherited from the European fin de siècle, specifically to authors as radical and diverse as Henri Bergson, Friedrich Nietzsche, Karl Marx, Sigmund Freud, H. G. Wells, D'Annunzio, Émile Verhaeren, and Sorel, Futurism spread in the space of twenty years, across the face of the entire globe from Argentina to Japan, from Mexico City to New York, from right to left, from the Italian Futurists to the Russian Cubo-Futurists. In terms of its universal impact it would remain unparalled as a radical cultural movement. Its indirect, long-term influence following its ultimate eclipse as an artistic force by the late twenties has hardly diminished. It was, first and foremost, the ne plus ultra of twentieth-century avant-gardism and it has remained so to this day.

[*See also* Avant-Garde; Collage; *and* Theater.]

BIBLIOGRAPHY

Apollonio, Umbro, ed. *Futurist Manifestos.* Translations by Robert Brain et al. London, 1973.

Banham, Reyner. *Theory and Design in the First Machine Age.* London, 1960.

Giedion, Sigfried. *Mechanization Takes Command: A Contribution to Anonymous History.* New York and Oxford, 1948.

Herbert, Robert L., ed. *Modern Artists on Art.* Englewood Cliffs, N.J., 1965.

Hulten, Pontus, ed. *Futurismo & Futurismi.* Milan, 1986.

Joll, James. *Three Intellectuals in Politics: Blum, Rathenau and Marinetti.* London, 1960.

Kirby, Michael. *Futurist Performance: Theory and Practice in the Drama, Scenography, Acting, Costumes, Film and Music of the Italian Futurists.* Translations by Victoria Nes Kirby, New York, 1971.

Marinetti, F. T. *The Futurist Cookbook.* Translated by Suzanne Brill, edited by Lesley Chamberlain. San Francisco, 1989.

Poggioli, Renato. *The Theory of the Avant-Garde.* Translated by Gerald Fitzgerald. Cambridge, Mass., 1968.

Schlemmer, Oskar, László Moholy-Nagy, and Farkas Molnar. *The Theater of the Bauhaus.* Edited by Walter Gropius, translated by Arthur S. Wensinger. Middleton, Conn., 1961.

Taylor, Joshua. *Futurism.* New York, 1961.

Stangos, Nikos, ed. *Concepts of Modern Art.* Harmondsworth, England, 1974.

Tisdall, Caroline, and Angelo Bozzolla. *Futurism.* London, 1977.

KENNETH FRAMPTON

G

GADAMER, HANS-GEORG. *[To explore the thought of the twentieth-century German philosopher Hans-Georg Gadamer, this entry comprises three essays:*

> Survey of Thought
> Gadamer and Aesthetics
> Gadamer and the Truth of Art

The first essay is an overview of Gadamer's philosophical hermeneutics (theory of interpretation). The second essay discusses his critique of aesthetics in his major work, Truth and Method. *The third essay analyzes his later critique of aesthetics, especially the notion of truth in art. For related discussion, see* Alienation, Aesthetic; Hermeneutics; Historicism; Interpretation; Literature; *and* Poetics.*]*

Survey of Thought

The academic career of Hans-Georg Gadamer (b. 1900) mirrors the major philosophical movements in Germany in the twentieth century. Gadamer encountered neo-Kantianism in Marburg, where he completed his doctoral dissertation in 1922; he studied phenomenology with Edmund Husserl and Martin Heidegger in Freiburg in 1923, and completed his *Habilitation* under Heidegger's direction in Marburg in 1929. Gadamer's doctoral dissertation and *Habilitation* in philosophy were on Plato. After teaching at the Universities of Marburg, Leipzig, and Frankfurt, he assumed the chair in philosophy in 1949 at the University of Heidelberg previously occupied by Karl Jaspers.

In the earliest stages of his philosophical journey, Gadamer became convinced that the experience of art was intimately connected to philosophy and to truth. *Truth and Method* (1960), his late and mature work, makes this connection explicit. His major contributions to aesthetics are to be found in part one of this book. Gadamer first explores the question of truth as it emerges in our experience of works of art and literature. Although he openly and gratefully acknowledges the pivotal role played by Heidegger's discussion of art and truth in *The Origin of the Work of Art* (1936), he titles his own project "transcending the aesthetic dimension" *(die Transzendierung der ästhetischen Dimension)* in contrast to "the overcoming of aesthetics" *(die Überwindung der Ästhetik),* the title Heidegger uses for his own project. Gadamer targets the "aesthetic dimension," that is, the dimension of an art or literary work that emerges for an

"aesthetic experience." Gadamer seeks to overcome not aesthetics but the abstraction required for an aesthetic experience *(a)* of the artwork or the literary work from its original world, and *(b)* of the subject's from her own concrete historical situation. Gadamer transcends this twofold abstraction by concentrating instead on the living presence we experience in our encounters with works of art and literature. Whether these be past or modern works, we experience their contemporaneity, their relevance, because of the truth claims they make on us.

Gadamer's second contribution to aesthetics in part 1 of *Truth and Method,* the elaboration of the ontology of the work of art and its hermeneutic significance, highlights an important difference between Gadamer and his teacher, Heidegger. Heidegger seeks to overcome aesthetics as part of his project of overcoming and ending metaphysics in order to initiate an altogether new kind of postmetaphysical "poetic thinking" *(Andenken).* Gadamer seeks instead to restore the dimension of truth that is missing from our "aesthetic experiences" of works of art (including literature) and thereby to begin anew the possibility of thinking philosophically about a different kind of experience of truth and the kind of "understanding" that results from this experience. In contrast to Heidegger, who finds the ontology of the work of art (its event structure that makes the artwork the advent of something altogether new) to call for a similar kind of "poetic" thinking, Gadamer discovers the ontology of the work of art (the work's potential to renew its claims to truth and to continue to provoke ever new interpretations) to hold true also for a wide range of historical events and texts that we renew and appropriate as our tradition through ever new efforts to understand them. Although *Truth and Method* is not finally a work in aesthetics but a hermeneutics, Gadamer's concepts of the relation between truth and art and of the ontology of the work of art are crucial to the larger project of *Truth and Method,* which is to understand our "understanding" of these things that matter to us as human beings.

The overall structure of *Truth and Method* is as follows. Part 1 begins by working out a different concept of "experience" and of the "experience of truth" by taking up and rethinking key elements of Immanuel Kant's aesthetics. The truth claims (truth) that Gadamer has in mind are such as to cause us to change our minds and ourselves. Truths that transform the way the self understands itself are not,

Gadamer insists, something that we can bring about by using a "method" such as René Descartes's method for ascertaining the truth of the proposition "I am, I exist," for these are truths about ourselves that happen "to us over and above our wanting and doing" (Gadamer, 1975, p. xxviii). Part 2 explains how this new concept of "experience" and of the "experience of truth" that requires us to change our presently existing self-understanding is also at work in our understanding the historical works, texts, and events that make up what we call our tradition. The truth claims made on us in these experiences of what has been handed down to us from tradition also transform our present self-understanding, that is, the way we understand ourselves concretely in our present historical situation. Part 3 turns to the medium of language and the "linguisticality" *(Sprachlichkeit)* of understanding in order to show how understanding the other, be it the work of art in its contemporaneity or the historically past work, text, or event, leads to increased self-understanding. Gadamer argues for a reciprocal relation between self-understanding and understanding the other. Understanding the other dialogically and undogmatically enriches both our understanding of our self and our understanding of the other.

Gadamer's hermeneutical philosophy productively appropriates and thereby preserves key insights found in the philosophy of Plato, Aristotle, Kant, G. W. F. Hegel, and Heidegger, among others. For example, part 1 of *Truth and Method* preserves Kant's insight into the enhancement of one's understanding provoked by works of art and literature. According to Gadamer, Kant fails to realize the full potential of his insight insofar as he subjectivizes aesthetics in his *Critique of Judgment* by narrowing the concept of aesthetic experience to the experience of the subject's state of mind (the free play of the mental faculties). This narrowing, Gadamer maintains, is in contrast to the role played earlier by judgment and taste in connection to a common culture and the formation of a *sensus communis*. By establishing instead a relation between taste and the individual who judges, Kant restricts aesthetic experience to the internal experience, *Erlebnis*, of the subject's pleasurable *state of mind*. Gadamer proposes that we think instead of our experience of works of art and literature on the model of *Erfahrung*, the concept of experience that is at work in Hegel's *Phenomenology of Spirit*. *Erfahrung* gives us something to understand that so challenges us that we must *change our minds*. Gadamer concludes that the subjective universality—the "disinterested interest" that Kant claims for aesthetic experience and the judgments we make about it—cuts us off from the interest we take in works of art and literature precisely because they challenge us to change our minds, not to mention our lives.

By retaining Hegel's insight into the negative moment of experience, Gadamer is able to develop a concept of understanding as an event that preserves in addition two crucial insights of Heidegger. Heidegger's first insight concerns the phenomenon of understanding. What Heidegger calls "understanding" *(Verstehen)* in section 32 of *Being and Time* (1927) differs from what Kant calls the "understanding" *(Verstand)* in his *Critique of Pure Reason*. For Kant, "understanding" *(Verstand)* is objectifying. It is the natural scientific understanding we have of objects that are other than ourselves and that exist in time and space. For Heidegger, "understanding" *(Verstehen)* is not objectifying, although it is also of something other than oneself, but in this case of other possible ways for human beings, including oneself, to be. Gadamer takes up and preserves Heidegger's insight that understanding is historically situated, that it is always already thrown into a tradition that presents us with other possible ways to be.

Heidegger's second insight is found in "The Origin of the Work of Art," an essay that Gadamer admits was key to his first formulating his own idea of a philosophical hermeneutics in the 1930s. In this essay, Heidegger claims that the truth that a work of art discloses emerges out of the "strife" *(Streit)* between "earth" *(Erde)* and "world" *(Welt)*. According to Gadamer, "The important insight that Heidegger's 'The Origin of the Work of Art' opened up is that 'earth' is a necessary determination of the Being of the work of art" (1994, p. 100). "World," the openness of being-there *(Dasein)*, was already worked out in and central to *Being and Time* (1927). With "earth," however, Heidegger introduces an altogether new conceptuality that is drawn from the poetry of Friedrich Hölderlin (1770–1843). "Earth" does not mean "matter" *(Stoff)* but that out of which everything emerges but which conceals itself. "Earth" becomes the counterconcept of "world" or that which continues to open up.

Beginning in the early 1930s, Heidegger increasingly resorts to the language and conceptuality of Hölderlin's poetry the more convinced he becomes that the language and conceptuality of metaphysics must be overcome. Gadamer rejects Heidegger's belief in the need to inaugurate a new, because German, origin of European history and proposes instead a creative renewal of the European tradition that began with the Greeks. Gadamer uses Heidegger's insight into the happening of truth in our encounter with works of art and poetry, but he rejects Heidegger's reading of the history of Western philosophy in terms of metaphysics and of the ever-increasing forgetfulness of Being. Gadamer is able to mine the history of Western philosophy by redirecting our attention to the tradition of practical philosophy and hermeneutics that become obscured when one's focus is primarily on metaphysics. This philosophical tradition is like "earth" in Heidegger's "The Origin of the Work of Art" in that it continues to conceal within itself resources. But for Gadamer these become resources for renewal instead of unmediated sources of something altogether new. The hermeneutical philosophy Gadamer presents in *Truth and Method* brings this philosophical tradition into the open, thereby renewing the history of Western philosophy starting

with the Greeks and, ironically, also creatively renewing Heidegger's thinking.

Hermeneutics becomes philosophical when it becomes a general theory of understanding that is also an ontology of ourselves as historically situated beings. Understanding is accordingly always mediated; it is always already interpretation. Four important concepts are at the core of Gadamer's hermeneutical philosophy: the hermeneutical circle, the principle of history of effect, play, and the fusion of horizons. The concept of the hermeneutical circle comes from sections 31 and 32 of *Being and Time,* where Heidegger maintains that understanding is circular in a productive and nonvicious sense. Heidegger's account of the circularity of understanding is in terms of an individual's understanding of her existence and of possible ways for her to be. Gadamer expands it to include our understanding of the other, be this a work of art or literature or a historical event or text. When we seek to understand something (for example, a text), we necessarily project wholes out of the parts of the text that we are interpreting as we anticipate the meaning of the text. Anticipations of meaning are really prejudgments or "prejudices" and they play a constitutive role in the process of coming to understand a text. But the whole, the meaning we anticipate based on the prejudices we import into our reading of the text, is not usually borne out by the parts as we continue to read and to interpret the text, and in this way the text leads us to confront and to question our prejudices. The circle of understanding continues—understanding the whole from the part and the part from the whole—as we continue to read and make sense of the text.

The second concept, the principle of history of effect *(Wirkungsgeschichte),* borrows and develops Heidegger's insight that we are thrown into a world and always already historically situated. According to Gadamer, the interpreter (for example, of a text) can never arrive at an unprejudiced, and in this sense immediate, starting point, an immediate intuition *(Anschauung),* like the kind sought by Descartes. We remain always "affected by history." Put in a more positive light, however, as historically situated beings we are always at the same time "open to the effects of history." When we seek to understand, for example, some text produced by another historically situated being, we put our own historically situated understanding of the matter discussed by the text at risk. Being open to the effects of history, therefore, brings with it the possibility that by understanding the text of another we will be challenged to change.

The third concept is that of play *(Spiel),* a concept that had been at the core of the aesthetics of Kant (the free play of the mental faculties) and of Schiller (the play impulse). Gadamer introduces the concept of play or game at two different places within *Truth and Method* to transcend the subjectivism that he finds characteristic of aesthetics and of modern philosophy in general. When Gadamer argues in

part 1 against Kant's subjectivization of aesthetics, he counters the abstraction inherent in Kant's aesthetic consciousness with the absorption of consciousness at play. An individual subject must lose herself to experience the truth claimed by the artwork, just as an individual player must lose herself in order to be absorbed in the game she is playing. Later, when Gadamer, focuses on language as the medium of hermeneutical experience in the transition from part 2 to part 3 of *Truth and Method,* he again turns to play and grounds language not in the consciousness of an individual subject but in the give-and-take of the language game *(Sprachspiel)* we call conversation.

Language in the form of conversation introduces us to the fourth concept, the fusion of horizons *(Horizontverschmelzung).* As mentioned earlier, understanding the text of another involves challenging one's own understanding of the matter at issue in the text. Insofar as the truth claimed by the text is at variance with what the interpreter believes to be the truth about the matter at issue, the interpreter's horizon of understanding at first excludes the truth claimed by the text. The interpreter next opens herself to a horizon other than her own when she allows the text to question her own prejudgments and claims to know the truth about the matter at issue in the text. This openness allows a dialogue of question and answer to take place during which the interpreter both questions the truth claimed by the text and allows herself and what she claims to be true to be put into question by the text. When the interpreter achieves a better understanding of the truth claimed by the text, as well as of her own prejudgments and claims to the truth, the horizons become fused and the dialogue comes to an end. Increased self-understanding on the part of the interpreter involves usually altering her prejudgments and claims to the truth because of what she has learned from her interpretation of the text. Increased self-understanding may also lead her, however, to reaffirm her initial prejudgments because they have survived the challenge.

The experience of the truth of art and literature, as we have seen, provides Gadamer a model for how we understand and interpret the works, texts, and events handed down to us as our tradition and history. We say that we understand a work, text, or historical event when we understand it to be making a true claim about how we are to understand something that is an issue for us in our own world and historical age. Like the experience we undergo with art, the event of understanding what is handed down to us challenges us to change our minds. In his "Reflections on My Philosophical Journey," Gadamer raises the question, "Can the example of art really provide a framework within which a universal hermeneutics could be developed?" (quoted in Hahn, 1997, p. 44). In the years following *Truth and Method,* Gadamer comes to recognize a potential problem with his concept of play or game. In conversation, the game of language that mediates understanding, one transcends

subjectivity dialogically by changing one's self and self-understanding along with one's changed understanding of the truth claimed by the other. In the experience of art, however, one seems to transcend subjectivity in a more one-sided way. The game we engage in with art challenges us to change our minds and ourselves without our apparently being able to reciprocate and challenge it.

Beginning in the 1960s, Gadamer takes up the question of the relation between philosophy and literature in general, and his philosophical hermeneutics and poetry in particular. In pursuing the question of the linguisticality *(Sprachlichkeit)* of interpretation, Gadamer ends up suggesting that the lyric poem is the model for the language game of the text to be interpreted and that the Platonic dialogue is the model for the language game of the interpretation of the text. These two paradigmatic forms of linguisticality help explain the reciprocity that for Gadamer characterizes the game that we engage in with the artwork or the literary work. Just as the work of art or literature challenges us to change our minds, so the very activity of interpreting it challenges it to provide answers to our questions. In interpreting the work, we do not reproduce the truth claimed by the artwork. We produce a different understanding of its claim to truth.

Ironically, once play became associated with poststructuralist theories of interpretation, especially that of Jacques Derrida ("Structure, Sign, and Play in the Discourse of the Human Sciences" [1978]), Gadamer's attempt to rethink his earlier concept of play risked being misunderstood as an affirmation of the very subjectivity in understanding that part 1 of *Truth and Method* set out to deny. In Paris in 1981, Gadamer attempted but failed to initiate a dialogue with Derrida on the difference between the hermeneutical and the deconstructive concepts of text and interpretation *(Dialogue and Deconstruction: The Gadamer-Derrida Encounter* [1989]). Even though Manfred Frank has taken up Gadamer's challenge to Derrida and raised additional questions about deconstruction from the perspective of hermeneutics in *What Is Neostructuralism?* (1988), the encounter between hermeneutics and deconstruction remains a project for the future.

[*See also* Heidegger; *and* Play.]

BIBLIOGRAPHY

Works by Gadamer

Wahrheit und Methode: Grundzüge einer philosophischen Hermeneutik. Tübingen, 1960.
Gesammelte Werke. 10 vols. Tübingen, 1985–1995. Vols. 1 and 2 on hermeneutics; vols. 3 and 4 on recent philosophy; vols. 5–7 on Greek philosophy; vols. 8 and 9 on aesthetics and poetics; vol. 10 on hermeneutics in retrospective.

Works by Gadamer in Translation

Dialogue and Dialectic: Eight Hermeneutical Studies on Plato. Translated by P. Christopher Smith. New Haven, 1980.
Hegel's Dialectic: Five Hermeneutical Studies. Translated by P. Christopher Smith. New Haven, 1976.
Heidegger's Ways. Translated by John W. Stanley. Albany, N.Y., 1994.
Philosophical Hermeneutics. Translated and edited by David E. Linge. Berkeley, 1976.
Plato's Dialectical Ethics: Phenomenological Interpretations Relating to the Philebus. Translated by Robert M. Wallace. New Haven, 1991.
Reason in the Age of Science. Translated by Frederick G. Lawrence. Cambridge, Mass., 1981.
The Relevance of the Beautiful and Other Essays. Translated by Nicholas Walker. Cambridge and New York, 1986.
Truth and Method. Translation by Sheed and Ward, edited by Garrett Barden and John Cumming. New York, 1975. 2d rev. ed., translation revised by Joel Weinsheimer and Donald G. Marshall (New York, 1989).

Other Sources

Bleicher, Josef. *Contemporary Hermeneutics: Hermeneutics as Method, Philosophy, and Critique.* London and Boston, 1980.
Grondin, Jean. *Introduction to Philosophical Hermeneutics.* Translated by Joel Weinsheimer. New Haven, 1994.
Hahn, Lewis Edwin, ed. *The Philosophy of Hans-Georg Gadamer.* Chicago, 1997.
Hirsch, E. D., Jr. *Validity in Interpretation.* New Haven, 1967.
Hoy, David Couzens. *The Critical Circle: Literature, History, and Philosophical Hermeneutics.* Berkeley, 1978.
Michelfelder, Diane P., and Richard E. Palmer, eds. *Dialogue and Deconstruction: The Gadamer-Derrida Encounter.* Albany, N.Y., 1989.
Palmer, Richard E. *Hermeneutics: Interpretation Theory in Schleiermacher, Dilthey, Heidegger, and Gadamer.* Evanston, Ill., 1969.
Wachterhauser, Brice R., ed. *Hermeneutics and Modern Philosophy.* Albany, N.Y., 1986.
Warnke, Georgia. *Gadamer: Hermeneutics, Tradition, and Reason.* Stanford, Calif., 1987.
Weinsheimer, Joel C. *Gadamer's Hermeneutics: A Reading of Truth and Method.* New Haven, 1985.
Wright, Kathleen, ed. *Festivals of Interpretation: Essays on Hans-Georg Gadamer's Work.* Albany, N.Y., 1990.

KATHLEEN WRIGHT

Gadamer and Aesthetics

In the first section of *Truth and Method*, Hans-Georg Gadamer warns readers: "The present conceptual analysis has to do not with theory of art but with ontology. Its first task, the criticism of traditional aesthetics, is only a stage on the way to acquiring a horizon that embraces both art and history" (Gadamer, 1989, p. 137). Although the book opens with intensive discussions of Immanuel Kant's *Critique of Judgment* (1790) and its successors, Gadamer's interest focuses not on aesthetics, ultimately, but on hermeneutic ontology, comprehending the nature of historical as well as artistic being. Because his exposition passes beyond the aesthetic, thus denying its autonomy, Gadamer's magnum opus offers not so much an aesthetics as an antiaesthetics. Not only does he show "how questionable the concept of the aesthetic is" (ibid., p. 129), but he argues that "aesthetics has to be absorbed into hermeneutics" (ibid., p. 164). Yet, precisely because Gadamer takes art as his starting

point, his insights mark a major innovation in the field of aesthetics.

Gadamer's critique of traditional philosophy of art begins by questioning Kant's "subjectification of aesthetics." Kant discovered that, despite taste's proverbial variability, it is nevertheless universally communicable because the beautiful is suited to engaging the perceiving subject's cognitive faculties, and this suitability is in principle the same for all subjects. The great significance of taste's claim to subjective universality is that it grounds the uniqueness and autonomy of the aesthetic. In Gadamer's view, however, the price that Kant paid for segregating the beautiful from the true and the good was exorbitantly high, because it distorted not only the idea of art but of scientific and moral knowledge as well. Grounding the autonomy of the aesthetic on subjective universality implies, negatively, that the judgment of artistic beauty has no truth value and that aesthetic judgment imparts no knowledge of the object. Correlatively, it implies, as Kant's successors inferred, that art is essentially a matter of subjective experience, for the artwork as conceived by subjectivist aesthetics is both produced out of the artist's life experiences and produced for the spectator's experience and disinterested enjoyment. In neither the process of production nor reception, however, can art make a legitimate truth claim, because taste is by nature precluded from offering objective conceptual knowledge.

For Gadamer, overcoming the subjectivization of aesthetics above all requires admitting that "what we experience in a work of art and what invites our attention is how true it is" (ibid., p. 114). This admission does not mean that we should ignore Kant's achievement in superseding rule-governed classicist aesthetics or reduce art once again to conceptual knowledge. On the contrary, for Gadamer the experience of art as truth shows that Kant subjectivized the aesthetic because he artificially limited truth to conceptual, determinant judgment. Once truth is restored to its full breadth, however, it is not the case that aesthetic judgment becomes dispensable. Just the opposite: we come to see that it is needed to discover nonconceptual truth not just in art but in the ethical and scientific realms as well. Broadened throughout what Plato called "the whole ocean of the beautiful," aesthetic judgment loses its unique identity and autonomy, but also acquires universal provenance.

"In order to do justice to art," Gadamer argues, "aesthetics must go beyond itself and surrender the 'purity' of the aesthetic" (ibid., p. 92). An "impure" aesthetics would recognize that the beautiful overlaps with the true and the good, as Gadamer shows, and it would also deny the double differentiation by which aesthetic purity is effected. First, aesthetic purism differentiates art as such from the world in which it originated. Just as purism views the artist as a self-expressing subject, free of the world or alienated from it, so also it views the pure artwork as abstractable from its original relationships and purposes—the celebratory occasion of

a court masque, say, or the meditative purpose of a religious icon. Severed from its particular past and rendered timeless, art becomes immediately accessible to all times. From the viewpoint of aesthetic purism, art's timelessness legitimates a second abstraction as well: the artwork as such is differentiated from all the mediations by which it becomes accessible to later ages. Now, what is negated, for example, is not the play's original occasion but rather the occasions of its subsequent presentation. The timeless work in itself is distinguished from the temporal performances of it.

In denying both aspects of aesthetic differentiation, Gadamer's impure aesthetics affirms, for example, that the play proper is to be seen onstage, not in the script. Nowhere does drama exist in a more authentic state; indeed, nowhere does it exist *an sich*. Similarly, the artist is never wholly the bohemian. The recent revaluation of allegory, no longer considered inimical to art, suggests to Gadamer that art cannot be understood merely as a function of the artist's free symbol-making power, for that power is never wholly free but always tied, like allegory, to the community's symbol-making traditions and conventions. Art, then, is impure in being inalienably tied to both its worlds—original and subsequent, past and present—and it can be severed from neither without loss.

Gadamer conceives the artwork's bond to the past world of its origin in terms of mimesis. This Platonic concept has become notoriously suspect because so much modern art is militantly nonrepresentational; but if no art, however little it depicts some recognizable thing, can entirely break with the world of its origin, then all art is, in this sense, inescapably mimetic: it always (re)presents a world that preexists it. That is not all it does, however. Like Socrates in the *Ion*, Gadamer speaks of a double mimesis, a (re)presentation of (re)presentation; for, just as a symphony presents its world, so also it is to be presented in subsequent performance, and the symphony is in principle no less itself when presented in this later world than when it first appeared, though the two may be (in fact, always are) very different.

From this discussion of double mimesis, Gadamer concludes, first, that the mode of being of art is production or presentation (*Darstellung;* presentation being not something the work suffers but what it is: an event of appearance), and second, that the work of art is best conceived as the mediation between worlds and times. If hermeneutics is understood as just this kind of mediation, then it follows that the ontological character of art is hermeneutical. More simply, presentation (mimesis) is interpretation. Artists interpret their worlds in the same way that actors interpret the plays that the artists write. Given that an artwork interprets a past to a present world, no interpretation, whether performative or scholarly, can ignore the work's history, the context of life in which it was originally rooted, for even a work physically removed from that context (e.g., a stained glass window transferred from a church to a museum) still retains a connection, however attenuated, to its original life-world.

Yet, for Gadamer, neither art in general nor the work in particular can be reduced (in Hegel's phrase) to "a thing of the past," which is to say that it cannot be treated merely historically. Historians cannot rightly conceive of their task as transporting themselves to the past by negating their present world. That would amount to forgetting that, just as a translator mediates between two languages, the artwork interprets two worlds. The criterion of correct interpretation therefore cannot be confined to just one of them. For example, the work's meaning cannot be limited, as some historicists would have it, to the intention of a finite author who did not and could not foresee how the work would be realized in potentially limitless future performances and interpretations. An interpretation is not only an interpretation of something but an interpretation to someone, and must be judged in terms of both. A true interpretation is like a fine metaphor: it gives insight. Gauging its success requires aesthetic judgment—not a hard rule or standard (such as intention) but tact and taste.

Asking for a definitive scholarly interpretation of a play makes no more sense than asking for a definitive performance of it. Indeed, art can have no definitive interpretation in Gadamer's view precisely because presentation defines the ontological character of the artwork: it *is* to be. No single moment, not even the moment of its inception, can claim to be final or complete in interpreting the artwork because it exists in a genuinely historical way, which means precisely the opposite of fixed and determinate. Gadamer discusses art as preparatory to history because art symbolizes a historic past that is never definitively past but always still to be presenced and presented. And this presentation is interpretation. The work of art, like everything historic, is always still to be interpreted and understood. Art exists hermeneutically, and it is for this reason that aesthetics must dissolve into hermeneutics.

There can be no doubt, then, that Gadamer believes that the artwork exists in the interpretations of it, and only there. That art exists hermeneutically does not at all imply, however, that the work means whatever the interpreter thinks it means. This is just as mistaken as saying that the work means whatever the author thinks it means, and for the same reason: it tries to judge a successful mediation in terms of only one of the things mediated, and it overestimates the power, freedom, and autonomy of the interpreter. To say that the artwork means whatever the interpreter makes of it is to assume that interpreting is something that the interpreter does to the interpreted, a free subject's action on a submissive object. This assumption, however, is a postulate of subjectivist aesthetics and its notion that the work exists in the subject's (here, the interpreter's) experience.

To overcome subjectivism in this form requires that Gadamer reconceive interpretation not as an action but as an event. Understanding is "not what we do," he writes, "but what happens to us over and above our wanting and doing" (ibid., p. xxvii). In this respect, Gadamer takes his cue from Hegel and the Greeks, who "did not conceive understanding as a methodic activity of the subject, but as something that the thing itself does and which thought 'suffers'" (ibid., p. 474). The paradigm of drama again helps in making this conception intelligible. The play exists only onstage—that is, only in the player's interpretation of it—but this is not the act of a free and disinterested subject detached from its object. Quite the contrary, subject and object here become indistinguishable. Although the play exists only in the interpretation of it and therefore is no object, the interpreter is part of the play and therefore no subject.

To be (presented or interpreted) is what the work is, and its interpretative realizations therefore belong to the work, not the interpreting subject. "Being understood belongs to the meaning of the text," Gadamer concludes; "understanding must be conceived as part of the event in which meaning occurs" (ibid., p. 164). "The experience induced by the work [of art] . . . does not leave him who has it unchanged" (ibid., p. 100), for to be caught up in the play that occurs beyond our wanting and doing, especially in what happens to us in the greater drama of history, is to lose one's free subjectivity—or, more exactly, to lose the illusion of having such freedom. The experience of art is in that sense an experience of truth, a truth that comes from belonging instead of distantiation and critical reflection.

In Gadamer's view, then, the artwork is its interpretations. The interpretations are *of* (i.e., belong to) the work itself. This self-interpreting or self-presenting quality, not incidentally, also characterizes the nature of beauty. "The beautiful of itself presents itself" (ibid., p. 481). It is precisely in the case of beauty that there can be no aesthetic differentiation, no distinction between noumenal and phenomenal or between what is and what merely appears beautiful, because it is in the nature of beauty to appear, to present itself. Because what appears beautiful is beautiful in fact, for Gadamer beauty becomes an apt symbol of truth—in particular, the self-evident, apparent truth that is its own evidence.

Art exemplifies not just the event of truth; it symbolizes the event of being. This does not mean only that the work exists nowhere apart from its interpretations. To say that the artwork is its interpretations means that through them it comes to be more and other than what it was. An interpretation is not a copy or mere repetition, for interpreting never merely duplicates the work or previous interpretations of it. Rather, difference necessarily belongs to the mediation performed by interpretation. If the fact that the work is *self*-interpreting bespeaks the self-identity of the work, the fact that it is self-*interpreting* indicates that the art is "an entity that exists only by always being something different" (ibid., p. 123). Differences of interpretation, Gadamer insists, belong no less than self-sameness to the work itself.

In the process of interpretation, the work that gives itself to be understood (otherwise) comes to be more and other than it had been previously. This is a process of hermeneutical emanation, to use Neoplatonic language, and in it an event (an increment) of being occurs, not just an extension; for interpretation is, by reason of its essential difference, productive, not merely reproductive. Through interpretation, something new is continually coming into being, though it is no less true that what is continually produced in the history of interpretations is the artwork itself.

[*See also* Kant.]

BIBLIOGRAPHY

Works by Gadamer

Ästhetik und Poetik. In *Gesammelte Werke,* vols. 8–9. Tübingen, 1993.
Philosophical Hermeneutics. Translated and edited by David E. Linge. Berkeley, 1976.
The Relevance of the Beautiful and Other Essays (1977). Translated by Nicholas Walker. Cambridge and New York, 1986.
Truth and Method (1960). 2d rev. ed. Translation revised by Joel Weinsheimer and Donald G. Marshall. New York, 1989.

Other Sources

Palmer, Richard E. *Hermeneutics: Interpretation Theory in Schleiermacher, Dilthey, Heidegger, and Gadamer.* Evanston, Ill., 1969.
Velkely, Richard L. "Gadamer and Kant: The Critique of Modern Aesthetic Consciousness in *Truth and Method.*" *Interpretation* 9 (1981): 353–364.
Weinsheimer, Joel. *Gadamer's Hermeneutics: A Reading of Truth and Method.* New Haven, 1985.
Weinsheimer, Joel. *Philosophical Hermeneutics and Literary Theory.* New Haven, 1991.
Wright, Kathleen, ed. *Festivals of Interpretation: Essays on Hans-Georg Gadamer's Work.* Albany, N.Y., 1990.

JOEL WEINSHEIMER

Gadamer and the Truth of Art

Gadamer devoted the first section of *Truth and Method* (1960) to the question of art for two basic reasons, one "negative," the other "positive." On the one hand, he delivered a sharp criticism of what he termed "aesthetic consciousness," which he saw as a hollow and disastrous modern abstraction. Yet, on the other hand, his entire project did follow the lead of something like the aesthetic experience of truth when it undertook to liberate the meaning of understanding and truth, as it is lived, for instance, in the human sciences and in everyday practical judgments, from the scientific, methodical model of truth. This might appear confusing: a severe rebuttal of aesthetics (which many aestheticians might deem excessive) coupled with a decisive reliance on the aesthetic experience of truth (which will, this time, irritate most epistemologists). Actually, the two movements go hand in hand: according to Gadamer, it is only by destroying the modern notion of an "aesthetic consciousness" and its disastrous consequences for the under-standing of understanding that one can hope to rediscover the truth-meaning function of art or aesthetic experience. This article will recall in broad strokes this twofold, yet unitary attitude toward aesthetics before addressing some of the inflections of Gadamer's later aesthetics.

The Overcoming of Aesthetic Consciousness and the Quest for the Hermeneutical Truth of Art. It is well known that the idea of "aesthetics" is a specifically modern phenomenon. Of course, the Ancients also had a notion of *aesthesis,* from which our idea of "aesthetics" is derived, but for them this notion covered the entire realm of sensory perception, which did not necessarily have what would today be considered an artistic or "aesthetic" value. They also reflected on the idea of beauty, but nothing indicates that they viewed it as an "aesthetic" affair, that is, as an experience that would somehow be different or isolated from the realm of cognitive and moral concerns.

It is precisely this idea that there is an "aesthetic" realm, an "aesthetic" beauty, and so on that is specifically modern. The first to use the notion in this sense was most probably the German philosopher Alexander Gottlieb Baumgarten, who hoped to dampen the rationalist urge of his time by developing the notion that there was something like an "aesthetic" cognition that would also embody rationalist ideas, yet without following the syllogistic ways of pure reason. There is reason in the senses, as it were. They provide a form of cognition that supplements and even enlarges the scope of reason. The first major philosopher to draw on this idea was Immanuel Kant. In the very first section of his *Critique of Pure Reason* (1781), he distanced himself from Baumgarten's enterprise and the meaning he gave to the notion of "aesthetics." In matters of taste, Kant argued, there is no such thing as a priori or rational principles. Here, everything is rather a matter of subjective judgment. This is why he discouraged the use of "aesthetics" for this realm, believing at the time that no science of it was possible. "Aesthetics," he contended, could only be the science of the a priori conditions of human receptivity, and this receptivity is by no means "aesthetic" in any modern sense of the word; it is, rather, purely cognitive, obeying the fundamental conditions of space and time. Aesthetics is thus, for Kant, the theory of the conditions of our cognitive receptivity. There is no such thing as a science of aesthetic taste, because here every judgment is precisely just a matter of taste and inclination.

Kant's attempt to claim back the ancient notion of aesthetics did not fly. Indeed, in his *Critique of Judgment,* written nine years later, Kant himself resorted to the use of the word *aesthetic* that was common at his time, but that he had earlier condemned. He now spoke of "aesthetic judgment" as a specific form of judgment that would start off with individual instances (*aesthesis* had always been aimed at the particular) and try to find the universal concept that corresponds to it, without, however, really succeeding. It is this

"play" of the human faculties, Kant likes to say, that induces something like an aesthetic pleasure, the sentiment of beauty or of the sublime. The details of Kant's complex analysis, which launched the career of aesthetics, cannot interest us here. What is important—and for Gadamer of momentous consequence—is that, according to Kant, this aesthetic judgment remains devoid of cognitive value. One can communicate and thus hope to share an aesthetic feeling, but there is no real objectivity to it, meaning by that a binding logical universality. Objective knowledge always follows this pattern of a universal judgment, or law, that determines the individual instances. In aesthetic judgment, which results from a play of our faculties, only the particular is given and the universal is only sought after and never reached. This is perhaps what delights us in aesthetic experience, but such a play does not amount to objective, hard knowledge, the kind that can be gained in science, but also, and most important for Kant, from the evidence of moral consciousness.

The consequence of Kant's treatment of aesthetic judgment was thus to create an independent realm for aesthetics, one that would be strictly separated from the spheres of knowledge (that of theoretical reason) and morality (practical reason). Whether Kant was responsible for this new invention of the aesthetic is a moot point. He could not have imagined the currency that the celebration of the artistic per se would enjoy in the next two centuries. Moreover, the evidence of his texts seems to show that he was also well aware of the moral underpinnings of any aesthetic judgment. Consequently, Gadamer's critique of the autonomy of aesthetic consciousness seems to waver somewhat in pinning the responsibility on Kant. In *Truth and Method*, Kant seems to be the immediate culprit, but in later works—for instance, in the 1980 essay "Anschauung und Anschaulichkeit" ("Intuition and Vividness," in the English collection *The Relevance of the Beautiful and Other Essays*, 1986)—one has the feeling that Gadamer is very sympathetic to Kant's moral understanding of the aesthetic experience, so that the responsibility for the isolation of an aesthetic consciousness must fall on his immediate Romantic successors.

Be that as it may, it cannot be denied that it was something like the Kantian paradigm that dominated the treatment of aesthetics for the next two centuries. According to Gadamer, the tacit presupposition was that aesthetic judgment had no cognitive import; it was "merely" aesthetic. One can, of course, see that as a good thing. It did in fact bestow "autonomy" (that is, self-regulation) on aesthetic experience, thus liberating it for itself. Ever since this autonomy has been conquered—and its conquest remains a constant struggle for most artists and philosophers—it has been possible to judge works of art independently (or so it seems) from the requirements of science or morality. An artwork does not aim at enlarging the scope of our knowledge; it is neither true nor false, neither right nor wrong in any narrow moral sense. It is a "piece of art" and must thus be judged by its own standards that are alleged to be intrinsically aesthetic.

Although Gadamer does not really want to revoke this autonomy, which is a modern fact that cannot be wished away, he does question whether it can be total. For him, the idea of a separate "aesthetic" consciousness is an abstraction because there is no such thing as an aesthetic experience that would disregard the *truth claim* raised by a work of art. An artwork always has something to say; it speaks to our cognitive and moral sensibilities and brings them into play. It is this truth claim that Gadamer will promise to sort out in his own theory of art. He will even seek to extend it to our larger experience of truth, that of cognitive truth.

But before getting to this "positive" aspect of Gadamer's aesthetics, one must grasp the far-reaching consequences of his genetic interpretation of aesthetic consciousness. For Gadamer, aesthetic experience did not spring out of itself; it did not even emerge out of the immediate contact with artworks. Rather, it was imposed by modern science in its exclusive claim on all matters of truth. For modernity, as it understands itself and as Gadamer also reads it, truth can only be ascertained by a methodical inquiry that remains the domain of science.

In this context, art and aesthetic experience had lost every cognitive legitimacy. They could only defend their legitimacy by relinquishing any sort of truth claim or practical purpose. Aesthetic experience was confined to the margins of truth and science and could only define itself through this predicament imposed on it by modern science. The "inferiority complex" of the aesthetic actually amounted to an "exteriority complex," because it was excluded from all matters of truth. It had to recognize that its predicament was "merely aesthetic," but it tried to put a positive twist on things by celebrating this aesthetic dimension. From now on, there would be such a thing as an independent aesthetic consciousness, with its own logic, requirements, and institutions. Indeed, art and aesthetics actually prospered once their autonomy was acknowledged (even if it was actually imposed on them by science). In the nineteenth and twentieth centuries, every respectable city had to have its "arts center" where museums and concert halls would be bundled together to create a space where one could enjoy and feel "aesthetically," a space that would be sharply demarcated from the rest of the "real" world dominated by the hard logic of science and economy. Soon, every newspaper would also have its "arts section" and every government its "art department" that would handle this allegedly distinct sphere of human activity. The premise of it all remains that this autonomous sphere has little to do with the real outside world, that it might to some extent seem superfluous and frivolous, but it is there and its unidentifiable autonomy requires that it be confined to some distinct places and institutions: an arts center, an arts sec-

tion, a museum, and so on. These institutions have become so evident to us that we might believe them to be natural. They are not. As Gadamer reminds us, prior to the nineteenth century, what we call artworks were usually found in temples or churches, where they were not works of art but of worship; secular paintings were mostly held in private collections according to substantial criteria; plays were not destined for "city theaters" but were staged by traveling groups, and so on. Gadamer's point is that art is never separated from the world but is always part of the world out of which it nurtures itself and that it helps to understand in return. "Aesthetic consciousness" is thus for Gadamer the result of an aesthetic separation or differentiation (of art and world, which is simply a false abstraction, and, furthermore, one that was imposed on art by methodical science). It is no wonder that this kind of aesthetic experience constantly has to struggle with this model of science when it seeks to define, and even fund itself. Gadamer thus calls into question the very premise of aesthetics, namely, the notion that there is something like an aesthetics.

There is, therefore, something like an "anti-aesthetic" in Gadamer's philosophy. But one can only be against something if one can provide a better description for what it covers up. Gadamer will turn the tables on aesthetic consciousness by claiming, in the simplest terms, that the experience of art is at its root an experience of truth. Not only is it one experience of truth among others, it is an experience that can also help us discover what truth is all about. It is this experience that methodical science expatriated when it grounded truth on method.

The Hermeneutical Understanding of Aesthetic Truth and Gadamer's Later Aesthetics. It can be argued that Gadamer focused mostly on the critique of aesthetic consciousness in his magnum opus *Truth and Method*. This is understandable because his aim was to free hermeneutics from an aesthetic reading of understanding that was actually imposed by modern science. He did aim, however, to reconquer a concept of truth out of the experience of art, but many critics, such as Käte Hamburger (1979), have found that his book did not really deliver on this promise. One of Germany's leading epistemologists, Karl Albert, wrote: "However one may judge Gadamer's inquiry into the 'experience of art' . . . , it did not achieve a clarification of the question of truth to the extent that art is concerned by it, nor did it demonstrate that it represents a specific mode of knowledge that would bring about any kind of 'essential knowledge'" (1994, p. 41).

There is some truth in these criticisms. Even if an important systematic ("positive") section is devoted to art in *Truth and Method*, it is striking that it hardly discusses the issue of truth in art. The focus in 1960 is truly on the *ontological* import of the art experience. To counter the Romantic notion that art deals with illusion, appearance, and thus frees us from reality, Gadamer argued that art, on the con-

trary, provides something like a *Seinszuwachs*, an overwhelming "increase" of being, which is akin to an addition or an excess of reality. It is striking that we are here in the presence of a "quantitative" notion. It must, of course, be understood from the backdrop of the *loss* of being and truth in the realm of aesthetic consciousness. But one must wonder whether such a quantitative notion was not itself also a by-product of methodical science and hence somewhat inadequate for the understanding of the specific truth experience of art. In *Truth and Method*, the principal notion that supported that of a quantitative "growth" of being was that of presentation *(Darstellung)*. It also insisted on the ontological import of the art event that went beyond the self-contained artwork itself. Presentation extended the horizon of art in two directions, including the upstream, but also the downstream of the work in the event of art itself. It refers *back* to a reality that is presented, yet transformed in the artwork, but it was also meant as a presentation *to* a spectator, whose response remains part of the work of art (in the Platonic sense of participation). It is illuminating that, for Gadamer, tragedy represented the paradigm of this all-important notion of presentation: a work of tragedy always throws us back to the tragedy of life, which is not suspended but is exacerbated in the artwork, and is always played out on a stage for spectators who will recognize themselves and their life tragedy in the play.

This ontological notion of presentation certainly has some Plotinian origins. It suggested the idea of the art event as a process or an emanation of being in which we would merely partake. By relying on such a Neoplatonic idea of emanation and presentation, it was surely Gadamer's aim in 1960 to liberate the experience of art from a too narrow subjectivist conception that would aesthetically trivialize its truth claim. But one could again wonder whether something was not lost in this ontological happening, namely, the response of the spectator in whom the truth of art accomplishes itself. It is this aspect of his aesthetics that Gadamer developed more thoroughly in his later work.

This conception is to be found in a recent collection of texts by Gadamer devoted to the question of literature and poetics. Published in 1993 as volume 8 of Gadamer's complete works edition under the title *Kunst als Aussage*, almost all of its thirty-six essays were written after *Truth and Method*. One can contend that it contains the most definitive formulation of Gadamer's positive aesthetics, which supplements the critique of aesthetic consciousness set forth in 1960. Out of this new book—one of the finest in the *Gesammelte Werke* edition because of its originality—one can also illuminate the aesthetic understanding of truth that Gadamer aimed at or presupposed in *Truth and Method*. It should also be noted that two volumes of the ten-volume complete works edition are devoted to aesthetics. Whereas volume 8 is concerned with philosophical or theoretical accounts of art experience, volume 9 offers for the most part

concrete interpretations of artworks. Disregarding arithmetics, Gadamer published volume 9 before volume 8. He no doubt wished thereby to convey the idea that the experience of the work of art precedes and overwhelms theoretical reflection on aesthetics.

The concise title of volume 8 already reveals its intention, but it is unfortunately difficult to render in English. *Kunst als Aussage* corresponds to something like "art speaks," in the sense that it has something to say and does so in a way no other medium can approximate. The best translation is perhaps "art as proposition," provided one hears the two senses of *proposition* in English: the idea that art says something in an almost propositional form (which it in fact surpasses) and that it proposes something, as a suggestion or an offering. One must always *respond* to an offering, to a proposition. When confronted with a work of art, something strikes us, invites us to rethink, rediscover our experience, yet we cannot translate its "proposition" in another language. Still, it is convincing, much more so in fact than a mere truth statement that could be objectively verified and isolated.

Why is it, then, that an artwork can be more convincing than a philosophical or scientific argument? A novel, an opera, a poem, or a movie leave an imprint on us and remain in our subconscious memory in a way that no arguments can equal. The names of Fyodor Dostoyevsky, Franz Kafka, Wolfgang Amadeus Mozart immediately mean something to us; they speak to us, revealing an infinite world of experience. This also means, of course, that we learn something from them, but what it is cannot be reduced to a specific expression or argument. It even suffices merely to evoke their names to apprehend this. Why is that so? The work of art does not really argue; it makes us see, it opens our eyes, in more ways than one (sensually, intellectually, attentively).

But what we see here is something that sees us too. What is experienced in a work of art—and that can be called truth since it reveals something that is there—is also a self-encounter, an encounter with oneself (or with one's self?). This is a precious indication of the truth experience that art can help us rediscover, and it is this experience that Gadamer will seek to extend to the entire realm of what can be understood. We are always intimately concerned by the truth that occurs in a piece of art. An artwork with no truth is one that does not speak to us. This hint is precious because it runs counter to the prevailing model of truth heralded by science, for which truth is something that is independent from the observer, where subjectivity does not come into play. Although this type of truth might be applicable in some spheres where objectivity is attainable (in the knowledge of nature, for instance), it is clearly out of place in the realm of art and in questions that pertain to meaning, where truth functions like an answer to questions in which we are immersed.

The experience of truth is thus one that implies our questioning selves. No metaphysical self is implied here, of course, only the notion that we are for ourselves a question, in Augustine's justly famous phrase. Art is the privileged testimony and reminder of this self-encounter. It requires participation, an engagement (imposed art, say, an obligatory visit to a museum or corporate season tickets for a concert series, will always be dreadful). Gadamer used to account for this participation through his quasi-Plotinian notion of presentation. In his later work, this ontological notion has for all intents and purposes been replaced by that of *Vollzug*, which points to the realization of meaning in its actual accomplishment by the reader or listener of a piece of art. Volume 9 of Gadamer's complete works edition has the title *Hermeneutik im Vollzug* (Hermeneutics in Its Actualization). The titles of the two volumes of Gadamer's later aesthetics need to be heard conjointly: *Art as Proposition* and *Hermeneutics in Its Actualization;* the proposition of art, its truth claim, can only be accounted for by its actualization. A poem is always an invitation to a dialogue (*Gedicht und Gespräch* [Poem and Dialogue] is also the title of a collection of poetic interpretations by Gadamer from 1990).

The response that the artwork invites corresponds to what Gadamer also calls "reading" *(lesen)*. The German *lesen* has a richness of meaning that cannot be perfectly conveyed in English, for in German, *lesen* also has the connotation of harvest, vintage, gathering. At the root of *lesen* lies the idea that a collection or recollection is taking place. This notion of recollection will enable Gadamer to bestow new meaning on the notion that art is a matter of mimesis or "reproduction." Yet, no "imitation of nature" is implied by this mimesis, but rather the notion that this reproduction actually takes place in our reading or rereading selves. (The similarity of this notion with the theory of mimesis developed by Paul Ricoeur in his three-volume work *Time and Narrative* [1983–1985] is striking. It also enables us, and perhaps for the first time, to discover the common hermeneutical ground shared by Gadamer and Ricoeur.) Such reading is requested for all forms of art, not only for literary works. To read a painting is to follow in one's own dialogue the lines it only sketches, the world it makes us enter into. To read a piece of music is to let oneself be carried by its rhythm. Its movement has something repetitive, but its unfolding seems to lead somewhere. Where this goes, no one can tell, but it takes us along. We also read architecture when we walk through a building or a temple and thus get a grasp of its presence, of its giving of space to those who partake in its world.

To read is also, and perhaps primarily, to hear, to let the presence of the work resound in our inner ear. This notion of the inner ear was unfortunately absent from the aesthetics of *Truth and Method*, but it is essential to the understanding of aesthetic or hermeneutical truth. Truth is al-

ways something that also goes through our inner ear, where it resounds and comes to be applied to our situation and our questions, that is, basically, to the question that we are for ourselves. One can say that we exist in this world through this capacity of reading or hearing, which is never only a taking up of something that is there independently of its being read or heard. It is the reading of the inner ear that confers truth to the encounter provided by the artwork. To read or to hear, is, in Martin Heidegger's terminology, to be "there" when truth occurs. Where does this truth take place? A false dichotomy would constrain us to locate it either in the object or in the subject. Nothing of the sort is evident here. The truth cannot be situated in the objective sphere, because it is only there when it is read, applied, heard, strided through as we go through a dance or an architectural masterpiece. It is not subjective or "merely" subjective either, because it is what is there, whatever it may be, which compels us to read, to listen, and to read further.

In his latest essays on the truth of art, such as "Word and Image—So True, So Being" and "On the Phenomenology of Ritual and Language" (1992), Gadamer steadfastly relies on the simple and almost tautological notion that in art "there is something there," that "it comes out": "so ist es, es kommt heraus," he writes, without ever specifying what this neutral "it" amounts to. It is there, this truth of art, yet only if we are also there. There is something reminiscent of Heidegger's fascination with expressions such as "es weltet," "es gibt" ("world is there," "there is") in those eloquently silent formulations of the later Gadamer. It is in such nakedness, with such truth and objectivity, that art addresses us, so much so that we do not know anymore who does the speaking and who does the reading.

The ancient notion of "contemplation" can also help to describe what is holding us here. We only have to hear out the notion of "temple" that is at the very heart of the event of contemplation. When we contemplate, we come to see or hear something in a state of awe, but that we can only read if we enter into its space and participate in its unfolding of meaning, one that we can never totally grasp.

[See also Kant; and Truth.]

BIBLIOGRAPHY

Works by Gadamer

Ästhetik und Poetik I: Kunst als Aussage. In *Gesammelte Werke,* vol. 8. Tübingen, 1993.

The Relevance of the Beautiful and Other Essays. Edited by Robert Bernasconi, translated by Nicholas Walker. Cambridge and New York, 1986.

Wahrheit und Methode: Grundzüge eines philosophischen Hermeneutik. Tübingen, 1960. 5th edition in *Gesammelte Werke,* vol. 1 (Tübingen, 1986). In English as *Truth and Method,* 2d rev. ed., translation revised by Joel Weinsheimer and Donald G. Marshall (New York, 1989).

Other Sources

Albert, Hans. *Kritik der reinen Hermeneutik: Der Antirealismus und das Problem des Verstehens.* Tübingen, 1994.

Grondin, Jean. "Das innere Ohr." In *Denken der Individualität: Festschrift für Josef Simon zum 65. Geburtstag,* edited by Thomas Soren Hoffmann and Stefan Majetschak. Berlin and New York, 1995.

Grondin, Jean. *Sources of Hermeneutics.* Albany, N.Y., 1995.

Hamburger, Käte. *Wahrheit und ästhetische Wahrheit.* Stuttgart, 1979.

Michelfelder, Diane P., Richard E. Palmer, eds. *Dialogue and Deconstruction: The Gadamer-Derrida Encounter.* Albany, N.Y., 1989.

JEAN GRONDIN

GARDENS. [*To explore the aesthetics of gardens, this entry comprises two essays:*

> Historical Overview
> Gardens as Art

The first essay examines the various ways in which gardens have been understood aesthetically and culturally. The second essay explores the historical debates within aesthetics and Western art theories about whether and how gardens can be viewed as art. For related discussion, see Environmental Aesthetics; Landscape; Nature; *and* Picturesque.]

Historical Overview

Gardens may best be viewed as a concentrated form of what since the late nineteenth century has usually been termed landscape architecture, though its professional proponents concern themselves with a wide range of tasks in which regional planning and environmental control can often occupy more of their time than garden making. Yet, it can be argued that the garden still represents in essence the products of the profession.

The sites of landscape architecture are contested territories. They are contested in their making, in the study and analysis of them, and in the way in which they accommodate paradox. Even the modern professional name of landscape architecture betrays the uneasy site of its work between building and field. All these contestations have a bearing on garden aesthetics.

Gardens are contested in their making because so many people contribute to them. If a poem or a novel has one author and a sonata one composer, gardens such as Villa d'Este, Versailles, or Stowe are the result of interventions by committee: patrons, their literary or cultural "advisers" (humanists in residence or clubs of *virtuosi* gathered for a country weekend), their estate managers, architects, hydraulic engineers, sculptors, gardeners (from supervisor down to laborer).

Nor have those who study these creations a single perspective; nobody really trains as a garden historian. Inquiry usually begins in other disciplines. If art history and litera-

ture, within the modern academy at least, provided some of the early garden historians, now cultural geography, social and economic history, philosophy, material culture, sociology, anthropology, archaeology, and history of religions have entered the territory; each necessarily arrives with certain predispositions, if not preconceptions. The more perspectives brought to bear on the garden, the more its origins in many contributions and skills are reaffirmed.

Finally, the garden is among the arts a site of conflict or dialogue. The most important of these is its accommodation of both nature and culture; it is the only art—with the possible exceptions of dance and body painting—that actively seeks to involve living materials in its creations; even the dry, stone garden associated with Zen Buddhism is subject to the changing conditions of light and climate under which it is experienced. Consequently, gardens are ever changing or never finished. One tendency, therefore, of garden aesthetics is to concentrate on hard elements—buildings and stonework, perhaps even the results of large-scale earthmoving or floating of lakes—that do not appreciably change once completed and allow the illusion of a stable and coherent object of study.

The nature/culture dualism has a further resonance; humans alone among the animals make gardens, and because humans are themselves an intricate mix of natural and cultural elements, the gardens they produce arguably come to represent that human condition more completely than other artifacts simply because of their involvement of cultural and natural materials (organic and inorganic); hence, the long-sustained and cherished desire for gardens among all societies and conditions of men and women, as well as the way in which larger aesthetic concerns were focused on and through the garden (the sublime with the André Le Nôtren garden, for instance, the picturesque with the later English landscape garden).

But that is not by any means the only duality that lurks in the garden (again, it yields perhaps a unique concatenation of such potential paradoxes). Product and process have already been noticed; additionally, there are visual and verbal meanings; the competing claims for the garden as a fine and a useful art; an insistent physicality along with the traditional promotion of metaphysics—the garden as a site of ideas; the same, haptic reality and what the eighteenth-century French garden maker Louis Carrogis, known as Carmontelle, called "pays d'illusion"; gardens endlessly renewing and reinventing themselves at every period, yet constituting an atavistic activity; they continuously reenact an idea of the garden, yet do so with plants materials not yet come to full maturity; finally, the full experience of garden art, in both its making and its reception, melds subject with object.

Such a richly contested site is not likely to accommodate aesthetics unproblematically. Indeed, the point at which gardenists (Horace Walpole's useful term for those who make and those who otherwise participate in gardens) argued most strenuously for the garden to be accepted in the pantheon of the fine arts was also the very point when its "naturalness" was acclaimed. It was also the point at which the modern tendency to distinguish between the practical and the beautiful, between gardens as a space of art and gardening as horticultural technology, between the *potager* or vegetable garden and the *jardin d'agrément*, first asserted itself clearly.

Conventional histories (and aesthetic narratives) concerning gardens oppose the formal to the informal, the geometric to the landscape or picturesque, the French to the English; they date a rupture between these modes somewhere during the late eighteenth century in England. Although not wholly wrong, this construction was devised largely in the interests of the English invention of what Walpole called a "more perfect perfection"; this account ignored a considerable Italian influence on estate layouts (Hunt, 1996b) so that the "modern," "natural," and above all "English" landscape garden could be promoted as the climax of all human intervention upon the land. Although this claim is explicable within the cultural history of the late eighteenth century in Britain, its large-scale export to other countries is much more puzzling. Landscape architecture studies have not yet adequately grappled with the way in which the "English" garden was subtly retooled for consumption in France (by Jean-Marie Morel, among others), Italy (by Ercole Silva), Germany (by Christian Cajus Lorenz Hirschfeld), or the United States (by Andrew Jackson Downing).

The watershed in garden practice of the late eighteenth century effectively ensured that a cluster of crucial theoretical perspectives on garden making would be lost. One had seen the garden as an integral and integrated part of a larger landscape. Another saw the garden as an art of representation. Yet another argued that intervention on a given site would be driven by local conditions (topography, economic resources, social conditions), with the further implication that there was no one particular format of design that was superior to any others. The loss, or at the very least the substantial playing down, of these ideas coincided with a period when general systems of value in the arts (including gardens) surrendered to personal adjudications; the picturesque, endlessly theorized, yet just as continuously experienced as private consumption of landscape, aptly characterizes this period.

Virtually all images and discussions from the Renaissance to the early eighteenth century see the garden as one of a series of interventions in the landscape (bird's-eye country house portraits around the year 1700 are a prime example); this scale, often focused along an axial vista, diminished in intricacy, concentration, and decorative intention as the eye is led away from the house. This commonplace phenomenon prompted two mid-sixteenth-century humanists, Ja-

copo Bonfadio and Bartolomeo Taegio, to invent the term *third nature (terza natura)* to describe the relationship of fine gardens to their surrounding in both agrarian and uncultivated land (see Hunt, 1996b). They almost certainly coined the phrase with an eye on Cicero's identification of the culturated landscape in his treatise *De natura deorum,* as "second nature"; in its turn, this world of fields, human habitations, and infrastructure was implicitly distinguished from the first nature of world of the gods (perhaps our "wilderness").

It was precisely from this idea and practice of a graduated landscape that the "English" landscape garden cut itself off. Treatises no longer bothered to discuss the relationship of the designed garden to its surroundings. A designer such as Lancelot "Capability" Brown was credited with sweeping away all immediate gardens and bringing the larger landscape of grass right up to the very walls of the house; the view outward, too, was supposed to be of uninterrupted scenery. In fact, the extent of the Brownian revolution has been exaggerated; moreover, its vogue was short-lived. The zoned and variegated landscape came again into favor during the early nineteenth century, but without any of the conceptual underpinnings that had characterized its earlier manifestations.

The major loss was an idea of the garden as representing in its own terms the worlds of second and first nature. The third nature had been understood as reflecting, abstracting, epitomizing the larger world beyond. This role of garden art was expressed and implied so frequently and yet so casually that one tends to miss its significance. John Worlidge stated it clearly, however, in *Systema Agriculturae* (London, 1669): "Gardens, orchards, parterres, avenues . . . represent unto us epitomized, the form and idea of the more ample and spacious pleasant fields, groves, and other rustick objects" (folio D1 verso). Beyond local manifestations of this representation—grottoes as caves, labyrinths as the wild wood, fountains as springs, lakes as the sea (complete with statues of Neptune), pergolas as forest walks—the garden was also seen to represent its surrounding topography and locality, its owner's status, and his/her ideology. "As is the gardener, so is the garden," wrote Thomas Fuller in 1732.

Fundamental to this perception of the garden was the belief that local circumstances would dictate the extent and modes of representation. Elaborate gardens—with *parterres de broderie* or topiary—were not necessarily the acme of design; some localities, some estate owners, would demand less intensive reworkings of their land. Andrea Palladio, for instance, virtually ignored the immediate garden in the interests of privileging an agrarian territory that better represented both his clients and their position in the Veneto landscape. This relativism became most pronounced around the year 1700, with such spokesmen as Anthony Ashley Cooper, third Earl of Shaftesbury (see Leatherbarrow, 1984), John Beale, or even A. J. Dezallier d'Argenville.

This openness to different ratios of artifice to nature according to the circumstances, along with ideas of a variegated or zoned landscape and of a garden's representative function, were effectively highjacked by the proponents of the so-called English landscape garden. They promoted one mode—the "natural"—at the expense of all others and worked vociferously to denigrate "French" (i.e., European) styles. This "new," "modern," and "natural" parkland was deemed to represent England and Englishness (above all English notions of liberty). Typified by the work of "Capability" Brown and conceptualized by Walpole and Thomas Whately, the landscape garden chose both to forget its role as only one of several natures and to blur the scope of its representation. For if, as Sir Joshua Reynolds shrewdly recognized in his thirteenth *Discourse,* the new garden was judged by its success at representing or imitating nature, its artfulness would effectively be disguised, even eliminated. The garden would then become, according to Sir William Chambers, no different than the common fields.

This démarche was crucial. What was lost was the compelling notion that art mediated natural (along with cultural) forms the better to instruct garden visitors in their understanding of the larger worlds outside the garden, a notion that had sustained garden thinking since the Renaissance. It was replaced by the essentially sterile debate (still flourishing in the discourses of landscape architecture) between those who privileged art in the garden and those who saw it as simply nature. Landscape architecture during the nineteenth century became an eclectic battle of styles that culminated in the strikingly artificial confrontation of "natural" gardenists such as William Robinson and exponents of the architectural, "formal" garden such as Sir William Blomfield.

The idea of the garden as a beautiful place, a *locus amoenus,* is ancient. What has been in question are the active, strategic ingredients of its special status, a debate that stems precisely from the contested nature of the garden where practical concerns (management, economic viability, horticultural needs) intersect with other values (the garden as sacred place, as glorification of its owner), as well as from the condition, antithetical to modernist aesthetics, that no garden is ever "complete" or coherent at any one time. It might be noted, however, that the emphasis on enclosure as endemic to the garden (and enshrined in its etymologies) may go some way toward promoting coherence.

The aesthetics of the garden has been most prominently in question when its proponents have urged its status as fine art. This occurs sometimes when garden artists have felt in full command of the scene of making (whatever style they worked in: Bernard Palissy in *La recette véritable* [1563], Jacques Boyceau in *Traité du jardinage* [1638], Walpole, or Whately). It also emerges whenever its artistry is in danger of being marginalized by appeals to nature—in the eighteenth century, or in the wake of the modern ecological

movement. The former produced the enormously popular picturesque taste that allowed the consumption of natural scenery as if it were art but often without physical intervention to change it. The latter, still under way, has produced a cluster of reactions: a professional as well as popular recourse to the garden once again; appeals to land art as cognate with landscape architecture; and arguments from both philosophers (Mara Miller, 1993) and historically orientated theorists (such as Hunt) that urge the status of the garden as art.

BIBLIOGRAPHY

Beale, John. "Draft Plans for 'A Physique Garden' and 'A Garden of Pleasure'" (1659). In *Culture and Cultivation in Early Modern England: Writing and the Land,* edited by Michael Leslie and Timothy Raylor, Appendix 3, pp. 226–231. Leicester, 1992.

Carrogis, Louis, called Carmontelle. *Jardin de Monceau, près de Paris.* Paris, 1779.

Hunt, John Dixon. "La représentation dans l'art du jardin." In *Hypothèse pour une troisième nature,* edited by Bernard Lassus. Paris, 1992.

Hunt, John Dixon. "Imitation, Representation, and the Study of Garden Art." In *The Art of Interpreting,* edited by Susan C. Scott, Papers in Art History from Pennsylvania State University, 9, pp. 198–215. University Park, Pa., 1995.

Hunt, John Dixon. "Paragone in Paradise: Translating the Garden." *Comparative Criticism* 18 (1996a).

Hunt, John Dixon. *Garden and Grove: The Italian Renaissance Garden in the English Imagination, 1600–1750.* New rev. ed. Philadelphia, 1996b.

Hunt, John Dixon, ed. *Garden History: Issues, Approaches, Methods.* Dumbarton Oaks Colloquium on the History of Landscape Architecture, 13. Washington, D.C., 1992.

Leatherbarrow, David. "Character, Geometry and Perspective: The Third Earl of Shaftesbury's Principles of Garden Design." *Journal of Garden History* 4 (1984): 332–358.

Miller, Mara. *The Garden as an Art.* Albany, N.Y., 1993.

Moore, Charles W., William J. Mitchell, and William Turnbull, Jr. *The Poetics of Gardens.* Cambridge, Mass., 1988.

Walpole, Horace. *The History of the Modern Taste in Gardening.* New York, 1995. See the introduction by John Dixon Hunt.

Weilacher, Udo. *Between Landscape Architecture and Land Art.* Translated by Felicity Gloth. Basel, 1996.

Whately, Thomas. *Observations on Modern Gardening.* Dublin, 1770; reprint, New York, 1982.

JOHN DIXON HUNT

Gardens as Art

Gardens have received very short shrift from philosophers. This is especially true of the past two hundred years, that is, since aesthetics has become a recognized field of inquiry. (Plato and Aristotle also ignored gardens in their writings on art, but in their case it is at least partially attributable to the fact that ancient Greece had no pleasure gardens.) Many factors have contributed to this dearth of philosophical attention, particularly the general preferences that philosophers have shown for arts that privilege sight and sound over touch and smell, for arts that are devoted to one

sense over *Gesamtkunstwerke* such as gardens, and for works that can be recognized in a single final form that can be taken for granted and referred to in writing, as opposed to a form that changes constantly and even unpredictably over time and in ways that are beyond the artist's control. Similarly, in spite of some interest in (anonymous) folk arts and "primitive" art, especially recently, the philosophical community has been more interested in works by single authors or artists, which gardens rarely are. Although a few gardens have been made by individuals, many others have been collaborative efforts; evidence regarding an individual's contributions and even the extent to which an individual's plans were carried out is often nonexistent or unreliable, and in even the most clear-cut cases, nature changes the gardenist's plans in ways that can be difficult to assess. In addition, although they may include the sorts of culture-specific stylistic shaping and the moments of individual genius that a Hegelian or Romantic theory looks for, gardens often seem to represent diffuse symbolic, even archetypal, schemes. Such schemes, often religious in origin and so pervasive as to transcend any individual's vision, serve as models of the cosmos, showing the difference between the original chaos and *maart*, the new correct order established at the time of creation, in ancient Egypt; illustrating a view of nature as unfailingly lovely, beguiling, and mirroring human feeling (even as its continuous change brings sadness), in Heian (tenth-century) Japan; revealing the Paradise to come in Islamic Persia and Mughal India and both Paradise and the Garden of Eden forsaken through sin in Europe and colonial Mexico.

Surely the main reason gardens have been ignored by philosophers, however, is the utter disparity between the sources of aesthetic value in gardens and the principal notion(s) of aesthetics as it has developed during the modern period: aesthetic distance or disinterest and the "autonomy of the work of art." Garden aesthetics is deeply grounded in a twofold relationship with environment, its and ours. For us, the environment is defined not only in reference to individuals, but to our species: the garden is that which promises to give us what we need as a species; its appeal is linked to our very survival. The beauty and formal regularity of a garden can be deeply utilitarian: obviously problematic for modern theories of disinterest are gardens like those of the Maya, elaborations of agriculture incorporating edible flowers and maize in raised beds surrounded by rectilinear canals used for irrigation, for raising fish to eat, and for the production of fertilizer.

Thomas Leddy (1988) has rightly pointed out that the same sort of recent moves in postmodern art and philosophy of art that have paved the way for environmental art and public art also bring gardens into the realm of art. Yet, gardens are more intimately and delicately dependent on the physical environment than any other art, even environmental art, even architecture. The aesthetic values of gar-

dens are largely, though not exclusively, a function of three separate kinds of relation to the environment: mutual comparison, dependence, and contrast. The three are brought into play both in terms of the garden's existence as a complex arrangement of physical and biological components with their own requirements and in terms of aesthetic significance as it is assigned to or discovered in features with survival value to human beings.

Gardens literally comprise a (relatively small) part of the landscape, while also being themselves comprised of a section of it. Garden and environment are contiguous, four-dimensional, and usually on the same scale, even when they appear not to be; that is, what appears to be a tree, a rock, or a pavilion usually *is* a tree, a rock, or a pavilion; if it looks like a pavilion we might be able to sit on, it usually is. (There are a few exceptions, most of them provided by China and Japan, to this generalization about the scale of the garden being identical to the scale of the real world: both in the form of miniature, tray-sized gardens, which are far smaller than human scale, and in the form of miniature landscapes modeled on Song-dynasty Chinese landscape paintings, such as that at Daitoku-ji in Japan, in which the appearance of a vast landscape at a great distance is imitated in small-scale rocks. In this latter case, however, unlike a painting, a human being can still enter the space physically, although to do so would be to break the illusion of the vastness.) Many garden styles take advantage of this in order to include scenery from beyond the garden precincts as part of the visual and conceptual effect. At the extreme are the gardened terraces of the Inca found at Ollantaytambo, from which the whole land seems to be a garden; a more moderate version is the Japanese *shakkei*, "borrowed scenery," but the principle is familiar from many eighteenth-century British gardens, especially after the adoption of the ha-ha, an abrupt ditch that serves as an invisible barrier between the garden proper and the pasture lands, allowing them to be included in an uninterrupted view of the garden but preventing the animals from wandering in and doing damage.

Many of the formal indicators of distinct and distinctive status and significance found in other artworks—the flatness of even a trompe l'oeil painting, the hardness or immobility of an otherwise realistic representational sculpture—are absent in the garden, so it can be much more difficult to recognize that one is encountering a work of art. Physically, visually, conceptually, one moves smoothly back and forth between the garden proper and its environs—so smoothly that it might be unnoticeable were it not for the walls, fences, and gateways, crucial features that must be erected in order to obstruct, and thus regulate, entry and exit, as well as to draw attention to the fact that there is an artificial construct, a work of art, there. Indeed, as Anne van Erp-Houtepen (1986) has shown, the very words for "garden" in the Indo-European languages (Eng. *garden, yard;* Ger. *Garten;* Fr. *jardin;* Lat. *hortus;* Sp. *huerto;* Itn. *giardino,* from

Indo-European *gher,* "fence," and *ghort,* "enclosure") come from one narrowly construed conception of control: fencing in or enclosing.

The walled garden is not exclusively a European phenomenon: the ancient "paradise" of King Cyrus described by Xenophon comes from Old Persian *pairidaeza* or *pardes,* "fence," and then "enclosure or garden," and traditional Chinese and Japanese gardens of all types—the imperial and the scholarly gardens of China, the private pleasure gardens of Japanese villas and the dry rock gardens of Japanese Buddhist monasteries—have also been enclosed. At least in the case of Japan it is probably the wall as much as the kind of planting and arrangement of space that distinguishes the garden from other kinds of sacred or symbolic space, such as shrine precincts. (Although some shrine precincts, such as Ise, employ concentric rings of fencing, others are open on all sides to the surrounding landscape, whether it be natural or urban.)

Cross-culturally, however, what is essential to the garden is not enclosure per se but control, and control is not merely a matter of walls permitting or denying access. (In fact, it is worth noting that what for us moderns seems to be virtually the antonym of the garden, namely, the concept "town," are, in Old Slavic *(gradu)* and Russian *(gorod,* "fence, enclosure, town," as in "Belgrade" and "Novgorod"), derived from the same Indo-European roots *gher* and *ghort,* and the English word *town* comes from the same concept, "fence," taking a different root (Old English *tun,* Old High German *tun*). This root, too, can lead to *yard,* as in the Scots dialects' term *tun* for "farmyard."

Gardens are often places that situate viewers in positions of visual control over vast landscapes, allowing an illusory symbolic control over the world external to the garden; this is seen in cultures as disparate as the Inca, the Italian Renaissance, and in at least one Chinese imperial garden, that of the Qing emperor Kang Xi at Jehol (although typically, Chinese gardens are enclosed and inward-turning). Control can also be expressed by geometric arrangement, such as that found in straightedged rectangular pools and canals of ancient Egyptian, Persian, Indian, French, and Dutch gardens and the regularly spaced plants that surround them.

But this overlapping and coexistence of gardens and surrounding landscape should not be taken to imply parity between them. Gardens and their human respondents are dependent on the larger environment for their physical survival, at the minimum for the air and light they require, and to a lesser extent (because human beings can intercede, adding, eliminating, amending, or substituting as needed or desired) for rain, pollinators, soil, and so on. This doubtless accounts for the peculiar fact that gardens do not appear in all cultures. Although there are traditions of grand gardens all over the world—most famously the Aztec, British, Chinese, Dutch, ancient Egyptian, French, German, Inca, Indian, Italian, Japanese, Mayan, Persian, and ancient

GARDENS: Gardens as Art. *Japanese Garden;* Daigoji Temple, Sanpoin, Japan. (Photograph courtesy of Bijutsu Shuppan-SHA/Art Resource, New York.)

Roman—and these have also given rise to vernacular offshoots and folk traditions, there are also many cultures without gardens: ancient Greece is perhaps the most surprising, and there are many others, including much of precolonial North America and virtually all of precolonial Africa other than Egypt. Unlike Paul Weiss's "Nine Basic Arts"—architecture, poetry, sculpture, painting, "musicry," story, music, theater, and dance—gardens evidently are not part of the fundamental artistic urge of humankind. An offshoot of agriculture, they require not only knowledge of planting, irrigation, and soil amendment techniques, but also permanent settlements, control over land of the kind that comes with private property, and certain habits of mind that agriculture induces—assumptions about permanence, deferred gratification, control over the natural environment, and the difference between domesticated and wild or chaotic spaces (*maart* and chaos, etc.). Such requirements make gardens inconceivable for nomadic societies of herders and hunters and gatherers, and for those for whom sufficient control of the physical environment is out of the question—peoples in regions with short growing seasons

(the Inuit) or extreme aridity, or, conversely, in jungles where vegetation grows too profusely.

More important from an artistic point of view, both gardens and the people responding to them find in the environment some of the basic (positive) qualities of the garden aesthetic—the particular quality of light, the intensity or vagueness of shadows and shade, the humidity and physical atmosphere, the degree of quiet and the kinds of sounds, the types of vegetation (lawns and tall deciduous trees or cactus and succulents). Much of what human beings find valuable in their actual environments is also valuable within gardens: terra firma, protection from extremes of weather, and the emotional and cognitive values identified by environmental psychologists Stephen and Rachel Kaplan (1978): complexity or diversity, coherence, mystery, and legibility. Each garden will reflect, even magnify, the strengths of its environment as positive values.

But not all of the qualities of the environment are positive. Many are disagreeable, and the garden takes on formal qualities that contrast with the environment and counteract or compensate for its perceived deficiencies: rocks from

Lake Tai are hauled hundred of miles to the gardens of Chinese scholars; marble for statues is imported to France; flowers from the Himalayas are cultivated in Sussex and from Turkey in Holland; gardens in hot, arid climates tend to highlight water, those in sunny ones, shade. Contrast is a special case of dependence on the environment, because it is logically defined in terms of the environment rather than within the work itself, as is the case with most other arts, like music, where contrast is established between different movements of a piece.

It goes without saying that these three relations of garden to environment—comparison, dependence, and contrast—are informed and shaped by national and local culture, historical period, individual taste, or artistic "genius," and so on. But the environmental impacts on garden aesthetics are so strong that gardens in similar climates (for instance) often have a great deal in common with each other in spite of tremendous artistic and cultural differences. The covered walkways surrounding courtyard gardens in New Orleans, Louisiana, the cloisters of France, and Malinalco, Mexico, may share common ancestors in Spain (the Alhambra) and ultimately Persia; the not entirely dissimilar walkways in the gardens of Suchou, in southern China have no such common influence, but share the goal—to get out of a sun that is often too hot, but that one wants nonetheless to see and, indirectly at least, to feel. As a result, they may provide experiences that are biologically and even aesthetically similar.

This intense and intimate—and inescapable—complex of relations to the environment and to the terms of human survival within it poses severe challenges for the long-entrenched aesthetic notions of distance and disinterest. Gardens are better handled by an aesthetic of engagement, such as that developed by Arnold Berleant (1991), and by theories that recognize the reciprocity between subjects and "objects" or environments.

Gardens and (Western) Theories of Art. Even so, there are many good reasons for considering gardens an art kind. They have beautiful and/or sublime qualities, they produce (or induce) aesthetic experience, they are the product of a forming human intention (at least to some extent), they have frequently been created by persons who are regarded as great artists (Kobori Enshu, William Kent, Alexander Pope), they have often been considered art (or in cultures lacking that concept, have been considered akin to other things that one would consider art, such as painting and architecture).

By more theoretical criteria, too, there should be no problem including gardens as an art kind; gardens fit easily into a number of contemporary theories. Monroe Beardsley's generous principle, for instance—"What establishes an art kind, on my view, is that a good many of its individual instances are created with the intention (perhaps among others) of making aesthetic experience obtainable"—would clearly include gardens. Similarly, George Dickie's "institu-

tional definition," according to which a work is a work of art if "some person or persons acting on behalf of a certain institution (the artworld)" designates it a "candidate for appreciation" (*Art and the Aesthetic: An Institutional Analysis,* 1974) and Arthur C. Danto's (1981) modification requiring a theory behind the work and a general sense of "aboutness," would also accommodate gardens, at least during some periods—although Dickie's theory would preclude most non-Western gardens in that most cultures, although they have certainly had artworks and artistic practices, have not had a category equivalent to "art" prior to the adoption of Western concepts. Donald Crawford's (1983) "dynamic interaction of natural forces and artifactual constructions" is exemplified in gardens. More pertinently, Leddy recommends the reintroduction of gardens into the category of art based on Rosalind Krauss's notion of an "expanded field" for sculpture in the postmodern period that would include "site construction," and points out that her reasoning "allows us . . . to think of objects that combine landscape and architecture as art, although these things were previously ideologically prohibited as art" (Leddy, 1988).

Earlier theories are more problematic for gardens. Because Kant's theory of the aesthetic recognized two basic sources of aesthetic experience, art and nature, one would expect that virtually all Kantian aesthetic theory would be especially interested in the garden, which combines them; why should not the fusion of art and nature that is the garden be the quintessential art kind? Yet, the association of many of even their aesthetic qualities with biological needs, and the fact that as whole environments they surround us and are therefore difficult to perceive as an object (in spite of some styles, such as the Picturesque, whose aims are precisely the making of landscape into an object), make it impossible to view them with strict disinterest; thus, they are problematic for any Kantian theory. With Hegelian theory, too, there are superficial reasons to include them, and deeper reasons to exclude them. Recent garden histories have begun to show how thoroughly European garden design was integrated with major artistic movements, how it reflected the "spirit of the time" just as did music, painting, and architecture. This may seem to suggest that gardens would fit into G. W. F. Hegel's theory. (This point has been difficult to recognize because garden history has lagged behind the history of the other arts, largely because of the fact that, as living things, gardens change so much and so fast that plans, notations, and other documentary evidence, where they exist, tend to be wholly inadequate.) Yet, the inescapable physicality of gardens thwarts their inclusion among the arts expressive of Spirit in its later stages—that is, precisely when gardens were becoming most artistic in Hegelian terms.

The dearth of philosophical reflection on gardens is revealed in the lack of appropriate vocabulary for discussing gardens. In spite of the great gardening tradition in English-

speaking lands, English, at any rate, has no word for the garden's equivalent of "audience"/"reader"/"viewer," each of which brings with it assumptions that are misleading in regard to gardens: assumptions of distance between the artistic "object" and the receptive subject, of separation of the aesthetic experience from everyday life and from survival value, and of prioritizing one or two senses, the visual and/or aural, a move that is inappropriate to gardens, which are visual, aural, kinesthetic, haptic, and often olfactory in equal measure.

Gardens and Other Arts. The kinds and degree of importance of the relations of gardens to other arts vary with different periods and styles of gardens. But, in virtually all cultures that have gardens, they are closely related, in both theory and practice, to other arts. These relations are of many kinds: resemblance and shared tasks or capabilities, making use one of the other, allusion of one to another, the deliberate modeling of one on the other. This is especially true of architecture. Styles otherwise as different as the Italian Renaissance, the English landscape garden, and the Japanese domestic garden take the house as the starting point and create realms of mediation—admittedly of quite different kinds!—between the house and the landscape beyond, while French formal gardens have plans laid out with reference to a point of view from a particular room within the palace or château, so that the garden ultimately makes sense only when one takes account of this privileged perspective. The Chinese literati garden, ancient Roman domestic courtyards, and the medieval cloister garden are literally defined by the walls of the building. Both gardens and architecture sometimes use the other for decorative purposes or as a thematic anchor; small, unusually decorative or otherwise visually distinctive buildings—follies, pergolas, pavilions, arbors, teahouses—are included in the gardens of Japan and China, most of Europe, and India. Gardens and buildings also have some of the same purposes—protection from the elements (in different ways and to different degrees), creation of distinctive protected spaces for particular kinds of socializing, and, in some cases, the affordance of spectacular views of surrounding landscape: the high rock terraces of the Incas and the Italian villa gardens overlooking steep cliffs are in this respect very like modern skyscrapers and the towers of ancient forts (whose views are not, however, designed with aesthetic experience in mind); and they use each other for protection from the weather: a hedge or row of trees may provide a windbreak for the house, the house a wall that warms the plants growing in front of it.

Gardens have complex relationships with painting, too. Both have the ability to conjure up ideal landscapes. "Picturesque" gardens of eighteenth-century England were alleged to be arranged according to the same principles as paintings (ut pictura hortus, though Stephanie Ross [1985] has shown that the situation is far more complex than is al-

lowed); some Japanese rock gardens of the Muromachi period were composed like Song-dynasty Chinese ink landscape paintings; the walls of Pompeiian gardens were painted with landscapes and images of gods. Gardens are important subjects for painting in Japan from the Heian period onward, in China especially from the Song onward, in the eighteenth and nineteenth dynasties of ancient Egypt, in Mughal India, in Renaissance Italy, in the portraits of landed gentry of eighteenth-century England, among the Impressionists, in Aztec/mission sites such as Malinalco. Sculpture frequently appears in gardens (in Italy, England, the United States, France), as do rocks chosen for either their aesthetic or their religio-sacred qualities (Chinese, Japanese, Inca). Poetry is carved into rock or wood, occasionally written out, and alluded to by visual forms, as are, also, other forms of literature such as history and legend. Gardens have been written about extensively in novels as far-flung as Johann Wolfgang von Goethe's *Elective Affinities,* the eighteenth-century Chinese *Dream of the Red Chamber,* and Murasaki Shikibu's eleventh-century *Tale of Genji;* the latter is said to have inspired the moon-viewing platform at Katsura Detached Imperial Villa in Kyoto. Gardens are well known for housing and inspiring theatrical productions and musical concerts all over the world; indeed, their ability to incorporate actual persons makes them very like theater themselves, while their sense of rhythm is shared with music.

Finally, the histories of gardens are often analogous to those of the other arts. The preoccupation with enclosure and physical safety evident in European medieval architecture is also seen in medieval gardens; in both cases, this gives way during the Renaissance to a fascination with the mentally and culturally expanding possibilities of views over vast territories. The symmetries and repetitions of seventeenth-century French decor, musical forms, and dances are also to be found in their parterres. The fascination with new information and with the codification of knowledge, as well as with perspective per se as a means of organizing information, is found not only in literature but in knot garden designs as well. In Muromachi Japan, gardens become more abstract and "ideal" at the same time as painting (both under the influence of specific forms of Buddhism); later, one finds a resurgence of the influence of the literary classics, especially *The Tale of Genji,* in garden design at the same time as these works are appearing as printed publications and assuming new importance as subjects for painting.

Yet, the history of gardens does not invariably flow seamlessly along with the histories of the other arts. Late eighteenth-century Britons do not, for example, express the same horror at allegedly tyrannical forms of poetry and painting that Horace Walpole expresses at "tyrannical" styles of gardens; gardens, in this case at least, carry a political weight in their very form that in literature and painting are found only on the level of content, their explicit views.

"Modern" art generated virtually no comparable developments in garden design. (Postmodernism seems more promising, although limited to a very small avant-garde.)

In spite of the variety and intimacy of their relations to the other arts, and their many similarities among them, there are other crucial differences between gardens and these other arts, in addition to the issues of distance/disinterest and their capacity to express Spirit. They are intimately related to, even dependent on, a number of areas of human endeavor that, although "artificial," and thus related to the arts in Plato's sense, nonetheless are the antithesis of art in the modern sense. They depend on the aforementioned agriculture, and engineering for bridges, pools, and fountains (which is not necessarily problematic, because architecture also needs it). As appealing arrangements of the actual physical environment that one can participate in, they resemble golf courses, parks and zoos, cemeteries, nature preserves, and hunting preserves. In their capacity to represent symbolically other realms of existence—paradise, the land of the immortals, and so on—they are akin to sacred spaces, and they may in some cases be meant to link us to these other realms, or at least to induce a mild alteration in state of consciousness.

In addition, there are a number of aesthetic issues that do not pertain to gardens. There are no forgeries of gardens, no mass production, no inauthenticity, no replicas. (In fact, there is one garden that is an exact replica, the Astor Court Garden at the Metropolitan Museum of Art in New York City, which is also, perhaps for that very reason, inauthentic, the exception that proves the rule.) The identity of a garden depends primarily on spatial location—to a much greater extent than is the case with gardens' closest artistic relatives, site-dependent sculpture, architecture, and contemporary environmental art, all of which can at least be recognized by their visual appearance even after being taken out of their original site. Gardens, on the other hand, even the self-enclosed ones that do not rely on "borrowed scenery" or views of the surrounding areas as do Inca, some Japanese, and some Italian villa gardens, are so dependent on their sites, and especially their microclimates, that they cannot in fact be duplicated, much less moved to a different site; not only that, most gardens change so much over time and so frequently that they have no single final form, no consistent set of formal features by which they may be identified. (Accurate notation, which on Nelson Goodman's [1976] theory would serve to identify a given work of art, while theoretically plausible, is extremely problematic in practice for actual gardens; given the rate of natural change in gardens, it is not clear that any plan is in fact adequate to the description of any living garden, at least over any significant length of time.) In this they are like improvisational performance. But whereas the latter derives the unity underlying its varying forms from the intention of the (human) performer (admittedly often in response to the audience and/or events occurring in the environment), human intention plays a far smaller role in the garden and is much less decisive in determining its effects. Although gardens are often spoken of as a "collaboration" with nature, the term is appropriate only if one recognizes that it implies in the garden a response or an interplay of a very different kind than is found among human collaborators, because when natural forces "collaborate," they rarely express intention, have goals, or exercise judgment, regardless of how actively they may affect—indeed, often counteract—the intentions, goals, and judgments of the gardenist. (Although "nature" may have goals for the garden, they pertain to components or materials of the garden, not to the work as a whole, nor to it as art; the squirrel nibbling tulips may be presumed oblivious both to the garden as a whole and to its specifically aesthetic effects.)

The difficulty in pinning down a single final form leads to another set of issues as well: gardens are less product than process, and less object than environment. As process, they thematize issues of temporality, control and response, subjectivity and objectivity; they resist the power and authority of the viewing/making subject; perhaps most important, they mediate in a dynamic way the tensions within a given culture between the oppositions or dichotomies with which it is preoccupied: polarities between human and divine, living and dead, order and chaos, public and private, wild and domesticated, natural and artificial, inner and outer, public and private, static and changing, animate and inanimate, subjective and objective, contingent and absolute, sacred and secular. As environment, they surround—literally helping to "de-fine"—the subject, even while offering a "grounding" that is at once personal, cultural, and natural.

[*See also* Japanese Aesthetics.]

BIBLIOGRAPHY

Berleant, Arnold. *Art and Engagement.* Philadelphia, 1991.

Berleant, Arnold. *The Aesthetics of Environment.* Philadelphia, 1992.

Carlson, Allen. "Appreciation and the Natural Environment." *Journal of Aesthetics and Art Criticism* 37.3 (Spring 1979): 267–275.

Chase, Isabel Wakelin Urban. *Horace Walpole: Gardenist: An Edition of Walpole's "The History of the Modern Taste in Gardening," with an Estimate of Walpole's Contribution to Landscape Architecture.* Princeton, N.J., 1943.

Chatterjee, Margaret. "Some Reflections on the Garden as an Art Object." *Diotima* 14 (1986): 20–27.

Coffin, David R. *The English Garden: Meditation and Memorial.* Princeton, N.J., 1994.

Crawford, Donald. "Nature and Art: Some Dialectical Relationships." *Journal of Aesthetics and Art Criticism* 42.1 (Fall 1983): 49–58.

Danto, Arthur C. *The Transfiguration of the Commonplace: A Philosophy of Art.* Cambridge, Mass., 1981.

du Prey, Pierre de la Ruffinière. *The Villas of Pliny from Antiquity to Posterity.* Chicago, 1994.

Garden Literature: An Index to Periodical Articles and Book Reviews (Boston) 1.1 (January–March 1992).

Ginz, Claude. "Neoclassical Rearmament: The Poetic-Philosophic Garden in Scotland Cultivated by Ian Hamilton Finlay." *Art in America* 75 (February 1987): 111–117.

Goodman, Nelson. *Languages of Art: An Approach to a Theory of Symbols*. 2d ed. Indianapolis, 1976.

Hunt, John Dixon. *The Figure in the Landscape: Poetry, Painting, and Gardening during the Eighteenth Century*. Baltimore, 1976.

Hunt, John Dixon, and Peter Willis, eds. *The Genius of the Place: The English Landscape Garden, 1620–1820*. London, 1975; reprint, Cambridge, Mass., 1988.

Kamal, Salim, and Ivan Gaskill, eds. *Landscape, Natural Beauty, and the Arts*. Cambridge and New York, 1993.

Kaplan, Stephen, and Rachel Kaplan, eds. *Humanscape: Environments for People*. North Scituate, Mass., 1978.

Keswick, Maggie. *The Chinese Garden: History, Art, and Architecture* (1978). 2d rev. ed. London and New York, 1986.

Leddy, Thomas. "Gardens in an Expanded Field." *British Journal of Aesthetics* 28.4 (Autumn 1988): 327–340.

Lippard, Lucy R. "Gardens: Some Metaphors for a Public Art." *Art in America* 69 (November 1981): 136–150.

MacDougall, Elisabeth B., and Richard Ettinghausen, eds. *Dumbarton Oaks Colloquium on the History of Landscape Architecture 4: The Islamic Garden*. Washington, D.C., 1976.

Miller, Mara. "Knot Gardens and the Genesis of the Scientific Attitude." *Comparative Civilizations Review* 18 (Spring 1988): 58–90.

Miller, Mara. *The Garden as an Art*. Albany, N.Y., 1993.

Morris, Edwin T. *The Gardens of China: History, Art, and Meanings*. New York, 1983.

Moynihan, Elizabeth B. *Paradise as a Garden: In Persia and Mughal India*. New York, 1979.

Peterson, Jeanette Favrot. *The Paradise Garden Murals of Malinalco: Utopia and Empire in Sixteenth-Century Mexico*. Austin, Tex., 1993.

Prest, John. *The Garden of Eden: The Botanic Garden and the Re-Creation of Paradise*. New Haven, 1981.

Protzen, Jean-Pierre. *Inca Architecture and Construction at Ollantaytambo*. New York and Oxford, 1993.

Ross, Stephanie. "*Ut Hortus Poesis*: Gardening and Her Sister Arts in Eighteenth-Century England." *British Journal of Aesthetics* 25 (Winter 1985): 17–32.

Ross, Stephanie. *What Gardens Mean*. Chicago, 1998 (forthcoming).

Slawson, David A. *Secret Teachings in the Art of Japanese Gardens: Design Principles, Aesthetic Values*. Tokyo and New York, 1987.

Stein, Rolf A. *The World in Miniature: Container Gardens and Dwellings in Far Eastern Religious Thought*. Translated by Phyllis Brooks. Stanford, Calif., 1990. Originally published as *Le monde en petit: jardin en miniature et habitations dans la pensée religieuse d'Extrême-Orient* (Paris, 1987).

Stone-Miller, Rebecca. *The Art of the Andes from Chavin to Inca*. London, 1995.

Strong, Roy. *The Renaissance Garden in England*. London, 1979.

van Erp-Houtepen, Anne. "Etymological Origin of the Garden." *Journal of Garden History* 6.3 (July–September 1986): 227–231.

MARA MILLER

GAY AESTHETICS. Gay aesthetics is an analytic category of recent origin. In 1969, in the wake of the civil rights and feminist movements, activists made the term *gay liberation* a rallying cry against the political and social stigmatization of homosexuals. Since that time, the term *gay* has come to be widely—though not exclusively—associated with male homosexuality. The novelty of the term *gay* does not mean, however, that the category of "gay aesthetics" is without history. Quite the contrary, the constitution of this category was itself an outcome of historical forces and was carried out largely in relation to history, with modern ideas being justified by reference to the past.

Historical Roots. Like the related category of lesbian aesthetics, the idea of gay aesthetics originated in the nineteenth century. Although homosexual activity is present throughout recorded history, only in the late nineteenth century did such activity come to be construed as a specific personality type. Drawing on broader efforts to understand behaviors by reference to biological or psychological type, scientists explained homosexual activity by proposing a third sex located between—or resulting from the inversion of—male and female. Among the supposed characteristics of the male "invert" were a feminine appearance and an ability or interest in feminine pastimes, which included the production and consumption of novels, music, and painting. (This history was influentially proposed in Michel Foucault's *History of Sexuality* [1976] and is rehearsed in more detail in David Greenberg's *The Construction of Homosexuality* [1988].) The same habits of mind prompted scientists—Sigmund Freud most prominently—to transform Romanticism's vague notions of art as a form of procreation into a concrete typology that explained art making as the product of diverted or excessive sexual impulses. By the turn of the century, therefore, artistry was linked to sexuality in general—and to homosexuality in particular—through medical discourses that focused not (as may be the case with lesbian aesthetics) on iconography, but on sensibility. The history of gay aesthetics, therefore, is not the history of homoerotic subject matter, but the history of how male homosexuality has been allied with certain styles of art production and consumption.

The scientific linkage of the artist and the invert prompted a diversity of responses by men who found themselves so labeled. Certain poets—the American Walt Whitman, along with such English followers as Edward Carpenter and John Addington Symonds—created a political identity around the idea that the intermediary perspective of "inverts" made them uniquely suited to effect a reconciliation between different classes of men and between men and women (on Whitman, see Martin, 1979; on the Englishmen who made Whitman's ideas more explicit, as both a political program and an affirmation of homosexuality, see Weeks, 1977). This conception of a gay aesthetic was eclipsed, however, in 1895 when the sensational trials of the critic and dramatist Oscar Wilde turned the figure of the refined aesthete into the stereotype of the homosexual.

After three trials, Wilde was convicted of "gross indecency" with another man and sentenced to two years at hard labor. To make their case, prosecutors quoted extensively from his own writings and from the poetry of other writers in the Aesthetic movement, cementing the link between homosexuality and this particular form of artistic sensibility. In the wake of this scandal, publishers in En-

gland and the United States shied away from any reference to homosexuality, canceling release of work even by more Whitmanesque writers, thus rendering the image of the homosexual as aesthete virtually unchallenged until the gay rights movements of the late 1960s.

The stereotype of the homosexual aesthete had enormous influence for both good and ill. On one hand, pejorative associations of artistry with effeminacy prevented many men from evincing interest in the arts, at the same time provoking within the art world a compensatory embrace of machismo—exemplified in the veneration of Pablo Picasso and Jackson Pollock—with all its attendant misogyny. On the other hand, the linkage of art and homosexuality provided other men with a way of understanding their outsider status, undergirding the careers of important authors (Marcel Proust, Jean Genet), painters (Charles Demuth, Duncan Grant), and photographers (George Platt Lynes, Cecil Beaton). The names cited here represent just a few examples of the men who, in the first half of the twentieth century, turned the idea of the artistic homosexual into something that, with historical hindsight, might be identified as a "gay aesthetic," characterized not just by homoerotic subject matter, but also by stylistic manifestations of such "feminine" attributes as sensitivity, sensuality, and refinement. The canonic status of some of these figures—the authors, in particular—sustains them as exemplars of the homosexual type for literary critics proposing theories of gay aesthetics as an "outlaw" sensibility rooted in identification with the "feminine" (Leo Bersani's *Homos* [1995] and Kaja Silverman's *Male Subjectivity at the Margins* [1992] exemplify this trend).

The linkage of homosexuality with aesthetics animated not only the careers of many well-known artists, but also more vernacular visual expressions of homosexual identity. After the Wilde trial, the persona of the "dandy"—the stylish, urban observer of modern life—shifted from its association with risqué heterosexuality to become a sign of homosexuality. Easily compatible with the attributes of aesthetic sensitivity, the dandy's association with fashion resulted in the stereotyping of certain art-allied professions as quintessentially gay, among them the fashion designer, the window dresser, and the interior decorator. As with all stereotypes, this one unwittingly served to concretize what it was intended to stigmatize, opening career paths for many gay men in fields rendered taboo to heterosexuals and giving visual definition to certain commodities and designs as symbols of gay identity. The image of Michelangelo's *David,* for instance—reproduced as signage for commercial establishments and as plaster casts for interior decor—became one such commodity, propelled both by its obvious embodiment of a standard of masculine beauty and by the association of Michelangelo's biography with the history of homosexuality.

The latter point is important to emphasize, for it demonstrates how the idea of a homosexual type, once invented, was projected back chronologically to propose historical evidence for contemporary ideas. Michelangelo's fame made him one of the first figures to be analyzed according to the emerging notions of homosexuality. As early as 1871, the Aesthete Walter Pater analyzed Michelangelo's genius as a fusion of masculine and feminine characteristics: "sweetness" and "strength," "energy of conception" and "a loveliness found usually only in the simplest natural things." In 1883, Symonds's biography of Michelangelo went so far as to report that, though chaste, the artist "was one of those exceptional, but not uncommon men, who are born with sensibilities abnormally deflected from the ordinary channel. He showed no partiality for women, and a notable enthusiasm for the beauty of young men" (James Saslow's *Ganymede in the Renaissance* [1986] offers an updated study of Michelangelo's sexuality in its Italian Renaissance context). Analogous efforts focused on art critics, most notably the eighteenth-century German Johann Joachim Winckelmann. For Pater, Winckelmann was the prototype of the aesthete, drawn to Greek sculpture and philosophy by an enthusiasm that "burns like lava." "That his affinity with Hellenism was not merely intellectual, that the subtler threads of temperament were interwoven in it, is proved by his romantic, fervent friendships with young men," Pater explained (1867). (Pater's idea of Winckelmann as the original aesthete/homosexual lingers—along with Victorian ideas of homosexuality as the blend of male and female—in Whitney Davis's [1994] analysis of the way Wincklemann's "dynamic of subjective, feminized 'witness' and objective, masculinized 'allusion'" became the model for the practice of art history.) Most influential among these historical projects was Freud's 1910 essay on Leonardo da Vinci as a case study in homosexuality. Drawing on extremely sketchy biographical information, Freud used Leonardo to exemplify his notion that homosexuality results when boys stall in what should be a temporary childhood stage of narcissistic eroticism where they identify with their mothers in loving themselves—this theory justified doctors in pathologizing homosexuals (and their mothers) through most of the twentieth century.

Whether intended as affirmative or pathologizing, however, the accumulation of such histories reinforced the linkage of the artistic and the homosexual personality, contributing to the belief in a gay aesthetic. Such ideas remained common both in mainstream representations of gay men and within the gay community through the 1980s. A study of gay newspapers' coverage of the secession of the gay and lesbian neighborhood of West Hollywood from the city of Los Angeles, California, identified numerous assertions that a gay-governed city would manifest a greater degree of aesthetic sensitivity, leading the researcher to conclude that the gay press "represented beauty as the rightful domain of gays" (discussed in Reed, 1996). Especially in anglophone cultures, this conflation of artistic personality and homosexual identity lingers in the popular imagination.

Identity Politics. The 1960s saw widespread changes in attitudes about both art and sexuality, changes that can be gauged by the career of Andy Warhol. Warhol began his career in the familiar mode of the homosexual artist, first making a name for himself as a fashion illustrator and then, around 1962 (and after his earlier overtly gay-themed work was rejected by galleries), achieving fame as an artist with the paintings of soup cans and soda bottles that became the paradigmatic imagery of the Pop art movement.

From the beginning, Pop was attacked by critics as, in the words of the New York *Village Voice* (1966), "a malicious fairy's joke." In 1968, a Time-Life book reviewing the sixties reported, "homosexual ethics and esthetics are staging a vengeful, derisive counterattack on what deviates call the 'straight' world. This is evident in 'pop,' which reduces art to triviality, and 'camp,' which pretends that the ugly and banal are fun" (for these and other references, see Joe Thomas's "What's Queer about Pop Art," 1996). What such remarks reveal—in addition to the prevailing homophobia of the era—is the grafting of a new characteristic onto the stereotype of the arty homosexual: a quality of subversive humor termed "camp."

First theorized in Susan Sontag's influential 1964 essay "Notes on Camp," this "sensibility" was defined as a "way of seeing the world as an aesthetic phenomenon," but with *aesthetic* defined in terms not of beauty but of artifice or style. Sontag's essay has been criticized for her insistence that camp be unwitting and concomitant blindness to its political potential (see David Bergman's introduction to *Camp Grounds*, 1993). It remains, nevertheless, a foundational document in efforts to describe gay aesthetics. Asserting that homosexuals, along with Jews, are "the two pioneering forces of modern sensibility," Sontag identified the influence of camp in a wide variety of cultural products, including Pop art. Ultimately, however, Sontag defines camp less as a mode of art production than of spectatorship: camp is a way of appreciating the spectacle of capitalism. "Dandyism in the age of mass culture," camp finds pleasure—often through humor—in the "failed seriousness" of products from Hollywood movies to Tiffany lamps. Tying this sensibility back to Wilde's humorous epigrams, Sontag connects camp to homosexuality by arguing that it "neutralizes" the "moral indignation" aroused in mainstream culture by sexual minorities. Later, Sontag (in "On Roland Barthes," 1982) and others (see especially R. K. Martin in Bergman, 1993) tied this Wildean tradition to the influential late twentieth-century literary and cultural critic Roland Barthes, extending the idea that this form of gay aesthetic is manifest primarily as critical spectatorship.

From this perspective, what threatened the art establishment about Warhol's work was the ironic light it cast on the Abstract Expressionist and Minimalist art of the era. Warhol's recycled news photos of automobile accidents and suicides, along with his rows of soup cans and Coke bottles, wittily implied that the fundamental concerns of this "high" art—from the metaphysics of violence to formal repetition—were at least as present in mass-cultural phenomena from newspapers to supermarket shelves. From this perspective, the aesthetic claims of high art appeared pretentious, while apparently frivolous aspects of mass culture became sources of aesthetic delight. The fact that Pop art, as befit its name, far exceeded the popularity of the museum-sponsored art discomfited artists and critics accustomed to imagining themselves the avant-garde of taste. In a climate where Americans were encouraged to perceive homosexuals—along with Communists—as a constant but secretive menace to American values, this sudden eruption of a subversive camp aesthetic was often characterized as a conspiracy of, in art critic Harold Rosenberg's words, "homosexual painters and their nonpainting auxiliaries."

The threat that camp posed to established values was reinforced by the simultaneous development in urban areas during the 1960s of subcultures that were openly "gay" rather than furtively homosexual. This too is registered in Warhol's work from the period—not his painting, which remained dependent on structures of exhibition and publicity associated with high art, but his films, which could be circulated in movie theaters catering to a gay clientele. It is not the explicit sexual content of Warhol's *Blow Job* (1963), *My Hustler* (1965), and *Flesh* (1968) that marks a shift in sensibility; many earlier artists had created private bodies of homoerotic imagery. Rather, it is the fact that by the midsixties his films took for granted—in both their narratives and their audience—an identifiable subculture just beginning to call itself "gay." The new term marked a self-conscious shift in the construction of sexual identity, which rejected pathologizing discourses of "homosexuality" imposed by medical authorities in favor of an affirmative embrace of a social identity whose very name implies a determination to be happy. The critical hostility aroused by Pop art cannot be separated from mainstream anxieties over the increasingly visible "gay" communities in and around major cities. One reaction among conservatives—ongoing to this day—was an effort to exert semantic control over homosexuality, as both an artistic and a social phenomenon, by defining it in ways that limit its scope of influence. Ironically, therefore, most of the literature proposing a coherent gay aesthetic arises from explicitly homophobic impulses to limit the creativity of gay people to, for example, the production of camp (the introduction to Hilton Kramer's *Revenge of the Philistines* [1985] is typical). Even where the motivation of delimiting a gay aesthetic is not homophobic, the effect is the same—if this is not obvious, try imagining its logical corollary: a single heterosexual aesthetic.

These tensions are clear in writings about art and homosexuality from the late 1960s to the early 1980s. Although mainstream opinion fretted that, in the words of the Time-Life author, "Increasingly, deviates are out in the open, par-

ticularly fashion and the arts," for many in the gay community of the 1970s the art world seemed to lag far behind vernacular culture in responding to the variety of new forms of gay identity. No museum approached the question of a possible gay aesthetic until 1982, when the New Museum in New York mounted an exhibition of work by gay and lesbian artists under the title *Extended Sensibilities: Homosexual Presence in Contemporary Art* (Cameron, 1982). While acknowledging that there is no such thing as a unitary gay lifestyle or attitude, the catalog proposed the existence of something called gay "sensibility," described as a kind of "aura" present, not in overt depictions of homosexual subject matter, but in "symbolic or allegorical" evocations of "ideas and feelings which represent gay expression"—and not necessarily recognizable even to other gays. Even this nebulous formulation of a gay aesthetic was contested by artists and critics who questioned the possibility—or the desirability—of formulating any coherent notion of a gay aesthetic. In a discussion responding to the exhibition, the novelist Edmund White insisted:

> If gay sensibility exists, it does not exist as a mystical attribute of a mythical identity—all gays, whenever and wherever . . . have nothing in common with homosexuals of classical Greece. . . . Indeed, I wonder what, if anything, we have in common with homosexuals in England or the United States of the last century.
>
> (as quoted in Ferguson)

The clear differences, noted by White, among formulations of homosexuality in different eras—not to mention the increasingly obvious diversity among those who identify as gay today—do seem to preclude any transhistorical or all-inclusive notion of a gay aesthetic. Nevertheless, as White went on to point out, particular historical contexts may give rise to certain people and certain styles—Wilde's dandyism, Warhol's films—as signifiers of gay identity. Although none of the participants in the debates over the *Extended Sensibilities* show in 1982 knew it, another episode in the history of gay aesthetics was about to begin.

In 1982, doctors in New York and Los Angeles were starting to codify the pattern of fatal diseases among gay men that would become widely known as Acquired Immune Deficiency Syndrome (AIDS). By 1987, 32,000 Americans—overwhelmingly gay men—had been diagnosed with AIDS, and more than 25,000 had died, with the prospect of worse to come (by 1995, diagnoses had risen to more than half a million, with 320,000 dead). Not until then—five years into the epidemic—did President Ronald Reagan acknowledge the existence of AIDS, while governmental initiatives to prevent and treat it were widely perceived to be floundering without any sense of direction or urgency. The response from the gay community, first in New York in 1987 and then in other cities, took the name of the AIDS Coalition to Unleash Power (ACT UP). Over the following years, ACT UP engaged in a series of highly publicized demonstrations de-

signed to change governmental and corporate AIDS policies and to educate the general public about the disease. ACT UP, which included among its members many in the arts and design professions, made sophisticated use of graphic design, producing posters, flyers, and T-shirts that made for striking images in television and newspaper reports about the group's activities. ACT UP's graphics (documented by Douglas Crimp and Adam Rolston, 1990) often exploited the history of gay aesthetics, quoting and reworking the styles of earlier gay artists, such as Warhol. Even as it mined this history, ACT UP became its next episode. By the end of the 1980s, the group's pastiche of provocative imagery and blunt text—the latter in blocky sans-serif typeface—become the vernacular look of an up-to-date, highly politicized gay community, angry and with specific demands. Nothing could be further from the ideas of vague symbols, allegories, and auras of gayness with which the decade began.

Under the influence of the AIDS emergency, many artists and critics who had earlier rejected ideas of gay aesthetics—and even hidden their homosexuality—began affirming their identity in their work (see Reed, 1994). What this outpouring of explicitly gay work confirms, however, is the diversity of aesthetics associated with gay identity. No identifiable characteristics link the work of all gay artists. Gay aesthetics cannot be imagined as a single sensibility, but rather must be seen as a series of historical instances in which artists and critics, in different places and at different times, have attempted to find expression for their experiences of homosexual identity.

[*See also* Camp; Lesbian Aesthetics; Politics and Aesthetics, *article on* AIDS, Aesthetics, and Activism; *and* Sexuality.]

BIBLIOGRAPHY

Bergman, David, ed. *Camp Grounds: Style and Homosexuality.* Amherst, Mass., 1993.

Bersani, Leo. *Homos.* Cambridge, Mass., 1995.

Bristow, Joseph. *Effeminate England: Homoerotic Writing after 1885.* New York, 1995.

Cameron, Dan. *Extended Sensibilities: Homosexual Presence in Contemporary Art.* New York, 1982. Transcripts of the panel discussions accompanying this New Museum exhibition are in Russell Ferguson et al., eds., *Discourses: Conversations in Postmodern Art and Culture* (Cambridge, Mass., 1990) pp. 130–153.

Cooper, Emmanuel. *The Sexual Perspective: Homosexuality and Art in the Last One Hundred Years in the West.* London and New York, 1986.

Crimp, Douglas, ed. *AIDS: Cultural Analysis, Cultural Activism.* Cambridge, Mass., 1988.

Crimp, Douglas, and Adam Rolston. *AIDS Demo Graphics.* Seattle, 1990.

Davis, Whitney. "Winckelmann Divided: Mourning the Death of Art History." In *Gay and Lesbian Studies in Art History,* edited by Whitney Davis. New York, 1994. Also issued as *Journal of Homosexuality* 27.1–2 (1994): 141–159.

Foucault, Michel. *The History of Sexuality,* vol. 1, *An Introduction* (1976). Translated by Robert Hurley. New York, 1978.

Freud, Sigmund. *Leonardo da Vinci and a Memory of His Childhood* (1910). Translated by Alan Tyson. New York, 1964.

Gagnier, Regenia. *Idylls of the Marketplace: Oscar Wilde and the Victorian Public.* Stanford, Calif., 1986.

Greenberg, David F. *The Construction of Homosexuality.* Chicago, 1988.

Kramer, Hilton. *Revenge of the Philistines: Art and Culture, 1972–1984.* New York, 1985.

Martin, Robert K. *The Homosexual Tradition in American Poetry.* Austin, Tex., 1979.

Meyer, Moe, ed. *The Politics and Poetics of Camp.* London and New York, 1994.

Pater, Walter, *Studies in the History of the Renaissance.* London, 1873. Includes "Winckelmann" (1867) and "Poetry of Michelangelo" (1871).

Reed, Christopher. "Postmodernism and the Art of Identity." In *Concepts of Modern Art*, 3d exp. updated ed., edited by Nikos Stangos, pp. 271–93. London and New York, 1994.

Reed, Christopher. "Imminent Domain: Queer Space in the Built Environment." *Art Journal* 55.4 (Winter 1996): 64–70.

Saslow, James M. *Ganymede in the Renaissance: Homosexuality in Art and Society.* New Haven, 1986.

Silverman, Kaja. *Male Subjectivity at the Margins.* New York and London, 1992.

Sinfield, Alan. *The Wilde Century: Effeminacy, Oscar Wilde, and the Queer Moment.* New York, 1994.

Sontag, Susan. "Notes on Camp" (1964). In *Against Interpretation and Other Essays.* New York, 1966.

Sontag, Susan. "On Roland Barthes." In *A Barthes Reader*, edited by Susan Sontag. New York, 1982.

Success Is a Job in New York: The Early Art and Business of Andy Warhol. New York, 1989. Exhibition catalog, Grey Art Gallery and Study Center, New York University.

Symonds, John Addington. *The Life of Michelangelo Buonarotti.* London, 1883.

Thomas, Joe. "What's Queer about Pop Art." *Decorum* 14 (December 1996): 43–48.

Weeks, Jeffrey. *Coming Out: Homosexual Politics in Britain from the Nineteenth Century to the Present.* London and New York, 1977.

CHRISTOPHER REED

GAZE. The notion of "the gaze" has played a pivotal role in the development of contemporary "theory" in at least two respects: as an epitome of human encounter, it has been central to the political turn that poststructuralist thought took in the late 1970s and 1980s, especially in the United Kingdom and the United States, and as an image of vision, it has been similarly central to a claimed visual (or antivisual) turn in the late 1980s and the 1990s.

The term *the gaze* translates the French *le regard*, a word that plays a major role in Jean-Paul Sartre's *Being and Nothingness* (1956)—but the standard English translation of that text renders this word as "the look," so that *the gaze* is widely taken to be a Lacanian term. This translational and conceptual game of hide-and-seek is a fair index to the difficulties of making out the term's real history and scope.

Lacan's most extended remarks on the gaze—four sessions from his seminar of 1964—have as their immediate theoretical context a discussion of chance and repetition in psychoanalysis, and they also have a clear external stimulus in the posthumous publication of Maurice Merleau-Ponty's *The Visible and the Invisible* (1968). At the outset, Lacan seems to intend a brief homage to the philosopher (on whom he had published an earlier piece *Les temps modernes*) but, on the evidence of the text, finds himself drawn into rereading Merleau-Ponty's later work and so extending his digression; as he does so, he finds himself also led into addressing Sartre's analysis of *le regard*. In general, the account Lacan offers is friendly to, if also critical of, Merleau-Ponty's work, which does not take *le regard* as its center, and markedly opposed to the Sartrean account. One mark of this is Lacan's persistent linking of his treatment of the gaze to questions of painting, a concern central to Merleau-Ponty and wholly foreign to Sartre. This is not to say that Lacan simply dismisses Sartre; indeed, there is enough of an overlap between the Sartrean account and elements of Lacan's that it is possible, through partial reading, to conflate them. Arguably, this has to a high degree happened in the derivation of the notion of the gaze that has been so influential in recent years.

To begin with the contrasts: Sartre's primary stories about vision are about the encounter of two persons, one of whom imposes vision on the other, thus rendering that subject object, mortifying and petrifying it through a radical theft of its freedom. Sartre's telling of these stories reflects recognizable asymmetries of power along lines of, for example, gender and class—people are spied on through keyholes, beggar women are glimpsed in the street, and so on. The stories that punctuate Lacan's meditation are significantly more various—they touch on such matters as the kinds of self-presence available to or impossible for the dreamer, the rivalry of Zeuxis and Parrhasios, and (moving beyond the immediate context of the 1964–1965 seminar) prisoners' dilemmas. But the most elaborate of them is clearly constructed in response to Sartre's: it gives us two subjects—Lacan and a fisherman—explicitly set in a situation of class difference and a presumed asymmetry of power; however, the action attributed to the gaze unfolds not between them but through the fisherman's evocation of their joint exposure or visibility to a glinting sardine can on the waves ("You see that? Well, it doesn't see you"). In setting matters this way, Lacan not only emphasizes a pictorial motif—a motif of human picturability as prior to human picturing—that informs all his stories, but also shifts the political burden away from questions of freedom and power toward questions of solidarity and acknowledgment (questions of the pictures we can take or admit of ourselves). He likewise shifts the tone from the Sartrean tragedy or quasi tragedy of visual oppression to the comedy of our mutual belongings, separations, and mistakings.

The role assigned to the sardine can is evidently crucial. It marks the limits of any purely human account of vision; read in terms of an underlying allegory of psychoanalytic development, it stands in for the Oedipal limitation of dual relations, measuring the determinative incidence of what Lacan calls the Symbolic on the Imaginary. If this latter were the sole field of human vision, that field would be essentially Sartrean; it is crucial to Lacan both that this field is developmentally limited by the Symbolic and logically dependent on it (were human being not overarchingly structured by the Symbolic order, no experience of the Imaginary could arise for it). This means that the sardine can marks no external limit on human vision but simply forces the acknowledgment of a limit internal to it; human vision does not begin with humans but is a mark of their belonging to or continuity with something other than themselves. It is just here that Lacan's account takes up and transforms Merleau-Ponty's effort to think human vision as, so to speak, a way the word touches itself and imagines that touch as distance. Like Merleau-Ponty, Lacan seems to see some part of the claim of painting on human beings as rooted in its ability to recall us from that distance (or to permit us to claim that distance as a form of contiguity).

This account necessarily minimizes the deep differences between Lacan and Merleau-Ponty, which can be summarized at a first level by noting that what Merleau-Ponty calls "Flesh," Lacan redescribes as language, as the Symbolic. But this offers too easy a trade-off between a concrete and palpable reality on Merleau-Ponty's side and a more abstract, semiotic fact on Lacan's—whereas what Lacan actually holds is that flesh cannot but turn itself immediately over to signification, the Real folding into the Symbolic and so finding what insistence it does only within the Symbolic's terms. In terms of the parable of the fisherman, this Real is well taken as figured by the sun in which the scene as a whole is bathed but which counts—is registered—only through its glinting reflection. Lacan's imagination of the gaze thus shares a certain ground with thoughts, more directly derived from Merleau-Ponty, informing such work as Gilles Deleuze's *The Fold* (1993).

As noted, both of these stories about vision have had a degree of influence on contemporary art and theory; the shape of this influence has been such as to muddy the difference between them, and thus much recent writing has tended to present an essentially Sartrean view of these matters under the name of Lacan. The real terms of the Lacanian view have only recently come strongly into view, as have their relations to various other strands of poststructuralist thought.

Laura Mulvey's 1975 essay "Visual Pleasure and Narrative Cinema" is frequently and properly taken as seminal for contemporary notions of the gaze, and the title of its third section, "Woman as Image, Man as Bearer of the Look," offers an epitome of the moral most often drawn from it. The actual analysis, keyed to the strategies that Mulvey argues inform the construction of the classic Hollywood film, knits this point to a considerably fuller account of point of view, montage, and the like, and ends by opposing the particular ideological mechanisms thus revealed to a possible materialist cinema that would operate otherwise. The argument is strongly psychoanalytic and its particular vocabulary implicitly Lacanian, with a heavy emphasis on Imaginary identification and misrecognition; in all these respects it both reflects and participates in a heavily Althusserian reading of Lacan associated with *Screen* magazine. To the extent that the essay has been taken up almost exclusively in terms of the Imaginary and its entwining with the Althusserian notion of ideology as interpellation, it has given a strong impetus to an image of ideology critique as critique of representation that has rapidly unfolded into a still more general critique of the visual as such. At this point it appears to lose any productive contact with the Symbolic constraints crucial to both Lacan and Mulvey's particular elaboration of his work, so that its underpinnings now appear essentially Sartrean and the reference to psychoanalysis becomes purely nominal.

Although Mulvey's emphases do clearly tend in this direction, the rapid generalization and assimilation of the shortened thesis seem to reflect a larger tendency within the community that formed around the notion of "theory" in the 1970s and 1980s and that would need its own separate accounting. If one were to examine simply the history of theory, its reception, and its elaboration, some notable features of this community might include its simultaneous fear and desire that "theory" should constitute a form of radical skepticism, and its deep orientation to the figures of prisonhouse and panopticon as descriptions both of "theory" and of one of its central historical objects. A more complex history might try to take into account the ways in which such an account of the visual found certain areas of strong resonance— for example, in avant-garde film (an area in which Mulvey is notably active; her 1976–1977 *Riddles of the Sphinx* is an effort at moving beyond the practices dissected in "Visual Pleasure") or in post-Minimalist artwork (the English artist Mary Kelly has worked particularly close to Mulvey's terms, but a wide range of recent art draws on the more generalized critique of representation). Finally, it should be noted that these very general terms play a large role in the theoretical positions that come to align themselves around the notion of the "anti-aesthetic" in the 1980s. Just as it is important to separate the two quite distinct views that circulate under the single title of "the gaze," it may also be important to separate this version of the critique of aesthetics from another version, rooted in Martin Heidegger, that finds itself extended and transformed in the wake of a very different, and significantly less visible, extension of the Lacanian problematic.

Exploring the further elaboration of Lacan's seminar formulations is a difficult matter, in no small part because the

most interesting further developments rarely happen under the explicit rubric of "the gaze." As Margaret Iversen (1994) has convincingly shown, Roland Barthes's *Camera Lucida* transcribes Lacan's notions with an extraordinary directness into his meditation on photography, and yet even in that text what Lacan calls "the gaze" is rewritten as the photographic feature that Barthes calls "punctum." As one moves further away from such direct appropriations, it becomes clear that Lacan's treatment of the gaze is best seen as but one moment of a more general effort to come to grips with certain aspects of Merleau-Ponty and, especially, Heidegger that are bound together by an overriding interest in the relation between how we are or are not before one another—the community or communities we form—and how we are before those things that might be said to assemble or picture us. Such an interest gets its start in Immanuel Kant and takes on the specific weight of community in G. W. F. Hegel's aesthetics, so taking it up as above all a Heideggerian issue underlines the way in which this general concern for the work of art becomes entangled with a certain critique of the aesthetic. As opposed to the currently more visible critique of the aesthetic as an ideological formation unjustifiably privileging certain objects and separating them off from the more general operations of history and culture, this is a critique made, as it were, in the name of art and its work insofar as the aesthetic appears as an attempt to circumscribe, limit, and ultimately render unrecognizable that very work. Such a critique markedly privileges the work of art as the (nonsacred) place that most clearly and forcefully displays the necessity and difficulty of acknowledging the Other as such—an act that presumably must neither fall into appropriation of that Other nor amount to a blank assertion of its (unreachable, untouchable, and finally empty or skeptical) alterity. In Heidegger and his major French inheritors and contesters (including Jacques Derrida, Maurice Blanchot, Jean-François Lyotard, Philippe Lacoue-Labarthe, and Jean-Luc Nancy, as well as at least moments of Michel Foucault, Deleuze, and Félix Guattari), this has seemed repeatedly to call for some kind of double move—first breaking down an apparent identity and then rebuilding it on the grounds of its dependence on an Otherness that it must necessarily cover over (such grounds are frequently described as "abyssal").

In this broad context, "the gaze" figures as the name for one effort among others to underline both the full weight or force of that modification of our participation in the world that is its visibility to us and so also our visibility in it and to one another, and the particular privilege of the things we recognize as works of art (things that show themselves in some peculiar crossing of the beautiful and the sublime) in rendering that. Taken this way, the notion of "the gaze" communicates with Nancy's (1991) discussion of "exposure" as a fundamental fact of community as well as such matters as the passage from the haptic to the optic that one might emblematically imagine to be at issue in Lascaux, with Lyotard on the play of discourse and figure, with Derrida on the turns of self-portraiture, and with Blanchot on lucidity ("the madness of the day") and darkness ("the other night") as they pass through the figure of Orpheus in his turning toward and away from Eurydice. Such references constitute no more than a partial and preliminary mapping of the complex field of "the gaze." Given the fluidity both of the current invocations of the gaze and the field sketched here, the accompanying bibliography can be only partial and suggestive.

[*See also* Feminism; Film; French Aesthetics, *article on* Contemporary Painting Theory; Lacan; Merleau-Ponty; *and* Sartre.]

BIBLIOGRAPHY

Barthes, Roland. *Camera Lucida: Reflections on Photography.* Translated by Richard Howard. New York, 1981.

Blanchot, Maurice. *The Gaze of Orpheus.* Translated by Lydia Davis, edited by P. Adams Sitney. Barrytown, N.Y., 1981.

Blanchot, Maurice. *The Space of Literature.* Translated by Ann Smock. Lincoln, Nebr., 1982.

Copjec, Joan. "The Orthopsychic Subject: Film Theory and the Reception of Lacan." *October* 49 (1989): 53–71.

Deleuze, Gilles. *The Fold: Leibniz and the Baroque.* Translated by Tom Conley. Minneapolis, 1993.

Derrida, Jacques. *Memoirs of the Blind: The Self-Portrait and Other Ruins.* Translated by Pascale-Anne Brault and Michael Naas. Chicago, 1993.

Iversen, Margaret. "What Is a Photograph?" *Art History* 17.3 (September 1994): 450–463.

Jay, Martin. *Downcast Eyes: The Denigration of Vision in Twentieth-Century French Thought.* Berkeley, 1993.

Krauss, Rosalind. "Cindy Sherman: Untitled." In *Cindy Sherman 1975–1993.* New York, 1993.

Lacan, Jacques. *Écrits: A Selection.* Translated by Alan Sheridan. New York, 1977.

Lacan, Jacques. *The Four Fundamental Concepts of Psychoanalysis.* Edited by Jacques-Alain Miller, translated by Alan Sheridan. New York, 1978.

Lyotard, Jean-François. *Discours, Figure.* Paris, 1971.

Melville, Stephen. "Division of the Gaze, or Remarks on the Color and Tenor of Contemporary 'Theory.'" In *Vision in Context: Historical and Contemporary Perspectives on Sight,* edited by Teresa Brennan and Martin Jay, pp. 103–116. New York and London, 1996.

Merleau-Ponty, Maurice. *The Visible and the Invisible.* Edited by Claude Lefort, translated by Alphonso Lingis. Evanston, Ill., 1968.

Mulvey, Laura. *Visual and Other Pleasures.* Bloomington, Ind., 1989.

Mulvey, Laura. *Fetishism and Curiosity.* Bloomington, Ind., 1996.

Nancy, Jean-Luc. *The Inoperative Community.* Edited and translated by Peter Connor. Minneapolis, 1991.

Nancy, Jean-Luc. *The Muses.* Translated by Peggy Kamuf. Stanford, Calif., 1996.

Newman, Michael. "The Trace of Trauma: Blindness, Testimony, and the Gaze in Blanchot and Derrida." In *Maurice Blanchot: The Demand of Writing,* edited by Carolyn Bailey Gill, pp. 153–173. London and New York, 1996.

Sartre, Jean-Paul. *Being and Nothingness.* Translated by Hazel Barnes. New York, 1956.

STEPHEN MELVILLE

GENDER. *See* Duchamp, *article on* En-Gendering of the Artistic Subject; Feminism; Gay Aesthetics; Lesbian Aesthetics; *and* Sexuality.

GENIUS. [*This entry comprises two essays on the history and significance of the concept of genius in aesthetics:*
 Conceptual and Historical Overview
 Genius and Feminism
The first essay traces this history through many of the key theorists who contributed to its development; the second essay offers a critique of this same history from a feminist perspective. For related discussion, see Artist; Autonomy; Creativity; Originality; *and* Romanticism.]

Conceptual and Historical Overview

Sometime in the late fourteenth century, the Latin word *genius* entered English as a name for a kind of inborn or indwelling spirit. *Genius* derives ultimately from the Greek word *gignesthai*, which has to do with coming into being and, more specifically, with begetting and being born. In its first uses, *genius* seems to follow in the track of the Greek *daimon*, and both were applied to the tutelary spirit or attendant god that was allotted to each person at birth. The idea of an innate spirit allows us to attribute genius to a language ("the genius of Italian"), to a people ("the English genius for public civility"), to a system of institutions ("the genius of the common law"), to a cuisine ("the genius of French cooking"), or to a place (the "genius loci").

In modern European and American culture, "genius" has come to designate two more particular conditions: (1) exceptional natural intelligence or creative powers or other aptitude (equivalent to the Latin *ingenium*), and (2) an individual human being, possessing such powers. In the latter usage, the sense of the extraordinariness of the powers that are being exercised is transferred to the individual that is exercising them. Very often, if not quite invariably, this transfer leads to the idea that the individual "genius" in question represents an altogether different order of human being. The point of transfer most often seems to be the idea that those who possess the *capacity* for extraordinary efforts and achievements must be in some sense extraordinary at all times and in all aspects of their existence. This idea of genius reached its highest tide in the Romantic period, but aspects of the idea continue to the present time. Such associations of genius with the concept of a superior level of human existence lend the concept a great deal of interest and allure. But they also make the concept a target of (frequently justified) suspicion and criticism.

In principle, a "genius" in the sense of an exceptionally gifted individual can occur in any field of human endeavor. Despite Immanuel Kant's (1987) claim that we ought to refer to scientists not as geniuses but as "great minds," we do in fact commonly refer to Isaac Newton, Albert Einstein, Charles Darwin, and Carl Friedrich Gauss as geniuses, perhaps less commonly so to Sigmund Freud, Friedrich Nietzsche, and Karl Marx. It is possible to refer to military or political genius, as in the case of Napoleon, Otto von Bismarck, Mao Zedong, Rosa Luxemburg, or Indira Gandhi. Other instances abound. Nevertheless, both the popular and the more critical use of *genius* have gravitated toward the field of the "creative arts." (The image of Einstein, and more generally of the "mad scientist," seems to represent a composite of the idea of the scientist and that of the artist.) Indeed, *genius* soon began to refer precisely to that region of the human spirit that enables the individual to produce the works that are received as creative and original.

Students of human culture must try to understand at least two shifts in the history of the concept of genius: (1) a shift away from a wider context of quasi-religious spirits and anthropomorphic guiding presences toward an idea of genius as the extraordinary capacity of single human being; and (2) a tendency to pull the term *genius* away from the full range of human cultural and intellectual activities and reserve it for the practitioners of what came to be called the "fine arts." These shifts are a bare indication of a more cataclysmic set of developments that occurred throughout the eighteenth and well into the nineteenth century.

Tracing these developments in the idea of genius is not merely a matter of tracing shifts in linguistic usages or in the ideological fashions of "high" culture. There are indications here of shifts in consciousness that go back to the Greek roots of European civilization. The concept of genius reaches down to the level at which human beings conceptualize the relation between what is perceived and felt and what is produced or accomplished. The student of genius may today no longer think of herself as a student of an exceptional *kind* of human being. But she is nevertheless a student of the ways in which human beings have put into words and concepts certain exceptional human *states* or *conditions* or *activities*. From time to time, human beings have found themselves in such states, and within such states they have seen things or heard things or made things that other human beings have found to be of interest and sometimes have come to see for themselves. Under the heading of *genius*, we are studying the myths, concepts, and explanations that human cultures have offered for these occurrences.

The various ideas and elements of genius are often so shrouded in mythology that it is difficult to command a clear view of their origins or development. It is striking that many of the elements of the modern or post-eighteenth-century views of genius are already to be found in texts of poetry and philosophy from the classical period. It is equally striking that these elements rarely come together, much less coalesce, in a unified conception of the artist as genius. It will prove helpful to take Plato as a representative

figure of classical Greek culture, as well as the first great philosopher to address the issues of art, mind, and creation. Plato's *Dialogues* contain at least five separate elements, each of which will later have something to contribute to the concept of genius. The fact that these elements do *not* coalesce into a single concept can teach us something about how the modern and more unified concept of genius and creativity came into existence.

1. *Mantike.* The word *mantike* refers to a state of being possessed by something divine and immortal, or to some other access to divine communication. In the *Phaedrus*, Plato has Socrates joke about the connection of the *manic*, which was (and is) the name for a kind of madness, and the *mantic*, which was a name for the condition that prophets and poets habitually found themselves in (*Phaedrus*, 244c–e). Plato seems to have taken the lead in understanding poets and their condition in relation to the *prophetes* and his condition (see Plato, *Apology*, 22b–c; *Ion*, 536c).

2. *Enthousiasmos.* The condition of *mantike* is not completely distinguishable from *enthousiasmos*. This refers more immediately to being possessed by a god and particularly to the *indwelling* of a god (*Apology*, 22c). Along with the possibility of a genuine communication from the gods through the medium of the mind and the body of the *prophetes* there goes the idea that the poet-prophet sees and hears things that the rest of humanity does not see. Taken together, the *mantic* and the *enthusiast* point to the idea that there are states in which the poet says things that are beyond his normal level of comprehension. Plato does not directly question the divinity of these utterances, but he focuses some irony on a consequence of this condition, namely, that the poets are not capable of understanding their own utterances.

3. *Technē.* The word *technē* is normally and directly translated as "art," though it can also be translated as "craft" or "skill." (It is the word from which are derived *technique* and *technology*.) Both Plato and Socrates rely on the idea of a "craft" in order to investigate the concept of virtue or excellence *(arete). Technē* is fundamental to Greek culture, and the term is very far from having any pejorative connotations such as "mere craft" or "mere technique." Nevertheless, Plato is very clear that *technē* is not enough to produce great poetry: "He who without the divine madness [*manias*] approaches the doors of the muses, confident that he will be a good poet by art [*ek technēs*], meets with no success, and the poetry of the sane man [*sophronountos*] vanishes into nothingness before that of the inspired madmen" (*Phaedrus*, 245a). There is an irony in Plato's locating *technē* on the side of sanity and poetic success on the side of madness. Nevertheless, one gets a glimpse of the fundamental distinction between what humans can produce according to knowledge that can be expressed in words and rules and what can only be produced in some other state of mind.

4. *Daimon.* The term *daimon* refers to a guardian spirit. It is the constellation of ideas around this term that was taken up into Roman mythology and was translated into Latin as *genius. Daimon* derives from *daio*, to distribute. *Daimon* thus emphasizes the awareness that destinies are distributed— portioned out—to every individual, who then has to bear the destiny received. The word occurs throughout Homer and in a number of places in Plato (e.g., at *Republic* 617e and 620d). Later versions of *genius* will merge the idea of being possessed by a divine order of existence with the idea of the artist-genius as possessing a peculiar fate or aptitude for such possession.

5. *Demiourgos.* The ordinary word for maker or craftsman is *demiourgos.* Plato does not often use the word in relation to the poets or other artists, presumably in part because the word *poetes* still retained a much larger share than the English word *poet* of the original meaning of *poein*, "to make." Moreover, if the poets are still thought of in Athens as "makers" and producers, Plato sometimes compares them unfavorably to the craftsmen and artisans, who are at least in the business of copying the Forms rather than copying the copies. The poets, Plato famously suggests, are only making copies of copies *(Republic).* The *Timaeus*, however, focuses close attention on the *demiourgos.* If the term there primarily refers to the operation of a divine principle, ultimately it became part of the process by which certain kinds of "production" and "craft" were elevated to a more than human level.

The case of Socrates is already a kind of transitional moment in the understanding of the exceptional individual. He names the voice that comes to him a *daimon* (e.g., at *Euthyphro* 9b), but the primary function of this voice is to say "No" to a particular course of action. When it finally tells him to "practice the arts" (*Phaedo* 60e), he is already in prison. He uses the moment to blur the very lines that Plato is often careful to draw: the lines between poetic composition and the methodical pursuit of the desire for the truth that they called philosophy. Socrates moves us toward a later sense of genius, in part by exemplifying the workings of a *daimon*, apart from any existing theology (*Apology* 26c–e). This moment marks one beginning of the somewhat paradoxical secularization of the divine, without which the idea of genius is impossible.

Seeing these separate elements laid out in this way should help to clarify the later history of the concept of genius. Each of the concepts above continues to circulate in the realms of Western thought and culture. After Plato and the classical period, the largest event in the history of Western culture is the arrival and establishment of Christianity. A thousand years of Christian philosophical speculation and cultural productiveness saturated Europe with the ideas and concepts of creation. The creation of the world out of nothing, by God's Word, gets mixed into the brew of Greek

ideas. In the *Timaeus,* the *demiourgos* or divine craftsman took on a new importance. Plotinus's reflections on the beauty and goodness of the creation were also circulating, perhaps especially in the forms in which they were taken up by Augustine. The intimate link between the beauty of creation by the divine Word and the *fact* of its createdness is the Christian preoccupation that most feeds the later streams of reflection on creativity and originality. The full confluence of these different streams seems to occur in the various artistic and theoretical phenomena of the Renaissance.

If the period from the Greeks to the Renaissance represents a slow accumulation and coalescence of terms for production and creation around the work of figures who are thought of as "artists," the period from the eighteenth to the middle of the nineteenth century is more like a sustained series of cultural upheavals and conceptual cataclysms. In this period, the concept of genius acquired its current constellation of meanings and a heightened place in the European discourses and ideologies concerning the arts. This is also the period in which European culture took on the bulk of its modern ideas about the arts. The rise of the figure of the genius is inseparable from other contemporaneous conceptual shifts in the European awareness of the arts.

Crucially, this period saw the consolidation of the idea of a sphere of "fine arts," separated from the more immediately practical human skills and endeavors. At the same historical moment, the subject of aesthetics was coming into existence as a separate, and to some extent autonomous, philosophical discipline. The concept of genius played a critical role in both of these developments. Like the concept of "the work" and the concept of "expression," "the genius" became deeply embedded in the artistic and compositional practices of Europe, along with the philosophical discourses that addressed those practices.

The conceptual shifts associated with the shift in the concept of genius must include (1) the very formation of the "system of the arts," including the idea that the fine arts, such as poetry, painting, and music, are superior to the skills and arts with a more immediately practical aim; (2) the shift from *imitation* as a principal goal and medium of poetic and painterly composition to *expression* and expressiveness as both goal and criterion for the success of a composition; (3) the priority of the creator of the composition over the forms, traditions, genres, and the rules by which such work had previously been accomplished; and (4) a new emphasis and conceptualization of the sorts of experiences of greatness, awe, and terror that came to be called collectively "the sublime."

No one of these shifts necessarily required the newly emerging concept of genius, and the concept of genius might have acquired much of its significance without all of these metamorphoses in the structure of the arts. But in each of these transformations, genius has a special role to play. The genius rises above the mundane concerns of habit and ordinary life to create something of more than ordinary spirit and power. The genius has something to express that goes beyond the range and ability of more ordinary humans to express or even to experience in the first place.

The genius will master the rules that contain the sum of tradition and experience in a particular field of creative endeavor. But he or she will also know how to break the rules to the best and the most powerful effect. That knowledge will be a necessary, though not a sufficient, sign that the genius is creating something original—most immediately, something original in relation to the very tradition whose rules are being broken. Even a neoclassical theorist and painter such as Sir Joshua Reynolds (1723–1792) could write, "Could we teach taste and genius by rules, they would not longer be taste and genius." Such a view is still consistent with the kind of position emblematized in Alexander Pope's praise of "what oft was thought, but ne'er so well-expressed." At some point in the eighteenth century, the more radical writers and thinkers abandoned this way of thinking about expression and sense. Greatness in the arts became a matter not just of finding original forms of expression but of finding something original to express.

The rise to theoretical preeminence of the concept of genius is inseparable from this demand for originality. What began as a series of fissures and tensions in the decorum of eighteenth-century art and criticism had become, by the end of the century, a force that would crack open the social bonds of taste and political ideology. The idea that the work of fine art must be the product of originality achieves its first enduring formulation in the work of Immanuel Kant (1724–1804) and is thereafter shared by many Romantic writers. The Romantic era stretches from at least the 1790s to the 1850s and embodies a tremendous cultural ferment, which was as much political as it was artistic and philosophical. Kant's work anticipates many Romantic themes and claims concerning genius, but it also constitutes a profound and intimate criticism of them. Together, they bring the concept of genius to some kind of culmination: they encompass almost every conceivable strand of genius that is still in existence, and quite a few that are now defunct. Neither for Kant nor for most of the Romantic writers is the praise of originality and genius to be construed as merely a praise of sheer newness or innovation.

For Kant, genius is the production of something for which no rule can be given. Hence, its "primary property" is originality. But this product of originality must be able to serve as a *model* for those who came after. This exemplary quality of genius is meant to combat the danger of what Kant explicitly calls "original nonsense" (1987, section 46). We are therefore entitled to think of the products of genius as making a kind of "original sense." If we take the exemplariness of the work of genius as making a new type of

sense, we may better understand why Kant thought that the work of genius had to be followed in the spirit in which it was created. The truest understanding of the work of genius is achieved by the one who is, in that very act of understanding, awakened to his own capacity for original creation. Only thus can we receive the work of originality without becoming imitators who thus lose the very element of genius they are trying to apprehend (ibid., section 49).

This demand for an exemplary new way of making sense helps explain Kant's conviction that the genius cannot explain how his ideas "arise and meet in his mind" (ibid., section 47). The way in which an original production makes sense cannot have been known *before* the production has been achieved: unlike the steps of a scientific proof or the directions for making a new invention, the steps by which an original work was achieved cannot be understood as making sense, in advance of the moment in which the new work emerges. The movements of the mind can only be regarded as *steps* toward the new work, when the new work is there to instruct us about the nature of its origins (ibid., section 32).

The poet William Wordsworth put a similar thought as follows: "Genius is the introduction of a new element into the intellectual universe: or if that not be allowed, it is the application of powers to objects on which they had not before been exercised, or the employment of them in such a manner as to produce effects hitherto unknown" (1966, p. 750). Kant shares with Romanticism a sense that the existing elements of our understanding and imagination are insufficient to our current resources of expression and making sense. Originality involves not just a new way of meeting an old need, but a new way of conceiving the relation between human need and human activity.

We can interpret the emergence of "spirit" and "freedom" in the discourse about genius as expressions of the demand for a reconception of the relationship between human activity and the more passive aspects of human being. Kant insists first of all on "spirit [*Geist*]" as the element by which the works of genius exceed works that merely satisfy our judgments of taste. "Spirit" then emerges within the concept of genius as the moment of animation by which human beings are enabled to make "exemplary" and "original" use of gifts that are theirs by nature (Kant, 1987, section 49). The emphasis on the spirit of genius now falls on the "free employment" of the mind's powers of cognition and representation. The awakening to such an original and free use of one's own powers must occur through the comparatively passive moment of reception. That audience that the work of genius requires for its most appropriate reception stands in need of the work of genius in order to arrive at the free use of its own powers.

The ideas of spirit and of the awakening to the free use of powers already in some sense present in the subject provide an important link to the post-Kantian problematic of freedom and autonomy. The exploration of the reciprocity of the freedom of the human spirit and the free production of the expressive works of that spirit become staples of Romantic and idealist art, criticism, and philosophy. Indeed, German idealism (e.g., in G. W. F. Hegel, Friedrich Wilhelm Joseph von Schelling, and Johann Gottlieb Fichte) goes beyond even the Romantics in constructing an ontology of the world as an expression of the highest capacities of human freedom and the originality of the human spirit.

Kant himself drew back from these Romantic and idealist dimensions of the liberating powers of art, which his own work had been the first to open. He seems to avoid the possibility that his work invites us to explore: the idea that the awakening of genius to a free use of its representational powers could provide symbols of the routes by which human beings awaken to moral autonomy as such. Indeed, Kant never explicitly abandoned the idea that the beauty of nature was superior to the beauty of artistic production, precisely because it was the unself-conscious beauty of nature that harbored the symbols of freedom.

None of these conceptions of genius entail a being who belongs to a higher order of existence: they are hence removed in principle (though unfortunately not always in practice) from any reduction of the concept of genius to mere biology. The kind of exceptional existence that a genius possesses cannot be regarded as a merely quantitative or qualitative deviation from some well-known baseline of "normal" qualities or abilities. The genius does not have *more* of some quality whose meaning for human existence has already been adequately understood.

In these conceptions of genius, the reception and acknowledgment of genius in someone else's work or mind demands a reevaluation and reconfiguration of our powers to apprehend and to produce new ways and forms of making sense. These views have in common the conception of "genius" as providing the point of access to a more intensive order of human significance. The genius possesses a more perfected capacity to participate in and to enrich that order of significance. The Romantic genius reenvisions and re-creates the conditions under which human beings express themselves and the elements of their basic constitution that they wish to communicate.

This conception was one reason why the concept of genius seemed to be most at home in the realm of the arts. But it also helps to explain the Romantic drive to merge the figure of the artist-genius with the artist-philosopher. The decline of the human powers of expression and feeling would be crippling to any philosopher seriously wishing to educate the majority of humankind. The starting point for any such education—and, indeed, the starting point for any human political reform or revolution—would have to involve the reshaping of the most basic of human perceptions and feelings.

John Stuart Mill (1806–1873) and Ralph Waldo Emerson (1803–1882) write the elegy for genius as Romantic origi-

nality. Mill characterizes genius as contributing to the elements of the individuality of all human beings: "the first service which originality has to render . . . is that of opening their eyes" (1859, part 3). But he also wrote its epitaph; for if everyone can approve of genius when it lets one write an exciting poem or paint a picture, no one really admires it when it gets too close to home: "Originality is the one thing which unoriginal minds cannot feel the use of. They cannot see what it is to do for them: how should they? If they could see what it would do for them, it would not be originality." In Emerson, even more than Mill, the idea of genius moves back out from the realm of art and singular individuals and into the basic elements of selfhood and autonomy. Emerson plays with the old idea of the guardian spirit: "Ah, that our Genius were a little more of a genius!" The spirit that has charge over our awakening to individuality has come to seem somewhat too dull for the job. The spirit with which we struggle to wake ourselves up seems to be allowing us to slumber at the most inopportune times. [*See* Emerson.]

The idea of genius continued to play a dominant role throughout the nineteenth century and into the twentieth century. Increasingly in the first half and then relentlessly in the latter half of the twentieth century, the idea has been subjected to thoroughgoing and even savage criticism. Many philosophers have argued that the term possesses little or no definite content. And indeed, *genius* is often used as an empty title, serving above all to place the work or the human being thus qualified beyond criticism. Others have suggested that what little content the term may possess has contributed to some of the more pernicious social and political ideologies of the West. The German critic and Marxist theorist Walter Benjamin condemned the use of concepts such as "genius and creativity, eternal value and mystery" as likely to lead us to "the processing of data in the Fascist sense" (1985).

More recently, feminist scholarship has taken stock of the peculiar affinity of the concept of genius for the masculine self-conception of the artists and of the apparent readiness of Western culture to endorse and even to underwrite such self-conceptions. This criticism is not invalidated by the fact that it was often through the concept of genius that some artists (primarily men) were able to avail themselves of creative sources that were otherwise culturally stigmatized as "feminine." Christine Battersby has argued that the majority of the "strands . . . of our modern usage of the term 'genius'" have been "utterly contaminated by past usage, and by the way that the male (still) provides the paradigm for both the normal and supernormal personality-types, consciousness-types, and energy-types" (1989, p. 156).

Despite the intellectual decline and disreputability into which *genius* has fallen, the term continues to flourish among some historians of the arts, reviewers, critics, and those who go to museums and concerts. Michel Foucault (1979) has characterized the concept of genius as part of the more general field of functioning defined by terms like *author* and *work*. Although this analysis has played a part in the general deflationary critique of *genius*, it also provides a glimpse of one way in which the concept may survive; for Foucault focuses attention on those moments where a new possibility of "discourse" opens up, possibilities associated with names like Marx, Freud, and even Anne Radcliffe (as the founder of the "Gothic novel"). In characterizing the discursive possibilities as going beyond the name and the specific works associated with the founder, Foucault gives a new inflection to the Kantian idea of an exemplary originality. As with the originality of genius, the founding of a new discourse is not secured until it is followed by other workers, whose contributions to the discourse may be marked precisely by what Foucault calls "divergence" (1979, pp. 154–155).

Nevertheless, it may be that the role of the word *genius* is destined to become largely ceremonial. It may even be that for particular segments of Western culture, giving up the concept of genius is no more significant and no harder to give up than the practice of applauding at the end of a concert piece. The decline of such a concept and its associated hopes and fantasies, however, is likely to be as fraught with consequences as was its ascendance.

[*See also* Kant; *and* Plato.]

BIBLIOGRAPHY

Battersby, Christine. *Gender and Genius: Towards a Feminist Aesthetics.* London, 1989.

Benjamin, Walter. "The Work of Art in the Age of Mechanical Reproduction." In *Film Theory and Criticism,* 3d ed., edited by Gerald Mast and Marshall Cohen, pp. 675–694. New York, 1985.

Coleridge, Samuel Taylor. *Biographia Literaria.* London, 1817.

Crawford, Donald. "Kant's Theory of the Creative Imagination." In *Essays in Kant's Aesthetics,* edited by Ted Cohen and Paul Guyer, pp. 151–178. Chicago, 1982.

Emerson, Ralph Waldo. *Essays.* First and Second Series. Boston, 1841–1844. See especially "History," "Art," "The Poet," and "Experience."

Foucault, Michel. "What Is an Author?" In *Textual Strategies: Perspectives in Post-Structuralist Criticism,* edited by Josúe V. Harari. Ithaca, N.Y., 1979.

Gilbert, Katherine, and Helmut Kuhn. *A History of Esthetics.* Rev. enl. ed. Westport, Conn., 1972.

Gould, Timothy. "The Audience of Originality: Kant and Wordsworth on the Reception of Genius." In *Essays in Kant's Aesthetics,* edited by Ted Cohen and Paul Guyer, pp. 179–193. Chicago, 1982.

Hazlitt, William. "Genius and Common Sense." In *Table Talk or Original Essays on Men and Manners.* London, 1821–1822.

Jamison, Kay Redfield. *Touched with Fire: Manic-Depressive Illness and the Artistic Temperament.* New York, 1993.

Kant, Immanuel. *Critique of Judgment.* Translated by Werner S. Pluhar. Indianapolis, 1987.

Longinus. *On Great Writing (On The Sublime).* Edited and translated by G. M. A. Grube. Indianapolis, 1957.

Mill, John Stuart. *On Liberty.* London, 1859.

Murray, Penelope, ed. *Genius: The History of an Idea.* Oxford and New York, 1989.

Peters, F. E. *Greek Philosophical Terms: A Historical Lexicon.* New York, 1967.

Plato. *Collected Dialogues.* Edited by Edith Hamilton and Huntington Cairns. New York, 1961. See especially *Apology, Republic, Ion, Phaedrus.*

Schelling, Friedrich Wilhelm Joseph. *The Philosophy of Art.* Edited and translated by Douglas W. Stott. Minneapolis, 1989.

Williams, Raymond. *Culture and Society, 1780–1950.* New York, 1958.

Wordsworth, William. "Essay, Supplementary." In *Poetical Works,* edited by Ernest de Selincourt, pp. 743–751. London, 1966.

Young, Edward. *Conjectures on Original Composition.* London, 1759.

Timothy Gould

Genius and Feminism

In Britain during the seventeenth and eighteenth centuries, two distinct concepts—Latin *ingenium* (meaning "inborn ability," "disposition," or "talent") and Latin *genius* ("begetting spirit" or "protective god")—gradually collapsed into each other as the framework of aesthetic and cultural values also changed. At the start of the eighteenth century there was talk of a person "having" genius, and genius could be great or small. By the late eighteenth century, "genius" no longer primarily meant "talent," and could even be opposed to talent. The genius was now a type of human being: one who was frequently credited with a begetting spirit, a god, or an inner daimon that drove him to the most awe-inspiring kinds of artistic production. As such, *genius* was extensively employed to mark the boundaries between the truly "original" artwork and mere novelty or imitation, and between "sublime" and merely inferior human types. For the Romantics, this superior mode of being was linked above all with creative production, as opposed to merely mimetic reproduction or skill. The genius was also associated with cultural leadership, however, and there was a qualitative ranking of human (and superhuman) achievements that generally masqueraded as mere description. The genius was described in metaphors that evoked the power of a Christian God who creates something out of nothing. In Romanticism, however, one can also find traces of older Greek and Roman metaphysics in which production is modeled on reproduction that is explained hylomorphically, in terms of the imposition of form on otherwise chaotic or inert matter.

In Roman mythology, *genius* was the begetting spirit in the family embodied in the paterfamilias and involved those aspects of (male) procreativity that ensured that the *gens* or male family clan continued, and that the reproductive line remained true to itself. The *genius* was a guardian spirit that protected the paterfamilias throughout his life, and that was passed down to the eldest son when the father died. Gradually, every male was endowed with a specific *genius* to whom yearly offerings were made, generally on the man's birthday. It is important to register that the *genius* was never associated with the more mundane physical aspects of reproduction (the penis); it was the male forehead that was honored in the Roman rites of *genius.* For the early Romans, a *genius*

watched over the lands that were in the ownership of the male clan or *gens*—and *genii* later came to be protectors of towns, particular places, and the Roman state itself. Indeed, under Augustus (27 BCE–14 CE) Rome itself was linked to the *genius* of the emperor, and to those aspects of his lineage that made him a god on earth. Thus, the notion of the "genius of the place" that was revived in Romantic poetry and by earlier writers such as Joseph Addison (1672–1719) drew on aspects of a pretheistic religion that associated formative force with those spiritualized aspects of a male's reproductive power that ensured immortality for the family, race, or species via continuity of the patrilineal line.

For the ancient Greeks, reproduction involved the shaping of preexistent matter. Matter remained chaotic and heterogeneous until unified and rendered harmonious by its relationship to form. Plato's account of sameness and difference in *Timaeus* (c.350 BCE) fits this model, and was interpreted by Plotinus in *The Enneads* (254–270) in ways that incorporated elements of Aristotle's metaphysics. In Aristotelianism and the subsequent Galenic biology that came to dominate Europe until the eighteenth century, the mother provided the "matter" of reproduction *(hyle);* but matter was conceptualized as inert or unruly, at least in respect to its own reproductive power. The material emerging from the mother's womb needed to be shaped by the defining formula (the *logos*) implicit in the male seed. Only by such a process of hylomorphic molding could an essence emerge from passive or chaotic matter, or continuity of form be passed on to subsequent generations. Putting together Aristotle and Plato, Plotinus compared the world to a fountain with higher levels spilling over (emanating) into those below them. Matter is the lowest order of reality and is compared by Plotinus to a shadow lying below soul and to a reflection in a mirror. Plotinus's mélange was profoundly influential on Christian thought, and Renaissance accounts of artistic production also frequently drew on his teaching. In these one sees that the artist can reestablish the connection of the Soul with Reason through the skillful mirroring of the truth (the form) that overflows (emanates) into nature and matter itself, with the contemplation of beauty also playing a key role in reaccessing the "Oneness" that constitutes the apex of being. [See Plotinus.]

During the seventeenth and early eighteenth centuries, art was also represented primarily in terms of mimesis, but the dominant analogies for the processes of artistic production were now those of mechanistic design. The ideal artist was compared to a watchmaker. As a craftsman who designed or copied in accordance with a blueprint or plan, he was ingenious, not fired by genius. Training the artist's *ingenium* or inherited talents remained the dominant concern. Rationality, craft, and skill were important values, because the product was fashioned according to procedures and principles that were rule-governed. Harmony, regularity, and "ease" were both artistic and intellectual virtues. By the late eighteenth

century, however, this model had broken down, and *genius* was no longer linked primarily to skill, reason, or the contemplation of beauty. Instead, genius was especially valued where it was "primitive," "natural," and "sublime"—an aesthetic category linked to terror and immensity, and opposed to "beauty." Much of the vocabulary of "original genius" developed in a British context, in order to explain the apparent lack of coincidence between the works of William Shakespeare and the rules pertaining to greatness in neoclassical drama. Cross-cultural influences were also strong, however, with the works of the French philosophes—especially Jean-Jacques Rousseau—contributing much to the lively attacks on established artistic traditions and authorities. In *Émile* (1762), however, Rousseau emphasized much more strongly than did the contemporary British advocates of "original genius" that even "natural" genius requires an appropriate education.

In German, the two terms *Genie* (from Latin *ingenium*) and *Genius* (from Latin *genius*) remain distinct. The collapse in meanings of the two words that occurred in English, however, has had a profound influence on German aesthetics, and is particularly pronounced in those writers who were most influenced by the British advocates of original genius. In particular, "Conjectures on Original Composition" (1759) by the Englishman Edward Young helped shaped the aesthetics of Johann Georg Hamann (1730–1788), Johann Gottfried von Herder (1744–1803) and other Sturm und Drang writers who drew on Young's account of genius as a kind of inner god, "the stranger within." In 1770, a meeting between Herder and Johann Wolfgang von Goethe was important in shaping Goethe's youthful enthusiasm for original genius. Thus, in *The Sorrows of Young Werther* (1774) the "torrent of genius" is said to "thunder forth" and "shake the depths" of the soul only where artistic tradition and rules are disregarded. Goethe's early enthusiasm for—and translations of—Shakespeare also show Herder's influence, as does his admiration for Ossian, James Macpherson's fake bard. Many other exponents of Ossian and "original genius" influential in Germany were Scottish, including Hugh Blair (1718–1800) and William Duff (1767). The latter's *Essay on Original Genius* ranked William Shakespeare as the only modern to stand alongside Homer and "Ossian," and described genius as taking nothing from tradition, and everything from the "ardour" of feeling and imagination.

Immanuel Kant (1724–1804) distanced himself from those who advocated original genius. Instead, he explicitly commended the far more moderate account of genius in *An Essay of Genius* (1774) by another Scot, Alexander Gerard, as the best work on the subject. Like Kant himself, Gerard allowed judgment a role alongside imagination, and emphasized the links between genius and talent both in this work and in an earlier work also translated into German, *An Essay on Taste* (1759). Gerard tied the imagination to the association of ideas in ways reminiscent of David Hume

(1711–1776), however, even though in lectures on *Anthropology, from a Pragmatic Point of View* (1802–1974). Kant is recorded by a student as crediting Gerard with the invention of a notion of "productive imagination" in which even simple ideas can be invented. The famous definition of genius in Kant's *Critique of Judgment* (1790) emphasizes the latter faculty, but in section 46 he is careful to start by linking genius *(Genie)* to "talent" (a gift of nature), and to specify the Latin word *ingenium* when he describes genius as "an innate mental predisposition *through which* nature gives the rule to art." Although, in the same section, Kant also registers the Latin use of the term *genius* as relevant, it is made subservient to the explanation of a talent that is both nonconceptual and rule-giving. Kant's genius is not a "natural" and "spontaneous" automaton, directed by a kind of inner god or *daimon* who acts as a "stranger within" in the manner of Young. Indeed, in *Reflexionen zur Anthropologie* no. 899, Kant explicitly denies genius the status of a daemon. Kant establishes the link between genius and nature via "inner gifts" that enable the genius to act as an example (for other geniuses) and as source of rules to be imitated (but not aped) within a new school or a tradition (Kant, 1987, section 49).

The Kantian genius is not, however, simply the product of hard work and talent; nor does he simply "copy" nature or follow the rules of others. As in the case of the transcendental schemata of the imagination, which act as preconceptual templates for an order that will emerge out of the heterogeneous material of the senses, the Kantian genius acts as an exemplar—as a "monogram" or "silhouette"—for new patterns of artistic order. Thus, in Kantian aesthetics the genius synthesizes art as the "I" also synthesizes the phenomenal world in the *Critique of Pure Reason* (1781–1787). The genius's productive imagination shapes materials ("intuitions") that come in through the senses. This material remains chaotic (a "manifold") until shaped by the "I" of the genius via the imposition of imaginative structures and space-time form. [*See* Kant.] Although matter has a different status than in Plato and Aristotle, Kant continues to refuse matter the capacity for self-formation. Kant's aesthetics had a profound influence on the German and British Romantics. Indeed, despite the apparent emphasis on organicism that developed in Romantic aesthetics, matter was still generally presented in hylomorphic terms, as chaotic and diverse until shaped into a unity by the "I" or by preconscious forces working within the self.

When Georg Wilhelm Friedrich Hegel comments in section 68 of the preface to his *Phenomenology of Spirit* (1807) that genius "was once all the rage in poetry, as it now is in philosophy," he is correctly remarking on a transition in aesthetic theory that had already occurred. Furthermore, Hegel's implicit attack on the cult of "genius" in philosophy is not directed at Kant, but at the kind of Romantic mysticism that valorized "intellectual intuition," and that Kant

also attacked in his late essays. Hegel's *Phenomenology* also correctly registers Denis Diderot's *Rameau's Nephew* (written c.1761–1779) as complicating the contemporary debate about the supposed "progress" of spirit that occurred during the Enlightenment period. Here Diderot's two voices—an "I" and a "He"—debate many aspects of genius, in particular whether there might not be genius of "wickedness." Like Goethe, who translated the dialogue into German for its first (posthumous) appearance in print (1805), Diderot places his "hero" in an ambiguous relationship to a spirit or daemon—Latin *genius*—although in Diderot's case the spirit is secular and made a part of the psyche of the pantomimic madman who would be a musical *génie,* like his uncle Jean-Philippe Rameau (1683–1764). As in *Faust, Part I* (1775–1808), Diderot's voice-drama highlights the acute moral dilemmas relating to the way that the vocabularies of "talent" *(ingenium)* and noble ancestry *(genius)* had merged. Genius was being presented as the bedrock of European culture, and as working for the benefit of the species rather than for the happiness or good of the genius himself.

Goethe's own attitude to genius changed sharply over his life as the dominant aesthetics in Germany also developed away from an aesthetics that privileged a passive response to nature, and also away from simple obedience to genius as "the stranger within." Already in December 1786 we find him writing to Herzogin Luise, in ways that show him registering that tradition plays a significant part in works of art. Indeed, when Goethe comes to review his life in his autobiography, he regrets that in his youth he was "seduced" by the aesthetic theories of Hamann and Herder. In his later works, Goethe is more important as an exemplar of genius than as a theorist of genius *(Genie),* and many of his apparent remarks on the topic of genius should more properly signal the use of the German term *Genius,* which brings with it connotations of patrilineal traditions. The end of *Faust, Part II* (published posthumously in 1832) is interesting in this respect. The closing lines famously evoke "the Eternal Feminine" that "draws us onwards"—and this was a theme that was taken up by such Romantic composers as Robert Schumann in 1844–1849, Franz Liszt in 1854/57, and Gustav Mahler in 1906 in their various musical settings of themes from *Faust.* However, although the figures of Faust and Goethe had by this time collapsed into each other as a kind of double-headed exemplar of genius, the "femininity" that draws Faust on by no means annuls the maleness of his character.

The work of genius required "femininity," but was generally debarred to female creators (such as Mahler's own wife, Alma) who were Muses to genius, but also represented as "unsexed" by genius. The many female authors who manipulated the vocabularies of genius found themselves trapped by the discrepancy between the vocabularies of embodi-ment ("male"/"female") and the vocabularies of cultural expectation of psychic and behavioral difference ("masculine"/"feminine"), which operated differentially in the case of genius. Madame de Staël tried hard in her influential novel *Corinne, or Italy* (1807) to portray a female genius. She also recorded her meetings with male geniuses such as Goethe, Johann Christoph Friedrich von Schiller (1759–1805), and August Wilhelm von Schlegel (1767–1845) in her equally influential *On Germany* (1810/13). Staël's essay on her travels through Germany established "Romanticism." She was the first to use the term—and gave it a German face. However, her attempts to represent her fictional alter ego, Corinne, as a genius who is both "beautiful" and "sublime," although profoundly important for a succession of women writers from Elizabeth Barrett Browning (1806–1861) to Emily Dickinson (1830–1886), failed to displace the figure of genius as a supernormal male. Napoleon—who hated Staël so much that he claimed he could still "sense her" in exile—came to represent genius, alongside Goethe (read through the character of Faust) as modes of "universal" genius that excluded the female.

Genius was presented as undertaking work that was dangerous, difficult, and itself "sublime," bringing the teleology of man (the best European males) to its ultimate conclusion. Shakespeare continued to be admired, but was no longer represented as producing his dramas as easily and effortless as a "bird on the bough." Sociological studies of the genius vocabulary show how by the mid-nineteenth century the figure of the genius was frequently linked with madness, and with those modes of artistic work that were valuable for humanity, but that were also dangerous for the individual. Arthur Schopenhauer's account of genius in *The World as Will and Representation* (1818/44) and *Parerga and Paralipomena* (1851) fits the model of genius as a male who transcends individuality. [*See* Schopenhauer.] Schopenhauer's influence can also be detected in such writers as Friedrich Nietzsche (1844–1900) and Sigmund Freud (1856–1939). Thus, for Freud, as much as Schopenhauer, artistic creation was a function of the not-I, "it," or "id" *(das Es)* that threatened the bounds of the "I" or "ego" *(das Ich).* Furthermore, for Freud the drives of the id that were sublimated into art by the genius were the predominantly sexual energies of libido—and all libido was explicitly masculine. [*See* Freud.] For Schopenhauer, the breakdown of (male) individuation has a value that it is lacking in Freud; but Schopenhauer also insists that it is only the male that has the kind of ego that can be transcended through genius. *Genius,* however, stands alongside *Genie* as a key term in Schopenhauer's texts. Indeed, he explicitly employs the ancient Roman myths of *genius* to buttress his metaphysics in chapter 44 of *The World as Will and Idea.*

The explicit racial prejudices that linked the superior being of the "genius" with European art—as opposed to the anonymous "craft" or handiwork of the supposedly inferior races—also fed on the Roman mythology of the *genius* cult that linked the best males of European descent in a contin-

ual spiritual lineage of patrilineal ancestors that stretched back to the nobility of Rome and also of Greece. It is this tradition that Nietzsche is drawing on in the recently rediscovered section of *Ecce Homo* in which he relates himself to his Roman and Greek ancestors: "The *greatest* individuals are the oldest: I do not understand it but Julius Caesar could be my father—*or* Alexander, this Dionysus in the flesh" (Nietzsche, 1980, vol. 6, p. 269). Nietzsche positions himself as Augustus, the son of Julius Caesar, who institutionalized worship of his *Genius* (and hence of his male ancestry) when he became emperor. Alexander (356–323 BCE) was also reputedly the son of Zeus. He died without designating a successor, and his death opened the anarchic age of the *Diadochi* (the "successors"). In positioning himself as son of Alexander, and hence as a "successor," Nietzsche reworks the metaphorics of the Roman emperor-genius. He is the son who will receive the divine reproductive power of the paterfamilias at the man-god's death.

Although the Romantics and their heirs also frequently represented the genius as androgynous, and as creating out of the "feminine" side of his personality, there was also a frequent insistence that the genius could not be a woman. The metaphorics of genius in the late eighteenth century and the early nineteenth century involve insistent reference to seeds, plants, and fertile soil, to spiders spinning webs out of their bellies, and to other procedures of plant and animal reproduction. There was, however, only cautious reference to the human processes of procreation, and then the links were, above all, with the male processes of fertilization, and to the economy of a body in which, as Samuel Taylor Coleridge put it in chapter 11 of his *Biographia Literaria* (1817), "certain secretions must be taken up again into the circulation, and be again and again re-secreted" in order to ensure the "healthful vigour, both to the mind and to its intellectual offspring." Whereas the archetypal analogy for organic growth must surely be that of the fetus in the womb, the Romantics in general either avoided the metaphors of human reproduction entirely or linked genius to a physiology that secretes and resecretes sperm.

Later in the century, there was more frequent reference to the androgynous genius as a male mother, who shapes material provided to him by his own imagination, and who thus combines "femininity" with spermatic energies capable of shaping dead or unruly matter. Writers like Nietzsche, however, who employ the "male mother" metaphor also carry on describing art as generated "by the cerebral system bursting with sexual energy" and "the distribution of semen" in the artist's blood (1967, section 805). Nietzsche's paradigm creator and artist remain male. He eulogizes Dionysus, a god who was himself male, but who is also allied with "woman" and with the ecstatic breakdown of the classical male subject position of autonomy and self-control. Thus, Nietzsche reserves all cultural production—and intellectual and spiritual "work"—for males. "Woman,"

by contrast, continues to be allied with reproductive processes that are "easy" and "natural." Thus, Nietzsche's "woman attains perfection in everything that is not a work: in letters, in memoirs, even in the most delicate handiwork, in short in everything that is not a *métier*"—or a work of genius (ibid., section 817). [*See* Nietzsche.]

Analogously, Carl Jung (1875–1961) describes the male as the creative and productive being who bringing forth involves what he calls the "*logos spermatikos,* the spermatic word." Whereas "the man brings forth his work as a complete creation out of his feminine nature," the analogous intellectual and creative "seeds" that derive from the "inner masculine side of a woman" have only "the power to fertilise the feminine side of the man" (*Collected Works*, vol. 13, p. 41 and vol. 7, p. 207). Thus, female "seeds" have an inferior formative force in comparison to those of the male. Woman remains linked with a matter that is incapable of self-formation. In accordance with this logic, the female genius was a freak—either not truly a genius, or not truly female. Not androgynous, but hermaphroditic, she was represented as neither male nor female, and as spiritually and physiologically sterile. By the end of the nineteenth century, it had become a cliché to assert with Cesare Lombroso in *The Man of Genius* (1863): "There are no women of genius; the women of genius are men."

From the late eighteenth century on, then, "genius" was conceptualized as both emerging from the (male) self and as involving a breakdown (or transcendence) of the normal boundaries that could contain the (male) self. The addition of Christian and Judaic understandings of creation in early modern Europe had transformed the "genius" god, and made him analogous to Yahweh—the self-creating, all-creating "I am that I am"—of Exod. 3.13–14. For the Romantics, it was the strength of the genius's will or his "I" that imposed the unity of artistic form on material that would otherwise have remained diverse, inert, chaotic, unruly, and heterogeneous. The genius confronted "the sublime": that infinity that has the capacity to overwhelm the "I" or the ego. The genius courted madness. For the early Romantics, however, at least the genius does not surrender to madness. His "I" is not completely overwhelmed by its confrontation with the infinite or indefinite "other." Fear is transcended, and infinity is contained in ways that mean that the indefinite, the chaotic, and the "other" lack the power to disturb.

The uniqueness of the genius's "I am" is reflected in every facet of his artwork—and indeed his whole oeuvre. For the Romantics, the great work of art "grew" or "burst forth": it was not a product of calm forethought or rational design but an overflow of burgeoning energy. This outflow, however, was commonly still represented in terms of an active force shaping an inert or unruly matter. Mental and biological "work" was exerted on material that was itself either unstructured or unruly—chaotic dreams, fantasies, instincts, drives, longings—as well as on heteronomous mat-

ter that comes from outside the self. It was intellectual, spiritual, and artistic "work" that was privileged. Romanticism turned against the artistic "gentleman" who never engaged in intellectual or spiritual toil. Romanticism also reinstituted a contrast between the cultural and intellectual "labor" of artistic production ("genius") and those (female) processes of reproduction that were supposedly "natural," "easy," passive, and without formative force.

It was the male self that encountered and regulated that infinite and unrepresentable in the dynamics of material excess and control, blockage and transcendence that constituted the "sublime." Woman, by contrast, was allocated to the sphere of the "beautiful": the delicate and charming that is loved, but that does not threaten the very boundaries of the self by an excess of power. Nineteenth- and early twentieth-century portraits of male "geniuses" typically throw light on or otherwise emphasize the forehead, in ways that rework Roman conceptions of divine male spirituality. The "work" of imaginative synthesis is now psychological. This is what explains Michel Foucault's acute definition of "madness" as "the absence of an oeuvre" in the 1972 edition of *Histoire de la folie*.

Until the work of feminist scholars in the 1980s, the sexual politics of the language of genius was simply not noticed by the writers of the hundred of monographs on this topic. Edgar Zilsel's German monograph *Die Entstehung des Geniebegriffes* (1926) was exceptional in even mentioning the fact that the term *genius* started off in Latin as a word referring to the vital forces of male sexuality. It is not that those scholars who attempted to tie our modern notion of genius to the Latin term *ingenium*, rather than to the Latin term *genius* itself, did so in a conscious attempt to sanitize our modern concept and free it from its gender implications. It is rather that the extraordinary gender logic of genius—which makes the genius both "male" and "feminine"—was obscured by the numerous psychological, philosophical, literary, and historical commentaries on genius that emerged prior to second-wave feminism.

Furthermore, since Freudian, Jungian, and Lacanian psychoanalysis drew on—and further mythologized—the tradition that represented art and culture in terms of the dream "work" and sublimated male energies, the emphasis on the "feminine" within deconstructive and psychoanalytic theory is also frequently insufficiently critical of the language of the Dionysian. In postmodern aesthetics, too, there is a frequent tendency to equate "modernity" with the rational, the certain, and the "I" that is transparent to itself, and to suppose that the fragmentation of the self into energy or drives is a move that allows us to move "beyond" gender. The accounts of "genius" in post-Kantian aesthetics, however, involved a reversal of the values of the previous "Enlightenment" age, but retained a model of an "I" that worked hylomorphically on the material presented by the senses, and in ways that nevertheless still take the male subject as norm.

Although in aesthetic theory, it is the Romantic concept of genius that was—and remains—the most potent, it should not be supposed that other traditions of theorizing genius no longer have any contemporary currency. Indeed, any cursory investigation of current psychological and sociological research reveals at least three competing notions in the psychological literature: first, that among a group of contemporaries within a particular specified field of creative activity, whoever is the most innovative is the genius; second, that the genius is the one who is historically the most influential or successful; third, that genius is simply to be equated with an excess of talent. In all these instances, the genius can be conceptualized as healthy and as supernormal, rather than as deviant and abnormal or demonic.

Although the tradition of discussing genius as a healthy and "supernormal" individual is found primarily in texts that use statistical means to measure creativity and analyze highly talented individuals, similar strands of thought can also be detected in everyday talk about gifted children and creativity. A genius on these models is not necessarily a personality type; nor does genius require intellectual "struggle" or "labor." Instead, genius can be something one "has," rather than something one "is." The "genius" on the "supersane" model does not necessarily live out of step with his society, but is a product of it and might, indeed, live in harmony with it. Thus, these psychological and everyday notions of genius keep alive early eighteenth-century, and hence pre-Romantic, uses of the term *genius* in which individuals are not geniuses, except insofar as they have a genius: a talent or disposition that enables them to produce a particular kind of artwork or make some valuable discovery. Historically, of course, women have fared badly in the supposedly "neutral" and "objective" assessments of supernormal talent that have been used to assess genius. At least on these models, however, women do not find themselves facing the double bind of a hidden metaphysics and aesthetics that can think human excellence only in terms of the supernormal—or even "feminine"—male.

Romantic "genius" required an overall and shaping "I" that works on the material present in the psyche until it is marked through and through by the very "essence" or individuality of the "genius." While being highly critical of any notion of "genius," it can be argued that it is strategically important to represent women writers and artists in terms that do not fragment their oeuvres into discontinuities (see Battersby, 1990). The move to "deconstruct" genius is problematic for those categories of humans who have been denied the status of "author," or "artist" with an oeuvre that can persist—and develop and mature—through change. Whereas for Foucault the lack of an oeuvre can be regarded as the definition of madness, for women in artistic and literary histories the absence of an oeuvre has been the cultural norm. This is also true of black and/or "Third World" artists whose works are consigned to "craft" galleries or to anthro-

pological and social history museums that analyze "national traditions" and "ethnic groupings," rather than to "art" galleries that focus on the work of "individuals." Where the works of women and nonwhite subjects are registered as the productions of an individual personality, these works are too often interpreted as the products of "tendencies" or "movements" or as autobiographical and emotional "responses" to a particular set of circumstances. Even the specific *genres* in which women and black artists have been deemed to excel have also been frequently graded as not revealing "genius" in its highest form. Thus, sketching, embroidery, watercolors, singing, ballet, pottery, and novel writing (for women), and dancing, jazz, and blues (for nonwhites), were historically dismissed as revealing expressive powers, but not the type of transformative force that was requisite for the most "sublime" modes of "genius." Although there has been a recent loosening in this grading of artistic activities, greater esteem (and differential amounts of public funding) are still allocated to "high" art such as sculpture, oil painting, and musical composition, as opposed to patchwork, pottery, or improvisational music. The "genius" vocabulary—and the related vocabularies of "creativity" and "originality"—continue to secure this divide.

The deconstructive move is particularly problematic where it is phrased in terms of the "femininity" of genius or in terms of an "Otherness" that transcends the masculinized self, because historically the psychic femininity of genius and an emphasis on the virile body and energy of the (male) genius were concepts that belonged together. Rather than seeking to "deconstruct" genius, one could argue that we need to return to a conception of genius that has its origins in Kant, while also subverting that account. Kant represents "genius" in terms of the founder of a new "tradition" when he defines the genius as the one who gives the rule to art. On this model, the genius is the one whose work does not rely on preexistent rules, but who is also not autonomous. The genius only emerges within a context (as the forerunner of a tradition). As such, "genius" is defined relationally, but what is unique in his or her work is not lost— and neither do we lose the importance of an artist in terms of traditions of art production. We need to start with Kant, but subvert his account of genius by registering that "traditions" are not "real" entities that are given to us by "nature" or by a unilinear history. Because there are not just patrilineal lines of inheritance, "traditions" are multiplicitous and open-ended. Genius does not establish its preeminence only by standing in relation to, and competition with, spiritual fathers—in Oedipal succession. On the contrary, "genius" needs to be thought relationally, and as establishing a (fluid and provisional) identity through complex patterns of relationship. But to think in this way would also involve thinking an artist's oeuvre—and hence also "genius"—as emergent properties, and not as a "form" that is imposed on matter by the "I."

In Kant, identity (the transcendental self) is only ever established in contradistinction from the other (the transcendental object) through the imposition of necessary rules (the categories) that form space and time in a "top-down" way. The question is what happens when we think artistic form from a position that has abandoned hylomorphic metaphysics, but not given up on notions of identity, self, or an "I." We need to think what is distinctive or "essential" about an artist in ways that do not implicitly value "form" as that which is "imposed on" a (passive) matter by a "self" (or an "unconscious") that "works" on the material presented to the senses in the manner of the Romantic "genius." In this respect, "genius" is not a topic that has been exhausted or is of merely historical or sociological interest. Instead, it offers important openings for philosophical critique, and clues also for reconstructing models of the "I" or the notion of an oeuvre in ways that are attuned to issues of racial and sexual difference (not the "feminine"). To this end, we can carry on employing the "genius" vocabulary, but only tentatively and strategically. This is because the language of "genius" cannot be completely cleansed of a hylomorphic metaphysics, nor of a metaphorics that links genius to the (spiritual) alter ego of the man-god who would control nature (and culture) through his patrilineal (European) noble bloodlines.

[*See also* Feminism.]

BIBLIOGRAPHY

Abrams, M. H. *The Mirror and the Lamp: Romantic Theory and the Critical Tradition.* New York and Oxford, 1953.

Battersby, Christine. *Gender and Genius: Towards a Feminist Aesthetics.* London, 1989; Bloomington, Ind., 1990.

Becker, George. *The Mad Genius Controversy: A Study in the Sociology of Deviance.* Beverly Hills, Calif., 1978.

Brown, Jane K., Meredith Lee, and Thomas P. Saine, in collaboration with Paul Hernandi and Cyrus Hamlin, eds. *Interpreting Goethe's Faust Today.* Columbia, S.C., 1994.

Fabian, Bernhard. Introduction to critical edition of Alexander Gerard, *An Essay on Genius, 1774.* Facs. ed. Munich, 1966.

Foucault, Michel. "La folie, l'absence d'œuvre" (1964). In *Histoire de la folie à l'âge classique: Suivi de mon corps, ce papier, ce feu et la folie et l'absence d'œuvre,* 2d ed., pp. 575–582. Paris, 1972.

Jung, Carl Gustav. *Collected Works.* 20 vols. Edited by Herbert Read, Michael Fordham, and Gerhard Adler; translated by R. F. C. Hull. London, 1953–1971.

Kant, Immanuel. *Critique of Judgment.* Translated by Werner S. Pluhar. Indianapolis, 1987.

Murray, Penelope, ed. *Genius: The History of an Idea.* Oxford and New York, 1989.

Nietzsche, Friedrich. *The Will to Power.* Edited by Walter Kaufman, translated by Walter Kaufman and R. J. Hollingdale. New York, 1967.

Nietzsche, Friedrich. *Sämtliche Werke: Kritische Studienausgabe in 15 Einzelbänden.* 15 vols. Edited by Giorgio Colli and Mazzino Montinari. Munich, 1980.

Nitzsche, Jane Chance. *The Genius Figure in Antiquity and the Middle Ages.* New York, 1975.

Schlapp, Otto. *Kants Lehre vom Genie und die Entstehung der 'Kritik der Urteilskraft.'* Göttingen, 1901.

Schmidt, Jochen. *Die Geschichte des Genie-Gedankens in der deutschen Literatur, Philosophie und Politik, 1750–1945.* 2 vols. Darmstadt, 1985.

Schopenhauer, Arthur. *The World as Will and Idea.* 3 vols. Translated by R. B. Haldane and J. Kemp. London, 1883–1886.

Weininger, Otto. *Sex and Character.* London, 1906.

CHRISTINE BATTERSBY

GENTILE, GIOVANNI (1875–1944), neo-idealist philosopher in whose philosophy aesthetics played a central role. Born in Castelventrano, Sicily, and assassinated in Florence, Gentile was a professor of the history of philosophy at Palermo, professor at Pisa and Rome, and minister of Education in Benito Mussolini's government in 1922, as well as the first president of the National Fascist Institute of Culture.

Gentile was attracted to philosophy by Jaja, a pupil of the influential Hegelian Bertrando Spaventa. Gentile was thereafter on the closest terms, personally and professionally, with Benedetto Croce until political and philosophical differences led to a bitter separation. Throughout his life, Gentile had a passionate concern with the theory and practice of education. In 1900, he published the important work *The Concept of Education,* followed in 1913–1914 by his two-volume *Summary of the Theory of Education as a Scientific Theory* in which his idealist philosophy is first expounded. In 1914, having acceded to Jaja's chair in Palermo, he published *The General Theory of the Spirit as Pure Act,* followed, after his move to Rome, by a more systematic statement in the first volume of his *System of Logic as a Theory of Knowing.* His aesthetics, which is integrated into and assumes a central role his general philosophy, and, indeed, his idealism, is best grasped through the remarkable, and remarkably neglected, *Philosophy of Art* (1931), a book of rare passion and eloquence that has been translated, with a valuable introduction, by Giovanni Gullace. His last work, posthumously published, is the short but important *The Genesis and Structure of Society.*

For Gentile, philosophy was the science of knowledge, the discipline that, though presupposing nothing, grounds, a priori, all the other sciences. René Descartes's method of doubt initiated this movement of thought, and George Berkeley's claim that *esse est percipi* was an important advance toward a solution. Immanuel Kant's and G. W. F. Hegel's advance on this was to examine the conditions under which the mind organizes its experiences, and the neo-idealism of Gentile and Croce is an attempt to develop further their approach. Thus, we can understand *neo*-idealism as continuing the tradition—to which, for example Johann Gottlieb Fichte and Hegel belong—of assenting to the motivations that prompted the Kantian revolution and, then, querying whether various attempts to reshape the story of

the nature of knowledge in accordance with that revolution had been properly undertaken in the light of those motivations.

Kant had talked of the way in which the mind organizes the data that impinge on it. To put it thus is to assume that there are mind-independent data to be organized, and that is, again, to allow entry to the very model of the mind as set against something independent of and different from it that, as we have seen, was meant to be avoided by the idealist project. Gentile shares with many of Kant's successors a skepticism as to whether Kant, or, in the particular case of Italian neo-idealism, Hegel, had gone far enough in making knowledge dependent on the mind's constructions.

Here is Gentile asserting the full-blooded idealist thesis that there is no reality that is not the reality of thought:

> We give the name *realism* to the method of thought that makes the whole of reality an independent object, abstracted from thought, to which thought as an activity should conform. By idealism, on the other hand, we mean the higher point of view, from which we discover the impossibility of conceiving a reality that is not the reality of thought itself. . . . The reality that is called material and external to thought is in some way illusory, and . . . true reality is just that which comes through the efforts of actual thinking. For, indeed, there is no way of thinking any reality, except by establishing thought as the basis of it. . . . According to this concept, there is, properly speaking, no reality; there is only the spirit that creates reality. . . . The realist speaks of something already existent, a world into which man comes and must adapt himself. The idealist knows only the world that the spirit creates, a world in which nature itself is always internal to the progress of the spirit and throbs within the soul of man, who continually remakes it with his intellect and his will in a restless activity that knows no pause, a world that is never complete. (1920, pp. 55–57)

For Gentile, then, the act of sensation (where the emphasis on what happens in the *act* of sensation explains why Gentile's version of idealism is called "actual idealism") is a process of self-constitution in which the subject preserves its own past and connects it to the present sensation, language being the instrument that gives this process form and expression. In that act there is a spontaneous self-affirmation that it can be the aim of the artist to capture in the disciplined expression of feeling. Thus, art is integrally linked to expression. The task of the interpreter (and Gentile wrote a number of pieces motivated by this conception) is to comprehend the motivation, the philosophy of the self-affirmed life, that is articulated in the expressed work.

Gentile was an influential figure. Much of R. G. Collingwood's *Philosophy of Art* may derive from Gentile's views, although, as Merle Brown argues in *Neo-Idealistic Aesthetics* (1966), Collingwood's work is an uneasy mélange of the very different and antagonistic views of both Croce and Gentile. Gentile, as Gullace makes clear in the preface to his

translation of Croce's *La Poesia* (1936), influenced Croce greatly. In the end, Croce's idealism was not thoroughgoing enough for Gentile, who subjected it to bitterly sarcastic rejoinders in his *Philosophy of Art*.

Gentile's philosophy understandably fell under a cloud because of his self-assumed role as the official theorist of Fascism, his intimate association with that movement being the direct cause of his assassination in the last days of the war in Italy. It is debatable whether idealism, as a philosophy, was especially suited to the task undertaken by those seeking to give an intellectual underpinning, coherence, and respectability to the somewhat bizarre ragbag of notions that constituted Mussolini's pseudophilosophical ramblings. Gentile, however, certainly believed that neo-idealism did justify Fascism (a matter on which he radically and bitterly differed with Croce, for whom that philosophy underpinned a radical liberalism). Whether or not idealism essentially entails a corporative political philosophy, it is obviously easy to take the step that Gentile took from believing that divisions between, say, mind and world are intellectually incoherent to believing that the elimination of divisions by the corporate state has an intellectual justification, particularly when that step was made at a time, in the 1920s, when a postwar fragmentation of political life threatened social chaos. It is equally easy to see why neo-idealists, opposed as they are to divisions, were tempted toward the more defensible notion of art as an organic unity.

At present, at least in Italy, Gentile's philosophy is being reexamined as a development of idealism. Certainly, he is worth reading to see what a thoroughgoing idealism might look like, and his views of art are interesting, if only because of the eloquence and passion with which he expresses and elaborates his distinctive version of the view that art is expression.

BIBLIOGRAPHY

Works by Gentile

The Theory of Mind as Pure Act (1919). Translated by W. Wildon Carr. London, 1922.
La riforma dell'educazione. Bari, 1920.
The Philosophy of Art (1931). Translated by Giovanni Gullace. Ithaca, N.Y., 1972.
Opere Complete. Florence, 1935–.
The Genesis and Structure of Society. Translated by H. S. Harris. Urbana, Ill., 1960.

Other Sources

Brown, Merle E. *Neo-Idealistic Aesthetics.* Detroit, 1966.
Croce, Benedetto. *Benedetto Croce's Poetry and Literature: An Introduction to Its Criticism and History.* Translated by Giovanni Gullace. Carbondale, Ill., 1981.
Harris, H. S. *The Social Philosophy of Giovanni Gentile.* Urbana, Ill., 1960.
Holmes, Roger W. *The Idealism of Giovanni Gentile.* New York, 1937.

Romanell, Patrick (Pasquale Romunelli). *Gentile: The Philosophy of Giovanni Gentile: An Inquiry into Gentile's Conception of Experience.* New York, 1938.
Romanell, Patrick. *Croce versus Gentile: A Dialogue on Contemporary Italian Philosophy.* Reprint, New York, 1982.

COLIN A. LYAS

GERARD, ALEXANDER (1728–1795), eighteenth-century scholar, philosopher, and theorist of taste, and a member of the Scottish, "commonsense" Aberdeen Philosophical Society (1758–1773) founded by Thomas Reid. A precocious scholar, Gerard was appointed professor of philosophy at Marischal College, Aberdeen, at age twenty-two. Ten years later, he assumed a professorship of divinity at Marischal, and in 1771 he left Marischal to become professor of divinity at King's College, Aberdeen. In aesthetics, he is primarily noted for *An Essay on Taste* (1759), based on a prizewinning essay of 1756. A second edition appeared in 1764 with translations of essays on taste by François-Marie Voltaire, Jean Le Rond d'Alembert, and Baron de Montesquieu. Gerard's *An Essay on Genius* appeared in 1774, and in 1780 a third edition of *An Essay on Taste* appeared without the French essays but with an added fourth part, "Of the Standard of Taste," replying to positions taken by Abbé Jean-Baptiste Dubos, David Hume, and Henry Home (Lord Kames) on deriving critical standards inductively from experience.

Like most eighteenth-century theorists prior to Immanuel Kant's popularization of the concept of the aesthetic, Gerard thought of human critical interest in art and nature as *fine taste*. In step with Francis Hutcheson's Lockean theory, Gerard held that fine or critical taste was an *internal* or *reflex sense*. As a sense, fine taste provided experiences that were simple in nature (distinctive and irreducible in feeling), immediately felt (rather than realized after the encounter), and nonvolitional (the content was unalterable by will). As a reflex sense, taste depended on the data of external sensation, reflecting on and subsequently responding to those sensations. In hearing harmony in music, one hears the sounds, but one pleasurably senses the harmony reflectively with one's taste. This mode of critical perception was analogically identified as a taste because of the predisposition of gustatory taste to respond hedonically to what one orally ingests. Gerard's most frequent term for critical appreciation is *relishing*: in this usage, he underlines the analogical parallel between critical experience and enjoyment of food and drink.

For Gerard, the pleasure in critical experience is not just an effect of stimulus sensory contact but involves the active operations of the imagination. Thus, taste is not a passive mode of experience; it is an aggregate sensibility combining

several powers of the imagination. Gerard discusses seven distinct kinds of imaginative sensibility: novelty, sublimity, beauty, imitation, harmony, ridicule, and virtue. These are reflexive operations because consciousness experiences pleasure or pain as a product of the particular mental activity engaged in. He claims that "Facility in the conception of an object, if it be moderate, gives us pleasure: the mind thinks well of itself, when it is able to form its conception without pain or labour." Perceiving an object as novel is pleasurable because of the "moderate difficulty" that the mind finds enjoyable in conceiving of a new object. The ridiculous or humorous is a pleasurable diversion created by mentally processing surprisingly incongruous objects or features. A beautiful figure or appearance, having the distinctive Hutchesonian qualities of uniformity and variety as well as proportion, initiates the pleasurable experience of "giving the mind at once the opposite gratifications of facility and active exertion."

These pleasurable mental exertions are activities of the imagination because they connect ideas according to rules of association. For Gerard, as for other eighteenth-century associationists such as Hume, the exercise of the imagination naturally and systematically relates ideas. Instead of being an erratic faculty, the imagination "commonly observes certain general rules, associating chiefly ideas of such objects as are connected by the simple relations of *resemblance, contrariety,* or *vicinity;* or by the more complex ties of *custom, co-existence, causation,* or *order*" (1764). The imagination makes these associations subtly, quickly, and in a universal or lawlike way. Thus, although taste is reducible to activities of the imagination, as an experience, taste has the distinctive and simple quality of a sense. It is, Gerard says, "compound in its *principles,* but perfectly simple in its *feeling*" (1764). In *An Essay on Genius,* Gerard develops an account of genius as employing these associative principles in creative inventions.

Gerard also acknowledges the importance that judgment, good sense, and the intellect play in taste; for fine taste, although dependent on certain natural gifts (one cannot be tone-deaf and relish musical harmony), must be cultivated and improved. Judgment plays a significant part in this education. Taste, he claims, improves to the extent that its delicacy of sensibility can be rendered more acute and refined; it is perfected by becoming correct and proportionate in its focus operations.

Judgment also undercuts that critical relativism according to which all taste is equal, no taste is mistaken, and reasons lack efficacy. Judgment's most prominent role in critical activity is to discover the exact nature of the object under critical scrutiny. Its goal is the identification of the qualities of objects responsible for the sensory relish of the critical experience. Knowledge of the object exhibited in one's judgment will allow one to see whether one has responded appropriately.

In "Of the Standard of Taste" (1780), Gerard (contra Hume) argues against allowing a broad inductive sampling of those qualities that produce pleasurable experience as the basis for determining critical standards. Only a judiciously criticized pairing of objective qualities and resulting sentiments can establish a valid critical standard. Even such a generalization, however, cannot aid in offering an immediate evaluation of a new work of art. For that, one must rely on critics whose "vigorous and improved taste" is rendered in evaluations supported by arguments.

The major problem in Gerard's theory is the tenuous linkage between taste as simple, immediate, and nonvolitional feeling and taste as discerning judgment—deliberate, complex in operation, and volitional because it is reasoned. Gerard acknowledges this unresolved split in his theory of taste by recognizing two "lights" of taste, either "as a species of *sensation,* or as a species of *discernment.*" At the root of this forced reconciliation is an overly simplistic conception of critical pleasure. Whereas Gerard acknowledges the variety of pleasurable, imaginative experiences, he insists on the simple unitary nature of critical pleasure as a sensation. He admits that there are a variety of pleasurable activities, but that pleasure per se is a simple feeling. He claims that "we cannot distinguish them [the feelings] into kinds." If he were to admit that pleasure could assume different forms in accordance with different critical interests, that would compromise his holding to the simple, immediate, and nonvolitional senselike character of critical experience.

[*See also* Taste.]

BIBLIOGRAPHY

Works by Gerard

An Essay on Taste. 2d ed. with three dissertations on taste by Voltaire, d'Alembert, and Montesquieu (1764). New York, 1970.
An Essay on Genius (1774). New York, 1970.
An Essay on Taste, 3d ed. with an added part 4, "Of the Standard of Taste" (1780). Delmar, N.Y., 1978.

Other Sources

Dickie, George. "The Association and Coalescence of Ideas: Alexander Gerard." In *The Century of Taste: The Philosophical Odyssey of Taste in the Eighteenth Century,* pp. 29–54, 69–84. New York and Oxford, 1996.
Grene, Marjorie. "Gerard's *Essay on Taste.*" *Modern Philology* 41 (August 1943): 45–58.
Hipple, Walter John. "Alexander Gerard." In *The Beautiful, the Sublime, and the Picturesque in Eighteenth-Century British Aesthetic Theory,* pp. 67–82. Carbondale, Ill., 1957.
Kallich, Martin. *The Association of Ideas and Critical Theory in Eighteenth-Century England.* The Hague, 1970. See pp. 167–198.
Kivy, Peter. "Recent Scholarship and the British Tradition: A Logic of Taste: The First Fifty Years." In *Aesthetics: A Critical Anthology,* 2d ed., edited by George Dickie, Richard Sclafani, and Ronald Roblin, pp. 254–268. New York, 1989.
Larsen, Elizabeth. "Re-Inventing Invention: Alexander Gerard and *An Essay on Genius.*" *Rhetorica* 11.2 (Spring 1993): 181–197.

McCosh, James. *The Scottish Philosophy, Biographical, Expository, Critical, from Hutcheson to Hamilton.* New York, 1890; reprint, Hildesheim, 1966. See pp. 227–229, 467–473.

Wiley, Margaret Lee. "Gerard and the Scots Societies." *University of Texas Publication Studies in English* 4026 (July 1940): 132–136.

KEVIN W. SWEENEY

GERMAN AESTHETICS. *The main theorists or artists in the history of aesthetics in the German-speaking world from the eighteenth century to the present are discussed in individual entries.* See Adorno; Auerbach; Baumgarten; Benjamin; Beuys; Bötticher; Brecht; Cassirer; Dilthey; Freud; Gadamer; Giedion; Gottsched; Habermas; Hamann; Hegel; Heidegger; Helmholtz; Herder; Herz; Heydenreich; Hölderlin; Hübsch; Husserl; Kant; König; Lessing; Lotze; Mannheim; Mattheson; Meier; Moritz; Nietzsche; Novalis; Paul; Riegl; Schleiermacher; Schmarsow; Schoenberg; Schopenhauer; Semper; Simmel; Sulzer; Wagner; Winckelmann; Wittgenstein; *and* Worringer.

GESTALT PSYCHOLOGY. *See* Arnheim; Attitude, *article on* Pictorial Attitude; Perception, *article on* Psychology of Perception; *and* Psychology of Art.

GIBSON, JAMES JEROME. *See* Attitude, *article on* Pictorial Attitude; Gombrich; *and* Perception.

GIEDION, SIGFRIED (1880–1968), Swiss architectural historian, critic, and secretary of CIAM (Congrès Internationaux d'Architecture Moderne, 1928–1956). For Giedion, aesthetic theory was rarely an explicit topic. Like others in architecture and design during the 1920s, he rejected "aesthetics" and "style" as socially codified impediments to historical change and instead embraced the potential of new technologies to produce a new architecture capable of fulfilling a utopian striving for the unmediated expression of collective will or desire. Yet, in mobilizing notions of pure means, form creation *(Gestaltung),* space-time, and transparency, Giedion drew on discourses in art history as well as art and architecture that were saturated with aesthetic theories—neorationalist theories of artistic means, neo-Kantian theories of mediation and artistic cognition, neo-Romantic theories of expression and empathy, and neo-Hegelian phenomenology. Preferring to write history over systematic expositions of architectural theory, Giedion subsumed aesthetic categories into a psychologized history of architecture that embraced the entire world of material culture from art, architecture, and urbanism to the mechanization of construction, agriculture, meat production, everyday domestic life, furniture, and locks. Giedion's im-

plicit aim with this expansive materialist history was nothing less than the articulation of a new system of freedom, whose characteristics and effects he presented in terms that were implicitly aesthetic.

Where Giedion's teacher in art history, Heinrich Wölfflin (1864–1945), as well as his more distant mentors Jacob Burckhardt (1818–1897) and Alois Riegl (1858–1905), had already introduced aesthetics into the historiography of art, Giedion went so far as to give his histories the mission of contributing to what he understood to be the progressive developmental trajectory of history, which he characterized in aesthetic terms. For this he was routinely criticized by other historians throughout his later career, most notably by Manfredo Tafuri, who rejected Giedion's mode of "operative criticism" as he developed his own approach to writing history critically engaged in contemporary issues.

Giedion's dissertation, published as *Spätbaroker und romantischer Klassizimus* in 1922, thematized the backward glance of history as "a mirror that reflects only the features of the observer" and applied Wölfflin's stylistic categories (linear-painterly, closed-open, multiplicity-unity) to draw a distinction within neoclassicism between the late Baroque and Romantic classicism that was to be of consequence for Giedion's own time. He portrayed the former as the last period of unity and Romantic fragmentation as both the sign of modernity and its challenge. Turning, during the following years, to the radical, technologically based architecture promoted by Le Corbusier, Walter Gropius, and László Moholy-Nagy, among others, Giedion's second book, *Bauen in Frankreich, Bauen in Eisen, Bauen in Eisenbeton* (1928), inaugurated a programmatic genre that he later called "contemporary history." Beginning with the proposition that the task of the historian was "to extract from the vast complexity of the past those elements that will be the point of departure for the future," Giedion proceeded to present the nineteenth-century origins and morphological development of the technical forms considered central to the new architecture. He showed how long-span iron structures of train stations, iron and glass skylights of arcades and department stores, glass curtain walls independent of structure, concrete frames, and cantilevered slabs had all contributed to the self-realization of new constructive means through the gradual transformation of earlier architectonic systems.

In his most extensive prehistory of architectural modernity, *Space, Time, and Architecture* (1941), Giedion rearticulated the role of the historian as revealing the hidden unity within the welter of seemingly heterogeneous productions of the era, "showing how our period has come to consciousness of itself." He later expanded this ambition to the sphere of everyday life in *Mechanization Takes Command* (1948). By applying the analytic tools of modern "scientific" art history to the present period of transition, Giedion hoped to contribute to the emergence of a self-conscious new tradi-

tion capable of harnessing "the potentialities at our disposal which are the key to a new and balanced life for enormous numbers of men." Giedion characterized the present as schizophrenic, marked by a profound split between the spheres of science and art, reason and emotion, as well as the tension between new and old. Taking up Charles Baudelaire's theory of *modernité* as finding the eternal and beautiful within the fleeting transitoriness and ugliness of modern life, Giedion sought to isolate the constitutive characteristics of the emerging era from transitory symptoms of the new destroying the old "ruling taste" as it struggled to realize itself from within it. Adapting the psychoanalytic theory of repression to historical analysis, he defined "constituent facts" as those that, "when they are suppressed, inevitably reappear." The recurrence of slabs and planes, undulating walls, column grids, and interpenetrating interior and exterior spaces, he held, "makes us aware that these are elements which, all together, are producing a *new tradition*"— a new architectonic language of expression, which was epitomized by Le Corbusier's "Five Points" of 1927.

In *Bauen in Frankreich*, Giedion designated the expressive and generative capacity of the collective unconscious with the notion of *Gestaltung*, meaning form creation or the organic lawfulness of self-construction. While *Gestaltung* could be associated with Conrad Fielder's theory of artistic creativity and autonomy (which he opposed to speculative aesthetics), on the one hand, and scientific theories of physiognomic expression (Johann Caspar Lavater) and plant morphology (Johann Wolfgang von Goethe), on the other, Giedion applied the term to works of industry and engineering. These appeared to him not only anonymous and unself-conscious but intensively enmeshed in the life process, and hence capable of displaying historical change before other fields. *Gestaltung* could break through the historicizing mask of architecture with "emanations of the highest impulse: LIFE!" His observation that "construction in the nineteenth century plays the role of the subconscious" came to play a strategic role in Walter Benjamin's approach to the prehistory of modernity in early artifacts of mass culture in nineteenth-century Paris.

Later, in *Space, Time, and Architecture,* Giedion subsumed the idea of *Gestaltung* within a historical analysis of what he took to be the emergence of a new postperspectival and nonanthropomorphic space conception. Like other historians of his time working within the broad influence of both neo-Kantian and Hegelian aesthetics, Giedion wanted to articulate the structural conditions, underlying categories, or state of consciousness of the historical period he was studying. August Schmarsow (1853–1936), for instance, had already suggested in 1894 that the history of architecture is the history of the sense of space, and thus consciously or unconsciously it is a basic constituent in the history of worldviews; and Erwin Panofsky (1892–1968) had, in the 1920s, borrowed Ernst Cassirer's theory of symbolic forms (such as language) to suggest that the science of perspective had served to structure, mediate, and regulate art, architecture, and urbanism in the Renaissance. [*See* Panofsky; *and* Schmarsow.] In this context, Giedion sought to articulate a system of constituent facts that could be understood as structuring the potentiality of architecture and urbanism in the modern period as it had emerged since the Baroque and struggled to realize a new universal in the twentieth century.

In *Space, Time, and Architecture,* Giedion argued that affinities between the means and spatial affects of engineering structures, post-Cubist art, modern mathematics, and the new architecture pointed to the emergence of a postperspectival and relational space conception, which he called "space-time." Elaborating on the use of the term in both art and science—citing Guillaume Apollinaire's interpretation of Cubism's fourth dimension as well as Hermann Minkowski's and Albert Einstein's theories of relativity—he distinguished space-time by its openness to relativity, incompleteness, and dynamism and assimilated it to the constitutive ambiguity of a phenomenology of spatial perception. In contrast to the graphic rules of perspective, he presented space-time as the mutual mediation of subject and object in the unrepresentable, yet equally scientific, fourth dimension in which inside and outside, subject and object, were considered interwoven. Informed by Schmarsow's theory of *Raumgestaltung* (space creation), which had been taken up in the 1920s by the artists Theo van Doesburg and Moholy-Nagy, Giedion's notion of space-time proposed that the cognitive status of the building had become contingent on the partial and shifting perceptions of observers moving through and around them. He conflated this with theories of image formation found in the post-Cubist discourse of "new optics" in which he was active during the late 1920s. He was optimistic that architecture could participate in realizing the new world of forms and experience suggested by the ability of scientific instruments, airplanes, photography, and film to expand vision. Rather than presuming to comprehend buildings definitively from a single viewpoint, Giedion deferred cognitive closure indefinitely and kept the process—like Le Corbusier's concrete Domino skeleton—"eternally open," the subject forming and perceiving space while being formed by and for it. In order to grasp the true nature of space, Giedion suggested that "the observer must project himself through it" and photographed the dizzying stairway of the upper levels of the Eiffel Tower as one of the first opportunities for this. Le Corbusier's Domino represented, for Giedion, the means for making that experience structural to mass society in the form of housing.

Giedion's enthusiasm for such self-alienating space-time experiences erupts at key moments in the text of *Space, Time, and Architecture,* invoking the ideal of hovering excitedly above the ground, dissolving the boundary be-

tween artifice and nature into the dynamic infinite beyond representation—in likening the dematerialization of Joseph Paxton's Crystal Palace to the atmospheric affects of J. M. W. Turner's paintings; the opened volumes and suspended planes of the Bauhaus at Dessau to Pablo Picasso's first double-faced portrait; and above all the simultaneous assertion and denial of volumes in Le Corbusier's Pessac housing and the hovering transparency of his Purist still lifes. The importance assigned by Giedion to these affects is revealed most clearly in the almost ecstatic prose of his earlier, more youthful *Building in France*. There he defended the paper-thinness of the buildings—their solid volumes eaten away with cubes of air, rows of windows passing suddenly into the sky, corners merging into one another, collapsing into two-dimensionality only to spring back into depth a few steps later. For him, they "create—as in a landscape of snow under certain light—that dematerialization of fixed borders, within which it is no longer possible to distinguish between rising and falling, and walking feels like being in the clouds." The developmental curve of Giedion's histories of modernity was aimed, then, at a state of artistic and experiential freedom whose sign was the self-conscious deployment of elementarized architectonic means in open space-time constructions. The affects of such works, at odds with representation and conceptually beyond or prior to it, were presented in terms of a sublime and ecstatically experienced amorphousness. Like his close friend Moholy-Nagy, Giedion imagined the immanent condition of freedom in terms of an aesthetics of transparency and ceaseless becoming, a dynamic and fluid process of simultaneous dissolution and formation in the virtuality of space-time. In taking up the new optics of Moholy-Nagy and Giedion, Benjamin subtly transformed their images into figures of an optical unconscious capable of dialectically exploding the phantasmagoric cultural production of industrial capitalism.

Although Giedion's space-time attempted to model the systemic form of modernity so that it could be generalized into the fabric of everyday life, he took individual works of art and architecture as working toward it and at the same time already embodying it proleptically. He considered these to be artistic symbols in the Romantic sense, capable of mediating the antinomies of modernity (inner/outer, ratio/vision, reason/feeling) to produce a momentary experience of harmony, of subject-object unity. They were vital, for without them the course of historical development remains "detached from our emotional life" and "the whole machinery runs down." Like the Romantics, Giedion pursued the philosophical aspiration to unity through art, rather than theory, or, more precisely, through *poiesie* or generative production, and aimed to operationalize the longed-for system through acts of individuation that strive to be absolutely self-positing. The making of art was thought to cojoin the critical dissolution of existing systems

with the relentless energy of formation, positing the exigency of a total closure that could never be perfected. By thematizing what is at work in the individual as the capacity to produce itself by means of its internal formative force, Romanticism set up a dialectical unity between artificial and natural production, understanding the organic as essentially autoformation. The operative criterion for this form (the mark of nature) was to be the perfection of character, which was announced through the exigency of a physical knowledge of the nature of the soul, for which the science of physiognomy was to serve as objective critic and judge. A reader of Friedrich Schlegel, the Romantic writer who attempted to achieve a new unity through the writing and assembly of fragments and ideas, as well as the idealist philosopher Friedrich Wilhelm von Schelling, who grappled for forty years with the problem of grounding an ungroundable system, Giedion approached the writing of history, like the making of architecture, as operationalizing the production of such a new system.

At the end of the 1940s, in *Mechanization Takes Command,* Giedion declared that the era of mechanization—of a rationality divorced from emotion, art, and the irrational—was coming to a close. Subsequently, during the 1950s and 1960s, he turned to prehistoric art, architecture, and anthropology to pursue the origins of modernity now conceived as the persistence and renewal of elemental forms of artistic expression and experience. In his two-volume study of the beginnings of art and architecture, *The Eternal Present,* published in 1962 and 1964, and his final book, *Architecture and the Phenomenon of Transition* (1971), Giedion tracked the origin of the problematic of representation to the birth of Western civilization and then followed its history once more into the twentieth century. In the postwar period, it seemed to Giedion that the reemergence of primitive forms, abstraction, transparency, and the interweaving of inner and outer space signaled the collapse of time into an "eternal present." Citing James Joyce's *Finnegans Wake,* a quartet by Béla Bartók, and a picture by Joan Miró, he described this in terms of a simultaneity in which "present, past, and future are not chopped off from one another but merge into one uninterrupted fabric" of continuity and change.

Despite the widespread appreciation of Giedion's research into the material culture of industrialization, mass society, and new media, his pursuit of a symbolic form beyond representation and his neo-Romantic approach to history as a form of artistic production were not generally understood among English-speaking readers. His erroneous reputation as the person who, more than anyone else, fixed the canon of modern architecture is one important indication of this. To avoid the codification and regulation associated with the idea of style, Giedion, like the architects with whom he aligned himself, posited the exigency of a higher law that would serve as a regulative ideal for a system of

freedom considered immanent yet only partially installed. This, however, harbored a paradox that undermined both the logic of his project and its reception; for a regulative ideal cannot, by definition, actually exist or be known, cannot have properties, and so cannot actually serve as a criterion against which to judge conformity to law. Consequently, Giedion had to project characteristics from examples already at hand, thereby collapsing claims for natural lawfulness and historical necessity into cultural conventions despite his intentions. It was not accidental that Giedion and his colleagues in CIAM found it necessary to undertake a moralizing campaign for what they thought would be automatic. The achievement of a new mediation between the human and the nonhuman under the conditions of industrialization became, in the end, contingent on the retraining and reculturation of the audience to an ethic of self-overcoming that no one was, after all, obliged to subscribe to. This paradox engendered the unresolved tension in Giedion's writings between completion and incompletion, being and becoming, that appeared to most readers as an unaccountable disparity between astute accounts of emerging tendencies and seemingly perplexing and extraneous aesthetic preoccupations.

Nevertheless, as Walter Benjamin's admiration suggests, Giedion's legacy remains significant. He mapped the regulatory matrix of architectural production and reception in the modern period more thoroughly and precisely than anyone had before; he scrutinized its internal contradictions, confused mixtures, and problematic transumption of the past into the system of modern production, and he struggled for an ethical relationship between self and Other under the sign of a transparency yet to come.

[*See also* Architecture.]

BIBLIOGRAPHY

Works by Giedion

Spätbarocker und romantischer Klassizismus. Munich, 1922.
Bauen in Frankreich, Bauen in Eisen, Bauen in Eisenbeton. Leipzig, 1928. Translated into English as *Building in France, Building in Iron, Building in Ferro-Concrete* by J. Duncan Berry (Santa Monica, Calif., 1995).
Space, Time, and Architecture: The Growth of a New Tradition. Cambridge, Mass., 1941; 2d ed., 1949; 3d ed., 1954; 4th ed., 1962; 5th ed., 1967.
Mechanization Takes Command: A Contribution to Anonymous History. New York and Oxford, 1948; reprint, New York, 1969.
Walter Gropius: Work and Teamwork. New York, 1954.
Architektur und Gemeinschaft: Tagebuch einer Entwicklung. Hamburg, 1956. In English as *Architecture, You and Me: The Diary of a Development* (Cambridge, Mass., 1958).
The Eternal Present, vol. 1, *The Beginnings of Art.* New York, 1962.
The Eternal Present, vol. 2, *The Beginnings of Architecture.* New York, 1964.
Architektur und das Phänomen des Wandels: Die drei Raumkonzeptionen in der Architektur. Tübingen, 1970. In English as *Architecture and the Phenomenon of Transition: The Three Space Conceptions in Architecture* (Cambridge, Mass., 1971).

Other Sources

Georgiadis, Sokratis. *Sigfried Giedion: Eine Intellektuelle Biographie.* Zurich, 1989. Translated into English as *Sigfried Giedion: An Intellectual Biography* by Colin Hall (Edinburgh, 1993).
Hays, K. Michael. "Introduction." In *Modernism and the Posthumanist Subject: The Architecture of Hannes Meyer and Ludwig Hilberseimer.* Cambridge, Mass., 1992.
Hommage à Giedion: Profil seiner Persönlichkeit. Basel, 1971.
Mertins, Detlef. "The Enticing and Threatening Face of Prehistory: Walter Benjamin and the Utopia of Glass." *Assemblage* 29 (April 1996): 6–23.
Mertins, Detlef. *Transparencies Yet to Come: Sigfried Giedion and the Prehistory of Architectural Modernity.* Ann Arbor, 1996. Princeton University Dissertation.
Mertins, Detlef. "Walter Benjamin and the Tectonic Unconscious." *ANY* 14 (1996): 29–35.
Mertins, Detlef. "Anything but Literal: Sigfried Giedion and the Reception of Cubism in Germany." In *Architecture and Cubism,* edited by Nancy Troy and Eve Blau. Cambridge, Mass., 1997.
Mertins, Detlef. "System and Freedom: Sigfried Giedion, Emil Kaufmann and the Constitution of Architectural Modernity." In *Autonomy and Ideology: The Origins of the Avant-Garde in America, 1923–1949,* edited by Robert Somol. New York, 1997.
Sigfried Giedion, 1888–1968: Der Entwurf einer modernen Tradition. Zurich, 1989. Exhibition catalog.
Sigfried Giedion: A History Project. Special issue of *Rassegna* 25 (March 1986).
Tafuri, Manfredo. *Teorie e storia dell'architettura.* Bari, 1968. Translated into English as *Theories and History of Architecture* by Giorgio Verrecchia (New York, 1980).
Wigley, Mark. *White Walls, Designer Dresses: The Fashioning of Modern Architecture.* Cambridge, Mass., 1995.

DETLEF MERTINS

GLASS. [*To explore glass as an artistic medium, this entry comprises two essays:*

Glass Art
Aesthetics of Glass

The first essay explains the relatively unknown history of glass art in relation to art history in general; the second essay examines the aesthetic issues that emerge in glass. For related discussion, see Mies van der Rohe; *and* Morris.]

Glass Art

Light and its meaning have preoccupied artists throughout time. The substance of glass has often been the vehicle for their studies—as a symbol, a tool, or the embodiment of luminosity. The aesthetics of glass is tied to both the inherent properties of the material and the stylistic and technological developments within the applied, industrial, and fine arts.

Glass is composed of silica and small amounts of other ingredients that are melted together at a high temperature. As it cools, it thickens, then becomes rigid while still retaining a liquid molecular (rather than crystalline) structure. Glass is extremely strong under compression; however, it is brittle under tension. Although rapidly chilled molten rock,

usually volcanic, can create naturally occurring glasses, the history of glass is essentially the story of a human-made material.

Glass makes a powerful aesthetic statement before the intervention of the artist. It can be transparent, translucent, or almost completely opaque. It reflects, transmits, and absorbs light and selective wavelengths of color. It can also bend or refract light. As light bounces back and forth between polished surfaces, its brilliance is enhanced. Glass can be made in virtually any color or value, and it may be homogeneous, striated, or layered. The hues of glass are often abnormally bright because light is transmitted through it rather than reflected solely off of the surface.

The earliest manufactured glass of the sixteenth and fifteenth centuries BCE was made to imitate the appearance of more valuable substances—a function that glass has maintained through the succeeding centuries. It was used as a less expensive alternative to gems, and glass pearls were made in Alexandria for the Roman market. The magical and protective powers associated with certain stones were believed to be transmitted to the glasses that imitated them.

The technique of blowing molten glass on the end of a hollow pipe was probably invented during the first century BCE in the Syro-Palestinian area. This revolutionary development enabled greater thinness and transparency, as well as increased scale. Hellenistic- and Roman-period glass was notable for its wide variety of techniques and decorative effects inspired by luxury objects of metal. Islamic glass of the tenth through the sixteenth centuries displayed sensitively carved surface reliefs and opaque enamel decoration in geometric, calligraphic, and plant and animal motifs.

The transformation of sand and ashes into a lustrous translucent material positioned glassmaking within the realm of alchemy, while its physical and optical qualities made it ripe for metaphorical allusions and supernatural associations. The abilities of glass to distort or to clarify reality caused it to assume an important role in Christian thought during the fourth century. During the Middle Ages, the colored light transmitted by stained glass was believed to enhance the spirituality of humanity; it was thought to emanate from God as the perfect source of light. Glass became a symbol of the Virgin Mary, and it was frequently used in her depictions.

Rock crystal is a mineral that has long symbolized purity, harmony, and transformation. It inspired glassmakers to produce deliberately decolorized glass by removing or counteracting natural impurities such as copper. The earliest examples of colorless glass date from the late eighth century BCE and are believed to have been made by Phoenician glassmakers. In his *Historia Naturalis*, published in 77 CE, Pliny the Elder wrote: "Still, however, the highest value is set upon glass that is entirely colorless and transparent, as nearly as possible resembling crystal, in fact."

The ability to produce colorless glass was apparently lost during the Middle Ages, then rediscovered in the sixteenth century, when the Venetians developed their *cristallo*. In a molten state, the glass mixture flowed easily and could be blown quickly. Much Venetian glass was valued for its lightness and complexity of form. During the second half of the seventeenth century, the elaborate Venetian styles declined in popularity. At that time, the English glass industry developed a colorless composition that included lead. This increased both the clarity and the brilliance of glass. Lead glass has a high index of refraction, which gives it a "fiery" quality. The introduction of significant amounts of lead has the additional effect of rendering the glass relatively soft—a characteristic ideal for cutting and engraving. By breaking the surface with different facets, prisms are formed and even more widely divergent colored light rays are produced.

The flamboyance of cut glass did not coincide with the strong aesthetic views expounded by the English reformer John Ruskin. He directed his attention to glass in his 1853 publication *The Stones of Venice*. In it Ruskin reveals his affinity for Venetian styles of the Renaissance, which were distinguished, in his opinion, by the two most significant characteristics of glass: its ductility when heated and its transparency when cold. Ruskin wrote, "All work in glass is bad which does not, with loud voice, proclaim one or the other of these qualities. Consequently, all cut glass is barbarous for the cutting conceals it ductility and confuses it with crystal. No workman ought to spend more than an hour in the making of any glass vessel" (1853, vol. 3).

The most intensely personal and symbolic glass was produced at the turn of the century by the firm owned by Émile Gallé in Nancy, France. An artist and student of botany and mineralogy, Gallé was one of the central figures of French Art Nouveau. He used nature as the language of his art and inscribed lines of Symbolist poetry on the vessels he called *verreries parlantes* (talking glasses). Gallé saw his creations as works of art communicating the moods and emotions emerging from the mysterious depths of nature and the unconscious. He used dark, smoky tones and deeply colored inlays to recall twilight, mist, microbes, flowers buried under ice, and ultimately dreamlike states. He also prized the accidental and unpredictable effects created during glassmaking.

In the United States at the end of the nineteenth century, Louis Comfort Tiffany's company was also producing blown glass. Again, it was nature that served as inspiration. For Tiffany and for the Johann Lötz-Witwe glassworks in Bohemia, however, organic forms were abstracted with little of the symbolism imparted by Gallé. Still, the tactile surfaces and brilliant color created a stunning expressiveness. Both Tiffany and Lötz glasses were renowned for their chemically induced metallic iridescence, which shimmered like the plumage of a tropical bird or the weathered surface of ancient glass.

Tiffany and his competitor, the painter John La Farge, had already revolutionized the aesthetics of stained glass during the late 1870s and 1880s. Although it had always been used to relate Bible stories, in the fifteenth century stained glass began to become increasingly pictorial in the manner of easel painting. Almost all details were painted on the glass with vitreous enamels or stains. Although Tiffany and La Farge worked independently, each wished to use glass to "paint with light" in a more illusionistic manner than had been previously attempted. With the help of technicians, they developed new kinds of glasses that imitated drapery, water, and the effect of light filtering through foliage. Opalescent glass produced a milky glow that was employed extensively by Tiffany's designers and craftspeople. Almost all painterly effects were produced in the glass, rather than applied with brushwork.

William Morris, leader of the English Arts and Crafts movement, had envisioned stained glass as one part of the harmonious architectural whole. This idea was elaborated on by the architect Frank Lloyd Wright, who rejected what he perceived as the suffocating banality of all stained glass since the Gothic period. Wright's designs were geometric abstractions dominated by colorless sections through which the outdoors could still be viewed. His windows modulated light levels and served as a transition between interior and exterior spaces.

For centuries, the potential of glass as a building material has appealed to architects on the basis of its functional practicality as well as its appearance. The glass "curtain wall" was made possible when Gothic engineering conceived of architecture supported by columns and buttresses. Walls were then freed to become colored glass that, in addition to providing light, was believed to facilitate communication between humanity and God. In 1851, Sir Joseph Paxton proved with the Crystal Palace that large-scale architecture of glass (three hundred thousand uniform panes) and iron was possible.

Bruno Taut was one of the progressive architects working in Germany and France after World War I who saw the future in glass and, in particular, colored light. Taut believed that glass architecture could be a vehicle for expression and would improve the ethics of humankind. He wrote, "The man who sees the splendors of glass every day cannot have ignoble hands." At the 1914 German Werkbund exhibition in Cologne, Taut erected a small prismatic-domed building of colored glass with cast concrete supports. Glass was used in the walls, floors, ceiling, and staircase. During the early 1920s, many architects, including Ludwig Mies van der Rohe, designed buildings clad in glass. Mies van der Rohe noted "the play of reflections" and their effectiveness as a screening device. The functionalists also found glass to be an ideal material for industrial design: it was inexpensive, it could be mass-produced into minimal shapes, it looked hygienic, and it was noncontaminative. Although Taut's romantic views were sidelined by the severity of German modernist ideology, decoration sustained a strong position within the advanced design of other countries. Rather than viewing semimechanized production techniques as an inferior alternative to handwork, René Lalique in France exploited its superior ability to produce geometricized repeating designs in high relief on blown and pressed glass.

During the 1920s and 1930s, a handful of artists were returning to Gallé's idea of the blown vessel as a sculptural form. His countryman, the painter Maurice Marinot, trained himself to blow glass. Marinot's thick-walled vessels were transparent, often incorporating brilliant color, masses of small bubbles, and deeply etched designs in keeping with the French *moderne* style. Glass that is created by hand is always imperfect, and no two pieces are exactly the same. For Marinot, the craft process was an intrinsic part of the artistry, and it was imperative that the artist fabricate the work himself or herself.

After Marinot ceased blowing glass in 1937, there were few successors to his artist/glassblower philosophy. Throughout the first sixty years of the twentieth century, however, there were a great number of artists worldwide using glass in a variety of approaches. Glass could be cold-worked the way a sculptor carves stone, shaped in a kiln-fired mold, cut apart and reassembled, painted or engraved, or used as a plate for printmaking.

Marcel Duchamp was one of the first artists to take glass beyond the vessel or the architectural window. His free-standing double panel *The Bride Stripped Bare by Her Bachelors, Even* or the *Large Glass* was completed in 1923. Duchamp forced the viewer to consider the colorless glass plane as part of the composition, not simply something to look through. When the glass cracked during transportation, Duchamp regarded it as part of the creative process.

Flat, almost colorless glass can be a neutral nonintrusive starting point for ideas. In the 1960s, the sculptor Robert Smithson used window glass in his conceptual works. Mario Merz and Therman Statom are among those making constructions that are fraught with the tension between precariousness and beauty. Windows torn out of buildings and still in their frames are assembled by Bruce Chao to form their own melancholy spaces.

Despite the variety of ways in which glass may be used outside of industry, molten glass remained an alluring and elusive material for the artist. During the late 1950s and early 1960s in the United States, a concerted effort was mounted to make hot glass directly accessible by melting and forming it in a studio rather than a factory. Experiments by individuals such as the potter Harvey Littleton (inspired by the German artist Erwin Eisch) and the glass scientist Dominick Labino changed the way hot glass was used and who was using it.

Much of the first decade of the American studio glass movement was devoted to acquiring manual and technical

skills. The viscous substance inspired methods somewhat analogous to the painting styles of the contemporary Abstract Expressionists. Glass artists dropped, dripped, slumped, sagged, and stretched the pliable medium to form sculpture. Vessels were influenced by contemporary ceramics, the field from which most of the new glass craftspeople emerged. There also began a tendency to incorporate the dramatic process of forming molten studio glass into the overall aesthetic—a phenomenon more akin to the "happenings" of the 1960s than to the private meditations of Marinot.

As knowledge and experience were acquired through the 1970s, artists could concentrate more on ideas than on mechanics. Glass art became increasingly sophisticated as different techniques made new visual effects possible. Scale increased as industrial adhesives permitted the successful bonding of parts. Glasswork aspiring to the status of sculpture emulated contemporary trends in other media, including the introduction of other materials and the construction of environments.

In Czechoslovakia, artists had been making sculpture from glass for decades, and the field was far more sophisticated than in the rest of the world. By the late 1960s, their predominating aesthetic aimed at exploiting recent scientific refinements as a means to intellectual abstraction. Optical glass, developed for science and industry to meet exceptionally high standards of purity, was the material of choice. Václav Cigler and many of his colleagues used the high-tech glass to change how the viewer experienced the environment.

Cigler's sculptures became precision-ground and polished lenses and prisms. In a lens, rays of light are bent and ideally brought to focus accurately at a single point. A convex lens causes rays of light to converge, whereas a concave surface causes them to diverge. In Cigler's words:

> I create glass which does not wish to be art, but a means of observation. I produce non-technical apparatuses which magnify, diminish, decompose, swallow, mirror, or partly or completely deform the outer environment. I use and misuse the optical properties of glass for visual games. Contrary to painters and sculptors who create a new reality, I leave this transformation to the glass article and the viewer. (as quoted in Adlerova, 1991)

The team of Stanislav Libenský and Jaroslava Brychtová also used optical glass. Their focus was on the effects taking place inside the glass as a result of the intersection of geometric shapes. The pellucid quality of the glass became a solid entity rather than a void. Over the following decades, Libenský and Brychtová would also employ dramatic contrasts in the thickness of the glass—ranging from blade-thin edges to solid masses—to "shape color and light in space."

Perhaps the most exaggerated optical effects are produced by coating glass with metals to heighten perspective.

The most familiar example is the common mirror. During the 1960s, Lucas Samaras built rooms and furniture completely sheathed with mirrors. An individual entering the boxlike space was completely surrounded by reflected images radiating into infinity. Larry Bell's sculptures are formed of sheet glass coated with vaporized metals. The appearance shifts between reflective and transparent as the viewer moves around them. Transparent dichroic coated glass, which changes color when subjected to transmitted and reflected light, has been effectively incorporated on an architectural scale by James Carpenter.

One of the most potent uses of glass has been as a metaphor for humankind, both physically and spiritually. Glass is so delicate that it can be shattered by a sound, yet it can resist the elements for thousands of years. As Vannoccio Biringuccio wrote in his *Pirotechnia,* first printed in Venice in 1540:

> Thus, in short, to anyone considering it well, all the effects of glass are marvelous. Considering its brief and short life, owing to its brittleness, it cannot and must not be given too much love, and it must be used and kept in mind as an example of the life of man and of the things of this world which, though beautiful, are transitory and frail. (1966)

The sculptor Christopher Wilmarth understood the analogy perfectly and used the contrast between glass and metal

GLASS: Glass Art. Emile Gallé, *Dragonfly Coupe* (1903), layered, inlaid, blown, and trailed glass with internal metal-foil decoration cut and engraved, 18.3 cm high; Corning Museum of Glass, Corning, New York. (Photograph copyright by the Corning Museum of Glass; used by permission.)

to evoke the human condition. By etching a membrane of blown or sheet glass with acid, the glass appeared to be marked by rain or tears. A head was suggested by a simple egg shape that, as Wilmarth noted, entrapped forever the breath of the blower. A gesture or a moment in time was captured by a curve or intersection between glass and dark bronze. The life-size figures of Czech artist Dana Zámečníková are cut window glass, built layer upon layer, like the subconscious, with each level of depth incorporating images from the artist's dreamworld and real world.

The sculptural vessel or "nonvessel" has also played an important role in twentieth-century glass. The bubble—a container—is the natural form assumed by blown glass. When the artist renders the vessel dysfunctional by deformation, covering it with a fugitive surface, or even altering the context, a change can occur that the philosopher Arthur Danto describes as "the transfiguration of the commonplace." The object may be "elevated" to the status of art. The British artist Diana Hobson fuses glass ground to the consistency of sand to produce fragile vessels too porous to entertain the traditional concept of containment. Toots Zynsky pulls glass into threads, then fuses them into irregular nestlike shapes. Dale Chihuly's nested and piled "Sea Form" vessels are blown so thinly that they seem ethereal. Howard Ben Tré's monumental works, like many of the nonfunctioning vessels, become symbols of the body or architecture. Their internal fissures, created as the glass cools, bring to mind age and decay.

Although brilliant luminescence, silky sheen, and optical virtuosity have historically drawn artists to glass, some view the natural beauty of the material as a hindrance to the serious consideration of their work. In response, inherent qualities may be diminished or denied by directing the viewer's attention to the shape and surface treatment rather than fixating on the material. In Germany, beginning in the 1950s, Erwin Eisch made blown-glass sculpture that was partly or completely covered with opaque color and deliberately ugly in terms of contemporary taste. Although, on first assessment of some work, it may seem that all transparency has been obscured, it is often subtly present to a sufficient degree to validate the choice of glass as the medium.

Debates regarding "truth to material" have circulated around glass ever since critics began venturing definitions concerning its proper use and appearance. In truth, because of its chameleonlike nature and its historical role of imitating other materials, there is no one fixed appearance that glass must legitimately assume.

Contemporary artists using glass remain essentially divided between those who see glass as a means to express an idea and those whose ideas are derived from—and dependent on—the glass. The attention-grabbing properties of the material, as well as the fascination of the process, can easily be mistaken for imagination and creativity on the part of the artist. Glass is a viable and limitless medium, but one that must be continuously scrutinized to determine if the light, the beauty, and the theatrics add up to a meaningful statement.

[*See also* Craft.]

BIBLIOGRAPHY

Adlerova, Alena. "Interview: Václav Cigler." *Glasswork* 10 (October 1991): 2–9.
Biringuccio, Vannoccio. *Pirotechnia.* Translated by Cyril Stanley Smith and Martha Teach Gnudi. Reprint, Cambridge, Mass., 1966.
Bostock, John, and H. T. Riley, eds. *The Natural History of Pliny,* vol. 6. London, 1898.
Corning Glass Works. *All about Glass.* Corning, N.Y., 1984.
Corning Museum of Glass and Museum of Decorative Arts, Prague. *Czechoslovakian Glass, 1350–1980.* Corning, N.Y., 1981.
Danto, Arthur C. *The Transfiguration of the Commonplace: A Philosophy of Art.* Cambridge, Mass., 1981.
Frank, Susan. *Glass and Archaeology.* London and New York, 1982.
Frantz, Susanne K. *Contemporary Glass: A World Survey from the Corning Museum of Glass.* New York, 1989.
Gombrich, E. H. *The Sense of Order: A Study in the Psychology of the Decorative Art.* 2d ed. Ithaca, N.Y., 1984.
Klein, Dan, and Ward Lloyd, eds. *The History of Glass.* London, 1984.
Polak, Ada. *Glass: Its Tradition and Its Makers.* New York, 1975.
Ruskin, John. *The Stones of Venice.* 3 vols. London, 1851–1853.
Sharp, Dennis, ed. *Glass Architecture by Paul Scheerbart and Alpine Architecture by Bruno Taut.* Translations by James Palmes and Shirley Palmer. New York, 1972.
Silverman, Debora L. *Art Nouveau in Fin-de-siècle France: Politics, Psychology, and Style.* Berkeley, 1989.
Tait, Hugh, ed. *Five Thousand Years of Glass.* London, 1991.

SUSANNE K. FRANTZ

Aesthetics of Glass

With the tremendous growth of the studio glass movement since the 1960s, glass has become a medium for making serious works of art. Because glass has often been used to produce merely pretty objects with little artistic or aesthetic merit, however, contemporary glass artists have had a difficult time in having their work taken seriously. When many think of glass, they think only of what may be purchased in gift stores rather than what is now beginning to be displayed in museums and contemporary galleries. Yet, workers in studio glass are artists, having learned how to exploit various properties of glass to produce works of significant artistic and aesthetic merit.

Glass and Some of Its Attributes. Six particularly important capacities or characteristics of glass have been manipulated by artists to produce works of artistic and aesthetic merit. First, glass has the capacity to manipulate light. It can transmit light without distortion, it can refract and reflect light, and it can appear to capture or retain light within its interior. Second, glass can be colored. Coloring may be absorbed into glass by various chemical means, or it may be applied onto glass by enameling or painting.

Third, glass may serve as a medium for images. These images may be applied to the surface of the glass, or placed

within the glass. Surface images may be made by using techniques such as painting and enameling that are commonly employed by artists working in nonglass media. Surface images may also be produced by sandblasting, etching, and engraving. Internal images may be constructed by using a wide variety of techniques ranging from gluing bits of glass together to manipulating bits of hot glass. Moreover, a surface image may be turned into an internal image by covering up a surface image with an added layer of glass.

Fourth, because of glass's transparency, it has the capacity to reveal internal form. The internal form may be real, illusory, or a combination of the two. Real internal form is produced by a variety of techniques ranging from gluing or fusing pieces of glass together, encasing glass within glass, forming internal surfaces by carving, drilling, sandblasting, or blowing, injecting tiny air bubbles into molten glass, to inserting nonglass materials into the glass. Illusory internal form is generated by the refractive and reflective qualities of glass. In such cases, the real internal form appears to change its shape because of the optical properties of the glass.

Fifth, glass has the ability to freeze shapes. Glass cools to a rigid condition without crystallization. When it is heated it loses much of its rigidity. If its rigidity is entirely or almost entirely lost, it flows in response to gravitational forces, and its shape may be completely controlled or predicted by casting the glass into a mold, or less severely controlled by slumping the glass in thermostatically controlled environments such as those provided by kilns and annealing ovens. If the glass is slightly more rigid, its shape may be altered by increasing the internal pressure within a piece of glass by blowing, and the blown object may be allowed to form freely producing a spherical bubble, or it may be forced into other shapes by rolling, pinching, denting, or blowing the hot glass into a mold. Glass may also be lampworked. Here the artist uses a torch to control the rigidity of the glass. The glass's response to the prevailing forces while in a nonrigid condition is frozen or fossilized into the glass once it cools, and the response becomes a permanent feature of the resultant object. Thus, glass flows in response to forces while in a nonrigid state, and fixes its response while in a rigid state.

Finally, the shape of glass can be manipulated even when it is in a rigid condition. Changes in shape may be brought about by decreasing or increasing the glass's volume. Highly controlled decreasing is brought about by cutting, drilling, carving, sandblasting, and etching. Less controlled ways to decrease the volume of glass are accomplished by chipping and smashing. The most common ways to increase the volume of a piece of glass are gluing and fusing pieces of glass together. These techniques lead to constructivist sculptures.

Exploitation of Properties for Aesthetic and Artistic Aims. The Czech husband and wife team, Stanislav Libenský and Jaroslava Brychtová, two of the foremost glass artists in the world, are particularly adept at exploiting the light-manipulating, color-retaining, shape-freezing, and shape-manipulating properties of glass. Since the 1970s, they have produced a number of nonvessel sculptural forms that are concerned with how light interacts with color and form. Beginning with a drawing, they sculpt a clay model that is used to prepare a mold. They cast their works out of monochromatic blocks of glass. Once the work is annealed, they grind and polish its surfaces. A recent series of sculptures, which range in height from one to four feet, are small monumental Minimalist works. They typically contain elements of tension brought about through breaking up simple pyramidal forms by slicing off corners, breaking up faces, and removing interior portions of glass to create regions of negative space. But the tension is countered by restricting their works to a single color, which they often mute by matting the surfaces of the works. The single color, even with its subtle changes in intensity, provides an overall unity to their works, which appear to emit a soft glow of color. Indeed, the glow of the cast glass makes it appear as if the works are still full of heat energy used to melt the glass. Another leading Czech artist, Frantisek Vizner, who usually works within the traditional vessel form, carves miniature monumental Minimalist works out of blocks of monochromatic glass. Vizner exploits the light-manipulating, color-retaining, and shape-manipulating properties of glass. Vizner's works have an inner serenity achieved through their simple form, matted finish, and restricted palette, where shifts in color are restricted to subtle alteration of saturation. His works are as peaceful as Mark Rothko's, and, like Rothko's better work, some of the lighter color fields appear to float within the surrounding darker colored areas.

The American glass artist Thomas Patti is adept at exploiting the ability of glass to reveal internal form. Patti views his works as essays about light, designing them to be seen under soft lighting conditions, with most of the light coming up from beneath the object. He places some of his works on a base containing an artificial light, and uses various techniques such as making the edges of his works as sharp as possible to diminish the amount of surface reflections. Patti wants viewers to think of the light moving up through his pieces. He also wants viewers to focus their attention on everyday manufactured materials that they take for granted, and he uses only commercially available glass. Patti works under extremely controlled conditions, and even invented a process that fuses plates of glass together but keeps the visual integrity of each plate. During the 1990s, he has worked on his Lumina Echo Series. Works in this series, like most of his works, are very small, about six inches by four inches by four inches high. Made up of about a dozen horizontal plates of alternating clear and colored glass, they are miniature monumental Minimalist sculptures. Patti creates an internal form by hollowing out simple, often repetitive shapes within most of the layers. The repetition of the alternating dark and clear layers of glass and internal shape provides a compositional echo. Patti's

careful attention to detail, repetitive forms, and use of industrial materials is reminiscent of Donald Judd.

The American glass artist Mary Shaffer has a particular fondness for the shape-freezing capacity of glass. Some of Shaffer's works are about this very property of glass and constitute studies on gravity. She also plays on the idea that glass is a frozen liquid by comparing it to water in both its solid and liquid states, and draws our attention to the fact that glass and bronze have the ability to freeze the shapes they acquire when in a liquid condition. Her work reminds us of the ubiquity of gravity and the shape-freezing property of material such as glass, water, and bronze. Her well-known work titled *Center Cube* contains two thirty-three-foot-high, identically shaped objects. One, made completely of bronze, is shaped like a sheet of water spilling out of the cube onto the ground, making it look like a bronze fountain with liquid bronze flowing out of it. In the second object, Shaffer substitutes a glass sheet for the bronze one, making the object look like a bronze fountain with water flowing out of it. Both glass and bronze flow as a response to gravity when in a nonrigid condition. The freestanding glass piece plays on the fact that the rigid piece of glass has sufficient internal strength to withstand the downward gravitational force created by the mass of the bronze cube. But, because she has slumped the glass to make it appear as if it were water falling from a fountain, our initial reaction is one of confusion. How can a column of falling water hold up a weighty piece of bronze? Surely, it is impossible; we are not on the moon. In some of Shaffer's other works she explicitly compares glass with ice, making the glass look like ice. Indeed, these works suggest an extension of the concept of trompe l'oeil. When applied to painting, the eye is deceived into thinking that a two-dimensional image is a three-dimensional object. In these sculptures, as with works by Duane Hanson, the eye is deceived into thinking that one kind of object or material is another. Shaffer is a sculptor who uses her wit and technical abilities to make amusing works that refer to her chosen material, especially how it is affected by gravity.

There is a long tradition of using stained glass to create images. Studio glass artists have also become adept at creating three-dimensional images. Consider, for example, works by the American Pike Powers and the Czech artist Dana Zámečníková, who utilize very different techniques in exploiting this property of glass for artistic purposes. Powers's piece, from her 1993 Coney Island Series, is known as *Dixie Cup High Diver*. It stands thirteen inches high with a nine-and-a-half-inch diameter, and looks like a glass half filled with water containing a little female doll, clad in a red bathing suit. The doll represents a diver ascending to the surface after completing a dive. The internal image, enameled onto an inner layer, is subsequently covered with gather of glass. The smiling diver has just completed her high dive. Exhilarated and relieved, she has not missed her target, which had appeared no larger than a Dixie cup from her high perch. Moreover, there is the added trompe l'oeil. Is the liquid in the glass water or glass?

Zámečníková typically uses multiple layers of transparent glass, painting, engraving, and sandblasting images onto each sheet of transparent glass. By splitting her images onto successive layers of glass she is able to give them an added sense of depth. For example, her piece titled *Theater* from the Saxe collection, made up of vertical glass sheets placed in front of each other, shows a group of dancers. Caught within a moment of time, the dancers appear to float above the dance floor.

One ingenious way in which glass artists have exploited the general shape-manipulating properties of rigid glass is by utilizing the fragility and brittleness of glass to enhance the artistic force of their work. Consider, for example, *Possessions*, a work by the American artist Dan Dailey. It depicts an unhappy former owner of one of his vases, frozen in time just after dropping the vase. Art glass is very fragile. But, there is more. Dailey, in depicting how a collector's happiness depends so much on her possessions, draws a parallel between the brittleness of glass, the ephemerality of worldly possessions, and the emotional fragility of those who place too much importance on their material goods. Moreover, the work also suggests a parallel between the broken vase and the eventual loss of the woman's physical beauty, which, again, places the work into a tradition of works about the concept of *vanitas*. But the work is even more interesting because the part that represents a shattered glass vase was actually a Dailey glass vase that he had shattered. Thus, Dailey represents glass's brittleness through exemplification, and the former owner's vanity through imagery.

The preceding discussion of artists and specific works demonstrates that contemporary glass artists are making works of aesthetic and artistic merit. Any adequate defense of the claim that glass is a medium for producing serious works of art must also address the work of Dale Chihuly, without question the best-known American glass artist. Chihuly is the recipient of the first National Living Treasure award given by the United States and has works housed in major museums throughout the world. His works overwhelm viewers with their sensuousness, as he skillfully exploits the light-manipulation, color-absorbing, and shape-freezing properties of glass—perhaps the three properties most responsible for its sensuousness. His works are so sensuous, however, that some viewers classify them as pretty instead of beautiful; others, mistakenly believing that less is always better than more, do not consider Chihuly to be a serious artist.

Nevertheless, many of Chihuly's works are aesthetically beautiful, exhibiting an overall harmony in shape and color. Many are about light: light interacting with color and form, and light as energy. In addition, they are about both the

forces that shaped them and the light that often appears to radiate from them. Thus, their beauty is enhanced through the harmonious relation between their color and shape and internal pressure. The pieces, blown and reblown, whirled about on the end of a spinning blowpipe, and reshaped with a wooden paddle, have frozen the resistance of glass to internal pressure and to centrifugal and gravitational forces. Chihuly also has created bundles of light energy; the vessels appear to be self-luminous bodies, radiating varying levels of light energy.

Just as works by Diego Velázquez or Jan Vermeer are not just paintings of a young princess having her portrait painted or a young woman reading a letter while standing at a window, but are about light, these works by Chihuly are essays about light (and color) as energy. Unlike Velázquez or Vermeer, however, Chihuly is not interested in narrative. His work is abstract, and he expresses his ideas by manipulating glass's shape and color; Chihuly is an Abstract Expressionist who has chosen glass as his medium of expression.

Concluding Remarks. Glass has tremendous aesthetic potential, a fact that is easily recognized once one exam-ines works in glass without being confined by glass's long history as a medium for making pretty, solely functional or highly decorative objects. Contemporary glass artists, fortified with a battery of techniques learned during the rise of the studio glass movement and gaining inspiration from twentieth-century artists working with traditional materials, have begun exploiting various properties of glass to produce works of high aesthetic and artistic merit.

BIBLIOGRAPHY

Contemporary American and European Glass from the Saxe Collection. Oakland, Calif., 1986. Exhibition catalog, Oakland Museum.

Dmoch, Mary, Ferdinand Hampson, John Lawson, Kathy Schey, and Lillian Zonars. *The 22nd Annual Glass Invitational: Habatat Galleries.* Riviera Beach, Fla., 1994.

Elliott, Kate. *Stanislav Libenský and Jaroslava Brychtová: Paintings, Drawings and Sculpture.* Seattle, 1995. Exhibition catalog, Elliott-Brown Gallery.

Frantz, Susanne K. *Contemporary Glass: A World Survey from the Corning Museum of Glass.* New York, 1989.

Frantz, Susanne K. "The Evolution of Studio Glass Collecting and Documentation in the United States." In *Contemporary Crafts and the Saxe Collection,* edited by Terry Ann R. Neff. New York, 1993. Exhibition catalog, Toledo Museum of Art.

Frantz, Susanne K., ed. *Stanislav Libenský/Jaroslava Brychtová: A 40-Year Collaboration in Glass.* Munich and New York, 1994.

Hampson, Ferdinand C. *Exhibition of Works by Mary Shaffer, November 7–28, 1992, at Habatat Galleries.* Pontiac, Mich., 1992.

Lewis, Albert. "Dan Dailey: Captured Phenomena." *New Work* (October/November 1979): 6–11.

Littleton, Harvey K. *Glassblowing: A Search for Form.* New York, 1971.

Taragin, Davira S. "Studio Craft Comes of Age." In *Contemporary Crafts and the Saxe Collection,* edited by Terry Ann R. Neff, pp. 11–19. New York, 1993. Exhibition catalog, Toledo Museum of Art.

HENRY FRANKEL

GOETHE, JOHANN WOLFGANG VON (1749–1832), poet, novelist, dramatist, and scientist of the German Romantic movement. Although not a philosopher in the usual sense, Goethe was strongly influenced by his reading of Baruch Spinoza, Gottfried Wilhelm Leibniz, Immanuel Kant, and Friedrich Wilhelm von Schelling, whose ideas he interpreted idiosyncratically and assimilated into his own thinking. Goethe's contributions to aesthetics are not contained in any single theoretical treatise, but are scattered throughout his critical essays and voluminous correspondence. Goethe had a tremendous impact on European literature through his portrayal of turbulent artistic temperaments in his poems, plays, and novels, and in his personal embodiment of the Romantic aesthetic ideal as a man of science, arts, and letters.

Life and Times. Goethe was born in Frankfurt am Main to an affluent Lutheran family. He studied at the University of Leipzig from 1765 to 1768, and wrote his first comedy, *Die Laune des Verliebten* (The Lover's Caprice) in 1767, and a verse tragedy, *Die Mitschuldigen* (The Fellow Culprits) in 1768. Goethe began his literary career in 1773 with the success of his drama *Götz von Berlichingen,* a work that typified the Sturm und Drang (storm and stress) era of German Romanticism, and that is sometimes regarded as the first German theater adaptation of Shakespearean tragedy.

Over the next several years, Goethe wrote poems, plays, and novels, notably a prose prototype of his later great masterpiece *Faust* (begun in 1775), *Egmont* (1788), and *Die Leiden des jungen Werthers* (1774) (The Sorrows of Young Werther). Following a brief period of illness, Goethe traveled in Switzerland and visited many of the German states, and on his return accepted a succession of political administrative posts. Through his friendship with Johann Gottfried von Herder, Goethe experienced a transition from Romanticism to neoclassicism that was to characterize his lifelong intellectual and artistic development. He combined diplomatic duties with botanical studies, travels in Italy, and various dramatic verse projects, including *Torquato Tasso* (1790), and *Iphigenie auf Tauris* (1787). He published *Versuch, die Metamorphose der Pflanzen* (The Metamorphosis of Plants) in 1789, two volumes of *Beiträge zur Optik* (Contributions to Optics) in 1791–1792, *Römische Elegien* (Roman Elegies) in 1795, and *Wilhelm Meisters Lehrjahre* (Wilhelm Meister's Apprenticeship) in 1795–1796. At the invitation of Prince Karl August of Weimar, Goethe assumed directorship of the royal theater in 1791, and in 1794 began his at first antagonistic and later collaborative relationship with Friedrich von Schiller. During the next thirteen years, Goethe wrote some of his most beloved fables or fairy tales *(Märchen),* completed the first part of *Faust* (1808), among many other poems and plays, and co-wrote with Schiller a series of satirical poems titled *Xenien* (1796), while contributing art criticism and art-history articles to

Schiller's journal *Die Horen*. In later years, Goethe published his novels *Achilleis* (1808) and *Die Wahlverwandtschaften* (Elective Affinities) (1809), *Farbenlehre* (Theory of Colors, 1810), a six-volume autobiography, *Aus meinem Leben: Dichtung und Wahrheit* (From My Life: Poetry and Truth, 1811–1822), *Italienische Reise* (Italian Journey, 1816), *Wilhelm Meisters Wanderjahre* (Wilhelm Meister's Travels, 1821–1929), *Novelle* (1826), and the second part of *Faust* (1832).

Goethe's Aesthetics and Art Criticism. As an essayist, Goethe's short studies of artworks and his appreciative descriptions and historical analyses of art trends and traditions illustrate his criteria of aesthetic excellence. Goethe's art criticism is widely dispersed throughout his correspondence with such contemporaries as Schiller, his lover the poet Charlotte von Stein, and Marianne von Willemer, and in numerous pamphlets and journal articles.

The principles by which Goethe judges beauty in art are similar to those by which he judges truth in science. Goethe's aesthetics is an expression of his theory of the unfolding of internal forms in nature according to an intuited balance and interplay of opposed cosmic forces. Goethe was deeply impressed by Kant's *Critique of Judgment* (1790), though his philosophy of art was markedly different. Where Kant distinguishes between the domains of art and science, Goethe sees no sharp division between aesthetic beauty and scientific truth. [*See* Kant.] It is important to observe, however, that Goethe takes an iconoclastic view of science that is quite different from the predominant ethos of eighteenth- and nineteenth-century thinkers. Goethe attempted to separate physics and the other natural sciences from mathematics and experimentation involving mechanical instruments. He understood science instead as the discernment and description of primal phenomena *(Urphänomena)* and primal polarities *(Urpolaritäten)* by intuitive insight. Goethe's commitment to intuitive methods is exemplified by his 1784 observation, by unaided visual comparison, that the human *os intermaxillare* is the anatomical analogue of ape and other mammal jawbone structures. In his *Metamorphose der Pflanzen*, Goethe similarly articulated some of his most basic metaphysical principles of science, explaining the transformations in plants by means of an intuitive grasp of alternating polarities and equilibriums of expansion *(diastole)* and contraction *(systole)*. In his investigations of optics and color, the *Beiträge zur Optik* and *Farbenlehre*, Goethe persistently but unconvincingly sought to refute Isaac Newton's mathematical and experimental theory as counterintuitive.

In commenting on the thought of Jean-Jacques Rousseau, Kant, and Goethe, Ernst Cassirer emphasizes the differences between Goethe's and Kant's attitudes toward the relation between science and aesthetics, and Goethe's concept of science as aesthetic and of aesthetics as scientific:

For Kant there are a priori principles of taste, as there are a priori principles of theoretical knowledge. Nevertheless nature and art, truth and beauty, remain divorced; they cannot be reduced to one and the same denominator. For Goethe, on the other hand, there is no sharp division between the two domains. His motto remains the words of [Anthony Ashley Cooper, earl of] Shaftesbury: "All Beauty is Truth." For him the beautiful is "a manifestation of secret natural laws, which without its appearance would have remained forever hidden from our view" [*Maximen und Reflexionen*, no. 183]. Laws of nature and laws of beauty cannot be set off from each other or in their meaning. The transition from the consideration of nature to the consideration of art is accomplished almost insensibly in Goethe's mind. He constantly alternates between the two and finds satisfaction only in this alternation. (Cassirer, 1945, p. 85)

If Goethe's aesthetics is "scientific," his science is equally if not more aesthetic. He is generally interested in the phenomenology of empirical experience. This makes it tempting to dismiss Goethe's natural philosophy with its implications for aesthetics as nothing more than poetic pseudoscience. Such criticisms are reinforced by the fact that in 1768, while recovering in Frankfurt from an illness contracted in Leipzig, Goethe began actively studying astrology, alchemy, occult hermetic philosophy, and religious mysticism. The scientific and the supernatural often play complementary roles in Goethe's literature. Thus, Goethe applies a principle of chemical attraction to human relationships, as the title of his novel *Elective Affinities* suggests, leaving it ambiguous whether the attribution is intended to be literal or metaphoric. But although Goethe's forays into physics are generally rejected today, other aspects of his empirical findings continue to be accepted even by contemporary scientists. His discovery of the isomorphism in ape and human intermaxillary bones is undisputed, and his single-handedly founding the discipline of plant morphology constitutes a lasting contribution to biological science. Goethe's theory of colors, based on direct observations and simple phenomenological experiments in opposition to Newton's mechanical-mathematical theory, was also of interest to Ludwig Wittgenstein in his *Remarks on Color* (1977). [*See* Color.] Moreover, the insight into the forms and relations of things that Goethe celebrates is arguably essential to, even if it is not the whole of, science. If there is a criticism of Goethe's idea of science, it is not that the method is incapable of leading to scientific advances but that it is too limited to embrace all of what has come to be defined as scientific knowledge. The connection between Goethe's understanding of science and aesthetics is epitomized in his model of an internal primitive botanical form or *Urpflanze*, like the essential form of artwork, on which all different types of plants are patterned, the variations among which ultimately explain their distinctive morphology, growth, and development.

Goethe explains his philosophy of the unity nature and art in a sonnet on the classical ideal, translated by Thomas H. Jappe in Paul Carus's study *Goethe:*

> Antagonistic art and nature seem,
> And yet are one without your knowing how;
> Gone is that feeling of aversion now,
> And equally attractive both I deem.
> Honest endeavor is no idle dream;
> And once in measured moments bound, I trow,
> To art with mental force and sweat of brow,
> Nature in us again will freely gleam
> True culture at all times demands the same:
> Untutored minds will ever vainly strive
> Perfection's height immaculate to scale.
>
> (1915, pp. 270–271)

The continuum of art and science in Goethe's thought further entails an objectivity in aesthetics and art criticism, and a scientific approach to art history that surpasses subjective preferences. Thus, in his 1824 article on Giotto di Bondone's *Last Supper,* Goethe maintains against the critic's prejudicial tastes:

> What we must do is recognize achievement in whatever shape or form it may appear, always remembering not to expect roses in winter and grapes in spring. That is, the fair, discriminating art critic does not praise or find fault with a work according to his personal likes or dislikes. Rather, his judgment is always based on an awareness of the history of art; he carefully considers place and time of origin, the state of the arts at the time and the taste of a particular school and a particular master. (1986, p. 60)

It is only through familiarization with these fine points of Goethe's art criticism that it is possible to acquire a sense of his intuition of the aesthetic ideal. Goethe offers rhapsodic eulogies to the great masterpieces of European art, and in the process finds opportunity to explain his underlying aesthetic principles. Goethe's topics include, among others, Gothic architecture, Leonardo da Vinci's *Last Supper* fresco in the monastery Santa Marie delle Grazie in Milan, landscape painting by Jakob van Ruisdael, Philipp Hackert, and Rembrandt van Rijn's etching of the biblical allegory, *The Good Samaritan.* Goethe presents an eloquent tribute to the genius of William Shakespeare, and comments in insightful detail on the poetry of George Gordon (Lord) Byron, on the plays and prose of Pedro Calderón de la Barca, Laurence Sterne, and Thomas Carlyle, and on the aesthetic literary theory of Plato and Aristotle, along with such contemporaries as Justus Möser.

Goethe has definite opinions about the purpose and value of art and of the difference between good and bad or mediocre art. He expounds his philosophy of art at length in his 1798 essay on the Laocoön statue group in a passage worth quoting as Goethe's manifesto of fine art as it applies to all the classical and neoclassical artists he admires:

All great works of art portray some aspect of man. The preferred subject of the plastic arts is the human body, and our present discussion is limited to them. Art has many levels, and on each one excellent artists may be found. However, a great work comprises all qualities that are normally found only separately.

The greatest works of art show us:

Highly Developed Human Forms. We expect above all a knowledge of the human body, of its parts, proportions, its internal and external functions, its form and movement in general.

Individuation. Knowledge of the variations in appearance and function of these parts. Particular qualities are isolated and highlighted; as a result, individual characters emerge. Through this, a significant relationship can be established among various works of art, and, in the case of a composite work, its individual components can exhibit a significant relationship among themselves. The subject can be:

At Rest or in Motion. The whole work of art or its components can either be represented complete within themselves, at rest, just existing, or they can be presented in motion, acting, expressing strong feelings.

The Ideal. To achieve ideal presentation, the artist must be profound, thorough, and perservering in his endeavors. Yet he must also possess the intellectual breadth to grasp his subject fully and to find the best moment to be portrayed. Such an artist can raise his subject above the limitations of the here and now and make it part of an ideal world, lending it the restraint, moderation, immediacy and dignity that befit this ideal world.

Grace. The subject and the manner in which it is portrayed must, however, obey the physical laws of art: order, clarity, symmetry, positioning, and so forth, which render it beautiful to the eye, that is, aesthetically graceful.

Beauty. The subject must also follow the law of spiritual beauty which results from restraint; an artist trained to portray or create beauty knows how to subject everything, even extremes, to this law.

These are the characteristics we expect of a great work of art. I do not believe I am overstating the case if I assert that the *Laocoön* group fulfills all these requirements, and that we can even derive all of the above laws from this particular work.

(1986, pp. 15–16)

Here is an entire dissertation on aesthetics condensed into a few paragraphs. Goethe endorses the humanist preoccupation with the ideal human form, from which reference point he derives all the other requirements of great art. In his "Introduction to the *Propylaea*" (1798), the art and literary journal he began at this time, Goethe similarly declares: "Man is the highest, indeed the only proper subject for the visual arts" (ibid., p. 81). The focus in Goethe's art criticism is specifically on the plastic arts. But it is not difficult to see Goethe's concept extended to include the depiction of human thought and action in masterpieces of poetry, tragedy, stories, and novels. What is methodologically interesting in Goethe's recitation of aesthetic ideals is his claim that they can all be derived intuitively from the contemplation of a single outstanding paradigm. In admiring

the *Laocoön*, Goethe abstracts from the experience the necessary elements that constitute great art, in much the same way that he intuits the essential internal organic forms of plant morphology in his perception of particular botanical specimens.

Goethe as Artist—from Sturm und Drang to Neoclassicism. The impact that Goethe exerted on aesthetics is as much a result of the example of his actual practice as of his abstract theory of the arts. Among his other extraordinary literary and scientific talents, Goethe struggled through much of his early life to perfect his skill as a painter, an ambition he did not relinquish until the age of forty. His surviving sketches, watercolors, and oil paintings, especially those executed during his travels in Italy, testify to a remarkable ability. But it seems that Goethe could not satisfy his own artistic standards, and so he eventually abandoned painting to concentrate on literature.

Goethe was an inspiration to generations of Romantic artists. As Dante is to Italy and Shakespeare to England, so Goethe is to Germany, for which the classic period of German letters is still known as the *Goethezeit*. Friedrich Nietzsche esteemed Goethe so highly as to consider him the ideal artist. This is not to say that Goethe's own artistic style and aesthetic ideology did not evolve. There is a movement in Goethe's thought from his early Sturm und Drang years of turbulent emotional expression toward a later mature phase in which he aspires to a kind of Stoic or Spinozist quieting of the perturbations of the soul by the soothing effects of classical aesthetic unity and harmony. The theme is manifested in a comparison of his early work *The Sorrows of Young Werther* and his later dramatic magnum opus *Faust*.

Werther is an epistolary novel that Goethe published anonymously in 1774. It recounts the psychological disintegration and suicide of a youthful artist through a disappointed, unconsummated love affair and the unbearable melancholy induced by his sensitivity to the woes of the world, compounded by the acute frustration of being unable adequately to express his most profound feelings. Werther is a portrait of Goethe's own life as an artist, and of his passionate involvement with Charlotte Buff in the provincial town of Wetzlar, where Goethe was appointed as an apprentice lawyer to the Supreme Court. The novel was written in four weeks, drawing on Goethe's correspondence with Charlotte's fiancé Johann Christian Kestner, and incorporating the later shocking news about the suicide of another mutual Wetzlar acquaintance, K. W. Jerusalem.

Young Werther is thus the personification of the German Romantic spirit of Sturm und Drang. It is easy to see in *Werther* the precursor of Arthur Schopenhauer's theory of the sufferings of aesthetic genius in suppressing the distractions of individual will in working to represent the intuited but strictly unrepresentable Platonic forms perceived in nature and perfected in imagination. In a series of letters to his friend Wilhelm, Werther describes his alarmingly rapid descent from exuberant joy into dark despair.

> I am so happy, my dear friend, so absorbed in the exquisite sense of tranquil existence, that I neglect my art. I could not draw a single line at the present moment; and yet I feel that I was never a greater painter than I am now. . . . Oh, would I could express it, could impress upon paper all that is living so full and warm within me, that it might become the mirror of my soul, as my soul is the mirror of the infinite God! O my friend—but it will kill me—I shall perish under the splendor of these visions!
>
> (1949, Letter of 10 May, p. 3)

By the end of the novel, as he loads his pistol and drinks a final glass of wine, Werther complains of being unable to express his feelings about the meaning of death. The melodrama of Werther's introspection remains moving despite its maudlin exaggeration, when Goethe has Werther exclaim:

> For the last, last time, I open these eyes. Alas! they will see the sun no more. . . . There is nothing like this feeling, Charlotte, and yet, it is most akin to a diffused dream, when you say to yourself—This is the last morning! The last! Charlotte, no word can adequately express this feeling.
>
> (Ibid., Letter of 20 December, p. 119)

Throughout, it is the artist's anguish that torments Werther. The ordinary vicissitudes of life are the pretext for Werther's disappointment in being unable to fully represent his intensity of emotion.

Goethe worked at his epic drama *Faust* intermittently over many years. The scholar-magician Faust was already a stock character of German fiction by the time Goethe adapted the story to his own purposes. Goethe is said to have known about the figure of Faust through the tradition of children's puppet theater. The tale of an avaricious man who makes a pact with the devil to surrender his soul in exchange for unlimited knowledge, wealth, and power is a subject of universal appeal. The treatment of the theme in Goethe's *Faust* has the added complication of taking as its central purpose the problem of representing the place of human existence in the physical universe. The play, although not obviously autobiographical, dramatizes Goethe's transition from the early Sturm und Drang period to his later neoclassicism. It is by renouncing the restless striving of will in its desire to dominate nature through an unholy alliance with Mephistopheles that Faust finally attains a state of harmony in which his excessive unnatural urges are resolved. By comparison with Christopher Marlowe's *The Tragedy of Dr. Faustus* (1589), Goethe's *Faust* depicts the protagonist as possessing an unquenchable thirst for knowledge rather than pleasure and wealth, and, unlike Marlowe's predecessor, is not damned and carried off to hell, but finds redemption by bringing his will into accord with the impersonal forces of nature. The transformation in Faust's attitude recalls Goethe's *Novelle*, in which music soothes and civilizes savage beasts, in agree-

ment with, if not actually inspired by, Schiller's program for the moral improvement of humanity in his *On the Aesthetic Education of Man* (1794). At the denouement, near the end of Goethe's play, Faust experiences a moment of peaceful epiphany just as he is about to die, in sharp contrast with Werther's final hours of misery. Faust achieves this state of bliss after a riot of satisfying overwhelming desires, fleeing from Mephistopheles to the temporary refuge of Gretchen's love, and undergoing many other trials and adventures, by immersing himself in everyday productive tasks. In his dying words a few scenes before the end of the play, Faust, like Goethe in his aesthetic theory and artistic practice, praises the natural sphere of activity proper to humankind.

[*See also* Poetics; Romanticism, *article on* Philosophy and Literature; *and* Theater.]

BIBLIOGRAPHY

Works by Goethe

Goethes Werke. Weimarer Ausgabe. 143 vols. Edited by Gustav von Loeper, Erich Schmidt, Herman Grimm, Wilhelm Scherer, Bernhard Seuffert, Bernhard Suphan, et al. Weimar, 1887–1919.

Goethes sämtliche Werke. Jubiläums Ausgabe. 40 vols. Edited by Konrad Burdach, Wilhelm Creizenach, et al. Berlin, 1902–1907.

Goethes Werke. Hamburger Ausgabe. 14 vols. Edited by Erich Trunz. Hamburg, 1948–1960.

Translations

Botanical Writings. Translated by Bertha Mueller. Honolulu, 1952.

Elective Affinities. Translated by Elizabeth Mayer and Louise Bogan. Chicago, 1963.

Essays on Art and Literature. Edited by John Gearey, translated by Ellen von Nardroff and Ernest H. von Nardroff. New York, 1986.

Goethe on Art. Edited and translated by John Gage. Berkeley, 1980.

Goethe's Faust. Translated by Barker Fairley. Toronto, 1970.

Goethe the Critic: A Selection of His Writings on the Arts. Edited by G. F. Senior, revised by C. V. Bock. Manchester, 1960.

Italian Journey, 1786–1788. Translated by W. H. Auden and Elizabeth Mayer (1962). San Francisco, 1982.

Selections from Goethe's Letters to Frau von Stein, 1776–1789. Edited and translated by Robert M. Browning. Columbia, S.C., 1990.

The Sorrows of Young Werther, The New Melusina, Novelle. Translated by R. D. Boylan, Victor Lange, and Jean Starr Untermeyer. New York, 1949.

Tales for Transformation. Translated by Scott Thompson. San Francisco, 1987.

Theory of Colours. Translated by Charles Lock Eastlake (1840). Cambridge, Mass., 1970.

Wilhelm Meister's Apprenticeship and Travels. 3 vols. Translated by Thomas Carlyle. London, 1874.

Other Sources

Angelloz, Joseph-François. *Goethe.* Translated by R. H. Blackley. New York, 1958.

Boyle, Nicholas. *Goethe: The Poet and the Age.* Oxford, 1991.

Carlyle, Thomas. *Critical and Miscellaneous Essays.* Boston, 1838.

Carus, Paul. *Goethe, with Special Consideration of His Philosophy.* Chicago, 1915.

Cassirer, Ernst. *Rousseau, Kant, Goethe: Two Essays.* Translated by James Gutmann, Paul Oskar Kristeller, and John Herman Randall, Jr. Princeton, N.J., 1945.

Grimm, Hermann. *The Life and Times of Goethe* (1880). Translated by Sarah Holland Adams. Freeport, N.Y., 1971.

Jacquette, Dale, ed. *Schopenhauer, Philosophy, and the Arts.* Cambridge and New York, 1996.

Lessing, Gotthold Ephraim. *Laocoön: An Essay on the Limits of Painting and Poetry* (1766). Translated by Edward Allen McCormick. Reprint, Baltimore, 1984.

Mann, Thomas. *Freud, Goethe, Wagner.* New York, 1937.

Nisbet, H. B. *Goethe and the Scientific Tradition.* London, 1972.

Robson-Scott, W. D. *The Younger Goethe and the Visual Arts.* Cambridge and New York, 1981.

Schiller, Friedrich. *On the Aesthetic Education of Man in a Series of Letters.* Translated by Reginald Snell. New York, 1965.

Schopenhauer, Arthur. *Schopenhauers sämtliche Schriften.* 7 vols. Edited by Arthur Hübscher. Wiesbaden, 1946–1958. See *Die Welt als Wille und Vorstellung,* vols. 1–2.

Schulze, Sabine, Friedmar Apel, et al., eds. *Goethe und die Kunst.* Stuttgart, 1994.

Sepper, Dennis L. *Goethe contra Newton: Polemics and the Project for a New Science of Color.* Cambridge and New York, 1988.

Simmel, Georg. *Kant und Goethe: Zur Geschichte der modernen Weltanschauung.* 4th ed. Leipzig, 1924.

Spaethling, Robert. *Music and Mozart in the Life of Goethe.* Columbia, S.C., 1987.

Stephenson, R. H. *Goethe's Conception of Knowledge and Science.* Edinburgh, 1995.

Swales, Martin. *Goethe: The Sorrows of Young Werther.* Cambridge and New York, 1987.

Wittgenstein, Ludwig. *Remarks on Colour.* Edited by G. E. M. Anscombe, translated by Linda L. McAlister and Margarete Schattle. Berkeley, 1977.

DALE JACQUETTE

GOMBRICH, ERNST HANS JOSEF

GOMBRICH, ERNST HANS JOSEF (b. 1909), theoretician and historian of art and culture. Gombrich began his long career in the immediate aftermath of the era when Alois Riegl, Heinrich Wölfflin, Aby Warburg, and assorted other polymath humanists forged the synthesis of methods and problems that became the modern study of art. The rich brew of philosophical and historical concerns that Gombrich drank in his student days sustained him in later years, for it was his fate to do his most innovative work in England and America from the 1940s through the 1970s, in contexts that were less than hospitable to the more speculative and transdisciplinary aspects of the founders' projects. His most important projects—his historiographical essays relocating the norms of modern art-historical explanation within the historical afterlife of the Renaissance, his investigations into the significance of reception theory for understanding pictorial meaning, and his lecture cycles on psychology, sign theory, and art, which became *Art and Illusion* (1960) and *The Sense of Order* (1979)—all carried on, by means of both elaboration and revision, the synthetic and self-reflexive inquiries of the generation in whose wake he studied. Despite the nearly overwhelming movement toward disciplinary stabilization from the 1940s onward, Gombrich persistently maintained a sense of art history as a

discipline whose boundaries remain perpetually up for grabs. At the present time, when the need for fundamental questioning is again being felt and embraced, Gombrich's preservation of his ties to those earlier thinkers should be a valuable treasure; that preservation is the living afterlife of the intellectual turmoil out of which was generated both the modern study of art and its contemporary tribulations.

In order to grasp the importance of Gombrich today, however, he needs to be read afresh and against the grain of much of the reception and criticism of his work, and even, in some measure, against his own retrospective depiction of it. Perhaps the foremost obstacle to such a rereading is the way in which his contributions have become scattered among the history of art, the philosophy of art history, the theory of culture, and philosophical aesthetics, fields that so often stand apart from, or are even hostile to, one another. Each of these disciplines has taken up Gombrich's ideas and arguments, but often stripped of their synthetic components and thus rendered suitable only for specialized disciplinary concerns. For instance, Gombrich's investigations into the psychology of perception, especially into the beholder's share in making the meaning of visual signs, have spurred creative work in representational and perspective theory by figures such as Nelson Goodman, Marx Wartofsky, and Richard Wollheim. In art history, to choose a quite discrepant example, themes developed in his iconological and historiographical analyses have continued pertinence in the work of Svetlana Alpers, Michael Ann Holly, and David Summers. These appropriations, adaptations, critiques, and revisions have been fruitful, a fact that need not be denied in the least in order to note also that the discrete points of view from which they have been pursued have also led to an obscuring of the overarching and underlying nature of Gombrich's project: what he calls, in a subtle and recuperative humanistic displacement of *Kulturwissenschaft,* "cultural history." The synthesis gestured toward by this concept has always informed Gombrich's more particular interests; it also represents a still largely untapped source of insight for our age of disciplinary instability.

Nowhere are Gombrich's cultural-historical concerns more vivid than in *Art and Illusion,* where he takes up the question of why different cultures, periods, and artists have mimetically represented the world in different and even incompatible ways that nonetheless can be organized into cohesive and explanatory historical sequences. Art history, he argues, cannot resolve this "riddle of style," or, at least, not without supplement. Art historians describe images, group them into historically and geographically significant types, and trace lines of influence and effacement between types. In short, they organize a field of phenomena in accord with the concept of style, but they do not address from within the field the question of why there is style at all. Many students of art who ponder this question do so in abstract and philosophical terms alone, as if it were a problem orthogonal to the real issues of art history. But such a segregation of conceptual and methodological from historical and empirical matters supposes that issues of how a field of study is constituted can be safely cordoned off from work within that field. Gombrich argues, to the contrary, that even undertaking to characterize, group, and explain images—the daily bread of art history—requires an answer to the riddle of style and that, indeed, the presumption that providing such an answer is extracurricular for art history merely permits tacit and distorting foundational views to remain unexamined and uncriticized. Further, Gombrich argues, the tacit answer too often has been a species of expressivism in which style is regarded as the simple surface appearance of some deeper impulse expressed through it. On such a view, style must be explained in terms of some other phenomenon that governs it—a zeitgeist, say, or a period or movement or mentality—and so, paradoxically, style is treated simultaneously as an organizing concept and as explanatorily idle. Expressivist positions subtly displace questions about the historicity of style outside the field of artifacts with style; rather than explain style, expressivism prevents issues about its historicity from being properly posed. Hence, only a reflectively enriched art history can execute adequately even its "primary" task.

The need to enrich art history motivates Gombrich's concrete historical studies throughout *Art and Illusion* (as well as other texts). To cite two instances: his analyses of the different functions of *sprezzatura* in European painting from the sixteenth through the nineteenth centuries, and his account of the interlocking roles of eyewitness observation and compliance with inherited motifs in the practice of drawing "from life," both aim to establish, over and above their particular interpretative judgments, that a nonexpressivist view of style is necessary in understanding the historical nature of the activities in question. In pursuit of both the more particular and the more general claim, the tools that Gombrich brings to bear, in addition to his formidable historical knowledge, are psychoanalysis, semiotics, and psychological and philosophical research in perception, in the veins of, respectively, Ernst Kris, the pragmatically inflected sign and information theories of Charles Morris and Roman Jakobson, and J. J. Gibson and Karl Popper. This tool kit may sound surprising to readers familiar only with Gombrich's later essays on perception and representation, mostly collected in *The Image and the Eye.* Many of the pieces in this collection are devoted to Gombrich's counterattack against what he regards as the improper Constructivist uses to which his work had been put. Goodman and Wartofsky, for instance, had used Gombrich's own argument that the ways we see the world are constructed by the ways we picture it to infer (to Gombrich's chagrin) that styles of mimetic visual representation are conventional through and through. In its conclusion that the visual facts to which mimesis aims to be faithful are themselves but ar-

tifacts of conventional representational systems, the Constructivist account of pictorial fidelity denies all canons of natural resemblance, a denial that Gombrich finds implausible. In order to press his rebuttal, Gombrich emphasizes throughout *The Image and the Eye* the realism he picked up from Popper (along with the latter's falsificationist theory of perception), in contrast to his own semantic, psychoanalytic, and pragmatic perspectives. But this polemical inflection ought not to obscure how, in his earlier works—in his review of Charles Morris's *Signs, Language, and Behavior,* in the essays collected in *Meditations on a Hobby Horse* (1963), in *The Sense of Order,* and, of course, in *Art and Illusion*—Gombrich brings all these elements together to argue:

1. that faithfully depicting the visual world is only one of the many human activities in which pictures play a role;
2. that specifically mimetic pictures are tools in the particular activity of faithful depiction (or, put differently, that mimetic pictures do not depict the world on their own, but rather we depict the world with them);
3. that the tools used in the interests of mimesis, as is true of tools in general, shape the ways in which the perceptual interests they serve are pursued;
4. that because the success of faithful depiction can only be measured by the perceptual achievements of beholders, pictures are also probes into the interest-saturated human perceptual apparatus; and that therefore
5. mimetic visual representation is a kind of image making whose styles are determined by the aims and interests embedded in the activity, the relative historical levels of achievement of the image makers, and the perceptual interests of beholders.

Now, this pragmatic-epistemic understanding of mimesis need not be shared by the image makers or beholders whose activities are being explained, as Gombrich demonstrates in his critique of the historically widely held view of pictorial fidelity as the reproduction of the vision of an innocent eye that is achieved by means of a stripping away of cultural accretions. Indeed, Gombrich reorients the problem of mimesis by shifting the problem from questions of objective correspondence between pictures and their motifs achieved by unlearning conceptual habits to questions of how pictorial techniques and habits of attention accumulate and are transmitted within and across historically and geographically diverse representational cultures. In a nutshell: the riddle of style is answered by picturing mimesis as a learning process.

Gombrich calls the learning process that structures the development of mimetic styles by different names: "schema and correction" initially, then "recall and recognition" in "Visual Discovery through Art" (in *The Image and the Eye*). Although the two names stress different features of the practice in question, they both emphasize the dialectical nature of mimesis: a picture is proposed as a faithful represen-

tation of a motif, is then tested against the element of the visual world it models, and then is modified in accord with perceived discrepancies. This dialectical cell is itself but a part of a larger dialectic, because the initial proposal is a residue of previous efforts and the modified outcome the starting point for later ones. Every mimetic practice is a kind of work on prior mimetic practices. Thus, Gombrich's answer to the riddle of style is that mimesis is an essentially historical experiment in learning.

Both the details and the overall adequacy of Gombrich's account of mimesis have been subjected to searching criticisms, albeit sometimes in ways that attribute to him mutually incompatible positions. Richard Wollheim (1973), for instance, argues in a review of *Art and Illusion* that without some "exit to the real" from the dialectic of mimesis, some nonpictorialized characterization of the visual world that mimesis aims to represent, all criteria for pictorial fidelity would become entirely internal to particular representational regimes; thus, there would be no way to prevent the history of styles from becoming entirely conventionalist and also, therefore, no way to understand such a history as a cumulative learning process. Because Wollheim interprets Gombrich as denying the existence of such an exit, he argues that the entire explanatory project of *Art and Illusion* is incoherent. Norman Bryson in *Vision and Painting* (1983), on the other hand, interprets Gombrich as availing himself too easily of an exit to the real and thus as taking insufficiently seriously the generative powers of different pictorial codes. This discrepancy in determining precisely what thesis Gombrich means to defend, even among those convinced that he is wrong about it, is rooted in ambivalences in the texts about mimesis themselves; indeed, the question to what degree, if at all, Gombrich is a conventionalist is still a hotly contested issue (see, for example, Murray Krieger's exchange with Gombrich about this issue [Krieger, 1985] initiated by Krieger's "The Ambiguities of Representation and Illusion: An E. H. Gombrich Retrospective" [1984]). Perhaps, though, more interesting questions can be posed about him by remembering the role of the account of mimetic styles in his larger project: to analyze cultural histories in general as dialectical learning processes. Only in this way can one make sense, and use, of Gombrich's intermixing of what seem to be separate disciplines, because learning processes involve perceptual, cognitive, developmental, and social components. It is just here that contemporary interdisciplinary studies of art and culture will find Gombrich most helpful, for his is one of the most powerful synthetic views of cultural processes and transmission that is not based on Hegelian historicism.

To say, however, that Gombrich's cultural history is not based on Hegelian historicism is too weak; his views are based, in fact, on an explicitly antagonistic relation to the Hegelian heritage in cultural explanation. It is a recurring theme in Gombrich that the direst obstacles to cogent cul-

tural history are the search for cultural totalities and the treatment of individual domains of culture—for example, religion, philosophy, and art—as expressions of those totalities. The model of the learning process offers the possibility of understanding cultural transmission, hence traditions, and cultural closures, hence discontinuities, without invoking nonexplanatory metaphysical entities such as "the spirit of the age." (This theme is most fully spelled out in *Ideals and Idols* [1979] and *Tributes* [1984]). It is undeniable that Gombrich's interpretation of Hegel is tendentious (and not surprisingly, because in this respect Gombrich is a too-close follower of Popper); he treats the doctrine of entelechy and the view of philosophy as a science of history as exhaustive of Hegel and neglects almost entirely Hegel's views on the centrality of negation and agon to culture and cultural transmission and his accounts of the historical embeddedness of understanding and interpretation. These Hegelian motifs are not really very far from Gombrich's own: in fact, to the degree that Gombrich's apology for writing cultural history is that it is a means to, and also an instance of, the heritability of the valuable achievements of a culture that writes its own history and worries about its afterlife, his ethical stance toward the past seems a genuine Hegelian inheritance. Still, the historicism that Gombrich does reject, with its belief in specific cultural totalities that the historian tries to reconstruct and its search for the ways those totalities shine through works of art, has exerted a strong pull on modern art history. Even more deleteriously, though, when, more typically, such a historicism is rejected *as if it were the only model of synthetic cultural history,* the study of art falls apart into an empiricist obsession with detail, on the one hand, and "theory," on the other. It is against this diremption that Gombrich has long taken a stand. For the many cultural and art historians now interested in developing nonexpressivist theories of "articulated" or "absent" totalities, Gombrich's work is fruitful and unexplored terrain.

It should be acknowledged that one other major obstacle does stand in the way of a fresh return to Gombrich. Because his most famous lectures, books, and essays have dealt with mimetic practices, Gombrich is sometimes thought to endorse the thesis that all visual art is mimetic. If he believes this, and if his theories of culture are developed in light of it, then clearly his approach would be nearly useless in understanding nonmimetic visual art. Gombrich nowhere defends this thesis, however. Indeed, *The Sense of Order,* his wide-ranging study of the psychology of ornament and decoration, is an obvious counterexample to it. Nonetheless, the false attribution of the thesis in question to Gombrich is not utterly without grounds. In "The Vogue of Abstract Art" (in *Meditations on a Hobby Horse*), *Art and Illusion,* and "Illusion and Visual Deadlock" (in *Meditations on a Hobby Horse*) Gombrich stretches the perfectly reasonable claim that mimesis is the eminent aim of Western art in order to apply it to modern abstractionism. Despite his remarkable sensitivity

to historical discontinuities in general, Gombrich is unaccountably—and, it must be said, mostly without argument—opposed to any discontinuous view of the art of his lifetime. Furthermore, in treating abstractionism as simply continuous with the preceding history of Western art, Gombrich finds it wan and idle by comparison; such a judgment is not surprising, because, considered as mimetic art, abstractionism is not very good. In other words, *The Sense of Order* notwithstanding, Gombrich's insight falters when confronted with nonmimetic art, and he appears drastically out of touch with the dynamics of modernism, as Arthur Danto argues in his 1983 *Grand Street* essay.

Because any theoretically informed history of art must be adequate for the full range of its objects, Gombrich's limited view of modernism might be thought to be a fatal flaw in his model of cultural history. Therefore, a robust engagement with his model would have to show that nothing in his theoretical position entails the limited view in question. Even more strongly: while recognizing that Gombrich himself provides no detailed pointers along the way, it is a hypothesis worth testing whether the enriched history of mimesis and symbolism effectuated by Gombrich's multidisciplinary approach to culture as learning processes might provide a suitable signpost for an enriched history of modernism. Many of the lacunae in the study of modernism—the contradiction between authenticity as an aesthetic norm and the historical compossibility of modernist forms, the incompatibility of autonomous artistic practices with their social functions, and so on—derive from conceptions of cultural history rooted in preconceived conflicts between art and history; thus, an alternative conception with a demonstrated effectiveness in resolving analogous difficulties in the study of mimesis has much to recommend it, at least as a model. Although Gombrich's own taste and historical judgment resist regarding modernism as sufficiently discontinuous with premodern art that it can be considered a distinct artistic domain, that resistance is not binding on other investigators. Of course, whether the model of a learning process will be adequate for understanding modernism awaits demonstration, but that waiting need not impede the long-delayed engagement with Gombrich's synthetic project.

[*See also* Mimesis; *and* Psychology of Art.]

BIBLIOGRAPHY

Works by Gombrich

Art and Illusion: A Study in the Psychology of Pictorial Representation. London, 1960; 2d rev. ed., reprint, Princeton, N.J., 1969.

Meditations on a Hobby Horse and Other Essays on the Theory of Art. London, 1963; 4th ed., Chicago, 1985.

Norm and Form: Studies in the Art of the Renaissance, vol. 1. London, 1966; 4th ed., Chicago, 1985.

Symbolic Images: Studies in the Art of the Renaissance, vol. 2. London, 1972; 3d ed., Chicago, 1985.

The Heritage of Apelles: Studies in the Art of the Renaissance, vol. 3., Oxford, 1976.

Ideals and Idols: Essays on Values in History and in Art. Oxford, 1979.

The Sense of Order: A Study in the Psychology of Decorative Art. Oxford, 1979, 2d ed., Ithaca, N.Y., 1984.

The Image and the Eye: Further Studies in the Psychology of Pictorial Representation. Oxford, 1982.

"Representation and Misrepresentation." *Critical Inquiry* 11.2 (December 1984): 195–201.

Tributes: Interpreters of Our Cultural Tradition. Oxford, 1984.

New Light on Old Masters: Studies in the Art of the Renaissance, vol. 4. Oxford, 1986.

Reflections on the History of Art: Views and Reviews. Oxford, 1987.

Topics of Our Time: Twentieth-Century Issues in Learning and in Art. London, 1991.

Other Sources

Alpers, Svetlana. *The Art of Describing: Dutch Art in the Seventeenth Century.* Chicago, 1983.

Bryson, Norman. *Vision and Painting: The Logic of the Gaze.* New Haven, 1983.

Danto, Arthur C. "E. H. Gombrich." *Grand Street* 2.2 (winter 1983): 120–132.

Goodman, Nelson. *Languages of Art: An Approach to a Theory of Symbols.* 2d ed. Indianapolis, 1976.

Holly, Michael Ann. "Past Looking." *Critical Inquiry* 16.2 (winter 1990): 371–396.

Krieger, Murray. "The Ambiguities of Representation and Illusion: An E. H. Gombrich Retrospective." *Critical Inquiry* 11.2 (December 1984): 181–194.

Krieger, Murray. "Optics and Aesthetic Perception: A Rebuttal." *Critical Inquiry* 11.3 (March 1985): 502–508.

Summers, David. "Real Metaphor: Towards a Redefinition of the Conceptual Image." In *Visual Theory: Painting and Interpretation,* edited by Norman Bryson, Michael Ann Holly, and Keith Moxey, pp. 231–259. New York, 1991.

Wartofsky, Marx. *Models: Representation and the Scientific Understanding.* Dordrecht and Boston, 1979.

Wollheim, Richard. "Reflections on *Art and Illusion.*" In *On Art and the Mind,* pp. 261–289. London, 1973; Cambridge, Mass., 1974.

Wollheim, Richard. *Art and Its Objects.* 2d ed., with six supplementary essays. Cambridge and New York, 1980.

GREGG HOROWITZ

GOODMAN, NELSON. [*This entry comprises two essays on the contemporary American philosopher and aesthetician Nelson Goodman:*

Survey of Thought

Art in Action

The first essay discusses Goodman's general philosophy and his aesthetic theory, especially in Languages of Art. *The second is an essay by Goodman himself, in which he further pursues his own earlier proposal to understand art in terms of the "When is Art?" (rather than "What is Art?") question.*]

Survey of Thought

Nelson Goodman (b. 1906), American philosopher of science, language, and art, was born in Somerville, Massachusetts, and was educated at Harvard, where he studied under the philosopher C. I. Lewis. His doctorate, completed in 1941, served as the basis for his great work *The Structure of Appearance* (1951). This book is a sketch of a constructed language that has at its disposal a minimum of primitive terms and uses them, together with the tools of modern logic, to define terms that make it possible to formalize the totality of human experience systematically and precisely. The central primitive term is the relation of matching: one experience matches another if there is no discernible difference between them.

The aim of Goodman's proposal is the aim of philosophy in general: to achieve understanding through clarity. One begins with a basis that is considered to be clear and uses it to explain what is less clear by means of definitions. It is thus of the utmost importance that the intuitively clear basis should be as small as possible.

Goodman does not believe that there is a criterion for distinguishing between what is clear and what is not. He considers the positivistic attempt to separate what is meaningful from what is meaningless through the possibility of verification to have been a failure—a heroic failure, but a failure nonetheless. What is clear and what is not is for Goodman an issue to be left to each philosopher's conscience. Goodman, who spent most of his life in New England (he taught at Tufts in 1945–1946, at Brandeis from 1964 to 1967, and at Harvard, where he is now professor emeritus, from 1968 to 1977) and Pennsylvania (where he taught from 1947 to 1963), developed the strictest New England Puritan conscience for clarity in philosophy.

Goodman's preference for a narrow, clear basis is also evident in his treatment of logic. In his logic, he commits himself to speak only about individuals and does not permit the use of sets because he does not consider the notion of a set to be sufficiently clear. On the other hand, Goodman's individuals can be abstract objects as well as material objects or sense data.

In *The Structure of Appearance,* Goodman proposes a philosophical reconstruction of language that is supposed to allow a reformulation of the scientific and the common-sense versions of the world. His later philosophical development brought him to the realization (which he shares with Ernst Cassirer) that there are a plurality of equally good versions of the world, even equally good literary or nonverbal fine-arts versions. In this view, there is no ready-made world independent of language to which these versions can be compared. Thus, our talk of a world amounts to talk of a right world-version.

A version is right if it does not offend our unyielding beliefs and principles. These beliefs are a mixed bag, including long-lived laws of logic and short-lived observations. But observations are by no means the sole arbiters of truth. A curve on a graph that summarizes observations but does not include even one observed point is right if it is a good approximation of the observed points and fulfills the criterion of choosing the simplest possible representation.

The next step in Goodman's philosophical development was an intensive concentration on artistic versions of the world. In Goodman's view, such versions do not lack cognitive status. He rejects the attempt to distinguish science and art by characterizing science as literal, cognitive, and emotionally neutral, while considering art metaphoric, expressive, and lacking in cognitive value. He claims that works of art, like science, present views of the world that can provide right insights, information, and knowledge.

The first stage in Goodman's philosophical development also includes his influential little book of 1955, *Fact, Fiction, and Forecast*. In this book, Goodman discusses a variety of topics, among them induction—that is, inference from the observed to the unobserved, or, crudely, inference from the past to the future. Goodman does not consider the "old problem of induction"—how to justify inference from the observed to the unobserved—to be particularly interesting or important. In its place he presents a new challenge: how to distinguish between valid and invalid inductive inferences. Goodman's inventiveness is apparent in the way he presents the problem. He devises an extremely artificial predicate, which he calls "grue," defining "x is grue" as "x is examined before time t (say, the year 2100) and is green, or is not examined before this time and is blue." Imagine, then, some emeralds that were examined before 2100 and were found to be green. These emeralds fit the definition of "grue." Any stones that are observed to support the claim that "all emeralds are green" by the same token support the rival claim that "all emeralds are grue." In that case, what is it rational for us to believe—that the first emerald we will come across after 2100 and which we have not previously examined will be green, or that it will be grue (that is, blue in our "natural" language)? We have no doubt whatsoever as to which of these we believe and which we reject. But why?

Goodman tries to convince us that the reason for this is not that "green" is a "natural" predicate that fits the nature of things in the world or our psychological makeup (for example, our judgments of similarity between colors), while "grue" is an unnatural one. Our justification for projecting the predicate "green" rather than "grue" on unexamined emeralds is, in the last analysis, grounded in our habits.

Goodman's rejection of the notion that some representation in art is more "realistic" than others is exactly parallel to his rejection of the "natural" justification for using "green" rather than "grue." In both cases, he presents instead a justification based on habits and familiarity. Goodman's claim is that we do not consider a picture realistic because the representation in the picture is similar to the thing represented in some natural way. Rather, we consider a mode of representation realistic only if we are used to it and familiar with it. Renaissance perspective is not more realistic than the reverse perspective of Chinese drawings; it is just the perspective we have gotten used to in our culture. In empiricist thought, the concept of similarity as the intersection of shared features was supposed to be the central explanatory notion for the creation of our concepts. Goodman denies the idea of similarity its explanatory force. Notions such as induction, metaphor, feature, or realism in art cannot be explained on the basis of similarity.

The basic work of the second phase of Goodman's philosophy, in which the artistic version of the world takes center stage, is his 1969 book *Languages of Art*. In contrast to Cassirer, who defended the claim of multiple symbolic forms that cannot be reduced to one another by presenting an eclectic and anecdotal description of the development of symbolic forms, Goodman proposes a systematic analytic theory of the different versions of the world (which are equivalent to Cassirer's "symbolic forms"). Here too there are a very small number of key concepts that are taken as understood. At the center is the concept of reference, which is meant to describe the relation between symbols and objects, and is systematically explored by the use of the pair of concepts "denotation" and "exemplification." The use of these concepts is not restricted to language, but also includes symbolic systems that contain diagrams, pictures, sculpture, musical notes, and other things. A symbol denotes the object or objects it applies to. The direction of fit is from the symbol to the object. The symbols "David," "Absalom's father," and Andrea del Verrocchio's David sculpture all denote David. In contrast, the direction of fit in exemplification is from the object serving as exemplar to the symbol. The tiny vials of perfume given out at airports are exemplars of the label "Chanel No. 5." The perfume vial refers to the label by exemplification.

An expression can denote an object either literally or metaphorically. The word *sapphire* literally denotes the precious stone, but it can also metaphorically denote a lake. Both cases involve denotation. The only difference between literal and metaphoric denotation is habit. We are more accustomed to apply the word *sapphire* to precious stones than to lakes. Goodman sees no reason not to speak of metaphoric truth. A lake is not literally a sapphire, but it is one metaphorically. On the other hand, the lake is not a ruby either literally or metaphorically.

Exemplification too can be either literal or metaphoric. A harp literally exemplifies a musical instrument, but it also metaphorically exemplifies David or Orpheus. The idea of metaphoric exemplification also gives us a way of talking about expression. Expression is metaphoric exemplification. The paintings of Picasso's "blue period" express sadness. It is not the case that the pictures are sad in the absurd sense that the canvases feel sadness. Moreover, we do not necessarily feel sad when we see these paintings; we may rather feel elated at having the opportunity to see them. Nor is it the case that Picasso must have been sad when he painted them. The pictures *express* sadness—that is, they exemplify "sadness" metaphorically.

Goodman also has a trick for using his limited number of concepts to solve the ancient problem of fictitious symbols—namely, how to describe the difference between names such as "Hercules" and "the Hydra" using only reference terms. After all, the two expressions denote the same thing—namely, nothing—because neither Hercules nor the Hydra has ever existed. On the other hand, there is all the difference in the world between Hercules and the Hydra, because Hercules killed the Hydra but the Hydra did not kill Hercules, and the Hydra had nine heads whereas Hercules had only one. Goodman's trick is to admit that there is no difference between what "Hercules" and "the Hydra" denote, but to assert that there is a difference between what a "Hercules-description" denotes and what a "Hydra-description" denotes, as there is between what a "Hercules-statue" and a "Hydra-statue" denote, or between what a "Hercules-picture" and a "Hydra-picture" denote. The term *Hydra* in the expression "Hydra-picture" has no independent meaning, just as the term *dwarf* in the expression "intellectual dwarf" has no independent meaning. The fact that there is a Hydra-picture does not mean that there is a Hydra whose picture it is, but only that there is a type of picture that is classified by the label "Hydra-picture." Fictitious symbols like "Hercules" and "the Hydra" do not differ in their primary reference, but they do differ in their secondary reference—for example, the references they get when they enter into different combinations.

Goodman considers the relation of symbolization to be the primitive relation of one thing "standing for" something else—that is, reference. As we have seen, he divides reference into two basic relations, denotation and exemplification, which operate in two opposite directions. But Goodman offers illuminating explications of much more complex reference relations in which there are references going in all directions—for example, allusions. Even such a simplistic example as the relation between the eagle as the symbol of a state and the state symbolized by the eagle requires a complex notion of reference. A picture of a bald eagle denotes a bird, which in turn exemplifies the label "daring, free, and in command." This label in its turn represents and is exemplified by the state whose symbol it is. Even more interesting is Goodman's use of multiple reference to explicate the important notion of variations as a special case of allusions. In Goodman's view, a variation must literally exemplify features it shares with the original, whereas metaphorically exemplifying features contrast with those of the original (as when a joyful passage serves as a variation on a plaintive one).

Symbols such as predicates in language come in families. A family of symbols, in Goodman's terminology, is a scheme, whereas the objects to which these symbols apply, and which are sorted by them, constitute a realm. The scheme and the realm together constitute a system. Goodman distinguishes between conditions applying to schemes

(syntactic conditions) and those applying to realms (semantic conditions). The conditions applying to the symbols involve their degree of disjointness, whereas the conditions applying to the realm of the objects involve both the degree of disjointness of the objects and their unambiguous fit with the symbols. The semantic conditions determine to what extent the relation between the symbols and the symbolized objects is free of vagueness and ambiguity. At one end of this continuum Goodman places notation, such as the Western system of musical notes in which the notes are disjointed—that is, clearly distinguished from one another—and the tones represented by the notes also fit them disjointly, without overlapping. A notation is thus a system in which all the elements are disjointed. At the other end of the continuum is painting, in which neither the symbols nor the objects are clearly distinguished from one another. In such a case, the symbols are said to be syntactically dense, whereas the objects are semantically dense. Between the two ends of the continuum are other systems, such as the linguistic one, which is used to write poems and stories, in which the symbols are disjointed but the objects symbolized are not. Such a system is thus syntactically disjointed but semantically dense, because the application of the linguistic expressions to the objects is subject to vagueness and ambiguity.

The disjointness conditions for symbols play an important role for Goodman in determining the identity conditions he proposes for different sorts of artistic works—the conditions we use to determine whether or not two items are instances of the same work. With literary works, such as *War and Peace,* every copy that preserves the spelling of the original is the same work. In contrast, a copy of a painting by Jan Vermeer, as similar to the original as it may be, is not the same work. Goodman distinguishes between two types of art—autographic and allographic. A type of art is autographic if the identification of works in that art necessarily depends on the history of their production. If this is not the case, then the type of art is allographic.

The identity conditions for works of art in terms of the disjointness of the symbols are also relevant, in Goodman's view, to the role of authenticity in such works. The difference between a Vermeer painting and a Han van Megeren forgery is an aesthetic one even if this is not noticeable at a particular time. Knowing the production history of the Vermeer should lead us to find differences between it and the van Megeren forgery. The syntactic density of paintings means that any difference can make a difference aesthetically.

Goodman diagnoses the aesthetic realm by characteristic symptoms. Because he considers this realm a symbol system, all its features are syntactic or semantic ones, rather than causal features such as the effect of the work on the observer.

None of Goodman's symptoms for belonging to the aesthetic realm is necessary—although it is necessary for at

least one of these symptoms to be present—and all of them together are sufficient. One symptom of the aesthetic is the syntactic density of the symbol system. A second symptom is semantic density—for example, the role of ambiguity in the language of poetry. The third symptom is the repleteness of the symbol—that is, its having rich and varied meaningful aspects. The fourth is the broad use of both literal and metaphoric exemplification. Even when a symbol does not play a referential role, as in Piet Mondrian's "nonfigurative" paintings, it exemplifies organizational principles that give it its meaning. In the same sense, a Bauhaus building that exemplifies organizational principles can be characterized as clearly aesthetic. The fifth symptom is the multiplicity and complexity of the reference relations in the symbol system, which are the outstanding signs of art.

Goodman's conceptual apparatus for discussing symbol systems also allows him to explicate the concept of style without recourse to the dubious distinction between form and content. He considers a particular feature of an artistic work to be characteristic of a style if it relates the work to other works of art—for example, other works of the same artist or the works of a particular school, period, or geographic area. A stylistic feature is thus a feature exemplified by the work, and it contributes to placing the work within a particular body of works. A work of art can also exemplify more than one style.

Not only does Goodman have much to say about style, but he also says it with style.

[*See also* Relativism; *and* Science and Aesthetics, *article on* Contemporary Thought.]

BIBLIOGRAPHY

Works by Goodman

The Structure of Appearance (1951). 3d ed. Dordrecht and Boston, 1977.
Fact, Fiction, and Forecast (1955). 4th ed. Cambridge, Mass., 1983.
Languages of Art: An Approach to a Theory of Symbols (1968). 2d ed. Indianapolis, 1976.
Problems and Projects. Indianapolis, 1972.
Ways of Worldmaking. Indianapolis, 1978.
Of Mind and Other Matters. Cambridge, Mass., 1984.
Reconceptions in Philosophy and Other Arts and Sciences. Written with Catherine Z. Elgin. Indianapolis, 1988.

Other Sources

Black, Max. "The Structure of Symbols." In *Caveats and Critiques: Philosophical Essays in Language, Logic, and Art*, pp. 180–215. Ithaca, N.Y., 1975.
Elgin, Catherine Z. *With Reference to Reference.* Indianapolis, 1983.
Robinson, Jennifer. "Some Remarks on Goodman's Language Theory of Pictures." *British Journal of Aesthetics* 19.1 (1979): 63–75.
Rudner, Richard, and Israel Scheffler, eds. *Logic and Art: Essays in Honor of Nelson Goodman.* Indianapolis, 1972.
Scheffler, Israel. *Beyond the Letter: A Philosophical Inquiry into Ambiguity, Vagueness, and Metaphor in Language.* London and Boston, 1979.
Snyder, Joel. "Picturing Vision." *Critical Inquiry* (1980): 499–526.
Wollheim, Richard. "Nelson Goodman's *Languages of Art.*" In *On Art and the Mind,* London, 1973; reprint, Cambridge, Mass., 1974.

Avishai Margalit

Art in Action

Consideration of seemingly scattered topics such as "What is art?" and the implementation of works of art, the role of conservation and restoration (or, for short, maintenance), the mission of the museum, and the prospects for an aesthetic economics has disclosed some fundamental interconnections between them. It has also focused attention on the importance of the working of works and on the following issues: the question "When is art?" is prior to "What is art?"; the implementation or activation of works matters no less than their execution; maintenance of works must certainly involve sustaining or reviving their functional capacities; and in aesthetics as in finance, both principal and yield have to be taken duly into account.

Activation. Some aspects of the functioning of a work of art now call for further attention. The working of a work consists of the response of an observer or audience coming to grasp the work—to understand it, and through it, other works and experiences. This is a subtle and complex phenomenon. It may be affected by whatever affects the object itself or the viewer or the circumstances of observation, and it is clearer from encounters with actual examples than from any general description. It may occur instantly and strikingly when a work is juxtaposed with another that compares and contrasts with it in unexpected ways, or when a shaft of sunlight hits a sculpture from a new angle, or more gradually with reading of relevant commentary or with increasing awareness of other works, other arts, or other cultures. A vast and motley miscellany of factors—from framing, lighting, and installation to publication, education, and advertising—may enter into how a work acts, into starting or enhancing or rectifying or blocking or inhibiting or disrupting its functioning. The problem and process of administering such positive and negative factors to optimize the functioning of a work can be called implementation; but because that may suggest utilizing an inert object, we may rather speak of *activation;* for works of art, like machines or persons, need to be thought of more as dynamic entities that often need to be started, restarted, and kept working.

Although activation of a work is typically not carried out by the artist and often involves routine and even menial procedures by technicians, it is not to be slighted as a purely practical matter adventitious to art; for what works *are* derives ultimately from what they *do.* The purpose in examining some questions about activation is theoretical, not technological. The philosophy of art aims no more at a manual of activation than at instructions or advice on how to paint.

To optimize the functioning of a work is not simply to maximize the quantity or intensity of activity, but rather to

maximize the functioning of the work as a work of art and as that particular work—that is, bringing about insight into and understanding of the work. An understanding of a work often has many aspects wanting different measures of activation. Furthermore, because there may be several different and right, yet not always compatible, understandings, we should speak of *a* rather than *the* proper functioning of a work—and there may be several alternative right ways of activating it.

Light against Life. Among factors in the activation of paintings, drawings, and prints, light is the most common, needed, and available—and it is variable and controllable. Light that is too dim or too bright blinds. Light of the wrong color or kind or from a wrong angle may distort. Under improved lighting, obscure details may emerge and the overall organization be better grasped, resulting in new insights. Whether a lighting is right, even whether it is too dim or too bright, will be affected by other features or the environment or installation, such as the coloring and contrast in surrounding works, the color and lighting of the background wall, of the ceiling and the floor, and of the room the spectator enters from. Of course, right lighting is not necessarily unique or final. Occasional abnormal lighting—as by shifting or flashing light, raking light, strobe light, or infrared or violet-ray—can be momentarily or even (in retrospect) lastingly effective. Plainly, lighting a work to optimize its functioning calls for a network of inquisitive experiments, delicate adjustments, and sensitive judgments by those in charge; and the complexity increases when other components of installation are taken into account, too. Luckily for the philosopher, all this is no more his task than is the execution of a work of art. With a clear conscience, he can leave it to the institutions or individuals and the experts responsible for the works.

But a deeper problem arises from a stubborn paradox: light, a powerful, adaptable, and virtually indispensable element in activating a drawing, watercolor, or Song painting, is also insidiously destructive. Even lighting too feeble for minimal visibility gradually fades or otherwise alters, and eventually eradicates, the work; and the stronger the lighting, the more damaging. To optimize the functioning of a work may be to curtail its survival; and measures taken for protecting a work may thwart its functioning. For a work of art, as for a person, the basic paradox is that living is dying. Contending demands of activation and conservation leave those responsible for works in a perpetual quandary. In each case, a decision has to be made to sacrifice one or the other aim or to reach a compromise between the two. One extreme (actual or proposed) instance of this is the locking away in darkness of an illuminated manuscript in order to preserve it.

We have arrived here at a problem of a different order from those already mentioned. Perceptive, experienced practitioners, with the help of instruments and experiments,

can arrive at fair judgments of the virtues and vices of a lighting. But a decision on how to compromise between effectiveness and destructiveness cannot be reached that way; it is a deep-rooted and far-reaching question of policy that calls for theoretical study. We philosophers cannot be expected to make the decisions in particular cases, but only to inquire into the underlying general problem, how to approach it, and what considerations are relevant in making these particular decisions. We shall soon encounter the same problem in a different guise.

Before leaving lighting, one final remark: obviously, lighting is a major factor in the performance of a play; but we must not jump to the conclusion that lighting has the same relation to a play as, say, to a painting. In painting, lighting is an activator of a completed work; execution and activation are clearly distinct here. On the other hand, in drama—a two-stage art—lighting and all other components of performance belong to the execution of a work; for a play consists of, and is only, completed in, its performances.

Care and Cure. A work can work only if in working condition. For instance, a painting requires light to work, but light will not make a too dirty painting work; cleaning it and lighting it are here partners in activation. Conservation and restoration have to be taken into account in any survey of art in action. Conservation includes protection not only from excess lighting but also from too much or too little heat or humidity, from harmful gases and rays, from subversive ingredients, from rust, rot, and vermin, as well as from physical damage by accident or vandalism. Restoration includes dealing with the result of neglect or ineffective conservation or mistaken prior attempts at restoration.

A decision to apply one or another means in order to retain or regain good condition or to refrain from doing anything has to be taken with recognition that any of the available means, and even doing nothing, is hazardous, risking constriction of the work's functioning and even destruction of the work itself. Moreover, we have no clear and comfortable idea of what constitutes good or right condition in general. Even so pat a maxim as that a work should, so far as possible, be kept in or returned to its condition on original completion is disgraced by any early Shang bronze with a glistening green patina gained from centuries of burial in earth. Removal of such a patina leaves a grim gray ghost. Such encrustations, the soft luster of some African wood carvings from decades of handling, some natural mellowing of materials in painting and sculpture, and some weathering of architectural works are not excrescences. Indeed, decisions about conservation and restoration and about rightness of condition are so highly variable with the particular work and circumstances that our best course now is to consider and compare some actual examples where a critical problem of what to do about the condition of a work has been confronted and dealt with.

Murals and Manuscript. Our first case must of course be the restoration of one of the world's greatest works of art: the frescoes on the ceiling of the Sistine Chapel in the Vatican in Rome, painted by Michelangelo about five hundred years ago. Not far into the twentieth century, the paintings had become so coated with grime and soot as to be nearly invisible. In contrast to the patina on an ancient bronze, this encrustation was nothing to be cherished, but an excrescence that almost kept the work from functioning at all, and had to be taken off. After intensive scientific study and exhaustive consultation and discussion on the effectiveness and risks of alternative available means of restoration on questions of policy, the authorities in charge decided to go ahead. They engaged a master craftsman who, with a small team, spent nine years of painstaking and knowing handwork to clear off the black pall and reveal the paintings that had been hidden.

Our second case has to do with another set of murals, also of major importance, but by unknown artists and some fourteen thousand years older than Michelangelo's, not on a ceiling but deep underground in the Lascaux caves in southern France, not on fresh plaster but on hard rock; and the painting is sometimes combined with engraving or carving. Because these works are not so familiar to the general public as the Sistine ceiling, let me quote a brief description by a scholar who knows them well. This is from *The Cave Artists* by Ann Sieveking:

> Lascaux is probably the most famous of the paleolithic caves and has the claim to be the most beautiful also. Originally it may have had rivals, but today its state of preservation makes it unique. It is near Montignac in the Dordogne. It was found by chance during the war, opened to the public in the late 1940's and closed nearly 20 years ago because the temperature and humidity changes caused by a great number of visitors threatened to ruin it. Extremely careful chemical treatment and the monitoring of its interior climate have saved the paintings, but there is no prospect that it will be opened again. . . . Lascaux is a small cave (hence the difficulty in ventilating it and controlling the temperature) consisting of a main hall that is approximately 30m long by 10 wide, with two further galleries beyond, and is more densely decorated than any known paleolithic sanctuary. Much of the wall's surface has a natural coating of white crystalline calcide and is covered with paintings in red, brown, black, and yellow, a polychrome combination of these producing a marvelous effect. The abundance of decoration is one reason for this visual impact and the other is, perhaps, the impression of movement that the paintings give. This is rare in paleolithic sanctuaries. (1979)

Some time after the caves were closed, a full-size replica was dug and painted nearby for the public to visit.

Our third case, already mentioned in passing, is the famous illuminated manuscript of the fifteenth-century known as *Les très riches heures du Duc de Berry*. Its functioning of course, depends on light, which also gradually destroys it. The common practice of libraries is to display such manuscripts seldom, in low light, open to a different page each day, thus reducing the rate of fading to the slowest compatible with the works' working at all, but storing them in darkness and controlled atmosphere most of the time. The decision concerning the *Très riches heures* at the Musée Condé in Chantilly went far beyond this: after having the finest possible reproductions made, the original would be locked away forever, never to be opened or seen again by anyone.

In comparing our first two cases, the ceiling and the caves, we note some obvious likenesses and differences. Both are large, immovable, and important works, susceptible to damage from atmospheric instability and impurities, especially those contributed by human breath. On the other hand, they differ greatly in age, situation, and available history; we know little about the artists of the cave murals, about the social background, interactions, motivations, or convictions of those who produced or viewed the work. More relevant to our present inquiry, though, is the dawning recognition that in the problem confronted and in response to it, the two cases are almost exact opposites of each other.

The primary problem with the ceiling was that it could hardly be seen. Its visibility, always somewhat below optimal because of the height of the ceiling, was steadily diminishing as the progressive accumulation of an opaque coating allowed less and less light to reach the paintings. Complete loss of function loomed. The indicated treatment was to remove the obstruction, thus admitting more light to the paintings and putting them back to work.

In contrast, the primary problem with the cave murals was not loss of function. Given some light, they could be seen fairly well; but because they were slowly fading and eroding, the decision was made to close the caves forever, entirely sacrificing function in favor of longer survival of the work.

Now compare the second and third cases. Although the cave murals and the illuminated manuscript are about as different as any two works can be, what is more to the point here are certain likenesses between the two cases: first, both were still in functioning condition, and the concern was for their future; and second, the response in both cases included not only therapeutic termination of the functioning but also an attempt to compensate by providing a substitute. In the case of the caves, producing a replica nearby seemed somewhat a gesture of apology for having taken the original work out of action. But in the case of the book of hours, plans were drawn in advance to have the finest possible reproductions made, apparently with the idea that they would serve as a fully equivalent replacement for the original. Ironically, the reproductions are sometimes taken to provide complete justification for the stoppage of function.

This calls attention to the need, in a study of art in action, for examining the nature and role of reproductions in the

functioning of a work. So far, we have considered only *direct* action, when a work and the agents and circumstances of activation are present to the observer; but when reproductions intervene, we have cases of indirect functioning.

Indirect Action. Attitudes toward reproductions vary over a wide range. At one extreme, as when facsimiles of *Les très riches heures* are taken as rendering the original superfluous or disposable thereafter, they are ranked as fully equivalent to the original. This reduces the museum's function to temporary storage of originals until perfect facsimiles can be produced. When the ideal of a "perfect facsimile" is recognized to be unrealizable, a reproduction *almost* indiscernible from the original is sometimes held to be good enough. But the most minute differences may have enormous effect, and indistinguishability is nontransitive and neither uniform nor constant. And Michael Camille (1990) has pointed out that what is judged to be a perfect facsimile at one time may later be seen as heavily influenced by the visual experience, cultural background, and interests of the period. Just as there is no innocent eye, there is no innocent process of mechanical reproduction.

All this tends to foster the opposite extreme, with reproductions being decried and dismissed as utterly deficient, irrelevant, or deceptive as surrogates for an original. But that attitude clashes with the overwhelming circumstance that what most of us, laypeople and experts alike, know or understand about most works of art, derives from reproductions: photographs, transparencies, slides, illustrations in books, and so on.

The collision here results from a derailment in our thinking. We started by looking on reproductions as substitutes for an absent original, taking over for it and performing all its functions. For this, exact duplication seemed to be required. But the reproductions that play an extensive and indispensable role in art education are seldom even approximate duplicates of the original; they differ from it in any of countless ways—some are black-and-white reproductions of colored works, two-dimensional prints of three-dimensional works, enlargements or reductions, details of complicated compositions, photographs under various abnormal kinds or angles of lighting. All these are a far cry from a facsimile. They by no means function as replacements, substitutes, surrogates, or proxies for the original, but rather as *instruments for activating it.* In many cases, their very contrasts with the original work bring out important features shared with it. In this, caricatures, parodies, and variations work like more commonplace reproductions. Reconceiving reproductions as means of activation rather than as replacements is highly consequential in other ways. It makes clear why two drastically different reproductions of the same work can be equally effective. The statement that destruction of the original or its permanent removal from display terminates its action no longer holds. *Direct* action ceases, but indirect functioning through reproductions

may continue; that is, action at a distance, or even posthumous action, is possible. Furthermore, what has been said of reproductions applies to verbal commentaries; these too are activators, and the illumination they provide usually depends heavily on characteristics other than descriptive faithfulness to the original work.

For an extant and present work, indirect action may supplement but can hardly supplant direct action. With the actual work readily available, some reproductions may provide additional understanding, but no one in his or her right mind with a drawing by Rembrandt van Rijn at hand will put it aside and concentrate exclusively on looking at reproductions and reading books about it instead. What accounts for the generally superior quality and force of direct action? Sometimes the difference is said to lie in that the original has an aura lacking in all reproductions, copies, and commentaries; but this tells us little until the notion of aura is clarified beyond the dictionary's "subtle emanation or exhalation." The aura has to be interpreted rather as a complex of the history of the work together with its associations, allusions, and other referential relationships.

This essay has laid a weight of words onto what is a subtle, in some ways fragile and often fleeting, yet momentous phenomenon, the aesthetic response to works of art. What has all this discussion accomplished? This essay has offered no solutions to any particular or general problem, nor any formulas or prescriptions for arriving at any decisions, but only consideration of a few aspects of art at work, and of some sample cases involving these aspects, with the hope that such a study might clear the ground for progress. Along the way, plenty of trouble has turned up. At every step, we have run into new and nasty difficulties: self-defeating activators, activators that are also destroyers, measures of preservation that vitiate or even prevent action. Of course, a study of art in action must embrace difficulties in as well as effective means of activating and sustaining function. The primary undertaking here has not been to improve the functioning of works but to improve our understanding of art in action and thus of art itself—because art is as much a matter of action as of objects.

On the other hand, we are not now able to deal here with a major problem that, as noted earlier, cannot be likewise unloaded as the responsibility of the practitioner rather than the philosopher: the question of how to arrive at decisions that require comparing kinds and force of action, or balancing action against survival. No calculus is to be expected, and no numerical values, but gross comparative judgments have to be and often are considerably reached. A sound account of the process of reaching them may depend on new psychological investigation—for example, into such informal processes as occur when some speakers of colloquial English say "I reckon" or "I figger," even when no figures are involved.

[*See also* Definition of Art.]

BIBLIOGRAPHY

Camille, Michael. "Les Très Riches Heures: An Illuminated Manuscript in the Age of Mechanical Reproduction." *Critical Inquiry* 17.1 (autumn 1990): 72–107.

Siveking, Anne. *The Cave Artists.* London, 1979.

NELSON GOODMAN

GOTHIC AESTHETICS. In the history of art, the Gothic ranks as the culminating phase of the Middle Ages in Europe (from the twelfth through the fifteenth centuries), characterized by technical daring and refinement of execution. Today, most scholars agree that Gothic began in architecture—with the activity of Abbé Suger (1081–1151) at Saint-Denis near Paris. Suger's rebuilding of the west and east ends of his venerable abbey church (1137–1144) yielded the first paradigm of Gothic architecture. This work started a trend leading away from the mural solidity of the previous Romanesque style, and toward the soaring grandeur of the cathedrals of Chartres (designed 1196), Reims (1211), and Amiens (1220). Typically, Gothic churches rejoiced in a wealth of sculpture, stained glass, and precious reliquaries.

Alongside art and architecture, analogous period phenomena have been detected in literature, music, philosophy, and other modes of cultural expression. During the time, Europe underwent major social transformations, including the rise of cities, growth of monarchical power (especially in France and England), increased trade and manufactures, and the flowering of universities. For many historians, this amalgam constitutes the Age of the Gothic.

Even more broadly, some writers employ the word *Gothic* in the sense of fantastic, extravagant, or imaginatively bizarre, whatever the medium or period. In this extended, evocative usage, *Gothic* overlaps with *Baroque* and *Mannerist*, also terms from art and architecture.

Term and Concept. Introduced in the sixteenth century (it was not used during the Gothic period itself), the meaning of the term *Gothic* has since changed in various ways.

A Germanic people, the Goths ravaged several provinces of the late Roman Empire. They split into two branches: the western group, the Visigoths, settled in Spain (412 and after), while their eastern cousins, the Ostrogoths, established themselves in Italy, beginning in 488. Although the heritage of the *Godos* played some role in later Hispanic culture, it was the historical memory of the Ostrogoths in Italy that fostered the proliferating senses the word *Gothic* engendered.

In fifteenth-century Italy, the Humanists, champions of the emerging Renaissance, addressed the puzzle of the Middle Ages (in their view the Dark Ages), a vast chronological wasteland separating them from their cherished Greco-Roman antiquity. For this catastrophe they blamed the Germans, whose incursions had given the Roman Empire its deathblow. As the Goths had been the first of these barbarians to ravage Italy, the Humanists collared them as the culprits in the perceived coarsening of arts and manners. In this pejorative sense, the term *Gothic* is analogous with *Vandal*. The Vandal stereotype, however, connotes only destruction, whereas the Gothic one implies both destruction and creation—though the latter was of an inferior order. In short, the word *Gothic* is an ethnic term turned into a disparaging aesthetic designation, equivalent to *barbarous*.

These hostile critiques of the Gothic are not always lacking in perceptiveness, for the cathedrals and public buildings populating European cities exacted a grudging admiration. This ambivalence is evident in the wry characterization of the artist-art historian Giorgio Vasari (1511–1574) found in his monumental *Lives of the Painters, Sculptors, and Architects,* first published in Florence in 1550. Vasari conceded the virtuosity of Gothic architecture, while stigmatizing its perverse indulgence in effects that were grotesquely profuse and untrue to materials.

Such was the magnetism of the new ideals of the Renaissance that the pejorative connotation of the word *Gothic* prevailed even in northern Europe, with its Germanic heritage. In France, the triumph of classic ideals during the reign of Louis XIV (1643–1715) meant the excoriation of the Gothic mode as the quintessence of all that was coarse, vulgar, and outmoded. This judgment applied across the board, to literature and social customs as much as to architecture and painting—ironically helping, in its comprehensiveness, to prepare for the later positive "unified field" concept of Gothic.

Reevaluation and Reenchantment. Exceptionally, England resisted the pattern of denigration. Antiquarians such as William Dugdale and Roger Dodsworth studied the outstanding buildings of the Middle Ages, distilling their labors in the three volumes of their *Monasticon anglicanum* (1654–1673). The concept of the Gothic also figured in political debates flowing from the English civil war of the seventeenth century. Defenders of the rights of Parliament liked to ascribe that venerable institution to the Angles, Saxons, and Jutes who came in the fifth century. As a national heritage, these writers discerned a special polity, labeling it the "Gothic balance" and "Gothic government." Ultimately, these ideas were to resurface in the moralistic interpretation propagated by advocates of the nineteenth-century Gothic revival.

A major theoretical advance was the publication, in London in 1756, of the treatise *A Philosophical Enquiry into the Origins of Our Ideas of the Sublime and the Beautiful* by Edmund Burke (1729–1797). Alongside the beautiful, long recognized as central to classicism, Burke posited a complementary principle, that of the sublime, which generated feelings of awe, dread, and magnificence. The sublime concept proved useful in reevaluating Gothic buildings. It also

fostered the emergence of the Gothic romance. The vogue for this literary subgenre, with its haunted castles and blasted moors, is generally considered to have been launched by Horace Walpole (1717–1797) in *The Castle of Otranto: A Gothic Story* (1765). Other novels of this kind are William Beckford's *Vathek* (1786), Matthew Gregory Lewis's *The Monk* (1796), and Charles Robert Maturin's *Melmoth the Wanderer* (1820). (Significantly, Walpole had Strawberry Hill, a small house he purchased west of London, transformed into an elaborate early prototype of Gothic revival architecture.) Fascinating as they are, these Gothic (or "Gothick") romances lie outside the main currents of the evolution of the Gothic concept.

The culminating, perhaps decisive step in what may be termed the reenchantment of medieval Gothic was taken by the young Johann Wolfgang von Goethe (1749–1832). In his pamphlet *Von deutscher Baukunst* (Of German Architecture, 1772), Goethe eulogized Strasbourg Cathedral. Likening the building to a "tree of God," he found the very profusion of the ornament a strength. The whole structure Goethe hailed as a manifestation of Germanic originality, contrasting with the servile imitations of classical models favored by the French and Italians.

In fact, the end of the eighteenth century and the first part of the nineteenth saw a concurrent vogue for both Gothic revival and neoclassical styles. This coexistence of two opposing styles, fostered in part by the philosophical distinction of the sublime and the beautiful, represented a momentous advance in the direction of aesthetic pluralism.

Gothic had started life as a term of denigration, but it was successfully detoxified. This redemption, though it was a lengthy one, foreshadowed the ascent of such modern stylistic labels as Impressionist and Fauve, which had begun as dismissive epithets.

During the nineteenth century, especially in northern Europe, the vogue of neo-Gothic structures signaled a renewal of national pride. Exceptionally, preference sprouted on the shores of the Mediterranean when, toward the end of the nineteenth century, the architects of Barcelona revived their region's Gothic as a token of glories of the Aragonese Middle Ages. Toward the middle of the century it became clear that, contrary to earlier expectations, Gothic architecture had actually begun in France, giving that nation its own motive for reviving it.

The heart of the Gothic revival lay, however, in Victorian Britain, where it gained the support of powerful writers. The first of these was the energetic Augustus Welby Northmore Pugin (1812–1852), who promoted the revival of Gothic building and design not solely on artistic grounds but as a vehicle for recapturing the superior ethos of the Christian Middle Ages. Pugin's precepts gained widespread circulation when they were adopted by John Ruskin (1819–1900). Ruskin promoted Gothic ideals as an antidote to the excesses of industrialism and commercialism in his

GOTHIC AESTHETICS. *View of North Spire* (c.1507–1513); Chartres Cathedral, France. (Photograph courtesy of Giraudon/Art Resource, New York.)

own day. Collectivism, rather than individualism, and craft production, rather than the machine, were his key precepts. Ruskin sought to promote these ideals in his Guild of St. George. Although this organization enjoyed little practical success, it provided the model for the arts and crafts associations that by the end of the century had sprung up in most English-speaking countries. The Arts and Crafts movement looked to the example of William Morris (1834–1896), a poet and prose writer who also designed all sorts of work in the applied arts. Morris's approach to design sought to break with the Renaissance distinction between the fine arts of architecture, painting, and sculpture, on the one hand, and the so-called applied arts (weaving, furniture, glass, wallpaper, etc.), on the other. In addition, Morris carried the broader aims of the Gothic revival further by his involvement in socialist politics. In this way, the preoccupation with the Gothic revival, so easy to caricature as nostalgic and even reactionary, fused with contemporary strivings for social change.

The Gothic revival assisted the rise of modern architecture, often indirectly. The Crystal Palace, erected in Lon-

don for the Great Exhibition of 1851, was not obviously Gothic in appearance, but showed deeper affinities in its reduction of load-bearing elements and their replacement with glass, together with the modular planning that allowed the vast hall to be dismantled and rebuilt. The architect Joseph Paxton, and others who put up glass structures of this type, exploited the most advanced engineering techniques of their own day, just as their medieval predecessors had done.

In the United States, the Gothic revival fused with the skyscraper as seen in Cass Gilbert's Woolworth Building (1913) in New York City. Rising to a height of almost eight hundred feet, the structure's vertical emphasis and filigree detailing earned its sobriquet as the "Cathedral of Commerce." Only in the 1930s did the soaring vertical lines of skyscrapers begin to yield to the horizontals of International Style architecture.

The 1919 brochure announcing the founding of the Bauhaus in Weimar, Germany, shows a cathedral-like structure, symbolizing the collectivist ideals embodied in the new modernist educational institution. As Sir Nikolaus Pevsner (1975) has shown, the Bauhaus concept is a lineal descendant of the ideals of Ruskin and Morris.

The 1920s in Germany acknowledged the similarity of the new style of Expressionism and the emotional intensity of the late Gothic. This affinity appears not only in fantastic designs for buildings, for the most part unexecuted, but also in the sets of such silent film classics as *The Cabinet of Dr. Caligari* (1919) and *The Golem* (1920).

Changing Foci in Scholarship. During the seventeenth and eighteenth centuries, French students of architecture noticed that medieval buildings fell into two distinct categories: an earlier type with massive walls and heavy vaults and a later type in which these features became lighter and more delicate. In the early nineteenth century, the former style of architecture (and the related arts) was termed *Romanesque,* permitting *Gothic* to be reserved for the later medieval building and visual arts of Europe.

This clarification of terms cleared the way for sustained efforts to understand the nature of Gothic itself. The foremost question, permeating all others, was whether Gothic architecture was primarily a symbol of the religious and philosophical commitments of the Christian religion or chiefly a manifestation of advancing technology. Broadly speaking, these two approaches may be denominated the idealist and the materialist theses.

The standard-bearers of idealism were the Gothic revivalists, such as Pugin and Ruskin, who preached the return to Gothic as a vehicle for the attainment of a better society. In their view, the greatest benefit of the change they were seeking lay not in adherence to higher standards of building and the crafts, though these would be essential, but in a fundamental change in human consciousness.

Less concerned than these advocates with social change in the present, cultural historians, committed to the principle of the inner unity of periods, looked for Gothic creativity in other spheres, such as poetry (the Grail and other romances) and music (polyphonic liturgical settings).

Following a nineteenth-century tradition, Erwin Panofsky (1951) posited a special affinity between Gothic architecture and scholastic philosophy. Such comparisons are seductive. Efforts to forge a unified field theory of Gothic civilization, however, meet an obstacle in castles and military architecture in general, which for practical reasons generally eschew the lightness and delicacy so characteristic of Gothic work in other types of building.

Other scholars have tried to link Gothic with a characteristically medieval type of social hierarchy. Attempts to ground Gothic in the "feudal system" must come to grips with the objections to the concept of feudalism that have surfaced with increasing insistence in recent years. Still, the ideal of Gothic totality enjoys considerable appeal—if only as a contrast to the fragmentation of late twentieth-century civilization.

The materialist thesis, opposing the idealist one just discussed, has reverberated outside the strict sphere of scholarship. The French architect and historian Eugène-Emmanuel Viollet-le-Duc (1814–1879), whose official assignments in restoring crumbling monuments had disclosed to him the finer details of Gothic structures, argued that Gothic resulted from the rational analysis of constructional problems; it is not perhaps too much of a simplification to say that for him the Gothic cathedrals represented the highest achievements of engineering prior to his own century. For this reason, Viollet-le-Duc's extensive writings were carefully studied by such modern architectural innovators as Frank Lloyd Wright and Le Corbusier (Charles-Édouard Jeanneret).

Close examination of the buildings along these lines led to an important series of debates about the essential constructional features of Gothic. For the early nineteenth century, the pointed arch ranked as the most important hallmark, so that some writers simply referred to Gothic building as "Pointed." It did not escape notice, however, that some Romanesque buildings employed pointed arches without forfeiting their characteristic massiveness. Attention shifted to the flying buttress. It is true that this feature, invented at the Cathedral of Paris about 1180, is a revealing engineering advance, but there are many smaller buildings, fully Gothic, that lack it. Then the rib vault was seen as the key feature. Finally, the lightening of the wall providing for huge expanses of stained-glass windows emerged as the essential element. These considerations encouraged a broader definition of Gothic architecture as driven by aspiration to elevation and attenuation. These characteristics may be interpreted, depending on the observer, either as primarily symbolic (the idealist thesis) or structural (the materialist).

Gothic architecture supplanted Romanesque, with the other arts following in relatively short order. Since the middle of the nineteenth century, it has been generally acknowledged that Gothic architecture arose in northern France, specifically at the Abbey of Saint-Denis, near Paris. The localization in Saint-Denis suggests that, unlike such styles as Egyptian and Romanesque, which arose gradually from a number of growing points, Gothic spread from a single place. In this diffusion it resembles the Baroque and Cubism, which began at Rome and Paris, respectively.

Erwin Panofsky and other scholars have argued that the new work at Saint-Denis was inspired by the mystical writings of Dionysius the Areopagite, a Syrian monk of the fifth century. Dionysius's emphasis on light, for example, would account for the great windows in the apse with their stained glass. Other scholars, scoffing at the Dionysian thesis, have failed to find traces of these writings in the apologetic texts of the founder Suger; they prefer to attribute his innovations to practical considerations. Yet others have stressed the link to the Capetian monarchy, for not only did the Gothic start in the French royal domain, at first it spread only there. These debates also show the interplay of the idealist and materialist theses.

At the opposite end of the medieval Gothic trajectory stands the conundrum of late Gothic. The elaborate vaulting and profuse decorative patterns of some late-Gothic buildings, especially in England and Germany, seem deliberately to contradict the rational approach of the earlier Gothic, centered in France. Perhaps anachronistically, some have attributed these adversarial trends to a nationalistic reaction against foreign (French) cultural hegemony. Late-medieval thought tended to shift from the certainties of Scholasticism to an emphasis on anomaly and mysticism. In the visual arts, the late Gothic witnessed a powerful surge of emotion as seen in the tortured body of Christ Crucified and the many grieving Madonnas. This emphasis on human vulnerability has made its appeal felt in particular periods, as in nineteenth-century Symbolism (the writings of Joris-Karl Huysmans) and twentieth-century Expressionism.

Aftermath. The United States, whose first European settlements came after the Gothic heyday, possesses no original works of Gothic architecture. Yet, it has shown a genuine affinity for the period, not yet fully explained. This affinity has flourished particularly in New England, where in fact the Puritans seem to have brought some remnants of late-medieval concepts. Two examples of scholars of this provenance are Henry Adams, author of *Mont Saint-Michel and Chartres* (1913), and the Harvard art historian Arthur Kingsley Porter.

Reflecting the roots of so many middle Americans in northern Europe, the Gothic has also enjoyed a certain resonance in the South and the Midwest. Appropriately or not, the writings of Tennessee Williams and Truman Capote have

sometimes been dubbed Gothic. In the Midwest, the 1930 painting *American Gothic* (Chicago, Art Institute), created by Grant Wood (1891–1942), has attained the status of an American icon. The rustic couple stands before a building with a pointed, traceried window, copied from a cottage (still existing) in Eldon, Iowa. Although the cottage was built in 1881–1905 as a private residence, such traceried windows are nowadays commonly inserted in the facades of meeting places of American Christian congregations that have been converted from warehouses or cinemas. The introduction of this semiotic marker suffices to "baptize" the structure, transferring it from the secular to the sacred realm.

Crumbling neo-Gothic architecture sometimes figures in Hollywood horror movies, an immensely profitable offshoot of the aesthetics of the sublime and its exploitation in Gothic novels. To be sure, only a few film treatments of this general type, such as *The Hunchback of Notre Dame*, stemming from the 1831 novel of Victor Hugo, are actually set in the Middle Ages. In fact, the Hugo story was especially popular; there were movie versions in 1923, 1939, 1957, 1982, and 1996.

The resonance of the Gothic is persistent and pervasive. Yet, it seems impossible to reduce the phenomenon to a single essence. Sometimes it has meant risk taking through technical daring and experiment, at others the reassuring comfort of tradition. It has signified romantic subversion and it has signified classical order. It has been limited to the Middle Ages and it has been stretched out to constitute a pervasive, ongoing heritage. It has figured as something remote and settled and something immediate and undefinable. As a component of emerging modernism, Gothic has provided both authorization and critique. The abiding legacy of the Gothic may lie in just this continuing flux.

[*See also* Architecture.]

BIBLIOGRAPHY

Bushart, Magdalena. *Der Geist der Gotik und die expressionistische Kunst: Kunstgeschichte und Kunsttheorie, 1911–1925.* Munich, 1990.

Clark, Kenneth. *The Gothic Revival.* 3d ed. London, 1962.

Frankl, Paul. *The Gothic: Literary Sources and Interpretations through Eight Centuries.* Princeton, N.J., 1960.

Germann, Georg. *Gothic Revival in Europe and Britain: Sources, Influences and Ideas.* Translated by Gerald Onn. London, 1972.

Halberstam, Judith. *Skin Shows: Gothic Horror and the Technology of Monsters.* Durham, N.C., 1995.

Mâle, Émile. *Religious Art in France: The Thirteenth Century.* Princeton, N.J., 1984.

Panofsky, Erwin. *Gothic Architecture and Scholasticism.* Latrobe, Pa., 1951; reprint, New York, 1976.

Pevsner, Nikolaus. *Pioneers of Modern Design: From William Morris to Walter Gropius.* 3d ed. London, 1975.

Robson-Scott, W. D. *The Literary Background of the Gothic Revival in Germany.* Oxford, 1965.

Wilson, Christopher. *The Gothic Cathedral: The Architecture of the Great Church, 1130–1530.* London and New York, 1990.

WAYNE R. DYNES

GOTTSCHED, JOHANN CHRISTOPH (1700–1766), German literary critic and follower of the Enlightenment. Gottsched pressed successfully for reforms in German theater (based on French models) as well as for uniformity in German grammar, orthography, and pronunciation while championing an enlightened poetry (one that consists in more than rhyme and meter) and criticizing the Baroque tastes of a preceding generation. He pursued these goals in monthly literary journals that he edited for more than four decades, in plays of his own, in numerous publications (including translations of Horace's *Ars poetica* and Gottfried Wilhelm Leibniz's *Theodicée*) and in an enormously influential grammar. Trained as a philosopher in his native Königsberg, Gottsched also wrote a substantial treatise on metaphysics, *Erste Gründe der gesamten Weltweisheit* (First Principles of Philosophy in its Entirety) that went through eight editions between 1734 and 1778. Gottsched's major contribution to aesthetics is his *Versuch einer critischen Dichtkunst vor die Deutschen* (Attempt at a Critical Art of Literary Composition for Germans). This work, of which four editions appeared between 1729 and 1751, constitutes the first attempt in Germany in the first half of the eighteenth century to give a systematic, philosophical foundation to literary criticism. Working out that philosophical foundation in terms of the thought of Leibniz and Christian Wolff, Gottsched set out to show both why criticism must be philosophically (that is, metaphysically) grounded and on what principles this grounding must be based.

Painfully aware of the absence in Germany of a poet of the stature of Jean Racine or Molière, Gottsched attributed this fact to the lack of competent critics of both poets and critics. To the question "Who criticizes the critics?" Gottsched had a straightforward answer: philosophers. Yet, it is not as though the task of poet, critic, and philosopher should fall on the shoulders of different individuals. Gottsched's aim was to elevate the level of the art of literary composition in Germany precisely through the formation of poets as critics and, if possible, as philosophers.

In concert with Johann Ulrich König, Gottsched rejects Jean-Baptiste Dubos's contention that public opinion is the supreme court of taste and that taste is a "sixth sense," in other words, a matter of sentiment or sensation rather than understanding. History, the existence of expert knowledge in other fields, and the undeniable presence of rules in poetry provide ample reason, Gottsched maintains, to suppose that the taste of a person who is knowledgeable about literary art is superior to that of a person who is not. While acknowledging the necessity, but not the sufficiency of natural talent, Gottsched argues further that, when good taste has disappeared from the cultural landscape, it is not enough for the poet merely to become a critic, that is, to learn the rules underlying criticism. At such a time a critical poet must also understand the reason for the rules, in other words, the poet must become a philosopher.

Gottsched's conception of the rule-based character of art and criticism and, as a consequence, of their subordination to philosophy can be traced to the rationalist epistemology underlying his analysis of taste. Taste, for Gottsched, is a matter of judgment, not sentiment or sensation, where what is judged is the beauty of an object. At the same time, however, it is a judgment in which sentiment or sensation plays an undeniable role. In the Wolffian terminology adopted by Gottsched and later refined in the aesthetic writings of Alexander Baumgarten and Moses Mendelssohn, sensation is defined generally as a clear but indistinct presentation or representation of something. For the aesthetic tradition that takes its bearings from this account of sensation, it is of paramount importance that being aware of things in this manner is inferior to the perfectly distinct knowledge of them that, indeed, exists in the divine mind. The sensation of red, for example, is clear insofar as it can be reidentified and differentiated from other sensations, but it is indistinct insofar as parts of it cannot be identified (allowing it to be reduced or analyzed). There is, moreover, an affective side to every sensation (hence, it is also a "sentiment"), depending on the degree of completeness or perfection in the subject, the object, and the relation obtaining between them. Beauty is a clear but indistinct presentation or representation of some perfection (that, once again, is in principle distinctly knowable because it is distinctly known by God). Gottsched accordingly defines good taste as "the understanding that judges rightly about the beauty of a thing in terms of the mere sentiment or sensation *(Empfindung)* of it, in matters of which one has no distinct and thorough knowledge" (Gottsched, 1973, vol. 6.1, p. 174). As this definition suggests, Gottsched does not subscribe to the view that all aesthetic judgments are equally valid (as though, for example, they were merely reports of different sentiments). There are right and wrong judgments of taste, based ultimately on the fact that beauty, the object of the judgment, is a presentation or representation of a perfection that is in principle distinctly knowable, even though it is only able to be approximated by human knowers. In the context of art, the nearest approximation to such distinct knowledge of the perfection involved is an understanding of the rules that make for such beauty. Thus, for Gottsched, good taste is nothing else but taste in conformity with rules.

Gottsched is aware that not all critics are philosophers and not all poets are critics. A beautiful poem can be written by a poet without the poet's knowledge of the rules employed that account for its beauty and the same poem can be adequately judged by a critic who must know the rules but may lack an understanding of their philosophical foundation. But Gottsched insists that only philosophy, that is to say, knowledge of the basis for criticism, can restore good taste in poets and criticism.

As for the rules themselves, Gottsched identifies the imitation of nature as the most general, theoretical rule, sub-

suming both general stylistic rules and genre-specific rules. The general rule about the imitation of nature is rooted in a metaphysical conception, adapted from Leibniz, of the relation obtaining between beauty and perfection, given the principle of sufficient reason and the so-called principle of the best. As a clear but indistinct representation of a perfection, beauty is based, according to Gottsched, on the nature of things themselves or, more precisely, on a harmony or agreement of things or parts of things according to a single set of general rules. This harmony or agreement, which itself presupposes a degree of similarity and uniformity, is a kind of perfection exhibited to a certain degree by every possible world in accordance with the principle of sufficient reason. In the case of our world (nature), moreover, the greatest possible perfection obtains, thanks to a divine decree, based on God's free choice to create the best possible world (in conformity with what Leibniz calls "the principle of the best"). By imitating the harmonious and, therefore, rule-governed arrangement of nature, the artist is accordingly capable of mirroring in an artifact the perfection that is the origin of natural beauty. [*See* Mimesis.]

In an audacious adaptation of Aristotle's claim about myth constituting the principle of tragedy, Gottsched applies the mimetic principle to literary art in general by claiming that its essence or at least highest form consists in telling a "fable" or "story." Yet, this claim, in addition to its apparent disregard for the stylistic side of literary art (to which a considerable part of Gottsched's work is, nonetheless, devoted), presents the traditional problem of differentiating literary art from history, on the one hand, and from mindless fantasy, on the other. A story is, Gottsched maintains, "nothing else but an imitation of nature," and yet it is the product not of an historian but of a poet. Gottsched deals with both difficulties by characterizing a story as a "possible world" or, in other words, once again a harmony in accordance with the principle of sufficient reason (ibid., p. 206). From this definition Gottsched derives a severe principle of poetic plausibility ("a fable's agreement with nature") that he uses to censure not only John Milton, but also Homer and Virgil (ibid., pp. 256ff.). Gottsched also applies the metaphysical principle of the best to his literary theory by demanding that the poet imitate the creator by crafting stories for the sake of moral instruction. Thus, according to Gottsched, the first step in writing a poem consists in "choosing an instructive moral principle" (ibid., p. 215).

As might be expected, the restriction of the scope of poetry and, particularly, poetic fantasy by Gottsched's metaphysically based and moralistic literary theory produced a strongly adverse reaction, namely, the celebrated controversy known as the *Dichterkrieg* (war of the poets) between him and the Swiss critics Johann Bodmer and Johann Breitinger. The controversy with the Swiss critics, who had the ear of several young writers and poets, contributed to the

diminishment of Gottsched's stature as a critic. But his reputation suffered irreparable damage when he became the object of Gotthold Ephraim Lessing's criticism and Johann Wolfgang von Goethe's sarcasm.

[*See also* Literature, *article on* Literary Aesthetics; *and* Taste.]

BIBLIOGRAPHY

Work by Gottsched

Ausgewählte Werke. 12 vols. Edited by Joachim Birke. Berlin, 1968–. See *Erste Gründe der gesamten Weltweisheit*, vol. 5, edited by Paul Mitchell (1983), and *Versuch einer critischen Dichtkunst vor die Deutschen*, vol. 6. 1–2, edited by Joachim Birke and Brigitte Birke (1973).

Other Sources

Birke, Joachim. *Christian Wolffs Metaphysik und die zeitgenössische Literatur- und Musiktheorie: Gottsched, Scheibe, Mizler.* Berlin, 1966.
Dahlstrom, Daniel O. "Die Aufklärung der Poesie: J. C. Gottsched, 1700–1766: Critische Dichtkunst, 1729." *Zeitschrift für Ästhetik und allgemeine Kunstwissenschaft* 31.1 (1986): 139–168.
Hasselbeck, Otto. *Illusion und Fiktion.* Munich, 1979.
Hohner, Ulrich. *Zur Problematik der Nachahmung in der Ästhetik des 18. Jahrhunderts.* Erlangen, 1976.
Wetterer, Angelika. *Publikumsbezug und Wahrheitsanspruch: Der Widerspruch zwischen rhetorischem Ansatz und philosophischem Anspruch bei Gottsched und den Schweizern.* Tübingen, 1981.

DANIEL O. DAHLSTROM

GRAMSCI, ANTONIO (1891–1937), founding member of the Italian Communist Party and its preeminent theoretician, political and cultural critic, best known for the concept of hegemony. Gramsci elaborated his concept of hegemony somewhat unsystematically in a voluminous body of wide-ranging and mostly fragmentary notes composed between 1929 and 1935 under extremely difficult conditions while he was confined by Mussolini's Fascist regime, first at the prison of Turi di Bari and later at a clinic in Formia. For a variety of reasons, but mainly because of the importance it attaches to the role of culture in its myriad manifestations within the complex relations of power at work in the social and political order, Gramsci's theory of hegemony has had considerable impact on several fields of inquiry, most notably history, political science, anthropology, sociology, and literary studies.

Gramsci arrived at the concept of hegemony through a series of detailed reflections on numerous aspects of the political, social, and cultural history of the Italian peninsula from antiquity to his own time. In spite of their specificity, these reflections touch on issues that are pertinent to the study of literature, art criticism, and aesthetics. To be sure, Gramsci does not provide a full-fledged aesthetic or even literary critical theory, nor did he set out to do so. Nevertheless, his observations on literature, drama, and criticism may still serve as fruitful starting points for critics and his-

torians who are interested in exploring the relations between artistic productions and politics, or between the history of art and the history of culture, but are also wary of conflating aesthetic with ideological judgment, or of separating considerations of content from analyses of form in the study of works of art. Gramsci's sketchy views on such critical approaches have thus far received relatively little attention within the enormous corpus of study that surrounds his work. Until now, it is the concept of hegemony that has had the strongest, most direct impact on the study of the arts (especially literature) and of culture in general.

Gramsci is very often cited and the concept of hegemony routinely employed in the arena of cultural studies. This sphere of research and criticism has no clear disciplinary boundaries, even though it grew, in large measure, out of the work in the 1970s of literary critics interested in the connections between literature (as well as other cultural productions, such as drama, film, the popular press, and television) and society and politics. The broad sweep of hegemony, in the Gramscian sense, with its far-reaching implications for the study of most aspects of culture, society, and politics, has been succinctly articulated by Raymond Williams, whose own work provided a major early stimulus to the development of cultural studies. Hegemony, Williams writes in *Marxism and Literature* (1977, p. 110),

> is a whole body of practices and expectations, over the whole of living: our senses and assignments of energy, our shaping perceptions of ourselves and our world. It is a lived system of meanings and values—constitutive and constituting—which as they are experienced as practices appear as reciprocally confirming. It thus constitutes a sense of reality for most people in the society, a sense of absolute because experienced reality beyond which it is very difficult for most members of the society to move, in most areas of their lives. It is, that is to say, in the strongest sense a 'culture', but a culture which has also to be seen as the lived dominance and subordination of particular classes.

The enormous amount of attention that Gramsci devoted to the analysis and critique of culture, in his mostly journalistic precarceral writings no less than in his prison notebooks, stemmed from a double motive. He wanted to understand and reveal the cultural processes that not only produce the ensemble of "meanings and values" that circulate in a given society (and, specifically, in the Italian society of his time), but also disseminate them so thoroughly that they become the unreflective common sense of the overwhelming majority of people in that society. By showing that meanings and values, as well as "perceptions of ourselves and our world," are neither the products of chance nor immutable facts of nature but, rather, are culturally shaped, he sought to foster among the subaltern classes whose cause he espoused a sense of critical independence and intellectual self-sufficiency that would enable them to develop a counterhegemonic, liberatory culture and to disseminate it on a mass scale. "Every revolution," he wrote in one of his early articles, "Socialism and Culture" (1916), "has been preceded by an intense labor of criticism, by the diffusion of culture and the spread of ideas among masses of men" (in 1985).

Gramsci does not provide anywhere in his writings a definition or description of culture that encompasses the entire range of meanings and nuances that are explicitly or implicitly attached to the term as he employs it in diverse contexts. Furthermore, it is virtually impossible to isolate Gramsci's concept of culture and to treat it independently of the rest of his concerns without at the same time producing a reductive version of it or obscuring its centrality in his work as a political activist and in the general thrust of his thinking. Thus, for example, his extensive critique of the philosophy of Benedetto Croce is an ineluctable part of his study of the role of intellectuals in society, which in turn is a crucial element of his deliberations on the social reasons for—and the political consequences of—the gulf separating high culture from popular culture. Likewise, his scathing treatments of the mostly Catholic, reactionary authors he labels "Brescianist" (after the nineteenth-century ultraconservative Jesuit novelist Antonio Bresciani) and of the Positivist social scientists he labels "Lorianists" (after the pseudosocialist political economist Achille Loria) lie at the heart of his chilling account of the impoverishment of Italian culture and the collapse of the "bulwarks of criticism" that, in his view, made the rise and triumph of Fascism possible.

If Gramsci were not so wary of the scientist tendencies of sociology, it would not be entirely off the mark to say that he was primarily concerned with the sociology of culture and, hence, was not especially preoccupied with questions of aesthetics in the narrow sense of the term. Nor do many aspects of cultural production that he dwelled on most frequently—among them, historiography, education, philosophy, religion, journalism, popular and popularizing literature of all kinds—lend themselves readily to aesthetic considerations. Unlike some of the best-known Western Marxists who evinced a special interest in cultural matters, such as György Lukács and Theodor Adorno, he did not engage in a systematic study of the relations between aesthetics and ideology. Nevertheless, Gramsci's writings on literature, literary history, and literary criticism provide ample evidence that he was attentive to certain aesthetic issues, and some of his observations on the subject, fragmentary and dispersed though they are, continue to merit serious attention.

Gramsci, obviously, did not share the formalist and aestheticist view that a literary work—or any work of art, for that matter—inhabited a totally autonomous sphere and that it could only be properly analyzed and evaluated in its own terms. At the same time, however, he did endorse the notion of the partial or limited autonomy of art, and he repeatedly pointed out that his views on the moral, social, and

political significance of a given novel or play did not constitute an aesthetic judgment per se on that work. Thus, for example, in a notebook devoted to "Literary Criticism," he concludes a set of observations on the popularity of Leo Nicolaevich Tolstoy and Alessandro Manzoni with a cautionary remark:

> This is a study of cultural history, not artistic criticism in the strict sense. . . . The study of the beauty of a work is made subordinate to the study of why it is 'read', 'popular', 'sought after', or, in the opposite sense, why it does not touch the people and arouse their interest, showing up the lack of unity in the cultural life of the nation. (Notebook 23, section 51)

It is possible, Gramsci maintained, to appreciate the aesthetic merits of a literary work even while repudiating the ideology that informs it. In one of his letters from prison (1 June 1931), he explained to his wife Julia that it is possible to admire the aesthetic achievement of a poet without "loving" that poet, for "'love' implies an adherence to the ideological content of the poetry." He added: "One can 'love his very own poet' while 'admiring' the artist in general. Aesthetic admiration can even be accompanied by a certain 'civil' scorn, as in the case of Marx's attitude towards Goethe." He returned to the same issue in another letter (5 September 1932), where he told Julia that he can "admire Tolstoy's *War and Peace* from an aesthetic viewpoint without agreeing with the ideological contents of the novel. If both factors coincided, Tolstoy would be my vade mecum, my *livre de chevet*. This holds also for Shakespeare, Goethe, and Dante."

In many of his notes on Marxism—or, the "philosophy of praxis," as he often called it—Gramsci inveighed against vulgar materialism and reductive ideologizing. Over and over again, he insisted on the need for critical rigor to replace reliance on dogma in order for Marxism to become a effective cultural force capable of generating radical social transformation. When he set out tentatively to articulate an approach to literature that was consistent with his philosophy of praxis, he was therefore particularly anxious lest ideological tendentiousness (or what one might call political correctness) interfere with and occlude aesthetic judgment. He addressed this question most directly in the opening paragraph of a relatively long note titled "Art and the Struggle for a New Civilization":

> The artistic relationship reveals, especially in the philosophy of praxis, the fatuous ingenuousness of those parrots who believe that they possess in a few brief and stereotyped formulas the key to open all doors (those keys are actually called 'picklocks'). Two writers may represent (express) the same socio-historical moment, but one might be an artist and the other just an unctuous individual. If the treatment of the question amounts to nothing more than a description of what the writers represent or express socially—that is, a more or less reliable summary of the characteristics of a specific socio-historical moment—it means that the

question of art has not even been broached. All this can be useful and necessary, indeed it certainly is but in a different field: in the field of political criticism, in the criticism of customs, in the struggle to destroy and overcome certain currents of feeling and belief, certain attitudes toward life and the world. This is not criticism and history of art, nor can it be presented as such—not without causing confusion and regression or the stagnation of scientific concepts; in other words, not without the failure precisely to pursue the intrinsic aims of the cultural struggle.
> (Notebook 23, section 3)

In order to convey the complexity of Gramsci's approach to the study of literature, however, this passage needs to be juxtaposed to another one from a note titled "Criteria of Literary Criticism":

> Given the principle that one should look only to the artistic character in a work of art, this does not in the least prevent one from investigating the mass of sentiments and the attitude towards life circulating in the work of art itself. . . . What is excluded is the notion that a work is beautiful because of its moral and political content and not for its form, with which the abstract content is fused and becomes one.
> (Notebook 15, section 38)

According to Gramsci, a work of art is never so cut off from history as not to embody even in its form an attitude toward life, a worldview. It is also noteworthy, in this context, that Gramsci sought to avoid the pitfalls of separating form from content, other than for heuristic purposes; he steered clear of stressing one in favor of the other. Indeed, arguing against the Parnassian claims of assorted formalists, aesthetes, and hermeticists, he maintained that "the so-called poetry of form is characterized by indifference toward content, that is, by moral indifference; but this, too, is content, the historical and moral vacuity of the writer" (Notebook 5, section 95). In another note that deals directly with the relation between form and content, he affirmed that "changes of content are also changes of form" and that "'content and form' besides having an aesthetic meaning have also a 'historical' meaning." He then went on to explain that to study literature is to study the history of literature, and "literary history is a part and an aspect of a much larger history of culture" (Notebook 14, section 72).

An extremely important aspect of Gramsci's fragmentary theory of literary criticism is its rebuttal of certain elements that lie at the core of Croce's aesthetics and general approach to literature. Croce insisted that the "history of art" had to be kept distinct from the "history of culture." To study the "history of art," for Croce, meant to study a series of individual works. The "history of culture," again according to Croce, could yield useful insights about intellectual currents, social mores, and so on, but contributed nothing to the evaluation and appreciation of individual works of art and their aesthetic quality. Because of Croce's bad influence, Gramsci maintained, "there is no concrete under-

standing of the fact that art is always tied to a definite culture or civilization" (Notebook 21, section 1). Furthermore, Gramsci was convinced that Croce's distinction between poetry and nonpoetry or "structure" (the conceptual, ideological, didactic, practical level of the text) was actually detrimental to textual analysis and interpretation—so much so that it led Croce to partially misread Dante. He set out to prove this point in a cluster of notes on canto 10 of the *Inferno,* where, among other things, he addresses the question: "Can one reconstruct and criticize a poem other than in the world of concrete expression and historically produced language?" (Notebook 4, section 79). The conclusions derived by Gramsci from his close reading of canto 10 of the *Inferno* are summarized succinctly by Robert Dombroski in his book, *Antonio Gramsci* (1989, p. 70): "Poetic form has its own historicity. It does not transport us into an atemporal sphere of pure lyricisim. Rather, it draws our attention to a specific manner of communicating, the modalities of which are largely determined by specific literary and social institutions. Structure and poetry are inextricably linked."

If the work of art is rooted in history, so of course is the critic. In his discussions on the role of intellectuals in society, Gramsci showed that, notwithstanding all claims to the contrary, the autonomy of intellectuals is an illusion or a deception. Literary critics are no exception and their claims of aesthetic disinterestedness are just as invalid. When the critic analyzes the work of literature as a work of literature, she or he is inevitably performing a cultural act and culture is not, in the final analysis, separable from politics. To illustrate this point, Gramsci contrasts the critical approaches of Francesco De Sanctis and Croce. Even though he did not share all his political views, Gramsci regarded De Sanctis as an exemplary figure because he did not divorce his literary criticism from his own cultural situation and from his participation in the cultural struggle of his day. Gramsci wrote (in a note that in its first version appears under the title "Historical Materialism and Practical Criteria or Canons of Historical and Political Interpretation"):

> The criticism of De Sanctis is militant, not "frigidly aesthetic": it belongs to a period of cultural struggle. The analyses of content, the criticism of the "structure" of works . . . are connected to the cultural struggle. . . . It is pleasing to feel in him the impassioned fervor of a committed person who has firm moral convictions and does not hide them or even try to.

In Croce, by contrast, one finds "a struggle for a refinement of culture, not for its right to exist; romantic passion and fervor have settled into a superior serenity and an indulgence full of bonhomie." For this reason, Gramsci concluded, "the type of literary criticism suitable to historical materialism is offered by De Sanctis, not by Croce" (Notebook 4, section 5). Gramsci considered Croce's aesthetic theory symptomatic of a much larger phenomenon: the posture of Olympian detachment cultivated by the priests of high culture who maintained a disdainful distance from what Gramsci called the "people-nation." The absence of a rapport or bond between the intellectuals and the people, in Gramsci's view, had left Italy culturally impoverished and fatally vulnerable to the onslaught of Fascist demagoguery.

From his prison cell, Gramsci had no opportunity to intervene in the events of his day. The impact of his thought was felt long after his death, following the publication of his writings in Italy in the late 1940s and early 1950s. Gramsci's posthumous influence pervaded virtually every level of political and cultural activity in Italy, including literary criticism, where it eclipsed the prevailing Crocean orthodoxy. It has been said that Gramsci contributed to the politicization of literary studies; this is true, but only to a certain extent, for it must not be taken to mean that Gramsci's work lends itself in any way to a reductive subordination of literary interpretation and aesthetic evaluation to political dogma. After all, Gramsci's abhorrence of dogmatism of all stripes is the hallmark of his work. Still, even when writing about literature and the arts, Gramsci remained first and foremost a social and political critic in the sense that—as Gianni Scalia pointed out in *La città futura* (1959, p. 336)—he looked at "the whole complex communicative exchange that constitutes the work of art understood in the fullness of its life, as a 'circuit' that is not only stylistic and psychological but also sociohistorical and cultural."

[*See also* Croce; Cultural Studies; *and* Marxism.]

BIBLIOGRAPHY

Work by Gramsci

Quaderni del carcere. 4 vols. Edited by Valentino Gerratana. Turin, 1975.

Works by Gramsci in English Translation

An Antonio Gramsci Reader: Selected Writings, 1916–1935. Edited by David Forgacs. New York, 1988.
Letters From Prison. 2 vols. Edited by Frank Rosengarten, translated by Raymond Rosenthal. New York, 1994.
Prison Notebooks. 5 vols. Edited by Joseph A. Buttigieg, translated by Joseph A. Buttigieg and Antonio Callari. New York, 1992–. The first and second volumes of this complete critical edition of Gramsci's notebooks were published in 1992 and 1996, respectively.
Selections from Cultural Writings. Edited by David Forgacs and Geoffrey Nowell Smith, translated by William Boelhower. London, 1985.
Selections from Political Writings, 1910–1920. Edited by Quintin Hoare, translated by John Matthews. London, 1977; reprint, Minneapolis, 1990.
Selections from Political Writings, 1921–1926. Edited and translated by Quintin Hoare. London, 1978; reprint, Minneapolis, 1990.
Selections from the Prison Notebooks. Edited and translated by Quintin Hoare and Geoffrey Nowell Smith. London, 1971.

Other Sources

Bové, Paul. "Dante, Gramsci, and Cultural Criticism." *Rethinking Marxism* 4.1 (spring 1991): 74–86.
Caracciolo, Alberto, and Gianni Scalia, eds. *La città futura.* Milan, 1959.

Davidson, Alastair. *Antonio Gramsci: Towards an Intellectual Biography.* London, 1977.

Dombroski, Robert S. *Antonio Gramsci.* Boston, 1989.

Entwistle, Harold. *Antonio Gramsci: Conservative Schooling for Radical Politics.* London and Boston, 1979.

Holub, Renate. *Antonio Gramsci: Beyond Marxism and Postmodernism.* London and New York, 1992.

Landy, Marcia. *Film, Politics, and Gramsci.* Minneapolis, 1994.

Rosengarten, Frank. "Gramsci's 'Little Discovery': Gramsci's Interpretation of Canto X of Dante's *Inferno.*" *Boundary 2* 14.10 (Spring 1986): 71–90.

Williams, Raymond. *Marxism and Literature.* Oxford and New York, 1977.

JOSEPH BUTTIGIEG

GREEK AESTHETICS. *See* Architecture, *article on* Early Greek Aesthetics; Aristotle; Classicism; Hellenistic Aesthetics; *and* Plato.

GREENBERG, CLEMENT (1905–1994), for some three decades a dominant voice in American art criticism and since the early 1980s (even after his death), a figure of central controversy in contemporary art and criticism. Greenberg's writings in defense of abstract art, particularly Abstract Expressionism and the postpainterly abstraction (or color field painting) that followed it, articulate a vision of modernist art that, although much contested, has yet to be supplanted, and the resistance to this vision that became evident in the 1960s with the emergence of Pop and Minimalism continues to fuel both contemporary criticism and a wide range of alternative accounts of modernism in art. Indeed, much of the current discussion of artistic and critical "postmodernism" is not fully comprehensible apart from an extended consideration of Greenberg's formalist criticism and the account of modernism he erected on the basis of its judgments.

Although Greenberg first appeared in print in 1939 as a literary critic, two of his earliest essays—"Avant-Garde and Kitsch" (1939) and "Towards a Newer Laocoön" (1940)—laid out with extraordinary authority the terms through which he would engage contemporary art. The first of these argues the role of the avant-garde to be the preservation of cultural standards in the face of capitalism's ongoing work of commodification, and understands the preservation of aesthetic value to be continuous with the cultivation and possession of one's experience over and against the tendency of capitalism to produce vicarious experience or mere effects. The second essay extends this argument by tying it to a historical justification for the emergence of an art at once "pure" and "abstract." For Greenberg, the argument for abstraction is in its essence an argument for integrity and uniqueness of a given artistic medium and the experience it subtends. In the modern period, this issues

above all in a demand that each art defeat its tendencies toward "literature" and toward meaning in favor of a foregrounding of its own material condition. Literature itself is not exempt from this demand:

> To deliver poetry from the subject and give full play to its true affective power it is necessary to free words from logic. . . . Poetry subsists no longer in the relation between words and meanings but in the relations between words as personalities composed of sound, history, and possibilities of meaning. . . . The poem still offers possibilities of meaning—but only possibilities. Should any of them be too precisely realized, the poem would lose the greatest part of its efficacy, which is to agitate the consciousness with infinite possibilities by approaching the brink of meaning and yet not falling over it. The poet writes, not so much to *express,* as to create a thing which will operate upon the reader's consciousness to produce the emotion of poetry.
> ("Towards a Newer Laocoön")

One recognizes in this the now-familiar demand that a poem should not mean but be. By sometime in the early 1950s, Greenberg gives up the underlying Marxist account in favor of a loosely Kantian understanding of aesthetic experience as radically disinterested, depending on no intellectual grasp of its object; subjective, involuntary, universal, and essentially a matter of a certain indeterminate but irreducible pleasure. As a matter of practice Greenberg maintains (or at least means to maintain) this position throughout his career, although by 1960 he has, in the synoptic essay "Modernist Painting," simplified its formulation down to a proposition that seems to derive more nearly from the preface to Immanuel Kant's first critique, the *Critique of Pure Reason,* than from the complexities of the third, the *Critique of Judgment,* which is the text in which Kant addresses questions of art and aesthetics:

> The essence of modernism lies, as I see it, in the use of characteristic methods of a discipline to criticize the discipline itself, not in order to subvert it but in order to entrench it more firmly in its area of competence. Kant used logic to establish the limits of logic, and while he withdrew much from its old jurisdiction, logic was left all the more secure in what there remained to it.
> ("Modernist Painting")

This simplification is not without its costs—the connection with experience is lost or misplaced, and Greenberg is left with an opposition between "self-criticism" and "criticism from the outside" that he has trouble sustaining throughout the remainder of the essay (it is also true, of course, that Kant's concern was with the limits of reason and not, strictly speaking, logic).

The difficulties Greenberg's writing runs into in "Modernist Painting" and elsewhere are themselves but symptoms of what is perhaps the underlying difficulty of his writing—his desire to hold at once to a radical Kantianism (radical in the sense that it derives its self-understanding and terms of criticism from the particular terms Kant de-

velops around the judgment of natural beauty with little theoretical quarter given to the complexities that Kant acknowledges to be bound up with products of human making) and his countervailing desire to be able to provide both historically justificatory accounts of the necessity of such practice and broad accounts of the kinds of plastic values (flatness, opticality) that give rise to aesthetic pleasure. Although Greenberg repeatedly insists (most notably in a postscript appended to "Modernist Painting" in 1978 and in the 1964 essay "Complaints of an Art Critic") on the separateness of the critical and theoretical or historical dimensions of his writing, his many critics can hardly be blamed for feeling them to be repeatedly run together in his writing ("This may," the 1978 postscript says, "be a fault of the writing or the rhetoric"). By the 1960s, he is widely perceived not to be judging on the case-by-case basis he claims but against a theoretically derived and historically justified norm.

The art-historical account embedded in Greenberg's writings insists on the centrality of Cubism to modernist style, over and against the "literary" impulses of Dada and Surrealism, and on the importance of the move beyond Cubism first achieved in Jackson Pollock's "painterly abstraction" (a term that Greenberg came to prefer to either Abstract Expressionism or Action Painting) and subsequently secured, along a path first opened by Barnett Newman and Mark Rothko, in the postpainterly abstraction of Morris Louis, Kenneth Noland, and Jules Olitski. By the 1960s, the values apparently inhering in this line of development—flatness, opticality, and an orientation to the essential conventions of painting—were increasingly under challenge from new tendencies that Greenberg could recognize only dismissively as "Novelty Art." Of these, two have proved particularly important to the current practice of visual art and the claims to a visual postmodernism—Pop, which appeared primarily as an exterior, neo-Dadaist assault on the terms of Greenberg's narrative, and Minimalism, which appeared to engage a more complex and internal argument with those terms, extending the apparent reductive and positivist lines of his account to the point of overrunning the distinction between painting and sculpture.

That these lines had, in a certain sense, already been blurred by Greenberg himself, albeit in a very different manner, in his support for the nonmonolithic, open, and distinctly pictorial constructed sculpture of David Smith and Anthony Caro—a sculpture he understood to derive directly from the example of Cubism—is a good indicator of something of the actual complexity of the internal crisis provoked by Minimalism. Although it is currently popular to link this crisis to Greenberg's clear postwar disillusionment with Marxism—a disillusionment that, on this account, left his formalist practice separated from its actual historical ground and so unable to respond to changing avant-garde strategies as they evolved within the new logic of global capitalism and its massive acceleration of commodification—it is also possible to continue to explore it on grounds that seem more nearly its own.

Here one would have to insist on the importance of Greenberg's refusal to allow any strong linkage between what he distinguished as the history of art as "a history of value" and the history of art as "the history of a phenomenon," even though his own writing repeatedly enacted and depended on just such a linkage. The form this refusal takes in Greenberg's own defenses of his position is that of a separation between the logic of his argument, on which he insists, and its rhetorical effects, which he disavows. The admission of "rhetoric," in art and criticism alike, can for Greenberg only point backward to the literary or the academic, the region of mere effects denounced in "Avant-Garde and Kitsch." He is thus unable to imagine the dimensions of artistic practice that he sometimes assimilates to "convention" and sometimes to "material conditions" as wholly embedded within a history or tradition, wholly exposed to interpretation. This is perhaps nowhere clearer than in his grudging and uneasy exemption of the early work of Jasper Johns from his general strictures on Pop art: Johns's work is acknowledged as a persuasive interpretation of the work of the painterly abstraction that preceded it, but the very terms of that acknowledgment—that it is an interpretation—entail its immediate qualification as "literary" and therefore "minor."

Concealed within the arguments of "Towards a Newer Laocoön" for the modernist orientation to medium specificity are elements of an argument that would open toward a broader construal of the terrain of art-historical action. In particular, Greenberg admits here a problematic of dominance and hierarchy among the arts that suggests that the specificity of a given art can never be fully thought or enacted in its own terms but must always show itself in some relation, compliant or resistant, to the other arts among which it necessarily finds itself. Although the explicit problematic of the system of the arts and its implicit hierarchies does not continue to figure in Greenberg's subsequent writings (except in his ongoing opposition to "literariness" in visual art), it is implicitly carried forward in, for example, his willingness to accept a mode of sculpture that takes its terms from painting while not accepting a mode of painting that takes its terms from sculpture. When these terms are made explicit, judgment—the pleasure one takes in a work and finds attributable to nothing other than that work as a realization of its medium—has to admit its dependence on a history of experience that cannot be anchored outside of the interpretative actions by which it is constituted. In 1967, Greenberg's strongest critical follower, Michael Fried, breaks with him along just these lines by asserting the ine-

liminability of specific histories from any construal of the "essential" conventions of an art:

> The crucial question is not what these minimal and, so to speak, timeless conditions are, but rather what at a given moment, is capable of compelling conviction, of succeeding as painting. This is not to say that painting *has no* essence; it *is* to claim that that essence—i.e., that which compels conviction—is largely determined by, and therefore changes continually in response to, the vital work of the recent past. (Fried, 1968)

Separating the threads within Greenberg's writing in this way can seem to bring out its inner Hegelianism—a Hegelianism that appears to enter his writing through the intermediary of Benedetto Croce, that is repeatedly dissembled by the uneasy dual reference to Kant and Karl Marx early on and the radical separation of value and phenomenon later, and that Greenberg would repudiate, inadequately, as a mere rhetorical effect. It may thus seem unsurprising that Greenberg's "Kantian formalism" should be succeeded in some measure by the more full-blown Hegelianism of Arthur C. Danto. It is perhaps equally unsurprising that a number of contemporary writers most intimately opposed to Greenberg have found themselves drawn to various strands of rhetorically oriented and distinctly anti-Hegelian French thought.

Since the late 1960s, "Greenberg bashing"—much of it extremely sloppy and distorting of Greenberg's actual positions—has been a leading feature of American critical and artistic discourse. Some writers, however, among whom one might include Michael Fried, Jeremy Gilbert-Rolfe, and Rosalind Krauss, have continued to explore (if in ways that Greenberg would find unacceptable) the inner dimensions and tensions of formalist theory and practice. More recently, a significant group of French writers, including Yves-Alain Bois and Thierry de Duve, have been willing to identify themselves with a (transformed) formalist practice. Still others, perhaps most prominently Tom Crow and T. J. Clark, have attempted to engage once again Greenberg's abandoned Marxist foundations.

This short summary, although accurate enough, does little to render the complexity and at least occasional urgency of the issues left in Greenberg's wake. Very roughly, those who find Greenberg to have been simply, and perniciously, wrong have tended to find themselves at home in the extraordinary variousness of post-Minimalist artwork as a no longer aesthetically determined field of visual culture whose primary structures and questions are essentially institutional—as if when Greenberg gives way there is no further barrier to a complete collapse into a particular version of Marcel Duchamp as revealer of "the art world" (Danto reroutes this toward philosophy by substituting Andy Warhol and theory for Duchamp). In general, such critics and theorists will feel free to use the word *postmodern* to describe this field and the art arising within it. Those who continue to feel some kind of obligation toward (or some of kind of power in) Greenberg's formalist concerns are more problematically placed—either condemned to simply extending and repeating Greenberg's strictures in ways that often appear critically unresponsive and artistically academic, or else obliged to rethink the foundations of formalist modernism in terms that can in some measure admit and adequately acknowledge work to which Greenberg's theoretical commitments blinded him. In the case of a writer such as de Duve, this has meant trying to rediscover a Greenbergian voice and formulate a quasi-Greenbergian theory on an explicitly Duchampian base; an achievement of this kind would essentially moot the question of any presumed "postmodernism." Still others, such as Michael Fried or Rosalind Krauss or—further from Greenberg and the specific history he informs—Jean-François Lyotard have been led toward more nuanced and more deeply divided accounts of modernism itself, such that *postmodernism* figures for them primarily as the name for a sort of internal fold or complexity to *modernism* rather than some asserted sequel to it. Although this field can be mapped in short compass by plotting the divagations of the term *postmodernism* within it, it should be clear that these are but an index to a much larger arena in which the achievements of such seminal figures as Gustave Courbet, Édouard Manet, Pablo Picasso, and Pollock (to take a not quite arbitrary few) are once again at stake. This is necessarily also an arena in which the historically complex passage from Kant to G. W. F. Hegel and beyond has taken on a particular cast and interest.

[*See also* Abstract Expressionism; Formalism; *and* Modernism, *overview article.*]

BIBLIOGRAPHY

Works by Greenberg

Art and Culture: Critical Essays. Boston, 1961.
"Counter Avant-Garde." *Art International* 15 (May 1971): 16–19.
"Can Taste Be Objective?" *Art News* 72 (February 1973): 22–23.
"Necessity of Formalism." In *Esthetics Contemporary,* edited by Richard Kostelanetz, pp. 191–194. Buffalo, 1989. First printed in *New Literary History* 3 (Autumn 1971).
"Seminar One." *Arts Magazine* 48 (November 1973): 44–46.
"Seminar Two." *Art International* 18 (Summer 1974): 73–74.
"Seminar Four." *Art International* 19 (January 1975): 16–17.
"Seminar Five." *Studio* 89 (May 1975): 191–192.
"Seminar Six." *Arts Magazine* 50 (April 1976): 25.
"Seminar Seven." *Arts Magazine* 52 (June 1978): 96–97.
"Seminar Eight." *Arts Magazine* 53 (January 1979): 84–85.
The Collected Essays and Criticism. Edited by John O'Brian. Chicago, 1986–1993. Vol. 1, *Perceptions and Judgments, 1939–1944;* vol. 2, *Arrogant Purpose, 1945–1949;* vol. 3, *Affirmations and Refusals, 1950–1956;* vol. 4, *Modernism with a Vengeance, 1957–1969.*

Other Sources

Bois, Yves-Alain. *Painting as Model.* Cambridge, Mass., 1990.
Clark, T. J. "Clement Greenberg's Theory of Art." *Critical Inquiry* 9.1 (September 1982): 139–156.

Clark, T. J. "More on the Differences between Comrade Greenberg and Ourselves." In *Modernism and Modernity,* edited by Benjamin H. D. Buchloh, Serge Guilbaut, and David Solkin, pp. 169–187. Halifax, Nova Scotia, 1983.

Crow, Thomas. *Modern Art in the Common Culture.* New Haven, 1996.

de Duve, Thierry. *Clement Greenberg between the Lines.* Translated by Brian Holmes. Paris, 1996. Includes a debate with Greenberg.

de Duve, Thierry. *Kant after Duchamp.* Cambridge, Mass., 1996.

Fried, Michael. "Art and Objecthood." In *Minimal Art: A Critical Anthology,* edited by Gregory Battcock, pp. 116–147. New York, 1968.

Fried, Michael. "How Modernism Works: A Response to T. J. Clark." *Critical Inquiry* 9.1 (September 1982): 217–234.

Fried, Michael. *Manet's Modernism: or, The Face of Painting in the 1860s.* Chicago, 1996.

Gilbert-Rolfe, Jeremy. *Beyond Piety: Critical Essays on the Visual Arts, 1986–1993.* Cambridge and New York, 1995.

Krauss, Rosalind E. *The Optical Unconscious.* Cambridge, Mass., 1993.

Kuspit, Donald B. *Clement Greenberg: Art Critic.* Madison, Wis., 1979.

Lyotard, Jean-François. *The Inhuman: Reflections on Time.* Translated by Geoffrey Bennington and Rachel Bowlby. Stanford, Calif., 1991.

Steinberg, Leo. "Other Criteria." In *Other Criteria: Confrontations with Twentieth-Century Art,* pp. 55–91. Oxford, 1972.

STEPHEN MELVILLE

GROPIUS, WALTER ADOLPH. *See* Bauhaus.

GROTESQUE. The grotesque describes a category of images that fits uneasily within the field of Western aesthetics and art history. The relative neglect of these images may be attributed in part to the classical foundations of these disciplines, because the grotesque presents the inverse of beauty and rationality. Where the creation of the beauty was exemplified by the legendary Greek artist Zeuxis, who combined the best features of several figures in order to create one of perfect and unified beauty, the grotesque was formed by combining parts of unlike creatures, seemingly without reason or purpose. Also weighing against the grotesque is its extreme variability, its extravagance and hyperbole, and its resistance to rules or fixed categories. As Victor Hugo observed, there was only one standard for ideal beauty, whereas the variations and combinations possible for the grotesque were limitless. One can go further and say that these combinations are also unpredictable, because they defy logic and deliberately break down established boundaries. If a fish tail can be attached to a woman's torso, or a child can sprout from a flower, what rules can be established for the grotesque?

Another factor to be considered is that the grotesque appears most often in the visual arts of the Western world in the decorative and graphic arts, media typically considered the "minor arts." Mikhail Bakhtin (c.1940) added a further insight, arguing that the grotesque emerged from folk and popular art forms, and developed especially from carnival, the riotous inversion of all norms and hierarchies. As the expression of the powerless and voiceless, the grotesque was

for Bakhtin a sort of brilliant subterfuge, cloaking serious messages within raucous laughter and earthy imagery. [*See* Bakhtin.] Perhaps most antithetical to aesthetic inquiry is the sheer concreteness of the grotesque. Associated with the grotto or cave, the earth, the corrupted flesh, the sexual, and the scatalogical, and just as likely to be constructed from animal or plant as from human forms, the grotesque exudes a physicality that seems to resist leaps to transcendental thoughts or logical arguments.

Definitions of the grotesque have shifted from one period to another, and the term has been frequently associated or conflated with caricatures, arabesques, emblems, and hieroglyphics. Further complicating the prospect of fixing a definition is the fact that the visual forms of the grotesque existed long before the term came into common use for this category of images. The efforts to define the grotesque in aesthetic terms cluster in classical or classicizing periods, where it was categorized as a subservient and marginal expression best kept in close check by the rational intellect. Yet, paradoxically, an inverse relationship exists between the aesthetic discussions of the grotesque and its stature as visual expression. For example, the grotesque was frequently used to express some of the most exalted of meanings in the medieval image tradition, but no critical aesthetic existed in regard to these images. Put another way, the aesthetic issues most often triggered by the grotesque, specifically those of artistic license, artifice, truth to nature, do not correspond to the meanings embodied by those grotesques that represent visionary or sacred meanings. Further, the grotesque should be recognized as an exclusively Western term, with issues and ideas generated from cultural contexts unrelated to non-Western traditions. Although there are divergent interpretations and incarnations of grotesque imagery from one Western period to another, such as from Greco-Roman to medieval, these styles and their cultural contexts overlap and build on one another. To describe a non-Western image, such as the Indian deity Śiva, as grotesque or monstrous because it has multiple arms imposes a foreign set of presumptions on the image that contradicts its intended meanings. Śiva's multiple aspects represent the pervasiveness and completeness of the deity, whose many hands hold many attributes.

Beginning with a visual description, there are at least three types of figures that typify the grotesque: the misshapen or deformed, the mixed or combinatory, and the metamorphic. Each of these types has a range of intensity from the comic to the horrible to the wondrous. The first of these types encompasses a range of deformity from the ugly to the monstrous and is most directly contrasted to the beautiful. This type of grotesque frequently overlaps with caricature and even with realism. The combinatory, or mixed, grotesque includes such figures as the centaur, griffin, and minotaur, all conjoined creatures that defy reason and nature, whereas the metamorphic grotesque describes

figures that appear to be in the process of transmuting from one form or substance to another. The metamorphic grotesque is in state of flux, whereas the combinatory grotesque presents a fixed relationship, however strange, and often carries a more stable meaning.

Grottesche was the term coined in the last quarter of the fifteenth century to describe the fanciful wall paintings uncovered in the excavation of the Domus Aurea (or, the Golden Palace of Nero). Although there was initial confusion as to whether the rooms of this Roman villa were built as grottos, because the site was beneath street level, linking this ornamental style of combinatory and metamorphic forms with the grotto emphasized its associations with earth, darkness, carnality. Debates about the grotesque emerge in Western aesthetics much earlier, however. One of the earliest arguments was set out by Horace in his *Ars poetica,* in which he established the seminal concept of "ut pictura poesis," and warned against excessive artistic license that led to the creation of wild fantasies: "a horse's neck with a man's head," or "a woman, lovely above, foully ended in an ugly fish below." Equally influential arguments against the grotesque in architecture and design were made by Vitruvius (27 CE), who condemned the "unreasonable fashion" of his time to "decorate the walls with monstrous forms. . . . Such things . . . never existed, do not now exist, and shall never come into being. For how can the stem of a flower support a roof, or a candelabrum pedimental sculpture? How can a tender shoot carry a human figure, and how can bastard forms composed of flowers and human bodies grow out of roots and tendrils?" The grotesque style in late antiquity was rejected by Horace and Vitruvius as ridiculous, contrary to reality and good sense, but it was not perceived as dangerous or threatening. Horace asked, "Could you restrain your laughter, my friends?" and Vitruvius pointed out the "falseness" of the grotesque image.

In the archaic, severe, and classical periods of Greek art and myth that preceded Horace's aesthetics, the grotesque embodied far more threatening forces. The human faculties of reason were expressed through idealized human figures with facial expressions of masklike calm, but these figures were depicted in ongoing conflict with an array of grotesque creatures, including centaurs, satyrs, and sphinxes. Although there were positive representations, such as the centaur Chiron, these combinatory grotesques more typically projected varieties of unreason, particularly violence and sexuality. In addition, apotropaic figures, such as the Medusa's head, appeared on shields and portals to ward off evil spirits. By the second and first centuries BCE, taste turned toward the grotesque, with a new delight in the ugly, in the deformed, and in the fantastic decorations that later drew the censure of Vitruvius.

In the medieval world, the meanings of the grotesque shifted. To be sure, grotesque and monstrous forms carved onto the tympana over church doorways or the drolleries painted in the margins of vellum manuscripts were still highly charged images representing the threat (and sometimes the delight) of all things bodied, carnal, and bestial, but now the threat was to the spirit, not the reasoning mind. The grotesque emerged as a major part of the image culture in the Romanesque period (c.1000 CE), where depictions of the struggle between good and evil were staged, most often in vivid representations of the Last Judgment. Satan and devilish accomplices were represented in terrifying mixtures of man and beast, such as those depicted in the huge tympanum of Autun cathedral (c.1130). The profusion and exuberance of these representations of sin and damnation in the twelfth century caused Bernard of Clairvaux to warn that excessive license could lead to licentiousness: "what is the point of such ridiculous monstrosity, the strange kind of shapely shapelessness?" Meyer Schapiro delved into the nature of these grotesques in chapter 1 of his monumental study *Romanesque Art* (1976). In *The Gothic Idol: Ideology and Image-Making in Medieval Art* (1989), Michael Camille explored the depiction of idols as grotesque or monstrous in medieval art in order to emphasize their associations with sexuality and other forms of carnality, as well as their material, false nature. Yet, the medieval era produced another kind of grotesque creature, one that embodied truths beyond the grasp of the rational mind. The four evangelists, the visions from the Book of Revelations, and prophecies of the Old Testament were frequently represented as combinatory grotesques. These fantastic creatures were not conceived within the realm of aesthetic inquiry, but were created within a ritual context as embodiments of spiritual belief.

The Gothic world was somewhat more restrained in the use of grotesque figures, perhaps because of the order and stability of the era. Gargoyles and other monsters inhabited the cathedrals but made fewer appearances among the rows of saintly figures lining the doorways. Instead, they were held at bay in the upper reaches of the structure. The practice of representing evil as deformed and horribly caricatured found a prolonged life in the northern Europe. Hieronymus Bosch's *Garden of Earthly Delights* (1505–1510) presents an encyclopedic array of debasement, deformity, and monstrous combinations in his misanthropic meditation on the wickedness of humankind.

An extensive critical debate concerning the grotesque emerged in Italy during the High Renaissance. This was influenced in part by the discovery of the "grotto" paintings in Rome, but also by the increasing premium put on the individual creativity of the artist. *Grottesche* became virtually synonymous with inspired invention, *fantasia,* and audacity, a development discussed by David Summers in his 1981 study of Michelangelo. Horace's linkage of the grotesque to artistic license was well known by the sixteenth century, but it was interpreted as artistic freedom and considered the marker of great genius. In his *Four Dialogues on Painting*

(1538–1540), Francisco de Hollanda attributed to Michelangelo this response to Horace's famous dictum regarding fantasy:

> I am happy to explain to you why it is customary to paint that which never is seen in the world, and how reasonable and right such liberty is; because they interpret badly who are used to saying that Horace, the lyric poet, wrote these verses in blame of painters. . . . he does in nowise blame painters but praises and favors them, since he says that poets and painters have license to dare to do . . . to dare what they choose.

Overstepping the boundaries of convention and good taste was justified by the claim that the creations of the artist-genius were the fruit of divine inspiration. This idea was repeated in the influential *Libro dell'arte* of Cennino Cennini and *Trattato dell'arte de la pittura* of Paolo Lomazzo (1584).

Michelangelo's art and thought were closely tied to these ideas and the tortuous and inspired movements of the *ignudi* in the Sistine ceiling frescoes exemplify his interpretation, as do the gigantism and exaggerated proportions of his architectural designs. A more literal revival of the Roman grotesque was created by Raphael, who toured the Domus Aurea with Giovanni da Udine and then planned and executed an ambitious decorative program, based on the Roman model, for the Vatican loggias, completed in 1519. Leonardo da Vinci mined another aspect of the grotesque with his savage caricatures and realistic drawings of deformity and decrepitude. The Mannerist period that followed on the heels of the High Renaissance pushed exaggerations and invention to extravagant degrees. Decorative grotesques quickly spread throughout Europe through engravings and printed books, and the profusion of sixteenth- and seventeenth-century exponents included Cornelius Floris, Agostino Veneziano, Marco Dente, Cornelius Bos, Lucas van Leyden, Étienne Delaune, and Christoph Jamnitzer.

In contrast to the extreme sophistication of Mannerist invention, the sixteenth century also saw the emergence of grotesques that were crude and earthy. In France, François Rabelais completed *Gargantua and Pantagruel* in 1532, a work so original and strange that it might be reckoned as a defining moment in the history of the grotesque similar to the discoveries of Nero's palace. In marked contrast to the classical conception of *grottesche* and its predominantly decorative context, Rabelais presented a world of coarse satire, bawdy and exuberant, mixed with a dark underlay of disorder and chaos. In his seminal study published in 1968, Bakhtin interpreted *Gargantua and Pantagruel* as "an encyclopedia of folk culture" that drew its imagery from the "unofficial" peasant culture of the medieval period, especially that of carnival. He saw the carnival and Rabelais's incorporation of it as an attempt to subvert dogma and official culture.

On one level, the eighteenth century simply expanded on issues and ideas about the grotesque established in earlier centuries. Archaeological excavations of Herculaneum and Pompeii sparked another round of decorative grotesques in architectural interiors and graphic design. Also folded into these decorative programs were designs inspired by Chinese imagery (chinoiserie), much of which struck European eyes as fantastic and grotesque. But the Age of Enlightenment also introduced a significant new interpretation of the grotesque by linking it to early, or "primitive," expression. Early in the eighteenth century, scholarly inquiry turned to the origins and development of human culture, augmented by a new wave of voyages of exploration and extensive archaeological excavations. One of the first of these studies was Giambattista Vico's *La scienza nuova*, published in 1725, which drew parallels between Native Americans, Homeric Greeks, and prehistoric Germans and Picts in an attempt to discern the characteristics shared by cultures presumed to be in the early stages of development. Vico paid particular attention to the development of language and postulated that the first written communication was not through words, but through images, or "poetic monsters," which were disparate objects combined in order to express ideas. In this same period, Jean-François Lafitau seconded this notion by comparing the masking practices of the Iroquois to satyrs and other hybrid creatures of classical mythology. Drawing on his five years among the Iroquois, his *Mœurs des sauvages ameriquains* (1724) proposed that the satyr was not a mythical creature at all but a poetic figure symbolizing the character of powerful leaders.

As Europeans moved into a more overtly imperialistic relationship with those peoples they defined as "primitive," they increasingly described various types of non-Western imagery as grotesque, and with far more negative connotations. In the nineteenth century, the Enlightenment notion of the grotesque as an emblematic language was discarded for a harsher use of the term to identify other image traditions as monstrous and deformed. Sculptures of sacred figures in African and Oceanic cultures were most frequently identified as grotesque and targeted for destruction because they seemed to embody the irrational superstitions and savagery of "uncivilized" peoples. Other commentators postulated that the "primitive" artist was simply incapable of producing anything but grotesque imagery. Thomas Wright, in *A History of Caricature and Grotesque* (1865), wrote that "Art itself, in its earliest forms, is caricature; for it is only by that exaggeration of features which belongs to caricature, that unskilled draughtsmen could make themselves understood."

It was through the rise of the Romantic sensibility in the late eighteenth and early nineteenth centuries that the grotesque gained legitimacy as a viable expressive vehicle and became a central element of aesthetic debate. It was of particular interest to artists and intellectuals, such as Victor Hugo and Charles Baudelaire, who rightly perceived that the grotesque was the exact contrary to classical beauty and

therefore a key weapon to use against it. The Romantics rejected the rational formulas of academic pedagogy as well, voicing the belief that artistic creativity originated in the figurative grotto, springing from the unfettered imagination. Thus, the grotesque was celebrated as emblematic of creative freedom and of individual expression. This was implicit in Hugo's contrast of the limitless possibilities for the grotesque with the narrow standard for ideal beauty. The Romantic exploration of the irrational, with an emphasis on madness, horror, and the diabolical, generated much of the grotesque imagery of this era, from Francisco Goya's graphic work and late paintings, to Charles Meryon's etchings of the gargoyles in the upper reaches of Notre-Dame de Paris. Notably, some of the most vivid grotesque images of the Romantic era were created by writers, specifically Edgar Allan Poe, E. T. A. Hoffmann, and Hugo. Medievalism and the Gothic Revival, subsets of Romanticism, provided other avenues for the exploration of the grotesque.

Although G. W. F. Hegel and Immanuel Kant each briefly discussed the role of the grotesque in their aesthetic universe, neither changed its marginal standing nor significantly revised the definition of its essential character. This work was done by intellectuals such as Friedrich Schlegel (*Gespräch über die Poesie,* 1800), Hugo (*Preface to "Cromwell,"* 1827), and most notably, Baudelaire and John Ruskin. Baudelaire's short but insightful essay *De l'essence du rire et généralement du comique dans les arts plastiques* (1855) established two categories: the *comique significatif* and the *comique absolu.* The former was a satirical form of comedy that stemmed from a sense of superiority, and took human frailties and foibles as its object of derision. In contrast, the absolute comic created works of powerful, even horrific expression, and with no intention of parody. While the significative comic worked in the realm of caricature, the absolute comic was grotesque, primitive, transgressive, raw—Rabelaisian, according to Bakhtin's definition. Although the Romantics looked primarily to medieval grotesques, Baudelaire invoked folk and children's art, as well as what he described as "carib" art. [*See* Baudelaire.]

Ruskin wrote on the nature of the grotesque all through his long career and his ideas are important, complex, and sometimes contradictory. Ruskin described grotesque expression as a continuum between the ludicrous and the fearful, with few grotesques being purely one or the other. Although he sanctioned grotesques resulting from the free play of a serious mind, such as Michelangelo's ornamental figures, Ruskin excoriated the majority of Renaissance *grottesche* as decadent, describing Raphael's loggia decoration as "an elaborate and luscious form of nonsense . . . a poisonous root; an artistic pottage, composed of nymphs, cupids, and satyrs, with shreddings of heads and paws of meek wild beasts" (1903–1912). But Ruskin reserved the designation of a "thoroughly noble grotesque" to those combined forms that expressed the exalted or wondrous:

"A fine grotesque is the expression, in a moment, by a series of symbols thrown together in bold and fearless connection, of truths which it would have taken a long time to express in any verbal way" (1903–1912). A stone griffin guarding and supporting the portal at San Zeno in Verona was taken by Ruskin as his prime example of the noble grotesque. Ruskin considered this griffin, carved by a medieval artisan, to be a creation of great artistic and spiritual imagination, free of the proscribed boundaries of aesthetic taste, ignorant of the rules of proportion and beauty. Adding another layer onto this interpretation, and anticipating Bakhtin, Ruskin also celebrated the grotesque as emblematic of the individual freedom accorded the medieval craftsman to create his part of the greater whole, contrasting this to the diminished role of the artisan in subsequent fine-art traditions where one artistic vision ruled. [*See* Ruskin.]

Since the early nineteenth century, it has no longer been possible to describe the grotesque as peripheral. The Romantic period marked the entrance of the grotesque into the mainstream of modern expression. In fact, it is arguable that the grotesque has become the normative expression of the modern era, but by the same token it has lost much of its transgressive power, its capacity to shock. Many factors in twentieth-century Western culture supported these developments. In many ways, modern culture has played out themes established in the Romantic era, especially the penchant for hyperindividualism and the belief that familiarity and routine mask a terrifying and chaotic reality. Primitivism, the deliberate emulation of "primitive" characteristics, played a major role in the proliferation of grotesque imagery in modern art. Because the grotesque was linked to "primitive" expression, European artists borrowed heavily from non-Western imagery that resembled Western grotesques. These efforts to recover a more vital expression led to ever more transgressive imagery, until a point was reached where the exotic was assimilated, and the strange became familiar. The representation of the other as grotesque found new life, however, in the exploration of outer space and weird, alien life forms will people our imaginations until fact finally crowds them into some other sphere of the unknown. Another path of regression intersecting with the grotesque was the incorporation of primitive life-forms into visual imagery, from insects and reptiles to microorganisms. In 1889, Joris Karl Huysmans published an essay, "Le monstre," suggesting that the microscope had opened up an entire new field of monstrosities, equal to those animating medieval art. Huysmans did not anticipate the impact of Sigmund Freud's theories of the subconcious. In relation to the interpretation of the grotesque, psychology, and particularly the concept of the id, mark another major shift. Those characteristics attached to the grotesque and then projected onto the primitive were now internalized, were recognized as the generative and primal force within each of us. In the visual arts, these revelations sparked the exploration of the imagery of

dreams, much of it grotesque, by artists and writers associated with the Surrealist movement. Finally, the grotesque moved into the mainstream of modern culture through the rise of the mass media. Certainly, the modern fine-arts tradition has been energized by its assimilation of popular culture, but the fine arts have been marginalized by the mass media, which draw primarily from popular culture. The popular traditions of the past, those celebrated by Bakhtin, were grotesque and burlesque, but always in relation to a dominant and elite culture. This is no longer the case. The mass media, particularly television, seem to democratize culture when in fact they monopolize it. Although popular traditions, such as carnival, were grotesque, profound meanings and deep traditions lay buried beneath the manic surface. In the unblinking realism of the camera and in its unflagging desire to entertain, the mass media transform popular tradition into mass culture, and make the grotesque and comic merely banal.

[*See also* Caricature.]

BIBLIOGRAPHY

Bakhtin, Mikhail. *Rabelais and His World* (c.1940). Translated by Helene Iswolsky. Cambridge, Mass., 1968; reprint, Bloomington, Ind., 1984.

Baltrusaitis, Jurgis. *Le moyen age fantastique: Antiquités et exotismes dans l'art gothique.* Paris, 1955.

Baltrusaitis, Jurgis. *Reveils et prodiges: Le gothique fantastique.* Paris, 1960.

Connelly, Frances S. *The Sleep of Reason: Primitivism in Modern European Art and Aesthetics, 1725–1907.* University Park, Pa., 1995.

Dacos, Nicole. *La découverte de la Domus Aurea et la formation des grotesques à la Renaissance.* London, 1969.

Farkas, Ann E., Prudence O. Harper, and Evelyn B. Harrison, eds. *Monsters and Demons in the Ancient and Medieval Worlds: Papers Presented in Honor of Edith Porada.* Mainz, 1987.

Fowler, Barbara Hughes. "The Grotesque." In *The Hellenistic Aesthetic,* pp. 66–78. Madison, Wis., 1989.

Harpham, Geoffrey Galt. *On the Grotesque: Strategies of Contradiction in Art and Literature.* Princeton, N.J., 1982.

Kayser, Wolfgang. *The Grotesque in Art and Literature.* Translated by Ulrich Weisstein. Bloomington, Ind., 1963; reprint, New York, 1981.

Kuryluk, Ewa. *Salome and Judas in the Cave of Sex: The Grotesque: Origins, Iconography, Technique.* Evanston, Ill., 1987.

Randall, Lillian. M. C. *Images in the Margins of Gothic Manuscripts.* Berkeley, 1966.

Ruskin, John. "Of the True Ideal: Thirdly, Grotesque." *Modern Painters* (1843–1860), vol. 3. In *The Works of John Ruskin* (Library Edition), edited by E. T. Cook and Alexander Wedderburn, vol. 5, pp. 130–148. London and New York, 1903–1912.

Ruskin, John. *The Stones of Venice* (1853). 3 vols. In *The Works of John Ruskin* (Library Edition), edited by E. T. Cook and Alexander Wedderburn, vols. 9–11. London and New York, 1903–1912.

Summers, David. *Michelangelo and the Language of Art.* Princeton, N.J., 1981.

Vico, Giambattista. *The New Science of Giambattista Vico* (1725). Translated by Thomas Goddard Bergin and Max Harold Fisch from 3d ed. (1774). Ithaca, N.Y., 1948.

Warncke, Carsten-Peter. *Die ornamentale Groteske in Deutschland, 1500–1650.* Berlin, 1979.

FRANCES S. CONNELLY

GUATTARI, FÉLIX. *See* Deleuze.

GURNEY, EDMUND (1847–1888), English music theorist. Gurney's *The Power of Sound* (1880) deserves far greater fame than it has achieved or is likely to achieve. It is a book more read about than read, and with good reason, even though it is one of the best, and most thorough, studies of musical aesthetics and philosophy ever undertaken—certainly the most powerful work of its kind to come out of Britain, or perhaps anywhere else, in the nineteenth century.

The Power of Sound is by no means user-friendly. It is extremely difficult to read, and provides little "readerly" pleasure. The reasons for this are not far to seek.

To begin with, Gurney's literary style is ponderous and ungracious. His favorite punctuation mark is the semicolon, by the excessive use of which he constructs clumsy sentences of Germanic length, strung together like the long chains of organic molecules. To compound the difficulty, the trail of argument is littered with frequent and frequently enormous footnotes, in double columns and small print, which, if collected together under separate covers, would constitute a substantial monograph in their own right. Finally, Gurney pursues relentlessly many arguments and issues of his day that, although of some intrinsic interest, must seem to the contemporary reader as distracting irrelevancies that continually postpone or lead away from central philosophical concerns.

Yet, in spite of these difficulties, Gurney's compendious work remains one of the most rewarding of its kind in the literature. No one who expects to make a contribution to the aesthetics and philosophy of music can afford not to master it. It contains an enormous amount of wisdom and good sense, whether or not one ultimately finds Gurney's overall view congenial.

The position propounded in *The Power of Sound* is certainly what can be described as a musical "purism" that, for all intents and purposes, although not entirely, turns out to be a musical "formalism" as well. For Gurney, one learns very early on, "the world of Beauty is preeminently the world of Form."

At the very center of Gurney's musical philosophy is the premise that melody is far and away the primary, the basic musical function: the sufficient and necessary condition for music: the *fons et origo* for greatness in music because, without great melody (on Gurney's view), there cannot be a great musical work.

Melody is the fundamental fact of Music, not only in the sense that homophony, or a succession of single unaccompanied notes is the primary form in the historical development of the art as well as in the tentative efforts of every individual, but in the sense that throughout the most highly developed music the melodic element, however adorned and supported by other elements, is displayed in full pre-eminence.

Before going on, it would be well to take the full measure of this complete commitment to the melodic in music. Its significance is not to be underestimated.

Ninety percent or more of Gurney's musical illustrations are taken from the works of Ludwig van Beethoven; Gurney makes it very plain that Beethoven represents for him the pinnacle of musical accomplishment. No composer before Franz Joseph Haydn is mentioned by name, at any length, in a seriously complimentary manner, except for (once or twice) George Frideric Handel and Johann Sebastian Bach. The only pre-Baroque composer discussed extensively is Giovanni Pierluigi da Palestrina, whose music, just because it cannot be understood on Gurney's melodic principles, is put down as inferior: "But however enchanting to the ear be Palestrina's harmonic tissues, there are many of them which it would be absurd to represent as arresting the attention and haunting the inward ear like the more definite and closely organic forms of later composers" (ibid.). Of music prior to Palestrina, Gregorian chants can only be enjoyed "in spite of their uncouthness, by venerable memories"; and as for pre-Tridentine polyphony: "In the later middle ages, when almost all serious music was written for Church purposes, the liturgical formed for the most part a purely mechanical framework, often most strangely strained and distorted, for the scientific ingenuity of the composer to build on." In short, Gurney's work is imbued with the idea of musical progress: "as a matter of undeniable fact, by far the greatest music the world has known appeared within the last two centuries, and by far the greater proportion of it in the latter half of that time." To the charge that his melody-oriented aesthetic can make little or no musical sense out of a motet by Josquin Des Prés, the reply would be that that is because such early, "primitive" examples of the art do, indeed, make little or no musical sense, as theory predicts. (That the theory could make little or no musical sense of a Bach fugue is true as well, but would, up to a point, be contested by Gurney, although Bach enters the scene long before Gurney's golden age, and the Well-Tempered Clavier is no match, in his eyes, for Beethoven's thirty-two piano sonatas.) In any case, it is with this historical blindness, born of the idea of progress, always in mind that one must temper one's judgment of Gurney's accomplishment, considerable though it may be.

Beauty is form and music is melody. What, then, is the form that makes a melody beautiful? It is what Gurney calls "ideal motion":

> We must now turn to the actual *process* by which Music is followed, to the facts connected with the evolution of melodic form moment after moment in time. The transition which this will involve of the phraseology of *form* into the phraseology of *motion* will make clearer the essential difference of this experience where form and motion are blended—where form is perceived by continuous advance along it—from perceptions both of visible form and physical motion. It is the *oneness of form and motion*

which constitutes the great peculiarity of melody and of the faculty by which we appreciate it. (Ibid.)

Two points can be made about the concept of ideal motion—one historical, the other philosophical. Gurney's form-in-motion—which is what ideal motion is—must strike anyone conversant with Eduard Hanslick's *Vom Musikalisch-Schönen* as at least in some way reminiscent of the latter's well-known precept: "The content of music is tonally moving forms." But whereas Gurney shows himself intimately familiar with the writings of Richard Wagner, Hermann von Helmholtz, and other, more obscure, German writers, the name of Hanslick never appears in *The Power of Sound*. Whether this is a forgotten influence or mere accident (form and apparent motion being, after all, fairly obvious musical characteristics) remains an open question.

More important is the philosophical point. To begin with, ideal motion is, for Gurney, a nonreductive property: it cannot be defined in terms of any formula. Any attempt to explain why a melody is beautiful by appeal to some identifiable feature is immediately met with Gurney's favorite argument: the feature in question may be present in a banal melody, or absent in a beautiful one, and hence is neither a necessary nor a sufficient condition for ideal motion.

Gurney's appeal to counterexamples, when facile explanations are offered for musical beauty, is doubtless a welcome and effective remedy. But it does grow tiresome to be continually told that the beauty or effectiveness of a passage is inherently inexplicable, thus putting an end to all thought or inquiry into the matter. Moreover, it by no means follows from the fact that some aspect of a particular melody is neither a necessary nor a sufficient condition for melodic beauty that it is not in that particular place contributing to melodic beauty. Furthermore, Gurney never seems to consider the possibility that although no *single* interesting feature of a melody is a necessary or sufficient condition for melodic beauty, a *cluster* of interesting features might be sufficient; for the more interesting features a beautiful melody has, the harder it is to find a banal melody that also has all of them.

In any event, ideal motion cannot be recognized, on Gurney's view, by any of the "normal" faculties of the human mind or sensory apparatus, just because, one assumes, the property is nonreductive. It is, rather, perceived by what Gurney characterizes as "the musical faculty [that] defies all explanation of its action and its judgments," and sometimes presents as a faculty of intuition, as when he writes, for example, of the ideal motion and the faculty that perceives it,

> we can but accept it as a fact of musical intuition that these two particular bits of melodic motion, resting on this simple harmonic foundation, do by their mutual reaction merge into an organism of incomparable beauty, in which, conformably to the true nature of an organism, each of them is indispensable to, and each reveals the necessity of, the other. (Ibid.)

The combined doctrine of a nonreductive property and a special, intuitive faculty to perceive it is very suggestive of philosophical things to come: Clive Bell's "significant form," for example, as well as G. E. Moore's "nonnatural" property of goodness. Perhaps it is no mere accident that Gurney, like Moore, was conversant with, and greatly impressed by, Henry Sidgwick's *Methods of Ethics* (1874). In any event, although Gurney, like Bell, associates the perception of his special aesthetic property with a special aesthetic emotion, he does *not*, like Bell, commit himself to the vicious circle of defining the property in terms of the emotion and the emotion in terms of the property. What role *does* emotion play in Gurney's system?

Gurney is insistent that the perception of ideal motion by the musical faculty is a deeply moving experience. His problem is how this can be so, given the separateness of the musical faculty, and the "purity" of the ideal motion. For the representational arts, and for architecture, connected, as they are, to our world and its necessities, the avenues to emotive arousal are clear. But how are we touched, emotively, by the beautiful in music? To answer this perplexing question, Gurney buys into Charles Darwin's evolutionary explanation of music's origin, based on the principle of sexual selection, which, during Gurney's lifetime, was in direct conflict with Herbert Spencer's theory that music had its origin in the impassioned tones of human speech. It was a conflict that, in the last forty years of the nineteenth century, generated a spirited quarrel in print to which Gurney contributed.

According to Darwin, the origin of music lies in such phenomena as the mating calls of primates, and the songs of birds, which serve the evolutionary function of attracting the opposite sex, and thus would have been preserved and accentuated through the process of sexual selection and mating. The emotive power of music, then, on the Darwinian view, is the result of its connection, though remote, with sexual passion, which it preserves in its deep intensity. "These ideas," Gurney avers, "may seem at first sight somewhat startling; but when we realize the extraordinary depth and indescribability of the emotions of music, the very remoteness of and far-reachingness of the explanation is in favour of its validity" (ibid.). Gurney goes on to say of this musical emotion:

> So far as it can be described, it seems like a fusion of strong emotions transfigured into a wholly new experience, whereof if we seek to bring out the separate threads we are hopelessly balked; for triumph and tenderness, desire and satisfaction, yielding and insistence, may seem to be all there at once, yet without any dubiousness or confusion in the result; or rather elements seem there which we struggle dimly to adumbrate by such words, thus making the experience seem vague only by our own attempt to analyse it, while really the beauty has the unity and individuality pertaining to clear and definite form. . . . it is

just this characteristic of *fused and indescribable emotion* which seems explicable on Mr. Darwin's view: for pleasure which was associated with the most exciting passions, would have correspondingly large opportunities not only for increase but for differentiation. . . . [These] "mighty emotions, by the extent to which they baffle analysis, might lead us to suspect their connection with the earliest and most universal of instincts." (Ibid.)

Gurney calls this deeply emotional effect of beautiful music its *impressiveness;* and he sharply distinguishes it from its *expressiveness,* by which he means its ability to express what might be called the "garden-variety" emotions: anger, melancholy, joy and the like—something he, unlike Hanslick, never doubts. For Gurney, music can be, and frequently is impressive without being expressive, and perhaps can be (although this is not altogether clear) expressive without being impressive.

The major operator, in giving music its emotive expressiveness, is, on Gurney's view, the analogy of musical "motion" with the real motions of human beings in space, with the major-minor contrast and timbre as secondary contributions:

> Now the Ideal Motion of Music gives us an aspect of physical movement ready-made; the aspect, namely, of pace and rhythm; which can be presented without any sound at all, by movements in space, and the correspondence of which with movements in space we perpetually exemplify in our own persons, as we follow and in any way keep time with Music. Moreover, the Ideal Motion regarded in its completeness, without such particular reference to the rhythmic element, will be found to present certain faint affinities to external movement and gesture. But while it is naturally in motion that we should look for the signs of definite emotional states, or, as we commonly say, the *expression* of such states, in Music there are three other features connected with expression, two of which belong especially to the tone- or pitch-element apart from peculiarities of motion; the use, namely, of the major or the minor "mode," and of occasional noticeable harmonies; the other being *timbre* or sound-colour. (Ibid.)

The recognition that music can be deeply moving through its beauty, and also be expressive of the garden-variety emotions, and that these are two distinctly different things, Gurney has achieved and expressed with great incisiveness. But he makes two ancillary points with regard to this crucial distinction, one of which must certainly be wrong, the other in need of the most careful and skeptical scrutiny.

The first of these ancillary points is that, according to Gurney, we should "relegate the definitely expressive element in Music to a much more subordinate position than that commonly assigned to it": "it is subordinate (1) in the sense that it is far from being a constant element; (2) in the sense that not in it but in the independently impressive aspect of Music, must be sought the explanation of the essential effect of the art" (ibid.) With this claim in itself one can

have no quarrel, particularly considering the preponderance of "expression" theories of music, many of them silly, with which Gurney was doing battle. But, as it turns out, Gurney is so anxious to deny to expression the whole or preponderant responsibility for music's effect that he seems to end up denying it any effect at all. This attitude is perhaps epitomized by the remark: "But clearly if I find twenty bits of music beautiful and indescribable, and one bit beautiful and pathetic, it is unreasonable to say I enjoy the last (which I don't enjoy at all more than the others) *because* it is pathetic" (ibid.).

The problem is both one of logic and, more important, one of aesthetics. Of course, it does not follow that because one finds much nonexpressive music beautiful, expressiveness has nothing to do with beauty when it does occur in beautiful music. But the problem goes far beyond merely catching Gurney on a point of logic. It would be intolerable, given the overpowering "presence" of expressive qualities in so much of the music we admire, that a philosophy of music should deny those qualities an important role in the aesthetic appreciation of music, when they importantly occur. A theory may perhaps safely ignore the froth on the wave, in explaining the tides; but when *consciousness* becomes the froth on the wave of behavior, the theory that holds *that* is justly condemned. Having perspicaciously observed that music is both expressive and deeply moving, Gurney, understandably but nevertheless unacceptably, spoiled his insight by making of expression merely an epiphenomenon.

The second of Gurney's ancillary points, closely related to the first, is that either (it is not altogether clear which) music cannot truly be expressive unless it is also impressive (even though they are logically separate) or (more moderately) that music cannot be *intensely* expressive without being impressive, which is to say, for example, that it cannot be intensely melancholy unless it is also beautiful and (so) emotionally moving. Thus, Gurney writes (suggesting the first alternative, and probably the one he inclines toward),

> music may be slow and quick and even and interrupted and furious and capricious and loud and soft and all the rest, and still be perfectly unbeautiful and unimpressive. But if it is musically impressive, particular features of motion will give it particular expressive characters: and then slowness becomes solemnity, and quickness becomes it may be merriness or it may be agitation, and fury becomes energy, and abruptness becomes passion, and so on. (Ibid.)

Suggestive of the latter alternative, Gurney writes:

> A hundred melodies might be written which would seem to embody the same features of motion, but there being no special beauty in them, there would be no *special* pathos. In the explanation of the expressive character of the phrases their musical value is, so to speak, presupposed: the reason why they fill up their place in the form to such divine issue lies hid in the instinctive perceptions of the musical faculty. (Ibid.)

The first alternative—which does seem the one most often suggested by what Gurney says—is plainly false; but the second must at least give us pause: there does seem to be plenty of banal music that is melancholy or happy, angry or yearning, and so forth; but can banal music be *intensely* melancholy or happy, angry or yearning?

There does seem to be some prima facie appeal to the notion that where we want to say a passage of music is *intensely* sad or happy or whatever, it is always with regard to a passage that is also "beautifully" so. Thus, it does seem to be true that beauty is a necessary, contributing condition for *intense* emotive properties in music. In any case, it is hard to come up with counterexamples. But, on the other hand, because the emotive character of a musical passage, where that character is describable in terms of the garden-variety emotions, is so much a matter of well-known musical conventions and materials, it does seem possible to have intense expression without musical beauty, although perhaps one has no particular reason to remark on it. Might not the "Royal Theme," for example, that Frederick the Great is supposed to have given Bach be an example of a somewhat clumsy theme, yet, because of the overload of "conventional" melancholy features in it, an *intensely* melancholy one? In any event, Gurney's recognition that the emotively moving—the musically impressive—in music is distinct from the emotively expressive is an extremely important one, no matter what its vagaries. Gurney was surely correct too in divining an intimate relation, at least in many cases, between the presence of the one and the intenseness of the other.

These, then, are the bare bones of Gurney's musical philosophy. But they can hardly give any idea of its richness and breadth; for, by the time *The Power of Sound* reaches its conclusion, Gurney has expressed his views on just about every important issue in the field up to that time. His views are always interesting, always powerfully argued, and frequently full of musical common sense, as opposed to the musical nonsense he is so adroit at debunking. Although Gurney's name (if not his doctrine) comes down to us as that of a musical "purist," a sworn and implacable enemy of "expression" theories of music, his own passion for the art must never on that account be doubted, as the following expression of it indicates:

> The activities of most lives are in themselves dull and trivial, having little or nothing of vivid and changing interest, and no power of satisfying the imagination, leaving the mind moreover but little leisure to explore unknown worlds; while on the spiritual side there may be more of struggle than of peace. To such lives love of beauty is like water in the desert; and among the arts it is more especially the privilege of Music, that, penetrating where Nature's face is veiled, she can open the springs of this love to the poorest dwellers in the dingiest cities. (Ibid.)

[*See also* Expression Theory of Art; Hanslick; *and* Music, *historical overview article*.]

BIBLIOGRAPHY

Work by Gurney

The Power of Sound. London, 1880; reprint, New York, 1966.

Other Sources

Hanslick, Eduard. *On the Musically Beautiful: A Contribution towards the Revision of the Aesthetics of Music.* Translated and edited by Geoffrey Payzant. Indianapolis, 1986.

Kivy, Peter. "Herbert Spencer and a Musical Dispute." *Music Review* 23.4 (November 1962): 317–329.

Levinson, Jerrold. "Edmund Gurney and the Appreciation of Music." *Iyyun: The Jerusalem Philosophical Quarterly* 42.1 (1993): 181–205.

Levinson, Jerrold. *Music in the Moment.* Ithaca, N.Y., 1997.

PETER KIVY

H

HABERMAS, JÜRGEN (b. 1929), contemporary German philosopher and public intellectual. Habermas is especially known for his notions of communicative action, rationality, and discourse ethics, and for the range, systematicity, and originality of his contributions to philosophy and social theory. In addition, he is a public intellectual who has tried to integrate social theory, theories of language, rationality, law, and morality into a synoptic vision of modernity—a vision that offers an account of the problems that plague modernity and suggestions as to how they might be overcome. Habermas's most-discussed and debated works are: *Knowledge and Human Interests* (1968), *The Theory of Communicative Action* (1984 [1984]), *The Philosophical Discourse of Modernity* (1985 [1987]), and *Between Facts and Norms* (1992). He is professor emeritus at the University of Frankfurt, Germany, where he taught for many years.

Throughout the course of his intellectual career Habermas has engaged in productive debates with representatives of philosophical and social-theoretical positions both close to and opposed to his own. Invariably, these critical exchanges substantially affected his own outlook as well as that of his interlocutors. In Germany, the most important exchanges involved the philosophers Hans-Georg Gadamer, Ernst Tugendhat, and Dieter Henrich, and the social theorist Niklas Luhmann. Exchanges with North American colleagues involved Richard Rorty, John Searle, Lawrence Kohlberg, Charles Taylor, and most recently John Rawls. The most polemical of these exchanges took place in the early 1980s and involved the major representatives of French poststructuralism, Michel Foucault and Jacques Derrida. Habermas's penchant for critical debate is not restricted to the academic world; as a public intellectual, he has also engaged in some important political debates in Germany—for example, the *Historikerstreit* (Historians' debate). In this debate, which took place in the pages of the *Frankfurter Allgemeine Zeitung* in the mid-1980s, Habermas challenged the attempts of conservative historians to "normalize" the Nazi past.

Although Habermas's work is largely focused on developing a fallible but universalistic conception of rationality and justice, his work deserves the serious attention of historians of art and culture, philosophers of art, and cultural theorists. Habermas is not just Germany's most important postwar philosopher, he is also the leading exponent of Frankfurt School "critical theory." Already with the publication of his first book in 1962, *The Structural Transformation of the Public Sphere* (1989), Habermas brought into focus the central theme of all of his subsequent inquiries: What are the necessary preconditions for the creation and preservation of communicatively structured, interlocking, yet autonomous public spheres through which (and only through which) members of modern societies can participate in democratic practices of collective self-determination and individual self-realization?

The historical and sociological analyses that organize *The Structural Transformation of the Public Sphere* track the emergence and eventual disintegration of the "bourgeois public sphere." The interpretative key to these analyses is Habermas's contention that the "bourgeois public sphere" introduced and simultaneously undermined a new understanding of reason—an understanding of reason conceived of not as an instrument of knowledge but as a public medium of communication. Habermas did not restrict his empirical analyses to the political, social, or economic conditions responsible for the emergence and disintegration of the public sphere; rather, he made clear that the new understanding of reason—the main focus of this book and Habermas's project as a whole—expressed an "idea of humanity" that was in crucial respects made possible by developments initiated by the emergence of the literary public sphere. To explain this change in the meaning of reason, Habermas singled out the development of a self-disclosing, "audience-oriented" subjectivity (which was itself facilitated by changes in the nature of family life and the separation of public and private spheres) and the institutionalization of an autonomous sphere of art and and literature with its own correlative modes of production, reception, and criticism. Thanks to these contributions of the literary public sphere, the idea of reason as a public medium of communication incorporated, if only for a brief historical moment, an aesthetically enriched conception of human flourishing and emancipation.

In the years following the publication of his first book, Habermas continued to develop the idea of reason as a public medium of "domination-free" communication in different arenas of inquiry; but almost twenty years would pass before he tried to recast the aesthetic component of reason that his sociological and historical analyses had made accessible into the philosophical terms of his mature theory of ra-

tionality. With the publication of *The Theory of Communicative Action* in 1981 (1984), Habermas presented the first systematic attempt since the time of German idealism to integrate art and aesthetic experience into a comprehensive theory of rationality. After the decline of German idealism, the idea that art and aesthetic experience contained an ineliminable moment of reason that it was philosophy's responsibility to incorporate into a comprehensive conception of reason simply lost its urgency—and, morever, its appeal. But it was an extremely urgent and appealing idea during the period of roughly forty years that spanned the publication of Immanuel Kant's *Critique of Judgment* (1790/93), Friedrich von Schiller's *On the Aesthetic Education of Humankind* (1794/95), Friedrich Wilhelm von Schelling's *System of Transcendental Idealism* (1800), and G. W. F. Hegel's lectures on aesthetics, given at the University of Berlin in the mid to late 1820s and published posthumously as, *Aesthetics: Lectures on Fine Art* (1835, 1842). It was in these lectures that Hegel (in)famously announced the "end of art." With that announcement, supported by a sweeping historical and conceptual analysis of developments in Western art from the ancient civilizations of the Near East, Greece, and Rome to Hegel's own time, this strikingly fecund period in aesthetics and the philosophy of art came to a close.

These diverse and various attempts to make a case for the rationality of art and aesthetic experience need to be understood as an integral part of a larger project: the reformulation of the meaning of reason through a critique of reason. Typically, the more radical the critique of reason, the more prominent the role art and aesthetic experience play, both in the critique of existing forms of reason and in the proposed reformulation. But no matter how radical the critique of reason, nor how "aesthetic" the proposed reformulation of reason, be it Schiller's or Schelling's, the goal of the whole undertaking was directed toward a more generous and capacious conception of theoretical and practical reason. Later in the nineteenth century, the critical and transformative power of art was reconceived in such a way that it was apparently rendered independent of the goal of reformulating the meaning of reason. Art and aesthetic experience continued to play a prominent role in the critique of reason, but the role was recast. They were neither the vehicle by which reason could become more expressive and less repressive, nor the very model on which a new conception of reason would be based; rather, art and aesthetic experience began to play the role of the "other of reason." This recasting of the role art and aesthetic experience were to play in the critique of reason was largely the result of the work of Arthur Schopenhauer, and above all, the work of Friedrich Nietzsche. The appeal of art's role as the "other of reason" continues to be felt in contemporary intellectual movements, especially poststructuralism and postmodernism.

But this was not the only historically influential attempt to redirect the critical and transformative power of art. The hermeneutical tradition from Wilhelm Dilthey to Hans-Georg Gadamer redirected this power toward practical contexts of life—toward the illumination of problems of individual and collective self-understanding. Habermas's understanding of how art and aesthetic experience are to be incorporated into a comprehensive theory of rationality is influenced most by this practical orientation. Nonetheless, he is also very conscious of the need to close the gap that had opened up in the German philosophical tradition between this orientation and the Nietzschean one. It is not by any means a matter of historical contingency that a German philosopher would once more take up the problem of systematically incorporating the whole domain of the aesthetic into a comprehensive theory of rationality. First of all, it is only in this tradition that such systematic and comprehensive conceptions have had any currency. In the Anglo-American tradition, minimalist accounts of rationality have been the exceptionless rule. Second, it is only in this tradition (with the exception of John Dewey's *Art and Experience* in the American pragmatist tradition) that art and aesthetic experience have enjoyed an unparalleled philosophical prominence. One need only add names such as Martin Heidegger, Walter Benjamin, and Theodor Adorno to the list of those already mentioned in order to appreciate how deeply embedded is the philosophical and cultural significance of art and aesthetic experience in the German tradition. Furthermore, as the most important exponent of Frankfurt School critical theory since Adorno, Habermas had to come to terms with the aesthetic turn that Adorno had proposed for critical theory.

Ever since *The Structural Transformation of the Public Sphere*, Habermas's entire philosophical project has been guided by the persistently defended idea that reason is a medium of public communication through which social actors achieve mutual understanding about something in their objective, social, or subjective worlds (formally, not ontologically distinguished). Habermas's theory of rationality is also a theory of modernity. He takes from Kant's three critiques the idea that with the emergence of modernity, reason has been differentiated into three aspects: the cognitive-instrumental, the moral-practical, and the aesthetic-practical. These three aspects of reason have been institutionalized into three corresponding cultural domains: the domains of science, morality, and art, each of which circumscribes independent logics of development, types of knowledge, justification, and criticism. Drawing from the sociological investigations of Max Weber, Habermas regards the differentiation of reason as the culmination of a historical process which, pace Weber, he regards as an irreversible process of social learning—that is, as a defensible history of cognitive, moral, and aesthetic progress. But the

social and cultural potentiality of these independent domains of social learning has been only selectively exploited. The cognitive-instrumental aspect of reason has been one-sidedly developed and institutionalized at the expense of the moral-practical and the aesthetic-practical aspects; and it is this one-sided predominance of the cognitive-instrumental dimension that has provoked the critique of reason (qua "instrumental reason" or "subject-centered reason"). What is required, on Habermas's view, is a "balanced institutional embodiment" of all three aspects of reason (1984, vol. 1, p. 273). Only through such a balanced embodiment in the institutions, practices, and life contexts of modern societies can the potential gain in reason that modernity represents be finally realized.

In *The Theory of Communicative Action*, Habermas proposed two ways to integrate the aesthetic into his theory of rationality. First, he made the case for a distinctly aesthetic validity claim (or "truth claim"), which he incorporated into his system of validity claims; and second, he made the case for a distinctive notion of aesthetic critique. The first way proved to be unsuccessful, forcing Habermas to abandon this approach altogether. The second proved to be more viable, but the results have been less than satisfactory when measured against the promise of restoring the rationality of art and aesthetic experience.

There are three irreducible validity claims in Habermas's system: the truth of propositions, the rightness of norms, and the authenticity or truthfulness of expressions. "A validity claim is equivalent to the assertion that the conditions for the validity of an utterance have been fulfilled" (ibid., p. 38). As such, any validity claim admits of criticism, of yes/no positions. The construction of the system of validity claims owes a good deal to speech act theory. Each of the basic validity claims has corresponding types of speech acts in which it is couched—constative speech acts for truth claims, regulative speech acts for rightness claims, and expressive speech acts for authenticity claims. Habermas is of the view that the criticizability of individual communicative utterances can be extended to cultural products such as works of art, moral and legal norms, and theories. These cultural products can be treated as complex types of speech acts that make "higher-level validity claims" (ibid., p. 40). An internal connection between the validity claims of individual utterances and those of cultural products allows works of art to be regarded as expressive speech acts, making validity claims (analogous to truth or rightness claims) whose rationality can be evaluated according to their capacity to authentically disclose an exemplary experience of "eye-opening" power.

But this connection turned out to be unsustainable. In an important paper on Adorno's aesthetics, Albrecht Wellmer (1991, p. 29) successfully argued that aesthetic validity claims could not be understood in the way that Habermas had proposed. First of all, works of art are not reducible to expressive speech acts claiming authenticity or truthfulness. This approach reverts back to an "aesthetics of genius" according to which the work of art is the "authentic" expression or disclosure of the artist's subjectivity. Second, as Habermas himself came to admit, the one-to-one relationship between the three basic validity claims and their corresponding speech acts is an inappropriate model for comprehending aesthetic validity claims. An aesthetic validity claim cannot be identified with, or connected to, any one of the basic validity claims to truth, rightness, or authenticity; rather, an aesthetic validity claim stands for a "truth potential" that can be released into the communicative practice of everyday life only through the interpretative activity and practical orientation of its addressees.

As a result of Wellmer's telling criticisms, Habermas gave up his attempt to integrate the claims of art and aesthetic experience into his system of validity claims. This is a very unfortunate circumstance, because the system of validity claims is the normative core of Habermas's communicative reformulation of reason. Although Habermas asserts that art's "truth potential" is such that it can affect our languages of evaluation, renew our need interpretations, and transform our cognitive and normative frameworks, he is unable to find a place for it in the basic concepts of his theory of rationality (Habermas, 1985, p. 202). Instead of rethinking these basic concepts, Habermas (mistakenly, it seems) preferred to accept a less comprehensive integration of the aesthetic realm into his reformulation of reason.

Habermas defines aesthetic critique as "a variation of a form of argumentation in which the adequacy of value standards, the vocabulary of our evaluative language generally, is made thematic" (1984, vol. 1, p. 20). It requires that one take a reflective attitude toward culturally established values in light of which one's needs and desires are interpreted and assessed. In the context of aesthetic experience, reasons can guide one's perceptions in such a way that they provide both the motivation and justification for transforming one's language of evaluation. But the justifications that issue from aesthetic critique do not have the generalizability, the universal reach, of those justifications proper to what Habermas calls discourse. A discourse can be theoretical, in which case it thematizes controversial truth claims, or it can be practical, in which case it thematizes normative rightness claims. The distinction between aesthetic critique and discourse is meant to mark the difference between justifications that can in principle transcend the horizons of culturally specific identities and forms of life and those that cannot. Because values cannot be universalized, the scope and effects of aesthetic critique remain bound to the culture in which they are situated—bound, in other words, to a shared preunderstanding of the values in question.

On the other hand, aesthetic critique can articulate new values, or bring about revaluations that have any innovative

character akin to novel works of art. Thus, it can initiate a process of normative change that is similar to aesthetic change, where norms, and not just perceptions, are affected. But Habermas never really develops in just what the rationality of this innovative dimension of aesthetic critique consists. His primary concern has been and continues to be the development of the moral and political implications of "discourse ethics." Nonetheless, Habermas was forced to return to the problematic of aesthetic critique later in the 1980s when he decided to respond to the growing challenge of the poststructuralist critique of modernity. This response appeared in a set of public lectures that were published under the title *The Philosophical Discourse of Modernity*. One of the primary aims of these lectures was to curtail and limit what he called the "aesthetic horizon" of critique (1987, p. 125).

In Habermas's historical account, this "aesthetic horizon" of critique was first opened up by Nietzsche when he set out to derive criteria for the critique of reason from the experience of art as an "other of reason." According to Habermas, Nietzsche's intentions were to render the critique of reason independent of the goal of reformulating reason in more capacious terms. It is from within this "aesthetic horizon" of critique, Habermas argued, that theorists such as Michel Foucault, Jacques Derrida, and Georges Bataille engaged in their deconstructive, rather than reconstructive, critique of reason. Thus, they follow Nietzsche in a form of critique that is rendered independent of any connection to reason, and that leads unavoidably to an aestheticization of critique. This is a process by which critique is modeled on the power of innovative works of art to create meaning and values.

Clearly, Habermas's criticism of Nietzsche and his poststructuralist and postmodernist "disciples" can easily backfire, for the rational status of aesthetic critique was left unsatisfactorily developed in Habermas's own theory of rationality: his criticism can be easily redirected at himself. Still, the force of his arguments against Nietzsche and neo-Nietzscheans are certainly compelling: aesthetic critique qua critique can be separated from reason only on pain of incoherence. It is an open question, however, whether any current conceptions of reason, including Habermas's, can make sense of the creative or transformative dimension of reason manifest in art and aesthetic critique. It is certainly to Habermas's credit that he has once again made this into an urgent and appealing problem. If he has failed to solve this problem, he has failed in an instructive way. The aesthetic realm remains in need of an unforced integration into a more capacious, generous conception of reason. Until then, no matter how warmly it is welcomed into the house of reason, the very design of the existing house will force the aesthetic to bow its head in shame: in shame of its particularity, and of its inability to universalize itself.

[*See also* Kant; *and* Marxism.]

BIBLIOGRAPHY

Works by Habermas

The Philosophical Discourse of Modernity: Twelve Lectures. Translated by Frederick G. Lawrence. Cambridge, Mass., 1987.
"Questions and Counter-Questions." In *Habermas and Modernity*, edited by Richard Bernstein. Cambridge, Mass., 1985.
The Structural Transformation of the Public Sphere: An Inquiry into a Category of Bourgeois Society. Translated by Thomas Burger with the assistance of Frederick Lawrence. Cambridge, Mass., 1989.
The Theory of Communicative Action, vol. 1. Translated by Thomas McCarthy. Boston, 1984.

Other Sources

Kompridis, Nikolas. *Crisis and Transformation: The Aesthetic Critique of Modernity from Hegel to Habermas.* Berkeley, forthcoming.
Menke-Eggers, Christoph. *Die Souveränität der Kunst: Ästhetische Erfahrung nach Adorno und Derrida.* Frankfurt am Main, 1988.
Rochlitz, Rainer. "De l'expression au sens: Perspectives esthétiques chez Habermas." *Revue Internationale de Philosophie* 49.194 (1995): 409–436.
Seel, Martin. *Die Kunst der Entzweiung: Zum Begriff der ästhetischen Rationalität.* Frankfurt am Main, 1985.
Warnke, Georgia. *Justice and Interpretation.* Cambridge, Mass., 1992. See chapters 5 and 7, pp. 87–110, 135–157.
Wellmer, Albrecht. "Truth, Semblance, and Reconciliation." In *The Persistence of Modernity: Essays on Aesthetics, Ethics, and Postmodernism*, translated by David Midgley. Cambridge, Mass., 1991.

NIKOLAS KOMPRIDIS

HAIKU. A major Japanese verse form, haiku consists of seventeen syllables and contains a word suggestive of the season. Prior to the twentieth century, it was called *hokku* (opening verse), because it served as the first verse of a poetic sequence known as *haikai*. Today it is composed as an autonomous poem, and even a hokku written in the premodern period is called a haiku when it is singled out and appreciated independently of other verses that follow it in the original text.

The early aesthetics of haiku centered around humor, for *haikai* as a poetic form started in reaction to the tradition of elegant court poetry that had dominated the Japanese poetic scene from the eighth through the fifteenth century. By introducing elements of humor, pioneers of *haikai* in the sixteenth and seventeenth centuries tried to liberate poetry from aristocratic court culture and make it more appealing to a large number of commoners who had become literate and begun participating in poetic activities. As the following example by Nishiyama Sōin (1605–1682) illustrates, the main ingredient of haiku humor was often incongruity arising from a surprising comparison, although wit, wordplay, parody, and slangy colloquialism were also employed as devices to elicit laughter:

> wisteria vines
> round the pine—an octopus
> climbing up the tree!

By and large, humor in early haiku was lacking in intellectual and emotional depth because it was derived more from clever manipulations of words or literary conventions than from keen observations of nature or human nature.

Matsuo Bashō (1644–1694) elevated haiku to a mature poetic form by setting up high aesthetic ideals and writing outstanding poems that embodied them. One such ideal was *sabi* (loneliness), originally a medieval concept connoting sad recognition of universal human loneliness. Bashō, as a postmedieval poet, added a more humanist meaning to the concept by believing in the pleasure of loneliness. Reportedly, he found *sabi* in the following haiku, in which the two old men, whose job is to "watch" fragile cherry blossoms, have learned to accept the mutability of life and are enjoying themselves in a chat.

> two blossom-watchmen
> with their white heads together
> in a chitchat

Bashō's second major aesthetic ideal, *karumi* (lightness), leans even more toward postmedieval humanism. *Karumi* is contrasted with "heavy" beauty, the type of beauty that emerges out of a poem dealing with a weighty philosophical theme. Yet, it is not the opposite of "heaviness"; rather, it is a dialectic transcendence of it. One of the haiku that Bashō thought had *karumi* presents a scene of falling blossoms:

> under the trees
> soup, fish salad and all
> in cherry blossoms

Although medieval poets would have made the scene into a "heavy" poem by striking home the transitoriness of life, this haiku focuses on the enjoyment of picnickers who are eating, drinking, and singing under the blossoms. With fallen petals all around them, these people are not unaware of their mortality, yet they accept it as part of the process of nature. Such earthly images as soup and fish salad, totally alien to the aesthetics of court poetry, appear side by side with the elegant image of cherry blossoms and help to create a new type of poetic beauty.

Unfortunately, less gifted poets who emulated Bashō's ideal of *karumi* often ended up with banal poems that reeked of earthly interests. Therefore, Yosa Buson (1716–1784), the leader of the haikai reform movement in the eighteenth century, came to emphasize the importance of *rizoku* (detachment from the mundane) in writing poetry. In his teaching, a person wishing to write a verse should first free his mind from earthly concerns by such means as reading classical poetry, drinking wine, and taking a stroll in the woods. The result would be a poem that aspires toward a realm higher than the physical world. The ethereal quality that Buson sought in poetry is discernible in the following example of his haiku:

> plum blossoms' scent
> rising and rising—a halo
> round the moon

The reader visualizes a scene of unearthly beauty in which thousands of plum blossoms are blooming in white, sending their fragrance toward the sky and seemingly creating a halo around the moon.

In sharp contrast to Buson, Kobayashi Issa (1763–1828) was strongly attached to mundane life with all its ugliness and to common humanity with all its idiocy. "There is no medicine for curing an idiot," he wrote one New Year's Day. "My only wish is to remain an idiot and live for many more New Years to come." Hampered by family and money problems during much of his life, he vented his frustrations and grief in numerous poems in his diaries. Even his haiku mourning his wife's death reveals the less noble side of himself:

> would there still be
> someone I could grumble to
> on this full-moon night!

In its honest presentation of the baser aspects of human nature, Issa's haiku foreshadowed the naturalistic realism that was to become the main current of Japanese literature in the early twentieth century.

Issa had no knowledge of Western-type realism because Japan was largely closed to the rest of the world during his lifetime. When the country opened up in the mid-nineteenth century, many literary and aesthetic thoughts began to flow in from the West. The first major haiku poetics formulated under Western influence came from Masaoka Shiki (1867–1902), who, impressed by realistic qualities of Western art, advocated the principle of *shasei* (sketch from life). In his theory, a haiku poet should observe his subject in nature objectively and sketch it as faithfully as possible in seventeen syllables, using no word extraneous to that subject. For example:

> flowering cockscombs—
> fourteen or fifteen of them
> there seem to be

Some critics praised this haiku by Shiki as a seminal example of the *shasei* principle, but others wondered whether such a plain description of a commonplace sight would deserve the name of poetry. Shiki himself became interested in sketching his own mind in his later years as he fought a losing battle with tuberculosis.

Some followers of the *shasei* principle, led by Ogiwara Seisensui (1884–1976) and Nakatsuka Ippekirō (1887–1946), pursued it to an extreme and began advocating freestyle haiku, a type of haiku not restricted by syllable count. In their view, a poet who uses no word extraneous to his subject matter should not be using a fixed pattern of syllables either, because it is an arbitrary requirement that ob-

structs the natural process of sketching from life. The form of each poem should be determined solely by its subject, they insisted. The rationale of freestyle haiku, then, seems identical with that of free verse. Seisensui, however, distinguished freestyle haiku from free verse in three major respects. First, free verse treats all manner of subjects, whereas freestyle haiku deals primarily with nature. Second, free verse can be of any length, whereas freestyle haiku has to be read out in one breath. Third, free verse has a linear structure and allows for elements of time, whereas freestyle haiku is centripetal in form and presents a momentary flash of insight. Seisensui's distinction did not convince everyone and caused some lively debates in the early twentieth century, helping to promote reflections on the basic nature of the haiku form. The freestyle haiku movement itself lost much of its momentum after the death of its two outstanding practitioners, Ozaki Hōsai (1885–1926) and Taneda Santōka (1882–1940).

Haiku theorists who defended the traditional seventeen-syllable form were also influenced in various ways by Western literary and aesthetic thoughts that entered Japan in the twentieth century. The European theory of symbolism, for instance, was one of the sources that inspired Ōsuga Otsuji (1881–1920) to conceive a metaphoric theory of haiku, a theory that identified the season word as the intermediary between the physical and the metaphoric world. Okazaki Yoshie (1892–1982) also considered haiku a type of symbolic poetry, observing that *sabi,* a product of Buddhism and Daoism, is what largely differentiates it from the kind of symbolism conceived by modern French poets. In the early 1930s, some radical haiku poets such as Kuribayashi Issekiro (1894–1991) and Hashimoto Mudō (1893–1974) became interested in Marxism and went on to formulate a proletarian theory of haiku. Hino Sōjō (1901–1956) and Saitō Sanki (1900–1962) promoted "modernist haiku" that reflected the European literary trends after World War I, whereas Takayanagi Shigenobu (1923–1983), under the influence of modern Western poetry, wrote "visual haiku" that utilized the visual effects of lineation and indentation.

There were also attempts to reinterpret traditional haiku poetics in the context of Western aesthetics. The most elaborate attempt was made by Ōnishi Yoshinori (1888–1959), a specialist in German aesthetics, who constructed a comprehensive aesthetic scheme that integrated both Japanese and Western ideas. The scheme was founded largely on Hermann Cohen's theory and centered on three basic types of beauty: "the beautiful," "the sublime," and "humor." Yet, Ōnishi went beyond Cohen when he conceived six derivative types of beauty: "the tragic," *yūgen* (mystery and depth), "the graceful," *aware* (pathos), "the comic," and *sabi.* The type of beauty known as *sabi,* he believed, would emerge when factors that generally produce humor are connected less with human life than with nature. In Ōnishi's view, humans have occasion to feel something like a shadow

of eternity in a natural phenomenon unaffected by humanity, and that feeling in turn gives them a heightened sense of spiritual freedom that enables them to look at human life with detachment and a kind of humor.

The poetics of haiku became known outside Japan in the early twentieth century and has merged with various theories of art in the West. It was one of the bases of the Imagist theory of poetry formulated by Ezra Pound, who considered haiku a "one-image poem" that utilized "a form of superposition." Sergei Eisenstein preferred to think of haiku as a "montage phrase," the principle of which he developed into the technique of montage in film. To the Beat generation poets, haiku seemed like a question with no answers, the contemplation of which would eventually lead to a Zen-like enlightenment. As more and more Westerners write haiku in their own languages, new views of haiku that incorporate Western aesthetic ideas will continue to appear.

[*See also* Japanese Aesthetics; *and* Poetics.]

BIBLIOGRAPHY

Beichman, Janine. *Masaoka Shiki.* Tokyo, 1986.

Higginson, William J., and Penny Harter. *The Haiku Handbook.* New York, 1985.

Hisamatsu, Sen'ichi. *The Vocabulary of Japanese Literary Aesthetics.* Tokyo, 1963.

Matsui, Toshihiko. *Kindai hairon shi* (A History of Modern Haiku Poetics). Tokyo, 1965.

Ōnishi, Yoshinori. *Bigaku* (Aesthetics). 2 vols. Tokyo, 1959–1960.

Shirane, Haruo. "Matsuo Bashō and the Poetics of Scent." *Harvard Journal of Asiatic Studies* 52.1 (June 1992): 77–111.

Ueda, Makoto. *Literary and Art Theories in Japan.* Cleveland, Ohio, 1967; reprint, Ann Arbor, 1991.

Ueda, Makoto. *Modern Japanese Poets and the Nature of Literature.* Stanford, Calif., 1983.

Yasuda, Kenneth. *The Japanese Haiku.* Tokyo, 1957.

Zolbrod, Leon. "Talking Poetry: Buson's View of the Art of Haiku." *Literature East and West* 15–16 (December 1971–June 1972): 719–734.

MAKOTO UEDA

HAMANN, JOHANN GEORG (1730–1788), German aesthetician and critic of the Enlightenment. Hamann was born in Königsberg, Germany. Before becoming a government employee in 1767, he worked as a private tutor, a merchant, a commercial traveler, and a secretary. He remained in government employment in Königsberg until 1784, when private patronage alone could support him. Hamann has been widely regarded as the leader of the Sturm und Drang, a literary movement that began in the 1760s in Germany, whose chief spokesmen were Johann Wolfgang von Goethe, Johann Gottfried von Herder, Friedrich von Schiller, and Justus Möser. The Sturm und Drang rebelled against the constraints of classical art and social convention, and it celebrated the values of self-expression, freedom, and genius. Hamann has been seen as the leader of this rebellion against

classicism, as the first champion of its faith in self-expression and genius.

This picture of Hamann is not, however, entirely accurate. He was not the first to celebrate genius or to revolt against classicism; and his influence on the Sturm und Drang was largely indirect, because his works were rare even in his lifetime. Still, there can be no question that his ideas were seminal. They were spread largely through Herder, his erstwhile student, and their importance was later acknowledged by some leading lights of the *Goethezeit*, especially Friedrich Schlegel, Friedrich Heinrich Jacobi, G. W. F. Hegel, and Goethe.

Hamann's most important work in aesthetics is his short tract *Aesthetica in nuce* (Aesthetics in a Nutshell), which first appeared in 1762 as part of his *Kreuzzüge des Philologen* (Crusades of a Philologist). This tract is primarily an attack on the rationalist hermeneutics of the Enlightenment and a defense of a more spiritual interpretation of the Bible. It is written in a rhapsodic and sibylline style, filled with obscure allusions, hidden polemics, and private jokes. Underneath all the difficult prose, however, lies the germ of an aesthetic theory, just as the title suggests.

Hamann's theory consists in two elements: a doctrine of expression and one of imitation. Each of these components deserves separate explanation. Hamann's originality and importance ultimately stem from his striking synthesis of both. This was an apparent squaring of the circle, given that these doctrines had been opposed in the eighteenth century.

Hamann's doctrine of expression consists in his belief that the artist should reveal the intensity, urgency, and depth of his feelings, and that he should portray the immediacy, freshness, and vitality of his perceptions. Passion, for Hamann, is not only the creative source of a work of art, but also the basis for its excellence. A work of art is valuable, he implies, to the extent that it expresses the personal feelings, and embodies the individual experience, of its creator. Thus, he says that "the fruitful womb of passion" is the germ not only of "new ideas and new expressions," but also of the "perfection of the plan and the strength of its execution."

Because the passions are the source of all creativity and excellence, Hamann urges the artist to know and satisfy them fully. This holds even for his apparently "lower desires," the need for food, wine, and sex. For the sake of art, Hamann does not hesitate to cast aside all moral and religious scruples. One should not bother with the "metaphysics of the fine arts," he says, until one initiates oneself into the "orgies and Eleusian mysteries." Of course, such passions can be the source of dishonor; but, Hamann asks, do they cease, for all that, to be creative "weapons of masculinity"?

Such a radical reappraisal of the role of the passions led Hamann to reject the fussy and fusty constraints of classicism. All its rules and conventions only repress the feelings of the artist and inhibit the development of his individuality. They are also abstract and artificial when it is the task of the artist to reveal the concreteness and immediacy of personal experience. For these reasons, Hamann scorns Johann Joachim Winckelmann's advice that the artist should follow the model of the Greeks. The source of artistic revival lay not in imitating an alien and dead culture but in having the daring and vitality to express the secret truths of the heart.

Of course, Hamann was not the first aesthetician of the eighteenth century to defend the rights of passion or to take a stand against the tyranny of classicism. That rebellion was already under way with the "Sensibility" or *Empfindsamkeit* movement of English and German literature, a tradition that influenced Hamann himself. Still, in two respects Hamann went much further than most of his contemporaries. First, he was skeptical of virtually *all* the prevalent rules, which were only so many obstacles in the path of self-expression. Second, he valued *all* the passions, especially sexual desires and energies, and not merely the *moral* sentiments so loved by eighteenth-century aestheticians. His emphasis on the creative role of sexuality and the depths of the subconscious anticipates even modern ideas.

The other component of Hamann's aesthetic, his doctrine of imitation, has to be placed within the context of his metaphysics. The basis of Haman's metaphysics was his religious conversion and mystical experience, which took place in London in 1759. At the heart of this experience was a remarkable vision: that the entire universe is a sacred text, the secret language of God. All the events of nature and history are words, symbols, or hieroglyphs, the means by which God speaks and reveals himself to humanity.

Although it appears fanciful and eccentric, Hamann's vision had a venerable legacy going back to the Middle Ages. According to Pseudo-Dionysus and Saint Bonaventure, for example, the universe is a holy book, whose words are the individual creatures. Like any sacred text, nature must be interpreted and explained, so that the will of its author is understood. This idea had been an inspiration to artists in the Middle Ages, providing them with not only a fund of symbolism, but also a divine calling: the interpretation of the sacred text of the universe. It had been largely forgotten during the Enlightenment, because of the rise of the new natural philosophy and its mechanical vision of the universe. It was Hamann's role to revive it in the eighteenth century, so that it could again be the source of aesthetic inspiration.

Armed with this mystical vision and a firm belief in the religious calling of art, Hamann reinterpreted the classical doctrine of imitation. He gave the concept a religious meaning, a mystical significance, that is absent from most eighteenth-century aesthetics. To imitate nature now means to reveal the secret language of God. All that appears to humans after the Fall, Hamann says, are scattered remnants and fragments of the divine message. It is the task of the scholar to collect

them, of the philosopher to interpret them, and of the poet to imitate them or bring them into order. The artist should therefore aspire to be like the ancient prophet, giving voice to divine inspirations and visions.

In retaining the doctrine of imitation, Hamann's aim was to reestablish the metaphysical significance of art, the classical equation of truth and beauty, which had been undermined by much eighteenth-century aesthetics. The predominant tendency of eighteenth-century aesthetics had been to accord art a merely subjective significance, so that it amounted to nothing more than pleasure. Thus, the British empiricists (Francis Hutcheson, David Hume, and Edmund Burke) reduced aesthetic experience to sensation, and the German rationalists (Christian Wolff, Alexander Baumgarten, and Moses Mendelssohn) demoted it to a confused representation of sensibility. Hamann explicitly attacked this trend, comparing its protagonists with Pontius Pilate: "Yes, you fine critics! You ask for what truth is and reach for the door because you cannot wait for an answer." He bluntly denied that art is a source of knowledge inferior to science and mathematics, as if it gives insight only into appearances. Rather, he stoutly maintained that it is a *superior* source of knowledge, indeed, the best instrument and criterion for acquiring knowledge of reality itself.

How could Hamann so boldly reinstate the metaphysical status of art? The answer lies largely in his epistemology, in his radical empiricism, and his critique of reason. A keen student of Francis Bacon, Hamann thinks that all knowledge arises from the five senses, and that perceptions are best expressed in images, which are "the whole store of human knowledge and happiness." What is the stock-in-trade of art, though, but images? Although its images are capable of capturing all the particularity, freshness, and richness of human experience, the concepts of reason are only pale and remote abstractions from it. Influenced by Hume and the nominalist tradition, Hamann affirms that all that exists is particular, and that the only direct means of knowing the particular is through our senses. Hence, art alone, which expresses our sense perceptions in images, gives us immediate insight into reality itself.

How does Hamann combine his doctrines of expression and imitation? On the face of it, any synthesis seems impossible. The artist cannot both express his passions and imitate nature, for self-expression and imitation seem to have different objects and different requirements. Self-expression has the self for its object and requires introspection; imitation has the world for its object and demands abstraction from the self. Such an opposition had in fact become commonplace in eighteenth-century aesthetics, because those writers who saw sentiment as the subject matter of art also denied it all metaphysical significance. The interest of Hamann's aesthetics is precisely that it breaks down such an opposition.

The key to Hamann's synthesis goes back to his metaphysics, and indeed to his mystical experience in London.

Part of Hamann's vision was his feeling for the omnipresence of God, for his guiding hand in all the events of nature and history. More important, he came to believe that God is in everyone, hidden deep within the heart, so that whatever one thinks and does is also a revelation of God. As he put it in one of his London notebooks: "The breath of life in our nose is also the exhaltation of God." According to this vision, then, there is no sharp opposition between the mind and nature, subject and object, the self and God. The self knows itself only through nature, only through its place within the universe and the divine economy. Hamann therefore denied the whole Cartesian legacy. Not "I think, therefore I am," he said, but "I am, therefore I think" should be the motto of philosophy.

This belief in the omnipresence of God, his presence deep within the soul, is the link between Hamann's doctrines of expression and imitation. If God is inside man, and if he reveals himself especially through man's senses and passions, then it follows that the artist only has to express his senses and passions to reveal God. What he senses and feels is only what God senses and feels through him, so that his self-expression becomes the revelation of God. Artistic creativity is not only the activity of the artist, then, but also of God himself. Thus, the artist does not simply decipher, but also creates the divine language. His images are its symbols, the revelation of God through his prophet.

It was Hamann's synthesis of these doctrines that anticipated the aesthetic credo of the *Stürmer und Dränger* and the whole Romantic generation. His aesthetic must have been very appealing to any budding writer in late eighteenth-century Germany; for Hamann not only defended the creativity of the artist—his right to express all his feelings and passions—but he also conferred the greatest metaphysical honors on him for doing so. Self-expression was not just self-indulgence; it was the revelation of God! This faith in the metaphysical significance of art reappears in some of the main works of Romanticism, in Friedrich Schleiermacher's *Reden über die Religion* (Speeches on Religion), Friedrich Hölderlin's *Hyperion*, Friedrich Wilhelm von Schelling's *System des transcendentalen Idealismus* (System of Transcendental Idealism), Novalis's *Lehrling zu Saïs* (The Apprentice at Saïs), and Friedrich Schlegel's *Vorlesungen über Transcendentalphilosophie* (Lectures on Transcendental Philosophy).

It is important to keep Hamann in mind to avoid one of the old clichés about Romanticism and post-Kantian aesthetics. Usually, the primacy of aesthetics in post-Kantian and Romantic philosophy is ascribed to the influence of Immanuel Kant's *Critique of Judgment* and Schiller's *Aesthetic Letters*. These works were indeed seminal; yet, they cannot explain the Romantic beliefs in self-expression and the metaphysical significance of art. Kant and Schiller swam against the Romantic current, with their classical taste and their relegation of art to the realm of mere appearance. Although Hamann was not the sole source of these Romantic

doctrines, he was one of the first thinkers to espouse them. The best testimony for his influence is the great honor that Goethe, Herder, Schlegel, and Hegel later bestowed on him.

[*See also* Classicism; Expression Theory of Art; Mimesis; *and* Romanticism, *article on* Philosophy and Literature.]

BIBLIOGRAPHY

Works by Hamann

Briefwechsel. 6 vols. Edited by Walter Ziesemer and Arthur Henkel. Wiesbaden, 1955–1979.
Sämtliche Werke. 6 vols. Edited by Josef Nadler. Vienna, 1949–1957.
Sokratische Denkwürdigkeiten: Aesthetica in nuce. Edited by Sven-Aage Jorgensen. Stuttgart, 1968.

Works by Hamann in English Translation

Hamann's Socratic Memorabilia: A Translation and Commentary. Translated by James C. O'Flaherty. Baltimore, 1967.
Aesthetica in nuce [Aesthetics in a Nutshell]: *A Rhapsody in Cabbalistic Prose.* Translated by Joyce P. Crick. In *German Aesthetic and Literary Criticism: Winckelmann, Lessing, Hamann, Herder, Schiller, Goethe,* edited by H. B. Nisbet. Cambridge and New York, 1985.

Other Sources

Alexander, W. M. *Johann Georg Hamann: Philosophy and Faith.* The Hague, 1966.
Berlin, Isaiah. *The Magus of the North: J. G. Hamann and the Origins of Modern Irrationalism.* London, 1993.
Blanke, Fritz. *Johann Georg Hamanns Hauptschriften Erklärt.* Gütersloh, 1959–1963.
German, Terence J. *Hamann on Language and Religion.* Oxford, 1981.
O'Flaherty, James. *Johann Georg Hamann.* Boston, 1979.
Unger, Rudolf. *Hamann und die Aufklärung.* 2 vols. Halle, 1925.

FREDERICK BEISER

HANSLICK, EDUARD (1825–1904), Czech music journalist. Hanslick had his musical training in Prague, where he was born, and studied law at the universities of Prague and Vienna, graduating from the latter in 1849 with a doctorate. In 1856, he took up a position teaching the history and aesthetics of music at the University of Vienna, a position created especially for him. He resided in Vienna from 1846 to the end of his life, but his literary career began in Prague in 1844, while he was still a student, and spanned more than fifty years.

In 1854, a little book appeared in Leipzig by Hanslick with the main title *Vom Musikalisch-Schönen.* Hanslick presented it as merely the groundwork for an aesthetics of music that he intended to write subsequently, but did not, and that would be concerned not with all music, but only with European music from Wolfgang Amadeus Mozart to Johannes Brahms. During its author's lifetime, the book appeared in ten German editions and several translations, including English in 1891. Despite its modest pretensions, *Vom Musikalisch-Schönen* has generated more discussion than any other writing on musical aesthetics.

In it Hanslick asks and answers two questions: What is the *purpose* of music? and What is the *content* of music? He provides various formulations of his answers to these questions, but in essence they are that music as such has no purpose outside itself, and the content of music as such is its form, or, which is the same thing, music's form is its content.

The former, the negative thesis that music has no purpose outside itself has received more critical attention than the latter, because it runs against the view prevailing in 1854 (and since) that the defining purpose of music is to express, arouse, or portray feelings *(Gefühle).* Hanslick calls this view the "feeling theory," and devotes his first two chapters to refuting it. The remainder of the book, chapters 3 to 7, is devoted mainly to Hanslick's positive thesis that the content of music is its form.

Here, paraphrased and much abbreviated, is a sampling from Hanslick's arguments against the feeling theory:

The purpose of arousing feelings in the beholder is not unique to music, but is common to all the arts, and is merely a secondary effect in each.

The connection between specific passages of music and specific changes of feeling is not one of strict causation; hence, the effect of music on feeling lacks the necessity that it would require to serve as an aesthetic principle.

In vocal music, one and the same melody can be heard as equally appropriate to words expressing one emotional state, or to words expressing another, or even the opposite, state.

The feelings are not so isolated from the other mental capacities that they can themselves be the defining feature of an art.

Even if it were possible for specific feelings to be represented in music, the degree of beauty in the music would not correspond to the degree of exactitude with which the music represented them.

The feelings, then, are too inclusive and inconstant to yield any theoretical principles, hence no aesthetical theory can be derived from them. Hanslick proposes instead a musical aesthetics based not on the subjective response to music but on the aesthetic object, namely, the musical artwork itself. His account of this musical aesthetics is so compact as to be a mere sketch, not to be further summarized; therefore, the present article will touch on just four of the book's main topics: the material of music; the creative act of the composer; the aesthetic perception of music; and form and content in music.

The Material of Music. Hanslick says that the material of music is not physical sounds or the instruments and voices that produce them, but is the whole system of "tones," the so-called diatonic system of tonal relationships, that is, all the major and minor scales, and their modifications, with their limitless capacity for melodic, harmonic, and rhythmic combination and variation.

This is "material" in the sense of "that out of which a thing is made," in this case a musical artwork. The material of music is nonphysical, that is, ideal; indeed, it is as volatile and evanescent as thought itself, with which it is for that reason uniquely compatible. The artistic laws governing the composer's use of this material are inscrutably linked to the laws of psychology and of nature, but the material of music does not exist in nature, because the diatonic system is a cultural artifact, as, for example, are languages. Hanslick insists that music is not a kind of language, but says that, like a language, the diatonic system is subject to historical change as the culture of which it is a part changes.

The Creative Act of the Composer. The composer, in Hanslick's view, is prompted not by an inner feeling, as the feeling theory maintains, but by an "inner singing." Hanslick's theory of the creative act of the composer is a two-stage theory in which this inner singing is the first stage. A distinctive musical motif presents itself in the composer's imagination, perhaps a whole melody or perhaps a brief melodic sequence of tones. Thanks to his training and cultivation, the composer immediately recognizes in this motif its potential for artistic development as the principal theme of the composition. In the second stage, the composer develops it, works out its combinations and variations. The composition thus has its origin in the composer's imagination, not in his feelings; it is destined not for the feelings but for the imagination of the listener.

The composer cannot will the motif into being, because its occurrence is spontaneous and inexplicable. Hanslick's argument for the autonomy of music, for music's essential purposelessness, depends on this spontaneity. The motif is the ultimate and irreducible nucleus determining every aspect of the completed composition.

At the second stage, however, the composer's task is voluntary, the deliberate exercise of artistic skills. Composing, at this stage, is a constructive activity, like building and sculpturing. It is a modeling in tonal relationships.

The Aesthetic Perception of Music. According to Hanslick, genuine music listening is itself an art, but most people are able to enjoy music only as pleasing auditory sensations, or as stimulus to the feelings, or as both. All musical perception depends in some degree on sensation *(Empfindung)* and feeling *(Gefühl)*, both of which are passive and require no special cultivation on the part of the hearer, but genuine music listening is an activity of the musically cultivated intellect. The physical and emotional responses Hanslick calls "elemental"; the intellectual response he calls "aesthetic." He describes the aesthetic response as follows:

> The most significant factor in the mental process which accompanies the comprehending of a musical artwork and makes it enjoyable . . . is the mental satisfaction which the listener finds in continuously following and anticipating the composer's designs, here to be confirmed in his expectations, there to be agreeably

led astray. It goes without saying that this mental streaming this way and that, this continual give and take, occurs unconsciously and at the speed of lightning. Only such music as brings about and rewards this mental pursuing . . . will provide fully artistic satisfaction. Without mental activity, there can be no aesthetical pleasure whatever.

Form and Content in Music. "The content of music is tonally moving forms," wrote Hanslick in his most striking formulation of his positive thesis that the content of music is its form. "Tonally," as used in this quotation, refers to the dynamic relationships between each degree and every other degree of the major and minor scales: differences of apparent tension, of attraction and repulsion among tones and groups of tones. The tonal (diatonic) system is exclusively a relational, that is, a formal system, which is to say that "tones" (in the relevant sense) exist only as related within the system and within musical artworks composed according to the artistic laws governing the system.

From all the foregoing it should be evident that "material" in the sense of "that out of which a thing is made," and "content" in the sense of "that which a thing contains within itself," are one and the same in the case of a musical composition. This is Hanslick's view. The material (= content) of music is inherently and already formed, not an inert "stuff" not yet given form by the artist (e.g., marble, pigment).

One sense of "content" is that of "subject," that is, what an artwork is about, its topic. In this sense (to borrow examples from Hanslick), a painting might have as its content the image of a flower girl, or, more abstractly, the idea of youthful innocence. Similarly, a poem might have as its content the depiction of the character and actions of a hero, or, more abstractly, the ideas of courage and magnanimity. Of "content" in this sense (in German, *Inhalt*) music has none. But to "content" in an obscurer sense, Hanslick devotes the concluding paragraphs of his book. This is "content" in the sense of "substance" (in German, *Gehalt*, sometimes *geistige Gehalt*, signifying something like "spiritual content"). According to Hanslick, this is a sense unique to music, because it originates in the first of the two stages of composing outlined earlier, to which nothing in the other arts corresponds.

In a musical composition, "substance" in this sense is identifiable only by musically cultivated listeners capable of attending actively to music in the manner described by Hanslick in the quotation in the preceding section. It is a salient aspect of a composition's principal theme, that is, the distinctive musical motif that presents itself in the composer's imagination as the first stage of composing. It is the motif's potential for artistic development. A motif possessing a high degree of this potential possesses substance in corresponding degree. At the second stage, in which the motif is developed by the composer, the substance of the composition is manifest accordingly, because there can be

nothing in the composition (i.e., no content) except what is implicit in the motif from the beginning.

"Substance," in another of its aspects, is that in a composition by which is revealed the spiritual depth and individuality of the composer as creative spirit. It marks the essential characteristics of each composition, that is, those characteristics that, to the musically cultivated intelligence, distinguish each substantial composition from every other. As Hanslick put it: "The forms which construct themselves out of tones are not empty but filled; they are not mere contours of a vacuum but mind giving external shape to itself from within." Hanslick does not use the jargon word *formalism* in *Vom Musikalisch-Schönen,* but that is the label most frequently attached to the book by commentators, with good reason. A more thoroughgoing formalism than Hanslick's would be hard to conceive, because, in his theory, the very *material* of music, namely, the whole system of tonal relationships, is itself immanently *formed.*

Nevertheless, *Vom Musikalisch-Schönen* is not, and does not include, a philosophical exposition and defense of formalism as an aesthetic theory. Hanslick's main purpose in writing the book was to proclaim and defend the autonomy of music. For this, a formalistic theory is well suited, because formalism demotes all extramusical considerations to at best a secondary role, though not necessarily a dispensable one. Beginning with his first published article in November 1844, and for fifty years after that, Hanslick, in his journalistic writings, campaigned against the subjection of music to any kind of political censorship, or to the requirements of any other art, or to those of an art of all the arts, as was proposed by Richard Wagner.

[*See also* Formalism; Gurney; *and* Music, *historical overview article.*]

BIBLIOGRAPHY

Works by Hanslick

Vom Musikalisch-Schönen: Ein Beitrag zur Revision der Ästhetik der Tonkunst. Leipzig, 1854.
The Beautiful in Music: A Contribution to the Revisal of Musical Aesthetics. Translated by Gustav Cohen from 7th edition (1885). London and New York, 1891; reprint, New York, 1974.
On the Musically Beautiful: A Contribution towards the Revision of the Aesthetics of Music. Edited and translated by Geoffrey Payzant from 8th edition (1891). Indianapolis, 1986.
Aus Meinem Leben. 2d ed. 2 vols. Berlin, 1894. Edited in one volume by Peter Wapnewski (Kassel, 1987).

Other Sources

Kivy, Peter. "Something I've Always Wanted to Know about Hanslick." *Journal of Aesthetics and Art Criticism* 46.3 (Spring 1988): 413–417.
Kivy, Peter. "What Was Hanslick Denying?" *Journal of Musicology* 8.1 (Winter 1990): 3–18.
Payzant, Geoffrey. "Hanslick, Heinse, and the 'Moral' Response to Music." *Music Review* 49.2 (May 1988): 126–133.
Payzant, Geoffrey. "Eduard Hanslick and Bernhard Gutt." *Music Review* 50.2 (May 1989): 124–133.
Payzant, Geoffrey. "Hanslick on Music as Product of Feeling." *Journal of Musicological Research* 9.2/3 (1989): 133–145.
Strauss, Dietmar. *Eduard Hanslick: Vom Musikalisch-Schönen: Ein Beitrag zur Revision der Ästhetik in der Tonkunst,* vol. 1, *Historisch-kritische Ausgabe;* vol. 2, *Eduard Hanslick's Schrift in textkritischer Sicht.* Mainz, 1990.

GEOFFREY PAYZANT

HAPPENINGS. *See* Performance Art; *and* Situationist Aesthetics.

HARDENBURG, FRIEDRICH FREIHERR VON. *See* Novalis.

HARLEM RENAISSANCE. The Harlem or Negro Renaissance movement was a period of cultural production dating from the end of World War I through the onset of the Great Depression (recent scholarship, though, is inclined to place the terminal date at the end of the 1930s, largely because one of the works best exemplifying the spirit of this era, Zora Neale Hurston's novel *Their Eyes Were Watching God,* did not appear until 1937). In his 1940 autobiography *The Big Sea,* poet and novelist Langston Hughes uncharitably looked back on the Harlem Renaissance as a rather peculiar moment when a small cadre of black intellectuals shared a fantasy that "the race problem had at last been solved through Art." Chiding his fellow intellectuals for believing that poets, singers, dancers, band leaders, and philosophers would lead the "New Negro" into "green pastures of tolerance," Hughes concluded both that this cultural movement had failed and that its failure could be attributed largely to the naive delusions of its chief proponents.

Whether or not one agrees with Hughes that the Harlem Renaissance failed, it is undeniable that many of the period's major figures—among them, Alain Locke, Jessie Fauset, W. E. B. Du Bois, Charles S. Johnson, and Claude McKay—did conceive of their roles in terms that explicitly or implicitly addressed the relation of cultural production to political practice. In fact, among the reasons that the Harlem Renaissance continues to command our attention is the way that it foregrounds the claim that artists and writers, precisely by functioning as artists and writers rather than as propagandists for political programs, could constitute the advance guard of a progressive social movement.

The Renaissance followed hard upon and further stimulated the first great migration of black Americans from the South to Northern urban centers. Motivated by a straitened Southern economy, a rising wave of Southern oppression, and the lure of more freedom and higher-paying jobs in the North, thousands upon thousands of Southern blacks made

their way North to cities like New York and Chicago during and after the war years. Several factors gave this demographic shift its aesthetic character. First and foremost was the way that the music produced by black Americans, particularly blues and jazz, came to express the sensibilities of the modern era. The words and music of early blues performances spoke to the angst and alienation that often attended urbanization while providing secular, often sexual or humorous, forms of solace and transcendence. In its turn, jazz, a dance music that thrived on improvisation and individual virtuosity, seemed the perfect form to express the sense of social and personal freedom that modern cities appeared to be making available in an unprecedented way. For black writers and artists, this period was the Harlem Renaissance; for a variety of Americans—black, white, straight, and gay—the decade of the 1920s was the Jazz Age.

Although black secular and popular music was not uniformly celebrated and embraced by black elites, cultural success of any sort had long been seized upon by black intellectuals as an effective weapon in their collective efforts to repudiate the racist claim that blacks had contributed little of value to world culture. The charge of black cultural inferiority, challenged as early as 1892 by Anna Julia Cooper in a work titled *A Voice from the South*, received a resounding refutation in Du Bois's 1903 *The Souls of Black Folk*, which included a brief on behalf of the artistry of black spirituals as already constituting a significant African-American contribution to the world. Du Bois's arguments, along with Franz Boas's anthropological studies demonstrating that human differences resulted from environmental rather than biological causes (Du Bois invited Boas to speak at Atlanta University in 1906), helped ensure that discussions of black progress would center on the question of culture.

The attempt to make political and social progress a function of aesthetic achievement received further elaboration in James Weldon Johnson's novel *The Autobiography of an Ex-Coloured Man* (1912), which celebrated not only the spirituals, but also ragtime music, folktales, and the popular dance called the "cakewalk." Widely regarded as a harbinger of the literature of the 1920s, Johnson's novel featured a light-skinned hero who considered himself a failure for abandoning his mission to transmute the raw material of black spirituals into high symphonic art, and opting instead for a life as a white businessman.

What the careers of Cooper, Du Bois, and Johnson also suggest is that an additional factor shaping the aesthetic response to the northward migration was the small but significant wedge that the higher education of blacks in the South and the North following Reconstruction had made in the ranks of the nation's literary and intellectual elites. Although the efforts of Booker T. Washington, the most influential black spokesperson of his day, to champion vocational rather than liberal education had enjoyed considerable political and financial success, the biographies of those who played key roles in framing the national debate about race appeared to give the lie to Washington's insistence that progress would proceed from an incremental mastery of agricultural, artisanal, and industrial skills. The key players in the Renaissance were not only liberally educated, but were also often conversant with the vanguard of thought in their various fields of endeavor. Charles S. Johnson, credited in David Levering Lewis's study of the period, *When Harlem Was in Vogue* (1981), with orchestrating the Renaissance, was an early proponent of and contributor to the University of Chicago's school of sociological thought under the leadership of Robert Park. (Park, significantly, had worked with Washington at Tuskegee Institute in Alabama.) Jessie Fauset earned her degree at Cornell University; Du Bois had been educated at Fisk, Harvard, and the University of Berlin; Alain Locke, who in 1925 edited the landmark collection *The New Negro*, which sought to define and describe the new movement, had been trained at Harvard, Oxford, and the University of Berlin. Jamaican-born poet and novelist Claude McKay, whose educational itinerary was less conventional (he was first apprenticed as a wheelwright and cabinetmaker, and later spent a brief, unfruitful time at Tuskegee), was, by the early 1920s, circulating in the upper echelons of Marxist political circles and addressed the Fourth Congress of the Communist International in 1923.

As the northward stream of blacks increased, these intellectuals and writers were, as Nathan Huggins, author of *Harlem Renaissance* (1971) has observed, strategically placed to be regarded as knowledgeable interpreters of the needs and desires of black people generally. Dramatizing the need for understanding and interpretation across racial lines was the racial violence marring the landscape of the United States as the war neared its end and the decade of the 1920s dawned. In 1917, East Saint Louis, Illinois, experienced the nation's worst race riot, a slaughter that left almost two hundred blacks dead and more than six thousand homeless. Two years later, during what is known as the Red Summer of 1919, race riots erupted in Longview, Texas, Washington, D.C., and Chicago. The human cost of racism could not be swept from sight, and the possibility that such violence might recur, perhaps on an even larger scale, provided the impetus for Alain Locke's warning in *The New Negro* that "the only safeguard for mass relations in the future must be provided in the carefully maintained contacts of the enlightened minorities of both race groups" (1992). Art and sociological knowledge administered by educated elites would work hand in hand to create a new era of peaceful and just race relations in the nation.

Rank-and-file black support for this leadership cadre was anything but assured, however, as indicated by the success of Jamaica-born Marcus Garvey's Universal Negro Improvement Association (UNIA), founded in 1914. Over the objections and criticisms of many of Harlem's most noted

intellectuals—particularly Du Bois—Garvey fashioned what is still the nation's largest mass-based black movement by yoking black nationalism, racial uplift, and black entrepreneurialism (the latter crystallized in his founding of his ill-fated shipping venture, the Black Star Line) into a symbol of racial pride and insurgency. Garvey's reign was brief: the shipping line, founded in 1919, was bankrupt by 1924, by which time the UNIA had fallen into disrepute because of internal mismanagement, betrayal by the Liberian government, and an ill-calculated meeting with the Ku Klux Klan. In his heyday, though, Garvey had discomfited many of the black elites who opposed him by exposing the thinness of their grassroots support. On the other hand, his movement provided apparent verification of the claims made by Du Bois, Locke, and others that blacks in Harlem could and should act on a worldwide scale.

Indeed, Locke's *New Negro* had begun as "Harlem: Mecca of the New Negro," a special issue of *Survey Graphic* magazine. *The New Negro* brought together essays, poems, short stories, and drawings by many of the writers and artists central to the Renaissance, including selections from Jean Toomer's *Cane* (1923)—a pastiche of poems, literary sketches, and a short play that Toomer described as a "swan song" to black Southern peasant culture. Also noteworthy in this volume was Countee Cullen's lyric poem "Heritage," which posed the haunting question, "What is Africa to me?" Taken together, the offerings by the thirty-four black and four white contributors to *The New Negro*, while not speaking in a single voice or from a single point of view, did highlight some of the significant questions with which Harlem writers would wrestle throughout the decade: What constituted the cultural legacy of Africa? (Locke viewed this legacy as consisting not in the emotionalism associated with the South but with the hard classicism of African masks that had inspired the work of Pablo Picasso.) Were black Americans significantly different from their white counterparts? (Melville Herskovits, who some fifteen years later would focus on the role of African retentions in shaping black American culture and behavior, maintained in *The New Negro* that on cultural matters black and white Americans were, by and large, the same.) Did the Renaissance portend a sea change in the status of black Americans generally or was this merely a false dawn? (James Weldon Johnson held that the vibrancy of Harlem rested on sound economic footing.)

The genesis of Locke's volume in a magazine not specifically identified with black issues (*Survey Graphic* covered sociology and social work), and the presence in it of white contributors, illustrate that the black intellectual and literary elite of the Harlem era could not have attained on its own the prominence that it did enjoy. The circulation figures for the major black-edited journals that published black literature—the National Association for the Advancement of Colored People's *Crisis,* the National Urban League's *Opportunity,* or Asa Philip Randolph and Chandler Owen's *The Messenger*—never topped 120,000 combined (the *Crisis* owned the lion's share of this number with a peak circulation of 95,000 during the troubled year of 1919). To be sure, through the means of well-publicized literary contests and through their effects on white readers, these journals exerted an influence that extended beyond their number of subscribers. Nonetheless, several other progressive and literary magazines—the *Nation,* the *Liberator,* the *New Masses,* and H. L. Mencken's *American Mercury*—supported and published writings by black authors.

Even here, though, it helps to step back and see that broad intellectual currents traversing the discourses of race contributed to the strategic importance of black writers and thinkers nationally during the 1920s. The modernist aesthetic that brought prestige to such figures as T. S. Eliot, Ezra Pound, and James Joyce played a crucial role in shaping assumptions about the epoch-making power of aesthetic endeavors. In the words of David Harvey, "Artists, writers, architects, composers, poets, thinkers, and philosophers had a very special position within this new conception of the modernist project. . . . the modern artist had a creative role to play in defining the essence of humanity" (1989). That black writers would be likewise expected to play a central role in defining the black race should come, then, as no surprise. To be sure, the precise effect of modernism on the Harlem Renaissance is still the subject of intense disagreement. Houston A. Baker, Jr.'s *Modernism and the Harlem Renaissance* (1987) holds that the tenets of European and American modernism played only a small role in black cultural production of the period, whereas George Hutchinson's *The Harlem Renaissance in Black and White* (1995) asserts that the American, anti-Eliot grain of modernism—fueled by William James's and John Dewey's pragmatist philosophy—constituted the most vital source of Harlem Renaissance thinking. In any case, the upshot of this discussion remains the same: the intellectual trends of the moment granted cultural producers an almost preternatural ability to determine the shape of modern society.

Precisely what the shape of this society was to be was up for debate. On racial issues in particular, critics and artists argued vociferously over whether a black or African difference would define black cultural contributions or whether black artists and writers would produce work exhibiting no discernible difference from the offerings of their white counterparts. Jessie Fauset, literary editor of the *Crisis* from 1919 to 1926, inclined to the latter view in both her novels and her critical writings. Never tendentious, however, she generously supported writers of different temperaments. Langston Hughes, whose aesthetic assumptions diverged significantly from Fauset's, nonetheless was one of the writers whom the *Crisis* cultivated. His signature poem, "The Negro Speaks of Rivers," appeared in the paper in 1919. Not all participants in this cultural debate were as generous as Fauset, however. In 1926, critic and novelist

George S. Schuyler published an essay in the *Nation* titled "The Negro-Art Hokum" in which he proclaimed provocatively that "the Aframerican is merely a lampblacked Anglo-Saxon." Accordingly, Schuyler continued, how "can the black American be expected to produce art and literature dissimilar to that of the white American?"

Aware that Schuyler's views were controversial, the *Nation* asked Langston Hughes to rebut; Hughes complied with "The Negro Artist and the Racial Mountain" (1926), an aesthetic manifesto that upbraided black artists for adopting tepid, white bourgeois standards, challenging them to embrace the cultural styles associated with the black peasant and urban working classes. Hughes intoned, "Let the blare of Negro jazz bands and the bellowing voice of Bessie Smith singing Blues penetrate the closed ears of the colored near-intellectuals until they listen and perhaps understand" (1926). Not merely issuing an aesthetic challenge, Hughes was also describing his own attempts in his poetry to explore and apply jazz and blues idioms.

The volatility of this debate was exacerbated by racial, class, and generational tensions. Despite attempts by editors such as Fauset to remain open to aesthetic differences, many of the younger black writers chafed under what they felt were the editorial restrictions of the established journals. Accordingly, in 1926, with Wallace Thurman as editor, a group of these writers and artists, among them Hughes, Gwendolyn Bennett, Aaron Douglas, and Hurston, brought out the short-lived *FIRE!! A Quarterly Devoted to the Younger Negro Artists*. The journal, which did not last beyond its first issue, is less noteworthy for any of the individual pieces collected in it than for the way it symbolized an attempted break from the critical expectations of writers such as Fauset. Thurman concluded the issue with an essay defending Carl Van Vechten, a white novelist, whose *Nigger Heaven* (1926) aroused controversy for having sensationalized blacks as primitive exotics. Thurman's defense of Van Vechten explicitly contrasted the white novelist's characters to those of Fauset.

Importantly, several novels of the late 1920s and 1930s effectively exploited this tension between black bourgeois striving for respectability and the desire to locate a rich and powerful difference in black identity. Nella Larsen's *Quicksand* (1928) followed its heroine, Helga Crane, from her role as a teacher in a black, Southern industrial college to Chicago, Harlem, and Denmark, before leaving her mired in a Southern landscape of failed uplift, unable to reconcile herself successfully to living in any black or white society. Claude McKay, whose *Home to Harlem* (1928) was published to critical censure for reveling is the seamier sides of black urban life, published another novel, *Banana Bottom* (1933), that chronicles a successful "return" home by its heroine, Bita Plant, against a narrative backdrop fully cognizant of Jamaica's colonial history and economy, and the perils of decolonization. But perhaps the most well known

of those novels addressing simultaneously the pull of the folk and the desires of the upwardly mobile was Hurston's *Their Eyes Were Watching God* (1937). Hurston, trained as an anthropologist by Boas—she published two books of folklore in the 1930s, *Mules and Men* (1935) and *Tell My Horse* (1938)—brought to her depictions of black Southern life an eye trained for cultural observation. *Their Eyes Were Watching God* rendered its heroine's journey to self-fulfillment through a narrative voice that exploited the verbal resources of folk expression.

As is the case with any aesthetic movement, assessing the influence or impact of the Harlem Renaissance is necessarily imprecise. Certainly, some of the preoccupations associated with that period—valorizing folk or vernacular expressions, asserting and celebrating ties to the African continent, insisting on the progressive dimension of aesthetic expression—remain in play in discussions of African-American literature. Even these themes, however, were contested and fell short of consensus during the 1920s, and all of them figure into black intellectual discourse in the decades before the Harlem Renaissance. Nonetheless, some straightforward indications of influence are undeniable. Many contemporary African-American women writers credit Hurston as a source of inspiration and as a model for novelistic craft. Earlier in the twentieth century the careers of African-Caribbean poets associated with the negritude movement were shaped by the poetry of Langston Hughes. With Hughes, one can cast the net even wider: Iranian poet Ahmad Shamlu credits his reading of Langston Hughes, among others, as opening his eyes to resources of the Persian poetic tradition. What seems consistent here is not so much a set of aesthetic protocols and themes deriving from that moment "when Harlem was in vogue" but an attitude suggesting the possibility of aesthetic rebirth when writers turn from established canons of taste in order to embrace sensibilities that seem to lie much closer to "home."

[*See also* Black Aesthetic; *and* Locke.]

BIBLIOGRAPHY

Baker, Houston A., Jr. *Modernism and the Harlem Renaissance*. Chicago, 1987.

Cooper, Anna Julia. *A Voice from the South* (1892). New York, 1988.

Cruse, Harold. *The Crisis of the Negro Intellectual*. New York, 1967.

Harvey, David. *The Condition of Postmodernity: An Enquiry into the Origins of Cultural Change*. Oxford and Cambridge, Mass., 1989.

Huggins, Nathan Irvin. *Harlem Renaissance*. New York and Oxford, 1971.

Hughes, Langston. "The Negro Artist and the Racial Mountain." *Nation* 122 (1926): 692–694.

Hughes, Langston. *The Big Sea: An Autobiography* (1940). New York, 1963.

Hurston, Zora Neale. *Mules and Men* (1935). New York, 1990.

Hurston, Zora Neale. *Their Eyes Were Watching God* (1937). New York, 1990.

Hurston, Zora Neale. *Tell My Horse: Voodoo and Life in Haiti and Jamaica* (1938). New York, 1990.

Hutchinson, George. *The Harlem Renaissance in Black and White*. Cambridge, Mass., 1995.

Johnson, James Weldon. *The Autobiography of an Ex-Coloured Man* (1912). New York, 1989.

Larsen, Nella. *Quicksand* (1928). Westport, Conn., 1969.

Lewis, David Levering. *When Harlem Was in Vogue*. New York, 1981.

Locke, Alain, ed. *The New Negro: Voices of the Harlem Renaissance* (1925). New York, 1992.

McKay, Claude. *Home to Harlem* (1928). Boston, 1987.

McKay, Claude. *Banana Bottom* (1933). New York, 1974.

Schuyler, George S. "The Negro-Art Hokum." *Nation* 122 (1926): 662–663.

Thurman, Wallace, ed. *FIRE!! A Quarterly Devoted to the Younger Negro Artists* (1926). Westport, Conn., 1970.

Van Vechten, Carl. *Nigger Heaven*. New York, 1926.

Wall, Cheryl A. *Women of the Harlem Renaissance*. Bloomington, Ind., 1995.

KENNETH W. WARREN

HAUTE COUTURE. *See* Fashion.

HEGEL, GEORG WILHELM FRIEDRICH. [*To clarify the significance of Hegel's philosophy for both aesthetics and art history, this entry comprises three essays:*

Survey of Thought
Hegel on the Historicity of Art
Hegel's Conception of the End of Art

The first essay is a survey of Hegel's philosophy in general, which provides the theoretical context of his aesthetic theory. The other essays concern two topics that link aesthetics (or, more accurately, the philosophy of art) and art history, a linkage that was internal to Hegel's own thought: art's "historicity" (that is, the idea that art is essentially historical) and "the end of art," that is, the argument that art has achieved its historical "end" in the sense of "full self-realization" or the complete historical embodiment of its concept. For related discussion, see Essentialism; Kant; *and* Marxism.]

Survey of Thought

Hegel (1770–1831), German Idealist philosopher. Not only have most of the major philosophical positions of the last 150 years—including Marxism, Søren Kierkegaard's existentialism, John Dewey's pragmatism, Bertrand Russell's logical atomism, Hans-Georg Gadamer's hermeneutics, and Derridean deconstruction—been developed in response to Hegel's thought (or in response to what was perceived to be "Hegelian" thought), but many of Hegel's own claims (for example, that human civilization develops historically, that philosophy is its own time comprehended in thought, and that the modern age is defined by the demand for freedom) have become a familiar part of the common stock of modern ideas. Hegel's thoughts on art and beauty are no exception in this regard. Many of the most important aesthetic theorists of the twentieth century—including Benedetto Croce, György Lukács, Ernst Bloch, and Theodor Adorno—have been profoundly influenced by Hegel's aesthetics, and the art historian Ernst Gombrich has even gone so far as to call Hegel the "father" of art history.

Hegel was born in Stuttgart into the family of a civil servant in the employ of the Duke of Württemberg in southern Germany. From 1788 to 1793 he studied theology and philosophy at the theological seminary or *Stift* in Tübingen, where he became close friends (indeed, shared a room) with Friedrich Hölderlin (1770–1843), who would become one of the greatest of German poets, and with Friedrich Wilhelm von Schelling (1775–1854), who would become the most important philosopher of nature of his generation. After working as a house tutor for three years in Bern, Switzerland, and for four years in Frankfurt am Main (where he was again in close contact with Hölderlin), Hegel became a *Privatdozent* or unsalaried lecturer at the University of Jena in 1801. In the same year, he published his first book, *Difference between the Systems of Fichte and Schelling*, and in 1802 he founded the *Critical Journal of Philosophy* with Schelling, who had been called to a professorship of philosophy in Jena in 1798. In 1807, shortly before he left Jena for Bamberg to edit the *Bamberger Zeitung*, Hegel's illegitimate son, Ludwig Fischer, was born to the wife of his landlord, and in the same year his monumental *Phenomenology of Spirit* was published, a work that sets out his understanding of the development of human consciousness and that contains extended and justly famous discussions of Greek tragedy (especially Sophocles' *Antigone*) and the Greek "religion of art."

In 1808, Hegel took over as rector of the Aegydiengymnasium in Nuremberg, where he stayed until 1816. The years in Nuremberg saw his marriage in 1811 to Marie von Tucher (b. 1791), the birth of his two sons, Karl (1813) and Immanuel (1814), and the publication of his *Science of Logic* (1812–1816). In 1816, Hegel was called to a full professorship at the University of Heidelberg, where he published the first edition of his *Encyclopedia of the Philosophical Sciences* (1817) and first gave lectures on aesthetics. In 1818, he accepted an offer of the chair of philosophy at the University of Berlin, where he taught until he died in 1831 of cholera. During his time in Berlin he published the *Elements of the Philosophy of Right* (1820), the second and third editions of the *Encyclopedia* (1827 and 1830), and several articles (including one on Karl W. F. Solger and one on Johann Georg Hamann) in the *Yearbooks for Scientific Criticism*. By the time he died, Hegel had become, in Johann Wolfgang von Goethe's view, "the most famous modern philosopher." He had been appointed rector of the University of Berlin in 1829, had been awarded the Red Eagle (Third Class) by King Friedrich Wilhelm III in 1831, and had become well acquainted with many of the literary and philosophical luminaries of the time, including Goethe, August

Wilhelm von Schlegel, and Alexander and Wilhelm von Humboldt.

In the years after Hegel's death, criticism of his philosophy proliferated and intensified, and after the 1830s Hegelianism never again recovered its position as the leading philosophy in Prussia or Germany. The profound influence of Hegel's thought, however,—both directly, through his own texts and edited lectures, and indirectly, through the writings of such figures as Karl Marx and Kierkegaard—continued well into the twentieth century. Indeed, there are signs—for example, in the emergence of neo-Hegelian social and political theory, in the growing interest in post-Kantian philosophy, and in Jacques Derrida's claim that his own thought occupies a position of "almost absolute proximity to Hegel"—that Hegel's thought is once more moving toward the center of philosophical concern.

Hegel wrote no book titled *Lectures on Aesthetics*. He did, however, hold lectures on art and beauty at the University of Heidelberg (in 1818) and four times at the University of Berlin (in the winter semester of 1820–1821, the summer semesters of 1823 and 1826, and the winter semester of 1828–1829). The text now known as *Lectures on Aesthetics* (or *Lectures on Fine Art*) was compiled by Hegel's student, H. G. Hotho, from Hegel's manuscripts and from transcripts of the 1823, 1826, and 1828–1829 lectures made by several of Hegel's students, including Hotho himself, K. G. J. von Griesheim, M. Wolf, and D. Stieglitz. The first edition of this text appeared in 1835, the second in 1842.

Even though Hegel did not produce a completed text on aesthetics to compare with the *Science of Logic,* this does not mean that he relegated art and beauty to the periphery of his philosophical system. Indeed, Hegel's aesthetic theory lies at the heart of his system, and can only be properly understood from the perspective of that system (though it should be noted that the lectures on aesthetics are much more accessible than other texts or lectures of Hegel and actually serve well as an introduction to his system).

In his *Science of Logic,* Hegel sets out his understanding of the fundamental character of being. At the end of that text, it becomes clear that being is not to be thought of as simple existence, or blind necessity, but as self-determining, developing reason, or what Hegel calls "Idea." Such self-determining, developing reason is, for Hegel, the ultimate—"divine"—source of all significant change and progress in the world. Such absolute, divine reason is not to be understood, however, as transcending creation and governing it, as it were, from the outside; rather, it is to be thought of (in the manner of Spinozist substance) as constituting nature itself—specifically, nature developing into more and more complex forms of organized matter (including life). The culmination of this development, Hegel claims, is the emergence of human consciousness and freedom. Hegel thus understands human consciousness and freedom not just to be

an accidental product of blind natural forces, but to be the result of the process whereby reason-as-nature develops into free, conscious existence and so becomes spirit or *Geist.*

Consciousness itself is understood by Hegel to be developing throughout history toward full *self*-consciousness and *self*-understanding. True self-understanding is thus not something that can be acquired through individual self-scrutiny at any time in history, but is something that has to be acquired by humanity as a whole through a long process of historical development. Hegel thinks that there are three main forms in which human beings come to understand themselves and their essential nature, and that each of these forms of self-understanding undergoes historical development.

The first of these forms is *philosophy* in which, Hegel thinks, we come to form a clear conception of our own nature as freely self-determining, conscious beings and of the process whereby being itself, or absolute reason, leads to the emergence of such freely self-determining, conscious beings. The second of these forms is *religion,* in which that process is represented or pictured in faith and worship as (for example, in the case of Judaism) the process whereby God "creates" human beings in his own image, or (in the case of Christianity) the process whereby God becomes incarnate as Jesus Christ and comes to be Holy Spirit in the community of believers constituting the church. The third form of human self-understanding Hegel identifies is *art,* in which the nature of human freedom and the indebtedness of human beings to God (or absolute reason) is brought before our minds by means of sensuous, external material, such as stone, paint, or sound. In art, therefore, natural objects, sounds, and colors are transformed by the creative talent of human beings into the expression of human character and freedom and of the relation between humanity and God (or the gods).

A number of things follow from this that make Hegel's account of art distinctive. Most important, perhaps, it is clear that Hegel does not regard art as a purely human phenomenon. Human beings give expression to the way they understand themselves and their world by painting pictures, writing poems, or composing music, but in so doing they fulfill the rational drive within being itself toward self-consciousness. In human self-consciousness, including the self-consciousness that human beings achieve through art, being itself becomes self-conscious. There is thus for Hegel, as there is for the Nietzsche of *The Birth of Tragedy,* a metaphysical drive within the very fabric of the world to produce art (though Nietzsche will claim that through art the world seeks to draw an Apollonian veil over the terrors of existence and dissolve individuality in Dionysian self-forgetfulness, rather than come to self-consciousness). [*See* Nietzsche.]

It is also clear from what has been said that Hegel does not see it as the primary purpose of art to give pleasure, but

to enhance our understanding of ourselves and our world. Hegel certainly acknowledges that we take sensuous pleasure in the colors in a painting or in the sounds in a symphony, and he agrees with Kant that we take pleasure in the free play of the imagination and the understanding that the beautiful form of an object affords. [*See* Kant.] But, like Aristotle and unlike Kant, he locates the main value of art in its disclosure of *truth*. Hegel's aesthetic theory thus concentrates less on the effect that a work of art might have on a subject (on the emotions it might arouse or the delight it might cause) and more on the way in which art itself communicates and articulates truth. [*See* Truth.]

The truth that art has to communicate, according to Hegel, is that human beings are fundamentally free, self-determining beings, that this freedom is realized and furthered by certain institutions such as the family and the state, and that such freedom is ultimately the result of the work of absolute reason (or the Idea) in and as nature and history. In contrast to many other aesthetic theorists, therefore, Hegel believes that there is a distinctive *content* that art has to present in order to count as genuine art, and that artistic talent or genius does not just consist in the formal ability to represent or imitate accurately and skillfully whatever the artist happens to encounter in the world (such as, in Aristotle's famous example, corpses). Hegel recognizes that the ability to represent things accurately in stone, paint, or words is an important skill for an artist to have, but he does not think that by itself such skill makes one into a true artist. For that, one must have a profound appreciation for human freedom and a perceptive understanding of its historical and ontological (or "divine") ground. This is why Hegel thinks that the greatest works of art, such as Aeschylus's *Oresteia* or Sophocles' *Antigone*, portray the conflict between noble, heroic individuals who freely promote the values of the state or the family, and the interaction between such individuals and the gods who stand behind those institutions.

Of course, not all genuine works of art depict the divine, either in the form of the Greek gods or of the Christian God incarnated in Jesus Christ. Hegel also sees it as the role of art to present human freedom by itself in all of its manifold forms, and he finds some of the most satisfying examples of such art in the supremely secular works of seventeenth-century Dutch painters such as Adriaen van Ostade and Jan Steen, and in Bartolomé Esteban Murillo's paintings of beggar boys (in the Alte Pinakothek in Munich), which, he says, give expression to an exuberant feeling of inner freedom, well-being, and delight in life even in the midst of poverty. Nevertheless, much of what Hegel regards as genuine art (rather than mere decoration or formal artistry) does involve depictions or descriptions of the divine, and this clearly indicates the close proximity in Hegel's mind between art and religion.

Whether the subject matter of art is divine freedom (or love) or secular human freedom, art is distinguished from philosophy and religion by the fact that it presents such freedom in the form of sensuous, material *externality*—for example, in wood, stone, paint, or sound—rather than in abstract concepts or in inner faith and feeling. The distinctive role of art is thus not to use sensuous material to point to or "symbolize" in an allegorical fashion an idea that is articulable in abstraction from its external, sensuous expression, but to express its content *in* the line of the stone or the arrangement of colors itself and so *in* the very physical forms that such colors and sculpted stone present. The distinctive character of art as art thus yields a criterion for judging what is to count as the best art: art achieves its perfection when the content expressed is indeed harmoniously fused with the material (and mode) of expression. This perfect harmony of human or divine freedom with the sensuous material of expression is what Hegel understands by beauty. Beauty is thus, as Hegel puts it, the *"sensuous* shining of *the idea" (das sinnliche* Scheinen *der Idee).* [*See* Beauty.]

Beauty does not, however, just lie in the formal harmony of a "spiritual" content and the external, sensuous material of expression, but consists in the sensuous expression of a content that is *in itself* in harmony with sensuous, material externality, in particular with the physical body. This is because only a form of freedom and spirit that is in itself in harmony with the realm of the sensuous can be given direct sensuous expression. For this reason, Hegel thinks that the forms of freedom that are most suited to aesthetic, beautiful expression are those found in classical Greece (in which, according to Hegel, freedom is explicitly conceived as the perfect harmony of the spirit and the body) and in Christianity (in which, according to Hegel, the inward freedom of Christian love is understood to be incarnated in, and to transform from within, the physical body). By contrast, the spiritual freedom that Hegel associates with Islam and that he takes to abstract from all bodily incarnation is judged by him to be incapable of genuinely beautiful expression.

Hegel by no means denies the importance of dissonance, negation, and pain in art; however, he thinks that the dominant tone of a work of art should always be that of harmony and reconciliation. Where there is pain, therefore, there should always be "smiling through tears." This "classical" preference for harmony and beauty in art, which puts Hegel at odds with those twentieth-century aestheticians who wish to "emancipate" art from beauty and harmony, is clearly influenced by the classicism of Johann Joachim Winckelmann (1717–1768). But it also derives directly from Winckelmann's systematic conception of art as the external, sensuous expression of human freedom, and so as the harmonious fusion of sensuous, material form and of a "spiritual" content that is itself in harmony with sensuousness. (Besides Winckelmann, other important influences on Hegel's

aesthetic theory are Johann Gottfried von Herder, Gotthold Ephraim, Lessing, Kant, and Friedrich von Schiller.)

Hegel regards art as a mode of human self-understanding, and he believes that, like the human self-understanding expressed in philosophy and religion, art undergoes *development* in history. The earliest art, Hegel argues, gives expression to a spirituality that is still seeking to free itself from nature and sensuous form and that is thus not yet reconciled with natural sensuous externality. Consequently, such art either has to distort natural forms in order to intimate through such forms the presence of nonnatural powers (as, for example, in Hindu depictions of the god Śiva with multiple arms), or it has to employ images that seem to point to spiritual meanings that remain hidden and mysterious (as Hegel believes is the case in Egyptian art). In either case, such art does not directly express its spiritual content *in* its sensuous material and form, but rather intimates or points to its content *through* its sensuous material and form, and so is understood as *symbolic* art.

The art of the Greek period gives expression to a spirituality that is itself in perfect harmony with the physical and the sensuous, and so achieves what Hegel regards as the height of classical beauty, above all in fifth- and fourth-century sculpture and in fifth-century tragic drama (in which Hegel believes the great tragic characters have a "sculptural" quality). Because Greek spirituality is so perfectly suited to aesthetic expression, there is, according to Hegel, the closest connection between religion and art among the Greeks; indeed, it is in their art that Hegel thinks the Greeks come to know and understand their gods (and themselves) most fully.

The spirituality of the Christian era is marked, according to Hegel, by a profound inwardness that finds its fullest expression in religious feeling and love and, ultimately, in philosophy. Such a spirituality withdraws out of the realm of the natural and the sensuous into the realm of inward faith and devotion, and so, unlike the spirituality of the Greeks, does not find its most perfect expression in the sensuous externality of art. Yet, Christian inwardness and love are also understood to be *incarnate* inwardness and love and, to that extent, such inwardness can be represented sensuously in art, as, for example, in Raphael's many paintings of the Madonna and child. (The aesthetic expression of incarnate inwardness finds its secular counterpart, for Hegel, in portrait painting, particularly after the Reformation.)

In the Christian era, art becomes the sensuous expression of inner freedom and love—just as the face gives concrete, visual expression to the character and spirit "behind" it—and is called *romantic* art. Yet, art also ceases to be the highest expression of truth, as Hegel understood it to be for the Greeks. This does not mean that Hegel thinks that art dies during the Christian era. "One can certainly hope," he says, "that art will always rise higher and perfect itself"; it is just

that, however fine we judge artistic representations of Mary and Jesus to be, "we bow our knee no longer."

Hegel's lectures on aesthetics end with his account of the five major arts: architecture, sculpture, painting, music, and linguistic art or poetry. Hegel understands architecture to be fundamentally a symbolic art, but he claims that it reaches perfection in Greek architecture and he was one of the first to acknowledge the distinctive merits of Gothic architecture. Sculpture, for Hegel, is the art most suited to the sensuous, material expression of freedom and so to classical beauty, whereas painting and music are clearly the most inward, romantic arts. Hegel regards poetry (including epic, lyric, and dramatic poetry) as the universal art that flourishes in various periods of history.

Hegel's is the last grand systematic theory of art and, in a culture that has grown suspicious of philosophical system building, it has found many detractors in the years since his death, even among those who have been most strongly influenced by it. Hegel's theory remains significant today, however, for five main reasons. First, Hegel shows that art provides a distinctive form of aesthetic freedom and experience that is not simply subservient to the pursuit of political and social goals (yet that can and does play an important role in the religious life of many cultures). Second, Hegel explores the expressive possibilities of different arts and forces us to consider whether particular arts may not be better suited to the expression of certain kinds of subject matter than others. Third, Hegel emphasizes the importance of a historical approach to art and is one of the first aesthetic theorists to take a serious interest in non-Western art. Fourth, unlike Kant (and some noted twentieth-century aesthetic theorists), Hegel always tries to breathe life into his ideas about art and beauty by showing how they are exemplified in individual artworks with which he is familiar (works by artists as varied as Sophocles, Aristophanes, Raphael, Michelangelo, John Milton, Rembrandt van Rijn, Goethe, Amadeus Mozart, and Gioacchino Rossini, though not, strangely, his contemporary, Ludwig van Beethoven). Hegel thus demonstrates that it is possible—indeed, essential—to integrate one's concrete experience of actual works of art into any systematic aesthetics. Fifth, not only is Hegel eclectic, but his theory of art offers, together with important philosophical insights, interpretations of individual artworks and artistic styles that have become some of the most famous and influential in history. Most notable, perhaps, are his interpretations of Greek tragedy, Gothic architecture, Dutch painting, and German Romantic drama and prose. Even if it were only because of his interpretation of Sophocles' *Antigone* (as presenting the clash between the ethical realms of the family and the state), Hegel's lectures on aesthetics would always remain essential reading for anyone seeking to understand the nature of art and beauty.

[*See also* Irony; *and* Romanticism, *article on* Philosophy and Literature.]

BIBLIOGRAPHY

Works by Hegel

Aesthetics: Lectures on Fine Art. 2 vols. Translated by T. M. Knox. Oxford, 1975. English translation of Hotho's text of Hegel's lectures.

Vorlesungen über die Ästhetik. In *Werke in zwanzig Bänden,* edited by Eva Moldenhauer and Karl Markus Michel, vols. 13–15. Frankfurt am Main, 1970. A modern edition of Hotho's text of Hegel's lectures.

Vorlesung über die Ästhetik: Berlin 1820–1821: Eine Nachschrift, vol. 1, Textband. Edited by Helmut Schneider. Frankfurt-am-Main, 1995. Transcript of Hegel's lectures of 1820–1821 made by Wilhelm von Ascheberg, but not used by Hotho.

Other Sources

Bungay, Stephen. *Beauty and Truth: A Study of Hegel's Aesthetics.* Oxford and New York, 1984.

Collenberg, Bernadette. "Hegels Konzeption des Kolorits in den Berliner Vorlesungen über die Philosophie der Kunst." In *Phänomen versus System: Zum Verhältnis von philosophischer Systematik und Kunsturteil in Hegels Berliner Vorlesungen über Ästhetik oder Philosophie der Kunst,* edited by Annemarie Gethmann-Siefert, pp. 91–164. Bonn, 1992.

Desmond, William. *Art and the Absolute: A Study of Hegel's Aesthetics.* Albany, N.Y., 1986.

Donougho, Martin. "Remarks on 'Humanus heisst der Heilige'." *Hegel-Studien* 17 (1982): 214–225.

Donougho, Martin. "The Woman in White: On the Reception of Hegel's Antigone." *The Owl of Minerva* (Biannual Journal of the Hegel Society of America) 21.1 (fall 1989): 65–89.

Etter, Brian. "The Sounds of the Ideal: Hegel's Aesthetic of Music." *The Owl of Minerva* (Biannual Journal of the Hegel Society of America) 26.1 (fall 1994): 47–58.

Gethmann-Siefert, Annemarie. *Die Funktion der Kunst in der Geschichte: Untersuchungen zu Hegels Ästhetik.* Bonn, 1984.

Gethmann-Siefert, Annemarie, and Otto Pöggeler, eds. *Welt und Wirkung von Hegels Ästhetik.* Bonn, 1986.

Henrich, Dieter. "The Contemporary Relevance of Hegel's Aesthetics." In *Hegel,* edited by Michael Inwood, pp. 199–207. Oxford and New York, 1985.

Houlgate, Stephen. *Hegel, Nietzsche and the Criticism of Metaphysics.* Cambridge and New York, 1986. See chap. 8, "Hegel and Nietzsche on Tragedy."

Houlgate, Stephen. *Freedom, Truth and History: An Introduction to Hegel's Philosophy.* London and New York, 1991. See chap. 4, "Art and Human Wholeness."

Kaminsky, Jack. *Hegel on Art: An Interpretation of Hegel's Aesthetics.* Albany, N.Y., 1962.

Pöggeler, Otto, ed. *Hegel in Berlin: Preussische Kulturpolitik und idealistische Ästhetik: Zum 150. Todestag des Philosophen.* Berlin, 1981.

Pöggeler, Otto, and Annemarie Gethmann-Siefert, eds. *Kunsterfahrung und Kulturpolitik im Berlin Hegels.* Bonn, 1983.

Winfield, Richard Dien. *Stylistics: Rethinking the Artforms after Hegel.* Albany, N.Y., 1996.

STEPHEN HOULGATE

Hegel on the Historicity of Art

That a particular art should come to be thought of as even having a history—a past, and perhaps a novel future—is itself a historical achievement, J. G. Droysen remarks in his *Historik.* Still more radical was the launching of art as such into the flux of history. That step was taken by the Romantic generation (Johann Gottfried von Herder, Friedrich von Schiller, August Wilhelm von Schlegel, Friedrich Schlegel, Novalis), which was Hegel's own, and which discovered—or better, invented—not just the classical tradition but also the normative traditions before and after. The preclassical Hegel borrowed from his friend Friedrich Creuzer (1771–1856) the name "symbolic," while following then current usage in calling the third worldview "romantic." This nomenclature dates from Hegel's Berlin lectures; in the 1828–1829 series, it appears in the second general division, between universal Ideal and singular arts. Yet, as Otto Pöggeler and others have shown, the triadic articulation goes back to the 1807 *Phenomenology of Spirit,* where it is framed as stages of religious consciousness and cultic activity. In the *Aesthetics,* this larger segmentation overlays the historical development discerned within each art (all save music, for some reason). Hegel suggests that the expressive capacity of each medium correlates with a particular worldview: architecture would be symbolic in its typical form and content, sculpture (and epic) would be classical, music (and lyric) romantic. Then lastly, art forms, individual arts, as well as particular artworks, are all understood by Hegel as signifying—if no more than obliquely—a truth about the culture in which they are embedded. Indeed, art is taken as offering privileged access to the meaning of a culture or society, a line of thought that may be traced in Karl Marx as well as in much subsequent history writing.

So pervasive has Hegel's influence been that it is difficult to discern how far the historicizing of art was itself a historical process. When the art historian E. H. Gombrich goes "in search of cultural history," it is in response to both orthodoxy and what is seen as the "crumbling" of its Hegelian foundations (Gombrich, 1969, p. 6). The revolt against Hegel as "the father of art history" (Gombrich, 1977) may offer no positive alternative, but it is at least clear about the father's ideological sins. These include "aesthetic transcendentalism," even if secularized as art worship; "historical collectivism," or making the work stand for the *Volk;* "historical determinism" and its twin, "metaphysical optimism," that is, construing history Whiggishly as inevitable progress; and "dialectic relativism," or taking art to express a deep spiritual value. It is easy to trace them in the outline just given: art expresses ideas, progresses necessarily in a quasi-natural evolution, and is measured holistically against competing worldviews. Yet, Gombrich lodges his complaint specifically against the movement called "historism" *(Historik),* which came to rule historiography; and in fact it proves difficult to sort out historically the roles played by Romantic aesthetics and by Hegelian philosophy. [*See* Gombrich.] Let us nevertheless try to get some perspective on events.

Around 1800, "the time was ripe" for art to become historicized, on at least two counts. First, various institutional

formations such as the museum, the concert hall, indeed, "literature" proper (now taken as something more than "literacy"), became part of the social fabric, ingredients in the so-called public sphere of civil society. Hegel's lectures respond to the thriving "art world" in Berlin. But this was not merely aesthetic appreciation or a nascent "high culture": what was on display in the art world was the artistic past. For example, a lecture (dated by the editor H. G. Hotho as 17 February 1829) takes note of Karl Friedrich Schinkel's new museum built just across from Hegel's own house, saluting its galleries for having made visible the historical evolution of painting. Hegel was present too at the first modern performance of Johann Sebastian Bach's St. Matthew Passion, invited by its conductor Felix Mendelssohn (an enthusiast of Hegel's lectures on aesthetics). Of course, the concert did not simply aim at putting the artwork back into history: it was an occasion at once religious, nationalistic, and celebrating a repeatable "classic." Yet the general point remains: in details and overall, this newly historicist art world informs Hegel's approach to his subject.

A second fundamental shift making possible the historicizing of art is more conceptual: baldly put, it is art itself. Only now might Art as such present itself as a unified practice under a systematic description, whereas previously "art" would have been used distributively, or as one of the "fine arts." Both as conceptual theory and as institutional practice, therefore, this singular Art could be taken to epitomize the creative potential of human imagination and culture. Art is henceforth systematically analyzable in a threefold way: within a new metaphysic of form—or rather of *forming*—as *formalist,* and as *self-reflexive* in character. Traits such as originality, creativity, or expressivity are simply beyond the reach of classical form or *poiēsis* (i.e., production according to given rules, the deliberate informing of given matter, via determinable means to given ends). In Romantic aesthetics, artistic activity becomes formal*ist* as well as formative: it imitates or articulates no essential content (perhaps the first to make explicit the radical autonomy of the artwork being Johann Wolfgang von Goethe's friend Karl P. Moritz). At its extreme, formalist autonomy verges on self-reflexivity. It falls within the broad shift during the late eighteenth century from assigned values and hierarchical structures to what Niklas Luhmann has called the "functional" operations of self-beholden systems. (History itself is one such system, alongside society, state, literature, and art generally.) Given the threefold semantic shift resulting from Romantic aesthetics, one may ask how it bears on Hegel's conceptualizing of art. But before that, one should ask how the Romantic transcendentalizing of art—Art as supreme value, once other norms have become relative—can be squared with its historicity; for an autonomous creativity seems on the face of it to operate outside of time altogether. How does Hegel manage to historicize a post-Romantic conception of Art?

Some Romantics tended to romanticize or monumentalize past ages, consecrating them as self-creating organic wholes: Herder, for example, but equally the youthful Hegel in his Grecomanic phase. A subtler transformation occurs when the historian frames his own belatedness as opening up a truth that is no longer ours. Thus, Johann Joachim Winckelmann advocated the superior value of imitating the Greeks where they had imitated nature. This self-reflective constitution of the past opens up three conceivable paths. One may (with Schiller or Friedrich Schlegel) trace out a narrative of declension, when irony becomes the end point, ambivalently inside and outside the historical stages it observes. Or one can propose that the historicist perspective is itself aesthetic, not just because it offers, as it were, imaginative picture galleries of the past, but also in operating under the aegis of artistic unification—the speculative stance of Friedrich Wilhelm von Schelling's "identity philosophy" around 1800, which saw the artwork as having fused together unconscious nature and cultural reflection. Hegel at first endorsed this position, though in his own manner: he brought out the communal basis of the *work* of art as well as its ethical and religious resonance (the *Natural Law* essay of 1802, the 1803 ethical system). By the time of the *Phenomenology,* however, Hegel had found still another way, one that (with modifications) he kept to for the rest of his life. It attempts to explicate history—including the history of art—philosophically and to situate philosophy historically, without, however, reducing one to the other (Gethmann-Siefert, 1984, p. 188). This attempt acknowledges the claims of historical ironism, on the one hand (the view that the historian is himself part of the process viewed), and, on the other, the demands of speculative system, discerning reason at work in cultural phenomena. From this complex position there follows a series of paradoxes. Paradox dissolves the conventional picture of Hegel's *Aesthetics* as beholden to a classical Ideal, or as forcibly "sublating" artistic intuition *(Anschauung)* into the eternal verities of an overriding Reason or some insistently dialectical machine, and so on.

First paradox: *essential pastness.* By that is not meant the notorious verdict of the end of art (talk about which seems never ending). Art is necessarily past, because subsisting only in retrospect. Hegel suggests (in a passage from the *Phenomenology* emblematic for the hermeneutics of Hans-Georg Gadamer or Paul Ricoeur) that the works of the muses

> are now what they are for us—beautiful fruit picked from the tree. A friendly fate presents them to us, as a maiden might offer that fruit. It does not give the actual life of their existence—the tree that bore them fate does not give us their world along with the works of that art . . . but only the veiled recollection [*Erinnerung*] of this actuality. (1977, p. 455)

Considered as cultic practice within the life-world, art is "always already" past; to paraphrase, it comes about only at the

falling of dusk, when a shape of life has grown old. Moreover, it is the *artist* who shapes and commemorates that world of substance, in bestowing the Judas kiss of his subjectivity (ibid., p. 426). The products of this complex transaction then become subsumed as "great art"—are dusted off and preserved philologically, celebrated ideologically, or interpreted philosophically. In all this, "our act," as Hegel puts it, remains purely external. Yet, in a further twist, fate prompts the reintegration of the *disjecta membra* of culture: not via the reviving touch of aesthetic "de-differentiation" or conscious "application" to our own situation, as Gadamer proposes, but through their cultural internalizing or recollection *(Erinnerung)* in the unifying pantheon of the museum, the university, and so on.

Indeed, the retrospective birth and rebirth of artworks answers Barnett Newman's citizen's complaint that a science of aesthetics affects the artist about as much as ornithology is "for the birds" (Newman, 1990, p. 247). Hegel admits the difficulty philosophy faces in saying anything about this most empirical field of inquiry, and admits that its conceptual talk cannot be justified, only taken up *"lemmatically,* so to say" (*Aesthetics,* vol. 1, pp. 24, 42), from an encyclopedic frame of reference. It is plausible to understand such resort as analogous to the "thinking approach" *(die denkende Betrachtung)* with which the *Lectures on the Philosophy of World History* characterize "philosophical history," accommodating specialist studies (with their mix of contingent fact and search for structure), while repudiating the Romantic projections of comparative historiography. But here one glimpses another resort, predicated on the assumption that art is itself a cultural construct or interpretation. Art is already a reflection on the past, including its own; art reflexively defines *itself,* by historical self-differentiation. One cannot deny that the process is drastic; it would be of little comfort to the birds that they are mounted under glass, in a museum culture. But works of culture are different: they speak to us, according to Hegel, in our own terms.

Second paradox: *the Ideal.* Hegel presents beauty as a symbiosis of human and divine, finite and infinite, a *Gestaltung* of sacred powers operative in the saeculum of history. Human shape mirrors divine, and art (more especially sculpture) reveals the transfiguration. Yet the secular is not the divine. As Hegel analyzes the "Ideal"—which the *Aesthetics* understands as the essential content of artistic presentation—the exemplum of its peculiar theomorphism is the category of "action" *(Handlung),* whether as performed deed or as narrated plot. While action is a simultaneous structure, it occurs sequentially; it embodies norms (for the deed should carry its ethical meaning on its face) while undercutting them (for it enacts the ambiguity by which opposing laws constitute *the* Law). In his *Theory of the Novel* (1971), Györgi Lukács captures the defining irony whereby epic unity has no outside

and discerns no horizon, whereas drama operates as self-delimiting, and hence confers meaning. Put another way, Hegel's Ideal comprises a union of union with disunion, of beauty with the sublime, as Schiller conceived of them: both the "shining forth" *(Scheinen)* of truth, and what must go unsaid, namely, an ethical value beyond sensuous shape *(Schein).* (That is why Hegel's sublime appears in the *Aesthetics* shorn of its Kantian, transcendental aura, as a mere figure of speech filed away under the rubric "Conscious Symbolics.") A dialectical tension pervades all aspects of an Ideal commonly (but wrongly) taken to be Hegel's own ideal. One need only think of Hegel's theorizing of Greek tragedy as at once a drama of reconciliation of opposite social norms, and a false—indeed, misrepresented—unity: tragic semantics and pragmatics just do not match. Already in 1806, in the *Gesammelte Werke,* Hegel could write of beauty as "the veil that hides the truth rather than its presentation [*Darstellung*]" (1976, p. 279).

Hegel's Ideal owes much to Winckelmann, not merely its divine status, but more subtly too its four-step development from rigid via "high" to "beautiful" styles, then degeneration into pleasing imitation. But where Winckelmann projected a natural, organic progression, Hegel's is also one of progressive reflexivity, as if each stage considers—at least to later eyes and ears—the conceptual viability of the previous. [*See* Winckelmann.]

Third paradox: *(neo)classicism.* Hegel, so often typed as a proponent of the ancient Greeks, in fact thinks little of their cold pastoral. More beautiful one cannot become—but nostalgia has no place in philosophy, he claims. Instead, Hegel's Berlin lectures display not merely his wide reading, but also a growing interest in both "symbolic" and "romantic" cultural forms. It is as if the classical Ideal exists only as a vanishing point in the historian's view from nowhere: something that can only be surmised between a before and an after. It might be thought that the classical still rules from its grave, as when Gadamer appeals to its authority, or formalism posits its parodic repetition. But Hegel thinks differently. As a supreme fiction, the classical is flattered with an ironic perspective: its economimesis (Jacques Derrida's term, whereby art imitates nature, which imitates art, etc.) is refigured as Romantic creativity. In philosophical retrospect, "we" can see that mimetic *poiēsis* is in fact creativity in disguise. Subjectivity is staged as in fact staging the entire show; and yet—Hegel's dialectical corrective—it is not Schlegelian irony that prevails. Whether as so-called "Romantic" irony or as the "beautiful soul," subjectivism appears on the stage of history—which leads to the most challenging paradox of all.

Fourth paradox: *ideality.* In Berlin, Hegel battled on two fronts. He distanced himself from Karl W. F. Solger, his predecessor at the University of Berlin, and equally from his academic rival, Friedrich Schleiermacher. Both offered

courses on aesthetics, but (in different ways) assumed a Romantic worldview that Hegel was concerned to keep in its historical place. Yet, Hegel had also to avoid the opposite danger, the empiricism exemplified by another Berlin contemporary, Carl Friedrich von Rumohr (1785–1843), often called the first scientific art historian. Rumohr polemicized against all "spiritualizing" tendencies, in particular the Neoplatonic allegorizing that he detected in Winckelmann and his school. For Rumohr, style was a purely formal colligation of empirical traits. Hegel in turn argued against this professed materialism, and its reduction of meaning to sensuous form. Scholars have barely begun to chart the shifts and accommodations made by the widely read Hegel between the several lecture series at Berlin, obscured as they may be in Hotho's editing. Yet, Hegel is constant in championing "the Idea" and its normative power. This constitutes the supreme paradox of Hegel's aesthetic idealism: he historicizes and so ironizes the Ideal, yet, in a further ironic shift, he defends the ideality of art, at one remove, so to speak. As one of the shapes of absolute spirit, Art escapes the finite measures of state and history (*Encyclopedia of Philosophical Sciences* 1817, Section 552). Nevertheless, state and society legitimate it as historical display—in the museum or its cultural equivalents.

Turning from the speculative overview of the past to current prospects—from museum curator to artist-critic—one might well ask what normative standards remain once full self-reflexivity is achieved and there is no longer any occulted content to be divined and presented. Dieter Henrich (1979) has argued that Hegel's diagnosis of art's loss of religious content prefigures humanist tendencies in modernism. Yet, if it is to avoid parody or else some kind of conceptual art, what *shape* does its productive activity assume? Others advocate muting Hegel's systemics in favor either of the promise of a "new mythology" found in Hegel's earlier thinking (Gethmann-Siefert), or of utopian protest against a prevailing "prose of life" (Theodor W. Adorno). Another possibility would see art as a communal *epigram* (see the remarks on this transitional figure in *Aesthetics*, pp. 608–609): it puts a spiritual mark on the shared objectivity of the past. But it is an open question whether and how the post-Hegelian artist could attain a speculative *tour d'horizon* of world art analogous to the Hegelian philosopher's perspective on past philosophy.

In general, much remains to be done in thinking through, on Hegelian terms, the status of art qua institution, that is, as part of society or the public sphere while nevertheless professing autonomy. Art and the arts reflect this divided heritage still: both *in* history and yet not wholly *of* it. Like Schinkel's museum, Art for Hegel serves the state and answers social interests, while supposedly escaping into an absolute empyrean.

[*See also* Historicism, *article on* Historicism and Philosophy.]

BIBLIOGRAPHY

Works by Hegel

Aesthetics: Lectures on Fine Art. 2 vols. Translated by T. M. Knox. Oxford, 1975.
Gesammelte Werke, vol. 8. Hamburg, 1976.
Natural Law. Translated by T. M. Knox. Philadelphia, 1975.
Phenomenology of Spirit. Translated by A. V. Miller. Oxford, 1977.
Vorlesungen über die Ästhetik. In *Werke in zwanzig Bänden,* edited by Eva Moldenhauer and Karl Markus Michel, vols. 13–15. Frankfurt am Main, 1970.

Other Sources

Gombrich, E. H. *In Search of Cultural History.* Oxford, 1969.
Gombrich, E. H. "The Father of Art History" (1977). In *Tributes: Interpreters of Our Cultural Tradition,* pp. 51–69. Ithaca, N.Y., 1984.
Gethmann-Siefert, Annemarie. *Die Funktion der Kunst in der Geschichte: Untersuchungen zu Hegels Ästhetik.* Bonn, 1984.
Henrich, Dieter. "Art and Philosophy of Art Today: Reflections with Reference to Hegel." In *New Perspectives in German Literary Criticism: A Collection of Essays,* edited by Richard E. Amacher and Victor Lange, Translated by David Henry Wilson et al., pp. 107–133. Princeton, N.J., 1979.
Lukács, György. *The Theory of the Novel: A Historico-Philosophical Essay on the Forms of Great Epic Literature.* Translated by Anna Bostock. Cambridge, Mass., 1971.
Newman, Barnett. *Selected Writings and Interviews.* Edited by John P. O'Neill. New York, 1990.
Szondi, Peter. *Poetik und Geschichtsphilosophie,* vol. 1. Frankfurt am Main, 1974.

MARTIN DONOUGHO

Hegel's Conception of the End of Art

Perhaps no aspect of Hegel's aesthetic theory is as well known, yet so frequently misunderstood, as his pronouncement that art has reached its end (Hegel, 1972, vol. 13, pp. 25, 142). Two potential misconceptions concerning this claim can be put aside straightaway. First, Hegel was not the first to hold the view that art had reached its limit. Immanuel Kant also held that art had progressed as far as it was capable (*Akademieausgabe* vol. 5, p. 309), but Kant's view lacks the inherent support that Hegel's receives from his teleological metaphysics and thus seems more an opinion than a philosophical position. Second, though often assumed to be a doctrine to which Hegel adhered throughout his career, the idea that art has realized its essential purpose is actually of relatively late vintage. The conceptual framework for the assertion that art has come to an end, in the sense that Hegel means, is not firmly in place until he develops his account of Absolute Spirit, which does not occur until the *Encyclopedia of Philosophical Sciences* of 1817 and the *Lectures on Aesthetics* delivered in the 1820s.

Hegel locates the end of art in the transition from the "classical" to "romantic" epochs of art. This claim that art has reached its end flows directly from his belief that classical art, at its pinnacle with Attic sculpture, is the histori-

cal period in which art is the most adequate expression of the self-consciousness of *Geist*. Because Hegel defines art as the purely sensuous appearance of *Geist* to itself, his claim is that the Greeks most fully realized the essential purpose of art—that is, its essence as a mode of self-consciousness. The content of the claim must be considered in the light of Hegel's general interpretation of the differences between the place of art in Athenian and Christian cultures. In some ways, Hegel's understanding of the Greeks is a function of his time—the cult of things Greek was prominent in late eighteenth- and early nineteenth-century Germanophone culture and, as such, was an idealized portrayal of Greek civil and cultural institutions. But, unlike some of his contemporaries (whose views he shared in his youthful Jena period), Hegel is not uncritically nostalgic about the ancients. In fact, he held modern, Christian (that is, German Protestant) culture to be superior to ancient life. The inferiority of Greek culture consists in its alleged comparative lack of reflective self-awareness. Specifically, Hegel felt that the Greeks lacked the highly developed, subjective sense of self indicative of the modern Christian era (which he claims began with Roman Stoicism). But this deficit is also the source of Greek art's aesthetic superiority. Unlike the modern world, engaged in ever-deepening representational and conceptual awareness of the self, Greek self-consciousness is organized aesthetically. This Greek aesthetic mode of life involves not only what would be retrospectively identified as the production and appreciation of art objects, but also religious customs and political institutions. In this connection, it is often remarked that the Greek *to kalon*, usually translated simply as "beauty," has an expanded meaning for the Greeks. For the Greek, *beauty* is a term that includes within its meaning other important sorts of excellences: physical, moral, and theoretical, as well as artistic (the Latin *pulchrum* also has something of this plasticity). Simply put, Hegel thinks that aesthetic awareness afforded the most exhaustive self-expression possible for the Greeks, given their particular level of self-consciousness.

Not so in the Romantic epoch: here art is no longer the most advanced vehicle for *Geist*'s self-consciousness. Religion and philosophy are in turn ascendant—which is to say that our self-conception has progressed to the point where we recognize ourselves as possessed of a quality of subjectivity that is representationally and, finally, conceptually comprehensible. The Romantic art aesthetically reflects this greater level of self-awareness. But the heightened abstract and subjective nature of Romantic self-understanding is only imperfectly modeled sensuously. Thus, Romantic art demonstrates its own limitations in its attempts to represent the definitive features of modern self-consciousness. The three arts that Hegel takes to achieve their perfection in the Romantic stage of art—painting, music, and poetry—all attempt to embody this increased subjectivity. Hegel inter-

prets these arts as progressive stages in the attempt to present what is known to be abstract conceptual matter by reducing the "physicality" of the artistic medium: from three-dimensionality characteristic of sculpture to the two-dimensionality of painting, the entire lack of spatial dimension in music, and the attempt to marry the two in poetry. Poetry, and particularly German poetry and novel writing of Hegel's time, is deemed to be the limit of the Romantic artistic expression of *Geist*. Because Hegel believes this poetry to be the final form of the sensuous expression of *Geist's* self-consciousness, it would not be incorrect to say that art has a second "end" at the conclusion of the Romantic era (1972, vol. 14, pp. 231–242).

Thus, art of the Romantic age is inferior to that of the classical epoch, when considered as the state-of-the-art expression of *Geist*'s self-consciousness. It is superior to earlier art in terms of the level of self-awareness it attempts to express, however. Romantic art is, from the viewpoint of religion and philosophy, superior to classical art. Hegel speaks of postclassical art as art "set free" from its aim, implying that postclassical art is free in a way that classical art is not. It is reasonably easy to state in what this freedom consists—Romantic art, and particularly its most developed poetic form, no longer shoulders the burden of being the primary mode of our self-consciousness and, therefore, need not function as an organizing mode of self-consciousness. Art of the Romantic era is open to mere reflection and the aesthetic theories characteristic of modernity—particularly those such as Kant's that attempt to identify an autonomy for aesthetic reflection divorced from other modes of assessing an object's value or nature—are results of this evolution. For Hegel, then, there is a definitive character, not only of art, but also of its criticism, that is only possible at the end stage of art.

Questions concerning the nature of the "art world" at the end stage can be divided into two groups: those having to do with the experience of past art and those concerning the production and assessment of current (and future) art. Let us first address the question of the relation of end-state consciousness to the experience of artworks of prior stages of consciousness. Hegel defines beauty as the sensuous appearance of truth, marking beauty as the canonical quality of art in terms of which aesthetic valuation proceeds. Any understanding of what is beautiful about a work of art is determined by the critical conception available at any stage of self-consciousness. This holds true for the critic positioned at the end state as well. We cannot judge classical art in the same way as a fifth-century Athenian. True, we can philosophically reconstruct *that* classical culture experienced art in the way that it did, but we cannot aesthetically recoup *how* that art was experienced in the culture. Our critical practice is committed to a set of attitudes about art that foil any attempt to inhabit the art world of another era aesthetically. To do so would be to inhabit an earlier

(and inferior) self-conception of *Geist,* so to speak "from the inside out." That cannot be accomplished for the simple reason that having such a (for us, limited) self-conception would require us to ignore the advances in our self-realization and this ignorance itself could not be part of the prior self-conception that we were attempting to inhabit.

Let us now turn to the question of what Hegel considers the final form of art to be. Hegel thinks that the art of his present is essentially romantic. It is often assumed on Hegel's behalf that, because we have reached an end state of self-consciousness, there will be no further development of art. Hegel need not think this, nor does he. In fact, he states that there can always be hope for progress in art (ibid., vol. 13, p. 142). Of course, he does not mean by this that there will be *essential* progress in art, but he does seem to allow that further refinements in the artistic expression of truth are possible. The fact that we have achieved perfect conceptual self-knowledge does not entail that there be advances in expressing this fact sensuously. Of course, such expressions will always be inferior to religious and philosophical thought, but the point is that Hegel's account permits him to say that there is in principle some further novelty, perhaps even perfectibility, in art. One does get the overall impression, however, that such novelty or perfectibility will be had in terms of the type of art that Hegel considers the aesthetic state of the art and the philosophical end state. For the sake of convenience, and in keeping with Hegel's own division of art, one can term this "late-Romantic." Such art will be, broadly speaking, "poetic" and have as its defining feature what Hegel calls "objective humor."

Although Hegel is far from specific about what objective humor consists in, one can glean some understanding of the meaning of the term by considering what he has to say about what he takes to be objective humor's main competitors for aesthetic allegiance in late-Romanticism: "subjective humor" and irony. Hegel associates subjective humor almost exclusively with the writings of Jean Paul Richter (ibid., vol. 14, pp. 229–231), although he also mentions Johann Georg Hamann as a practitioner (ibid., vol. 11, p. 336). The essence of humor for Hegel, and what makes it the essential crowning form of art of the Romantic type, is that it consists in the expression of the personality of the artist in its particularity. This introduces an essential subjective contingency into the work in that all that initially seems "objective" to the reader (and here Hegel seems to mean both the accepted formal narrative forms of fiction and the content of the story) is later revealed as entirely contingent and empty of content. It is this annihilation *(Vernichtung)* of that which can be taken objectively and seriously that is the sole end of subjective humor. It is, as Jean Paul himself insisted, entirely negative in this sense. Such humor is completely one-sided and invariably introduces into the artwork a fundamental absurdity. All elements of the work are put at the disposal of the author's wit, which attacks even the artifice and accepted narrative conventions of the work. In Jean Paul at least, the attempt to flaunt conventions of art making introduces a Baroque or even grotesque element into the structure of the work, obscuring its aesthetic appeal.

Irony also relies on subjectivity to achieve its effect. Although Hegel notes that Solger and Ludwig Tieck are often cited as ironists, he is careful to qualify the ways in which they are so considered (ibid., vol. 11, pp. 233–234; vol. 13, p. 99), focusing his critique of ironism at Friedrich Schlegel. This critique is quite complex; a thumbnail sketch of it will suffice here. Hegel associates irony very closely with Fichteanism—in fact, it would not be inaccurate to say that irony is for Hegel "aestheticized" Fichte. Hegel objects to Johann Gottlieb Fichte's reliance on the *Ich* as the foundation of his account of the relation of subjects and objects of experience as entirely subjective and formal. All content is defined as that which stands in opposition to the subject, yet the subject can be nothing more than that which is opposed by an object. Considered as a living individual, such a subject's activity is limited to constructing its identity in terms of what is "other" to it. Formal and empty itself, such a subject treats the world merely as a receptacle for its own striving (ibid., vol. 13, p. 94). When that subject is an artist, this means living through one's art products by imprinting one's individual personality on those works. Hegel states that such a Fichtean artist lives *künstlerlich*, exploiting the double sense of the adjective: such an artist lives "as an artist," but also "artificially." The artificiality resides in the fact that, because of the empty character of the identity of the *Ich*, the artwork can be nothing more than a shell. The essence of irony for Hegel is that the work at once poses as a substantial object for contemplation and undercuts itself at all turns by revealing its lack of substance; that is, an ironic work, in Friedrich Schlegel's sense, is one that simultaneously asserts and takes back its content. Such work is not comical, in the true sense. Humor only negates the merely accidental, Hegel states, and therefore does not question seriousness as such (at least it does not necessarily amount to a claim that *nothing* can be held rightly to be worthy of seriousness). Irony, however, is universal negation and mocks all claims. Hegel puts the point slightly differently when he says that irony involves the attitude of divine ingenuity *(göttliche Genialität)* (ibid., vol. 13, p. 95). The phrase is meant to invoke two rival, perhaps antithetical, traditional ways of explaining of creative power that are married in romantic irony. One way is the conception of creativity current in the French neoclassicism of Nicolas Boileau and Charles de Batteaux where the artist's talent to produce praiseworthy works of art relies on *génie*—ingenuity in manipulating stock rhetorical and poetic devices with the aim of creating novel and surprising effects while remaining within the prescribed rules of taste. A competing tradition, beginning with

the radical views of Edward Young and including the Sturm und Drang and the more moderate position of Kant, emphasizes the inexplicable and free power of genius in the creative process—a capacity not bound by rules, at least not by rules issuing from external authority. The creative character proper to the ironic poet is an unholy mixture of these seeming disparate traits ascribed to genius—of the ingenuity indicative of a heightened ability to play with poetic form and meter and the unfettered divinity usually associated with theories of genius. The result is that the ironist is something of a monster, possessed of or by an entirely unbounded power to play. Irony shares with subjective humor its negative function, though irony is the more genuinely philosophical position because it extends that negation globally. In Hegelian terms, what both lack is an understanding that univocal negation must also be negated. Such an understanding is what Hegel thinks is the essence of objective humor; put another way, objective humor is dialectical. It allows the substantial to win through contingency, the negation of contingency revealing a yet deeper level at which things are united. Although a bit overly schematic, it would not be wrong to say that a work of objective humor contains within it the three dialectical moments: (1) a situation, claim, or character trait is presented as generally accepted, universal, or objectively true; (2) expectations about it are dashed as it is revealed to be merely contingent; (3) that revelation opens up a deeper sense of order in which one can situate the contingency and feel the necessity for its being revealed as such (as well as the reason why it was initially taken to be otherwise) (ibid., vol. 14, p. 231). Hegel mentions as examples of this sort of literature Laurence Sterne, Johann Wolfgang von Goethe (the *West-östlicher Divan*), the poetry of Friedrich Rückert (ibid., vol. 14, p. 242), and T. G. Hipple (ibid., vol. 11, p. 336).

[*See also* Danto.]

BIBLIOGRAPHY

Works by Hegel

Werkeausgabe. 20 vols. Edited by Eva Moldenhauer and Karl Markus Michel. Frankfurt am Main, 1972–.

Other Sources

Desmond, William. *Art and the Absolute: A Study of Hegel's Aesthetics.* Albany, N.Y., 1986.
Gombrich, E. H. "Hegel und die Kunstgeschichte." *Neue Rundschau* 88 (1977): 202–219.
Harries, Karsten. "Hegel on the Future of Art." *Review of Metaphysics* 27 (1973–1974): 677–696.
Henrich, Dieter. "Kunst und Natur in der idealistischen Ästhetik." In *Nachahmung und Illusion,* edited by H. R. Jauss, pp. 128–134. Munich, 1964.
Henrich, Dieter. "Art and Philosophy of Art Today: Reflections with Reference to Hegel." In *New Perspectives in German Literary Criticism: A Collection of Essays,* edited by Richard E. Amacher and Victor Lange, translated by David Henry Wilson et al., pp. 107–133. Princeton, N.J., 1979.
Henrich, Dieter. "The Contemporary Relevance of Hegel's Aesthetics." In *Hegel,* edited by Michael Inwood, pp. 199–207. Oxford and New York, 1985.
Kuhn, Helmut. *Die Vollendung der klassischen deutschen Ästhetik durch Hegel.* Berlin, 1931.
Oelmüller, Willi. "Hegels Satz vom Ende der Kunst." *Philosophisches Jahrbuch* 73 (1965): 75–94.
Rosen, Michael. *Hegel's Dialectic and Its Criticism.* Cambridge and New York, 1982.
Taylor, Charles. *Hegel.* Cambridge and New York, 1975.

FRED L. RUSH, JR.

HEIDEGGER, MARTIN. [*This entry comprises five essays on Heidegger's philosophy and its significance for contemporary art theory and aesthetics:*

Survey of Thought
Heidegger's Confrontation with Aesthetics
Heidegger's Displacement of the Concept of Art
Philosophical Heritage in Heidegger's Concept of Art
Heidegger and Hölderlin

The first essay is an overview of Heidegger's philosophy, both his early "fundamental ontology" (a metaphysical distinction between "Being" and "beings"—with humans among the latter) and his later notion of poetic thinking. The second essay explains how Heidegger challenges the tradition of aesthetics, particularly Hegel. The third essay analyzes Heidegger's attempts to overcome the concept of "art." The fourth essay continues the discussion of the preceding two essays by showing how much Heidegger is indebted to the philosophical aesthetic tradition he tries to overcome. The fifth essay reveals in more depth the centrality of Hölderlin's poetry to Heidegger's thinking. For related discussion, see Gadamer; Hegel, article on Hegel's Conception of the End of Art; Hermeneutics; Hölderlin; and Phenomenology.]

Survey of Thought

Martin Heidegger (1889–1976), twentieth-century German philosopher. Heidegger's magnum opus, *Being and Time* (1927), was inspired by but also radically transformed the phenomenology of his elder colleague and mentor, Edmund Husserl. Heidegger's early philosophy exerted enormous influence on the French existentialists Jean-Paul Sartre and Maurice Merleau-Ponty, and on theologians such as Rudolf Karl Bultmann and Paul Tillich, and in addition sparked renewed interest in philosophical hermeneutics, for example in the work of Hans-Georg Gadamer. More recently, Heidegger has found a critical audience among English-speaking philosophers influenced by the American pragmatists and the later work of Ludwig Wittgenstein. Finally, the influence of Heidegger's later writings has been felt especially among French poststructuralists such as Michel Foucault and Jacques Derrida.

Born in the small town of Messkirch in Baden, Heidegger received a Catholic education at the Bertholdgymnasium in

Freiburg and the Theological Seminary at the University of Freiburg. While preparing for a life in the priesthood, however, he read works by Franz Brentano and Husserl, and subsequently redirected his studies toward logic, mathematics, and philosophy. Working under the neo-Kantian philosopher Heinrich Rickert, Heidegger wrote a dissertation on psychologism and the theory of judgment, began lecturing in Freiburg on ancient and medieval philosophy, and published his *Habilitationsschrift* on a scholastic text then commonly attributed to Duns Scotus (it was later shown to have been written by Thomas of Erfurt).

In 1916, Husserl succeeded Rickert at the University of Freiburg. Before long, intermittent with a few months of uneventful military service, Heidegger established himself as Husserl's most sympathetic and promising assistant. By 1919, under Husserl's influence, Heidegger had openly renounced his Catholicism and resumed lecturing on phenomenology and the history of philosophy. His early lectures reveal a commitment to the methodological spirit of phenomenology and at the same time a deep and persistent dissatisfaction with the substance of Husserl's views. Still, with Husserl's enthusiastic support Heidegger secured a professorship in Marburg in 1923, and five years later succeeded Husserl himself at Freiburg.

In 1927, Heidegger published *Being and Time* and dedicated it to Husserl "in admiration and friendship." After arriving in Freiburg, however, Heidegger broke off almost all contact with Husserl and within a few years the two had become completely estranged. In 1933, Heidegger was elected rector of the university and officially joined the Nazi Party, remaining a member until the end of the war. As rector, Heidegger sought to bring academic life at Freiburg into conformity with the spirit of National Socialism as he then conceived it, while at the same time providing Nazism itself a kind of philosophical guidance and support drawn from his own thought. Heidegger's idiosyncratic conception both of the movement and of his own role in it, however, met with little sympathy or success and he resigned the rectorate the next year, in 1934. Contrary to popular legend, Heidegger did not personally ban Husserl, who was Jewish, from access to the university library. Rather, a nationwide law against Jewish civil servants came into effect just prior to Heidegger's assumption of the rectorate; indeed, his immediate predecessor had resigned after only a few days in office to protest a similar local decree against "non-Aryan" emeritus professors, notably Husserl. In assuming office, Heidegger certainly complied with but did not instigate the measures taken against Jewish faculty.

During the 1930s, Heidegger lectured extensively on Friedrich Hölderlin and Friedrich Nietzsche and wrote the essay "The Origin of the Work of Art" (1935–1936). After the war, until 1952, he was barred from teaching in Germany, though by 1949 he was lecturing privately and again enjoying considerable acclaim, especially in France. He re-

mained philosophically productive throughout the 1950s, and even into the 1960s, during which time he turned his thoughts increasingly to questions concerning language and technology. He died in 1976.

Heidegger's early philosophy rests on an application of his own original conception of phenomenological method to the question concerning the "meaning of being." Heidegger maintains that the question of being has been systematically neglected and obscured throughout the history of Western philosophy and that phenomenology offers a new way of making it explicit again, as it once was for the pre-Socratic philosophers of ancient Greece. The question of being is not, strictly speaking, a metaphysical question, for whereas traditional metaphysics concerns the general nature or cause of things, that is—*what* or *why* something is—Heidegger's question has to do with what it *means* for something to be, that is to say, what we understand when we understand *that* something is.

Heidegger therefore draws a sharp distinction between entities *(Seiende)* and being *(Sein)*. Heidegger's own "fundamental ontology," then, consists in a phenomenological interpretation of our understanding of *being*, which conditions our understanding of *entities*, including ourselves. To avoiding begging the question of being, Heidegger argues, we must avoid all metaphysical and even commonsense presuppositions about the kind of entities we are in order to secure a firm hold on our understanding of being as it presents itself in everyday life. This is why Heidegger eschews all traditional philosophical conceptions of human beings as rational animals, creatures, thinking substances, the will to power, even persons or subjects, and instead enlists "Dasein," an ordinary German word meaning existence (literally, *being there*), as a technical term for those entities who have an understanding of being. This manner of transforming ordinary language into a new, unfamiliar philosophical vocabulary is characteristic of the unique style one finds in all of Heidegger's writings.

Being and Time, then, prepares the ground for the question of being by advancing a phenomenological account of *Dasein's* everyday understanding of being. Contrary to Husserl's approach to phenomenology, however, which seeks to bracket the world of mundane experience and so provide a rigorous description of "pure" consciousness, or trancendental subjectivity, Heidegger's phenomenology is avowedly "hermeneutical," or interpretative; that is, Heidegger aspires not to a theoretically pure description of mental states but to an interpretation of the pretheoretical understanding that *Dasein* always already has of itself in everyday life.

Consequently, Heidegger rejects what he calls Husserl's "worldless" or Cartesian conception of the subject as a transcendental ego and instead describes *Dasein* as "being-in-the-world," as being "thrown" into a historically constituted world of moods and situations while at once "project-

ing" into a future constituted by practical possibilities or options. *Dasein's* being, understood in this way as "thrown projection" *(geworfener Entwurf),* is thus essentially temporal. *Dasein* cannot be understood as an object present in the world at a single moment; rather, it must interpret itself in terms of its relation to its own future possibilities in light of its given historical context.

Heidegger maintains that *Dasein's* understanding is always conditioned by a kind of anonymous social authority that he calls *das Man,* or "the one" (a neologism alluding to expressions such as "One doesn't talk about such things," standardly translated as "the They," as in "What will they say!"). *Das Man* is not any particular person or group; it is rather the often tacit normative demands inherent in all aspects of social life. *Dasein* never escapes the normalizing influence of *das Man,* because conformity to norms is constitutive of *Dasein's* understanding of being. *Dasein* has an "authentic" *(eigentlich)* or "inauthentic" *(uneigentlich)* relation to *das Man,* however, depending on whether it owns up to its groundlessness and finitude as "being-unto-death" or instead flees its facticity into the prejudices of conventional wisdom and received opinion. At the same time, authenticity and inauthenticity have to do with the stand one takes on one's own temporality, either committing oneself in a "resolute" *(entschlossen)* way to a particular contingent future or else tarrying alongside things in the present and conforming passively to the normative pressures of *das Man.*

The conclusion of *Being and Time* is not any direct pronouncement concerning the meaning of being, except to say that we always understand being in terms of time— hence the title of the book. *Dasein's* own temporality is the horizon against which we have any understanding of being at all. Metaphysics has always understood being in terms of the present, or the now, so that the past means "no longer now" whereas the future means simply "not yet now." On the traditional view, then, things lying in the past and future strictly speaking *are not.* But Heidegger urges that no such ontology of presentness can make sense of *Dasein* as a temporal entity existing as thrown projection. Heidegger carries out his "destruction of traditional ontology" by drawing attention to the temporal horizons for any understanding of being.

Whereas *Dasein* is fundamentally future-directed in its temporality, the being of other entities lies in the various ways in which they show up for us in the present. Here Heidegger famously distinguishes between objects, which are merely "occurrent" *(vorhanden),* and the environment and artifacts that are "available" *(zuhanden)* for use. Contrary to the tradition, Heidegger argues that the things we make use of in daily life cannot be understood as so many occurrent objects with so many utility predicates added on in thought specifying their functions. Rather, *Dasein* orients itself in the world through preconceptual skills and abilities that embody an understanding of the availability of things. Practical

competence is therefore more basic than cognition or propositional knowledge: knowing-*that* presupposes knowing-*how.* Our understanding of objects consists in our ability to abstract from the pragmatic contexts of everyday life and reflect on ourselves as subjects with mental attitudes toward things occurrent.

Being and Time remains an unfinished work; the published portion of the text amounts to only a third of the entire project as Heidegger originally conceived it. Fundamental difficulties lay in the way of its completion, specifically the transition Heidegger anticipated from the analytic of *Dasein* to a more general inquiry into the meaning of being. That transition proved elusive, and eventually Heidegger abandoned his commitment to an account of any particular entity, for example *Dasein,* as a necessary prerequisite for asking the question of being as such. Being is always the being of some entity, but the entities he later takes up include such things as the peasant's shoes in a Vincent van Gogh painting, a Greek temple, a hydroelectric plant on the Rhine, and a simple jug or pitcher. Part of Heidegger's dissatisfaction with *Being and Time* no doubt lay in the apparent narrowness of the categories of availability and occurrentness as definitive of nonhuman entities.

Consequently, in "The Origin of the Work of Art," Heidegger describes artworks as cultural paradigms, which are no more reducible to their practical availability than useful things are reducible to their occurrentness or objectivity; that is, just as a useful thing is not a mere object with functional properties added on, neither is a work of art simply a useful thing with aesthetic qualities in addition. Whereas useful things fit more or less neatly into the practical intelligibility of the daily world, works of art make manifest what is recalcitrant, anomalous, and resistant to ordinary understanding— and yet what sustains it. That combination of recalcitrance and support is what Heidegger calls "earth," and by initiating a strife between the intelligibility of the world and the resistance of the earth, works of art have a unique capacity to disclose understandings of being and constitute or inaugurate truth for historical cultures. Other such history-making, truth-constituting paradigms, according to Heidegger, include the founding of a political state, the felt presence of the gods or God, sacrificial events, and philosophical thinking. Science, by contrast, never founds truth, but only cultivates a domain of truth already disclosed in some other way.

Heidegger criticizes theories of art, and indeed aesthetics at large, particularly in its Kantian form, for trying to understand artworks in terms of the inner experiences of perceiving subjects, and for presupposing a hylomorphism that obscures what is really "at work" in works of art, namely, the happening of truth and the founding of history. Heidegger accepts the identification of beauty with form, precisely because an emphasis on form underlies the metaphysical interpretation of being, to which aesthetics itself belongs, for example, in Plato's theory of forms. But form, beauty,

and aesthetic experience indicate only partial aspects of the work of art and typically go hand in hand with historically specific interpretations of truth and being. If, for example, we understand being as creation, the work of art will appear as a reflection of the divine, whereas if we understand being with reference to subjectivity, the work of art will reveal itself as the self-expression of creative genius. Yet, it is art itself, Heidegger argues, that articulates and defines those understandings of being and truth to begin with. "The Origin of the Work of Art" therefore concludes with the identification of the essence of art as "poetry" *(Dichtung)*, that is, a primordial saying or establishing of truth and being. The "origin" of the work of art, then, is not the object or the artist, but art itself.

The work of art that Heidegger regarded as definitive of contemporary German history lay in the poetry of Hölderlin. More than any other artist, he thought, Hölderlin had articulated the historical identity and destiny of the German people in their remembrance of Greek antiquity and their confrontation with modern technology. As Heidegger became increasingly disillusioned with German politics before World War II, however, he came to regard National Socialism itself not as a manifestation of the spirit of Hölderlin's poetry but as a symptom of the technological manipulation and domination of the earth. Heidegger understood technology to have grown out of the metaphysical interpretation of being, the last great philosophical expression of which occurs in Nietzsche's concepts of the will to power and the eternal return. Heidegger's reading of Nietzsche is controversial not least of all because of its reliance on the posthumous notes published and presented misleadingly as Nietzsche's magnum opus under the title *The Will to Power*. Heidegger's two-volume *Nietzsche* (1961), drawn from his lectures of the 1930s and 1940s, interprets Nietzsche's philosophy as the culmination of the subjectivism of modern thought and the nihilism of Western metaphysics in general. In conceiving of entities in terms of the will to power and the eternal return, Nietzsche gives expression to what Heidegger would later call the "technological" understanding of being.

Heidegger's Nietzsche lectures thus demonstrate his ambivalence toward mainstream Nazism and mark a shift in his conception of technology. He had earlier conceived of technology as the principal challenge of modernity, to which National Socialism offered the only adequate response in what Ernst Jünger called the worker's "total mobilization" of the world. At some point in the late 1930s, Heidegger came to regard that activist mentality as part of the problem of technology, rather than as part of the solution. In his 1953 lecture "The Question concerning Technology," for example, Heidegger describes technology as "enframing" *(Gestell)*, that is, as an understanding of being that sets upon and challenges forth the earth for the sake of maximizing order, organization, and efficiency; and where being is understood enframing, entities in turn show up as mere resource material or "standing reserve" *(Bestand)*, which is constantly available for use and at the same time transparent in its instant accessibility. Heidegger therefore comes to regard technology as a uniquely dangerous and decisive epoch in the history of metaphysics because he thinks that the sheer immanence and efficiency of things at once blinds us to the fact that we live in an understanding of being and opens our eyes to the totalizing syndrome inherent in that understanding. Only by resisting that totalizing syndrome, he thinks, can we gain a free relation to the technological artifacts and institutions of modern society and preserve within it a place for art, that is, for the happening of truth.

BIBLIOGRAPHY

Works by Heidegger

Prolegomena zur Geschichte des Zeitbegriffs (Marburg lectures, summer 1925). In *Gesamtausgabe,* vol. 20, edited by Petra Jaeger. Frankfurt am Main, 1979. Translated as *History of the Concept of Time: Prolegomena* by Theodore Kisiel (Bloomington, Ind., 1985).

Grundprobleme der Phänomenologie (Marburg lectures, summer 1927). In *Gesamtausgabe,* vol. 24, edited by Friedrich-Wilhelm von Hermann. Frankfurt am Main, 1975. Translated as *The Basic Problems of Phenomenology* by Albert Hofstadter (Bloomington, Ind., 1982; rev. ed., 1988).

Sein und Zeit (1927). 15th ed. Tübingen, 1984. Translated as *Being and Time* by John Macquarrie and Edward Robinson (New York, 1962) and Joan Stambaugh (Albany, N.Y., 1996).

Metaphysische Anfangsgründe der Logik im Ausgang von Leibniz (Marburg lectures, summer 1928). In *Gesamtausgabe,* vol. 26, edited by Klaus Held. Frankfurt am Main, 1978. Translated as *The Metaphysical Foundations of Logic* by Michael Heim (Bloomington, Ind., 1984).

Kant und das Problem der Metaphysik. Bonn, 1929; 4th enl. ed., Frankfurt am Main, 1973. Translated as *Kant and the Problem of Metaphysics* by Richard Taft (Bloomington, Ind., 1990).

"Vom Wesen des Grundes" (1929). In *Wegmarken,* 2d enl. ed. Frankfurt am Main, 1978. Translated as *The Essence of Reasons* by Terrence Malick (Evanston, Ill., 1969).

Die Grundbegriffe der Metaphysik: Welt, Endlichkeit, Einsamkeit (Freiburg lectures, winter 1929–1930). In *Gesamtausgabe,* vols. 29–30, edited by Friedrich-Wilhelm von Hermann. Frankfurt am Main, 1983. Translated as *The Fundamental Concepts of Metaphysics: World, Finitude, Solitude* by William McNeill and Nicholas Walker (Bloomington, Ind., 1995).

Hegels Phänomenologie des Geistes (Freiburg lectures, Winter 1930–1931). In *Gesamtausgabe,* vol. 32, edited by Ingtraud Görland. Frankfurt am Main, 1980. Translated as *Hegel's Phenomenology of Spirit* by Parvis Emad and Kenneth Maly (Bloomington, Ind., 1988).

Aristoteles, Metaphysik Θ 1–3: Von Wesen und Wirklichkeit der Kraft (Freiburg lectures, summer 1931). In *Gesamtausgabe,* vol. 33, edited by Heinrich Hüni. Frankfurt am Main, 1981. Translated as *Aristotle's Metaphysics Θ 1–3: On the Essence of Actuality and Force* by Walter Brogan and Peter Warnek (Bloomington, Ind., 1995).

Hölderlins Hymnen "Germanien" und "Der Rhein" (Freiburg lectures, winter 1934–1935). 2d ed. Frankfurt am Main, 1989.

"Der Ursprung des Kunstwerkes" (1936). In *Holzwege,* 6th ed. Frankfurt am Main, 1980. Translated as "The Origin of the Work of Art" by Albert Hofstadter, in *Poetry, Language, Thought* (New York, 1971), pp. 15–87.

Erläuterungen zu Hölderlins Dichtung (1936–1944). 5th ed. Frankfurt am Main, 1981.

Grundfragen der Philosophie: Ausgewählte "Probleme" der "Logik" (Freiburg lectures, winter 1937–1938). In *Gesamtausgabe,* vol. 43, edited by Friedrich-Wilhelm von Herrmann. Frankfurt am Main, 1984. Translated as *Basic Questions of Philosophy: Selected "Problems" of "Logic"* by Richard Rojcewicz and André Schuwer (Bloomington, Ind., 1994).

Hölderlins Hymne "Andenken" (Freiburg lectures, winter 1941–1942). In *Gesamtausgabe,* vol. 52, edited by Curd Ochwadt. Frankfurt am Main, 1982.

Hölderlins Hymne "Der Ister" (Freiburg lectures, summer 1942). In *Gesamtausgabe,* vol. 53, edited by Walter Biemel. Frankfurt am Main, 1984.

Einführung in die Metaphysik. Tübingen, 1953. Translated as *An Introduction to Metaphysics* by Ralph Manheim (New Haven, 1959).

"Die Frage nach der Technik" (1953). In *Vorträge und Aufsätze,* 6th ed. Pfullingen, 1990. Translated in *The Question concerning Technology and Other Essays* by William Lovitt (New York, 1977), pp. 3–35.

Was heisst Denken? Tübingen, 1954. Translated as *What Is Called Thinking?* by Fred D. Wieck and J. Glenn Gray (New York, 1968).

Identität und Differenz. Pfullingen, 1957. Translated as *Identity and Difference* by Joan Stambaugh (New York, 1969).

Der Satz vom Grund. Pfullingen, 1957. Translated as *The Principle of Reason* by Reginald Lilly (Bloomington, Ind., 1991).

Gelassenheit. Pfullingen, 1959. Translated as *Discourse on Thinking* by John M. Anderson and E. Hans Freund (New York, 1966).

Unterwegs zur Sprache. Pfullingen, 1959. Translated as *On the Way to Language* by Peter D. Hertz and Joan Stambaugh (New York, 1971).

Nietzsche. 2 vols. Pfullingen, 1961. Translated as *Nietzsche* by David Furrell Krell in 4 vols. (New York, 1979–1987).

Zur Sache des Denkens. Tübingen, 1969. Translated as *On Time and Being* by Joan Stambaugh (New York, 1972).

Works on Heidegger

Biemel, Walter, and Friedrich-Wilhelm von Hermann, eds. *Kunst und Technik: Gedächtnisschrift zum 100. Geburtstag von Martin Heidegger.* Frankfurt am Main, 1989.

Dreyfus, Hubert L. *Being-in-the-World: A Commentary on Heidegger's "Being and Time," Division 1.* Cambridge, Mass., 1991.

Dreyfus, Hubert L., and Harrison Hall, eds. *Heidegger: A Critical Reader.* Oxford and Cambridge, Mass., 1992.

Gadamer, Hans-Georg. *Heidegger's Ways.* Translated by John W. Stanley. Albany, N.Y., 1994.

Guignon, Charles, ed. *The Cambridge Companion to Heidegger.* Cambridge and New York, 1993.

Hermann, Friedrich-Wilhelm von. *Heideggers Philosophie der Kunst.* 2d ed. Frankfurt am Main, 1994.

Kockelmans, Joseph J. *Heidegger on Art and Art Works.* Dordrecht and Boston, 1985.

Ott, Hugo. *Martin Heidegger: A Political Life.* Translated by Allan Blunden. New York and London, 1993.

Pöggeler, Otto. *Die Frage nach der Kunst: von Hegel zu Heidegger.* Freiburg, 1984.

Pöggeler, Otto. *Martin Heidegger's Path of Thinking.* Translated by Daniel Magurshak and Sigmund Barber. Atlantic Highlands, N.J., 1987.

Schürmann, Reiner. *Heidegger on Being and Acting: From Principles to Anarchy.* Translated by Christine-Marie Gros with the author. Bloomington, Ind., 1987.

Tugendhat, Ernst. *Der Wahrheitsbegriff bei Husserl und Heidegger.* Berlin, 1967.

Tugendhat, Ernst. "Heidegger's Idea of Truth." In *The Heidegger Controversy: A Critical Reader,* edited by Richard Wolin, pp. 245–263. New York, 1991.

TAYLOR CARMAN

Heidegger's Confrontation with Aesthetics

Although Martin Heidegger has discussed art and works of art in numerous places, "The Origin of the Work of Art" (1935) remains the key text and it seems appropriate that the two most ambitious presentations of Heidegger's "philosophy of art," Friedrich-Wilhelm von Herrmann's *Heideggers Philosophie der Kunst* (1980) and Joseph Kockelmans's *Heidegger on Art and Art Works* (1985), should offer commentaries on that text. All such attempts, however, are called into question by Otto Pöggeler's (1994) insistence that to attribute to Heidegger a "philosophy of art" is to do violence to his understanding of art and philosophy as equal partners in the attempt to open up a path to a postmodern age.

Undeniable is Heidegger's critical and questioning confrontation with aesthetics, especially with its most comprehensive elaboration: G. W. F. Hegel's *Vorlesungen über die Ästhetik* (1975). If interest in Heidegger's reflections on art has been steadily growing, not only among philosophers, but also among literary critics and art historians, artists, and architects, this reflects increasing dissatisfaction with the aesthetic approach that ever since the eighteenth century has shaped not just reflection on art and artworks, but also the way artists have come to understand and pursue their vocation, tending toward the progressive marginalization of art. Heidegger thus links the rise of aesthetics to the dying of art: "Aesthetics takes the work of art as an object, the object of *aisthesis,* of sensuous apprehension in the widest sense. Today we call this apprehension experience. . . . Yet perhaps experience is the element in which art dies. The dying occurs so slowly that it takes a few centuries" (1971). Such remarks bring to mind Hegel's thesis of the death of art in its highest sense. We moderns, Hegel insists,

> are beyond the stage of reverence for works of art as divine and objects deserving our worship. The impressions they produce is one of a more reflective kind, and the emotions which they arouse require a higher test and a further verification. Thought and reflection have taken their flight above fine art. (1975)

The rise of aesthetics is part of this development. Challenging Hegel, Heidegger would have us at least keep open the possibility that art might still be "an essential and necessary way in which that truth happens which is decisive for our historical existence" (1971).

For Hegel the death of art in its highest sense is of a piece with humanity's coming-of-age, where, as Heidegger points out, Hegel did not mean to assert that art would no longer be created. On the aesthetic approach, what Hegel calls the death of art may indeed be welcomed as the emancipation of art from what is external to it. The history of Western art can thus be written as a history of the progressive emancipation of art from everything foreign to its essence, where Hegel, too, is thinking first of all of religion. As Heidegger puts it in the *Spiegel* interview (1990): today's art "no longer

points out a path" and has become uncertain of its place. To be sure, this tendency has not gone unchallenged and, in recent decades, such challenges have become so powerful that they invite talk of the death of art for art's sake, a death that can be understood as a precondition for art's recovery of what was once its highest function. Such challenges can draw strength from Heidegger's reflections.

Heidegger is well aware of all that argues against such romantic attempts to move art once again toward the sacred, and although he does not interpret history as the triumphant progress of the spirit, he finds it difficult to dispute Hegel's judgment as far as the present is concerned; that judgment is said to have the support of the entire history of metaphysics, which issues in the understanding of the being of things that rules modern science and technology and the world it has shaped. Aesthetics depends on metaphysics—to challenge the latter is also to call the former into question. As Heidegger points out in the "Addendum" (1956) to "The Origin of the Work of Art": "Reflection on what *art* may be is completely and decidedly determined only in regard to the question of *Being*"; and because that question is one that permits no definitive answer, "What art may be is one of the questions to which no answers are given in the essay. What gives the impression of such an answer are directions for questioning."

Hegel understands the death of art in its highest sense as but one manifestation of the spirit's progress. Heidegger links that death to the progress of metaphysics. Metaphysics is determined by the attempt to lay hold of the essence of whatever is. Greek metaphysics thus opposed what truly is and abides to the shadowy fleeting world that presents itself to our senses. Becoming is declared an inferior kind of being: mere appearance, dependent for its being on true being. Yet, this inferior being retains a certain autonomy. Plato therefore insists on the resistance offered by sensible matter to the forms. Our attempts to understand things inevitably meet with a certain opacity. Heidegger will name what thus must remain "undisclosed and unexplained" and "shatters every attempt to penetrate into it" the "earth."

The essential opacity of the earth is further obscured when in the Middle Ages metaphysics allies itself with the Christian understanding of an omnipotent, all-seeing creator God: everything lies open to his gaze. Metaphysics enters a third stage, decisive for our situation, with René Descartes: the knowing subject is made the foundation of the being of reality, which comes to be understood as what in principle can become an object of clear and distinct knowledge. With this reality is given its measure in the human subject's ability to comprehend and to manipulate. At least since Descartes there thus have been dreams that science and technology would give us ever-increasing mastery, not just over things, but over our bodies, over disease, over

ourselves—perhaps even over death, although this remains a dream. To thus make reason the measure of reality is to lose sight of what Heidegger named the earth.

With his talk of the "earth" Heidegger gestures in the same direction as Immanuel Kant with his "thing-in-itself," which is present to us only as appearance. What invites us to supplement appearance with the thing-in-itself is the fact that, even if constituted by our concepts or language and as such appearance, what thus appears is not created by our understanding, but given. Inseparable from our experience of things is this sense of their gift, some awareness that our understanding is finite, and that means also that the reach of our concepts is limited: everything real resists full translation into words; like Kant's "aesthetic idea," it is "inexponible." To be sure, language opens human beings to reality; yet, as Heidegger emphasizes, language conceals even as it reveals. Where this essential concealment is neglected or forgotten, words and concepts threaten to replace reality with a false, merely linguistic reality. To be sure, human being is essentially a dwelling in language; but the house of language is not a prison. When Heidegger understands the work of art as a presentation of the earth, he invites us to understand art as a way of opening windows in that house. Poetry thus should not be understood as a speaking that offers particularly effective descriptions of things, but rather as speaking that re-presents the essence of language in such a way that it becomes conspicuous, and that means a speaking that opens up the rift between word and thing that is essential to language.

What puts us in touch with the earth, this transcendence within the visible, within the sensible, is first of all the body, where Heidegger insists that the embodied self is also a caring, desiring self. What presents itself is not just an assemblage of mute facts, but an inevitably meaningful configuration of objects of desire or things to be avoided. To be in the world is to be claimed in countless different ways by persons and things. What Heidegger names the earth may thus not be reduced to the mute presence of things. To be open to the earth is inevitably to be affected, moved, claimed. Heidegger's talk of the earth thus gestures also toward the affective base without which all our talk of values or divinities is ultimately groundless: idle talk, even as it helps us to understand why metaphysics must progress toward nihilism. The step beyond nihilism is possible only as a recovery of the earth. Philosophical reflection, by its very nature, lacks the strength to effect such a recovery. The presentation of the earth requires art, and because it does, Heidegger's postmetaphysical thinking does not lead to a philosophy of art but seeks to enter into dialogue with artists and poets, especially with Friedrich Hölderlin.

The work of art is said by Heidegger not only to "let the earth be an earth," but also, indeed first of all, "to set up a world," where "world" does not mean a collection of mute

facts but a meaningful order that assigns to persons and things their proper places. In this sense, Heidegger understands a Greek temple as world-establishing work:

> In setting up the work the holy is opened up as holy and the god is invoked into the openness of his presence. . . . Dignity and splendor are not properties beside and behind which the god, too, stands as something distinct, but it is rather in the dignity, in the splendor, that the god is present. In the reflected glory of this splendor there glows, that is, there lightens itself what we called world. (1971)

The world establishing of the work of art is here understood as a response to a god, to an integrating power. Letting the god be present the temple gathers human beings into a genuine community.

But, like Hegel, Heidegger too wonders whether ours is an age when "great art, together with its nature, has departed from among men," had to depart, because the ruling sense of reality prevents us from granting art such authority: no art today can claim the integrative power needed to establish a world in Heidegger's sense. Our modern world has no room for gods. Heidegger's discussion of art in "The Origin of the Work Art" invites thus characterization as a nostalgic longing for lost unity. Such romantic longing remained with him to the end and finds expression in his famous proclamation in the *Spiegel* interview that "Only a god can still save us." To open up even the possibility for such salvation, we first must learn to let the earth be an earth. Much art today struggles to keep human beings open to this elusive dimension, without claiming the integrative power needed to establish a world in Heidegger's sense. That such art is deficient, given the understanding of art developed in "The Origin of the Work of Art," is evident.

But what supports that understanding? Does art really betray its essence when it comes under the sway of the aesthetic approach? How are we to understand that essence? To ask for the essence of art is to ask for what it is that makes something a work of art. But, as Heidegger has noted, the search for essence is by its very nature circular. Where are we to look for the essence of art? To works of art, of course. But how do we know what to count as a work of art unless we already know what art is? The truth or falsity of a particular determination of art cannot be decided by looking at art. Is it even meaningful to speak here of truth and falsity? When we claim to determine the essence of art, we express what strikes us as essential and that will inevitably be bound up with the concrete way in which we exist in the world, including our encounters with art, including also what we would like art and our world to be. Heidegger's reflections on art stem from his experience and understanding of the neediness of the age, an understanding intertwined with his postmetaphysical understanding of the history of Being, mediated also by his encounters with

poetry and art, especially with Hölderlin: "the poet who points toward the future" and "expects the god."

BIBLIOGRAPHY

Works by Heidegger

Erläuterungen zu Hölderlins Dichtung. Frankfurt am Main, 1944.

Holzwege. Frankfurt am Main, 1950. See especially "Der Ursprung des Kunstwerkes," translated with Addendum as "The Origin of the Work of Art," by Albert Hofstadter, in *Poetry, Language, Thought* (New York, 1971), pp. 15–87.

Vorträge und Aufsätze. Pfullingen, 1954. See especially "Bauen, Wohnen, Denken," translated by Albert Hofstadter as "Building, Dwelling, Thinking," in *Poetry, Language, Thought* (New York, 1971), pp. 143–161; "Das Ding," translated by Albert Hofstadter as "The Thing" in *Poetry, Language, Thought,* pp. 163–186; ". . .dichterisch wohnet der Mensch. . . ." translated by Albert Hofstadter as ". . . Poetically Man Dwells . . . , in *"Poetry, Language, Thought,* pp. 211–229.

Unterwegs zur Sprache. Pfullingen, 1959. Translated by Peter D. Hertz and Joan Stambaugh as *On the Way to Language* (New York, 1971). The first essay, "Die Sprache," is translated by Albert Hofstadter as "Language," in *Poetry, Language, Thought* (New York, 1971), pp. 187–210.

Nietzsche. 2 vols. Pfullingen, 1961. See especially the first part of vol. 1, translated by David Farrell Krell as *Nietzsche: The Will to Power as Art* (New York, 1979).

Beiträge zur Philosophie: Vom Ereignis. In *Gesamtausgabe,* vol. 65, edited by Friedrich-Wilhelm von Herrmann. Frankfurt am Main, 1989.

"Der Spiegel Interview with Martin Heidegger." In *Martin Heidegger and National Socialism: Questions and Answers,* edited by Günther Neske and Emil Kettering, translated by Lisa Harries, pp. 41–66. New York, 1990.

Other Sources

Biemel, Walter. *Philosophische Analysen zur Kunst der Gegenwart.* The Hague, 1969.

Derrida, Jacques. *The Truth in Painting.* Translated by Geoff Bennington and Ian McLeod. Chicago, 1987.

Hegel, Georg Wilhelm Friedrich. *Aesthetics: Lectures on Fine Art.* Translated by T. M. Knox. New York, 1975.

Herrmann, Friedrich-Wilhelm von. *Heideggers Philosophie der Kunst.* Frankfurt am Main, 1980.

Kockelmans, Joseph J. *Heidegger on Art and Art Works.* Dordrecht, 1985.

Pöggeler, Otto. "Heidegger on Art." In *Martin Heidegger: Politics, Art, and Technology,* edited by Karsten Harries and Christoph Jamme, pp. 107–110. New York and London, 1994.

KARSTEN HARRIES

Heidegger's Displacement of the Concept of Art

Martin Heidegger's writings on art and poetry in the 1930s and 1940s are dedicated to the overcoming of the aesthetic tradition, a task that is integral to the overcoming of the metaphysical tradition. Just as the overcoming of metaphysics is supposed to be accomplished with reference to the thinking of the so-called pre-Socratics who preceded philosophy, so the overcoming of aesthetics takes place by recalling the great art of the Greeks that flourished prior to

the introduction of aesthetics by Plato and Aristotle. Some commentators have recognized that Heidegger's accounts of the temple, the sculpture of the god, the poems of Pindar, and the tragedies of Sophocles in "The Origin of the Work of Art" (1936) would be distorted if aesthetic considerations intervened, but there has been relatively little scrutiny of Heidegger's attempt to free the concept of art itself from its status as an aesthetic category. It seems that Heidegger's claim was that by offering an account of the truth of the work of art he had overcome the aesthetic reduction of the artwork to the realm of the beautiful and that, by presenting truth in terms of the play of concealment and unconcealment, he had surpassed the metaphysical conception of truth. But in the background lurks the question of whether and how the concept of art can be freed from aesthetics. What made this so difficult for Heidegger to negotiate was the fact that the concept of art was, in his terms, a late arrival. This meant that he could not adopt his usual strategy of rehabilitating a questionable term on the basis of the way it had been used by the early Greeks.

"The Origin of the Work of Art" almost fails to get started because of doubts as to what art is. The potential impasse is bypassed only under the pretext that works of art are familiar to everyone, a suggestion that seems less and less plausible as the essay proceeds and it emerges that Heidegger was concerned only with "great art." Heidegger understood "great art" as the fashioner and preserver of what G. W. F. Hegel called the Absolute, which in Heidegger's language corresponds to the "truth of Being." Corresponding to Hegel's famous assertion that "art is and remains for us, on the side of its highest vocation, something past," Heidegger acknowledged that great art is not familiar today. What differentiates Heidegger from Hegel is that Heidegger allowed for the possibility that there might be great art in the future, but the Addendum to "The Origin of the Work of Art" written in 1956 concedes that "What art may be is one of the questions to which no answers are given in the essay." Heidegger's difficulty was that any delimitation of what future art might be in terms of what art has been in the past has to pass through the concept of art, a concept that arose in the context of aesthetics, and that he seemed to lack the resources to displace from that context.

"The Origin of the Work of Art" culminates in questions about the future possibility of art. The title of the essay sounds as if Heidegger was asking where art comes from, but in fact his question was whether art could again function as an origin, as it had done for the Greeks. For art to be an origin means for it to found a people; hence Heidegger's insistence that the question of whether or not Hölderlin's poetry was great art had not yet been decided and that it would only be decided if and when the German public became a people through it. Even though Heidegger was less inclined after World War II to pose the question of art in terms of "the peo-ple," a term that he had already recognized in the 1930s as problematic, it is clear that art was still being asked to play a world-historical role. This is especially clear at the end of "The Question concerning Technology," where Heidegger addressed the danger that arose from the dominance of technology by having recourse to art as a mode of revealing that exceeded that of production. What underlay the recourse to art was the ambiguity of *poiēsis* as both poetry and production and of *technē* as encompassing both art and craft. Heidegger developed the resources offered by these words as far as he could, but their very indeterminacy left him unable to follow his usual procedure of appealing to the Greek language for guidance as to what art was and might be.

The fact that the Greeks did not readily distinguish art from craft was particularly significant given that Heidegger had, since *Being and Time* (1927) explored the distortions that resulted from philosophy's overreliance on the model of production even for understanding nature. The analysis of the work of art in "The Origin of the Work of Art" was supposed not only to clarify these distortions, but also to open up a source of access to the question of Being that overreliance on the model of production had left closed. When, however, in "The Question concerning Technology," Heidegger asked why the Greeks gave "the modest name *technē*" to art, his answer—that it was because art shared in the revealing that the Greeks called *poiēsis*—made it clear that the difficulty of attributing to the Greeks a recognition of art as such left him with an acute problem. Heidegger's appeal to "the *poiēsis* of the fine arts" at the end of "The Question concerning Technology" was not only a mark of respect to his audience at the Bavarian Academy of Fine Arts; it was also an indication that the lack of reflection on art among the early Greeks, which had earlier been judged by Heidegger as a precondition of their great art, meant that he could not group together all of the various forms of *poiēsis* that he sought to identify as "art" without anachronism. One might have expected Heidegger to be as dismissive of the idea of Greek "art" as he was of Greek "religion," and on the same grounds that the Greeks lacked these concepts. "Technology and Art—Ge-stell" was less conclusive than "The Question concerning Technology," but on this issue more coherent. To the suggestion that art might dissolve at the end of metaphysics, Heidegger responded that it would be followed by "something more original than art—*technē*." This proposal was in much better conformity with Heidegger's standard strategy of looking to the most fundamental words of Greek thought to provide the basis for another beginning. Juxtaposition of these two texts suggests that Heidegger alternated between trying to rehabilitate the concept of art and seeking a more original name with which to replace it. Heidegger pursued this second, less recognized, option largely in the context of his attempts to negotiate the classification of the fine arts.

Nothing in antiquity corresponds to the modern system of the arts. Poetry was not associated in a systematic way with painting, sculpture, music, and especially architecture until the eighteenth century. Nevertheless, Heidegger usually gave the impression that he accepted this way of classifying the arts in spite of its dependence on the aesthetic tradition. For example, in "The Origin of the Work of Art," Heidegger not only attributed to the Greeks the view that poetry in the conventional sense has a certain preeminence over the other arts, but he also claimed in his own right that all art is essentially poetry *(Dichtung),* with explicit reference to the arts of architecture, painting, sculpture, and music. Heidegger's uniquely broad use of the concept of poetry as a projective saying that prepares what is sayable and unsayable for a historical people represents a tentative attempt to offer an alternative to the aesthetic concept of art. It seems as if Heidegger had it in mind to displace the concept of art into that of poetry, not just on the basis that poetry best exemplified the *poiēsis* or revealing that characterized all the arts, but also because its dominance corresponded to the pervasiveness of language.

The direction of this reading is confirmed most clearly in Heidegger's 1942–1943 lectures on Parmenides, when, in the context of a discussion of the concluding myth of Plato's *Republic,* he returned to the problem of the ranking of and the relation among "the classes of art"—architecture, sculpture, poetry. On the face of it, Heidegger was attempting to counter the view that for the Greeks architecture and sculpture were of a higher rank than poetry and thinking, a view that had arisen because they were made of more stable material than mere words, of which they were independent. Heidegger argued the reverse by focusing on the relation to the divine: "only if we recognize the original relation between the word and the essence of Being will we be capable of grasping why, for the Greeks and only for them, saying *(ho muthos)* must correspond to the divine." It is apparent that Heidegger continued even here to measure the priority of poetry with reference to the other arts, as if it had been as obvious to the ancient Greeks as it was to the aestheticians of the eighteenth century and their modern counterparts that poetry, architecture, sculpture, and perhaps even painting and music as well, belong together. Beneath the surface of Heidegger's argument that all of the arts rely on "the silent word" for the divine, however, it is possible to identify an underlying dissatisfaction with the concept of art that led him to seek to displace it into *muthos* and the related word for "word," *epos.*

Until recently, readers of Heidegger might have had the general impression that if the turn to the *muthos* was intended as a step toward the displacement of the concept of art, it was left relatively undeveloped by him and eventually aborted. With the publication of the 1957 lecture series, "Principles of Thinking," however, it is possible to correct this impression. Toward the end of the final lecture of "Prin-

ciples of Thinking," Heidegger drew attention to the relation of saying *(die Sage)* to the Homerian *epos,* as he had done in the Parmenides lecture course, where *die Sage* was the word Heidegger used to translate *muthos.* Of itself, this reference does not amount to much, but it is the missing link that connects the displacement of the concept of art in the 1940s with the writings on language from the late 1950s. The essays in *On the Way to Language* are often neglected in discussions of Heidegger's account of art, although Gerald L. Bruns's *Heidegger's Estrangements* (1989) is a notable exception. They can now be read, however, particularly in their clarification of the relation between poetry and thinking, as the culmination of his efforts to explore what he had earlier expounded as the task of great art, with the focus now on showing, gathering, and granting. By giving the name "saying" to the realm of the belonging together of thinking, being, and language, Heidegger explored another, more penetrating, way to express his claim from the 1930s that art was an origin, a way that was no longer obliged to rely on the concept of art as such.

Heidegger never resolved the basis on which he extended his discussion of poetry to architecture, painting, sculpture, and music. His argument that all the arts, even the nonlinguistic arts, were dictated by language was insufficient. It left unexplained why he tended to place so much faith in the aesthetic classification of the fine arts when he had rejected the basis on which it was made. In "The Origin of the Work of Art," however, seemingly just in passing, Heidegger expressly left open "whether art, in all its modes from architecture to poesy, exhausts the nature of poetry *[Dichtung],*" thereby acknowledging the possibility of a future in which poetry would again no longer be circumscribed by the concept of art. Heidegger's attempts to replace the concept of art by a broad concept of poetry, by myth, and by saying were as important a part of his efforts to negotiate the aesthetic concept of art as his attempts to twist the concept of art free from aesthetics. In other words, by turning away from the concept of art in search of something more original, and also by rehabilitating it as if it had its own originality, Heidegger pursued two formally inconsistent strategies. Future investigation of this topic needs to ask whether this was vacillation on Heidegger's part or, as deconstruction would insist, the evidence of a necessity that contemporary reflection on the arts is unwise to ignore.

BIBLIOGRAPHY

Works by Heidegger

"Der Ursprung des Kunstwerkes." In *Gesamtausgabe*, vol. 5, edited by Friedrich-Wilhelm von Hermann, *Holzwege*, pp. 1–74. Frankfurt am Main, 1977. Translated by Albert Hofstadter as "The Origin of the Work of Art," in *Poetry, Language, Thought* (New York, 1971), pp. 15–87.

"Die Frage nach der Technik." In *Vorträge und Aufsätze*, pp. 13–44. Pfullingen, 1954. Translated by William Lovitt as "The Question

concerning Technology," in *The Question concerning Technology and Other Essays* (New York, 1977), pp. 3–35.

"Grundsätze des Denkens." In *Gesamtausgabe*, vol. 79, edited by Friedrich-Wilhelm von Herrmann. *Bremer und Freibürger Vorträge*, pp. 79–176. Frankfurt am Main, 1994.

Parmenides. In *Gesamtausgabe*, vol. 54, edited by Manfred S. Frings. Frankfurt am Main, 1982. Translated by André Schuwer and Richard Rojcewicz as *Parmenides* (Bloomington, Ind., 1992).

"Technik und Kunst—Ge-stell." In *Kunst und Technik*, edited by Walter Biemel and Friedrich-Wilhelm von Herrmann, pp. xiii–xiv. Frankfurt am Main, 1989.

Unterwegs zur Sprache. In *Gesamtausgabe*, vol. 12, edited by Friedrich-Wilhelm von Herrmann. Frankfurt am Main, 1985. Translated by Peter D. Hertz as *On the Way to Language* (New York, 1971).

Other Sources

Bernasconi, Robert. "The Greatness of the Work of Art." In *Heidegger in Question: The Art of Existing*, pp. 99–116. Atlantic Highlands, N.J., 1993.

Bruns, Gerald L. *Heidegger's Estrangements: Language, Truth, and Poetry in the Later Writings.* New Haven, 1989.

Gadamer, Hans-Georg. "The Truth of the Work of Art." In *Heidegger's Ways*, translated by John W. Stanley, pp. 95–109. Albany, N.Y., 1994.

Kristeller, Paul Oskar. "The Modern System of the Arts." In *Renaissance Thought and the Arts*, pp. 163–227. Princeton, N.J., 1980.

ROBERT BERNASCONI

Philosophical Heritage in Heidegger's Conception of Art

Martin Heidegger's philosophical interest in the work of art appears relatively late in his writings. One searches in vain for the terms *art* and *work of art* in the otherwise careful index of the English translation of *Sein und Zeit* (Being and Time). Furthermore, philosophical reflections on the work of art are practically nonexistent in the Marburg lectures that mark the genesis of his 1927 masterpiece. Even those such as Hans-Georg Gadamer who were familiar with the first phases of Heidegger's writings were surprised to receive an announcement in 1935 of a Heidegger philosophy conference in Freiburg about the origin of the work of art. Yet, it would be wrong to conclude, on the one hand, that the themes discussed at this 1935 lecture, and clarified in a series of three lectures held in Frankfurt in 1936, created a discontinuity in Heidegger's itinerary, or that these themes were not at all anticipated in Heidegger's earlier works. On the other hand, the fact that the texts actually delivered at the 1935–1936 lectures did not refer explicitly to predecessors in German philosophy (with the exception of allusions made to G. W. F. Hegel in an introduction written a little later) could lead one to think that Heidegger's concept of art had no ties to the heritage of German philosophy about the subject. In both cases, the first impression is false because the Heideggerian approach to the work of art has both an ontogenesis and a phylogenesis.

Ontogenesis. *Being and Time* introduces a theoretical project that Heidegger calls "Fundamental Ontology." This project, which is centered on the question of Being, maintains that it inheres in the question that it be thrown back at the one who poses it, and that a rigorous and synthetic investigation of the different meanings of the word *Being*—such polysemy already having been mentioned by Aristotle—first requires a transcendental analysis of the mode Being of the human being [*d'être de l'existant humain*]. This mode is what Heidegger calls *Dasein* precisely because to exist is a question that concerns *Dasein* intimately. In such an analysis, Heidegger selects a certain number of transcendental factors constitutive of human existence (the "existentialia"), and regroups them into a synthetic structure that is called *Sorge* (Care). Care is a condition whereby *Dasein* is thrown into its own Being and, at the same time, into the project of the possibility of its own Being. This condition of Care is finite insofar as its own end, death, determines it intrinsically; and Care is, at the same time, fundamentally temporal but has a temporality that is itself intrinsically finite. Fundamental ontology aims to show that such finite temporality is the only transcendental horizon of intelligibility for the diverse meanings of the term *Being*.

The earliest ontogenesis of the Heideggerian meditation on the work of art is found in the works that set the background for the analysis of *Dasein* and, more generally, for "fundamental ontology." This ontology places the human being in between two poles: authenticity *(Eigentlichkeit)* and inauthenticity *(Uneigentlichkeit)*. The way of Being is authentic when *Dasein* cares about its own possibility, that is about Being-unto-death. The way of Being is inauthentic, by contrast, when *Dasein* is concerned with possibilities (for example, objectives to meet, works to produce) that are actualized in everyday life that are shared with others. The origin of this distinction between authentic and inauthentic, in the formal sense of a contrast between the proper and the improper, is found in the phenomenological works of Edmund Husserl during the time that the young Heidegger was his assistant. But Heidegger's Marburg lectures, which took place during the gestation of *Being and Time*, demonstrate that his concrete formulation of the polarity between authenticity and inauthenticity owes a great deal to a long meditation on Aristotle's *Nicomachean Ethics*, a work in which Heidegger recognizes a genuine ontology of *Dasein*.

The term *art* appears in Aristotelian discussions. The Greeks only had one term—*technē*—to designate the art of artists (painters, sculptors, epic and tragic poets), of artisans (potters, carpenters), and of experts (doctors and strategists). All those who had an art, that is, a know-how, formed a group called *technitēs*, however diverse their activities. In the same way, there was only one term, *poiēsis* "production," to designate the activity of all the producers with a particular capacity; little did it matter whether the capacity was to produce works, utensils, or results. The *Nicomachean Ethics* discusses *technē* by inquiring into levels of excellence or virtue of which human life is capable. This in-

quiry establishes a hierarchy between active life and contemplative or theoretical life, and sees art, in the broad sense of the term *technē*, among two types of excellence that active life can attain. At the heart of the active life, art is an excellence of an inferior type, because any practical know-how presides over an activity that produces only ends or goods exterior to the agent. By contrast, the superior excellence of active life, what Aristotle refers to as *phronēsis*, governs an activity that intrinsically concerns the agent, because this activity, known as *praxis*, is the very conduct of life.

Heidegger's analysis of the *Nicomachean Ethics* while he was at Marburg and before *Being and Time* was published, above all in the long introduction to his 1924 manuscript on Plato's *Sophist* (*Gesamtausgabe*, vol. 19), shows that the future formulation of the analytic of *Dasein* between the polarity of inauthenticity and authenticity is indebted largely in its tenor to Aristotle's distinction between the activity preoccupied with the exterior aims of *poiēsis* and the activity of *praxis* that is *hou eneka* (self-intentional) in the sense that the agent cares about living well. This Heideggerian analysis of Aristotle, however, precisely because it is oriented toward a bipolar formulation of an analysis of *Dasein* and fundamental ontology, indicates that during this phase of the elaboration of *Being and Time,* art in general, including art by the artists, belongs to the order of inauthenticity and, given the limits of this order, would not be able to have a fundamental and authentic status. At the same time, it is incontestable that the Marburg lectures revealed a major theme of Heidegger's future meditation on the origin of the work of art. The theme is truth understood in terms of unveiling or *alētheia.* In effect, the Heideggerian lecture on Aristotle emphasizes that art, understood in the sense of a know-how presiding over production, owes its excellent "dianoetic" character to its ability for discovering materials, the means and the ends that are required for production. Art is a way of *alētheuein,* of unveiling or discovering, and a capability to put *alētheia* into a work. In continuation of this lecture in Aristotle, the 1935–1936 lecture argues that the origin of the work of art is art understood as truth setting itself to work. In order for the meditation on the work of art to grant this happening of truth in art an eminently ontological status, however, Heidegger had in the meantime to introduce a clear distinction between, on the one hand, a *technē* and *poiēsis* directly tied to everyday life and destined to fall away from authenticity and, on the other, a *technē* and *poiēsis* rescued from the inauthentic. This distinction does not exist in the 1927 formulation of *Dasein*, which entirely reduces *technē*—the technical know-how as well as the art of artists—to the level of practical circumspection immersed in everyday life. This distinction between two levels of *technē* will reemerge forcefully in the sadly famous Rectorat discourse of 1933.

The 1933 discourse is manifestly inspired by Plato's *Republic.* For those who are interested in the ontogenesis of the Heideggerian meditation about the origin of the work of art, the question becomes how the meditation is indebted to Heidegger's long association with Platonic texts in his previous writings. This question is all the more necessary because in his lecture on Aristotle, while Heidegger is teaching in Marburg, he appears not to pay any attention to Aristotle's *Poetics,* a text where he would have been able to find arguments in favor of a clear distinction between the *technē* and *poiēsis* of artisans and experts and the *technē* and *poiēsis* of artists and poets specializing particularly in tragedy. On the other hand, if Aristotle cannot inspire the young Heidegger's lack of philosophical interest in the work of art, because the *Poetics* claims that there is a philosophical teaching to be found in works of art and above all in tragedy, Heidegger's lack of interest could certainly have derived from Plato's criticism of poetry.

By looking at the Heideggerian approach to the work of art in 1935–1936 from this angle, one has to recognize that the ensemble of works written just before and after *Being and Time* tended not only to erase the differences between Plato and Aristotle, but to take from Aristotle that which helps clarify and confirm, in terms of existence, the absolute priority Plato gives to contemplative life, namely, life devoted to the understanding of Being. It is probably because this ensemble of works concentrates exclusively on this priority that Aristotle's *Poetics* is ignored; the *Poetics* has as its axes not the presence of early philosophy but rather, and more important, the specific dignity of human fragile interaction within a world of appearances. In any case, it is in terms of the Platonic predominance of the view of Being that Heidegger could, soon after completing *Being and Time,* during a course on "The Essence of Truth" dated 1931–1932 (*Gesamtausgabe*, vol. 34), give his own conception of truth through a detailed analysis of the parable of the cave, a key passage in Plato's *Republic.*

This analysis interprets the phases of the parable as steps toward the light, that is, toward the unveiling of being in its Being *(de l'étant dans son être).* In this context, Heidegger introduces his account of the Platonic concept of the philosopher-king; Heidegger believes that the one who is dedicated to the contemplation of Being is the only authentic guardian of the "polis," the one who is capable of establishing its rules (1962, p. 100). In the same context, he says for the first time that the fine arts (more precisely, what he calls "great poetry"—in particular, that of Sophocles) have the power to see the light of Being. He adds that in order to understand this, one must stop considering the problem of art as an aesthetic problem (ibid., pp. 63–64). In this double gesture in which he celebrates both the ontological bearing of great art and what he calls the transcendence of aesthetics, it is Plato who inspires Heidegger. He claims that, in the *Phaedrus,* the Beautiful is one of the names of Being (ibid., p. 198). This role of the Beautiful suggests implicitly that Plato's diatribes against the artists are not the result of an

objection to art but of an objection to works that set out to please immediately, an objection to a low level of art made in the name of a higher *technē*, which has a superior position in the unveiling of Being. In the 1933 Rectorat discourse, this eminent "technē," qualified as "a supreme setting into work" *(mise-en-œuvre-suprême)*, is given a superior position in the ontological unveiling of Being, attributing to it a Promethean grandeur. It is *technē* that forms the dominant theme in philosophical discussions about art that appear, even before the essay on the origin of the work of art, in Heidegger's first lecture course on Friedrich Hölderlin (1934–1935) and in the *Introduction to Metaphysics* (1935). It is during this period that the ontogenesis outlined above joins the phylogenesis through the inspiration emanating from the German philosophical heritage, in particular that of Hegel and Friedrich Nietzsche, both fervent admirers of the Greeks.

Phylogenesis. The structure of the first Heideggerian interpretation of Hölderlin (*Gesamtausgabe*, vol. 39) follows from fundamental ontology, in the sense that the structure shows the contrast between the decadence of everyday life and the resolute authenticity of *Dasein*. But *Dasein* is no longer simply an individual entity, because, according to Heidegger, it is the German people that the poet addresses, calling them to await the return of the gods that fled. Rather than on the finite temporality of the individual, the focus now is placed on the destiny of a people. Heidegger believes that only a few creators are capable of confronting such a destiny, namely, the poet who will unveil it, the thinker who will conceptually articulate the unveiling initiated by the poet, and the founder of a state who will lead the nation *(Volk)* toward its destiny. The topic of creation, as origin or as new beginning, insofar as it institutes or unveils the Being of a collective *Dasein*, moves to the forefront *(avant-plan)* of Heideggerian thought in this period. Now, since the Being of a collective *Dasein* is, like that of individual *Dasein*, polarized between fallenness and authenticity, the theme of creation is associated here with that of conflict, a theme dear to Heracleitus.

The association of these two themes becomes an important role, and gives rise to diverse appropriations. Because Hölderlin and Hegel had both been inspired by Heracleitis when they were young friends, Heidegger lets Hegel's *Phenomenology of Spirit* overlap onto his own reading of Hölderlin. Heidegger does this in three points: the idea that the life of Spirit depends on the ability that one has to face the negative and death; the notion of "unhappy conscience," which is tied to the flight of the gods; and, finally, the idea that it is the state that permits the Spirit to fulfill itself historically.

Moreover, the same association of the theme of creation with that of Heraclitean *polemos* allows Heidegger, in his lecture on the poet, to appropriate the writings of Nietzsche. Heidegger focuses on three points: the Dionysian;

the distress stemming from the flight of the gods; and the struggle for the *Übermensch* (overman) in opposition to those who are superfluous.

In this analysis, the question of complicity with the National Socialist revolution can hardly be contested. The text of the lecture course on the *Introduction to Metaphysics* was written shortly thereafter, the same year that Heidegger drafted "The Origin of the Work of Art." In the *Introduction to Metaphysics*, to introduce oneself to metaphysics means to reconquer actively, as the destiny of a people, the German people, the fundamental experience that brought the beginning of philosophy to the pre-Socratic thinkers. But Heidegger argues that this reconquest has a German antecedent: German Idealism as it marks the effigy of a "world of Spirit" in which the philosopher celebrates "grandeur, amplitude, original authenticity." In this way, the *Introduction to Metaphysics* does not hide the inspiration drawn from the Hegelian notion that a work of Spirit consists of a triple modality: political, artistic, and philosophical. But, on the other hand, the inspiration drawn from Nietzsche, which was already evident in the lectures on Hölderlin, is present in terms of "violence" and "combat" and "power" where Heidegger proclaims the necessity of recapturing the pre-Socratic fundamental experience. According to Heidegger, this experience would have been about the struggle between the unveiling and the veiling of Being, and the strength of the creators *(créateurs)* in the Greek world would have been to face—in poetry, in philosophy, and in politics—that struggle to whose importance the weak caught up in everyday life are blind. Besides, the antagonism between the unveiling and the veiling of Being in sheer appearance was, according to Heidegger, the dominant theme not only in pre-Socratic thought but also in the poetry of the Greek tragedies, of Sophocles in particular. On this point, again, the double heritage of Hegel and Nietzsche appears to guide from afar the Heideggerian interpretation of the Greek world. In effect, as in Hegel's *Phenomenology of Spirit*, Heidegger interprets the Greek world in light of Sophocles' *Antigone*. Like his predecessor, Heidegger attributes a metaphysical tendency to tragedy as well as to other works, and sees the crowning of Greek art therein. On the other hand, like Nietzsche during the time of *The Birth of Tragedy*, Heidegger underscores the importance of the tragic chorus and insists on a profound accord between the meaning of tragedy and the dialogue of Heracleitus, yet without aligning himself with the anti-Parmenidean views of Nietzsche. Given that the writings discussed above include a number of criticisms of Plato, notably for having begun to conceive truth more as rectitude than as unveiling, one might be tempted to think that Heidegger's revival of this heritage of German philosophy marks a rupture with the dominant Platonist ontogenesis retraced earlier. Such a belief is wrong: despite the incontestable metamorphosis that this belief represents in relation to the work of the preceding

decade, Heidegger's thinking between 1933 and 1936 remains Platonist in the sense that it does not cease to proclaim the excellence of the *bios theoretikos,* which is devoted to the contemplation of Being. Moreover, it does not cease to proclaim, in agreement with Plato, that the Beautiful is one of the names for Being.

It is the junction of the ontogenesis described earlier and the ontogenesis just outlined that is discussed in "The Origin of the Work of Art." In light of the ontogenesis, Heidegger's reflection on a Vincent van Gogh painting of shoes allows him, in this essay, to integrate two new levels of *technē:* a pragmatic, everyday level and a new ontological level achieved in the work of art. The painting by van Gogh reveals that Being-in-the-world is both presupposed and hidden by the use of everyday work shoes; it also shows the essence of the work of art; in such a work, truth has set itself to work. Some particular being, a pair of peasant shoes, has come in the world to stand in the light of its Being.

Regarding phylogenesis, on the other hand, it is in the line of the Hegelian heritage, and more precisely the chapter "Religion in the Form of Art" in Hegel's *Phenomenology of Spirit,* that Heidegger claims to be only interested in what he calls "great art," exemplified in his eyes by the Greek temple, as far as the plastic arts are concerned, and in tragedy, as far as the spoken word is concerned. It is in the wake of Hegelian reflection on the successive figures of Spirit since Greece that Heidegger associates art history with a history of truth, and insists on the irrevocable character of the collapse of the worlds that were revealed by the works created at the time. It is also in the wake of Hegel that Heidegger asks in a postscript whether art can still be an essential and necessary accession to truth in the modern period of the reign of aesthetics. This Hegelian inspiration, however, is not solely exclusive in the sense that Hegel's philosophy of art marks the commemoration of a path already taken by the Spirit, whereas Heidegger's meditation claims to prepare the future and calls for a decision. But it is on this point, no doubt, that he listens to Nietzsche, whose second "Untimely Meditation" seems to put the Heideggerian characterization of History on the alert about returning to the past, and insists on the necessity for finding caretakers for creative works. On the other hand, when Heidegger confirms that truth irrupts in the work of art in the form of combat between the World and the Earth, and emphasizes that the combat at stake is not only between concealment and unconcealment, but also between measure and excess, one cannot refrain from suspecting here an echo of Nietzsche's meditation about Apollo and Dionysus. This echo is confirmed by the fact that during the winter semester of 1936–1937, just a few months after the conferences on the origin of the work of art, Heidegger comes back to many of these themes during a course devoted to "The Will to Power as Art," a course in which, far from distancing himself from Nietzsche, he expresses his thoughts through Nietzsche. It

is important to note that this course on Nietzsche included certain pages related to Immanuel Kant's *Critique of Judgment,* which only confirms a posteriori the Platonist tendency that is present in the ontogenesis and phylogenesis above. In fact, these pages do not address what in the Kantian analysis of aesthetic judgment concerns the debate with others inherent in the *sensus communis.* Only one Kantian topic is discussed, namely, favor *(Gunst),* carefully described in terms of "letting be," ontological contemplation. Everything takes place as though only the metaphysician is competent in the eyes of Heidegger to speak about the work of art. This thesis was without doubt that of Plato in ancient times, and of Hegel and Nietzsche in modernity, but certainly not of Aristotle or Kant.

[*See also* Aristotle; *and* Plato.]

BIBLIOGRAPHY

Works by Heidegger

An Introduction to Metaphysics. Translated by Ralph Manheim. New Haven, 1959.
Being and Time. Translated by John Macquarrie and Edward Robinson. New York, 1962.
Gesamtausgabe. Frankfurt am Main. 1976–.
"The Origin of the Work of Art." In *Poetry, Language, Thought,* translated by Albert Hofstadter, pp. 15–87. New York, 1971.

Other Sources

Aristotle. *Complete Works of Aristotle.* 2 vols. Translated and edited by Jonathan Barnes. Princeton, N.J., 1984.
Hegel, G. W. F. *Phenomenology of Spirit.* Translated by A. V. Miller. Oxford, 1977.
Nietzsche, Friedrich. *The Birth of Tragedy* (with *The Case of Wagner*). Translated and edited by Walter Kaufmann. New York, 1967.
Nietzsche, Friedrich. *Untimely Meditations.* Translated by R. J. Hollingdale. Cambridge and New York, 1983.
Plato. *The Collected Dialogues of Plato.* Corr. ed. Edited by Edith Hamilton and Huntington Cairns, translated by Lane Cooper et al. Princeton, N.J., 1963.

JACQUES TAMINIAUX
Translated from French by Annabel Manning

Heidegger and Hölderlin

It is widely accepted that the tragic poetry of Sophocles is the key to Aristotle's *Poetics,* which is, arguably, the founding text of aesthetics as a philosophical discipline. Until recently, it has not been appreciated how important the poetry of Friedrich Hölderlin (1770–1843) has been in shaping the ideas about art found in Martin Heidegger's "The Origin of the Work of Art" (1936), a work that aims to overcome aesthetics. Nor has it been understood to what extent Heidegger mistakenly identifies the views expressed in the poetry of Hölderlin with the ideas of Friedrich Nietzsche (1844–1900), whose *Birth of Tragedy* also calls into question the philosophical discipline of aesthetics. Like Hölderlin before him, Nietzsche offers a theory of the origin of tragic

poetry that challenges the picture of Sophoclean tragic poetry that Aristotle presents in his *Poetics*. Both Hölderlin and Nietzsche are enlisted by Heidegger in the mid-1930s to assist him in overcoming aesthetics.

The title of Heidegger's essay, "The Origin of the Work of Art," is meant to call into question two beliefs that are fundamental to aesthetics: first that an individual subject, the author or artist, is the cause of an object that is the work of art, and second that the object, the work of art, causes an effect on the feelings of an isolated subject, the reader or viewer. Heidegger's strategy in "The Origin of the Work of Art" is to lead readers to abandon their belief in artistic genius and in an aesthetic experience. Both beliefs assume a subject in isolation either as the cause of or as affected by the work of art. Aesthetics, according to Heidegger, must be overcome so that art can begin to be a community-building and history-making force for the German people.

Recently published archival material shows that Heidegger drafted a version of "The Origin of the Work of Art" as early as 1931. He first lectured publicly on this topic in November 1935, and a year later, in November and December of 1936, expanded this into the three lectures that were published together under the final title in 1950. Most famous for its much-debated discussion of Vincent van Gogh's painting of a pair of peasant shoes, Heidegger's essay is also widely known for its discussion of a Greek temple. Until recently, little attention has been paid to the complex role played in this essay by the poetry of Hölderlin, which Heidegger still reads in 1936 through the lenses of Nietzsche's philosophy. Prior to his first public lecture on the origin of the work of art in 1935, Heidegger had already devoted a lecture course in the winter semester of 1934/35 to a reading of Hölderlin's poems "Germania" and "The Rhine." In addition, Heidegger had delivered a lecture titled "Hölderlin and the Essence of Poetry" in April 1936, just before he expanded his 1935 lecture into three lectures. This background helps to explain why "The Origin of the Work of Art" (the three lectures) concludes that "The nature of art is poetry" (Heidegger, 1971, p. 75) and why it ends with a poem, "The Journey," by Hölderlin.

Hölderlin's poetry provides more than just the background for Heidegger's essay. It also provides concepts. When Hans-Georg Gadamer introduced the popular 1960 Reclam edition of the essay, he wrote of the surprise he felt at reading about Heidegger's new concept of earth that suddenly is said to be in a relationship of strife with world, a concept familiar to readers of Heidegger's *Being and Time* (1927). Gadamer notes correctly that this new concept of earth comes from the poetry of Hölderlin. But there is more. In the 1980s, Heidegger's three lecture courses on poems by Hölderlin from the winter semesters of 1934/35 and 1941/42 and from the summer semester of 1942 were published. They show the mark of Hölderlin on the very concept of the strife *(Streit)* between earth and world, a

strife that Heidegger takes—or rather mistakes—to be of history-making importance for the German people in the Adolf Hitler years.

Until these lecture courses were published in the 1980s, it was difficult to establish any direct connection between Heidegger's writings about art and poetry and his membership in the National Socialist Party. Now, however, one finds Heidegger claiming that Hölderlin's poetry contains a formulation of the distinction between the Apollonian and the Dionysian that both predates the distinction offered by Nietzsche in *The Birth of Tragedy* and reformulates this Greek distinction for Germany and its poetry (Heidegger, 1980, pp. 293–294). This distinction is important for Heidegger because he believes that it explains the "essential opposition" between Greece and Germany (ibid., p. 290), which, according to Heidegger, accounts for Germany's unique history-shaping destiny among modern nations. Heidegger believes that Sophocles' tragic poetry (as opposed to Homer's or Hesiod's epic poetry) inaugurated the epoch of Western history that begins with the Greeks and he is convinced that Hölderlin's hymns will inaugurate a new history of the West to be identified with Germany and no longer with Greece. It is therefore not at all by chance that virtually all of Heidegger's writings on Hölderlin's poetry and Nietzsche's philosophy take place during the Hitler years when Germans called for a radical break with the last two thousand years of Western history.

Heidegger's artwork essay with its emphasis on the necessary strife between world and earth has, until now, been read independently of his commitment to National Socialism, and many, especially postmodernists, have valorized Heidegger's idea that the "work-being of the [art]work consists in the fighting of the battle *[Streit]* between world and earth" (1971, p. 49). The historical parallel, however between Heidegger's belief in strife *(Streit)* between earth and world and the essential opposition *(Wesensgegensatz)* between Germany and Greece suggests a different way to understand the progress of the three parts of "The Origin of the Work of Art," as well as the significance of the specific artworks that Heidegger discusses in each of the essay's three parts. The relation of necessary strife between earth and world worked out in Heidegger's essay nevertheless remains a significant breakthrough independent of Heidegger's political views about the essential opposition between Germany and Greece. Gadamer, for example, retrieves what is valuable from the essay in *Truth and Method* (1960) (part 1) by connecting it with an altogether different understanding of history, including German history (part 2). Gadamer's interpretation of "the conversation that we are" (Hölderlin) as a dialogue in search of consensus (part 3) is his alternative to Heidegger's interpretation of this conversation in terms of strife and essential opposition.

How does Heidegger's commitment to Nazism at the time of his writing "The Origin of the Work of Art" affect

what he says in this work? The first part of the essay, titled "Thing and Work," aims at shattering the way of asking about the work of art that is customary in aesthetics (1971, p. 39). It and the example of van Gogh's painting of a pair of peasant shoes have received the most attention, for example, by Meyer Schapiro and Jacques Derrida. Van Gogh's painting is the occasion for the introduction of the concept of earth: "This equipment [the peasant shoes] belongs to the *earth,* and it is protected in the *world* of the peasant woman. From out of this protected belonging the equipment itself rises to its resting-within-itself" (ibid., p. 34). When one reexamines Heidegger's discussion of van Gogh's painting in relation to his commitment at that time to Nazism, one notices two things. First, he does not claim that there is any strife between earth and world in van Gogh's painting. Instead, everything is at rest *(Ruhe).* Second, he chooses to discuss an artwork that presents the ordinary and everyday world of work. As the daily world of a peasant woman, this work world is far removed from any kind of history-making politics.

Heidegger discusses a Greek temple, presumably the temple at Olympia dedicated to Zeus, in the second part of "The Origin of the Work of Art," titled "The Work and Truth." Again one finds earth and world, but this time they are said to be at strife with one another:

> In setting up a world and setting forth the earth, the work is an instigating of this strife *[Streit].* . . . The work-being of the work consists in the fighting of the battle *[die Bestreitung des Streites]* between world and earth. . . . The fighting of the battle is the continual self-overreaching gathering of the work's agitation *[Bewegtheit].* The repose *[Ruhe]* of the work that rests in itself thus has its presencing in the intimacy of strife.
> (Ibid., pp. 49–50; the author has replaced "striving," a mistranslation of *Streit,* with "strife."—K. W.)

Heidegger emphasizes strife because the temple is historically decisive: "The temple-work, standing there, opens up a world and at the same time sets this world back again on earth, which itself *only thus emerges as a homeland [der heimatliche Grund]*" (ibid., p. 42; emphasis added; translation modified). The temple celebrates the extraordinary event of Greece's emergence as a history-making force that gives direction to the next two thousand years of Western history.

The strife between earth and world is necessary, according to Heidegger, if the earth is to become a homeland in which a "historical people *[Volk]* . . . returns to itself for the fulfillment of its vocation" (ibid.). Heidegger employs his concept of truth as unconcealedness *(alētheia)* to characterize the fulfillment of a people's, in this case the Greek people's, vocation: "Setting up a world and setting forth the earth, the work [the Greek temple] is the fighting of the battle in which the unconcealedness of beings as a whole, or truth, is won" (ibid., p. 55). Heidegger believes that in and

through Greek art what "was in the future to be called Being was set into work, setting the standard" (ibid., p. 77). He also believes that the Germans are like the ancient Greeks in that they too are a historical people with a historically decisive vocation. The vocation of the Germans, however, requires of them that they turn against all that goes back to ancient Greece and overcome, for example, aesthetics. Heidegger will soon add to this the requirement of overcoming metaphysics.

The third part of "The Origin of the Work of Art," titled "Truth and Art," connects the question of what the work of art originates, the existence of a people that decides how Western history will be defined, with the readiness of the German people in the mid-1930s to allow the poetry of Hölderlin to make of them a history-making people. This last part of Heidegger's artwork essay does not feature a discussion of a particular work of art, as was the case in the first part (van Gogh's painting) and the second part (the Greek temple). Instead, it brims over with explicit and implicit references to poems, letters, and other writings of Hölderlin that speak to the history-making decision facing the German people. Heidegger phrases this decision as follows: "Are we in our existence historically at the origin? . . . Or, in our relation to art, do we still merely make appeal to a cultivated acquaintance with the past?" (ibid., p. 78). Heidegger's own decision is clear from the end of his lecture course in 1934/35 on the poems "Germania" and "The Rhine": "The hour of our history," he pronounces, "has struck" (1980, p. 294). "The Origin of the Work of Art" ends in 1936 with the following line from Hölderlin: "With difficulty *[Schwer]* / that which dwells near its origin departs" (1971, p. 78; "The Journey," verses 18–19; translation modified). This line evokes, now in translation, the well-known Greek adage "All beginnings are difficult" ("arche depou pantos ergou kalepoteron esti"), thereby challenging the Germans of Heidegger's own time (1936) to make the decision to do the difficult and to be the origin of a new, because German, history.

BIBLIOGRAPHY

Works by Heidegger

"De l'origine de l'œuvre d'art." Translated by Emmanuel Martineau. Paris, 1987.

"Der Ursprung des Kunstwerkes." In *Gesamtausgabe,* vol. 5, *Holzwege,* edited by Friedrich-Wilhelm von Herrmann, pp. 1–74. Frankfurt am Main, 1977.

Erläuterungen zu Hölderlins Dichtung. In *Gesamtausgabe,* vol. 4, edited by Friedrich-Wilhelm von Herrmann. Frankfurt am Main, 1981.

Hölderlins Hymne "Andenken." In *Gesamtausgabe,* vol. 52, edited by Curd Ochwadt. Frankfurt am Main, 1982.

Hölderlins Hymne "Der Ister." In *Gesamtausgabe,* vol. 53, edited by Walter Biemel. Frankfurt am Main, 1984.

Hölderlins Hymnen "Germanien" und "Der Rhein." In *Gesamtausgabe,* vol. 39, edited by Suzanne Ziegler. Frankfurt am Main, 1980.

"The Origin of the Work of Art." In *Poetry Language, Thought,* translated by Albert Hofstadter, pp. 15–87. New York, 1971. The Epilogue and Addendum were not part of the 1936 manuscript.

"Vom Ursprung des Kunstwerkes." *Heidegger Studies* 5 (1989): 5–22. This version was edited by Hermann Heidegger.

Other Sources

Biemel, Walter. "Elucidations of Heidegger's Lecture 'The Origin of the Work of Art' and the Destination of Thinking." In *Reading Heidegger: Commemorations,* edited by John Sallis, pp. 370–382. Bloomington, Ind., 1993.

Bruns, Gerald L. *Heidegger's Estrangements: Language, Truth, and Poetry in the Later Writings.* New Haven, 1989.

Dreyfus, Hubert L. "Heidegger on the Connection between Nihilism, Art, Technology, and Politics." In *The Cambridge Companion to Heidegger,* edited by Charles Guignon, pp. 289–316. Cambridge and New York, 1993.

Foti, Veronique M. *Heidegger and the Poets.* Atlantic Highlands, N.J., 1992.

Gethmann-Siefert, Annemarie. "Heidegger and Hölderlin: The Over-Usage of 'Poets in an Impoverished Time'." *Research in Phenomenology* 19 (1989): 59–88.

Harries, Karsten. "Poetry as Response: Heidegger's Step beyond Aestheticism." *Midwestern Studies in Philosophy* 16 (1991): 73–88.

Harries, Karsten, and Christoph Jamme, eds. *Martin Heidegger: Politics, Art, and Technology.* New York, 1994.

Pöggeler, Otto. "Philosophy in the Wake of Hölderlin." *Man and World* 7 (May 1974): 158–176.

Taminiaux, Jacques. "The Origin of 'The Origin of the Work of Art'." In *Reading Heidegger: Commemorations,* edited by John Sallis, pp. 392–404. Bloomington, Ind., 1993.

Wright, Kathleen. "Heidegger's Hölderlin and the Mo(u)rning of History." *Philosophy Today* 37.4 (winter 1993): 423–435.

Wright, Kathleen. "Heidegger's Authorization of Hölderlin's Poetry." In *Martin Heidegger: Politics, Art, and Technology,* edited by Karsten Harries and Christoph Jamme, pp. 164–174. New York, 1994.

Zimmerman, Michael. *Heidegger's Confrontation with Modernity: Technology, Politics, and Art.* Bloomington, Ind., 1990.

KATHLEEN WRIGHT

HELLENISTIC AESTHETICS. [*To treat the aesthetics of the Hellenistic Age (336 BCE to 31 BCE), this entry comprises two essays:*

Visual Arts

Philosophers and Literary Critics

The first essay examines the aesthetics of the art that is characteristic of the period. The second essay analyzes the relationship between the philosophy and literary theories developed in this same age. For related discussion, see Comparative Aesthetics; Roman Aesthetics; *and* Sculpture.]

Visual Arts

Variability, stylistic ambivalence, a rapidly expanding artistic repertory, and the progressive diffusion of motifs and forms beyond the Greek homeland characterize Hellenistic art from Alexander the Great, (336 BCE) to the Roman conquest (31 BCE). The concomitant uncertainty, even the probable absence, of a singular aesthetic tradition, usually traced from Aristotle to Longinus, provides a similarly confused picture of the attitudes and taste of the very diverse patrons, creators, and consumers of artworks over three centuries of intense activity. To say that diversity is the ultimate characteristic of Hellenistic art is to acknowledge the difficulty that art historians of the period have had, and continue to have, in isolating and defining its most salient patterns of development. In part, this critical historiographical problem reflects the physical expansion of the Greek world and its interactive encounter with other mature cultures, the gradual displacement of the classical polis by national or supranational kingdoms, and the rise into prominence of a distinctly metropolitan visual experience, exemplified by Alexandrian topoi. Two other factors must also be considered: the decline of Athens as a dominant, creative center of artistic activity that once, but no longer, established a normative, aesthetic standard, and the development of a new patronage and an art market increasingly devoted to the service of private rather than public interests. Indeed, even the integrity of the historical dimension and character of the very concept "Hellenistic"—meaning, essentially, not quite "Greek"—can be problematic, given its extension to late-republican Rome in one direction and to Central Asia in another.

The complexity and shapelessness of the historical situation cannot, or should not, be subsumed within a terminology dependent on analogous but anachronistic tags, such as the "Hellenistic Baroque." Undeniably, the dearth of an ancient aesthetic literature that responded to and analyzed contemporary artistic practice and goals has left a critical void, inadequately filled by the tantalizing survivals preserved, more or less authentically, in works by Cicero, Vitruvius, Dionysius of Halicarnassus, Quintilian, and Pliny the Elder. References to treatises on sculpture, painting, and architecture, written by Xenokrates of Athens, Antigonus of Karystos, Hermogenes of Alabanda, and Pasiteles from Magna Graecia—all of them practitioners—indicate that these writers were conscious of the nature and objectives of their art (or craft), had a sense of the history and achievements of their masterful predecessors, and wished to convey their views and knowledge to others. Indeed, the precedent for doing so had been established in the fifth century by the sculptor Polykleitos, whose treatise, the *Canon,* incorporated both the principles of his practice and his ideal of meaningful form. The *Canon* does not survive, nor do the subsequent treatises, yet enough is known of the latter to understand their engagement in an ongoing debate about the nature of fifth-century classicism as a normative or, alternatively, repressive standard of aesthetic performance.

In the eighteenth century, Johann Joachim Winckelmann's enthusiastic preference for Greek works of art originating in the fifth and fourth centuries reflected an attitude previously cultivated in antiquity by the Hellenistic scholar-savants who conceived of the "classical tradition" as a canonical collection of masterworks, created by great artists and endowed with an idealizing virtue, representative of

Greek culture at its most noble. It was, of course, a concept put forward in the third and second centuries largely by the poets, philosopher-scientists, and textual critics associated with the great libraries of Alexandria and Pergamon. Their effort to identify and maintain the past glories of artistic endeavor—the works of the sculptors Phidias and Polykleitos, of the architect, Iktinos, of the painters, Polygnotos and Zeuxis—tacitly acknowledged that the Hellenistic present lacked the same ennobling distinction, or failed to reach the high aesthetic standard evident in works of art produced during Athens's golden age.

Beginning in the latter part of the fourth century, especially around the protean figure of the sculptor Lysippus, the tension between the estimable fifth-century style and imagery and the demands of a new age and of an unrepressed creative imagination could not be denied. Looking over his shoulder at Polykleitos, Lysippus nevertheless invented a new scheme for representing the human body that was more elegant and more active than his predecessor's, effectively enhancing the apparent naturalism of the sculptures' form. Contemporary painters, such as Apelles and Philoxenus of Eretria, intensified the spatiality of their pictorial illusions, which, like the new sculpture, began to break down the apparent separation between the viewer's own space and that of the artwork. The viewer's psychology and sensibility were, thereby, more directly engaged in the act of visual perception than ever before—for example, the *Alexander Mosaic* after a lost painting by Philoxenus.

Ultimately, the canonized classical past, the weighty precedents of its idealizing and subsequently idealized models, the cool modalities of its emphatic insistence on moderation and harmony, would be perceived either as a burden to be rejected by artists or as a standard to be constantly reaffirmed. The intensity of its rejection informs the alleged extravagances of Pergamene and Rhodian works—such as the great frieze from the Altar of Zeus at Pergamon, the pathetic *Laocoön*, or the theatrical excesses of the *Farnese Bull*—typical instances of the anticlassical current in the formal language of Hellenistic art. Conversely, the fifth century resurfaced in neoclassical and eclectic revivals associated with Damophon of Messene and other archaizing artists of the second century; they created new works with an "old" gloss or (re-)created earlier types, as if through seeming replication the essence of that past art could once again be recaptured. It is no accident that the contrast between floridity and austerity, familiar characterizations of these two divergent Hellenistic modes, should also be implicated in the criticism of rhetorical practice under the rubrics "Asiatic" (Pergamene) and "Attic" (Athenian).

The critics, ensconced in the libraries, even at Pergamon, might not have gotten their way in shaping the formal appearance of artworks when their tradition-bound concept of beauty failed to obtain support, but they often prevailed in determining the works' content or program. Artistic culture at the highest level was becoming intellectualized, the servant of a cultivated in-group, in part because the role of the intellectual per se had achieved significant codification in the portraiture of philosophers and other revered thinkers, established in the third century and displayed in portrait galleries or in ensembles such as *The Seven Sages.* Beyond the domain of popular self-representation, the savants required an increasingly abstract, even preciously sophisticated, imagery to visualize the rarefied flights of their imagination, the *phantasiae* that rendered visible what had before only existed in their minds. So the academicians of the Pergamon Library participated in the invention of an abstruse iconography for the gigantomachy represented in the program of the great frieze of the Altar of Zeus and dependent on Hesiodic echoes, accessible to so few that identifying labels had to be applied to the individual figures, still visible today in Berlin. Representations of the Muses as vehicles of artistic inspiration became commonplace, as did many other abstract personifications, products of a specialist iconography in the service of an educated class exploiting its special knowledge and its influence on the exercise of royal patronage—the libraries were royal foundations.

Allegorical compositions, informed by dense, abstruse, textual references, were typical products of the hothouse atmosphere of the libraries. They often supported the claims of the classicizing traditionalists, perhaps never so completely as in the late second-century relief representing the *Apotheosis of Homer* in the British Museum. The relief combines Homer, the Olympian gods, the Ptolemaic rulers of Egypt, the Muses, and possibly a contemporary poet, all staged on the rocky slopes of a hill—Parnassus or Helicon—anticipating, as it were, Raphael's greater fresco on a similar theme in the Vatican Stanze. The presence of the Ptolemaic rulers and their full participation in the advancement of the classical tradition, most completely personified in the figure of Homer, constituted but a small portion of the evolving imagery of rule, deemed beneficial to their subjects, and of the widely disseminated portraits of the royal dynasts. But, if Ptolemaic, Seleucid, Attalid, and Antigonid patronage and imagery constituted a new, grand repertory of representation, subsequently influential on the budding imperialists of Rome, they also provided apt models, when used discreetly, by the emerging class of wealthy private patrons who began to enter the art market, employ and commission artists to decorate their houses, fill them with objects, even portraits of themselves, and, eventually, induce artists to create new repertories reflective of their own interests, their own experiences of the world.

After Alexander, the physical expansion of the Greek horizon led not only to the gradual diffusion of Greek art into areas unfamiliar with its principles, but also encouraged Greek scientists, historians, geographers, and ethnographers to observe, record, classify, and assimilate their growing knowledge of a larger, often strange, world. The fa-

HELLENISTIC AESTHETICS: Visual Arts. *Combat between Amazons and Giants,* relief; Pergamon Museum, Staatliche Museen, Berlin. (Photograph courtesy of Giraudon/Art Resource, New York.)

mous mosaic in the Sanctuary of Fortuna at Palestrina, probably a copy of a third-century choragraphic painting representing the journey of Ptolemy II up the Nile, exemplifies the impact of this new knowledge and its transformation into novel artistic motifs: topographical observations, replete with the representation of features of the local terrain, architecture, Greeks and Egyptians, and indigenous plant and animal life, constitute, in effect, an encyclopedic landscape, which only in the "upper reaches of the Nile" trails off into fantasy, no less explicitly described. Landscape and bucolic imagery, often associated with Theocritan escapist poetics, here takes a realistic turn, evidence of a great interest in the visual transcription of the ordinary, and extraordinary, in the knowable world. Furthermore, the congruence of space and time, implicit in the composition of the *Palestrina Mosaic,* reflects the elaborate staging or contextualization of narrative episodes exploited in Menander's comedies, Hellenistic travel literature, the Greek novel, and, eventually, Roman historical reliefs. Whether or not these descriptive motifs were "real" or invented, a mimetic impulse brought much of Hellenistic art down to earth.

Hellenistic artists did not have to go to strange lands to seek out novel subjects. Many of them were to be found in the great commercial centers of the Hellenistic world, espe-

cially in Alexandria, whose urban themes, aspects of "life on the street" in all its particulars, gave rise to the concept of "Alexandrianism," a repertory of genre types that was the very antithesis of Attic classicism. The repertory included unflattering representations of the old and the infirm, of poor peasants and fishermen, of the deformed by birth or disease, particularly dwarfs and hunchbacks, and drunks—in effect, the detritus of society, often caricatured. At the same time, Hellenistic genre subjects included more favorable representations of racial types, of slaves, of young, fashionable women, often rapt in private conversation or at play, of pupils and their teachers, and a great variety of portraits ranging from the high-and-mighty rich to more ordinary folk, both Greek and non-Greek.

Many of these urban types populate the new comedies and constitute the stock characters of the mime with its burlesque tone. In the particularities of their physical appearance, these types also fulfill the theories of the physiognomists, articulated by Theophrastus, the Pseudo-Aristotle, and their scientific followers—primitive psychologists all—who believed that a person's character could be read in the configuration of face and body. If a Stoic moralism also permeated the judgmental reception of many of these genre subjects, equally significant in their generation and recep-

tion is sentiment. Sentiment does not lead to Aristotelian *katharsis,* but rather to a "feeling" that could be intellectualized, controlled, contemplated at some distance, and made into an object of representation. All of this is consistent with Stoic attitudes about the importance of worldly knowledge and the necessity of separating oneself from its sordidness and vulgarity.

Yet, consciously limiting the viewer's empathetic response to an image that demanded attention—*Laocoön,* struggling in the serpent's grasp, *Niobe,* embracing the last of her dying daughters, the *Ludovisi Gaul,* suicidally defiant to the last—often failed when a powerful visual rhetoric was at work, an aesthetic of irresistible spectacle. Artistic virtuosity and psychological insight in the service of dramatic effects elicited the viewer's participation in the mise-en-scène, as if the voyeur were inescapably implicated in the artwork's outreach and thus constituted an essential part of its inclusive composition, a factor readily exploited in the growing use of ekphrastic narratives. Accordingly, illusionistic painting found its willing, reactive beholder, the imagery of the hermaphrodite a viewer who shared the gendered confusion of the subject, erotic motifs of rape and sexual congress a ready public, and the spatially and emotionally aggressive forms of the Pergamon altar frieze a responsive spectator, whose submission to its demands the Christian fathers recognized when they called the altar the "Throne of Satan."

BIBLIOGRAPHY

Fowler, Barbara Hughes. *The Hellenistic Aesthetic.* Madison, Wis., 1989.

Goldhill, Simon. "The Naive and Knowing Eye: Ecphrasis and the Culture of Viewing in the Hellenistic World." In *Art and Text in Ancient Greek Culture,* edited by Simon Goldhill and Robin Osborne, pp. 197–223. Cambridge and New York, 1994.

Gordon, R. L. "The Real and the Imaginary: Production and Religion in the Graeco-Roman World." *Art History* 2.1 (1979): 5–34.

Himmelmann, Nikolaus. *Alexandria und der Realismus in der griechischen Kunst.* Tübingen, 1983.

Laubscher, Hans-Peter. *Fischer und Landleute: Studien zur hellenistischen Genre-Plastik.* Mainz, 1982.

Onians, John. *Art and Thought in the Hellenistic Age: The Greek World View, 350–50 BC.* London, 1979.

Pollitt, Jerome J. "Professional Art Criticism in Greece." *Gazette des Beaux-Arts* 64 (1964): 317–327.

Pollitt, Jerome J. *Art in the Hellenistic Age.* Cambridge and New York, 1986.

Schweitzer, Bernhard. "Xenokrates von Athen." *Schriften der königsberger gelehrten Gesellschaft, Geisteswissenschaftliche Klasse* 9 (1932): 1–52.

Trimpi, Wesley. "The Meaning of Horace's *Ut Pictura Poesis.*" *Journal of the Warburg and Courtauld Institutes* 36 (1973): 1–34.

RICHARD BRILLIANT

Philosophers and Literary Critics

Hellenistic aesthetic theory has its roots in Plato and Aristotle, but diverges significantly from these origins. Building on the work of Plato, the Stoics developed a unique aesthetics that postulates the sensory perception of moral harmony. The other main new philosophical school, Epicureanism, distinguished aesthetic beauty from moral goodness. Apart from the philosophical schools, there is a large body of evidence, albeit scattered, for literary aesthetics. The best single source for Hellenistic aesthetics is the Epicurean Philodemus (c.110–40 BCE), whose prose writings are preserved in fragmentary papyri.

Stoic Aesthetics. The Stoic school (founded by Zeno of Citium, c.335–263 BCE) identified beauty *(to kalon)* with moral goodness *(to agathon)*. Aesthetics is thus a form of ethics. Like every branch of Stoic philosophy, aesthetics is closely integrated within a general worldview. The world is a harmoniously functioning system, regulated in every detail by divine reason *(logos)*, or God. The whole is perfectly good and beautiful. It is the task of human beings to perfect their own reason so as to integrate it within this order.

The Stoics laid down a number of aesthetic principles, which may be regarded as the foundation of a theory of aesthetics. In the first place, they defined beauty as the "perfectly symmetrical." This consists in the perfect adjustment of the parts to each other and to the whole. Whereas Plato, followed later by the Neoplatonists, held that perfect beauty consists in a unity without parts, the Stoics proposed that it is a beauty combined out of parts. In addition, beauty is identical with utility, because moral goodness is identical with utility.

Further, all beauty is natural, as well as artistically crafted. Because nature is an artist (literally, a craftsman, *technitēs*), all works of nature are also works of art *(technē)*. Cosmic nature is identical with god, who pervades every part of the world. The human artist emulates this divine craftsman.

Finally, beauty exists in material form; for everything that exists, including the mind, is corporeal. In a strict sense, a work of art is a mental creation, having the same corporeality as the mind itself. The external product, such as the spoken or written text, has a different kind of corporeality, which is extrinsic to the work of art in the strict sense. Mental matter does not resist beauty but is the medium in which beauty is realized. Nothing, therefore, prevents a human artist from achieving perfect beauty and so sharing in the work of god.

The Stoics are heirs to Plato's vision of art as an expression of moral beauty. Whereas Plato, however, restricted the human artist to an imperfect imitation of moral beauty, the Stoics grant the artist the freedom to soar to the level of divinity.

Although these general principles apply to all arts, we have detailed information only about music and poetry. Following Plato, the Stoic Diogenes of Babylon (first half of the second century BCE) argued that music has a moral character that influences the moral character of the hearer, as well as stimulates the hearer to action (Philodemus, *On Music,* book 1, fragment 28, and book 4, cols. 15–16). When music

accompanies the words of a poem, it has the power to reinforce the moral message of the poem.

The Stoics took up Plato's challenge of reconciling poetry with philosophy. They welcomed poetry both as a preliminary to wisdom and as a form of wisdom. While recognizing errors made by the poets of the past, they credited them with anticipating truths uncovered by philosophy. The Stoics also gave new meaning to the traditional view of poets as inspired by god. Only the wise person, they held, is strictly a poet. Evidence of their high esteem for poetry is that Cleanthes (331–232 BCE) chose to teach Stoic doctrine by writing poems himself.

One summary of a complete Stoic poetic system has come down to us. It is presented, with criticism, by Philodemus in the fifth book of his *On Poems*. Only the final letters -*on* of the author's name are preserved, suggesting that he may have been Aristo of Chios (third century BCE). Despite textual difficulties, Philodemus's report provides a key to Stoic poetics. It also reinforces the reputation of the Stoics for writing in a dry, unattractive style.

The entire theory consists of a highly schematic system of distinctions couched in a highly technical vocabulary. Its basis is a dichotomy between thought *(dianoia)* and verbal composition *(synthesis)*. A poem is good only if both thought and diction are good. In Stoic terminology, both must be "fine" ("urbane," *asteion*), "worthwhile," and "useful." These terms are coextensive; all signify aspects of moral goodness. A poem is bad when either thought or diction is bad in any way. Poems that are neither good nor bad form an intermediate category. As a bearer of moral attributes, a poem in the strict sense is not the text that is spoken or heard, or written or read, but the mental activity of the poet. The poet is thus in perfect control of his or her work. The way the poem is written or performed makes no difference to its intrinsic worth.

The external product lies between poet and hearer, who perform reciprocal functions. As a mental state, goodness belongs not only to the poet's creative act, but also to the hearer's interpretative response. The poet is good or bad in creating a poem; the hearer is good or bad in responding to the external product. Just as the poet has control over the creation, so the hearer has control over the reception of the poem.

This emphasis on the hearer is an important innovation of the Stoics. Before the Hellenistic period, the hearer was regarded primarily as a passive recipient. The Stoics originated what might be called a reader-response theory. As a consequence, they are in a position to welcome traditional poetry. Unlike Plato, who condemned traditional poetry because of its moral distortions, the Stoics are free to accept all kinds of poetry on the ground that the hearer can make an autonomous, moral use of it. Because the hearer is not defenseless against external impressions, bad poetry need not be avoided. A good hearer can appropriate a bad poem by compensating mentally for any deficiencies.

According to Philodemus's Stoic theorist, the thought and diction of a poem are judged by two distinct faculties: thought is judged by reason, and verbal composition is judged by the perceptual faculty of hearing. In both cases, the fundamental criterion is moral worth. This view is especially problematic with respect to diction. How can verbal expression be morally good, and how can moral goodness be judged the faculty of hearing? The Stoics answered that the linguistic composition of a poem reflects the moral quality of the thought. Spoken language is thought expressed in voice; therefore, the moral goodness of words corresponds to that of the underlying thought. The faculty of hearing judges the moral quality of the words by its own rational ordering; for all perceptions are centered in a single, rationally ordered mind, which judges in response to a sensory stimulus. All perceptual judgments are ultimately judgments of the moral beauty of a sensory object.

Like Plato, the Stoics disapproved of poems that excite the emotions. In opposition to Aristotle, they banned the "pleasure" *(hēdonē)* of pity or fear from both the poet's creative act and the hearer's response. In their view, the emotions (divided into pleasure, desire, distress, and fear) are irrational impulses that must be replaced by a rationally ordered disposition. The Stoics did not eliminate feeling altogether, for they replaced the irrational impulse of pleasure by "joy" *(chara)*, a serene contentment arising from a harmoniously ordered soul.

The Stoics elaborated Plato's requirement that diction must follow content. They held that, by nature, letter sounds correspond to things in the world. It is possible, therefore, to combine letters and syllables in such a way that they naturally reflect the meaning of the words. Certain sounds, rhythms, and melodies express moral goodness, whereas others do the opposite. For this reason, Cleanthes exalted poetic form as especially appropriate to divinity (Philodemus, *On Music*, book 4, col. 28). Because of its special acoustic properties, he proposed, poetry can convey the truth about God better than prose, even philosophical prose.

The Stoic Posidonius (c.135–50 BCE) defined "poem" as verse and "poetry" as a "poem with meaning, containing an imitation of divine and human things" (Diogenes Laertius 7.60). By imitating divine creation through verse, the poet acquires a special affinity with god. He also acts as a companion to the philosopher. Whereas the philosopher explains the work of god, the poet displays it with a lucidity and persuasiveness of his own. The poet—and this is of course the ideal poet, the true poet—imitates god by harmonizing all parts of his creative work.

The Epicureans. Epicurus (341–270 BCE), founder of Epicureanism, identified goodness and utility with pleasure. All moral or physical beauty, he held, is simply a means to the end of pleasure. He offers no aesthetic theory as such. Indeed, he repudiates literary and musical theory, along

with all other scholarly enterprises, as a waste of time, because it contributes nothing to pleasure or happiness.

At the same time, Epicurus offered some guidance on aesthetic enjoyment. He claimed that the wise person is a "lover of sights and enjoys hearing and seeing Dionysiac performances as much as anyone" (Plutarch, *A Pleasant Life* 1095e–1096c; cf. Diogenes Laertius 10.120). This is a response to Plato's denunciation of the vulgar "lover of sights" and "lover of sounds" (*Republic* 475d–476b). Unlike Plato's lover of wisdom, the Epicurean philosopher does not give up a love of sensible particulars for the love of a transcendental moral beauty.

Epicurus's followers, most prominently Philodemus and his teacher Zeno of Sidon (c.155–75 BCE), elaborated and modified Epicurus's views. They agreed with the basic premise that what makes poetry and music good is the enjoyment they provide. Against the Stoic view that music can instill moral virtue, Philodemus proclaims that music is "useful" simply "for pleasure" (*On Music*, book 4, col. 18). Poems produce pleasure by the thoughts expressed in them rather than by the sound or the words simply. These thoughts need not be morally useful, but they must provide relaxation and good cheer. This kind of poetry is illustrated by Philodemus's own elegant epigrams, which deal with topics of ordinary life, including love, youth and old age, and parties.

The classificatory zeal of the Hellenistic period provoked, in the case of Philodemus, a new assertion of artistic unity. Philodemus developed a new conception of the relationship between language and thought. Words and thought cannot be separated, but thoughts are fashioned by the way they are expressed. We respond to a good poem with a single, undivided experience, which may be called aesthetic pleasure. This pleasure is primarily intellectual, for it consists in a recognition of artistically created ideas.

Other Hellenistic Literary Theorists. In addition to the philosophers, Hellenistic poets and literary critics made significant contributions to literary aesthetics. Indeed, the Hellenistic period is the era in which the concept of literature first arose, along with an investigation of literary aesthetics.

Poets expressed their views both in their own poems and in prose treatises. Callimachus (c.305–240 BCE), who is considered by many the greatest Hellenistic poet, is the chief spokesman for a new poetics. In the preface to his *Aitia* (Explanations), and in his *Hymn to Apollo*, Callimachus contrasts his own "slender" Muse with the endless epic verses composed by others, and the "tiny, pure" stream of a "holy" spring with the murky waters of the broad Euphrates. Callimachus calls for a poetry that is unpretentious, limpid, concise, and innovative. In his demand for meticulous craftsmanship, he comes close to enunciating a position of "art for art's sake." His contemporary, the geographer, poet, and astronomer Eratosthenes, specifically rejected a didactic goal for poetry. He proclaimed that the aim of poetry is "to move the soul," not to teach (Strabo, *Geography* 1.2.3).

Neoptolemus of Parium, a poet and literary critic of the third century BCE, developed the most comprehensive classification of poetry known from antiquity. As Philodemus reports (*On Poems*, book 5, cols. 14–16), he divided the art of poetry into three parts: the ability of the poet (called "poet" simply), verbal composition ("poem"), and theme ("poetry"). This classification is based on a distinction between art as a mental capacity and the implementation of this capacity in a work of art. The artwork is divided in turn into the medium ("poem") and the message ("poetry"). Neoptolemus demanded that the poet both benefit the hearers (by saying what is "useful") and move the soul.

Next to Neoptolemus, probably the most important author of a poetic system in the Hellenistic period is Crates of Mallos (first half of the second century BCE). Crates called himself a "critic" to distinguish his vocation from that of the grammarian (Sextus Empiricus, *Against the Grammarians*, section 79). He is traditionally called a Stoic. But although he was strongly influenced by Stoic linguistics and metaphysics, he differed radically from the Stoics by rejecting a moral or didactic purpose for poetry. Crates proposed that what makes a poem good is good sound (Philodemus, *On Poems*, book 5, cols. 24–30). Good sound is not determined by rules, but inheres in a poem naturally as a correspondence between the sound of the words and the underlying thought. Crates contributed to the notion of an aesthetic faculty by proposing that the faculty of hearing can judge the goodness of poem by recognizing the principles of good sound that inhere in a poem.

[*See also* Literature, *article on* Literary Aesthetics.]

BIBLIOGRAPHY

Works by Philodemus

Philodemus: Über die Musik IV. Buch. Edited by Annemarie Jeanette Neubecker. Naples, 1986.
Filodemo: Il quinto libro della Poetica: PHerc. 1425 e 1538. Edited by Cecilia Mangoni. Naples, 1993.

Other Sources

Asmis, Elizabeth. "The Poetic Theory of the Stoic 'Aristo.'" *Apeiron* 23.3 (1990): 147–201.
Asmis, Elizabeth. "Epicurean Poetics." In *Proceedings of the Boston Area Colloquium in Ancient Philosophy,* edited by John J. Cleary, pp. 69–93. New York, 1991.
Asmis, Elizabeth. "Crates on Poetic Criticism." *Phoenix* 46.2 (summer 1992): 138–169.
Asmis, Elizabeth. "Neoptolemus and the Classification of Poetry." *Classical Philology* 87 (July 1992): 206–231.
Grube, G. M. A. *The Greek and Roman Critics.* London, 1965.
Kennedy, G. A., ed. *The Cambridge History of Literary Criticism.* Cambridge, 1989.
Obbink, Dirk, ed. *Philodemus and Poetry: Poetic Theory and Practice in Lucretius, Philodemus, and Horace.* Oxford, 1995. See pp. 69–96.
Russell, D. A. *Criticism in Antiquity.* Berkeley, 1981.

ELIZABETH ASMIS

HELMHOLTZ, HERMANN VON (1821–1894), German physiologist and physicist, also known for his explanations of the scientific foundations of music and painting. At mid-century, while holding a series of professorships in physiology at Königsberg, Bonn, and Heidelberg, Helmholtz was above all prized for his application of physical methods and principles to understanding the physiology of muscles, nerves, and senses (especially the eye and the ear). His principal accomplishments here included the discovery (in 1850–1851) of the finite speed of nerve impulses and the invention of the ophthalmoscope. More generally, his studies of (human) color and space perception made him into a leading theorist; during the 1850s, he helped to initiate the neo-Kantian movement, even as he rejected Kant's claim of a priori knowledge of space. Helmholtz consistently advocated an empiricist (as opposed to nativist) explanation for human spatial perception.

Along with his physiological researches of mid-century, Helmholtz pursued his passion for physics. In 1847, he (co)discovered the principle of the conservation of energy; and in 1858, he proposed a pathbreaking set of equations for representing hydrodynamic phenomena. After 1871, when he was appointed professor of physics at the University of Berlin, he devoted himself largely to assessing various electrodynamic theories; to providing a theory of chemical thermodynamics; and to proposing a mechanical foundation for thermodynamics and for physics as a whole. He also spent much of the final seven years of his life (1888–1894) as the founding president of the Physikalisch-Technische Reichsanstalt in Berlin, an institution devoted to setting physical units and standards, and to applied physics and technology.

During childhood, Helmholtz acquired an extraordinarily strong love of nature and the fine arts, one that he nourished throughout his life. He played the piano, viewed paintings, read poetry, and attended plays and concerts. From the 1850s onward, his multiple passions to understand the laws of physical nature and of visual and aural perception led him to offer explanations of the scientific foundations of music and painting as well as to portray the epistemological differences between the scientist and the artist. He presented his views in four popular, occasional lectures and in a semipopular book.

On Music. First in a lecture of 1857, and then in his musicological masterpiece, *On the Sensations of Tone as a Physiological Basis for the Theory of Music* (first edition, 1862), Helmholtz undertook to explain the physics and physiology of musical tones, and their import for musical aesthetics. In the first of the book's three parts, Helmholtz was principally concerned with harmonic upper partial tones, as well as with the composition of vibrations and tone quality. He defined and characterized musical tone physically, distinguished between fundamental and harmonic upper partial tones, and showed how tones could be understood in terms of Ohm's acoustical law and Fourier analysis. Similarly, he explained the phenomenon of resonance and its role in analyzing musical tones, above all showing how the ear analyzes such tones and apprehends differences in their qualities. In part 2, Helmholtz analyzed the interruptions of harmony, above all explaining the physical and physiological nature of combinational tones and beats and their various roles in producing consonance or dissonance. Here he sought to relate the neurology of the ear to musical rules.

In part 3, Helmholtz turned from a scientific to an aesthetic analysis of tones. He showed the historical development of tonality and its role in defining different musical styles and tastes. He viewed aesthetic taste as a matter of historical development and national differences rather than of transcendent laws. Indeed, he considered music to be constituted by both unchanging natural laws and changing aesthetic principles. Yet, he did not believe that music (or any other art) was arbitrary in nature; rather, he thought that in any given historical epoch musical rules formed a well-developed system that neither the artist nor the listener could consciously articulate. Both worked on a trial-and-error basis to discover what was most pleasing to them.

On the Sensations revolutionized musicology during the final third of the nineteenth century and has remained a classic ever since. Yet, neither in Helmholtz's day nor subsequently has it been without its critics. Helmholtz himself was keenly aware of the book's limitations. Although he believed that he had shown much about the scientific foundations of modern tonal (and other) musical systems, he also declared that he had done no more than analyze the elements of music (i.e., the sensations of tone). He did not attempt to discuss the theory of rhythm or composition, nor did he enter into the deeper psychological problems of musical expression. Yet, he did maintain that music does not and cannot imitate nature. He thought that it "expresses the kind of mental transition which is due to [a] feeling," that it did "not represent feelings and situations, but only frames of mind." Different listeners, he noted, often adduce different causes for different mental states. Music cannot directly represent any specific feeling because, for example, different listeners may have experienced different feelings while hearing one and the same musical passage. In contrast to the other arts, Helmholtz argued, music does not create images of external objects or events; in music alone, "the sensations of tone are the material of the art" and the source of enjoyment.

On Painting. Helmholtz complemented his musicological analysis with a series of lectures during the early 1870s on the scientific foundations of painting. The painter's aim, he claimed, was to use the colored canvas to elicit images of objects in the viewer's mind, to create a sort of optical illusion of (nonexistent) objects before the viewer. As in his analysis of music, Helmholtz did not think that he, or any scientist, could explain qua scientist "the ultimate aims and purposes

of art." Yet, here too he thought that an understanding of the more elementary (i.e., physiological) parts of painting, like their counterparts in music, could contribute to understanding painting's deeper (i.e., aesthetic) aims.

Helmholtz divided his lectures into four parts. In part 1, on form, he argued that the artist cannot and does not seek to copy objects; rather, he uses various materials and arrangements to give an idealization of an object, to portray that which is characteristic of it. He grounded his argument against copying on two physiological facts: first, three-dimensional external objects had to be imaged onto the viewer's one-dimensional retina; and second, (most) observers see the world with two eyes (binocularly), giving a sense of depth perception, something that painters can only hint at. Hence, the artist could hope only to represent "a true perspective view" of the external world, not copy it. The painter is thus forced to select a proper perspective (position and aspect), lighting, and shading in order to obtain fitting images and to enhance a true representation.

In part 2, on shading, Helmholtz argued that the painter cannot use colors in the same proportion as they are to be found in the natural world. The physiology of the eye requires that the painter use colors having only moderate sensitivity on the eye so as to elicit the same impression on the viewer of a picture as natural light would elicit on viewing the same scene in nature. Here again, the painter cannot simply copy an object but instead must translate its impression into that of a different scale, namely, that of the impression of light made on the eye of an observer of a painting. What counted was not the color used but the effect of light on the eye.

In part 3, on color, Helmholtz further argued that everything depended on the eye's nervous apparatus and the degree to which it is stimulated by incident light. He discussed the threefold physiological basis of colors, as well as the different types and roles of contrast phenomena. Because a different scale of sensitivity reigned when a viewer gazed on a painting than when he or she gazed at objects in the natural world, the painter could not seek to copy light and colors as seen in the natural world but had to translate them onto a different scale of sensitivity. The key, again, was the impression made by the light or color on the eye, not the light or color as experienced in the natural world.

In part 4, on harmony of color, Helmholtz argued that there are naturally complementary colors and limits to color mixing that are dependent on the eye's physiology. He argued, too, that moderation of light and of colors in painting has its advantages. In painting, truth to nature really meant "ennobled fidelity to nature." The impressions of a great painting, he thought, give insight into the essence of its objects, yet they do so without tiring or injuring the eye. By translating, rather than transcribing, nature, the artist does not fatigue the eye but instead "ennobles" the impressions made on the senses. This allows the viewer to contemplate the painting longer and more intensely than he or she could possibly do in the natural world.

On Goethe and on Art in General. Along with his writings on painting and music, Helmholtz's other statements of his aesthetic views came in addresses (of 1853 and 1892) on Johann Wolfgang von Goethe as a scientist and an artist. Goethe's accomplishments as a scientist, Helmholtz thought, lay in his comparative botanical and anatomical studies; his physics, by contrast, was a failure. Helmholtz maintained that Goethe's artistic intuition had greatly helped him in imagining a common structural plan that lay behind various animal and plant forms, but his unwillingness or inability to think about nature conceptually and abstractly, and ultimately to frame scientific laws, doomed his theory of colors. Helmholtz thought that Goethe had failed to observe basic optical facts, had misunderstood Isaac Newton's optical theory, and had misunderstood the sources of instrumental error in his prism. Goethe's strength as a scientist and artist, Helmholtz argued, came from his use of images; his weakness lay in his inability to use the inductive method. In his second address on Goethe, Helmholtz showed a more positive view of the poet; indeed, he now suggested that scientists and artists had much in common. He now declared that science was not the only way to gain insight into the workings of nature and humanity; art can do so as well. Both science and art were ultimately dependent on experience and both represent truth, he thought; both sought after the "lawlike behavior" of phenomena, even if they did so in very different ways.

In his most general aesthetic statements, Helmholtz averred that there are laws and rules of beauty and that they are "dependent on the nature of human intelligence." Yet, for both artists and their audiences such laws remain in the unconscious realm of human experience. There is always design in art, Helmholtz argued, even though art must always appear as if it were without design. One measure of great art, he maintained, is that the listener or viewer, like the artist, is aware (however vaguely) of some pattern or regularity, yet he or she cannot ever quite articulate the precise nature of that pattern or regularity. Conversely, if the pattern or regularity is too obvious, the art is of lesser value. All art, Helmholtz believed, must be "reasonable," it must contain some harmony of its parts; the more harmonious or reasonable it is, the greater is its value as a work of art. The critical listener or viewer seeks time and again to make conscious the (unconscious) pattern on which the work of art rests, and in so doing comes to appreciate (if only in part) the work's beauty and so to glimpse its reasonableness. To be sure, Helmholtz also believed that art could be understood and enjoyed by a cultivated listener or viewer immediately upon being heard or seen, without any critical search for its reasonableness. Yet, the critical appreciation of a work of art, he maintained, elevates the viewer or listener morally, be-

cause he or she thereby gains a sense of order in the universe and in other minds.

[*See also* Goethe; *and* Science and Aesthetics.]

BIBLIOGRAPHY

Works by Helmholtz

"On Goethe's Scientific Researches" (1853). In *Science and Culture: Popular and Philosophical Essays,* edited by David Cahan, pp. 1–17. Chicago, 1995.

"On the Physiological Causes of Harmony in Music" (1857). In *Science and Culture: Popular and Philosophical Essays,* edited by David Cahan, pp. 46–75. Chicago, 1995.

"On the Relation of Optics to Painting" (1871–1873). In *Science and Culture: Popular and Philosophical Essays,* edited by David Cahan, pp. 279–308. Chicago, 1995.

On the Sensations of Tone as a Physiological Basis for the Theory of Music. 2d English ed. Translated from the 4th German ed. (1877) by Alexander J. Ellis (1885). Reprint, New York, 1954.

"Goethe's Presentiments of Coming Scientific Ideas" (1892). In *Science and Culture: Popular and Philosophical Essays,* edited by David Cahan, pp. 393–412. Chicago, 1995.

Other Sources

Bailhache, Patrice. "Valeur actuelle de l'acoustique musicale de Helmholtz." *Revue d'histoire des sciences* 39.4 (1986): 301–324.

Cahan, David. "Helmholtz and the Civilizing Power of Science." In *Hermann von Helmholtz and the Foundations of Nineteenth-Century Science,* edited by David Cahan, pp. 559–601. Berkeley, 1993.

Dahlhaus, Carl. "Hermann von Helmholtz und der Wissenschaftscharakter der Musiktheorie." In *Über Musiktheorie: Referate der Arbeitstagung 1970 in Berlin,* edited by Frieder Zaminer, pp. 49–58. Cologne, 1970.

Hatfield, Gary. "Helmholtz and Classicism: The Science of Aesthetics and the Aesthetics of Science." In *Hermann von Helmholtz and the Foundations of Nineteenth-Century Science,* edited by David Cahan, pp. 522–558. Berkeley, 1993.

Hiebert, Elfrieda, and Erwin Hiebert. "Musical Thought and Practice: Links to Helmholtz's Tonempfindungen." In *Universalgenie Helmholtz: Rückblick nach 100 Jahren,* edited by Lorenz Krüger, pp. 295–311. Berlin, 1994.

DAVID CAHAN

HERBART, JOHANN FRIEDRICH. *See* Kant.

HERDER, JOHANN GOTTFRIED (1744–1803), German philosopher and literary critic. Herder was born in Mohrungen, East Prussia. He studied at Königsberg, and in 1764 became a teacher and assistant pastor in Riga. He was appointed court preacher at Bückeberg in 1770, and in 1776 he moved to Weimar, where he held the position of first preacher until his death in 1803.

Although Herder's fame now rests on his role in the development of modern nationalism and historicism, his main philosophical concern throughout his life was aesthetics. Most of his major works were devoted to its themes and problems, from his first publication, the early 1767 *Frag-* *mente,* to his last, the 1800 *Kalligone.* Herder's grand ambition was to be the Isaac Newton of the aesthetic world. He wanted to do for aesthetics what Immanuel Kant had tried to do for metaphysics: place it on a firm foundation.

One might be skeptical of Herder's ideal, but his success can be measured by how many of his ideas are commonplace today: that a writer should explore and exploit the strengths of his native language; that the critic should judge a work according to its intentions and context; that literature should be seen as an integral part of a culture; and that one should not imitate the literature of the past when its language and culture are dead. Herder's influence on his age was immense: Jean Paul, Johann Wolfgang von Goethe, Friedrich von Schiller, Friedrich Schlegel, G. W. F. Hegel, and Wilhelm von Humboldt were all in his debt.

Herder's aesthetics has to be placed in the context of the German cultural revival of the late eighteenth century. The aim of his first major work, the *Fragmente,* was to provide a justification for this revival. Like Gotthold Ephraim Lessing, Johann Georg Hamann, and Justus Möser, Herder advocated the development of German as a literary language and rejected French classicism. The essence of his argument was that German, like all languages, has its unique advantages, and that writers should exploit them rather than imitate the classics.

Herder's aesthetics has frequently been interpreted as a defense of the Sturm und Drang, a rationale for its cult of genius, contempt for classicism, and love of folk poetry. There is some justification for this reading because Herder shared, and indeed shaped, some of the values of the Sturm und Drang. But, in other respects, he deliberately distanced himself from this movement. He questioned the authority of sentiment; he denied that genius alone produces art; and he stressed the value of rules. It is telling that Herder wrote a sharp critique of Hamann's *Aesthetica in nuce,* a Sturm und Drang classic. In his attempt to provide a rational foundation for aesthetics, Herder firmly belongs in the tradition of the Aufklärung.

The main influences on Herder's aesthetics were Johann Joachim Winckelmann, Lessing, Hamann, Alexander Gottlieb Baumgarten, and Kant. Winckelmann made Herder aware of the cultural context of art; Lessing taught him the need to distinguish genres; Hamann drew his attention to the importance of language; Baumgarten convinced him of the need to provide a foundation for aesthetics; and Kant, his teacher from 1762 to 1764, showed him the proper methodology for it. Yet, if Herder acknowledged his debt to all these thinkers, he also criticized them severely. His aesthetics was a creative and critical synthesis of them, not an eclectic potpourri.

Herder's concern with the foundations of aesthetics is apparent in several of his early works, the first collection of his *Fragmente* (1767), the fourth of his *Kritische Wälder,* and his early unpublished writings on Baumgarten, Hamann, and

Winckelmann. Following in the footsteps of Kant's *Prize Essay* (1763), Herder argues that the proper method of aesthetics is the same as that for metaphysics: the "analytic method," which had been such a success in the physics of Newton and Galileo Galilei. This method began with an analysis of some phenomenon, and then reconstructed it by determining the relationships between the parts. In the case of aesthetics, this method would analyze the experience of beauty into its basic components—the sensations of touch, sight, and sound—and then show how it arose from their combination.

Herder developed his method in reaction to two of the predominant trends of eighteenth-century German aesthetics, the rationalism of Baumgarten and the Wolffian school, and the sentimentalism of C. J. Crusius and F. J. Riedel. The problem with the rationalists lay with their deductive method: they began with abstract definitions of the beautiful and the sublime when they should have ended with them. Rather than tracing how these concepts arose from experience, they simply analyzed the meaning of the terms as they are found in ordinary language, so that their systems consisted in little more than nominal definitions. The difficulty with the sentimentalists was that they reduced aesthetic experience down to simple feelings, so that it became impossible to analyze them. They failed to see that the experience of beauty is complex, the result of many mental activities, such as comparison, abstraction, and appraisal.

Of course, there is little new in Herder's methodology, which goes back to the "plain, historical method" of John Locke's *Essay concerning Human Understanding*. This method had been applied to aesthetics by Francis Hutcheson and David Hume in England, and by Abbé Jean-Baptiste Du Bos and Étienne Bonnot de Condillac in France. Indeed, Herder was simply following Kant, who, in his 1763 *Beobachtungen über das Gefühl des Schönen und Erhabenen,* used Burke's methods of observation to analyze the feelings of the sublime and the beautiful.

What, then, if anything, is new to Herder's aesthetics? Not the analytic method, but his transformation of it. Under the influence of Hamann, Winckelmann, and Baron de Montesquieu, Herder stresses the role of culture—the language, traditions, and history of a nation—in the formation of aesthetic experience. The poetry, painting, sculpture, and music of a people have to be seen as one part of its general way of life, as inseparable from its language, politics, religion, and history. Herder therefore goes a step beyond his predecessors. He does not think that analysis should stop with such apparent primitives as taste, aesthetic sense, or the simple ideas of sensation—the stock-in-trade of most eighteenth-century aesthetics. Rather, he insists that it should place these ideas in their cultural and historical context. Hence, in Herder's hands, the analytic method becomes synthetic: the terms of aesthetic analysis are seen as parts of a whole, the culture and history of a nation.

The most important factor in a culture, Herder believed, was its language. Under Hamann's influence, Herder stressed the importance of language for thought. He insisted that an idea or concept is inseparable from its sensible manifestation or "expression" in sounds and letters, and indeed from the characteristics of a specific language. This simple point led Herder to a weighty conclusion: "The genius of a language is also the genius of the literature of a nation." In other words, the best qualities of a nation's literature are unique to its language, and thus untranslatable.

Herder's emphasis on history, culture, and language soon brought him into conflict with the leading critics of his day, especially the Berlin *Aufklärer,* Friedrich Nicolai, Moses Mendelssohn, Johann Georg Sulzer, Thomas Abbt, and Lessing. The *Fragmente* is a sustained critique of the Berliners' journal, the *Briefe, die neueste Literatur betreffend* (Letters concerning the Latest Literature), which continued to defend, despite its rejection of Johann Christoph Gottsched, a neoclassical aesthetics. Herder argued that the Berliners' classicism ignored the fact that each language has its unique qualities and strengths, deriving from its characteristic sounds, idioms, syntax, and meter. What is good poetry and prose in one language will then not be so in another. This means that it is pointless to imitate ancient authors—be they Jewish, Greek, or Roman—or to evaluate modern writers by classical standards. Modern writers can no more write Jewish psalms, Greek dithyrambs, or Roman satires than they can revive the whole way of life of these peoples.

The apparent implication of Herder's cultural and historical approach is that there are no universal standards of criticism. If good literature depends on a specific language and culture, then standards will vary from one nation to another. Herder warned critics that their seemingly universal standards are often only those of their age, and that they should evaluate each work according to the intentions of its author. Yet, he resisted drawing relativist conclusions from his approach, and insisted that, despite great variations, there were still some natural laws of beauty.

Herder made some attempt to explain these laws in the first and fourth parts of his early *Kritische Wälder* (Critical Forests, 1769). In the first part, Herder developed his own classification of the fine arts through a critique of Lessing's famous distinction between poetry and painting in his *Laokoön* (1766). According to Lessing, painting and poetry have different media, and therefore distinct objects: the signs of painting are color and shapes in space, and thus its objects are physical bodies; the signs of poetry are sounds in time, so that its objects are actions. Herder made two objections to Lessing's distinction. First, it does not give a sufficient characterization of poetry, because music also works through time, a succession of sounds. Second, there is a fundamental difference between poetry and music and painting: poetry uses artificial signs, words that have a meaning, whereas music and painting use natural signs, col-

ors, shapes, and sounds. This distinctive feature of poetry—that it works on the soul in virtue of the meaning of its words—raises the question of how it affects the soul, by what medium it produces its effect. Herder's answer is that it does so through its own sui generis medium, namely, power or energy *(Kraft)*. Unfortunately, Herder did not fully explain the meaning of this concept, which has been much criticized for its vagueness.

In the fourth part of the *Kritische Wälder,* Herder applied his analytic method to aesthetic experience, tracing its various forms to their origins in sensation. He attempted to find the laws for each art form from its specific object and sense. Painting is the art of sight, having colors and surfaces for its object; music is the art of hearing, having sounds for its object; and sculpture is the art of touch, with shapes and forms for its object. The advantage of this threefold classification, Herder believed, is that it reinstates sculpture to its proper status and does not treat it simply as a form of painting. He argued that sculpture is sui generis because touch is a fundamental form of aesthetic experience. It is only by touch that one perceives shapes and forms, because one *sees* only colors and surfaces. Herder later developed these ideas about sculpture into a full treatise, *Plastik: Einige Wahrnehmungen über Form und Gestalt aus Pygmalions bildendem Träume,* which appeared in 1778.

In two works of the 1770s, *Ursache des gesunkenen Geschmacks* (Causes of the Decline in Taste) and *Wirkung der Dichtkunst* (Effects of the Poetic Art), Herder applied his cultural and historical standpoint to two classic questions: What are the causes of the rise and fall of taste? and What are the consequences of poetry for morals? He argued that those who complain about the decline in taste, or the moral consequences of poetry, simply do not know their history. They do not recognize that taste and poetry were central to Jewish, Greek, Arab, and Nordic culture. Whether there is good taste, or whether poetry has good consequences, Herder maintained, ultimately depends on the general state of a culture. He then suggested that the key to a flourishing culture—and thus to ensuring good taste and moral poetry—lies in its political constitution. If a government only permits its people sufficient freedom, then their natural powers will blossom into a healthy culture. Behind Herder the critic, then, there stood the political reformer.

Herder's last major work on aesthetics, *Kalligone,* is an attack on Kant's *Kritik der Urteilskraft* (Critique of Judgment). Still a loyal follower of the empiricism of the early Kant, Herder dismissed the transcendental method of the later Kant as barren scholasticism. Just as he once criticized Baumgarten, he now reproaches his old teacher for his a priori approach. Although much of Herder's polemic misses its target, some of his criticisms are noteworthy. True to his historicism and naturalism, he contends that Kant places art in an artificial intellectual realm, which abstracts it from history, society, and nature. Human beings cannot separate art from history, he argues, because we *make* history through art. It is through art that we develop our characteristic human powers, and create our culture. It is art that summarizes the whole spirit of an age, and that represents the pinnacle of a nation's achievement. We also cannot abstract art from nature, Herder further insists, because it is the highest organization and development of all nature's powers. What is beauty, he asks, but the expression of the well-being of a creature, the perfection of all its organic powers? It is a leitmotif of *Kalligone* that art is the product of the natural development of humanity, as essential to its development and well-being as food, clothing, and shelter. The main problem with the third *Kritik,* in Herder's view, is that it treats art as if it were a frivolous intellectual pastime, a mere matter of the "play of the faculties." Whatever the merits of its polemics, *Kalligone* contains many useful summaries of Herder's views, and its naturalism and historicism pose an interesting alternative to Kant's aesthetics.

[*See also* Baumgarten; Hamann; Historicism, *article on* Historicism and Philosophy; Kant, *article on* History of Kantian Aesthetics; *and* Lessing.]

BIBLIOGRAPHY

Works by Herder

Briefe: Gesamtausgabe, 1763–1803. Edited by Wilhelm Dobbek and G. Arnold. Weimar, 1977–.
Sämtliche Werke. 33 vols. Edited by Bernhard Suphan. Berlin, 1881–1913. The standard edition.
Werke. 10 vols. Edited by Ulrich Gaier et al. Frankfurt am Main, 1994–. The latest edition with notes and commentary.

Works by Herder in English Translation

Essay on the Origin of Languages. In *On the Origin of Language,* translated by John Moran and Alexander Gode. New York, 1966; reprint, Chicago, 1986.
Johann Gottfried Herder on Social and Political Culture. Translated and edited by Frederick Barnard. Cambridge, 1969.
Johann Gottfried Herder: Selected Early Works, 1764–1767. Edited and translated by Ernest Menze, Karl Menges, and Michael Palma. University Park, Pa., 1992.

Other Sources

Berlin, Isaiah. *Vico and Herder: Two Studies in the History of Ideas.* London, 1976.
Clark, Robert T. *Herder: His Life and Thought.* Berkeley, 1955.
Haym, Rudolf. *Herder nach seinem Leben und seinen Werken dargestellt.* Berlin, 1877–1885.
Norton, Robert E. *Herder's Aesthetics and the European Enlightenment.* Ithaca, N.Y., 1991.

FREDERICK BEISER

HERMENEUTICS. The term *hermeneutics* refers to a family of questions, ranging across many different cultural traditions and intellectual disciplines, about what it is to understand and interpret anything, whether a text, a human action, an alien culture, or oneself. Much of the history of

interpretation has centered on the problem of justifying the recontextualization of writings that originate outside the interpreter's historical and cultural situation. How can non-philosophical texts be squared with philosophical norms of reason? How can texts within a religious culture be squared with the rule of faith? The allegorization of Homer and the moralization of Virgil, the typological appropriation of the Hebrew Scriptures by the church fathers, the application of general rules and legal principles to unprecedented situations—in each case the hermeneutical problem is how to make the refractory text speak to the present moment and the issues at hand. During the Reformation, this problem was formulated for the first time in terms of the individual reader. Understanding a biblical narrative, for example, means understanding how its meaning applies to oneself—or, in other words, understanding oneself in light of the text.

Arguably, modern hermeneutics begins with the attempt to square this principle of *applicatio* with increasing awareness that the Scriptures are, after all, historical documents whose meanings are internal to the time and place of their composition. In other words, how to bring together both the subjective and objective dimensions of interpretation? In the early nineteenth century, the theologian Friedrich Schleiermacher (1786–1834) gave a complex turn to this question when he said that the task of interpretation is "to understand a text at first as well as and then even better than its author" (1977). As stated, this is an old philological principle, but Schleiermacher formulated it as a philosophical problem of how to understand the thoughts and experiences of other people. Here there is no longer any question of redescribing another's text in one's own language. The idea now is to enter into the standpoint of another person. At a minimum, and as a practical matter of historical criticism and philological rigor, this means reconstructing the time and place of a text's composition, but for Schleiermacher, understanding must have an inner dimension as well, and so he added that the interpreter must also transform himself into the author by means of empathy *(Einfühlung)*. Understanding presupposes a certain intimacy; it remains an event of the present. Thus, for Wilhelm Dilthey (1833–1911), the task of historical reason is nothing less than to reexperience *(nacherleben)* the subjective life of those whose expressions have come down to us from the past. The goal of interpretation, he says, is to "reproduce an alien life in oneself." The question is: How can this subjective process have an objective validity?

Much of twentieth-century hermeneutical inquiry has tried to neutralize the difficulty of this question by reconceptualizing it in nonsubjectivist language. For Edmund Husserl (1859–1938), understanding is not of minds but of ideal objects called meanings. For Martin Heidegger (1889–1976), however, understanding is not a theoretical project at all but a mode of practical involvement with the world around us. Heidegger's "hermeneutics of facticity"

means knowing what things are for, what they are doing here; it means knowing how to inhabit a world made up of such things. There is, at any rate, no getting outside our own historical and cultural environment and the schemes and practices by which we make sense of things. Unfortunately, this leaves unanswered the originary problem of hermeneutics. How can we understand what is not of our world? For Heidegger, understanding can only mean appropriation and integration of the other into our scheme of things, finding a place for it within our hermeneutical circle. This, indeed, is very much in keeping with ancient practices such as allegory. What happens, however, when we encounter something that refuses appropriation and resists the forms and practices of interpretation that underlie the intelligibility of our cultural experience?

According to Hans-Georg Gadamer (b. 1900), this is the fundamental question of a philosophical hermeneutics that attempts to clarify the experience of human finitude. Basic to hermeneutical experience, Gadamer says, is the moment when we are "brought up short by the text" and discover that the text cannot be made sense of in our terms. Gadamer thinks of this as a moment of critique. It is an encounter with "the limit that is implicit in every hermeneutical experience of meaning." It is what cultural materialists have in mind when they speak of the prison house of our language and culture. For Gadamer, however, the limits of understanding are flexible and porous rather than ideological fixtures rooted in the unconscious. In our encounter with the refractory text, we can extend the limits of our understanding through the revision and reconstruction not of the text, but of ourselves and our world. In order to understand the text of another, we must change. This does not mean merely an alteration in perspective, merely expanding our horizon so as to assimilate the other into our view of things. Gadamer also means changing in the strong sense of intervening in our world and altering the course of its future. Our relation to the text is one of responsibility as well as one of interpretation. Understanding is not just making sense; it also means learning how to inhabit a new world. Indeed, Gadamer has always thought of understanding as a mode of practical wisdom or *phronēsis,* and in his later writings he characterizes hermeneutics as "practical philosophy."

Gadamer's hermeneutical analysis has taken two avenues of inquiry. In *Truth and Method* (1960 [1989b]), his concern was to understand what happens in our encounter with all that comes down to us in tradition. Here the main question is: What is it to understand a text that has a claim on us? Within the ethical context of the claim, understanding can never be thought of simply in terms of meanings or conceptual agreement with other minds. It always entails a judgment vis-à-vis our relation with the world; hence, Gadamer's insistence that "understanding always involves something like applying the text to be understood to the interpreter's present situation" (1989b). Ultimately, under-

standing always shows itself in action. In such later writings as "The Relevance of the Beautiful" (1975 [1986]), however, Gadamer has tried to clarify the hermeneutical nature of aesthetic experience, especially within the context of twentieth-century art. In particular, how do we stand with respect to the avant-garde work that sets absolute limits to our capacity to render things intelligible? Gadamer has in mind such things as the Cubist painting, the hermetic poem of Stéphane Mallarmé among other instances of literary modernism, and Arnold Schoenberg's development of serial composition in music. How are we to respond, for example, to Marcel Duchamp's provocation when he buys a snow shovel in a hardware store and sets it up in his studio as his latest composition? Gadamer's position is that none of this is to be brushed aside as nonsense. On the contrary, modern art has a claim on us that is equal to the claims of tradition. How can this be so?

Here hermeneutics situates aesthetics at the intersection with recent ethical theory and its critique of a subject-centered rationality that seeks conceptual determination and control. For the French philosopher Emmanuel Lévinas (1906–1995), our encounter with other people shows the limits of such rationality, for the other confronts us in its irreducible singularity. It cannot be brought under the rule of identity or the legislation of our concepts. The other is refractory to the categories by which we make the world intelligible. For Gadamer, this refractoriness is an essential feature of our experience of art. A recurring theme in Gadamer's writings is the way in which art exposes us to our human finitude. The avant-garde text, like the action of tragedy, shows us that our relation to reality is not so much one of cognition and control as one of recognition and habitation. Here Gadamer is close to the American philosopher Stanley Cavell (b. 1928), who thinks of our relation to the world and to others—and to art—in terms of acknowledgment and responsibility. Understanding Schoenberg's "new music," Cavell says, is a matter of "naturalizing ourselves to a new form of life, a new world" (1976). [*See* Cavell.] Or, as Arthur C. Danto (b. 1926) puts it in a famous essay (1964), it means entering into the history and theory—the hermeneutical circle—of the "artworld."

Like Lévinas, Gadamer tries to clarify our relation to "the otherness of the other" in dialogical terms. Our relation to language, for example, is not simply that of speaking subjects. On the contrary, the experience of language is fundamentally an experience of listening. "Understanding begins," Gadamer says, "when something addresses us." Lévinas would say that this is the beginning of ethics. For Lévinas, the model of being addressed is biblical; it is the prophetic experience of being exposed to a transcendent alterity whose claims cannot be evaded. For Gadamer, the model is that of the Platonic dialogue where our mode of existence takes the form of being in question. For Lévinas, there is no such thing as understanding the other on the model of cognition, and this is also true in Gadamer's analysis, where understanding the other only begins when we enter into the play of question and answer that our experience of alterity or the refractoriness of the other opens up. But in this event the other does not give up its alterity; it is not laid bare. On the contrary, understanding can now only be clarified in terms of "being-with" the other *(Mitsein)*. Understanding means learning how to live with differences without reducing them to the same. This is perhaps the ultimate point of Gadamer's principle that "understanding is more being than consciousness."

The first part of Gadamer's *Truth and Method* is a critique of aesthetic consciousness understood as a mode of disinterested contemplation. Unlike Heidegger, however, whose "Origin of the Work of Art" (1936) was intended as an overcoming of aesthetics, Gadamer has always tried to understand what it is that actually happens when we come into contact with the work of art in all of its irreducible strangeness and irreplaceability. Indeed, part of what is implied in "hermeneutical claim to universality" is the need to come to terms with the dissonance of modern art as that which resists exegesis and the construction of meaning. As Theodor W. Adorno (1903–1969) puts it in his *Aesthetic Theory*, "Aesthetics cannot hope to grasp works of art if it treats them as hermeneutical objects. What at present needs to be grasped is their unintelligibility" (1997). [*See* Adorno.] Accordingly, in his later writings on art and poetry—particularly in his essays and commentaries on the poetry of Paul Celan (1920–1970)—Gadamer emphasizes that our relation with the work of art is to be understood in terms of "being-with" rather than in terms of "grasping" by means of formal analysis and historical explanation. "All interpretation," he says, "comes to share in the being of the poem." This sharing in being does not mean the possession of common ground or beliefs. It is not a species of agreement; it is not governed by any rule of identity or logic of exclusion. It means sharing a dwelling. Or perhaps it is close to what the French philosopher Jean-Luc Nancy (b. 1942) means when he speaks of the "sharing *[partage]* of voices" (1990), which is not a speaking in unison but a passing from one to another, sharing or distributing voices in the manner of the gift. As Nancy thinks of it, "Hermeneutics is a gift." Hermeneutically speaking, the mode of being that we come to share with the work of art, as indeed with other people, is the mode of being of the gift, a movement of being-with that passes one to another and allows a multiplicity of voices their endless syncopation.

It follows that understanding shows itself most fully in action and conduct in the world. Paul Ricoeur (b. 1913) has applied this notion very fruitfully to the interpretation of texts, especially narrative texts. For Ricoeur (1981), the meaning of a text does not lie behind it in the form of an originating intention that interpretation then tries to reproduce. Rather, the text projects its meaning in front of itself

in the form of a possible world. Ricoeur asks us to think about our relation to this world existentially as if it were a projection of our own possibilities; that is, to understand a text means to understand oneself in terms of the possibilities opened up by the text. The world of the text, in this respect, is continuous with one's own as an enlargement of it or as a movement into the future. In any case, understanding takes the form of appropriation, making the text our own in the way we live. Arguably, this has been, from the beginning, the most fundamental truth of hermeneutics.

[*See also* Dilthey; Gadamer; Heidegger, *article on* Survey of Thought; Interpretation; Reception Aesthetics; Ricoeur; *and* Schleiermacher.]

BIBLIOGRAPHY

Adorno, Theodor W. *Aesthetic Theory.* Edited by Gretel Adorno and Rolf Tiedemann, translated by Robert Hullot-Kentor. Minneapolis, 1987.

Bruns, Gerald L. *Hermeneutics: Ancient and Modern.* New Haven, 1992.

Cavell, Stanley. "Aesthetic Problems of Modern Philosophy." In *Must We Mean What We Say?* Reprint, Cambridge and New York, 1976.

Danto, Arthur. "The Artworld." *Journal of Philosophy* 61.19 (October 1964): 571–582.

Dilthey, Wilhelm. "The Construction of the Historical World in the Human Sciences." In *Dilthey: Selected Writings,* edited and translated by H. P. Rickman. Cambridge and New York, 1976.

Dilthey, Wilhelm. *Hermeneutics and the Study of History.* Edited by Rudolf A. Makkreel and Frithjof Rodi. Princeton, N.J., 1996.

Gadamer, Hans-Georg. *Reason in the Age of Science.* Translated by Frederick G. Lawrence. Cambridge, Mass., 1981.

Gadamer, Hans-Georg. *The Relevance of the Beautiful and Other Essays.* Translated by Nicholas Walker, edited by Robert Bernasconi. Cambridge and New York, 1986.

Gadamer, Hans-Georg. "Text and Interpretation." Translated by Dennis Schmidt and Richard Palmer. In *Dialogue and Deconstruction: The Gadamer–Derrida Encounter,* edited by Diane P. Michelfelder and Richard E. Palmer, pp. 21–54. Albany, N.Y., 1989a.

Gadamer, Hans-Georg. *Truth and Method.* 2d rev. ed. Translation revised by Joel Weinsheimer and Donald G. Marshall. New York, 1989b.

Gadamer, Hans-Georg. *Hans-Georg Gadamer on Education, Poetry, and History: Applied Hermeneutics.* Edited by Dieter Misgeld and Graeme Nicholson, translated by Lawrence Schmidt and Monica Reuss. Albany, N.Y., 1992.

Gadamer, Hans-Georg. *Literature and Philosophy in Dialogue: Essays in German Literary Theory.* Translated by Robert H. Paslick. Albany, N.Y., 1994.

Gadamer, Hans-Georg. *Who Are You and Who Am I: A Commentary on Paul Celan's "Atemkristal."* Translated by Bruce Krajewski. Albany, N.Y., 1997.

Heidegger, Martin. *Being and Time.* Translated by John Macquarrie and Edward Robinson. New York, 1962.

Heidegger, Martin. *Poetry, Language, Thought.* Translated by Albert Hofstadter. New York, 1971.

Nancy, Jean-Luc. "Sharing Voices." Translated by Gayle L. Ormiston. In *Transforming the Hermeneutic Context: From Nietzsche to Nancy,* edited by Gayle L. Ormiston and Alan D. Schrift, pp. 211–259. Albany, N.Y., 1990.

Redding, Paul. *Hegel's Hermeneutics.* Ithaca, N.Y., 1996.

Ricoeur, Paul. *Hermeneutics and the Human Sciences.* Edited and translated by John B. Thompson. Cambridge and New York, 1981.

Schleiermacher, Friedrich. *Hermeneutics: The Handwritten Manuscripts.* Edited by Heinz Kimmerle, translated by James Duke and Jack Forstman. Missoula, Mont., 1977.

GERALD L. BRUNS

HERZ, MARCUS (1747–1803), physician, philosopher, and a member of the Jewish intelligentsia of late eighteenth-century Berlin. Herz is today best remembered as the respondent at the defense of Immanuel Kant's inaugural dissertation in 1770, as the author of his own *Betrachtungen aus der spekulativen Weltweisheit* (Considerations from Speculative Philosophy) in 1771, a work that is partly a popularization of Kant's dissertation and partly a defense of the rationalism that Herz learned from Moses Mendelssohn, and as the recipient of Kant's letters documenting his progress on the *Critique of Pure Reason* during the 1770s. During his lifetime, however, Herz was widely known for his medical practice and writings, although he destroyed his medical reputation shortly before his death by his polemic against inoculation.

Herz's contribution to aesthetics was his *Versuch über den Geschmack und die Ursachen seiner Verschiedenheit* (Essay on Taste and the Causes of Its Variation), first published in 1776 and in a revised edition in 1790—thus, at the same time as Kant's *Critique of Judgment.* Kant praised the work when he thanked Herz for a copy of the first edition, but never mentioned it in his own publications. Kant's discussion of the "empirical interest in the beautiful" in the *Critique of Judgment,* however, is clearly a criticism of Herz's way of connecting sociability and taste, and perhaps Kant's aesthetic theory can be seen more broadly as an attempt to show how an a priori principle of taste could be built on foundations that Herz thought could support only an empirical theory. Thus, Herz's work, now little known, had a considerable, even if indirect, influence on the history of aesthetics.

The essay on taste, which shows the influence of diverse sources including Alexander Gottlieb Baumgarten, Mendelssohn, Johann Georg Sulzer, Alexander Gerard, Baron de Montesquieu, and the Kant of the first *Critique,* has the form of a typical eighteenth-century essay on the possibility of a standard of taste. On the surface, its three main parts present an initial analysis of beauty and of taste as the capacity for responding to beauty, an analysis of the causes of diversity of taste among individuals, and an analysis of the causes of diversity of taste among peoples or nations. At the same time, the second part adds an account of fine art to the initial analysis of beauty, and the third part adds an account of the connection between taste and society to the initial account of the pleasures of taste for the individual. It is in these additions that the interest of Herz's work lies. In part 1, Herz expands on Francis Hutcheson's analysis of beauty by arguing that beauty consists not only in the variety (or "manifoldness") of the parts and the unity of the form in an object, but

also in the proportion *(Haltung)* of the parts to their causes and to the end of the object; he then argues that taste consists in the capacities of imagination, for apprehending manifoldness, reason, for judging of unity, and the "feeling of proportion" *(Haltungsgefühl)*, for responding to proportion. In part 2, Herz ostensibly discusses grounds for individual variations in imagination, and reason, and the feeling of proportion. But to explain how artists can attain a proportion in their works that can earn agreement in the face of human diversity, he includes an extensive discussion of fine art. Here he argues that, as products of human activity, works of art must have an ultimate end *(Endzweck)*, that this can only be maximal happiness, and that such happiness does not consist in the mere accumulation of pleasant sensations but in the fullest possible development and exercise of one's active powers. From this he argues further that "the artist must exhibit something to us that has an interest for us through its influence on what is inmost in us," or our emotional life, for it is precisely what stirs our emotions and passions that most exercises our active powers. In part 3, Herz sets out to explain differences in national tastes as the result of differences in freedom of thought, religion, morality, resources, climate, and form of government. Under the heading of morality, he offers a complex argument that morality requires the development of sociability, or inclinations to other human beings, and that art contributes to the development of morality by producing objects that stimulate feelings of sociability, and in so doing "set our activity into motion in the greatest and most manifold ways." The connection between art and the development of sociability creates an interest *(Interesse)* in art that goes beyond individual pleasure in beauty of form, and our interest in sociability is so central to our pleasure in beauty that such pleasure is significantly diminished by solitude.

Herz's account of our social interest in fine art and even natural beauty was clearly a rejection of Hutcheson's conception of disinterestedness, although it was based on a conception of morality closer to that of David Hume than of Kant. Kant's criticism of an interest in beauty based on its contribution to feelings of sociability as merely empirical *(Critique of Judgment,* section 41) was clearly a repudiation of Herz's account, although one dependent on Kant's own assumption that the principles of both moral and aesthetic judgment must be entirely a priori, an assumption that has itself always been highly controversial.

[*See also* Kant; *and* Taste.]

BIBLIOGRAPHY

Works by Herz

Betrachtungen aus der spekulativen Weltweisheit. Königsberg, 1771. Modern edition edited by Elfriede Conrad, Heinrich P. Delfosse, and Birgit Nehren (Hamburg, 1990).
Versuch über den Geschmack und die Ursachen seiner Verschiedenheit. Leipzig, 1776.
Briefe an Aerzte. 2 vols. 2d ed. Berlin, 1784.
Versuch über den Schwindel. Berlin, 1784.
Versuch über den Geschmack und die Ursachen seiner Verschiedenheit. 2d enl. corr. ed. Berlin, 1790.

Other Source

Davies, Martin. *Identity or History? Marcus Herz and the End of the Enlightenment.* Detroit, 1995.

PAUL GUYER

HEYDENREICH, KARL HEINRICH (1764–1801), professor of philosophy in Leipzig from 1789 to 1798 who wrote voluminously on aesthetics, philosophy of religion, and moral and political philosophy. After publishing a number of articles on aesthetics in journals between 1786 and 1790, Heydenreich published his *System der Ästhetik* in 1790, a translation of a commentary titled *Archibald Alison über den Geschmack* (1792), a four-volume *Ästhetisches Wörterbuch über die bildenden Künste, nach Watelet und Levesque* (1793–1995), *Grundsäzte der Kritik des Lächerlichen* (1797), and numerous further essays on aesthetics, including one on Immanuel Kant's *Critique of Judgment,* in the three volumes of his *Originalideen* (1793–1786). Outside of aesthetics, Heydenreich's many publications included *Betrachtungen über die Philosophie der natürlichen Religion* (1790–1791), *Propädeutik der Moralphilosophie* (1794), and a *System des Naturrechts nach kritischen Principien* (1794–1795).

Heydenreich's *System der Ästhetik,* which was labeled *Erster Band* (first volume) although no sequel was ever published, was the most widely noted of his works in aesthetics during his life and is the only one to have been reprinted in modern times. The work was published after the appearance of the *Critique of Judgment* in March 1790, and Kant's work is cited in the introduction; but obviously the bulk of the work was completed before Heydenreich had been able to read Kant's contribution to aesthetics. It is interesting in part to see how a writer strongly influenced by Kant's two previous critiques developed a position in aesthetics in many ways quite different from Kant's own. Heydenreich attacks the rationalist aesthetics of Alexander Gottlieb Baumgarten and Moses Mendelssohn, arguing that their conception of sensuously represented perfection is too vague to be of any use, and also rejects Karl Philip Moritz's attempt to ground the autonomy of art while retaining the rationalist conception of perfection: he mercilessly lampoons Moritz's idea that a work of art contains a self-referential perfection in itself, arguing instead that "every work of art obviously has its end outside itself, in a being who is to be affected by it, whether this be the artist himself . . . or other persons beside him" (1978). Heydenreich himself argues that it is futile to try to reduce all beauty to a single explanation, and instead recognizes four kinds of beauty: beauties connected with the "immediate impressions of objects on our senses," beauties con-

nected with the "contingent associations of certain images and representations with certain objects," beauties whose "efficacy rests on an essential relation of certain forms and tones with certain conditions of the human being as a being sensitive to well-being and misery," and beauties that "arouse enjoyment through relation to certain objects, images, representations, thoughts and actions to the laws of the understanding or speculative or practical reason" (ibid.). In the case of the fine arts, however, Heydenreich is more than willing to countenance a reduction of all these forms of beauty to a single principle: works of fine art please because of their satisfaction of the human need for feeling or sentiment. His main claim is thus that,

> as a cognitive being, the human being possesses a necessary drive to expand his knowledge and communicate it to his fellow humans, as a sensitive being, a drive to exhibit and communicate his feelings *[Empfindungen]*. . . . Every work of fine art is thus the exhibition of a determinate condition of sensitivity *[Empfindsamkeit]*. (Ibid.)

On this basis, Heydenreich offers a detailed classification of the fine arts, focusing on the ways in which different media and genres exhibit and communicate feelings. He planned also to enumerate detailed principles of taste, which would apparently have used the Baumgartian distinction among matter, form, and sign or vehicle to evaluate the various dimensions of our response to the representation of feeling. This part of his system, however, was never written.

Heydenreich considered the need to develop our capacity for the representation and communication of feeling essential to human well-being, and based his defense of the moral and political significance of the arts on this assumption. In this he anticipated, though in much less dramatic prose, the theme that Friedrich von Schiller was to develop five years later in his *Letters on the Aesthetic Education of Mankind;* and although Heydenreich was not mentioned by name in that work, his influence on Schiller's argument was no secret to his contemporaries.

[*See also* Kant.]

BIBLIOGRAPHY

Work by Heydenreich

System der Ästhetik. Texte zum literarischen Leben um 1800, vol. 5. Reprint, Hildesheim, 1978.

Other Sources

Adickes, Erich. *German Kantian Bibliography*. Reprint, New York, 1970. See pp. 111–115.

PAUL GUYER

HISTORICISM. [*To examine the roles of historicism in various fields related to aesthetics, this entry comprises four essays:*

The first essay explains the general philosophical meaning of historicism, that is, the view that all phenomena, such as art, can be understood and interpreted only in historical terms. The other three essays concern the different ways historicism has developed and been practiced in the fields of literary theory, art history, and architecture. See also Hegel; *and* Heidegger]

Historicism and Philosophy

Until recently, *historicism* was an epithet reserved for certain anti-Hegelian theories of history, most prominently those of Leopold von Ranke (1795–1886) and like-minded thinkers. Thus restricted, the doctrine is sometimes termed "classical historicism." More recently, it has been used as a term of opprobrium by Karl Popper (1902–1994) to signify theories such as Karl Marx's, which, according to Popper, wrongly hold that there are laws of history distinct from the laws of nature. Historicism fails, Popper insists, because there cannot be any "historical prediction" under would-be laws of "historical evolution." Apart from Popper's objection, there may indeed not be any *laws* of history; but that is not to say, or to demonstrate, that there *are* laws of nature (if there are no laws of history) or to deny that natural science is itself inseparable from its history—that the would-be laws of nature are themselves abstracted and idealized from uniformities discerned under historicized conditions. One sees here that objections to historicism are often veiled assertions of some strong realism, that is, a realism requiring exceptionless lawlike necessities in physical nature. In Anglo-American philosophy, the issue has been rendered problematic by the publication of Thomas S. Kuhn's *The Structure of Scientific Revolutions* (1970). Also, in our own time, largely with the appearance of post-Heideggerian hermeneutics, as in the work of Hans-Georg Gadamer (b. 1900), and of poststructuralism, as in the work of Michel Foucault (1926–1989), historicism has been taken to signify the implicit theme common to all three doctrines, namely, the historied or historicized nature of thought itself. In both Gadamer and Foucault, the theme bears on the analysis of human existence more than on the methodology of science or history.

Given these complex options, we gain a considerable economy by treating historicism simply as the thesis that thinking is inherently historicized, that is, formed or preformed by internalizing (in infancy, living among apt adults) the local conceptual resources and general orientation of the particular historical culture in which persons or selves, "subjects" or "rational agents" (ourselves), first emerge. Nevertheless, Gadamer speaks of historicism in Ranke's sense of the term—hence, as making a mystery of

the mutual intelligibility of different historical cultures—and so rejects it; and Foucault does not speak of the doctrine directly. It is also clear that to favor the sense just improvised is, emphatically, to favor one or another form of relativism; and, indeed, in Ranke, Gadamer, and Foucault, relativistic leanings cannot be denied, though they are characteristically muted or called into question. The issue is important for the largest theories of culture, art, morality, and reason; for, however historicism is consensually defined, the theme of the historicity of thinking has continued to attract diverse and more and more adherents.

In Ranke, historicism signifies the repudiation of G. W. F. Hegel's (1770–1831) advocacy of an all-encompassing world history (as Ranke reads Hegel, misleadingly). Ranke holds that every culture, or age, has its distinct *Geist,* its ideally unifying sense of the meaning of the events of its separate world. The historian's project is to capture "objectively" that individual *Geist* in interpreting the documents and other archival remains of one age or another ("wie es eigentlich gewesen"). Ranke only vaguely grasps the self-referential paradox that threatens his theory, namely, that the historian, practicing his discipline within the horizon of one *Geist,* may not be able to recover the *Geist* of another age. God, apparently, knows each age as it is; and the fortunate historian, somehow favored in a cognate way, objectively grasps the *Geist* of another age. Apart from the dubious metaphysics and epistemology of Ranke's approach, one finds an excellent vindication in such a work as Wilhelm Worringer's *Abstraction and Empathy* (1907 [1953]). Worringer was able to demonstrate compellingly, within the terms of an idealist conception of history, that the hieratic forms of Old Kingdom figures (in Egypt) could not be construed as signifying an incapacity to model the human form naturalistically: it must signify instead (Worringer argued) a certain artistic intention (or "will") on the part of ancient sculptors, who plainly produced (at the same time that they favored the archaic forms in representing the pharaoh) fully modeled figures of scribes and the like. [*See* Worringer.] The point is that classical historicism from Ranke to Worringer construed relativism in the benign sense of admitting the plurality of historical cultures *and* the competence of historians and historically minded commentators to fix the objective facts about each such culture. Characteristically, it treated cultures as *geistlich* (intrinsically interpretable but historicized) artifacts but not, to ensure the objectivity of history, the interpreting competence of the historian.

Gadamer and Foucault risk the more radical possibilities of this generic "idealism" (the doctrine that the very reality of the human world is no more than its historically changing interpretation), in that, in different ways—Gadamer: proximately, through Martin Heidegger; Foucault: more distantly, through Friedrich Nietzsche; both: ultimately, through Hegel—they treat the interpreting mind (the human self) as itself an artifact of diverging histories. Gadamer's *Truth and Method* (1960 [1989]) is very much occupied with a running polemic against Romantic hermeneutics, which, like classical historicism (with which it ultimately converges, as in Worringer), treats the literary and plastic artifacts of a culture as manifestations of an original "authorial [or *geistlich*] intent" but fails to address the historicity of every effort to recover authorial intent itself. It would not be unreasonable to claim, though the argument cannot be entirely straightforward, that this older mix of objectivism and historicity is ultimately grounded in something like Giambattista Vico's (1668–1744) formula "verum = factum" (the invariant truths of history reveal what human history has truly produced): the ultimate artifactuality of human nature rests with God's art, not merely man's.

This Vichian vision is what Gadamer's and Foucault's "historicisms" sweep away. Gadamer, as it happens, attempts to recover, rather mysteriously, the constancy of the "universal horizon" of humanity (apparently never quite lost in the divergent formations of historicized selves)—in the "classical," the abiding Western tradition of Hellenic Greece. Foucault is more consistent, but also more paradoxical, in that he never satisfactorily resolves the aporia of the radical historicity he advocates. His own term for the historicizing process (or, rather, its narrative recovery) is *genealogy,* borrowed and deepened from Friedrich Nietzsche's *Genealogy of Morals* and applied in somewhat post-Marxian and poststructuralist terms, that is, by abstracting from all conceptions of the underlying causal engine of history. Notably in the essay "Nietzsche, Genealogy, History" (1971 [1977]), Foucault implicates the central historicist paradox: that truth itself is an artifact of history, or a historically contingent "discursive regime" or "episteme," while, at the same time, we apparently are able to fix "objectively" the function of truth in any such regime by way of genealogical reflection. Foucault never discusses how the paradox is actually resolved—though it is benign enough.

By the end of the 1990s, historicism promised (or threatened) to be global—in the sense that the entire intelligible world, not merely the world of human history and human culture (to take up and eclipse Vico's thesis) is socially constituted or constructed. Historicism, metonymically in Gadamer and Foucault, is, then, a radical form of social constructionism (the doctrine that the intelligible world is, in part, constituted by the powers of historical understanding), though constructivism need not be a form of historicism. It may well be that this characteristic orientation originates with Nietzsche (1844–1900). [*See* Nietzsche.] In any case, construed epistemologically and metaphysically, historicism is the plausible consequence of embracing a certain series of philosophical gains: proceeding, first, by accepting the "constructive" lesson of Immanuel Kant's (1724–1804) critical or transcendental repudiation of the naive objectivisms espoused by classical empiricism and rationalism; second, by recognizing the threatening incoherence of Kant's own con-

fidence about discerning (under contingent conditions) the necessary, transcendental, possibly not even human constituting functions of "pure" understanding—which the German Idealists put in question; and third, by integrating the social, historical, and postcritical correction of Kant, the Idealists, and Edmund Husserl (1859–1938)—which later phenomenologists, hermeneuts, existentialists, pragmatists, genealogists, and poststructuralists hoped to supply, by insisting on the essential formative role of actual human history. The result is that the naïveté of a Ranke, the implicit dualism of a Popper, the unexplained reserve of a Gadamer, and the default of a Foucault converge on a global historicism that is also a Constructivism.

On the Constructivist side, the argument leads to what has been termed "symbiosis," the thesis—shared by such disparate thinkers as Nietzsche, Martin Heidegger (1889–1976) in *Being and Time* (1926 [1962]), John Dewey (1857–1952), Gadamer, and Foucault—that cognizing subjects and cognizable world are themselves artifactually defined within a holist space and that determinate questions of truth, knowledge, and realism arise only in that space. (See, particularly, Margolis, 1993, 1995.) From this vantage, it is instructive to think of Ludwig Wittgenstein (of the *Philosophical Investigations* [1953]) and W. V. Quine (of *Word and Object* [1960]) as nonhistoricizing Constructivists manqués (that is, Constructivists who do not historicize human understanding but who yield somewhat in the historicized direction); for Wittgenstein never assigns the Constructivism of his epistemology (within what he calls *Lebensformen* [forms of life] and "language games") a full metaphysical function, and Quine reserves a privileged sensory access to the real world (beyond the range of his "analytic hypotheses") that is not Constructivist at all. [*See* Wittgenstein.] These comparisons confirm the profound divide between Anglo-American and Continental European philosophies; for, until recently, there has been very little interest in (and almost no support for) historicism in the Anglo-American tradition. The best-known exceptions are afforded by Dewey and Kuhn. Dewey's historicism is, however, more atmospheric than determinate; and Kuhn's is plainly abortive, because it is meant to be subsumed under the invariances of a theory that favors the vision of the unity of science program, though not its usual inductivism or falsificationism.

Historicism is now characteristically tolerated, but not widely favored, in the theory of interpretation, and more when it is applied to literature and the arts than when it is applied to human conduct in moral and political terms or in the physical sciences. On arguments in accord with post-Kantian thinking—refusing any principled disjunction between cognizing subjects and cognized world—any such division risks conceptual incoherence. But it is clear enough that Western moral philosophy is profoundly antihistoricist and that historicized and Constructivist accounts of science have, until recently, been almost nonexistent. The reason,

perhaps, is that the advocacy of historicism has been, and has been perceived to be, destructive of standard philosophical claims. This is clearly not true of Kuhn's theory, though it is true enough, for example, of Paul Feyerabend's regarding science and scientific methodology and of Roberto Mangabeira Unger's regarding political and legal liberalism.

Historicism focuses, minimally, on the twin concerns of metaphysics and epistemology. Cultural phenomena are seen to be historically constituted, and the human condition itself is construed as a contingently and variably formed aptitude keyed to specific enabling cultural practices. These themes have been regularized in the hermeneutic literature by theorists such as Gadamer, who hold both that the "nature" of persons and literary texts can only be imputed, changeably, under changing horizons (of description and interpretation) and that the aptitude for pertinent description and interpretation is itself altered within the changing tradition in which it is effectively exercised. The hermeneuts, who are perhaps the most easily recognized historicists, have pointedly analyzed, in this respect, (1) the "double hermeneutic" of interpreting texts as "possessing" historically alterable meanings and of interpreting our aptitude for such interpretation as similarly alterable; and (2) the sense in which the "hermeneutic circle" (the doctrine that the meaning of a text's parts depends on the meaning of the whole text and that the meaning of the whole depends on the meaning of its parts) is itself subject to historicized change. [*See* Hermeneutics.] The poststructuralists (including Foucault most prominently) tend not to oppose (but also not to advance) such themes, because they affirm more frontally the sheer (historical) contingency of "normalized" thought within one or another "discursive regime" (or episteme)—within which the division between cognizing subjects and cognized objects is already of little interest; that is, they emphasize the unperceived, radical, historical contingency of even the most settled categories of understanding. [*See* Poststructuralism.]

Behind these variant theories, there remains the deep puzzle of "intentionality"—more in the collective sense the classical historicists implicitly featured than the solipsistic (now canonical) sense preferred by Franz Brentano (1838–1917) and Edmund Husserl. Thus construed, historicism would make no sense at all if cultural realism (the admission of the genuine reality of the phenomena of the human world), primarily regarding traditions, institutions, social practices, or objectivity regarding the interpretation of cultural artifacts—persons, artworks, histories—could not be convincingly sustained. Both realism and objectivity require a realist reading of intentionality; also, the admission of intentionality (collective and historicized) tends to undermine the older forms of context-free objectivisms, which of course Western philosophy has regularly favored (see Margolis, 1995).

The matter is complicated, for the ultimate plausibility of historicism depends on what should count as an adequate theory of language—the resources of reference and predication especially. Thus, if natural-language reference and predication cannot be accounted for apart from the enabling practices of something like Wittgensteinian *Lebensformen,* and if such practices change (as Gadamer and Foucault claim) in the historicized way without disabling our linguistic fluency, then a global historicism appears quite plausible. In effect, this means that language is intentional in the sense suggested, whether it is applied to physical nature or human culture. The success of reference must then be inherently informal—consensual but only holistically, not criterially. The same would hold true for the extension of general predicates beyond the initial paradigms by which they are introduced. Predicates would (then) be intentional in a double way: first, because their generality would be inseparable from *lebensformlich* practices, no matter what predicates were involved (physical or *geistlich*); second, because *geistlich* (intentional, significative, interpretable) predicates or predicables would themselves signify the real features of cultural phenomena. The natural sciences would then be seen to be cultural artifacts abstracted from more encompassing human sciences.

In general, the argument favoring historicism contends that it is both coherent and conceptually more resourceful than its denial. For one thing, it would not oblige us to insist on necessary invariances of any kind (particularly regarding would-be laws of nature); for another, any such insistence would implicate some form of cognitive privilege (foundational sources ensuring certainty in knowledge); and, for a third, having abandoned modal necessities (necessities of thought or reality impossible to deny), historicism would be free to acknowledge salient regularities of whatever strength might be required (as Foucault admits, in speaking of the "historical a priori"). Historicism, therefore, is a radical thesis capable of absorbing (coherently) all the standard work of philosophy, science, art, history, criticism, technology, morality, and the like.

In sum: historicism now signifies: (1) the historicity of thought; (2) a symbiosis of subjects and objects along post-Kantian lines; (3) a global Constructivism; and (4) an accommodation of relativism. Its paradigm arguments center on the analysis of reference and predication in natural languages. Historicism holds that the effective fluency of such resources is inescapably collective, consensual, informal, intentional; and that the enculturing processes by which pertinent aptitudes are transmitted from one generation to another (thus formed) are themselves subject to the flux of history. The single most distinctive consequence of accepting the historicist thesis is to concede the ineliminability of interpretive *tertia,* that is, theories, conceptual schemes, "forms of life," "life-worlds," "cognitive perspectives," "discursive regimes," "paradigms," "horizons" mediating between belief and reality. Such *tertia* function tacitly in forming our own distinctive powers, which are altered in a historied way as a result of continuous cultural work. Hence, the recovery of the past, whether in ordinary reference and predication or in interpreting what may be perceived in (say) Diego Velázquez's *Las Meninas*—since Foucault, the best-known emblem of historicism—cannot fail to be formed and re-formed by the historicing powers of our interpretive *tertia.*

[*See also* Definition of Art; Essentialism; Foucault; Gadamer; Herder; *and* Relativism.]

BIBLIOGRAPHY

Foucault, Michel. "Nietzsche, Genealogy, History." In *Language, Counter-Memory, Practice,* edited by Donald F. Bouchard, translated by Donald F. Bouchard and Sherry Simon. Ithaca, N.Y., 1977.

Gadamer, Hans-Georg. *Truth and Method.* 2d rev. ed. Translation revised by Joel Weinsheimer and Donald G. Marshall. New York, 1989.

Heidegger, Martin. *Being and Time.* Translated from 7th ed. by John Macquarrie and Edward Robinson. New York, 1962.

Kuhn, Thomas S. *The Structure of Scientific Revolutions.* 2d enl. ed. Chicago, 1970.

Margolis, Joseph. *The Flux of History and the Flux of Science.* Berkeley, 1993.

Margolis, Joseph. *Interpretation Radical but Not Unruly: The New Puzzle of the Arts and History.* Berkeley, 1995.

Quine, W. V. *Word and Object.* Cambridge, Mass., 1960.

Wittgenstein, Ludwig. *Philosophical Investigations.* 3d ed. Translated by G. E. M. Anscombe. New York, 1958.

Worringer, Wilhelm. *Abstraction and Empathy: A Contribution to the Psychology of Style.* Translated by Michael Bullock. New York, 1953.

JOSEPH MARGOLIS

New Historicism in Literary Theory

Although he was not the first to use the term *New Historicism* to describe a mode of literary analysis, Stephen Greenblatt gave the phrase its current valence. In his introduction to a special issue of the journal *Genre* published in 1982, Greenblatt declared that New Historicism places literature and history in a new relation to each other, "challeng[ing] the assumptions that guarantee a secure distinction between 'literary foreground' and 'political background' or, more generally, between artistic production and other kinds of social production" (1982, p. 6). As his coinage indicates, Greenblatt's quarrel was in part with formalism or New Criticism, dominant in American literary criticism until the 1960s, which described each literary work as a self-contained, organic expression of unitary meaning—a "well-wrought urn," in Cleanth Brooks's phrase appropriating John Keats's image. But primarily Greenblatt objected to an earlier historicism that placed the literary work against its historical "background," assumed to be "a single political vision, usually identical to that said to be held by the entire literate class or indeed the entire population" (ibid., p. 5). That political vision in the Renaissance, according to this

"old historicism," was one of a stable, ordered society, governed by its belief in a hierarchical Great Chain of Being ruled on earth by the monarch and in heaven by God. For both old historicism and New Criticism, then, literature is an entity separate from history that operates according to unique principles, at best able to reflect this stable cultural background without either distortion or anxiety.

Greenblatt had already begun to formulate an alternative to old historicism in his landmark *Renaissance Self-Fashioning: From More to Shakespeare* (1980), which described the Renaissance self: one who "fashioned" himself or herself not, as in the Middle Ages, solely in imitation of Christ, but rather in response to a newly conceived and increasingly emphasized set of social postures and civilizing habits. (This "invention of manners" has been described foremost by historian Norbert Elias [1982]). In Greenblatt's view, Renaissance literature is analogous to this Renaissance self: "Literature functions . . . as a manifestation of the concrete behavior of its particular author, as itself the expression of the codes by which behavior is shaped, and as a reflection upon those codes" (1980, p. 4). The literary work is not to be regarded as different in kind from any other product—physical or ideological—of the author's culture, because all such products represent interpretative responses to the culture's customs and institutions. Greenblatt thus clearly shows his debt to Michel Foucault, who defines every human linguistic act, including the literary text, as a product wholly embedded within its own historical period's peculiar epistemological configuration.

But Greenblatt, less committed to an idea of the all-pervasive episteme than Foucault, also allies himself with cultural anthropologists such as Clifford Geertz (1973), who seeks to explore, through a technique of "thick description," how humans construct their own social systems through their interpretative responses. Hence, although the term *New Historicism* remains the one most often used to describe his approach, Greenblatt himself has abandoned it in favor of "a poetics of culture," which more nearly expresses his sense of the interdependence of historical circumstance and aesthetic creation. In particular, Greenblatt rejects a purely Marxist sense of individuals' helpless absorption within an all-encompassing economic and ideological superstructure. In this way, Greenblatt draws on the British historian Raymond Williams and the French philosopher Louis Althusser, both of whom revise a Marxist notion of "material history" to include aesthetic productions. [*See* Williams.] Greenblatt views the individual in culture according to French sociologist Pierre Bourdieu's concept of the habitus, a "system of structured, structuring dispositions . . . which is constituted in practice and is always oriented towards practical functions" (Bourdieu, 1980, p. 52). [*See* Bordieu.] Both the individual and the literary work, then, take part in an endlessly reciprocal relation to culture—or, as Greenblatt's influential collection of essays on Shakespeare has it, a relation of "negotiation," "a subtle, elusive set of exchanges, a network of trades and trade-offs, a jostling of competing representations" (1988, p. 7). The individual author, then, is both a product of competing cultural discourses and a remaker of those discourses, reflecting them in the distorted glass of his or her own social place and constellation of desires.

In particular, the Renaissance author, in Greenblatt's view, seeks to express and to manage his or her sense of "anxiety," a condition to which, historian William Bouwsma (1990) contends, late-medieval and early modern society was newly prone as it progressively abandoned once-secure boundaries: between the country and the growing city; between clergy and laity; between aristocrat and peasant; between God's design and humans' apprehension of it. This anxiety, eventually expressed in Blaise Pascal's terror of the "eternal silence of these infinite spaces," has been most studied by New Historicist scholars of the Renaissance in the context of the individual subject's relation to monarchy. In an influential essay on Shakespeare's *A Midsummer Night's Dream,* for example, Louis Montrose delineates the masculine anxiety generated by Queen Elizabeth I's anomalous place on the English throne, arguing that the play airs that anxiety in order to neutralize it, thus "altering the lines of force" within Elizabethan culture (1986, p. 69).

The term *anxiety* highlights several, often unacknowledged, debts that most New Historicists share. First, their model of the self resembles that of Sigmund Freud, who describes individual neurosis as a complex (if ultimately damaging) design for managing one's incompletely understood familial configuration. Similarly, New Historicism argues that the peculiar features of a literary text highlight its author's attempts to manage the incomprehensible weight of his or her uncertain cultural position. That management, as in Freud's family romance, can take the form of responses ranging from the oblique to the overt, from submission to outright rebellion. Second, New Historicists owe certain habits of reading to the poststructuralist practices of Jacques Derrida and Paul de Man, who, like Freud, expose the instabilities of a seemingly stable textual meaning by putting pressure on the text's most fissured—its most "anxious"—moments. Third, New Historicism's sense that material history is accessible only as a text shares something with Jacques Lacan's theory that the "real" is what is elided whenever representation takes place.

But unlike psychoanalysis or deconstruction, which in their usage by American literary critics have hardened into a fairly programmatic series of principles and practices, New Historicism has never been assigned a definitive or systematic methodology by its practitioners. This evasion of method has elicited a number of different critiques of New Historicism. First, historians have taken issue with New Historicists' inattention to facts: not only their mistaking or ignoring the current state of historical knowledge—

especially that offered by the "new history," which studies local events and social structures rather than national or movements and personages—but also their taking bizarre or anecdotal evidence as offering access to the general state of a historical period's episteme. Montrose's essay, for example, begins with an anecdote about the late sixteenth-century dream reported by Simon Forman, a "professional astrologer and physician, amateur alchemist, and avid play-goer," in which Forman flirts with the (already elderly) Queen Elizabeth I and expresses his desire to impregnate her (ibid., p. 65). Whereas Montrose classifies Forman's dream as expressing "the cultural contours of a psyche that is both distinctively male and distinctively Elizabethan" (ibid., p. 66), a historian might question whether the average Elizabethan gave much thought to the monarch's gender, as opposed to, say, the monarch's edicts on taxes or religious conformity.

Second, whereas one of the continuing objects of New Historicist Renaissance study has been Europe's cultural clash with the newly explored African and American continents, in its emphasis on cultural oddities New Historicism, like the anthropology to which Greenblatt compares his practice, can be accused of a sort of cultural imperialism of its own: an artifact or anecdote of the past is viewed out of context, like an African fetish or a Peruvian headdress, as simultaneously totemically privileged and perfectly symptomatic of its culture. As Joel Fineman (1989) puts it, the characteristic New Historicist anecdote serves first to introduce a rupture into a Hegelian, teleological history, but then to reinscribe the *grand récit* of cultural narrative, of inescapable cause and effect. New Historicism's primary fields of application, the distant yet much-studied late-medieval and Renaissance periods, serve to keep this sense of simultaneous strangeness and familiarity alive. The more contemporary field to which New Historicism has been most applied, American studies, has similarly, if less imperiously, profited from reading the United States' racial, ethnic, and religious heterogeneity—which a predominantly homogeneous academy still tends to view as eccentric and extremist—as central to American culture.

Third, in its evasion of method the New Historicism has proved to have a "portmanteau quality," as H. Aram Veeser terms it, capable of absorbing any number of issues and agendas (1989, p. xi); thus, it has drawn charges of being apolitical to the point of being amoral. (Critics of the academy from the right, in contrast, perceive New Historicism as dangerously political in its reversal of Ben Jonson's formula that authors such as Shakespeare are figures not "of an age, but for all time.") In this regard, New Historicism may be contrasted to the otherwise similar "cultural materialism" practiced by Marxist-influenced critics such as Jonathan Dollimore and Alan Sinfield (1985), who extend their analyses of Renaissance culture, for instance, into a study of the modern British classroom's oppressive calcifi-

cation of Shakespeare as Great English Poet. New Historicist readings that operate under what has come to be called the "subversion-containment model" have particularly come under fire. Greenblatt's essay "Invisible Bullets," reprinted in *Shakespearean Negotiations* (1988), has been taken as paradigmatic of this model. In this essay, Greenblatt juxtaposes the reports of Thomas Harriot, one of the first Virginia colonists, with Shakespeare's *Henry IV* plays and *Henry V* in order to demonstrate how both authors air, and indeed produce, subversive notions: in Harriot's case that religion, at least as the Algonquians practice it, is merely a social construction; in Shakespeare's, that Prince Hal's ascent to kingship is merely a matter of theatrical dissembling. In both instances, asserts Greenblatt, the author raises a potentially devastating challenge to a social order that depends on belief in God or king, only to demonstrate that such a challenge does not undermine that order. Greenblatt argues that this operation replicates that of early modern English society as a whole, in which seemingly subversive sites of cultural production such as the public theater can only contribute to dominant institutions' desire for stability. From this reading, one might surmise that there was no authentic stance in early modern English culture from which subversion or alterity might be proposed; and one might wonder how, then, that culture went so far in the seventeenth century as to behead its monarch.

Several rejoinders have been issued to the charges of political conservatism raised by readings such as "Invisible Bullets" and by New Historicism's general refusal to define its methodological allegiances. The eighteenth- and nineteenth-century literary scholar Catherine Gallagher, associated with New Historicism through her coeditorship (with Greenblatt and others) of the Berkeley, California-based journal *Representations*, argues that New Historicism actually has an agenda superior to that of demonstrating either social institutions' hegemony or their subversion by literature: "The contribution of the new historicism has been to identify a third alternative in which the very antagonism between literature and ideology becomes, in specific historical environments, a powerful and socially functional mode of constructing subjectivity" (1989, p. 44). Montrose, in contrast, has responded by agreeing that New Historicists need to reveal themselves as complicit in a project of social containment: "to speak today of an historical criticism must be to recognize that not only the poet but also the critic exists in history . . . and that our comprehension, representation, interpretation of the texts of the past always proceeds by a mixture of estrangement and appropriation" (1989, p. 24).

The final kinds of attack on New Historicist studies of Renaissance literature are more specific. Given their Foucauldian emphasis on the equivalence and interrelation of all texts within a culture, New Historicists have been taken to task for their reliance on canonical literary works, and especially the epitome of canonical works, the plays of Shake-

speare. Marjorie Garber (1990), for example, accuses New Historicists such as Greenblatt and Montrose of fashioning Shakespeare into the transcendent truth that they find so problematic in New Criticism and old historicism. As well, some New Historicists have been criticized for their unexamined assumption that early modern culture is intertextually constituted entirely by means of the viewpoints of white, Christian males. Feminist critics, for example, have interrogated New Historicists' readiness to agree with historian Thomas Laqueur's thesis that early modern medicine, following Galen's model, believed there was really only one sex: men, and imperfect, ill-formed men—that is, women.

A second generation of critics influenced by New Historicism, however, is engaged in reversing these trends. A number of analyses have been published, of which three studies of Renaissance England might be singled out for comment. In *Dangerous Familiars* (1994), Frances Dolan addresses early modern England's anxiety over domestic household subordinates—servants, wives, and mothers—and their potential for causing social disorder. Dolan's analysis extends not only to Shakespearean plays that represent crime within the household, such as *Macbeth* and *Othello*, but also to a vast range of other texts that share this concern: for example, ballads, pamphlets, and trial transcripts, as well as other once-popular but less well remembered plays. Kim Hall's *Things of Darkness* (1995) studies how Renaissance literature and art conjoin images of the black man and the white woman in order to define white masculinity in opposition to them. Finally, James Shapiro's *Shakespeare and the Jews* (1996) examines how England constituted its fledgling ideas of "Englishness" by means of contrast to its fabricated ideas of "Jewishness"; Shapiro's readings of Shakespeare are interspersed among a wealth of other literary and nonliterary early modern representations of Jews. The value of these studies is not only that they expand New Historicism's topical and textual frames of reference, but also that they treat sources of cultural anxiety that, unlike the Elizabethan subject's relation to monarchy, have hardly been comfortably resolved. In this way, these new New Historicists make Renaissance strangeness all too familiar, and thus erase any lingering notions that the Renaissance era is a secure and safely distant "background" to the "foreground" of our own time.

BIBLIOGRAPHY

Althusser, Louis. "Ideology and Ideological State Apparatuses." In *Lenin and Philosophy, and Other Essays*, pp. 127–186. Translated by Ben Brewster. New York, 1971.

Bourdieu, Pierre. *The Logic of Practice* (1980). Translated by Richard Nice. Stanford, Calif., 1990.

Bouwsma, William J. "Anxiety and the Formation of Early Modern Culture." In *A Usable Past: Essays in European Cultural History*, pp. 157–189. Berkeley, 1990.

Dolan, Frances E. *Dangerous Familiars: Representations of Domestic Crime in England, 1550–1700*. Ithaca, N.Y., 1994.

Dollimore, Jonathan, and Alan Sinfield. *Political Shakespeare: Essays in Cultural Materialism* (1985). 2d ed. Ithaca, N.Y., 1994.

Elias, Norbert. *The Civilizing Process* (1939). 2 vols. Translated by Edmund Jephcott. Reprint, New York, 1982.

Fineman, Joel. "The History of the Anecdote: Fiction and Fiction." In *The New Historicism*, edited by H. Aram Veeser, pp. 49–76. New York and London, 1989.

Foucault, Michel. *The Order of Things: An Archaeology of the Human Sciences* (1966). New York, 1970.

Gallagher, Catherine. "Marxism and the New Historicism." In *The New Historicism*, edited by H. Aram Veeser, pp. 37–48. New York and London, 1989.

Garber, Marjorie. "Shakespeare as Fetish." *Shakespeare Quarterly* 41.2 (Summer 1990): 242–250.

Geertz, Clifford. *The Interpretation of Cultures*. New York, 1973.

Greenblatt, Stephen. *Renaissance Self-Fashioning: From More to Shakespeare*. Chicago, 1980.

Greenblatt, Stephen. "Introduction." In *The Power of Forms in the English Renaissance*, edited by Stephen Greenblatt, pp. 3–6. Norman, Okla., 1982. Reprint of *Genre* 15.1–2 (Summer 1982).

Greenblatt, Stephen. *Shakespearean Negotiations: The Circulation of Social Energy in Renaissance England*. Berkeley, 1988.

Greenblatt, Stephen. "Towards a Poetics of Culture." In *The New Historicism*, edited by H. Aram Veeser, pp. 1–14. New York and London, 1989.

Hall, Kim F. *Things of Darkness: Economies of Race and Gender in Early Modern England*. Ithaca, N.Y., 1995.

Laqueur, Thomas. *Making Sex: Body and Gender from the Greeks to Freud*. Cambridge, Mass., 1990.

Montrose, Louis A. "*A Midsummer Night's Dream* and the Shaping Fantasies of Elizabethan Culture: Gender, Power, Form." In *Rewriting the Renaissance: The Discourses of Sexual Difference in Early Modern Europe*, edited by Margaret W. Ferguson, Maureen Quilligan, and Nancy J. Vickers, pp. 65–87. Chicago, 1986.

Montrose, Louis A. "Professing the Renaissance: The Poetics and Politics of Culture." In *The New Historicism*, edited by H. Aram Veeser, pp. 15–36. New York and London, 1989.

Shapiro, James. *Shakespeare and the Jews*. New York, 1996.

Veeser, H. Aram. "Introduction." In *The New Historicism*, edited by H. Aram Veeser, pp. x–xvi. New York and London, 1989.

Williams, Raymond. *Culture and Society 1780–1950* (1958). Reprint, New York, 1983.

KATHERINE EGGERT

Historicism in Art History

Historicism can be considered central to all of the discourses and institutions concerned with historical studies before poststructuralism, but none more so than art history because of its focus on the material object or artifact and its formation as an academic discipline in the German-speaking context in the nineteenth century. Since the beginning of modernity, historicism has been deeply implicated in both popular and philosophical conceptions of art because art-historical criticism has used historicizing methods of understanding the past and brought them to bear on the interpretation of the art object. These methods allow the interpreter of art to understand that art objects have a history without questioning the underlying assumptions of that position. Historicism in art history places those objects from

the past called art in the narrative context of history writing. In the text this is accomplished through naming and description. By being represented in the discourse of art history in these ways, art objects could be seen to provide access to a historical consciousness through their connection to specific places, events, audiences, and makers.

With historicism in art history came the possibility of conceiving representation—in the visual and material form of the art object—as endowed with historical understanding. With this historical understanding implicit within it, the art object could be said to be the material embodiment of both ethical and ideological dimensions of history. The value of history and the value of the art object are thus mutually enhancing and codependent positions in discourse and culture. The implications for material productions of all sorts of the understanding that history resides in the object contribute to historicism's power in art history and visual studies in general. By examining historicism in art history, one can see how interpreters have constructed and contested the meaning and value of history in art.

It should be noted that although the methods associated with German historicism can be found in much art history since Johann Joachim Winckelmann (1717–1768), art history lacks in its discourse any explicit debate or extended dialogue on historicism, whether as a philosophical or a historical problem. Unlike the disciplines of history and literature, where historicist approaches have been criticized and thematized, they have been in the main taken for granted in art history. The historicist turn that Fredric Jameson (1988) has named "historical and cultural aestheticism," or "existential historicism," has not been criticized in the discipline whose foundations it established. E. H. Gombrich's influential *Art and Illusion* (1960), for example, disclaims historicism's significance absolutely. The important exception to this pattern, Arnold Hauser's *The Philosophy of Art History* (1958), recognized that historicism in terms of the subject, "art history," led immediately to considerations about history, art, and the historiography of art history. Donald Preziosi (1989), Keith Moxey (1994), and others have also said that such a comprehensive critical project of historicism's influence will include the relatively recent history and construction of art history as a discipline in the academy, although, again, no sustained disciplinary critiques have thus far emerged. One is tempted to attribute the unexamined assumptions in the discipline to the very conjunction of art and history initiated with historicist methods.

The writing about what later came to be called "art" began with the Greeks, but this discourse has never been considered history writing proper, in part because only partial written accounts have survived. These fragments do not possess the rhetorical structures now associated with history writing. Rather, they consist in the main of anecdotal material, not properly speaking history because not part of a larger narrative context or schema. The Italian Renais-

sance, beginning with Antonio Manetti's *Life of Brunelleschi* (c.1470), erected an account of "modern" art based primarily on the life of the individual artist, the vita, and on the biographical anecdote. Giorgio Vasari's series of *Lives* (1550 and 1568) remain the best-known example of this genre. This writing was not called history at the time. The term *history* used in conjunction with art to mean a narrative about the past (but not about art) and distinguished from *magistra vitae* appeared first early in the seventeenth century in the Italian context with Agostino Mascardi's (1591–1640) *Dell'arte historica* (1636).

The early Italian model of the vita of the artist was the most prevalent form of writing about art throughout Europe until the publication of Winckelmann's *The History of Ancient Art* (1764). The origin of a historicist approach to the history of art begins here with Winckelmann's writing on Greek sculpture, hence his appellation as "the first art historian." Enlightenment discussions of history, contemporaneous with Winckelmann, attempted to distinguish history from science and to establish its laws as separate from but equal to the laws of nature, as a result of the scientific revolutions of the seventeenth century. In such endeavors, history often called on art as an alternative to the available scientific models in seeking to distinguish itself from them. Winckelmann's *History of Ancient Art* appeared in a German context, just when the first discussions of the nature of history and art came together in German philosophical discourse, for example, in Alexander Baumgarten, Johann Gottfried von Herder, and Immanuel Kant. Although important differences exist among these thinkers, they established aesthetics as an area of knowledge and inquiry pertaining to all of the arts and separate from other human activities. Commentators on historicism cite works by these writers as central to the later debates of the concept. The same philosophers also dominate, although not usually explicitly, art history's traditional understanding of art and its history.

Winckelmann idealized the Greek world and its art, particularly its marble sculpture, in his history writing. Unlike the earlier authors in the biographical tradition, he tied particular art objects (i.e., marble sculptures) and the styles of individual makers (i.e., artists with names) to a systematic and developmental account of the rise of Greek art from simple and debased beginnings to an apogee of beauty and advancement. His argument asserted that particular historical conditions pertaining uniquely to the Greek world and its climate were responsible for the artistic ideals on which Greek sculpture was based. In Winckelmann, one finds a consideration of the past and its applicability to a *present* appreciation of ancient sculpture. He uses methods such as lengthy description, connoisseurship, and antiquarian archaeology that came to be commonly used in art history. One also finds a consideration of the relationship of art to the politics of the state. The *History of Ancient Art* exhibits

virtually all of the negative characteristics of historicism that have been seen by later historiographers to be typical of that approach. These include a developmental or teleological narrative of the past, the attribution of predictability to past events, and a nationalistic ideology.

From its very beginnings, then, the historicist approach to art contained within its compass an ideal model—classical sculpture—for the art object that, when inserted in the historical account, justified such an idealist conception of art. It has been said of historicism that it regards all intellectual life to be permeated by history. As early as Winckelmann's book on Greek art, one also sees that this approach invests history and its theoretical presuppositions, such as chronology, in art. Thus, historicism locates art and, eventually, culture in history and gives art history its relevance to aesthetics. This understanding of the relationship of art and history can be said to permeate the writing on art known since the nineteenth century as art history. Art historian David Freedberg (1994) argues that the figuration of any material object determines its status as art and its placement in a narrative about the past. Rather, as Winckelmann's text clearly demonstrates, the linking of art to history leads to the ontic status of art. The a priori figuration of the material object only matters to art and history inasmuch as it becomes art when inserted in the supportive historical narrative. Art theorists of Winckelmann's time held the ideal model for art, antique sculpture, to be universal and timeless, while at the same time insisting that art could be incorporated into a historical narrative based on chronology. The conflict in the two positions has been noted, notably by Jeffrey Morrison (1996), but more often has been ignored in art history.

In the context of the foregoing discussion, the ethical consequences of Winckelmann's canonical status deserve elaboration. Not only did Winckelmann precede other historicists, but his subject was Greek art. From the time of the earliest Renaissance humanists, the Greeks had been considered the historical model for progress. From the High Renaissance, Greek sculpture (or its Roman copies) had served as the standard for visual progress in the European mimetic theory of art. The antique works were used comparatively with contemporary painting and sculpture to invoke the ideal to which art should aspire. With Winckelmann, Greek sculpture became the material embodiment of a concept of historical *and* visual perfection. The classical model, essential to the beginnings of the history of art, insisted on what Fredric Jameson has described as "not merely the Identity of these formal languages and sign systems with our own aesthetic values and ideals, but rather also, and through the symbolic medium of the aesthetic experience, a whole political analogy between two forms of social life" (1988)—the present and the past. The objects of the past served as lessons for the present.

Nineteenth-century German art history followed Winckelmann in most ways: art objects functioned as primary sources for the historicist approach, much as church documents did for the historicist history supposedly initiated by Leopold von Ranke (1795–1886), the "father of German historicism." Contemporary documents, texts, and art objects were seen as providing excellent representations of specific historical circumstances. The retrieval and interpretation of these were essential to the operation and aims of reconstructing an earlier society or culture. The art historian Aby Warburg's "Lecture on Serpent Ritual" (1939) is a good example by an Italian Renaissance specialist at understanding native Americans according to the views of a historicist social consciousness achieved through such means.

Ranke's student Jacob Burckhardt accepted literary texts, that is, biographies of poets and painters, as historical documents in *The Civilization of the Renaissance in Italy* (1860). This allowed him to include the "historicized" visual arts in the context of his reconstruction of a cultural history, an approach known as *Kulturwissenschaft*. In this approach one finds a much-expanded view of the role of art in history in comparison to that previously held by Winckelmann. In this strand of German historicism, Burckhardt and others argued that the history represented in art objects of all kinds could serve as a moral exemplar for present political behavior. This ethical position of the interpreted object was enhanced by the dominant theoretical concepts of mimesis in representation. Art theorists in the early modern period saw an ideal simulacrum of nature exemplified in Greek sculpture, and later art historians found it in Renaissance painting, sculpture, and architecture. Nineteenth-century German art historians held that the moral authority inherent in the art object and revealed by the interpreter conferred its weight onto the interpretational project of art historians. Burckhardt, to take an early and extreme example, found the ethical life mirrored in the writing of art history. E. H. Gombrich (1986) found such a life mirrored in the writing of Aby Warburg, if somewhat tendentiously at times. By the beginning of the twentieth century, both ethical and national considerations were implied in an art-historical consciousness.

Nevertheless, Burckhardt's methods came under attack in the early part of the twentieth century by art historians, such as Warburg, who realized that his predecessor's "sources" in form and content spoke only of an artistic elite and neglected important factors in the history of art such as change in forms, motifs, and visual theories. Warburg and his followers, such as Edgar Wind, Erwin Panofsky, and many others, were keenly aware of the relativism of an art history based only on particular kinds of sources. They sought to base their explanations of objects on a variety of sources. Still, they relied on a concept of art as it had been theorized by the Enlightenment philosophers of aesthetics.

This assertion is particularly apparent in the approach found in the Viennese school of art history, usually cited for its philological approach to art history. The influence of the

idealist philosophy of Benedetto Croce's "humanist historicism" is crucial to this school after 1902, the date of the publication of Croce's book on aesthetics. Using G. W. F. Hegel's view that the ends of history are in art, Croce's arguments were first made in terms of a philosophy of aesthetics that held out for historical absolutes (understood to consist of a rational structure) as opposed to purely formal, stylistic ones in approaching the work of art. Here change occurs because change is historically determined: the artist is the agent of change; the work of art the representative of change. Croce's philologically minded adherents in the Viennese school of art history found in the early sources and documents of the discourse on art (i.e., from the Italian humanist context) justification for assuming a consistency in the history of writing on art and the artist. In the discourse on art, the privileging of the artist has, in fact, not changed since the Italian Renaissance biography of the artist, although the form of the discourse in which one finds that artist also constructs the view one has of the individual maker, as Winckelmann's work shows. The Viennese school's approach could be called idealist-historicist and its influence can be seen, for example, in Gombrich's emphasis on the absolute agency of the "making and matching" individual maker in *Art and Illusion*, even as he reacted against other aspects of historicism (discussed later in this essay).

Virtually contemporaneously with Warburg, Heinrich Wölfflin and Alois Riegl sought a way out of the privileging of the individual artist that Burckhardt's approach imposed by turning to a study of art based on formal analysis and consistencies, such as intention, that were observable and observably repetitious in the works alone. Hauser (1958) has called this an "Art History without Names," or what purports to be a history of style pure and simple. In fact, it refuses to abandon individualism and an absolute historicism based on the incomparability in the historical structure, that is, art, to other historical and philosophical structures. This is where Burckhardt, Croce, Gombrich, Warburg, Wölfflin, and Riegl agree and where art history, as Norman Bryson (1983) has argued, attempts a mimesis of the absolute idealism of art, as the latter had been defined since early aesthetics. As already noted, this mimetic priority for art history begins with Winckelmann and his use of classical sculpture.

In the mid-twentieth century, a vehement rejection of historicism on the grounds of its predictability thesis came from the neopragmatist Karl Popper. Popper's influence on his friend, the art historian Gombrich, can be seen above all in Gombrich's *Art and Illusion* (delivered as lectures in 1956, published in 1960), which attempts to salvage a progressivist account of "why styles change" while at the same time rejecting historicism. He posited a "making and matching" visible in all art "based on individual efforts to criticize and to improve on predecessors' achievements" rather than on inexorable laws of historical development.

According to Gombrich, in the European mimetic system of representation, change occurs through the agency of the individual maker working in response to earlier artistic models. Gombrich's conflicted and contradictory views toward historicism have characterized Anglo-American art history until recently, when a critique based on semiotics has emerged. [*See* Gombrich.] Furthermore, the implicit notion of value inherent in the concept of mimesis and its function in the interpretation of art according to historicist methods lies at the heart of Walter Benjamin's critique of both art and historicism in "Theses on the Philosophy of History" (written c.1939–1940, published in 1950) and "The Work of Art in the Age of Mechanical Reproduction" (1936). [*See* Benjamin.] This critique of historicism is most relevant to the problems with historicism in art history because it questions the status of the art object and refuses to give history the status of art. Yet, in spite of important exceptions, particularly in the study of postmodern visual culture, art history has resisted critiques such as Hauser's and Benjamin's.

Georg Iggers (1983) has argued that historicism, particularly when understood as the self-conscious consideration of history writing, was a distinctly German phenomenon and therefore tied to the German views of themselves as a nation and a people. In these terms, the significance of the preeminence of *Kulturwissenschaft* in pre–World War II art history becomes clear: for Burckhardt and virtually all of the early art historians mentioned here, the art object became the document of civilization, because of the common mimetic standard on which representation in the European tradition was based. If the interpreters of that civilization were ethically endowed as individuals, they were also recognizable as Germans, the founders of the academic discipline of the history of art. When Iggers referred to "the decline of the German Idea of history" after World War II and the fall of the Third Reich, he did not consider, however, the simultaneous persistence of German historicism in the context of the dispersal of German art historians and the growth of the discipline in the English-speaking context.

Art history was born and developed as an academic discipline in Germany, Austria, and Switzerland long before it did elsewhere: beginning in the 1870s, the major figures in the practice of *Kulturwissenschaft* held important and early chairs in departments of art history or as directors of museums in the largest universities and cities of these countries, such as Berlin, Vienna, and Basel. By the turn of the twentieth century, art history had come to be incorporated in most German universities, and museums, especially in Austria and Berlin, had become major public institutions. When World War II led to the diaspora of German-Jewish intellectuals to the English-speaking countries, art historians took the methods of historicism with them, but the critique did not follow, except, as already noted, in the form in which it is found in Popper. The influence of these immigrant art

historians on the shape of art history both as discipline and discourse in America and Britain since World War II cannot be overestimated.

For apparent and understandable reasons, the immigrants were reluctant to grapple publicly with critiques of a method that had led to the establishment of their discipline in Germany and had aided them and their predecessors in their assimilation as Jews into (German) culture. Similarly in America, art history gave these immigrants entry into American high culture in the universities and museums. Because their existence as intellectuals and professionals both in Europe and America was centered on the object, they had to believe that the same German nation that had attempted to eradicate them had also been responsible for their salvation through the tradition and the training that had led to their vocations. How could the atrocities committed by Germany during the war be related to the Western cultural artifacts they chose to study? The postwar critiques of historicism in the discipline of history foreground issues of language, ethnicity, and the nation-state; they were barely felt in a field that has refused the interrelatedness of these with art's history and a situated history of art historians.

A critique of historicism in art history would depend on Benjamin's analysis of the art object and Michel Foucault's premise that an institution's (or discipline's) discourse is inseparable from institutional (or disciplinary) history and power. The so-called crisis in art history that began in 1973 and brought about the "New Art History" recognized that much was amiss in both discourse and discipline, but its most important effect has been to engender a number of critiques of art and the discipline based on a doomsday, "end of art" history. Just as Friedrich Nietzsche at the end of the nineteenth century responded to the nation-state by totally rejecting the historicist concept of history, a nihilistic version of historicism on his part, these art historians and critics have displayed a hostility to art history at the end of the twentieth century without fully incorporating a sustained critique of the methods that support the closeness of the concepts of art and history.

Studies of art history's founders, such as those by Hans Dilly (1979) and Silvia Feretti (1989), as well as translations of some of their important preimmigration writing, signal a growing awareness on the part of art historians of the intellectual history of its founders that must be considered central to the understanding of historicism in art history because of the role of ethics in the tradition of historicist art history. In addition, since the early 1980s, the most compelling critiques about the position of art in culture have come from outside the discipline, for example, Martin Bernal (1987), James Clifford (1988), Kobena Mercer (1994), and Marianna Torgovnick (1990). These writers, trained in disciplines that either have successfully critiqued historicism or never used it, have raised questions about the primacy of the art object and the ways that history is employed in the interpre-

tation of visual culture. Instead of understanding history and art as universal and ideal concepts, these writers, each in a distinct way, attempt to stress the significance of their present location in the interpretation of the past and its artifacts. Historicism's place in these discussions has shifted or disappeared entirely and with it art history's relative primacy in discussions of the history of art. Interestingly, the "New Historicism" in literary studies, in part a critique of old historicist positions, has used visual culture and artifacts among its appropriations of nonliterary texts as comparative material in interpretation. Thus far, this rather undefined method has had little obvious effect on the practice of art historians, particularly those who deal with the art of the distant past. In any case, since its beginnings, art history has invariably incorporated literary texts and documents as comparative materials in the interpretation of art. Whether the discipline has theorized it or not, art history is now on the sidelines of a multidisciplinary debate about art, culture, and meaning. Contemporary discussion of these issues centers on issues of identity (nationalities, race, ethnicity, sexuality, gender) in "visual culture," including film, video, digitized media, and popular culture. These discussions exclude from the debate methods that have refused to recognize that historicism lurks behind every art object.

[*See also* Formalism, *article on* Formalism in Art History; *and* Winckelmann.]

BIBLIOGRAPHY

Belting, Hans. *The End of the History of Art?* Translated by Christopher S. Wood. Chicago, 1987.

Benjamin, Walter. *Illuminations.* Edited by Hannah Arendt, translated by Harry Zohn. New York, 1968.

Bernal, Martin. *Black Athena: The Afroasiatic Roots of Classical Civilization,* vol. 1. New Brunswick, N.J., 1987.

Bryson, Norman. *Vision and Painting: The Logic of the Gaze.* New Haven, 1983.

Burckhardt, Jacob. *The Civilization of the Renaissance in Italy.* 2 vols. Translated by S. G. C. Middlemore. New York, 1958.

Carrier, David. *Principles of Art History Writing.* University Park, Pa., 1991.

Clifford, James. *The Predicament of Culture: Twentieth-Century Ethnography, Literature, and Art.* Cambridge, Mass., 1988.

Croce, Benedetto. *Aesthetic as Science of Expression and General Linguistic.* 2d ed. Translated by Douglas Ainslie. London, 1922.

Dilly, Heinrich. *Kunstgeschichte als Institution: Studien zur Geschichte einer Disziplin.* Frankfurt am Main, 1979.

Ferretti, Silvia. *Cassirer, Panofsky, and Warburg: Symbol, Art and History.* Translated by Richard Pierce. New Haven, 1989.

Freedberg, David. "The Object of Art History: Context, Visuality, and the Objects of Art History." *Art Bulletin* 76 (September 1994): 394–396.

Gombrich, E. H. *Art and Illusion: A Study in the Psychology of Pictorial Representation.* (1960). 2d rev. ed. Reprint, Princeton, N.J., 1969.

Hauser, Arnold. *The Philosophy of Art History.* New York, 1958.

Holly, Michael Ann. *Panofsky and the Foundations of Art History.* Ithaca, N.Y., 1984.

Iggers, Georg G. "Historicism." In *Dictionary of the History of Ideas,* edited by Philip P. Wiener, vol. 2, pp. 456–464. New York, 1973.

Iggers, Georg G. *The German Conception of History: The National Tradition of Historical Thought from Herder to the Present*. Rev. ed. Middletown, Conn., 1983.

Jameson, Fredric. "Marxism and *Historicism*." In *The Ideologies of Theory: Essays, 1971–1986*, vol. 2, *The Syntax of History*, pp. 148–177. Minneapolis, 1988.

Kultermann, Udo. *The History of Art History*. New York, 1993.

Manetti, Antonio di Tuccio. *The Life of Brunelleschi*. Edited by Howard Saalman, translated by Catherine Enggass. University Park, Pa., 1970.

Mascardi, Antonio. *Dell'arte historia*. Florence, 1859.

Mercer, Kobena. *Welcome to the Jungle: New Positions in Black Cultural Studies*. New York, 1994.

Moxey, Keith. *The Practice of Theory: Poststructuralism, Cultural Politics, and Art History*. Ithaca, N.Y., 1994.

Panofsky, Erwin. *Perspective as Symbolic Form*. Translated by Christopher S. Wood. New York, 1991.

Podro, Michael. *The Critical Historians of Art*. New Haven, 1982.

Preziosi, Donald. *Rethinking Art History: Meditations on a Coy Science*. New Haven, 1989.

Soussloff, Catherine M. *The Absolute Artist: The Historiography of a Concept*. Minneapolis, 1997.

Soussloff, Catherine M., ed. *Jewish Identity in Art History: Ethnicity and Discourse*. Berkeley, forthcoming.

Torgovnick, Marianna. *Gone Primitive: Savage Intellects, Modern Lives*. Chicago, 1990.

Vasari, Giorgio. *Lives of the Most Eminent Painters, Sculptors, and Architects*. 10 vols. Translated by Gaston du C. de Vere. London, 1912–1914.

Warburg, Aby. "A Lecture on Serpent Ritual." *Journal of the Warburg and Courtauld Institutes* 2 (1939): 277–292.

White, Hayden. "Historicism, History, and the Figurative Imagination." In *Tropics of Discourse: Essays in Cultural Criticism*, pp. 101–120. Baltimore, 1978.

Williams, Raymond. *Keywords: A Vocabulary of Culture and Society*. Rev. ed. New York, 1985.

Winckelmann, Johann Joachim. *The History of Ancient Art*. 4 vols. Translated by G. Henry Lodge. Reprint, New York, 1969.

Wölfflin, Heinrich. *The Principles of Art History: The Problem of the Development of Style in Later Art*. Translated by M. D. Hottinger. Reprint, New York, 1950.

CATHERINE SOUSSLOFF

Historicism and Architecture

In current architectural discourse, the term *historicism* carries two meanings. The more common one generally refers to the use of old formal motifs and devices in current practice. It can be called eclectic historicism. It arose early in the nineteenth century, it dominated official and establishment circles on the Continent, and it brought a revival of styles ranging from the Egyptian through Greek, Roman, and Gothic to the only recently outmoded Baroque. Each new revival sought to invest architectural form with a legitimacy or justification that had been lost or to associate a building with some highly regarded past or sentiment about that past.

There were dissenters. In his *Discourses on Architecture* (1863–1872), Eugène-Emmanuel Viollet-le-Duc (1814–1879) said that this architecture resulted from a compromise among officials, "amateurs, academicians, professors of art who profess nothing, archaeologists, partisans of Classical Antiquity or of the Middle Ages, men of science, political economists and enthusiasts." At the end of the day, "no one is satisfied, and all the city exclaims, 'Why do not our architects produce an architecture characteristic of the age?'"

The idea that there is an architecture characteristic of its age is the other, much richer, meaning of the term *historicism*. Each age has a unique and distinct culture, and among its cultural products is architecture. Architecture is good to the extent that it embodies the unique characteristics of unique, successive ages. The architecture of one age is unsuitable for any other age. Viollet-le-Duc's lament at the height of eclectic historicism asks, Why is our age false to itself?

Viollet-le-Duc's observations refer to the world built atop the ruins of the French Revolution and the industrial revolution. In the older world, regimes and their architects worked with rules that were not the product of human intention or will but were discovered in nature, and buildings were good according to their ability to embody enduring standards and norms within traditional formal conventions and functional uses. Because society and architecture shared similar origins, a discussion of, for example, proportional harmony had social and political significance. Aesthetic quality revealed and preserved social harmony.

In this view of the past, Renaissance architecture and its development into the architecture current at the time of the French Revolution was not a revival of Roman architecture but a restoration of good architecture. Differences in architectural form revealed different circumstances and varying understandings of the nature of architecture, not different, equally valid architectures. An aberrant architecture such as that of the Gothic of the medieval cathedrals was interesting only insofar as it represented a departure from a fixed, metahistorical norm of good architecture that the ancients had "discovered" in nature and that the Renaissance had rediscovered.

This secure world was undermined by philosophes and others even during its most optimistic period. As the eighteenth century drew to a close, architects changed their intentions. Rather than seeing their art as being primarily that of constructing well-proportioned structures fitting social purposes, architecture became, in the words of the architect and theorist Étienne-Louis Boullée (1728–1799), "the art of presenting images through the disposition of bodies." Boullée's images were extracted from the then-current conventional architecture, but soon a more rigorous challenge would disconnect even that link with nature. Giambattista Vico (1668–1744) argued that history was the product of men. Johann Gottfried von Herder (1744–1803) asserted that society was in constant evolution. Jean-Jacques Rousseau (1712–1778) pronounced society to be an artifice created by social convention, and the revolutionary acts of Georges-Jacques Danton (1759–1794) proved that this was indeed the case.

The guillotine severed the old architectural connection between "nature" and "society." The old architecture was no longer useful. Buildings now had to refer to society not through the intermediary of their common origin in nature but directly. They had to serve society's needs, and they had to embody in their form the technical, political, economic, and material means that society used to satisfy those needs. As E. H. Gombrich has made clear, in the aftermath of the formulations of Georg Wilhelm Friedrich Hegel (1770–1831) the character of a society or a nation was embodied in the style of its cultural products—its art, religion, constitution, morality, law, customs, science, and technology—and this style had to be the authentic manifestation of the spirit of the age. The problem now became, and remains, how is the architect to "produce an architecture characteristic of the age"?

Because societies evolve and because architecture reflects society, changes in architecture reveal not a movement away from or toward a standard of good architecture but merely changes in society. Architectural knowledge no longer turned on aesthetic judgments linked with general, enduring metahistorical questions about the good and the beautiful but instead consisted in knowing how architecture serves the ends sought by societies. Any judgments that were made were based on assessments about how symptomatic buildings are of the society that produced them. Collecting the information required to formulate these assessments became the program of the modern discipline of architectural history.

This historicist program was, and remains, not only relativistic but also moralistic. One of its early preoccupations was the extent of freedom available to the architect and workman within the newly emerged technologically based means of production. Augustus Welby Northmore Pugin (1812–1852) agreed that just as the history of society was the product of the human spirit, so too the history of architecture embodied certain spiritual qualities. His polemic *Contrasts, or, A Parallel between the Noble Edifices of the Middle Ages and Corresponding Buildings of the Present Day; Shewing the Present Decay of Taste* (1836) argued that the Renaissance revival of classical architecture had restored "corrupt ideas of pagan sensuality" and had reinstated a fundamentally pagan glorification of the individual. He believed that a restored Gothic architecture would embody the faith of Christianity and promote its practices. John Ruskin (1819–1900), who abhorred the increasing alienation of the worker in the expanding system of mass production based on the division of labor, also sought a reformation in architecture and building. The Renaissance idea that the architect should and could determine in advance the architectural design in all its details had reinstated a slavery much worse than that of ancient Greece, Egypt, and Assyria. Yet, "it is, perhaps, the principal admirableness of the Gothic schools of [Christian] architecture, that they

. . . receive the results of the labor of inferior minds; and out of fragments full of imperfection, and betraying that imperfection in every touch, indulgently raise up a stately and unaccusable whole" (Ruskin, 1981).

French historicists saw history as epic, an epic clearly embodied in the powerful metaphors of the Gothic cathedrals—Laon, Paris, Bourges, Chartres, Reims, Amiens, Beauvais, and the more than a hundred less-known ones: Sens, Soissons, Troyes, Le Mans, and so on. *Notre-Dame de Paris* (1830) of Victor Hugo (1802–1885) presented the cathedral, which was built over many centuries, as a complex and hybrid aesthetic composite. "The greatest products of architecture are not so much individual as they are social works. . . . The man, the artist and the individual are obliterated in these huge anonymous piles; they represent the sum total of human intelligence. Time is the architect, the nation is the mason." But in his time architecture was

irrevocably dead; killed by the printed book, killed because it was less enduring, killed because it was more costly. . . . [A]rchitecture will never again be the social art, the collective art, the dominant art. The great poem, the great edifice, the great work of humanity, will no longer be built; it will be printed. (1978)

Viollet-le-Duc presented his reading of history in *History of a Town Hall and a Cathedral* (1878). Set in the imaginary French town of Clusy, architectural history served as a metaphor of social-political history by tracing and illustrating the architectural development of the town hall and the cathedral from Gallo-Roman times to the French Revolution. Not surprisingly, as the town experienced its greatest unity, wealth, and glory around 1200, its architectural ambitions and achievements reached their zenith.

In his *Discourses on Architecture,* Viollet-le-Duc explained that architectural form represented historical development. Roman construction had been based on slave labor and could therefore afford to use inefficient, inert masses of masonry to support vaults, but structurally efficient Gothic architecture, having been built by free men, was a monument to a republican spirit and expressed "enfranchised labor" and "the triumph of an intellect which feels its power, which acts, which is asserting its independence, while ironically concealing its secrets from blind and indifferent masters, and which is conscious that it will one day become in its turn the ruling power" (1959). Gothic structural systems embodied, to put it differently, the secular program of the nineteenth-century bourgeoisie.

Like Ruskin, Viollet-le-Duc found that the way buildings were put together had profound social consequences. Both believed that the principles embedded in medieval architecture were universal, as valid in the nineteenth century as in the thirteenth. In his program for architectural education, Viollet-le-Duc argued that the historian should investigate a building as an anatomist does when dissecting bodies and

thereby discover the "hidden methods and reasons that have determined its form" (ibid.). The architect learns not through inquiry into nature but only in the study of precedent based on a knowledge of architectural history. This study will provide the knowledge about the union of technical means and functional ends that is the prerequisite to enhancing the professional position of the architect, salvaging architecture from the scrap heap of history, and investing architects with a professional expertise not shared by the "amateurs, academicians, professors of art who profess nothing" and others mentioned at the beginning of this essay. A knowledgeable, technically sound, socially useful architecture could restore moral content to a free society.

But the movement of the spirit made the world, and architecture, more technical, not more moral. In the revolutionary fervor attending the search for order after World War I, architects exorcised the study of history because it sanctioned eclectic historicism. The architect now sought the knowledge of the engineer who could build railroads, roads, sewers, waterworks, automobiles, and airplanes, of the public health official who guarded against disease and epidemics, and of others harnessing the science of the new age to serve social ends directly. But the architect was no mere technician. He was also the guardian of the aesthetic spirit of art and the vessel through which the world-historical spirit spoke in aesthetic, that is, spiritual, terms. In addition, he was charged with responsibility for reaching social ends and with maintaining sincerity when he exercised his art. By the 1920s, eclectic historicism had been defeated and replaced by a new architecture void of any visible, eclectic, historicist link with the past. As Henry-Russell Hitchcock and Philip Johnson put it in concluding *The International Style* (1932), the book that authenticated the new European modernism for use in the United States:

> We have, as the Egyptians had or the Chinese, as the Greeks and our own ancestors in the Middle Ages before us, a style which orders the visible manifestation of a certain close relationship between structure and function. . . . Those who have buried architecture, whether from a thwarted desire to continue the past or from an over-anxiety to modify and hurry on the future, have been premature: We have an architecture still. (1932)

By the middle of the twentieth century, this new architecture, embedded in the historical moment of the modern world, had achieved hegemony in all official, educational, institutional, corporate, and aesthetic circles. But when, early in the 1970s, it became obvious that the new architecture had failed to deliver on its promises, two important consequences followed.

One was the recognition, as clearly explained in Alan Calquhoun's essay from 1983, that historicism was the basis of modernism and that modernism was simply another historicist style. This recognition led a few individuals such as Allan Greenberg, Demetri Porphyrios, and Leon Krier to seek to reconstruct architecture outside historicism; for them this would occur within classicism, which Porphyrios (1982) explained is not a style. The second, and dominant and pervasive, consequence was to bring architects to a simple failure of nerve. Eschewing their earlier commitment to improving society or changing human nature, they isolated themselves within the avant-garde, where they contended among one another for fame and linked architecture to a succession of other, similarly esoteric artistic endeavors.

To signal the newness of the era they were entering, they declared that modernism had given way to postmodernism. In the subsequent activity, many old modernist themes persisted (a dominance of technology or of architecture's alliance with art and aesthetics) and new ones emerged (a rationalism pushed to absurd extremes and a formalism that used architectural forms to illustrate literary theory). Now void of millennialism, eclectic historicism stepped in to supply formal devices that architects distorted or parodied and hung on modernism's reductionist frameworks. Because these forms now bore no content external to architecture, they could be selected purely on the basis of taste, caprice, the internal logic of the particular building, or the ideological position of the architect. Forms from the heroic modernists of earlier in the century might signal a commitment to newness in art; some sort of historical argument could justify using some style of classicism; and "contextualism," or basing a design on the existing stylistic character of a new building's setting, would declare one's commitment to rejecting a vapid modernism and honoring traditional stylistic and urban values. But, as Robert A. M. Stern makes clear, these were choices justified merely by pluralism and ungrounded in any principled connection between buildings and larger metaphysical, civil, or moral positions.

Architectural theory now flourished but avoided using the term *historicism*. Instead, critics used the term as a synonym for *contextualism* or, along with historians, interchanged it with *eclecticism* or *revivalism*. Thus, a historicist architecture hatched in the spirit of revolution and seen as the means of righting the wrongs of the past by calling principles from the past for use in the present now serves an avant-garde of nihilism and narcissism or the self-indulgence of caprice and irony.

[*See also* Architecture.]

BIBLIOGRAPHY

Boullée, Étienne-Louis. "Considérations sur l'importance et l'utilité de l'architecture." In *L'architecture visionnaire et néoclassique*, edited by Jean-Marie Pérouse de Montclos, pp. 29–40. Paris, 1993.

Calquhoun, Alan. "Three Kinds of Historicism" (1983). In *Theorizing a New Agenda for Architecture: An Anthology of Architectural Theory, 1965–1995*, edited by Kate Nesbitt, pp. 202–209. New York, 1996.

Gombrich, E. H. *In Search of Cultural History*. Oxford, 1969.

Hitchcock, Henry-Russell, and Philip Johnson. *The International Style: Architecture since 1922*. New York, 1932.

Hugo, Victor. *Notre-Dame de Paris* (1830). New York, 1978.

Porphyrios, Demetri. "Classicism Is Not a Style." *Architectural Design* 52.5–6 (1982): 50–57.

Pugin, Augustus Welby Northmore. *Contrasts, Or, A Parallel between the Noble Edifices of the Middle Ages and Corresponding Buildings of the Present Day; Shewing the Present Decay of Taste.* n.p., 1836.

Ruskin, John. "The Nature of Gothic." In *Stones of Venice* (1851–1853), edited by Jan Morris, chap. 7, pp. 118–139. Boston, 1981.

Stern, Robert A. M. "New Directions in Modern American Architecture: Postscript at the Edge of the Millennium" (1977). In *Theorizing a New Agenda for Architecture: An Anthology of Architectural Theory, 1965–1995,* edited by Kate Nesbitt, pp. 100–108. New York, 1996.

van Pelt, Robert Jan, and Carroll William Westfall. *Architectural Principles in the Age of Historicism.* New Haven, 1991.

Viollet-le-Duc, Eugène-Emmanuel. *Discourses on Architecture* (1863–1872). 2 vols. Translated by Benjamin Bucknall. New York, 1959.

Viollet-le-Duc, Eugène-Emmanuel. *Dictionnaire raisonné de l'architecture française du XI^e au XVI^e siècle.* 10 vols. Paris, 1858–1875.

Viollet-le-Duc, Eugène-Emmanuel. *Histoire d'un hôtel de ville et d'une cathédrale.* Paris, 1878.

ROBERT JAN VAN PELT and
CARROLL WILLIAM WESTFALL

HOFFMANN, ERNST THEODOR WILHELM.

See Romanticism, *article on* Philosophy and Literature.

HÖLDERLIN, JOHANN CHRISTIAN FRIEDRICH

(1770–1843), German lyric poet who modeled his poetry on classical Greek verse. Little known in the nineteenth century, Hölderlin is generally considered an important poet of modernity. His importance for the development of German Idealism has also become increasingly clear. Like G. W. F. Hegel and Friedrich Wilhelm von Schelling, his fellow students at the Tübingen theological academy, Hölderlin welcomed the French Revolution, studied the new transcendental philosophies of Immanuel Kant and Johann Gottlieb Fichte in the context of a renewed controversy over the interpretation of Baruch Spinoza, and lamented the loss of Greek ideals of beauty as they were conceived by Johann Joachim Winckelmann and Friedrich von Schiller. Unlike Hegel and Schelling, however, Hölderlin decided that poetry alone could properly present, and thus help to resolve, the conflicts and antagonisms to which each of these phenomena gave witness. Rarely, if ever, has a poet of Hölderlin's stature devised philosophical and "poetological" projects that are as intricately constructed and systematically developed as his theory of the alteration of tones, his reflections on tragedy, and his remarks on the relation between the ancients and the moderns.

Hölderlin's Thought and Poetry. Many of Hölderlin's early remarks on poetic technique are concerned with the latent conflict within the apparent naïveté of the Homeric epics. Although the *Iliad* appears "natural" and thus gives rise to the Winckelmannian notion of "noble simplicity," the motivating impulse of the poem is anything but the simplicity of nature. Underlying the naive mode of presentation is a heroic mood—Achilles' striving to separate himself from all others. The *Iliad* had to appear simple and moderate in order for this heroic distinctiveness to be recognized in the course of the poem: "so as to maintain the individuality of the presented character, which is his principal occupation, the epic poet is so thoroughly moderate" ("On the Different Ways to Write Poetry"). By showing that Homer's poem only appears to be natural, Hölderlin can overcome Schiller's distinction between "naive and sentimental poetry" and assert that naturalness in poetry is always a matter of art. In his first mature poetological remarks, Hölderlin thus disavows the basic principle of classicism in favor of a dialectical exposition of a poem's internal dynamics. In order for the "basic mood" of a poem to be recognized or "felt," it must first present itself in the guise of an opposing tendency.

Hölderlin's analyses of Homer are not so much concerned with correcting a mistaken view of the poet as working himself out of a technical impasse that he encountered when writing poetry in an exalted manner. His most ambitious early poems bear titles such as "Hymn to the Goddess of Harmony" and "Hymn to Freedom" (1792–1793), and for all the pathos of their revolutionary ideals, they are marked by a twin failure: the mood is too subjective, and the objects to which they are dedicated (harmony, freedom, humanity, friendship) are too abstract. As Hölderlin's early poetological reflections make clear, the two failures are closely connected: the subjectivity of the hymnic voice generates the abstractness of the ideals it seeks to praise. Hölderlin indicates that Homer could serve as a model when he names him "the poet of all poets," but this phrase must be understood in a precise manner: Hölderlin has no intention of imitating the Homeric *poems;* instead, he recognizes Homer's poetological *procedures* as a valid solution to his own impasse. If a poet wishes to praise an ideal, he cannot do so in an "ideal" manner; on the contrary, he must choose "real" objects or common situations from which the poet would then take leave. Only in this way can the sublimity of the ideal be felt. Even before Schiller and Johann Wolfgang von Goethe admonished him for shying away from everyday objects and losing himself in ecstatic reveries, Hölderlin had overcome this tendency and had embarked, in turn, on a concentrated investigation into the philosophical foundation for a poetic practice that would not be caught in the unproductive alternative between ecstatic visions of the whole or sober descriptions of particular objects, events, and situations.

Hölderlin's critique of Johann Gottlieb Fichte and his revision of Spinoza's notion of *natura naturans* are motivated by the same concerns that led him to reject Weimar classicism. In a brief text published under the title "Judgment

and Being" (c.1795), Hölderlin shows the inadequacy of Fichtean idealism:

> When I say "I am I," the subject ("I") and the object ("I") are not so unified that absolutely no separation can be made without injuring the being *[Wesen]* of that which is supposed to be separated; on the contrary, the "I" is only possible through this separation of the "I" from itself.

The unification prior to self-consciousness cannot have the character of identity. Because, according to Hölderlin's specious but trenchant etymology, judgment *(Urtheil)* consists in primordial separation *(Ur-theilung),* this unification cannot be named in accordance with the paradigm of the subject-predicate sentence. Instead of using Spinoza's term "substance," Hölderlin speaks of "being" *(Seyn)* or "intellectual intuition" (the absolute union of subject and object). In both his poems and his poetological reflections, he then poses this question: how is it possible for "being" or "intellectual intuition" to be felt? Only parts separated from "being" can feel and be felt, and what these parts feel is precisely their separation from the prior unity. "Being" can thus be felt only if it breaks apart in the form of separate—and suffering—self-conscious beings.

This conception of the relation between "being" and self-consciousness gives shape to all of Hölderlin's mature work. His one novel, *Hyperion* (1796–1798), is an epistolary account of the dissonances to which an "elegiac character" is exposed when he no longer enjoys childlike oneness with nature: "To be one with all—that is the life divine. . . . But an instant of reflection hurls me down." Hyperion is the one who goes beyond or transcends *(hyper-ion)* his experience of separation. Each station on his "eccentric path"—naive nature worship, heroic struggle for political freedom, undivided love—ends in failure; but the course of his failure and the self-reflective account of his suffering allow him to become aware of himself as a poet who could present (and not simply live) the "life divine." The novel concludes in mid-course with Hyperion quoting himself. By separating itself from the life of the speaker, his voice is able to formulate the law of separation: "all things that are separated find one another again."

During the years in which Hölderlin completed the final version of *Hyperion* and thus achieved a certain self-clarification concerning philosophical principles, he trained himself in the constraints of poetic composition. Following the letter, if not the spirit, of Schiller's "clever advice" to concentrate on less exalted topics, Hölderlin eschewed the expansive horizons of his earlier hymns and dedicated himself to the task of writing short odes according to rigid metrical rules. Hölderlin adopted Friedrich Gottlieb Klopstock's manner of integrating classical meter into German prosody, but he was much more circumspect in his selection of appropriate metrical patterns. Almost without exception, his odes conform either to alcaic or (the third) asclepiadic meter.

Some of these odes consist of only a single strophe, and many others are composed of two antithetical ones. This extensive exercise in obeying a narrow set of metrical rules within a limited space is responsible in part for certain characteristic features of Hölderlin's mature poetry, including, for instance, his unusual employment of conjunctions, his ability to control complicated inversions, and the facility with which he devises gnomic-like utterances. But these odes are more than merely formal exercises; many of them explore in a condensed form the same conflicts that *Hyperion* seeks to articulate and resolve: the opposition between the unity of nature and the freedom of human subjectivity, the antithetical relation between a poetic vocation and political action, and the paradoxical dynamics of love. Hölderlin often returned to these formal and thematic experiments, rewriting and expanding them as he gained greater and greater awareness of the tasks poets must solve if they are not only to gain insight into the law of separation but also to give it a "secure" linguistic expression.

What happens "when the poet is in control of spirit" is then the subject of Hölderlin's most systematic poetological investigation, an incomplete treatise to which his editors have given the title "On the Procedures of Poetic Spirit" (c.1800). The opening sentence of this labyrinthine text is almost two pages long and describes in bewildering detail the conflicting demands that a poet must already have recognized and fulfilled before he comes to choose a subject matter ("material") for his poem, and the next sentence proceeds to describe in an equally complex fashion the structure of the "material" that allows for the presentation of the "life divine." Hölderlin's reflections thus achieve a remarkable density: each phrase is balanced by opposing as well as complementary ones, and each sentence condenses the results of the previous one as it simultaneously transforms the significance of the terms through which the argument is structured. Near the end, Hölderlin speaks of a "poetic prose," and in a text as relentlessly self-reflective as this one, it is impossible not to apply this phrase to the treatise itself.

The extraordinary complexity of Hölderlin's exposition of the procedures whereby a poet not only gains insight into the "life divine" but also grants this insight a secure linguistic structure arises from an implicit injunction: the genuine poet can never lay claim to immediacy. The "divine moment" in which "the infinite is made present" can be given only to a "poetic individuality," but the latter "cannot appear, or only in the character of a positive nothingness, an infinite standstill." The "poetic I" is unable to recognize itself by itself and must therefore choose an "external object" to which it stands in "harmonious opposition." Yet, the choice of an objective sphere allows this "I" to attain only mediated *knowledge* of itself. Another stage of reflection is necessary for the poet to create a *language* in which the "transcendental sensation" of an all-embracing "unity in

harmonious opposition"—a sensation to which this knowledge alone grants access—can find expression. Hölderlin formulates the relation between philosophical insight and poetic language in the following terms: "Knowledge intimates the language, whereas the language remembers the knowledge." The poet cannot proceed without philosophical insight; but the latter is only a propaedeutic for poetic language, which, by retaining and completing the merely "ideal" result of "mental" reflection, calls the mind back to "life." Only a second, "creative" reflection whose product is the language of poetry can at once secure and enliven the "transcendental sensation" for which philosophical reflection strives.

As a "specific act of creative reflection," poetic language enlivens purely philosophical insight. But the "lively sense" of such language can be recognized only if the poem follows a "calculable law" ("Remarks on 'Oedipus'"). The law of each poem determines the precise sequence in which its formal features and subject matter modulate. Although Hölderlin may have drawn certain aspects of his theory of the alteration of tones from classical notions of poetic *metabolē* and mystical versions of the Pythagorean "harmony of the spheres," he never makes any tone into a rigid category. In accordance with the etymology of the word, he treats each tone as a tension and a tendency: the "naive" or "natural" tone is characterized by the simplicity of sensation, the "heroic" or "energetic" by unfettered passion, and the "ideal" by the unifying power of "intellectual intuition" or "fantasy." The naive tone is restful, the heroic tone active, and the ideal tone actively contemplative. Poetic language has a dynamic character because every well-articulated work, such as the *Iliad,* is rooted in the opposition between its underlying impulse and its overall appearance. The progress of the poem consists in the "balancing out" of this opposition.

The three modes of poetic composition that Hölderlin sets out to analyze—lyric, epic, and tragic—correspond to the three basic tones: naive, heroic, and ideal, respectively. Every genuine poem is a "metaphor" or translation of an otherwise inarticulate "meaning" or "basic mood" into an objective sphere, which, as the "appearance" or "art-character" of the poem, expresses an opposing mood. The naive is opposed to the ideal because of its sensuousness; the heroic to the naive because of its "aorgic" or unbounded character; the ideal to the heroic because it recognizes the fundamental unity of "everything living." Hölderlin's most extensive discussion of the rules for translating an underlying meaning into an articulate poem, "On the Difference of Poetic Modes" (c.1800), thus begins with these categorical declarations:

> The lyric poem—which appears to be idealist—is naive in its meaning; it is a continuous metaphor of a feeling. The epic poem—which appears to be naive—is heroic in its meaning; it is the metaphor of a great striving. The tragic poem—which appears to be heroic—is idealist in its meaning; it is the metaphor of an intellectual intuition.

Hölderlin's conception of the alteration of tones is less concerned with the nature of different genres than with the inner dynamics of individual poems. The complexity and systematicity of this conception give Hölderlin the chance to analyze specific kinds of poetic tasks and develop guidelines for the completion and revision of his own work. Far from asserting the purity of any particular genre, Hölderlin maintains that each of the three ways to write poetry must include the basic features of the other two. Indeed, the initial conflict between the basic tone and its opposing art-character can be resolved only if another tone comes to "mediate and unite" them. He calls this tone "the spirit of the poem" and locates it in the tone excluded from the poem's initial definition: for the lyric, it lies in the heroic; for the epic, in the ideal; for the tragic, in the naive. He thus recommends that the lyric poet study epics (naive tone), the epic poet tragedies (ideal tone), and the tragedian lyric poetry (naive tone). Hölderlin's poetological system is even more complex and capable of even greater precision because the defining feature of a poem, its initial "art-character," can be further modulated by a subordinate tone. The lyric poem thus breaks down into the lyric-naive (lyric proper), the lyric-heroic (which characterizes some of Hölderlin's finest odes), and the lyric-ideal.

The lawful progress of a poem consists in explicating the implicit tone: the basic tone of a poem, which at first recedes in favor of its opposing art-character, becomes the art-character of the following section and is accordingly opposed by a new basic tone; the latter then becomes the next art-character of the poem, which in turn conceals another basic tone. This sequence does not proceed indefinitely, however. The conflicts that give rise to a poem remain unresolvable unless its progress is interrupted. One of the most important tasks of Hölderlin's conception of tonal modulation is to locate the moment of interruption. After the "catastrophe" occurs, the course of the poem is reversed: the explicit tones turn into implicit ones in a regular fashion; the basic tone of the poem can be felt in a mediated fashion as its "effect," and the poem returns to its initial art-character. By securing a "lawful" succession of its interacting tones, Hölderlin shows how a poem can and does interrupt itself. Without the "positive nothingness" of this "catastrophic" moment, the "lively sense" of a poem could not make itself known.

The tragic character of this "catastrophe" does not escape Hölderlin's notice. The very definition of the tragic poem within the system of tonal modulations indicates its importance not only for his poetological reflections but also for his poetic practice. As the "metaphor" of intellectual intuition, tragedy has the paradoxical task of making "original matter," as the unification of "being" prior to any separation into self-conscious beings, appear to consciousness ("The Significance of Tragedies"). The tragic poem assumes a heroic appearance because this unity presents itself at first as a boundless striving for separation, the unavoidable and unconscious event for which Hölderlin invented a memorable

phrase: "the necessary arbitrariness of Zeus." But speaking of a "necessary arbitrariness" in the context of a theoretical program is considerably easier than presenting the relation between necessity and arbitrariness in a dramatic form. Hölderlin had no interest in following Gotthold Ephraim Lessing and writing a tragedy of middle-class conflicts in which the quotidian character of the setting guarantees the particularity and thus the arbitrariness of its sub-ject matter. After momentarily considering the death of Socrates as a suitable subject matter for a tragedy, Hölderlin decided to devote his energy toward someone who is more active in his self-destruction: Empedocles. Hölderlin's commitment to "higher tragedy" as the "most rigorous of all poetic forms" then gave rise to an impasse not unlike the one that marked his early efforts at a sublime hymnic style: in order to show the divine moment of "intellectual intuition," he had to re-move the events from the accidents of its locality, purify the situation of its particularities, and deny the peculiarities of the tragic character through whom the divine moment works. Hölderlin thus runs into an aporia of tragedy, if not a tragic aporia: in order to show the rigorous necessity of Empedocles' suicide, he had to purge the events of their par-ticularity; but by so doing, the motivating ground for the ac-tion becomes increasingly less comprehensible and, as a re-sult, appears increasingly more arbitrary.

In response to this aporia Hölderlin interrupted his work on the tragedy and renewed his investigation into the "Ground of Empedocles," which means, at the very least, the motivating basis of the hero's self-immolation. These notes for the revision of *Empedocles* also constitute one of Hölder-lin's most powerful diagnoses of the historico-philosophical condition of his age. Empedocles is a suitable subject matter for tragedy, it seems, because certain characteristics of the modern era can be correlated with those of his time; more specifically, the conflict between boundless nature (the "aor-gic") and historically instituted art forms (the "organic") has once again become so extreme that it can be resolved only in a tragic manner. When human beings become too au-tonomous in their self-reflective activities, nature "tran-scends into . . . the incomprehensible, the nonsensuous." This conflict can be resolved only when humanity becomes more aorgic and nature more organic: "In the middle there lies the struggle and death of the individual." At this moment of individuated mediation, the conflict assumes its most ex-treme form and thus appears resolved—but only appears so in the organic-aorgic intensity *(Innigkeit)* of the hero, who, by proclaiming himself divine, commits an irreparable crime. As long as the hero remains apart (that is, alive), no resolu-tion can take place, and so he falls victim to the very unifica-tion he inwardly accomplishes.

After much philological labor, the editors of Hölderlin's writings have discerned three drafts of the Empedocles tragedy, none of which brings the tragic event to its conclu-sion. The final draft breaks off just after Empedocles seems

to break down in response to the question of an enig-matic foreigner, "Who are you and who am I?" But, in the course of composing the final draft, Hölderlin had arrived at another solution to the aporia he encountered in translat-ing his conception of a tragic poem, as the "metaphor of in-tellectual intuition," into a tragic drama, and one of the keys to this solution was his decision to translate Pindar's odes into German. Instead of celebrating a divine moment by means of "foreign, analogical material" (the tragic hero), Hölderlin could learn from Pindar and take his own poetic predicament as the "material" for a holy celebration. Be-cause this predicament was in every way bound up with a specific locality—Hesperia, Germany, or "the fatherland"—Hölderlin could escape from the dilemma into which his work on *Empedocles* issued. The purification of the event goes hand in hand with the determination of its particular-ity, for the event in this case is the very composition of the poem. Hölderlin could thus strive to be "the poet of poets" in an entirely new sense: a poet who never expresses his own subjectivity but, instead, allows the objective condi-tions and necessary tasks of poetry to come to light. Of par-ticular importance for the execution of this new program is Hölderlin's first mature attempt to write in a Pindaric man-ner, "As on a Holiday . . ." (c.1800), and near the end of this incomplete poem, he formulates in a striking manner this new conception of the poetic vocation: "Yet it is up to us, under the thunderstorms of God, / Fellow poets, to stand bareheaded, / to grasp the Father's ray, this itself, with our own hands, / and to offer unto the people / the heavenly gift, wrapped in song."

The lengthy and wide-ranging poems that Hölderlin wrote between 1800 and 1806 are the ones through which he achieved renown in the twentieth century. Among them are the elegy "Bread and Wine," the river poems "The Rhine" and "The Ister," and works such as "Celebration of Peace," "Patmos," and "Remembrance," which defy generic de-scription. The extent to which these poems conform to Hölderlin's conception of tonal modulation remains a matter of dispute among scholars, but at least this much is clear: having completed his novel and having abandoned his ef-forts at writing a tragedy, Hölderlin dedicated himself en-tirely to lyric poetry; but the underlying intention of this po-etry is no longer correlated with "basic mood" to which he had previously assigned lyric poetry, and none of his great lyric poems should therefore be understood as a "continu-ous metaphor of a feeling." The central stanza of "The Rhine," for example, responds to an implicit question why there need be feeling (suffering) at all by presenting a daring version of intellectual intuition: "Because, therefore, / the blessed ones feel nothing of themselves / another [a mortal] must indeed, if to say such things / is allowed, in the name of the gods / taking part, feel; this other / they need." Yet, "The Rhine" is not a tragedy, nor even a tragic poem. Only in the poem itself can one find an adequate word to describe it.

Hölderlin generally preferred the term *song* when speaking of these poems, but this does not imply that he conceived of his poems as "heavenly gifts" that he "wrapped" for "the people." Indeed, each of his late poems (c.1802–1806) is a critical interrogation of this very conception of the poetic vocation. Thus, the celebration of the "bridal feast" in "The Rhine" as the day in which mortals and immortals momentarily come to a reconciliation is related in only a tenuous way with an act of poetic composition, and the linguistic events to which the day of celebration is most closely bound in the course of the poem—Rousseau's silence and Socrates' discourse—cannot be unambiguously understood as poetic acts. Like the generic description of the poem, the poetic vocation to which the poet binds himself can be discerned only in the experience of reading it—and perhaps not even there.

The centrality of intellectual intuition to "The Rhine" indicates that, even if he declined to write a tragedy, Hölderlin had not ceased to reflect on the implications of tragic poetry for his own poetic practice, and he pursued his study of Greek tragedy into his last lucid days. Indeed, the only major poetological texts that Hölderlin himself published were the remarks he appended to his translations of Sophocles' *Oedipus Tyrannos* and *Antigone* (1804); and here, too, the "catastrophic" moment is decisive. Hölderlin calls it the "caesura, the pure word, the counterrhythmic interruption." When the caesura takes place (figured in the Sophoclean plays by Teiresias's speeches), "no longer the alteration of representation but representation itself appears," and man is henceforth torn out of "his life-sphere . . . and [sent] into the eccentric sphere of the dead." From this moment onward—and this moment determines the time of tragedy as one "in the middle"—there is no turning back: "man forgets himself and the god and turns around like a traitor, of course in a holy manner." Tragedy thus enacts the rigorous separation of the human sphere from the divine one; but it does so for the sake of a certain continuity: "so that the course of the world not show any rupture and the memory of the heavenly ones not expire." Mortal and immortal beings turn away from one another, and this "divine infidelity" *(göttliche Untreue)* allows for no reconciliation. But for this very reason it can be remembered or, as Hölderlin also writes, "maintained." The task of tragedy turns out to be the opposite of Platonic anamnesis: not to recall the things themselves but to maintain in memory the loss of all relation to a divine sphere.

Because the relation between the human and divine spheres determines the course of history for Hölderlin, tragedy assumes an eminently historical function. In his "Remarks on 'Antigone,'" he explicitly poses the philosophical-historical question to which his concept of "divine infidelity" and much of his late poetry refer: "how it goes from the Greek to the Hesperian." Hölderlin's discovery of a dynamic relation between basic mood and art-character in the *Iliad* grants him access to this movement. Because the basic

mood of the Greeks is "holy pathos," their appearance (thanks to Homer) is sober precision. "With us it is the reverse," Hölderlin writes in a famous letter to Böhlendorff: Hesperian (Western) poets must learn how to "use freely" their native "clarity of presentation" and thus avoid indulging in the opposing tendency to which they are otherwise inclined, namely, enthusiastic revelry (characteristic of Klopstock, much of Schiller, and almost all of Hölderlin's early poems). The free use of native clarity demands that Hesperian poets must precisely locate and localize their poems: instead of subordinating their poetry to the "ideal" tone of unifying "fantasy," they should freely coordinate moments in an "epic" manner (as Böhlendorff does, according to Hölderlin). By disclosing a historical dimension to the dynamics of basic mood and art-character, Hölderlin radicalizes his theory of tonal alteration and thereby does away with its founding telos: the creation of a *whole* poem the progress of which is interrupted in the middle. To invoke one of Hölderlin's late remarks, the course of his poetological reflections has shown him "the apriority of the individual over the whole."

The short poems that Hölderlin wrote in the last years of his life (c.1825–1843) confine themselves to a strict set of thematic and rhythmic formulas. The poems and fragments he wrote during the years in which he struggled with mental breakdown (c.1802–1806), by contrast, coordinate the three tonal moments so freely that they gain almost total independence from one another: images, narratives, and gnomic utterances do not so much follow each other in "The Only One," the last versions of "Patmos," and "Mnemosyne" as happen one upon the other. Although—or precisely because—these works do not place the question of "the fatherland" at their center, their form can perhaps best be understood in terms of the "patriotic reversal" *(vaterländische Umkehr)* that Hölderlin describes in his "Remarks on 'Antigone.'" Whereas "divine infidelity" leaves no room for the presentation of the whole except through a commemoration of its absence, "patriotic reversal" does not even hold out this possibility. Such a reversal takes place in *Antigone* in the form of a rebellion, according to Hölderlin, and in other, unspecified forms it gives shape to the Hesperian poetry that *Antigone* helps inaugurate. Hölderlin contrasts "patriotic reversal" with "total reversal," which is "forbidden to man as a cognitive being." Because the former is almost total—"the entire form of things changes"—it is even more damaging to the project of presenting the infinite unification of "being" than total reversal, for this unification can no longer be presented in the guise of its exact opposite: "such a form is relevant precisely because the infinite, like the spirit of states and of the world, cannot be grasped other than from an awkward *(linkischen)* perspective." Hölderlin's last poetological thoughts before his long decades of almost complete silence, like many of his poetic fragments of the same period, are incomparably dense

meditations on separations that are no longer of a heroic cast and are no longer based on an opposing ideal tone in which a unifying whole is envisaged. "Infidelity" and "reversal" take precedence over "separation" and "unification" as terms through which Hölderlin presents the course of poetry, history, and thought.

Reception of Hölderlin's Work. Under the influence of a poorly edited version of Hölderlin's writings, Nietzsche planned to write a drama about Empedocles. His youthful attachment to Hölderlin was so unusual, however, that it prompted Richard and Cosima Wagner to express concern for his sanity. During the nineteenth century, Hölderlin's work was rarely read, but his "madness" did attract some attention. Contrasting his work with that of Jean Paul, Wilhelm Windelband and Friedrich Vischer attributed this madness to a lack of humor. Wilhelm Dilthey made an effort to rescue him from narrow diagnoses of his "case" and included a study of his life and work in *Experience and Poetry* (1905). But it was not until Stefan George and his "circle" took up his cause and presented him as a poet whose fate testified to the purity of his poetic mission and therefore to his ability to assume the position of spiritual leadership that he was accorded serious attention. Of undeniable importance for the appreciation and reconstruction of his poetic practice was the work of Norbert von Hellingrath, a young member of the "George circle." With Hellingrath's discussion of the Pindar translations and his edition of the poetry from 1800 to 1806, Hölderlin's poetry began to occupy an exalted position in the literary history of modernity. Along with George, Rainer Maria Rilke and Georg Trakl soon came under his spell. In a critical essay whose density rivals that of Hölderlin's own poetological writings, Walter Benjamin attempts to wrench the poet loose not only from his admirers among the George circle but also from all psychologically inclined interpreters. Disregarding the poet's life and instead concentrating on what he calls "the poetized" *(das Gedichtete)*, Benjamin's essay of 1915 shows how Hölderlin radically changed his conception of a true poetic vocation from the defiant heroism of "The Poet's Courage" to the sober passivity of "Timidity."

In the first few years after the end of World War I, the scholarly edition begun by Hellingrath was completed, and Hölderlin's work became widely known. His reflections on the relation between Germany and Greece, on the one hand, and Christ and the classical gods, on the other, soon became the starting point for an enormous amount of interpretive activity. Because Hölderlin had been an enthusiastic supporter of the French Revolution, had turned the penultimate chapter of *Hyperion* into an unparalleled invective against the Germans, and had shown no interest in the German nationalist cause against Napoleon, he would seem to have been an unpromising candidate for induction into the National Socialist pantheon of heroes, but inducted he was: soldiers were given a "battle-ready" collection of

his poems in which "Death for the Fatherland"—an ode that Hölderlin wrote on the model of the "Marseillaise"—occupied a prominent position. On the one hundredth anniversary of his death in 1943, a large celebration was orchestrated in his honor, and the first volume of a newly edited complete works was published. After the war, Hölderlin's work became the focal point of intense and often highly technical philological debates. Beginning in the 1960s, scholarly discussion began to focus on his contacts with certain radical political activists of his day, including Isaak von Sinclair, a friend and benefactor to whom "The Rhine" is dedicated. Scholars and writers began to portray Hölderlin as a Jacobin who, unlike Hegel and Schelling, would sooner retreat into obscurity and silence than abandon his revolutionary goals. His writings were also translated into many languages, and he became better known outside of Germany. Certain features of contemporary European and American literature can be understood in part as responses to the challenge of his poetry.

Because of his close association with Hegel and Schelling, Hölderlin's work has solicited more interest among twentieth-century philosophers than would otherwise be expected of a poet. When Franz Rosenzweig published "The Oldest System-Program of German Idealism" (1917), he initiated a scholarly industry whose goal consists in determining which of the three friends contributed which sentences to the fragmentary document. Efforts to understand Hölderlin's place in the development of German Idealism have culminated in a massive project of philosophical-historical reconstruction initiated and directed by Dieter Henrich. On the basis of meticulous research, Henrich shows that Hölderlin pushed Schelling and Hegel to go beyond the limits of both Kantian critique and Fichtean "theory of science" toward a philosophy no longer founded solely on the principle of subjectivity. Henrich's massive *Ground in Consciousness* (1992) summarizes the results of his inquiries and proposes to develop certain Hölderlinian themes into a form of philosophical reflection that would no longer take it for granted that such reflection must issue into an absolute system.

The most imposing philosophical exposition of Hölderlin—and the one to which Henrich implicitly responds—is, however, that of Martin Heidegger. In a series of lectures and essays on Hölderlin's work, Heidegger not only sets out to distinguish it from the philosophical systems of Schelling and Hegel but also to present it as a singular and still uncomprehended occurrence in the history of the West. In the preface to his first lecture series on the poet, the date of which is itself significant (1934), Heidegger lays down the stakes of his inquiry:

One takes Hölderlin 'historically' and fails to perceive the one essential matter: that his work, which is still without a space and a time, has already overcome our historical fussing-about and

has founded the inception of another history, the history that begins with the struggle over the decision concerning the arrival or flight of God.

Because Hölderlin's work is the inception of "another history," it allows for the exposition of the "essence" *(Wesen)* of poetry. This "essence" is not a universal that particular poems all have in common. On the contrary, it is a way in which "Being"—as the open horizon in which things can manifest themselves—takes place, and the singular event of Being occurs, according to Heidegger, when the poet names the "holy." Holiness is not to be understood as an attribute of God or the gods (Heidegger, like Hölderlin, hesitates between singular and plural), but as the space of divine manifestation and withdrawal. By naming the holy, the poet grants the otherwise self-concealing openness of Being a certain constancy. In this way, Heidegger interprets the last line of "Remembrance": "what remains, however, poets found *(stiften)*." Hölderlin is exceptional, for Heidegger, because he alone names the holy.

Philosophers and literary scholars as well as poets have been drawn to, and repelled by, Heidegger's commentaries on Hölderlin. Beda Allemann's *Heidegger and Hölderlin* (1954) set the stage for much of the discussion of the relation between the two when he asserted that they engage in comparable "reversals": Hölderlin turns away from Greece and returns to "the fatherland" in his late poems, just as Heidegger turns away from the analytic of human existence undertaken in *Being and Time* toward a thinking of Being as such. After documenting the comparison and explicating the direction of their respective "turns," Allemann asks whether Heidegger, basing himself on a philologically unsupportable conjecture about the meaning of "spirit" *(Geist)* in a late fragment of "Bread and Wine," draws Hölderlin into the idealist project of demonstrating that spirit must return home. Allemann insists on keeping Hölderlin's "patriotic reversal" absolutely distinct from the Hegelian circle of "spiritual" homecoming. Subsequent analyses of the relation between Heidegger and Hölderlin tend to accentuate the differences between their trajectories and to emphasize the intractable difficulties to which Heidegger's philological procedures give rise. Heidegger's interpretation of "As on a Holiday . . .," in particular, has attracted a number of important critical engagements with his commentaries as a whole. With great philological acumen, Peter Szondi shows in "The Other Arrow" (1963) that the version of the hymn on which Heidegger bases his interpretation is faulty. The poem does not conclude with an invocation of the "bareheaded" poets receiving the divine ray; it proceeds to speak of "another arrow" striking the poem, an arrow that the poet himself, as a "false priest," casts. Drawing on the careful editorial work of Friedrich Beissner, Szondi presents the poem as a transitional one in which the poet learned to give up his elegiac mood, overcame his grief, and, as the final

line of the incomplete poem says, "sing the warning song to those able to learn."

At the opening of his essay "Heidegger's Exegesis of Hölderlin" (1955), Paul de Man catalogs the various ways in which Heidegger violates proper philological procedures; but he goes on to insist that philology is not in a position to decide on the significance of this exegesis. De Man then takes issue with Heidegger's interpretation of "As on a Holiday . . ." more directly. According to de Man, Heidegger turns to Hölderlin because his thought needs a "witness" to "Being"—someone, that is, who has an immediate relation to Being and who, unlike all philosophers in the Western tradition, no longer represents it in an occluded manner. The decisive proof that Heidegger assigns this role of witness to Hölderlin is found in his reading of certain dramatic lines in "As on a Holiday . . .": "But now day breaks! I waited and saw it come, / And what I say, the holy be my word." The holiness of the word allows for an unmediated relation to the dawning of the horizon in which things come to light—or, in other words, an immediate relation to Being. Heidegger, according to de Man, thus discovers the central "concern" of Hölderlin's poetry, but he pays dearly for his discovery: "Hölderlin says exactly the opposite of what Heidegger makes him say." As de Man notes, the word "be" *(sei)* in "the holy be my word" is in the subjunctive mood; it does not affirm that the word of the poet is holy but prays that it be so: "this hymn suggests a conception of the poetic as an essentially open and free act, a pure intention, a mediated and conscious prayer that achieves self-consciousness in its failure; in short, a conception diametrically opposed to Heidegger's." The poetic word cannot be the "holy word" it begs to be, because the very nature of the word as mediation undermines its holiness. De Man thus challenges Heidegger's contention that Hölderlin escapes the metaphysical tradition that finds its completion in Hegel, and at the end of the essay he calls for a reassessment of the relation between Hölderlin, as the poet of necessarily failed immediacy, and Hegel, as the philosopher of inconsolable mediation.

An implicit point of reference for de Man's critique is Maurice Blanchot's collection of essays, *The Work of Fire* (1949). Many of the essays in this collection bring out certain resonances and dissonances in the works of Hölderlin, Hegel, and Heidegger. "The 'Sacred' Speech of Hölderlin," in particular, does not so much take issue with Heidegger's commentary on "As on a Holiday . . ." as draw out its consequences to a point where the poetic word discloses its own impossibility. Reflecting on the same line as de Man— "the holy be my word"—Blanchot emphasizes the paradoxical character of this prayer for a word in and to which prayer speaks:

> How can the holy, which is "unexpressed," "unknown," which is what opens provided only that it is not discovered . . . fall into speech . . .? In truth, that cannot really be, that is the impossi-

ble. And the poet is nothing but the existence of this impossibility, just as the language of the poem is nothing but the retention, the transmission of its own impossibility. (Blanchot, 1995)

Because silence is not an option for the poet—every poetic silence being a mode of speaking—this impossibility is ineluctable, and the poet can exist only insofar as he is "not yet" the poet he intends to be. Blanchot thus arrives at a formulation of the poetic vocation that indicates why Hölderlin chose to remain under the guardianship of a carpenter for the last decades of his life: "Impossible, the reconciliation of the holy with speech demanded that the poet's existence come nearest to nonexistence."

[See also Heidegger; Poetics; and Romanticism, article on Philosophy and Literature.]

BIBLIOGRAPHY

Editions of Hölderlin's Works

Sämtliche Werke: Historisch-kritische Ausgabe. 6 vols. Edited by Friedrich Seebass and Norbert von Hellingrath. Munich, 1913–1923.
Sämtliche Werke: Stuttgarter Ausgabe. 8 vols. Edited by Friedrich Beissner. Stuttgart, 1943–1985.
Sämtliche Werke: Frankfurter Ausgabe. 20 vols. planned. Edited by D. E. Sattler and Wolfram Goddeck. Basel, 1975–.
Sämtliche Werke und Briefe. 3 vols. Edited by Jochen Schmidt. Frankfurt am Main, 1992.
Sämtliche Werke und Briefe. 3 vols. Edited by Michael Knaupp. Munich, 1992–1993.

Translations of Hölderlin's Works

Essays and Letters on Theory. Translated by Thomas Pfau. Albany, N.Y., 1988.
Hymns and Fragments. Translated by Richard Sieburth. Princeton, N.J., 1984.
Hyperion and Selected Poems. Edited by Eric Santner. New York, 1990.
Poems and Fragments. Translated by Michael Hamburger. Reprint, Cambridge and New York, 1980.

Critical Writings

Adorno, Theodor. "Parataxis: On Hölderlin's Late Poetry." In Notes to Literature, edited by Rolf Tiedemann, translated by Shierry Weber Nicholsen, vol. 2, pp. 109–149. New York, 1991.
Alemann, Beda. Hölderlin und Heidegger. Zurich, 1954.
Benjamin, Walter. "Two Poems by Friedrich Hölderlin." Translated by Stanley Corngold. In Selected Writings, vol. 1, 1913–1926, edited by Marcus Bullock and Michael W. Jennings, pp. 18–36. Cambridge, Mass., 1996.
Blanchot, Maurice. "Hölderlin's Itinerary." In The Space of Literature, translated by Ann Smock, pp. 269–276. Lincoln, Nebr., 1982.
Blanchot, Maurice. "The 'Sacred' Speech of Hölderlin." In The Work of Fire, translated by Charlotte Mandell. Stanford, Calif., 1995.
Böschenstein, Bernhard, and Ulrich Gaier, eds. Hölderlin-Jahrbuch. Tübingen, 1947–.
de Man, Paul. "Heidegger's Exegesis of Hölderlin." Translated by Wlad Godzich. In Blindness and Insight, 2d rev. ed., pp. 246–266. Minneapolis, 1983.
de Man, Paul. "The Image of Rousseau in Hölderlin." In The Rhetoric of Romanticism. New York, 1984.
Dilthey, Wilhelm. Das Erlebnis und die Dichtung: Lessing, Goethe, Novalis, Hölderlin. 2d rev. ed. Leipzig, 1905.

Fioretos, Aris, ed. "The Solid Letter": Essays on Hölderlin. Stanford, Calif., 1997.
Heidegger, Martin. ". . . Poetically Man Dwells. . . ." In Poetry, Language, Thought, translated by Albert Hofstadter, pp. 211–229. New York, 1971.
Heidegger, Martin. Hölderlins Hymnen: "Germanien" und "Der Rhein." Frankfurt am Main, 1980.
Heidegger, Martin. Erläuterungen zu Hölderlins Dichtung. Frankfurt am Main, 1981. Partially translated in Existence and Being, edited by Werner Brock (reprint, London, 1968).
Heidegger, Martin. Hölderlin's Hymn "The Ister." Translated by William McNeill and Julia Davis. Bloomington, Ind., 1996.
Henrich, Dieter. Der Grund im Bewusstsein. Stuttgart, 1992.
Jamme, Christoph, and Otto Pöggeler, eds. Jenseits des Idealismus. Bonn, 1988.
Kommerell, Max. "Hölderlins Empdockles-Dichtungen." In Geist und Buchstaben der Dichtung. Frankfurt am Main, 1940.
Lacoue-Labarthe, Philippe. "The Caesura of the Speculative." In Typography, edited by Christopher Fynsk. Cambridge, Mass., 1989.
Lukács, Györgi. "Hölderlin's Hyperion." In Goethe and His Age, translated by Robert Anchor. London, 1968.
Nägele, Rainer. Hölderlin: Text, Geschichte und Subjektivität. Stuttgart, 1985.
Rosenzweig, Franz. Das älteste Systemprogramm des deutschen Idealismus. Heidelberg, 1917.
Ryan, Lawrence. Hölderlins Lehre vom Wechsel der Töne. Stuttgart, 1960.
Sattler, D. E. Friedrich Hölderlin: 144 fliegende Blätter. 2 vols. Darmstadt, 1981.
Szondi, Peter. Hölderlin-Studien. Frankfurt am Main, 1967. Partially translated by Harvey Mendelsohn in On Textual Understanding and Other Essays (Minneapolis, 1986).
Warminksi, Andrzej. Readings in Interpretation: Hölderlin, Hegel, Heidegger. Minneapolis, 1987.

PETER FENVES

HOME, HENRY (Lord Kames; 1696–1782), Scottish lawyer who was a figure in the Scottish Enlightenment. Home's early writings on the subjects of law and equity, which sought to discern the general principles informing the individual cases, prepared the ground for his approach to social and aesthetic philosophy. In these early works, Home dedicated himself to a reform of the Scottish legal system that was intended to replace the feudal system with a modern civil society. This enterprise continued to inform his two main works in social philosophy and the philosophy of art: the Essays on the Principles of Morality and Natural Religion (1751) and the Elements of Criticism (1762), each of which was informed by a search for principles, whether of society or of criticism, that would ensure harmony and order in the new civil society.

Home's two main works are related to each other in a variety of ways. The Essays on the Principles of Morality defends Francis Hutcheson's account of the moral sense against David Hume's objections, extending the former's moral sense or "feeling" for the good into "feelings" for the sanctity of property and the respect for contracts that formed the basis of the emergent commercial society. These are the basic feelings necessary to ensure social harmony in

the new social order, and are supplemented by Home with the pleasures of the feeling for beauty. In the *Essays,* the feeling of beauty is introduced as a desirable addition or an ornament to social order, but one that in the last analysis is irrelevant to securing social order; it is only when "strict obligation ends [that] pleasure is employed to reward the performance" (p. 381).

In the *Elements of Criticism,* the pleasures of the feeling of beauty are accorded a more positive role in ensuring the harmony of civil society. In the dedication to George III, Home argues that the fine arts have a "beneficial influence in society" by virtue of "uniting different ranks in the same elegant pleasures," promoting benevolence, cherishing love of order, enforcing submission to government, as well as "inspiring a delicacy of feeling" that effectively makes such submission pleasurable (vol. 1, p. 5). By reason of the new role envisaged for the pleasures of the beautiful in the social order, Home considers it necessary to secure the principles of criticism by proposing a standard or a number of principles "that ought to govern the taste of every individual" (ibid., p. vi). The standard of taste informing the principles is derived from a "conviction common to the species" that is experienced as a sense of what is beautiful or ugly, high or low, proper or improper, proportioned or disproportioned. Thus, Home derives the discrete principles of criticism such as "dignity" from a shared human nature consisting of senses and passions, which give rise to a sense of discrimination that is also a rational "standard of taste." For this reason, the *Elements* begins with an extended discussion of human nature (two hundred pages) from which is derived, somewhat speciously, the host of "Principles of the Arts" that are discussed at length (three hundred pages) in volume 1, and that are supplemented in volume 2 by a recasting of neoclassical rhetorical poetics (five hundred pages).

Perhaps the most interesting feature of the *Elements of Criticism* is the tension it exposes and attempts to resolve between the arguments for a civil society based on property and contract and the "opulence and luxury" that are generated by such a society but that are morally corrosive and a threat to social order. Home argues, in genealogical fashion, that a "flourishing commerce begets opulence," which begets an inflamed "appetite for pleasure," which, through being "vented on luxury, and on every sensual gratification," begets selfishness, which infects the ranks and destroys *amor patriae* and public spirit (ibid., vol. 1, p. vii). What distinguishes Home from other moralists, however, is that, rather than denying the cause of social disorder—commerce—he proposes to regulate its effects. The *Elements* are thus an attempt to regulate the "appetite for pleasure" provoked by commerce and wealth through a principled enjoyment of art. In this way, art and the pleasures of the beautiful will break the sequence from commerce, through pleasure, to sensual gratification, selfishness, and disorder by inserting in the sequence the sensual pleasures of art,

which, because based on a shared human nature, will lead to benevolence and social harmony. In this way, a principled beauty will tend to reconcile the opposed forces of wealth and virtue.

The *Elements of Criticism* proved extremely influential, even though contemporary opinion of the book was mixed: one reviewer considered it to rival Aristotle, whereas Adam Smith commented, "They are all bad, but this is the worst." Perhaps surprisingly for such a gargantuan tome, the *Elements* went through eight editions between 1762 and 1807, with selections from it being translated in the *Encyclopaedia Britannica.* It was also translated into German by J. N. Meinhard a year after publication in 1763 and went through several German editions, one of which fells into the hands of Immanuel Kant and was to play a minor role in the development of his aesthetic theory in the *Critique of Judgment.*

[*See also* Criticism; *and* Pleasure.]

BIBLIOGRAPHY

Works by Home

Essays on the Principles of Morality and Natural Religion. Edinburgh, 1751; reprint, New York, 1976.
Elements of Criticism. 3 vols. Edinburgh, 1762; reprint, New York, 1967.

Other Sources

Lehmann, William C. *Henry Home, Lord Kames, and the Scottish Enlightenment.* The Hague, 1971.
Randall, Helen Whitcomb. *The Critical Theory of Lord Kames.* Northampton, Mass., 1944.

HOWARD CAYGILL

HORACE, QUINTUS HORATIUS FLACCUS. *See* Roman Aesthetics; *and* Ut Pictura Poesis.

HÜBSCH, HEINRICH (1795–1863), German architect and architectural theorist. Hübsch was born in Weinhem, near Mannheim, Germany. At the age of twenty, after a brief university career, Hübsch entered the private architectural academy of the celebrated neoclassicist and educator Friedrich Weinbrenner. After two years, Hübsch moved to Rome and continued his education in the immediate presence of Roman, medieval, Renaissance and—through travel—Greek and Byzantine models. In 1822, Hübsch published his first book, *Über griechische Architektur* (On Greek Architecture). Following a brief stint on the faculty of the Städelsches Kunstinstitut in Frankfurt, Hübsch was called to Karlsruhe in 1826 to assume the duties of the recently deceased Weinbrenner. Six years later he was named head of the architecture faculty at Karlsruhe's Polytechnische Hochschule, by which time he was active as a practicing architect. His most engaging designs include the Karlsruhe Kunsthalle and the Trinkhalle in Baden-Baden, both dating

from the mid-1830s; each building is a representative example of the round-arch style *(Rundbogenstil)* that he advocated. The neo-Renaissance surge of the 1840s and beyond left Hübsch in a professional cul-de-sac, and despite a rewarding and full career, little notice was taken at the time of his death in 1863.

Although Hübsch's primary contribution to the history of architecture was the theory and practice of the *Rundbogenstil*, his contribution to aesthetics devolves around the concept of style. Born to a generation that promoted the architectural agenda of a new Romantic medievalism above the claims of earlier neoclassicists, Hübsch and his German peers enjoined a deep and enduring debate over the nature and direction of contemporary architecture. The essence of this debate is neatly summarized in Hübsch's most important tract, and the work that triggered the debate, *In welchem Style sollen wir bauen?* (In What Style Should We Build?) of 1828.

It is important to note that both sides agreed that the vitality and authenticity of architectural style had been called into question since the turn of the century, and that as a result the profession was incapable of anything but derivative creations. It is this search for an authentic style as an expression of one's age that sparked the emergence of a historicist aesthetic consciousness that, in one form or another, persists to this day.

The basis for Hübsch's critique of contemporary architectural practice was the concept of imitation. For Hübsch, such an attack struck at the core of the aesthetic hegemony of architectural neoclassicism. The mimetic basis of all classical theory was the notion that its forms were derived from natural models that, although remote in time, remained immutable for all human experience. The classical doctrine of mimesis, beginning with Vitruvius, claims that all architectural forms descend from an archetypal primitive wooden hut. Over the course of generations, these forms were perfected by the ancients and gradually translated into (or imitated in) stone for the sake of permanence and the pursuit of higher aesthetic aims. To Hübsch, this amounted to an unfounded belief in the priority of wood architectural prototypes that he dubbed "Xylomania" and represented the "original sin" of classical doctrine (Hübsch, 1992, p. 71 and note). This conceptual error represented the root of the problem: the absolutism of the antique claim on style led to the "unfailingly empty sophisms" of those who "go to enormous and self-deceiving lengths to construct a system of specious arguments" about the nature of the "specific detail of architectural forms" (ibid., p. 64). Indeed, Hübsch's remedy was to abstain from any and all aesthetic speculation about the matter. Before one could even begin an enterprise of aesthetic analysis, let alone pronounce a remedy for the malaise of the profession, one needed a sure understanding of the material and historical realities of the concept of style.

Hübsch's *In What Style Should We Build?* is made up of eighteen sections with a brief concluding coda. The bulk of the sections (8–18) constitute a thematic summary of the history of Western architecture beginning with the Greeks and ending with the Renaissance. The themes that unite this brisk account are set forth in the introductory sections (1–8) and present a rich and revealing theoretical nugget offering, in many cases for the first time, a pattern of recurring theoretical issues that remained vital until well into the twentieth century.

For Hübsch, the concept of style "means something general," and is not to be attached exclusively to classical architecture. Here is one of the most strident, programmatic declarations of stylistic relativism in Germany and a touchstone for ensuing debates on architectural style. In an additionally provocative declaration, Hübsch claimed that the "most general requirement for all kinds of buildings [regardless of style] is the enclosure of a specific space" (ibid., p. 66). This emphasis on the spatial nature of architectural experience was discussed further by Gottfried Semper around mid-century and perhaps most definitively by August Schmarsow near the end of the century. Given the conceptual and actual priority that Hübsch places on space enclosure, the elements of any building's style are thus defined by the handling of a few specific architectonic elements: walls, supports, spanning members, doors, and windows. Further, the uniform and regular variation of these elements that contribute to a national or a period style is a function of the interplay of three critical factors: climate, material, and what Hübsch dubbed "technostatic proportion," that is, a feeling for structural elegance and economy. These three factors alone were said to provide the basis for the only two valid styles of architecture: "one with straight, horizontal stone architraves; the other with curved vaults and arches" (ibid., p. 68).

These elements also furnished the conditions for architectural progress, whether by innovative material technique, the introduction of daring structural solutions, or the employment of more economical means resulting in the reduction of the built mass. Drawing on a topos of ancient and Renaissance art theory, Hübsch claimed that "technostatic judgement by eye" enabled progressively "lighter technostatic proportions" in newer buildings; in this and in his strictures against "decorative overloading" (ibid., 69 and 70), Hübsch thus employs the neoclassical aesthetic's moral penchant for simplicity and economy against the neoclassical taste for antique style.

All this is offered to buttress Hübsch's primary claim: that Greek forms are not suited to German needs. In this, Hübsch offers several practical, pre-aesthetic desiderata for his assertion. First, Greek trabeation could not be reproduced satisfactorily in the local German sandstone, whose spanning capacity was rather limited. Arcuated forms, based on smaller lithic units, were thus required to handle necessarily larger interior spans of contemporary buildings. Second, the Nordic climate required steeper roof pitches to enable efficient

runoff and drainage. Classical cornice profiles could not be relied on to withstand the exigencies of climate and would diminish the durability of any building so employing them.

Hübsch's penchant for the bold technostatic experience of medieval architecture made vaulting the inevitable solution. But the Romans, who invented vaulting, produced a "sham architecture" by trying to wed the formal vocabulary of Greek trabeation (that is, the Orders), to a more plastic, arcuated vaulting system of construction. In this, the Romans exhibited the "utmost poverty of imagination" (ibid., p. 79). Renaissance architects managed to compound the error by applying temple facades to vaulted basilicas, forms that, to Hübsch, possessed no necessary, organic connection to one another. Hübsch's answer was not the soaring vaults of Gothic, or pointed-arch *(Spitzbogen)*, architecture but the forms of what he called "Byzantine" architecture. Here, there was none of the decorative "overloading" of the Gothic. Indeed, Hübsch believed that the *Rundbogenstil* offered all the best principles that made Greek architecture perpetually appealing: (1) simplicity of composition, (2) moderation of ornament, and (3) an even development of forms (ibid., p. 77).

Hübsch presents the *Rundbogenstil* as a reasonable, even inevitable, solution to the moribund state of architectural production in the third and fourth decades of the nineteenth century in Germany. Indeed, his claims for its validity were allegedly superior to mere scholarly, aesthetic analyses because of the "influence of reality"; in fact, he claimed that such a solution would "no longer have a historical or conventional character" (ibid., p. 99).

Although Hübsch's views attracted early attention, they were not uniformly accepted. The architect-journeyman Rudolf Wiegmann reviewed Hübsch's tract and criticized its materialist tenor (Wiegmann, 1992, pp. 103–112). Another critical architect, Carl Albert Rosenthal, referred to the same problem in his book *The Origin and Significance of the Architectural Forms of the Greeks* (1830). The young art historian Franz Kugler made similar claims in an article on religious architecture (1834).

In the years and decades that followed, the debate on the nature of architectural style gained both in momentum and in seriousness among German architects. By the mid-1840s, theoretical works by Eduard Metzger, Hubert Stier, Johann Heinrich Wolff, Carl Gottlieb Wilhelm Bötticher, and Rosenthal, while nominally confronting one another, nevertheless addressed themselves directly to Hübsch's foundational question. By this time, taste had changed and an awareness of the aesthetic potential of iron construction had become an increasingly urgent issue of debate. It was Bötticher who, in 1846, finally posited the likelihood of iron as the material generator of a new style altogether (Bötticher, 1992, p. 158).

Debates on style characterize several of the key episodes in the history of nineteenth-century architectural theory.

Rudolf Redenbacher, a neglected but original and important theorist of the 1880s, declared his pronounced debt to Hübsch, along with Bötticher and Semper, for the formation of his ideas about architecture and style (Redenbacher, 1881). Finally, at the outset of the 1890s, Albert Hofmann, coeditor of the *Deutsche Bauzeitung,* the leading organ of the profession in German, brought up the issue once again in a rival periodical, taking his title and theme directly from Hübsch (Hofmann, 1890).

Hübsch's primary contribution to the history of aesthetic thought consists in (1) his definitive separation of style from ornament, and (2) his rejection of the Rousseauesque cult of feeling so prevalent among other German Romantic architects and theorists (Hübsch, 1992, pp. 84 and 172). The subsequent elaboration of the concept of style, in architectural theory, art history, hermeneutics, and aesthetics, owes an enormous debt to Hübsch's bold stance of 1828.

[*See also* Architecture; *and* Style.]

BIBLIOGRAPHY

Works by Hübsch

In What Style Should We Build? The German Debate on Architectural Style. Translated by Wolfgang Herrmann. Santa Monica, Calif., 1992.
Über griechische Architektur. Heidelberg, 1822.

Primary Sources

Bötticher, Carl Gottlieb Wilhelm. "The Principles of the Hellenic and Germanic Ways of Building with Regard to Their Application to Our Present Way of Building." In Heinrich Hübsch, *In What Style Should We Build? The German Debate on Architectural Style,* trans-lated by Wolfgang Herrmann, pp. 147–167. Santa Monica, Calif., 1992.
Hofmann, Albert. "Im welchem Style sollen wir bauen?" *Allgemeine Bauzeitung* 55 (1890): 81–84, 89–92.
Kugler, Franz. "Ueber den Kirchenbau und seine Bedeutung für unsere Zeit." *Museum: Blätter für bildende Kunst* 2 (1834): 1–8.
Redenbacher, Rudolf. *Tektonik: Prinzipien der künstlerischen Gestaltung der Gebilde und Gefüge von Menschenhand welche den Gebieten der Architektur, der Ingenieurfächer und der Kunst-Industrie angehören.* Vienna, 1881.
Rosenthal, Carl Albert. *Über die Entstehung und Bedeutung der architektonischen Formen der Griechen.* Berlin, 1830.
Schmarsow, August. *Das Wesen der architektonischen Schöpfung: Antrittsvorlesung, gehalten in der Aula der k. Universität am 8. November 1893.* Leipzig, 1894. English translation as "The Essence of Architectural Creation," in *Empathy, Form, and Space: Problems in German Aesthetics, 1873–1893,* edited and translated by Harry Francis Mallgrave and Eleftherios Ikonomov, pp. 281–297. Santa Monica, Calif., 1994.
Semper, Gottfried. *Der Stil in den technischen und tektonischen Künsten oder praktische Aesthetik.* 2 vols. Munich, 1861–1863.
Wiegmann, Rudolf. "Remarks on the Treatise *In What Style Should We Build?*" In Heinrich Hübsch, *In What Style Should We Build? The German Debate on Architectural Style,* translated by Wolfgang Herrmann, pp. 103–112. Santa Monica, Calif., 1992.

Other Sources

Döhmer, Klaus. *"In welchem Style sollen wir bauen?"* In *Architekturtheorie zwischen Klassizismus und Jugendstil.* Munich, 1976.

Kruft, Hanno-Walter. *A History of Architectural Theory from Vitruvius to the Present*. Translated by Ronald Taylor, Elsie Callander, and Antony Wood. London, 1994.

Muthesius, Stefan. *Das englische Vorbild: Eine Studie zu den deutschen Reformbewegungen in Architektur, Wohnbau und Kunstgewerbe im späteren 19. Jahrhundert*. Studien zur Kunst des neunzehnten Jahrhunderts, vol. 26. Munich, 1974.

J. DUNCAN BERRY

HUME, DAVID. [To analyze Hume's contribution to aesthetics, *this entry comprises two essays:*
> Survey of Thought
> "Of the Standard of Taste"
The first is a survey of Hume's philosophy and the place of aesthetics in it. The second essay is an in-depth analysis of his major essay "Of the Standard of Taste." For Hume's views on tragedy, see Tragedy. *For related discussion, see* Taste.]

Survey of Thought

The Scottish philosopher David Hume (1711–1776) was born and died in Edinburgh. He entered the university there at the age of twelve, but left after three, or possibly four, years without graduating, as was usual for the time. He then embarked on a long and private study of philosophy. Between 1734 and 1737 he lived in France, where he wrote *A Treatise of Human Nature*, which was published anonymously in 1739–1740. To boost disappointing sales of the work, he then published an anonymous summary of its main theses. This pamphlet, now known as the *Abstract*, singled out precisely the tenets he selected for discussion in rewriting the original three volumes for publication as three separate works between 1748 and 1752. These are now known as the *An Enquiry concerning Human Understanding*, (1748), *An Enquiry concerning the Principles of Morals* (1751), and *A Dissertation on the Passions* (1757).

Hume did not obtain full-time employment until he was over forty. He twice failed to become a professor of philosophy, first in Edinburgh in 1745, and then in Glasgow in 1752; on both occasions, hostility toward his overtly skeptical and antireligious views combined with local political ambitions to prevent his appointment. In 1752, however, he became Keeper of the Advocates' Library in Edinburgh, which later became the National Library of Scotland. Using the magnificent collection, Hume was able to publish between 1754 and 1762 the six volumes of his *History of England*. This work was immensely successful during Hume's lifetime and into the following century. Earlier, in 1741–1742, Hume had published anonymously two volumes of *Essays Moral, Political, and Literary,* and in his own name another set of essays on political and economic topics, in 1752, titled *Political Discourses.*

Hume served in Paris from 1763 to 1766 as secretary and then chargé d'affaires at the British embassy. He became friendly with most of the great writers and philosophers of the day, such as Denis Diderot, Jean Le Rond d'Alembert, and Jean-Jacques Rousseau, but to their mutual regret he never met Voltaire. Hume returned to London, and in 1767 became undersecretary of state, handling diplomacy with foreign powers to the north of France, including the Russia of Catherine the Great. Hume retired to Edinburgh in 1769 and played host to all the important visitors to town, including Benjamin Franklin. His closest friend was Adam Smith, whose *Wealth of Nations* appeared in 1776, shortly before Hume died. Smith adopted a great deal from Hume's philosophical work, including emphasis in moral philosophy on imagination and sympathy. Moreover, like Hume, Smith embedded his economic theory in his account of moral philosophy and the nature of society.

The scientific, political, and social contexts in which Hume was writing were quite different from today, and the full force of many of his claims and concepts cannot easily be translated into twentieth-century forms. Hume himself was most influenced by the writings of the Roman politician Marcus Tullius Cicero, and the seventeenth-century French philosopher Nicolas de Malebranche. He read widely, however, and expected his equally learned audience to recognize his other allusions and debts. Hume's letters reveal how much his public life, which began *after* publication of almost all his works, influenced his views about what mattered most—moral conduct in civil society; his published writings must therefore be carefully anchored in the contexts in which they were conceived and produced. The only works he revised after his government service were new editions of his *History* and his posthumously published *Dialogues concerning Natural Religion*.

Hume accepted the traditional view that the only way to plan for the future is to understand how and why things change; only a knowledge of causes can help us dispel our fears, explain the past, and influence what happens. But he held that many philosophers had misunderstood the nature of causal knowledge and had thus failed to benefit from it. To start with, an examination of the natural world must be grounded on a study of the investigator himself; a study of the nature of man should reveal the nature and extent of his knowledge, and enable us all better to conduct our lives. Hume argues that we are all governed less by reason than philosophers have typically claimed, and that we are motivated essentially by our passions; moreover, our capacities for interpreting experience, from which all our ideas ultimately derive, are extremely varied. This means both that we should not overintellectualize human activities and that we must learn to live with probabilities rather than certainties. Such philosophical views yield intensely practical benefits, as is proper, because the goal of philosophy is to achieve understanding that enhances life.

Only after Hume's death did aesthetics achieve recognizably modern form, as a result of many intellectual and social factors, such as the spread of wealth, which hastened the

emergence of an informed, but nonpracticing, "public" for the arts with the leisure to travel, buy, and comment. Hume's own scattered reflections on art are firmly anchored in his conception of social life, which is why he embeds his account of both making and responding to art within his general analysis of other human actions. Almost all his claims are based on literature, in which his entirely orthodox tastes are anchored in classical, and modern neoclassical, works. Even so, his interest in drama and poetry is markedly less than his interest in history and philosophy, and he shows almost no knowledge of the other arts.

Hume derived his views about the arts from a single book, and he saw his own task as that of underpinning those claims by reference to his own theory of human nature. The Abbé Jean-Baptiste Du Bos was secretary of the French Academy from 1723 until his death in 1742; he had been a friend of the Huguenot scholar Pierre Bayle, whose encyclopedic and skeptical works had inspired Hume. In 1719, Du Bos published *Réflexions critiques sur la poésie et sur la peinture*, and the book dominated European thinking about the arts for the next fifty years. Hume frequently referred to the work from the 1730s onward, drawing from it extensively in his essay "Of the Standard of Taste." Unlike Du Bos, however, who signaled his awareness of important differences between the arts, deriving in part from their various mediums, Hume usually ignored such differences and generalized his remarks on the basis of literature alone. [*See* Du Bos.]

In *A Treatise of Human Nature*, Hume discusses the nature of judgment and disinterested evaluation on which he grounds his later views about literary criticism. His passing remarks on beauty are also of interest, however. He holds that beauty is an indefinable "power" in objects that causes a pleasurable sentiment. Beauty is not a sentiment, nor even a property discernible by the five senses, but rather a property whose presence is *felt* (by a sixth or seventh sense, as Du Bos and Francis Hutcheson, respectively, said) only when objects with certain detectable properties causally interact, under specifiable conditions, with minds having certain properties (1978, p. 299; 1987, pp. 164–166; 1975, p. 292). Discussion of something in which a person takes delight can alter his perception of it, and thereby set off a new causal chain that results in a new sentiment.

Hume argues that two important principles operate when we make judgments about beauty: comparison functions in our classification of objects, and sympathy operates whenever we think of things in association with people. Beauty is species-relative and culture-relative, and we can justify our judgments only by reference to the species and nature of the object claimed to be beautiful. We must know precisely what kind of thing we are talking about before we can decide whether it is beautiful. Although beauty is detected by a sentiment, it is as "real" as color and other supposedly secondary qualities. Three factors secure the objectivity of judgments in such contexts: the conventions of language,

the universal psychological makeup of human beings, and the possibility of publicly shareable viewpoints. It is crucial to recognize, Hume argues, that human beings cannot agree in the absence of either a literal or metaphoric point of view: that alone can ground the general standards that are necessary for general evaluations. Strictly speaking, a general point of view should be understood as a cognitive attitude in which we think of things in certain ways, whatever we may privately feel to the contrary; and although the standards formulated in this frame of mind are general, they are also revisable, because they serve the needs of the community and those needs may change. It is thus contingent which standards are accepted within a particular context, but it is necessary that there be some standards, in the sense that agreed anchors and conventions are the only guarantees of mutual understanding and communication.

Such views demonstrate that Hume understood the central importance of context to critical judgments, and that he was one of the first British writers to do so. In the finer arts, he insists, "it is requisite to employ much reasoning" (1975, p. 173). Here Hume followed Du Bos closely, agreeing with him both that we can derive sustained pleasure from a work only if we understand it in some way, and that the role of reason is to justify the verdict of sentiment by picking out those features of a work that have affected us.

Hume's most influential essay on aesthetics, "Of the Standard of Taste" (1987, pp. 226–249), was published in 1757, in a context of intense debate by his friends in Edinburgh and further afield. Hume seeks to find in human nature as well as in social practices a resolution for a problem that Allan Ramsay, the painter, and to some extent Adam Smith, located only in social practices. Hume holds that some so-called judgments of taste are palpably foolish and indefensible. Accordingly, he tries to show that criticism can be a factually based, rational, and social activity, capable of integration within the rest of intelligible discourse.

A purely passive attitude toward things is inadequate for detecting their most subtle properties: an observer must self-consciously attend to the object in question, and be in a proper state of mind when doing so. A merely causal reaction will be replaced by an appropriate causal interaction, to which the observer crucially contributes. Hume adopts Du Bos's list of the "causes" why people fail to respond appropriately to works of art: they are all technical notions from seventeenth-century French debates. A person may lack "delicacy," or sensory discrimination; may lack "good sense," or the capacity to make true sensory rather than true intellectual judgments; and may be "prejudiced." The point about prejudice is that, having made up one's mind in advance, one does not fully attend to the matter at hand, and may even fail to perceive that one's prior judgment was verified. Two steps are necessary for proper judgment: practice and comparison. Practice is necessary in order to overcome any superficial or incomplete first impressions, and to ensure that all relevant

aspects of a complex work have been discerned. Comparison is necessary in order to determine not only the boundaries of a work, but the similarities and differences between it and anything else, as well as its relations with other things.

Hume's discussion trades on a deep ambiguity in the French and English term *sentiment,* which covers both physical feelings or sensations, and mental thoughts, opinions, and judgments, but his general model of aesthetic response and judgment emerges clearly. Someone becomes conscious of pleasure in a certain object, and this induces him to pay closer perceptual and intellectual attention to it, with the aim of sustaining the pleasure. If the spectator can make sense, in some way, of what he perceives, he will experience new sentiments that may loosely be described as enhancing the original ones. In the broadest sense, works of art are pleasurable means of communication between human beings, and so the preconditions of effective communication apply to art as much as to other means of communication. Works please because of the particular properties they possess, and one of our tasks is to identify these causes in order to enable others to share in our enjoyment. We cannot comprehend a work, however, merely by being in its presence. Two properties that belong to human actions, and that are goals of our comprehension, are meaning and value: neither is discernible by the five ordinary senses alone. The mind must be actively involved. Just as inference beyond present data is necessary for all factual reasoning, so interpretation is necessary to establish the meaning of what another has done, and for that an understanding of context is essential. The cognitive engagement of an artist requires a reciprocal engagement by the audience.

The importance of Hume's essay "Of the Standard of Taste" has been fully grasped only in the late twentieth century. It is the first sustained attempt to give an experientially grounded account of critical judgment that escapes the accusation of being merely causal and recognizes the ineliminable role of the mind in interpretation of the selected data.

[*See also* Beauty, *conceptual and historical overview article.*]

BIBLIOGRAPHY

Works by Hume

An Abstract of a Book Lately Published: Entitled A Treatise of Human Nature, *&c. Wherein the Chief Argument of That Book Is Farther Illustrated and Explained* (1740). In *A Treatise of Human Nature* (1739–1740), edited by L. A. Selby-Bigge and P. H. Nidditch, 2d ed. Oxford, 1978.
A Treatise of Human Nature (1739–1740). Edited by L. A. Selby-Bigge and P. H. Nidditch. 2d ed. Oxford, 1978.
Essays Moral, Political, and Literary (1741–1742). Edited by Eugene F. Miller. Rev. ed. Indianapolis, 1987.
A Letter from a Gentleman to His Friend in Edinburgh: Containing Observations on . . . the Principles . . . said to be maintain'd in . . . A Treatise of Human Nature (1745). Edited by Ernest C. Mossner and John V. Price. Facs. ed., Edinburgh, 1967.
An Enquiry concerning Human Understanding (1748) In *Enquiries concerning Human Understanding and concerning the Principles of Morals,* edited by L. A. Selby-Bigge and P. H. Nidditch, 3d ed. Oxford, 1975. First published as *Philosophical Essays concerning Human Understanding.*

An Enquiry concerning the Principles of Morals (1751). In *Enquiries concerning Human Understanding and concerning the Principles of Morals,* edited by L. A. Selby-Bigge and P. H. Nidditch, 3d ed. Oxford, 1975.
The History of England (1754–1762). 6 vols. Indianapolis, 1983.
A Dissertation on the Passions (1757). In *David Hume: The Philosophical Works,* edited by Thomas Hill Green and Thomas Hodge Grose, vol. 4. London, 1882–1886; reprint, Darmstadt, 1964.
The Natural History of Religion (1757). In *David Hume: The Philosophical Works,* edited by Thomas Hill Green and Thomas Hodge Grose, vol. 4. London, 1882–1886; reprint, Darmstadt, 1964.
Dialogues concerning Natural Religion (1778). Edited by Norman Kemp Smith. Oxford, 1935; 2d ed., London, 1947.
The Letters of David Hume. 2 vols. Edited by J. Y. T. Greig. Oxford, 1932.
New Letters of David Hume. Edited by Raymond Klibansky and Ernest C. Mossner. Oxford, 1954.

Other Sources

Brunius, T. *David Hume on Criticism.* Stockholm, 1951.
Crow, Thomas E. *Painters and Public Life in Eighteenth-Century Paris.* New Haven, 1985.
Dubos, Jean-Baptiste. *Réflexions critiques sur la poésie et sur la peinture* (1719). 3 vols. New ed. Utrecht, 1732–1736.
Fleming, John. *Robert Adam and His Circle in Edinburgh and Rome.* London, 1962.
Fontenelle, Bernard le Bovier de. *Digression sur les anciens et les modernes.* Paris, 1688.
Gerard, Alexander. *An Essay on Taste.* London, 1759.
Harris, Eileen. *British Architectural Books and Writers, 1556–1785.* Cambridge and New York, 1990.
Hipple, Walter John, Jr. *The Beautiful, the Sublime, and the Picturesque in Eighteenth-Century British Aesthetic Theory.* Carbondale, Ill., 1957.
Home, Henry, Lord Kames. *Elements of Criticism.* London, 1762.
Hunter, J. Paul. *Before Novels: The Cultural Contexts of Eighteenth-Century English Fiction.* New York, 1990.
Johnson, David. *Music and Society in Lowland Scotland in the Eighteenth Century.* New York and Oxford, 1972.
Jones, Peter. *Hume's Sentiments: Their Ciceronian and French Context.* Edinburgh, 1982.
Jones, Peter. "Hume and the Beginnings of Modern Aesthetics." In *The Science of Man in the Scottish Enlightenment,* edited by Peter Jones, pp. 54–67. Edinburgh, 1989.
Jones, Peter. "The Aesthetics of Adam Smith." In *Adam Smith Reviewed,* edited by Peter Jones and Andrew S. Skinner, pp. 56–78. Edinburgh, 1992.
Kivy, Peter. *The Seventh Sense: A Study of Francis Hutcheson's Aesthetics and Its Influence in Eighteenth-Century Britain.* New York, 1976.
Mossner, Ernest Campbell. *The Life of David Hume.* Edinburgh, 1954.
Pears, Iain. *The Discovery of Painting: The Growth of Interest in the Arts in England, 1680–1768.* New Haven, 1988.
Rykwert, Joseph. *The First Moderns: The Architects of the Eighteenth Century.* London, 1980.
Smith, Adam. *The Theory of Moral Sentiments* (1759). Edited by D. D. Raphael and A. L. Macfic. Oxford, 1976.
Summerson, John. *Georgian London.* London, 1945.
Turnbull, George. *A Treatise on Ancient Painting.* London, 1740.

PETER JONES

"Of the Standard of Taste"

In his Introduction to the *Treatise of Human Nature* (1739) David Hume announces his intention to deal systematically with "Logic, Morals, Criticism and Politics." "In these four

sciences," he wrote, "is comprehended almost every thing, which it can in any way import us to be acquainted with, or which can tend either to the improvement or ornament of the human mind." As it turned out, criticism was the one science that he did not get around to, either in the *Treatise* or in his subsequent publications. It was not that Hume lost interest: observations about the arts and about critical judgment occur in all his principal works as well as in his essays and in his letters, but his one and only formal statement is the essay "Of the Standard of Taste," which was published in 1757 when Hume was forty-six years old. Frequently reprinted, respectfully anthologized, the essay has not been exactly neglected, but until fairly recently, it was regarded as being of greater interest to the historian of ideas than to the philosopher. It does precede the recognition of aesthetics as an autonomous field (or "subject")—the Germans get credit for that—and it does endorse a number of commonplaces (some would say clichés) of neoclassical criticism. Peter Jones, one of the most learned and most sympathetic of Hume scholars has remarked: "It is unfortunate, perhaps, that in the absence of alternative texts, a single, condensed, derivative essay of under twenty pages should be taken as representative of Hume's considered views on art and criticism" (Jones, 1982). It is true, as has been demonstrated by Jones and others, that taken individually, not one of Hume's main points in the essay is original with him. In addition to his British sources—Joseph Addison, Francis Hutcheson, Anthony Ashley Cooper, earl of Shaftesbury—Hume borrowed copiously, sometimes word for word, from the French author Abbé Jean-Baptiste Du Bos (*Réflexions critiques sur la poésie and sur la peinture*, 1719): (Philosophers in those days did not go in for acknowledgments or citing their sources; think of René Descartes or Baruch Spinoza. It was not plagiarism, we are told, because the assumption was that educated readers would recognize borrowings and did not need footnotes.) To identify and trace the history of Hume's ingredients is interesting and helpful but should not keep us from appreciating the brilliant and innovative way that he organizes these familiar materials.

Immanuel Kant's *Critique of Judgment* (1790) is the first modern work in philosophical aesthetics and Kant in our own day seems to be overshadowed by G. W. F. Hegel. Aesthetics, as it is taught, is linked to the history of art, to the social sciences, to culture criticism and an interest in the spirit of the age. Such questions as whether the judgment of taste stands in need of a transcendental deduction are not in the forefront. Hume's ideas, in particular those that inspire the essay "Of the Standard of Taste," foreshadow Kantian doctrine on a number of central points. Although Kant scholars might protest the claim, it can be argued that it was Hume, in this "condensed, derivative essay of under twenty pages," who set the stage for discussion of the central issue for philosophical aesthetics. The present essay summarizes Hume's arguments and adds some brief comments.

Hume begins by observing that it is obvious even to "men of the most confined knowledge" that tastes differ within "the narrow circle of their acquaintance" and as, through travel and education, they become less provincial, they are increasingly struck by the "inconsistence and contrariety" of tastes. Initially, we tend to dismiss as "barbarous" the preferences of "distant nations and remote ages," but when we find that they return the compliment, we become less dogmatic and more tolerant.

The opening paragraph is an epitome. It sums up the conflict that the essay is designed to resolve. In the sort of situation that Hume is imagining, one's first thought might be, "If you think *that's* beautiful, you're out of your mind!" But people differ from one another in many ways and one person can admire what another finds boring. Tastes differ; that is a fact of life.

Actually, Hume continues, there are greater differences both in taste and in morals than appear on the surface. "There are certain terms in every language which import blame and others praise." Everyone agrees that elegance, propriety, and simplicity are what one looks for in a writer, just as heroism and prudence are virtues in a man of action; conflicts arise when it comes to deciding who is entitled to claim such merits. (Homer's heroes might not strike us as good role models.)

True enough, although Hume goes on to draw a contrast with "matters of opinion and science," which is very unclear.

Hume continues: "It is natural for us to seek a *Standard of Taste;* a rule, by which the various sentiments of men may be reconciled; at least, a decision, afforded, confirming one sentiment, and condemning another."

If one begins by consulting our everyday intuitions, one finds an inconsistency between what Hume calls "two species of common sense." The first species, as it happens, echoes the "skeptical" views of philosophers "in holding that there is a big difference between "judgment and sentiment." Judgments "have reference to something beyond themselves" and are either true or false, but

a thousand different sentiments, excited by the same object, are all right: Because no sentiment represents what is really in the object. It only marks a certain conformity or relation between the object and organs or faculties of the mind; and if that conformity did not really exist, the sentiment could never possibly have being. . . . To seek the real beauty or real deformity is as fruitless an inquiry as to pretend to ascertain the real sweet or real bitter; According to the disposition of the organs, the same object may be both sweet and bitter; and the proverb has justly determined it to be fruitless to dispute concerning tastes.

The second species of common sense points in the opposite direction:

Whoever would assert an equality of genius between Ogilby and Milton, or Bunyan and Addison, would be thought to defend no

less an extravagance than if he had maintained a mole-hill to be as high as Tenerife or a pond as extensive as the ocean. Though there be found persons who give preference to the former authors, no one pays attention to such a taste and we pronounce without scruple, the sentiment of these pretended critics to be absurd and ridiculous.

Hume's task will be to tell a story that allows a place for both species of common sense, which are (although he does not quite admit it) flatly inconsistent. Something has to go. What will it be? Hume can hardly jettison the distinction between sentiment and judgment because from the time of the *Treatise* it has been central to his whole system. (It has, however, been noted that for Hume, as for his French forbears, the term *sentiment* encourages ambiguities because it can mean not only feeling or affect but also opinion.) The real sticking point is the first species' claim that "a thousand sentiments, excited by the same object, are all right." That cannot be if the person who prefers Ogilby to John Milton is not merely mistaken but "absurd and ridiculous." The trick will be to show that the sentiment-judgment distinction does not entail commitment to what Hume calls "the natural equality of tastes." Does Hume in the end pull it off? Opinions differ. Some think that a causal theory in which a judgment of taste articulates a response to a candidate work of art precludes the possibility of a judgment's being mistaken, or "absurd and ridiculous." That thesis, however, requires supporting argument; otherwise, it is mere dogma and has little to recommend it. An aesthetic judgment—what Hume calls a "pronouncement"—is an event and hence may be presumed to have causes. But it is also an assertion—a speech act—and to ask whether or not it is true is to raise a completely different question.

Hume suddenly changes gears and lays it down as "evident" that the "rules of composition" are not established a priori but are based on observation of "what has universally been found to please in all countries and in all ages." He speaks as if everyone knew what the "rules" are; he gives no examples; to anyone inclined, following neoclassical convention, to assume, for example, a rule of verisimilitude—that art "holds the mirror up to nature" and so forth—Hume observes that "many of the beauties of poetry and even of eloquence are founded on falsehood and fiction, on hyperboles, metaphors and an abuse of perversion of terms from their natural meaning." He then goes on to what amounts to an attack on the whole idea of rules: a poem either works or it does not, and, if it does, then the fact that it breaks the rules is irrelevant. Ariosto pleases, not by his faults (e.g., his "monstrous improbable fictions") but by "the force and clearness of his expression, the readiness and variety of his inventions." (Note that these are examples of those "general terms" that "import praise.")

And however his faults may diminish our satisfaction, they are not able entirely to destroy it. Did our pleasure really arise from

those parts of his poem, which we denominate faults, this would be no objection to criticism in general; it would only be an objection to those particular rules of criticism, which would establish such circumstances to be faults, and would represent them as universally blameable. If they are found to please, they cannot be faults; let the pleasure, which they produce, be ever so unexpected and unaccountable.

The putative "rules of composition" are supposed to be grounded in laws of taste, that is, in generalizations to the effect that certain qualities are causes of pleasure to everyone in every age. Here Hume is in a bind: he is going to claim that a judgment of taste is justified or vindicated when a critic correctly identifies the features of a poem that account for the pleasure that reading the poem affords; and Hume's conception of causality is that it depends on the observation of repeated sequences. This suggests that the critic is seeking a causal law on the order of "A poem that manifests 'force and clearness of expression' has some claim to beauty." On the other hand, it would have been obvious to Hume that such a law holds only because it is definitional and based "on reasonings a priori", and that for reasons he has indicated, a substantive law (or rule) has no teeth; a violation (or disconfirmation) does not unseat the judgment. ("If they are found to please, they cannot be faults.")

A stroke of genius on Kant's part was to recognize that if there were rules (or laws) of taste, then a judgment of taste could be supported in the one case deductively, and in the other by induction, and that either possibility is absurd. (For one thing, it would follow that one could assess the beauty of a poem without ever reading it, just by ascertaining that it had or lacked the relevant properties.) It would be nice to be able to credit Hume with having anticipated Kant, but even with special pleading, there is too much that tells against it. One can say this much for Hume: he recognized that there was something fishy about the notion of laws, principles, or rules of taste and nothing, it can be argued, in his positive theory depends on the supposition that there are any such. Hume very much wanted to be famous as a man of letters and the critical establishment of his day was committed to aesthetic generalizations, usually of a rather bland and innocuous sort. The French, whom he admired, tended to go on and on about "propriety" and why Jean Racine is more correct than William Shakespeare. It is to Hume's credit that although he did not, like Kant, flatly dismiss this talk as hot air, he did not trade on it either.

Next Hume moves to an interesting and rather tricky step in his argument. In his summary of the first species of common sense, he had said that "beauty is no quality of things in themselves: It exists merely in the mind which contemplates them, and each mind perceives a different beauty." This is (perhaps intentionally) a sloppy way of putting a point that he had made quite clearly in the *Treatise* and in subsequent works: beauty is not a perceptible property ei-

ther of works of art or of a critic's mind. We *ascribe* beauty to a poem when we find that not just once, but on repeated readings, it is a cause of pleasure for us *and* if we suppose, with or without reason, that it will be a cause of pleasure to others. Beauty, as Hume says repeatedly in the *Treatise,* is to be thought of as a "power" (what one would call a dispositional property) and, as such, it is a supervenient property. Sometimes Hume likens it to color, a so-called secondary property, and is eager to insist that being supervenient is not being somehow illusory or unreal. (Actually, in Lockean terms, beauty should be listed as a *tertiary* property because secondary properties—for example, color, at least in the visual arts—are among those on which beauty supervenes.)

We pronounce a work beautiful when our encounter with it is a cause of pleasure, but the causal relation here is not simple: it is not as if the poem were a magnet and the pleasurable response iron filings. The "power" of a poem may escape notice if the test conditions are not exactly what they should be.

> Those fine emotions of the mind are of a very tender and delicate nature, and require the concurrence of many favorable circumstances to make them play with facility and exactness, according to their general and established principles. The least exterior hindrance to such small springs, or the least internal disorder, disturbs their motion, and confounds the operation of the whole machine. When we would make an experiment of this nature, and would try the force of any beauty or deformity, we must choose with care a proper time and place, and bring the fancy to a suitable situation and disposition. A perfect serenity of mind, a recollection of thought, a due attention to the object; if any of these circumstances be wanting, our experiment will be fallacious, and we shall be unable to judge of the catholic and universal beauty.

Hume's idea, as it starts to emerge here, is that although the sentence "*O* is beautiful" (where *O* is the name or description of an individual) cannot be true or false, it may or may not be justified. It is justified if the speaker is correct in identifying the cause of his pleasure and if he has reason to suppose that *O* will be a cause of pleasure to anyone who satisfies the test conditions. This is a plausible view, but then why say that it is an expression of sentiment and not a judgment? If "*O* is soluble" or "*O* is green" qualify, why not "*O* is beautiful"?

What follows the provisos for reliable test procedures is a panegyric.

> The same HOMER who pleased at ATHENS and ROME two thousand years ago, is still admired at PARIS and at LONDON. All the changes of climate, government, religions and language, have not been able to obscure his glory. Authority or prejudice may give a temporary vogue to a bad poet or orator. . . . On the contrary, a real genius, the longer his works endure, and the more wide they are spread, the more sincere is the admiration which he meets with. Envy and jealousy have too much place in a narrow circle; and even familiar acquaintance with his person

may diminish the applause due to his performances: But when these obstructions are removed, the beauties, which are naturally fitted to excite agreeable sentiments, immediately display their energy and while the world endures, they maintain their authority over the minds of men.

Hume takes seriously the notion that time will tell, that genuine poetic excellence is notably durable. In part this may be a reflection of Hume's own taste, which was old-fashioned and somewhat conventional. (It was also limited: Hume actively disliked music, and had little interest in the visual arts: literature, which for him included history and philosophy, was the major art form and his preferences were for classical authors.)

His celebration of Homer does suggest an idea that is consistent with his own theory and that would get him off the hook on the topic of laws and rules of taste. It goes as follows: the beauties of Homer, those qualities that are "naturally fitted to excite agreeable sentiments," are not qualities that the *Iliad* has in common with the *Aeneid;* they are qualities that are distinctive to the *Iliad*—to the original and perhaps to every adequate translation. This would mean that if we praise both the *Iliad* and the *Aeneid* for, for example, "clearness of expression," that term will have to be flagged as indexical. A reader could still attribute his pleasure in the *Iliad* to its "force and clearness of expression" without committing himself at all to admiring the *Aeneid* and might well insist that (in its own way) the *Aeneid* has "force and clearness of expression."

Throughout Hume's essay, the reader is left wondering about "the standard"? What is it that it is "natural for us to seek"? What are we looking for? At times it seems to be a criterion and one that has explicit conditions of application. In the *Euthyphro,* Socrates pretends to be seeking instruction and asks Euthyphro to tell him what feature it is that makes every pious act pious—just giving examples will not do. It sounds as if what he is asking for is a definition. What Socrates says is that, for his own purposes, he needs a "standard"—which is a plausible translation of the Greek *paradigma.* But a paradigm in Greek, as in English, need not be a rule: it can be a clear instance, a central case. Now, Hume may be thinking that it is the great works that set our standards. If asked to evaluate someone's epic poem, and one takes the request seriously, one reads the poem in the light of the *Iliad*—and similarly for novels, string quartets, and other art forms that Hume does not consider. That would not be a bad idea.

What about the appeal to posterity? There certainly is something impressive about real perennial favorites. Highbrow critics are apt to dismiss popular art as junk, which it mostly is, but it is *ephemeral* junk. Hume thinks of Homer; we would probably think of Shakespeare, who has always been popular. (Hume, incidentally, regards Shakespeare as a genius but lacking in taste—no doubt a further instance of

the French influence.) The question—one that could not have failed to occur to Hume—is, How we are to be guided in our assessment of our contemporary poets? What are we supposed to be doing while waiting for the verdict of posterity? Hume's answer—and the final major theme of the essay is that we should be guided by the opinion of those critics we have learned to trust—what Hume calls the "true judges."

Hume lists the requirements for the good critic and they turn out to be a good deal more complicated than his earlier suggestion about "perfect serenity of mind" and so forth. (This is the section in which Hume follows most closely the pattern set by Du Bos.) The first is "delicacy of imagination," a term used (rather oddly) to mean the capacity to make fine analytic discriminations. A good critic not only can say what makes for excellence in a poem as a whole but can explain in detail what each line contributes to the overall effect. By way of illustration, Hume cites from *Don Quixote* the story of Sancho Panza's kinsmen; they were laughed at for claiming to detect the taste of iron and of leather in wine submitted for their approval but were vindicated when a key with a leather thong was discovered at the bottom of the barrel. Hume finds a parallel with the good critic who can distinguish qualities of the work that are inconspicuous and escape the attention of others. (The paragraphs in which Hume draws what he sees as the moral of this tale are thoroughly confused. It is a good exercise to try to see where and why he went wrong.) The second thing a critic needs is practice. First impressions are apt to be vague and sentiments uncertain, but if one works hard, "the mist dissipates." Competence is no more easily come by for a critic than for a creative artist: "the same address and dexterity which practice gives to the execution of any work is also acquired by the same means, in the judging of it."

Practice develops skill at making comparisons, the third virtue of a critic and an important one, because it is only by comparison that one is able to determine the correct degree of praise or blame that a poem merits. (This harks back to the opening section where Milton's "elegance and genius" are highlighted when his work is contrasted with Ogilby's.) But, for comparisons to be illuminating (and this is the fourth virtue), the critic must free his mind of prejudice. If I am asked to judge the work of "a different age and nation," or if I have some personal feelings for the author in question, "I must depart from this situation; and considering myself as a man in general, forget, if possible, my individual being and my peculiar circumstances."

Finally, the critic must have good sense, which means that he must be able to take account of an author's intentions and to appreciate in detail the steps the author takes to achieve them. Moreover, the critic must have a developed feeling for logical and dramatic coherence: "every kind of composition, even the most poetical, is nothing but a chain of propositions and reasonings; not always, indeed, the justest and most exact, but still plausible and specious, however disguised by the colouring of the imagination."

Hume's summary: "Strong sense, united to delicate sentiment, improved by practice, perfected by comparison, and cleared of all prejudice, can alone entitle critics to this valuable character; and the joint verdict of such, wherever they are to be found is the true standard of taste and beauty."

Is this the final answer to the question raised at the outset of the essay? Does the "joint verdict" give us a rule for deciding among "the various sentiments of men" which to confirm and which to condemn? Surely not: whether or not a poem is any good is not something to be decided by appeal to the experts, if for no other reason than that such an appeal would mean that the only valid judgments of taste are secondhand—which is absurd. The only available recourse is to become oneself a "true judge," and that, as Hume correctly observes, may be a lifetime's vocation.

On the other hand, the role of authority in critical as in moral debates is extremely murky. If one cannot in good faith assert, "This has *got* to be right (beautiful) because *he* says it is," one is nonetheless swayed by the opinions of those whose judgment one has come to trust. Here there are many issues that need to be explored.

Two points, as if afterthoughts, are raised at the conclusion of the essay. The first is that, although (as Hume sees it) "the general principles of taste are uniform in human nature," there are preferences that have to do with personal idiosyncrasies and should simply be accepted; so there *is* one limited field in which it is useless to dispute about tastes. The second point raises weightier issues and certainly deserves more attention than Hume is willing to allot. It is that, despite the need to free ourselves from prejudice in considering the work of another age or nation, we cannot and should not suspend our moral conviction. It would be silly to take exception to characters from the historical past because of their quaint costumes.

> But where the ideas of morality and decency alter from one age to another, and where vicious manners are described without being marked with the proper characters of blame and disapprobation, this must be allowed to disfigure the poem, and to be a real deformity. I cannot, nor is it proper I should, enter into such sentiments; and however I may excuse the poet on account of the manners of his age, I can never can relish the composition.

Hume's claim is reminiscent of Plato at his least enlightened. The only poets or playwrights allowed in the Republic are those who are careful to depict the good guys in glowing colors and the villains as despicable. It is hard to associate such a view with Hume's own temperamental and moral commitments; the truth is that he does not say enough in the essay to give a clear picture of what he really has in mind.

The issues raised, if not settled, in "Of the Standard of Taste" are central to aesthetics. The essay is a remarkable performance and deserves careful study.

[*See also* Taste, *article on* Modern and Recent History.]

BIBLIOGRAPHY

Work by Hume

"Of the Standard of Taste." In *Essays Moral, Political, and Literary,* rev. ed., edited by Eugene F. Miller, pp. 226–249. Indianapolis, 1987.

Other Sources

Carroll, Noël. "Hume's Standard of Taste." *Journal of Aesthetics and Art Criticism* 43.2 (Winter 1984): 181–194.

Cohen, Ted. "Partial Enchantments of the Quixote Story in Hume's Essay on Taste." In *Institutions of Art: Reconsiderations of George Dickie's Philosophy,* edited by Robert J. Yanal. University Park, Pa., 1994.

Dickie, George. "Taste and Attitude: The Origin of the Aesthetic." *Theoria* 39.1–3(1973): 153–170.

Guyer, Paul. "The Standard of Taste and the Most Ardent Desire of Society." In *Pursuits of Reason: Essays in Honor of Stanley Cavell,* edited by Ted Cohen, Paul Guyer, and Hilary Putnam. Lubbock, Tex., 1993.

Jones, Peter. "Hume's Aesthetics Reassessed." *Philosophical Quarterly* 26.102 (January 1976): 48–62.

Jones, Peter. *Hume's Sentiments: Their Ciceronian and French Context.* Edinburgh, 1982.

Kivy, Peter. "Hume's Standard of Taste: Breaking the Circle." *British Journal of Aesthetics* 7.1 (January 1967): 57–66.

Korsmeyer, Carolyn W. "Hume and the Foundations of Taste." *Journal of Aesthetics and Art Criticism* 35.2 (Winter 1976): 201–215.

MacAchlan, Christopher. "Hume and the Standard of Taste." *Hume Studies* 12 (1986): 18–38.

Mooij, J. J. A. "Hume on Is-Ought and the Standard of Taste." *Journal of Value Inquiry* 14.3–4 (Fall–Winter 1980): 319–332.

Mothersill, Mary. *Beauty Restored.* Oxford, 1984. See chap. 7.

Mothersill, Mary. "Hume and the Paradox of Taste." In *Aesthetics: A Critical Anthology,* 2d ed., edited by George Dickie, Richard Sclafani, and Ronald Roblin, pp. 269–286. New York, 1989.

Potkay, Adam. *The Fate of Eloquence in the Age of Hume.* Ithaca, N.Y., 1994.

Rose, Mary Carman. "The Importance of Hume in the History of Western Aesthetics." *British Journal of Aesthetics* 16.3 (Summer 1976): 218–229.

Savile, Anthony. "Progress and Common Sense: Two Approaches to a Problem in Criticism." *British Journal of Aesthetics* 17.4 (Autumn 1977): 305–319.

Wieand, Jeffrey. "Hume's Two Standards of Taste." *Philosophical Quarterly* 34.135 (April 1984): 129–142.

MARY MOTHERSILL

HUSSERL, EDMUND GUSTAV ALBRECHT

(1859–1938), German philosopher and founder of phenomenology, a movement in twentieth-century thought that attempts to deal with traditional issues in philosophy through analysis of human consciousness. Husserl came to philosophy as a trained mathematician and psychologist, convinced of the inadequacy of "psychologism," his contemporaries' endeavor to account for mathematical knowledge solely in terms of empirical descriptions of mental states. Husserl's first major work, *Logical Investigations* (1900–1901), aims at providing a description of the acts of consciousness involved in logical understanding and knowing, without reducing the validity of the content of logic to the empirical status of those acts. The success of this work gained Husserl a position at the University of Göttingen, but a penchant for self-criticism soon produced a fundamental change in his thinking. Although he was convinced that the analysis of consciousness in the *Logical Investigations* was not psychologistic, he became dissatisfied with its overemphasis on acts of consciousness and its consequent failure to investigate the "senses" and "essences" of things—that make the relation between consciousness and the world possible at all—as well as how those senses and essences can be availed. Taking up this sort of investigation into his analysis of consciousness and appropriating a traditional label for it, Husserl develops a "transcendental phenomenology" in his second major work, *Ideas Pertaining to a Pure Phenomenology and to a Phenomenological Philosophy* (1913). Roughly paralleling Husserl's move to the University of Freiburg in 1916, where he would retire in 1929, his conception of phenomenology takes a further turn as he probes limitations of his previously published analyses: notably, their privileging of acts of judgment and their neglect of the temporal nature and passive and intersubjective aspects of consciousness. In the last two decades of his life, Husserl accordingly develops a "genetic phenomenology," concerned with the way the senses of things come to be constituted—ultimately, in the consciousness of a "life-world"—prior to explicit acts of perception and judgment. One fruit of these labors, most of which remained unpublished during Husserl's lifetime, is his final major work, *Crisis of European Sciences and Transcendental Phenomenology* (1936).

Although traditional issues in aesthetics and the philosophy of art appear only on the periphery of Husserl's published work, he was no stranger to them, nor did he fail to recognize their import for his project of establishing the nature as well as the necessity of phenomenology as the study of pure consciousness. In his unpublished studies—especially his investigations of pictorial consciousness and fantasy—insightful treatments of such issues recur with a telling regularity. In addition, he assigns at least some of those issues a particular place within a projected system of philosophical sciences. Because Husserl's philosophy is bent on determining the senses and essences of things, as something disclosable only through an analysis of consciousness and the ways in which things are originally given to consciousness, the relevance of his philosophy to aesthetics (and vice versa) comes as no surprise. Indeed, Husserl's initial and innovative attempts to demonstrate the objectivity of logical form—without reducing the latter to either the psychological processes of the logician (psychologism) or

the relations among the logical signs (logicism)—provided early students of his such as Waldemar Conrad, Moritz Geiger, Roman Ingarden, and Fritz Kaufmann (all of whom eventually made names for themselves in aesthetics), later students such as Martin Heidegger and Oskar Becker, sympathetic contemporaries such as Wilhelm Dilthey and Max Scheler, and prominent philosophers of art in the twentieth century such as Maurice Merleau-Ponty, Hans-Georg Gadamer, Mikel Dufrenne, and Wolfgang Iser with a valuable, even indispensable, model for the analysis of the objectivity claimed for aesthetic consciousness.

Clarification of Husserl's thought on aesthetic issues is complicated by the lack of a self-contained treatment of such issues in his published works and by the fact that the views expressed on aesthetic matters in his unpublished writings change. Any attempt to sort out Husserl's conception of aesthetics must also contend with a distinct, though ultimately not unrelated, use of the term in what he calls "transcendental aesthetics" as well as with the fact that the issue of aesthetic value forms a part of the axiology (the "doctrine of pure value") projected by him as a separate undertaking in the division of philosophical disciplines. The following exposition of Husserl's views on aesthetics accordingly begins with his notions of "transcendental aesthetics" and "axiology" before turning to his specific conceptions of an aesthetic attitude and art. In conclusion, Husserl's influence will be indicated by briefly tracing the early development of a "phenomenological aesthetics" by some of his contemporaries and students.

Transcendental Aesthetics and Axiology. In the early 1920s, as Husserl begins to work out the details of "genetic phenomenology" (described earlier), he speaks of a "transcendental aesthetic," the aim of which is to unpack the essential structures of the passively undergone experiences that underlie our more active experiences, such as judging and inferring (the stuff of logic). A transcendental aesthetic, as Husserl originally conceives it, thus focuses on perception and intuition exclusively, leaving out all knowledge in the form of judgments, "the entire sphere of determinative and predicative thinking" (Husserl, 1966, p. 295). Although Husserl is quite deliberately harking back to Kantian terminology, a transcendental aesthetic is not for him, as it was for Kant, a mere complement to transcendental logic, but rather the foundation for the latter. Thus, perception and intuition, as he puts it here, make up "the first basic shapes of consciousness that come into question for the construction of the specifically logical consciousness" (ibid., p. 319n). At the end of the decade, he regards a transcendental aesthetic as "the extraordinarily huge complex of various kinds of research in regard to the primordial world" (Husserl, 1960, p. 173). The aim of such an investigation is to give an essential description of "the world of pure experience," in advance of all intersubjectivity and science, without which "objects could not appear in a unified way" and the very unity of the

world could not constitute itself as a "passive synthetic unity," that is to say, a unity that takes shape before we make judgments—"active synthetic unities"—in regard to it (cf. Husserl, 1969, p. 297). Shortly thereafter, however, the notion of a "transcendental aesthetic" no longer excludes intersubjectivity, but instead encompasses the synthesis of agreement *(Einstimmigkeit)* among transcendental subjects constituting the same world (cf. Husserl, 1973, pp. 234ff.). "In transcendental intersubjectivity," as Husserl puts it, "I find the real world to be 'aesthetic'" and it is in this context that artworks explicitly become a subject of a transcendental aesthetic (cf. Husserl, 1989, p. 157).

But not only artworks. In regard to this intersubjectively aesthetic world, Husserl singles out first an "aesthetic morphology" or "aesthetic of nature" as a descriptive science, the task of which is to identify and describe the natural types of things and properties, forms and causal connections, as they present themselves and are constituted strictly within "the realm of intersubjective aesthetic normality," presupposed by mathematics and mathematical natural science (ibid., p. 159). There is also a corresponding morphology of the psychological domain, that is to say, the "soulful aesthetic" *(seelisch Ästhetische)*, which includes, Husserl adds, "works of art" (ibid., pp. 160ff.). What distinguishes these two aesthetics is their accessibility. The aesthetic of nature admits of a distinction between aesthetically primary and aesthetically secondary qualities, where the former alone are "absolutely accessible to each person by means of perception, whether he be normal or not" (ibid., p. 161). Although there is not the same distinction of qualities in the sphere of the "psychologically aesthetic," an analogous distinction must be made between the types of psychological expression revealed in the primary experience, made by every ego, of its specific bodiliness and historically as well as geographically situated types of self-understanding. "Aesthetic objectivity" requires the former as something able to be experienced by everyone. Yet, it is the latter self-understanding with its expanded and stratified conception of human experience that makes it possible to understand the objective accomplishments of human beings, from the most universal types of those accomplishments—among which Husserl singles out "the category artwork"—to more differentiated (national, regional, even individual) types. Husserl mentions "German art" as an example of the latter type. He adds, however, that all such types, in contrast to the types of expression disclosed in the primary experience of bodiliness, are not unqualifiably accessible to everyone. Only those born and raised in a particular historical environment and who have made it their own can "concretely experience" and "actually understand" the practices and products of that environment:

The Chinese who comes to Europe and becomes acquainted with our music, literature, and so forth does not become ac-

quainted with them in their actual sense. He must first build up a European within himself . . . must first constitute a European ego in himself. . . . Only by the detour of such a self-transformation that is only imperfectly realizable . . . is actual objectivity thinkable. (Ibid., p. 162)

Even then, Husserl adds, the objectivity remains an idea and, moreover, an idea fundamentally different from the idea of nature.

For Husserl's views on traditional aesthetics and philosophy of art, there are two striking features to this final development of his notion of a transcendental aesthetic. In the first place, it underscores a constant theme in his thinking about aesthetic experience proper, namely, that the experience, whether directed at nature or art, is based on some foregoing discernment of an object, a discernment that provides a necessary, albeit insufficient, foundation for the objectivity of the aesthetic experience. In the second place, it demonstrates his cognizance of the bounds that enable some and disable others in matters of aesthetic objectivity in regard to a specific genre or work. Those bounds are determined by the fact that understanding of a specific genre or work requires, in his view, some level of identification—historical, geographic, or individual—with the artistic accomplishment.

As far as Husserl's own systematic pretensions are concerned, he considered aesthetic values (together with ethical and other sorts of values) the object of axiology, though in his opinion the constitution of axiological disciplines remained in his day a "theoretical desideratum." Nevertheless, axiology is construed by him as a separate, complementary discipline alongside theoretical and practical philosophy (in keeping with the neo-Kantian division of philosophy). At the same time, axiology, like practical philosophy, is a "material discipline," ranging over a particular region of objects (namely, values) and thus subordinate to the formal ontology of objectivity in general, the provenance of theoretical philosophy (cf. Husserl, *Ideas*, 1980a, sections 147ff.). This subordination is the systematic counterpart to the aforementioned theme that aesthetic experience is founded on a more basic level of experience.

That theme is echoed in Husserl's account of the way in which an evaluation or appreciation of something is based on the perception of it. In *Ideas*, this relation between appreciation and perception is invoked by Husserl to introduce the crucial distinction between an intentional object and a grasped object. Although an object must be grasped in some measure in order to be evaluated or appreciated, the thing insofar as it is merely grasped is not "the full, intentional correlate" of the act of evaluation or appreciation (ibid., section 37). The relation sketched here between an aesthetic act of appreciation and an act of perception bears an obvious kinship to the way in which Husserl, in his phenomenological analyses of knowing, relates perceptions of individual objects to higher-level perceptions of essences

(*Wesen*, understood as individual objects' ways of being) or states of affairs (*Sach-verhalte*, that is to say, whole-part or part-part relations within an individual object or between it and other objects). Those essences and states of affairs are not directly apparent in the straightforward perception of an individual object, but they also are not created or constructed by the act of perceiving them. Instead, they are themselves perceived, that is to say, they present themselves and are constituted in the act of perceiving them (and, typically, as a consequence of imagining an arbitrary assortment of instances of them). So, too, an object of appreciation such as an artwork, both as an artwork and as an embodiment of values, is founded on an objectivity and corresponding perception—without being reduced to either. In the posthumously edited *Experience and Judgment* (cf. sections 12, 65), Husserl further exploits this grounded character of artworks in an attempt to clarify a "precise concept of reality." Artworks and geometrical essences, he maintains, are alike "irreal," that is, they are based on specifically real things as far as their spatiotemporal appearances are concerned and yet they are able to appear in different realities, not merely as similar, but as identical.

The Aesthetic Attitude and Art. The direct counterpart to the theme of the preceding paragraphs—the distinctive, yet nonetheless grounded character of acts of appreciation and their intentional correlates—can be found in Husserl's occasional treatments of the constitution of an aesthetic object. His most sustained treatment of this topic is in a collection of texts, encompassing a mere eighteen pages written between 1906 and 1918, titled "Aesthetics and Phenomenology" (classified in the Husserl Archives as *Sammelkonvolut A VI 1;* cf. Gabriele Scaramuzza and Karl Schuhmann, 1990). In "Aesthetic Objectivity," a text from this collection, Husserl distinguishes between a "grounding (theoretical) objectivity" and the constitution of an "axiologically aesthetic objectivity." The former is composed of certain primary contents as well as their intuitive and symbolic meanings. Although founded on those contents and meanings, the "axiologically aesthetic objectivity" is essentially constituted by something further, namely, a manner in which the object presents itself, its way of appearing—even if, Husserl adds, the appearance is only imaginary or the result of symbolic representation (as in the case of literature). An array of moods and thoughts radiates from this appearance, constituting a complex that underpins aesthetic feelings in which, he maintains, the essential character of the aesthetic objectivity emerges. Much like Theodor Conrad and Roman Ingarden, each of whose aesthetic theories bears the stamp of Husserl's earliest phenomenological investigations, Husserl appears to differentiate two senses of this essential character, namely, as an artwork and as the embodiment of values.

The aesthetic attitude, as Husserl construes it, has, he acknowledges, a close (and potentially quite misleading) affin-

ity with the phenomenological attitude that he makes a requisite for philosophy. The phenomenological attitude is achieved by canceling every doxic modality (for example, belief, supposition, negation) in order to reflect directly on the respective consciousness and what is given to it. Husserl calls this modification a "neutralization" of consciousness but warns that it is to be distinguished from the sort of neutralization that can take place in some acts of imagination, fantasy, or aesthetic contemplation. In the aesthetic contemplation of a picture, for example, a normal perception and discrimination of the figures formed must be presupposed and yet neutralized in order to focus, not on the figures as objects, but on "the realities displayed or, more precisely, portrayed in the picture" by the figures. Indeed, any object—and not just a picture—can be contemplated in a purely aesthetic manner insofar as it is taken as a "sheer image or picture" without regard for its existence or nonexistence, possibility or plausibility, and so on (cf. Husserl, *Ideas*, section 111). On the other hand, as long as the objective context of things and not their appearance matters to us, our attitude is not aesthetic. "Aesthetic evaluation is essentially connected with the distinction between consciousness of an object in general and the *manner of appearance* of the object" (cf. Husserl, 1980b, p. 388).

Husserl exploits a corresponding distinction between the picture and its subject in some early reflections on pictorial art. That distinction, he avers, is "the essential foundation of the possibility of aesthetic feeling," though clearly not in the sense that the picture is used as a mere external sign, immediately diverting attention away from itself and to the subject (ibid., p. 41). In the aesthetic contemplation of a picture, attention is precisely paid to the "picture-object" *(Bild-objet)*, that is, to the immanent pictorial character of the picture, rather than to the "subject matter of the picture" as is normally the case in a nonaesthetic perception of a picture. Thus, for example, a picture in a newspaper often has the function of awakening a new intention, an act of memory or imagination, directed at the subject (and not through a picture of the subject matter). In an aesthetic contemplation of a picture, by contrast, one sees the subject *in the picture*. To be sure, new intentions are likely to be awakened as fantasy is set in motion, but the aesthetic interest, Husserl insists, is principally directed at the picture-object and "clings internally to it, finding the enjoyment in the *manner* of its picturing" (ibid., p. 37). It is, of course, possible, he adds, for an individual to enjoy even his or her own fantasies aesthetically. "But then, in the contemplation of that picture, we do not merely look to the subject but are interested in *how* the subject displays itself there, what pictorial manner of appearance it shows, and perhaps how aesthetically pleasing the appearance is" (ibid.). Indeed, in some cases of pictorial art (as an example, Husserl cites a Titian painting "representing heavenly and earthly love to me"), the conception of the work as a portrait or picture of

a likeness, an inauthentic or indirect representation of something, is completely inappropriate (ibid., pp. 154ff.).

In two crucial and interrelated—indeed, intentional—respects, Husserl's depiction of aesthetic contemplation and its "object" is reminiscent of the Kantian aesthetics so dominant among his contemporaries. According to Immanuel Kant himself, what matters in an aesthetic judgment of beauty is not the existence of an object but the sheer form (spatial and/or temporal structure) of its appearance, the ascertainment of which is the work of imagination (cf. Kant, *Critique of Judgment*, sections 2, 11, and Husserl's reference to Kant's doctrine in Husserl, 1980, p. 145 n. 1). The same can be said for Husserl, at least for some of his early accounts of aesthetic experience and art (though even Husserl's early doctrine of imagination, it bears noting, differs markedly from that of his predecessor). Not incidentally, Husserl characterizes art as "the realm of shaped fantasy" *(das Reich der gestalteter Phantasie)* (ibid., p. 514). Furthermore, in Husserl's, as in Kant's, aesthetic theory, one finds an effort to distinguish aesthetic pleasure from intellectual pleasures or the pleasures of various accomplishments (including those of the imagination) and a refusal to restrict the aesthetic attitude and its analysis to the experience of art. "If I am in the aesthetic attitude," Husserl observes, "I do not abandon it if I pass over into consciousness of the actuality of nature." Aesthetic consciousness can still have thematic preference, in which case "the theme of my consciousness is not the actuality [of a thing], but the beauty of its manner of appearance" (ibid., p. 392).

Nevertheless, there remain substantial differences between Husserl's and Kant's accounts of the aesthetic experience, differences that become more pronounced as Husserl increasingly refines his aesthetic views. The difference is evident, in the first place, in the elaboration of the subjective side of that experience. Far from characterizing the aesthetic attitude as disinterested, Husserl attributes to it an interest in manners of appearance and, indeed, not just any interest (because there can be a theoretical interest in appearances) but an interest in appearance solely for the sake of enjoying that appearance. Moreover, according to Husserl, aesthetic feelings are, as noted earlier, the very medium in which aesthetic objectivity emerges. In order to experience a drama aesthetically, a certain emotive stance toward it is imperative. It is not sufficient merely to entertain the characters and the story in fantasy; it is also necessary, Husserl insists, to have certain feelings awakened by the reflection on the depicted characters (or, in general, on the manner of presentation or appearance of an object).

The difference between Husserl's and Kantian aesthetics can also be found in the former's account of the role played in the aesthetic experience by the object and the object's content. He insists that the object of the aesthetic experience must be chosen or crafted in such a way that it or its functions not only do not conflict with its form or the

"characteristic placement" of its parts, but above all contribute to the expressiveness of the appearance. "Aesthetic appearances are exclusively appearances that just express and exhibit something, and not in the manner of an empty sign." Moreover, it is only in the wake of this expressiveness, according to Husserl, that an aesthetic differentiation of the beautiful, the less beautiful, and even the ugly can be considered. In this way, by reference to expressiveness and not mere form, Husserl makes a plea for coherence and unity in the constitution of "a beautiful aesthetic object" and, at the same time, clearly countenances the possibility of ugliness in the form—a possibility that presents Kantian aesthetics with a distinct difficulty (ibid., pp. 145ff.). The appearance of every object has its "aesthetic coloring," as Husserl later puts it (ibid., p. 389). Herein lies a difference in the two theories of art as well. Unlike Kant, Husserl does not limit fine art to the beautiful. "All art is 'aesthetic,' it is joy at what is seen *in concreto*. But not all art is callistic," that is to say, occupied with beautiful artifacts (ibid., p. 542).

Further differences with Kantian aesthetics are evident in Husserl's observation that "the content of the object itself is not aesthetically meaningless" (ibid., p. 390). This last remark is meant, in the first place, to call attention to the fact that aesthetic enjoyment can be "enhanced," as Husserl puts it, by being combined with nonaesthetic enjoyment—in other words, with the pleasure we might actually take in the object itself. (The implications of this possibility of enhancement for Husserl's conception of art are discussed later in this essay.) At the same time, the remark is intended to introduce an expanded sense of "manners of appearance," encompassing all the ways that we can become conscious of objects insofar as they form the basis of distinctive feelings and we adopt a corresponding stance toward them. It bears noting that when Husserl makes these observations about aesthetics (around 1912), he is struggling to develop an accurate conception of fantasy or imagination. Originally, he conceived imagination as a kind of mental picturing, a conception motivated in part by the fact that the German term for imagination *(Einbildungskraft),* like the English, contains within it the word for image or picture *(Bild).* Eventually, however, aided by further reflection on what is involved in enjoying works of art, he saw the need to distinguish between imagining in the sense of picturing something and in the sense of presenting something to ourselves in its absence without picturing it. When we speak of imagining (or remembering) something, we are often using the term in the latter sense; that is to say, we imagine (or remember) the object without first forming a mental picture of it. Husserl's expanded conception of the aesthetic realm corresponds to this new conception of imagination. Eventually, however, he dispenses altogether with the appeal to fantasy in defining the aesthetic attitude. In the early 1920s, he observes: "What is essential for the aesthetic attitude . . . is not fantasy, but the attitude toward what is aesthetically interesting, the objectivity in the way [it presents itself]" (ibid., p. 591).

As this last remark suggests, the enjoyment of any of several ways that things can be presented to us apparently displaces the enjoyment of appearances and fantasy—or at least pictorial fantasy—as the center of gravity for Husserl's conception of aesthetic experience and art. Criticizing his earlier views, he observes:

> It cannot be said that art must move in the sphere of observability. I used to think that it was part of the essence of art to display things in a picture and I used to understand this displaying in the sense of picturing a likeness *[Abbilden].* But on a closer look, this is not correct. (Ibid., p. 514)

For, as is more obviously the case in the experience of some forms of art (e.g., drama and fiction) than others (e.g., painting), the aesthetic attitude of neutralization is often carried out, as Husserl puts it, "from the very outset" of the experience (ibid., p. 516).

Like any object of aesthetic experience, a work of art is something to be appreciated and hence its constitution presupposes an individual's experience of it (and, indeed, a certain level of pre-aesthetic experience of it). In addition, a work of art has a value as someone's creation. Applying to art the two senses of imagination elaborated by him, Husserl specifies the two extremes within which the artist moves: pictorial art (typified by some painting and literature, for example, "Schnitzler's Vienna" or "Fontane's Berlin") and the art of pure fantasy (typified by music and literature that suffuse the world with value). As an apparent complement to this distinction, he distinguishes between realistic art and idealistic composition. In the former, beauty plays at best a minor role and, although the depictions in this kind of art can be quite wonderful, the joy comes from observing the concrete world with understanding, in a manner very much akin to theory. In the context of this form of art, furthermore, it is the poem and not the poet that is to be understood. "Realistic art is a type of biography of the age, of strata of the age. It depicts by means of characteristic 'images.' It constructs fictions in which characteristic types of the age present themselves. It is art, not science and yet still communicating knowledge" (ibid., p. 541). The idealistic writer has, on the other hand, a normative attitude. "And he doesn't merely present values and the battle of good and evil, he wants to enkindle in our souls with a love for the good, without moralizing or preaching. He transfigures that love in the medium of beauty" (ibid., p. 542). At an even higher level, Husserl adds, art can be philosophical, elevating people to the divine and uniting them with it.

Apart from the individual appreciation of a work of art and its value as someone's creation, Husserl also attributes an "objectivity" and "intersubjective existence" to a work of art and its beauty. This objectivity consists in its iterability in repeated acts of seeing, speaking, reading, and so on. "In

fantasy I identify, I posit the quasi form *[quasi-Gebilde]* as the same. But the form is actually the same and the value of its beauty is the same" (ibid., p. 543). Thus, a poem, Husserl continues, is "an individual 'objective' idea" that can be realized more or less perfectly in an endless variety of ways.

Along with several fundamentally aesthetic notions (e.g., "myth," "heroism," "naïveté"), a concept of art plays a central role in Husserl's maturest meditations in *Crisis of European Sciences and Transcendental Phenomenology*. The crisis, which in Husserl's eyes is by no means confined to science, is traced by him in large measure to the confusion of art in a broad sense *(technē)*—"the Galilean art that is called physics" (1970, p. 51)—with science. In the nineteenth century in particular, Husserl observes, there was a proliferation of "theoretical arts," developed from positive sciences, that he ranks between fine arts (also architecture) and lower-level arts (ibid., p. 197). The way out of the crisis is through the reassertion of Europe's Greek birthright: the subjection of all traditional (i.e., local) norms—among which Husserl mentions the norm "beauty"—to the idea of a universal truth and the wholesale replacement of a "mythical-practical" attitude with the theoretical attitude of philosophy (ibid., pp. 328–334). In the process, it is the philosopher's task, according to Husserl, to hold in abeyance all traditional prejudices in order, in strictly theoretical and scientific fashion, to pose—along with the questions of being, goodness, value, and utility—the questions of what it means "to be beautiful" and "to be art" (ibid., pp. 155, 180).

While working on the *Crisis* in the summer of 1935, Husserl further elaborates the difference between art and science by emphasizing the greater finality of the former. Each type of cultural formation, such as art and science, has its own distinctive historical character. Given its aim of "objective truth," however, a work of science is not even relatively complete until it is referred to "the universal horizon of other scientists as actual and possible co-workers" whose task is to test the work and construe it as material on which to build. By contrast, Husserl maintains, a work of art is "something complete in itself within its category"; "it is not and should not be material for the construction of new works of art. Each begins for itself and ends for itself" (ibid., p. 506). Although Husserl maintains that "the original use of an artwork" is to edify or uplift a person as a human being through enjoyment of the work, he apparently attributes the distinctiveness of art in part to its inherently parochial character. Even when the oneness and spirit of a people are expressed in an artwork, it is done "according to their own sense of being" (ibid.). Thus, for Husserl, an artwork is more complete and self-contained than a work of science, presumably because the former is more localized and, within that locale, addressed to a wider group. In this connection, Husserl rejects the position of those artists who, under the banner of *l'art pour l'art*, create works of art only for one another. After emphasizing the "intersubjective existence" of artworks as a "common good," Husserl asserts that "an art that intends to be art for artists, is an anomalous art" (ibid., p. 507).

The Influence of Husserlian Phenomenology on Aesthetics. Husserl's influence on the course of aesthetics in the early twentieth century must be traced predominantly to others' adaptation of his methods of analyzing consciousness to the investigation of aesthetic experience. Perhaps the first instance of this application is to be found in Waldemar Conrad's 1908–1909 essay "Der ästhetische Gegenstand" (The Aesthetic Object). Conrad attempts to construe aesthetic objects, namely, works of art, as ideal objects in deliberate imitation of Husserl's account of ideal objects in *Logical Investigations*. Geiger published his "Beiträge zur Phänomenologie des ästhetischen Genusses" (Contributions to the Theme of Aesthetic Enjoyment) in the very first volume of Husserl's journal, *Jahrbuch für Philosophie und phänomenologische Forschung* (Yearbook for Philosophy and Phenomenological Research) in 1913. According to Geiger, as he puts it in his 1925 essay "Phänomenologische Ästhetik" (Phenomenological Aesthetics), what distinguishes phenomenological aesthetics is the fact that (1) it remains steadfastly with the phenomena, (2) it strives to understand these phenomena, not individually and contingently, but essentially, and (3) it grasps that essence neither through induction nor through deduction but through the intuition of the essence in an individual example. The emphasis on this third feature was not universally accepted; Werner Ziegenfuss, for example, eschews the notion of eidetic intuition in his own draft of a "phenomenological aesthetics" (*Die phänomenologische Ästhetik*, 1927).

Ingarden and Kaufmann, on the other hand, continued to develop the more orthodox Husserlian conception of aesthetics initiated by Geiger. In "Phenomenological Aesthetics: An Attempt at Defining Its Range" (Ingarden, 1973), after criticizing traditional tendencies to collapse aesthetics into its objective or subjective pole, Ingarden asserts that he rejected the method of empirical generalization in favor of an eidetic analysis of the idea of a literary work of art or a work of art in general. His method in aesthetics, moreover, takes its bearing from the Husserlian understanding of the artwork as a distinctive intentional object. "This method," as he puts it, "aims above all at bringing its objects of enquiry to an immediate givenness in a suitably shaped experience and to a faithful description of the data of that experience" (1973). Similarly, in "Art and Phenomenology" (1940), Kaufmann describes the import of Husserlian phenomenology for aesthetics as follows: "The spirit of free acknowledgement of the qualitative richness of true experience on the one side and the insistence on the qualitative irreducibility of its essential character on the other proved to be congenial to art."

Husserl's influence was also evident in musical theory, where, by the early 1930s, a "musical phenomenology" had emerged, aimed at determining the essential features of sounds. In keeping with Husserl's approach, it was presumed, on the one hand, that these essential features were in principle independent of the existence of the sensations of any particular subject. Investigating those features is, on the other hand, something completely different from explaining their causes and effects acoustically, physiologically, and psychologically (cf. Schäfke, 1934, p. 410).

This application of Husserlian phenomenology to music exemplifies the irreducible yet grounded character of the aesthetic object that is the hallmark of Husserl's thinking on aesthetics and art. Aesthetics and art encompass activities and objects of the mind that are based on, but cannot be reduced to, a host of acts of a different order (perceiving, remembering, imagining, signifying, knowing, etc.). At the same time, it is a centerpiece of Husserlian phenomenology that any act of the mind can be neutralized so that the sense of what presents itself in that act can become the subject of a preeminently theoretical reflection, with a view toward determining what is essential about it. So, too, for all the distinctiveness of the aesthetic attitude and its objects, the determination of their sense and essence lies within the purview of the science of pure consciousness that Husserl calls "phenomenology."

[*See also* Phenomenology.]

BIBLIOGRAPHY

Works by Husserl

Logische Untersuchungen (1900–1901). 2 vols. 2d rev. ed. Halle, Germany, 1913; reprint, Tübingen, 1968. Translated as *Logical Investigations* by J. N. Findlay (London, 1970).

Ideen zu einer reinen Phänomenologie und phänomenologischen Philosophie (1913). 4th ed. Tübingen, 1980. Translated as *Ideas Pertaining to a Pure Phenomenology and to a Phenomenological Philosophy* by F. Kersten (The Hague, 1980a).

Formale und transzendentale Logik (1929). Edited by Paul Janssen. The Hague, 1974. Translated as *Formal and Transcendental Logic* by Dorion Cairns (The Hague, 1969).

Cartesianische Meditationen (1931). Edited by Stephen Strasser. The Hague, 1950. Translated as *Cartesian Meditations* by Dorion Cairns (The Hague, 1960).

Die Krisis der europäischen Wissenschaften und die Transzendentale Phänomenologie (1936). Edited by Walter Biemel. The Hague, 1954. Translated as *Crisis of European Sciences and Transcendental Phenomenology* by David Carr (Evanston, Ill., 1970).

Erfahrung und Urteil (1939). Edited by Ludwig Landgrebe. Hamburg, 1954. Translated as *Experience and Judgment* by James S. Churchill and Karl Ameriks (Evanston, Ill., 1973).

Analysen zur passiven Synthesis. Edited by Margot Fleischer. The Hague, 1966.

Zur Phänomenologie der Intersubjektivität, vol. 3, *1929–1935*. Edited by Iso Kern. The Hague, 1973.

Phantasie, Bildbewusstsein, Erinnerung. Edited by Eduard Marbach. The Hague, 1980b.

Aufsätze und Vorträge, 1922–1937. Edited by Thomas Nenon and Hans Rainer Sepp. Dordrecht, 1989.

Other Sources

Conrad, Theodor. *Definition und Forschungsgehalt der Ästhetik.* Bergzabern, 1909.

Conrad, Theodor. *Über Wahrnehmung und Vorstellung.* Leipzig, 1911.

Conrad, Waldemar. "Der ästhetische Gegenstand." *Zeitschrift für Ästhetik und allgemeine Kunstwissenschaft* 3–4 (1908–1909).

Dufrenne, Mikel. *The Phenomenology of Aesthetic Experience.* Translated by Edward S. Casey. Evanston, Ill., 1973.

Geiger, Moritz. "Beiträge zur Phänomenologie des ästhetischen Genusses." *Jahrbuch für Philosophie und phänomenologische Forschung* 1.2 (1913): 567–684.

Geiger, Moritz. "Phänomenologische Ästhetik." *Zeitschrift für Ästhetik und allgemeine Kunstwissenschaft* 19 (1925): 29–42.

Ingarden, Roman. *The Literary Work of Art.* Translated by George G. Grabowicz. Evanston, Ill., 1973.

Ingarden, Roman. *Selected Papers in Aesthetics.* Edited by Peter J. McCormick. Washington, D.C., 1985.

Kaufmann, Fritz. "Art and Phenomenology." In *Philosophical Essays in Memory of Edmund Husserl*, edited by Marvin Farber, pp. 187–202. Cambridge, Mass., 1940.

Scaramuzza, Gabriele, and Karl Schuhmann. "Ein Husserlmanuskript über Ästhetik." *Husserl Studies* 7 (1990): 165–177.

Schäfke, Rudolf. *Geschichte der Musikästhetik in Umrissen.* Berlin, 1934.

Ziegenfuss, Werner. *Die phänomenologische Ästhetik.* Leipzig, 1927.

DANIEL O. DAHLSTROM

HUTCHESON, FRANCIS (1694–1746), British philosopher. Hutcheson was born in Ireland and educated at Glasgow. He returned to Ireland, where he was licensed to preach by the Presbyterian Church of Ireland, though his moderate and humanist sentiments were not approved by the church or his parishioners. He operated an academy and engaged in correspondences and periodical controversies, including one with Gilbert Burnet. In 1725, Hutcheson published *An Inquiry into the Original of Our Ideas of Beauty and Virtue*, in which he undertook to advance the ideas of Anthony Ashley Cooper, earl of Shaftesbury, against the positions of Thomas Hobbes and Bernard de Mandeville. This was followed in 1728 by *An Essay on the Nature and Conduct of the Passions and Affections*. His correspondence with Burnet was published in 1735. At the end of 1729, Hutcheson was elected to the faculty at Glasgow University, where he lectured on natural religion, morals, jurisprudence, and government and published in Latin works on metaphysics and moral philosophy. He died in 1746 on a visit to his wife's family in Dublin. His son published his *System of Moral Philosophy* posthumously in 1754.

Hutcheson began his inquiry into beauty with a description of ideas and their relation to sense that is based on John Locke's *An Essay concerning Human Understanding*. There are significant differences in detail, but it is likely that Hutcheson took himself to be following Locke closely. On Hutcheson's view, all knowledge begins with simple ideas

provided by sense. The external senses are distinguished by their fundamental differences and correspond, except for touch, to different organs. In addition, Hutcheson appealed to what he called internal or reflex senses. A concept of internal sense had classical roots—in Augustine's writings, for example—but in that context it was a consciousness of ideal form. Locke adopted the term to describe the reflective powers of the mind. For Locke, because all ideas came originally from experience, some way was required to account for the mind's ideas of its own powers. Those ideas were gained, Locke held, by the mind's experience of its own operations through reflection, and they constituted an internal sense.

Hutcheson abandoned Locke's doctrine of reflection and treated internal sense as an independent source of ideas on a par with external senses. Hutcheson was careful to state that the ideas produced by internal sense were not innate, so he remained firmly within the Lockean, empiricist camp in disputes with both scholastics and Cartesian rationalists. Although the internal sense depends on a causal relation to objects that also stimulate the external senses, however, its ideas are different and substantially independent.

In *An Essay on the Nature and Conduct of the Passions and Affections,* Hutcheson referred to five classes of sense: the external senses, the sense of beauty or pleasure, the public sense, the moral sense, and the sense of honor (1969, pp. 5–6). One is justified in calling all of these "senses" by the immediacy and independence of the ideas they provide. One need not know the causal source of an idea provided by a sense, and the power of perception of such ideas is antecedent to custom, education, or example (1725, p. 79). The production of ideas of sense is independent of our will. The sense of beauty produces ideas of order, harmony, and regularity, and those ideas are a source of unreflective pleasure. Hutcheson sought to place the pleasure derived from nature and art on the same empirical footing as the ideas from the five external senses (though Hutcheson objected that the analysis of even the external senses was much too careless and that limiting them to five senses obscured subtle differences within each sense). His goal was ultimately moral. The pleasure of beauty was a fortunate coincidence of our sensitive nature and the teleological order provided by the deity. We could have been constituted differently, and for all we know, some other beings may take pleasure from other causes. We are fortunate, however, in that our sense of beauty responds to just those principles of order, regularity, and harmony that the deity has built into the cosmos, so our moral sense, our sense of beauty, and our ordained ends coincide.

The result of Hutcheson's move to an internal sense was to place discussions of beauty and art on the same footing as discussions in the sciences. During the eighteenth century, Isaac Newton exercised tremendous influence not just as a scientist but also as a thinker who provided a model for all knowledge. Scientific investigations were having great successes. Those successes produced a widespread sense of optimism that the Newtonian model could solve all problems. Hutcheson sought the same kind of authority for beauty that experimental science found in the senses by trying to fit our experience of beauty to an internal sense.

Hutcheson's doctrine of internal sense led him to describe the sense of beauty and the moral sense as disinterested. His principal opponent was Hobbes, who was widely opposed for reducing all virtue to self-love. Hutcheson advanced two arguments. First, because both beauty and virtue are perceived by senses, that perception is antecedent to any calculation or rational control. Even if on reflection we conclude that our interest is otherwise, our sense will already have responded positively to beauty and virtue. "This Love then is antecedent to the Conjunction of Interest, and the Cause of it, not the Effect: this Love then must be disinterested" (ibid., p. 144). Here the argument is from the order of production. A sense responds independently of our interest. Second, the ideas of beauty and virtue are produced when interest is kept out of the process. In particular, property and ownership are not required and interfere with the perception of beauty and virtue: "It is necessary to keep the Sense free from foreign Ideas of Property, and the Desire of Distinction, as much as possible. If this can be done, we may receive Pleasure from every Work of Nature or Art around us" (1969, pp. 173–174). This argument sees some forms of interest as interfering with pleasure. There are certainly hints in this use of disinterestedness of later Kantian and attitudinal theories, but Hutcheson stopped short of attributing to disinterestedness a power to produce aesthetic emotions. Hutcheson meant by disinterested "benevolent" and "unpossessing." The senses themselves produce the ideas, and as long as they are allowed to operate normally, no special conditions are required. Their existence provides the evidence to refute claims that humans act only through calculation and self-love.

The sense of beauty is thus a direct causal response to properties of the perceptual object. One need not be able to identify those properties to experience their effects any more than one need be able to identify the optical properties of objects in order to see them as colored. Nevertheless, even if one cannot identify the microscopic, mechanical chain of causes, one can classify perceptions. Hutcheson bridged the gap between classical and modern concepts of beauty by identifying the perceptual responses of our sense of beauty as arising from uniformity amid variety. This retained the classical definition of beauty as harmony and perfection while shifting its ground from ideal forms to perceptual experience. Our internal sense of beauty is adapted to respond to uniformity amid variety in objects. One can enter into comparisons on the basis of greater degrees of

uniformity in cases of equal variety, and greater variety in cases of equal uniformity. At the same time, the response is still that of a sense, and it is antecedent to any knowledge of the cause. Moreover, we could have been constituted so as to respond differently. Beauty is located in the experience of sense, not in the harmony and perfection of the object.

Hutcheson was aware that experience showed a greater diversity than his theory of internal senses would account for. Most people agree on which objects are red, and those who do not can be easily identified as color-blind. But the sense of beauty provides no such uniform response. Hutcheson appealed to the association of ideas to account for the disagreements. Subsequently, association was given prominence by David Hartley in his *Observations on Man* (1749) and used by writers on both morals and beauty to account for the production of specifically moral and aesthetic ideas and pleasures. For Hutcheson, however, association was essentially negative. Left alone, the internal senses would produce ideas in accord with the way we have been made to receive them. But association corrupts the process by attaching to the ideas of sense other emotional qualities derived from our accumulated experience. Thus, if one has an aversion to heights from having fallen, the beauty of a precipice will be overcome by the fear associated with it. Such associations are adventitious and can be corrected by training and rational reflection. Thus association of ideas is always secondary, and it functions primarily to interfere with the senses themselves: "The common Effect of these Associations of Ideas is this, that they raise the Passions into an extravagant Degree, beyond the proportion of real Good in the Object" (1969, p. 95). Association of ideas may produce pleasure or pain, but its ideas are different in kind from those provided by sense.

Although association of ideas is not productive of ideas of beauty, Hutcheson does distinguish two classes of ideas of beauty within the realm of sense. Original or absolute beauty arises from the internal sense's response directly to some object that possesses uniformity amid variety. Thus, absolute beauty is associated with the object itself even though it is the idea produced by the sense that constitutes the beauty and not the properties of the object. But relative or comparative beauty requires more than one object. Its source is imitation or likeness. Again, Hutcheson was able to incorporate the classical theories of realism and imitation into his empiricist reconstruction of beauty. Internal sense recognizes when one thing successfully imitates another and responds to the comparative likeness. Thus, even if the forms of a painting are not themselves beautiful in an original or absolute way, they may still stimulate the internal sense of beauty by their successful imitation of their object. Hutcheson thought of this process as still falling within the scope of a sense because the response is immediate and independent of one's will, even though it involves comparison.

Hutcheson acknowledged the influence of Lord Shaftesbury in his preface to the *Inquiry*. Like Lord Shaftesbury, Hutcheson relied on sentiment as the ultimate evidence for the classical virtues and harmonies: "So propose the whole World as a Reward, or threaten the greatest Evil, to make us approve a deform't Object, or disapprove a beautiful one; . . . our Sentiments of the Forms, and our Perceptions, would continue invariably the same" (1725, 11). Hutcheson goes much farther than Lord Shaftesbury did, however, in treating sentiment as independent of reflective control or education. With Hutcheson, persons of taste are defined finally in terms of their ability to take pleasure in certain sensations. At the same time, Hutcheson's moral scheme is teleological and utilitarian. That teleology carries over to the sense of beauty as well. Hutcheson enters into moral and aesthetic calculations based on his confidence that our senses are guided by a higher order that makes them reliable indicators of god's will and the natural right built into the order of things by the deity. It was left to David Hume (1711–1776) to free sentiment from presuppositions of order and let it stand on its own. Hume pushed a reliance on sentiment to its logical conclusion. In the process, both Hume and his Scottish opponents largely abandoned Hutcheson's strict reliance on an internal sense. Hutcheson had already begun to multiply senses. As that process continued, the analogy to perception was replaced by a metaphoric sense that is essentially linguistic rather than organic. Alexander Gerard (1728–1795) and Archibald Alison (1757–1839) are among the most prominent aesthetic theorists of that continuing tradition.

Hutcheson's scheme is significant not so much for the actual results it produced as for the shift in thinking that it fixes. After Hutcheson, it is much more difficult, and eventually it becomes impossible, to take beauty to be some substantial relation built into the structure of the cosmos. Hutcheson's conclusions are that beauty is first of all some feeling raised in us, and second that it is traceable, at least as our senses are constructed, to certain causal uniformities in our perception of objects. Thus, it is one's response that is the real test. That the combination of uniformity and variety is instrumental in producing the response of beauty by means of an internal sense is an empirical observation.

The fact that some of Hutcheson's empirical observations about what produces beauty are more than a little problematic is less important than the basis he offered. We are unlikely to accept the comparisons of the beauty of simple figures that Hutcheson laid down. He hoped to develop a kind of mechanics of beauty like Newton's mechanics of objects. That hope proved illusory. But his insistence that the experience provided by sense forms the basis for hypotheses fundamentally changed the basis for aesthetics.

[*See also* Beauty; Cooper; *and* Taste, *article on* Modern and Recent History.]

BIBLIOGRAPHY

Works by Hutcheson

An Essay on the Nature and Conduct of the Passions and Affections, with Illustrations on the Moral Sense (1728; 3d ed., 1742). Gainesville, Fla., 1969.

An Inquiry into the Original of Our Ideas of Beauty and Virtue. London, 1725.

Collected Works of Francis Hutcheson. 7 vols. Facsimile edition prepared by Bernhard Fabian. Hildesheim and New York, 1969–1971.

On Human Nature (Reflections on Our Common Systems of Morality and on the Social Nature of Man). Edited by Thomas Mautner. Cambridge and New York, 1993.

Other Sources

Alison, Archibald. *Essays on the Nature and Principles of Taste.* London, 1790.

Dickie, George. "Taste and Attitude: The Origin of the Aesthetic." *Theoria* 39.1–3 (1973): 153–170.

Dickie, George. *The Century of Taste: The Philosophical Odyssey of Taste in the Eighteenth Century.* New York and Oxford, 1996.

Gerard, Alexander. *An Essay on Taste* (1759). Facsimile of 3d ed. (1780). Gainesville, Fla., 1963.

Kivy, Peter. *The Seventh Sense: A Study of Francis Hutcheson's Aesthetics and Its Influence in Eighteenth-Century Britain.* New York, 1976.

Korsmeyer, Carolyn. "Relativism and Hutcheson's Aesthetic Theory." *Journal of the History of Ideas* 36.2 (April–May 1975): 319–330.

Locke, John. *An Essay concerning Human Understanding.* Edited by P. H. Nidditch. Oxford, 1975.

Michael, Emily. "Francis Hutcheson on Aesthetic Perception and Aesthetic Pleasure." *British Journal of Aesthetics* 24.3 (Summer 1984): 241–255.

Townsend, Dabney. "From Shaftesbury to Kant." *Journal of the History of Ideas* 48.2 (April–June 1987): 287–305.

DABNEY TOWNSEND

HYPERTEXT. The term *hypertext,* coined by Theodor H. Nelson in the 1960s, refers to a form of electronic text, a radically new information technology, and a mode of publication. By his term, Nelson meant a kind of "text that branches and allows choices to the reader, best read at an interactive screen . . . a series of text chunks connected by links which offer the reader different pathways" (1987). The idea of hypertext, of which the World Wide Web provides a flat, somewhat primitive approximation, derives from Vannevar Bush's ideas of the Memex—essentially a microfilm device that would permit readers to create a new form of writing in which chains of associations overlaid already existing books. To a certain extent, hypertext theorists, authors, and system designers divide into those who see this medium chiefly as a form of Nelsonian atomized, electronically linked textuality and those who emphasize Bush's reading paths.

Whichever emphasis one takes, hypertext is an information technology in which a new element—the link—plays a major part. All its chief practical, cultural, educational, and aesthetic characteristics derive from the fact that linking creates a new kind of multisequential or multilinear (rather than nonlinear) writing. Until digital computing, writing had always been a matter of physical marks on physical surfaces. Computer-based information technology, however, transforms writing into electronic codes that can be rapidly duplicated, transmitted, and searched or otherwise manipulated. This electronic coding also permits the creation of the link.

Reading a fully hypertextualized version of the present essay, one encounters it on (or by means of) some electronic device—usually a computer monitor or projector—and, at points where links indicate the availability of additional information, one finds some sign of that fact: differently colored text, underlining, icons, changes in the on-screen appearance of one's pointing device, or the like. Using a touch screen, computer mouse, or other such tool, one commands the virtual page to display indicated material, which can take the form of annotations, images, or the entire text to which a note might refer. For example, in the hypertext version of this essay one might find links to material on Nelson, Bush, and their writings, and these in turn might link to other texts.

Although the hypertext experience to an extent resembles that of reading a scholarly article with footnotes or endnotes, there are major differences. First, although the quasi hypertext created by an annotated essay does permit readers to pursue different reading paths—some choose to neglect the notes, others glance at them, and a few abandon the original article to consult materials it discusses—the structure remains fundamentally axial. In any large hypertext environment, such as the World Wide Web, each text exists within an enormous network. Second, linking tends to erase the manuscript- and print-based distinction between main and subsidiary text, and annotations can link to other notes, thus creating a kind of layered text, or to many points in the same text, thus producing glossary-like functions.

Some rudimentary hypertext systems, such as early Hypercard or most World Wide Web viewers, replace the current text by the linked text, but more sophisticated ones make such annotations appear in a second window or document. Some systems, such as Intermedia, Storyspace, or Microcosm, permit authors to attach two or more links to particular words, phrases, or images.

Hypertext exists both as a congeries of actually existing systems and as an ideal of textuality with certain political, cultural, and aesthetic assumptions. For example, because electronic links can create multiple reading paths, the pioneering theorists of hypertext have always emphasized that some of the author's power and responsibilities shift from author to reader. In fact, they also urge that readers should have power to add their own links and even texts as well. All forms of literary hypertext do not, however, necessarily share authorial power with readers. Michael Joyce's *Afternoon* (1987), which is generally recognized as the first im-

portant novelistic hyperfiction, uses its multiple pathways to create multiple versions of its story, all under the author's control. This and other examples suggest that informational and fictional hypertexts might represent fundamentally different modes and instantiate differing degrees of democratization.

Hypertext redefines print-based conceptions of authorship, intellectual property, textuality, and related matters. For example, electronic links tend to blur the boundaries of individual texts because they can connect passages "external" to a work—say, commentary on it by another author or parallel or contrasting texts—as well as passages within it. Like devices of information retrieval, links permit readers to move as rapidly between the beginning of *Paradise Lost* to a parallel passage in the *Divine Comedy* as between that passage and one that appears on the next page of a printed edition.

Manuscript and printed versions of the Bible, the Talmud, and other heavily annotated works also provide partial analogies for such reader-directed information technology, as does the modern newspaper. Nonetheless, one must be careful about analogies that can misdescribe this developing form of textuality as much as illuminate it. For example, those who are trying to develop a new rhetoric and stylistics of writing in large networked nonspatial electronic environments have drawn on spatial analogies involving physical movement: orientation, departure, navigation, and arrival. Similarly, the unfortunate metaphor of "home page" seems inappropriate in the gigantic, networked, definitively non-book environment of the World Wide Web. Describing a hypertext document as an "electronic book" also seems unhelpful: electronifying a printed book involves translating its alphanumeric characters into code, preserving its hierarchical organization, and perhaps adding a search tool. Fully hypertextualizing a print book—an action obviously appropriate for reference works such as dictionaries, encyclopedias, airline schedules, and parts catalogs—requires placing it in an environment of other books that in some way opens it to them. Creating something equivalent to a book in hypertext (rather than hypertextualizing an already existing one) demands that one begin with a networked, rather than a hierarchical, structure to take advantage of what Nelson calls the digital "docuverse." The appropriate paradigm for the hypertext web, therefore, is not the electronic book but the electronic library.

Before proposing to locate any putative universal aesthetic qualities or implications of hypertext, one must realize that the limitations and constraints of existing systems mean that each thus far provides only a partial approximation of the powerful Bush–Nelson vision. Although all hypertext systems permit readers to make choices, only a few, such as Intermedia and Microcosm, permit readers to share the author's power more fully and create materials that other readers can see. Similarly, hypertext environments

that work only on individual computers, such as Hypercard, Toolbook, and Storyspace, can realize only a small part of the ideal, which requires the availability of a large number of texts. The kinds of linking possible also prove to have major effects on the text and hypertext created. For example, although true hypertext would seem to require that multiple links can attach to any single word or phrase, thus promoting reader choice, many early systems, including current World Wide Web viewers and their underlying hypertext markup language (HTML) code, do not permit this crucial feature.

Although the wide variety of both systems and authorial practice makes difficult proscribing rules, this new information medium does seem to have certain discernible aesthetic implications. For example, like other forms of digital text, hyperfiction, poetry, and informational works readily combine visual and verbal elements, and they do so not only because they so easily include images and sound but also because elements of graphic design become increasingly important. Writing hypermedia, in other words, promises to require increased visual literacy. One can also see that hypertexts tend toward network structures that emphasize fragmentation and collage, multiplicity, and forms of argumentation or narration featuring multiple climactic moments or resolutions. As one might expect of an information technology based on linking, hypertext encourages abrupt juxtapositions and border crossing, thereby calling manuscript- and print-based notions of genre into question. It also permits and may encourage placing major emphasis on irony and the grotesque.

The ideas associated with postmodernism and poststructuralism have proved especially appropriate to theorizing this new information technology, and much of the creative work produced within and by it embodies their qualities. Ideal descriptions of hypertext, for example, share many qualities with Gilles Deleuze and Félix Guattari's rhizome, and they also approximate the "ideal" writerly text of Roland Barthes, whose

> networks are many and interact, without any one of them being able to surpass the rest; this text is a galaxy of signifiers, not a structure of signifieds; it has no beginning; it is reversible; we gain access to it by several entrances, none of which can be authoritatively declared to be the main one. (Barthes, 1974, pp. 5–6)

Similarly, Mikhail Bakhtin's multivocality describes hypertexts far better than it does print-based works, and poststructuralist pronouncements on the death of the author appear more convincing when applied to the ways in which linked electronic environments reconfigure our experience of authorship and intellectual property. Jacques Derrida's ruminations on texts, deconstruction, and meaning, particularly in *Dissemination* (1981), provide a valuable means of understanding hypertextuality and its relation to texts created in the world of print. His notions of decentering and

the difficulties of defining the limits of individual texts seem especially helpful.

Although much literary and critical theory since the 1960s illuminates both the ideal and the experience of hypertext, practical criticism faces major difficulties when confronting literary and artistic texts that readers can enter or leave in multiple places. Even when not considered as an open-ended, perpetually unfinished textuality, hypertext in the form of individual webs causes problems for the critic, in part because they refuse to be mastered. Because many webs are simply too large and complex to follow every possible reading path, critics can no longer pretend to master the documents about which they write. Because each reading of a web produces an individual performance, critics will consequently have to work in ways resembling those of music or drama critics, who must base their work on a limited number of performances. Indeed, as some have already pointed out, perhaps the only appropriate form in which the critic can work involves working in hypertext itself, adding links and commentary that readers may or may not wish to pursue.

[*See also* Artificial Intelligence; Computer Art; Cyberspace; Digital Media; Multimedia; Text; *and* Virtual Reality.]

BIBLIOGRAPHY

Barthes, Roland. *S/Z.* Paris, 1970. Translated by Richard Miller. New York, 1974.

Bolter, Jay David. *Writing Space: The Computer, Hypertext, and the History of Writing.* Hillsdale, N.J., 1991.

Bush, Vannevar. "As We May Think." *Atlantic Monthly* 176 (July 1945): 101–108.

Landow, George P. *Hypertext: The Convergence of Contemporary Critical Theory and Technology.* Baltimore, 1992. Revised edition published as *Hypertext 2.0* (Baltimore, 1997).

Landow, George P., ed. *Hyper/Text/Theory.* Baltimore, 1994.

Nelson, Theodor H. *Computer Lib/Dream Machines.* Rev. ed. Redmond, Wash., 1987.

Nielsen, Jakob. *Multimedia and Hypertext: The Internet and Beyond.* Boston, 1995.

Nyce, James M., and Paul Kahn, eds. *From Memex to Hypertext: Vannevar Bush and the Mind's Machine.* Boston, 1991.

Yankelovich, Nicole, Norman Meyrowitz, and Andries van Dam. "Reading and Writing the Electronic Book." *IEEE Computer* 18 (October 1985): 15–30.

GEORGE P. LANDOW

I

IBN RUSHD (1126–1198), Islamic philosopher. Abū al-Walīd al-Ḥafīd Ibn Rushd (known in Latin as Averroës) was born in Córdoba (now in Spain), then ruled by the Almohads. He trained first in Islamic law, serving as a *qāḍī* or judge in Córdoba before turning to an examination of Aristotelian philosophy. Under the patronage of the Almohad prince, he produced short, middle-length, and long commentaries on Aristotle's works. In the case of the *Poetics,* he seems to have produced only short and middle-length commentaries. Perhaps that is just as well, since the logical character of even these texts has usually been severely misunderstood by some contemporary European scholars, who have tended to conflate the logical and philosophical thrust of Ibn Rushd's texts with their own concerns with rhetoric and politics.

Perhaps the most important later development in Arabic of the logical theory of poetical validity occurs with Ibn Rushd, whose *Middle Commentary on Aristotle's Poetics* harks back to the intervening presence of Abū Ḥamid al-Ghazālī to reappropriate the tradition begun by Muḥammad al-Fārābī and Ibn Sīnā. Ibn Rushd's conception of philosophy and truth emerges in his book *The Incoherence of Incoherence* (1978), in which he defends logic and our capacity for knowledge against al-Ghazālī's attack on *The Incoherence of Philosophy.* He argues that truth and knowledge form the most secure community because they can claim and sustain subjects' agreement. In *On the Harmony of Religion and Philosophy* he extends that conception of reason by arguing that philosophy and religion are conterminous, and in *Kitāb al-kashf ʿan manāhij al-adilla* he uses al-Ghazālī's account of "existences" to set out the role of metaphors in their scriptural use. When the standard ways of arriving at truth prove inadequate, and in order to explain the meanings of the metaphors and allegories contained in religious texts, we rely on the validity of figurative language. Accordingly, considerations of metaphor occur within the context of a search for truth, and to examine the force, scope, and validity of metaphors Ibn Rushd turns to the *Poetics.*

Both al-Fārābī, in his *Commentary on the Poetics* and other writings on the syllogistic validity of poetic discourse, and al-Ghazālī, in the outline of poetics that he includes in *Miʿyar al-ʿilm,* gave prominence to arguments by example and associationist psychology. Although in instances Ibn Rushd seems to follow al-Fārābī and al-Ghazālī (rather than Ibn Sīnā) in his construal of poetic syllogisms, his conception of poetic validity remains obscure: perhaps aware of the limitations of arguments by example, in *Three Short Commentaries on Aristotle's "Topics," "Rhetoric," and "Poetic"* and in his *Commentary on Rhetoric,* Ibn Rushd carefully qualifies the role of syllogisms in poetic discourse, saying that even though the poetic art is considered syllogistic, it neither actually uses syllogisms nor depends on a distinctive kind of syllogism (1977, p. 86). Yet, in other contexts, such as his commentary on Aristotle's *Rhetoric,* he relies on the power of syllogisms to explain figurative language in terms of images and their access to meanings (p. 96).

Further, in the *Middle Commentary on Aristotle's Poetics,* in his analysis of the validity of poetics, Ibn Rushd recognizes that many parts of the Greek context of Aristotle's thought are inaccessible to him, "for much of its contents are either rules particularly characteristic of [classical Greek] poems and their customs therein are not found in the speech of the Arabs but are found in other languages." Nevertheless, he expects "to comment upon those universal rules in Aristotle's *Poetics* that are common to all or most nations" (p. 59). These principles he applies to examples of Arabic poetry; indeed, of the Aristotelian Arabic philosophers, Ibn Rushd in his *Commentary* displays the most thorough familiarity with examples of poetry and their analysis.

In any case, Ibn Rushd's turn to the *Poetics* to explain the nature of metaphors and their underlying concern for truth is not always clear, for he often distinguishes metaphors in "scriptural texts" from poetry (e.g., pp. 77–78, 92–94, 135); however, that cannot be the whole story. If poetic statements are implausible, exaggerated, incoherent, loose, and other cognates of truth, they will lack validity, he maintains (pp. 82–86), thus suggesting that truth is vital to poetic validity. We might explain this as follows.

As the structure of a poem is made up of representations, imitations, likenesses, and their development, any inadequacy in them occurs as an incoherence in the poem. This incoherence may be caused, say, by trying to use impossible means to present objects or events, distorting the imitation, or perhaps using irrational things to represent rational beings (par. 104–111). But its occurrence makes a poem inadequate because it misrepresents or misexplains its actions, motivations, and development. The poem will be disrupted in the sense that the sequence of representations or imita-

tions will appear inconsistent, ambiguous, or unexplained in some measure. This fault has other consequences, for any implausibility in the representations affects the unity of their development, where the latter make up the structure of the poem. Moreover, "unity" or "harmony" is an aesthetic criterion also, and its absence is a reason for finding a work aesthetically unsatisfactory. Accordingly, the aesthetic evaluation of a work by reference to the presence or absence of harmony or unity is explained ultimately by the truths embodied in the imitations, representations, and likenesses that make up the poem. It suggests why we will find the poem better aesthetically if its imitations and likenesses are not erroneous at the same time as their development has a harmonious unity in the structure of the poem.

Further, the depth of a poem will clearly be better the more truths it makes available, and the more successful the harmony of elements of our deeper access to the truth—that is, the deeper the truth and the more diverse the elements held in harmony in the story—the better we shall think it aesthetically. In other words, its truth is essential to the aesthetic character and value of the poem, and this yields what we wanted to infer: the aesthetic value of a work, far from precluding it from gaining access to the truth, depends on such access. Where a poem is less than truthful—where it is implausible, exaggerated, one-sided, incoherent—there it is shallow and lacks unity. Thus, a poem that misrepresents love is shallow and unlikely to satisfy anyone possessing or wanting a deeper understanding and experience of the subject—whether this understanding is philosophical, dialectical, or native to the masses. The poem will not be generally admired and will fail to generate a common response because it does not get at the real matter of love. Further, by contrast with that better understanding, where a poem's structure depends on a superficial understanding of love, there we will find it incoherent because it is implausible, and therefore lacking in unity and, as such, ill-constructed. The untruthfulness of the poem, then, determines its lack of unity and renders it aesthetically inadequate.

These references to truth allow Ibn Rushd to affirm the close relation between philosophy and poetry, by contrast with "the art of inventing parables," and therefore between truth and beauty (p. 84). Moreover, Ibn Rushd can argue that, as a part of the logical organon of philosophy, poetics tells us of the sorts of demands we may make of each other on the basis of the truth-seeking and affective validity of poetry. Although the account he gives in the *Middle Commentary* provides the tools to explain the social features of poetics, he explicitly refers to the latter in the *Short Commentary*.

Within the political and intellectual ferment that occurred in twelfth- and thirteenth-century Arabic culture, Ibn Rushd's commentaries did not have the impact they were to have in medieval Latinate Europe. His commentary on the *Poetics* was translated into Latin by Hermannus Alemannus

and influenced the reception of Aristotle's work in that context. The translation from the Greek of the *Poetics* produced by William Moerbeck, an associate of Thomas Aquinas, did not especially disrupt this influence. Ibn Rushd's influence shows itself, for example, in Dante Alighieri's finding reason to place Averroës among members of the First Circle, which was as high as an infidel could go, or, for example, in Michelangelo's sonnets. Arguably, the conception of poetic validity worked out by the Arabic philosophers, and transmitted to Europe through translations of their work into Latin, provide analogues for theories such as Alexander Gottlieb Baumgarten's on "the logic of the senses."

[*See also* Arab Aesthetics; Islamic Aesthetics; *and* Poetics.]

BIBLIOGRAPHY

Works by Ibn Rushd

Averroes' Middle Commentary on Aristotle's Poetics. Translated and edited by Charles E. Butterworth. Princeton, N.J., 1986.
Averroes' Tahāfut al-tahāfut: The Incoherence of the Incoherence. Translated by Simon Van den Bergh. Reprint, London, 1978.
Averroes' Three Short Commentaries on Aristotle's "Topics," "Rhetoric," and "Poetics." Edited and translated by Charles E. Butterworth. Albany, N.Y., 1977.
On the Harmony of Religion and Philosophy. Translated by George F. Hourani. Reprint, London, 1976. A translation, with introduction and notes, of Ibn Rushd's *Kitab fasl al-maqal*, with its appendix *(Damima)* and an extract from *Kitāb al-kashf 'an manāhij al-adilla.*
The Philosophy and Theology of Averroes. Translated by Mohammad Jamil ur-Rehman. Baroda, 1921.

Other Source

Summers, David. *The Judgment of Sense: Renaissance Naturalism and the Rise of Aesthetics.* Cambridge and New York, 1987.

SALIM KEMAL

IBN SĪNĀ (980–1037), Islamic philosopher. One of the few medieval Arabic philosophers to have written an autobiography, Abū 'Alī al-Ḥusayn Ibn Sīnā (known in Latin as Avicenna) was born near Bukhara, the capital of the Samanid dynasty. In his adult life he was enormously productive, despite the unstable political conditions he lived in, and the bibliography of his philosophical works is estimated to contain up to 250 works. His major work is the *Kitāb al-Shifā*, which includes a commentary on Aristotle's *Poetics*. As with other Arabic Aristotelians, Ibn Sīnā saw the *Poetics* as part of logic, and consequently sought to explain the logical nature of poetic thought by working out an account of poetic syllogisms in order to justify the validity of poetic and literary language and discourse. In doing so, he relies on Aristotle's conception of demonstrative syllogisms to argue for the formal role of pleasure in constituting poetic syllogisms and for their moral value. His work thus advances beyond the problems that beset Muḥammad al-Fārābī's account.

Ibn Sīnā presents his theory in numerous works, including the *Commentary on the Poetics of Aristotle, Remarks and Admonitions,* and *Kitāb al Qiyās.* In *Remarks and Admonitions,* he argues that poetic syllogisms are "composed of imagined propositions," or "premises [having] a certain disposition and composition" (1984, p. 148). Poetic and literary utterances have a logical form that parallels demonstrative reasoning although poetic syllogisms use "premises inspired by emotion"—our imaginative assent to a poetic syllogism depends on the pleasure and wonder that we feel in response to understanding and appreciating the harmony of its elements.

For Aristotle and Ibn Sīnā logical necessity is obvious and primitive. It is expressed in the first-figure syllogistic, consisting of four perfect and self-evident conclusions, and does not need any more basic underlying principles to justify its validity. Logicians test the validity of statements, arguments, and examples of informal reasoning by trying to reduce them to the first figure without loss of meaning and sense. Where such reduction proves impossible they involve meanings and connotations that are ill-formed, and if subjects find them plausible it must be because of their own psychology rather than on the basis of the rules for rational thought. While primitiveness and obviousness may work for demonstration, however, they are not clearly appropriate to justifying poetic syllogisms. In poetic statements meanings are cumulative and synthetic, and their nuances and connotations depend on complex constructions that would be lost on being translated into simpler first-figure terms. Their openness to infinite interplay, based on their complex construction, suggests that they are irreducible to first-figure terms. In turn, that means that the validity of their syllogism cannot be tested; consequently, their acceptance would depend on the psychology of the subject rather than the rules of rational thought.

To answer this doubt, Ibn Sīnā relies on a proposal with a long history that predates his work and is usual even in contemporary thought: the construction of meanings in figurative language involves a relation between terms that, when it is harmonious, occasions pleasure. This proposal blunts the threat of incoherence and psychologism contained in the possibility of an infinite interplay. The occurrence of pleasure from understanding the meanings of terms shows that the mix of terms is not infinite because a harmony between an infinite number of elements is an implausible event: harmony presupposes some sense of a whole, of elements held in a known relation and found to have a harmony. The possibility of an infinite number of elements would disrupt any claim to harmony by always leaving open the possibility of unruly elements. The presence of pleasure and harmony suggests a meaningful order—that meanings are not open to the infinite interplay that threatens incoherence.

In *Commentary on the Poetics of Aristotle* (pp. 66–68), Ibn Sīnā limits the interplay of meanings also through the theme guiding the deployment of figurative language. These themes include tragedy, satire, and the motifs usual to Arabic poetry such as the *nasība,* the caravan site and journey. The themes, once established, will exhaust all figurative language. They are open, however, to rational argument and defense: perhaps one theme becomes outdated or the need for a new form makes itself felt, and so on. In any case, the theme and the pleasure in a harmony of terms provide a meaningful order and relation of terms in poetic discourse.

For Ibn Sīnā pleasure is a part of the formal structure of the poetic syllogistic since it is essential to the validity of the syllogistic, where it establishes which meanings are relevant by restricting the interplay of associations to those that form a harmonious relation. The feeling is not a factor added externally or arbitrarily, but shows how poetic syllogisms work. It shows that poetic discourse is rule governed; and while it cannot claim the certainty of demonstrative arguments, its emotive and ratiocinative quality is still valid because it is a pleasure in a harmonious relation of terms with given meanings.

The role of feeling also shows that the formal validity of poetry involves an essential reference to the subject and its participation in the aesthetic activity *and also* bars morally unjustifiable content from poetry. Poetic discourse imitates subjects in that it treats them as the end of the process of production of aesthetic discourse because their experience of pleasure forms the ultimate ground for appreciating a poem. Only the occurrence of this feeling in the subject will validate aesthetic discourse, and his or her response therefore becomes crucial to the success of poetic discourse.

If subjects play this role, then, by implication, evil poems cannot be aesthetically valuable for the following reasons. Ibn Sīnā follows Aristotle in thinking of virtue as a balanced individual; in "Healing: Metaphysics X," he also adds that just political relations are partnerships "only achieved through reciprocal transactions" between individuals: when they contain a balance between individuals. Moreover, the existence of a community is necessary for individual existence, for we recognize ourselves as human beings only through interaction with others like ourselves. An unjust or evil society, then, excludes individuals from reciprocal transactions, especially, perhaps, when they maintain a balanced and virtuous individual life. As an imbalance, evil stresses some parts of ourselves over others, and cannot account for the qualities whose possession makes us virtuous human beings. By this argument, an evil but beautiful poem would be impossible because it would have to be morally imbalanced yet procure an aesthetically pleasing harmony between terms. This is impossible because pleasure has to be given by a subject; however, the partial stresses in the evil poem connote a divisive society that excludes the virtuous individual and moral mean. But those people excluded by the evil poem are also the ones who must appreciate it and constitute its aesthetic value by grasping its meaning and

giving assent. To find the evil poem beautiful, not only would they have to thwart their own participation in the community just when they gave assent to a poem, but their assent would also be vacuous because they are excluded by the society subtended by the poem. Yet, only their pleasurable response can validate the claim that the evil poem is beautiful. In other words, the poem's aesthetic value cannot be constituted except through their participation, but if their participation is serious, it will restore the balanced community. Similarly, arguably an evil poem, even if found beautiful by the evil community (assuming that there can be such a thing), will not be beautiful in any serious sense because its aesthetic value has not been tested, so to speak, since the criteria for distinguishing the community are not defeasible. In either case, poetic discourse that is structured by the kind of syllogistic Ibn Sīnā proposes will also have moral connotations.

With this unity between poetic validity and moral value Ibn Sīnā brings to fruition the initial steps taken by al-Fārābī to bring logic to bear on poetic thought. They share in the classical understanding of beauty. Perhaps the most important next development of this theory of logical poetical validity occurs with Ibn Rushd, whose *Middle Commentary on Aristotle's Poetics* harks back to the intervening presence of al-Ghazālī to reappropriate the tradition begun by al-Fārābī and Ibn Sīnā.

[*See also* Arab Aesthetics; Islamic Aesthetics; *and* Poetics.]

BIBLIOGRAPHY

Works by Ibn Sīnā

"Healing: Metaphysics X." In *Medieval Political Philosophy: A Sourcebook,* edited by Muhsin Mahdi and Ralph Lerner. New York, 1963.
Avicenna's Commentary on the Poetics of Aristotle: A Critical Study with an Annotated Translation of the Text. Translated by Ismail M. Dahiyat. Leiden, 1974.
Remarks and Admonitions. Translated by Shams Constantine Inati. Toronto, 1984.

SALIM KEMAL

ICON. [*This essay discusses the original, that is, religious, meaning of* icon. *Secular notions of icon are discussed in the various entries related to art history, as well as in* Semiotics; *for cases of modern cultural icons, see* Pop Art *and* Popular Culture.]

The icon (from the Greek *eikōn,* "image, representation") is one of the main phenomena of Orthodox, and in particular Russian Orthodox, culture. It plays an important part in Orthodox aesthetic consciousness and is one of the primary categories of Russian religious aesthetics. The icon emerges as a ritual image during the early Byzantine period (fourth to sixth century) and develops into its classical form in the ninth to the eleventh century in Byzantium, after the final victory of iconolatry, and later in the fourteenth to the fif-

teenth century in medieval Rus. The great significance of the icon for the Orthodox mentality is marked by a special feast in honor of the victory of iconolatry, which is referred to as the "Triumph of Orthodoxy" and has been celebrated by the church since 843 on the first Sunday of Lent.

The icon as a liturgical object is central to Byzantine and medieval Russian culture. It assumed the sacred functions of late Egyptian funeral portraits and images of the emperor in the late Roman Empire. Artistic-aesthetic intuitions that were highly developed in the Greco-Roman culture are also of great importance in the emergence of the cult of icons. Christianity connects the icon (above all, the image of Christ) with the most important dogma: that of the Incarnation of the Word of God—one of the hypostases of the transcendent God. In Byzantium, the icon assumes its place of honor both in the church—fulfilling the function of presenting the ever-existent spiritual world during liturgy—and in the dwelling of each Christian: as an image for worship that replaces the images of pagan gods. In the church, the icon is prominent in the liturgical synthesis of the arts. Orthodox iconography itself was formed under the strong influence of church liturgy, in particular, hymnography. The use of icons for liturgical purposes has for many centuries, however, provoked objections on the part of extremely conservative Christian leaders (especially monks), who see in the icon a continuation of idolatry. This attitude results in rancorous discussions that last for many years, as well as the appearance of diverse heretical movements, of which the iconoclasm of the eighth and the ninth century was the strongest and most enduring.

In Byzantium, the main contributions to the theory of the icon were made by Pseudo-Dionysius the Areopagite, John of Damascus, the church fathers of the Seventh Ecumenical Council (787), the patriarch Nikephoros, and Theodore of Studios. In medieval Rus, their ideas were eagerly received and interpreted (sometimes in quite opposite ways) by Iosif Volotsky, Maxim Grek, Zinovij Otensky, the hegumen Artemij, the participants of the church councils of 1551 *(Stoglav)* and 1554, the deacon Ivan Viskovaty, Evfimij Chudovsky, the archpriest *(protopop)* Avvakum, Simon Ushakov, Iosif Vladimirov, Simeon Polotsky, and other thinkers and icon painters. The results of many centuries of reflection in theology, metaphysics, and the aesthetics of the icon in the Orthodox regions are summed up by Russian religious thinkers of 1900–1930: Eugenij N. Trubetskoy, Pavel A. Florensky, and Sergij N. Bulgakov. On the whole, there now exists a rather complex, multifaceted theory of the icon that reflects the essence of the phenomenon of Orthodox aesthetics.

For the Orthodox mentality, the icon is first of all a narrative about the events of sacred history, or the life of a saint in pictures (using the phrase of Basil the Great, which has become a sort of theological formula, a "book for the illiterate")—that is, a realistic representation or an illustration. In

this idea the expressive-psychological function of the icon comes to the foreground: not only does it relate events of ancient times, it also arouses a whole range of emotions in the beholder—compassion, pity, affection, admiration, and consequently the desire to emulate those who are represented. Hence the moral function of the icon: the arousal of feelings of love and compassion, the softening of human souls that have been hardened by daily concerns. Therefore, the icon is the expression and carrier of the main moral principle of Christianity: humanism, the all-embracing love for others as a consequence of God's love for them and their love for God.

The icon is a story in images, though it is not a story about the events of daily life, but of events that are unique, miraculous, and significant for the whole of humankind. It contains nothing contingent, transient, or insignificant: it is a generalized, laconic image. Moreover, the icon is the atemporal *eidos* (form, idea) of an event that has taken place in history, or of a concrete historical person. It is an everlasting countenance, or a visual image in which the person has been made by the Creator, and which was lost as a result of the Fall but must be regained after the resurrection from the dead (or even in this life, as has happened to some renowned ascetics who had transformed their psychophysical essence in the process of ascetic life).

The icon is an imprint of the divine seal on human destiny. The ultimate example of this seal, the main Icon, was the Word of God himself, who became flesh. The icon is a materialized copy of the divine face. From this follows a particular illusionist, or, as we might say today, photographic character of the icon, for, according to the church fathers, it bears witness to the truth of the Incarnation. The canons of the Seventh Ecumenical Council state that the icon confirms the ancient tradition of making painted representations of Jesus Christ; it "serves as a confirmation of the fact that the Word of God has become flesh in truth, and not in appearance." The icon acts as a mirror that captures only material objects: if there is a reflection in the mirror, or any other imprint, the material object itself—in this case, the human being in the flesh, Jesus Christ—therefore does, or did, truly exist.

The icon is not, however, simply the representation of the earthly face of the historical Jesus that was subject to temporal changes: it is the imprint of the ideal, preexistent face of the *Pantokrator* and Savior. According to Theodore of Studius, the icon reveals this face, the original "visible image," even more distinctly than did the face of the historical Jesus Christ himself. The icon is the symbol, for it not only represents but also expresses that which can scarcely yield to representation. In the iconic image of Jesus, who lived and acted almost two thousand years ago, the spiritual vision of the believer intuits in reality the Person of God become human who possesses two natures "joined without conflation" and "indivisibly divisible"—divine and human—a feat that is

ICON. *Mother of God of the Passion* (c.1450–1500), Greek (Crete, attributed to Andreas Ritsos); Ikonen-Museum Recklinghausen. (Photograph copyright by the Ikonen-Museum Recklinghausen, Germany; used by permission.)

in principle impossible for the human intellect, but is symbolically manifested to our spirit through mediation on the icon.

By pointing to the spiritual phenomena of the celestial world, which are beyond representation, the icon lifts the human mind and spirit to that world, uniting them with it and allowing them to share in the infinite delight of the spiritual creatures that surround the throne of the Lord. The icon thus has a contemplative and anagogic function. As the object of prolonged and deep contemplation, it helps initiate spiritual concentration and leads to meditation and spiritual ascent. The icon depicts the past, present, and future of the Orthodox world. It is in its essence beyond time and space. In the icon, believers find an eternal spiritual cosmos, participation in which is the life's goal for a member of the Orthodox community. In the icon, the unity of the heavenly and the earthly, and the communion *(sobor)* of all creatures before the face of God, is accomplished. The icon is the symbol and embodiment of *sobornost* (companionship).

The icon is a special kind of symbol. Uplifting the soul of the believer to the spiritual spheres, the icon presents to our transitory world that which it depicts. It is a sacred, or litur-

gical, symbol that is endowed with the power, energy, and holiness of the represented sacred personage or event. The gracious power of the icon rests on the very likeness, or similitude of the image with its prototype (hence, once again, the tendency toward illusionism in icon painting), as well as on the naming, or name, of the icon (hence follows, on the contrary, the conventionality and symbolism of the image). The icon is antinomical in its essence, just as its original divine Prototype: it is the expression of something beyond expression, and the representation of the non-representable. The ancient antithetical archetypes of the mirror, as really presenting its prototype (the Hellenic tradition), and the name, as the carrier of the essence of the named (the Middle Eastern tradition), find in the icon their antinomical unity.

Because the icon manifests its prototype, it must be worshiped and has the ability to work miracles. Believers love and worship the icon as they would what it depicts—"the honor paid to the image extends to the prototype," according to the church fathers)—and receive spiritual aid from the icon, just as they would from its archetype. The icon is for this reason an object of prayer. Believers pray before it, and open up their souls in trusting confession, supplication, or thanksgiving.

The icon is also a beautiful image that serves, with its bright colors, as an ornament for the church and brings spiritual joy to the beholder. "The color of painting," John of Damascus writes about church art, "draws me toward contemplation and, delighting my sight as a meadow, pours the glory of God into my soul."

In the icon, church tradition lives on in its artistic form. The main carrier of tradition, the iconic canon, preserves, as if in a specific inner norm of artistic process, the principles, methods, and peculiarities of the artistic language of icon painting that have accumulated over centuries of Orthodox spiritual and artistic practice. The canon does not impede the creative will of the icon painter but instead disciplines it, and facilitates, for artistic thought, the breakthrough into the realm of absolute spirituality, as well as the expression of the acquired spiritual experience in the language of icon painting. The icon is therefore an outstanding work of representational art that conveys its deepest spiritual content by exclusively artistic means: color, composition, line, and shape. According to Bulgakov, the icon embodies, to the ultimate degree of manifestedness, the "spiritual, sacred corporeity" (*dukhotelesnost'*). The corporeal entelechy, which is the object of intuitive striving for every true art, is realized in the icon most fully and, for the Orthodox mentality, in the best way possible. In the icon, the eternal antinomy of culture between the spiritual and corporeal is eliminated, for it is in the icon (we have in mind the classical examples of the icon from its "golden age" in Russia around 1400) that spirituality receives its ultimate realization in the created world, and reveals its visually perceptible beauty. Finally, all this bears witness to the wisdom (*sophijnost'*) of the icon. The icon possesses *sophijnost'* for it contains in itself, in a certain incomprehensible unity, all the things described here, together with many other ineffable essences, a fact that makes clear the involvement of Sophia, the Wisdom of God herself, in its creation.

[*See also* Byzantine Aesthetics; Iconoclasm and Iconophobia; Iconography and Iconology; Religion and Aesthetics; *and* Russian Aesthetics, *article on* Religious Aesthetics.]

BIBLIOGRAPHY

Bulgakov, Sergij N. *Ikona i ikonopochitanie* (The Icon and Iconolatry). Paris, 1931.

Bychkov, Victor V. *Malaya istoriya vizantiyskoy estetiki* (The Concise History of Byzantine Aesthetics). Kiev, 1991.

Bychkov, Victor V. *Russkaya srednevekovaya estetika, XI–XVII veka* (Medieval Russian Aesthetics). Moscow, 1992.

Bychkov, Victor V. *Dukhovno-esteticheskiye osnovy russkoj ikony* (The Spiritual and Aesthetic Foundations of the Russian Icon). Moscow, 1995.

Filosofia russkogo religioznogo iskusstva, XVI–XX vv., Antologia (The Philosophy of Russian Religious Art, an Anthology). Edited by Nikolaj K. Gavryushin. Moscow, 1993.

Florensky, Pavel A. *Ikonostasis*. Moscow, 1994.

Ouspensky, Leonid A., and Vladimir N. Lossky. *The Meaning of Icons.* 2d ed. Translated by G. E. H. Palmer and E. Kadloubovsky. Crestwood, N.Y., 1982.

Ouspensky, Leonid A. *Bogoslovie ikony pravoslavnoj Tserkvi* (The Theology of the Icon of the Orthodox Church). Paris, 1989.

Trubetskoy, Eugenij N. *Tri ocherka o russkoj ikone* (Three Essays on the Russian Icon). Moscow, 1991.

VICTOR V. BYCHKOV
Translated from Russian by Oleg V. Bychkov

ICONOCLASM AND ICONOPHOBIA. The Renaissance humanists believed that the arts had their origins in the praise and invocation of the gods. But the beauty of the poems and songs, they admitted, sometimes led people to worship the works of art rather than the gods; or worse, to worship the wrong gods altogether. The problems are only more acute when the work represents the visible form of the gods. Hostility to carved and painted images, then, had its origins in religious disputes.

Theory. The most serious iconophobic accusation is that the supposedly sacred image in fact represents a false god. But this charge does not necessarily call into question the accuracy of the representation. Accuser and accused may even share the same theory of representation. For example, everyone might agree that images make supernatural spirits present through sympathetic magic. In that case, the image of the true god could be distinguished from the idol by practical testing: the true image should be powerful and beneficial. Or everyone might agree that the image is only a conventional sign for divinity. In that case, the idol can be converted into a true image simply by changing the convention, as medieval

Christians did when they converted Roman statues of Venus into Virgin Marys by relabeling them. The charge of *inauthenticity* seldom involves a true iconophobia.

True iconophobia does surface in the accusation that the image incompletely or falsely represents the gods, even the true gods. The fear is that some *inadequacy* of the representation will result in a misleading account of the divine nature. The accusation of inadequacy can fix on either the production of the image or its reception. Thus, it is connected to issues in epistemology, semiotics, psychology, poetics, and aesthetics.

A mimetic image, apparently, represents its object less conventionally than language does. Because an image lacks grammar and a conventionally established vocabulary, it cannot formulate statements. Instead, the image represents objects by resemblance. Resemblance is a psychological fact. The image offers surrogate stimuli for perceptions of real things. Even without being literally mistaken for the real things, mimetic images nevertheless work powerful effects on beholders. They are often used to alter behavior in the realms of religion, politics, and consumerism.

Because the image appears to bypass convention, its effectiveness seems to transcend particular cultures and communities. This is a great practical advantage.

But the psychological fact of resemblance is not so easy to account for logically. Resemblance, or an "iconic" relationship of sign to object, is based on shared physical properties. But an icon may share many properties with any number of things in the world, and it is not necessarily obvious which ones a beholder is meant to focus on. The match between image and world cannot be shown to be either intrinsic or universally manifest.

Iconophobia, then, is the fear of the uncontrolled nature of iconic representation. A theologian, for example, will naturally mistrust an image because there is no guarantee that it will capture and convey the appropriate aspects of divine reality. On the contrary, the materiality of the image might dangerously mislead worshipers into thinking, for instance, that their gods have material substance and are limited by space-time. The image delivers at best an incomplete list of properties, and at worst a list of false properties. The theologian will prefer words because they can articulate more complicated relationships among things and ideas, and because the materiality of words is in principle a merely contingent attribute.

Theologians also prefer words because of their "inscriptional" quality, that is, their status as tokens irrevocably severed from an original, authentic performance or moment of utterance. This means that they can more easily be assigned a divine provenance and so invested with scriptural or oracular authority. Images, too, can be "revealed," but this involves more awkward deceptions.

Semiotics has attempted to circumvent such critiques by arguing for the full conventionality of iconic representation.

According to this argument, communities agree to read images in certain ways, and to judge some images as more realistic than others. But the conventions of iconic representation are never as finely articulated as the conventions of linguistic exchange. The image has no truth-value worthy of the name.

In the face of skepticism about iconicity, iconophiles can also take the opposite tack. They can push the image even further in the direction of nonconventionality. For example, they point out that iconic representation often involves imprints, traces, or other signs that claim physical contiguity with their objects (what are known as indexical signs). Cultic traditions will offer homomaterial samples, or parts meant to stand for an absent whole (in other words, relics). Finally, portraits can be said to operate not only by resemblance but also simply by naming their objects. That is, they manage to designate individuals successfully even with minimal effort to establish commonality of properties. Effigies, for instance, can be very effective referential pointers, inciting violent or reverential behavior, without even bothering to reproduce unique features of the individual subject. But such extreme efforts to universalize the power of iconic representation only invite still more pointed charges of inaccuracy or inadequacy.

An important set of iconophobic critiques focuses on what can be called the rhetoric of the image rather than its logic. Such accusations do not necessarily call into doubt the adequacy of the representation. Instead, they worry about the psychological effects the material sign might work after it has successfully stood in for its absent object. Apparently contingent attributes of its material form might suddenly get seen as constitutive attributes, that is, as expressions in their own right capable of conveying new contents. These new, unforeseen sign-content relationships will not be governed by the old code that the recipient was using to read the original sign.

The conversion of contingent materiality into a constitutive property tends to introduce the issue of beauty, or the formal properties of a sign that provoke pleasure and yet are logically unrelated to the possible pleasurable features of the signified real object. Beauty is a highly effective rhetorical device, because it can draw attention to or prejudice the beholder in favor of the signified. Beauty even plays a role in traditional defenses of the religious image, in that the sensual appeal of the image is said to stimulate holy thoughts and emotions. Materiality and even beauty might also be read as metaphors for some immaterial idea, again greatly augmenting the power of the image. This apology has been pushed still further in the paradoxical argument that the uncoded and uncontrolled aspects of aesthetic textuality actually make for the most adequate representation of a transcendental and unknowable divinity.

All these advantages can of course be considered disadvantages. Beauty is no reliable guide to the facts of the mat-

ter. Metaphoricity grants the recipient of the image too much responsibility for hypothesizing a code.

The extreme case of iconophobia on the grounds of inadequacy is a general breakdown in confidence in all forms of representation, and not only iconic representation. This representational pessimism, or Puritanism, dismisses virtually any claims that an object might make to stand for anything else. Such a demystifying critique is apt to see an idol in any object or idea whose function and range have been metaphorically extended. Attacks on monarchy and other political institutions; on the cultic aspects of erotic love; on the arts in general; and even on metaphysics and on ideology, can all be understood as varieties of iconoclasm.

Because the materiality of the image seems to initiate semiotic processes posterior to some primary communicative function, these processes have been called "excessive" or "supplementary." The most recent semiotic theory tends to mistrust this distinction between primary and supplementary semiosis. Materiality and metaphoricity are seen as inevitable conditions of textuality. This removes the sting of the iconophobic critique.

History. Any of the iconophobic accusations just discussed can lead to the literal destruction of the offending images, or iconoclasm in the strict sense. But such destructions are political acts, with meanings and repercussions in the realm of things and events. They do not necessarily register people's beliefs about representation with much precision. Iconoclasm may not even reflect a conflict of beliefs about representation. For example, when an army burns the flag of the defeated enemy, it is not contesting the flag's claim to represent the enemy's nation and its ideals; on the contrary. Likewise, it is difficult in moments of political crisis to distinguish iconophobic iconoclasm from mere vandalism. The history of iconoclasms is not equivalent to the history of iconophobic theory.

Many in ancient Greece and Rome objected to the worship of false gods. The educated elite mocked the follies of popular piety and doubted the representational claims of visual mimesis. But a thoroughgoing doctrine of iconophobia emerged only with monotheistic religion. The Jewish tribes encoded their hostility to "strange gods" or idols in the Decalogue, which prohibits the fashioning of any similitude of the deity, or indeed of any living thing. The Jews have stood by this teaching throughout their history. Medieval rabbis like Maimonides extended the skepticism to language and refused to authorize even linguistic representations of the deity. (This resembled the attitude of some medieval Christian mystics.)

The earliest Christians inherited both the Judaic prohibition and the Platonic mistrust of fictions. But many of the first Christians were converted from paganism, and it proved difficult and inexpedient to abandon the habit of image worship altogether. The texts of the church fathers generally disapproved of images. Sculpture was especially suspect; the Second Commandment had mentioned "graven" images, and the experience of pagan worship showed that supernatural power was more readily attributed to statues than to flat images. But archaeological data show that already by the sixth century many Christians were using pictures as the focus of worship. Theologians like John of Damascus and ecclesiastics like Pope Gregory the Great mounted theoretical justifications of the image. At the very least, images were defended as the bibles of the illiterate. At the most, they might serve as channels or vessels for divine grace. The Incarnation provided a model for such a penetration of the transcendental into the material. The Incarnation—the doctrine that the deity took the form of a human being—also circumvented the traditional Jewish argument that the deity was unrepresentable.

The painting of sacred figures flourished especially under the auspices of the Eastern or Byzantine Church. Between 726 and 842 the icons fell victim to a series of violent iconoclasms. The war on the holy portraits was motivated partly by doctrinal objections, partly by local imperial politics. In the end, the images won the right to existence, but they remained subject to strict formal and antimimetic aesthetic conventions designed to ward off any animistic identification of the image with its sacred model and to conceal the traces of human fabrication.

There is some evidence that medieval Christian iconophobes were aware of the Islamic ban on images, which was well established by the eighth century. Muslims shunned mimetic images in part to avoid competition with the already complex system of Christian sacred imagery. Islam never produced a real theory of representation. In the absence of figuration, however, decorative art thrived.

The Western Church, meanwhile, was continually relaxing its restrictions on imagery. By the eleventh century even sculpture was widespread. The cult of images was popular and profitable. The Scholastic theologian Thomas Aquinas, following John of Damascus, legitimated images by drawing fine distinctions between the "worship" due to the cross, the "extreme reverence" due to the figure of Christ, and the "reverence" due to the saints. Some images were justified through connections to legendary "natural" images of Christ, such as the Veil of Veronica, or to the authoritative portrait of the Virgin painted by Saint Luke. Other images were physically linked to corporeal relics, in other words, material samples of the holy personages and not mere signs.

Medieval Christians were quick to accuse heathens and even Muslims of idolatry. But a few reform-minded theologians turned this critique back on Christianity. Bernard of Clairvaux, the twelfth-century leader of the Cistercians, railed against the sensuality and gratuitous artifice of church sculpture. His diatribe stands out as an early acknowledgment of aesthetic effects at odds with the safer, pedagogical function of Christian images. The Lollards and the Hussites, the radical reforming sects that flourished in

England and in Bohemia around 1400, also objected to any reliance on the senses as an avenue to the divine. Traditionalists had a ready response, however: by stimulating the emotions, the beautiful image actually encouraged and enhanced proper devotion.

The Protestant Reformation, initiated by Martin Luther in 1517, brought iconophobia to the forefront of contemporary politics. Following the lead of classically educated humanists, the reformers subjected popular superstitions about the power of images to rational critique. They dismissed the paintings and statues in churches as seductive but false foci for religious devotion. They argued that the ubiquitous images of the saints encouraged unsound doctrine, for example, by distracting from the centrality of Christ's sacrifice.

Beginning in the 1520s, and throughout the century in many lands, reformers incited or persuaded people to destroy the paintings and statues. Some Protestant iconoclasms were spontaneous and violent, others official and orderly. Some involved mere confiscation of the images—perhaps to resurface in a private home—rather than literal destruction. Iconoclasm made good political theater. The whitewashed church wall became a public symbol of the doctrinal triumph of the Reformation.

The iconoclasms were often inspired by popular resentment against the clergy or the wealthy citizens who had paid for the images, rather than by a theory of representation. Often the iconoclasts made no better argument than an appeal to the authority of the Old Testament. Luther realized this and distanced himself from the issue. He conceded that metaphors of the divine were necessary, precisely because the divine was radically unknowable. But if pictures were mere conventional signs, then they were theologically uninteresting. Luther's rival, the Swiss reformer Huldrych Zwingli, was more fearful that these signs were being abused by worshipers and so supported outright destruction.

Iconophobia was pushed to its extreme in the teachings of John Calvin. Calvin relentlessly correlated idolatry with other, more insidious forms of false consciousness. Radical iconophobia in the sixteenth century was often exploited as a political weapon, for example, by Henry VIII of England when he wanted to assert his power over the clergy, or by the Netherlanders when they rebelled against their Spanish overlords. The result was that Calvin and his followers effectively eradicated the cult of images in much of northern Europe.

Protestant iconophobia had a huge and not exclusively negative impact on aesthetics and the history of art. It permanently affected the ways images were made, exhibited, and judged.

The need to ward off the charge of idolatry inspired artists to devise various insulating strategies. German painters and printmakers, in some cases under direct supervision from theologians, illustrated Protestant teachings through schematic diagrams. There was no danger of illegitimately worshiping such images. Other painters masked sacred subject matter with allegory. In Netherlandish art especially, landscape, still life, and scenes of everyday life were conventionally encoded with doctrinal and moralizing content. Such works made no attempt to represent transcendental objects through the technology of pictorial mimesis. No beholder would confuse the picture with the proper object of devotion. Even Catholic Europe responded to the accusation of idolatry. Reform-minded theologians, ultimately supported by the decrees of the Council of Trent (1563), urged suppression of profane and worldly elements in sacred painting, curtailing the cult of the saints, and a purification of the illustration of doctrine.

The risk of substitution was further diminished by adjustments in format, medium, and style. Sculpture of the human figure was virtually abandoned in the Protestant countries, except on very small scale. Painters were less likely to seduce the eye with brilliant colors and imitations of costly materials (exceptions were made when the represented object was deserving of worship after all, for instance, in the iconic portraits of Elizabeth I of England). Mistrust of the eloquence of color was brought to a pitch in seventeenth- and eighteenth-century Idealist art criticism, and indeed has remained a constant in Western aesthetics. Reformers and humanists often preferred black-and-white prints to paintings. Artists strove for plain, straightforward styles, stripped of conspicuous artifice and deceptive pictorial rhetoric.

The Reformation changed the way paintings were displayed. The institution of the altarpiece, an assemblage of paintings or sculptures mounted on an altar, collapsed in the Protestant north. Painters switched to small, horizontal formats suitable for hanging on the walls of private dwellings. In Holland the oil painting became a standard element of interior decoration. Indeed, the styles, subject matters, and functions of Dutch painting had a lasting impact, especially on French and English art and taste.

In some ways, paradoxically, Protestant iconophobia enhanced the prestige of art. Once the image relinquished its claims to represent the divine object, beholders were licensed to enjoy it frankly as a work of art. The image was now physically disengaged from the church and from ritual. Allegorical strategies discouraged beholders from reading images literally and instead taught them to take them, like literary texts, as fictions. Long before aesthetics was given a philosophical basis, beholders began valuing images not for the expensive materials they were made from, nor for the practical purposes they served, nor even for their content, but for their quality, ingenuity, and beauty. To be sure, this development was in the long run independent of the Reformation. Nowhere was the culture of art appreciation more advanced than in Catholic Italy. Here the persistence of traditional iconography only testifies to its irrelevance to connoisseurs

of art. Consistency of subject matter may even have made it easier to compare artists on the basis of quality and style.

The suspicion that beholders were admiring the image for its formal qualities, rather than looking through it to the object it signified, lay at the heart of Protestant iconophobia. Zwingli, for example, chided believers for preferring a finely crafted crucifix to a poorly made one.

It is sometimes argued that the modern work of art never abandoned the old representational ambitions of the sacred image. The modern masterpiece gives access not to an elusive deity, but allegedly to the ineffable genius of an artist. According to this so-called secularization theory, the supernatural aura of the sacred image was transferred to the profane work of art. The modern idea of a masterpiece was freely projected backward onto premodern artifacts. Inevitably, the work of art's new aura was subjected to a fresh iconophobic critique. Modernist aesthetics either denied the sign-status of the image altogether, or insisted on the pure conventionality of its signification. Pictorial mimesis, the cult of genius, and the idea of "art" itself were successively challenged by the practices of Abstraction, Dada, Pop, and any number of movements and trends of the 1960s, 1970s, and 1980s.

But because the modern work of art has been released from its burden of representing remote sources of religious and political power, its success or failure no longer seems so important. Modern works of art have seldom been the objects of literal iconoclasm—unless the relegation of art to the sanctuary of the museum is itself understood as a disabling tactic. Some art has been destroyed because it symbolizes the wealth, power, or effete values of an enemy, as, for example, in the French and Russian revolutions and under German National Socialism. A uniquely modern phenomenon is the lone vandal who publicly inflicts damage on famous works in museums. Such acts appear to signal only the vandal's overall nonparticipation in the ideals of the society. The true echo of the old debates over the power of sacred images is perhaps heard only in modern disagreements over the adequacy of advertising images, especially images of political candidates.

[*See also* Icon; Iconography and Iconology; *and* Religion and Aesthetics.]

BIBLIOGRAPHY

Aston, Margaret. *England's Iconoclasts*, vol. 1, *Laws against Images*. Oxford, 1988.

Belting, Hans. *Likeness and Presence: A History of the Image before the Era of Art.* Translated by Edmund Jephcott. Chicago, 1994.

Besancon, Alain. *L'image interdite: Une histoire intellectuelle de l'iconoclasme.* Paris, 1994.

Bevan, Edwyn. *Holy Images: An Inquiry into Idolatry and Image-Worship in Ancient Paganism and in Christianity.* London, 1940.

Bredekamp, Horst. *Kunst als Medium sozialer Konflikte: Bilderkämpfe von der Spätantike bis zur Hussitenrevolution.* Frankfurt am Main, 1975.

Eire, Carlos M. N. *War against the Idols: The Reformation of Worship from Erasmus to Calvin.* Cambridge and New York, 1986.

Freedberg, David. *Iconoclasm and Painting in the Revolt of the Netherlands, 1566–1609.* New York, 1988.

Goux, Jean-Joseph. *Les iconoclastes.* Paris, 1978.

Grabar, André. *L'iconoclasme byzantin* (1957). 2d ed. Paris, 1984.

Halbertal, Moshe, and Avishai Margalit. *Idolatry.* Translated by Naomi Goldblum. Cambridge, Mass., 1992.

Michalski, Sergiusz. *The Reformation and the Visual Arts: The Protestant Image Question in Western and Eastern Europe.* London and New York, 1993.

Stirm, Margarete. *Die Bilderfrage in der Reformation.* Quellen und Forschungen zur Reformationsgeschichte, vol. 45. Gütersloh, 1977.

Warnke, Martin, ed. *Bildersturm: Die Zerstörung des Kunstwerks.* Munich, 1973.

CHRISTOPHER S. WOOD

ICONOGRAPHY AND ICONOLOGY. In the recent attempts to rethink the history of art history, the question of iconography and iconology has been raised with repeated insistence. This insistence stems from the ways in which the coupling of iconographic and iconological readings of works of art have worked both to justify the object domain for art history and to secure that domain methodologically. This double achievement is above all the work of Erwin Panofsky, the art historian whose writings not only reworked various traditions of iconography and iconology into a systematic and synthetic method but seem to have single-handedly wrested art history out of its dependent position in the early decades of the century, establishing art history on its own disciplinary ground.

The claims Panofsky made for art history are clearly evident in the clinically demarcated levels of description and interpretation through which the work of art is rendered the meaningful object of a humanistic discipline. In "Iconography and Iconology," his seminal introductory essay to *Studies in Iconology: Humanistic Themes in the Art of the Renaissance,* an introduction in which Panofsky is attempting to distinguish between a traditional definition of iconography as a pictorial representation of a subject through a figure and iconology as a larger understanding of iconic representations, he outlines three specific levels: (1) the *preiconological* level of description of the primary and natural subject matter, the "motifs" or "pure forms" that are "carriers of primary and natural meanings," motifs that define for Panofsky a "history of style"; (2) the *iconographic* level that recognizes the conventionality of images and their themes, demanding knowledge of their literary and textual sources and defining a "history of types"; (3) the *iconological* level that discloses the "intrinsic meaning" or symbolical values of a work, demanding a "synthetic intuition" of the work, familiarity with "the essential tendencies of the human mind," and conditioned by "personal psychology" and "Weltanschauung." This third level discovers "the unifying principles which reveal the basic attitude of a nation, a

period, a class, a religious or philosophical persuasion—qualified by one personality and condensed into one work." If iconography thus constitutes for Panofsky "a method applicable to the history of art," and if iconology is then "art history turned interpretive," and if the three levels of description, analysis, and interpretation come together to form a "synthesis," then Panofsky's writings can be said to offer the discipline a sense of rigorous scholarship, a justification for archival and historical research, a clear integration of such research with larger cultural terms and issues, and an eminently reproducible and pedagogically effective method. The stakes implied in recent revisionist rereadings of Panofsky's writings are thus inseparable from a rethinking of iconography and iconology in their relations to one another, to the various traditions of iconography and iconology from which Panofsky draws, to the Renaissance works of art through which they find their interpretative measure in his writings, and to the institutional contexts in which they find their effective force.

In keeping with their debts to much of contemporary theory, these alternative rereadings have tended to downplay or criticize those elements in Panofsky's orientation to normative and quasi-scientific knowledge in favor of a more discursive or semiotic understanding of the work of art's signifying practices, a clearer understanding of the issues of gender and cultural politics that Panofsky's work tends to efface, and a more Heideggerian emphasis on the groundlessness of interpretation over which Panofsky erects his iconographic and iconological project for art history. Following both structuralist and poststructuralist readings, these alternative rereadings thus emphasize Panofsky's gradual remove from his early aspirations for a "tranzcendentale Kunsttheorie," his folding over of hermeneutics into a form of positivism, as well as his reworking of the "vicious circle" of interpretation into a more securely grounded method and totalizing synthesis. A barely concealed stake here is the relation of art to knowledge, specifically, the relation of art history to various terms made available by Immanuel Kant and significantly reworked in the intellectual traditions close to Panofsky. As several commentators have recently suggested, if Panofsky's writings claim their origins in a reading of Kant, as well as in a conspicuous neglect of the writings of Friedrich Nietzsche, then Panofsky is especially indebted to the neo-Kantian tendencies of Ernst Cassirer rather than the radical revision of Kant and Nietzsche that was also available at the time in the writings of Martin Heidegger.

The relation of Panofsky to Cassirer has thus constituted an important feature of many of the recent rereadings of his essays, with particular attention given to the important and long-untranslated "Perspective as Symbolic Form." More recently, the French art historian Georges Didi-Huberman has usefully highlighted the contrast between Panofsky's early flirtation with Heidegger on the necessary "violence" of interpretation and the developed and more influential position of "Iconography and Iconology," while Hubert Damisch has clearly demonstrated how iconography can be said to have its roots deeply entrenched in the metaphysics of the sign. And yet it can also be remarked that the specific implications of the logocentric and metaphysical underpinnings to Panofsky's writings on iconography and iconology are as difficult to pin down at their roots as his relation to Kant. Panofsky's formulations are not simply a matter of methodological "choice" (indeed, that image is continuous with the use he makes of Cassirer) nor even of his belonging to one or another philosophical tradition. For "iconography" and "iconology" also both preexist Panofsky's distinction and theoretical reformulation of them, and in such a manner that Panofsky is usefully seen as inheritor of a long tradition of iconographic and iconological practices, practices that are by no means confined to Renaissance art. Indeed, beyond Panofsky's seminal and synthetic reworking of these traditions for art history, it would thus also be necessary to rethink iconography and iconology back into a wider series of contexts and critical practices. These contexts would include (1) classical definitions of mimesis and theories of the sign from Plato and Aristotle onward; (2) a long tradition of "ut pictura poesis" and rhetoric with which Panofsky's writings on iconography and iconology are closely tied, notably writings stemming from Aristotle's *Poetics* and Horace's *Ars poetica;* (3) a radical rethinking of this same tradition of "ut pictura poesis" in which, from Gotthold Ephraim Lessing to Clement Greenberg, the relation or "system" of the "sister" arts is critically rethought beyond, or prior to, iconographic readings and image-text relations, a rethinking in which the arts are now thought in their specificity; (4) a long tradition of iconographic and hieroglyphic manuals and emblem books—from Horapollon to Valeriano, Cartari to Andrea Alciati—that expound on various moral, social, educational, and psychological values and practices within specific discursive contexts; (5) medieval allegory and symbolism and Christian iconography and iconology, bound to representations of scriptural scenes and figures and foregrounding the relation between iconographic interpretation and biblical exegesis; (6) Reformation and Counter-Reformation contexts for pictorial representations and conventions; (7) the survival of pagan iconographic traditions within a Christian context; (8) the iconographic representation of the Incarnation, of the figurability, legibility, and representability of the unrepresentable, and of the figural aspects of the work that are irreducible to iconographic identification, decoding, and meaning; (9) the history of deciphering and translating hieroglyphic inscriptions; (10) Renaissance *Imprese* and Baroque emblematics; (11) representational traditions that privilege description over narration and which thereby disclose Panofsky's debts to southern over northern Renaissance pictorial traditions; (12) a seventeenth-century re-

configuring of rhetoric and imitation in the French Academy in which "the eloquence of color" is seen as irreducible to the more literary claims made by iconographic meanings; and (13) all the subsequent renewals and reworkings of visual allegory up to the twentieth century. Indeed, it could be suggested that only by passing back through Panofsky, or by situating their insights against his pervasive influence, scholars have been able to reopen and rethink the specificities and complexities of these various histories.

In light of this complex history, it becomes exemplary to turn to the most important source for Panofsky's iconology, the work of Cesare Ripa. First published in 1593 and reprinted in 1603 with woodcut illustrations, Ripa's *Iconologia* is not merely a classification and taxonomy of iconology as an effective manual for artists and scholars but a manual that supplies the necessary rules for the representation and figuration of certain values that are at once philosophical and moral, pertaining to the arts as well as the sciences, to psychological dispositions as well as human ideas and ideals. Through the specific attributes of each of the personified and illustrated figures, the orientation of Ripa's text would thus be toward rigorous definitions, governed by a certain claim for rhetoric and the guiding premises of "ut pictura poesis." In this sense, the *Iconologia* offers a meticulous description of images that, in Ripa's words, "are made to signify something different from what the eye sees." In other words, the *Iconologia* is determined by the rendering visible and legible of concepts, its images thus always in excess of purely contemplative vision. As a compendium of definitions that are at once meticulously defined and fully legible, Ripa's *Iconologia* would thereby find its natural place within Panofsky's general reading of the movement from Platonism to the Neoplatonism of the Renaissance, an index of the transformation of a metaphysical definition to a classic-phenomenal definition, from the Idea concept to the object of allegorical-symbolical representation. Ripa's *Iconologia* would be situated not only as a fundamental response to the status of images raised by the Council of Trent but as the theoretical outcome of a long history of Aristotelian and Neoplatonic thinking that would appear to find, for Panofsky, its ultimate corroboration in the very philosophy of Kant himself. Indeed, it is clear that the guiding term that grounds Ripa's *Iconologia*, a "ragionamenti d'imagini," translated invariably as a "science" or "logic of images," could come to define the very terms in which Panofsky could claim a scientific basis for art history as a discipline read simultaneously through both Kant *and* Ripa.

Whatever the status of Ripa's text for Panofsky—and it was only late in his life that Panofsky acknowledged its importance, precisely at the point at which he seemed to be renouncing the claim to iconology "at a deeper level" in favor of iconography as a more modest methodological project—the *Iconologia*, as Damisch has suggested, can also open questions of the sign, representation, and mimesis in ways

that suggest a divergence between semiotic and hermeneutic concerns. The question raised here, on the basis of the iconographic tradition itself, is thus not that of the grounds on which the interpretative circuit of form, iconography, and Panofskyan iconology is to be articulated, but that of the actual possibility of passage from iconography, understood in Ripa's sense, to the larger project of iconological synthesis as Panofsky defines it.

This criticism also offers an opportunity to expand our sense of the field of possibilities in and against which Panofsky's seminal work emerged by reference to the early writings of Walter Benjamin, writings that disclose a radical reconfiguration of the history of iconography and iconology. Benjamin cites Panofsky's and Fritz Saxl's work on Albrecht Dürer at some length in his study of the Baroque *Trauerspiel*, a book we know Panofsky to have read in an attempt by others to secure Benjamin a position at the Warburg Institute. Panofsky's unpublished review—"a cool, resentment-ridden response" as Benjamin himself suggested—seems to confirm that Benjamin's book explicitly takes as one of its principal targets the "undialectical" neo-Kantian modes of thinking and humanism that characterize the work of both Panofsky and his mentor Cassirer. Indeed, the opening paragraph to Benjamin's last chapter on the *Trauerspiel* appears to be clearly aimed at the writings of Panofsky and Cassirer, notably in the suggestion that the introduction of a "distorted conception of the symbol into aesthetics was a romantic and destructive extravagance which preceded the desolation of modern art criticism" (Benjamin, 1977). The difference is clearly figured in Benjamin's "Epistemo-Critical Prologue," in which the significance of what he terms the "idea" is at radical odds with Panofsky's book of the same name published four years earlier.

More marked in the context of our reading on iconography and iconology, however, is the manner in which Benjamin's reading of the *Trauerspiel* is continually traversed by his reading of the iconography of emblem books and allegorical printed engravings, these "authentic documents" of the modern allegorical way of looking at things. The modern concept of allegory to which Benjamin refers passes beyond Christian and didactic exegesis and turns on the development of humanist scholars to decipher hieroglyphic inscriptions. In other words, he foregrounds the reading of emblem books that are based on historico-philosophical issues of intertwined pictorial languages, "pictorial signs" rather than "phonetic signs" that are woven out of the allegorical constellations of iconic representations and written texts. As Benjamin remarks, the translation of sentences word for word by special "pictorial signs" and enclosed in books of iconology or lexica is clearly exemplified by Ripa's *Iconologia*.

The "new type of writing" revealed in the emblem books resists any form of cognitive apprehension and possession. For Benjamin, there is no intuitive and symbolic relation or

correspondence between the pictorial sign and the determination of its intelligible or spiritual meaning. Rather, he claims that these images are defined by the distance that separates the image and its reflection, signifier and signified. Working in the divergence between semiotics and hermeneutics opened up in such texts as Ripa's *Iconologia,* he therefore suggests that this new kind of pictorial writing can be considered in terms of a figural force, occupying a middle ground between the capacity to conceive an idea or concept and the capacity to present it, between the conceptual understanding of the idea and its sensible presentation. The relation between the image and the text should now be seen as a play of "nonsensuous similarities" or "correspondences" analogous to what he terms a rebus, a kind of "script" that flashes open as an "idea," understood along the same lines as what he later terms a "dialectical image." In other words, as in the paradox of what he also refers to as the theological symbol—"the unity of the material and the transcendental object"—this new kind of pictorial writing found in the emblem books constitutes a fictioning force, a "density" or "intensity" that is inseparable from a certain materiality of the medium. Reading the figural presentations of these hieroglyphic inscriptions is thus not secured by a method for decoding the enigmas but demands a continual task of translation. Instead of positing a conventional relationship between an illustrative image and its abstract meaning, their translation involves the removal from one language into another through "a continuum of transformations," the opening of alterity and not the positing of abstract areas of identity and similarity held together by the symbol. In short, the emblematist's understanding of iconography does not present the essence implicitly behind the image, as if such emblematic or iconographical codes constitute a series of symbolic or enigmatic codes standing in need of decoding and unveiling in order to be understood and culturally meaningful, notions that dovetail with the interpretative logic of Panofsky's writings. Benjamin suggests on the contrary that the emblematist "rips"—or, figuratively speaking, disfigures or disassembles—the essence of what is depicted out before the image, as a form of allegorical or figural writing: allegory, he claims, is not essentially different from writing. It remains without closure or synthesis, for the temporal repetition of allegory necessitates that "no form can repose closed on itself" in the manner of a symbol. As "the intermittent rhythm of an incessant interruption," in which the originally temporal data is transposed into "a figurative spatial simultaneity," the allegorical reading of iconography exposes us not only to the melancholic gaze of Dürer's famous engraving but to the essence of history and tradition itself.

Benjamin's approach to iconography and allegory stems not only from a sustained rethinking of German Romanticism within a philosophical tradition but from a detailed knowledge of the iconographic and iconological work undertaken by scholars in the Warburg school. It also points ahead to more recent reappraisals of the relation between allegory and the symbol as well as a renewed thinking concerning questions of rhetoric in the visual arts, notably, a rhetoric that remains irreducible to literary terms or sources. Above all, Benjamin's reading of iconography urges a recognition of the autonomy of the visual emblem and the tradition in which it figures; it is, one might say, more accepting of the kind of confusion between Salome and Judith that Panofsky is at such pains to straighten out in the exemplifying sections of "Iconography and Iconology," and to that extent offers an alternative to the essential literariness of Panofsky's claims for art history. Or rather, if this literariness guarantees the passage between iconography and iconology for Panofsky, then Benjamin renews and extends the ways in which such texts as Ripa's *Iconologia* permanently displace this guarantee at one of its most privileged origins. One effect of this is to minimize what in Panofsky's hands appeared a relatively arbitrary distinction between the artist-emblematist and the art historian-iconographer in ways that may be said to lead, as it were, back from Panofsky toward Aby Warburg's complex and incomplete, quasi-pictorial projects. At the same time, Benjamin's reading of allegory and iconography forces a recognition that Panofsky's foregrounding of symbolic forms of thinking is deeply inscribed in a totalizing and fundamentally Hegelian synthesis and closure, a tradition that Benjamin's work and reading of allegory is equally intent on exposing and rethinking in the work of mourning it discloses. In short, Benjamin's writing is open to a thought of modernity in ways that radically displace Panofsky's more circumscribed art-historical and methodological claims in which the work of art is secured as a meaningful form of human and symbolic practice. [*See* Benjamin.]

It is also worth stressing that Benjamin is thinking through allegory and iconography together as that which exists *prior* to the word-image opposition and not, as some recent arguments have tended to suggest, a dialectical movement *beyond* this opposition. It simultaneously reconfigures recent claims that Panofsky's preiconological level contains, in spite of Panofsky, the measure through which the work of art can find its renewed sense of figural force against the claims of art history to define itself in terms of the symbolic and discursive forms of cultural meanings and reason. While such resurgent "formalist" approaches often take a central impetus from the writings of Maurice Merleau-Ponty, it was precisely the same philosopher who once claimed that it might be possible to find in paintings themselves a figured philosophy of vision *and its iconography.* His claim bears repeating in the iconographic and iconological traditions it reopens.

[*See also* Icon; Iconoclasm and Iconophobia; Interpretation; *and* Panofsky.]

BIBLIOGRAPHY

Bätschmann, Oskar. "Logos in der Geschichte: Erwin Panofskys Ikonologie." In *Kategorien und Methoden der Deutschen Kunstgeschichte, 1900–1930,* edited by Lorenz Dittmann, pp. 89–112. Stuttgart, 1985.

Benjamin, Walter. *The Origin of German Tragic Drama.* Translated by John Osborne. London, 1977.

Bonnet, Jacques, ed. *Erwin Panofsky.* Cahiers pour un temps. Paris, 1983.

Damisch, Hubert. *Théorie de nuage: Pour une histoire de la peinture.* Paris, 1972.

Damisch, Hubert. "Semiotics and Iconography." In *The Tell-Tale Sign: A Survey of Semiotics,* edited by Thomas Sebeok, pp. 27–36. Lisse, the Netherlands, 1975.

Didi-Huberman, Georges. *Devant l'image: Question posée aux fins d'une histoire de l'art.* Paris, 1990.

Ferretti, Silvia. *Cassirer, Panofsky, and Warburg: Symbol, Art, and History.* New Haven, 1989.

Holly, Michael Ann. *Panofsky and the Foundations of Art History.* Ithaca, N.Y., 1984.

Klein, Robert. "Thoughts on Iconography." In *Form and Meaning: Essays on the Renaissance and Modern Art,* translated by Madeline Jay and Leon Wieseltier, pp. 143–160. New York 1979; reprint, Princeton, N.J., 1981.

Lichtenstein, Jacqueline. *La couleur eloquente: Rhétorique et peinture à l'age classique.* Paris, 1989.

Melville, Stephen. "Art History and the Temptation of New Perspectives." *October* 52 (Spring 1990).

Merleau-Ponty, Maurice. "Eye and Mind." In *The Primacy of Perception and Other Essays,* edited by James M. Edie, pp. 159–190. Evanston, Ill., 1964.

Panofsky, Erwin. *Studies in Iconology: Humanistic Themes in the Art of the Renaissance* (1939). Reprint, New York, 1962.

Ripa, Cesare. *Iconologia* (1593). 3d ed. Rome, 1603.

PHILIP ARMSTRONG

IDEALISM. *See* Collingwood; Croce; Gentile; Hegel; Kant; *and* Schelling.

IDENTITY. *See* Feminism; Ontology of Art; *and* Portraiture.

IDEOLOGY. The term *ideology* was invented by the Enlightenment philosopher and aristocrat-turned-revolutionary Antoine Destutt de Tracy to name a new materialist science of ideas. Following John Locke and Étienne Bonnot de Condillac, Destutt de Tracy held that the rational content of ideas could be identified by working out their origins in sensation, thus facilitating the resolution of moral and political disputes. Within Destutt's lifetime, however, *ideology* had already acquired a derogatory meaning. Napoleon himself used the term to criticize thinkers like its coiner for their commitment to abstract doctrine, instead of pragmatic engagement with social and political facts. Throughout its complex history *ideology* has continued to signify in both of these directions, which share the idea of a conceptual system to be characterized, understood, or criticized by reference to its relations to some underlying, nonideological reality.

Its most important modern uses derive from the writings of Karl Marx and Friedrich Engels. Their manuscript *The German Ideology,* written in 1845–1846 but left unpublished to (in Marx's words) "the gnawing criticism of the mice," was an attack on the Young Hegelian circle of German philosophers in which they themselves had moved. In accusing their compatriots of ideology they mocked them for imagining that the ills of society can be traced to mistaken ideas and that therefore Enlightenment, by curing theoretical error, would bring social liberation.

Thinking, Marx and Engels asserted, is a part of the "life-process" that people carry on in ways constrained by specific social-historical conditions. Characteristic of the ideological point of view is the appearance of the autonomy of theory; "people and their relations appear upside-down as in a *camera obscura*": the historical process appears to be regulated by the history of ideas, when in reality the latter is only an element of a process that, therefore, cannot be adequately understood or explained by reference to theoretical developments:

> Morality, religion, metaphysics, and all the rest of ideology as well as the forms of consciousness corresponding to these . . . have no history, no development; but people, developing their material production and their material intercourse, alter, along with this their actual world, also their thinking and the products of their thinking.
>
> (Marx and Engels, 1976, pp. 36–37; translation amended)

Ideology here applies particularly to more or less systematic complexes of ideas, to which individuals' actual consciousness "corresponds."

The existence of ideology itself must be social-historically explained: the inverted primacy granted to thought arises from the "historical life-process." In an adaptation of the early Romantic critique of the division of labor in society, Marx and Engels traced the origin of ideology to the appearance of professional thinkers. As the activity of a particular group of people, consciousness ceases to look like a necessary aspect of all social activity and appears as an autonomous domain. In reality, theorists (e.g., German liberal philosophers) are only putting into a particular form conceptions that, articulating elements of the mode of social life in which they share, characterize the consciousness of the people generally with whom they share it. These elements are to be found crystallized in language, described by Marx as "practical, real consciousness" (p. 44), consciousness in its specifically human, historical, and social dimension. Theory is the everyday "language of real life" (p. 36) raised to a higher power; the vagaries of both are to be accounted for by reference to their relation to the practices of social life.

While Marx came to abandon his early focus on the division of labor as the root of social evil, he never discarded the conception of thought as "conscious being," and this remained the starting point both of his theory of revolution and of the critique of ideology. Thus, Marx introduced the *Contribution to the Critique of Political Economy,* the first installment of his study of economic theory, with the assertion that "it is not the consciousness of people that determines their existence, but their social existence that determines their consciousness" (Marx, 1987, p. 263; translation amended). He describes the "relations of production"—the social relationships within which people conduct the activities on which their continued existence as biological beings depends—as "the real foundation, on which arises a legal and political superstructure and to which correspond definite forms of social consciousness." This is a metaphor for an explanatory relation: it is only by reference to people's actual social activities that we can critically evaluate the terms in which they understand their doings. In particular, in analyzing revolutionary social change,

> it is always necessary to distinguish between the material transformation of the economic conditions of production, which can be determined with the fidelity of natural science, and the legal, political, religious, artistic [*künstlerisch*] or philosophic—in short, ideological—forms in which people become conscious of this conflict and fight it out. (p. 263; translation amended)

The self-understanding of those whose lives are given shape by a particular social order cannot be assumed to describe or explain their activity adequately, especially when it obscures their relation to historical reality. Instead, that self-understanding must itself be explained by reference to a scientifically constructed description of the social order. Such a redescription of social reality, explaining while indicating the limits of its ideological conceptualization, was what Marx meant by "critique."

Alongside his continuing engagement with political questions—*The Eighteenth Brumaire of Napoleon Bonaparte,* written in 1851–1852, contains a stunning analysis of political ideology in mid-century France—economics was the ideology to which Marx chiefly dedicated his analytic energies. The methodological standpoint from which he studied it might be described as a sort of cultural anthropology, applied by a tremendous effort of intellectual distancing from his own culture to the analytic and historical investigation of that culture. Thus, he discussed economic theory, in both its more scientific and its more purely apologetic varieties, as a systematized interpretation of the everyday consciousness of people in capitalist society, a consciousness produced and reproduced not in the first place thanks to the educative efforts of economists but through the learning and enactment of the social forms of production and distribution represented by the buying and selling of commodities.

The rise of socialist and communist movements from the later nineteenth century on brought the conversion of Marx's critical social science into various versions of Marxism, which as doctrine, philosophy, or worldview provided ideological support for those movements. Marx's metaphor of "base" and "superstructure" was transformed into a "theory" of causal connection between distinct levels or subsystems of social reality. This led to insoluble theoretical problems about the nature of this supposed causal connection, to which different writers sought solutions in concepts of "dialectical causality," "totality," and various sorts of "mediation" between the theoretically abstracted extremes of class relations and ideology (Eagleton, 1991, contains a useful survey of these developments).

A late version of such theorizing was the conception of ideology elaborated by Louis Althusser at the end of the 1960s, which has been influential among some art historians and cultural theorists. Like others before him (starting with Engels), Althusser attempted to reconcile the base-superstructure metaphor with the apparent autonomy of culture by invoking both "a 'reciprocal action' of the superstructure on the base" and the "determination in the last instance" of the former by the latter (Althusser, 1971, p. 135). In the absence of any specification of the "last instance," of course, this idea is explanatorily empty, which robs the metaphor, understood causally, of theoretical power, and leaves superstructural ideology floating as an independent historical force (see Korsch, 1938, pp. 233–236; this work remains the best introduction to Marx's treatment of ideology). Indeed, Althusser insisted that ideology, like the Freudian unconscious, is "trans-historical and therefore immutable in form throughout the extent of history" (p. 161). Ideology, for Althusser, is the set of basic assumptions about social relations that define what it is to be an individual subject in any particular social system; consequently, while (Marxist-Leninist) theory may formulate scientific, nonideological accounts of reality, people's lived activity—including the activity of class conflict—must be ideologically governed. Where theory comes from is from this point of view unclear.

For Marx, in contrast, just as ideology is a historical product, it can be expected to lose its grip as the development of capitalism forces people to come to grips with social reality. Correlatively, there could be no such thing as a general theory of ideology, thanks to the variability of culture; Marx insisted that "to examine the connection between intellectual production and material production it is above all necessary to grasp the latter itself not as a general category but in *definite historical* form" (Marx, 1989, p. 182). A mode of production can be understood only in terms of the particular cultural categories invoked by people in the course of reproducing it. Commodity exchange, for instance, involves historically particular conceptions of alienable private property, on the one hand, and of a specifically economic value

regulating exchange, on the other. Since no "base" exists independently of its "superstructure," critique amounts to a demonstration of the way in which ideology structures the processes of social reproduction, setting limits to behavior whose large-scale violation would constitute revolutionary change (see Mattick, 1986).

Despite Marx's success at escaping the boundaries of his own cultural horizon in relation to economic and political categories, his occasional remarks on artistic matters leave the transhistorical status of the modern concept of art unquestioned. This difficulty with escaping the thrall of artistic ideology has been characteristic of the Marxist tradition. To take a recent instance, Terry Eagleton's effort to confront aesthetic theory as ideology speaks the vocabulary of that ideology when he writes of the "debasement" of art as a branch of commodity production (Eagleton, 1990, p. 2). Eagleton's argument is that aesthetics, the product of a system that both places its highest value on human subjectivity and requires subjectivity's submission to class oppression, at once expresses basic ideological themes of modern society and provides a powerful challenge to those themes. In its freedom from social and economic utility—threatened by commodification—art provides "a utopian glimpse of an alternative to this sorry condition" (p. 65), in principle, shareable by everyone.

Such an argument, despite its author's wishes, restates fundamental elements of the aesthetic ideology against which it is directed; in particular, the idea of a polarity between creative freedom and the compulsions of the market. A true critique of aesthetic ideology would have to go beyond the discovery of contradictory or utopian elements in aesthetic theorizing to a consideration of the place of aesthetics in the history and continued functioning of present-day society. The result would be not a materialist or radical aesthetics but a form of theory taking aesthetics itself as an object, as Marx's study of capitalism offered a contribution not to political economy but to its critique.

It is significant that Marx included *künstlerisch* in his list of ideological forms, for only in relatively modern times could art be considered an aspect of ideology, part of the domain of the mind alongside law, morality, religion, and politics, as opposed to that of productive labor or quotidian life generally. This peculiarity of the modern idea of art cannot be explained within the terms set by that idea. Art developed along with the commercialized mode of production that became capitalism, and it is only by understanding art as an aspect of this mode of production that the supposed antagonism between them so central to aesthetics can be understood.

Renaissance artists laid the groundwork for the modern ideology of art when they struggled for social status by insisting that they practiced not a craft but a liberal art, the object-making hand merely fulfilling the dictates of the imaginative mind. The nineteenth-century modernization of art that replaced working to the order of religious, state, and private patrons with producing on speculation for the market redefined it as the expression of individual genius. Here, as ever, ideology provides a representation of a reality: artworks are produced by independent entrepreneurs (or, latterly, professionals employed by nonprofit cultural or educational institutions) rather than by wageworkers. Art can therefore incarnate free individuality, validating the social dominance of those who collect and enjoy it, and signifying a cultural end to which the making of money becomes only a means. The freedom of the artist, including his or her freedom to starve, provides a model for that of the ruling elite (who have, for example, the education and leisure necessary for the appreciation of art) purchased by the unfreedom of the many. Thus, it is precisely its distance from market considerations, its "noneconomic" character, that gives art its social meaning—and its market value.

To take the most striking example of this, the ideological conflict between "bohemian" and "bourgeois" characteristic of the post-1800 history of artistic activity—a conflict, as just suggested, located inside the limits of bourgeois society—helped to legitimate the bourgeois claim that aesthetic sensitivity is a mark of natural distinction rather than a mark of social privilege (see Bourdieu, 1984, p. 57). Perhaps it was only to be expected, therefore, that the acceptance of once avant-garde art as the quasi-official art of bourgeois society has been accompanied, at the end of the twentieth century, by the decline of the idea of art's transcendence of its social context and uses.

Aesthetics, along with the artistic ideologies at work in critical and pedagogical theory and in such fields as the history and psychology of art, consists of theoretical constructions open like other discursive products to critique in the Marxian manner. But can art objects themselves embody ideology? A negative answer would fall within the terms of the modern ideology of art, for it would imply that artworks have meaning independently of their places in social practice (as in Clive Bell's idea of a mysterious "significant form" of visual art). It is not in principle difficult, though it may take ingenious and scholarly work, to identify ideological elements in the aspects of artworks that can be given linguistic representation, such as Émile Zola's biologism or the vision of a fruitful natural order crowned by aristocratic ownership presented by some English landscape painting around 1800. But since the nineteenth century the question of artistic meaning has increasingly been addressed in terms of a contrast between the "content"—statable in words—of artworks and their nondiscursive "form." Can ideology be identified in artistic form?

This question provides a meeting point for two important problems: the relation between experience and what is said about it in words; and the means and nature of the production of meaning in nondiscursive modes of signification, such as gesture, sound, and imagery. The first of these arises

as soon as ideology is understood as a systematic rendering of social practices that might be lived without much conceptual articulation. What is decisive in social life, as Raymond Williams says, "is not only the conscious system of ideas and beliefs, but the whole lived social process as practically organized by specific and dominant meanings and values" (Williams, 1977, p. 109). Williams wrote of "structures of feeling," meaning "not feeling against thought, but thought as felt and feeling as thought: practical consciousness of a present kind" (p. 132). [See Williams.] This may be compared to Bourdieu's concept of "habitus," dispositional schemata of action and perception, learned in the family and reshaped as individuals move through social institutions like school and workplace. Habitus includes, for instance, the unconscious details of carriage, tone of voice, vocabulary, and differentiated response—reactions of enjoyment, displeasure, or indifference—to objects and activities that allow people to sort each other out by social class. It involves for some a sense of being at home with works of art, and a felt assumption of a high place for art in the scale of social values. [See Bourdieu.] We can think of ideology as a systematizing (and simplifying, since abstracting) presentation of such structures of feeling and action as natural forms of experience. The abstractions of economics theorize as inevitable the felt necessity of subordination to practices of commodity production; the doctrine of "aesthetic experience" defines art, a cultural practice, as the natural producer of a particular psychological response (if only on the part of certain, properly sensitive individuals).

But why should language be seen as the only medium for such systematization? Even within the linguistic domain, the plot summary of a novel leaves out much that readers might look for in the work, and that a writer might have labored to put into it; no description of a painting is a substitute for the visual experience of the picture itself; and the question of the "meaning" of music antedated the development of abstract composition, in eighteenth-century debates about the relation between music and text in opera. Yet it is hard to see how a piano sonata or an abstract painting can be understood as exhibiting the features of ideology. Can the meanings inherent in such works, or identifiable in the formal aspects of narrative or descriptive art, be capable of ideology, presenting people's experience of their social existence in ways that occlude the historical specificity of that experience?

Theodor W. Adorno argued that it was the very irreducibility of an artwork to its description—a version of the Kantian idea of the autonomy of art, its independence as a mode of meaning and value from other modes of truth—that constituted its social significance. Music, the most abstract art, provided the clearest case. Adorno saw the music of Viennese classicism as ideological by virtue of its submission to formal laws of composition, by which "it closes itself off against the manifest portrayal of society in which it has its enclaves," hiding class conflict with harmonically structured wholeness (Adorno, 1985, p. 129). He believed that the new music of the second Viennese school, in contrast, was "no longer an ideology," because in its hermeticism and refusal to please an audience it "surrendered the deception of harmony" and made the alienation of the oppressive class system in which art music has its being audible in the rigors of serial technique (p. 131). [See Adorno.]

Despite the brilliance of Adorno's writing the relation he discovers between Arnold Schoenberg's liberation of dissonance and the avowal of social disharmony is in the last analysis only a suggestive analogy. Elsewhere, he compares serial composition to bureaucratic rationalization and the relation between theme and variation in sonata form to the dialectic of individual and society. Such analogies or allegories can be stimulating and even revealing, but they can also be arbitrary or mechanical. At best they point to further, deeper questions, about the origin of the seeming similarity between such disparate orders of social reality as economic organization and compositional technique.

The relation between the two tends to be mediated in Marxist cultural theory by some conception of "worldview" or "class outlook." Such conceptions demand further exploration of the relation between artistic activity and the social class to whose outlook it supposedly gives formal definition. One path art historians have taken into this territory is the study of patronage, ranging from examination of the constraints on earlier artistic activity by the religious or courtly commissioners of work to more recent examples such as the effect on Abstract Expressionist painting of its utilization by the American ruling class as a propaganda weapon in the cold war. Serge Guilbaut, for instance, concluded with regard to the latter case that American "[a]vant-garde art succeeded because the work and the ideology that supported it, articulated in the painters' writings as well as conveyed in images, coincided fairly closely with the ideology that came to dominate American political life after the 1948 presidential elections" (Guilbaut, 1983, p. 3). Guilbaut was not in fact able to demonstrate the ideological content of the images, as opposed to interpretations of them. But this sort of study raises the issue of the contribution of art to the construction of social viewpoints, and so returns us to the question of how form in art can constitute ideology.

This can only be because art does not exist in a world of its own, sealed off from the conceptualizing performed in language. In Meyer Schapiro's words, "there is no 'pure art,' unconditioned by experience; all fantasy and formal construction, even the random scribbling of the hand, are shaped by experience and by nonaesthetic concerns" (Schapiro, 1937a, p. 196). The mute experience of an art object is no different from any other lived event. Just as all language is an articulation of nonverbal as well as verbal practices, so nondiscursive form—visual, aural, and other—shares its world of meaning

with that constructed in speech. Not only can a mode of depiction or mark making be, for instance, described as literally "free," meaning ungoverned by convention or a definite idea of an image's final configuration; images and sounds can also metaphorically exemplify (to use Nelson Goodman's terminology) the same descriptions as other things (giving us gloomy colors, happy tunes, or mechanical shapes), in this way establishing links to them.

Schapiro gives an example, the appeal of machines to modernist painters after World War I:

> The older categories of art were translated into the language of modern technology; the essential was identified with the efficient, the unit with the standardized element, texture with new materials, representation with photography, drawing with the ruled or mechanically traced line, color with the flat coat of paint, and design with the model or the instructing plan. The painters thus tied their useless archaic activity to the most advanced and imposing forms of modern production; and precisely because technology was conceived abstractly as an independent force with its own inner conditions, and the designing engineer as the real maker of the modern world, the step from their earlier Expressionist, Cubist or Suprematist abstraction to the more technological style was not a great one.
>
> (1937a, p. 210)

Ideology can be identified in such artistic work in the location of "modernity" in engineering (and indeed in what might be analyzed as ideological forms in the presentation of machine-made things), ignoring the historical specificity of the ways in which the mechanization of production was being accomplished. The advance of capitalist production—including, in Russia, its state-directed analogue—was equated visually with the progress of universal norms of rationality and efficiency. [*See* Schapiro.]

We can say, then, that ideology appears in artistic form where the latter can be conceptually linked, by maker or receiver, to other areas of social practice. Ideological content, in form and subject matter alike, is for this reason not univocal, as Schapiro pointed out in a discussion of Diego Rivera's Mexican murals: "in so far as the revolutionary work of art projects slogans, phrases and their counterpart images, in so far as it forms a spectacle rather than determines an action, its effect in stirring the imagination may be manipulated in contrary ways" (Schapiro, 1937b, p. 465). Artistic ideology is a matter of the uses to which a work of art is open.

Ideology in form is as vulnerable to unmasking as its theoretical relatives. In fact, it seems that the verbal explication that renders its ideological content explicit tends to rob art of some of its potential power. If ideology is cultural production experienced with insufficient understanding of its location in social life, sociohistorical consciousness—including experiencing the present as history—is the decisive step toward critique.

[*See also* Art Market; Film, *article on* Film and Ideology; *and* Marxism.]

BIBLIOGRAPHY

Adorno, Theodor W. *Philosophy of Modern Music*. Translated by Anne G. Mitchell and Wesley V. Blomster. New York, 1973.

Althusser, Louis. "Ideology and Ideological State Apparatuses (Notes towards an Investigation)" (1969). In *Lenin and Philosophy*, pp. 127–186. New York, 1971.

Bourdieu, Pierre. *Distinction: A Social Critique of the Judgement of Taste*. Translated by Richard Nice. Cambridge, Mass., 1984.

Bourdieu, Pierre. *The Rules of Art: Genesis and Structure of the Literary Field*. Translated by Susan Emanuel. Stanford, Calif., 1996.

Clark, T. J. *Image of the People: Gustave Courbet and the 1848 Revolution*. Princeton, N.J., 1982.

Eagleton, Terry. *The Ideology of the Aesthetic*. Oxford and Cambridge, Mass., 1990.

Eagleton, Terry. *Ideology: An Introduction*. London and New York, 1991.

Guilbaut, Serge. *How New York Stole the Idea of Modern Art: Abstract Expressionism, Freedom, and the Cold War*. Translated by Arthur Goldhammer. Chicago, 1983.

Korsch, Karl. *Karl Marx*. London, 1938.

Marx, Karl. *A Contribution to the Critique of Political Economy* (1859). In Karl Marx and Frederick Engels, *Collected Works*, vol. 29, pp. 257–417. New York, 1987.

Marx, Karl. *Theories of Surplus Value* (1861–1863). In Karl Marx and Frederick Engels, *Collected Works*, vol. 31. New York, 1989.

Marx, Karl, and Frederick Engels. *The German Ideology*. In *Collected Works*, vol. 5, pp. 19–539. New York, 1976.

Mattick, Paul, Jr. *Social Knowledge: An Essay on the Nature and Limits of Social Science*. Armonk, N.Y., 1986.

Mitchell, W. J. T. *Iconology: Image, Text, Ideology*. Chicago, 1986.

Schapiro, Meyer, "The Nature of Abstract Art" (1937a). In *Modern Art: 19th and 20th Centuries*, pp. 185–226. New York, 1978.

Schapiro, Meyer. "The Patrons of Revolutionary Art." *Marxist Quarterly* 1.3 (October–December 1937b): 462–466.

Williams, Raymond. *Marxism and Literature*. Oxford and New York, 1977.

PAUL MATTICK

ILLUSION. *See* Mimesis; Realism; *and* Theories of Art.

IMAGERY. [*This entry reflects two distinct perspectives on a long-standing debate in the history of aesthetics about the nature of imagery:*

> Visual Imagery in Reading
> Reading Images

The first essay concerns the discussion, connected though not confined to cognitive science, about the role of imagery in reading literary texts. The second essay analyzes the flip side, as it were, of the first problem: the sense in which we read pictures as we see them. For related discussion, see Discourse; Ecphrasis; Psychology of Art; Representation; Semiotics; Text; *and* Ut Pictura Poesis.]

Visual Imagery in Reading

The significance of visual imagery to reading and to literary experience has shifted from one extreme to another over

the centuries. Whereas visual images were long regarded as integral not only to the mental act of reading but also to thought itself, during the modern period they were judged to hold no merit for adult reading experiences of literature. In recent decades, however, the tide has turned. Although visual imagery has not been reinstated as the privileged vehicle of thought and feeling, visual images are coming to be recognized as significant to the cognitive and affective process of reading and to literary experience.

The changing fortunes of visual imagery might be characterized as the product of three elements: one is a psychological view about how the mind works; the second is an aesthetic judgment about what visual imagery can contribute to literary experience, and the third is an institutional decision regarding proper objects of literary study. For centuries philosophers and scientists who explored the mind argued about the role of images in thought and memory. The deathblow for the strong claims about visual images began at the opening of the twentieth century with the rejection by early behaviorist psychologists of subjective mental states and continued through mid-century with the philosophers' move to conceptualize mental experience as an exclusively linguistic process.

The models of aesthetic value and literary study also changed radically during this time period. At the turn of the twentieth century literature was praised for its abilities to evoke vivid visual images in the minds of its recipients, as it was in the days of the Roman orators, and texts were discussed for their capacities to provoke such responses. By mid-century the aesthetic experience of literature had become a linguistic one—a savoring of sound and meaning—and these were studied by close textual analysis. Even in the 1970s when scholars became dissatisfied with the exclusive textual focus of previous criticism and began to attend to the subjective response of the reader, the hypostatized reader was one who responded only through internalized sounds and meanings. In more recent times, exploration of the reader has come to be associated less with an examination of aesthetic experience produced through mental states and more with the impact of cultural differences on the pragmatic and ideological reception of texts.

In the contemporary period some of the most incisive, aesthetic criticisms of visual imagery have come from William H. Gass. In essays from the 1950s through the 1970s, Gass has argued that aesthetic experience requires that the reader remain within the verbal constructions given by the author. For example, if a writer does not specify whether or not a character is bald, the reader should maintain the indeterminacy of the textual meaning. According to Gass, visual images are detrimental to reading because they would demand, by their very nature, that the reader visualize the character as either bald or hairy and, thus, violate the text's aesthetic indeterminacy.

Perhaps the strongest challenge to Gass' anti-imagery aesthetic is to question his claim regarding the preeminent value of textual indeterminacy. Margaret Atwood's *The Handmaid's Tale* provides a compelling test case. In one passage the powerful Commander secretly gives his female servant, Offred, an object forbidden by the strict fundamentalist sect that governs the land. It is a feathery garment, with sequins over the breasts, like a theatrical nightclub costume. Offred reflects on the childlike enticement of the garment. Suppose that a female reader visualizes the white breast cups of the garment's inside. In visualizing the costume's white cups from the inside, this reader imaginatively participates in the familiar childhood experience of trying on an exotic costume or the adolescent girl's unfamiliar experience of pressing one's youthful body into the cold, white breast cups of a formal dress or swimsuit.

Such an image would reduce the indeterminacy of the text insofar as the breast cups are not noted as being white or being observed from the inside. Indeed, the image also involves a shift of perspective—from the garment's outside to its inside, which would also be inappropriate according to the spirit of Gass's critique. If one judges that such visually evocative experiences are integral to the aesthetics of Atwood's work and that being integral is not the mark of an aesthetically inferior text, then one would be forced to reject Gass's anti-imagery argument.

Granting the aesthetic value of this particular experience does not, however, answer the broader, more important question as to what general contributions imagery might make to the aesthetics of reading. Although an answer to this question certainly involves aesthetic judgments, the central issue is psychological. What does visual imagery do?

After years of ignoring visual imagery, researchers since the 1970s have begun to study the process of image production. Cognitive and neuropsychologists have ascribed to the visualizing process a range of properties that would be inconsistent with the characterization of imagery as purely epiphenomenal, that is, as a wholly subjective experience without causal consequences. Such empirical findings converge with certain therapeutically oriented theories stemming from psychoanalysis. As demonstrated from the convergence of these theories and research, visual images have cognitive and affective consequences for reading.

The best-researched effects of visual imagery pertain to cognition, in particular, memory. Literary experience requires complex operations of memory because the reader's response to a text, even during a first reading, is not merely a processing of the words on the page but a recognition of these words in terms of what has occurred in the past, both formally and semantically, and what is thus anticipated to occur in the future. Thus, by guiding expectations and focusing attention on sound, sense, and image, memory partakes in shaping the reader's interpretation and aesthetic experience of a text.

In terms of memory for narrative texts, the forming of visual imagery is most useful when the text from which it is drawn has a discontinuous narrative (Marschark and Hunt, 1989). Visual images serve as the reader's means of organizing complex verbal material within paragraphs. The psychologists Allan Paivio and James M. Clark (1991) suggest that visual images might also function as overarching representations that tie together large segments of text. For example, Italo Calvino's *Invisible Cities* (1974) is a literary work built on short, discontinuous narrative segments. A strong advocate of the visual imagination, Calvino crafted this discontinuous text and others on the visual principles of mnemonic construction that stretch back to antiquity.

The reader who visualizes complex spatial relationships within a text will also find them easier to grasp and recollect (Denis, 1982). Dante Alighieri's *Inferno*, a text that seems to benefit from one's visualization of its intricate, symbolic topography, might have been designed to be held in memory in a visual format in order to serve as an educating icon for its audience. Historians have documented a long tradition of such visually oriented memory constructions.

The discipline that has most directly addressed affect is psychoanalysis. In its theory, research, and practice psychoanalysis has long recognized the affective dimension of imaging, although linguistically oriented, psychoanalytic literary scholars have not generally attended to visualization during reading. Sigmund Freud (1953) and Jacques Lacan (1977) acknowledge the early, primary role of imaging in human development, as do other psychoanalytic theorists. For Freud, these include such visual pleasures as the early, pregenital fascination with the visual world, the oral devouring of visual stimuli, and the voyeuristic savoring of bodies and objects. Lacan's best-known discussions of imaging occur in connection with his notion of the mirror stage of development.

Despite their different perspectives, psychoanalytic theorists would agree that individuals manage their contacts with affect-laden imagery in different ways. Some people use imagery to heighten affective response, whereas others repress imagery as a means of dampening affect. Empirical research has shown that the different ways subjects visualize in response to affectively charged verbal stimuli are consistent with Freud's theories of how repression operates (Moses and Reyher, 1985; and Kuzendorf 1985–1986).

The affective dimension of visual imagery has received scant attention by scientists working outside the psychoanalytic tradition, although they have increasingly acknowledged its significance. When cognitive psychologists pursue the subject of affective imagery, they often seek to identify how it functions, rather than how it is produced. For example, Marguerite Cocude (1988) asked subjects to visualize in response to different words and determined that visual imagery rated by the subject as highly emotional lasted longer than neutral imagery and that visual images with ag-

gressive themes required a longer time to generate than neutral imagery. Such studies of functional properties will likely be complemented by future research on the neurological architecture of affective imaging.

It is important to note that the cognitive and affective consequences of imaging described above do not necessarily occur every time a reader takes up a text. Although one might expect some variability of reader responses to textual stimuli that is potentially affective, responses vary also when the textual stimuli are sheerly cognitive. Three factors influence whether a reader will visualize and experience the kinds of consequences described above: reader, context, and text.

Readers appear to differ greatly as to their skills in visualizing and their choices of situations in which to exercise it. Some readers, like literary scholar Geoffrey Hartman, claim not to be able to form images, while others, like writer John Hawkes, claim not to be able to read without visualizing (Esrock, 1994, p. 94). Psychological tests that compare those who are instructed to visualize when performing a task with those who are instructed to perform the same task using a variety of nonvisual activities sometimes reveal no significant differences in performance until one distinguishes, among those who are instructed to visualize, between high imagers—those who habitually use imagery—and low imagers—those who use it infrequently. In such cases, only high imagers use their visual imagery in ways that facilitate their cognitive performances relative to individuals who do not use visual imagery. The designation of low and high has itself been the subject of experiments, for researchers have been trying to find accurate ways of characterizing the different styles of imagery use.

For some readers, visualization seems to be the way of becoming empathically involved in the text (Hilgard, 1979). Differences between readers appear also according to gender. For instance, gender differences have been demonstrated in experiments concerned with the speed and efficiency of image generation, factors that might account for cognitive and affective behavior. Women have been shown to generate static images faster and more efficiently than men, whereas men excel when dealing with moving imagery (Paivio and Clark, 1991). Moreover, gender differences in how a reader visualizes pronoun references suggest various ways in which men and women identify with fictional characters (Martyna, 1980).

Context is a broad factor. It would include the physical conditions in which readings occur. The particular setting would not only affect the anxiety of an experimental test subject and, perhaps, her imaging style but would also affect the comfort of an ordinary reader, perhaps, the reader's propensity to move into an affectively charged kind of textual reading. The context might also refer to the reader's expectations for the kind of reading that is appropriate to the occasion. One might read the same text differently for the

comprehension section of the Scholastic Aptitude Test, in which the assimilation of textual themes is anticipated, than for pleasure reading at home, where savoring small details is often central. Expectation is a particularly important factor to investigate further because these expectations are socially generated through our educational institutions and, as such, are open to change.

The impact of the text has been less understood than might be expected, given the long rhetorical tradition concerned with the use of texts to evoke visual imagery in their audience. Note, however, that all texts are not necessarily constructed in visually evocative ways. Throughout the centuries injunctions against imagery have appeared sporadically, though they have been articulated on diverse grounds. The paucity of description in the Hebrew Bible conveys its writers' religious commitment to the biblical prohibition against the creation of graven images. Some twentieth-century fiction also discourages its readers from imaging. Texts by Ivy Compton Burnette and some by Samuel Beckett are oriented toward sound and not vision, whether it be through a foregrounding of the subtleties of character dialogues and inner thoughts, through the abstract, schematic portrayal of characters, or through rapidly shifting spatial perspectives. In many cases, writers concerned with creating new forms of verbal reality reject visual imagery because it has long been associated with traditional, representational theories of literature.

For other writers, equally concerned with creating new verbal forms, the evocation of the visual lies at the heart of their creative efforts. Judging from the construction of their prose, writers like Gabriel García Márquez, Hawkes, and Atwood demonstrate strong interest in the visual responses of their readers. Writers of visually evocative texts are often attentive not only to the impact of their prose on the reader's imagination but also to the impact of their own visualizations on the writing of their texts. Calvino, for example, explains at length how visual images often function as the generative impulse for his stories and ultimately fuse with words in complex, textual networks (Calvino, 1988).

Visually evocative texts do not necessarily use lavish visual descriptions. Indeed, visually rich descriptions can be structured so as to minimize or maximize the reader's visualizations. Imaging takes time, and the text that rapidly changes physical perspective does not permit the reader time to construct and sustain a visual image while still maintaining the flow of reading. Furthermore, texts that are visually evocative generate widely divergent imagery. Although images are generally thought to represent textual referents, like dark-eyed characters or lush scenery, images can be responses to textual connotations and, as such, need not depict mentioned or implied textual entities. Images might be created as abstract representations of plot developments, as overarching textual symbols, or as various other forms of spatial or temporal configuration. The reader's imagery need not even be representational. Responding to a description of tulips, a reader might visualize a disembodied swatch of the color red. When imagery does depict explicit, textual referents, such referents might have occurred in the past and be reawakened in the reader's memory in imagistic form, or they might be imagistic anticipations of the future. To identify these many dimensions of imagery, current models of textual analysis need to be developed beyond their exclusively verbal orientations (Long, Winograd, and Bridge, 1989).

That imagery effects are subject to factors of reader, context, and text suffices to curtail any easy, universalizing description of the role of visual imagery in reading and aesthetic experience. What can be shown, however, is that visualization is a process that can have powerful cognitive and affective consequences. Scholars have begun to explore how to integrate the reader's visual imaging responses into their interpretations of literary works, how to use visual imagery to understand a writer's allusions (Collins, 1991), and how to introduce visual imaging strategies into the classroom to enhance comprehension and literary appreciation (Esrock, 1994). As we learn more about the operation of the brain in relationship to consciousness, no doubt the functions of imaging and other cognitive operations will become clearer. Nonetheless, the full incorporation of imaging into literary discourse depends, as it always has, not only on our models of the mind but also on our judgments about the value of the imaging generated by this mind and on the practices of literary study and teaching that are fostered within our institutions.

BIBLIOGRAPHY

Atwood, Margaret. *The Handmaid's Tale.* Boston, 1986.

Calvino, Italo. *Invisible Cities.* Translated by William Weaver. New York, 1974.

Calvino, Italo. *Six Memos for the Next Millennium.* Translated by Patrick Creagh et al. Cambridge, Mass., 1988.

Cocude, Marguerite. "Generating and Maintaining Visual Images: The Incidence of Individual and Stimulus Characteristics." In *Cognitive and Neuropsychological Approaches to Mental Imagery,* edited by Michel Denis, Johannes Engelkamp, and John T. E. Richardson, pp. 213–222. Dordrecht, 1988.

Collins, Christopher. *The Poetics of the Mind's Eye: Literature and the Psychology of Imagination.* Philadelphia, 1991.

Denis, Michel. "Imaging While Reading Text: A Study of Individual Differences." *Memory and Cognition* 10 (1982): 540–545.

Esrock, Ellen J. *The Reader's Eye: Visual Imaging as Reader Response.* Baltimore, 1994.

Freud, Sigmund. *Three Essays on the Theory of Sexuality.* In *The Standard Edition of the Complete Psychological Works of Sigmund Freud,* edited by James Strachey, vol. 7. London, 1953.

Gass, William H. *Fiction and the Figures of Life.* New York, 1970.

Hilgard, Josephine R. *Personality and Hypnosis: A Study of Imaginative Involvement.* 2d ed. Chicago, 1979.

Kosslyn, Stephen Michael. *Image and Brain: The Resolution of the Imagery Debate.* Cambridge, Mass., 1994.

Kunzendorf, Robert G. "Repression as the Monitoring and Censoring of Images: An Empirical Study." *Imagination, Cognition, and Personality* 5 (1985–1986): 31–39.

Lacan, Jacques. "The Mirror Stage as Formative of the Function of the I." In *Ecrits: A Selection,* translated by Alan Sheridan, pp. 1–7. New York, 1977.

Long, Shirley, Peter Winograd, and Connie Bridge. "The Effects of Reader and Text Characteristics on Reports of Imagery during and after Reading." *Reading Research Quarterly* 24 (1989): 353–372.

Marschark, Marc, and R. Reed Hunt. "A Reexamination of the Role of Imagery in Learning and Memory." *Journal of Experimental Psychology: Learning, Memory, and Cognition* 15 (1989): 710–720.

Martyna, Wendy. "The Psychology of the Generic Masculine." In *Women and Language in Literature and Society,* edited by Sally McConnell-Ginet, Ruth Borker, and Nelly Furman, pp. 69–78. New York, 1980.

Moses, Ira, and Joseph Reyher. "Spontaneous and Directed Visual Imagery: Image Failure and Image Substitution." *Journal of Personality and Social Psychology* 48 (1985): 233–242.

Paivio, Allan, and James M. Clark. "Static Versus Dynamic Imagery." In *Imagery and Cognition,* edited by Cesare Cornoldi and Mark A. McDaniel, pp. 221–245. New York, 1991.

Yates, Frances A. *The Art of Memory.* Chicago, 1966.

ELLEN ESROCK

Reading Images

It was probably the seventeenth-century French painter Nicolas Poussin who spoke for the first time of actually "reading" a picture, in a letter to his patron Chantelou: "I beg you to read both the narrative *and* the picture" ("Lisez l'histoire et le tableau"; Poussin, 1964, p. 36). For the first time, an artist referred to an actual "reading" dissociated from the virtual reading of the narrative that underpins a picture. Almost a century after Poussin, the eighteenth-century *philosophe* and inventor of modern art criticism, Denis Diderot, both in his *Salons* and in his *Pensées détachees sur la peinture* (see especially 1957, pensée no. 142), also described his viewing pictures as a sort of "reading," although he did not actually use the term. The twentieth-century Swiss painter Paul Klee followed in their footsteps when he asserted, in his *Pedagogical Sketchbook,* "The eye travels along the paths cut out for it in the work" (Klee, 1963, p. 33). For Klee, an artist was above all a guide who "cut out" scanning paths on the surface of a picture for the viewer to follow.

Iconological versus Semiotic "Reading." Whereas the first half of the twentieth century had been dominated by the "history of ideas" *(Geistesgeschichte)* approach (Anton Dvořák, Julius von Schlosser) and the iconographical studies of French and German scholars such as Émile Mâle, Walter Friedländer, Wilhelm Pinder, and then between the world wars, by Erwin Panofsky with his "iconological" research (i.e., the actual identifying and reading of the texts forming the philosophico-religious underpinnings of a picture), the second part of the twentieth century saw the rise of scholars who approached art from the point of view of linguistics and semiotics and wanted to deal concretely, not metaphorically, with the question of "reading" pictures.

Thus, in a period still dominated by iconographical studies, the late 1930s, the American art historian Meyer Schapiro discovered that medieval illuminations had to be read according to a specific pictorial "syntax" (Schapiro, 1973). Later, E. H. Gombrich, in his *Art and Illusion: A Study in the Psychology of Pictorial Representation* (1969), examined "the beholder's share," that is, the viewer's "reading." Yet, even earlier, at the end of the nineteenth century, the Austrian art historian, Alois Riegl, had the idea of a "grammar" and a "syntax" of pictorial styles. During the hegemony of iconographic research, at the turn of the century, Riegl had defined a dichotomy of human vision that implied the possibility of the concrete "reading" of a flat surface, just as one reads a text. According to this great precursor of the semiotic approach to the visual arts, *optics* and *haptics* were the two fundamental categories. Thus, a certain type of artistic procedure was based on the scanning of objects according to their outlines. The gaze that follows lines virtually (or implicitly) inscribed over the representation of things, and follows the outlines of figures, Riegl called "optical." The opposite type of vision, which focuses on surface and texture, and emphasizes the value of the surface of objects, Riegl called "haptic"—from the Greek verb *haptein,* to grasp, and from *haptikos,* capable of touch. Thus, on the level of artistic creation the "optical" gaze of a painter will produce linearity and angularity. In contrast, "haptic" creativity will emphasize surface and have the gaze focus on surfaces. According to Riegl's formula, all forms of art can be grouped under the heading "Outline and/or color in plane or volume *(Umriss und/oder Farbe in Ebene oder Raum)*" (Riegl, 1992). The "optical" eye merely brushes the surface of things, whereas the "tactile" eye penetrates "in depth" and finds its pleasure in texture and grain. In the 1920s and 1930s Riegl's dichotomy was rewritten by another pioneer of visual semiotics, the Swiss art historian Heinrich Wölfflin, in terms of another dichotomy—"the linear" as opposed to the "pictorial."

Riegl and Wölfflin died without knowing that their dichotomy of vision corresponded to the findings of modern physiology, when it was discovered that there are, indeed, two types of vision, according to whether it is the "cones" or the "rods" in our retina—the instruments of our apperception of things—that do the "seeing."

The Scanning of Pictures. Our contemporary period has seen the flourishing of so much research on the question of visual apperception that the Rieglian/Wölfflinian categories of optics and haptics have been made more tangible. In the former Soviet Union, the physiologist A. L. Yarbus (1967), and in the United States a team of researchers, David Noton and Lawrence Stark (1971), devoted illuminating studies to the question of "eye movements" in the apperception of flat pictures. Yarbus showed the results of the "linear" ("optic") traces of eyes looking at

a picture by Ilya Repin in seven scanning episodes of twenty-five seconds each.

The picture by Repin that was "scanned" by Yarbus's subjects is titled *The Unexpected.* It represents the return of a political prisoner, a 1905 "Decembrist," from Siberia after a long exile. His children seated at the table do not recognize him. No matter what the content of the "scanned" picture might be, the seven linear diagrams are formed by straight lines (called by the French term *saccades*). It is the movements of the fovea that Yarbus recorded here, that is, the part of the retina endowed with the greatest concentration of photosensitive cells. By placing each recorded saccade over a transparent paper, and subsequently applying this paper over the surface of the illustration, it is possible to see in detail how seven persons have "read" this picture. In Rieglian/Wölfflinian terms, the saccades represent the "optical" vision of eyes that quickly follows the outlines or "jumps" over things to establish relations between the persons and the objects in the painting—indeed, many saccades here follow the lines of the floorboards, the door and window frames, and so forth. They are the fixed traces of "optical" readings. Another example from Yarbus shows how an eye reads "optically" the photograph of a sculpture, the head of the Egyptian queen Nefertiti, during a period of three minutes.

It is remarkable that during a fairly long period of time, the scanning begins to resemble (to "mimic") the thing it scans, whereas in a rapid scanning of twenty-five seconds, it is a mere "reading trace." "Haptic" vision—concentrated in the points of fixation between the saccades—can also be observed. This comes graphically to the fore in the study by the Americans Noton and Stark, as in the scanning of a Klee drawing by a viewer's eye.

Here, the trace of the scanning shows the importance of the points of fixations in relation to the saccades, which link one to the other. These saccades are seen in their true, hesitating trajectory, which is different from the straight lines in the Yarbus diagrams. Moreover, the "fixation points" are fairly large blots. Their size is due to the length of the "in-depth" fixation of the observer's gaze. These small pools of darkness are the imprint of the "touch" of the viewer's eye on the canvas. They show "the haptic" element in the observer's gaze while the *fovea* focuses on the surfaces, on the texture, and on the coloring. Such a "reading trajectory" resembles the "drippings" of Jackson Pollock's Action Painting rather than the "seismographic" saccades in the Yarbus diagram. This type of haptic eye-touch reads a picture "nonsemantically," as a penetration into a texture, and as an index of the painter's hand—that is, as texture and *signature* rather than as *representation.*

In contrast, the "optical," "seismographic" saccades Yarbus recorded establish a kind of "bridge" between semantic elements: from the floor to the feet, then to the hand, then to the eye, and so on. This type of reading is a "seman-

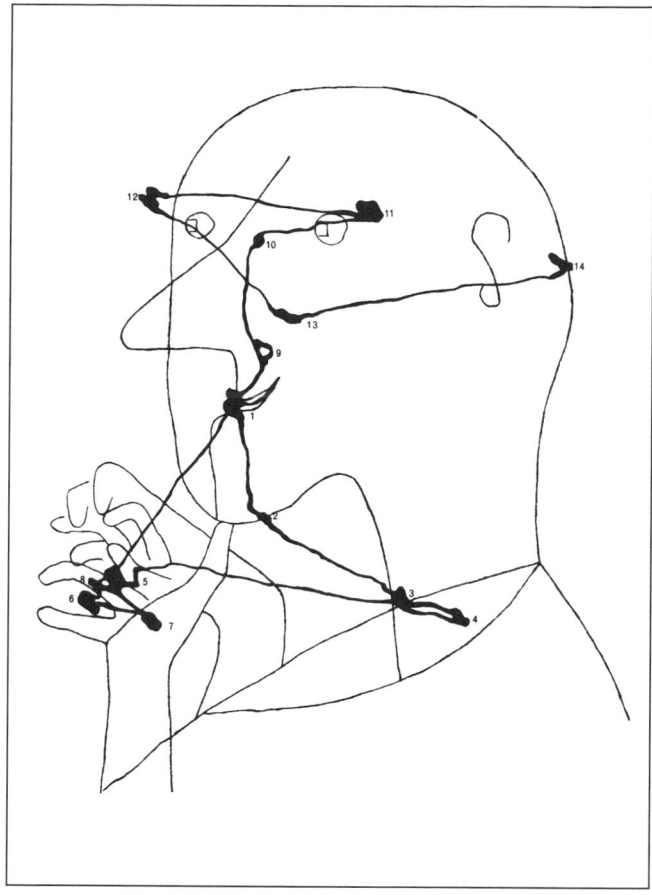

IMAGERY: Reading Images. Lorelle A. Raboni, *Redrawn for Scientific American Magazine after Paul Klee's Old Man Figure* (originally from *Scientific American,* June 1971). (Copyright by Lorelle A. Raboni; used by permission.)

tic" reading of the picture. The painting is read, here, as an *icon* in the sense of C. S. Peirce. The eye produces a diagrammatic reproduction of the thing it scans.

A Quattrocento "Scan Path": The Gesture of Demonstration. One of the protosemioticians who most influenced the "reading" approach to pictures was certainly the German linguist Karl Bühler, whose three functions of language, *expression, representation,* and *appeal* were later to be recast by the linguist Roman Jakobson as *emotive, referential,* and *conative* functions. It was Jakobson who suggested the transfer of these three functions to all objects of communication. Extending the "appeal function" to pictures, the Israeli semiotician Claude Gandelman (1986a, 1986b, 1991) isolated an "appeal structure" inscribed in pictures of the Italian quattrocento, before the Italian Renaissance was in full swing. Appeal structures are present in the work of such painters as Fra Angelico, Masaccio, and Piero della Francesca, who frequently represented a human figure pointing toward an object located within the picture frame while looking orthogonally "outside the picture plane" as

though addressing the viewer. This gesture by a demonstrator who "seeks out" the gaze of the outside observer is a gaze-directing device. In the quattrocento, Leon Battista Alberti promoted such a device: "I like to see in a painting a figure which admonishes us and designates with his finger that which is taking place inside the picture, or one which makes us a sign with its hand for us to watch something that is occurring inside" (Alberti, 1966, p. 123). Gandelman renamed this "outward looking gaze" "the gesture of demonstration," with reference to the dramaturgy of Bertolt Brecht. By practicing the gesture of demonstration *(der Gestus des Zeigers)* the Brechtian actor meant to foreground the "showing" element in his acting instead of the mimetic-illusionistic element. He always showed that he was showing, or demonstrating, social possibilities but never tried to embody a part. In painting, the gesture of demonstration is a rhetorical gesture directed toward the external viewer: it is an "appeal" signal inviting him or her to look in a specific direction or at a specific object located "inside" the representation. Thus, a specific scan path for the eye of the outside "reader" to follow is always to be found in pictures containing the "gesture of demonstration."

The Picture as "Scanning Matrix." The idea of "reading" pictures in a literal sense refutes Gotthold Ephraim Lessing's famous eighteenth-century dichotomy between the temporal arts—literature, music—and the spatial ones—painting, sculpture, architecture (Lessing, 1965). As Paul Klee wrote in his "Creative Credo,"

> In Lessing's *Laokoön* . . . , there is much ado about the difference between time and space in art. Once we examine it more closely, this is really just a bit of erudite hair-splitting, for space, too, implies the concept of time. It takes time for a dot to start moving and to become a line, or for a line to shift its position so that a plane is formed. . . . The act of viewing the work of art, too, is in essence a function of time. The beholder focuses on one section and then on the other. . . . The work of art, then, results from physical movement; it is a record of such movement, and it is perceived through the movement [of the eye muscles].
> (Quoted by Grohmann, 1981)

Similarly, the French scholar Louis Marin described the "built-in" scan path for the viewer's eye as a "matrix of meaning":

> The picture forms an organic totality of eye movements and the finished picture is the impregnation of the surface texture by a compound of signs both topical and dynamic in character whose aim it is to differ the final unity of the scene as a structured totality both in time and space, in opposition to the view that claims that a picture can be "seized" at a glance—Lessing in his *Laocoön*. This circuitous route inscribed on the surface for the gaze of the spectator is not a fixed one. The points of fixation imply possibilities of choice. Yet, the picture forms a "matrix" of trajectories [for the eye to follow] from which each viewer—and each generation of viewers—generates the global meaning of the picture.
> (Marin, 1968, p. 863)

BIBLIOGRAPHY

Alberti, Leon Battista. *On Painting* (1435). Rev. ed. Translated by John R. Spencer. New Haven, 1966.

Diderot, Denis. *Salons.* 4 vols. Edited by Jean Seznec and Jean Adhémar. Oxford, 1957–1967.

Diderot, Denis. *Essais sur la peinture/Pensées détachees sur la peinture* (1795). Paris, 1955.

Gandelman, Claude. "The Scanning of Pictures." *Communication and Cognition* 19 (1986): 3–24.

Gandelman, Claude. *Le Regard dans le texte: Image et écriture du quattrocento au XXe siècle.* Paris, 1986.

Gandelman, Claude. *Reading Pictures/Viewing Texts.* Bloomington, Ind., 1991.

Gombrich, E. H. *Art and Illusion: A Study in the Psychology of Pictorial Representation* (1960). 2d. rev. ed. Reprint, Princeton, N.J., 1969.

Grohmann, Will. *Paul Klee.* Translated by Norbert Guterman. Reprint, New York, 1981.

Klee, Paul. *Pedagogical Sketchbook* (1925). Translated by Sibyl Moholy-Nagy. New York, 1953.

Lessing, Gotthold Ephraim. *Laokoön: or, The Limits of Painting and Poetry* (1766). Edited by Dorothy Reich. Oxford, 1965.

Marin, Louis. "Semiologie de l'art." In *Encyclopaedia universalis,* vol. 14, p. 863. Paris, 1968.

Marin, Louis. *Études semiologiques.* Paris, 1971.

Noton, David, and Lawrence Stark. "Eye Movements and Visual Perception." In *Perception: Mechanisms and Models, Readings from Scientific American,* pp. 219–230. San Francisco, 1971.

Poussin, Nicolas. *Lettres et propos sur l'art.* Edited by Anthony Blunt. Paris, 1964. See especially p. 36.

Riegl, Alois. *Problems of Style: Foundations for a Theory of Ornament* (1893). Translated by Evelyn Kain. Princeton, N.J., 1992.

Schapiro, Meyer. *Words and Pictures: On the Literal and the Symbolic in the Illustration of a Text.* The Hague, 1973.

Wölfflin, Heinrich. *Principles of Art History: The Problem of the Development of Style in Later Art* (1921). Translated by M. D. Hottinger. New York, 1950.

Yarbus, A. L. *Rol Dvizhenij Glaz V-Protsesse Zrenya.* Moscow, 1965. Translated as *Eye Movements and Vision* by Basil Haigh, edited by Lorrin A. Riggs (New York, 1967).

CLAUDE GANDELMAN

IMAGINATION. [*This entry comprises two essays on a notion central to the creation of art and to aesthetics:*

The Imaginary

Contemporary Thought

The first essay is a survey of the history of philosophical and literary treatments of the "imaginary," using Jean-Paul Sartre's term for speaking about the imagination. The second essay reflects contemporary thinking about the imagination, particularly within Anglo-American philosophy. For related discussion, see Creativity; Epistemology and Aesthetics; Fiction; Perception; Play; *and* Sartre.]

The Imaginary

The *imaginary*—a relatively modern term—has gained currency in the face of mounting skepticism concerning the "true nature" of imagination or fantasy. Jean-Paul Sartre in-

troduced the term *l'imaginaire* into the critical discussion, not least as this human potential manifests itself in rather different ways: with flights of fancy it can wander off into worlds of its own, or, as imagination, it can conjure up images, or summon the absent into presence. In ordinary experience, the imaginary tends to manifest itself, in a somewhat diffuse manner, as fleeting impressions, projections, daydreams, and other reveries. Small wonder that there have been a great many attempts in human history to come to grips with this protean potential.

Zachary Mayne remarked in his *Two Dissertations concerning Sense and the Imagination* (1728) that the imagination is "like the Chameleon, of which Creature it is reported, that it changes its Hue according to the Colour of the Place where it happens to be." This was by no means the first attempt to gain a cognitive hold on the imagination. Already in the seventeenth century, increasing attention to it had once more focused on Aristotle's conception of fantasy as lying halfway between perception and thought. Thomas Hobbes considered the Greek *phantasia* as the imagining of an object no longer present, and thus constituting a "decaying sense." Up to the sixteenth century the imagination had occupied a lower rank, not least because, through its link to the senses and memory, it was present as a latent subversion, if not an actual defiance, of a reason-dominated hierarchy. But since then, imagination began its advance, and in the eighteenth century, it gained prominence thanks to its multiple uses. When it became a matter of trying to take possession of the empirical world, knowledge had to be wrested from experience, which meant processing the data that found their way into the mind. Joseph Addison suggested that "the exercise or pleasure of the imagination depends on a full and directed interplay and integration of as many faculties and operations of the mind as possible." Samuel Johnson, unlike Hobbes, considered imagination no longer a "decaying sense" but a central guide, linking past, present, and future so as to enable one to hold a steady course in the midst of all the changes one undergoes.

So long as the eighteenth-century view of the imagination prevailed that it acted as a means of combining—and this applies particularly to the first generation of association psychologists, of whom David Hartley was the most prominent representative—it was conceived, in accordance with the guidelines of empirical cognition, as a mechanistic operation linking existing data. As such, it was used as a foundation for cognition without any satisfactory account of how it could achieve such a status. This lack preoccupied the second generation of association psychologists, who revealed the imagination as a synthetic power by describing it as a process of dynamic flow that fuses heterogeneous elements into a unity.

These diverse descriptions of the imagination as a combining agent did not, however, provide a definition of what it actually is. David Hume used the terms *fancy* and *imagina-*

tion synonymously; he saw the imagination as "kind of magical faculty of the soul, which, tho' it be always most perfect in the greatest geniuses, and is properly what we call a genius, is however inexplicable by the utmost efforts of human understanding." He is far from being alone in this view. In the late 1780s Johann Gottfried von Herder described the imagination as one of the least explored of all the human faculties, although it represented the real link between body and mind. Johann Gottlieb Fichte, generally following Immanuel Kant, took it to be the bridge between reason and the senses—a power "of which we are seldom or never conscious," although in his *Wissenschaftslehre* (1794/95) he stresses: "Here . . . it is taught that all reality is brought forth solely through the imagination."

Thus, in both idealist and empiricist philosophy the imagination was on its way to becoming the ground of all cognition, although the various functional descriptions led only to its unfathomableness. A function whose basis is inexplicable and a grounding that dwindles into a *regressus* points to a fundamental ambiguity of the imagination. Is it an evidential experience that defies conceptualization by discourse, or is it discourse that makes it ambiguous because imagination cannot be contained within its demarcations? What is obvious is that imagination changes according to the direction from which it is approached, and since no approach can ever encompass it in its entirety, it turns into a stimulant for an unceasing endeavor to define it. What, then, is it as itself?

The perennial answer to this question from Aristotle right through to the end of the eighteenth century was: "a faculty" that, on close inspection, however, was divided into several faculties as undertaken by Johannes Nikolaus Tetens, who provided the most detailed analysis of the imagination as a faculty. According to the activities exercised, the imagination is split into three different faculties: "the faculty of absorbing ideas into itself *(facultas percipiendi)*, a power of comprehension, but also a faculty of bringing them forth again, a power of re-conception, which we normally call fantasy or imagination . . . insofar as it renews pictorial ideas of sensation," and finally "the faculty of poetry," which is one of "developing, dissolving and reuniting . . . bringing together and merging" (1979). This was a highly progressive insight at the time. Everything in the imagination has come from elsewhere, and ultimately this means that the imagination is not self-activating but requires activating stimuli from outside itself. Since even the "poetic power" needs given circumstances to bring it into operation, the inescapable impression remains that the imagination as a faculty is a blank that resists filling by whatever faculty psychology has proposed.

This was brought out into the open by Samuel Taylor Coleridge, who made the last significant attempt to grasp the imagination as a faculty, although he changed the inherited notion to such an extent that the time-honored faculty con-

cept ceased to be the overarching view of the imagination. This can be seen from a remark he made in 1818: "As every faculty, with every the minutest organ of our nature, owes its whole reality and comprehensibility to an existence incomprehensible and groundless, because the ground of all comprehension [*sic*]." If a faculty originated from a knowable ground, then it would, first and foremost, represent that ground and could not, therefore, ramify into a plurality of activities, which Coleridge specified by his division of that "faculty" into "primary imagination," "secondary imagination," and "fancy." If these graduated definitions of the imaginative faculties have no fathomable ground, then their groundlessness cannot be equated with a hidden intentionality. Activities, according to Coleridge, are powered not by the faculties themselves but by the subject in its different attempts to relate to itself and to the world in an equal manner. Within the finite mind the infinity of self-constitution is repeated (primary imagination); the conscious will guides the operation of undoing the world of objects in order to enable the subject to recreate it anew in such a way that it will become present to itself (secondary imagination); and finally empirical choices—made by fancy—facilitate the combining of data to meet the needs of the situation.

So long as the imagination was linked to the subject, it could be regarded as a faculty that, like all the other faculties, appeared to represent the natural equipment of human beings. Even the prevalent concern for the self-constitution of the subject only modified the inherited concept. The loss of certainty as regards the ground of the faculties led Coleridge, however, to view the imagination not only as divided but also as a potential that has to be pricked into action from outside itself.

The nature of "faculty" is bound to change when the agent that activates it changes. Coupling the imagination with the self-constitution of the subject considerably altered the traditional concept of the faculties; but finally, the classification of the imagination as a faculty became virtually obsolete when it ceased to be linked to the subject. Such a shift caused a change in the designation of that human potential: imagination as a faculty gave way to the rather diffuse notion of the imaginary. This was the step taken by Sartre in his book *L'imaginaire* (1940; The Psychology of Imagination, 1948), in which he "aims to describe the great functions of the consciousness to create a world of unrealities or 'imagination,' and its noetic correlative the imaginary." If the imaginary is closely allied to the activities of consciousness, the categorization of the imagination as a faculty is invalidated. Instead of being an unplumbable attribute of human beings, the imaginary is a mental image that indicates "the relation of consciousness to the object; in other words, it means a certain manner in which the object makes its appearance to consciousness." These images present their object "as being given-as-absent." Forming a mental image entails capturing the absent or nonexistent through an ana-

logue that feeds on knowledge, memory, experience, information, desire, and so on. There is evidently no consciousness without image making, for only through image making is consciousness able to fulfill its intention that is concretized to the degree to which its target becomes determinate. As it seeks to open up this target, what is must be animated by mental images—not least in order to bridge the difference between intention and object.

The imaginary, then, is not a self-activating potential but has to be brought into play from outside itself. It follows that the imaginary has no intentionality of its own but has intentions imposed on it by the demands of its activator. Precisely because it is without intentionality, it appears to be open to all intentions that will always be tied to what they trigger, so that something will "happen" to the activator. Therefore, consciousness may be flooded by its images if attention lapses, or it may slide into the mental image, and thus undergo modification that inscribes itself into intentionality by splitting it into various modes of positing, perceiving, ideating, and realizing, as Sartre had it. Thus, the imaginary can never be identical to its intention-led mobilization (otherwise, it would constitute the intentional activation of itself). Instead, the imaginary discloses itself in interplay with its activators. This interplay is identical neither to the intentions nor to the imaginary gestalt that brings it about, although that gestalt could not come into being without the intention-led mobilization of the imagery.

The activation of the imaginary is by no means confined to consciousness. There is no doubt that the imaginary plays a role in perception, and that imaginary components increase in ideas, dreams, daydreams, and hallucinations. They are all different manifestations through which the imaginary takes on an experiential existence. As an idea, the imaginary gives presence to what is absent; as a dream, it is the captivity of the dreamer to his or her imagery; as a daydream, it is a fading of gestalts into featurelessness; and as hallucination, it is a consciousness overwhelmed. Perhaps it is only in madness that the imaginary takes on a comparatively pure presence; but even madness has forms that might either ossify the imagination through vagrant consciousness or drive intentionality out of consciousness. For madness, like all other phenomena, has the nature of a product, and the imaginary always bears the imprint of its contextually conditioned result.

Again, this fact implies that the imaginary does not produce its own salience, which comes about through interplay with the factors that have mobilized it. The factors that influence the manifestation of the imaginary will vary in their complexity. In perception, it is visual anticipation governed by intentional projections that release the imaginary. In the idea, the imaginary is directed by memory-laden, cognitive factors aiming at presenting the absent. The dream is dominated by a consciousness that the imaginary is nothing but imagistic, although consciousness has no control over the

images. In the daydream, the imagery plays with our preoccupations, producing projected gestalts and then snuffing them out. In the hallucination, the imaginary triumphs over consciousness, which is present only as a mutilated intentionality. The factors mentioned all endow the imaginary with its shape, but they, in turn, also undergo changes when mobilizing such a potential.

Historically, the imaginary, from its humble beginnings as imagination, embarked on an unforeseeable career, advancing to a kind of *materia prima* in spheres such as psychoanalysis, anthropology, and, in recent times, social theory. Concepts like *ur*-fantasy or the primordial *état imaginaire*, which play a role in present-day discussion, show that the imaginary is on its way to becoming the be-all and end-all. The problem with *ur*-fantasy and the primordial *état imaginaire* is that if they are meant to be ultimates, they are frequently something other than what we may understand by the *imaginary*. This is evident in certain psychoanalytical and anthropological discourses. With Sigmund Freud, for instance, fantasy is the meeting point for ideas of expectation and fulfillment. Jacques Lacan conceives the imaginary as an oscillation between a specular self and a core self that looks at itself in the mirror with fascination and aggression. For Donald Woods Winnicott, the imaginary is a "transitional object" that allows the infant to bridge the gap after separation from the mother. Arnold Gehlen regards the *état imaginaire* as a state of transference, indicative of deficient instincts, which human beings can repair only by projecting themselves into a featureless future, thus countering a challenging reality by overlaying it with self-projections. Projecting entails transposing oneself into something other, and incorporating it into oneself to such a degree that one can cope with the rift between deficient instincts and the pressure of the external world. And finally Cornelius Castoriadis proposes the "radical imaginary" as the unfathomable precondition for the institutionalization of society.

Owing to the urge of getting to know what it is, investigators have continually forced the imaginary into pigeonholes that do not fit. The massive revaluation, however, that it has enjoyed since Romanticism has been more concerned with its operational nature, which involves the interplay with the factors mobilizing it. Discourse remains unavoidable, but its exploration will be aimed less at a definition of the imaginary than at a clarification of what happens to it when pricked into action.

In this respect, the fictive component of literature is bound to mobilize the imaginary in a different manner, for it has less of the pragmatic orientation required by the subject, the consciousness, or the exigencies of the social world, all of which channel the imaginary in quite specific directions. The fictive compels the imaginary to take on form while at the same time acting as the medium for its manifestation. As the fictive is intentional, it has to activate the imaginary, since the fulfillment of intentions requires imaginings, for intentionality cannot itself bring forth what it targets. At best, it adumbrates an empty idea that requires substantiation, which is supplied by the imaginary on which the fictive has imposed form.

BIBLIOGRAPHY

Bachelard, Gaston. *On Poetic Imagination and Reverie.* Translated by Colette Gaudin. Indianapolis, 1971.

Caillois, Roger. *La Pieuvre.* Paris, 1973.

Casey, Edward S. *Imagining: A Phenomenological Study.* Bloomington, Ind., 1976.

Costa Lima, Luíz. *Control of the Imaginary: Reason and Imagination in Modern Times.* Translated by Ronald W. Sousa. Minneapolis, 1988.

de Laurentis, Teresa et al., eds. *The Technological Imagination: Theories and Fictions.* Madison, Wis., 1980.

Engell, James. *The Creative Imagination: Enlightenment to Romanticism.* Cambridge, Mass., 1981.

Harari, Josué V. *Scenarios of the Imaginary: Theorizing the French Enlightenment.* Ithaca, N.Y., 1987.

Irwin, W. R. *The Game of the Impossible: A Rhetoric of Fantasy.* Urbana, Ill., 1976.

Iser, Wolfgang. "The Aesthetic and the Imaginary." In *The States of 'Theory': History, Art, and Critical Discourse,* edited by David Carroll, pp. 201–220. New York, 1990.

Kamper, Dietmar. *Zur Soziologie der Imagination.* Munich, 1986.

Kearney, Richard. *The Wake of Imagination: Ideas of Creativity in Western Culture.* London, 1988.

McFarland, Thomas. *Originality and Imagination.* Baltimore, 1985.

Sartre, Jean Paul. *The Psychology of Imagination.* New York, 1948.

Schwab, Gabriele. "Genesis of the Subject. Imaginary Functions, and Poetic Language." *New Literary History* 15 (1984): 453–474.

Tetens, Johann Nicolas. *Die philosophischen Werke.* Hildesheim and New York, 1979.

Warnock, Mary. *Imagination.* Berkeley, 1976.

WOLFGANG ISER

Contemporary Thought

Imagination may refer to a mental faculty or capacity, that is, a state or condition of a person allowing the performance of a certain activity. Or it may refer to the exercise of a capacity or ability, that is, the activity itself. Imagination, usually to its detriment, may be contrasted with knowledge, with reasoning, or with perception. What is imagined may be propositional (I imagine that something is true), objectual ("Imagine an airplane painted with stripes like a zebra"), or experiential (what it is like to experience something).

What follows is an analysis of experiential imagination: imagining what it would be like to *do* this or that, how others feel, what it is like to be a certain kind of person (including situations where one encounters and judges other sorts of people), and what it would be like for the imaginer to be in a certain kind of situation. In all of these cases imagining something is a temporally extended process, rather than a momentary psychological state—not merely a matter of imagining *that* something is the case, but imagining doing, being, or experiencing something. There is also an affective

or experiential component to these imaginings that cannot be captured by a linguistic description in the way the content of a thought is captured by a sentence.

Aesthetically, the relevance of this set of imaginings is far-ranging. Imagining what it is like to live in various fictional worlds can stimulate artistic creativity. Imagining what it is like to be a certain way may be part of performance technique for acting, dancing, playing music, or singing. Such imaginings are also often essential to appreciating an artwork. One may imagine what it is like to be a given fictional character, or to live at a particular time, or to move or act in a given way. In response to an artwork one may imagine having various emotions, or have emotions or other affective experiences by virtue of engaging in imaginal activity.

This essay discusses the psychological processes that underlie experiential imaginings as described above. Explained first is how these imaginings depend on our cognitive and affective malleability, so that ideas, thoughts, and perceptions may play different cognitive and affective roles for the same person. Certain perceptions and beliefs go "off-line," and sensitivities are mobilized so that one experiences things differently from how one normally would. Second, the concept of simulation is described. If the mental role of one's perceptions, ideas, and thoughts simulates another person's mental activity, then one imagines what it would be like to be that sort of person.

Human beings are cognitively and affectively malleable in that perceptual stimuli, ideas, and thoughts may all play a variety of cognitive and affective roles for the same person at different times. Sometimes a shift in cognitive or affective role is under voluntary control. A silhouetted figure is seen now as a vase and later as the profiles of two people facing each other. A painting is seen now as a picture of a boat, later as blotches of contrasting colors, and still later as daubs of paint. A novel is read first with empathy for the narrator, a second time experiencing things from the point of view of one of the narrator's victims, and a third time from a perspective that sees both of the others as narrow and selfish.

Context, past experiences, relatively automatic mental shifts analogous to the perceptual kind, and concerted voluntary effort all affect the kind of mental role perceptions, ideas, and thoughts play, and also whether they figure in generating some kind of emotion or affect. A passage in a novel, painting, or film may prompt the thought of a neglected child. Depending on what other thoughts and desires—imagined or actual—play a role in producing other thoughts, ideas, and feelings, the thought might in turn generate empathy with or compassion for the child, anger with the parents, or frustration with the political regime. Value-laden language or style of writing in a novel, the colors and spatial arrangements of images in a painting, music and camera angle in a film, will prompt some trains of imaginings and discourage others, provided one has a sensitivity to these sorts of properties of an artwork. It is possible to encourage and develop these sensitivities along with the ability to shift into a variety of psychological gears. One might shift into the first person, seeing things from the protagonist's point of view, or one might be sympathetic while not actually sharing his or her feelings, or one might loathe the protagonist and exult in his or her downfall.

Imagining what it would be like to *do* this or that, how others feel, what it is like to be a certain kind of person, and what it would be like to be in a certain kind of situation all require using psychological flexibility. One may use it to *simulate* the mental activities of a person or the activities a person would perform in the situation envisioned. When one imagines how fictional characters feel, or what it would be like to live during a certain period in history, one makes use of this capacity to simulate. A simulation instantiates the structurally significant relationships among processes and subprocesses of what is simulated. One in effect uses one's own mind to model someone else's, that is, how another person's mind would function, psychologically, under the relevant set of circumstances. A simulation may become *so* like what it simulates that it no longer merely simulates, but actually instantiates those mental processes. Actors have been known to have difficulty distinguishing between being "in character" and being (what has become) themselves. Plato's worry that audiences at tragedies will acquire the emotional traits of characters in a play is also accounted for by imagination as simulation.

The capacity to simulate depends on having the ability to make perceptual, cognitive, and affective shifts as described above. It requires being able to put some psychological states—such as some of our beliefs and values—"off-line" and make new "inputs"—for example, the forms and content of the work—function in ways appropriate to the simulating. It is not always possible to do this; the imagination may not be up to the task. Suppose the character and the person attempting to simulate have different character and personality traits: the simulator is timid and reticent, and the character is irascible and pugnacious. If the stimulator imagines how the *character* feels when a bully enters the room, the simulator would imagine being angry and prepared to fight, but if the simulator imagines how *she* or *he* would feel she or he imagines being fearful and prepared to escape. The simulator may think he or she is imagining how a character in a novel feels when he or she is really imagining how *he* or *she* would feel under those circumstances (which is a case of projection, not empathy).

With sufficient imagination one can also shift into a point of view or attitude of someone whose values or desires one does not share. It is sometimes argued that it is not possible to imagine having feelings and attitudes deriving from values and desires one does not actually have. But if it were not possible, there would have been no need for anyone—from Plato to Marxists to feminists of various stripes, and other ethically and politically oriented criticism—to worry about it. Words,

sounds, and images are powerful, and may subtly or radically shift one's psychological orientation in ways one does not even realize. Artworks that present phenomena of which one disapproves in a favorable light, or phenomena of which one approves in an unfavorable light, can be disturbing because they challenge one's values. Imagining things from a contrary point of view is potentially threatening: one might find it so persuasive and natural to see things from that perspective that one's values and desires actually shift. Nevertheless, an inability to imagine an opposing point of view evidences a lack of imagination. *Control* over one's imagination, so that one can shift into a different point of view, enables one to take advantage of human mental flexibility without being controlled *by* whatever prompts the imagining.

BIBLIOGRAPHY

Elliott, R. K. "Aesthetic Theory and the Experience of Art." *Proceedings of the Aristotelian Society* 67 (1966–1967): 111–126. Reprinted in *Philosophy Looks at the Arts: Contemporary Readings in Aesthetics,* edited by Joseph Margolis, rev. ed. (Philadelphia, 1978), pp. 45–57.

Goldman, Alvin I. *Philosophical Applications of Cognitive Science.* Boulder, Colo., 1993. See pp. 63–97, 125–154.

Morton, Adam. *Frames of Mind: Constraints on the Common-Sense Conception of the Mental.* Oxford, 1980.

Scruton, Roger. *Art and Imagination: A Study in the Philosophy of Mind.* London, 1974.

Walsh, Dorothy. "The Non-Delusive Illusion of Literary Art." *British Journal of Aesthetics* 23 (Winter 1983): 53–60.

Wollheim, Richard. *On Art and the Mind.* London, 1973; reprint, Cambridge, Mass., 1974. 54–83.

Wollheim, Richard. *The Thread of Life.* Cambridge, Mass., 1984. See pp. 62–96.

SUSAN L. FEAGIN

IMPRESSIONISM. The term *Impressionism* designates a movement in French painting that emerged in Paris during the late 1860s and 1870s, attained great international renown by 1900, and—despite the inevitable cycles of approbation and disapproval from critics and historians over the years—now undoubtedly stands as the most popular and most widely reproduced body of paintings of all time. Artists participating in the movement include Gustave Caillebotte, Mary Cassatt, Edgar Degas, Claude Monet, Berthe Morisot, Camille Pissarro, Pierre-Auguste Renoir, and Alfred Sisley. The name of the movement derives from the quip of a critic prompted by Monet's *Impression, Sunrise* of 1873 when the painting was shown in 1874 at the first independent exhibition held in Paris by the group; seven more such independent exhibitions would follow, extending until 1886. Éduoard Manet, though somewhat older than most of these artists and by his own choice not a participant in their independent exhibitions, is often considered an Impressionist owing to his close personal and stylistic affinities with various members of the group. Similarly, Paul Cézanne occupies a place at the margins of Impressionism:

while active in two of the early group exhibitions and in age quite close to Monet and Renoir, Cézanne has since frequently been treated as a member of the generation of French artists following the Impressionists owing to his virtual invisibility in the public eye preceding his enthusiastic discovery by critics during the 1890s.

The term Impressionism, more generally, refers to the manner of painting exemplified by this circle of French artists, which painters in other countries subsequently adopted, giving rise to such late-century classifications as "American Impressionism" and "the German Impressionists." In its adjectival forms, the term may be applied even more widely across time and across media to indicate a general looseness of handling; hence, for instance, Jan Vermeer has been characterized as the most "impressionistic" of the Dutch genre painters (and the revival of interest in Vermeer during the latter half of the nineteenth century is not unrelated to the rapid success of Impressionism), whereas Auguste Rodin, a contemporary of the Impressionist painters, has been labeled an "impressionist" sculptor.

What was French Impressionist painting about? The answer to this question—much like the manner of Impressionist painting itself—presents a superficial simplicity behind which lurks a marked epistemic complexity. Impressionism, in the most unadorned formulation of its program, concerned itself with the capturing, in paint, of "impressions"; or, as the critic Jules Castagnary phrased it in 1874 in his since oft-cited definition: "They are *impressionists* in the sense that they render not the landscape, but the sensations produced by the landscape." It seems these painters had, as if through an act of phenomenological reduction, bracketed the hard reality of the external world in order to devote their attention to the phenomena of impressions offered to the senses of perception.

Yet this assessment—served up over and over in various platitudinous guises within the critical and historical literature—only defers an account of the movement. For where, precisely, were these impressions supposed to reside? Richard Shiff (1984) has analyzed the consequent interpretative uncertainty with great precision and insight.

On one hand, the impression could locate itself in the "effect" (*effet,* a term frequently used by contemporary critics) generated by the natural forces of an externalized real world—a collection of light rays bouncing off certain material objects and passing into the eye, for example. Impressions of this sort could claim a certain objectivity since any observer, presumably, could step up to a given spot and receive the same basic stimuli as had the artist; the artist's skill lay solely in his or her technical capacity to record accurately in paints that constant prospect. This interpretative tack, articulated by many writers in a long-standing historiographic tradition, regards the meaning of Impressionist painting to derive principally from the close *iconic* match it establishes between its own surface and the sensations emit-

IMPRESSIONISM. Claude Monet, *Boulevard des Capucines* (1873–1874), oil on canvas, 31 5/8 × 23 3/4 in. (80.4 × 60.3 cm); Nelson Atkins Museum of Art, Kansas City, Missouri (Purchase: the Kenneth A. and Helen F. Spencer Foundation Acquisition Fund). (Photograph copyright by the Nelson-Gallery Foundation; used by permission.)

ted by some external reality, and thus, by extension, from the constituent elements of that reality. Many critics and art historians, accordingly, have dealt with Impressionism by cataloging and analyzing its iconography of modern life, that distinctively contemporary collection of things and their settings ostensibly present at the actual time of painting and thus providing the artist with his or her impressions of them: cityscapes of a newly Haussmannized Paris, scenes of bourgeois domesticity and of more risqué forms of urban entertainment, depictions of new practices of leisure—boating, strolling, and the like—to be found in the suburbs surrounding the capital, landscapes of more distant sites rapidly being transformed into tourist destinations in this age of increased rail mobility.

On the other hand, the impression can equally well be seen to dwell in the individual artist's personal experience of such external stimuli, and that experience potentially derives much more from the particular subjective temperament of the artist than from any objective sensory data flowing from the external world into that artist's eye. Here the artist's gift stems, not from a basic reportorial competence, but rather from his or her heightened sensibility—sensitivity both to that which is given to the senses and, more important, to that which is felt by the emotions. The painting spoke of (and to) that individual sensibility. In line with this second interpretive predilection, a string of writers—as lengthy as those of the first historiographic tradition—have plumbed Impressionist paintings for signs, perforce *indexical,* of the personal presence of the creator at the moment of creation before the canvas: large oily brushstrokes registering the act of painting, a bright palette suggesting a high-pitched temperament, the avoidance of the learned (and thus communally held) academic practices of composition, modeling, and drawing.

Whereas a practitioner of the first interpretive tendency might look at Monet's *Boulevard des Capucines* of 1873 and declare, "it's a fresh new boulevard in the most fashionable *quartier* of Paris," a writer in the second tradition would ascertain before this same canvas, "it's a Monet." Whereas the first practitioner would tend to subordinate the stylistic characteristics of the painting to the demands of the scene portrayed—a light and airy city calls for a light and airy technique—the second writer could treat the contemporary subject matter as merely the necessary consequence of an autonomous temperament that refuses any imposed or inherited corpus of themes. Nevertheless, it would be misleading to imagine a rigid division of semiotic labor between icon and index in, respectively, the first and second lines of interpretation. Just as the iconic sign can register the sensibility of the individual artist—he or she, after all, chose to stand before this particular scene, chose to frame it in this particular manner—the indexical trace can make reference to the physical world giving rise to the effect: this cottony patch of paint evokes a puff of cloud or of steam from a train, this dry pastel smear the blur of breeze-tossed foliage, this unctuous glob the glitter of sunlight off the ripple across a river's surface. Impressionist painting and its critical interpretation, in short, have constantly conflated and confused both the means of signification and the competing entities thus signified as it moves between the representation of the scene portrayed and of the artist portraying it.

It is possible to regard the efficacy, and astounding success, of Impressionism as deriving from precisely its incapacity to resolve the uncertainty over the proper location of the impression, its failure to settle these ambiguities of signification and meaning. By productively plying the middle ground between the objective and subjective senses of the term *impression,* Impressionism and its subsequent interpretation not only have found an inexhaustible source of material for a sustained debate that keeps the movement, to its enormous benefit, perpetually before the public. Art and criticism together have also posited a relation—an arrangement, a matrix—that inexorably links the subject and the

objects of modern perception and places them together within the same spatial array, the same field of action. This is quite similar to the manner in which Hubert Damisch, in *The Origin of Perspective* (1987), describes the operation of traditional perspective:

> *Construzione legittima* proposes, in its very configuration, a formal apparatus which, with its viewpoint, vanishing point, and "distance point"—organizing itself as it does around the position of a "subject" taken to be the origin of the perspectival construction, the index of what is *here,* what *there,* and what *over there*—is equivalent to a network of spatial adverbs, if not personal pronouns: in other words, to what linguists call a *dispositif d'énonciation.*

Damisch, expanding on this linguistic exposition of the visual field, insists on the paradigmatic character (in the strictly technical sense) of perspective: like a language, perspective provides for the inflection of its figures, according to a set of syntactical rules, for the sake of articulating the relative positioning of elements, including the perceiving subject, in space. In the broadest meaning of perspective, Impressionism can be seen as an ingenious extension of the perspectival *dispositif d'énonciation,* in which painting continues the activity of locating subject and objects relative to each other—in this case, by installing sensitive artist in intimate proximity to external effect, separated only by the thin membrane of the impression.

When, however, one adds to this account the nonlinguistic connotations brought to the concept of paradigm by Thomas Kuhn in *The Structure of Scientific Revolutions* of 1962 (connotations explicitly dismissed by the structuralist Damisch), Impressionism potentially emerges as something other than a continuation of a static enunciative structure: it represents instead a "paradigm shift" away from traditional perspective toward a new ordering of subject and object. This is, in its essence, the claim voiced by numerous writers who have identified Impressionism as the first true instance of modern art (a formulation that begs the question—not broached here, for that matter—of what properly constitutes the modern). To characterize that shift in paradigm: traditional perspective annuls a crucial means of differentiating artist from viewer—either can occupy the viewpoint at the apex of the perspectival pyramid—while opening up a space (quite literally) between that shared subject position and the objects depicted; but it does these two things only to collapse that distance again by turning the objects depicted into, precisely, a view—now ostensibly as accessible to subsequent viewers as it once was to the original artist (the final section of Erwin Panofsky's *Perspective as Symbolic Form* of 1927 (translated in 1991) remains unrivaled in its brilliant exposition of this point). Impressionism, in contrast, might be described as reversing this process: it annuls the distance between artist and world by conflating the two in the impression, while opening up a gap between a highly sensitive artist and a viewer, who only sees the product of that sensibility; but it does these two things only to collapse that gap again by also, at moments, allowing the viewer, like the artist, full access to the effect.

The body of critical and historical literature on Impressionism, which in practice has collectively held in abeyance any resolution to the question of whether the viewer is viewing an effect (the first interpretive tradition, above) or is viewing the vision of the gifted artist viewing the effect (the second), operates into perpetuity this mechanism of oscillating identification and differentiation between viewer and artist. Meanwhile, the immediacy, the proximity—definitive of the movement—between artist and world, or at least between that artist's perception and that world's effect, has attained the status of received truth, attained it through the indeterminate device of the impression. Indeed, the very concept of impression recasts the distance potentially separating the elements matched by the iconicity of vision into the proximity of an indexical relation, as the effect seemingly presses itself—as if in a tactile manner—directly into eye, or consciousness, of the artist.

It proves productive to assay the history of Impressionist painting as well as the historiography of its reception according to this tension between the affirmation and the conflation of the paradigms of spatial disposition; which is to say, to apprise painting and writing along this axis that locates world, effect, artist, and viewer in positions of relative distance or propinquity. Productive, because this approach—unlike a taxonomy, such as the one mooted above, that bifurcates painting and scholarship into artifacts that place emphasis on the scene portrayed and those that stress the artist's creativity—never loses sight of the crucial central relation, that constructive ambiguity between effect and sensibility, activated by the concept of the impression. Ultimately, all Impressionist paintings, as well as subsequent written accounts of them, explicitly or implicitly posit some version of this relation as they inflect the four constituent parts of Impressionist representation.

The interplay begins among the Impressionist artists themselves: choice of subject matter and stylistic rendition suggest a rough division of painters into two competing yet complementary camps. (This division would eventually manifest itself in the eight independent exhibitions, where control of and participation in the installations tended to alternate irregularly between one group and the other. Morisot and Pissarro, for various reasons, fell either between or beyond these two camps; fittingly, they were able to sidestep the partisanship and participate with both coteries in the later exhibitions.) In the first camp, Manet (although not a participant in the exhibitions) and Degas, as well as others, developed a manner of painting that—despite local patches of paint that refused the task of carving out spatial recession through modeling and compositional tactics—nevertheless opened up a stagelike space into

which could enter the appropriate dramatis personae to act out scenes of a cosmopolitan and sophisticated Paris; Degas's many portrayals of the ballet at the opera literalize the metaphor. Frequently, these works strongly indicate an identity for the viewer, quite close in character to that of these patently worldly artists themselves, that allows him (nominatively male) to have seeming direct access to the scene portrayed: the *abonné* backstage at the opera, the flaneur dispassionately strolling the boulevards. In the other camp, artists such as Monet and Renoir, who devoted more of their attentions to the modern landscape in and around Paris, tended to render the surfaces of their canvases in a flat and uniform manner that seemingly drew near the scenes portrayed, as if the effects had reconfigured themselves into an opaque wall of sensation pressing directly against the receptive artist's consciousness. Yet, even as the contemporary world thus became selfsame with the artist's cognition of it, viewers—with but a painted surface in front of them—find themselves inevitably somewhat removed from the perceptual intimacy ostensibly enjoyed by the artist directly before the motif. Early critics writing at the time of the group exhibitions capture well the differing dynamics of distance. Edmond Duranty (1876), with his thinly disguised preference for Degas, gives short shrift to the painters' stylistic idiosyncrasies ("even the most learned physicist," he maintained, "could find nothing to criticize") in order to permit artist and viewer alike, through these images, to embark on a nuanced classification of the "special characteristics of the modern individual," whose "faultless rendition" in paint conveyed "temperament, age, and social position." In contrast, Théodore Duret (1878) lauded the manner in which the Impressionist landscapist, by painting in bright and unconventional colors for the sake of capturing the nuances of light and "the accidents of atmosphere," demonstrated a "complete honesty, declar[ing] that he only reproduces that which he sees"; that this honesty left the vast majority of convention-bound viewers "indignant" with the picture led to an estrangement from his or her audience that the sensitive artist simply had to bear.

Over the coming decades through the turn of the century, however, Impressionism actually garnered not antipathy but great approbation from its ever-growing audiences, both in France and internationally. Shiff, while insisting on the essential continuity of the basic interpretive machinery, has traced the accompanying shift in emphasis after 1880—in the works of the artists and in the critical writing—away from the transcription of things and toward the visible mediations of artistic temperament. And despite—or perhaps because of—the movement's eclipse by Cubism during and after World War I as the art movement serving in the critical and historical commentary as the principal rhetorical figure for progress and innovation, Impressionism had by the 1930s firmly established itself as an indispensable corner-stone in the edifice of the French national tradition in the arts. Only in the years surrounding World War II did Impressionism once again give rise to intriguingly new interpretive gambits—on both sides of the Atlantic.

In a few short but highly influential paragraphs appearing in an article published in *Marxist Quarterly,* Meyer Schapiro (1937; rpt. 1978) stretched the ties of Impressionist intimacy shared by artist and effect to posit a similar direct contact, through them, uniting a contemporary audience of viewers—the bourgeois class—to the world that they, like the artist, inhabited. The early Impressionist artists discovered "a constantly changing phenomenal outdoor world of which the shapes depended on the momentary position of the causal or mobile spectator." Yet, just as that "causal or mobile spectator"—in the first instance, the artist him- or herself—came to embody as well the values of "individual enjoyment" and "freedom" that defined the class character of the "enlightened bourgeois," the "changing phenomenal world" portrayed in Impressionist painting corresponded directly to the "mobility of environment, the market and of industry"—real substructural forces, in other words—that gave rise to that class in the first place. In Impressionism a specific class finds the perfect, seemingly transparent, cultural mechanism for the realization of its own particular world. (Estrangement, in Schapiro's account, characterizes the period following 1880, and takes much the form of economic alienation.)

Writing quite soon after the war, Maurice Merleau-Ponty (1945, translated in 1964)—in perhaps the most significant account of Impressionism from the perspective of philosophy to date—found the scope of Impressionism much more constrained. The Impressionist painters activated a series of technical devices (devices here described as the mediating procedures of skilled artists, rather than simply taken, as in many earlier accounts, for the direct transcription of effects) in order to capture and represent the bright "atmosphere" of the outdoor landscape on canvases hung within the "dim light" of contemporary parlors. As a consequence, however, the painting lost its hold on the depicted object's "proper weight," and thereby on the object beyond the atmosphere itself. In contrast, Cézanne's painting (for Merleau-Ponty's essay is a championing of Cézanne over and against Impressionism, and has thus played its part in widening the historiographic divide separating Cézanne from his Impressionist comrades) counters the dissolution of the object's "solidity and material substance" for the sake of "find[ing the object] again behind the atmosphere." World and effect coalesce in that moment when Cézanne brackets "the human organization of ideas and sciences" in order to allow his "natural vision" to perceive "an object in the act of appearing, organizing itself before our eyes." The result is a painting that "penetrates right to the root of things beneath the imposed order of humanity" and that

thereby offers forth "the presence, the insurpassable [*sic*] plenitude which is for us the definition of the real." This profound phenomenological perception is first and foremost Cézanne's, but it is not his alone. Other sensitive souls—the proper audience for Cézanne's paintings—can grasp this same unity by looking through Cézanne's eyes: "The artist is the one who arrests the spectacle in which most men take part without really seeing it and who makes it visible to the most 'human' among them." Merleau-Ponty, in essence, has Cézanne's pictures do what Schapiro's Impressionism did (but his own Impressionism did not): unite viewer—in this case more the existentialist hero, to be sure, than a class actor—to the world, by way of the seemingly unmediated route through painter and effect.

After another period of relative historiographic quiescence, Impressionist scholarship returned to prominence during the period from the late 1960s until the mid-1980s with the rise in the "social history of art." It might seem, at first, that the call for a social history of Impressionism constituted a return to a strictly realistic reading of the movement: sailboats, for instance, documenting the emergence of new habits of leisure; smokestacks in the distance hinting at the forces of industrialization. Yet, in the hands of its best practitioners the social history of Impressionism always maintained a commitment to the more complex task of exploring with some rigor the nature of the relation between artist and viewer, on one hand, and effect and world, on the other. Two leading accounts, by Robert Herbert (1988) and by T. J. Clark (1985), each in its own manner exemplifies the approach.

Herbert argues that the Impressionist painters did indeed directly perceive and, often through quite sophisticated technical means that only appear spontaneous, capture effects—most significantly, the effects generated by Baron Georges-Eugène Haussmann's bad-faith transformations of Paris for the sake of perpetuating Louis-Napoleon's despotic rule. Yet in spite of this capacity for exacting perception, the Impressionists did not render exactly that which they saw: through careful compositional framing and the stylistic effacement or exaggeration of certain distinct features of the depicted social scene, these artists allowed for a mild celebration or (more frequently) performed a subtle critique of the new Imperial Paris. In a certain sense, Herbert returns the analysis of Impressionism to the terms of Duret: the artist enjoyed his or her closeness to the effect, while the viewer of the resulting canvas remains somewhat distanced from that intimacy—in Herbert's case owing not to the viewer's incomprehension but rather to the artist's artful editorializing. Clark, in contrast, faults the likes of Monet and Renoir for succumbing to the lure of the effect: these artists, Clark suggests, blithely contributed to the false ideology of Louis-Napoleon's "spectacle" (the Situationists are never far from Clark's analysis) by repeating its obfuscating precepts. Manet emerges as the hero in Clark's ac-

count: his paintings (and, to a lesser extent, Degas's) cut through the illusions of Impressionist painting and commercial spectacle alike to reveal the social fissures and class contradictions that actually sundered France during the second half of the nineteenth century. The structure (if not the content) of Clark's argument thus perhaps most closely resembles that of Merleau-Ponty (even as Clark shared with Duranty a desire to have his favored artist reveal the truth of "social position"): in each, the perspicacious artist—respectively, Manet or Cézanne—glimpses the real world across the gap of the Impressionist artist tied too closely to the effect, while the exceptional discriminating viewer—respectively, the attuned phenomenologist and the demystifying critic of ideology—establishes intimate kinship with that perspicacious artist.

Interest in Impressionism among modernist art historians has waned somewhat over the last decade, as a discipline influenced by the insights of poststructural literary criticism and intrigued by the promise of semiotic analysis has shifted its attentions toward other movements in modern art—Cubism and Surrealism, for instance—seemingly more explicit about the mediating role of their means of signification. Yet, much remains for scholars to accomplish in the task of explicating the semiotic devices through which Impressionist painting—in a manner virtually unquestioned during the hundred-plus years of its critical and historical reception, and even as analysts have pried open distance between it and the viewer, or it and the world—has managed, repeatedly, to represent an unmediated proximity between artist and effect.

BIBLIOGRAPHY

Clark, T. J. *The Painting of Modern Life: Paris in the Art of Manet and His Followers.* New York, 1985.

Damisch, Hubert. *The Origin of Perspective.* Translated by John Goodman. Cambridge, Mass., 1994.

A Day in the Country: Impressionism and the French Landscape. Los Angeles, 1984. Exhibition catalog.

Duranty, Edmond. *La nouvelle peinture: À propos du groupe d'artistes qui expose dans les galeries Durand-Ruel.* Paris, 1876.

Duret, Théodore. *Les peintres impressionnistes.* Paris, 1878.

Herbert, Robert L. *Impressionism: Art, Leisure, and Parisian Society.* New Haven, 1988.

Merleau-Ponty, Maurice. "Cézanne's Doubt" (1945). In *Sense and Non-Sense,* translated by Hubert L. Dreyfus and Patricia Allen Dreyfus, pp. 9–25. Evanston, Ill., 1964.

The New Painting: Impressionism, 1874–1886. San Francisco, 1986. Exhibition catalog.

Rewald, John. *The History of Impressionism.* 4th ed. New York, 1973.

Schapiro, Meyer. *Impressionism: Reflections and Perceptions.* New York, 1997.

Schapiro, Meyer. "The Nature of Abstract Art." *Marxist Quarterly* 11 (January–March 1937). Reprinted in *Modern Art: 19th & 20th Centuries* (New York, 1978) pp. 185–211.

Shiff, Richard. *Cézanne and the End of Impressionism: A Study of the Theory, Technique, and Critical Evaluation of Modern Art.* Chicago, 1984.

JAMES D. HERBERT

IMPROVISATION. [*To explore the roles of improvisation in art, this entry comprises two essays. The first essay analyzes the aesthetic implications of improvisation. The second continues and expands this analysis while discussing jazz music in particular.*

An Overview
Jazz Improvisation

For related discussion, see Intention; Jazz; *and* Music, *overview article.*]

An Overview

To improvise is to do or produce something on the spur of the moment. There is a sense in which all human action is improvisatory, and in that sense, all art, as the result of human action, has some improvisatory element.

Beyond that general point, improvisation has a special place in aesthetic theory and in artistic practice. For a start, improvisation may be assigned an originative role in at least two important senses. Aristotle makes an art-historical claim, speculating that it was the improvisations of the authors of dithyrambs and phallic songs that gave birth to poetry and its two proper types, tragedy and comedy.

But improvisation may also be thought to be originative in a more fundamental sense, to the extent that improvisation is regarded as an anticipatory phase, an indispensable element, or even the primal core of artistic creative activity. Artists know well the importance of improvisation in their creative endeavors and this aspect of improvisation has not escaped the attention of philosophers. Aristotle himself suggests that the prepoetic improvisations of ancient Greece grew out of fundamental and natural human proclivities, above all, the instinct for imitation, the sense of harmony and rhythm, and the natural desire to know. R. G. Collingwood, although he does not use the term *improvisation* explicitly, captures the sense of its importance for artistic activity when he opposes the essentially expressive activity he calls "art proper" to "craft," an activity in which a preconceived result is produced by means of consciously controlled and directed action. Suppose a sculptor were simply playing about with clay, Collingwood asks, and found the clay under his fingers turning into a little dancing man. Would this not be not a work of art because it was done without being planned in advance? To the contrary, Collingwood argues: a planned work may indeed be a work of art, but artistic expression is necessarily an exploratory activity whose end cannot be foreseen and preconceived.

Of course, the extent to which improvisation may involve consciously controlled and directed action or distinctions between means and ends and planning and execution is a complex issue. The originative claim about the role of improvisatory activity with respect to artistic creation thus finds itself taking up issues, not only in philosophical anthropology, but also in psychology, action theory, and the philosophy of mind, not to mention the actual conventions and histories of individual artistic genres and practices.

In any case, it is certainly safe to say that improvisation in the arts has seemed to many to feature certain qualities on which the arts have traditionally placed a high premium: invention, transformation, expressive freedom, spontaneity, and creativity. And it is not surprising that improvisational activity has been explicitly highlighted in a wide array of artistic practices, ranging from ancient poetic recitation, medieval minstrelsy, musical performance in the Baroque era, to modern jazz and blues, Beat poetry, rap music, folk music, drama, dance, and stand-up comedy.

Artistic improvisation brings something into being, but what that something is and how we are to understand its mode of becoming are questions of great theoretical interest. Artistic styles and artworks in a variety of arts can be said to be "improvisatory" insofar as certain artistic features (bold brush strokes, frequency of embellishment and repetition, freely varying metrical schemes) might bring this or that aspect of improvised actions to mind. But the most difficult ontological questions are raised by those artistic practices that explicitly feature improvisational activity. In this connection, improvisation can be profitably understood in relation to prevailing views about artistic production, performance, and presentation. In the case of much Western classical music practice of the past two hundred years or so, for example, it is not implausible to distinguish more or less clearly between the stages of composition and performance: one commonly thinks of the composer's activity as concerned with creating at least the defining outlines of the musical work of art and capturing those outlines in the notated score, and one thinks of the performer as presenting and interpreting that work for an audience. In such a framework, much hangs on the notion of a "work," a reasonably well articulated, enduring thing that, in the minds of many, figures as the main focus of the composer's activity, the performer's interpretive efforts, and the audience's proper object of attention.

Some theorists understand the activity of improvisation, on the other hand, as a spontaneous activity in which the improviser simultaneously practices the interdependent functions of composition and performance, a position that itself raises many interesting questions. It is a point of contention, for example, whether all improvisations bring works into existence. Clearly, some improvisations are more in the spirit of embellishments and can legitimately be seen as interpretations of preexisting works whose themes and patterns provide the organizing structure around which improvisatory activity is focused. In such cases, one can reasonably appreciate improvisations as if they were primarily performances of a work to which the twin values of fidelity and creativity would appropriately apply.

But there is no a priori principle governing either what aspects of a work might or might not be improvised on or

how radical improvisations might be within a given domain. Consider the range of improvisational practice in modern jazz, often thought to be the paradigmatic case of improvisatory art. One might have thought that one indispensable feature of jazz has been improvisation around more or less fixed harmonic patterns or structures. It is certainly true that many jazz performances do involve a clear statement of a melodic theme in the context of an explicit or implied harmonic structure followed by improvised choruses on "the changes," that is, the harmonic structure of the piece. Additionally, some harmonic patterns (those in George Gershwin's "I've Got Rhythm," for example) have become canonical to the point of being regarded virtually as works themselves. But it must also be remembered that jazz musicians have improvised on works precisely by reharmonizing them, often in ways that depart radically from the original. Indeed, the degree of departure from works can become so high that some jazz improvisations "of" previous works cannot be readily identified as such, even by those thoroughly immersed in the tradition. There is also, of course, the existence of "free jazz" in which not only does harmonic patterning cease to be a central structuring principle but so, too, does the normative ideal of obedience to a preexisting work. To further complicate the ontological issue, one might even go so far as to claim that neither improvisations nor performances qualify as works, just insofar as they are activities, events, or processes, whereas artworks, presumably, are enduring, repeatable things.

This last position, however, rests on a notion of a work that seems both conceptually contestable and unduly narrow in the face of artistic practice. A more reasonable position would be that, certainly in the context of artistic improvisation, the notion of a "work" is transformed from something conceived of primarily as a product, as something made, to something more along the lines of a process, of an experience, or of something in the making.

Improvisations themselves can be considered as aesthetic objects in their own right, as objects suitable for aesthetic contemplation, whether the improvisations are live or recorded and whether or not improvisations are presented or understood as deriving from preexisting works. Experienced audiences can and do attend to the formal, expressive, and mimetic features of the sounding structures of musical improvisations, for example, just as they can in the case of performances of previously composed works. Those familiar with the musical culture and traditions in which improvisations are produced know that extraordinarily high levels of complex, subtle, and expressive achievement are possible, a fact to which the improvisations of Charlie Parker and John Coltrane, for example, readily attest.

But one must not lose sight of the fact that improvisations also feature productive activity as such. One great attraction of improvised performances is precisely the opportunity to witness, as it were, the shaping activity of the improviser who creates an artistic utterance unmediated by another human being. It is as though we are able to gain access to the artist's mind at the moment of artistic creation. This sense is heightened by the risk and excitement inherent and often evident in spontaneous creative acts and by the innovation, inventiveness, technical agility, and imagination that such activity demands. Improviser and audience share in a sense of immediacy. In improvisation, the rough draft is also necessarily the final product. One has only one go at it, without having the luxury of rewriting, erasing, replaying, and so on. One feels that one's abilities are far more exposed and obvious to the experienced audience than is the case for nonimprovisatory arts. One might well say, then, that in improvisation, as perhaps in no other form of art, are the experience of the artist and that of the audience so thoroughly intermingled. The evaluation of improvisations must be made within this context and it might not be going too far to suggest that, insofar as improvisation in art broadens the notion of the "work," so does it deepen our understanding of art in general.

BIBLIOGRAPHY

Alperson, Philip. "Improvisation in Music." *Journal of Aesthetics and Art Criticism* 43.1 (Fall 1984): 17–29.

Aristotle. *Poetics.* Translated by Richard Janko. Indianapolis, 1987.

Collingwood, R. G. *The Principles of Art.* Oxford, 1938.

Goehr, Lydia. *The Imaginary Museum of Musical Works: An Essay in the Philosophy of Music.* Oxford, 1992.

Hamilton, Andy. "The Aesthetics of Imperfection." *Philosophy* (July 1990): 323–340.

Kivy, Peter. *Authenticities: Philosophical Reflections on Musical Performance.* Ithaca, N.Y., 1995.

Thom, Paul. *For an Audience: A Philosophy of the Performing Arts.* Philadelphia, 1994.

PHILIP A. ALPERSON

Jazz Improvisation

Jazz music, and particularly jazz improvisation, is a little-discussed topic within the larger field of aesthetics. This is regrettable, not only because of the diminished coverage of the arts that this omission implies, but also because the art of jazz improvisation presents a rather large number of aesthetic practices that would revise or unsettle prominent aesthetic theories. A few of these tensions between jazz improvisational practice and aesthetic theory will be discussed here under four main headings: improvisation and the definition of a musical work; the relation between improvisation and language; critical evaluation and creative aspiration in jazz improvisation; and conceptions of intention and execution in improvisational contexts.

Improvisation and the Definition of a Musical Work. It will be helpful to begin by briefly characterizing some of the central improvisational practices in jazz. Many jazz compositions are written with underlying recurrent

harmonic structures (often referred to as "changes") on which the performance is constructed, and they are also usually composed with a melodic line that is played at the beginning and the close of the piece. Although these musical building materials are written down, they are almost never played as written. First, the harmonic structures will be voiced by chordal instruments (e.g., piano, guitar, vibraphone) differently as each interpretation of the piece develops; it is extremely unlikely that these structures will be realized exactly the same way in any two performances (and if they were this would be cause for aesthetic suspicion, not congratulation). Second, the melody (often referred to as the "head" of the piece) will rarely, if ever, be played the same way on any two occasions; the melody as written is itself perennially open to interpretation, and the interpretive license implicitly granted within this musical culture covers a very broad range of possibilities. Thus, successful innovative and creative performances depart substantially from the musical information given in the score in ways idiosyncratic to the individual performer or in ways stylistically demanded by the genre of the piece, so the improvisational aspect of a jazz performance extends into the written as well as the unwritten sections of the piece.

Moving from improvisational practice to aesthetic theory, it is undeniably clear that the problem of the definition of the musical work has been central to musical aesthetics. If the work is not reducible to any particular performance of the work, and if it is the kind of object that can be given a very large number of interpretations, then it seems that it must in some sense be an abstract object, a fully articulated mental object, originating in the mind of the composer, against which all performances, or instantiations, of the work, must be judged. On this view, often called musical idealism or Platonism, criticism is a matter of the comparison of the particular sonic event against its ideal progenitor. This view is a close relative of the theory of musical works constructed on the type-token distinction derived from the philosophy of C. S. Peirce; here the many works are tokens of the single type constituted by the composition. Another definitional theory suggests that the musical work is a sound pattern, exemplified in its performances.

But even the preceding cursory glance at the nature of jazz compositions, and particularly the performances of those compositions, renders such theories of musical definition implausible. These definitional theories—Platonistic, type-token, or fixed sound-pattern conceptions of the musical work—progress from the realm of the implausible to that of the utterly inapplicable when we move beyond the "head" to the body of the piece (and is meant primarily to small ensembles in jazz, e.g., trios, quartets, and quintets), where the individual players in the ensemble "take solos" over the "changes," the recurrent harmonic structures. Very frequently, the improvising soloists begin building a solo with motivic material derived from the head; this could be a melodic or rhythmic fragment taken from the written melody, or, as often happens, it could be a fragment taken from the melody as played by the performer on that one occasion. Frequently, soloists will investigate the permutational possibilities of these motifs as the solo unfolds, and usually this melodically integrative material will be supplemented with freely improvised new material (i.e., melodic strains not thematically related to the head), which in turn can be motivically fragmented and reemployed in turn. That is one of the many ways a jazz solo can develop "organically"; there are countless others. But it is clear even at this preliminary point that the musical work being undertaken in an improvised solo is as close (or closer) to composition as it is to a preexisting musical work. To put it Platonically, these musical practices give us sonic embodiments without mental ideals behind those embodiments; to put it in terms of the type-token distinction, one has tokens without types; and to put it in terms of sound patterns, one may indeed have patterns of sound, but the patterns do not predate the sounds. Jazz improvisation, as a body of practices, constitutes a powerful challenge to a good deal of music-definitional theory that has not been satisfactorily met.

The Relation between Improvisation and Language. Since much speech is undeniably improvisatory in nature, and since music, and particularly improvisatory music, is often construed as a kind of language, this relationship seems clearly to warrant investigation. Here, too, the practices of musical improvisation present further facts interestingly at odds with received opinion.

Despite the revolutionary work of Ludwig Wittgenstein in the philosophy of language, linguistic dualism is still a powerfully attractive conception of linguistic meaning to many; that is, for every outward or physical utterance there is an inner mental predecessor to that utterance of a nonlinguistic nature that gives the utterance its meaning. Such alleged prelinguistic ideas predate the utterance just as Platonic musical-ideational works allegedly predate their outward expressions in sound. Thus, linguistic dualism is closely related to its aesthetic counterpart, and just as linguistic dualists believe we need to look behind the utterance for its inner meaning, so musical dualists believe we need to look behind the melody for its meaning. The actual artistic labors, in context, of jazz improvisers quickly reveal the misleading nature of this dualistic conception of musical meaning. As in any improvisatory activity, in jazz soloing one very frequently discovers in midstream where one "wants to go": that is, the melodic trajectory, or the developmental impetus of a motif, reveals itself in the performance itself *to the performer*. This plain fact is strange and counterintuitive only to those who attend more closely to the dictates of dualistic theory than to the facts of practice; interestingly, this fact becomes by contrast intuitively plausible as soon as one attends more closely to the above-mentioned improvisatory nature of conversations and of

the frequently improvised specific linguistic utterances contained within them.

Indeed, it might be suggested that listeners find jazz improvisation familiar in an intuitively deep way precisely because they already possess, and use extensively, a parallel improvisatory capacity in language. Highly skilled musical improvisers—Sonny Rollins's extended improvisations constitute one case in point—are often extraordinarily *attentive* improvisers, finely attending to the nuances of directional significance, to the developmental implications of the musical or thematic logic as it unfolds in the music as they are making it. Here musical creation is contemporaneous with critical discovery; as a soloist is creating a musical structure, that soloist is simultaneously functioning as an analytic critic of that structure. Moving to the concept that ties music and language together at the deepest level, if we are to speak of meaning in these contexts, then it becomes apparent, given a healthy respect for the details of these musical practices and a healthy disrespect for the mischaracterizations of those practices that are motivated by theory, that the meanings of improvisational phrases are discovered and attended to as (and just after) they are created, and not—as the dualistic conception appropriated from linguistic dualism would suggest—before. Thus, jazz improvisation severely destabilizes a familiar conception of musical meaning that has been generated by a prior conception of linguistic meaning, and there is every reason to believe that a philosophical inquiry into the details of improvisation could shed new light on the nature of linguistic meaning.

Critical Evaluation and Creative Aspiration in Jazz Improvisation. Several problems emerge from the previous sections. If we first inquire into the evaluational "logic" of our critical judgments of particular improvisational performances, we see at once that criticism can rarely if ever itself be dualistic in structure, that is, it cannot proceed in terms of the comparison of the sonic event with an ideal as indicated by a score or any other extramusical type. It may be true that there are cases that initially appear to approximate dualistic criticism, as in the critical evaluation of new performances of Duke Ellington's or Count Basie's big-band arrangements, but in such cases, first, the comparison is made not to an abstract type or score, but to the recordings of the original *sonic* events; second, the improvised solos in these neoclassical performances blur the correspondence between score and sound as discussed above. Despite the impossibility of modeling critical evaluation in dialectic terms, however, there certainly are critical standards that are employed in the evaluation of jazz improvisations, but such standards are extremely resistant to generalization or summarization; they are in any case *not* imaginary ideals. Here the historical dimension of this art form comes to the fore: very frequently (1) the levels of technical facility, (2) improvisational resourcefulness, (3) imaginative reach, (4) improvisational-compositional ability conjoined to spontaneous analytic attentiveness, (5) capacities to maintain a sense of form and organic growth within an improvised solo, (6) ranges of tonal and dynamic shadings, nuances of attack and vibrato and so forth, and (7) the employment of vocalization and speech patterns instrumentally, along with countless other critical criteria, will be employed in making historically informed judgments in terms of the standards of a given instrument. There is, in fact, an instructive sense in which these critical standards are doubly resistant to generalization: it is not just that these criteria are numerous and of manifold kinds; it is also true that although the great performers in the history of improvisation frequently display many or all of these critical excellences, beyond a certain level it is often considered a failing rather than an achievement to sound too great an extent like, for example, John Coltrane or Sonny Stitt, Miles Davis or Clifford Brown, Charlie Christian or Django Reinhardt. A fundamental creative aspiration in jazz is thus very often to find one's own improvisational voice, but to do it in a way that at once both does and does not acknowledge precedent: the aspiration is often to find a style that is both unique and simultaneously deeply historically rooted. Aesthetic theory has labored extensively on the problem of critical evaluation and evaluative canons; it is clear that when it turns to the criticism of jazz improvisation as a philosophical problem it must, in a fashion that is seemingly paradoxical, allow for a kind of critical standard that is simultaneously historical and sui generis, simultaneously generalizable and unique.

Intention and Execution in Improvisational Contexts. Jazz improvisation can be connected to the philosophy of mind. The improvising soloist is, to be sure, engaged in intentional activity, but the role that intention plays within the various contexts of this activity is every bit as complicated as the role played by intention in literary aesthetics. For example, a saxophonist might well intend to play a passage within an improvised solo exactly as Charlie Parker played it (a common practice called "quoting"), and in doing so simultaneously intend to acknowledge the master's insight at that juncture, intend to entertain the Parker connoisseurs in the audience, and intend to sound, for a fleeting moment, unlike himself. This improviser may well also expand the harmonic significance of the melodic line he has, until the quotation, been playing, but while the Parker quotation achieves this expanded significance it may be the farthest thing from his mind. Yet, it is insufficiently nuanced to say that because the expanded significance was not preconceived, it was a *mistake*. It was not, and is surely acceptable as one aspect of the performance in the critical description of this solo. Thus, like much-investigated literary cases, the meaning of the phrase is not circumscribed wholly by authorial (or player's) intention. On a lower level of intentional analysis, the improviser might intend to play a C-natural but accidentally play a C-sharp; here there may

be in use a clear criterion of one-to-one correspondence between intention and execution, but such a simple conception of intentional activity as a dualistic match of outward execution to inner intention applies only to the most rudimentary cases. Suffice to say for the present that when aesthetic theory moves into the area of improvisational intention it will need a vastly more sophisticated—and multifaceted—conception of intentional activity than that offered by the conventional twin categories of imaginative preconception and technical execution. Beyond this intentional problem, it seems likely that the very nature of musical thought, of presonic ideation, will be illuminated by an investigation into improvisational thought; as indicated above, we could certainly stand to be substantially clearer than we are at present about the relations between composition (meaning the very large number of musical practices that are encompassed by that generic term) and improvisation. Indeed, it is arguable that all composition has, if in a restricted sense, its genesis in improvisation, and it seems undeniable that one of the central virtues of improvisation is that it provides access to musical thought directly and immediately. In any case, "improvisation" is also a word that encompasses a very broad range of musical deeds, and the elucidation of its many multiform senses, as it appears in problems of musical definition, in analogies with language, in critical evaluation, in problems of musical intention, and in many further issues beyond those mentioned here, is one of the philosophical achievements to which we might look forward within the larger field of aesthetics.

BIBLIOGRAPHY

Alperson, Philip. "Improvisation in Music." *Journal of Aesthetics and Art Criticism*. 43.1 (Fall 1984): 17–29.

Gioia, Ted. *The Imperfect Art: Reflections on Jazz and Modern Culture.* New York and Oxford, 1988.

Hodeir, André. *Jazz: Its Evolution and Essence.* Translated by David Noakes. New York, 1956.

Howard, V. A. *Artistry: The Work of Artists.* Indianapolis, 1982.

Kivy, Peter. *Authenticities: Philosophical Reflections on Musical Performance.* Ithaca, N.Y., 1995.

Roberts, Howard, and Garry Hagberg. *The Guitar Compendium: The Praxis System: Technique, Improvisation, Musicianship, Theory.* 3 vols. Tübingen, 1989.

Schuller, Gunther. *Early Jazz: Its Roots and Musical Development.* New York and Oxford, 1968.

GARRY HAGBERG

INDIAN AESTHETICS. [*To explore the history of Indian aesthetics, this entry comprises two essays:*
Historical and Conceptual Overview
Indian Theory of Interpretation
The first essay explains the various aesthetic traditions in Indian art history. The second essay discusses the different theories of interpretation, mostly of literary texts, in this same history.

For further discussion, see Abhinavagupta; Comparative Aesthetics; Interpretation; *and* Rasa.]

Historical and Conceptual Overview

Notions of the meaning and ideals of the arts in India have emerged from the various civilizations that have flourished, contended, and blended over the course of a four-thousand-year history in the South Asian subcontinent. These civilizations include Hinduism, Buddhism, Jainism, and Islam, as well as the distinctive regional cultures of the subcontinent. Aesthetic notions have in some cases been formalized as theories, such as the concept of *rasa*, and have attained an ongoing prestige in the classical Sanskrit discourses of philosophy, theology, and cultural analysis traditionally maintained by the brahman (priestly, literate) castes. In Tamil literary culture as well, self-consciously articulated theories (such as the *akam/puram* distinction, and the five landscapes) have powerful hermeneutic value with respect to much of the literary expression throughout India. Yet, important artistic values are also expressed in the characteristic forms of performance praxis, many of which share similar contexts of patronage and connoisseurship across religious, civilizational, and even social boundaries. Still another source for aesthetic ideals may be discovered in the hundreds of technical treatises on the production and performance of artistic works known as *śāstra*s in the Hindu and other non-Islamic traditions; these texts on music, dance, painting, sculpture, and architecture generally do not approach questions of beauty or the goals of art directly, but are indispensable in coming to terms with the phenomenon of representation in India. Since World War II, South Asian culture has been intensely scrutinized by folklorists and anthropologists; as a result there has been important work on the study of the aesthetics of art in the nonelite sphere, and a growing realization that there has always been a dynamic interplay between the forms and ideals of court culture and that of the village, or more recently, the street. High culture, folk culture, and popular culture: Hindu, Buddhist, and Islamic civilizations have been shown historically to have permeable membranes rather than rigid walls between them. Finally, much of the framing of artistic products and praxes in India has been accomplished by non-South Asians and South Asians familiar with European aesthetic theory during the colonial (1757–1947) and postindependence eras (1947–present); these more recent frames have themselves become part of the ongoing conversation about art in India, even among native traditionalists.

Although the various aesthetic conceptions of India bear detailed examination, the most effective way to understand how the aesthetic realms have come to be demarcated is through social and cultural history rather than through consideration of schools and movements. The term *India* should be understood as a kind of shorthand for that re-

gion that historically has included India, Pakistan, and Bangladesh, as well as, to a slightly lesser extent, Sri Lanka, Afghanistan, Nepal, and Burma. Especially during the first millennium CE, the cultural forms and aesthetic ideals of Indian civilization were transmitted to the lands of Southeast Asia through colonization and trade, giving rise to the notion of Greater India. The early Indus Valley civilization (2300–1700 BCE) left a rich legacy of visual, architectural, and religious forms, but the characteristic forms understood today as "Indian" began to be set in place with the coming of the Indo-European peoples who called themselves Aryans (1300–1200 BCE). The priestly, literate stratum of this society, known as brahmans, controlled ritual observance, reflections on which eventually yielded theories of linguistics (as well as epistemology), which in turn gave rise to basic notions of poetics. Likewise, speculations in the Upanishads on the fundamental elements of the cosmos—heat, breath, and ultimately the ineffable *brahman*—that endured beyond the ephemeral sacrifice yielded the first notions of a transcendent realm, access to which could be gained through meditative trance. Throughout the history of Indian aesthetics, accordingly, brahman involvement in describing and theoretically prescribing the arts has sustained two models of experience that are apprehended in the West as "religious": On the one hand, there is the notion of art making as ritual, that music, dance, sculpture, or architecture recapitulates a cosmic creative process. Performance and execution therefore are attended with preparatory rituals, including benedictions for success; even rehearsal practice is regarded as a kind of rite. As investment in the Vedic ritual recedes into orthodox nostalgia, and devotional, temple-centered theism emerges to become a major part of the synthesis known as Hinduism in the early centuries of the common era, this ritual quality of art making is accompanied or replaced by a mode of worship. On the other hand, the savoring and appreciation of art are from the very beginning viewed as an experience of "transport," the notion that profound and knowledgeable communion with art, especially the performing arts, lifts one out of the particularity of one's own time and place. The model seems to be very much that of the phenomenal personality giving way to the knowledge of the Self *(ātman)*, which is, eternally, pure consciousness and bliss, enduring through the cycles of rebirth that involvement in the world of ethical causality and consequences, or *karma*, propels. At times the aesthetic experience has been declared to be an analogue of spiritual transcendence, while at other times it is identified with it. Even in this century when sympathetic art historians such as Ananda Kentish Coomaraswamy began to rescue the study of the Indian visual arts from earlier colonial evaluations of their grotesqueness, they did so by following the elite Sanskrit theories that stressed a transcendent referent *(brahman)* for all artistic production. It should be remembered that in the Sanskrit tradition a sign of the independence of art (except for overtly religious or devotional art) is its tendency to assert the independence, and even superiority, of the aesthetic rapture disclosed through intense communion with the arts over the Creator's mundane world.

It seems that early on brahmans expanded their role as keepers of ritual order to that of arbiters of many aspects of the social order through formulation of law texts and counsel to rulers. From the beginning of the common era, in fact, brahmans engaged in the production of vast numbers of treatises on all aspects of human activity, from astronomy to sexual love, from drama and dance to sculpture. These treatises and the project of constructing a normalizing discourse are known by the term *śāstra*, and they were crucial in the formulation of aesthetic ideals in ancient India.

Debates about the relationship between the texts, the self-understanding of the work process by the artists, and the artifacts produced take a different shape in the various arts. The Sanskrit law texts provide a clue as to the nature of this most important of Sanskrit genres: they are neither completely descriptive of actual practices, nor were they ever regarded as strictly prescriptive. Rather, they represent a brahmanic "dream of order." In the arts, they are naturally infused with a ritual quality of production that marks them as brahminic, yet paradoxically they represent more a legitimated way of seeing rather than a legitimated way of making. It is very tempting to view these texts as repositories of lore, but often the recommended proportions for a figure or the astronomically proliferated catalogs of musical modes have little or no counterpart in reality. It has been suggested that, at least for the visual arts, the texts are most usefully regarded as a *complement* to the actual paintings, sculptures, and buildings; their use to the art historian or the aesthetician accustomed to stylistic analysis or cultural history would be in constructing a comprehensive *epistēmē* of the premodern Indian view of art (Maxwell, 1989). It would certainly be inappropriate to imagine that the master of a sculptural workshop would use the *śāstra* in teaching, or that an artisan would have the texts, virtually all of which were unillustrated, before him when carving a figurative column. In literature, the situation is quite different, since the theoreticians and producers were often the same people, or at least people of the same class with identical training. No matter how elaborate a scheme of figures of speech of varieties of nuance a poetician would care to develop, he could easily provide examples asserted to illustrate the feature.

By and large the practitioners of arts other than literature in premodern India were not brahmans. Musicians, actors, dancers, painters, and sculptors were understood as artisans; references in literature show that their work was admired for its technical skill, artistic sensitivity, and power to evoke aesthetic delight, but as in virtually every other premodern hierarchical society they were regarded as the ser-

vants of those with the means to engage them. (It must be emphasized, however, that the revival/reconstitution of the performing arts as part of the nationalist reawakening in the late nineteenth and twentieth centuries was mostly undertaken by Brahmans and other literate high-caste communities.) Some received their training as a caste heritage. Others seem to have been drawn from a variety of caste communities, but were apprenticed to workshops of artisans organized in guilds, schools, or lineages. Nonetheless, their lore was formalized in *śāstric* texts in the Sanskrit language. In part this must have been due to the meeting of the world of the artisans and that of the brahmans at the courts of their royal patrons. Although it may seem puzzling that an elite literary format was used to describe the training and work of "entertainers" and "workmen," one can imagine that the members of these professions probably welcomed the prestige that *śāstras* conferred. It is also quite possible that brahmans were hired to produce the texts, much as royalty of low-caste backgrounds and newly powerful mercantile communities engaged brahmans to produce noble genealogies with roots in legendary divine ancestors. Some have seen the incongruence between the arid seriousness of the theoretical *śāstric* descriptions and the exuberant sensuality of all the arts as reflecting a distinct socioritual disjunction between brahman authors schooled in the texts of ritual and transcendence, and non-brahman artisans who drew their inspiration from the lush environment. A more historically and socially precise formulation of this hypothesis may remove it from the realm of stereotypes, as well as explain the richly textured and sensual quality of Sanskrit poetry and drama, a great deal of which was produced by brahmans.

The older professional *śāstras* often begin with a foundational myth in which the gods establish the profession to ward off the demonic forces of chaos for the benefit of all classes of humanity. A divine prototype (such as the architect Viśvakarman) or a superhuman sage with divine parentage may transmit the professional knowledge from the gods to human artists. In later texts the mythic preamble is often replaced or supplemented by a theological disquisition on the original and inherent divinity of the fundamentals (sound, dance movement, etc.) of the artistic activity.

Indian speculation on aesthetic experience is usually understood to have originated in such a *śāstric* text, the famous *Nāṭyaśāstra* of Bharata (200 BCE–200 CE). In this book are contained not only myths of origin for theater and rituals pertaining to performance, but technical instructions for the arts that contribute to theatrical production—playhouse architecture, dramatic language, but especially dance and dramatic movement, character portrayal, and musical forms. In the *Nāṭyaśāstra* as in other *śāstras*, drama is conceived as *anukaraṇa*—"imitation," though not even the *laukika* (mundane) mode of representation should be understood as anything like realism or naturalism. Although

the text of the *Nāṭyaśāstra*, corrupt and disjointed, was not discovered until 1865, embedded in the much more elegant (in language as well as thought) commentary of Abhinavagupta, many of its ideas were known through other sources. The discussion of *rasa*, in particular, as the organizing principle in artistic production and the goal of aesthetic experience is considered a foundational moment in the Indian speculation on art.

The original meaning of this term is "flavor" or "taste" and was first used in early dramaturgical manuals to describe the primary emotional experiences available in the theater—the comic, the heroic, the erotic, and so on. The application of this scheme first to dramatic situations, then to dramatic language, and finally to all poetic language eventually yielded a comprehensive theory about the values and techniques involved in the production and apprehension of art. That this theory should come to dominate all philosophical discussions about art has to do not only with its rich explanatory power but also with the sociology of knowledge in premodern India: although Indian theater per se was never associated with the ritually higher castes before the colonial-influenced styles arose, the production, ordering, and criticism of Sanskrit texts were a virtual Brahman monopoly. (It was also, naturally, through this literate community and their texts that Europeans first approached Indian literature, including dramatic literature.) Hence, a classification scheme best suited to the overall theatrical experience came to be, by means of the infusion of *dhvani* (a complex theory about the various suggestive powers of language) a "poetics" broadly applied to many forms of art. With the transition in Hinduism from ritual performance to devotional theism in the early Middle Ages, *rasa* theory expanded to include the notion that certain rarefied experiences of art could resemble union with the divine (whether conceived of abstractly or personally), or even that they could stimulate this union, permanently.

Despite the general division between systematic aesthetics in literature and, as it were, practical aesthetics in the visual and musical arts, the original dramaturgic setting of *rasa* helps us trace the evolution of an aesthetic discourse that came to be applied to nearly all forms of artistic expression. In the *Nāṭyaśāstra* itself, *rasa*, "aesthetic flavor" seems to be an intensification of *bhāva*, the fundamental emotional quality of a performance. The theory that eventually attained the greatest prestige, however, posited a *disjunction* between worldly emotions *(bhāva)* and transcendent aesthetic moods *(rasa)*; this was clearly due to the theological commitments of the most brilliant and prestigious theorist, the twelfth-century Kashmiri philosopher Abhinavagupta. Although Sanskritists have until recently been less interested in other interpretations, it is clear that in all the performance arts, the aesthetic discourse centers on *bhāva*, the means of its production, and the conditions of its apprehension, just as in the *Nāṭyaśāstra* (and in the every-

day expression of vernacular Indian languages), *rasa* is seen as existing along the same experiential continuum as *bhāva,* not as a radical departure from it. It is no wonder that the frequent subject matter of the visual arts and music, when it is not human erotic passion or military heroism, is a sensual devotion to very unabstract gods and goddesses.

Although the *Nāṭyaśāstra* is organized in such a way as to frame its varied contents as a single discourse, its uneven textuality indicates that it is a compendium from sources of differing provenances. While it is unlikely that *rasa* had achieved the crucial and exalted role of later theorizing, it is clear that attitudes of aestheticism and cultures of elite connoisseurship were already in place by the beginning of the first millennium CE, attitudes and cultures that have attended the performing arts till this day. This does not mean that nonelite audiences did not witness plays or enjoy dance and music concerts. On the contrary, professional guilds, merchants, and local governing bodies sponsored performances for large audiences by artists who shared their styles, techniques, and source material with those who played before kings and nobles. Rather the elite cult of connoisseurship meant that (1) drama, music, and dance were intended to elevate the experience of self rather than provoke questions about the social order or what the gods intended for the human community; (2) the ideal audience was composed of men and women whose knowledge of the technical requirements of performance was virtually identical with that of the performers; and (3) the ideal audience saw itself as one in spirit with the creative goals of the author or artist: the spectator was called a *sahṛdaya,* one whose heart is "one with" the artist, or a *rasika,* one who experiences *rasa.* Even today at performances of Indian music, drama, and dance, the shivers of delight, the grunts and ejaculations of pleasure—these indicate a profoundly informed involvement; in the art forms that involve improvisation, the performers derive further creative energy from their "with-hearted" audiences. Thus, connoisseurship was and is a discipline—a Yoga. In this way, the aesthetic sociology of artist and audience much more closely resembles at end one of the spectrum, the informed, worshipful adulation of balletomane for dancer, and at the other, the symbiosis between jazz musicians and their fans. The aesthetic rapture, the transport is itself the goal, not the revelation or transformation of a worldview, such as the political recontextualization found in the Greek theater.

There are reasons to believe that even by the era of the Gupta dynasty (320–500 CE) the fundamental pattern of a community of artists and connoisseurs brought together in an assembly by a wealthy or royal patron, both equally familiar with a body of technical requirements, both focused on the goal of a gradual elevation to raptures of aesthetic delight, would gather for a relaxed and very long session of performance. In music it is immediately obvious that the essential form is improvisation that explores the possibilities

of a composition that "realizes" a particular mode, and a great deal of the skill involved in creating the music has to do with producing innovation while honoring the tones of the *rāga* and remaining with the extended rhythmic format called *tāla.* As with all forms of art that involve improvisation, sustaining coherence while introducing surprise is the goal. Yet, it cannot be emphasized too strongly that the performer must have so thoroughly absorbed the technical discipline that the *rāga*'s melodic sequence and the long, sometimes asymmetrical rhythm line must be as natural as breathing before he or she can begin to improvise. Except in the case of a prodigy, the artist cannot begin to create until long years of practice and apprenticeship bring him or her to a state of mastery. This practice is considered not only a matter of technical mastery, but also a spiritual discipline of *memory, imitation,* and *internalization* that is part of the process of transmission from guru to disciple. The student is expected to simply reproduce what the teacher does without recourse to the aid of notebooks or the *śāstra*s. Of course, at first the student simply cannot do this, but personal instruction ideally takes place on a daily basis and gradually the student catches on. The model has its roots in the oral transmission of the ancient scriptural texts of Hinduism, the Vedas (Reck, 1989)—except that in the arts the student will at some point take flight and begin to be able to improvise creatively. (This was probably true of Vedic poesy before the texts were frozen and declared ahistorically transcendent.) Because of the element of improvisation in the performing arts, Indian classical music has often been compared to jazz. This is somewhat misleading sociologically, and ignores the features of ritual and tradition that are inseparable from Indian performance traditions, but the comparison between those styles of jazz that explore the melodic implications of a "standard" and the development of a *rāga* is not entirely misplaced.

This performance model finds expression in poetry as well. In the Sanskrit texts, poetry is classified as *śravya* and *dṛśya,* that which is to be heard (recited verse) and that which is to be seen (drama, which is a medium of mixed verse and prose). Reading, silent or otherwise, is a secondary form of transmission. Descriptions in texts such as the *Sāhitya-darpana* (The Mirror of Composition) indicate that the poet-in-training, as well as the would-be connoisseur, is expected to be thoroughly versed in the many syllabic and moric meters of Sanskrit, not to mention the extensive catalogs of figures *(alaṃkāra),* and to have memorized the standard lexica. Court annals reveal that kings would not only engage poets as panegyrists to praise them and their exploits, they would also invite poets to engage in competitions, improvising on a theme suggested by the king or other members of the royal assembly.

Aesthetic ideals, like all cultural models, have been disseminated through language, religious traditions, and political hegemony. It is perhaps not surprising that these three

have often been found together. Although the Vedic language of the ritual and speculative texts evolved into the classical Sanskrit adopted for the many varieties of *śāstra*s, for many centuries it did not achieve hegemony, largely because of the rise of Buddhism from about 200 BCE until approximately the fifth century CE. The greatest of the Maurya dynasts, Aśoka (c.268–233 BCE), favored Buddhism, early forms of which rejected brahmanical ritualism and ideas of transcendent permanence, as well as their language, Sanskrit.

Paradoxically, the beginning of the permanent record of many of the Indian arts begins with artifacts of a religious civilization that no longer exists in India. Buddhist sculpture, initially borrowing from Hellenic models in the post-Alexandrian cultures of the Northwest, developed into a graceful articulation of the human figure that formed the basis of the proportions and stance of later Indian iconography; bodhisattvas transmuting into Hindu gods. Carvings of playful and voluptuous mythological creatures on Buddhist gateways are the ancestors of the grinning demigods and impossibly gorgeous dancing girls that emerge from the walls of later Hindu temples. The earliest stone architecture of South Asia consists of these gateways, as well as pillars and great halls that sheltered the mendicant monks during the rainy season. The cave paintings of Ajanta reveal a splendid, luxurious court life. Frequently embedded in the painted and sculpted scenes of Buddhist life and worship is the figure of a patron, sometimes a noble, but often a wealthy merchant, or the female head of a merchant family. The integration of patronage into artistic representation becomes an abiding feature of architecture, painting, and the literary arts. Similarly, the renunciant, nonviolent religious tradition of Jainism has made tremendous contributions to what are now apprehended as characteristically "Hindu" forms of literature, painting, sculpture, and architecture.

The three later dynasties that were able to consolidate power over a great deal of the subcontinent each promulgated a language and associated set of aesthetic ideals: the Guptas sponsored a revival of Sanskrit learning and fostered a royal theism that was expressed in the confident classicism in sculpture and architecture. The middle ages saw shifting patterns of power in which smaller kingdoms tried for periods of time to emulate Gupta grandeur on a smaller scale. This was a period during which earlier Sanskrit technical compendia gave way to systematic philosophical treatises. For example, monistic, semimonistic, and dualist philosophical trends in the Upanishads emerged as full-blown theologies, reorienting realms of knowledge developed earlier, including aesthetic theory. It was during this period, for example, that theories about the expressive powers of language (*dhvani,* literally "resonance") were synthesized with the *rasa,* so that what was originally a theory about the emotional effects of drama became an explanation for the ways in which poetic language reaches into each individual's experience. Following Ānandavardhana's treatise on *dhvani,* the *Dhvanyāloka* (ninth century CE), Abhinavagupta (c.1000 CE) built on the superstructure of Kashmiri Śaiva theology to produce his great treatises in the form of commentaries on both the *Nāṭyaśāstra* and the *Dhvanyāloka* itself, ensuring that thenceforward aesthetic discourse would be dominated by the same transcendent reference points that brahman textuality had been developing in most other śāstras. Although debates have continued about the applicability of the *rasa* theory to the visual arts and music, it is incontrovertible that from the time of Abhinavagupta, *rasa* attained to a status of theologically legimated transcendence, and the *rasika* might not only be an enthusiastic lover of the arts, but a votary in an elite cult of connoisseurship.

During the first five centuries of the common era, the literature of the Tamil region of the deep South attained a magnificence of expression in the quasi-legendary assemblies *(cankams,* pronounced *sangams)* of wandering bards sponsored by royal patrons. While the Dravidian tongues of the South are linguistically distinct from the Indo-Aryan Sanskrit and its derivative vernaculars, the theoretical taxonomy of Tamil poetry not only affords an insight into the aesthetic realm of the *cankam* poets, it also provides a template for understanding literary topoi in all the regional languages and Sanskrit itself. Poetry was thought to be of two types—*akam* (the domestic, including matters of kinship and family as well as love and sexuality) and puṟam (the public, including kingship, warfare, and politics). Within the former, the varieties and especially the stages of passionate love were correlated with an idealized division of the Tamil country into five landscapes. Although the dramatis personae and scenarios are found in a variety of genres in literatures throughout the subcontinent, it is quite possible that the original formulation does derive from this material. The naive young woman, sometimes married, sometimes under the watchful eye of her elders, sends her friend to her lover as a go-between, and eventually steals away to him under cover of night or the black monsoon clouds. Their illicit meeting is brief, for the lover must depart to fulfill obligations in military service or mercantile affairs. The characteristic poetic expression of love therefore is the yearning of love-in-separation rather than the fulfillment of love-in-union.

After the fifth century CE, the poetics of *bhakti,* devotional theism, appropriate this mode to express longing for the absent lord, but the scenario of human love described above filters into the poetry of Prakrit, a cluster of languages sometimes described as vernacular forms of Sanskrit but equally characterizable as literary versions of prototypes for the emergent vernaculars of the northern regions of the subcontinent. The poetic texts in Sanskrit often exemplify the various modes and genres with literary Prakrits, and most characters who are not royal males or brahmans in so-

called Sanskrit dramas speak Prakrit. Thus, it is probable that this striking scenario of passionate love ubiquitous in Sanskrit poetry and drama, depicted hundreds of years later in the paintings of North India, and evoked in the wild desolation of North Indian musical *rāgas* even today began its life in the villages of the Tamil country. Beyond its echoes in the more or less chaste devotional hymns to the elusive incarnations of Vishnu, Shiva, or the Goddess, the lover-beloved-go-between story is expressed in the stories and plays of western India about the stolen love of Krishna, who as a rustic cowherd lures the swooning village girls into a moonlit tryst with his sensual flute playing, and disappears after bestowing his divine passion on the favored cowherd girl, Rādhā. The devotees of the charming Krishna are invited to imagine themselves in the dramatic scenario radiating from the lord; the Gosvāmins (sixteenth century CE), medieval theologians of *bhakti* from Bengal, appropriate the aesthetic psychology of *rasa* as a devotional poetics to describe the divine erotic rapture of the devotee with Kṛṣṇa (Krishna). Rather than following the intellectual lineage of Abhinavagupta, in which the transcendent is abstracted in a disjunct fashion from human experience, the Gosvāmins followed a more popular and less cerebral interpretation of *rasa*, articulated by the aesthetician Bhoja of Dhāra (eleventh century CE), which (as mentioned above) survives to this day in its colloquial usage on the subcontinent—*rasa* is an intensification of worldly emotion rather than a departure from it, just as the love of the lord builds on human experience rather than marking a departure from it. A related formulation of the aesthetic psychology of *rasa* along these lines is that *rasa* is the experience of aesthetic delectation itself, which brings in its train the flood of emotion (*bhāva*) on which it is based.

Indian classical music has often been understood within the framework of the *rasa* aesthetic, but music theory and aesthetics present a number of additional paradigms that came to be applied to other cultural forms as well. Literary evidence such as the plays of Kālidāsa (fifth century CE) reveal a great deal of information about the development of aesthetics in music during the Gupta period, that is, before the production of the extant musical *śāstras*: music, like drama, is understood as having the power to evoke deep layers of emotional memory, perhaps even from former lives, refined and universalized as *rasa;* there appears to be an elaborate grammar of musical language as well as a consensus about critical norms in the connoisseur culture of court musical life. It was probably about this period or slightly later that the first texts describing the microtones of the modal scales or *rāgas* were produced. But the seminal encyclopedic text for the theory and practice of music, the *Saṃgītaratnākara* of Śārṅgadeva, is not dated until the early 1200s. The text speaks in terms of distinctions that have been cultural commonplaces in the subcontinent for millennia. By dividing music into *sāman* (ritual chanting), *deśī*

(folk or rural music), and *mārga* (high culture, court, or urban music), a template was established in which the archaic sacred was revered as the primal font of the fundamental medium and safely placed out of range for relevance to performance and the ongoing development of creative musical theory, while serious, sophisticated, connoisseur-worthy art was clearly set apart from the rustic and untutored. *Mārga* versus *deśī* remains an important distinction in India's self-representation of its cultural artifacts.

As a prelude to presenting the great "Hindu" treatise on music, it is necessary to describe a characteristic of brahmanic discourses produced during India's many periods of cultural turmoil and synthesis. First challenged by the indigenous but nonorthodox movements of Buddhism and Jainism, both of which wielded considerable influence between about 500 BCE and 800 CE, and the political and cultural hegemony of Islamic rule (800 to 1600 CE), brahmanism responded by refining its own ideas and incorporating theological conceptions and worship practices from popular religion as well as from its rival faiths. This syncretism resulted in a dynamic and complex skein of traditions called "Hinduism" by Muslims and later the British. At the courts and in the streets there was a great deal of exchange between those who by default fell within the Vedic tradition (ordinary "Hindus") and others, whether native or alien, who did not follow that tradition; this has obvious ramifications for artistic practice. Yet, in the world of brahmanic texts, especially epics and *śāstras*, India was largely a land dominated by sages and royal orthopraxy. Foreign nationalities were acknowledged, but not competing ideologies. One would never know from the *Mahābhārata*, for example, that Buddhism was powerful during most of the millennium of the epic's composition. One would never know from the *Saṃgītaratnākara* that Indian music was undergoing a defining transformation under the influence of Perso-Islamic models. A persistent brahmanic strategy was to assert the eternality and therefore the ahistoricity of its texts, attempting to remove them from critique on the grounds of incoherence, obscurity, or irrelevance. Thus, the *Saṃgītaratnākara* begins not with a general description of music, but a disquisition on the first vibration of sound emanating from Brahmā, the most mythless of the Hindu creator gods. More than a thousand years before, brahmans faced with the Buddhist critique of the dubious efficacy of the sacrificial ritual and the hierarchical caste system that maintained it, had developed the startling idea that the Vedic hymns were not actually created by the brilliant priest-poets whose family names were attached to the books of the Veda but were eternal; the Vedic priests rather "saw" and "heard" the poetry that has come down to us as the earliest specimen of Indian literature.

The rise and splendor of Indo-Islamic civilization from the thirteenth through the eighteenth century forced a retrenchment of orthodox Hindu values, reflected in aesthetic

practice, even as it stimulated the hybrid forms of art and literature in the North. Throughout its history India had been subject to invasions and immigrations, and had largely absorbed and blended the foreign influences. The serene Greco-Bactrian images of the Buddha, draped in a toga and displaying distinctly Mediterranean features, attest to an early example of this blending, and there are those who wish to counter the harsh characterization of the onslaught of Muslim rule in India with reminders that the previous Hindu culture was created out of diverse strands, indigenous and foreign. It is true that even before Maḥmūd of Ghazna's invasion in about 1000 CE, Arab traders had settled in many places along the western coast. On the other hand, the great Indo-Islamic dynasties had the military power to overwhelm rather than blend with the existing culture. Maḥmūd and his descendants established a lavish court and inaugurated the production of visual and performing arts in the style of their Turkish ancestors, whereas their successors, the Mughals, were entranced with the Persian culture of their previous homeland. The importation of these models by the Mughal kings and ministers not only established an imperial preference, it also joined the cultural production of Indo-Islamic rule with other great centers of Islamic culture to the west and north or the subcontinent. The "mission to India" was, in fact, part of the public discourse of Islamic rule: although individual rulers were not without respect for Hindu cultural achievements, they regarded the Islamic civilizations of Isfahan, Samarkand, and Baghdad as far superior, especially as they glorified the one God. Elite Muslims were impressed by monistic strains in Hindu philosophical thought, but horrified at the riot of idolatrous images that played so exuberantly on Hindu temples. Yet, just as we can now see striking points of contact between indigenous Indian lyrical poetry and the Islamic literature that drew on Persian models, features such as the love of ornamentation (in clothing and architecture, for example) in retrospect can be viewed as productive points of convergence between the two cultures.

The Persian models beloved by the Mughals were to have their greatest influence in the public arts of music, architecture, and painting. Even the untrained eye can see some immediate similarities between the jewel-like colors, sweeping line, and delicate brushwork in Persian and Mughal miniature painting (Chandra, 1983, quoting Stchoukine, p. 92). Yet it soon becomes apparent that indigenous South Asian patterns of representation, such as royal and divine groupings and mythical narratives, as well as European traditions of naturalism in portraiture, have been incorporated in the Mughal pieces, making the style a truly Indian one rather than an Indian variation on Persian models. The dynamic and evolving Mughal synthesis in turn stimulated a revival of "Hindu" painting in the courts of the Rajput rulers of western India, some of whom continued to resist Mughal

imperial consolidation and some of whom thrived as vassal kings in its service.

Doubtless the most obvious feature of pan-Islamic architecture introduced into India was the dome. The early Buddhists had erected moundlike *stūpa*s as memorials and reliquaries around which the faithful moved in ritual procession to enact reverence to the Buddha and his teaching. These, however, were solid structures, not true domes. Elsewhere in pre-Islamic India are found various kinds of vaulted ceilings or drumlike towers; it is a technical deficiency that domes are absent on the subcontinent. Nonetheless it was during the five hundred years of Muslim rule that domed structures transformed many South Asian cityscapes and royal centers into variants of those in the Middle East and Central Asia, effectively aligning their realm on the grandest visual scale with the Islamic world. Not only in the *masjid*s (mosques) large and small, memorial tombs and monuments (such as the Taj Mahal), and public buildings of all kinds made use of this "Islamic" dome. So powerful was the association of this feature with rule that Hindu princes also incorporated them into their palaces. For the centers of colonial administration and the great buildings emblematic of the "gift" of colonial infrastructure (post offices, railway stations, etc.), the British imperial architects expanded the generalized exotic, tropical architectural vocabulary with the Islamic dome—suggestive at once of British possessions in other parts of the Islamic world, as well as of the British will to carry on the magnificence of the Mughals in their own style.

Even before the Islamic incursion, India had been characterized by a variety of regional differences roughly equivalent to the national differences within Europe, but the division of North and South was most pronounced owing to the moderate barrier of the Vindhya mountains and the broad linguistic separation of Indo-Aryan languages in the North and Dravidian tongues (such as Tamil) in the South. The strength of Muslim rule in the North, especially after the dominance of the Mughals, made of the earlier geographic divide a cultural one, with all the arts more or less influenced by Indo-Islamic sensibilities in the North, and a more pervasively brahmanic Hindu quality found in the South. It was long thought that in the realm of music especially, the South had remained "pure," free from Islamic "taint," and therefore retained a greater degree of archaic Hindu musical theory and style. But if this is true, it is only part of the story.

Patterns of patronage stimulated the development of regional musical styles, the more secular (and erotic), more emotionally indulgent and musically volatile style of the North (Hindustani), and the more chaste (and devotional), rhythmic, and classically virtuosic style of the South (Carnatic). Both forms doubtless represent significant creative evolutions from the pre-Muslim eras and the devotional

texts in Carnatic music are not of ancient but late medieval origin. It is true that the Muslim emperors of the North favored the sensually intoxicating, and often mystical explorations of the *rāgas,* while southern Hindu kings were highly impressed by the compositions in praise of Hindu deities by Tyāgarāja and others that formed the basis for Carnatic vocal and instrumental performance, but this cannot be an argument for the antiquity of Carnatic music. After all, the man called the "father of Carnatic music," Purandara Dāsa, lived in the seventeenth century. Further, there were at different times significant Muslim rulers at various locations in the South, which meant that the general process of borrowing back and forth really took place all over the subcontinent, despite the construction of "Hindu" and "Muslim" sociocultural realms by the British colonial administrators and the guardians of religious orthodoxy on both sides.

The great Mughal ruler Akbar encouraged a synthesis of Islam and Hinduism by bringing together sages and philosophers from both traditions and likewise encouraged artists and musicians from Persia to share the assembly halls with native artisans and performers. Although his successors Jahangir and Shah Jahan, builder of the Taj Mahal, were more assiduous in developing the Persian cultural models, it was precisely during this period that indigenous forms began to alter the elite Persian world that the Mughals reproduced in courts extending all over the North and well into the southern plain of the Deccan. Fundamentally, Islamic architecture begins to show decorative and structural features borrowed from Hindu temples; on the other hand, the great structures built by both the British and Hindu rulers during the decline and extinction of the Mughals exhibit a distinctly Islamic quality with their domes and minaret-like towers evocative of mosques. Court music adopts the classical *śāstric* grammar of the *rāga* and *tāla* system. Most important, a new pan-Indian language, Urdu, arises out of a blend of a late medieval Indo-Aryan dialect and Persian.

The significance of language in the aesthetic life of South Asia cannot be overestimated. From the long historical view, the periods in which large portions of the subcontinent have been brought under a single political entity have been relatively few, but they have been cultural and aesthetic watersheds owing to the promulgation of a link language by the ruling powers. Despite Aśoka's rejection of the brahmanic Sanskrit, that language attained such pan-Indian hegemony for the dissemination of elite texts, both *śāstric* and belletristic, that the Buddhists themselves were obliged to employ it in their writings. The Mughals naturally required the use of Persian for all administrative and public literary occasions, but the new language, Urdu, became within the space of two hundred years both the language of the bazaar and the language of poets. Eventually, it became the language of most of India's Muslims, even those who

lived outside of its northern Indian homeland where other regional languages were the vernacular (excepting those who lived in Bengal and the deep South), the language of Hindus and Muslims alike in a great swath across northern India, the unofficial lingua franca of all of northern and central India, the main language of Pakistan, and in its Sanskritized version as Hindi, one of two sanctioned national languages (along with English) in the modern Indian state.

Almost all of the nouns in Urdu are Persian, which means that many of them are in origin Arabic: not only did Urdu literature begin with Persian genre models, but a Perso-Arabic Islamic sensibility infuses the literature. The distinctive Urdu verse form, the *ghazal,* abundantly illustrates this aesthetic, yet paradoxically shows the many points of contact with indigenous Indic aesthetic practices. The love poetry embodied in the *ghazal* couplet imports the Persian imagery of the rose and the nightingale, the jinni and the peri. It speaks of the exquisite torture of doomed illicit passion for an unattainable woman or youth, the intoxication of wine and the intoxication of sexual allure, which can become indistinguishable from or an instance of the spiritual intoxication with God sought by Sufi mystics, whose love is also illicit in terms of orthodox Islam.

Yet, certain persistent images and situations, with somewhat different associations and contexts, are shared with the Tamil, Prakrit, and Sanskrit lyric poems—the cover of night hiding the illicit tryster, the lightning flash that like the sudden erotic surge reveals the beloved and emblematizes his fugitive appearance, and disappearance. For as in many traditional societies in which marriage was a contractual alliance between families for mutual economic satisfaction, love was usually a dangerous, unsanctioned emotion between two people who could never hope to be united, or even to meet. Both literatures rely on intensely conventionalized traditions in which expert manipulation of familiar tropes and exaggeration of the "natural" are not at all considered a sign of lack of originality, but rather a sublime elevation of a commonly shared truth. The apparent similarities of style, once taken as a sign of the unimportance of individual creative personae, actually betrays distinct authorial preferences to the serious student, whether the specimens are from ancient times when work was unsigned or during the Islamic period, when names of artists are regularly attached to painting and poetry. In both literatures, memory is more than a documentary revisiting of the past; rather, memory serves as a creative power, defeating the pain of separation (in the Tamil or Sanskrit lyric) or the impossibility of union, and out of the slenderest event calling into life all the phases of a love affair and their concomitant emotional changes. Although the Sanskrit poems are primarily known as written text now, the original performance context of the assembly *sabhā* at which verses would be spontaneously thrown up in playful competition was not

INDIAN AESTHETICS: Overview. *Śiva Naṭarāja* (eleventh century); National Museum, Madura, Tamil Nadu, India. (Photograph courtesy of Giraudon/Art Resource, New York.)

hundreds of niches containing representations of the dance postures of the *Nāṭyaśāstra,* and these postures form the basis of much of Indian sculptural iconography. Goddess Sarasvatī is depicted with a vīnā, prototype of string instruments. Lord Śiva, the erotic ascetic, is Naṭarāja, lord of the dance, patron of the performing arts. Possibly the best-known image, along with exemplar of the most sublime Indo-Islamic architecture, the Taj Mahal, is Śiva serenely stamping down the untroubled chubby demon dwarf of spiritual ignorance, dancing his apocalyptic dance of destruction and rebirth, the dance that generates a circular halo of flames, the emblem of a brilliantly glowing and ever-expanding sacred universe.

BIBLIOGRAPHY

Chandra, Pramod. *On the Study of Indian Art.* Cambridge, Mass., 1983.
Coomaraswamy, Ananda Kentish. *The Transformation of Nature in Art.* Reprint, New York, 1956.
Coomaraswamy, Ananda Kentish. *Essays in Architectural Theory.* Edited by Michael W. Meister. Delhi, 1995.
Dallapiccola, Anna Libera, ed. *Śastric Traditions in Indian Arts.* 2 vols. Stuttgart, 1989.
Delmonico, Neal G. "Sacred Rapture: A Study of the Religious Aesthetic of Rūpa Gosvāmin." Ph.D. diss., University of Chicago, 1989.
Goswamy, B. N. *Essence of Indian Art.* San Francisco, 1986.
Hart, George L. *The Poems of Ancient Tamil: Their Milieu and Their Sanskrit Counterparts.* Berkeley, 1975.
Maxwell, T. S. "Śilpa vs. Śāstra." In *Śastric Traditions in Indian Arts,* edited by Anna Libera Dallapiccola, vol. 1, pp. 5–16. Stuttgart, 1989.
Miller, Barbara Stoler, ed. *Theater of Memory: The Plays of Kālidāsa.* Translated by Edwin Gerow, David Gitomes, and Barbara Stoler Miller. New York, 1984.
Miller, Barbara Stoler, ed. *The Powers of Art: Patronage in Indian Culture.* Delhi, 1992.
Neuman, Daniel M. *The Life of Music in North India: The Organization of an Artistic Tradition.* Reprint, Chicago, 1990.
Ramanujan, A. K. "Two Realms of Kannaḍa Folklore." In *Another Harmony: New Essays on the Folklore of India,* edited by Stuart H. Blackburn and A. K. Ramanujan, pp. 41–75. Berkeley, 1986.
Ramanujan, A. K., ed. *The Interior Landscape: Love Poems from a Classical Tamil Anthology.* Reprint, Delhi, 1994.
Reck, David B. "The Invisible Śāstra: A Personal Memoir of Musical Study in India." In *Śastric Traditions in Indian Arts,* edited by Anna Libera Dallapiccola, vol. 1, pp. 407–414. Stuttgart, 1989.
Richmond, Farley P., Darius L. Swann, and Phillip B. Zarrilli, eds. *Indian Theatre: Traditions of Performance.* Honolulu, 1990.
Russell, Ralph. *The Pursuit of Urdu Literature: A Select History.* Bombay, 1992.
Sivaramamurti, C. *Naṭarāja in Art, Thought and Literature.* New Delhi, 1974.

DAVID L. GITOMER

dissimilar from that of the Urdu poetry—the *mushaira*s that still take place in India, Pakistan, London, and New York.

The *ghazal* found further life as a text in Hindustani music, and lives today in the poignant lyrics of the best of the so-called Hindi (really Urdu) film songs. But the interweaving of the arts has always played a major part in the Indian aesthetic sensibility. The revival of Hindu painting in the sixteenth, seventeenth, and eighteenth centuries, much stimulated by the ateliers of the Mughal courts, chose for a standard subject the series of Rāgamāla pictures, in which the various musical modes were depicted as personifications or distinct emotional scenarios. Over a thousand years earlier, Kālidāsa filled his plays with poems about connoisseurship in painting and dance, as well as, of course, about the power of poetry. With his flute song Lord Krishna calls the cowherd girls, and all his devotees, to dance with him in the dark-clouded monsoon night, and the scene becomes the subject of countless paintings, and of classical dance itself. Hindu temples through the medieval period were girded with friezes depicting musicians at work, and dancers serenely basking in their own beauty. The walls of the Chidambaram temple in South India are lined with

Indian Theories of Interpretation

As in the Judeo-Christian tradition, Indian theories of interpretation originated in scholastic debates about the correct determination of the meanings of the scriptures—the Hindu Vedas and, in the case of Buddhism, the spoken words of Buddha. All philosophical schools, both pro-Vedic and anti-

Vedic (mainly Buddhist) contributed to the discussions, and although primarily concerned with metaphysical and soteriological matters, they formulated a general philosophy of language and devised a sophisticated system of textual exegesis. The literary critics generally remained detached from metaphysical discussions, but borrowed their conceptual apparatus from these schools—particularly from logic, grammar, and Vedic exegesis (called Mīmāṃsā)—while they also made significant contributions to semantic analysis. Of all these schools, it is the Vedic exegetes (Mīmāṃsakas), headed by Jaimini (c.200 BCE) and his commentators Śabara (c. first century BCE), Kumārila, and Prabhākara (seventh century CE), who developed a full-fledged methodology of textual interpretation, which came to enjoy wide acceptance in spite of sectarian differences reflecting the individual epistemological and ontological stances of rival schools.

The science of interpretation, as developed by the Mīmāṃsā school, was treated as a separate discipline, distinct from grammar or the study of language that concerned word forms and their syntactic relations. It typically involved the larger investigation of the import of whole passages or texts and of their contextually generated meanings as well as the settling of dubious or problematic passages in the light of the overall purport or intent of the Vedic texts. In short, it comprehended the whole process of reasoning that is required for making coherent sense of a verbal text. A common assumption of all schools of thought was that the same set of interpretive principles or the same metacritical approach applied to all linguistic texts—scriptural, legal, literary, or other—and that in all cases the linguistic convention and the well-known canons of reasoning provided the sole criteria for testing the validity of an interpretation. The rules of interpretation must be derived from the rules of ordinary language. There was, however, no conception of language in its diachronic or sociohistorical dimensions. The language of a text—Vedic or other—was treated synchronically by the norms of standard usage, established by the grammarian Pāṇini (c.400 BCE).

There are two major presuppositions of the Mīmāṃsā theory of interpretation: (1) the autonomy of verbal meaning and its impersonality, and (2) the unity and determinacy of meaning.

A fundamental premise of the theory of language derived from grammar is that the linguistic sign (as in Sausserian linguistics) is a binary entity consisting of the word form (*śabda* or the acoustic form) and a mentally perceived sense or object content (*artha* or signified) and that meaning is objective reference; that is, it involves a reference to what lies outside the sign—whether the referent is an actually existing thing or a pure mental entity. Again, both grammar and Mīmāṃsā take the relation between the word and its meaning to be uncreated and primordially given, in opposition to the logicians, for whom it is arbitrary and created by human convention. In its most basic sense, what this implies

is that the word-meaning connection must be present as a given before language use could occur and that changes or deviations in usage, too, can be judged only against the background of a "streamlike continuity" of the linguistic system. But, in either case, all Indian schools agree that the words of a text, once they are put together in a certain formal order, remain fixed in their particular meanings and that verbal understanding can result only from a stable and self-same text. For all but the Buddhists and the Vaiśeṣikas verbal cognition is an independent epistemic category not to be conflated with sense perception or inference. What the word signifies has its being only as a concept or mental representation, and it is directly conveyed by it, not inferred.

There is another sense in which language as a sign system is autonomous. Whether it was created by a convention or whether it has been there from the beginning of time, the word does not depend for its meaning on any particular author/speaker—human or divine—because of its self-revealing character. A linguistic text is a self-sufficient structure capable of delivering its own meaning. Here some logicians and the followers of Prabhākara argue for the personal derivation of verbal meanings and of their validity. A sentence has meaning because it has been put together by an intelligent agent to cause a certain understanding in the hearer. The validity of a statement likewise is dependent on the trustworthiness of the speaker. But the Mīmāṃsakas of the Bhaṭṭa school, the grammarians, and the Vedāntists maintain that although a sentence is composed by the free will of a speaker, the speaker does not create its meaning, nor for that matter does the interpreter constitute it. They both participate in an already existing impersonal apparatus of language. The validity of human utterances cannot be tested by the speaker's reliability, since the speaker's reliability itself cannot be authenticated except by some other corroborating evidence.

The Vedic exegetes were, of course, dealing with texts of unknown authorship, and hence they had to defend the authority of those texts purely in "intrinsic" terms. But their theory of impersonality extends even to texts whose authors are known. In their view, even such texts should be treated as anonymous because, in their own meaning, they are not dependent on the will or intention of any author. The Mīmāṃsā maxim says that the word itself is the speaker. Similarly, Buddhist hermeneutics adopts an impersonal approach to Buddha's teachings. It is guided by Buddha's dictum in regard to his own doctrines: "Rely on the teaching, not the teacher." Such being the case, authorial intention or utterer's meaning is not a separable element, distinct from utterance meaning, nor is the knowledge of it a precondition *(hetu)* for verbal understanding. Some logicians insist that since all human utterances are intentional acts the speaker's intention *(tātparya)* must be appealed to especially for determining the purport of ambiguous or paronomastic expressions. But the Mīmāṃsaka replies that the linguistic

and contextual setting of the utterance is alone sufficient to disambiguate such expressions. Moreover, when the speaker's intention has itself to be deduced from the meaning of the text, how can that be said to be the cause of verbal understanding?

This is not, however, to say that the anti-intentionalist argument ignores the fact that human speech occurs in a communicative situation involving a speaker, the sign, and a hearer who decodes the sign. Nor does it abolish the notion of intentionality or intention ascription. It is fully recognized that human intentionality is the condition of all speech, a necessary presupposition of all meaning. The Mīmāṃsaka himself, in spite of his apparent reification of the text, constantly invokes human intentions, motivations, and speech contexts in fixing the intent of the text and in drawing out its implications and unstated meanings. But he makes no appeal to an actual speaker. What he appeals to are paradigms of human speech situations in which speakers intend certain meanings and express them according to the public conventions of meaning. This approach in terms of the "authorless" text has traditionally characterized all branches of textual interpretation in India, including literary criticism.

The unity and determinacy of meaning constitute another axiom of interpretation accepted by all schools, and they follow as a corollary to the theory of verbal signs outlined above. Human utterances normally express meanings that are both definite and unified; indefiniteness and multiplicity of meanings destroy linguistic discourse and cut at the very root of the convention of verbal signs. Even where a text is seemingly ambiguous and admits of many constructions, a unity of idea or purpose is to be imposed by some means in the interests of intelligibility. The Mīmāṃsaka protest that it is manifestly unjust to cultivate deliberate ambiguity. They do recognize that words are tricky instruments and behave in a bewildering variety of ways. Even connected utterances can be ambiguous, incongruous, or incomplete, or the connections between sentences may not be apparent. Hence, they have devised a whole armory of maxims *(nyāyas)* and interpretive procedures designed to avoid multivocity of meaning and secure interpretative certainty. One such maxim, the "nonsimultaneity" of multiple meanings, requires that not all the meanings of a multiple-meaning word should be taken as operative or as being of equal importance in the same instance of use: they should either be forced into relation, as subject and modifier, or all but the one that is demanded by the context should be excluded. A word or a sentence cannot be taken both in its literal and its metaphorical senses, as being "twice true," in any single instance of speech act. When the metaphorical sense is invoked, the context of the sentence will nullify the literal sense. Similarly, the etymological sense of a word should be invoked only when the conventional sense is unsuitable in the context. Mythical

and allegorical meanings are to be demythicized or rationalized to conform to literal norms. Another important maxim, which is common to both Vedic and Buddhist hermeneutics, stipulates that the definitive or explicitly stated meaning of a text (*śruti*, or *nītārtha* in Buddhist terminology) is more authoritative than that which is deduced metaphorically or from the context (*(liṅga*, or *neyārtha* in Buddhism). The one, in fact, excludes the other. Even the literary critics, who set great store by the suggestive or implicative function of poetic language, felt obliged to abide by these time-honored principles.

The unity of an utterance is understood pragmatically as the unity of a single consistent motive or purpose (illocutionary force, in speech act terminology). But such a unity is to be looked for only at the level of the sentence or a sequence of sentences constituting a discourse. The Mīmāṃsā texts analyze various sentence types according to the nature of the purpose they serve in the context of the Vedas—such as declarations, descriptions, imperatives, commendations, and deprecations. But in the interests of unity of discourse, these types are hierarchically subordinated to one central type. Thus, the Vedas, considered as one macro speech act, are shown to be injunctive in their overall purport; their illocutionary force is prompting *(bhāvanā)* men to perform certain ritual acts.

It is recognized by all schools that verbal meaning is context bound. Words may sometimes fail to convey definite meanings in isolation and need the larger context of a sentence. A sentence may often, in turn, require other sentences and sometimes a whole text or discourse to complete, clarify, or confirm its meaning. The Mīmāṃsakas believe, however, that words do possess core meanings, although they are also capable of extended or deviant uses, and that the sense of the sentence is made up of the senses of the words (the atomistic view). For the grammarian Bhartṛhari, on the other hand, word meanings are an unreal abstraction, and only the indivisible sentence or speech act should be endowed with the designative power (the holistic theory of *sphoṭa*). In any case, both schools acknowledge that the meaning of a sentence extends beyond its syntactic boundaries and requires the supplying of connections, missing links, and implications from the immediate textual/linguistic context and sometimes from the nonverbal/situational context, as in situations involving live speech. Context is defined as the mutual requirement of one sentence for another; it is any piece of knowledge that is required for the completion of the sense of a sentence or for its justification in case of some obstruction to it.

But context is what is demanded by the text itself and not anything extraneous. Naturally, every meaning will give rise to a train of associated meanings, so much so that a sentence may never come to a stop (the fault of endlessness). But in the interests of unity and determinacy, one has to set boundaries to its operations, guided strictly by what the ex-

egetes call "semantic expectancy" (*ākāṅkṣā*) or the logical need of one meaning for another. The limits of the meaning of a sentence are therefore set by the twin criteria of (1) apparent inconsistency or incompatibility of the sentence meaning with the context of the given discourse, which accounts for all secondary or metaphorical meaning; and (2) logical or psychological incompleteness (*ākāṅkṣā*). But for these two conditions, no other meaning need be understood from the sentence. The sentence must be deemed to rest in itself if it possesses a minimum of logical consistency (*yogyatā*). Commenting on the limits of contextualization, Kumārila says that there are no limits to the extent to which one may have to go in search of the required meaning. If the required meaning is not found within the sentence, then, one should go to another sentence in the text or to another context. And if it is not found there either, one should look for it in the world. But in all this, we are to be guided by what is actually given in the text and warranted by it.

Consistently with their premise that the verbal text—oral or written—is a stable and reidentifiable object with a unitary meaning, Indian theorists, even like their classical counterparts in the West, would not allow incompatible interpretations or multiplicity of interpretive choices. Divergences, they insist, must be settled through further discussion; they should in no case result in loss of definition. In practice, however, their monistic position did not prevent the spawning of radically different readings of the same body of scriptures, from different metaphysical standpoints. The theological schools systematically manipulated the texts to conform to sectarian viewpoints. But this is explained by the fact that the texts, which they all appealed to, were by no means unified compositions, but a vast body of accretions, often antinomic in character and capable of diverse constructions according to differing purposes, attitudes, or interests. In contrast, in the area of literary criticism, where one was dealing with unified structures, there was over the centuries a remarkable consensus both on the meaning and the value of the texts. In any case, the goal of interpretation remained the same, namely, the recovery of what is given in the text—its objective content. Free play and open-ended pluralism were never considered as viable alternatives.

BIBLIOGRAPHY

Chari, V. K. *Sanskrit Criticism.* Honolulu, 1990.
Gachter, Othmar. *Hermeneutics and Language in Pūrvamīmāṃsā.* Delhi, 1983.
Iyer, K. A. Subramania. *Bhartṛhari: A Study of the Vākyapadīya in the Light of the Ancient Commentaries.* Poona, 1965.
Lopez, Donald S., Jr., ed. *Buddhist Hermeneutics.* Honolulu, 1988; Indian ed., Delhi, 1993.
Matilal, Bimal Krishna. *The Word and the World: India's Contribution to the Study of Language.* Delhi and New York, 1990.
Raja, K. Kunjunni. *Indian Theories of Meaning.* 2d ed. Madras, 1969.

V. K. CHARI

INEFFABILITY. [*This entry explores a basic problem in all aesthetic theories: that some part of any work of art is ineffable; that is, it is "unspeakable" and so cannot be captured by aesthetic theory.*]

The word *ineffable* literally means *unspeakable,* but the two words are fringed in different auras of connotation. *Unspeakable* suggests the forbidden, even the monstrous—unspeakable practices, unnatural acts; but *ineffable* suggests a divine power too great for words. In the seventeenth century, *ineffable* was sometimes used in a quite literal manner, as a synonym for *unpronounceable:* a writer on the Chinese language, for example, noted that certain single brush strokes represented ineffable letters. But throughout the history of the English language, *ineffable* has typically been used as a word of mystification, appropriate to the sacred matters beyond the range of language. In Hebrew, the name of God was represented by a sacred formula that could only be written, never spoken: the tetragrammaton, traditionally given in English as *Jehovah.* Through such taboos, the notion of cannot-be-pronounced easily shades into the notion of must-not-be-pronounced.

Because *ineffable* seems somewhat occult, theological, or quaint, it is perhaps used most frequently in the jocular sense. In the nineteenth century, people sometimes referred to underwear as *ineffables,* and the word *unmentionables* retains, faintly, a similar sense today. The fact that *ineffable* is (like *disgruntled*) a word in a negative form, but without any positive in modern use, may add to its potential for humor. *Ineffable* is a poetic sort of word, and here is the best-known case of its use in a poem:

> When you notice a cat in profound contemplation
> Of the thought, of the thought, of the thought of his name:
> His ineffable effable
> Effanineffable
> Deep and inscrutable singular name.

This is from the beginning of T. S. Eliot's *Old Possum's Book of Practical Cats* (1963), where Eliot discusses the three names of every cat: the everyday, the fancy, and the ineffable, known by no human being. But the notion that names vary in importance between the casual and the absolute is potentially serious; and possibly the word *ineffable* can be rehabilitated and made useful for philosophy.

Cratylus. Homer mentions that the gods use a system of nomenclature different from the human: for example, in Troy there is a river called Xanthus by gods, and Scamander by men (*Iliad* 20.74). In Plato's *Cratylus,* Socrates speculates at length concerning this discrepancy between human and divine speech. In this dialogue, Hermogenes claims that the relation between word and thing is conventional and arbitrary, while Cratylus disagrees, arguing that things may be rightly or wrongly named. Socrates, meanwhile, listens to a summary of their debate, taking no firm position; he seems first on Cratylus's side, then ambigu-

ously on Hermogenes's. He argues against Hermogenes by pointing out that some names are more telling, more satisfactory, than other names, and that a name is an imitation of the essence of a thing, just as a painting is an imitation of the outline of the thing (423e). Then he turns to Cratylus and argues that he is wrong, too: a name cannot be an inviolable aspect of a thing's reality, because if a thing is to be rightly, irrefutably named, there needs to be a fundamental resemblance, indeed, a perfect correspondence, between the name and what it refers to; but the idea that a word can be equivalent to an object is ludicrous: "how ridiculous would be the effect of names on things, if they were exactly the same with them! For they would be doubles of them, and no one would be able to determine which were the names and which were the realities" (432d). Any real name, Socrates suggests, is a more or less faulty representation of its correspondent thing. He asks himself if the original "legislator"—the fellow who invented names and attached them to things—was extraordinarily wise or knowledgeable; no, Socrates answers, the legislator was more ignorant than we are today—so there is no special prestige to be discovered in the origins of Greek or of any other language.

This dialogue presents a clear formulation of the two principal models of the ineffable. The linguistics of Hermogenes—which affirms that names are just labels slapped at random on the surfaces of things—promotes a user-friendly model of the ineffable: the ineffable is simply the body of objects and ideas to which a sign has not yet been affixed. The linguistics of Cratylus—which affirms that names *are* things, in their mode of extension into the domain of speech—promotes mystery. What does it mean, to Cratylus, if a thing does not have a name? Either the name is lost, and probably lost forever, except insofar as a supersensitive Homer may arise, capable of hearing the inaudible frequencies of such divine names as *Xanthus;* or there is no name at all, in which case the thing must always remain nameless, since a nameless thing is not likely to acquire a whole new mode of reality in the realm of names— an *invented* name is, for Cratylus, inevitably a *wrong* name.

Roughly speaking, the Christian tradition of the West has tended to follow the lines of Cratylus: the Scholastic realist positions that "nomina sunt consequentiae rerum," that God is an abyss of such unfathomable mysteries as the concept of trinity, all tend to draw a firm boundary between the securely nameable and the ineffable. But the nominalist, somewhat more secular tradition of the West has followed Hermogenes: the boundary between the named and the nameless is provisional and shifting, as the ineffable steadily gets displaced by the advance of human knowledge.

Older Art. Whatever criticism the concept of ineffability has received from philosophers, it has always had a home among artists. The traditional name for the sort of art that explores, not carefully delineated areas of human experi-

ence, but levitation, dizziness, rapture, is the sublime. In Plato's *Ion,* Socrates describes the inspired rhapsode as connected to heaven as if by a chain of magnets, "seized with the Bacchic transport . . . a light and winged thing, and holy" (534a–b); and in the Bible, the apostles of Christ, at Pentecost, are crowned with "cloven tongues as of fire . . . And they were all filled with the Holy Ghost, and began to speak with other tongues" (Acts 2.3–4). It is possible that divine language can at times find a direct conduit to the human organs of speech; the ineffable can sound out, having simultaneous aspects of nonsensical babbling and of devastatingly intense meaning. The same age that produced the Acts of the Apostles also produced the first known treatise on the sublime, ascribed to Longinus, which presents the highest art as a kind of liquidation of the normal world and normal human selves: "our soul is uplifted by the true sublime; it takes a proud flight, and is filled with joy and vaunting, as though it had itself produced what it has heard"; "[it] breathes into dead things a kind of living voice" (*On the Sublime,* chap. 7, 30). The sublime assaults the boundary between writer and reading, even the boundary between the living and the dead; the universe is reconstituted as a field of dazzlingly unstable metaphors, in which anything can be compared to anything else. Sublime art tends to regard the ineffable as an ascription of a kind of sense to many sorts of verbal constructions that assault ordinary models of speech acts. Perhaps it is fitting, after all, that the word *ineffable* should have found a home in Eliot's nonsense verse.

But the rhapsodic or Pentecostal model is not the only model by which the ineffable may be vented into the domain of speech. It is also possible to make experiments with a language not composed of sounds or syllabic written characters, a language that thereby might hope to epiphanize ineffabilities. For example, Plotinus, in the third century of the common era, thought that Egyptian hieroglyphics—a language of pictograms—might contain wisdom beyond the range of discursive languages:

> The wise of Egypt . . . indicated the truth where, in their effort towards philosophical statement, they left aside the writing-forms that take in the detail of words and sentences—those characters that represent sounds and convey the propositions of reasoning—and drew pictures instead For each manifestation of knowledge and wisdom is a distinct image, an object in itself, an immediate unity, not an aggregate of discursive reasoning and detailed willing. Later from this wisdom there appears, in another form of being, an image, already less compact, which announces the original in terms of discourse.
>
> (Plotinus, 1956, 5.8.6)

Plotinus used the same word (*agalma,* or form) to refer to (1) a statue, (2) the intelligible form that guided the sculptor's hand, and (3) Egyptian sacred writing. An ineffability, then, may be imagined indifferently as a sculpture or as a

hieroglyph, an instantaneous and triumphant autotely; it will suffer dilution and corruption if translated, compelled to channel itself into some consecutive, time-bound language.

The history of literature is full of fascinating experiments in finding ways to speak of those matters that cannot be spoken of; to overcome language's discursivity, its mechanical beat of subject and predicate, in favor of some seizure of transcendental meaning. There are a number of rhetorical tricks by which language can pretend to hieroglyphic immediacy or sublime involuntariness. There is a certain style of writing, sometimes called revelation discourse, that can be found in a number of ancient texts all over the world. Here is an example from Old Irish, "The Mystery," attributed to the mythical Milesian colonist Amergin, as translated by Douglas Hyde:

> I am the wind which breathes upon the sea,
> I am the wave of the ocean,
> I am the murmur of the billows,
> I am the ox of the seven combats,
> I am the vulture upon the rocks,
> I am a beam of the sun . . .

Elsewhere in this catalog Amergin speaks of himself as a "word of science"; and he is indeed a poet who has become language—the whole dictionary is nothing but a predicate of his ego. A poem like this is obviously a fragment of an infinitely long poem, since the rhetorical form (I am *x*, for any *x*) cannot be exhausted until every object in the universe is mentioned. In this way Amergin includes the ineffable, since the poem operates on a field larger than speech.

More Recent Art. In 1757, Edmund Burke tried to redefine for modern sensibilities Longinus's term *sublime*. He considered that there existed two categories of aesthetic effect, exactly opposite to one another: the sublime and the beautiful. The beautiful consists of small, intricate, fragile things, from glass flowers to lyric poems; the sublime, however, is an effect of astonishment or terror caused by overstress of the sense organs, provoked by the attempt to grasp the starry sky or the infinite ocean or the receding colonnade of a Gothic cathedral or the roar of a waterfall. Burke was one of the finest politicians of his age, and his theory offers a Realpolitik of aesthetic phenomena: what you have power over is beautiful; what has power over you is sublime.

Before Burke's time, the artistic prestige of the vague and inconceivable had not always been great. At the bottom of hell, only a few feet from Satan himself, Dante Alighieri placed Nimrod, a giant sunk to his waist in the frozen ground, condemned to babble incomprehensible words because he had built the impious Tower of Babel and forever shattered the integrity of language (Dante Alighieri, 1977, p. 31). But for Burke the incomprehensible had a sort of majesty, and he cited John Milton's description of Death to prove it:

> The other shape,
> If shape it might be call'd that shape had none
> Distinguishable in member, joint, or limb,
> Or substance might be call'd that shadow seem'd,
> For each seem'd either; black it stood as Night,
> Fierce as ten Furies, terrible as Hell,
> And shook a dreadful Dart; what seem'd its head
> The likeness of a Kingly Crown had on.
>
> (*Paradise Lost* [1674] II 666–73)

Death is simply a personified inkblot, described through procedures of semantic vertigo. Burke thought that this intellectual reeling showed the highest state of sublimity; and such gropings after things that strain and hurt the sense organs, that dance out of the grip of language, must heighten the status of the ineffable over the merely speakable and writable and thinkable.

The Romantic movement was devoted to the ineffable in a variety of ways: as Demogorgon, the master of the universe, remarks in Percy Bysshe Shelley's *Prometheus Unbound* (1819), "the deep truth is imageless" (2.4.116). Perhaps the chief Romantic poet of the ineffable was Samuel Taylor Coleridge, who defined imagination in chapter 13 of *Biographia Literaria* (1817)—just after quoting Milton's description of Death—as "the living Power and prime Agent of all human perception . . . a repetition in the finite mind of the eternal act of creation in the infinite I AM. . . . It dissolves, diffuses, dissipates in order to re-create." In an 1812 lecture on William Shakespeare's *Romeo and Juliet*, Coleridge was more specific about how we can know when we are in the presence of supreme operations of the imagination: such a text must show a "hovering between images. As soon as it is fixed on one image, it becomes understanding; but while it is unfixed and wavering between them, attaching itself permanently to none, it is imagination." To demonstrate this hovering, Coleridge cited, first, Romeo's oxymoronic description of his predicament in love: "Misshapen chaos of well-seeming forms! / Feather of lead, bright smoke, cold fire, sick health!" (1.1.185–186); and then, once again, Milton on Death. A string of oxymorons provides a continuous texture of self-contradiction, an aporia by which language deletes itself as quickly as it sets itself down; it is a useful device for alluding to the ineffable, which is more likely to be found in erasures and other sorts of verbal limbos than in precise settled description.

Twentieth-Century Art. This century has taken to remarkable lengths the Romantic desire to dissolve, dissipate, and diffuse the fixed artistic image; there is pride in intimacy with the abyss. The linguistics of the followers of Ferdinand de Saussure—a school of Hermogenes—has tended to extend the kingdom of the ineffable to include, not simply God, Fate, and the exact value of pi, but the ordinary furniture of one's daily life. If a word is a sign that points, straight as an arrow, at another sign, and if these sign-vectors never terminate in real objects, then the cushioned but firm thing

on which one sits is, strictly speaking, ineffable. Immanuel Kant believed that physical objects existed on a plane of supersensible reality, beyond real knowledge; but, in the dereferentialized sign systems of modern linguistics, even the sensible is beyond the reach of language. If such theories are correct, Helen Keller, a blind and deaf woman, should have been able to converse in English (with the aid of a Morse code dictionary and grammar book) as meaningfully as anyone else, even if Miss Sullivan had never taught her to associate the word *water* with the actual liquid gushing from a pump: for language's efficacy is in no way predicated on the richness of the relation between words and the outer world. We delude ourselves in our inevitable but naive belief that we stride through an effable world.

But, while modern linguistics has extended the domain of the ineffable, it has also desecrated it, made it commonplace. And modern science joined this project when Werner Heisenberg set forth his famous uncertainty principle: the act of observing alters the object observed, since light, or gamma rays, or any other medium for making an object visible, necessarily deforms the object. But twentieth-century art, not twentieth-century linguistics or science, has labored to resacralize the ineffable. Wassily Kandinsky devised his style of abstract painting precisely in order to render art more spiritual, less bound to the fallen and corrupt world of finite objects:

> The more abstract is form, the more clear and direct is its appeal. In any composition the material side may be more or less omitted in proportion as the forms are more or less material, and for them substituted pure abstractions, or largely dematerialized objects. The more an artist uses these abstracted forms, the deeper and more confidently will he advance into the kingdom of the abstract. (Kandinsky, 1977, chap. 6)

But perhaps the most influential consecration of what is beyond the reach of our senses—the infrared and ultraviolet of art—lies in the Expressionist movement.

In our age, the ineffable is centered on the Nazi Holocaust. Even the extraordinary experiments in carrying artistic expressivity to its limits that occurred *before* Adolf Hitler took power have seemed, in retrospect, like clairvoyant gesturings toward the gas chambers. In early Modernist music, one potent metaphor for the ineffable was the shriek made by playing many of the twelve notes of the chromatic scale at the same time: at the climax of the adagio of Gustav Mahler's Tenth Symphony (1911), a piercing, overwhelming chord sounds out, containing nine of the twelve notes; and at the end of Arnold Schoenberg's *Erwartung* (1909, 1924), there is a remarkable shudder in which various glides up and down the chromatic scale get faster and quieter until they vanish into silence—an acoustic smear and erasure that point to the ineffable simply by exhausting and undoing the resources of musical speech. For Schoenberg's pupil, the Communist composer Hanns Eisler, Schoen-

berg's challenging, "atonal" music forced the middle class to contemplate horrors beyond its complacent existence:

> When his music is heard in the concert hall of the bourgeoisie they are no longer charming and agreeable centers of pleasure . . . but places where one is forced to think about the chaos and ugliness of the world . . . He is the lyric composer of the gas chambers of Auschwitz, of the Dachau concentration camp, of the complete despair of man in the street under the heel of fascism. That is his humanity. It is proof of Schoenberg's genius and instinct that he gave expression to all these emotions at a time when the world seemed safe.

It is as if Schoenberg had perfected means of expressing terrors that did not yet exist; almost as if the twentieth century responded to Schoenberg's music by finding an event for which his method of *chromatic saturation* (to use Charles Rosen's term) was a sufficient correlative.

The actual music composed inside Terezin (Theresienstadt), the "model" Nazi concentration camp and the only concentration camp in which artistic pursuits were feasible, approaches the ineffable in a manner quite different from the scream-expressionism of Edvard Munch or the Schoenberg of *Erwartung*. Perhaps the masterpiece of Terezin is *Der Kaiser von Atlantis* (1943), by Viktor Ullmann (also a pupil of Schoenberg's): it is a parable of Fascism in which Death punishes the emperor for his war crimes, by refusing to let anybody die, until the emperor himself volunteers to die. Eventually the emperor does agree to die, to an astonishingly gentle aria: "It falls, the snow, where you are not. It streams down, the summer rain, where you are not. Where you are not, are many things . . . For faraway things aren't worthy of being mourned, much more worthy are nearby things, that lie in eternal shadow." Instead of the brain convulsions of Schoenberg's more radical works, Ullmann offers delicacy, learning, parody, grace: the ineffable is, for him, best captured by a psychic dissociation from terror, by conspicuous espousal of the cultural values that the Nazis have destroyed. *Der Kaiser von Atlantis* is a great hymn to "where you are not."

Perhaps the earliest significant postwar artistic response to the Holocaust was Thomas Mann's *Doktor Faustus* (1947). By Mann's own admission, the central chapter, in which the (rather Schoenberg-like) composer Adrian Leverkühn sells his soul to the devil in return for effortless spasms of musical inspiration, was based on accounts of the gas chambers. Here is the devil on the nature of hell:

> This is the secret joy and security of Hell, that it can't be denounced by informants, that it is hidden away from all speech, that it merely is and doesn't get into the newspapers, can't be brought into criticizing consciousness through any word-forms, for which the words *subterranean, cellar, thick walls, soundlessness, forgottenness, incapacity for salvation,* are weak symbols. One has to be satisfied with symbols, my good fellow, when one speaks of Hell, for here everything stops—not only the indicating word, but everything whatsoever stops.

Since our eyes turn away from the photographs taken at Auschwitz, since our minds turn away from visualizing what happened there, the Holocaust exists for us—as it exists in Mann's novel—as allegories, parables, foreshadowings, vague memories, and tongueless gestures, all of which self-consciously confess their impotence. In the presence of the Holocaust, art and philosophy alike have had to acquire a certain modesty, a certain deferentiality. The ineffable has converged, at last, with the unspeakable: the transcendental with the monstrous.

Ineffable is derived from the Latin word *fari,* also the source of *fable* and *fate;* and perhaps the notion of ineffability requires a rejection, not only of the speakable, but of fable and fate as well; perhaps artists concerned with the ineffable need to write anti-fables that attempt to escape the dicta of fate. This sort of speech that rejects the speakable seems as necessary to the human race as the more normal kind of speech. We continually try to prove Ludwig Wittgenstein wrong, to prove that, whereof one cannot speak, thereof one must not be silent.

[*See also* Death and Aesthetics; Difficulty, Aesthetics of; Discourse; Ecphrasis; *and* Romanticism.]

BIBLIOGRAPHY

Burke, Edmund. *A Philosophical Enquiry into the Origin of our Ideas of the Sublime and Beautiful.* New York, 1971.

Coleridge, Samuel Taylor. *Biographia Literaria.* Edited by James Engell and W. Jackson Bate. Princeton, N.J., 1983.

Coleridge, Samuel Taylor. *Complete Works,* edited by W. G. T. Shedd. vol. 4, *Lectures upon Shakespeare and Other Dramatists.* New York, 1884.

Dante Alighieri. *The Divine Comedy.* 3 vols. corr. ed. Edited by Giorgio Petrocchi, translated by Charles S. Singleton. Princeton, N.J., 1977.

Dieckmann, Liselotte. *Hieroglyphics: The History of a Literary Symbol.* Saint Louis, 1970.

Eliot, T. S. *Collected Poems, 1909–1962.* New York, 1963.

Kandinsky, Wassily. *Concerning the Spiritual in Art.* Translated by M. T. H. Sadler. New York, 1977.

Mann, Thomas. *Doktor Faustus: Das Leben des deutschen Tonsetzers Adrian Leverkühn erzählt von einem Freunde.* Hamburg, 1971.

Milton, John. *Complete Poems and Major Prose.* Edited by Merritt Y. Hughes. New York, 1957.

Plato. *The Collected Dialogues of Plato.* Edited by Edith Hamilton and Huntington Cairns. New York, 1961; reprint, Princeton, N.J., 1969.

Plotinus. *The Enneads.* 2d rev. ed. Translated by Stephen MacKenna. London, 1956.

Rhys, Ernest. *Lyric Poetry.* London and New York, 1913.

Richter, David, ed. *The Critical Tradition: Classic Texts and Contemporary Trends.* New York, 1989.

Rosen, Charles. *Arnold Schoenberg.* New York, 1975.

Sandow, Gregory. *Hanns Eisler/Bertolt Brecht Collaboration.* Tomato CD 2696382, 1989.

Ullmann, Viktor. *Der Kaiser von Atlantis.* Text by Peter Kien. London CD 440 854-2, 1994.

Wittgenstein, Ludwig. *Tractatus Logico-Philosophicus.* Translated by D. F. Pears and B. F. McGuinness. London, 1961.

DANIEL ALBRIGHT

INGARDEN, ROMAN WITOLD (1893–1970), Polish philosopher. Roman Ingarden was born in Austrian-occupied Kraków on 5 February 1893. In 1912 he went to Göttingen to study under Edmund Husserl, following him to Freiburg in 1916. He published his dissertation in 1921, completing his Habilitationsschrift in 1923. In 1931, he published *Das literarische Kunstwerk* (The Literary Work of Art), and from 1933 to 1939 he held the position of Extraordinary Professor at Jan Kazimierz University in Lwów. He was subsequently appointed professor in the Department of German Philology at that institution, holding the position until June 1941, when the outbreak of the German-Soviet conflict closed the universities. In 1946 he was appointed Ordinary Professor and Chair of the Department of Ordinary Philosophy at Jagiellonian University in Kraków. In 1949 he was denounced by the Communists, removed from his position, forbidden to teach, and officially transferred to the University of Warsaw. In 1957 the Department of Philosophy was reestablished at Jagiellonian and Ingarden was appointed Chair, a position he held until his retirement in 1963. He died of a cerebral hemorrhage in Kraków on 14 June 1970.

Ingarden's contribution to contemporary thought becomes clear when his work in aesthetics is located within its proper context. In the preface to the first edition of *The Literary Work of Art* he writes:

> Although the main subject of my investigation is the literary work, or the literary work of art, the ultimate motives for my work on this subject are of a general philosophical nature, and they far transcend this particular subject. They are closely connected to the problem of idealism-realism, with which I have been concerning myself for many years
>
> (Ingarden, 1973, p. LXXII)

Like many other students of Husserl's, Ingarden disagreed with his teacher's turn to transcendental idealism. Unlike the other students, however, Ingarden devoted most of his academic life to proving that Husserl was wrong. In 1918 he sent Husserl a lengthy letter (he had to stuff it into two large envelopes) in which he discussed the problems involved in the idealism/realism debate and expressed reservations about Husserl's idealist solutions, and his published works offer a record of the development of his realist response. According to Ingarden, Husserl's transcendental idealism is basically like all the other forms of idealism that make the world, or *being,* or *reality,* dependent on the activity of some mind or consciousness; Husserl's idealism attempts "to conceive the real world and its elements as purely intentional objectivities which have their ontic and determining basis in the depths of the pure consciousness that constitutes them" (ibid.). In Husserl's phenomenology, the external world, to the extent that one can be conscious of it, is a construction of consciousness—everything is reduced to the status of a purely intentional objectivity. In order to at-

tack that position, Ingarden had to present an ontological analysis of something that was indisputably an "intentional object" but that owed its existence to more than just the act of consciousness alone. He chose the literary work of art. His ontological analyses in *The Literary Work of Art* demonstrated that for such a work to exist, various nonintentional, ontically autonomous entities must also exist, both real, material objects, like paper and ink, and such purely ideal objectivities as concepts, ideas, and essences.

The Literary Work of Art, itself a superb model of literary analysis, deftly employed a new method of analyzing not just the literary work of art, but all sorts of artworks. The fundamental conception was that the work of art is a stratified formation *(ein schichtliches Gebilde)* whose various strata contain elements that are concretized, or "fleshed out," by the audience experiencing the work; the aesthetic object arises as the object of this aesthetic experience. The metaphysical implications of the ontological distinction between artistic and aesthetic object are immense, and the distinction between artistic and aesthetic value (and value qualities) promises a similarly consequential impact on contemporary art criticism. Ingarden's work has only recently been recognized by the general philosophical audience, however, so the full extent of its influence remains to be seen. The method employed in *The Literary Work of Art* was, however, successfully employed by Ingarden and others in the ontological analysis of other sorts of artworks, including the dramatic work, the musical work, the picture, and the architectural work. These ontological studies—in conjunction with Ingarden's *Controversy over the Existence of the World* (1947)—continue to provide the basis for a good deal of the current research in realist phenomenology.

Ingarden's analyses establish that the most basic ontological features of the literary work of art are common to all artworks. Most fundamentally, the work of art is neither a real nor an ideal object. A real object comes into being at some point in time, perhaps undergoes change throughout time, and finally ceases to exist, whereas an ideal object neither comes into being nor passes away, and does not experience change. A work of art seems strangely to partake of both real and ideal features. William Shakespeare's *Hamlet*, for example, came into being at a particular time, has continued to exist since that time, and may perhaps cease to exist sometime in the future. But it also consists of nothing other than a determinately ordered manifold of sentence structures. So while *Hamlet* might seem to be a real object on the one hand, on the other hand it is ideal, for a sentence is constructed of purely ideal elements—namely, word meanings. In order to avoid positing the existence of an aesthetic object—a move that would have required drastic revision of their basic ontological and metaphysical assumptions—theoreticians offered two different solutions to this ontological problem: the psychologistic solution and the performance solution.

Ingarden revealed the shortcomings of each as follows. It might be suggested that the written characters that constitute the literary work are no more than the vehicle of its communication and cognition, and that the work itself is what the author experienced during its production. Alternatively, one might maintain that the written characters are nothing more than conventional signs by means of which readers represent to themselves the experience of the author. But each of these psychologistic solutions is absurd. It is not possible to identify the work with the psychic experience of the author, for that experience ended as soon as the work was completed. Nor is it possible to identify the work with the psychic experience of the reader, for then there would be as many *Hamlet*s as there are readers. The performance solution fails for the same reason: if the work were identical with its performance, then there would be as many *Hamlet*s as there have been performances of it—yet one speaks meaningfully of a self-identical "it" and of its different performances. Ingarden's solution to the problem of the identity of the work of art is given with the acknowledgment of its mode of being—that is, *intentional* being. The work of art, that is, is an intentional object: it is a stratified formation intentionally constructed out of both real and ideal elements. Ingarden's painstaking analyses in *The Literary Work of Art*, together with those he provides in *The Cognition of the Literary Work of Art*, describe the basic elements and intentional activities involved in this construction.

The literary work is a formation *(Gebilde)* constructed of four heterogeneous strata that differ from one another with respect to the material with which they are constructed and the role that each of them plays in the overall structure of the work, which is assembled so as to produce a polyphonic harmony among the strata and the various elements belonging to the strata. These strata are as follows:

1. the stratum of word sounds and the higher-order phonetic formations built on them;
2. the stratum of different orders of meaning units;
3. the stratum of represented objects (or "objectivities"); and
4. the stratum of manifold schematized aspects.

It is by virtue of the stratum of word sounds that the work enjoys such features as melody, rhythm, tempo, rhyme, and assonance. But the stratum of word sounds also contributes significantly to the operation of the stratum of meaning units. A word or combination of words may have several different meanings, and it is the word sound that determines which of these meanings is the appropriate one in the given context. The most basic meaning units are names and finite verbs, which are distinguished by two types of intentionality. Whenever a name or verb is used, it is said to "determine" its intended referent, which is called its "intentional object." Whereas the name determines its intentional

object as a complete whole, the finite verb determines something that is incomplete—namely, an activity—so it is said to "point back" to the subject of the sentence. The contribution of the stratum of word sounds may be discerned when one notices that whenever a certain name is used, it is being actualized (or "made real") a part of the (ideal) meaning contained in the (ideal) concept that corresponds to the intentional object. (This occurs as part of the process of concretization discussed below.) The various intentional objects that are determined during the course of reading a literary work belong to the stratum of represented objects (or "objectivities"). This stratum comprises everything that is nominally and verbally projected—that is, intended—in the work: activities, persons, objects, events, states of affairs, and so on. In a literary work of art the represented objectivities are related to one another so as to constitute a unified ontic sphere in which particular things are portrayed as really existing. During the reading of the work, certain aspects of the represented objectivities that are not explicitly revealed or developed—in other words, "determined"—by the author are concretized by the reader. These are called schematized aspects, and they may take the form of discernible "spots of indeterminacy."

This can be summarized as follows. The literary work of art may be regarded as a potential aesthetic object; that is, the aesthetic object may be said to exist "in potentiality." Reading the text is a process of actualizing, or "realizing," that potential. In a literary work of art the appearance of a word with a determinate corresponding word sound is determined by the intended meaning of the word; the individual word meaning is determined with respect to the sentence as a whole. The sentence represents an objectivity; for example, a state of affairs, an event, or a person. While reading the text, one concretizes the schematized aspects of these represented objectivities by filling in spots of indeterminacy and actualizing features of the text that had been "held in readiness"—that is, that had previously existed only potentially. The end product of this act of concretization—which is a central function of the aesthetic experience—is the literary work of art as aesthetic object. The work of art, in short, is different from, and is a condition of the possibility of, the aesthetic object.

A distinction between artistic and aesthetic value may now be made. Artistic value derives immediately from the artist and has to do with skill and technical ability. The artistic value of a work of art lies in the extent to which the work succeeds as a foundation for the construction of the intended aesthetic object. When the represented objectivities are related to one another so as to constitute a unified ontic sphere in which particular things are portrayed as really existing, a successful portrayal will enable the reader to concretize the schematized aspects of the work in such a way as to construct an intentional object that enjoys the character

of reality. Artistic value, then, belongs to the work of art and derives from the artist's mastery of the art. Aesthetic value, on the other hand, belongs not to the work of art but to the aesthetic object, that is, the intentional object of the aesthetic experience. Aesthetically valuable qualities—for example, emotional qualities, like "sad," "sublime," and "tragic," and intellectual qualities like "witty," "clever," and "profound"—are actualized (realized) only in the aesthetic experience.

Although analysis of the aesthetic experience is not an ontological but an epistemological task and, with the notable exception of *Controversy over the Existence of the World* (1947), Ingarden's focus was generally on the ontology of the work of art, his work has exercised considerable influence on subsequent treatments of the aesthetic experience, both in Europe and in North America. In *The Phenomenology of Aesthetic Experience*, Mikel Dufrenne adopted Ingarden's analyses of the ontology of the work of art as the starting point for his own epistemological analyses of the aesthetic experience, and, while a chief task of Dufrenne's work was, in a manner reminiscent of that of Maurice Merleau-Ponty, critically to move beyond Ingarden's radical separation of the two spheres of (conceptual) meaning and the perceived, he employed many of Ingarden's central concepts (e.g., concretization) in his attempt to demonstrate that the aesthetic object "bears this meaning within itself in order to present it to perception" (Dufrenne, 1973, p. 212). The reception theory advanced by Hans Robert Jauss and the related reader-response theory elaborated by Wolfgang Iser have also drawn extensively from Ingarden's analyses. During the course of his construction of a literary communication theory in *The Act of Reading: A Theory of Aesthetic Response*, for example, Iser refers to Ingarden at a number of crucial steps, citing his treatment of, among other things, the "quasi-judgmental" character of literary sentences and the concretization of schematized aspects and "spots of indeterminacy," and "his classical concept of the work of art as polyphonic harmony" (Iser, 1966, p. 112).

Ingarden's treatment of the literary work of art offered only a *first* step in his preparatory analysis, and, immediately on completion of *The Literary Work of Art*, he undertook similar investigations into the ontology of other sorts of works of art—namely, the musical work, the picture, the architectural work, and the film—intending to include these studies as an appendix to *The Literary Work of Art*. He explains in his foreword to *Untersuchungen zur Ontologie der Kunst* (1962), however, that as he "prepared that book [*The Literary Work of Art*] for the press in 1930, it became apparent that the volume had become too bulky, so that [he] had to forego publication of the appendix." Ingarden published these studies separately in 1946 and 1947 and together in 1962 in the German text just cited, which was published in English in 1989 as *The Ontology of the Work of Art*, but they

were never as widely read as *The Literary Work of Art*, and until quite recently had received little critical attention. The question of the *identity* of the work of art guides Ingarden's inquiry throughout these analyses, all of which follow basically the same procedure as that used in the analysis of the dramatic work in *The Literary Work of Art*, section 57, "The Stage Play," to which this discussion is restricted. Ingarden regards the dramatic work as a "borderline case" of the literary work, so the question of its identity—the question, that is, of whether the play read is identical to the one seen on stage—is especially provocative:

> The question is only whether what is counterposed to the individual performance is the corresponding "written" literary work or a "stage play" that is different from it. If it is the latter, we will have to contrast literary works of a certain kind ("dramatic" works) with stage plays, which will then be seen as heterogeneous to them; if it is the former, we will have to accept only a special kind of concretization of "dramatic" works, namely, the kind that is realized in a "performance." (p. 318)

Ingarden opts for a version of the latter alternative, distinguishing the written text from the performance *Aufführung)* and placing the "stage play" *(Schauspiel)* between them. The written text is structurally different from that of a novel or a poem, for it comprises both the "main text," consisting of sentences that are actually spoken by the characters, and the "side text," which consists of stage directions. We can best grasp the ontology of the dramatic aesthetic object by attending to the four steps involved in the creation and staging of the work, thinking again in terms of actuality and potentiality. To begin with, the author composes the main written text (text A) and the side written text (text B). Each text consists of undetermined objectivities that exist as potentially determined (i.e., these objectivities may be determined—or concretized, or "fleshed out"—in one of a given number of ways). Next, the director, actors, and others involved in the actual production, or "staging," of the work actualize the potential objectivities of text B and, in so doing, partially actualize the potentialities of text A, producing the stage play. The actors then further actualize the potentialities of text A in and as the performance. The audience further actualizes the potentialities of text A in their aesthetic experience of the performance as aesthetic object. The question of the identity of a particular play is thus far more complicated than previous theorists and critics had assumed. The dramatic work of art presents us with a progressive ontological hierarchy: from the initial stage of indeterminate potentiality one proceeds through stages of determination and actualization until one reaches the final stage of the fully concretized aesthetic object, which is the intentional object constituted through the conscious activity of the person engaged in the aesthetic experience of actually seeing the play performed.

[*See also* Phenomenology.]

BIBLIOGRAPHY

Works by Ingarden

Intuition und Intellekt bei Henri Bergson: Darstellung und Versuch einer Kritik (Intuition and Intellect in Bergson: Exposition and Attempt at a Self-Criticism). *Jahrbuch für Philosophie und phänomenologische Forschung* 5 (1922): 285–461. Ingarden's Ph.D. dissertation.

"Essentiale Fragen: Ein Beitrag zum Wesensproblem" (Essential Questions: A Contribution regarding the Problem of Essence). *Jahrbuch für Philosophie und phänomenologische Forschung* 7 (1925): 125–304. Ingarden's Habilitationsschrift.

Das literarische Kunstwerk: Eine Untersuchung aus dem Grenzgebiet der Ontologie, Logik und Literaturwissenschaft. Halle, 1931. Translated by George G. Grabowicz as *The Literary Work of Art* (Evanston, Ill., 1973).

Spór o istnienie Świata (Controversy over the Existence of the World). vol. 1. Warsaw, 1947.

Spór o istnienie Świata, vol. 2. Warsaw, 1948.

Studia z estetyki (Studies in Aesthetics). 3 vols. Warsaw, 1957–1970.

Untersuchungen zur Ontologie der Kunst. Tübingen, 1962. Translated by Raymond Meyer with John T. Goldthwait as *The Ontology of the Work of Art* (Athens, Ohio, 1989).

Über die kausale Struktur der realen Welt: Der Streit um die Existenz der Welt (On the Causal Structure of the World: Controversy over the Existence of the World), vol. 3. Tübingen, 1974.

Selected Papers in Aesthetics. Edited by Peter J. McCormick. Washington, D.C., 1985.

Einführung in die Phänomenologie Edmund Husserls: Osloer Vorlesungen 1967. Edited by Gregor Haefliger: In *Gesammelte Werke*, edited by Rolf Fieguth and Guido Küng, vol. 4. Tübingen, 1992.

Vom Erkennen des literarischen Kunstwerks. Edited by Rolf Fieguth and Edward N. Swiderski. In *Gesammelte Werke*, edited by Rolf Fieguth and Guido Küng, vol. 13. Tübingen, 1997.

Other Sources

Dufrenne, Mikel. *The Phenomenology of Aesthetic Experience*. Translated by Edward S. Casey, Albert A. Anderson, Willis Domingo, and Leon Jacobson. Evanston, Ill., 1973.

Dufrenne, Mikel. *In the Presence of the Sensuous: Essays in Aesthetics*. Edited and translated by Mark S. Roberts and Dennis Gallagher. Atlantic Highlands, N.J., 1987.

Dziemidok, Bohdan, and Peter J. McCormick, eds. *On the Aesthetics of Roman Ingarden: Interpretations and Assessments*. Dordrecht and Boston, 1989.

Gierulanka, D. "The Philosophic Work of Roman Ingarden (A Systematic Outline)." *Dialectics and Humanism* 4 (1977): 117–128.

Graff, Piotr, and Slaw Krzemien-Ojak, eds. *Roman Ingarden and Contemporary Polish Aesthetics*. Translated by Piotr Graff et al. Warsaw, 1975.

Hanneborg, Kurt. "New Concepts in Ontology: A Review Discussion of Roman Ingarden: *Der Streit um die Existenz der Welt*." *Inquiry* 9 (1966): 401–409.

Iser, Wolfgang. *The Act of Reading: A Theory of Aesthetic Response*. Baltimore, 1978.

Lech-Piwowarczyk, Ewa. "Language and the Definition of Art: Analytic and Continental Discussions of the Nature of Art." Ph.D. diss., University of Ottawa, 1992.

Mitscherling, Jeff. *Roman Ingarden's Ontology and Aesthetics*. Ottawa, 1996.

Rudnick, Hans H. ed. *Ingardeniana II: New Studies in the Philosophy of Roman Ingarden. Analecta Husserliana*, vol. 30, pp. 225–296. Dordrecht and Boston, 1990. Contains thorough bibliography of works by and on Ingarden.

Wellek, René. *Four Critics: Croce, Valéry, Lukács, and Ingarden*. Seattle, 1981.

JEFF MITSCHERLING

INSANE, ART OF THE. In 1921 the Swiss psychiatrist Walter Morgenthaler published his book *Ein Geisteskranker als Künstler* (A Mental Patient as Artist), in which he presented both a clinical case history and an informed aesthetic analysis of the graphic work of his patient Adolf Wölfli, who had been hospitalized at the Waldau Asylum near Bern since 1895. This landmark study, the first monograph devoted to the artistic productions of a psychotic person, was followed a year later by the work of a German psychiatrist, Hans Prinzhorn, *Bildernei des Geisteskranken* (The Artistry of the Mentally Ill). Prinzhorn, who had been trained as an art historian, had assembled at the Psychiatric Clinic of the University of Heidelberg from hospitals all over Europe what remains today the largest and most important collection of the "art of the insane." As his title suggests, he, like Morgenthaler, was interested primarily in the artistic aspects of the works he illustrated and described, and offered only as much clinical data as seemed necessary to illuminate the work. Both the book and the collection itself became major sources of inspiration to avant-garde artists all over Europe.

The nearly simultaneous appearance of these two volumes was no accident. They represented the culminating events in a process of inquiry that had begun long before and is carefully detailed in John MacGregor's *The Discovery of the Art of the Insane.* The question of the relationship between art and mental illness can be traced at least to Plato who, like Aristotle, addressed the "divine madness" of creative individuals. More recently, much was made of Michelangelo's "terribilita," and Romantically tinged formulations about the nature of Vincent van Gogh's disorder and its effects on his art continue to flow unabated. There has also been a significant body of art about mental illness, systematically discussed in Sander Gilman's *Seeing the Insane.*

Interest in the artistic productions of persons who are mentally ill is, however, of more recent vintage. As MacGregor points out, it can be seen in the final plate of William Hogarth's *Rake's Progress,* where one of the madmen in Bedlam is seen drawing an elaborate cosmic chart on the wall of the hospital. For Hogarth, this activity would seem to have been merely one of the variety of mad behaviors represented in this remarkably perceptive engraving. But in 1812 Benjamin Rush, often called the "father of American psychiatry," made particular note of the graphic productions of his patients:

> Two instances of a talent for drawing, evolved by madness, have occurred within my knowledge. The disease which thus evolves these new and wonderful talents and operations of the mind may be compared to an earthquake which, by convulsing the upper strata of our globe, throws upon its surface precious and splendid fossils, the existence of which was unknown to the proprietors of the soil in which they were buried.
>
> (MacGregor, 1989, p. 30)

One is struck by the modernity of Rush's imagery, anticipating as it does the psychoanalytic construction of the nature of psychosis and of psychotic productions.

It was in the nexus of the Romantic movement, however, that art by and about the insane became an object of interest in both artistic and popular circles. Both the young genius Théodore Géricault and the old genius Francisco José de Goya produced remarkable paintings of patients in psychiatric hospitals, while William Blake, Henry Fuseli, and others produced powerful images of imagined psychotics. At the same time the Royal Bethlem Hospital was host to a succession of inspired madmen such as Jonathan Martin and Richard Dadd, whose work, much of it now preserved in the Bethlem archives, was taken up by the art world at the time and became the focus of extensive public attention. Indeed, it was in this era that the link between madness and genius was, if not forged, certainly reinforced, as writers such as Lord Byron, Johann Wolfgang von Goethe, Charles Lamb, and Samuel Taylor Coleridge at times romanticized madness and glorified intoxication as the sovereign route to creative work. Denis Diderot, for one, was quite explicit in his equation of madness and creativity: "Oh, how near are genius and madness" (MacGregor, p. 76).

The "scientific" study of the art of the insane has been shown by MacGregor to have begun with the work of the French psychiatrist Paul-Max Simon, who, beginning in 1876, published a series of essays on the *dessins des alienès.* His approach was largely descriptive and classificatory, as was the dominant tendency in French psychiatry of that era. Simon was, however, acutely sensitive to the artistry that could at times be seen in these drawings, appraising their formal characteristics and attempting, within the methods available to him, to grasp their meanings. His objectivity was shaken only by the erotic images that he often found in his examples; these conflicted with the sense of propriety that governed both his work and his private life.

Simon's pioneering work was, however, overshadowed by that of the more flamboyant Italian alienist and criminologist Cesare Lombroso, who vigorously proclaimed in his voluminous publications the equation of genius with madness and attributed both to a degenerative psychosis which took the form of "moral insanity." Lombroso developed a sizable collection of *arte nei pazzi,* and formulated a set of criteria that, in his view, distinguished such works from those of "normal" artists; most of these served to buttress his notion of the "degeneracy" he ascribed to artists, criminals, and the mentally ill alike.

Neither Morgenthaler nor Prinzhorn followed this path. Both were actuated by a profound sympathy for the *geisteskranken* whose work they studied, and both were, to some degree, influenced by the new tendencies in European psychiatry initiated by Sigmund Freud. Morgenthaler was able, with the aid of the rich store of autobiographical data and elaborate fantasy supplied by his subject, to demonstrate

the developmental and historical origins of much of Wölfli's imagery and to approach much of its formal character within the terms of then-prevalent aesthetic theories. He treated Wölfli, that is, not as a madman who made drawings but as a gifted artist who happened to be psychotic. Similarly, Prinzhorn sought to minimize distinctions between psychotic and "normal" artists; indeed, he insisted that there was no "art of the insane" and that there were no specific characteristics that differentiated it from the art of the museums, but that the work of the mentally ill reflected universal drives for "configuration," form making, emotional expression, and communication.

As noted earlier, Prinzhorn's book was eagerly taken up by avant-garde artists of the period—among them Paul Klee, Alfred Kubin, Max Ernst, and Hans Bellmer, who pronounced it the most important educational experience of his life. Above all, it was the Surrealists who were most captivated by "the artistry of the mentally ill"; to André Breton and his colleagues it seemed the very actualization of their goals of spontaneity, automatism, and the direct expression of the "unconscious." In his characteristically militant vein, Breton said, "We will not cease until justice has corrected the blind and intolerant prejudice under which works of art produced in asylums have suffered for so long, and so long as we have not freed them from the aura of worthlessness which has been created around them" (MacGregor, p. 1).

The Surrealists never succeeded (nor could they have done so) in their aim of eliminating conscious control and obliterating the boundary between their art and that of the insane; indeed, it was after a visit from Salvador Dalí that Freud commented, "In classic paintings I look for the unconscious; in a surrealist painting, for the conscious" (Dalí, 1993, p. 397). Nonetheless, the Surrealists presented, as MacGregor states, "a completely coherent and convincing argument for the acceptance of the art of the insane within the boundaries of art" (p. 291).

The most vigorous and the most effective recent proponent of the "art of the insane" was Jean Dubuffet, who with his concept of *art brut* (raw art) brought to the issue the fullest measure of energy and understanding. A relentless advocate, he sought, like Prinzhorn, to demolish the boundaries between *art brut* and what he called the "art of culture." "Can you imagine," he wrote, "an art that is not mad? . . . There is no art of the insane any more than there is an art of dyspeptics or an art of people with knee complaints" (Dubuffet, 1959, n.p.).

Despite the fact that the efforts of Lombroso and others to define specific formal characteristics of psychotic art were unsuccessful, there are certain features that are frequently found in such work. One of these is the use of unusual or untraditional materials—often, whatever comes to hand or is made available, such as bread crumbs or sea shells. The eccentric English mystic Madge Gill produced endless ink drawings on long rolls of muslin rigged up in a special contraption by her son; she maintained that the drawings were actually the creation of the spirit she called Myrninerest, of which she was only the mediumistic vehicle. Another, and most typical, feature is the compulsive tendency to fill every inch of space on the sheet—the horror *vacui*. Yet another is an extraordinary sureness of line through which the untrained artist creates incredibly complex graphic structures without correction and with a sense of intense determination. Further, there is often what both Prinzhorn and the Austrian psychiatrist Leo Navratil have called "physiognomization"—the practice of assigning human facial features to nonhuman and inanimate objects (even to landscape), suggesting a regressive dedifferentiation of the animate and inanimate worlds and the attribution to the latter of the frightening aspects of people. And finally, there is a tendency to perseveration both in formal structure and content themes without the development one finds in the work of "normal" artists.

It is perhaps in the works of the best known of psychotic artists, the Swiss child molester and paranoid Adolf Wölfli, that these characteristics can best be seen. Wölfli, who lived from 1864 to 1930, was hospitalized at age thirty after his third episode of child molestation. He was so violent and assaultive that he spent almost twenty years in solitary confinement, and it was in that setting that he began to draw, ultimately devoting all his waking hours to creating drawings accompanied by lengthy and, at times, incomprehensible "texts" and to composing music in an idiosyncratic notation. Roger Cardinal has written of him,

> The quality of monstrous exactitude may be found in the drawings. Wölfli never planned in advance, and never hesitated. Normally he would begin drawing at the edge of the sheet, and having drawn the border device would add successive layers moving inward toward the center and stopping only when the whole space was filled in—at which he would turn over and start his commentary . . . he left not a single empty space, if necessary adding disjointed sentences in the gaps in the picture.
>
> (Cardinal, 1972, p. 59)

For the most part, the works that we know of the "art of the insane" were produced by patients who were hospitalized for many years—often for their entire adult lives. Usually, they undertook their creative activity after the acute phase of their illness had subsided, often as an antidote to boredom and inactivity. It has been estimated that no more than 2 percent of hospitalized patients did so spontaneously, before (or apart from) the introduction of formal art therapy programs. For many of them, the compulsion to create constituted efforts at integration and restructuring of the personality and at both representing and ordering their chaotic and frightening experience of the world. Current tendencies toward brief hospitalizations and the use of chemotherapeutic agents and intensive activity programs in

the treatment of the mentally ill are likely to spell the end of this phenomenon. A few gifted eccentrics and ambulatory psychotics like the aforementioned Madge Gill and the Chicago hospital janitor Henry Darger may continue to surface, but the psychiatric hospital as a breeding ground for spontaneous creative effort seems to have run its course.

[*See also* Outsider Art; *and* Surrealism.]

BIBLIOGRAPHY

Allderidge, Patricia. *The Late Richard Dadd, 1817–1886.* London, 1974.

La beauté insensée: Collection Prinzhorn, Université de Heidelberg, 1890–1920. Charleroi, 1995.

Cardinal, Roger. *Outsider Art.* New York, 1972.

Dalí, Salvador. *The Secret Life of Salvador Dalí.* Translated by Haakon M. Chevalier. London, 1948; reprint, New York, 1993.

Dubuffet, J. "Art brut." In *Catalogue: Collection de l'art brut.* Lausanne, 1959.

Gilman, Sander L. *Seeing the Insane.* New York, 1982; reprint, Lincoln, Nebr., 1996.

MacGregor, John M. *The Discovery of the Art of the Insane.* Princeton, N.J., 1989.

Morgenthaler, Walter. *Madness and Art: The Life and Works of Adolf Wölfli.* Edited and translated by Aaron H. Esman and Elka Spoerri. Lincoln, Nebr., 1992.

Prinzhorn, Hans. *The Artistry of the Mentally Ill: A Contribution to the Psychology and Psychopathology of Configuration* (1921). Translated by Eric von Brockdorff. New York, 1972.

Spoerri, Elka, and Jürgen Glaesemer, eds. *Adolf Wölfli.* Translated by Radka Donnell et al. Bern, 1976.

Thevoz, Michel. *Art brut.* Translated by James Emmons. New York, 1976.

AARON H. ESMAN

INSTALLATION ART. The term *installation art* has been used increasingly since the 1960s to denote temporary, site-specific artworks designed to surround or interact with the spectator and/or extant architecture in a given exhibition space. Formal and conceptual precedents are most often traced to Dadaist and Surrealist exhibitions of the early twentieth century or to the "environments" and "Happenings" of the late 1950s and early 1960s when artists in western Europe and the United States sought to redefine the role and function of the work of art, rejecting the formalist paradigms and market-oriented production governing more traditional media such as easel painting and freestanding sculpture and embracing the possibilities that ephemeral materials and staged performances could offer. Indeed, the words *environment* and *installation* have been indexed synonymously in reference guides for almost twenty years though they carry somewhat different connotations. In environment art, it is implied, the audience literally enters *into* the work of art, becoming a functional part of the art itself or at least physically encompassed by it. Installation art is usually taken to mean the partial transformation of a given exhibition space, often in conjunction with an explicit exploration of the artwork's production, exhibition, and reception. Both terms nevertheless rely on conceptual art's rejection of the notion that art must take the form of discrete and durable objects that bear the mark of fine artisanship. For the conceptual artist of the 1960s and 1970s, art was no longer to be defined as an artisanal activity, but was rather to be seen as a set of operations performed in a field of signifying practices, perhaps centered on a medium, but not bounded by it (Burgin, 1986). Emphasizing or reproducing this field of signification in the work itself, environments and installations disrupted the traditional semiotic and somatic boundaries assumed to exist between the audience, the work of art, the site of exhibition, and the world beyond. This disruption of boundaries may mark the contribution of installation art to a contemporary aesthetics.

Claims made about aesthetics and installation art have for the most part been implicit rather than explicit. As late as 1989 critics were asking whether installation art was a genre, a medium, or a mode of conceptual practice (Handy, 1989). At the date of this publication the answers to such questions are still under debate. The term *installation* is itself merely a formal designation—derived from curatorial practice—that reveals little about the artwork it defines save for the fact that it has been "installed" at a given site. It is no doubt because its definition and usage are far from fixed that there have been numerous attempts to link the idea of "installation art" to a set of functional connotations and, as mentioned above, to establish a developmental history that traces both formal and conceptual precedents. For some, nothing less than the political agenda of a new genre is at stake; for others, the term serves only to indicate a passing trend. The brief overview that follows is intended as a provisional selection of perspectives on installation art taking into account (1) historical precedents and contemporary examples and (2) the ways in which aesthetic theories have intersected with the practices of installation art broadly defined.

In *Installation Art,* Nicholas De Oliveira, Nicola Oxley, and Michael Petry chart a historical continuum from the Wagnerian *Gesamtkunstwerk* (or "total" work of art), to nineteenth-century world fairs, to the later performances, exhibitions, and working philosophies particular to Futurism, Dada, Constructivism, and the Bauhaus (each of which called for the collaboration and interdependence between art forms such as painting, sculpture, and architecture), to Minimalism and conceptual art's dematerialization of the aesthetic object. The authors suggest that the origins of what today is called installation art can be found in this ideological lineage of popular arts and avant-garde practices that emphasize material and spatial experimentation. They prefer this account of origins to any that relies on merely formal antecedents such as the complex interior design histories of cathedrals, houses, or gardens. There can be no doubt, of course, that both histories have had an important function in determining the shape of installation

art. Still, the authors argue that any definition of installation art must minimally include a consideration of location, site, gallery, public, environment, space, time, and duration. In this way the term can be used to link artworks that have very little in common formally, but that share a commitment to certain conceptual goals.

Others have argued that installation art derives in part from a hybridization of art media that can be traced to Pablo Picasso's and Georges Braque's incorporation of collaged elements onto the Cubist canvas. According to artist Allan Kaprow and curator William C. Seitz, this defining moment in the "destruction" of pictorial space would lead to two-dimensional relief, three-dimensional assemblage, and finally the extension of the work of art from the wall into whole rooms and architectural sites (Kaprow, 1966; Seitz, 1961). Kurt Schwitters's autobiographical and theatrical room-size assemblages—or *Merzbau*—of the 1920s and 1930s are often read, for example, as an extension of collage into architecture; just as El Lissitzky's *Proun Space* (1923) made up of reliefs in painted wood is read as an early emergence of painting from its traditional frame into a carefully choreographed extension of the exhibition site. Gallery and museum spaces became not only an institutional framework but also a formal component in works of art such as Marcel Duchamp's *1,200 Bags of Coal* at the International Exhibition of Surrealism (1938) or his *Mile of String* (consisting of a web of string that entangled both gallery and artworks) in the First Papers of Surrealism exhibition (1942), or in Frederick Kiesler's early environmental sculpture *Galaxies* at the Museum of Modern Art (1952). In such works the context of art's display becomes integral to the work of art itself, and the gallery space is transformed into the primary material of aesthetic encounter. Such interrogation and metamorphosis of what Brian O'Doherty calls "the white cube" have become the motivating principle of many installation projects since.

It is possible—provisionally, if somewhat artificially—to read installation projects according to the primary mode by which the works articulate the exhibition site; either as a system of physical constraints to be integrated into a formal composition, as the setting for a tableau or theatrical event, or as a social institution to be examined for its own cultural politics. Although these modes are not mutually exclusive, most installations tend to emphasize one or another approach. In 1958, for example, in his quest for immateriality, Yves Klein opened *Le vide* (The Void), which was nothing more than the empty white interior of a gallery. One year later, Arman, another member of the Nouveau Réalistes collective in France, commented ironically on this work by exhibiting *Le plein* (Fullness), which consisted of filling the same gallery space (or at least blocking windows and doors) with rubbish.

Certainly, in putting forth a conceptual play with existential metaphors and an ironic critique of both exhibition space and art production, the artists here also redefine the gallery space as a simple container that is to be made empty or full. For Kaprow—who saw environment art as indebted to the heritage of painting in its concern for color, space, surface, and the disruption or questioning of the "field" of representation—visitors were more than spectators in such works as *Tire-environment* (1958–1961); they were also to be seen as formal elements in the work itself, as moving, colored shapes that could add to the overall composition of car tires piled into the gallery. Minimalists of the 1960s attended to the floors and walls of the gallery space. Richard Serra's *Splashing* (1968) consisted of molten lead thrown against the baseboard of the gallery forming a cast impression of the meeting of floor and wall. Carl Andre's *Spill* (1966) was the random scattering of eight hundred small plastic blocks across the gallery floor. Working in a visual vocabulary drawn from large-scale earthworks Robert Smithson produced carefully arranged piles of rocks, gravel, and dirt offset by mirrors in the gallery space called *Non-Sites* (1968). Synecdoches of an absent wilderness, the nonsites also reflected the urbanity of the immediate environment. In an escape from the gallery, Gordon Matta-Clark transformed abandoned buildings by carving through floors, walls, and ceilings to create a sculptural labyrinth of passages through the structure of urban life. In the early 1970s, Robert Irwin and James Turrell created spatial and optical illusions with projected colored light using or rebuilding the architecture of the site as necessary. Chris Jenning's *Vault* (1992) used metal rods to produce a series of arcs inside the gallery walls. Eve Laramee covered the floor of a room with deep blue cobalt glass in *Requiem for a Blue Field* (1991). Employing diverse media, these artists each focused on a transformation of the exhibition site by incorporating the physical properties of the site as formal elements of the work.

In the 1960s, artists in Europe and the United States began to collapse distinctions between high and low art, turning to signs from mass media culture in the production of Pop Art, to the detritus of consumption in Assemblage, and to the participatory performance of "Happenings." Many of the installations produced during this period used everyday objects to restage familiar domestic or public spaces. In Claes Oldenburg's *Store* (1961), for example, painted sculptural replicas of food were exhibited and "sold" in a gallery, rearticulating the commercial function of the site. In Edward Kienholz's *Beanery* (1965) and in George Segal's *Subway* (1968) the space of the gallery, filled with everyday objects and life-size human figures, became a critical staging of other "nonart" sites. Illustrating the banality or the desperation of everyday existence, or the anonymity of human relations, such works confront the audience on a corporal as well as conceptual level.

More recently, artists such as Christian Boltanski, Amalia Mesa-Bains, and Ilya Kabokov have brought the material

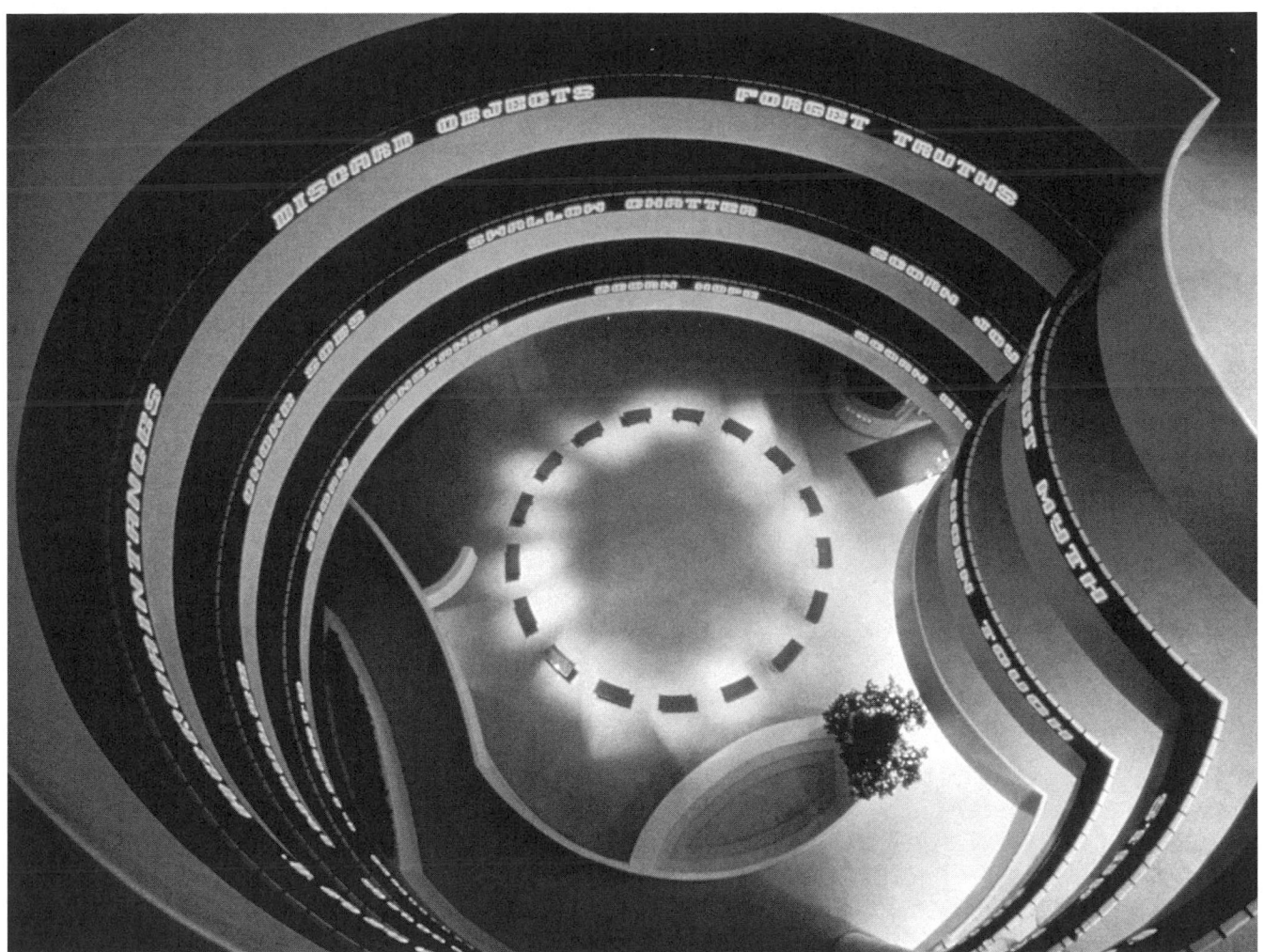

INSTALLATION ART. Jenny Holzer, *Various Texts* (1989), LED sign, Indian Red granite benches; installation at the Solomon R. Guggenheim Museum, New York, 12 December 1989–11 February 1990. (Photograph by David Heald, courtesy of Jenny Holzer Studio, New York. Copyright 1989 by Jenny Holzer; used by permission.)

culture of everyday life into the space of museum or gallery by using found objects and traditional forms of display to produce memorials to the past, recreating shrines for the dead or private collections of memorabilia in their room-size installations. In other instances, artists have made the performative aspect of the installation explicit by openly requesting audience participation or by including live actors. In 1973, Betye Saar encouraged her audience to contribute personal objects as offerings to an installation entitled *Mti,* which took the form of an altar. Many of the works of Joseph Beuys, such as *I Like America and America Likes Me* (1974) were also carefully staged installations in which the artist as actor performed the role of shaman, mediating between his audience and the symbolic objects he had gathered as his props. The contemporary artist Ann Hamilton has populated her large-scale, labor-intensive installation works, such as *dominion*

(1990) with human or animal inhabitants—witnesses who participate silently to transform the function of the site. Rather than responding formally to the material space of the gallery, these works reconstitute a social, theatrical space, situating the audience within both uncanny and familiar fields of signs. In such works the audience is understood to be a conduit in a system of meaning that is no longer confined to the closed circuit of contemplation elicited by abstraction (art object/viewer) but that also includes the material traces of social institutions and collective practices that exist beyond the walls of the gallery.

Other artists have sought to explore the social, historical, or economic function of the gallery or museum itself. Feminist artworks of the 1970s, such as Mierle Laderman Ukeles's *Washing, Tracks, Maintenance* (1973) drew attention to the invisible work of those who clean and maintain, not only

exhibition spaces, but also other cultural and domestic sites. Andy Warhol's *Raid the Icebox* (1970) at the Rhode Island School of Design was one of the earliest works to treat the collection of an art museum as the artistic medium itself. Given access to the museum's storage vaults, Warhol chose to exhibit those things that most pleased him—shoes, jars, parasols, chairs—and insisted on a thorough catalog of the exhibition. In a somewhat more critical mode, Marcel Broodthaers's imaginary *Musée d'Art Modern, Département des Aigles* (1972) was a mock collection of artifacts organized under a single, overdetermined, nationalistic rubric— the image of the eagle. A museum within a museum, Broodthaers's installation also criticized the fixed pattern of interpretive modes and taxonomies which such institutions of culture systematically follow. In Daniel Spoerri's *Musées Sentimentals,* which exhibited at the Centre National d'Art Georges Pompidou in Paris (1977) and the Kunstverein in Cologne (1979), objects both common and rare, found inside and outside the museum, were grouped in evocative combinations to produce a series of personal and historical connotations concerning the institutions that housed them. In a similar effort, Michael Asher transferred a bronze sculpture from the exterior to the interior of the Art Institute of Chicago in 1979, treating the institution, according to Claude Gintz, as a "closed system" shaped temporarily by the artist but leaving no material trace.

It has now become common for artists to perform such interrogations, highlighting not only their own personal fascinations with objects but also the institutional functions of museums as arbiters of aesthetics or history. Contemporary installation artists such as Louise Lawler, Judith Barry, Hans Haacke, James Luna, and Fred Wilson have taken a critical look at galleries and museums in terms of their role in the contextualization and decontextualization of art, in the production of meaning for objects, in the commodification and protection of art, in the perpetuation of racist histories, in the promotion of nationalistic agendas, and as the public relations department for large corporations. As a metadiscursive commentary on the apparatus of display, such work attempts to make visible the systems of power that underlie all exhibition practices.

As mentioned earlier, installation art has been closely associated with other art forms—such as conceptual and performance art—that since the 1960s have rejected traditional aesthetic criteria that emphasize the formal qualities or material properties of discrete objects, and instead seek to (1) change the conditions of art's reception and (2) mediate between the audience and other nonart domains of culture. A not insignificant inspiration for this new approach to art making were the social movements—student protests, civil rights movements—that sought to overturn the conservative politics of the cultural institutions of the time. As early as 1958, in the French journal *Internationale situationiste,* art was redefined as a "situation," a moment of life concretely and deliberately constructed, and in some cases a "psychogeographical" study of the effects of the social and political environment on the emotions and behavior of individual subjects.

The conceptual residue of this earlier work can be found in the writings of the artist and theorist Victor Burgin, who in 1969 argued for a new way of reading the concerns of conceptual, temporary, and site-specific or event-specific works of art in an essay titled "Situational Aesthetics." Art understood as a material, fabricated object is here replaced, suggests Burgin, by the idea of art as a set of *conditions* for the demonstration of concepts, or *systems* that might generate objects—in short, *situations.* In such work the specific nature of any art object is largely contingent on the situation for which it is designed, and moreover, the situation is itself determined not only by the material signs it produces, but by the psychological activity of the perceiver who encounters this situation. By recognizing art production as an intervention in a continuous perceptual field, the artist may be less interested in the creation of new material forms, Burgin suggests, than in the coordination or reorganization of existing forms. Indeed, the *subtraction* of objects from a given domain or site may take the place of their *production* or presentation. But what will then distinguish art making from other cultural practices? What will signal its aesthetic status? The identification of art by an audience, Burgin argues, relies on "the recognition of cues which signal that the type of behavior termed aesthetic appreciation is to be adopted" (Burgin, 1969). Such "situational cues" contextualize the art object, informing the audience that "this is art." For Burgin, however, this art object may take the form of a recollection that links discrete points in a continuum of perception that takes place over time. Works that emphasize temporality, whether by focusing on a process or by being temporarily installed, are read as counterconservative to the degree that they reveal "permanence" to be a relationship between people and not an attribute of an object. The status of the art object is therefore not a matter of material forms, but is rather the combined result of the behavior of the perceiver in the context of given "situational cues."

In his recent book *Kant after Duchamp,* Thierry de Duve offers a rereading of Kantian aesthetics taking into account the rupture that Marcel Duchamp's readymade created in the definition of aesthetic practice in the twentieth century. The readymade is seen to mark the shift from a classical to a modern aesthetics, requiring a substitution, de Duve argues, of the sentence "this is beautiful" in Immanuel Kant's *Critique of Judgment* with the sentence "this is art." Recognizing that art making has become no longer a matter of proper proportions and correct techniques but rather the effect of a series of social relations, de Duve argues that it is worth recuperating the notion of aesthetic judgment in Kant for the reason that it marks the domain of artistic practice based on a revised notion of the *sensus communis.*

What it allows, in other words, is a continuing exploration of a collective definition of art practice, while rejecting the universalist criteria of beauty that Kant's definition requires. Burgin's *Situational Aesthetics* argues that art is only recognizable *as art* through situated cues, read as legible signs of art's context. While written at different moments and from within divergent conceptual paradigms, both perspectives require that art *objects* be reconceived as collective semiotic *processes*.

The term *situational aesthetics* also appears—more narrowly defined—in the writings of Michael Asher as an aesthetic system that juxtaposes predetermined elements that occur within an institutional framework, that are recognizable and identifiable to the public because they are drawn from the institutional context itself. In Asher's work, Claude Ginz argues, attention normally devoted to the artist is displaced onto the elements that make up the institutional context of the installation. The artist thereby avoids creating new objects and also rejects the authorial function and fetishism of the artistic signature. At the same time, Asher's "situational aesthetic" calls on audience members to decipher their own roles or positions within the social institutions under examination.

Installation art is seen to take part in a "situational aesthetics" to the degree that it forms the context for art's reception and to the degree that it deemphasizes the fabricated object in favor of the coordination of already formed elements of signification. Defined as a temporary event, installation art also adheres to the rejection of "permanence" and to the repositioning of the spectator/participant (both physically and psychologically) that Burgin claims are central to situational aesthetics. The production of an environment or situation thus decenters the eye, or the "I," of the spectator, who is no longer located in a transcendental role of metadiscursive and solitary contemplation, but is seen as a culturally situated subject who both constitutes, and is constituted by, the work of art. While not referring to a situational aesthetic, other theorists have made observations about installation art that in many cases parallel this paradigm.

The philosopher Paul Crowther argues that installation art shares with assemblage an ability to transform the complex excess of images and signs in contemporary industrial societies into a finite and rational form—an activity he likens to no less than the Kantian sublime. He sees installation art as an inherently "postmodern" phenomenon, and postmodernism, for Crowther, is both an outgrowth and recognition of contemporary culture's semiotic excesses and a parallel sense that "meaning" is not simply "there in the world" to be perceived but must rather be decoded from within an always shifting domain of social signification. Installation art is seen as a particularly appropriate form for the representation of this postmodern condition precisely because it is temporary, site specific, and, Crowther claims,

it more overtly marks the inescapable fact that art production and its reception are determined by a field of cultural relations. For Crowther, Kant's sublime hinges on a relation between perceptual and imaginative excess and rational containment. The audience of installation art, he claims, is no longer lost in an excess of images and signs because it is able to enjoy a sensory overload that is sublime precisely because sensory and imaginative excess can be comprehended as an idea (Crowther, 1995). Crowther qualifies his own usage of the term by suggesting that the "sublime" in installation art depends on humor and irony as an exemplar of the triumph of reason over the excess of banality; that through testing perceptual or imaginative limits, installation art makes the scope of rational comprehension more vivid. Aesthetics in this case is dependent on a structural relation of signs, a mechanism of temporary order over an excess of signs, rather than a fixed set of qualities or methods of production.

Benjamin Weil makes a case surprisingly similar to Crowther's insofar as he observes in installations a method of ordering a chaotic context of signs. Weil argues, however, that there are important differences between the installations of the 1970s and 1990s, the primary difference being a reorganization of time frames and conceptions of geographic space in everyday life. Like Crowther he argues that a constant flow of images released by media advertising, information technologies, and music videos has generated a feeling of sensory saturation (Weil, 1992). He suggests that current artistic production has been developing mainly in the form of environments that require the spectator to go beyond the solitary viewing experience to develop a consciousness of art as bodily experience, as a surface upon which audience members project their own collective history. Weil argues that installation art places the audience members in the position of explorers invited to use their own interpretations as a source of meaning. Of course, this observation only makes explicit that which is implicit in all art production. The installation artist is thus read as an information editor seeking to generate a new approach to the environment, a mediation between current social practices and subjectivities, rather than a collection of new objects.

If the installation artist is put in the position of one who selects and combines preexisting materials, can the installation become a semantic system to be read as language or speech? The artist Joseph Kosuth has commented that installations return the "language of art" to something like "speech acts" and away from the tradition of "gilded illuminations" (Kosuth, 1993). Each installation constructs what Kosuth calls its own "event context" that establishes the parameters of the work's legibility and makes evident, as Weil points out, the integral role of the viewer in the construction of the meaning of the work. Of course, this emphasis on speech act theory tends to leave little room for ambiguity in interpretation. Rather than setting rigid parameters, the in-

stallation may simply restructure the familiar architecture of gallery or museum to promote new ways of reading art. Rolf Wedewer noted in 1970 that the point of early environments and installations was to create a dialogue between the object and/or the room and the public (Wedewer, 1970). The artist is no longer seen as the creator of eternal values; rather, the presentation of a temporary fixed situation assumes the visitor will become an active performer in the work and the production of its meaning.

Of course, the participation on the part of the audience, whether in installation art or any other art practice, cannot be prescribed. It relies on signs that are legible in the space of exhibition, though the meanings of even familiar signs will inevitably change in the process of their repositioning within the installation. For Andrew Benjamin, installation art falls conceptually between sculpture and the readymade, as each is the transposition of matter from one site to another, a process that also consists in a transformation of meaning. The interruption of the signifying system in which the object was originally located disrupts the space in which it comes to be placed (Benjamin, 1993). Installation art is here once again read as a mode of production that relies on the disruption of systems of situated signs, an alternative spatial domain in which to reconsider the subject's relation to objects.

To the degree that installations link art practice to other "nonart" areas of life, they disrupt the boundaries that have been traditionally used to distinguish art as a unique domain of aesthetic experience. Theorists Peter Bürger and Arthur C. Danto have each separately read this disruption—the moment at which a work of art has become indistinguishable from "mere reality"—as "the end of art"; yet both recognize that art making can and must continue despite its supposed impossibility (Bürger, 1991; Danto, 1997). Installation art appears to be one such impossible form. In the above accounts of installation art, aesthetics as a set of abstract principles or universally recognizable qualities is replaced by the view that art is to be defined as a mediating phenomenon. An "aesthetics" of installation art characterizes the means by which this mediation takes place, whether it is through the representation of recognizable objects and appropriated signs or through the creation of new environments. It is an aesthetics that recognizes that art is a set of practices that "situate" the viewing subject in relation to the social domains that lie beyond the walls of gallery or museum.

[See also Bourgeois; Conceptual Art; Contemporary Art; Performance Art; and Situationist Aesthetics.]

BIBLIOGRAPHY

Asher, Michael. "Accommodating Art." *Art Monthly* (September 1996): 7–9.
Benjamin, Andrew, ed. *Installation Art.* London, 1993.
Bürger, Peter. "Aporias of Modern Aesthetics." In *Thinking Art: Beyond Traditional Aesthetics,* edited by Andrew Benjamin and Paul Osborne. London, 1991.
Burgin, Victor. "Situational Aesthetics." *Studio International* 178.915 (October 1969): 118–121.
Burgin, Victor. *The End of Art Theory: Criticism and Post-Modernity.* Atlantic Highlands, N.J., 1986.
Corrin, Lisa. "Installing History." *Art Papers* 18 (July/August 1994): 6–14.
Crowther, Paul. "The Postmodern Sublime: Installation and Assemblage Art." *Art and Design* 10.1/2 (February 1995): 9–17.
Danto, Arthur C. *After the End of Art: Contemporary Art and the Pale of History.* Princeton, N.J., 1997.
de Duve, Thierry. *Kant after Duchamp.* Cambridge, Mass, 1996.
Deitcher, David. "Art on the Installation Plan." *Artforum* (January 1992): 79–84.
De Oliveira, Nicholas, Nicola Oxley, and Michael Petry, eds. *Installation Art.* London, 1994.
Foote, Nancy. "Situation Esthetics: Impermanent Art and the Seventies Audience." *Artforum* (January 1980): 26–33.
Handy, Ellen. "Installations and History." *Arts Magazine* (February 1989): 62–65.
Kaprow, Allan. *Assemblage, Environments, and Happenings.* New York, 1966.
Kosuth, Joseph. *eine grammatische bemerkung = a grammatical remark.* Stuttgart, 1993.
O'Doherty, Brian. *Inside the White Cube: The Ideology of the Gallery Space.* Santa Monica, Calif., 1986.
Seitz, William C. *The Art of Assemblage.* New York, 1961.
Wedewer, Rolf. "Environments and Rooms." *Art and Artists* 5.4 (July 1970): 16–19.
Weil, Benjamin. "Remarks on Installations and Changes in Time Dimensions." *FlashArt* (January/February 1992): 104–109.

JENNIFER GONZÁLEZ

INSTITUTIONAL THEORY OF ART. The first institutional theory of art is outlined in a 1964 essay by Arthur Danto, "The Artworld," which ruminates on the paradox that Andy Warhol's *Brillo Boxes* is art even though any of its perceptually indistinguishable twins—any stack of Brillo boxes in a grocery store—is not. Danto offers this solution to the paradox: "To see something as art requires something the eye cannot descry—an atmosphere of artistic theory, a knowledge of the history of art: an artworld" (Danto, 1964). Ultimately, however, it is "art theory" that makes Warhol's stack of silk-screened plywood boxes into art.

How does this "making" occur? Danto unpacks "art theory" in terms of what he calls "artrelevant predicates," which are predicates that apply uniquely to artworks. Let it be that predicates P, Q, and R are, at a certain point in history, the only art-relevant predicates in critical use. It will follow that any artwork is some or all of P, Q, or R. "But suppose," Danto continues, "an artist determines that H shall henceforth be artistically relevant for his paintings." Then objects to which H applies can now and in the future stand as artworks. How does H become an art-relevant predicate? Danto says only that an artist "determines" that it be so. His theory leaves unexplained, however, why someone's decision can render a certain predicate art-relevant. In fact, at this point one wonders what work the art-relevant predi-

cates are doing, if in the end making something art comes down to someone's decision. Why not, for example, allow decisions to make a thing art directly, unmediated by art-relevant predicates?

In later works Danto turned away from an institutional view, and toward a theory that defined art in terms of "aboutness" (or "semantic character"). In *The Transfiguration of the Commonplace*, Danto explicitly abjured George Dickie's institutional theory that was motivated by his "Artworld" essay. The influence of that essay is nonetheless considerable. It was the death knell for aesthetic definitions of art, because Danto forced into philosophical attention the paradox (which I'll refer to as "Danto's paradox") of two materially identical objects, only one of which is a work of art. If two objects alpha and beta are materially identical, then alpha and beta share all material properties. Alpha and beta are then perceptually, hence aesthetically, equivalent. Each, that is, is equally beautiful (or ugly), equally serene (or garish), equally balanced (or unbalanced), and so on. Further, if alpha but not beta is a work of art, then the arthood of alpha cannot lie in its aesthetic features, for it shares these features exactly with beta, which is not art. Danto's essay also spawned other institutional theories by pointing to one way of solving his paradox, namely, the possibility that something is art not by virtue of any properties of *it* but rather by virtue of a relation it bears to some larger context. Danto, and after him, Dickie, calls this larger context "the artworld."

The best-known institutional theorist is George Dickie, who has taken Danto's paradox seriously, and has drawn out in detail the view that Danto, perhaps inadvertently, fell into at the end of his artworld essay, namely, that it is the decisions of persons, paradigmatically though not exclusively artists, that make objects into art. Dickie's first version of the institutional theory of art appeared in a 1969 essay, "Defining Art," which was slightly modified into the more familiar version that appears in his 1974 book, *Art and the Aesthetic: An Institutional Analysis:* "A work of art in the classificatory sense is (1) an artifact (2) a set of the aspects of which has had conferred upon it the status of candidate for appreciation by some person or persons acting on behalf of a certain social institution (the artworld)" (Dickie, 1974). We shall set aside both the artifactuality condition, since this is not specific to the institutional theory, and Dickie's claim that it is "a set of aspects" of something that obtains arthood status, since this introduces a complication that also has little to do with the institutional theory (it means, for example, that the status of candidate for appreciation is typically conferred on the painted sides of paintings—that "set of their aspects"—and not on their unpainted backs).

Two features of Dickie's institutional theory embrace a value neutrality. The classificatory sense of art is supposed to be a nonevaluative use of the term *art* that picks out a thing of a certain kind without thereby attributing to it any value (or disvalue). The classificatory sense of *art* is thus

like *bicycle* or *water* and unlike *durable* or *brackish*. "Appreciation," Dickie says, means that "in experiencing the qualities of a thing one finds them worthy or valuable." To be a candidate for appreciation is, then, to be an object for evaluation, not an already evaluated object.

The most prominent institutional aspect of the definition is the conferral of status by someone on behalf of the artworld. Dickie likened the conferring of the status of candidate for appreciation on objects to such things as a university's bestowing a Ph.D. degree on a person and Congress's making a piece of geography a national park. He acknowledged that the artworld has no codified procedures, lines of authority, or conditions for membership. Although there are "core personnel" in the artworld, such as artists, museum curators, and art critics, ultimately, "every person who sees himself as a member of the artworld is thereby a member." The institutional theory, if correct, neatly solves Danto's paradox. For it accepts that there is no material difference between Marcel Duchamp's notorious urinal-artwork *Fountain* and any of its plumbing store twins; and accordingly that the arthood of *Fountain* must derive from some institutional status it has come to have. Institutional status is not a material but a relational property. Accordingly, *is art*, on analysis, is revealed to be not a one-place predicate but a multiplace relation.

Although Dickie was influenced positively by Danto's essay, it should be said that he was also reacting against the view that art is an open concept, a view rooted in the later Ludwig Wittgenstein that was most famously expressed by Morris Weitz in his 1956 essay, "The Role of Theory in Aesthetics." Dickie took that essay to deny the possibility of defining *art*, and in advancing his definition took himself to have refuted the open-concept theorists. It should be noted, however, that there is an ambiguity in the claim that art is an open concept. It may mean (1) that there is no true statement of the form "For any x, x is art if and only if x meets finitely statable conditions thus and such." But it may also mean (2) that for all things that are art, x, y, z, . . . there need be no property they all share. The open-concept theorists took (1) and (2) to entail each other, hence to be equivalent. Dickie may have followed suit. Now, Dickie's definition, at least if true, shows (1) to be false. But this did not entail that (2) is thereby false. Indeed, on the institutional theory there can be two works of art—perhaps Leonardo da Vinci's *Mona Lisa* and Duchamp's *Fountain* are examples—which share no single property, where *property* is understood as nonrelational. Thus, the institutional theory still admits a sense, namely (2), in which art is an open concept, but it nonetheless provides a definition.

The institutional theory has been the object of much philosophical target practice. Here are five arguments against it.

1. Ted Cohen (1973) points out that something can be appreciated only if it is appreciable; and that some things—

he offers as examples "ordinary thumbtacks, cheap white envelopes, the plastic forks given at some drive-in restaurants," and possibly even Duchamp's *Fountain*—are unappreciable, hence cannot be candidates for appreciation, hence cannot be art.

2. Monroe C. Beardsley (1976) claims that it is incoherent to say that an artwork is made by a certain practice (conferring the status of candidate for appreciation) on behalf of a certain institution (the artworld) when neither the rules of the practice nor the limits of the institution can be further specified.

3. Danto (1981) objects to Dickie's concept of appreciation. In spite of Dickie's denial that he intends "a special kind of *aesthetic* appreciation," Danto thinks Dickie nonetheless implies aesthetic appreciation, given that an appreciation of "the qualities of a thing" suggests perceptible aspects of the object rather than the artist's act or the ideas the thing conveys. He criticizes Dickie's brief art-critical appreciation of *Fountain*, which praised that work for "its gleaming white surface . . . its pleasing oval shape," as inappropriately aesthetic, and this because the appeal of Duchamp's witty work lies, Danto would have it, in its expression of an idea, not in its curves and color.

4. Richard Wollheim (1987) tries to saddle the institutionalist with this dilemma. Either the representatives of the artworld have reasons for making a certain object a work of art or they do not. If they do, then their reasons for conferring the status ought to have been part of the theory. If they do not have reasons for making that object art, then the institutional theory embraces an unconvincing irrationality.

5. Noël Carroll (1994) objects that the institutional theory does not meet the challenge set by the open-concept theorists, for it does not explain at all what a work of art *is*, only that a work of art, whatever it is, fits into a certain social context.

Meeting Cohen's objection would mean giving up on the value neutrality—the classificatory thrust—of the institutional theory. For Cohen in effect calls for a kind of minimal or ground-floor "appreciability" condition that an object would have to meet to be qualified as an artwork. Dickie wants to keep open the possibility that literally anything could be a work of art; hence, he cannot search for such a minimal condition without becoming apostate to his own theory. If one had such a ground-floor condition—one discovers (decides?) that something *o* is art only if it has quality *G*—then it becomes uncertain what relevance the further decisions of persons have in determining the arthood of *o*. In any event, Cohen himself has not supplied any such ground-floor condition.

Both Wollheim and Carroll, in different ways, wish to resurrect the challenge thrown down by the open-concept theorists in the 1950s, namely, that a definition of art must say something about the qualities of the art *object* in virtue of which it is art. But this is something the institutional theory does *not* do. Hence, the institutional theory is incomplete or irrational (Wollheim) or not really a definition at all (Carroll). It is, of course, true that if the representatives of the artworld conferred the status of candidate for appreciation on objects, say, in virtue of their being beautiful—they had beauty as their reason for conferring arthood—then it should be expected that beauty ought to be part of a complete theory of art. Indeed, one should wonder (as with Danto's theory) what the decisions of anyone have to do with art making, if the criterion on which it is decided to make art is the beauty of the object. In that case, one should promote a beauty theory, and any institutionalism would become irrelevant. In response, however, one should keep in mind that Dickie embraces the open-concept view to this extent: he agrees that there is no set of properties that all artworks—past, present, and future—must share. Hence, there exists no closed set of reasons that representatives of the artworld must, as a matter of definition, always present to justify their decision to let something be art. Wollheim himself embraces a meaning theory of art; but it is clearly not the case that every artwork is endowed with meaning—think of a Bach fugue or a visual arabesque. Carroll is correct to the extent that the institutional theory has not told us what a work of art is in terms of the qualities or functions such works have. These are serious and interesting issues. Yet, it does not follow that if a theory does not mention qualities or functions of artworks it has not thereby given a definition. For a definition of *F* can be, minimally, a finite and stable set of conditions that something must meet in order to qualify as *F*. If *art* can only be defined in terms of its role in social practice, then this may just be a brute fact about the concept of art, not an objection to the institutional theory. It may be that *being* art is more like *being enclosed* than like being *bumpy*.

It was primarily to meet Beardsley's reservations about the unfounded legalisms of the 1969/1974 theory and Danto's objections to its concept of appreciation that Dickie produced another version. In his 1984 book, *The Art Circle: A Theory of Art,* "A work of art is an artifact of a kind created to be presented to an artworld public." "An artist is a person who participates with understanding in the making of a work of art." The artist's understanding must include the belief that there is an artworld and that the artworld contains options—Dickie calls them "frameworks"—for the presentation of artworks. "The artworld is the totality of all artworld systems," and "An artworld system is a framework for the presentation of a work of art by an artist to an artworld public." The *Art Circle* theory abandons the quasi legalisms of the earlier version. It no longer makes reference to a status conferred by people acting on behalf of an institution. Thus, Beardsley's reservation would seem to be met. Moreover, it exchanges the near-aesthetic notion of "appreciation" for the more neutral one of "presentation." Thus, Danto's objection is met.

The circularity of the definition, about which Dickie has been entirely open, may appear a crippling defect. Dickie defends himself by pointing out that some circles are informative, and his is one. In fact, there are families of concepts that seem impossible to define except in terms of each other. Dickie thinks *art, artist,* and *artworld* to be one such family. For another, consider the concepts *premise, conclusion,* and *argument.* It seems one is forced to say that an argument is by definition composed of at least one premise and at least one conclusion; and a premise is that part of the argument that is intended to provide evidence for the conclusion, which is that part for which evidence in the form of the premise is provided. (Some honest logician should title his or her introductory text *The Logic Circle.*) It is possible to provide a noncircular definition of *art*—for example, defining art as the expression of emotion—but it is doubtful whether we can have a noncircular *institutional* definition of art.

What, ultimately, is an institutional theory? To qualify as an institutional theory of *F*-hood, such a theory must do three things. First, it must acknowledge that there are a relatively small number of statable conditions for determining whether any object *o* is or is not *F.* Thus *F*-hood is not an "open concept," if being an open concept denies the existence of such statable and determining conditions. Second, any theory institutional in structure must deny that the *F*-hood of any *o* is grounded in or limited by any constellation of material properties of *o.* In this sense, anything could be *F,* and in this sense an institutional theory accepts half of an open-concept view. Third, an institutional theory must posit as a necessary condition for the truth of "*o* is *F*" some institutionality: perhaps certain practices or certain intentions of individuals as grounded in some social or cultural circumstance. In sum, an institutional theory of *F*-hood treats *F*-hood as an objectively determinable and socially determined multiplace concept, with some social institution as one of the relata. "Something is a painting if and only if it is hung in a gallery" counts as a little institutional theory of arthood. We can easily see, however, that institutional theories can be offered of concepts other than art. "A being is a person if and only if that being is given rights in the U.S. Constitution" would be an institutional theory of personhood.

There are other institutional theories by other philosophers. Three will be discussed here. As will be seen, Dickie's still seems the most viable of the institutional theories.

T. J. Diffey, in "The Republic of Art," published in the same year as Dickie's first institutional theory (though Dickie wrote in *Art and the Aesthetic* that he did not know of the piece at that time), speaks of art as a "status" that is "conferred" on things by a group he refers to as the "republic of art." Diffey's, however, is an *evaluative* institutional definition. "To say that something is a work of art is to imply that it is a thing of interest and of worth." It is not the purely private taste of a single individual, however, that "honors" something sufficiently to make it art. Neither is it the decision of the general public, but rather the joint verdict of the republic of art, which has as "first-stage" members creators, performers, spectators, and critics; and ultimately includes "anyone involved with the arts." In Diffey's example, this republic confers the status of art on Jane Austen's novels, because it sees value in them, and withholds the status from Mrs. Ann Radcliffe's, because it deems them inferior. Of course, the republic may change its mind and expel things from the status of art; it may also be inconsistent, and even irrational. Diffey's theory is remarkably like Dickie's, except that Dickie wants to keep evaluation out of the definition of art. The main problem with Diffey's theory is that it makes bad art, contrary to the facts, a definitional impossibility.

Timothy Binkley (1977) has perhaps the most minimal institutional theory of art: "To be a piece of art, an item need only be indexed as an artwork by an artist. . . . To make art is, basically, to isolate something (an object, an idea . . .) and say of it, 'This is a work of art,' thereby cataloguing it under 'Artworks.'" Art is what artists index as art; and artists, presumably, are those who index objects as art. Binkley embraces a theory somewhat like that in Dickie's *Art Circle,* but Binkley's theory exhibits a very tight and therefore uncomfortable circularity.

Marcia Muelder Eaton (1983) maintains that "*x* is a work of art if and only if (1) *x* is an artifact and (2) *x* is discussed in such a way that information concerning the history of production of *x* directs the viewer's attention to properties which are worth attending to." Eaton's view invites one of the criticisms directed at Dickie's, for she tells us that by "property" she means something perceivable, and by "discussion" something aesthetic, which makes her definition unable to accommodate such conceptual works as Duchamp's *Fountain.* Eaton also swallows whole one implication of her definition, namely, that a Coke bottle can be a work of art if its perceivable qualities are discussed aesthetically. But this also admits many automobiles, garments, hair styles, and plates of food as works of art, since these are discussed aesthetically in such publications as *Car and Driver, Vogue,* and *Gourmet.* (It would beg the question for Eaton to stipulate that these are not art forums, and the discussants not art critics.) It would appear, then, that Eaton's definition is both too narrow and too broad.

The institutional theory may be in decline among aestheticians. Historical theories (e.g., Carroll 1994) and meaning theories (Danto, 1981, and Wollheim 1987) seem to be in the ascendance. Perhaps people have simply tired of discussing the theory. No one has in any way refuted the institutional theory, however, and Stephen Davies in his fine book (1991) has recently defended it.

[*See also* Art World; Artifact; Danto; Definition of Art; Dickie; Museums; *and* Sociology of Art.]

BIBLIOGRAPHY

Beardsley, Monroe C. "Is Art Essentially Institutional?" In *Culture and Art,* edited by Lars Aagaard-Mogensen, pp. 194–209. Atlantic Highlands, N.J., 1976.

Binkley, Timothy. "Piece: Contra Aesthetics." *Journal of Aesthetics and Art Criticism* 35.3 (Spring 1977): 265–277.

Carroll, Noël. "Identifying Art." In *Institutions of Art: Reconsiderations of George Dickie's Philosophy,* edited by Robert J. Yanal, pp. 3–38. University Park, Pa., 1994.

Cohen, Ted. "The Possibility of Art: Remarks on a Proposal by Dickie." *Philosophical Review* 82.1 (January 1973): 69–82.

Danto, Arthur. "The Artworld." *Journal of Philosophy* 61.19 (October 15, 1964): 571–584.

Danto, Arthur C. *The Transfiguration of the Commonplace.* Cambridge, Mass., 1981.

Davies, Stephen. *Definitions of Art.* Ithaca, N.Y., 1991.

Dickie, George. "Defining Art." *American Philosophical Quarterly* 6.3 (July 1969): 253–256.

Dickie, George. *Art and the Aesthetic: An Institutional Analysis.* Ithaca, N.Y., 1974.

Dickie, George. *The Art Circle: A Theory of Art.* New York, 1984.

Diffey, T. J. "The Republic of Art." *British Journal of Aesthetics* 9.2 (April 1969): 145–156.

Eaton, Marcia Muelder. *Art and Nonart.* Rutherford, N.J., 1983.

Weitz, Morris. "The Role of Theory in Aesthetics." *Journal of Aesthetics and Art Criticism* 15.1 (September 1956): 27–35.

Wollheim, Richard. *Painting as an Art.* Princeton, N.J., 1987.

ROBERT J. YANAL

INTENTION.

[*This entry comprises three essays on the role of artists' (or authors') intentions in the interpretation of artworks:*

An Overview
Intentional Fallacy
Speech Act Theory and the Interpretation of Images

The first essay explores the history of the notion of intention in philosophical aesthetics. The second essay explains the fallacy that some theorists think is present whenever art interpretation relies on artistic intention. The third essay shows the relevance of speech act theory (a philosophy of language that emphasizes the act of speech over any of the act's component parts—words and sentences—in the interpretation of linguistic meaning) to the interpretation of intentions in art; this discussion reveals one way that intentions may still be relevant in art interpretation after the critique of earlier attempts to rely on them.]

An Overview

Works of art are generally viewed as objects requiring interpretation, but there is no general agreement about the nature and aim of the interpretative process. A large part of the history of the concept of intention in aesthetics can be told by tracing the significance theorists have attributed to the artist's intentions in the ongoing debate about the nature and aim of our interpretive responses to art. (*Art* and *artist* are here being used generically to include works and practitioners in all the arts.) Much of the material dealt with in what follows focuses on the case of literature—a major locus of theorizing in the latter half of the twentieth century—though many of the mentioned themes and arguments are generalizable to the other arts. The discussion is largely confined to developments in the twentieth century, since the concept of intention has only assumed a central place in aesthetics since the rise of Romanticism.

Granted that works of art are objects of interpretation, what kind of interpretation is involved? That the concern here is with the interpretation of *meaning* rather than with the kind of interpretive activity employed, say, in trying to explain how something works, is suggested by certain facts about art—that works of art are artifacts, that they are usually made for contemplation, and that they typically possess a representational or expressive significance. If this is so—if the aim is the interpretation of meaning—it is a short step to what will be called *intentionalism:* the view that the success of an interpretation is dependent on grasping the intended meaning conveyed by the artist through the work.

A highly influential view, which in some versions claims precisely this, is the *expression theory of art.* The expression theory originated with the Romantic movement in the nineteenth century. William Wordsworth espoused it, as did Leo Tolstoy. It was defended in a philosophically sophisticated way by, among others, Benedetto Croce, R. G. Collingwood, and John Dewey. According to the closely related theories defended in Croce's *Aesthetic* (1964) and Collingwood's *Principles of Art* (1938), the process of artistic expression is one in which an inchoate feeling or emotion finds articulate embodiment in the work of art. Associated with this view is a cognitive function for art: that of providing the artist with self-understanding. But Croce and Collingwood conceive the work as an inner mental object or private "intuition"; according to them, the tangible objects ordinarily associated with art are inessential to the process of artistic expression—mere vehicles for communicating the artist's "intuition" to a public. In Dewey's *Art as Experience* (1934), the process of expression is, perhaps more plausibly, identified with the artist's efforts to produce a public embodiment of feeling through engagement with a medium, and the resulting sense of insight acquired by the artist is enjoyed by the public as well. On Tolstoy's account in *What Is Art?* (1930), expression is explicitly viewed as a process of communication: one gets whatever enjoyment and value one derives from art by interpreting it in such a way as to retrieve the original feeling state which it was the artist's intention to reveal when she or he embodied it in the work.

In Romantic aesthetics (as illustrated by the expression theory) artists occupied a central place of authority in criticism. Their intentions were taken to fix the meaning of a work, and there was a presumption that critics should defer to whatever elucidations artists offered about their aims and

preoccupations during the composition of their works. Biographical criticism flourished. (A prominent example is John Livingston Lowes's *Road to Xanadu* [1927], a massively researched account of the writing practices, techniques, and artistic aims that inspired the poetry of Samuel Taylor Coleridge.) The development of critical theory in the second half of the twentieth century can be seen as a process in which criticism has looked for emancipation from the constraining influence of the artist, and where the critic has been accorded an increasingly active, creative role in interpreting the meaning of art. While the Romantic influence resulted in an expectation that a work has a single meaning and that all valid interpretations will be consistent and convergent, subsequent theory has accommodated the idea that works of art can have more than one meaning, and that incompatible interpretations can be true of a work. In this dispute the question of the restraints imposed on interpretation by the artist's intentions has been central.

A seminal contribution to this issue, originally published in 1946, was Monroe Beardsley and William Kurtz Wimsatt's "The Intentional Fallacy" (1976). [*See the article on* Intentional Fallacy, *below.*] Beardsley and Wimsatt were closely associated with the New Criticism, a movement that maintained that the scrutiny of the text is the central activity of literary criticism, and that literary meaning is entirely the outcome of the particular combination of words comprising the work itself. A diverse group of writers, including John Crowe Ransom and Cleanth Brooks in the United States and Ivor Armstrong Richards and Frank Raymond Leavis in Britain, argued for the self-sufficiency of the work as a source of meaning, or endeavored to demonstrate this self-sufficiency through close textual analyses of poems, novels, and plays. Rejecting earlier emphases on historical and biographical investigations in criticism, the New Critics—the term comes from the title of Ransom's book *The New Criticism* (1979)—practiced a method of "scrutiny" premised on the idea that sufficiently alert readers can grasp the meaning of a literary work by attending exclusively to the text, and that they are not required to defer to, and may override, historical and biographical investigations concerning a work's origin—including writers' avowals about their works' intended meanings. To accommodate the kind of ineliminable references to the author implicit in judgments about a work's perceptiveness, profundity, sincerity, and the like, Brooks's *Well-Wrought Urn: Studies in the Structure of Poetry* (1968) introduced the notion of the "implied author" to refer to the controlling intelligence visible in a work. The implied author is distinct from the actual author, being a figure conjured entirely by a particular work, or a group of works by the same author; it is the figure the author appears to be when she is approached through her writing. By means of this device, even references to the author's intentions can be reinterpreted as referring to the "intentions" of a notional authorial figure conjured by the writing itself. (In recent theory, the

notion of the implied author has continued to exercise an influence—for example, through Alexander Nehamas's related but distinct notion of the "postulated author" [Nehamas, 1981–1982].) Freed from the boundaries set by the actual intentions of the flesh-and-blood author, more flexible criteria for the acceptability of an interpretation were introduced: that it should make coherent sense of the work's perceptible features, that it should cast the work in an aesthetically satisfying light, and so on. It is possible that incompatible interpretations should fare equally well on criteria such as this, and rank among the best available.

The notion of the self-sufficiency of a literary work gets a degree of support from the idea that language is a system embodying shared semantic and syntactic rules that are independent of, and presupposed by, the purposes to which words are put by individual speakers and writers. In "The Intentional Fallacy," Beardsley and Wimsatt assert that language is "public," drawing the conclusion that for those fluent in the relevant language, the meanings of literary works are available through the text—an inference that feeds, in a self-explanatory way, into New Criticism. This account of literary meaning faces the difficulty that, although an unambiguous sentence like "God save the Queen" indeed has a "propositional" meaning dictated by its grammar and the meanings of its constituent words, this kind of meaning cannot fully account for the literary meaning it might have in different instances of its use. If, for example, the sentence occurred ironically in one poem and nonironically in another, its propositional meaning, being tied to its semantic and syntactic features, would remain unchanged. These features would therefore not account for the difference in literary meaning. The insight on which this objection is based— that meaning arises partly from the context in which a work is produced—has in recent years resulted in a distinction being drawn between *texts* and *works*. It is possible to imagine the same text being produced by different authors in different contexts; if, as seems likely, their meaning was affected by their different origins, the two texts, since they differed in meaning, would have to be regarded as different works. (This argument was first advanced in connection with music, by Jerrold Levinson [1990]: identical musical scores composed in different circumstances may differ in meaning, implying distinct works.) Thus, context emerges as a factor contributing to literary meaning. Intentionalists would argue that the author's intentions are a crucial contextual feature.

The idea that meaning is a function of a preestablished linguistic system is also central to the structuralist movement in literary theory (henceforward referred to simply as structuralism), but here, with the introduction of the notion of codes that determine meaning in the various literary genres, the idea was carried much further. Structuralism has some of its roots in structural anthropology, and others in Russian Formalism. A more direct influence is the work of

the Swiss linguist, Ferdinand de Saussure, especially his *Course in General Linguistics* (1959). Saussure argued that words are arbitrary signs whose meanings are formal in origin, arising from their relative position in the linguistic system taken as a whole. Although he observed a distinction between the established system of language *(langue)* and the use of language in particular utterances *(parole)*, the focus of his work was on the former, and it is here that his influence on structuralism was decisive. Thus, a feature of structuralism in literary theory is the way it treats texts almost exclusively as products of a general system—one that dictates the character of particular literary works. In Roland Barthes's *S/Z* (1974), a reading of Honoré de Balzac's *Sarrasine* is constructed by Barthes on the basis of five identified narrative codes, conceived as independent conventions that explain the meaning of the text. In general, structuralist readings of literary works are little concerned with authorship, their focus being on the way literary works display features of the overarching system of which they are the product. In some varieties, structuralism tends toward determinism—a tendency reflected in Barthes's "death of the author" doctrine (propounded in an essay of that name collected in the volume *Image, Music, Text* [1977]). In these varieties, the writer's consciousness is taken not as the source of the work but as itself the product of the system in which the writer is enmeshed. Any concern with the writer's motivations is now seen as misplaced.

Marxist literary theory shares some common ground with structuralism, but is in other respects at odds with it. What it shares is the tendency toward determinism, for Marxist theorists, too, are prone to be dismissive of the contribution to literature of individual authors, seeing art and other cultural artifacts as products of the underlying structure of society. But unlike the structuralist, whose approach is essentially ahistorical, the Marxist takes society to be in a state of progressive change determined by historical forces, and so Marxist criticism pays careful attention to the historical context of art—especially prevailing class structures and economic conditions. There are many varieties of Marxism, and a corresponding variety of ways in which theorists have applied Marxist ideas to literature. Marxists vary in their construal of how far the activities of individuals are determined by sociological laws, and by the ideologies generated by the social "infrastructure." While Karl Marx himself (especially in his early writings) made provision for individual agency, and ascribed—for instance, in the introduction to the *Grundrisse* (1973)—a degree of autonomy to artists, the more general tendency in the Marxist tradition is to view human consciousness as determined by ideology. Under the influence of Louis Althusser, recent Marxist theorists like Pierre Macherey and Terry Eagleton, whose *Criticism and Ideology: A Study in Marxist Literary Theory* (1976) is an accessible introduction to this style of theorizing, have propounded a kind of criticism in which literary works are viewed as the outcome of economic forces, and the personalities and intentions of individual authors are excluded from literary analyses of their works.

In *Of Grammatology* (1976) Jacques Derrida, the founder of deconstructionism, takes from Saussure the idea that the meaning of a word is determined by its relation to other words in the language, but rejects what he regards as the static conception of language implicit in the Saussurean view. Derrida emphasizes the repeatable ("iterable") nature of linguistic signs, asserting that they accrue meaning in each different context of use. There are thus, on his account, two ways in which meaning outruns the intentions of the language user: (1) the meaning of a sign is, à la Saussure, a function of its place in the system, hence independent of its use by a speaker or writer on a particular occasion; and (2) meaning is in flux, being a function of its past and future contexts as well as of the present context of its use. These claims issue in the idea that meaning is indeterminate: because the meaning of one sign is linked to others in an expanding circle, and because present usages are linked to past and future ones, any attempt to achieve finality of interpretation is forever "deferred."

Another theme in Derrida concerns his emphasis on the distinction between speech and writing. A feature of writing is that we typically encounter it in the absence of the writer. In his essay "Signature, Event, Context" (collected in *Margins of Philosophy* [1982]), Derrida argues that even when we know nothing at all about the context in which a piece of writing was produced, it remains readable. Derrida concludes that understanding is independent of a grasp of the author's motivations and other states of consciousness. This argument has produced mixed reactions. John Searle (1977) has resolutely rejected it, pointing out, among other objections, the possibility that an isolated text may by itself offer access to the author's intentions, since "a meaningful sentence is just a standing possibility of the corresponding (intentional) speech act." On the other hand, Stanley Fish (1982) has attempted to endorse and expand Derrida's argument, deducing that authorial intentions are not so much discerned by readers as manufactured by them—an antirealist conclusion that takes the notion of the creativity of the reader to its logical terminus. He takes up Derrida's claim that the interpretation of writing regularly proceeds without any evidential basis for inferring the author's intentions, and also asserts (somewhat dogmatically) that meaning and intention "are the same thing." Fish reasons that since the reading process, insofar as it grasps meaning at all, views writing from the outset as having an intended meaning, the writing furnishes no evidence of the author's intentions, and these must therefore be viewed as projections of the reader. Fish's premise—that all meaning is intended meaning—stands at the opposite extreme to the claim that the text is self-sufficient. Whereas the latter takes all meaning to derive from syntactic and semantic features of the linguistic

system, Fish's view in effect denies that these features make any contribution whatever to the meaning of a piece of writing.

Since the publication of Beardsley and Wimsatt's "Intentional Fallacy" in the mid-century, the question of intention and its relevance for interpretation has shown a continuing capacity to generate controversy. Analytic aesthetics—aesthetics, that is, in the Anglo-American tradition, represented by journals such as the *Journal of Aesthetics and Art Criticism* and the *British Journal of Aesthetics*—has maintained a philosophical distance from developments in literary theory since structuralism, although contributing a salutary element of skepticism, with regular calls for arguments and clarity. In analytic aesthetics intentionalism has remained a vigorously debated philosophical option. The arguments both in favor of and against it have been fed by developments in the philosophy of mind, and their implications explored through extensive discussions of particular artworks. Using ideas of the later Ludwig Wittgenstein, writers like Frank Cioffi (1963–1964) and Colin Lyas (1973, 1992) have drawn attention to the fact that artists' intentions can assume a public form in their work, offering a way of reconciling intentionalism with the public character of artistic meaning and the autonomy of criticism. Speech act theory, developed by writers like John Langshaw Austin, Paul Grice, and John Searle, has been productively applied—for example, by William E. Tolhurst—to the analysis of literary meaning. Drawing on a distinction made by Grice, Tolhurst argues (1979) that literary meaning is a variety of utterance—as distinct from utterer's—meaning. In direct conflict with this view is that of E. D. Hirsch. In his book *Validity in Interpretation* (1967), which takes inspiration from the hermeneutic tradition, Hirsch defends the claim that literary meaning is identical with the author's intentions—or utterer's meaning. Richard Wollheim (1980, 1987) and Michael Baxandall (1985) have defended sophisticated versions of intentionalism while eruditely exploring works of visual art from an intentionalist perspective. Edited volumes of essays in which the arguments for and against intentionalism in its various forms have been debated have regularly appeared—for example, David Newton-de Molina (1976) and Gary Iseminger (1992). The concept of intention has also played a role in other debates in analytical aesthetics—for example, in discussions of the definition of art, the ontology of art, and the nature of metaphor.

[*See also* Expression Theory of Art; Improvisation; New Criticism; *and* Originality.]

BIBLIOGRAPHY

Barthes, Roland. *S/Z.* Translated by Richard Miller. New York, 1974.
Barthes, Roland. *Image, Music, Text.* Edited and translated by Stephen Heath. New York, 1977.
Baxandall, Michael. *Patterns of Intention: On the Historical Explanation of Pictures.* New Haven, 1985.

Beardsley, Monroe, and William Kurtz Wimsatt. "The Intentional Fallacy." In *On Literary Intention,* edited by David Newton-de Molina, pp. 1–13. Edinburgh, 1976.
Brooks, Cleanth. *The Well-Wrought Urn: Studies in the Structure of Poetry.* Rev. ed. London, 1968.
Cioffi, Frank. "Intention and Interpretation in Criticism." *Proceedings of the Aristotelian Society* 64 (1963–1964): 85–106.
Collingwood, R. G. *The Principles of Art.* Oxford, 1938.
Croce, Benedetto. *Aesthetic: As Science of Expression and General Linguistic.* 2d ed. Translated by Douglas Ainslie. London, 1922.
Derrida, Jacques. *Of Grammatology.* Translated by Gayatri Chakravorty Spivak. Baltimore, 1976.
Derrida, Jacques. "Signature, Event, Context." In *Margins of Philosophy,* translated by Alan Bass, pp. 307–330. Chicago, 1982.
Dewey, John. *Art as Experience.* New York, 1934.
Eagleton, Terry. *Criticism and Ideology: A Study in Marxist Literary Theory.* London, 1976.
Fish, Stanley Eugene. "With the Compliments of the Author: Reflections on Austin and Derrida." *Critical Inquiry* 8.2 (1982): 693–721.
Hirsch, E. D., Jr. *Validity in Interpretation.* New Haven, 1967.
Iseminger, Gary, ed. *Intention and Interpretation.* Philadelphia, 1992.
Levinson, Jerrold. "What Musical Works Are." In *Music, Art and Metaphysics: Essays in Philosophical Aesthetics.* Ithaca, N.Y., 1990.
Lowes, John Livingston. *The Road to Xanadu: A Study in the Ways of the Imagination.* Boston, 1927.
Lyas, Colin. "Personal Qualities and the Intentional Fallacy." In *Philosophy and the Arts,* edited by Godfrey Vesey. London, 1973.
Lyas, Colin. "Wittgensteinian Intentions." In *Intention and Interpretation,* edited by Gary Iseminger. Philadelphia, 1992.
Macherey, Pierre. *A Theory of Literary Production.* Translated by Geoffrey Wall. London and Boston, 1978.
Marx, Karl. *Grundrisse: Foundations of the Critique of Political Economy.* Translated by Martin Nicolaus. Harmondsworth, 1973.
Nehamas, Alexander. "The Postulated Author: Critical Monism as a Regulative Ideal." *Critical Inquiry* 8.2 (1981–1982): 133–149.
Newton-de Molina, David, ed. *On Literary Intention.* Edinburgh, 1976.
Ransom, John Crowe. *The New Criticism.* Reprint, Westport, Conn., 1979.
Saussure, Ferdinand de. *Course in General Linguistics.* Edited by Charles Bally and Albert Reidlinger, translated by Wade Baskin. New York, 1959.
Searle, John Rogers. "Reiterating the Differences: A Reply to Derrida." *Glyph* 1 (1977): 198–208.
Tolhurst, William E. "On What a Text Is and How It Means." *British Journal of Aesthetics* 19.1 (Winter 1979): 3–14.
Tolstoy, Leo. *What Is Art? and Essays on Art.* Translated by Aylmer Maude. Oxford, 1930.
Wollheim, Richard. "Criticism as Retrieval." In *Art and Its Objects,* 2d ed. with six supplementary essays, pp. 185–204. Cambridge and New York, 1980.
Wollheim, Richard. *Painting as an Art.* Princeton, 1987.

PAUL TAYLOR

Intentional Fallacy

The intentional fallacy was first discussed in print in 1943 by the philosopher Monroe C. Beardsley, and the literary critic William K. Wimsatt, Jr., although apparently it was not so christened until the appearance in 1946 of their article with that title (later collected in a volume of essays by Wimsatt).

A fallacy is presumably not just any mistake, but specifically a mistake in reasoning (although not all philosophers identifying their favorite "fallacies" have been scrupulous on this point). So interpreted, the intentional fallacy may be regarded as an inference appealing to, among others, a premise providing "external evidence" (e.g., diaries, first drafts, correspondence) of the meaning that the author intended to express in the work (what Wimsatt and Beardsley call the "intentional meaning," as contrasted, significantly, with the "actual meaning") and concluding with a judgment of the value of the work as proportional to its success in carrying out those intentions. In 1946, Wimsatt and Beardsley summarized their earlier position thus: "the design or intention of the author is neither available nor desirable as a standard for judging the success of a work of literary art." Committers of the fallacy were Romantic "biographical" critics, and Wimsatt and Beardsley's argument was one of the major theoretical underpinnings of the New Criticism, which saw itself as displacing "biographical" criticism and came to dominate American literary studies around the middle of the twentieth century.

Wimsatt and Beardsley's diagnosis of what is fallacious about the intentional fallacy depends crucially on the distinction between actual meaning and intentional meaning and on the claim that the literary work (along with its actual meaning) "is detached from the author at birth and goes about the world beyond his power to intend about it or control it . . ., belongs to the public, [and] . . . is embodied in language, the peculiar possession of the public." Thus, ontological questions about the nature of "the poem itself" and its relations to the author's intentions became a focus of discussion in their own right; someone with no particular interest in defending "biographical" criticism nonetheless might still be an intentionalist in the sense that he or she might think that there was a closer connection between actual meaning and intentional meaning than that allowed by Wimsatt and Beardsley. Thus, the literary critic E. D. Hirsch, Jr., writing in 1967, in one of the main texts of what Beardsley called the "Intentionalist Backlash," set out to defend "the sensible belief that a text means what its author meant" (Hirsch, 1967); his views have been the starting point for most recent discussions.

Hirsch's main argument is adumbrated in the following passage:

> A determinate verbal meaning requires a determining will. Meaning is not made determinate simply by virtue of its being represented by a determinate sequence of words. . . . But if a determinate word sequence does not in itself represent one, particular, self-identical, unchanging complex of meaning, then the determinacy of verbal meaning must be accounted for by some other discriminating force. . . . That discriminating force must involve an act of will, since unless one particular complex of meaning is *willed* . . . , there would be no distinction between what an author does mean by a word sequence and what he could mean by it.

Beardsley takes Hirsch to be defending what he calls "the Identity Thesis," the thesis that "what a literary work means is identical to what its author meant in composing it," and, although he does not deal directly with the above argument by Hirsch, Beardsley presents three arguments of his own that he claims "conclusively refute" that thesis. These arguments are based on the existence of authorless works, on the claim that works may change their meaning after their author's death, and on the fact that a work can have meanings of which its author is not aware.

Other writers have objected to various of the premises of Hirsch's argument. One obvious and crucial premise is that meaning is, in fact, determinate, that a poem, for example, does indeed "represent one, particular, self-identical, unchanging complex of meaning." Hirsch is clear that this is not meant to rule out ambiguity, but it still seems to be a fairly ambitious claim. Why suppose, for example, that poems cannot change? (Note that Beardsley assumes that they can.)

What is required of the concept of determinacy at a minimum, if it is to function in Hirsch's argument, is that it support the idea that "the poem itself" (as distinguished, say, from the poem relative to its being brought into existence in a certain way by its author or to its having certain effects on its readers) definitely either does or does not have the meaning one might think of attributing to it; determinacy, in a word, entails a kind of realism about poems and their semantic properties.

Here the most direct challenge has been issued by the philosopher Joseph Margolis, who argues that poems, in common with other "culturally emergent" entities, precisely are not determinate entities even in this minimal sense. Although in some hands a view like this might degenerate into mystery mongering, Margolis argues at length that his "Robust Relativism" does not violate any of "the usual logical constraints." (Wimsatt and Beardsley, of course, would not be inclined to attack Hirsch's argument at this particular point.)

Even if, with Hirsch, one accepts realism rather than relativism about meaning and argues that linguistic norms are insufficient to underwrite it without adding the "discriminating force" of a human mind to the mix, one still has to say whose mind does the job. (The passage quoted earlier from Hirsch does not speak to this point.) Stanley Fish's claim that readers acting as members of "interpretive communites" constitute meaning, and, in a different way, the tradition of what has been called "ludic" interpretation, defended, for example, by Roland Barthes, show that it cannot be taken for granted that the author, rather than the critic or reader, will win the nomination.

Emphasis on the reader or critic, while strictly compatible with realism is, however, more likely to issue in relativism. For the realist the choice of the author as the person whose intentions render the work determinate is most plausibly defended by assimilating literature to ordinary communica-

tive utterances, which we are less inclined to think of as issuing in entities that are "detached from their authors at birth" and thus have their essential properties either independent of further human intervention or at most in virtue of the agency of people other than their speakers. (Correspondingly, an important part of the anti-intentionalist strategies of Beardsley, Fish, and Barthes, different as each of them is from the others, is an emphasis on the distinction between ordinary language and literature.)

A point that Hirsch would surely concede to Beardsley is that one cannot, by willing or intending, make a sequence of words mean whatever one wants it to mean (Humpty-Dumpty famously claims to mean by "glory" a "nice knock-down argument"; Ludwig Wittgenstein, on the other hand, asks rhetorically, "Can I say 'bububu' and mean 'If it doesn't rain I shall go for a walk'?") The literary critics Steven Knapp and Walter Benn Michaels, in a widely discussed series of articles, argue that Hirsch's intentionalism is insufficiently radical. Using well-chosen and ingeniously developed examples, they argue that there is no reason to suppose that there is "a moment of interpretation" prior to intention, as the notion that the public norms of language at least limit what we can mean by what we say would have us do. Their extreme version of intentionalism thus amounts to a defense of Humpty-Dumpty's implied theory of meaning.

Recently the philosopher Jerrold Levinson has tried to accommodate the conflicting intuitions that separate a Hirsch from a Beardsley by defending "hypothetical intentionalism." The answer to the question of what "discriminating force" makes a mere sequence of words into a determinate poem is that no actual intention is crucial; rather, it is the intention that can most reasonably be attributed to the author:

> The core of literary meaning, as with any piece of discourse publicly presented, is not the meaning (the many meanings) of the words taken in abstraction from the author, or precisely of necessity the meaning that the author actually intended to put across, but our best hypothetical attribution of such, formed from the position of the intended audience. (Levinson, 1992)

Clearly the answers given by Levinson's theory and Hirsch's will often coincide. Crucial cases where they do not can be illustrated by an example due to Hirsch: a man standing beside an apparently disabled car holding a fuel can says to a motorist who stops, "My car ran out of gas." Can his utterance mean that a caboose that he owns just dashed from a cloud of argon? Perhaps our intuitions are not clear enough to be decisive, but even if hypothetical intentionalism's answer (presumably in the negative) seems persuasive, one can wonder whether this "compromise between intentionalism and anti-intentionalism," as Levinson calls it, sufficiently honors the basic impulses behind both views. Is the independence from the author that the work gains on this view sufficient to satisfy the anti-intentionalist, and is reference to hypothetical rather than actual intentions

sufficient to satisfy the intentionalist's vision of literature as communicative utterance? Perhaps any reference to the author's intentions will be too much for the anti-intentionalist, whereas reference to (merely) hypothetical intentions will not be enough for the intentionalist.

[*See also* Beardsley.]

BIBLIOGRAPHY

Barnes, Annette. "Half an Hour before Breakfast." *Journal of Aesthetics and Art Criticism* 34.3 (Spring 1976): 261–271.

Barthes, Roland. "The Death of the Author." In *Image, Music, Text,* edited and translated by Stephen Heath, pp. 142–148. New York, 1977.

Beardsley, Monroe C. *The Possibility of Criticism.* Detroit, 1970.

Beardsley, Monroe C. "Intentions and Interpretations: A Fallacy Revived." In *The Aesthetic Point of View,* edited by Michael J. Wreen and Donald M. Callan, pp. 188–207. Ithaca, N.Y., 1982.

Beardsley, Monroe C., and W. K. Wimsatt. "The Intentional Fallacy." In *On Literary Intention,* edited by David Newton-de Molina, pp. 1–13. Edinburgh, 1976.

Carroll, Noël. "Art, Intention, and Conversation." In *Intention and Interpretation,* edited by Gary Iseminger, pp. 97–131. Philadelphia, 1992.

Fish, Stanley. *Is There a Text in This Class? The Authority of Interpretive Communities.* Cambridge, Mass., 1980.

Hirsch, E. D., Jr. *Validity in Interpretation.* New Haven, 1967.

Iseminger, Gary. "A Revised Identity Thesis." In *Text, Literature, and Aesthetics,* edited by Lars Aagaard-Mogensen and Luk de Vos, pp. 67–71. Amsterdam, 1986.

Iseminger, Gary. "Actual Intentionalism vs. Hypothetical Intentionalism." *Journal of Aesthetics and Art Criticism* 54.2 (Fall 1996): 319–326.

Iseminger, Gary, ed. *Intention and Interpretation.* Philadelphia, 1992.

Knapp, Steven, and Walter Benn Michaels. "Against Theory." In *Against Theory: Literary Studies and the New Pragmatism,* edited by W. J. T. Mitchell, pp. 11–30. Chicago, 1985.

Levinson, Jerrold. "Intention and Interpretation: A Last Look." In *Intention and Interpretation,* edited by Gary Iseminger, pp. 221–256. Philadelphia, 1992.

Margolis, Joseph. *Art and Philosophy.* Atlantic Highlands, N.J., 1980.

Newton-de Molina, David, ed. *On Literary Intention.* Edinburgh, 1976.

Tolhurst, William E. "On What a Text Is and How It Means." *British Journal of Aesthetics* 19.1 (Winter 1979): 3–14.

Wimsatt, W. K. "Genesis: A Fallacy Revisited." In *On Literary Intention,* edited by David Newton-de Molina, pp. 116–138. Edinburgh, 1976.

GARY ISEMINGER

Speech Act Theory and the Interpretation of Images

The analogy between speech acts and images is drawn on the basis that both are intentional actions. Just as a speaker issues an utterance, so too an artist produces an image. Moreover, we can understand images made by an agent in the same manner that we understand utterances made by an agent.

Formulating the foundations of speech act theory, J. L. Austin stated that "the occasion of an utterance matters se-

riously, and . . . the words used are to some extent to be 'explained' by the 'context' in which they are designed to be or have actually been spoken in a linguistic interchange" (Austin, 1975). In interpreting the utterances of another, we want to understand not only what was said (word sequence meaning), but also the purpose for which it was said (force). The context, including the specific situation within which the utterance occurs as well as who the speaker is, aids us in this project.

The image act model sets out four components that are part of the interpretive process: the artist (speaker), the viewer (hearer), the image (utterance), and the context. It is the conjunction of these four components that comprise an image act. The artist, working within a social, historical, and cultural context, creates an image through an intentional action. The viewer who wants to understand what is done by the image act needs to understand something about the person who made it. Part of constructing an interpretation includes making hypotheses about the artist's aims, desires, beliefs, fears, and so on. It is within this framework that one can speculate about what the artist was doing. Familiarity with elements of the context of production as well as knowledge about the artist can contribute to interpretation construction. For instance, knowing that a monumental statue of a goddess was created by Audrey Flack in twentieth-century America rather than by Phidias in fourth-century Greece will influence the aims and beliefs attributed to the artist. We may suppose that Phidias believed in the existence of the deity depicted. Hence, his sculpture functions as an homage to the goddess. In Flack's case, however, we may not assign this belief. Rather than being an homage to a deity, her sculpture may function as a talisman for women.

The context of presentation also offers clues as to how to interpret what the artist was doing. The kind of context in which it is presented—church, museum, billboard, refrigerator, park—as well as specific features of the context tell us something about what is being done with the image. The suppositions we make about a gallery show titled "AIDing Vision" may be quite different from those we make for a "Celebration of Women's History Month" show.

In addition to the context and the artist, the viewer is a component of the image act. The consequences of an action are part of that action. Thus, the effect on, or response of, the audience is part of what is accomplished through the action. A stand-up comedian's utterance on stage may be taken as a joke or as an insult depending on who the hearer is. Installation work explicitly acknowledges the role of the viewer as a participant. Vito Acconci's image acts are not complete as works of art until the viewer walks over or crawls through them.

Images in the context of speech act theory are to be treated as utterances. Although the most thoroughly analyzed speech acts are single utterances such as "It is raining," there is a considerable body of literature that treats texts (from grocery lists to novels) as speech acts. There is no predetermined way to decide where to draw the parallel between a text and an image. Some complex images, such as Diego Rodríguez de Silva y Velázquez's *Las Meninas*, may be analogous to a book, whereas Marcel Duchamp's *In Advance of a Broken Arm* may seem more like a single sentence.

The import of considering an image as an utterance is that its interpretation will include the components of meaning and force. The meaning of utterance, often designated sentence meaning, is given by truth conditions. The force indicates how the sentence is used. Images can also be seen as having meaning and force. The force of an image is construed as the purpose the image serves, how it is to be taken, or what is done by the image. For example, to take one aspect of an image, we may decide that the meaning of a crown is, "I am royalty." In Jean-Michel Basquiat's *CPRKR* (1982), the crown appears with Charlie Parker's initials. Here it functions as a tribute to "Charles the First." Alternatively, the zirconium crown on Justine Colette's head in her piece, *Justine of Colette is Dead Co., Ltd. Posing as the Doll in Living Environment* (1970–1980) carries an ironic force. Adorned in lace and gemstones, Colette displays herself in a Wall Street storefront. Her performance mocks the royalty of Miss America, the crowning of Barbie.

Meaning and force are logically independent in the sense that the meaning of an image does not determine its force, nor does the force of an image determine its meaning. This principle of the autonomy of meaning suggests that the same image can evoke different forces and thus accounts for one source of divergent interpretations. Meaning and force are interdependent, however, in the sense that an interpretation incorporates both. The viewer assigns meaning and force conjointly within the framework of beliefs and aims attributed to the artist. The viewer continues revising these elements to make sense of what was created by the artist.

We can identify four different kinds of force that are involved in image interpretation. The first two originate in Austin's work. Designated as the illocutionary force, an image can be an assertion, warning, command, request, tribute, or question. The perlocutionary force encompasses the effect or consequences on the viewer. An image may have the perlocutionary force of annoying, amusing, irritating, educating, or prompting action on the part of the audience.

A third kind of force can be found in Donald Davidson's treatment of speech acts. An image can be taken humorously, ironically, sarcastically, or metaphorically. Finally, Marcia Eaton and Søren Kjørup address a fourth kind of force, namely, whether an image serves the purpose of an advertisement, a work of art, a map, or a comic.

Illocutionary force is illustrated in Eaton's discussion of how an image can be seen as an assertion. One may judge that an image is true or false, which "necessitates the viewer's taking an active role in which he or she first formulates a statement which he or she believes to be a possible

interpretation of the picture and then relates that statement to his or her beliefs" (Eaton, 1980). In her example, one might say that Norman Rockwell's paintings say something true about American life. Looking at *Freedom from Want,* a viewer may formulate the proposition "The United States is a land with a high standard of living where most people are free from want." If the viewer believes this proposition is true and that it is "said" in or by the painting, she or he may say that the painting is true. Unpacking the proposition that can be expressed in the painting paves the way for treating it as an assertion.

Eaton's discussion also effectively distinguishes between the meaning (application of truth conditions) and the force of the image. If we decide that the proposition above gives the meaning of the Rockwell image, we must still decide what force it evokes. Whether it is taken as an assertion made by Rockwell or a send-up will change our interpretation of the image. This question focuses on the third kind of force, asking if we should take Rockwell seriously or treat the image as a satire.

Carolyn Korsmeyer's discussion of pictorial assertion further develops the difficulties in deciding what is being asserted if one takes an image as an assertion made by the artist. Using Paul Revere's engraving of the Boston Massacre, Korsmeyer states that a number of propositions from the specific ("The British army is an unjust occupying force") to the general ("Military force tragically may be used against the legitimate protests of innocent civilians") can be postulated. Holding the force constant (assertion), the viewer who wants to know what assertion is being made by Revere will decide this within the context of what is known about Revere—his beliefs, aims, attitudes, role in the confrontation.

The Revere engraving also provides a good example of the function of perlocutionary force. Presented in a newspaper, the image may function as an assertion meant to inform readers of an event. It was also distributed at the head of a broadside. The assertion may have the perlocutionary force of making viewers angry or urging them to rally against the British occupation. As Korsmeyer points out, some viewers (Tories) took it as propaganda while other viewers (colonial patriots) took it as a call to protest against injustice. The perlocutionary force is assessed in reference to the audience and may vary with particular viewers.

This brings us to the fourth kind of force involved in image actions. Kjørup points out that the same image can be used for alternate purposes as the context changes. George Inness's *Lackawanna Valley* may be used by the president of the railroad firm to extol the efficiency of his trains or by a curator in a museum exhibition on American painting of the nineteenth century. In the first case it functions as an advertisement; in the second it functions as a work of art. A different agent using the image to accomplish an alternate purpose performs a new image act.

The artist's intentions will be relevant to an interpretation of an image act. This interpretative model is supported by views on intention that carve a middle road between intentionalism (the correct interpretation of the image is determined by the artist's intentions) and anti-intentionalism (the artist's intentions are irrelevant to the interpretation of the image). Such accounts are found in the work of Noël Carroll, Nan Stalnaker, and William E. Tolhurst.

The recommendation here is that image acts be taken as particular cases of intentional behavior. Like all intentional behavior, interpreting it will require the viewer to make hypotheses about the ends the agent aims to fulfill. This does not mean, however, that the artist's stated intentions determine the correct interpretation of a work. With the availability of more than one interpretation of an action, it is not always the case that the agent's interpretation is the most important one. As Colin Lyas states,

> It is equally important to realize that a description of an action in terms of the agent's conception of it, though it tells us something about how the action is properly to be characterized, may not tell us everything about what was done, nor even the most important things about what was done . . . an artist, the agent producing the work, may not be the best judge of what is important and interesting about it.

Similarly, Nan Stalnaker argues that "formal relationships are created by the artist for expressive and communicative purposes, and . . . it is in terms of these purposes that formal qualities have meaning—are interpretable." It is not the artist's expressed intention that will determine the meaning of a work, however.

Additionally, an action can have consequences unintended by the agent that are still part of what was done and thus properly part of its interpretation. Advertising that employed caricatures of African Americans to emphasize the whiteness their products could provide (Carter's Inky Racer, Rinso's Whitewashing power) count as racist even if that was not intended by their creators. That they count as racist, however, stems from their desire to present a recognizable stereotype of the pickaninny. Similarly, whatever Richard Serra intended the effect on the audience to be, the irritated and outraged response of people who had to live with *Tilted Arc* is part of the image act.

Speech act theory can be usefully applied to images when one's interpretative aim is to understand what the artist has done through the image. Instead of unveiling a universal truth, an interpretation makes sense of an artist's action on an occasion of viewing by incorporating contextual elements and the effect on the viewer.

[*See also* Hermeneutics; *and* Interpretation.]

BIBLIOGRAPHY

Austin, J. L. *How to Do Things with Words.* 2d ed. Edited by J. O. Urmson and Marina Sbisa. Cambridge, Mass., 1975.

Beardsley, Monroe. "Intentions and Interpretations: A Fallacy Revived." In *The Aesthetic Point of View: Selected Essays,* edited by Michael J. Wreen and Donald M. Callen, pp. 188–207. Ithaca, N.Y., 1982.

Carroll, Noël. "Art, Intention, and Conversation." In *Intention and Interpretation,* edited by Gary Iseminger, pp. 97–131. Philadelphia, 1992.

Carroll, Noël. "Anglo-American Aesthetics and Contemporary Criticism: Intention and the Hermeneutics of Suspicion." *Journal of Aesthetics and Art Criticism* 51.2 (Spring 1993): 245–252.

Davidson, Donald. "Moods and Performances." In *Inquiries into Truth and Interpretation,* pp. 109–121. Oxford, 1984.

Davidson, Donald. "Communication and Convention." In *Inquiries into Truth and Interpretation,* pp. 265–280. Oxford, 1984.

Davidson, Donald. "A Nice Derangement of Epitaphs." In *Truth and Interpretation: Perspectives on the Philosophy of Donald Davidson,* edited by Ernest Lepore. Oxford and New York, 1986.

Davies, Stephen. "The Aesthetic Relevance of Authors' and Painters' Intentions." *Journal of Aesthetics and Art Criticism* 41.1 (Fall 1982): 65–76.

Eaton, Marcia. "Truth in Pictures." *Journal of Aesthetics and Art Criticism* 39.1 (Fall 1980): 15–26.

Kjørup, Søren. "George Inness and the Battle at Hastings, or Doing Things with Pictures." *Monist* 58 (April 1974): 216–235.

Korsmeyer, Carolyn. "Pictorial Assertion." *Journal of Aesthetics and Art Criticism* 53.3 (Spring 1985): 257–265.

Lyas, Colin. "Wittgensteinian Intentions." In *Intention and Interpretation,* edited by Gary Iseminger, pp. 132–151. Philadelphia, 1992.

Matthews, Robert J. "Describing and Interpreting a Work of Art." *Journal of Aesthetics and Art Criticism* 36.1 (Fall 1977): 5–14.

Margolis, Joseph. "Meaning, Speakers' Intentions, and Speech Acts." *Review of Metaphysics* 19 (1973): 1007–1022.

Margolis, Joseph. "Art as Language." *Monist* 58.2 (April 1974): 175–186.

Meiland, Jack W. "The Meanings of a Text." *British Journal of Aesthetics* 21.3 (Summer 1981): 195–203.

Searle, John R. *Speech Acts: An Essay in the Philosophy of Language.* Cambridge, 1969.

Stalnaker, Nan. "Intention and Interpretation: Manet's *Luncheon in the Studio.*" *Journal of Aesthetics and Art Criticism* 54.2 (Spring 1996): 121–134.

Stecker, Robert. "The Role of Intention and Convention in Interpreting Artworks." *Southern Journal of Philosophy* 31.4 (Winter 1993): 471–489.

Stecker, Robert. "Art Interpretation." *Journal of Aesthetics and Art Criticism* 52.2 (Spring 1994): 193–206.

Tolhurst, William E. "On What a Text Is and How It Means." *British Journal of Aesthetics* 19.1 (Winter 1979): 3–14.

BETH ANN DOBIE

INTERPRETATION. [*This entry comprises two essays on different modes of interpretation in aesthetics:*

Interpretation in Art
Analytic Interpretation

Once interpretation becomes an integral part of aesthetics, a serious philosophical issue arises. Insofar as interpretation is subjective (that is, an interpretation of a work of art may say more about the subject doing the interpreting than it does about the work interpreted), can there be more than one correct interpretation of a single work of art? What is the standard for choosing the "correct" interpretation? The first essay here explains the contrast between those who hold that there can be a single correct interpretation of a work of art ("singularism") and those who argue that interpretation is always multiple ("multiplism"). The second essay explains in some depth the question whether the view that there is a plurality of interpretations of a single work of art is compatible with the idea that critics can argue and even defeat one another's interpretations of that same work. Other themes of interpretation, as well as the various types of interpretation, are discussed in separate entries (see, for example, Hermeneutics; Indian Aesthetics; *and* Theories of Art). *For related discussion, see* Appreciation; Criticism; Epistemology and Aesthetics; Evaluation; Iconography and Iconology; Intention; Law and Art, *article on* Law and Literature; *and* Value.]

Interpretation in Art

A theory of artistic interpretation may address one of a number of interrelated questions. For example, can an interpreted art object answer to more than one admissible interpretation? What constraints should there be for the range of admissible interpretations? What sort of thing is an interpreted art object? Are all the features of such art objects autonomous of the activity of interpretation, or are some features imputable in the very activity of interpretation? Further, how should one discern the meaning(s) or significance(s) of an interpreted art object? And what is the relation between a creator's intention(s) and the meaning(s) of the interpreted art object?

Since the answer to one of these questions may affect those of another, it is difficult to say in a non–question begging way what interpretation as such is. Correspondingly, the idea of interpretation resists strict rules of application. As Stuart Hampshire has suggested, interpretation is not a distinct concept, and so our starting point must be the typical circumstances in which the word is put to use (Hampshire, 1966, pp. 101–102). Yet, here we may indicate certain points of strategy.

First, we should distinguish between global and piecemeal approaches to interpretation. According to the global approach, one attempts to formulate a general theory of interpretation to range over all possible objects of interpretation, including those in all human and nonhuman sciences. These would include middle-sized objects, entites in formal studies like mathematics, the natural sciences, the social sciences, and objects in the arts. According to the global approach, interpretation in art, specifically, would be a special case of a global theory of interpretation. Such an approach would proceed from certain views about the nature of interpreted objects generally construed as well as about interpretation as such. One who offers a global theory of interpretation would likely offer a unified ontology of objects of interpretation (realist, constructionist, or some alternative).

Alternatively, according to the piecemeal approach, one attempts to formulate a theory of interpretation specific to a designated domain, leaving open the question of what commonalties there might be between accounts of different objects of interpretation in different domains. Here we shall leave the question open whether there is an adequate global theory of interpretation that captures the nature of interpretation as such or the nature of objects of interpretation generally.

One question about interpretation concerns its aims. Some theorists (notably Monroe C. Beardsley and E. D. Hirsch) urge that the aim of interpretation is always to fix upon *the* singular right interpretation (we may call this thesis "singularism"). Singularism assumes that whatever practical difficulties there might be in settling upon such an interpretation, it is the ideal aim of the activity of interpreting to reach for that one presumed interpretation. In contrast, some theorists (notably Joseph Margolis and Michael Krausz) suggest that more than one interpretation of a pertinent object is characteristically admissible, and that the singularist approach mistakens the nature of at least cultural objects and their interpretive practices. (We may call this thesis "multiplism.") The multiplist holds that such a mistake does not derive from a practical difficulty about actually settling on the presumed single right interpretation.

Distinct strategies may be deployed by proponents of each of these views. When confronted with a case in which it appears that an object of interpretation reasonably answers to two or more admissible interpretations, the singularist might attempt to install the singularist condition by *pluralizing* or bifurcating the object of interpretation. That is, the initially offered object of interpretation may be redescribed in such a way that there should result two or more objects of interpretation; and for each of those redescribed objects of interpretation there results one and only one interpretation of it. Consider such a pluralizing maneuver where the multiplist initially offers several putatively admissible interpretations of, say, Vincent van Gogh's *Potato Eaters*. The multiplist may urge several competing interpretations, for example, formalist, psychological, Marxist, and feminist. In response, the singularist might attempt to pluralize the object of interpretation by saying that these interpretations really are not interpretations of the same thing. In fine, they are interpretations of different intentionalized objects bearing different meanings or significances. The singularist might urge that what the formalist, for example, takes as the intentional object will be significantly different from what the Marxist takes as the intentional object, notwithstanding the fact that such intentionalized objects may be embodied in the same physical canvas.

The multiplist may counter that sometimes invoking this pluralizing maneuver is misplaced. Indeed, on occasion the multiplist may urge that one ought to deploy the contrary strategy of *aggregating* objects of interpretation. That is, he might urge that the intentionalized objects should be aggregated into one intentional object. Formalist and Marxist objects might be aggregated into a formalist-Marxist object. Or, Marxist and feminist objects may be aggregated into a Marxist-feminist object. Doing so raises questions about possible strains that would then be inherent in the resultant object, and the tolerances of such strains.

The strategies of pluralizing or aggregating an intentional object find their analogue in corresponding strategies of pluralizing or aggregating interpretations themselves. That is, confronting a multiplist condition, a singularist may seek to aggregate two or more interpretations to install a singularist condition. For example, a Marxist and feminist interpretation of the *Potato Eaters*, taken separately, may initially be taken to exemplify a multiplist condition, but the singularist might bring these together as one Marxist-feminist interpretation, claiming that the intentional object answers to that one interpretation. There seems to be no general rule-governed procedure for determining whether such maneuvers should be invoked. The appropriateness of doing so needs to be determined on a case-by-case basis.

In turn, prior ideological commitments may well motivate the choice of types of descriptions of intentional objects or their interpretations. For example, initially on finding no Marxist intentional object (that is, one articulated in terms of Marxist concepts) a Marxist may be tempted to reconstruct such an object in accordance with the general conviction that the truth of Marxism is ubiquitous at a level deeper than that initially exhibited. Such ideological commitments may well motivate the search for many different object types or interpretation types, such as feminist, liberal, conservative, Catholic, Protestant, Jewish, Hindu, Buddhist, African-American, Hispanic, or white.

As a strategy for identifying the single right interpretation the singularist might be tempted to appeal to the original intention of the creator of an art object. This suggestion runs into several problems. First, there are epistemological difficulties in recovering original intentions of creators. Second, the individual creator might well have had essentially ambiguous or incongruous intentions. Third, for those art objects that were collectively rather than individually created, it is unclear where one should look for intentions. Fourth, a work may embody emergent features that the creator, individual or collective, never intended. Indeed, art objects characteristically embody features that are unintended by their creator. So, the intentions of a creator need not locate or exhaust admissible interpretations of an art object. The original intention of a creator does not guarantee the singularist's requirement.

Does one impute or change certain features of an art object by interpreting it? The imputationalist view holds that the art object is the class of its interpretations, and that there is no art object independent of interpretation as such; it has no structure independent of interpretation. What is inter-

preted is not constituted independently of interpretation; interpretation constitutes art objects. Yet this need not mean that any particular interpretation on any given occasion may fully constitute its art object. Such a radical kind of imputationalism would result in arbitrariness. But according to a more moderate imputationalism, a given art object may be constituted within webs of interpretations with more or less complex histories. Art objects are altered by the histories of their interpretations. Since there may be an infinite number of interpretive moments, there may also be an infinite number of interpretations. But according to the moderate imputationalist view, this does not entail that there are no constraints on the range of admissible interpretations at any particular historical moment. There may be just a handful of such interpretations at any given historical moment. The moderate imputationalist holds that there are constraints on what is reasonably imputable, and they may be provided within the context of pertinent interpretive practices.

In contrast, the anti-imputationalist view of interpretation holds that art objects are fully constituted independently of interpretation as such. They cannot be constituted by interpretation. Whichever interpretations are admissible must be constrained by something other than interpretation itself. The anti-imputationalist view holds that if an interpretation fully constitutes its art object, there would be a population explosion of art objects. Consequently, no two interpretations of a given art object could compete. Art objects would multiply by the number of interpretations offered. Further, the anti-imputationalist holds that it must be possible for an interpretation to be wrong or inadmissible, and this implies that interpretations cannot impute their own art objects. Anti-imputationalists hold that it must be possible to identify and reidentify art objects on different occasions, and they charge that imputationalists are unable to satisfy this requirement. For interpretations to explain art objects and not be just responsive to them, art objects must be autonomous of interpretations. The admissible interpretation should capture the work's real nature, independent of interpretation. Anti-imputationalists may well concede that every act of interpretation is situated in a historical and cultural setting, and that interpretations of art objects have changed throughout history. But they are concerned that, on the imputationalist account, the art object is reduced to the status of a Rorschach inkblot, into which interpreters may read anything that might be entertained. Without some account of art objects independent of interpretation, one could not say whether an interpretation was faithful to the art object.

Moderate imputationalists, in turn, may well concede that their thesis entails that, over time, there may be an infinity of interpretations. But this does not amount to anarchism. Rather their thesis is consistent with the claim that, at any given time, there may be reasonable constraints on what in-

terpretations are admissible, and there may be reasonable constraints on what is imputable. It is only from a more global point of view that the claim of infinite interpretations is made to start with. From a more local point of view, the range of ideally admissible interpretations is not infinite. Structurally, the situation would look rather like this. For any set, there are a limited number of points in it—and there are an infinite number of sets. It follows, therefore, that globally there are an infinite number of points. Correspondingly, in the context of actually settling on admissible interpretations in a given historical moment, the range is not infinitely open. Rejecting the view that there must be one single right interpretation of an art object does not entail anarchism. There are intermediary stages that may constrain what is reasonable at a particular time. So, the moderate imputationalist and the anti-imputationalist may well agree that at a certain historical moment some interpretations of a given art object are inadmissible.

To provide pertinent constraints on the range of admissible interpretations, the moderate imputationalist may invoke the function of an interpretive practice, which may be said to supply the terms in which interpretations are admissible. They orient practitioners as to what features of its objects are significant or salient. Their rules, guidelines, values, or procedures indicate appropriate methods and maneuvers to be pursued. Further, the character of an interpretive practice is set within the context of its pursuit, which involves the realization of its aims as well as its continual reformulations in light of its achievements. Being a function of the consensual agreement of pertinent practitioners, the range of admissible interpretations is socially constituted. While there must always be a collection of individual persons constituting such a group, the rules, guidelines, values, or procedures of a practice are not embodied in any one or number of particular individuals in the group. What, in the last resort, makes an interpretation admissible is, as José Ferrater-Mora says, "the consensus, or agreement, that it is, indeed, acceptable or not. This consensus functions within the rules laid down, implicitly or explicitly, by the community of researchers by virtue of habits engendered by a multitude of common experiences" (Ferrater-Mora, 1976, pp. 51–52). Since the range of admissible interpretations is a function of its place in a given practice, and since a practice is consensually defined, so the range of admissible interpretations is consensually defined. In sum, the very idea of a range of admissible interpretations is predicated on a practice as a social rather than a person-specific or occasion-specific condition.

The differences between the imputationalist and the anti-imputationalist views may be brought out by considering the fact that, with the introduction of new theories, lexica, and idioms, postdictively one characteristically comes to see the plausibility of certain new interpretations. This would have been true of Marxist or Freudian interventions, for example. Anti-imputationalists might well concede that peo-

ple do come to see the same things in different ways under different historical circumstances. But they would hold that has nothing to do with the nature of what is seen. In contrast, imputationalists hold that, in light of the fact that the interpreted object may be come to be seen differently in virtue of intervening interventions, the object is intentionalized differently and so its nature is different.

The anti-imputationalist and the imputationalist might well agree that generally one should be historically responsible in settling on admissible interpretations. But the very construal of history itself is implicated in the present controversy. The question whether history is imputed or not remains open, and so the appeal to historical authenticity as such does not advance the discussion.

Finally, it should be noted that the imputationalist view does not entail multiplism, for the constraints that an imputationalist might invoke can still be so restrictive as to narrow the range of admissible interpretations to a limit of one. In turn, the anti-imputationalist view does not entail singularism, for the features of the interpreted object—which are thought to be autonomous of interpretation—may, for independent reasons, still answer to more than one admissible interpretation. The controversy between singularists and multiplists, on the one hand, and between imputationalists and anti-imputationalists, on the other hand, are therefore distinct and logically detachable.

BIBLIOGRAPHY

Barnes, Annette. *On Interpretation: A Critical Analysis.* Oxford and New York, 1988.

Beardsley, Monroe C. *The Possibility of Criticism.* Detroit, 1970.

Ferrater-Mora, José. "On Practice." *American Philosophical Quarterly* 13 (January 1976): 51–52.

Gadamer, Hans-Georg. *Truth and Method.* 2d rev. ed. Translation revised by Joel Weinsheimer and Donald G. Marshall. New York, 1989.

Hampshire, Stuart. "Types of Interpretation." In *Art and Philosophy,* edited by Sidney Hook. New York, 1966.

Hirsch, E. D., Jr. *Validity in Interpretation.* New Haven, 1967.

Iseminger, Gary, ed. *Intention and Interpretation.* Philadelphia, 1992.

Krausz, Michael. *Rightness and Reasons: Interpretation in Cultural Practices.* Ithaca, N.Y., 1993.

Krausz, Michael, ed. *The Interpretation of Music: Philosophical Essays.* Oxford, 1993.

Margolis, Joseph. *Art and Philosophy.* Atlantic Highlands, N.J., 1980.

Margolis, Joseph. *Interpretation: Radical but Not Unruly: The New Puzzle of the Arts and History.* Berkeley, 1995.

MICHAEL KRAUSZ

Analytic Interpretation

Is it possible that toleration of a plurality of sometimes incompatible interpretations of artworks is a permanent feature of critical practice, while, nevertheless, genuine confrontation and significant defeat take place between critics?

In order to answer this question one needs to address two others: What is it for artworks to tolerate a plurality of sometimes incompatible interpretations? What is it for genuine confrontation and significant defeat to take place between critics?

Some argue that the toleration in question requires of each interpretation that it be equally acceptable. Moreover, confrontation and defeat between critics require defeat of one interpretation by another, the ultimate aim being to establish one of the interpretations as true and interpretations incompatible with it as false. On such an understanding, the answer to the original question would seem to be no. Critical defeat of the kind specified and plurality of sometimes incompatible interpretations would seem to be opposing ideals.

It is not difficult to see why. Imagine that critics offer genuinely incompatible interpretations. The ideal of confrontation and defeat at issue would require critics to directly challenge and defeat interpretations incompatible with their own. But if the critical aim is to eliminate rival interpretations, it does not seem that a permanent feature of critical practice is to tolerate a plurality of sometimes incompatible interpretations. Rather, the ideal would be to make such tolerance a transient feature of critical practice.

If, on the other hand, we suppose that critical practice tolerates a plurality of interpretations, some of which seem incompatible, it is difficult to see how critical interpretations can genuinely confront and significantly defeat one another. If toleration of a plurality of interpretations requires that the interpretations so tolerated be equally acceptable, and among the acceptable interpretations are ones that would be incompatible if interpretations were true and false, then acceptability, it would seem, cannot be a matter of truth. If, however, interpretations are not true, but merely plausible or interesting, then alternative interpretations are not genuinely incompatible. It would, therefore, not be possible for critics to defeat critical interpretations incompatible with their own, for there would be no interpretations that fit this description. The tendency for those who accept the kind of multiplicity in question to deny defeasibility and for those who affirm defeasibility to deny this kind of multiplicity would appear to be justified.

We seem to be faced with a simple choice: either the incompatibility between apparently incompatible interpretations is only apparent or not of a worrisome kind—as it would be if interpretive remarks were, for example, hypotheses aiming at plausibility or interest; or, if interpretations were prescriptions or were reports of seeings-as—or the incompatibility between such interpretations is real—as it would be if interpretive remarks were ordinary statements taking the values true and false. If the incompatibility is only apparent, interpretive remarks do not really attribute incompatible features to the work and, hence, seemingly incompatible interpretations can coexist, but one interpretation does not confront and significantly defeat another. If the incompatibility is real, then incompatible interpretive

remarks do actually attribute incompatible features to the work and significant confrontation and defeat are possible, but long-term coexistence of genuinely incompatible interpretations is not.

We are faced with this choice, however, only if we accept certain assumptions, about what it is for critical practice to tolerate a plurality of sometimes incompatible interpretations, and what it is for genuine confrontation and significant defeat to occur between critics. If, for example, a kind of genuine confrontation and significant defeat between critics were possible that did not require defeat of one interpretation by another, did not require critics to establish the falsity of rival critical interpretations, would it be possible for critical practice to tolerate a plurality of sometimes incompatible interpretations of artworks while genuine confrontation and significant defeat took place between critics? That is, would it be possible for the incompatibility between interpretations to be real while critical practice tolerated a plurality of sometimes incompatible interpretations?

What kind of confrontation and defeat between critics would not involve the defeat of one interpretation by another? Would such confrontation and defeat be significant?

Suppose a critic writes: "*Wuthering Heights* is a work of sober realism." If we can understand the critic to be stating that *Wuthering Heights* is a work of this kind and also to be claiming to know that it is, then the critic's utterance can be subject to a certain kind of defeat, defeat by a relevant counterpossibility. This kind of defeat, however, challenges only one element of the utterance, namely, the utterance qua knowledge claim; it does not challenge the utterance qua interpretive claim.

A relevant counterpossibility is directed at the critic's claim to know, in the case at hand, that *Wuthering Heights* is a work of sober realism. Showing that a critic has not ruled out a relevant counterpossibility is one way of showing that the critic is not justified in his or her knowledge claim—it is one way of defeating the knowledge claim.

"*Wuthering Heights* is a perverse fantasy," a claim entailed by a rival critic's claim to know, would, for example, be a relevant counterpossibility to the claim that "*Wuthering Heights* is a work of sober realism" if and only if: (1) if it is not logically possible for these claims jointly to be true; and (2) the critic has adequate warrant for thinking that "*Wuthering Heights* is a work of sober realism" is true only if the critic has adequate warrant for thinking that "*Wuthering Heights* is a perverse fantasy" is false. One could consider those two claims to be relevant counterpossibilities to each other if they cannot be jointly true, and the critic would have adequate warrant for thinking one true only if the critic had adequate warrant for thinking the other false.

Showing that a critic has not ruled out a relevant counterpossibility directly challenges the critic's claim to know and can defeat that claim. That challenge is, however, compatible with the legitimate presence in criticism of multiple, even incompatible, interpretations, since, even if successful, it only defeats the critic's claim to know; it does not defeat the other, interpretive, element of the utterance. Two competing interpretations may thus be tolerated even if at most only one of them can be true. Given what is known, the interpretations are equally acceptable. Either of them, but not both of them, could be true.

This sort of defeat requires that the critic who writes "*Wuthering Heights* is a work of sober realism" is both stating that it is such a work and claiming to know that it is. If we understand the critic to be doing something other than this, we do not get defeat by a relevant counterpossibility.

For example, the critic who claims that it is plausible that *Wuthering Heights* is a perverse fantasy is not typically challenging and defeating the critic who claims that it is plausible that *Wuthering Heights* is a work of sober realism, assuming here that both interpretations have roughly the same textual support. The critic who reports that he or she experiences *Wuthering Heights* as a work of sober realism would not be challenged or defeated by the critic who experiences it as a perverse fantasy. Nor would the critic who prescribes that we view it as a work of sober realism ordinarily be challenged and defeated by the critic who prescribes that we view it a perverse fantasy.

If defeat by a relevant counterpossibility is a possible mode of defeat with regard to critics' interpretive utterances, is it a significant mode of defeat? That is, is such defeat an important, and a typical kind of, defeat within critical practice?

If defeat by a relevant counterpossibility exists, critics must not only be making certain kinds of statements about the work, they must be claiming to know that their statements are true. If they claim to know that a work is a given way, and their claim is correct, then an interpretive claim incompatible with theirs would be false. Critics who claim to know seem, therefore, not to be aiming in their critical confrontations at *merely* presenting a relevant counterpossibility but rather to be aiming at defeating rival interpretations by showing them to be false, for critics who attempt to establish their interpretations as true can plausibly be understood to be attempting to establish the falsity of interpretations incompatible with their own. If critics succeed in their attempts, however, they would automatically succeed at presenting a relevant counterpossibility. Critics' actual attempts must in a sense contain the attempt to provide a relevant counterpossibility; that is, a rational critic could not claim to have meant *not* to present such a counterpossibility.

If defeat by a relevant counterpossibility is a typical mode of defeat, it is not because critics typically aim at just this defeat. It would be a typical mode of defeat, however, if despite their aims, it turned out that presenting a relevant counterpossibility was often all that critics did. If this were

all they did, then critics, while directly challenging other critics, would, when successful, defeat only the opposing critics' knowledge claim and not the interpretive claim.

Presenting relevant counterpossibilities against other critics would be all that critics frequently did if critics were not often able to rule out alleged relevant counterpossibilities against themselves—if they did not frequently have adequate warrant for thinking that an interpretive claim, incompatible with theirs, was false. But is it not true that critics are often able to rule out such counterpossibilities? Do they not frequently have adequate warrant for thinking a rival interpretive claim false?

Whether or not they have the requisite warrant and ability, or have it frequently, is subject to debate. Those arguing against the importance of a defeat by a relevant counterpossibility might argue for the common availability of the requisite warrant and ability.

The ability to establish truth and falsity depends in a problematic domain (e.g., regarding interpretation) on it being possible to establish criteria for truth and falsity and on these criteria being satisfied. Criteria have, of course, been proposed. For example, some hold that an interpretive claim is true if the work not only could have the properties the interpretation attributes to it (the interpretation is compatible with the work if the work, say, is identified syntactically), but having such properties also does not conflict with successfully realized artistic intention. If, for example, given these criteria, it could be ascertained that Emily Brontë successfully realized her intention to make *Wuthering Heights* a work of sober realism, then it would be true that it was such a work. If this were true, then it would be false that it was a perverse fantasy. *Wuthering Heights* would not tolerate incompatible interpretations even though it might tolerate multiple interpretations that were compatible with the true one. If critics commonly knew that interpretive claims, such as this, were true, then defeat by a relevant counterpossibility would be a transient option in critical practice.

If the criteria for truth and falsity operating within critical practice were uncontroversial and if the evidence available about artworks were often sufficient for establishing, on the basis of those criteria, the truth of one interpretation over its rivals, then defeat by a relevant counterpossibility would be a transient, and not a significant, mode of defeat in the critical realm. But are either of these conditions met? The criteria for establishing the truth and falsity of an interpretive claim are contested. The ongoing debates between intentionalists and anti-intentionalists suggest that critics do not agree on the criteria. Even within each of the opposing camps, there are divergent views as to what would be required. Further, if there were agreement on criteria, it is not clear that the evidence available about individual artworks often would be sufficient for establishing which of two incompatible interpretations was true, which false. If the evidence were not sufficient, then critics, although wanting more than defeat of a rival critic's knowledge claim, might not be entitled to more. Defeat by a relevant counterpossibility would be a standard mode of defeat. Tolerating a plurality of sometimes incompatible interpretations would be a permanent feature of critical practice.

There are critics for whom the truth of an interpretation, if truth is linked in some crucial way to artistic intention, is not of paramount importance. These critics want interpretations that make the work a significant work or a relevant work or an effective work, assuming, of course that there is some legitimate way of identifying the work that is independent of how the artist intended the work to be understood. Such critics may be prepared to acknowledge that their interpretations are not true, if truth requires, for example, not conflicting with successfully realized artistic intent, but, nevertheless, claim an acceptability for them based on their making the work a more effective or relevant or significant work. Their claim is that a work, identified in the appropriate way, is a work of such and such a kind, that is, being of such and such a kind is possible for this text in this language, even though being of such and such a kind is in conflict with the way the work is if artistic intent is considered. Acceptability is made independent of truth, when truth is linked to artistic intent.

If one stipulates that a work can only be ways that do not conflict with what the artist successfully realized in his or her work—although the artist could have written a perverse fantasy using the same text in the same language, she wrote a work of sober realism—then even if it would have been a more significant work if she had written a perverse fantasy, since it is not such a work, the interpretation of it as a perverse fantasy cannot make the work we have a more relevant work. On this view an interpretation genuinely incompatible with the true interpretation would not be acceptable and hence not tolerated.

If the work were identified in a different way, so that one could claim that the interpretation of the work that conflicted with successfully realized intent was an acceptable one, then would the acceptable, but not true, interpretation be genuinely incompatible with the true interpretation? Would the truth of one necessitate the falsity of the other? If it would not, the presence of alternative criteria for acceptability would show only that tolerating multiple interpretations is a permanent feature of critical practice; it would not show that tolerating a plurality of sometimes incompatible interpretations was such a feature.

It can be true, for example, that a work is a perverse fantasy if this means that reading it as a perverse fantasy is a natural way of reading it and a way, moreover, that makes the work a more significant work than do other ways of reading it. It could also be true that if it is a work of sober realism, the writer successfully realized her intention to make

it such a work. Viewing the interpretive assertions in these ways makes them compatible rather than incompatible claims about the work. This suggests that whether tolerating a plurality of sometimes incompatible interpretations is a permanent feature of critical practice depends not on whether there are alternative criteria for acceptability but rather on whether there is an uncontroversial standard for establishing truth and falsity, a standard that, moreover, often enough yields unequivocal results.

BIBLIOGRAPHY

Barnes, Annette. *On Interpretation: A Critical Analysis.* Oxford and New York, 1988.
Beardsley, Monroe. *The Possibility of Criticism.* Detroit, 1970.
Carroll, Noël. "Anglo-American Aesthetics and Contemporary Criticism: Intention and the Hermeneutics of Suspicion." *Journal of Aesthetics and Art Criticism* 51.2 (Spring 1993): 245–252.
Derrida, Jacques. "Signature Event Context." *Glyph* 1 (1977): 172–197.
Fish, Stanley. *Is There a Text in This Class? The Authority of Interpretive Communities.* Cambridge, Mass., 1980.
Levinson, Jerrold. "Intention and Interpretation: A Last Look." In *Intention and Interpretation,* edited by Gary Iseminger, pp. 221–256. Philadelphia, 1992.
Margolis, Joseph. *Art and Philosophy.* Atlantic Highlands, N.J., 1980.
Stecker, Robert. "Incompatible Interpretations." *Journal of Aesthetics and Art Criticism* 50:4 (Fall 1992): 291–298.

ANNETTE BARNES

INTUITION. *See* Husserl; *and* Novalis.

IRIGARAY, LUCE (b. 1932), Belgian psychoanalyst and philosopher. Since 1985, when two major works by Luce Irigaray were published in English, her name has become familiar within anglophone feminist theory, critical theory, and associated fields. She has published extensively, often in volumes of collected essays. Transcriptions of spoken presentations, with open and digressive structures, and the use of poetics, particularly in meditations on the mediation present in subject-subject relationships, are integral to her published oeuvre. Her feminist, left-wing politics can be evidenced not only in the content of her writing, but importantly for this discussion, in its structures. For example, she eschews formal academic style in favor of a writing that will "unpick" the authoritative and patriarchal structures within academic thought: "It is surely not a matter of interpreting the operation of discourse while remaining within the same type of utterance as the one that guarantees discursive coherence" (Irigaray, 1985b, p. 78). This activist approach— her practice of theory—is consistent and is not restricted to her writing. It can be seen, for example, in her engagement with various women's groups and struggles, and with the Italian Communist Party. It can also be seen in her attention to women and women's relationships with men, rather than to qualities of femininity that can be applied to—or appro-

priated by—men, once more excluding women (this marks a clear difference between Irigaray and Hélène Cixous). Some critics in the United States have seen this as evidence of essentialism, but this criticism ignores, first, structural political differences between the United States and Europe, and second, the evidence of Irigaray's own writing on the need to attend to philosophical structures in order to effect cultural, religious, legal, and sexual change.

Although Irigaray frequently works across and between disciplines in her approach to a given subject, it is the aspects of her writing that touch on philosophy (particularly ethics), psychoanalysis, linguistics, and female subjectivity and sexuality that can be most easily located within her writings and worked on by others. This has happened within literary criticism, in particular, where Irigaray's insights have been used to analyze the approaches of women writers and the characters they create. Surprisingly, however, an Irigarayan aesthetic is rarely discussed, either as a product of her own work or as a possibility for other writers and artists. There are three main reasons for this: first, Irigaray infrequently mentions art or attends to aesthetics as a distinct discipline; second, most critics draw on the overt *content* of her work, glossing over or bypassing its *form,* and thereby render a focus on or deep analysis of an Irigarayan aesthetic unlikely; and third, Irigaray's strategies for the form her writing should take are multiple, political, and highly resistant to critical or academic containment or categorization. Indeed, if anything can, this latter point summarizes Irigaray's writerly "style." From the loosely conversational, through the tightly structured, to the acute, poetic evocation of liminal experience, Irigaray chooses the means appropriate to her focus. This strategy can, where necessary, extend to sentence structures, grammatical usage, and the employment of punctuation.

These three points have had direct effects on critical evaluation of an Irigarayan aesthetic. For example, within art history, theory, or criticism there are few writers who provide anything but a tangential mention or superficial understanding of Irigaray's work. Two recent major anthologies of feminist writings on aesthetics have only passing references to Irigaray. More pointedly, some of her critics (most prominently, Martin Jay) consider her antivisual, a criticism based almost exclusively on readings of *Speculum of the Other Woman* (Irigaray, 1985a) and *This Sex Which Is Not One* (Irigaray, 1985b), and bypassing work such as "Flesh Colors" (in Irigaray, 1993b, pp. 153–165) or the intense visual pleasure evoked in *Elemental Passions* (Irigaray, 1992). Those who have written on Irigaray's style restrict their attention to her interrogation of other philosophers and psychoanalysts, and their discussion to her use of mimesis. (For example, Weed, 1994, underestimates the politics of Irigaray's approaches to writing and ignores the more lyrical texts, with the result that Irigaray's choices do seem—in Weed's account—to be more about surface style than polit-

ical strategy.) All commentators on Irigaray's concept of the "morphology" of women take it to refer to biology alone, ignoring the use of the term in linguistics (the field of one of Irigaray's doctorates) to refer to the structure of words, thus moving the debate toward essentialism and away from aesthetics. These slippages make the critical consideration of an Irigarayan aesthetic less likely.

Of the writers brought together under the somewhat misleading anglophone designation "new French feminism," it is Irigaray who is most clearly working toward a developing feminist aesthetic (Battersby, 1991, p. 35). It would, however, be a mistake when considering her approach to aesthetics to concentrate only on her writings on art or artists: first, they are often mentioned for a particular reason; and second, her broader writing can be highly suggestive for discussions of aesthetics (as I shall discuss below). In "A Natal Lacuna" (Irigaray, 1994a), Irigaray is looking at artist Unica Zürn's work symptomatically and explicitly from the point of view of an analyst whose prime interest is in the psychoanalytic structures demonstrated by the artist, in order to shed light on Zürn's state of dereliction and ultimate suicide. In "Flesh Colors," Irigaray discusses the importance of painting in the analytic scenario, as a therapeutic process. In *Je, Tu, Nous: Toward a Culture of Difference* (Irigaray, 1993a, p. 47) and *Thinking the Difference: For a Peaceful Revolution* (Irigaray, 1994b, p. 12), she advocates placing "attractive" images of mothers and daughters in public places, and says that Christian mothers should place images of the Virgin Mary and Anne, Mary's mother, in their daughters' bedrooms. These references and suggestions can appear to be highly restrictive or naive in their understandings of visual representations. When regarded, however, as points on a web of understandings that spread across her writing, their strategic purpose becomes clearer.

"How Can We Create Our Beauty?" (Irigaray, 1993a, pp. 107–111) provides signals for such a web. A short, polemical essay, it begins in terms that may seem utopian to those involved with late twentieth-century visual art:

> Very often, when looking at women's works of art, I have been saddened by the sense of anguish they express, an anguish so strong it approaches horror. . . . The experience of art, which I expected to offer a moment of happiness and repose, of compensation for the fragmentary nature of daily life, of unity and communication or communion, would become yet another source of pain, a burden. (p. 107)

She then outlines why she thinks women make images of pain, and how they could create beauty. She discusses the cathartic nature of such work, and the distinction made between creativity and procreativity. The result is that "as women, we have thus been enclosed in an order of forms inappropriate to us. In order to exist, we must break out of these forms" (p. 109)—an enterprise fraught with danger, as well as the possibility of rediscovering our flesh, identity,

colors, and forms. Intrinsic to this is a concept of the Divine for women and its representation.

From this essay, three salient points emerge: (1) the general issue of flesh, body, female morphology, and its representation; (2) the nature of women's creativity, and in close relation, subjectivity; and (3) the representation of what Irigaray terms the "female Divine" and its inevitable adjuncts, universality and transcendence. Running through these discussions, as they fold into Irigaray's wider writings, are two other points: (4) the necessity for productive acknowledgment of female genealogies (two-way interchange between mother and daughter, and its concomitant, exchange between women); and (5) in very close relation to this, a notion of fulfillment or "becoming" for women. Without any of these, women's beauty is not possible: indeed, these in conjunction would be productive of and allow for the performativity of women's beauty. The Irigarayan discussion of beauty, therefore, is not about defining a new aesthetic or new criteria by which evaluative judgments can be made, nor is it an essentialist notion of a female aesthetic that has been overshadowed by a male aesthetic and that only requires illumination in order to become visible: "In other words, the issue is not one of elaborating a new theory of which woman would be the *subject* or the *object*, but of jamming the theoretical machinery itself, of suspending its pretension to the production of a truth and of a meaning that are excessively univocal" (Irigaray, 1985b, p. 78). For Irigaray, beauty for women is a potential state of being that can only come about as a result of rethinking political and cultural discourse. Any Irigarayan aesthetic would be located within this rethinking. Her discussion of beauty is about making possible an order of discourse that would in and of itself, and inevitably, be productive of beauty. It is a discussion that requires the reader to think "differently" (to think *in* her or his state of difference)—to reconceptualize what might be productive of beauty, and what might constitute the transcendental and the universal: "It is more a question of breaking out of the autological and tautological circle of systems of representation and their discourse so as to allow women to speak their sex. . . . Man's language separates her from her mother and from other women, and she speaks it without speaking in it" (Irigaray, 1991, pp. 97, 101).

It should be clear by now that an "Irigarayan aesthetic" will not be found in the reproduction of certain methodologies in the studio, or adherence to one or another "style" of imagery or production. Irigaray is scornful of an unreconsidered, disengaged aesthetics, and of the thought that art might simply attempt to illustrate, without attempting to resolve through its discourses the ethics of sexual difference:

> For a dialectic of the couple to occur, we need an art of perception that cannot be reduced either to a pure innerness or a pure outerness but passes ceaselessly from one to the other. This . . . is lacking in our culture. . . . When art does show an awareness

of the flesh this does not mean that the ethics of the couple has been resolved. It gives the nod to the importance of the issue and to its valuable aesthetic potential. There is a danger that ethics should become a part of aesthetics and seen as secondary to the life of the people, pleasant but not essential to spiritual development. This avoids the need to go beyond contradictions and oppositions and achieve a truly sexualized thought.

(Irigaray, 1993b, pp. 144–145)

There is no possibility therefore of discussing what an Irigarayan aesthetic looks like, or how certain artists might be working within one. Rather, it is necessary to search for the places within particular artworks (the practices that form them as well as how they function) that are performative of Irigarayan thinking and that will thus begin to be productive of beauty. Only with this approach do her comments on art (such as those on images of mothers and daughters, mentioned above) function within her web of meanings, rather than achieve the overvaluation they might otherwise be granted in a more conservatively academic structure. They can then be seen as suggestions for developing languages, from within present possibilities that are capable of the cultural articulation of mediation between women. Her concept of the "dialectic of the couple" (including its mediation through vision) appears at its most advanced in *Elemental Passions*. This work addresses the need to progress from subject-object relationships to subject-subject relationships, which allow for transcendence and in which all senses are engaged:

> The most secret part of your face offered without anything held back. Fearlessly welcoming its being revealed in another's gaze. Where your most impenetrable part appeared uncovered to you, where your most unapproachable part returned to you. . . . And you did not fix your gaze on the nearest or the furthest point, but proximity was seen through you. (Irigaray, 1992, p. 99)

Reading across Irigaray's work also allows for the drawing in of "extra-art" concepts. For example, Irigaray's insights into the psychoanalytic implications of the gestures of little girls when they miss their mothers (including whirling dances and playing with dolls; Irigaray, 1993b, pp. 91–104) have rich implications for feminist understandings of some women artists' use of space or some of their representations of women's bodies (Robinson, 1996). Such revaluation of terms (e.g., gesture) extant within art writing can play richly into a radical rethinking of art practices. Irigaray's discussion of painting in the analytic scenario broadens in a subtle and complex move to discuss the colors of the body (what she terms "chromo-soma") and art as an enabler of affective relationships:

> Creativity is a goal only for someone who gives priority to *making an object*, to anality(?). For the sublimation of genitality—a dimension that is still unknown to us, particularly because it has been buried beneath the idea that reproduction is our duty as women and the sole way we can achieve our sexuality—artistic

creation represents a means, not an end. This is the indispensable road to take not only for psychoanalysis but, more generally, in every relationship, if we are to realize an art of the sexual that respects the colors, the sounds, and the forms proper to each sex.

(Irigaray 1993b, pp. 164–165)

Likewise, Irigaray's discussions of women's mucus and its function as mediation between women and between a woman and herself (Irigaray, 1991, pp. 105–117) can be seen to offer potential for the analysis of the use of paint by women, marking a clear gendered difference from the popular psychoanalytic understanding of paint as feces. Add this to Irigaray's insights into gesture, her explication of the universal as mediation (Irigaray, 1993b, pp. 127–149), her complex outlining of the relation between flesh and color, within the framework laid down by "How Can We Create Our Beauty?" (in Irigaray, 1993a), and there is a rich conceptual network for a developing feminist aesthetic of painting based on a radical, "sexuate" (as Irigaray terms it) rethinking of cultural structures and their effects on production.

Reading across Irigaray's writing is important for another reason. Throughout, there are moments, frequently implicit, of the "revaluing" of terms, where she not only reevaluates certain concepts but rebuilds the terms with shifted meanings:

> This brings to mind the political stake—in the restricted or generalised sense—of this work. The fact that women's "liberation" requires transforming the economic realm, and thus necessarily transforming culture and its operative agency, language. Without such an interpretation of a general grammar of culture, the feminine will never take place in history, except as a reservoir of matter and of speculation. (Irigaray, 1985b, p. 155)

Thus, where a concept such as "the universal," as it is generally used, has been exposed as being partial, patriarchal, and applicable by and large only to men, Irigaray will explore a viable concept of the universal for women; she will shift what it is that is meant when the word *universal* is used, in order to demonstrate a concept of universality that is productive for women. Such shifts may become explicit only at particular points in her writing, and remain implicit in her uses of the words elsewhere. So, "The Universal as Mediation" (Irigaray, 1993b, pp. 127–149) accounts for her use of those terms elsewhere. Other essays such as "The Limits of Transference" (Irigaray, 1991, pp. 105–117) focus on aspects of mediation between women; writings like *Elemental Passions* (Irigaray, 1992) are a representation of this form of becoming in the couple—the subject-subject relationship. It is only when reading across the whole of her work that the full revolutionary extent of her cultural and political analysis can be comprehended. It is also only then that her own aesthetic—the continuum between the overtly present analyses (the "content" of her writings) and the suggestively productive form that they take—becomes apparent.

Irigaray does not offer her autobiography as a matter of import, but she has claimed French and Italian ancestry. She was born in Belgium. After gaining a master's degree and teaching, she moved to Paris in 1964, where she earned a doctorate in linguistics in 1968 (title: "Le langage des déments") and another in philosophy in 1974 ("Speculum de l'autre femme"). The latter was the result of her studies at the École freudienne under Jacques Lacan. Despite the thesis getting high honors, its attack on the phallocentric nature of Freudian psychoanalysis led to her being barred from teaching at the University of Paris VIII and from the École freudienne. In this professional rejection and its lasting ramifications, we can see a crucial reason for the relative lack of academic credence granted Irigaray until recently in France when compared with her standing in the United Kingdom, Ireland, North America, Italy, Holland, and Australia. In the 1960s she was granted a position with the Centre national de la recherche scientifique, where she is now director of research. (See Stephenson, 1991, for fuller professional details.)

[*See also* Feminism; Freud, *survey article; and* Lacan.]

BIBLIOGRAPHY

Works by Irigaray

Elemental Passions. Translated by Joanne Collie and Judith Still. London, 1992.

The Irigaray Reader. Edited by Margaret Whitford, translated by David Macey et al. Oxford and Cambridge, Mass., 1991.

Je, Tu, Nous: Toward a Culture of Difference. Translated by Alison Martin. London and New York, 1993a.

"A Natal Lacuna." Translated by Margaret Whitford. *Women's Art Magazine* 58 (May/June 1994a): 11–13.

Sexes and Genealogies. Translated by Gillian C. Gill. New York, 1993b.

This Sex Which Is Not One. Translated by Catherine Porter with Carolyn Burke. Ithaca, N.Y., 1985b.

Speculum of the Other Woman. Translated by Gillian C. Gill. Ithaca, N.Y., 1985a.

Thinking the Difference: For a Peaceful Revolution. Translated by Karin Montin. London and New York, 1994b.

Other Sources

Battersby, Christine. "Situating the Aesthetic: A Feminist Defence." In *Thinking Art: Beyond Traditional Aesthetics,* edited by Andrew Benjamin and Peter Osborne, pp. 31–43. London, 1991.

Berg, Maggie. "Luce Irigaray's 'Contradictions': Poststructuralism and Feminism." *Signs: Journal of Women in Culture and Society* 17.1 (Autumn 1991): 50–70.

Kruse, Felicia E. "Luce Irigaray's *Parler Femme* and American Metaphysics." *Transactions of the Charles S. Peirce Society* 27.4 (1991): 451–464.

Robinson, Hilary. "Irigaray's Imaginings." *Women's Art Magazine* 61 (November/December 1994): 20.

Robinson, Hilary. "Louise Bourgeois' 'Cells': Gesturing towards the Mother." In *Museum of Modern Art Papers,* vol. 1, *Louise Bourgeois,* edited by Ian Cole, pp. 21–30. Oxford, 1996.

Stephenson, Katherine. "Luce Irigaray." In *French Women Writers: A Bio-Bibliographical Source Book,* edited by Eva Martin Sartori and Dorothy Wynne Zimmerman, pp. 229–243. New York, 1991.

Weed, Elizabeth. "The Question of Style." In *Engaging with Irigaray: Feminist Philosophy and Modern European Thought,* edited by Carolyn Burke et al., pp. 79–109. New York, 1994.

Whitford, Margaret. *Luce Irigaray: Philosophy in the Feminine.* London and New York, 1991.

Whitford, Margaret. "Woman with Attitude." *Women's Art Magazine* 60 (September/October 1994): 15–17.

Xu, Ping. "Irigaray's Mimicry and the Problem of Essentialism." *Hypatia* 10.4 (Fall 1995): 76–89.

HILARY ROBINSON

IRONY. Irony has had, since at least Socrates, a general philosophical significance, and contemporary conceptions of irony have infiltrated an imposing array of disciplines, including not only literary criticism and aesthetics, but also sociology, psychology, political theory, ethics, and even epistemology. Colloquial uses of the term now cover a similarly wide spectrum of meanings ranging from the domain of sarcasm to that of chance, from traditional dictionary definitions ("saying something different from what is meant"), through "dramatic irony" (Oedipus vows to bring the culprit to justice for crimes he, Oedipus, has committed), to so-called situational ironies consisting of almost any perceived incongruity or coincidence of events ("Isn't it ironic that . . . ?").

The Greek *eirōnia (eirōn)* is of uncertain origin and etymology, and even more etymologies have been proposed throughout the centuries than definitions of the term. First extant in Aristophanes' *Clouds, eirōn* seems to have been primarily a term of abuse meaning dissembler or deceiver. Plato's Socrates is similarly criticized by his opponents in the dialogues as an *eirōn,* although it is widely held that the personality of Socrates transformed the meaning of the term into the less pejorative definition it soon found in Aristotle. In the absence of the latter's treatise on comedy, we can summarize his *Rhetoric, Eudemian Ethics, Magna Moralia,* and *Nicomachean Ethics* by saying that Aristotle situates truthfulness as a mean between the self-depreciating *eirōn* and the boaster *(alazon).* While not losing its pejorative association with deception, and said to be liable to provoke ire in its audience, irony is also said by Aristotle to be more suitable to the gentleman than buffoonery. Indeed, in those who use understatement and avoid ostentation, like Socrates, irony in moderation is said to be desirable, and preferable to boasting. Yet too great or obvious an understatement of one's worth, Aristotle specifies, is also potentially boastful.

So-called blame-by-praise, and vice versa, became firmly associated with irony only with the classical Latin *ironia.* In general, a tendency to emphasize *either* the ironist's scornful superiority *or* the ironic self-depreciation—his or her role as either victimizer or victim—may be regarded as defining some important features of its history. For example, although self-depreciating irony was certainly known in the Middle Ages, in theory through Aristotle and in practice

more ubiquitously, it was more common to stress its derisive and mocking effects, and only in the Renaissance was specifically Socratic rhetoric more generally resurrected from its Greek origins to stand as a different, more sustained, and often "philosophical" mode alongside the medieval trope.

After Aristotle, the figure of the ironist seems to have reverted for a time to a pejorative usage. The term was applied by Demosthenes to those who avoided military service and taxation, and by Theophrastus in his *Characters* to lovers of deception. Both Epicureans and Stoics held irony in disdain, and even the Cynics, according to Zoja Pavloskis, "were too harsh to take a kindly view of the refined weapon of irony, preferring to it sardonic laughter" (Pavloskis, 1968). Anaximenes of Lampsacus (c.380–320 BCE) proposed what was to become the familiar definition of irony as a trope in which sense was contrary to surface meaning, inherited as *ironia* by Cicero, Quintilian, and others, and transmitted by grammarians such as Donatus (fourth century CE) and Isidore of Seville (c.570–636 CE) to medieval scholars. Although medieval authors rarely, if ever, seem concerned with more extensive "Socratic" definitions, Quintilian himself distinguished, beyond the trope and the figure (an extended trope, related to the trope as allegory is to metaphor), a meaning of irony associated with "a way of life," for which Socrates served as exemplum. Horace also adopted in his *Satires* a highly conscious self-depreciating irony connected, either directly or indirectly, to Plato and Aristotle, whereas Cicero explicitly praised Scipio as an ironist in the Socratic vein. At this point, then, we may observe an at least provisional rehabilitation of irony, noting further that what was first achieved for its dignity by association with Socrates would later be achieved by association with the Bible.

By contrast both with Aristotelian and with later post-Romantic definitions, discussions of *ironia* in the Middle Ages were characteristically restrictive in their understanding of the word. Whatever Romantic and twentieth-century authors might say about the irony of medieval poets, definitions from the period generally concentrated on well-circumscribed tropes and figures, and—although the element of deception was still relevant to much ironic practice—were often concerned rather minutely with the signs (either intonation or gesture) by which the ironic meaning of a given trope might be clearly revealed. As such, the ancient concern with ironic deception appeared in the Middle Ages mainly by default, in a rhetorical *punctilio* that elaborately classified signals of transparency. By contrast, it is perhaps worth mentioning what Dilwyn Knox calls "*ironia* as an etymological principle"—the medieval classification of supposed etymological antiphrasis (such as *bellus* for *bellum,* or the biblical substitution of *benedicere* for *maledicere*) as originating either in *ironia* or euphemism. Far from being transparent, this kind of *ironia* tended toward genuine ambiguity and obscurity as demonstrated by the conflicting etymological theories advanced for such cases.

In spite of their familiarity with Aristotle's definition of Socratic irony as self-depreciation, medieval authors often interpreted such *ironia* as straightforwardly modest, and clearly distinguished it from *ironia* as trope. Renaissance authors such as Francesco Filelfo (1398–1481) and Alberico Gentili (1552–1608), debated the extent to which the Socratic profession of ignorance was held by Aristotle to be mendacious or immoral. More commonly, Renaissance interpretations of Socratic irony stressed its derisive and humiliating character in a manner common to medieval discussions of the trope. Desiderius Erasmus, for example, while defending Socratic irony as a modest means of rebutting others' arrogance, suggested like Aristotle that it was likely to produce hostility in its audience. Although embracing a classical tradition reaching back at least to Cicero which regarded irony, not so much in terms of its victims (whether self or other), but as a kind of sustained, urbane humor, Erasmus nevertheless held that irony rather than humor belonged to the Scriptures. Recognizing irony's tendency toward rhetorical comprehensiveness, Erasmus remarked further that irony is like a "stroll" through all the rest of the tropes and figures. Elsewhere in medieval and Renaissance rhetoric, irony appears as one of the four main tropes, second only to dominant metaphor.

From the Latin Middle Ages to the seventeenth century, discussions of *ironia* remained largely within the confines of scholarly circles. A relative absence of colloquial use in English is suggested by George Puttenham's *Arte of English Poesie* (1589), which defines *ironia* as "what we call the *drye mock*," while Richard Perceval's English-Spanish *Dictionarie* of 1623 gives *ironia* in the Spanish section, but no *irony* in the English counterpart. A similar absence is characteristic also of seventeenth-century France, with the notable exception of Blaise Pascal, whom the Jesuits reproached for using irony in his *Provincial Letters* (1656). In answering the reproach, Pascal appealed to the authority of the Scriptures, at once using irony and arguing for its use as a means of establishing divine truth by refuting the false opinions of human beings. With respect to irony, Pascal's polemic with the Jesuits is reminiscent of medieval and Renaissance debates in which some theologians considered irony to be sinful through its association with lying, while others vindicated irony by referring to its usage in the Bible. As regards the polemical use of irony, a comparable approach was taken in the French Enlightenment by François-Marie Arouet de Voltaire. For Voltaire, as for Pascal, *ironie* was synonymous with *raillerie* (mockery), and throughout his illustrious career, Voltaire used irony as a means of provoking public laughter at the expense of his opponents. For all its apparent lightness, irony was given a privileged role in discussions of the most serious subjects: whereas Pascal employed irony in the defense of religion, Voltaire directed

his irony against religion, which he considered antagonistic to the cause of philosophy.

Although both the English and French word were customary in erudite parlance by the latter half of the seventeenth century, *irony* only fully entered colloquial English as well as its ordinary literary discussion in the early part of the eighteenth century. Jonathan Swift's claim that he was the first writer to introduce irony into English letters *(On the Death of Dr. Swift),* though literally untrue, is nonetheless indicative of his influence on a period notable for ironic works such as those of Alexander Pope and Henry Fielding—the latter having defined the novel as "comic epic" in prose. While Swift, Pope, and Fielding appealed to classical models, it was perhaps Daniel Defoe's defense of his *Shortest Way with Dissenters,* as Norman Knox observes, "that began the process which was to bring the word down from its pedestal."

During the eighteenth century, the term found its way into more and more dictionaries and encyclopedias, though the reflection on irony at first remained conservative relative to the audacity of its usage. César Chesnau Dumarsais (1730) defines irony in typical fashion as a "figure by which one is given to understand the opposite of what is said," suggesting at the same time the potential absorption of all figural language into irony, since "the words used in irony are not taken in the literal and proper sense." In a comparison recalling Quintilian's classic definition of irony as a species of allegory, Ephraim Chambers's *Cyclopaedia* (1728) observes the following fundamental difference: "*allegory* imports a similitude between the thing spoken and intended; *irony* a contrariety between them."

The eighteenth century is considered to be a turning point in the history of the concept of irony. Such, at any rate, was the post-Romantic view of G. W. F. Hegel and Søren Kierkegaard, for whom Enlightenment "reflection on reflection" (exemplified by Immanuel Kant and Johann Gottlieb Fichte) was bound to give rise to a world-historical ironic phase, an irony "to the second power" (in comparison with the inaugural irony of Socrates). The German Romantics, notably Friedrich Schlegel, conceived Romantic literature as a dissolving of genres into a Romantic and ironic hybrid exemplified by the *roman,* the novel—which Schlegel further defined as "the modern form of Socratic dialogue" (an equation adopted by Friedrich Nietzsche). Although the Romantics emphasized the relation between irony and *Witz,* they tended to distance Romanticism and irony from the satirical and "classical" wit of Voltaire or Swift, emphasizing rather their antecedents in more "baroque" or "arabesque" authors, such as Miguel de Cervantes, François Rabelais, William Shakespeare, and their near contemporaries Denis Diderot and Laurence Sterne. Sterne's *Tristram Shandy* also alludes to the first three of these authors, and comically defines "IRONY" not only as something the French do not understand, but also by allusion to the classical ironical figure of the ass. Although

Sterne himself did not avow his irony, one of Sterne's anonymous eighteenth-century imitators stressed the genuinely deceptive element in his rhetoric: "I write, Sir, that I may not be understood." We should mention also that Sterne's ironic ass is connected both to the traditional "Socratic" gesture of cognitive modesty and to a gesture of *erotic* modesty (specifically, in *Tristram Shandy,* a refusal to say the word *ass*). The connection between the ironist and eroticist has a significant history deriving in part from Plato's *Symposium.*

"To study Romantic irony," writes D. C. Muecke in *The Compass of Irony* "is to discover how modern Romanticism could be, or if you like, how Romantic Modernism is" (Muecke, 1969). It is certainly difficult, if not impossible, to keep modern accretions from sticking to the Romantic "original," whether this be located in Friedrich Schlegel's seemingly pathbreaking fragmentary aphorisms and notebook entries concerning irony (most of which have only this century been assimilated outside German scholarship), or whether the term is extended to include the likes of August Wilhelm von Schlegel, Johann Wolfgang Goethe, Novalis, Karl Solger, Ludwig Tieck, and others who used it in a philosophical manner that would have been unfamiliar abroad. Hegel's assimilation of Romantic irony to the negative element in dialectic, for example, is especially tempting for an encyclopedist, since the formulations of the Romantics regarded en masse do seem to detach irony from any particular rhetorical strategy, let alone trope as such, reducing (or magnifying) it to the level of an intellectual abstraction—what Hegel called "infinite absolute negativity."

Beyond his formulation that "irony is the form of paradox," Friedrich Schlegel produced the following pre-Hegelian definition in the *Athenaeum* fragments (1798): "An idea is a concept perfected to the point of irony, an absolute synthesis of absolute antitheses, a constantly self-producing fluctuation of two conflicting thoughts. An idea is at the same time an idea and a fact." Elsewhere, Schlegel defined irony as "logical beauty," specifying that philosophy rather than rhetoric is its proper home, and that only by "the divine spirit of irony" can poetry rise to the level of "the Socratic muse." Goethe wrote comparably of the "objectivity" (for example, in Shakespeare) that could be achieved through irony—which he associated, by virtue of its critical and self-critical character, with the scientific spirit. Nevertheless, Romantic irony is generally perceived as having been defined by its major proponents in terms of the contradictory and creative unity of opposites: subjective and objective, art and science, finite and infinite, freedom and necessity, seriousness and jest. With such putative synthetic power, it is no wonder that the influence of German Romanticism could pass into English literature, through such authors as Samuel Taylor Coleridge, in the form of an affirmation of the creative and "organic" unity of the imagination, ostensibly stripped of skeptical caveat.

Hegel's critique of Romantic irony, directed mainly against Friedrich Schlegel, took particular exception to the erotic allegories of Schlegel's novel *Lucinde*. In philosophical terms, Hegel accused Schlegel of arbitrarily appropriating the dialectic of "reflection on reflection," and in particular Fichte's metaphysical ego, in the service of unwarranted subjectivity and egoism. Although Hegel treated Socratic irony as a necessary dialectical moment in the emergence of philosophical reflection (a destructive assertion of subjectivity against the *doxa* of false objectivities), his highest praise of Romantic irony was reserved for Solger, regarded essentially as prototype for "the negative" in Hegelian dialectic (or since Solger died young, best regarded, as Kierkegaard wittily put it, as a sacrifice to the Hegelian system). Solger also generalized irony as a principle of art anticipating, mutatis mutandis, much twentieth-century literary criticism, including the Anglo-American New Criticism.

In *The Concept of Irony: With Constant Reference to Socrates* (1841), although he followed much of the Hegelian critique of Romanticism, Kierkegaard attempted to play off Hegelians and Romantics against each other in a kind of ironic mutual destruction. Arguing from Hegel's definition of irony as "infinite absolute negativity" that "the substantial aspect of Socrates' existence was irony" (an intentional contradiction), Kierkegaard made Schlegelian paradox and Hegelian dialectic into ironic bedfellows, offering Hegelians a more "Romantic" Socrates than they might have liked, while subjecting Romantics to "the master's voice" of Hegel. Although Kierkegaard seemed to follow Hegel in insisting against the Romantics that irony must be firmly "mastered" to be legitimate, there was a difference between mastery of irony by, as Lee Capel put it, "for Hegel the concrete universal, for Kierkegaard the concrete individual" (Kierkegaard, 1968). Kierkegaard's quasi-Hegelian account of the mastery of irony by historical dialectic was doubtless itself ironic (as his contradictory definition of Socrates suggests), inasmuch as he did not accept that any purely rational system could achieve such a feat. In any event, Kierkegaard's assertion that Socrates, in contradistinction to Plato, "did not merely use irony, but was so dedicated to irony that he himself succumbed to it" (Kierkegaard, 1968) emphasized the "unmastered" character of irony *sensu eminentiori* (in Kierkegaard's phrase) attributed to Socrates. Indeed, insofar as he portrayed Socratic irony as not only intellectually unmastered but unmasterable, Kierkegaard formalized the epistemological uncertainty or "incomprehensibility" associated with Schlegelian irony that has also become standard in many twentieth-century derivatives. The official position of the later Kierkegaard—or rather his pseudonyms—was that in the Christian world (based on faith rather than knowledge) irony must give way to humor, related to each other as doubt (an epistemological category) relates to consciousness of sin (an ethical or religious one).

Alongside its epistemological dimension, however, Kierkegaard also stressed irony's "sociological" one. As Hegel mocked Schlegel's supposed claims to nobility of spirit, Kierkegaard likened the ironist's putative elevation to that of king over commoner. Kierkegaard also adapted Solger's sacrificial formulation of the metaphysical relation between divine infinitude and material finitude into a full-blown sacrificial account of irony's psychological, social, and historical dynamics. Although Socrates is said to be skeptical of sacrifice, and famously, of death, it is said to be an irony at his expense as well of that of the Greek state that "he falls for such a ridiculous argument as a death sentence." Irony, in Kierkegaard, resembles the dialectic of world history, which destroys every given actuality, including ultimately the ironist, in service of the new. The ironist is "a sacrifice required by the world process," since "the past shows it is still justified by demanding a sacrifice, the new by providing a sacrifice."

Kierkegaard's sacrificial account thus emphasizes what D. C. Muecke calls "the close resemblance of ironist and victim." On the one hand, this may recall earlier tendencies to regard Socratic irony as genuinely self-depreciating rather than mockingly so; but on the other, Kierkegaard emphasizes irony's psychology of exclusive ressentiment: "Irony is not so much apathy, divested of all tender emotions of the soul; instead it is more like vexation over the fact that others also enjoy what it desires for itself." Or again: "Irony . . . fashions itself in similitude with the divine envy, yet which is not only envy of the great and eminent but even of the small and insignificant, thoroughly envious of all finitude." While the highest polemical reach of irony is said to be *nil admirari*—a kind of psychological equivalent of the claim that "irony in a strict sense can never set forth a thesis"—the ironist's envy ironically turns out to be highly imitative. Mimesis is similarly stressed in Kierkegaard's treatment of Aristophanes' *Clouds*, said to have come "very close to the truth" in its portrayal of Socratic irony, and the potential symmetry of ironist and victim is well expressed in a formula from Kierkegaard's journals: "the most terrible fight is not when there is one opinion against another, but when two men say the same thing."

Although both Kierkegaard and Hegel considered Schlegel's grand pretensions as an ironist to have been sufficiently unmasked by his later conversion to Catholicism, and although many subsequent doubts have been expressed as to whether Romantic definitions of irony succeed in defining any definitive aspect of Romantic practice, there can be no doubt that the Romantic use of the term has had powerful, if often curiously indirect and delayed effects on the history of aesthetics and criticism. It is perhaps no minor irony that from early twentieth-century Formalism to much contemporary theory, it might seem that the path to various versions of Romantic irony were being beaten all over again. Wherever one looks—whether at the develop-

ment of Slavic literary theory, French structuralism or post-structuralism, Anglo-American New Criticism, rhetorical criticism as practiced by Kenneth Burke or Paul de Man, Marxist theory, psychoanalysis, or indeed analytic philosophy in the throes of epistemological skepticism—one finds irony both by its own name and disguised under various pseudonyms. One may suspect, to be sure, that the main consistency of post-Romantic ironic terminologies lies in their inconsistency—a symptom of what Joseph Dane has called "the critical mythology of irony"—yet the very existence of such a mythology attests to the critical importance of the term. Assuming we bear in mind the element of genuine deception that definitively returned to the concept with Romanticism and Kierkegaard, a genuine modern history would have to distinguish between modern authors who use the term *irony* ironically and those who do not—not to speak of those who are ironic without using the term itself in any special sense.

Although such a history lies beyond the scope of this article, a twentieth-century sampler would surely mention György Lukács's influential *Theory of the Novel* (1920), which inherits the Romantic linkage of irony and the novel via Hegelian and Marxist dialectic. Superficially very different from this, the Russian Formalist theory of parody (exemplified by Viktor Shklovsky's essay on *Tristram Shandy*) can also, on the one hand, be compared to the later emergence of irony as a key term for the New Criticism in Britain and the United States, and, on the other hand, linked to Mikhail Bakhtin, whose overtly Marxist philosophy of language and concern with the performative effects of "the double-voiced word" resulted in a theory of novelistic "heteroglossia" and carnival that indirectly echoed both Schlegel's collapse of genre in the *roman* and Kierkegaard's comments on the ironic significance of the inversions of carnival—the Feast of the Ass, the Feast of Fools, Easter humor, and so on. The German novelist Thomas Mann stands out as providing perhaps the most extensive treatment of the term itself in modern fiction, a treatment whose elaboration of Romantic and Nitzschean themes (for instance, in *Doktor Faustus*) has been highly influential. In French, Vladimir Jankélévitch (1936) pursued Romantic and Kierkegaardian lines, while French structuralism and post-structuralism (which reestablished links with the Slavic tradition in linguistics and literary theory) later provided such formulations as Roland Barthes's in *Critique et vérité* (1966), which defines criticism itself in terms of "irony" suitably redefined for structuralist purposes as "a question put to language by language."

While the French literary and philosophical tradition is surely rich in ironic sensibility in several senses—whether in the poetry and critical writings of Charles Baudelaire and Paul Valéry, the novels of Gustave Flaubert, or the philosophy of Jean-Paul Sartre—it nevertheless provides no more consistent a usage or theorization of the term than the British and American. Similarly, while Romantic irony's joy in wreaking epistemological havoc makes it an easy label for French deconstruction, whose philosophical inventor, Jacques Derrida, exploits such classically ironic images as the intellectual midwifery of Socrates or the Shandyan monstrous birth, Derrida does not himself privilege the term. We have to turn, rather, to Paul de Man (Belgian émigré to the United States) to find "irony" itself at center stage of deconstructivist discourse, for instance, in his strategic citation of Schlegel's definition of irony as "permanent parabasis" (parabasis being the part of classical comedy in which the poet interrupts the fiction to speak in his own name) at the conclusion of *Allegories of Reading: Figural Language in Rousseau, Nietzsche, Rilke, and Proust* (1979). The seeming opacity of de Man's style, not to mention his apparently exalted claims to ignorance, make him a possible candidate for ironist as genuine deceiver; thus, he might be seen as playing the *eirōn*, say, to such conservative studies as Wayne Booth's *The Rhetoric of Irony* (1974), concerned with showing how "stable" ironic interpretation may be safeguarded from Romantic and deconstructive instabilities. But, if we are tempted to embrace the equation of irony and deconstruction as a result of de Man, as many literary critics have been, de Man himself, on the contrary, makes irony into "the undoing of the deconstructivist allegory of all tropological cognitions, the systematic undoing, in other words, of understanding"—in other words, the deconstruction of deconstruction? Similarly, while a popular view of deconstructive irony regards it as suspending "the referent" in favor of poststructuralist "textuality," "reality" in favor of "fiction," Schlegel's definition of irony as "permanent parabasis" seems rather to emphasize not so much that "reality" but even more that "fiction" is fiction—"permanently interrupted" by the real.

Whatever the merits of *his* irony, de Man's prominence in Britain and America has done much to close the gap that divided New Critical uses of the term, including William Empson's *Seven Types of Ambiguity* (1930), Northrop Frye's *Anatomy of Criticism* (1957), and Cleanth Brooks's *The Well-Wrought Urn* (1947), both from the Continental tradition and from American literary and social theorists such as Kenneth Burke. In *A Grammar of Motives*, Burke offers a "literal" or "realistic" application of the four tropes (metaphor, metonymy, synecdoche, irony) in which "for *irony* we could substitute *dialectic*" (Burke, 1969). On familiar terms with the Hegelian-Kierkegaardian sense of ironic dialectic, and notably with the sacrificial horizon of irony in Kierkegaard, Burke also drew on social science, including Sigmund Freud and Émile Durkheim, and anticipated such authors as Erving Goffman in sociology and René Girard in literary theory and anthropology. In Burke, irony is a rhetoric of miscommunication whose structure of victimage aligns it with his more general theory of the construction of social and rhetorical hierarchy.

Besides irony's seepages from the languages of philosophy and art into those of the social sciences, we should also mention Leo Strauss in the domain of political philosophy. Usually overlooked in books on modern irony, Strauss's claims for the irony of many political philosophers usually read unironically (such as John Locke) has had considerable influence in American political circles, both practical and academic. Besides regarding deceptive irony as a means by which philosophers concealed subversive or otherwise dangerous ideas from the general public or from the authorities, Strauss elaborated in *The City and Man* a more general classical sense of Socratic irony as a kind of dissimulation—a "graceful vice" dissimulating virtue rather than vice—structuring the form of Plato's work that says "different things to different people." Socratic irony, so conceived, is an important part of "the literary question" which "properly understood is the question of the relation between society and philosophy" (Strauss, 1964). A certain "radical irony" is therefore a meeting ground of political and literary theory.

For all their differences, Burke and Strauss both conceive irony in terms of social (mis)communication rather than aesthetics. This is also the case for Richard Rorty in *Contingency, Irony, and Solidarity* (1989), where irony is connected not only to his assertion of epistemological contingency (through his critique of the epistemological projects of analytic philosophy) and thus the possible "incommensurability" of discourses, but also to the possibility of solidarity and communication in such circumstances. In *The Critical Mythology of Irony* (1991), Joseph Dane notes that recent studies of Romantic irony have also stressed the "community" supposedly implied by ironic communication, although Dane himself emphasizes the hierarchical, even elitist constitution of Romantic and post-Romantic conceptions of irony, as well as the critical "struggle for mastery" that seems to be central to the modern history of the term. The tendency of post-Romantic irony to establish a hierarchy of "higher" and "lower" kinds of ironists might recall Burke's linkage of irony, social hierarchy, and the scapegoat principle. In Burke's intertwining of ironic dialectic with the scapegoat mechanism, as in Girard's later theory of sacrifice and its representation in literature, irony functions on the one hand as a name for the "dialectical coincidence of opposites" that characterizes sacrificial phenomena (what Burke calls "such reversible pairs as disease-cure, hero-villain, active-passive"), and on the other hand, especially in Girard, as a name for the poetic lucidity, for example, in Shakespearean tragedy, which both perceives and dissembles the sacrificial character of the drama. This sense of irony, which unmasks while simultaneously reenacting a delusion, has also informed such recent studies as Helene Foley's *Ritual Irony: Poetry and Sacrifice in Euripides* (1985).

If Burke, Strauss, and de Man may be taken as representing a wide spectrum of the most provocative theorizations of irony since the 1950s, we may conclude that all three situate the term at the intersection of literature, philosophy, and the social sciences, and that all three associate irony with the possibility of genuine deception—including that of their readers—without excluding the sense of irony as a "drye mock" (with a different kind of victim or mock victim). Although the ironist's "victims" or "inferiors" may be others, as in Strauss's contention that Plato's Socrates is never shown talking to equals in the dialogues, "dialectical irony (or humility)," may, according to Burke, provide a kind of "technical equivalent for the doctrine of original sin"—placing the ironist on a level with everyone else. For Burke, "irony as approached through either drama or dialectic moves us into the area of 'law' and 'justice,'" and the symmetry of the *lex talionis*.

Northrop Frye particularly emphasized the arbitrariness of victimage revealed by tragic irony: "Irony isolates from the tragic situation the sense of arbitrariness. . . . Thus the figure of a typical or random victim begins to crystallize in domestic tragedy as it deepens in ironic tone. We may call this typical victim the *pharmakos* or scapegoat." A scene of victimage is also the context of de Man's now famous invocation of Schlegelian irony at the conclusion of *Allegories of Reading*, a context in which de Man defines fiction in terms of arbitrariness, generalizing from his analysis of the famous denunciation of the servant Marion in Rousseau's *Confessions*. Indeed, Schlegel's definition of irony as "permanent parabasis"—or continual entrance of the real into fiction—seems in this instance, in de Man's treatment, identical with the reality of the victimage itself.

Readers may be forgiven for regarding this somber conclusion as a terminal development of the post-Romantic insistence that we theorize irony, not merely in dead earnest, but out of ordinary existence. Such versions of irony as Frye's, Burke's, and de Man's, of course, reflect only certain aspects of recent rhetorical thinking, however representative, and the term *irony* may very well fade in the future from critical and theoretical prominence. It may be that we shall eventually be cured of theoretical irony, as Aristophanes was cured of his hiccups in the *Symposium:* by a fit of philosophical sneezing.

[*See also* Hegel; Kierkegaard; *and* Romanticism.]

BIBLIOGRAPHY

Booth, Wayne C. *A Rhetoric of Irony.* Chicago, 1974.

Burke, Kenneth. *A Grammar of Motives.* Reprint, Berkeley, 1969.

Dane, Joseph A. *The Critical Mythology of Irony.* Athens, Ga., 1991.

de Man, Paul. *Allegories of Reading: Figural Language in Rousseau, Nietzsche, Rilke, and Proust.* New Haven, 1979.

Kierkegaard, Søren. *The Concept of Irony: With Constant Reference to Socrates.* Translated by Lee M. Capel. Reprint, Bloomington, Ind., 1968.

Knox, Dilwyn. *Ironia: Medieval and Renaissance Ideas on Irony.* Leiden and New York, 1989.

Knox, Norman. *The Word Irony and Its Context, 1500–1755.* Durham, N.C., 1961.

Markantonatos, G. "On the Origin and Meanings of the Word *Eironia*." *Rivista di Filologia e di Istruzione Classica* 103.1 (1975): 16–21.

Muecke, D. C. *The Compass of Irony.* London, 1969.

Pavloskis, Zoja. "Aristotle, Horace and the Ironic Man." *Classical Philology* 53.1 (1968): 22–41.

Strauss, Leo. *The City and Man.* Chicago, 1964.

Rorty, Richard. *Contingency, Irony, and Solidarity.* New York, 1989.

JOHN VIGNAUX SMITH and REGINALD MCGINNIS

ISLAMIC AESTHETICS. [*To explore the history of Islamic Aesthetics, this entry comprises three essays:*

> Historical and Conceptual Overview
> Islamic Music
> Visual Experience in Islamic Culture

The first essay is an overview of the question whether there is a distinct Islamic aesthetic theory to capture the multiple traditions of Islamic art. Since the focus in this history tends to be on particular art forms rather than on general theory, the second and third essays discuss music and the visual arts. All three essays discuss mainly, if not only, medieval Islamic aesthetics. For related discussion, see Arab Aesthetics; Aristotle; *and* Comparative Aesthetics.]

Historical and Conceptual Overview

"God is beautiful and He loves beauty." This celebrated saying from the Muslim tradition of statements attributed to the prophet Muḥammad clearly identifies beauty as a desirable but not necessary attribute of all that can be perceived, heard, or seen. The saying itself is unlikely to have come from the religious and cultural milieu of seventh-century Arabia. It was derived, at some unknown time, from the transfer to the theological realm of philosophical and spiritual values developed in the great urban centers of learning in medieval Islam.

The recognition of an aesthetic component does not, however, ensure the existence of an aesthetic doctrine derived from revealed or philosophical arguments. The religion of Islam contains features like its opposition to idolatry and its proclamation of the absolute unicity of God as Creator, which had an impact on the arts (on aniconism, for instance), but these features led not to aesthetic judgments of value, but at best only to moral judgments of praise or of condemnation. On the other hand, Arabic poetry and later Persian and the Persianate Turkish and Urdu poetry, whether recited or written down, were judged according to aesthetic criteria. The study and knowledge of this poetry led eventually to written manuals of judgment and interpretation, but there were certainly, long before, unwritten ways of criticism according to rules that were understood but never spelled out. How much these older ways affected the later theories is still unknown. Music was also provided with such criteria, often under the impact of classical philosophy

and mathematics. Less is known, on the theoretical level, about the visual arts, because an awareness of these arts did not enter into formal statements of what it meant to be a person of culture, to possess what was known as *adab*. We shall see that there are in fact significant exceptions to this generality, which seems valid on a philosophical level. An exception should be made about the Platonic and Aristotelian theme of *mimēsis* understood as an imitation of visible experience, which appears regularly in philosophical writing and in poetic imagery. For, even though the initial Muslim revelation did not deal with the arts, the relationship it established between God, on the one hand, and creation and human beings, on the other, remained throughout as a central concern of Muslim thought and incorporated most antique discussions of this matter. It acquired a particularly powerful expression in mystical thought and in mystical imagery, where the formal considerations so central to any aesthetic theory or attitude often overwhelmed content.

Next to a theoretical aesthetic thought dealing with beauty, which, as befits most of the Middle Ages, Christian or Muslim, protrudes occasionally from lengthy metaphysical, logical, and even moral discussions, there was what may be called a practical or cultural aesthetic—that is, the passing of judgments about things made or used by men and women and decisions that affected their taste, their opinion and practice of the arts and crafts of their own or earlier times, and the ways in which they actually reacted or should have reacted to whatever surrounded them. Most of the data about this aesthetics derive from the thousands of stories and anecdotes, learned or derived from folklore, imaginatively inventive or allegedly truthful, which characterize much of *adab* literature in Arabic and Persian. It is not, at this stage of research, possible to sketch out the components or the characteristics of this practical aesthetics, which must have evolved considerably over the centuries and varied somewhat from one area to the other. It is particularly difficult to do so because too much uncertainty remains about the history of meanings to be attributed to key words or to the major roots (e.g., *hsn* or *jml*) used to circumscribe beauty. Yet, it is by gathering, collating, and classifying masses of stories with implied or expressed aesthetic judgments (as was initiated by F. Rosenthal) that, little by little, our understanding of the operative vocabulary of aesthetic judgment will be enhanced and general hypotheses can be proposed about the nature of that judgment.

A few examples of such practical judgments will explain the wealth and importance of data that were not meant to deal with beauty, but end up by providing elements for an eventual theory and practice of cultural aesthetics.

In the middle of the ninth century, one of the Abbasid caliphs in Baghdad is credited with having introduced a type of construction technique that altered the character of buildings and a new type of textile that changed the appearance of everyone at the court. A century later, in the same

milieu and always according to written sources, writing spreads as the most common means of decorating objects and clothes. In the tenth century, again in Baghdad, mathematicians elaborate all sorts of ways to apply geometric theory and irrational numbers to the decoration of architectural facades and to the composition of buildings, thereby asserting in practice, if not in theory, the primacy of geometry in the making of effective forms; while the historical fact of this mathematical thinking is not at issue, much is still unclear about the aesthetic principles to be derived from it. More or less at the same time, what will eventually be known as calligraphy acquires its first abstract definitions and the beginning of an aesthetic judgment that has been very difficult to translate into contemporary terms. In the tenth to the eleventh century the great physicist and optician Ibn al-Haytham (who lives in Andalusia) is led to a theory of beauty (using the root *hasana* rather than the more common *jamala*) in order to explain the ways in which the eye perceives. He develops a psychology of vision in which proportion and harmony play an essential part in providing beauty and pleasure. Proportion and harmony are the terms used as well in literary and musical criticism. Also in the twelfth century, al-Jurjānī, a native of northeastern Iran, is one of several writers to develop a theory of literary form in which technical definitions of images and metaphors *(majaz* or *istiara)* acquire a prescriptive quality. Much later, in the fifteenth century, the Persian writer Sharaf al-Dīn Rami composes a similar, although much less profound, book on the "good" and proper images to be used in poetry.

None of these examples is in itself an attempt to define an aesthetic theory in any formal sense, as did happen among philosophers and as is often implied among mystics. What they do suggest is that a careful and thorough analysis of the most unexpected sources—technical manuals and even historical chronicles—can bring to light the many facets of most sophisticated and complex judgments made by traditional Muslim culture.

[*See also* Beauty; al-Fārābī; Ibn Rushd; Ibn Sīnā; *and* Poetics.]

BIBLIOGRAPHY

Abu Dib, K. *Al-Jurjani's Theory of Poetic Imagery.* Warminster, 1979.

Ahsan, M. M. *Social Life under the Abbasids.* London, 1979.

Grabar, Oleg. *The Mediation of Ornament.* Princeton, N.J., 1992.

Melikian-Chirvani, A. S. "Khwaja Miraq Naqqash." *Journal Asiatique* 276 (1988).

Necipoğlu, Gulru. *The Topkapi Scroll: Geometry and Ornament in Islamic Architecture.* Santa Monica, Calif., 1995.

Rami, Sharaf al-Din. *Traité des termes figurés relatifs à la description de la beauté.* Translated by C. Huart. Paris, 1885.

Rosenthal, Franz. *Four Essays on Art and Literature in Islam.* Leiden, 1971.

Sabra, A. I., ed. *The Optics of Ibn al-Haytham: Books I-III: On Direct Vision.* 2 vols. London, 1989. See esp. book II, pp. 97–102.

Schimmel, Annemarie. *A Two-colored Brocade: The Imagery of Persian Poetry.* Chapel Hill, N.C., 1992.

OLEG GRABAR

Islamic Music

In medieval Islamic civilization, just as in antiquity and the Christian Middle Ages, there was no independent subject called "aesthetics." Aesthetic questions were discussed, often tangentially, in philosophical treatises, including cosmological, and in some theological works. They were also touched on in specialized writings on music and poetry.

Coincidentally, what we nowadays call the fine arts were not carved out as a separate group. Ya'qūb al-Kindī, Muḥammad al-Fārābī, and Ibn Sīnā all classified music as a theoretical inquiry, and among the mathematical sciences, along with arithmetic, geometry, and astronomy. 'Abd al-Raḥmān Ibn Khaldūn discusses music as a performing art together with the "crafts," which include medicine, carpentry, weaving, tailoring, midwifery, architecture, and calligraphy. The discussion of poetry is usually appended to the linguistic sciences. It is also referred to in treatises on singing.

One could advance three reasons for the nonprominence and nonseparateness of the aesthetic: (1) according to predominant views, the functions of art were religious, or moral, or therapeutic (especially in the case of music), but rarely purely aesthetic; (2) while the pleasure that results from the beauty of the object was recognized by all, it was welcomed only by some, and suspected by many, especially the religious authorities, for fear that it would lure the listener away from the path of righteousness; and (3) it was believed that each of the various arts does indeed belong to the category to which it was assigned, and this differs from the views of the modern aesthetician. If so, it is equally legitimate to ask why *did* later centuries separate and elevate the fine arts, as it would be for a modern to ask why in antiquity and the Middle Ages they were *not*.

In the philosophies of music in Islam, the thinkers were essentially following the agenda set by the Greeks, notably by the Pythagoreans, Plato and Aristotle, and by Aristoxenus, mentioned in Arabic encyclopedic works variously as Aristokas and Arisoxene.

One finds adherents to Pythagorean ideas on music as late as the sixteenth century, despite the rejection of the basic tenets of that school by such major philosophers as al-Fārābī (d. 950) and Ibn Sīnā (Avicenna, d. 1037). The most important advocates of Pythagoreanism are al-Kindī (d. 870) and the Brethren of Purity, a group of thinkers who lived in the second half of the tenth century.

According to al-Kindī, the task of the philosopher seeking theoretical knowledge about music is to understand the place of music in the total scheme of things. There is an affinity between certain features of music and other aspects of the cosmos. Al-Kindī lays less stress on arithmetical ratios than do the Brethren, who came after him, but he offers an extensive correlation between music, on the one hand, and each of the following, on the other: astronomy, astrology, meteorology, the four natural elements, the ages of hu-

mankind, the mental and bodily powers, the humors, and human character traits. Besides theoretical knowledge, the philosopher has a more practical aim. Knowing all the external connections of music, the philosopher, now doubling as physician, can choose the right rhythm and mode to use in the right season, at the right time of day, and so on, and thus tailor the best therapy for the individual physical or mental ailment at hand.

The Brethren of Purity are more Pythagorean than al-Kindi in two respects. They make much fuller use of ratios in their discussion of music, and they restore the religious tone of their Greek masters. The ratio affinity that binds the sublunar with the celestial world, together with the belief that human music was devised in emulation of celestial music in order to soften the hearts of humans for religious devotion—these facilitate the ascent of the human spirit to its destiny in the higher world. And in this is the highest fulfillment of our music.

The Aristotelian rejection of Pythagorean astronomy has its echo in the thought of al-Fārābī. The latter maintains that it has been shown in natural science that the celestial spheres do not make sound of any kind, musical or otherwise. Furthermore, à la Aristoxenus, the primary substances of music, which are the subject matter of theoretical study are not arithmetical ratios, but tones, intervals, melodies, and rhythm. Finally, human music originated from our natural disposition. We need to express what goes on within us, be it joyful or sad. We also need to divert ourselves from the toil, fatigue, and boredom of our daily lives.

Ibn Sīnā continues the rejection of Pythagoreanism with a more explicit declaration of the independence of the theoretical study of music from all externalities. To bind music to other sciences such as arithmetic and astronomy is the mistake of those who do not distinguish, among the sciences, the one science from the other. Moreover, to base the understanding of music on similarities with arithmetical ratios and human character traits is to treat as essential what is only accidental.

Ibn Sīnā's aesthetics of music stresses two complementary points. The first is that music is a source of enjoyment, but not accidentally and as a bonus. Rather, enjoyment is the telos of music. Ibn Sīnā makes a similar point about the lyric poetry of the Arabs, contrasted with the dramatic poetry of the Greeks, which aims to instill character.

The second idea in Ibn Sīnā's aesthetics concerns what it is that we enjoy in music. There are three aspects: (1) the characteristics of musical sound as sound (musical sounds by their quality, their volume, and their intensity can be pleasing or offensive); (2) the order and arrangement of notes and intervals into melody and rhythm, and the occurrence and recurrence of these in a composition; and (3) the existence of similarities between music, on the one hand, and natural processes (buildup, acceleration, etc.) and human emotions (joy, sadness, etc.), on the other. But these

are accidental facts about music. Nonetheless, we do derive enjoyment from recognizing in music what we experience outside it. Of the three aspects, the greatest source of enjoyment, according to Ibn Sīnā, is the second: the order.

Ibn Sīnā agrees with al-Fārābī that music has a natural, not a supramundane origin. It is an outgrowth of the functional use of sound by the animal kingdom, whether in calling for mates, alerting about danger, or expressing what goes on within oneself. When we are listening to composed music, however, we focus our discerning faculty on the three aspects of music that are the sources of pleasure. We are no longer in the natural state in which meaningful sounds are simply functional.

Two other figures are of special interest, al-Hasan al-Katib and Ibn Khaldūn, who, although they do not stand completely outside the orbit of Greek influence, do not simply fit into the Pythagorean-Aristoxenian controversies.

Al-Katib (d. eleventh century) is highly eclectic. He was greatly influenced by Kindi and al-Fārābī, among others. He states the interdependence of the theoretical and practical aspects of music in the tighest terms. To understand and enjoy music one must acquire the theoretical knowledge that qualifies one as a listener, for music has meaning just as language does. One would then have to know the elements of good composition, good structure, and the proper matching of melody and poetic text. The listener must further know the basic principles and techniques of performance, although it is not necessary for that listener to be a performer. On the other hand, the performer needs to have theoretical knowledge much as the builder needs to have the plans drawn by the architect.

All this is required for the higher art music. But then there is populist music, which only has surface qualities and no underlying meaning. Al-Katib inveighs against the popular and vulgar taste prevalent in his day and the music on which it feeds. Such populist music has only surface qualities, it requires no knowledge on the part of the listener, and it can be enjoyed equally by humans and animals. In the case of what al-Katib calls serious music, there is a goal that is beyond enjoyment, and that is to affect the spirit, uplift its qualities, and lead to noble deeds.

Ibn Khaldūn (d. 1406) is a unique figure in world intellectual history. He brings a broad vision from his perspective as a philosopher of history. Music, as one of the "crafts," gets a fair amount of attention in his monumental *Introduction to History*. Music comes on the scene in a given society when a level of affluence and luxury has been attained. It is then that people seek the pleasure and the diversion that music can provide. Late to come, music is also among the first to go when a civilization begins to distintegrate.

As to the analysis of musical pleasure itself, Ibn Khaldūn offers a relational theory. Harmony is the quality in the object that, when it affects the listener in an agreeable manner,

is experienced as pleasure. In such a case, beauty is attributed to that object. Beauty, then, is the capacity of an object to affect us in an agreeable way when we are in the condition to be so affected.

What is significant about all these thinkers who make pleasure the object of listening to music is the contrast this makes with other prevailing views of the day. These ranged from the prohibitionist view that saw no good purpose served from listening to music; on the contrary, it is a lure that can pull humans away from God. There was also the view, inherited from antiquity, that music should ennoble the soul, or that it should help it to raise to the higher world of the Spirit. And then there was the Sufis' view, which relied on their music, performed in special sessions, to inflame the seekers' love of God and propel them in their return journey to the ultimate Source. Many orthodox religious authorities opposed the Sufis and their music for providing an alternative way to God.

[*See also* Music, *historical overview article.*]

BIBLIOGRAPHY

Al-Farabi. *Grand traité de la musique.* French translation in *La musique arabe,* edited and translated by Rodolphe d'Érlanger, vols. 1 and 2. Paris, 1936.

Farmer, Henry George. *A History of Arabian Music to the Thirteenth Century.* London, 1930.

Ibn Khaldūn, ʿAdu I-Rahman. *The Muquaddimah: An Introduction to History.* 3 vols. Translated by Franz Rosenthal. New York, 1958. See vol. 2.

Ibn Sina, Abu ʿAli (Avicenna). *Kitab al-Shifaʾ* (The Book of the Healing). French translation of chap. 12, *The Book on Music,* in *La musique arabe,* edited and translated by Rodolphe d'Érlanger, vol. 2. Paris, 1936.

Ikhwān al-Ṣafā (The Brethren of Purity). *The Epistle on Music of the Ikhwan Al-Safa.* Translated by Amnon Shiloah. Tel Aviv, 1978.

Al-Kātīb, al-Hasan ibn Ahmad. *La perfection des connaissances musicales.* Translated into French by Amnon Shiloah. Paris, 1972.

Nasr, Seyyed Hossein. *An Introduction to Islamic Cosmological Doctrines.* Rev. ed. London, 1978.

Perkuhn, Eva Ruth. *Die Theorien zum arabischen Einfluss auf die europäische Musik des Mittelalters.* Waldorf-Hessen, 1976.

Randel, Don M. "Al-Fārābī and the Role of Arabic Music Theory in the Latin Middle Ages." *Journal of the American Musicological Society* 29 (1976): 173–183.

Rosenthal, Franz. "Al-Kindi and Ptolemy." In *Studi orientalistici in onore di Giorgio Levi della Vida,* vol. 2, pp. 436–456. Rome, 1956.

Shehadi, Fadlou. "Art and Imitation: Plato and Ibn Sina." In *Of Scholars, Savants, and Their Texts: Studies in Philosophy and Religious Thought: Essays in Honor of Arthur Hyman,* edited by Ruth Link-Salinger. New York, 1989.

Shehadi, Fadlou. *Philosophies of Music in Medieval Islam.* Leiden and New York, 1995.

Shiloah, Amnon. *The Theory of Music in Arabic Writing, c. 900–1900.* Munich, 1979.

Shiloah, Amnon. *The Dimension of Music in Islamic and Jewish Culture.* London, 1993.

Werner, Eric, and Isaiah Sonne. "The Philosophy and Theory of Music in Judaeo-Arabic Literature." *Hebrew Union College Annual* 16 (1941): 251–319; 17 (1942–1943): 511–572.

FADLOU SHEHADI

Visual Experience in Islamic Culture

Although there is no exact equivalent to the term *aesthetics* in medieval Islamic languages, visual beauty is mentioned in several contexts that cumulatively provide an insight into aesthetic attitudes in the medieval Islamic Near East. The form taken by this appreciation was shaped by Islamic culture itself, but the assimilation and transformation of Greek thought in ninth- and tenth-century Iraq provided an important catalyst for its articulation. Reservations voiced about the representational arts in religious literature such as legal texts or Quranic commentaries, in which painters and sculptors were criticized for their presumption in emulating God's creative powers, probably both inhibited popular development of those arts and lowered the social status of artists. Figural wall paintings are known, however, to have existed in the palaces of the Abbasid caliphs during the ninth and tenth centuries. At the same time, classical ideas that gained currency in intellectual circles within the Islamic world, linked physical beauty and moral good, and paradoxically supported the view that a painter's emulation of nature also reflects or mimics an archetype of superhuman or divine origin—eventually these concepts would be used to justify the representational arts despite the continued opposition of some religious leaders.

The medieval Islamic world made no distinction between fine and applied arts, but viewed calligraphy, architecture, painting, carpentry, metalworking, and weaving primarily as useful crafts that depended on the acquistion of skills that were only secondarily vehicles for aesthetic expression. Texts that comment on beauty or aesthetic values include collections of aphorisms or anecdotes, accounts of the teachings of masters in a given field, and essays on subjects ranging from the physical attributes of young female slaves to calligraphy, geometry, painting, and music.

Comments on the curative value of beautiful paintings by the physician Muḥammad ibn Zakarīyā al-Rāzī (864–925) stress the practical benefits of aesthetic pleasure. Al-Rāzī recommends looking at wall paintings or paintings in books in order to dispel gloom and keep the humors in their proper balance. He claims that earlier (i.e., Greek) scholars had discovered the importance of using wall paintings in baths that depict animal combats, the embrace of lovers, as well as the trees and flowers of a garden, because each theme revives a different faculty among the bathers. Although it is unknown whether al-Rāzī's comments affected the practice of art in his own time, an eighth-century bathhouse in Jordan known as Quṣayr ʿAmra has wall paintings that exemplify al-Rāzī's categories.

References to the visual arts that occur in collections of aphorisms have been collected by Franz Rosenthal in his essay on Greco-Arabic wisdom literature. These range from descriptions of statues and paintings to comments about the ethical and spiritual value of beauty and are included in

compilations by tenth- and eleventh-century authors such as the *Jāwidhān-Khirad* of Abū ʿAlī Aḥmad (Ibn) Misk-awayh (d. 1030), the *Ṣiwān al-Ḥikmah* of Abū Sulaymān al-Manṭiqī al-Sijistānī (d. 985), and the *Mukhtār al-Ḥikam* of Abū al-Wafāʾ al-Mubashshir ibn Fātik. These remarks also provided a model for new compositions such as the treatise on calligraphy by Abū Ḥayyān al-Tawḥīdī (d. 1023). This text combines practical advice from practitioners on the craft of the scribe with aphorisms on the moral and aesthetic value of calligraphy.

Observations on beauty in early Arabic literary texts reflect an idealization and abstraction of the human body. Physical aspects of the body that arouse aesthetic appreciation are considered in the *Kitāb al-Qiyān,* an essay by ʿAmr b. Baḥr, known as al-Jāḥiẓ (d. 869), concerning the attributes and delights of "Singing Slave-Girls." He defines how their beauty *(jamāl)* should be evaluated by a prospective purchaser. For him beauty consists of an equilibrium *(wazn)* in the shape and arrangement of the constituent parts of the human body and the features of the face. These comments probably reflect Greek ideas about the human beauty, but Jāḥiẓ also applies the same criteria to buildings, carpets, textiles, clothing, and even irrigation channels.

An appreciation of formal regularity and harmonious balance is also central to the aesthetics of Islamic calligraphy. Literary sources such as the *Fihrist* of Abū al-Faraj Muḥammad Ibn al-Nadīm (d. 990) mention scribes active during the Umayyad era (661–750) who attained renown for the beauty of their Quranic or epigraphic scripts, but the articulation of specific aesthetic criteria for the evaluation of writing probably occurred only at the Abbasid court in Baghdad. It was there that *khaṭṭ al-mansūb* (measured writing) was defined and established as a standard of excellence. Ibn al-Nadīm suggests that by the reign of the caliph al-Maʾmūn (813–833) there was a link between the scale of a script and the size of paper on which it was written and between a text's content and the style of writing employed.

Subsequently, the concept of "measured writing" was applied also to the shapes and proportions of individual letters. It has become traditional to credit an Abbasid official, Abū ʿAlī b. Muḥammad, known as Ibn Muqla (885–940), with the invention of this system. In "measured writing" a dot made by the point of the pen is used as a module to define the height of the alphabet's first letter, the *alif,* which also functions as the diameter of a circle. The shape and size of the remaining letters of the alphabet were then created in reference either to the vertical *alif* or to the curvature of the circle drawn around it. Although Ibn Muqla is usually mentioned as the inventor of this system, later calligraphers also created their own versions of it; the one used by a later ʿIraqī calligrapher, ʿAlī b. Hilāl known as Ibn al-Bawwāb (d. about 1022), attained particular renown. A manual on calligraphy compiled by Abū al-ʿAbbās Aḥmad al-Qalqashandī (1355–1418) compares the proportional

system attributed to Ibn Muqla with those proposed by other scribes.

An explanation of "measured writing" that puts greater stress on its aesthetic implications is contained in an encyclopedic treatise entitled *Rasāʾil* (Epistles) compiled circa 980 by a group of scholars known by the pseudonym Ikhwān al-Ṣafāʾ (the Brethren of Purity). This treatise blends Neoplatonic, neo-Pythagorean and Near Eastern concepts into a hierarchical cosmological system, and had the didactic purpose of guiding people toward salvation through a stage-by-stage mastery of religious and philosophical concepts.

According to the *Rasāʾil,* the skilled craftsman and his work occupies the first stage in a spiritual and religious hierarchy that culminates with the prophets and their divinely inspired revelations. Humans are viewed as microcosms of the universe, so that their actions and reactions mirror its fundamental structure. Thus, the painter who creates a painting, the weaver who produces cloth, or the calligrapher who copies a text can reflect the principles of harmonious proportion that underlie the universe as a whole. This underlying structure is also expressed through models or archetypes that could be rendered verbally, visually, or numerically. When viewed from this perspective, the geometrical and proportional structure of "measured writing" exemplifies how man's actions are linked to the broader universal order. The Ikhwān al-Ṣafāʾ explain how each letter can be given its proper proportion and shape by comparing it to the *alif* or its circumscribed circle, but they stress that mastery of the letters is only the first stage in becoming a calligrapher.

The Ikhwān al-Ṣafāʾ also present figure painting as one of the skills that reflect the harmony and beauty of the universe. They stress the importance of mixing colors in a proper proportion and the need to depict the human figure with a harmonious ratio among its constituent parts. Their view that the skill of the painter in depicting the human body provides a reflection of the underlying beauty and harmony of the universe must have had a classical, probably Neoplatonic, source. Their discussion is significant because it offers a counterargument to the criticism voiced in religious circles that the mimetic skill of the painter was presumptuous or even blasphemous because it sought to compete with God's creative power. Although it is unclear whether these views, which gave a new respectability and even a degree of religious merit to the occupation of figure painter, had an immediate impact, they did provide a rationale used by later supporters of the painter's craft.

Harmonic proportion and visual balance are also central to the observations on beauty *(al-ḥusn)* contained in the *Kitāb al-Manāẓir,* a treatise on optics by Abū ʿAlī ibn al-Ḥasan, known as Ibn al-Haytham (c.965–1039). It is in keeping with his stress on rationality and his distrust of dogmatic positions that he draws examples of beauty from his

own personal observations. He states that beauty arises from the process of vision itself and can be produced by qualities such as light, color, shape, size, roughness, and smoothness, either singly or in conjunction with each other. Many of Ibn al-Haytham's examples concern the impact of light on the eye. He explains the attraction of the full moon by noting the appeal of large luminous bodies and regular shapes, and of the contrast between its radiance and the darkness of the sky around it. He also notes the appeal of clusters of stars or candles that are sufficiently dispersed to be individually perceptible.

Ibn al-Haytham's choice of examples also reflects the values of his epoch. He makes no mention of painting, sculpture, or architecture but returns repeatedly to the question of what produces beauty in writing and in the human form. With respect to scripts, he notes the cumulative impact of proportional harmony and visual balance. Letters must be regularly shaped and joined in a harmonious manner, and yet have sufficient "dissimilarity" in the width of their contours to avoid ugliness. With respect to human beauty, he praises a particular physical type with a well-rounded visage, flat yet delicately shaped cheeks, large almond-shaped eyes, and a small mouth with thin to moderately thick lips.

While Ibn al-Haytham's observations on the beauty of harmonic proportion is in basic agreement with many of the observations of the Ikhwān al-Ṣafā', he does not connect those perceptions to an underlying numerical or cosmological system. Others, however, chose to emphasize those factors and to diminish the role of the senses in perceiving beauty. The philosopher Abū Ḥāmid al-Ghazālī identifies two different types of beauty, one recognized through sense perception and the other comprehended only through an inner vision that he calls the "eye of the heart." The nature of the beauty apprehended by the senses differs from case to case. The features that make a horse beautiful are distinct from those that create beauty in a landscape, a human body or voice, a garment, a vessel, or handwriting, yet God's power is revealed through all the works of his creation. Al-Ghazālī's emphasis on the parallels between divine and human creativity may reflect the esteem accorded in his day to both the visual arts and their creators. During the twelfth and thirteenth centuries finely executed metalwork, ceramics, and manuscripts from Central Asia, Iran, and Iraq were increasingly inscribed with laudatory texts and with the names of their makers.

The popularity of figural wall painting and manuscript illustration appears to have broadened during the fourteenth and fifteenth centuries, particularly in the area of the former Abbasid Empire. However, a full synthesis of theory and practice occurs only in the sixteenth and seventeenth centuries, when apologists for painting, who were usually connected with princely courts, used the dichotomy between inner significance (ma'nā) and outer form (ṣūra) stressed by al-Ghazālī and his followers to explain the relationship between an artist's visual imagination and his paintings. These efforts are exemplified by Dost Muḥammad's essay on painting and calligraphy that served as the preface to an album compiled in 1544 for the Safavid prince Bahrām Mīrzā b. Shāh Ismāʿīl (1517–1549). He explains that the legendary paintings of Mānī were drawn from the "mirror of his mind" and the "eye of his imagination." Another album preface by Mīr Sayyid Aḥmad Mashhadī, written in 1564–1565, praises the painter's ability to depict the prototype underlying the outer reality of a thing or person. He links that skill with the vision of beauty that is formed in the painter's visual imagination or "the tablet of painter's mind." Ṣādiqī Beg Afshār's Qānūn al-ṣuwar (The Canon of Painting), written between 1576 and 1602, offers practical advice on technique for the aspiring painter. It also opens with a defense of hunar (art) in which he states that the painter's work can reveal the underlying qualities of things stripped of their transitory features because his creations derive from his creative imagination (khiyāl), which registers the intrinsic reality of things. These formulations create implicit parallels between the painter's creative role, and that of God, a view expressed even more directly by the Mughal emperor Akbar (reigned 1556–1605). His biographer, Abū al-Faẓl ʿAllāmī, quotes him as saying that the painter's skill in depicting living creatures has religious merit because it recalls God's creative power.

BIBLIOGRAPHY

Abū al-Faẓl ʿAllāmī. *The Āʾīn-i Akbarī.* Translated by H. Blochmann, edited by D. C. Phillott. Delhi, 1989. See vol. 1, pp. 113–115.

Abū al-ʿAbbās Aḥmad Qalqashandī. *Ṣubḥ al-aʿshā.* 3 vols. Cairo, 1914. See vol. 3, pp. 24–49.

Dost-Muhammad. "Preface to the Bahram Mirza Album." In *A Century of Princes: Sources on Timurid History and Art,* edited and translated by W. M. Thackston, pp. 335–352. Cambridge, Mass., 1989.

Ettinghausen, Richard. "Al-Ghazzālī on Beauty." In *Art and Thought: Issued in Honor of Dr. Ananda K. Coomaraswamy on the Occasion of His 70th Birthday,* edited by K. Bharatna Iyer, pp. 160–165. London, 1947.

[Ibn al-Haytham]. *The Optics of Ibn al-Haytham: Books I–III: On Direct Vision.* 2 vols. Translated and edited by A. I. Sabra. London, 1989. See vol. 1, pp. 200–206; vol. 2, pp. 97–102.

Ikhwān al-Ṣafā'. *Rasā'il.* 4 vols. Beirut, 1957. See esp. vol. 1, pp. 219–222, 242–245, 252–253.

Marquet, Yves. "La place du travail dans la hiérarchie ismailienne." *Arabica* 8.3 (1961): 225–237.

Mīr Sayyid Aḥmad Mashhadī. "Preface to the Amir Gayb Beg Album." In *A Century of Princes: Sources on Timurid History and Art,* edited and translated by W. M. Thackston, pp. 353–356. Cambridge, Mass., 1989.

Pellat, Charles. "Les esclaves-chanteuses de Ǧāḥiẓ." *Arabica* 10 (1963): 121–147. See esp. 135–136.

Rice, David S. *The Unique Ibn al-Bawwāb Manuscript in the Chester Beatty Library.* Dublin, 1955. See pp. 5–9.

Rosenthal, Franz. "Abū Haiyān al-Tawḥīdī on Penmanship." *Ars Orientalis* 13–14 (1948): 1–30.

Rosenthal, Franz. "On Art and Aesthetics in Greco-Arabic Wisdom Literature." In *Four Essays on Art and Literature in Islam,* pp. 1–19. Leiden, 1971.

Rosenthal, Franz, ed. *The Classical Heritage in Islam*. Translated by Emile Marmorstein and Jenny Marmorstein. London, 1975. See esp. pp. 265–266.

Ṣādiqī Bek. "The Canons of Painting." In *The Houghton Shahnameh*, edited by Martin Bernard Dickson and Stuart Cary Welch, vol. 1, pp. 259–269. Cambridge, Mass., 1981.

Shiloah, Amnon. "L'épitre sur la musique des Ikhwan al-Ṣafāʾ: Traduction annotée." *Revue des études islamiques* 32 (1964): 125–167; 34 (1966): 160–193. See esp. vol. 34, pp. 166–172.

PRISCILLA SOUCEK

ITALIAN AESTHETICS. See Alberti; Aquinas; Architecture, *article on* Italian Renaissance Aesthetics; Croce; Eco; Futurism; Gentile; Gramsci; Origins of Aesthetics; Pareyson; Renaissance Italian Aesthetics; Roman Aesthetics; Vasari; *and* Vico.

J

JAKOBSON, ROMAN. *See* Metonymy; Russian Aesthetics, *article on* Russian Formalism; *and* Structuralism.

JAMESON, FREDRIC (b. 1934), American philosopher and cultural critic. Fredric Jameson is generally considered to be one of the foremost contemporary English-language Marxist literary and cultural critics. Over the past three decades, he has published a wide range of works analyzing literary and cultural texts, while developing his own neo-Marxist theoretical perspectives. In addition, Jameson produced many important critiques of opposing theoretical schools and positions. A prolific writer, he has assimilated an astonishing number of theoretical discourses into his project, while intervening in many contemporary debates and analyzing a diversity of cultural texts, from the novel to video, from fairy tales to postmodernism.

Jameson's first published book analyzed the literary theory and work of Jean-Paul Sartre (1961; 2d ed., 1984). Written as a doctoral dissertation at Yale University, *Sartre: The Origins of a Style* was influenced by Jameson's teacher Erich Auerbach and by the stylistics associated with Leo Spitzer. The text focused on Sartre's style, narrative structures, values, and vision of the world. It is devoid of the Marxian categories and political readings characteristic of Jameson's later work, but read in the context of the stifling conformism and banal business society of the 1950s, Jameson's choice of subject matter (Sartre) and his choice of an intricate literary-theoretical writing style (already the infamous Jamesonian sentences appear full-blown) can be seen as an attempt to create himself as a critical intellectual against the conformist currents of the epoch. One also sees him already turning against the literary establishment, against the dominant Anglo-American modes of literary criticism, seeking alternatives in European theory. Indeed, all of Jameson's subsequent works constitute critical interventions against the hegemonic forms of literary criticism and modes of thought regnant in the Anglo-American world, and throughout his career he has drawn heavily on Continental European methods, modes of thought, and discourses to provide alternatives to the more prosaic literary and cultural theories dominant in the United States.

After intense study of Marxian literary theory in the 1960s, when he was influenced by the New Left and the antiwar movement, Jameson published *Marxism and Form: Twentieth-Century Dialectical Theories of Literature* (1971), which introduced a tradition of dialectical neo-Marxist literary theory to the English-speaking world. After engaging the structuralist project in *The Prison House of Language: A Critical Account of Structuralism and Russian Formalism* (1972), Jameson concentrated on developing his own literary and cultural theory in a series of works that includes *Fables of Aggression: Wyndham Lewis, the Modernist as Fascist* (1979), *The Political Unconscious: Narrative as a Socially Symbolic Act* (1981), and *Postmodernism; or, The Cultural Logic of Late Capitalism* (1991). He has also published several volumes of essays, including *The Ideologies of Theory* vol. 1, *Situations of Theory,* and vol. 2, *Syntax of History,* both published in 1988). There followed *Signatures of the Visible* (1990), which collects some essays on film and visual culture; *The Geopolitical Aesthetic: Cinema and Space in the World System* (1992), a collection of essays on American and world cinema; and *The Seeds of Time* (1994), which collected his recent articles in cultural criticism. Jameson has also published a book on Theodor W. Adorno, *Late Marxism: Adorno; or, The Persistence of the Dialectic* (1990), in which he presents this great cultural critic of the Frankfurt School as a key theorist for our time.

Jameson has characteristically appropriated into his theory a wide range of theoretical positions ranging from structuralism to poststructuralism, from psychoanalysis to postmodernism, producing a highly eclectic and original brand of Marxian literary and cultural theory. Yet, throughout his work Jameson privileges Marxism as the most all-encompassing and comprehensive theoretical framework, within which other methods function as local or factional tools. In *The Political Unconscious,* Jameson argues for the primacy of Marxism on the grounds that its horizon—history and the socioeconomic totality—provides the most comprehensive framework in which gender, race, class, sexuality, myth, symbol, allegory, and other more limited concerns can be explored and interpreted. In his later works, which reach out to grasp the elusive world system as a whole, Jameson argues that it is capitalism and its processes of commodification and reification that provide the motor and matrix of today's world system, especially after the collapse of Soviet communism. Consequently, his work as a whole should be read as a series of attempts

to provide a Marxian method of interpretation and aesthetic theory.

Jameson's Marxism is far from conventional. Influenced by the Marxist theorist Ernst Bloch, Jameson utilizes a dual hermeneutic of ideology and utopia to criticize the ideological components of cultural texts while setting forth their utopian dimension, which includes those visions of a better world that provide perspectives from which to criticize existing society. For Jameson, even conservative texts often project visions of a better world and thus criticize the organization of existing society and its values.

Jameson's focus on the utopian dimension interests him in fantasy, science fiction, and other forms of popular culture that he believes contain utopian and critical moments. Yet, he is also deeply attracted to realist texts that he believes provide both knowledge and criticisms of existing capitalist society. György Lukács's work on realism and on the historical novel strongly influenced Jameson's way of seeing and situating literature. Although he has never accepted Lukács's polemics against modernism, Jameson appropriated key Lukácsian categories such as reification, the process through which human beings and their products become thinglike, to describe the fate of culture and human beings in contemporary capitalism. The Lukácsian-Hegelian markers of Jameson's work include the contextualizing of cultural texts in history, broad historical periodizing, and the use of Hegelian-Marxian dialectical categories and methods.

Dialectical criticism for Jameson involves the attempt to synthesize competing positions and methods into a more comprehensive theory, as he does in *The Prison House of Language*, in which he incorporates elements of French structuralism and semiotics, as well as Russian Formalism, into his theory. Dialectical criticism also involves for Jameson thinking that reflexively analyzes categories and methods, while carrying out concrete analyses and studies. Categories for Jameson articulate historical content and thus must be read in terms of the historical environment out of which they emerge. Dialectical criticism thus involves thinking that reflects on categories and methods while carrying out concrete analyses and studies; relational and historicizing thinking that contextualizes the object of study in its historical environment; utopian thinking that compares the existing reality with possible alternatives and finds utopian hope in literature, philosophy, and other cultural texts; and totalizing, synthesizing thought that provides a systematic framework for cultural studies and a theory of history within which dialectical criticism can operate.

Jameson thus develops in his early 1970s work a totalizing, synthesizing thought which provides a systematic framework for cultural studies and a theory of history within which dialectical criticism can operate. All of these aspects continue to be operative in Jameson's later works with the totalizing element coming more prominently (and

controversially) to the fore as his work evolved. It is in *The Political Unconscious* that Jameson's theoretical synthesis finds its most systematic articulation. The text contains an articulation of Jameson's literary method, a systematic inventory of the history of literary forms, and a hidden history of the forms and modes of subjectivity itself, as it traverses the field of culture and experience. Jameson employs a Lukácsian-inspired historical narrative to tell the story of how cultural texts contain a "political unconscious"—buried narratives and social experiences that require sophisticated literary hermeneutics to decipher. One particular narrative of *The Political Unconscious* concerns, in Jameson's striking phrase, "the construction of the bourgeois subject in emergent capitalism and its schizophrenic disintegration in our own time" (Jameson, 1981, p. 9). Key stages in the odyssey of the disintegrating bourgeois subjectivity are articulated in George Gissing, Joseph Conrad, and Wyndham Lewis, a story that will find its culmination in Jameson's account of postmodernism.

Indeed, Jameson's studies on postmodernism are a logical result of his theoretical project. In *Postmodernism; or, The Cultural Logic of Late Capitalism* (1991), which updates his studies of the 1980s, Jameson interprets postmodernism in terms of the Marxian theory of capitalism as a new "cultural dominant." Within his analysis, Jameson situates postmodern culture in the framework of a theory of stages of society—based on a neo-Marxian model of stages of capitalist development—and argues that postmodernism is part of a new stage of capitalism. In developing his concept of postmodernism, Jameson (1984 and 1991) tends to underscore the rupture with modernism, claiming that postmodern culture implodes high culture and mass culture, while exhibiting a "new depthlessness" that resists interpretation; involves pastiche and not parody; reduces history to historicity, to a stereotyped and clichéd set of images that pander to nostalgia rather than promote genuine historical understanding; and ruptures narrative and subjects in a "schizophrenic" dispersal of fragments. This postmodern culture loses its critical distance and is an embellishment to existing society. Jameson claims that postmodernism is a new "cultural logic of capitalism," that its fragmented image culture and aestheticization are part of a shift to a new global capitalism and that postmodernism is therefore not just another aesthetic style beside modernism but rather a new cultural dominant.

Jameson's analysis of postmodernism has been submitted to many sharp criticisms. Linda Hutcheon (1988 and 1989), for example, argues that he has a one-sided conception of postmodern culture and misses its engagement with history, parody, and double-coded distance and complicity with mainstream culture. Others object that postmodernism is not, as Jameson claims, a cultural dominant, but one aesthetic style among many that is novel and provocative, though far from dominant. In this light, many critics say

that Jameson's conception is too totalizing and object to his correlation of stages of culture with stages of capitalism. Others contend that he excessively privileges class and does not pay sufficient attention to race and gender. Moreover, almost every assertion that Jameson makes in his now classic analysis of postmodernism has been subject to detailed polemic and discussion (see the studies in Kellner, 1989).

In an age marked by poststructuralism and postmodern skepticism, Jameson's totalizing mode of thought has been under attack. Yet, he attempts to incorporate key aspects of poststructuralist and postmodern theory into his work and to strengthen Marxism by bringing into it modes of thought that are often taken as antithetical to Marxian dialectic. This in turn has generated great controversy, with more orthodox Marxists claiming that Jameson has illicitly diluted Marxian thought, while poststructuralists and others complain about tensions between different modes of thought in Jameson's work (see the discussion in Best and Kellner, 1991).

Jameson thus emerges as a synthetic and eclectic Marxian cultural theorist who attempts to preserve and develop the Marxian theory while analyzing the politics and utopian moments of a stunning diversity of cultural texts. His work expands literary analysis to include popular culture, architecture, theory and other texts, and thus can be seen as part of the movement toward cultural studies as a replacement for canonical literary studies. Nothing cultural is alien to Jameson's synoptic grasp, and his restless and wide-ranging studies continue to illuminate our contemporary culture and to map its future, while elaborating valuable methods for the study of culture and society.

BIBLIOGRAPHY

Works by Jameson

Sartre: The Origins of a Style. New Haven, 1961; 2d.ed., New York, 1984.

Marxism and Form: Twentieth-Century Dialectical Theories of Literature. Princeton, N.J., 1971.

The Prison House of Language: A Critical Account of Structuralism and Russian Formalism. Princeton, N.J., 1972.

Fables of Aggression: Wyndham Lewis, the Modernist as Fascist. Berkeley, 1979.

The Political Unconscious: Narrative as a Socially Symbolic Act. Ithaca, N.Y., 1981.

"Postmodernism: The Cultural Logic of Late Capitalism." *New Left Review* 146 (1984).

"Third-World Literature in the Era of Multinational Capitalism." *Social Text* 15 (1986).

"Cognitive Mapping." In *Marxism and the Interpretation of Culture,* edited by Larry Grossberg and Cary Nelson. Urbana, Ill., 1988.

Late Marxism: Adorno; or, The Persistence of the Dialectic. London and New York, 1990.

Signatures of the Visible. New York and London, 1990.

Postmodernism: or, The Cultural Logic of Late Capitalism. Durham, N.C., 1991.

The Geopolitical Aesthetic: Cinema and Space in the World System. Bloomington, Ind., 1992.

The Seeds of Time. New York, 1994.

Other Sources

Best, Steven, and Douglas Kellner. *Postmodern Theory: Critical Interrogations.* New York, 1991.

Dowling, William C. *Jameson, Althusser, Marx: An Introduction to "The Political Unconscious."* Ithaca, N.Y., 1984.

Hutcheon, Linda. *The Poetics of Postmodernism: History, Theory, Fiction.* New York and London, 1988.

Hutcheon, Linda. *The Politics of Postmodernism.* London and New York, 1989.

Kellner, Douglas, ed. *Postmodernism/Jameson/Critique.* Washington, D.C., 1989.

Sprinker, Michael. *Imaginary Relations: Aesthetics and Ideology in the Theory of Historical Materialism.* London and New York, 1987.

Diacritics 12 (Fall 1982); *Critical Exchange* 14 (Fall 1983); and *New Orleans Review* (1984) have special issues on Jameson's work.

DOUGLAS KELLNER

JAPANESE AESTHETICS. [*To explore the artistic and aesthetic traditions in Japan, this entry comprises two essays on this distinct artistic and aesthetic tradition:*

Historical Overview

Kire and Iki

The first essay surveys the history of Japanese aesthetics and the ways in which it is similar to and different from Western aesthetics. The second essay examines the particular concepts of kire *and* iki, *or "cutting" and "style." For other essays on Japanese aesthetics, see* Haiku; *and* Nature, *article on* Japanese Aesthetic Appreciation of Nature. *For related discussion, see* Beauty; Chinese Aesthetics, *historical overview article;* Comparative Aesthetics; Disinterestedness; *and* Landscape.]

Historical Overview

The aesthetic traditions of Japan have become established over its long history. The process of Westernization began in the late nineteenth century, and Japan has rapidly transformed itself into an industrialized, technologically advanced culture. It may appear that the traditional (i.e., pre-Westernization) Japanese aesthetic ideas, values, and practices have all but disappeared amidst the skyscrapers and high-tech industries. The indigenous Japanese aesthetic sensibility still exists, however, not only in classical art forms such as the tea ceremony, nō drama, and haiku, but also in daily activities such as packaging, cooking, and even etiquette. While Japanese aesthetics by no means designates a uniform development, several common themes persist throughout its diverse historical and artistic manifestations. Such themes, discussed here, together comprise what may be termed a uniquely or characteristically Japanese aesthetic tradition.

Aesthetic Egalitarianism. Perhaps the most prominent aspect of Japanese aesthetics is that aesthetic concern affects every area of people's lives. Both in producing and in appreciating, the object of an aesthetic experience is not

limited to what is generally referred to as fine arts, such as painting, music, and literature. Other objects and phenomena, whether made by human beings or natural, are equally celebrated for giving rise to rewarding aesthetic experiences in the Japanese tradition. So are numerous daily practices such as cooking, bathing, and gift wrapping, as well as physical exercises, such as archery and swordsmanship. Instead of comprising a special realm divorced from everyday life, aesthetic matters are thoroughly integrated into Japanese life, past and present.

This inclusivity, which may be called aesthetic egalitarianism, extends further. The Japanese tradition does not make the Western distinction between the so-called higher and lower senses, the former generally considered as more legitimate vehicles for conveying aesthetic values. A typical aesthetic experience, according to the Japanese tradition, engages several senses, which together inform the aesthetic value.

The best example of such an experience that does not privilege any particular realm or senses can be found in an art medium unique to Japan: the tea ceremony. Established during the sixteenth century, this art form elevates the otherwise mundane activities of serving and drinking tea and eating snacks to a ritualized event. No limit is set on what can become its aesthetic content. Participants are encouraged to attend not only to the tea itself but also to the surrounding environment—the garden, tea hut, privy, and weather, as well as objects such as the tea bowl, flowers, the scroll decorating the alcove, and various utensils for making the tea. Equally important are the host's elegant and seamless bodily movements in preparing and serving the tea, as well as the content of the conversation that ensues spontaneously among the guests.

Furthermore, the participant's experience of each item engages several different senses. For example, the entire body is engaged in walking through the garden, kneeling to cleanse one's hands and mouth in the water basin, crawling into the tea hut through a small entrance, taking one's seat, and lifting the tea bowl. The activity of drinking tea itself involves not only the senses of taste and smell, but also the sound of slurping, the tactile sensations of the bowl's surface texture and the warmth of its contents, as well as its visual effect. The tea hut is also appreciated for its visual and spatial factors, the sound of the water boiling and the leaves rustling in the garden, and the tactile sensations of the wood grain and the weave of the straw mat against one's feet. These elements are valued all the more because of the awareness, encouraged by the philosophy behind this art form, that this specific occasion will never repeat itself. It is literally a once-in-a-lifetime event *(ichigo ichie).*

While the tea ceremony is an explicitly participatory art form, many other Japanese aesthetic activities both entice and challenge. The preferred mode of indirect communication and artistic minimalism (to be explained later) requires that the audience take an active role in facilitating an aesthetic experience.

Several factors, both religious and historical, have contributed to this egalitarianism. Chronologically, the first contribution can be traced to the indigenous Japanese religion, Shintō, according to which the dynamic, ever-changing spirit, *kami*, resides in every natural object. Neither an abstract entity nor an entity existing in the other world, deity is always embodied in concrete objects. That is, this world is viewed as constituting reality itself rather than as a faint shadow or a veil covering reality that transcends this world. The consecration of objects making up this world, nature in particular, provided the foundation for the subsequent Japanese tendency to find aesthetic value in everything. The Shintō worldview also determined the decidedly "this-worldly" characteristic of Japanese representational art. While some Japanese paintings do depict other-worldly mandalas, the esoteric Buddhist metaphysical worldview, the subject matter of Japanese art is predominantly nature—landscapes, flowers, and animals. Similarly, Japanese literary works generally shun abstract metaphysical speculations and instead, deal primarily with concrete objects and events.

This attention to every aspect of experience became firmly established among the nobility during the Heian court period (794–1185), thus setting the standard for cultural sophistication. Supported by landed wealth, the Heian aristocracy could afford to indulge in cultivating sensibility through literature, calligraphy, nature appreciation, music, fashion, courtship ritual, and even incense connoisseurship. In courtship ritual, consisting primarily of exchanging poems, aesthetic sophistication was regarded as the sole criterion for one's worthiness as a lover, a cultured person, and sometimes even as an ethical person. The relevant considerations included not only the poem itself, its sensitivity to the season, the weather, and the occasion, but also the calligraphy, the selection of paper, the attachment of a flower or a twig suitable for the poem, and the appropriate fragrance of incense.

Although these specific rituals did not survive beyond the Heian period, this legacy of the keen sensitivity to every aspect of an object or phenomenon has persisted throughout Japanese history. It was first embraced by the warriors who seized political power after the decline of the Heian aristocracy. Commoners by birth, these warriors made a conscious effort to emulate the cultural accomplishments of the aristocrats by training themselves in the arts in addition to their military pursuits. It was under their patronage that many Japanese art forms, such as the tea ceremony, Nō theater, screen painting, and architecture, flourished.

The spiritual foundation for the medieval warriors' preoccupation with culture and the arts was Zen Buddhism,

introduced from China to Japan toward the end of the twelfth century. The warriors embraced Zen Buddhism primarily for its teaching concerning mental and physical discipline, a requirement for a successful military career, as well as for its accompanying cultural and artistic sophistication. It was Zen metaphysics that furthered aesthetic egalitarianism. Zen metaphysics consists of transcending ordinary discriminations and categorizations in its enlightened experience of the ultimate reality, Buddha nature. All objects and phenomena are rendered equally real, valuable, and worthy of appreciation simply by virtue of "being such" or "being thus." The commonly held distinctions between sentient and nonsentient, noble and ignoble, valuable and unvaluable are overcome by a thoroughgoing Zen egalitarianism, resulting in the absolute affirmation of the facticity of everything. The thirteenth-century Zen priest Dōgen (1200–1253) hence describes the prevalence of Buddha nature by referring to a donkey's jaw, a horse's mouth, the sound of breaking wind, and the smell of excrement, as well as by pointing to grasses, trees, mountains, and streams.

This uncompromising Zen commitment to appreciating the "thusness" of everything is manifested in several Zen-inspired art forms. For example, Matsuo Bashō (1644–1694), the most noted haiku master, sometimes took animals' bodily functions, fleas, and lice as his subject matter and elevated them to aesthetic heights. Similarly, a superb garden design is created by arranging otherwise ordinary rocks and pebbles or by reusing "found objects" such as discarded millstones and abandoned temple foundations in the construction of stepping-stones. The same observation can be made regarding flower arrangement, an art form that predates the introduction of Zen Buddhism but one to which Zen subsequently gave a theoretical foundation. The materials for arrangement need not be opulent flowers; in fact, withered twigs or branches with simple, unassuming flowers are often favored.

By the time the common townsfolk started commissioning or supporting popular arts such as *kabuki* theater, woodblock prints, and genre literature during the Edo period (1600–1867), aesthetic sensitivity to every aspect of daily life had become thoroughly entrenched in all segments of Japanese society. This prevalence of the aesthetic sensitivity and its integration with life continue even today amidst the thoroughly modernized, high-tech life. In addition to participating in traditional arts such as the tea ceremony, flower arrangement, and calligraphy, contemporary Japanese lives are still permeated by aesthetic concerns. Seasonal haiku are featured on the front page of major newspapers; both at home and in restaurants, food is carefully arranged on a plate specifically chosen to suit both the food and the season; a package is meticulously wrapped with the appropriate material; and even business letters invariably contain some reference to the beauty of the season.

The possibility of aestheticizing every object and phenomenon, however, raises a question about the feasibility of any criteria for aesthetic value. That is, if anything whatsoever can provide an aesthetic experience, what makes some qualities aesthetically more positive than others? What are some specific aesthetic values, according to Japanese aesthetics?

The Aesthetics of Implication, Suggestion, Imperfection. One such set of criteria of aesthetic value in the Japanese tradition is given by a cluster of concepts: *yojō* (emotional aftertaste in poetry), *wabi* (the beauty of poverty in the tea ceremony), *sabi* (lonely beauty in haiku), and *yūgen* (mystery and depth, primarily in Nō theater).

Yojō refers to the quality of a poem in which the words do not fully reveal the emotion the poet is trying to express. Frequently likened to the moon hidden behind clouds or autumnal mountain scenery shrouded in mist, this suggestion of lingering emotion through understatement is used as the evaluative criterion for the aesthetic value of a poem.

Landscape imagery also appears in the definition of *wabi* given by tea masters. Takeno Jōō (1502–1555), for example, cites the following poem by Fujiwara no Teika (1162–1241), for illustrating *wabi*:

> Looking about
> Neither flowers
> Nor scarlet leaves,
> A bayside reed hovel
> In the autumn dusk.

A little later Sen no Rikyū (1522–1591) refers to a poem by another poet, Fujiwara no Ietaka (1158–1237), for the same purpose:

> To those who wait
> Only for flowers,
> Show them a sprig
> Of grass amid the snow,
> In a mountain village.

Both poems depict rather desolate and colorless landscapes in contrast to a more typical autumnal landscape with its brilliantly colored leaves or a spring landscape with the lustrous pink of cherry blossoms.

Yūgen, although applicable to poetry, was primarily developed as the aesthetic ideal for Nō theater by Zeami (sometimes Seami) Motokiyo (1362–1443), as well as for linked verse *(renga)* by Shinkei (1401–1475). The Nō theater consists of a small number of actors, at least one of whom wears a mask, conveying a relatively well known story of historical figures or supernatural events through minimal, stylized, and usually slow body movements. The drama takes place on a sparse stage with few props, accompanied by both instrumentalists and singers seated on the stage. Linked verse started out as a kind of parlor game involving a number of people, and it later developed into an art medium and be-

came a precursor of haiku. Each participant takes turns improvising a short verse that builds on the verse made by the previous person by maintaining some continuity of imagery while introducing a new element.

Yūgen, according to Zeami, is the highest and most ineffable beauty achievable in a Nō play, which is illustrated by, among other things, a lonely snowless peak among many snow-clad mountain tops, and a silver bowl filled with snow, suggesting chillingly exquisite and severe beauty achieved without opulence. Shinkei gives the examples of the glimpse of moon through clouds as well as a single white blossom amidst bamboo, reminding us of *yojō* and *wabi*.

Derived both from *sabishi* (loneliness) and *sabi* (rust), *sabi* refers to the austere, stark, and rustic atmosphere conjured up primarily in haiku, but also in other literary expressions, the tea ceremony, and Nō theater. Because of its association with rust, it also refers to agedness.

Although initially put forward to describe the genre-specific aesthetic value, these concepts subsequently become more widely applicable. Merging with each other, they indicate some uniform preferences that characterize the Japanese aesthetic tradition. First, they all favor artistic minimalism. For example, the two most celebrated poetic forms in Japanese literature are extremely short: *waka* and *haiku*, which are 5-7-5-7-7- and 5-7-5-syllable verses, respectively. Their brevity both necessitated and encouraged the expression of the poetic content through implication and suggestion. An explicit and comprehensive account is neither possible nor desirable. Similarly, the Nō play conveys deep and rich emotions through restricted body movements, subtle tilts of the mask, a sparsely set stage, and the monotonous, drone-like chanting of the chorus. As for rock gardens and brush ink paintings, they avoid colorfulness and consist of monochrome shading. Their respective spaces are filled only with minimal elements, although their empty spaces are as significant and fertile in expressive power as their filled spaces. Some ink paintings, by Zen priests in particular, consist only of a circle. Flower arrangement, as mentioned previously, shuns showy bouquets of flowers, instead favoring sparse elements such as a single branch. Finally, while affording multisensory aesthetic experiences through various elements, everything about the tea ceremony underscores simplicity, from the tea hut to the utensils, from the tea bowls to the food accompanying the tea.

This taste for minimalism gives rise to a further preference for indirect communication. This standard was established in the first work of Japanese literary criticism, the Preface to *Kokinshū* (The Anthology of Ancient and Modern Poems), compiled in 905. Its editor, Ki no Tsurayuki (868?–946), claims that in a good poem the emotive content is conveyed in terms of or by way of nature, evoking a certain mood or atmosphere. A straightforward outpouring of feelings and ideas, while it may provide optimal, unambiguous communication, is considered crude and uncouth,

hence aesthetically undesirable. Indeed, such directness is still considered aesthetically and even ethically inferior in Japanese discourse, which consistently prefers implication, allusion, and innuendo.

When applied to the visual realm, this preference for suggestion and implication results in appreciating things that are hidden or obscured, rather than fully exposed. If the optimal view of the moon requires a clear sky and unobstructed view, the moon's beauty is heightened by a cloud covering it, or a tree blocking its view. Similarly, the allure of distant mountains is enhanced by mist. Hence, when constructing a garden for strolling, one should not make a scenic view visible all at once; rather, the view should be partly hidden, inviting the stroller to follow the path, which will provide a succession of unfolding scenes. This aesthetic sensibility is also manifested in the Japanese fondness for wrapping gifts or packaging objects. Inordinate aesthetic considerations go into engaging people's senses in undoing a package one layer at a time.

The historical root of this preference for implied or hidden things can again in part be traced to the Heian sensibility that was developed for the courtship ritual. At the initiation of the courtship, the lady is always situated inside a room adjacent to an open veranda with a shade or curtain drawn inbetween. This arrangement gives the suitor a glimpse of the bottom of the lady's many-layered kimono peeking out beneath the curtain, a glimpse that is sufficient to display her taste and pique his interest. What began as a means of seduction subsequently became a norm that favors implication and obscurity.

The reason for this preference is stimulation of the imagination. Implication, obscurity, and hiddenness are considered more attractive than a straightforward expression or unhindered view, because a certain mysteriousness on the surface entices further exploration by engaging subjects' imaginations. Appreciation is based not only on what is said or seen, but also on what is not said or not seen, as well as on the manner in which the two are contrasted.

The allure of implication and obscurity, together with artistic minimalism, poses a challenge to the recipient of the aesthetic experience. On the one hand, it is demanded that one be satisfied with whatever is given, no matter how seemingly incomplete or insufficient it may be. On the other hand, participation and the engagement of one's imagination are required to complete the experience. One needs to feel the intense emotion suggested by the subtle tilt of a nō mask, to explore fully and appreciate the poetic content only hinted at by a poet, and to celebrate the essence of a specific season represented by a single branch in a vase.

Another aspect of aesthetic preference stemming from *wabi*, *sabi*, and *yūgen* is the celebration of objects or phenomena falling short of their optimal condition, things not normally appreciated because of their failure to entertain us with opulence and perfection. If cherry blossoms are in

prime condition at full bloom, their aesthetic value is increased when they are in decline; that is, when they start falling. Or, if a perfect tea bowl consists of an unblemished surface, a regular contour, and an orderly pattern, an even higher aesthetic value is accorded a chipped and cracked surface, discoloration, irregular shape, and a disorderly pattern brought about either by accident, aging, or design. The tea masters' pursuit of this studied imperfection extends to other implements used in the tea ceremony. One master is said to have replaced the luxurious cloth framing of a scroll painting with shabby material, while another is known for intentionally breaking off the handle of a flower vase. Finally, many other Japanese objects are made with materials that age gracefully. Unpainted wood is used in architecture to show prominent grain resulting from use and weathering, clay pottery exhibits crackles with repeated handling, and lacquerware reveals its inner layer of a different color when rubbing wears off the surface lacquer.

The appreciation of the aged and the imperfect is again a legacy of Zen philosophy. Aging and accidents, universal circumstances governing both objects and human beings, are not cause for lamentation but rather are fully embraced in Zen because every state of an object, including its decline and destruction, expresses its "thusness." It should be noted that what starts out as paying equal homage to those aesthetic values normally un- or underappreciated results in overcompensating the common tendency by undervaluing more popular aesthetic values.

This emphasis on appreciating the defective and the decaying underscores, however, the prevalence of what it sets out to overcome: the preference for beautiful objects in perfect condition. This contrast in taste gave rise to one of the few debates in the Japanese aesthetic tradition. On one side are the advocates for the more challenging aesthetic qualities of impoverishment, obscurity, and agedness, represented most notably by Yoshida Kenkō (1283–1350), a retired Buddhist priest whose collection of essays titled *Tsurezuregusa* (Essays in Idleness, c.1340) is often regarded as the manifesto of this viewpoint. In discussing the beauty of falling cherry blossoms and the moon hidden by clouds, he contrasts such aesthetic appreciations with a more prevalent and "natural" sentiment of people who "*commonly* regret that the cherry blossoms scatter or that the moon sinks in the sky" and claims that a person who only derives a negative aesthetic experience from such a situation is "exceptionally insensitive." That is, while recognized as natural and prevalent, the penchant for opulence and perfection is criticized as the taste of the uneducated, unsophisticated common folk. Connoisseurs, according to the advocates of *wabi*, will not give in to such easy attractions; a refined and cultivated sensitivity is required for appreciating what may be regarded as more challenging aesthetic values.

Several centuries later, however, a noted philologist and literary critic, Motoori Norinaga (1730–1801), criticizes the

JAPANESE AESTHETICS: Historical Overview. *Shino Style White-Glazed Chawan*, possibly Hagi (seventeenth–eighteenth century). (Photograph courtesy of Art Resource, New York.)

sensibility advocated by Kenkō and his followers by saying that Kenkō's view "does not accord with human feeings but is a *fabricated* aesthetic taste formed in the impertinent mind of a man. . . . what that monk said can be described . . . as *contrived* only to make what does not accord with human wishes a refined taste."

Despite their opposing conclusions, however, both seem to agree that it is more "natural" for humans to long for clarity, perfection, and abundance. The aesthetics of imperfection, desolation, and agedness advocates transforming what would normally be a disappointing experience, such as viewing an obscured moon or a shabby scroll, into a positive experience. The appreciation of the imperfect can then be interpreted as an end product of a dialectic movement, a resolution to the disappointment or dissatisfaction in the ordinary context. However, why challenge the "natural" aesthetic attraction and advocate what may be considered a subversive aesthetics? Is it simply to exercise one's imagination more?

It is important to note that this aesthetic preference for the obscured, imperfect, and insufficient presupposes certain cultural and economic conditions. For example, the possibility and effectiveness of indirect expression require some degree of culturally shared associations and allusions, such as cherry blossoms symbolizing transience and elegance or an autumn dusk evoking desolation and loneliness. Or, the appreciation of objects not altogether visible is based on the possibility of an unobstructed view of the object, such as the moon not covered by clouds. Otherwise, the experience will simply result in frustration and disappointment.

The aesthetics of *wabi* similarly presupposes a ready access to beautiful and expensive objects; it is an aesthetics born out of wealth and privilege. Being powerful and

wealthy, the advocates of *wabi* aesthetics—shoguns and tea masters—could afford to emulate the impoverished look of peasant life by creating rustic tea huts with stark interiors and ordinary, sometimes defective, tea bowls. For poor peasants who had no choice but to make do with such objects, the look of poverty would merely symbolize ever-unrequited longing. The Japanese penchant for simplicity and insufficiency was thus cultivated as part of an elite aesthetics, first by aristocrats, then by the military class. Both groups were educated in the arts and shared a common cultural literacy; at the same time they were wealthy and powerful enough to indulge in taking pleasure in the appearance of impoverishment and imperfection.

This aesthetics sometimes took on a political undertone of appeasing the dissatisfaction of the poor through aestheticizing the look of poverty. This is most notably illustrated by the development of the tea ceremony. Its increasing imitation of poverty and emphasis on being satisfied with one's lot, no matter how inadequate, paralleled the increasing power of its patron shoguns.

Although initially cultivated as rather indulgent preoccupations supported by sophistication and wealth, the aesthetics of insufficiency came to be entrenched in every segment of the society, forming one of the consistent characteristics of Japanese aesthetics. Its appeal to the general populace, both wealthy and poor, stems from its aestheticization of the conditions of life in general—the impermanence and imperfection of existence. As in many other world cultures and traditions, the transience of everything, particularly of human life, poses a challenge to humans because this predicament is difficult to accept. The initial Japanese response to this irrevocable fact of life ranged from regret and resignation to seeking salvation in the other world, the Pure Land.

The aestheticization of the aged and the imperfect in the aesthetics of *wabi, sabi,* and *yūgen* became the prevalent mode of affirming one's existential predicament in the Japanese tradition. The aesthetic celebration of perishability and imperfection makes the difficulties of life more palatable and even attractive. Death itself becomes aestheticized, as seen in the aura surrounding the medieval ritual of suicide by disembowelment. This *seppuku* (or *harakiri*) is considered not only an ethical act of carrying out one's duty according to the warrior ethic and an act of bravery, indicating the strength of one's physical and mental will, it is also the final aesthetic act showing how elegantly and gracefully one can take leave of this world, evidenced not only by the physical act itself but also by the calligraphic composition of a parting poem. This aestheticization of death also appears in this century. The wartime loyalty to the emperor was often inspired by the parallel drawn between death and falling cherry blossoms, a traditional symbol for beauty marked by transience and poignancy.

Respect for the Essential Characteristics of the Object. Another criterion for aesthetic evaluation in the Japanese tradition, specifically of designed objects, is the attention to and respect for the indigenous characteristics of the subject matter or the material. This principle is best illustrated by a comparison between European formal gardens and Japanese gardens. The geometric pattern of the former is primarily executed with shrubs and plants, irrespective of their native growth patterns, characteristics, or the genius loci. On the other hand, Japanese garden design takes advantage of what the site and the materials offer. A distant mountain view, for example, is frequently incorporated in a Japanese garden through the technique called *shakkei* (borrowed scenery). A tree is shaped to articulate its native characteristics through elimination of its branches and leaves, considered to be adventitious and to detract from its essential features.

This principle of garden design was first established in the eleventh-century manual for garden making, *Sakuteiki* (The Book of the Garden). Primarily concerned with rock arrangement, its author, Tachibana no Tsunayoshi (1028–1094), reiterates the importance of "obeying the request" in designing a garden. That is, the native characteristic of each rock must be attended to and enhanced in making an arrangement, as if each rock is requesting that its voice be heard and respected.

A further theoretical basis of this respect for each object was supplied by Zen Buddhism. The Zen identification of Buddha nature, or ultimate reality, with the thusness or being-such-ness of an object cultivates an attentive and respectful attitude toward the characteristics essential to defining its individuality. This Zen celebration of the thusness of each object encourages the artist to lend sympathetic eyes and ears to what the object is prepared to articulate and to assist in distilling its quintessential features.

Art forms other than garden making that also deal directly with nature as the material share the same design principle. The arts of flower arrangement and bonsai consist of discerning the unique characteristics of each plant and shaping it to highlight such individuality, by eliminating leaves and flowers, bending stems, and training branches with wires and ropes. Japanese cooking is also marked by its methods, presentation, and choice of vessels that accentuate the characteristics of the ingredients. In using materials such as Japanese paper and bamboo for creating various artifacts, ranging from packaging and ornaments to architectural elements, the characteristics of the materials are fully utilized. Furthermore, manufactured materials, such as concrete for architecture, plastics and synthetic fabric for clothing design, and artificial flowers for flower arrangements are wholeheartedly embraced by contemporary Japanese designers. Following the time-honored principle of respect for the material, they give homage to, rather than

shun, hide, or contradict, the unique features of these new materials. Thus, this design canon is flexible in accommodating and making creative use of new challenges.

The same principle of respecting the voice of the material applies to representational arts as well. For example, Bashō instructs his disciples to "listen to the pine tree" and "enter into the bamboo" when composing a haiku about them. When successful, he claims, the resulting haiku reads as if it "became" rather than having been "composed." If one captures the heart of bamboo into the paint brush, the nineteenth-century painter Tsubaki Chinzan (1801–1854) claims, the painting of bamboo will take form by itself. A similar suggestion is given with respect to acting in a Nō play: if the actor discerns and identifies with the essence of the character to be portrayed, the outward expression of imitation will follow naturally. Indeed, the actor's preparation before the performance includes a period during which he meditates on the character of the mask he is to don.

On the other hand, this emphasis on respecting the object's characteristics tends to result in a deemphasis of the artist's unique contribution to the creative process. Consequently, the success of an artwork, judged by the degree to which it conveys the essential features of the object depicted or used, is frequently described by epithets such as "natural" or "spontaneous." A successful work of art should facilitate the audience's appreciation of the material used and the subject matter depicted. In contrast, a design composed primarily as a vehicle for the artist's self-expression, and imposed on the material irrespective of its characteristics, is criticized as appearing "artificial" or "forced."

While these terms do refer to the features of the objects, in the Japanese tradition they also refer to some aspects of the artists and their creative acts. In one sense, therefore, the Japanese tradition accepts what in Western aesthetics would be considered an intentional fallacy. What, then, is the relationship between spontaneity as a feature of the artwork, and the spontaneity of the artist's creative act?

Sincerity and Devotion as Criteria for Intentionalist Criticism. In the Western aesthetic tradition, the controversial nature of the relationship between the artist's state of mind and the artwork is discussed as the problem of the intentional fallacy. According to Japanese aesthetics, the inseparability of artists and their artworks is assumed and requires that the evaluation of a work of art refer to the artist's state of mind.

As a powerful and persistent legacy of Zen Buddhism, the Japanese aesthetic tradition considers artistic endeavor to be a secular means of experiencing enlightenment. This is indicated by the fact that most of the artistic disciplines can be referred to as "the way" *(dō): sadō* for the tea ceremony, *gadō* for painting, *kadō* for flower arrangement, *shodō* for calligraphy, *kyūdō* for archery, and *kendō* for swordsmanship. Zen enlightenment requires, among other things,

transcendence of oneself because ordinary egocentric concerns render one's experience of the world, mediated by discrimination, conceptualization, and valuation, useful for satisfying self-interest. These, however, hamper direct experience of the Buddha nature of everything.

Thoroughgoing devotion and commitment to artistic discipline is one way of continuously transcending oneself. Masterful artists are distinguished from novices not only by their mastery and internalization of techniques and skills but also by their dedication to the process of art making. Successful artistic training culminates in freedom from one's conscious intent to achieve artistic success.

The "disinterested" attitude in the Western tradition is often proposed as a necessary condition for having an aesthetic experience as a viewer. In contrast, the same notion in the Japanese tradition refers to a requirement for the artist in creating a successful artwork. This highest stage of "no intent" or "no self," which is attained through rigorous training, facilitates an approach to the subject matter or material with a humble, open attitude, allowing the object to evoke its inner essence to which the artists listen and submit themselves. All of this will in turn be reflected in the "effortless," "spontaneous," "natural," or "unconstrained" appearance of the resultant object or performance. A potter, for example, is expected to form a vessel based on an interaction between the particular clay and his or her hands rather than forcing the clay to conform to a preconceived shape.

Furthermore, the potter must submit to accepting unexpected reactions of the glaze to the firing. The harmonious atmosphere in the tea ceremony, according to Rikyū, must be generated by the participants' spontaneous cooperation; a conscious effort to create such harmony will fail. The Nō actor's performance invoking *yūgen* is characterized by "effortless proficiency" and "perfect freedom," which are achieved only after long, arduous, and rigorous training. The highest compliment paid to an artist can only be expressed paradoxically: the most artful achievement is indicated when the art becomes thoroughly artless. Hence, aesthetic evaluation in the Japanese tradition does engage in an intentionalist evaluation. The gauge of artistic merit, however, is the degree of the absence (not a mere lack but rather the absence achieved through transcendence) of an artistic intention, thereby lending an air of spontaneity and artlessness to the product.

The art of artlessness, it is important to note, also results from the thorough integration of the body and the mind. The devotion to and rigorous training in a particular medium will lead the artist to experience the complete harmony between what the mind dictates and what the body executes—sometimes through a certain tool: the mind and the hand of a painter or a calligrapher will need to become one with the brush; no separation can creep in between the sword, the body, and the mind of a warrior; body movement

and its rhythm help determine the design of an object, ranging from pottery to sword, from woven textile to carved wood.

The Challenge to Western Aesthetics. Studying Japanese aesthetics is valuable both for multicultural explorations of aesthetic issues as well as for a comprehensive understanding of Japanese culture, which is dominated by aesthetic concerns. It also offers areas for comparative inquiry—for example, between the concepts of *wabi* and *sabi* and the eighteenth-century British cult of the picturesque. Both traditions celebrate obscurity and irregularity, the creation of an aesthetic value through iconoclastic acts, an emphasis on the stimulation of the imagination, and a fascination with the passage of time.

Another comparison can be made between the Japanese means of justifying every aspect of one's life through aestheticization (particularly of those aspects that are difficult to accept) and Friedrich Nietzsche's aesthetic justification of life by which every contingency is given necessity as an ingredient in forming an organic whole. Both propose a total affirmation of what exists and happens through aesthetic means.

In addition to offering prospects for these inquiries, Japanese aesthetics also challenges some important assumptions in Western aesthetics. For example, the Japanese aesthetics of egalitarianism challenges the Western paradigm, which primarily involves experiencing a fine-art object through one of the higher senses. We are compelled to question whether rich aesthetic experiences outside of the fine arts, such as ordinary artifacts, the environment, nature, and daily activities, can be appropriately and adequately explained by applying this model.

The Japanese notions of good works of art and successful artists also suggest alternatives to Western parallels. In particular, the design principle of respecting the subject matter and material could be translated into an ecologically desirable attitude that begins with overcoming the view of nature as mere raw material to be used for various human purposes, both practical and aesthetic. This practice may suggest a model of partnership between humans and nature that encourages careful attention to nature's dictates.

The evaluative criteria of spontaneity, both of the works of art and of the artists, can also challenge the competing Western criteria of formal unity and expressive power of the object. Is it sometimes possible that a work of art is valued to the extent that the creator submits his or her individuality and originality to the demands of the subject matter or the material? Can we sometimes describe successful artists as those who faciliate the experiencing agents' communion with the subject matter represented or the material used in the art objects?

These alternative ideas suggested in Japanese aesthetics should not be regarded as invalidating or replacing Western aesthetic ideas. The challenge they pose can, however, shed light on both the nature and the limitations of Western aesthetic theories and concepts for explaining the rich and diverse content of human aesthetic experience.

[*See also* Appreciation; *and* Gardens, *article on* Gardens as Art.]

BIBLIOGRAPHY

Primary Sources

de Bary, Wm. Theodore, Ryusaku Tsunoda, and Donald Keene, eds. *Sources of Japanese Tradition.* 2 vols. New York, 1964.

Dōgen. *Shōbōgenzō: Zen Essays by Dōgen* (1231–1253). Translated by Thomas Cleary. Honolulu, 1986.

Fujiwara, Teika. *Maigetsushō* (c.1219). In *The Theory of Beauty in the Classical Aesthetics of Japan,* translated by Toshihiko Izutsu and Toyo Izutsu, pp. 79–96. The Hague, 1981.

Hattori, Dohō. *The Red Booklet* (1776). In *The Theory of Beauty in the Classical Aesthetics of Japan,* translated by Toshihiko Izutsu and Toyo Izutsu, pp. 159–167. The Hague, 1981.

Hayashiya, Tatsusaburo, ed. *Kodai chūsei geijutsuron* (Ancient and Medieval Theories of Art). Tokyo, 1973.

Herrigel, Eugen. *Zen in the Art of Archery* (1948). Translated by R. F. C. Hull. New York, 1953.

Kawabata, Yasunari. *Japan the Beautiful and Myself.* Translated by Edward G. Seidensticker. Tokyo, 1969.

Keene, Donald, ed. *Anthology of Japanese Literature from the Earliest Era to the Mid-Nineteenth Century.* New York, 1955.

Nanbō, Sōkei. *A Record of Nanbō* (1690). In *The Theory of Beauty in the Classical Aesthetics of Japan,* translated by Toshihiko Izutsu and Toyo Izutsu, pp. 135–158. The Hague, 1981.

Nihon koten bungaku taikei (Outline of Classical Japanese Literature). 100 vols. Tokyo, 1958–1968.

Nihon koten bungaku zenshū (Complete Works of Classical Japanese Literature). 51 vols. Tokyo, 1973.

Sei Shōnagon. *The Pillow Book of Sei Shōnagon* (c.1002). Translated by Ivan Morris. Harmondsworth, 1971.

Tanizaki, Jun'ichiro. *In Praise of Shadows.* Translated by Thomas J. Harper and Edward G. Seidensticker. New Haven, 1977.

Yamamoto, Tsunetomo. *Hagakure: The Book of the Samurai* (1716). Translated by William Scott Wilson. Tokyo, 1979.

Yasuda, Ayao, ed. *Nihon no geijutsuron* (Theories of Art in Japan) (1957). Rev. ed. Tokyo, 1990.

Yoshida, Kenkō. *Essays in Idleness: The Tsurezuregusa of Kenkō* (1330–1332). Translated by Donald Keene. New York, 1967.

Secondary Sources

Hisamatsu, Shin'ichi. *Zen and the Fine Arts* (1957). Translated by Gishin Tokiwa. Tokyo, 1971.

Hume, Nancy G., ed. *Japanese Aesthetics and Culture: A Reader.* Albany, N.Y., 1995.

Ienaga, Saburo. *Japanese Art: A Cultural Appreciation* (1959). Translated by Richard L. Gage. New York, 1979.

Izutsu, Toshihiko, and Toyo Izutsu. *The Theory of Beauty in the Classical Aesthetics of Japan.* The Hague, 1981.

Kasulis, T. P. *Zen Action/Zen Person.* Honolulu, 1981.

Katō, Shūichi. *Form, Style, Tradition: Reflections on Japanese Art and Society.* Translated by John Bester. Berkeley, 1971.

Keene, Donald. *Landscapes and Portraits: Appreciations of Japanese Culture.* Tokyo, 1981.

Lee, O-Young. *The Compact Culture: The Japanese Tradition of "Smaller Is Better."* Translated by Robert N. Huey. Tokyo, 1991. Originally published as *Smaller Is Better: Japanese Mastery of the Miniature* (Tokyo, 1984).

Marra, Michele. *The Aesthetics of Discontent: Politics and Reclusion in Medieval Japanese Literature.* Honolulu, 1991.

Minami, Hiroshi. *Psychology of the Japanese People* (1953). Translated by Albert R. Ikoma. Tokyo, 1971.

Morris, Ivan. *The World of the Shining Prince: Court Life in Ancient Japan* (1964). Tokyo, 1994.

Odin, Steve. "Intersensory Awareness in Chanoyu and Japanese Aesthetics." *Chanoyu Quarterly* 53 (1988): 35–44.

Oka, Hideyuki. *How to Wrap Five Eggs: Japanese Design in Traditional Packaging.* New York, 1967.

Okakura, Kakuzō. *The Book of Tea.* 5th ed. New York, 1931.

Pilgrim, Richard B. *Buddhism and the Arts of Japan.* Chambersburg, Pa., 1981.

Reeve, John, ed. *Living Arts of Japan.* London, 1990.

Richie, Donald. *A Lateral View: Essays on Culture and Style in Contemporary Japan.* Berkeley, 1992.

Saito, Yuriko. "The Japanese Aesthetics of Imperfection and Insufficiency." *Journal of Aesthetics and Art Criticism* 55.4 (Fall 1997): 377–385.

Sandrisser, Barbara. "On Elegance in Japan." In *Aesthetics in Perspective,* edited by Kathleen M. Higgins, pp. 628–633. Fort Worth, Tex., 1996.

Sokoloff, Garret. "By Pausing Before a *Kicho.*" In *Aesthetics in Perspective,* edited by Kathleen M. Higgins, pp. 620–627. Fort Worth, Tex., 1996.

The Sun: 100 Key Words for Understanding Japan. Tokyo, 1993.

Tanahashi, Kazuaki. *Brush Mind.* Berkeley, 1990.

Ueda, Makoto. *Literary and Art Theories in Japan.* Cleveland, Ohio, 1967.

Varley, H. Paul. *Japanese Culture* (1973). 3d ed. Honolulu, 1984.

Yanagi, Sōetsu. *The Unknown Craftsman: A Japanese Insight into Beauty.* Adapted by Bernard Leach. Tokyo, 1973.

YURIKO SAITO

Kire and Iki

As a term in Japanese aesthetics *kire* ("cut" or "cutting") is a basic element not only in artistic creation but also in a stance toward life; it takes on new meaning in modern art, and indeed in modernity generally. Let us begin with a general sense of the term. The word *kire* is first found as a basic concept of Japanese poetry in the so-called *kire-ji,* or "cut-syllable." A cut-syllable has the function of cutting off the flow of expression in order to create space for a new phase. An example of this is given below. The reader stops at a cut-syllable and awaits the opening of a new perspective. This kind of cut is not a breaking off or severing, but rather a connection to a new flow. The word *kire* thus constitutes together with another term, *tsuzuki* ("continuance"), the expression "cut-continuance," which is a key term in Japanese poetics. This can be made clearer by adducing as an example a well-known poem by the haiku master Matsuo Bashō (1644–1694):

An ancient pond—	*Furuike ya*
a frog plops in,	*kawazu tobikomu*
sound of the water.	*mizu no oto*

The cut-continuance in this poem is the *ya,* which is rendered in the English translation by the long dash after the word *pond.* The ancient pond, which would otherwise remain inconspicuous, is brought to our attention by means of this cut-continuance. The cut-syllable *ya* opens up a new phase in which a frog plops in. The stillness of the pond is emphasized and made prominent by the sound of the water. An everyday event in nature is "cut out" by the poem and brought to awareness. By means of this cut the everydayness of nature or the naturalness of the everyday suddenly attains expression in a new way. The poem is thus itself a kind of cutting.

The original meaning of the "cutting" is a technical operation through which the immediate naturalness of nature, as yet untouched by human hand, is "cut off"—but in such a way that nature is "cut out" of its immediate naturalness and presents itself in its inner naturalness. This cut-continuance is a basic Japanese technique that conditions the whole of Japanese art in its traditional sense.

"Cutting" is not only a basic technique in artistic activity but also, and primarily, an art and way of *living.* For Bashō, the best-loved poet in Japan, "all syllables in language are cut-syllables," and to be a poet meant for him a decisive cut in the sense of a "cutting away" from his everyday life. (See the account in Mukai Kyorai, *Kyorai-shō* [Kyorai's Writings], 1955.) He wandered as a poet, and on his travels he devoted himself to poetry, since traveling was for him precisely the locus in which mortal human beings experience their own mortality. The cut as poetic art was for Bashō ultimately the art of life.

The cut finds expression in the plastic arts of the middle ages and modern period in Japan in many ways. Let us adduce as an example the art of flower arrangement (*ikebana*—literally: "making [flowers] live"). A flower in *ikebana* has been cut off from natural life. Such a flower is still able to suck up water in the vase in order to preserve its life. But since it no longer has any roots it is no longer able to propagate itself. It is precisely through its being cut off at the root from its unconditioned will to life that its authentic finitude and naturalness attains pure expression. (The philosopher Nishitani Keiji draws attention to this function of the cut, 1995.)

Another example of the cut is the so-called dry garden (*karesansui*—literally: "withered mountains and waters") in the Japanese art of horticulture. The consummate form of this garden art can be seen in the famous rock garden in the temple of Ryōanji in Kyoto. In this plain garden, whose rectangular area of 336.6. square meters is surrounded by a simple wall, fifteen rocks lie on gravel—nothing more. No vegetation or sculptures. The rocks and gravel are to depict mountains and the water of rivers. The question is: Why "withered" mountains and waters—why lifeless rocks and cold gravel instead of living trees and flowing water? The garden artist and Zen master Musō Soseki deliberately let the natural elements of mountains and waters "wither," in that he cut off the naturalness of nature. (It used to be thought

that the rock garden at Ryōanji was designed by Sōami, but recent research attributes it to Musō, 1275–1351.) In his short poem "Dry Garden Poem," Musō writes:

> Without a single particle of dust,
> High mountains are made to rise;
> With no drop of water to be found,
> Torrents pour into the valleys.

When organic life is cut off and withered in this way, the rest and motion, rigor and play of the world of nature come to the fore. The rock garden at Ryōanji is an outstanding example of the art of cutting. (For a comprehensive discussion of this garden, see part 2 of Ōhashi, 1994.)

A third example is to be found in the so-called way of tea. The house used in the tea ceremony is cut off from the secular world by a path. This path, or *roji*, symbolizes a place of purity in the Buddhist sense. Whoever passes through this place on the way to the teahouse, according to the idea of the way of tea, is purified of the dust of the secular world. The events in the tearoom consist of "nothing more than boiling water and making tea" and "drinking tea," as the well-known tea master Sen no Rikyū (1523–1591) puts it—just as one does in everyday life. (See the chapter titled "Metsugo" [After Death] in Rikyū, 1995.) Only one thing separates the two realms from each other: the cut, which is spatially manifested in the place of purity that is the *roji*. In the tearoom, which is cut off from everyday life in this way, everydayness is reproduced in such a way that the mortality and "this-one-time-only" quality of life, which are generally forgotten, are experienced explicitly. The appurtenances of the tea ceremony—the implements, movements of the body, the flowers on display, the hanging calligraphy scroll, and so forth—are all expressions of the cut-continuance between non-everyday and everyday life.

In all these examples there is expressed a certain conception of life and death. According to this conception, which has its roots in the spiritual-intellectual world of medieval-modern Japan, a world informed by Buddhism, life and death do not stand in stark opposition. Life itself—as in the inhalation-exhalation of respiration, or in the human gait in which the right and left legs work in intermittent opposition—is not a mere continuum but is rather a cut-continuance in each and every moment. The finitude and mortality of life manifest themselves here. To experience truly the reality of life is to become aware of the cut-continuance in life. Artistic cutting was thus often executed, as in the case of Bashō, in concert with and as the fruit of Zen practice. The spirit of art informed by the cut in medieval-modern Japan is actually rooted in Zen Buddhism. Religious practice itself counts as a cut-continuance in a distinctive sense, insofar as it is a cutting away from the everyday, in which—especially when practice leads to an experience of *satori* (awakening)—life is experienced anew and illumined in its basic sources.

It may be helpful at this point to compare the idea of the cut with another basic aesthetic concept, namely, that of *iki*. This idea may be familiar to some readers, since it is discussed in an essay by Martin Heidegger titled "A Dialogue on Language" (Heidegger, 1971). It may help to begin by presenting the more general concept.

During the 1830s, a period marked by the final flowering and maturation of Edo-period culture in Japan, there arose a style of life and art that was called *iki* and that normatively informed the aesthetic consciousness of that era. It had to do with a refined erotic and graceful style of expression that manifested itself in speaking, walking, sitting, and other movements and gestures, as well as in such fields as design, architecture, and music. Heidegger's text makes mention of the ideas of the Japanese philosopher "Count Kuki" (Kuki Shūzō), who in his book *"Iki" no kōzō* (The Structure of "Iki," 1981) tried to articulate these aesthetic expressions of *iki* with the help of European concepts.

According to Kuki, *iki* comprises, as the basic constituent of these aesthetic expressions, three elements. The first is "idealized coquetry." Since the tension of the relationship between man and woman in the form of coquetry culminates in the sexual act, idealized coquetry demands moral-ascetic restraint. Kuki discerned behind the idealized coquetry of *iki* the moral idealism of the "way of the warrior" *(bushidō)* as well as the practice of renunciation in Buddhism. These two are the second and third elements of *iki*, respectively.

Kuki saw a typical manifestation of *iki* in the resolute and proud, but at the same time restrained and gentle, behavior of the *yūjo* (literally: "play-woman"). The popular but misleading name for this occupation these days is *geisha*. The *yūjo* was a woman who entertained men with the practice of such arts as music and dance, but who was not a prostitute in the common sense. She lived in a special community informed by a highly refined aesthetic style, which was an important element of the aesthetic life of modern Japan. For this reason such women often figure in woodblock prints from the period. Many *yūjo*, however, were forced to enter the profession because their parents were poor and had no option other than selling their daughters. Such circumstances obliged them to renounce worldly happiness in their lives. For Kuki the study of *iki* was ultimately a matter of explicating the specific way of being of the Japanese people as a whole, in a kind of "hermeneutics of existence as a people" (Kuki, 1981, p. 78).

For our purposes the issue is the relationship between *iki* and *kire* (cutting). The first thing to notice is that both the idealism of *bushidō* and the renunciation of Buddhism involve some kind of cutting. The way of the warrior is the mode of life of a human being whose task is to wage war. A warrior must be constantly ready to die in battle. In the way of the warrior this readiness for death, which everyday life chooses to forget, itself becomes everyday life. Everyday-

ness is "cut off," but precisely through this cutting off there arises a new daily life in which one is aware of one's own mortality. What Buddhism calls "renunciation" is another decisive cutting. Kuki explains the idea by way of the Japanese term *akirame* (literally: "illumine"). Renunciation is illuminating insight into the processes of the world and into one's own fate. Egoistic wishes and desires are cut off by means of this insight, and one attains purified life. In this way the cut-continuance of life is realized.

It is clear that the principal condition of the possibility of *iki* lies in a *kire:* the question is now what kind of cutting leads to the attitude of *iki*. It is apparently not a matter of the same radical and severe cut we saw exemplified in the rock garden of Ryōanji. The epoch during which *iki* came to prominence as a life-ideal had reached a stage of cultural overripeness, made possible by a long period of peace and economic prosperity. The way of the warrior originated in a context of two centuries without war, in which warriors had no opportunity to practice their martial arts. In the absence of actual experience of combat, readiness for death was idealized in the way of the warrior and refined into an ideal mode of expression. The art of the sword as the art of killing became an art of practicing living, and at the same time a "way" (Chinese, *dao*) as a mode of life.

The same is true for Buddhism, which as a result of the economic and cultural blossoming of feudalistic society in Japan lost the momentum from the tensely creative Kamakura period (1185–1333) or the Muromachi period (1392–1573), during which people found themselves constantly exposed to real death. The forces of moral influence had long since passed into the hands of Confucianism, which took over the ideology of the feudalistic social system with its ideas of order and the task of being human. It is not hard to understand how the illuminating insight of renunciation *(akirame)* was an echo of the true sharpness of the cut—whereby precisely the attitude of *iki* arose.

[*See also* Style.]

BIBLIOGRAPHY

Heidegger, Martin. "A Dialogue on Language." In *On the Way to Language,* translated by Peter D. Hertz, pp. 1–54. New York, 1971.

Hume, Nancy G., ed. *Japanese Aesthetics and Culture: A Reader.* Albany, N.Y., 1995.

Isutzu, Toshihiko, and Toyo Isutzu. *The Theory of Beauty in the Classical Aesthetics of Japan.* The Hague and Boston, 1981.

Kuki Shūzō. *"Iki" no kōzō* (The Structure of "Iki"; 1930). In *Kuki Shūzō zenshū* (Complete Works of Kuki Shūzō), vol. 1, pp. 1–186. Tokyo, 1981.

Mukai Kyorai. *Kyorai shō* (selections). In *An Anthology of Japanese Literature, from the Earliest Era to the Mid-Nineteenth Century,* edited by Donald Keene, pp. 377–383. New York, 1955.

Nishitani Keiji. "The Japanese Art of Arranged Flowers." In *World Philosophy: A Text with Readings,* edited by Robert C. Solomon and Kathleen M. Higgins, pp. 23–27. New York, 1995.

Ōhashi Ryōsuke. *Kire—Das "Schöne" in Japan: Philosophisch-ästhetische Reflexionen zu Geschichte und Moderne.* Cologne, 1994.

Rikyū. *Nampōroku* (selections). In *Wind in the Pines: Classic Writings of the Way of Tea as a Buddhist Path,* edited by Dennis Hirota, pp. 215–240. Fremont, Calif., 1995.

ŌHASHI RYŌSUKE
Translated from Japanese by Graham Parkes